KU-672-576

REFERENCE COPY

The Condensed Chemical Dictionary

The Condensed Chemical Dictionary

NINTH EDITION

Revised by

GESSNER G. HAWLEY

Coeditor, Encyclopedia of Chemistry
Coauthor, Glossary of Chemical Terms

VNR

VAN NOSTRAND REINHOLD COMPANY

NEW YORK CINCINNATI ATLANTA DALLAS SAN FRANCISCO
LONDON TORONTO MELBOURNE

Van Nostrand Reinhold Company Regional Offices:
New York Cincinnati Atlanta Dallas San Francisco

Van Nostrand Reinhold Company International Offices:
London Toronto Melbourne

Library of Congress Catalog Card Number: 76-19024
ISBN: 0-442-23240-3

Manufactured in the United States of America

Published by Van Nostrand Reinhold Company
450 West 33rd Street, New York, N.Y. 10001

Published simultaneously in Canada by Van Nostrand Reinhold Ltd.

15 14 13 12 11 10 9 8 7 6 5 4 3 2 1

Library of Congress Cataloging in Publication Data
Main entry under title:

The Condensed chemical dictionary.

1. Chemistry—Dictionaries. I. Hawley, Gessner Goodrich, 1905–
QD5.C5 1976 540'.3 76-19024
ISBN 0-442-23240-3

Contents

Publisher's Preface

With the appearance of the Ninth Edition, "The Condensed Chemical Dictionary" rounds out half a century of continuous service to the chemical and process industries, and to the many thousands of people throughout the world whose work and interests have brought them into contact with chemistry and its commercial products. The First Edition appeared in 1919, when the chemical industry in the United States was just entering on a huge expansion program as a result of World War I. At that time, the urgent need for such a reference book became apparent to Francis M. Turner, then Vice-President of the Chemical Catalog Company, predecessor of Reinhold Publishing Corporation. He remained Editorial Director of the Dictionary until his death.

The Second and Third Editions were prepared with the assistance of Thomas C. Gregory. The Fourth and following three editions underwent considerable expansion under the capable editorship of Professor and Mrs. Arthur Rose, of State College, Pennsylvania; their competent and effective work improved the Dictionary in many ways. The Eighth Edition was published in 1971, and the Publishers entertain the hope that the Ninth Edition will be equally well received.

VAN NOSTRAND REINHOLD COMPANY

Introduction

The first edition of the *Condensed Chemical Dictionary* appeared in 1919, when the chemical industry in the United States was entering on a huge expansion program as a result of World War I. The urgent need for such a reference book became apparent to Francis M. Turner, President of the Chemical Catalog Company, predecessor of the Reinhold Publishing Corporation. Under his supervision a succession of Editors developed and expanded the *Condensed Chemical Dictionary* to meet the growing needs of the chemical industries. Since his death this development has continued, with the result that the work has achieved world-wide recognition in its field.

The *Condensed Chemical Dictionary* is a unique publication. It is not a dictionary in the usual sense of an assemblage of brief definitions, but rather a compendium of technical data and descriptive information covering many thousand chemicals and chemical phenomena, organized in such a way as to meet the needs of those who have only minutes to devote to any given substance or topic.*

Three distinct types of information are presented: (1) technical descriptions of chemicals, raw materials, and processes; (2) expanded definitions of chemical entities, phenomena, and terminology; and (3) description or identification of a wide range of trademarked products used in the chemical industries. Supplementing these are listings of accepted chemical abbreviations used in the literature, short biographies of chemists of historic importance, and descriptions or notations of the nature and location of many American technical societies and trade associations. In special cases editorial notes have been supplied where it was felt necessary to clarify or amplify a definition or description. A few entries written by specialists are acknowledged by use of the author's name.

In a work of this nature, selection of topics for inclusion can hardly fail to be influenced by current interests and developing concerns within the topic area. The growing importance to chemists—and to the general public as well—of environmental and health hazards, which came to the forefront so quickly in the 1960's, was reflected in the Eighth Edition, which greatly increased its coverage of this aspect of chemistry. Now, nearly ten years later, the magnitude of the energy problem has been uppermost in the thinking of a broad spectrum of engineers, chemists, and physicists, since it is certainly the No. 1 technical problem confronting this country. Thus the present edition, while retaining its emphasis on environmental considerations, has been expanded in the area of energy and its sources, as far as permitted by presently available information. The effort has been to provide condensed, authoritative, factually oriented statements and descriptions, and to resist prognostications as to the future potential of any particular energy source. At the same time, continuing attention has been devoted to common hazards, such as flammable and explosive materials, poisons, pesticides, carcinogens, corrosive agents, etc., in line with the practice followed in earlier editions, and with the increasing public concern over these matters.

In connection with certain classifications of substances, particularly pesticides and carcinogens, which have occasioned the most controversy in recent years, the statement "Use may be restricted" indicates that a state or local regulation may exist even though the product has not been officially banned, or that a definitive ruling on its use is pending. A number of disputed cases have arisen recently; though some have been definitely settled, others are still being evaluated or are in the process of litigation. Typical is the instance of diethylstilbestrol (DES), use of which in cattle feeds for rapid fattening was first prohibited, later conditionally approved, and finally once again prohibited. Where a spe-

*More detailed and sophisticated treatment is presented in the *Encyclopedia of Chemistry*, Third Edition, edited by C. H. Hampel and G. G. Hawley (Van Nostrand Reinhold, 1973). An elementary volume designed primarily for student use, entitled *Glossary of Chemical Terms*, by the same authors and publisher, is also available (1976).

cific ruling has been made by a Federal agency limiting the use of a product to certain purposes, this is indicated by the statement "Use has been restricted," e.g., for DDT. When a product has been banned outright, the statement "Use has been prohibited" is used. In such a work as this, in view of the many materials in various stages of evaluative testing, court procedures, appeals, hearings, etc., it is impossible to keep abreast of every development. The user should check the current status of any questionable products before making decisions that involve them. (See also paragraph on Hazards, below.)

Arrangement of Entries

The entries are listed in strict alphabetical order; that is, those comprised of two or more words are alphabetized as if they were a single word, e.g., "acidimetry" precedes "acid value," and "waterproofing agent" precedes "water softener." The many prefixes used in organic chemistry are disregarded in alphabetizing, since they are not considered an integral part of the name; these include ortho-, meta-, para-, alpha, beta-, gamma-, sec-, tert-, sym-, as-, uns-, cis-, trans-, endo-, exo-, d-, 1-1, dl-, n-, N-, O-, as well as all numerals denoting structure. However, there are certain prefixes that are an integral part of the name (iso-, di-, tri-, tetra-, cyclo-, bis-, neo-, pseudo-), and in these cases the name is placed in its normal alphabetical position, e.g., dimethylamine under D and isobutane under I. The same is true of mono- (used as little as possible) and of ortho-, meta-, and para- in inorganic compounds such as sodium orthophosphate.

Chemicals and Raw Materials

The information in the categories listed below is given for each substance in the sequence indicated; where entries are incomplete, it may be presumed that no reliable data were provided by the reference systems utilized.

Name: The commonly accepted name is the key entry. Terminological variations are indicated where necessary. In virtually all cases, the name is given in the singular number. A name having initial caps and enclosed in quotes is a trademark; the superior numbers refer to the name of the manufacturer given in Appendix I.

Synonym: Alternate names (IUPAC and others), as well as trivial names, are indicated. Obsolete and slang names have been eliminated as far as possible. Most synonyms are entered independently and cross-referenced, but space limitation has not permitted complete consistency in this regard.

Formula: The molecular (or atomic) formula is regularly given; structural formulas are used in special cases of unusual importance or interest.

Properties: The properties typically given are: physical state; atomic number; atomic weight; valence; isotopes; odor; taste; specific gravity; boiling point (at 760 mm Hg unless otherwise stated); melting point (freezing point); refractive index; solubility; or miscibility. Various other properties are given where pertinent: flash point; autoignition point; electrical properties; tensile strength; hardness; or expansion coefficient.

Source or Occurrence: Geographical origin of metals, ores, essential oils, vegetable oils, and other natural products.

Derivation: The chemical reactions or other means of obtaining the product; current industrial methods are emphasized; obsolete and "curiosity" methods have been largely eliminated.

Grades: Recognized grades as reported in the industrial literature, including technical, C.P., U.S.P., refined, reactor, semiconductor, and the like.

Containers: Unit types; tank cars; tank trucks; carlots; bulk; barges; pipelines.

Hazard: This category includes flammability, toxicity, tissue irritation, explosion risks, etc., based on authoritative data. Also given are the tolerance ratings (Threshold Limit Values) for workroom exposures established by the American Conference of Government Industrial Hygienists; various rulings of the Food and Drug Administration; as well as reference to chemical safety data sheets prepared by and available from the Manufacturing Chemists Association, 1825 Connecticut Ave., N.W., Washington, D.C., 20009. It was not considered practicable to include exposure standards established by the Federal regulatory bodies NIOSH and OSHA; those that have been determined have been published periodically in the *Federal Register* over the last few years.

The toxicity ratings are intended to be used only as indications of the industrial hazard presented by a given material; qualified toxicologists or physicians should be consulted for specific evaluations, dosages, exposure times and concentrations. For further information regarding these hazards, the reader is referred to the following entries: combustible material; flammable material; dust, industrial; corrosive material; oxidizing material; poison (1); toxicity; toxic materials; carcinogen.

Uses: These are primarily large-scale applications. Because of the rapidity of change in the chemical industries and the difficulty of obtaining reliable current data, no attempt has been made to list uses in the order of their tonnage consumption. The patent literature is not specifically represented.

Shipping Regulations: The requirements for transportation labelling for both rail and air are given under this heading; they are based on those established by the official regulatory agencies at the time this book was in prepartion. These agencies are the Department of Transportation (Graziano's Tariff No. 25), Washington, D.C., for rail and water transport, and the Federal Aviation Authority, which observes the regulations of the International Air Transport Association, Geneva, Switzerland (P.O. Box 160). These schedules are periodically updated—indeed the latter issues a revised tariff annually.

In view of the frequent and numerous changes in labelling specifications, the designations "Rail" and "Air" are used in the Shipping Regulations entry, rather than the initials of the agencies (DOT and IATA), as in previous editions. This has been done to avoid the impression that such regulations are currently definitive. It should not be assumed that any given label specification in this book is ipso facto the latest ruling of either agency. For this reason, **this Dictionary is not to be cited as a final authority for any transportation label.** All manufacturers and shippers should obtain the official revised tariffs on a regular basis to ensure proper compliance, not only in respect to labeling but to the many other shipping specifications for hazardous materials contained therein.

General Entries

It is likely that no two editors would completely agree about what general subjects should be included in a dictionary of this kind. The major subdivisions of matter directly involved with chemical reactions, the various states of matter, and the more important groups of compounds would almost certainly be regarded as essential; but beyond these, the area of selectivity widens rapidly. The topics either added or expanded by the present Editor were chosen chiefly because of their interest and importance, both industrial and biochemical, and secondarily because of the terminological confusion evidenced in the literature and in industrial practice. Regarding the latter, the reader is referred to the entries on gum, resin, pigment, dye, filler, extender, reinforcing agent, homogeneous, and combustible materials. In some cases a position has been taken which may not be accepted by all, but which is defensible and certainly not arbitrary. Even editors must acknowledge that the meanings and uses of terms often change illogically, and that such changes are usually irreversible.

Among the general entries are: important subdivisions of chemistry; short biographies of outstanding chemists of the past; numerous group definitions (barbiturate, peroxide); major chemical and physicochemical phenomena (polymerization, catalysis); functional names (antifreeze, heat-exchange agent, sequestrant); terms describing special material forms (aerosol, foam, fiber); energy sources (solar cell, fuel cell, fusion); and the more important chemical processes. No general entry is intended to be encyclopedic or definitive, but rather a condensation of essential information, to be supplemented by reference to specialized sources. To present all this in useful and acceptably complete form has been a challenging, though often frustrating task, which the Editor leaves with the uneasy feeling that, like the breadcrumbs in the Mad Hatter's butter, some mistakes may have got in as well.

Trademarks

Continuing the policy of previous editions, an essential component of the Dictionary comprises descriptions of several thousand proprietary industrial products. The information was either provided by the manufacturers of these materials or taken from announcements or advertisements appearing in the technical press. Each proprietary name is enclosed in quotation marks, is stated to be a trademark (or brand name), and is followed by a superscript number referring to the Numerical List of Manufacturers on page 948. From this, the address of the manufacturer can easily be found in the Alphabetical List of Manufacturers. We wish to thank the owners of these trademarks for making the information available. The space devoted to these is necessarily limited, as the constant proliferation of trademarked products makes it impossible to list more than a small fraction of them in a volume such as this.

The absence of a specific trademark designation does not mean that proprietary rights may not exist in a particular name. No listing, description, or designation in this book is to be construed as affecting the scope, validity, or ownership of any trademark rights that may exist therein. Neither the Editor nor the Publisher assumes any responsibility for the accuracy of any such description, or for the validity or ownership of any trademark.

A Request

Many corrections and suggestions have been received from readers during the long history of the earlier editions; the Editors have always tried to acknowledge these to the best of their ability, and they have welcomed this correspondence, for it has been an important source of information about the acceptance of the Dictionary by its market. The present Editor and Publisher wish to encourage this reaction from the field, not only to permit corrections to be made in reprinted issues, but also as a basis for preparing future editions. All letters addressed to the Publisher will be forwarded.

THE EDITOR

Abbreviations

ACS American Chemical Society (q.v.)
ASTM American Society for Testing and Materials (q.v.)
b.p. boiling point
°C degrees centigrade (Celsius)
cc cubic centimeter
C.P. chemically pure: a grade designation signifying a minimum of impurities, but not 100% purity.
CNS central nervous system
C.C. closed cup
COC Cleveland Open Cup
C.I. "Colour Index" (a standard British publication giving official numerical designations to colorants).
cp centipoise
cu cubic
°F degrees Fahrenheit
FCC "Food Chemicals Codex," First Edition and supplements. Published by National Research Council, Washington, D.C. Second Edition available 1972.
FDA Food and Drug Administration, the Federal agency responsible for enforcing the Food, Drug, and Cosmetic Act. It also has authority to establish tolerances for pesticides on food products, to regulate food additives, including colors, and to require proof of the efficacy and safety of drugs. In recent years its authority has been extended to cover flammable products, dangerous chemicals, packaging and labeling.
f.p. freezing point
FTC Federal Trade Commission (a consumer protection agency).
g gram
gal gallon
lb pound
ml milliliter
m.p. melting point
MCA Manufacturing Chemists Association (q.v.)
mg milligram
mm millimeter
N.D. "New Drugs" (an annual publication).
N.F. "National Formulary" (a publication).
O.C. Open Cup
ppm parts per million
psi(a) pounds per square inch (absolute)
q.v. quod vide (which see)
sec. second
sp. gr. specific gravity
TCC Tagliabue Closed Cup
TOC Tagliabue Open Cup
USAN United States Adopted Name: a nonproprietary name approved by the American Pharmaceutical Association, American Medical Association, and the U.S. Pharmacopeia. Such names applied to pharmaceutical products do not imply endorsement; their use in advertising and labeling is required by law.
USDA U.S. Department of Agriculture.
U.S.P. United States Pharmacopeia: a standard publication of authorized drugs and materia medica, now in its 18th edition. Published by Mack Publishing Co., Easton, Pa.
wt/gal weight per gallon

The
Condensed Chemical Dictionary

A

A (1) Abbreviation for absolute temperature; (2) abbreviation for Angstrom (q.v.); (3) symbol for mass number (AEC).

abaca (Manila hemp). The strongest vegetable fiber, obtained from the leaves of a tree of the banana family. The fibers are 4–8 ft long, light in weight, soft, lustrous, nearly white, and do not swell or lose strength when wet. Denier ranges from 300 to 500. Combustible, but self-extinguishing.
Sources: Philippines, Central America, Sumatra.
Grades: Available in 18 grades based on color and length.
Uses: Heavy cordage and twine, especially for marine use; manila paper.
See also hemp.

ABC salt. Abbreviation for 4′,4‴ azobis-4-biphenylcarboxylic acid, disodium salt used as a chemical intermediate.

abherent. Any substance that prevents adhesion of a material to itself or to another material. It may be in the form of a dry powder (talc, starch); a suspension (bentonite-water); a solution (soap-water); or a soft solid (stearic acid, tallow). Abherents are used as dusting agents and mold washes in the adhesives, rubber and plastics industries. Fats and oils are used as abherents in the baking industry. Fluorocarbon-resin coatings on metals are widely used on cooking utensils.

Abies Siberica oil. See fir needle oil.

abietic acid (abietinic acid; sylvic acid) $C_{19}H_{29}COOH$ (having a phenanthrene ring system). A major active ingredient of rosin, where it occurs with other resin acids. The term is often applied to these mixtures, separation of which is not achieved in technical grade material.
Properties: Yellowish resinous powder; m.p. 172–175°C; optical rotation −106°; soluble in alcohol, ether, chloroform, and benzene; insoluble in water. Combustible. Low toxicity.
Derivation: Rosin, pine resin; tall oil.
Method of purification: Crystallization.
Grades: Technical.
Containers: Kegs; drums; multiwall paper bags.
Uses: Abietates (resinates) of heavy metals as varnish driers; esters in lacquers and varnishes; fermentation industries; soaps.

"Abitol."[266] Trademark for a colorless, tacky, very viscous liquid; mixture of tetra-, di-, and dehydroabietyl alcohols made from rosin.
Uses: Plasticizers, tackifiers, adhesive modifiers.

ablation. The rapid removal of heat (5000 to 10,000°F) from a metallic substrate by pyrolysis of a material of low thermal conductivity, which is able to absorb or dissipate the heat while being decomposed to gases and porous char. Ablative materials applied to the exterior of temperature-sensitive structures isolate them from hyperthermal effects of the environment. Interaction of a high-energy environment with the exposed ablative material results in a small amount of sacrificial erosion of the surface material. The attendant energy-absorption processes control the surface temperature and greatly restrict the flow of heat into the substrate. Ablative materials are usually composed of a ceramic or glass-reinforced plastic.

abrasive. A finely divided, hard, refractory material, ranging from 6 to 10 on the Mohs scale, used to reduce, smooth, clean, or polish the surfaces of other, less hard substances, such as glass, plastic, stone, wood, etc. Natural abrasive materials include diamond dust, garnet, sand (silica), corundum (aluminum oxide, emery), pumice, rouge (iron oxide), and feldspar; the more important synthetic types are silicon carbide, boron carbide, cerium oxide, and fused alumina. Abrasives in powder form are used in several ways: (1) applied directly to the surface to be treated by mechanical pressure or compressed-air blast, as in cleaning building stone; (2) affixed to a paper or textile backing after the particles have been coated with an adhesive; or (3) mixed with a bonding agent such as sodium silicate or clay, the particles being compressed into a wheel rotated by a power-driven shaft.

abrasive, coated. See abrasive (2).

ABS. Abbreviation for (1) alkyl benzene sulfonate (detergent); (2) acrylonitrile-butadiene-styrene copolymer. See ABS resin.

"Absafil."[539] Trademark for an acrylonitrile-butadiene-styrene copolymer reinforced with glass fiber.
See also reinforced plastic.

abscisin. A plant hormone which promotes the aging process in plants, by inducing the dropping of leaves and fruit. Occurs naturally in many plants; the synthetic product is commercially available as abscisic acid. See also plant growth regulator.

absolute. (1) Free from admixture of other substances; pure. Example: absolute alcohol is dehydrated ethyl alcohol, 99% pure.
(2) The pure essential oil obtained by double solvent extraction of flowers in the manufacture of perfumes. See concrete (2).
(3) Absolute temperature (q.v.).

absolute temperature. The fundamental temperature scale used in theoretical physics and chemistry, and in certain engineering calculations such as the change in volume of a gas with temperature. Absolute temperatures are expressed either in degrees Kelvin or in degrees Rankine, corresponding respectively to the centigrade and Fahrenheit scales. Degrees Kelvin are obtained by adding 273 to the centigrade temperature, while degrees Rankine are obtained by adding 460 to the Fahrenheit temperature. The nearest practical approach to absolute zero is about −272°C.

Superior numbers refer to Manufacturers of Trade Mark Products. For page number see Contents.

absorbent. (1) Any substance exhibiting the property of absorption, e.g., absorbent cotton, so made by removal of the natural waxes present. See absorption (1). (2) A material that does not transmit certain wavelengths of incident radiation. See absorption (2).

absorption. (1) In chemical terminology, the penetration of one substance into the inner structure of another, as distinguished from adsorption, in which one substance is attracted to and held on the surface of another. Physicochemical absorption occurs between a liquid and a gas or vapor, as in the operation known as scrubbing (q.v.) in which the liquid is called an absorption oil; sulfuric acid, glycerol, and some other liquids absorb water vapor from the air under certain conditions. Physiological absorption takes place through porous tissues such as the skin and intestinal walls, which permit passage of liquids and gases into the bloodstream. See also adsorption; hygroscopic.

(2) In physical terminology, retention by a substance of certain wavelengths of radiation incident upon it, followed by either an increase in temperature of the substance or by a compensatory change in the energy state of its molecules. The ultraviolet component of sunlight is absorbed as the light passes through glass and some organic compounds, the radiant energy being transformed into thermal energy. The radiation-absorptive capacity of matter is utilized in analytical chemistry in various types of absorption spectroscopy (q.v.).

(3) In physical chemistry, the ability of some elements to pick up or "capture" thermal neutrons produced in nuclear reactors as a result of fission. This is due to the large capture cross section of their atoms, which is measured in units called barns; elements of particularly high neutron absorption capability are cadmium and boron.

absorption band. The range of wavelengths absorbed by a molecule; for example, absorption in the infrared band of 2.3 to 3.2μ indicates the presence of OH and NH groups, while the 3.3 to 3.5 band indicates aliphatic structure. Atoms absorb only a single wavelength, producing lines, such as the sodium D line. See also spectroscopy; resonance (2); ultraviolet absorber; excited state.

absorption oil. See absorption (1).

absorption spectroscopy. An important technique of instrumental analysis involving measurement of the absorption of radiant energy by a substance as a function of the energy incident upon it. Absorption processes occur throughout the electromagnetic spectrum, ranging from the gamma region (nuclear resonance absorption or the Mossbauer effect) to the radio region (nuclear magnetic resonance). In practice they are limited to those processes that are followed by the emission of radiant energy of greater intensity than that which was absorbed. All absorption processes involve absorption of a photon by the substance being analyzed. If it gives off the excess energy by emitting a photon of less energy than that absorbed, fluorescence or phosphorescence occurs, depending on the lifetime of the excited state.

The emitted energy is normally studied. If the source of radiant energy and the absorbing species are in identical energy states, i.e., in resonance, the excess energy is often given up by the nondirectional emission of a photon whose energy is identical with that absorbed.

Either absorption or emission may be studied, depending upon the chemical and instrumental circumstances. If the emitted energy is studied, the term "resonant fluorescence" is often used. However, if the absorbing species releases the excess energy in small steps by the process of inter-molecular collision or some other mode, it is commonly understood that this phenomenon falls within the realm of absorption spectroscopy. (The terms absorption spectroscopy, spectrophotometry, and absorptimetry are often used synonomously.)

Most absorption spectroscopy is done in the ultraviolet, visible, and infrared regions of the electromagnetic spectrum. See also emission spectroscopy; infrared spectroscopy.

ABS resin. Any of a group of tough, rigid thermoplastics deriving their name from the three letters of the monomers which produce them; *A*crylonitrile-*B*utadiene-*S*tyrene. Most contemporary ABS resins are true graft polymers consisting of an elastomeric polybutadiene or rubber phase, grafted with styrene and acrylonitrile monomers for compatibility, dispersed in a rigid styrene-acrylonitrile (SAN) matrix. Mechanical polyblends of elastomeric and rigid copolymers, e.g., butadiene-acrylonitrile rubber and SAN, historically the first ABS resins, are also marketed.

Varying the composition of the polymer by changing the ratios of the three monomers and use of other comonomers and additives results in ABS resins with a wide range of properties.

Properties: Dimensional stability over temperature range from –40 to +160°F. Attacked by nitric and sulfuric acids, and by aldehydes, ketones, esters, and chlorinated hydrocarbons. Insoluble in alcohols, aliphatic hydrocarbons, mineral and vegetable oils. Processed by conventional molding and extrusion methods. Sp. gr. 1.04; tensile strength about 6500 psi; flexural strength 10,000 psi; good electrical resistance; water absorption 0.3–0.4%. Nontoxic. Combustible, but slow-burning; flame retardants may be added. Can be vacuum-metallized or electroplated.

Grades: High-, medium-, and low-impact; molding and extrusion.

Uses: Automobile body parts and fittings; telephones; bottles; heels; luggage; packaging; refrigerator door liners; plastic pipe (subject to local building codes); building panels (ditto); shower stalls; boats; radiator grills; machinery housings; business machines.

Note: Several trademarked types are "Cycolac," "Abson," "Kralastic," "Lustran." For further information refer to Society of Plastics Industry, 250 Park Ave., New York.

abundance. The relative amount (% by weight) of a substance in the earth's crust, including the atmosphere and the oceans.

(a) The abundance of the elements in the earth's crust is:

Rank	Element	% by wt.
1	Oxygen	49.2
2	Silicon	25.7
3	Aluminum	7.5
4	Iron	4.7
5	Calcium	3.4
6	Sodium	2.6
7	Potassium	2.4
8	Magnesium	1.9
9	Hydrogen	0.9
10	Titanium	0.6
11	Chlorine	0.2
12	Phosphorus	0.1

Rank	Element	% by wt.
13	Manganese	0.1
14	Carbon	0.09
15	Sulfur	0.05
16	Barium	0.05
	all others	0.51

(b) The percentages of inorganic compounds in the earth's crust, exclusive of water, are:

(1) SiO_2 55	(2) Al_2O_3 15	(3) $CaCO_3$ 8.8
(4) MgO 1.6	(5) Na_2O 1.6	(6) K_2O 1.9

(c) The most abundant organic materials are cellulose and its derivatives, and proteins.
Note: On the universal scale, the most abundant element is hydrogen.

Ac Symbol for actinium; abbreviation of acetate.

AC. Abbreviation for allyl chloride.

acacia gum. See arabic gum.

acaricide. A type of pesticide effective on mites and ticks (acarides).

"Accel."[123] Trademark for a lactic acid starter culture for use in food processing.

accelerator. (1) A compound, usually organic, that greatly reduces the time required for vulcanization of natural and synthetic rubbers, at the same time improving the ageing and other physical properties. Organic accelerators invariably contain nitrogen, and in some cases both nitrogen and sulfur. The latter type are called ultra-accelerators because of their greater activity. The major types include amines, guanidines, thiazoles, thiuram sulfides, and dithiocarbamates. The amines and guanidines are basic, the others acidic. The introduction of organic accelerators in the early twenties was largely responsible for the successful development of automobile tires and mechanical products for engineering uses. A few inorganic accelerators are still used in low-grade products, e.g., lime, magnesium oxide, and lead oxide. See also vulcanization, rubber.

(2) A compound added to a photographic developer to increase its activity, such as certain quaternary ammonium compounds and alkaline substances.

(3) A particle accelerator (q.v.).

acceptor. See donor.

"Acele."[28] Trademark for a cellulose acetate fiber.
Properties: Sp. gr. 1.32; tensile strength (psi), 18,000–24,000; break elongation 28%; moisture regain 6%; soluble in glacial acetic acid, acetone, acetonitrile, butyrolactone, dimethyl formamide, dioxane-1,4. Combustible.
Containers: Tubes and cones in cases; beams.
Uses: Textiles.
See also acetate fiber.

acenaphthene (1,8-dihydroacenaphthalene; ethylenenaphthalene) $C_{10}H_6(CH_2)_2$, a tricyclic compound.
Properties: White needles; sp. gr. 1.024 (99/4°C); freezing point 93.6°C; b.p. 277.5°C; refractive index (100°C) 1.6048. Soluble in hot alcohol; insoluble in water. Combustible.
Derivation: From coal tar.
Grades: Technical; 98%.
Containers: Fiber drums.
Hazard: Irritating to eyes and skin.
Uses: Dye intermediate; pharmaceuticals; insecticide; fungicide; plastics.

acenaphthenequinone (1,2-acenapthenedione) $C_{10}H_6(CO)_2$, a tricyclic compound.
Properties: Yellow needles; m.p. 261–263°C; insoluble in water; soluble in alcohol.
Derivation: By oxidizing acenaphthene, using glacial acetic acid and sodium or potassium dichromate.
Grades: Technical.
Uses: Dye synthesis.

acenocoumarin (3-(alpha-acetonyl-4-nitrobenzyl)-4-hydroxycoumarin) $C_{19}H_{15}NO_6$.
Properties: White crystalline powder, tasteless and odorless; m.p. 197°C. Slightly soluble in water and organic solvents. Low toxicity.
Use: Medicine (anticoagulant).

acetal (diethylacetal; 1,1-diethoxyethane; ethylidenediethyl ether) $CH_3CH(OC_2H_5)_2$.
Properties: Colorless, volatile liquid; agreeable odor; nutty after-taste. Stable to alkalies but readily decomposed by dilute acids. Forms a constant-boiling mixture with ethyl alcohol. Soluble in alcohol, ether, and water. Sp. gr. 0.831; b.p. 103–104°C; vapor pressure 20.0 mm (20°C); flash point (closed cup) −5°F; specific heat 0.520; refractive index 1.38193 (20°C); wt (lb/gal) 6.89; autoignition temp. 446°F.
Derivation: Partial oxidation of ethyl alcohol, the acetaldehyde first formed condensing with the alcohol.
Grades: Technical.
Containers: Cans; carboys.
Hazard: Highly flammable. Dangerous fire risk. Explosive limits in air 1.65 to 10.4%. Moderately toxic and narcotic in high concentrations.
Uses: Solvent; cosmetics; organic synthesis; perfumes; flavors.
Shipping regulations. (Rail) Flammable liquid, n.o.s., Red label. (Air) Flammable Liquid label.
See also acetal resin.

acetaldehyde (acetic aldehyde; aldehyde; ethanal; ethyl aldehyde) CH_3CHO.
Properties: Colorless liquid; pungent, fruity odor. Sp. gr. 0.783 (18/4°C); b.p. 20.2°C; m.p. −123.5°C; vapor pressure 740.0 mm (20°C); flash point −40°F (open cup); specific heat 0.650; refractive index 1.3316 (20°C); wt. 6.50 lb/gal (20°C); miscible with water, alcohol, ether, benzene, gasoline, solvent naphtha, toluene, xylene, turpentine and acetone.
Derivation: (a) oxidation of ethylene; (b) vapor-phase oxidation of ethanol; (c) vapor-phase oxidation of propane and butane; (d) direct conversion of ethylene.
Grades: Technical 99%.
Containers: 5-, 10-, 55-, and 110-gal steel drums; tank cars.
Hazard: Highly flammable and toxic. Dangerous fire and explosion risk. Explosive limits in air 4 to 57%. Tolerance, 100 ppm in air. Safety data sheet available from Manufacturing Chemists Assn., Washington, D.C.
Uses: Manufacture of acetic acid and acetic anhydride, *n*-butanol, 2-ethylhexanol, peracetic acid, pentaerythritol, pyridines, chloral, 1,3-butylene glycol, and trimethylolpropane; also as intermediate.
Shipping regulations: (Rail) Red label. (Air) Flammable Liquid label.)

acetaldehyde ammonia. See aldehyde ammonia.

acetaldehyde cyanohydrin. See lactonitrile.

acetaldol. See aldol.

acetal resin (polyacetal). A polyoxymethylene thermoplastic polymer obtained by ionically initiated polymerization of formaldehyde (CH_2O) to obtain a linear molecule of the type -O-CH_2-O-CH_2-O-CH_2-. Single molecules may have over 1500 -CH_2O- units. As the molecule has no side chains, dense crystals are formed. Acetal resins are hard, rigid, strong, tough and resilient; dielectric constant 3.7; dielectric strength 1200 volts/mil (20-mil), 600 volts/mil (80-mil); dimensionally stable under exposure to moisture and heat, resistant to chemicals, solvents, flexing and creep, and have a high gloss and low friction surface. Can be chromium-plated, injection-molded, extruded, and blow-molded. Not recommended for use in strong acids or alkalies. They may be homopolymers or copolymers.
Properties: Sp. gr. 1.425; thermal conductivity 0.13 Btu/hr/sq ft/°F/ft; coefficient of thermal expansion 4.5×10^{-5} per °F; specific heat 0.35 Btu/lb/°F; water absorption 0.41%/24 hr; tensile strength 10,000 psi; elongation 15%; hardness (Rockwell) R120; impact strength (notched) 1.4 ft-lb/in.; flexural strength 14,100 psi; shear strength 9500 psi. Combustible, but slow burning.
Uses: An engineering plastic, often used as substitute for metals, as oil and gas pipes; automotive and appliance parts; industrial parts; hardware; communication equipment; aerosol containers for cosmetics.
See also "Delrin"; "Celcon."

acetamide (acetic acid amine, ethanamide) CH_3CONH_2.
Properties: Colorless deliquescent crystals. Mousy odor. Soluble in water and alcohol; slightly soluble in ether. Sp. gr. 1.159; m.p. 82°C; b.p. 223°C; refractive index 1.4274 (78.3°C). Low toxicity. Combustible.
Derivation: Interaction of ethyl acetate and ammonium hydroxide.
Method of purification: Crystallization.
Grades: Technical; C.P. (odorless); intermediate; reagent.
Containers: Fiber cartons.
Uses: Organic synthesis (reactant, solvent, peroxide stabilizer); general solvent; lacquers; explosives; soldering flux; hygroscopic agent; wetting agent; penetrating agent.

acetamido-. Prefix indicating the group CH_3CONH-. Also called acetamino- or acetylamino-.

5-acetamido-8-amino-2-naphthalenesulfonic acid (acetyl-1,4-naphthalenediamine-7-sulfonic acid; acetylamino-1,6-Cleve's acid)
$C_{10}H_5(NHCOCH_3)(NH_2)(SO_3H)$. A reddish brown paste.
Hazard: May be toxic.
Use: Chemical intermediate.

8-acetamido-5-amino-2-naphthalenesulfonic acid (acetyl-1,4-naphthalenediamine-6-sulfonic acid; acetylamino-1,7-Cleve's acid)
$C_{10}H_5(NHCOCH_3)(NH_2)(SO_3H)$. A paste.
Hazard: May be toxic.
Use: Chemical intermediate.

para-**acetamidobenzenesulfonyl chloride.** See N-acetylsulfanilyl chloride.

acetamidocyanoacetic ester. See ethyl acetamidocyanoacetate.

8-acetamido-2-naphthalenesulfonic acid magnesium salt (acetyl-1,7-Cleve's acid)
$[C_{10}H_6(CH_3CONH)(SO_3)]_2Mg$.
Properties: Brownish-gray paste containing approximately 80% solids.
Use: Intermediate.

para-**acetamidophenol.** See para-acetylaminophenol.

"Acetamine."[28] Trademark for a group of azo dyes and developers made for application to acetate yarn, and especially suited to nylon.

acetamino-. See acetamido-.

acetaminophen. See para-acetylaminophenol.

acetanilide (N-phenylacetamide) $C_6H_5NH(COCH_3)$.
Properties: White, shining crystalline leaflets or white, crystalline powder; odorless; stable in air; slightly burning taste; sp. gr. 1.2105; m.p. 114–116°C; b.p. 303.8°C; soluble in hot water, alcohol, ether, chloroform, acetone, glycerol and benzene. Flash point 345°F; combustible; autoignition temp. 1015°F.
Derivation: Acetylation of aniline with glacial acetic acid.
Grades: Technical; C.P.
Containers: 1-lb cartons; bottles; fiber drums; multiwall paper bags.
Hazard: Moderately toxic by ingestion.
Use: Rubber accelerator; inhibitor in hydrogen peroxide; stabilizer for cellulose ester coatings; manufacture of intermediates (para-nitroaniline, para-nitroacetanilide; para-phenylenediamine); synthetic camphor; pharmaceutical chemicals; dyestuffs; precursor in penicillin manufacture; medicine (antiseptic).

acetanisole. See para-methoxyacetophenone.

acetate. (1) A salt of acetic acid in which the terminal H atom is replaced by a metal, as in copper acetate, $Cu(CH_3COO)_2$.
(2) An ester of acetic acid where the substitution is by a radical as in ethyl acetate, $CH_3COOC_2H_5$. In cellulose acetate the hydroxyl radicals of the cellulose are involved in the esterification. See also cellulose acetate; vinyl acetate.

acetate dye. One group comprises water-insoluble azo or anthraquinone dyes that have been highly dispersed to make them capable of penetrating and dyeing acetate fibers. A second class consists of water-insoluble amino azo dyes that are made water-soluble by treatment with formaldehyde and bisulfite. After absorption by the fiber the resulting sulfonic acids hydrolyze and regenerate the insoluble dyes.

acetate fiber. Generic name for a manufactured fiber in which the fiber-forming substance is cellulose acetate. Where not less than 92% of the hydroxyl groups are acetylated, the term triacetate may be used as a generic description of the fiber (Federal Trade Commission). This fiber was formerly called "acetate rayon" or "acetate silk." The term "rayon" is no longer permissible for this type.
Properties: Thermoplastic; becomes tacky at 350°F. Soluble in acetone and glacial acetic acid; decomposed by concentrated solutions of strong acids and alkalies. Moisture absorption 6%. Tenacity (dry) 1.4 g/denier; (wet) about 1 g/denier. Elongation 50% dry, 40% wet. Combustible.
Uses: Wearing apparel; industrial fabrics.
See also "Acele" (acetate); "Arnel" (triacetate); and following entry.

acetate fiber, saponified. Regenerated cellulose fibers obtained by complete saponification of highly oriented cellulose acetate fibers. Available in continuous filament form having a high degree of crystallinity and great strength.

Properties: Tensile strength (psi) 136,000–155,000; elongation 6%; sp. gr. 1.5–1.6; moisture regain 9.6–10.7%; decomposes about 300°F. Similar to cotton in chemical resistance, dyeing, and resistance to insects and mildew. Combustible.
Uses: Cargo-parachutes; typewriter ribbons; belts; webbing; tapes; carpet backing.

acetate film. A durable, highly transparent film with nondeforming characteristics, produced from cellulose acetate resin. It is grease-, oil-, dust-, and airproof and hygienic. Combustible.
Available forms: Rolls and cut-to-size sheets.
Uses: Laminates; support for photographic film; document preservation; pressure-sensitive tape; magnetic sound recording tape; window cartons and envelopes; packaging.

acetate of lime. Commercial term for calcium acetate made from pyroligneous acid and milk of lime. There are brown and gray acetates of lime. For further data see calcium acetate.

acetate process. See cellulose acetate.

acethydrazidepyridinium chloride. See Girard's "P" reagent.

acetic acid (ethanoic acid, vinegar acid, methanecarboxylic acid) CH_3COOH. Glacial acetic acid is the pure compound (99.8% min.), as distinguished from the usual water solutions known as acetic acid. It is the end-product of oxidation of fermentation alcohols, as in vinegar (q.v.). 36th highest-volume chemical produced in U.S. (1975).
Properties: Clear, colorless liquid; pungent odor. M.p. 16.63°C; b.p. 118°C (765 mm), 80°C (202 mm); sp. gr. 1.0492 (20/4°C); wt/gal (20°C) 8.64 lb; viscosity (20°C) 1.22 cps; flash point (open cup) 110°F; refractive index 1.3715 (20°C). Miscible with water, alcohol, glycerin, and ether; insoluble in carbon disulfide; autoignition temp. 800°F. Combustible.
Derivation: (a) liquid- and vapor-phase oxidation of petroleum gases (with catalyst); (b) oxidation of acetaldehyde; (c) reaction of methanol and carbon monoxide; (d) directly from naphtha.
Grades: U.S.P. (glacial, 99.4 wt %, and dilute, 36–37 wt %), C.P.; technical (80; 99.5%); commercial (6, 28, 30, 36, 56, 60, 70, 80 and 99.5%); N.F. (diluted; 6.0 g/100 ml).
Containers: 5-lb bottles; 6-, 13-gal carboys; 8.5-, 13-, 55-, 100-gal barrels and drums; tank cars.
Hazard: Moderate fire risk. Pure acetic acid is toxic by ingestion and inhalation, but dilute material is approved by FDA for food use. Strong irritant to skin and tissue. Tolerance, 10 ppm in air. Safety data sheet available from Manufacturing Chemists Assn., Washington, D.C.
Uses: Manufacture of acetic anhydride, cellulose acetate, and vinyl acetate monomer; acetic esters; chloroacetic acid; production of plastics, pharmaceuticals, dyes, insecticides, photographic chemicals, etc.; food additive (acidulant); latex coagulant; oil-well acidizer; textile printing.
Shipping Regulations: Glacial or over 80% acid solution: (Rail) Corrosive liquid, n.o.s., White label. (Air) Corrosive label.
See also vinegar.

acetic acid amine. See acetamide.

acetic acid, glacial. See acetic acid.

acetic aldehyde. See acetaldehyde.

acetic anhydride (acetyl oxide; acetic oxide) $(CH_3CO)_2O$.
Properties: Colorless, mobile, strongly refractive liquid; strong odor; sp. gr. 1.0830 (20/20°C); b.p. 139.9°C; f.p. −73.1°C; flash point 121°F (C.C.). Autoignition temp. 732°F; wt/gal (20°C) 9.01 lbs. Miscible with alcohol, ether, and acetic acid; soluble in cold water; decomposes in hot water to form acetic acid. Combustible.
Derivation: (a) oxidation of acetaldehyde with air or oxygen with catalyst; (b) by catalyzed thermal decomposition of acetic acid to ketone.
Grades: C.P., technical (75, 85, 90–95%).
Containers: Bottles; carboys; aluminum drums; tank cars.
Hazards: Strongly irritating and corrosive; may cause burns and eye damage. Tolerance, 5 ppm in air. Moderate fire risk. Safety data sheet available from Manufacturing Chemists Assn., Washington, D.C.
Uses: Cellulose acetate fibers and plastics; vinyl acetate; dehydrating and acetylating agent in production of pharmaceuticals, dyes, perfumes, explosives, etc.; aspirin. Esterifying agent for food starch (5% max.).
Shipping Regulations: (Rail) Corrosive liquid, n.o.s., White label. (Air) Corrosive label.

acetic ester. See ethyl acetate.

acetic ether. See ethyl acetate.

acetic oxide. See acetic anhydride.

acetin (monoacetin; glyceryl monoacetate) $C_3H_5(OH)_2OOCCH_3$. Acetin may also refer to glyceryl di- or triacetate, also known as diacetin and triacetin (q.v.).
Properties: Colorless, thick liquid; hygroscopic; sp. gr. 1.206 (20/4°C); b.p. 158°C (165 mm); 130°C (3 mm); soluble in water, alcohol; slightly soluble in ether; insoluble in benzene; combustible.
Derivation: By heating glycerol and strong acetic acid, distilling off the weak acetic acid formed and again heating with strong acetic acid and distilling. Method of purification: Rectification.
Hazard: Moderately toxic; irritant.
Uses: Tanning; solvent for dyes; food additive; gelatinizing agent in explosives.

acetoacetanilide (acetylacetanilide) $CH_3COCH_2CONHC_6H_5$.
Properties: White, crystalline solid; m.p. 85°C. Resembles ethyl acetoacetate in chemical reactivity. Slightly soluble in water, soluble in dilute sodium hydroxide, alcohol, ether, acids, chloroform, and hot benzene; density 25 lb/cu ft. Flash point 325°F; combustible.
Derivation: By reacting ethyl acetoacetate with aniline, eliminating ethyl alcohol. Acetoacetanilide may also be prepared from aniline and diketene.
Grades: Technical.
Containers: Fiber drums.
Uses: Organic synthesis; dyestuffs (intermediate in the manufacture of the dry colors generally referred to as Hansa and benzidine yellows).

acetoacet-ortho-**anisidide** $CH_3COCH_2CONHC_6H_4OCH_3$.
Properties: White crystalline powder; m.p. 86.6°C; sp. gr. 1.1320 (86.6/20°C); flash point (open cup) 325°F. Combustible.
Containers: 1-gal cans; 5-, 55-gal drums.
Uses: Intermediate for azo pigments.

acetoacet-ortho-**chloranilide** $CH_3COCH_2CONHC_6H_4Cl$.
Properties: White crystalline powder; m.p. 107°C; b.p. decomposes; sp. gr. 1.1920 (107/20°C); flash point (open cup) 350°F. Almost insoluble in water. Combustible.
Containers: 1-gal cans; 5-, 55-gal drums.
Hazard: Moderately toxic, by ingestion.
Use: Intermediate for azo pigments.

acetoacet-para-**chloranilide** $CH_3COCH_2CONHC_6H_4Cl$.
Properties: White crystalline powder; m.p. 133°C; b.p. decomposes; flash point (open cup) 320°F. Combustible. Very slightly soluble in water.
Containers: 200-lb drums.
Hazard: Moderately toxic by ingestion.
Use: Intermediate for azo pigments.

acetoacetic acid (acetylacetic acid; diacetic acid; acetone carboxylic acid) CH_3COCH_2COOH.
Properties: Colorless oily liquid; soluble in water, alcohol, and ether; decomposes below 100°C into acetone and carbon dioxide.
Hazard: Irritant to eyes and skin.
Uses: Organic synthesis.

acetoacetic ester. See ethyl acetoacetate.

acetoacet-para-**phenetidide** $CH_3COCH_2CONHC_6H_4OCH_2CH_3$.
Properties: Crystalline powder; m.p. 108.5°C; b.p. decomposes; sp. gr. 1.0378 (108.5/20°C); flash point (open cup) 325°F. Combustible.
Containers: 1-gal cans; 5-, 55-gal drums.
Hazard: Moderately toxic by ingestion.
Use: Intermediate for azo pigments.

acetoacet-ortho-**toluidide** $CH_3COCH_2CONHC_6H_4CH_3$.
Properties: Fine, white granular powder; m.p. 106°C; b.p. decomposes; density 1.062 g/ml (106°C); slightly soluble in water. Flash point 320°F; combustible.
Grades: Technical.
Hazard: Moderately toxic.
Uses: Intermediate in the manufacture of Hansa and benzidine yellows.

acetoacet-para-**toluidide** $CH_3COCH_2CONHC_6H_4CH_3$.
Properties: White crystalline powder; m.p. 93.0–96.0°C; purity, 99% min.
Hazard: Moderately toxic.
Uses: Light-fast yellow pigment intermediate; diazo coupler.

acetoacet-meta-**xylidide** (AAMX) $(CH_3)_2C_6H_3NHCOCH_2COCH_3$.
Properties: White to light yellow crystalline solid; m.p. 89–90°C; sp. gr. (20°C) 1.238; solubility in water (25°C) 0.5%; flash point 340°F; combustible.
Use: Intermediate for yellow pigments.

acetoaminofluorene. A pesticide. May not be used in food products or beverages (FDA).
Hazard: Toxic by ingestion.

para-**acetoanisole.** See para-methoxyacetophenone.

acetoglyceride. Usually an acetylated monoglyceride although commercial acetoglycerides will contain di- and triglycerides. See acetostearin.

acetoin. See acetylmethylcarbinol.

acetol. See hydroxy-2-propanone.

acetomeroctol $CH_3COOHgC_6H_3(OH)C(CH_3)_2CH_2C(CH_3)_3$.
Properties: White solid; m.p. 155–157°C; freely soluble in alcohol; soluble in ether or chloroform; sparingly soluble in benzene; insoluble in water.
Hazard: Toxic by ingestion.
Use: Medicine (antiseptic, solution 1:1000)
Shipping regulations: (Rail, Air) Mercury compounds, solid, n.o.s., Poison label.

acetone (dimethylketone; 2-propanone) CH_3COCH_3. 40th highest-volume chemical produced in U.S. (1975).
Properties: Colorless volatile liquid; sweetish odor. M.p. –94.3°C; b.p. 56.2°C; refractive index (20°C) 1.3591; sp. gr. (20/20°C) 0.792; wt/gal (15°C) 6.64 lb; flash point (open cup) 15°F. Autoignition temp. 1000°F. Miscible with water, alcohol, ether, chloroform and most oils.
Derivation: (a) Oxidation of cumene (q.v.); (b) dehydrogenation or oxidation of isopropyl alcohol with metallic catalyst) (c) vapor-phase oxidation of butane; (d) by-product of synthetic glycerol production.
Grades: Technical; C.P.; N.F.; electronic; spectrophotometric.
Containers: Cans; drums; tank cars; tank trucks.
Hazard: Flammable; dangerous fire risk. Explosive limits in air 2.6 to 12.8%. Tolerance, 1000 ppm in air. Safety data sheet available from Manufacturing Chemists Assn., Washington, D.C.
Uses: Chemicals (methyl isobutyl ketone, methyl isobutyl carbinol; methyl methacrylate; bisphenol-A); paint, varnish and lacquer solvent; cellulose acetate, especially as spinning solvent; to clean and dry parts of precision equipment; solvent for potassium iodide and permanganate; delusterant for cellulose acetate fibers; specification testing of vulcanized rubber products.
Shipping regulations: (Rail) Red label. (Air) Flammable Liquid label.

acetone bromoform. See tribromo-*tert*-butyl alcohol.

acetonecarboxylic acid. See acetoacetic acid.

acetone chloroform. See chlorobutanol.

acetone cyanohydrin (alpha-hydroxyisobutyronitrile; 2-methyllactonitrile) $(CH_3)_2COHCN$.
Properties: Colorless liquid; b.p. 82°C (23 mm); m.p. –20°C; density 0.932 (19°C); refractive index n 20/D 1.3996; flash point 165°F; soluble in water, alcohol, and ether. Combustible; autoignition temp. 1270°F.
Derivation: Condensing acetone with hydrocyanic acid.
Grades: Technical (97–98% pure).
Containers: 6-gal carboys; 380-lb drums.
Hazard: Toxic. Readily decomposes to hydrocyanic acid and acetone.
Uses: Insecticides; intermediate for organic synthesis, especially methyl methacrylate.
Shipping regulations: (Rail, Air) Poison label.

acetonedicarboxylic acid. See beta-ketoglutaric acid.

acetone oxime. See acetoxime.

acetone semicarbazone $(CH_3)_2CNNHCONH_2$. White powder; m.p. 188°C. A chemical intermediate.

acetone sodium bisulfite. See sodium acetone bisulfite.

acetonitrile (methyl cyanide) $CH_3C \equiv N$.
Properties: Colorless, limpid liquid; aromatic odor; sp. gr. 0.783; m.p. –41°C; b.p. 82°C; flash point 42°F. Soluble in water and alcohol; high dielectric constant; high polarity; strongly reactive.
Derivation: By-product of propylene-ammonia process for acrylonitrile.
Grades: Technical; nanograde; spectrophotometric.
Containers: Drums; tank cars; tank trucks.
Hazard: Toxic. Flammable, dangerous fire risk. Tolerance, 40 ppm in air.

Uses: Solvent in hydrocarbon extraction processes, especially for butadiene; specialty solvent; intermediate; separation of fatty acids from vegetable oils; manufacture of synthetic pharmaceuticals.
Shipping regulations: (Rail) Red label. (Air) Flammable Liquid label.

acetonylacetone (1,2-diacetylethane; hexanedione-2,5; 2,5-diketohexane) $CH_3COCH_2CH_2COCH_3$.
Properties: Colorless liquid. Soluble in water; sp. gr. 0.9734 (20/20°C); b.p. 192.2°C; vapor pressure 0.43 mm at 20°C; freezing point -5.4°C; flash point 185°F; wt 8.1 lb/gal (20°C). Combustible; autoignition temp. 920°F.
Derivation: By-product in the production of acetaldehyde from acetylene.
Grades: Technical.
Containers: 1-gal cans; 5-, 55-gal drums.
Hazard: Irritant to eyes and skin.
Uses: Solvent for cellulose acetate, roll-coating inks, lacquers, stains; intermediate for pharmaceuticals and photographic chemicals; electroplating.

acetonyl alcohol. See hydroxy-2-propanone.

3-(alpha-**acetonylbenzyl)-4-hydroxycoumarin.** See warfarin.

3-(alpha-**acetonylfurfuryl)-4-hydroxycoumarin** (sodium salt also used). A rodenticide.
Hazard: Toxic by ingestion and inhalation.

3-(alpha-**acetonyl-4-nitrobenzyl)-4-hydroxycoumarin.** See acenocoumarin.

acetophenetidin (para-acetylphenetidin; acetophenetidide; phenacetin; para-ethoxyacetanilide) $CH_3CONHC_6H_4OC_2H_5$.
Properties: White crystals or powder; odorless and stable in air. Soluble in alcohol, chloroform and ether; slightly soluble in water; has slightly bitter taste. M.p. 135°C.
Derivation: By the interaction of para-phenetidin and glacial acetic acid, or of ethyl bromide and para-acetaminophenol.
Method of purification: Crystallization.
Grades: Technical; U.S.P., as phenacetin.
Containers: 100-, 200-, 1000-lb drums.
Hazard: Moderately toxic by ingestion.
Uses: Medicine (analgesic); veterinary medicine.

acetophenone (phenyl methyl ketone; hypnone; acetylbenzene) $C_6H_5COCH_3$.
Properties: Colorless liquid with sweet, pungent odor and taste. B.p. 201.7°C; f.p. 19.7°C; sp. gr. (20/20°C) 1.030; wt/gal (20°C) 8.56 lb; refractive index (20°C) 1.5363; flash point (COC) 180°F. Slightly soluble in water; soluble in organic solvents. Combustible. Low toxicity.
Derivation: (a) Friedel-Crafts process with benzene and acetic anhydride or acetyl chloride; (b) by-product from the oxidation of cumene (c) oxidation of ethylbenzene.
Method of purification: Distillation and crystallization.
Grades: Technical; refined; perfumery.
Containers: Glass bottles; cans; drums; tank cars; tank trucks.
Hazard: Narcotic in high concentrations.
Uses: Perfumery; solvent; intermediate for pharmaceuticals, resins, etc.; flavoring.

"Acetoquat."[400] Trademark for a series of quaternary ammonium salts.

"Aceto-Slip."[400] Trademark for a refined oleamide (q.v.).

acetostearin.
$CH_3(CH_3)_{16}COOCH_2CHOHCH_2OOCCH_3$. Acetylated glyceryl monostearate. Solid with peculiar combination of flexibility and nongreasiness. Derived from glyceryl monostearate or mixed glycerides by acetylation with acetic anhydride. Used as a protective coating for food and as a plasticizer. Nontoxic.

acetotoluidide. See acetyl-ortho- or -para-toluidine.

acetoxime (acetone oxime; 2-propanone oxime) $(CH_3)_2CNOH$.
Properties: Colorless crystals; both basic and acidic in properties. Volatilizes in air. Chloral-like odor. Sp. gr. 0.97 (20/20°C); b.p. 136.3°C; m.p. 61°C. Fairly readily hydrolyzed by dilute acids. Soluble in alcohols, ethers, and water. Combustible.
Grades: Technical.
Uses: Organic synthesis (intermediate); solvent for cellulose ethers; primer for diesel fuels.

ortho-**acetoxybenzoic acid.** See aspirin.

acetoxylation. A recently developed method of synthesizing ethylene glycol in which ethylene is reacted with acetic acid in the presence of a catalyst such as tellurium-bromine, resulting in formation of mixed mono- and diacetates; this is followed by hydrolysis to ethylene glycol and acetic acid, with up to 95% yield of the glycol. It is thus considerably more efficient than the ethylene oxide method.

4-(para-**acetoxyphenyl)-2-butanone.** See Q-lure.

acetozone. See acetylbenzoyl peroxide.

"Acetulan."[493] Trademark for a special fraction of acetylated lanolin alcohols.
Properties: Anhydrous, pale straw color, practically odorless, low-viscosity liquid. Miscible with ethanol, mineral oil, and other common formulating materials; insoluble in water. Combustible, nontoxic.
Uses: Hydrophobic penetrant; emollient; plasticizer; cosolvent; pigment dispersant; cosmetics.

aceturic acid (N-acetylglycine; acetylaminoacetic acid). $CH_3CONHCH_2COOH$.
Properties: Long needles; m.p. 206–208°C; soluble in water and alcohol; slightly soluble in acetone, chloroform; glacial acetic acid; practically insoluble in ether, benzene. Forms stable salts with organic bases.
Use: Medicine.

acetylacetanilide. See acetoacetanilide.

acetylacetic acid. See acetoacetic acid.

acetylacetone $CH_3COCH_2OCCH_3$ (diacetylmethane; pentanedione-2,4).
Properties: Mobile, colorless liquid. Unpleasant odor. When cooled, solidifies to lustrous, pearly spangles. The liquid is affected by light, turning brown and forming resins. B.p. 140.5°C sp. gr. (20/20°C) 0.9753; wt 8.1 lb/gal; freezing point -23.5°C; flash point 105°F (TOC). Soluble in water (acidified by hydrochloric acid); fairly soluble in neutral water; soluble in alcohol, chloroform, ether, benzene, acetone, and glacial acetic. Combustible.
Derivation: By condensing ethyl acetate with acetone.
Containers: 5-gal cans; 55-gal drums.
Hazard: Moderate fire risk; moderately toxic.

Uses: Solvent for cellulose acetate; intermediate; metal chelates; paint driers; lubricant additives; pesticides.

N-acetyl-L(+)-alanine. See alanine.

acetylamino-. See acetamido-.

acetylaminoacetic acid. See aceturic acid.

para-**acetylaminobenzenesulfonyl chloride.** See N-acetylsulfanilyl chloride.

ortho-**acetylaminobenzoic acid.** See acetylanthranilic acid.

para-**acetylaminophenol** (APAP; N-acetyl-para-aminophenol; acetaminophen; para-acetamidophenol; para-hydroxyacetanilide) $CH_3CONHC_6H_4OH$.
Properties: Crystals; odorless; slightly bitter taste; sp. gr. 1.293 (21/4°C); m.p. 168°C; slightly soluble in water and ether; soluble in alcohol; pH saturated aqueous solution 5.5–6.5
Derivation: Interaction of para-aminophenol and an aqueous solution of acetic anhydride.
Grade: N.F.
Containers: Drums.
Uses: Intermediate in making pharmaceuticals; stabilizer for hydrogen peroxide.

para-**acetylaminophenyl salicylate** (phenetsal) $C_6H_4(NHCOCH_3)OOCC_6H_4OH$.
Properties: Fine, white, crystalline scales; odorless, tasteless. Soluble in alcohol, ether, and hot water; insoluble in light hydrocarbon solvents; decomposed by strong alkalies. M.p. 187–188°C.
Derivation: By reducing para-nitrophenol salicylate to para-aminophenol salicylate and acetylating the latter.
Containers: Tins; glass bottles.
Hazard: See aspirin.
Use: Medicine (analgesic).

para-**acetylanisole.** See para-methoxyacetophenone.

N-**acetylanthranilic acid** (ortho-acetylaminobenzoic acid) $CH_3CONHC_6H_4COOH$.
Properties: Needles, plates, rhombic crystals (crystallized in glacial acetic acid); m.p. 185°C; slightly soluble in water; soluble in hot alcohol, ether and benzene. Combustible.
Derivation: Oxidation of ortho-acetyltoluidine with potassium permanganate in the presence of magnesium sulfate or potassium chloride.
Grades: Technical.
Uses: Chemical (organic synthesis, anthranilic acid).

acetylation. Introduction of an acetyl radical (CH_3CO-) into the molecule of an organic compound having OH or NH_2 groups. The usual reagents for this purpose are acetic anhydride or acetyl chloride. Thus ethyl alcohol C_2H_5OH may be converted to ethyl acetate ($C_2H_5OCOCH_3$). Cellulose is similarly converted to cellulose acetate by treatment with a mixture containing acetic anhydride. Acetylation is commonly used to determine the number of hydroxyl groups in fats and oils (see acetyl value).

acetylbenzene. See acetophenone.

acetyl benzoyl peroxide (acetozone; benzozone) $C_6H_5CO \cdot O_2 \cdot OCCH_3$.
Properties: White crystals; decomposed by water, alkaloids, organic matter and some organic solvents; decomposes slowly and evaporates when gently heated, and instantaneously (possibly explosively) if quickly heated, ground or compressed. M.p. 36.6°C; b.p. 130°C (19 mm); moderately soluble in ether, chloroform, carbon tetrachloride and water; slightly soluble in mineral oils and alcohol. The commercial product is mixed with a neutral drying powder and contains 50% acetyl benzoyl peroxide.
Hazard: Moderately toxic by ingestion. Strong irritant to skin and mucous membranes. Strong oxidizing agent; dangerous in contact with organic materials. Moderate explosion risk when shocked or heated.
Uses: Medicine (active germicide); disinfectant.
Shipping regulations: (Rail) Solid, not accepted; solution, Yellow label. (Air) Solid, or solutions containing more than 40% by weight of peroxide. Not acceptable. Solutions containing not more than 40% by weight of peroxide in a nonvolatile solvent, Organic Peroxide label. Not acceptable on passenger planes.

acetyl bromide CH_3COBr.
Properties: Colorless, fuming liquid; turns yellow in air; reacts violently with water or alcohol; soluble in ether, chloroform and benzene. B.p. 81°C; m.p. −96°C; sp. gr. 1.663 (16/4°C).
Derivation: Interaction of acetic acid and phosphorus pentabromide.
Grades: Technical.
Containers: Metal bottles; iron drums.
Hazard: Toxic by ingestion and inhalation; strong irritant to eyes and skin.
Uses: Organic synthesis; manufacture of dyes.
Shipping regulations: (Rail) Corrosive liquid, n.o.s., White label. (Air) Corrosive label.

N-**acetyl-N-bromodiethylacetyl urea.** See acetylcarbromal.

alpha-**acetylbutyrolactone** (gamma-lactone) $\overline{CH_2CH_2CH(OCCH_3)C(O)O}$.
Properties: Liquid with ester-like odor. Sp. gr. 1.18–1.19 (20°C); b.p. (30 mm) 142–144°C; soluble in water. Combustible.
Derivation: Sodium acetoacetate and ethylene oxide in absolute alcohol.
Use: Intermediate.

acetyl carbinol. See hydroxy-2-propanone.

acetylcarbromal (N-acetyl-N-bromodiethylacetylurea) $(C_2H_5)_2CBrCONHCONHCOCH_3$.
Properties: Crystals, slightly bitter taste; m.p. 109°C. Slightly soluble in water; freely soluble in alcohol, and ethyl acetate.
Hazard: Overdoses may be fatal.
Use: Medicine (sedative).

acetyl chloride (ethanoyl chloride) CH_3COCl.
Properties: Colorless, highly refractive, fuming liquid; strong odor; sp. gr. 1.1051; m.p. −112°C; b.p. 51–52°C; flash point 40°F (closed cup); soluble in ether, acetone, acetic acid.
Derivation: By mixing glacial acetic acid and phosphorus trichloride in the cold and heating a short time to drive off hydrochloric acid. The acetylchloride is then distilled.
Containers: Iron drums; 110-lb carboys. Protect from moisture.
Hazard: Toxic; strong irritant to eyes; flammable, dangerous fire risk. Reacts violently with water and alcohol.
Uses: Organic preparations (acetylating agent); dyestuffs; pharmaceuticals.
Shipping regulations: (Rail) White label. (Air) Corrosive label.

acetylcholine (acetylethanoltrimethylammonium hydroxide) $CH_3COOCH_2CH_2N(CH_3)_3OH$. A derivative

of choline important because it acts as the chemical transmitter of nerve impulses in the autonomic system. It has been isolated and identified in brain tissue. The enzyme cholinesterase (q.v.) hydrolyzes acetylcholine into choline and acetic acid, and is necessary in the body to prevent acetylcholine poisoning. See nerve gas.
Uses: Medicine (as bromide and chloride); biochemical research.
See also cholinesterase inhibitor.

acetylcholinesterase. See cholinesterase.

N-acetyl-L-cysteine (USAN)
$HSCH_2CH(NHCOCH_3)COOH$. The N-acetyl derivative of the naturally occurring amino acid, L-cysteine. Used in medicine.
Grade: N.D.

alpha-**acetyldigitoxin (anhydrous)** $C_{43}H_{66}O_{14}$.
Properties: Crystals; sparingly soluble in water, ether, petroleum ether; soluble in most organic solvents.
Derivation: Obtained by enzymatic hydrolysis of a digitalis extract.
Hazard: Moderately toxic.
Use: Medicine (heart disease).

acetylene (ethyne) $HC \equiv CH$.
Properties: Colorless gas; ethereal odor; sp. gr. 0.91 (air = 1); m.p. -81.8°C (890 mm); b.p. -84°C; soluble in alcohol, acetone; slightly soluble in water; flash point (closed cup) 0°F; autoignition temp. 635°F. An asphyxiant gas.
Derivation: (a) by the action of water on calcium carbide; (b) By cracking petroleum hydrocarbons with steam (Wulff process), or natural gas by partial oxidation (BASF process); (c) from fuel oil by modified arc process; (d) from coal by ARC-Coal process (experimental).
Grades: Technical, containing 98% acetylene and not more than 0.05% by volume of phosphine or hydrogen sulfide; 99.5%.
Containers: Steel cylinders; pipeline (see Hazard).
Hazard: Highly flammable; dangerous fire risk; burns with intensely hot flame. Explosive limits in air 2.5 to 80%. Forms explosive compounds with silver, mercury, and copper, which should be excluded from contact with acetylene in transmission systems. Copper alloys may be used with caution. Piping used should be electrically bonded and grounded. Safety data sheet available from Manufacturing Chemists Assn., Washington, D.C.
Uses: Vinyl chloride and vinylidene chloride; vinyl acetate; welding and cutting metals; acrylonitrile; acrylates; per- and trichloroethylene, cyclooctatetraene; 1,4-butanediol; carbon black.
Shipping regulations: (Rail) Red Gas label. (Air) Flammable Gas label. Not acceptable on passenger planes.

acetylene black. The carbon black resulting from incomplete combustion or thermal decomposition of acetylene. See also carbon black.
Properties: High liquid adsorption, retention of high bulk volume; high electrical conductivity.
Containers: Bags; cases.
Uses: Dry cell batteries; component of explosives; reinforcing agent in rubber and in thermal and sound insulation; gloss suppressor in paints; carburizing agent in hardening of steel; pigment in printing inks; filler in electroconductive polymers.

acetylene dichloride. See sym-dichloroethylene.

acetylene hydrocarbon (alkyne). One of a class of unsaturated hydrocarbons of the homologous series having the generic formula C_nH_{2n-2} and a structural formula containing a triple bond.

acetylene polymer. The best known polymer of acetylene is cyclooctatetraene, developed by Reppe in Germany during World War II. Another is made with a Ziegler catalyst. Both conjugated and unconjugated types have been synthesized, but no use for them on an industrial scale has yet been found.

acetylene tetrabromide (sym-tetrabromoethane) $CHBr_2CHBr_2$.
Properties: Yellowish liquid. Soluble in alcohol and ether; insoluble in water. Sp. gr. 2.98 to 3.00; b.p. 239–242°C with decomposition; 151°C (54 mm); m.p. 0.1°C; refractive index 1.638. Combustible.
Derivation: Interaction of acetylene and bromine, and subsequent distillation.
Method of purification: Rectification.
Grades: Technical.
Container: Steel drums.
Hazard: Toxic and irritant by inhalation. Tolerance, 1 ppm in air.
Uses: Separating minerals by specific gravity; solvent for fats, oils, and waxes; fluid in liquid gauges; solvent in microscopy.

acetylene tetrachloride. See sym-tetrachloroethane.

acetylenogen. See calcium carbide.

N-acetylethanolamine (hydroxyethylacetamide) $CH_3CONHC_2H_4OH$.
Properties: Brown viscous liquid, soluble in alcohol, ether, and water; sp. gr. 1.122 (20/20°C); boiling range 150–152°C (5 mm), decomposes (10 mm); lb/gal 9.34 (20°C); refractive index n 20/D 1.4730; flash point 350°F; f.p. 15.8°C. Combustible; low toxicity.
Grades: Technical (75% solution in water).
Uses: Plasticizer for polyvinyl alcohol and for cellulosic and proteinaceous materials; humectant for paper products, glues, cork and inks; high-boiling solvent for fountain-pen inks; textile conditioner.

acetylethanoltrimethylammonium hydroxide. See acetylcholine.

acetyleugenol. See eugenol acetate.

acetylferrocene (ferrocenyl methyl ketone) $C_5H_5FeC_5H_4COCH_3$. Orange crystalline solid; m.p. 85–86°C. Used as an intermediate. See also ferrocene.

acetylformic acid. See pyruvic acid.

N-acetylglycine. See aceturic acid.

acetylide. A salt-like carbide formed by the reaction of acetylene and an alkali or alkaline-earth metal in liquid ammonia, or with silver, copper and mercury salts in aqueous solution. The latter are explosive when shocked or heated. See also carbide; calcium carbide.

acetyl iodide CH_3COI.
Properties: Colorless, transparent, fuming liquid, turning brown on exposure to air or moisture; soluble in ether and benzene; decomposed by water and alcohol. Sp. gr. 1.98; b.p. 105–108°C.

Derivation: Interaction of acetic acid, iodine, and phosphorus.
Method of purification: Distillation.
Grades: Technical.
Containers: Glass bottles.
Hazard: Toxic; strong irritant to eyes and skin.
Uses: Organic synthesis.
Shipping regulations: (Rail) Corrosive liquid (n.o.s.). White label. (Air) Corrosive label.

acetylisoeugenol (isoeugenol acetate) $C_6H_3(CHCHCH_3)(OCH_3)(OCOCH_3)$.
Properties: White crystals; spicy clove-like odor; congealing point 77°C; soluble 1 part in 27 parts of 95% alcohol. Combustible. Low toxicity.
Method of purifiation: Crystallization.
Grades: Technical.
Containers: Tin cans.
Uses: Perfumery, particularly for carnation-type odors; flavoring.

acetyl ketene. See diketene.

acetylmethylcarbinol (acetoin; 3-hydroxy-2-butanone; dimethylketol) $CH_3COCHOHCH_3$.
Properties: Slightly yellow liquid or crystalline solid (dimer); oxidizes gradually to diacetyl on exposure to air; sp. gr. 1.016; b.p. 140–148°C; m.p. 15°C; soluble in alcohol; miscible with water in all proportions; slightly soluble in ether. Combustible. Low toxicity.
Derivation: Reduction of diacetyl.
Grades: Technical; F.C.C. (as acetoin).
Containers: Polyethylene-lined cartons.
Uses: Aroma carrier; preparation of flavors and essences.

acetyl oxide. See acetic anhydride.

acetyl peroxide (diacetyl peroxide) $(CH_3CO)_2O_2$.
Properties: Colorless crystals; m.p. 30°C; slightly soluble in cold water; soluble in alcohol and ether. Marketed as a 25% solution in dimethyl phthalate; flash point (open cup) 113°F; f.p. (approx.) –8°C; sp. gr. (20°C) 1.18. Combustible.
Containers: 65-lb carboys.
Hazard (25% solution): Moderate fire risk. Strong oxidizing agent. Moderately toxic and irritant. The pure (100%) material is a severe explosion hazard; should not be stored after preparation, nor heated above 30°C.
Uses: Initiator and catalyst for resins.
Shipping regulations: (Rail) (Solid) Not accepted. (Solution) Yellow label. (Air) Solid, or solutions containing more than 25% by weight of peroxide, Not acceptable. Solutions containing not more than 25% by weight of peroxide in a nonvolatile solvent, Organic Peroxide label. Not acceptable on passenger planes.

acetylphenetidin. See acetophenetidin.

acetylphenol. See phenyl acetate.

N-acetyl-para-**phenylenediamine.** See para-aminoacetanilide.

acetylpropionic acid. See levulinic acid.

acetyl propionyl (2,3-pentanedione; methyl ethyl glyoxal; methyl ethyl diketone) $CH_3COCOCH_2CH_3$.
Properties: Yellow liquid; m.p. –52°C; b.p. 106–110°C; sp. gr. (15/4°C) 0.955–0.959; partly soluble in water. Combustible. Nontoxic.
Grade: 99%.
Use: Flavors of butterscotch and chocolate type.

4-acetylresorcinol (2,4-dihydroxyacetophenone) $C_6H_3(OH)_2COCH_3$.
Properties: Light tan crystals; m.p. 146–148°C. High absorptivity in ultraviolet light region. Slightly soluble in water; soluble in most organic solvents except benzene and chloroform.
Uses: UV absorber in plastics; dye intermediate; fungicide; plant growth promoter.

acetylsalicylic acid. See aspirin.

N-acetylsulfanilyl chloride (para-acetamidobenzenesulfonyl chloride; para-acetylaminobenzenesulfonyl chloride) $(CH_3CONH)C_6H_4(SO_2Cl)$.
Properties: Light tan to brownish powder or fine crystals. M.p. 149°C; soluble in benzene, chloroform, and ether.
Containers: 150-lb steel drums.
Hazard: May be toxic.
Use: Intermediate in the manufacture of sulfa drugs.

2-acetylthiophene (methyl 2-thienyl ketone) $CH_3COC_4H_3S$.
Properties: Yellowish oily liquid; m.p. 10–11°C; b.p. 213.5°C; 88–90°C (10 mm); very slightly soluble in ether.
Use: Organic intermediate.

acetyl-ortho-**toluidine** (ortho-acetotoluidide) $CH_3CONHC_6H_4CH_3$.
Properties: Colorless crystals; m.p. 110°C; b.p. 296°C; sp. gr. 1.168 (15°C); soluble in alcohol, ether, benzene, chloroform, glacial acetic acid; slightly soluble in cold water; insoluble in hot water.
Derivation: By boiling glacial acetic acid with ortho-toluidine and distilling the product.
Grade: Technical.
Use: Organic synthesis.

acetyl-para-**toluidine** (para-acetotoluidide) $CH_3CONHC_6H_4CH_3$.
Properties: Colorless needles; m.p. 153°C; b.p. 307°C; density 1.212 (15/4°C); slightly soluble in water; soluble in alcohol, ether, ethyl acetate, glacial acetic acid.
Grades: Technical.
Containers: Wooden barrels or fiber drums.
Uses: Dyes.

acetyl triallyl citrate $CH_3COOC_3H_4(COOCH_2CH{:}CH_2)_3$.
Properties: Liquid; boiling range 142–143°C (0.2 mm); sp. gr. 1.140 (20°C); refractive index n 25/D 1.4665; flash point 365°F. Combustible.
Use: Cross-linking agent for polyesters; monomer for polymerization.

acetyl tributyl citrate $CH_3COOC_3H_4(COOC_4H_9)_3$.
Properties: Colorless, odorless liquid. Distillation range (1 mm) 172–174°C; pour point –75°F; sp. gr. (25°C) 1.046; wt/gal (25°C) 8.74 lb; refractive index (25°C) 1.4408; viscosity (25°C) 42.7 cps. Insoluble in water. Flash point 400°F. Combustible; low toxicity.
Derivation: Esterification and acetylation of citric acid.
Grade: Technical.
Containers: Metal drums; tanks.
Uses: Plasticizer for vinyl resins.

acetyl triethyl citrate $CH_3COOC_3H_4(COOC_2H_5)_3$.
Properties: Colorless, odorless liquid. Distillation range (1 mm) 131–132°C; pour point –45°F; sp. gr. (25°C) 1.135; wt/gal (25°C) 9.47 lb; refractive index (25°C) 1.4386; viscosity (25°C) 53.7 cps. Slightly

soluble in water. Flash point 370° F; combustible; low toxicity.
Derivation: Esterification and acetylation of citric acid.
Grades: Technical.
Containers: Metal drums; cans.
Uses: Plasticizer for cellulosics, particularly ethyl cellulose.

acetyl tri-2-ethylhexyl citrate $CH_3COOC_3H_4(COOC_8H_{17})_3$.
Properties: Liquid; b.p. 225°C (1 mm); insoluble in water. Flash point 430° F. Combustible; low toxicity.
Grades: Technical.
Use: Low-volatility plasticizer for vinyl resins.

N-acetyltryptophan. Available commercially as N-acetyl-L-tryptophan; m.p. 185–186° C; N-acetyl-DL-triptophan, m.p. 205°C.
Use: Nutrition and biochemical research; medicine.

acetyl valeryl (heptadione-2,3) $CH_3COCOC_4H_9$. Yellow liquid, 92% pure. Combustible. Nontoxic.
Uses: Cheese, butter, and miscellaneous flavors.

acetyl value. The number of milligrams of potassium hydroxide required for neutralization of acetic acid obtained by the saponification of one gram of acetylated fat or oil sample. Acetylation is carried out by boiling the sample with an equal amount of acetic anhydride, washing and drying. Saponification values on the acetylated and on untreated fat are determined. From the results the acetyl value is calculated. It is a measure of the number of free hydroxyl groups in the fat or oil.

ACGIH. Abbreviation for American Conference of Governmental Industrial Hygienists. See threshold limit values.

"Achromycin."[315] Trademark for tetracycline hydrochloride.

acicular. Needle-shaped; used in describing crystals or the particles in powders.

acid. One of a large class of chemical substances whose water solutions have one or more of the following properties: sour taste, ability to make litmus dye turn red and to cause other indicator dyes to change to characteristic colors, ability to react with and dissolve certain metals to form salts, and ability to react with bases or alkalies to form salts. All acids contain hydrogen. In water, ionization or splitting of the molecule occurs, so that some or most of this hydrogen forms H_3O^+ ions (hydronium ions), usually written more simply as hydrogen ions, H^+.

Acids are referred to as strong or weak according to the concentration of hydrogen ion that results from ionization. Hydrochloric acid (HCl), nitric acid (HNO_3), and sulfuric acid (H_2SO_4), are strong or highly ionized acids; acetic acid (CH_3COOH) and carbonic acid (H_2CO_3) are weak acids. Tenth normal HCl is 100 times as acid (pH = 1) as tenth normal acetic acid (pH = 3). pH range of acids is from 6.9 to 1. See also pH.

When dealing with chemical reactions in solvents other than water it is sometimes convenient to define an acid as a substance that ionizes to give the positive ion of the solvent. The common definitions of acid have been extended as more detailed studies of chemical reactions were made. The Lowry-Bronsted definition of an acid as a substance that can give up a proton is more useful in connection with an understanding of bases (see base). Perhaps the most significant contribution to the theory of acids was the electron-pair concept introduced by G. N. Lewis about 1915. See Lewis electron theory.

A brief outline of the major groups of acids is as follows:

Inorganic
Mineral acids: sulfuric, nitric, hydrochloric, phosphoric.
Hazard: All mineral acids are highly irritating and corrosive to human tissue.

Organic
Carboxylic (contain –COOH group)
aliphatic: acetic, formic
aromatic: benzoic, salicylic
Dicarboxylic (contain 2 –COOH groups)
oxalic, phthalic, sebacic, adipic
Fatty acids (contain –COOH group)
aliphatic: oleic, palmitic, stearic
aromatic: phenylstearic
Amino acids: nitrogen-containing protein components
For further details see Lewis acid, carboxylic acid, fatty acid, amino acid, and specific compounds.

1, 2, 4 acid. See 1-amino-2-naphthol--4-sulfonic acid.

acid amide. See amide.

acid ammonium tartrate. See ammonium bitartrate.

acid anhydride. An oxide of a nonmetallic element of an organic radical which is capable of forming an acid when united with water, or which can be formed by the abstraction of water from the acid molecule, or which can unite with basic oxides to form salts.

acid butyl phosphate. See *n*-butyl acid phosphate.

acid calcium phosphate. See calcium phosphate, monobasic.

acid dye. An azo, triarylmethane or anthraquinone dye with acid substituents such as nitro-, carboxy-, or sulfonic acid. They are most frequently applied in acid solution to wool and silk, and no doubt combine with the basic groups of the proteins of those animal fibers. Orange II (C.I. 15510), black 10B and acid alizarine blue B are examples.

acid ethylsulfate. See ethylsulfuric acid.

"Acidex."[205] Trademark for an epoxy coating.

acid fungal protease (fungal protease enzyme). Highly off-white powder. Used as replacement for pepsin; chill-proofing agent for beer; in cereal treatment; feed supplement for baby pigs; rennet extender.

acid glaucine blue. See peacock blue.

acid, hard. See Lewis electron theory.

acidic oxide. An oxide of a nonmetal, e.g., SO_2, CO_2, P_2O_5, SO_3, which forms acids when combined with water. See also acid anhydride.

acidimetry. The determination of the concentration of acid solutions or of the quantity of acid in a sample or mixture. This is usually done by titration with a solution of base of known strength (standard solution) and an indicator is used to establish the end point. See also pH.

acid lining. Silica brick lining used in steel-making furnaces.

acid magnesium citrate. See magnesium citrate, dibasic.

acid magnesium phosphate. See magnesium phosphate, monobasic.

acid methyl sulfate. See methylsulfuric acid.

acid mine drainage (AMD). Water from both active and inactive coal mines which has become contaminated with sulfuric acid as a result of hydrolysis of ferric sulfate, the oxidation product of pyrite. This is a factor in water and stream pollution, which can be corrected by use of appropriate ion-exchange resins. See also "Amberlite."

"Acidol."[162] Trademark for betaine hydrochloride. (q.v.)

"Acidolene."[244] Trademark for a series of sulfonated oils made from neatsfoot, sperm, cod, fish and coconut oil.
Uses: Leather industry, as fatliquor. Also used wherever an oil emulsifiable in water is needed for plasticizing or softening.

acid phosphatase. An enzyme found in blood serum which catalyzes the liberation of inorganic phosphate from phosphate esters. Optimum pH 5; is less active than alkaline phosphatase.
Uses: Biochemical research.

acid phosphate. An acid salt of phosphoric acid such as NaH_2PO_4, $CaHPO_4$, etc. Also used to refer specifically to calcium phosphate monobasic, $Ca(H_2PO_4)_2$, or superphosphate of lime.

acid potassium oxalate. See potassium binoxalate.

acid potassium sulfate. See potassium bisulfate.

acid, soft. See Lewis electron theory.

acidulant. Any of a number of acids (mostly organic) used in food processing as flavor intensifiers, preservatives, buffers, meat-curing agents, viscosity modifiers, etc. Those most commonly used are citric, fumaric, propionic, sorbic, lactic, succinic, adipic, malic, and tartaric acids.

"Acidulin."[100] Trademark for glutamic acid hydrochloride (q.v.).

acid value. The number of milligrams of potassium hydroxide neutralized by the free acids present in one gram of oil, fat, or wax. The determination is made by titrating the sample in hot 95% ethyl alcohol, using phenolphthalein as indicator.

Acid Yellow 9. See 4-aminoazobenzene-3,4′-disulfonic acid.

"Acintene."[252] Trademark for a group of terpene products from tall oil.

"Acintol."[252] Trademark for a series of tall oils, crude and distilled, and tall oil derivatives, such as fatty acids, rosins, tall oil heads, pitch. The derivatives are usually obtained by fractional distillation. Combustible.
Uses: Adhesives; cement; intermediates; degreasing compounds; emulsifiers; flotation agents; ink vehicles; leather chemicals; lubricants; metallic soaps; oil-well drilling muds; paints and varnishes; rubber chemicals; soaps and cleaners.

"Acitrol."[320] Trademark for a series of pickling inhibitors for acid baths.

"ACL."[55] Trademark for a series of solid, organic chlorine-liberating compounds; used in bleaches, cleansers, sanitizers, etc.

"Aclar."[50] Trademark for a series of fluorohalocarbon films. Retail useful properties from −320°F to +390°F. Used in packaging applications, where a transparent, vapor and/or gas barrier is required, as in packaging of foods for astronauts. Used in electronic and electrical applications because of insulating and heat-resistant properties. Extreme chemical resistance and ability to seal make it useful as a tank lining, etc.

"A-C-M."[299] Trademark for a balanced mixture of ascorbic acid (vitamin C) and citric acid, used as an antioxidant that protects flavor and prevents browning of fruits exposed to air. Used in home freezing and canning of fresh fruits.

"Acofor."[79] Trademark for distilled tall oil fatty acids.
Properties: Sp. gr. 0.907 (25/25°C); refractive index (20°C) 1.471; flash point (open cup) 380°F; acid number 192; saponification number 194; unsaponifiable matter 2.5%; rosin acids 4.5%. Combustible.
Uses: Paint and varnish; inks; soaps; disinfectants; textile oils; core oils, etc. See also "Aconew."

acoin. See di-para-anisyl-para-phenetylguanidine hydrochloride.

"Aconew."[79] Trademark for pale distilled tall oil fatty acids. See also "Acofor."

aconite (monkshood; wolfsbane; friar's cowl).
Derivation: Dried tuberous root or leaves of the perennial herbaceous plant Aconitum napellus.
Occurrence: Mountainous regions of Europe, Asia and North America.
Grades: Technical.
Containers: Burlap bags and bales.
Hazard: Toxic by ingestion and skin absorption.
Uses: Medicine and veterinary medicine.

aconitic acid (propene-1,2,3-tricarboxylic acid) $H(COOH)C{:}C(COOH)CH_2(COOH)$.
Properties: White to yellowish crystalline solid; m.p. (about) 195°C with decomposition; soluble in water and alcohol. Nontoxic. Combustible.
Derivation: (a) By dehydration of citric acid with sulfuric acid; (b) extraction from sugar cane bagasse, Aconitum napellus and other natural sources.
Uses: Preparation of plasticizers and wetting agents; antioxidant; organic syntheses; making itaconic acid.

"Acousti-Celotex."[351] Trademark for sound-absorbing tile or panels made from cellulose fibers or mineral wool fibers with an organic binder.

"A-C" Polyethylene.[175] Trademark for a series of low molecular weight polyethylenes. Available in both emulsifiable and nonemulsifiable grades.
Containers: 50-lb multiwall paper bags.
Uses: Coating containers and paper; printing inks, paints and paste polishes; rubber lubricants and mold release agents; emulsifiable grades in various liquid polishes, textile finishes, paper sizes.

acraldehyde. See acrolein.

"Acrawax."[73] Trademark for a series of synthetic waxes supplied in solid and powdered forms. Melting range from 83 to 143°C. Used as antitack agents; flatting agents in paint; lubricant and mold release agents for butyl and neoprene elastomers; adhesives; rubber; plastics.

acridine $C_{13}H_9N$ (tricyclic).
Properties: Small colorless needles. Soluble in alco-

hol, ether or carbon disulfide, sparingly soluble in hot water. Sublimes at 100°C; m.p. 111°C; b.p. above 360°C.

Derivation: (a) By extraction with dilute sulfuric acid from the anthracene fraction from coal tar and adding potassium dichromate. The acridine chromate precipitated is recrystallized, treated with ammonia and recrystallized. (b) Synthetically.

Hazard: Carcinogenic agent. Tolerance, 0.2 mg per cubic meter in air.

Uses: Manufacture of dyes; derivatives, especially acriflavine, proflavine; analytical reagent.

acriflavine $C_{14}H_{14}N_3Cl$. A mixture of 3,6-diamino-10-methylacridinium chloride and 3,6-diaminoacridine.

Properties: Brownish or orange, odorless, granular powder. Soluble in 3 parts of water; incompletely soluble in alcohol; nearly insoluble in ether and chloroform; the aqueous solutions fluoresce green on dilution. Also available as the hydrochloride.

Uses: Antiseptic and bacteriostat.

"Acrilan."[58] Trademark for a synthetic acrylic fiber.

Properties: Tenacity (wet and dry) 3 g/denier; softened by heat at 455°F; resistant to common solvents, mineral acids and weak alkalies. Can be satisfactorily wet-laundered. Ignites readily; not self-extinguishing.

Derivation: A solution of polymerized acrylonitrile is forced through minute holes of a spinneret, the solvent is removed, and the resulting fiber is stretched.

Uses: Woven and knitted clothing fabrics; carpets; drapes; upholstery; electrical insulation; laminates.

"Acrite" 100.[125] Trademark for a durable, non-nitrogenous textile reactant; a cross-linking agent for cellulose.

Containers: 55-gal drums.

Hazard: Strong irritant to eyes and skin.

"Acrival."[496] Trademark for an aqueous acrylic emulsion. Used as a fabric finish, hand modifier, anticrock agent for pigment prints, non-woven binder, and upholstery backing.

"Acriviolet."[243] Trademark for dye mixture used as oral antiseptic.

acroleic acid. See acrylic acid.

acrolein (2-propenal; acrylaldehyde; allyl aldehyde; acraldehyde) CH_2CHCHO.

Properties: Colorless or yellowish liquid; disagreeable choking odor. Soluble in water, alcohol and ether. Polymerizes readily unless inhibitor (hydroquinone) is added. Very reactive. B.p. 52.7°C; m.p. -87.0°C; sp. gr. (20/20°C) 0.8427; wt/gal (20°C) 7.03 lb; flash point (COC) below 0°F. Autoignition temp. 532°F.

Derivation: (a) Oxidation of allyl alcohol or propylene; (b) by heating glycerol with magnesium sulfate; (c) from propylene with bismuth-phosphorus-molybdenum catalyst.

Method of purification: Rectification.

Grades: Technical.

Containers: Up to tanks.

Hazard: Toxic by inhalation and ingestion; strong irritant to eyes and skin. Tolerance, 0.1 ppm in air. Flammable, dangerous fire risk. Explosive limits in air 2.8 to 31%. Safety data sheet available from Manufacturing Chemists Assn., Washington, D.C.

Uses: Intermediate for synthetic glycerol, polyurethane, and polyester resins, methionine, pharmaceuticals; herbicide; tear gas.

Shipping regulations: (inhibited) (Rail) Red label. (Air) Flammable Liquid label. Not acceptable on passenger planes. (uninhibited) Not acceptable.

acrolein dimer (2-formyl-3,4-dihydro-2H-pyran)

$\overline{OCH:CHCH_2CH_2C}HCHO$.

Properties: Liquid; sp. gr. 1.0775 (20°C); b.p. 151.3°C; freezing point -100°C; flash point (open cup) 118°F; wt/gal (20/20°C) 8.96 lb; soluble in water. Combustible.

Containers: 55-gal drums.

Hazard: Moderate fire risk.

Uses: Intermediate for resins, pharmaceuticals, dyestuffs.

"Acronal."[440] Trademark for dispersions, solutions, and solids of acrylate homo- and copolymers.

"Acrylafil."[539] Trademark for styrene-acrylonitrile polymer with glass fiber reinforcement. Available with 35 and 40% glass fiber content.

acrylaldehyde. See acrolein.

acrylamide $CH_2CHCONH_2$.

Properties: Colorless, odorless crystals; m.p. 84.5°C; b.p. (25 mm) 125°C; sp. gr. 1.122 (30°C); soluble in water, alcohol, acetone; insoluble in benzene, heptane. The solid is stable at room temperature but may polymerize violently on melting.

Derivation: Hydration of acrylonitrile with sulfuric acid (84.5%) and neutralization.

Grade: Technical (approximately 97% pure).

Containers: Fiber drums.

Hazard: Toxic by skin absorption. Tolerance, 0.3 mg per cubic meter of air.

Uses: Synthesis of dyes, etc.; cross-linking agent; adhesives, paper and textile sizes, soil conditioning agents; flocculants; sewage and waste treatment; ore processing; permanent press fabrics.

acrylate. (1) Any of several monomers used for the manufacture of thermosetting acrylic surface coating resins, e.g., 2-hydroxyethyl acrylate (HEA) and hydroxypropyl acrylate (HPA).

(2) Polymer of acrylic acid or its esters, used in surface coatings, emulsion paints, paper and leather finishes, etc.

See also acrylic acid; acrylic resin.

"Acrylene."[265] Trademark for a modified acrylic latex particularly useful as an exterior paint vehicle.

acrylic acid (acroleic acid; propenoic acid)

$$H_2C{=}\overset{\displaystyle H}{\overset{|}{C}}{-}\overset{\displaystyle O}{\overset{\|}{C}}{-}OH$$

Properties: Colorless liquid; acrid odor. Polymerizes readily. Miscible with water, alcohol and ether. B.p. 140.9°C; m.p. 12.1°C; sp. gr. (20/20°C) 1.052; vapor pressure (20°C) 3.1 mm; wt/gal (20°C) 8.6 lb; refractive index (20°C) 1.4224; flash point (open cup) 130°F. Combustible.

Derivation: (a) Condensation of ethylene oxide with hydrocyanic acid followed by reaction with sulfuric acid at 320°F; (b) acetylene, carbon monoxide and water, with nickel catalyst; (c) propylene is vapor-oxidized to acrolein, which is oxidized to acrylic acid at 300°C with molybdenum-vanadium catalyst.

Grades: Technical (esterification and polymerization grades); glacial (97%).

Containers: Bottles; drums; tank cars.

Hazard: Irritant and corrosive to skin. May polymerize with explosive violence. Moderate fire risk.
Use: Monomer for polyacrylic and polymethacrylic acids and other acrylic polymers. See acrylic resin.
Shipping regulations: (Rail) Corrosive liquid, n.o.s., White label. (Air) (inhibited) Corrosive label; (uninhibited) Not acceptable.

acrylic fiber. Generic name for a manufactured fiber in which the fiber-forming substance is any long-chain synthetic polymer composed of at least 85% by weight of acrylonitrile units $—CH_2CH(CN)—$ (Federal Trade Commission).
Properties: Tensile strength, 2 to 3 g/denier; water absorption 1.5 to 2.5%; sp. gr. about 1.17.
Hazard: Readily combustible; fumes may be toxic.
Uses: Modacrylic fibers; blankets; carpets.
See also modacrylic fiber; acrylic resin.

acrylic resin. Thermoplastic polymers or copolymers of acrylic acid, methacrylic acid, esters of these acids, or acrylonitrile. The monomers are colorless liquids that polymerize readily in the presence of light, heat, or catalysts such as benzoyl peroxide; they must be stored or shipped with inhibitors present to avoid spontaneous and explosive polymerization. See also acrylic acid, acrylonitrile, methyl methacrylate; acrylic fiber, nitrile rubber.
Properties: Acrylic resins vary from hard, brittle solids to fibrous, elastomeric structures to viscous liquids, depending on the monomer used and the method of polymerization. A distinctive property of cast sheet and extruded rods of acrylic resin is ability to transfer light through the solid material as exemplified in "Lucite" and "Plexiglas."
Uses: Bulk-polymerized: Hard, shatterproof transparent or colored material (glass substitute, decorative illuminated signs, contact lenses, dentures, medical instruments, specimen preservation, furniture components). Suspension-polymerized: beads and molding powders (headlight lenses, adsorbents in chromatography, ion-exchange resins). Solution polymers: coatings for paper, textiles, wood, etc. Aqueous emulsions: adhesives, laminated structures, fabric coatings, nonwoven fabrics. Compounded prepolymers: exterior auto paints, applied by spray and baked. Acrylonitrile-derived acrylics are extruded into synthetic fibers and are also the basis of the nitrile family of synthetic elastomers.

"Acryloid" Coating Resins.[23] Trademark for acrylic ester polymers in organic solvent solutions or 100% solids form, water-white and transparent. Films range from very hard to very soft.
Use: Exceptionally resistant surface coatings, such as heat-resistant and fumeproof enamels, vinyl and plastic printing, fluorescent coatings, clear and pigmented coatings on metals.

"Acryloid" Modifiers.[23] Trademark for thermoplastic acrylic polymers in powder form. Various grades facilitate processing or improve physical properties of rigid or semi-rigid poly (vinyl chloride) formulations.

"Acryloid" Oil Additives.[23] Trademark for acrylic polymers supplied in special oil solution or in diester lubricant.
Use: Viscosity-index improvement, pour-point depression of lubricating oils and hydraulic fluids, sludge dispersancy in lubricating and fuel oils

"Acrylon."[65] Trademark for a group of acrylic rubbers outstanding in resistance to oil, grease, ozone, and oxidation.
Uses: Gaskets and rubber parts for contact with oils and diester lubricants.

acrylonitrile (propenenitrile; vinyl cyanide) $H_2C{:}CHCN$.
Properties: Colorless, mobile liquid; mild odor; freezing range -83° to -84°C; b.p. 77.3–77.4°C; sp. gr. 0.8004 (25°C); flash point (TOC) 32°F. Soluble in all common organic solvents; partially miscible with water.
Derivation: (a) From propylene, oxygen, and ammonia with either bismuth phosphomolybdate or a uranium-based compound as catalyst; (b) addition of hydrogen cyanide to acetylene with cuprous chloride catalyst; (c) dehydration of ethylene cyanohydrin.
Containers: Tank cars; tank trucks.
Hazard: Toxic by inhalation and skin absorption. Flammable, dangerous fire risk. Explosive limits in air 3 to 17%. Tolerance, 20 ppm in air. Safety data sheet available from Manufacturing Chemists Assn., Washington, D.C.
Uses: Monomer for acrylic and modacrylic fibers and high-strength whiskers; ABS and acrylonitrile-styrene copolymers; nitrile rubber; cyanoethylation of cotton; synthetic soil blocks (acrylonitrile polymerized in wood pulp); organic synthesis; grain fumigant; monomer for a semiconductive polymer that can be used like inorganic oxide catalysts in dehydrogenation of *tert*-butyl alcohol to isobutylene and water; bottles for soft drinks.
Shipping regulations: (Rail) Red label. (Air) Flammable Liquid label.

acrylonitrile-butadiene rubber. See nitrile rubber.

acrylonitrile-butadiene-styrene resin. See ABS resin.

acrylonitrile dimer. See methylene glutaronitrile.

acrylonitrile-styrene copolymer. A thermoplastic blend of acrylonitrile and styrene monomers having good dimensional stability and suitable for use in contact with foods. Among its numerous applications is that of bottles for soft drinks (trademarked "Cyclesafe"). FDA regulations limit the amount of acrylonitrile monomer that can migrate from the container to the contents to 0.3 ppm.

acryloyl chloride (acrylyl chloride) $H_2CCHCOCl$. Liquid; b.p. 75°C.
Containers: Glass containers, up to 55 gal.
Uses: Monomer; intermediate.

"Acrysol."[23] (1) Trademark for aqueous solutions of sodium polyacrylate or other polymeric acrylic salts.
Use: Thickeners in paints, fabric coatings and backings; adhesives.
(2) Trademark for polyacrylic acid and copolymer products in aqueous solutions or dispersions. Some grades are solutions of sodium polyacrylate.
Use: Warp size for synthetic fibers, cotton and rayon; modifier of starch sizes.

A.C.S. Abbreviation for American Chemical Society (q.v.)

"Actafoam."[511] Trademark for an activator-stabilizer for vinyl foams containing azodicarbonamide. Lead-free.

"Actane."[142] Trademark for fluoride-containing additive supplied in powder form for acid pickling solutions to dissolve siliceous films on metals as well as to assist in etching aluminum and other metals.

ACTH (adrenocorticotropic hormone; corticotropin). One of the hormones secreted by the anterior lobe of the pituitary gland. It stimulates an increase in the secretion of the adrenal cortical steroid hormones. It is a polypeptide consisting of a 39-unit chain of amino acids, the sequence varying in certain positions with the biological species. ACTH was synthesized in 1960.

Properties: White powder; freely soluble in water; soluble in 60–70% alcohol or acetone. Solutions are stable to heat. Molecular weight is approximately 3500.
Source: Extracted from whole pituitary glands of swine, sheep and ox. Normally isolated from swine.
Grades: Pure; U.S.P., as corticotropin injections.
Units: Based on comparison with U.S.P. Corticotropin Reference Standard.
Hazard: May have damaging side effects.
Uses: Medicine; biochemical research; veterinary medicine.

"Acti-dione."[519] Trademark for antibiotic cycloheximide, an agricultural fungicide.

"Actidip."[204] Trademark for a mildly alkaline powder having the property of refining crystal size in phosphate baths when used just prior to the phosphate. Commercially available in 100-lb fiber drums.

"actif • 8."[1] Trademark for a phosphate leavening system consisting of sodium aluminum phosphate and anhydrous monocalcium phosphate. (F.C.C.)

actinide series (actinoid series). The group of radioactive elements starting with actinium (q.v.) and ending with Element 105. All are classed as metals. Those with atomic number higher than 92 are called transuranic elements. The series includes the following elements: actinium, 89; thorium, 90; protoactinium, 91; uranium, 92; neptunium, 93; plutonium, 94; americium, 95; curium, 96; berkelium, 97; californium, 98; einsteinium, 99; fermium, 100; mendelevium, 101; nobelium, 102; lawrencium, 103; rutherfordium, 104; and hahnium, 105. The isotopes of several of these elements are under study for possible uses in such fields as radiography, neutron activation analysis, hydrology, and geophysics.

actinium Ac A radioactive metallic element; first member of the actinide series (q.v.)
Properties: Atomic number 89; atomic weight 227 (most stable isotope); silvery white metal; m.p. 1050°C; b.p. (est) 3200°C; oxidation state +3. Eleven radioactive isotopes; 227 has half-life of 21.8 years.
Derivation: Uranium ores; neutron bombardment of radium. Several compounds have been prepared. Available commercially at 98% minimum purity.
Hazard: Radioactive poison, radiation being of the bone-seeking type.
Uses: Radioactive tracer (225 isotope).
See also lanthanum, to which actinium is closely similar.

actinomycin. A family of antibiotics produced by Streptomyces; reported to be active againse E. coli, other bacteria, fungi and to have cytostatic and radiomimetic activity. There are many forms of actinomycin; two of commercial importance are cactinomycin and dactinomycin.

activated alumina. See alumina, activated.

activated carbon. See carbon, activated.

activated sludge. See sewage sludge.

activation. The process of treating a substance or a molecule or atom by heat or radiation or the presence of another substance so that the first mentioned substance, atom or molecule will undergo chemical or physical change more rapidly or completely. Common types of activation are:

(1) Processing of carbon black, alumina and other materials to impart improved adsorbent qualities. Subjecting the material to steam or carbon dioxide at high temperatures is the method usually used. See alumina, activated; carbon, activated.

(2) Heating or otherwise supplying energy to a substance (for example, ultraviolet or infrared radiation) to attain the necessary level of energy for the occurrence of a chemical reation, or for emission of desired light waves, as in fluorescence or chemical lasers. The term excitation is also used.

(3) An important variation of (2) is the process of making a material radioactive by bombardment with neutrons, protons, or other nuclear particles. See also activation analysis.

(4) Catalysis (q.v.), basically a process by which energy of activation for occurrence of a reaction is lowered by the presence of a non-reacting substance.

activation analysis. An extremely sensitive technique for identifying and measuring very small amounts of various elements. A sample is exposed to neutron bombardment in a nuclear reactor, for the purpose of producing radioisotopes from the stable elements. The characteristics of the induced radiations are sufficiently distinct that different elements in the sample can be accurately identified. The technique is particularly useful when concentrations of the elements are too small to be measured by ordinary means. Trace elements have thus been determined in drugs, fertilizers, foods, fuels, glass, minerals, dusts, water, toxicants, etc.

activator. (1) A metallic oxide that makes possible the cross-linking of sulfur in rubber vulcanization. By far the most widely used is zinc oxide; in rubber mixes where no organic accelerator is used, oxides of magnesium, calcium or lead are effective.

(2) A fatty acid that increases the effectiveness of acidic organic accelerators; stearic acid is generally used, especially with thiazoles.

(3) A substance necessary in trace quantities to induce luminescence in certain crystals. Silver and copper are activators for zinc sulfide and cadmium sulfide.

See also initiator.

active amyl alcohol. See 2-methyl-1-butanol.

active carbon. See carbon, activated.

activity. (1) Chemical activity (thermodynamic activity): A quantity replacing actual molar concentration in mathematical expressions for the equilibrium constant so as to eliminate the effect of concentration on equilibrium constant. (2) Activity coefficient: A fractional number which when multiplied by the molar concentration of a substance in solution, yields the chemical activity. This term gives an idea of how much interaction exists between molecules at higher concentrations. (3) Activity of metals or elements: An active element will react with a compound of a less active element, to produce the latter as the free element, and the active element ends up in a new compound. Thus, magnesium, an active metal, will displace copper from copper sulfate to form magnesium sulfate and free metallic copper; chlorine will liberate iodine from sodium iodide, and sodium chloride is formed. See electromotive series. (4) Activity product: The number resulting from the multiplication of the activities of slightly soluble substances. This is frequently called the solubility product. (5) Catalytic activity: See catalysis. (6) Optical activity: The existence of optical rotation (q.v.) in a substance. (7) Radioactivity (q.v.).

activity coefficient. See activity (2).

activity series (displacement series; electromotive series). An arrangement of the metals in the order of their tendency to react with water and acids, so that each metal displaces from solution those below it in the series and is displaced by those above it. The arrangement of the more common metals is: potassium, sodium, magnesium, aluminum, zinc, iron, tin, lead, hydrogen, copper, mercury, silver, platinum, gold.

"Acto."[51] Trademark for refined petroleum sodium sulfonate. Used as an oil-soluble emulsifier and surface-active agent.

"Actol."[243] Trademark for a series of polyoxypropylene diols, triols, and polyols. These vary in molecular weight from about 1,000 to 3,600; the diols and triols are almost insoluble in water, but the polyols are completely miscible with it.
Containers: 55-gal drums.
Uses: Urethane foams, elastomers, and coatings.

acyl. An organic acid group in which the OH of the carboxyl group is replaced by some other substituent (RCO—). Examples: acetyl, CH_3CO—; benzoyl, C_6H_5CO—.

ADA. Abbreviation for acetonedicarboxylic acid. See beta-ketoglutaric acid.

"Adalin."[162] Trademark for carbromal or bromodiethylacetylurea (q.v.).

1-adamantanamine hydrochloride. See amantadine hydrochloride.

adamantane (diamantane; sym-tricyclodecane) $C_{10}H_{16}$. Has unique molecular structure consisting of four fused cyclohexane rings. White crystals; m.p. 205–210°C (subl); about 99% pure. Derivatives recently developed (alkyl adamantanes) have potential uses in imparting heat-, solvent-, and chemical resistance to many basic types of plastics. Synthetic lubricants and pharmaceuticals are also based on adamantane derivatives. See also "Symmetrel."

Adams, Roger (1889–1971). American chemist, born in Boston; graduated from Harvard, where he taught chemistry for some years. After studying in Germany he moved to the University of Illinois in 1916, where he later became Chairman of the Dept. of Chemistry (1926–1954). During his prolific career he made this department one of the best in the country, and strongly influenced the development of industrial chemical research in the U.S. His executive and creative ability made him an outstanding figure as a teacher, innovator and administrator. Among his research contributions were development of platinum hydrogenation catalysts, and structural determinations of chaulmoogric acid, gossypol, alkaloids, and marijuana. He held many important offices, including presidency of the ACS and AAAS, and was a recipient of the Priestley medal.

adamsite. See diphenylamine chloroarsine.

addition polymer. A polymer formed by the direct addition or combination of the monomer molecules with one another. An example is the formation of polystyrene by stepwise combination of styrene monomer units (about 1000 per macromolecule). See also polymerization.

additive. This general term is conventionally defined as any substance incorporated into a base material, usually in low concentrations, to perform a specific function, e.g., antioxidants, stabilizers, colorants, inhibitors, preservatives, thickeners, softeners, driers, dispersing agents, plasticizers, viscosity-index improvers, pour-point depressants, biodegrading agents, and numerous others. They are used in many fields (petroleum products, foods, paints, plastics, rubber, textiles, and cosmetics).

Note: This definition does not make a necessary distinction between ingredients that are essential to the existence of the end product and those that are merely helpful and thus of secondary importance. It is questionable whether ingredients without which the end product could not exist should be considered additives. For example, leavening agents are essential to the existence of bread and other bakery products, as are blowing agents to cellular plastics; emulsifers are essential to mayonnaise; and sulfur or its equivalent is essential to vulcanized rubber. These are in fact base materials rather than additives, since they are essential to a satisfactory end product.
See also food additive.

additive, unintentional. See food additive.

adduct. See inclusion complex.

adenine (6-aminopurine) $C_5H_5N_5$.
Properties: White, odorless microcrystalline powder with sharp salty taste. M.p. 360–5°C (dec.). Very slightly soluble in cold water; soluble in boiling water, acids and alkalies; slightly soluble in alcohol; insoluble in ether and chloroform. Aqueous solutions are neutral.
Occurrence: Ribonucleic acids and deoxyribonucleic acids, nucleosides, nucleotides, and many important coenzymes.
Derivation: By extraction from tea; by synthesis from uric acid; prepared from yeast ribonucleic acid.
Use: Medicine and biochemical research.

adenosine (adenine riboside; 9-beta-D-ribofuranosyladenine) $C_5H_4N_5{\cdot}C_5H_9O_4$. The nucleoside composed of adenine and ribose.
Properties: White, crystalline, odorless powder with mild, saline or bitter taste. M.p. 229°C. Quite soluble in hot water; practically insoluble in alcohol.
Derivation: Isolation following hydrolysis of yeast nucleic acid.
Use: Biochemical research.

adenosine diphosphate (5′-adenylphosphoric acid; ADP; adenosine 5′-pyrophosphate; adenosine diphosphoric acid) $C_{10}H_{15}N_5O_{10}P_2$. A nucleotide found in all living cells and important in the storage of energy for chemical reactions.
Derivation: (a) From adenosine triphosphate by hydrolysis with the enzyme adenosinetriphosphatase from lobster or rabbit muscle; (b) by yeast phosphorylation of adenosine.
Use: Biochemical research.
Commercially available as the sodium or barium salt.

adenosinediphosphoric acid. See adenosine diphosphate.

adenosine monophosphate. See adenylic acid.

adenosine phosphate. USAN for 5′-adenylic acid.

adenosine triphosphate (5′-adenyldiphosphoric acid; ATP) $C_{10}H_{16}N_5O_{13}P_3$. A nucleotide which serves as a source of energy for biochemical transformations in plants (photosynthesis) and also for many chemical reactions in the body, especially those associated with muscular activity, and with replication of cell components.
Properties: White amorphous powder; odorless; faint

sour taste. Soluble in water; insoluble in alcohol, ether, and organic solvents; stable in acidic solutions; decomposes in alkaline solution.
Derivation: Isolation from muscle tissue; yeast phosphorylation of adenosine.
Use: Biochemical research.
Commercially available as the disodium, dipotassium, and dibarium salts.

adenylic acid (adenosine monophosphate; AA; adenosine phosphate; adenosinephosphoric acid; AMP) $C_{10}H_{14}N_5O_7P$. The monophosphoric ester of adenosine; i.e., the nucleotide containing adenine, D-ribose and phosphoric acid. Adenylic acid is a constituent of many important coenzymes. Cyclic adenosine-3′,5′,-monophosphate is designated by biochemists as cAMP (q.v.).
Properties (muscle adenylic acid): Crystalline solid; m.p. 196–200°C. Readily soluble in boiling water. Gives only traces of furfural when boiled with 20% hydrochloric acid.
(yeast adenylic acid monohydrate): Long crystalline rods. Decomposes 195°C. Anhydrous form decomposes at 208°C. Almost insoluble in cold water; slightly soluble in boiling water. Gives quantitative yield of furfural when distilled with 20% HCl.
Derivation: Yeast adenylic acid by precipitation from yeast nucleic acid. Muscle adenylic acid by precipitation from tissues; by hydrolysis of ATP with barium hydroxide; by enzymatic phosphorylation of adenosine.
Uses: Medicine and biochemical research.

adhesion. The state in which two surfaces are held together by interfacial forces, which may consist of valence forces or interlocking action, or both. (ASTM)

adhesive. Any substance, inorganic or organic, natural or synthetic, that is capable of bonding other substances together by surface attachment. A brief classification by type is as follows:

I. Inorganic
  1. Soluble silicates (water glass)
  2. Phosphate cements
  3. Portland cement (calcium oxide-silica)
  4. Other hydraulic cements (mortar, gypsum)
  5. Ceramic (silica-boric acid)
  6. Thermosetting powdered glasses ("Pyroceram")

II. Organic
  1. Natural
    (a) Animal
      Hide and bone glue; fish glue
      Blood and casein glues
    (b) Vegetable
      Soybean, starch, cellulosics, rubber latex and rubber-solvent (pressure-sensitive).
      Gums, terpene resins (rosin), mucilages
    (c) Mineral
      Asphalt, pitches, hydrocarbon resins
  2. Synthetic
    (a) Elastomer-solvent cements
    (b) Polysulfide sealants
    (c) Thermoplastic resins (for hot-melts)
      Polyethylene, isobutylene, polyamides, polyvinyl acetate
    (d) Thermosetting resins
      Epoxy, phenolformaldehyde, polyvinyl butyral, cyanoacrylates.
    (e) Silicone polymers and cements

See also following entries. For further information refer to Adhesives Manufacturers Association, 441 Lexington Ave., New York.

adhesive, high-temperature. (1) Organic polymers, e.g., polybenzimidazoles, that retain bonding strength up to 500°F for a relatively long time (500–1000 hours); above 500°F, strength drops rapidly, 80% being lost after 10 minutes at 1000°F.
(2) Inorganic (ceramic), e.g., silica-boric acid mixtures or cermets produce bonds having high strength above 2000°F; adhesive lap-bond strengths can be over 2000 psi at 1000°F. These adhesives are used largely for aerospace service, and metal/metal and glass/metal seals.

adhesive, hot-melt. A solid, thermoplastic material which quickly melts upon heating, and then sets to a firm bond on cooling. Most other types of adhesives set by evaporation of solvent. Hot-melt types offer the possibility of almost instantaneous bonding, making them well-suited to automated operation. In general, they are low-cost, low-strength products, but are entirely adequate for bonding cellulosic materials. Ingredients of hot-melts are polyethylene, polyvinyl acetate, polyamides, hydrocarbon resins, as well as natural asphalts, bitumens, resinous materials, and waxes.
Uses: Rapid and efficient bonding of low-strength materials, e.g., bookbinding, food cartons, side-seaming of cans, miscellaneous packaging applications.
See also sealant.

adhesive, rubber-based (cement, rubber). (1) A solution of natural or synthetic rubber in a suitable organic solvent, without sulfur or other curing agent; (2) a mixture of rubber (often reclaimed), filler, and tackifier (pine tar, liquid asphalt) applied to fabric backing (pressure-sensitive friction tape); (3) a room-temperature curing rubber-solvent-curative mixture, often made up in two parts, which are blended just before use; (4) rubber latex, especially for on-the-job repairing, such as conveyor belts; (5) silicone rubber cement (see "RTV" and silicone (uses).
Hazard: Those containing organic solvents, (1) and (3) above, are flammable.
Shipping Regulations: Cement, liquid, n.o.s., (Rail) Red label. (Air) Flammable Liquid label.

adiabatic. A process, condition, or operation during which there is no gain or loss of heat from the environment.

adipic acid (hexanedioic acid; 1,4-butanedicarboxylic acid) $COOH(CH_2)_4COOH$.
Properties: White, crystalline solid. M.p. 152°C; b.p. (100 mm) 265°C; sp. gr. (20/4°C) 1.360; flash point (closed cup) 385°F. Slightly soluble in water; soluble in alcohol and acetone. Relatively stable. Combustible; low toxicity.
Derivation: Oxidation of cyclohexane, cyclohexanol, or cyclohexanone with air or nitric acid.
Grades: Technical; F.C.C.
Containers: Glass bottles; tins; 50-lb multiwall paper bags; drums.
Uses: Manufacture of nylon and of polyurethane foams; preparation of esters for use as plasticizers and lubricants; food additive (neutralizer and flavoring agent); adhesives.

"Adipol."[55] Trademark for a series of adipate plasticizers.

adiponitrile (1,4-dicyanobutane) $NC(CH_2)_4CN$.

Properties: Water-white, odorless liquid; m.p. 1–3°C; n (20/D) 1.4369; b.p. 295°C; flash point 200°F (open cup); combustible. Slightly soluble in water; soluble in alcohol and chloroform.

Derivation: Chlorination of butadiene to dichlorobutylene, which is reacted with 35% sodium cyanide solution to yield 1,4-dicyanobutylene; this is hydrogenated to adiponitrile. Also by electroorganic synthesis from acrylonitrile.

Hazard: Toxic by ingestion.

Uses: Intermediate in the manufacture of nylon; organic synthesis.

"Adiprene."[28] Trademark for a polyurethane rubber, the reaction product of diisocyanate and polyalkylene ether glycol. In its raw polymer form it is a liquid of honey-like color and consistency which is compounded chemically (to polymerize it further) and converted into products by casting and other techniques. See also polyurethane rubber.

"Admerol."[221] Trademark for a long oil, oxidizing copolymer for zinc- and lead-free stains and blister-resistant paints; binder for aluminum paints and varnish oils.

"Admex."[221] Trademark for a series of plasticizers consisting variously of epoxidized soybean oil, tallate esters, monomeric esters and polyesters. Used in vinyl plastics.

admiralty metal. A nonferrous alloy containing 70–73% copper, 0.75–1.20% tin, remainder zinc. It offers good resistance to dilute acids and alkalies, sea water, and moist sulfurous atmospheres. Sp. gr. (20°C) 8.53; liquidus temperature 935°C; solidus temperature 900°C.

Uses: Condenser, evaporator, and heat exchanger tubes, plates, and ferrules.

"Adofoam."[544] Trademark for oil-field additive used specifically in air-drilling and hydraulic fracturing to produce stable foam in fresh water, salt water, acid/water solutions, sulfur/water, and oil/water mixtures.

"Adogen."[221] Trademark for a series of fatty nitrogen chemicals including amines, amides, amine acetates and quaternary ammonium compounds. Available in various grades for specific applications in fabric softeners, ore separation, detergents, petroleum additives, corrosion inhibitors, bactericides, printing inks, antiblock and slip agents, water-proofing formulations and chemical intermediates.

"Adol."[221] Trademark for a series of industrial fatty alcohols, specifically cetyl, stearyl, and oleyl, available in a variety of grades for specific applications.

"Adomall."[544] Trademark for multi-value water-fracturing additive for oil-field operations; kills bacteria, inhibits corrosion, lowers surface tension, reduces permeability damage, removes drilling mud from bore and fracture areas, and produces stable foam to return fracturing water from wells.

"Adomite."[544] Trademark for oil-field additives to reduce fluid loss during hydraulic fracturing. "Adomite Mark II" functions in oil-based fracturing fluids; "Adomite Aqua" in water-based fracture fluids.

"Adoquat."[544] Trademark for a quaternary ammonium salt used in waterflooding operations for secondary recovery of petroleum. It increases efficiency by inhibiting bacterial growth and reducing microbial plugging.

"Adoxal."[227] Trademark for 2,6,10-trimethyl-9-undecen-1-al. (q.v.)

ADP. Abbreviation for (1) adenosine diphosphate (q.v.); (2) ammonium dihydrogen phosphate. See ammonium phosphate, monobasic.

"Adrenalin."[330] Trademark for epinephrine (q.v.).

adrenaline. A hormone having a benzenoid structure with the formula $C_9H_{13}O_3N$ (also called epinephrine). It is obtained by extraction from the adrenal glands of cattle, and is also made synthetically. Its effect on body metabolism is pronounced, causing an increase in blood pressure and rate of heart beat. Under normal conditions its rate of release into the system is constant, but emotional stresses such as fear or anger rapidly increase the output and result in temporarily heightened metabolic activity.

adrenocorticotropic hormone. See ACTH; corticoid hormone.

adriamycin (doxorubicin). An antibiotic drug reported to be effective against such types of cancer as leukemia and cancers of the breast and bladder. It is made by fermentation of a soil fungus. Approved by FDA for clinical research, but is said to have deleterious side effects. Synthetic routes to adriamycin and its analogs have been developed.

adsorbent. A substance which has the ability to condense or hold molecules of other substances on its surface. Activated carbon, activated alumina, and silica gel are examples.

"Adsorbosil."[425] Trademark for adsorbents for use in chromatography. "Adsorbosil"-1 is a mixture of specially purified silica gel and 10% calcium sulfate designed as a thin-layer chromatography powder; "Adsorbosil" CAB is designed for column chromatography.

adsorption. Adherence of the atoms, ions or molecules of a gas or liquid to the surface of another substance, called the adsorbent (q.v.) The best-known examples are gas/solid and liquid/solid systems. Finely divided or microporous materials presenting a large area of active surface are strong adsorbents, and are used for removing colors, odors, and water vapor (activated carbon, activated alumina, silica gel). The attractive force of adsorption is relatively small, of the order of van der Waal's forces (q.v.). When molecules of two or more substances are present, those of one substance may be adsorbed more readily than those of the others. This is called preferential adsorption.

See also absorption, chemisorption.

adsorption indicator. A substance used in analytical chemistry to detect the presence of a slight excess of another substance or ion in solution as the result of a color produced by adsorption of the indicator on a precipitate present in the solution. Thus a precipitate of silver chloride will turn red in a solution containing even a minute excess of silver ion (silver nitrate solution), if fluorescein is present. In this example, fluorescein is the adsorption indicator.

"Advabrite."[230] Trademark for a series of optical brighteners giving a bright, blue-white fluorescence in very dilute solutions either in daylight or ultraviolet light.

"Advacar."[230] Trademark for a series of water-dispersible paint driers. Supplied in high metal concentrations (calcium 4%, cobalt 6%, lead 24%, manganese 6%).

"Advacide."[230] Trademark for a series of fungicides and mildewcides for paints, etc., wood and fabric preservatives. "Advacide" TPLA is a free-flowing powder of triphenyl lead acetate containing 10% of a liquid aromatic hydrocarbon mixture, and used in antifouling paints.

"Advalite."[230] Trademark for a series of organic and organometallic compounds, specifically designed for the stabilization of vinyl coatings systems against heat and light degradation.

"Advawax."[230] Trademark for a series of wax modifiers.
P. High molecular weight polybutene and paraffin wax.
M. High molecular weight polybutene and amber-colored microcrystalline wax.
280. A hard, high-melting point synthetic wax for use in lacquers, varnishes, as a plastic lubricant and anti-blocking agent, as an extender or substitute for carnauba wax.

"Advawet."[230] Trademark for a series of wetting agents for emulsion paints, latex paints and vinyl plastisols.

AEC. Abbreviation of Atomic Energy Commission, now replaced by Energy Research and Development Agency (q.v.).

AEH. Abbreviation for anhydroenneaheptitol.

AEPD. Abbreviation for 2-amino-2-ethyl-1,3-propanediol.

aerate. To impregnate or saturate a material (usually a liquid) with air, or some similar gas. This is usually achieved by bubbling the air through the liquid, or by spraying the liquid into air.

aerobic. Requiring air or oxygen. See bacteria.

"Aerocarb."[57] Trademark for mixtures containing salts of sodium, potassium, and barium used as a molten bath for case-hardening and heat treatment of steel.
Containers: Steel drums.
Hazard: Toxic by ingestion.

"Aerocat."[57] Trademark for a series of synthetic silica-alumina catalysts used for refining of petroleum in fluid catalytic cracking units.

"Aero" Cyanamid.[57] Trademark for an agricultural chemical.
Properties: Gray to black dry material of various finenesses.
Containers: 50-and 100-lb bags.
Uses: For direct application to soil; formulating complete fertilizers and special uses such as leaf removal. vine killing, wood control and disease control.

"Aerofloat."[57] Trademark for aryl dithiophosphoric acids (liquids) and alkyl dithiophosphoric acid salts (solids) used as mining flotation reagents to promote frothing properties.

"Aerofloc"[57] Trademark for a group of synthetic water-soluble polymers used as flocculating agents to improve solid-liquid separations by thickening and filtration.

"Aerofroth" Frothers.[57] Trademark for a group of surface-active agents used primarily as foaming agents or frothers in flotation processing of ores and minerals.

aerogel. Dispersion of a gas in a solid or a liquid. The reverse of an aerosol; flexible and rigid plastic foams are examples. See foam; aerosol.

"Aeroheat" Heat Treating Salts.[57] Trademark for various mixtures containing alkaline chlorides, carbonates, nitrates and nitrites. Used in the form of molten baths in the hardening, quenching, annealing and tempering of ferrous and non-ferrous metals.
See also salt bath.

"Aeromet."[57] Trademark for metallurgical additive in steel making where nitrogen content is desired; also applied to iron and steel for desulfurization.
Containers: Bulk freight cars, moisture-resistant paper bags in fiber drums.

"Aeromine"[57] Trademark for a group of cationic flotation reagents used in froth flotation processing of ores and minerals, primarily silica and silicates.

"Aeroprills."[57] Trademark for an ammonium nitrate fertilizer containing 33.5% nitrogen (16.75% nitrate nitrogen; 16.75% ammonia nitrogen).
Properties: Light cream-colored pellets.
Containers: 80 or 100 lb bags.
Hazard: See ammonium nitrate.
Uses: For direct application and in fertilizer mixtures.

"Aerosil."[562] Trademark for submicroscopic pyrogenic silica made at 1100° C; high surface area.
Uses: Reinforcing and dispersing agent in rubber and plastics; thickening agent in lacquers; anticaking agent.

"Aerosize"[57] Trademark for a series of aqueous high free-rosin emulsions used for treatment of paper and paper board to impart water, ink, and lactic acid resistance. May be used to give a surface sizing effect as on floor coverings.

aerosol. A suspension of liquid or solid particles in a gas, the particles often being in the colloidal size range. Fog and smoke are common examples of natural aerosols; fine sprays (perfumes, insecticides, inhalants, anti-perspirants, paints, etc.) are man-made aerosols. Suspensions of various kinds may be formed by placing the components, together with a compressed gas, in a container (bomb). The pressure of the gas causes the mixture to be released as a fine spray (aerosol) or foam (aerogel) when a valve is opened. This technique is used on an industrial scale to spray paints and pesticides. It is also used in consumer items such as perfumes, deodorants, shaving cream, whipped cream, and the like. The propellant gas may be a hydrocarbon (propane, isobutane), a chlorofluorocarbon, carbon dioxide, or nitrous oxide. About half of the propellants are chlorofluorocarbons, 45% are hydrocarbons, the other 5% being nitrous oxide and carbon dioxide. Most paint sprays utilize hydrocarbons, which have an excellent safety record in spite of their flammability, and are not environmentally damaging. See also chlorofluorocarbon.
Note: The use of chlorofluorocarbons for aerosol propellants is under extensive investigation because of their possible deleterious effect on atmospheric ozone; as a result such propellants are being discontinued by some aerosol manufacturers until positive determinations are completed. See also propellant, and ozone (note).

"Aerosol" Wetting Agent.[57] Trademark for neutral wetting agents obtainable in various types. They do not decompose on heating and are soluble in practically all nonaqueous media as well as in water.
Uses: Antiseptics; cosmetics; detergents; inks; cleaning, degreasing metals; case-hardening processes for

metals; paints, varnishes, and lacquers; paper; shampoos; toilet preparations; acid treatment of oil wells; embalming fluids; settling operations; breaking mineral oil emulsions; shoe cleaners and polishes.

"Aerosporin."[301] Trademark for polymyxin B sulfate.

"Aerotex"[57] Trademark for a group of synthetic resins used in finishing textile fabrics.

"Aerotex" Water Repellent S.[57] Trademark for a silicone emulsion for producing water repellency on textile fabrics.

"Aerothene."[233] Trademark for a group of chlorinated solvents used as vapor-pressure depressants, or with compressed gases to replace fluorocarbon propellant systems.

aerozine. A 50-50 mixture of hydrazine and unsdimethylhydrazine (UDMH); one of most used of bipropellant rocket fuels.
Hazard: Flammable and explosive. See also hydrazine.

AET. See aminoethylisothiourea dihydrobromide.

"Afaxin."[162] Trademark for synthetic oleo-vitamin A.

affinity. The tendency of an atom or compound to react or combine with atoms or compounds of different chemical constitution. For example, paraffin hydrocarbons were so named because they are quite unreactive, the word "paraffin" meaning "very little affinity." The hemoglobin molecule has a much greater affinity for carbon monoxide than for oxygen. The free energy decrease is a quantitative measure of chemical affinity.

aflatoxin. $C_{17}H_{10}O_6$. A polynuclear substance derived from molds; a known carcinogen (q.v.). Produced by a fungus occurring on many vegetables, especially those with high moisture content. It is highly toxic.

after-chromed dye. A dye that is improved in color or fastness by treatment with sodium dichromate, copper sulfate or similar materials, after the fabrics are dyed.

Ag Symbol for silver.

agar (agar-agar). A phycocolloid derived from red algae such as Gelidium and Gracilaria; it is a polysaccharide mixture of agarose and agaropectin.
Properties: Thin, translucent, membranous pieces, or pale buff powder. Strongly hydrophilic, it absorbs 20 times its weight of cold water with swelling; forms strong gels at about 40°C. Nontoxic.
Grades: Technical; U.S.P.; F.C.C.
Containers: Bales (strip); fiber drums (powder).
Uses: Microbiology and bacteriology (culture medium); antistaling agent in bakery products; confectionery, meats, and poultry; gelation agent; desserts and beverages; protective colloid in ice cream; pet foods; health foods; laxative; pharmaceuticals; dental impressions; laboratory reagent; photographic emulsions.
See also algae; alginic acid.

age-resister. See antioxidant.

"Agerite."[119] Trademark for a series of antioxidants for rubber.
Alba, para-Benzyloxyphenol.
DPPD. N,N'-Diphenyl-para-phenylenediamine.
Gel. A mixture of alkylated diphenylamines with selected petroleum wax.
Geltrol. Phosphited polyalkyl polyphenol.
Hipar. Mixture of phenyl-beta-naphthylamine, isopropoxydiphenylamine, and diphenyl-para-phenylenediamine.
HP. A blend of phenyl-beta-naphthylamine and N,N'-diphenyl-para-phenylenediamine.
Iso. para-Isopropoxydiphenylamine.
Powder. Phenyl-beta-naphthylamine.
Resin. Aldol-alpha-naphthylamine.
Resin D. A polymerized 1,2-dihydro-2,2,4-trimethylquinoline.
Spar. Mixed mono-, di-, and tristyrenated phenols.
Stalite and Stalite S. Mixture of alkylated diphenylamines.
Superflex. Diphenylamine-acetone reaction product.
Superlite. A mixture of polybutylated bisphenol A.
White. syn-Di-beta-naphthyl-para-phenylenediamine.

agglomeration. See aggregation.

agglutination. The combination or aggregation of particles of matter under the influence of a specific protein. The term is usually restricted to antigen-antibody reactions characterized by a clumping together of visible cells, such as bacteria or erythrocytes. The antigen is called an agglutinogen and the antibody an agglutinin, because of an apparent gluing or sticking action. See also aggregation.

aggregation. A general term describing the tendency of large molecules or colloidal particles to combine (coalesce) in clumps, especially in solution. When this occurs, usually as a result of removal of the electric charges by addition of an appropriate electrolyte, by the action of heat, or by mechanical agitation, the aggregates precipitate or separate from the dissolved state. Included in this term are the more specific terms agglutination, coagulation, flocculation, and peptization. See also these entries.

aglucone. The nonsugar-like portion of a glucoside molecule. See glycoside.

aglycone. A nonsugar hydrolytic product of a glycoside. See glycoside.

agricultural chemical (agrichemical). A chemical compound or mixture used to increase the productivity and quality of farm crops. They include fertilizers, soil conditioners, fungicides, insecticides, herbicides, nematocides, and plant hormones. For further information, refer to National Agricultural Chemicals Association, Washington, D.C.

A.I.Ch.E. Abbreviation for American Institute of Chemical Engineers (q.v.).

"Al Polymer."[216] Trademark for an aromatic polymer that can be cured at moderate temperatures to produce a poly(amide-imide). It is a single polymer system based on trimellitic anhydride (q.v.). Supplied as a completely solid resin.
Uses: As a complete 220° insulation system in wire enamels, dipping varnishes, impregnated cloth, glass and asbestos laminates, film and sleeving.

air. A mixture (or solution) of gases, the composition of which varies with altitude and other conditions at the collection point. The composition of dry air at sea level is:

Substance	% by Wt.	% by Vol.
Nitrogen	75.53	78.00
Oxygen	23.16	20.95
Argon		0.93
Carbon dioxide		0.03
Neon		0.0018
Helium		0.0005
Methane		0.0002
Krypton		0.0001
Nitrous oxide		0.000,05

Substance	% by Vol.
Hydrogen	0.000,05
Xenon	0.000,008
Ozone	0.000,001

The density of dry air is 1.29 g/liter (at 0°C and 760 mm Hg). It is noncombustible, but will support combustion.

Liquid air is air which has been subjected to a series of compression, expansion and cooling operations until it liquefies.

Containers: Steel cylinders or tanks, under high pressure.

Uses: Source of oxygen, nitrogen, and rare gases; coolant; power source (compressed); cryogenic agent (liquid); particle classification; blowing agent (asphalt, soap, ice-cream mixes, whipped cream, etc.); floatation.

Shipping regulations: Air, compressed: (Rail) Green label. (Air) Nonflammable compressed gas label. Air, liquid: Not acceptable on passenger planes; (nonpressured) Not acceptable.

See also air pollution.

air classification. The separation of solid particles according to weight and/or size, by suspension in and settling from an air stream of appropriate velocity, as in air-floated clays and other particulate products.

air floatation. See air classification.

"Air Floate."[177] Trademark for a technical grade of barium carbonate with improved flow properties.

Hazard: Toxic by ingestion.

Use: Ceramic industry.

"Air-Flo Green."[147] Trademark for a mosquito larvicide containing copper meta-arsenite.

Hazard: Highly toxic by ingestion.

"Airocel PK Foam Liquid."[270] Trademark for a protein-based liquid concentrate used to produce lightweight concrete, low density refractories, foamed ceramics and other rigid inorganic foams. Also used as an air entraining agent to facilitate the pumping of concrete through pipe and flexible hose.

air pollution (atmospheric pollution). Introduction into the atmosphere of substances that are not normally present therein, and that have a harmful effect on man, animals, or plant life. Important among these are sulfur dioxide, which forms sulfuric acid on contact with water vapor; automotive exhaust emission products (carbon monoxide, lead compounds, polynuclear hydrocarbons, nitrogen oxides); toxic metal dusts from smelters, coal smoke, and other particulates; formaldehyde and acrolein; and radioactive emanations. Control of these is exercised by the Environmental Protection Agency. As conventionally used, the term does not apply to interior air spaces such as industrial workrooms. Tolerances for the latter are established by the American Conference of Governmental and Industrial Hygienists (ACGIH). See threshold limit values.

Note: EPA requirements for automotive exhaust emission standards have been postponed until 1980, since it has been found that the present catalytic converters cause formation of sulfuric acid and sulfates from the sulfur in the gasoline, and are also inefficient in other respects.

"Akineton."[9] Trademark for biperiden, employed in medicine as the hydrochloride and lactate salts.

"Akon."[116] Trademark for a colloidal iron product used for removing oxygen from boiler feed water to reduce corrosion.

"Akroflex" C.[28] Trademark for a rubber antioxidant containing 35% diphenyl-para-phenylenediamine $C_6H_4(NHC_6H_5)_2$, and 65% phenyl-alpha-naphthylamine ("Neozone" A).

Uses: To improve the aging and service life of rubbers; anticrosslinking agent for SBR (styrene-butadiene rubber).

"Aktoflo-S"[236] Trademark for a mixed oxyethylated phenol derivative with an optimum proportion of defoamant added. Added to emulsified-surfactant drilling muds, to reduce clay dispersion, and viscosity.

"Aktone."[285] Trademark for a rubber accelerator; density 1.39. Used as a secondary accelerator, deodorizer and secondary blowing agent in sponge compounds.

"Akweons."[152] Trademark for metal-processing oils.

674 and 700. Wetting agents and corrosion inhibitors for use in acids; also exhibit fume-depressant properties.

250. An acid fume depressant and pickling activator.

"Akwilox."[152] Trademark for high specific gravity, brominated vegetable oils manufactured for use in the soft drink industry.

Al Symbol for aluminum.

alabaster. A fine-grained, compact gypsum.

"Alacsan."[542] Trademark for a series of quaternary ammonium compounds. Used in cosmetics, germicides, algaecides, disinfectants, deodorants, and sanitizers.

"Alacstat" C-2.[542] Trademark for N,N-bis(2-hydroxyethyl alkylamine); used as antistatic agent for polyolefins.

"Alamac."[259] Trademark for acetate salts of primary amines. Water-soluble or dispersible.

Uses: Non-metallic ore flotation; corrosion inhibition.

"Alamide."[259] Trademark for a series of high molecular weight, aliphatic amides, such as palmitamide and stearamide, produced by reacting ammonia with fatty acids.

Uses: Intermediates for durable water repellents and finishes for textiles and paper; mold release and anti-blocking agents.

"Alamine."[259] Trademark for a series of primary, secondary, and tertiary aliphatic amines, organic substituted ammonia derivatives. Soluble in variety of organic solvents. Not appreciably soluble in water. Chain length from $C_{12}-C_{18}$ with varying degrees of unsaturation.

Uses: Corrosion inhibitors, ore flotation agents, textile finishing agents, asphalt antistripping agents, rubber compounding, color particle dispersion, petroleum product additives, chemical intermediates.

"Alanap."[248] Trademark for N-1-naphthylphthalamic acid; used as a herbicide.

alanine (alpha-alanine; alpha-aminopropionic acid; 2-aminopropanoic acid) $CH_3CH(NH_2)COOH$. A nonessential amino acid.

Properties: Colorless crystals; soluble in water; slightly

soluble in alcohol; insoluble in ether; optically active; nontoxic.
DL-alanine, m.p. 295°C with decomposition; sublimes at 200°C.
L(+)-alanine, m.p. 297°C with decomposition.
D(−)-alanine, m.p. 295°C with decomposition.
L(+)-alanine hydrochloride: prisms; m.p. 204°C with decomposition.
L(+)-alanine, N-acetyl; crystals; m.p. 116°C.
L(+)-alanine, N-benzoyl; crystals; m.p. 152–154°C.
Derivation: Hydrolysis of protein (silk, gelatin, zein); organic synthesis.
Grades: Reagent; technical.
Containers: Drums (DL-form).
Uses: Microbiological research; biochemical research; dietary supplement.

beta-**alanine** (3-aminopropanoic acid; beta-aminopropionic acid). $NH_2CH_2CH_2COOH$. A naturally occurring amino acid not found in protein.
Properties: White prisms; m.p. 198°C with decomposition. Soluble in water; pH (50% solution) 6.0–7.3, slightly soluble in alcohol; insoluble in ether. Hydrochloride: plates and leaflets; m.p. 122.5°C; platinichloride: yellow leaflets; m.p. 210°C (dec). Nontoxic.
Derivation: Addition of ammonia to beta-propiolactone; other processes based on the reaction of ammonia with acrylonitrile, etc.
Containers: Fiber drums.
Uses: Biochemical research; organic synthesis; production of calcium pantothenate (q.v.); buffer in electroplating.

beta-**alanylhistidine.** See carnosine.

"Alar."[248] Trademark for a plant growth regulator (succinic acid 2,2-dimethylhydrazide) which improves the color and texture of apples, grapes, and tomatoes, and prevents premature dropping.

"Alathon."[28] Trademark for a polyethylene resin. "Alathon" G-0530, designated as a reinforced polyethylene, contains 30% by weight of glass-fiber treated with a proprietary coupling agent that optimizes its reinforcing properties.

"Albacer."[73] Trademark for a hard, white synthetic wax; m.p. 99–104°C; soluble in some organic solvents; compatible with a range of synthetic resins, mineral and vegetable oils, and other waxes.
Uses: Coatings for paper; mold lubricant; polishes.

"Albamycin."[327] Trademark for novobiocin. (q.v.).
Hazard: May have damaging side effects.

"Albaoil."[244] Trademark for series of sulfonated castor oils varying in percentage of oil.
Uses: Fatliquor and plasticizer in leather; also in textile and other industries.

"Alberene Stone."[427] Trademark for a natural soapstone (magnesium silicate) for chemical laboratory benches, sinks, fume hoods, etc. Highly resistant to acids and alkalies.

"Albigen."[440] Trademark for a water-soluble polymer used in the textile industry for stripping vat and other dyestuffs. Has no affinity for the fiber; promotes the stripping effect of alkaline hydrosulfite solutions.

"Albone."[28] Trademark for a series of hydrogen peroxide solutions, which vary in hydrogen peroxide content from 35% to 90% by weight.
Hazard: See hydrogen peroxide.

albumin. Any of a group of water-soluble proteins of wide occurrence in such natural products as milk (lactalbumin), blood serum, and eggs (ovalbumin). They are readily coagulated by heat and hydrolyze to to alpha-amino acids or their derivatives. See following entries.

albumin, blood.
Properties: Brown amorphous lumps soluble in water and alcohol.
Derivation: Ox-blood is allowed to coagulate and the serum separated by centrifuging. The decanted liquor is filtered, decolorized and subsequently evaporated at low pressure. Nontoxic.
Grades: Technical, light and dark.
Containers: 100-, 332-lb drums; 300-lb barrels.
Uses: Photographic papers; textile printing; dye preparations; clarifying agent; adhesive; pesticides.

albumin, egg (ovalbumin). Principal protein found in egg white.
Preparation: Fresh white separated from the yolk, diluted with water, beaten to froth and subsequently filtered and evaporated. Nontoxic. Molecular weight 43,000.
Grades: Technical; edible.
Containers: Cases; barrels.
Uses: Leather industry; foodstuffs; clarifying agent; photography; adhesives; sugar refining.

albumin, milk. See lactalbumin.

albumin, serum (normal human).
Properties: Light yellow to cream colored lumps; practically odorless. About 96% of the total protein is albumin. Molecular weight 66,000.
Derivation: Fractionation of human blood and careful drying.
Grades: U.S.P., as sterile solution.
Use: Medicine.

A.L.C.A. Abbreviation for American Leather Chemists' Association.

alchemy. The predecessor of chemistry, practiced from as early as 500 B.C. through the 16th century. Its two principal goals were transmutation of the baser metals into gold and discovery of a universal remedy. Modern chemistry grew out of alchemy by gradual stages, but a few of the old ideas persisted almost into the 19th century, e.g., phlogiston theory.

Alcoa smelting process. A new and more efficient method for production of aluminum metal from bauxite announced in 1973 by the Aluminum Co. of America. It requires about one-third less electric power than the Hall technology, the only large-scale process now in use. In the new method, alumina made in the usual manner (Bayer process) is reacted with chlorine, the resulting aluminum chloride yielding the metal and chlorine on electrolysis. Pilot-plant production is under way.

alcohol. A broad class of hydroxyl-containing organic compounds occurring naturally in plants and made synthetically from petroleum derivatives such as ethylene. Many are manufactured in tonnage quantities; their major uses are in organic synthesis, as solvents, detergents, beverages, pharmaceuticals, plasticizers, and fuels. The many types may be summarized as follows:

I. Monohydric (1 OH group)
  1. Aliphatic
    (a) paraffinic (ethyl alcohol)
    (b) olefinic (allyl alcohol)
  2. Alicyclic (cyclohexanol)
  3. Aromatic (phenol, benzyl alcohol)
  4. Heterocyclic (furfuryl alcohol)
  5. Polycyclic (sterols)

II. Dihydric (2 OH groups): glycols and derivatives (diols)
III. Trihydric (3 OH groups): glycerol and derivatives
IV. Polyhydric (polyols) (3 or more OH groups)
For further information see monohydric, dihydric, trihydric, polyol, and specific alcohol.

alcohol, absolute. See ethyl alcohol.

alcohol dehydrogenase. An enzyme found in animal and plant tissue which acts upon ethyl alcohol and other alcohols producing acetaldehyde and other aldehydes.
Use: Biochemical research.

alcohol, denatured. Ethyl alcohol to which another liquid has been added to make it unfit to use as a beverage (chiefly for tax reasons). In the U.S. it may be either Completely Denatured (CDA), usually by the use of the toxic methyl alcohol, or Specially Denatured (SDA). The number of formulations authorized officially for making SDA have been reduced to four (Nos. 40, 40A, 40B and 40C). They include the following denaturants: SDA 40B must contain brucine, brucine sulfate, or quassin, plus tert-butyl alcohol; SDA 40A must contain sucrose octa-acetate plus tert-butyl alcohol; SDA 40B must contain Bitrex and tert-butyl alcohol; SDA40C must contain only tert-butyl alcohol. For exact formulas consult the Alcohol and Tobacco Tax Division of IRS, Washington, D.C.
Properties: See ethyl alcohol.
Hazard: Flammable, dangerous fire risk; may be toxic by ingestion.
Uses: Manufacture of acetaldehyde and other chemicals; solvents; antifreeze and brake fluids; fuel.
Shipping Regulations: (Rail) Red label. (Air) Flammable Liquid label.

alcohol, grain. Synonym for ethyl alcohol, applying to that made from grain.

alcohol, industrial. A mixture of 95% ethyl alcohol and 5% water, plus additives for denaturing or special solvent purposes. See also alcohol, denatured.

alcoholysis. A chemical reaction between an alcohol and another organic compound, analogous to hydrolysis. The alcohol molecule decomposes to form a new compound with the reacting substance, the other reaction product being water. Both hydrolysis and alcoholysis may be considered as forms of solvolysis. See also solvolysis.

alcohol, wood. See methyl alcohol.

"Aldactone."[70] Trademark for spironolactone (q.v.).

"Aldaromes."[188] Trademark for compositions of aromatic chemicals and essential oils used in embalming fluids and sprays to mask or cover the unpleasant odor of formaldehyde. Will not decompose when in contact with formaldehyde; give clear solutions in 40% formaldehyde. Not water-soluble.

aldehyde. A broad class of organic compounds having the generic formula RCHO, and characterized by an unsaturated carbonyl group (C=O). They are formed from alcohols by either dehydrogenation or oxidation, and thus occupy an intermediate position between primary alcohols and the acids obtained from them by further oxidation. Their chemical derivation is indicated by the name: *al*(cohol) + *dehyd*(rogenation). Aldehydes are reactive compounds, participating in oxidation, reduction, addition, and polymerization reactions. For specific properties, see individual compound.

aldehyde ammonia (acetaldehyde ammonia; 1-aminoethanol) $CH_3CHOHNH_2$.
Properties: White crystalline solid; stable in closed containers, resinifies on long exposure to air. Very soluble in water and alcohol.
Constants: M.p. 97°C (partly decomposed).
Derivation: Action of acetaldehyde on ammonia.
Hazard: Moderately toxic by ingestion and inhalation, strong irritant. Moderate fire and explosion risk.
Uses: Accelerator for vulcanization of thread rubber; organic synthesis; source of acetaldehyde and ammonia.

aldehyde collidine. See 2-methyl-5-ethylpyridine.

aldehydine. See 2-methyl-5-ethylpyridine.

"Aldo."[73] Trademark for a series of waxes, glyceryl monofatty acid esters, in technical, water-dispersible, and edible grades.
Uses: Foods; plastics and rubber; cosmetics; lubricants; plasticizers; emulsion stabilizers.

aldol (acetaldol; beta-hydroxybutyraldehyde) $CH_3CHOHCH_2CHO$.
Properties: Water-white to pale yellow syrupy liquid. Decomposes into crotonaldehyde and water on distillation under atmospheric pressure. Miscible with water, alcohol, ether, organic solvents. Sp. gr. 1.1098 (15.6/4°C); b.p. 83°C (20 mm); vapor pressure < 0.1 mm (20°C); specific heat 0.737; weight 9.17 lb/gal (20°C). Flash point 150°F (open cup); freezing point below 0°C. Autoignition temp. 530°F. Combustible.
Derivation: By condensation of acetaldehyde.
Grades: Technical (98%).
Hazard: Moderately toxic and irritant. Moderate fire risk.
Containers: Cans; drums; tank cars.
Uses: Synthesis of rubber accelerators and age resisters; perfumery; engraving; ore flotation; solvent; solvent mixtures for cellulose acetate; fungicides; organic synthesis; printer's rollers; cadmium plating; dyes; drugs; dyeing assistant; synthetic polymers.

aldolase (zymohexase). An enzyme present in muscle involved in glycogenolysis and anaerobic glycolysis. It catalyzes production of dihydroxyacetone phosphate and phosphoglyceric aldehyde from fructose 1,6-diphosphate.
Use: Biochemical research.

aldol condensation. A reaction between two aldehyde or two ketone molecules in which the position of one of the hydrogen atoms is changed in such a way as to form a single molecule having one hydroxyl and one carbonyl group. Since such a molecule is partly an alcohol (OH group) and partly an aldehyde (CO group) and represents a union of two smaller molecules, the reaction is called an ald-ol condensation. It can be repeated to form molecules of increasing molecular weight. The condensation of formaldehyde to sugars in plants, which on repetition builds up the more complex carbohydrate structures such as starch and cellulose, is thought to be a reaction of this type. It occurs most effectively in an alkaline medium.

aldo-alpha-naphthylamine condensate.
Properties: Orange to dark red solid with characteristic odor; softens at 64°C min; sp. gr. 1.16; insoluble in water, gasoline; slightly soluble in alcohol and petroleum hydrocarbons; soluble in acetone, benzene, chloroform and carbon disulfide.
Uses: Antioxidant in tire carcasses, tubes, insulating tape, black soles.

aldose. Any of a group of sugars whose molecule contains an aldehyde group and one or more alcohol groups. An example is glyceraldehyde ($HOCH_2 \cdot CHOH \cdot CHO$), specifically called an aldotriose because it contains three carbon atoms.

"Aldosperse."[73] Trademark for a series of polyethylene glycol monofatty acid esters.
Properties: yellow liquids to soft white solids (S-9); sp. gr. about 1.0; m.p. $< 2°C$ (O-9) to 24–29°C (S-9). Soluble in most organic solvents. Combustible; nontoxic.
Uses: Antistatic agents; emulsifiers and detergents; wetting agents; dye assistants; thickening agents; viscosity stabilizers.

aldosterone (electrocortin) $C_{21}H_{28}O_5$. An adrenal cortical steroid hormone which is the most powerful mineralocorticoid. Probably the chief regulator of sodium, potassium, and chloride metabolism, approximately 30 times as active as deoxycorticosterone.
Properties: Crystals; m.p. 108–112°C.
Derivation: Isolated from adrenals; has been synthesized.
Use: Medicine.

"Aldox."[204] Trademark for an acidic powdered compound used to deoxidize aluminum prior to spot welding or to desmut aluminum subsequent to etching. Commercially available in 400-lb fiber drums.

aldrin (HHDN) $C_{12}H_8Cl_6$. The assigned common name for an insecticidal product containing 95% or more of 1,2,3,4,10,10-hexachloro-1,4,4a,5,8,8a-hexahydro-1,4,5,8-endo-exodimethanonaphthalene. See also dieldrin and endrin.
Properties: Brown to white crystalline solid; m.p. 104–105.5°C; insoluble in water; soluble in most organic solvents. Not affected by alkalies or dilute acids; compatible with most fertilizers, herbicides, fungicides, and insecticides.
Grade: Technical.
Containers: Fiber drums.
Uses: Insecticide.
Hazard: Highly toxic by ingestion and inhalation; absorbed by skin. Tolerance, 0.25 mg per cubic meter of air. Said to be carcinogenic. Use restricted to nonagricultural applications.
Shipping regulations: (Rail, Air) Liquid mixtures with more than 60% aldrin, Poison label; dry mixtures with more than 65% aldrin, Poison label. See regulations for further details.

"Aldyl."[28] Trademark for plastic pipe made of "Delrin" acetal resin clad in a jacket of yellow "Alathon" polyethylene resin.

ale. See brewing.

"Alert"[151] Trademark for a series of mercaptan type gas odorants, recommended for warning odorization of natural gas.

"Alfane."[41] Trademark for an acid-, solvent-resistant, synthetic-resin cement of the epoxy type used as a mortar cement up to 212°F.

alfin. A catalyst obtained from alkali alcoholates derived from a secondary alcohol which is a methyl carbinol and olefins possessing the grouping $—CH=CH—CH_2$, which may be part of a ring, as in toluene. The interaction of the alkali alcoholate (sodium isopropoxide) with the olefin halide (allyl chloride) gives a slurry of sodium chloride on which sodium isopropoxide and allyl sodium are adsorbed. This slurry is a typical alfin catalyst used to convert olefins into polyolefins. The elastomers produced are called alfin rubbers, and are now in commercial production.

"Alfol."[544] Trademark for synthetic primary straight-chain alcohols made by Ziegler-type reaction of aluminum alkyls, ethane, and hydrogen. Lower alcohols (C_6 to C_{12}) are chiefly used in phthalate and other esters for use as vinyl plasticizers. Higher alcohols (C_{12} to C_{18}) are intermediates producing biodegradable surfactants, alcohol sulfates, alcohol ethoxylate nonionics, alcohol ether sulfate anionics, and amine-derived cationics.

"Alfonic."[544] Trademark for ethoxylate primary straight-chain alcohols and alcohol ether sulfate salts for use as active components in biodegradable detergent compounds.

"Alfrax."[280] Trademark for a series of refractory products composed principally of electrically fused aluminum oxide grain. Available as bonded refractories and refractory cements. Used as furnace linings.

algae. Chlorophyll-bearing organisms occurring in both salt and fresh water; they have no flowers or seeds, but reproduce by unicellular spores. They range in size from single cells to giant kelp over 100 ft long, and include most kinds of seaweed. There are four kinds of algae: brown, red, green, and blue-green. Their photosynthetic activity accounts for the fact that over two-thirds of the world total of photosynthesis takes place in the oceans. Algae are harvested and used as food supplements (see carrageenan, agar), soil conditioners, animal feeds, and as a source of iodine; they also contain numerous minerals, vitamins, proteins, lipids, and essential amino acids. Alginic acid (q.v.) is another important derivative. Blue-green algae are water contaminants and are toxic to fish and other aquatic life. Phosphorus compounds in detergent wastes stimulate the growth of algae to such an extent that overpopulation at the water surface prevents light from reaching many of the plants; these decompose, removing oxygen and releasing carbon dioxide, thus making the water unsuitable for fish. Algae are being used in treatment of sewage and plant effluent in a proprietary flocculation process. See also: eutrophication; agar.

"Algepon."[300] Trademark for series of dyeing, stripping and discharge printing assistants for various applications in textile processing. Several types are each formulated for a specific function.

algicide. Chemical agent added to water to destroy algae. Copper sulfate is commonly employed for large water systems.

"Algimaster."[426] Trademark for a composition containing the active ingredients alkyl quaternary ammonium bromides, organic polyamine, amine hydrobromides.
Containers: 1-qt polyethylene bottles.
Uses: Control of algae in swimming pool water.

algin. A hydrophilic polysaccharide (phycocolloid or hydrocolloid) found exclusively in the brown algae. It is analogous to agar (q.v.). The seaweed (giant kelp) is sea-harvested, water-extracted and refined. U.S. (California) and Great Britain are the chief producers. See also alginic acid; alginate.

alginate. Any of several derivatives of alginic acid (e.g., calcium, sodium or potassium salts or propylene glycol alginate). They are hydrophilic colloids (hydrocolloids) obtained from seaweed. Sodium alginate is water-soluble but reacts with calcium salts to form insoluble calcium alginate.
Uses: Food additive (thickener, stabilizer); yarns and fibers; medicine (first-aid dressings); meat substitute; high-protein food analogs.

alginic acid $(C_6H_8O_6)_n$, a polysaccharide composed of beta-D-mannuronic acid residues linked so that the carboxyl group of each unit is free, while the aldehyde group is shielded by a glycosidic linkage. It is a linear polymer of the mannuronic acid in the pyranose ring form.
Properties: White to yellow powder, possessing marked hydrophilic colloidal properties for suspending, thickening, emulsifying and stabilizing. Insoluble in organic solvents; slowly soluble in alkaline solutions. Nontoxic.
Grades: Refined (food); technical (commercial); N.F. (sodium alginate); F.C.C.
Containers: Drums; fiber containers; bottles; multiwall paper bags.
Uses: Food industry as thickener and emulsifier; protective colloid; tooth paste; cosmetics; pharmaceuticals; textile sizing; coatings; waterproofing agent for concrete; boiler water treatment; oil-well drilling muds; storage of gasoline as a solid.

alicyclic. A group of organic compounds characterized by arrangement of the carbon atoms in closed ring structures sometimes resembling boats, chairs, or even birdcages. These compounds have properties resembling those of aliphatics, and should not be confused with aromatic compounds (q.v.) having the hexagonal benzene ring. Alicyclics are comprised of three subgroups: (1) cycloparaffins (saturated), (2) cycloolefins (unsaturated, with two or more double bonds), and (3) cycloacetylenes (cyclynes) with a triple bond. The best-known cycloparaffins (sometimes called naphthenes) are cyclopropane, cyclohexane, and cyclopentane; typical of the cycloolefins are cyclopentadiene and cyclooctatetraene. Most alicyclics are derived from petroleum or coal tar. Many can be synthesized by various methods. See also subgroups referred to above.

"Alidase."[70] Trademark for hyaluronidase. See hyaluronic acid.

"Alipal."[307] Trademark for a series of anionic surfactants. CO-433. Sodium salt of sulfated nonylphenoxypoly(ethyleneoxy)ethanol; 28% active.
Uses: Detergents, shampoos, scrub soaps, lime soap dispersant; emulsifier; antistatic agent.

aliphatic. One of the major groups of organic compounds characterized by straight- or branched-chain arrangement of the constituent carbon atoms. Aliphatic hydrocarbons are comprised of three subgroups: (1) paraffins (alkanes), all of which are saturated and comparatively unreactive, the branched-chain types being much more suitable for gasoline than the straight-chain; (2) olefins (alkenes or alkadienes), which are unsaturated and quite reactive; (3) acetylenes (alkynes), which contain a triple bond and are highly reactive. In complex structures the chains may be branched or cross-linked. See also alicyclic; aromatic; chain.

"Aliquat."[259] Trademark for a group of fatty quaternary ammonium chlorides, stable in both acidic and alkaline media.
Uses: Cationic textile softeners, corrosion inhibitors, and flow control agents.

aliquot. A part which is a definite fraction of a whole; as, aliquot samples for testing or analysis.

"Alite."[326] Trademark for a series of sintered metallic oxides.

"Alitrile."[259] Trademark for a series of high molecular weight, aliphatic nitriles produced from fatty acids from tallow, hydrogenated tallow, coconut and soybean oils, and palmitic and stearic acids.
Uses: Lubricating oil additives; plasticizers.

"Alizachrome."[134] Trademark for a series of mordant acid dyes.

alizarin (1,2-dihydroxyanthraquinone) $C_6H_4(CO)_2C_6H_2(OH)_2$. Parent form of many dyes and pigments, including mordants.
Properties: Orange-red crystals; brownish-yellow powder. Soluble in alcohol and ether; sparingly soluble in water. C.I. No. 58000. M.p. 289°C; b.p. 430°C (sublimable). Low toxicity. Combustible.
Derivation: Anthracene is oxidized to anthraquinone, the sulfonic acid of which is then fused with caustic soda and potassium chlorate; the melt is run into hot water and the alizarin precipitated with hydrochloric acid. Occurs naturally in madder root.
Grades: Technical.
Containers: Barrels; kegs; fiber drums.
Uses: Manufacture of dyes; production of lakes; both dye and intermediate.

alizarin oil. Turkey red oil (q.v.).

alizarin yellow R (para-nitrophenylazosalicylate sodium) $O_2NC_6H_4NNC_6H_3OHCOONa$. C.I. 14030.
Properties: Yellow brown powder; soluble in water.
Use: Acid-base indicator in pH range 10.1 (yellow) to 12.0 (violet); biological stain.

"Alizarol."[243] Trademark for a group of acid mordant dyes.

alkadiene. See diolefin.

"Alkalate."[244] Trademark for a compound consisting of sodium sesquisilicate base intimately combined with sodium phosphate. Total Na_2O content 38.2%; per cent of total Na_2O in active form 32.4%; per cent of total Na_2O in inactive form 5.8%; compounded to balance the effect of bicarbonates and soil conditions.
Uses: General free-rinse cleaning in high bicarbonate areas, such as laundries and dairies.

alkali. Any substance which in water solution is bitter, more or less irritating or caustic to the skin, turns litmus blue, and has a pH value greater than 7.0. See also base; pH; alkali metal.

The alkali industry produces sodium hydroxide, sodium carbonate (soda ash), sodium chloride, salt cake, sodium bicarbonate, and corresponding potassium compounds.
Hazard: Strong alkalies in solution are corrosive to the skin and mucous membranes.

alkali blue. Class name for a group of pigment dry powders prepared by the phenylation of pararosaniline or fuchsine, followed by drowning in hydrochloric acid, washing and sulfonating. Alkali blue, on a weight basis, has the highest tinting strength of all blue pigments. The presscake may be "flushed" with vehicle to replace the water in the pulp.
Uses: Printing inks; interior paints.

alkali cellulose. The product formed by steeping wood pulp with sodium hydroxide, the first step in the manufacture of viscose rayon and other cellulose derivatives. See also carboxymethylcellulose.

alkali metal. A metal in Group 1A of the periodic system; lithium, sodium, potassium, rubidium, cesium, and francium. Except for francium (q.v.), the alkali metals are all soft, silvery metals, which may be readily fused and volatilized, the melting and boiling points becoming lower with increasing atomic weight. The specific gravity increases with, but less rapidly than, the atomic weight, the atomic volume therefore becoming greater as the series is ascended. They are the most strongly electropositive of the metals. They react vigorously, at times violently with water; within the group itself the basicity increases with atomic weight, that of cesium being the greatest.
Hazard: Dangerous fire risks, especially in contact with moisture, acids, or oxidizing materials.

alkalimetry. The measurement of the concentration of bases or of the amount of free base present in a solution by titration or some other means of analysis.

alkaline earth. An oxide of an alkaline earth metal (lime).

alkaline-earth metals. Calcium, barium, strontium, and radium (Group II A of Periodic Table). In general they are white, and differ by shades of color or casts, malleable, extrudable, and machinable, and may be made into rods, wire, or plate; less reactive than sodium and potassium, and have higher melting and boiling points. See also specific entry.

alkaloid. A basic nitrogenous organic compound of vegetable origin. Usually derived from the nitrogen ring compounds: pyridine, quinoline, isoquinoline, pyrrole; designated by the ending -ine. Though some are liquids, they are usually colorless, crystalline solids, having a bitter taste, which combine with acids without elimination of water. They are soluble in alcohol; insoluble, or sparingly soluble in water. Examples are atropine, morphine, nicotine, quinine, codeine, caffeine, cocaine and strychnine.
Hazard: Most alkaloids are toxic by ingestion.
Uses: Medicine.
See specific entry.

"Alkamortar H-W."[446] Trademark for a ceramic air-setting mortar, resistant to caustic solutions of moderate concentrations at atmospheric temperatures.
Uses: Bonding acid-resistant brick and in limited installations under alkaline conditions.

"Alkam" 4T.[505] Trademark for a polyaliphatic amine.
Properties: Clear pale yellow liquid; b.p. 211.1°C (160 mm Hg); refractive index (n 20/D); 1.453; heat of vaporization, 66.8 cal/g; completely miscible with methanol, benzene and hexane.
Uses: Curing agent for epoxy resins; also as an acid gas absorber and as an antiozonant synergist, dispersant and emulsifier.

alkane. See paraffin.

alkanesulfonic acid, mixed RSO_3H (R is methyl, ethyl, propyl, mixed). Trade designation for a mixture of methane-, ethane-, and propanesulfonic acids. A strong nonoxidizing, nonsulfonating liquid acid which is thermally stable at moderately high temperatures.
Properties: Light amber liquid with sour odor; very corrosive; miscible with water, and saturated fatty acids. M.p. below -40°C; b.p. 120–140° (1 mm); sp. gr. 1.38 (20°C); pH (1% solution) 1.15.
Containers: 5-, 25-, 70-lb carboys.
Hazard: Strong irritant to eyes and skin.
Uses: Catalyst; intermediate; reaction medium.
Shipping regulations: (Rail) Corrosive liquid, n.o.s., White label. (Air) Corrosive label.

"Alkanol."[28] Trademark for a series of fatty alcohol-ethylene oxide condensation products used as nonionic surface-active agents in detergents, dispersing and emulsifying agents in paper, leather, and textiles. These include grades OA, OE, OJ, OP, and HC. 189-S is a saturated hydrocarbon sodium sulfonate. B and BG are sodium alkylnaphthalene sulfonates. S is tetrahydronaphthalene sodium sulfonate.

alkanolamine (alkylolamine). A compound such as ethanolamine, $HOCH_2CH_2NH_2$, or triethanolamine, $(HOCH_2CH_2)_3N$, in which nitrogen is attached directly to the carbon of an alkyl alcohol. See specific compound.

"Alkaterge."[319] Trademark for a group of oil-soluble surface-active agents.
Containers: Up to 55-gal drums.
Uses: Corrosion inhibitor; salts and soap; paper, textiles, metal cleaners; emulsion stabilizer; acid acceptor; pigment grinding and dispersion; antifoam agent; antioxidant.

"Alkazene" 42.[233] Trademark for dibromoethylbenzene. ($C_6H_3C_2H_5Br_2$). Colorless liquid, f.p. -40°C; b.p. 262°C; sp. gr. 1.744; refractive index 1.587 at 25°C; 14.51 lb/gal at 25°C. Insoluble in water; soluble in methanol and ether. Used in the synthesis of pharmaceuticals, dyes, quaternary ammonium compounds and other organic compounds.

alkene. See olefin.

"Alkeran."[301] Trademark for melphalan (q.v.).

"Alkolene."[244] Trademark for series of slightly alkaline oils emulsified with soap; used in the leather industry.

"Alkophos."[58] Trademark for aluminum acid phosphates in liquid form.
Uses: High-temperature water-insoluble binder, coating, or adhesive.

"Alkor."[41] Trademark for a synthetic, furan-type resin cement which is acid- and alkali-proof and used as a mortar cement where temperatures do not exceed 380°F.

alkoxyaluminum hydrides (H_nAlOR_{3n}). A group of reducing agents especially useful in converting epoxides to alcohols. Derived by reaction of aluminum hydride with the corresponding alcohol in tetrahydrofuran.

"Alkyd Molding Compound."[175] Trademark for a thermosetting plastic comprised of an unsaturated polyester (usually formulated with a diallyl phthalate cross-linking monomer), inorganic mineral fillers (clay, glass fiber, asbestos, etc.), and other modifiers.
Properties: High dimensional stability, good electrical resistivity, ease of molding at low pressures. Glass-reinforced type has high mechanical strength and impact resistance.
Forms available: Granular, putty (soft), glass-reinforced.

Uses: Components of electrical systems, encapsulating compounds, vacuum tube bases; electrical insulation.
See also alkyd resin.

alkyd resin. The thermosetting reaction product of a dihydric or polyhydric alcohol (ethylene glycol or glycerol) and a polybasic acid (phthalic anhydride) in the presence of a drying oil (linseed, soybean) which acts as a modifier. Alkyds are actually a type of polyester resin, which has a similar derivation, but is not oil-modified. Alkyd resins may be produced by direct fusion of glycerol, phthalic anhydride and drying oil at from 410–450°F. Solvents are then added to adjust the solids content. The amount of drying oil varies depending on the intended use. The solutions are used as vehicles in exterior house paints, marine paints and baking enamels. Molded alkyd resins are used for electrical components, distributor caps, encapsulation, and a variety of similar applicatoins. See also "Alkyd Molding Compound."

alkyl. A paraffinic hydrocarbon group which may be derived from an alkane by dropping one hydrogen from the formula. Examples are methyl, CH_3-, ethyl, C_2H_5-; propyl, $CH_3CH_2CH_2$-; isopropyl, $(CH_3)_2CH$-; etc. Corresponding aromatic groups are known as aryl.

alkylaryl polyethyleneglycol ether. See isooctylphenoxypolyoxyethylene ethanol for a typical example of this class of compound. Used as surface-active agents, as in detergents.

alkylaryl sulfonate. An organic sulfonate of combined aliphatic and aromatic structure, e.g., alkylbenzene sulfonate (q.v.).

alkylate. (1) A product of alkylation (q.v.). (2) A term used in the petroleum industry to designate a branched chain paraffin derived from an isoparaffin and an olefin, e.g., isobutane reacts with ethylene (with catalyst) to form 2,2-dimethylbutane (neohexane). The product is used as a high-octane blending component of aviation and civilian gasolines.
(3) In the detergent industry, the term is applied to the reaction product of benzene or its homologs with a long-chain olefin to form an intermediate, e.g., dodecylbenzene, used in the manufacture of detergents. It also designates a product made from a long-chain normal paraffin, which is chlorinated to permit combination with benzene to yield a biodegradable alkylate. The adjectives "hard" and "soft" applied to detergents refer to their ease of decomposition by microorganisms. See biodegradability; detergent.

alkylation. (1) The introduction of an alkyl (q.v.) radical into an organic molecule. This was one of the early chemical processes used in Germany to furnish intermediates for improved dyes, e.g., dimethylaniline. Other alkylation products are cumene, dodecylbenzene, ethylbenzene, and nonylphenol.
(2) A process whereby a high-octane blending component for gasolines is derived from catalytic combination of an isoparaffin and an olefin. See also alkylate (2); neohexane.

alkylbenzene sulfonate (ABS). A branched-chain sulfonate type of synthetic detergent, usually a dodecylbenzene or tridecylbenzene sulfonate. Such compounds are known as "hard" detergents because of their resistance to breakdown by microorganisms. They are being replaced by linear sulfonates. See alkyl sulfonate, linear; detergent. See also sodium dodecylbenzene sulfonate.

alkyldimethylbenzylammonium chloride. General name for a quaternary detergent. See, for example, benzalkonium chloride.

alkyl fluorophosphate. See diisopropyl fluorophosphate.

alkylolamine. See alkanolamine.

alkyl sulfonate, linear (linear alkylate sulfonate; LAS). A straight-chain alkylbenzene sulfonate; a detergent specially tailored for biodegradability (q.v.). The linear alkylates may be normal or iso (branched at the end only), but are C_{10} or longer. See sodium dodecylbenzene sulfonate.

alkyne. See acetylene hydrocarbon.

allantoin (glyoxyldiureide; 5-ureidohydantoin) $C_4H_6N_4O_3$. The end product of purine metabolism in mammals other than man and other primates; it results from the oxidation of uric acid.
Properties: White to colorless, odorless, tasteless powder or crystals; m.p. 230°C (dec.). One gram is soluble in 190 cc water or 500 cc alcohol; readily soluble in alkalies. Optically active forms are known.
Preparation: Produced by oxidation of uric acid. Also present in tobacco seeds, sugar beets, wheat sprouts.
Uses: Biochemical research; medicine.

allene (propadiene; dimethylenemethane) $H_2C{:}C{:}CH_2$.
Properties: Colorless gas; unstable; m.p. -136.5°C; b.p. -34.5°C. Can be readily liquefied.
Derivation: (a) Action of zinc dust on 2,3-dichloropropene; (b) pyrolysis of isobutylene; (c) electrolysis of potassium itaconate.
Containers: Cylinders.
Hazard: Flammable; dangerous fire risk.
Use: Organic intermediate.
Shipping regulations: (Rail) Red Gas label. (Air) Flammable Gas label. Not acceptable on passenger planes.

allergen. Any substance that acts in the manner of an antigen on coming into contact with body tissues by inhalation, ingestion or skin adsorption. The allergen causes a specific reagin to be formed in the bloodstream; the ability to produce reagins in response to a given allergen is an inherited characteristic that differentiates an allergic from a non-allergic person. A reagin is actually an antibody. The specificity of the allergen-reagin reaction and its dependence on molecular configuration is similar to the antigen-antibody reaction (q.v.).

The allergen molecule (often a protein such as pollen or wool) may be regarded as a key which precisely fits the corresponding structural shape of the reagin molecule. Allergies in the form of contact dermatitis can result from exposure to a wide range of plant products, some metals, and a few organic chemicals. Though they are alike in some ways, antigen-antibody reactions protect the individual, whereas allergen-reagin reactions are harmful. See also antigen-antibody.

allethrin $C_{19}H_{26}O_3$ Generic name for 2-allyl-4-hydroxy-3-methyl-2-cyclopenten-1-one ester of chrysanthemummonocarboxylic acid. A synthetic insecticide structurally similar to pyrethrin and used in the same manner. For other synthetic analogs see barth-

rin, cyclethrin, ethythrin, furethrin. Pyrethrin I differs in having a 2,4-pentadienyl group in place of the allyl of allethrin.
Properties: Clear, amber-colored, viscous liquid. Sp. gr. (20/20°C) 1.005–1.015; refractive index (20°C) 1.5040. Insoluble in water; incompatible with alkalies; soluble in alcohol, carbon tetrachloride, kerosine, and nitromethane. Low toxicity. Combustible.
Derivation: Synthetically (glycerine, acetylene, and ethyl acetoacetate are the major raw materials).
Grades: 90% technical (about 90% pure, with 10% of isomers or related compounds); 20% technical, 2.5% technical.
Containers: Drums.
Uses: Insecticides; synergist.

"Allexcel."[342] Trademark for allethrin-containing insecticidal concentrates.

allicin ($C_6H_{10}OS_2$). An antibacterial substance extracted from garlic (allium).
Properties: Colorless oily liquid; sharp garlic odor; unstable, decomposing rapidly when heated; slightly soluble in water; very soluble in alcohol, benzene and ether.
Hazard: Skin and eye irritant.
Use: Medicine.

"Allmul."[455] Trademark for a firebrick having a melting point of 3335°F, and a chemical analysis of 76% alumina, 20% silica. Recommended for most severe furnace applications.

allo-. A prefix designating the more stable of two geometrical isomers (Chemical Abstracts).

allomaleic acid. See fumaric acid.

allomerism. A constancy of crystalline form or structure with a variation in chemical composition. See also polyallomer.

alloocimene (2,6-dimethyl-2,4,6-octatriene) $(CH_3)_2C(CH)_3CCH_3CHCH_3$.
Properties: Clear, almost colorless liquid. Boiling range (5–95%) 89–91°C (20 mm); sp. gr. (15/15°C) 0.824; refractive index (20°C) 1.5278. Polymerizes and oxidizes readily. Combustible; low toxicity.
Derivation: Pyrolysis of alpha-pinene.
Uses: Component of varnishes and a variety of polymers; fragrance.

allophanamide. See biuret.

allophanate. An unsaturated nitrogenous product made by reaction of an alcohol with two moles of isocyanic acid (a gas). Usually crystalline, high-melting products that are easily isolated. Acid-sensitive and tertiary alcohols can be converted into allophanates.

"Alloprene."[206a] Trademark for a type of chlorinated rubber.

allothreonine. See threonine.

allotrope. One of several possible forms of a substance. See allotropy.

allotropy (polymorphism). The existence of a substance in two or more forms, which are significantly different in physical or chemical properties. The difference between the forms involves either (1) crystal structure, (2) the number of atoms in the molecule of a gas, or (3) the molecular structure of a liquid. Carbon is a common example of (1), occurring in several crystal forms (diamond, carbon black, graphite) as well as several amorphous forms. Diatomic oxygen and ozone (O_3) are instances of (2); and liquid sulfur and helium of (3). Uranium has 3 crystalline forms, manganese 4, and plutonium no less than 6. A number of other metals also have several allotropic forms, which are often designated by Greek letters, e.g., alpha-, gamma- and delta-iron.

alloxan (mesoxalylurea) $C_4H_2O_4N_2 \cdot H_2O$ and $4H_2O$.
Properties: White crystals, become pink on exposure to air; colorless, aqueous solution imparts pink color to skin; m.p. 170°C (dec.) (various m.p. in literature); soluble in water and alcohol.
Derivation: Oxidation of uric acid in acid solution.
Uses: Biochemical research; cosmetics; organic synthesis.

alloy. A solid or liquid mixture of two or more metals; or of one or more metals with certain nonmetallic elements, as in carbon steels. The properties of alloys are often greatly different from those of the components. The purpose of an alloy is to improve the specific usefulness of the primary component—not to adulterate or degrade it. Gold is too soft to use without a small percentage of copper; the corrosion and oxidation resistance of steel is markedly increased by incorporation of from 15 to 18% of chromium and often a few percent of nickel (stainless steel). The presence of up to 1.5% carbon profoundly affects the properties of steels. Similarly, a low percentage of molybdenum improves the toughness and wear resistance of steel. The hundreds of special alloys available are instances of the tailor-made nature of alloys to meet specific operating conditions. See alloy, fusible; amalgam; superalloy.

alloy, fusible (low-melting alloy; fusible metal). An alloy melting in the range of about 125–500°F, usually containing bismuth, lead, tin, cadmium, or indium. Eutectic alloys are the particular compositions that have definite and minimum melting points, compared with other mixtures of the same metals. The compositions of a few fusible alloys are given below:

System	Composition	Eutectic Temperature
Cd—Bi	60 Bi 40 Cd	144
In—Bi	33.7 Bi 66.3 In	72
	67.0 Bi 33 In	109
Pb—Bi	56.5 Bi 43.5 Pb	125
Sn—Bi	58 Bi 42 Sn	139
Pb—Sn—Bi	52 Bi-16 Sn-32 Pb	96
Pb—Cd—Bi	52 Bi-8 Cd-40 Pb	92
Sn—Cd—Bi	54 Bi-20 Cd-26 Sn	102
In—Sn—Bi	58 Bi-17 Sn-25 In	79
Pb—Sn—Cd—Bi	*50Bi-10 Cd-13.3 Sn-26.7 Pb	70
In—Pb—Sn—Bi	49.4 Bi-11.6 Sn-18 Pb-21 In	57
In—Cd—Pb—Sn—Bi	44.7 Bi-5.3 Cd-8.3 Sn-22.6 Pb-19.1 In	47

*Wood's metal.

alloy steel. A steel containing up to 10% of elements such as chromium, molybdenum, nickel, etc., usually with a low percentage of carbon. These added elements improve hardenability, wear resistance, toughness, and other properties. This term includes low-alloy steels, in which the alloy content does not exceed 5%, but does not include stainless steel. See also steel.

allylacetone (5-hexene-2-one) $CH_2CHCH_2CH_2COCH_3$.
Properties: Colorless liquid; sp. gr. (20/4°C) 0.846; wt/gal (20°C) 6.99 lb; 5% to 95% distills between 127–129°C; soluble in water and organic solvents.

Containers: 5-gal steel drums; 55-gal resin steel drums.
Hazard: Probably highly toxic by ingestion and inhalation.
Uses: Intermediate in pharmaceutical synthesis; perfumes, fungicides; insecticides; fine chemicals.

allyl acrylate $CH_2CHCOOCH_2CHCH_2$. Liquid; b.p. 122–124°C. Used as monomer for resins.
Hazard: Probably highly toxic by ingestion and inhalation.

allyl alcohol (2-propen-1-ol; propenyl alcohol) $CH_2{=}CHCH_2OH$.
Properties: Colorless liquid with pungent mustardlike odor. B.p. 96.9°C; m.p. –129°C; sp. gr. (20/4°C) 0.8520; wt/gal (20°C) 7.11 lb; refractive index n 20/D 1.4131; flash point (TOC) 90°F. Miscible with water, alcohol, chloroform, ether, petroleum ether.
Derivation: (a) Hydrolysis of allyl chloride (from propylene) with dilute caustic; (b) isomerization of propylene oxide over lithium phosphate catalyst at 230–270°C; (c) dehydration of propylene glycol.
Containers: 55-gal drums; tank cars.
Hazard: Toxic by ingestion and inhalation; strong irritant to eyes and skin. Flammable, moderate fire risk. Tolerance, 2 ppm in air.
Uses: Esters for use in resins and plasticizers; intermediate for pharmaceuticals and other organic chemicals; manufacture of glycerol and acrolein; military poison gas; herbicide.
Shipping regulations: (Rail, Air) Poison label.

allylamine (2-propenylamine) $C_3H_5NH_2$.
Properties: Colorless to light yellow liquid; strong ammoniacal odor; attacks rubber and cork; b.p. 55–58°C; sp. gr. 0.759–0.761 (20/20°C); refractive index n 22/D 1.4194. Soluble in water, alcohol, ether, and chloroform. Combustible.
Grades: C.P.; technical.
Containers: 55-gal drums; tank cars.
Hazard: Fumes are strongly irritant to eyes and skin.
Uses: Pharmaceutical intermediate; organic synthesis.

allyl bromide (3-bromopropene; bromoallylene) C_3H_5Br.
Properties: Colorless to light yellow liquid; irritating, unpleasant odor. Sp. gr. 1.398 (20/4°C); m.p. –199°C; b.p. 71.3°C; refractive index 1.4654; soluble in alcohol, ether, chloroform, carbon tetrachloride, carbon disulfide; insoluble in water. Flash point 30°F; autoignition temp. 563°F.
Grades: Technically pure (95% minimum purity by bromine titration).
Containers: 5-gal carboys.
Hazard: Strong irritant to skin and eyes. Flammable, dangerous fire risk.
Uses: Organic synthesis; preparation of resins and perfume intermediates.
Shipping regulations: (Rail) Red label. (Air) Flammable Liquid label. Not acceptable on passenger planes.

allyl caproate (allyl hexanoate; 2-propenyl hexanoate) $CH_3(CH_2)_4COOCH_2CHCH_2$.
Properties: Colorless to pale yellow liquid; pineapple odor. Insoluble in water; soluble in 1 volume of 80% alcohol. Sp. gr. 0.885–0.888; refractive index 1.424–1.426; b.p. 186–188°C.
Hazard: May be toxic.
Uses: Perfumery; flavors.

allyl chloride (3-chloropropene; alpha-chloropropylene) CH_2CHCH_2Cl.
Properties: Colorless liquid with unpleasant pungent odor; b.p. 45.0°C; f.p. –134.5°C; sp. gr. (20/4°C) 0.9382; wt/gal (20°C) 7.83 lb; refractive index 1.416; flash point (TOC) –20°F. Autoignition temp. 737°F. Insoluble in water; miscible with alcohol, chloroform, ether, and naphtha.
Derivation: By gas-phase direct chlorination of propylene at 15 psi and 400–500°C.
Method of purification: Distillation.
Containers: 55-gal drums; tank cars.
Hazard: Strong irritant to tissue. Toxic by inhalation. Highly flammable, dangerous fire risk. Explosive limits in air 3.3 to 11.2%. Tolerance, 1 ppm in air.
Uses: Preparation of allyl alcohol and other allyl derivatives; thermosetting resins for varnishes, plastics, adhesives; synthesis of pharmaceuticals, glycerol, and insecticides; precursor of epichlorohydrin.
Shipping regulations: (Rail) Flammable liquid, n.o.s. Red label. (Air) Flammable Liquid label. Not acceptable on passenger planes.

allyl chlorocarbonate (allyl chloroformate) C_3H_5OOCCl.
Properties: Colorless liquid; pungent odor. B.p. 106°C; sp. gr. 1.14; flash point 88°F (closed cup).
Hazard: Toxic; strong irritant to tissue. Flammable, moderate fire risk.
Shipping regulations: (Rail) White label. (Air) Corrosive label. Not acceptable on passenger planes.

allyl chloroformate. See allyl chlorocarbonate.

allyl cyanide (3-butenenitrile; vinylacetonitrile.) $CH_2{:}CHCH_2CN$.
Properties: Liquid; agreeable onion-like odor; sp. gr. 0.8341; m.p. –87°C; b.p. 119°C. Slightly soluble in water.
Derivation: Treating dry cuprous cyanide with allyl bromide.
Hazard: Toxic; absorbed by skin. Tolerance (as CN): 5 mg per cubic meter of air.
Uses: Crosslinking agent in polymerization.

allyl diglycol carbonate. See diethylene glycol bis (allyl carbonate).

4-allyl-1,2-dimethoxybenzene. See methyl eugenol.

allylene. See methylacetylene.

2-allyl-2-ethyl-1,3-propanediol $C_3H_5C(CH_2OH)_2CH_2CH_3$. Off-white solid; f.p. 31°C; density (35°C) 0.976 g/cc. Suggested as polymer additive and chemical intermediate.

allyl hexanoate. See allyl caproate.

1-allyl-4-hydroxybenzene. See chavicol.

***l*-N-allyl-3-hydroxymorphinan bitartrate.** See levallorphan tartrate.

allyl-alpha-**ionone.** $C_{16}H_{24}O$. Yellow liquid with fruity odor; sp. gr. (25/25°C) 0.928–0.935. Stable; soluble in 70% alcohol. Made synthetically. Used as perfume and flavoring. Combustible; low toxicity.

5-allyl-5-isobutylbarbituric acid $C_{11}H_{16}N_2O_3$.
Properties: White crystalline powder; odorless; slightly bitter taste; soluble in alcohol, ether and chloroform; almost insoluble in water; m.p. 138–139°C.
Hazard: Toxic by ingestion.

Use: Medicine (sedative).
See also barbiturate.

allyl isocyanate C_3H_5NCO.
Properties: Colorless liquid that turns yellow on standing; sp. gr. 0.935–0.945 (20°C); flash point 110°F; b.p. 85.5–86°C; f.p. < -80°C. Combustible.
Grades: Purity, 98% min.
Hazard: Probably toxic. Moderate fire risk.
Uses: Organic intermediate; cross-linking agent; polymer modifier.

allyl isothiocyanate (allyl isosulfocyanate; mustard oil; 2-propenyl isothiocyanate) CH_2CHCH_2NCS.
Properties: Colorless to pale yellow, oily liquid; pungent, irritating odor; sharp, biting taste. Sp. gr. 1.013–1.016 (25°C); b.p. 152°C; flash point 115°F; refractive index 1.527; optically inactive. Soluble in alcohol, ether, carbon disulfide; slightly soluble in water. Combustible.
Derivation: Distillation of sodium thiocyanate and allyl chloride, or of seed of black mustard.
Grades: Technical; F.C.C.
Containers: 5-lb bottles.
Hazard: Toxic by ingestion, inhalation and skin contact. Moderate fire risk.
Uses: Fumigant; ointments and mustard plasters.
Shipping regulations: (Air) Poison label. Not acceptable on passenger planes.

allyl mercaptan. See allyl thiol.

allyl methacrylate $CH_2C(CH_3)COOC_3H_5$. Used as monomer and intermediate.

4-allyl-2-methoxyphenol. See eugenol.

5-allyl-5-(1-methylbutyl) barbituric acid. See secobarbital.

4-allyl-1,2-methylenedioxybenzene. See safrole.

allyl pelargonate $C_3H_5OOC(CH_2)_7CH_3$. Liquid; fruity odor; b.p. (3 mm) 87–91°C; refractive index 1.4332 (20.5°C). Used in flavors and perfumes; polymers. Combustible.

para-**allylphenol.** See chavicol.

allyl resin. A special class of polyester resins derived from esters of allyl alcohol and dibasic acids. Common monomers are allyl diglycol carbonate, also known as diethylene glycol bis(allyl carbonate), diallyl chlorendate, diallyl phthalate, diallyl isophthalate, and diallyl maleate. Polymerization occurs through the unsaturated allyl double bond to form thermosetting resins which are highly resistant to chemicals, moisture, abrasion, and heat. They have low shrinkage and good electrical resistivity.
Uses: As laminating adhesives and coatings, especially by impregnation of layered materials with prepolymers (called prepregs); allylic glass cloth; varnishes; applications requiring microwave transparency; encapsulation of electronic parts; vacuum impregnation of metal castings of ceramics; molding compositions; heat-resistant furniture finishes.

allyl sulfide (diallyl sulfide; thioallyl ether) $(CH_2CHCH_2)_2S$.
Properties: Colorless liquid with garlic odor. Combustible. B.p. 139°C; sp. gr. (27/4°C) 0.888; refractive index n 27/D 1.4877. Insoluble in water; miscible with alcohol, ether, chloroform, and carbon tetrachloride.
Hazard: Strong irritant to eyes and skin.
Uses: Component of artificial oil of garlic.

allyl thiol (allyl mercaptan; 2-propene-1-thiol). CH_2CHCH_2SH.
Properties: Water-white liquid (darkens on standing); strong garlic odor; sp. gr. 0.925 (23/4°C); b.p. 67–68° (90°C); insoluble in water; soluble in ether and alcohol. Combustible.
Containers: Bottles.
Hazard: May be toxic.
Uses: Pharmaceutical intermediate; rubber accelerator intermediate.

allylthiourea (allylsulfocarbamide; thiosinamine; allyl sulfourea). $C_3H_5NHCSNH_2$.
Properties: White crystalline solid; slight garlic odor; bitter taste; sp. gr. 1.22; m.p. 78°C. Soluble in water, ether, and solutions of borax, benzoates, urethane; insoluble in benzene; slightly soluble in 70% alcohol.
Derivation: Warming a mixture of equal parts of allyl isothiocyanate and absolute alcohol with an equal amount of 30% ammonia.
Hazard: May be toxic. A strong allergen.
Uses: Medicine; corrosion inhibitor; organic synthesis.

allyltrichlorosilane $CH_2CHCH_2SiCl_3$.
Properties: Colorless liquid; pungent, irritating odor; b.p. 117.5°C; sp. gr. 1.217 (27°C); refractive index n 20/D 1.487; flash point (COC) 95°F. Readily hydrolyzed by moisture, with the liberation by hydrochloric acid; polymerizes easily.
Derivation: Reaction of allyl chloride with silicon (copper catalyst).
Hazard: Toxic; strong irritant to tissue. Flammable, moderate fire risk.
Use: Intermediate for silicones; glass fiber finishes.
Shipping regulations: (Rail) White label. (Air) Corrosive label. Not acceptable on passenger planes.

4-allyl veratrole. See methyl eugenol.

almond oil, bitter. See oil of bitter almond.

almond oil, sweet. See oil of sweet almond.

"Alnico." Trademark for an alloy containing chiefly aluminum, nickel, and cobalt; it has outstanding properties as a permanent magnet.

aloin (barbaloin). A mixture of active principles obtained from aloe. Varies in properties according to variety used.
Properties: Yellow crystals with bitter taste; darkens on exposure to air; odorless or slight odor of aloe; soluble in acetone, alkalies, water, and alcohol; slightly soluble in ether.
Grades: Technical.
Containers: Barrels; drums; kegs.
Uses: Medicine; proprietary laxatives; electroplating baths; fermentation.

"Alon."[275] Trademark for a fumed alumina.
Properties: Powder with surface area of 100 square meters per gram. Isoelectric point at pH 9.1.
Derivation: Flame hydrolysis of aluminum chloride.
Uses: Spinning aid for textiles, especially wool; paper making; antistatic agent; armature cores; hair sprays; reinforcement of elastomers; hydrophilic and thickening agent.

"Alox."[117] Trademark for a series of oxygenated hydrocarbons derived from the controlled, liquid phase, partial oxidation of petroleum fractions. Each consists of mixtures of organic acids and hydroxy acids, lactones, esters, and unsaponifiable matter. Esters, amines, amides and metal soap derivatives are also available.
Containers: Up to tank car lots.
Uses: Corrosion inhibitors; film-forming rust preventives; lubricity agents; emulsifiers.

"Aloxite."[280] Trademark for aluminum oxide made by fusing materials high in alumina, such as bauxite, and for articles made therefrom.
Containers: Multiwall paper sacks.
Uses: Abrasive grains and powders; grinding wheels; stones; razor hones; refractory cements; filter plates and tubes; diffuser plates and tubes; porous undergrain plates and coated abrasive products.

alpaca. A natural fiber obtained from a South American animal similar to the llama. Properties resemble those of wool. Used for specialty clothing and also blended with rayon. Combustible.

"Alpco."[271] Trademark for a series of high-melting mineral waxes and resins, used for carbon paper, inks, polishes, paper, plasticizers, surfactants, dispersants, casting waxes and surface coatings.

"Alperox" C.[154] Trademark for lauroyl peroxide (96.5% min.).

alpha (α) (1) A prefix denoting the position of a substituting atom or group in an organic compound. The Greek letters alpha, beta, gamma, etc., are usually not identical with the IUPAC numbering system, 1, 2, 3, etc., since they do not start from the same carbon atom. However, alpha and beta are used with naphthalene ring compounds to show the 1 and 2 positions, respectively. Alpha, beta, etc. are also used to designate attachment to the side chain of a ring compound.
(2) Both a symbol and a term used for relative volatility in distillation.
(3) Symbol for optical rotation (q.v.).
(4) A form of radiation consisting of helium nuclei. See alpha particle.
(5) The major allotropic form of a substance, especially of metals, e.g., alpha-iron.

alpha particle. A helium nucleus emitted spontaneously from radioactive elements, both natural and man-made. Its energy is in the range of 4 to 8 MeV, and is dissipated in a very short path, i.e., in a few centimeters of air or less than 0.005 mm of aluminum. It has the same mass (4) and positive charge (2) as the helium nucleus. Accelerated in a cyclotron, alpha particles can be used to bombard the nuclei of other elements. See also helium; decay, radioactive.

"Alphazurine."[243] Trademark for triphenylmethane acid blues.

"Alsilox."[468] Trademark for a fusion product of 1% alumina, 65% litharge, and 34% silica; used in ceramics. Available in various particle sizes.

"Altax."[265] Trademark for benzothiazyl disulfide.
Properties: Cream to light yellow powder; sp. gr. 1.51 ± .03; moderately soluble in benzene, carbon disulfide, chloroform; insoluble in water, dilute caustic, gasoline.
Uses: Primary accelerator in natural and nitrile rubber and SBR; plasticizer and vulcanization retarder in neoprene Type G; cure modifier in neoprene Type W; oxidation cure activator in butyl. For extruded and molded goods, tire and tubes, wire and cable, sponge.

altheine. See asparagine.

"Aludur-11."[326] Trademark for an alumina ceramic of comparable chemical resistance to porcelain, but with superior mechanical properties.

alum. See aluminum ammonium sulfate, aluminum potassium sulfate, aluminum sulfate.

"Alumalith."[250] Trademark for amblygonite (q.v.).

alum, burnt (alum, dried) $AlNH_4(SO_2)_2$ or $AlK(SO_4)_2$. Aluminum ammonium sulfate or aluminum potassium sulfate heated just sufficiently to drive off the water of crystallization.
Properties: White, odorless powder; sweetish taste. Absorbs moisture on exposure to air. Soluble in hot and slowly soluble in cold water; insoluble in alcohol.
Grades: N.F.
Use: Medicine (astringent).

alum, chrome. See chromium-potassium sulfate.

alum, chrome ammonium. See chromium-ammonium sufate.

"Alumel."[166] Trademark for an alloy consisting of about 94% nickel, with small, carefully controlled amounts of silicon, aluminum and manganese. It is chiefly used in thermocouples and lead wire.

"Alumex."[285] Trademark for a china clay; particle size 55–60% minus 2 microns, 20–25% plus 5 microns; oil absorption 30 cc/100 g clay; pH 4.5–5.5

alumina. See aluminum oxide.

alumina, activated. A highly porous, granular form of aluminum oxide having preferential adsorptive capacity for moisture and odor contained in gases and some liquids. When saturated, it can be regenerated by heat (350–600°F). The cycle of adsorption and reactivation can be repeated many times. Granules range in size from powder to pieces about 1½" diam. Average weight about 50 lb/cu ft. An effective desiccant (q.v.) for gases and vapors in the petroleum industry. It is also used as a catalyst or catalyst carrier; in chromatography; and in water purification.

alumina, fused. See aluminum oxide; "Alundum."

alumina gel. See aluminum hydroxide gel.

alumina-silica fiber.
Properties: Amorphous structure; excellent resistance to all chemicals except hydrofluoric and phosphoric acids and concentrated alkalies. Available in both short and long staple. Low heat conductivity; high thermal shock resistance. Tensile strength 400,000 psi; elastic modulus 16 million psi; upper temperature limit in oxidizing atmosphere 800°C. Noncombustible.
Derivation: The short fiber type is made by blasting a stream of molten alumina and silica with a steam jet. Long staple is spun from a molten mixture of alumina and silica modified with zirconium.
Forms: Fibers; sheets; blankets.
Uses: Nonwoven fabrics (short staple); woven fabric structures; cordage; thermal insulation; repair of furnace linings; piping molten metals; welding insulation (re-usable); insulation for rocket and space applications.
Note: In finely divided form, alumina-silica is also used as a catalyst (q.v.).

alumina trihydrate (aluminum hydroxide; aluminum hydrate; hydrated alumina; hydrated aluminum oxide) $Al_2O_3 \cdot 3H_2O$ or $Al(OH)_3$.
Properties: White crystalline powder, balls or granules; sp. gr. 2.42; insoluble in water; soluble in mineral acids and caustic soda. Noncombustible.

Derivation: From bauxite; the ore is dissolved in strong caustic and aluminum hydroxide precipitated from the sodium aluminate solution by neutralization (as with carbon dioxide) or by autoprecipitation (Bayer process, q.v.).
Grades: Technical; C.P.
Containers: Bags; drums; tonnage lots.
Uses: Glass, ceramics, iron-free aluminum and aluminum salts; manufacture of activated alumina; base for organic lakes; flame retardant. Finely divided form (0.1–0.6 microns) used for rubber reinforcing agent, paper coating, filler, cosmetics.

aluminium. British spelling of aluminum.

aluminosilicate. A compound of aluminum silicate with metal oxides or other radicals. Used as catalyst in the refining of petroleum; to soften water. See also zeolite.

aluminum (aluminium). Al Metallic element of atomic number 13; Group IIIA of the Periodic Table. Atomic weight 26.9815. Valence 3; no stable isotopes; monovalent in high-temperature compounds AlCl and AlF. Most abundant metal in earth's crust; third most abundant of all elements. Does not occur free in nature. Nontoxic.
Properties: Silvery white, crystalline solid. Tensile strength (annealed) 6800 psi, cold-rolled 16,000 psi. Sp. gr. 2.708; m.p. 660°C; b.p. 2450°C. Forms protective coating of Al_2O_3 about 50 A thick, which makes it highly resistant to ordinary corrosion. Attacked by conc. and dilute solutions of HCl, hot. conc. H_2SO_4 and perchloric acid, also violently by strong alkalies. Rapidly oxidized by water at 180°C. Unattacked by dilute or cold conc. H_2SO_4 and conc. nitric acid. Nontoxic and noncombustible except in powder form. Electrical conductivity about 2/3 that of copper. Aluminum qualifies as both a light metal and a heavy metal, according to their respective definitions.
Derivation: From bauxite by Bayer process (q. v.) and subsequent electrolytic reduction by Hall process (q.v.). There are several processes for obtaining ultrapure aluminum: (a) electrolytic (three-layer); (b) zone refining (q.v.); and (c) chemical refining. Purities up to 0.2 ppm are possible.
Two new and more efficient processes in the advanced experimental stage are the Alcoa and Toth processes (q.v.).
Forms available: Structural shapes of all types, plates, rods, wire, foil, flakes, powder (technical and U.S.P.). Aluminum can be electrolytically coated and dyed by the anodizing process (see anodic coating); it can be foamed by incorporating zirconium hydride in molten aluminum; and it is often alloyed with other metals, or mechanically combined (fused or bonded) with boron and sapphire fibers or whiskers. Strengths up to 55,000 psi at 500°C have been obtained in such composites. A vapor-deposition technique is used to form a tightly adherent coating from 0.2 to 1 mil thick on titanium and steel.
Hazard: Fine powder forms flammable and explosive mixtures in air. Tolerance, 50 million particles per cubic foot of air.
Uses: Building and construction; corrosion-resistant chemical equipment (desalination plants); die-cast auto parts; electrical industry (power transmission lines); photoengraving plates; permanent magnets; cryogenic technology; machinery and accessory equipment; containers for fissionable reactor fuels; tubes for ointments, tooth paste, shaving cream, etc. As powder, in paints and protective coatings, as rocket fuel, ingredient of incendiary mixtures (thermite); as a catalyst; foamed concrete; vacuum metallizing and coating. As foil, in packaging, cooking, decorative stamping. As flakes, for insulation of liquid fuels.
Shipping regulations (powder): (Rail) Flammable solid, n.o.s. Yellow label. (Air) Flammable Solid label.
See also aluminum alloy. For further information refer to Aluminum Association, New York, N.Y.

aluminum acetate. (a) Normal $Al(C_2H_3O_2)_3$; (b) basic $Al(C_2H_3O_2)_2OH$.
Properties: (a) Known only in solution; (b) white powder; insoluble in water; decomposed by heat.
Derivation: Interaction of aluminum hydroxide and acetic acid. The product is recovered by crystallization.
Grades: Technical; C.P.; also sold in solution (U.S.P.).
Containers: Boxes; drums; carboys (solutions).
Uses: Waterproofing cloth; mordant in textile-dyeing; preparation of lakes; embalming fluids; calico printing; treatment of dermatitis.
See also mordant rouge.

aluminum acetylacetonate $Al(C_5H_7O_2)_3$.
Properties: Solid; m.p. 189°C; b.p. 315°C. Soluble in benzene and alcohol.
Uses: Deposition of aluminum; catalyst.

aluminum acetylsalicylate (aluminum aspirin) $[C_6H_4(OCOCH_3)(COO)]_2AlOH$.
Properties: White to off-white granules or powder; odorless or slight odor; m.p., decomposes. Insoluble in water and organic solvents; soluble with decomposition in alkali hydroxides and carbonates.
Derivation: Reaction of aluminum hydroxide with acetylsalicylic acid.
Grade: N.F.
Uses: Medicine.

aluminum alkyl (aluminum trialkyl). Catalyst used in the Ziegler process.
Hazard: Pyrophoric liquid.
Shipping regulations: (Rail) Red label. (Air) Not acceptable.
See triethylaluminum and triisobutylaluminum.

aluminum alloy. Aluminum containing variable amounts of manganese, silicon, copper, magnesium, lead, bismuth, nickel, chromium, zinc, or tin. A wide range of uses and properties is possible. Alloys may be obtained for casting or working, heat-treatable or non-heat-treatable, with a wide range of strength, corrosion resistance, machinability and weldability. See also duralumin.

aluminum ammonium chloride (ammonium aluminum chloride) $AlCl_3 \cdot NH_4Cl$.
Properties: White crystals; soluble in water. M.p. 304°C.
Uses: Fur treatment.

aluminum ammonium sulfate (ammonium alum; alum N.F.) $Al_2(SO_4)_3(NH_4)_2SO_4 \cdot 24H_2O$ or $AlNH_4(SO_4)_2 \cdot 12H_2O$.
Properties: Colorless crystals; odorless; strong astringent taste. Soluble in water, glycerine; insoluble in alcohol. Sp. gr. 1.645; m.p. 94.5°C; b.p., loses $20H_2O$ at 120°C.
Derivation: By crystallization from a mixture of ammonium and aluminum sulfates.
Method of purification: Recrystallization.
Grades: Technical, lump, ground, powdered; C.P.; N.F.; F.C.C.
Containers: Bags; drums.
Hazard: Moderately toxic by ingestion.
Uses: Medicine; mordant in dyeing; water and sew-

age purification; sizing paper; retanning leather; clarifying agent; ingredient in baking powder; food additive.

aluminum, anodized. See anodic coating.

aluminum arsenide AlAs. A semiconductor used in rectifiers, transistors, thermistors.
Hazard: Highly toxic by ingestion.
(Rail, Air) Arsenical compounds, solid. Poison label.

aluminum borate $2Al_2O_3 \cdot 3H_2O$ (approx.)
Properties: White, granular powder; decomposed by water.
Derivation: Interaction of aluminum hydroxide and boric acid.
Grades: Technical; C.P.
Use: Glass and ceramic industries.

aluminum boride
Properties: Powder; apparent bulk density (fully settled): light, 0.6–0.8 g/cc; dense, 1.2–1.4 g/cc; high neutron absorption.
Uses: Nuclear shielding.

aluminum borohydride $Al(BH_4)_3$.
Properties: Volatile pyrophoric liquid; b.p. 44.5°C; m.p. -64.5°C.
Derivation: (1) By reaction of sodium borohydride and aluminum chloride in the presence of a small amount of tributyl phosphate; (2) by reaction of trimethylaluminum and diborane.
Hazard: Ignites spontaneously in air; reacts violently with water.
Uses: Intermediate in organic synthesis; jet fuel additive.
Shipping regulations: (Rail) Pyrophoric liquid, n.o.s., Red label. (Air) Pyrophoric liquid n.o.s. Not acceptable.

aluminum brass. An alloy containing 76% copper, 21.5 to 22.25% zinc, 1.75 to 2.50% aluminum.
Uses: Condenser, evaporator and heat exchanger tubes and ferrules.

aluminum bromide (a) $AlBr_3$; (b) $AlBr_3 \cdot 6H_2O$.
Properties: White to yellowish, deliquescent crystals; exists as double molecules Al_2Br_6 in the vapor; soluble in water, alcohol, carbon disulfide or ether. (a) Sp. gr. 3.01; m.p. 97.5°C; b.p. 265°C; (b) sp. gr. 2.54; m.p. 93°C (decomposes).
Derivation: (a) By passing bromine over heated aluminum; (b) reaction of hydrobromic acid with aluminum hydroxide.
Containers: Glass jars; air-tight drums.
Hazard: The anhydrous form reacts violently with water and is corrosive to skin.
Uses: Bromination, alkylation, and isomerization catalyst in organic synthesis.
Shipping regulations: (anhydrous) (Air) Corrosive label.

aluminum bronze. An alloy containing 88–96.1% copper, 2.3–10.5% aluminum, and small amounts of iron and tin. Characterized by high strength, ductility, hardness and resistance to shock, fatigue, most chemicals, and sea water.
Powder: Also called gold bronze powder. An alloy of 90% copper and 10% aluminum reduced from leaf form to powder, polished mechanically, and coated with stearic acid. Available in the following grades: litho, molding, printing-ink, and radiator. Used as a pigment in paints and inks.

aluminum n-butoxide $Al(OC_3H_9)_3$.
Properties: Yellow to white crystalline solid; m.p. 101.5°C (pure) and 88–96°C (commercial); density 1.0251 (20°C); b.p. 290–310°C (30 mm). Soluble in aromatic, aliphatic and chlorinated hydrocarbons.
Uses: Ester exchange catalyst; defoamer ingredient; hydrophobic agent intermediate.

aluminum carbide Al_4C_3.
Properties: Yellow crystals or powder; decomposes in water with liberation of methane. Sp. gr. 2.36. Stable to 1400°C.
Derivation: By heating aluminum oxide and coke in an electric furnace.
Grades: Technical.
Containers: Iron drums.
Uses: Generating methane; catalyst; metallurgy; drying agent.
Hazard: Dangerous fire risk. Keep dry.
Shipping regulations: (Rail) Flammable solid, n.o.s., Yellow label. (Air) Flammable Solid label.

aluminum carbonate. A basic carbonate of variable composition; formula sometimes given as $Al_2O_3 \cdot CO_2$. White lumps or powder, insoluble in water, dissolves in hot hydrochloric acid or sulfuric acid. Formerly used as mild astringent, styptic. Normal aluminum carbonate $Al_2(CO_3)_3$ is not known as an individual compound.

"Aluminum Chelate."[134] Trademark for a group of compounds based on aluminum.
BEA-1: Chemically modified aluminum secondary butoxide.
Properties: Pale yellow liquid; sp. gr. 1.030 (21°C); aluminum content 8.9–9.1%.
PEA-1: Chemically modified aluminum isopropylate.
Properties: Pale yellow liquid; sp. gr. 1.035 (25°C); aluminum content 9.3–10.0%.
PEA-2: Chemically modified aluminum isopropylate.
Properties: Pale yellow; soluble in aromatic, aliphatic and chlorinated hydrocarbons; aluminum content 7.8–7.9%.
Uses: Curing of epoxy, phenolic, castor oil alkyds and high molecular weight polymers which are hydroxyl or carboxyl bearing; textile hydrophobing, in solvent based systems; adhesion promotion.

aluminum chloride, anhydrous $AlCl_3$.
Properties: White or yellowish crystals; sp. gr. (25°C) 2.44; m.p. 190°C (2.5 atm); sublimes readily at 178°C; the vapor consists of double molecules, Al_2Cl_6. Soluble in water.
Derivation: (a) By reaction of purified gaseous chlorine with molten aluminum; (b) by reaction of bauxite with coke and chlorine at about 1600°F. This product is used to make the hydrate.
Impurities: (a) Ferric chloride; free aluminum; insolubles.
Grades: Technical; reagent.
Containers: Drums; car lots.
Hazard: Toxic by ingestion and inhalation. Strong irritant to tissue. Reacts violently with water, evolving HCl. Safety data sheet available from Manufacturing Chemists Assn., Washington, D.C.
Uses: Ethylbenzene catalyst; dyestuff intermediate; detergent alkylate; ethyl chloride; pharmaceuticals and organics (Friedel-Crafts catalyst); butyl rubber; petroleum refining; hydrocarbon resins; nucleating agent for titanium dioxide pigments.
Shipping regulations: (Rail) Not listed. (Air) Corrosive label.

aluminum chloride hydrate $AlCl_3 \cdot 6H_2O$.
Properties: White or yellowish deliquescent crystalline powder; nearly odorless;; sweet, astringent taste; sp. gr. 2.4; m.p., decomposes. Soluble in water and alcohol. The water solution is acid.
Derivation: By crystallizing the anhydrous form from hydrochloric acid solution.
Grades: Technical; C.P.; N.F. See also aluminum chloride solution.
Hazard: May be toxic.
Uses: Pharmaceuticals and cosmetics; pigments; roofing granules; special papers; photography; textile (wool).

aluminum chloride solution 32° Bé. Special grade of a solution containing only 0.005% iron as impurity, and having an acid reaction but containing no free acid.
Containers: Carboys; tank cars; tank trucks.
Uses: Antiperspirants; roofing granules.

aluminum chlorohydrate. $[Al_2(OH)_5Cl]_x$. An ingredient of commercial antiperspirant and deodorant preparations. Also used for water purification and treatment of sewage and plant effluent.

aluminum diethyl monochloride. See diethylaluminum chloride.

aluminum diformate (aluminum formate, basic) $Al(OH)(CHO_2)_2 \cdot H_2O$.
Properties: White or gray powder. Soluble in water. Low toxicity.
Derivation: Aluminum hydroxide is dissolved in formic acid and spray-dried. Solutions are also prepared by treating aluminum sulfate with formic acid, followed by lime.
Grades: Technical solutions (12–20° Bé).
Containers: Carboys; carlots.
Uses: Waterproofing; mordanting; antiperspirants; tanning leather; improving wet strength of paper.

aluminum distearate $Al(OH)(C_{18}H_{35}O_2)_2$.
Properties: White powder; m.p. 145°C; sp. gr. 1.009. Insoluble in water, alcohol, ether. Forms gel with aliphatic and aromatic hydrocarbons. Low toxicity.
Uses: Thickener in paints, inks, and greases; water repellent; lubricant in plastics and ropes; in cement production.

aluminum ethylhexoate (aluminum octoate). A metallic salt of 2-ethylhexoic acid, used as a gelling agent for liquid hydrocarbons as gasoline and common petroleum fractions used in coating thinners.

aluminum fluoride, anhydrous AlF_3.
Properties: White crystals. Sublimes about 1260°C without melting; sp. gr. 2.882. Slightly soluble in water; insoluble in most organic solvents.
Derivation: (1) Action of hydrogen fluoride gas on alumina trihydrate; (2) reaction of hydrofluoric acid on a suspension of aluminum trihydrate, followed by calcining the hydrate formed; (3) reaction of fluosilicic acid on aluminum hydrate.
Grades: Technical
Containers: Multiwall paper sacks; fiber drums.
Hazard: Moderately toxic; strong irritant to tissue. Tolerance (as F): 2.5 mg per cubic meter of air.
Uses: Production of aluminum to lower the melting point and increase the conductivity of the electrolyte; flux in ceramic glazes and enamels; manufacture of aluminum silicate.

aluminum fluoride hydrate $AlF_3 \cdot 3\frac{1}{2}H_2O$.
Properties: White crystalline powder. Slightly soluble in water.
Derivation: Action of hydrofluoric acid on alumina hydrate and subsequent recovery by crystallization.
Grades: Technical; C.P.
Containers: Bags; drums.
Hazard: Moderately toxic.
Uses: Ceramics (production of white enamel).

aluminum fluosilicate (aluminum silicofluoride) $Al_2(SiF_6)_3$.
Properties: White powder. Slightly soluble in cold water; readily soluble in hot water.
Grades: Technical.
Hazard: Moderately toxic.
Uses: Artificial gems, enamels, glass.

aluminum formate. See aluminum triformate, and aluminum diformate.

aluminum formate, basic. See aluminum diformate.

aluminum formate, normal. See aluminum triformate.

aluminum formoacetate $Al(OH)(OOCH)(OOCCH_3)$. White powder; soluble in water and alcohol; used in textile water repellents.

aluminum hydrate. See alumina trihydrate.

aluminum hydride AlH_3.
Properties: White to gray powder; decomp. 160°C (100°C if catalyzed). Evolves hydrogen on contact with water.
Hazard: Dangerous fire and explosion risk.
Uses: Electroless coatings on plastics, textiles, fibers, other metals; possible rocket fuel.
Shipping regulations: (Rail) Flammable solid, n.o.s., Yellow label. (Air) Flammable Solid label. Not acceptable on passenger planes.

aluminum hydroxide. See alumina trihydrate.

aluminum hydroxide, anhydrous. See alumina trihydrate.

aluminum hydroxide gel (hydrous aluminum oxide; alumina gel) $Al_2O_3 \cdot xH_2O$.
Properties: White, gelatinous precipitate. Constants variable with the composition; sp. gr. about 2.4. Insoluble in water and alcohol; soluble in acid and alkali. Nontoxic; noncombustible.
Derivation: By treating a solution of aluminum sulfate or chloride with caustic soda, sodium carbonate or ammonia; by precipitation from sodium aluminate solution by seeding or acidifying (carbon dioxide is commonly used).
Grades: Technical; C.P.; U.S.P. (containing 4% Al_2O_3); N.F. (dried, containing 50% Al_2O_3).
Containers: Fiber drums.
Uses: Dyeing mordant; water purification; waterproofing fabrics; manufacture of lakes; filtering medium; chemicals (aluminum salts); lubricating compositions; manufacture of glass; sizing paper; ceramic glaze; medicine.

aluminum hydroxystearate $Al(OH)[OOC(CH_2)_{10}CHOH(CH_2)_5CH_3]_2$.
Properties: White powder; m.p. 155°C; sp. gr. 1.045. Less soluble in nonpolar compounds than other aluminum stearates, and more soluble in polar compounds.
Uses: Waterproofing of leather and cements; lubricant for plastics and ropes; paints and inks.

aluminum iodide AlI_3 (anhydrous).
Properties: Brown-black crystalline pieces (white when pure); m.p. 191°C; b.p. 385°C; sp. gr. 3.9825. Soluble in water, alcohol, ether, carbon disulfide.
Derivation: Heating aluminum and iodine in a sealed tube.

Method of purification: Crystallization.
Grades: Technical.
Containers: Bottles.
Hazard: Moderately toxic. Reacts violently with water.
Uses: Organic synthesis.

aluminum isopropoxide. See aluminum isopropylate.

aluminum isopropylate (aluminum isopropoxide) $Al(OC_3H_7)_3$.
Properties: White solid; sp. gr. 1.035 (20°C); m.p. 128–132°C; b.p. 138–148°C (10 mm). Soluble in alcohol, benzene; decomposes in water.
Derivation: From isopropanol and aluminum.
Grades: Distilled (purity approximately 100%).
Containers: Waterproof drums.
Uses: Dehydrating agent; catalyst; waterproofing textiles.

aluminum metaphosphate $Al(PO_3)_3$.
Properties: White powder; insoluble in water; m.p. approx. 1537°C.
Uses: As a constituent of glazes, enamels and glasses, and as a high temperature insulating cement.

aluminum monobasic stearate. See aluminum monostearate.

aluminum monopalmitate. See aluminum palmitate.

aluminum monostearate (aluminum monobasic stearate) $Al(OH)_2[OOC(CH_2)_{16}CH_3]$.
Properties: Fine, white to yellowish white powder; faint characteristic odor; m.p. 155°C; sp. gr. 1.020. Insoluble in water, alcohol and ether. Forms a gel with aliphatic and aromatic hydrocarbons. Low toxicity.
Derivation: Mixing solutions of a soluble aluminum salt and sodium stearate.
Grades: U.S.P., which describes it as a mixture of the monostearate and monopalmitate containing 14.5–16.0% Al_2O_3.
Uses: Medicine; manufacture of paints, inks, greases, waxes; thickening lubricating oils; waterproofing; gloss producer; stabilizer for plastics.

aluminum naphthenate.
Properties: Yellow substance of rubbery consistency with high thickening power. Combustible.
Derivation: Reaction of an aluminum salt with an alkali naphthenate in aqueous solution.
Uses: Paint and varnish drier and bodying agent; detergent in lube oils; the solution in organic solvents has been proposed for insecticides and siccatives.
See also soap (2).

aluminum nitrate $Al(NO_3)_3 \cdot 9H_2O$.
Properties: White crystals. Soluble in cold water; decomposes in hot water. Soluble in alcohol and acetone. M.p. 73°C; decomposes at 150°C.
Derivation: Formed by the action of nitric acid on aluminum and crystallization.
Grades: Technical; C.P.; 99.75%.
Containers: Multiwall paper sacks; drums.
Hazard: Strong oxidizing agent. Do not store near combustible materials.
Uses: Textiles (mordant); leather tanning; manufacture of incandescent filaments; catalyst in petroleum refining; nucleonics.
Shipping regulations: (Rail) Yellow label. (Air) Oxidizer label.

aluminum oleate $Al(C_{18}H_{33}O_2)_3$.
Properties: Yellowish-white viscous mass. Insoluble in water; soluble in alcohol, benzene, ether, oil and turpentine. Combustible. Nontoxic.
Derivation: By heating aluminum hydroxide, water and oleic acid. The resultant mixture is filtered and dried.
Uses: Waterproofing; drier for paints, etc.; thickener for lubricating oils; medicine; as lacquer for metals; lubricant for plastics; food additive.

aluminum orthophosphate. See aluminum phosphate.

aluminum oxide (alumina) Al_2O_3. The mineral corundum (q.v.) is natural aluminum oxide, and emery, ruby, and sapphire are impure crystalline varieties. The mixed mineral bauxite is a hydrated aluminum oxide.
Properties: Vary according to the method of preparation. White powder, balls or lumps of various mesh. Sp. gr. 3.4–4.0; m.p. 2030°C; insoluble in water; difficultly soluble in mineral acids and strong alkali. Noncombustible; nontoxic. See also alumina trihydrate; aluminum hydroxide gel.
Derivation: (a) Leaching of bauxite with caustic soda followed by precipitation of a hydrated aluminum oxide by hydrolysis and seeding of the solution. The alumina hydrate is then washed, filtered and calcined to remove water and obtain the anhydrous oxide. See Derivation under alumina trihydrate. (b) Coal mine waste waters are used to obtain aluminum sulfate, which is then reduced to alumina.
Grades: Technical; C.P.; fibers; high purity; fused; calcined.
Containers: Multiwall paper sacks; drums; barrels.
Uses: Production of aluminum; manufacture of abrasives, refractories, ceramics, electrical insulators, catalyst and catalyst supports; paper; spark plugs; crucibles and laboratory wares; absorbing gases and water vapors (see alumina activated); chromatographic analysis; fluxes; light bulbs; artificial gems; heat-resistant fibers.

aluminum oxide, hydrated. See alumina trihydrate.

aluminum oxide, hydrous. See aluminum hydroxide gel.

aluminum palmitate (aluminum monopalmitate) $Al(OH)_2(C_{16}H_{31}O_2)$.
Properties: White powder; m.p. 200°C; sp. gr. 1.072. Insoluble in alcohol and acetone; forms gel with hydrocarbons. Combustible; nontoxic.
Derivation: By heating aluminum hydroxide and palmitic acid and water. The resultant mixture is filtered and dried.
Uses: Waterproofing leather, paper, textiles; thickening for lubricating oils; thickening or suspending agent in paints and inks; production of high gloss on leather and paper; ingredient of varnishes; lubricant for plastics; food additive.

aluminum paste. Aluminum powder ground in oil; used for aluminum paints.

aluminum phenolsulfonate (aluminum sulfocarbolate) $Al_2(C_6H_4OHSO_3)_6$.
Properties: Reddish-white powder with slight phenol odor. Strongly astringent taste. Soluble in water and alcohol, glycerin.
Containers: Drums.
Use: Medicine.

aluminum phosphate (aluminum orthophosphate) $AlPO_4$.
Properties: White crystals. Insoluble in water and alcohol, soluble in acids and alkalies. Sp. gr. 2.566; m.p. 1500°C.
Derivation: Interaction of solutions of aluminum sulfate and sodium phosphate.
Containers: Bags, drums.
Hazard: Solutions are corrosive to tissue.
Uses: Ceramics, dental cements; cosmetics; paints and varnishes; pharmaceuticals; pulp and paper.
Shipping regulations: (solution) (Air) Corrosive label.

aluminum phosphide AlP.
Properties: Dark gray or dark yellow crystals; sp. gr. 2.85. Evolves phosphine.
Hazard: Toxic; flammable, dangerous fire risk.
Uses: Insecticide; fumigant.
Shipping regulations: (Rail) Flammable solid, n.o.s., Yellow label. (Air) Corrosive label.

aluminum picrate $Al[(NO_2)_3C_6H_2O]_3$.
Properties: Strong oxidizing agent.
Hazard: Toxic by ingestion and inhalation. Dangerous fire risk in contact with combustible materials; severe explosion risk when shocked or heated.
Uses: Explosive compositions.
Shipping regulations: (Rail) Consult regulations (explosive compositions); (Air) Compositions, explosive, n.o.s. Not acceptable.

aluminum potassium sulfate (potash alum; alum, N.F.; potassium alum) $Al_2(SO_4)_3 \cdot K_2SO_4 \cdot 24H_2O$, sometimes written $AlK(SO_4)_2 \cdot 12H_2O$.
Properties: White odorless crystals having an astringent taste; sp. gr. 1.75; m.p. 92°C; b.p., loses $18H_2O$ at 64.5°C; anhydrous at 200°C; soluble in water; insoluble in alcohol. Solutions in water are acid. Noncombustible; nontoxic.
Derivation: (a) From alunite, leucite or similar mineral. (b) Also derived by crystallization from a solution made by dissolving aluminum sulfate and potassium sulfate.
Grades: Technical; lump; ground; powdered; N.F.; F.C.C.
Containers: Bags and drums.
Uses: Dyeing (mordant); paper; matches; paints; tanning agent; waterproofing agent; purification of water; aluminum salts; food additive; medicine.

aluminum resinate.
Properties: Brown solid. Insoluble in water; soluble in oils.
Derivation: Heating soluble aluminum salts and rosin.
Grades: Technical (fused, precipitated).
Uses: Drier for varnishes.
Hazard: Flammable; dangerous fire risk.
Shipping regulations: (Powder) (Rail) Flammable solid, n.o.s., Yellow label. (Air) Flammable Solid label.

aluminum rubidium sulfate (rubidium alum) $AlRb(SO_4)_2 \cdot 12H_2O$.
Properties: Colorless crystals. Soluble in water (hot); insoluble in alcohol. Sp. gr. 1.867; m.p. 99°C.

aluminum salicylate $Al(C_6H_4OHCOO)_3$.
Properties: Reddish-white powder. Odorless. Soluble in dilute alkalies; insoluble in water, alcohol. Decomposed by acid.
Use: Medicine.

aluminum silicate. Any of the numerous types of clay (q.v.), which contain varying proportions of Al_2O_3 and SiO_2. Made synthetically by heating aluminum fluoride at 1000–1200°C with silica and water vapor; the crystals or whiskers obtained are up to 1 cm long, have high strength, and are used in reinforced plastics. See also mullite; kaolin.
Uses: Same as for clay.

aluminum silicofluoride. See aluminum fluosilicate.

aluminum silicon. Ingots and powder.
Hazard (powder): Flammable, dangerous fire risk.
Shipping regulations: (Powder): (Rail) Flammable solid, n.o.s., Yellow label. (Air) Flammable Solid label.

aluminum soaps. See aluminum oleate, aluminum palmitate, aluminum resinate, aluminum stearate. See also soap (2).

aluminum sodium sulfate (SAS; sodium aluminum sulfate; soda alum; alum; porous) $Al_2(SO_4)_3 \cdot Na_2SO_4 \cdot 24H_2O$, or $AlNa(SO_4)_2 \cdot 12H_2O$.
Properties: Colorless crystals; saline, astringent taste; effloresces in air. Soluble in water; insoluble in alcohol. Sp. gr. 1.675; m.p. 61°C. Nontoxic; noncombustible.
Derivation: By heating a solution of aluminum sulfate and adding sodium chloride. The solution is allowed to cool, with constant stirring. The alum meal deposited is washed with water and centrifuged.
Method of purification: Recrystallization.
Grades: Pure crystals; technical; C.P.; F.C.C.
Containers: Bulk in cars; bags; barrels.
Uses: Textiles (mordant, waterproofing); dry colors; ceramics; tanning; paper size precipitant; matches; inks; engraving; sugar refining; water purification; medicine; confectionery; baking powders; food additive.

aluminum stearate (aluminum tristearate) $Al(C_{18}H_{35}O_2)_3$ (approx.).
Properties: White powder; sp. gr. 1.070; m.p. 115°C. Insoluble in water, alcohol, ether; soluble in alkali, petroleum, turpentine oil. Forms gel with aliphatic and aromatic hydrocarbons. Nontoxic.
Derivation: Reaction of aluminum salts with stearic acid.
Grade: Technical.
Containers: 25-, 50-lb bags; cartons; drums.
Uses: Paint and varnish drier; greases; waterproofing agent; cement additive; lubricants; cutting compounds; flatting agent; cosmetics and pharmaceuticals; additive for chewing gums.

aluminum sulfate (trade term for alum; pearl alum; pickle alum; cake alum; filter alum; paper makers' alum; patent alum). (a) $Al_2(SO_4)_3$; (b) $Al_2(SO_4)_3 \cdot 18H_2O$. Noncombustible; nontoxic.
Properties: White crystals; soluble in water (has a sweet taste); insoluble in alcohol. Stable in air. Sp. gr. (a) 2.71, (b) 1.62 m.p. (a) decomposes at 770°C, (b) decomposes at 86.5°C.
Derivation: (1) By treating pure kaolin or aluminum hydroxide or bauxite with sulfuric acid. The insoluble silicic acid is removed by filtration and the sulfate is obtained by crystallization; (2) similarly from waste coal-mining shale and sulfuric acid.
Grades: Iron-free; technical; C.P.; U.S.P.; F.C.C. A liquid form (49.7 H_2O) is also available.
Containers: Bags; fiber drums; multiwall paper sacks; bulk in carloads; solution in rubberlined tank cars or trucks.
Uses: Sizing paper; lakes; alums; mordant for dyeing; foaming agent in fire foams; fireproofing cloth; tannage of white leather; catalyst in manufacture of ethane; pH control in paper industry; waterproof-

ing agent for concrete; clarifying agent for fats and oils; lubricating compositions; deodorizer and decolorizer in petroleum refinery processes; precipitating agent in sewage treatment and water purification; food additive.

aluminum sulfide Al_2S_3.
Properties: Yellowish-gray lumps. Odor of hydrogen sulfide gas. In moist air, decomposes and forms a gray powder; decomposed by water to evolve hydrogen sulfide. Sp. gr. 2.02; m.p. 1100°C.
Hazard: Moderately toxic by ingestion. Irritant to skin and tissue.
Uses: Preparation of hydrogen sulfide.

aluminum sulfocarbolate. See aluminum phenolsulfonate.

aluminum thiocyanate (aluminum sulfocyanate) $Al(SCN)_3$.
Properties: Yellowish powder. Soluble in water; insoluble in alcohol and ether. Probably low toxicity.
Uses: Textile industry; manufacturing pottery.

aluminum trialkyl. See triethylaluminum and triisobutylaluminum.

aluminum tributyl. See tributyl aluminum.

aluminum triformate (aluminum formate, normal) $Al(HCOO)_3 \cdot 3H_2O$.
Properties: White, crystalline powder. Soluble in hot water; slightly soluble in cold water.
Uses: Textile (delustering rayon, mordanting, waterproofing, after-treatment of dyeings); paper (sizing); fur dyeing (mordant); and medicine.

aluminum trimethyl. See trimethyl aluminum.

aluminum triricinoleate $Al(C_{17}H_{32}OHCOO)_3$.
Properties: Yellowish to brown plastic mass. Limited solubility in most organic solvents. M.p. 95°C. Combustible. Low toxicity.
Derivation: Castor oil.
Uses: Gelling agent; waterproofing; solvent-resistant lubricants.

aluminum tristearate. See aluminum stearate.

alum, N.F. May be either aluminum ammonium sulfate or aluminum potassium sulfate.

alum, papermakers'. See aluminum sulfate.

alum, pearl. Specially prepared aluminum sulfate for the paper making industry.

alum, pickle. Aluminum sulfate prepared to meet specifications of packers and preservers.

alum, porous. See aluminum sodium sulfate.

alum, potash. See aluminum potassium sulfate.

alum, rubidium. See aluminum rubidium sulfate.

alum, soda. See aluminum sodium sulfate.

"Alundum."[249] Trademark for a series of fused-alumina refractory and abrasive products. Fusion point 2000–2050°C.
Grades: Grains, cements and refractory shapes.
Uses: (grains) Electrical insulation in radio and television tubes; (cements) embedding electrical resistors, metal melting applications, refractory brick setting; (refractory shapes) high-temperature work, as in furnaces, bricks, plates, muffles, tunnel kilns, tubes, laboratory ware.

alunite (alum stone) $KAl_3(OH)_6(SO_4)_3$. A naturally occurring basic potassium aluminum sulfate, usually found with volcanic and other igneous rocks.
Occurrence: Utah, Arizona, California, Colorado, Nevada, Washington; Italy; Australia.
Uses: Production of alum potassium compounds, millstones; substitute for bauxite in aluminum manufacture; decolorizing and deodorizing agent; fertilizer.

"Alusite" D.[446] Trademark for a 70% alumina brick with relatively low porosity, good resistance to mechanical abrasion and penetration by molten slags. Used in rotary lime, lime sludge and dolomite kilns, soaking pit curb walls, gas regenerators, non-ferrous metallurgical refining furnaces, in reverberatory and brass melting furnaces, and various lead furnaces.

"Alwax" Sizes.[57] Trademark for series of aqueous emulsions of paraffin and microcrystalline waxes.
Uses: Coating paper to resist water, lactic acid, blood serum, and other organic liquids. Improves paper pliability, gives smoother surface, added scuff resistance to container board, and improved printing qualities.

Am Symbol for americium.

"AM."[433] Trademark for brands of selenium dioxide, tellurium dioxide, and germanium dioxide.

amalgam. A mixture or alloy of mercury with any of a number of metals or alloys, including cesium, sodium, tin, zinc, lithium, potassium, gold and silver, as well as with some nonmetals. Dental amalgams are mixtures of mercury with a silver-tin alloy. A sodium amalgam is formed in the preparation of pure sodium hydroxide by electrolysis of brine.
Uses: Dental fillings; silvering mirrors; catalysis; analytical separation of metals; to facilitate application of active metals such as sodium, aluminum, and zinc in the preparation of titanium, etc., or in reduction of organic compounds. See also sodium amalgam.

amantadine hydrochloride (1-adamantanamine hydrochloride) $C_{10}H_{17}N \cdot HCl$. A derivative of adamantane (q.v.). An anti-viral drug. Also used in treatment of Parkinson's disease. See "Symmetrel."

amaranth (FDC Red No. 2; Red Dye No. 2). $NaSO_3C_{10}H_5N{=}NC_{10}H_4(SO_3Na)_2OH$. An azo dye derived from naphthionic and R acids.
Properties: Dark red to purple powder; sp. gr. about 1.50; soluble in water, glycerin, propylene glycol; insoluble in most organic solvents.
Hazard: A suspected carcinogen. Use in foods, drugs and cosmetics restricted.
Uses: Formerly a certified food and drug colorant. Replaced for some applications by FDC Red No. 40.

amatol. An explosive mixture of ammonium nitrate and T.N.T. The 50-50 mixture can be melted and poured for filling small shells; the 80% ammonium nitrate mixture is granular.
Hazard: A high explosive. Moderately toxic by ingestion, inhalation, and skin absorption. Strong irritant.
Shipping regulations: (Rail) Consult regulations. (Air) Not acceptable.

"Amax."[69,265] Trademark for N-oxydiethylene benzothiazole-2-sulfenamide.
Uses: Accelerator for natural rubber and SBR.

amber. A polymerized fossil resin derived from an extinct variety of pine. Readily accumulates static electrical charge by friction; good electrical insulator.

ambergris.
Properties: Irregular, gray, waxy, opaque masses; peculiar odor. Insoluble in water; soluble in alcohol, chloroform, ether, fats, and oils. M.p. about 60°C; sp. gr. 0.80–0.92.
Chief constituents: Cholesterol, ambrein, benzoic acid.
Derivation: Intestinal tract of the sperm whale. Found on beaches and afloat in the ocean.
Uses: Fixative in perfumes; also provides animal note to cosmetics. Its popularity has increased in recent years.

"Amberlac."[23] Trademark for modified alkyd-type resins for quick-drying lacquers. Some grades are resin-modified; some oxidizing and baking grades.
Uses: Metal primers; bottle cap coatings; food can coatings; appliance coatings.
"Amberlac" 165. Trademark for synthetic water-soluble polymer. Colorless additive for film formers; improves scuff resistance; prevents sticking of thermoplastic coatings.
Uses: High gloss paper coatings; coating of book covers, decorative papers, paperboard.

"Amberlite."[23] Trademark for several types of ion-exchange resins. Insoluble crosslinked polymers of various types in minute bead form. Strong acid, weak acid, strong base, and weak base forms, each having various grades differing in exchange capacity and porosity, for removing simple and complex cations and anions from aqueous and non-aqueous solutions. Reversible in action, can be regenerated.
Grades: Laboratory; liquid; nuclear, mixed bed; pharmaceutical.
Uses: Water conditioning (softening and complete deionization); recovery and concentration of metals, antibiotics, vitamins, organic bases; catalysis; decolorization of sugar; manufacture of chemicals; neutralization of acid mine water drainage; analytical chemistry; water treatment in nuclear reactors; pharmaceuticals.

"Amberol."[23] Trademark for maleic-resin and resin-modified and unmodified phenolformaldehyde-type polymers in solid form. They react with various oils to produce fast drying, high gloss protective coatings and vehicles for printing inks.
Uses: Varnishes; enamels; can liners; nitrocellulose sanding sealers; printing inks; tackifying and vulcanization of butyl rubber.

"Ambidex."[458] Trademark for a dark-colored oil-soluble, cationic surface-active agent, containing both an amide group and an amine salt. Used to make oil-in-water and water-in-oil emulsions.

ambient temperature. The temperature of the environment in which an experiment is conducted or in which any physical or chemical event occurs. See also room temperature.

"Ambiflo."[233] Trademark for polyglycol lubricants. Used in brake fluids, rubber mold release formulations, metalworking, compressor and textile fiber lubricants.

"Ambitrol."[233] Trademark for coolants used in stationary industrial engines.

amblygonite $Li(AlF)PO_4$ or $AlPO_4 \cdot LiF$. A natural fluorophosphate of aluminum and lithium.
Properties: White to grayish-white. Contains up to 10.1% lithia, sometimes with partial replacement by sodium. Sp. gr. 3.01–3.09 Mohs hardness 6.
Occurrence: California, Maine, Connecticut, South Dakota; Germany; Norway; France.
Use: A source of lithium; used in glazes and coatings.

"Ambodryl."[330] Trademark for bromodiphenhydramine hydrochloride [beta-(para-bromobenzhydryloxy) ethyl dimethylamine hydrochloride]
$C_{17}H_{21}BrNO \cdot HCl$
Properties: White to off-white odorless amorphous or crystalline powder possessing a bitter taste; melting range 148–150°C. Soluble in water and alcohol; insoluble in ether.
Use: Medicine.

ambomycin (USAN). An antibiotic produced by Streptomyces ambofaciens.

"Ambraloy-687."[324] Trademark for an aluminum-brass alloy containing arsenic as an inhibitor to increase its resistance to dezincification. Its nominal composition is copper 77%, zinc 20.96%, aluminum 2%, arsenic 0.04%. Used principally as a condenser tube alloy where cooling water is salt or brackish, and where it is resistant to the impinging action of turbulently flowing sea water containing air bubbles.

ambrettolide (omega-6-hexadecenlactone; 6-hexadecenolide; 16-hydroxy-6-hexadecenoic acid, omega lactone) $C_{16}H_{28}O_2$. Colorless liquid, having powerful musk-like odor. Found in ambrette-seed oil.
Uses: Flavoring; perfume fixative.

"Amcrom."[433] Trademark for copper-chromium alloys.

"Amerchol."[493] Trademark for a series of surface-active lanolin derivatives; most are soft solids.
Uses: Emulsifiers and stabilizers for water and oil systems; emollients in pharmaceuticals and cosmetics.

American Carbon Society. The present name of the American Carbon Committee, a group incorporated in 1964 to operate the Biennial American Carbon Conferences. The committee also has sponsored the international journal *Carbon*.

American Chemical Society (ACS). The nationally chartered professional society for chemists in the United States. One of the largest scientific organizations in the world, it was started in 1876, and now has over 110,000 members. Its publications include the world-famous *Chemical Abstracts*, *Journal of the American Chemical Society*, a number of high-level journals devoted to major subdivisions of chemistry, as well as technical books, including the ACS Monograph Series. It has 28 Divisions and many regional Sections. The Priestley Medal, considered the highest award in chemistry, was established by the ACS in 1923. The Society celebrated its 100th anniversary at its Centennial Meeting in New York City, April 5–8, 1976. Its offices are at 1155 16th St., N.W., Washington, D.C.
See also Chemical Abstracts Service.

American Conference of Governmental Industrial Hygienists (ACGIH). An official group of scientists established for the purpose of determining standards of exposure to toxic and otherwise harmful materials in workroom air. The standards are revised periodically and are available from the Secretary, Box 1937, Cincinnati, Ohio. See also threshold limit value.

American Institute of Chemical Engineers (AIChE). Founded in 1908, the AIChE is the largest society in the world devoted exclusively to the advancement and development of chemical engineering. Its official publication is *Chemical Engineering Progress*. It has over 50 local sections and many committees working in a wide range of activities. Its offices are located at 345 East 47th Street, N.Y.

American Institute of Chemists. Founded in 1923, the AIC is primarily concerned with chemists and chemical engineers as professional people, rather than with chemistry as a science. Special emphasis is placed on the scientific integrity of the individual and on a code of ethics adhered to by all its members. It publishes a monthly journal, *The Chemist*. Its offices are at 60 East 42nd Street, N.Y.

American Petroleum Institute (API) Incorporated in 1919 under the laws of the District of Columbia; membership about 15,000. The objects of the Institute as stated in its charter, are: (1) to afford a means of cooperation with the government in all matters of national concern; (2) to foster foreign and domestic trade in American petroleum products; (3) to promote, in general, the interests of the petroleum industry in all its branches; (4) to promote the mutual improvement of its members and the study of the arts and sciences connected with the petroleum industry.

The activities of the Institute include the fields of standardization, design, care, and correct practice in the use of equipment; engineering and technology; fundamental research; safety and fire protection; industrial health; product labeling; waste disposal; testing methods and specifications; measuring, sampling, and testing; nomenclature; metallurgy; corrosion prevention; pipeline, highway, waterway, and railroad transportation; radio facilities; fuels and lubricants; agriculture; highways; aviation; vocational and personnel training; education and public relations; finance and accounting; statistics; petroleum reserves; taxation, legislation, and regulation. Its offices are located at 1271 Avenue of the Americas, New York, N.Y. 10020.

American Society for Metals (ASM). Formally organized in 1935, this society actually had been active under other names since 1913, when the need for standards of metal quality and performance in the automobile became generally recognized. ASM publishes *Metals Review* and the famous "Metals Handbook," as well as research monographs on metals. It is active in all phases of metallurgical activity, metal research, education and information retrieval. Its headquarters is at Metals Park, Ohio 44073.

American Society for Testing and Materials (ASTM). This society, organized in 1898, and chartered in 1902, is a scientific and technical organization formed for "the development of standards on characteristics and performance of materials, products, systems, and services; and the promotion of related knowledge." It is the world's largest source of voluntary consensus standards. The society operates through more than 125 main technical committees which function in prescribed fields under regulations that ensure balanced representation among producers, users, and general interest participants. Headquarters of the society is at 1916 Race St., Philadelphia, Pa. 19103.

americium Am A synthetic radioactive element of atomic number 95; a member of the actinide series (q.v.). Atomic weight 241; 14 isotopes of widely varying half-life.

Properties: Normal valence +3, but divalent, tetravalent and higher valencies exist. Alpha and gamma emitter; forms compounds with oxygen, halides, lithium, etc. Metallic americium is silver-white, crystalline, sp. gr. 13.6, m.p. about 100°C. Half-life of Am 241 is 458 years.

Derivation: Multiple neutron capture in plutonium in nuclear reactor; Pu isotopes yield Am 241 and Am 243 on beta decay. The 241 isotope is available in 100-gram quantities, the 243 isotope only in milligram quantities. The metal is obtained by reduction of Am trifluoride with barium in a vacuum at 1200°C.

Hazard: A radioactive poison.

Uses: Gamma radiography; radiochemical research; electronic devices.

Shipping regulations: (Rail, Air) Consult regulations.

"Ameripol" CB.[561] Trademark for synthetic cis-polybutadiene rubber.

"Ameripol" SN.[561] Trademark for a synthetic natural rubber (*cis*-polyisoprene).

"Ameripol" 4502 and 4503.[561] Adhesive polymers based on SBR and produced in free-flowing crumb form to reduce dissolving time and eliminate need for milling equipment.

"Amerlate" P.[493] Trademark for the isopropyl ester of hydroxy, normal and branched chain acids of lanolin. A light yellow soft solid that liquefies on contact with skin. A hydrophilic emollient; moisturizer; conditioning agent; lubricant; pigment dispersant; nonionic auxiliary w/o emulsifier.

"Amersol."[319] Trademark for denatured alcohols for all industrial uses.

amethopterin. See methotrexate.

amiben. Generic name for 3-amino-2,5-dichlorobenzoic acid; $C_6H_2NH_2Cl_2COOH$.

Use: Herbicide or plant growth regulator.

Hazard: May be toxic.

amide. A nitrogenous compound related to or derived from ammonia. Reacton of an alkali metal with ammonia yields inorganic amides, e.g., sodium amide ($NaNH_2$). Organic amides are characterized by an acyl group ($-CONH_2$), usually attached to an organic group ($R{=}CONH_2$). They are closely related to organic acids. Acetamide (CH_3CONH_2) and formamide ($HCONH_2$) are common examples. Carbamide (urea) is $CO(NH_2)_2$. See also polyamide.

4-amidino-1-(nitrosamino-amidino)-1-tetrazene. See tetrazene.

amidol. See 2,4-diaminophenol hydrochloride.

aminacrine hydrochloride. USAN name for 9-aminoacridine hydrochloride.

amination. The process of making an amine (RNH_2). The methods commonly used are (a) reduction of a nitro compound and (b) action of ammonia on a chloro-, hydroxy-, or sulfonic acid compound.

amine. A class of organic compounds of nitrogen that may be considered as derived from ammonia (NH_3) by replacing one or more of the hydrogen atoms with alkyl groups. The amine is primary, secondary, or tertiary depending on whether one, two, or three of the hydrogen atoms are replaced. All amines are basic in nature, and usually combine readily with hydrochloric or other strong acids to form salts. See also fatty amine.

amine 220. 2-(8-Heptadecenyl)-2-imidazoline-1-ethanol, $C_{17}H_{33}C{:}NC_2H_4NC_2H_4OH$.
Properties: Sp. gr. 0.9330 (20/20°C); lb/gal 7.76 (20°C); b.p. 235°C (1 mm); flash point 465°F. Combustible.
Containers: 1-gal can; 5- and 55-gal drums.
Uses: Demulsifier used particularly in the recovery of tar from water-gas process emulsions. A powerful cationic wetting agent. Useful in flotation processes involving siliceous minerals and the formation of emulsions and dispersions under acidic conditions.

amine 248 Dark-colored liquid or paste consisting of a non-volatile amine mixture with bis-(hexamethylene)triamine and its homologs as principal components. Disperses readily in water.
Containers: Up to tank cars.
Uses: Coagulant and flocculating agent; ingredient of antistripping agents for asphalt; corrosion inhibitor; anchoring agent; for making cationic amine salts.

"Amine 750."[266] Trademark for technical grade dehydroabietylamine with low secondary amine content.

amine absorption process. See Girbotol absorption process.

"Amine C, O and S."[219] $CH_3(CH_2)_nCNC_2NC_2N_4OH$. Trademarks for high molecular weight imidazolines soluble in organic solvents but sparingly soluble in water at alkaline pHs. React with acids such as hydrochloric to form water-soluble positively charged colloids with unusual tolerance for electrolyte.
Uses: Fungicides; antistatic treatment for upholstery; water-repellent treatment for plaster; detergents; rust-preventive oils.

"Amine D."[266] Trademark for a technical grade of dehydroabietylamine.
Properties: Pale yellow viscous liquid; oil soluble.
Uses: Production of bactericides; fungicides; corrosion inhibitors; asphalt additives; flotation agents.

"Amine D Acetate."[266] Trademark for acetic acid salt of "Amine D."

aminimide. Any of a group of nitrogenous compounds derived by reaction of 1,1-dimethylhydrazine with an epoxide in the presence of an ester of a carboxylic acid. A number of types of epoxides and esters may be used, providing a wide variety of products, including short- and long-chain aliphatics and aromatics. This family of compounds was announced in 1973; their major uses are considered to be in tire-cord dips to increase adhesion of cord to rubber, in soil-removing detergents (nonionic), in coatings formulations, and in cosmetic creams and shampoos. They are stated to be biodegradable and without toxic hazard. They also have elastomer and cross-linking applications from adhesives, caulks and sealants, to foams and mechanical goods. As isocyanate precursors, aminimides can be prepared in a large number of structural variations. Bis-aminimides are especially valuable for producing stable, single-package prepolymer compositions for *in situ* generation of isocyanates in polyurethane applications.

para-**aminoacetanilide** (4′-aminoacetanilide; N-acetyl-para-phenylenediamine) $NH_2C_6H_4NHCOCH_3$.
Properties: Colorless or reddish crystals; soluble in alcohol and ether, slightly soluble in water. M.p. 162°C; b.p. 267°C. Combustible.
Derivation: Acetylation of para-phenylenediamine.
Containers: Drums.
Hazard: Probably moderately toxic by ingestion.
Uses: Intermediates; azo dyes.

aminoacetic acid. See glycine.

aminoacetophenetidide hydrochloride. See phenocoll hydrochloride.

amino acid. An organic acid containing both a basic amino group (NH_2) and an acidic carboxyl group (COOH); thus they are amphoteric and exist in aqueous solution as dipolar ions. The 25 amino acids that have been established as protein constituents are alpha-amino acids (i.e., the $-NH_2$ group is attached to the carbon atom next to the $-COOH$ group). Many other amino acids occur in the free state in plant or animal tissue. 22 amino acids with structures identical with those that exist today have been identified in pre-Cambrian sedimentary rock, indicating an age of at least 3 million years.

Amino acids can be obtained by hydrolysis of a protein; or they can be synthesized in various ways, especially by fermentation of glucose. An essential amino acid is one which cannot be synthesized by the body and is necessary for survival, namely, isoleucine, phenylalanine, leucine, lysine, methionine, threonine, tryptophan, and valine. Non-essential amino acids (alanine, glycine, and about a dozen others) can be synthesized by the body in adequate quantities. Arginine and histidine are essential during periods of intensive growth.

All the essential and most of the non-essential amino acids have one or more asymmetric carbon atoms and are optically active.

Research in molecular biology has established the fact that heredity characteristics are to a large extent determined by the sequence of amino acids in the genes, which is programmed by deoxyribonucleic acid. This substance is composed of four amino acids, and is itself a protein. (See genetic code, deoxyribonucleic acid, chromatin, protein.)

Note: Use of amino acids as fortification additives to foods restricted by FDA to foods containing proteins.

amino acid oxidase. See amino oxidase.

5-aminoacridine. See 9-aminoacridine.

9-aminoacridine (5-aminoacridine) $C_{13}H_{10}N_2$.
Properties: Sulfur-yellow crystals; m.p. 241°C; moderately strong base. Freely soluble in alcohol.
Use: Medicine (also as hydrochloride).

aminoamylene glycol. See 2-amino-2-ethyl-1,3-propanediol.

5-amino-2-aniliobenzenesulfonic acid. See 4-aminodiphenylamine-2-sulfonic acid.

ortho-**aminoanisole.** See ortho-anisidine.

para-**aminoanisole.** See para-anisidine.

aminoanthraquinone $C_6H_4(CO)_2C_6H_3NH_2$ (tricyclic) (a) 1-amino; (b) 2-amino.
Properties: (a) Red, iridescent needles. (b) Red or orange-brown needles. Soluble in alcohol, chloroform, benzene, and acetone; insoluble in water. M.p. (a) 252°C; (b) 302°C; b.p. sublime (both a and b).
Derivation: By reduction of nitroanthraquinones, or by the substitution of the amino radical direct for the sulfonic acid.

Hazard: May be toxic.
Use: Dye intermediates.

4-aminoantipyrine (4-amino-1,5-dimethyl-2-phenyl-3-pyrazolone; 1,5-dimethyl-2-phenyl-4-aminopyrazolone) $C_{11}H_{13}N_3O$.
Properties: M.p. 107–109°C.
Uses: Medicine; analytical reagent.

para-aminoazobenzene (aniline yellow; phenylazoaniline) $C_6H_5NNC_6H_4NH_2$.
Properties: Yellow to tan crystals. Soluble in alcohol and ether; slightly soluble in water. M.p. 126–128°C; b.p. above 360°C.
Derivation: (a) Heating diazoaminobenzene with aniline hydrochloride as catalyst. (b) Diazotization of a solution of aniline and aniline hydrochloride with hydrochloric acid and sodium nitrite.
Uses: Dyes (chrysoidine, induline, solid yellow and acid yellow); insecticide.

4-aminoazobenzene-3,4′-disulfonic acid (Acid Yellow 9) $C_6H_4(SO_3H)NNC_6H_3NH_2(SO_3H)$. Bright, violet needles. Used for synthesizing dyes; for dyeing wool.

para-**aminoazobenzene hydrochloride** (aminoazobenzene salt) $C_6H_5NNC_6H_4NH_2 \cdot HCl$.
Properties: Steel-blue crystals. Soluble in alcohol; slightly soluble in water.
Derivation: By passing dry hydrogen chloride gas into a solution of aminoazobenzene.
Uses: Dyes; coloring lacquers; intermediate.

aminoazobenzenemonosulfonic acid $NH_2C_6H_4NNC_6H_4SO_3H$.
Properties: Yellowish-white, microscopic needles. Barely soluble in water; almost insoluble in alcohol, ether and chloroform.
Derivation: By sulfonating aminoazobenzene.
Use: Dyestuff manufacture.

ortho-**aminoazotoluene** (Solvent Yellow 3; 2-amino-5-azotoluene; toluazotoluidine) $CH_3C_6H_4N_2C_6H_3NH_2CH_3$.
Properties: Reddish-brown to yellow crystals; soluble in alcohol, ether, oils, and fats; slightly soluble in water; m.p. 100–117°C.
Derivation: From ortho-toluidine by treatment with nitrite and hydrochloric acid.
Uses: Dyes; medicine.

6-para-(para-aminobenzamido)benzamido-1-naphthol-3-sulfonic acid $H_2NC_6H_4CONHC_6H_4CONHC_{10}H_5(OH)(SO_3H)$. Gray paste containing approx. 35% solids. Used as an intermediate.

aminobenzene. See aniline.

para-**aminobenzenearsonic acid.** See arsanilic acid.

2-amino-para-**benzenedisulfonic acid** (aniline-2,5-disulfonic acid) $C_6H_3NH_2(SO_3H)_2 \cdot 4H_2O$.
Properties: Crystals; very soluble in water and alcohol.
Derivation: Boiling sodium salt of 4-chloro-3-nitrobenzene sulfonate with sodium sulfite, resulting in formation of sodium 2-nitrobenzene disulfonate, which is reduced with iron and acetic acid to aniline-2,5-disulfonic acid.
Uses: Intermediate.

4-amino-meta-**benzenedisulfonic acid** (aniline-2,4-disulfonic acid) $C_6H_3NH_2(SO_3H)_2 \cdot 2H_2O$.
Properties: Needles decomposing when heated over 120°C; very soluble in water and alcohol.
Derivation: By heating sulfanilic acid with fuming sulfuric acid at 170–180°C.
Use: Dye intermediate.

meta-**aminobenzenesulfonic acid.** See metanilic acid.

para-**aminobenzenesulfonic acid.** See sulfanilic acid.

meta-**aminobenzoic acid** $C_6H_4NH_2CO_2H$. Yellowish or reddish crystals; sublimes easily; sweet taste. Slightly soluble in water, alcohol, and ether; m.p. 173°C to 174°C.
Use: Dye intermediate.

ortho-**aminobenzoic acid**. See anthranilic acid.

para-**aminobenzoic acid** (PABA) $NH_2C_6H_4CO_2H$. Required by many organisms as a vitamin for growth; active in neutralizing the antibacteriostatic effect of some sulfonamide drugs.

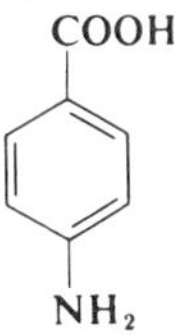

Properties: Light buff odorless crystals; white when pure; discolors on exposure to light and air; m.p. 186–187°C. Sparingly soluble in cold water; soluble in hot water, solutions of alkali hydroxides or carbonates. Unstable to ferric salts and oxidizing agents.
Derivation: Reduction of para-nitrobenzoic acid. Commercially available as the calcium, potassium, and sodium salts.
Food source: Widely distributed, especially in yeast.
Grades: Technical; N.F.
Containers: Bottles; drums.
Uses: Dye intermediate; pharmaceuticals; nutrition; UV absorber in suntan lotions.

2-aminobenzothiazole $C_6H_4NC(NH_2)S$ (bicyclic). Used for chemical intermediates; photographic chemicals.

meta-**aminobenzotrifluoride** $CF_3C_6H_4NH_2$.
Properties: Colorless to oily yellow liquid.
Grade: Technical (88% min); purified (98% min).
Use: Pharmaceutical intermediate.

ortho-**aminobenzoylformic anhydride.** See isatin.

N-(para-**aminobenzoyl)glycine.** See para-aminohippuric acid.

ortho-**aminobiphenyl** (ortho-phenylaniline; ortho-biphenylamine) $C_6H_5C_6H_4NH_2$.
Properties: Colorless or purplish crystals; m.p. 49.3°C; b.p. 299°C; insoluble in water.
Derivation: Reduction of ortho-nitrobiphenyl.
Containers: Drums; tank cars.
Hazard: May be toxic.
Uses: Intermediate for organic synthesis (carbazole); resins; synthetic rubbers.

1-aminobutane. See n-butylamine.

2-aminobutane. See sec-butylamine.

2-amino-1-butanol $CH_3CH_2CHNH_2CH_2OH$.
Properties: Colorless liquid; sp. gr. 0.944 at 20/20°C; m.p. −2°C; b.p. 178°C; (10 mm) 79–80°C; flash point 164°F; wt 7.85 lb/gal (20°C); pH (0.1M aqueous solution) 11.11; refractive index 1.453 at 20°C. Completely miscible in water at 20°C; soluble

in alcohols; corrosive to copper, brass, aluminum. Combustible. Low toxicity.
Containers: 1-gal cans; 5- and 55-gal drums.
Uses: Emulsifying agent (in soap form) for oils, fats, and waxes; absorbent for acidic gases; chemical synthesis.

aminobutyric acid. $CH_3CH_2CH(NH_2)COOH$. A rare amino acid, having three major modifications; optically active; soluble in water.

alpha-**aminocaproic acid.** See norleucine.

aminocaproic lactam. See caprolactam.

aminochlorobenzene. See chloroaniline.

2-amino-4-chlorophenol (para-chloro-ortho-aminophenol) $C_6H_3OHNH_2Cl$.
Properties: Light brown crystals; m.p. 138°C (dec.).
Derivation: Reduction of para-chloro-ortho-nitrophenol.
Use: Intermediate.

2-amino-4-chlorotoluene [5-chloro-2-methylaniline (NH_2 = 1); 4-chloro-ortho-toluidine (CH_3 = 1)] $ClNH_2C_6H_3CH_3$. Off-white solid or light brown oil which tends to darken on storage; m.p. 20–22°C.
Use: Intermediate.

2-amino-5-chlorotoluene $ClNH_2C_6H_3CH_3$. Crystalline solid; m.p. 26–27°C; sparingly soluble in water; soluble in dilute acids. Used as an intermediate.

2-amino-6-chlorotoluene [6-chloro-ortho-toluidine (CH_3 = 1); 3-chloro-2-methylaniline (NH_2 = 1)] $ClNH_2C_6H_3CH_3$. Liquid; m.p. 0–2°C.
Use: Intermediate.

4-amino-2-chlorotoluene [2-chloro-para-toluidine (CH_3 = 1); 3-chloro-4-methylaniline (NH_2 = 1)] $ClNH_2C_6H_3CH_3$. Liquid; m.p. 21–24°C.
Use: Intermediate.

4-amino-2-chlorotoluene-5-sulfonic acid (Brilliant Toning Red Amine; Permanent Red 2B Amine) $ClNH_2CH_3C_6H_2SO_3H$.
Properties: White to buff powder; essentially insoluble as free acid; soluble as sodium or ammonium salt.
Grades: 98.5% min purity.
Containers: Fiber drums.
Uses: Intermediate for azo pigments.

5-amino-2-chlorotoluene-4-sulfonic acid (Lake Red C Amine). $ClNH_2CH_3C_6H_2SO_3H$.
Properties: White to pink powder; essentially insoluble as free acid; soluble as sodium or ammonium salt.
Grades: 98.5% min purity.
Containers: Fiber drums.
Uses: Intermediate for azo pigments.

meta-**amino**-para-**cresol methyl ether.** See 5-methyl-ortho-anisidine.

aminocyclohexane. See cyclohexylamine.

L-amino-dehydrogenase. See amino oxidase.

3-amino-2,5-dichlorobenzoic acid. See amiben.

para-**aminodiethylaniline** (N,N-diethyl-para-phenylenediamine; diethylaminoaniline) $(C_2H_5)_2NC_6H_4NH_2$.
Properties: Liquid; b.p. 260–2°C; insoluble in water; soluble in alcohol and ether.
Derivation: Treatment of diethylaniline with nitrous acid and subsequent reduction.
Hazard: Probably toxic. See aniline.
Uses: Dye intermediate; source of diazonium compounds in diazo copying process.

para-**aminodiethylaniline hydrochloride** $C_{10}H_{16}N_2 \cdot HCl$.
Properties: Colorless needles; soluble in water, alcohol; insoluble in ether.
Hazard: Probably toxic. See aniline.
Use: Color photography.

2-amino-4,6-dihydroxypteridine. See xanthopterin.

para-**aminodimethylaniline** (dimethylaminoaniline; dimethyl-para-phenylenediamine) $(CH_3)_2NC_6H_4NH_2$.
Properties: Colorless, asbestos-like, needles; stable in air when pure. If impure, the crystals liquefy. Soluble in water, alcohol and benzene. M.p. 41°C; b.p. 257°C.
Derivation: By reduction of para-nitrosodimethylaniline with zinc dust and hydrochloric acid.
Method of purification: Recrystallization from mixture of benzene and ligroin.
Hazard: Toxic by ingestion or inhalation of vapor; irritant to skin and eyes.
Uses: Base for production of methylene blue; photodeveloper; reagent for detection of hydrogen sulfide; reagent for cellulose; organic synthesis; reagent for certain bacteria.

aminodimethylbenzene. See xylidine.

1-amino-2,3-dimethylbenzene. See 2,3-xylidine.

1-amino-2,4-dimethylbenzene. See 2,4-xylidine.

1-amino-2,5-dimethylbenzene. See 2,5-xylidine.

1-amino-2,6-dimethylbenzene. See 2,6-xylidine.

4-amino-1,5-dimethyl-2-phenyl-3-pyrazolone. See 4-aminoantipyrine.

2-amino-4,6-dimethylpyridine $(CH_3)_2C_5H_2NNH_2$.
Properties: Solid; m.p. 65.2–68.5°C; b.p. 235°C. Water-soluble.
Derivation: Prepared from 2-aminopyridine.
Use: Organic intermediate.

2-amino-4,6-dinitrophenol. See picramic acid.

para-**aminodiphenyl.** $C_6H_5C_6H_4NH_2$.
Properties: Colorless crystals. M.p. 53°C; b.p. 302°C.
Hazard: Toxic by ingestion, inhalation, skin absorption. A carcinogenic agent.
Uses: Organic research.

para-**aminodiphenylamine** (N-phenyl-para-phenylenediamine) $NH_2C_6N_4NHC_6H_5$.
Properties: Purple powder; m.p. 75°C; insoluble in water; soluble in alcohol and acetone.
Derivation: Reduction of the coupling product of diazotized sulfanilic acid and diphenylamine.
Uses: Dye intermediate; pharmaceuticals; photographic chemicals.

4-aminodiphenylamine-2-sulfonic acid (5-amino-2-anilinobenzenesulfonic acid) $C_6H_5NHC_6H_3NH_2SO_3H$.
Properties: Needle-like crystals. Barely soluble in water.
Derivation: From para-nitrodiphenylamine-ortho-sulfonic acid by reduction with iron and hydrochloric acid.
Use: Synthesis of dyestuffs.

aminodithioformic acid. See dithiocarbamic acid.

aminoethane. See ethylamine.

2-aminoethanesulfonic acid. See taurine.

2-aminoethanethiol (cysteamine; mercaptamine; thioethanolamine) $HSCH_2CH_2NH_2$.
Properties: Crystals with unpleasant odor; oxidizes on contact with air; m.p. 97°C; soluble in water. Combustible.

Uses: Medicine (believed to offer protection against radiation).

1-aminoethanol. See aldehyde ammonia.

2-aminoethanol. See ethanolamine.

2-(2-aminoethoxy)ethanol ("Diglycolamine"; DGA). $NH_2CH_2CH_2OCH_2CH_2OH$.
Properties: Colorless, slightly viscous liquid with a mild amine odor. Miscible with water and alcohols; b.p. 221°C; sp. gr. (20/20°C) 1.0572; flash point 260°F; f.p. -12.5°C. Combustible.
Hazard: Strong irritant to tissue.
Uses: Removal of acid components from gases, especially CO_2 and H_2S from natural gas; intermediate.
Shipping regulations: (Air) Corrosive label.

aminoethylethanolamine. See hydroxyethyl-ethylenediamine.

4-aminoethylglyoxaline. See histamine.

alpha-(**1-aminoethyl**)-meta-**hydroxybenzyl alcohol bitartrate.** See metaraminol bitartrate.

4-(2-aminoethyl)imidazole. See histamine.

beta-**aminoethylisothiourea dihydrobromide** (2-(2-aminoethyl)-2-thiopseudourea dihydrobromide; AET) $C_3H_9N_3S \cdot 2HBr$.
Properties: Crystals; hygroscopic; m.p. 194–195°C.
Derivation: Thiourea is refluxed with 2-bromoethylamine hydrobromide in isopropanol.
Use: Enzyme activator; free radical detoxifier (believed to offer protection against radiation).

para-beta-**aminoethylphenol.** See tyramine.

N-aminoethylpiperazine
$H_2NC_2H_4NCH_2CH_2NHCH_2CH_2$. High boiling triamine combining a primary, secondary and tertiary amine in one molecule.
Properties: Liquid; sp. gr. 0.9837; b.p. 222.0°C; flash point 200°F; freezing point 17.6°. Soluble in water. Combustible.
Containers: 55-gal drums; tank cars.
Hazard: Strong irritant to tissue.
Uses: Epoxy curing agent; intermediate for pharmaceuticals; anthelmintics, surface-active agents, synthetic fibers.
Shipping regulations: (Air) Corrosive label.

2-amino-2-ethyl-1,3-propanediol (AEPD; aminoamylene glycol) $CH_2OHC(C_2H_5)NH_2CH_2OH$.
Uses: Emulsifying agent (in soap form) for oils, fats, and waxes; absorbent for acidic gases; chemical synthesis; paints and other coatings.

3-(2-aminoethyl)pyrazole dihydrochloride. See betazole hydrochloride.

2-aminoethylsulfuric acid $NH_2CH_2CH_2OSO_3H$.
Properties: White, noncorrosive crystalline powder; m.p. 274–280°C; sinters at 274°C and darkens without complete melting at 280°C; sp. gr. 1.782; bulk density 1.007 g/cc. Soluble in water; insoluble in most organic solvents; pH of 1% aqueous solution 4.0 (20°C), 5% aqueous solution 3.3 (39°C).
Use: Organic synthesis of ethyleneimine and various other compounds; amination of cotton.

2-(2-aminoethyl)-2-thiopseudourea dihydrobromide. See beta-aminoethylisothiourea dihydrobromide.

3-amino-alpha-**ethyl-2,4,6-triiodohydrocinnamic acid.** See iopanoic acid.

4-aminofolic acid. See aminopterin.

aminoform. See hexamethylenetetramine.

amino-G acid. See 2-naphthylamine-6,8-disulfonic acid.

alpha-**aminoglutaric acid.** See glutamic acid.

amino-4-guanidovaleric acid. See arginine.

aminohexamethyleneimine.
$NH_2NCH_2CH_2CH_2CH_2CH_2CH_2$.
Properties: Liquid; b.p. 170°C; soluble in water and most organic solvents. Combustible.
Uses: Intermediate for dyes, pharmaceuticals and photographic chemicals.

2-aminohexanoic acid. See norleucine.

6-aminohexanoic acid. See aminocaproic acid.

para-**aminohippuric acid** [N-(para-aminobenzoyl)glycine; PAHA] $NH_2C_2H_4CONHCH_2COOH$.
Properties: White, crystalline powder. Discolors on exposure to light. Slightly soluble in water, alcohol, and most organic solvents. Very soluble in dilute hydrochloric acid and alkalies. Forms a water-soluble sodium salt. M.p. 197–199°C.
Grade: U.S.P.
Uses: Medical diagnostic reagent; intermediate.

aminohydroxybenzoic acids. See aminosalicylic acids.

alpha-**amino**-beta-**hydroxybutyric acid.** See threonine.

2-amino-2-hydroxymethyl-1,3-propanediol. See tris(hydroxymethyl)aminomethane.

alpha-**amino**-beta-**hydroxypropionic acid.** See serine.

alpha-**amino**-beta-**imidazolepropionic acid.** See histidine.

alpha-**aminoisocaproic acid.** See leucine.

alpha-**aminoisovaleric acid.** See valine.

amino-J acid. See 2-naphthylamine-5,7-disulfonic acid.

aminomercuric chloride. See mercury, ammoniated.

aminomethane. See methylamine.

3-amino-4-methoxybenzanilide
$CH_3OC_6H_3(NH_2)CONHC_6H_5$. Gray powder.
Hazard: May be toxic.
Uses: Dyes, pharmaceuticals and other organic chemicals.

1-amino-2-methoxy-5-methylbenzene. See 5-methylorthoanisidine.

meta-(**4-amino-3-methoxyphenylazo) benzenesulfonic acid** $H_2NC_6H_3(OCH_3)NNC_6H_4SO_3H$.
Properties: Maroon paste containing approximately 38% solids.
Use: Intermediate.

para-**aminomethylbenzenesulfonamide hydrochloride.** See mafenide hydrochloride.

l-alpha-(**aminomethyl)-3,4-dihydroxybenzyl alcohol.** See lavarterenol.

4-amino-4'-methyldiphenylamine-2-sulfonic acid (aminotoluidinobenzenesulfonic acid) $CH_3C_6H_4NHC_6H_3NH_2SO_3H$.

Properties: Light to dark gray paste with characteristic odor.
Use: Intermediate.

4-amino-10-methylfolic acid. See methotrexate.

4-amino-2-methyl-1-naphthol. See "Synkamin."

2-amino-3-methylpentanoic acid. See isoleucine.

2-amino-2-methyl-1,3-propanediol (AMPD; aminobutylene glycol; butanediolamine) $CH_2OCH(CH_3)NH_2CH_2OH$. Corrosive to copper, brass, aluminum.
Uses: Emulsifying agent (in soap form) for oils, fats, and waxes; absorbent for acidic gases; chemical synthesis; cosmetics.

2-amino-2-methyl-1-propanol (isobutanolamine; AMP) $CH_3C(CH_3)NH_2CH_2OH$.
Uses: Emulsifying agent (in soap form) for oils, fats, and waxes; absorbent for acidic gases; chemical synthesis.

2-amino-3-methylpyridine (2-amino-3-picoline) N:C(NH₂)C(CH₃):CHCH:CH.
Properties: Liquid. B.p. 221°C; freezing point 29.5–33.3°C; soluble in water.
Derivation: From 2-aminopyridine.
Hazard: May be toxic. See 2-aminopyridine.
Use: Intermediate.

2-amino-4-methylpyridine (2-amino-4-picoline) $N{:}C(NH_2)CH{:}C(CH_3)CH{:}CH$.
Properties: Crystals; b.p. 230.9°C; 115–117°C (11 mm); melting point 96–99.0°C. Sublimes on slow heating; soluble in water, lower alcohols.
Derivation: Prepared from 2-aminopyridine.
Hazard: May be toxic. See 2-aminopyridine.
Use: Intermediate; medicine.

2-amino-5-methylpyridine (2-amino-5-picoline) $N{:}C(NH_2)CH{:}CHC(CH_3){:}CH$.
Properties: Crystals. B.p. 227.1°C; melting point 76.6°C; soluble in water.
Derivation: Prepared from 2-aminopyridine.
Hazard: May be toxic. See 2-aminopyridine.
Uses: Intermediate.

2-amino-6-methylpyridine (2-amino-6-picoline) $N{:}C(NH_2)CH{:}CHCH{:}C(CH_3)$.
Properties: Crystals. B.p. 214.4°C; melting point 43.7°C; soluble in water.
Derivation: Prepared from 2-aminopyridine.
Hazard: May be toxic. See 2-aminopyridine.
Use: Intermediate.

2-amino-4-(methylthio)butyric acid. See methionine.

alpha-**amino**-beta-**methylvaleric acid.** See isoleucine.

alpha-**amino**-gamma-**methylvaleric acid.** See leucine.

3-amino-1,5-naphthalenedisulfonic acid. See 2-naphthylamine-4,8-disulfonic acid.

6-amino-1,3-naphthalenedisulfonic acid. See 2-naphthylamine-5,7-disulfonic acid.

7-amino-1,3-naphthalenedisulfonic acid. See 2-naphthylamine-6,8-disulfonic acid.

1-aminonaphthalene-4-sulfonic acid. See naphthionic acid.

2-amino-1-naphthalenesulfonic acid. See 2-naphthylamine-1-sulfonic acid.

4-amino-1-naphthalenesulfonic acid. See naphthionic acid.

5-amino-1-naphthalenesulfonic acid. See 1-naphthylamine-5-sulfonic acid.

5-amino-2-naphthalenesulfonic acid. See 1-naphthylamine-6-sulfonic acid.

6-amino-2-naphthalenesulfonic acid. See 2-naphthylamine-6-sulfonic acid.

8-amino-1-naphthalenesulfonic acid. See 1-naphthylamine-8-sulfonic acid.

8-amino-2-naphthalenesulfonic acid. See 1-naphthylamine-7-sulfonic acid.

8-amino-1,3,6-naphthalenetrisulfonic acid. See 1-naphthylamine-3,6,8-trisulfonic acid.

1-amino-8-naphthol-2,4-disulfonic acid. See 8-amino-1-naphthol-5,7-disulfonic acid.

1-amino-8-naphthol-3,5-disulfonic acid. (B acid; 8-amino-1-naphthol-4,6-disulfonic acid) $C_{10}H_4NH_2OH(SO_3H)_2$.
Derivation: By sulfonating 1-amino-8-naphthol-3-sulfonic acid.
Use: Dye intermediate.

1-amino-8-naphthol-3,6-disulfonic acid (H acid; 8-amino-1-naphthol-3,6-disulfonic acid) $C_{10}H_4NH_2OH(SO_3H)_2$.
Properties: Gray powder. Soluble in water, alcohol and ether.
Derivation: From alpha-naphthylamine trisulfonic acid by soda fusion.
Use: Dye intermediate.

1-amino-8-naphthol-4,6-disulfonic acid (K acid; 8-amino-1-naphthol-3,5-disulfonic acid) $C_{10}H_4NH_2OH(SO_3H)_2$.
Derivation: By soda fusion of a naphthylamine trisulfonic acid.
Use: Dye intermediate.

2-amino-8-naphthol-3,6-disulfonic acid (2R acid; RR acid; 7-amino-1-naphthol-3,6-disulfonic acid) $C_{10}H_4NH_2OH(SO_3H)_2$.
Derivation: By soda fusion of a naphthylamine trisulfonic acid.
Use: Dye intermediate.

7-amino-1-naphthol-3,6-disulfonic acid. See 2-amino-8-naphthol-3,6-disulfonic acid.

8-amino-1-naphthol-3,5-disulfonic acid. See 1-amino-8-naphthol-4,6-disulfonic acid.

8-amino-1-naphthol-3,6-disulfonic acid. See 1-amino-8-naphthol-3,6-disulfonic acid.

8-amino-1-naphthol-4,6-disulfonic acid. See 1-amino-8-naphthol-3,5-disulfonic acid.

8-amino-1-naphthol-5,7-disulfonic acid. Chicago acid; SS acid, 1-amino-8-naphthol-2,4-disulfonic acid) $C_{10}H_4NH_2OH(SO_3H)_2$.
Properties: Gray paste; white when pure. Soluble in water and sodium hydroxide solution.
Derivation: Fusion of 1,8-naphthosultam-2,4-disulfonic acid with caustic potash.
Uses: Dye intermediate.

1-amino-2-naphthol-4-sulfonic acid (1,2,4 acid) $C_{10}H_5NH_2OHSO_3H$.
Properties: Pinkish-white to gray needles. Soluble in hot water; almost insoluble in cold water.
Derivation: Beta-naphthol is changed to the 1-nitroso-beta-naphthol which is treated with sodium bisulfite.

Upon acidification the free sulfurous acid effects simultaneous reduction and sulfonation.
Uses: Aniline dyes.

1-amino-2-naphthol-6-sulfonic acid $C_{10}H_5NH_2OHSO_3H$.
Properties: Needles or prisms. Slightly soluble in hot water, alcohol; insoluble in ether.
Derivation: By reducing 1,2,6-nitrosonaphtholsulfonic acid.
Uses: Dye intermediate and chemical intermediate.

1-amino-5-naphthol-7-sulfonic acid (M acid; 5-amino-1-naphthol-3-sulfonic acid) $C_{10}H_5NH_2OHSO_3H$.
Properties: Gray needles. Slightly soluble in cold water; soluble in hot water and alcohol.
Use: Dye intermediate.

1-amino-8-naphthol-4-sulfonic acid (S acid; 8-amino-1-naphthol-5-sulfonic acid)
Properties: Gray needles; white when pure. Slightly soluble in water; insoluble in alcohol and ether.
Derivation: Fusion of 1-naphthylamine-4,8-disulfonic acid with caustic soda.
Use: Dye intermediate.

1-amino-8-naphthol-5-sulfonic acid $C_{10}H_5NH_2OHSO_3H$.
Properties: Small needles; slightly soluble in water.
Derivation: Fusion of 1-naphthylamine-5,8-disulfonic acid with caustic soda.
Use: Dye intermediate.

2-amino-1-naphthol-4-sulfonic acid $C_{10}H_5NH_2OHSO_3H$.
Properties: White needles; insoluble in water, alcohol, ether, benzene.
Derivation: By heating 2-nitroso-1-naphthol with sodium sulfite.
Use: Dye intermediate.

2-amino-3-naphthol-6-sulfonic acid (R acid) $C_{10}H_5NH_2OHSO_3H$.
Properties: Needles; slightly soluble in water.
Derivation: Fusion of 2-naphthylamine-3,6-disulfonic acid with caustic soda.
Uses: Dyes.

2-amino-5-naphthol-7-sulfonic acid (J acid; 6-amino-1-naphthol-3-sulfonic acid) $C_{10}H_5NH_2OHSO_3H$.
Properties: Gray needles, white when pure. Soluble in hot water; sparingly soluble in cold water.
Derivation: Fusion of beta-naphthylamine-5,7-disulfonic acid with caustic.
Hazard: May be toxic.
Use: Dyes.

2-amino-8-naphthol-6-sulfonic acid. See 7-amino-1-naphthol-3-sulfonic acid.

5-amino-1-naphthol-3-sulfonic acid. See 1-amino-5-naphthol-7-sulfonic acid.

6-amino-1-naphthol-3-sulfonic acid. See 2-amino-5-naphthol-7-sulfonic acid.

7-amino-1-naphthol-3-sulfonic acid (gamma acid; 2-amino-8-naphthol-6-sulfonic acid) $C_{10}H_5NH_2OHSO_3H$.
Properties: White crystals. Soluble in alcohol and ether; very slightly soluble in water.
Derivation: By heating caustic soda and 2-naphthylamine-6,8-disulfonic acid in an autoclave.
Use: Azo dye intermediate.

8-amino-1-naphthol-5-sulfonic acid. See 1-amino-8-naphthol-4-sulfonic acid.

2-amino-5-nitrothiazole $C_3H_3N_3O_2S$.
Properties: Greenish yellow to orange-yellow fluffy powder; slightly bitter taste. Decomposes 202°C; very sparingly soluble in water; soluble in dilute acids.
Use: Veterinary medicine.

amino oxidase (L-amino acid oxidase; D-amino acid oxidase; L-amino dehydrogenase). An enzyme which catalyzes the deamination of alpha-amino acids by dehydrogenation to keto acids and ammonia. Two types are recognized, acting on the D- and L-amino acids. Recent emphasis has been on characterization of the D-amino oxidase, which is known to contain the flavin isoalloxazine as coenzyme. Both types are found in animal tissue, especially in liver and kidney, as well as in snake venom and certain bacteria.

2-amino-6-oxypurine. See guanine.

1-aminopentane. See n-amylamine.

2-aminophenetole. See ortho-phenetidine.

4-aminophenetole. See para-phenetidine.

meta-**aminophenol** (meta-hydroxyaniline) $C_6H_4NH_2OH$.
Properties: White crystals; m.p. 122°C; soluble in water, alcohol, and ether.
Derivation: Fusion of meta-sulfanilic acid with caustic soda and subsequent extraction of the melt with ether.
Hazard: Moderately toxic; an allergen and skin irritant.
Uses: Dye intermediate; intermediate for para-aminosalicylic acid.

ortho-**aminophenol** (ortho-hydroxyaniline) $C_6H_4NH_2OH$.
Properties: White crystals; turn brown with age; m.p. 172–173°C; sublimes on further heating. Soluble in cold water, alcohol, benzene; freely soluble in ether.
Derivation: By reduction of ortho-nitrophenol mixed with aqueous ammonia by means of a stream of hydrogen sulfide. Also available as the hydrochloride.
Grades: Technical; 99% min.
Containers: Fiber cans and drums.
Hazard: Moderately toxic by ingestion.
Uses: Dyeing furs and hair; dye intermediate for azo and sulfur dyes; pharmaceuticals.

para-**aminophenol** (para-hydroxyaniline) $C_6H_4NH_2OH$.
Properties: White or reddish yellow crystals; turn violet on exposure to light; m.p. 184°C, with decomposition; soluble in water and alcohol.
Derivation: (a) By reduction of para-nitrophenol with iron filings and hydrochloric acid. (b) By electrolytic reduction of nitrobenzene in concentrated sulfuric acid and treatment with an alkali to free the base. Also available as the hydrochloride.
Grades: Technical; photographic.
Containers: Glass bottles; fiber drums.
Hazard: Moderately toxic by ingestion.
Uses: Dyeing textiles, hair, furs, feathers; photographic developer; pharmaceuticals; antioxidants, oil additives.

4-amino-1-phenol-2,6-disulfonic acid $C_6H_2OHNH_2(SO_3H)_2$.
Properties: Fine needles. Soluble in water; slightly soluble in alcohol; insoluble in ether.

Derivation: Action of sulfur dioxide on para-nitrophenol.
Uses: Dyes.

2-amino-1-phenol-4-sulfonic acid (ortho-aminophenol-para-sulfonic acid) $C_6H_3OHNH_2SO_3H$.
Properties: Brown crystals. Fairly soluble in hot water; very soluble in alkaline solution. No melting point. Decomposes on heating.
Derivation: (a) Sulfonation and nitration of chlorobenzene followed by hydrolysis to phenol with caustic soda with subsequent reduction by sodium sulfide. (b) Sulfonation of ortho-aminophenol. (c) Sulfonation of phenol followed by nitration and reduction.
Containers: 100-lb kegs; 250-lb wooden barrels.
Use: Intermediate for dyes.

ortho-**aminophenol**-para-**sulfonic acid.** See 2-amino-1-phenol-4-sulfonic acid.

6-(D-alpha-**aminophenylacetamido)penicillanic acid.** See ampicillin.

para-**aminophenylarsonic acid.** See arsanilic acid.

1-amino-2-phenylethane. See beta-phenylethylamine.

ortho-**aminophenylglyoxalic lactim.** See isatin.

para-**aminophenylmercaptoacetic acid** $NH_2C_6H_4SCH_2CO_2H$.
Properties: M.p. 186–187°C; insoluble in water, alcohol, benzene, chloroform; soluble in aqueous acid or alkali solutions.
Uses: Synthetic intermediate for dyes and pharmaceuticals.

2-(para-**aminophenyl)-6-methylbenzothiazole.** See dehydrothio-para-toluidine.

meta-**aminophenyl methyl carbinol** $NH_2C_6H_4CH(OH)CH_3$.
Properties: Solid; sp. gr. 1.12; b.p. 217.3°C (100 min); melting point 66.4°C; soluble in water; flash point 315°F; combustible.
Uses: Carrier for dyeing synthetic fibers; intermediate for perfume chemicals and pharmaceuticals.

1-(meta-**aminophenyl)-3-methyl-5-pyrazolone** $NH_2C_6H_4NNC(CH_3)CH_2CO$.
Properties: Light tan paste containing approximately 45% solids.
Uses: Intermediate.

alpha-**amino**-beta-**phenylpropionic acid.** See phenylalanine.

aminophylline (theophylline ethylenediamine) $C_{16}H_{24}N_{10}O_4 \cdot xH_2O$. Contains 84–86% anhydrous theophylline; 14–15% ethylenediamine.
Properties: White or slightly yellowish granules or powder with slight ammoniacal odor and a bitter taste. Soluble in water; insoluble in alcohol and ether. Deteriorates slowly on exposure to air. Solutions are alkaline to litmus.
Grade: U.S.P.
Containers: Bottles; 25-, 100-lb drums.
Use: Medicine; veterinary medicine.

aminopicolines. See aminomethylpyridines.

aminoplast resin (amino resin). A class of thermosetting resins made by the reaction of an amine with an aldehyde. The only aldehyde in commercial use is formaldehyde, and the most important amines are urea and melamine.
Uses: Molding, adhesives, laminating, textile finishes, permanent-press fabrics, wash-and-wear apparel fabrics, protective coatings, paper manufacture; leather treatment, binders for fabrics, foundry sands, graphite resistors, plaster of paris fortification, foam structures and ion-exchange resins.
See dimethylol urea; methylol urea; melamine resins; urea-formaldehyde resins.

2-aminopropane. See isopropylamine.

2-aminopropanoic acid. See alanine.

3-aminopropanoic acid. See beta-alanine.

1-amino-2-propanol. See isopropanolamine.

2-amino-1-propanol (2-aminopropyl alcohol; beta-propanolamine) $CH_3CH(NH_2)CH_2OH$.
Properties: Colorless to pale yellow liquid. Both *l* and *dl* forms are available. *dl*-form: fishy odor; boils at 173–176°C; freely soluble in water, alcohol, ether. *l*-form: refractive index 1.4480–1.4495 at 25°C; distillation range about 114°C at 100 mm. Combustible.
Uses: Organic synthesis and chemical intermediate.

3-amino-1-propanol (propanolamine) $H_2NCH_2CH_2CH_2OH$.
Properties: Colorless liquid; m.p. 12.4°C; b.p. 184–186°C, 168°C (500 mm); flash point 175°F; sp. gr. 0.9786 (30°C). Miscible with alcohol, water, acetone and chloroform. Combustible.
Grade: 99% pure.
Hazard: Moderately toxic by ingestion; moderate irritant to tissue.
Uses: Organic intermediate.

2-aminopropyl alcohol. See 2-amino-1-propanol.

4-(3-aminopropyl)morpholine. See N-aminopropylmorpholine.

N-aminopropylmorpholine (4-(3-aminopropyl)morpholine) $CH_2CH_2OCH_2CH_2NC_3H_6NH_2$.
Properties: Colorless liquid; sp. gr. 0.9872 (20/20°C); b.p. 224.5°C; flash point 220°F (O.C.). f.p., -15°C; soluble in water and alcohol. Combustible.
Hazard: Strong irritant to tissue.
Use: Fiber synthesis; chemical intermediate.
Shipping regulations: (Air) Corrosive label.

para-**(2-aminopropyl) phenol.** See hydroxyamphetamine.

gamma-**aminopropyltriethoxysilane.** $NH_2(CH_2)_3Si(OC_2H_5)_3$.
Properties: Liquid; b.p. 217°C; sp. gr. 0.94 (25°C).
Uses: Sizing of glass fibers for making laminates.

gamma-**aminopropyltrimethoxysilane** $NH_2(CH_2)_3Si(OCH_3)_3$. 100% active; water-white liquid; sp. gr. 1.01; refractive index 1.42
Uses: Glass fabric sizing; binder; adhesion promoter.

aminopterin (4-aminofolic acid; aminopteroylglutamic acid) $C_{19}H_{20}N_8O_5 \cdot 2H_2O$. Differs slightly in structure from folic acid (q.v.) and antagonizes the utilization of folic acid by the body; an antimetabolite.
Properties: Occurs as clusters of yellow needles which are soluble in aqueous sodium hydroxide solutions.
Hazard: Toxic by ingestion.
Uses: Medicine; rodenticide.

aminopteroylglutamic acid. See aminopterin.

6-aminopurine. See adenine.

2-aminopyridine (alpha-pyridylamine) $C_5H_4NNH_2$.
Properties: White leaflets or large, colorless crystals; freezing point 58.1°C; b.p. 210.6°C; soluble in water, alcohol, benzene, ether.

Hazard: Toxic by ingestion and inhalation. Tolerance, 0.5 ppm in air.
Use: Intermediate for antihistamines, and other pharmaceuticals.

3-aminopyridine (beta-pyridylamine) $C_6H_4NNH_2$.
Properties: White crystals; m.p. 64°C; b.p. 250–252°C; soluble in water, alcohol, benzene, ether.
Hazard: May be toxic.
Use: Intermediate in preparation of drugs and dyestuffs.

4-aminopyridine $C_5H_4NNH_2$.
Properties: Crystals; m.p. 158.9°C; b.p. 273.5°C; soluble in water.
Derivation: From 2-aminopyridine.
Grade: 95% (minimum).
Hazard: May be toxic.
Used: Intermediate.

amino R acid. See 2-naphthylamine-3,6-disulfonic acid.

4-aminosalicylic acid (PASA; PAS; para-aminosalicylic acid; 4-amino-2-hydroxybenzoic acid) $NH_2C_6H_3(OH)COOH$.
Properties: White or nearly white bulky powder. Odorless or has slight acetous odor. M.p. 150–151°C (decomp.). Affected by light and air. Darkened solutions should not be used. Soluble in sodium bicarbonate and phosphoric acid; somewhat soluble in alcohol; slightly soluble in ether or acetone; practically insoluble in benzene and water; pH (saturated aqueous solution) 3.0–3.7. Low toxicity.
Derivation: From meta-aminophenol and potassium bicarbonate solution under pressure.
Grade: U.S.P.
Use: Medicine; industrial preservative.

5-aminosalicylic acid (meta-aminosalicylic acid; 5-amino-2-hydroxybenzoic acid) $NH_2C_6H_3(OH)COOH$.
Properties: White crystals; sometimes pinkish; m.p. 260–280°C with decomposition; soluble in hot water or alcohol.
Derivation: From the corresponding nitrosalicylic acid by reduction.
Uses: Dyes: intermediate.

alpha-**aminosuccinamic acid.** See asparagine.

aminosuccinic acid. See aspartic acid.

2-aminothiazole (2-thiazylamine) $SCHCHNCNH_2$.
Properties: Light yellow crystals; m.p. 90°C; distills at 3 mm without decomposition. Slightly soluble in cold water, alcohol and ether; soluble in hot water and dilute mineral acids. Low toxicity.
Derivation: Chlorination of vinyl acetate and condensation with thiourea.
Use: Intermediate in synthesis of sulfathiazole.

alpha-**amino**-beta-**thiolpropionic acid.** See cysteine.

aminothiourea. See thiosemicarbazide.

aminotoluene. See meta-, ortho-, or para-toluidine. See also benzylamine.

6-amino-s-triazine-2,4-diol. See ammelide.

3-amino-1,2,4-triazole (amitrole) $NHNC(NH_2)NCH$.
Properties: White crystalline solid; m.p. 156–159°C; soluble in water; slightly soluble in alcohol.
Hazard: Use on food crops restricted; a carcinogen.
Uses: Herbicide and plant growth regulator.

4-amino-3,5,6-trichloropicolinic acid ("Tordon"). $C_5NCl_3(COOH)(NH_2)$. Herbicide and defoliant, especially for perennials, broad-leaf weeds, etc.; considered relatively nontoxic to animals and humans.

3-(3-amino-2,4,6-triiodophenyl)-2-ethylpropanoic acid. See iopanoic acid.

aminourea hydrochloride. See semicarbazide hydrochloride.

"Aminox."[248] Trademark for a low-temperature reaction product of diphenylamine and acetone.
Properties: Light-tan powder; sp. gr. 1.13; m.p. 85–95°C; soluble in acetone, benzene, and ethylene dichloride; insoluble in water and gasoline.
Use: Antioxidant for nylon and light-colored rubber products.

aminoxylene. See xylidine.

amitrole. Generic name for 3-amino-1,2,4-triazole (q.v.).

"Amizyme."[114] Trademark for a series of enzyme preparations, high in dextrinizing or starch-liquefying properties. Available in tablet, powdered or liquid form.
Derivation: Produced by growing pure microbial cultures on select media.
Uses: Conversion of starch; used in paper coatings, adhesives, and textile sizes.

"Ammate."[28] Trademark for ammonium sulfamate in various grades.
Hazard: See ammonium sulfamate.
Use: Non-selective herbicide.

ammelide (6-amino-s-triazine-2,4-diol; cyanuramide) $NC(OH)NC(OH)NC(NH_2)$. Crystalline solid; m.p., decomposes; insoluble in alcohol; slightly soluble in hot water. Similar to melamine and suggested for melamine-type (amino) resins.

ammeline (4,6-diamino-s-triazine-2-ol; cyanurdiamide) $NC(OH)NC(NH_2)NC(NH_2)$. Crystalline solid; m.p., decomposes; insoluble in water and alcohol. A compound similar to melamine and used in melamine-type resins and in special high temperature lubricants.

ammine. A coordination compound formed by the union of ammonia with a metallic substance, in such a way that the nitrogen atoms are linked directly to the metal. See cobaltammine and coordination compound. Note the distinction from amines, in which the nitrogen is attached directly to the carbon atom.

ammonia, anhydrous. NH_3. Third highest-volume chemical produced in U.S. (1975).
Properties: Colorless gas (or liquid); sharp, intensely irritating odor; lighter than air; easily liquefied by pressure; b.p. -33.5°C freezing point -77.7°C; vapor pressure of liquid 8.5 atm (20°C); sp. volume 22.7 cu ft/lb (70°C); sp. gr. of liquid 0.77 at 0°C; 0.6819 at b.p.; very soluble in water, alcohol, and ether. Autoignition temp. 1204°F. Combustible.
Note: Ammonia is the first complex molecule to be identified in interstellar space; it has been observed in galactic dust clouds in the Milky Way, and is believed to constitute the rings of the planet Saturn.
Derivation: Natural gas is the basic raw material for most of the ammonia made in the U.S.; it can also be made from coal. A mixture of natural gas, air, and

steam is converted to CO_2, H_2, and N_2; the CO_2 is absorbed and the mixture is adjusted to a composition of 1 part nitrogen (from the air) to 3 parts hydrogen. See also synthesis gas. Synthesis is carried out at 1000 atm or less and at temperatures up to 700°C. A number of different catalysts have been used, the most effective being produced by fusion of iron oxide (Fe_3O_4) containing aluminum oxide and potassium oxide as promoters, followed by reduction of the oxide. Chemisorption of the nitrogen on the catalyst surface is the rate-controlling step. Ammonia is formed as the end-product of animal metabolism by decomposition of uric acid.
Grades: Commercial 99.5%; refrigerant 99.97%. See also ammonium hydroxide.
Containers: Steel cylinders; barges; unit trains; ocean tankers; special tank cars at -28°F; pressurized tank cars; pipelines.
Hazard: Toxic and irritant by inhalation; tolerance, 25 ppm in air. Inhalation of concentrated fumes may be fatal. Moderate fire risk; explosive limits in air 16 to 25%. Forms explosive compounds in contact with silver or mercury. Safety data sheet available from Manufacturing Chemists Assn., Washington, D.C.
Uses: Fertilizers, either as such or in form of compounds, e.g., ammonium nitrate; manufacture of nitric acid, hydrazine hydrate, hydrogen cyanide, urethane, acrylonitrile, and sodium carbonate (by Solvay process); refrigerant; nitriding of steel; condensation catalyst; synthetic fibers; dyeing; neutralizing agent in petroleum industry; latex preservative; sulfite cooking liquors; fuel cells; rocket fuel; yeast nutrient; developing diazo films.
Shipping regulations: (Rail) Green label. (Air) Nonflammable Compressed Gas label. Not acceptable on passenger planes. (For solutions consult regulations.)
See also Haber, Fritz; synthesis gas.

ammonia, aromatic spirits. A mixture of 10% of ammonia in alcohol. Strong, suffocating odor.
Hazard: Flammable; keep tightly closed; irritating to mucous membranes.
Uses: Medicine; veterinary medicine.

ammonia-soda process. See Solvay process.

ammoniated mercury chloride. See mercury, ammoniated.

ammoniated ruthenium oxychloride. See ruthenium red.

ammoniated superphosphate. Fertilizer produced by mixing ammonia with superphosphate in the ratio of 5 parts to 100.

ammonio-cupric sulfate. See copper sulfate, ammoniated.

ammonio-ferric oxalate. See ferric ammonium oxalate.

ammonio-ferric sulfate. See ferric ammonium sulfate.

"Ammo-nite."[532] Trademark for a mixture of ammonium nitrate (98%) and coating agents (2%). Used as a fertilizer and blasting agent.
Hazard: Dangerous. See ammonium nitrate.

ammonium acetate $NH_4(C_2H_3O_2)$.
Properties: White, deliquescent, crystalline mass. Soluble in water, alcohol. M.p. 114°C; sp. gr. 1.073.
Derivation: By the interaction of glacial acetic acid and ammonia. Combustible.
Uses: Reagent in analytical chemistry; drugs; textile dyeing; preserving meats; foam rubbers; vinyl plastics; explosives.

ammonium acid carbonate. See ammonium bicarbonate.

ammonium acid fluoride. See ammonium bifluoride.

ammonium acid methanearsonate $CH_3AsO(OH)ONH_4$. A post-emergent herbicide; available as "Ansar" 157, a clear solution containing 9.54% elemental arsenic.
Hazard: Highly toxic by ingestion and inhalation.
Shipping regulations: (Rail, Air) Arsenical compounds, n.o.s., Poison label.

ammonium acid phosphate. See ammonium phosphate, monobasic.

ammonium alginate (ammonium polymannuronate) ($C_6H_7O_6 \cdot NH_4$). A hydrophilic, colloid substance. See also algin.
Properties: Filamentous, grainy, granular, or powdered; colorless or slightly yellow and may have a slight characteristic smell and taste. Slowly soluble in water forming a viscous solution. Insoluble in alcohol. Low toxicity.
Grades: Technical; F.C.C.
Uses: Thickening agent and stabilizer in food products.

ammonium alum. See aluminum ammonium sulfate.

ammonium aluminum chloride. See aluminum ammonium chloride.

ammonium arsenate $(NH_4)_2HAsO_4$.
Properties: White crystals or powder efflorescing in air with loss of ammonia; sp. gr. 1.99; Soluble in water; decomposes in hot water.
Hazard: Highly toxic by ingestion or inhalation.
Shipping regulations: (Rail, Air) Poison label.

ammonium aurin tricarboxylate
$C(C_6H_3OHCOONH_4)_2(C_6H_3(O)COONH_4)$. Forms colored lakes with aluminum, chromium, iron and beryllium.

ammonium benzenesulfonate (ammonium sulfonate) $C_6H_5SO_3NH_4$. M.p. 271°C (decomp.); sp. gr. 1.34.
Grade: 35% solution in kerosene.
Hazard: Probably toxic by ingestion. Flammable.

ammonium benzoate $C_6H_5COONH_4$.
Properties: White crystals or powder. Soluble in water, alcohol and glycerol. Decomposes at 198°C; sp. gr. 1.260; sublimes at 160°C.
Uses: Medicine; food preservative.

ammonium biborate. See ammonium borate.

ammonium bicarbonate (ammonium acid carbonate; ammonium hydrogen carbonate) NH_4HCO_3.
Properties: White crystals. Soluble in water; insoluble in alcohol. Sp. gr. 1.586; m.p. decomposes at 36° to 60°C. Noncombustible. Nontoxic.
Derivation: By heating ammonium hydroxide with an excess of carbon dioxide, and evaporation.
Impurities: Ammonium carbonate.
Grades: Technical; C.P.; F.C.C.
Containers: 100-lb drums; carloads.
Hazard: Evolves irritant fumes on heating to 95°F.
Uses: Production of ammonium salts; dyes; leavening agent for cookies, crackers, cream puff doughs; fire-extinguishing compounds; pharmaceuticals; degreasing textiles; blowing agent for foam rubber.

ammonium bichromate. See ammonium dichromate.

ammonium bifluoride (ammonium acid fluoride; ammonium hydrogen fluoride) NH_4HF_2.
Properties: White crystals, deliquescent; sp. gr. 1.211; soluble in water and alcohol.

Derivation: Action of ammonium hydroxide on hydrofluoric acid with subsequent crystallization.
Hazard: Toxic by inhalation. Corrosive to skin. Tolerance (as F): 2.5 mg per cubic meter of air.
Uses: Ceramics; chemical reagent; etching glass (white acid); sterilizer for brewery, dairy and other equipment; electroplating; processing beryllium.
Shipping regulations: (Rail) (liquid) Corrosive liquid, n.o.s. White label. (Air) (solid or solution) Corrosive label.

ammonium binoxalate $(NH_4)HC_2O_4 \cdot H_2O$.
Properties: Colorless crystals. Soluble in water. Sp. gr. 1.556; decomposes on heating.
Derivation: Action of ammonium hydroxide on oxalic acid with subsequent crystallization.
Hazard: Toxic by ingestion and inhalation.
Uses: Analytical reagent; ink removal from fabrics.

ammonium biphosphate. See ammonium phosphate, monobasic.

ammonium bisulfide. See ammonium sulfide.

ammonium bitartrate (acid ammonium tartrate) $(NH_4)HC_4H_4O_6$.
Properties: White crystals. Soluble in water, acids and alkalies; insoluble in alcohol. Sp. gr. 1.636. Low toxicity.
Derivation: By the action of ammonium hydroxide on tartaric acid.
Use: Baking powder.

ammonium borate (ammonium biborate) $NH_4HB_4O_7 \cdot 3H_2O$.
Properties: Colorless crystals; efflorescent with loss of ammonia. Soluble in water. Sp. gr. 2.38–2.95. Noncombustible.
Derivation: Action of ammonium hydroxide on boric acid with subsequent crystallization.
Hazard: Evolves irritant fumes, especially when heated.
Uses: Medicine; fireproofing compounds; electrical condensers; herbicide.

ammonium bromide NH_4Br.
Properties: Colorless crystals or yellowish white powder; soluble in water and alcohol; somewhat hygroscopic. Sp. gr. 2.43; m.p., sublimes. Noncombustible.
Derivation: Action of hydrobromic acid on ammonium hydroxide with subsequent crystallization.
Grades: Technical; pure; C.P.; N.F.
Hazard: Moderately toxic by ingestion and inhalation.
Uses: Precipitating silver salts for photographic plates; medicine (for its bromide ion); analytical chemistry; process engraving; textile finishing; fire retardant.

ammonium cadmium bromide. See cadmium ammonium bromide.

ammonium carbamate $NH_4CO_2NH_2$.
Properties: White, crystalline rhombic powder; exceedingly volatile; forms urea on heating. Soluble in water and alcohol. Sublimes at 60°C. Decomp. in air evolving ammonia.
Derivation: Interaction of dry ammonia gas and carbon dioxide; recovered from ammonia liquor with ammonia and ammonium carbonate.
Grades: Technical.
Hazard: Evolves irritant fumes, especially when heated.
Use: Fertilizer.

ammonium carbazotate. See ammonium picrate.

ammonium carbonate (crystal ammonia; ammonium sesquicarbonate; hartshorn). A mixture of ammonium acid carbonate and ammonium carbamate. $(NH_4)HCO_3 \cdot (NH_4)CO_2NH_2$.
Properties: Colorless crystal plates or white powder; unstable in air, being converted into the bicarbonate. Strong odor of ammonia, sharp ammoniacal taste. Soluble in water; decomposes in hot water, yielding ammonia and carbon dioxide. Noncombustible.
Derivation: Ammonium salts are heated with calcium carbonate.
Method of purification: Sublimation.
Grades: Technical; lumps; cubes; powder; C.P.; N.F.; F.C.C.
Containers: Bottles; barrels; drums.
Hazard: Evolves irritant fumes, especially when heated.
Uses: Ammonium salts; medicine; baking powders; smelling salts; fire extinguishing compounds; pharmaceuticals; textiles (mordant, washing fabrics); fermentation accelerator in wine manufacture; also, organic chemicals, ceramics, in washing wool.

ammonium chlorate. NH_4ClO_3.
Properties: Colorless or white crystals. Water-soluble.
Derivation: Reacton of ammonium chloride with sodium chlorate in solution.
Hazard: Spontaneous chemical reaction with reducing agents. Powerful oxidizer. When contaminated with combustible materials, it may ignite. Shock-sensitive; may detonate when exposed to heat or vibration, especially when contaminated.
Uses: Explosives.
Shipping regulations: (Rail) Chlorates, n.o.s., Yellow label. (Air) Not acceptable.

ammonium chloride (sal ammoniac) NH_4Cl.
Properties: White crystals; cool, saline taste; somewhat hygroscopic. Soluble in water and glycerol; slightly soluble in alcohol. Sublimes 350°C; sp. gr. 1.54.
Derivation: (a) As a by-product of the ammonia-soda process; (b) reaction of ammonium sulfate and sodium chloride solutions.
Grades: Technical (lumps or granulated); C.P.; U.S.P.; F.C.C.
Containers: Barrels; multiwall paper sacks.
Hazard: Toxic by inhalation. Tolerance (fume), 10 mg per cubic meter of air.
Uses: Dry batteries; mordant (dyeing and printing); soldering flux; manufacture of various ammonia compounds; fertilizer; pickling agent in zinc coating and tinning; electroplating; washing powders; snow treatment; resins and adhesive of urea-formaldehyde; medicine; food industry.

ammonium chloroplatinate. See ammonium hexachloroplatinate.

ammonium chloroplatinite (platinous-ammonium chloride; platinum ammonium chloride) $(NH_4)_2PtCl_4$. Dark red ruby crystals; decomposes 140–150°C; soluble in water; insoluble in alcohol. Used in photography.

ammonium chloroosmate. See ammonium hexachloroosmate.

ammonium chromate $(NH_4)_2CrO_4$.
Properties: Yellow crystals; soluble in cold water; insoluble in alcohol. Sp. gr. 1.866; m.p., decomposes.
Derivation: Addition of ammonium hydroxide to a

solution of ammonium bichromate; recovery by crystallization.
Impurities: Dichromates.
Grades: Technical; C.P.
Hazard: Toxic by inhalation; strong irritant.
Uses: Mordant in dyeing; photography (sensitizer for gelatin coatings); analytical reagent; catalyst; corrosion inhibitor.

ammonium chrome alum. See chromium ammonium sulfate.

ammonium chromium sulfate. See chromium ammonium sulfate.

ammonium citrate, dibasic $(NH_4)_2HC_6H_5O_7$.
Properties: White granules; soluble in water; very slightly soluble in alcohol. Low toxicity.
Uses: Pharmaceuticals; rustproofing; cotton printing; plasticizer; analytically in determination of phosphate in fertilizer.

ammonium decaborate. See ammonium pentaborate.

ammonium dichromate (ammonium bichromate) $(NH_4)_2Cr_2O_7$.
Properties: Orange needles; soluble in water and alcohol. Sp. gr. 2.152 (25°C); m.p., decomposes with slight heating.
Derivation: Action of chromic acid on ammonium hydroxide with subsequent crystallization.
Containers: Bottles; cartons; drums; multiwall paper bags.
Hazard: Dusts and solutions are toxic; irritant to eyes and skin; dangerous fire risk. Strong oxidizing agent; may explode in contact with organic materials. Safety data sheet available from Manufacturing Chemists Assn., Washington, D.C.
Uses: Mordant for dyeing; pigments; manufacturing of alizarin; chrome alum; oil purification; pickling; manufacture of catalysts; leather tanning; synthetic perfumes; photography; process engraving, and lithography (sensitizer for photo-chemical insolubilizatin of albumin, etc.) chromic oxide; pyrotechnics.
Shipping regulations: (Rail) Yelow label. (Air) Flammable Solid label.

ammonium dihydrogen phosphate. See ammonium phosphate monobasic.

ammonium dimethyldithiocarbamate $(CH_3)_2NCS_2NH_4$.
Properties: Yellow crystals; soluble in water; decomp. in air.
Grade: 42% solution in water.
Containers: 5-gal cans, 55-gal drums.
Use: Fungicide.

ammonium dinitro-ortho-**cresolate.**
Hazard: Dangerous in contact with combustible materials. Strong oxidizing agent. Flammable.
Use: Herbicide.
Shipping regulations: (Rail) Flammable solid, n.o.s., Yellow label. (Air) Flammable Solid label. Not accepted on passenger planes.

ammonium fluoride NH_4F.
Properties: White crystals; sp. gr. 1.31; decomposed by heat; soluble in cold water.
Derivation: Interaction of ammonium hydroxide and hydrofluoric acid with subsequent crystallization.
Method of purification: Recrystallization.
Grades: Technical; C.P.
Containers: 1-lb waxed bottles; 250-lb barrels.
Hazards: Toxic by inhalation. Corrosive to tissue. Tolerance (as F): 2.5 mg per cubic meter of air.
Uses: Fluorides; analytical chemistry; antiseptic in brewing; etching glass; textile mordant; wood preservation.

ammonium fluosilicate (ammonium silicofluoride) $(NH_4)_2SiF_6$.
Properties: White, crystalline powder; sp. gr. 2.01; soluble in alcohol and water.
Containers: Drums.
Hazard: Toxic by inhalation. Strong irritant to eyes and skin. Tolerance (as F): 2.5 mg per cubic meter of air.
Uses: Laundry sours; moth proofing; disinfectant in brewing industry; glass etching; light metal casting; electroplating.

ammonium gluconate $NH_4C_6H_{11}O_7$.
Properties: White powder; soluble in water; insoluble in alcohol. Optical rotation +11.6° (in water). Nontoxic.
Preparation: From gluconic acid by neutralization with ammonia.
Use: Emulsifying agent for cheese and salad dressings.

ammonium glutamate (monoammonium glutamate). See sodium glutamate.

ammonium hexachloroosmate (ammonium chlorosmate; osmium ammonium chloride) $(NH_4)_2OsCl_6$.
Properties: Red powder. Contains 43.5% osmium. Soluble in alcohol, water.
Hazard: Moderately toxic. See osmium.

ammonium hexachloroplatinate (ammonium chloroplatinate; platinic-ammonium chloride; platinic sal ammoniac; platinum ammonium chloride) $(NH_4)_2PtCl_6$.
Properties: Orange-red crystals, or yellow powder. Slightly soluble in water; insoluble in alcohol. Sp. gr. 3.06; m.p., decomposes. Low toxicity.
Grades: Technical; C.P.
Containers: Glass bottles.
Uses: Plating; platinum sponge.

ammonium hexafluorogermanate $(NH_4)_2GeF_6$.
Properties: White crystalline solid; m.p. 380°C (sublimes); soluble in cold water; insoluble in alcohol; sp. gr. 2.564 g/cc.

ammonium hexanitratocerate. See ceric ammonium nitrate.

ammonium hydrate. See ammonium hydroxide.

ammonium hydrogen carbonate. See ammonium bicarbonate.

ammonium hydrogen fluoride. See ammonium bifluoride.

ammonium hydrosulfide. See ammonium sulfide.

ammonium hydroxide (ammonia solution; aqua ammonia; ammonium hydrate) NH_4OH.
Properties: Colorless liquid; strong odor. Concentrations of solutions range up to about 30% ammonia.
Grades: Technical; C.P.; 16°; 20°; 26°; N.F. (strong); F.C.C.
Containers: Steel tank trucks, tank cars, barges.
Hazard: Toxic by ingestion; Both liquid and vapor extremely irritating, especially to eyes. Safety data sheet available from Manufacturing Chemists Assn., Washington, D.C.
Uses: Textiles; manufacture of rayon, rubber, fertilizers; refrigeration; condensation polymerization; photography (development of latent images); pharmaceuticals; ammonia soaps; lubricants; fireproofing wood; ink manufacture; explosives; ceramics; ammonium compounds; saponifying fats and oils; organic synthesis; detergent; food additive; household cleanser.
Shipping regulations: (Air) Solutions containing free

ammonia (pressurized), Nonflammable Gas label. Not accepted on passenger planes; containing 10% or more ammonia (nonpressurized), No label required.

ammonium hypophosphite $NH_4N_2PO_2$.
Properties: Colorless crystals or white powder; decomposes when heated. Soluble in water and alcohol.
Hazard: Evolves flammable and toxic fumes on heating. See phosphine.

ammonium ichthosulfonate. See ichthammol.

ammonium iodate. NH_4IO_3. White, odorless, granular powder.
Hazard: Fire risk in contact with organic materials. May be toxic.
Use: Oxidizing agent.

ammonium iodide NH_4I.
Properties: White, hygroscopic crystals or powder; soluble in water or alcohol. Sp. gr. 2.56; m.p., sublimes with decomposition. Affected by light.
Hazard: Moderately toxic.
Uses: Iodides; medicine; photography.

ammonium iron tartrate. See ferric-ammonium tartrate.

ammonium lactate $NH_4C_3H_5O_3$.
Properties: Colorless to yellowish syrupy liquid. Slight odor of ammonia. Decomposes when hot. Sp. gr. 1.19–1.21 (15°C). Soluble in water and alcohol.

ammonium laurate, anhydrous $C_{11}H_{23}COONH_4$.
Properties: Tan, wax-like material, free from ammonia odor. Soluble in ethyl alcohol, methyl alcohol, cottonseed oil, and mineral oil. Soluble (hot) in naphtha, toluene, and vegetable oil. Sp. gr. (25°C) 0.88; pH (5% dispersion) 7.6–7.8; m.p. 42–56°C; neut. value 120–125. Nontoxic; combustible.
Uses: Production of oil-in-water emulsions with high oil content; cosmetics.

ammonium lignin sulfonate. See lignin sulfonate.

ammonium linoleate ($C_{17}H_{31}COONH_4$).
Properties: Yelow paste with ammoniacal odor. Soluble in water, ethyl alcohol, methyl alcohol. Emulsifies in naphtha, toluene, mineral oil, and vegetable oil. Sp. gr. 1.1; pH (5% dispersion) 9.5–9.8; total solids 82%. Nontoxic; combustible.
Grades: Technical, 80%.
Uses: Emulsifying agent for oils, waxes, and hydrocarbon solvents; surface tension reducer; detergent; water-repellent finishes.

ammonium metatungstate $(NH_4)_6H_2W_{12}O_{40} \cdot xH_2O$.
Properties: White, crystalline powder; very soluble in water. Absolute density 4g/cc; m.p., decomp. at 200–300°C to relatively stable anhydrous form, $(NH_4)_6H_2W_{12}O_{40}$, which decomp. above 300°C to form tungstic oxide, WO_3.
Containers: Leverpak drums.
Uses: Preparation of high-purity tungsten metal, alloys and chemicals; tungsten catalysts; corrosion-inhibiting solutions.

ammonium metavanadate (ammonium vanadate) NH_4VO_3.
Properties: White crystals; insoluble in saturated ammonium chloride solution; slightly soluble in cold water. Sp. gr. 2.326; m.p., breaks up at 210°C. Low toxicity. Nonflammable.
Derivation: Alkali solutions of vanadium pentoxide and precipitation with ammonium chloride.
Uses: For catalyst as vanadium pentoxide; dyes; varnishes; drier for paints and inks; photography.

ammonium molybdate (molybdic acid 85%) $(NH_4)_6Mo_7O_{24} \cdot 4H_2O$.
Properties: White crystalline powder; soluble in water; insoluble in alcohol. Sp. gr. 2.27; m.p. decomposes. Nonflammable.
Derivation: Dissolving molybdenum trioxide in aqueous ammonia.
Grades: Technical; C.P.; reagent (contains 85% MoO_3).
Hazard: Moderately toxic and irritant.
Uses: Analytical reagent; pigments; catalyst for dehydrogenation and desulfurization in petroleum and coal technology; production of molybdenum metal; source of molybdate ions.

ammonium 12-molybdophosphate (ammonium phosphomolybdate) $(NH_4)_3PO_4 \cdot 12MoO_3 \cdot 3H_2O$, or $(NH_4)_3PMo_{12}O_{40} \cdot xH_2O$.
Properties: Yellow crystalline powder; soluble in alkali; insoluble in alcohol and acids; very slightly soluble in water. Nonflammable.
Derivation: Interaction of ammonium molybdate and phosphoric and nitric acids.
Grade: 91% MoO_3.
Hazard: May be irritant to skin.
Uses: Reagent; ion exchange columns; photographic additives; imparting water resistance.

ammonium 12-molybdosilicate (ammonium silicomolybdate) $(NH_4)_4SiMo_{12}O_{40} \cdot xH_2O$.
Properties: Crystalline, yellow granules; thermally stable. Only slightly soluble in water, ethyl alcohol, and ethyl acetate. Nonflammable.
Grade: Technical; reagent.
Hazard: May be irritant to skin.
Uses: Catalysts; reagents; in atomic energy as precipitants and inorganic ion-exchangers; in photographic processes as fixing agents and oxidizing agents; in plating processes as additives; and in plastics, adhesives, and cement industries for imparting water resistance.

ammonium nickel chloride. See nickel ammonium chloride.

ammonium nickel sulfate. See nickel ammonium sulfate.

ammonium nitrate (Norway saltpeter) NH_4NO_3. Twelfth highest-volume chemical produced in U.S. (1975).
Properties: Colorless crystals; soluble in water, alcohol, and alkalies. Sp. gr. 1.725; m.p. 169.6°C; b.p. decomposes at 210°C, with evolution of oxygen. Low toxicity.
Derivation: Action of ammonia vapor on nitric acid.
Grades: Usually expressed in percent of nitrogen, as 20.5% N; 33.5% N. FGAN is a fertilizer grade, prilled and usually coated with kieselguhr. Also available as an 83% solution. A temperature-stabilized grade is also available which inhibits breakdown of prills due to crystalline changes.
Containers: Bags; carloads; tank cars; tank trucks.
Hazard: May explode under confinement and high temperatures, but not readily detonated. Ventilate well. To fight fire use large amounts of water. The material must be kept as cool as possible and removed from confinement and flooded with water in event of fire. Explodes more readily if contaminated with combustible materials. Strong oxidizing agent. May be made resistant to flame and detonation by proprietary process involving addition of 5 to 10% ammonium phosphate. The Fertilizer Institute states that the fertilizer grade of ammonium nitrate is not explosive.

Uses: Fertilizer; explosives, especially as prills/oil mixture; pyrotechnics; herbicides and insecticides; manufacture of nitrousoxide; absorbent for nitrogen oxides; ingredient of freezing mixtures; oxidizer in solid rocket propellants; nutrient for antibiotics and yeast; catalyst.
Shipping regulations: (Rail) Yellow label. (Air) Oxidizer label. Regulations apply to mixtures, fertilizers, etc.
See also explosive, high.

ammonium nitrate-carbonate mixtures. See calcium ammonium nitrate.

ammonium nitroso-beta-phenyl hydroxylamine. See cupferron.

ammonium oleate $C_{17}H_{33}COONH_4$. An ammonium soap.
Properties: Yellow to brownish, ointment-like mass; ammonium odor; decomposes on heating. Soluble in water and alcohol. Combustible.
Uses: Emulsifying agent; cosmetics.

ammonium oxalate $(NH_4)_2C_2O_4 \cdot H_2O$.
Properties: Colorless crystals; soluble in water. Sp. gr. 1.502; decomposed by heat.
Hazard: Toxic by ingestion.
Uses: Analytical chemistry; safety explosives; manufacture of oxalates; rust and scale removal.
Shipping regulations: (Rail) Poisonous solids, n.o.s., Poison label. (Air) Oxalic salts, solid, Poison label.

ammonium paratungstate. See ammonium tungstate.

ammonium pentaborate (ammonium decaborate) $(NH_4)_2B_{10}O_{16} \cdot 8H_2O$.
Properties: Crystals or powder; soluble in water.
Uses: Intermediate for boron chemicals; as a power-level control in atomic submarines.

ammonium perchlorate (AP; APC) NH_4ClO_4.
Properties: White crystals. Soluble in water. Sp. gr. 1.95; m.p., decomposes on heating.
Derivation: Interaction of ammonium hydroxide, hydrochloric acid, and sodium chlorate. Recovery by crystallization.
Hazard: Strong oxidizing agent; ignites violently with combustibles. Shock-sensitive; may explode when exposed to heat or by spontaneous chemical reaction. Sensitive high explosive when contaminated with reducing materials. Skin irritant.
Uses: Explosives; pyrotechnics; analytical chemistry; etching and engraving agent; smokeless rocket and jet propellant.
Shipping regulations: (Rail) Yellow label. (Air) Oxidizer label.

ammonium permanganate NH_4MnO_4.
Properties: Crystal or powder form, having a metallic sheen in rich violet-brown or dark purple shades; a powerful oxidizing material; soluble in water.
Hazard: Strong oxidizing agent; may explode on shock or on exposure to heat. Toxic by ingestion and inhalation of dust and fume.
Shipping reglations: (Rail) Yellow label. (Air) Not acceptable.

ammonium perrhenate NH_4ReO_4.
Properties: Colorless, only weakly oxidizing solid; stable to heat; density 3.55 g/cc; slightly soluble in cold water but moderately soluble in hot water; decomp. at 365°C.
Derivation: Liquid ion-exchange.
Hazard: Moderate fire risk by reaction with reducing materials.

ammonium persulfate $(NH_4)_2S_2O_8$.
Properties: White crystals; strong oxidizing agent; soluble in water. Sp. gr. 1.98; m.p., decomposes.
Derivation: Electrolysis of a concentrated solution of ammonium sulfate. Recovered by crystallization.
Hazard: Oxidizing agent; fire risk in contact with organic materials.
Uses: Oxidizing agent; bleaching agent; photography; etchant for printed circuit boards; etching copper; electroplating; manufacture of other persulfates; deodorizing and bleaching oils; aniline dyes; preserving food; depolarizer in batteries; washing infected yeast.

ammonium phosphate, dibasic (ammonium phosphate, secondary; diammonium hydrogen phosphate; diammonium phosphate; DAP) $(NH_4)_2HPO_4$.
Properties: White crystals or powder; sp. gr. 1.619; mildly alkaline in reaction; soluble in water; insoluble in alcohol. Noncombustible. Low toxicity.
Derivation: Interaction of ammonium hydroxide and phosphoric acid in proper proportions.
Grades: Technical; C.P.; fertilizer; feed; dentifrice; highly purified, for phosphors; F.C.C.
Uses: Flameproofing of wood, paper, and textiles; coating vegetation to retard forest fires; to prevent afterglow in matches, and smoking of candlewicks; fertilizer (high analysis phosphate type); plant nutrient solutions; manufacture of yeast, vinegar, and bread improvers; feed additive; flux for soldering tin, copper, brass, zinc; purifying sugar; in ammoniated dentrifices; halophosphate phosphors.

ammonium phosphate, hemibasic $NH_4H_2PO_4 \cdot H_3PO_4$.
Properties: White crystalline material; somewhat hygroscopic. Strongly acid in reaction. Soluble in water. Noncombustible. Low toxicity.
Uses: Nutrient for truck gardens; yeast food; buffer for adjustment of pH values; metal cleaning.

ammonium phosphate, monobasic (ammonium acid phosphate; ammonium biphosphate; ammonium dihydrogen phosphate; ammonium phosphate, primary) $NH_4H_2PO_4$.
Properties: Brilliant white crystals or powder. Mildly acid in reaction. Moderately soluble in water; sp. gr. 1.803. Noncombustible. Low toxicity.
Derivation: Interaction of phosphoric acid and ammonia in proper proportions.
Grades: Technical; C.P.; F.C.C.; single crystals.
Uses: Fertilizers, flameproofing agent; to prevent afterglow in matches; plant nutrient solutions; manufacture of yeast, vinegar, yeast foods, and bread improvers; food additive; analytical chemistry.

ammonium phosphite (neutral ammonium phosphite) $(NH_4)_2HPO_3 \cdot H_2O$.
Properties: Colorless, crystalline mass. Hygroscopic. Soluble in water.
Grades: Technical.
Use: Chemical (reducing agent).

ammonium phosphomolybdate. See ammonium 12-molybdophosphate.

ammonium phosphotungstate (ammonium phosphowolframate) $2(NH_4)_3PO_4 \cdot 24WO_3 \cdot xH_2O$.
Properties: White powder. Soluble in alkali; insoluble in acid; slightly soluble in water.
Derivation: Interaction of ammonium tungstate, ammonium phosphate and nitric acid.
Use: Chemical reagent.

ammonium phosphowolframate. See ammonium phosphotungstate.

ammonium picrate (ammonium carbazoate; ammonium picronitrate) $C_6H_2(NO_2)_3ONH_4$.
Properties: Yellow crystals; sp. gr. 1.72; m.p., decomposes; slightly soluble in water and alcohol.
Hazard: A high explosive when dry; flammable when wet.
Uses: Pyrotechnics; explosive compositions.
Shipping regulations: (Rail) Consult regulations. (Air) Dry, or with less than 10% water. Not acceptable. Wet with not less than 10% water. Flammable Solid label.

ammonium polymannuronate. See ammonium alginate.

ammonium polyphosphate. See urea-ammonium polyphosphate.

ammonium polysulfide $(NH_4)_2S_x$.
Properties: Known only in solution; yellow, unstable; hydrogen sulfide odor; decomposed by acids with evolution of hydrogen sulfide.
Derivation: Passing hydrogen sulfide into 28% ammonium hydroxide and dissolving an excess of sulfur in the resulting solution.
Hazard: Evolves toxic and flammable gas on contact with acids.
Uses: Analytical reagent; insecticide spray.

ammonium reineckate. See Reinecke salt.

ammonium ricinoleate $C_{17}H_{32}OHCOONH_4$.
Properties: White paste. Combustible.
Grades: Technical.
Uses: Detergent; emulsifying agent.

ammonium saccharin $C_7H_8N_2O_3S$.
Properties: White crystals or a white crystalline powder; freely soluble in water. About 500 times as sweet as sucrose in dilute solutions. See saccharin.
Grade: F.C.C.
Use: Nonnutritive sweetener.

ammonium salicylate $C_6H_4OHCOONH_4$.
Properties: Colorless crystals and white powder with pink tinge; odorless; stable in dry air, but affected by light. Soluble in water and alcohol.
Use: Medicine.

ammonium salt. A salt formed on direct union of ammonia or neutralization of ammonium hydroxide with acids. They are in general white and are soluble in water. Usually decomposed by heat into ammonia and the corresponding acid, which may also be decomposed. All ammonium salts liberate ammonia (NH_3) when heated with a strong base, e.g., sodium hydroxide or calcium hydroxide.

ammonium selenate $(NH_4)_2SeO_4$.
Properties: Colorless crystals; sp. gr. 2.194; soluble in water; insoluble in alcohol.
Hazard: Moderately toxic.
Use: Mothproofing agent.

ammonium selenite $(NH_4)_2SeO_3 \cdot H_2O$.
Properties: Colorless or slightly reddish crystals. Keep away from dust or light! Soluble in water.
Grades: Technical.
Hazard: Moderately toxic.
Uses: Analysis (test for alkaloids); glass colorant.

ammonium sesquicarbonate. See ammonium carbonate.

ammonium silicofluoride. See ammonium fluosilicate.

ammonium silicomolybdate. See ammonium 12-molybdosilicate.

ammonium soap. A soap resulting from the reaction of a fatty acid with ammonium hydroxide. Has an appreciable vapor pressure of ammonia and decomposes on continued exposure leaving the fatty acid residue. Usually not sold as detergents but used in toilet preparations and emulsions.

ammonium stearate $C_{17}H_{35}COONH_4$.
Properties: Tan, wax-like solid, free from ammonia odor. Dispersible in hot water. Soluble (hot) in toluene; partially soluble (hot) in butyl acetate and ethyl alcohol. Sp. gr. (22°C) 0.89; pH (3% dispersion) 7.6; m.p. 73–75°C; neutralization value 70–80. Low toxicity. Combustible.
Grades: Available as anhydrous solid or as paste.
Uses: Vanishing creams, brushless shaving creams, other cosmetic products; waterproofing of cements, concrete, stucco, paper and textiles, etc.

ammonium sulfamate $NH_4OSO_2NH_2$.
Properties: White crystalline solid; m.p. 130°C, decomposes at 160°C. Soluble in water. Nonflammable.
Derivation: Hydrolysis of the product obtained when urea is treated with fuming sulfuric acid.
Containers: Bags, carlots, truck loads.
Hazard: Hot acid solutions when enclosed may explode. Moderately toxic by ingestion. Tolerance, 10 mg/cubic meter of air.
Uses: Flameproofing agent for textiles and certain grades of paper; weed and brush killer; electroplating; generation of nitrous oxide.

ammonium sulfate $(NH_4)_2SO_4$.
Properties: Brownish-gray to white crystals according to degree of purity. Soluble in water; insoluble in alcohol and acetone. Sp. gr. 1.77; m.p. 513°C, with decomposition. Low toxicity. Nonflammable.
Derivation: (a) Ammoniacal vapors from destructive distillation of coal react with sulfuric acid, followed by crystallization and drying. (b) Synthetic ammonia is neutralized with sulfuric acid. (c) By-product of manufacture of caprolactam (q.v.); (d) from gypsum, by reaction with ammonia and carbon dioxide.
Method of purifying: Recrystallization or sublimation.
Grades: Commercial; technical; C.P.; enzyme (no heavy metals); F.C.C.
Containers: Bags, drums; bulk in railroad cars and ships.
Uses: Fertilizers; water treatment; fermentation; fireproofing compositions; viscose rayon; tanning; food additive.

ammonium sulfate nitrate.
Properties: A double salt of approximately 60% ammonium sulfate and 40% ammonium nitrate; 26% nitrogen content. White to light gray granules; soluble in water.
Hazard: Oxidizing material; dangerous in contact with organic materials.
Shipping regulations: Nitrates, n.o.s. (Rail) Yellow label. (Air) Oxidizer label.

ammonium sulfide $(NH_4)_2S$. The true sulfide is stable only in the absence of moisture and below 0°C. The ammonium sulfide of commerce is largely ammonium bisulfide or hydrosulfide, NH_4HS.
Properties: Yellow crystals. Soluble in water, alcohol, and alkalies. M.p., decomposes. Evolves hydrogen sulfide on contact with acids.
Grades: Technical; C.P.; liquid, 40–44%.

Containers: Iron drums; tins; glass bottles; tanks.
Hazard: Strong irritant to skin and mucous membranes. Toxic by skin absorption.
Uses: Textile industry; photography (developers); coloring brasses, bronzes; iron control in soda ash production.

ammonium sulfite $(NH_4)_2SO_3 \cdot H_2O$.
Properties: Colorless crystals; acrid, sulfurous taste. Hygroscopic; sublimes at 150°C with decomposition; soluble in water. Sp. gr. 1.41.
Hazard: Moderately toxic by ingestion and inhalation.
Uses: Chemical (intermediates, reducing agent); medicine; permanent wave solutions; photography.

ammonium sulfocyanate. See ammonium thiocyanate.

ammonium sulfocyanide. See ammonium thiocyanate.

ammonium sulfonate. See ammonium benzenesulfonate.

ammonium sulforicinoleate.
Properties: Yellow liquid; soluble in alcohol. Combustible.
Grades: Technical.
Containers: 400-lb barrels.
Use: Medicine; furniture polish.

ammonium tartrate $(NH_4)_2C_4H_4O_6$.
Properties: White crystals; soluble in water and alcohol. Sp. gr. 1.601; decomposes on heating.
Uses: Textile industry; medicine.

ammonium tetrathiocyanodiammonochromate. See Reinecke salt.

ammonium tetrathiotungstate $(NH_4)_2WS_4$.
Properties: Orange-colored, crystalline powder; sensitive to heat; hydrogen sufide odor; m.p., decomposes. Soluble in water, ammoniacal and amine solutions.
Uses: Source of high purity tungsten disulfide for catalysts, lubricants, semiconductors.

ammonium thiocyanate (ammonium sulfocyanide; ammonium sulfocyanate) NH_4SCN.
Properties: Colorless, deliquescent crystals; soluble in water, alcohol, acetone, and ammonia. Sp. gr. 1.3057; m.p. 149.6°C, decomposes at 170°C.
Derivation: By boiling an aqueous solution of ammonium cyanide with sulfur or polysulfides, or by the reaction of ammonia and carbon disulfide.
Grades: Technical; C.P.; 50–60% solution.
Containers: 50-, 100-lb bags; plastic-lined kegs; drums; solution in tank cars.
Uses: Analytical chemistry; chemicals (thiourea); fertilizers; photography; ingredient of freezing solutions, especially liquid rocket propellants; fabric dyeing; zinc coating; weed killer and defoliant; adhesives; curing resins; pickling iron and steel; electroplating; temporary soil sterilizer; polymerization catalyst; separator of zirconium and hafnium, of gold and iron.

ammonium thioglycollate $HSCH_2COONH_4$.
Properties: Colorless liquid, repulsive odor. Evolves hydrogen sulfide. Combustible.
Containers: 55-gal drums, commercial grade.
Uses: Solutions of various strengths are used for hair waving and for hair removal.

ammonium thiosulfate $(NH_4)_2S_2O_3$.
Properties: White crystals, decomposed by heat; very soluble in water; pH of 60% solution 6.5–7.0. Low toxicity.
Grades: Pure crystals (97%); 60% photographic solution.
Uses: Photographic fixing agent, especially for rapid development; analytical reagent; fungicide; reducing agent; brightener in silver plating baths; cleaning compounds for zinc-base die-cast metals; hair-waving preparations; fog screens.

ammonium tungstate (ammonium wolframate; ammonium paratungstate) $(NH_4)_6W_7O_{24} \cdot 6H_2O$. See also ammonium metatungstate.
Properties: White crystals; soluble in water; insoluble in alcohol.
Derivation: Interaction of ammonium hydroxide and tungstic acid with subsequent crystallization.
Containers: Polyethylene bags; Leverpak drums.
Uses: Preparation of ammonium phosphotungstate and other tungsten compounds.

ammonium uranium carbonate. See uranyl ammonium carbonate.

ammonium uranium fluoride. See uranyl ammonium fluoride.

ammonium vanadate. See ammonium metavanadate.

ammonium wolframate. See ammonium tungstate.

ammonium zirconifluoride. See zirconium ammonium fluoride.

ammonium zirconyl carbonate $(NH_4)_3ZrOH(CO_3)_3 \cdot 2H_2O$.
Properties: Available in aqueous solution; sp. gr. 1.238 (24°C). Stable up to about 68°C; decomposes in dilute acids, alkalies.
Containers: 500-lb drums.
Uses: Ingredient in water repellents for paper and textiles; catalyst; stabilizer in latex emulsion paints; ingredient in floor wax to aid in resistance to detergents; lubricant in fabrication of glass fibers.

ammonobasic mercuric chloride. See mercury, ammoniated.

"Ammonyx."[328] Trademark for a series of quaternary ammonium chloride derivatives, in which the substituents are methyl, ethyl, benzyl, stearyl, lauryl, phenyl and cetyl groups. Some items are similar isoquinolinium and pyridinium salts. These are all cationics. The "O" series are alkylamine oxides, and are nonionics.
Uses: Softeners; wetting agents; emulsifiers; some grades are germicides, fungicides, disinfectants, and slimicides.

"Ammo-Phos."[84] Trademark for high analysis ammonium phosphate-containing fertilizers.

"Amnic."[433] Trademark for copper-nickel alloy.

amobarbital (5-ethyl-5-isoamylbarbituric acid) $C_{11}N_{18}N_2O_3$.
Properties: White, crystalline powder; odorless with bitter taste. M.p. 156–161°C; solutions are acid to litmus. Very slightly soluble in water; soluble in alcohol.
Grade: U.S.P.
Hazard: Toxic by ingestion; may be habit-forming.
Use: Medicine (also as sodium salt).

amodiaquine hydrochloride $ClC_9H_5NNHC_6H_3(OH)CH_2N(C_2H_5)_2 \cdot 2HCl \cdot 2H_2O$.
Properties: Yellow, odorless bitter, crystalline solid. M.p. 150–160°C (dec). Soluble in water; sparingly soluble in alcohol; very slightly soluble in benzene, chloroform, and ether; pH (1% solution) 4.0–4.8.
Grade: N.F.
Use: Medicine (antimalarial).

amorphous. Noncrystalline; having no molecular lattice structure, which is characteristic of the solid state.

All liquids are amorphous; some materials that are apparently solid, such as glasses, or semisolid, such as some high polymers, rubber and sulfur allotropes, also lack a definite crystal structure and a well-defined melting point. They are considered as high-viscosity liquids. The cellulose molecule contains amorphous as well as crystalline areas. Carbon derived by thermal decomposition or partial combustion of coal, petroleum and wood is amorphous (coke, carbon black, charcoal), though other forms (diamond, graphite) are crystalline.

AMP.
(1) Abbreviation for 2-amino-2-methyl-1-propanol.
(2) Abbreviation for adenosine monophosphate. See adenylic acid.

A5MP. Abbreviation for adenosine-5-monophosphoric acid. See adenylic acid (muscle adenylic acid)

"Ampco."[407] Trademark for a series of aluminum-iron-copper alloys containing 6–15% aluminum, 1.5–5.25% iron, balance copper. Resistant to fatigue, corrosion, erosion, wear and cavitation-pitting. Used for bushings, bearings, gears, slides, etc.

"Ampcoflex."[41] Trademark for normal-impact rigid sheet and also pipe and fittings of Type I unplasticized ployvinyl chloride used to fabricate structures where optimum corrosion resistance is desired.

"Ampcoloy."[407] Trademark for a series of industrial copper alloys including low iron-aluminum bronzes, nickel-aluminum bronzes, tin bronzes, manganese bronzes, lead bronzes, beryllium-coppers and high-conductivity alloys.

"Ampco-Trode."[407] Trademark for a series of aluminum-bronze, arc-welding electrodes and filler rod, containing 9.0–15.0% aluminum, 1.0–5.0% iron, balance copper, for joining like or dissimilar metals and overlaying surfaces resistant to wear, corrosion, erosion and cavitation-pitting.

AMPD. Abbreviation for 2-amino-2-methyl-1,3-propanediol.

amphetamine (1-phenyl-2-aminopropane; methylphenethylamine; "Benzedrine").
$C_6H_5CH_2CH(NH_2)CH_3$.
Properties: Colorless, volatile liquid; characteristic strong odor and slightly burning taste; b.p. 200–203°C (dec); flash point 80°F; soluble in alcohol and ether; slightly soluble in water.
Grades: Dextro-, dextrolevo-.
Containers: Glass bottles.
Hazard: Flammable, moderate fire risk. Basis of a group of hallucinogenic, habit-forming drugs which affect the CNS. Sale and use restricted to physicians. Production limited by law.
Use: Medicine. Also available as phosphate and sulfate.

amphibole. A type of asbestos. See asbestos.

ampholyte. A substance that can ionize to form either anions or cations, and thus may act as either an acid or a base. An ampholytic detergent is cationic in acid media and anionic in basic media. Water is an ampholyte. See also amphoteric.

"Amphos."[433] Trademark for phosphorized oxygen-free copper in cast or wrought form.

amphoteric. Having the capacity of behaving either as an acid or a base; thus aluminum hydroxide neutralizes acids with the formation of aluminum salts: $Al(OH)_3 + 3HCL \longrightarrow AlCl_3 + 3H_2O$, and also dissolves in strongly basic solutions to form aluminates: $Al(OH)_3 + 3NaOH \longrightarrow Na_3AlO_3 + 3H_2O$. Amino acids and proteins are amphoteric, i.e., their molecules contain both an acid group (COOH) and a basic group (NH_2). Thus wool can absorb both acidic and basic dyes.

amphotericin B. A polyene antifungal antibiotic.
Properties: Pale yellow semicrystalline powder; m.p. gradual decomposition above 170°C. Insoluble in water; slightly soluble in methanol. Somewhat more soluble in dimethylsulfoxide.
Derivation: Fermentation with Streptomyces nodosus. Commercially available as a deoxycholate complex.
Grade: U.S.P.
Hazard: May have side effects.
Use: Medicine (meningitis treatment).

ampicillin (USAN) (6-(D-alpha-aminophenyl-acetamido)-penicillanic acid) $C_{16}H_{19}N_3O_4S$. An antibiotic; used in medicine. Grade: N.D.

"Amprol."[123] Trademark for amprolium (q.v.).

amprolium. 1-[(4-Amino-2-propyl-5-pyrimidinyl)-methyl]-2-picolinium chloride hydrochloride. A coccidiostat.

amprotropine phosphate (phosphate of the *dl*-tropic acid ester of 3-diethylamino-2,2-dimethyl-1-propanol) $C_{18}H_{29}NO_3 \cdot H_3PO_4$.
Properties: Bitter crystals. M.p. 142–144°C. Soluble in water, slightly soluble in alcohol.
Use: Medicine (antispasmodic).

"Ampvar."[41] Trademark for synthetic-resin metal conditioner of the vinyl-phosphoric acid-zinc chromate type used to prepare metal surfaces for the application of corrosion-proof coatings.

"Amthio."[50] Trademark for an ammonium thiosulfate solution.
Properties: A reddish liquid fertilizer and soil conditioner; used before planting. Contains 12% nitrogen and 26% sulfur. Sp. gr. 1.33; weighs approx 11.1 lb/gal. Can be mixed and applied with many liquid fertilizer solutions, or alone in irrigation water.
Precautions: Avoid contact with skin or eyes.

amygdalic acid. See mandelic acid.

amygdalin (mandelonitrile beta-gentiobioside; amygdaloside) $C_6H_5CHCNOC_{12}H_{21}O_{10}$. A glycoside found in bitter almonds.
Properties: White crystals; bitter taste. Anhydrous form m.p. 214–216°; soluble in water and alcohol; insoluble in ether.
Note: "Bitter almonds contain amygdalin, together with an enzyme that catalyzes its hydrolysis. When the kernels are ground and moistened, a volatile oil produced by the hydrolysis can be distilled from them, consisting mainly of benzaldehyde and hydrocyanic acid. This is the oil of bitter almond, used in pharmacy as a food flavor after removal of the hydrocyanic acid." (Eckey, "Vegetable Fats and Oils.") See also oil of bitter almond.

amyl. The five-carbon aliphatic group C_5H_{11}, also known as pentyl. Eight isomeric arrangements (exclusive of optical isomers) are possible. In addition to this theoretical source of confusion, the amyl compounds occur (as in fusel oil), or are formed (as from the petroleum pentanes) as mixtures of several isomers, and since their boiling points are close and their other properties similar, it is neither easy nor usually necessary to purify them. As used in this dictionary, amyl means a mixture of isomers, unless a specific isomer is designated. Several entries are under isoamyl. See also amyl alcohol.

amyl acetate (amylacetic ester; banana oil; pear oil) $CH_3COOC_5H_{11}$. Commercial amyl acetate is a mixture of isomers, the composition and properties depending upon the grade and derivation. The principal isomers are isoamyl, normal, and secondary amyl acetates (q.v.).
Properties: Colorless liquid; flash point from 65 to 95°F (C.C.), depending on grade; persistent banana-like odor. Autoignition temp. about 714°F.
Derivation: Esterification of amyl alcohol (often fusel oil) with acetic acid and a small amount of sulfuric acid as catalyst.
Method of purification: Rectification.
Grades: Commercial (85–88%), high test (85–88%), technical (90–95%), pure (95–99%); special antibiotic grade. Amyl acetate is also sold by original source, as from fusel oil, pentane, or Oxo process.
Containers: Drums; tank cars.
Hazard: Flammable, dangerous fire risk. Explosive limits in air 1.1 to 7.5%. Moderately toxic by inhalation and ingestion.
Uses: Solvent for lacquers and paints; extraction of penicillin; photographic film; leather polishes; nail polish; warning odor; flavoring agent; printing and finishing fabrics; solvent for phosphors in fluorescent lamps.
Shipping regulations: (Rail) Red label. (Air) Flammable liquid label.

n-**amyl acetate** $CH_3COOCH_2CH_2CH_2CH_2CH_3$.
Properties: Colorless liquid; b.p. 148.4°C; m.p. -70.8°C; sp. gr. (20/20°C) 0.879; wt/gal (20°C) 7.22 lbs; flash point (closed cup) 77°F. Very slightly soluble in water; miscible with alcohol and ether. Vapor heavier than air. Autoignition temp. 714°C.
Derivation: Esterification of *n*-amyl alcohol with acetic acid.
Hazard: Flammable, dangerous fire risk. Moderately toxic. Tolerance, 100 ppm in air.
Uses: See amyl acetate.
Shipping regulations: (Rail) Red label. (Air) Flammable liquid label.

sec-**amyl acetate** $CH_3CO_2C_5H_{11}$.
Properties: Colorless liquid. May be a mixture of secondary isomers. Distillation range 123–145°C; odor mild, nonresidual; purity, ester content as amyl acetate, 85–88%; sp. gr. 0.862–0.866 (20/20°C); flash point 89°F (C.C.), wt/gal (20°C) 7.19 lbs (approximate).
Derivation: Esterification of sec-amyl alcohol and acetic acid.
Grades: Technical.
Containers: Drums; tank cars.
Hazard: Flammable, moderate fire risk. Moderately toxic. Tolerance, 125 ppm in air.
Uses: Solvent for nitrocellulose and ethyl cellulose; cements; coated paper; lacquers; leather finishes; nail enamels; plastic wood; textile sizing and printing compounds.

amylacetic ester. See amyl acetate.

amyl acid phosphate $(C_5H_{11})_2HPO_4$ and $C_5H_{11}H_2PO_4$. A mixture of primary and amyl isomers. Water-white liquid; density 1.070–1.090; flash point (COC) 245°F. Insoluble in water; soluble in alcohol; petroleum naphtha. Combustible.
Containers: Up to 55-gal drums.
Hazard: Strong irritant to tissue.
Uses: Curing catalyst and accelerator in resins and coatings; stabilizer; dispersion agent; lubricating and antistatic agent in synthetic fibers.
Shipping regulations: (Air) Corrosive label.

amyl alcohol (amyl hydrate). Eight isomers of amyl alcohol, $C_5H_{11}OH$, are possible (exclusive of several optical isomers) and six are offered commercially. In addition, definite mixtures of the isomers are sold under a variety of names (unfortunately some of them identical with the names of the pure isomers) as well as fusel oil (q.v.), a natural fermentation product. For descriptive data on the pure isomers, see
(1) *n*-amyl alcohol, primary
(2) 2-methyl-1-butanol (active amyl alcohol, from fusel oil)
(3) isoamyl alcohol, primary
(4) 2-pentanol
(5) 3-pentanol
(6) tert-amyl alcohol
The other two isomers, not described in detail, are
(7) sec-isoamyl alcohol
(8) 2,2-dimethyl-1-propanol
(1), (2), (3) and (8) are primary alcohols, (4), (5) and (7) are secondary alcohols, and (6) is a tertiary alcohol. (1), (4) and (5) are normal, and (2), (3), (6), (7) and (8) are branched-chain compounds. (2), (4) and (7) are asymmetric, and have optically active forms.

amyl alcohol, primary. A mixture of primary amyl alcohols made from normal butenes by the Oxo process is sold under this name. It consists of 60% primary *n*-amyl alcohol, 35% 2-methyl-1-butanol, and 5% 3-methyl-1-butanol. Used as a solvent.
Hazard: Flammable, moderate fire risk. Probably toxic.

n-**amyl alcohol,** primary (1-pentanol; n-butyl carbinol) $CH_3(CH_2)_4OH$.
Properties: Colorless liquid; mild odor; b.p. 137.8°C; freezing point -78.9°C; sp. gr. (20/20°C) 0.812–0.819; wt/gal (20°C) 6.9 lbs; flash point (open cup) 123°F. Autoignition temp. 572°F. Slightly soluble in water; miscible in all proportions with alcohol, benzene, and ether. Combustible.
Derivation: Fractional distillation of the mixed alcohols resulting from the chlorination and alkaline hydrolysis of pentane.
Grades: Technical; C.P.; 98%.
Containers: 55-gal drums; tank cars.
Hazard: Moderately toxic by ingestion and inhalation. Lower explosive limit in air, 1.2% by volume. Moderate fire risk.
Uses: Raw material for pharmaceutical preparations; organic synthesis; solvent.

amyl alcohol, primary, active. See 2-methyl-1-butanol.

n-sec-**amyl alcohol.** See 2- and 3-pentanol.

sec-**amyl alcohol,** active. See 2-pentanol.

tert-**amyl alcohol** (dimethyl ethyl carbinol; 2-methyl-2-butanol; amylene hydrate; tert-pentanol). $(CH_3)_2C(OH)CH_2CH_3$.
Properties: Colorless liquid; camphor odor and burning taste; sp. gr. 0.81 (20/20°C); freezing point -11.9°C; b.p. 101.8°C; refractive index 1.4052 (20°C); wt/gal 6.76 lbs; flash point (open cup) 70°F. Slightly soluble in water; miscible with alcohol and ether; solutions neutral to litmus. Autoignition temp. 819°F.
Derivation: Fractional distillation of the mixed alcohols resulting from the chlorination and alkaline hydrolysis of pentanes.

Grades: Technical; C.P.; N.F.
Containers: 55-gal drums.
Hazard: Flammable, dangerous fire risk; moderately toxic by ingestion, inhalation, skin absorption.
Uses: Solvent; flotation agent; organic synthesis; medicine (sedative).
Shipping regulations: (Rail) Flammable liquid, n.o.s., Red label. (Air) Flammable Liquid label.

amyl alcohol, fermentation. See fusel oil.

amyl aldehyde. See *n*-valeraldehyde.

n-**amylamine** (pentylamine; 1-aminopentane) $C_5H_{11}NH_2$.
Properties: Colorless liquid; sp. gr. 0.75 (20/20°C); f.p. -55.0°C; b.p. 104.4°C; flash point 45°F (open cup). Soluble in water, alcohol and ether.
Derivation: Reaction of ammonia and amyl chloride, which gives a mixture of mono-, di-, and triamyl amines.
Grade: Technical.
Containers: 55-gal drums; tank cars.
Hazard: Flammable, dangerous fire risk. Strong irritant.
Uses: Chemical intermediate; dyestuffs; rubber chemicals; insecticides; synthetic detergents; flotation agents; corrosion inhibitors; solvent; gasoline additive; pharmaceuticals.
Shipping regulations: (Rail) Flammable liquid, n.o.s., Red label. (Air) Flammable Liquid label.

sec-**amylamine** (2-aminopentane) $CH_3(CH_2)_2CH(CH_3)NH_2$.
Properties: Colorless liquid; b.p. 198°F; sp. gr. 0.7; flash point 20°F.
Hazard: Flammable, dangerous fire risk.
Uses: See n-amylamine.
Shipping regulations: (Rail) Flammable liquid, n.o.s., Red label. (Air) Flammable Liquid label.

amylase. A class of enzymes which convert starch into sugars. Fungal and bacterial amylases, from specific fungi and bacteria, have been suggested for commercial fermentation processes. See also amylopsin, diastase, and ptyalin.
Uses: Textile desizing; conversion of starch to glucose sugar in syrups (especially corn syrups); baking (to improve crumb softness and shelf life); dry cleaning (to attack food spots and similar stains).

n-**amylbenzene** (1-phenylpentane) $C_6H_5CH_2(CH_2)_3CH_3$. Water-white liquid; mild odor; f.p. -75°C; b.p. 205°C; sp. gr. (20/4°C) 0.8585. Flash point 150°F (open cup). Insoluble in water; soluble in hydrocarbons and coal-tar solvents. Combustible.
Hazard: Probably irritant to skin and eyes; narcotic in high concentrations. Moderate fire risk.

sec-**amylbenzene** (2-phenylpentane). $CH_3CH(C_6H_5)CH_2CH_2CH_3$.
Properties: Clear liquid; f.p. -75°C; b.p. 190.3°C; sp. gr. 0.861 (20/4°C); flash point 155°F. Combustible.
Grades: Pure, 99.0 mole %; technical, 95.0 mole %.
Hazard: Probably irritant to skin and eyes; narcotic in high concentrations. Moderate fire risk.
Use: Weed control; chemical intermediate.

amyl benzoate. See isoamyl benzoate.

amyl butyrate. See isoamyl butyrate.

amyl carbinol. See 1-hexanol.

n-**amyl chloride** (1-chloropentane) $CH_3(CH_2)_3CH_2Cl$.
Properties: Colorless liquid. B.p. 107.8°C; freezing point -99°C; sp. gr. 0.883 (20/4°C); refractive index n 20/D 1.4128; flash point 54°F (open cup). Miscible with alcohol and ether; insoluble in water. Autoignition temp. 650°F.
Derivation: (a) Distillation of amyl alcohol with salt and sulfuric acid; (b) addition of hydrogen chloride to alpha-amylene.
Grades: Technical.
Hazard: Flammable, dangerous fire risk. Explosive limits in air 1.4 to 8.6%. May be narcotic in high concentrations.
Use: Chemical intermediate.
Shipping regulations: (Rail) Red label. (Air) Flammable Liquid label.

amyl chlorides, mixed.
Properties: Straw to purple-colored liquid; sp. gr. (20°C) 0.88; 95% distills between 85 and 109°C; wt/gal 7.33 lb; refractive index (20°C) 1.406; insoluble in water; water azeotrope at 77–82°C, 90% $C_5H_{11}Cl$ (approximate); miscible with alcohol and ether. Flash point 38°F (open cup).
Components: 1-chloropentane, b.p. 107.8°C; 2-chloropentane, b.p. 96.7°C; 3-chloropentane, b.p. 97.3°C; 1-chloro-2-methylbutane, b.p. 99.9°C; 1-chloro-3-methylbutane, b.p. 98.8°C; 3-chloro-2-methylbutane, b.p. 93.0°C; and 2-chloro-2-methylbutane, b.p. 86.0°C.
Derivation: Vapor phase chlorination of mixed normal pentane and isopentane.
Containers: Drums; tank cars.
Hazard: Flammable, dangerous fire risk. Explosive limits in air 1.4 to 8.6%. May be narcotic in high concentrations.
Uses: Synthesis of other amyl compounds; solvent; rotogravure ink vehicles; soil fumigation.
Shipping regulations: (Rail) Red label. (Air) Flammable Liquid label.

alpha-**amylcinnamic alcohol** (2-benzylidene-1-heptanol) $C_6H_5CH{:}C(CH_2OH)C_5H_{11}$.
Properties: Yellow liquid; floral odor; sp. gr. 0.954–0.962 (25/25°C); soluble in 3 parts 70% alcohol. Combustible.
Derivation: Synthetic.
Use: Perfumery; flavoring

alpha-**amylcinnamic aldehyde** (jasmine aldehyde; alpha-pentylcinnamaldehyde) $C_6H_5CH{:}C(CHO)C_5H_{11}$.
Properties: Clear, yellow, oily liquid. Jasmine-like odor. Aldehyde content 97%. Soluble in 6 volumes of 80% alcohol. Sp. gr. 0.962 to 0.968; refractive index 1.554 to 1.559. Combustible.
Derivation: Synthetic.
Grades: Technical; F.C.C.
Uses: Perfumery, flavoring.

amyl para-**dimethylaminobenzoate.** See "Escalol 506."

alpha-n-**amylene.** See 1-pentene.

beta-n-**amylene.** See 2-pentene.

amylene dichloride. See dichloropentane.

amylene hydrate. See tert-amyl alcohol.

amyl formate $HCOOC_5H_{11}$.
Properties: Colorless liquid composed of a mixture of isomeric amyl formates with isoamyl formate in predominance. Plum-like odor. Less odoriferous and more active solvent than amyl acetate. It also has

both a lower boiling point and a higher rate of evaporation. Miscible with oils, hydrocarbons, alcohols, ketones. Slightly soluble in water. Sp. gr. 0.880 to 0.885; b.p. 123.5° C; flash point 80° F.
Grades: Technical; F.C.C.
Hazard: Flammable, dangerous fire risk; moderately toxic by ingestion and inhalation.
Uses: Solvent for cellulose esters, resins; solvent mixtures; films and coatings; perfume for leather; flavoring.
Shipping regulations: (Rail) Flammable liquid, n.o.s., Red label. (Air) Flammable Liquid label.

n-**amyl furoate** (amyl pyromucate) $C_4H_3OCO_2C_5H_{11}$.
Properties: Colorless oil, decomposes on standing. Sp. gr. 1.0335 (20/4° C); b.p. 233° C. Insoluble in water; soluble in alcohol. Combustible; low toxicity.
Derivation: Esterification of furoic acid.
Uses: Perfumes; lacquers.

amyl hydrate. See amyl alcohol.

amyl hydride. See pentane.

amyl hydrosulfide. See amyl mercaptan.

"Amyliq."[173] Trademark for starch-liquefying enzyme for sizes and adhesives.

amyl mercaptan. Legal label name for pentanethiol (q.v.).

tert-**amyl mercaptan.** See 2-methyl-2-butanethiol.

amyl nitrate (mixed isomers) $C_5H_{11}NO_3$.
Properties: Colorless liquid; b.p. 145° C; flash point 118° F; sp. gr. at 20° C 0.99; ethereal odor. Combustible.
Hazard: Oxidizing agent; moderate fire risk. Moderately toxic.
Use: Additive to increase cetane number of diesel fuels.
Shipping regulations: Nitrates,n.o.s., (Rail) Yellow label. (Air) Oxidizer label.

amyl nitrite (isoamyl nitrite) $(CH_3)_2CHCH_2CH_2NO_2$.
Properties: Yellowish liquid, of peculiar ethereal, fruity odor and pungent aromatic taste. Soluble in alcohol; almost insoluble in water. Decomposes on exposure to air, light, or water. Sp. gr. 0.865–0.875 at 25° C; b.p. 96–99° C.
Derivation: Interaction of amyl alcohol with nitrous acid.
Grades: N.F. (95% min); technical.
Containers: Dark amber glass bottles.
Hazard: Flammable, dangerous fire risk, a strong oxidizing material. Vapor may explode if ignited. Moderately toxic.
Uses: Medicine; perfumes; diazonium compounds.
Shipping regulations: (Rail) Red label (Air) Flammable Liquid label.

amyloglucosidase. An enzyme used commercially to convert starches to dextrose.

"Amylon."[53] Trademark for a high-amylose starch derived from high-amylose corn.

amylopectin. The outer, almost insoluble portion of starch granules. It is a hexosan, a polymer of glucose, and is a branched molecule of many glucose units. It stains violet with iodine and forms a paste with water.

amylopsin (animal diastase). The starch-digesting enzyme of pancreatic juice, the most powerful enzyme of the digestive tract. It is an amylase which converts starches through the soluble-starch stage to various dextrins and maltose.It acts in neutral, slightly acid and slightly alkaline environments with an optimum pH of 6.3–7.2. It requires the presence of certain negative ions for activation.
Uses: Biochemical research.

amylose. The inner, relatively soluble portion of starch granules. Amylose is a hexosan, a polymer of glucose, and consists of long straight chains of glucose units joined by a 1,4-glycosidic linkage. It stains blue with iodine. Microcrystalline amylose is available, chiefly as a food ingredient and dietary energy source.

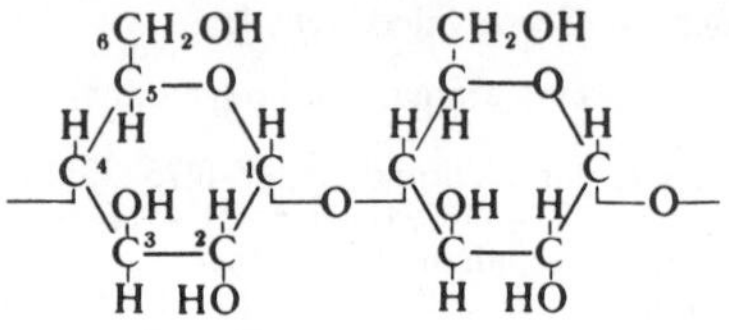

```
      6CH2OH              CH2OH
       |                   |
       C5—O                C—O
   H/H      \H         H/H     \H
   C4        1C        C        C
 —|\OH   H/|—O—|\OH    H/|—O—
    C3——2C               C——C
    |     |              |   |
    H    HO              H   HO
```

ortho-sec-**amylphenol** $C_5H_{11}C_6H_4OH$.
Properties: Clear, straw-colored liquid; sp. gr. (30/30° C) 0.955–0.971; initial b.p. not below 235.0°C; final b.p. not above 250.0°C; wt/gal 8.0 lbs; very slightly soluble in water; soluble in oil and organic solvents. Flash point (open cup) 219° F. Combustible.
Uses: Dispersing and mixing agent for paint pastes; antiskinning agent for paint, varnish and oleoresinous enamels; organic synthesis.

para-tert-**amylphenol** $(CH_3)_2(C_2H_5)CC_6H_4OH$.
Properties: White crystals; m.p. 93° C; b.p. 265–267° C (138°C at 15 mm); slightly soluble in water; soluble in alcohol and ether. Flash point 232° F (open cup). Combustible.
Containers: Drums; carlots.
Uses: Manufacture of oil-soluble resins; chemical intermediate.

para-tert-**amphenyl acetate** $C_5H_{11}C_6H_4OOCCH_3$.
Properties: Colorless liquid. Sp. gr. (20° C) 0.996; boiling range 253–272° C; odor, fruity. Flash point 240° F. Combustible.
Use: Perfumes; flavorings.

amyl propionate $CH_3CH_2COOC_5H_{11}$. Probably the isoamyl isomer.
Properties: Colorless, high-boiling liquid; apple-like odor; sp. gr. (20/20° C) 0.869–0.873; wt/gal (20° C) 7.25 lbs (approximate); distillation range 135–175° C; flash point 106° F (open cup); autoignition temp. 712° F; miscible with most organic solvents. Combustible.
Derivation: By reacting amyl alcohol with propionic acid in the presence of sulfuric acid as a catalyst, followed by neutralization, drying and distillation.
Hazard: Moderately toxic; moderate fire hazard.
Uses: Perfumes; lacquers; flavors.

amyl pyromucate. See n-amyl furoate.

amyl salicylate. See isoamyl salicylate.

amyl sulfide. See diamyl sulfide.

amyltrichlorosilane $C_5H_{11}SiCl_3$. A mixture of isomers.
Properties: Colorless to yellow liquid. B.p. 168° C; sp. gr. 1.137 (25/25° C); refractive index n 20/D 1.4152; flash point (COC) 145° F. Combustible. Readily hydrolyzed by moisture with the liberation of hydrochloric acid.
Derivation: By Grignard reaction of silicon tetrachloride and amylmagnesium chloride.
Hazard: Highly toxic and corrosive.
Use: Intermediate for silicones.
Shipping regulations: (Rail) White label. (Air) Corrosive label. Not acceptable on passenger planes.

amyl valerate. See isoamyl valerate.

amyl valerianate. See isoamyl valerate.

"Amytal."[100] Trademark for amobarbital, U.S.P.

"Amzirc."[433] Trademark for zirconium-containing copper-base alloys.

anabasine (neonicotine; 2-(3-pyridyl)piperidine) $C_{10}H_{14}N_2$. A naturally occurring alkaloid.
Properties: Colorless liquid; darkens on exposure to air. B.p. 105°C; freezing point 9°C; sp. gr. 1.046 (20/20°C); refractive index n 20/D 1.5430. Miscible with water; soluble in alcohol and ether.
Derivation: (a) Extraction from Anabasis aphylla and Nicotiana glauca; (b) synthetic.
Hazard: Toxic by ingestion, inhalation, and skin absorption.
Use: Insecticide.

anacardic acid. $C_{22}H_{32}O_3$. The chief component of cashew nutshell oil (q.v.).

anacardium gum. See cashew gum.

anaerobic. Descriptive of a chemical reaction or a microorganism that does not require the presence of air or oxygen; examples are the fermentation of sugars by yeast and the decomposition of sewage sludge by anaerobic bacteria. It also applies to certain polymers that solidify when kept out of contact with air. See also "Loctite"; bacteria.

"Anaesthesin."[162] Trademark for ethyl para-aminobenzoate.

analcite (analcime) $Na_2O \cdot Al_2O_3 \cdot 4SiO_2 \cdot 2H_2O$. A mineral; one of the zeolites (q.v.).
Properties: Colorless, white; sometimes greenish-grayish, yellowish or reddish white; hardness 5–5.5; sp. gr. 2.22–2.29.
Occurrence: Europe; United States; Nova Scotia.

analytical chemistry. The subdivision of chemistry concerned with identification of materials (qualitative analysis) and with determination of the percentage composition of mixtures or the constituents of a pure compound (quantitative analysis). The gravimetric and volumetric (or "wet") methods (precipitation, titration, and solvent extraction) are still used for routine work; indeed, new titration methods have been introduced, e.g., cryoscopic, pressuremetric (for reactions that produce a gaseous product), redox methods, and use of a fluoride-sensitive electrode. However, faster and more accurate techniques (collectively called instrumental) have been developed in the last few decades. Among these are infrared, ultraviolet, and x-ray spectroscopy, where the presence and amount of a metallic element is indicated by lines in its emission or absorption spectrum; colorimetry, by which the percentage of a substance in solution is determined by the intensity of its color; chromatography of various types, by which the components of a liquid or gaseous mixture are determined by passing it through a column of porous material, or on thin layers of finely divided solids; separation of mixtures in ion-exchange columns; and radioactive tracer analysis. Optical and electron microscopy, mass spectrometry, microanalysis, nuclear magnetic resonance (NMR), and nuclear quadrupole resonance (NQR) spectroscopy all fall within the area of analytical chemistry. See also spectroscopy, nuclear magnetic resonance, NQR spectroscopy. chromatography.

anatase (octahedrite). A natural crystallized form of titanium dioxide (q.v.). Sp. gr. 3.8; refractive index 2.5; m.p. 1560°C.

"Anattene."[342] Trademark for annatto derivatives for use in coloring food-stuffs, i.e., cheeses, oranges.

andalusite Al_2OSiO_4. A natural silicate of aluminum.
Properties: Gray, greenish, reddish, or bluish in color. Sp. gr. 3.1–3.2; hardness, 7–7.5.
Occurrence: Massachusetts, Connecticut, California, Nevada; Europe; South Africa; Australia.
Use: Constituent of sillimanite refractories; spark plug insulators; laboratory ware; superrefractories.

androgen. A male sex hormone. The androgenic hormones are steroids and are synthesized in the body by the testis, the cortex of the adrenal gland, and, to slight extent, by the ovary.

androsterone. $C_{19}H_{30}O_2$. An androgenic steroid; metabolic product of testosterone (q.v.). The international unit (I.U.) of androgenic activity is defined as 0.1 mg androsterone.
Properties: Crystalline solid; m.p. 185–185.5°C; sublimes in high vacuum; dextrorotatory in solution; not precipitated by digitonin; practically insoluble in water; soluble in most organic solvents.
Derivation: Isolation from male urine; synthesis from cholesterol.
Use: Medicine; biochemical research.

anesthetic. A chemical compound that induces loss of sensation in a specific part or all of the body. A brief classification of the more important agents is as follows:

A. General
1. Hydrocarbons
(a) Cyclopropane (U.S.P.). Effective in presence of substantial proportions of oxygen. Flammable.
(b) Ethylene (U.S.P.). Anesthesia quickly induced and recovery rapid. Flammable.
2. Halogenated hydrocarbons
(a) Chloroform. Nonflammable but toxic. Its use is being abandoned because of its high toxicity.
(b) Ethyl chloride. A gas at room temperature; liquefies at relatively low pressure. Applied as a stream from container directly on tissue. Sometimes used in gaseous form as inhalation type general anesthetic. Flammable.
(c) Trichloroethylene. Toxic and flammable. Used as general anesthetic since 1934.
3. Ethers
(a) Ethyl ether (U.S.P.). First anesthetic used in surgery (1842). Now largely replaced with less dangerous types. Highly flammable; explodes in presence of spark or open flame.
(b) Vinyl ether. A liquid having many of the physiological properties of ethylene and ethyl ether. Highly flammable.
4. Miscellaneous
(a) Tribromoethanol. Basal anesthetic, supplemented by an inhalation type when general anesthesia is needed. Ingredient of "Avertin."
(b) Nitrous oxide. Originally prepared by Priestley in 1772 (laughing gas); first used

as anesthetic by Humphry Davy in 1800. Used (with oxygen) largely for dental surgery. Nonflammable
(c) Barbiturates (q.v.).

B. Local
(a) alkaloids (cocaine)
(b) synthetic products (procaine group, e.g., "Novocain," alkyl esters of aromatic acids (topical).
(c) quinine hydrochloride.

anethole (anise camphor; para-methoxypropenylbenzene; para-propenylanisole) $CH_3CH:CHC_6H_4OCH_3$.
Properties: White crystals with sweet taste; melt at room temperature; odor of oil of anise. Affected by light. Soluble in 8 volumes of 80% alcohol, 1 volume of 90% alcohol. Almost immiscible with water. Sp. gr. 0.983–0.987; refractive index 1.557–1.561; optical rotation 0.08°; m.p. 22–23°C; distillation range 234–237°C.
Derivation: By crystallization from anise or fennel oils; synthetically from para-cresol.
Grades: U.S.P. technical; F.C.C.
Uses: Perfumes, particularly for dentifrices; flavors; synthesis of anisic aldehyde; licorice candies; color photography (sensitizer in color-bleaching process); microscopy.

ANFO. A high explosive based on ammonium nitrate. See also explosive, high.

angelic acid (cis-2-methyl-2-butenoic acid; alpha-methylcrotonic acid) $CH_3CH:C(CH_3)COOH$. The cis isomer of tiglic acid.
Properties: Colorless needles or prismatic crystals; spicy odor. Soluble in alcohol, ether, and hot water. Sp. gr. 0.9539 (76/4°C); m.p. 45°C; b.p. 185°C; refractive index n 47/D 1.4434. Low toxicity.
Derivation: From the root of Angelica archangelica or from the oil of Anthemis nobilis by distillation.
Uses: Medicine; flavoring extracts.

angelica oil.
Properties: Essential oil; strong aromatic odor; spicy taste. Soluble in alcohol. Sp. gr. 0.853–0.918; optical rotation +16° to +41°. Chief known constituents: Phellandrene; valeric acid. Combustible.
Derivation: Distilled from the roots and seeds of Angelica archangelica, found principally in Europe.
Grades: Technical; F.C.C.
Uses: Medicine; preparation of liqueurs; perfumery.

"Angio-Conray."[329] Trademark for an 80% solution of sodium iothalamate; used in diagnostic medicine.

angiotensin (angiotonin; hypertensin). A peptide found in the blood, important in its effect on blood pressure. Both a decapeptide and an octapeptide are known. Their amino acid sequences, and hence the complete structures, have been established.

angiotensin amide. USAN name for 1-L-asparaginyl-5-L-valyl angiotensin octapeptide; $C_{49}H_{70}N_{14}O_{11}$. Described in N.D. Used in medicine.

angiotonin. See angiotensin.

angstrom. A unit of length almost one one-hundred-millionth (10^{-8}) centimeter. The ångstrom (abbreviated Å.) is now defined in terms of the wave length of the red line of cadmium (6438.4696 Å.). Used in stating distances between atoms, dimensions of molecules, wavelengths of short-wave radiation, etc. See also nanometer.

anhydride. A chemical compound derived from an acid by elimination of a molecule of water. Thus sulfur trioxide, SO_3, is the anhydride of sulfuric acid (H_2SO_4); carbon dioxide, CO_2, is the anhydride of carbonic acid, H_2CO_3; phthalic acid, $C_6H_4(CO_2H)_2$, minus water gives phthalic anhydride, $C_6H_4(CO_2)O$. This term should not be confused with anhydrous (q.v.).

anhydrite $CaSO_4$. A natural calcium sulfate usually occurring as compact granular masses and resembling marble in appearance. Differs from gypsum in hardness and lack of hydration.

anhydroenneaheptitol (AEH; 4-hydroxy-2H-pyran-3,3,5,5(4H,6H)tetramethanol)
$OCH_2(CH_2OH)_2CHOHC(CH_2OH)_2CH_2$ (ring).
Uses: Alkyd resins; rosin esters; urethane coatings and foams; surfactants; lubricating oil additives.

"Anhydrol."[214] Trademark for a gasoline-free solvent composed of 100 gal S.D. ethanol denatured with 10 gal isopropanol (90%) and 1 gal methyl isobutyl ketone.
Properties of anhydrous grade: B.p. 76.2–79.5°C; sp. gr. 0.7895–0.7935 (20/20°C); lb/gal 6.6 (20°C); flash point 54°F.
Grades: Anhydrous to 190 proof.
Containers: 1-gal can; 5- and 55-gal drums; tank cars.
Hazard: Flammable, dangerous fire risk.
Uses: Solvent in manufacture of printing inks, textile dyestuff solutions, window cleaners, synthetic detergents, aircraft de-icing fluids, and photographic chemicals.

"Anhydron."[100] Trademark for cyclothiazide (q.v.).

anhydrous. Descriptive of an inorganic compound that does not contain water either adsorbed on its surface or combined as water of crystallization. Do not confuse with anhydride (q.v.).

"Anhydrox."[236] Trademark for a compound to prevent or overcome anhydrite or gypsum contamination in drilling mud, by pretreatment of the mud to remove calcium and sulfate ions.

anidex. A synthetic fiber designated by the FTC as a cross-linked polyacrylate elastomer.

anileridine (ethyl 1-(para-aminophenethyl)-4-phenylisonipecotate) $C_{22}H_{28}N_2O_2$.
Properties: White, odorless, crystalline powder. Oxidizes, darkens in air and on exposure to light. Exhibits polymorphism, and of two crystalline forms observed, one melts at about 80°C and the other at about 89°C. Soluble in alcohol and chloroform. Very slightly soluble in water.
Grade: N.F.
Hazard: Toxic by ingestion. Causes addiction.
Use: Medicine (narcotic).
Shipping regulations: Subject to Federal narcotic restrictions.

aniline (aniline oil; phenylamine; aminobenzene) $C_6H_5NH_2$. One of the most important of the organic bases; the parent substance for many dyes and drugs.
Properties: Colorless oily liquid; characteristic odor and taste; rapidly becomes brown on exposure to air and light. Vapors will contaminate foodstuffs and damage textiles. Soluble in alcohol, ether, and benzene; soluble in water. Sp. gr. 1.0235; m.p. −6.2°C; b.p. 184.4°C; wt/gal (20°C) 8.52 lbs; refractive index n 20/D 1.5863; flash point (closed cup) 158°F. Combustible. Autoignition temp. 1418°F.
Derivation: (a) By catalytic vapor-phase reduction of nitrobenzene with hydrogen; (b) Reduction of nitrobenzene with iron filings (hydrochloric acid as catalyst); (c) Catalytic reaction of chlorobenzene and

aqueous ammonia; (d) ammonolysis of phenol (Japan).
Grades: Commercial; C.P.
Containers: Bottles; drums; tank cars.
Hazard: Toxic by ingestion, inhalation, and skin absorption. An allergen. Tolerance, 5 ppm in air. Safety data sheet available from Manufacturing Chemists Assn., Washington, D.C.
Uses: Rubber accelerators and antioxidants; dyes and intermediates; photographic chemicals (hydroquinone); isocyanates for urethane foams; pharmaceuticals; explosives; petroleum refining; diphenylamine; phenolics; herbicides, fungicides.
Shipping regulations: (Rail) Poison label. (Air) Poison label. Not acceptable on passenger planes.

aniline acetate $C_6H_5NH_2 \cdot CH_3COOH$.
Properties: Colorless liquid; becomes dark with age; on standing or heating is converted gradually to acetanilide; sp. gr. 1.070–1.072; miscible with water and alcohol. Combustible.
Use: Organic synthesis.

aniline black. A black color developed on cotton and other textiles from a bath containing aniline hydrochloride, an oxidizing agent (usually chromic acid) and a catalyzer (usually a vanadium or copper salt).

aniline chloride. See aniline hydrochloride.

aniline-2,4-disulfonic acid. See 4-amino-meta-benzenedisulfonic acid.

aniline-2,5-disulfonic acid. See 2-amino-para-benzenedisulfonic acid.

aniline dye. Any of a large class of synthetic dyes made from intermediates based upon, or made from, aniline. Most are somewhat toxic and irritating to eyes, skin, and mucous membranes; they are generally much less toxic than the intermediates from which they are derived.

aniline hydrochloride (aniline salt; aniline chloride) $C_6H_5NH_2 \cdot HCl$.
Properties: White crystals; commercial product frequently greenish in appearance; darkens in light and air. Soluble in water, alcohol, and ether. Sp. gr. 1.2215; m.p. 198°C; b.p. 245°C.
Derivation: (a) By passing a current of dry hydrochloric acid gas into an ethereal solution of aniline; (b) neutralizing aniline at 100°C with concentrated hydrochloric acid and subsequent crystallization.
Hazard: Toxic. See aniline.
Uses: Dyes; intermediates; dyeing and printing; aniline black.
Shipping regulations: (Rail) Poisonous solids,n.o.s., Poison label. (Air) Poison label.

aniline ink. A fast-drying printing ink used on kraft paper, cotton fabric, cellophane, polyethylene, etc. The name is due to the fact that original inks for this purpose were solutions of coal-tar dyes in organic solvents. Modern inks usually employ pigments rather than dyes and are of two types: spirit inks, containing organic solvent as the vehicle, and emulsion inks, in which water is the main vehicle.

aniline point. The lowest temperature at which equal volumes of aniline and the test liquid are miscible. Cloudiness occurs on phase separation. Used as a test for components of hydrocarbon fuel mixtures.

aniline salt. See aniline hydrochloride.

para-**anilinesulfonic acid.** See sulfanilic acid.

aniline yellow. See para-aminoazobenzene.

1-anilino-4-hydroxyanthraquinone
$C_6H_4(CO)_2C_6H_2(OH)NHC_6H_5$(tricyclic). A chemical intermediate.
Hazard: Probably toxic.

6-anilino-1-naphthol-3-sulfonic acid. See phenyl-2-amino-5-naphthol-7-sulfonic acid.

7-anilino-1-naphthol-3-sulfonic acid. See phenyl-2-amino-8-naphthol-6-sulfonic acid.

anilinophenol. See para-hydroxydiphenylamine.

animal black. See bone black, ivory black.

animal diastase. See amylopsin.

animal oil. See bone oil.

animal starch. See glycogen.

anion. An ion having a negative charge; anions in a liquid subjected to electric potential collect at the positive pole or anode. Examples are hydroxyl, OH^-; carbonate, CO_3^{--}; phosphate, PO_4^{---}

ortho-**anisaldehyde** (ortho-methoxybenzaldehyde; ortho-anisic aldehyde) $C_6H_4(OCH_3)CHO$.
Properties: White to light tan solid, burned, slightly phenolic odor; b.p. 238°C; m.p. (2 crystalline forms) 38–39°C and 3°C; sp. gr. (liquid) 1.1274 (25/25°C); (solid) 1.258 (25/25°C); refractive index (n 20/D) 1.5608; flash point 244°F; insoluble in water. Soluble in alcohol. Combustible.
Grade: 95% (min.).
Use: Intermediate.

para-**anisaldehyde** (aubepine; para-anisic aldehyde; para-methoxybenzaldehyde) $C_6H_4(OCH_3)CHO$.
Properties: Colorless to plae yellow liquid, having odor of hawthorn. Insoluble in water. Soluble in 5 volumes of 50% alcohol. Sp. gr. 1.119–1.122; refractive index 1.570–1.572; m.p. 0°C; b.p. 248°C. Combustible.
Derivation: Obtained from anethole or anisole by oxidation.
Grades: Liquid and crystals, latter being the disulfite compound.
Uses: Perfumery; intermediate for antihistamines; electroplating; flavoring.

anise alcohol. See anisic alcohol.

anise camphor. See anethole.

anise oil (anise seed oil; aniseed oil). See anethole.

anisic acid (papr-methoxybenzoic acid) $CH_3OC_6H_4COOH$.
Properties: White crystals or powder; sp. gr. 1.385 (4°C); m.p. 184°C; b.p. 275–280°C; soluble in alcohol and ether; almost insoluble in water. Low toxicity.
Derivation: Oxidation of anethole.
Use: Medicine; repellent and ovicide.

anisic alcohol (anisyl alcohol; anise alcohol; para-methoxybenzyl alcohol) $CH_3OC_6H_4CH_2OH$.
Properties: Solidifies at room temperature; has a floral odor suggesting hawthorn. Soluble in 1 volume of 50% alcohol. Sp. gr. 1.111–1.114; refractive index 1.541–1.545; boiling range 255–265°C. Combustible.
Derivation: Obtained from anisic aldehyde, by reduction.
Grades: Technical; F.C.C.

Uses: In perfumery, for light floral odors; pharmaceutical intermediate; flavoring.

anisic aldehyde. See anisaldehyde.

ortho-**anisidine** (ortho-methoxyaniline; ortho-aminoanisole) $CH_3OC_6H_4NH_2$.
Properties: Reddish or yellowish colored oil, becomes brownish on exposure to air; volatile with steam; sp. gr. 1.097 (20°C); b.p. 225°C; f.p. 5°C; soluble in dilute mineral acid, alcohol, and ether; slightly soluble in water.
Derivation: (a) Reduction of ortho-nitroanisole with tin (or iron) and hydrochloric acid; (b) heating ortho-aminophenol with potassium methyl sulfate.
Method of purification: Steam distillation.
Grades: 99% (1% maximum moisture).
Containers: 55-, 110-gal drums; tank cars.
Hazard: Toxic by inhalation. Absorbed by skin; strong irritant. Tolerance, 0.5 mg per cubic meter of air.
Use: Intermediate for azo dyes and for guaiacol.

para-**anisidine** (para-methoxyaniline; para-aminoanisole) $CH_3OC_6H_4NH_2$.
Properties: Fused, crystalline mass; crystallizing point 57.2°C min; sp. gr. 1.089 (55/55°C); b.p. 242°C; soluble in alcohol, and ether; slightly soluble in water.
Derivation: (a) Reduction of para-nitroanisole with iron filings and hydrochloric acid; (b) methylation of para-aminophenol.
Grades: Technical.
Containers: 500-, 800-lb drums.
Hazard: Toxic by inhalation. Absorbed by skin; strong irritant. Tolerance, 0.5 mg per cubic meter of air.
Uses: Azo dyestuffs, intermediate.

anisindione (2-para-anisyl-1,3-indandione) $C_{16}H_{12}O_3$.
Properties: Pale yellow crystals; m.p. 156–157°C.
Grade: N.D.
Use: Medicine.

anisole (methylphenyl ether; methoxybenzene) $C_6H_5OCH_3$.
Properties: Colorless liquid; aromatic odor; soluble in alcohol and ether; insoluble in water. Sp. gr. 0.999 (15/15°C); f.p. -37.8°C; b.p. 155°C; refractive index (n 20/D) 1.5150–1.5170; flash point 125°F (open cup). Combustible.
Derivation: From sodium phenate and methyl chloride; heating phenol with methyl alcohol.
Uses: Solvent; perfumery; vermicide; intermediate; flavoring.

anisotropic. Descriptive of crystals whose index of refraction varies with the direction of the incident light. This is true of most crystals, e.g., calcite (Iceland spar); it is not true of isometric (cubic) crystals, which are isotropic (q.v.).

anisoyl chloride $CH_3OC_6H_4COCl$.
Properties: Clear crystals or amber liquid. M.p. 22°C; b.p. 262–263°C. Soluble in acetone and benzene; decomposed by water or alcohol. Fumes in moist air.
Containers: Drums.
Uses: Intermediate for dyes and medicines.
Hazard: Toxic; solutions are corrosive to tissue.
Shipping regulations: (Rail) White label. (Air) Corrosive label.

anisyl acetate (para-methoxybenzyl acetate) $CH_3OC_6H_4CH_2COCH_3$.
Properties: Colorless liquid, lilac-odor. Soluble in 4 vols. of 60% alcohol. Sp. gr. 1.104–1.107; refractive index 1.514–1.516. Combustible.
Derivation: Reaction of anisic alcohol with acetic anhydride.
Grades: Technical; F.C.C.
Use: Perfumery; flavoring.

anisylacetone. Generic name for 4-(para-methoxyphenyl)-2-butanone, $CH_3OC_6H_4C_2H_4COCH_3$. A colorless to pale yellow liquid; m.p. 8°C. Used in insect attractants; organic synthesis; flavoring. Combustible.

anisyl alcohol. See anisic alcohol.

anisyl formate (para-methoxybenzyl formate) $CH_3OC_6H_4CH_2OCOH$.
Properties: Colorless liquid, lilac odor. Soluble in 5.5 vols. of 70% alcohol. Sp. gr. 1.139–1.141. Combustible.
Uses: Perfumery; flavoring.

2-para-**anisyl-1,3-indandione.** See anisindione.

annatto. Vegetable dyestuff containing coloring principle called bixin, $C_{25}H_{30}O_4$, the mono-methyl ester of a dibasic acid having 24 carbons.
Properties: Soluble in alcohol and water.
Derivation: From the seeds of Bixa orellana.
Occurrence: South America; West Indies; India.
Uses (as extract): Coloring margarine, sausage casings, etc., food-product marking inks. For details see regulations of Meat Inspection Division, USDA, and FDA regulations.

annealing. Maintenance of glass or metal at a specified temperature for a specific length of time (at least three days for plate glass), and then gradually cooling it at a predetermined rate. This treatment removes the internal strains resulting from previous operations, and eliminates distortions and imperfections; a clearer, stronger and more uniform material results. See also tempering.

annellation. A chemical reaction in which one cyclic or ring structure is added to another to form a polycyclic compound.

"Ano."[307] Trademark for a series of dyestuffs used for coloring anodized aluminum. See anodic coating.

anode. The positive electrode of an electrolytic cell, to which negatively charged ions travel when an electric current is passed through the cell. Such anodes are usually made of graphite or other form of carbon, though titanium has been successfully introduced in the chlor-alkali industry. In a primary cell (battery or fuel cell) the anode is the negative electrode. See also cathode; electrode.

anode mud. Residue obtained from the bottom of a copper or other plating bath. In the electrolytic refining of copper the anode mud contains the relatively inert metals platinum, silver and gold and is usually collected and treated for the recovery of these metals and other rare elements.

anode process. See electrophoresis.

anodic (anodized) coating. The electrolytic treatment of aluminum, magnesium, and a few other metals as a result of which heavy, stable films of oxides are formed on their surfaces. A thin oxide film will form on an aluminum surface without special treatment on exposure of the metal to air. This provides excellent resistance to corrosion. This fact led to the development of electrochemical processes to produce much thicker and more effective protective and decorative coatings. The chief electrolytes used are sulfuric, oxalic and chromic acids. The metal acts as the anode. Such anodic coatings are hard and have good electrical insulating properties. Their ability to absorb dyes and pigments makes it possible to obtain finishes in a complete range of colors, including black. The luster of the underlying metal gives them a metallic sheen; colorants can be used to reproduce the color of any metal with which the aluminum might be used. Anodized coatings can be used as preparatory treatments to electroplating; copper, nickel, cadmium, silver and iron have been successfully deposited over oxide coatings. See also aluminum.

anomer. A specific kind of diastereoisomer (or epimer) occurring in some sugars and other substances having asymmetric carbon atoms.

"Anoplex."[85] Trademark for the sodium bisulfite complex of anisaldehyde [$C_6H_4(OCH_3)HC(OH)SO_3Na$].
Properties: Dry, white powder; slightly soluble in water.
Containers: 100-lb fiber drums.
Uses: Compounding of brighteners requiring a uniform dry solid bisulfite complex for electroplating.

ANPO. Abbreviation for alpha-naphthylphenyloxazole (q.v.).

"Ansar."[548] Trademark for a series of herbicides.

"Ansco Preventol."[307] Trademark for a series of fungicides and bactericides.

"Ansol" Solvent.[192] Trademark for ethyl alcohol, with high ester and hydrocarbon content designed to increase solvency. Used in resin-solvent systems.
Containers: Up to tank cars and trucks.
Hazard: Flammable, dangerous fire risk.

"Anstac-2M."[238] Trademark for an antistatic and cleaning agent for plastics, such as methyl methacrylates, vinyls, and polystyrenes.
Containers: 1-qt bottles; 1-, 5-gal drums.
Uses: In aircraft, sign, novelty, electrical, photographic, optical, and other industries.

antacid. Any mildly alkaline substance, such as sodium bicarbonate, taken internally in water solution to neutralize excess stomach acidity.

antagonist, structural (antimetabolite). An organic compound that is structurally related to a biologically active substance (enzyme, nucleic acid, amino acid, etc.) and which acts as an inhibitor of its growth and development. Such biological antagonism exists between sulfa drugs and *p*-aminobenzoic acids, and also between histamine and a group of compounds collectively called antihistamines (q.v.). One of the most important of these from an agricultural standpoint is imidazole (q.v.) which, together with similar compounds, is used to "antagonize" the metabolism of insects, especially those attacking fabrics; it is also being used in irrigation waters to protect plants from deleterious pests. Structural antagonists have important medical applications in the complex field of allergic disease. See also chemotherapy; antigen-antibody.

"Antara."[307] Trademark for a series of high-molecular-weight organic phosphate esters offering lubricity, corrosion inhibition, and surface-active properties in oil- or water-based systems including automotive engine oils, hydraulic fluids, cutting oils, and automatic transmission fluids. Available as free acids and neutralized salts.

"Antaron FC-34."[307] Trademark for a high foaming, water soluble, amphoteric surfactant with soap-like qualities; a complex fatty amido compound; 40% active.
Uses: Fulling agent and detergent for woolen and worsted fabrics, effective under neutral, acid and alkaline conditions; recommended for use in bubble baths, detergents; in soaps for dedusting purposes.

"Antarox."[307] Trademark for a series of surfactants.
Uses: Viscose spin bath additive, preventing accumulation of sludge in pipe lines, reels, etc.; prevents clogging of spinnerets; in manufacture of cellophane, prevents deposits from forming on extrusion slits and rollers; in the steel industry, to obtain cleaner sheets in the final wash of the reduced sheet during cold reduction.

antazoline $C_{17}H_{19}N_3$.
Properties: White, odorless crystalline powder with bitter taste. M.p. 237–241° (dec). Sparingly soluble in alcohol and water; practically insoluble in benzene and ether.
Use: Medicine (antihistamine). Available as hydrochloride and phosphate.

anteiso-. Prefix denoting an isomer (usually a fatty acid or derivative) which has a single, simple branching attached at the third carbon from the end of a straight chain, in distinction to an iso- compound, where the attachment is to the second carbon from the end. For example, isododecanoic acid would be $CH_3CH(CH_3)(CH_2)_8COOH$, while anteisododecanoic acid would be $CH_3CH_2CH(CH_3)(CH_2)_7COOH$.

"Antepar."[301] Trademark for anhydrous piperazine citrate.

anthelmycin. USAN name for an antibiotic substance produced by Streptomyces longissimus.

anthocyanin. A flavonoid plant pigment which accounts for most of the red, pink, and blue colors in plants, fruits and flowers. Water-soluble. See also flavonoid.

"Anthomine."[300] Trademark for a dying assistant primarily for use in wool dyeing.

anthopyllite $(Mg,Fe)_7Si_8O_{22}(OH)_2$. A natural magnesium-iron silicate. See asbestos.

anthracene $C_6H_4(CH)_2C_6H_4$.
Properties: Yellow crystals with blue fluorescence. Soluble in alcohol and ether; insoluble in water. Sp. gr. 1.25 (27/4°C); m.p. 217°C; b.p. 340°C; flash point 250°F (closed cup). Combustible. It has semiconducting properties.

Derivation: (a) By salting out from crude anthracene oil, and draining. The crude salts are purified by pressing and finally, by the use of various solvents, phenanthrene and carbazole are removed; (b) by distilling crude anthracene oil with alkali carbonate in iron retorts, the distillate containing only anthracene and phenanthrene. The latter is removed by carbon disulfide.
Method of purification: By sublimation with superheated steam, or by crystallization from benzene followed by sublimation; for very pure crystals, by zone melting of solid anthracene.

Impurities: Phenanthrene, carbazole and chrysene.
Grades: Commercial (90 to 95%); pure crystals.
Hazard: Carcinogenic agent. Tolerance, 0.2 mg per cubic meter in air.
Uses: Dyes; alizarin; phenanthrene; carbazole; anthraquinone; calico printing; as component of smoke screens; as scintillation counter crystals; organic semiconductor research.

anthracene oil. A coal-tar fraction boiling in the range 270–360°C; a source of anthracene and similar aromatics. Also used as a wood preservative, and pesticide, except on food crops.

1,8,9-anthracenetriol. See anthralin.

anthracite. See coal.

anthragallic acid. See anthragallol.

anthragallol (1,2,3-trihydroxyanthraquinone; anthragallic acid) $C_6H_4(CO)_2C_6H(OH)_3$. Tricyclic.
Properties: Brown powder. Soluble in alcohol, ether, glacial acetic acid; slightly soluble in water and chloroform. Sublimes at 290°C.
Derivation: Product of the reaction of benzoic, gallic, and sulfuric acids.
Use: Dyeing.

"Anthragen."[307] Trademark for a series of lake colors. Used for printing inks, wallpaper, coated paper, paint, rubber, and organic plastics.

"Anthralan."[307] Trademark for a series of acid dyestuffs. Used on wool.

anthralin (1,8,9-anthracenetriol; 1,8-dihydroxyanthranol) $C_{14}H_{10}O_3$.
Properties: Odorless, tasteless, yellow powder. M.p. 176–181°C. Filtrate from water suspension is neutral to litmus. Soluble in chloroform, acetone, benzene, and in solutions of alkali hydroxide; slightly soluble in alcohol, ether, and glacial acetic acid; insoluble in water. Combustible.
Derivation: By catalytic reduction of 1,8-dihydroxyanthraquinone with hydrogen at high pressure.
Grade: N.F. (95%).
Hazard: Moderately toxic by ingestion, inhalation and skin absorption. Do not use on scalp or near eyes.
Use: Medicine (treatment of psoriasis).

anthranilic acid (ortho-aminobenzoic acid) $C_6H_4(NH_2)(CO_2H)$.
Properties: Yellowish crystals; sweetish taste; soluble in hot water, alcohol, and ether. M.p. 144–146°C. Combustible. Sublimes.
Derivation: Treatment of phthalimide with an alkaline hypobromite solution.
Grades: Technical (95–98%); 99% or better.
Hazard: May be toxic by ingestion.
Uses: Dyes, drugs, perfumes and pharmaceuticals.

anthranol (9-hydroxyanthracene) $C_{14}H_9OH$.
Properties: Crystals, m.p. 120°C; soluble in organic solvents with a blue fluorescence. Changes in solution to anthrone. Combustible.
Use: Dyes.

anthranone. See anthrone.

"Anthrapole."[300] Trademark for a group of dye carriers or assistamts, for use in dyeing polyester fibers and blends. Active ingredients are aromatic esters, chlorinated hydrocarbons and phenol derivatives. Emulsifying agents are incorporated.

anthrapurpurin (1,2,7-trihydroxyanthraquinone; isopurpurin; purpurin red) $C_6H_3OH(CO)_2C_6H_2(OH)_2$. Tricyclic.
Properties: Orange-yellow, crystalline needles. Soluble in alcohol and alkalies; slightly soluble in ether and hot water. M.p. 369°C; b.p. 462°C. Probably low toxicity.
Derivation: By fusion of anthraquinonedisulfonic acid with caustic soda and potassium chlorate; the melt is run into hot water and the anthrapurpurin precipitated by hydrochloric acid.
Uses: Dyeing; organic synthesis.

anthraquinone $C_6H_4(CO)_2C_6H_4$.
Properties: Yellow needles. Soluble in alcohol, ether, and acetone; insoluble in water. Sp. gr. 1.419–1.438; m.p. 286°C; b.p. 379–381°C; flash point (closed cup) 365°F. Combustible. Low toxicity.
Derivation: (a) By heating phthalic anhydride and benzene in the presence of aluminum chloride and dehydrating the product; (b) by condensation of 1,4-naphthoquinone with butadiene.
Method of purification: Sublimation.
Grades: Sublimed; 30% paste (sold on 100% basis); electrical 99.5%.
Containers: Bags; drums.
Uses: Intermediate for dyes and organics; organic inhibitor; bird repellent for seeds.
See also anthraquinone dye.

anthraquinone-1,5- and 1,8-disulfonic acids (rho acid, chi acid respectively) $C_{14}H_8O_8S_2$.
Properties: In their pure state, slightly yellow to white crystals. The technical grade is grayish-white. Soluble in water and strong sulfuric acid. The 1,8-isomer is the more soluble. The 1,5-disulfonic acid melts with decomposition at 310–311°C. The 1,8-isomer melts with decomposition at 293–294°C. Probably low toxicity.
Derivation: Anthraquinone is sulfonated with fuming sulfuric acid in the presence of mercury or mercuric oxide to a mixture of the 1,5- and 1,8-disulfonic acids which are separated by fractional crystallization.
Method of purification: Fractional crystallization from strong sulfuric acid or in form of their alkali salts from either acid or alkaline solutions.
Grades: Technical.
Use: Dyes.

anthraquinone dye. A dye whose molecular structure is based on anthraquinone ($C_6H_4(CO)_2C_6H_4$). The chromophore groups are $>C{=}O$ and $>C{=}C<$; the benzene ring structure is important in the development of color. C.I. numbers from 58000 to 72999. These are acid or mordant dyes when OH or HSO_3 groups respectively are present. Those anthraquinone dyes that can be reduced to an alkaline soluble leuco (vat) derivative that has affinity for fibers, and which can be reoxidized to the dye, are known as anthraquinone vat dyes. They are largely used on cotton, rayon, and silk, and have excellent properties of color and fastness, and relatively low toxicity.

anthrarufin. See 1,5-dihydroxyanthraquinone.

anthrone (anthranone; 9,10-dihydro-9-oxoanthracene) $C_{14}H_{10}O$. The keto, more stable form of anthranol.
Properties: Colorless needles; m.p. 156°C; insoluble in water; soluble in alcohol, benzene, and hot sodium hydroxide.
Derivation: Reduction of anthraquinone with tin and hydrochloric acid.
Use: Rapid determination of sugar in body fluids, and of animal starch in liver tissue; general reagent for carbohydrates; organic synthesis.

anti-. (1) A prefix used in designating geometrical isomers in which there is a double bond between the carbon and nitrogen atoms. This prevents free rotation so that two different spatial arrangements of substituent atoms or groups are possible. When a given pair of these are on opposite sides of the double bond, the arrangement is called *anti-*; when they are on the same side it is called *syn-*, as indicated below:

$$\begin{array}{cc} C_6H_5-\underset{\|}{C}-H & C_6H_5-\underset{\|}{C}-H \\ HO-N & N-OH \\ \text{anti} & \text{syn} \end{array}$$

These prefixes are disregarded in alphabetizing chemical names.
(2) A prefix meaning "against" or "opposed to," as in antibody, antimalarial, etc.

antianxiety agent. See psychotropic drug.

antibiotic. A chemical substance produced by microorganisms that has the capacity, in dilute solutions, to inhibit the growth of other microorganisms or destroy them. Only about 20 out of several hundred known have proved generally useful in therapy; those that are used must conform to FDA requirements. The most important groups of antibiotic-producing organisms are the bacteria, lower fungi or molds, and actinomycetes. These antibiotics belong to very diverse classes of chemical compounds. Most of the antibiotics produced by bacteria are polypeptides (such as tyrothricin, bacitracin, polymyxin). The penicillins are the only important antibiotics produced by fungi. Actinomycetes produce a wide variety of compounds (actinomycin, streptomycin, chloramphenicol, tetracycline). The antimicrobial activity (antibiotic spectrum) of antibiotics varies greatly; some are active only upon bacteria, others upon fungi, still others upon bacteria and fungi; some are active on viruses, some on protozoa, and some are also active on neoplasms. An organism sensitive to an antibiotic may, upon continued contact with it, develop resistance and yet remain sensitive to other antibiotics. *Note:* A number of antibiotics have been restricted by FDA, both for direct use by humans and as additives to animal feeds. Among those that are in question are streptomycin, chloramphenicol, tetracycline and some forms of penicillin.
See also penicillin; cephalosporin.

antibody. See antigen-antibody.

antiblock agent. A substance, e.g., a finely divided solid of mineral nature, which is added to a plastic mix to prevent adhesion of the surfaces of films made from the plastic to each other or to other surfaces. They are of particular value in polyolefin and vinyl films. The hard, infusible particles tend to roughen the surface and thus maintain a small air space between adjacent layers of the film, thus preventing adhesion. Silicate minerals are widely used for this purpose. Another type of antiblock function is performed by high-melting waxes, which bloom to the surface and form a layer that is harder than the plastic.

anticaking agent. An additive used primarily in certain finely divided food products that tend to be hygroscopic to prevent or inhibit agglomeration and thus maintain a free-flowing condition. Such substances as starch, calcium metasilicate, magnesium carbonate, and magnesium stearate are used for this purpose in table salt, flours, sugar, coffee, whiteners, and similar products.

anticancer agent. Any of several antibiotic drugs reported to be effective in treatment of various types of cancer. One of these is adriamycin, produced by fermentation of a mold fungus; approved for clinical research by FDA, it is said to be effective against leukemia, breast cancer and bladder cancer; it has deleterious side effects involving the heart. Other such agents are daunorubicin and carminomycin.

anticholinesterase. See cholinesterase inhibitor; nerve gas; insecticide.

antichlor. Any product which serves to neutralize and remove hypochlorite or free chlorine after the bleaching operation. For many years the term was considered synonymous with sodium thiosulfate, but it may equally be applied to sodium disulfite or any other product used for the purpose.

anticoagulant. A complex organic compound, often a carbohydrate, that has the property of retarding the clotting or coagulation of blood. The most effective of these is heparin (q.v.), which acts by interfering with the conversion of prothrombin to thrombin and by inhibiting the formation of thromboplastin. In addition to their specific clinical uses, anticoagulants have been applied with limited success to rodenticides (see warfarin). They are regarded as cumulative poisons, requiring multiple ingestions to be lethal. One type (diaphenadione) has been found to reduce blood cholesterol in experimental animals.

antidepressant. See psychotropic drug.

antidote. Any substance that inhibits or counteracts the effects of a poison which has entered the body by any route. Mild acids or alkalies (except sodium bicarbonate) exert neutralizing action if corrosive materials have been swallowed; for noncorrosive poisons warm salt water or milk may be given to cause vomiting. Activated charcoal in water is effective in protecting the throat and stomach linings, except for corrosive poisons. Nothing should be administered by mouth if the subject is unconscious. Artificial respiration may be necessary. In no case should alcohol be used as an antidote. Atropine is known to be an antidote for poisoning by cholinesterase inhibitors.

antiemetic agent. A compound, usually classed as a pharmaceutical, that inhibits or prevents nausea. A well-known type is dimenhydrinate ("Dramamine") which is useful to counteract motion sickness, nausea due to pregnancy, etc. A more recent antiemetic is a benzoquinolizine derivative used for nausea resulting from anesthesia ("Emete-con."[299])

antienzyme. A substance present in the substrate which restricts or negates the catalytic activity of the enzyme on that substrate.

antifertility agent. A synthetic steroid sex hormone of the type normally produced by the body during pregnancy. Said to act by simulating the conditions of pregnancy and thus suppressing ovulation, which automatically prevents conception. The hormones used are basically of two kinds: (1) progestin (synthetic progesterone) and (2) synthetic estrogen. There are also several other derivatives, all of which are much stronger than natural progesterone and estrogen when taken orally. The chemical modifications are necessary for oral potency.

Injectable contraceptives that remain effective over much longer periods than the oral type are also available. The active ingredient is medoxyprogesterone acetate. Experimental work is also being done with N,N'-octaethylenediamine bis(dichloroacetamide) ("Fertilysin"). An estrogen-free type is available in England (chloromadinone acetate); it is said to be less effective, but with less tendency to blood clotting, than estrogenic types.

Hazard: FDA requires that oral contraceptives carry a label warning of the tendency of these agents to form blood clots. There is also a possibility that they have other adverse side effects.

Note: Proteins and peptides which act as enzyme inhibitors have been identified in the semen of some mammals.

"Antifoam A."[149] Trademark for a silicone defoamer used to prevent or suppress foams in a wide variety of aqueous and nonaqueous systems. Generally effective at concentrations in the range of 1 to 200 ppm. Permissible in food processing up to 10 parts per million. Also available as a water-dilutable emulsion.

"Antifoam 3, 5, 7, 9, 44, 60, 66."[245] Trademark for silicone emulsions and fluids designed for the prevention or suppression of foam in aqueous and nonaqueous systems. Antifoams 3 and 44 used in food-grade applications.

"Antifoam C-1 and HP."[108] C-1 is a light-colored water-soluble liquid; HP a dry, powdered, water dispersible blend containing organic dispersing agents and sodium sulfite.

Uses: Foaming and carryover control in steam boilers. Improves steam purity; disperses sludge; minimizes blowdown.

antifoam agent. See defoaming agent.

antifouling paint. An organic coating formulated especially for use on the hulls and bottoms of ships, boats, buoys, pilings, and the like, to protect them from attack by barnacles, teredos, and other marine organisms. The chief specific ingredient is a metallic naphthenate, e.g., copper naphthenate; mercury compounds are also used.

antifreeze. (1) Water additive. Any compound that lowers the freezing point of water. Both sodium chloride and magnesium chloride were once used, but their extreme corrosive properties made them a liability in automotive cooling systems. Methyl alcohol requires only 27% by volume for protection to 0°F. Due to its tendency to evaporate rapidly at operating temperatures, its flammability and low boiling point (147°F), it has been replaced by glycol derivatives, which are relatively noncorrosive, nonflammable, have very low evaporation rate, and are effective heat-exchange agents. A concentration of about 35% protects against freezing to 0°F. Ethylene and propylene glycol antifreezes can be carried in an automotive cooling system for several years without damage, and are satisfactory coolants at summer operating temperatures. Methoxy propanol has been introduced as an antifreeze-coolant for diesel engines. See also coolant.

(2) Gasoline additive. A proprietary preparation trademarked "Drygas."[580] consisting of methyl alcohol, isopropyl alcohol, or mixtures of these, which lower the freezing point of water enough to inhibit ice formation in feed lines and carburetors. It is added directly to the gasoline. It is highly toxic and flammable.

"Anti-Fume S Solution."[28] Trademark for a durable protective agent for inhibiting and fading of dyed acetate by atmospheric gases.

antigen-antibody. An antibody is a blood serum protein of the globulin fraction, which is formed in response to introduction of an antigen. It has a molecular weight of about 160,000. An antigen is an infective organism (protein) with molecular weight of at least 10,000; it is able to induce formation of an antibody in an organism into which it is introduced (by injection). Thus an animal is able to resist infections to which it has previously been exposed. The entire science of immunology is due to antigen-antibody reactions, the most outstanding feature of which is their specificity.

The antibodies produced in the bloodstream can react only with the homologous antigen, or with those of a similar molecular structure. As a result, the animal can destroy a particular virus or bacterium and become immune. The specificity of antigens is due not so much to the molecules as a whole as to its configuration; certain radicals (polar and quaternary ammonium groups) seem to "mate" with corresponding complementary structures in the antibody molecule. A precipitate or agglutinate (q.v.) is formed by the reaction, which is analogous to colloidal (catalytic) reactions in some respects, e.g., surface configuration. See also allergen; immunochemistry; antagonist, structural.

"Anti-Germ 500."[299] Trademark for a highly concentrated and potent disinfectant, sanitizer and deodorizer with residual bacteriostatic action. Effective against a broad range of bacteria and fungi. Contains two quaternary ammonium compounds and isopropyl alcohol.

antiglobulin. An agent used to coagulate globulin.

antihistamine. A synthetic substance structurally analogous to histamine whose presence in minute amounts prevents or counteracts the action of excess histamine formed in body tissues. These compounds are usually complex amines of various types, and also have other physiological effects and medical uses. Examples are, chlorpheniramine maleate, dimenhydrinate, diphenhydramine hydrochloride, imidazole, pheniramine maleate, pyrilamine maleate, thonzylamine hydrochloride, tripelennamine hydrochloride. See also antagonist, structural.

antihypertensive agent. An organic compound having the property of lowering blood pressure in animals and man; among the better-known types are the alkaloid reserpine and its derivative, syrosingopine; guanethidine sulfate; alpha-methyl dopa (α-methyl-3,4-dihydroxyphenylalanine); and hydralazine. They function by various nerve-blocking mechanisms involving structural antagonism. They should be taken only by prescription.

antiknock agent. Any of a number of organometallic compounds, most important of which are lead alkyls, that increase the octane number of a gasoline when added in low percentages by reducing knock, especially in high-compression engines. Knock is caused by spontaneous oxidation reactions in the cylinder head, resulting in loss of power and characteristic ignition noise. Branched-chain hydrocarbon gasolines ameliorate this problem, and antiknock additives virtually eliminate it. Tetraethyllead, the most effective of these, has been used for many years, but its contribution to air pollution has necessitated considerable reduction of its use in automotive fuels. Lead-free gasolines (which contain only 0.05 gram per gallon) are available for use in engines equipped with catalytic converters. A rare-earth compound (beta-diketoenolate) is claimed to have substantial antiknock properties; cerium compounds are also being researched.

antilymphocytic serum (ALS). An immunological suppressant for use in organ transplants. It acts to control the build-up of rejection factors in the blood which result from introduction of foreign organs into the body. See also immunochemistry.

"Antilac."[165] Trademark for liquid antimony lactate containing 15% available antimony oxide. Recommended as a replacement for technical tartar emetic. See "Mordantine."

antimalarial agent. A natural or synthetic drug of the alkaloid type that is specific to combating malaria, a disease of the tropics. Most of the synthetic types are derivatives of 8-aminoquinoline (nitrogen-containing heterocyclic compounds) developed by research teams before and during the Second World War. Quinine (q.v.) has been the standard natural antimalarial drug for centuries. The first synthetic was pamaquin (1926), but it proved too toxic for more than limited use. Mepacrin ("Atabrin") was developed in 1932; it was used in the war and was found more effective then quinine, though it discolored the skin. Chloroquine (q.v.) and pentaquine followed, the former being preferable to mepacrin with no skin discoloration. A more recently discovered non-heterocyclic called proguanil (1945) has the advantages of lower cost and toxicity than other antimalarials. See also chemotherapy.

antimatter. See antiparticle.

antimetabolite. See antagonist, structural; antihistamine.

antimonial lead alloy (hard lead). Lead containing from about 6 to 28% antimony. Common grades are as follows: (a) 15% antimony; resistant to sulfuric acid; used in type metal; (b) national stock pile specification, 10.7–11.3% antimony; (c) battery grids; 5–11% antimony; (d) battery terminals; 4% antimony; (e) cable sheaths; 1% antimony.

antimonic. The variation of the name antimony used for compounds in which the antimony has a valence of five, as antimony pentachloride, pentasulfide, etc.

antimonic acid. See antimony pentoxide.

antimonic anhydride. See antimony pentoxide.

antimonite. See stibnite.

antimonous (antimonious). The variation of the name antimony used for compounds in which the antimony has a valence of three, as in antimony tribromide, antimony trichloride, antimony trioxide, antimony trisulfide.

antimonous sulfide. See antimony trisulfide.

antimony Sb (from Latin stibnium). Element of atomic number 51; Group VA of the Periodic Table. 2 stable isotopes.

Properties: Atomic weight 121.75; valences 3, 4, and 5; silver-white solid; m.p. 630.5°C; b.p. 1635°C; low thermal conductivity; Mohs hardness 3 to 3.5. Oxidized by nitric acid; unattacked by HCl in absence of air; reacts with H_2SO_4. Nontoxic as solid. Combustible. A semiconductor.

Forms: Besides the stable metal there are two allotropes—yellow crystalline and amorphous black modifications.

Ores: Stibnite, kermasite, tetrahedrite, livingstonite, jamisonite.

Occurrence: Algeria, Bolivia, China, Mexico, South Africa, Peru, Yugoslavia.

Derivation: Reduction of stibnite with iron scrap; direct reduction of natural oxide ores. About half the antimony used in U.S. is recovered from lead-base battery scrap metal.

Grades: Up to 99.999% pure; technical; powder; commercial grade in 55-lb cakes 10 × 10 × 2½".

Containers: 224-lb cases.

Hazard: Toxic as fume. Tolerance (as Sb) 0.5 mg per cubic meter of air. Use with adequate ventilation.

Uses: Hardening alloy for lead, especially storage batteries and cable sheaths; bearing metal; type metal; solder; collapsible tubes and foil; sheet and pipe; semiconductor technology (99.999% grade); pyrotechnics.

antimony 124. Radioactive antimony isotope.

Properties: Half-life, 60 days; radiation, beta and gamma. The chemical form used is often antimony trichloride and oxychloride in hydrochloric acid solution.

Uses: As a tracer (q.v.) especially in solid state studies, and marker of interfaces between products in pipe lines; the gamma ray has sufficient energy to eject neutrons from beryllium. Convenient portable neutron sources, which may be reactivated in a nuclear reactor, are made by irradiation of an antimony pellet encased in a beryllium shell.

Hazard: Toxic; radioactive material.

Shipping regulations: (Rail, Air) Consult regulations.

antimony 125. Radioactive antimony isotope. Half-life 2.4 years; emits beta and gamma rays.

Hazard: Radioactive poison.

Shipping regulations: (Rail, Air) Consult regulations.

antimony black. Metallic antimony in the form of a fine powder produced by electrolysis or chemical action on an antimony salt solution. See also antimony trisulfide.

antimony bromide. See antimony tribromide.

antimony, caustic. See antimony trichloride.

antimony chloride. See antimony trichloride.

antimony chloride, basic. See antimony oxychloride.

antimony fluoride. See antimony trifluoride.

antimony hydride (stibine) SbH_3.

Properties: Colorless gas; m.p. −88°C; b.p. −17°C.

Derivation: Action of HCl on antimony/metal compounds such as Zn_3Sb_2; also released by reduction of antimony compounds in HCl solutions with zinc or other reducing metal.

Hazard: Toxic. Tolerance, 0.1 ppm in air; 0.5 mg per cubic meter of air (as Sb).

antimony iodide. See antimony triiodide.

antimonyl. The radical or group SbO, which occurs commonly in formulas of antimony compounds. Thus, SbOCl is often named antimonyl chloride, and numerous other antimony compounds are sometimes named in a similar manner.

antimony lactate $Sb(C_3H_5O_3)_3$.

Properties: Tan-colored mass. Solution in water.

Grades: Technical.

Containers: 500-lb barrels.

Hazard: Toxic. Tolerance (as Sb), 0.5 mg per cubic meter of air.
Uses: Mordant in fabric dyeing.

antimony oxide. See antimony trioxide.

antimony oxychloride (antimony chloride, basic; antimonyl chloride) SbOCl.
Properties: White powder; m.p. 170°C (decomposes); soluble in hydrochloric acid and alkali tartrate solutions; insoluble in alcohol, ether, and water.
Derivation: Interaction of water and antimony chloride.
Hazard: Toxic. Tolerance (as Sb) 0.5 mg per cubic meter of air.
Uses: Antimony salts; flameproofing textiles.

antimony pentachloride (antimony perchloride) $SbCl_5$.
Properties: Reddish-yellow, oily liquid. Offensive odor. Hygroscopic. Solidifies by absorption of moisture. Decomposed by excess water into hydrochloric acid and antimony pentoxide. Soluble in an aqueous solution of tartaric acid, in hydrochloric acid, and chloroform. M.p. 2.8°C; sp. gr. 2.34; b.p. 92°C (30 mm).
Derivation: Action of chlorine on antimony powder.
Containers: Drums.
Hazard: Toxic and corrosive. Fumes in moist air and reacts strongly with organic materials. Tolerance (as Sb), 0.5 mg per cubic meter of air.
Uses: Analysis (testing for alkaloids and cesium); dyeing; intermediates; as chlorine carrier in organic chlorinations.
Shipping regulations: (Rail) White label. (Air) Corrosive label.

antimony pentafluoride SbF_5.
Properties: Liquid; sp. gr. 2.99 (23°C); m.p. 7°C; b.p. 149.5°C; hydrolyzed by water; soluble in potassium fluoride, liquid sulfur dioxide.
Derivation: Antimony pentachloride and anhydrous hydrogen fluoride.
Containers: Steel cylinders.
Hazard: Toxic; corrosive to skin and tissue. Tolerance (as Sb), 0.5 mg per cubic meter of air.
Use: Catalyst and/or source of fluorine in fluorination reactions.
Shipping regulations: (Rail) White label. (Air) Corrosive label. Not acceptable on passenger planes.

antimony pentasulfide (antimony red; antimony persulfide; antimony sulfide golden, Sb_2S_5.
Properties: Orange yellow powder; odorless; insoluble in water; soluble in concentrated hydrochloric acid with evolution of hydrogen sulfide; soluble in alkali. Decomposes on heating.
Hazard: Flammable, dangerous fire risk near oxidizing materials. Toxic. Tolerance (as Sb), 0.5 mg per cubic meter of air.
Uses: Red pigment; rubber accelerator.
Shipping regulations: (Rail) Flammable solid,n.o.s., Yellow label. (Air) Flammable Solid label.

antimony pentoxide (antimonic anhydride; antimonic acid; stibic anhydride) Sb_2O_5.
Properties: White or yellowish powder; sp. gr. 5.6; m.p. 450°C; loses oxygen above 300°C; insoluble in water; soluble in strong bases forming antimonates; insoluble in acids except concentrated hydrochloric.
Derivation. Action of concentrated nitric acid on the metal or the trioxide.
Hazard: Oxidizing material. Moderate fire risk. Tolerance (as Sb), 0.5 mg per cubic meter of air. Toxic by inhalation.
Use: Preparation of antimonates and other antimony compounds.

antimony perchloride. See antimony pentachloride.

antimony persulfide. See antimony pentasulfide.

antimony potassium tartrate (tartar emetic; potassium antimonyl tartrate; tartrated antimony) $K(SbO)C_4H_4O_6 \cdot \frac{1}{2}H_2O$.
Properties: Transparent, odorless crystals, efforescing on exposure to air, or white powder; sweetish, metallic taste. Sp. gr. 2.6; at 100°C loses all its water. Soluble in water, glycerol; insoluble in alcohol. Aqueous solution is slightly acid.
Derivation: By heating antimony trioxide with a solution of potassium bitartrate and subsequent crystallization.
Grades: Technical; crystals; powdered; C.P.; U.S.P.
Containers: Drums; barrels.
Hazard: Toxic by ingestion and inhalation. Tolerance (as Sb), 0.5 mg per cubic meter of air.
Uses: Textile and leather mordant; medicine; insecticide.

antimony salt (deHaens salt). Mixture of antimony trifluoride and either sodium fluoride or ammonium sulfate.
Properties: White crystals; soluble in water.
Hazard: Toxic. Tolerance (as Sb), 0.5 mg per cubic meter of air.
Use: Dyeing and printing textiles.

antimony sodiate. See sodium antimonate.

antimony sodium thioglycollate $C_4H_4O_4NaS_2Sb$.
Properties: White or pink powder; odorless or with a faint mercaptan odor; freely soluble in water; insoluble in alcohol.
Hazard: Toxic. Tolerance (as Sb), 0.5 mg per cubic meter of air.
Use: Medicine.

antimony sulfate (antimony trisulfate) $Sb_2(SO_4)_3$.
Properties: White powder or lumps. Deliquescent. Decomposes in water. Sp. gr. 3.62 (4°C).
Hazard: Toxic and flammable; tolerance (as Sb), 0.5 mg per cubic meter of air.
Uses: Matches; pyrotechnics.

antimony tribromide (antimony bromide) $SbBr_3$.
Properties: Yellow, deliquescent, crystalline mass. Soluble in carbon disulfide, hydrobromic acid, hydrochloric acid, ammonia. Decomposed by water. Sp. gr. 4.148; m.p. 96.6°C; b.p. 280°C.
Hazard: Toxic. Tolerance (as Sb), 0.5 mg per cubic meter of air.
Uses: Analytical chemistry; mordant; manufacturing antimony salts.

antimony trichloride (antimonous chloride; antimony chloride; caustic antimony) $SbCl_3$.
Properties: Colorless, transparent, crystalline mass. Very hygroscopic. Fumes slightly in air. Soluble in alcohol, acetone, acids; with water forms antimony oxychloride. Sp. gr. 3.14; b.p. 223.5°C; m.p. 73.2°C.
Derivation: Interaction of chlorine and antimony or by dissolving antimony sulfide in hydrochloric acid.
Grades: Technical; C.P.
Containers: Crystals: bottles; pails. Liquid: bottles, jugs; demijohns; carboys.
Hazard: Corrosive liquid or solid. Strong irritant to eyes and skin. Safety data sheet available from Manufacturing Chemists Assn., Washington, D.C.
Uses: Antimony salts; bronzing iron; mordant; manufacturing lakes; chlorinating agent in organic synthesis; pharmaceuticals; fireproofing textiles.
Shipping regulations: (Rail) Corrosive liquid, n.o.s., White label. (Air) Corrosive label.

antimony trifluoride (antimony fluoride) SbF_3.
Properties: White to gray crystals, hygroscopic. M.p. 292°C; sp. gr. 4.58; soluble in water. Grade: 99–100%.
Hazard: Toxic; strong irritant to eyes and skin. Tolerance (as Sb), 0.5 mg per cubic meter of air.
Uses: Porcelain; pottery; dyeing; fluorinating agent.

antimony triiodide (antimony iodide) SbI_3.
Properties: Red crystals. Volatile at high temperatures. Soluble in carbon disulfide, hydrochloric acid, and solution of potassium iodide; insoluble in alcohol and chloroform, decomposes in water with precipitation of oxyiodide. Sp. gr. 4.768; m.p. 167°C; b.p. 401°C.
Derivation: Action of iodine on antimony.
Hazard: Toxic. Tolerance (as Sb), 0.5 mg per cubic meter of air.

antimony trioxide (antimony white; antimony oxide). Sb_2O_3. Occurs in nature as valentinite.
Properties: White, odorless, crystalline powder. Sp. gr. 5.67; m.p. 655°C. Insoluble in water; soluble in concentrated hydrochloric and sulfuric acids, strong alkalies; amphoteric.
Derivation: Burning antimony in air; adding ammonium hydroxide to antimony chloride; directly from low-grade ores.
Grades: Technical; pigment.
Containers: Bags; barrels.
Hazard: Toxic. Tolerance (as Sb), 0.5 mg per cubic meter of air; may cause dermatitis.
Uses: Flameproofing of textiles, paper, and plastics (chiefly polyvinyl chloride); paint pigments; ceramic opacifier; catalyst; intermediate; staining iron and copper; phosphors; mordant; glass decolorizer.

antimony trisulfate. See antimony sulfate.

antimony trisulfide (antimony orange; black antimony; antimony needles; antimonous sulfide; antimony sulfide) Sb_2S_3.
Properties: (a) Black crystals; (b) orange-red crystals. Insoluble in water; soluble in concentrated hydrochloric acid, and sulfide solutions; Sp. gr. 4.562; m.p. 546°C.
Derivation: (a) Occurs in nature as black crystalline stibnite (q.v.). (b) As precipitated from solutions of salts of antimony, the trisulfide is an orange-red precipitate, which is filtered, dried and ground.
Grades: Technical.
Containers: Barrels; bags.
Hazard: Explosion risk in contact with oxidizing materials. Toxic. Tolerance (as Sb), 0.5 mg per cubic meter of air.
Uses: Vermilion or yellow pigment; antimony salts; pyrotechnics; matches; percussion caps; camouflage paints (reflects infrared radiation in same way as green vegetation).

antimycin A ($C_{28}H_{40}O_9N_2$). An antibiotic substance said to have strong fungicidal properties.
Properties: Crystals, M.p. 139–140°C; soluble in alcohol, ether, acetone, and chloroform; slightly soluble in benzene, carbon tetrachloride, and petroleum ether; insoluble in water.
Derivation: From Streptomyces.
Use: Active against a large group of fungi, but in general not against bacteria; possible insecticide and miticide.

antioxidant. An organic compound added to rubber, natural fats and oils, food products, gasoline and lubricating oils to retard oxidation, deterioration, rancidity, and gum formation, respectively. Rubber antioxidants are commonly of an aromatic amine type, such as di-beta-naphthyl-para-phenylenediamine and phenyl-beta-naphthylamine; a fraction of a percent affords adequate protection. Many antioxidants are substituted phenolic compounds (butylated hydroxyanisole, di-tert-butyl-para-cresol, and propyl gallate). Food antioxidants are effective in very low concentrations (not more than 0.01% in animal fats) and not only retard rancidity but protect the nutritional value by minimizing the breakdown of vitamins and essential fatty acids. Sequestering agents, such as citric and phosphoric acids, are frequently employed in antioxidant mixtures to nullify the harmful effect of traces of metallic impurities.
Note: Maximum concentration of food antioxidants approved by FDA is 0.02%.

antiozonant (antiozidant). A substance used to reverse or prevent the severe oxidizing action of ozone on elastomers, both natural and synthetic. Among antiozonant materials used are petroleum waxes, both amorphous and microcrystalline, secondary aromatic amines such as N.N-diphenyl-para-phenylenediamine, quinoline, and furan derivatives. See also ozone.

antiparticle (antimatter). A subatomic particle identical in mass with electrons, protons, and neutrons but opposite in electrical charge or (in the case of the neutron) in magnetic moment. Thus a positron is an electron with a positive charge; an antiproton is a proton with a negative charge; an antineutron has no charge but has a magnetic moment opposite to that of a neutron. A photon has no antiparticle, since it has no property except motion. When an antiparticle collides with its opposite particle (that is, a collision of an electron and a positron), both particles are annihilated and their masses are converted to photons of equivalent energy. The same is true of other subatomic particles (neutrinos, mesons, etc.), which are fantastically short-lived (of the order of billionths of a second). The quantum-mechanical concepts that led to the discovery of antiparticles include that of "strangeness," which is expressed numerically.

antiperspirant. Any substance having a mild astringent action which tends to reduce the size of skin pores and thus restrain the passage of moisture on local body areas. The most commonly used antiperspirant agent is aluminum chlorohydrate; use of zirconium compounds in antiperspirant sprays has been virtually discontinued because of their suspected carcinogenicity, though they are permissible in creams. Antiperspirants exert a neutralizing action which gives them deodorant properties. The FDA classifies them as drugs rather than as cosmetics.

antipsychotic agent. See psychotropic drug.

antipyrine (phenazone; 2,3-dimethyl-1-phenyl-3-pyrazolin-5-one). $(CH_3)_2(C_6H_5)C_3HN_2O$.
Properties: Colorless crystals or powder; odorless; slightly bitter taste; sp. gr. 1.19; m.p. 110–113°C; b.p. 319°C; soluble in water, alcohol, and chloroform; slightly soluble in ether.
Derivation: Condensation of methylphenylhydrazine and ethyl acetoacetate.
Method of purification: Crystallization.
Grades: Technical; N.F.
Containers: Drums; barrels.
Hazard: Moderately toxic by ingestion.
Uses: Medicine; an analytical reagent for nitrous acid, nitric acid, and iodine number.

antipyrine chloral hydrate. See chloral hydrate antipyrine.

antiscorbutic. Tending to prevent scurvy. See ascorbic acid.

antiseptic. A substance applied to humans or animals that retards or stops the growth of microorganisms without necessarily destroying them, e.g., alcohol; boric acid and borates; certain dyes, as acriflavine; menthol; hydrogen peroxide; hypochlorites; iodine; mercuric chloride; and phenol. Many of these are corrosive and poisonous and should be used with great caution. Among the newer antiseptics are hexachlorophene, which also has some degree of toxicity, and some quaternary ammonium compounds.
See also disinfectant; sanitizer; fumigant.

antiskinning agent. A liquid antioxidant used in paints and varnishes to inhibit formation of an oxidized film on the exposed surface in cans, pails, or other open containers.

antistatic agent. The marked tendency of thermoplastic polymers to accumulate static charges, which result in adherent particles of dust and other foreign matter, has required study of possible means of eliminating or reducing this property. The following have been tried. (1) Development of more electrically conductive polymers, e.g., tetracyanoquinodimethane. (2) Incorporation of additives which migrate to the surface of the plastic or fiber and modify its electrical properties; examples of these are fatty quaternary ammonium compounds, fatty amines, and phosphate esters; other types of antistatic additives are hygroscopic compounds, such as polyethylene glycols, and hydrophobic slip additives that markedly reduce the coefficient of friction of the plastic. (3) Copolymerization of an antistatic resin with the base polymer.

"Antistine" Phosphate.[305] Trademark for antazoline phosphate.

"Antox."[28] Trademark for rubber antioxidant; a condensation product of butyraldehyde-aniline. Amber liquid.

"Antozite."[69,119] Trademark for a series of antiozonants for use in rubber and synthetic rubber.
1. N,N′-Di(2-octyl)-para-phenylenediamine.
2. N.N′-Di-3(5-methylheptyl)-para-phenylenediamine.
MPD. N,N′-Bis(1,4-dimethylpentyl)-para-phenylenediamine.

"Antron."[28] Trademark for nylon textile fibers in the form of continuous filament yarns and staple.

ANTU. Abbreviation for alpha-naphthylthiourea (q.v.).

AP. Abbreviation for ammonium perchlorate (q.v.).

"Apamide."[272] Trademark for acetaminophen. See para-acetylaminophenol. Used in medicine.

APAP. Abbreviation for acetyl-para-amino-phenol. See para-acetylaminophenol.

"Apasol" W-1345.[309] Trademark for a completely sulfonated ester of a high molecular weight alcohol; used in many textile operations.

apatite. A natural calcium phosphate (usually containing fluorine) occurring in the earth's crust as phosphate rock. It is also the chief component of the bony structure of teeth. See also fluoridation.
Properties: Color variable; sp. gr. 3.1–3.2; hardness 5; frequently in hexagonal crystals.
Occurrence: Eastern U.S., California; U.S.S.R.; Canada; Europe.
Uses: Source of phosphorus and phosphoric acid; manufacture of fertilizers; laser crystals.

APC. Abbreviation for ammonium perchlorate.

"Apco."[200] Trademark for a series of petroleum solvents with carefully graded boiling ranges and known flash points; some are deodorized. Available in up to tank car lots. Combustible.
Uses, according to specific properties: Dry cleaning; degreasing; extraction; paint and varnish thinner; wood- and weed-treating solvent.

apholate. Generic name for 2,2,4,4,6,6-hexakis(1-aziridinyl)-2,2,4,4,6,6-hexahydro-1,3,5,2,4,6-triazatriphosphorine. An insect sterilant. Prevents reproduction of certain insects by inhibiting formation of DNA in eggs.
Hazard: Probably toxic by ingestion.

aphrodine. See yohimbine.

API. Abbreviation for American Petroleum Institute. (q.v.).

API gravity. A scale of measurement adopted by the American Petroleum Institute. It runs from 0.0 (equivalent to specific gravity 1.076) to 100.0 (equivalent to specific gravity 0.6112). The API values as used in the petroleum industry decrease as specific gravity values increase.

"Apiezon."[431] Trademark for a series of hydrocarbon oils, greases and waxes that are produced by molecular distillation and characterized by very low vapor pressures and good thermal stability. Used as lubricants and seals in high vacuum equipment and operations; as a stationary phase in gas chromatography. Combustible.

apo-. A prefix denoting formation from, or relationship to, another compound, e.g., apomorhine.

APO. See triethylenephosphoramide.

apocarotenal. Food color; supplied in dark purplish-black beadlets. Vitamin A activity 120,000 units/gram. Dispersible in warm water. Approved for food use by FDA.

"Apodol."[412] Trademark for anileridine (q.v.).

Appert, Francois (1750–1840). French pioneer in the science of food preservation. Though not a chemist, his work on application of heat to food products was in effect a form of home preserving, which eventually developed into the canning industry. The idea of destroying bacteria by heat treatment was later applied more exhaustively by Pasteur.

apple acid. See malic acid.

apple oil. See isoamyl valerate.

applied research. The experimental investigation of a specific practical problem for the immediate purpose of creating a new product, improving an older one, or evaluating a proposed ingredient. The experimental program is set up to answer the question "What happens?" rather than "Why does it happen?" Examples might be the determination of the value of a new rubber antioxidant; the substitution of one drying oil for another in paint formulation; or the development of a new synthetic product.

"Appramine."[42] Trademark for cream-colored, cationic fatty amides. Disperse readily in water at temperatures above 60°C. Used as softening agents for all types of textile fibers.

"Appretole."[42] Trademark for a cream-colored anionic fatty amide dispersion. Disperses readily in water above 60°C. Used as a softener for cotton and rayon textile fabrics.

"Apresoline."[305] Trademark for hydralazine hydrochloride (q.v.).

aprotic solvent. A type of solvent which neither donates nor accepts protons. Examples: dimethylformamide; benzene; dimethyl sulfoxide.

"APW."[244] Trademark for a compound consisting of a balanced blend of buffered alkalies and a surface-active agent.
Properties: Soluble in water; total Na_2O content 39.9%; percent of total Na_2O in active form 24.0%; percent of total Na_2O in inactive form 15.9%; white, granular dedusted mechanical mix.

aqua ammonia. See ammonium hydroxide.

"Aquablak."[133] Trademark for carbon black and bone black aqueous dispersions used in latex paints, latex compounding, paper coatings, leather finishing, etc.

"Aquabond."[553] Trademark for dry and liquid resinous adhesives.

"Aquace."[266] Trademark for vinyl acetate-ethylene copolymer emulsions.
Properties: 55% to 57% solids; 200 to 600 cps; 4.5 to 6.0 pH; 0.13μ to 0.17μ average particle size; nonionic; 8.9 to 9.1 lbs per gal.
Containers: Bulk; 55-gal lined steel or fiber drums.
Uses: Binder for pigments in paper coatings; exterior or interior paints; adhesives and textiles.

"Aquadag."[46] Trademark for a dispersion of colloidal graphite in water. Used as a lubricant for dies, tools, and molds for metalworking, glassmaking, etc.; conductive coating.

"Aquagel."[236] Trademark for a gel-forming colloidal bentonite clay used in drilling muds. See also bentonite.

"Aqualin."[125] Trademark for a herbicide, biocide, and slimicide; contains 92% acrolein.
Hazard: See acrolein.
Uses: Herbicide used to control aquatic weeds; biocide to control microorganisms in oil well injection water; slimicide to control slime-forming organisms in pulp and paper mills.

"Aquapel."[266] Trademark for a series of alkylketene dimers prepared from long chain fatty acids. Used for paper sizing.

"Aquaprint."[293] Trademark for a resin-bonded pigment color for printing on textiles. The vehicle, an oil-in-water emulsion, contains a water-insoluble binder which adheres to the fibers and anchors the color permanently to the cloth.

aqua regia (nitrohydrochloric acid; chloronitrous acid; chloroazotic acid).
Properties: Fuming yellow, suffocating, volatile liquid.
Derivation: A mixture of nitric and hydrochloric acids, usually one part of nitric acid to three or four parts of hydrochloric acid.
Grades: Technical.
Containers: Glass bottles.
Hazard: Toxic; a powerful oxidizing agent; corrosive liquid.
Uses: Metallurgy; testing metals; dissolving metals (platinum, gold, etc.).
Shipping regulations: (Rail) White label. (Air) Corrosive label. Not acceptable on passenger planes. Legal label name: nitrohydrochloric acid.

"Aquaresin."[73] Trademark for glycol bori-borate.
Properties: Water-white viscous liquid which is nondrying. Odorless. Soluble in water, methyl alcohol, glycerine, and diethylene glycol; insoluble in ethyl alcohol and toluene. Sp. gr. (25°C) 1.375; pH (5% dispersion) 8.0.
Uses: Textile lubricant and softener; softener or plasticizer for glues, gelatins, gums; adhesive for cellophane, glassine; sealing joints in systems carrying oils, hydrocarbons; fire retardant for treating paper, leather, textiles; prevention of caking in pigments in water suspensions.

"Aquarex."[28] Trademark for a series of wetting agents for elastomers. They act as stabilizers and mold lubricants.

"Aquarol."[300] Trademark for water repellents of the wax-multivalent metal salt type for textiles.

"Aquasol."[57] Trademark for a highly sulfonated castor oil.

"Aquasorb AR."[329] Trademark for a phosphorus pentoxide (P_2O_5)-based desiccant.
Hazard: May be flammable and caustic.

"Aqua-Tone."[29] Trademark for a series of polyvinyl acetate emulsion coatings that are easily applied, quick drying, completely washable and resistant to the effects of sunlight, mildew, and most industrial fumes.

"Aquazinc."[354] Trademark for an aqueous dispersion of zinc stearate containing a wetting agent and designed to replace powdered zinc stearate for many applications in order to eliminate the dust, fire hazard, and other difficulties encountered with the dry material. It will volatize at or below 100°C.
Uses: Manufacture of butyl rubber, neoprene adhesives, various types of rubber latex, etc.

"Aquet."[204] Trademark for a product used to raise the pH of swimming pool water.

Ar Symbol for argon approved by IUPAC.

"A.R."[329] Trademark for chemical products for laboratory and industrial use which are specially produced and controlled to meet critical purity and uniformity requirements.

arabic gum (acacia gum). The dried water-soluble exudate from the stems of Acacia Senegal or related species.
Properties: Thin flakes, powder, granules or angular fragments; color white to yellowish white, almost odorless, and have a mucilaginous taste. Completely soluble in hot and cold water, yielding a viscous solution of mucilage; insoluble in alcohol. The aqueous solution is acid to litmus. Nontoxic; combustible.
Occurrence: Sudan, West Africa, Nigeria.
Composition: A carbohydrate polymer, complex and highly branched. The central core or nucleus is D-galactose and D-glucuronic acid (actually the calcium, magnesium, and potassium salts), to which are attached sugars such as L-arabinose and L-rhamnose.
Grades: U.S.P.; F.C.C. (both grades as acacia).

Containers: Bags; multiwall paper sacks; barrels.
Uses: Pharmaceuticals; adhesives; inks; textile printing; cosmetics; thickening agent and colloidal stabilizer in confectionery and food products.

arabinogalactan. A water-soluble polysaccharide extracted from timber of the western larch trees. It is a complex highly branched polymer of arabinose and galactose.
Properties: Dry, light tan powder; readily soluble in hot and cold water; both powder and solutions relatively stable. Nontoxic; combustible.
Uses: Dispersing and emulsifying agent; lithography.

arabinose (pectin sugar; gum sugar) $C_5H_{10}O_5$. Both the D- and L-enantiomers occur naturally. L-Arabinose is common in vegetable gums, especially arabic.
Properties: White crystals. Soluble in water and glycerin; insoluble in alcohol and ether. M.p. 158.5°C; sp. gr. 1.585 (20/4°C). Nontoxic. Combustible.
Uses: Medicine; culture medium.

Ara-C. Abbreviation for cytosine arabinoside (q.v.).

arachidic acid (eicosanoic acid) $CH_3(_2)_{18}COOH$. A widely distributed but minor component of the fats of peanut oils and related plant species.
Properties: Shining, white, crystalline leaflets. Soluble in ether; slightly soluble in alcohol; insoluble in water. M.p. 75.4°C; sp. gr. 0.8240 (100/4°C); b.p. 205°C (1 mm), 328°C (decomposes); refractive index 1.4250 (100°C). Nontoxic; combustible.
Derivation: From peanut oil.
Grades: Technical 99%.
Uses: Organic synthesis; lubricating greases, waxes, and plastics; source of arachidyl alcohol; biochemical research.

arachidonic acid (5,8,11,14-eicosatetraenoic acid). $CH_3(CH_2)_4(CH{:}CHCH_2)_4(CH_2)_2COOH$. A C_{20} unsaturated fatty acid. Nontoxic; combustible. An essential fatty acid.
Properties: Liquid; m.p. −49.5°C; iodine value 333.50.
Source: Liver; brain; lecithin.
Grade: 99% methyl ester.
Uses: Medicine; biochemical research.

arachidyl alcohol (1-eicosanol) $CH_3(CH_2)_{18}CH_2OH$. A long-chain saturated fatty alcohol, much like stearyl alcohol.
Properties: White, wax-like solid. M.p. 66.5°C; b.p. 220°C (3 mm); 369°C at 760 mm. Refractive index 1.455; soluble in hot benzene. Combustible; low toxicity.
Derivation: Ziegler synthesis (trialkylaluminum process).
Grades: Technical; 99%.
Containers: Up to tankcars.
Uses: Lubricants, rubber, plastics, textiles; research.

arachin (arachine). A protein from peanut, a globulin containing arginine, histidine, lysine, cystine. Yellow green syrup. Soluble in water and alcohol; insoluble in ether. Nontoxic; combustible.

aragonite. A form of calcium carbonate appearing in pearls. See nacre.

aralkyl. See arylalkyl.

aramid. Generic name for a distinctive class of aromatic polyamide fibers. It differs from nylon in chemical structure and physical properties. See polyamide.

"Aramite."[248] Trademark for 2-(para-tert-butylphenoxy) isopropyl 2-chloroethyl sulfite $(CH_3)_3CC_6H_4OCH_2CH(CH_3)OSOOC_2H_4Cl$.
Properties: Clear, light-colored oil; sp. gr. 1.148–1.152 (20°C); b.p. 175°C (0.1 mm); very soluble in common organic solvents; insoluble in water; noncorrosive.
Grades: Technical (90% min.); wettable powder; emulsifiable concentrates; dusts.
Hazard: Irritant to eyes and skin; toxic by ingestion. Use may be restricted.
Uses: Acaricide on a wide variety of vegetation.

"Aranox."[248] Trademark for para-(para-tolysulfonylamido)-diphenylamine. $CH_3C_6H_4SO_2NHC_6H_4NHC_6H_5$.
Properties: Gray powder; sp. gr. 1.32; m.p. 135°C min; soluble in acetone, benzene and ethylene dichloride; insoluble in gasoline and cold water; slightly soluble in hot water or hot alkaline solutions which extract it from thin rubber sheets after considerable exposure.
Uses: Antioxidant for light-colored rubber products.

"Arapahoe Pre-Mix."[109] Trademark for mixtures of liquid scintillation-counting materials.

"Arasan."[28] Trademark for seed disinfectants based on thiram (q.v.).
Hazard: Moderately toxic by ingestion.

"Arazate."[248] Trademark for zinc dibenzyl dithiocarbamate (q.v.).

arbutin (ursin). Available commercially in both natural and synthetic forms. Pure synthetic arbutin is hydroquinone-beta-D-glucopyranoside, $C_{12}H_{16}O_7$.
Properties: (pure synthetic): White powder; m.p. 199–200°C; soluble in water and alcohol; stable in storage.
Derivation: A glucoside found in the leaves of the cranberry and blueberry and in the roots, trunks and leaves of most pear species. Pure arbutin can be prepared synthetically from acetobromoglucose and hydroquinone in presence of alkali.
Uses: Medicine; possible industrial oxidation inhibitor, polymerization inhibitor, color stabilizer, intermediate.

archeological chemistry. Application of the techniques of analytical chemistry to identification of materials found in excavations. Among those used are microanalytical methods, spectroscopic analysis, x-rays, and other types of nondestructive tests. For age determination, C-14 measurement (chemical dating) has been a valuable tool. See also chemical dating.

"Arcturus Red."[141] Trademark for azo red pigments derived from beta-naphthol.
Grades: Resinated and nonresinated.
Uses: Printing inks, especially flexographic inks for printing on foil, rubber, plastics.

arecoline (arecaidine methyl ester; methyl 1,2,5,6-tetrahydro-1-methylnicotinate; methyl arecaidinate) $C_8H_{13}O_2N$. Alkaloid obtained from Areca catechu or betel nut.
Properties: Colorless, odorless, oily liquid; strongly alkaline; optically inactive; b.p. 209°C; sp. gr. 2.02; soluble in water, alcohol, chloroform, ether; volatile with steam. Combustible.
Hazard: Moderately toxic.
Use: Medicine (also as hydrobromide)

arene. See aromatic.

"Areskap."[58] Trademark for sodium butyl-orthophenylphenolsulfonate in liquid or dry powder form.
Uses: Wetting, penetrating and spreading agent for insecticides, embalming fluids.

"Aresket."[58] Trademark for sodium butylbiphenylsulfonate in dry powder form.
Uses: Wetting, penetrating, and spreading agent.

argentite (silver glance) Ag_2S. Lead-gray to black or blackish-gray mineral. A natural silver sulfide. Contains 87.1% silver. Differs from other soft black minerals in cutting like wax. Soluble in nitric acid. Sp. gr. 7.2–7.36; Mohs hardness 2–2.5.
Occurrence: Nevada, Colorado, Montana; Mexico; Chile; Canada.
Use: An important ore of silver.

argentum. The Latin name for silver, hence the symbol Ag in chemical nomenclature.

arginase. An enzyme producing ornithine and urea by splitting arginine. It is found in liver.
Use: Biochemical research.

arginine (guanidine aminovaleric acid; amino-4-guanidovaleric acid) $NHC(NH_2)NH(CH_2)_3CH(NH_2)COOH$. An essential amino acid for rats, occurring naturally in the L(+) form. Available as the glutamate and hydrochloride.
Properties: Prisms from water containing 2 molecules of H_2O, anhydrous plates from alcohol solution; dehydrates at 105°C; decomposes at 244°C; sparingly soluble in alcohol; insoluble in ether. Nontoxic.
Derivation: Widely found in animal and plant proteins. It is precipitated as the flavianate from gelatin hydrolyzate in industry.
Containers: Drums.
Uses: Biochemical research; medicine; pharmaceuticals; dietary supplement.

argon Ar. Inert element of atomic number 18, in noble gas group of the Periodic System. Atomic weight 39.948; occurs in air to extent of 0.94% by volume.
Properties: Colorless, odorless, tasteless, monatomic gas; it is not known to combine chemically with any element, but forms a stable clathrate (q.v.) with beta-hydroquinone; m.p. −189.3°C; b.p. −185.8°C; sp. gr. 1.38 (air = 1); sp. vol. 9.7 cu ft/lb (70°F, 1 atm). Slightly soluble in water. Noncombustible; an asphyxiant gas.
Derivation: (a) By fractional distillation of liquid air. (b) By the treatment of atmospheric nitrogen with certain metals such as magnesium and calcium to form nitrides. (c) Recovery from natural gas oxidation bottoms steam in ammonia plant. (d) Originally formed by radioactive decay of potassium 40.
Methods of purification: (a) Highly purified argon is obtained by passing the gas through a bed of titanium at 850°C. (b) Synthetic zeolite molecular sieves separate oxygen from argon to give high purity gas.
Grades: Technical; highest purity (99.995%).
Containers: Steel cylinders (technical); hermetically sealed glass flasks (highest purity); tube trailers.
Uses: Inert gas shield in arc welding; furnace brazing; plasma jet torches (with hydrogen); electric and specialized light bulbs (neon, fluorescent, sodium vapor, etc.); titanium and zirconium refining; flushing molten metals (steel) to remove dissolved gases; in Geiger-counting tubes, lasers; inert gas or atmosphere in miscellaneous applications.
Shipping regulations: (Rail) (gas) Green label; (pressurized liquid) Green label. (Air) (gas) Nonflammable Compressed Gas label; (liquid, nonpressurized) No label required; (liquid, low-pressure) Not acceptable on passenger planes; (liquid, pressurized) Nonflammable Compressed Gas label. Not acceptable on passenger planes.

"Argo" Steepwater.[30] Trademark for a liquid concentrate of proteins, soluble nutrients, unidentified growth factors (U.G.F.); 53.0–56.0% solids, pH 3.7–4.2. Used in manufacture of enzymes and for drugs and pharmaceuticals. Shipped in tank cars. See also corn steep liquor.

"Argyrol." Trademark for an organic compound of silver and a protein, used in medicine for its specific antiseptic and bacteriostatic action. Low toxicity.

"Aridex."[28] Trademark for renewable water repellents used in textile industry.

"Aridye."[293] Trademark for a product and process for printing colors on textiles using permanent and insoluble pigments suspended in an organic vehicle, into which water is emulsified to give printing consistency. The vehicle contains a water-insoluble binder which adheres to the cloth and anchors the color permanently to the fibers.

"Ariperm."[300] Trademark for a gas-fading inhibitor for application to dyed acetate fabrics. Protects the dyes from fading due to acid gases in the atmosphere.

"Aristocort."[315] Trademark for triamcinolone (q.v.).

"Arizole."[252] Trademark for a group of terpene products, including anethole and sulfate pine oil. Used in perfumes, soaps, etc.

"Arkolene."[300] Trademark for textile wetting agents of the alkylarylsulfonate type, sodium or ammonium salts.

"Arkolube."[300] Trademark for textile lubricating and softening agents, based on silicones, polyethylenes, and waxes.

"Arko Stat-Ex."[300] Trademark for antistatic agent for use on synthetic fibers.

"Arkotan."[300] Trademark for a synthetic tanning agent for leather. Ammonium salt of a naphthalenesulfonic acid complex.

"Arlacel."[89] Trademark for each of a series of nonionic emulsifiers for use in cosmetics and pharmaceuticals. They are fatty acid partial esters of polyols or polyol anhydrides.

"Arlex."[89] Trademark for noncrystallizing industrial humectant solution, containing 83% solids consisting of sorbitol and related polyhydric materials.
Use: For flexibilizing and moisture-conditioning in industrial applications, including tobacco, glue compositions, cellulose products, etc.

"Armac."[15] Trademark for a series of amine acetate salts derived from primary, secondary or tertiary amines. The alkyl groups range from C_8 to C_{18}.
Uses: Anticaking agents; biocides; cationic emulsifiers; corrosion inhibitors; mineral flotation agents; pigment dispersants; pigment flushing agents; textile assistants.

"Armalon."[28] Trademark for TFE-fluorocarbon fiber felt and also for TFE-fluorocarbon resin-coated glass fabrics, tapes, and laminates

"Armazide."[204] Trademark for an algaecide for swimming pools.

"Armeen."[15] Trademark for a series of high-molecular-weight aliphatic amines: primary, secondary and tertiary. Available in chain lengths ranging from C_8 to C_{18} and up to 92% purity of a single homolog, also in natural mixtures of these chains derived from coco, soya or tallow fatty acids.

Uses: Biocides; catalysts; chemical intermediates; corrosion inhibitors; curing agents; detergents; emulsifiers; mineral flotation agents; mold release agents; petroleum additives; pigment grinding aids; rubber reclaiming agents.

"Armid."[15] Trademark for a series of high-melting, wax-like amides derived from fatty acids.
Uses: Antiblocking agents in plastic formulations; antitack and antiscratch agents in printing inks; dye solubilizers; mutual solvents for waxes and plastics.

"Armofos."[1] Trademark for sodium tripolyphosphate, anhydrous $Na_5P_3O_{10}$.
Uses: Sequestering agent for iron, calcium and magnesium ions; soap builder; detergent mixtures; deflocculator in drilling muds; paper, ceramics and textiles.

"Armosul."[15] Trademark for alpha-sulfonated alkyl acids, C_{16} and C_{18}. Biodegradable and suitable for detergent preparations. Used in mineral flotation.

"Armotan."[15] Trademark for a series of sorbitan and polysorbate esters of fatty acids.
Uses: Bodying agents; corrosion inhibitors; defoamers; dispersants; emulsifiers; lubricants; opacifiers; solubilizers; stabilizers; wetting agents. Approved for many food uses.

Armstrong's acid. See naphthalene-1,5-disulfonic acid.

"Arnel."[352] Trademark for an acetate fiber made from cellulose triacetate. It has a higher melting point and is less soluble than cellulose acetate.
Properties: Staple or continuous filament. Tensile strength (psi) 20,000–26,000; elongation 22–28%; sp. gr. 1.3; moisture regain 3.2%; m.p. approximately 300°C; soluble in methylene chloride, glacial acetic acis, chloroform; swollen by acetone and trichloroethylene. Combustible.
Uses: Knitted and woven fabrics for wearing apparel, alone and blended with other fibers; laundry pads; electrical insulation; laminated papers.

arnica. Medication derived from a plant; usually in tinctures of various strengths. Used topically for relief of bruises, strains, etc.
Hazard: Toxic by ingestion; 1 ounce of 20% tincture may cause severe poisoning.

"Arochem."[221] Trademark for a series of hard coating resins, including modified phenolics, modified alcohols, ester gums, modified maleics.

"Aroclor."[58] Trademark for a series of polychlorinated polyphenols: liquids, resins, or solids.
Properties: Water-white mobile liquids and pale yellow viscous oils to light amber resins and opaque crystalline solids. Some types are combustible.
Hazard: May be toxic by ingestion and inhalation; absorbed by skin.
Uses: As dielectric liquids; as impregnating agents for electrical apparatus; thermostat fluids; swelling agents for transmission seals; additives or base for lubricants, oils and greases; plasticizers for cellulosics, vinyls, and chlorinated rubber.

"Aroflat."[221] Trademark for a series of oxidizing alkyds in a petroleum solvent for flat wall coatings.

"Aroflint."[221] Trademark for a two-component system for industrial and marine finishes.

"Arolast."[221] Trademark for a set of components for urethane elastomer.

"Aromaldehydes."[188] Trademark for a series of simulated terpeneless oils.

aromatic (arene). A major group of unsaturated cyclic hydrocarbons containing one or more rings; these are typified by benzene, which has a six-carbon ring containing three double bonds. The vast number of compounds of this important group, derived chiefly from petroleum and coal tar, are rather highly reactive and chemically versatile. The name is due to the strong and not unpleasant odor characteristic of most substances of this nature. Certain 5-membered cyclic compounds such as the furan group (heterocyclic) are analogous to aromatic compounds.
Note: The term "aromatic" is often used in the perfume and fragrance industries to describe essential oils, which are not aromatic in the chemical sense.

aromaticity. A stable electron shell configuration in organic molecules, especially those related to benzene. See resonance, orbital theory.

aromatization. See hydroforming.

"Aromin."[51] Trademark for a highly aromatic solvent widely used as a carrier for chemical pesticides.

"Aromox."[15] Trademark for a series of methylated or ethoxylated amine oxides derived from high-molecular-weight aliphatic amines.
Uses: Foam and suds stabilizers in detergent and cosmetic formulations; surfactants.

"Aroset."[221] Trademark for a thermosetting acrylic resin solution to be used as a coating.

"Arox."[51] Trademark for specially compounded fluid lubricants for all air tools.

"Arquad."[15] Trademark for a series of quaternary ammonium salts containing one or two alkyl groups ranging from C_8 to C_{18}.
Uses: Corrosion inhibitors; emulsifiers; germicides and sanitizing agents; textile fabric softeners.

Arrhenius, Svante (1859–1927)
Native of Sweden, Nobel prize in chemistry, 1903. Best known for his fundamental investigations on electrolytic dissociation of compounds in water and other solvents, and for his basic equation stating the increase in the rate of a chemical reaction with rise in temperature:

$$\frac{d \ln k}{dT} = \frac{A}{RT^2}$$

in which k is the specific reaction velocity, T is the absolute temperature, A is a constant usually referred to as the energy of activation of the reaction, and R is the gas law constant.

arsacetin (sodium acetylarsanilate; sodium para-acetylaminophenylarsonate)
$CH_3CONHC_6H_4AsO(OH)ONa$.
Properties: White, crystalline powder; odorless, tasteless, free of arsenous or arsenic acid; solutions will admit of thorough sterilization. Soluble in cold, but more so in warm water.
Hazard: Highly toxic by ingestion.
Use: Medicine.
Shipping regulations: (Rail, Air) Arsenical compounds, n.o.s. Poison label.

arsanilic acid (atoxylic acid; para-aminobenzenearsonic acid; para-aminophenylarsonic acid)
$C_6H_4NH_2 \cdot AsO(OH)_2$.
Properties: White, crystalline powder; practically odorless; soluble in hot water; slightly soluble in cold water, alcohol, and acetic acid; insoluble in acetone, benzene, chloroform, and ether. M.p. 232°C.
Derivation: By condensing aniline with arsenic acid,

removing the excess of aniline by steam distillation in alkaline solution and setting the acid free by hydrochloric acid.
Hazard: Highly toxic. Yields flammable vapors on heating above melting point.
Uses: Arsanilates; manufacture of arsenical medicinal compounds, such as arsphenamine, etc.; veterinary medicine; grasshopper bait.
Shipping regulations: (Rail, Air) Arsenical compounds, n.o.s. Poison label.

arsenic As. A nonmetallic element of atomic number 33; group VA of Periodic Table; atomic weight 74.9216; valence 2,3,5; no stable isotopes.
Properties: Silver-gray brittle, crystalline solid that darkens in moist air. Allotropic forms: black, amorphous solid (beta-arsenic); yellow, crystalline solid. Sp. gr. 5.72 (commercial product ranges from 5.6 to 5.9); m.p. 814°C (36 atm); sublimes at 613°C (1 atm); Mohs hardness 3.5; insoluble in water, caustic and nonoxidizing acids. Attacked by HCl in presence of oxidant. Reacts with nitric acid. Low thermal conductivity; a semiconductor.
Derivation: Flue dust of copper and lead smelters, from which it is obtained as white arsenic (arsenic trioxide) in varying degrees of purity. This is reduced with charcoal. The commercial grade is not made in U.S.
Grades: Technical; crude (90–95%); refined (99%); semiconductor grade 99.999%; single crystals.
Containers: Bottles; drums; barrels.
Hazard: Highly toxic by ingestion and inhalation. The ACGIH tolerance for arsenic and its compounds for workroom exposures is 0.5 mg per cubic meter of air (as As), but OSHA has proposed that this limit be lowered to 0.004 mg for 8-hour exposure, with a maximum limit of 0.01 for short exposures. There is said to be considerable evidence that arsenic is carcinogenic, and that all inorganic arsenic compounds are occupational carcinogens.
Uses (metallic form): Alloying additive for metals, especially lead and copper (shot, battery grids, cable sheaths, boiler tubes). High-purity (semiconductor) grade: used to make gallium arsenide for dipoles and other electronic devices; doping agent in germanium and silicon solid state products; special solders; medicine.
Shipping regulations: (Rail, Air) Poison label.
See also arsenic trioxide.

arsenic acid (orthoarsenic acid) $H_3AsO_4 \cdot ½H_2O$.
Arsenic pentoxide is also sometimes called arsenic acid.
Properties: White, translucent crystals. Soluble in water, alcohol, alkali, glycerin. Sp. gr. 2–2.5; m.p. 35.5°C; b.p. loses water at 160°C.
Derivation: By digestion of arsenic with nitric acid.
Grades: Pure; technical; C.P.
Containers: Glass bottles; barrels.
Hazard: Highly toxic by ingestion.
Uses: Manufacture of arsenates; glass making; wood treating process; defoliant (under special regulations); desiccant for cotton; soil sterilant.
Shipping regulations: (Rail, Air) Poison label.

arsenical Babbitt. See Babbitt metal.

arsenical nickel. See niccolite.

arsenic anhydride. See arsenic pentoxide.

arsenic, black (beta-arsenic). See arsenic.

arsenic bromide. Legal label name for arsenic tribromide. (q.v.).

arsenic chloride. See arsenic trichloride.

arsenic disulfide (arsenic monosulfide; ruby arsenic; red arsenic glass; red arsenic sulfide; red arsenic) As_2S_2 or AsS. Occurs as mineral, realgar (q.v.).
Properties: Orange-red powder. Soluble in acids and alkalies; insoluble in water. Sp. gr. 3.4–2.6; m.p. 307°C.
Derivation: By roasting arsenopyrite and iron pyrites and sublimation.
Grades: Technical.
Containers: Steel drums.
Hazard: Highly toxic by ingestion and inhalation.
Uses: Leather industry, depilatory agent; paint pigment; shot manufacture; pyrotechnics; rodenticide.
Shipping regulations: (Rail, Air) Arsenical compounds, n.o.s. Poison label. Legal label name: arsenic sulfide.

arsenic hydride. See arsine.

arsenic pentafluoride AsF_5. A gas; b.p. −52.8°C.
Grade: About 98%.
Containers: 50-g, 1-lb cylinders.
Hazard: Highly toxic by inhalation.
Shipping regulations: (Rail, Air) Arsenical compounds, n.o.s. Poison label.

arsenic pentasulfide As_2S_5.
Properties: Yellow or orange powder. Soluble in nitric acid and alkalies; insoluble in water; decomposes to sulfur and the trisulfide when heated.
Derivation: By precipitation from arsenic acid in a hydrochloric acid solution with hydrogen sulfide. It is filtered, then dried.
Grades: Technical.
Containers: Barrels; boxes.
Hazard: Highly toxic by inhalation.
Uses: Paint pigments; blue fire; other arsenic compounds.
Shipping regulations: (Rail, Air) Arsenical compounds, n.o.s. Poison label.

arsenic pentoxide (arsenic oxide; arsenic anhydride; arsenic acid) As_2O_5.
Properties: White, amorphous solid; deliquescent. Forms arsenic acid in water. Soluble in water, alcohol. Sp. gr. 4.086; m.p. 315°C.
Derivation: By action of oxidizing agent such as nitric acid on arsenious oxide.
Containers: Drums; cartons; boxes; tins.
Hazard: Highly toxic.
Uses: Arsenates; insecticides; dyeing and printing.
Shipping regulations: (Rail, Air) Poison label.

arsenic sesquioxide. See arsenic trioxide.

arsenic sulfide. Legal label name for arsenic disulfide (q.v.).

arsenic thioarsenate $As(AsS_4)$
Properties: Dry, free-flowing yellow powder; stable; high-melting. Insoluble in water and organic solvents but soluble in aqueous caustics.
Hazard: Highly toxic.
Uses: Scavenger for certain oxidation catalysts and thermal protectant for metal-bonded adhesives and coating resins.
Shipping regulations: (Rail, Air) Arsenical compounds, n.o.s. Poison label.

arsenic tribromide (arsenic bromide; arsenious bromide; arsenous bromide) $AsBr_3$.

Properties: Yellowish-white hygroscopic crystals; sp. gr. 3.54 (25°C); m.p. 33°C; b.p. 221°C. Decomposed by water.
Derivation: Direct union of arsenic and bromine.
Hazard: Highly toxic.
Uses: Analytical chemistry; medicine.
Shipping regulations: (Rail, Air) Arsenical compounds, n.o.s. Poison label. Legal label name: arsenic bromide.

arsenic trichloride (arsenic chloride; arsenious chloride; arsenous chloride; caustic arsenic chloride; fuming liquid arsenic) $AsCl_3$.
Properties: Colorless or pale yellow, oily liquid. Soluble in concentrated hydrochloric acid and most organic liquids; decomposed by water. Fumes in moist air. B.p. 130.5°C; m.p. -18°C; sp. gr. 2.163 (14/4°C); noncombustible.
Derivation: (a) By action of chlorine on arsenic; (b) by distillation of arsenic trioxide with concentrated hydrochloric acid.
Grades: Technical.
Containers: Bottles; 20-, 55-gal drums.
Hazard: Highly toxic; strong irritant to eyes and skin.
Uses: Intermediate for organic arsenicals (pharmaceuticals, insecticides).
Shipping regulations: (Rail, Air) Poison label.

arsenic triiodide (arsenous iodide; arsenious iodide; arsenic iodide) AsI_3.
Properties: Orange-red shining crystalline scales or powder; unstable in sunlight or moisture; sp. gr. 4.70 (25°C); m.p. 146°C; sublimes when heated slowly. Soluble in alcohol, ether, carbon disulfide, chloroform, and benzene; soluble in water with hydrolysis.
Derivation: By the direct union of arsenic and iodine.
Hazard: Highly toxic.
Uses: Analytical chemistry; medicine.
Shipping regulations: (Rail, Air) Arsenical compounds, n.o.s. Poison label.

arsenic trioxide (crude arsenic; white arsenic; arsenious acid; arsenious oxide; arsenous anhydride) As_2O_3.
Properties: White, odorless, tasteless powder; slightly soluble in water; soluble in acids and alkalies; soluble in glycerin; sp. gr. 3.865; sublimes at 193°C.
Derivation: Smelting of copper and lead concentrates. Flue dust, to which pyrite or galena concentrates are added, yields As_2O_3 vapor. Condensation gives product of varying purity called crude arsenic (90–95% pure). A higher-purity oxide called white arsenic is obtained by resubliming the crude As_2O_3 (99+% pure).
Containers: Drums; barrels; carloads.
Hazard: Highly toxic by ingestion or dust inhalation. May be carcinogenic. Safety data sheet available from Manufacturing Chemists Assn., Washington, D.C.
Uses: Pigments, ceramic enamels, aniline colors; decolorizing agent in glass; insecticide; rodenticide; herbicide; sheep and cattle dip; hide preservative; wood preservative; preparation of other arsenic compounds.
Shipping regulations: (Rail, Air) Poison label.

arsenic trisulfide (arsenious sulfide; arsenic sulfide, yellow; arsenous sulfide; arsenic tersulfide) As_2S_3.
Properties: Yellow crystals or powder, changes to a red form at 170°C; sp. gr. 3.43; m.p. 300°C; insoluble in water and hydrochloric acid; dissolves in alkaline sulfide solutions and in nitric acid.
Derivation: Occurs in nature as the mineral orpiment. May be precipitated from arsenious acid solution by the action of hydrogen sulfide.
Grades: Technical; pigment; single crystals.
Hazard: Highly toxic.
Uses: Pigment; reducing agent; pyrotechnics; glass used for infrared lenses.
Shipping regulations: Arsenical compounds, n.o.s. Poison label.

arsine (arsenic hydride) AsH_3.
Properties: Colorless gas; m.p. -113.5; b.p. -55°C; decomposes 230°C; soluble in water; slightly soluble in alcohol, alkalies.
Derivation: Reaction of aluminum arsenide with water or HCl; electrochemical reduction of arsenic compounds in acid solutions.
Grades: Technical; 99% pure or in mixture with other gases.
Containers: Steel cylinders.
Hazard: Highly toxic. Tolerance, 0.05 ppm in air.
Use: Organic synthesis; military poison gas; doping agent for solid state electronic components.
Shipping regulations: (Air) Not acceptable. (Rail) Poison gas, n.o.s. Poison Gas label. Not accepted by express.

arsphenamine. A specific for syphilis originally developed by Ehrlich, but no longer in use. It was a derivative of arsenic and benzene. See Ehrlich.

arsthinol $C_{11}H_{14}NO_3S_2As$. Cyclic 3-hydroxypropylene ester of 3-acetamido-4-hydroxydithiobenzenearsonous acid.
Properties: White, odorless, microcrystalline powder; sparingly soluble in alcohol; very slightly soluble in water.
Hazard: Toxic in high concentrations.
Use: Medicine.
Shipping regulations: (Rail, Air) Arsenical compounds, n.o.s. Poison label.

***l*-arterenol.** See levarterenol.

"Artic."[28] Trademark for refrigeration grade of methyl chloride (q.v.).

artificial cinnabar. See mercuric sulfide, red.

artificial snow. A copolymer of butyl and isobutyl methacrylate, often dispersed from an aerosol bomb or other atomizing device; used in decorative window displays, etc. Man-made snow is crystallized water vapor made by mechanical means.

"Arubren CP."[470] Trademark for a highly chlorinated aliphatic hydrocarbon compound used in rubber compounds to decrease flammability of vulcanizates.

arylalkyl. A compound containing both aliphatic and aromatic structures, e.g., alkyl benzenesulfonate. Also called aralkyl.

aryl. A compound whose molecules have the ring structure characteristic of benzene, naphthalene, phenanthrene, anthracene, etc., i.e., either the six-carbon ring of benzene or the condensed six-carbon rings of the other aromatic derivatives. For example, an aryl group may be phenyl, C_6H_5 or naphthyl ($C_{10}H_9$).

As Symbol for arsenic.

as.- Abbreviation for asymmetrical; same as uns- (q.v.).

ASA. Abbreviation for acrylic ester-modified styrene-acrylonitrile terpolymer. See also "Luran S."

asarone. See 2,4,5-trimethoxy-1-propenylbenzene.

asbestos. A group of impure magnesium silicate minerals which occur in fibrous form. Colors: white, gray, green, brown. Sp. gr. 2.5. Noncombustible.
(1) Serpentine asbestos is the mineral chrysotile, a magnesium silicate. The fibers are strong and flexible

so that spinning is possible with the longer fibers. A microcrystalline form trademarked "Avibest" has been developed.

(2) Amphibole asbestos includes various silicates of magnesium, iron, calcium, and sodium. The fibers are generally brittle and cannot be spun but are more resistant to chemicals and to heat than serpentine asbestos.

(3) Amosite.

Occurrence: Yukon, Quebec, Vermont, Mexico, Arizona, California, North Carolina, Africa, Italy.

Hazard: Toxic by inhalation of dust particles. An active carcinogen. Tolerance, 5 fibers/cc of air more than 5 microns in length.

Uses: Fireproof fabrics; brake lining; gaskets; roofing compositions; electrical and heat insulation; paint filler; chemical filters; reinforcing agent in rubber and plastics; component of paper dryer felts; diaphragm cells.

ascaridole $C_{10}H_{16}O_2$ 1,4-Peroxido-para-menthene-2.

Properties: A liquid, naturally occurring peroxide; B.p. 84°C (5 mm); sp. gr. 1.011 (13/15°C); refractive index n 20/D 1.4743.

Derivation: By vacuum distillation of chenopodium oil.

Hazard: Strong oxidizing agent; explodes on heating to 130°C or in contact with organic acids. Toxic by ingestion.

Uses: Initiator in polymerization; medicine.

Shipping regulations: (Rail) Peroxides, organic, liquid, Yellow label. (Air) Not acceptable.

"Ascarite."[16] A trademark for sodium hydrate-asbestos absorbent, for rapid and quantitative absorption of carbon dioxide, in the determination of carbon in iron and steel by direct combustion method, and other analyses.

Grades: Mesh 8–20, 20–30.

ascorbic acid (L-ascorbic acid; vitamin C)

OCOCOH:COHCHCHOHCH$_2$. A dietary factor which must be present in the diet of man to prevent scurvy. It cures scurvy and increases resistance to infection. Ascorbic acid presumably acts as an oxidation-reduction catalyst in the cell. It is readily oxidized; citrus juices should not be exposed to air for more than a few minutes before use.

Properties: White crystals (usually plates, sometimes needles); m.p. 192°C; soluble in water; slightly soluble in alcohol; insoluble in ether, chloroform, benzene, petroleum ether, oils, and fats; stable to air when dry. Nontoxic.

Sources: Food source: citrus fruits, tomatoes, potatoes, green leafy vegetables, other fruits and vegetables. Commercial source: Synthetic product made by fermentation of sorbitol.

Units: One international unit is equivalent to 0.05 mg of L-ascorbic acid.

Grades: U.S.P.; F.C.C.

Containers: Glass bottles; fiber cans; multiwall paper drums.

Uses: Medicine; nutrition; color fixing, flavoring, and preservative in meats and other foods; oxidant in bread doughs; abscission of citrus fruit in harvesting; reducing agent in analytical chemistry. The ferric, calcium and sodium salts are available for biochemical research.

ascorbic acid oxidase. An enzyme found in plant tissue which acts upon ascorbic acid in the presence of oxygen to produce dehydroascorbic acid.

Use: Biochemical research.

ascorbyl palmitate $C_{22}H_{38}O_7$. A white or yellowish-white powder having a citrus-like odor. M.p. 116–117°; soluble in alcohol, animal and vegetable oils; slightly soluble in water.

Derivation: Palmitic and L-ascorbic acids.

Grade: F.C.C.

Uses: Antioxidant for fats and oils; source of vitamin C; stabilizer; emulsifier.

-ase. A suffix characterizing the names of many enzymes, e.g., diastase, cellulase, cholinesterase, etc. However, the names of some enzymes end in -in (pepsin, rennin, papain).

"Aseptoform."[19] Trademark for esters of para-hydroxybenzoic acid, such as "Aseptoform" methyl, "Aseptoform" propyl, "Aseptoform" butyl.

ash. Mineral content of a material remaining after complete combustion.

askarel. A generic descriptive name for synthetic electrical insulating (dielectric) material which when decomposed by the electric arc evolves only nonexplosive gases or gaseous mixtures. Examples are chlorinated aromatic derivatives, particularly pentachlorodiphenyl and trichlorobenzene, but also including pentachlorodiphenyl oxide, pentachlorophenylbenzoate, hexachlorodiphenylmethane, pentachlorodiphenyl ketone, and pentachloroethylbenzene. Nonflammable.

Uses: Insulating medium in transformers; dielectric fluid.

See also dielectric; transformer oil.

A.S.M. Abbreviation for American Society for Metals (q.v.).

asparagic acid. See aspartic acid.

L-asparaginase. An enzyme used in the treatment of certain types of leukemia. Produced by biochemical activity of certain bacteria. Yields are now in excess of 3500 units per gram of source.

asparagine (alpha-aminosuccinamic acid; beta-asparagine; althein; aspartamic acid; aspartamide) $NH_2COCH_2CH(NH_2)COOH$. The beta amide of aspartic acid, a nonessential amino acid, existing in the D(+)- and L(−)-isomeric forms as well as the DL-racemic mixture. L(−)-asparagine is the most common form. Nontoxic.

Properties L(−)-asparagine monohydrate: White crystals; m.p. 234–235°C; acid to litmus; nearly insoluble in ethanol, methanol, ether, and benzene; soluble in acids and alkalies.

Derivation: Widely distributed in plants and animals, both free and combined with proteins.

Uses: Biochemical research; preparation of culture media; medicine.

asparaginic acid. See aspartic acid.

1-L-asparaginyl-5-L-valyl angiotensin octapeptide. See angiotensin amide.

aspartamic acid. See asparagine.

aspartamide. See asparagine.

aspartic acid (asparaginic acid; asparagic acid; aminosuccinic acid) $COOHCH_2CH(NH_2)COOH$. A natu-

rally occurring nonessential amino acid. The common form is L(+)-aspartic acid. Nontoxic.
Properties: Colorless crystals; soluble in water; insoluble in alcohol and ether; optically active.
DL-aspartic acid: M.p. 278–280°C with decomposition; sp. gr. 1.663 (12/12°C).
L(+)-aspartic acid: M.p. 251°C.
D(−)-aspartic acid: M.p. 269–271°C with decomposition; sp. gr. 1.6613.
Source: Young sugar cane; sugar beet molasses.
Derivation: Hydrolysis of protein; reaction of ammonia with diethyl fumarate.
Uses: Biological and clinical studies; preparation of culture media; organic intermediate; dietary supplement; detergents; fungicides; germicides; metal complexation; synthetic sweetener base (L-form).
Available commercially as D(−)-, L(+)-, and DL-aspartic acid.

aspartocin. USAN for antibiotic produced by Streptomyces griseus.

aspergillic acid $C_{12}H_{20}N_2O_2$. 2-Hydroxy-3-isobutyl-6-(1-methylpropyl)pyrazine 1-oxide. An antibiotic from strains of Aspergillus flavus. Nontoxic.
Properties: Solid. M.p. 97°C; insoluble in cold water; soluble in common organic solvents and dilute acids. Hydrochloride melts at 178°C and is soluble in water.

asphalt (petroleum asphalt, Trinidad pitch, mineral pitch). A dark-brown to black cementitious material, solid or semisolid in consistency, in which the predominating constituents are bitumens, which occur in nature as such or are obtained as residua in petroleum refining (ASTM). It is a mixture of paraffinic and aromatic hydrocarbons and heterocyclic compounds containing sulfur, nitrogen, and oxygen.
Properties: Black solid or viscous liquid; sp. gr. about 1.0; soluble in carbon disulfide. Flash point 450°F; autoignition temp. 900°F; solid softens to viscous liquid at about 200°F; penetration value (paving) 40–300 (roofing) 10–40. Good electrical resistivity. Combustible; low toxicity.
Occurrence: California, Trinidad, Venezuela, Cuba, Canada (Athabasca tar sands).
Containers: Drums, barrels, tank trucks, tank cars.
Uses: Paving and road-coating; roofing; sealing and joint filling; special paints; adhesive in electrical laminates and hot-melt compositions; diluent in low-grade rubber products; fluid loss control in hydraulic fracturing of oil wells; medium for radioactive waste disposal; pipeline and underground cable coating; rust-preventive hot-dip coatings; base for synthetic turf; water-retaining barrier for sandy soils; supporter of rapid bacterial growth in converting petroleum components to protein.
See also bacteria; protein; asphalt, blown.

asphalt, blown (mineral rubber, oxidized asphalt, condensed asphalt). Black, friable solid obtained by blowing air at high temperature through petroleum-derived asphalt, with subsequent cooling. Penetration value 10–40; softening point 185–250°F. Combustible. Uses are primarily roofing, as diluent in low-grade rubber products, and as thickener in oil-based drilling fluids. Shipped in 55-gal. metal drums. For further information on asphalt, refer to the Asphalt Institute, 1270 Avenue of the Americas, New York, N.Y.

asphalt, cut-back. A liquid petroleum product, produced by fluxing an asphaltic base with suitable distillates. (A.S.T.M.)
Properties: Flash point (open cup) 50°F. Solubility of residue from distillation in carbon tetrachloride 99.5%.
Hazard: Flammable, dangerous fire hazard.
Use: Road surfaces.
Shipping regulations: (Rail) Red label. (Air) Flammable Liquid label.

asphaltene. A component of the bitumen in petroleums, petroleum products, malthas, asphalt cements, and solid native bitumens, soluble in carbon disulfide but insoluble in paraffin naphthas. (A.S.T.M.)

asphalt, liquid. See residual oil; asphalt.

asphalt, oxidized. See asphalt, blown.

asphalt paint. Asphaltic base in a volatile solvent with or without drying oils, resisns, fillers, and pigments. Ground asbestos is frequently used as a component of heavy asphaltic paints for roofing and waterproofing purposes.
Hazard: Flammable; dangerous fire risk.

asphyxiant gas. A gas which has little or no positive toxic effect but which can bring about unconsciousness and death by replacing air and thus depriving an organism of oxygen. Among the so-called asphyxiant gases are carbon dioxide, nitrogen, helium, methane, and other hydrocarbon gases.

aspidospermine $C_{22}H_{30}O_2N_2$.
Properties: White to brownish-yellow crystalline alkaloid. M.p. 108°C; b.p. 200°C (1 to 2 mm); sublimes at 180°C under reduced pressure. Soluble in fats and fixed oils; sparingly soluble in absolute alcohol and ether. Its sulfate and hydrochloride are soluble in water.
Hazard: Moderately toxic by ingestion.
Use: Medicine.

aspirin (acetylsalicylic acid; ortho-acetoxybenzoic acid). $CH_3COOC_6H_4COOH$.
Properties: White crystals or white, crystalline powder. Odorless; slightly bitter taste. Stable in dry air; slowly hydrolyzes in moist air to salicylic and acetic acids. Soluble in water, alcohol, chloroform, and ether; less soluble in absolute ether. Dissolves with decomposition in solutions of alkali hydroxides and carbonates. M.p. 132–136°C; b.p. 140°C (dec.).
Derivation: Action of acetic anhydride on salicylic acid.
Method of purification: Crystallization.
Grades: Technical, U.S.P.
Containers: 25-lb boxes; 25-, 100-, 250-lb drums.
Hazard: An allergen; may cause local bleeding, especially of the gums; 10-gram dosage may be fatal. Dust dispersed in air is serious explosion risk.
Use: Medicine (anodyne).

"Aspon."[1] Trademark for a concentrate of tetra-n-propyl dithionopyrophosphate, a liquid insecticide.

assay. Determination of the content of a specific component of a mixture, with no evaluation of other components. Such determinations are made on ores of various metals (especially precious metals), on pharmaceutical products to validate the amount of drug present in a given unit, and on organisms (bacteria) to determine their reactions to an antibiotic or insecticide. The latter procedure is called bioassay. Ores are assayed by heat fractionation, organic materials by solvent extraction and chemical separation.

assistant. A term loosely used in the textile industry for any chemical compound that aids in a processing step, e.g., scouring, dyeing, bleaching, finishing, etc. See also auxiliary; dyeing assistant.

association. A reversible chemical combination due to any of the weaker classes of chemical bonding forces. Thus the combination of two or more molecules due to hydrogen bonding, as in the union of water molecules with one another, or of acetic acid molecules with water molecules, is called association; also, combination of water or solvent molecules with molecules of solute or with ions, i.e., hydrate formation, or solvation. Formation of complex ions or chelates, as copper ion with ammonia or copper ion with 8-hydroxyquinoline, are other examples. Aqueous solutions of soaps or synthetic detergents are often called association colloids.

A-stage resin (resole; one-step resin). An alkaline catalyzed thermosetting phenol-formaldehyde type resin consisting primarily of partially condensed phenol alcohols. At this stage the product is fully soluble in one or more common solvents (alcohols, ketones) and is fusible at temperatures below 150° C. On further heating and without use of a catalyst or additive, the resin is eventually converted to the insoluble, infusible cross-linked form (C-stage). The A-stage resin is a constituent of most commercial laminating varnishes, and is also used in special molding powders. See B-stage resin, C-stage resin, novolak; phenol-formaldehyde resin.

astatine At Element of atomic number 85. Group VIIA of Periodic Table; atomic weight 211. Heaviest number of the halogen family. It has 20 isotopes, all radioactive; derived by alpha bombardment of bismuth. The two most stable isotopes have half-lives of about 8 hours. Astatine occurs in nature to the extent of about one ounce in entire earth's crust. Like iodine, it concentrates in the thyroid gland. Its use in medicine is still experimental.

"Asterol" Dihydrochloride.[190] Trademark for a brand of diamthazole dihydrochloride (q.v.).

A.S.T.M. Abbreviation for American Society for Testing and Materials (q.v.).

"Aston 108."[328] Trademark for a thermosetting polyamine, 20% active.
Uses: Durable antistatic agents; softeners.

Aston, Francis William (1877–1945). This noted English chemist and physicist carried out much of his work with J. J. Thomson at Cambridge. He was the pioneer investigator of isotopes and his method of separating the lighter from the heavier atomic nuclei provided the technique that later developed into the mass spectroscope, which utilizes a magnetic field for this purpose. Aston received the Nobel prize for this discovery in 1922, just three years after Rutherford performed the first transmutation of elements. Aston also correctly estimated the energy content of a hydrogen atom and predicted the controlled release of this energy.

"Astracel."[307] Trademark for a group of fast dyes for union shades on polyester-carbon blended fabrics.

astrochemistry. Strictly, the application of radioastronomy (microwave spectroscopy) to determination of the existence of chemical entities in the gas clouds of interstellar space.

Astrochemistry also includes the occurrence of elements and compounds in celestial bodies, including their atmospheres. Such data have been obtained from spectrographic study of the light from the sun and stars, from analysis of meteorites, and from actual samples from the moon. The most abundant element is hydrogen. Over 25% of the elements, including carbon, have been identified, as well as water, carbon dioxide, ammonia, and the following organic compounds: ethane, methane, acetylene, formaldehyde, formic acid, methyl alcohol, hydrogen cyanide, carbon monoxide and acetonitrile. When applied to the planets only, the science is called chemical planetology (q.v.).

"Astrol."[307] Trademark for a group of fast alizarin direct blues.

"Astrosil."[540] Trademark for a flexible silica fabric (over 99% SiO_2).
Properties: High-temperature resistance, 2000° F; low expansion coefficients; excellent ablation characteristics; resembles fiber glass fabrics in appearance; compatible with phenolics, polyesters, epoxies and can be impregnated on either horizontal or vertical coating equipment.
Uses: Nose cones, heat shields, nozzle exit cones, blast shields, liquid propulsion ablative chambers; furnace insulation; diffusion bonding blankets; brazing applications; liquid, gas and molten metal filtration; die lubricant for metal extrusion; conformable seal for soaking pit covers.

asymmetry. A molecular structure in which an atom having four tetrahedral valences is attached to four different atoms or groups. The most common cases involve the carbon atom, though they may exist also with other elements such as nitrogen and sulfur. An example of this structure is lactic acid, which contains one asymmetric carbon (indicated by *). In such cases two optical isomers (L and D enantiomers) result which are nonsuperposable mirror images of each other:

```
    COOH          COOH
     |             |
HO—C*—H       H—C*—OH
     |             |
    CH3           CH3
     L             D
```

Amino acids are also characterized by asymmetric carbons. Many compounds have more than one asymmetric carbon, e.g., tartaric acid, sugars, terpenes, etc. This results in the possibility of many optical isomers, the number being determined by the formula 2^n, where n is the number of asymmetric carbons. See also optical isomer; enantiomer; glyceraldehyde.

At Symbol for astatine.

"Atabrine" Hydrochloride.[162] Trademark for quinacrine hydrochloride (q.v.).

atactic. A type of polymer molecule in which substituent groups or atoms are arranged randomly above and below the backbone chain of atoms, when the latter are all in the same plane. See polymer, stereospecific.

"Atarax."[299] Trademark for hydroxyzine hydrochloride (q.v.).

-ate. A suffix having two different meanings. (1) In inorganic compounds it indicates a salt whose metal or radical is in the highest oxidation state, as calcium sulfate, ammonium nitrate, etc. (2) In engineering terminology it means "result of," as in precipitate, condensate, alkylate, distillate, etc.

ATE. Abbreviation for aluminum triethyl. See triethyl aluminum.

"Atlac."[89] Trademark for a series of polyester resins for use in reinforced plastics.

"Atlacide."[147] Trademark for herbicide containing 58% sodium chlorate.

"Atlas."[147] Trademark for a series of pesticides containing various amounts of sodium arsenite.
Hazard: Highly toxic by ingestion.

"Atlastavon."[41] Trademark for synthetic-resin sheet lining of the plasticized polyvinyl chloride type used to protect steel tanks at temperatures up to 160°F without brick sheathing. Outstanding resistance to oxidizing acids.

"Atlastic 31."[41] Trademark for a hot-melt asphaltic lining material used as a resilient membrane between concrete and acid brick sheathing.

"Atlox."[89] Trademark for series of emulsifiers developed for use with agricultural pesticides.

ATM. Abbreviation for aluminum trimethyl. See trimethyl aluminum.

atmosphere. (1) The pressure exerted by the air at sea level (14.696 psi), which will support a column of mercury 760 mm high (about 30 in.). This is standard barometric pressure, though it varies slightly with local meteorological conditions. It is often used to indicate working pressures of steam. The accepted abbreviation is atm.

(2) Any environmental gas or mixture of gases, e.g., an atmosphere of nitrogen, or an inert atmosphere.

(3) The air itself, as atmospheric oxygen.

atmosphere, controlled. As used in the technology of food preservation and storage, a gaseous environment in which the concentrations of oxygen, carbon dioxide and nitrogen are held constant at a specific level, the temperature also being controlled. CA storage techniques were introduced on a commercial scale in the U.S. in 1940.

atmospheric pollution. See air pollution.

"Atmul."[89] Trademark for a series of mono- and diglyceride emulsifiers used in baked goods and other food products.

atom. The smallest possible unit of an element, consisting of one or more protons and (except hydrogen) two or more neutrons located in the nucleus, and one or more electrons which revolve around it. The protons are positively charged; the neutrons have no charge; the electrons are negatively charged. As each atom contains the same number of protons as electrons, the atom is electrically neutral. Atoms of the various elements differ in mass (weight), that is, in the number of neutrons and protons, and also in the number of electrons. Atoms of a given element are identical, but an element may have atoms of slightly different masses, called isotopes (q.v.). Individual atoms of uranium and thorium have been resolved at 5 angstrom units in the scanning electron microscope. Motion pictures of uranium atoms at magnification of 7.5 million times have been made at the Enrico Fermi Institute at University of Chicago.

Atoms of the same or different elements combine to form molecules; when the atoms are of two or more different elements, these molecules are called compounds (q.v.). Atoms remain essentially unchanged in chemical reactions, except that some of the outermost electrons may be removed, shared, or transferred, as occurs in oxidation, ionization, and chemical bonding. A few atomic species disintegrate as a result of nuclear changes, and thus become radioactive. Heavy, unstable atoms such as uranium-235 and plutonium can be split by bombardment with high-energy particles, yielding tremendous energy. See also electron, proton, bonding, orbital theory, ionization, radioactivity, fission.

atomic energy. See nuclear energy.

Atomic Energy Commission. See Energy Research and Development Administration.

atomic hydrogen welding. A method of welding in which hydrogen gas is passed through an arc between two tungsten electrodes. The arc breaks down the molecules to form atomic hydrogen. The recombination of the atoms to form molecules and the combustion of the molecular hydrogen in atmospheric oxygen produce a flame temperature of 4000–5000°C.

atomic number. The number of positive charges carried by the nucleus of an atom, which is equal to the number of negative charges represented by the electrons. With a few exceptions, it is the number of the element in the sequence obtained by arranging the elements in the order of increasing atomic weight. See also Periodic Table; mass number.

atomic theory. See Dalton, John.

atomic volume. The atomic weight of an element divided by its density.

atomic weight. The mass (q.v.) of an atom of an element compared with the mass of the carbon-12 isotope taken as the standard at 12. The carbon-12 standard results in atomic weights for natural carbon of 12.011 and for oxygen of 15.9994, because of the slight difference involved in averaging the weights of the isotopic forms. Official atomic weight designations are released periodically by the IUPAC.

ATP. Abbreviation for adenosine triphosphate.

"Atpet."[89] Trademark for a series of emulsifiers used in conjunction with oil as corrosion inhibitors, solubilizers for production of soluble cutting oils, and water-block removal agents in oil well drilling.

atrazine. Generic name for 2-chloro-4-ethylamino-6-isopropylamino-s-triazine; used as a herbicide, plant growth regulator and weed-control agent for corn etc., and for noncrop and industrial sites.

atropamine. See apoatropine.

atropine (daturine) $C_{17}H_{23}NO_3$, An alkaloid obtained from species of Atropa, Datura or Hyoscyamus.
Properties: White crystals or white crystalline powder; optically inactive (but usually contains levorotatory hyoscyamine). Soluble in alcohol, ether, chloroform, and glycerol. Slightly soluble in water. M.p. 114–116°C.
Derivation: By extraction from Datura stramonium, or synthetically.
Grades: Technical; N.F.
Hazard: Toxic. An allergen.
Use: Medicine; antidote for cholinesterase-inhibiting compounds (organophosphorus insecticides, nerve gases); artificial respiration is also necessary.

attapulgite. $(Mg,Al)_5Si_8O_{22}(OH)_4 \cdot 4H_2O$. A hydrated aluminum-magnesium silicate, the chief ingredient of fuller's earth (q.v.). See also clay.
Uses: Drilling fluids; decolorizing oils; filter medium.

attar (otto) An essential oil (fragrance) made by steam distillation of flowers, especially roses. See essential oil; perfume; rose oil.

atto-. Prefix meaning 10^{-18} unit (symbol a). E.g., 1 ag = 1 attogram = 10^{-18} gram.

Au Symbol for gold, from Latin *aurum*.

"Aura."[108] Trademark for a powdered, chlorinated, alkaline polyphosphate detergent for mechanical dishwashing.

auramine $(CH_3)_2NC_6H_4(C{:}NH)C_6H_4N(CH_3)_2 \cdot HCl$.
Properties: Yellow flakes or powder; soluble in water, alcohol, and ether.
Uses: Yellow dye for paper, textiles, leather; also an antiseptic; fungicide.

"Aurantiol."[227] $C_{18}H_{27}O_3N$. Trademark for hydroxycitronellal-methyl anthranilate Schiff base (methyl N-3,7-dimethyl-7-hydroxyoctylidene-anthranilate).

aureolin. See Indian yellow.

"Aureomycin."[315] Trademark for chlortetracycline hydrochloride. An antibiotic. Must conform to FDA requirements.

"Auric."[28] Trademark for a ferric oxide brown pigment.

auric and aurous compound. See corresponding gold compound.

aurin (para-rosolic acid) $(C_6H_4OH)_2CC_6H_4O$. A triphenylmethane derivative.
Properties: Reddish-brown pieces with greenish metallic luster; easily powdered; insoluble in water, benzene, and ether; soluble in alcohol.
Hazard: May be toxic.
Uses: Indicator; dye intermediate.

austenite. A component of steel; a nonmagnetic solid solution of carbon or ferric carbide in gamma iron. Very unstable below its critical temperature, but may be obtained in high carbon steels by rapid quenching from high temperatures. Addition of manganese and nickel lowers critical transition temperature and stable austenite may be obtained at room temperature. Characterized by a face-centered cubic lattice.

austenitic alloys (austenitic steels). Alloys of iron, chromium, nickel noted for their resistance to corrosion.

Australian bark. See wattle bark.

autocatalysis. A catalytic reaction induced by a product of the same reaction. This occurs in some types of thermal decomposition, in autoxidation, and in many biochemical systems, as when an enzyme activates its own precursor. See also autoxidation.

autoclave. A chamber, usually of cylindrical shape, provided with a door or gate at one end which can be securely closed during operation. It is built heavily enough to accommodate steam pressures of considerable magnitude. It is used to effect chemical reactions requiring high temperature and pressure, such as open-steam vulcanization of rubber. Sizes vary from laboratory units to production size, which may be over 50 ft long and 3 or more feet in diameter; the latter are provided with baffles to ensure equal distribution of the entering steam. Autoclaves are also used in certain sterilization processes.

"Autoset."[65] Trademark for urea-formaldehyde resin for use as a binder in the manufacture of particle board.

autoignition point (ignition point). The minimum temperature required to initiate or cause self-sustained combustion in any substance in the absence of a spark or flame. Ignition temperature varies, depending upon the test method. Some approximate ignition temperatures are the following: acetone 1000°F, amyl acetate 750°F, aniline 1000°F, butane 806°F, carbon disulfide 212°F, ethyl ether 356°F, phenol 1319°F, toluene 1000°F, white pine shavings 507°F, cotton batting 446°F, magnesium powder (fine) 883°F, nitrocellulose film 279°F. (National Fire Protection Assn.) See also flash point.

automotive exhaust emission. See air pollution.

autoxidation. A spontaneous, self-catalyzed oxidation occurring in the presence of air; it usually involves a free-radical mechanism. It is initiated by heat, light, metallic catalysts, or free-radical generators. Industrial processes such as manufacture of phenol and acetone from cumene are based on autoxidation. Other instances are the drying of vegetable oils, the spoilage of fats, gum formation in lubricating oils, and the degradation of high polymers exposed to sunlight for long periods. See also autocatalysis.

auxiliary. Any of a number of chemical compounds used in some phase of textile processing. They may be classified as follows: (1) fats, oils, and waxes; (2) starches, gums, and glues (sizing); (3) soaps and detergents; (4) inorganic chemicals (bleaching, mercerizing); (5) organic solvents; (6) special-purpose products (flameproofing, mildew-proofing, repellent and decorative coatings, and permanent press resins). They are sometimes also called assistants.

auxin. A natural or synthetic plant growth hormone that regulates longitudinal cell structure so as to permit bending of the stalk or stem in phototropic response. The natural materials are formed in small amounts in the green tips of growing plants, in root tips, and on the shaded side of growing shoots. 3-Indoleacetic acid (q.v.), is the most important natural auxin. See also plant growth regulator.

auxochrome. A radical or group of atoms whose presence is essential in enabling a colored organic substance to be retained on fibers. The best examples are the groups -COOH, $-SO_3H$, -OH and $-NH_2$.

"AV20."[507] Trademark for a vitreous, 85% pure alumina, used in ceramics. Impervious to gas; compressive strength 250,000 psi; max. service temperature 2550°F; hardness (Mohs) 8.5. Fabricated by dry press or extrusion methods.

"Avadex."[58] Trademark for a series of liquid or granular herbicides containing 2,3-dichloroallyl diisopropylthiolcarbamate. Widely used to control growth of wild oats in agricultural crops.
Hazard: May be toxic by ingestion and inhalation.

"Availaphos."[62] Trademark for a mineral supplement supplying phosphorus and calcium in readily available form for animal and poultry feeds.

"Avertin."[162] Trademark for tribromoethanol solution of amylene hydrate. An anesthetic (q.v.).

aviation gasoline. See gasoline.

"Avibest."[55] Trademark for a microcrystalline form of asbestos.

"Avicel."[261] Trademark for microcrystalline cellulose, a highly purified particulate form of cellulose; particle size ranges from <1 to 150 microns, average varies with grade; density 1.55 (bulk density 0.3–0.5). Insoluble in dilute acid, organic solvents, oils; swells in dilute alkali. Dispersible in water to form stable gels or pourable suspensions. Adsorbs oily and syrupy materials.
Uses: Aid to stabilization and emulsification; ingredient in foods; suspending agent; binder and hardening agent in tableting; separatory medium in column and thin-layer chromatography; pure cellulose raw material.

avidin. A protein occurring in egg white, where it comprises about 0.2% of the total protein. It has the property of combining firmly with biotin (q.v.) and rendering it unavailable to organisms, since proteolytic enzymes do not destroy the avidin-biotin complex. Avidin loses its ability to combine with biotin when subjected to heat; hence cooked egg white does not lead to biotin deficiency.
Hazard: Toxic by ingestion, but inactivated by heat.

"Avilon."[443] Trademark for metal-complex dyes for polyamide fibers.

"Avisun."[429] Trademark for polypropylene resin; molecular weight 100,000–500,000; sp. gr. 0.90–0.92; m.p. 347°F. Available in a variety of molding and film grades.

"Avitene."[55] Trademark for a microcrystalline form of collagen (q.v.).

"Avitex."[28] Trademark for a group of textile softeners, lubricants, and antistatic agents. Both anionic and cationic types.

"Avitone."[28] Trademark for a group of chemical compounds based on hydrocarbon sodium sulfonates that are used principally as softening, lubricating, and finishing agents for textiles, leather, and paper.

avocado oil. An edible oil high in unsaturated fatty acids.
Properties: Sp. gr. 0.91; acid value 1–7; saponification value 177–198; iodine value 71–95; solidification temp. 7–9°C; refractive index (40°C), 1.461–1.465. Faint odor, bland taste, greenish color. Nontoxic.
Source: Pulp of the avocado, Persea americana; occurs naturally in Central America; cultivated in Brazil, California, Florida, West Indies, Hawaii.
Chief constituents: Oleic, palmitic and linoleic acids.
Containers: Drums.
Uses: Cosmetic creams; hair conditioners; suntan preparations; salad oils.

Avogadro's law. A principle stated in 1811 by the Italian chemist Amadeo Avogadro (1776–1856), namely, that equal volumes of gases at the same temperature and pressure contain the same number of molecules, regardless of their chemical nature and physical properties. This number (Avogadro's number) is 6.023×10^{23}; it is the number of molecules of any gas present in a volume of 22.41 liters, and is the same for the lightest gas (hydrogen) as for a heavy gas such as CO_2 or Br_2. Avogadro's number is one of the fundamental constants of chemistry. It permits calculation of the amount of pure substance (mole), which is the basis of stoichiometric relationships. It also makes possible determination of how much heavier a simple molecule of one gas is than that of another, as a result of which the relative molecular weights of gases can be ascertained by comparing the weights of equal volumes. Avogadro's number (conventionally represented by **N** in chemical calculations) is now considered to be the number of atoms present in 12 grams of the carbon-12 isotope (one mole of carbon-12), and can be applied to any type of chemical entity. See also mole.

"Avolin."[188] $C_6H_4(COOCH_3)_2$. Trademark for a special perfume grade of dimethyl phthalate (q.v.).

"Avril."[261] Trademark for a high wet modulus rayon fiber, usually marked in various blends with other fibers.

A.Z. See 1,1′-azobisformamide.

aza-. Prefix indicating the presence of nitrogen in a heterocyclic ring.

3-azabicyclo(3,2,2)nonane $C_8H_{15}N$.
Properties: White-tan solid; m.p. 180°C (sublimes); partially soluble in water; solubility decreases with an increase in temperature. Readily soluble in alcohol; density 4.67 lb/gal (20°C).
Use: Intermediate for the preparation of pharmaceuticals and rubber chemicals.

"Azak."[266] Trademark for an herbicide, 2,6-di-tert-butyl-p-tolyl methylcarbamate (q.v.).

azathioprine (Imuran). An immunosuppressive drug administered for the purpose of inhibiting the natural tendency of the body to reject foreign tissues by one or more types of immunizing reactions, i.e., formation of leucocytes or antibodies. It has been used with some success in cases of kidney and liver transplants.

6-azauridine (6-azauracil riboside; as-triazine-3,5(2H, 4H)dione riboside).
Derivation: Microbiological fermentation.
Use: Research on cell formation and cancer.

azelaic acid (nonanedioic acid; 1,7-heptanedicarboxylic acid). $COOH(CH_2)_7COOH$.
Properties: Yellowish to white crystalline powder; m.p. 106°C; b.p. 365°C (decomp.); soluble in hot water, alcohol, and organic solvents. Low toxicity.
Derivation: Oxidation of oleic acid by ozone.
Grades: Technical.
Uses: Organic synthesis; lacquers; production of hydrotropic salts; alkyd resins; polyamides; polyester adhesives; low-temperature plasticizers; urethane elastomers.

azelaoyl chloride $ClOC(CH_2)_7COCl$.
Properties: B.p. 125–130°C (3 mm). Slowly decomposes in cold water; soluble in hydrocarbons and ethers.
Uses: Organic synthesis.

azeotrope. See azeotropic mixture.

azeotropic distillation. A type of distillation in which a substance is added to the mixture to be separated in order to form an azeotropic mixture with one or more of the components of the original mixture. The azeotrope or azeotropes thus formed will have boiling points different from the boiling points of the original mixture and will permit greater ease of separation.

azeotropic mixture (azeotrope). A liquid mixture of two or more substances which behaves like a single substance in that the vapor produced by partial evaporation of liquid has the same composition as the liquid. The constant boiling mixture exhibits either a maximum or minimum boiling point as compared with that of other mixtures of the same substances.

azide. Any of a group of compounds having the characteristic formula $R(N_3)_x$. R may be almost any metal atom, a hydrogen atom, a halogen atom, the ammonium radical, a complex ($[Co(NH_3)_6]$, $[Hg(CN)_2M]$ with M = Cu, Zn, Co, Ni) an organic radical (e.g., methyl, phenyl, nitrophenol, dinitrophenol, *p*-nitrobenzyl, ethyl nitrate, etc.), and a variety of other groups or radicals. The azide group has a chain structure (N=N=N) rather than a ring structure. All the heavy metal azides, hydrogen azide, and most if not all of the light metal azides (under appropriate conditions) are explosive; they should be handled with utmost care and protected from light, shock, and heat. Many of the organic azides are also explosive. See also lead azide, hydrogen azide.

aziminobenzene. See 1,2,3-benzotriazole.

azine dye. A class of dyes derived from phenazine $(C_6H_4)N_2(C_6H_4)$ (tricyclic). The chromophore group may be >C=N—; but the color is probably due to the characteristic unsaturation of the benzene rings. The members of the group are quite varied in application. The nigrosines (Colour Index 50415–50440) and safranines (Colour Index 50200–50375) are examples of this group. See also dye, synthetic.

azinphos. See methyl azinphos.

1-aziridineethanol (N-(2-hydroxyethylethyleneimine) $C_2H_4NCH_2CH_2OH$. Colorless liquid; b.p. 167.9°C; flash point 185°F (open cup). Combustible.
Handling: Inhibited with 1-3% dissolved sodium hydroxide.
Hazard: Toxic. Irritant to skin and eyes.
Use: Chemical intermediate.

aziridine. A compound based on the ring structure

$\overline{CH_2NHCH_2}$. See ethyleneimine, which is aziridine itself; ethylethyleneimine; propyleneimine; polypropyleneimine; 1-aziridineethanol.

azlon. Generic name for a manufactured fiber in which the fiber-forming substance is composed of any regenerated naturally occurring protein (Federal Trade Commission). Proteins from corn, peanuts and milk have been used. Azlon fiber has a soft hand, blends well with other fibers, and is used, in general, like wool. Combustible.

azobenzene (diphenyldiimide; benzeneazobenzene) $C_6H_5N_2C_6H_5$.
Properties: Yellow or orange crystals; m.p. 68°C; b.p. 297°C; sp. gr. 1.203 (20/4°C); soluble in alcohol and ether; insoluble in water. Combustible.
Derivation: Reduction of nitrobenzene with sodium stannite.
Hazard: Moderately toxic.
Use: Manufacture of dyes, and rubber accelerators; fumigant; acaricide.

azobenzene-para-sulfonic acid $C_{12}H_{10}O_3N_2S$.
Properties: Orange crystals; m.p. 129°C.
Hazard: Moderately toxic.
Uses: Intermediate and reagent chemicals.

1,1'-azobisformamide (AZ; azodicarbonamide) $H_2NCONNCONH_2$.
Properties: Yellow powder; sp. gr. (20/20°C) 1.65; m.p., above 180°C (dec.); insoluble in common solvents; soluble in dimethyl sulfoxide. Nontoxic. Hydrolyzes at high temperatures to nitrogen, carbon dioxide, and ammonia.
Derivation: From hydrazine.
Grades: Technical, F.C.C.
Uses: Blowing agent for plastics and rubbers; maturing agent for flours.

azobisisobutyronitrile $(CH_3)_2C(CN)NNC(CN)(CH_3)_2$.
Properties: White powder; m.p. 105°C (dec); insoluble in water; soluble in many organic solvents and in vinyl monomers.
Containers: Fiber drums.
Hazard: May be toxic by ingestion.
Uses: Catalyst for vinyl polymerizations and for curing unsaturated polyester resins; blowing agent for plastics.

"Azocel."[552] Brand name for 1,1'-azobisformamide (q.v.).

azodicarbonamide. See 1,1'-azobisformamide.

azodine (benzeneazonaphthylethylenediamine) $C_{18}H_{18}N_4$.
Properties: Red crystals; m.p. 107–108°C.
Use: Reagent for rapid determination of penicillin in blood, urine and other media.

"Azodrin" Insecticide.[125] Trademark for dimethyl phosphate of 3-hydroxy-N-methyl-cis-crotonamide, $(CH_3O)_2P(O)OC(CH_3){:}CHC(O)NHCH_3$.
Properties: Reddish brown solid with a mild ester odor; b.p. 125°C; soluble in water and alcohol; very slightly soluble in kerosine and diesel fuel. Commercially available as a water-miscible solution.
Hazard: Toxic by ingestion, inhalation, and skin absorption. Flammable, dangerous fire risk. Use may be restricted.
Uses: Controls certain insects which attack cotton plants.

azo dye. A synthetic dye that has —N=N— as a chromophore group in its molecular structure and is produced from amino compounds by the processes of diazotization and coupling. Over half of the commercial dyestuffs are in this general category. By varying the chemical composition it is possible to produce acid, basic, direct, or mordant dyes. This general group is subdivided as monoazo, disazo, trisazo, and tetrazo according to the number of —N=N— groups in the molecule. Examples are Chrysoidine Y. Bismarck Brown 2R and Direct Green B. See also dye, synthetic.

azoic dye. An insoluble azo dye made on or in the fiber, except direct dyes that are further developed on the fiber. Formerly called developed, ice, or ingrain colors. See also dye, synthetic.

azophenylene. See phenazine.

"Azosol."[307] Trademark for a series of dyestuffs; soluble in organic solvents.
Uses: Coloring spirit lacquers and spirit inks.

azosulfamide (disodium 2-(4'-sulfamylphenylazo)-7-acetamido-1-hydroxynaphthalene-3,6-disulfonate) $C_{18}H_{14}N_4Na_2O_{10}S_3$.
Properties: Dark red, odorless, tasteless powder. Soluble in water with an intense red color; practically insoluble in organic solvents.
Use: Medicine.

azote. The French word for nitrogen (*a* = not, plus *zoo* = alive, as in *zoon* = animal). Nitrogen-bearing compounds can be recognized by this root word as in such terms as azo, azide, azobenzene, carbazole, thiazole, etc. The derivation is due to the chemical inertness of nitrogen.

azotic acid. See nitric acid.

azoxytoluidine. See diaminoazoxytoluene.

azulene $C_{16}H_{26}O$. The blue coloring matter of chamomile, wormwood and other essential oils. A terpene.
Properties: Blue, oily liquid; b.p. 170°C; sp. gr. 0.987; insoluble in water.
Use: Cosmetics.

azure blue. See cobalt blue.

azuresin (azure A carbacrylic resin).
Properties: Moist, irregular, dark blue or purple granules. Slightly pungent odor.
Derivation: Carbacrylic cation-exchange resin, in reversible combination with 3-amino-7-dimethylaminophenazathionium chloride (azure A dye).
Grade: N.F.
Use: Medicine (diagnostic test).

B

B Symbol for boron.

Ba Symbol for barium.

Ba-137. See cesium 137.

babassu oil. A nondrying, edible oil expressed from the kernels of the Babassu palm which grows in profusion in Brazil.
Composition: 44% lauric acid, 15% myristic acid, 16% oleic acid, balance mixed acids. Usable in foods and soap-making, but supply is limited by cost of exploitation of large quantites potentially available. Combustible; nontoxic.

Babbitt metal. One of a group of soft alloys used widely for bearings. They have good bonding characteristics with the substrate metal, maintain oil films on their surfaces and are nonseizing and antifriction. Used as cast, machined or preformed bimetallic bearings in the form of a thin coating on a steel base; the main types are lead base, lead silver base, tin base, cadmium base, and arsenical. The latter contains up to 3% arsenic.
Hazard: Dust may be toxic by inhalation.

Babcock test. A rapid test for butterfat in milk introduced by Stephen M. Babcock in 1890 and now in world-wide use in the dairy industry.

B acid. See 1-amino-8-naphthol-3,5-disulfonic acid.

"Baciferm."[319] Trademark for zinc bacitracin antibiotic feed supplement.

bacitracin. An antibiotic, of polypeptide structure, produced by the metabolic processes of Bacillus subtilis. It is effective against many Gram-positive bacteria, but ineffective against most aerobic Gram-negative bacteria.
Properties: White to pale buff, hygroscopic powder; odorless or with slight odor; bitter taste. Powder is stable to heat but solutions deteriorate at room temperature. Freely soluble in water; soluble in alcohol, methanol, and glacial acetic acid; insoluble in acetone, chloroform, and ether. Solutions are neutral or slightly acid to litmus. Nontoxic.
Grade: U.S.P., having a potency of about 40 to 50 units per milligram.
Use: Medicine; feed supplement.

bacitracin methylene disalicylate.
Properties: White to gray-brown powder. Slight unpleasant odor; less bitter than bacitracin. Soluble in water, pyridine, ethanol; less soluble in acetone, ether, chloroform, benzene; pH of saturated aqueous solution 3.5–5.0. Available also as the sodium salt.
Use: Antibiotic; food additive.

bacteria. A type of microorganism often composed of a single cell, in the form of straight or curved rods (bacilli), spheres (cocci), or spiral structures. Their chemical composition is primarily protein and nucleic acid. Some types, called anaerobic, are able to live and reproduce in the absence of air or oxygen; aerobic types require oxygen. Filamentous bacteria are related to blue-green algae; molds that yield antibiotics (Actinomycetes) are of this type. Pathogenic bacteria are infectious organisms that cause such diseases as pneumonia, tuberculosis, syphilis, and typhus. The staining of bacteria for microscopic identification was originated by Koch. Bacteria are often classified as Gram-positive or Gram-negative (q.v.). Food spoilage is often induced by bacterial contamination; but there are many beneficial types of bacteria in the body, e.g., intestinal flora, that aid in metabolism. Bacteria rich in proteins can be produced by fermentation of animal wastes for feed supplements.
Uses: (1) Fermentation processes used in baking and the manufacture of alcohol, wine, vinegar, beer (yeast), and antibiotics (molds). See also (3).
(2) Fixation of atmospheric nitrogen in the soil.
(3) Reaction with hydrocarbons (methane and other paraffins) to yield proteins (yeasts).
(4) Purification of sewage sludge, activated by bacteria (see sewage sludge).
(5) Reaction with cellulose to form biopolymers (q.v.) and high-protein foodstuffs.
(6) Reaction with waste materials (coal and cement dusts, gasworks effluent) to release plant nutrients for inexpensive fertilizers (U.S.S.R.).
(7) Precipitation and concentration of uranium and some other metals by compounds obtained from bacteria grown on carbonaceous materials such as lignin and cellulose.
(8) Formation of azo compounds in soil treated with the herbicide propanil.
(9) Miscellaneous reactions, e.g., oxidation of pentaerythritol to tris (hydroxymethyl) acetic acid; conversion of the sulfur in gypsum to elemental sulfur via hydrogen sulfide.
See also fermentation; virus; enzyme.

bactericide (germicide). Any agent that will kill bacteria, especially those causing disease. Bactericides vary greatly in their potency and specificity. They may be other organisms (bacteriophages), chemical compounds, or short-wave radiation. See also virus; antibiotic.

bacteriophage. A type of virus which attacks and destroys bacteria by surrounding and absorbing them.

bacteriostat. A substance which prevents or retards the growth of bacteria. Examples are quaternary ammonium salts and hexachlorophene. See also antiseptic.

baddeleyite (zirconia). ZrO_2. A natural zirconium oxide.
Properties: Color, black, brown, yellow, to colorless; streak white; luster submetallic to vitreous to greasy. M.p. 2500–2950°C. Highly resistant to chemicals. Sp. gr. 5.5–6.0.
Grades: Crude (53%, 73–75%); purified (98%).
Occurrence: Brazil; Ceylon.

Uses: Corrosion- and heat-resistant applications; source of zirconium.

Badische acid. See 2-naphthylamine-8-sulfonic acid.

Baekeland, L. H. (1863–1944). Born in Ghent, Belgium, did early research in photographic chemistry and invented Velox paper (1893). After working for several years in electrolytic research he undertook fundamental study of the reaction products of phenol and formaldehyde which culminated in his discovery in 1907 of phenol-formaldehyde polymers, originally called "Bakelite." The reaction itself had been investigated by Bayer in 1872, but Baekeland was the first to learn how to control it to yield dependable results on a commercial scale. The Bakelite Co. (now a division of Union Carbide) was founded in 1910. (See phenolformaldehyde resins).

bagasse. A form of cellulose derived as a by-product of the crushing of sugar-cane. Contains a high proportion of hemicellulose (q.v.). After pulping with either soda or kraft cooking liquor, it can be made into a low grade of paper. It is also used in compressed form as an insulating board in construction; medium for growth of nutritive bacteria; animal feeds; manufacture of furfural. See "Celotex".

Hazard: Toxic by inhalation of dust. Dust is flammable in high concentrations in air.

"Bairstat" 14E.[440a] Trademark for an internal antistatic agent (q.v.) especially adapted for polyethylene and polypropylene. Chemically it is N,N-bis(2-hydroxyethyl)-alkyl amine. Approved for use in packaging films for food products.

bait. An insecticide or rodenticide placed in such a way as to attract the pest. Arsenic compounds and Bordeaux mixture are typical insect baits. All types are highly toxic. See pesticide.

"Bake Aid."[1] Trademark for hydrated monocalcium phosphate.

Properties: White, free-flowing, crystalline powder. F.C.C. grade.

Use: "Rope" preventive in bread baking.

"Bakelite."[214] Trademark for polyethylene, polypropylene, epoxy, phenolic, polystyrene, phenoxy, parylene, polysulfone, ethylene copolymers, ABS, acrylics, and viny resins and compounds.

baking finish. A paint or varnish that requires baking at temperatures above 150° F for the development of desired properties (ASTM). Such finishes are based on oil-modified alkyd, melamine, epoxy, nitrocellulose or urea resins, or combinations of these; baking is often done by infrared radiation, producing high molecular weight coatings that are dense and tough.

baking powder. A synthetic leavening agent widely used in the baking industry. There are several types, all of which are composed of a carbonate, a weak acid or acidic compound, and a filler. A typical composition is sodium bicarbonate, tartaric acid or monobasic calcium phosphate, and cornstarch. Ingredients sometimes used are ammonium carbonate and potassium bitartrate. Upon contact with moisture and heat, the active ingredients react to evolve carbon dioxide, which "raises" the dough in the early minutes of heat exposure, thus producing a stable solid foam. Wheat flour gluten is sufficiently elastic to retain the bubbles of carbon dioxide.

baking soda. See sodium bicarbonate.

BAL. Abbreviation for British Anti-Lewisite. See 2,3-dimercaptopropanol.

"Balan."[530] Trademark for benefin; a selective herbicide.

balance. Exact equality of the number of atoms of various elements entering into a chemical reaction and the number of atoms of those elements in the reaction products. For example, in the reaction $NaOH + HCl \longrightarrow NaCl + H_2O$, the atoms in the input side are hydrogen 2, sodium 1, oxygen 1, and chlorine 1; each of these is also present in the products, though in different combination. The atoms of catalysts (when present) do not enter into reactions, and therefore are not involved. The balance of chemical reactions follows the law of conservation of mass.

The term "material balance" is used by chemical engineers in designing processing equipment; it denotes a precise list of all the substances to be introduced into a reaction and all those that will leave it in a given time, the two sums being equal.

balata. See rubber, natural.

"Balco."[155] Trademark for an alloy of 70% nickel and 30% iron.

Properties: Density, 8.46 g/cm³; resistivity, 20 ohm-μcm; tensile strength, 70,000 psi.

Uses: Square-loop laminated cores for magnetic amplifiers.

"Bali-Hi."[79] Trademark for a limed special dark wood resin.

Properties: Color "B"; m.p. (cap. tube) 105°C; s.p. (Ball & Ring) 127°C; acid number 42; 5% lime (approx.); sp. gr. (25/25°C) 1.13.

Use: Black and dark colored news inks.

"Ballast"[155] Trademark for an alloy composed of nickel 99.70%; cobalt, 0.08%; iron, 0.17%.

Properties: M.p. 1455°C; sp. gr. 8.90; tensile strength at 20°C 55,000 psi; resistivity at 20°C 7.90 microhm cm.

Uses: Electron tubes.

ball clay. A clay that has good plasticity, strong bonding power, high refractoriness and fires to a white or cream colored product. Used as bonding and plasticizing agents, or chief ingredients of whiteware, porcelains, stoneware, terra cotta, glass refractories; floor and wall tile. See also clay.

balsam. A resinous mixture of varying composition obtained from several species of evergreen trees or shrubs. Contains oleoresins, terpenes, and usually cinnamic and benzoic acids. All types are soluble in organic liquids and insoluble in water. Some have a penetrating, pleasant odor; they are combustible and in general non-toxic. The best-known types are:

(1) Peru balsam, from Central America, is a thick, viscous liquid (sp. gr. 1.15) containing vanillin. Used in flavoring, chocolate manufacture, as an ingredient in expectorants and cough sirups, and as a fragrance in shampoos and hair conditioners. A mild allergen. Shipped in drums.

(2) Tolu balsam, a plastic solid, is derived from a related tree in Colombia; its uses are similar to Peru balsam. Shipped in cans. Source of tolu oil. Odorless.

(3) Copaiba balsam, from Brazil and Venezuela (sp. gr. 0.94–0.99), is a viscous liquid used in varnishes and lacquers, as an odor fixative, and in medicine. It is the source of copaiba oil. Shipped in cans or drums.

(4) Balm of Gilead, from a Middle Eastern shrub, is used in perfumery and medicine. Shipped in bags.

(5) Canada balsam, from the North American balsam fir, is a liquid of sp. gr. 0.98; used in microscopy, in fine lacquers, as a flavoring, and as a fragrance.

(6) Benzoin resin (Benjamin gum). See benzoin resin.

bamboo. A grass or plant native to southeast Asia having a rather high cellulose content which makes possible its use for specialty papers. Its fibers are longer than those of most other plants of this type, and are comparable to those of coniferous woods. Its composition is: total cellulose 58%, alpha-cellulose 35%, pentosans 28%, lignin 23%. Also used for making light furniture, fishing rods, etc. Combustible.

"Bamca."[505] Trademark for a brand of substituted alpha-methylhydrocinnamic aldehyde.
Properties: Clear, slightly yellow liquid; sp. gr. 0.942; refractive index (n 20/D), 1.503; acid value 5.0; flash point T.C.C., 204°F. Combustible.
Use: Floral-type perfume.

banana oil.
(1) (banana liquid). A solution of nitrocellulose in amyl acetate or similar solvent, so termed because of its penetrating banana-like odor.
(2) Synonym for amyl acetate.

band, absorption. See absorption band.

"Bandane."[316] Trademark for polychlorodicyclopentadiene isomers; used as an herbicide.

"Ban-Kal."[204] Trademark for a nonfoaming liquid acid cleaner for dairies and food-processing plants.

"Banox."[108] Trademark for a series of dry, powdered, phosphate-type corrosion inhibitors. No. 1 is artificially colored. No. 1-P and WT are colorless.
Uses: Refrigerator cars; refrigeration brine; cooling towers and small water systems.

"Banthine."[70] Trademark for methantheline bromide.

Banting, Sir Frederick (1891–1941). A native of Ontario, Banting did his most important work in endocrinology. His brilliant research culminated in the preparation of the antidiabetic hormone which he called insulin (q.v.), derived from the "isles of Langerhans" in the pancreas. He received the Nobel prize in medicine for this work, together with MacLeod of the University of Toronto. In 1930 the Banting Institute was founded in Toronto. He was killed in an airplane crash during World War II.

"Banvel" D.[316] Trademark for an herbicide containing 2-methoxy-3,6-dichlorobenzoic acid (dimethylamine salt).

BAP. Abbreviation for benzyl-para-aminophenol (q.v.).

"Barafene."[293] Trademark for coatings used to prevent permeation of volatile ingredients, oils and oxygen through polyolefin containers.

"Barafos."[236] Trademark for a polyphosphate compound for the treatment of drilling mud to reduce viscosity and gel strength.

"Baragel."[236] Trademark for a compound of purified bentonite and an organic base.
Uses: Gelling agent for lubricating oils to prepare nonmelting greases.

"Barak."[28] Trademark for dibutylammonium oleate, $(C_4H_9)_2NH_2COO(C_{17}H_{33})$.
Properties: Translucent, light brown liquid. Combustible.
Use: To activate accelerators and improve processing of rubber and synthetic rubbers.

barban. Generic name for 4-chloro-2-butynyl metachlorocarbanilate, $C_6H_4(Cl)NHCOOCH_2C{:}CCH_2Cl$; herbicide and plant growth regulator.

barberite. A nonferrous alloy containing 88.5% copper, 5% nickel, 5% tin, 1.5% silicon. Sp. gr. 8.80; m.p. 1070°C. It offers good resistance to sulfuric acid in all dilutions up to 60%, sea water, moist sulfurous atmospheres, and mine waters.

barbital (diethylmalonylurea; diethylbarbituric acid; "Veronal"). $C_8H_{12}N_2O_3$.
Properties: White crystals or powder; bitter taste; odorless; stable in air; m.p. 187–192°C; soluble in hot water, alcohol, ether, acetone, and ethyl acetate.
Derivation: By the interaction of diethyl ester or diethylmalonic acid and urea.
Grades: Technical; C.P.
Hazard: See barbiturate.
Use: Medicine (sedative); stabilizer for hydrogen peroxide.
See also barbiturate.

barbiturate. A derivative of barbituric acid (q.v.) which produces depression of the central nervous system and consequent sedation. Used (by prescription only) for sedative and anesthetic purposes.
Hazard: Toxic and habit-forming. Several types, including amo-, seco-, and pentabarbital are under government restriction.

barbituric acid (malonylurea, pyrimidinetrione, 2,4,6-tri-oxohexahydro pyrimidine).

$\overline{OCNHCOCH_2CONH} \cdot 2H_2O$.

Properties: White crystals, efflorescent; odorless; m.p. 245°C with some decomposition; slightly soluble in water and alcohol; soluble in ether. Forms salts with metals.
Derivation: By condensing malonic acid ester with urea.
Grades: Technical.
Containers: 1-, 5-lb glass bottles; tins.
Hazard: See barbiturate.
Uses: Preparation of barbiturates; polymerization catalyst; dyes.

"Barden".[285] Trademark for a group of hydrous aluminum silicates (sedimentary kaolins) from South Carolina.
Properties: Sp. gr. 2.60; bulk density, aerated, 18–20 lbs/cu ft, packed, 35–40 lb/cu ft; creamy white; pH 4.5–5; air-floated; particle size 90% less than 2 microns.
Containers: 50-lb multiwall bags or bulk.
Uses: In pesticides; boxboard, flooring and tile adhesives, fertilizers, roofing granules, putties, caulking compounds, etc.

"Bardol"[175] Trademark for a coal-tar oil with aromatic content.
Properties: Dark-colored liquid; sp. gr. 1.07–1.12 (25/25°C); distillation at 300°C, 60% max; low viscosity at 40°F. Combustible.
Containers: 55-gal steel drums; tank trucks; tank cars.
Uses: Swelling agent for natural and synthetic elastomers; dispersing agent for blacks and mineral fillers; tackifier; plasticizer.

barite ($BaSO_4$). Natural barium sulfate (q.v.); barytes; heavy spar.

barium Ba Alkaline-earth element of atomic number 56, Group IIA of Periodic Table. Atomic weight 137.34. Valence 2. Seven stable isotopes.
Properties: Silver-white, somewhat malleable metal. Sp. gr. 3.6; values for melting and boiling points are reported ranging from 704°C to 850°C for m.p. and from 1140°C to 1637°C for b.p. The most acceptable values based on reliable original work appear to be m.p. 710°C and b.p. 1500°C. Extremely reactive; reacts readily with water, ammonia, halogens, oxygen, and most acids. Gives green color in flame. Extrudable and machinable.
Occurrence: Ores of barite and witherite (Georgia, Missouri, Arkansas, Kentucky, California, Nevada, Canada, Mexico).
Derivation: Reduction of barium oxide with aluminum or silicon in a vacuum at high temperature.
Forms: Rods, wire, plate, powder.
Grades: Technical; pure.
Hazard: Toxic. Flammable (pyrophoric) at room temperature in powder form; store under inert gas, petroleum, or other oxygen-free liquid. When heated to about 200°C in hydrogen, barium reacts violently, forming BaH_2. Tolerance for all soluble barium compounds is 0.5 mg per cubic meter in air.
Uses: Getter alloys in vacuum tubes; deoxidizer for copper; Frary's metal; lubricant for anode rotors in x-ray tubes; spark-plug alloys.
Shipping regulations: (Rail) Flammable solid, (n.o.s.) Yellow label. (Air) Flammable Solid label.

barium 137. See cesium 137.

barium acetate $BA(C_2H_3O_2)_2 \cdot H_2O$.
Properties: White crystals. Soluble in water; insoluble in alcohol. Sp. gr. 2.02; m.p., decomposes.
Derivation: Acetic acid is added to a solution of barium sulfide. The product is recovered by evaporation and subsequent crystallization.
Grades: Technical; C.P.
Hazard: Toxic by ingestion. Tolerance, see under barium.
Uses: Chemical reagent; acetates; textile mordant; catalyst manufacture; paint and varnish driers.
Shipping regulations: (Air) Barium compounds, n.o.s., water- or acid-soluble: Poison label.

barium aluminate $3BaO \cdot Al_2O_3$.
Properties: Gray pulverized mass, soluble in water, acids.
Hazard: Toxic by ingestion. Tolerance, see under barium.
Uses: Ceramics; water treatment.
Shipping regulations: See barium acetate.

barium azide $Ba(N_3)_2$. Crystalline solid; sp. gr. 2.936; loses nitrogen at 120°C; soluble in water; slightly soluble in alcohol.
Hazard: Toxic. Explodes when shocked or heated. Tolerance, see under barium.
Uses: High explosives.
Shipping regulations: (Air) Not acceptable when dry, or wet with less than 50% water. Wet with more than 50% water. Flammable Solid label. Not acceptable on passenger planes. (Rail) 50% or more water-wet, Yellow label.

barium binoxide. See barium peroxide.

barium borotungstate (barium borowolframate) $2BaO \cdot B_2O_3 \cdot 9WO_3 \cdot 18H_2O$.
Properties: Large, white crystals. Effloresces in air. Keep well stoppered! Soluble in water.
Hazard: Toxic by ingestion. Tolerance, see under barium.
Use: Making borotungstates.
Shipping regulations: See barium acetate.

barium borowolframate. See barium borotungstate.

barium bromate $Ba(BrO_3)_2 \cdot H_2O$.
Properties: White crystals or crystalline powder. Slightly soluble in water; insoluble in alcohol. Sp. gr. 3.820; decomposes at 260°C.
Derivation: By passing bromine into a solution of barium hydroxide, barium bromide and barium bromate being formed which are separated by crystallization.
Grades: Pure, reagent.
Hazard: Moderate fire risk in contact with organic materials. Toxic by ingestion. Tolerance, see under barium.
Uses: Analytical reagent; oxidizing agent.
Shipping regulations: See barium acetate.

barium bromide $BaBr_2 \cdot 2H_2O$.
Properties: Colorless crystals. Soluble in water and in alcohol. Sp. gr. 3.852, m.p. anhydrous $BaBr_2$ 847°C.
Derivation: Interaction of barium sulfide and hydrobromic acid, with subsequent crystallization.
Grades: Technical; C.P.
Hazard: Toxic by ingestion. Tolerance, see under barium.
Uses: Manufacturing bromides; photographic compounds.
Shipping regulations: See barium acetate.

barium carbonate $BaCO_3$.
Properties: White powder; found in nature as the mineral witherite. Insoluble in water; soluble in acids (except sulfuric). Sp. gr. 4.275; m.p. 174°C at 90 atm; 811°C at 760 mm.
Derivation: Precipitated $BaCO_3$ is made by reaction of $NaCO_3$ or CO_2 with barium sulfide.
Grades: Technical; C.P.; reagent 99.5%.
Containers: Bags; carload (bulk).
Hazard: Toxic by ingestion. Tolerance, see under barium.
Uses: Treatment of brines in chlorine-alkali cells to remove sulfates; rodenticide; barium salts; ceramic flux; optical glass; case-hardening baths; ferrites; in radiation-resistant glass for color television tubes.
Shipping regulations: See barium acetate.

barium chlorate $Ba(ClO_3)_2 \cdot H_2O$.
Properties: Colorless prisms or white powder. Soluble in water. Sp. gr. 3.179; m.p. 414°C. Combustible.
Derivation: Electrolysis of barium chloride.
Grades: Technical; C.P.; reagent.
Hazard: Toxic by ingestion. Tolerance, see under barium. Strong oxidizing agent; fire risk in contact with organic materials.
Uses: Pyrotechnics; explosives; textile mordant; manufacture of other chlorates.
Shipping regulations: (Rail) Yellow label. (Air) (dry or wet), Oxidizer label.

barium chloride $BaCl_2 \cdot 2H_2O$.
Properties: Colorless flat crystals. Soluble in water; insoluble in alcohol. Sp. gr. 3.097; m.p. 960°C (anhydrous). Combustible.
Derivation: (a) By the action of hydrochloric acid on barium carbonate or barium sulfide; (b) by heating a mixture of $BaSO_4$, carbon, and calcium chloride.
Grades: Technical (crystals or powdered) 99%; crystalline; powdered; C.P.
Containers: Bottles; multiwall paper sacks; drums; carloads.
Hazard: Toxic by ingestion; 0.8 gram may be fatal. Tolerance, see under barium.

Uses: Chemicals (artificial barium sulfate, other barium salts); reagent; lube oil additives; boiler compounds; pigments; manufacture of white leather.
Shipping regulations: See barium acetate.

barium chromate (lemon chrome; ultramarine yellow; baryta yellow; Steinbuhl yellow) $BaCrO_4$.
Properties: Heavy, yellow, crystalline powder. Soluble in acids; insoluble in water. Sp. gr. 4.498. Combustible.
Derivation: Interaction of barium chloride and sodium chromate. The precipitate is washed, filtered and dried.
Grades: Technical; C.P.
Hazard: Toxic. Tolerance, see under barium.
Uses: Safety matches; pigment in paints, ceramics; fuses; pyrotechnics; metal primers; ignition control devices.
Shipping regulations: See barium acetate.
See also chrome yellow.

barium citrate $Ba_3(C_6H_5O_7)_2 \cdot 2H_2O$.
Properties: Grayish white crystalline powder. Soluble in water, hydrochloric and nitric acids.
Hazard: Toxic; tolerance, see barium.
Uses: Manufacture of barium compounds; stabilizer for latex paints.
Shipping regulations: See barium acetate.

barium cyanide $Ba(CN)_2$.
Properties: White, crystalline powder. Soluble in water and alcohol.
Derivation: By the action of hydrocyanic acid on barium hydroxide with subsequent crystallization.
Hazard: Toxic by ingestion. Tolerance, see under barium.
Uses: Metallurgy; electroplating.
Shipping regulations: (Rail, Air) Poison label.

barium cyanoplatinite (platinum barium cyanide; barium platinum cyanide) $BaPt(CN)_4 \cdot 4H_2O$.
Properties: Yellow or green crystals; m.p. 100°C ($-2H_2O$); sp. 2.08; soluble in water, insoluble in alcohol.
Grades: C.P.
Hazard: Toxic by ingestion. Tolerance, see under barium.
Uses: X-ray screens.
Shipping regulations: See barium acetate.

barium dichromate (barium bichromate) $BaCr_2O_7 \cdot 2H_2O$.
Properties: Brownish-red needles or crystalline masses. Soluble in acids, decomposed by water.
Hazard: Toxic by ingestion. Tolerance, see under barium.
Shipping regulations: See barium acetate.

barium diorthophosphate. See barium phosphate, secondary.

barium dioxide. See barium peroxide.

barium diphenylamine sulfonate $(C_6H_5NHC_6H_4SO_3)_2Ba$.
Properties: White crystals, soluble in water.
Hazard: Toxic by ingestion. Tolerance, see under barium.
Use: Indicator in oxidation-reduction titrations.
Shipping regulations: See barium acetate.

barium dithionate (barium hyposulfate) $BaS_2O_6 \cdot 2H_2O$.
Properties: Colorless crystals. Soluble in water; slightly soluble in alcohol. Sp. gr. 4.536.
Derivation: Action of manganese dithionate on barium hydroxide.
Hazard: Toxic by ingestion. Tolerance, see under barium.
Shipping regulations: See barium acetate.

barium diuranate. See uranium-barium oxide.

barium ethylsulfate $Ba(C_2H_5SO_4)_2 \cdot 2H_2O$.
Properties: Colorless crystals. Soluble in water and alcohol. Combustible.
Derivation: Interaction of barium hydroxide and ethylsulfuric acid.
Hazard: Toxic by ingestion. Tolerance, see under barium.
Use: Organic preparations.
Shipping regulations: See barium acetate.

barium fluoride BaF_2.
Properties: White powder. Sparingly soluble in water. Sp. gr. 4.828; m.p. 1354°C.
Derivation: Interaction of barium sulfide and hydrofluoric acid followed by crystallization.
Grades: Technical; C.P.; single pure crystals, 99.98%.
Hazard: Toxic by ingestion. Tolerance. See under barium.
Uses: Ceramics; manufacture of other fluorides, crystals for spectroscopy, electronics, dry-film lubricants.
Shipping regulations: See barium acetate.

barium fluosilicate (barium silicofluoride) $BaSiF_6H$.
Properties: White, crystalline powder. Insoluble in water.
Grades: Technical.
Hazard: Toxic by ingestion. Tolerance, see under barium.
Uses: Ceramics; insecticidal compositions.
Shipping regulations: See barium acetate.

barium fructose diphosphate. See fructose diphosphate, calcium and barium salts.

barium hexafluorogermanate $BaGeF_6$. White crystalline solid; m.p. about 665°C; dissociates to barium fluoride and germanium fluoride; sp. gr. 4.56 g/cc.
Hazard: Probably toxic.

barium hydrate. See barium hydroxide.

barium hydrosulfide $Ba(SH)_2$.
Properties: Yellow crystals. Hygroscopic. Soluble in water.
Hazard: Toxic by ingestion. Tolerance, see under barium.
Shipping regulations: See barium acetate.

barium hydroxide, anhydrous $Ba(OH)_2$. Available commercially. See barium hydroxide hydrates, following.

barium hydroxide, monohydrate (barium monohydrate) $Ba(OH)_2 \cdot H_2O$.
Properties: White powder; soluble in dilute acids; slightly soluble in water.
Hazard: Toxic by ingestion. Tolerance, see under barium.
Uses: Manufacture of oil and grease additives; barium soaps and chemicals. Refining of beet sugar; alkalizing agent in water softening; sulfate removal agent in treatment of water and brine; boiler scale removal; dehairing agent; catalyst in manufacture of phenol-formaldehyde resins; insecticide and fungicide; sulfate controlling agent in ceramics; purifying agent for caustic soda; steel carbonizing agent; glass.
Shipping regulations: See barium acetate.

barium hydroxide, octahydrate (barium hydrate; barium octahydrate; caustic baryta) $Ba(OH)_2 \cdot 8H_2O$.
Properties: White powder or crystals; absorbs carbon dioxide from air. Keep well stoppered! Soluble in water, alcohol and ether. Sp. gr. 2.18; m.p. 78°C, losing its water of crystallization (m.p. anhydrous $Ba(OH)_2$ 408°C).
Derivation: (a) By dissolving barium oxide in water with subsequent crystallization. (b) By precipitation from an aqueous solution of the sulfide by caustic soda. (c) By heating barium sulfide in earthenware retorts into which a current of moist carbonic acid is passed after which superheated steam is passed over the resulting heated carbonate.
Impurities: Iron and calcium in commercial grades.
Grades: Technical (crystals or anhydrous powder); C.P.; ACS reagent.
Containers: Barrels; multiwall paper sacks.
Hazard: Toxic by ingestion. Tolerance, see under barium.
Uses: Organic preparations; barium salts; analytical chemistry. See also the monohydrate.
Shipping regulations: See barium acetate.

barium hydroxide pentahydrate (barium pentahydrate) $Ba(OH)_2 \cdot 5H_2O$.
Properties: Translucent free-flowing white flakes; density 65 lb/cu ft (approx.).
Containers: 100-lb paper bags; 400-lb fiber drums.
Hazard: Toxic by ingestion. Tolerance, see under barium.
Uses: Same as the octahydrate.
Shipping regulations: See barium acetate.

barium hypophosphite $BaH_4(PO_2)_2$.
Properties: White, crystalline powder; odorless; soluble in water; insoluble in alcohol.
Hazard: Toxic by ingestion. Tolerance, see under barium.
Use: Medicine.
Shipping regulations: See barium acetate.

barium hyposulfate. See barium dithionate.

barium hyposulfite. See barium thiosulfate.

barium iodate. $Ba(IO_3)_2$.
Properties: White, crystalline powder. Slightly soluble in water, hydrochloric and nitric acids; insoluble in alcohol. Sp. gr. 5.23; m.p., decomposes at 476°C.
Hazard: Toxic by ingestion. Tolerance, see under barium.
Shipping regulations: See barium acetate.

barium iodide $BaI_2 \cdot 2H_2O$.
Properties: Colorless crystals; decomposes and reddens on exposure to air. Soluble in water; slightly soluble in alcohol. Sp. gr. 5.150; m.p., loses $2H_2O$ and melts at 740°C.
Derivation: Action of hydriodic acid on barium hydroxide or of barium carbonate on ferrous iodide solution.
Hazard: Toxic by ingestion. Tolerance, see under barium.
Use: Preparation of other iodides.
Shipping regulations: See barium acetate.

barium manganate (manganese green; Cassel green) $BaMnO_4$.
Properties: Emerald-green powder. Insoluble in water; decomposed by acids. Sp. gr. 4.85.
Hazard: Toxic by ingestion. Tolerance, see under barium.
Use: Paint pigment.

barium mercury bromide. See mercuric barium bromide.

barium mercury iodide. See mercuric barium iodide.

barium metaphosphate $Ba(PO_3)_2$.
Properties: White powder; slowly soluble in acids; insoluble in water.
Hazard: Toxic by ingestion. Tolerance, see under barium.
Uses: Glasses, porcelains and enamels.
Shipping regulations: See barium acetate.

barium metasilicate. See barium silicate.

barium molybdate $BaMoO_4$.
Properties: White powder; slightly soluble in acids and water. Absolute density 4.7 g/cc; approximate m.p. 1600°C.
Grade: Crystal, 99.84% pure.
Hazard: Toxic by ingestion. Tolerance, see under barium.
Uses: Electronic and optical equipment; pigment in paints and other protective coatings.
Shipping regulations: See barium acetate.

barium monohydrate. See barium hydroxide, monohydrate.

barium monosulfide. See barium sulfide.

barium monoxide. See barium oxide.

barium nitrate $Ba(NO_3)_2$.
Properties: Lustrous, white crystals. Soluble in water; insoluble in alcohol. Sp. gr. 3.244; m.p. 575°C. Strong oxidizing agent.
Derivation: By the action of nitric acid on barium carbonate or sulfide.
Grades: Technical; crystals; fused mass or powder; C.P.
Containers: Barrels; multiwall paper sacks; drums.
Hazard: Dangerous fire risk in contact with organic materials. Toxic. Tolerance, see under barium.
Uses: Pyrotechnics (gives green light); incendiaries; chemicals (barium peroxide); ceramic glazes; rodenticide; electronics.
Shipping regulations: (Rail) Yellow label. (Air) Oxidizer label.

barium nitrite $Ba(NO_2)_2 \cdot H_2O$.
Properties: White to yellowish, crystalline powder. Soluble in alcohol, water. Sp. gr. 3.173; decomposed at 217°C.
Hazard: Toxic by ingestion. Tolerance, see under barium.
Uses: Diazotization.
Shipping regulations: See barium acetate.

barium octahydrate. See barium hydroxide, octahydrate.

barium oxalate $BaC_2O_4 \cdot H_2O$.
Properties: White crystalline powder. Sp. gr. 2.66. Insoluble in water; soluble in dilute nitric or hydrochloric acids.
Hazard: Toxic by ingestion. Tolerance, see under barium.
Use: Analytical reagent; pyrotechnics.
Shipping regulations: See barium acetate.

barium oxide (barium monoxide; barium protoxide; calcined baryta) BaO.
Properties: White to yellowish-white powder; absorbs carbon dioxide readily from air. Soluble in acids; reacts violently with water to form the hydroxide. Sp. gr. 5.72; m.p. 1923°C.
Derivation: Decomposition of carbonate at high temperature in presence of carbon; oxidation of barium nitrate.
Grades: Technical regular grind (208 lb/cu ft); tech-

nical fine grind (175 lb/cu ft); porous, carbide-free; 97%.
Containers: Glass bottles; hermetically sealed cans.
Hazard: Toxic by ingestion. Skin irritant. Tolerance, see under barium.
Uses: Dehydrating agent; detergent for lubricating oils.
Shipping regulations: (Air) No label required.

barium pentahydrate. See barium hydroxide pentahydrate.

barium perchlorate $Ba(ClO_4)_2 \cdot 4H_2O$.
Properties: Colorless crystals. Soluble in alcohol and water. Sp. gr. 2.74; m.p. 505°C.
Hazard: Toxic by ingestion. Oxidizing material; fire and explosion risk in contact with organic materials. Tolerance, see under barium.
Use: Manufacture of explosives; experimentally in rocket fuels.
Shipping regulations: (Rail) Yellow label. (Air) Oxidizer label.

barium permanganate $Ba(MnO_4)_2$.
Properties: Brownish-violet crystals. Soluble in water.
Hazard: Toxic by ingestion. Oxidizing material. Fire and explosion risk in contact with organic materials. Tolerance, see under barium.
Uses: Strong disinfectant; manufacture of permanganates.
Shipping regulations: (Rail) Yellow label. (Air) Oxidizer label.

barium peroxide (barium binoxide; barium dioxide; barium superoxide) BaO_2 and $BaO_2 \cdot 8H_2O$.
Properties: Grayish-white powder. Slightly soluble in water. Sp. gr. 4.96; m.p. 450°C; decomposes 800°C. Oxidizing agent.
Derivation: By heating BaO in oxygen or air at about 1000°F.
Grades: Technical; reagent.
Hazard: Toxic by ingestion. Oxidizing material. Fire and explosion risk in contact with organic materials. Tolerance, see under barium. Skin irritant. Keep cool and dry.
Uses: Bleaching; decolorizing glass; thermal welding of aluminum; manufacture of hydrogen peroxide.
Shipping regulations: See barium perchlorate.

barium phosphate, secondary (barium diorthophosphate) $BaHPO_4$.
Properties: White powder. Soluble in nitric acid (dilute), hydrochloric acid (dilute); slightly soluble in water.
Hazard: Toxic by ingestion. Tolerance, see under barium.
Shipping regulations: See barium acetate.

barium platinum cyanide. See barium cyanoplatinite.

barium potassium chromate (Pigment E) $BaK(CrO_4)_2$.
Properties: Pale yellow pigment. As compared with other chromate pigments, it has a low chloride and sulfate content and forms stronger, more elastic paint films. Sp. gr. 3.65.
Derivation: By a kiln reaction at 500°C between potassium dichromate and barium carbonate.
Hazard: Toxic by ingestion. Tolerance, see under barium.
Uses: Component of anticorrosive paints for use on iron, steel, and light metal alloys.
Shipping regulations: See barium acetate.

barium protoxide. See barium oxide.

barium pyrophosphate $Ba_2P_2O_7$.
Properties: White powder, soluble in acids and ammonium salts; very slightly soluble in water.
Hazard: Toxic by ingestion. Tolerance, see under barium.
Shipping regulations: See barium acetate.

barium silicate (barium metasilicate) $BaSiO_3$. Colorless powder; sp. gr. 4.4; b.p. 1604°C. Insoluble in water; soluble in acids. Used in ceramics.
Hazard: Toxic by ingestion. Tolerance, see under barium.
Shipping regulations: See barium acetate.

barium silicofluoride. See barium fluosilicate.

barium-sodium niobate. A synthetic electro-optical crystal used to produce coherent green light in lasers; also to make such devices as electro-optical modulators and optical parametric oscillators. The crystal undergoes no optical damage from laser irradiation at high power levels.

barium stannate $BaSnO_3 \cdot 3H_2O$.
Properties: White crystalline powder, sparingly soluble in water, readily in hydrochloric acid.
Hazard: Toxic by ingestion. Tolerance, see under barium.
Use: Production of special ceramic insulations requiring dielectric properties.
Shipping regulations: See barium acetate.

barium stearate $Ba(C_{18}H_{35}O_2)_2$.
Properties: White crystalline solid; insoluble in water or alcohol; m.p. 160°C; sp. gr. 1.145. Combustible.
Hazard: Toxic. Tolerance, see under barium.
Uses: Waterproofing agent; lubricant in metalworking, plastics and rubber; wax compounding; preparation of greases; heat and light stabilizer in plastics.

barium sulfate [barytes (natural); blanc fixe (artificial, precipitated); basofor]. $BaSO_4$.
Properties: White or yellowish, odorless, tasteless powder. Soluble in concentrated sulfuric acid; sp. gr. 4.25–4.5; particle size, 2–25 microns; m.p. 1580°C. Nontoxic. Noncombustible.
Derivation: (a) By treating a solution of a barium salt with sodium sulfate (salt cake). (b) By-product in manufacture of hydrogen peroxide. (c) Occurs in nature as the mineral barite (Arkansas, Missouri, Georgia, Nevada, Canada, Mexico).
Grades: Technical, dry, pulp, bleached, ground, floated, natural; C.P.; U.S.P.; x-ray.
Containers: 5-lb bottles; 100-, 250-lb drums; wooden barrels; multiwall paper sacks.
Uses: Weighting mud in oil-drilling; paper coatings; paints; filler and delustrant for textiles, rubber, linoleum, oilcloth, plastics, and lithograph inks; base for lake colors; x-ray photography; opaque medium for gastrointestinal radiography; in battery plate expanders.

barium sulfide (barium monosulfide; black ash) BaS.
Properties: Yellowish-green or gray powder or lumps. Soluble in water, decomposes to the hydrosulfide. Sp. gr. 4.25.
Derivation: Barium sulfate (crude barite) and coal are roasted in a furnace. The melt is lixiviated with hot water, filtered and evaporated.
Impurities: Iron, arsenic.
Hazard: Toxic by ingestion. Tolerance, see under barium.

Uses: Dehairing hides; fireproofing agent; barium salts; generating pure hydrogen sulfide for analytical purposes.
Shipping regulations: See barium acetate.

barium sulfite $BaSO_3$.
Properties: White powder, decomposed by heat. Soluble in hydrochloric acid (dilute); insoluble in water.
Grades: Technical; C.P.
Hazard: Toxic by ingestion. Tolerance, see under barium.
Uses: Analysis; paper manufacturing.
Shipping regulations: See barium acetate.

barium sulfocyanide. See barium thiocyanate.

barium superoxide. See barium peroxide.

barium tartrate. $BaC_4H_4O_6$
Properties: White crystals; sp. gr. 2.98; soluble in water; insoluble in alcohol.
Hazard: Toxic by ingestion. Tolerance, see under barium.
Uses: Pyrotechnics.
Shipping regulations: See barium acetate.

barium thiocyanate (barium sulfocyanide) $Ba(SCN)_2 \cdot 2H_2O$.
Properties: White crystals. Soluble in water and in alcohol. Deliquescent.
Derivation: By heating barium hydroxide with ammonium thiocyanate and subsequent crystallization.
Hazard: Toxic by ingestion. Tolerance, see under barium.
Uses: Making aluminum or potassium thiocyanates; dyeing; photography.
Shipping regulations: See barium acetate.

barium thiosulfate (barium hyposulfite) $BaS_2O_3 \cdot H_2O$.
Properties: White, crystalline powder; slightly soluble in water. Insoluble in alcohol. Sp. gr. 3.5; decomposed by heat.
Hazard: Toxic by ingestion. Tolerance, see under barium.
Uses: Explosives; luminous paints; matches; varnishes.
Shipping regulations: See barium acetate.

barium titanate $BaTiO_3$.
Properties: Light gray-buff powder; m.p. 3010°F; sp. gr. 5.95; insoluble in water and alkalies; slightly soluble in dilute acids; soluble in concentrated sulfuric acid and hydrofluoric acid.
Hazard: Tolerance, see under barium.
Uses: Ferroelectric ceramic. Single crystals, either pure or doped with iron, are used in storage devices, dielectric amplifiers, and digital calculators.
Shipping regulations: See barium acetate.

barium tungstate (barium wolframate; barium white; tungstate white; wolfram white) $BaWO_4$.
Properties: White powder. Insoluble in water. Sp. gr. 5.04. Low toxicity.
Uses: Pigment; in x-ray photography for manufacture of intensifying and phosphorescent screens.

barium uranium oxide. See uranium barium oxide.

barium white. See barium tungstate.

barium wolframate. See barium tungstate.

"Barium XA."[554]Trademark for a product used by manufacturers of high quality tool steels. Eliminates chain type occlusions and degasifies the steel.

barium zirconate $BaZrO_3$.
Properties: Light gray-buff powder; sp. gr. 5.52; density 140 lb/cu ft; m.p. 4550°F; insoluble in water and alkalies; slightly soluble in acid.
Uses: Manufacture of a white, easily colored silicone rubber compound having good heat stability at temperatures up to 500°F; electronics.

barium zirconium silicate A complex of BaO, ZrO_2, SiO_2.
Properties: White powder. Density 118 lb/cu ft; m.p. 2800°F; insoluble in water, alkalies. Slightly soluble in acids; soluble in hydrofluoric acid.
Containers: 80-lb paper bags; 500-lb drums.
Hazard: Toxic by ingestion.
Uses: Production of electrical resistor ceramics; glaze opacifiers; stabilizer for colored ground coat enamels.
Shipping regulations: See barium acetate.

barn. A unit of measurement, equal to 10^{-24} square centimeter, for the cross-section target area of the nucleus of an atom.

"Barnesite."[88] Trademark for a special rare earth for instrument lens polishing.

"Baroco."[236] Trademark for a high yield clay, compounded for the preparation of drilling muds for use in formations containing moderate quantities of salt or other oil field electrolytes that may flocculate ordinary drilling muds.

"Baroid."[236] Trademark for a weighting material manufactured from selected barytes (barium sulfate ore). Added to drilling muds to increase unit weight, thus increasing the hydrostatic head on the formations being drilled in deep wells, to prevent the walls from caving.

barometric pressure. The pressure of the air at a particular point on or above the surface of the earth. At sea level this pressure is sufficient to support a column of mercury about 29.9 in. in height (760 mm), equivalent to 14.7 pounds per square inch absolute (psia) or 1 atmosphere.

"Baron."[233] Trademark for a weed-killing composition containing 2-(2,4,5-trichlorophenoxy)ethyl-2,2-dichloropropionate.

"Bar-O-Sil."[304] Trademark for a complex barium silicate vinyl stabilizer.
Properties: Fine white powder; sp. gr. 2.5, refractive index 1.5.
Containers: Fiberboard drums containing 50 lb.
Uses: Supplementary stabilizer for barium-cadmium and/or zinc types. Useful in film, sheeting, extrusions and dispersion resin systems. Controls plating, hazing, crocking and fogging.

"Barosperse."[329] Trademark for a special barium sulfate formulation used in radiographic examinations of the gastro-intestinal tract.

barrier, moisture. Any substance that is impervious to water or water vapor. Most effective are high-polymer materials like vulcanized rubber, phenolyformaldehyde resins, polyvinyl chloride, and polyethylene, which are widely used as packaging films. The chief factors involved are polarity, crystallinity, and degree of cross-linking. Water-souble surfactants and protective colloids increase the susceptibility of a film to water penetration. Any pigments and fillers must be completely wetted by the polymer. Properly formulated paints are effective moisture barriers.

barthrin. Generic name for a synthetic analog of allethrin described as the 6-chloropiperonyl ester of chrysanthemummonocarboxylic acid. Used as insecticide with applications similar to allethrin and other analogs as furethrin, ethythrin and cyclethrin. Relatively nontoxic to humans.

"Bartyls."[12] Trademark for a series of compounded antiskinning agents. Developed for use in low concentrations in printing inks, paints and varnishes.

baryta, calcined. See barium oxide.

baryta, caustic. See barium hydroxide.

baryta water. A solution of barium hydroxide.

baryta yellow. See barium chromate.

barytes. See barium sulfate.

"Basacryl."[440] Trademark for a series of cationic dyestuffs for the dyeing and printing of polyacrylonitrile fiber.

basal metabolism. See metabolism.

"Basazol."[440] Trademark for dyes used in printing and dyeing fabrics composed of cellulosic fibers.

base. Any of a large class of compounds with one or more of the following properties: bitter taste, slippery feeling in solution, ability to turn litmus blue and to cause other indicators to take on characteristic colors, ability to react with (neutralize) acids to form salts. Included are both hydroxides and oxides of metals.

Water-soluble hydroxides, such as sodium, potassium, and ammonium hydroxide, undergo ionization to produce hydroxyl ion (OH—) in considerable concentration, and it is this ion that causes the previously mentioned properties common to bases. Such a base is strong or weak according to the fraction of the molecules which break down (ionize) into positive ion and hydroxyl ion in the solution. Base strength in solution is expressed by pH (q.v.). Common strong bases (alkalies) are sodium and potassium hydroxides, ammonium hydroxide, etc. These are caustic and corrosive to skin, eyes, and mucous membranes. The pH range of bases is from 7.1 to 13.

Modern chemical terminology defines bases in a broader manner. A Lowry-Bronsted base is any molecular or ionic substance that can combine with a proton (hydrogen ion) to form a new compound. A Lewis base is any substance that provides a pair of electrons for a covalent bond with a Lewis acid (q.v.). Examples of such bases are hydroxyl ion and most anions, metal oxides, compounds of oxygen, nitrogen, sulfur with nonbonded electron pairs (such as water, ammonia, hydrogen sulfide).

For hard and soft bases, see Lewis electron theory.

BASF process. A process for producing acetylene by burning a mixture of low molecular weight hydrocarbons (as, natural gas) with oxygen to produce a 2700°F temperature. The combustion products and cracked gases are quickly chilled by scrubbing with water, and the acetylene is separated by distillation and solvent extraction from ethylene, carbon monoxide, hydrogen and other reaction products. The Sachsse process is similar.

basic. Descriptive of a compound that is more alkaline than other compounds of the same name. See, e.g., lead carbonate, basic; basic salt.

basic dichromate. See bismuth chromate.

basic lining. A furnace lining containing basic compounds that decompose under furnace conditions to give basic oxides. The usual basic linings contain calcium and magnesium oxides or carbonates.

"Basicol."[188] Trademark for a series of essential oils intended as replacements for oils of lavender, geranium, lemon, pine, ylang ylang, neroli and orris root.

basic oxide. An oxide which is a base or which forms a hydroxide when combined with water, and/or which will neutralize acidic substances. Basic oxides are all metallic oxides, but there is a great variation in the degree of basicity. Some basic oxides such as those of sodium, calcium and magnesium combine with water with vigor or with relative ease, and also neutralize all acidic substances rapidly and completely. The oxides of the heavy metals are only weakly basic, do not dissolve or react with water to any extent, and neutralize only the more strongly acidic substances. There is a gradual transition from basic to acidic oxides and certain oxides, as aluminum oxide, show both acidic and basic properties. See also base.

basic research. See fundamental research.

basic salt. A compound belonging in the category of both salt and base because it contains OH (hydroxide) or O (oxide) as well as the usual positive and negative radicals of normal salts. Among the best examples are bismuth subnitrate, often written $BiONO_3$, and basic copper carbonate, $Cu_2(OH)_2CO_3$. Most basic salts are insoluble in water and many are of variable composition.

basic slag. A slag produced in the manufacture of steel. It contains a variable amount of tricalcium phosphate, calcium silicate, lime and oxides of iron, magnesium and manganese. Used as a fertilizer for its phosphorus and lime. See also slag.

basis metal. In electroplating, the metal that is being coated; it constitutes the cathode and may be any of a large number of metals.

"Basogal P."[440] Trademark for leveling agent for vat dyeing.

batch distillation. Distillation in which the entire sample of the material to be distilled, the charge, is placed in the still before the process is begun and product is withdrawn only from the condenser of the apparatus.

bating. In leather processing, the treatment of delimed skins with pancreatin or other tryptic enzyme to give a softer and smoother-grained product. The extent of bating varies from none for sole leather to ten hours or more for soft kid skins. The chemical mechanism is not clearly defined.

batrachotoninin A ($C_{24}H_{33}NO_4$). An isomeric component of batrachotoxin, the strongest neurotoxin among venoms. It is a steroidal alkaloid; the A form is only 1/500th as strong as the complete venom, but is still as toxic as strychnine. It is found in the so-called poison dart frog of Colombia. Its structure has been elucidated; when synthesized it may prove useful in medicine. See also, snake venom.

battery. An electrochemical device that generates electric current by converting chemical energy to electrical energy. Its essential components are positive and negative electrodes made of more or less electrically conductive materials, a separating medium, and an electrolyte. There are four major types: (1) primary batteries (dry cells), which are not reversible and in which the anode (zinc) is the negative plate and the cathode (graphite) is the positive plate, with ammonium chloride as electrolyte; (2) secondary or storage batteries, which are reversible and can be recharged, and in which lead sponge is the negative plate (anode) and lead oxide the positive plate (cathode), with sulfuric acid as electrolyte; (3) nuclear and solar cells, or energy converters; and (4) fuel cells. For further information see dry cell; storage battery; voltaic cell; fuel cell; solar cell.

battery acid (electrolyte acid). Sulfuric acid of strengths suitable for use in storage batteries. Water-white, odorless and practically free from iron.
Derivation: By diluting high-grade commercial sulfuric acid with distilled water to standard strengths.
Containers: Glass carboys; tank cars.
Hazard: Corrosive to skin and tissue. See also sulfuric acid.
Use: Storage batteries.
Shipping regulations: (Rail) White label. (Air) Corrosive battery fluid, not over 47% strength, Corrosive label. Legal label name, electrolyte acid.

battery limits. That portion of a chemical plant in which the actual processes are carried out, as distinguished from storage buildings, offices and other subordinate structures, called offsites.

batu. A variety of East India copal resin. See East India.

Baumé (Bé). An arbitrary scale of specific gravities devised by the French chemist Antoine Baumé and used in the graduation of hydrometers. The relations to specific gravity (at 60/60°F) are as follows: °Bé = 145 −145/sp. gr. for materials heavier than water; °Bé = 140/sp. gr. −130 for materials lighter than water.

bauxite. A natural aggregate of aluminum-bearing minerals, more or less impure, in which the aluminum occurs largely as hydrated oxides. It is usually formed by prolonged weathering of aluminous rocks. Contains 30–75% Al_2O_3, 9–31% H_2O, 3–25% Fe_2O_3, 2–9% SiO_2, and 1–3P TiO_2. Nontoxic; noncombustible.
Properties: White cream, yellow, brown, gray or red; sp. gr. 2–2.55; Mohs hardness 1–3. Insoluble in water; decomposed by hydrochloric acid.
Occurrence: Australia, Jamaica, France, Guiana, Guinea, U.S. (Arkansas), Brazil.
Uses: Most important ore of aluminum; aluminum chemicals; abrasives; aluminous cement; refractories; decolorizing and deodorizing agent; catalysts; filler in rubber, plastics, paints, cosmetics.
See also Bayer process; Hall process.
Note: Due to increasing cost of bauxite, the use of other Al-containing minerals is under active investigation.

"Baycovin."[470] Trademark for diethylpyrocarbonate.

Bayer process. Process for making alumina from bauxite. The main use of alumina is in the production of metallic aluminum. Bauxite is mixed with hot concentrated sodium hydroxide, which dissolves the alumina and silica. The silica is precipitated and the dissolved alumina is separated from the solids, diluted, cooled and then is crystallized as aluminum hydroxide. The aluminum hydroxide is calcined to give anhydrous alumina, which is then shipped to reduction plants. See also Hall process.

Bayer's acid. See 2-naphthylamine-7-sulfonic acid; also crocein acid.

"Baygon."[181] Trademark for ortho-isopropoxyphenyl methylcarbamate (q.v.).
Uses: Molluscicide.

"Baymal."[28] Trademark for a colloidal alumina whose chemical formula is AlOOH.
Properties: White powder; dispersible in water to form a stable, positively charged, colloidal sol. Ultimate particles are fibrillar crystals approximately 100 millimicrons long by 5 millimicrons in diameter.
Grades: Technical and drug.
Containers: 5-, 50-lb fiber drums.
Uses: High temperature binders and adhesives; coatings for metals and ceramics; emulsifying and thickening agent; flocculating agent for clarification and purification; chromatographic separations.

bay oil. See myrcia oil.

"Bayol."[51] Trademark for technical-grade white mineral oils that are widely used where U.S.P. or N.F. quality is not required.

"Baytex."[181] Trademark for O,O-dimethyl (O-(4-(methylthio)-meta-tolyl)phosphorothioate. See fenthion.

"Baytown."[110] Trademark for a series of "cold", styrene-butadiene copolymers (SBR), many of which contain carbon black and/or extending oil.

BBO. See 2,5-dibiphenylyloxazole.

BBP. Abbreviation for butyl benzyl phthalate (q.v.).

BCWL. Abbreviation for basic carbonate white lead. See lead carbonate, basic.

"BDA."[233] Trademark for inhibited hydrochloric acid solution containing surfactants. Used in limestone and dolomite formations, in oil-well fracturing and acidizing.

Be Symbol for beryllium.

Bé. Abbreviation for Baumé (q.v.).

"Beaconol."[354] Trademark for a group of sulfonated fatty alcohols used as detergents, wetting agents, and dispersing agents in the paper, textile, paint, and rubber industries.

"Bearflex."[500] Trademark for a group of oils used in plasticizers and as extenders for light-colored compounds.

"Beatad."[266] Trademark for a series of pregelatinized starches, either alone or with selected cellulose gums, for use as paper and paperboard additives.

"Beckacite."[36] Trademark for fumaric, maleic and modified phenolic resins.

"Beckamine."[36] Trademark for urea-formaldehyde resins (q.v.).

Beckmann rearrangement. The conversion of a ketone oxime to a substituted amide by an intermolecular rearrangement brought about by a catalyst. For example, the oxime of cyclohexanone is converted into caprolactam, with sulfuric acid as catalyst.

"Beckolin."[36] Trademark for synthetic drying oils. Two grades available: Light and dark. Nonvolatile, 100%; color 4–7 (Hellige-Klett); acid number 6–16; viscosity X to Y (Gardner-Holdt).
Uses: Paint, varnish, and enamel products.

"Beckosol."[36] Trademark for phenolated, phthalic-free, rosin-modified, pure drying and pure non-drying alkyd resins.

Becquerel, Henri (1851–1908). French physicist who shared the Nobel Prize in physics with the Curies for the discovery of the radioactivity of uranium salts. He also discovered the deflection of electrons by a magnetic field as well as the existence and properties of gamma radiation.

beer. See brewing.

beerstone. A deposit occurring during brewing operations on containers and consisting of calcium oxalate and organic material.

beeswax. Wax from the honeycomb of the bee. Beeswax consists largely of myricyl palmitate, cerotic acid and esters, and some high-carbon paraffins.
Properties: Brown or white (bleached) solid with faint odor. Sp. gr. 0.95; melting range 62–65°C; insoluble in water; slightly soluble in alcohol; soluble in chloroform, ether, and oils. Combustible; nontoxic.
Grades: Technical; crude; refined; N.F.; F.C.C.; U.S.P. (white).
Containers: Bags; bricks and slabs in 100-lb cartons.
Uses: Furniture and floor waxes; shoe polishes; leather dressings; anatomical specimens; artificial fruit; textile sizes and finishes; church candles; cosmetic creams; lipsticks; adhesive compositions.

beet molasses. See molasses.

beet sugar. See sucrose.

behenic acid (docosanoic acid) $CH_3(CH_2)_{20}COOH$. A saturated fatty acid; a minor component of the oils of the type of peanut and rapeseed.
Properties: Liquid; m.p. 80.0°C; b.p. 306°C (60 mm), 265°C (15 mm); sp. gr. 0.8221 (100/4°C); refractive index 1.4270 (100°C). Combustible; low toxicity.
Derivation: Peanut oil; occurs in ben oil, hydrogenated mustard oil and rapeseed oil.
Grades: Technical; 99%.
Uses: Cosmetics; waxes; plasticizers; chemicals; stabilizers.

behenone $C_{22}H_{44}O$. An aliphatic ketone. Insoluble in water. Inert; compatible with high-melting waxes, fatty acids. Incompatible with resins, polymers, and organic solvents at room temperature, but compatible with them at high temperature. Used as antiblocking agent.

behenyl alcohol (1-docosanol) $CH_3(CH_2)_{20}CH_2OH$. A long-chain, saturated fatty alcohol.
Properties: Colorless waxy solid; m.p. 71°C; b.p. 180°C (0.22 mm). Insoluble in water; soluble in ethanol and chloroform. Low toxicity. Combustible.
Derivation: Reduction of behenic acid with lithium aluminum hydride as catalyst.
Grades: Technical; 99%.
Uses: Synthetic fibers; lubricants; evaporation retardant on water surfaces.

Beilstein, F. K. (1838–1906). A German chemist noted for his compilation "Handbuch der Organischen Chemie," the first edition of which appeared in 1880. A multi-volume compendium of the properties and reactions of organic compounds, it has been revised several times and remains a unique and fundamental contribution to chemical science.

BEK. See butyl ethyl ketene.

belladonna (deadly nightshade; banewort). An herbaceous perennial bush (Atropa belladonna) of which the leaves and roots are used for their content of hyoscyamine and atropine.
Occurrence: Southern and central Europe; Asia Minor; Algeria; cultivated in North America, England, France.
Grades: Belladonna leaf, U.S.P.; belladonna root.
Containers: Boxes; bales.
Hazard: Highly toxic.
Use: Medicine (gastrointestinal relaxant).

bemberg. A cuprammonium rayon fiber. Flammable, not self-extinguishing.

"Bemol."[456] Trademark for a series of solid lubricants for extreme temperature/high vacuum use, such as various forms of molybdenum disulfide, niobium diselenide, tungsten disulfide.

"Bemul."[345] Trademark for a nontoxic, practically odorless emulsifying agent; a pure white, edible glycerol monostearate in bead form; m.p. 58–59°C; completely dispersible in hot water; completely soluble in alcohols and hot hydrocarbons.
Uses: Pharmaceuticals, cosmetics, and foodstuffs; protective coating for edible hygroscopic powders, tablets, and crystals; pour-point depressant for lubricating oils; textile sizes; etc.

"Benadryl."[330] Trademark for diphenhydramine hydrochloride.

"Ben-A-Gel."[304] Trademark for highly beneficiated hydrous magnesium silicate for aqueous systems. Used for thickening or gelling water systems, or as a suspension agent, and an emulsion stabilizer in oil-in-water emulsions.

bench gas. See coal gas.

Benedict solution. A water solution of sodium carbonate, copper sulfate and sodium citrate. The blue color changes to a red, orange, or yellow precipitate or suspension in the presence of a reducing sugar such as glucose, and is therefore used in testing for such materials, especially for urinalysis in the treatment of diabetes. See Fehling's solution.

beneficiation. A process used in extractive metallurgy whereby an ore, either metallic or nonmetallic, is concentrated in preparation for further processing (smelting). Calcination (q.v.) is often an important step in beneficiation; others are physical separation of high-grade ore from impurities (gangue) by screening, washing, milling or magnetic means. A process for removing sulfur from coal by chemical comminution has been developed. This is a type of beneficiation.

benefin (N-butyl-N-ethyl-alpha, alpha, alpha-trifluoro-2,6-dinitro-para-toluidine) $C_6H_2(NO_2)_2CF_3NC_6H_{14}$.
Properties: Yellow-orange solid; m.p. 65–66.5°C; b.p. 121–122°C (0.5 mm). Slightly soluble in water; readily soluble in acetone and xylene.
Hazard: May be toxic.
Use: Herbicide.

"Benemid."[123] Trademark for probenecid; used in medicine.

Benjamin gum. See benzoin resin.

"Benodaine."[123] Trademark for piperoxan hydrochloride.

benomyl (methyl-1-(butylcarbamoyl)-2-benzimidazolecarbamate). Generic name for a post-harvest fungicide for peaches, apples, etc. Toxic. Use may be restricted.

"Bentolite."[471] Trademark for a series of white bentonites from Texas.
Grades: H and L, high and low gelling.
Uses: Suspending agent; thixotropic agent; adhesives; ceramic bonding and plasticizing agent; desiccant; medicated powders.

"Bentone."[304] Trademark for organic derivatives of hydrous magnesium aluminum silicate minerals.
Uses: Gelling and pigment-suspending agents.

bentonite. A colloidal clay (aluminum silicate) composed chiefly of montmorillonite. There are two varieties: (1) sodium bentonite (Wyoming or western), which has high swelling capacity in water; and (2) calcium bentonite (southern), with negligible swelling capacity.
Properties: (Wyoming) Light to cream-colored impalpable powder; forms colloidal suspension in water, with strongly thixotropic properties. Nontoxic; noncombustible.
Occurrence: Wyoming; Mississippi; Texas; Canada; Italy; U.S.S.R.
Containers: Paper bags; drums; bulk carloads.
Uses: Oil-well drilling fluids; cement slurries for oil-well casings; bonding agent in foundry sands and pelletizing of iron ores; sealant for canal walls; thickener in lubricating greases and fireproofing compositions; cosmetics; decolorizing agent; filler in ceramics, refractories, paper coatings; asphalt modifier; polishes and abrasives; food additive; catalyst support. See also clay.

"Benzahex."[147] Trademark for a group of insecticides containing gamma-benzene hexachloride.
Hazard: See benzene hexachloride.

benzalacetone. See benzylidene acetone.

benzalazine (benzylidene azine) $C_6H_5CH{:}NN{:}CHC_6H_5$.
Properties: Yellow crystals; m.p. 91–93°C; insoluble in cold water; soluble in benzene and hot alcohol.
Uses: Stabilizer; polymerization catalyst; ultraviolet absorbent; reagent and intermediate.

benzal chloride. See benzyl dichloride.

benzaldehyde (benzoic aldehyde; synthetic oil of bitter almond) C_6H_5CHO.
Properties: Colorless or yellowish, strongly refractive, volatile oil with odor resembling oil of bitter almond, and burning aromatic taste; oxidizes readily; miscible with alcohol, ether, fixed and volatile oils; slightly soluble in water. Sp. gr. 1.0415 (25/4°C; refractive index (20°C) 1.5440–1.5464; f.p. -56°C; b.p. 178°C. Flash point 145°F (C.C.). Oxidizes in air to benzoic acid. Combustible. Autoignition temp. 377°F.
Derivation: (a) Air oxidation of toluene with uranium or molybdenum oxides as catalysts; (b) chlorination of toluene with hydrolysis by acid or alkali.
Impurities: Usually chlorine derivatives.
Method of purification: Rectification.
Grades: Technical; N.F.. Note: The specifications, especially regarding impurities, vary considerably for the grades used for dye manufacture from those used in perfumery.
Containers: Tins; carboys; drums; tank cars.
Hazard: Moderately toxic by ingestion.
Uses: Chemical intermediate for dyes, flavoring materials, perfumes, and aromatic alcohols; solvent for oils, resins, some cellulose ethers, cellulose acetate and nitrate; flavoring compounds; synthetic perfumes; manufacture of cinnamic acid, benzoic acid; pharmaceuticals; photographic chemicals.
See also oil of bitter almond.

benzaldehyde cyanohydrin. See mandelonitrile.

benzaldehyde green. See malachite green.

benzalkonium chloride. A mixture of alkyl dimethylbenzylammonium chlorides of general formula $C_6H_5CH_2N(CH_3)_2RCl$ in which R is a mixture of the alkyls from C_8H_{17} to $C_{18}H_{37}$. It is a typical quaternary ammonium salt.
Properties: White or yellowish-white, amorphous powder or gelatinous pieces. Aromatic odor and very bitter taste; soluble in water, alcohol or acetone; almost insoluble in ether; slightly soluble in benzene. Water solutions foam strongly when shaken and are alkaline to litmus.
Grade: U.S.P.
Hazard: Toxic by ingestion and skin absorption.
Uses: Cationic detergent; surface antiseptic; fungicide.

benzamide (benzoylamide) $C_6H_5CONH_2$.
Properties: Colorless crystals; m.p. 130°C; b.p. 288°C; sp. gr. 1.341. Soluble in hot water, hot benzene, alcohol, and ether. Combustible.
Derivation: From benzoyl chloride and ammonia or ammonium carbonate.
Grades: Technical.
Uses: Organic synthesis.

benzaminoacetic acid. See hippuric acid.

benzanilide (benzoylaniline; phenylbenzamide) $C_6H_5NH(COC_6H_5)$.
Properties: White to reddish crystals and powder, closely related to acetanilide, containing benzoyl in place of acetyl radical. Sp. gr. 1.306; m.p. 160–162°C. Soluble in alcohol; insoluble in water; slightly soluble in ether.
Derivation: From benzoic anhydride and aniline with caustic soda.
Uses: Intermediate in the synthesis of dyes, drugs and perfumes.

benzanthrone $C_{17}H_{10}H$, a four-ring system.
Properties: Pale yellow needles; soluble in alcohol and other organic solvents. M.p. 170°C.
Derivation: (a) From anthranol and glycerol by condensation by means of sulfuric acid (anthranol is made from anthraquinone); (b) from anthracene in sulfuric acid solution by addition of glycerol and heating to 100–110°C until the anthracene disappears. The reaction mass is then diluted with water, salted out and purified.
Method of purification: Crystallization from toluene.
Use: Dyes.

benzathine penicillin G (N,N^1-dibenzylethylenediamine dipenicillin G) $2C_{16}H_{18}N_2O_4S \cdot C_{16}H_{20}N_2 \cdot 4H_2O$.
Properties: White, odorless, crystalline powder; slightly soluble in alcohol; practically insoluble in water; pH of a saturated solution is 4.5–7.5.
Grade: U.S.P.
Use: Medicine (antibiotic).

benzazimide. See 4-ketobenzotriazine.

"Benzedrine."[71] Trademark for amphetamine sulfate.

benzene C_6H_6. Thirteenth in order of high-volume chemicals produced in U.S. (1975).

CH
HC CH
HC CH
CH
I

II

Structure: I. Complete ring showing all elements.
II. Standard ring showing double bonds only.
III. Simple ring without double bonds, with numerals indicating position of carbon atoms to which substituent atoms or groups may be attached (2 = ortho, 3 = meta, 4 = para).
IV. Generalized structure, with enclosed circle suggesting the resonance of this compound.

These structures are also referred to as the benzene nucleus.

Properties: Colorless to light-yellow, mobile, nonpolar liquid of highly refractive nature; aromatic odor; vapors burn with smoky flame; b.p. 80.1°C; m.p. 5.5°C; sp. gr. 0.8790 (20/4°C); wt/gal 7.32 lb; refractive index (n 20/D) 1.50110; flash point (closed cup) 12°F; surface tension 29 dynes/cm. Autoignition temp. 1044°F. Miscible with alcohol, ether, acetone, carbon tetrachloride, carbon disulfide, acetic acid; slightly soluble in water.

Derivation: (a) Hydrodealkylation of toluene or of pyrolysis gasoline (q.v.); (b) transalkylation of toluene by disproportionation reaction; (c) catalytic reforming of petroleum; (d) fractional distillation of coal tar.

Grades: Crude; straw color; motor; industrial pure (2°C); nitration (1°C); thiophene-free; 99 mole %; 99.94 mole %; nanograde.

Containers: Drums; tank cars; barges.

Hazard: Flammable, dangerous fire risk. Explosive limits in air, 1.5 to 8% by volume. Toxic by ingestion, inhalation, and skin absorption. Tolerance, 25 ppm in air. Safety data sheet available from Manufacturing Chemists Assn., Washington, D.C.

Uses: Ethylbenzene (for styrene monomer); dodecylbenzene (for detergents); cyclohexane (for nylon); phenol; nitrobenzene (for aniline); maleic anhydride; dodecylbenzene; chlorobenzene; diphenyl; benzene hexachloride; benzene-sulfonic acid; solvent; antiknock gasoline.

Shipping regulations: (Rail) Red label. (Air) Flammable Liquid label.

See also aromatic.

benzene azimide. See 1,2,3-benzotriazole.

benzeneazoanilide. See diazoaminobenzene.

benzeneazobenzene. See azobenzene.

benzeneazo-para-**benzeneazo**-beta-**naphthol** ("Sudan" III; tetraazobenzene-beta-naphthol) $C_6H_5NNC_6H_4$-4-NN-1-$C_{10}H_6$-2-OH. A red dye; C.I. 26100.

Properties: Brown powder; m.p. 195°C; insoluble in water; soluble in alcohol and oils.

Uses: Coloring oils red; biological stain.

benzeneazonaphthylethylenediamine. See azodine.

benzenecarboxylic acid. See benzoic acid.

benzenediazonium chloride $C_6H_5N(N)Cl$.

Properties: Ionic salt. Very soluble in water; insoluble in most organic solvents.

Hazard: Explodes on heating.

Use: Dye intermediate.

Shipping regulations: Not listed. Consult authorities.

benzene dibromide. See dibromobenzene.

benzene-ortho-**dicarboxylic acid.** See phthalic acid.

benzene-para-**dicarboxylic acid.** See terephthalic acid.

benzene hexachloride (BHC). A commercial mixture of isomers of 1,2,3,4,5,6-hexachlorocyclohexane (q.v.). insecticide. The gamma isomer is toxic. Use may be restricted. See also lindane.

benzenemonosulfonic acid. See benzenesulfonic acid.

benzenephosphinic acid (phenylphosphinic acid) $C_6H_5H_2PO_2$.

Properties: Colorless crystals; m.p. 82–84°C; sp. gr. 1.376 (29°C). Decomposes at 200°C. Stable in air. Soluble in water, alcohol, acetone. Slightly soluble in ether; insoluble in benzene, hexane, carbon tetrachloride. Combustible.

Containers: 100-lb fiber drums.

Uses: Antioxidant; intermediate for metallic salt formation; accelerator for organic peroxide catalysts.

benzenephosphonic acid (phenylphosphonic acid) $C_6H_5H_2PO_3$.

Properties: Colorless crystals. M.p. 158°C; sp. gr. 1.475 (4°C); decomposes at 275°C; soluble in water, alcohol, carbon tetrachloride. Combustible.

Containers: 100-lb fiber drums.

Uses: Intermediate in antifouling paint agents; catalyst in organic reactions.

benzenephosphorus dichloride $C_6H_5PCl_2$.

Properties: Highly reactive colorless liquid. M.p. -51°C; b.p. 224.6°C; sp. gr. 1.315 (25°C); refractive index 1.5958 (n 25/D). Soluble in common inert organic solvents; fumes in air; hydrolyzes in water.

Containers: 55-gal stainless steel drums.

Hazard: Flammable; corrosive to skin and tissue.

Uses: Organic synthesis, for derivation of plasticizers, polymers, antioxidants; oil additives.

Shipping regulations: (Rail) Red label. (Air) Corrosive label. Not acceptable on passenger planes.

benzenephosphorus oxydichloride $C_6H_5POCl_2$.

Properties: Reactive colorless liquid. M.p. 3.0°C; b.p. 258°C; sp. gr. 1.197 (25°C); refractive index 1.5585 (n 25/D). Soluble in common inert organic solvents; hydrolyzes in water. Combustible.

Containers: 55-gal nickel drums.

Hazard: Strong irritant to skin.

Uses: Organic synthesis, for derivation of plasticizers, polymers, antioxidants, oil additives.

benzenesulfonic acid (benzenemonosulfonic acid; phenylsulfonic acid) $C_6H_5SO_3H$.

Properties: Fine, deliquescent needles or large plates; m.p. 65–66°C when anhydrous; with 1.5 molecules water, m.p. is 43–44°C; soluble in water and alcohol; slightly soluble in benzene; insoluble in ether and carbon disulfide.

Derivation: By reacting benzene with fuming sulfuric acid.

Uses: Manufacture of phenol, resorcinol and other organic syntheses, and as a catalyst.

benzene-1,3-5-tricarboxylic acid chloride. See trimesoyl trichloride.

benzenoid. Any organic compound containing or derived from the benzene ring structure, e.g., phenol, nitrobenzene, anthracene, styrene. This large array of unsaturated compounds, derived chiefly from petroleum and coal tar, provides a broad base for the synthesis of polymers, dyes, and intermediates. See also aromatic.

benzenyl trichloride. See benzotrichloride.

benzethonium chloride. A synthetic quaternary ammonium compound, $C_{27}H_{42}ClNO_2$.
Properties: Colorless, odorless plates. Very bitter taste; m.p. 164–166°C. Soluble in water, alcohol, and acetone. Aqueous solution yields flocculent white precipitate with soap solutions.
Grade: N.F.
Hazard: Highly toxic by ingestion; 1 gram may be fatal.
Use: Antiseptic; cationic detergent.

benzhydrol (benzohydrol; diphenylmethanol; diphenylcarbinol) $(C_6H_5)_2CHOH$.
Properties: Needlelike colorless crystals; m.p. 69°C; b.p. 298°C, 176°C (13 mm). Slightly soluble in water, easily soluble in alcohol, ether, chloroform and carbon disulfide. Combustible.
Derivation: Reduction of benzophenone with magnesium or zinc dust.
Use: Preparation of other organic compounds, including certain antihistamines; insecticide.

benzhydryl bromide. See diphenylmethyl bromide.

benzhydryl chloride $(C_6H_5)_2CHCl$.
Properties: Water-white to light straw-colored liquid; refractive index 1.596; m.p. 16°C; b.p. 140°C (3mm). Combustible.
Uses: Organic synthesis.

2-(benzhydryloxy)-N,N-dimethylethylamine hydrochloride. See diphenhydramine hydrochloride.

benzidine (benzidine base; para-diaminodiphenyl) $NH_2(C_6H_4)_2NH_2$.
Properties: Grayish-yellow, white or reddish gray crystalline powder; m.p. 127°C; b.p. 400°C; soluble in hot water, alcohol, and ether; slightly soluble in water. Combustible. Also available as the hydrochloride.
Derivation: (a) By reducing nitrobenzene with zinc dust in alkaline solution followed by distillation; (b) by electrolysis of nitrobenzene, followed by distillation; (c) nitration of diphenyl followed by reduction of the product with zinc dust in alkaline solution, with subsequent distillation.
Grades: Technical (paste; powder 80–85%).
Hazard: Toxic by ingestion, inhalation, and skin absorption. A carcinogen.
Uses: Organic synthesis; manufacture of dyes, especially of Congo red; detection of blood stains; stain in microscopy; reagent; stiffening agent in rubber compounding.
Shipping regulations: (Air) Poison label.

benzidinedicarboxylic acid. See diaminodiphenic acid.

benzidine dye. Any of a group of azo dyes derived from 3,3′-dichlorobenzidine; they include yellow and orange colors claimed to be light-fast and alkali-resistant. Congo Red is derived from benzidine and naphthionic acid. These compounds are active carcinogens and physical contact with them should be avoided.

benzidine sulfate. $C_{12}H_{12}N_2 \cdot H_2SO_4$.
Properties: White crystalline powder. Soluble in ether; sparingly soluble in water, alcohol, dilute acids.
Derivation: Action of sulfuric acid or sodium sulfate on benzidine with subsequent recovery by precipitation.
Hazard: Toxic by ingestion and skin absorption. A carcinogen.
Use: Organic synthesis.

benzil (dibenzoyl) $C_6H_5CO \cdot COC_6H_5$.
Properties: Yellow needles. Soluble in alcohol and ether; insoluble in water. M.p. 95°C; b.p. 346–348°C; sp. gr. 1.521.
Derivation: From benzoin by oxidation with nitric acid.
Use: Organic synthesis; insecticide.

benzilic acid (diphenylglycolic acid) $(C_6H_5)_2C(OH)COOH$.
Properties: White to tan powder with a characteristic odor; m.p. 148–151°C; soluble in hot water and alcohol. Combustible.
Containers: Bottles; drums.
Uses: Chemical intermediate.

benzine. The name "benzine" is archaic and misleading and should not be used. (A.S.T.M. Petroleum Definitions D-288). Do not confuse with benzene.
Shipping regulations: (Rail) Red label. (Air) Flammable Liquid label.
See also ligroin.

benzocaine. See ethyl para-aminobenzoate.

benzodihydropyrone (dihydrocoumarin) $C_9H_8O_2$ (bicyclic).
Properties: White to light yellow oily liquid with a sweet odor; congeals at 23°C. Insoluble in water; soluble in alcohol, chloroform and ether. Combustible.
Uses: Perfumery.

"Benzoflex."[316] Trademark for a series of plasticizers which are dibenzoate esters of dipropylene glycol or any of several polyethylene glycols.
Uses: Primary plasticizer for vinyl resins; adhesive formulations; some grades in food packaging adhesives.

benzofuran. See coumarone.

benzoglycolic acid. See mandelic acid.

benzoguanamine (2,4-diamino-6-phenyl-s-triazine) $C_6H_5C_3N_3(NH_2)_2$.
Properties: Crystals; sp. gr. 1.40 (d 25/4); m.p. 227–228°C. Soluble in methyl "Cellosolve," alcohol, and dilute hydrochloric acid; partially soluble in dimethylformamide and acetone; practically insoluble in chloroform, ethyl acetate; insoluble in water, benzene, and ether. Combustible.
Derivation: Prepared from benzonitrile and dicyandiamide in the presence of sodium and liquid ammonia.
Containers: Drums; tank cars.
Uses: Thermosetting resins, resin modifiers; chemical intermediate for pesticides, pharmaceuticals and dyestuffs.

benzohydrol. See benzhydrol.

benzoic acid (carboxybenzene; benzenecarboxylic acid; phenylformic acid) C_6H_5COOH. It occurs naturally in benzoin resin (q.v.).
Properties: White scales or needle crystals with odor of benzoin or benzaldehyde; sp. gr. 1.2659; m.p. 121.25°C; b.p. 249.2°C; sublimes at 100°C; freely volatile in steam; flash point (closed cup) 250°F; soluble in alcohol, ether, chloroform, benzene, carbon disulfide, carbon tetrachloride, and turpentine; slightly soluble in water. Combustible.
Derivation: (a) Decarboxylation of phthalic acid by steam, in the presence of catalysts; (b) chlorination

of toluene to yield benzotrichloride, which is hydrolyzed to benzoic acid; (c) oxidation of toluene; (d) from benzoin resin (q.v.).
Method of purification: Sublimation.
Grades: Technical; C.P.; U.S.P.; F.C.C.
Containers: Barrels; multiwall paper sacks; drums; tanks.
Hazard: May cause nausea and gastroenteric disturbance. Possible allergic reactions. Use restricted to 0.1% in foods.
Uses: Sodium and butyl benzoates; plasticizers; benzoyl chloride; alkyd resins; food preservative; seasoning tobacco; flavors; perfumes; dentifrices; medicine (germicide).

benzoic aldehyde. See benzaldehyde.

benzoic trichloride. See benzotrichloride.

benzoin (bitter almond-oil camphor; benzoylphenyl carbinol; 2-hydroxy-2-phenylacetophenone; phenylbenzoyl carbinol) $C_6H_5CHOHCOC_6H_5$.
Properties: White or yellowish crystals; slight camphor odor; m.p. 137°C; slightly soluble in water and ether; soluble in acetone and hot alcohol. Optically active. Combustible.
Derivation: Condensation of benzaldehyde in an alkaline cyanide solution.
Uses: Organic synthesis; intermediate; photopolymerization catalyst.
Note: Do not confuse with benzoin resin. (q.v.).

alpha-**benzoin oxime** (benzoin antioxime) $C_6H_5CHOHC{:}NOHC_6H_5$. Solid; m.p. 150–152°C.
Uses: Organic intermediates and photographic chemicals.

benzoin resin (gum benzoin; Benjamin gum).
Properties: Reddish-brown globules; balsamic, vanilla-like odor; brittle at room temperature, but softened by heat. Soluble in warm alcohol and carbon disulfide; insoluble in water.
Source: Obtained from the Styrax tree in Southeast Asia and Sumatra. The Sumatran grade is higher-melting and only 75% soluble in alcohol.
Grades: Technical; tincture U.S.P.
Constituents: Benzoic acid, cinnamic acid, vanillin.
Uses: Source of benzoic acid; perfumery; cosmetics; medicine (antiseptic and expectorant).
Note: Do not confuse with benzoin (q.v.).

benzol. Obsolete name for benzene, no longer in approved use.

benzonitrile (phenyl cyanide) C_6H_5CN.
Properties: Colorless oil; almond-like odor; sharp taste; viscosity (100°F) 1.054 centistokes; refractive index 1.5289; soluble in boiling water, alcohol, and ether; slightly soluble in cold water. Sp. gr. 1.0051; b.p. 190.7°C; m.p. -13.1°C.
Derivation: From benzoic acid by heating with lead thiocyanate.
Hazard: Toxic and flammable.
Uses: Manufacture of benzoguanamine; intermediate for rubber chemicals; solvent for nitrile rubber, specialty lacquers, and many resins and polymers, and for many anhydrous metallic salts.
Shipping regulations: (Rail, Air) Not listed. Consult authorities.

benzophenol. See phenol.

benzophenone (diphenylketone) $(C_6H_5)_2CO$.
Properties: White prisms, with sweet, rose-like odor. Soluble in alcohol and ether; insoluble in water; m.p. 47.5°C; b.p. 305°C. Combustible.
Purification: Crystallization from alcohol.
Grades: Free from chlorine (FFC); also F.C.C.
Containers: Tin cans; fiberboard containers; drums.
Uses: Organic synthesis; odor fixative; derivatives are used as ultraviolet absorbers; flavoring; soap fragrance; pharmaceuticals; polymerization inhibitor for styrene.

benzophenone oxide. See xanthone.

3,3′,4,4′-benzophenone tetracarboxylic dianhydride (BTDA) $C_{17}H_6O_7$.
Properties: Free-flowing powder; m.p. 228°C.
Uses: Epoxy curing agent; heat-resistant polymers, specialty alkyd resins, polyesters and plasticizers.

benzopyrene (benzypyrene) $C_{20}H_{12}$. A polynuclear (five-ring) aromatic hydrocarbon. Found in coal tar, cigarette smoke, and in the atmosphere as a product of incomplete combustion. Occurs as benzo[a]pyrene and benzo[e]pyrene.
Properties (benzo[a]pyrene): Yellowish crystals; m.p. 179°C; b.p. 310–312°C (10 mm). Insoluble in water; slightly soluble in alcohol; soluble in benzene, toluene, xylene.
Hazard: An active carcinogen. Toxic by inhalation.

benzopyrone. See coumarin.

benzoquinone. See quinone.

benzosulfimide. See saccharin.

benzothiazole C_6H_4SCHN (bicyclic).
Properties: Yellow liquid with unpleasant odor; b.p. 227°C; slightly soluble in water; soluble in alcohol. Combustible.
Hazard: May be toxic by ingestion.
Use: Rubber accelerators.

benzothiazolyl disulfide. See 2,2′-dithiobisbenzothiazole.

benzothiazyl 2-cyclohexylsulfenamide. See N-cyclohexyl-2-benzothiazolesulfenamide.

2-benzothiazyl-N,N-diethylthiocarbamyl sulfide (diethyldithiocarbamic acid, 2-benzothiazoyl ester) $(C_6H_4SCN)SSCN(C_2H_5)_2$.
Properties: Free-flowing, light yellow to tan powder; sp. gr. 1.27; m.p. 69°C (min).
Uses: Rubber accelerator.

benzothiazyl disulfide. See 2,2′-dithiobisbenzothiazole.

1,2,3-benzotriazole (aziminobenzene; benzene azimide) $C_6H_4NHN_2$.
Properties: White to light tan, odorless, crystalline powder; boiling range 201–204°C (15 min); very stable toward acids and alkalies, and toward oxidation and reduction. Its basic characteristics are very weak but it forms stable metallic salts. Can exist in 2 tautomeric forms. Soluble in alcohol and benzene; slightly soluble in water.
Containers: Bottles; fiber drums.
Hazard: Moderately toxic by ingestion.
Uses: Photographic restrainer; chemical intermediate; derivatives used as ultraviolet absorbers.

benzotrichloride (toluene trichloride; benzenyl trichloride; benzoic trichloride; phenylchloroform) $C_6H_5CCl_3$.
Properties: Colorless to yellowish liquid; fumes in air; hydrolyzes in presence of water; penetrating odor.

Soluble in alcohol and ether; insoluble in water. Sp. gr. 1.38; b.p. 213–214°C; m.p. –5°C; refractive index 1.5584.
Derivation: Chlorination of boiling toluene.
Method of purification: Rectification.
Hazard: Moderately toxic; fumes highly irritant.
Use: Synthetic dyes; organic synthesis.

benzotrifluoride (toluene trifluoride; trifluoromethylbenzene) $C_6H_5CF_3$.
Properties: Water-white liquid with aromatic odor. B.p. 102.1°C; m.p. –29.1°C; sp. gr. 1.1812 (25/4°C); refractive index 1.4146. Flash point (closed cup) 54°F. Miscible with alcohol, acetone, benzene, carbon tetrachloride, ether, and *n*-heptane; insoluble in water.
Containers: 55-gal drums.
Hazard: Toxic by inhalation. Flammable, dangerous fire risk. Tolerance (as F), 2.5 mg per cubic meter of air.
Uses: Intermediate for dyes, and pharmaceuticals; as a solvent and dielectric fluid; vulcanizing agent; insecticides.
Shipping regulations: (Rail) Flammable liquid, n.o.s., Red label. (Air) Flammable liquid, n.o.s., Flammable Liquid label.

trans-beta-**benzoylacrylic acid** $C_6H_5COCH{:}CHCOOH$.
Properties: Straw yellow needles or plates; m.p. 99°C; soluble in most solvents but only slightly soluble in cold water and ligroin. Combustible.
Containers: Polyethylene-lined fiber drums.
Uses: Reagent for characterizing phenols; intermediate in the manufacture of bactericides, insecticides, surface-active agents and the upgrading of drying oils.

N-benzoyl-L(+)-alanine. See alanine.

benzoylamide. See benzamide.

benzoylaminoacetic acid. See hippuric acid.

benzoylaniline. See benzanilide.

benzoyl chloride C_6H_5COCl.
Properties: Transparent, colorless liquid; pungent odor; vapor causes tears. Sp. gr. 1.2188; m.p. –0.5°C; b.p. 197.2°C; refractive index (n 20/D) 1.5536; flash point 162°F. Soluble in ether and carbon disulfide; decomposes in water. Combustible.
Derivation: (a) Interaction of benzoic acid and sulfuryl chloride; (b) benzotrichloride and water in the presence of zinc chloride; (c) phosphorus tri- or pentachloride and benzoic acid.
Containers: Drums; tank trucks.
Grades: Technical; C.P.
Hazard: Strong irritant to skin and tissue. Toxic by ingestion and inhalation.
Uses: Medicine; intermediate for introduction by benzoyl groups; intermediate for other organics.
Shipping regulations: (Rail) White label. (Air) Corrosive label.

benzoyl-2,5-diethoxyaniline $C_6H_5CONHC_6H_3(OC_2H_5)_2$.
Properties: Gray pellets; m.p. 83–84°C. Used as an intermediate for pharmaceuticals, dyestuffs and other organic chemicals.
Hazard: May be toxic.

benzoylferrocene (phenyl ferrocenyl ketone) $C_5H_5FeC_5H_4COC_6H_5$.
Properties: Dark red crystalline solid; m.p. 107–108°C.
Hazard: May be toxic.
Uses: Intermediate.

benzoyl fluoride. C_6H_5COF.
Containers: Available in quantities up to 5 lb in aluminum bottles.
Hazard: Toxic. Tolerance (as F), 2.5 mg per cubic meter of air.
Uses: Manufacture of acyl and other fluorides.

benzoylglycin. See hippuric acid.

benzoylglycocoll. See hippuric acid.

benzoyl peroxide (dibenzoyl peroxide) $(C_6H_5CO)_2O_2$.
Properties: White, granular, crystalline solid; tasteless; faint odor of benzaldehyde. Active oxygen, about 6.5%. Soluble in nearly all organic solvents; slightly soluble in alcohols and vegetable oils; slightly soluble in water. M.p. 103–105°C; decomposes explosively above 105°C. Autoignition temp. 176°F; sp. gr. 1.3340 (25°C).
Grades: Technical, wet or dry; F.C.C.
Containers: 1-lb net fiber containers or polyethylene-lined bags; standard cases contain 5, 25, and 50 containers.
Hazard: Flammable and explosive. May explode spontaneously when dry (1% of water). Never mix unless at least 33% water is present. Caution!! Safety data sheet available from Manufacturing Chemists Assn., Washington, D.C.
Uses: Bleaching agent for flour, fats, oils, and waxes; polymerization catalyst; drying agent for unsaturated oils; pharmaceutical and cosmetic purposes; rubber vulcanization without sulfur; burn out agent for acetate yarns; production of cheese; embossing vinyl flooring (proprietary).
Shipping regulations: (Rail) Yellow label. (Air) Organic peroxide label. Not acceptable on passenger planes.

benzoylphenyl carbinol. See benzoin.

2-benzoylpyridine $C_6H_5COC_5H_4N$.
Properties: Colorless liquid. Freezing point 42.7°C; insoluble in water.
Grade: 98% (minimum).
Use: Organic synthesis.

4-benzoylpyridine $C_6H_5COC_5H_4N$.
Properties: Colorless liquid. Freezing point 71.4°C; insoluble in water.
Grade: 98% (minimum).
Use: Organic synthesis.

benzoylsulfonic imide. See saccharin.

benzozone. See acetyl benzoyl peroxide.

1,2-benzphenanthrene. See chrysene.

benzyl abietate $C_{19}H_{29}COOCH_2C_6H_5$.
Properties: Nonvolatile, viscous liquid which resembles Canada balsam. Soluble in most anhydrous solvents. See also balsam.

benzyl acetate (phenylmethyl acetate) $C_6H_5CH_2OOCCH_3$.
Properties: Water-white liquid; floral odor. Soluble in alcohol and ether. Slightly soluble in water; soluble in alcohol. Flash point 216°F (closed cup). Sp. gr. 1.059–1.062 (15°C); b.p. 212°C; refractive index 1.5015–1.5035. Combustible. Autoignition temp. 862°F.
Derivation: (a) By treating benzyl chloride with sodium acetate in various solvents; (b) by esterification of benzyl alcohol with acetic anhydride or acetic acid.
Method of purification: Distillation.
Grades: Free-from-chlorine grade which should have an ester content of 97% but for which lower grade material is sometimes substituted; technical grade which is not free from chlorine and for which ester content varies considerably; F.C.C.
Containers: 55-gal drums; carboys; cans; bottles.

Hazard: Moderately toxic; irritant to skin.
Uses: Artificial jasmine and other perfumes; soap perfume; flavoring; solvent and high boiler for cellulose acetate and nitrate, natural and synthetic resins; oils; lacquers; polishes; printing inks; varnish removers.

benzyl alcohol (alpha-hydroxytoluene; phenylmethanol; phenylcarbinol) $C_6H_5CH_2OH$.

H
—C—OH
H

Properties: Water-white liquid; slight odor; sharp, burning taste. B.p. 206°C; flash point 213°F; sp. gr. 1.040–1.050 (25/25°C); refractive index (20°C) 1.5385–1.5405. Somewhat soluble in water; miscible with alcohol, ether and chloroform. Combustible, low toxicity. Autoignition temp. 817°F.
Derivation: (a) By hydrolysis of benzyl chloride; (b) from benzaldehyde by catalytic reduction or Cannizzaro reaction.
Method of purification: Distillation and chemical treatment.
Grades: Free from chlorine (F.F.C.); technical; N.F.; textile; photographic reagent; F.C.C.
Uses: Perfumes and flavors; photographic developer for color movie films; dyeing nylon filament, textiles and sheet plastics; solvent for dyestuffs, cellulose esters, casein, waxes, etc.; heat-sealing polyethylene films; intermediate for benzyl esters and ethers; local anesthetic; cosmetics, ointments, emulsions; ball point pen inks; stencil inks.

benzylamine (aminotoluene) $C_6H_5CH_2NH_2$.
Properties: Light amber liquid; strongly alkaline reaction. Soluble in alcohol, ether, and water. Sp. gr. 0.9813; b.p. 184.5°C; refractive index (n 20/D) 1.540. Combustible.
Derivation: From benzyl chloride and ammonia.
Use: Chemical intermediate for dyes, pharmaceuticals, and polymers.

N-benzyl-para-**aminophenol** (BAP) $C_6H_5CH_2NHC_6H_4OH$.
Properties: Light brown powder, melts between 84–90°C; 96–99% pure; solubility 50% in anhydrous methanol, 50% in 95% ethyl alcohol, 0.06% in water; 0.1–0.5% in gasoline, varying with chemical nature of gasoline.
Hazard: Carcinogenic agent. Tolerance, 0.2 mg per cubic meter in air.
Use: In cracked gasoline, in concentration of 0.001–0.004% by weight to prevent gum formation.

2-benzylamino-1-propanol $C_6H_5CH_2NHCH(CH_2OH)CH_3$.
Properties: White to yellow solid. Both *l* and *dl*-forms are available. M.p. (*dl*-form) 70–73°C. Specific rotation (*l*-form) +38° to +44° (1.0% solution in alcohol) at 25°C. Combustible.

benzylaniline $C_6H_5NHCH_2C_6H_5$.
Properties: Colorless prisms. Soluble in alcohol and ether; insoluble in water. M.p. 33°C; b.p. 310°C.
Use: Organic synthesis.

benzylbenzene. See diphenylmethane.

benzyl benzoate $C_6H_5CH_2OOCC_6H_5$.
Properties: Water-white liquid; readily freezes. Sharp, burning taste and faint aromatic odor. Supercools easily. Insoluble in water and glycerin; soluble in alcohol, chloroform and ether. Sp. gr. 1.116–1.120 (25/25°C); b.p. 325°C; m.p. 18.8°C; refractive index 1.568–1.569 (20°C). Flash point, 298°F. Combustible; low toxicity.
Derivation: (a) By a Cannizzaro reaction from benzaldehyde; (b) by esterifying benzyl alcohol with benzoic acid; (c) by treating sodium benzoate with benzyl chloride.
Method of purification: Distillation and crystallization.
Grades: Technical; U.S.P.; F.C.C.
Containers: 55-gal drums (tinned); 100-lb aluminum drums; cans; bottles.
Uses: Fixative and solvent for musk in perfumes and flavors; medicine (external); plasticizer; miticide.

benzyl bromide (alpha-bromotoluene) $C_6H_5CH_2Br$.
Properties: Clear, refractive liquid. Pleasant odor. Not easily hydrolyzed. Soluble in alcohol, benzene, ether, insoluble in water. A lachrymator. Sp. gr. 1.438 at 16°C; b.p. 198–199°C; m.p. -39°C; vapor density 5.8.
Derivation: (a) Bromination of toluene; (b) interaction of benzyl alcohol and hydrobromic acid.
Hazard: Corrosive to skin and tissue.
Uses: Making foaming and frothing agents; organic synthesis.
Shipping regulations: (Rail) White label. (Air) Corrosive label. Not acceptable on passenger planes.

benzyl butyrate $C_3H_7COOCH_2C_6H_5$.
Properties: Liquid; fruity odor; b.p. 240°C; density 1.016 (17.5°C); soluble in alcohol. Combustible.
Grades: Technical; F.C.C.
Uses: Plasticizer; odorants and flavoring.

benzyl carbinol. See phenethyl alcohol.

benzyl "Cellosolve." See ethylene glycol monobenzyl ether.

benzyl chloride (alpha-chlorotoluene) $C_6H_5CH_2Cl$.
Properties: Colorless liquid; pungent odor. A lachrymator. Sp. gr. 1.090–1.111 (25/25°C); m.p. -43°C; b.p. 179°C; n 25/D 1.5365; flash point 153°F (O.C.). Combustible. Autoignition temp. 1085°F. Soluble in alcohol and ether; insoluble in water.
Derivation: By passing chlorine over boiling toluene until it has increased 38% in weight. The product is washed with water and separated by fractional distillation.
Grades: Technical; C.P.; 95%; redistilled.
Forms: Anhydrous; stabilized (with aqueous sodium carbonate solution).
Containers: Anhydrous (unstabilized); 475-lb nickel drums; 100-lb carboys. Stabilized: 475-lb steel drums; tank trucks; tank cars.
Hazard: Intensely irritating to eyes and skin. Safety data sheet available from Manufacturing Chemists Assn., Washington, D.C.
Uses: Dyes; intermediates; benzyl compounds; synthetic tannins; perfumery; pharmaceuticals; manufacture of photographic developer; gasoline gum inhibitors; penicillin precursors; quaternary ammonium compounds.
Shipping regulations: (Rail) White label. (Air) Corrosive label.

benzyl chlorocarbonate (benzyl chloroformate) $C_6H_5CH_2OCOCl$.
Properties: Oily liquid, with lachrymatory properties; odor of phosgene; decomposes over 100°C; reacts with water to form hydrochloric acid.
Hazard: Irritant to eyes. Emits highly toxic phosgene fumes when heated to 100°C.
Use: Peptide synthesis.
Shipping regulations: (Rail) White label. (Air) Corrosive label. Not acceptable on passenger planes.

benzyl chloroformate. See benzyl chlorocarbonate.

ortho-**benzyl**-para-**chlorophenol** (chlorophene, USAN; "Santophen"; septiphene; 4-chloro-alpha-phenyl-ortho-cresol) $C_6H_5CH_2C_6H_3OHCl$.
Properties: White to light tan or pink flakes; crystallizing point 45°C min; sp. gr. 1.202–1.206 (55/55°C); odor slight phenolic. Insoluble in water; highly soluble in alcohol and other organic solvents; dispersible in aqueous media with the aid of soaps or synthetic dispersing agents; noncorrosive to most metals. Combustible.
Containers: 22- and 55-gal lacquer-lined drums.
Hazard: Moderately toxic and irritant.
Uses: Active principle or enhancing agent for disinfectants.

benzyl cinnamate (cinnamein) $C_9H_7O_2 \cdot C_7H_7$.
Properties: White crystals; aromatic odor; m.p. 39°C; congealing point min 34°C; b.p. 244°C (25 mm); insoluble in water; soluble in alcohol. Low toxicity.
Grades: Technical; F.C.C.
Uses: Perfumery and flavors.

benzyl cyanide (phenylacetonitrile; alpha-tolunitrile) $C_6H_5CH_2CN$.
Properties: Colorless oily liquid; aromatic odor; soluble in alcohol and ether; insoluble in water. Sp. gr. 1.0157; m.p. –24°C; b.p. 230°C; refractive index (n 25/D) 1.5211.
Derivation: Interaction of benzyl chloride and potassium cyanide.
Grades: Technical.
Containers: Metal drums; tins; bottles.
Hazard: Highly toxic.
Use: Organic synthesis; especially penicillin precursors.
Shipping regulations: (Rail) Poisonous liquid, n.o.s., Poison label. (Air) Poison label.

benzyl dichloride (benzylidene chloride; benzal chloride; chlorobenzal; alpha, alpha-dichlorotoluene) $C_6H_5CHCl_2$.
Properties: Colorless oily liquid; faint aromatic odor; sp. gr. 1.295 (16°C); m.p. –16.1°C; b.p. 207°C; refractive index 1.5502 (20°C); soluble in alcohol, ether, and in dilute alkali; insoluble in water. Combustible.
Derivation: Chlorination of toluene, until two formula weights of chlorine are absorbed, in absence of catalysts but presence of light.
Hazard: Toxic; strong irritant and lachrymator.
Uses: Dyes.
Shipping regulations: (Air) Poison label. (Rail) Poisonous liquids, n.o.s., Poison label. Legal label name (Air) benzylidine chloride.

N-benzylidiethanolamine $C_6H_5CH_2N(C_2H_4OH)_2$.
Properties: Colorless to light yellow liquid; completely miscible with water. Sp. gr. 1.073; refractive index 1.5345–1.5375; distilling range 240–255°C. Combustible.
Uses: Corrosion inhibitor; intermediate.

N-benzylidimethylamine $C_6H_5CH_2N(CH_3)_2$.
Properties: Colorless to light yellow liquid. Sp. gr. 0.894 (27°C); refractive index 1.4985–1.5005 (25°C); b.p. 180–182°C; distilling range 65–68oC (18 mm). Combustible.
Uses: Intermediate, especially for quaternary ammonium compounds; a dehydrohalogenating catalyst; corrosion inhibitor; acid neutralizer.

benzyl disulfide. See dibenzyl disulfide.

N-benzylethanolamine $C_6H_5CH_2NH(C_2H_4OH)$.
Properties: Colorless to light yellow liquid. Sp. gr. 1.044 (27°C); refractive index 1.5400–1.5430; distillation range 240–255°C. Combustible.
Uses: Corrosion inhibitor; intermediate.

benzyl ethyl ether $C_6H_5CH_2OC_2H_5$.
Properties: Colorless, oily liquid; aromatic odor; volatile in steam; insoluble in water; miscible with alcohol and ether. B.p. 185°C; sp. gr. 0.949; refractive index 1.4955 at 20°C. Combustible.
Derivation: By boiling benzyl chloride with either sodium or potassium ethylate.
Hazard: Narcotic in high concentrations; may be skin irritant.
Uses: Organic synthesis; flavoring.

benzyl ethylsalicylate $C_6H_5CH_2OOCC_6H_4OC_2H_5$.
Used as fixative and solvent in perfumes.

benzyl fluoride $C_6H_5CH_2F$.
Properties: Colorless liquid. Forms acicular crystals on prolonged cooling. Sp. gr. 1.022 at 25°C; b.p. 139.8°C (753 mm); m.p. -35°C.
Derivation: By decomposing benzyltrimethylammonium fluoride.
Hazard: Toxic and irritant. Tolerance (as F): 2.5 mg per cubic meter of air.
Use: Organic synthesis.

benzyl formate $C_6H_5CH_2OOCH$.
Properties: Colorless liquid; fruity-spicy odor. Resembles benzyl acetate in many respects but differs in its greater volatility. Sp. gr. 1.083–1.087; refractive index, 1.511–1.513; b.p. 203°C; miscible with alcohols, ketones, oils, aromatic, aliphatic and halogenated hydrocarbons; insoluble in water.
Hazard: May be narcotic in high concentrations.
Uses: Perfumes; flavoring; solvent for cellulose esters.

benzylhexadecyldimethylammonium chloride. See cetalkonium chloride.

benzylhydroquinone. See para-benzyloxyphenol.

benzylidene acetone (benzalacetone; methyl styryl ketone; 4-phenyl-3-buten-2-one) $C_6H_5CHCHCOCH_3$.
Properties: Colorless crystals; odor of coumarin. Soluble in alcohol, ether, benzene, and chloroform; insoluble in water. M.p. 42°C; congealing point 39°C (min); b.p. 260–262°C; refractive index 1.5836 at 46°C; density (15/15°C) 1.0377. Combustible.
Derivation: Condensation of benzaldehyde and acetone.
Use: Organic synthesis; perfumery; fixative, flavoring.

benzylidene azine. See benzalazine.

benzylidene chloride. Legal label name for benzyl dichloride.

2-benzylidene-1-heptanol. See alpha-amylcinnamic alcohol.

benzyl iodide $C_6H_5CH_2I$.
Properties: Colorless crystals or liquid. Soluble in alcohol, carbon disulfide and ether; insoluble in water. Sp. gr. 1.7335; m.p. 34.1°C; b.p., decomposes.

Derivation: Interaction of benzyl chloride and hydriodic acid.
Hazard: Moderately toxic and irritant.

benzyl isoamyl ether. See isoamyl benzyl ether.

benzyl isobutyl ketone. See 4-methyl-1-phenyl-2-pentanone.

benzyl isoeugenol (1-alpha-phenyl-4-propenylveratrole) $CH_3CHCHC_6H_3(OCH_3)OCH_2C_6H_5$.
Properties: White crystalline solid; floral odor of the carnation type. Soluble in alcohol and ether. Combustible.
Use: Perfumery; fixative.

N-benzylisopropylamine $C_6H_5CH_2NH(CH_3CHCH_3)$.
Properties: Colorless to yellow liquid; sp. gr. 0.895 (25°C); refractive index 1.4995–1.5015 (25°C). Combustible.
Uses: Rust inhibitor; intermediate.

benzyl isothiocyanate $C_6H_5CH_2NCS$. Colorless to slightly yellow liquid; a lachrymator.
Hazard: Toxic; irritating to tissues.
Use: Chemical intermediate.

benzyl mercaptan. See benzyl thiol.

benzylmethylamine $C_6H_5CH_2NHCH_3$.
Properties: Colorless to light yellow liquid. Sp. gr. 0.936 (25°C); refractive index 1.5185–1.5220 (25°C); distillation range 183–188°C. Combustible.
Uses: Organic synthesis.

N-benzyl-N,N-methylethanolamine $C_6H_5CH_2NCH_3(C_2H_4OH)$.
Properties: Colorless to light yellow liquid; sp. gr. 1.006 (27°C); refractive index 1.5250–1.5270 (25°C); distillation range 95–105°C (2 mm). Combustible.
Uses: Corrosion inhibitor; intermediate.

3-benzyl-4-methyl umbelliferone $C_6H_5CH_2CH_3C_9H_4O_3$.
Properties: Tan crystalline powder; m.p. 255°C min; slightly soluble in ethyl alcohol; insoluble in water.
Use: Optical whitening agent; intermediate.

para-**benzyloxyphenol** (benzylhydroquinone; "Agerite Alba") $C_6H_5CH_2OC_6H_4OH$.
Properties: Light tan powder; sp. gr. 1.26; faint odor; m.p. 121–122°C. Slightly soluble in water; practically insoluble in petroleum hydrocarbons; very soluble in benzene and alkalies. Low toxicity. Combustible.
Use: Rubber antioxidant; stabilizer; polymerization inhibitor; chemical intermediate.

benzyl pelargonate $C_6H_5CH_2OOCC_8H_{17}$. Liquid; sp. gr. 0.962 (15.5/15.5°C); b.p. 315°C; mild odor. Used in flavors and perfumes; bactericides and fungicides; organic synthesis.

para-**benzylphenol** (4-hydroxydiphenylmethane) $C_6H_5CH_2C_6H_4OH$.
Properties: White crystals from ethyl alcohol; m.p. 84°C; b.p. 320–322°C. Soluble in ethyl alcohol, ether, chloroform, benzene, acetic acid, caustic alkalies; moderately soluble in hot water. Combustible.
Hazard: Toxic by ingestion.
Uses: Antiseptic and germicide; organic synthesis.

benzyl phenylacetate $C_6H_5CH_2COOCH_2C_6H_5$.
Properties: Colorless liquid; honeylike odor. Soluble in alcohol; sp. gr. 1.097–1.099; refractive index 1.554–1.556. Combustible.
Uses: Perfumery and flavors.

benzyl phenyl ketone. See deoxybenzoin.

benzyl propionate $C_2H_5COOCH_2C_6H_5$. Similar to benzyl acetate but has sweeter odor.
Properties: Liquid. B.p. 220°C; sp. gr. 1.036 (17.5°C); insoluble in water. Combustible.
Grades: Technical; F.C.C.
Uses: Perfumes; flavoring.

benzylpyridine $C_6H_5CH_2C_5H_4N$.
Properties: Liquid. Boiling point 276.8°C; m.p. 13.6°C; sp. gr. (20°C) 1.061; refractive index 1.5797. Insoluble in water. Combustible.

benzyl salicylate $C_6H_4(OH)COOCH_2C_6H_5$.
Properties: Colorless liquid; faint sweet odor. Soluble in 9 vols. of 90% alcohol. Sp. gr. 1.176–1.179; refractive index 1.580–1.581; m.p. min 24°C; b.p. (26 mm) 208°C. Combustible. Low toxicity.
Grades: Technical; F.C.C.
Uses: Perfume fixative; solvent for synthetic musk; sun-screening lotions; soap odorant.

benzyl succinate (dibenzyl succinate) $C_6H_5CH_2OOCCH_2CH_2COOCH_2C_6H_5$.
Properties: White crystalline powder, almost tasteless. Soluble in alcohol, ether, chloroform, also in fixed and volatile oils; insoluble in water. M.p. 45°C. Combustible.

benzyl sulfide $(CH_2C_6H_5)_2S$.
Properties: Colorless plates. Soluble in alcohol and ether; insoluble in water. Sp. gr. 1.0712; m.p. 49°C.
Derivation: Action of potassium sulfide on benzyl chloride and subsequent distillation.
Use: Organic synthesis.

benzyl thiocyanate $C_6H_5CH_2CNS$.
Properties: Colorless crystals; m.p. 41°C; b.p. 230°C; insoluble in water; soluble in alcohol and ether. Low toxicity.
Containers: 5-25 lb drums.
Hazard: Moderate fire hazard.
Use: Insecticide.

benzyl thiol. (benzyl mercaptan; alpha-toluenethiol). $C_6H_5CH_2SH$.
Properties: Colorless liquid; b.p. 195°C; insoluble in water; soluble in alcohol and carbon disulfide; flash point (closed cup) 158°F; sp. gr. 1.05; strong odor; combustible.
Hazard: Toxic by inhalation; irritant to tissue.
Uses: Odorant; flavors.

2-benzyl-6-thiouracil. $C_6H_5CH_2C_4H_3N_2OS$. A drug intermediate.

benzyltrimethylammonium chloride $C_6H_5CH_2N(CH_3)_3 \cdot Cl$. A quaternary ammonium salt.
Properties: Colorless crystals; stable up to 135°C, above which benzyl chloride and trimethylamine are formed. Readily soluble in water, ethyl alcohol, and butanol; slightly soluble in butyl phthalate and tributyl phosphate. Properties of 60% solution: Sp. gr. (20/20°C) 1.07; wt/gal 8.90 lb; f.p. less than -50°C.
Grades: 60–62% aqueous solution.
Containers: Bottles; carboys; 50-gal coated drums.
Uses: Solvent for cellulose; gelling inhibitor in polyester resins; intermediate.

benzyltrimethylammonium hexafluorophosphate $C_6H_5CH_2N(CH_3)_3PF_6$. Crystals; m.p. 160°C.
Hazard: Probably toxic by ingestion, and irritant to skin.

4-benzyltrimethylammonium methoxide $C_6H_5CH_2(CH_3)_3NOCH_3$. A quaternary ammonium salt.
Properties: Yellow liquid; decomposes on distillation.
Hazard: Probably toxic by ingestion, and irritant to skin.
Uses: Catalyst; organic-soluble strong base.

benzyne C_6H_4. An unsaturated cyclic hydrocarbon with a structure similar to benzene, in which one of the double bonds is replaced by a triple bond. It may be prepared from benzenediazonium-2-carboxylate or from isatoic anhydride.

"Beraloy."[155] Trademark for beryllium-copper alloys supplied in two grades: "Beraloy" A (1.80–2.05% beryllium) and "Beraloy" D (1.60–1.80% beryllium) "Beraloy" A meets A.S.T.M. Specifications B-194-51T and B-197-51T.
Properties: High electrical conductivity; high resistance to fatigue; very low hysteresis or drift; easily formed when annealed; high strength and rigidity when heat treated; corrosion resistant.
Forms: Strip, round wire, flat wire.
Uses: Diaphragms; springs; fabrication of lightweight, intricate parts.

berberine $C_{20}H_{19}NO_5$.
Properties: White to yellow crystals; m.p. 145°C (anhydrous); insoluble in water; soluble in ether and alcohol. Salts of berberine are: berberine bisulfate, berberine sulfate, and berberine hydrochloride. All three are yellow crystals, slightly soluble in water.
Derivation: From the root of Berberis vulgaris or Hydrastis canadensis.
Hazard: Toxic by ingestion, inhalation, and skin absorption.
Use: Medicine, in form of alkaloid, its sulfate or hydrochloride.

bergamot oil.
Properties: An essential oil. Brownish-yellow to green liquid; agreeable odor; bitter taste. Linalyl acetate content 34–45%. Soluble in alcohol and glacial acetic acid; affected by light. Sp. gr. (25/25°C) 0.875–0.880; optical rotation +8° to +24°; refractive index 1.4650–1.4675 (20°C). Chief constituents: Limonene; linalyl acetate; linalool.
Derivative: By expression from the fruits of citrus bergamia Risso et Poiteau.
Grades: N.F.; Italian.
Containers: Cans.
Uses: Perfumery; flavoring.

berkelium Bk Synthetic radioactive element with atomic number 97 first produced (1949) as the 243 isotope by bombarding americium with helium ions in a cyclotron. The chemical properties of berkelium have been studied by tracer techniques and are similar to those of the other transuranium elements. Its oxidation behavior is similar to that of the rare earth cerium. It has a m.p. of 986°C. There are 8 isotopes ranging from 243 to 250; the 249 isotope has been made by neutron bombardment of curium 244. Atomic weight uncertain; 249 is generally accepted. The following compounds have been identified by x-ray diffraction; berkelium dioxide (BkO_2), berkelium sesquioxide (Bk_2O_3), berkelium trifluoride (BkF_3), berkelium trichloride ($BkCl_3$), and berkelium oxychloride (BkOCl).

Berlin blue. Any of a number of the varieties of iron blue pigments. See iron blue.

Berlin red. A red pigment consisting, essentially, of red iron oxide.

Berthelot, M. P. E. (1827–1917). French chemist who did fundamental work on the organic synthesis of hydrocarbons, fats, and carbohydrates. Opposed the then current idea that a "vital force" is responsible for synthesis. Did important work on explosives for French government. He was one of the first to prove that all chemical phenomena depend on physical forces that can be measured.

Berthollet, Claude Louis (1748–1822). French chemist. Followed Lavoisier, but did not accept the latter's contention that oxygen is the characteristic constituent of acids. He was the first to propose chlorine as a bleaching agent. His essay on chemical physics (1803) was first attempt to explain this subject. His speculations on stoichiometry, especially as regards relative masses of reacting atoms, profoundly affected later theories of chemical affinity.

beryl. $Be_3Al_2(SiO_3)_6$ sometimes with replacement of Be by Na, Li, Cs. A natural silicate of beryllium and aluminum. See beryllium.

beryllia. See beryllium oxide.

beryllides. Intermetallic compounds made by chemically combining beryllium with such metals as zirconium and tantalum.

beryllium Be Element of atomic number 4, group IIA of the periodic system. Atomic weight 9.0121. Valence 2; no stable isotopes.
Properties: A hard, brittle, gray-white metal; sp. gr. 1.85; m.p. 1280°C; soluble in acids (except nitric) and alkalies. Resistant to oxidation at ordinary temperatures. High heat capacity and thermal conductivity. It is the lightest structural metal known; can be fabricated by rolling, forging, and machining. Joining is chiefly by shrink-fitting; brazing and welding are difficult. Highly permeable to x-rays.
Occurrence: Beryl, the ore of beryllium, is found chiefly in South Africa, Rhodesia, Brazil, Argentina and India. Principal sources in U.S. are Colorado, Maine, New Hampshire, and South Dakota. There are undeveloped deposits in Canada.
Derivation: The ore is converted to the oxide or hydroxide, then to the chloride or fluoride. The halide may be (a) reduced in a furnace by magnesium metal, or (b) reduced by electrolysis. (c) Liquid-liquid extraction with an organophosphate chelating agent can be used as a method of purification, or as an alternate process on the ore itself.
Grades: Technical; over 99.5% pure.
Forms: Hot-pressed or cold-pressed and sintered blocks; sheet (0.04 inch); tube; rods; wire; powder.
Hazard: Highly toxic, especially by inhalation of dust. Tolerance, 0.002 mg per cubic meter of air.
Uses: Structural material in space technology; moderator and reflector of neutrons in nuclear reactors; special windows for x-ray tubes; in gyroscopes, computer parts, inertial guidance systems; additive in solid propellant rocket fuels; Be-Cu alloys (q.v.).
Shipping regulations: (flake or powder) (Air) Poison label. (Rail) Poisonous solids, n.o.s., Poison label.

beryllium acetate, basic $Be_4O(C_2H_3O_2)_6$.
Properties: White crystals; m.p. 285–286°C; b.p. 330–331°C; insoluble in water; hydrolyzed by hot water and dilute acids. Soluble in chloroform and other organic solvents. Can be crystallized from hot acetic acid in very pure form.
Hazard: Highly toxic by inhalation and ingestion.
Use: Source of pure beryllium salts.
Shipping regulations: (Rail, Air) Beryllium compounds, n.o.s., Poison label.

beryllium acetylacetonate $Be(C_5H_7O_2)_2$. Crystalline powder; slightly soluble in water; resistant to hydrolysis. A chelating nonionizing compound. M.p. 108°C; b.p. 270°C; freely soluble in alcohol and ether; slightly soluble in water.
Hazard: Highly toxic by inhalation and ingestion.
Shipping regulations: (Rail, Air) Beryllium compounds, n.o.s., Poison label.

beryllium carbide Be_2C.
Properties: Fine hexagonal, hard, refractory crystals; attacked vigorously by strong, hot alkali solutions forming methane gas and alkali beryllate. Sp. gr. 1.91; decomposes above 2100°C.
Derivation: By direct interaction of elemental Be and carbon; by reduction of beryllium oxide with carbon above 1500°C.
Hazard: Highly toxic by inhalation.
Use: Atomic energy applications; experimental rocket fuels.
Shipping regulations: See beryllium acetate.

beryllium carbonate (basic beryllium carbonate) $BeCO_3 + Be(OH)_2$.
Properties: White powder. Variable composition. Soluble in acids; insoluble in water.
Hazard: Highly toxic by inhalation and ingestion.
Shipping regulations: See beryllium acetate.

beryllium chloride $BeCl_2$.
Properties: White or slightly yellow, deliquescent crystals; sweetish taste. M.p. 440°C; b.p. 520°C; sp. gr. 1.90. Very soluble in water; soluble in alcohol, benzene, ether, carbon disulfide. Readily hydrolyzed.
Derivation: By passing chlorine over a mixture of beryllium oxide and carbon.
Hazard: Highly toxic by inhalation and ingestion.
Shipping regulations: (Rail, Air) Poison label.

beryllium copper. A precipitation-hardenable alloy; often also contains nickel or cobalt, and has relatively high electrical conductivity, high strength, and high hardness.
Properties: Specific gravity 8.22. Tensile strength of heat-treated sheet 175,000 psi, elongation 5% in 2 inches; Brinell hardness; 350; good electrical conductivity. Typical analysis: copper 97.4; beryllium 2.25; nickel 0.35.
Hazard: Moderately toxic as powder. Avoid inhalation.
Uses: In electrical switch parts; watch springs; optical alloys; electronic equipment; valves and parts; spot-welding electrodes; nonsparking tools; springs and diaphragms; shims; cams; and bushings.
Note: A comparatively recent development is an 85 copper, 9 nickel, 6 tin alloy reported to be 15% stronger than Be-Cu.

beryllium fluoride BeF_2.
Properties: Hygroscopic solid; m.p. 800°C; sp. gr. 1.986. Readily soluble in water; sparingly soluble in alcohol.
Derivation: By the thermal decomposition (at 900–950°C) of ammonium beryllium fluoride.
Hazard: Highly toxic by inhalation and ingestion. Tolerance (as F), 2.5 mg per cubic meter of air.
Use: Production of beryllium metal by reduction with magnesium metal.
Shipping regulations: See beryllium acetate.

beryllium hydrate. See beryllium hydroxide.

beryllium hydride BeH_2. White solid.
Properties: Reacts with water, dilute acids, methyl alcohol, liberating hydrogen. When heated to 220°C it liberates hydrogen rapidly.
Hazard: Fire risk when exposed to water, organic materials, and heat. Highly toxic.
Uses: Experimentally in rocket fuels.
Shipping regulations: See beryllium acetate.

beryllium hydroxide (beryllium hydrate) $Be(OH)_2$.
Properties: White powder; decomposed to the oxide at 138°C; soluble in acids, alkalies; insoluble in water.
Derivation: By precipitation with alkali from pure beryllium acetate, basic.
Grades: Technical.
Hazard: Highly toxic by inhalation and ingestion.
Shipping regulations: (Rail, Air) Poison label.

beryllium metaphosphate $Be(PO_3)_2$.
Properties: White porous powder or granular material; has a high melting point; insoluble in water.
Hazard: Highly toxic by inhalation and ingestion.
Uses: Raw material for special ceramic compositions; catalyst carrier.
Shipping regulations: See beryllium acetate

beryllium nitrate $Be(NO_3)_2 \cdot 3H_2O$.
Properties: White to faintly yellowish, deliquescent mass; m.p. 60°C; decomposes 100–200°C; soluble in water, alcohol.
Derivation: Action of nitric acid on beryllium oxide, with subsequent evaporation and crystallization; reaction of beryllium sulfate with barium nitrate.
Grades: Technical; C.P.
Uses: Chemical reagent; gas mantle hardener.
Hazard: Highly toxic by inhalation. Oxidizing material; dangerous fire risk.
Shipping regulations: (Rail) Nitrates, n.o.s., Yellow label. (Air) Oxidizer label.

beryllium nitride Be_3N_2.
Properties: Hard, refractory white crystals; m.p. 2200 ± 40°C; reacts with mineral acids to form the corresponding salts of beryllium and ammonia. Readily attacked by strong alkali solutions, liberating ammonia.
Derivation: By heating beryllium metal powder in a dry, oxygen-free nitrogen atmosphere at temperatures of 700–1400°C.
Hazard: Highly toxic by inhalation.
Uses: Atomic energy; production of the radioactive carbon isotope C^{14} for tracer uses; experimental rocket fuels.
Shipping regulations: (Rail, Air) Beryllium compounds, n.o.s., Poison label.

beryllium oxide (beryllia) BeO.
Properties: White powder. A unique ceramic material. Sp. gr. 3.016; m.p. 2570°C. Hardness (Mohs) 9. Soluble in acids and alkalies; insoluble in water. High electrical resistivity and thermal conductivity; transparent to microwave radiation; undamaged by nuclear radiation. High heat-stress resistance. Can be fabricated into finished shapes.
Derivation: By heating beryllium nitrate or hydroxide.
Grades: Technical; C.P.; pure; single crystals.
Hazard: Highly toxic by inhalation. Keep container tightly closed and flush out after use (MCA).
Uses: Electron tubes; resistor cores; windows in klystron tubes; transistor mountings; high-temperature

reactor systems; additive to glass, ceramics and plastics; preparation of Be compounds.
Shipping regulations: (Rail, Air) Poison label.

beryllium potassium fluoride (potassium beryllium fluoride) $BeF_2 \cdot 2KF$.
Properties: White, crystalline masses. Soluble in water; insoluble in alcohol.
Hazard: Highly toxic by inhalation and ingestion. Tolerance (as F), 2.5 mg per cubic meter of air.
Shipping regulations: See beryllium acetate.

beryllium silicate Be_2SiO_4. Single crystals available, spherical and prismatic.
Shipping regulations: See beryllium acetate.

beryllium sodium fluoride (sodium beryllium fluoride) $BeF_2 \cdot 2NaF$.
Properties: White, crystalline mass. Soluble in water. M.p. about 350°C.
Hazard: Highly toxic by inhalation and ingestion. Tolerance (as F), 2.5 mg per cubic meter of air.
Use: Making pure beryllium metal.
Shipping regulations: See beryllium acetate.

beryllium sulfate $BeSO_4 \cdot 4H_2O$.
Properties: Colorless crystals. Soluble in water; insoluble in alcohol. Sp. gr. 1.713; decomposes at 540°C.
Hazard: Highly toxic by inhalation and ingestion.
Shipping regulations: (Rail, Air) Poison label.

Berzelius, J. J. (1779–1848). A native of Sweden, Berzelius was one of the foremost chemists of the 19th century. He made many contributions to both fundamental and applied chemistry: he coined the words *isomer* and *catalyst*; classified minerals by chemical composition; recognized organic radicals which maintain their identity in a series of reactions; discovered selenium and thorium and isolated silicon, titanium and zirconium; did pioneer work with solutions of proteinaceous materials which he recognized as being differnt from "true" solutions.

"Besk."[204] Trademark for a high-foaming heavy duty manual cleaner for dairies and food-processing plants.

"Be Square."[128] Trademark for grades of petroleum microcrystalline wax. Melting range 170-175°, 180–185°, 190–195°F.

BET. Abbreviation for Brunauer, Emmett, and Teller, applied to an equation and method for determining the surface area of an adsorbent such as carbon.

beta- (β). A prefix having meanings analogous to those of alpha (q.v.).

(1) It indicates (a) the position of a substituent atom or radical in a compound; (b) the second position in a naphthalene ring; or (c) the attachment of a chemical unit to the side-chain of an aromatic compound.

(2) It refers to a secondary allotropic modification of a metal or compound.

(3) It designates a type of radioactive decay. See beta particle.

"Betacote"[532a] Trademark for a series of urethane prepolymers specifically designed for flame-proofing coatings on wood. Coatings 6 mils thick on Douglas fir have a flame-spread rating of 55.

"Betaflux."[250] Trademark for petalite (q.v.).

betaine hydrochloride (lycine hydrochloride) $C_5H_{11}O_2N \cdot HCl$.
Properties: Colorless crystals; m.p. 227–228°C (dec.); soluble in water and alcohol; insoluble in chloroform and ether. Aqueous solutions are strongly acid. Liberates hydrogen chloride at the melting point. Low toxicity.
Grades: Technical.
Containers: Drums.
Uses: Source of hydrogen chloride in solders and fluxes; organic synthesis; medicine (muscular degeneration).

betaine phosphate $C_5H_{11}O_2N \cdot H_3PO_4$.
Properties: White, odorless granules; acid taste; m.p. 198–200°C; very soluble in water.
Grades: Technical.
Uses: Source of phosphoric acid.

"Betalight."[529] Trademark for a self-luminous light source consisting of a phosphor-lined glass capsule into which tritium gas has been introduced under pressure. Available in numerous configurations, colors, and luminosities. Used for illumination of panels, meters, dials, instruments, etc.

"Betalin" S.[100] Trademark for thiamine hydrochloride U.S.P. (q.v.).

betamethasone (9-fluoro-11β, 17,21-trihydroxy-16β-methylpregna-1,4-diene-3,20-dione) A corticosteroid hormone.
Properties: White, odorless, crystalline powder; m.p. 240°C with some decomposition. Sparingly soluble in acetone and alcohol; insoluble in water.
Grade: N.D.; N.F.
Uses: Medicine.

"Betanol."[354] Trademark for a group of dispersing agents consisting of high molecular weight esters. They form dispersions which are stable over a wide pH range in the presence of acids and mineral salts. They can be used to prepare both water-in-oil and oil-in-water emulsions.
Uses: Cosmetics, pharmaceuticals, textiles, paints.

"Betanox" Special.[248] Trademark for low-temperature reaction product of phenyl-beta-naphthylamine and acetone.
Properties: Tan powder; sp. gr. 1.16; m.p. above 120°C; soluble in acetone, benzol, and ethylene dichloride; insoluble in water and gasoline.
Uses: Antioxidant for wire insulation, tire treads, carcass, inner tubes, dark-colored footwear, proofing, and mechanical goods.

beta particle. A charged particle emitted from a radioactive atomic nucleus, either natural or manmade. Their energies range from 0 to 4 MeV. They carry a single charge; if this is negative, the particle is identical with an electron; if positive; it is a positron. Beta rays (streams of these particles) may cause skin burns, and are harmful if they enter the body. Protection can be afforded by a thin sheet of metal. See also electron; decay, radioactive.

"Betasan"[1] (N-beta-O, O-diisopropyl dithiophosphoryl ethyl benzene sufonamide). Trademark for a concentrated herbicide, both liquid and granular.
Hazard: May be toxic.

betazole hydrochloride (3-(2-amioethyl)pyrazole dihydrochloride) $C_5H_9N_3 \cdot 2HCl$.
Properties: White, crystalline, nearly odorless powder; pH of 5% solution 1.5; m.p. not higher than 240°C. Soluble in water; practically insoluble in chloroform. Low toxicity.
Grades: U.S.P.
Use: Medicine (substitute for histamine).

betel. See arecoline.

bethanechol chloride (carbamylmethylcholine chloride) $H_2NCOOCH(CH_3)CH_2N(CH_3)_3Cl$.
Properties: Colorless hygroscopic crystals with amine-like odor. Exhibits polymorphism. Of two crystalline forms, one melts about 211° and the other 219°C. Very soluble in water; freely soluble in alcohol; practically insoluble in chloroform, benzene, and ether. Stable in air; pH (1% solution) 5.5–6.5.
Derivation: Propylene chlorohydrin is treated with phosgene and then with ammonia in ether. The product is heated with trimethylamine.
Grade: U.S.P.
Hazard: Toxic in overdose.
Use: Medicine (intestinal disorders).

Bettendorf's reagent. A reagent used for the detection of arsenic in presence of bismuth and antimony compounds. It consists of a concentrated solution of stannous chloride in fuming hydrochloric acid.
Hazard: Highly toxic; strong irritant to tissue.

Betterton-Kroll process. A process for obtaining bismuth and purifying desilverized lead that contains bismuth. Metallic calcium or magnesium is added to the molten lead to cause formation of high melting intermetallic compounds with bismuth. These separate as a surface scum and are skimmed off. The excess calcium and magnesium are removed from the lead by use of chlorine gas as mixed molten chlorides of lead or zinc. Bismuth of 99.995% purity is produced in this way.

Betts process. An electrolytic process for removing impurities from lead, in which pure lead is deposited on a thin cathode of pure lead, from an anode containing as much as 10% of silver, gold, bismuth, copper, antimony, arsenic, selenium, and other impurities. The electrolyte is lead fluosilicate and fluosilicic acid. The scrap anodes and the residues of impurities associated with them are either recast into anodes or treated to recover antimonial lead, silver, gold, bismuth, etc.

betula oil. See methyl salicylate.

"Beutene."[248] Trademark for a butyraldehyde-aniline reaction product.
Properties: Reddish-brown, free-flowing liquid; sp. gr. 0.95; soluble in acetone, benzol, and ethylene dichloride; slightly soluble in gasoline; insoluble in water.
Uses: Rubber accelerator.

BFE. Abbreviation for bromotrifluoroethylene (q.v.).

BF_3-ether complex. See boron trifluoride etherate.

BF_3-MEA. See boron trifluoride monoethylamine.

BF_3-MeOH. See boron trifluoride-methanol.

BFPO. Abbreviation for bis(dimethylamino)fluorophosphine oxide. See dimefox.

BHA. Abbreviation for butylated hydroxyanisole (q.v.).

BHC. Abbreviation for benzene hexachloride (q.v.).

BHT. Abbreviation for butylated hydroxytoluene. See 2,6-di-tert-butyl-para-cresol.

Bi Symbol for bismuth.

bi- Prefix meaning two; di- is preferred in chemical nomenclature; exceptions are bicarbonate, bisulfate, bitartrate, in which it indicates the presence of hydrogen in the molecule, i.e., $NaHCO_3$ (sodium bicarbonate). See also bis-.

bialamicol hydrochloride (USAN) (6,6'-diallyl-alpha,-alpha'-bis(diethylamino-4,4'-bi-ortho-cresol dihydrochloride) $C_{28}H_{440}N_2O_2 \cdot 2HCl$. Crystals; m.p. 209–210°C; soluble in water.
Use: Medicine.

bibenzyl. See diphenylethane, symmetrical.

bicalcium phosphate. See calcium phosphate, dibasic.

bicarburetted hydrogen. See ethylene.

"Bicillin."[24] Trademark for benzathine penicillin G.

bicyclic. An organic compound in which two (only) ring structures occur. They may or may not be the same type of ring. See naphthalene.

bicyclohexyl (dicyclohexyl) $C_{12}H_{22}$.
Properties: Colorless, mobile liquid with pleasant odor. B.p. 238.5°C; f.p. 1 to 3°C; sp. gr. 0.883 (25/16°C); wt/gal 7.37 lb; refractive index (n 20/D) 1.480; flash point 165°F. Combustible. Autoignition temp. 471°F.
Derivation: Hydrogenation of diphenyl (q.v.).
Hazard: May be toxic.
Uses: High-boiling solvent and penetrant.

"Bidrin."[125] Trademark for dimethyl phosphate of 3-hydroxy-N,N-dimethyl-cis-crotonamide. $(CH_3O)_2P(O)OC(CH_3){:}CHC(O)N(CH_3)_2$.
Properties: Brown liquid with a mild ester odor; b.p. 400°C; miscible in water and xylene; slightly soluble in kerosine and diesel fuel. Commercially available water-miscible solution.
Uses: Insecticide.

biformin. An antibiotic produced by the fungus Polyporus biformis, reported to be active against various bacteria and fungi.

"B-I-K."[248] Trademark for a surface-coated urea.
Properties: White powder; sp. gr. 1.32; melting range 129–134°C; soluble in water. Surface coating not soluble in water, but is soluble in rubber. Slightly soluble in acetone; insoluble in benzene, gasoline and ethylene dichloride.
Uses: Promoter for azodicarbonamide, a nitrogen blowing agent; activator for thiazoles, sulfenamides, and thiurams; odor reducer when used with nitrosoamine-type blowing agents.

"Bikalith."[88] Trademark for a series of lithium silicate ores including lepidolite, petalite, spodumene, and eucryptite. Used in glass-making and ceramics.

bile acid. An acid found in bile (the secretion of the liver). Bile acids are steroids, having a hydroxyl group, and a five-carbon-atom side chain terminating in a carboxyl group. Cholic acid is the most abundant bile acid in human bile. Others are deoxycholic and lithocholic acids. The bile acids do not occur free in bile but are linked to the amino acids, glycine and taurine. These conjugated acids are water-soluble; their salts are powerful detergents and as such, aid in the absorption of fats from the intestine.

bilirubin (bilifulvin) $C_{32}H_{36}O_6N_4$. Red coloring matter of bile. Chemical structure related to hemoglobin.
Properties: Orange-red powder; m.p. 192°C; soluble in acids, alkalies, chloroform and benzene; insoluble in water; very slightly soluble in alcohol and ether.
Derivation: From bile pigment.
Use: Analytical chemistry.

bimetal. A combination of two metals in which one metal is clad on another. Bimetallic couples of metals with different expansion coefficients are used as temperature indicators and control instruments.

binapacryl. Generic name for 2-sec-butyl-4,6-dinitrophenyl 3-methyl-2-butenoate, $CH_3CH_2CH(CH_3)C_6H_2(NO_2)_2OOCCHC(CH_3)CH_3$.
Hazard: Probably toxic by ingestion and inhalation.
Use: Acaricide and fungicide.

binary. Descriptive of a system containing two and only two components. Such a system may be a chemical compound composed of two elements, an element and a group (hydroxyl, methyl, etc.), or two groups e.g., oxalic acid; it may also be a two-component solution or alloy.

bind. To exert a strong physiochemical attraction, as often occurs between various proteins and water in hydrophilic gels, between organic dyes and fabrics, or between acids or bases and various chemical complexes.

binder. The film-forming ingredient in paint, usually either a drying oil or a polymeric substance; a resin used in foundry casting.
See also paint.

binding energy. The binding energy of a nucleus is the minimum energy required to dissociate it into its component neutrons and protons. Neutron or proton binding energies are those required to remove a neutron or a proton, respectively, from a nucleus. Electron binding energy is that required to remove an electron from an atom or a molecule. (Atomic Energy Commission). See also mass defect; fission; ionization.

bioassay. See assay.

"Biobate."[173] Trademark for an enzymatic preparation for use in bating in the leather industry.

"Biocheck."[108] Trademark for a family of biocides, fungicides, and slimicides.
Containers: 55 gal. drum.
Uses: Controlling and eliminating microbiological growth in pulp and paper mill water systems. Also for antibacterial papers.

biochemical oxygen demand (B.O.D.). A standardized means for estimating the degree of contamination of water supplies, especially those which receive contamination from sewage and industrial wastes. (See also sewage sludge, biodegradability). It is expressed as the quantity of dissolved oxygen (in mg/liter) required during stabilization of the decomposable organic matter by aerobic biochemical action. Determination of this quantity is accomplished by diluting suitable portions of the sample with water saturated with oxygen and measuring the dissolved oxygen in the mixture both immediately and after a period of incubation, usually five days. See also dissolved oxygen (D.O.) and oxygen consumed (C.O.D.) as related terms.

biochemistry. Originally a subdivision of chemistry, but now an independent science, biochemistry includes all aspects of chemistry that apply to living organisms. Thus photochemistry is directly involved with photosynthesis (q.v.) and physical chemistry with osmosis (q.v.)—two phenomena that underlie all plant and animal life. Other important chemical mechanisms that apply directly to living organisms are catalysis, which takes place in biochemical systems by the agency of enzymes (q.v.); nucleic acid and protein constitution and behavior, which is known to control the mechanism of genetics; colloid chemistry, which deals in part with the nature of cell walls, muscles, collagen, etc.; acid-base relations, involved in the pH of body fluids; and such nutritional components as amino acids, fats, carbohydrates, minerals, lipids, and vitamins, all of which are essential to life. The chemical organization and reproductive behavior of microorganisms (bacteria and viruses) and a large part of agricultural chemistry are also included in biochemistry. See also nutrition, cell, nucleic acid.

biocide. General name for any substance that kills or inhibits the growth of microorganisms such as bacteria, molds, slimes, fungi, etc. Many of them are also toxic to humans. Biocidal chemicals include chlorinated hydrocarbons, organometallics, halogen-releasing compounds, metallic salts, organic sulfur compounds, quaternary ammonium compounds, and phenolics. See also antiseptic, disinfectant, fungicide.

bicolloid. An aqueous colloidal suspension or dispersion produced by or within a living organism. Blood, milk, and egg yolk are examples.

bioconversion. Utilization of animal manures, garbage and similar organic wastes for production of fuel gases by digestion or destructive distillation. See also biogas.

biocytin $C_{16}H_{28}N_4O_4S$ (epsilon-N-biotinyl-L-lysine). A naturally occurring complex of biotin isolated from yeast. Water-soluble crystals; m.p. 228.5°C. It is believed to be an intermediate in the utilization of biotin by animal organisms.

biodegradability. The susceptibility of a substance to decomposition by microorganisms; specifically, the rate at which detergents and pesticides and other compounds may be chemically broken down by bacteria and/or natural environmental factors. Branched chain alkylbenzene sulfonates (ABS) are much more resistant to such decomposition than are linear alkylbenzene sulfonates (LAS), in which the long straight alkyl chain is readily attacked by bacteria (q.v.). If the branching is at the end of a long alkyl chain (isoalkyls), the molecules are about as biodegradable as the normal alkyls. The alcohol sulfate anionic detergents and most of the nonionic detergents are biodegradable. Among pesticides, the organophosphorus types, while highly toxic are more biodegradable than DDT and its derivatives. Tests on a number of compounds gave results as follows: Easily biodegraded: n-propanol, ethanol, benzoic acid, benzaldehyde; ethyl acetate. Less easily biodegraded: ethylene glycol, isopropanol, orthocresol, diethylene glycol, pyridine, triethanolamine. Resistant to biodegradation: aniline, methanol, monoethanolamine, methyl ethyl ketone, acetone. Additives that accelerate biodegradation of polyethylene, polystyrene, and other plastics are available.

bioethics. An interdisciplinary science for which research facilities were established in 1971; it encompasses the ethical and social issues resulting from advances in medicine and the biosciences; its scope includes a number of areas of importance to chemistry, e.g., reproductive and genetic phenomena, organ transplants, gerontology and antiaging techniques, biological warfare, contraception, etc. The Kennedy Institute at Georgetown University, Washington, D.C., is the chief center for information about this developing aspect of biomedical science.

bioflavonoid. A group of naturally occurring substances thought to maintain normal conditions in the walls of the small blood vessels. The bioflavonoids are widely distributed among plants, especially citrus fruits, black currants and rose hips (hesperidin, rutin, quercitin). They have little or no medicinal value.

biogas. Methane produced by digestion of manure in small-scale units on farms. Pilot-scale production is planned (from cow dung) in Texas and Oklahoma for use in home heating.

biogeochemistry. A branch of geochemistry dealing with the interactions between living organisms and their mineral environment. It includes among other studies that of the effect of plants on the weathering of rocks; of the chemical transformations that produced petroleum and coal; of the concentration of specific elements in vegetation at some time in the geochemical cycle (iodine in sea plants, uranium in some forms of decaying organic matter); and of the organic constituents of fossils.

bioluminescence. See chemiluminescence.

biomas. (1) As used by fuel and energy technologists, this term refers to the total uncontrolled plant growth in a given area; it is considered as a possible source of energy by chemical, bacterial, and enzymic conversion, after present supplies of fossil fuels have been exhausted. It is estimated that 5% of world biomass (which totals about 145 billion tons) could be converted to enough synthetic fuel to satisfy world requirements, and that those of the U.S. could be met by conversion of the biomass obtained by intensive cultivation of about 6% of its continental land area. The term also includes water-growing plants such as the various forms of algae.

(2) A more specific definition, used by marine technologists, is the amount of living matter, both plant and animal, that is present in a given segment of ocean floor, expressed in grams per square meter. This is called the standing crop. By far the highest values occur in shallow water, e.g., in a Pacific coastal area (littoral) biomass has been determined to be from 1000 to 5000 g/sq m, compared with 200 g/sq m at a depth of 200 m and 5 g/sq m at 4000 m.

biomaterial. Any material used as a surgical implant within the body to replace or support tissue; these include metals, ceramics, plastics, elastomers, carbon in various forms, bone derivatives, and the like. They must be entirely compatible with the interior environment, be nondegradable, and completely nontoxic; they must also duplicate as closely as possible the properties of the tissues they replace. Numerous materials used in dental surgery are included.

"bioMeT 12."[288] Trademark for a rodent-repellent coating for cables in which the active ingredient is a mixture of tributyl tin salts. It has proved 95% effective in preventing destruction of telephone cables by rats and other rodents. Flexible, transparent, and effective for six months or more, it is applied mechanically over the plastic cable sheathing. Can also be used to protect other types of wiring, shipping containers, and similar products.

"Bionol."[307] Trademark for a series of cationic bactericidal agents composed of 50% alkyldimethyl benzyl ammonium chloride.
Uses: Disinfectant; germicide; deodorant.

"Biopal" VRO-20.[307] Trademark for an iodophor, consisting of a solution of 20% available iodine in alkylphenoxypoly(ethyleneoxy)ethanol.
Properties: Brownish-black liquid; readily soluble in water. Slowly decreases in activity in the presence of light, and should be packed in dark amber bottles or in suitably lined metal containers. Has both detergent and germicidal properties.
Hazard: Toxic by ingestion.
Uses: Bactericide, sporocide, fungicide, and protozoacide in hard or soft water and at high or low temperatures.

biopolymer. A water-soluble polymer resulting from the action of bacteria (genus Xanthomonas) on carbohydrates. The viscosity is almost as low as that of water; chromium ion can be added to increase viscosity, if desired. Such polymers are being used as viscosity builders in oil-well drilling muds, and as thickeners and gel strength additives in aqueous media. See also fermentation.

biophyl. A highly refined form of verxite (q.v.), in excess of 99% pure hydrobiotite (magnesium-iron-aluminum silicate). Used in foods and pharmaceuticals.

biosynthesis. Synthesis of edible single-cell proteins by a fermentation process; the basic ingredients of one process are a specially developed yeast strain, oxygen or air, ethyl alcohol or other carbonaceous substance such as acetic acid or paraffin hydrocarbons, nitrogen source, and an inorganic salt solution. These are fermented at 40°C, centrifuged, and dried. Other methods are also under development. See also amino acid; protein.

biotin (vitamin H) $C_{10}H_{16}N_2O_3S$. 2′-Keto-3,4-imidazolido-2-tetrahydrothiophene-n-valeric acid. Biotin, frequently referred to as a member of the vitamin B complex, is necessary for the maintenance of health in animals and for growth of many microorganisms. It influences fat metabolism, decarboxylation and carbon dioxide fixation and deamination of some amino acids. It is closely related metabolically to pantothenic acid and folic acid. A biotin deficiency may be induced by ingestion of avidin (q.v.), a raw-egg protein, because of the formation of a nonabsorbable biotin-avidin complex. Biotin is synthesized in the intestinal tract of humans; therefore, normally not required.
Properties: White crystals; m.p. 230–232°C; soluble in water and alcohol; insoluble in naphtha and chloroform; stable to heat; stable in neutral or acid solutions; destroyed by strong alkali or oxidizing agents.
Sources: Egg yolk, kidney, liver, yeast, milk, molasses.
Units: Amounts are expressed in milligrams or micrograms of biotin.
Grades: Practical; F.C.C.
Containers: Bottles.
Uses: Medicine; nutrition.

N-biotinyl-L-lysine. See biocytin.

biotite. A component of igneous rocks and of soil, similar to mica. It is a silicate of magnesium, iron, potassium and aluminum.

biphenyl. See diphenyl.

ortho-**biphenylamine.** See ortho-aminobiphenyl.

ortho-**biphenyl biguanide** $NH_2(CNHNH)_2C_6H_4C_6H_5 \cdot H_2O$.
Properties: White to faintly pink powder; m.p. above 150°C on dried material; ash not over 0.5%. Soluble in alcohol, "Carbitol" and "Cellosolve"; very slightly soluble in water.
Containers: 50-lb paper bags; 150-lb fiber drums.
Use: Soap antioxidant.

birefringent. Descriptive of a type of crystal which separates an impinging light ray into two components that are at right angles to each other; as a result, two images appear, each of which is caused by a light ray vibrating in only one direction (plane-polarized light). Such anisotropic crystals (Iceland spar) are used in nicol prisms. See also nicol; anisotropic.

bis-. Prefix meaning "twice" or "again." Used in chemical nomenclature to indicate that a chemical grouping or radical occurs twice in a molecule, e.g., bisphenol A, where two phenolic groups appear $(CH_3)_2C(C_6H_5OH)_2$. See also following entries.

2,2-bis(acetoxymethyl)propyl acetate $CH_3C(CH_2OOCCH_3)_2CH_2OOCCH_3$. Colorless liquid; n(20/D) 1.4359. Used as a plasticizer.

bisacodyl (4,4′-(2-pyridylmethylene)diphenol diacetate; bis-(para-acetoxyphenyl)-2-pyridylmethane) $(CH_3COOC_6H_4)_2CHC_5H_4N$.
Properties: Tasteless crystals; m.p. 38°C. Practically insoluble in water and alkaline solutions. Soluble in acids, alcohol, acetone, propylene glycol and other organic solvents.
Grade: N.D.
Use: Medicine.

bisamides. General formula RCONHR′NHCOR. When R and R′ have high molecular weights they are hard, light-colored waxes.

bis(2-aminoethyl)sulfide $(H_2NCH_2CH_2)_2S$. An ethyleneimine derivative.
Properties: Colorless liquid; f.p. 2.6°C; b.p. 238°C; density 8.7 lb/gal; refractive index 1.5277; flash point 246°F; very soluble in water, benzene and ethanol. Combustible.
Uses: See ethyleneimine.

N,N-bis(3-aminopropyl)methylamine $CH_3N(C_3H_6NH_2)_2$.
Properties: Liquid; sp. gr. 0.9307 (20/20°C); b.p. 240°C; f.p. −29.6°C; flash point 220°F. Completely soluble in water. Combustible.
Hazard: Moderately toxic and irritant.
Uses: Chemical intermediate.

1,3-bis(2-benzothiazolylmercaptomethyl)urea (1,3-bis[2-benzothiazolylthio)methyl]urea) $(C_6H_4NCS \cdot SCH_2NH)_2CO$.
Properties: Buff to light tan powder; m.p. 220°C; sp. gr. 1.38 (25°C.
Use: Rubber accelerator.

para-**bis[2-(5-para-biphenylyloxazoyl)]benzene** (BOPOB) $C_{36}H_{25}O_2N_2$.
Properties: Shiny yellow flakes; m.p. 327–328°C; fluorescence peak 4400A; sparingly soluble in toluene.
Grade: Purified.
Use: Scintillation counter; wave length shifter in liquid scintillators.

2,2-bis(para-**bromophenyl)-1,1,1-trichloroethane.** $C_{14}H_9Br_2Cl_3$. The bromine analog of DDT.
Hazard: Probably toxic by skin absorption.
Use: Insecticide.

bis(tert-**butylperoxy)-2,5-dimethylhexane** $CH_3C(CH_3)(OOC_4H_9)CH_2CH_2C(CH_3) \cdot (OOC_4H_9)CH_3$. A cross-linking agent for polymers. See "Varox".

bis(2-chloroethoxy)methane. See dichloroethyl formal.

4-[para-**[bis(2-chloroethyl)amino]phenyl]butyric acid.** See chlorambucil.

5-[bis(2-chloroethyl)amino]uracil (uracil mustard) CONHCONHCHCN$(C_2H_4Cl)_2$ (ring from first C to last C). A cream-white, odorless, crystalline compound; moderately soluble in methanol and acetone. Used in medicine.

bis(2-chloroethyl) ether. See dichloroethyl ether.

bis(chloromethyl)ether.
Properties: Reported to form spontaneously from formaldehyde and chloride ions in moist air.
Hazard: Carcinogenic to animals in low concentrations; a known occupational carcinogen.
Uses: Intermediate for ion-exchange resins; laboratory reagent.

1,1′-bischloromercuriferrocene [1,1-di(chloromercuri)-ferrocene]). $(ClHgC_5H_4)_2Fe$.
Hazard: Highly toxic by ingestion.
Uses: Inorganic polymers.
Shipping regulations: (Rail, Air) Mercury compounds, solid (n.o.s.), Poison label.

3,3-bis(chloromethyl)oxetane. See "Penton."

bis(para-**chlorophenoxy)methane** $(ClC_6H_4O)_2CH_2$.
Properties: Solid; m.p. 65°C; insoluble in water and oils; soluble in ether and acetone.
Hazard: Moderately toxic by ingestion.
Use: Acaricide.

2,2-bis(para-**chlorophenyl)-1,1-dichloroethane.** See TDE.

1,1-bis(para-**chlorophenyl)ethanol.** See di(para-chlorophenyl)ethanol.

1,1′-bis(para-**chlorophenyl)-2,2,2-trichloroethanol** (4,4′-dichloro-alpha-trichloromethylbenzhydrol) (Kelthane) $CCl_3C(C_6H_4Cl)_2OH$. An alcohol analog of DDT.
Hazard: Toxic by inhalation and ingestion.
Use: Miticide.

bis-cyclopentadienyliron. See dicyclopentadienyliron.

bis(2,6-diethylphenyl)carbodiimide $(CH_2H_5)_2C_6H_3NCNC_6H_3(C_2H_5)_2$.
Properties: Light yellow to red-brown liquid with faintly acrid odor; sp. gr. 1.007 (20/20°C); b.p. 192–194°C (0.4 mm); n(23/D) 1.591. Soluble in inert organic solvents. Combustible.
Hazard: Toxic by inhalation. Damaging to eyes.
Uses: Stabilizers in polyester and urethane systems; intermediate for textile chemicals and pharmaceuticals.

2,6-bis(dimethylaminomethyl)cyclohexanone $[(CH_3)_2NCH_2]_2C_6H_8O$. Sp. gr. 0.95 (68°F).
Uses: Preservative for aqueous paint systems, casein, pigment dispersions and adhesives.

2,6-bis(dimethylaminomethyl)cyclohexanone dihydrochloride $C_{12}H_{24}N_2O \cdot 2HCl$. Free-flowing, white to off-white crystalline salt.
Uses: Preservative for aqueous systems, latex paints, adhesives, coatings, wax emulsions, casein and starch solutions.

bis(1,3-dimethylbutyl)amine $[(CH_3)_2CH_2CH_2CH(CH_3]_2NH$.
Properties: Liquid; sp. gr. 0.772–0.778 (20/20°C); distillation range 179.0–205°C; 6.5 lb/gal; flash point 160°F. Combustible.

N,N′-bis(1,4-dimethylpentyl)-para-**phenylenediamine** (diheptyl-para-phenylenediamine) $C_6H_4(NHC_7H_{15})_2$.
Properties: Amber to red liquid; sp. gr. 0.90; solidification temperature 45°F. Combustible.
Uses: Gasoline antioxidant and sweetener.

1,3-bisethylaminobutane $C_2H_5NHCH_2CH_2CHNH(C_2H_5)CH_3$. Water-white amine, boiling range 179–185°C. Combustible.

N,N′-bis(1-ethyl-3-methylpentyl)-para-**phenylenediamine** $C_6H_4(NHC_8H_{17})_2$.
Properties: Dark reddish brown liquid; sp. gr. 0.90 approx; wt/gal 7.5 lb; combustible.
Use: Antioxidant for polyunsaturated elastomers.

2,2-bis(para-**ethylphenyl)-1,1-dichloroethane.** See 1,1-dichloro-2,2-bis(para-ethylphenyl)ethane.

bisethylxanthogen $(C_2H_5OCSS)_2$.
Properties: Yellow needles; onion-like odor; m.p. 28–32°C. Insoluble in water; freely soluble in benzene, ether, petroleum fractions.
Grades: 58% solution in oil.
Containers: 5-gal cans; 30-gal drums.
Uses: Weed control; rubber accelerator; fungicide.

1,3-bis(3-glycidoxypropyl)tetramethyldisiloxane
$[\overline{OCH_2CHC}H_2O(CH_2)_3Si(CH_2)_2]_2O$.
Properties: Liquid; sp. gr. 0.99 (25°C); n(25°C) 1.4500; b.p. about 185°C (2 mm). Soluble in acetone and benzene; insoluble in water. Flash point 300°F. Combustible.
Containers: Up to 55-gal drums.
Use: Chemical intermediate.

bishydroxycoumarin [3,3′-methylenebis(4-hydroxycoumarin)dicoumarol] $C_{19}H_{12}O_6$.
Properties: White or creamy-white, crystalline powder; faint pleasant odor and slightly bitter taste; m.p. 287–293°C. Readily soluble in solutions of fixed alkali hydroxides; slightly soluble in chloroform; practically insoluble in water, alcohol and ether.
Derivation: (a) Originally, extracted from spoiled sweet clover; (b) synthetically from methyl acetylsalicylate, sodium, and formaldehyde.
Grade: U.S.P.
Hazard: High dosage may cause hemorrhage.
Use: Anticoagulant for blood.

bis(1-hydroxycyclohexyl) peroxide $C_6H_{10}OH)_2O_2$.
Properties: Fine white powder; m.p. 66–68°C. Active oxygen 6.6% min.
Hazard: Dangerous fire risk in contact with organic materials. Strong oxidizing agent.
Use: Catalyst for polymerization of polyester resins.
Shipping regulations: (Rail) Oxidizing material, n.o.s., Yellow label. (Air) Peroxide, organic, solid, n.o.s., Organic Peroxide label.

N,N-bis(2-hydroxyethyl)alkylamine. Clear liquid used as an antistatic for blow molding applications for polyolefins. Approved for use in food packaging films.

bis(hydroxyethyl) butynediol ether $HO(CH_2)_2OCH_2C \cdot CCH_2O(CH_2)_2OH$.
Properties: Dark brown liquid; sp. gr. 1.136 (25/15°C); solidifies at < -15°C; distillation range 116–235°C (10 mm).
Hazard: May decompose with explosive violence under alkaline conditions at high temperature.
Uses: Intermediate for polyesters, plasticizers, and plastics; nickel brightener in electroplating; corrosion inhibitor; pickling inhibitor prior to copper plating.

bis(hydroxyethyl) cocoamine oxide. A derivative of coconut oil (q.v.) claimed to be useful as a gasoline additive to inhibit rust formation and iceing of carburetors; also used as foaming agent in shampoos, detergents, tooth pastes and the like.

N,N-bis(hydroxyethyl)oleamide $CH_3(CH_2)_7HC{:}CH(CH_2)_7CON(CH_2CH_2OH)_2$.
A technical grade containing 25% excess amine. Light-amber liquid with faint odor. Used as a surface-active agent.

beta-**bishydroxyethyl sulfide.** See thiodiglycol.

1,3-bishydroxymethylurea. See dimethylolurea.

4,4-bis(4-hydroxyphenyl)pentanoic acid (diphenolic acid; DPA) $CH_3C(C_6H_4OH)_2CH_2CH_2COOH$.
Properties: Light tan granules; m.p. 170–173°C; sp. gr. 1.30–1.32. Soluble in acetic acid, acetone, and ethanol; insoluble in benzene, carbon tetrachloride, and xylene. Slightly soluble in water.
Uses: Paint formulations; coatings and finishes.

bishydroxyphenyl sulfone. See dihydroxydiphenyl sulfone.

1,4-bis(2-hydroxypropyl)-2-methylpiperazine $C_{11}H_{24}O_2N_2$.
Properties: Liquid; b.p. 145°C at 3 mm Hg; sp. gr. 1.0013 (25/25°C); n(20/D) 1.4803; flash point 300°F; miscible in water. Combustible; odor-free.
Uses: Catalyst; chemical intermediate.

2,4-bis(isopropylamino)-6-methoxy-s-triazine $[(CH_3)_2CHNH]_2C_3(OCH_3)$. White extruded pellets.
Hazard: Toxic by ingestion.
Use: Industrial herbicide.

bis-keto-triazine ("Permafresh" 110). Water-white liquid; 40% active. Can be cured with magnesium chloride to provide chlorine resistance; compatible with optical brighteners. Used as a wash/wear finish for cellulosics and blends of cellulosics.

"Bismanol" (MnBi). An alloy or compound of bismuth and manganese which has an exceptionally high coercive force (3400 oersteds) and a high energy product. Produced by U.S. Naval Ordnance Laboratory by powder metallurgy methods, and used as a permanent magnet.

Bismarck Brown R
$CH_3C_6H_3[NNC_6H_2(CH_3)(NH_2)_2]_2 \cdot 2HCl$.
Toluene-2,4-diazo-bis-meta-toluylenediamine hydrochloride.
Properties: Dark brown powder; soluble in water and alcohol.
Derivation: Action of nitrous acid on toluylene diamine.
Use: Dye for wool and leather; biological stain.

Bismarck Brown Y (4,4′-[meta-phenylenebis(azo)]bis-(meta-phenylenediamine) dihydrochloride) C_6H_4-1,3-$[NNC_6H_3$-2,4-$(NH_2)_2 \cdot HCl]_2$.
Used as a dye; biological stain.

"Bismate."[69] Trademark for bismuth dimethyldithiocarbamate. (q.v.).

bis(2-methoxyethoxy)ethyl ether. See dimethoxytetraglycol.

N,N′-bis(1-methylheptyl)-para-phenylenediamine $C_6H_4[NHCH(CH_3)C_6H_{13}]_2$.
Properties: Dark reddish brown liquid; sp. gr. approx. 0.90 (80°F); wt/gal, 7.5 lb. Combustible.
Use: Antiozonant in polyunsaturated elastomers.

1,4-bis[2-(4-methyl-5-phenyloxazolyl)]benzene (dimethyl-POPOP) $C_6H_4[C_3NO(CH_3)C_6H_5]_2$. Yellow needles with a bluish fluorescence; m.p. 231–234°C. Used as a scintillation phosphor.

bismite. See bismuth trioxide.

bismuth Bi Metallic element of atomic number 83, Group VA of the periodic system. Atomic weight 208.9804. Valences 2, 3, 4. No stable isotopes; 5 naturally radioactive isotopes.
Properties: Crystalline, brittle metal, with reddish tinge. Soluble in nitric and hydrochloric acids. Highly diamagnetic (mass susceptibility -1.35×10^6). Expands 3.3% on solidification. Electrical resistivity higher in solid than in liquid state. Extrudable at 437°F; not fabricable at room temperature. Sp. gr. 9.8 at 20°C; m.p. 271°C; b.p. 1560°C; Brinell hardness 7. Thermal conductivity (0.018 cal/sec/cc at 250°C) is lowest of all metals except mercury. On heating it burns to form the oxide.
Source: (1) Metallurical by-products (often lead bullion) obtained chiefly from smelting ores of lead, silver, copper and gold; (2) ores used chiefly for their bismuth and one or two other metals, as tin and tungsten. See also bismuthinite, and cosalite.
Derivation: Debismuthizing of lead bullion by (a) fractional crystallization, (b) electrolytic (Betts) refining, or (c) addition of calcium or magnesium (Betterton-Kroll process) which removes bismuth.
Purification: By addition of molten caustic, zinc, and finally chlorine (to make removable chlorides of the impurities).
Impurities: Lead, iron, copper, arsenic, antimony, selenium.
Forms: Rods, wire, lump, powder.
Grades: 99.5 + % pure; high purity (less than 10 ppm impurities) (single crystals).
Hazard: Flammable in powder form.
Uses: Pharmaceuticals and medicine; cosmetics (eye polish, lipstick, etc.);component of low-melting (fusible) alloys; catalyst in making acrylonitrile; additive to improve machinability of steels and other metals; coating selenium; thermoelectric materials; permanent magnets; semiconductors.

bismuth ammonium citrate
Properties: Pearly, shining, transparent scales or white powder; slightly acid, metallic taste; composition varies. Soluble in water; slightly soluble in alcohol.
Derivation: Interaction of bismuth subnitrate, citric acid and ammonium hydroxide.
Use: Medicine.

bismuth antimonide BiSb. Single crystals used as semiconductors.

bismuth bromide (bismuth tribromide) $BiBr_3$.
Properties: Yellow, crystalline powder. Hygroscopic. Decomposed by water with formation of bismuth oxybromide. Soluble in ether, hydrochloric acid (dilute), solutions of potassium iodide, potassium bromide and potassium chloride; insoluble in alcohol. Sp. gr. 5.7; b.p. 453°C; m.p. 218°C.

bismuth carbonate, basic. See bismuth subcarbonate.

bismuth chloride (bismuth trichloride) $BiCl_3$.
Properties: White, very deliquescent crystals; volatilized by heat. Soluble in acids; insoluble in alcohol; decomposes in water to the oxychloride. Sp. gr. 4.56; m.p. 227&°C; b.p. decomposes at 300°C.
Derivation: Action of hydrochloric acid on bismuth.
Containers: Bottles; drums.
Use: Bismuth salts.

bismuth chloride, basic. See bismuth oxychloride.

bismuth chromate (basic dichromate) $Bi_2O_3 \cdot {}_2CrO_3$.
Properties: Orange-red amorphous powder; soluble in alkalies and acids; insoluble in water.
Derivation: Interaction of bismuth nitrate and potassium chromate.
Hazard: Toxic by ingestion and inhalation. Tolerance (as CrO_3), 0.1 mg per cubic meter of air.
Use: Pigment.

bismuth citrate $BiC_6H_5O_7$
Properties: White powder; sp. gr. 3.458; soluble in ammonia or alkali citrates; insoluble in water; slightly soluble in alcohol; m.p. decomposes.
Derivation: Boiling bismuth subnitrate with citric acid.
Use: Medicine.

bismuth dimethyldithicarbamate $Bi[(CH_3)_2NC(S)S]_3$.
Properties: Lemon yellow powder; sp. gr. 2.04; melts above 230°C with decomposition; soluble in chloroform; slightly soluble in benzene, carbon disulfide; insoluble in water.
Use: Accelerator for natural rubber and SBR, especially in cable covers and mechanical items.

bismuth ditannate. See bismuth tannate.

bismuth ethyl camphorate $C_{36}H_{57}BiO_{12}$.
Properties: Solid; faint aromatic odor; m.p. 61–67°C. Insoluble in water; soluble in ether, chloroform and oils.
Use: Medicine.

bismuth ethyl chloride $BiHC_2H_5Cl$. White powder.
Hazard: Ignites spontaneously in air; dangerous fire risk.
Shipping regulations: (Rail, Air) Not listed. Consult authorities.

bismuth gallate, basic. See bismuth subgallate.

bismuth glance. See bismuthinite.

bismuth hydrate. See bismuth hydroxide.

bismuth hydroxide (bismuth hydrate; bismuth oxyhydrate; bismuth trihydroxide; bismuth trihydrate; hydrated bismuth oxide) $Bi(OH)_3$.
Properties: White amorphous powder. Soluble in acids; insoluble in water. Sp. gr. 4.36.
Derivation: Action of sodium hydroxide on a solution of bismuth nitrate.
Containers: Glass bottles; tins; drums.
Use: Bismuth salts.

bismuthinite (bismuth glance) Bi_2S_3, may contain copper or iron.
Properties: Lead-gray mineral, often with yellow tarnish, metallic luster. Contains 81.2% bismuth, 18.8% sulfur. Soluble in nitric acid. Sp. gr. 6.4–6.5; Mohs hardness 2.
Occurrence: Utah; Bolivia; Mexico.
Use: Ore of bismuth.

bismuth iodide (bismuth triiodide) BiI_3.
Properties: Grayish-black, metallic, glistening crystals. Soluble in alcohol, hydriodic acid and potassium iodide; insoluble in water; decomposes in hot water; sp. gr. 5.778 (15°C); m.p. 408°C.
Derivation: By the interaction of bismuth and iodine.
Hazard: Moderately toxic by ingestion.
Uses: Analytical chemistry; bismuth oxyiodide.

bismuth beta-**naphthol** $Bi_2O_2(OH) \cdot C_{10}H_7O$.
Properties: Brown to gray powder, almost insoluble in water or other solvents.
Derivation: By treating a solution of sodium beta-naphtholate with acetic acid solution of bismuth nitrate, and adding caustic soda solution to neutralize excess acid.
Use: Medicine.

bismuth nitrate (bismuth ternitrate: bismuth trinitrate) $Bi(NO_3)_3 \cdot 5H_2O$.
Properties: Lustrous, clear, colorless, hygroscopic crystals; acid taste. Soluble in dilute nitric acid, alcohol and acetone; slowly decomposed by water to the subnitrate. Sp. gr. 2.83; b.p., decomposes 75–80°C.
Derivation: Action of nitric acid on bismuth with subsequent recovery by evaporation and crystallization.
Containers: Bottles; tins; drums; multiwall paper sacks.
Hazard: Oxidizing material; fire risk near organic materials.
Uses: Preparation of other bismuth salts; bismuth luster on tin; luminous paints and enamels; medicine.
Shipping regulations: (Rail) Nitrates, n.o.s., Yellow label. (Air) Nitrates, n.o.s., Oxidizer label.

bismuth nitrate, basic. See bismuth subnitrate.

bismuth oleate.
Properties: Yellowish-brown, soft, granular mass. Soluble in ether; insoluble in water.
Derivation: A mixture of bismuth trioxide and oleic acid.
Use: Medicine.

bismuth oxide. See bismuth trioxide; bismuth tetraoxide.

bismuth oxide, hydrated. See bismuth hydroxide.

bismuth oxycarbonate. See bismuth subcarbonate.

bismuth oxychloride (bismuth chloride, basic; bismuth subchloride) BiOCl.
Properties: White, lustrous crystalline powder. Sp. gr. 7.717. Soluble in acid; insoluble in water.
Derivation: By action of water on bismuth chloride; interaction of dilute nitric acid solution of bismuth nitrate with sodium chloride.
Containers: Drums.
Uses: Medicine; cosmetics; pigment.

bismuth oxyhydrate. See bismuth hydroxide.

bismuth oxynitrate. See bismuth subnitrate.

bismuth pentafluoride BiF_5.
Properties: Crystals; sublimes at 550°C.
Hazard: Reacts violently with water and petrolatum above 50°C. Toxic; strong irritant to eyes and skin. Tolerance, 2.5 mg per cubic meter of air.
Uses: Fluorinating agent.

bismuth pentoxide Bi_2O_5. An acid anhydride; its salts have not been prepared in pure state. Made by oxidation of bismuth trioxide, giving a scarlet precipitate.

bismuth phenate (bismuth phenolate; bismuth phenylate) $C_6H_5O \cdot Bi(OH)_2$.
Properties: Grayish-white powder; odorless and tasteless. Insoluble in water, alcohol and ether.
Derivation: Interaction of bismuth nitrate and sodium phenolate.
Grades: Technical, 80% Bi_2O_3.
Use: Medicine.

bismuth phenosulfonate (bismuth sulfocarbolate).
Properties: Pale, reddish powder. Slightly soluble in water.
Derivation: By the interaction of bismuth hydroxide and phenolsulfonic acid.
Uses: Medicine; antiseptic.

bismuth phenylate. See bismuth phenate.

bismuth potassium iodide $BiI_3 \cdot 4KI$.
Properties: Red crystals. Decomposed by water. Soluble in potassium iodide solution.
Use: Medicine.

bismuth-potassium tartrate. See potassium-bismuth tartrate.

bismuth pyrogallate (basic bismuth pyrogallate) $C_6H_3(OH)_2OBiOH$.
Properties: Yellowish-green amorphous powder; odorless; tasteless. Soluble in dilute acid and alkaline solutions; insoluble in water and alcohol. Decomposed by acids.
Derivation: Action of pyrogallic acid on bismuth carbonate.
Grades: Technical, 60% Bi_2O_3.
Use: Medicine.

bismuth salicylate, basic. See bismuth subsalicylate.

bismuth stannate $Bi_2(SnO_3)_3 \cdot 5H_2O$.
Properties: Light colored crystalline powder. Insoluble in water. Approximate temperature of dehydration is 140°C.
Hazard: Tolerance (as Sn) 2 mg/cu m of air.
Uses: Component of ceramic capacitors, especially useful with barium titanate.

bismuth subcarbonate (bismuth oxycarbonate; bismuth carbonate, basic) $(BiO)_2CO_3$ or $Bi_2O_3 \cdot CO_2$, with one-half H_2O.
Properties: White, odorless powder; tasteless; insoluble in water and alcohol; soluble in nitric or hydrochloric acid with effervescence. Sp. gr. 6.86. Stable in air, but slowly affected by light.
Derivation: By adding ammonium carbonate to a solution of a bismuth salt.
Grades: Technical; C.P.; U.S.P. (90% Bi_2O_3 min).
Uses: Bismuth compounds; cosmetics; treatment of gastric ulcer; opacifier in x-ray diagnosis; enamel fluxes and ceramic glazes.

bismuth subchloride. See bismuth oxychloride.

bismuth subgallate (basic bismuth gallate) $C_6H_2(OH)_3COOBi(OH)_2$.
Properties: Saffron-yellow powder; odorless and tasteless. Soluble in dilute alkalies; insoluble in water, alcohol and ether. Stable in air, but affected by light.
Derivation: Interaction of bismuth nitrate, glacial acetic acid, and gallic acid in aqueous solution.
Use: Medicine (treatment of alimentary canal).

bismuth subnitrate (basic bismuth nitrate; bismuth oxynitrate) $4BiNO_3(OH)_2 \cdot BiO(OH)$.
Properties: White, heavy, slightly hygroscopic powder. which shows acid to moistened litmus paper.

Soluble in acids; insoluble in water and alcohol. Sp. gr. 4.928; m.p. decomposes at 260°C.
Derivation: Hydrolysis of bismuth nitrate, filtering and drying.
Grades: Technical; C.P.; N.F.
Containers: Multiwall paper bags; 5- to 250-lb drums.
Hazard: See bismuth nitrate.
Uses: See bismuth subcarbonate.

bismuth subsalicylate (basic bismuth salicylate) $Bi(C_7H_5O_3)_3Bi_2O_3$.
Properties: White, bulky crystalline powder; tasteless; odorless; soluble in acids and alkalies; insoluble in water, alcohol and ether. Stable in air but affected by light.
Derivation: By treating freshly prepared bismuth hydroxide with salicylic acid.
Use: Medicine.

bismuth sulfate $Bi_2(SO_4)_3$.
Properties: White needles or powder. Contains 68.5% (approx.) bismuth. Sp. gr. 5.08; decomposes at 405°C. Soluble in hydrochloric acid (dilute), nitric acid (dilute); insoluble in alcohol, water.
Use: Medicine.

bismuth sulfide (bismuth trisulfide) Bi_2S_3.
Properties: Blackish-brown powder. Soluble in nitric acid; insoluble in water. Sp. gr. 7.6–7.8; m.p., decomposes.
Derivation: (a) By melting bismuth and sulfur together. (b) By passing hydrogen sulfide into a solution of a bismuth salt. (c) Occurs as the mineral bismuthinite.
Use: Bismuth compounds.

bismuth sulfocarbolate. See bismuth phenolsulfonate.

bismuth tannate (bismuth ditannate).
Properties: Light brownish-yellow powder containing about 36% bismuth. Insoluble in water and alcohol; soluble in mineral acids.
Derivation: From freshly prepared bismuth hydroxide and tannin.
Use: Medicine.

bismuth tartrate $Bi_2(C_2H_4O_6)_3 \cdot 6H_2O$.
Properties: White powder. Decomposes with loss of water at 105°C. Sp. gr. 2.59 at 20°C. Soluble in water.
Use: Medicine (treatment of syphilis), often with potassium or sodium as part of compound.

bismuth telluride (bismuth tritelluride) Bi_2Te_3.
Properties: Gray hexagonal platelets. M.p. 573°C; sp. gr. 7.642.
Derivation: Stoichiometric combination of the elements.
Grades: Ingots, single crystals.
Hazard: Toxic; tolerance, 10 mg per cubic meter of air; (if selenium-doped, 5 mg per cubic meter of air).
Use: Semiconductors for thermoelectric cooling and power generation applications.

bismuth tetraoxide Bi_2O_4.
Properties: Heavy, yellowish-brown powder. Soluble in acids; insoluble in water. Sp. gr. 5.6; m.p. 305°C.
Derivation: By further oxidation of bismuth trioxide.
Use: Bismuth salts.

bismuth tribromide. See bismuth bromide.

bismuth trichloride. See bismuth chloride.

bismuth trihydrate. See bismuth hydroxide.

bismuth trihydroxide. See bismuth hydroxide.

bismuth triiodide. See bismuth iodide.

bismuth trinitrate. See bismuth nitrate.

bismuth trioxide (bismuth oxide; bismuth yellow; bismite). Bi_2O_2.
Properties: Heavy, yellow powder. Soluble in acids; insoluble in water. Sp. gr. 8.8; m.p. 820°C.
Derivation: Heating bismuth nitrate in air; ignition of bismuth hydroxide.
Uses: Enameling cast iron; ceramic and porcelain colors.

bismuth trisulfide. See bismuth sulfide.

bismuth tritelluride. See bismuth telluride.

bismuth yellow. See bismuth trioxide.

bismuth zirconate Bi_2ZrO_5. Used in electronics.

para**bis[2-5(5-alpha-naphthyloxazolyl)]benzene** (NOPON) $C_{32}H_{20}O_2N_2$.
Properties: Crystals; m.p. 215–217°C.
Grade: Purified.
Use: Scintillation counter.

bis(3-nitrophenyl) disulfide. See nitrophenide.

bisphenol A (4,4′-isopropylidenediphenol; 2,2-bis(4-hydroxyphenol)propane). $(CH_3)_2C(C_6H_4OH)_2$.
Properties: White flakes with a mild phenolic odor. B.p. 220°C (4 mm); freezing point 153°C; sp. gr. 1.195 (25/25°C); flash point 175°F; insoluble in water; soluble in alcohol and dilute alkalies; slightly soluble in carbon tetrachloride. Combustible.
Derivation: Condensation reaction of phenol and acetone catalyzed by HCl at 150°F.
Containers: 50-lb bags; 400-lb metal drums.
Hazard: May be toxic by ingestion.
Uses: Intermediate in manufacture of epoxy, polycarbonate, phenoxy, polysulfone and certain polyester resins; flame retardants; rubber chemicals.

1,4-bis-2-(5-phenyloxazoyl)benzene (POPOP). $(C_6H_5C_3HNO)_2C_6H_4$.
Properties: Light yellow, cottony needles; m.p. 245–246°C; fluorescence max. 4200 A; solubilities g/100 g at 25°C: water 0.00; 95% ethanol 0.00; toluene 0.12; hexane 0.02. Combustible.
Grade: Purified.
Containers: Glass bottles.
Use: Band-shifter in scintillation counting.

bis(tetrachloroethyl) disulfide $C_4H_2Cl_8S_2$.
Properties: Liquid; sp. gr. (23.3°C) 1.785; b.p. 185°C (3 mm). Soluble in benzene, hexane, ethanol. Combustible.
Hazard: May be toxic by ingestion and inhalation.
Uses: Agricultural chemicals; additives.

bis(tri-n-butyltin) oxide$(C_4H_9)_3SnOSn(C_4H_9)_3$.
Properties: Slightly yellow liquid; b.p. 180°C (2 mm); f.p. less than −45°C; sp. gr. 1.17 (25°C); flash point, above 212°F (TCC); viscosity 4.8 centistokes at 25°C; practically insoluble in water; miscible with organic solvents. Forms compounds with cellulosic and lignin-containing materials not easily decomposed or dissolved in water. Combustible.
Derivation: Hydrolysis of tributyl tin chloride.
Hazard: Toxic by ingestion and inhalation. Tolerance (as Sn), 0.1 mg per cubic meter of air.
Uses: Fungicide and bactericide, especially in underwater and antifouling paints (q.v.); pesticide.

bis(trichlorosilyl)ethane (1,1,1,4,4,4-hexachloro-1,4-disilabutane). $Cl_3SiCH_2CH_2SiCl_3$.
Properties: Colorless liquid. B.p. 202.9°C); flash point (COC) 190°F. Readily hydrolyzed with liberation of hydrochloric acid. Combustible.
Derivation: Reaction of acetylene and trichlorosilane in presence of a peroxide catalyst.

Grades: Technical.
Hazard: Corrosive when exposed to moisture.
Use: Intermediate for silicones.

bistridecyl phthalate $C_6H_4(COOC_{13}H_{27})_2$.
Properties: Liquid; sp. gr. 0.9497 (25°C); boiling range 280–290°C (4 mm); f.p. -35°C; flash point 485°F; refractive index 1.483 (25°C). Combustible.
Use: Primary plasticizer for most PVC resins.

bithionol [bis(2-hydroxy-3,5-dichlorophenyl) sulfide; 2,2′-thiobis(4,6-dichlorophenol)] $HOCl_2C_6H_2SC_6H_2Cl_2OH$.
Properties: White or grayish-white crystalline powder, m.p. 187°C. Odorless or with slight aromatic or phenolic odor. Insoluble in water; freely soluble in acetone, alcohol and in ether; soluble in chloroform and dilute solutions of fixed alkali hydroxides.
Grade: N.F.
Containers: 50-, 200-lb drums.
Hazard: Skin irritant; may not be used in cosmetics (FDA).
Uses: Medicine; deodorant; germicide; fungistat; pharmaceuticals.

4,4′-bi-ortho-**tolylene diisocyanate.** See 3,3′-dimethyl-4,4′-biphenylene diisocyanate.

bitter almond oil. See oil of bitter almond.

bittern. The solution of bromides, magnesium, and calcium salts that remains after sodium chloride has been crystallized by concentration of sea water or brines.

"Bitumastic."[11] Trademark for refined coal tar-based water-resistant protective coatings including hot-applied impermeable enamels and industrial coatings.

bitumen. A mixture of hydrocarbons occurring both in the native state and as residue from California petroleum distillation. Soluble in carbon disulfide. Solid to viscous, semisolid liquids. Used in hot-melt adhesives, coatings, paints, sealants, roofing and road-coating. Bitumens are found in asphalt, mineral waxes and in lower grades of coal. Combustible. See also asphalt, gilsonite, glance pitch.

bituminous coal. See coal.

biuret (allophanamide; carbamylurea) $NH_2CONHCONH_2 \cdot H_2O$.
Properties: White needles; odorless; m.p. 190°C with decomposition; soluble in water and alcohol; very slightly soluble in ether. Loses water of crystallization at about 110°C.
Derivation: From urea by heat.
Methods of purification: Crystallization.
Use: Analytical chemistry.

bixin. See annatto.

Bk Symbol for berkelium.

"B-K Chlorine-Bearing Powder."[204] Trademark. Calcium hypochlorite powder containing 50% available chlorine.
Use: Dairy bactericide in plants and on farms, restaurants; disinfectant and deodorant.

"B-Kleer."[204] Trademark for a heavy-duty caustic alkali; contains polyphosphate and wetting agent in fused flake form.
Uses: Machine bottle washing; H.T.S.T. Pasteurizer; vacuum pan cleaning.

"BL-60."[1] Trademark for a modified form of sodium aluminum phosphate specifically designed for use in cakes made with lactylated type shortening.
Grade: Technical; F.C.C.
Uses: Leavening agent in prepared cake mixes.

black. Any of several forms of finely divided carbon, either pure or admixed with oils, fats, or waxes. See acetylene black, animal black, bone black, carbon black, etc.

black, aniline. See aniline black.

black antimony. See antimony trisulfide.

black ash. (1) (papermaking) The product obtained by heating black liquor in furnaces. The organic material is reduced to carbon. The alkaline components are leached out and used again in papermaking. The carbon may be treated to obtain activated carbon. (2) See barium sulfide.

blackbody. In radiation physics, an ideal blackbody is a theoretical object that absorbs all the radiant energy falling upon it and emits it in the form of thermal radiation. The power radiated by a unit area of a blackbody is given by Planck's radiation law, and the total power radiated is expressed by the Stefan-Boltzmann law.

black, bone. See bone black.

black cyanide. A mixture containing 45% calcium cyanide, made from calcium cyanamid by heating it with salt and carbon.
Hazard: Highly toxic.
Shipping regulations: See calcium cyanide.

black lead. See graphite.

black liquor. (1) The liquor resulting from cooking pulpwood in an alkaline solution in the soda or sulfate (kraft) papermaking process. It is a source of lignin and tall oil, and is said to be effective in removal of mercury from industrial effluents (USDA).
(2) Iron acetate liquor (black mordant).

black oil. See residual oil.

black phosphorus. See phosphorus.

black plate. Thin sheet steel obtained by rolling and usually used for containers. It is not coated with any metal but a special lacquer or baked enamel finish is usually applied by the manufacturer.

black, platinum. See platinum black.

black powder (blasting powder). A low explosive composed of potassium nitrate, charcoal and sulfur. In some cases sodium nitrate is substituted for potassium nitrate. Typical proportions are 75, 15 and 10%. Gunpowder is probably the oldest variety.
Hazard: Sensitive to heat, and deflagrates rapidly. Does not detonate, but is a dangerous fire and explosion hazard.
Uses: Time fuses for blasting and shell; in igniter and primer assemblies for propellants; pyrotechnics; mining and blasting.
Shipping regulations: (Rail) Not accepted by rail express. Consult regulations. (Air) Not accepted.

black rouge. See iron oxide, black.

black sand. A deposit of dark minerals with a high specific gravity found in stream beds and on beaches. Magnetite and ilmenite are usually present, as also monazite and other minerals.

blackstrap. See molasses.

blanc fixe. Precipitated barium sulfate (q.v.).

"Blancol."[307] Trademark for an anionic dispersing agent, composed of the sodium salt of a sulfonated naphthalene condensate; 90% active.
Uses: Dispersing agent for pigments, earths, and solids in water; peptizing agent in insecticide formulations; in the paper industry for slime control, preventing coagulation of pitch, reducing two-sidedness, to improve sizing, etc.; in the leather industry as a bleaching, dispersing, leveling, and neutralizing agent.

"Blancophor."[307] Trademark for optical whitening agents.
FFG. A coumarin derivative. Used as whitening agent for wool, nylon, acetate rayon and mixed fibers.
HS Brands. Stilbene derivatives; used on cellulosic fibers, cotton and rayon fabrics, paper, and in household and industrial detergents.

"Blandol."[45] Trademark for white mineral oil, N.F.
Uses: Pharmaceutical and cosmetic formulations; plasticizers; paper penetrants; foam depressants.

"Blankit I."[307] Trademark for a bleaching and reducing agent composed of a stabilized, buffered hydrosulfite product; 67% active.
Containers: Must be stored in air-tight, dry containers protected from heat and moisture.
Uses: Bleaching agent for wool, jute, linen and hemp; textile stripping agent.

blast-furnace dust. Dust deposited by or recovered from blast-furnace gases. It contains a variable amount of potash which renders it valuable.

blast-furnace gas. By-product gas from smelting iron ore, obtained by the passage of hot air over the coke in the blast furnaces. A typical gas will analyze 12.9% carbon dioxide, 26.3% carbon monoxide, 3.7% hydrogen, 57.1% nitrogen.
Hazard: Dangerous by inhalation. See carbon monoxide.
Uses: Heating blast-furnace stoves; boiler- or gas-engine fuel.

blasting agent. See black powder; ammonium nitrate; explosive, high, permissible, and low.

blasting gelatin. (SNG) A type of dynamite containing about 7% of nitrocellulose.
Hazard: High explosive.
Shipping regulations: (Air) Not accepted. (Rail) Consult regulations.

blasting powder. See black powder.

"B-L-E."[248] Trademark for high-temperature reaction product of diphenylamine and acetone.
Properties: Dark-brown viscous liquid; sp. gr. 1.087; soluble in acetone, benzene, and ethylene dichloride; insoluble in gasoline and water. Combustible.
Uses: General-purpose rubber antioxidant.

bleach. To whiten a textile or paper by chemical action; also the agent itself. Bleaching agents include hydrogen peroxide (the most common), sodium hypochlorite, sodium peroxide, sodium chlorite, calcium hypochloride, hypochlorous acid, and many organic chlorine derivatives. Chlorinated lime is a bleaching powder used on an industrial scale; household bleaching powders are sodium perborate and dichlorodimethylhydantoin.
Hazard: Some bleaching agents are toxic and strong oxidizing agents. See calcium hypochlorite; lime, chlorinated.

bleaching assistant. A material added to bleaching baths to secure more rapid and complete penetration of the bleach or improved regulation of the bleaching action, e.g., compounds of sulfonated oils and solvents, soluble pine oils, fatty alcohol salts, sodium silicate, sodium phosphate, magnesium sulfate, and borax.

bleach liquor. Calcium hypochlorite solution.

bleed. (1) Of a dye, to run. (2) To release pressure gradually, as from a valve.

"Blendene."[73] Trademark for a terpene-soap composition.
Properties: Oily liquid having a pine odor; dispersible in water, in toluene, mineral spirits, and mineral oil. Soluble (hot) in methyl alcohol, ethyl alcohol, and naphtha; partly soluble in cottonseed oil. Sp. gr. (25°C) 0.947; titer below 6°C; pH (10% dispersion) 9.2. Combustible.
Uses: Manufacture of fluid emulsions, oils, solvents, etc., for water paint base, emulsion sprays, furniture and automobile polishes.

"Blendex."[525] Trademark for synthetic resinous products prepared from a variety of copolymer combinations. They are used to modify other polymers to attain a wide range of properties.

"Blendtex."[228] Trademark for a flame-retardant, water-resistant nonwoven fabric composed of nylon, rayon, and cellulose fibers mechanically compacted into a single sheet or layer. Used chiefly for disposable hospital sheeting, clothing items, etc. See also nonwoven fabric.

blister copper. Copper (96–99% purity) produced by the reduction and smelting of copper ores. It has a blistered appearance probably caused by gas pockets. It is usually further refined electrolytically.

blister packaging. A type of packaging used widely in the food and pharmaceutical industries consisting of a hollow cavity of various shapes and capacities in which the material is enclosed. Polyester and polyethylene resins are often used.

"BLO."[307] Trademark for a solvent, gamma-butyrolactone.

block polymer. A high polymer whose molecule is made up of alternating sections of one chemical composition separated by sections of a different chemical nature, or by a coupling group of low molecular weight. An example might be blocks of polyvinyl chloride interspersed with blocks of polyvinyl acetate. Such polymer combinations are made synthetically; they depend on the presence of an active site on the polymer chain which initiates the necessary reactions. See also graft polymer; stereoblock polymer.

block polymers

blood. A complex, liquid tissue of sp. gr. 1.056 and pH of 7.35–7.45. It is comprised of erythrocytes (red cells), leucocytes (white cells), platelets, plasma proteins and serum. The plasma fraction (55–70%) is whole blood from which the red and white cells and the platelets have been removed by centrifuging. Hemoglobin (q.v.) is a protein found in the erythrocytes; it contains the essential iron atom, and functions as the transport agent for oxygen to the heart (artery) and of carbon dioxide from the heart (vein). Experimental work has been reported on the effectiveness of fluorocarbon compounds in carrying out the essential transport functions of blood, especially of the red cells.

Uses: Plasma is used to restore liquid volume and thus osmotic pressure in the body where blood loss has been extensive. Animal blood is used as a component of adhesive mixtures; in dried or powder form it is a component of fertilizers, poultry feeds, and deer repellents.

See also hemoglobin; plasma; platelet.

bloom. (1) A thin coating of an ingredient of a rubber or plastic mixture that migrates to the surface, usually within a few hours after curing or setting. Sulfur bloom in vulcanized rubber products is most common; it is harmless, but impairs the eye appeal of the product. Paraffin wax is often included purposely; when it migrates to the surface it provides an efficient barrier to sunchecking and oxidation.

(2) A piece of steel made from an ingot.

(3) An arbitrary scale for rating the strength of gelatin gels. When so used the word is capitalized.

blowing agent. A substance incorporated in a mixture for the purpose of producing a foam. One type decomposes when heated to processing temperature to evolve a gas, usually carbon dioxide, which is suspended in small globules in the mixture. Typical blowing agents of this kind are baking powder (bread and cake); sodium bicarbonate or ammonium carbonate (cellular or sponge rubber); halocarbons and methylene chloride in urethane, pentane in expanded polystyrene, and hydrazine in various types of foamed plastics. Another type is air used at room temperature as a blowing agent for rubber latex; it is introduced mechanically by whipping, after which the latex is coagulated with acid. Air is also used for this purpose in ice cream, whipped cream, and other food products, as well as in blown asphalt, and blown vegetable oils. See also foam.

blow molding. A technique originally developed in the 1950's for production of hollow thermoplastic products. It involves placing an extruded tube of the thermoplastic in a mold and applying sufficient air pressure to the inside of the tube to cause it to take on the conformation of the mold. Polyethylene is usually used, but a number of other materials are adaptable to this method, e.g., cellulosics, nylons, polypropylene and polycarbonates. It is an economically efficient process and is especially suitable for production of toys, bottles and other containers, as well as air-conditioning ducts and various industrial items. The method is not limited to hollow products; for example, housings can be made by blowing a unit and sawing it along the parting line to make two housings.

blown asphalt. See asphalt, blown.

blown oil (oxidized oil; base oil; thickened oil; polymerized oil). Vegetable and animal oils which have been heated and agitated by a current of air or oxygen. They are partially oxidized, deodorized and polymerized by the treatment, and are increased in density, viscosity and drying power. Common blown oils are castor, linseed, rape, whale, and fish oils.

Uses: Paints, varnishes, lubricants, and plasticizers.

blue copperas. See copper sulfate.

blue gas. See water gas.

blue lead. See lead sulfate, blue basic.

blueprint. See Turnbull's Blue.

blue verdigris. See copper acetate, basic.

blue vitriol. See copper sulfate.

blush. Precipitation of water vapor in the form of colloidal droplets on the surface of a varnish or lacquer film, caused by lowering of the temperature immediately above the coated surface due to solvent evaporation. This results in unsightly graying of the dried film and can be avoided by use of a less volatile solvent.

"B-Nine."[248] Trademark for succinic acid, 2,2-dimethylhydrazide; used as a growth retardant.

B.O.D. See biochemical oxygen demand.

boghead coal. A variety of bituminous or subbituminous coal resembling cannel coal in appearance and behavior during combustion. It is characterized by a high percentage of algal remains and volatile matter. Upon distillation it gives exceptionally high yields of tar and oil (ASTM).

"Bogol."[48] Trademark for a crude tall oil (q.v.).

boiled oil. See linseed oil, boiled.

boiler scale. A rocklike deposit occurring on boiler walls and tubes in which hard water has been heated or evaporated. Consists largely of calcium carbonate, calcium sulfate or similar materials, depending on the mineral content of the water. Boiler scale decreases the rate of heat transfer through the boiler and tube walls resulting in increased heating costs and shortening of boiler life. Most boiler feed water is softened (treated to remove calcium and magnesium ions) before being used. See also water, hard; zeolite.

boiling point. The temperature of a liquid at which its vapor pressure is equal to or very slightly greater than the atmospheric pressure of the environment. For water at sea level it is 212°F (100°C).

bois de rose oil. See oil bois de rose.

Boltzmann, Ludwig (1844–1906). Born in Vienna, Boltzmann was interested primarily in physical chemistry and thermodynamics. His work has importance for chemistry because of his development of the kinetic theory of gases and the rules governing their viscosity and diffusion. His mathematical expression of his most important generalizations is known as Boltzmann's Law, still regarded as one of the cornerstones of physical science.

bombardment. Impingement upon an atomic nucleus of accelerated particles such as neutrons or deuterons for the purpose of inducing fission or of creating unstable nuclei which will thus become radioactive. This operation was first accomplished with positively charged particles in the cyclotron in the early 1940's, and subsequently in nuclear reactors. Neutrons are commonly used in reactors because their lack of electrical charge permits easier penetration of the target nucleus. See also radioisotope; fission.

"Bomyl."[50] Trademark for dimethyl 3-hydroxyglutaconate dimethyl phosphate ($C_9H_{15}O_8P$); used as an insecticide in control of flies.
Properties: Insoluble in water and kerosine. Corrosive to iron, steel, and brass. Stable when stored in glass. Nonflammable.
Grades: Granular, technical, and concentrated.

BON. Abbreviation for beta-oxynaphthoic acid. See 3-hydroxy-2-naphthoic acid.

"Bonaril."[233] Trademark for a hydrolyzed polyacrylamide for use in foundry sands.

bond, chemical. An attractive force between atoms strong enough to permit the combined aggregate to function as a unit. A more exact definition is not possible because attractive forces ranging upward from zero to those involving more than 250 kcal per mole of bonds are known. A practical lower limit may be taken as 2-3 kcal per mole of bonds, the work necessary to break about 1.5×10^{24} bonds by separating their component atoms to infinite distance.

All bonds appear to originate with the electrostatic charges on electrons and atomic nuclei. Bonds result when the net coulombic interactions are sufficiently attractive. Different principal types of bonds recognized include metallic, covalent, ionic, and bridge.

Metallic bonding is the attraction of all the atomic nuclei in a crystal for the outer shell electrons which are shared in a delocalized manner among all available orbitals. Metal atoms characteristically provide more orbital vacancies than electrons, for sharing with other atoms.

Covalent bonding results most commonly when electrons are shared by two atomic nuclei. Here the bonding electrons are relatively localized in the region of the two nuclei, although frequently a degree of delocalization occurs when the shared electrons have a choice of orbitals. The conventional *single* covalent bond involves the sharing of 2 electrons. There may also be *double* bonds with 4 shared electrons, *triple* bonds, with 6 shared electrons, and bonds of intermediate multiplicity.

Covalent bonds may range from *nonpolar*, involving electrons evenly shared by the two atoms, to extremely *polar*, where the bonding electrons are very unevenly shared. The limit of uneven sharing occurs when the bonding electron(s) spends full time with one of the atoms. This makes this atom into a negative ion and leaves the other atom in the form of a positive ion. Ionic bonding is the electrostatic attraction among oppositely charged ions.

Bridge bonds involve compounds of hydrogen in which the hydrogen bears either a positive or negative charge. When hydrogen is attached by a polar covalent bond to one molecule, it may attract another molecule, bridging the two molecules together. If the hydrogen is positive, it may attract an electron pair of the other molecule. This is called a *protonic bridge*. If the hydrogen is negative, it may attract, through a vacant orbital, the nucleus of an atom of a second molecule. This called a *hydridic bridge*. Such bridges are at the lower range of bond strength, but may be very significant in their effect on the physical properties of condensed states of those substances in which they are possible. (See also hydrogen bond).

R. T. Sanderson

"Bonderite."[62] Trademark for chemical compositions for producing a corrosion-inhibiting finish on metals, preparing metal surfaces for the subsequent application of finish coats, conditioning metal surfaces to facilitate metal deformation operations.

"Bonderizing."[343] Trademark for a process for furnishing a corrosion-resisting base for paint finishes on steel, aluminum, zinc, and their alloys and diecastings. Also used as an aid in deep drawing steel and aluminum.

"Bonderlube."[62] Trademark for soap-like chemical composition for treating metal surfaces which have been pretreated with phosphatizing coating chemicals, to form a lubricant layer adapted to cold forming and to retard the formation of rust.

"Bondogen."[69,496] Trademark for oil soluble sulfonic acid of high, molecular weight with a high-boiling hydrophilic alcohol and a paraffin oil.
Properties: Dark mahogany liquid; sp. gr. 0.93; acid number 40–42. Combustible.
Uses: Plasticizer and processing aid for elastomers.

bone ash. An ash composed principally of tribasic calcium phosphate, but containing minor amounts of magnesium phosphate, calcium carbonate and calcium fluoride. Noncombustible.
Derivation: By calcining bones. A synthetic product is also available.
Uses: Cleaning and polishing; ceramics; animal feeds. The better grades are used in coating molds for copper wire, bar, slabs, and other metals.

bone black (bone char; bone charcoal). Black pigment made by carbonizing bones. Carbon content is usually about 10%. Nonflammable in bulk; nontoxic.
Containers: Multiwall paper sacks; fiber drums.
Hazard: Flammable as suspended dust.
Uses: Manufacturing activated carbon; decolorizing agent and filtering medium; cementation reagent; adsorptive medium in gas masks; paint and varnish pigment; clarifying shellac; water purification.

bone china. Ceramic tableware of high quality in which a small percentage of bone ash is incorporated. Made chiefly in England.

bone meal. A product made by grinding animal bones. Raw meal is made from bones that have not been previously steamed; if pressure-steaming has been used, the meal is called "steamed." The fertilizer grade contains from 43–55% tricalcium phosphate, 20–25% phosphoric acid, and 4–5% ammonia. The feed grade, according to Bureau of Animal Industry specifications, must contain from 65–75% tricalcium phosphate and only about 2% ammonia. Much of the latter grade is imported.
Uses: Fertilizer (raw); animal feeds (steamed).

bone oil (animal oil; Dippel's oil; hartshorn oil).
Properties: Dark brown, fixed oil; repulsive odor. Soluble in water. Sp. gr. 0.900–0.980. Combustible.
Chief constituents: Hydrocarbons, pyridine bases, and amines.
Derivation: Destructive distillation of bones or other animal substance. Note distinction from bone tallow.
Grades: Technical.
Containers: Drums; tank cars.
Hazard: Evolves ammonium cyanide when heated to 180°C.
Uses: Organic preparations; source of pyrrole; denaturant for alcohol.

bone phosphate (BPL) Phosphoric acid occurring in bones in the form of tribasic calcium phosphate (q.v.).

bone seeker. An element or radioisotope that tends to lodge in the bones when introduced into the body. Examples are fluorine, calcium, and strontium 90.

bone tallow (bone fat). Fat obtained from animal bones by boiling in water, by treating with steam under pressure, or by solvent extraction. It is a glyceryl ester.

"Bonine."[299] Trademark for meclizine hydrochloride.

BON maroon. A calcium or manganese precipitated compound of 3-hydroxy-2-naphthoic acid and 2-naphthylamine-1-sulfonic acid. See BON red.

BON red. Class name for a group of organic azo pigments made by coupling 3-hydroxy-2-naphthoic acid to various amines and forming the barium, calcium, strontium or manganese salts. They have bright shades ranging from yellow red to deep maroon; good light and heat resistance, non-bleeding in vehicles and solvents, and good opacity. They are widely used in printing inks, paints, enamels, lacquers, rubber, plastics, wall paper, textiles, floor coverings, and crayons.

boort (bort). See diamond, industrial.

BOPOB. See para-bis[2-(5-para-biphenylyloxazolyl)]-benzene.

boracic acid. See boric acid.

boral. A composite material consisting of boron carbide crystals in aluminum, with a cladding of commercially pure aluminum. Concentrations of up to 50% boron carbide can be obtained.
Uses: Reactor shields; neutron curtains; shutters for thermal curtains; safety rods; containers for fissionable material.
See also composite.

"Boralloy."[521] Trademark for boron nitride in the form of a polycrystalline material produced by a proprietory vapor deposition process.

"Boran."[169] Trademark for diaminochrysazin used in the colorimetric determination of boron.

borane. One of a series of boron hydrides (compounds of boron and hydrogen). The simplest of these, BH_3, is unstable at atmospheric pressure and becomes diborane (B_2H_6) (q.v.), a gas at normal pressures. This is converted to higher boranes (penta-, deca-, etc.) by condensation; this series progresses through a number of well-characterized crystalline compounds. Hydrides up to $B_{20}H_{26}$ exist. Most are not very stable and readily react with water to yield hydrogen; many react violently with air. As a rule, they are highly toxic. Their properties have suggested investigation for rocket propulsion, but they have not proved satisfactory for this purpose. There are also a number of organoboranes used as reducing agents in electroless nickel-plating of metals and plastics. Some of the compounds used are di- and triethylamine borane and pyridine borane. See also carborane, diborane, pentaborane, hydride.

borax (tincal; borax decahydrate) $Na_2B_4O_7 \cdot 10H_2O$. A natural hydrated sodium borate, found in salt lakes and alkali soils; also the commercial name for sodium borate (q.v.).

borax, anhydrous (borax, dehydrated; sodium borate, anhydrous) $Na_2B_4O_7$.
Properties: White, free-flowing crystals; hygroscopic; forms partial hydrate in damp air; m.p. 741°C; sp. gr. 2.367; slightly soluble in cold water; noncombustible; low toxicity.
Grades: Technical (99% $Na_2B_4O_7$); standard; fine granular form; glass, or fused.
Containers: 100-lb paper bags; boxcars.
Uses: Manufacture of glass, enamels, and other ceramic products; herbicide.

borax pentahydrate $Na_2B_4O_7 \cdot 5H_2O$.
Properties: Free-flowing powder; begins to lose water of hydration at 122°C; sp. gr. 1.815. Noncombustible; low toxicity.
Grades: Crude; technical (99.5% $Na_2B_4O_7 \cdot 5H_2O$).
Containers: 100-lb bags; bulk.
Uses: Weed killer and soil sterilant; fungus control on citrus fruits (FDA tolerance 8 ppm of boron residue).

borazole (borazine) $B_3N_3H_6$. Inorganic analog of benzene.
Properties: Colorless liquid; m.p. −58°C; b.p. 53°C; sp. gr. 0.824 at 0°C. Hydrolyzes to evolve boron hydrides.
Hazard: Toxic by ingestion and inhalation; strong irritant to tissue. Dangerous fire risk.

borazon. A boron nitride (q.v.) formed at very high pressures (85,000 atm) and temperatures (1800°C) from mixtures of boron and nitrogen, or from ordinary hexagonal boron nitride in the presence of catalysts such as lithium, calcium, magnesium, or their nitrides. Mohs hardness 10; density 3.48 g/cc. Reacts extremely slowly with water; dissolves in fused sodium carbonate. For uses, see under boron nitride.

Borcher's metal. A group of alloys of chromium with nickel and cobalt, or of chromium and iron with a small proportion of molybdenum and/or silver or gold. Heat- and corrosion-resistant.
Uses: Chemical apparatus; crucibles; pyrometer tubes; heat treating or annealing pots.

Bordeaux mixture. A fungicide and insecticide mixture made by adding slaked lime to a copper sulfate solution. It is either made by the user or bought as a powder ready for dissolving. Stabilizing agents are sometimes added to delay settling. Used especially for potato bugs and similar garden pests.
Hazard: Irritant and corrosive by ingestion.

Borden's "38."[65] Trademark for ureaform, a plant nutrient containing 38% nitrogen.

"Borester."[441] Trademark for a series of boric acid esters.

"Bor-guard."[1] Trademark for a borax-propylene glycol-butynediol condensate. A corrosion inhibitor particularly suited for brake fluids of high boiling point; also for antifreeze and hydraulic fluids.

boric acid (boracic acid; orthoboric acid) H_3BO_3.
Properties: Colorless, odorless scales or white powder; stable in air. Sp. gr. 1.4347 (15°C); m.p., indeterminate, since it loses water in stages, through metaboric acid, HBO_2, pyroboric acid, $H_2B_4O_7$, and to the oxide, B_2O_3. Soluble in boiling water, alcohol and glycerin. Noncombustible.
Derivation: (a) By adding hydrochloric acid or sulfuric acid to a solution of borax and crystallizing; (b) From weak borax brines, by extraction with a kerosine solution of a chelating agent such as 2-ethyl-1,3-hexanediol, or other polyols. Borates are stripped from the chelate by sulfuric acid.

Method of purification: Recrystallization.
Hazard: Toxic by ingestion. Use only in weak solution. Irrirant to skin in dry form.
Grades: Technical, 99.9%; C.P.; U.S.P.
Uses: Heat-resistant glass (borosilicate glass); glass fibers; porcelain enamels; boron chemicals; metallurgy (welding flux, brazing copper); fireproofing compositions for textile products such as cotton batting in mattresses; fungus control on citrus fruits (FDA tolerance 8 ppm boron residue); ointment and eye wash (water solution only); nickel electroplating baths.

boric acid ester (borate ester). Trimethyl, tri-*n*-butyl, tricyclohexyl, tridodecyl, tri-*p*-cresyl borates; trihexylene glycol biborate; compounds which are readily hydrolyzed to boric acid and the respective alcohols.
Properties: Colorless to yellow liquids; b.p. 230–350°C. Low toxicity. Combustible.
Uses: Dehydrating agents; catalysts; sources of boric oxide; special solvents; stabilizers; plasticizers or adhesion additives to latex paints; ingredients of soldering and brazing fluxes.

boric acid, ortho-. See boric acid.

boric anhydride. See boric oxide.

boric oxide (boric anhydride; boron oxide) B_2O_3.
Properties: Colorless powder or vitreous crystals; slightly bitter to taste. Soluble in alcohol and hot water; slightly soluble in cold water. Sp. gr. 1.85; m.p. approx. 450°C; b.p. 1500°C. Noncombustible.
Derivation: By heating boric acid.
Grades: Glass or fused form; powder, technical or high purity 99.99+%.
Hazard: Toxic. Tolerance, 10 mg per cubic meter of air.
Uses: Production of boron; heat-resistant glassware; fire-resistant additive for paints; electronics; liquid encapsulation techniques; herbicide.

boride. An interstitial compound of boron and another metal (transition, alkaline-earth, or rare-earth); such compounds are not stoichiometric, the boron atoms being linked together in zigzag chains, two-dimensional nets, or three-dimensional structures throughout the crystal.
Properties: Highly refractory, with m.p. from 2000 to 3000°C; Mohs hardness from 8 to 10; thermally and electrically conductive; high chemical stability; do not react with HCl or HF but are attacked by hot alkali hydroxides; color varies from gray (transition-metal) to black (alkaline-earth) to blue (rare-earth).
Derivation: (1) Sintering mixtures of metal powder and boron at 2000°C; (2) reduction of mixture of the metal oxide and boric oxide with aluminum, silicon, or carbon; (3) fused salt electrolysis; (4) vapor-phase deposition.
Uses: High-temperature service as rocket nozzles, turbines, etc.

borneol (bornyl alcohol; 2-camphanol; 2-hydroxycamphane). $C_{10}H_{17}OH$.
Properties: White, translucent lumps; sharp, camphor-like odor; burning taste; soluble in alcohol and ether; insoluble in water. Optically active in natural form; racemic form made synthetically. Sp. gr. 1.011; m.p. 208°C; b.p. 212°C. Low toxicity.
Derivation: Natural form from a species of tree in Borneo, Sumatra, etc. Synthesized from camphor by hydrogen reduction, or from alpha-pinene.
Trades: Technical.
Containers: Cans; boxes.
Hazard: Fire risk in presence of open flame.
Uses: Perfumery; esters.

Shipping regulations: (Rail) Flammable solid, n.o.s., Yellow label. (Air) Flammable Solid label.

bornyl acetate $C_{10}H_{17}OOCCH_3$.
Properties: Colorless liquid, solidifying to crystals at about 50°F; piny-camphoraceous odor. Soluble in 3 volumes of 70% alcohol; miscible with 95% alcohol and with ether. Sp. gr. 0.980–0.984; refractive index 1.463–1.465; m.p. 29°C. Combustible.
Derivation: Interaction of borneol and acetic anhydride in the presence of formic acid.
Grades: Technical; F.C.C.
Uses: Perfumery; flavoring; nitrocellulose solvent.

bornyl alcohol. See borneol.

bornyl formate $C_{10}H_{17}OOCH$.
Properties: Colorless liquid, having a piny odor. Sp. gr. 1.007–1.009. Combustible.
Grades: Technical.
Uses: Perfuming of soaps, disinfectants, and sanitary products; flavoring.

bornyl isovalerate $C_{10}H_{17}OOC_5H_9$. A constituent of valerian oil.
Properties: Limpid fluid, aromatic, valerian-like odor. Soluble in alcohol and ether; insoluble in water. Sp. gr. 0.951 at 20°C; b.p. 255–260°C. Combustible.
Use: Medicine; essential oil intermediate; flavoring.

boroethane. See diborane.

boron. B Nonmetallic element of atomic number 5; Group IIIA of the Periodic Table. Atomic weight 10.81. Valence 3; 2 stable isotopes, 11 (about 81%) and 10 (about 19%). See also boron 10.
Properties: Black, hard solid; brown, amorphous powder; crystals. Highly reactive. Soluble in concentrated nitric acid and sulfuric acid; insoluble in water, alcohol, and ether. High neutron absorption capacity. Low toxicity. Amphoteric. A plant micronutrient. Sp. gr. 2.45; m.p. 2300°C; Mohs hardness 9.3.
Sources: Borax; kernite; colemanite; ulexite.
Derivation: (a) by heating boric oxide with powdered magnesium or aluminum; (b) by vapor-phase reduction of boron trichloride with hydrogen over hot filaments (800–2000°C); (c) electrolysis of fused salts.
Forms: Filament, powder, whiskers, single crystals.
Grades: Technical (90–92%); 99% pure; high-purity crystals.
Hazard: Dust ignites spontaneously in air; severe fire and explosion hazard. Reacts exothermally with metals over 900°C; explodes with hydrogen iodide.
Uses: Special-purpose alloys; cementation of iron; neutron absorber in reactor controls; oxygen scavenger for copper and other metals; fibers and filaments in composites with metals or ceramics; semiconductors; boron-coated tungsten wires; rocket propellant mixtures; high-temperature brazing alloys. See also boron alloy; boron fiber.

boron 10. Nonradioactive isotope of boron of mass number 10.
Properties: Has marked capacity for absorbing slow neutrons, emitting a high-energy alpha particle in the process.
Derivation: Constitutes about 19% of natural boron.
Forms available: Crystalline powder; dry amorphous powder; colloidal suspension of dry amorphous powder in oil; in boron trifluoride-calcium fluoride; in potassium borofluoride; in boron trifluoride ethyl etherate; in boric acid.
Uses: Neutron counter; radiation shielding (in the form of boral, q.v.); medicine.

boron alloy. A uniformly dispersed mixture of boron with another metal or metals. Ferroboron (q.v.) usu-

ally contains 15–25% boron; manganese boron usually 60–65% manganese.
Uses: Degasifying and deoxidizing agents; to harden steel (in trace quantities); to increase conductivity of copper; turbojet engines.

boron bromide. See boron tribromide.

boron carbide B_4C.
Properties: Hard black crystals; sp. gr. 2.6; Mohs hardness 9.3; m.p. 2350°C; b.p. 3500°C. Soluble in fused alkali; insoluble in water and acids. High capture cross-section for thermal neutrons.
Derivation: Heating boron oxide with carbon in an electric furnace.
Forms: Powder, crystals, rods, fibers, whiskers.
Hazard: Avoid inhalation of dust or particles.
Uses: Abrasive powder; abrasion resister and refractory; control rods in nuclear reactors; reinforcing agent in composites for military aircraft and other special applications. See also boral.

boron chloride. See boron trichloride.

boron fiber. A vapor-deposited filament made by deposition of boron on a heated tungsten wire. Filament is 0.004 inch in diameter, while the wire is only 0.0005 inch. Tensile strength 350,000 to 450,000 psi; elastic modules 55 million psi; upper temperature limit in oxidizing atmosphere 250°C. Used in composites with epoxy resins for aircraft and space applications. They can be woven into fabrics.

boron fluoride. See boron trifluoride.

boron fuel. See rocket fuel.

boron hydride. See borane, diborane, pentaborane.

boron nitride BN.
Properties: White powder, particle size about 1 micron. M.p. 3000°C (sublimes). Graphite-like hexagonal plate structure. High electrical resistance. Compressed at a million psi, it becomes hard as diamond. Excellent heat-shock resistance; low mechanical strength; hygroscopic. Noncombustible. Low toxicity.
Derivation: Heating a mixture of boric acid and tricalcium phosphate in ammonia atmosphere in an electric furnace. See also borazon.
Forms: Powder; compressed solid; fibers; whiskers.
Uses: Refractory; furnace insulation; crucibles; rectifying tubes; dielectric; chemical equipment; self-lubricating bushings; molten metal pump parts; transistor and rectifier mounting wafers; heatshield for plasma; nose-cone windows; heat-resistant fibers stable to 1600°F in oxidizing atmosphere for military composites; metalworking abrasive; high-temperature insulator; high-strength fibers.

boron nitride, pyrolytic. See "Boralloy."

boron oxide. See boric oxide.

boron phosphate (sometimes called borophosphoric acid) BPO_4.
Properties: White, non-hygroscopic crystals; sp. gr. 1.873. Soluble in water; pH (1% solution) 2.0.
Containers: 400-lb barrels.
Use: Special glasses; ceramics; acid cleaner; dehydration catalyst.

boron phosphide BP. A refractory maroon powder; noncorrosive; Mohs hardness 9.5. Inert to corrosion.
Derivation: Direct union of boron and phosphorus at about 1000°C in a reducing atmosphere.
Hazard: Ignites spontaneously at 390°F. Evolves toxic fumes in contact with water and acids.

boron steel. See ferroboron.

boron tribromide (boron bromide) BBr_3.
Properties: Colorless, fuming liquid. Decomposed by alcohol and by water. Sp. gr. 2.69 at 15°C; b.p. 90°C; m.p. -46°C.
Derivation: (high purity). Direct bromination of boron, followed by rectification in quartz columns.
Grades: Technical; high purity.
Hazard: May explode when heated. Corrosive to tissue. Fumes are highly toxic. Tolerance, 1 ppm in air.
Use: Catalyst in organic syntheses.
Shipping regulations: (Rail) Corrosive liquid, n.o.s., White label. (Air) Corrosive label. Not acceptable on passenger planes.

boron trichloride (boron chloride) BCl_3.
Properties: Colorless, fuming liquid. Decomposed by alcohol and by water. Sp. gr. 1.35 (15°C); b.p. 12.5°C; m.p.—107°C. Reacts with hydrogen at 1200°C.
Derivation: (a) Heating boric oxide and carbon with chlorine; (b) combining boric oxide with phosphorus pentachloride.
Grades: Technical (99%); C.P. (99.5%).
Containers: Up to 50-lb pressure cylinders; tank cars.
Hazard: Strong irritant to tissue. Fumes are corrosive and highly toxic.
Uses: Catalyst in organic syntheses; source of many boron compounds; refining of alloys; soldering flux; making electrical resistors; extinguishing magnesium fires in heat-treating furnaces.
Shipping regulations: (Rail) White label. (Air) Corrosive label. Not acceptable on passenger planes.

boron trifluoride (boron fluoride) BF_3.
Properties: Colorless gas; 2.3 times as dense as air; m.p. -126.8°C; b.p. -101°C; does not support combustion; soluble in cold water; hydrolyzes in hot water; soluble in concentrated sulfuric acid; decomposes in alcohol. Easily forms double compounds such as that with ether, known as boron trifluoride etherate, or BF_3 - ether complex.
Derivation: From borax and hydrofluoric acid or from boric acid and ammonium bifluoride. The complex formed is then treated with cold fuming sulfuric acid.
Grades: Pure (99% min.).
Containers: Steel cylinders.
Hazard: Toxic by inhalation; corrosive to skin and tissue. Tolerance, 1 ppm in air.
Uses: Catalyst in organic synthesis; instruments for measuring neutron intensity; silver soldering fluxes; gas brazing.
Shipping regulations: (Rail) Green label. (Air) Nonflammable gas label. Not acceptable on passenger planes.

boron trifluoride etherate (BF_3-ether complex). $CH_3CH_2O(BF_3)CH_2CH_3$. A relatively stable coordination complex formed by the combination of diethyl ether with boron trifluoride, in which the boron atom is bonded to the oxygen of the ether. Colorless liquid. B.p. 259°F; flash point 147°F (open cup). Combustible.
Hazard: Toxic by inhalation; corrosive to skin and tissue.
Shipping regulations: (Rail, Air) Not listed. Consult authorities.

boron trifluoride-methanol (BF_3-MeOH). A solution of boron trifluoride in methanol; sp. gr. 0.90 (20°C); concentration 14 grams/ 100 cc. Used as an esterification reagent for fats and oils. Combustible.
Hazard: Highly toxic; moderate fire risk.
Shipping regulations: (Rail, Air) Poisonous liquid, n.o.s., Poison label.

boron trifluoride monoethylamine (boron fluoride monoethylamine; BF_3-MEA) BF_3-$C_2H_5NH_2$.
Properties: White to pale tan flakes. Sp. gr. 1.38; m.p. 88–90°C. Soluble in furfuryl alcohol, polyglycol, acetone. Releases boron trifluoride above 110°C. Combustible.
Hazard: Highly toxic; moderate fire risk.
Use: Elevated temperature cure of epoxy resins.
Shipping regulations: (Rail, Air) Not listed. Consult authorities.

borophosphoric acid. See boron phosphate.

borosilicate glass. See glass, heat-resistant.

"Boro-Spray."[88] Trademark for a crystalline product consisting chiefly of sodium pentaborate. Used for spray applications to tree fruit and truck crops where boron deficiency is indicated.

borotungstic acid (borowolframic acid) Various formulas and properties given.
Grades: Technical.
Containers: Glass bottles.
Use: Mineralogic assay.

"Botran."[519] Trademark for an agricultural fungicide, 2,6-dichloro-4-nitroaniline.
Hazard: Probably toxic.

bound water. Water molecules that are tightly held by various chemical groups in a larger molecule. Carboxyl, hydroxyl and amino groups are usually involved; hydrogen-bonding is often a factor. Proteins tend to bind water in this way, and in meats it will remain unfrozen as low as −40°F.

Boyle's Law. The volume of a sample of gas varies inversely with the pressure, if the temperature remains constant. The relation is strictly true only for an imaginary perfect or ideal gas, but the law is satisfactory for practical calculations except when pressures are high, or temperatures are approaching the liquefaction point. Van der Waal's equation is a refinement to take care of the inherent inaccuracy of Boyle's Law.

"BPIC."[177] Trademark for technical grade of tertbutyl peroxy isopropyl carbonate, a polymerization initiator for acrylic, ethylene, styrene and other monomers, and a cross-linking agent for silicone and ethylene propylene elastomers.

BPL. (1) Abbreviation for bone phosphate of lime. See bone phosphate.
(2) Abbreviation for beta-propiolactone.

"BPN."[248] Trademark for phenyl-beta-naphthylamine (q.v.).

"BP Pyro No. 5."[1] Trademark for a grade of sodium acid pyrophosphate possessing a controlled slow rate of reaction.
Derivation: Dehydration of sodium phosphate, monobasic.
Grade: F.C.C.
Uses: Leavening agent in prepared biscuit doughs; baking powder.

"BPR."[55] Trademark for insecticidal mixture containing varying proportions of pyrethrin, piperonyl butoxide and rotenone; in liquid or dust base.

Br Symbol for bromine.

BRA. Abbreviation for beta-resorcylic acid.

"Bradosol" Bromide.[305] Trademark for domiphen bromide.

brake fluid. See hydraulic fluid.

branched chain. See chain.

bran oil. See furfural.

brasilin (brazilin; Brazilwood extract) $C_{16}H_{14}O_5$. The crystalline colorizing principle of Brazilwood.
Properties: White or pale yellow rhombic needles from alcohol; turns orange in air or light. Soluble in water, alcohol, ether and in alkali hydroxide solution with a carmine-red color. Decomposes above 130°C.
Uses: Dyeing red and purple shades of wood, ink, textiles, etc.; acid-base indicator, turning yellow in acid and carmine red in alkali; biological stain.

brass. Copper-zinc alloys of varying composition. Low-zinc brasses (20% or less) are resistant to stress-corrosion cracking and are easily formed. Red Brass (15% zinc) is highly corrosion-resistant. Yellow brasses contain from 34 to 37% zinc and have good ductility and high strength, and can withstand severe cold-working. Cartridge brass contains 30–33% zinc. Muntz metal (40% zinc) is primarily a hot-working alloy, used where cold-forming operations are unnecessary. Some brasses also contain low percentages of other elements, e.g., manganese, aluminum, silicon, lead, and tin (admiralty metal, naval brass).
Hazard: Flammable in powder or finely divided form.
Uses: Condenser tube plates; piping; hose nozzles and couplings; marine equipment; jewelry; fine arts; stamping dies.
See also admiralty metal, aluminum brass, naval brass, red brass, yellow brass, Muntz metal, cartridge brass. For additional information refer to Copper and Brass Fabricators Council, 225 Park Ave., New York.

brassidic acid $CH_3(CH_2)_7CHCH(CH_2)_{11}COOH$. (*trans*-13-Docosenoic acid). An isomer of erucic acid.
Properties: White crystals. M.p. 61–62°C; b.p. 282°C (30 mm); sp. gr. 0.859; refractive index (n 57/D) 1.448. Insoluble in water; slightly soluble in alcohol; soluble in ether. Combustible.
Derivation: By treating erucic acid with nitrous acid (catalyst).

"Braze Bonding Agent."[69] Trademark for halogenated rubber derivatives and selected modifiers in solvent.
Properties: Deep red liquid; sp. gr. 1.01 ± 0.02; total solids 20–22%; flash point 93°F.
Hazard: Flammable; moderate fire risk.
Uses: Cement to bond natural rubber, SBR or neoprene to steel.

Brazil wax. See carnauba wax.

brazing. A method of joining metals that involves the use of a filler metal or alloy, at above 800°F. It is an irreversible process, as the filler cannot be remelted. The filler alloys may be silver/copper/zinc (1200–1600°F) or nickel/boron/silicon/chromium, suitable for superalloys in the range above 2000°F; silver-cadmium alloy is also used. The filler forms an intermetallic solution with the metal being joined. See also soldering.

breeder. A particularly efficient type of nuclear reactor, now undergoing development both in the U.S. and abroad, which is able to utilize the tremendous energy latent in U-238. This cannot be exploited in conventional nuclear reactors, which are fueled with enriched uranium or plutonium, for these eventually become depleted and must be replaced. The fuel used for the breeder reactor is a mixture of nonfissionable U-238 and plutonium-239 sealed in long, thin hexagonal metal tubes, which are in turn contained in cans called subassemblies. These constitute the reactor core. Around it are placed several layers of U-238, also in subassemblies. When criticality is reached, the neutron flux from the core permeates the entire system, and thus "breeds" fissionable U-239 (plutonium) in the surrounding U-238. The amount of fissionable material thus made available is about 100 times as great as that obtainable with a conventional reactor, since all the energy potential of the U-238 can be released. Twenty pounds of uranium has the potential of delivering about 52 million kilowatt hours of electricity; only a small fraction of this would be extractable without breeding.

The breeder utilizes fast neutrons, which are much more efficient than the slow (thermal) neutrons used in conventional reactors. Liquid sodium is the coolant in breeder reactors, as it has no retarding effect on the neutrons; 2.9 neutrons per fission are produced in the breeder, compared with only 2.4 in water-moderated reactors. It is this excess of neutrons that makes it possible for the fast breeder to produce more fuel than it consumes. A large-scale demonstration breeder is under construction at Oak Ridge, but it will be some years before breeders will produce commercial power in the U.S., though France achieved this in 1973.

breeze. Coke particles less than one-half inch in diameter. This occurs to the extent of about 100 pounds per ton of coal processed.

"Brellin."[342] Trademark for gibberellic acids as plant growth-stimulating compounds.

"Bretol."[430] Trademark for cetyldimethylethyl ammonium bromide, a quaternary ammonium compound used in dental preparations and soldering fluxes.

"Brevon."[448] Trademark for vinyl resin films; used for cocoon-type storage coatings.

brewing. The production of beer, ale, and malt liquors by a process involving a complex series of enzymatic reactions. The most important of these is the conversion of starch to a malt extract (wort), which in turn is fermented with yeast. Mashing is the preparation of wort from malt and cereals by enzymatic hydrolysis, after which the product is boiled with hops, which impart the characteristic taste and aroma of beer. The malt extract must contain the nutrients required for yeast growth. Mashing involves a complex interplay of chemical and enzymatic reactions that are not fully understood. Few changes have been made in the basic brewing processes for over a century; increased automatic operation and quality control techniques ensure a consistently good product. See also fermentation, yeast, wort.

Brewster process. A method for the extraction of acetic acid from the acid distillate of the destructive distillation of wood. Isopropyl ether is used as the solvent for the acetic acid. See Coahran process.

"BRH" 2 Rubber Softener.[50] Trademark for an asphaltic product.

Properties: Viscous fluid; sp. gr. 1.0 (25/25°C); flash point, 400°F; max. evaporation loss of 1% in 5 hours on heating 50 g at 163°C. Combustible.

Uses: Friction, adhesive, and electrical tapes because of excellent aging and tack characteristics; reclaiming oil for rubber scrap, especially in pan process.

brick, refractory. A highly heat-resistant and nonconductive material used for furnace linings, as in the glass and steel industries, and other applications where temperatures above 1600°C are involved. Some types are made of quartzite or high-silica clays (see firebrick), others of metallic ores such as chromite, magnesite, and zirconia. See also refractory.

bridged system. A cyclic organic compound in which more than two atoms are common to two rings. In the structure illustrated, the carbon atoms that connect the rings (1 and 4) are called bridgehead atoms. The groups of atoms that span the bridgeheads are called bridging groups, or bridges (2, 3, 5, 6, and 7).

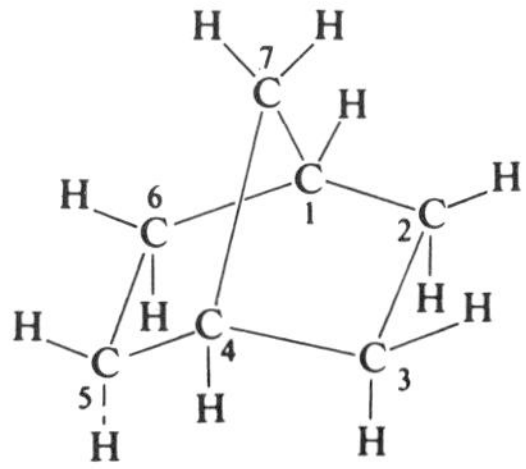

brightener. A compound which, when added to a nickel plating formulation of the Watts type (nickel sulfate and nickel chloride in a 6 to 1 ratio, plus boric acid) will yield a bright, reflective finish. There are two types: (1) naphthalene disulfonic acids, diphenyl sulfonates, aryl sulfonamides, etc., which give bright deposits on polished surfaces; and (2) metal ions having high hydrogen overvoltage in acid solution (zinc, cadmium, selenium, etc., and unsaturated organic compounds such as thiourea, acetylene derivatives, azo dyes, etc.) which give mirror brightness as a result of their "leveling" action. Usually both types are used for maximum effectiveness. See also leveling (2); optical brightener.

bright stock. Lubricating oil of high viscosity, obtained from residues of petroleum distillation by dewaxing and treatment with fuller's earth or similar material. Sometimes also applied to viscous petroleum distillates.

Use: For blending with neutral oils in preparing automotive engine lubricating oils.

"Brij."[206] Trademark for a series of emulsifiers and wetting agents developed for use in emulsions of high alkalinity or acidity. They are polyoxyethylene ethers of higher aliphatic alcohols. Soluble in water and lower alcohols. Insoluble in coal-tar hydrocarbons.

Brilliant Crocein (Crocein Scarlet MOO) $C_6H_5N_2C_6H_4N_2C_{10}H_4OH(SO_3Na)_2$.

Properties: Light brown powder; cherry red solution in water. C.I. 27290.

Use: To dye wool and silk red from acid solution, and cotton and paper with aid of a mordant; also used for red lakes; biological stain.

"Brilliant Toning Red."[141] Trademark for Permanent Red 2B azo pigments derived from betahydroxynaphthoic acid.

Uses: Printing inks; paints; enamels; lacquers; rubber; plastics; wallpaper; textiles; floor coverings; crayons; paper coatings.

Brilliant Toning Red Amine. See 4-amino-2-chlorotoluene-5-sulfonic acid.

brine. Any solution of sodium chloride and water, usually containing other salts also. The most industrially important brines are (1) in subterranean wells, as in Michigan; (2) in desert lakes, such as Great Salt, Searles, Salton Sea, and Dead Sea; and (3) the ocean. These are the sources of many inorganic chemicals such as soda ash, sodium sulfate, potassium chloride, bromine, chlorine, borax, etc. Concentration range from 3% (ocean) to 20% or more.

Large areas of sand and shale containing brines under high pressure exist along the Gulf Coast; these are reported to be an important undeveloped source of natural gas and other hydrocarbons suitable for fuel or petrochemical feedstocks.

See also desalination; demineralization.

Brinell hardness. See hardness.

brisance. The shattering power of an explosive measured by the ratio of the weight of graded sand shattered when a charge of the test explosive is detonated in a standard manner to the weight of sand shattered by TNT detonated in the same manner.

britannia metal. See pewter and white metal

"Britecarb."[155] Trademark for carbonized nickel with the carbon completely removed from one side.

"BriteSil."[531] Trademark for a series of soluble, clarified sodium silicates, with controlled Na_2O and SiO_2 contents and densities. Available in nine grades.

"BriteSorb."[531] Trademark for a synthetic, precipitated magnesium silicate having the empirical formula $MgO \cdot 2.5SiO_2 \cdot 1.5H_2O$. A white, finely divided, free flowing powder with large surface area; insoluble in water and most organic solvents; pH (10% slurry) 9.5. Available in various particle sizes.

Uses: Purification; clarification; adsorption; catalyst base.

British Anti-Lewisite. See 2,3-dimercaptopropanol.

British thermal unit. See Btu.

"Britone Reds."[141] Trademark for resinated type lithol reds. See "Graphic Red."

Brix degree. A measure of the density or concentration of sugar solutions. The degrees Brix equals percent by weight of sucrose in the solution and is related empirically to the specific gravity.

Broenner's acid. See 2-naphthylamine-6-sulfonic acid.

"Brom 55."[225] Trademark for dibromodimethylhydantoin (q.v.).

bromacil. Generic name for 5-Bromo-3-sec-butyl-6-methyluracil, an herbicide. Substitute for some uses of 2,4,5-T, approved by EPA.

bromal. See tribromoacetaldehyde.

"Bromat."[430] Trademark for cetyl trimethyl ammonium bromide, a quaternary ammonium compound with high germicidal activity.

bromcresol green. Tetrabromo-meta-cresolsulfonphthalein, an acid base indicator, showing color change from yellow to blue over the range pH 3.8–5.4. Yellow crystals; m.p. 218°C; slightly soluble in water, soluble in alcohol. See also indicator.

bromcresol purple. Dibromo-ortho-cresolsulfonphthalein, an acid base indicator, changes from yellow to purple between pH 5.2–6.8. Yellow crystals; m.p. 241°C; insoluble in water, soluble in alcohol. See also indicator.

bromelin (bromelain). A milk-clotting, protein-digesting enzyme. It is precipitated from pineapple juice with alcohol or ammonium sulfate.

Use: Biochemical research; meat-tenderizing formulations; texturizer in baking.

bromeosin. See eosin.

bromic acid $HBrO_3$.

Properties: Colorless or slightly yellow liquid; turns yellow on exposure; unstable except in very dilute solution; soluble in water. Sp. gr. 3.18; b.p. decomposes at 100°C.

Derivation: Sulfuric acid is added to a solution of barium bromate and the product recovered by subsequent distillation and absorption in water.

Hazard: Toxic by ingestion and inhalation. Strong irritant to tissue.

Uses: Dyes; intermediates; pharmaceuticals.

brominated camphor. See camphor bromate.

bromine. Nonmetallic, halogen element of atomic number 35, group VIIA of the Periodic Table. Atomic weight 79.904. Valences 1, 3, 5 (valence of 7 also reported). There are two stable isotopes.

Properties: Dark, reddish-brown liquid; irritating fumes. Soluble in common organic solvents; very slightly soluble in water. Attacks most metals, including platinum and palladium; aluminum reacts vigorously, and potassium explosively. Dry bromine does not attack lead, nickel, magnesium, tantalum, iron, zinc, and sodium (the latter below 300°C). B.p. 58.8°C; m.p. -7.3°C; sp. gr. (20/4°) 3.11; vapor density (air = 1) 3.5; wt. per gal 25.7 lb; specific heat 0.107 cal/gram; refractive index 1.647; dielectric constant 3.2.

Derivation: From sea water and natural brines by oxidation of bromine salts with chlorine; solar evaporation (Great Salt Lake); from salt beds at Stassfurt; Dead Sea.

Method of purification: Distillation.

Grades: Technical; C.P.; 99.8%; 99.95%.

Containers: Bottles; drums; tank cars; tank trucks.

Hazard: Highly toxic; severe skin irritant. Tolerance, 0.1 ppm in air; 0.7 mg per cubic meter of air. Strong oxidizing agent; may ignite combustible materials on contact. Safety data sheet available from Manufacturing Chemists Assn., Washington, D.C.

Uses: Manufacture of ethylene dibromide (antiknock gasoline); organic synthesis; bleaching; water purification; solvent; intermediate for fumigants (methyl bromide); fire-extinguishing fluid; analytical reagent; fire-retardant for plastics; dyes; pharmaceuticals; photography; shrink-proofing wool.

Shipping regulations: (Rail) White label. (Air) Corrosive label. Not acceptable on passenger planes.

bromine azide (bromoazide). BrN_3.

Properties: Crystals or red liquid. M.p. approx. 45°C; b.p. explodes! A strong oxidizing agent.

Hazard: Toxic; explosive when heated or shocked. Will ignite combustible materials on contact.

Uses: Detonators and other explosive devices.

Shipping regulations: (Rail, Air) Not listed. Consult authorities.

bromine-chloride (chlorine-bromide) BrCl.

Properties: Reddish-yellow, mobile liquid. M.p. -66°C; decomposes with evolution of chlorine at 10°C. Soluble in carbon disulfide, ether, water.

Hazard: Highly toxic; irritant. A powerful oxidizer; reacts vigorously with combustible materials.
Use: Industrial disinfectant.

bromine cyanide. See cyanogen bromide.

bromine iodide. See iodine monobromide.

bromine pentafluoride BrF_5.
Properties: Colorless liquid; sp. gr. 2.466 (25°C); m.p. −61°C; b.p. 40.5°C; vapor pressure (70°F) 7 psia; decomposes in water. Reacts with every known element except inert gases, nitrogen and oxygen.
Derivation: By reacting bromine diluted with nitrogen with fluorine in a copper vessel at 200°C.
Grade: 98% min.
Containers: Cylinders.
Hazard: Highly toxic; corrosive to skin and tissue. Tolerance, 0.1 ppm in air. Explodes on contact with water.
Uses: Synthesis; oxidizer in liquid rocket propellants.
Shipping regulations: (Rail) White label. (Air) Corrosive label. Not acceptable on passenger planes.

bromine trifluoride BrF_3.
Properties: Colorless liquid; sp. gr. (135°C) 2.49; m.p. 9°C; b.p. 135°C; vapor pressure (70°F) 0.15 psia; decomposed violently by water.
Derivation: See bromine pentafluoride.
Grade: 98% min.
Containers: Cylinders.
Hazard: Highly toxic; corrosive to skin. Extremely reactive and dangerous. Tolerance (as F), 2.5 mg per cubic meter of air.
Use: Fluorinating agent; electrolytic solvent.
Shipping regulations: (Rail) White label. (Air) Corrosive label. Not acceptable on passenger planes.

"Brominex."[152] Trademark for a series of technical grade brominated oils. Bromine content ranges from 20 to 45%.

"Brominol"[3] Trademark for a brominated olive oil used to adjust specific gravity of citrus oils to provide cloud in beverages.

bromisovalum. See bromoisovaleryl urea.

N-bromoacetamide (NBA) $CH_3CONHBr$.
Properties: White powder with bromine odor. M.p. 105–108°C. Contains about 57% active bromine; decomposes appreciably above 80°F.
Containers: Glass bottles; polyethylene-lined fiber drums.
Hazard: Emits toxic fumes of bromine on heating.
Uses: Brominating and oxidizing agent in organic synthesis.

bromoacetic acid $CH_2BrCOOH$.
Properties: Colorless, deliquescent crystals. Keep from air and moisture. Soluble in water, alcohol and ether. M.p. 51°C; b.p. 208°C; sp. gr. 1.93.
Derivation: By heating acetic acid and bromine.
Hazard: Strong irritant to skin and tissue.
Uses: Organic synthesis; abscission of citrus fruit in harvesting.
Shipping regulations: (Air) Corrosive label (solid and solution).

bromoacetone $CH_2BrCOCH_3$.
Properties: Colorless liquid when pure; rapidly becomes violet even in absence of air. Soluble in acetone, alcohol, benzene, and ether; slightly soluble in water. Sp. gr. 1.631 at 0°C; b.p. 136°C (partial decomposition); m.p. −54°C; vapor density 4.75; vapor tension 9 mm (20°C).
Derivation: By treating aqueous acetone with bromine and sodium chlorate at 30–40°C.
Grades: Technical.
Containers: Lead-lined cylinders.
Hazard: Highly toxic by inhalation and skin contact. A lachrymating gas; strong irritant.
Uses: Organic synthesis; tear gas.
Shipping regulations: (Rail) Poison gas label; not accepted by rail express. (Air) Not acceptable.

bromoacetone cyanohydrin $CH_2BrC(OH)(CN)CH_3$.
Properties: Colorless liquid. Soluble in alcohol, ether, and water. Sp. gr. 1.584 at 13°C; b.p. 94.5°C (5 mm).
Derivation: Interaction of bromoacetone and hydrocyanic acid at (approximate) 0°C.
Hazard: May be toxic.
Use: Organic synthesis.

5-(2-bromoallyl)-5-sec-**butylbarbituric acid.** See butallylonal.

bromoallylene. See allyl bromide.

bromoauric acid (gold tribromide, acid) $HAuBr_4 \cdot 5H_2O$.
Properties: Dark, red-brown needle crystals or granular masses; odorless; metallic and acid taste. Stable in air if pure but deliquescent if chloride is present. Soluble in water and alcohol. M.p. 27°C.
Derivation: By dissolving auric bromide in hydrobromic acid, concentrating and crystallizing.

para-**bromobenzaldehyde** BrC_6H_4CHO. Solid; m.p. 58°C. Used as a chemical intermediate.

bromobenzene (phenyl bromide) C_6H_5Br.
Properties: Heavy, mobile, colorless liquid. Pungent odor. Sp. gr. 1.499 (15°C); wt/gal 12.51 lb; b.p. 156.6°C; freezing point −30.5°C, flash point 124°F; refractive index 1.5625. Miscible with most organic solvents; insoluble in water. Autoignition temp. 1051°F. Combustible.
Derivation: Bromination of benzene in presence of iron.
Grades: Technical; pure.
Hazard: Moderate fire risk. Skin irritant.
Uses: Solvent; motor fuels; top-cylinder compounds; crystallizing solvent; organic synthesis.

para-**bromobenzenesulfonic acid** $BrC_6H_4SO_3H$.
Properties: Crystallizes in needles. M.p. 102–103°C; b.p. 155°C (25 mm); soluble in hot water and hot alcohol.

para-**bromobenzoic acid** $C_6H_4BrCOOH$.
Properties: Colorless or reddish crystals. Soluble in alcohol and ether; very slightly soluble in water. M.p. 254°C.
Derivation: From para-bromotoluene by oxidation.
Use: Organic synthesis; detection of strontium.

ortho-**bromobenzyl cyanide** (ortho-bromophenylacetonitrile; 2-bromo-alpha-cyanotoluene) $BrC_6H_4CH_2CN$.
Properties: Colorless liquid; sp. gr. 1.519; m.p. 25.5°C; b.p. 242°C (decomposes). Nonflammable.
Hazard: Strong lachrymator; irritant to tissue.
Shipping regulations: (Rail) Tear gas label. (Air) Poison label. Not acceptable on passenger planes.

1-bromobutane. See n-butyl bromide.

2-bromobutane. See sec-butyl bromide.

5-bromo-3-sec-**butyl-6-methyluracil.** See bromacil.

alpha-**bromobutyric acid** $CH_3CH_2CHBrCOOH$.
Properties: Clear, colorless, oily liquid. Soluble in alcohol and ether; sparingly soluble in water. Sp. gr. 1.54; b.p. 181°C at 250 mm, 214–217°C with decomposition; m.p. -4°C. Combustible.
Derivation: By heating bromine and butyric acid.
Use: Organic synthesis.

bromocarnallite. An artificial carnallite (q.v.) in which chlorine is replaced by bromine.

3-bromo-1-chloro-5,5-dimethylhydantoin $BrCl(CH_3)_2C_3N_2O_2$.
Properties: Free-flowing white powder; faint halogen odor; m.p. 163–164°C; soluble in benzene, methylene dichloride, chloroform. Active bromine, 33% min; active chlorine, 14% min.
Hazard: May be toxic. See bromine, chlorine.
Uses: Germicide and fungicide in treatment of water; disinfectant; halogenating agent; catalyst of ionic type; selective oxidant.

sym-**bromochloroethane** (ethylene chlorobromide) CH_2BrCH_2Cl.
Properties: Colorless, volatile liquid. Chloroform-like odor; soluble in alcohol and ether; insoluble in water. Sp. gr. 1.70; b.p. 107–108°C; wt/gal 14.9 lb (0°C); f.p. -16.6°C. Nonflammable.
Derivation: By action of bromine and chlorine on ethylene gas.
Hazard: Avoid inhalation of fumes.
Uses: Solvent, especially for cellulose esters and ethers; organic synthesis; fumigant for fruits and vegetables.

bromochloromethane (methylene chlorobromide) $BrCH_2Cl$.
Properties: Clear, colorless, volatile liquid with chloroform-like odor; sp. gr. 1.93 (25°C); b.p. 67°C; m.p. -86.5°C; refractive index (n 25/D) 1.48. Nonflammable. Soluble in organic solvents; insoluble in water.
Containers: Drums; tankcars.
Hazard: Avoid inhalation of fumes.
Uses: Fire extinguishers; organic synthesis.

1-bromo-3-chloropropane (trimethylene chlorobromide) $BrCH_2CH_2CH_2Cl$.
Properties: Colorless liquid. Freezing point below -50°C; b.p. 143–145°C; sp. gr. 1.594 (25/25°C); lb/gal 13.27 (25°C); refractive index 1.484 (n 25/D). Insoluble in water. Soluble in methanol and ether. Nonflammable.
Hazard: Avoid inhalation of fumes.
Uses: Organic synthesis; pharmaceuticals.

2-bromo-2-chloro-1,1,1-trifluoroethane. See halothane.

2-bromo-alpha-**cyanotoluene.** See bromobenzyl cyanide.

bromocyclopentane. See cyclopentyl bromide.

bromodiethylacetylurea (carbromal) $C(C_2H_5)_2BrCONHCONH_2$.
Properties: White, crystalline powder; odorless; tasteless. Soluble in chloroform, ether, alcohol, concentrated mineral acids and alkali hydroxide solutions; almost insoluble in cold water; slightly soluble in hot water. M.p. 116–117°C.
Use: Medicine.

1-bromododecane. See lauryl bromide.

bromoethane. See ethyl bromide.

2-bromoethyl alcohol. See ethylene bromohydrin.

2-bromoethylamine hydrobromide $BrCH_2CH_2NH_2 \cdot HBr$.
Uses: Intermediate; suggested as a soldering flux.

bromoethyl chlorosulfonate $BrCH_2CH_2OSO_2Cl$.
Properties: Liquid. B.p. 100–105°C (18 mm).
Derivation: Interaction of sulfuryl chloride and glycol bromohydrin.
Hazard: Irritant to skin and tissue.

para-**bromofluorobenzene** C_6H_4BrF.
Properties: Colorless liquid; b.p. 151–152°C; freezing point -17.4°C; sp. gr. (15°C) 1.593; refractive index (n 25/D) 1.5245; insoluble in water.
Uses: Intermediate; production of para-fluorophenol.

bromoform (tribromomethane; methyl tribromide) $CHBr_3$.
Properties: Colorless, heavy liquid; odor and taste similar to those of chloroform. Soluble in alcohol, ether, chloroform, benzene, solvent naphtha, fixed and volatile oils; slightly soluble in water. Sp. gr. 2.887; m.p. 9°C; b.p. 151.2°C; wt/gal 24 lb; boiling range 150.3–151.2°C; freezing point 9°C; surface tension 41.53 dynes/cm (20°C); dielectric constant 4.5 (20°C). Refractive index 1.6005. Nonflammable.
Derivation: By heating acetone or ethyl alcohol with bromine and alkali hydroxide, and recovery by distillation. (Similar to acetone process for chloroform.)
Grades: Technical; pharmaceutical; spectrophotometric.
Hazard: Toxic by ingestion, inhalation and skin absorption. Tolerance, 0.5 ppm in air.
Uses: Intermediate in organic synthesis; geological assaying; solvent for waxes, greases, and oils.

"Bromofume."[88] Trademark for a soil fumigant composition, consisting of ethylene dibromide in volatile solvent.
Use: Control of wireworms and rootknot nematodes.

1-bromohexane. See n-hexyl bromide.

2-bromoisovaleryl urea (bromisovalum) $(CH_3)_2CHCHBrCONHCONH_2$.
Properties: White needles; slightly bitter in taste. Soluble in hot water, alcohol, and ether. M.p. 147–149°C.
Use: Medicine.

bromol (2,4,6-tribromophenol) $C_6H_2Br_3OH$.
Properties: Soft, white needles; sweet taste; penetrating bromine odor. Soluble in alcohol, chloroform, ether, and caustic alkaline solutions; almost insoluble in water. M.p. 96°C, sublimes; sp. gr. 2.55 (20/20°C); b.p. 244°C.
Derivation: Action of bromine on phenol.
Hazard: May be toxic.

bromomethane. See methyl bromide.

bromomethylethyl ketone $BrCH_2COC_2H_5$.
Properties: Colorless to pale-yellowish liquid. Affected by light. Soluble in alcohol, benzene, ether; insoluble in water. Sp. gr. 1.43; b.p. 145–146°C (decomposes).
Derivation: Interaction of sodium bromide and methyl ethyl ketone in the presence of sodium chlorate.
Hazard: May be irritant to skin and eyes.
Uses: Organic synthesis.

alpha-**bromonaphthalene** $C_{10}H_7Br$.
Properties: Colorless, thick liquid; pungent odor; sp. gr. 1.4870; m.p. 6.2°C; b.p. 279°C; refractive index 1.6601; soluble in alcohol, ether, and benzene; slightly soluble in water.
Derivation: Bromination of naphthalene.
Uses: Organic synthesis; microscopy; refractometry.

1-bromo-2-naphthol $BrC_{10}H_6OH$. Solid; m.p. 121–125°C. Used as a dye intermediate.

2-bromopentane $CH_3CH_2CH_2CHBrCH_3$.
Properties: Colorless to yellow liquid; strong odor. Sp. gr. 1.1850 (25/25°C).

para-**bromophenol** $HO(C_6H_4)Br$.
Properties: Crystals; sp. gr. 1.840 (15°C), 1.5875 (80°C); m.p. 64°C; b.p. 238°C, slightly soluble in water; soluble in alcohol, chloroform, ether, and glacial acetic acid. Used as a disinfectant.

bromophenol blue. Tetrabromophenolsulfonphthalein, an acid-base indicator, showing color change from yellow to purple over the range pH 3.0 to 4.6.

ortho-**bromophenylacetonitrile.** See bromobenzyl cyanide.

2-bromo-4-phenylphenol $C_6H_5C_6H_3BrOH$.
Properties: Light-colored solid with faint characteristic odor; sp. gr. (25/4°C) 1.536; m.p. 93.6–95.6°C; flash point 207°C; b.p. with decomposition (18 mm) 195–200°C. Soluble in alkalies, most organic solvents; insoluble in water. Combustible.
Hazard: May be toxic by ingestion and inhalation.
Use: Germicide.

bromophosgene (carbonyl bromide; carbon oxybromide) $COBr_2$.
Properties: Heavy, colorless liquid. Strong odor. Hydrolyzed by water and is decomposed by light and heat. Sp. gr. 2.5 (approx.) (15°C); b.p. 64–65°C.
Derivation: Action of sulfuric acid on carbon tetrabromide.
Hazard: Highly toxic by ingestion and inhalation.
Uses: Military poison gas (toxic suffocant); making crystal violet-type coloring agents.
Shipping regulations: (Rail) Poison gas label. Not accepted by express. (Air) Not acceptable.

bromopicrin (nitrobromoform; tribromonitromethane) CBr_3NO_2.
Properties: Prismatic crystals; decomposes with explosive violence if heated rapidly. Soluble in alcohol, benzene, and ether; slightly soluble in water. Sp. gr. 2.79 at 18°C; b.p. 127°C (118 mm Hg); m.p. 103°C.
Derivation: Action of picric acid on an aqueous solution of bromine and calcium oxide, followed by distillation under reduced pressure.
Hazard: Highly toxic; powerful irritant. Severe explosion hazard when heated.
Uses: Organic synthesis; military poison gas.
Shipping regulations: (Rail, Air) Not listed. Consult authorities.

3-bromopropene. See allyl bromide.

alpha-**bromopropionic acid** (2-bromopropionic acid) $CH_3CHBrCOOH$.
Properties: Colorless liquid. Sp. gr. 1.69; m.p. 24.5°C; b.p. 203°C, with decomposition; soluble in water, alcohol, and ether. Probably low toxicity.
Derivation: By heating propionic acid with bromine.
Method of purification: Distillation.

3-bromo-1-propyne. See propargyl bromide.

2-bromopyridine C_5H_4NBr.
Properties: Liquid; b.p. 195°C; sp. gr. 1.627 (20°C); refractive index 1.5714 (n 20/D); solubility in 100 g water 2.08 (20°C). Probably low toxicity.
Use: Synthesis of pyridine compounds.

3-bromopyridine C_5H_4NBr.
Properties: Needles;ˆ b.p. 174.4°C; sp. gr. 1.628 (20/20°C); refractive index 1.5710 (n 20/D). Difficulty soluble in water, readily soluble in common organic solvents. Probably low toxicity.

5-bromosalicylhydroxamic acid
$BrC_6H_3(OH)CONHOH$. Crystals; decomposes at 232°C. Very sparingly soluble in water; forms a water-soluble sodium salt. Used in medicine.

beta-**bromostyrene** (bromostyrol) $C_6H_5CHCHBr$.
Properties: Yellowish liquid; strong floral odor. Soluble in 4 volumes of 90% alcohol. Sp. gr. 1.395–1.424; refractive index 1.602–1.608; m.p., min -2°C.
Use: Perfumery.

bromosuccinic acid $HOOCCH_2CHBrCOOH$.
Properties: Colorless crystals; sp. gr. 2.073; m.p. 159–161°C; soluble in water and alcohol; insoluble in ether.
Derivation: By heating bromine and succinic acid.
Use: Organic synthesis.
Note: The above are the properties of the *dl* form. Optically active forms are also known.

N-bromosuccinimide (NBS) $(CH_2CO)_2NBr$.
Properties: Fine crystals, white to cream in color; melting range 172–178°C (decomposes). Soluble in carbon tetrachloride; 44.5% min. active bromine.
Hazard: Respirators should be used in handling dry material, which evolves toxic fumes of bromine.
Use: For controlled low-energy bromination.

4-bromothiophenol BrC_6H_4SH.
Properties: White solid; f.p. 73°C; b.p. 239°C; almost insoluble in water; soluble in methanol, alkaline solutions (with which it reacts).
Hazard: May be toxic by ingestion.
Use: Intermediate.

bromothymol blue. Dibromothymolsulfonphthalein, an acid-base indicator, showing color change from yellow to blue over the range pH 6.0–7.6.

alpha-**bromotoluene.** See benzyl bromide.

para-**bromotoluene** (*p*-tolyl bromide). $BrC_6H_4CH_3$.
Properties: Crystals; m.p. 28.5°C; b.p. 184–185°C; sp. gr. 1.3898 (20°C); refractive index 1.5490; flash point 185°F. Combustible. Insoluble in water, soluble in alcohol, ether and benzene.
Hazard: May be toxic by ingestion.
Use: Intermediate.

bromotrichloromethane (trichlorobromomethane) CCl_3Br.
Properties: Colorless heavy liquid with chloroform-like odor. Miscible with many organic liquids; sp. gr. 2.0; b.p. 104°C; n 20/D 1.5051.
Hazard: Moderately toxic by ingestion and inhalation of fumes.
Use: Organic synthesis.

bromotrifluoroethylene (BFE) $BrFC{:}CF_2$. Gas (monomer) or liquid (polymer). The latter are usually clear oils at room temperature and noncracking solids at -65°F. Viscosities and densities vary widely.
Monomer: High purity gas (97%). Shipped in cylinders.
Hazard: Flammable (gas or liquid); dangerous fire risk.
Uses (BFE polymers): Flotation fluids for gyros or accelerometers used in inertial guidance systems. BFE polymers can also be used like CFE polymers.

Shipping regulations: (Rail) Compressed gases, n.o.s., Red label. (Air) Flammable Gas label. Not acceptable on passenger planes.

bromotrifluoromethane $CBrF_3$.
Properties: Colorless gas; noncorrosive, nontoxic, nonflammable; f.p. −168°C; b.p. −58°C; density at b.p. 8.71 grams/liter.
Derivation: Bromination of fluoroform or perfluoropropane in nonmetallic reactor.
Uses: Chemical intermediate; refrigerant; metal hardening; fire extinguishment.
Shipping regulations: (Air) Nonflammable compressed gas label. (Rail) Compressed gases, n.o.s., Green label.

bromopheniramine maleate (2-[para-bromo-alpha-(2-dimethylaminoethyl)benzyl]pyridine bimaleate) $C_{16}H_{19}BrN_2 \cdot C_4H_4O_4$.
Properties: Crystals. Rather soluble in water; less soluble in alcohol; m.p. 130–135°C.
Grade: N.F.
Use: Medicine.

"Bromsulphalein."[348] Trademark for sodium sulfobromophthalein (q.v.).

"Bromural."[9] Trademark for bromoisovalerylurea (bromisovalum).

"Bromvegol."[342] Trademark for brominated vegetable oils used for weighting soft-drink emulsions.

bronze. An alloy of copper and tin, usually containing from 1 to 10% tin; special types contain from 5 to 10% aluminum (aluminum bronze); fractional percentages of phosphorus (phosphor bronze) as deoxidizer; or low percentages of silicon (silicon bronze).
Hazard: Powder is flammable.
Uses: Spark-resistant tools; springs; fourdrinier wire; paint; cosmetics (as powder); electrical hardware; architecture; fine arts.
See also brass; phosphor bronze.

bronze blue. Any of a number of varieties of iron-blue pigments (q.v.).

bronze orange. See red lake C.

bronzing liquid.
(1) A solution of pyroxylin in amyl acetate together with a bronze powder, usually aluminum bronze.
(2) Gloss oils and aluminum bronze.
(3) Spirit varnishes and aluminum bronze.
Shipping regulations: (Rail) Red label. (Air) Flammable Liquid label.

brosylate ester. An ester of para-bromobenzenesulfonic acid.

Brownian movement. The continuous zigzag motion of the particles in a colloidal suspension, visible in an optical microscope. The motion is caused by impact of the molecules of the liquid upon the colloidal particles. Named after the British botanist Robert Brown who first noted this phenomenon.

browning reaction (Maillard reaction). A complicated and not completely evaluated sequence of chemical changes occurring without the involvement of enzymes during heat exposure of foods containing carbohydrates (usually sugars) and proteins, as well as during storage. It is responsible for the surface color change of bakery products and meats. It begins with an aldol condensation reaction involving the carbonyl groups of the carbohydrates and the amino groups of the proteins, and ends with formation of furfural, which produces the dark brown coloration. Besides color change, the reaction is accompanied by alterations in flavor and texture, as well as in nutritive value. It was first noted by the French chemist Maillard.

"Brozone."[233] Trademark for a liquid formulation of methyl bromide in solvent; used as soil fumigant to control weeds, nematodes and fungi.
Hazard: See methyl bromide.

brucine (dimethoxystrychnine) $C_{23}H_{26}O_4N_2 \cdot 2H_2O$ or $4H_2O$.
Properties: White crystalline alkaloid. Very bitter taste; loses water at 100°C; m.p. 178°C. Soluble in alcohol, chloroform, and benzene; slightly soluble in water, ether, glycerin, and ethyl acetate. Forms brucine sulfate, hydrochloride, and nitrate (m.p. 230°C). Also available as the sulfate.
Derivation: By extraction and subsequent crystallization from nux vomica or ignatia seeds.
Hazard: Highly toxic by ingestion and inhalation.
Use: Medicine; denaturing alcohol; lubricant additive; separation of racemic mixtures.
Shipping regulations: (Rail, Air) Poison label.

brucite $Mg(OH)_2$. Natural magnesium hydroxide.
Properties: Color white, gray, greenish; luster pearly or waxy. Sp. gr. 2.39; Mohs hardness 2.5.
Occurrence: Nevada, Washington; Canada.
Uses: Refractories.

"Brush-Blitz."[58] Trademark for a highly concentrated formulation of 88% isooctyl ester of 2,4,5-trichlorophenoxyacetic acid; used as a brush killer.

"Brush-Rhap."[266] Trademark for butyl and 2-ethylhexyl esters or amine salts of 2,4,5-trichlorophenoxyacetic acid. Available in various concentrations of active ingredient and in combination with esters of 2,4-dichlorophenoxyacetic acid. Used as a herbicide.

"BRV."[50] Trademark for a heavy, high-boiling coal-tar distillate.
Properties: Dark, coal-tar oil; sp. gr. 1.14–1.18 (25/25°C); Engler specific viscosity, 5–10 (50°C); distillation, 26% max at 355°C. Combustible.
Containers: 55-gal steel drums; tank trucks; tank cars.
Uses: Rubber plasticizer, softener, and reclaiming oil; dispersing agent.

"Brymul."[51] Trademark for an emulsifiable grade of cleaner for general use on metals, etc. Contains Stoddard-type solvent.
Hazard: Moderate fire risk.

"Bryton."[544] Trademark for a series of oil-soluble petroleum sulfonates used as additives to motor oils and diesel fuels.
Uses: Detergent dispersants, rust-inhibiting agents, and alkaline carriers.

"BS."[1] Trademark for a series of seven trimethylolpropane ester base stocks for custom formulating of high-temperature lubricants. Also used as additives in lubricating oils and greases.

B-stage resin (resitol). A thermosetting phenolformaldehyde type resin which has been thermally reacted beyond the A-stage so that the product has only partial solubility in common solvents (alcohols, ketones), and is not fully fusible even at 150–180°C. The B-stage resin has limited commercial use.

BT (Bacillus thuringiensis). A species of bacteria used as a pesticide for agricultural crops. It is of the stomach-poison type, and has been approved for commercial use.

"BTC."[328] Trademark for a series of cationic quaternary ammonium chlorides, generally alkyldimethylbenzylammonium chloride.
Uses: Disinfectant; deodorant; germicide; fungicide; algaecide; slimicide.

BTDA. See 3,3′,4,4′-benzophenone tetracarboxylic dianhydride.

Btu (British thermal unit). The quantity of heat required to raise the temperature of one pound of water one degree Fahrenheit (usually from 39 to 40°F). This is the accepted standard for the comparison of heating values of fuels. For example, fuel gases range from 100 (low producer gas) to 3200 (pure butane) Btu per cu ft. The usual standard for a city gas is about 500 Btu.

BTX. Commercial abbreviation for benzene, toluene, xylene, the three major aromatic compounds.

Bu Informal abbreviation for butyl.

buclizine hydrochloride $C_{28}H_{33}ClN_2 \cdot 2HCl$. 1-para-chlorobenzhydryl-4-(para-(tert)-butylbenzyl) piperazine dihydrochloride.
Uses: Medicine (antihistamine).

"Budene."[265] Trademark for a cis-1,4-polybutadiene synthetic dry rubber manufactured by solution polymerization utilizing a stereospecific catalyst. This elastomer has a relatively high cis content, and is protected with a non-staining, non-discoloring antioxidant.

"Budium."[28] Trademark for a polybutadiene finish for application to tin plate.

buffer. A solution containing both a weak acid and its conjugate weak base whose pH changes only slightly on the addition of acid or alkali. The weak acid becomes a buffer when alkali is added, and the weak base becomes a buffer on addition of acid. This action is explained by the reaction

$$A + H_2O \rightleftharpoons B + H_3O_+$$

in which the base B is formed by the loss of a proton from the corresponding acid A. The acid may be a cation such as NH_4^+, a neutral molecule such as CH_3COOH, or an anion such as $H_2PO_4^-$. When alkali is added, hydrogen ions are removed to form water; but, as long as the added alkali is not in excess of the buffer acid, many of the hydrogen ions are replaced by further ionization of A to maintain the equilibrium. When acid is added, this reaction is reversed as hydrogen ions combine with B to form A.

The pH of a buffer solution may be calculated by the mass law equation

$$pH = pK' + \log \frac{C_B}{C_A}$$

in which pK' is the negative logarithm of the apparent ionization constant of the buffer acid and the concentrations are those of the buffer base and its conjugate acid.

bufotenine [3-(2-dimethylaminoethyl)-5-indolol] $C_{12}H_{16}N_2O$.
Properties: Colorless prisms; insoluble in water; soluble in alcohol; slightly soluble in ether; soluble in dilute acids and alkalies.
Derivation: From toads and toadstools; also made synthetically.
Hazard: A hallucinogenic agent.
Use: Medicine (experimental).
See also hallucinogen.

builder, detergent. A substance that increases the effectiveness of a soap or synthetic surfactant by adding to its detergent power. Builders act as water softeners and as sequestering and buffering agents. Phosphate-containing compounds were formerly used, but their use has been prohibited because of their tendency to eutrophy lakes and rivers. Replacing phosphates, though less effective, are sodium carbonate, sodium silicate and ethylenediaminetetraacetic acid. Pilot plant production of a product containing trisodium 2-oxy-1,1,3-propane tricarboxylate has been reported. Nitrilotriacetic acid is an effective builder, but its use has been avoided because of possible carcinogenicity.

bunamiodyl $C_3H_7CONHC_6HI_3CH{:}C(C_2H_5)COONa$ 3(3-Butyrylamino-2,4,6-tri-iodophenyl)-2-ethyl sodium acrylate. Used in medicine (radiopaque contrast media, diagnostic aid).

bunker fuel oil. A heavy residual oil used as fuel by ships, industry, and for large-scale heating installations. Combustible.

"Bunnatol-G."[354] Trademark for a plasticizer for synthetic and reclaimed rubber. Insoluble in mineral and vegetable oils; imparts to rubber a high resistance to greases and oils.

Bunsen, Robert Wilhelm (1811–1899). Born in Germany, Bunsen is remembered chiefly for his invention of the laboratory burner named after him. He engaged in a wide range of industrial and chemical research, including blast-furnace firing, electrolytic cells, separation of metals by electric current, and production of light metals by electric decomposition of their molten chlorides. He also discovered two new elements—rubidium and cesium.

Burgundy pitch. A resin obtained from Norway spruce or European silver fir. Other types, e.g., that from various species of pines, are also offered under this name. Characterized by extreme tackiness; soluble in acetone and alcohol. Used to some extent in surgeon's tape and various special adhesive compositions.

burnable poison. A neutron absorber (or poison), such as boron, which, when purposely incorporated in the fuel or fuel cladding of a nuclear reactor, gradually "burns up" (is changed into nonabsorbing material) under neutron irradiation. This process compensates for the loss of reactivity that occurs as fuel is consumed and fission-product poisons accumulate, and keeps the overall characteristics of the reactor nearly constant during use. (Atomic Energy Commission).

burnt lime. See calcium oxide.

burnt sienna. See iron oxide red.

burnt umber. See umber.

"Buromin."[108] Trademark for sodium hexametaphosphate in glass plate form for boiler water conditioning.

"Burosil."[108] Trademark for a granular, alkaline, phosphate-silicate compound used in boiler water conditioning to precipitate calcium and magnesium as a loose sludge.

"Burundum."[326] Trademark for wear- and corrosion-resistant high-alumina ceramic bodies used in cylindrical form as grinding media for particle size reduction.

bushy stunt virus. A viral protein present in tomato plant infections; pH 4.1; molecular weight 7,600,000. See virus.

"Butacite."[28] Trademark for polyvinyl butyral resin; available as soft pliable sheeting, in 750-ft rolls, 10–84 in. wide. See polyvinyl acetal resins.

1,3-butadiene (vinylethylene; erythrene; bivinyl; divinyl) $H_2C{:}CHHC{:}CH_2$. 31st in order of high-volume chemicals produced in U.S. (1975).
Properties: Colorless gas with mild aromatic odor; easily liquefied; b.p. –4.41°C; sp. gr. 0.6211 (liquid at 20°C); f.p. –108.9°C; flash point –105°F; specific volume 6.9 cu ft/lb (700°F); autoignition temp. 804°F; vapor pressure 17.65 psia (0°C). Soluble in alcohol and ether; insoluble in water. The material polymerizes readily, particularly if oxygen is present, and the commercial material contains an inhibitor to prevent spontaneous polymerization during shipment or storage.
Derivation: (a) Catalytic dehydrogenation of butenes or butane; (b) oxidative dehydrogenation of butenes.
Methods of purification: Extractive distillation in the presence of furfural, absorption in aqueous cuprous ammonium acetate, or use of acetonitrile.
Grades: Technical, 98.0%; C.P. 99.0%; instrument 99.4%; research 99.8%.
Containers: Cylinders; pressure tank trucks and tank cars; ocean tankers.
Hazard: Highly flammable gas or liquid; flammable limits in air 2–11.5%; may form explosive peroxides on exposure to air. Must be kept inhibited during storage and in shipment; inhibitors often used are di-n-butylamine or phenyl-beta-naphthylamine. Storage is usually under pressure or in insulated tanks below 35°F. Moderately toxic; irritant to eyes. Tolerance, 1000 ppm in air. Safety data sheet available from Manufacturing Chemists Assn., Washington, D.C.
Uses: Synthetic elastomers (styrene-butadiene, polybutadiene, neoprene; nitriles); ABS resins; chemical intermediate.
Shipping regulations: (inhibited) (Rail) Red gas label; (Air) Flammable Gas label; not acceptable on passenger planes; (uninhibited) (Rail) Consult authorities; (Air) Not acceptable.

butadiene-acrylonitrile copolymer. See nitrile rubber.

butadiyne. See diacetylene.

butaldehyde. See butyraldehyde.

butallylonal (5-(2-bromoallyl)-5-sec-butylbarbituric acid) $C_{11}H_{15}BrN_2O_3$.
Properties: Fine, white crystals or crystalline powder; slightly bitter taste; m.p. 130–134°C; soluble in alcohol or ether; slightly soluble in cold water; insoluble in paraffin hydrocarbons.
Use: Medicine.

"Butamer."[416] Trademark for a process for the isomerization of normal butane to isobutane in the presence of hydrogen and a solid, noble metal catalyst of undisclosed composition. By recycling unconverted normal butane, ultimate yield of isobutane on a volumetric basis exceeds 100%.

butanal. See butyraldehyde.

butane (*n*-butane) C_4H_{10}.
Properties: Colorless gas; natural-gas odor; extremely stable; has no corrosive action on metals; does not react with moisture; very soluble in water; soluble in alcohol and chloroform; b.p. –0.5°C; f.p. –33°C; condensing pressure (approx.) 30 lb at 90°F; sp. gr. of liquid at 0°C, 0.599; sp. gr. of vapor at 0°C (air = 1) 2.07; critical temperature 153.2°C; critical pressure (absolute) 525 psi; heating value (77°F) 3266 Btu/cu ft; specific volume (70°F) 6.4 cu ft/lb; flash point –76°F; autoignition temp. 761°F. An asphyxiant gas.
Derivation: A by-product in petroleum refining or natural gasoline manufacture.
Grades: Research, 99.99 mole %; pure, 99 mole %; technical 95 mole %; also available in various mixtures with isobutane, propane, pentanes, etc.
Containers: Steel cylinders; tank cars; tank trucks; ocean tankers.
Hazard: Highly flammable, dangerous fire and explosion risk. Explosive limits in air 1.9 to 8.5%. Tolerance, 500 ppm in air.
Uses: Organic synthesis; raw material for synthetic rubber and high-octane liquid fuels; fuel for household and for many industrial purposes; manufacture of ethylene; solvent; refrigerant; standby and enricher gas; propellant in aerosols; pure grades used in calibrating instruments; food additive.
Shipping regulations: (Rail) Liquefied petroleum gas, Red gas label; (Air) Flammable Gas label. Not accepted on passenger planes.
Note: Butane in liquid form may be stored both above and below ground. Besides storage in liquefied form under its vapor pressure at normal atmospheric temperatures, refrigerated liquid storage at atmospheric pressure may be used. Such systems are closed and insulated, and the LP gas vapor is circulated through pumps and compressors to serve as the refrigerant for the system. Butane may be stored in pits in the earth capped by metal domes, and in underground chambers. (Compressed Gas Association). The foregoing also applies to propane (q.v.).

butanedial. See succinaldehyde.

1,4-butanedicarboxylic acid. See adipic acid.

butanedioic anhydride. See succinic anhydride.

1,3-butanediol. See 1,3-butylene glycol.

1,4-butanediol. See 1,4-butylene glycol.

2,3-butanediol. See 2,3-butylene glycol.

butanediolamine. See 2-amino-2-methyl-1,3-propanediol.

butanedione. See diacetyl.

2,3-butanedione oxime thiosemicarbazone $CH_3C(NOH)C(CH_3)N_2HCSNH_2$. A test reagent for manganese in very dilute solutions made from dimethylglyoxime and thiosemicarbazide.

butane dioxime. See dimethylglyoxime.

butanenitrile. See butyronitrile.

1-butanethiol (*n*-butyl mercaptan) C_4H_9SH.
Properties: Colorless liquid; sp. gr. 0.8412 (20/4°C); refractive index 1.4427 (20/D); flash point 35°F; b.p. 97.2–101.7°C; strong, obnoxious odor. Slightly soluble in water; very soluble in alcohol and ether.
Grades: 95%.
Hazard: Flammable, dangerous fire risk. Toxic by inhalation. Tolerance, 0.5 ppm in air.
Uses: Intermediate; solvent.

Shipping regulations: (Rail) Red label. (Air) Flammable Liquid label.

2-butanethiol (*sec*-butyl mercaptan) $C_2H_5CH(SH)CH_3$.
Properties: Colorless liquid; obnoxious odor. Boiling range 73–89°C; sp. gr. (20/4°C) 0.8288; refractive index (20/D) 1.4363; flash point -10°F.
Grades: 95%.
Containers: Cars and drums.
Hazard: Flammable, dangerous fire risk. Toxic by inhalation. Tolerance, 0.5 ppm in air.
Shipping regulations: (Rail) Red label. (Air) Flammable Liquid label.

1,2,4-butanetriol $HOCH_2CHOHCH_2CH_2OH$.
Properties: Almost colorless, odorless liquid; completely miscible in water and ethyl alcohol; hygroscopic. B.p. 312°C (extrap); sp. gr. 1.184; refractive index 1.473; flash point 332°F; combustible; nontoxic.
Derivation: Reaction of 2-butyne-1,4-diol with water, followed by reduction.
Grades: Technical; nitration.
Containers: Drums.
Uses: Intermediate for alkyd resins and explosives; cellulose plasticizer; emulsifier for cosmetics, inks, finishes, paper, cork, textiles.

butanoic acid. See butyric acid.

1-butanol. See n-butyl alcohol.

2-butanol acetate. See sec-butyl acetate.

2-butanone. See methyl ethyl ketone.

butanoyl chloride. See butyroyl chloride.

"Butaprene."[5] Trademark for synthetic rubbers, latexes, and resins comprising copolymers of butadiene with other monomers except those copolymers of butadiene with styrene which are classified as general purpose synthetic rubbers. See "FR-S."

"Butaprene PL."[35] Trademark for a complete series of latexes comprising copolymers of butadiene with styrene, acrylonitrile, acrylate esters, etc.
Properties: Air dries; forms continuous films capable of carrying high pigment and/or filler loadings. Keep from freezing.
Containers: 55-gal drums; tank trucks; tank cars.
Uses: Interior and exterior water-based paints; metal primers; adhesives; paper coating; grease-resistant coatings and saturants.

"Butaprene SL."[5] Trademark for a series of high styrene-butadiene copolymer resins used as reinforcing agents for rubber.
Grades: "Butaprene SL," "Butaprene SL/AB," "Butaprene SL-1" covering a range of hardness and dispersibility. All grades are supplied as white, friable, resinous crumbs.
Containers: 50-lb paper bags or suitably sized fiber drums.
Use: For imparting stiffness, strength, and abrasion resistance to rubber and synthetic rubber compounds, especially shoe soles, floor tile and similar compounds.

"Butarez" CTL.[303] Trademark for a liquid polybutadiene.

"Butasan."[58] Trademark for zinc dibutyldithiocarbamate (q.v.).

"Butazate."[248] Trademark for zinc dibutyldithiocarbamate (q.v.).

"Butazolidin."[219] Trademark for phenylbutazone (q.v.).

2-butenal. See crotonaldehyde.

butene-1 (ethylethylene; alpha-butylene) $CH_2{:}CHCH_2CH_3$. A liquefied petroleum gas.
Properties: Colorless gas; b.p. -6.3°C; sp. gr. 0.5951 (20/4°C); f.p. about -185°C. Specific volume 6.7 cu ft/lb (70°F); flash point -110°F; soluble in most organic solvents; insoluble in water. Autoignition temperature near 700°F.
Derivation: (a) Gases containing appreciable content of butene-1, along with other butene and butane hydrocarbons, are obtained by fractional distillation of refinery gas. (b) Can be produced directly from ethylene.
Grades: Technical, 95%; C.P. 99.0%; research 99.4%.
Containers: Cylinders, tanks.
Hazard: Highly flammable; flammable limits in air 1.6–9.3% by volume; dangerous fire and explosion risk. Asphyxiant gas.
Uses: Polymer and alkylate gasoline; polybutenes; butadiene; intermediate for C_4 and C_5 aldehydes, alcohols, and other derivatives; production of maleic anhydride by catalytic oxidation.
Shipping regulations: (Rail) Liquefied petroleum gas, Red Gas label. (Air) Flammable Gas label. Not acceptable on passenger planes.

cis-**butene-2** (dimethylethylene; beta-butylene; also called the "high-boiling" butene-2). $CH_3CH{:}CHCH_3$. Cis- and trans-butene-2 are geometric structural isomers. A liquefied petroleum gas.
Properties: Colorless gas; b.p. 3.7°C; sp. gr. 0.6213 (20/4°C); freezing point -139°C; specific volume 6.7 cu ft/lb (70°F); flash point -100°F; soluble in most organic solvents. Insoluble in water. Autoignition temp. 615°F.
Derivation: Gases containing appreciable content of cis-butene-2, along with other butene and butane hydrocarbons, are obtained by fractional distillation of refinery gas.
Grades: Technical 95%; C.P. 99%; research 99.8%.
Containers: Cylinders; tanks.
Hazard: Highly flammable. Flammable limits in air 1.8–9.7% by volume. Dangerous fire and explosion risk. Asphyxiant gas.
Uses: Solvent; cross-linking agent; polymer gasoline; butadiene synthesis; synthesis of C_4 and C_5 derivatives.
Shipping regulations: Same as butene-1.

trans-**butene-2** (dimethylethylene; beta-butylene; also called the "low-boiling" butene-2) $CH_3CH{:}CHCH_3$. Cis- and trans-butene-2 are geometrical structural isomers. A liquefied petroleum gas.
Physical properties: Colorless gas; b.p. 0.88°C; freezing point -105.8°C; sp. gr. 0.6042 (20/4°C); specific volume 6.7 cu ft/lb (70°F); flash point -100°F; soluble in organic solvents. Insoluble in water. Autoignition temp. 615°F.
Derivation: Gases containing appreciable content of trans-butene-2, along with other butene and butane hydrocarbons, are obtained by fractional distillation of refinery gas.
Grades: Technical, 95%; C.P. 99.0%; research 99.9%.
Hazard: Highly flammable. Flammable limits in air 1.8–9% by volume. Dangerous fire and explosion risk. Asphyxiant gas.

Uses: Same as for cis-butene-2.
Shipping regulations: Same as butene-1.

trans-**butenedioic acid.** See fumaric acid.

2-butene-1, 4-diol $HOCH_2CH:CHCH_2OH$.
Properties: Almost colorless, odorless liquid; very soluble in water, ethyl alcohol and acetone, sparingly soluble in benzene. Technical butenediol is predominantly the cis isomer. F.p. range 4.0–7.0°C; b.p. range 232–235°C; sp. gr. 1.067–1.074; refractive index (n 25/D) 1.476–1.478; flash point 263°F; combustible.
Derivation: By reduction of 2-butyne-1,4-diol; by high pressure synthesis from acetylene and formaldehyde.
Containers: Lined steel drums.
Hazard: Primary skin irritant.
Uses: Intermediate for alkyd resins, plasticizers, nylon, pharmaceuticals; cross linking agent for synthetic resins; fungicides.

3-butenenitrile. See allyl cyanide.

butenoic acid. See crotonic acid.

3-buten-2-one. See vinyl methyl ketone.

butethal (5-butyl-5-ethylbarbituric acid) $C_{10}H_{16}N_2O_3$.
Properties: White crystals or powder; odorless; bitter taste; m.p. 124–127°C; fairly soluble in alcohol or ether; practically insoluble in water.
Use: Medicine.
See also barbiturate.

butethamine hydrochloride [2-(Isobutylamino)ethyl para-aminobenzoate hydrochloride] $NH_2C_6H_4COOCH_2CH_2NHCH_2CH(CH_3)_2 \cdot HCl$.
Properties: White, odorless, crystals or crystalline power with bitter taste and local anesthetizing effects on tongue. M.p. 192–196°C; sparingly soluble in water; slightly soluble in alcohol and chloroform; very slightly soluble in benzene; practically insoluble in ether. pH (1% solution) about 4.7; stable in air.
Grade: N.F.
Use: Medicine.

"Buton."[29] Trademark for butadiene-styrene copolymers for surface coatings and thermosetting plastics. Characteristics include chemical resistance, hardness, flexibilty, adhesion, abrasion resistance, and gloss range. Cure temperatures range from 70° to 1100°F.
Grades and Uses:
100—Basic all-hydrocarbon resin for can coatings, metal sheet primers, and chemical intermediates.
200—Polar modification for baked metal primers, tank, and drum linings.
300—Higher polarity gives greater compatibility and reactivity for resistant lacquers and lowbake coatings.
A-500—High molecular weight resin. Excellent for laminating, potting and encapsulation.

butonate $(CH_3O)_2P(O)CH(CCl_3)OOCC_3H_7$. Generic name for O,O-dimethyl-2,2,2-trichloro-1-*n*-butyryloxyethyl phosphonate.
Properties: Colorless, somewhat oily liquid with slight ester odor; miscible with most organic solvents; stable in neutral or acid aqueous solutions; unstable in aqueous alkali. Sp. gr. 1.3742; refractive index 1.4707; wt/gal 11.5 lb.
Hazard: Toxic. Use may be restricted. Cholinesterase inhibitor.
Use: Insecticide.

butopyronoxyl (butyl mesityl oxide) $C_{12}H_{18}O_4$. n-Butyl 3,4-dihydro-2,2-dimethyl-4-oxo-1,2H-pyran-6-carboxylate.
Properties: Yellow to pale reddish-brown liquid with aromatic odor. Reasonably stable in air; slowly affected by light. Insoluble in water; miscible with alcohol, chloroform, ether, glacial acetic acid. Sp. gr. 1.052–1.060 (25/25°C); refractive index (n 25/D) 1.4745–1.4755. Distilling range 256–270°C. Low toxicity.
Derivation: Condensation of mesityl oxide and dibutyl oxalate in the presence of sodium ethoxide.
Grade: Technical.
Use: Insect repellent.

"Butoxone."[147] Trademark for a selective hormone-type weedkiller based on 2,4-dichlorophenoxybutyric acid (2,4-DB). Available as the isooctyl ester and dimethylamine salt.

2-butoxyethanol. See ethylene glycol monobutyl ether.

2-(2-butoxyethoxy)ethyl thiocyanate. See beta-butoxy-beta'-thiocyanodiethyl ether.

1-butoxyethoxy-2-propanol $CH_3CHOHCH_2OC_2H_4OC_4H_9$.
Properties: Colorless liquid; sp. gr. 0.9310 (20/20°C); b.p. 230.3°C; freezing point -90°C. Soluble in water; wt/gal 7.8 lb; flash point 250°F; combustible.
Uses: Solvent; hydraulic fluid components; anti-stall additive for automotive fuels; plasticizer intermediate.

butoxyethyl laurate. See ethylene glycol monobutyl ether laurate.

butoxyethyl oleate. See ethylene glycol monobutyl ether oleate.

butoxyethyl stearate. See ethylene glycol monobutyl ether stearate.

"Butoxyne."[307] Trademark for a series of specialty chemicals.
Properties: Nonionic, ashless, 100% active materials differing in alkyl substituent. Effective at high temperatures and in concentrated electrolyte solutions. Stable to acids and alkalies.
Uses: Emulsifiers for water-in-oil systems; corrosion and rust inhibitors, gelling and thickening agents for a wide variety of organic liquids.

para-**butoxyphenol** $HOC_6H_4OC_4H_9$.
Properties: White to faint yellow crystalline powder; m.p. 61–65°C. Soluble in alcohol, acetone, ether, benzene, aqueous alkali; insoluble in water. Combustible.
Grades: 93% pure.
Containers: Fiber drums.
Use: Synthesis.

butoxy polypropylene glycol. Generic name for polypropylene glycol monobutyl ether, probably $CH_3CHOH(CH_2OCHCH_3)_n\ CH_2OC_4H_9$. Colorless liquid; used as an insect repellent. Low toxicity.

n-**butoxypropanol.**
Properties: Colorless liquid; sp. gr. 0.8801 (20/20°C); b.p. 170.2°C; f.p. -80°C (sets to glass); soluble in water. Flash point 154°F. Combustible.
Uses: Solvent for water-based enamels.

beta-**butoxy**-beta'-**thiocyanodiethyl ether** [2-(2-butoxyethoxy)ethyl thiocyanate] $CH_3(CH_2)_3OCH_2CH_2OCH_2CH_2SCN$.
Hazard: Moderately toxic. Skin irritant.
Use: Insecticide.

butoxytriglycol (triethylene glycol monobutyl ether) $C_4H_9O(C_2H_4O)_3H$.
Properties: Liquid; sp. gr. 1.0021 (20/20°C); b.p. de-

composes; f.p. –47.6°C. Completely soluble in water. Flash point 290°F. Combustible. Low toxicity.
Uses: Plasticizer intermediate.

buttercup yellow. See zinc yellow.

butter. (1) A colloidal system (emulsion) in which the continuous phase is composed of liquid fat from fat globules disintegrated by mechanical agitation, and the disperse phase is comprised of finely divided water droplets and undamaged fat globules. (2) Outmoded term for hygroscopic metallic chlorides of viscous consistency, e.g., butter of zinc, etc.

butter fat. The oily portions of the milk of mammals. Composition is largely glycerides of oleic, stearic, and palmitic acids with smaller amounts of the glycerides of butyric, caproic, caprylic and capric acids. Sp. gr. range 0.910–0.914. Cow's milk contains about 4% butter fat. See also milk.

butter yellow. See dimethylaminoazobenzene.

"Butvar." Trademark for polyvinyl butyral resins. See polyvinyl acetal resin.
Uses: Coatings (metal, textile, wood, etc.); film; adhesives; sealers; molded materials; insulation and safety glass.

butyl. (1) The group C_4H_9. (2) butyl rubber.

n-**butyl acetate** $CH_3COOCH_2CH_2CH_2CH_3$.
Properties: Colorless liquid; fruity odor. Soluble in alcohol, ether, and hydrocarbons; slightly soluble in water. Vapor is heavier than air. Sp. gr. 0.8826 (20/20°C); b.p. 126.3°C; vapor pressure 8.7 mm Hg (20°C; freezing point –75°C; refractive index 1.3951 (20°C); wt/gal 7.35 lb (20°C); flash point 98°F. (TOC). Autoignition temp. 790°F.
Derivation: Esterification and then distillation, after contact of butyl alcohol with acetic acid in the presence of a catalyst such as sulfuric acid.
Containers: 1-gal cans; 5-, 55-gal drums; tank cars.
Hazard: Flammable; moderate fire risk. Moderately toxic; skin irritant. Tolerance, 150 ppm in air.
Uses: Solvent in nitrocellulose lacquers, leather dressings, perfumes, flavoring extracts; solvent for natural gums and synthetic resins; dehydrating agent.
Shipping regulations: (Air) Flammable Liquid label.

sec-**butyl acetate** $CH_3COOCH(CH_3)(C_2H_5)$. (2-butanol acetate.)
Properties: Colorless liquid; b.p. 112.2°C; sp. gr. 0.8905 at 0/4°C, 0.870 at 20/4°C; refractive index 1.389 (20°C); wt/gal 7.21 lb; flash point (open cup) 88°F. Miscible with alcohol and ether; insoluble in water.
Derivation: Esterification of sec-butyl alcohol.
Containers: 1-gal cans; 55-gal drums; tank cars.
Hazard: Flammable, moderate fire risk. Moderately toxic. Tolerance, 200 ppm in air.
Uses: Solvent for nitrocellulose lacquers; thinners; nail enamels; leather finishes.
Shipping regulations: (Air) Flammable Liquid label.

tert-**butyl acetate** $CH_3COOC(CH_3)_2$.
Properties: Colorless liquid, b.p. 96°C; sp. gr. 0.896 (20°C). Insoluble in water; soluble in alcohol and ether.
Hazard: Flammable, moderate fire risk. Moderately toxic. Tolerance, 200 ppm in air.
Uses: Solvent.

butyl acetoacetate $CH_3COCH_2COOCH_2CH_2CH_2CH_3$.
Properties: Colorless liquid; insoluble in water; soluble in alcohol and ether. Sp. gr. 0.9694 (20/20°C); b.p. 213.9°C; vapor pressure 0.19 mm (20°C); flash point 185°F; wt/gal 8.1 lb (20°C). Combustible.
Grades: Technical.
Use: Intermediate in synthesis of metal derivatives, dyestuffs, pharmaceuticals; flavoring.

butyl acetoxystearate
$CH_3(CH_2)_5CH(CH_3COO)(CH_2)_{10}COOC_4H_9$.
Properties: See butyl acetyl ricinoleate.
Derivation: From castor oil, butyl alcohol, and acetic anhydride, with hydrogenation.
Uses: Plasticizer; textile oils; adhesives.

butyl acetylene. See 1-hexyne.

butyl acetyl ricinoleate
$CH_3(CH_2)_5CH(CH_3CO_2)CH_2(CH)_2(CH_2)_7CO_2C_4H_9$.
Properties: Yellow, oily liquid; mild odor; miscible with most organic solvents. Sp. gr. 0.940 (20/20°C); saponification number 235; f.p., indefinite, becomes cloudy at –32°C, solidifies at –65°C; flash point 230°F (open cup); refractive index 1.4614 (20°C); Saybolt viscosity 123 secs at 100°F; wt/gal 7.8 lb (68°F); practically insoluble in water. Combustible. Autoignition temp. 725°F.
Derivation: From castor oil, butyl alcohol and acetic anhydride.
Grades: Technical.
Containers: 1-gal cans; 5-, 55-gal steel drums; tank cars.
Uses: Plasticizer; emulsifier; lubricant; detergent; protective coatings; special cleansing compounds; quick-breaking emulsions.

n-**butyl acid phosphate** (n-butylphosphoric acid; acid butyl phosphate).
Properties: Water-white liquid; sp. gr. 1.120–1.125 (25/4°C); refractive index (n 25/D) 1.429; flash point (COC) 230°F. Soluble in alcohol, acetone, and toluene. Insoluble in water and petroleum naphtha. Combustible.
Containers: Up to 55-gal drums.
Hazard: Strong irritant to skin and tissue.
Uses: Esterification catalyst and polymerizing agent; curing catalyst and accelerator in resins and coatings; special detergents.
Shipping regulations: (Air) Corrosive label.

N-tert-**butylacrylamide** $H_2C{:}CHCONHC(CH_3)_3$.
Properties: White crystalline solid; m.p. 128–130°C; sp. gr. 1.015 (30°C). Soluble in methanol, ethyl alcohol, chloroform, and acetone. Combustible.
Hazard: Moderately toxic by ingestion and inhalation. Moderately irritant to skin.
Uses: Monomer; organic intermediate.

n-**butyl acrylate** $CH_2{:}CHCOOC_4H_9$.
Properties: Colorless liquid; m.p. –64°C; boiling range 145.7–148.0°C (760 mm); polymerizes readily on heating; vapor pressure (20°C) 3.2 mm; sp. gr. 0.9015 (20/20°C); wt/gal 7.5 lb (20°C); flash point 120°F (open cup); nearly insoluble in water. Combustible.
Derivation: Reaction of acrylic acid or methyl acrylate with butyl alcohol.
Grades: Technical (inhibited).
Containers: 1-gal cans; 5-, 55-gal drums; tank cars.
Hazard: Moderate fire risk; moderately toxic and irritant.

Uses: Intermediate in organic synthesis; polymers and copolymers for solvent coatings, adhesives, paints, binders; emulsifier.
See also acrylic resin.

tert-**butyl acrylate** $CH_2:CHCOOC(CH_3)_3$.
Properties: Liquid; b.p. 120°C; sp. gr. 0.879 (25°C); refractive index n 25/D 1.4080; flash point 66°F (TOC). Commercial grade contains 100 ppm hydroquinone monomethyl ether as stabilizer.
Hazard: Flammable, dangerous fire risk.
Use: Monomer for acrylic resins.
Shipping regulations: (Rail) Flammable liquid, n.o.s., Red label. (Air) Flammable liquid, n.o.s., Flammable Liquid label.

n-**butyl alcohol** (1-butanol; butyric alcohol) $CH_3(CH_2)_2CH_2OH$.
Properties: Colorless liquid; vinous odor. B.p. 117.7°C; f.p. -89.0°C; sp. gr. (20/20°C) 0.8109; wt/gal (20°C) 6.76 lb; refractive index (n 20/D) 1.3993; flash point 95°F. Solubility in water (20°C) 7.7 wt %; solubility of water in n-butyl alcohol 20.1%. Miscible with alcohol and ether. Autoignition temp. 689°F.
Derivation: (a) hydrogenation of butyraldehyde, obtained in the Oxo process; (b) condensation of acetaldehyde to form crotonaldehyde, which is then hydrogenated (aldol condensation).
Containers: 1-, 5-, 55-gal drums; tank cars; tank trucks.
Hazard: Toxic on prolonged inhalation; irritant to eyes and skin. Tolerance, 100 ppm in air. Flammable, moderate fire risk.
Uses: Preparation of esters, especially butyl acetate; solvent for resins and coatings; plasticizers; dyeing assistant; hydraulic fluids; detergent formulations; dehydrating agent (by azeotropic distillation); intermediate; "butylated" melamine resins; glycol ethers; butyl acrylate.
Shipping regulations: (Air) Flammable Liquid label.

sec-**butyl alcohol** (SBA; 2-butanol; methylethylcarbinol) $CH_3CH_2CHOHCH_3$.
Properties: Colorless liquid; strong odor. B.p. 99.5°C; f.p. -114.7°C; sp. gr. (20/4°C) 0.808; wt/gal (20°C) 66.74 lb; refractive index (n 25/D) 1.3949; flash point (closed cup) 75°F. Autoignition temp. 763°F. Moderately soluble in water; miscible with alcohol and ether.
Derivation: Absorption of butene, from cracking petroleum or natural gas, in sulfuric acid with subsequent hydrolysis by steam.
Grades: Technical.
Containers: Drums; tank cars; tank trucks.
Hazard: Toxic on prolonged inhalation; irritant to eyes and skin. Flammable, dangerous fire risk. Tolerance, 150 ppm in air.
Uses: Preparation of methyl ethyl ketone; solvent; organic synthesis.
Shipping regulations: (Rail) Red label. (Air) Flammable Liquid label.

tert-**butyl alcohol** (2-methyl-2-propanol; trimethyl carbinol) $(CH_3)_3COH$.
Properties: Colorless liquid or crystals; camphor odor. f.p. 25.5°C; b.p. 82.9°C; sp. gr. (liquid, 26°C) 0.779; refractive index (n 20/D) 1.3878; flash point (closed cup) 52°F. Autoignition temp. 892°F. Miscible with water, alcohol and ether.
Derivation: Absorption of isobutene, from cracking petroleum or natural gas, in sulfuric acid with subsequent hydrolysis by steam.
Grades: Technical.
Containers: Drums; tanks.
Hazard: Irritant to eyes and skin. Flammable, dangerous fire risk. Tolerance, 100 ppm in air. Toxic on prolonged inhalation.
Uses: Alcohol denaturant; solvent for pharmaceuticals; dehydration agent; perfumery; chemical intermediate.
Shipping regulations: (Rail) Red label. (Air) Flammable Liquid label.

n-**butyl aldehyde.** See butyraldehyde.

n-**butylamine** (1-aminobutane) $C_4H_9NH_2$.
Properties: Colorless, volatile liquid with amine odor. B.p. 77.1°C; f.p. -49.1°C; sp. gr. (20/20°C) 0.7385; wt/gal (20°C) 6.2 lb; refractive index (n 20/D) 1.401; flash point (open cup) 30°F. Miscible with water, alcohol, and ether.
Derivation: Reaction of butanol or butyl chloride with ammonia.
Grades: Technical.
Containers: Drums; tank cars.
Hazard: Flammable, dangerous fire risk. Skin irritant. Toxic; tolerance, 5 ppm in air.
Uses: Intermediate for emulsifying agents, pharmaceuticals, insecticides, rubber chemicals, dyes, tanning agents.
Shipping regulations: (Rail) Flammable liquid, n.o.s., Red label. (Air) Flammable Liquid label.

sec-**butylamine** (2-aminobutane) $CH_3CHNH_2C_2H_5$.
Properties: Colorless liquid; sp. gr. (20°C) 0.725; boiling range 63–68°C; refractive index (20°C) 1.395; solidification point -104.5°C; odor, amine; flash point 15°F; wt/gal (20°C) 6.0 lb.
Hazard: Flammable, dangerous fire risk. Moderately toxic.
Shipping regulations: See *n*-butylamine.

tert-**butylamine** $(CH_3)_3CNH_2$.
Properties: Colorless liquid; b.p. 44–46°C; sp. gr. 0.700 (15°C); refractive index (n 18/D) 1.3794; flash point about 50°F. Miscible with water; soluble in common organic solvents.
Grades: Technical.
Containers: Drums; tank cars.
Hazard: Moderately toxic; skin irritant. Flammable, dangerous fire risk.
Uses: Intermediate for rubber accelerators, insecticides, fungicides, dyestuffs, pharmaceuticals.
Shipping regulations: See *n*-butylamine.

butyl ortho-**aminobenzoate.** See butyl anthranilate.

n-**butyl** para-**aminobenzoate** $H_2NC_6H_4COOC_4H_9$.
Properties: White, crystalline powder, odorless, tasteless; m.p. 57–59°C; b.p. 174°C (8 mm). Soluble in dilute acids, alcohol, chloroform, ether, and fatty oils. Almost insoluble in water.
Grade: N.F.
Containers: Polyethylene-lined fiber drums.
Use: Medicine (local anesthetic); treatment of burns; ointments; UV absorber in suntan preparations.

N-n-butylaminoethanol $C_4H_9NHC_2H_4OH$.
Properties: Liquid; sp. gr. 0.88–0.99 (20/20°C); distillation range 192–210°C; wt/gal 7.4 lb; flash point 170°F. Combustible.

tert-**butylaminoethyl methacrylate** $CH_2:C(CH_3)COOCH_2CH_2NHC(CH_3)_3$.
Properties: Liquid; b.p. 100–105°C (12 mm); density 0.914 g/ml (25°C); 7.61 lb/gal; refractive index 1.4440 (25°C); flash point 205°F (COC). Combustible.
Uses: Coatings; textile chemicals; dispersing agent for nonaqueous systems; antistatic agent; stabilizer for chlorinated polymers; ion exchange resins; emulsifying agent; cationic precipitating agent.

N′-n-**butyl-3-amino-4-methoxybenzenesulfonamide** $CH_3OC_6H_3(NH_2)SO_2NHC_4H_9$. Pink powder; m.p. 96–100°C; insoluble in water. Partially soluble in alcohol and acetone. Used as an intermediate.

butyl-para-**aminophenol.** See Gasoline Antioxidant No. 5.

N-n-**butylaniline** $C_6H_5NHC_4H_9$.
Properties: Amber liquid. Sp. gr. (20°C) 0.932; boiling range 236–242°C; refractive index 1.534 (20°C); odor aniline; very soluble in alcohol and ether; insoluble in water. Flash point 225°F. Combustible.
Uses: Organic synthesis; dyes.
Hazard: Probably toxic.

butyl anthranilate (butyl ortho-aminobenzoate) $C_4H_9OOCC_6H_4NH_2$. Used in flavoring.

2-tert-**butylanthraquinone** $C_{18}H_{16}O_2$.
Properties: Yellow powder. M.p. 102–104°C; soluble in alcohol and acetone. Combustible; low toxicity.
Grades: Technical (98%).
Use: Organic synthesis.

butylated hydroxyanisole (BHA) $(CH_3)_3CC_6H_3OH(OCH_3)$. A mixture of 2- and 3-*tert*-butyl-4-methoxyphenol.
Properties: White or slightly yellow, waxy solid having a faint characteristic odor; melting range 48–63°C. Not naturally water-soluble, but can be made so by special treatment. Freely soluble in alcohol and propylene glycol. Combustible.
Grade: F.C.C.; water-soluble.
Hazard: Moderately toxic by ingestion. Use in foods restricted; consult FDA regulations.
Uses: Antioxidant for fats and oils; food packaging.

butylated hydroxytoluene. See 2,6-di-tert-butyl-para-cresol.

n-**butylbenzene** (1-phenylbutane) $C_6H_5C_4H_9$.
Properties: Colorless liquid; b.p. 183.2°C; f.p. –87.9°C; sp. gr. 0.860 (20°C); refractive index (n 20/D) 1.489; flash point 160°F (open cup). Combustible. Autoignition temp. 774°F.
Grades: Technical; pure; research.
Hazard: Moderately toxic by ingestion.
Uses: Organic synthesis, especially of insecticides.

sec-**butylbenzene** (2-phenylbutane) $C_6H_5C(CH_3)C_2H_5$.
Properties: Colorless liquid; b.p. 170.65°C; vapor pressure 15 mm Hg (60°C; f.p. –75.68°C; sp. gr. 0.8618 (204°C; wt/gal 7.2 lb (20°C); refractive index (n 20/D) 1.4901; flash point 145°F (open cup). Combustible. Autoignition temp. 784°F.
Grades: Technical, 95%; pure; research.
Containers: Drums; tank trucks and tank cars.
Hazard: Moderately toxic by ingestion.
Use: Solvent for coating compositions and organic synthesis; plasticizer; surface-active agents.

tert-**butylbenzene** (2-methyl-2-phenylpropane) $C_6H_5C(CH_3)_3$.
Properties: Colorless liquid; insoluble in water; soluble in alcohol; b.p. 169.1°C; f.p. –57.8°C; sp. gr. 0.866 (20°C); refractive index (n 20/D) 1.492; flash point 140°F (open cup). Combustible. Autoignition temp. 842°F.
Grades: Technical; pure; research.
Hazard: Moderately toxic by ingestion.
Use: Organic synthesis; polymerization solvent; polymer linking agent.

butylbenzenesulfonamide (N-n-butyl benzenesulfonamide) $C_6H_5SO_2NHC_4H_9$.
Properties: Liquid; pleasant odor; amber to straw color; sp. gr. 1.148 (25/25°C); refractive index 1.5235 (25°C); b.p. 189–190°C (4.5 mm).
Hazard: Moderately toxic by ingestion.
Uses: Synthesis of dyes, pharmaceuticals, and other organic chemicals; in resin manufacturing; and as plasticizer for some synthetic polymers.

butyl benzoate (n-butyl benzoate) $C_6H_5COOC_4H_9$.
Properties: Colorless oily liquid; insoluble in water; miscible with alcohol or ether. Sp. gr. 1.00 (20°C); b.p. 247.3°C; m.p. –22°C; flash point 225°F (open cup). Combustible. Low toxicity.
Grades: Technical.
Uses: Solvent for cellulose ether; plasticizer; perfume ingredient; dyeing of textiles.

N-tert-**butyl-2-benzothiazolesulfenamide** $C_6H_4NCS(SNHC_4H_9)$.
Properties: Light buff powder or flakes; (sometimes colored blue); m.p. 104°C min; sp. gr. 1.29 at 25°C. Soluble in most organic solvents. Combustible.
Use: Rubber accelerator.

butyl benzyl phthalate (BBP) $C_4H_9OOCC_6H_4COOC_7H_7$.
Properties: Clear, oily liquid. Slight odor. Sp. gr. 1.113–1.121 (25/25°C). Flash point 390°F. Combustible. Low toxicity.
Grades: Technical.
Containers: 5- and 55-gal drums; tank cars.
Uses: Plasticizer for polyvinyl and cellulosic resins; organic intermediate.

butyl benzyl sebacate $C_4H_9OOC(CH_2)_8COOC_7H_7$. Ester used as plasticizer.
Properties: Light straw colored liquid. B.p. (10 mm) 245–285°C; sp. gr. 1.023 (25°C); wt/gal 8.6 lb; flash point 395°F. Combustible; low toxicity.
Uses: Plasticizer for resins.

butyl borate. See tributyl borate.

n-**butyl bromide** (1-bromobutane) C_4H_9Br.
Properties: Colorless liquid; sp. gr. (20/20°C) 1.279; b.p. 101.6°C; f.p. –112.4°C. Flash point 75°F (open cup). Autoignition temp. 509°F. Insoluble in water; soluble in alcohol and ether.
Grade: 99.7%.
Hazard: Probably toxic. Flammable; dangerous fire risk.
Use: Alkylating agent.
Shipping regulations: (Rail) Flammable liquid, n.o.s., Red label. (Air) Flammable Liquid label.

sec-**butyl bromide** (2-bromobutane) $CH_3CHBrCH_2CH_3$.
Properties: Clear, colorless liquid with pleasant odor; boiling point 91.2°C; sp. gr. 1.2425 (25/25°C); refractive index at 25°C 1.4320–1.4344; flash point 70°F (open cup); soluble in alcohol and ether; insoluble in water.
Containers: Steel drums.
Hazard: Narcotic in high concentrations. Flammable; dangerous fire risk.
Use: Synthesis; alkylating agent.
Shipping regulations: (Rail) Flammable liquid, n.o.s., Red label. (Air) Flammable liquid, n.o.s., Flammable Liquid label.

butyl butanoate. See n-butyl butyrate.

n-butyl butyrate (butyl butanoate) $CH_3(CH_2)_2COOC_4H_9$.
Properties: Colorless liquid; sp. gr. 0.8721 (20/20°C); refractive index (n 20/D) 1.4059; f.p. -91.5°C; b.p. 165.7°C (736 mm); flash point 128°F. (o.c.); insoluble in water; soluble in alcohol and ether. Combustible.
Grade: F.C.C.
Hazard: Moderately toxic; irritant and narcotic; moderate fire risk.
Use: Flavoring.

butylcarbamoylsulfanilamide. See 1-butyl-3-sulfanilylurea.

n-**butyl carbinol.** See n-amyl alcohol, primary.

sec-**butyl carbinol.** See 2-methyl-1-butanol.

butyl "Carbitol."[214] Trademark for diethylene glycol monobutyl ether (q.v.).

butyl "Carbitol" acetate.[214] Trademark for diethylene glycol monobutyl ether acetate.

para-tert-**butylcatechol** (4-tert-butyl-1,2-dihydroxybenzene) $(CH_3)_3CC_6H_3(OH)_2$.
Properties: Colorless crystals; m.p. 56–57°C; sp. gr. 1.049 (60/25°C); b.p. 285°C; flash point 265°F. Combustible; soluble in ether, alcohol, and acetone; slightly soluble in water at 80°C.
Containers: Glass bottles; lined drums.
Hazard: Skin irritant.
Use: Polymerization inhibitor for styrene-butadiene and other olefins.

butyl "Cellosolve."[214] Trademark for ethylene glycol monobutyl ether (q.v.).

butyl "Cellosolve" acetate.[214] Trademark for ethylene glycol monobutyl ether acetate (q.v.).

butyl chloral hydrate (trichlorobutyraldehyde hydrate) $CH_3CHClCCl_2CH(OH)_2$.
Properties: Colorless leaflets; sp. gr. 1.693 (20/4°C); m.p. 78°C. Slightly soluble in water; soluble in alcohol and ether.
Derivation: Action of chlorine on paraldehyde.
Use: Medicine. (hypnotic, anticonvulsant)

n-**butyl chloride** (1-chlorobutane) $CH_3CH_2CH_2CH_2Cl$ or C_4H_9Cl.
Properties: Colorless liquid. Insoluble in water. Miscible with alcohol and ether. Sp. gr. 0.8875 (20/20°C); b.p. 78.6°C; wt/gal 7.35 lb (20°C); refractive index 2.4015 (20°C); vapor pressure 80.1 mm (20°C); f.p. -122.8°C; viscosity 0.0045 poise (20°C); flash point 15°F (open cup). Autoignition temp. 860°F.
Grades: N.F.; technical.
Containers: Drums; tanks.
Hazard: Flammable; dangerous fire risk. Toxic on prolonged inhalation.
Uses: Organic synthesis (alkylating agent); solvent; anthelmintic.
Shipping regulations: (Rail) Flammable liquid, n.o.s., Red label. (Air) Flammable Liquid label.

sec-**butyl chloride** $CH_3CHClCH_2CH_3$. Clear liquid; sp. gr. 0.875 (20/4°C); refractive index 1.3963 (20°C); 99% pure.
Hazard: May be flammable.
Uses: Organic synthesis.

butyl citrate. See tributyl citrate.

6-tert-**butyl**-meta-**cresol** (MBMC; 6-tert-butyl-3-methylphenol) $(CH_3)_3CC_6H_3(OH)CH_3$.
Properties: Clear liquid; solidifies slightly below room temperature; f.p. 23.1°C; b.p. 244°C; sp. gr. 0.922 (80°C). Soluble in organic solvents and aqueous potassium hydroxide. Flash point 116°F. Combustible.
Containers: 5- and 55-gal drums; tank cars.
Hazard: Moderate fire risk. May be toxic and irritant to skin.
Uses: Germicide; disinfectant; synthesis of antioxidants and rubber-processing chemicals; additives to lubricating oils; synthetic resins; perfumes (fixative).

butyl crotonate $CH_3CH{:}CHCOOC_4H_9$.
Properties: Water-white liquid; persistent odor. Sp. gr. 0.9037 (20/20°C); b.p. 180.5°C; wt/gal 7.52 lb (20°C); soluble in alcohol and ether; insoluble in water. Combustible.

butyl cyclohexyl phthalate $C_4H_9OOCC_6H_4COOC_6H_{11}$.
Properties: Clear liquid; very mild odor; sp. gr. 1.078; saponification number 369; acidity (as phthalic acid), 0.01% max.; miscible with most organic solvents. Probably low toxicity. Combustible.
Containers: Drums; tank trucks; tank cars.
Use: Plasticizer for polymers and elastomers; nitrocellulose lacquers.

butyl decyl phthalate $C_4H_9OOCC_6H_4COOC_{10}H_{21}$.
Properties: Clear, oily liquid; sp. gr. 0.977 to 0.987 (25/25°C). Probably low toxicity. Combustible.
Containers: Drums; tank cars and trucks.
Uses: Primary plasticizer for polyvinyl chloride and copolymer resins.

n-**butyldiamylamine** $C_4H_9N(C_5H_{11})_2$.
Properties: Straw-colored liquid. Sp. gr. (20°C) 0.788; boiling range 229–241°C; odor amine; flash point 200°F. Combustible.

n-**butyldichloroarsine** $C_4H_9AsCl_2$.
Properties: Oily liquid. Somewhat agreeable odor. Decomposed by water. B.p. 192–194°C.
Hazard: Highly toxic by inhalation and ingestion.
Uses: Military poison gas.
Shipping regulations: (Rail) Poison gas label. Not accepted by rail express. (Air) Poisonous gases, n.o.s. Not acceptable.

butyl dichlorophenoxyacetate. See 2,4-D.

1-n-butyl-3-(3,4-dichlorophenyl)-1-methylurea (neburon) $Cl_2C_6H_3NHCONCH_3(C_4H_9)$.
Properties: White crystalline solid; m.p. 102°C; very low solubility in water and hydrocarbon solvents. Stable towards oxidation and moisture.
Use: Weed killer.

n-**butyl diethanolamine** $C_4H_9N(CH_2CH_2OH)_2$.
Properties: Liquid; sp. gr. (20°C) 0.97; b.p. 272°C; color very light straw; odor faint amine; wt/gal 8.08 lb (20°C); flash point 245°F. Combustible; low toxicity.

n-**butyl diethyl malonate** $C_4H_9CH(COOC_2H_5)_2$.
Properties: Colorless liquid with a fruity odor; sp. gr. 0.972–0.974 (25/25°C); refractive index 1.420–1.422 (n 25/D); soluble in alcohols, ketones and esters. Combustible.
Containers: Bottles; tins; drums.
Use: Intermediate.

butyl diglycol carbonate (diethylene glycol bis(n-butylcarbonate) $(C_4H_9OCO_2CH_2CH_2)_2O$.
Properties: Colorless liquid of low volatility; insoluble in water (very stable to hydrolysis); widely soluble in organic solvents; compatible with many resins and plastics. Sp. gr. 1.07 (20/4°C); boiling range 164–166°C (2 mm); flash point 372°F; Saybolt viscosity 21 cps (20°C); refractive index 1.435 (20°C);

evaporation rate 0.59 mg/sq cm/hr (100°C). Combustible.
Uses: Plasticizer; high-boiling-point solvent and softening agent; manufacture of pharmaceuticals and lubricant compositions.

4-tert-**butyl-1,2-dihydroxybenzene.** See para-tert-butylcatechol.

Butyl "Dioxitol."[125] Trademark for diethylene glycol monobutyl ether (q.v.).

4-butyl-1,2-diphenyl-3,5-pyrazolidinedione. See phenylbutazone.

butyldithiocarbonic acid. See butylxanthic acid.

butyl dodecanoate. See butyl laurate.

"Butyl Eight."[69] Trademark for an ultra-accelerator for rubber of the dithiocarbamate type.
Properties: Dark red liquid; odor distinct; sp. gr. 1.01; low toxicity. Partly soluble in water; soluble in acetone, benzene, carbon disulfide, chloroform, and gasoline.

butylene (butene). One of the liquefied petroleum gases butene-1, *cis*-butene-2, *trans*-butene-2, and isobutene. See specific entry for details.

butylene dimethacrylate
$CH_2{:}C(CH_3)COOCH_2CH_2C(CH_3)HOOCC(CH_3){:}CH_2$.
Properties: Liquid; boiling range 110°C (3 mm); density 1.011 g/ml (25/15.6°C); refractive index 1.4502 (25°C); flash point > 150°F. Combustible.
Use: Monomer for resins.

1,3-butylene glycol (1,3-butanediol)
$HOCH_2CH_2CH(OH)CH_3$. Can exist in optical isomeric forms.
Properties: Practically colorless, viscous liquid; hygroscopic; sp. gr. 1.0059 (20/20°C); 8.4 lb/gal (20°C); b.p. 207.5°C; vapor pressure 0.06 mm (20°C); refractive index 1.4401 (20°C); flash point (COC) 250°F; completely soluble in water and alcohol; slightly soluble in ether. Combustible. Autoignition temp. 741°F. Low toxicity.
Derivation: Reduction of aldol.
Containers: Drums; tank cars.
Uses: Polyesters; polyurethanes; surface active agents; plasticizers; humectant; coupling agent; solvent; food additive and flavoring.

1,4-butylene glycol (1,4-butanediol; tetramethylene glycol) $HOCH_2CH_2CH_2CH_2OH$.
Properties: Colorless, oily liquid. B.p. 230°C; m.p. 16°C; sp. gr. 1.020 (20/4°C). Flash point over 250°F. Miscible with water; soluble in alcohol; slightly soluble in ether. Combustible.
Derivation: From acetylene and formaldehyde by high-pressure synthesis.
Grades: Technical.
Containers: Drums; tank cars.
Uses: Solvents; humectant; intermediate for plasticizers, pharmaceuticals; cross-linking agent in polyurethane elastomers; manufacture of tetrahydrofuran; terephthalate plastics.

2,3-butylene glycol (2,3-dihydroxybutane; 2,3-butanediol; pseudobutylene glycol; sym-dimethylethylene glycol). Can exist in optical isomeric forms.
$CH_3CHOHCHOHCH_3$.
Properties: Nearly colorless crystalline solid or liquid; hygroscopic; sp. gr. 1.045 (20/20°C); m.p. 23–27°C; b.p. 179–182°C; refractive index 1.438 (20°C); flash point 185°F (open cup); soluble in alcohol and ether; miscible with water in all proportions. Combustible; low toxicity.
Derivation: From corn sugar by acid hydrolysis; also from fermentation of sugar beet molasses.
Grades: 99%.
Uses: Resins; solvent for dyes; intermediate; blending agent.

1,2-butylene oxide (1,2-epoxybutane)
$H_2\underline{COC}HCH_2CH_3$.

Properties: Colorless liquid; sp. gr. 0.8312 (20/20°C); b.p. 63°C; sets to a glass below -150°C; flash point (closed cup) near 0°F. Soluble in water and miscible with most organic solvents.
Grades: About 97.5% purity.
Hazard: Highly flammable, dangerous fire risk. Toxic concentrations occur at room temperature.
Uses: Intermediate, especially for various polymers; stabilizer for chlorinated solvents.
Shipping regulations: (Rail) Flammable liquid, n.o.s., Red label. (Air) Flammable liquid, n.o.s., Flammable Liquid label.

2,3-butylene oxide (2,3-epoxybutane)
$CH_3H\underline{COC}HCH_3$.

Two forms, cis and trans, are known.
Properties: Cis: f.p. -80°C; b.p. 59.7°C (742 mm); sp. gr. 0.8266 (25/4°C). Trans: f.p. -85°C; b.p. 53.5°C (742 mm); sp. gr. 0.8010 (25/4°C). Flash point near 0°F. Very soluble in ether, benzene and organic solvents; decomposes in water.
Hazard: Highly flammable; dangerous fire risk. Toxic concentrations occur at room temperature.
Use: Intermediate.
Shipping regulations: See 1,2-butylene oxide.

butyl 9,10-epoxyoctadecanoate. See butyl epoxystearate.

butyl epoxystearate (butyl 9,10-epoxyoctadecanoate)

$CH_3(CH_2)_7\overline{CHOC}H(CH_2)_7COOC_4H_9$.
Properties: Colorless liquid with mild slightly fatty, slightly fruity odor; sp. gr. (20°C) 0.910; wt/gal 7.59 lb. Combustible.
Containers: 1-, 5-gal cans; 55-gal drums; tank cars.
Use: Plasticizer for low-temperature flexibility improvement of vinyl resins.

n-**butylethanolamine** $C_4H_9NHCH_2CH_2OH$.
Properties: Colorless liquid. Sp. gr. (20°C) 0.892; boiling range 194–204°C; odor, very faint amine type; flash point 170°F. Combustible. Low toxicity.

butyl ether (n-dibutyl ether) $C_4H_9OC_4H_9$.
Properties: Colorless liquid; stable; mild, ethereal odor; sp. gr. 0.7694 (20/20°C); b.p. 142.2°C; vapor pressure 4.8 mm (20°C); flash point (TOC) 92°F; f.p. -95.2°C; latent heat of vaporization 67.8 cal/g at 140.9°C; refractive index 1.3992 (20°C); wt/gal 6.4 lb (20°C); viscosity 0.0069 poise (20°C). Autoignition temp. 382°F. Miscible with most common organic solvents; immiscible in water.
Grades: Technical; spectrophotometric.
Containers: 1-gal cans; 5- and 55-gal drums; tankcars.
Hazard: Flammable, moderate fire risk. May form explosive peroxides, especially in anhydrous form. Toxic on prolonged inhalation.
Uses: Solvent for hydrocarbons, fatty materials; ex-

tracting agent, used especially for separating metals; solvent purification; organic synthesis (reaction medium).

butylethylacetaldehyde. See 2-ethylhexaldehyde.

5-butyl-5-ethylbarbituric acid. See butethal.

n-**butyl ethyl ether.** See ethyl n-butyl ether.

butyl ethyl ketene (BEK) $(C_4H_9)(C_2H_5)CCO$.
Properties: Yellow liquid with pungent odor; sp. gr. 0.8266 (20/4°C); b.p. 36°C (12 mm); f.p. below −80°C; refractive index 1.4224 (n 20/D); flash point (TOC) 64°F.
Grade: Available as 20% solution in hexane.
Hazard: Flammable, dangerous fire risk; fumes are toxic and irritant. Store solution under nitrogen.
Use: Reactive chemical intermediate.
Shipping regulations: Same as 1,2-butylene oxide.

2-butyl-2-ethylpropanediol-1,3. See 2-ethyl-2-butylpropanediol-1,3.

tert-**butylformamide** $(CH_3)_3CNHCOH$. Colorless, highboiling liquid; soluble in water and common hydrocarbon solvents; used as a solvent and in petroleum additives.

butyl formate $HCOOC_4H_9$.
Properties: Colorless liquid. Sp. gr. 0.885–0.9108; b.p. 107°C; f.p. −90°C; flash point 64°F (closed cup); autoignition temp. 612°F; miscible with alcohols, ethers, oils, hydrocarbons; slightly soluble in water.
Grades: Technical.
Hazard: Flammable, dangerous fire risk. Narcotic and irritant in high concentrations.
Uses: Solvent for nitrocellulose, some types of cellulose acetate, many cellulose ethers, many natural and synthetic resins; lacquers; perfumes; organic synthesis (intermediate); flavoring.
Shipping regulations: (Rail) Flammable liquid, n.o.s., Red label. (Air) Flammable Liquid label.

n-**butyl furfuryl ether** $C_4H_9OCH_2C_4H_3O$.
Properties: Colorless liquid, turning dark on exposure to air; extremely hygroscopic, unstable in presence of moisture. Sp. gr. 0.955 (20/0°C); b.p. 189–190°C (765 mm); refractive index (n 20/D) 1.4522.

n-**butyl furoate** $C_4H_3OCO_2C_4H_9$.
Properties: Colorless oil; decomposes on standing. Sp. gr. 1.055 (20/f°C); b.p. 83–84°C (1 mm), 118–120°C at 25 mm. Insoluble in water; soluble in alcohol and ether.

n-**butyl glycol phthalate.** See dibutoxyethyl phthalate.

tert-**butyl hydroperoxide** $(CH_3)_3COOH$. A highly reactive peroxy compound.
Properties: Water-white liquid; m.p. −8°C; decomposes at 75°C; sp. gr. 0.896 (20/4°C); refractive index 1.396 (25°C) (90% pure); flash point (90%) 130°F. Moderately soluble in water; very soluble in organic solvents and alkali metal hydroxide solutions. Combustible.
Grades: 70%, 90% pure.
Containers: Polyethylene bottles.
Hazard: Moderate fire risk. Oxidizing agent. Moderately toxic.
Uses: Polymerization, oxidation, and sulfonation catalyst; bleaching; deodorizing.
Shipping regulations: (Rail) Oxidizing material, n.o.s., Yellow label. (Air) Solution exceeding 70%, Not acceptable; solution not exceeding 70% in nonvolatile solvent, Organic Peroxide label.

tert-**butylhydroquinone** $C_6H_3(OH)_2C(CH_3)_3$. Intermediate; m.p. 125°C; insoluble in water; soluble in alcohol, acetone and ethyl acetate.

4,4′-butylidenebis (6-tert-**butyl-**meta-**cresol)** $[(CH_3)_3CC_6H_2(OH)(CH_3)]_2CHC_3H_7$.
Properties: White powder; m.p. 209°C min; sp. gr. 1.03 (25°C).
Use: Antioxidant for rubber, dry or latex.

butyl isocyanate C_4H_9NCO. Colorless liquid; b.p. 115°C; sp. gr. 0.88 at 20°/4°C. Intermediate for pesticides; herbicides, pharmaceuticals.
Hazard: Toxic. Strong irritant to eyes and skin.
Shipping regulations: (Air) Poison label; not acceptable on passenger planes. (Rail) Poisonous liquids, n.o.s., Tear gas label.

butyl isodecyl phthalate.
Properties: Clear liquid; mild odor; color (Hazen) 50 max; sp. gr. 0.9 (20/20°C); saponification number 310; acidity (as phthalic acid) 0.01% max.
Containers: Drums; tankcars; tank trucks.
Use: Plasticizer; especially for polyvinyls.

tert-**butylisopropylbenzene hydroperoxide.** White crystalline solid.
Hazard: Dangerous fire risk; reacts strongly with reducing materials. Oxidizing agent.
Shipping regulations: (Rail) Oxidizing material, n.o.s., Yellow label. (Air) Solution exceeding 60%, Not acceptable; solution not exceeding 60% in a nonvolatile solvent, Organic Peroxide label.

"Butyl Kamate."[69] Trademark for an aqueous solution of potassium di-n-butyl dithiocarbamate; 50% minimum assay.
Uses: Ultra-accelerator for natural and synthetic latexes.

n-**butyl lactate** $CH_3CHOHCOOC_4H_9$.
Properties: Water-white, stable, liquid. Mild odor. Miscible with many lacquer solvents, diluents, and oils; slightly soluble in water; hydrolyzed in presence of acids and alkalies. Sp. gr. 0.974–0.984 (20/20°C); flash point 168°F (TOC); m.p. −43°C; wt per U.S. gal 8.15 lb (68°F); b.p. 188°C; refractive index 1.4216 (20°); vapor pressure 0.4 mm Hg (20°C); latent heat of vaporization 77.4 cal/g (20°C). Autoignition point 720°F; combustible; low toxicity.
Grades: Technical, 95% min.
Containers: 1-gal cans; 5- and 55-gal steel drums; tank trucks; tankcars.
Uses: Solvent for nitrocellulose, ethyl cellulose, oils, dyes, natural gums and many synthetic polymers; lacquers; varnishes; inks; stencil pastes; antiskinning agent; chemical (intermediate); perfumes; drycleaning fluids; adhesives.

N-n-**butyl lauramide** $C_{11}H_{23}CONHC_4H_9$.
Properties: White solid. Boiling range 200–225°C at 2 mm; odor lauric acid; flash point 375°F. Combustible. Low toxicity.

butyl laurate (butyl dodecanoate) $C_{11}H_{23}COOC_4H_9$.
Properties: Liquid with sp. gr. (25°C) 0.855; b.p. (5 mm) 130–180°C; m.p. −10°C; insoluble in water. Combustible; nontoxic.
Derivation: Alcoholysis of coconut oil with butyl alcohol followed by fractional distillation.
Containers: Drums.
Use: Plasticizer; flavoring.

butyllithium. n-Butyllithium $CH_3CH_2CH_2CH_2Li$, sec-butyllithium $CH_3CHLiCH_2CH_3$, and tert-butyllithium $(CH_3)_3CLi$, are available, usually in solution in one of the C_5 to C_7 hydrocarbons, in which

they are quite stable. sec-Butyllithium solutions must be kept at or below 60°F.
Derivation: Reaction of lithium ribbon and one of the butyl chlorides in pentane or hexane.
Grades: Sold by percentage butyllithium in the solution.
Hazard: Solid and solutions highly flammable, ignite on contact with moist air. Toxic and irritant. Safety data sheet available from Manufacturing Chemists Assn., Washington, D.C.
Uses: Polymerization of isoprene and butadiene; intermediate in preparation of lithium hydride; rocket fuel component; metalating agent.
Shipping regulations: (Rail, Air) Not specified. Consult authorities.

n-**butylmagnesium chloride** C_4H_9MgCl.
Properties: Liquid; sp. gr. 0.88.
Derivation: From magnesium and butyl chloride.
Grade: Available in solution in ethyl ether, or in tetrahydrofuran.
Containers: Glass bottles; 5-gal drums.
Hazard: Flammable, dangerous fire risk. Probably toxic.
Use: Grignard reagent, as an alkylating agent.
Shipping regulations: (Rail, Air) Not specified. Consult authorities.

n- and sec-**butyl mercaptan.** See 1- and 2-butanethiol.

tert-**butyl mercaptan.** See 2-methyl-2-propanethiol.

butyl mesityl oxide. See butopyronoxyl.

n-**butyl methacrylate** $H_2C:C(CH_3)COOC_4H_9$.
Properties: Colorless liquid; b.p. 163.5–170.5°C; f.p. below -75°C; sp. gr. 0.895 (25/25°C); flash point 130°F (open cup); refractive index 1.4220; readily polymerized; insoluble in water. Combustible.
Derivation: Reaction of methacrylic acid or methyl methacrylate with butyl alcohol.
Grades: Technical (inhibited).
Containers: Drums; tanks.
Hazard: Toxic by ingestion. Moderate fire risk.
Uses: Monomer for resins, solvent coatings, adhesives, oil additives; emulsions for textiles, leather and paper finishing. See also acrylic resin.

tert-**butyl methacrylate** $H_2C:C(CH_3)COOC(CH_3)_3$.
Properties: Colorless liquid; b.p. 66°C (57 mm); sp. gr. 0.877 (25°C); refractive index 1.4124 (n 25/D); flash point 92°F (TOC).
Grade: Technical, containing 100 ppm hydroquinone monomethyl ether as inhibitor.
Hazard: Flammable, moderate fire risk. Toxic by ingestion.
Use: Monomer for resins. See also acrylic resin.

tert-**butyl-4-methoxyphenol.** See butylated hydroxyanisole.

sec-**butyl 6-methyl-3-cyclohexene-1-carboxylate.** See siglure.

para-tert-**butyl**-alpha-**methylhydrocinnamaldehyde** (alpha-methyl-beta-(para-tert-butylphenyl)-propionaldehyde) $(CH_3)_3CC_6H_4CH_2CH(CH_3)CHO$.
Properties: Light-yellow liquid; strong floral odor; sp. gr. 0.942–0.949 (25/25°C); refractive index 1.503–1.510 (20°C); flash point (TCC) 204°F. Soluble in 1 part 90% alcohol. Stable, non-discoloring. Combustible. Low toxicity.
Grades: 93%, 85% purity.
Use: Perfume.

2-tert-**butyl-4-methylphenol.** See 2-tert-butyl-para-cresol.

6-tert-**butyl-3-methylphenol.** See 6-tert-butyl-meta-cresol.

n-**butyl myristate** $CH_3(CH_2)_{12}COOC_4H_9$. The butyl ester of myristic acid.
Properties: Water-white oily liquid; saponification number 193–203; f.p. 1–7°C; boiling range 167–197°C at 5 mm; sp. gr. (25°C) 0.850–0.858; insoluble in water; soluble in acetone, castor oil, chloroform, methanol, mineral oil, and toluene. Combustible; nontoxic.
Derivation: Alcoholysis of coconut oil with butyl alcohol followed by fractional distillation.
Containers: Tins; drums.
Uses: Plasticizer; lubricant for textiles, paper stencils; cosmetic preparations.

"Butyl Namate."[69] Trademark for an aqueous solution of sodium di-n-butyldithiocarbamate; 47% minimum assay.
Uses: Ultra-accelerator for natural and synthetic latices.

n-**butyl nitrate** $C_4H_9NO_3$.
Properties: Liquid. Sp. gr. (20°C) 1.103; b.p. 123°C; water-white; odor ethereal. Insoluble in water; soluble in alcohol and ether; flash point 97°F.
Hazard: Flammable, moderate fire risk in contact with reducing materials. Oxidizing agent; may explode on shock or heating.
Shipping regulations: (Rail) Nitrates, n.o.s., Yellow label. (Air) Nitrates, n.o.s., Oxidizer label.

butyl nonanoate. See butyl pelargonate.

butyl octadecanoate. See butyl stearate.

butyl octyl phthalate $C_4H_9OOCC_6H_4COOC_8H_{17}$.
Properties: Water-white liquid; mild characteristic odor; sp. gr. 0.991–0.997 (20/20°C); saponification number 298–308. Miscible with most organic solvents. Combustible; low toxicity.
Containers: Drums; tank trucks; tankcars.
Uses: Plasticizer for vinyl resins.

butyl octadecanoate. See butyl stearate.

butyl oleate $CH_3(CH_2)_7CH:CH(CH_2)_7COOC_4H_9$.
Properties: Light-colored, oleaginous liquid; mild odor; insoluble in water; miscible with alcohol, ether, vegetable and mineral oils. Sp. gr. 0.873 (20/20°C); iodine value 76.8; f.p. opaque at 12°C, solid at -26.4°C; wt/gal 7.26 lb (20°C); boiling range 173–227°C (2 mm). Flash point 356°F. Combustible; low toxicity.
Derivation: Alcoholysis of olein or esterification of oleic acid with butyl alcohol.
Containers: 7-, 35-lb tins; 380-lb drums.
Uses: Plasticizer, particularly for PVC; solvent; lubricant; water-resisting agent; coating compositions; polishes; water-proofing compounds.

Butyl "Oxitol."[125] Trademark for ethylene glycol monobutyl ether (q.v.).
Containers: 1-, 5-, 55-gal pails or drums; tank cars and tank trucks.
Hazard: Slightly irritating to skin and eyes.
Uses: Solvent in various types of surface coating formulations, to improve gloss and leveling; component of hydraulic fluids; coupling agent in various types of cleaning and cutting oils.

n-**butyl pelargonate** (n-butyl nonanoate) $C_4H_9OOCC_8H_{17}$. Lqiuid with fruity odor; combustible; sp. gr. 0.865 (15.5/15.5°C); b.p. 270°C. Used in flavors and perfumes; as chemical intermediate.

sec-**butyl pelargonate** $C_2H_5CH(CH_3)OOCC_8H_{17}$. Liquid; sp. gr. 0.8608 (20/4°C); b.p. 123°C (15 mm); combustible; refractive index 1.4220. Shows optical activity. Used as chemical intermediate.

tert-**butyl peracetate.** See tert-butyl peroxyacetate.

tert-**butyl perbenzoate.** See tert-butyl peroxybenzoate.

tert-**butyl perisobutyrate.** See tert-butyl peroxyisobutyrate.

tert-**butyl permaleic acid.** See tert-butyl peroxymaleic acid.

tert-**butyl peroctoate.** See tert-butyl peroxy(2-ethylhexanoate).

tert-**butyl peroxide.** See di-tert-butyl peroxide.

tert-**butyl peroxyacetate** (tert-butyl peracetate) $(CH_3)_3COOOCCH_3$. Available as a 72–76% solution in benzene. Flash point below 80°F (COC); sp. gr. 0.923.
Hazard: Flammable, dangerous fire risk. Oxidizing material.
Uses: Polymerization initiator for vinyl monomers; in manufacture of polyethylene and polystyrene.
Shipping regulations: Peroxides, organic, (liquid or solution), n.o.s. (Rail) Yellow label. (Air) Not acceptable.

tert-**butyl peroxybenzoate** (tert-butyl perbenzoate) $(CH_3)_3COOOCC_6H_5$.
Properties: Colorless liquid with mild aromatic odor; sp. gr. 1.04 (25/25°C); f.p. 8.5°C; vapor pressure 0.33 mm (50°C); flash point 200°F. Soluble in alcohols, esters, ethers, ketones; insoluble in water.
Grade: 98% min.
Hazard: Oxidizing material; do not store near combustible materials.
Uses: Polymerization initiator for polyethylene; polystyrene, polyacrylates and polyesters; chemical intermediate.
Shipping regulations: (Rail) Peroxides, organic, liquid, Yellow label. (Air) Not acceptable.

tert-**butyl peroxy-2-ethylhexanoate** (tert-butyl peroctoate) $(CH_3)_3COOOCCH(C_2H_5)C_4H_9$.
Properties: Colorless liquid with a faint odor; sp. gr. 0.895 (25/25°C); f.p. below -30°C; refractive index 1.426 (25°C) decomposes at 89°C; flash point 190°F. Insoluble in water; miscible with most organic solvents.
Hazard: Oxidizing material. Do not store near combustible materials.
Use: Polymerization catalyst.
Shipping regulations: Peroxides, organic (liquid or solution), n.o.s., (Rail) Yellow label; (Air) Not acceptable.

tert-**butyl peroxyisobutyrate** (tert-butyl perisobutyrate) $(CH_3)_3COOOCCH(CH_3)CH_3$. Available as a 72–75% solution in benzene. Flash point below 80°F.
Hazard: Flammable, dangerous fire risk. Oxidizing agent.
Use: Polymerization catalyst.
Shipping regulations: Peroxides, organic (liquid or solution), n.o.s., (Rail) Red label; (Air) Not acceptable.

tert-**butylperoxy isopropyl carbonate** (BPIC) $(CH_3)_3COOOCOCH(CH_3)_2$.
Properties: Liquid; f.p. -3°C; sp. gr. 0.945 g/cc; refractive index (n 20/D) 1.4050; flash point (TOC) 112–118°F. Almost insoluble in water; miscible with hydrocarbons, esters and ethers. Relatively stable under ordinary conditions. Combustible.
Grade: Technical (8.6% active oxygen).
Hazard: Moderate fire risk. Oxidizing agent.
Uses: Polymerization initiator; cross-linking agent.
Shipping regulations: Peroxides, organic, liquid, n.o.s., (Rail) Yellow label; (Air) Not acceptable.

tert-**butyl peroxymaleic acid** (tert-butyl permaleic acid) $(CH_3)_3COOOCCH:CHCOOH$. An unsaturated peroxide.
Properties: White crystalline solid; m.p. 114–116°C (dec); slightly soluble in water, cool 5% alkaline solutions and alcohols; moderately soluble in oxygenated organic solvents, polyester monomers; slightly soluble in petroleum ether, carbon tetrachloride, and chloroform; insoluble in benzene.
Grade: 95% pure.
Hazard: Oxidizing agent. Do not store near combustible materials.
Uses: Polymerization catalyst; bleaching; pharmaceuticals.
Shipping regulations: (Rail) Oxidizing materials, n.o.s., Yellow label. (Air) Peroxides, organic, solid, n.o.s., Organic Peroxide label.

tert-**butyl peroxyphthalic acid** (tert-butyl perphthalic acid) $(CH_3)_3COOOCC_6H_4COOH$.
Properties: White crystalline solid; m.p. 96–99°C; insoluble in water; soluble in cool 5% alkaline solutions and in alcohols; moderately soluble in oxygenated organic solvents, chlorinated hydrocarbons, polyester monomers; slightly soluble in petroleum hydrocarbons.
Grades: 95% pure.
Hazard: Oxidizing agent. Do not store near combustible materials.
Uses: Polymerization catalyst and oxidizing agent.
Shipping regulations: (Rail) Oxidizing materials, n.o.s., Yellow label. (Air) Peroxides, organic, solid, n.o.s., Organic Peroxide label.

tert-**butyl peroxypivalate** $(CH_3)_3COOOCC(CH_3)_3$.
Properties: Colorless liquid; sp. gr. 0.854 (25/25°C); f.p. below -19°C; refractive index 1.410 (25°C); decomposes at 70°C; flash point 155–160°F. Insoluble in water and ethylene glycol; soluble in most organic solvents.
Grade: Available as 75% solution in mineral spirits.
Hazard: (solution) Flammable, dangerous risk. Oxidizing agent. May explode on heating.
Use: Polymerization initiator.
Shipping regulations: (Rail) Peroxide, organic, liquid, n.o.s., Yellow label; (solution) Red label. (Air) Not acceptable.

tert-**butyl perphthalic acid.** See tert-butyl peroxyphthalic acid.

ortho-tert-**butylphenol** $(CH_3)_3CC_6H_4OH$.
Properties: Light yellow liquid; freezing point -7°C; sp. gr. 0.982 (20°C); b.p. 224°C; flash point (open cup) 230°F. Soluble in isopentane, toluene, and ethyl alcohol; insoluble in water. Combustible.
Hazard: Moderate irritant to eyes and skin.
Uses: Chemical intermediate for synthetic resins, plasticizers, surface-active agents, perfumes, and other products; a permissible antioxidant for aviation gasoline (A.S.T.M. D910-64T).

para-tert-**butylphenol** $(CH_3)_3CC_6H_4OH$.
Properties: White crystalline solid, with a distinctive odor; sp. gr. (crystals) 1.03; sp. gr. (molten) 0.908 (114/4°C); b.p. 239°C; m.p. 100°C. Combustible.
Derivation: Catalytic alkylation of phenol with olefins.
Grades and Containers: Flake in 50-lb bags; molten in insulated tank cars and tank trucks.
Hazard: Moderate irritant to eyes and skin.
Uses: Plasticizer for cellulose acetate; intermediate for antioxidants, special starches, oil soluble phenolic resins; pour-point depressors and emulsion breakers for petroleum oils and some plastics; synthetic lubricants; insecticides; industrial odorants.

2(para-tert-**butylphenoxy)isopropyl 2-chloroethyl sulfite.** See "Aramite."

n-**butyl phenylacetate** $C_4H_9OOCCH_2C_6H_5$.
Properties: Colorless liquid with rose-honey odor; sp. gr. 0.991–0.994 (25/25°C); b.p. 135–141°C (18 mm) refractive index 1.488–1.490 (20°C). Soluble in 2 volumes 80% alcohol. Combustible. Low toxicity. Made synthetically.
Grade: 98% min.
Uses: Perfumes; flavoring.

n-**butyl phenyl ether** $C_4H_9OC_6H_5$.
Properties: Liquid. Sp. gr. (20°C) 0.929; boiling range 202–212°C; color water-white; odor aromatic; flash point 180°F. Combustible. Low toxicity.

4-tert-**butylphenyl salicylate**
$(CH_3)_3CC_6H_4OOCC_6H_4OH$.
Properties: Off-white, odorless crystals; m.p. 62–64°C; soluble in alcohol, ethyl acetate, toluene; insoluble in water.
Use: Light absorber, best at 290–330 μ.

n-**butylphosphoric acid.** See n-butyl acid phosphate.

butyl phthalylbutyl glycolate
$C_4H_9OOCC_6H_4COOCH_2COOC_4H_9$.
Properties: Colorless, odorless liquid; sp. gr. 1.093–1.103 (25/25°C); b.p. 219°C (5 mm); f.p. below -35°C; darkens on heating above 290°C. Flash point 390°F (open cup). Combustible. Low toxicity. Insoluble in water; extremely light-stable.
Uses: Plasticizer for polyvinyl chloride. FDA-approved for use in vinyl food wrappings.

n-**butyl propionate** $C_2H_5CO_2C_4H_9$.
Properties: Water-white liquid. Apple-like odor. Soluble in alcohol and ether; miscible with all coaltar and petroleum distillates; very slightly soluble in water. Sp. gr. 0.875 (20°C); 0.874 (15.5°C); wt/gal 7.3 lb; b.p. 146°C (commercial grades boil over a range of 130–150°C due to presence of butyl alcohol and esters); f.p. -89°C; flash point 90°F; dilution ratio (nitrocellulose solution method) with toluene 2.1, with petroleum naphtha 1.2. Autoignition temp. 800°F.
Derivation: Esterification of propionic acid with butyl alcohol, with sulfuric acid catalyst.
Grades: Technical (85–90% to 95% ester content).
Containers: Cans; steel drums; tank cars.
Hazard: Flammable, moderate fire risk. Skin and eye irritant.
Uses: Solvent for nitrocellulose; retarder in lacquer thinner; lacquers; ingredient of perfumes, flavors.

butyl ricinoleate $C_{17}H_{32}(OH)COOC_4H_9$.
Properties: Yellow to colorless oleaginous liquid; soluble in alcohol and ether; insoluble in water. Sp. gr. 0.916 (20/20°C); b.p. approximately 275°C (13 mm); flash point 220°F. Saybolt viscosity 112 (100° F); freezing point indefinite, slightly opaque at -30°C, and very viscous at -50°C; wt/gal 7.62 lb (20°C). Combustible. Low toxicity.
Derivation: Castor oil and butyl alcohol.
Containers: 5-gal cans; 55-gal drums.
Uses: Plasticizer; lubricant.

butyl rubber. A copolymer of isobutylene (97%) and isoprene (3%). Polymerized below -140°F, with aluminum chloride catalyst.
Properties: Sp. gr. 0.92. Vulcanizates have tensile strength up to 2000 psi (unreinforced) and 3000 psi (reinforced). Service temperature range -70 to +400°F. Good abrasion resistance; excellent impermeability to gases; high dielectric constant; excellent resistance to aging and sunlight; superior shock absorbing and vibration damping qualities. Resistance to oils and greases only fair. Will support combustion.
Grades: Stabilized; latex; chlorine-containing elastomer; low molecular weight (liquid).
Containers: 55-lb blocks or bales, with or without polyethylene film wrapper.
Uses: Tire carcases and linings, especially for tractors and other out-size vehicles; electric wire insulation; encapsulating compounds; steam hose and other mechanical rubber goods; pond and reservoir sealant. Latex is used for paper coating, textile and leather finishing, and adhesive formulations; air bags for tire vulcanization; self-curing cements; pressure-sensitive adhesives; tire cord dips; sealants.

butyl sebacate. See dibutyl sebacate.

butyl sorbate $C_4H_9OOCC_5H_7$. An insect attractant for the European chafer.

n-**butylstannoic acid** $[(C_4H_9)Sn(OH)_3]$ or $[(C_4H_9)SnOOH]_n$.
Properties: White, infusible and insoluble free-flowing powder. Exists as a polymer of undetermined chemical structure. Further dehydration results in polymers having molecular weights of 1000–5000 which are soluble in organic solvents.
Hazard: May be toxic by ingestion and skin absorption.
Uses: Polymerization catalyst, antioxidant, and heat stabilizer for PVC (not approved by FDA for food containers); electrically conducting tin oxide coatings on glass, vitreous and metal surfaces; activator for alkaline earth metal phosphates in fluorescent light bulbs; intermediate for silicones.

butyl stearamide. $C_{17}H_{35}CONHC_4H_9$.
Properties: Light straw-colored liquid; sp. gr. (20/20°C) 0.869; boiling range 195–200°C (2 mm); flash point 430°F; amide odor. Combustible. Low toxicity.
Uses: Plasticizer and intermediate for the synthesis of insecticides, surface-active agents, pharmaceuticals, and textile assistants.

butyl stearate (butyl octadecanoate) $C_{17}H_{35}COOC_4H_9$.
Properties: Colorless, stable, oleaginous liquid solidifying at about 19°C. Practically odorless, sometimes with faint fatty odor. Sp. gr. 0.855–0.860 (25/25°C); m.p. 19.5–20°C; flash point 320°F (closed cup); combustible. Low toxicity. wt/gal 7.14 lb (68°F); refractive index 1.4430 (20°C); b.p. 350°C. Miscible

with mineral and vegetable oils; soluble in alcohol and ether; insoluble in water.
Derivation: Alcoholysis of stearin or esterification of stearic acid with butyl alcohol.
Grades: Technical; cosmetic; chemically pure.
Containers: Vinyl-coated drums.
Uses: Ingredient of polishes, special lubricants and coatings; lubricants for metals, and in textile and molding industries; in wax polishes as dye solvent; plasticizer for laminated fiber products, rubber hydrochloride, chlorinated rubber, and cable lacquers; carbon paper and inks; emollient in cosmetic and pharmaceutical products; lipsticks; damp-proofer for concrete; flavoring.

1-butyl-3-sulfanilylurea (N-)(butylcarbomoyl) sulfanilamide; N-sulfanilyl-N-butylurea; carbutamide) $H_2NC_6H_4SO_2NHCONH(CH_2)_3CH_3$.
Properties: Crystals; m.p. 144–145°C. Soluble in water (pH 5–8).
Derivation: Prepared from butylurea and sulfanilamide.
Use: Medicine.

butyl sulfide. See dibutyl sulfide.

4-tert-**butyl**-ortho-**thiocresol** (2-methyl-4-tert-butylthiophenol) $(CH_3)_3CC_6H_3(CH_3)SH$.
Properties: Colorless liquid with mild (non-mercaptan) odor; sp. gr. 0.983 (25°C); f.p. -4°C; b.p. 250°C; refractive index (n 25/D) 1.546. Insoluble in water; soluble in hydrocarbons. Combustible.
Grades: Available as 98% pure, supplied under nitrogen atmosphere; as 55% solution in hydrocarbon.
Hazard: May be toxic. Solution probably flammable.
Uses: Peptizer for rubbers; polymer modifier; lube oil additive.

4-tert-**butylthiophenol** $(CH_3)_3CC_6H_4SH$.
Properties: Colorless liquid with mild (non-mercaptan) odor; sp. gr. 0.986 (25°C); f.p. -11°C; b.p. 238°C; refractive index (n 25/D) 1.546. Insoluble in water; soluble in hydrocarbons.
Grade: 98%, supplied under nitrogen atmosphere.
Hazard: May be toxic.
Uses: Lube oil additives; polymer modifiers; antioxidants; dyes.

n-**butyltin trichloride** $C_4H_9SnCl_3$.
Properties: Colorless liquid; fumes in contact with air; sp. gr. 1.71 (25/4°C); b.p. 102°C (12 mm); refractive index (n 25/D) 1.5190. Soluble in organic solvents; sparingly soluble in water with partial hydrolysis.
Hazard: Toxic by ingestion. Strong irritant to skin. Do not expose to liquids or vapors. Tolerance, (as Sn), 0.1 mg per cubic meter of air.
Uses: Intermediate; catalyst; stabilizer.

butyl titanate. See tetrabutyl titanate.

para-tert-**butyltoluene** 1-methyl-4-tert-butylbenzene) $(CH_3)_3CC_6H_4CH_3$.
Properties: Colorless liquid. Sp. gr. (20/20°C), 0.857–0.863; b.p. 192.8°C. Combustible. Insoluble in water.
Grade: Technical.
Containers: Tank cars; drums.
Hazard: Toxic by inhalation, ingestion, and skin absorption. Tolerance, 10 ppm in air.
Uses: Solvent; intermediate.

n-**butyltrichlorosilane** $C_4H_9SiCl_3$.
Properties: Colorless liquid; b.p. 142°C; sp. gr. 1.1608 (25/25°C); refractive index (n 25/D); 1.4363; flash point (COC) 126°F. Readily hydrolyzed, with liberation of hydrochloric acid. Soluble in benzene, ether, heptane.
Derivation: Grignard reaction of silicon tetrachloride and n-butylmagnesium chloride.
Grades: Technical, 95%.
Containers: Drums.
Hazard: Moderate fire risk; corrosive to skin and tissue.
Use: Intermediate for silicones.
Shipping regulations: (Rail) White label. (Air) Corrosive label. Not accepted on passenger planes.

tert-**butyltrimethylmethane.** See hexamethylethane.

N-n-**butylurea** $C_4H_9HNCONH_2$.
Properties: White solid; decomposes on heating; odorless; m.p. 96°C; soluble in water; alcohol, and ether.

n-**butyl vinyl ether.** See vinyl n-butyl ether.

butyl xanthate. See butylxanthic acid.

butylxanthic acid (butyl xanthate; butyldithiocarbonic acid) $CH_3(CH_2)_3O \cdot CS \cdot SH$. See xanthic acids.

"Butyl Zimate."[69] Trademark for zinc dibutyldithiocarbamate. (q.v.).

1-**butyne.** See ethylacetylene.

2-**butyne.** See crotonylene.

butynediol $HOCH_2C{:}CCH_2OH$.
Properties: White orthorhombic crystals; m.p. 58°C; b.p. 238°C; refractive index 1.450 (n 25/D); soluble in water, aqueous acids, alcohol and acetone. Insoluble in ether and benzene. Combustible.
Derivation: High-pressure synthesis from acetylene and formaldehyde.
Grades: Crystalline solid, 97%; aqueous solution 35%.
Containers: Drums; tank cars; tank trucks.
Hazard: Toxic; strong irritant. May explode on contamination with mercury salts, strong acids and alkali earth hydroxides and halides at high temperatures.
Uses: Intermediate; corrosion inhibitor; electroplating brightener; defoliant; polymerization accelerator; stabilizer for chlorinated hydrocarbons; cosolvent for paint and varnish removal.

3-butyn-1-ol (beta-ethynyl ethanol) $HC{:}CCH_2CH_2OH$.
Properties: Water-white liquid with characteristic odor; sp. gr. 0.9257 (20/4°C); refractive index 1.4409 (20°C); b.p. 128.9°C; f.p. -63.6°C. Combustible.
Uses: Preparation of perfume bases, acetylenic esters, plastics, plasticizers, pharmaceuticals, wetting agents, medicinals, and organic synthesis.

butyraldehyde (butaldehyde; n-butanal; n-butyl aldehyde; butyric aldehyde) $CH_3(CH_2)_2CHO$.
Properties: Water-white liquid; characteristic, pungent, aldehyde odor. Sp. gr. 0.8048 (20/20°C); b.p. 75.7°C; vapor pressure 91.5 mm (20°C); flash point 20°F; wt/gal 6.7 lb (20°C); coefficient of expansion 0.00114 (20°C); freezing point -99°C; viscosity 0.0043 poise (20°C). Autoignition temp. 446°F. Slightly soluble in water; soluble in alcohol and ether.
Derivation: (a) Oxo process (q.v.); (b) dehydrogenating butanol vapors over a catalyst, the butyraldehyde being separated by distillation; (c) partial reduction of crotonaldehyde.
Grades: Technical (93% min.).
Containers: Drums; tank cars.
Hazard: Flammable; dangerous fire risk. Moderately toxic. Safety data sheet available from Manufacturing Chemists Assn., Washington, D.C.
Shipping regulations: (Rail) Red label. (Air) Flammable Liquid label.

butyric acid (n-butyric acid; butanoic acid; ethylacetic acid; propylformic acid). $CH_3CH_2CH_2COOH$.
Properties: Colorless liquid; penetrating and obnox-

ious odor; refractive index 1.3981 (20°C); sp. gr. 0.9583 (20/4°C); m.p. -5.0 to -8°C; b.p. 163.5°C (757 mm), 75°C (25 mm); vapor pressure 0.84 mm (20°C); flash point 170°F; viscosity 1.61 cps (20°C); autoignition temp. 846°F. Completely soluble in water; miscible with alcohol and ether. Combustible.
Derivation: Occurs as glyceride in animal milk fats. Produced as a by-product in hydrocarbon synthesis, by oxidation of butyraldehyde, and by butyric fermentation of molasses or starch.
Grades: 90%; 95%; 99%; edible; synthetic; reagent; technical; F.C.C.
Containers: Drums; tank cars.
Hazard: Strong irritant to skin and tissue.
Uses: Synthesis of butyrate ester perfume and flavor ingredients; pharmaceuticals; deliming agent; disinfectants; emulsifying agents; sweetening gasolines.
Shipping regulations: (Air) Corrosive label.

butyric alcohol. See n-butyl alcohol.

butyric aldehyde. See butyraldehyde.

butyric anhydride $(CH_3CH_2CH_2CO)_2O$.
Properties: Water-white liquid. Hydrolyzes to butyric acid. Sp. gr. 0.9681 (20/20°C); m.p. -75°C; b.p. 199.5°C; vapor pressure 0.3 mm (20°C); flash point 190°F; wt/gal 8.1 lb (20°C). Combustible. Low toxicity.
Grades: Technical; 98%.
Containers: Drums; tanks.
Uses: Manufacture of butyrates, drugs, and tanning agents.

butyrin. See glyceryl tributyrate.

butyrolactam. See 2-pyrrolidone.

butyrolactone (gamma-butyrolactone) $OCH_2CH_2CH_2CO$.
Properties: Colorless liquid with pleasant odor. B.p. 204°C; m.p. -44°C; sp. gr. 1.144; flash point 209°F. (open cup). Miscible with water, alcohol and ether. Combustible.
Derivation: High-pressure synthesis from acetylene and formaldehyde.
Grades: Technical.
Containers: Drums; tank cars.
Hazard: Moderately toxic by ingestion.
Uses: Intermediate for butyric and succinic acids; solvent for resins; paint removers; petroleum processing.

butyrone. See dipropyl ketone.

butyronitrile (propyl cyanide; butanenitrile) $CH_3(CH_2)_2CN$.
Properties: Colorless liquid; sp. gr. (15°C) 0.796; f.p. -112.6°C; b.p. 116–117.7°C. Flash point 79°F (open cup). Slightly soluble in water; soluble in alcohol and ether.
Containers: Drums and tank cars.
Hazard: Toxic. Flammable, dangerous fire risk.
Uses: Basic material in industrial, chemical and pharmaceutical intermediates and products; poultry medicines.

butyroyl chloride (butyryl chloride; butanoyl chloride) C_3H_7COCl.
Properties: Colorless liquid with pungent acid chloride odor. Reacts with alcohol and water; infinitely miscible with ether. Freezing point -89°C; distillation range 100–110°C; sp. gr. 1.028 (15°C); refractive index 1.4121 (n 20/D).
Hazard: Toxic by ingestion and inhalation. Strong irritant to tissue.
Use: Organic synthesis.

N-**butyryl**-para-**aminophenol.** See "Suconox."

butyryl chloride. See butyroyl chloride.

"Buxine."[227] Trademark for amylcinnamic aldehyde.

"Buxinol."[227] Trademark for n-amylcinnamic alcohol.

BVE. Abbreviation for butyl vinyl ether. See vinyl n-butyl ether.

"B-X-A."[248] Trademark for a diarylamine-ketone-aldehyde reaction product.
Properties: Brown powder; sp. gr. 1.10; melting range, 85–95°C; store in a cool place. Soluble in acetone, benzene and ethylene dichloride; insoluble in water and gasoline.
Use: Antioxidant for rubber and nylon.

"BxDC."[177] Trademark for butoxyethyl diglycol carbonate.

"B.Y."[319] Trademark for riboflavin feed supplement.

BZ. A nonlethal gas which causes temporary disability. It is a derivative of lysergic acid.

C

C Symbol for carbon.

C^{14} (carbon-14). The naturally occurring radioactive isotope of carbon, used in chemical dating, tracer studies, etc.

Ca Symbol for calcium.

CA. Abbreviation for cellulose acetate and cortisone acetate; also for controlled atmosphere.

C_3A. Abbreviation for tricalcium aluminate, as used in cement. See cement, Portland.

CAB. Abbreviation for cellulose acetate butyrate.

"Cab-O-Brade."[275] Trademark for garnet material in granular form for use as an abrasive.

"Cab-O-Cure."[275] Trademark for cross-linking peroxide agents for elastomers and polyolefins.
Hazard: May be fire risk. See peroxide.

"Cab-O-Lime."[275] Trademark for granular oxide, primarily of calcium and magnesium, for reacting with or fluxing certain undesirable impurities in molten steel or iron.

"Cab-O-Sil."[275] Trademark for colloidal silica particles sintered together in chain-like formations. Surface area ranges from 50 to 400 square meters per gram, depending on grade.
Grades: Standard; M-5; L-5; SD-20.
Uses: Thickening and emulsifying agent for oil/water systems; drilling muds; cattle-feed supplements; tile cleaners; dispersion of oil slicks on sea water; plastics.

"Cab-O-Sperse."[275] Trademark for aqueous dispersions of pyrogenic silica for use in the paper and textile industries.

cacao butter (cocoa butter). See theobroma oil.

C acid. See 2-naphthylamine-4,8-disulfonic acid.

cacodylic acid (dimethylarsinic acid) $(CH_3)_2AsOOH$.
Properties: Colorless, odorless, deliquescent crystals. M.p. 200°C; soluble in water, alcohol, and acetic acid; insoluble in ether.
Derivation: By distilling a mixture of arsenic trioxide and potassium acetate and oxidizing the resulting product with mercuric oxide.
Hazard: Highly toxic.
Uses: Herbicide, especially for control of Johnson grass on cotton; soil sterilant; chemical warfare; timber thinning.
Shipping regulations: (Rail, Air) Poison label.

cactinomycin. USAN name for an antibiotic produced from Streptomyces which is 10% dactinomycin and 90% two kinds of actinomycin C.

"Cadalume" L.[288] Trademark for a bright cadmium electroplating process for a protective coating on iron and steel. Materials used are cadmium oxide, sodium cyanide, and addition agents.

"Cadalyte."[28] Trademark for a series of compound for cadmium electroplating.

cadaverine (1,5-diaminopentane; pentamethylenediamine) $NH_2(CH_2)_5NH_2$. A ptomaine formed in the decay of animal proteins after death; also made synthetically.
Properties: Syrupy, colorless, fuming liquid; m.p. +9°C; b.p. 178–179°C; soluble in water and alcohol; slightly soluble in ether.
Hazard: Highly toxic by ingestion; absorbed by skin; irritant.
Uses: Preparation of high polymers; intermediate; biological research.

"Caddy."[48] Trademark for a liquid cadmium fungicide used on turf grass.
Hazard: See under cadmium.

cadinene $C_{15}H_{24}$. See sesquiterpene.

"Cadmate."[69] Trademark for cadmium diethyldithiocarbamate.
Hazard: See under cadmium.

"Cadminate."[329] Trademark for a turf fungicide containing 60% cadmium succinate and 40% inert matter.
Hazard: See under cadmium.

cadmium Cd Metallic element of atomic number 48, group IIB of the periodic table. Atomic weight 112.4. Valence 2. There are eight stable isotopes.
Properties: Soft, blue-white, malleable metal or grayish-white powder. Tarnishes in moist air; corrosion resistance poor in industrial atmospheres. Becomes brittle at 80°C. Resistant to alkalies; high neutron absorber. Sp. gr. 8.642; m.p. 320.9°C; b.p. 767°C; refractive index 1.13; Mohs hardness 2.0. Soluble in acids, especially nitric, and in ammonium nitrate solutions. Lowers melting point of certain alloys when used in low percentage. Combustible.
Occurrence: As greenockite (CdS) ore containing zinc sulfide; also with lead and copper ores containing zinc. Canada, central and western U.S., Peru, Australia, Mexico, Congo.
Derivation: (1) Dust or fume from roasting zinc ores is collected, mixed with coal or coke and NaCl or ZnCl and sintered; the cadmium volatilized is collected in an electrostatic precipitator, leached, fractionally precipitated, and distilled. (2) By direct distillation from cadmium-bearing zinc. (3) By recovery from electrolytic zinc process (about 40%).
Hazard: Highly toxic, especially by inhalation of dust or fume. May be fatal. Cadmium plating of food and beverage containers has resulted in a number of outbreaks of gastroenteritis (food poisoning). Flammable in powder form. Tolerance (dust and soluble compounds), 0.2 mg per cubic meter of air; (oxide fume, as Cd), 0.1 mg per cubic meter of air. Soluble compounds of cadmium are highly toxic; however, ingestion usually induces a strong emetic action which minimizes the risk of fatal or severe poisoning. Use as fungicide may be restricted.
Grades: Technical; powder; pure sticks; ingots; slabs; high-purity crystals (less than 10 ppm impurities).
Uses: Electrodeposited and dipped coatings on metals; bearing and low-melting alloys; electrical equipment; brazing alloys; fire-protection systems; nickel-

cadmium storage batteries; power transmission wire; TV phosphors; basis of pigments used in ceramic glazes, machinery enamels, baking enamels; Weston standard cell; control of atomic fission in nuclear reactors; fungicide; photography and lithography; selenium rectifiers.
Shipping regulations: (metal and soluble compounds) (Rail, Air) Not listed. Consult current regulations.

cadmium acetate (a) $Cd(OOCCH_3)_2 \cdot 3H_2O$; (b) $Cd(OOCCH_3)_2$.
Properties: Colorless crystals; soluble in water and alcohols. (a) Sp. gr. 2.01; m.p., loses water at 130°C. (b) Sp. gr. 2.341; m.p. 256°C.
Derivation: Interaction of acetic acid and cadmium oxide.
Hazard: Highly toxic. See under cadmium.
Uses: Ceramics (iridescent glazes); manufacture of acetates; assistant in dyeing and printing textiles; laboratory reagent.
Shipping regulations: See under cadmium.

cadmium ammonium bromide (ammonium-cadmium bromide) $CdBr_2 \cdot 4NH_4Br$.
Properties: Colorless crystals. Soluble in alcohol and water.
Hazard: Highly toxic. See under cadmium.
Shipping regulations: See under cadmium.

cadmium antimonide. A semiconductor used in thermoelectric devices.
Hazard: Highly toxic. See under cadmium.
Shipping regulations: See under cadmium.

cadmium-base Babbitt. See Babbitt metal.

cadmium borotungstate $Cd_5(BW_{12}O_{40}) \cdot 18H_2O$.
Properties: Yellow, heavy crystals; m.p. 75°C. Soluble in water. Solution yellow or light brown.
Grades: Technical.
Hazard: Highly toxic. See under cadmium.
Use: Separating minerals.

cadmium bromate $Cd(BrO_3)_2 \cdot H_2O$.
Properties: White crystals or crystalline powder; sp. gr. 3.758; m.p., decomposes. Soluble in water; insoluble in alcohol.
Derivation: By adding cadmium sulfate to a solution of barium bromate.
Hazard: Highly toxic. Strong oxidizing agent; dangerous in contact with organic materials.
Uses: Analytical reagent.
Shipping regulations: Oxidizing material, n.o.s., (Rail) Yellow label. (Air) Oxidizer label.

cadmium bromide $CdBr_2$ or $CdBr_2 \cdot 4H_2O$.
Properties: White to yellowish, efflorescent crystalline powder; sp. gr. 5.192; m.p. (anhydrous) 568°C; b.p. 863°C; soluble in water, acetone, alcohol, and acids.
Derivation: By heating cadmium in bromine vapor.
Grades: Technical; reagent.
Containers: Bottles; cans; cases.
Hazard: Highly toxic. See under cadmium.
Uses: Photography; process engraving; lithography.
Shipping regulations: See under cadmium.

cadmium carbonate $CdCO_3$.
Properties: White, amorphous powder; sp. gr. 4.258; decomposes below 500°C. Soluble in acids (dilute) and in concentrated solutions of ammonium salts; insoluble in water.
Hazard: Highly toxic. See under cadmium.
Grades: Reagent.
Shipping regulations: See under cadmium.

cadmium chlorate $Cd(ClO_3)_2 \cdot 2H_2O$.
Properties: Colorless, prismatic crystals. Hygroscopic. Sp. gr. 2.28 (18°C); m.p. 80°C. Soluble in alcohol, water, and acetone.
Grades: Technical.
Hazard: Highly toxic. Dangerous in contact with organic materials.
Shipping regulations: Chlorates, n.o.s., (Rail) Yellow label. (Air) Oxidizer label.

cadmium chloride (a) $CdCl$; (b) $CdCl_2 \cdot 2\frac{1}{2}H_2O$.
Properties: Small white crystals, odorless; sp. gr. (a) 4.05, (b) 3.327; m.p. (a) 568°C; b.p. (a) 960°C. Soluble in water, alcohol, and acids.
Derivation: Action of hydrochloric acid on cadmium with subsequent crystallization.
Grades: Technical; reagent.
Containers: Bottles; tins; drums.
Hazard: Highly toxic. See under cadmium.
Uses: Preparation of cadmium sulfide; analytical chemistry; photography; dyeing and calico printing; ingredient of electroplating baths; addition to tinning solutions; manufacture of special mirrors; vacuum tube industry.
Shipping regulations: See under cadmium.

cadmium cyanide $Cd(CN)_2$. Obtained as a white precipitate when potassium or sodium cyanide is added to a strong solution of a cadmium salt. A complex ion is formed when it is dissolved in an excess of the precipitating agent, and a solution of this complex ion is used as an electrolyte for electrodeposition of cadmium.
Hazard: Highly toxic. See under cadmium.
Shipping regulations: (Rail, Air) Cyanides or cyanide mixtures, n.o.s., Poison label.

cadmium diethyldithiocarbamate $Cd[SC(S)N(C_2H_5)_2]_2$.
Properties: White to cream colored rods; sp. gr. 1.39; melting range 68–76°C; mostly soluble in benzene, carbon disulfide, chloroform; insoluble in water and gasoline.
Hazard: Highly toxic. See under cadmium.
Uses: Accelerator for butyl rubber.
Shipping regulations: See under cadmium.

cadmium fluoride CdF_2. Available as pure crystals, 99.89%; density 6.6 g/cc; m.p. approx. 1110°C. Soluble in water and acids; insoluble in alkalies.
Hazard: Highly toxic. See under cadmium.
Uses: Electronic and optical applications; high-temperature dry-film lubricants; starting material for crystals for lasers.
Shipping regulations: See under cadmium.

cadmium hydroxide (cadmium hydrate) $Cd(OH)_2$.
Properties: White, amorphous powder; sp. gr. 4.79; m.p., loses H_2O (300°C); soluble in ammonium hydroxide and in dilute acids; insoluble in water and alkalies; absorbs carbon dioxide from air.
Derivation: By the action of sodium hydroxide on a cadmium salt solution.
Grades: Technical; C.P.
Containers: Glass bottles; boxes.
Hazard: Highly toxic, see under cadmium.
Use: Cadmium salts; cadmium plating.
Shipping regulations: See under cadmium.

cadmium iodate $Cd(IO_3)_2$.
Properties: Fine, white powder; sp. gr. 6.48; m.p., decomposes; slightly soluble in water; soluble in nitric acid or ammonium hydroxide.
Grades: Technical.
Hazard: Highly toxic; fire risk in contact with organic materials.
Use: Oxidizing agent.
Shipping regulations: (Rail) Oxidizing material, n.o.s., Yellow label. (Air) Oxidizing material, n.o.s., Oxidizer label.

cadmium iodide CdI_2.
Properties: White, flaky, crystals; odorless; becomes yellow on exposure to air and light. Occurs in two allotropic forms. Sp. gr. 5.67 (alpha), and 5.30 (beta); m.p. (alpha) 388°C, m.p. (beta) 404°C; b.p. (alpha) 796°C; soluble in water, alcohol, ether, acetone, ammonia, and acids.
Derivation: By the action of hydriodic acid on cadmium oxide.
Hazard: Highly toxic. See under cadmium.
Uses: Photography; medicine; process engraving and lithography; analytical chemistry.
Shipping regulations: See under cadmium.

cadmium molybdate $CdMoO_4$.
Properties: Yellow crystals; sp. gr. 5.347; m.p. approx. 1250°C. Slightly soluble in water; soluble in acids.
Grades: Technical; crystals 99.98% pure.
Hazard: Highly toxic. See under cadmium.
Uses: Electronic and optical applications.
Shipping regulations: See under cadmium.

cadmium nitrate (a) $Cd(NO_3)_2 \cdot 4H_2O$; (b) $Cd(NO_3)_2$.
Properties: White, amorphous pieces or hygroscopic needles. Soluble in water, ammonia, and alcohol; (a) sp. gr. 2.455; m.p. 59.5°C; b.p. 132°C; (b) m.p. 350°C.
Derivation: Action of nitric acid on cadmium or cadmium oxide and crystallization.
Grades: Technical; reagent.
Containers: Tins; glass bottles, 400-lb drums.
Uses: Coloring glass and porcelain; laboratory reagent; cadmium salts.
Hazard: Highly toxic. Dangerous fire and explosion hazard.
Shipping regulations: Nitrates, n.o.s., (Rail) Yellow label. (Air) Oxidizer label.

cadmium oxalate $Cd(COO)_2 \cdot 3H_2O$.
Properties: White, amorphous powder; soluble in dilute acids, ammonium hydroxide; insoluble in alcohol and water; sp. gr. 3.32 (dehydrated); m.p., decomposes at 340°C.
Hazard: Highly toxic.
Shipping regulations: See under cadmium.

cadmium oxide CdO.
Properties: (a) Colorless, amorphous powder; sp. gr. 6.95. (b) brown or red crystals; sp. gr. 8.15. Both decompose on heating at 900°C; the crystals are soluble in acids and alkalies and insoluble in water.
Derivation: Cadmium metal is distilled in a retort, the vapor reacted with air, and the oxide collected in a baghouse.
Hazard: Highly toxic; inhalation of vapor or fume may be fatal. Tolerance (fume, as Cd): 0.1 mg per cubic meter of air.
Uses: Cadmium plating baths; electrodes for storage batteries; cadmium salts.
Shipping regulations: See under cadmium.

cadmium pigment. A family of pigments based on cadmium sulfide or cadmium selenide, used chiefly where high color retention is required. They are light-fast and have good alkali resistance. Red shades are obtained with cadmium selenide, yellow with cadmium sulfide. Used in paints and high-gloss baking enamels; often extended with barium sulfate, and are then called cadmium lithopone.
Hazard: Highly toxic.
Shipping regulations: See under cadmium.

cadmium potassium iodide $CdI_2 \cdot 2KI \cdot 2H_2O$.
Properties: White powder, becomes yellowish with age; deliquescent; soluble in water, alcohol, ether, and acid; sp. gr. 3.359; m.p. 76°C, with decomposition.
Derivation: By combining cadmium iodide and potassium iodide in solution, in proportion of their combining weights, and subsequent crystallization.
Hazard: Highly toxic. See under cadmium.
Uses: Analytical chemistry; medicine.
Shipping regulations: See under cadmium.

cadmium propionate $Cd(OOCC_2H_5)_2$. A solid; used in scintillation counters.
Shipping regulations: See under cadmium.

cadmium ricinoleate
$Cd[CH_3(CH_2)_5CHOHCH_2CH{:}CH(CH_2)_7CO_2]_2$.
Properties: Odorless, fine, white powder derived from castor oil. M.p. 104°C; sp. gr. 1.11.
Hazard: Highly toxic. See under cadmium.
Uses: Solutions used to stabilize polyvinyl chloride and copolymers against light and heat.
Shipping regulations: See under cadmium.

cadmium selenide CdSe. Red powder. Sp. gr. 5.81 (15/4°C); m.p. above 1350°C; insoluble in water; stable at high temperatures. The red powder form is used as a paint pigment. Pure cadmium selenide is used as a semiconductor.
Shipping regulations: See under cadmium.

cadmium selenide lithopone. See cadmium pigments.

cadmium stearate. Used as a lubricant and stabilizer in plastics.
Hazard: Highly toxic. See under cadmium.
Shipping regulations: See under cadmium.

cadmium succinate $Cd(OOCCH_2\text{-})_2$. A white powder; slightly soluble in water; soluble in alcohol. Used in insecticides; turf fungicide.
Hazard: Highly toxic. See under cadmium.
Shipping regulations: See under cadmium.

cadmium sulfate (a) $CdSO_4$; (b) $3CdSO_4 \cdot 8H_2O$; (c) $CdSO_4 \cdot 4H_2O$.
Properties: Colorless, odorless crystals. Sp. gr. (a) 4.69; (b) 3.09; (c) 3.05; m.p. (a) 1000°C. Soluble in water; insoluble in alcohol.
Derivation: By the action of dilute sulfuric acid on cadmium or cadmium oxide.
Grades: Technical; C.P.
Hazard: Highly toxic. See under cadmium.
Uses: Pigments; medicine; vacuum tubes; fluorescent screens; electrolyte in Weston standard cell.
Shipping regulations: See under cadmium.

cadmium sulfide (orange cadmium; see also cadmium pigment) CdS.
Properties: Yellow or brown powder. Sp. gr. 4.82; m.p. 1750°C (100 atm.); sublimes in nitrogen 980°C. Insoluble in cold water; forms a colloid in hot water; soluble in acids and ammonia. Can be polished like a metal. It is an *n*-type semiconductor.
Derivation: (a) By passing hydrogen sulfide gas into a solution of a cadmium salt acidified with hydrochloric acid. The precipitate is filtered and dried. (b) Occurs naturally as greenockite.

Grades: Technical; N.D.; high purity (single crystals).
Hazard: Highly toxic. See under cadmium.
Uses: Pigments and inks; ceramic glazes; pyrotechnics; phosphors; fluorescent screens; scintillation counters; rectifiers; photoconductor in xerography; transistors; photovoltaic cells; solar cells.
Shipping regulations: See under cadmium.

cadmium telluride CdTe.
Properties: Brownish-black, cubic crystals. Oxidizes on prolonged exposure to moist air. Insoluble in water and mineral acids, except nitric, in which it is soluble with decomposition. M.p. 1090°C; sp. gr. 6.2 (15/4°C).
Derivation: Fusion of the elements; reaction of hydrogen telluride and cadmium chloride.
Grades: High purity crystals, 99.99+%.
Hazard: Toxic by inhalation.
Use: Semiconductors.
Shipping regulations: See under cadmium.

cadmium tungstate $CdWoO_4$.
Properties: White or yellow crystals or powder. Soluble in ammonium hydroxide, alkali cyanides; very slightly soluble in water.
Derivation: By the interaction of cadmium nitrate and ammonium tungstate.
Forms: Single crystal rods; broken crystals (crackle).
Hazard: Toxic by inhalation.
Uses: Fluorescent paint; x-ray screens; scintillation counters; catalyst.
Shipping regulations: See under cadmium.

"Cadmofixe."[470] Trademark for co-precipitated red, orange and yellow cadmium pigments containing barium sulfate.
Hazard: May be toxic. See under cadmium.

"Cadmolith."[296] Trademark for a series of yellow and red cadmium-lithopone pigments.
Hazard: May be toxic. See under cadmium.
Uses: Automotive finishes, textile coatings, printing inks, lacquer and rubber.

"Cadmopur."[470] Trademark for pure red, orange and yellow cadmium pigments.
Hazard: May be toxic. See cadmium.

"Cadox."[419] Trademark for a series of organic peroxide catalysts.
Hazard: May be fire risk in contact with organic materials.

"Cadoxen."[50] Brand name for cadmium ethylenediamine hydroxide complex; a colorless, stable cellulose solvent.
Hazard: May be toxic. See under cadmium.

caesium. See cesium.

C_4AF. Abbreviation for tetracalcium aluminoferrate, as used in cement. See cement, Portland.

caffeine (theine; methyltheobromine; 1.3.7-trimethylxanthine) $C_8H_{10}N_4O_2 \cdot H_2O$ (bicyclic compound).
Properties: White, fleecy masses or long, flexible, silky crystals; an alkaloid; loses H_2O at 80°C. Efflorescent in air. M.p. 236.8°C; soluble in chloroform, slightly soluble in water and alcohol, very slightly soluble in ether. Odorless; bitter taste; solutions neutral to litmus.
Derivation: By extraction of coffee beans, tea leaves, or kola nuts; also synthetically. Much of the caffeine of commerce is a by-product of decaffeinized coffee manufacture.
Method of purification: Recrystallization.
Grades: Technical; U.S.P.; F.C.C.
Containers: 1-lb bottles; 5-, 10-, 25-lb cans; 100-150-lb drums.
Hazard: Moderately toxic in dosage of 1 grain or more. Concentration of 200 micrograms per ml has been found to inhibit activity of the enzyme DNA polymerase. Use in soft drinks not to exceed 0.02%.
Uses: Beverages; medicine.

caffeine bromide. See caffeine hydrobromide.

caffeine hydrobromide (caffeine bromide) $C_8H_{10}O_2N_4 \cdot HBr \cdot 2H_2O$.
Properties: Colorless, efflorescent crystals; become brownish on exposure to air; decompose at 80–100°C. Soluble in water or alcohol (with decomposition).
Hazard: Moderately toxic.
Use: Medicine.

caffeine sodium benzoate. A mixture of caffeine and sodium benzoate containing 47–50% anhydrous caffeine and 50–53% sodium benzoate. More water-soluble than caffeine.
Properties: White, odorless powder with slightly bitter taste. Slightly soluble in chloroform; soluble in alcohol and water.
Grades: U.S.P.
Uses: See caffeine.

cage compound. See inclusion complex; clathrate compound.

"Cairox."[459] Trademark for potassium permanganate (q.v.).

cake alum. See aluminum sulfate.

calabarine. See physostigmine.

"CAL"[323] Trademark for a decolorizing carbon designed for use in fixed or moving beds. Total surface area (N_2, BET Method) 1000–1100 sq m/g; particle size 12 × 40 mesh; apparent density 27.5 lb/cu ft. Used especially in beet sugar refining.

"Calade."[108] Trademark for a powdered, alkaline sodium hexametaphosphate compound containing wetting agents and an aluminum corrosion inhibitor.
Uses: Detergent for dishwashing and general cleaning.

calamine. (1) A hydrated zinc silicate containing 67.5% zinc oxide. Sp. gr. 3.5; Mohs hardness 4.5 to 5. Pyroelectric. Occurs in U.S. and Europe. Source of metallic zinc.
(2) Zinc oxide with low percentage of ferric oxide; ZnO must be 98%. Soluble in mineral acids; insoluble in water. Pharmaceutical preparation (U.S.P.).

calamus oil.
Properties: Yellow to brownish-yellow essential oil. Sp. gr. 0.959–0.970 (15°C); refractive index 1.503–1.510; saponification value 6–20. Slightly soluble in water.
Derivation: By steam distillation of calamus, the stem or root of the sweet flag. Chief known constituents; asarone (see 2,4,5-trimethoxy-1-propenylbenzene) and eugenol.
Containers: Bottles.
Uses: Perfumery; flavoring agent.

"Calatac 1250."[325] Trademark for a polyvinyl acetate emulsion. Produces durable, stiff finish on textiles. Used for backfilling, bodying, binding.

"Calatac" S.[325] Trademark for a borated resin used as a water-soluble resin binder and plasticizer for textile sizing.

calaverite $AuTe_2$. One of the gold telluride group of minerals. Corresponds to the same general formula as sylvanite and krennerite. Pale bronze-yellow color or tin-white, tarnishing to bronze yellow on exposure. Metallic luster. Contains 40–43% gold, 1–3% silver. Sp. gr. 9.0; Mohs hardness 2.5.
Occurrence: United States (California, Colorado); Australia; Canada.
Use: Important source of gold.

"Calcene."[177] Trademark for a specially prepared precipitated calcium carbonate for use in compounding rubber, paints, and plastics. The particles of TM-grade are coated with stearic acid to aid in dispersion. Grade NC is not coated.

calciferol. See ergocalciferol.

calcimine (kalsomine). Essentially chalk and glue in powdered form ready to mix with water. Used as temporary decoration for interior plastic walls. Will not withstand washing.

calcination. Heating of a solid to a temperature below its melting point to bring about a state of thermal decomposition or a phase transition other than melting. Included are the following types of reactions: (1) thermal dissociation, including destructive distillation of organic compounds, e.g., concentration of aluminum by ignition of bauxite; (2) polymorphic phase transitions, e.g., conversion of anatase to rutile form of TiO_2; (3) thermal recrystallization; e.g., devitrification of glass. Calcination is often used in the beneficiation of ores. See also destructive distillation; pyrolysis.

calcite $CaCO_3$. The most common form of natural calcium carbonate. Dogtooth spar, Iceland spar, nailhead spar, and satin spar are varieties of calcite. Essential ingredient of limestone, marble, and chalk (q.v.).
Properties: Colorless, white, and various colored crystals; vitreous to earthy luster; good cleavage in 3 directions. May contain small amounts of magnesium, iron, manganese and zinc. Reacts with acids to evolve carbon dioxide. Sp. gr. 2.72; Mohs hardness 2.
Uses: Phosphor; Iceland spar is used in optical instruments.

calcitonin. A thyroid hormone controlling the proportion of calcium in circulating blood; may be used in calcium balance control and possibly also in treatment of bone fractures, hypervitaminosis, and other calcium-related diseases. It is obtained in purified form from pig thyroid, and research on synthesis is active.

calcium Ca Alkaline-earth element of atomic number 20, group IIA of the periodic system. Atomic weight 40.08. Valence 2. Six stable isotopes.
Properties: Moderately soft, silver-white, crystalline metal; oxidizes in air to form adherent protective film; can be machined, extruded, or drawn. Soluble in acid; decomposes water to liberate hydrogen. Sp. gr. 1.57; m.p. 845°C; sublimes below its m.p. in vacuum; b.p. 1480°C.
Derivation: Electrolysis of fused $CaCl_2$; by thermal process under high vacuum from lime reduced with aluminum. Does not occur free in nature.
Forms: Crowns, nodules, ingots; crystals up to 99.9% pure.
Containers: Air-tight tins; well-stoppered glass bottles.
Hazard: Evolves hydrogen on contact with moisture. Flammable in finely divided state; fire and explosion hazard when heated or on contact with strong oxidizing agents. Nontoxic.
Uses: Alloying agent for aluminum, copper, and lead; reducing agent for beryllium; deoxidizer for alloys; dehydrating oils; decarburization and desulfurization of iron and its alloys; getter in vacuum tubes; separation of nitrogen from argon; reducing agent in preparation of chromium metal powder, thorium, zirconium, and uranium; fertilizer ingredient.
Shipping regulations: (Rail) Yellow label. (Air) Flammable Solid label. Crystalline form not accepted on passenger planes.
Note: Calcium is an essential component of bones, teeth, shells and plant structures. It occurs in milk in trace percentages, and is necessary in animal and human nutrition. Vitamin D aids in the deposition of calcium in bones.

calcium 45. Radioactive calcium of mass number 45.
Properties: Half-life 164 days; radiation, beta.
Derivation: By reactor irradiation of calcium carbonate, by neutron bombardment of scandium, or as a by-product of the irradiation of calcium nitrate for the preparation of carbon 14.
Forms available: Calcium chloride in hydrochloric acid solution and solid calcium carbonate.
Hazard: Dangerous radiation hazard; allowable concentration in air 3×10^{-8} microcurie per ml. This isotope is a boneseeker, and may cause damage to the blood-forming organs.
Uses: Research aid for studying water purification, calcium exchange in clays, detergency, surface wetting and other surface phenomena, calcium uptake and deposition in bone, soil characteristics as related to soil utilization of fertilizer and crop yield, diffusion of calcium in glass, etc.
Shipping regulations: (Rail, Air) Consult regulations.

calcium abietate $(C_{20}H_{29}O_2)_2Ca$. Product of the action of lime on rosin or resin acids. See calcium resinate.

calcium acetate (vinegar salts; gray acetate; lime acetate; calcium diacetate) $Ca(CH_3COO)_2 \cdot H_2O$.
Properties: Brown, gray or white (when pure) powder; amorphous or crystalline; slightly bitter taste; slight odor of acetic acid; decomposes on heating. Soluble in water; slightly soluble in alcohol. Combustible. Nontoxic.
Derivation: Action of pyroligneous acid on calcium hydroxide; the solution being filtered and evaporated to dryness; yielding gray acetate of lime.
Grades: Technical (80% basis); reagent; C.P.; pure; brown; gray; F.C.C.
Containers: 100-lb Fiberpak; 350-lb Leverpak drums.
Uses: Manufacture of acetone, acetic acid, acetates; mordant in dyeing and printing of textiles; stabilizer in resins; additive to calcium soap lubricants; food additive, as antimold agent in bakery goods, in sausage casings; catalyst manuacture; medicine.

calcium acetylsalicylate (aspirin, soluble) $Ca(CH_3COOC_6H_4COO)_2 \cdot 2H_2O$.
Properties: White powder. Aqueous solutions are unstable. Soluble in water.
Derivation: (a) Action of acetylsalicylic acid upon calcium carbonate in the presence of a small amount of water. (b) By passing carbon dioxide into an aqueous solution of calcium carbonate and acetylsalicylic acid.
Use: Medicine.

calcium acid sulfite. See calcium hydrogen sulfite.

calcium acrylate $(H_2C{:}CHCOO)_2Ca$.
Properties: Free-flowing white powder; soluble in water; deliquescent. Solutions polymerize to form hydrophilic resin.
Uses: Ion-exchange clay soil stabilizer, binder and sealer. (laboratory scale only).

calcium alginate.
Properties: White or cream-colored powder, or filaments, grains or granules. Slight odor and taste. Insoluble in water; insoluble in acids, but soluble in alkaline solutions. Nontoxic.
Grade: F.C.C.
Hazard: Flammable, but self-extinguishing.
Uses: Pharmaceutical products; food additive; thickening agent and stabilizer in ice cream, cheese products, canned fruits and sausage casings; synthetic fibers.
See also algin; alginic acid.

calcium aluminate (tricalcium aluminate) $3CaO \cdot Al_2O_3$. Crystals or powder; sp. gr. (25°C) 3.038; m.p., decomposes 1535°C; soluble in acids. A refractory, and an important ingredient of cements, especially of aluminous cement. Fused calcium aluminate (a glass) can be used for infrared transmission and detection. Noncombustible.

calcium ammonium nitrate. A uniform mixture of about 60% ammonium nitrate and 40% limestone and/or dolomite. A fertilizer containing about 20% nitrogen.
Hazard: Oxidizing material; fire risk in contact with organic materials.
Shipping regulations: Nitrates, n.o.s., (Rail) Yellow label. (Air) Oxidizer label.

calcium arsenate (tricalcium ortho-arsenate) $Ca_3(AsO_4)_2$.
Properties: White powder. Slightly soluble in water, soluble in dilute acids. Decomposes on heating.
Derivation: Interaction of calcium chloride and sodium arsenate.
Grades: Technical; C.P.
Hazard: Highly toxic by ingestion and inhalation. Tolerance, 1 mg per cubic meter of air.
Uses: Insecticide; germicide.
Shipping regulations: (Rail, Air) Poison label.

calcium arsenite $CaAsO_3H$.
Properties: White, granular powder. Insoluble in water; soluble in acids.
Grades: Technical.
Hazard: Highly toxic.
Uses: Germicides; insecticides.
Shipping regulations: (Rail, Air) Poison label.

calcium ascorbate $Ca(C_6H_7O_6)_2 \cdot 2H_2O$.
Properties: A white to slightly yellow, odorless, crystalline powder. Soluble in water; slightly soluble in alcohol; insoluble in ether. The pH of a 10% solution is between 6.8 and 7.4. Nontoxic.
Grade: F.C.C.
Uses: Food preservative.

calcium biphosphate. See calcium phosphate, monobasic.

calcium bisulfide. See calcium hydrosulfide.

calcium bisulfite. See calcium hydrogen sulfite.

calcium bromate $Ca(BrO_3)_2 \cdot H_2O$.
Properties: White crystalline powder; sp. gr. 3.329; loses water at 180°C. Very soluble in water. Low toxicity.
Grade: F.C.C.
Hazard: Oxidizing agent. Fire risk in contact with organic materials.
Uses: Maturing agent; dough conditioner.
Shipping regulations: Oxidizing material, n.o.s., (Rail) Yellow label. (Air) Oxidizer label.

calcium bromide (a) $CaBr_2 \cdot 6H_2O$; (b) $CaBr_2$.
Properties: White powder or crystals; odorless; sharp saline taste; very deliquescent; very soluble in water. Also soluble in alcohol and acetone. (a) Sp. gr. 2.295 (25°C); m.p. 38°C; b.p. 149°C (decomposes). (b) Sp. gr. 3.353 (25°C); m.p. 730°C, with slight decomposition; b.p. 806–812°C.
Derivation: By the action of hydrobromic acid on calcium oxide, carbonate, or hydroxide and subsequent crystallization.
Grades: Technical: C.P.
Uses: Photography; medicine; dehydrating agent; food preservative; road treatment; freezing mixtures; sizing compounds; wood preservative.

calcium carbide CaC_2.
Properties: Grayish-black, irregular hard solid; must be kept dry. Sp. gr. 2.22; m.p. about 2300°C. Garlic-like odor. Decomposes in water, with formation of acetylene and calcium hydroxide and evolution of heat.
Derivation: Interaction of pulverized limestone or quicklime with crushed coke or anthracite in an electric furnace.
Grades: Technical; lumps; powder.
Containers: 2 lb to 5 tons; metal packages, water- and airtight.
Hazard: Forms flammable and explosive gas and corrosive solid on exposure to moisture. Safety data sheet available from Manufacturing Chemists Assn., Washington, D.C.
Uses: Generation of acetylene gas for welding; manufacture of neoprene; chloroethylenes; vinyl acetate monomer; acetylene chemicals.
Note: Rapidly being replaced by hydrocarbons for most chemical uses.
Shipping regulations: (Rail) Flammable solid, n.o.s., Yellow label. (Air) Flammable Solid label. Not acceptable on passenger planes.

calcium carbonate $CaCO_3$.
Properties: White powder or colorless crystals, odorless, tasteless; sp. gr. 2.7–2.95; decomposes at 825°C; noncombustible; nontoxic; very slightly soluble in water; soluble in acids with evolution of carbon dioxide.
Occurrence: Calcium carbonate is one of the most stable, common, and widely dispersed materials. It occurs in nature as aragonite, oyster shells, calcite, chalk, limestone, marble, marl, and travertine; especially in Indiana (structural limestone), Vermont (marble), Italy (travertine), and England (chalk).
Derivation: (a) Mined from natural surface deposits. (2) Precipitated (made synthetically) by reaction of calcium chloride and sodium carbonate in water solution, or by passing carbon dioxide through a suspension of hydrated lime ($Ca(OH)_2$) in water.
Uses: Source of lime; neutralizing agent; filler and extender in rubber, plastics, paints; opacifying agent in paper; fortification of bread; putty; tooth powders; antacid; whitewash; Portland cement; SO_2 re-

moval from stack gases; metallurgical flux; analytical chemistry; CO_2 generation (laboratory).
See also chalk; calcite; marble; limestone; whiting.
Note: Calcium carbonate is a major cause of boiler scale when hard water is used in heating systems.

calcium caseinate.
Properties: White or slightly yellow, nearly odorless powder. Insoluble in cold water; forms a milky solution when suspended in water, stirred and heated. Nontoxic.
Containers: 100-lb bags.
Uses: Medicine; special foods.
See also casein.

calcium chlorate $Ca(ClO_3)_2 \cdot 2H_2O$.
Properties: White to yellowish crystals. Keep well stoppered. Melts when rapidly heated at 100°C; m.p. (anhydrous) 340°C. Soluble in water and alcohol. Hygroscopic. Sp. gr. 2.711.
Derivation: By the action of chlorine on hot calcium hydroxide slurry.
Hazard: Oxidizing agent; dangerous fire risk; forms explosive mixtures with combustible materials.
Uses: Photography, pyrotechnics; dusting powder to kill poison ivy.
Shipping regulations: (Rail) Yellow label. (Air) Oxidizer label.

calcium chloride (a) $CaCl_2$; (b) $CaCl_2 \cdot H_2O$; (c) $CaCl_2 \cdot 2H_2O$; (d) $CaCl_2 \cdot 6H_2O$.
Properties: White, deliquescent crystals, granules, lumps or flakes. (a) Sp. gr. 2.15 (25°C); m.p. 772°C; b.p. > 1600°C. (b) melt. p. 260°C. (c) U.S.P. grade; sp. gr. 0.835 (25°C). (d) Sp. gr. 1.71 (25°C); b.p., loses $4H_2O$ at 30°C and $6H_2O$ at 200°C. All forms soluble in water and alcohol. Water solution is neutral or slightly alkaline. Low toxicity.
Derivation: (1) Action of hydrochloric acid on calcium carbonate and subsequent crystallization. (2) Commercially obtained as a by-product in the Solvay soda and other processes. (3) Recovery from brines.
Grades: Technical; C.P.; F.C.C.; U.S.P. (the dihydrate); various forms and purities; solutions.
Uses: De-icing and dust control of roads; drilling muds; dustproofing, freezeproofing, and thawing coal, coke, stone, sand, ore; concrete conditioning; paper and pulp industry; fungicides; refrigeration brines; drying and desiccating agent; sequestrant in foods; firming agent in tomato canning; tire weighting; pharmaceuticals; electrolytic cells. For further information refer to Calcium Chloride Institute, Ring Bldg., Washington, D.C.

calcium chlorite. $Ca(ClO_2)_2$. White crystals; sp. gr. 2.71; decomposes in water.
Hazard: Strong oxidizing agent; fire risk in contact with organic materials.
Shipping regulations: (Rail) Yellow label. (Air) Oxidizer label. Not acceptable on passenger planes.

calcium chromate $CaCrO_4$; $CaCrO_4 \cdot 2H_2O$. Bright yellow powder; hydrate loses water at 200°C. Anhydrous; sp. gr. 2.89. Soluble in dilute acids and alcohols; slightly soluble in water.
Containers: Bags; drums.
Grades: Technical.
Hazard: May be toxic; a suspected carcinogen.
Uses: Pigment; corrosion inhibitor; oxidizing agent; depolarizer for batteries; coating for light metal alloys.

calcium citrate (lime citrate; tricalcium citrate) $Ca_3(C_6H_5O_7)_2 \cdot 4H_2O$. A by-product in the manufacture of citric acid.
Properties: White odorless powder; loses most of its water at 100°C and all of it at 120°C. Almost insoluble in water; insoluble in alcohol.
Grades: Reagent; technical; F.C.C.
Use: Dietary supplement; sequestrant, buffer and firming agent in foods; medicine.

calcium cyanamide (lime nitrogen; calcium carbimide) $CaCN_2$.
Properties: Colorless crystals or powder; sp. gr. 1.083; m.p. 1300°C; sublimes > 1150°C. Decomposes in water, liberating ammonia and acetylene.
Derivation: Calcium carbide powder is heated in an electric oven, into which nitrogen is passed (24 to 26 hrs.). Any uncombined calcium carbide is leached out after removal.
Grades: Fertilizer, 21% N; industrial.
Containers: 200-lb bags; drums; bulk in cars.
Hazard: Moderately toxic. Skin irritant. Fire risk if exposed to moisture or combined with calcium carbide.
Uses: Fertilizer; nitrogen products; hardening iron and steel.
Shipping regulations: (Air) Containing more than 0.5% calcium carbide: Flammable Solid label.

calcium cyanide $Ca(CN)_2$.
Properties: Colorless crystals or white powder; gray-black (technical); decomposes in moist air liberating hydrogen cyanide. Dissolves in water and very weak acid, with evolution of hydrogen cyanide. Decomposes above 350°C.
Containers: 4-oz to 100-lb metal boxes or drums.
Hazard: Highly toxic. Safety data sheet available from Manufacturing Chemists Assn., Washington, D.C.
Uses: Rodenticide; fumigant for greenhouses, flour mills, grain, seed, citrus trees under tents for control of scale insects; leaching of gold and silver ores; other cyanides.
Shipping regulations: (Rail, Air) Poison label.

calcium cyclamate (calcium cyclohexylsulfamate; calcium cyclohexanesulfamate) $(C_6H_{11}NHSO_3)_2Ca \cdot 2H_2O$.
Properties: White, crystalline, practically odorless powder with very sweet taste. Freely soluble in water (solutions are neutral to litmus); practically insoluble in alcohol, benzene, chloroform and ether; pH (10% solution) 5.5–7.5. Sweetening power approximately, 30 times that of sucrose.
Grades: N.F.; F.C.C.
Containers: 100-lb drums.
Hazard: Not permitted for use in foods and soft drinks due to suspected carcinogenicity.
Uses: Nonnutritive sweetener.
See also cyclamate.

calcium cyclobarbital (calcium cyclohexenylethylbarbiturate) $Ca(C_{12}H_{15}N_2O_3)_2$.
Properties: Whitish powder. Soluble in water; insoluble in alcohol, chloroform, and ether.
Grade: N.F.
Hazard: May be toxic. See barbiturate.
Use: Medicine.

calcium cyclohexanesulfamate. See calcium cyclamate.

calcium cyclohexylsulfamate. See calcium cyclamate.

calcium dehydroacetate $(C_8H_7O_4)_2Ca$. See also dehydroacetic acid.
Properties: White to cream powder. Almost insoluble in water and organic solvents.
Grades: 96% minimum.
Use: Fungicide.

calcium diacetate. See calcium acetate.

calcium dibromobehenate $(C_{22}H_{41}O_2Br_2)_2Ca$.
Properties: White to yellow powder; odorless; tasteless. Protect from light. Soluble in ether, chloroform, acetone, carbon tetrachloride, and benzene; insoluble in water or alcohol.
Use: Medicine.

calcium dichromate $CaCr_2O_7 \cdot 3H_2O$, or 4.5 H_2O.
Properties: Brownish-red crystals; deliquescent; sp. gr. (4.5 H_2O variety) 2.136. Soluble in water.
Grades: Technical; C.P.
Hazard: May be toxic.
Use: Corrosion inhibitor.

calcium dihydrogen sulfite. See calcium hydrogen sulfite.

calcium dioxide. See calcium peroxide.

calcium disodium edetate (USAN) (calcium disodium EDTA; edathamil calcium disodium; calcium disodium ethylenediaminetetraacetate) $CaNa_2C_{10}H_{12}N_2O_8 \cdot xH_2O$. For other variations of the name see ethylenediaminetetraacetic acid. The calcium disodium salt is a mixture of the dihydrate and trihydrate.
Properties: White, odorless powder or flakes; slightly hygroscopic; faint saline taste. Stable in air. Soluble in water; insoluble in organic solvents. It acts as a chelating agent for heavy metals.
Grades: U.S.P.; F.C.C.
Uses: Medicine (antidote in heavy-metal poisoning); in foods to "complex" trace heavy metals, as a preservative, and to retain color and flavor; antigushing agent in fermented malt beverages. (For restrictions on food uses see FDA regulations.)

calcium disodium ethylenediaminetetraacetate. See calcium disodium editate.

calcium ethylhexoate. See soaps (2).

calcium ferrocyanide $Ca_2Fe(CN)_6 \cdot 12H_2O$.
Properties: Yellow crystals; decomposes on heating. Soluble in water; insoluble in alcohol; sp. gr. 1.68. Low toxicity.
Use: Removal of metallic impurities in the manufacture of citric, tartaric, and other acids.

calcium fluoride CaF_2.
Properties: White powder, occurring in nature as fluorite (pure form) or fluorspar (mineral). M.p. 1402°C, b.p. 2500°C (approx.); sp. gr. 3.18. Reacts with hot concentrated sulfuric acid to liberate hydrofluoric acid. Insoluble in water; soluble in ammonium salts.
Derivation: (a) By powdering pure fluorite or fluorspar; (b) by the interaction of a soluble calcium salt and sodium fluoride.
Hazard: Toxic and irritant. Tolerance (as F), 2.5 mg per cubic meter of air.
Grades and Uses: See fluorspar. Single pure (99.93%) crystals of calcium fluoride are also produced for use in spectroscopy and electronics, lasers, high-temperature dry-film lubricants.

calcium fluorophosphate (fluoroapatite, FAP). Recently developed laser crystal said to have lowest energy threshold of any room-temperature crystal.

calcium fluosilicate (calcium silicofluoride (a) $CaSiF_6$; (b) $CaSiF_6 \cdot 2H_2O$.
Properties: White, crystalline powder. Sp. gr. (a) 2.662 (17.5°C), (b) 2.254. Very slightly soluble in water.
Derivation: By the action of fluosilicic acid on calcium carbonate and subsequent crystallization.
Hazard: Toxic and irritant. Tolerance (as F), 2.5 mg per cubic meter of air.
Use: Ceramics.

calcium folinate (calcium leucovorin) $C_{20}H_{21}CaN_7O_7 \cdot 5H_2O$. The calcium salt of folinic acid, formerly called the citrovorum factor.
Properties: Yellowish-white or yellow, odorless microcrystalline powder. Very soluble in water. Practically insoluble in alcohol.
Use: Medicine.

calcium formate $Ca(OOCH)_2$. White powder; m.p. $>$ 300°C; sp. gr. 2.015. Soluble in water; insoluble in alcohol. Used in the chrome tanning of leather.

calcium gluconate $Ca(C_6H_{11}O_7)_2 \cdot H_2O$.
Properties: White, odorless, practically tasteless, fluffy powder or granules. Stable in air. Loses H_2O at 120°C. Soluble in hot water; less soluble in cold water; insoluble in alcohol, acetic acid, and other organic solvents; specific rotation (20/D) about +6°. Solutions neutral to litmus. Nontoxic.
Derivation: Neutralization of gluconic acid with lime or calcium carbonate.
Grades: Technical; U.S.P.; F.C.C.; special for ampules.
Containers: Cans; fiber drums; barrels.
Uses: Medicine and veterinary medicine; food additive, buffer and sequestering agent; vitamin tablets.

calcium glutamate. Similar to sodium glutamate (q.v.).

calcium glycerophosphate (calcium glycerinophosphate) $CaC_3H_7O_2PO_4$.
Properties: White, crystalline powder; odorless; almost tasteless; slightly hygroscopic; decomposes above 170°C; slightly soluble in water; insoluble in alcohol. Nontoxic.
Derivation: By esterification of phosphoric acid with glycerol and conversion of glycerophosphoric acid to the calcium salt.
Grades: Technical; pure; F.C.C.
Uses: Medicine and veterinary medicine; stabilizer for plastics; nutrient and dietary supplement.

calcium glycolate $(CH_2OHCOO)_2Ca$.
Properties: White solid.
Grades: Technical.
Use: Source of glycolic acid and of the glycolic acid radical in chemical synthesis.

calcium hydrate. See calcium hydroxide.

calcium hydride CaH_2.
Properties: Grayish-white lumps or crystals. Acted upon by moist air with formation of calcium hydroxide and evolution of hydrogen. Sp. gr. 1.7; decomposes at 675°C. Decomposed by water, acids, and lower alcohols.
Grades: Technical; 94% pure.
Containers: Drums.
Hazard: Evolves highly flammable hydrogen when wet; solid product is slaked lime, irritating to skin.
Uses: Reducing agent; drying agent; analytical reagent in organic chemistry; easily portable source of hydrogen; cleaner for blocked-up oil wells.
Shipping regulations: (Air) Flammable Solid label. Not acceptable on passenger planes. (Rail) Flammable solids, n.o.s., Yellow label.

calcium hydrogen sulfite (calcium bisulfite; calcium dihydrogen sulfite; calcium acid sulfite) $Ca(HSO_3)_2$. Exists only in solution and is really a solution of calcium sulfite in an aqueous sulfur dioxide solution.
Properties: Yellowish liquid with strong sulfur-dioxide odor; sp. gr. 1.06; corrosive to metals. Soluble in water and acids.
Derivation: Action of sulfur dioxide on calcium hydroxide solution.
Hazard: Irritant and corrosive to skin and tissue.
Uses: Antichlor in bleaching textiles; paper pulp (dissolving lignin); preservative; bleaching sponges; hydroxylamine salts; germicide; disinfectant.
Shipping regulations: (Air) Corrosive label. (Rail) Corrosive materials, n.o.s., White label.

calcium hydrosulfide (calcium bisulfide; calcium sulfhydrate) $Ca(HS)_2 \cdot 6H_2O$.
Properties: Colorless, transparent crystals. Soluble in alcohol and water. Decomposes in air (15–18°C).
Use: Leather industry.

calcium hydroxide (calcium hydrate; hydrated lime; caustic lime; slaked lime) $Ca(OH)_2$.
Properties: Soft, white crystalline powder with alkaline, slightly bitter taste. Sp. gr. 2.34; m.p., loses water at 580°C; pH of water solution (25°C) 12.4. Very slightly soluble in water; soluble in glycerin, syrup, and acids; insoluble in alcohol. Absorbs carbon dioxide from air.
Derivation: Action of water on calcium oxide.
Impurities: Calcium carbonate, magnesium salts, iron.
Grades: Technical; chemical lime (insoluble matter under 2%, Mg under 3%); building lime; U.S.P., C.P., F.C.C.
Containers: Wooden barrels; multiwall paper sacks; bulk.
Hazard: Skin irritant; avoid inhalation.
Uses: Mortar; plasters; cements; calcium salts; causticizing soda; hydrogen; depilatory; unhairing of hides (leather industry); lime paints; medicine; agriculture (to "sweeten" acid soil); ammonia recovery in gas manufacture; disinfectant; water softening; purification of sugar juices; accelerator for low-grade rubber compounds; water paints; soil stabilizers, petrochemicals; food additive, as buffer and neutralizing agent.

calcium hypochlorite (calcium oxychloride) $Ca(OCl)_2$. See also lime, chlorinated.
Properties: White crystalline solid; sp. gr. 2.35; decomposes at 100°C; decomposes in water and alcohol; not hygroscopic; practically clear in water solution. Stable chlorine carrier. An oxidizing material.
Derivation: Chlorination of a slurry of lime and caustic soda with subsequent precipitation of calcium hypochlorite dihydrate, dried under vacuum.
Grades: Commercial (70%); high purity (99.2% available chlorine as calcium hypochlorite).
Containers: Steel drums and cans; plastic and glass bottles.
Hazard: Toxic by ingestion, skin contact and inhalation. Dangerous fire risk in contact with organic materials.
Uses: Algicide; bactericide; deodorant; potable water purification; disinfectant for swimming pools; fungicide; bleaching agent (paper, textiles).
Shipping regulations: (Dry, containing more than 39% available chlorine) (Rail) Yellow label. (Air) Oxidizer label.

calcium hyposulfite. See calcium thiosulfate.

calcium iodate $Ca(IO_3)_2$.
Properties: White crystals or powder; odorless; sp. gr. 4.5 (15°C); decomposes 540°C. Soluble in water and nitric acid; insoluble in alcohol. Oxidizing material.
Grades: Technical; C.P.; F.C.C.
Hazard: Fire risk in contact with organic materials.
Uses: Deodorant; medicine; mouth washes; feed additive; food additive; dough conditioner (up to 0.0075 part per 100 lb flour used).

calcium iodide $CaI_2 \cdot 6H_2O$.
Properties: Yellowish-white crystals; deliquescent; decomposes in air by absorption of carbon dioxide. Soluble in water, ethyl alcohol, and amyl alcohol; sp. gr. 2.55 (anhydrous 4.0 at 25°C); loses 6 H_2O at 42°C; m.p. 783°C; b.p. ca. 1100°C.
Derivation: Action of hydriodic acid on calcium carbonate.
Hazard: Moderately toxic by ingestion and inhalation.
Use: Photography; medicine.

calcium iodobehenate $Ca(OOCC_{21}H_{42}I)_2$.
Properties: White or yellowish powder containing approximately 24% iodine; unctuous to the touch; odorless, or slight fatty odor. Soluble in warm chloroform; only slightly soluble in alcohol and ether; insoluble in water.
Hazard: Moderately toxic.
Uses: Medicine; feed additive.

calcium ipodate (calcium 3-dimethylaminomethyleneamino)-2,4,6-triiodohydrocinnamate) $Ca[OOCCH_2CH_2C_6H(I_3)N{:}CHN(CH_3)_2]_2$. Listed in N.D. Used as a radiopaque agent in medicine.

calcium lactate $Ca(C_3H_5O_3)_2 \cdot 5H_2O$.
Properties: White, almost tasteless powder; almost odorless. Somewhat efflorescent. Soluble in water, practically insoluble in alcohol. Loses H_2O at 120°C. Nontoxic.
Derivation: By neutralizing dilute lactic acid with calcium carbonate and evaporating the solution.
Grades: N.F.; F.C.C.
Uses: Medicine, veterinary medicine; manufacture of foods; beverages.

calcium laurate $Ca(OOCC_{11}H_{23})_2 \cdot H_2O$. A soap (crystals when pure) produced by precipitation of a soluble laurate with calcium chloride. For uses, see soap (2).

calcium leucovorin. See calcium folinate.

calcium levulinate $Ca[CH_3CO(CH_2)_2COO]_2 \cdot 2H_2O$.
Properties: White powder; faint odor like burnt sugar; bitter salty taste. Freely soluble in water; slightly soluble in alcohol; insoluble in ether and chloroform; m.p. 119–125°C.

calcium lignosulfonate.
Uses: Binder for nonmagnetic ores; retards setting rate of Portland cement in oil-well casings. See also lignin sulfonate.

calcium linoleate $Ca(C_{18}H_{31}O_2)_2$.
Properties: White powder. Soluble in alcohol and ether; insoluble in water.
Derivation: Interaction of solutions of calcium chloride and a soluble linoleate.
Containers: 100-, 200-, 400-lb drums.
Uses: Waterproofing compounds; emulsifying agent and stabilizer for flat paints, fillers, and enamels.

calcium magnesium aconitate. See dicalcium magnesium aconitate.

calcium magnesium carbonate. See dolomite.

calcium magnesium chloride (magnesium calcium chloride) $CaCl_2 \cdot MgCl_2$.
Properties: White, deliquescent crystals; soluble in water and acids; insoluble in alcohol and ether. Nonflammable. Low toxicity.
Derivation: (a) A by-product in the salt industry; (b) by the action of hydrochloric acid on dolomite.
Use: Manufacture of intermediates, dyes, fireproof paints, paper and textile sizing; preservative; laboratory reagent; dehydrating starch.

calcium mandelate $Ca(C_8H_7O_3)_2$.
Properties: White, odorless powder. Insoluble in alcohol; slightly soluble in cold water.
Use: Medicine and pharmaceuticals.

calcium metasilicate. $CaSiO_3$.
Properties: White powder; sp. gr. 2.9. Insoluble in water.
Containers: Bags.
Hazard: Irritant dust. Use in foods restricted to 5% in baking powder, 2% in table salt.
Uses: Absorbent; antacid; adhesives; filler for paper and paper coatings; cosmetics; food additive (anticaking agent).
See also dicalcium silicate.

calcium molybdate $CaMoO_4$.
Properties: White, crystalline powder; m.p. ca. 1250°C; sp. gr. 4.35; soluble in mineral acids; insoluble in alcohol, ether, or water. Noncombustible.
Derivation: Fusion of calcium oxide and a molybdenum ore.
Grades: Technical; single crystals, 99.97%.
Use: Molybdic acid; alloying agent in production of iron and steel; crystals in optical and electronic applications.

calcium naphthenate.
Properties: Sticky, tenacious mass. Insoluble in water; soluble in ethyl acetate, carbon tetrachloride, gasoline, benzene and ether. Combustible.
Derivation: Precipitation from aqueous solution of calcium salts and sodium naphthenate.
Uses: Waterproofing compositions; adhesives; driers; wood fillers; grafting waxes; cements; varnishes; color lakes.
See also soap (2).

calcium nitrate (lime nitrate; nitrocalcite; lime saltpeter; Norwegian saltpeter)
(a) $Ca(NO_3)_2 \cdot 4H_2O$; (b) $Ca(NO_3)_2$.
Properties: White, deliquescent mass. Soluble in water, alcohol and acetone. Sp. gr. (a) 1.82, (b) 2.36; m.p. (a) 42°C, (b) 561°C.
Hazard: Strong oxidizer; dangerous fire risk in contact with organic materials; may explode if shocked or heated.
Grades: Technical; pure; C.P.; reagent.
Uses: Pyrotechnics; explosives; matches; fertilizers; other nitrates; source of carbon 14 by nuclear irradiation.
Shipping regulations: (Rail) Yellow label. (Air) Oxidizer label.

calcium nitride Ca_3N_2.
Properties: Brown crystals; sp. gr. 2.63 (17°C); m.p. 1195°C. Soluble in water with evolution of ammonia, an irritant gas; soluble in dilute acids; insoluble in absolute alcohol.

calcium nitrite $Ca(NO_2)_2 \cdot H_2O$.
Properties: Colorless or yellowish crystals. Hygroscopic. Soluble in water; slightly soluble in alcohol. Sp. gr. 2.23 (34°C, anhydrous); m.p., loses water at 100°C.
Grades: Technical.

calcium novobiocin. See novobiocin.

calcium octoate. See soap (2).

calcium orthophosphate. See calcium phosphate, tribasic.

calcium orthotungstate. See calcium tungstate.

calcium oxalate CaC_2O_4.
Properties: White, crystalline powder; soluble in dilute hydrochloride acid, dilute nitric acid; insoluble in acetic acid and water. Sp. gr. 2.2.
Grades: Technical; C.P.
Hazard: Toxic and irritant.
Use: Making oxalic acid and organic oxalates.

calcium oxide (lime; quicklime; burnt lime; calx; unslaked lime; fluxing lime) CaO. Second in order of high-volume chemicals produced in U.S. (1975).
Properties: White, or grayish-white hard lumps, sometimes with a yellowish or brownish tint, due to iron; odorless. Crumbles on exposure to moist air; sp. gr. 3.40; m.p. 2570°C; b.p. 2850°C. Soluble in acid; reacts with water to form calcium hydroxide, with evolution of heat.
Derivation: Calcium carbonate (limestone) is roasted in kilns until all the carbon dioxide is driven off.
Impurities: Calcium carbonate; magnesium, iron, and aluminum oxides.
Grades: N.F.; technical; chemical lime; agricultural lime; building lime; F.C.C.; single crystals.
Containers: Wooden barrels; bags; freight cars; multiwall paper sacks.
Hazard: Strong irritant; evolves heat on exposure to water. Dangerous near organic materials. Tolerance, 5 mg per cubic meter of air.
Uses: Refractory; flux in steel manufacture; pulp and paper; mfg. of calcium carbide; SO_2 removal from stack gases; sewage treatment (phosphate removal, pH control); poultry feeds; neutralization of acid waste effluents; insecticides and fungicides; dehairing of hides; food processing; sugar refining.

calcium oxychloride. See calcium hypochlorite.

calcium palmitate $Ca(C_{15}H_{31}CO_2)_2$. White or pale-yellow powder produced by reacting a soluble palmitate with a soluble calcium salt. Insoluble in water; slightly soluble in alcohol or ether. Combustible.
Uses: Waterproofing agent; thickener for lubricating oils; manufacture of solidified oils. Available only as technical grade. See also soap (2).

calcium pantothenate $(C_9H_{16}NO_5)_2Ca$. The calcium salt of pantothenic acid, available in either the dextro- or racemic forms; only the dextro- form has vitamin activity.
Properties: (both forms identical) White, slightly hygroscopic, odorless powder; sweetish taste; stable in air; solutions have a pH of 7–9; soluble in water and glycerol; insoluble in alcohol, chloroform, and ether. M.p. 170–172°C; dec. 195–196°C; specific rotation (5% aqueous solution) + 28.2° (25°C). Nontoxic.
Source: Same as pantothenic acid.
Grades: U.S.P. (both forms); F.C.C.

Uses: Medicine; animal feeds; dietary supplement. See also pantothenic acid.

calcium pectate. A material developed in plants such as beans, peas, potatoes, which enables them to seal off fungus-infected areas. See pectic acid; pectins.

calcium perborate $Ca(BO_3)_2 \cdot 7H_2O$.
Properties: Gray-white lumps or powder. Soluble in acids; also in water with evolution of oxygen.
Uses: Medicine; bleach; tooth powders.

calcium perchlorate $Ca(ClO_4)_2$. White crystals; sp. gr. 2.651; decomposes at 270°C. Soluble in water and alcohol.
Hazard: Strong oxidizer; dangerous fire risk in contact with organic materials.
Shipping regulations: (Rail) Oxidizing material, n.o.s., Yellow label. (Air) Oxidizer label.

calcium permanganate $Ca(MnO_4)_2 \cdot 4H_2O$.
Properties: Violet crystals, deliquescent; sp. gr. 2.4. Soluble in water and ammonia; decomposed by alcohol.
Grades: Technical; pure.
Hazard: Moderately toxic. Strong oxidizer; dangerous fire risk in contact with organic materials.
Uses: Textile industry; sterilizing water; dentistry; disinfectant; deodorizer; an additive (with hydrogen peroxide) in liquid rocket propellants.
Shipping regulations: (Rail) Yellow label. (Air) Oxidizer label.

calcium peroxide (calcium superoxide; calcium dioxide) CaO_2.
Properties: White or yellowish, odorless, almost tasteless powder. Decomposes about 200°C. Practically insoluble in water; soluble in acids with formation of hydrogen peroxide. Available oxygen 22.2% (min. 13.3% in technical grade).
Derivation: Interaction of solutions of a calcium salt and sodium peroxide, with subsequent crystallization.
Grades: 60–75%; F.C.C.
Hazard: Strong oxidizing agent. Dangerous fire risk in contact with organic materials. Irritant in concentrated form.
Uses: Seed disinfectant; dentrifices; dough conditioners; medicine; bleaching of oils; modification of starches; high-temperature oxidations.
Shipping regulations: (Rail) Yellow label. (Air) Oxidizer label.

calcium phenolsulfonate (calcium sulfocarbolate) $Ca(C_6H_4OHSO_3)_2 \cdot H_2O$.
Properties: White, crystalline powder; odorless. Soluble in water and alcohol.
Derivation: Action of phenolsulfonic acid on calcium hydroxide.
Uses: Disinfectant; medicine.

calcium phosphate. See calcium phosphate, dibasic; calcium phosphate, monobasic; or calcium phosphate, tribasic.

calcium phosphate, dibasic (dicalcium orthophosphate; bicalcium phosphate; secondary calcium phosphate) $CaHPO_4 \cdot 2H_2O$ and $CaHPO_4$.
Properties: White, tasteless, crystalline powder; odorless; soluble in dilute hydrochloric, nitric, and acetic acids; insoluble in alcohol; slightly soluble in water. (Hydrate) sp. gr. 2.306; loses water at 109°C. Nontoxic. Nonflammable.
Derivation: Interaction of fluorine-free phosphoric acid with milk of lime.
Grades: U.S.P.; F.C.C.; dentrifice grade; feed grade, 18½ or 21% P.
Uses: Animal feed supplement; food supplement; dentifrice; medicine; glass; fertilizer; stabilizer for plastics; dough conditioner; yeast food.

calcium phosphate, monobasic (calcium biphosphate; acid calcium phosphate; calcium phosphate, primary; monocalcium phosphate) $CaH_4(PO_4)_2 \cdot H_2O$. See also superphosphate.
Properties: Colorless, pearly scales or powder, deliquescent in air. Soluble in water and acids. Aqueous solutions are acid. M.p., loses H_2O at 100°C, decomposes at 200°C; sp. gr. 2.20. Nontoxic. Nonflammable.
Derivation: By dissolving either dicalcium or tricalcium phosphates in phosphoric acid and allowing the solution to evaporate spontaneously.
Grades: F.C.C.; ceramic; anhydrous; hydrated.
Uses: Baking powders; fertilizers; mineral supplement; stabilizer for plastics; to control pH in malt; glass manufacture; buffer in foods; firming agent.

calcium phosphate, precipitated. See calcium phosphate, tribasic.

calcium phosphate, primary. See calcium phosphate, monobasic.

calcium phosphate, secondary. See calcium phosphate, dibasic.

calcium phosphate, tertiary. See calcium phosphate, tribasic.

calcium phosphate, tribasic (calcium orthophosphate; tricalcium phosphate; precipitated calcium phosphate; tricalcium orthophosphate; tertiary calcium phosphate) $Ca_3(PO_4)_2$. True $Ca_3(PO_4)_2$ can be prepared thermally but is rare. Precipitated 'tricalcium phosphate' is a hydroxyapatite (q.v.) with the approximate formula $Ca_5OH(PO_4)_3$. (See also bone ash).
Properties: White, odorless, tasteless crystalline powder. Sp. gr. 3.18; m.p. 1670°C; refractive index 1.63. Soluble in acids; insoluble in water, alcohol, and acetic acid. Nontoxic, nonflammable.
Derivation: (a) Phosphate rock, apatite, and phosphorite (q.v.). (b) By the interaction of solutions of calcium chloride and sodium triphosphate with excess of ammonia. (c) By interaction of hydrated lime and phosphoric acid.
Grades: Granular; technical; C.P.; N.F.; pure precipitated; F.C.C.
Uses: Ceramics; calcium acid phosphate; phosphorus and phosphoric acid; polishing powder; cattle foods; clarifying sugar syrups; medicine; mordant (dyeing textiles with Turkey red); fertilizers; dentifrices; stabilizer for plastics; in meat tenderizers; in foods as anticaking agent, buffer, nutrient supplement; can remove strontium-90 from milk.

calcium phosphide (photophor) Ca_3P_2 (or Ca_2P_2).
Properties: Red-brown crystals or gray granular masses; sp. gr. 2.51 (15°C); m.p. about 1600°C; insoluble in alcohol and ether.
Derivation: By heating calcium phosphate with aluminum or carbon; by passing phosphorus vapors over metallic calcium.
Grades: Technical.
Hazard: Dangerous fire risk; decomposed by water to phosphine, which is highly toxic and flammable.
Uses: Signal fires; torpedoes; pyrotechnics.
Shipping regulations: (Rail) Yellow label. (Air) Flammable Solid label. Not acceptable on passenger planes.

calcium phosphite (dicalcium orthophosphite) $CaHPO_4 \cdot 2H_2O$.
Properties: White powder; loses its water at 200–300°C (with decomposition). Slightly soluble in water; insoluble in alcohol.

calcium phytate (hexacalcium phytate) $C_6H_6(CaPO_4)_6$.
Properties: Free-flowing white powder; slightly soluble in water. pH of saturated solution is neutral. Low toxicity.
Derivation: Corn steep liquor.
Containers: 50-lb bags.
Uses: To remove excess metals from wine and vinegar; source of calcium in pharmaceuticals and nutrition; source of phytic acid and its salts.

calcium plumbate Ca_2PbO_4.
Properties: Orange to brown crystalline powder; decomposed by hot water or carbon dioxide; sp. gr. 5.71. Soluble in acids (with decomposition); insoluble in cold water.
Hazard: Fire risk in contact with organic materials. May be toxic by ingestion.
Uses: Oxidizing agent; pyrotechnics and safety matches; glass; storage batteries.

calcium polysilicate $CaO \cdot 12SiO_2$. A powder, used as an anticaking agent.

calcium propionate $Ca(OOCCH_2CH_3)_2$. (Occurs also with one H_2O.) White powder, soluble in water; slightly soluble in alcohol. Low toxicity.
Grade: F.C.C.
Uses: Mold-inhibiting additive in bread, other foods, tobacco, pharmaceuticals.

calcium propyl arsenate $C_3H_7AsO_3Ca$. Crystals, soluble in water. Used for preemergence control of crab grass.
Hazard: Highly toxic by ingestion.
Shipping regulations: (Rail, Air) Arsenical compounds, n.o.s., Poison label.

calcium pyrophosphate $Ca_2P_2O_7$.
Properties: White powder. Soluble in dilute hydrochloric and nitric acids; insoluble in water. Sp. gr. 3.09; m.p. 1230°C. Nontoxic.
Grade: F.C.C.
Uses: Polishing agent in dentrifices; mild abrasive for metal polishing; nutrient and dietary supplement.

calcium resinate.
Properties: Yellowish-white, amorphous powder or lumps. Rosin odor. Soluble in acid; insoluble in water; soluble in amyl acetate, butyl acetate, ether amyl alcohol.
Derivation: By boiling calcium hydroxide with rosin and filtering; fusion of hydrated lime and melted rosin.
Grades: Technical; fused.
Hazard: Flammable; dangerous fire risk; spontaneous heating.
Uses: Waterproofing; manufacturing paint driers, porcelains, perfumes, cosmetics, enamels; coating for fabrics, wood, paper; tanning leather. See also soap (2).
Shipping regulations: (Rail) Yellow label. (Air) Flammable Solid label. Not acceptable on passenger planes.

calcium ricinoleate $Ca[CH_3(CH_2)_5CHOHCH_2CHCH(CH_2)_7CO_2]_2$.
Properties: White powder with a slight odor of fatty acids. Derived from castor oil. M.p. 84°C; sp. gr. 1.04. Combustible; low toxicity.
Uses: Greases and lubricants; stabilizer for polyvinyl chloride.

calcium D-saccharate $CaC_6H_8O_8 \cdot 4H_2O$.
Properties: White, crystalline powder; odorless; tasteless; insoluble in water and alcohol; soluble in calcium gluconate solutions. Nontoxic.
Derivation: Oxidation of D-gluconic acid and neutralization with lime.
Use: Medicine.

calcium saccharin $Ca(C_6H_4COSO_2N)_2 \cdot 3\frac{1}{2}H_2O$.
Properties: White, crystalline powder. Odorless or faint, aromatic odor; intensely sweet taste even in dilute solutions. 10 mg is equivalent in sweetening power to approximately 5 g of sucrose. Soluble in water. Nontoxic.
Grade: N.F.; F.C.C.
Use: Non-nutritive sweetener in medicine and foods. See also saccharin.

calcium salicylate $Ca(C_7H_5O_3)_2 \cdot 2H_2O$.
Properties: White powder; odorless, tasteless. Loses all water at 120°C. Soluble in water and alcohol. Nontoxic.
Grades: Purity 99+%.
Use: Medicine.

calcium silicate. See calcium metasilicate; wollastonite; cement, Portland.

calcium silicide $CaSi_2$.
Properties: Solid; sp. gr. 2.5. Insoluble in cold water; decomposes in hot water; soluble in acids and alkalies.
Hazard: Flammable; may ignite spontaneously in air.
Shipping regulations: (Rail) Flammable solid, n.o.s., Yellow label. (Air) Flammable Solid label.

calcium silicofluoride. See calcium fluosilicate.

calcium-silicon alloy. Contains 30% calcium.
Hazard: Flammable; may ignite spontaneously in air.
Shipping regulations: (Rail) Flammable solid, n.o.s., Yellow label. (Air) Flammable Solid label.

calcium sorbate $Ca(OOCC_5H_7)_2$. Used as a chemical preservative in foods. Nontoxic.

calcium stannate $CaSnO_3 \cdot 3H_2O$. White crystalline powder; insoluble in water; approximate temperature of dehydration 350°C.
Hazard: Toxic by ingestion and inhalation. Tolerance, 2 mg per cubic meter of air.
Uses: Additive in ceramic capacitors; production of ceramic colors.

calcium stearate $Ca(C_{18}H_{35}O_2)_2$.
Properties: White powder; m.p. 179°C. Insoluble in water; slightly soluble in hot alcohol. Decomposed by many acids and alkalies. Low toxicity.
Derivation: Interaction of sodium stearate and calcium chloride; then filtration.
Grades: Technical; F.C.C.
Containers: Fiber drums; cartons; multiwall paper sacks.
Uses: Water repellent; flatting agent in paints; lubricant in making tablets; emulsions; cements; wax crayons; stabilizer for vinyl resins; anticaking agent in foods; cosmetics.

calcium strontium sulfide $CaSrS_2$ or $CaS \cdot SrS$. Used as a phosphorescent pigment; a phosphor.

calcium succinate $CaC_4H_4O_4 \cdot 3H_2O$.
Properties: Colorless crystals. Slightly soluble in water; soluble in dilute acids.
Use: Medicine.

calcium sulfamate $Ca(SO_3NH_2)_2 \cdot 4H_2O$.
Properties: White, crystalline solid. Soluble in water. Aqueous solution is stable on boiling. Nonflammable; low toxicity.
Grades: Technical.
Use: Flameproofing agent for textiles and certain grades of paper.

calcium sulfate $CaSO_4$ or $CaSO_4 \cdot 2H_2O$. Occurs in nature as anhydrite, and in hydrated form as gypsum (plaster of paris).
Properties (pure anhydrous): White odorless powder or crystals; sp. gr. 2.964; m.p. 1450°C. Slightly soluble in water. (Dihydrate, pure precipitated) sp. gr. 2.32; loses 1½ H_2O at 128°C; becomes anhydrous at 163°C. Noncombustible; nontoxic. Neither the anhydrous nor the dehydrate forms can set with water.
Derivation: From natural sources and as a by-product in many chemical operations.
Grades: Technical; pure precipitated (as the dihydrate); F.C.C.
Containers: Bags; bulk shipments in freight cars.
Uses: Portland cement retarder; tile and plaster; source of sulfur and sulfuric acid; polishing powders; paints (white pigment, filler, drier); paper (size, filler, surface-coating); dyeing and calico printing; metallurgy (reduction of zinc minerals); drying industrial gases, solids, and many organic liquids; nutrient supplement; in granulated form as soil conditioner; quick-setting cements, molds and surgical casts; wallboard.

calcium sulfhydrate. See calcium hydrosulfide.

calcium sulfide CaS.
Properties: Yellow to light-gray powder with odor of hydrogen sulfide in moist air; unpleasant alkaline taste. Gradually decomposes in moist air or in weak acids. Soluble in acids; slightly soluble in water with partial decomposition; insoluble in alcohol. Sp. gr. 2.8. Low toxicity.
Derivation: Strong heating of pulverized calcium sulfate and charcoal.
Hazard: Irritant to skin and mucous membranes.
Uses: Luminous paint; depilatory; preparation of arsenic-free hydrogen sulfide; veterinary medicine; ore dressing and flotation agent.

calcium sulfite $CaSO_3 \cdot 2H_2O$.
Properties: White powder; loses water at 100°C. Soluble in sulfurous acid; slightly soluble in water. Low toxicity.
Derivation: Action of sulfurous acid on calcium carbonate.
Uses: Textiles (antichlor); disinfectant in sugar industry, brewing; biological cleansing; food preservative and discoloration retarder; paper manufacture.

calcium sulfocarbolate. See calcium phenolsulfonate.

calcium sulfocyanate. See calcium thiocyanate.

calcium superoxide. See calcium peroxide.

calcium tannate.
Properties: Yellowish-gray powder. Soluble in dilute acids; slightly soluble in water.
Uses: Pharmaceuticals; adhesives.

calcium tartrate $CaC_4H_4O_6 \cdot 4H_2O$.
Properties: White, crystalline powder. Soluble in dilute acids; slightly soluble in water or alcohol. Low toxicity.
Derivation: Interaction of a calcium salt and crude cream of tartar.
Grades: Technical; C.P.
Use: Tartaric acid.

calcium thiocyanate (calcium sulfocyanate) $Ca(SCN)_2 \cdot 3H_2O$.
Properties: White hygroscopic crystals or powder. Soluble in water and alcohol. Low toxicity.
Uses: Solvent for cellulose and polyacrylate; for parchmentizing; stiffening and swelling of textiles.

calcium thioglycolate $Ca(SCH_2COO)_2 \cdot 3H_2O$. Stable white powder; loses water at 100°C; decomposes 250°C. Slightly soluble in water; insoluble in alcohol. Used in cosmetic depilatories.

calcium thiosulfate (calcium hyposulfite) $CaS_2O_3 \cdot 6H_2O$.
Properties: White crystals; effloresces at 40°C; sp. gr. 1.872. Soluble in water and alcohol.
Use: Medicine.

calcium titanate $CaTiO_3$.
Properties: Powder; sp. gr. 3.98; m.p. 1800°C. Low toxicity.
Use: Electronics.

calcium trisodium pentetate (USAN) (calcium trisodium{[(carboxymethyl)imino]bis(ethylenenitrolo)} tetraacetate) $CaNa_3C_{14}H_{18}N_3O_{10}$. A chelating agent; antidote for lead poisoning.

calcium tungstate (calcium orthotungstate; calcium wolframate, normal) $CaWO_4$.
Properties: White crystalline powder; sp. gr. 6.062. Soluble in ammonium chloride; slightly soluble in water; decomposed by hot acids.
Derivation: (a) Interaction of calcium chloride and sodium tungstate. (b) Occurs in nature as scheelite; Nevada, California, Arizona, Utah, Colorado; New Zealand; Europe.
Method of purification: A slurry of powdered scheelite is treated with soda ash to form the soluble sodium tungstate. Insoluble impurities are filtered off and calcium tungstate is precipitated with lime.
Uses: Luminous paints; fluorescent lamps; photography; x-ray pictures; medicine.
Note: Synthetic crystals are available for use as scintillation counters, and possible application in lasers.

calcium undecylenate $(CH_2CH(CH_2)_8COO)_2Ca$. A fine, white powder of limited solubility. M.p. 155°C. Nontoxic.
Uses: Bacteriostat and fungistat in cosmetics and pharmaceuticals.

calcium zirconate $CaZrO_3$.
Properties: Solid; m.p. 2550°C; sp. gr. 4.78. Soluble in nitric and other acids. Noncombustible.
Uses: Refractory metal.

calcium zirconium silicate $CaZrSiO_5$ or CaO, ZrO_2, SiO_2.
Properties: White solid; m.p. 2900°F; insoluble in water, alkalies; slightly soluble in acids; soluble in hydrofluoric acid. Noncombustible.
Uses: Electrical resistor ceramics; glaze opacifier.

"Calcochrome."[57] Trademark for a series of chrome colors applied in the dyeing of finished goods.

"Calcocid."[57] Trademark for a series of acid dyestuffs used in the dyeing of wool and worsted goods, natural silk, jute, and in coloring diversified materials.

"Calcodur."[57] Trademark for a series of direct colors applied in the dyeing of cotton, rayon, and miscellaneous vegetable fibers.

"Calcofast."[57] Trademark for a series of metallized dyes containing chemically combined chromium used for dyeing wool. They can also be applied to leather, nylon, etc.

"Calcofluor."[57] Trademark for a series of direct dyeing dyes which possess fluorescent properties. Used for dyeing cotton, linen, viscose, acetate, nylon, wool and certain synthetics. Used also in soaps as a brightener for textile use.

"Calconyl."[57] Trademark for a series of coloring matters which are in substance stabilized combinations of a diazotized color base and a naphthol. They are used for dyeing or printing of fast shades on cotton and rayon.

"Calcophen."[57] Trademark for a series of oil dyes with non-subliming properties.
Uses: Varnish stains, foil coating, plastics.

"Calcosyn."[57] Trademark for a series of direct dyes for the dyeing of certain synthetic fibers such as cellulose acetates and nylon.

"Calcotone."[57] Trademark for a series of highly dispersed pigment pastes used whenever water suspensions of pigments are indicated.

calcspar. See calcite.

calendering. Passing a material between revolving metal rolls for any of several purposes: (1) to convert it into a sheet of uniform thickness, (2) to cause it to impregnate a textile fabric, or (3) to increase its surface gloss and hardness. Calenders (i.e., cylinders) are composed of from three to as many as 10 or 12 hollow, cast-iron rolls up to 84 inches in width, set vertically in a frame. The standard ruber or plastics calender has three steam-heated rolls which turn in opposing directions, either at the same speed or at different speeds. When moving at the same speed, they deliver the mixture fed between the top and center rolls in a smooth sheet, which can be as thin as 0.005 inch, from between the center and bottom rolls. The sheet is usually adhered to a fabric fed from the rear between the center and bottom rolls. If fabric impregnation is desired, the center roll runs faster than the other two, thus pressing the soft and tacky mixture into the textile material (tire carcases, electrical tape, etc.). In paper manufacturing, a high-speed calender "stack" or supercalender imparts a smooth finish to the sheet as it leaves the drying unit. Laboratory sizes are available.

"Calginate."[322] Trademark for calcium alginate (q.v.).

"Calgolac."[108] Trademark for a powdered alkaline, sodium hexametaphosphate detergent.
Uses: Cleaning bars, fountains, and laboratory glassware and equipment.

"Calgon."[108] Trademark for a sodium phosphate glass, commonly called sodium hexametaphosphate. It has a molecular ratio of 1.1 Na_2O: 1 P_2O_5 with a guaranteed minimum of 67% P_2O_5.
Derivation: From food-grade phosphoric acid and commercial soda ash by a thermal process.
Forms: Powder, agglomerated particles, and broken glassy plates, either pure or adjusted with mild alkalies.
Properties: Completely soluble in water in all proportions but is insoluble in organic solvents. It possesses sequestering, dispersing and deflocculating properties and precipitates proteins. In very low concentrations, it inhibits corrosion of steel and prevents the precipitation of slightly soluble, scale-forming compounds such as calcium carbonate and calcium sulfate.
Uses: Softening water without precipitate formation as in dyeing, laundering, textile processing and washing operations; corrosion inhibitor in deicing salt preparations; frozen desserts; pretanning hides in the manufacture of leather; dispersing clays and pigments; threshold treatment for scale and corrosion prevention.

"Calgon, Composition T."[108] Trademark for a complex glassy phosphate produced by a thermal process. White powder passing an 80 mesh sieve. Differs from "Calgon" sodium hexametaphosphate as part of the sodium has been replaced by other cations, predominantly zinc.
Containers: 100-lb bags; 100-lb drums.
Uses: Dispersing calcium carbonates, titanium dioxide, and other pigments used in the pulp and paper mill for coating and filling.

"Calgon, Composition TG."[108] Trademark for colorless glass platelets of sodium zinc hexametaphosphate.
Uses: Corrosion protection in recirculating cooling water and municipal water systems, and after mechanical cleaning of water mains where rapid film formation is important.

"Calgonite."[108] Trademark for alkaline detergent compositions containing complex phosphate as a principal ingredient. Generally recommended for spray-type washing operations where an alkaline cleaner is necessary and superior detergency with freedom from lime deposits is required.

"Calgosil."[108] Trademark for a metaphosphate-silicate compound; used to provide corrosion reduction in low hardness waters.

caliche. See sodium nitrate.

"Califlux."[500] Trademark for a series of oils composed principally of nitrogen bases and acidaffins. Used in plasticizers, extenders, as reclaiming agents for dark colored compounds.

californium Cf A synthetic radioactive element of the actinide group, with atomic number 98 and atomic weight 252. It has several isotopes, two of which (Cf 252 and Cf 249) are available in milligram amounts. The pure metal has not yet been obtained. Several compounds are known: the trioxide, the trifluoride the trichloride, and the sesquioxide. The 252 isotope has potential uses in neutron activation analysis for continuous materials testing, mineral prospecting, oil-well logging, etc. Biologically, it is a bone-seeking element, and has specialized applications in medicine.

"Calktite."[326] Trademark for an acid- and alkali-proof caulking compound. Used in protective coating for masonry, acid tanks and floors.

"Calo-Clor."[329] Trademark for a mercurial turf fungicide containing 73% mercury in chemical combination (principally mercuric and mercurous chlorides).
Hazard: Highly toxic by ingestion.

"Calocure."[329] Trademark for a mercurial turf fungicide having a mercury content of 36%; principally mercuric and mercurous chorides.
Hazard: Highly toxic by ingestion.

"Calodorant."[151] Trademark for a group of gas odorants for use in odorization of natural gas.

"Calogreen."[329] Trademark for a turf fungicide composed of extremely finely divided form of mercurous chloride. Contains 85% mercury (insoluble in water).
Hazard: Highly toxic by ingestion.

calomel. See mercurous chloride.

caloric. The amount of heat necessary to raise one gram of water one degree centigrade (at 1 atm); a kilogram calorie is the amount of heat required to raise a kilogram of water one degree centigrade. In the latter case, the word Calorie is capitalized when used alone, or the abbreviation kcal may be used. In connection with foods and beverages, kilogram calories are referred to.

calorizing. The process by which steel is coated with aluminum by heating it in aluminum powder. The aluminum forms an alloy with the steel surface and produces a thin, tightly adherent coating. See also cementation.

"Calsolene Oil HSA."[325] Trademark for a sulfated ester; anionic wetting and dispersing agent. Used primarily for wet processing of hydrophobic fibers.

"Cambrelle."[206] Trademark for a nonwoven melded fabric used for carpet backing, road reinforcement, upholstery, interlinings, tablecloths and other household applications. It is said to be composed of two different polymers, one lying within the other. Upon heating to the melting point of the external polymer, the fibers soften and unite to form a fabric. The term "melded" refers to this type of fusion. See also nonwoven fabric.

"Camoform."[330] Trademark for bialamicol hydrochloride.

camomile oil. See chamomile oil.

"Camoquin."[330] Trademark for amodiaquine hydrochloride.

cAMP. Biochemical designation for cyclic adenosine-3′,5′-monophosphate, an activator of hormones and initiator of prostaglandin synthesis.

2-camphanol. See borneol.

2-camphanone. See camphor.

camphene $C_{10}H_{16}$. A terpene.
Properties: Colorless crystals. Soluble in ether; slightly soluble in alcohol; insoluble in water. M.p. 48–51°C; b.p. 159–162°C.
Derivation: (a) By heating pinene hydrochloride with alkalies, aniline, or alkali salts, such as sodium acetate. (b) A constituent of certain essential oils.
Grades: Technical (46°C m.p.).
Containers: Tins; drums; tanks.
Hazard: Gives off flammable vapors when heated.
Uses: Medicine; manufacture of synthetic camphor; camphor substitute.

camphor (gum camphor; 2-camphanone) $C_{10}H_{16}O$. A ketone occurring naturally in the wood of the camphor tree (Cinnamomum camphora).
Properties: Colorless or white crystals, granules or easily broken masses; penetrating aromatic odor; sp. gr. 0.99; m.p. 174–179°C; sublimes slowly at room temperature; slightly soluble in water; soluble in alcohol, ether, chloroform, carbon disulfide, solvent naphtha, and fixed and volatile oils. Flash point 150°F (closed cup); autoignition temp. 871°F. Combustible.
Derivation: Steam distillation of the camphor-tree wood and crystallization. This product is called natural camphor and is dextrorotatory. Synthetic camphor, most of which is optically inactive, may be made from pinene, which is converted into camphene, (q.v.), which by treatment with acetic acid and nitrobenzene becomes camphor; turpentine oil is also used.
Containers: Tins; drums.
Grades: Technical (synthetic, m.p. 163–168°C); U.S.P. (m.p. 174–179°C).
Hazard: Evolves flammable and explosive vapors when heated; explosive range 0.6 to 3.5% in air; moderately toxic. Tolerance, 2 mg per cubic meter of air.
Uses: Medicine (internal and external); plasticizer for cellulose nitrate, other explosives, and lacquers; insecticides, moth and mildew preventives; tooth powders; flavoring; embalming; pyrotechnics; intermediate.
Note: A liquid form (camphor oil) is produced almost exclusively in Taiwan; formerly used in the manufacture of sassafras oil, the available supply is now used chiefly as a fragrance or flavoring material, and to some extent as a pharmaceutical product.

camphor bromate (alpha-bromo-1-camphor; brominated camphor) $C_{10}H_{15}BrO$.
Properties: Colorless crystals with slight camphor odor and taste. Also available as powder. Discolors in light and should be stored in cool, dark place. M.p. 76°C; b.p. 274°C; sp. gr. 1.449. Soluble in alcohol, ether, chloroform, and oils; insoluble in water.
Derivation: By heating camphor with bromine.
Use: Medicine; camphor derivatives.

camphoric acid $C_8H_{14}(COOH)_2$.
Properties: Colorless, odorless needles or scales. Soluble in alcohol, ether, fatty oils, and water; insoluble in chloroform. Sp. gr. 1.186 (20/4°C); m.p. 186–188°C.
Derivation: By oxidizing camphor with nitric acid.
Uses: Pharmaceuticals; medicine.

camphor, Malayan. See borneol.

camphor, peppermint. See menthol.

Canada balsam. See balsam.

cananga oil.
Properties: Yellowish essential oil; floral odor similar to ylang ylang (q.v.). Sp. gr. 0.908–0.925; optical rotation −15° to −30°; refractive index, 1.495–1.503. Combustible. Low toxicity.
Grades: Regular (native); rectified (the latter being lighter in color, has better solubility in alcohol, and is more stable); F.C.C.
Derivation: By distillation from the flowers of the Javanese variety of Cananga odorata.
Containers: Cans.
Use: Perfumery, particularly for floral types; flavoring agent in foods.

canavanine $NH_2CNHNHOCH_2CH_2CHNH_2COOH$. A non-protein amino acid obtained from jackbean meal. It is found naturaly in the L(+) form.
Properties: Crystals, from dilute alcohol; m.p. 184°C (dec); soluble in water; nearly insoluble in alcohol.
Sulfate: Crystals from dilute alcohol; m.p. 172°C (dec); soluble in water.
Use: Biochemical research.

candelilla wax.
Properties: Yellowish-brown, opaque to translucent solid. Soluble in chloroform, turpentine, carbon tetrachloride, trichloroethylene, toluene, hot petroleum ether and alkalies; insoluble in water. Sp. gr.

0.983; m.p. 67–68°C; saponification value 65; iodine number 37; refractive index 1.4555. Combustible; nontoxic.
Occurrence: Mexico, Texas.
Grades: Crude; refined; powdered.
Containers: Bags; boxes.
Uses: Leather dressing; polishes; cements; varnishes; candles; electric insulating composition; sealing wax; waterproofing and insect-proofing containers; paint removers; dentistry; paper sizes; rubber and rubber substitutes; stiffener for soft waxes.

candicidin (USAN). An antifungal antibiotic produced by Streptomyces griseus.

candidin. An antifungal antibiotic produced by Streptomyces viridoflavus.

canescine. See deserpidine.

cane sugar. See sucrose.

cannabis (marijuana). Its principle, tetrahydrocannabinol, can be made synthetically.
Derivation: Dried flowering tops of pistillate plants of Cannabis sativa.
Habitat: Iran, India; cultivated in Mexico and Europe.
Hazard: Controversial. Hallucinatory and possibly habit-forming by ingestion and inhalation.
Uses: Formerly as sedative. No present legitimate use.

cannel coal. A variety of bituminous or subbituminous coal of uniform and compact fine-grained texture. Dark gray to black, has a greasy luster. It is noncaking; yields a high percentage of volatile matter; and burns with a luminous, smoky flame. (ASTM definition, ASTM D493-39.) Combustible.
Hazard: Explosion risk in form of dust.

Cannizzaro, Stanislao (1826–1910). Born in Italy, he extended the research of Avogadro on the molecular concentrations of gases, and thus was able to prove the distinction of between atoms and molecules. His investigations of atomic weights helped to make possible the discovery of the Periodic Law by Mendeleev. His research in organic chemistry led to the establishment of the Cannizzaro reaction involving the oxidation-reduction of an aldehyde in the presence of concentrated alkali.

"Cantaxin."[162] Trademark for ascorbic acid.

"Cantrece."[28] Trademark for a self-crimping nylon yarn; designed specifically for hosiery.

CAP. Abbreviation for chloramphenicol.

capillarity. The attraction between molecules, similar to surface tension (q.v.), which results in the rise of a liquid in small tubes or fibers, or in the wetting of a solid by a liquid. It also accounts for the rise of sap in plant fibers and of blood in capillary (hair-like) vessels.

"Capracyl."[28] Trademark for a group of neutral-dyeing, premetalized acid colors that produce the highest possible degree of light fastness on nylon. Also suitable for dyeing wool, particularly in blends with cellulosic fibers.

capraldehyde. See decanal.

"Capran."[50] Trademark for a transparent, nylon 6 thermoplastic film used for food packaging.

capreomycin. (USAN) Antibiotic produced by streptomyces capreolus. Used in medicine.

capric acid (decanoic acid; decoic acid; decylic acid) $CH_3(CH_2)_8COOH$. Occurs as a glyceride in natural oils.
Properties: White crystals. Unpleasant odor; soluble in most organic solvents, and dilute nitric acid; insoluble in water. Sp. gr. 0.8858 (40°C); b.p. 270°C, 172.6°C (30 mm); m.p. 31.5°C; refractive index 1.4288 (40°C); acid number 308–315. Combustible; nontoxic.
Derivation: Fractional distillation of coconut oil fatty acids.
Grades: Technical; 90%; F.C.C.
Containers: Drums; tanks, carload lots.
Uses: Esters for perfumes and fruit flavors; base for wetting agents; intermediates; plasticizer; resins; intermediate for food-grade additives.

caproic acid (hexanoic acid; hexylic acid; hexoic acid) $CH_3(CH_2)_4COOH$. Present in milk fats to extent of about 2%.
Properties: Oily, colorless or slightly yellow liquid; odor of limburger cheese. Soluble in alcohol and ether; slightly soluble in water. Sp. gr. 0.9276 (20/4°C); f.p. -4.0°C; b.p. 205°C; 119°C (30 mm); refractive index 1.4168 (20°C); wt/gal 7.7 lb; viscosity 0.031 poise (20°C). Flash point 215°F (open cup). Combustible.
Derivation: From crude fermentation of butyric acid; fractional distillation of natural fatty acids.
Grades: Technical; reagent to 99.8%; F.C.C.
Containers: Drums; carload lots.
Hazard: Strong irritant to tissue.
Uses: Analytical chemistry; flavors; manufacture of rubber chemicals, varnish driers, resins and pharmaceuticals.
Shipping regulations: (Air) Corrosive label. Legal label name hexanoic acid.

caproic aldehyde. See n-hexaldehyde.

caprolactam (aminocaproic lactam; 2-oxohexamethyleneimine)

$\overline{CH_2(CH_2)_4NHCO}$.
Properties: White flakes or fused; m.p. 68–69°C; sp. gr. 70% solution 1.05; refractive index (40°C) 1.4935, (31°C) 1.4965; soluble in water, chlorinated solvents, petroleum distillates and cyclohexene; heat of fusion 29 cal/g; heat of vaporization 116 cal/g; viscosity 9 cps at 78°C; vapor pressure 3 mm Hg at 100°C, 50 mm Hg at 180°C.
Derivation: (1) Catalytic oxidation of cyclohexane to cyclohexanol, reacting with peracetic acid to form caprolactone, and further reaction with ammonia; (2) catalytic hydrogenation of phenol to cyclohexanone, reaction with ammonia to cyclohexanone oxime, with Beckmann rearrangement with sulfuric acid catalyst; (3) catalytic oxidation of cyclohexane to cyclohexanone, reaction with hydroxylamine sulfate and ammonia to cyclohexanone oxime, followed by sulfuric-acid catalyzed Beckmann rearrangement; (4) UV-catalyzed reaction of cyclohexane with nitrosyl chloride to cyclohexanone oxime hydrochloride, followed by Beckmann rearrangement. Method (1) was never used commercially. Recent modifications of method (3) have been made in order to minimize formation of by-product ammonium sulfate.
Forms: Flake; molten.
Containers: 80-lb paper bags; 300-lb fiber drums; tanks (molten).
Hazard: Toxic by inhalation. Tolerance, 5 ppm in air.
Uses: Manufacture of synthetic fibers (especially

nylon 6), plastics, bristles, film, coatings, synthetic leather, plasticizers, and paint vehicles; cross-linking agent for curing polyurethanes; synthesis of amino acid lysine.

caprolactone.
Derivation: Reaction product of peracetic acid and cyclohexanone; an intermediate product in manufacture of caprolactam (q.v.).
Containers: Tank cars; tank trucks.
Uses: Intermediate in adhesives, urethane coatings and elastomers; solvent; diluent for epoxy resins; synthetic fibers; organic synthesis.

"Caprolan."[523] Trademark for a polyamide fiber made from polymerized caprolactam. Has excellent dyeability and a wide variety of end-uses; maintains a superior level of dimensional stability after heat setting and has outstanding mechanical qualities. See also caprolactam; nylon.

capryl compounds. The term "capryl" is generally but erroneously used in the trade to refer to octyl compounds. Thus the definition of capryl and caprylic compounds will be found under the corresponding octyl entry, e.g., for capryl alcohol, see 2-n-octanol; for caprylic halides see corresponding octyl halide; for caprylic acid see octanoic acid.

caprylyl peroxide. Legal label name for octyl peroxide (q.v.).

"Captan."[183] Trademark for a gas odorant built on a mercaptan base. Clear liquid, lighter than water, similar in odor to butyl mercaptan.
Use: To add odor to odorless gases both for safety and for the detection of leaks.

captan (N-trichloromethylmercapto-tetrahydrophthalimide) $C_9H_8Cl_3NO_2S$.
Properties: White to cream powder; m.p. 158–164°C; sp. gr. 1.5. Practically insoluble in water; partially soluble in acetone, benzene and toluene; slightly soluble in ethylene dichloride and chloroform.
Derivation: Reaction product of tetrahydrophthalamide and trichloromethylmercaptan.
Containers: 50-lb fiber drums; 5-lb bags.
Hazard: Toxic and irritant. Avoid contamination of feed and foodstuffs. Avoid inhalation of dust or spray mist. Avoid prolonged or repeated contact with skin.
Uses: Seed treatment; fungicide in paints, plastics, leather, fabrics, and fruit preservation.

"Captax."[265] Trademark for 2-mercaptobenzothiazole. Cream-colored to buff powder; characteristic odor. An acid accelerator of vulcanization in tire treads, mechanical rubber goods, wire insulation, footwear, clothing, drug sundries; requires stearic acid as activator. Imparts excellent ageing and heat resistance.

capture. The process in which an atomic or nuclear system acquires an additional particle; for example, the capture of electrons by positive ions, or capture of electrons or neutrons by nuclei. (AEC). See also cross-section (1); neutron; fission.

"Capyco."[481] Trademark for pyrophyllite.

caramel. A sugar-based food colorant made from liquid corn syrup by heating in the presence of catalysts to about 250°F for several hours, cooling to 200°F, and filtering. The brown color results from either Maillard reactions, true caramelization, or oxidative reactions. Caramels are colloidal in nature, the particles being held in solution by either positive or negative electric charges. See also caramelization; browning reaction.

caramelization. A type of nonenzymic browning reaction occurring during exposure of food products to heat when the products contain no nitrogen compounds, e.g., sugars. See also browning reaction.

carbachol (carbamylcholine chloride; choline chloride carbamate) $NH_2COOCH_2CH_2 \cdot N(CH_3)_3Cl$.
Properties: White or faintly yellow crystals or as crystalline powder, odorless, or with faint amine odor, and hygroscopic. Soluble in water and alcohol; almost insoluble in chloroform and ether. M.p. 201–205°C (with decomposition).
Grade: U.S.P.
Use: Medicine.

carbamate. A compound based on carbamic acid, NH_2COOH, which is used only in the form of its numerous derivatives and salts.

carbamide. See urea.

carbamide peroxide. See urea peroxide.

carbamide phosphoric acid (urea phosphoric acid) $CO(NH_2)_2 \cdot H_3PO_4$.
Properties: White rhombic crystals. Very soluble in water and alcohol.
Containers: 350-lb drums.
Hazard: Evolves toxic fumes when heated.
Uses: Catalyst for acid-setting resins; flameproofing compositions; cleaning compounds; acidulant.

carbamidine. See guanidine.

carbamite. See sym-diethyldiphenylurea.

carbamylcholine chloride. See carbachol.

carbamylguanidine sulfate. See guanylurea sulfate.

carbamylhydrazine hydrochloride. See semicarbazide hydrochloride.

carbamylmethylcholine chloride. See bethanechol chloride.

carbamylurea. See biuret.

carbanil. See phenyl isocyanate.

carbanilide. See diphenylurea.

carbanion. A negatively charged organic ion such as H_3C—, RC—, having one more electron than the corresponding free radical. Carbanions are short-lived but important intermediates in base-catalyzed polymerization and alkylation reactions. See carbonium ion; carbene; free radical.

"Carbanthrene."[243] Trademark for anthraquinone vat dyes.

carbaryl. Generic name for 1-naphthyl-N-methylcarbamate. $C_{10}H_7OOCNHCH_3$.
Properties: Solid; m.p. 142°C.
Derivation: Synthesized directly from 1-naphthol and methyl isocyanate or from naphthyl chloroformate (1-naphthol and phosgene) plus methylamine.
Hazard: Toxic by ingestion, inhalation, and skin absorption. Irritant. A reversible cholinesterase inhibitor. Tolerance, 5 mg per cubic meter of air. Use may be restricted.
Use: Insecticide.

"Carbasols."[223] Trademark for products consisting of natural resin esters condensed with modifying agents of the polyhydric alcohol-polybasic acid type. Available in wide variation as to physical properties such as color, softening point, and acid number.

Uses: Cellulose-ester lacquers; oleoresinous varnishes; printing ink vehicles.

carbazide. See carbodihydrazide.

carbazole (dibenzopyrrole; diphenylenimine) $(C_6H_4)_2NH$ (tricyclic).
Properties: White crystals with characteristic odor; m.p. 244–246°C; b.p. 352–354°C; soluble in alcohol and ether; insoluble in water. Low toxicity.
Derivation: (a) From crude anthracene cake by selective solution of the phenanthrene with crude solvent naphtha, removal of the anthracene by conversion into a sulfonic derivative and extraction by means of water. (b) Synthetically, from ortho-aminobiphenyl.
Grades: Technical, 97%.
Containers: Barrels; fiber drums.
Uses: Manufacture of dyes; reagent; explosives; insecticides; lubricants; rubber antioxidants; odor inhibitor in detergents.

carbazotic acid. See picric acid.

carbene (methylene). An organic radical containing divalent carbon. Some divalent carbon derivatives where the carbon is multiple-bonded to oxygen or nitrogen are stable compounds (carbon monoxide); most are highly reactive units that are known only as reaction intermediates. Carbenes, carbonium ions, carbanions, and free radicals are the four most important classes of organic reaction intermediates containing carbon in an unstable valence state. A typical synthesis involving a carbene is that of cyclopropanes by the addition of carbenes to olefins.

2-carbethoxycyclohexanone $OC_6H_9COOC_2H_5$.
Properties: Colorless liquid with a characteristic ester odor; b.p. 106–107°C (11 mm); sp. gr. 1.074 (25°C); refractive index n (17.5/D) 1.4750. Soluble in dilute alkali; insoluble in water. Combustible.
Use: Intermediate.

2-carbethoxycyclopentanone (ethyl cyclopentanone-2-carboxylate; ethyl-2-oxocyclopentanecarboxylate) $OC_5H_7COOC_2H_5$.
Properties: Colorless liquid with characteristic ester odor; b.p. 122–124°C (25 mm); flash point 191°F; refractive index 1.451 (25°C); sp. gr. 1.0976 (0°C); soluble in equimolecular amounts of dilute alcohol; insoluble in water. Combustible.
Hazard: Avoid inhalation of vapors and contact with skin.
Uses: Intermediate; pharmaceutical intermediate.

beta-**carbethoxyethyltriethoxysilane** $C_2H_5OOC(CH_2)_2Si(OC_2H_5)_3$.
Properties: Colorless liquid. B.p. 246°C. Combustible.
Hazard: May be toxic and irritant.
Use: Intermediate.

N-carbethoxypiperazine $C_7H_{14}N_2O_2$.
Properties: Colorless, somewhat viscous liquid; slight odor; b.p. 116–117°C (12 mm); 237°C (760 mm); refractive index 1.4756 (25°C). Miscible with water and common organic solvents.
Containers: 55-gal drums.
Use: Intermediate.

beta-**carbethoxypropylmethyldiethoxysilane** $C_2H_5OOC(C_3H_6)CH_3Si(OC_2H_5)_2$.
Properties: Colorless liquid. B.p. 228°C. Combustible.
Hazard: May be toxic and irritant.
Use: Intermediate.

carbide. A binary solid compound of carbon and another element. The most familiar carbides are those of calcium tungsten, silicon, boron, and iron (cementite). Two factors have an important bearing on the properties of carbides: (1) the difference in electronegativity between carbon and the second element, and (2) whether or not the second element is a transition metal. Salt-like carbides of alkali metals are obtained by reaction with acetylene; those obtained from silver, copper and mercury salts are explosive. See acetylide. See also carbide, refractory and carbide, cemented.

carbide, cemented. A powdered form of refractory carbide (q.v.) united by compression with a bonding material (usually iron, nickel, or cobalt), followed by sintering. Tungsten carbide (q.v.) is bonded with cobalt at 1400°C; from 3 to 25% of cobalt is used depending on the properties desired. Used chiefly in metal cutting tools, which are hard enough to permit cutting speeds in rock or metal up to 100 times that obtained with alloy steel tools.

carbide, refractory. A carbide (q.v.) characterized by great hardness, thermal stability, high melting point, and chemical resistance. Decomposed by fusion with alkali, and attacked by mixtures of nitric and hydrofluoric acids. The best known refractory carbides are those of silicon, boron, tungsten, molybdenum, and tantalum. Used as abrasives, furnace linings, and other high-temperature applications. Some types are bonded.

carbinol
(1) Synonym for methyl alcohol, CH_3OH.
(2) Hence, any compound of similar structure retaining the COH radical, and in which hydrocarbon radicals may be substituted for the hydrogen originally attached to the carbon. Thus isopropyl alcohol $(CH_3)_2CHOH$, and benzyl alcohol, $C_6H_5CH_2OH$, may be named dimethylcarbinol and phenylcarbinol respectively.

carbinoxamine maleate
$ClC_6H_4CH(C_5H_4N)OCH_2CH_2N(CH_3)_2 \cdot C_4H_4O_4$.
(2-[para-chloro-alpha-(2-dimethyl-amino-ethoxy)-benzyl]pyridinemaleate.
Properties: White, odorless, bitter crystalline powder; m.p. 116–121°C. Very soluble in water; freely soluble in alcohol and chloroform; very slightly soluble in ether; pH (1% solution) 4.6–5.1.
Grade: N.F.
Use: Medicine.

"Carbitol."[214] Trademark for a group of mono- and dialkyl ethers of diethylene glycol and their derivatives; specialized solvents with a wide variety of properties and uses. Specific types are as follows:
butyl "Carbitol"
 See diethylene glycol monobutyl ether.
butyl "Carbitol" acetate
 See diethylene glycol monobutyl ether acetate.
"Carbitol" acetate
 See diethylene glycol monoethyl ether acetate.
"Carbitol" solvent
 See diethylene glycol monoethyl ether.
dibutyl "Carbitol"
 See diethylene glycol dibutyl ether.
diethyl "Carbitol"
 See diethylene glycol diethyl ether.
N-hexyl "Carbitol"
 See diethylene glycol monohexyl ether.

methyl "Carbitol"
See diethylene glycol monomethyl ether.

methyl "Carbitol" acetate
See diethylene glycol monomethyl ether acetate.

"Carbium."[244] Trademark for a precipitated calcium carbonate filler for PVC resins. Used in extruded stocks, calendered stocks, floor tile, plastisols and organosols.

"Carbocaine" Hydrochloride.[162] Trademark for mepivacaine hydrochloride.

carbocyclic. Any organic compound whose "skeleton" is in the form of a closed ring of carbon atoms. This includes both alicyclic and aromatic structures.

carbodihydrazide (carbazide) $CO(NHNH_2)_2$.
Properties: Colorless crystals; m.p. 154°C; sp. gr. 1.1616 (-5°C). Very soluble in water and alcohol.
Uses: Organic intermediate and photographic chemical.

carbodiimide. See cyanamide (1).

"Carbo-Dur."[184] Trademark for an adsorbent of granular activated carbon used in taste, color, and odor removal.

"Carbofrax."[280] Trademark for bonded refractory bricks for furnace linings, muffle walls, etc.; also the cement or mortar used to install them. Contains 85% or more silicon carbide. Porosity about 13%. See also refractory; silicon carbide.

carbohydrase. An enzyme whose catalytic activity is directed toward the breaking down of complex carbohydrates to simpler units. Illustrations are amylase; invertase; maltase (q.v.).

carbohydrate. A compound of carbon, hydrogen and oxygen that contains the saccharose grouping

```
     |
  H—C—C—
     |  ||
    OH O
```

or its first reaction product, and in which the ratio of hydrogen to oxygen is the same as in water. Carbohydrates are the most abundant class of organic compounds, constituting three-fourths of the dry weight of all vegetation. They are also widely distributed in animals and lower forms of life. They comprise (1) monosaccharides: simple sugars, such as fructose (levulose) and its isomer glucose (dextrose), both having the formula $C_6H_{12}O_6$; (2) disaccharides: sucrose ($C_{12}H_{22}O_{11}$), maltose, cellobiose, and lactose; and polysaccharides (high polymeric substances). The last group includes the entire starch and cellulose families, as well as pectin, the seaweed products agar and carrageenan, and natural gums. The simple sugars are crystalline and water-soluble, with a sweet taste; starches are water-soluble, tasteless and amorphous; cellulose is insoluble in water and organic solvents, and is only partially crystalline. Galactose, sorbose, xylose, arabinose, and mannose are constituents of more complex sugars. The natural gums are water-soluble plant products composed of monosaccharide units joined by glycosidic bonds (arabic, tragacanth). See also sugar; starch; cellulose; gums, natural; photosynthesis; nutrient.

"Carbo-Korez."[41] Trademark for a carbon-filled, synthetic-resin, acid-proof cement of the phenolformaldehyde type used as a mortar cement where temperatures do not exceed 370°F. Especially good for high concentrations of sulfuric acid.

"Carbolac."[275] Trademark for high color channel blacks; used for paint, varnish, and lacquer.

carbolfuchsin (Ziehl's stain). A staining solution of fuchsin in alcohol and aqueous phenol used in the study of microorganisms.

carbolic acid. Legal label name for phenol (q.v.).

carbolic oil (middle oil). Comprises the fraction having a boiling range of about 190–250°C obtained from distillation of coal tar, and containing naphthalene, phenol, and cresols.

3-carbomethoxy-1-methyl-4-piperidone hydrochloride $C_8H_{13}NO_3 \cdot HCl$. White crystalline solid; m.p. 165°C. Soluble in water, alcohols; insoluble in ether, hydrocarbons.
Use: Pharmaceutical intermediate.

2-carbomethoxy-1-methylvinyl dimethyl phosphate. For the alpha isomer, see mevinphos. The beta isomer is also a pesticide.

carbomycin. An antibiotic isolated from products of Streptomyces halstedii when grown in suitable media by deep culture method. It inhibits growth of certain gram-positive bacteria such as staphylococci, pneumococci and hemolytic streptococci.
Properties: White, odorless, bitter powder; m.p. 195–220°C (dec). Freely soluble in chloroform; very slightly soluble in water; slightly soluble in alcohol and ether. Stable when protected from moisture; pH (saturated solution) 5.5–8.0.
Use: Medicine.

carbon. C. A nonmetallic element; atomic number 6; atomic weight 12.0111; Group IVA of the Periodic Table; normal valence 4, but divalent forms are known (carbenes). Carbon has two stable and four radioactive isotopes. The C^{12} isotope, which comprises 99% of the element, is the standard to which atomic weights of all other elements are referred (i.e., C = 12.00 exactly). One mole of carbon atoms (6.02×10^{23}) is contained in 12 grams of C^{12}. Carbon has two crystalline allotropes (diamond and graphite) and several amorphous allotropes (coal, coke, carbon black, charcoal).

Carbon is present in all organic and in a few inorganic compounds (carbon oxides, carbon disulfide and metallic carbonates such as calcium carbonate). It is the active element in photosynthesis and thus occurs in all plant and animal life. The radioisotope C^{14} is used in tracer research and chemical dating. Carbon is a strong reducing agent and is used as such in purifying metals. It is the only element capable of forming four covalent bonds. Its strong electrical conductivity is used to advantage in electrodes and other electrical devices. Its presence in small proportions in steel has a pronounced effect on the properties of the metal.

Carbon forms binary compounds called carbides with many metals and some nonmetals. A few compounds are known which contain divalent carbon (carbenes or methylenes).

Since its major properties and uses vary widely with its form, the following entries should be consulted: diamond; graphite; carbon, activated; carbon black; carbon, industrial; charcoal; coke; steel; carbon cycle.

carbon 13. Stable, nonradioactive carbon isotope used for special analytical research. Commercially available in gram quantities.

carbon 14 (radiocarbon). Naturally occurring radioactive carbon isotope of mass number 14; a special

case of radioactivity induced by cosmic rays. Half-life, 5580 years; beta radiation. Can be made by reactor irradiation of calcium nitrate.
Uses: Radiation source in thickness gauges and other instruments; elucidation of mechanisms in organic chemistry, metallurgy, and biochemical reactions; radiocarbon dating in geology and archaeology.

carbon, activated (active carbon; activated charcoal). An amorphous form of carbon characterized by high adsorptivity for many gases, vapors, and colloidal solids. The carbon is obtained by the destructive distillation of wood, nut shells, animal bones, or other carbonaceous material; it is "activated" by heating to 800–900°C with steam or carbon dioxide, which results in a porous internal structure (honeycomb-like). The internal surface area of activated carbon averages about 10,000 square feet per gram. The specific gravity is from 0.08 to 0.5. It is not effective in removing ethylene.
Grades: Technical: U.S.P., as activated charcoal.
Containers: Bags; fiber drums.
Hazard: Moderate, by inhalation of dust. Flammable.
Uses: Decolorizing of sugar; water and air purification; solvent recovery; waste treatment; removal of sulfur dioxide from stack gases and "clean" rooms; deodorant; removal of jet fumes from airports; catalyst; natural gas purification; brewing; chromium electroplating; air conditioning.
Shipping regulations: (Rail) Yellow label. (Air) Flammable Solid label.

carbonate. A compound resulting from the reaction of either a metal or an organic compound with carbonic acid. The reaction with a metal yields a salt (calcium carbonate), and that with an aliphatic or aromatic compound forms an ester, e.g., diethyl carbonate, diphenyl carbonate. The latter are liquids used as solvents and in synthesizing polycarbonate resins.
See also carbonic acid.

carbon bisulfide. See carbon disulfide.

carbon black. Finely divided forms of carbon made by the incomplete combustion or thermal decomposition of natural gas or petroleum oil. The principal types, according to the method of production, are channel black (also called impingement black), furnace black (q.v.), and thermal black.

Channel black is characterized by lower pH, higher volatile content, and less chainlike structure between the particles. It has the smallest particle size (largest specific surface area) of any industrial material; the particles are in the colloidal range. Surface area runs to about 18 acres per pound. Its chief use is as reinforcing agent for rubber (tire treads); it increases both abrasion and oil resistance. Stearic acid is required for optimum dispersion. Channel black is no longer used extensively.

Thermal black consists of relatively coarse particles and is used principally as a pigment. Furnace black produced from natural gas has an intermediate particle size while that produced from oil can be made in a wide range of controlled particle sizes and is particularly suitable for reinforcing synthetic rubber. Furnace black is by far the most important form of carbon black, greatly exceeding the other two in volume used.
Properties: The raw product is extremely fine, smoke-like powder which remains suspended in air and penetrates thin paper and textiles (channel black). Thermal blacks do not "fly" as readily. Sp. gr. 1.8–2.1; b.p. 4200°C. Insoluble in all solvents.
Forms: Powder; pellets; pastes.
Grades: (ASTM) N660, N550, N330, N110, N220, N761, N762, N601, 5300 (channel), 5301.
Containers: Multiwall paper bags; lined barrels.
Uses: Tire treads, belt covers, and other abrasion-resistant rubber products; plastics, as reinforcing agent, opacifier, electrical conductivity, UV light absorber; colorant for printing inks; carbon paper; typewriter ribbons; paint pigment; nucleating agent in weather modification; expanders in battery plates.

carbon, combined. A metallurgical term for carbon which has combined chemically with iron to form cementite, as distinct from graphitic carbon in iron or steel. See also pearlite; ferrite.

carbon cycle.

(1) The progress of carbon from the air (carbon dioxide) to plants by photosynthesis (sugar and starches) then through the metabolism of animals to decomposition products which ultimately return it to the atmosphere in the form of carbon dioxide.

(2) One of the processes by which the sun and other self-luminous astronomical bodies are thought to derive their energy. The net process is the combination (fusion) of four hydrogen atoms to form helium. One mechanism, called the carbon cycle, involves successive additions of hydrogen atoms, followed by beta decay, to an initial carbon-12 atom, until a final step is reached in which the new nucleus breaks down to a helium atom and a carbon-12 is regenerated. The carbon thus functions as a catalyst for the process. At the temperatures prevailing in the sun all atoms are stripped of their electrons and the reaction is between the nuclei of the atoms (thermonuclear reaction). Symbolically, the set of reactions is written.

$$C^{12} + H^{1} \rightarrow N^{13};\ N^{13} \rightarrow C^{13} + e;$$
$$C^{13} + H^{1} \rightarrow N^{14};\ N^{14} + H^{1} \rightarrow O^{15};$$
$$O^{15} \rightarrow N^{15} + e;\ N^{15} + H^{1} \rightarrow C^{12} + He^{4}.$$

See also fusion (2).

carbon dichloride. See perchloroethylene.

carbon dioxide CO_2. 29th highest-volume chemical produced in the U.S. (1975).
Properties: (a) gas: colorless, odorless, sp. gr. 1.53 (air = 1.00), (b) liquid: volatile, colorless, odorless, sp. gr. (–37°C) 1.101, sp. volume 8.76 cu ft/lb (70°F). (c) solid (dry ice): white, snow-like flakes or cubes, sp. gr. 1.56 (–79°C), m.p. –78.5°C (sublimes). All forms are noncombustible. Soluble in water (carbonic acid, H_2CO_3), hydrocarbons, and most organic liquids. Nontoxic (solid); an asphyxiant gas.
Derivation: (a) gas—CO_2 is present in air to extent of 0.03% by volume and 0.0474% by weight, but is not economically recoverable. Its chief sources are as follows: (1) end product of combustion and respiration, (2) cracking of hydrocarbons, (3) fermentation of carbohydrates, (4) action of heat or acid on limestone, marble, and other carbonates, (5) natural springs or wells. (b) liquid—by compressing and cooling the gas to about –37°C. (c) solid (dry ice)—by expanding the liquid to vapor and snow in presses that compact the product into blocks. The vapor is recycled.
Grades: Technical; U.S.P.; commercial and welding 99.5%.
Containers: Liquid—steel cylinders, tank cars and

trucks. Solid—ten-inch cubes wrapped in paper, insulated corrugated board or other special types. Shipped in insulated refrigerator cars or trucks.
Hazard: (solid) Damaging to skin and tissue; keep away from mouth and eyes. Tolerance (gas): 5000 ppm in air.
Uses: Refrigeration; carbonated beverages; aerosol propellant; chemical intermediate (carbonates, synthetic fibers, paraxylene, etc.); low-temperature testing; fire extinguishing; inert atmospheres; municipal water treatment; medicine; enrichment of air in greenhouses; fracturing and acidizing of oil wells; mining (Cardox method); miscellaneous pressure source; hardening of foundry molds and cores; shielding gas for welding; cloud seeding; moderator in some types of nuclear reactors; immobilization for humane animal killing; special lasers; blowing agent.
Shipping regulations: (Rail) (liquid) Green label. (Air) (Gas or liquid) Nonflammable Gas label.
Note: Carbon dioxide is the source of the carbon utilized by plants to form organic compounds in the photosynthetic reaction, catalyzed by chlorophyll. See photosynthesis; carbon cycle (1).

carbon disulfide (carbon bisulfide) CS_2.
Properties: Clear, colorless or faintly yellow liquid, almost odorless when pure; strong disagreeable odor. Sp. gr. 1.260 at 25/25°C; b.p. 46.3°C; freezing point -111°C; wt/gal 10.48 lb (25°C); refractive index 1.6232 (25°C); flash point -22°F; autoignition temp. 212°F. Soluble in alcohol, benzene, and ether; slightly soluble in water. Classed as an inorganic compound.
Derivation: (a) Reaction of natural gas or petroleum fractions with sulfur. (b) From natural gas and hydrogen sulfide at extremely high temperature (plasma process). (c) By heating sulfur and charcoal and condensing the carbon disulfide vapors.
Method of purification: Distillation.
Impurities: Sulfur compounds.
Grades: 99.9%; spectrophotometric.
Containers: Drums; tank cars.
Hazard: Highly flammable; dangerous fire and explosion risk; can be ignited by friction. Explosive limits in air 1 to 50%. Poisonous. Tolerance, 20 ppm in air. Absorbed by skin. Safety data sheet available from Manufacturing Chemists Assn., Washington, D.C.
Uses: Viscose rayon; cellophane; manufacture of carbon tetrachloride and flotation agents; veterinary medicine; solvent.
Shipping regulations: (Rail) Red label. Not accepted by express. (Air) Not acceptable.

carbon fiber. See graphite fiber.

carbon fluoride CF_x, C_4F. A solid nonconductive material formed on carbon anodes during electrolysis of molten potassium fluoride-hydrogen fluoride mixtures to yield elemental fluorine. C_4F is unstable at 60°C and higher temperatures. (CF) forms only at higher temperatures.

carbon, graphitic. A metallurgical term referring to practically pure carbon which forms in pig iron during cooling, because the absorbing power of iron for carbon decreases as its temperature falls. It exists in the iron in the form of tiny flakes distributed throughout the mass. The tendency of graphitic carbon is to weaken the metal, while combined carbon up to the limit of about 0.90% strengthens it. See also pearlite; cementite; ferrite.

carbon hexachloride. See hexachloroethane.

carbonic acid.
H_2CO_3, or $\frac{HO}{HO}{>}C{=}O$.
A weak acid formed by reaction of carbon dioxide with water. Both organic and inorganic carbonates are formed from it by reaction with organic compounds or metals, respectively. Thus inorganic carbonates ($CaCO_3$, KCO_3, Na_3CO_3, etc.) are salts of carbonic acid and organic carbonates are esters of carbonic acid.

carbonic anhydrase. An enzyme in red blood cells which catalyzes the production of carbon dioxide and water from carbonic acid.
Use: Biochemical research.

carbon, industrial. Any form of pure carbon used for industrial purposes, exclusive of fuel. Coke is one of the most important. Besides its use (combined with coal-tar pitch) for refractories, furnace linings, electrodes, fibers, etc., it has tremendous volume consumption for reduction of iron in blast furnaces (see coke). Graphite in its many applications is another form; activated carbon for decolorizing and solvent recovery, carbon black for rubber and printing inks, industrial diamonds as abrasives and drilling bits, compressed carbon for electrodes and other electrical uses, and carbon fibers and whiskers are all included in this term.

carbonium ion. A positively charged organic ion such as H_3C^+, H_2RC^+, $R_3C^+{=}O$, etc., having one less electron than the corresponding free radical and acting in subsequent chemical reactions as though the positive charge was localized on the carbon atom. Such ions can exist only when corresponding negative ions are also present. An electron-deficient carbon atom is extremely reactive and has only a transitory existence in most cases, but many organic rearrangement and replacement reactions are effectively explained in terms of a carbonium ion intermediate, including acid-catalyzed polymerization of propylene and other olefins. In this case propylene and hydrogen ion form a carbonium ion as follows: $H_3C{-}HC{=}CH_2 + H^+ \rightarrow H_3C{-}HC^+{-}CH_3$. The latter then combines with another molecule of $HC_3{-}HC{=}CH_2$ to start chain growth.
The difference between a carbonium ion, a free radical and a carbanion may be illustrated as follows:

$$\left[\begin{array}{c} R \\ R : \ddot{C} \\ \ddot{R} \end{array}\right]^{+} \quad \left[\begin{array}{c} R \\ R : \ddot{C}\cdot \\ R \end{array}\right]^{0} \quad \left[\begin{array}{c} R \\ R : \ddot{C}: \\ \ddot{R} \end{array}\right]^{-}$$

carbonium ion — free radical — carbanion

carbonization. See destructive distillation.

carbon monoxide CO. Discovered by Priestley in America in 1799.
Properties: Colorless gas or liquid; practically odorless. Burns with a violet flame. Slightly soluble in water; soluble in alcohol and benzene. Sp. gr. 0.96716; b.p. -190°C; solidification point -207°C; specific volume 13.8 cu ft/lb (70°F). Autoignition temp. (liquid) 1128°F. Classed as an inorganic compound.
Derivation: (a) Obtained almost pure by placing a mixture of oxygen and carbon dioxide in contact with incandescent graphite, coke or anthracite. (b) Action of steam on hot coke or coal (water gas) or on natural gas (synthesis gas). In the latter case, carbon dioxide is removed by absorption in amine solutions, and the hydrogen and carbon monoxide sep-

arated in a low-temperature unit. (c) By-product in chemical reactions. (d) Combustion of organic compounds with limited amount of oxygen as in automobile cylinders. (e) Dehydration of formic acid.
Grades: Commercial (98%); C.P. (99.5%).
Containers: Cylinders; (liquid) special insulated double-walled trucks.
Hazard: Highly toxic by inhalation; highly flammable, dangerous fire and explosion risk; lower explosion limit in air 12.5% by volume. Tolerance, 50 ppm (industrial workrooms); U.S. standard, 35 ppm. Note: CO has an affinity for blood hemoglobin over 200 times that of oxygen. A major air pollutant.
Uses: Organic synthesis (methanol, ethylene, isocyanates, aldehydes, acrylates; phosgene); fuels (gaseous); metallurgy (special steels, reducing oxides, nickel refining); zinc white pigments.
Shipping regulations: (Rail) Red Gas label. (Air) Flammable Gas label. Not acceptable on passenger planes. See also synthesis gas (1).

"Carbonox."[236] An organic humic acid material used for treatment of drilling mud to reduce viscosity and gel strength. Also used to prepare emulsion muds characterized by low filtration rates, stability, and easy maintenance.

carbon oxybromide. See bromophosgene.

carbon oxychloride. See phosgene.

carbon oxycyanide. See carbonyl cyanide.

carbon oxyfluoride. See carbonyl fluoride.

carbon oxysulfide. See carbonyl sulfide.

carbon steel. See steel.

carbon tetrabromide (tetrabromomethane) CBr_4. A brominated hydrocarbon.
Properties: Colorless crystals; sp. gr. 3.42; m.p. 90.1°C; b.p. 189.5°C; insoluble in water; soluble in alcohol, ether, and chloroform. Noncombustible.
Hazard: Moderately toxic; narcotic in high concentrations.
Use: Organic synthesis.

carbon tetrachloride (tetrachloromethane; perchloromethane) CCl_4. A chlorinated hydrocarbon.
Properties: Colorless liquid; yields vapors 5.3 times heavier than air; sweetish, distinctive odor. Sp. gr. 1.585 (25/4°C); b.p. 76.74°C; freezing point -23.0°C; refractive index 1.4607 at 20°C; vapor pressure 91.3 mm (20°C); wt/gal 13.22 lb (25°C); flash point, none. Miscible with alcohol, ether, chloroform, benzene, solvent naphtha, and most of the fixed and volatile oils; insoluble in water. Noncombustible.
Derivation: (a) Interaction of carbon disulfide and chlorine in presence of iron; (b) chlorination of methane or higher hydrocarbons at 250–400°C.
Method of purification: Treatment with caustic alkali solution to remove sulfur chloride, followed by rectification.
Grades: Technical; C.P.; electronic.
Containers: 55-gal drums; tank cars; tank trucks.
Hazard: Highly toxic by ingestion, inhalation, and skin absorption. Narcotic. Tolerance, 10 ppm in air. Decomposes to phosgene at high temperature. Do not use to extinguish fire. Safety data sheet available from Manufacturing Chemists Assn., Washington, D.C.
Uses: Refrigerants and propellants, especially the chlorofluorohydrocarbons; metal degreasing; agricultural fumigant; chlorinating organic compounds; production of semiconductors.
Shipping regulations: (Air) Poison label. (Rail) Poisonous liquids, n.o.s., Poison label.
Note: Not permitted in products intended for home use.

carbon tetrafluoride. See tetrafluoromethane.

carbon trichloride. See hexachloroethane.

carbonyl group. The divalent group C=O, which occurs in a wide range of chemical compounds. It is present in aldehydes, ketones, organic acids and sugars, as well as in the carboxyl group, i.e., $C\genfrac{}{}{0pt}{}{\nearrow O}{\searrow OH}$. In combination with transition metals it forms coordination compounds which are highly toxic, as they decompose to release carbon monoxide when absorbed by the body, e.g., nickel carbonyl. Several metal carbonyls have antiknock properties. The carbonyl group is also found in combination with nonmetals, as in phosgene (carbonyl chloride); these compounds are also poisonous.

N,N′-carbonylbis(4-methoxymetanilic acid) disodium salt (sodium methoxymetanilate urea) $[C_6H_3(OCH_3)(SO_3Na)NH]_2CO$.
Properties: Gray paste; solids approx. 70%.
Grades: Technical.
Hazard: May be toxic.
Use: Intermediate.

carbonyl bromide. See bromophosgene.

carbonyl chloride. See phosgene.

carbonyl cyanide (carbon oxycyanide) $CO(CN)_2$.
Properties: Colorless liquid. Unstable in the presence of water. B.p. -83°C; sp. gr. 1.139 (-114°C); m.p. 114°C.
Hazard: Highly toxic. Tolerance (as CN), 5 mg per cubic meter in air.
Uses: Organic synthesis.
Shipping regulations: (Rail) (Air) Cyanide solutions, n.o.s., Poison label.

carbonyl fluoride (fluoroformyl fluoride; carbon oxyfluoride) COF_2.
Properties: Colorless gas. Unstable in the presence of water. B.p. -83°C; sp. gr. 1.139 (-114°C); f.p. -114°C. Minimum purity, 97 mole %. Nonflammable.
Derivation: Action of silver fluoride on carbon monoxide.
Containers: Liquefied gas in steel cylinders.
Grades: Technical.
Hazard: Highly toxic. Use may be restricted. Tolerance (as F), 2.5 mg per cubic meter of air.
Uses: Organic synthesis.
Shipping regulations: (Rail) Poisonous gases, n.o.s., Poison Gas label. Not accepted by express. (Air) Not acceptable.

carbonyl sulfide (carbon oxysulfide) COS.
Properties: Colorless gas with typical sulfide odor except when pure. Sp. gr. gas (air = 1) 2.1; f.p. -138.8°C; b.p. -50.2°C (1 atm). Soluble in water and alcohol.
Derivation: Hydrolysis of ammonium or potassium thiocyanate.
Containers: Steel cylinders.
Hazard: Flammable; explosion limits in air 12 to 28.5%. Highly toxic; narcotic in high concentrations.
Uses: Synthesis of thio organic compounds.

Shipping regulations: (Rail) Red gas label. (Air) Flammable Gas label. Not accepted on passenger planes.

carbophenothion Generic name for (S-[{(para-chlorophenyl)thio}methyl]-O,O-diethyl phosphorodithioate; O-O-diethyl S-(para-chlorophenylthiomethyl) phosphorodithioate. $(C_2H_5O)_2P(S)SCH_2S(C_6H_4)Cl$.
Properties: Amber liquid; b.p. 82°C (0.1 mm); sp. gr. 1.29 (20°C). Essentially insoluble in water; miscible in common solvents.
Hazard: Toxic. Use may be restricted. Cholinesterase inhibitor.
Uses: Insecticide; acaricide.

"Carbopol."[119] Trademark for a group of water-soluble resins having excellent suspending, thickening, and gel-forming properties.
Uses: Suspensions of glass fibers, graphite, powdered metals; gel formation in hydrocarbons; emulsifier for creosote, tars, and asphalts.

carborane. A crystalline compound comprised of boron, carbon, and hydrogen. It can be synthesized in various ways, chiefly by the reaction of a borane (penta- or deca-) with acetylene, either at high temperature in the gas phase or in the presence of a Lewis base. Alkylated derivatives have been prepared. Carboranes have different structural and chemical characteristics, and should not be confused with hydrocarbon derivatives of boron hydrides. The predominant structures are the cage type, the nest type and the web type, these terms being descriptive of the arrangement of atoms in the crystal. Active research on carborane chemistry has been conducted under sponsorship of the U.S. Office of Naval Research.

"Carbortam."[337] Trademark for a metallurgical processing alloy containing 15–20% titanium, 6–8% carbon, 2.4–4.0% silicon, 1.75–2.25% boron, and the balance iron except with traces of phosphorus and sulfur. Used to deoxidize and harden steel.

"Carborundum."[280] Trademark for abrasives and refractories of silicon carbide, fused alumina and other materials.
Properties: For silicon carbide—Crystalline form ranges from small to massive crystals in the hexagonal system, the crystals varying from transparent to opaque with colors from pale green to deep blue or black; hardness 9.17 (Mohs); sp. gr. 3.06–3.20; noncombustible; not affected by acids; slowly oxidizes at temperatures above 1000°C; good heat dissipator; highly refractory. For fused alumina—See properties under the trademark "Aloxite."
Uses: Abrasive grains and powders for cutting, grinding and polishing; valve-grinding compounds; grinding wheels; stones; bones; rubbing bricks; coated abrasive products; antislip tiles and treads; refractory grains.

carbosand. Fine sand that has been treated with an organic solution and roasted to produce a material that can be sprayed onto oil slicks to aid in sinking or dispersing them. See also oil spill treatment.

"Carbose."[203] Trademark for various grades of sodium carboxymethylcellulose.
Uses: Detergency promotion; paint, paper and textile coatings; and drilling muds.

"Carboseal."[214] Trademark for hygroscopic liquid compositions for joint sealing.
Uses: Swelling and moistening agent for jute and other packing in cast-iron gas mains, to correct joint leakage and lay dust.

"Carbo-Sour."[244] Trademark for a product consisting of highly soluble fluorine compounds. Used as a laundry sour, when high solubility is desired.

"Carbotronic."[280] Trademark for self-bonded silicon carbide in granular form; used in lightning arresters and for electrical purposes.

"Carbo-Vitrobond."[41] Trademark for a carbon-filled, sulfur-based compound for use as a hot-pour acid-proof cement where temperatures do not exceed 200°F. Has excellent resistance to hydrofluoric acid and fluoride salts.

"Carbowax."[214] Trademark for polyethylene glycols and methoxypolyethylene glycols.
Grades: Available in various numbered grades, i.e., 200, 400, 1000, 4000, 6000. Usually designated by approximate molecular weight of polymer.
Uses: Water-soluble lubricants; solvents for dyes, resins, proteins; plasticizers for casein and gelatin compositions, glues, zein, cork, and special printing inks; solvent and ointment bases for cosmetics and pharmaceuticals; intermediates for nonionic surfactants and alkyd resins.

"Carboxide."[214] Trademark for proprietary fumigantsterilant mixture of ethylene oxide and carbon dioxide. Ethylene oxide is the active agent.
Uses: Fumigant and sterilizing agent to eliminate insects such as beetles, moths, cockroaches, insect larvae and insect eggs. There is definite reduction in thermophylic bacteria, and destruction of mold and fungi, including spores.

carboxybenzene. See benzoic acid.

2-carboxy-2′-hydroxy-5′-sulfoformazylbenzene (ortho-[alpha-(2-hydroxy-5-sulfophenyl)-azobenzylidene]-hydrazino benzoic acid)
$HO_3SC_6H_3(OH)N{:}NC(C_6H_5){:}NNC_6H_4COOH$.
Uses: Reagent used for the colorimetric determinaton of zinc and copper.

carboxyl group. The chemical group characteristic of carboxylic acids, which include fatty acids and amino acids. It usually occupies the terminal position in the molecule and is capable of assuming a negative charge which makes the end of the molecule water-soluble. Though it is customarily shown as either COOH or CO_2H, the structure of the group is $C{\lessgtr}^{O}_{OH}$; thus it is composed of a carbonyl group and a hydroxyl group.

carboxylase. A decarboxylase enzyme found in plant tissues which acts upon pyruvic acid, producing acetaldehyde and carbon dioxide.
Use: Biochemical research.

carboxylic acid. Any of a broad array of organic acids comprised chiefly of alkyl (hydrocarbon) groups (CH_2, CH_3), usually in a straight chain (aliphatic), terminating in a carboxyl group (COOH). Exceptions to this structure are formic acid (HCOOH) and oxalic acid (HOOCCOOH). The number of carbon atoms ranges from 1 (formic) to 26 (cerotic), the carbon of the terminal group being counted as part of the chain. Carboxylic acids include the large and important class of fatty acids (q.v.), and may be either saturated or unsaturated. A few contain halogen atoms (chloroacetic). There are also some natural aromatic carboxylic acids (benzoic, salicylic) as well as alicyclic types (abietic, chaulmoogric). See also amino acid.

carboxymethylcellulose (CMC; sodium carboxymethylcellulose; CM cellulose). A semisynthetic, water-soluble polymer (q.v.) in which CH_2COOH groups

are substituted on the glucose units of the cellulose chain through an ether linkage. Molecular weight ranges from 21,000 to 500,000. Since the reaction occurs in an alkaline medium, the product is the sodium salt of the carboxylic acid R—O—CH_2COONa.
Properties: Colorless, odorless, nontoxic, water-soluble powder or granules; pH (1% solution) 6.5–8.0; stable in pH range 2–10. Sp. gr. 1.59; refractive index 1.51; tensile strength 8000–15,000 psi. Viscosity of 1% solution varies from 5 to 2000 centipoises, depending on the extent of etherification. Insoluble in organic liquids. Reacts with heavy-metal salts to form films that are insoluble in water, transparent, relatively tough, and unaffected by organic materials. Many of its colloidal properties are superior to those of natural hydrophilic colloids. It also has thixotropic properties (q.v.) and functions as a polyelectrolyte (q.v.). See also cellulose, modified.
Derivation: By reaction of alkali cellulose and sodium chloroacetate.
Grades: Crude; technical (about 75% pure); high viscosity; low viscosity; semirefined; refined (99.5+% pure); U.S.P.; F.C.C.
Containers: 13-, 51-gal drums; 50-lb bags.
Uses: Detergents, soaps, food products (especially dietetic foods and ice cream), where it acts as water binder, thickener, suspending agent, and emulsion stabilizer; textile manufacturing (sizing); coating paper and paper board to lower porosity; drilling muds; emulsion paints; protective colloid; pharmaceuticals; cosmetics.

carboxymethylmercaptosuccinic acid $HOOCCH_2SCH(COOH)CH_2(COOH)$.
Properties: White powder; melting range 136–138°C. Water solubility, 147 g/100 g at 25°C; ethanol solubility, 76 g/100 g at 25°C.
Uses: Heavy-metal chelator and deactivator.

carboxymethylpyridinium chloride hydrazide. See Girard's "P" reagent.

carboxymethyltrimethyl ammonium chloride hydrazide. See Girard's "T" reagent.

carboxypeptidase. A proteolytic enzyme found in the pancreas which catalyzes the hydrolysis of native food proteins. It acts upon polypeptides producing simpler peptides and amino acids.
Use: Biochemical research.

4-carboxyresorcinol. See beta-resorcylic acid.

6-carboxyuracil. See orotic acid.

carbromal. See bromodiethylacetylurea.

carbutamide. See 1-butyl-3-sulfanilylurea.

"Carbyne."[533] Trademark for an herbicide containing 4-chloro-2-butynyl meta-chlorocarbanilate (barban).

carcinogen. Any substance that causes the development of cancerous growths in living tissue. Such substances are usually grouped in two classifications: (1) Those that are known to induce cancer in man or animals either by operational exposure in industry or by ingestion in feedstuffs, e.g., asbestos particulates, nickel carbonyl, trichloroethylene, benzidine and compounds, vinyl monomer, diethylstilbestrol, benzopyrene, aflatoxin, chloromethyl ether, betanaphthylamine, as well as anthracene, phenanthrene, chrysene, and other polynuclear hydrocarbons of coal-tar origin. (2) Those that have been found to cause cancer in animals under experimental conditions, namely, by external application or feeding of extremely high concentrations of the substance; among these are dimethyl sulfate, cyclamate compounds, ethyleneimine, nitrilotriacetic acid, and 4-dimethylaminoazobenzene. (The validity of the tests made on cyclamates has been questioned). The Delany amendment to the Food, Drug and Cosmetic Act forbids the use in human foods of any substance falling in this group.

The substances mentioned above by no means comprise a complete list of carcinogens. There are several hundred known and suspected carcinogenic compounds, and new ones are constantly being discovered. The Environmental Protection Agency has stated that a substance will be considered a presumptive cancer risk when it causes a statistically excess incidence of benign or malignant tumors in humans or animals. See also oncogenic.

cardamom oil.
Properties: Colorless or pale-yellow essential oil; strongly aromatic, camphoraceous odor and taste. Sp. gr. 0.917–0.947 (25/25°C); refractive index 1.4630–1.4660 (20°C); optical rotation +22° to +44° in 100 mm tube at 25°C. Insoluble in water, soluble in alcohol and ether. Keep well closed, cool, and protected from light. Combustible; nontoxic.
Chief known constituents: Terpinene, dipentene, limonene, eucalyptol; borneol.
Derivation: Distilled from the seeds of Elettaria cardamomum from Malabar and Ceylon.
Grades: Technical; N.F.; F.C.C.
Uses: Flavoring; liqueurs; medicine.

"Cardanol."[158] Trademark for a mixture of 3-pentadecenyl and less saturated C_{15} phenols, as, $C_6H_4OH \cdot C_{15}H_{27}$.
Properties: Amber liquid, boiling range 180–230°C (1 mm); soluble in oils, waxes, and all organic solvents; insoluble in water, glycerin, and aqueous alkalies.
Uses: Aldehyde-reactive plasticizer for phenol-aldehyde resins, particularly for laminating purposes.

"Cardanol Bis-Phenol."[158] Stated to be 1,8-di(hydroxyphenyl)pentadecane $C_6H_4OH \cdot C_{15}H_{30} \cdot C_6H_4OH$.
Properties: Brown viscous liquid (10,000 cps at 25°C).
Uses: Base for epoxy resins and for synthetic resins of phenol-formaldehyde type useful as wire enamel.

"Cardene."[458] Trademark for technical grade of N,N-bis(hydroxyethyl)oleamide containing about 25% excess amine.

"Cardilate."[301] Trademark for erythritol tetranitrate (q.v.).

"Cardinal."[488] Trademark for phthalic anhydride.

"Cardio-Green."[348] Trademark for indocyanine green, a diagnostic dye used in medicine.

"Cardolite."[158] Trademark for friction components based upon a phenolic-type liquid found in the fibrous outer shell of the cashew nut. These materials are produced in the form of friction-fortifying particles, and binding resins in liquid or powdered form.
Uses: Brake linings, brake blocks, clutch facings and other friction units which have high heat resistance, and outstanding friction and wear characteristics.

"Cardosol" Brand Resin.[158] Trademark for a water-soluble ketone formaldehyde condensate which can be gelled and cured by alkali or heat.
Use: Water-resistant adhesives for box board, coatings, for glass fibers and as a ceramic binder.

"Cardura" E Ester.[125] Trademark for glycidyl ester of "Versatic" 911 Acid.
Hazard: Irritant to skin on prolonged contact.
Use: Modifier for alkyd resins and thermosetting acrylic systems; reactive diluent for epoxy resins.

delta-3-**carene** $C_{10}H_{16}$. 3,7,7-Trimethylbicyclo[4.1.0]-hept-3-ene, a terpene hydrocarbon having both a six-member and a three-member ring.
Properties: Clear, colorless liquid; sp. gr. 0.8668 (15°C); b.p. 170°C; refractive index (n 20/D) 1.4723. Stable to about 250°C. Resinifies with oxygen. Combustible.
Derivation: From wood turpentine.
Uses: Solvent; intemediate.

"Carmethose."[305] Trademark for sodium carboxymethylcellulose.

carmine. An aluminum lake of the pigment from cochineal (q.v.). Bright red pieces, easily powdered. Soluble in alkali solutions, borax; insoluble in dilute acids; slightly soluble in hot water.
Grades: Technical.
Uses: Dyes, inks, indicator in chemical analysis; coloring food materials, medicines, etc.

carminic acid $C_{22}H_{20}O_{13}$, a tricyclic compound. The essential constituent of carmine.
Properties: Dark, purplish-brown mass or bright-red powder. M.p., decomposes at 136°C; pH 4.8 yellow; pH 6.2 violet. Soluble in water, alcohol, alkali hydroxide solutions; insoluble in ether, benzene, chloroform. Combustible.
Derivation: By extraction from the insects, Coccus cacti (cochineal).
Uses: Stain in microscopy; indicator in analytical chemistry; coloring proprietary medicines; pigment for fine oil colors; color photography; dyeing.

carnallite $KCL \cdot MgCl_2 \cdot 6H_2O$ or $KMgCl_3 \cdot 6H_2O$.
Properties: A natural hydrated double chloride of potassium and magnesium. White, brownish and reddish; streak, white; shining, greasy luster; strongly phosphorescent; bitter taste; deliquescent. Sp. gr. 1.62; Mohs hardness 1.
Occurrence: West Germany; Alsace; New Mexico.
Use: A commercial source of manufactured potash salts.

carnauba wax (Brazil wax). The hardest and most expensive commercial wax.
Properties: Hard solid in form of yellow to greenish brown lumps; slight odor. Sp. gr. 0.995 (15/15°C); m.p. 84–86°C; acid number 2–9; iodine number 13.5. Soluble in ether, boiling alcohol and alkalies; insoluble in water. Combustible. Nontoxic.
Derivation: Exudation from leaves of the wax palm, Copernica cerifera (Brazil).
Grades: By numbers and sources; crude and refined; powdered; F.C.C.
Containers: Bags; boxes.
Uses: Shoe polishes; leather finishes; varnishes; electric insulating compositions; furniture and floor polishes; carbon paper; waterproofing; to prevent sun-checking of rubber and plastic products; confectionery.

carnosine (beta-alanylhistidine; ignotine) $C_9H_{14}N_4O_3$. An amino acid occurring in muscle of many animals and man. Occurs naturally in the L(+)-form.
Properties: M.p. 245–250°C (dec.); soluble in water. Nitrate: crystals; m.p. 222°C (dec.); soluble in water. Hydrochloride: crystals; m.p. 245°C (dec.); soluble in water. D(–)-carnosine: crystals; m.p. 260°.
Use: Biochemical research.

carnotite $K_2(UO_2)_2(VO_4)_2 \cdot 3H_2O$. A natural hydrated vanadate of uranium and potassium, usually found in sandstones and other sedimentary rocks.
Properties: Bright lemon yellow; luster dull or earthy, pearly or silky when coarsely crystalline. Soluble in acids. Radioactive. Usually occurs as a powder or in fine-grained aggregates.
Occurrence: Colorado, Utah, Arizona, New Mexico, South Dakota; Australia; Congo; U.S.S.R.
Hazard: A radioactive poison.
Use: Ore of uranium; source of radium.
Shipping regulations: (Rail, Air) Consult regulations.

Carnot's reagent. A reagent for the determination of potassium; an alcoholic solution of sodium bismuth thiosulfate, made from sodium thiosulfate and bismuth subnitrate.

carob-seed gum (locust bean gum). A polysaccharide plant mucilage which is essentially galactomannan (carbohydrate). Molecular weight about 310,000. Swells in cold water, but viscosity increases when heated. Insoluble in organic solvents. Combustible; nontoxic on basis of present data. See note.
Containers: Bags.
Derivation: Extracted from carob seeds, from the tree Ceratonia siliqua.
Grades: Technical; F.C.C. (as locust-bean gum).
Uses: In foods as stabilizer, thickener, emulsifier, and packaging material; cosmetics; sizing and finishes for textiles; pharmaceuticals; paints.
Note: Restricted by FDA to level of use current in 1973.

Caro's acid (peroxysulfuric acid; persulfuric acid) H_2SO_5 or $HOSO_2OOH$.
Properties: White crystals; m.p. 45°C (decomposes).
Derivation: Action of hydrogen peroxide on concentrated sulfuric acid; action of 40% sulfuric acid on potassium persulfate.
Hazard: Strong irrirant to eyes, skin and mucous membranes. Strong oxidizer; may explode in contact with organic materials.
Use: Caro's reagent, a pasty mass of great oxidizing power for testing aniline, pyridine, and alkaloids.

carotene (provitamin A) $C_{40}H_{56}$. A precursor of vitamin A occurring naturally in plants. It consists of 3 isomers; about 15% alpha, 85% beta, and 0.1% gamma. Carotene is a member of a large class of pigments called carotenoids. It has the same basic molecular structure as vitamin A and is transformed to the vitamin in the liver.
Properties: Ruby-red crystals, easily oxidized on contact with air; m.p. (alpha) 188°C, (beta) 184°C, (gamma) 178°C; insoluble in water; slightly soluble in alcohol and ether; soluble in chloroform, benzene, and oils. Nontoxic.
Source: Orange-yellow pigment in plants, algae, fungi, and some marine animals, especially in leaves, vegetation, and root crops, in trace concentrations. Notably present in butter and carrots.
Derivation: By extraction from carrots and palm oil concentrates; by a chromatographic process from alfalfa, beta-Carotene is also made by a microbial fermentation process from corn and soybean oil.
Grades: According to U.S.P. units of vitamin A. Solid as pure crystals; as solutions in various oils; as colloidal dispersion. Also F.C.C.
Uses: Pharmaceuticals; coloring margarine and butter; feed and food additive.

carotenoid. A class of pigments occurring in the tissues of higher plants, algae, and bacteria, as well as in fungi. Also present in some animals, as squalene

in shark liver oil. They include the carotenes and xanthophylls (q.v.).
Properties (general): Yellow to deep red crystalline solids; soluble in fats and oils; insoluble in water; high-melting; stable to alkali but unstable to acids and to oxidizing agents; color easily destroyed by hydrogenation or by oxidation; some are optically active.

Carothers, Wallace H. (1896–1937). Born in Iowa, Carothers obtained his doctorate in chemistry at University of Illinois. He joined the research staff of DuPont in 1928 and began the development of several of the greatest achievements of synthetic organic chemistry. One of these was polychloroprene, the first successful synthetic rubber, now known as neoprene. Further research in the detailed chemistry of polymerization culminated in the synthesis of nylon, the reaction product of hexamethylenetetramine and adipic acid. Carothers' work in the polymerization mechanisms of fiber-like synthetics of cyclic organic structures was brilliant and productive, and he is regarded as one of the most original and creative American chemists of the early 20th century.

carrageenan. A phycocolloid. The aqueous, usually gel-forming cell-wall polysaccharide mucilage found in the red algae (Chondrus crispus and several other species). Water-extracted from a seaweed called carrageen or Irish moss (east coast of southern Canada, New England, and south to New Jersey). It is a mixture of polysaccharide fractions; (1) the lambda fraction is cold-water soluble, contains D-galactose and 35% esterified sulfate, and does not gel; (2) the kappa fraction contains D-galactose and 3,6-anhydro-D-galactose (1.4 to 1 ratio) and 25% esterified sulfate. Kappa form does not gel without addition of a solute; the properties of the gel depend on the amount and nature of the added solute. Another species of seaweed produces 100% kappa form (N. Carolina to tropics). Carrageenan is a hydrophilic colloid which absorbs water readily and complexes with milk proteins.
Forms: Dehydrated, purified powder.
Grades: Technical; F.C.C.
Uses: Emulsifier in food products, especially chocolate milk; toothpastes; cosmetics; pharmaceuticals; protective colloid; stabilizing aid in ice cream (0.02%).

"Carrene 16."[54] Trademark for a 54% aqueous solution of lithium bromide containing an additive for corrosion inhibition in absorption refrigeration systems.

"Carrene 500."[54] Trademark for an azeotropic mixture of 73.8% dichlorodifluoromethane and 26.2% unsymmetrical difluoroethane boiling at −28°F and used as a refrigerant. Nonflammable; low toxicity.

carrier. (1) A neutral material such as diatomaceous earth used to support a catalyst in a large-scale reaction system. (2) A gas used in chromatography to convey the volatilized mixture to be analyzed over the bed of packing which separates the components. (3) An atomic tracer carrier: a stable isotope or a normal element to which radioactive atoms of the same element have been added for purposes of chemical or biological research. See also tracer.

"Carstan."[458] Trademark for a series of stannous carboxylate salts used as catalysts in flexible and rigid urethane foams and elastomers.

carthamin (carthamic acid; safflor carmine; safflor red) $C_{21}H_{22}O_{11}$.
Properties: Dark-red powder with green luster. Slightly soluble in water; soluble in alcohol; insoluble in ether; solutions rapidly decompose.
Derivation: A glucoside coloring matter from Carthamus tinctorius.

"Carum."[51] Trademark for grease-type lubricants prepared for use in chemical processing industries where insolubility in the material being processed is essential. Intended for valves and pumps handling such materials. One grade is permissible under FDA Regulations for lubrication where incidental contact with food might occur.

carvacrol (isopropyl-ortho-cresol; 2-methyl-5-isopropylphenol; 2-hydroxy-para-cymene) $(CH_3)_2CHC_6H_3(CH_3)(OH)$. An alcohol.
Properties: Thick, colorless oil; thymol odor; sp. gr. 0.976 (20/4°C); b.p. 237°C; m.p. 0°C; refractive index 1.523 (20°C). Insoluble in water; soluble in alcohol, ether, and alkalies. Combustible. Low toxicity.
Derivation: From para-cymene by sulfonation, followed by alkali fusion.
Uses: Perfumes; fungicides; disinfectant; flavoring.

carvone $CH_3\overline{C{:}CHCH_2CH[C(CH_3){:}CH_2]CH_2C}O$. A ketone derived from the terpene dipentene. It is optically active, occurring naturally in both *d*- and *l*-forms.
Properties: Pale-yellowish or colorless liquid with a strong characteristic odor. Sp. gr. 0.960 (20°C); b.p. 227–230°C; refractive index (n 18/D) 1.4999. Soluble in alcohol, ether, chloroform, propylene glycol, and mineral oils; insoluble in glycerin and in water. Combustible; low toxicity.
Derivation: The *d*- form is the main constituent of caraway and dill oils; the *l*- form occurs principally in spearmint oil and may be synthesized from *d*-limonene.
Method of purification: Rectification.
Grades: F.C.C. (both *d*- and *l*- forms); technical.
Uses: Flavoring; liqueurs; perfumery.

"Carwinate."[520] Trademark for isocyanates used to make urethane elastomers, coatings, foams, adhesives, rigid plastics, sealants, and one-shot flexible, semi-flexible and semi-rigid foams.

caryophyllene $C_{15}H_{24}$. A mixture of sesquiterpenes occurring in many essential oils. It forms the chief hydrocarbon component of clove oil (q.v.).

caryophyllic acid. See eugenol.

"Cascamite."[65] Trademark for a powdered urea-formaldehyde resin glue; water-resistant, moldproof, stainfree.

cascarilla oil. The volatile oil obtained by steam distillation of the dried bark of Croton eleuteria Bennet. Light yellow to brown amber liquid having a pleasant spicy odor. Soluble in most fixed oils, and in mineral oil; practically insoluble in glycerin and propylene glycol.
Use: Flavoring agent in foods, medicine, tobacco.

case-hardening compound. A material used to impart a hard surface to steel while the interior remains soft and tough. This is acocmplished by heating the steel out of contact with air while packed in carbonaceous material, cooling it to black heat, reheating to a high temperature, and quenching. The materials are usually wood charcoal with sodium, potassium, or barium carbonates, cyanides, etc.

casein. Though commonly regarded as the principal protein in milk (about 3%), casein is actually a colloidal aggregate composed of several identifiable proteins, together with phosphorus and calcium. It occurs in milk as a heterogeneous complex called calcium caseinate, which can be fractionated by a number of methods. It can be precipitated with acid at pH 4.7, or with the enzyme rennet (rennin). The product of the latter method is called paracasein, the term being applied to any of the casein fractions involved, i.e., alpha, beta, kappa, etc.
Properties: White, tasteless, odorless, nontoxic, solid; sp. gr. 1.25–1.31; hygroscopic; stable when kept dry but deteriorates rapidly when damp. Soluble in dilute alkalies and concentrated acids; almost insoluble in water; precipitates from weak acid solutions.
Derivation: Acid casein; warm skim milk is acidified with dilute sulfuric, hydrochloric, or lactic acid, the whey drawn off, the curd washed, pressed, ground and dried. Rennet casein: warm skim milk is treated with an extract of the enzyme rennin (rennet). The curd contains combined calcium and calcium phosphate.
Sources: Midwestern U.S.; Australia; Argentina; New Zealand; Poland.
Grades: Acid-precipitated (domestic edible, imported inedible); paracasein.
Uses: Cheese making; plastic items; paper coatings; water-dispersed paints for interior use; adhesives, especially for wood laminates; textile sizing; foods and feeds; textile fibers.

casein-sodium. See sodium caseinate.

cashew gum (anacardium gum). The exudation from the bark of the cashew-nut tree, Anacardium occidentale. Hard, yellowish-brown gum, partly soluble in water. Used for inks, insecticides, pharmaceuticals, mucilage, tanning agent, natural varnishes, bookbinders' gum.

cashew nutshell oil (cashew nutshell liquid). The oil obtained from the spongy layer between the inner and outer shells of cashew nuts. The raw liquid contains about 90% anacardic acid, $C_{22}H_{32}O_3$, and a blistering compound containing sulfur. Most of the liquid used in commerce has been heated or treated with chemicals to make it safe to handle. The liquid is non-drying, but can be made drying by proper treatment. It polymerizes on heating and forms condensation products with aldehydes.
Hazard (untreated): Toxic; strong irritant.
Uses: Varnishes and impregnating materials; modifier for phenol-based resins; plasticizers; germicides and insecticides, coloring materials and indelible inks, lubricants, and preservatives.

casinghead gasoline. See gasoline.

cassava. See tapioca.

Cassel brown. See Van Dyke brown.

Cassel green. See barium manganate.

Cassella's acid. See 2-naphthol-7-sulfonic acid; 2-naphthylamine-4,8-disulfonic acid.

Cassella's F acid. See 2-naphthylamine-7-sulfonic acid.

cassia oil (Chinese cinnamon oil; cinnamon, cassia oil; cinnamon oil, U.S.P.).
Properties: See cinnamic aldehyde, its chief constituent.
Derivation: Distilled from leaves and twigs of Cinnamomum cassia.
Grades: U.S.P. (as cinnamon oil); redistilled; technical; lead-free.
Containers: Bottles; tins; cans.
Uses: Flavoring; perfumery; medicine; soaps.

cassiterite (tinstone, wood tin, stream tin) SnO_2. Natural tin dioxide, usually in igneous rocks.
Properties: Color brown, black, yellow, white; luster adamantine or dull submetallic; streak white; Mohs hardness 6–7; sp. gr. 6.8–7.1.
Occurrence: Malaya; Bolivia; Indonesia; Africa.
Use: Principal ore of tin.

cast iron. Generic term for a group of metals that basically are alloys of carbon and silicon with iron. Relative to steel, cast irons are high in carbon and silicon, carbon ranging from 0.5 to 4.2% and silicon from 0.2 to 3.5%. All these metals may contain other alloys added to modify their properties.

Iron castings are produced in an exceptionally wide range of sizes and weight, from piston rings a fraction of an inch in diameter and weighing less than 1 oz to steam-turbine bases 20 ft long and weighing 180,000 lb.

Most cast iron is manufactured by melting a mixture of steel scrap, cast-iron scrap, pig iron, and alloys in a cupola using coke as a fuel. A small percentage is melted in electric furnaces. It is poured into molds of silica sand bonded with bentonite, fireclay, and water. A small percentage is cast into metal molds or into molds of baked or fired ceramics. Internal cavities are formed by hard but collapsible cores of sand bonded with drying oils or synthetic resins. Small molds and cores usually are made by machine, using patterns of wood or metal.

"Castolast H-W."[446] Trademark for a 93% high-alumina castable cement bonded with low iron calcium aluminate cement. Shipped dry; with water addition, develops and maintains high strength, through 3200°F. Resistant to abrasion and impact.
Uses: Petrochemical unit liner; burner blocks and other high temperature applications. Can be cast, trowelled or applied with an air placement gun.

castor.
Derivation: Dried preputial follicles with their secretions of the common beaver (Castor fiber). Solid unctuous masses contained in pairs of sacs, each about 2 in. in length. Characteristic irritating odor.
Contains 40–70% resin.
Grades: Canadian or American; Russian.
Uses: Medicine; perfume fixative; flavoring. A synthetic castor is also marketed.

"Castordag."[46] Trademark for a dispersion of colloidal graphite in castor oil. Used as an assembly and maintenance lubricant.

castor oil (ricinus oil).
Properties: Pale-yellowish or almost colorless, transparent, viscid liquid; faint, mild odor; nauseating taste. A non-drying oil. Sp. gr. 0.945–0.965 (25/25°C); saponification value 178; iodine value 85; solidifies at −10°C. Flash point 445°F; autoignition temp. 840°F. Combustible. Soluble in alcohol, benzene, chloroform and carbon disulfide; dextrorotatory. Nontoxic.
Derivation: From the seeds of the castor bean, Ricinus communis (Brazil, India, U.S.S.R., U.S.). They are cold pressed for the first grade of medicinal oil and hot pressed for the common qualities, about 40% of the oil content of the bean being obtained. Re-

sidual oil in the cake is obtained by solvent extraction.
Chief constituent: Ricinolein (glyceride of ricinoleic acid).
Grades: U.S.P. No. 1; No. 3; refined; F.C.C.
Containers: Bottles; drums; tanks.
Hazard: Develops heat spontaneously.
Uses: Plasticizer in lacquers and nitrocellulose; production of dibasic acids; lipsticks; polyurethane coatings, elastomers and adhesives; fatty acids; surface-active agents; hydraulic fluids; pharmaceuticals; industrial lubricant; electrical insulating compounds; mfg. of Turkey Red oil (q.v.); source of sebacic acid and of ricinoleates; medicine (laxative).
See also castor oil, dehydrated; blown oil.

castor oil, acetylated. See glyceryl triacetylricinoleate.

castor oil acid. See ricinoleic acid.

castor oil, blown. See blown oil.

castor oil, dehydrated (DCO). A castor oil from which about 5% of the chemically combined water has been removed, and which as a result, has drying properties similar to those of tung oil. Dehydration is carried out commercially by heating the oil in the presence of catalysts, such as sulfuric and phosphoric acids, clays, and metallic oxides. The commercial product is offered in a wide range of viscosities and analytical constants. Used in protective coatings and alkyd resins.

castor oil, hydrogenated. Principally glyceryl tri-12-hydroxystearate (q.v.). A hard, waxy product used in making hydroxystearic acid.

castor oil, polymerized. A rubber-like polymer results from combination of castor oil with sulfur or diisocyanates; this can be blended with polystyrene to give a tough, impact-resistant product.

castor oil, sulfonated. See Turkey red oil.

castor seed oil meal (castor cake; castor meal). The residue from extraction of oil from the castor seed (ricinus). The normal product contains 29.5% crude protein; 35.8% crude fiber; 13.2% nitrogen-free extract and 1.0% crude fat. The total digestible nutrients approximate 25%. The ash content of 7.5% is high in potash and phosphate.
Containers: Bag or bulk; carlots.
Hazard: Contains highly toxic ricin which must be removed before internal use.
Use: Animal feeds (after removal of toxic ingredients); fertilizer.

"Castorwax."[202] Trademark for hydrogenated castor oil, the triglyceride of 12-hydroxystearic acid.
Properties: White, hard, synthetic wax; m.p. 85°C; sp. gr. 0.9990 (25°C); acid value 2; iodine value 3; saponification value 180; insoluble in most organic solvents at room temperature and compatible with ethylcellulose, cellulose acetate butyrate, polyethylene (up to 25%), polymethacrylate, rosin, shellac, natural and synthetic rubbers, insect and vegetable waxes.
Uses: Potting compounds, greases, sealants and impregnating compositions; wax blends where increase in grease and solvent resistance, hardness and melting point is desired; blending agent and viscosity reducer in hot melts; direct application to paper for resistance to moisture and fat.

"Castung."[202] Trademark for dehydrated castor oil. See castor oil, dehydrated.

catalase. An oxidizing enzyme occurring in both plant and animal cells. It decomposes hydrogen peroxide. It can be isolated and is used in food preservation (removing oxygen in packaged foods) and in decomposing residual hydrogen peroxide in bleaching and oxidizing processes.

"Catalin."[353] Trademark for a phenolformaldehyde resin. Nonflammable.
Uses: Miscellaneous molded articles.

catalysis. One of the most important phenomena in nature, catalysis is the "loosening" of the chemical bonds of two (or more) reactants by another substance, in such a way that a fractionally small percentage of the latter can greatly accelerate the rate of the reaction, while remaining unconsumed. (See catalyst). Thus one part by volume of catalyst can activate thousands of parts of reactants. Though the mechanism of their action is not completely known, the electronic configuration of the surface molecules of the catalyst is often the critical factor. The surface irregularities give rise to so-called "active points," at which intermediate compounds can form. Most industrial catalysis is performed by finely divided transition metals or their oxides.

Solid catalysts may combine chemically (bond) at the surface with one or more of the reactants. This is known as chemisorption, and occurs on only a small portion of the catalyst surface (i.e., at the active points); it results in changing the chemical nature of the chemisorbed molecules. Catalysis of chemical reactions by surfaces must proceed by chemisorption of at least one of the reactants.
See also catalyst; chemisorption; hydrogenation; enzyme.

catalysis, heterogeneous. A catalytic reaction in which the reactants and the catalyst comprise two separate phases, e.g., gases over solids, or liquids containing finely divided solids as a disperse phase.

catalysis, homogeneous. A catalytic reaction in which the reactants and the catalyst comprise only one phase, e.g., an acid solution catalyzing other liquid components.

catalyst. Any substance of which a fractional percentage notably affects the rate of a chemical reaction without itself being consumed or undergoing a chemical change. Most catalysts accelerate reactions, but a few retard them (negative catalysts or inhibitors). Catalysts may be inorganic, organic, or a complex of organic groups and metal halides (see catalyst, stereospecific). They may be either gases, liquids, or solids. In some cases their action is destructive and undesirable, as in the oxidation of iron to its oxide, which is catalyzed by water vapor, and similar types of corrosion. The life of an industrial catalyst varies from 1000 to 10,000 hours, after which it must be replaced or regenerated.
Note: Though it is not a "substance," light in the visible wavelengths can act as a catalyst, as in photosynthesis and other photochemical reactions.

Following is a partial list of catalysts; the asterisk indicates a destructive effect.

Substance	*Reaction type*
aluminum chloride	condensation (Friedel-Crafts)
aluminum alkyl + titanium chloride	Ziegler catalyst for stereospecific polymers

Substance	*Reaction type*
aluminum oxide	hydration; dehydration
ammonia	condensation (polymers)
chromic oxide	methanol synthesis; aromatization; polymerization
cobalt	hydrocarbon synthesis; Oxo process
copper salts	oxidation (of rubber)*
hydrogen fluoride	alkylation; condensation; dehydration; isomerization
iron	ammonia synthesis; hydrocarbon synthesis
iron oxide	dehydrogenation (oxidation)
manganese dioxide	oxidation
molybdenum oxide	dehydrogenation; polymerization; aromatization; partial oxidation
nickel	hydrogenation (oils to fats)
phosphoric acid	polymerization; isomerization
platinum metals	hydrogenation; aromatization
silica-alumina	cracking hydrocarbons
silver	hydration; oxidation
sulfuric acid	isomerization; corrosion*
vanadium pentoxide	oxidation (sulfuric acid)
water (esp. + NaCl)	oxidation (corrosion)*

Catalysts are highly specific in their application; they are widely used in the petroleum-refining, synthetic organic chemical and pharmaceutical industries. For details of application, see the preceding list. Since the activity of a solid catalyst is often centered in a small fraction of its surface, the number of active points can be increased by adding promoters (q.v.), which increase the surface area in one way or another, e.g., by increasing porosity. Catalytic activity is decreased by substances that act as poisons (q.v.), which clog and weaken the catalyst surface, e.g., lead in the catalytic converters used to control exhaust emissions.

Besides inorganic substances, there are many organic catalysts that are vital in the life processes of plants and animals. These are called enzymes (q.v.) and are essential in metabolic mechanisms, e.g., pepsin in digestin. Vitamins and hormones also have catalytic properties. See also catalysis; catalyst, stereospecific; enzyme.

catalyst, negative. See inhibitor.

catalyst, organic. See enzyme.

catalyst, stereospecific. An organometallic catalyst which permits control of the molecular geometry of polymeric molecules. Examples are Ziegler and Natta catalysts derived from a transition metal halide and a metal alkyl, or similar substances. There are many patented catalysts of this general type, most of them developed in connection with the production of polypropylene, polyethylene, or other polyolefins. See also polymer, stereospecific; Natta catalyst; Ziegler catalyst.

catalyst, thermonuclear. See carbon cycle (2).

catechol. See pyrocatechol.

catenane. A compound with interlocking rings, which are not chemically bonded, but which cannot be separated without breaking at least one valence bond. The model would resemble the links of a chain.

"Cat-Floc"[108] (diallyldimethylammonium chloride). Trademark for a quaternary ammonium polymer.

Derivation: Monomer in water solution is mixed with catalytic amount of butylhydroperoxide and kept at 50–75°C for 48 hours; the solid formed is taken up in water, precipitated, and washed with acetone.

Uses: Flocculating agent; textile spinning aid; antistatic agent; wet-strength improvers in paper; rubber accelerators; curing epoxy resins; surfactants; bacteriostatic and fungistatic agents.

cathode. The negative electrode of an electrolytic cell, to which positively charged ions migrate when a current is passed, as in electroplating baths. The cathode is the source of free electrons (cathode rays) in a vacuum tube. In a primary cell (battery), the cathode is the positive electrode. See also anode; electrode.

cathode sputtering. See sputtered coating.

cation. An ion having a positive charge. Cations in a liquid subjected to electric potential collect at the negative pole or cathode.

cation exchange. See ion exchange.

cationic reagent. One of several surface-active substances in which the active constituent is the positive ion. Used to flocculate and collect minerals that are not flocculated by oleic acid or soaps (in which the surface-active ingredient is the negative ion). Reagents used are chiefly quaternary ammonium compounds e.g., cetyl trimethyl ammonium bromide.

catlinite (pipestone). A fine-grained silicate mineral related to pyrophyllite which is easily compressible, has high surface friction, and is used for gaskets in very high-pressure equipment.

"Cato."[53] Trademark for a cationic derivative of starch, available in ungelatinized or gelatinized (cold water soluble) form. Used in manufacture of paper, warp sizing, etc.

"Cat-Ox."[58] Proprietary catalytic oxidaton process for removing sulfur dioxide and fly ash from stack gases and converting the former to sulfuric acid. It involves passing the flue gases at 900°F (1) through an electrostatic precipitator, (2) through a converter where sulfur dioxide is oxidized to the trioxide with vanadium pentoxide, (3) through heat exchangers to cool the gases so that the SO_3 combines with water vapor, and (4) through an absorbing tower which condenses the sulfuric acid. Commercial acid (70%) is made in this way. See also sulfuric acid.

caulking compound. See sealant.

caustic. (1) Unqualified, this term usually refers to caustic soda (sodium hydroxide). (2) As an adjective, it refers to any compound chemically similar to sodium hydroxide, e.g., caustic alcohol (C_2H_5ONa). (3) Any strongly alkaline material which has a corrosive or irritating effect on living tissue.

caustic baryta. See barium hydroxide.

causticized ash. Combinations of soda ash and caustic soda in definite proportions marketed for purposes where an alkali is needed ranging in causticity between the two materials. Causticized ash is usually designated by its caustic soda content and the range of standard marketed products embraces 7%, 10%, 15%, 25%, 36%, 45%, and 67% of caustic soda.

caustic lime. See calcium hydroxide.

caustic potash. See potassium hydroxide. Caustic potash, liquid, is the legal label name for potassium hydroxide solution.

caustic soda. Legal label name for sodium hydroxide solution.

"Cavalite."[28] Trademark for a series of fiber-reactive dyes for cellulose.

cavitation. Formation of air bubbles in a liquid such as salt water when subjected to intense vibration, resulting severe mechanical damage to the surfaces of metals exposed to it, e.g., ship propellers, steam condensers, pumps and piping systems. The erosive effect is due to the shock waves created by collapse of the bubbles. The pressures exerted by cavitation have been calculated to be in the range of 30,000 psi. This phenomenon plays a part in corrosion of metals as well as in emulsion formation. See also corrosion; homogenization.

Cb Symbol for columbium, an obsolete name for niobium.

CBM. Abbreviation for chlorobromomethane (see bromochloromethane); also for constant boiling mixture. See azeotropic mixture.

cc. Abbreviation for cubic centimeter. See also milliliter.

Cd Symbol for cadmium.

CDA. Abbreviation for completely denatured alcohol. See alcohol, denatured.

CDAA. See alpha-chloro-N,N-diallylacetamide.

"CDB-59."[55] Trademark for potassium dichloroisocyanurate (q.v.).

"CDB-60"[55] Trademark for sodium dichloroisocyanurate (q.v.).

"CDB-85."[55] Trademark for trichloroisocyanuric acid (q.v.).

CDEC. See 2-chloroallyl diethyldithiocarbamate.

CDP. (1) Abbreviation for cytidine diphosphate. See cytidine phosphates. (2) Abbreviation for cresyl diphenyl phosphate.

"CDP-GA."[1] Trademark for a phosphorus-containing deposit modifier for automotive gasolines.

CDTA. See trans-1,2-diaminocyclohexanetetraacetic acid.

Ce Symbol for cerium.

"Cebicure."[123] Trademark for ascorbic acid for meat curing.

"Cebitate."[123] Trademark for sodium ascorbate for meat curing.

cedar leaf oil. True cedar leaf oil is distilled from the leaves of Juniperus virginiana. The name has also been used as a synonym for thuja oil (q.v.) (from Thuja occidentalis). The properties of the two oils are quite different.
Properties: Colorless liquid; sp. gr. 0.870–0.890; optical rotation +55° to 65° (20°C); soluble in alcohol and ether. Combustible; nontoxic.
Chief constituents: Limonene, cadinene, borneol, and bornyl esters.
Containers: Drums; cans.
Uses: Medicine; microscopy; perfume; flavoring.

cedrol $C_{15}H_{26}O$, a tertiary terpene alcohol.
Properties: Colorless crystals; cedarwood odor. M.p. 86°C. Soluble in 11 parts of 95% alcohol. Combustible.
Uses: Perfumery, for woody and spicy notes; odorant for disinfectants.

cedryl acetate $CH_3COOC_{15}H_{25}$.
Properties: Colorless liquid, having a light cedar odor. Sp. gr. 0.975–0.995; refractive index 1.496–1.510. Soluble in one volume of 90% alcohol. Combustible.
Use: Perfumery.

"Celanar."[352] Trademark for a polyester film made from polyethylene terephthalate, (q.v.).
Properties: Transparent; biaxially oriented; density 1.395; tensile strength 30,000 psi; dielectric strength 7000 volts/mil; melting point 260°C; service temperature -60 to 150°C; outstanding dimensional stability and chemical resistance.
Uses: Magnetic recording tape; drafting and engineering reproduction materials; metallic yarn; roll leaf; pressure-sensitive tapes; packaging; dielectric material in capacitors, wire and cable, motors, generators, transformers, and oils.

"Celanese CL."[352] Trademark for a series of polyvinyl acetate emulsions. Available as:
102: Fine particle size, water resistant homopolymer emulsion.
202: Fine particle size, water resistant copolymer emulsion.
203: Vinyl-acrylic copolymer emulsion.
204: Vinyl copolymer emulsion.
Uses: Paints, adhesives and paper-coating specialties.

"Celanese Solvent."[352] Trademark for a series of special solvents. Available as:
203: Replacement for normal butyl alcohol in nitrocellulose lacquers, alkyd resin formulations and thinners; distillation range 115–120°C; flash point 100°F (open cup).
601: Replacement for methyl ethyl ketone in vinyl and nitrocellulose applications; distillation range 74–84°C; flash point 10°F (open cup).
Hazard: Flammable, dangerous fire risk.
901H: Replacement for butanol and methylisobutyl carbinol in lacquers and brake fluids; distillation range 125–155°C; flash point 120°F (open cup).

"Celanthrene."[28] Trademark for a group of anthraquinone disperse dyes designed especially for acetate; also suitable for application to nylon.

"Celatom."[468] Trademark for a group of diatomaceous silicas (diatomite) of high quality and uniformity.
Uses: Filter aid, including foods and beverages; absorbents, as in insecticides and fertilizers; catalyst supports; fillers for paper, paints, explosives, concrete and asphalt; chromatography.

"Cel-Bak."[515] Trademark for latex foam used in carpeting; high-density and regular grades.

"Celcon."[352] Trademark for a highly crystalline acetal copolymer based on trioxane. See acetal resin.

celestine blue. A dye used for staining biological specimens.

celestite $SrSO_4$. Natural strontium sulfate, usually found in sedimentary rocks.

Properties: Colorless, white, pale blue, or red; luster vitreous to pearly. Resembles barite (q.v.). Sp. gr. 3.95; Mohs hardness 3–3.5.
Occurrence: United States; Canada; Europe, Mexico.
Uses: Strontium chemicals; oil-well drilling mud; sugar refining; ceramics; production of very pure caustic soda.

"Celestone."[321] Trademark for a brand of betamethasone.

"Celite."[247] Trademark for diatomaceous earth and related products.
Properties: Color, white to pale brownish white, depending on grade and processing, calcined grades being pink to buff; sp. gr. 0.24–0.34; porous, capable of absorbing 300–400% water by weight; poor conductor of heat, sound, and electricity; resistant to acids except hydrofluoric; slowly soluble in hot alkali.
Grades: Filtration; mineral filler.
Containers: Drums; multiwall paper bags; bulk.
Uses: Filtering; fillers; absorbent; abrasive in glass and metal polishing; catalyst carrier; ingredient in cements, flameproofing agents, and other products.

"Celkate."[247] Trademark for finely divided hydrated synthetic magnesium silicates having high absorption properties; light tan in color; density 10–18 lb/cu ft; surface area 150–250 sq m/gram.
Grades: Available in various grades of purifying of petroleum-base solvents, chemical and drug solutions, and for decolorizing of animal, fish, and vegetable oils.
Uses: Filter agent to remove solids and free fatty acids; absorptive carrier of liquids; conditioning agent to improve flow properties of dry powders.

cell. (1) The fundamental unit of biological structure, comprised of (a) an outer membrane, or wall, about 100 A thick which, being semipermeable, maintains by osmosis (q.v.) the biochemical equilibrium of the intracellular fluids; (b) the cytoplasm (q.v.) containing mitrochondria, ribosomes, and other structures; and (c) the nucleus (q.v.), in which lie the chromosomes and genes. An extremely complex biochemical organization, the cell is the dynamic unit of all life; its ability to reproduce itself and to control its functions systematically is of basic importance to maintenance of life and growth. All organic matter is either found in cells or is produced by cellular activity, and was ultimately derived from photosynthesis (q.v.). Cells are comparatively large units which can be resolved in an optical microscope; the largest single cells are represented by the eggs of oviparous animals. See also mitosis.

(2) Any self-contained unit having a specific functional purpose, as follows: (a) voltaic cell (battery) to generate electric current; (b) electrolytic cell to effect electrolysis; (c) fuel cell to convert chemicals into electricity; and (d) solar cell to capture heat from sunlight. All except the last involve use of electrodes and electrolytes. See also specific entry.

(3) Any completely enclosed hollow unit, as in a honeycomb or cellular plastic.

cell, dry. See dry cell.

cell, electrolytic. See electrolytic cell.

"Cellex."[236] Trademark for sodium carboxymethylcellulose. Used as an organic colloidal compound for providing low filtrate loss in drilling mud.

cell, fuel. See fuel cell.

"Cellidor."[470] Trademark for cellulose acetate and acetate butyrate plastics.

"Cellitazol."[307] Trademark for a series of developed acetate dyestuffs.

"Celliton."[307] Trademark for a series of disperse dyestuffs characterized by good fastness to light, washing, etc. Used for dyeing and printing acetate fibers.

cellobiose $C_{12}H_{22}O_{11}$. The product of the partial hydrolysis of cellulose; composed of two D-glucose molecules.
Properties: Colorless crystals; m.p. 225°C (dec); soluble in water; slightly soluble in alcohol; nearly insoluble in ether; insoluble in acetone.
Use: Bacteriology.

"Cello-Flex."[116] Trademark for a special kraft paper electrical insulation used mainly for transformers.

celloidin. A pure form of pyroxylin, used for imbedding sections in microscopy.

"Cellolyn."[266] Trademark for a series of pale, low-melting hydroabietyl ester resins. Used for lacquers, inks and adhesives.

cellophane (regenerated cellulose). Film produced from wood pulp by the viscose process.
Properties: Transparent, strong, flexible, and highly resistant to grease, oil, and air. The base cellulose film is modified by softeners, flame-resisting materials and dyes; also by coating with other materials. On exposure to heat the untreated film loses strength at 300°F, decomposes 350°–400°F; does not melt; burns readily and is not self-extinguishing.
Hazard: Flammable, moderate fire risk.
Uses: Wrapper or protective package for fabricated articles and industrial applications.
See also rayon, viscose.

"Cellosize."[214] Trademark for hydroxyethyl cellulose (q.v.) and carboxymethylcellulose (q.v.). See also cellulose, modified.

"Cellosolve."[214] Trademark for mono- and dialkyl ethers of ethylene glycol and their derivatives, widely used as industrial solvents.
butyl "Cellosolve"
See ethylene glycol monobutyl ether.
butyl "Cellosolve" acetate
See ethylene glycol monobutyl ether acetate.
"Cellosolve" acetate
See ethylene glycol monoethyl ether acetate.
"Cellosolve" solvent
See ethylene glycol monoethyl ether.
dibutyl "Cellosolve"
See ethylene glycol dibutyl ether.
n-hexyl "Cellosolve"
See ethylene glycol monohexyl ether.
methyl "Cellosolve"
See ethylene glycol monomethyl ether.
methyl "Cellosolve" acetate
See ethylene glycol monomethyl ether acetate.
phenyl "Cellosolve"
See ethylene glycol monophenyl ether.

cell, photovoltaic. See solar cell.

cell, solar. See solar cell.

"Cellu-Aid."[218] Trademark for diatomite used as extender for paper.

"Celluflex" 112.[1] Trademark for cresyl diphenyl phosphate (q.v.). Used for flame-resistance in vinyl formulations.

"Celluflex" 179.[1] Trademark for tricresyl phosphate (q.v.).

"Celluflux" TPP.[1] Trademark for triphenyl phosphate. (q.v.). Used as plasticizer in cellulose acetate and phenolic plastics.

"Celluguard."[1] Trademark for a water glycol fire-resistant hydraulic fluid.

cellular plastic. A thermosetting or thermoplastic foam composed of cellular cores with integral skins having high strength and stiffness. The cells result from the action of a blowing agent, either at room temperature or during heat treatment of the plastic mixture. The resulting product may be either flexible or rigid, the latter being machinable. The foaming action in some cases may occur *in situ* (foamed-in-place plastics). Cellular plastics are combustible. For details see foam, plastic.

Uses (flexible): Furniture, automobile interiors, mattresses, etc. where softness and resiliency are desired. (Rigid) Insulating material; boat building and similar light construction; salvage of water-logged ships.

See also foam, plastic; rubber sponge. For further information refer to Cellular Plastics Division, Society of the Plastics Industry, 250 Park Ave., New York.

cellulase. An enzyme complex produced by the fungi Aspergillus niger and Trichoderma viride which is capable of decomposing cellulosic polysaccharides into smaller fragments, primarily glucose. It has been used as a digestive aid in medicine and in the brewing industry. Research has been devoted to experimental applicatoin of cellulase to disposal of cellulosic solid wastes. The resulting glucose can be fermented to ethyl alcohol, used to grow yeast for animal feed proteins, or as a chemical feedstock.

"Celluloid."[352] Trademark for a plastic consisting essentially of a solid solution of cellulose nitrate and camphor or other plasticizer with or without the presence of pigments and coloring matter. Available in sheets, rods, tubes, films. Also called pyroxylin.

Hazard: Flammable, dangerous fire risk.

Shipping regulations: (Rail) Yellow label. (Air) Flammable Solid label. Scrap: Not acceptable. Consult regulations for details.

See also nitrocellulose; Hyatt.

cellulose $(C_6H_{10}O_5)_n$. A natural carbohydrate high polymer (polysaccharide) consisting of anhydroglucose units joined by an oxygen linkage to form long molecular chains that are essentially linear. It can be hydrolyzed to glucose. The degree of polymerization (q.v.) is from 1000 for wood pulp to 3500 for cotton fiber, giving a molecular weight from 160,000 to 560,000.

CH₂OH H OH CH₂OH H OH
H O O H H O O H
H OH H H OH H
OH H H O O OH H H
HO H H O H H O OH
H OH CH₂OH H OH CH₂OH
X

Cellulose is a colorless solid, specific gravity about 1.50, insoluble in water and organic solvents; it will swell in sodium hydroxide solution and is soluble in Schweitzer's reagent (q.v.). It is the fundamental constituent of all vegetable tissues (wood, grass, flowers) and is thus the most abundant organic material in the world. Cotton fibers are almost pure cellulose; wood contains about 50%.

The physical structure of cellulose is unusual in that it is not a single crystal, but consists of crystalline areas embedded in amorphous areas. Chemical reagents penetrate the latter more easily than the former. Cellulose is virtually odorless and tasteless, but it is combustible, with an ignition point of about 450°F; in some forms it is flammable. For example, railroad shipping regulations require a yellow label on such odd items as "burnt fiber," "burnt cotton," "wet waste paper" and "wet textiles." Fires have been known to occur in warehouses in which telephone books were stored. These were undoubtedly due to heat build-up in the paper caused by microbial activity and self-sustaining oxidation. See also flammable material.

The most important uses of cellulose as such are in the form of bulk woods of many kinds; paper, most of which is made from wood pulp; cotton products (clothing, sheeting, industrial fabrics); and packaging, ranging from wooden barrels to candy pats. Specialized uses include nonwoven fabrics, medical equipment (artificial kidney), insulation and soundproofing, sausage casings, etc.

There are manay chemical modifications of cellulose, including its esters (cellulose acetate), ethers (methylcellulose), the nitrated product (nitrocellulose), as well as rayon and cellophane (from cellulose xanthate). Thus it is the basis of many plastics, fibers, coatings, lacquers, explosives, and emulsion stabilizers. Alkali cellulose is an intermediate made by the action of sodium hydroxide solution on cellulose and is used for making cellulose ethers and viscose. See also cellulose, modified.

Cellulose exists in three forms—alpha, beta, and gamma. Alpha-cellulose has the highest degree of polymerization and is the chief constituent of paper pulp. The beta and gamma forms have much lower D.P. and are known as hemicellulose (q.v.). Methods of determining the alpha content of pulps are detailed in TAPPI Method T203 and ASTM D-588-42. See also pulp, paper.

Cellulose has been prepared in microcrystalline form. See "Avicel." Cellulose can be decomposed to glucose by the enzyme cellulase (q.v.).

Hazard: Cellulosic materials (paper, cotton and textile wastes) when wet with water are a dangerous fire hazard. Yellow label is required for rail freight; not accepted by express.

See also rayon; cellophane; nitrocellulose; carboxymethylcellulose; and following entries.

cellulose acetate (CA). A cellulose ester in which the cellulose is not completely esterified by acetic acid.

Properties: White flakes or powder. A thermoplastic resin, softening about 60–97°C and melting about 260°C. Sp. gr. 1.27–1.34; soluble in acetone, ethyl acetate, cyclohexanol, nitropropane, ethylene dichloride. Notable for toughness, high impact strength, and ease of fabrication. Subject to dimensional change due to cold flow, heat, or moisture absorption (1–7%). Fibers weaken above 80°C and are difficult to dye uniformly; not attacked by microorganisms. Nontoxic.

Derivation: Reacting cellulose (wood pulp or cotton linters) with acetic acid or acetic anhydride, with sulfuric acid catalyst. The cellulose is fully acetylated (three acetate groups per glucose unit) and at the same time the sulfuric acid causes appreciable degradation of the cellulose polymer so that the product contains only 200–300 glucose units per polymer chain. At this point in the process the cellulose acetate ordinarily is partially hydrolyzed by the addition of water until an average of 2–2.5 acetate groups per glucose unit remain. This product is thermoplastic and soluble in acetone. Fibers are produced by forcing an acetone solution through orifices of the spinneret into a stream of warm air, which evaporates the solvent. Fibers are also produced in a similar manner from cellulose triacetate, which is insoluble in acetone but soluble in methylene chloride. See acetate fiber; acetate film.

Forms: Sheet, film, or fiber; molded items.

Grades: Filtered and unfiltered. Also graded by percent combined acetic acid content: plastic 52–54%; lacquer 54–56%; film 55.5–56.6%; water-resisting 56.5–59%; triacetate 60.0–62.5%.

Containers: Cartons, drums, multiwall paper sacks.

Hazard: Flammable; not self-extinguishing. Moderate fire risk.

Uses: Acetate fiber, lacquers, protective coating solutions, photographic film, transparent sheeting, thermoplastic molding composition, cigarette filters, magnetic tapes, osmotic cell membrane.

cellulose acetate butyrate (CAB; cellulose acetobutyrate).

Properties: White pellets or granules. A thermoplastic resin. Sp. gr. 1.2. Soluble in ketones, organic acetates, lactates, methylene, ethylene, and propylene chlorides and high-boiling solvents. Exellent weathering properties, low water absorption, low heat conductivity and high dielectric strength; high resistance to oil and grease. Combustible. Nontoxic.

Derivation: Reaction of purified cellulose with acetic and butyric anhydrides with sulfuric acid as catalyst and glacial acetic acid as solvent. The ratio of acetic and butyric components may be varied over a wide range.

Grades: According to butyryl content (17, 27, 38, 50%).

Containers: Fiber cartons and drums.

Uses: Molding compositions; film and sheet; photographic film, lacquers, protective coating solutions; tail-light lenses; water desalination membrane; piping and tubing; covering for aluminum fibers; toys; packaging; brush handles; hydrometers; miscellaneous consumer products.

cellulose acetate phthalate. A reaction product of phthalic anhydride and cellulose acetate, used for coating of tablets and capsules.

cellulose acetate propionate (cellulose propionate). Similar to cellulose acetate butyrate but made with propionic anhydride instead of butyric anhydride. Unusually stable; requires less plasticizer and is compatible with more plasticizers than the butyrate.

cellulose acetobutyrate. See cellulose acetate butyrate.

cellulose ether. See ethylcellulose; methylcellulose.

cellulose gum. A purified grade of carboxymethylcellulose (q.v.).

cellulose, hydrated (hydrocellulose). Cellulose that has been caused to react with water (about 8–12%), forming a gelatinous mass. Combustible.

Derivation: By mechanical pulverization and agitation with water, by the action of strong salt solutions, alkalies, or acids.

Use: In the manufacture of paper, vulcanized fiber, mercerized cotton, viscose rayon.

See also hydration.

cellulose methyl ether. See methylcellulose.

cellulose, modified. One of many derivatives of cellulose, formed by substitution of appropriate radicals (carboxyl, alkyl, acetate, nitrate, etc.) for hydroxyl groups along the carbon chain. Such reactions are usually not stoichiometric. Some of these products (carboxymethyl and hydroxyalkyl cellulose) are water-soluble ethers; others are organosoluble esters (cellulose acetate), nitrates (nitrocellulose), or xanthates (viscose). See also modification.

Biodegraded cellulose has been used as a microbial growth medium for protein formation; a tobacco substitute based on a cellulose modification has been developed. See "Cytrel".

cellulose nitrate. See nitrocellulose.

cellulose, oxidized (cellulosic acid). Derivative of cotton cellulose produced by treatment with nitrogen dioxide. Is soluble in alkali but may be made to retain original form of the cellulose and much of its tensile strength.

Properties: Slightly off-white gauze, lint, or powder. Slight charred odor; acid taste; soluble in aqueous organic bases, in dilute alkali, and in ammonium hydroxide, forming salts and esters. It is insoluble in water, acids, and common organic solvents. It slowly degrades at room temperatures and should be kept cool. Combustible.

Grades: U.S.P. technical.

Containers: Glass bottles; fiber cans.

Uses: Surgery and medicine; ion-exchange medium; thickening agent.

cellulose propionate. See cellulose acetate propionate. See also "Forticel."

cellulose, regenerated. See cellophane; rayon.

cellulose sponge. A sponge of regenerated cellulose, highly absorbent, soft and resilient when wet, long-lasting. It will not scratch, can be sterilized by boiling and is not affected by ordinary cleaning compounds. The pores vary in size form coarse pore (the size of a pea) to fine pore (the size of a pinhead). Yarn made of cellulose sponge consists of cotton fiber covered with the sponge product. Combustible.

Uses: Washing automobiles and trucks, walls and painted surfaces, windows, etc.; general cleaning; sponge used in photographic laboratories; wet mops, cleaning pads, etc.

cellulose triacetate. A cellulose resin in which the cellulose is completely esterified by acetic acid.

Properties: White flakes; sp. gr. 1.2; soluble in chloroform, methylene chloride, tetrachloroethane. Combustible; nontoxic.

Derivation: Reaction of purified cellulose with acetic anhydride in the presence of sulfuric acid as catalyst and glacial acetic acid as solvent, followed by very slight hydrolysis.

Grade: Flake.

Containers: Fiber cartons or drums.

Use: Protective coatings resistant to most solvents; textile fibers; base for magnetic tape.

cellulose xanthate. See viscose process.

cellulosic plastic. One of a number of semisynthetic polymers based on cellulose. See cellophane; cellulose acetate; cellulose, modified; nitrocellulose; rayon, viscose; carboxymethylcellulose.

Note: Development and use of cellulosics may be expected to increase due to the shortage of petroleum.

cellulosic thiocarbonate. A reactive intermediate in the graft polymerization of certain synthetic polymers to cellulosic fibers. The latter are treated with sodium hydroxide and a sulfur-containing compound, and the cellulose thus activated is placed in an emulsion or solution of monomers. Polymerization at 50°C occurs with a peroxide catalyst to form the graft. See also graft polymer.

"Cellulube."[1] Trademark for a series of functional fluids (phosphate esters) combining fire-resistance and lubricating qualities. Available in controlled viscosities for industrial hydraulic and lubricant applications.

"Celluphos 4."[1] Trademark for tributyl phosphate.
Uses: Plasticizer for nitrile elastomers; foam depressant for emulsion systems; solvent, extractant and complexing agent for inorganics.

"Cellutherm."[1] Trademark for a series of synthetic lubricants based on trimethylolpropane esters.

cell, voltaic. See voltaic cell.

"Celogen."[248] Trademark for a series of blowing agents.
AZ. Azodicarbonamide.
OT. Para, para-Oxybis-(benzenesulfonylhydrazide).
RA. para-Toluene sulfonyl semicarbazide.

"Celoron."[281] Trademark for macerated canvas or paper-based industrial laminated or molded plastics.
Properties: Mottled brown or black; sp. gr. 1.35; high impact strength; unaffected by rapid temperature changes; resistant to heat, oil, water, and many chemicals; may be used continuously at 225–250°F. Combustible.
Forms: Sheets; cut pieces; blanks; rings; molded parts.
Uses: Timing gears for automobile industry; electrical insulation; structural parts.

"Celotex."[351] Trademark for structural building and insulation board produced in large sheets. Made from bagasse (q.v.) or wood fiber and treated to be resistant to fungi, termites, and water penetration. The name also includes roofing products, gypsum, wallboard, lath, plasters, mineral wool, and hard board.

Celsius, A. (1701–1744). A Swedish physicist who proposed the use of the centigrade temperature scale. His name is now generally applied to this scale (degrees centigrade = degrees Celsius). See centigrade.

cement, aluminous (high alumina cement). A hydraulic cement which contains at least 30 to 35% alumina (in contrast to Portland cement, which contains less than 5%). The alumina is usually supplied by inclusion of bauxite. Aluminous cement attains its maximum strength more rapidly than Portland cement. It is also more resistant to solutions of sulfates. It exists in two modifications, sintered and fused.

cementation. A process in which steel or iron objects are coated with another metal by immersing them in a powder of the second metal and heating to a temperature below the melting point of any of the metals concerned. Zinc, chromium, aluminum, copper and other metals are applied to iron or steel in this fashion. The process is basically diffusion of one metal into the other, so that intermetallic alloy layers are formed at the interface of the basis and coating metals. See also sherardizing.

cemented carbide. See carbide, cemented.

cement, hydraulic. Any mixture of fine-ground lime, alumina, and silica that will set to a hard product by admixture of water, which combines chemically with the other ingredients to form a hydrate. See also cement, aluminous and cement, Portland.

cementite Fe_3C. A carbide of iron formed in the manufacture of pig iron and steel. Composed of 93.33% iron and 6.67% carbon, it is very hard and brittle and will scratch glass and feldspar, but not quartz. It is about two-thirds as magnetic as pure iron under an exciting current. It occurs in ordinary steels of more than 0.85% carbon and takes its name from cement steel, made by the cementation process, which contains a great deal of this carbide. See also carbide.

cement, organic. Any of various types of rubber cement, silicone adhesives, deKhotinsky cement. See adhesive, rubber-based; silicone; deKhotinsky cement.

cement, Portland. A type of hydraulic cement in the form of finely divided gray powder composed of lime, alumina, silica and iron oxide as tetracalcium aluminoferrate ($4CaO \cdot Al_2O_3 \cdot Fe_2O_3$), tricalcium aluminate ($3CaO \cdot Al_2O_3$), tricalcium silicate ($3CaO \cdot SiO_2$), and dicalcium silicate ($2CaO \cdot SiO_2$). These are abbreviated respectively as C_4AF, C_3A, C_3S and C_2S. Small amounts of magnesia, sodium, potassium, and sulfur are also present. Hardening does not require air, and will occur under water. Sand is often added as a diluent. Cement may be modified with various plastic latices in proportions up to 0.2 part latex solids to 1 part cement to improve adhesion, strength, flexibility, and curing properties. Water evaporation can be retarded by adding such resins as methylcellulose and hydroxyethyl cellulose.

For further information refer to the Portland Cement Association, Chicago, Illinois.

cement, rubber. See adhesive, rubber-based.

"CE Methyl Esters."[487] Trademark for a series of methyl esters of straight chain (normal) even numbered fatty acids ranging from C_8 (octanoate) to C_{18} (octadecanoate) and including mixtures of these.
Properties: Colorless to light yellow liquids or white solids. Sp. gr. about 0.87.
Containers: Up to tank cars and trucks.
Uses: Chemical intermediates; lubricants; cosmetic ingredients; formulating aids (rubber, wax, etc.).

"Centifoliol."[188] Trademark for a replacement for otto of rose and rose absolute.
Uses: Perfume and cosmetic compositions.

centigrade. The internationally used scale for measuring temperature, in which 100° is the boiling point of water at sea level (1 atmosphere), and 0° is the freezing point. A temperature given in centigrade degrees may be converted to the corresponding Fahrenheit temperature by multiplying it by 9/5 (or 1.8), and adding 32 to the product. A temperature given in Fahrenheit degrees is converted to the corresponding centigrade temperature by subtracting 32, and multiplying the remainder by 5/9. The centigrade scale was devised by the Swedish scientist Celsius, and his name is now increasingly used in reference to it.

centigrade heat unit. See chu.

centipoise (cp). One one-hundredth of a poise. The poise is the metric system unit of viscosity, and has the dimensions of dyne-second per square centimeter or grams per centimeter-second.

centistoke (cs). One one-hundredth stoke, the kinematic unit of viscosity; it is equal to the viscosity in poises divided by the density of the fluid in grams per cubic centimeter, both measured at the same temperature.

centrifugation. A separation technique based on the application of centrifugal force to a mixture or suspension of materials of closely similar densities. The smaller the difference in density, the greater is the force required. The equipment used (centrifuge) is a chamber revolving at high speed to impart a force up to 17,000 times that of gravity; the materials of higher density are thrown toward the outer portion of the chamber, while those of lower density are concentrated at or near the inner portion. This technique is used effectively in a number of biological and industrial operations, such as separation of the components of blood, concentration of rubber latex, and of fat particles from other milk components. Separation of isotopes, especially those of uranium, by this method is now practicable for producing enriched uranium; this method is economically superior to the gaseous diffusion process.

"Century."[189] Trademark for a series of stearic acids, oleic acids, special fatty acids and hydrogenated fatty acids.
Uses: Cosmetics, pharmaceuticals, textile finishes, wetting agents, carbon paper, cutting oils, plastics, esters, polishes, rubber compounding, soaps and lubricating greases.

"Cenwax."[189] Trademark for a series of hydrogenated castor oil products. Used as lubricants, coatings for leather, paper and textiles, candles, crayons, sealing compounds, polishes and wax extenders.

cephalin (kephalin; phosphatidyl ethanolamine; phosphatidyl serine)
$CH_2OR_1CHOR_2CH_2OP(O)(OH)OR_3$. A group of phospholipids in which two fatty acids (R_1 and R_2) form ester linkages with the two hydroxyl groups of glycerophosphoric acid (q.v.) and either ethanolamine or serine (R_3) forms an ester linkage with the phosphate group. Cephalins are, therefore, either phosphatidyl ethanolamine or phosphatidyl serine. They are associated with lecithins found in brain tissue, nerve tissue, and egg yolk.
Properties: Yellowish, amorphous substance; characteristic odor and taste; insoluble in water and acetone; soluble in chloroform and ether; slightly soluble in alcohol.
Uses: Medicine; biochemical research.

cephalosporin. Any of a family of antibiotics related to penicillin discovered in 1953; an important member of this group was synthesized by Woodward in 1966. Several cephalosporins are used clinically (cephalothin, cephaloridine, and cephalexin). The molecule contains a fused beta-lactam-dihydrothiazine ring system with an N-acyl side chain and an acetoxy group attached to the dihydrothiazine ring. Cephalosporins are reported to be free from the allergic reactions common with penicillin. Development of new cephalosporin derivatives is being actively pursued by pharmaceutical research chemists. See also penicillin; antibiotic.

ceramic. A product manufactured by the action of heat on earthy raw materials, in which the element silicon with its oxide and complex compounds known as silicates occupy a predominant position (American Ceramic Society). The chief major groups of the ceramics industry are as follows: (1) structural clay products (brick, tile, terra cotta, glazed architectural brick); (2) whitewares (dinnerware, chemical and electrical porcelain, e.g., spark plugs, sanitary ware, floor tile); (3) glass products of all types; (4) porcelain enamels; (5) refractories (materials that withstand high temperatures); (6) Portland cement, lime, plaster and gypsum products; (7) abrasive materials such as fused alumina, silicon carbide and related products; (8) aluminum silicate fibers. A wide range of ceramics is now available as ultra-fine particles (10–150 microns); and ceramic foams are offered commercially. See also specific entries. For further information refer to the American Ceramic Society, 4055 North High St., Columbus, Ohio.

ceramic, ferroelectric. A unique type of polycrystalline ceramic having properties that make possible the production of reliable, high-density optical memories for computers that are more efficient than conventional types. Lead zirconate titanate (q.v.), heated and pressed into thin plates, is one of the compounds used. As a result of its ferroelectric properties, an applied voltage aligns the electric charges in the molecules of ceramic in the direction of the field, and the polarization so induced remains indefinitely, until changed. Thus the material accommodates itself to the requirements of the digital system, namely, binary 0 and binary 1. See also ferroelectric.

ceramic, glass. See glass ceramic.

"Ceramix."[177] Trademark for a technical grade of barium carbonate used in ceramic industry.
Hazard: May be toxic by ingestion.

"Ceramol."[400] Trademark for a blend of cetyl and stearyl alcohols and higher alcohol sulfates. Melts from 50–60°C; acid number 1.0 max; saponification value 3 max; iodine number 5 max; acetyl value 185–195. Not alkaline to phenolphthalein. Combustible.
Uses: Emulsifier for cosmetic creams, ointments and lipsticks.

"Ceramvar."[155] Trademark for an iron, nickel, cobalt alloy designed for ceramic-to-metal sealing.

"Ceraphyl."[10] Trademark for a series of lactate emollients designed for use in cosmetics and pharmaceutical applications.

"Ceratak."[128] Trademark for a grade of petroleum microcrystalline wax. Minimum m.p. 165°F.

"Cerathane" 63-L.[128] Trademark for an emulsifiable microcrystalline wax; minimum m.p. 200°F.

"Ceraweld."[128] Trademark for a grade of petroleum microcrystalline wax. Minimum m.p. 165°F.

"Cercor."[20] Trademark for thin-walled, cellular ceramic structures which can be used for a wide range of high temperature applications.
Uses: Gaseous heat exchangers; burner plates; acoustics; flame arresting; filtering and insulation.

cerebrosides. Derivatives of sphingosine (q.v.) in which the amino group is connected in an amide linkage to a fatty acid and the terminal hydroxyl group is connected to a molecule of sugar usually galactose, in glycosidic linkage. They are found in brain and nervous tissue, usually in association with sphingomyelin (q.v.).

"Cerelose."[30] Trademark for a white, crystallized, refined, dextrose (pure monosaccharide); 100% fermentable. Available in hydrate, anhydrous and liquid forms.
Containers: Up to tank cars and trucks.
Uses: Adhesives; chemicals; drugs and pharmaceuticals; foundry processes; and plastics.

"Ceresan."[28] Trademark for a series of mercury compounds used as seed disinfectants.
Hazard: Highly toxic.

ceresin wax (purified ozocerite; earth wax; mineral wax; cerosin; cerin).
Properties: White or yellow waxy cake; white is odorless; yellow has a slight odor. Sp. gr. 0.92–0.94; m.p. 68–72°C. Soluble in alcohol, benzene, chloroform, naphtha; insoluble in water. Combustible. Nontoxic.
Derivation: Purification of ozocerite by treatment with concentrated sulfuric acid and filtration through animal charcoal.
Grades: White; yellow.
Containers: Bags; cartons.
Uses: Candles; sizing; bottles for hydrofluoric acid; electrical insulation; shoe and leather polishes; impregnating and preserving agent; lubricating compounds; wood filler; floor polishes; antifouling paints; waxed papers; cosmetics; ointments; matrix compositions; waterproofing textile fabrics.

ceria. See ceric oxide; rare earth.

ceric ammonium nitrate (cerium-ammonium nitrate; ammonium hexanitratocerate) $Ce(NO_3)_4 \cdot 2NH_4NO_3$.
Properties: Small prismatic, yellow crystals. Soluble in water and alcohol; almost insoluble in concentrated nitric acid; soluble in other concentrated acids.
Derivation: By electrolytic oxidation of cerous nitrate in nitric acid solution, and subsequently mixing solutions of cerium nitrate and ammonium nitrate, followed by crystallization.
Hazard: Strong oxidizing agent. Dangerous fire risk in contact with organic materials.
Uses: Analytical chemistry; oxidant for organic compounds; scavenger in the manufacture of azides.
Shipping regulations: (Rail) Nitrates, n.o.s., Yellow label. (Air) Oxidizer label.

ceric hydroxide (ceric oxide, hydrated; cerium hydrate) $CeO_2 \cdot xH_2O$.
Properties: Whitish powder when pure; a hydrated oxide containing 85–90% ceric oxide. Soluble in concentrated mineral acids; insoluble in water.
Derivation: By treating a solution of a ceric salt with strong alkali. Reagent grade is prepared by adding a saturated solution of ceric ammonium nitrate to an excess of ammonium hydroxide.
Grades: Commercial; high purity; reagent.
Containers: Cans; bottles; fiber drums.
Uses: Production of cerium salts and ceric oxide; opacifier in glasses and enamels (imparts yellow color); shielding glass.

ceric oxide (cerium dioxide; cerium oxide; ceria) CeO_2.
Properties: Pale yellow, heavy powder (white when pure). Commercial product is brown. Sp. gr. 7.65; m.p. 2600°C. Soluble in sulfuric acid; insoluble in water and dilute acid; requires reducing agent with acid to dissolve the anhydrous oxide. Noncombustible.
Derivation: By decomposing cerium oxalate by heat. Hardness depends on firing temperature.
Grades: Technical; high purity (99.8%).
Containers: Fiber drums; car lots.
Uses: Ceramics; abrasive for glass polishing; opacifier in photochromic glasses; retarder of discoloration in glass, especially radiation shielding and color TV tubes; catalyst; enamels and ceramic coatings; phosphors; cathodes; capacitors; semiconductors; refractory oxides; diluent in nuclear fuels.

ceric sulfate (cerium sulfate) $Ce(SO_4)_2 \cdot 4H_2O$.
Properties: White or reddish-yellow crystals; sp. gr. 3.91; soluble in water (decomposes); soluble in dilute sulfuric acid. Strong oxidizing agent.
Derivation: Action of sulfuric acid on cerium carbonate.
Hazard: Fire risk in contact with organic materials.
Uses: Dyeing and printing textiles; analytical reagent; waterproofing; mildewproofing.

ceric sulfide (cerium sulfide). A high-temperature thermoelectric material that is stable and efficient up to 1100°C.

cerin. See ceresin wax.

cerite. A rare-earth ore found chiefly in Sweden. A minor source of cerium.

cerium Ce A rare-earth element of the lanthanide group of the Periodic Table. Atomic number 58; atomic weight 140.12; valences 3,4. Four stable isotopes.
Properties: Gray, ductile, highly reactive metal; sp. gr. 6.78; m.p. 795°C; b.p. 3257°C. Attacked by dilute and concentrated mineral acids and by alkalies. Readily oxidizes in moist air at room temperature. It has four allotropic forms. It is the second most reactive rare-earth metal. Cerium forms alloys with other lanthanides (see misch metal); it also forms a nonmetal with hydrogen, as well as carbides and intermetallic compounds. Decomposes water; nontoxic.
Ores: Cerite (Sweden), bastnasite (California, New Mexico), monazite (beach sands in Florida, Brazil, India; South Africa).
Derivation: Chemical processing and separation of ores.
Grades: Granules; ingots; rods (99.9% pure).
Hazard: May ignite on heating to 300°F. Strong oxidizing agent.
Uses: Cerium salts; cerium-iron pyrophoric alloys; ignition devices; military signalling; illuminant in photography; reducing agent (scavenger); catalyst; alloys for jet engines; solid state devices; rocket propellants; getter in vacuum tubes; diluent in plutonium nuclear fuels. See also misch metal.
Note: Cerium compounds have been found to have antiknock properties, e.g., cerium (2,2,6,6-tetramethyl-3,5-heptanedionate)$_4$, or Ce(thd)$_4$.

cerium 141. Radioactive cerium of mass number 141.
Properties: Half-life 32.5 days; radiation, beta and gamma.
Derivation: From cerium-140 by capture of a neutron and emission of a gamma photon.
Form available: $CeCl_3$ in hydrochloric acid solution.
Hazard: Radioactive poison.
Uses: Biological and medical research.
Shipping regulations: (Rail, Air) Consult regulations.

cerium-ammonium nitrate. See ceric ammonium nitrate.

cerium carbonate. See cerous carbonate.

cerium chloride. See cerous chloride.

cerium dioxide. See ceric oxide.

cerium hydrate. See ceric hydroxide or cerous hydroxide.

cerium naphthenate.
Properties: A rubbery material, very difficult to dry. Almost insoluble without small quantities of organic stabilizers.

Derivation: By saponifying naphthenic acids and treating the sodium naphthenate formed with a suitable cerium salt. The commercial product is a mixture of rare earth soaps.
Uses: See soap (2).

cerium nitrate. See cerous nitrate.

cerium oxalate. See cerous oxalate.

cerium oxide. See ceric oxide.

cerium sulfate. See ceric sulfate; cerous sulfate.

cermet (**cer**amic + **met**al). A semisynthetic product consisting of a mixture of ceramic and metallic components having physical properties not found solely in either one alone, e.g., metal carbides, borides, oxides, and silicides. They combine the strength and toughness of the metal with the heat and oxidation resistance of the ceramic material. The composition may range from predominantly metallic to predominantly ceramic, e.g., SAP sintered aluminum contains 85% aluminum and 15% aluminum oxide. The most important industrial cermets are titanium carbide-based, aluminum oxide-based, and special uranium dioxide types. Cermets are made by powder metallurgy techniques, involving use of bonding agents such as tantalum, titanium and zirconium. High stress-to-rupture properties; operate continuously at 1800°F, for short periods at 4000°F.
Uses: Gas turbines, rocket motor parts, turbojet engine components, nuclear fuel elements; coatings for high-temperature resistance; sensing elements in instruments; seals, bearings, etc. in special pumps and other equipment.

"Cer-O-Cillin."[327] Trademark for an antibacterial substance which differs from penicillin G in that the benzyl group is replaced by an allylmercaptomethyl group. Soluble in concentrations up to 500,000 units per cc. in sterile water for injection, sterile sodium chloride for injection, or sterile 5% dextrose.
Uses: Medicine.

cerosin. See ceresin wax.

cerotic acid (hexacosanoic acid; cerinic acid) $CH_3(CH_2)_{24}COOH$. A fatty acid obtained from beeswax, carnauba wax or Chinese wax.
Properties: White odorless crystals or powder; sp. gr. 0.8198 (100/4°C); m.p. 87.7°C; refractive index 1.4301 (100°C). Insoluble in water; soluble in alcohol, benzene, ether, acetone. Combustible; low toxicity.

cerous carbonate (cerium carbonate) $Ce_2(CO_3)_3 \cdot 5H_2O$.
Properties: White powder; soluble in mineral acids (dilute); insoluble in water. Low toxicity.
Derivation: By adding an alkali carbonate to a solution of a cerous salt.
Containers: Fiber drums; car lots.

cerous chloride (cerium chloride) $CeCl_3 \cdot xH_2O$.
Properties: White crystals; deliquescent; sp. gr. 3.88 (anhydrous); m.p. 848°C (anhydrous); b.p. 1727°C. Soluble in water, alcohol, and acids. Low toxicity.
Derivation: Action of hydrochloric acid on cerium carbonate or hydroxide.
Containers: Fiber drums; car lots.
Uses: Incandescent gas mantles; spectrography; preparation of cerium metal.

cerous fluoride $CeF_3 \cdot xH_2O$.
Properties: Off-white powder, insoluble in water and acids. Sp. gr. (anhydrous) 6.16; m.p. 1460°C; b.p. 2300°C.
Derivation: By treating cerous oxalate with hydrofluoric acid.
Hazard: Toxic and irritant. Tolerance (as F), 2.5 mg per cubic meter of air.
Uses: In arc carbons to increase their brilliance; preparation of cerium metal.

cerous hydroxide (cerium hydrate). Approximate formula $Ce(OH)_3$.
Properties: White gelatinous precipitate; yellow, brown or pink when impurities are present. Soluble in acids; insoluble in water and alkali.
Derivation: Chief source is monazite sand.
Grades: Pure; crude.
Use: Pure form: To produce cerium salts; impart yellow color to glass; opacifying agent in glazes and enamels. Crude form: flaming arc lamp.

cerous nitrate (cerium nitrate) $Ce(NO_3)_3 \cdot 6H_2O$.
Properties: Colorless crystals; deliquescent. Soluble in water, alcohol, and acetone.
Constants: M.p., loses $3H_2O$ at 150°C; b.p. decomposes at 200°C.
Derivation: Action of nitric acid on cerous carbonate.
Containers: Fiber drums; car lots.
Hazard: Strong oxidizing agent. Fire risk in contact with organic materials.
Uses: Incandescent gas mantles; medicine; reagent.
Shiping regulations: Nitrates, n.o.s., (Rail) Yellow label. (Air) Oxidizer label.

cerous oxalate (cerium oxalate) $Ce_2(C_2O_4)_3 \cdot 9H_2O$.
Properties: Yellowish white, odórless, tasteless, crystalline powder; decomposes upon heating; soluble in dilute sulfuric and hydrochloric acid; very slightly soluble in water; insoluble in oxalic acid solution, alkalies, alcohol and ether.
Derivation: By extraction from monazite sand with oxalic or hydrochloric acid and conversion into the oxalate, followed by crystallization.
Grades: Pure; the commercial product is a complex mixture of oxalates of cerium, lanthanum and didymium.
Hazard: Toxic; strong irritant.
Uses: Medicine; isolation of cerium metals.

cerous sulfate (cerium sulfate) $Ce_2(SO_4)_3 \cdot 8H_2O$.
Properties: White crystals or powder; soluble in water and in acids. M.p. 630°C (dehydrated); sp. gr. 2.886. Low toxicity.
Derivation: Reagent grade is prepared by reducing a solution of ceric sulfate in sulfuric acid with hydrogen peroxide.
Grades: Technical and purified (reagent).
Uses: Developing agent for aniline black.

"Cerox."[455] Trademark for a series of high alumina refractories with resistance to thermal shock, corrosion, abrasion, erosion, and reducing atmospheres. Fusion temperature up to 3600°F. Available in prefired shapes. Used in production of steel, of electronic ceramic components, of abrasives, in solid state ore reduction, metal powder sintering, in high-temperature gas reactors.

"Cerrobase."[60] Trademark for the eutectic alloy of bismuth and lead. M.p. 255°F. Shrinks slightly after solidifying, later expands.
Uses: Proof casting forging dies; master patterns; mandrels for electroforming; heat-transfer medium in autoclaves; liquid seal in bright annealing and nitriding furnaces; molds for plastics; fusible foundry cores; filler for bending large diameter tubing.

"Cerrobend."[60] Trademark for the eutectic alloy of bismuth, lead, tin and cadmium, m.p. 158°F. Expands during and after solidification.
Uses: Filler in thin-walled tubing; assembly, check-

ing, drilling, spotting fixtures in aircraft and automotive tooling; anchoring medium in precision machining jet engine components (buckets and blades).

"Cerrocast."[60] Trademark for a non-eutectic alloy of bismuth and tin. Melting range 281–338°F. Exhibits negligible volume change during and after solidification.
Uses: Soft metal dies for "lost wax" patterns; engraving machine patterns; split jaw chucks; molds for plastics; mandrels for electroforming.

"Cerromatrix."[60] Trademark for a bismuth alloy with melting range 217–440°F, pouring at 250°F, expanding during and after solidification.
Uses: Die making to anchor punches; fastening bearings, bushings, and non-moving parts in machinery; nests in drill jigs and dial feeding stations; sheet-metal forming dies, etc.

"Cerrosafe."[60] Trademark for a bismuth, lead, tin, and cadmium alloy with melting range 158–190°F. Shrinks 15 minutes after solidification, then expands.
Uses: Accurate duplicate patterns; proof casting cavities, such as gun chambers, bullet molds; toy castings and hobby models; sprayed-on protective coating on wood patterns and core boxes.

"Cerrotru."[60] Trademark for a non-shrinking bismuth-tin eutectic alloy,m.p. 281°F.
Uses: Anchoring shafts in Alnico rotors, forming blocks in stretch presses; engraving machining models; special jaws in tangent tube bending; soft metal dies in "lost wax" process; mandrels in electroforming; molds by "dip" casting.

certified color. See food color; FD&C color.

cerulean blue. A light blue pigment essentially cobaltous stannate $CoO \cdot n(SnO_2)$.

cerussite $PbCO_3$. Natural lead carbonate, found in the upper zone of lead deposits.
Properties: Colorless, white, gray; luster adamantine; Mohs hardness 3–3.5; sp. gr. 6.55. Effervesces in nitric acid.
Occurrence: Colorado, Arizona, New Mexico, Idaho; Australia; Europe.
Uses: An ore of lead.

"Cer-Vit."[191] Trademark for a glass ceramic (q.v.) having linear expansion coefficient near zero. Used for specialty products such as telescope mirrors, where minimum distortion is essential.

ceryl alcohol $C_{26}H_{53}OH$. An alcohol obtained from Chinese wax.
Properties: Colorless crystals; m.p. 79°C; insoluble in water; soluble in alcohol and ether. Combustible.

ceryl cerotate $C_{26}H_{53}OOCC_{25}H_{51}$. The chief constituent of Chinese wax and typical of natural waxes. Colorless crystals with m.p. 84°C.

CES. Abbreviation for cyanoethyl sucrose.

cesium (caesium) Cs An alkali-metal element of Group IA of the Periodic Table; atomic number 55; atomic weight 132.9055; valence 1. No stable isotopes.
Properties: Liquid at slightly above room temperature; soft solid below melting point; highly reactive. Decomposes water, with evolution of hydrogen, which ignites instantly. Also reacts violently with oxygen, the halogens, sulfur, and phosphorus, with spontaneous ignition and/or explosion. Sp. gr. 1.90; m.p. 28°C; b.p. 705°C. Mohs hardness 0.2. Cesium has highest position in the electromotive series; it also has the lowest m.p. of any alkali metal, and the lowest ionization potential of any element. Soluble in acids and alcohol. Low toxicity.
Derivation: By thermochemical reduction of cesium chloride with calcium, or by electrolysis of the fused cyanide. Its chief ore is pollucite, found in Maine, So. Dakota, Manitoba, Elba, So. Africa.
Grades: Technical; 99.9%.
Containers: Due to its reactivity, cesium must be handled in sealed glass ampules with special break seals, or in stainless steel cylinders.
Hazard: Dangerous fire and explosion risk. Ignites spontaneously in moist air; may explode in contact with sulfur or phosphorus; reacts violently with oxidizing materials; causes burns in contact with skin.
Uses: Photoelectric cells; getter in vacuum tubes; hydrogenation catalyst; ion propulsion systems; plasma for thermoelectric conversion; atomic clocks; rocket propellant; heat transfer fluid in power generators; seeding combustion gases for magnetohydrodynamic generators.
Shipping regulations: (Rail) Flammable solid, n.o.s., Yellow label. (Air) Flammable Solid label. Not acceptable on passenger planes unless in cartridges.

cesium 137. Radioactive cesium of mass number 137.
Properties: Half-life 33 years; radiation, beta.
Hazard: Radioactive poison. The beta decay of Cs-137 produces barium 137, which in turn is radioactive, emitting a 0.662 mev gamma ray, with a 2.6 minute half-life.
Uses: Most applications of Cs-137 depend on the fact that any Cs-137 preparation has an equivalent amount of the gamma-emitting Ba-137 daughter. Approved by FDA as gamma radiation source for certain foods.
Shipping regulations: (Rail, Air) Consult regulations.

cesium alum. See cesium aluminum sulfate.

cesium aluminum sulfate (cesium alum) $CsAl(SO_4)_2 \cdot 12H_2O$.
Properties: Colorless crystals; sp. gr. 2.0215; m.p. 117°C; soluble in water; insoluble in alcohol.
Derivation: By adding a solution of cesium sulfate to a solution of potassium alum, concentrating and crystallizing.
Grades: Pure.
Uses: Mineral waters; purification of cesium by fractional crystallization; preparation of cesium salts.

cesium antimonide. Used as a high-purity binary semiconductor.
Hazard: Toxic by ingestion.

cesium arsenide. Used as a high-purity binary semiconductor.
Hazard: Toxic by ingestion.
Shipping regulations: Arsenical compounds, n.o.s., (Rail, Air) Poison label.

cesium bromide CsBr.
Properties: Colorless crystalline powder; sp. gr. 4.44; m.p. 636°C; b.p. 1300°C; soluble in water; slightly soluble in alcohol.
Grades: Technical; single crystals.
Uses: Medicine; crystals for infrared spectroscopy; scintillation counters.

cesium carbonate Cs_2CO_3.
Properties: White, hygroscopic, crystalline powder; very stable; can be heated to high temperature with-

out loss of CO_2. Soluble in water, alcohol, and ether.
Derivation: By passing carbon dioxide into a solution of cesium hydroxide and subsequent crystallization.
Uses: Brewing; mineral waters; ingredient of specialty glasses.

cesium chloride CsCl.
Properties: Colorless crystals; sp. gr. 3.972; m.p. 646°C; sublimes at 1290°C. Soluble in water and alcohol; insoluble in acetone.
Derivation: Action of hydrochloric acid on cesium oxide and crystallization.
Grades: Pure; low-optical density; single crystals.
Uses: Medicine; brewing; preparation of other cesium compounds; mineral waters; evacuation of radio tubes (positive ions supplied at surface of filament); for a density gradient in ultracentrifuge separations.

cesium dioxide Cs_2O_2.
Properties: Yellow needles; sp. gr. 4.25; m.p. 400°C; decomposes at 650°C; soluble in water and acids.
Grades: Technical; pure.
Use: Cesium salts.

cesium fluoride CeF.
Properties: Deliquescent crystals; sp. gr. 4.115; m.p. 682° C; b.p. 1251° C. Very soluble in water; soluble in methanol; insoluble in dioxane and pyridine.
Grades: 99% min; single crystals.
Hazard: Toxic and irritant. Tolerance (as F), 2.5 mg per cubic meter of air.
Uses: Optics; catalysis; specialty glasses.

cesium hexafluorogermanate Cs_2GeF_6.
Properties: White crystalline solid; m.p. approx. 685°C; density 4.10 g/cc. Slightly soluble in cold water and acids; very soluble in hot water.
Hazard: Toxic and irritant.

cesium hydrate. See cesium hydroxide.

cesium hydroxide (cesium hydrate) CsOH.
Properties: Colorless or yellowish, fused, crystalline mass. Strong alkaline reaction. Hygroscopic. Keep well stoppered. Sp. gr. 3.675; m.p. 272.3°C. Extremely soluble in water; the strongest base known.
Derivation: By adding barium hydroxide to an aqueous solution of cesium sulfate.
Grade: Technical; 50% aqueous solution.
Use: Recommended as electrolyte in alkaline storage batteries at subzero temperatures.

cesium iodide Csl.
Properties: Colorless, crystalline powder; deliquescent. Sp. gr. 4.510; m.p. 621°C; b.p. 1280°C; soluble in alcohol and water.
Grades: Technical; single crystals.
Hazard: Moderately toxic.
Uses: Crystals for infrared spectroscopy; scintillation counters.

cesium nitrate $CsNO_3$.
Properties: Crystalline powder; saltpeter taste; sp. gr. 3.687; m.p. 414°C; b.p. (decomposes); soluble in water and acetone; slightly soluble in alcohol.
Derivation: Action of nitric acid on cesium oxide and crystallization.
Grades: Pure, 99.0% min.
Hazard: Dangerous; may ignite organic materials on contact.
Use: Cesium salts.
Shipping regulations: Nitrates, n.o.s., (Rail) Yellow label. (Air) Oxidizer label.

cesium oxide Cs_2O.
Properties: Orange-red crystals; sp. gr. 4.36; m.p. decomposes 360–400°C; very soluble in water; soluble in acids.
Grades: Technical; pure.
Use: Cesium salts.

cesium perchlorate $CsClO_4$.
Properties: Crystalline solid; sp. gr. 3.327 (4°C); m.p. (dec.) 250°C; soluble in water (much more in hot than cold); slightly soluble in alcohol and acetone.
Grade: 99% min.
Hazard: Dangerous fire risk; strong oxidizing agent; may ignite organic materials on contact.
Uses: Optics; catalysis; specialty glasses; power generation.
Shipping regulations: Perchlorates, n.o.s., (Rail) Yellow label. (Air) Oxidizer label.

cesium peroxide. See cesium tetroxide.

cesium phosphide. Used as a high-purity binary semiconductor.
Hazard: Fire risk by decomposition to phosphine.

cesium sulfate Cs_2SO_4.
Properties: Colorless crystals; soluble in water; insoluble in alcohol. Sp. gr. 4.2434; m.p. 1010°C. Low toxicity.
Derivation: Action of sulfuric acid on cesium carbonate.
Grades: Pure; low optical density.
Uses: Brewing; mineral waters; for density gradient in ultracentrifuge separation.

cesium tetroxide (cesium peroxide) Cs_2O_4.
Properties: Yellow crystals; sp. gr. 3.77; m.p. 600°C; decomposes violently in water; soluble in acids. Strong oxidizing agent.
Hazard: Fire risk in contact with organic materials.
Shipping regulations: Oxidizing material, n.o.s., (Rail) Yellow label. (Air) Oxidizer label.

cesium trioxide Cs_2O_3.
Properties: Chocolate-brown crystals; sp. gr. 4.25 (0°C); m.p. 400°C; decomposes in water; soluble in acids.

cetalkonium chloride (USAN) (cetyldimethylbenzylammonium chloride; benzylhexadecyldimethylammonium chloride) $C_6H_5CH_2N(CH_3)_2(C_{16}H_{33})Cl$. A quaternary ammonium germicide.
Properties: Colorless, odorless, crystalline powder; m.p. 58–60°C. Soluble in water to form colorless, odorless solution having pH 7.2. Compatible with alkalies and antihistamines. Soluble in alcohol, acetone, esters, carbon tetrachloride.
Uses: Medicine; germicide; fungicide; surface-active agent.

"Cetalon SZ"[309] Trademark for a combination of formaldehyde sulfoxylate in the form of a white powder, easily soluble in water.
Uses: Stripping agent for wool, nylon, acetate, etc.

cetane. See hexadecane.

cetane number. A rating for Diesel fuel comparable to the octane number rating for gasoline. It is the percentage of cetane ($C_{16}H_{34}$) which must be mixed with heptamethylnonane to give the same ignition performance, under standard conditions, as the fuel in question.

"Cetats."[430] Trademark for cetyltrimethylammonium tosylate (q.v.).

cetene. See 1-hexadecene.

cetin (cetyl palmitate; palmitic acid, cetyl ester) $C_{15}H_{31}COOC_{16}H_{33}$.
Properties: White crystalline wax-like substance. Chief constituent of commercial purified spermaceti. M.p. 50°C; b.p. 360°C; sp. gr. 0.832; refractive index (n 70/D) 1.4398. Soluble in alcohol and ether; insoluble in water. Combustible; nontoxic.
Derivation: By solution from spermaceti.
Uses: Base for ointments, cerates, and emulsions; manufacture of candles, soaps, etc.

"Cetol."[430] Trademark for cetyldimethylbenzylammonium chloride (cetalkonium) (q.v.).

"Cetone Alpha."[227] Trademark for alpha-isomethylionone.

"Cetone V."[227] Trademark for allyl-alpha-ionone.

cetyl alcohol (alcohol C-16; cetylic alcohol; 1-hexadecanol; normal primary hexadecyl alcohol; palmityl alcohol) $C_{16}H_{33}OH$. A fatty alcohol. Combustible; nontoxic.
Properties: White, waxy solid; faint odor; sp. gr. 0.8176 (49.5°C); m.p. 49.3°C; b.p. 344°C; refractive index (n 79/D) 1.4283; partially soluble in alcohol and ether; insoluble in water. Combustible.
Derivation: By saponifying spermaceti with caustic alkali; reduction of palmitic acid.
Method of purification: Crystallization; distillation.
Grades: Technical; cosmetic; N.F.
Containers: Tins; cartons; drums; bags; tank cars.
Uses: Perfumery; emulsifier; emollient; foam stabilizer in detergents; face creams, lotions, lipsticks, toilet preparations; chemical intermediate; detergents; pharmaceuticals; cosmetics; base for making sulfonated fatty alcohols; to retard evaporation of water, when spread as a film on reservoirs, or sprayed on growing plants.

cetyl bromide $C_{16}H_{33}Br$.
Properties: Dark yellow liquid. Freezing point 15°C; b.p. 186–197°C (10 mm); sp. gr. 0.991 (25/25°C); lb/gal 8.25 (25°C); refractive index 1.460 (n 25/D); flash point 350°F. Soluble in ether; very slightly soluble in water, methanol. Combustible.
Use: Synthesis.

cetyldimethylbenzylammonium chloride. See cetalkonium chloride.

cetyldimethylethylammonium bromide $C_{16}H_{33}(CH_3)_2H_5NBr$. A quaternary ammonium salt (q.v.).
Properties: Paste.
Uses: Disinfectant; deodorant; germicide; fungicide; detergents.

cetyldimethylethylammonium chloride $C_{16}H_{33}(CH_3)_2C_2H_5NCl$. A quaternary ammonium salt (q.v.).

cetylic acid. See palmitic acid.

cetylic alcohol. See cetyl alcohol.

cetyl lactate. See "Ceraphyl."

cetyl mercaptan (hexadecyl mercaptan) $C_{16}H_{33}SH$.
Properties: M.p. 18°C; b.p. 185–190°C (7 mm); strong odor; sp. gr. 0.8474 (20/4°C); refractive index 1.474 (n 20/D); flash point 275°F. Combustible.
Grades: 95% (min.) purity.
Containers: Up to tank cars.
Hazard: Probably toxic.
Uses: Intermediate; synthetic rubber processing; surface-active agent; corrosion inhibitor.

cetyl palmitate. See cetin.

cetyl pyridinium bromide $C_{16}H_{33}C_5H_5NBr$.
Properties: Cream-colored waxy solid. Soluble in acetone, ethanol and chloroform.
Uses: Surface-active agent; germicide.

cetylpyridinium chloride (1-hexadecylpyridinium chloride). The monohydrate of the quaternary salt of pyridine and cetyl chloride; $C_{16}H_{33}C_5H_5NCl\cdot H_2O$.
Properties: White powder with slight odor. M.p. 80–84°C. Very soluble in alcohol, chloroform and water; very slightly soluble in benzene and ether; pH (1% soln) 6.0–7.0.
Grades: Technical; N.F.
Use: Medicine.

cetyltrimethylammonium bromide (hexadecyltrimethylammonium bromide) $C_{16}H_{33}(CH_3)_3NBr$. A quaternary ammonium salt.
Properties: White powder; soluble in water, alcohol and chloroform.
Grade: Technical.
Uses: Surface-active agent; germicide.

cetyltrimethylammonium chloride $C_{16}H_{33}(CH_3)_3NCl$. A quaternary ammonium salt.

cetyltrimethylammonium tosylate $[C_{16}H_{33}(CH_3)_3N]OSO_2C_6H_4CH_3$. A high-temperature stable quaternary ammonium compound.
Uses: Germicide; surfactant.

cetyl vinyl ether (vinyl cetyl ether) $C_{16}H_{33}OCH{:}CH_2$.
Properties: Colorless liquid; sp. gr. 0.822 (27°C); m.p. 16°C; b.p. 142°C (1 mm), 173°C (5 mm); flash point (open cup) 325°F; refractive index (n 25/D) 1.444. Combustible.
Grades: 97%.
Containers: 350 lb drums.
Hazard: Toxic by inhalation; skin irritant. Reacts strongly with organic materials.
Use: Reactive monomer which may be copolymerized with a variety of unsaturated monomeric materials, including acrylonitrile, vinyl chloride, vinylidene chloride, and vinyl acetate, to yield internally plasticized resins.

Cf Symbol for californium.

CF. Abbreviation for citrovorum factor. See folinic acid.

"C- Fatty Acids."[487] Trademark for fatty acids derived from coconut oil. The major component acids are lauric and myristic. Differ primarily in amount of unsaturated acid components and color. Light-yellow solids which liquefy at about 25°C; obtained from naturally occurring triglycerides; nontoxic; combustible.
Containers: Tank cars and trucks.
Uses: Intermediates, rubber compounding, cosmetic ingredients, buffing compounds, alkyd resins, emulsifiers, grease manufacture and candles.

CFE. Abbreviation for chlorotrifluoroethylene. Also used for polychlorotrifluoroethylene resins.

cgs. Abbreviation of centimeter gram second, the system of measurement used internationally by scientists.

chabazite $CaAl_2Si_4O_{12} \cdot 6H_2O$. Essentially a natural hydrated calcium aluminum silicate, usually containing some sodium and potassium. A zeolite (q.v.).
Properties: Color white, reddish, yellow, brown; luster vitreous; sp. gr. 2.1; Mohs hardness 4-5.
Occurrence: New Jersey, Colorado, Oregon; Europe.
Use: Water treatment.

Chadwick, Sir James (1891-). A British physicist who was awarded the Nobel Prize in 1935 for his discovery of the neutron (1932), the existence of which had been predicted by Rutherford. This was an immensely important advance in the knowledge of subatomic particles. See also neutron.

chain. A series of atoms of a particular element directly connected by chemical bonds which constitutes the structural configuration of a compound. Such chains are usually composed of carbon atoms, often shown without their accompanying hydrogens. Carbon chains may be of the following types:

(1) Open or straight chain: a sequence of carbor atoms extending in a direct line; this is characteristic of paraffins and olefins, the former being saturated:

$$\begin{array}{ccccccccccc} & & H & & H & & H & & H & & \\ & & | & & | & & | & & | & & \\ H & - & C & - & C & - & C & - & C & - & H \\ & & | & & | & & | & & | & & \\ & & H & & H & & H & & H & & \end{array}$$

and the latter unsaturated:

$$\begin{array}{ccccccc} H & & H & & H & & H \\ | & & | & & | & & | \\ C & = & C & - & C & = & C \\ | & & & & & & | \\ H & & & & & & H \end{array}$$

(2) Branched chain: a paraffinic structure isomeric with its straight-chain counterpart, e.g., butane, but having a subordinate chain comprised of one or more carbon atoms:

$$\begin{array}{ccccccccc} & & H & & & H & & & H & & \\ & & | & & & | & & & | & & \\ H & - & C & & — & C & — & & C & - & H. \\ & & | & & & | & & & | & & \\ & & H & & H - & C & - H & & H & & \\ & & & & & | & & & & & \\ & & & & & H & & & & & \end{array}$$

Such compounds are designated by the prefix iso-, and their properties are often notably different from the straight-chain isomer. For example, octane has a very low antiknock rating, whereas that of isooctane is high.

(3) Closed chain, or ring: a cyclic arrangement of carbon atoms giving a closed geometric structure, that is, a ring, pentagon or other form; these are characteristic of alicyclic, aromatic, and heterocyclic compounds. See cyclic compound.

(4) Side-chain: a group of atoms attached to one or more of the locations in a cyclic or heterocyclic compound, e.g., tryptophan:

$CH_2CH(NH_2)COOH$

N

H

chain mechanism. See free radical.

chain reaction. See fission, nuclear; nuclear energy.

chalcocite (copper glance) Cu_2S. Natural cuprous sulfide, occurring with other copper minerals.
Properties: Color lead gray, tarnishing dull black; luster metallic; sp. gr. 5.5–5.8; Mohs hardness 2.5–3.
Occurrence: Montana, Arizona, Utah, Nevada; Alaska; Chile; Mexico; Europe.
Use: Important ore of copper.

chalcopyrite (copper pyrites, yellow copper) $CuFeS_2$. Natural copper-iron sulfide, found in metallic veins and igneous rocks.
Properties: Color brass yellow, frequently tarnished bronze or iridescent; luster metallic; streak greenish black; sp. gr. 4.1–4.3; Mohs hardness 3.5–4. May carry gold or silver or mechanically intermixed pyrite.
Occurrence: Montana, Utah, Arizona, Tennessee, Wisconsin; Europe; Chile; Canada.
Use: Important ore of copper.

chalk. A natural calcium carbonate composed of the calcareous remains of minute marine organisms. Decomposed by acids and heat. Odorless, tasteless, nontoxic. See also calcite; calcium carbonate; whiting; chalk, prepared.

chalk, drop. See chalk, prepared.

chalk, French. A variety of soapstone or steatite. See talc.

chalking. A natural process by which paints develop a loose, powdery surface formed from the film. Chalking results from decomposition of the binder, due principally to the action of ultraviolet rays.

chalk, precipitated. See calcium carbonate, precipitated.

chalk, prepared (drop chalk; calcium carbonate, prepared).
Properties: Fine, white to grayish-white impalpable powder, often formed in "conical drops." Odorless, tasteless and stable in air. M.p., decomposes at 825°C with evolution of carbon dioxide; decomposed by acids; practically insoluble in water; insoluble in alcohol. Nontoxic; noncombustible.
Derivation: By grinding native calcium carbonate to a fine powder, agitating with water, allowing the coarser particles to settle, decanting the suspension and allowing the fine particles to settle slowly.
Containers: Fiber cans; tins; glass bottles; multiwall paper sacks.
Uses: Medicine (antacid); tooth powders; calcimine; polishing powders; silicate cements. For other uses, see whiting and calcium carbonate.

chamber process. An obsolete method for manufacturing sulfuric acid from sulfur dioxide, air, and steam in the presence of nitrogen oxides as catalysts. It is no longer used in the U.S.

chamois. A very soft, flexible leather made from the flesh layer of a split sheepskin by treating with fish oils, piling in contact with similarly treated skins, and allowing the fish oils to oxidize. Used chiefly for fine cleaning of smooth surfaces and specialty leather items.

chamomile oil (camomile oil). An essential oil, usually blue in color, used in flavoring and in medicine. There are several varieties, including Roman (English), German, Hungarian, etc. It has a strong aromatic odor and bitter taste.

channel black. See carbon black.

charcoal, activated. See carbon, activated.

charcoal, animal. See bone black; ivory black.

charcoal, bone. See bone black.

charcoal, vegetable. See vegetable black; carbon, activated.

charcoal, wood. An amorphous allotropic form of carbon.
Derivation: Destructive distillation of wood.
Grades: Technical, in lumps; powdered; briquettes.
Containers: Barrels; multiwall paper sacks.
Hazard: Dangerous fire risk in briquette form or when wet; may ignite spontaneously in air.
Uses: Chemical (precipitant in the cyanide process, precipitant of iodine and lead salts from their solutions, catalyst, calcium carbide); decolorizing and filtering medium, gas adsorbent; component of black powder and other explosives; fuel; arc light electrodes; decolorizing and purifying oils; solvent recovery; deodorant.
Shipping reguations: (Rail) (ground, crushed or pulverized) Yellow label; (wet) Not accepted. (Air) (ground, crushed or pulverized) Flammable Solid label; (wet) Not acceptable.

Chardonnet (1839–1924). A native of France, he has been called the father of rayon because of his successful research in producing what was then called artificial silk from nitrocellulose. He was able to extrude fine threads of this semisynthetic material through a spinneret-like nozzle and the textile product was made on a commercial scale in several European countries. He was awarded the Perkin medal for this work in 1914, only a few years before the discovery of rayon.

Charles' Law. At constant volume, the pressure of a confined gas is proportional to its absolute temperature. See also Gay-Lussac's Law.

chaulmoogra oil (gynocardia oil; hydnocarpus oil).
Properties: Brownish-yellow oil or soft fat; repulsive odor; somewhat acrid taste. Soluble in ether, chloroform, benzene, solvent naphtha; sparingly soluble in cold alcohol; almost entirely soluble in hot alcohol, carbon disulfide. Sp. gr. 0.940. Iodine value 85–105, depending on type; optically active.
Chief constituents: Glycerides of chaulmoogric and hydnocarpic acids.
Derivation: Expressed from the seeds of Taraktogenos kurzii or Hydnocarpus anthelminthicus or wightianus.
Uses: Medicine (treatment of leprosy and other infective skin diseases).

chaulmoogric acid (hydnocarpyl acetic acid)

$\overline{CH_2CH_2CHCHC}H(CH_2)_{12}COOH$. A cyclic fatty acid.
Properties: Colorless shiny leaflets; m.p. 68.5°C; soluble in ether, chloroform, and ethyl acetate.
Source: Chaulmoogra oil.
Use: Medicine; biochemical research.

chavicol (para-allylphenol; 1-allyl-4-hydroxybenzene) $C_3H_5C_6H_4OH$. Liquid; m.p. 16°C; b.p. 230°C; sp. gr. 1.033 (18/4°C); soluble in water and alcohol. Occurs in many essential oils.

CHDM. See 1,4-cyclohexanedimethanol.

"Cheelox."[307] Trademark for a series of organic chelating and sequestering agents, consisting of polycarboxylic acid derivatives of amines or polyamines or their salts, as, for example, ethylenediaminetetraacetic acid (q.v.).

"Chel."[219] Trademark for a series of chelating agents.

chelate. The type of coordination compound (q.v.) in which a central metal ion such as Co^{2+}, Ni^{2+}, CU^{2+}, of Zn^{2+} is attached by coordinate links to two or more nonmetal atoms in the same molecule, called ligands. Heterocyclic rings are formed with the central (metal) atom as part of each ring. Ligands offering two groups for attachment to the metal are termed bidentate (two-toothed); three groups, tridentate, etc.

A common chelating agent is ethylenediaminetetraacetic acid (EDTA). Nitrilotriacetic acid $N(CH_2COOH)_3$ and ethyleneglycol-bis(beta-aminoethyl ether)-N,N-tetraacetic acid $(HOOCCH_2)_2N$-$CH_2CH_2OCH_2CH_2OCH_2CH_2N(CH_2COOH)_2$ are used in analytical chemical titrations and to remove ions from solutions and soils. Metal chelates are found in biological systems, e.g., the iron-binding porphyrin group of hemoglobin and the magnesium-binding chlorophyll of plants. Medicinally, metal chelates are used against Gram-positive bacteria, fungi, viruses, etc.
See also ammine; sequestration; complex; cobaltammine.

"Chelon."[428] Trademark for a series of chelating agents.

"Chemactant."[460] Trademark for a series of industrial products derived from lanolin.
Uses: Softener and conditioner for leather, fur, textiles; inks; plastics; rubber; paper; mold release compounds; resin polymerization; waxes, polishes, and cleaners; greases; emulsions; paints; coatings; quick breaking aerosol foams.

"Chemglaze"[547] Trademark for clear and pigmented, single package, moisture-curing, pure polyurethane coatings.

"Chem-Hoe."[177] Trademark for isopropyl N-phenylcarbamate (IPC), a selective herbicide.

Chemical Abstracts Service (CAS). A systematic, computerized chemical information source developed by the CAS Division of the American Chemical Society (q.v.). *Chemical Abstracts* (1907) has become the largest scientific abstract journal in the world and the repository of all significant chemical research information reported in the international literature. An extensive, in-depth systems approach to the storage and handling of chemical information has enabled CAS to assemble a vast computerized body of knowledge, typified by its Chemical Compound Registry—a collection of unique, structure-based "addresses," or fingerprints, which now contains the structural record of about 2.5 million substances. The ACS chemical publications are one form of the output of this information system.

chemical bond. See bond, chemical.

chemical change. Rearrangement of the atoms, ions, or radicals of one or more substances resulting in the formation of new substances often having entirely different properties. Such a change is called a chemical reaction (q.v.). In some cases energy in the form of heat, light, or electricity is required to initiate the change; this is known as an endothermic reaction. When energy is given off, as a result of rupture of chemical bonds, the change is said to be exothermic.

Chemical changes should be distinguished from physical changes, in which only the state or condition of a substance is modified, its chemical nature remaining the same. A physico-chemical change has some of the characteristics of both. Examples of the three types are:

Chemical Changes

fuel + oxygen ⟶ carbon dioxide + water + heat (exothermic)

water + carbon dioxide + energy ⟶ sugar + oxygen (endothermic)

Physical Changes

water to ice or steam — coagulation of latex
distillation processes

Physicochemical Changes

cooking of food
vulcanization of rubber
tanning of leather
drying of oil- or plastic-based paints

chemical dating. Estimation of the age of geologic structures and events by measuring the amount of radioactive decay products in existing samples. The age of a uranium-containing material can be determined by measuring the percentage of lead (or helium) formed as a result of disintegration of the uranium. Uranium decays to both helium and the 206 lead isotope, but measurement of helium content is inaccurate because of its strong tendency to escape. By determining the ratio of the percentage of lead in a sample to the percentage of uranium, the age can be calculated. A more recent method, applicable to events within about 10,000 years, involves the use of the natural radioactive carbon isotope (C-14); the percentage of this isotope determined in a carefully prepared sample is an index of its age, based on the half-life of C-14 (5700 years), which was present in the atmospheric CO_2 absorbed by plants centuries ago. This method has yielded valuable results in the study of archaeological specimens, deep-sea sedimentation and dates of volcanic and glacier activity.

chemical economics. The principles and practical application of industrial economics in particular reference to the manufacture of chemicals and chemically derived products. Economic factors that apply especially to the chemical industries are: (1) heavy capital investment; (2) vast range of raw materials and products; (3) complex and varied production methods; (4) high-level R & D for new products, including extensive testing for safety and acceptability; (5) market research to develop new product uses, etc.; (6) substitution of synthetic for natural products or (in some cases) the reverse; (7) utilization of by-product materials; (8) waste reclamation and pollution control. See also chemical process industry.

chemical engineering. That branch of engineering concerned with the production of bulk materials from basic raw materials by large-scale application of chemical reactions worked out in a laboratory. The unit operations of chemical engineering are those common to all chemical processing: fluid flow, heat transfer, filtration, evaporation, distillation, drying, mixing, adsorption, solvent extraction and gas absorption. Underlying these are the fundamentals involved in every technical problem in chemical engineering; conservation, equilibrium, kinetics, and control.

Conservation involves the laws of the conservation of matter and energy. The limits of any chemical process are established by its equilibrium conditions; the principles underlying the equilibrium concept (such as the Law of Mass Action) are of fundamental importance (see chemical laws). By kinetics is meant time-dependent processes or rate processes (momentum transfer, thermal transfer, mass transfer and chemical kinetics). Control involves either the systems approach or feedback or closed-loop methods, as well as considerations of the stability of the system. See also equilibrium constant; stoichiometry; kinetics, chemical.

Chemical Industry Institute of Toxicology. An industry-supported organization whose purpose is investigation of environmental and occupational health concerns involving widely used basic chemicals. It will study the toxicity of commodity chemicals and the pertinent testing procedures. The Institute has 20 company members (1976).

chemical kinetics. See kinetics, chemical.

chemical laws. A group of basic principles governing the combining power and reaction characteristics of elements. Among the more important are:

(1) Law of Mass Action: The rate of a homogeneous (uniform) chemical reaction at constant temperature is proportional to the concentrations of the reacting substances.

(2) Law of Definite or Constant Composition: Any chemical compound invariably contains the same elements in the same fixed proportions by weight. A number of exceptions to this law occur in solid compounds such as silicates, which are known as nonstoichiometric compounds.

(3) Law of Multiple Proportions: When two elements unite to form two or more compounds (e.g., nitrogen and oxygen can form five different oxides), the weights of one element that combine with a given weight of the other are in the ratio of small whole numbers. Hydrogen and oxygen unite in the ratio of 1 to 8 in water and of 1 to 16 in hydrogen peroxide. Thus the weights of oxygen that unite with one gram of hydrogen are in the ratio of 1 to 2.

(4) Law of Conservation of Mass: Any chemical reaction between two or more elements or compounds leaves the total mass unchanged, the reaction products having exactly the same mass that was present in the reactants, regardless of the extent to which their other properties are changed.

(5) Law of Avogadro: Equal volumes of gases at constant temperature and pressure contain the same number of molecules, whether the gases are the same or different. 22.4 liters of any gas contain 6.02×10^{23} molecules. See also Avogadro's law; mole.

Chemical Marketing Research Association (CMRA). A professional organization whose primary functions are concerned with market research in the chemical and process industries. Techniques for this comparatively new research are still being evaluated. Major purposes of the Association are to provide opportunities for members to keep in touch with the latest consumer and industrial research concepts; to encourage the general welfare of the chemical industries; to improve the service of these industries to the public; and to cooperate with Government officials in furthering the national welfare. The offices of the organization are at 100 Church St., New York, N.Y.

chemical microscopy. See microscopy, chemical.

chemical milling. The process of producing metal parts to predetermined dimensions by removing metal from the surface with chemicals. Acid or alkaline pickling or etching baths are used for this purpose. Immersion of a metal part will result in uniform removal of metal from all surfaces exposed to the solution. This process is used by the aircraft industry for weight reduction of large parts; it is also used in the manufacture of instruments and other components where exact tolerances are required.

chemical nomenclature. The origin and use of the names of elements, compounds, and other chemical entities, both individually and as groups, as well as the various proposals for systematizing them. It may be considered to include three major aspects: (1) the gradual sporadic development of these names, many of which go back to the alchemists of the Middle Ages; (2) the proliferation of terminology due to the rapid extension of organic chemistry in the mid-nineteenth century, which led to the recommendations of the Geneva System in 1892; (3) the additional reforms adopted by the International Union of Pure and Applied Chemistry in 1930. There is still no clear-cut elimination of the older names in spite of the changes introduced in these reformed systems. Thus the present nomenclature is in some respects a hybrid; for example, the earlier terms paraffin and olefin are still widely used instead of more modern alkane and alkene, and methyl alcohol is as acceptable as methanol.

A comparatively recent development in the nomenclature of inorganic and complex compounds is use of the Stock system, in which roman numerals indicate the oxidation state or coordination value. For example, iron II chloride stands for ferrous chloride ($FeCl_2$), and iron III chloride for ferric chloride ($FeCl_3$).

See also: Geneva System; benzene.

chemical oxygen demand. See oxygen consumed.

chemical planetology. Application of various branches of chemistry (analytical, physical and geochemistry) to study of the composition of the surface and atmosphere of the planets, particularly Venus, Mars and Jupiter. Much informatoin has been obtained by spectrographic methods, and valuable additional data have resulted from space probes. See also astrochemistry.

chemical process industry. An industry whose product(s) results from (a) one or more chemical or physicochemical changes; (b) extraction, separation, or purification of a natural product, with or without the aid of chemical reactions; (c) the preparation of specifically formulated mixtures of materials, either natural or synthetic. Examples are as follows (with allowance for a certain amount of overlapping); (a) the plastics, rubber, leather, food, dye, and synthetic organic industries; (b) the petroleum, paper, textile, and perfume industries; (c) the glass, cement, fertilizer, soap, and paint industries. Many of these involve one or more unit operations of chemical engineering (q.v.), as well as such basic processes as polymerization, oxidation, reduction, hydrogenation, etc., usually with the aid of a catalyst. This definition could be interpreted to include ore processing, separation, and refinement, as well as the manufacture of metal products; however, these are usually considered to comprise the metal and metallurgical industries.

chemical reaction. A chemical change (q.v.) that may occur in several ways, e.g. by combination, by replacement, by decomposition, or by some modification of these. Reactions are endothermic when heat is necessary to maintain them, and exothermic when they evolve heat. All chemical reactions are in balance, that is, the number of atoms in the reacting substances is invariably equal to the number of atoms in the reaction products. Common types of reactions are oxidation, reduction, ionization, combustion, polymerization, hydrolysis, condensation, enolization, saponification, rearrangement, etc. Chemical reactions involve rupture of only the bonds which hold the molecules together, and should not be confused with nuclear reactions, in which the atomic nucleus is involved. A reversible reaction is one in which the reaction product is unstable and thus changes back into the original substance spontaneously. In a complete reaction the activity goes to the right and is indicated by an arrow$\rightarrow$; if heat or a catalyst is used this is indicated by a symbol or word above the arrow, as $\xrightarrow{\Delta}$, $\xrightarrow{\text{catalyst}}$; a reversible reaction is shown by either $\rightleftharpoons$ or $\leftrightarrow$.

chemical smoke. Chemically generated aerosols used primarily for military purposes. They are of 4 types:

(a) FS, a mixture of sulfuric anhydride and chlorosulfonic acid; used in shells and bombs and sprayed from airplanes.

(b) FM, titanium tetrachloride; same as FS, but brilliant white and will drop like a curtain when sprayed.

(c) HC, a mixture of hexachloroethane, aluminum and zinc oxide; burns to yield a white cloud.

(d) WP, a white phosphorus; burns to form white cloud of phosphoric acid; an excellent smoke producer.

See also fog; smoke; chemical warfare.

chemical stoneware (brick, chemical). A clay pottery product widely employed to resist acids and alkalies. It is used for utensils, pipes, stopcocks, ball mills, laboratory sinks, etc.

chemical technology. A general term covering a broad spectrum of physicochemical knowledge of the materials, processes, and operations used in the chemical process industries. It includes (1) basic phenomena such as activation, adsorption, oxidation, catalysis, corrosion, surface activity, polymerization, etc.; (2) the properties, behavior and handling of industrial materials and products (plastics, textiles, coatings, soap, foods, metals, paper, pharmaceuticals, etc.); and (3) their formulation, fabrication and testing (compounding, extruding, molding, assembly, and the like). See also chemical process industry.

chemical thermodynamics. That aspect of thermodynamics concerned with the relationship of heat, work, and other forms of energy to equilibrium in chemical reactions and changes of state. See also thermodynamics; thermochemistry; kinetics, chemical; equilibrium constant.

chemical warfare. The employment of a chemical agent directly for military purposes, i.e., to cause casualties by irritation, burning, asphyxiation, or poisoning; to contaminate ground; to screen action by smoke; or to cause incendiary damage. It includes use of all forms of toxic or irritant gases, including nerve gases (q.v.), smoke-inducing agents, flammable gels such as napalm, and such incendiary materials as magnesium and thermite. Biological warfare involves the use of bacteria and other infective agents against enemy forces or populations.

Chemico process. A technique used for extracting sulfur from low-grade ores (25 to 50% sulfur) by means of hot water.

"Chemigum."[265] Trademark for synthetic acrylonitrile/butadiene dry elastomers and latices produced by emulsion polymerization. They have exceptional resistance to oils and aromatic fuels. They range from high to low nitrile content polymers.

Uses: (dry) Automotive gaskets and hose, refrigerator hose, fuel cell interliners, oil-resistant shoe soles and heels, belt covers, hard packing compounds, adhesives, kitchen and drug sundries, tubing, textile coatings, frictioning compounds, oil-resistant auto parts, rubber rolls, printers' supplies, flooring. (Latex): Saturated and beater-impregnated paper; nonwoven fabric binders, carpet backings, textile and paper coating, foam, dipped goods, textile inks, leather finishes.

"Chemigum AC." A polyacrylic elastomer intended for high temperature, oil resistant service—applications where temperature requirements exceed 275°F.

"Chemigum 200 series." Latices produced with both high and medium nitrile contents. Require conventional cure cycle.

"Chemigum 500 series." Carboxylic-modified NBR latices with medium nitrile content. Their use eliminates the conventional sulfur cure. Other properties: mechanical and chemical stability, low foam formation, advantageous film color and aging properties.

chemiluminescence. The emisson of absorbed energy (as light) due to a chemical reacton of the components of the system. It includes the subclassifications bioluminescence and oxyluminescence, in which light is produced by chemical reactions involving organisms and oxygen, respectively. Chemiluminescence occurs in thousands of chemical reactions covering a wide variety of compounds, both organic and inorganic. Emission of light by fireflies is a common example of bioluminescence. See also luminescence.

"Chemipen."[412] Trademark for potassium phenethicillin.

chemisorption. The formation of bonds between the surface molecules of a metal (or other material of high surface energy) and another substance (gas or liquid) in contact with it. The bonds so formed are comparable in strength to ordinary chemical bonds, and much stronger than the van der Waals type characterizing physical adsorption (q.v.). Chemisorbed molecules are almost always altered, e.g., hydrogen is chemisorbed on metal surfaces as hydrogen atoms; chemisorption of hydrocarbons may result in the formation of chemisorbed hydrogen atoms and hydrocarbon fragments. Even when dissociation does not occur, the properties of the molecules are changed by the surface in important ways. This mechanism is the activating force of catalysis.

A practical example of chemisorption is the boundary lubrication of moving metal parts in machinery. A film of oil forms a chemisorbed layer at the interface (q.v.) and averts the high frictional forces that would otherwise exist. Solids with high surface energies are necessary for chemisorption to occur, e.g., nickel, silver, platinum, iron.

See also catalysis.

"Chemivic."[265] Trademark for vinyl-reinforced nitrile dry polymers, produced by thoroughly fluxing "Pliovic" PVC resin into "Chemigum" NBR rubber.

Uses: Integral molded shoe soles and mechanical goods, gaskets, solvent resistant hole and tubing, wire and cable jacketing.

"Chemlok"[547] Trademark for a series of adhesives, primarily for bonding rubber to metal.

"Chemlok" 205. A mixture of polymers, organic compounds and mineral fillers in a methyl isobutyl ketone and "Cellosolve" solvent system. Used as a primer with "Chemlok" 220 when the bond must have exceptional resistance to adverse environmental conditions or when surface preparation must be minimized. Also used as a single coat adhesive for bonding nitrile elastomers.

"Chemlok" 220. Dissolved organic polymers and dispersed fillers in a xylene and perchloroethylene solvent system. Used as a versatile one-coat adhesive for bonding uncured elastomers to metals during the vulcanization.

chemodynamics. A comprehensive, interdisciplinary study of chemicals in the environment, with special reference to pesticides.

chemonite (copper arsenite, ammoniacal). A wood-preservative solution prescribed by Federal Specification TT-W-549 to contain copper hydroxide, $Cu(OH)_2$, 1.84%; arsenic trioxide, As_2O_3, 1.3%; ammonia, NH_3, 2.8%; acetic acid, 0.05%; water, as necessary to 100.0%.

Hazard: Highly toxic by ingestion.

Shipping regulations: (Rail) Arsenical compounds, n.o.s., Poison label. (Air) Poison label. Arsenites, liquid, n.o.s., Poison label.

chemonuclear production. Manufacture of chemicals using the energy of a nuclear reactor. Feasibility studies carried out on making hydrogen cyanide from nitrogen and methane, using fission fragments as the energy source for the heat of reaction, have indicated that this process cannot yet compete economically with standard methods. In the case of hydrazine the economic aspect is more favorable, because of the high cost of conventional methods.

chemosterilant. A term coined by the USDA for materials or processes which sterilize insects, usually the males, thus preventing their reproductoin. Gamma radiation is one method used. The males are brought to the radiation by sex attractants.

chemotherapy. The treatment or prevention of a disease by administration of a chemical. The term was first used by Paul Ehrlich (q.v.), discoverer of the arsphenamine treatment for syphilis (1910); he stated that chemotherapy results from the interaction of chemically reactive groups on drugs and of chemically reactive receptor groups on parasitic cells, and that an effective drug must be of quite low molecular weight. His achievement was one of the great triumphs of biomedical science. More recent important achievements have been the development of antimalarials (q.v.); the synthesis of sulfa drugs; the discovery and proliferative development of antibiotics (penicillin, streptomycin, etc.); and the synthesis of cortisone. Much research effort has been devoted to the chemotherapeutic investigation of cancer; polycyclohydrocarbons such as anthracene, phenanthrene, benzopyrene, and their derivatives have been exhaustively studied in animals without conclusive results. Some of these compounds have been identified in cigarette tars. See also immunochemistry; adriamycin.

CHEMTREC. Abbreviation of Chemical Transportation Emergency Center; it is a division of the Manufacturing Chemists Association established as an emergency information source for transportation accidents involving flammable, toxic or explosive materials.

"Chemtree."[461] Trademark for a group of metallic mortars used for various kinds of nuclear shielding, where a combination of formability and high attenuation values is required. They are dry powders which are mixed with water before use.

chemurgy. The linking of agricultural, scientific, and industrial efforts to improve active cooperation between these branches of a nation's economy for their mutual benefit (National Farm Chemurgic Council). The chief objectives of chemurgy are: (1) new, nonfood uses for farm crops, their residues and byproducts; (2) new and profitable uses for previously unused agricultural wastes of both plant and animal origin; (3) new crops that farmers may grow profitably; (4) more valuable uses for presently used crops through chemurgic upgrading.

chenopodium oil (wormseed oil, American; goosefoot oil).
Properties: Colorless or yellowish oil; characteristic penetrating odor; bitterish, burning taste. Soluble in 3 to 10 vols of 70% alcohol (inferior and adulterated oils do not yield a clear solution). Sp. gr. 0.965 to 0.990 (15°C); optical rotation -4° to -8° 50′; refractive index 1.4740-1.4790 (20°C).
Chief known constituents: Ascaridole, $C_{10}H_{16}O_2$; para-cymene; *l*-limonene.
Derivation: Distilled from the seeds and leaves of Chenopodium ambrosioides anthelminticum.
Containers: Cans.
Hazard: Toxic by ingestion.
Use: Medicine (antihelminthic).

chi acid. See anthraquinone-1,8-disulfonic acid.

Chicago acid. See 8-amino-1-naphthol-5,7-disulfonic acid.

chicle. A thermoplastic, gumlike substance obtained from the latex of the sapodilla tree native to Mexico and Central America. Softens at 90°F. Insoluble in water; soluble in most organic solvents. Chief use is as chewing gum, after incorporation of sugar and specific flavoring. Nontoxic, but ingestion should be avoided.

Chilean nitrate. See sodium nitrate.

Chilean saltpeter. See sodium nitrate.

China clay. See kaolin.

China-wood oil. See tung oil.

"Chip-Cal."[147] Trademark for low-lime calcium arsenate. Available in granular (48% tricalcium arsenate) and powder (85% tricalcium arsenate) form.
Hazard: Highly toxic by ingestion.

"Chipcote."[147] Trademark for a series of organic mercury seed treatments based on methyl mercury nitrile.
Hazard: Highly toxic by ingestion.

chiral. In chemistry this term describes asymmetric molecules that are mirror-images of each other, i.e., they are related to each other optically as right and left hands. Such molecules are also called enantiomers, and are characterized by optical activity. See also optical isomer; enantiomer.

chlophedianol hydrochloride (USAN) (alpha-(2-dimethylaminoethyl) ortho-chlorobenzhydrol hydrochloride) $C_{17}H_{20}ClNO \cdot HCl$. Listed in N.D. Used in medicine.

chlor-. See chloro-.

chloral (trichloroacetaldehyde) CCl_3CHO.
Properties: Colorless, mobile, oily liquid; penetrating odor. Sp. gr. 1.505 (25/4°C); f.p. -57.5°C; b.p. 97.7°C; vapor pressure 35 mm (20°C); index of refraction (n 20/D) 1.4557; latent heat of vaporization 97.1 Btu/lb. Soluble in water, alcohol, ether and chloroform; combines with water forming chloral hydrate.
Derivation: (a) By the chlorination of ethyl alcohol, addition of sulfuric acid, and subsequent distillation; (b) by the chlorination of acetaldehyde.
Grades: Technical, 94% min.
Containers: Drums; glass bottles; tank cars.
Uses: Manufacture of chloral hydrate and DDT.

chloralamide. See chloral formamide.

chloral formamide (chloralamide; chloramide) $CCl_3CHOHNHOCH$.
Properties: Colorless, lustrous crystals; odorless; slightly bitter taste. Soluble in water (hydrolyzes at 60°C), alcohol, ether and glycerol. M.p. 114-115°C; decomposes at higher temperatures.
Use: Medicine.

chloral hydrate ("knockout drops"; trichloroacetaldehyde, hydrated; trichloroethylidene glycol) $CCl_3CH(OH)_2$.
Properties: Transparent, colorless crystals; aromatic, penetrating, slightly acrid odor and slightly bitter, sharp taste. Slowly volatilizes when exposed to air. Soluble in water, alcohol, chloroform, and ether; also soluble in olive oil and turpentine oil. Sp. gr. 1.901; m.p. 52°C; b.p. 97.5°C.
Derivation: Action of 1/5 of its volume of water on chloral.
Grades: Technical; U.S.P.
Hazard: Hypnotic drug; dangerous to eyes; toxic in overdose.
Uses: Medicine (sedative); manufacture of DDT; liniments.

chloral hydrate antipyrine (antipyrine chloral hydrate) $C_{11}H_{12}N_2OCl_3CH(OH)_2$.
Properties: Colorless crystals; moderately soluble in water; soluble in alcohol; m.p. 67°C.
Hazard: Probably toxic.
Use: Medicine (sedative).

chlor-alkali cell. See electrolytic cell.

chlorambucil (4-(para[bis(2-chloroethyl)amino]phenyl) butyric acid) $(ClC_2H_4)_2NC_6H_4(CH_2)_3COOH$. A nitrogen mustard derivative.
Properties: Off-white powder; m.p. 65-69°C. Slightly soluble in water; soluble in acetone and ether.
Grade: U.S.P.
Uses: Medicine; insect sterilant.

chloramid. See chloral formamide.

chloramine NH_2Cl. A colorless, unstable, pungent liquid; soluble in water; decomposes (slowly in dilute solution) to form nitrogen, hydrochloric acid, and ammonium chloride. M.p. -66°C; soluble in alcohol and ether. (Do not confuse with chloramine-T). Chloramine is an intermediate in the Raschig process for hydrazine (q.v.).

chloramine-B. $C_6H_5SO_2NClNa$ (sodium benzenesulfochloramide)
Properties: White powder with faint chlorine odor; soluble in water.
Uses: Medicine (antiseptic).

chloramine-T. (sodium para-toluenesulfochloramine). $CH_3C_6H_4SO_2NNaCl \cdot 3H_2O$. See also dichloramine-T.
Properties: White or slightly yellow crystals or crystalline powder. Contains not less than 11.5 nor more than 13% active chlorine. Slight odor of chlorine. Decomposes slowly in air, liberating chlorine. (Not to be confused with NH_2Cl, which is also termed chloramine). Soluble in water, insoluble in benzene, chloroform, ether; decomposed by alcohol.
Derivation: Reaction of ammonia and paratoluenesulfochloride under pressure. The latter is reacted with sodium hypochlorite in the presence of an alkali and the chloramine produced by crystallinization.
Use: Medicine (antiseptic).

chloramphenicol
$NO_2C_6H_4CH(OH)CH(CH_2OH)NHCOCHCl_2$.
D(-)Threo-1-(para-nitrophenyl)-2-dichloroacetamido-1,3-propandiol. An antibiotic derived from Streptomyces venezuelae or by organic synthesis. It was the first substance of natural origin shown to contain an aromatic nitro group.
Properties: Fine, white to grayish-white or yellowish-white, needlelike crystals or elongated plates. Bitter to taste, neutral to litmus, and reasonably stable in neutral or slightly acid solutions. M.p. 149–153°C; alcoholic solution is dextrorotatory while ethyl acetate solution is levorotatory. Very slightly soluble in water; freely soluble in alcohol, propylene glycol, acetone and ethyl acetate.
Grade: U.S.P.
Hazard: May have deleterious and dangerous side effects. Must conform to FDA labelling requirements; use is closely restricted.
Uses: Medicine (antibiotic); antifungal agent.

chloramphenicol sodium succinate $C_{15}H_{15}Cl_2N_2NaO_8$.
Properties: Light yelow, crystalline powder. Freely soluble in water and alcohol.
Grade: U.S.P.
Use: Medicine. (See chloramphenicol).

chloranil (tetrachloroquinone; tetrachloro-para-benzoquinone) $C_6Cl_4O_2$.
Properties: Yellow leaflets; m.p. 290°C; sp. gr. 1.97; soluble in alcohol, ether, and benzene; insoluble in water; good storage stability.
Derivation: From phenol, para-chlorophenol, or para-phenylenediamine by treatment with potassium chlorate and hydrochloric acid.
Hazard: Skin irritant.
Uses: Agricultural fungicide; dye intermediate; electrodes for pH measurements; vulcanizing agent.

chloranthrene yellow. See flavanthrene.

chlorapatite. See apatite.

"Chlorasol."[214] Trademark for a fumigant composition; 70.3% ethylene dichloride, 29.7% carbon tetrachloride by weight. Boiling range 75–78°C; no flash; nonflammable.
Uses: Fumigant for meal, grain and clothes moths; grain weevils; grain, flour, and carpet beetles; the rice weevil; book lice; penetrates stored grain, rolled rugs, upholstered furniture, cartons, sacks, and stacked material.

chlorauric acid. See gold trichloride.

chlorazine. Generic name for 2-chloro-4,6-bis(diethylamino)-s-triazine.
$[(C_2H_5)_2N]_2\overline{CNC(Cl)NCN}$.
Properties: Solid; m.p. 15–18°C; sp. gr. 1.096 (20°C). Soluble in acetone, chloroform, ethyl alcohol, and xylene.
Hazard: May be toxic.
Use: Herbicide.

chlorbenside (par-chlorobenzyl para-chlorophenyl sulfide) $ClC_6H_4CH_2SC_6H_4Cl$. Generic name for an agricultural toxicant.
Properties: Crystals with almond-like odor (technical grades); m.p. 75–76°C; insoluble in water; soluble in most organic solvents; resistant to acid and alkaline hydrolysis.
Grades: Technical.
Hazard: Toxic; skin irritant.

chlor- compounds. Most organic compounds of chlorine retain the letter "o" in accepted chemical terminology, i.e., chlorobenzene, chloroacetic, etc., although the form without the "o" is sometimes used. Therefore, for chlor- compounds, see also chloro-.

chlorcyclizine hydrochloride
$ClC_6H_4CH(C_6H_5)C_4H_8N_2CH_3 \cdot HCl$. 1-(para-Chloro-alpha-phenylbenzyl)-4-methylpiperazine hydrochloride.
Properties: White, odorless, crystalline solid with bitter taste; m.p. 222–227°C; very soluble in water; soluble in chloroform and alcohol; practically insoluble in benzene and ether. Solutions acid to litmus; pH (1 in 100 solution) 4.8–5.5.
Grade: U.S.P.
Use: Medicine (antihistamine, q.v.).

chlordane $C_{10}H_6Cl_8$ (1,2,4,5,6,7,8,8-octachloro-4,7-methano-3a,4,7,7a-tetrahydroindane).
Properties: Colorless, odorless, viscous liquid; sp. gr. 1.57–1.67 (60/60°F); viscosity SSU 100 seconds (38°C); organic chlorine 64–67% by wt, purity 98%; b.p. 175°C (2 mm); refractive index (n 25/D) 1.56–1.57; soluble in many organic solvents; insoluble in water; miscible in deodorized kerosine; decomposes in weak alkalies.
Grades: Technical and pure.
Containers: Aluminum, aluminum-clad or phenolic enamel-lined metal containers; pails; drums.
Hazard: Toxic by ingestion, inhalation, and skin absorption. Use has been restricted. Tolerance, 0.5 mg per cubic meter of air.
Uses: Insecticides; oil emulsions, dusts and dispersible liquids.

chlordiazepoxide hydrochloride (USAN) (7-chloro-2-methylamino-5-phenyl-3-H-1,4-benzodiazepine-4-oxide hydrochloride) $C_{16}H_{14}ClN_3O \cdot HCl$.
Properties: Crystals; m.p. 212–218°C; soluble in water; sparingly soluble in alcohol; insoluble in ether and chloroform.
Hazard: Toxic in high concentrations. Central nervous system depressant. Manufacture and dosage controlled by law.
Use: Medicine (tranquilizer).

chlordimeform [N'-(4-chloro-*o*-tolyl)-N,N-dimethylformamine].
Ovicide, insecticide and miticide designed for use on cotton and vegetable crops. Available in a concentrated emulsion form, it is stated to be less toxic than organophosphates and to be biodegradable. It is trademarked "Galecron."

chlorendic anhydride (hexachloroendomethylenetetrahydrophthalic anhydride) $C_6H_2Cl_6O_3$. 1,4,5,6,7,7-Hexachlorobicyclo-(2,2,1)-5-heptene-2,3-dicarboxylic anhydride.
Properties: Fine, white, free-flowing crystals. M.p. 239–240°C; sp. gr. 1.73. Readily soluble in acetone, benzene, toluene; slightly soluble in water, n-hexane, and carbon tetrachloride. Nonflammable.
Derivation: By Diels-Alder reaction of maleic anhydride and hexachlorocyclopentadiene.
Grades: Technical; pure.
Uses: Flame-resistant polyester resins; hardening epoxy resins; chemical intermediate; source of chlorendic acid.

"Chloretone."[330] Trademark for chlorobutanol (q.v.).

"Chlorex."[214] Trademark for 2,2′-dichloroethyl ether (q.v.).

"Chlorextol."[116] Trademark for a synthetic, nonflammable, non-sludging liquid of high dielectric strength; used for insulating and cooling transformers and other electrical equipment.

"Chlorhydrol."[15a] Trademark for an aluminum chlorohydroxide antiperspirant. Offered in liquid or powder form for use in lotions, creams, and gels. Noncorrosive and nonirritant to skin. Does not damage fabrics.

chloric acid $HClO_3 \cdot 7H_2O$. Known only in aqueous solution. Decomposes at 40°C.
Hazard: Toxic; strong oxidizer. Ignites organic materials on contact.
Shipping regulations: (Rail) Oxidizing material, n.o.s., Yellow label. (Air) Not acceptable.

chloridizing. Heating in the presence of chlorine, as a step in the recovery of certain metals from their oxides or other compounds.

"Chlorimets."[47] Trademark for a series of nickel-base cast alloys.
"Chlorimet" 2 contains 32% molybdenum, 3% iron max, 1% silicon and 0.10% carbon.
"Chlorimet" 3 contains 18% molybdenum, 18% chromium, 3% iron max, 1.0% silicon and 0.07% carbon.

chlorinated acetone. See chloroacetone.

chlorinated camphene. See toxaphene.

chlorinated hydrocarbon. See hydrocarbon, halogenated.

chlorinated isocyanuric acid. See dichloro- or trichloroisocyanuric acid, and potassium or sodium dichloroisocyanurate. Used as dry bleaches.

chlorinated lime. See lime, chlorinated.

chlorinated naphthalene (chloronaphthalene). $C_{10}H_7Cl$. A product of the chlorination of naphthalene. The physical state varies from mobile liquids to crystalline solids depending on the extent of chlorination.
Hazard (tri- and higher): Toxic by ingestion, inhalation, and skin absorption. Strong irritants. Tolerance (tetra compound) 2 mg per cubic meter of air.
See also oil, chloronaphthalene; wax, chloronaphthalene.

chlorinated paraffin. See paraffin, chlorinated.

chlorinated para red. A modification of para red that contains some chlorine. Much lighter than para or toluidine red and has excellent brilliance but poorer heat resistance.

chlorinated polyether. See "Penton."

chlorinated polyolefin. See rubber, chlorinated; polypropylene, chlorinated.

chlorinated rubber. See rubber, chlorinated.

chlorinated trisodium phosphate. See trisodium phosphate, chlorinated.

chlorine Cl. Nonmetallic, halogen element of atomic number 17; Group VIIA of the Periodic Table. Atomic weight 35.453. Valences 1,3,4,5,7. Two stable isotopes Cl-35 (75.4%) and Cl-37 (24.6%). Seventh highest-volume chemical produced in U.S. (1975).
Properties: (1) A diatomic gas, which is heavy, noncombustible, greenish-yellow; pungent irritating odor. Liquefaction pressure 7.86 atm (25°C), 1 atm at −35°C. Water solubility 0.64 gram Cl_2 per 100 grams water. Sp. gr. 2.49 (0°C) (air = 1). Thermodynamic properties: (a) critical temperature 144.0°C; (b) critical pressure 78.525 atm absolute; (c) critical volume 1.763 liters per kg. Strongly electronegative.
(2) Liquid: Clear, amber color, irritating odor; sp. gr. 1.56 (−35°C); f.p. −101°C; 1 liter of liquid = 456.8 liters of gas at 0°C and 1 atm. Very low electrical conductivity. Soluble in chlorides and alcohols. Extremely strong oxidizing agent. Slightly soluble in cold water.
Occurrence: Not found free in nature; component of minerals halite (rock salt), sylvite, and carnallite; chloride ion in sea water.
Derivation: (1) Electrolysis of sodium chloride brine in either diaphragm or mercury cathode cells; chlorine is released at the anode. (2) Fused salt electrolysis of sodium chloride or magnesium chloride; (3) Electrolysis of hydrochloric acid. (4) Oxidation of hydrogen chloride with nitrogen oxide as catalyst and absorption of steam with sulfuric acid ("Kel-Chlor" process); no by-product caustic is produced.
Grades: Technical (gas and liquid); pure (99.9%).
Containers (liquid): 100 and 150-lb steel cylinders; single-unit tank cars; multi-unit tank cars; motor trucks in cylinders and containers; tank barges (up to 600 tons); pipelines (liquid and gas).
Hazard: Toxic as irritant and by inhalation. Tolerance, 1 ppm in air. Moderate fire risk in contact with turpentine, ether, ammonia, hydrocarbons, hydrogen, powdered metals, and other reducing materials. Safety data sheet available from Manufacturing Chemists Assn., Washington, D.C.
Uses: Manufacture of carbon tetrachloride, trichloroethylene, chlorinated hydrocarbons, polychloroprene (neoprene), polyvinyl chloride, hydrogen chloride, ethylene dichloride, hypochlorous acid, metallic chlorides, chloroacetic acid, chlorobenzene, etc. Also in water purification, shrinkproofing wool, in flame-retardant compounds and (with lithium) in special batteries; processing of meat, fish, vegetables and fruit.
Shipping regulations: (Rail) Green label. (Air) Nonflammable Gas label. Not accepted.
For further details, refer to the Chlorine Institute, 342 Madison Ave., N.Y.

chlorine 36. Radioactive chlorine of mass number 36. Half-life about 440,000 years; radiation, beta.

Derivation: Separated from various isotopes produced during the pile irradiation of potassium chloride.
Forms available: As hydrochloric acid solution and as solid potassium chloride.
Hazard: Moderately toxic. Permissible level 4×10^{-7} microcurie per milliliter in air.
Uses: Tracer in studying the salt water corrosion of metals, expecially steel; reaction mechanism of chlorinated hydrocarbons; location and flow of salt waters in porous media; etc.
Shipping regulations: (Rail, Air) Consult regulations.

chlorine-bromide. See bromine-chloride.

chlorine dioxide ClO_2.
Properties: Red-yellow gas; f.p. -59.5°C; b.p. 10°C; very reactive, unstable. Strong oxidizing agent. Decomposes in water. Dissolves in alkalies forming a mixture of chlorite and chlorate.
Derivation: Usually made at point of consumption from sodium chlorate, sulfuric acid and methanol, or from sodium chlorate and sulfur dioxide. Concentration of gas is limited to 10% to reduce explosion hazard.
Grades: Sold as hydrate, in frozen form.
Hazard: Irritant to skin and mucous membranes. Explodes when heated or by reaction with organic materials. Tolerance 0.1 ppm in air.
Uses: Bleaching wood pulp, fats and oils; maturing agent for flour (nontoxic in percentage used); water treatment (purification and taste removal); swimming pools; odor control.
Shipping regulations: (Rail) (frozen hydrate): Yellow label. Not accepted by express. (Air) Not listed. Consult authorities.

chlorine trifluoride ClF_3.
Properties: Nearly colorless gas, pale green liquid, or white solid. B.p. 11.3°C; f.p. -76.3°C; density of gas (air = 1) 3.14. Extremely reactive, comparable to fluorine.
Derivation: By reaction of chlorine and fluorine at 280°C and condensation of the product at -80°C. Obtained 99.0% pure.
Containers: Steel cylinders.
Hazard: Corrosive to skin; dangerous fire risk. Explodes in contact with organic materials or with water. Tolerance, 0.1 ppm in air.
Uses: Incendiary; fluorinations; cutting oil well tubes; oxidizer in propellants.
Shipping regulations: (Rail) White label. (Air) Corrosive label. Not acceptable on passenger planes.

chlorine water. Clear yellowish liquid; deteriorates on exposure to air and light. Made by saturating water with approximately 0.4% chlorine.
Use: Deodorizer, disinfectant, antiseptic.

chloriodized oil. Chlorinated and iodized vegetable oil. Contains 26.0–28.0% iodine in organic combination.
Properties: Pale yellow, viscous, oily liquid with faint, bland taste. Practically insoluble in water; slightly soluble in alcohol; freely soluble in benzene, chloroform and ether.
Derivation: Formed by chemical addition of iodine monochloride to a vegetable oil.
Hazard: May be toxic by ingestion.
Use: Medicine.

chlorisondamine chloride $C_{14}H_{20}Cl_6N_2$. 4,5,6,7-Tetrachloro-2-(2-dimethylaminoethyl) isoindoline dimethylchloride. A quaternary ammonium compound. Listed in N.D.
Properties: Crystals; decompose 258–265°C. Soluble in water and alcohol.
Use: Medicine.

chlormerodrin ([3-chloromercuri)-2-methoxypropyl]-urea) $ClHgCH_2CH(OCH_3)CH_2NHCONH_2$.
Properties: White, odorless powder with bitter metallic taste. Very soluble in sodium hydroxide; very slightly soluble in chloroform; slightly soluble in alcohol, methanol, and water. Stable to light and air; pH (0.5% solution) 4.3–5.0.
Grade: N.F.
Hazard: Highly toxic in overdose.
Use: Medicine.
Shipping regulations: (Rail, Air) Mercury compounds, n.o.s., Poison label.

chlormethazanone $C_{11}H_{12}ClNO_3S$. 2-(4-Chlorophenyl)-3-methyl-4-meta-thiazanone-1-dioxide.
Properties: Crystals; m.p. 117°C; insoluble in water; slightly soluble in alcohol.
Use: Medicine (tranquilizer and muscle relaxant.)

chloroacetaldehyde $ClCH_2CHO$.
Properties: (of 40% aqueous solution): Clear, colorless liquid with pungent odor. Boiling range 90–100°C; f.p. -16.3°C; sp. gr. 1.19 (25/25°C); refractive index 1.397 (25°C); wt/gal 9.9 lb (25°C). Soluble in water, acetone, methanol. At concentrations in water above 50% it forms an insoluble hemihydrate. Pure substance has flash pt. 190°F. Combustible.
Hazard: Corrosive to skin and mucous membranes; tolerance, 1 ppm in air.
Uses: Intermediate; fungicide.

chloroacetaldehyde dimethyl acetal. See dimethyl chloroacetal.

chloroacetamide (alpha-chloroacetamide; 2-chloroethanamide) $ClCH_2CONH_2$.
Properties: Colorless to pale yellow crystals; characteristic odor; m.p. 117–119°C; soluble in water and alcohol; insoluble in ether.
Use: Intermediate.

chloroacetic acid (chloracetic acid; MCA; monochloroacetic acid) $CH_2ClCOOH$.
Properties: Colorless to light-brownish crystals; deliquescent. Sp. gr. 1.370 (70°C); crystallizing point, alpha form, 61.0–61.7°C; beta form, 55.5 to 56.5°C; gamma form, 50°C. The commercial material melts at 61 to 63°C; boiling range 186 to 191°C. Soluble in water, alcohol, ether, chloroform, carbon disulfide. Nonflammable.
Derivation: Action of chlorine on acetic acid in the presence of acetic anhydride, phosphorus, or sulfur.
Grades: Technical; medicinal; 99.5% pure.
Containers: Drums.
Hazard: Use in food products prohibited by FDA. Irritating and corrosive to skin.
Uses: Herbicide; preservative; bacteriostat; intermediate in production of carboxymethylcellulose, ethyl chloroacetate, glycine, synthetic caffeine, sarcosine, thioglycolic acid, EDTA, 2,4-D, 2,4,5-T.
Shipping regulatons: (Rail) (liquid), White label. (Air) (solid and solution), Corrosive label. Legal label name (Rail) monochloroacetic acid.

chloroacetic anhydride (chloroethanoic anhydride; sym-dichloroacetic anhydride) $(ClCH_2CO)_2O$.
Properties: Colorless to slightly yellow crystals with pungent odor; m.p. 51–55°C; soluble in chloroform and ether; hydrolyzes with water to chloroacetic acid.
Hazard: Irritating to skin and eyes; moderately toxic by inhalation.
Use: Intermediate.

ortho-**chloroacetoacetanilide** $CH_3COCH_2CONHC_6H_4Cl$.
Properties: White, crystalline solid. Resembles ethyl acetoacetate in chemical reactivity. M.p. 107°C; vapor pressure 0.1 mm (20°C); nonflammable. Insoluble in water.
Uses: Organic synthesis; dyestuffs.

chloroacetocatechol. See chloroacetopyrocatechol.

alpha-**chloroaceto-3,4-dihydroxybenzene.** See chloroacetopyrocatechol.

chloroacetone (monochloroacetone; 1-chloro-2-propanone; chloracetone; chlorinated acetone) CH_3COCH_2Cl. A lachrymator.
Properties: Colorless liquid; pungent, irritating odor. Sp. gr. 1.162 (16°C); b.p. 119°C; f.p. -44.5°C. Soluble in alcohol, ether and chloroform and water.
Derivation: Chlorination of acetone.
Hazard: Strong irritant to tissue, eyes, and mucous membranes.
Uses: Couplers for color photography; enzyme inactivator; insecticides; perfumes; antioxidant intermediate; medicine; organic synthesis; tear gas.
Shipping regulations: (Rail) (stabilized) Tear Gas label; (unstabilized) Not accepted. Legal label name, monochloracetone. (Air) (stabilized) Poison label. Not acceptable on passenger planes; (unstabilized) Not acceptable.

chloroacetonitrile (chloroethane nitrile; chloromethyl cyanide) $ClCH_2CN$.
Properties: Colorless liquid, with pungent odor; sp. gr. 1.2020–1.2035 (25/25°C); refractive index 1.4210–1.4240 (n 25/D); 5–95% distils between 124–129°C; soluble in hydrocarbons, alcohols; insoluble in water.
Hazard: Probably toxic and irritant.
Uses: Fumigant; intermediate.

chloroacetophenone (chloracetophenone, CN; phenacylchloride; phenyl chloromethyl ketone) $C_6H_5COCH_2Cl$. This formula is for the omega (or alpha) isomer, which is a powerful lachrymator. The para form, $ClC_6H_4COCl_3$, is also available.
Properties: White crystals; floral odor; m.p. 56°C; b.p. 247°C. The para isomer has m.p. 20°C; b.p. 273°C. Insoluble in water; soluble in acetone, benzene, carbon disulfide.
Derivation: From chloroacetylchloride, benzene and aluminum chloride.
Hazard: Strong irritant to eyes and tissue as gas or liquid. Tolerance (alpha), 0.05 ppm in air.
Uses: Pharmaceutical intermediate; riot control gas.
Shipping regulations: (Rail) Not listed. (Air) (omega isomer) Poison label. Not acceptable on passenger planes.

chloroacetopyrocatechol (chloroacetocatechol; alpha-chloroaceto-3,4-dihydroxybenzene) $(HO)_2C_6H_3COCH_2Cl$. Melting range 171–178°C. Pharmaceutical intermediate.

chloroacetyl chloride (chloracetyl chloride) $ClCH_2COCl$. A lachrymator.
Properties: Water-white liquid; pungent odor. Sp. gr. 1.495 (0°C); b.p. 105 to 110°C; decomposes in water. Nonflammable.
Derivation: (a) Action of chlorine on acetyl chloride in sunlight. (b) Dropping phosphorus trichloride on chloroacetic acid.
Hazard: Irritating to eyes; corrosive to skin.
Uses: Preparation of chloroacetophenone; intermediate; tear gas.
Shipping regulations: (Rail) Corrosive liquid, n.o.s. White label. (Air) Corrosive label. Not acceptable on passenger planes.

chloroacetylurethane $ClCH_2CONHCOOC_2H_5$.
Properties: Crystals; soluble in alcohol; sparingly soluble in water. M.p. 129°C.
Derivation: By interaction of a urethane derivative and ethyl chloroacetate.

chloroacrolein $H_2C{:}CClCHO$.
Properties: Colorless liquid. Sp. gr. 1.205 at 15°C; b.p. 29 to 31°C (17 mm).
Derivation: Chlorination of acrolein.
Hazard: Toxic; irritating to eyes and skin.
Uses: Tear gas.
Shipping regulations: (Rail, Air) Not listed. Consult regulations.

alpha-**chloroacrylonitrile.**
Properties: Activated carbon-carbon double bond and reactive chlorine and metal groups. Readily polymerizes, and copolymerizes with other unsaturated monomers. High cross-linking ability.
Derivation: Chlorination of acrylonitrile and dehydrohalogenation by cracking.
Uses: Synthetic fibers, coatings, and films; acrylic polymers; treatment of cotton fiber; intermediate.

2-chloroallyl diethyldithiocarbamate (CDEC). $(C_2H_5)_2NCSSCH_2CCl{:}CH_2$.
Properties: Amber liquid; b.p. 128–130°C (1 mm); very slightly soluble in water; soluble in benzene, alcohol, acetone, chloroform, and ether.
Forms available: Liquid and granular.
Hazard: Dry preparations are irritating to eyes and skin.
Uses: Herbicide; pesticide.

1-(3-chloroallyl)-3,5,7-triaza-1-azoniaadamantane chloride $C_6H_{12}N_4(CH_2CHCHCl)Cl$. White to cream-colored powder; soluble in water and methanol; almost insoluble in acetone.
Uses: Bactericide used as preservative in latexes, paints, floor polishes, joint cements, adhesives, inks, starches, etc.

chloroaluminum diisopropoxide $[(CH_3)_2CHO]_2AlCl$. White crystalline solid; m.p. 160°C (dec). Soluble in most organic solvents; hydrolyzes.
Uses: Catalyst and intermediate.
Hazard: May be irritant and toxic.

chloroamino- See aminochloro-.

para-**chloro**-ortho-**aminophenol.** See 2-amino-4-chlorophenol.

2-chloro-4-tert-**amylphenol** $C_5H_{11}C_6H_3ClOH$.
Properties: Water-white liquid with aromatic odor; sp. gr. 1.11 (20°C); boiling range 253–265°C; flash point 225°F. Combustible.
Hazard: Probably toxic and irritant.

meta-**chloroaniline** (meta-aminochlorobenzene) $ClC_6H_4NH_2$.
Properties: Colorless to light amber liquid; tends to darken during storage. Boiling range 228–231°C; f.p. -10.6°C. Insoluble in water; soluble in organic solvents.
Grades: Technical.
Containers: Drums; tank cars.

Hazard: Toxic by ingestion, inhalation and skin absorption.
Uses: Intermediate for azo dyes and pigments; pharmaceuticals; insecticides; agricultural chemicals.
Shipping regulations: (Rail), Not listed. (Air) Poison label.

ortho-**chloroaniline** (ortho-aminochlorobenzene) $ClC_6H_4NH_2$.
Properties: Amber liquid; amine odor; darkens on exposure to air. Distillation range 208–210°C; f.p. –2.3°C; sp. gr. 1.213 (20/4°C); refractive index 1.5896 (n 20/D). Miscible with alcohol and ether; insoluble in water.
Grades: Technical.
Containers: Drums; tank cars.
Hazard: Toxic by ingestion, inhalation, and skin absorption.
Use: Dye intermediate; standards for colorimetric apparatus; manufacture of petroleum solvents and fungicides.
Shipping regulations: (Rail) Not listed. (Air) Poison label.

para-**chloroaniline** (para-aminochlorobenzene) $ClC_6H_4NH_2$.
Properties: White or pale yellow solid. M.p. 69.5°C; distilling range 229–233°C. Soluble in hot water and organic solvents.
Grades: Technical.
Containers: Drums.
Hazard: Toxic by inhalation and ingestion.
Use: Dye intermediate; pharmaceuticals; agricultural chemicals.
Shipping regulations: (Rail) Not listed. (Air) Poison label.

4-chloroaniline-3-sulfonic acid $HSO_3C_6H_3ClNH_2$.
Properties: White to light-grey powder.
Containers: Fiber kegs; polyethylene-lined steel drums.
Hazard: Probably toxic.
Use: Intermediate.

2-chloroanthraquinone $C_{14}H_7ClO_2$.
Properties: See anthraquinone and 2-methylanthraquinone. M.p. 208–211°C. Insoluble in water; soluble in hot benzene. Probably low toxicity.
Derivation: Condensing phthalic anhydride and chlorobenzene in the presence of anhydrous aluminum chloride to form para-chlorobenzoylbenzoic acid. Ring closure of the intermediate acid is brought about by heating in sulfuric acid solution.
Use: Starting material for certain vat dyes.

chloroauric acid. See gold trichloride.

chloroazotic acid. See aqua regia.

chlorobenzal. See benzyl dichloride.

chlorobenzaldehyde C_6H_4CHOCl.
Properties: Colorless to yellowish liquid (ortho-) or (powder (para-); boiling range 209–215°C; f.p. 8.0°C (min); sp. gr. 1.240–1.245 (25/25°C). Soluble in alcohol, ether and acetone. Insoluble in water. Combustible.
Hazard: May be toxic.
Uses: Intermediate in the preparation of triphenyl methane and related dyes; organic intermediate.

ortho-**chlorobenzalmalononitrile** (CS).
Properties: Off-white, crystalline powder. Available both unground, and ground with 5% silica aerogel or treated "Cab-O-Sil" (q.v.).
Hazard: Strong irritant to eyes and mucous membranes.
Use: Riot control aerosol.

3-chloro-4-benzamido-6-methylaniline $ClC_6H_2NH_2CH_3(NHCOC_6H_5)$.
Properties: White solid; m.p. 198–199°C.
Hazard: Probably toxic.
Uses: Azoic dyes; pigments.

chlorobenzanthrone $C_{17}H_9ClO$.
Properties: All isomers: yellow needles. Soluble in alcohol, benzene, toluene, acetic acid.
Derivation: From benzanthrone by treatment with chlorine.

chlorobenzene (monochlorobenzene; phenyl chloride) C_6H_5Cl.
Properties: Clear, volatile liquid. Almond-like odor. Sp. gr. 1.105 (25/25°C); b.p. 131.6°C; f.p. –45°C; wt/gal 9.19 lb (25°C); refractive index 1.5216 (25°C); flash point 85°F. (closed cup). Autoignition temp. 1180°F. Miscible with most organic solvents; insoluble in water.
Derivation: By passing dry chlorine into benzene with a catalyst.
Grades: Technical.
Containers: 55-gal drums; tank cars; tank trucks.
Hazard: Flammable, moderate fire risk; avoid inhalation and skin contact. Tolerance, 75 ppm in air. Explosive limits 1.8 to 9.6% in air.
Uses: Phenol; chloronitrobenzene; aniline; solvent carrier for methylene diisocyanate; solvent; pesticide intermediate.

para-**chlorobenzenesulfonamide** $ClC_6H_4SO_2NH_2$.
Properties: White, odorless powder; m.p. 145–148°C; soluble in alcohol.
Grade: 98–99% purity.
Hazard: May be toxic.
Use: Intermediate for pharmaceuticals and resins.

para-**chlorobenzenethiol.** See para-chlorothiophenol.

para-**chlorobenzhydrol** (para-chlorobenzohydrol) $ClC_6H_4C(C_6H_5)HOH$.
Properties: White to off-white, crystalline powder; m.p. 57–61°C; insoluble in water; soluble in ether, alcohol, and benzene.
Use: Organic synthesis.

chlorobenzilate (ethyl 4,4′-dichlorobenzilate; ethyl 2-hydroxy-2,2-bis(4-chlorophenyl acetate) $(C_6H_4Cl)_2C(OH)COOC_2H_5$.
Properties: Viscous yellow liquid; b.p. 141–142°C (0.06 mm). Sp. gr. (technical, 90%) 1.2816 (20/4°C). Slightly soluble in water; soluble in acetone, benzene, methanol.
Hazard: Reported to cause tumors in mice.
Use: Pesticides.

"Chlorobenzilate."[219] Trademark for ethyl 4,4′-dichlorobenzilate. Available in emulsifiable solution or 25% wettable powder. Used as a miticide.

para-**chlorobenzohydrol.** See para-chlorobenzhydrol.

para-**chlorobenzoic acid** ClC_6H_4COOH.
Properties: Nearly white coarse powder; m.p. (ortho-) 137°C, (para-) 238°C. Soluble in methanol, absolute alcohol, and ether; insoluble in water, 95% alcohol, and toluene.
Uses: Intermediate for the preparation of dyes, fungicides, pharmaceuticals and other organic chemicals.

para-**chlorobenzophenone** $ClC_6H_4COC_6H_5$.
Properties: White to off-white crystalline powder; m.p. 73–78°C; b.p. 332°C; soluble in acetone, benzene, carbon tetrachloride, ether and hot alcohol. Insoluble in water.
Use: Intermediate.

chlorobenzotriazole $ClC_6H_4NHN{:}N$.
Properties: White solid; m.p. 157–159°C.
Uses: Intermediate and photographic chemical.

ortho-**chlorobenzotrichloride** $ClC_6H_4CCl_3$.
Properties: Colorless liquid or solid. M.p. 29.37°C; b.p. 264.3°C; sp. gr. 1.5131 (25/4°C); refractive index (n 20/D) 1.5836. Soluble in alcohol, ether, and acetone; decomposes in water. Low toxicity.
Uses: Intermediate for pharmaceuticals, dyes, and other organic chemicals.

para-**chlorobenzotrichloride** $ClC_6H_4CCl_3$.
Properties: Water-white liquid; boiling range 248–257°C; f.p. (approx.) 3.8°C; sp. gr. 1.480–1.490 (25/25°C). Soluble in alcohol, ether, and acetone. Insoluble in water. Low toxicity.
Uses: Same as ortho-chlorobenzotrichloride.

meta-**chlorobenzotrifluoride** (meta-chlorotrifluoromethylbenzene; meta-chloro-alpha, alpha, alpha-trifluorotoluene) $ClC_6H_4CF_3$.
Properties: Water white aromatic liquid; b.p. 138°C; freezing point -56°C; refractive index 1.446 (n 20/D); flash point 122°F (closed cup); sp. gr. 1.351 (15.5/15.5°C). Combustible.
Hazard: Moderate fire risk. Toxic by ingestion and inhalation. Tolerance (as F), 2.5 mg per cubic meter of air.
Uses: Intermediate in dyes and pharmaceuticals; dielectrics; insecticides.

ortho-**chlorobenzotrifluoride** (ortho-chlorotrifluoromethylbenzene; ortho-chloro-alpha, alpha, alpha-trifluorotoluene) $ClC_6H_4CF_3$.
Properties: Colorless liquid with aromatic odor; sp. gr. 1.379 (15.5/15.5°C); refractive index (1.456 (20°C); b.p. 152°C; f.p. -7.4°C; flash point (closed cup) 138°F; wt/gal 11.50 lb (15.5°C). Combustible.
Hazard: Moderate fire risk. Toxic by ingestion and inhalation. Tolerance (as F), 2.5 mg per cubic meter of air.
Uses: Dye intermediate, chemical intermediate, solvent and dielectric fluid.

para-**chlorobenzotrifluoride** (para-chlorotrifluoromethylbenzene; para-chloro-alpha, alpha, alpha-trifluorotoluene) $ClC_6H_4CF_3$.
Properties: Water-white liquid; aromatic odor; b.p. 139.3°C; f.p. -36°C; refractive index (n 20/D) 1.446; flash point (closed cup) 116°F; sp. gr. (15.5/15.5°C) 1.353; wt/gal 11.28 lb (15.5°C). Combustible.
Hazard: Moderate fire risk. Toxic by ingestion and inhalation. Tolerance (as F), 2.5 mg per cubic meter of air.
Uses: Same as ortho-chlorobenzotrifluoride.

chlorobenzoyl chloride ClC_6H_4COCl.
Properties: Colorless liquid; boiling range 227–239°C; f.p. -4 to -6°C; f.p. (ortho-) -4 to -6°C, (para-) 10 to 12°C; sp. gr. 1.374–1.376 (25/15°C). Soluble in alcohol, ether, and acetone. Insoluble in water.
Use: Intermediate for pharmaceuticals, dyes, and other organic chemicals.

para-**chlorobenzoyl peroxide** $(ClC_6H_4CO)_2OO$.
Properties: White, odorless powder; decomposes violently on heating or contamination. Insoluble in water; soluble in organic solvents.
Hazard: Dangerous fire and explosion risk; explodes when heated to 38°C; strong oxidizer; will ignite on contact with organic materials. Store in dark, cool locality.
Uses: Bleaching agent; polymerization catalyst.
Shipping regulations: (Rail) Yellow label. (Air) Organic Peroxide label.

chlorobenzyl chloride $ClC_6H_4CH_2Cl$.
Properties: Colorless liquid; boiling range 216–222°C; f.p. (ortho-) below -30°C, (para-) 27°C, sp. gr. 1.270–1.280 (25/15°C). Soluble in alcohol, ether, and acetone; insoluble in water.
Hazard: Probably irritating to skin and eyes.
Use: Intermediate for organic chemicals, pharmaceuticals and dyes.

para-**chlorobenzyl** para-**chlorophenyl sulfide.** See chlorbenside.

para-**chlorobenzyl cyanide** $ClC_6H_4CH_2CN$.
Properties: Colorless to pale yellow solid. M.p. 27°C. Soluble in acetone and alcohol.
Containers: 25-lb cans; 500-lb drums.
Hazard: Highly toxic.
Use: Organic synthesis.
Shipping regulations: (Air) Cyanide solutions, n.o.s., Poison label.

para-**chlorobenzyl** para-**fluorophenyl sulfide.** See fluorbenside.

2-(para-**chlorobenzyl)pyridine** (2-(4-chlorobenzyl)pyridine) $ClC_6H_4CH_2C_5H_4N$.
Properties: Liquid; b.p. 310.5°C; f.p. 8.4°C; sp. gr. 1.168 (25°C); refractive index 1.5865 (n 20/D); insoluble in water.
Use: Organic synthesis.

chlorobromo-. See bromochloro-.

2-chlorobutadiene-1,3. See chloroprene.

1-chlorobutane. See n-butyl chloride.

chlorobutanol (trichloro-tert-butyl alcohol; 1,1,1-trichloro-2-methyl-2-propanol; acetone chloroform) $Cl_3CC(CH_3)_2OH$.
Properties: Colorless to white crystals with characteristic odor and taste. Soluble in alcohol and glycerol; slightly soluble in water; readily soluble in ether, chloroform, and volatile oils. M.p. (anhydrous form) 97°C; m.p. (hemihydrate) 78°C; b.p. 167°C; sublimes easily. Combustible.
Derivation: By action of potassium hydroxide on a solution of chloroform and acetone.
Grade: U.S.P.
Hazard: Toxic; action similar to chloral hydrate (q.v.).
Uses: Plasticizer for cellulose esters and ethers; preservative for biological fluids and solutions.

4-chloro-2-butynyl meta-**chlorocarbanilate.** See barban.

chlorocarbon. A compound of carbon and chlorine, or or carbon, hydrogen, and chlorine, such as carbon tetrachloride, chloroform, tetrachloroethylene, etc.

chlorocarbonyl ferrocene (ferrocenoyl chloride) $C_5H_5FeC_5H_4COCl$. Orange-red solid; m.p. 48–49°C.
Hazard: Probably toxic.
Use: Intermediate.

chlorochromic anhydride. See chromyl chloride.

chlorocosane. See paraffin, chlorinated.

3-chlorocoumarin $C_9H_5O_2Cl$.
Properties: Slightly yellow crystalline solid. Melting point, 118°C.
Grade: Technical.
Use: Tin-plating solutions.

para-**chloro**-meta-**cresol.** See 4-chloro-3-methylphenol.

alpha-**chloro-N,N-diallylacetamide** (CDAA) $ClCH_2CON(CH_2CH{:}CH_2)_2$.
Properties: Amber liquid or granules; b.p. 74°C (0.3 mm); slightly soluble in water; soluble in alcohol, hexane, and xylene.
Hazard: Dry formulations are irritating to eyes and skin.
Use: Herbicide.

1-chloro-2-dichloroarsinoethene. See beta-chlorovinyldichloroarsine.

2-chloro-1-(2,4-dichlorophenyl)vinyl diethyl phosphate. See diethyl-1-(2,4-dichlorophenyl)-2-chlorovinyl phosphate.

chlorodifluoroacetic acid $CClF_2 \cdot COOH$.
Properties: Colorless, pungent liquid; b.p. 122°C; f.p. 23°C. Completely soluble in water; miscible with most organic solvents. Strong acid, dissolving cellulose and proteins.
Uses: Catalyst, particularly for esterification and condensation reactions; herbicides; intermediate.

1,1,1-chlorodifluoroethane (1,1,1-difluorochloroethane; difluoromonochloroethane) CH_3CClF_2.
Properties: Colorless, nearly odorless gas. B.p. -9.2°C; f.p. -130.8°C; sp. gr. 1.194 (-9°C). Insoluble in water.
Derivation: Chlorinating 1,1-difluoroethane in ultraviolet light.
Grade: Technical.
Containers: Cylinders.
Hazard: Flammable; asphyxiant in high concentrations.
Uses: Refrigerant; solvent; intermediate.
Shipping regulations: (Rail) Red Gas label. (Air) Flammable Gas label. Not acceptable on passenger planes. Legal label name (Rail) difluoromonochloroethane.

chlorodifluoromethane (monochlorodifluoromethane; difluorochloromethane; difluoromonochloromethane; propellant 22; refrigerant 22). $CHClF_2$.
Properties: Colorless, nearly odorless gas; density (gas at its b.p.) 4.82 g/l; b.p. -40.8°C; f.p. -160°C; partly soluble in water; nonflammable.
Derivation: Reaction of chloroform with anhydrous hydrogen fluoride with antimony chloride catalyst.
Grades: Technical; 99.9% pure.
Hazard: Asphyxiant in high concentrations.
Uses: Refrigerant; low-temperature solvent; fluorocarbon resins, especially tetrafluoroethylene polymers.
Shipping regulations: (Rail) Green label. (Air) Nonflammable Gas label. Legal label name monochlorodifluoromethane (Rail).
See also chlorofluorocarbon.

1-chloro-2,4-dinitrobenzene (dinitrochlorobenzene) $C_6H_3(NO_2)_2Cl$.
Properties: Pale yellow needles; almond odor; soluble in alcohol; insoluble in water; sp. gr. 1.69; m.p. 27–53°C; b.p. 315°C. Flash point 382°F; combustible; upper explosion limit 22%.
Derivation: Chlorination of dinitrobenzene.
Grades: Technical; fused.
Containers: Drums; tank cars.
Hazard: Toxic by ingestion, inhalation, and skin absorption; skin irritant. Moderate explosion hazard.
Uses: Dyes; organic synthesis.
Shipping regulations: (Rail, Air) Poison label. Legal label name (Rail) dinitrochlorobenzol.

chlorodiphenyl.
Properties: Colorless, mobile liquid; b.p. 340–375°C; flash point 383°F (open cup). Combustible. Resistant to acids and alkalies.
Grades: 54% chlorine; 42% chlorine.
Hazard: Toxic by ingestion, inhalation, and skin absorption. Strong irritant. Tolerance (54%) 0.5 mg/cubic meter of air; (42%) 1 mg/cubic meter of air.
Uses: Plasticizer for cellulosics, vinyl resins, and chlorinated rubbers.
See also diphenyl.

4-chlorodiphenyl sulfone. See para-chlorophenyl phenyl sulfone.

2-chloroethanamide. See chloroacetamide.

chloroethane. See ethyl chloride.

chloroethane nitrile. See chloroacetonitrile.

chloroethanoic anhydride. See chloroacetic anhydride.

chloroethene. See vinyl chloride.

chloroethyl acetate. See ethyl chloroacetate.

beta-**chloroethylchloroformate** CH_2ClCH_2OOCCl.
Properties: Colorless liquid. Decomposed by alkaline solutions and hot water. Insoluble in cold water. Sp. gr. 1.3825 (20°C); b.p. 152.5°C (752 mm).
Derivation: By bubbling gaseous phosgene into ethylene chlorohydrin at 0°C.
Hazard: Irritating to eyes and skin. Moderately toxic by ingestion and inhalation.

ɔeta-**chloroethyl chlorosulfonate** $ClCH_2CH_2OSO_2Cl$.
Properties: Colorless liquid. Chloropicrin-like odor. Darkens on long storage and decomposes with evolution of hydrogen chloride. B.p. 101°C (23 mm).
Hazard: Toxic by ingestion and inhalation. Strong irritant to tissue.
Derivation: Interaction of sulfuryl chloride and ethylene chlorohydrin. Also from action of sulfur trioxide on ethylene chloride below 45°C.

chloroethylene. See vinyl chloride.

2-chloroethyl methyl sulfide. See hemisulfur mustard.

(2-chloroethyl)trimethyl ammonium chloride. $[ClCH_2CH_2N(CH_3)_3]Cl$. A plant growth regulator.

2-chloroethyl vinyl ether. $CH_2ClCH_2OCHCH_2$.
Properties: Liquid; b.p. 104–108°C.
Hazard: Moderately toxic by ingestion and inhalation; moderate fire and explosion hazard.

chlorofluorocarbon. Any of several compounds comprised of carbon, fluorine, chlorine and hydrogen, the best known of which are trichlorofluoromethane and dichlorodifluoromethane (numbers 11 and 12, respectively). They are used chiefly as refrigerants and as aerosol propellants. Chlorofluorocarbons are often erroneously called fluorocarbons. Though safer and more efficient than other propellant gases, their use for aerosol sprays has diminished because of their suspected depleting effect on stratospheric ozone. Though this has not been proved, there is a considerable body of evidence that it does occur. *Note:* Use of these substances for aerosol propellant sprays will be restricted as of 1977.

chloroform (trichloromethane, methenyl trichloride) $CHCl_3$.
Properties: Clear, colorless, highly refractive, heavy, volatile liquid; characteristic odor; sweet taste. Keep from light. Miscible with alcohol, ether, benzene, solvent naphtha, fixed and volatile oils; slightly soluble in water. Sp. gr. 1.485 (20/20°C); b.p. 61.2°C; freezing point -63.5°C; wt/gal 12.29 lb (25°C); refractive index 1.4422 (25°C); nonflammable, but will burn on prolonged exposure to flame or high temperature.
Derivation: (a) Reaction of chlorinated lime with acetone, acetaldehyde, or ethyl alcohol; (b) by-product from the chlorination of methane.
Method of purification: Extraction with concentrated sulfuric acid and rectification.
Grades: Technical; C.P.; A.C.S.; N.F.
Containers: Drums; tank cars; tank trucks.
Hazard: Toxic by inhalation; narcotic. Prolonged inhalation or ingestion may be fatal. Tolerance, 25 ppm in air; 120 mg per cubic meter of air. Safety data sheet available from Manufacturing Chemists Assn., Washington, D.C. The National Cancer Institute has found that chloroform causes cancer in rats and mice; it has been prohibited by FDA from use in drugs, cosmetics and food packaging, including cough medicines, toothpastes, etc.
Uses: Fluorocarbon refrigerants and propellants; fluorocarbon plastics; solvent; analytical chemistry; fumigant; insecticides.

chloroformoxime ClHCNOH.
Properties: Needles. Odor resembles that of hydrocyanic acid. Stable at 0°C; unstable at normal temperature. Small quantities volatilize. Large quantities decompose. Aqueous solutions slowly decompose. Soluble in water, alcohol, ether, benzene; slightly soluble in carbon disulfide.
Derivation: Interaction of hydrochloric acid and sodium cyanate.
Hazard: Toxic by inhalation; strong irritant to tissue.
Use: Organic synthesis; tear gas and vesicant.

chloroformyl chloride. See phosgene.

chlorogenic acid (3-(3,4-dihydroxycinnamoyl)quinic acid) $(HO)_2C_6H_3CHCHCOOC_6H_7(OH)_3COOH$. Occurs in many plant tissues, especially in diseased tissue. Said to act as a growth inhibitor in plants.
Properties: Crystals; m.p. 208°C. Slightly soluble in cold water; soluble in hot water, alcohol, acetone.

chloroguanide hydrochloride (1-(para-chlorophenyl)-5-isopropylbiguanide hydrochloride) $C_{11}H_{16}ClN_5 \cdot HCl$.
Properties: Crystals with bitter taste. M.p. 248–252°C. Soluble in alcohol; slightly soluble in water; insoluble in chloroform and ether; pH of saturated aqueous solution 5.8–6.3.
Use: Medicine.

chlorohydrin (alpha-chlorohydrin; 1-chloropane-2,3-diol; glyceryl alpha-chlorohydrin) $CH_2OHCHOHCH_2Cl$.
Properties: Colorless heavy liquid. Unstable; hygroscopic. The commercial grade is a mixture of the two isomers, alpha and beta, of which alpha is in a greater proportion. Sp. gr. 1.326 (18°C); b.p. 213°C (decomposes); wt/gal 11.012 lb; freezing point -40°C; viscosity 2.388 poise (20°C). Miscible with some organic solvents and water; immiscible with oils. Nonflammable.
Derivation: By passing hydrochloric acid gas into glycerol containing 2% acetic acid.
Grades: Technical.
Uses: Solvent for cellulose acetate, glyceryl phthalate resins; partial solvent for gums; intermediate in organic synthesis.

chlorohydrin rubber. An elastomer made from epichlorohydrin (q.v.). Both a homopolymer and a copolymer with ethylene oxide are available. See also "Hydrin."

chlorohydroquinone (2-chloro-1,4-dihydroxybenzene; 2,5-dihydroxychlorobenzene [Cl = 1] $ClC_6H_3(OH)_2$.
Properties: White to light-tan fine crystals; m.p. 100°C; b.p. 263°C. Very soluble in water and alcohol; slightly soluble in ether.
Grades: Photographic; commercial.
Uses: Photographic developer; organic intermediate; dyestuffs; bactericide.

chlorohydroxybenzene. See chlorophenol.

5-chloro-2-hydroxybenzophenone $C_6H_5COC_6H_3OHCl$.
Properties: Yellow crystals; nearly odorless; m.p. 93–95°C; soluble in alcohol, ethyl acetate, methyl ethyl ketone; insoluble in water.
Use: Light absorber, best at 320–380μ.

2-chloro-4-(hydroxymercuri)phenol. See hydroxymercurichlorophenol.

4-chloro-1-hydroxy-3-methylbenzene. See 4-chloro-3-methylphenol.

6-chloro-3-hydroxytoluene. See 4-chloro-3-methylphenol.

chloro-IPC (isopropyl N-(3-chlorophenyl) carbamate; isopropyl 3-chlorocarbanilate; CIPC) $C_6H_4ClNHCOOC_3H_7$.
Properties: Light tan powder; m.p. 41.4°C; vapor pressure (149°C) 2 mm; sp. gr. 1.18 (30°C); very slightly soluble in water.
Containers: 60-, 250-, 450-lb drums; for solution: 55-gal drums; tank cars.
Hazard: Toxic by ingestion.
Use: Pre-emergence herbicide; prevents sprouting of potatoes.

2-chloro-N-isopropylacetanilide (N-isopropyl-alpha-chloroacetanilide) $C_6H_5N[CH(CH_3)_2]COCH_2Cl$.
Properties: Light tan powder or granules; m.p. 67–76°C; b.p. 110°C (0.03 mm). Very slightly soluble in water; soluble in acetone, alcohol, benzene, xylene, and carbon tetrachloride.
Hazard: May be toxic.
Uses: Herbicide.

chloroisopropyl alcohol. See propylene chlorohydrin.

6-chloro-4-isopropyl-1-methyl-3-phenol. See chlorothymol.

chloromadinone acetate. A nonestrogenic sex hormone used in oral contraception. See also antifertility agent.

chloromaleic anhydride $\overline{CH:CClC(O)OC}(O)$.
Properties: Yellow liquid; sp. gr. 1.5; m.p. 10–15°C; b.p. 192°C.
Uses: Catalyst for epoxy resins; intermediate.

chloromercuriferrocene $C_5H_5FeC_5H_4HgCl$. Orange crystalline solid; m.p. 193–194°C. Used as an intermediate, and for inorganic polymers.

Hazard: Highly toxic by ingestoin and inhalation.
Shipping regulations: (Rail, Air) Mercury compounds, n.o.s., Poison label.

1-[3-(chloromercuri)-2-methoxypropyl] urea. See chlormerodrin.

ortho-**chloromercuriphenol.** See ortho-hydroxyphenylmercuric chloride.

chloromethane. See methyl chloride.

chloromethapyrilene citrate. See chlorothen citrate.

3-chloro-2-methylaniline. See 2-amino-6-chlorotoluene.

3-chloro-4-methylaniline. 4-amino-2-chlorotoluene.

5-chloro-2-methylaniline. See 2-amino-4-chlorotoluene.

chloromethylated diphenyl oxide $C_6H_5OC_6H_5$, with up to three $-CH_2Cl$ radicals substituted for H's. Straw-colored liquids or white solids; m.p. up to 55°C; sp. gr. 1.19–1.30 (25/25°C); flash point 307°F+. Insoluble in water; very soluble in ether. Combustible.
Uses: Intermediate; resins; plasticizers.

chloromethylbenzene. See chlorotoluene.

1-chloro-3-methylbutane. See isoamyl chloride.

chloromethylchloroformate $ClCOOCH_2Cl$.
Properties: Mobile, colorless liquid. Penetrating, irritating odor. Hydrolyzed by hot and cold water. Decomposed by alkalies. Sp. gr. 1.465 at 15°C; b.p. 106.5–107°C; vapor density 4.5 (air = 1); vapor tension 5.6 mm (20°C). Soluble in most organic solvents.
Hazard: Toxic by ingestion and inhalation.

chloromethylchlorosulfonate $ClCH_2OClSO_2$.
Properties: Colorless liquid. Sp. gr. 1.63; b.p. 49–50°C (14 mm).
Derivation: By protracted boiling of chlorosulfonic acid with chloromethylchloroformate; also from paraformaldehyde and chlorosulfonic acid.
Hazard: Toxic by ingestion and inhalation. Strong irritant to tissue.

chloromethyl cyanide. See chloroacetonitrile.

chloromethyl ether.
Hazard: A carcinogen.
Uses: Ion-exchange resins.

1-chloromethylethylbenzene. See ethylbenzyl chloride.

1-chloromethylnaphthalene (alpha-naphthylmethyl chloride) $C_{10}H_7CH_2Cl$.
Properties: Colorless to greenish-yellow liquid with sharp pungent odor; sp. gr. 1.182 (25/25°C); coagulation point 23°C; refractive index 1.6354–1.6360 (25/D); insoluble in water; soluble in usual organic solvents. Very reactive.
Hazard: Vapor irritating to eyes. Toxic.
Use: A lachrymator; intermediate.
Shipping regulations: (Rail, Air) Not listed.

4-chloro-3-methylphenol (4-chloro-1-hydroxy-3-methylbenzene; 6-chloro-3-hydroxytoluene; 4-chloro-meta-cresol; so-called para-chloro-meta-cresol) $C_6H_3CH_3OHCl$.
Properties: White or slightly pink crystals with phenolic odor; m.p. 64–66°C; b.p. 235°C; volatile with steam; solubility 1:250 in water at 25°C; soluble in alkalies, organic solvents, fats and oils.
Hazard: May be toxic and irritant to skin.
Uses: External germicide; preservative for glues, gums, paints, inks, textile and leather goods.

4-chloro-2-methylphenoxyacetic acid. See MCP.

chloromethylphosphonic acid $ClCH_2PO(OH)_2$.
Properties: White hygroscopic solid; m.p. 85–95°C.
Uses: Intermediate for flameproofing agents, resins, lubricants, additives, plasticizers.

chloromethylphosphonic dichloride $ClCH_2POCl_2$.
Properties: Water-white to light straw liquid, highly reactive. Sp. gr. 1.638 (25°C); refractive index 1.4960–1.4970 (n 25/D).
Hazard: Toxic by inhalation; irritant to eyes, lungs and mucous membranes.
Uses: Intermediate for flameproofing agents, resins, lubricants, additives, and plasticizers.

3-chloro-2-methyl-1-propene. See methylallyl chloride.

"Chloromycetin."[330] Trademark for a proprietary form of chloramphenicol, an antibiotic. Must comply with FDA requirements. See chloramphenicol.

chloronaphthalene. See chlorinated naphthalene.

alpha-**chloro**-meta-**nitroacetophenone** $NO_2C_6H_4COCH_2Cl$.
Properties: Off-white, free-flowing granules; approx. m.p. 95–100°C; soluble in chlorinated solvents; insoluble in water.
Uses: Bacteriostat and fungistat in cutting oils, water systems, paint, plastics, textiles; chemical intermediate.

2-chloro-4-nitroaniline (ortho-chloro-para-nitroaniline) $C_6H_3ClNO_2NH_2$.
Properties: Yellow needles. Soluble in alcohol, benzene, ether; slightly soluble in water and strong acids. M.p. 107°C.
Derivation: (a) From 1,2-dichloro-4-nitrobenzene by heating with alcoholic ammonia. (b) From the chlorination of para-nitroaniline in acid solution.
Hazard: Toxic by ingestion and inhalation.
Use: Intermediate in manufacture of dyes.

4-chloro-2-nitroaniline (para-chloro-ortho-nitroaniline) $C_6H_3ClNO_2NH_2$.
Properties: Orange, crystalline powder; m.p. 163°C; insoluble in water; soluble in methanol and ether.
Hazard: Toxic by ingestion and inhalation.
Use: Dye and pigment intermediate.

4-chloro-3-nitroaniline $C_6H_3ClNO_2NH_2$.
Properties: Yellow to tan powder; m.p. 95–97°C. Soluble in alcohol and acetone. Partially soluble in hot water.
Hazard: Toxic by ingestion and inhalation.
Uses: Intermediate in the manufacture of azo dyes, pharmaceuticals and other organic compounds.

meta-**chloronitrobenzene** $C_6H_4ClNO_2$.
Properties: Yellowish crystals; sp. gr. 1.534; m.p. 44°C; b.p. 236°C; soluble in most organic solvents; insoluble in water. Combustible.
Derivation: By chlorinating nitrobenzene in the presence of iodine and recrystallizing.
Containers: Drums.
Hazard: Highly toxic by inhalation and ingestion.
Use: Intermediate.
Shipping regulations: (Rail, Air) Poison label. Legal label name (Rail) meta-nitrochlorbenzene.

ortho-**chloronitrobenzene** $C_6H_4ClNO_2$.
Properties: Yellow liquid; sp. gr. 1.368; b.p. 245.5°C; m.p. 325°C; soluble in alcohol and benzene; insoluble in water. Flash point 261°F; combustible.
Derivation: By nitrating chlorobenzene and purifying by rectification.
Containers: Drums; tank cars.
Hazard: Highly toxic by inhalation and ingestion.

Uses: Intermediate, especially for dyes.
Shipping regulations: (Rail, Air) Poison label. Legal label name (Rail) ortho-nitrochlorbenzene.

para-**chloronitrobenzene** $C_6H_4ClNO_2$.
Properties: Yellowish crystals; sp. gr. 1.520; m.p. 83°C; b.p. 242°C; soluble in organic solvents; insoluble in water. Combustible.
Derivation: Nitration of chlorobenzene and recrystallization.
Containers: Drums.
Hazard: Highly toxic by inhalation and ingestion.
Use: Intermediate, especially for dyes; manufacture of *p*-nitrophenol, from which parathion is made; agricultural chemicals; rubber chemicals.
Shipping regulations: (Rail, Air) Poison label. Legal label name (Rail) para-nitrochlorbenzene.

2-chloro-5-nitrobenzenesulfonamide $ClNO_2C_6H_3SO_2NH_2$.
Properties: Grayish-white solid. Insoluble in water; soluble in benzene.
Hazard: Probably toxic by ingestion.
Uses: Dye and pharmaceutical intermediates.

6-chloro-3-nitrobenzenesulfonic acid, sodium salt $NaSO_3C_6H_3NO_2Cl$.
Properties: Off-white moist crystals.
Hazard: Probably toxic by ingestion.
Use: Intermediate.

4-chloro-3-nitrobenzoic acid $ClC_6H_3NO_2COOH$.
Properties: Light gray or white powder; m.p. 170–174°C.
Hazard: Probably toxic by ingestion.
Use: Intermediate.

4-chloro-3-nitrobenzotrifluoride (para-chloro-meta-nitrotrifluorotoluene) $C_6H_3CF_3NO_2Cl$.
Properties: Thin, oily liquid; sp. gr. 1.542 (15.5/15.5°C); f.p. -7.5°C; flash point 275°F. Combustible. Refractive index (n 20/D) 1.491; b.p. 222°C; soluble in organic solvents; insoluble in water.
Grade: 97.5%.
Hazard: Toxic by inhalation and ingestion. Tolerance (as F), 2.5 mg per cubic meter of air.
Use: Intermediate for dyestuffs; agricultural chemicals; pharmaceuticals.

4-chloro-2-nitrophenol, sodium salt $ClC_6H_3NO_2ONa$.
Properties: Red needle crystals, with one molecule of water of crystallization. Soluble in hot water.
Derivation: Nitration of para-dichlorobenzene followed by hydrolysis.
Grades: 90% anhydrous sodium salt, containing 80% base.
Hazard: Probably toxic.
Uses: Dye intermediate; manufacture of 2-amino-4-chlorophenol.

1-chloro-1-nitropropane. See korax.

2-chloro-6-nitrotoluene $ClNO_2C_6H_3CH_3$. Solid: m.p. 36.5–40°C; insoluble in water. Intermediate.
Hazard: Toxic; dangerous fire risk.

4-chloro-2-nitrotoluene $ClNO_2C_6H_3CH_3$. Solid, m.p. 35–37°C; insoluble in water; soluble in alcohol and ether; intermediate.
Hazard: Toxic; dangerous fire risk.

para-**chloro**-meta-**nitrotrifluorotoluene.** See 4-chloro-3-nitrobenzotrifluoride.

chloronitrous acid. See aqua regia.

chloropentafluoroacetone ClF_2CCOCF_3.
Properties: Colorless, nonflammable, hygroscopic, highly reactive gas. B.p. 7.8°C; f.p. -133°C; liquid density 1.43 g/ml (25°C).
Hazard: Evolves heat on contact with water.
Use: Intermediate.

chloropentafluoroethane (monochloropentafluoroethane; fluorocarbon 115; propellant 115) $CClF_2CF_3$.
Properties: Colorless gas; b.p. -37.7°F; f.p. -106°C; insoluble in water; soluble in alcohol and ether. Has good thermal stability. Nonflammable, nontoxic.
Containers: Cylinders up to 1800-lb tanks.
Uses: Refrigerant and dielectric gas.
Shipping regulations: (Rail) Green label. (Air) Nonflammable Gas label. Legal label name (Rail) monochloropentafluoroethane.

1-chloropentane. See n-amyl chloride.

chlorophene. USAN for ortho-benzyl-para-chlorophenol (q.v.).

meta-**chlorophenol** (3-chloro-1-hydroxybenzene) C_6H_4OHCl.
Properties: White crystals with odor similar to phenol; discolors on exposure to air; sp. gr. 1.245; m.p. 33°C; b.p. 214°C.
Soluble in alcohol, ether and aqueous alkali; slightly soluble in water.
Derivation: From meta-chloroaniline through the diazonium salt.
Hazard: Toxic by skin absorption, inhalation or ingestion. Strong irritant to tissue.
Use: Intermediate in organic synthesis.

ortho-**chlorophenol** (2-chloro-1-hydroxybenzene) C_6H_4OHCl.
Properties: Colorless to yellow brown liquid with unpleasant penetrating odor. Very soluble in water; soluble in alcohol, ether, and aqueous sodium hydroxide. Volatile with steam. B.p. 175°C; freezing point 7°C; sp. gr. 1.265 (15.5°C); flash point 225°F. Combustible.
Derivation: Chlorination of phenol.
Hazard: Toxic by skin absorption, inhalation or ingestion. Strong irritant to tissue.
Use: Organic synthesis (dyes).

para-**chlorophenol** (4-chloro-1-hydroxybenzene) C_6H_4OHCl.
Properties: White crystals (yellow or pink when impure) with unpleasant penetrating odor. Soluble in water; soluble in benzene, alcohol, and ether. Volatile with steam; b.p. 217°C; m.p. 42–43°C; sp. gr. 1.306; refractive index (n 40/D) 1.5579. A 1% solution is acid to litmus. Flash point 250°F. Combustible.
Derivation: Chlorination of phenol; from chloroaniline through the diazonium salt.
Grades: N.F.; technical.
Containers: 55-gal drums.
Hazard: Toxic by skin absorption, inhalation or ingestion. Strong irritant to tissue.
Uses: Intermediate in synthesis of dyes and drugs; denaturant for alcohol; selective solvent in refining mineral oils.

chlorophenothane. U.S.P. name for DDT (q.v.).

para-**chlorophenyl benzenesulfonate.** See fenson.

para-**chlorophenyl** para-**chlorobenzenesulfonate.** See ovex.

4-chloro-alpha-**phenyl**-ortho-**cresol.** See ortho-benzyl-para-chlorophenol.

3-para-**chlorophenyl-1,1-dimethylurea.** See monuron.

meta-**chlorophenyl isocyanate** ClC_6H_4NCO.
Properties: Clear water-white to light yellow liquid. F.p. -4.4°C; flash point (COC) 215°F. Combustible.
Hazard: Strong irritant to skin, eyes, and mucous membrane.
Use: Intermediate for pharmaceuticals, herbicides, and pesticides.

para-**chlorophenyl isocyanate** ClC_6H_4NCO.
Properties: Colorless to slightly yellow liquid or white crystalline solid. F.p. 29.0°C; flash point (COC) 230°F. Combustible.
Hazard: Strong irritant to skin, eyes, and mucous membranes.
Use: Intermediate for pharmaceuticals, herbicides, and pesticides.

1-(para-**chlorophenyl)-5-isopropylbiguanide hydrochloride.** See chloroguanide hydrochloride.

3-(para-**chlorophenyl)-5-methylrhodanine** $C_{10}H_8ClNOS_2$. Yellow crystals, m.p. 106–110°C. Insoluble in water; soluble in acetone.
Hazard: Toxic by ingestion.
Uses: Fungicide and for nematode control.

chloro-ortho-**phenylphenol** (chloro-2-phenylphenol) $C_6H_3(OH)ClC_6H_5$.
Properties: Clear, colorless to straw colored, viscous liquid with faint characteristic odor; sp. gr. 1.228 (20/4°C); freezing point less than -20°C; boiling range 5–95% 146–158.7°C (5 mm); flash point 273°F; readily soluble in most organic solvents. Combustible.
Composition: 80% 4-Chloro-2-phenylphenol; 20% 6-chloro-2-phenylphenol.
Hazard: Toxic by ingestion.
Uses: Fungicide.

para-**chlorophenyl phenyl sulfone** (4-chlorodiphenyl sulfone, sulphenone). $ClC_6H_4SO_2C_6H_5$.
Properties: Dimorphic crystals; slight aromatic odor; tasteless; insoluble in water; soluble in most organic solvents. Relatively stable to acids and alkalies.
Uses: Insecticide and acaricide (harmful to most grapes and pears).

chlorophenyltrichlorosilane $ClC_6H_4SiCl_3$. A mixture of isomers.
Properties: Colorless to pale yellow liquid. B.p. 230°C; sp. gr. 1.439 (25/25°C); refractive index (n 20/D) 1.5414; flash point (COC) 255°F. Readily hydrolyzed, with liberation of hydrochloric acid. Combustible.
Derivation: By Grignard reaction of silicon tetrachloride and chlorophenylmagnesium chloride.
Grades: Technical.
Hazard: Highly toxic. Corrosive to skin when wet.
Use: Intermediate for silicones.
Shipping regulations: (Rail) Corrosive material, n.o.s., White label. (Air) Corrosive label. Not acceptable on passenger planes.

4-chlorophthalic acid $C_6H_3Cl(COOH)_2$.
Properties: Colorless crystals; m.p. 150°C; decomposes on further heating; soluble in alcohol and ether; insoluble in water.
Derivation: Chlorination of phthalic acid.

chlorophyll. The green plant pigment essential to photosynthesis (q.v.). It occurs in three forms (a, b, and c), all of which are magnesium-centered porphyrins containing a hydrophilic carbocyclic ring with a lipophilic phytyl tail. Chlorophyll is a photoreceptor up to wavelengths of 700 mμ; it can readily transfer radiant energy to its chemical environment, and thus acts as a transducer in photosynthesis. It is structurally analogous to the red blood pigment hemin (q.v.). Chlorophyll a has been synthesized by two different routes. Its derivatives are relatively unstable to light, oxidizing agents, and chemical reagents. It is nontoxic.
Properties:
Chlorophyll a: $C_{55}H_{72}MgN_4O_5$.

```
                CH3—C————C—CH=CH2
                    ‖        ‖
         CH3  CH—C         C——CH   CH3
          |    ‖     \   /      ‖    |
         HC——C        N        C——C
          |      \ +   |   + /        ‖
          |        N=Mg=N             ‖
          |       //   |    \\        ‖
         HC——C         N        C——C
          |     |    /   \      |    |
         CH2   C==C       C==CH   C2H5
          |     |   |       |
         CH2   CH  C=======C—CH3
          |     |    \    /
C20H39OC=O      |      C
                |      ‖
                |      O
               COOCH3
```

Blue-green microcrystalline wax. About three times as plentiful as b. M.p. 117–120°C. Freely soluble in ether, ethanol, acetone, chloroform, carbon disulfide, benzene; sparingly soluble in cold methanol; insoluble in petroleum ether. The alcoholic solution is blue-green with a deep-red fluorescence.

Chlorophyll b: $C_{55}H_{70}MgN_4O_6$. Yellow-green microcrystalline wax. Sparingly soluble in absolute alcohol, ether. Ether solution has a brilliant green color. Solutions with other organic solvents are usually green to yellow-green with red fluorescence.

Chlorophyll c: Occurs in marine organisms, and may be as important as chlorophyll b.
Derivation: Alcoholic extraction of green plants; isolation by chromatography.
Grades: Aqueous, alcoholic or oil solutions; water solutions are prepared by saponification of oil-soluble chlorophyll.
Containers: Glass bottles.
Uses: Colorant for soaps, oils, fats, waxes, liquors, confectionery, preserves; cosmetics; perfumes; dentistry; source of phytol; sensitizer for color film; toothpaste additive; medicine.
See also photosynthesis; porphyrin.
Note: An experimental use of chlorophyll to act as energy converter in a synthetic photovoltaic cell has aroused interest in connection with solar energy research. The cell is termed a "synthetic leaf", as it is an attempt to approximate the energy-trapping function of a natural leaf.

chlorophyllin. Reaction product of alcoholic potassium or sodium hydroxide and alcoholic leaf extracts. The methyl and phytyl groups are replaced by alkali but the magnesium is not replaced. Used in food coloring, dyes, deodorants, and medicine.

chloropicrin (chloropicrin; nitrotrichloromethane; trichloronitromethane; nitrochloroform) CCl_3NO_2.
Properties: Pure product slightly oily, colorless, refractive liquid. Relatively stable; not decomposed by water or mineral acids. Sp. gr. 1.692 (0°C); b.p. 112°C; f.p. -69.2°C. Soluble in alcohol, benzene, carbon disulfide; slightly soluble in ether and water. Nonflammable.

Derivation: (a) Action of picric acid on calcium hypochlorite. (b) Nitrification of chlorinated hydrocarbons.
Hazard: Highly toxic by ingestion and inhalation; strong irritant. Tolerance, 0.1 ppm in air.
Uses: Organic synthesis; dye-stuffs (crystal violet); fumigants; fungicides; insecticides; rat exterminator; poison gas.
Shipping regulations: (Rail) Poison label. Legal label name chlorpicrin. (Air) Not acceptable.

chloroplatinic acid $H_2PtCl_6 \cdot 6H_2O$.
Properties: Red-brown crystals. Soluble in water, alcohol and ether. Sp. gr. 1.431; m.p. 60°C.
Derivation: By solution of platinum in aqua regia, evaporation and crystallization.
Containers: Glass bottles.
Hazard: See platinum chloride.
Uses: Electroplating; platinizing pumice and the like for catalysts; etching zinc for printing; platinum mirrors; indelible ink; ceramics (producing fine color effects on high-grade porcelain); microscopy.

chloroprene (2-chlorobutadiene-1,3) $H_2C{:}CHCCl{:}CH_2$.
Properties: Colorless liquid; b.p. 59.4°C; sp. gr. 0.9583 (20/20°C); soluble in alcohol; slightly soluble in water. Flash point -4°F.
Derivation: Addition of cold hydrogen chloride to vinylacetylene; chlorination of butadiene.
Grades: Pure, 95% min.
Containers: Cylinders; tank trucks.
Hazard: Flammable, dangerous fire risk; explosive limits in air, 4.0 to 20%. Toxic by ingestion, inhalation, and skin absorption. Tolerance, 25 ppm in air. May be carcinogenic.
Use: Manufacture of neoprene (q.v.).
Shipping regulations: Flammable liquid, n.o.s., (Rail) Red label (Air) Flammable Liquid label.

1-chloropropane. See propyl chloride.

3-chloropropane-1,2-diol. See chlorohydrin.

1-chloro-2-propanol. See propylene chlorohydrin.

1-chloro-2-propanone. See chloroacetone.

1- or **3-chloropropene.** See allyl chloride.

2-chloropropene (isopropenyl chloride) $CH_3CCl{:}CH_2$.
Properties: Colorless gas or liquid; sp. gr. 0.918 (9°C); f.p. -137.4°C; b.p. 22.65°C (1 atm).
Derivation: By treating propylene dichloride with alcoholic potassium hydroxide and fractionating from the simultaneously formed 1-chloropropene.
Grade: 95% purity.
Containers: Steel cylinders.
Hazard: Flammable, dangerous fire risk; may be toxic.
Uses: Intermediate in organic synthesis; formulation of copolymers.
Shipping reuglations: (Rail) Flammable liquid, n.o.s., Red label. (Air) Flammable Liquid label. Not accepted on passenger planes.

2-chloropropionic acid (alpha-chloropropionic acid) $CH_3CHClCOOH$.
Properties: Crystals; sp. gr. 1.260-1.268 (20°C); b.p. 183-187°C. Soluble in water. Combustible.
Use: Intermediate for weed killers.

3-chloropropionic acid (beta-chloropropionic acid) CH_2ClCH_2COOH.
Properties: Crystals; m.p. 41°C; b.p. 200°C; soluble in water, alcohol, chloroform. Combustible.
Use: Intermediate.

3-chloropropionitrile $ClCH_2CH_2CN$.
Properties: Colorless liquid; f.p. -51°C; flash point (closed cup) 168°F; refractive index (n 25/D) 1.4341; sp. gr. 1.1363 (25°C); b.p. 176°C (dec.). Miscible with acetone, benzene, carbon tetrachloride, alcohol, and ether. Combustible.
Hazard: Toxic by ingestion, inhalation and skin contact.
Uses: Intermediate.

alpha-**chloropropylene.** See allyl chloride.

chloropropylene oxide. See epichlorohydrin.

3-chloropropyl mercaptan $ClCH_2CH_2CH_2SH$.
Properties: Liquid; distillation range 141.1-145.6°C; sp. gr. 1.131 (15.5°C); refractive index 1.492 (20°C); flash point approx. 110°F. Combustible.
Hazard: Moderate fire risk.
Use: Intermediate.

3-chloro-1-propyne. See propargyl chloride.

2-chloropyridine ClC_5H_4N.
Properties: Oily liquid; sp. gr. 1.205 (15°C); b.p. 170°C. Slightly soluble in water; soluble in alcohol and ether.
Uses: Production of antihistamines, germicides, pesticides, and agricultural chemicals.

6-chloroquinaldine $C_9H_5N(CH_3)Cl$.
Properties: Brownish-black oily crystalline mass.
Grade: Technical.
Use: Intermediate.

chloroquine $C_9H_5NClNHCH(CH_3)(CH_2)_3N(C_2H_5)_2$. 7-Chloro-4-(4-diethylamino-1-methylbutylamino)-quinoline.
Properties: Colorless crystals. Bitter taste; m.p. (free base) 87°C. Soluble in water; insoluble in alcohol, benzene, chloroform, ether.
Derivation: Condensation of 4,7-dichloroquinoline with 1-diethylamino-4-aminopentane.
Hazard: Moderately toxic.
Use: Medicine (antimalarial). Usually dispensed as the phosphate.

5-chlorosalicylanilide $ClC_6H_3OHCONHC_6H_5$.
Properties: White crystals; m.p. 209-211°C; slightly soluble in water; soluble in alcohol, ether, chloroform and benzene.
Hazard: Toxic by ingestion.
Use: Fungicide; antimildew agent; intermediate for pharmaceuticals, dyes, pesticides.

5-chlorosalicylic acid $ClC_6H_3OHCOOH$.
Properties: White crystals; m.p. 174-176°C; slightly soluble in water; soluble in alcohol, ether, chloroform and benzene.
Hazard: May be toxic by ingestion.
Use: Mothproofing; insecticide; intermediate for pharmaceitcals, dyes, pesticides.

N-chlorosuccinimide (NCS) $COCH_2CH_2CONCl$.
Properties: White crystalline powder; m.p. 148-149°C; soluble in water; sparingly soluble in chloroform and carbon tetrachloride.
Hazard: Strong skin irritant.
Uses: Chlorinating agent; disinfectant for swimming pools; bactericide.

chlorosulfonic acid (sulfuric chlorohydrin) $ClSO_2OH$.
Properties: Colorless to light yellow, fuming, slightly cloudy liquid; pungent odor; sp. gr. 1.76-1.77 (20/20°C); f.p. -80°C; b.p. 158°C. Decomposes in water

to sulfuric and hydrochloric acids. Decomposed by alcohol and acids.
Derivation: By treating sulfur trioxide or fuming sulfuric acid with hydrochloric acid.
Grades: Technical.
Containers: Carboys; drums; tank cars.
Hazard: Highly toxic by inhalation. Strong irritant to eyes and skin. Can ignite combustible materials; evolves hydrogen on contact with most metals. Safety data sheet available from Manufacturing Chemists Assn., Washington, D.C.
Uses: Synthetic detergents; pharmaceuticals; dyes; pesticides; intermediates; ion-exchange resins; anhydrous hydrogen chloride and smoke-producing chemicals.
Shipping regulations: (Rail) White label (Air) Corrosive label.

4-chlorosulfonylbenzoic acid $ClSO_2C_6H_4COOH$.
Properties: Light tan powder; soluble in benzene; slightly soluble in ether.
Use: Intermediate.

chlorosulfuric acid. See sulfuryl chloride.

chlorotetracycline. See chlortetracycline.

chlorotetrafluoroethane (monochlorotetrafluoroethane) C_2HF_4Cl.
Properties: Nonflammable, nontoxic, odorless and colorless gas. Much heavier than air.
Containers: Cylinders.
Shipping regulations: (Rail) Green label. (Air) Nonflammable Gas label. Legal label name (Rail) monochlorotetrafluoroethane.

chlorothen citrate (chloromethapyrilene citrate) $C_{14}H_{18}ClN_3S \cdot C_6H_8O_7$.
Properties: White, practically odorless crystalline powder. Slightly soluble in alcohol and water; practically insoluble in chloroform, ether and benzene. 1% solution is clear and colorless. pH (1% solution) 3.9–4.1. Melts at 112–116°C; on further heating solidifies and remelts at 125–140°C (dec).
Grade: N.F.
Use: Medicine (antihistamine).

"Chlorothene."[233] Trademark for a series of chlorinated organic solvents.
Grades: NU and Industrial are both inhibited 1,1,1-trichloroethane. Used for cold cleaning of metal parts and as industrial solvents.

para-**chlorothiophenol** (para-chlorobenzenethiol) ClC_6H_4SH.
Properties: Moist white to cream crystals; m.p. 52–55°C; b.p. 205–207°C. Soluble in most organic solvents.
Hazard: May be toxic by ingestion.
Uses: Oil additives; agricultural chemicals; plasticizers; rubber chemical; dyes; wetting agents and stabilizers.

chlorothymol (6-chloro-4-isopropyl-1-methyl-3-phenol) $CH_3C_6H_2(OH)(C_3H_7)Cl$.
Properties: White crystals or crystalline granular powder; characteristic odor; aromatic, pungent taste; becomes discolored with age; affected by light; m.p. 59–61°C. Soluble in benzene, chloroform, dilute caustic soda, alcohol; insoluble in water.
Derivation: Action of sulfuryl chloride on thymol in a solution of carbon tetrachloride.
Grades: N.F.
Hazard: Irritating to skin and mucous membranes in concentrated solution.
Use: Bactericide; component of antiseptic solutions.

alpha-**chlorotoluene.** See benzyl chloride.

meta-**chlorotoluene.** (3-chloro-1-methylbenzene) $CH_3C_6H_4Cl$.
Properties: Colorless liquid; sp. gr. 1.07218 (20/4°C); b.p. 161.6°C; f.p. –48.0°C.
Derivation: Diazotization of meta-toluidine followed by treating with cuprous chloride.
Hazard: Narcotic in high concentrations. Avoid inhalation.
Uses: Solvent; intermediate.

ortho-**chlorotoluene** (2-chloro-1-methylbenzene). $CH_3C_6H_4Cl$.
Properties: Colorless liquid. B.p. 159.2°C; f.p. –35.1°C; sp. gr. 1.0776 (25/4°C); refractive index (n 20/D) 1.5268. Miscible with alcohol, acetone, ether, benzene, carbon tetrachloride, and n-heptane; slightly soluble in water.
Derivation: By catalytic chlorination of toluene.
Hazard: Toxic by inhalation. Tolerance, 50 ppm in air.
Uses: Solvent and intermediate for organic chemicals and dyes.

para-**chlorotoluene** (4-chloro-1-methylbenzene) $CH_3C_6H_4Cl$.
Properties: Colorless liquid; boiling range 162–166°C; f.p. approx. 6.5°C; sp. gr. 1.065–1.067 (25/15°C); refractive index 1.5184 (22°C). Soluble in alcohol, ether, acetone, benzene, and chloroform. Insoluble in water.
Hazard: Avoid inhalation.
Uses: Solvent and intermediate for organic chemicals and dyes.

2-chlorotoluene-4-sulfonic acid (ortho-chlorotoluene-para-sulfonic acid) $CH_3C_6H_3(SO_3H)Cl$.
Properties: White glistening plates. Soluble in hot water.
Derivation: Chlorination of toluene-para-sulfonic acid.
Method of purification: Recrystallization from water.
Uses: Dye intermediate.

2-chloro-para-**toluidine.** See 4-amino-2-chlorotoluene.

4-chloro-ortho-**toluidine.** See 2-amino-4-chlorotoluene.

6-chloro-ortho-**toluidine.** See 2-amino-6-chlorotoluene.

4-chloro-ortho-**toluidine hydrochloride** $CH_3C_6H_3(Cl)NH_2 \cdot HCl$.
Hazard: Toxic by ingestion, inhalation and skin absorption.

2-chloro-5-toluidine-4-sulfonic acid (6-chloro-meta-toluidine-4-sulfonic acid) $CH_3C_6H_2(NH_2)(SO_3H)Cl$.
Properties: Fine white crystals. Soluble in dilute caustic solution.
Derivation: From ortho-chlorotoluene-para-sulfonic acid by nitration and subsequent reduction.
Hazard: Probably toxic.
Use: Intermediate.

chlorotrianisene (chlorotris(para-methoxyphenyl) ethylene; tri-para-anisylchloroethylene) $(CH_3OC_6H_4)_2CCCl(C_6H_4OCH_3)$. A synthetic nonsteroid estrogen.
Properties: White, odorless, crystalline powder. M.p. 115–117°C. Freely soluble in acetone, benzene and chloroform; slightly soluble in ether; very slightly soluble in alcohol and water.
Grade: N.F.
Use: Medicine.

chlorotriazinyl dye. A fiber-reactive dye (q.v.); both mono- and di-derivatives have been developed. They react readily and permanently with cellulose under alkaline conditions with 70–80% efficiency.

chlorotrifluoroethylene (CFE; CTFE; trifluorochloroethylene) $ClFC{:}CH_2$.
Properties: Colorless gas with faint ethereal odor; b.p. -27.9°C; f.p. -157.5°C; sp. gr. (liquid) (20°C) 1.305; flammable limits in air 8.4–38.7% by volume; decomposes in water.
Derivation: From trichlorotrifluoroethane and zinc.
Grades: Technical; 99.0%.
Containers: Cylinders; tank cars; tank trucks. Shipped with inhibitor.
Hazard: Moderately toxic; flammable; dangerous fire risk.
Uses: Intermediate; monomer for chlorotrifluoroethylene resins.
Shipping regulations: (Rail) Red Gas label. (Air) (Inhibited) Flammable Gas label; Not accepted on passenger planes. (Uninhibited) Not acceptable. Legal label name (Rail) trifluorochloroethylene.

chlorotrifluoroethylene polymer (polytrifluorochloroethylene resin; fluorothene). A polymer of chlorotrifluoroethylene, usually including vinylidine fluoride, characterized by the repeating structure (CF_2-CFCl).
Properties: Colorless; sp. gr. 2.10–2.15; refractive index 1.43. Impervious to corrosive chemicals; resistant to most organic solvents; heat-resistant; nonflammable; thermoplastic. Zero moisture absorption; high impact strength; transparent films and thin sheets.
Uses: Chemical piping, gaskets, tank linings, connectors, valve diaphragms, wire and cable insulation; electronic components.
See also "Kel-F"; fluorocarbon polymer.

chlorotrifluoromethane (monochlorotrifluoromethane; trifluorochloromethane) $CClF_3$.
Properties: Colorless, nonflammable gas; ethereal odor; b.p. -81.4°C; f.p. -181°C; heavier than air.
Derivation: From dichlorodifluoromethane in vapor phase with aluminum chloride catalyst.
Grade: 99.0% min. purity.
Containers: Cylinders.
Hazard: Moderately toxic by inhalation; slightly irritant.
Uses: Dielectric and aerospace chemical; refrigerant and coolant; hardening of metals; pharmaceutical processing.
Shipping regulations: (Rail) Green label. (Air) Nonflammable Gas label. Legal label name (Rail) monochlorotrifluoromethane.
See also chlorofluorocarbon.

2-chloro-5-trifluoromethylaniline $C_6H_3ClCF_3NH_2$.
Properties: Amber colored oil; f.p. 6–8°C; insoluble in water; soluble in alcohol and acetone. Forms water-soluble salts with mineral acids.
Hazard: Toxic. Irritant to eyes and skin.
Use: Intermediate for dyes and other organic chemicals.

chlorotrifluoromethylbenzene. See chlorobenzotrifluoride.

chloro-alpha, alpha, alpha-**trifluorotoluene.** See chlorobenzotrifluoride.

chlorotris (para-**methoxyphenyl**)**ethylene.** See chlorotrianisene.

chlorovinylarsinedichloride. See beta-chlorovinyldichloroarsine.

beta-**chlorovinyldichloroarsine** (1-chloro-2-dichloroarsinoethene; dichloro(2-chlorovinyl)arsine; chlorovinylarsinedichloride; lewisite). Two isomers, probably cis and trans, are known. $ClCH{:}CHAsCl_2$.
Properties: Colorless liquid when pure. Impurities influence a color change ranging from violet to brown. Geranium-like odor. Decomposed by water and alkalies. Inactivated by bleaching powder. Antidote is dimercaptopropanol (q.v.).
Derivation: Condensation of arsenic trichloride with acetylene in the presence of aluminum or copper or mercury chloride. The mixed arsines are separated by fractionating.
Hazard: Highly toxic vesicant gas.
Use: Poison gas; skin blistering agent.
Shipping regulations: (Rail) Poison Gas label. Not accepted by express. (Air) Not acceptable. Legal label name: lewisite.

beta-**chlorovinyl ethyl ethynyl carbinol.** See ethchlorvynol.

beta-**chlorovinylmethylchloroarsine**
$ClCH{:}CHAsClCH_3$.
Properties: Liquid; decomposed by water. B.p. 112–115°C (10 mm).
Derivation: Interaction of acetylene and methyldichloroarsine in the presence of aluminum chloride.
Hazard: Highly toxic; strong irritant; absorbed by skin.
Shipping regulations: (Rail, Air) Arsenical compounds, n.o.s., Poison label.

"Chlorowax."[244] Trademark for a series of liquid and resinous chlorinated paraffins containing from 40% to 70% chlorine by weight. They are odorless, nonflammable, insoluble in water, and have low toxicity. See also paraffin, chlorinated.

para-**chloro**-meta-**xylenol** (4-chloro-3,5-dimethylphenol). $C_6H_2(CH_3)_2OHCl$. Crystals with phenolic odor.
Hazard: Toxic by ingestion; strong irritant; absorbed by skin.
Uses: Active ingredient in germicides, antiseptics, etc.; fungistat.

6-chloro-3,4-xylyl methylcarbamate (6-chloro-3,4-dimethylphenyl N-methylcarbamate) $ClC_6H_2(CH_3)_2OCONHCH_3$.
Properties: Solid; m.p. 120–133°C.
Hazard: May be toxic by ingestion and inhalation.
Use: Pesticide.

chlorpheniramine maleate (chlorprophenpyridamine maleate) $C_{16}H_{19}ClN_2 \cdot C_4H_4O_4$. 1-(para-Chlorophenyl)-1-(2-pyridyl)3-dimethylaminopropane maleate.
Properties: White odorless crystals. M.p. 130–135°C. Slightly soluble in ether; soluble in alcohol, chloroform, and water. pH (1% solution) about 4.8.
Grade: U.S.P.
Use: Medicine (antihistamine).

chlorphenol red (dichlorosulfonphthalein)
$C_6H_4SO_2OC(C_6H_3ClOH)_2$. An acid-base indicator (q.v.), showing color change from yellow to red over the pH range 5.2 to 6.8.

chlorpicrin. Legal label name (Rail) for chloropicrin (q.v.).

chlorpromazine (2-chloro-10-(3-dimethylaminopropyl)-phenothiazine) $C_{17}H_{19}ClN_2S$.
Properties: Oily liquid; amine odor; alkaline reaction; b.p. 200–205°C, (0.8 mm).

Hazard: May have toxic side effects; central nervous system depressant.
Use: Medicine (antipsychotic drug).

chlorprophenpyridamine maleate. See chlorpheniramine maleate.

chlorquinaldol (5,7-dichloro-8-hydroxyquinaldine; 5,7-dichloro-2-methyl-8-quinolinol) $CH_3C_9H_3N(OH)Cl_2$.
Properties: Yellow, crystalline, tasteless powder with a pleasant medicinal odor; m.p. 114°C; soluble in alcohol, chloroform; insoluble in water.
Use: Medicine (bactericide and fungicide).

chlortetracycline (CTC; chlorotetracycline) $C_{22}H_{23}ClN_2O_8$. An antibiotic produced by the growth of Streptomyces aureofaciens in submerged cultures. It has a wide antimicrobial spectrum including many Gram-positive and Gram-negative bacteria, rickettsiae and several viruses. Its chemical structure is that of a modified naphthacene molecule. Low toxicity. Also available as the hydrochloride.
Properties: Golden-yellow crystals. M.p. 168–169°C. Slightly soluble in water; very soluble in aqueous solutions above pH 8.5; freely soluble in the "Cellosolves," dioxane and "Carbitol"; slightly soluble in methanol, ethanol, butanol, acetone, ethyl acetate, and benzene; insoluble in ether and naphtha.
Derivation: By submerged aerobic fermentation, filtration, solvent extraction, and crystallization.
Use: Medicine (antibiotic); feed supplement; preservative for raw fish.

chlorthion. Generic name for O,O-dimethyl-O(3-chloro-4-nitrophenyl) thiophosphate. Phosphoric acid ester containing chlorine. Related chemically to parathion (q.v.) and malathion (q.v.). Cholinesterase inhibitor.
Hazard: Highly toxic by inhalation and ingestion; absorbed by intact skin. Use may be restricted.
Use: Insecticide.
Shipping regulations: (Rail, Air) Not listed.

"Chlor-Trimeton Maleate."[321] Trademark for chlorpheniramine maleate.

cholaic acid. See taurocholic acid.

cholecalciferol (5,7-cholestadien-3-beta-ol; 7-dehydrocholesterol, activated; vitamin D_3) $C_{27}H_{44}O$. A free vitamin D_3, isolated in crystalline state from the 3,5-dinitrobenzoate; produced by irradiation and equivalent in activity to the vitamin D_3 of tuna liver oil.
Properties: Colorless crystals. Unstable in light and air. Insoluble in water. Soluble in alcohol, chloroform and fatty oils. Melting range 84–88°C. Specific rotation +105 to +112°.
Grade: U.S.P.; F.C.C.
Package: Hermetically sealed under nitrogen.
Use: Medicine (antirachitic vitamin).

choleic acid. A general term applied to the coordination complexes formed by deoxycholic acid (a bile acid) with fatty acids or other lipids, and with a variety of other compounds, including such aromatics as phenol and naphthalene. These complexes are similar to those used in separation processes, such as the urea adducts, for large-scale purification. See also cholic acid.

cholesterol (cholesterin; 5-cholesten-3-beta-ol) $C_{27}H_{15}OH$. The most common animal sterol; a monohydric secondary alcohol of the cyclopentenophenanthrene (4-ring fused) system, containing one double bond. It occurs in part as the free sterol and in part esterified with higher fatty acids as a lipid in human blood serum. The primary precursor in biosynthesis appears to be acetic acid or sodium acetate. Cholesterol itself in the animal system is the precursor of bile acids, steroid hormones and provitamin D_3.

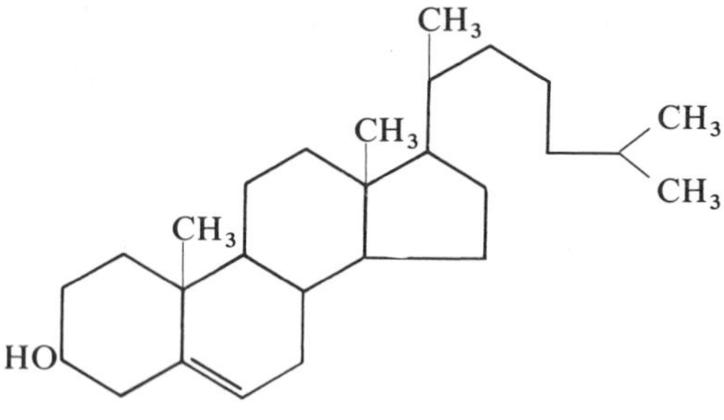

Properties: White, or faintly yellow, almost odorless, pearly granules or crystals; affected by light; m.p. 148.5°C; b.p. 360°C (dec); sp. gr. 1.067 (20/4°C); levorotatory; specific rotation (25°C) −34 to −38°; insoluble in water; slightly soluble in alcohol; soluble in fat solvents, vegetable oils and aqueous solutions of bile salts.
Occurrence: Egg yolk, liver, kidneys, saturated fats and oils.
Source: Prepared from beef spinal cord by petroleum ether extraction of the nonsaponifiable matter; purification by repeated bromination.
Grades: Technical; U.S.P.; S.C.W., standard for clinical work.
Containers: 1- and 5-lb bottles.
Uses: Medicine; emulsifying agent in cosmetic and pharmaceutical products; source of estradiol (q.v.).
Note: Regarding the part played by cholesterol in heart disease (atheroma), Dr. W. Stanley Hartroft of Toronto has stated "It still has not been shown that lowering the cholesterol in the blood by this amount [20%] will have any protective effect for the heart and vessels against the development of atheroma and the onset of its serious complications."
See also sterol.

cholic acid $C_{24}H_{40}O_5$. The most abundant bile acid. In bile it is conjugated with the amino acids glycine and taurine as glycocholic acid and taurocholic acid, respectively, and does not occur free.
Properties: The monohydrate crystallizes in plates from dilute acetic acid; bitter taste with sweetish aftertaste; anhydrous form, m.p. 198°C. Not precipitated by digitonin. Soluble in glacial acetic acid, acetone, and alcohol; slightly soluble in chloroform; practically insoluble in water and benzene.
Derivation: From glycocholic and taurocholic acids in bile; organic synthesis.
Grade: F.C.C.
Use: Biochemical research; pharmaceutical intermediate; emulsifying agent in foods, up to 0.1%.

choline (choline base; (beta-hydroxyethyltrimethylammonium hydroxide) $(CH_3)_3N(OH)CH_2CH_2OH$. Member of the vitamin B complex. Essential in the diet of rats, rabbits, chickens, and dogs. In man it is required for lecithin formation and can replace methionine in the diet. There is no evidence of disease in man due to choline deficiency. It is a dietary factor important in furnishing free methyl groups for transmethylation; has a lipotrophic function.
Source: Egg yolk, kidney, liver, heart, seeds, vegetables and legumes; synthetic preparation from tri-

methylamine and ethylene chlorohydrin or ethylene oxide.
Units: Amounts are expressed in milligrans of choline.
Uses: Medicine; nutrition; feed supplement; catalyst; curing agent; control of pH; neutralizing agent; solubilizer.

choline bicarbonate. See 2-hydroxyethyltrimethylammonium bicarbonate.

choline bitartrate ($C_5H_{14}NO \cdot C_4H_5O_6$).
Properties: White crystalline powder; odorless or faint trimethylamine-like odor; acid taste; hygroscopic; soluble in water and alcohol. Insoluble in ether, chloroform and benzene.
Grade: F.C.C.
Use: Medicine; dietary supplement; nutrient.

choline chloride $(CH_3)_3N(Cl)CH_2CH_2OH$. Animal feed additive derived from agricultural waste or made synthetically. Available as 50% dry feed-grade and 70% solution.

choline chloride carbamate. See carbachol.

cholinesterase
(1) (acetylcholinesterase). Enzyme specific for the hydrolysis of acetylcholine to acetic acid and choline in the body. It is found in the brain, nerve cells and red blood cells and is important in the mechanism of nerve action. See nerve gases; parathion; insecticide.
Derivation: From bovine erythrocytes.
Uses: Biochemical research; determination of phosphorus in insecticides and poisons.
(2) "Pseudo" or nonspecific cholinesterase, prepared from horse serum. This esterase hydrolyzes other esters as well as choline esters. It occurs in blood serum, pancreas and liver.

cholinesterase inhibitor. A chemical compound which deactivates the enzyme cholinesterase, thus preventing or retarding hydrolytic breakdown of the highly toxic acetylcholine formed in the body by the nervous system. Nerve gases act in this way, and so do a number of insecticides. The compounds involved are usually organic esters of phosphoric acid derivatives. Serious poisoning and death may occur on ingestion or inhalation of such compounds. See also parathion; nerve gas.

"Chulograffin Sodium."[412] Trademark for sodium iodipamide (q.v.).

cholytaurine. See taurocholic acid.

chondroitin sulfate. A major constituent of the cartilaginous tissue in the body.

chondrus. See carrageenan.

chorionic gonadotropin (HCG). A hormone isolated from blood and urine of pregnant women; is secreted by the placenta. It is a glycoprotein containing about 11% galactose and having a molecular weight of about 100,000.
Properties: Rods or needle-like crystals. Soluble in water and glycols. Relatively unstable in aqueous solution; stable in dry form. It enhances estrone and progesterone production.
Units: One international unit equals the activity of 0.1 mg of a standard preparation.
Use: Medicine; veterinary medicine.

"Chromacyl."[28] Trademark for a group of dyes that contain chromium in the molecule. Suitable for wool and nylon.

chlorinated zinc chloride. See zinc chloride chromated.

chromatin. A deoxyribonucleoprotein complex consisting of (1) double-stranded DNA molecules, (2) a basic protein called histone, and (3) other proteins. The latter protect the DNA from attack by enzymes. Chromatin occurs in the cell nucleus (q.v.), where it forms chromosomes, the carriers of genes. Its name is derived from its sensitivity to biological stains. See also deoxyribonucleic acid; chromosome; gene.

chromatography. A laboratory analytical technique for the separation and identification of chemical compounds in complex mixtures. Basically, it involves the flow of a mobile (gas or liquid) phase over a stationary phase (which may be a solid or a liquid). Liquid chromatography is used for soluble substances and gas (vapor-phase) chromatography for volatile substances. As the mobile phase moves past the stationary phase, repeated adsorption and desorption of the solute occurs at a rate determined chiefly by its ratio of distribution between the two phases. If the ratio is large enough, the components of the mixture will move at different rates, producing a characteristic pattern, or chromatograph, from which their identity can be determined. See also liquid chromatography; gas chromatography; paper chromatography; thin-layer chromatography; ion-exchange chromatography; gel filtration.

"Chromax" Castings.[350] Trademark for ferrous alloys containing 20–30% nickel and 15–20% chromium.

chrome alum. See chromium potassium sulfate.

chrome ammonium alum. See chromium ammonium sulfate.

chrome cake. A green form of salt cake (sodium sulfate) containing a low percentage of chromium. A by-product of sodium dichromate manufacture used in the paper industry.

chrome dye. A mordant dye, most frequently one in which sodium dichromate is used as the mordant.

chrome green. See chrome pigment.

"Chromekill 4A."[142] Trademark for a nearly neutral material composed of organic and inorganic reducing agents designed for the reduction of hexavalent chromium to trivalent chromium when it is present in low percentages in alkaline cleaning solutions and electroplating baths.

"Chromel."[166] Trademark for a series of nickel-chromium alloys. There are a number of different compositions ranging from 35% to 80% nickel and 16% to 20% chromium, some containing iron. The 8ONi-2OCr types are useful up to 2100–2200°F.

chrome-molybdenum steel. Steel made by any accepted method of quality steel-making containing both chromium and molybdenum, usually in the ranges of chromium 0.35 to 1.10% and molybdenum 0.08 to 0.35%.

chrome-nickel steel. See steel, stainless.

chrome pigment. An inorganic pigment (q.v.) containing chromium. The most important types are: (1) chrome oxide green, one of the most permanent and stable pigments known, the pure grade consisting of 99% Cr_2O_3, used in paints applied to cement and lime-containing surfaces; (2) chrome green, chrome yellow, and chrome red, consisting chiefly of lead

chromate, and used in paints, rubber, and plastic products; (3) miscellaneous pigments such as molybdate orange and zinc yellow, based on lead and zinc compounds of chromium, respectively. All these are more stable to sunlight, weathering, and chemical action than the brighter organic dyes.
Hazard: See lead chromate.

chrome potash alum. See chromium potassium sulfate.

chrome red. See chrome pigment.

chrome steel. A steel made by any accepted method of quality steel-making containing chromium as alloying element usually in the range 0.20 to 1.60%, although the percentage may be as high as 25% in specialized heat-resistant and wear-resistant steels and in stainless steels. See also steel, stainless.

chrome tanning. See tanning.

chrome-vanadium steel. A steel made by any accepted method of quality steel-making containing both chromium and vanadium, usually in the ranges of chromium 0.50 to 1.10% and vanadium 0.10 to 0.20%.

chrome yellow. See lead chromate; chrome pigment.

chromia. See chromic oxide.

chromic acetate (chromium acetate) $Cr(C_2H_3O_2)_3 \cdot H_2O$.
Properties: Grayish-green powder or bluish-green, pasty mass. Soluble in water; insoluble in alcohol.
Derivation: Action of acetic acid on chromium hydroxide. The solution is evaporated and crystallized.
Hazard: Toxic by ingestion.
Uses: Textile mordant; tanning.

chromic acid (chromium trioxide; chromic anhydride) CrO_3. The name is in common use, although the true chromic acid, H_2CrO_4, exists only in solution.
Properties: Dark-purplish red crystals; soluble in water, alcohol and mineral acids. Deliquescent. Sp. gr. 2.67–2.82; m.p. 196°C.
Derivation: (a) Sulfuric acid is added to a solution of sodium dichromate and the product is crystallized out; (b) chromite is fused with soda ash and limestone and then treated with sulfuric acid; (c) electrolysis.
Grades: Technical; C. P.
Containers: 100-, 400-lb Drums.
Hazard: Highly toxic, corrosive to skin. Tolerance (as CrO_3), 0.1 milligram per cubic meter of air. Powerful oxidizing agent; may explode on contact with reducing agents; may ignite on contact with organic materials. Safety data sheet available from Manufacturing Chemists Assn., Washington, D.C.
Uses: Chemicals (chromates; oxidizing agent, catalysts); chromium plating; intermediate; medicine (caustic); process engraving; anodizing; ceramic glazes; colored glass; metal cleaning; inks; tanning; paints; textile mordant.
Shipping regulations: (Rail) (solid) Yellow label; (solution) White label. (Air) (solid) Oxidizer label; (solution over 35%) Corrosive label.

chromic chloride (chromium chloride; chromium trichloride; chromium sesquichloride) (a) $CrCl_3$ or (b) $CrCl_3 \cdot 6H_2O$.
Properties: (a) Violet crystals. Sp. gr. 2.76; m.p. 1150°C; sublimes about 1300°C. Insoluble in water and alcohol. (b) Greenish-black or violet deliquescent crystals depending on whether or not chlorine is coordinated with the chromium. Sp. gr. 1.76; m.p. 83°C. Soluble in water; soluble in alcohol; insoluble in ether.
Derivation: (a) By passing chlorine over a mixture of chromic oxide and carbon. (b) By the action of hydrochloric acid on chromium hydroxide.
Hazard: Toxic. Tolerance (as Cr), 0.5 mg per cubic meter in air.
Uses: Chromium salts; intermediates; textile mordant; chromium plating including vapor plating; preparation of sponge chromium; catalyst for polymerizing olefins.

chromic fluoride (chromium fluoride; chromium trifluoride) $CrF_3 \cdot 4H_2O$ or $CrF_3 \cdot 9H_2O$.
Properties: Fine green crystalline powder. Sp. gr. (anhydrous) 3.8; m.p. over 1000°C; b.p. (sublimes) 1100–1200°C. The hydrates are soluble in water and acids; insoluble in alcohol.
Derivation: Interaction of chromium hydroxide and hydrofluoric acid.
Grades: Technical; high purity (CrF_3).
Hazard: Toxic. Irritating to skin and eyes, especially in solution. Tolerance (as F), 2.5 mg per cubic meter of air.
Uses: Printing and dyeing woolens; mothproofing woolen fabrics.
Shipping regulations: (Solid and solution): (Rail) Corrosive material, n.o.s., White label. (Air) Corrosive label.

chromic formate $Cr(OH)(HCOO)_2$.
Properties: Dark green liquid. Sp. gr. (20°C) 1.237; wt/gal 10.3 lb.
Hazard: Toxic by ingestion; strong irritant.
Uses: Tanning agent; textile mordant; synthesis.

chromic hydrate. See chromic hydroxide.

chromic hydroxide (chromic hydrate; chromium hydroxide; chromium hydrate) $Cr(OH)_3$.
Properties: Green, gelatinous precipitate; decomposed to chromic oxide by heat. Insoluble inwater; soluble in acids and strong alkalies.
Derivation: By adding a solution of ammonium hydroxide to the solution of a chromium salt.
Uses: Chromium salts and chromites; paint pigment; wool treatment.

chromic nitrate (chromium nitrate) $Cr(NO_3)_3 \cdot 9H_2O$.
Properties: Purple crystals; soluble in alcohol and water. M.p. 60°C; decomposes 100°C.
Derivation: By the action of nitric acid on chromium hydroxide.
Hazard: Strong oxidant; may ignite organic materials on contact. May be explosive when shocked or heated.
Shipping regulations: Nitrates, n.o.s., (Rail) White label (Air) Oxidizer label.

chromic oxide (chromium oxide; chromia; chromium sesquioxide; green cinnabar) Cr_2O_3.
Properties: Bright-green, crystalline powder; sp. gr. 5.04; m.p. 2435°C; b.p. 4000°C; insoluble in water, acids, and alkalies.
Derivation: (a) By heating chromium hydroxide; (b) by heating dry ammonium dichromate; (c) by heating sodium dichromate with sulfur and washing out the sodium sulfate.
Uses: Metallurgy; green paint pigment; ceramics; catalyst in organic synthesis; green granules in asphalt roofing; component of refractory brick.

chromic phosphate (chromium phosphate) (a) $CrPO_4 \cdot 6H_2O$; (b) $CrPO_4 \cdot 4H_2O$.
Properties: (a) Violet crystals; sp. gr. 2.12 (14°C); m.p. 100°C. (b) Green crystals. Soluble in acids; insoluble in water.
Derivation: (a) Interaction of solutions of chromium chloride and sodium phosphate; (b) by mixing chrome alum and disodium hydrogen phosphate.

Violet amorphous powder (not the hexahydrate) is formed which becomes crystalline on contact with water. On boilng it is converted into green crystalline hydrate.
Use: Paint pigment.

chromic sulfate (chromium sulfate) (a) $Cr_2(SO_4)_3$; (b) $Cr_2(SO_4)_3 \cdot 15H_2O$; (c) $Cr_2(SO_4)_3 \cdot 18H_2O$.
Properties: (a) Violet or red powder; (b) dark-green amorphous scales; (c) violet cubes. Sp. gr. (a) 3.012; (b) 1.867; (c) 1.70. (a) Insoluble in water and acids; (b) soluble in water; insoluble in alcohol; (c) soluble in water and alcohol.
Derivation: Action of sulfuric acid on chromium hydroxide, with subsequent crystallization.
Uses: Textile industries; green paints and varnishes; green ink; ceramics (glazes, green effects); the basic form (reduction of sodium dichromate) is used in tanning (q.v.).

chromite (chrome iron ore) $FeCr_2O_4$. A natural oxide of ferrous iron and chromium, sometimes with magnesium and aluminum present. Usually occurs in magnesium- and iron-rich igneous rocks.
Properties: Color iron-black to brownish-black; streak dark brown; luster metallic to submetallic; sp. gr. 4.6; Mohs hardness 5.5. Low toxicity.
Grades: Metallurgical; refractory; chemical.
Occurrence: U.S.S.R.; South Africa; Rhodesia; Philippines; Cuba; Turkey.
Uses: Only commercial source of chromium and its compounds.

chromium Cr Metallic element of atomic number 24, group VIB of the Periodic Table; atomic weight 51.996; valences 2, 3, 6; four stable isotopes. Name derived from Greek for color.
Properties: Hard, brittle, semi-gray metal. Sp. gr. 7.1; m.p. 1900°C; b.p. 2200°C. Compounds have strong and varied colors. Cr ion forms many coordination compounds. Exists in active and passive forms, the latter giving rise to its corrosion resistance, due to a thin surface oxide layer that passivates the metal when treated with oxidizing agents. Active form reacts readily with dilute acids to form chromous salts. Soluble in acids (except nitric) and strong alkalies; insoluble in water.
Occurrence: USSR, So. Africa, Turkey, Philippines, Rhodesia.
Derivation: From chromite (q.v.), by direct reduction (ferrochrome); by reducing the oxide with finely divided aluminum or carbon; and by electrolysis of chromium solutions.
Grades (ore): Chromium ores are classified as (1) metallurgical, (2) refractory, and (3) chemical, and their consumption in the U.S. is in that order. (1) must contain a minimum of 48% Cr_2O_3 and have chromium-iron ratio of 3 to 1; (2) must be high in Cr_2O_3 and Al_2O_3 and low in iron; (3) must be low in SiO_2 and Al_2O_3 and high in Cr_2O_3.
Forms available: (1) Chromium metal as lumps, granules, or powder; (2) high- or low-carbon ferrochromium (q.v.). Single crystals. High-purity crystals or powder run 99.97% pure.
Hazard: Elemental chromium and trivalent compounds are relatively nontoxic, but the hexavalent compounds have an irritating and corrosive effect on tissue, resulting in ulcers and dermatitis on prolonged contact. Tolerance for chromium dust and fume is 1 mg per cubic meter of air. It is a suspected carcinogen.
Uses: Alloying and plating element on metal and plastic substrates for corrosion resistance; chromium-containing and stainless steels (q.v.); protective coating for automotive and equipment accessories; nuclear and high-temperature research; constituent of inorganic pigments.

chromium 51. Radioactive chromium of mass number 51.
Properties: Half-life 26.5 days; radiation, gamma (0.32 MeV).
Grade: U.S.P. (as sodium chromate Cr 51 injection).
Hazard: Radioactive poison.
Uses: Diagnosis of blood volume, blood cell life and cardiac output; etc.
Shipping regulations: (Rail, Air) Consult regulations.

chromium acetate. See chromic acetate.

chromium acetylacetonate
$[CH_3COCHC(CH_3)O]_3Cr$.
Properties: Purple powder or red-violet crystals; m.p. 216°C; b.p. 340°C, insoluble in water; soluble in acetone and alcohol.
Derivation: Reaction of chromium chloride, acetylacetone and sodium carbonate.
Use: Reduction of detonation of nitromethane.

chromium ammonium sulfate (ammonium chromium sulfate; chrome ammonium alum)
$CrNH_4(SO_4)_2 \cdot 12H_2O$.
Properties: Green powder or deep violet crystals; sp. gr. 1.72; m.p. 94°C. Soluble in water; slightly soluble in alcohol. The aqueous solution is violet when cold; green when hot.
Grades: Technical.
Uses: Mordant; tanning.

chromium boride One of several compounds of chromium and boron, e.g., CrB, CrB_2, and Cr_3B_2. They have high melting points, are very hard and corrosion-resistant, and may be suitable for use in jet and rocket engines.
Properties: CrB, may be crystalline; sp. gr. 6.2; m.p. 1550°C; Mohs hardness 8.5; resistivity 67μ-ohm cm (20°C). CrB_2, sp. gr. 5.15; m.p. 1850°C; hardness 2010 (Knoop); resists oxidation up to 1100°C. Cr_3B_2, may be crystalline; sp. gr. 6.1; Mohs hardness 9+; resistivity 116μ-ohm cm (20°C).
Uses: Metallurgical additives; high temperature electrical conductors; cermets; refractories; coatings resistant to attack by molten metals.

chromium bromide. See chromous bromide.

chromium carbide Cr_3C_2.
Properties: Orthorhombic crystals; sp. gr. 6.65; microhardness, 2700 kg/sq mm (load 50 g); m.p. 1890°C; b.p. 3800°C; resistivity 95μ-ohm cm (room temperature). Highest oxidation resistance at high temperatures of all metal carbides; also resistant to acids and alkalies.
Uses: Gage blocks and hot extrusion dies; in powder form, as spray coating material; components for pumps and valves.

chromium carbonate. See chromous carbonate.

chromium carbonyl. See chromium hexacarbonyl.

chromium chloride. See chromic chloride and chromous chloride.

chromium copper. A copper-chromium alloy containing 8–11% chromium. Used in the manufacture of hard steels for increased elasticity.

chromium dioxide Cr_2O_2.
Properties: Black, semiconducting material. Sp. gr. 4.9. Acicular needle-like crystals having strong magnetic properties. Also in powder form.
Derivation: By heating chromic acid.
Use: Magnetic component in recording tapes (claimed to be superior to iron oxide for this purpose).

chromium fluoride. See chromic fluoride.

chromium hexacarbonyl (chromium carbonyl) $Cr(CO)_6$
Properties: White crystalline solid; sp. gr. 1.77; m.p. 150–151°C (inert atmosphere); dec. 210°C (explodes). Slightly soluble in iodoform and carbon tetrachloride; insoluble in water, alcohol, ether or acetic acid. More stable than most metal carbonyls; decomposed by chlorine and fuming nitric acid but resists bromine, iodine, water, and cold conc. nitric acid. Decomposes photochemically when solutions are exposed to light.
Grade: 98% pure.
Hazard: Highly toxic by inhalation and ingestion. Explodes at 400°F.
Use: Intermediate.
Shipping regulations: (Rail, Air) Not specified. Consult authorities.

chromium hydrate. See chromic hydroxide.

chromium hydroxide. See chromic hydroxide.

chromium manganese antimonide. Brittle gray solid having magnetic properties when above a definite temperature. If the composition of the compound is intentionally changed, the temperature required also changes.

chromium naphthenate
Properties: Dark-green liquid or violet powder.
Derivation: By addition of chromium salts to solution of sodium naphthenate and recovery of the precipitate.
Grades: 6% chromium.
Hazard; Toxic by ingestion.
Use: Paints (anti-chalking agent).

chromium nitrate. See chromic nitrate.

chromium oxide. See chromic oxide.

chromium oxychloride. See chromyl chloride.

chromium phosphate. See chromic phosphate.

chromium potassium sulfate (chrome alum; potassium chromium sulfate; chrome potash alum) $CrK(SO_4)_2 \cdot 12H_2O$.
Properties: Dark, violet-red crystals; efflorescent; sp. gr. 1.813; m.p. 89°C; loses $10H_2O$ at 100°C. Soluble in water.
Derivation: By reducing potassium dichromate in dilute sulfuric acid with sulfurous acid.
Hazard: Toxic by ingestion.
Uses: Tanning (chrome-tan liquors); textile dye (mordant); photography (fixing bath); ceramics.

chromium sesquichloride. See chromic chloride.

chromium sesquioxide. See chromic oxide.

chromium steel. See steel, stainless; iron, stainless.

chromium sulfate. See chromic sulfate.

chromium trichloride. See chromic chloride.

chromium trifluoride. See chromic fluoride.

chromium trioxide. See chromic acid.

chromogen. See chromophore.

"Chromegene."[307] Trademark for mordant dyestuffs used on wool and leather. Characterized by very good fastness to light, fulling, etc.

"Chromol."[244] Trademark for a series of raw oils emulsified with a non-ionic emulsifier and therefore stable to alum and salt. Used in the leather industry.

"Chromolan."[243] Trademark for metalized acid dyes.

chromophore. A chemical grouping which when present in an aromatic compound (the chromogen) gives color to the compound by causing a displacement of, or appearance of, absorbent bands in the visible spectrum. Dyes are sometimes classified on the basis of their chief chromophores, e.g.; —NO, nitroso dyes; —NO_2, nitro dyes; —N=N—, azo dyes, etc.

chromosome. The heredity-bearing gene carrier of the living cell, derived from chromatin (q.v.) and consisting largely of nucleoproteins (DNA), together with other protein components (histones). See also deoxyribonucleic acid; gene.

"Chromosorb."[247] Trademark for a series of screened calcined and flux-calcined diatomite aggregates. Available in non-acid washed, acid-washed, and acid-washed dimethyldichlorosilane treatments.
Grades: Chromosorb P, W, G for analytical use; Chromosorb A for preparative chromatography.
Containers: Glass and plastic bottles of various sizes dependent on grade.
Use: Supports in gas or liquid chromatography.

chromotropic acid. See 4,5-dihydroxy-2,7-naphthalenedisulfonic acid.

chromous bromide (chromium bromide) $CrBr_2$.
Properties: White crystals; changes to yellow on heating. Oxidizes in moist air but stable in dry air. Sp. gr. 4.356; m.p. 842°C; soluble in water (blue color).
Hazard: Toxic by ingestion; irritant to skin and tissue. Tolerance (as Cr), 0.5 mg per cubic meter in air.

chromous carbonate (chromium carbonate) $CrCo_3$.
Properties: Grayish-blue amorphous mass; sp. gr. 2.75. Soluble in mineral acids; slightly soluble in water containing carbon dioxide; insoluble in alcohol.
Hazard: Toxic by ingestion; irritant to skin and tissue. Tolerance (as Cr), 0.5 mg per cubic meter in air.

chromous chloride (chromium chloride) $CrCl_2$.
Properties: White, deliquescent needles; sp. gr. 2.878 (25°C); m.p. 824°C; active reducing agent; very soluble in water; insoluble in alcohol and ether.
Derivation: Reaction of the metal with anhydrous hydrogen chloride.
Hazard: Toxic. Tolerance (as Cr), 0.5 mg per cubic meter in air.
Uses: Reducing agent; oxygen absorbent.

chromous oxalate $CrC_2O_4 \cdot H_2O$.
Properties: Yellow crystalline powder; sp. gr. 2.468; soluble in water; active reducing agent.
Hazard: Toxic by ingestion; irritant to skin and tissue. Tolerance (as Cr), 0.5 mg per cubic meter in air.

"Chromoxane."[307] Trademark of mordant dyestuffs. Used on wool; characterized by fairly good fastness to light, very good fastness to fulling, etc., and by relatively bright shade.

"Chromspun."[256] Trademark for a solution-dyed acetate fiber.

chromyl chloride (chromium oxychloride; chlorochromic anhydride) CrO_2Cl_2.
Properties: Mobile, dark-red liquid. B.p. 116°C; freez-

ing point -96.5°C; sp. gr. 1.911. Fumes in air; reacts vigorously with water to form chromic acid, chromic chloride, hydrochloric acid, and chlorine. Miscible with carbon tetrachloride, tetrachloroethane, carbon disulfide.
Derivation: By heating sodium dichromate and sodium chloride with sulfuric acid.
Grades: Technical.
Hazard: Toxic. Corrosive to tissue.
Uses: Organic oxidations and chlorinations; solvent for chromic anhydride; chromium complexes and dyes.
Shipping regulations: (Rail) White label. (Air) Corrosive label. Not acceptable on passenger planes.

chrysamine G $C_{26}H_{16}N_4O_6Na_2$.
Properties: Yellowish-brown powder. Very sparingly soluble in water.
Derivation: By coupling diazotized benzidine with salicylic acid.
Use: Yellow direct cotton dyestuff.

chrysanthemummonocarboxylic acid, ethyl ester $(CH_3)_2CCH(COOC_2H_5)CHCH{:}C(CH_3)_2$.
Properties: Colorless to pale yellow, with pleasant ester odor; sp. gr. 0.924–0.927 (25/25°C); insoluble in water; soluble in alcohols and ketones.
Containers: Tins; drums.
Hazard: May be toxic by ingestion.
Uses: Synthesis of perfumes, pharmaceuticals; insecticides; other organic chemicals.
See also cinerin I, pyrethrin I.

chrysarobin.
Properties: Microcrystalline orange-yellow powder; slight odor or odorless; tasteless. Soluble in chloroform and benzene; slightly soluble in ether and alcohol; very slightly soluble in water. Combustible.
Derivation: Extraction from araroba (goa powder) obtained from the Brazilian tree, Andira araroba.
Hazard: Irritating to eyes and mucous membranes.
Use: Medicine (external).

chrysazin. See 1,8-dihydroxyanthraquinone.

chrysene (1,2-benzphenanthrene) $C_{18}H_{12}$ (a tetracyclic hydrocarbon)
Properties: Crystals; sp. gr. 1.274 (20/4°C); m.p. 254°C; b.p. 448°C; sublimes easily in a vacuum. Slightly soluble in alcohol, ether, glacial acetic acid. Insoluble in water.
Derivation: Distillation of coal tar.
Hazard: Carcinogenic agent. Tolerance, 0.2 mg per cubic meter in air.
Use: Organic synthesis.

chrysoidine (meta-diaminoazobenzene hydrochloride) $C_6H_5NNC_6H_3(NH_2)_2 \cdot HCl$.
Properties: Red-brown powder or large black shiny crystals with a green luster. Soluble in alcohol and water giving orange-brown solutions. Insoluble in ether; m.p. 117°C.
Uses: Orange dye for cotton and silk.

chrysolite. See olivine.

chrysophanic acid (1,8-dihydroxy-3-methylanthraquinone; chrysophanol) $C_{15}H_{10}O_4$.
Properties: Yellow crystals; m.p. 196°C; sublimes. Slightly soluble in water and cold alcohol; soluble in hot alcohol, chloroform and ether.
Source: Rhubarb root, cascara sagrada, senna leaves, goa powder.
Derivation: Oxidation of chrysarobin and other synthetic methods.
Use: Medicine.

chrysotile. See asbestos.

chu. Abbreviation for centigrade heat unit. It is the amount of heat required to raise the temperature of one pound of water one centigrade degree from 15°C to 16°C. It is sometimes called a pcu (pound centigrade unit).

chymosin. See rennin.

chymotrypsin. An enzyme found in the intestine which catalyzes the hydrolysis of various proteins (especially casein) and protein digestion products to form polypeptides and amino acids.
Properties: White to yellowish-white, odorless, crystalline or amorphous powder. Soluble in water or saline solution.
Derivation: Crystallized from an extract of the pancreas gland of the ox.
Grade: N.F.
Use: Medicine.

chrmotrypsinogen. A crystallizable enzyme occurring in the pancreas, which gives rise to chymotrypsin.
Use: Biochemical research.

ci. Abbreviation for curie (q.v.).

"Cibalan."[443] Trademark for metal-complex dyes for wool and polyamide fibers.

"Cibanone."[443] Trademark for anthraquinone vat dyes.

"Cibaphasol."[443] Trademark for coacervate-forming colloidal product used in textile dyeing.

cigarette tar. The comparatively non-volatile residue from the burning of cigarette tobacco which appears in finely divided form in the smoke. Cigarette tar is known to contain minute traces of various aromatic ring compounds (especially benzo[a]pyrene) found in coal tar, some of which are known carcinogens (q.v.). Various forms of activated carbon are used in filters to adsorb the toxic components of the tars that find their way into the smoke. See also smoke (5).

ciguatoxin $C_{28}H_{52}NO_5Cl$. A complex toxic principle in bony fishes; has both fat- and water-soluble fractions. Ninhyrdrin test positive. It is a type of quaternary ammonium compound and one fraction is said to be an irreversible anticholinesterase. The pharmacology is unknown.

CIIT. Abbreviation for Chemical Industry Institute of Toxicology (q.v.).

"Cinaryl."[233] Trademark for antimicrobial agents containing 1-(3-chloroallyl)-3,5,7-triaza-1-azoniaadamantane chloride.

cincholepidine. See lepidine.

cinchona bark. The bark of one of several species of Cinchona trees native to South America and cultivated in Indonesia, Peru, Ecuador, and western Africa. The best-known types are calisaya, loxa, and succirubra. It is used primarily as the source of natural quinine (q.v.), quinidine, (q.v.), cinchonidine, cinchonine, and related alkaloids formerly used as antimalarials (q.v.).

cinene. See dipentene.

cineol. See eucalyptol.

cinerin I $C_{20}H_{28}O_3$. One of the four primary active insecticidal principles of pyrethrum flowers. It is the 3-(2-butenyl)-4-methyl-2-oxo-3-cyclopenten-1-yl ester of chrysanthemummonocarboxylic acid. See also cinerin II, pyrethrin I and II.
Properties: A viscous liquid, quickly oxidized in air; b.p. 200°C (0.1 mm). Insoluble in water; soluble in organic solvents; incompatible with alkalies.
Hazard: Toxic by ingestion.
Use: Household insecticide.

cinerin II $C_{21}H_{28}O_5$. One of the four primary active insecticidal principles of pyrethrum flowers. It is the 3-(2-butenyl)-4-methyl-2-oxo-3-cyclopenten-l-yl ester of chrysanthemumdicarboxylic acid monomethyl ester. See also cinerin I, pyrethrin I and II.
Properties: A viscous liquid, quickly oxidized in air; b.p. 200°C (0.1 mm). Insoluble in water; soluble in organic solvents.
Hazard: Toxic by ingestion.
Use: Household insecticide.

cinnabar (natural vermilion; liver ore) HgS. Natural mercuric sulfide, occurring in veins near recent volcanic rocks and hot springs.
Properties: Red, scarlet, reddish brown to blackish solid; streak scarlet; luster adamantine to dull earthy when impure; sp. gr. 8.10; hardness 2.5. Soluble in aqua regia. Has greater optical rotation than any other substance (+325°).
Occurrence: California, Nevada; Spain; Italy; Mexico; Yugoslavia.
Use: The only important ore of mercury.

cinnamaldehyde. See cinnamic aldehyde.

cinnamein. See benzyl cinnamate.

cinnamene. See styrene.

cinnamic acid (beta-phenylacrylic acid; 3-phenylpropenoic acid; cinnamylic acid) $C_6H_5CH{:}CHCOOH$.
Properties: White, crystalline scales; soluble in benzene, ether, acetone, glacial acetic acid, carbon disulfide, oils; insoluble in water. Congealing point 133°C (min); b.p. 300°C. Combustible.
Derivation: By heating benzaldehyde with sodium acetate in presence of a dehydrating agent (acetic anhydride) or by heating benzyl chloride with sodium acetate in an autoclave. Occurs naturally in tobacco and some balsams.
Grades: Technical; refined.
Containers: Glass bottles; cans; drums.
Uses: Medicine; perfumes; intermediate.

cinnamic alcohol (cinnamyl alcohol; phenylallylic alcohol; 3-phenyl-1-propen-1-ol; styryl carbinol) $C_6H_5CH{:}CHCH_2OH$.
Properties: White needles or crystals; hyacinth-like odor. Soluble in 3 volumes of 50% alcohol. Congealing point 33°C (min): (pure), as low as 24°C (tech.); b.p. 257°C. Combustible.
Derivation: (a) From oil of cassia or oil of cinnamon. Occurs as an ester. (b) Reduction of cinnamic aldehyde.
Method of purification: Recrystallization.
Grades: Technical; F.C.C.
Containers: 1-, 2-, 5-lb bottles; tin cans.
Uses: Perfumery, particularly for lilac and other floral scents; flavoring agent; soaps; cosmetics.

cinnamic aldehyde (cinnamaldehyde; 3-phenylpropenal; cinnamyl aldehyde) $C_6H_5CH{:}CHCHO$.
Properties: Yellowish oil; cinnamon odor, sweet taste. Thickens on exposure to air. Soluble in 5 volumes of 60% alcohol; very slightly soluble in water. Sp. gr. 1.048–1.052; refractive index 1.618–1.623; f.p. -8°C; b.p. 248°C. Combustible.
Derivation: (a) From Ceylon and Chinese cinnamon oils (b) By condensation of benzaldehyde and acetaldehyde.
Method of purification: Rectification.
Uses: Flavors; spice perfumes.
See also cassia oil.

cinnamic ether. Incorrect name for ethyl cinnamate.

cinnamon oil. See cassia oil.

cinnamoyl chloride (phenylacrylyl chloride) $C_6H_5CH{:}CHCOCl$.
Properties: Yellow crystals; m.p. 35°C; b.p. 170°C (58 mm); sp. gr. 1.1617 (45/4°C). Decomposes in water; soluble in hydrocarbons and esters.
Hazard: Skin irritant.
Use: Reagent for determination of water; chemical intermediate.

cinnamyl acetate $C_6H_5CH{:}CHCH_2OOCCH_3$.
Properties: Colorless liquid having a floral-spicy odor. Soluble in 4 volumes of 70% alcohol. Sp. gr. 1.048–1.052; refractive index 1.539–1.542. Combustible.
Uses: Perfumery (fixative); flavoring.

cinnamyl alcohol. See cinnamic alcohol.

cinnamyl aldehyde. See cinnamic aldehyde.

cinnamyl cinnamate (styracin) $C_9H_7O_2C_9H_9$.
Properties: Rectangular prismic crystals; m.p. 40°C (min). Soluble in alcohol, ether, benzene.
Derivation: Esterification of cinnamic acid with cinnamic alcohol.
Use: Perfumery; flavoring.

cinnamylic acid. See cinnamic acid.

"Ciodrin"[125] Trademark for dimethyl phosphate of alpha-methylbenzyl 3-hydroxy-cis-crotonate. $(CH_3O)_2P(O)OC(CH_3){:}CHC(O)OCH(CH_3)C_6H_5$.
Properties: Light straw liquid with a mild ester odor; b.p. 275°F (0.03 mm). Miscible with xylene; soluble in ethanol, or acetone; slightly soluble in kerosine and water.
Hazard: Toxic by ingestion, inhalation, and skin absorption.
Use: Insecticide for external parasites on livestock.

CIPC. See chloro-IPC.

"Circosol."[494] Trademark for various grades of naphthenic oils which provide overall balance of nonstaining and processing properties for most polymers. Aid processing and breakdown of rubber or polymer.

cis- (Latin, on this side). A prefix used in designating geometrical isomers in which there is a double bond between two carbon atoms. This prevents free rotation and thus two different spacial arrangements of substituent groups or atoms are possible. When a given atom or radical is positioned on one side of the carbon axis or backbone, the isomer is called *cis-*; when it is in the opposite location relative to the carbon axis, the isomer is called *trans-* (Latin, on the other side). This is indicated in the following formulas:

```
CH3          COOH      CH3           H
   \        /             \        /
    C ═ C                   C ═ C
   /        \             /        \
  H          H           H          COOH

       cis                    trans
```

This is often called *cis-trans* isomerism. These prefixes are disregarded in alphabetizing chemical names.

"Cisdene."[454] Trademark for a polybutadiene rubber.

citraconic acid (methylmaleic acid) $CH_3C(COOH){:}CH(COOH)$.
Properties: Hygroscopic colorless crystals. M.p. 91°C (dec); sp. gr. 1.62. Soluble in water, alcohol, ether; insoluble in benzene and naphtha. Low toxicity.
Derivation: By carefully heating citric acid.
Grades: Technical; C.P.

citraconic anhydride (methylmaleic anhydride) $C_5H_4O_3$.
Properties: Colorless liquid. M.p. 7–8°C; b.p. 213–214°C; sp. gr. 1.25 (15/4°C). Soluble in ether. Low toxicity.
Grades: Reagent.
Containers: Sealed tubules.
Uses: Reagent for alkalies, alcohols, and amines.

citral (geranial; geranialdehyde; 3,7-dimethyl-2,6-octadienal) $(CH_3)_2CCHC_2H_4C(CH_3)CHCHO$. Commercial material is a mixture of alpha and beta isomers.
Properties: Mobile pale yellow liquid; strong lemon odor; sp. gr. 0.891–0.897 (15°C); refractive index 1.4860–1.4900 (20°C); not optically active. Soluble in 5 volumes of 60% alcohol; soluble in all proportions of benzyl benzoate, diethyl phthalate, glycerin, propylene glycol, mineral oil, fixed oils, and 95% alcohol; insoluble in water. Combustible; nontoxic.
Derivation: Principal constituent of lemon grass oil and can be isolated by fractional distillation. Obtained synthetically by oxidation of geraniol, nerol, or linalool by chromic acid.
Grades: Technical; pure; F.C.C.
Containers: Glass bottles; tins; (synthetic) drums.
Uses: Perfumes; flavoring agent; intermediate for other fragrances.

citric acid (2-hydroxy-1,2,3-propanetricarboxylic acid) $HOOCCH_2C(OH)(COOH)CH_2COOH \cdot H_2O$.
Properties: Colorless translucent crystals or powder; odorless; strongly acid taste; hydrated form is efflorescent in dry air. Sp. gr. 1.542; m.p. 153°C (anhydrous form); decomposes before boilng. Very soluble in water and alcohol; soluble in ether. Combustible. Nontoxic.
Occurrence: In living cells, both animal and plant. See TCA cycle.
Derivation: By mold fermentation of carbohydrates, including deep fermentation; from lemon, lime, pineapple juice, molasses.
Grades: Both hydrous (hydrated) and anhydrous; technical; C.P.; U.S.P.; F.C.C.
Containers: Various, including bags, cartons, barrels, drums.
Uses: Preparation of citrates, flavoring extracts, confections, soft drinks, effervescent salts; acidifier; dispersing agent; medicines; antioxidant in foods (for details see regulations of Meat Inspection Division of USDA); sequestering agent; water-conditioning agent and detergent builder; cleaning and polishing stainless steel and other metals; alkyd resins; mordant; removal of sulfur dioxide from smelter waste gases; abscission of citrus fruit in harvesting; cultured dairy products.

citric acid cycle. See TCA cycle.

citrinin $C_{13}H_{14}O_5$. An antibiotic yellow pigment produced by Penicillium citrinin Thom and Aspergillus niveus; m.p. 170–171°C (decomposes). Yellow crystals; insoluble in water; soluble in alcohol, dioxane, dilute alkali solutions.

"Citroflex."[299] Trademark for a series of organic citrates.

"Citronel 'B' and 'C'."[188] Trademark for lemon oil substitutes for technical use.

citronellal $C_9H_{17}CHO$. 3,7-Dimethyloct-6(or 7)-enal. Has both *d-* and *l-*isomers. The *d-*isomer is described.
Properties: Colorless liquid; intense lemon-like odor; sp. gr. 0.847–0.850; optical rotation +8° to +11°; b.p. 205°C; refractive index (n 20/D) 1.4566. Slightly soluble in water; soluble in alcohol and ether. Combustible; nontoxic.
Derivation: From lemon grass, citronella oil and other oils.
Containers: Bottles; drums.
Uses: Soap perfumery; manufacture of hydroxycitronellal; flavoring.

citronellal hydrate. See hydroxycitronellal.

citronella oil.
Properties: Light yellowish essential oil, with rather pungent, citrus-like odor. Soluble in 80% alcohol. Sp. gr. 0.887–0.906; refractive index 1.468–1.483; solutions are levorotatory. Combustible.
Derivation: Steam-distilled from the grass of Cymbopogon nardus.
Constituents: Geraniol 60%, citronellol 15%, camphene and dipentene 15%; also linalool and borneol.
Grades: Ceylon; Java; Taiwan.
Containers: Tins; glass bottles; drums.
Hazard: Moderately toxic.
Uses: Medicine; insect repellent; perfumes for soaps and disinfectants; manufacture of citronellal, geraniol; denaturant for alcohol; pesticide, but may not be used on food crops.

citronellol (3,7-dimethyl-6(or 7)-octen-1-ol) $CH_2{:}C(CH_3)(CH_2)_3CH(CH_3)CH_2CH_2OH$, the 7-octene form, which predominates. See also rhodinol.
Properties: Colorless liquid, having a somewhat rosy odor; sp. gr. 0.849–0.859; refractive index 1.456–1.458; optical rotation −1°30′ to +5°; b.p. 244°C (760 mm). Soluble in two or more volumes of 70% alcohol; soluble in most oils. Combustible; low toxicity.
Occurrence: Citronella, geranium, rose, savin and other essential oils.
Derivation: From the oils above, or by reduction of citronellal or geraniol.
Containers: Bottles; drums.
Use: Perfumery (floral odors, mainly rose types); flavoring.

citronellyl acetate $C_{10}H_{19}OOCCH_3$.
Properties: Colorless liquid. Odor somewhat like that of bergamot oil. Sp. gr. 0.884–0.891; b.p. 119–121°C (15 mm); optical rotation usually slightly dextro, up to +1°; refractive index 1.450–1.452. Soluble in 9 volumes of 70% alcohol. Combustible; low toxicity.
Derivation: Action of acetic anhydride upon citronellol.
Uses: Perfumery; flavoring.

citronellyl acetic ether. Incorrect name for citronellyl acetate.

citronellyl formate $C_{10}H_{19}OOCH$.
Properties: Colorless liquid; floral odor. Sp. gr. 0.890–

0.903 (25/25°C); refractive index 1.4430–1.4490 (20°C). One volume dissolves in 3 volumes 80% alcohol; soluble in most oils. Combustible; low toxicity.
Grade: F.C.C.
Use: Flavoring.

citron yellow. See zinc yellow.

citrovorum factor. See folinic acid.

citrulline $NH_2CONH(CH_2)_3CHNH_2COOH$. An arginine derivative; an amino acid found in watermelon juice in the L(+) form.
Properties: Crystals from methanol-water mixture; m.p. 222°C; soluble in water; insoluble in methanol and ethanol.
Use: Biochemical research.

citrus peel oils. Edible oils expressed from the peel or rind of grapefruit, lemon, lime, orange, and tangerine.
Properties: Color, odor, and taste characteristic of source. Sp. gr. 0.84–0.89; refractive index 1.473–1.478; optically active; unsaponifiable. Soluble in alcohol, vegetable oils, mineral oil; orange and lemon are soluble in glacial acetic acid. Combustible. Nontoxic.
Constituents: Limonene, citral, and terpenes in varying percentages.
Containers: Cans, drums.
Uses: Flavoring agents in desserts, soft drinks, ice cream; odorants in perfumery and cosmetics; furniture polish (lemon oil).
Note: Terpene-free grades are available at much higher concentrations (from 15–30 times) than original oil. These grades have much lower optical rotation and less tendency to spoil on storage, since terpenes tend to oxidize to undesirable components such as carvone and *p*-cymene. See terpeneless oil.

citrus seed oils. Edible oils expressed from seeds of grapefruit, orange, lemon, lime, and tangerine reclaimed from cannery processing.
Properties: Nondrying; odor, color, and taste characteristic of source; bitter baste removed by alkali refining. Sp. gr. 0.91–0.92; saponification value 190–195; iodine no. 100–110; optically inactive; combustible. Nontoxic. May be bleached, deodorized or hydrogenated.
Constituents: Chiefly palmitic, oleic, and linoleic acids.
Containers: Cans, drums, tank cars.
Uses: Flavoring; food products; cosmetics; odorants in special soaps.

civet (zibeth).
Derivation: Unctuous secretion from the civet cat. An artificial type is also available.
Habitat: Malacca Islands; Ethiopia; East Indies.
Properties: Yellow to brown. Semi-solid. Strong unpleasant odor. Soluble in hot alcohol and ether; insoluble in water.
Containers: Bottles.
Use: Perfumery (fixative); cosmetics, to provide animal note which has become popular.

civettal. See 1,2,3,4-tetrahydro-6-methylquinoline.

Cl Symbol for chlorine; the molecular formula is Cl_2.

"Cladkote."[41] Trademark for a modified polyester composite of resins and siliceous reinforcing material which cures to a tough, chemical-resistant topping; suggested for use in food plants under acid conditions.

cladding. The process in which two metals are bonded by being rolled together at suitable pressure and temperature. Controlled explosion is also used. At the interface each metal diffuses sufficiently into the other to form an alloy. Cladding is generally from 5 to 20% of total thickness, but may be heavier, depending on the properties desired. The base metal is usually carbon or low-alloy steel clad with stainless steel, nickel, or other protective metal. Nonferrous metals are also clad; copper clad with cupronickel is used for coinage.

"Clalite."[159] Trademark for a buffered sodium hydrosulfite used to improve whiteness of coating clay in paper industry.

Clapeyron equation. The equation $dp/dT = \Delta H/T\Delta V$, derived from thermodynamics. It states that the change of vapor pressure of a liquid (expressed in ergs) with absolute temperature (°K) equals the heat of vaporization (in calories) per gram divided by the product of the absolute temperature and the increase in volume (V, in cubic centimeters when a specified quantity (a gram) of the liquid changes to vapor. Other consistent units may be used. The approximate Clapeyron equation $d \ln p/dT = \Delta H/RT^2$, expresses the same relation in a less exact form because in its derivation, it is assumed that ΔV is equal to the volume of vapor. This assumption (that the volume of liquid is negligibly small) is usually true within a few percent under ordinary conditions of temperature and pressure.

"Clarase."[212] Trademark for a fungal alpha-amylase with strong liquefying and dextrinogenic activity together with considerable saccharogenic action.
Uses: Fruit juices and pectin; chocolate syrup; brewing.

"Claricol."[354] Trademark for a crystal inhibitor and winterizing aid for edible and non-edible oils.

"Clarolite."[471] Trademark for a series of active clays used for bleaching of oils and removal of color bodies.

clary sage oil. See sage oil, clary.

classification. In chemical engineering parlance, the mechanical process of separating subdivided solids (crushed stone, cement, mineral aggregate, and the like) into two (or more) "classes," each containing a specific size range. The latter may vary from an inch or more in diameter to powders of considerable fineness. Customary classifying equipment includes so-called grizzlies, perforated metal and vibrating screens, sifters, sieves, and similar devices. A magnetic separator is often used in conjunction with the screen to remove tramp metal. A similar method is used in the food industry for size grading of certain fruits and vegetables, e.g., peas, berries, etc.

clathrate compound. An inclusion complex in which molecules of one substance are completely enclosed within the other, as argon within hydroquinone crystals. Urea adducts are inclusion complexes of the channel or canal type. In these, the complexing urea crystals wrap around the molecule of the other substance, usually a straight-chain unbranched aliphatic hydrocarbon. Similar complexes are formed with thiourea. See also inclusion complex.

Claude system. A process for the production of liquid air in which the compressed gas is made to perform work in an expansion engine and thus cool itself.

clay. A hydrated aluminum silicate. Generalized formula $Al_2O_3SiO_2 \cdot xH_2O$. Component of soils in varying percentages.
Properties: Fine, irregularly shaped crystals ranging from 150 microns to less than 1 micron (colloidal); color, reddish-brown to pale buff, depending on iron oxide content; sp. gr. about 2.50; insoluble in water

and organic solvents; odorless; absorbs water to form a plastic, moldable mass and in some cases forms a thixotropic gel (bentonite); refractory material; strong ion-exchange capability, important in soil chemistry and construction engineering. Nontoxic.
Derivation: Weathering of rocks.
Occurrence: Southeastern U.S.; Wyoming, Texas, Canada; England; France; U.S.S.R.
Types: Kaolinite, montmorillonite, attapulgite, illite, bentonite, halloysite.
Grades: Natural; refined; air-floated.
Containers: Multi-wall paper bags; tank cars.
Hazard: Dusts may be irritant to nose and throat; suspensions of dust are a fire hazard.
Uses: Ceramic products; refractories; colloidal suspensions; oil-well drilling fluids; filler for rubber and plastic products; films; paper coating; decolorizing oils; temporary molds; filtration; carrier in insecticidal sprays; catalyst support.
See also fullers earth, bentonite, ceramic, refractory, kaolin, slip clay; polyorganosilicate graft polymer.

cleaning agents, chemical. See detergent; solvent; trichloroethylene; abrasive.

"Clearate."[48] Trademark for a high grade soya lecithin. Used in foods, inks, cosmetics and paints.

cleave. (1) Of a crystal, to break or separate along definite planes defined by the crystal structure; it may cleave in one direction, as in mica, or in several.
(2) Of an alkene molecule, to divide into two compounds (aldehydes or ketones) at the double bond; this is usually done by ozone, followed by hydrolysis in the presence of powdered zinc.

Cleland's reagent. See dithiothreitol.

Cleveland Open Cup. See COC.

Cleve's acid. See 1-naphthylamine-6-sulfonic acid; or 1-naphthylamine-7-sulfonic acid; known also as Cleve's 1,6 acid and Cleve's 1,7 acid, respectively.

cliffstone Paris white. A special grade of whiting (q.v.) made from a hard grade of English chalk.

"Climelt."[56] Trademark for arc-cast molybdenum and tungsten metals and their respective alloys.

clinical chemistry. A subdivision of chemistry that deals with the behavior and composition of all types of body fluids, including the blood, urine, perspiration, glandular secretions, etc. It involves analysis and testing of these for content of numerous metabolic constituents as well as foreign materials; thus it also includes toxicological factors.

clinoptilolite. A natural inorganic zeolite (q.v.) used as a selective ion-exchange medium for removal of ammonia from plant waste water.

"Clopane Hydrochloride."[100] Trademark for cyclopentamine hydrochloride.

"Clorafin."[266] Trademark for a series of chlorinated nonflammable paraffins. Used as plasticizers and lubricant additives.

"Cloran."[505] Trademark for a reactive difunctional chlorinated carboxylic acid anhydride.
Uses: Reactive intermediate to produce flame retardant polymers.

chlorphene (USAN). See ortho-benzyl-para-chlorophenol.

"Clor-Tabs."[204] Trademark for 70% available chlorine in tablet form.

cloud point. In petroleum technology, the temperature at which a waxy solid material appears as a diesel fuel is cooled. This material is harmful to engine performance.

clove oil (caryophyllus oil).
Properties: Pale-yellow liquid; darkens and thickens with age and exposure; strong aromatic odor; pungent and spicy taste. Soluble in ether and chloroform; soluble in 1 to 2 vols. and more of 70% alcohol; fresh, so-called extra-light oils in 2.5 to 3 vols. of 60% alcohol. Sp. gr. 1.038–1.060; b.p. 250–260°C; refractive index 1.5270–1.5350; optical rotation to −1° 10′. Nontoxic.
Derivation: Distilled from cloves, the unexpanded flowers of Eugenia aromatica (Eugenia caryophyllata).
Method of purification: Rectification.
Grades: Technical; U.S.P.; F.C.C.
Containers: Drums.
Uses: Medicine (local); flavoring; dentistry; perfumery; confectionery; soaps.

Cm Symbol for curium.

"CM."[28] Trademark for a flame-retardent composition based on ammonium sulfamate and modified to prevent afterglow and to improve penetration.
Properties: Fine, white granules; soluble in water; insoluble in dry cleaning solvents.
Uses: Treatment of fabrics, paper, paper products, and other cellulosic materials.

CMC. See carboxymethylcellulose.

CM-cellulose. Abbreviation for carboxymethylcellulose, used especially by biochemists.

CMHEC. Abbreviation for carboxymethyl hydroxyethyl cellulose.

CMP. Abbreviation for cytidine monophosphate. See cytidylic acid.

CMPP. Abbreviation for 2-(4-chloro-2-methylphenoxyl) propionic acid. See mecoprop.

CMRA. Abbreviation for Chemical Marketing Research Association (q.v.).

CMU. Abbreviation for chlorophenyldimethylurea. See monuron.

CNS. Abbreviation for central nervous system (as applied to the action of certain drugs).

Co Symbol for cobalt.

coacervation. An important equilibrium state of colloidal or macromolecular systems. It may be defined as the partial miscibility of two or more optically isotropic liquids, at least one of which is in the colloidal state. For example, gum arabic shows the phenomenon of coacervation when mixed with gelatin.

coagulant. A substance that induces coagulation (q.v.). Coagulants are used in precipitating solids or semisolids from solution, as casein from milk, rubber particles from latex, or impurities from water. Compounds that dissociate into strongly charged ions are normally used for this purpose. The blood contains the natural coagulant thrombin.

"Coagulant Aid."[108] Trademark for a series of polyelectrolytes and combinations of polyelectrolytes with other materials.
Uses: Clarification of water for municipal and industrial uses.

coagulation. Irreversible combination or aggregation of semisolid particles such as fats or proteins to form a clot or mass. This can often be brought about by addition of appropriate electrolytes, for example, by the addition of an acid to milk or of aluminum sulfate to turbid water. Mechanical agitation and removal of stabilizing ions, as in dialysis, also cause coagulation. The clotting of blood by thrombin (q.v.), the coagulation of rubber particles in latex by acetic acid and of egg-white by heat are additional instances. See also flocculation.

Coahran process. Recovery of acetic acid from pyroligneous acid by extracting with ether. It is an improved version of the Brewster process (q.v.), but is basically the same.

coal. A natural solid combustible material consisting chiefly of amorphous elemental carbon with low percentages of hydrocarbons, complex organic compounds and inorganic materials. Coal was formed from prehistoric plant life and occurs in layers or veins in sedimentary rocks. It is far more plentiful in the U.S. than petroleum and is an important source of heat and energy; its use as a fuel is increasing. In the U.S. it occurs chiefly in West Virginia and Kentucky, as well as in Wyoming and other western states. Much of it is too high in sulfur content to meet present pollution standards.

Coal is classified according to its heating value, expressed in Btu/lb, and its fixed carbon content.

	Fixed Carbon	Btu/lb
Anthracite	86–98%	13,500–15,600
Low to Medium Volatile Bituminous	69–86%	14,500–15,600
High Volatile Bituminous	46–69%	11,000–15,000
Subbituminous	46–60%	8,300–13,000
Lignite	46–60%	5,500– 8,300

Coal is most frequently specified in terms of its proximate analysis, giving the percentages of moisture, volatile combustible matter, fixed carbon, and ash. An ultimate analysis gives the percentages of the various elements present (C, H, O, N, and S).

Coal is also an important source of chemical raw materials: pyrolysis (destructive distillation) yields coal tar and hydrocarbon gases, which can be upgraded by hydrogenation or methanation to synthetic crude oil and fuel gas, respectively; catalytic hydrogenation yields hydrocarbon oils and gasoline; gasification produces carbon monoxide and hydrogen (synthesis gas) from which ammonia and other products can be made. Numerous processes for adapting these reactions to large-scale production of fuel oil and gasoline are in the pilot-plant stage, though none has yet been commercially developed. Methane is being produced from coal mines on a commercial scale.
See also gasification; hydrogenolysis; hydrosolvation.
Note: Pending the further development of alternative energy sources, coal is probably the safest and most abundant fuel available in the U.S. A large percentage can be obtained by surface or strip mining, though the necessary restoration of the environment (soil and plant life) is expensive. Improvements in safety are needed in order to realize full and safe production from deep mines.

coal gas (bench gas; coke-oven gas). A mixture of gases produced by the destructive distillation of bituminous coal in highly heated fire-clay or silica retorts or in by-product coke ovens.
Hazard: Flammable; highly toxic.
Uses: Directly in open hearth furnaces.
Shipping regulations: (Rail) Red Gas label. (Air) Flammable Gas label. Not accepted on passenger planes.

coalescence. The union of two or more droplets of a liquid to form a larger droplet, brought about when the droplets approach one another closely enough to overcome their individual surface tensions. The combination of fine mercury droplets is an example. Coalescing agents are used to remove solid contaminants from hydrocarbons; coalescence may also be brought about by mechanical means in centrifuges.

coal, gasification. See gasification.

coal, hydrogenation. See hydrogenolysis; gasification.

coal oil. The crude oil obtained by the destructive distillation of bituminous coal; or the distillate obtained from this oil.
Hazard: Flammable, moderate fire risk.

coal tar.
Properties: Black, viscous liquid (or semi-solid), naphthalene-like odor; sharp burning taste; obtained by destructive distillation of bituminous coal, as in cokeovens; one ton of coal yields 8.8 gallons of coal tar. Combustible. Sp. gr. 1.18 to 1.23 (60/60°F). Soluble in ether, benzene, carbon disulfide, chloroform; partially soluble in alcohol, acetone, methanol, and benzene; only slightly soluble in water.

Coal tar may be hydrogenated under pressure to form a petroleum-like fuel suitable for residual use.

Coal-tar fractions obtained by distillation and the chemicals found in each are as follows: (1) light oil (up to 200°C); benzene, toluene, xylenes, cumenes, coumarone, indene; (2) middle oil (200–250°C) and (3) heavy oil (250–300°C): naphthalene, acenaphthene, methylnaphthalenes, fluorene, phenol, cresols, pyridine, picolines; (4) anthracene oil (300–350°C): phenanthrene, anthracene, carbazole, quinolines; and (5) pitch (q.v.). Typical yields: 5% light oil; 17% middle oil; 7% heavy oil; 9% anthracene oil; 62% pitch. Treatment with alkalies, acids, and solvents is necessary to separate the individual chemicals.
Grades: Crude, refined; U.S.P.
Containers: Tank cars; barrels.
Hazard: Toxic by inhalation.
Uses: Raw material for plastics, solvents, dyes, drugs, and other organic chemicals. The crude or refined product or fractions therof are also used for waterproofing, paints, pipecoating, roads, roofing, insulation, as pesticides and sealants, and in medicine.

coal-tar distillate. The lighter fractions of coal tar. The terms coal tar light oil, coal tar naphtha, and coal tar oil are loosely defined and are sometimes regarded as synonymous.
Hazard: Flammable, dangerous fire risk. Toxic by inhalation and skin absorption.
Shipping regulations: (Rail) Red label. (Air) Flammable Liquid label.
See also entries under naphtha.

coal-tar dye. A dye produced from the coal-tar hydrocarbons or their derivatives such as benzene, toluene, xylene, naphthalene, anthracene, aniline, etc. See also dye, synthetic.
Shipping regulations: (Air) Coal-tar dye, liquid, n.o.s., Corrosive label.

coal-tar light oil. See coal-tar distillate and light oil.

coal-tar naphtha. See coal-tar distillate and naphtha, solvent.

coal-tar oil. See coal-tar distillate.

coal-tar pitch. Dark brown to black amorphous residue left after coal tar is redistilled. It is composed almost entirely of polynuclear aromatic compounds and constitutes 48–65% of the usual grades of coal tar. Different grades have different softening points: roofing pitch softens at 65°C, electrode pitch at 110–115°C. Combustible.
Hazard: Volatile components (anthracene, phenanthrene, acridine) are carcinogenic agents. Tolerance, 0.2 mg per cubic meter in air.
Uses: Binder for carbon electrodes; base for paints and coatings; impregnation of fiber pipe for electrical conduits and drainage; foundry core compounds; briquetting coal; tar-bonded refractory brick; paving and roofing; plasticizers for elastomers and polymers; extenders; saturants; impregnants; sealants.

coating. A film or thin layer applied to a base material called a substrate. The coatings most commonly used in industry are metals, alloys, resin solutions, and solid/liquid suspensions on various substrates (metals, plastics, wood, paper, leather, etc.). They may be applied by electrolysis, vapor deposition, vacuum, or mechanical means such as brushing, spraying, calendering, and roller coating. Products such as cables and power cord are coated by extrusion.

A comparatively recent development in coatings technology involves the production in powder form of such thermosetting resins as acrylics, epoxies, and polyesters appropriately compounded; these are applied to auto bodies, machinery, and other industrial products by electrostatic spraying techniques. This process is claimed to minimize the pollution problems and waste encountered with solvent-based sprayed coatings.
See also protective coating; film; paint; vacuum deposition.

"Coat" Series.[236] Trademark for a series of corrosion inhibitors that have the ability to attach themselves to the metal surface of oil production equipment. Selected polar organic salts, they are specially formulated to have the degree of oil solubility, water dispersibility and wetting needed for corrosion protection.

cobalamin (vitamin B_{12}). The antipernicious anemia vitamin. All vitamin B_{12} compounds contain the cobalt atom in its trivalent state. There are at least three active forms: cyanocobalamin, hydroxocobalamin, nitrocobalamin (q.v.). Vitamin B_{12} is a component of a coenzyme which takes part in the shift of carboxyl groups within molecules. As such it has an influence on nucleic acid synthesis, fat metabolism, conversion of carbohydrate to fat, and metabolism of glycine, serine, methionine and choline.
Source: Food source: Liver, eggs, milk, meats, and fish. Commercial source: Produced by microbial action on various nutrients (spent antibiotic liquors, sugar beet molasses, whey, etc.).
Uses: Medicine (blood and nerve treatment); nutrition; animal feed supplements.

cobalt Co Metallic element of atomic number 27, group VIII of the Periodic Table. Atomic weight 58.9332. Valences 2,3; no stable isotopes; there are several artificial radioactive isotopes, the most important being Co-60.
Properties: Steel-gray, shining, hard, ductile, somewhat malleable metal; ferromagnetic, with permeability two-thirds that of iron; has exceptional magnetic properties in alloys. Sp. gr. 8.9; m.p. 1943°C; b.p. 3100°C. Attacked by dilute hydrochloric and sulfuric acids; soluble in nitric acid. Corrodes readily in air. Hardness: cast, 124 Brinell; electrodeposited, 300 Brinell. An important trace element in soils and necessary for animal nutrition. Cobalt has unusual coordinating properties, especially the trivalent ion (coordination number 6). Noncombustible except as powder. Low toxicity.
Occurrence: Principal ores are smaltite, cobaltite, chloanthite, linnaeite. (Canada; Congo; Rhodesia; North Africa).
Derivation: From ore concentrates by roasting followed (a) by thermal reduction by aluminum, or (b) by electrolytic reduction of solutions of metal; (c) by leaching, with either ammonia or acid in an autoclave under elevated temperatures and pressures and subsequent reduction by hydrogen.
Forms available: Rondels (1 in. × ¾ in.); shot; anodes; 150 and finer mesh powder, to 99.6% purity; ductile strips (95% cobalt, 5% iron) and high purity strips, 99.9% pure; single crystals.
Containers: 500-lb kegs; drums.
Hazard: Dust is flammable and toxic by inhalation. Tolerance, 0.1 milligram per cubic meter of air.
Uses: Chemical (cobalt salts, oxidizing agent); electroplating; ceramics; lamp filaments; catalyst (sulfur removal from petroleum, Oxo process, organic synthesis); trace element in fertilizers; glass; drier in printing inks, paints and varnishes; colors; cermets. Principal use in alloys, especially cobalt steels for permanent and soft magnets and cobalt-chromium high-speed tool steels; cemented carbides; jet engines; coordination and complexing agent (see cobaltammine).

cobalt 57. Radioactive cobalt of mass number 57.
Properties: Half-life 267 days; radiation, gamma and K x-ray.
Derivation: Bombardment of a nickel target with protons.
Forms available: Cobaltous chloride in hydrochloric acid solution; cyanocobalamin cobalt-57 solution (U.S.P.).
Hazard: Radioactive poison.
Use: Medicine (diagnostic aid).
Shipping regulations: (Rail, Air) Consult regulatons.

cobalt 58. Radioactive cobalt of mass number 58.
Properties: Half-life 72 days; radiation: positron, gamma, K x-ray.
Derivation: Bombardment of nickel 58 with neutrons.
Forms available: Cobaltous chloride in hydrochloric acid solution.
Hazard: Radioactive poison.
Use: Biological and medical research.
Shipping regulations: (Rail, Air) Consult regulations.

cobalt 60. Radioactive cobalt of mass number 60. One of the most common radioisotopes.
Properties: Half-life 5.3 years; radiation, beta and gamma. Radiocobalt is available in larger quantities and is cheaper than radium.
Derivation: Pile irradiation of cobalt oxide, Co_2O_3, or of cobalt metal.

Forms available: Cobalt metal pellets or wire needles; cobaltous chloride in hydrochloric acid solution; solid cobaltic oxides; labeled compounds such as cyanocobalamin (U.S.P.).
Hazard: Radioactive poison.
Uses: Radiation therapy (cancer); radiographic testing of welds and castings; as a source of ions in gas-discharge devices; irradiation of meats; as the radiation source in liquid-level gages; for locating buried telephone and electrical conduits; portable radiation units; gamma irradiation for wheat and potatoes; as a research aid in studying the permeability of porous media to flow of oil, wearing quality of floor wax, oil consumption in internal combustion engines, wool dyeing, etc.
Shipping regulations: (Rail, Air) Consult regulations.

cobalt acetate. See cobaltous acetate.

cobaltammine. A coordination or complexing compound containing the group $[Co(NH_3)_6]^{3+}$ or its derivatives in which some of the ammonia has been replaced by other groups or ions. The names hexammine, pentammine, etc., are used to indicate the number of ammonia groups present in any case. Prepared by adding excess ammonia to a cobaltous salt, exposing to air so that oxygen is absorbed and boiling to oxidize the cobalt.
These compounds show none of the ordinary properties of cobalt. Different types of salts with various acid radicals are known. The ammonia in the ammines may be replaced, molecule for molecule, by other nitrogen compounds such as hydroxylamine or ethylene diamine; by water; by ions such as hydroxyl, chloride, nitrate, etc.; or by groups such as nitro (NO_2). See coordination compound.

cobalt ammonium sulfate. See cobaltous ammonium sulfate.

cobalt arsenate. See cobaltous arsenate.

cobalt black. See cobaltic oxide.

cobalt bloom. See erythrite 2.

cobalt blue (Thenard's blue; cobalt ultramarine; azure blue).
Properties: Blue to green pigment of variable composition, consisting essentially of mixtures of cobalt oxide and alumina, approximating cobaltous aluminate, $Co(AlO_2)_2$. Cobalt blue is said to be the most durable of all blue pigments, being resistant to both weathering and chemicals.
Derivation: By heating alumina with (a) cobaltous oxide, or a material yielding this oxide on calcination; (b) cobalt phosphate; (c) cobalt arsenate. Greenish shades may be made by incorporating zinc oxide.
Grades: Technical (called genuine, to distinguish it from the imitation, which is ultramarine blue).
Containers: 250-lb barrels.
Uses: Pigments in oil or water; cosmetics (eyeshadows, grease paints).

cobalt bromide. See cobaltous bromide.

cobalt carbonate. See cobaltous carbonate, basic.

cobalt carbonyl. See cobalt tetracarbonyl.

cobalt chloride. See cobaltous chloride.

cobalt chromate. See cobaltous chromate.

cobalt chromate, basic. See cobaltous chromate.

cobalt difluoride. See cobaltous fluoride.

cobalt 2-ethylhexoate (cobalt octoate). Probably the cobaltous salt of 2-ethylhexoic acid, $C_4H_9CH(C_2H_5)COOH$.
Properties: Blue liquid, sp. gr. 1.013 (25°C).
Uses: Paint drier; whitener; catalyst.

cobalt-gold alloy. Made by a vapor deposition technique into magnetic films. Compositions range from 25–60% gold.

cobalt hydrate. See cobaltic hydroxide and cobaltous hydroxide.

cobalt hydroxide. See cobaltic hydroxide and cobaltous hydroxide.

cobaltic acetylacetonate $Co[CH_3COCHC(CH_3)O]_3$.
Properties: Dark green or black crystals; sp. gr. 1.43; m.p. 241°C.
Derivation: Reaction of cobaltous carbonate with acetylacetone and peroxide.
Use: Vapor plating of cobalt.

cobaltic boride CoB.
Properties: Crystalline prisms; sp. gr. 7.25 (18°C); m.p. > 1400°C. Decomposes in water; soluble in nitric acid.
Use: Ceramics.

cobaltic fluoride. See cobalt trifluoride.

cobaltic hydroxide (cobalt hydroxide; cobalt hydrate) $Co(OH)_3$, actually considered to be $Co_2O_3 \cdot 3H_2O$.
Properties: Dark-brown powder; soluble in cold concentrated acids; insoluble in water and alcohol. Sp. gr. 4.46. Loses water at 100°C.
Derivation: Addition of sodium hydroxide to a solution of a cobaltic salt; action of chlorine on a suspension of cobaltous hydroxide; action of sodium hypochlorite on a cobaltous salt.
Use: Cobalt salts.

cobaltic oxide (cobalt oxide; cobalt black) Co_2O_3. Sometimes incorrectly called cobalt peroxide.
Properties: Steel-gray or black powder. Soluble in concentrated acids; insoluble in water. Sp. gr. 4.81–5.60; m.p., decomposes at 895°C.
Derivation: By heating cobalt compounds at low temperature with excess of air.
Uses: Pigment; coloring enamels; glazing pottery.

cobalt iodide. See cobaltous iodide.

cobaltite (cobalt glance) CoAsS. Silver-white to gray mineral; metallic luster. Contains 35.5% cobalt. Sp. gr. 6–6.3; Mohs hardness 5.5.
Occurrence: Canada; Congo; Sweden.
Hazard: Dust or powder toxic when inhaled.
Use: An important cobalt ore; ceramics.

cobalt linoleate. See cobaltous linoleate.

cobalt molybdate. A molybdenum catalyst (a gray-green powder) used in petroleum technology, in reforming and desulfurization.

cobalt monoxide. See cobaltous oxide.

cobalt naphthenate. See cobaltous naphthenate.

cobalt neodecanoate. Deep blue paste containing 12% cobalt. Used as a drying additive for printing inks.

cobalt nitrate. See cobaltous nitrate.

cobaltocene (dicyclopentadienylcobalt) $(C_5H_5)_2Co$. A metallocene (q.v.).
Properties: Purple crystals; m.p. 172–173°C; soluble in hydrocarbons, highly reactive compound which is readily oxidized by air, water and dilute acids.

Containers: Supplied in aromatic solvents in bottles and steel drums.
Uses: Polymerization inhibitor of olefins up to 200°C; Diels-Alder reaction; catalyst; paint drier; oxygen stripping agent.

cobalto-cobaltic oxide (tricobalt tetraoxide) Co_3O_4.
Properties: Steel-gray to black in anhydrous form. Insoluble in water, hydrochloric acid and nitric acid; soluble in sulfuric acid and fused sodium hydroxide; hygroscopic. Sp. gr. 6.07; changes to cobaltous oxide at 900–950°C.
Derivation: By heating strongly other cobalt oxides in air. Thus, the commercial oxides contain a substantial quantity of Co_3O_4.
Uses: Ceramics; pigments; catalyst; preparation of cobalt metal; electronic chemicals.

cobalt octoate. See cobalt 2-ethylhexoate. See also soap (2).

cobalt oleate. See cobaltous oleate.

cobaltous acetate (cobalt acetate) $Co(C_2H_3O_2)_2 \cdot 4H_2O$.
Properties: Reddish-violet, deliquescent crystals. Soluble in water, acids, and alcohol; sp. gr. 1.7043; m.p., loses H_2O at 140°C.
Derivation: Action of acetic acid on cobalt carbonate with subsequent crystallization.
Grades: Technical; pure crystalline; C.P.
Hazard: May not be used in food products (FDA).
Uses: Sympathetic inks; paint and varnish driers; catalyst; anodizing; mineral supplement in feed additives; foam stabilizer.

cobaltous aluminate. See cobalt blue.

cobaltous ammonium sulfate (cobalt ammonium sulfate) $CoSO_4 \cdot (NH_4)_2SO_4 \cdot 6H_2O$.
Properties: Ruby-red crystals; soluble in water; insoluble in alcohol; sp. gr. 1.902.
Derivation: Crystallization of cobaltous sulfate with ammonium sulfate.
Uses: Ceramics; cobalt plating; catalyst.

cobaltous arsenate (cobalt arsenate) $Co_3(AsO_4)_2 \cdot 8H_2O$.
Properties: Violet-red powder. Soluble in acids; insoluble in water. Sp. gr. 3.178 (15°C).
Derivation: Interaction of solutions of sodium arsenate and of a cobalt salt.
Grades: Technical.
Containers: Wooden kegs; boxes.
Hazard: Highly toxic.
Uses: Painting on glass and porcelain in light blue colors; coloring glass.
Shipping regulations: (Rail) Arsenical compounds, n.o.s., Poison label. (Air) Arsenates, Poison label.
See also erythrite.

cobaltous bromide (cobalt bromide) $CoBr_2 \cdot 6H_2O$.
Properties: Red violet crystals. Soluble in water, alcohol, and ether. Anhydrous crystals are red. Sp. gr. 2.46; m.p. 47–48°C; loses $6H_2O$ at 130°C.
Derivation: By the action of bromine on cobalt, or of hydrobromic acid on cobaltous hydroxide or carbonate, followed by crystallization.
Grades: Technical; C.P.
Containers: Glass bottles.
Uses: In hygrometers; catalyst.

cobaltous carbonate $CoCO_3$.
Properties: Red crystals; insoluble in water and ammonia; soluble in acids. Sp. gr. 4.13; m.p., decomposes. The cobalt carbonate of commerce is usually the basic salt (see following entry).
Derivation: By heating cobaltous sulfate with a solution of sodium bicarbonate.
Uses: Ceramics; trace element added to soils and animal feed; temperature indicator; catalyst.

cobaltous carbonate, basic $2CoCO_3 \cdot 3Co(OH)_2 \cdot H_2O$. The cobalt carbonate of commerce.
Properties: Red violet crystals; soluble in acids; insoluble in cold water; decomposes in hot water. M.p., decomposes.
Derivation: By adding sodium carbonate to a solution of cobaltous acetate, followed by filtration and drying.
Uses: Manufacturing cobaltous oxide; cobalt pigments; cobalt salts; intermediate.

cobaltous chloride (cobalt chloride) (a) $CoCl_2$; (b) $CoCl_2 \cdot 6H_2O$.
Properties: (a) blue, (b) ruby-red crystals. Soluble in water and alcohol; also soluble in acetone. Sp. gr. (a) 3.348, (b) 1.924; m.p. (a) sublimes, (b) 86.75°C.
Derivation: By the action of hydrochloric acid on cobalt, its oxide, hydroxide or carbonate. Concentration gives (b) and dehydration (a).
Hazard: May not be used in food products (FDA).
Uses: Absorbent for ammonia; gas-masks; electroplating; sympathetic inks; hygrometers; in soils and animal feeds; vitamin B_{12}; flux for magnesium refining; solid lubricant; dye mordant; catalyst; barometers; laboratory reagent; foam stabilizer.

cobaltous chromate (basic cobalt chromate; cobalt chromate). $CoCrO_4$.
Properties: Brown or yellowish-brown powder. Variable composition. (Pure cobaltous chromate is $CoCrO_4$; gray-black crystals.) Soluble in mineral acids, in solution of chromium trioxide; insoluble in water.
Hazard: May be toxic by inhalation.
Use: Ceramics (tinting).

cobaltous citrate $Co_3(C_6H_5O_7)_2 \cdot 2H_2O$.
Properties: Rose-red amorphous powder; m.p. 150°C ($-2H_2O$). Slightly soluble in water; soluble in dilute acids.
Uses: Vitamin preparations; therapeutic agents.

cobaltous cyanide (a) $Co(CN)_2 \cdot 2H_2O$, (b) $Co(CN)_2$.
Properties: (a) buff crystals, (b) blue-violet powder; sp. gr. (b) 1.872; m.p. (b) 280°C. Insoluble in water; soluble in potassium cyanide, hydrochloric acid, ammonium hydroxide.
Hazard: Highly toxic.
Shipping regulations: (Rail) Cyanides, dry, Poison label. (Air) Poison label.

cobaltous ferrite $CoOFe_2O_3$. A constituent of magnetically soft ferrites, which have high permeability, low coercive force, low magnetic saturation, and high resistivity. See also ferrite.

cobaltous fluoride (cobalt difluoride) CoF_2.
Properties: Rose-red crystals or powder; sp. gr. 4.46; m.p. ca 1200°C; b.p. 1400°C. Soluble in cold water, hydrofluoric acid. Decomposes in hot water. Ammine complexes can be prepared from the hydrate.
Hazard: Toxic and irritant. Tolerance (as F), 2.5 mg per cubic meter of air.

cobaltous formate $Co(CHO_2)_2 \cdot 2H_2O$.
Properties: Red crystals; sp. gr. 2.129 (22°C) m.p.

140°C (–$2H_2O$); decomposes 175°C. Soluble in cold water.

cobaltous hydroxide (cobalt hydroxide; cobalt hydrate) $Co(OH)_2$.
Properties: Rose-red powder. Soluble in acids and ammonium salt solutions; insoluble in water and alkalies. Sp. gr. 3.597.
Derivation: Addition of sodium hydroxide to a solution of cobaltous salt.
Use: Cobalt salts; in preparation of paint and varnish driers; catalyst.

cobaltous iodide (cobalt iodide) $CoI_2 \cdot 6H_2O$.
Properties: Brownish-red crystals; loses iodine on exposure to air; sp. gr. 2.90; loses $6H_2O$ at 27°C. Soluble in water and alcohol. Anhydrous cobaltous iodide, CoI_2, is in form of black crystals, sp. gr. 5.68; or in a yellow beta-modification which gives a colorless aqueous solution.
Derivation: Heating cobalt powder with hydrogen iodide. Anhydrous cobaltous iodide is prepared by heating cobalt in iodine vapor.
Hazard: Moderately toxic.
Use: In hygrometers; determination of water in organic solvents.

cobaltous linoleate (cobalt linoleate) $Co(C_{18}H_{31}O_2)_2$.
Properties: Brown, amorphous powder. Soluble in alcohol, ether and acids; insoluble in water. Combustible.
Derivation: By boiling a cobalt salt and sodium linoleate.
Use: Paint and varnish drier, especially enamels and white paints.
See also soap (2).

cobaltous naphthenate.
Properties: Brown, amorphous powder or bluish-red solid. Insoluble in water. Soluble in alcohol, ether, oils. Composition indefinite. Combustible.
Derivation: By treating cobaltous hydroxide or cobaltous acetate with naphthenic acid.
Uses: Paint and varnish drier; bonding rubber to steel and other metals.

cobaltous nitrate (cobalt nitrate) $Co(NO_3)_2 \cdot 6H_2O$.
Properties: Red crystals; deliquescent in moist air. Soluble in most organic solvents. Sp. gr. 1.88 (25°C); m.p. 56°C.
Derivation: Action of nitric acid on metallic cobalt, cobalt oxide, hydroxide, or carbonate, with subsequent crystallization.
Uses: Sympathetic inks; cobalt pigments; catalysts; additive to soils and animal feeds; vitamin preparations; hair dyes; porcelain decoration.
Hazard: Moderately toxic; oxidizing agent; dangerous fire risk in contact with organic materials.
Shipping regulations: Nitrates, n.o.s. (Rail) Yellow label. (Air) Oxidizer label.

cobaltous oleate (cobalt oleate) $Co(C_{18}H_{33}O_2)_2$.
Properties: Brown, amorphous powder. Soluble in alcohol and ether; insoluble in water. M.p. 235°C. Combustible.
Derivation: By heating cobaltous chloride and sodium oleate, followed by filtration and drying.
Use: Paint and varnish driers.
See also soaps (2).

cobaltous oxalate CoC_2O_4.
Properties: Reddish-white crystals; sp. gr. 3.021; insoluble in water; soluble in conc-ammonia.
Hazard: Toxic by ingestion.
Uses: Temperature indicator; catalysts (hydrated form).

cobaltous oxide (cobalt oxide; cobalt monoxide) CoO.
Properties: Grayish powder under most conditions; can form green-brown crystals. Soluble in acids and alkali hydroxides; insoluble in water and ammonium hydroxide. Sp. gr. 6.45; m.p. 1935°C. Low toxicity.
Derivation: Calcination of cobalt carbonate or its oxides at high temperature in a neutral or slightly reducing atmosphere.
Grades: Technical; ceramic.
Uses: Pigment in paints and ceramics; preparation of cobalt salts; catalyst; porcelain enamels; coloring glass; feed additive.

cobaltous perchlorate $Co(ClO_4)_2$.
Properties: Red needles; sp. gr. 3.327; soluble in water, alcohol, acetone.
Hazard: Fire and explosion risk in contact with organic materials. Strong oxidizing agent. Moderately toxic.
Use: Chemical reagent.
Shipping regulations: Perchlorates, n.o.s., (Rail) Yellow label. (Air) Oxidizer label.

cobaltous phosphate (cobalt phosphate) $Co_3(PO_4)_2 \cdot 8H_2O$.
Properties: Reddish powder; sp. gr. 2.769 (25°C); loses $8H_2O$ at 200°C; slightly soluble in cold water; soluble in mineral acids; insoluble in alcohol. Low toxicity.
Derivation: Interaction of solutions of cobalt salts and sodium phosphate.
Uses: Cobalt pigments; coloring glass; painting porcelain; animal feed supplement.

cobaltous resinate (cobalt resinate); principally cobalt abietate. See abietic acid.
Properties: Brown-red powder; insoluble in water; soluble in oils.
Derivation: See soap (2). The precipitated grade is higher in cobalt, more expensive, and more effective.
Grade: Fused; precipitated.
Use: Varnish drier.
Hazard: Spontaneously flammable in air; reacts strongly with oxidizing materials.
Shipping regulations: (precipitated) (Rail) Yellow label. (Air) Flammable Solid label. Not acceptable on passenger planes.

cobaltous silicofluoride $CoSiF_6 \cdot 6H_2O$.
Properties: Pale red crystals; sp. gr. 2.087; soluble in water.
Hazard: Toxic; irritant to tissue.
Use: Ceramics.

cobaltous succinate $Co(C_4H_4O_4) \cdot 4H_2O$.
Properties: Violet crystals; slightly soluble in cold water; soluble in alkalies; insoluble in alcohol.
Use: Vitamin preparations; therapeutic agents.

cobaltous sulfate (cobalt sulfate) (a) $CoSO_4$; (b) $CoSO_4 \cdot 7H_2O$.
Properties: Red powder; soluble in water and methanol. Sp. gr. (a) 3.472; (b) 1.948; m.p. (a) 735°C; (b) loses $7H_2O$ at 420°C.
Derivation: Action of sulfuric acid on cobaltous oxide.
Hazard: May not be used in food products (FDA).
Uses: Ceramics; pigments; glazes; in plating baths for cobalt; additive to soils and animal feeds; catalyst; foam stabilizer.

cobaltous tungstate (cobalt tungstate; cobalt wolframate) $CoWO_4$.
Properties: Reddish-orange powder; insoluble in water; soluble in hot concentrated acids; sp. gr. 8.42. Low toxicity.

Derivation: By adding a sodium tungstate solution to a solution of a cobalt salt.
Use: Pigment; drier for enamels, inks, paints; electronic devices; antiknock agents.

cobalt oxide. See cobaltic oxide, cobaltous oxide, cobalto-cobaltic oxide. The commercial cobalt oxides are not usually definite chemical compounds but are mixtures of two or more cobalt oxides.

cobalt phosphate. See cobaltous phosphate.

cobalt potassium cyanide $K_3Co(CN)_6$.
Properties: Yellow crystals; sp. gr. 1.878 (25°C); m.p., decomposes; soluble in water; slightly soluble in alcohol.
Grades: Pure; electronic.
Hazard: Highly toxic.
Uses: Electronic research.
Shipping regulations: (Rail, Air) Cyanides, dry, Poison label.

cobalt potassium nitrite (cobalt yellow; potassium cobaltinitrate; Fischer's salt; potassium hexanitrocobaltate III) $K_3Co(NO_2)_6$.
Properties: Yellow, microcrystalline powder. Slightly soluble in water; insoluble in alcohol; m.p., decomposes at 200°C.
Derivation: By adding potassium nitrate and acetic acid to a solution of a cobalt salt.
Uses: Medicine; yellow pigment; painting on glass or porcelain.

cobalt resinate. See cobaltous resinate.

cobalt selenite $CoSe_2O_3 \cdot 2H_2O$. A blue-red powder; insoluble in water.

cobalt silicide. A semiconductor reported to have as much as 15% efficiency in converting heat to electricity in the temperature range 20–800°C.

cobalt soap. See cobaltous linoleate, cobaltous naphthenate, cobaltous oleate, cobaltous resinate; cobalt tallate.

cobalt sulfate. See cobaltous sulfate.

cobalt tallate. Cobalt derivative of refined tall oil; of varying composition. Used as a drier in paints and varnishes. Combustible. See soap (2).

cobalt tetracarbonyl (cobalt carbonyl; dicobalt octacarbonyl) $Co_2(CO)_8$.
Properties: Orange or dark brown crystals. White when pure. Sp. gr. 1.78; m.p. 51°C, decomposing above this temperature. Insoluble in water; soluble in organic solvents as alcohol, ether, and carbon disulfide.
Derivation: Combination of finely divided cobalt with carbon monoxide under pressure.
Hazard: Toxic by ingestion and inhalation.
Uses: Anti-knock gasoline; catalyst in Oxo process; high-purity cobalt salts.

cobalt titanate Co_2TiO_4.
Properties: Greenish-black crystals; sp. gr. 5.07–5.12. Soluble in concentrated hydrochloric acid.

cobalt trifluoride (cobaltic fluoride) CoF_3.
Properties: Light brown, free-flowing powder; sp. gr. 3.88 (25°C); no odor, except HF odor developed in moist air; stable in sealed containers; reacts readily with moisture in the atmosphere to form a dark, almost black powder; reacts with water to form a black, finely divided precipitate (cobaltic hydroxide). Insoluble in alcohol and benzene. As a fluorinating agent, yields one atom of fluorine and reverts to the difluoride. The spent cobalt difluoride may be regenerated with elemental fluorine.
Hazard: Strong irritant to tissue. Tolerance (as F), 2.5 mg per cubic meter of air.

cobalt tungstate. See cobaltous tungstate.

cobalt ultramarine. See cobalt blue.

cobalt wolframate. See cobaltous tungstate.

cobalt yellow. See cobalt potassium nitrite.

"Cobanic."[155] Trademark for an alloy composed of nickel 55.0% and cobalt 45.0%.
Properties: M.p. 1500°C; sp. gr., 8.84; resistivity at 20°C, 11.0 microhm cm; tensile strength at 20°C, 85,000 lb/sq. in.
Uses: Electron tubes.

"Cobenium."[155] Trademark for a heat-treatable, high-cobalt alloy. Comprising cobalt, 40%; chromium, 20%; nickel, 15%; molybdenum, 7%; manganese, 2%; beryllium, 0.04%; carbon, 0.15%; iron, balance.
Properties: Corrosion-resistant; non-magnetic; resistant to set and fatigue; heat-treatable, high strength; high elasticity.
Uses: Springs; corrosion-resistant products.

"alpha **Cobione."**[123] Trademark for hydroxocobalamin, crystalline.

"Coblac."[123] Trademark for a series of carbon black-nitrocellulose dispersions, which make it possible to produce black lacquers without milling or grinding. Available in several types for pigmenting automotive lacquers, industrial lacquers, leather finishes, etc.

"Cobon."[169] Trademark for 2-nitroso-1-naphthol used for the colorimetric determination of cobalt.
Sensitivity: 0.005 ppm cobalt.

COC. Abbreviation for Cleveland Open Cup, a standard method of flash point determination.

cocaine (methylbenzoylepgonine) $C_{17}H_{21}NO_4$. An alkaloid.
Properties: Colorless to white crystals, or white crystalline powder. Soluble in alcohol, chloroform, and ether; slightly soluble in water (solution is alkaline to litmus). The hydrochloric acid solution is levorotatory. M.p. 98°C.
Derivation: By extraction of the leaves of Erythroxylon coca with sodium carbonate solution, treatment of the latter with dilute acid and extraction with ether, evaporation of the solvent; re-solution of the alkaloid and subsequent crystallization. Also synthetically from the alkaloid ecgonine.
Method of purification: Recrystallization.
Grades: Technical; N.F.
Hazard: Highly toxic; a habit-forming drug. Use restricted to physicians.
Use: Local anesthetic (medicine, dentistry), usually as the hydrochloride.

cocarboxylase (TPP; thiamine pyrophosphate chloride) $C_{12}H_{19}ClN_4O_7P_2S \cdot H_2O$. The coenzyme of the yeast enzyme carboxylase. It is the key substance in decarboxylation, an energy-producing reaction in the body.
Properties: Crystallizes from alcohol containing some

hydrochloric acid; m.p. 240–244°C (dec); soluble in water; dry substance very stable.
Use: Biochemical research; medicine.

cocculin. See picrotoxin.

cochineal (coccus). A red coloring matter consisting of the dried bodies of the female insects of Coccus cacti. The coloring principle is carminic acid $C_{22}H_{20}O_{13}$.
Grades: Technical; N.F.; silver grain; black grain.
Uses: Coloring food, medicinal products, toilet preparations; manufacture of red and pink lakes and carmine; indicator in analytical chemistry; inks; dyeing.

cocoa (cacao). Powder prepared from roasted cured kernels of ripe seed of Theobroma cacao. See also cacao bean.
Grade: Commercial; U.S.P.
Uses: Flavoring agent for pharmaceuticals; food products; beverage.

cocoa butter. See theobroma oil.

cocoa oil. See theobroma oil.

"Cocoloid."[322] Trademark for an algin-carrageenan composition; a hydrophilic colloid.
Uses: Stabilizer for chocolate-milk products, sterilized cream, and other milk products.

coconut acid. Mixture of fatty acids derived from hydrolysis of coconut oil. Acid chain lengths vary from 6–18 carbons but are mostly 10, 12, and 14. Combustible; nontoxic.
Grades: Distilled; double-distilled.
Containers: Drums; tank cars.
Uses: Soaps; detergents; source of long-chain alkyl groups.

coconut cake (coconut palm cake; copra cake). The residual product from expression of oil from the seed of the coconut. See coconut oil meal.

coconut oil.
Properties: White, semi-solid fat containing C_{12} to C_{15}; slight odor. Sp. gr. 0.92; saponification value 250–264; iodine value 7–10; sharp m.p. at 25°C (76°F). Soluble in alcohol, ether, chloroform, carbon disulfide; immiscible with water. Highly digestible; resists oxidative rancidity but is susceptible to that induced by molds and other microorganisms. Nondrying; nontoxic; combustible.
Chief constituents: Glycerides of lauric acid, as well as of capric, myristic, palmitic and oleic acids.
Derivation: Hydraulic press or expeller extraction from coconut meat, followed by alkali-refining, bleaching, and deodorizing.
Grades: Crude; refined; Ceylon; Manila.
Containers: Barrels; casks; tank cars.
Uses: Food products (margarine, hydrogenated shortenings); synthetic cocoa butter; soaps; cosmetics; emulsions; cotton dyeing; lubricating greases; synthetic detergents; medicine; dietary supplement; source of fatty acids, fatty alcohols, and methyl esters; base for laundering and cleaning preparations for soft leathers.

coconut oil meal. The dried and crushed form of coconut cake recovered from the hydraulic or expeller process of extraction of oil from the meat. The usual product of commerce contains 24.2% crude protein; 13.3% crude fiber; 35.7% nitrogen-free extract; 7.4% ether soluble (fat) and 6.0% ash. The total digestible nutrients approximate 72%.
Containers: Bulk or bags.
Uses: Animal feeds; fertilizer ingredient.

cocoonase. A proteolytic enzyme derived from silk moths; closely similar to trypsin; isoelectric point about 9.5, slightly less than for trypsin; pH about 8.0; molecular weight about 25,000. As secreted by the insect it is 80% pure. Identical with trypsin in catalytic specificity and mechanism of action.

cocoyl sarcosine. $C_{11}H_{23}CON(CH_3)CH_2COOH$. Yellow liquid; sp. gr. 0.970; m.p. 22–28°C. Combustible; nontoxic. Used as a detergent emulsifier.

"C-O-C-S."[55] Trademark for insecticides containing copper oxychloride sulfate. Used on fruits and vegetables.

C.O.D. Abbreviation for chemical oxygen demand. See oxygen consumed, biochemical oxygen demand (B.O.D.); dissolved oxygen (D.O.).

codehydrogenase I. See nicotinamide adenine dinucleotide.

codeine (methylmorphine) $C_{18}H_{21}NO_3 \cdot H_2O$. A narcotic alkaloid.
Properties: Colorless or white crystals or powder. Effloresces slowly in dry air; affected by light; m.p. 154.9°C. Slightly soluble in water; soluble in alcohol and chloroform; levorotatory in acid and alcohol solutions.
Derivation: From opium by extraction and subsequent crystallization; also by the methylation of morphine.
Grades: Technical; N.F.
Hazard: Habit-forming narcotic; sale legally restricted.
Use: Medicine (analgesic); cough sirups (usually as the hydrochloride, phosphate, or sulfate).

cod-liver oil (morrhua oil)
Properties: Pale yellow viscous liquid; fixed, nondrying oil; slightly fishy odor and taste. Soluble in ether, chloroform, ethyl acetate, and carbon disulfide; slightly soluble in alcohol. Sp. gr. 0.918–0.927; saponification value 180–192; iodine value 145–180; maumene test 102–113; acid value 204–207. Combustible; nontoxic.
Chief constituents: Glycerides of palmitic, stearic acids; cholesterol, butyl alcohol esters, etc.
Derivation: From the livers of codfish (Gadus morrhua) and other species of Galidae. These are rendered by steam heat and the oil separated and chilled until the stearin solidifies, when it is pressed and the clear oil collected.
Method of purification: Filtration.
Grades: Pale; light-brown; dark-brown; N.F.
Uses: Medicine (for its vitamin A and D content; now largely replaced by synthetic products); chamois-leather tanning.

coenzyme. A comparatively low molecular weight organic substance which can attach itself to, and thus supplement, a specific protein to form an active enzyme system. Generally synonymous with the term prosthetic group. See also following entries.

coenzyme I. See nicotinamide adenine dinucleotide.

coenzyme A (CoA) $C_{21}H_{36}O_{16}N_7P_3S$. Essential for the formation of acetylcholine and for acetylation reactions in the body. It has been synthesized, and is built up from pantothenic acid, cysteamine, adenosine, and phosphoric acid.

coenzyme Q
$CH_3C_6(O)_2(OCH_3)_2[CH_2CH{:}C(CH_3)CH_2]_nH$. Found in animal organs and yeast. Is active in the citric acid cycle in carbohydrate metabolism. The n in the formula varies according to the source.

"CO Fatty Alcohols."[487] Trademark for a series of primary straight chain fatty alcohols. "TA-1618" is principally C_{16} and C_{18}; "Umbrex" is principally C_6, C_8 and C_{10}.
Properties: Clear colorless liquids to waxy white solids. Sp. gr. 0.81–0.88. Combustible.
Containers: Tank cars and trucks.
Uses: Cosmetic ingredients, foamers, evaporation retardant, lubricants; chemical intermediates in cosmetics, detergents, dispersants, emulsifiers, lube oil additives, nonionic surfactants, plasticizers, solvents, wetting agents.

coffearine. See trigonelline.

coffinite $U(SiO_4)_{1-x}(OH)_{4x}$ (or $USiO_4$, with appreciable $(OH)_4$ in place of some SiO_4). A naturally occurring uranium mineral. Color black; sp. gr. 5.1; luster adamantine; commonly fine-grained and mixed with organic matter and other minerals.
Occurrence: Colorado, Utah, Wyoming, Arizona.
Use: Ore of uranium (Colorado).

cognac oil, green (wine yeast oil). Volatile oil obtained by steam distillation from wine lees. A green to bluish-green liquid with the characteristic aroma of cognac. Soluble in most fixed oils and in mineral oil. It is very slightly soluble in propylene glycol and insoluble in glycerine. Combustible. Low toxicity.
Grade: F.C.C.
Use: Flavoring agent.

"Coherex."[499] Trademark for a dust inhibitor, consisting of a stable, concentrated emulsion, based on natural petroleum resins.

cohune oil. An edible nondrying oil, with properties similar to coconut and babassu oils. Its composition is 46% lauric acid, 16% myristic acid, and 10% oleic acid, balance mixed acids. Obtained from a palm native to Mexico and Central America. Combustible; nontoxic.

"Coilife."[308] Trademark for special epoxy resin encapsulation of random wound stators utilizing solventless epoxy resin formulations and rotational seasoning process.

coke. The carbonaceous residue of the destructive distillation (carbonization) of bituminous coal, petroleum, and coal-tar pitch. The principal type is that produced by heating bituminous coal in chemical recovery or beehive coke ovens (metallurgical coke), one ton of coal yielding about 0.7 ton of coke. It is used chiefly for reduction of iron ore in blast furnaces, and as a source of synthesis gas. Petroleum yields coke during the cracking process. Coke derived from petroleum residues and coal-tar pitch is used for refractory furnace linings in the electrorefining of aluminum and other high-temperature service; also for electrodes in electrolytic reduction of alumina to aluminum, as well as in electrothermal production of phosphorus, silicon carbide, and calcium carbide.

cola (kola; kola nuts; kola seeds; Soudan coffee; guru). Contains caffeine, theobromine.
Derivation: Seeds of Cola nitida or other species of Cola.
Habitat: West Africa; West Indies; India.
Containers: Bags.
Hazard: Moderately toxic.
Use: Soft drinks.

colamine. See ethanolamine.

colchicine $C_{22}H_{25}NO_6$. An alkaloid plant hormone.
Properties: Yellow crystals or powder; odorless or nearly so. Soluble in water, alcohol, and chloroform; moderately soluble in ether; affected by light; m.p. 135–150°C. Solutions are levorotatory.
Derivation: From Colchicum autumnale by extraction and subsequent crystallizatoin. Has been synthesized.
Grades: Technical; U.S.P.
Hazard: Highly toxic; 0.02 gram may be fatal if ingested.
Use: Medicine; to induce chromosome doubling in plants.

cold flow. The permanent deformation of a material that occurs as a result of prolonged compression or extension at or near room temperature. Some plastics and vulcanized rubber exhibit this behavior; in metals it is known as creep.

cold rubber. Synthetic rubber produced by polymerization at relatively low temperatures; specifically, SBR or butadiene-styrene elastomers produced by polymerization at about 40°F compared with usual temperature of about 120°F. A special catalyst system is required.

colemanite. The ore of calcium borate ($Ca_2B_6O_{11} \cdot 5H_2O$). Sp. gr. 2.26–2.48. Used to replace boric acid in the manufacture of glass fibers. Mined in Turkey, it began to be imported into the U.S. in large volume in 1965 and is competitive with domestically produced B_2O_3, derived from kernite.

"Colex."[1] Trademark for a finely powdered bone glue used for adhesion in water paints.

colistin $C_{45}H_{85}N_{13}O_{10}$. Antibiotic produced by a soil microorganism. Probably identical to polymyxin E and closely related chemically to polymyxin B, since it is a polypeptide composed of amino acids and a fatty acid. See polymyxin.

collagen. A fibrous protein comprising most of the white fiber in the connective tissues of animals and man, especially in the skin, muscles and tendons. The most abundant protein in the animal kingdom, it is rich in proline and hydroxyproline. The molecule is analogous to a three-strand rope, in which each strand is a polypeptide chain; it has a molecular weight of about 100,000. Glue made from the collagen of animal hides and skins is still widely used as an adhesive. So-called "soluble" collagen is that first formed in the skin; upon aging it becomes increasingly crosslinked and less hygroscopie. "Soluble" collagen is being used in the cosmetic industry as the basis for face creams, lotions and hair-dressing preparations. Special forms of collagen have been developed for dialysis membranes. Microcrystalline collagen is being used in prosthetic devices and other medical and surgical applications. Regenerated collagen, used in sausage casings, is made by neutralizing with acid collagen that has been purified by alkaline treatment. Collagen is converted to gelatin by boiling in water, which causes hydrolytic cleavage of the protein to a mixture of degradation products. See also gelatin.

2,4,6-collidine (2,4,6-trimethylpyridine) $(CH_3)_3C_5H_2N$.
Properties: Colorless liquid. B.p. 170.4°C; freezing point −44.5°C; sp. gr. 0.913 (20/20°C); refractive index (n 20/D) 1.4981. Soluble in alcohol; slightly soluble in water. Combustible.
Grades: Technical (97.5% purity).

Uses: Chemical intermediate; dehydrohalogenating agent.

collodion. A solution of pyroxylin (nitrocellulose) in ether and alcohol. U.S.P. specifications are pyroxylin 40 g, ether 750 ml and alcohol 250 ml.
Properties: Pale yellow, syrupy liquid; odor of ether. Immiscible with water. Flash point approx. 0° F.
Grades: Technical; U.S.P.
Uses: Cements; coating wounds and abrasions; solvent for drugs; corn removers; process engraving and lithography.
Hazard: Flammable; dangerous fire risk.
Shipping regulations: (Rail) Red label. (Air) Flammable Liquid label.
See also nitrocellulose.

colloid chemistry. A subdivision of physical chemistry (q.v.) comprising the study of phenomena characteristic of matter when one or more of its dimensions lie in the range between 1 millimicron (nanometer) and 1 micron (micrometer). The science thus includes not only finely divided particles but also films, fibers, foams, pores, and surface irregularities. It is the dimension that is critical, rather than the nature of the material. Colloidal particles may be gaseous, liquid, or solid, and occur in various types of suspensions (imprecisely called solutions), e.g., solid/gas (aerosol), solid/solid, liquid/liquid (emulsion), gas/liquid (foam). In this size range, the surface area of the particle is so much greater than its volume that unusual phenomena occur; for example, the particles do not settle out of the suspension by gravity, and are small enough to pass through filter membranes. Macromolecules (proteins and other high polymers) are at the lower limit of this range; the upper limit is usually taken to be the point at which the particles can be resolved in an optical microscope. The first specific observations were made by Thomas Graham about 1860, and were extended by Ostwald, Hatchek, and Freundlich. Though the term is often used synonymously with surface chemistry, in a strict sense it is limited to the size range mentioned in at least one dimension, whereas surface chemistry is not. Natural colloidal systems include rubber latex, milk, blood, egg-white, etc. See also surface chemistry; colloid, protective; emulsion.

colloid, association. See association.

colloid mill. See homogenization.

colloid, protective. Any surface-active substance that prevents the dispersed phase of a suspension from coalescing by forming a thin layer on the surface of each particle. Thus in milk, the surface of the fat particles is coated with albumin, and in rubber latex the globules are kept from coagulating by a layer of protein.

"Colloisol."[440] Trademark for a series of vat dyes for dyeing and printing textiles of cellulosic fibers.

cologne (toilet water). A scented alcohol-based liquid used as a perfume, after-shave lotion, or deodorant. Combustible.

Cologne brown. See Van Dyke brown.

colorant. Any substance that imparts color to another material or mixture. Colorants are either dyes or pigments. A colorant may either be (1) naturally present in a material (chlorophyll in vegetation), (2) admixed with it mechanically (dry pigments in paints), or (3) applied to it in a solution (organic dyes to fibers).

Note: A valid distinction between dyes and pigments is almost impossible to draw. Some have established it on the basis of solubility, others on physical form and method of application. Most pigments are insoluble, inorganic powders, the coloring effect being a result of their dispersion in a solid or liquid medium; most dyes, on the other hand, are soluble synthetic organic products which are chemically bound to, and actually become part of, the material to which they are applied. Organic dyes are usually brighter and more varied than pigments, but tend to be less stable to heat, sunlight and chemical effects. The term colorant applies to black and white, as well as to actual colors. Instruments for measuring, comparing and matching the hue, tone and depth of colors are called colorimeters. See also dye; pigment; colorimetry, food color; FD&C color.

"Colorex"[1] Trademark for titanium trichloride in aqueous solution with zinc chloride. Dark violet to black liquid.
Uses: Powerful reducing agent; dye stripper for textiles.

colorimetry. An analytical method based on measuring the color intensity of a substance or a colored derivative of it. For example, the yellow carotene content of butter can be determined by saponifying a sample of butter in an alkaline solution, extracting the carotene with ether, and measuring the intensity of yellow color in the ether extract. Colorimetric methods can be used for determining very minute amounts. They are applied in hospital laboratories for blood and urine analysis, food laboratories for determination of vitamins, preservatives, coloring matter, etc. In metallurgical laboratories, traces of metals in raw materials and finished products are determined.

colorless dye. Synonym for optical brightener (q.v.).

"Colorundum."[205] Trademark for a balanced mixture of abrasive aggregates, mineral oxide color, and stearate for surfacing concrete floors, in a choice of colors.

"Columbia."[214] Trademark for activated carbons.
Properties: Hard, durable, inert pellets and granules. Ash content 0.5–6%. Unaffected by most chemical agents and conditions. Numerous grades available.
Uses: Solvent recovery; gas purification; air conditioning; gas separation; gas masks; catalyst carriers.

columbite. An ore of tantalum. See tantalite.

columbium Cb Alternate name for the element niobium. The latter name became official in 1949; columbium is still used by metallurgists.

"Coly-Mycin."[546] Trademark for colistin in form of sulfate or methane sulfonate (colistimethate sodium).

"Combatex."[443] Trademark for a cationic resin textile-finishing agent.

combination. A chemical reaction in which two substances unite to form a third; the reacting substances may be either elements or compounds, but the product is always a compound. Examples:
$Cu + Cl_2 \longrightarrow CuCl_2$; $2NH_3 + H_2SO_4 \longrightarrow (NH_4)_2SO_4$.
Polymerization is a special case of combination; here complex organic molecules of the same kind unite to form chains or clusters of high molecular weight. See also polymerization.

combining number. See valence.

combining weight. See equivalent weight.

combustible material. Any substance that will burn, regardless of its autoignition point (q.v.), or whether it is a solid, liquid, or gas. Although this definition necessarily includes all flammable materials (q.v.) as well, this fact is disregarded in official classifications. As usually defined, the term "combustible" refers to solids that are comparatively difficult to ignite and that burn relatively slowly, and to liquids having a flash point above 80°F (above 100°F, according to National Fire Protection Assn.).

It is difficult to generalize about the combustibility of solids, for the rate and ease of combustion may depend as much on their state of subdivision as on their chemical nature. Many metals in powder or flake form will ignite and burn rapidly, whereas most are noncombustible as bulk solids. Cellulose is combustible in the form of a textile fabric or as paper, but is flammable as fine fibers (cotton linters). A plastic that will burn at flame temperature will be a greater fire hazard as a foam than as a bulk solid because of the large surface area exposed to air and the thinness of the cell walls. Some polymers, such as nylon and polyvinylidene chloride, will melt and burn, but will not propagate flame; others, e.g., polyvinyl chloride and polyurethane, ignite at high temperature and evolve toxic fumes. Acrylics and cellulose-derived plastics, such as rayon and cellulose acetate, are readily combustible. This may be partially offset by use of fire-retardant chemicals. Glass is noncombustible in all forms.

Several definitions of combustible liquids are offered by various regulatory authorities. The National Fire Protection Association considers them as liquids having a flash point above 100°F (Tag Closed Cup). The Dept. of Transportation (DOT) defines them as those with a flash point above 80°F. The International Air Transport Authority (IATA) has defined them as liquids flashing from 80 to 101°F (Tag Open Cup). The Manufacturing Chemists Association (MCA) uses the range from 80 to 150°F (Tag Open Cup).

See also flammable material; hazardous material.

combustion. An exothermic oxidation reaction in which the heat evolved results from rupture of chemical bonds. In the case of organic fuels (wood, coal, petroleum, and their derivatives) the energy released was originally due to photosynthesis. Combustion may occur in an atmosphere of chlorine as well as of air. In pure oxygen combustion occurs with greatly increased intensity, and some substances may ignite spontaneously at room temperature. The products of combustion of organic compounds are carbon dioxide and water. One gallon of gasoline yields on complete burning approximately one gallon of water. Organic compounds differ greatly in their combustibility, ranging from those of almost explosive nature (carbon disulfide) to those which barely support combustion and are difficult to ignite (rubber). Oxygen itself is not combustible, but actively supports combustion. See also oxidation, respiration, ignition point, flash point. For further information consult National Fire Protection Institute, 470 Atlantic Ave., Boston, Mass.

comminution. Size reduction of materials by any of several means, e.g., grinding, cutting, shredding, chopping, etc. Solids can be reduced to a particle size approaching 1 micron in special fine-grinding equipment. Comminution of coal by chemical means is possible by use of a low-molecular weight compound such as sodium hydroxide or anhydrous ethanol, which penetrates the natural fault system of the coal, resulting in fragmentation without mechanical crushing. This permits removal of sulfur from coal without burning or grinding. About 100 pounds of chemical are needed per ton of coal.

compatibility. The ability of two or more materials to exist in close and permanent association for an indefinite period. Liquids (solvents) are compatible if they are miscible in all proportions and do not undergo phase separation on standing. Thus water is compatible with alcohol but not with gasoline. Liquids and solids are compatible if the solid is soluble in the liquid, but not otherwise. Solids are compatible if they can exist in intimate contact for long periods with no adverse effect of one on the other.

"Compazine."[71] Trademark for a brand of prochlorperazine, as the maleate or the edisylate.

complex compound. See coordination compound.

complexing agent. See ligand; chelate; ethylenediaminetetraacetic acid.

complex ion. An ion which has a molecular structure consisting of a central atom bonded to other atoms by coordinate covalent bonds. See coordination compound.

component. One of the minimum number of substances required to state the composition of all phases of a system. In the absence of chemical reaction, any of the substances in a mixture. See also constituent.

composite. A mixture or mechanical combination on a macro scale of two or more materials that are solid in the finished state, are mutually insoluble, and differ in chemical nature. The major types are: (1) laminates (q.v.) of paper, fabric or wood (veneer) and a thermosetting material (resin, rubber, or adhesive); examples are tire carcases, plywood, and electrical insulating structures. (2) Reinforced plastics (q.v.) principally of glass fiber and a thermosetting resin; other types of fibers such as boron, aluminum silicate, and silicon carbide may be used (see also whiskers). (3) Cermets (q.v.), which are mixtures of ceramic and metal powders, heat-treated and compressed. (4) Fabrics, i.e., woven combinations of wool or cotton and a synthetic fiber. (5) Filled composites in which a bonding material such as linseed oil, resin, or asphalt is loaded with a filler in the form of flakes or small particles; examples are linoleum, glass/flake-plastic mixtures for battery cases, and asphalt/gravel road-surfacing mixtures.

composting. Aerobic bacterial decomposition of solid organic wastes, both agricultural and urban, including sewage sludge. As much as 500 tons a day can be handled in the larger installations, the waste degrading quickly at low temperatures. The product can be used as a soil conditioner and for landfill. The waste is piled and turned frequently to provide aeration, while maintaining a high enough temperature in the pile to destroy pathogenic organisms. Volume of composted waste is from 20 to 60% of original volume.

compound. (1) A substance composed of atoms or ions of two or more elements in chemical combination. The constituents are united by bonds (q.v.) or valence forces. A compound is a homogeneous entity in which the elements have definite proportions by weight, and are represented by a chemical formula (q.v.). A compound has characteristic properties quite different from those of its constituent elements; it can be decomposed by energy in the form of a chemical reaction (q.v.), of heat, or of an electric current. Example: Water is a *liquid* formed by chemical combination of two *gases*; it can be separated into hydrogen and oxygen by an electric current (electrolysis); in certain reactions it is split into its constituent ions (H and OH) (hydrolysis); it is not chemically changed by heat or cold. See also mixture; homogeneous; reaction.

(2) Loosely, a product formula (often proprietary) of various types, e.g., pharmaceuticals (a vegetable compound), rubber (a fast-curing compound), etc.

(3) Having two sets of lenses (compound microscope).

Compound 1080. Toxic. Use may be restricted. See sodium fluoroacetate.

"Compound 4072."[125] Trademark for 2-chloro-1-(2,4-dichlorophenyl)-vinyl diethyl phosphate.

compreg. A hardwood impregnated with a phenolformaldehyde resin under heat and pressure.

"Compregnite."[65] Trademark for a phenol-formaldehyde liquid resin used for impregnation of wood to improve density and physical properties.

compressed gas. Any material or mixture that, when enclosed in a container, has an absolute pressure exceeding 40 psi at 70°F or, regardless of the pressure at 70°F, has an absolute pressure exceeding 140 psi at 130°F; or any flammable material having a vapor pressure exceeding 40 psi absolute at 100°F. (Vapor pressure determined by Reid method (ASTM)). Compressed gases include liquefied petroleum gases (q.v.), as well as oxygen, nitrogen, anhydrous ammonia, acetylene, nitrous oxide and fluorocarbon gases. Some of these are shipped in tonnage volume. For details on properties, containers and shipping regulations, see specific gas. For additional information consult Compressed Gas Association, 500 Fifth Avenue, New York, N.Y.

Shipping regulations: (Flammable) (Rail) Red Gas label. (Air) Flammable Gas label. Not acceptable on passenger planes. (Nonflammable) (Rail) Green label. (Air) Nonflammable Gas label.

compression molding. Formation of a rubber or plastic article to a desired shape, either by placing the raw mixture in a specially designed cavity or bringing it into contact with a contoured metal surface. After the material is in place, heat and pressure are supplied by a hydraulic press, the time and temperature varying with the nature of the material. In the case of rubber products, vulcanization occurs simultaneously. Most plastic molding is now done by the injection method, which is more economically efficient. See also injection molding.

"Compresto."[84] Trademark for an electrical conductor consisting of layers of shaped pure aluminum wires concentrically stranded about a single round core wire.

Compton effect. One of the principal processes by which high energy electromagnetic radiation, or gamma rays, interact with or are absorbed by matter. In the Compton process, the gamma ray frees an electron in matter as if the electron was unbound, dividing the momentum of the gamma ray betewen the ejected electron and a new gamma ray of lower energy going off in a new direction.

"Conac S."[28] Trademark for N-cyclohexyl-2-benzothiazole sulfenamide (q.v.). Used to accelerate the vulcanization of natural and synthetic rubber.

"Conac T." Cyclic thiuram; yellow to greenish-yellow powder; sp. gr. 1.26; m.p. 96°C. Used as a delayed action accelerator for SBR.

"Conastron."[464] Trademark for a high temperature and chemical-resistant coating, containing titanium metal powder and stainless steel flake in high temperature silicones and boron polymers.

"Concentals."[325] Trademark for a series of highly sulfated organic compounds. Used in cotton softening and finishing operations.

concentration. The amount of a given substance in a stated unit of a mixture, solution, or ore. Common methods of stating concentration are per cent by weight or by volume; normality; or weight per unit volume, as grams per cubic centimeter or pounds per gallon. The concentration of an atom, ion or molecule in a solution may be indicated by square brackets, as $[Cl^-]$. For radioactivity, the concentration is usually expressed as millicuries per milliliter (mc/ml), or millicuries per millimole (mc/mM).

conchiolin $C_{32}H_{98}N_2O_{11}II$. A natural bonding agent in the calcium carbonate structure of pearls.

conchoidal. A term adopted from mineralogy by chemists to describe a type of surface formed by fracturing a hard solid by impact. Certain materials present involutely curved fracture surfaces suggestive of the shape of the shells of bivalves (conch), from which the term is derived. Examples are glass, blown asphalt, and numerous minerals.

concrete.

(1) A conglomerate of gravel, pebbles, broken stone, blast-furnace slag or cinders, termed the aggregate, embedded in a matrix of either mortar or cement, usually standard Portland cement in the U.S. Reinforced concrete and ferro-concrete contain steel in various forms. Further information can be obtained from the American Concrete Institute, Detroit, Michigan. See cement, Portland.

Uses: Building and road construction; radiation shielding.

(2) A waxy solid obtained from roses by extraction with non-polar solvents (benzene) after trace quantities of solvent have been removed. When the concrete is dewaxed by a properly chosen second solvent (alcohol), the desired essential oil remains. This is called an absolute. See also perfume.

concrete, cellular. A light-weight concrete foam which may be made in several ways:

(a) By addition of aluminum powder to the concrete mix and applying heat, which releases hydrogen;

(b) By whipping air into the mix containing an entraining agent.

(c) By adding preformed foam to the mix. Such foams are made from a foaming agent such as dried blood, a stabilizer, organic solvents, and a germicide.

See also foam.

condensation. (1) A chemical reaction in which two or more molecules combine, with the separation of water or some other simple substance. If a polymer is formed, the process is called polycondensation (ASTM). Phenolformaldehyde resin is an example of a condensation polymer. (2) The change of state of a substance from the vapor to the liquid (or solid) form.

conductance. The conductivity of an electrolytic solution, defined as the reciprocal of the resistance. It is usually used in connection with electrolytic solutions. See conduction (2).

"Conductex."[133] Trademark for a group of carbon blacks designed to provide high electrical conductivity where required in rubber, plastics, coatings etc.

conduction. (1) Transference of heat through a substance or from one substance to another when the two substances are in physical contact (thermal conduction). Crystalline solids (especially metals and alloys) are good thermal conductors because of their high density; liquids, such as water and glass, and high polymers, such as rubber and cellulose, usually are not.

(2) Transference of an electric current through a solid or liquid (electrical conduction). In metallic or electronic conductors, the current is carried by a flow of electrons from atom to atom, the atomic nuclei remaining stationary. This type of conduction is common to all metals and alloys, carbon and graphite, and certain solid compounds (manganese dioxide, lead sulfide). In electrolytic or ionic conductors, the current is carried by ions, as in solutions of acids, bases, and salts and in many fused compounds. In electrolytic conduction, as in metallic conduction, heat is generated, and a magnetic field is formed around the conductor; a transfer of matter also occurs. In a few materials, as solutions of alkali and alkaline earth metals in anhydrous liquid ammonia, both types of conduction take place simultaneously; such conductors are called mixed conductors. See also semiconductor; transference number.

conductivity. The property of a substance or mixture that describes its ability to transfer heat or electricity. It is the reverse of resistivity. See also conduction.

"Conflex."[537] Trademark for a composite combining the electrical conductivity of copper and the elastic properties of spring steel. Used in fabricating rotating switches.

configuration. In an organic molecule, the location or disposition of substituent atoms or groups around asymmetric carbon atoms. This can be changed only by severing single covalent bonds.

conformation. The shapes or arrangements in three-dimensional space that an organic molecule can assume by rotating carbon atoms or their substituents around single covalent bonds. The conformation of a molecule is not fixed; though one or another shape may be more likely to occur, the number of conformational isomers is infinite. Conformational analysis involves the study of the preferred (or most likely) conformations of a molecule in the ground, transition, and excited states. Research on the conformations of cyclohexane and various sugars has contributed much to this aspect of stereochemistry.

Congo resin. A variety of copal fossil resin. The natural product is insoluble in organic solvents, but forms transparent gels with some alcohols and hot solvents. When thermally processed (cracked) it is soluble in all organic solvents, fatty acids, and vegetable oils. Its high acid number prevents its use in paints containing reactive pigments.

Uses: High-gloss varnishes; paints for metal surfaces; wrinkle finishes.

Congo red. Sodium diphenyl-bis-alpha-naphthylamine sulfonate $C_{32}H_{22}O_6N_6S_2Na_2$.

Properties: Brownish-red powder; soluble in water and alcohol; insoluble in ether; odorless; decomposes on exposure to acid fumes.

Derivation: Combination of tetraazotized benzidine and naphthionic acid.

Hazard: May have severe allergic reaction (anaphylactic shock). Possible carcinogen.

Uses: Dye; medicine (antidote); indicator; biological stain.

coniferin $C_{16}H_{22}O_8$. A glucoside contained in pine bark and other conifers. When decomposed it yields coniferyl alcohol which can be oxidized to vanillin. Used as a raw material for manufacture of synthetic vanillin.

coning oil. Usually an emulsified mineral oil used as lubricant for textile fibers in processing to the finished yarn. Fatty acid esters are often used for the oil-in-water emulsions.

conjugated double bonds. Two or more double bonds which alternate with single bonds in an unsaturated compound, as in the formula for butadiene-1,3 ($H_2C{=}CH{-}CH{=}CH_2$) or maleic acid (the $O{=}C{-}C{=}C{-}C{=}O$ skeleton).

conjugate layers. Two layers of a liquid system each composed of a different ternary mixture and in equilibrium with one another.

"Conjulin" Fatty Acids.[550] Trademark for a fatty acid mixture containing principally oleic, linolenic, and 9,11-octadecadienoic acids. Used in alkyd resins, epoxy esters, driers.

"Conpernik."[308] Trademark for an alloy of approximately equal proportions of iron and nickel having constant permeability over a range of low flux densities. It is used where constant inductance cores are required over a low range of inductions.

conservation of energy, law. (Also known as the first law of thermodynamics). See energy.

consistency. Resistance to flow of a material, usually a liquid. For Newtonian liquids, consistency and viscosity are synonymous. For non-Newtonian liquids, it qualitatively represents plastic flow. The term is used primarily by food technologists.

"Consol C."[1] Trademark for a chemically modified protein colloid derived from bones and hides. Used in the paper industry to improve interfiber bonding and filler retention.

constantan. Generic name for an alloy containing from 40 to 45% nickel and from 55 to 60% copper. Used in thermocouples and specialized heat-measuring devices.

constant-boiling mixture. See azeotropic mixture.

constant composition, law. See chemical laws (2).

constituent. Any of the elements or subgroups in a molecule of a compound. For example, nitrogen is a constituent of proteins; the carboxyl group is a constituent of fatty acids. See also component.

contact acid. Sulfuric acid made by the contact process (q.v.).

contact process. A process for manufacture of sulfuric acid and oleum, in which the sulfur dioxide from the burning of sulfur, pyrites, or other sulfur sources is oxidized with air to sulfur trioxide by contact with a vanadium pentoxide catalyst. The sulfur trioxide is absorbed in high-strength sulfuric acid to give a product of 98 or 100% acid, or fuming sulfuric acid higher in sulfur trioxide content. Almost all the sulfuric acid produced in the U.S. is made by this process.

contact resin (impression resin, low-pressure resin). A synthetic thermosetting resin characterized by cure at relatively low pressure. The usual components are an unsaturated high molecular weight monomer such as an allyl ester, or a mixture of styrene or other vinyl monomer with an unsaturated polyester or alkyd. Cure requires heat and a catalyst as well as some pressure. The curing does not result in water formation as with phenol-formaldehyde resins.

container. This term refers not only to units used to pack chemical products for shipment but also to transport them. The factors that dictate their selection include size of shipment, compatibility of the product with container material, ease of storage and handling, and cost. Common containers are:

tank cars, tank trucks, hopper cars (bulk chemicals)
barges (petroleum products, petrochemicals, other chemicals)
tankers (crude oil, refined products)
pipelines (gases, liquid chemicals, etc.)
55-gal drums of steel, plastic, or fiberboard, with or without polyethylene lining.
5-, 15-, 30-gal metal cans and pails
5-gal plastic carboys
1-gal metal cans
steel cylinders (compressed gases)
multiwall paper bags (dry powders)
glass and plastic bottles, vials, etc.
wooden kegs, barrels, boxes, bales; burlap bags (bulk imports, such as gums, etc.)
76-lb flasks (mercury)

contaminant. Any substance accidentally or unwillingly introduced into air, water, or food products which has the effect of rendering them toxic or otherwise harmful. Examples are sulfur dioxide resulting from combustion of high-sulfur fuels; pesticide residues in vegetables, fish, or other food products; industrial dusts; and radioactive materials resulting from nuclear explosions. See also waste disposal; fallout; decontamination; food additive.

Note: Pesticide residues in foods are often referred to as unintentional additives.

"Continental."[104] Trademark for a series of channel blacks used in natural and synthetic rubber, paints, inks and plastics.

"Continex."[104] Trademark for furnace blacks. Used in rubber, plastics, paints, paper.

continuous distillation. Distillation in which a feed, usually of nearly constant composition, is supplied continuously to a fractionating column and the product is continuously withdrawn at the top, bottom, and sometimes at intermediate points.

continuous phase. See phase (2).

control. (1) In any chemical or other scientific experiment, the reference base with which the results are compared. This base is invariably a sample of identical constitution and prepared under the same conditions, from which all experimental variations are omitted. Thus the control represents known values, as far as any specific experiment is concerned. Such a sample is often called a "blank." The use of a control is vital to significant interpretation of experimental data.

(2) Automatic: Maintenance of uniform physical conditions (temperature, pressure, power level, etc.) in a system by means of a sensing device operating on the feedback principle.

controlled atmosphere. See atmosphere, controlled; storage.

controlled-release. Descriptive of a compound manufactured in such a way that its effect will be kept uniform over an extended time period; this not only provides more effective control but also reduces the waste involved in using unnecessarily high concentrations. This principle has been applied successfully to fertilizers, pesticides, pharmaceuticals, flavors, and fragrances. It may involve (1) encapsulation of the agent, (2) incorporating it into a neutral matrix such as a rubber or plastic, (3) coating the particles with sulfur (fertilizers), (4) absorbing it into substrates of various types.

convection. The transfer of heat from one place to another by a moving gas or liquid. Natural convection results from differences in density caused by temperature differences. Thus warm air is less dense than cool air; the warm air rises relative to the cool air, and vice versa. Forced convection involves motion caused by pumps, blowers, or other mechanical devices.

converting. See fabrication.

"Convertit."[173] Trademark for a liquid invertase preparation.

cooking. (1) Conversion of a foodstuff from the raw, inedible state to a palatable and more readily digestible condition by application of heat, as in boiling, frying, roasting, or baking. One or more chemical and physical changes occur: (1) hydrolysis of collagen in the connective tissue of meats; (2) softening and partial hydrolysis of cellulose and starches to sugars in fruits and vegetables; (3) denaturation and coagulation of proteins in meat, resulting in a less tightly ordered structure and improved texture; (4) coagulation of egg albumin and wheat gluten (bakery products); (5) modification or "shortening" of wheat flour by added fats, which coat the particles to form a laminar structure, thus preventing formation of chains of gluten; (6) evolution of carbon dioxide by the action of leavening agents (yeast, baking powder), which causes "rising" or gas bubbles that are retained by the gluten to form a stable solid foam; (7) browning of meats and bakery products due to reaction of sugars and amino acids (nonenzymatic); (8) deactivation of enzymes and destruction of bacteria; (9) loss or deactivation of water-soluble vitamins in meat and vegetables.

(2) In the paper industry, digestion of wood pulp with such compounds as sodium sulfate, sodium hydroxide, and sodium sulfide to separate the lignin content from the cellulose. See also pulp, paper.

"Coolanol."[58] Trademark for coolant-dielectric fluids for electronic equipment.

coolant (heat-transfer medium; thermofor). Any liquid or gas having the property of absorbing heat from its environment and transferring it effectively away from its source. Coolants are used in all types of automobiles, as well as in chemical processing and nuclear engineering equipment. One of the most effective and cheapest coolants is water, which is almost universally used in automotive and ordinary reaction equipment. Air is also used. Where intense heating requires a more efficient medium, special coolants are used; liquid sodium in nuclear reactors, liquid hydrogen in high-thrust nuclear rocket engines; carbon dioxide, propylene glycol, and "Dowtherm" in chemical processing reactors. Methoxy propanol has been introduced for diesel engines. Some coolants provide antifreeze protection. See also antifreeze.

Coolidge, William D. (1873–1975). American physical chemist, born in Massachusetts. Received degree in electrical engineering at M.I.T. (1896) and doctorate in physics at Leipzig (1899). In 1905 he joined the General Electric Research Laboratory which had been established five years earlier. Here he invented the ductile tungsten filament and developed the use of tungsten in electrical switches and medical x-ray tubes. He also did pioneer research in experimental metallurgy and powder metallurgy. He also had a prominent part in evaluating uranium research (1941) and in setting up the Manhattan project. He was the recipient of many honors and awards, not the least of which was induction into the National Inventors Hall of Fame, in 1975.

coordination compound (complex compound). A compound formed by the union of a metal ion (usually a transition metal) with a nonmetallic ion or molecule called a ligand or complexing agent. The ligand may be either positively or negatively charged (such ions as Cl^- or $NH_2NH_3^+$), or it may be a molecule of water or ammonia. The most common metal ions are those of cobalt, platinum, iron, copper, and nickel, which form highly stable compounds. When ammonia is the ligand, the compounds are called ammines. The total number of bonds linking the metal to the ligand is called its coordination number; it is usually 2, 4, or 6, and often depends on the type of ligand involved. All ligands have electron pairs on the coordinating atom, e.g., nitrogen, that can be either donated to or shared with the metal ions. The metal ion acts as a Lewis acid (electron acceptor) and the ligand as a Lewis base (electron donor). The bonding is neither covalent nor electrostatic, but may be considered intermediate between the two types. The charge on the complex ion is the sum of the charges on the metal ion and the ligands; for example, $4NH_3 + 2Cl^- + Co^{+3}$ forms the complex $[Co(NH_3)_4Cl_2]^+$. The brackets enclose the metal ion and the coordinated ligands. See also chelate; sequestration; metallocene.

coordination number (CN). The number of points at which ligands (q.v.) are attached to the metal ion in a complex. Common coordination numbers are 2, 4, and 6, exemplified by the ions $[Ag(NH_3)_2]^+$, $[Ni(CN)_4]^{-2}$, and $[PdCl_6]^{-2}$

copaiba oil.
Properties: Essential oil; colorless to yellowish or bluish; characteristic copaiba-balsam, pepperlike odor; bitter taste. Sp. gr. 0.88–0.91 (15°C); optical rotation −7 to −33°; refractive index 1.494–1.500; insoluble in water; soluble in 5–6 vols 95% alcohol; soluble in ether, carbon disulfide. Combustible.
Chief constituent: Caryophyllene, a sesquiterpene.
Derivation: Distilled from copaiba balsam.
Grades: Technical; F.C.C.
Hazard: Moderately toxic.
Use: Medicine; flavoring.

copaiba resin. See balsam (3).

copaivic acid $C_{20}H_{30}O_2$. A monobasic acid derived from copaiba balsam.

copal. A group of fossil resins still used to some extent in varnishes and lacquers. Insoluble in oils and water. Most important types are Congo, kauri, and manila.

"Copeenblak."[133] Trademark for a series of carbon black dispersions in polyethylene for pigmenting cables and jackets, film and pipe stocks.
701. Pelletized polyethylene carbon black concentrate.
703. High quality colloidal dispersion of carbon black in polyethylene resin, in concentrated pellet form.

"Copel."[166] Trademark for 55-45 copper-nickel alloy used as a resistor material in the construction of electrical instruments where temperature coefficient of resistance must be very low.

copolymer. An elastomer produced by the simultaneous polymerization of two or more dissimilar monomers; as SBR synthetic rubber, from styrene and butadiene.

copper Cu Metallic element of atomic number 29, of group IB of the periodic system; atomic weight 63.546; valences 1, 2; two stable isotopes.
Properties: Distinctive reddish color; sp. gr. 8.96; m.p. 1083°C; b.p. 2595°C; ductile; excellent conductor of electricity. Complexing agent, coordination numbers 2 and 4. Dissolves readily in nitric and hot concentrated sulfuric acid; in hydrochloric and dilute sulfuric acid slowly but only when exposed to the atmosphere. More resistant to atmospheric corrosion than iron, forming a green layer of hydrated basic carbonate. Readily attacked by alkalies. A necessary trace element in human diet; a factor in plant metabolism. Essentially nontoxic in elemental form. Noncombustible except as powder.
Ores: Azurite, azurmalachite, chalcocite, chalcopyrite (copper pyrites), covellite, cuprite, malachite. Michigan, Arizona, Utah, Montana, New Mexico, Nevada; Mexico; Chile; Peru; Canada; Africa; U.S.S.R.
Derivation: Varies with the type of ore. With sulfide ores the steps may be (1) concentration (of low grade ores) by flotation and leaching, (2) roasting, (3) formation of copper "matte" (40–50% Cu), (4) reduction of matte to "blister" copper (96–98%), (5) electrolytic refining to 99.9+% copper.
Forms available: Ingots, sheet, rod, wire, tubing, shot, powder; high purity (impurities less than 10 ppm) as single crystals or whiskers.
Hazard: Toxic and flammable in finely divided form. Tolerance (fume) 0.1 mg per cubic meter; (dusts and mists) 1 mg per cubic meter.
Uses: Electric wiring; switches, plumbing, heating, roofing and building construction; chemical and pharmaceutical machinery; alloys (brass, bronze, Monel metal, beryllium-copper); electroplated protective coatings and undercoats for nickel, chromium, zinc, etc.; cooking utensils; corrosion-resistant piping; insecticides; catalyst; antifouling paints. Flakes used as insulation for liquid fuels. Whiskers used in thermal and electrical composites. For further information refer to Copper and Brass Fabricators Council, 225 Park Ave., New York, N.Y.

copper abietate (cupric abietate) $Cu(C_{20}H_{29}O_2)_2$.
Properties: Green scales; soluble in alcohol, and in oils, with fine green color; insoluble in water.
Derivation: Heating copper hydroxide with abietic acid.
Hazard: Moderately toxic.
Uses: Preservative metal paint; fungicide.

copper acetate (cupric acetate; crystals of Venus; verdigris, crystallized) $Cu(C_2H_3O_2)_2 \cdot H_2O$.
Properties: Greenish-blue, fine powder. Soluble in water, alcohol and ether. Sp. gr. 1.9; m.p. 115°C; decomposes at 240°C.
Derivation: Action of acetic acid on copper oxide and subsequent crystallization.
Hazard: Moderately toxic.
Use: Pesticide; catalyst (synthetic rubber).

copper acetate, basic (copper subacetate; verdigris; verdigris, blue; verdigris, green).
Properties: Masses of minute silky crystals either pale green or bright blue in color. Blue variety, approximate formula $(C_2H_3O_2)_2Cu_2O$. Green variety, approximate formula $CuO \cdot 2Cu(C_2H_3O_2)_2$. Coppery taste. The green rust with which uncleaned copper vessels become coated and which is commonly termed verdigris is a copper carbonate and must not be confused with true verdigris. Apart from its impurities, verdigris is a variable mixture of the basic copper acetates.
Soluble in acids; insoluble in alcohol; very slightly soluble in water.
Derivation: Action of acetic acid on copper in the presence of air.
Hazard: Moderately toxic.
Uses: Paint pigment; insecticide; fungicide; mildew preventive; mordant in dyeing and printing fabrics; copper acetoarsenite.

copper acetoarsenite (cupric acetoarsenite; Paris green; king's green; Schweinfurt green; imperial green) $(CuO)_3As_2O_3 \cdot Cu(Cu_2H_3O_2)_2$.
Properties: Emerald-green powder. Soluble in acids; insoluble in alcohol and water. Toxic to many plants.
Derivation: By reacting sodium arsenite with copper sulfate and acetic acid.
Hazard: Highly toxic by ingestion.
Uses: Wood preservative; larvicide.
Shipping regulations: (Rail, Air) Poison label.

copper acetylacetonate (copper 2,4-pentanedione) $Cu(C_5H_7O_2)_2$. Crystalline powder; slightly soluble in water and alcohol; soluble in chloroform. M.p. >230°C. Resistant to hydrolysis. A chelating nonionizing compound.

copper amalgam.
Properties: Hard, brown leaflets. Contain 74% (approx) mercury and 24% (approx) copper. Soluble in nitric acid.
Use: Dental cement.

copper aminoacetate. See copper glycinate.

copper aminosulfate. See copper sulfate, ammoniated.

copper ammonium acetate (cuprous acetate, ammoniacal). Used in an absorption process for separating butadiene from C_4 streams at refineries, or from byproducts of ethylene production.

copper arsenate. Probably the basic cupric arsenate.
Properties: Light blue, blue, or bluish-green powder. Variable composition. Contains 33% (approx) copper and 29% (approx) arsenic. Soluble in dilute acids, ammonium hydroxide; insoluble in alcohol, water.
Hazard: Highly toxic.
Uses: Insecticide; fungicide.
Shipping regulations: (Rail) Arsenical compounds n.o.s., Poison label. (Air) Arsenates, n.o.s., Poison label.

copper arsenite (cupric arsenite; copper orthoarsenite; Scheele's green) $CuHAsO_3$; or, $Cu_3(AsO_3)_2 \cdot 3H_2O$; variable.
Properties: Fine, light-green powder. Soluble in acids; insoluble in water and alcohol. M.p., decomposes.
Hazard: Highly toxic.
Uses: Insecticide; fungicide.
Shipping regulations: (Rail, Air) Poison label.

copper arsenite, ammoniacal. See chemonite.

copperas. See ferrous sulfate.

copperas, blue. See copper sulfate.

copperas, green. See ferrous sulfate.

copperas, white. See zinc sulfate.

copper benzoate $(C_6H_5COO)_2Cu \cdot 2H_2O$.
Properties: Blue, crystalline, odorless powder; slightly soluble in cold water, acids, and alcohol. M.p. (loses H_2O) 110°C.
Derivation: Interaction of solutions of a benzoate and copper salt.

copper-beryllium. See beryllium-copper.

copper, blister. See blister copper.

copper blue. See mountain blue.

copper borate. See copper metaborate.

copper bromide (cupric bromide) $CuBr_2$.
Properties: Black crystalline powder or crystals; deliquescent. Soluble in acetone, alcohol, water. M.p. 498°C; sp. gr. 4.77 (25°C).
Uses: Photogaphy (intensifier); organic synthesis (brominating agent).

copper carbonate (cupric carbonate; copper carbonate, basic; artificial malachite; mineral green. For the native mineral see malachite) $Cu_2(OH)_2CO_3$.
Properties: Green powder. Soluble in acids; insoluble in water. Sp. gr. 3.7–4.0; decomposes at 200°C.
Derivation: By adding sodium carbonate to a solution of copper sulfate, filtering and drying.
Grades: Technical; C.P.
Hazard: Toxic by ingestion.
Uses: Pigments; pyrotechnics; insecticides; copper salts; coloring brass black; astringent in pomade preparations; antidote for phosphorus poisoning; smut preventive; fungicide; feed additive (in small amounts).

copper chloride (cupric chloride) (a) $CuCl_2$; (b) $CuCl_2 \cdot 2H_2O$.
Properties: (a) Brownish-yellow powder; hygroscopic; (b) green, deliquescent crystals. Soluble in water and alcohol. Sp. gr. (a) 3.386 (25°C); (b) 2.54 (25°C). (a) M.p. 620°C; at 993°C decomposes to cuprous chloride. (b) Loses $2H_2O$ at 100°C.
Derivation: (a) By the union of copper and chlorine. (b) Copper carbonate is treated with hydrochloric acid and the product crystallized.
Grades: Technical; C.P.; reagent.
Hazard: Moderately toxic.
Uses: Isomerization and cracking catalyst; mordant in dyeing and printing fabrics; sympathetic ink; disinfectant; pyrotechnics; wood preservation; fungicides; metallurgy; preservation of pulpwood; de-

odorizing and desulfurizing petroleum distillates; photography; water purification; feed additive.
See also cuprous chloride.

copper chromate (cupric chromate, basic) $CuCrO_4 \cdot 2CuO \cdot 2H_2O$.
Properties: Light chocolate-brown powder. Loses water at 260°C. Soluble in nitric acid; insoluble in water.
Derivation: Action of chromic acid on copper hydroxide.
Hazard: Toxic; a suspected carcinogen. Tolerance, 0.1 mg per cubic meter of air.
Uses: Mordant in dyeing; wood preservative.

copper cyanide (cupric cyanide($Cu(CN)_2$.
Properties: Green powder. Keep well stoppered! Soluble in acids and alkalies; insoluble in water.
Derivation: Addition of potassium cyanide to a solution of copper sulfate; cupric cyanide is precipitated. This can be dried, but is not stable.
Grade: Technical.
Containers: Glass bottles; special drums.
Hazard: Toxic. Tolerance (as CN), 5 mg per cubic meter of air.
Uses: Electroplating copper on iron; intermediates (introduction of the cyanide group in place of the amino radical in aromatic organic compounds).
Shipping regulations: (Air) Poison label. (Rail) Consult regulations.
See also cuprous cyanide.

copper, deoxidized (copper, oxygen-free) Copper metal specially treated (as by addition of phosphorus) to remove all or a part of the 0.05% oxygen normally present. It is more ductile than ordinary copper.

copper dichromate (cupric dichromate) $CuCr_2O_7 \cdot 2H_2O$.
Properties: Brown-red crystals; soluble in water, alcohol, and ammonium hydroxide. Hygroscopic. Sp. gr. 2.286; loses $2H_2O$ at 100°C.
Hazard: Probably toxic.

copper dihydrazinium sulfate $CuSO_4(N_2H_4)_2 \cdot H_2SO_4$.
Properties: Bluish powder; m.p. above 300°C, starts to decompose at 140°C; very slightly soluble in water, 250 ppm at 80°C.
Hazard: Moderately toxic by ingestion or inhalation. Skin and eye irritant.
Use: Foliage fungicide.

copper dimethyldithiocarbamate. See "Cumate."

copper, electrolytic. Copper refined by electrolysis. The purest form of copper available commercially.

copper ethylacetoacetate $Cu(C_6H_9O_3)_2$.
Properties: Blue-green powder; m.p. 192–193°C. Insoluble in water; soluble in most organic solvents.
Hazard: Moderately toxic.
Use: Fungicide intermediate.

copper ferrocyanide (cypric ferrocyanide) $Cu_2Fe(CN)_6 \cdot 7H_2O$.
Properties: Reddish-brown powder; insoluble in water and acids; soluble in NH_4OH and KCN solutions. Low toxicity.
Uses: Pigment in paints and enamels; analytical test for traces of copper; inorganic osmotic membranes.

copper fluoride (cupric fluoride) $CuF_2 \cdot 2H_2O$. The anhydrous form, CuF_2, is also available.
Properties: Blue crystals. Slightly soluble in water; soluble in acids and alcohol. Sp. gr. 2.93 (25°C).
Derivation: (1) By decomposing copper carbonate with hydrofluoric acid and subsequent crystallization; (2) fluorination of copper hydroxyfluoride at 525°C.
Grade: Technical.
Hazard: Toxic. Tolerance (as F), 2.5 mg per cubic meter of air.
Uses: Ceramics; enamels; welding and brazing fluxes; additive to cast iron; high-energy batteries.

copper fluosilicate (copper silicofluoride; cupric fluosilicate; cupric silicofluoride) $CuSiF_6 \cdot 4H_2O$.
Properties: Blue, hygroscopic crystals. Soluble in water; slightly soluble in alcohol. Sp. gr. 2.158; decomposed by heat.
Derivation: Interaction of copper hydroxide and hydrofluosilicic acid.
Method of purification: Crystallization.
Grades: Technical.
Hazard: Toxic. Tolerance (as F), 2.5 mg per cubic meter of air.
Uses: Dyeing and hardening white marble; treating grape vines for "white disease."

copper glance. See chalcocite.

copper gluconate (cupric gluconate) $[CH_2OH(CHOH)_4COO]_2$ Cu.
Properties: Odorless, light blue, fine, crystalline powder. Soluble in water; insoluble in acetone, alcohol, and ether. Low toxicity.
Grades: Pharmaceutical; F.C.C.
Containers: Cans; 25-lb fiber drums.
Uses: Medicine; feed additive; nutrient; dietary supplement.

copper glycinate (copper aminoacetate) $(NH_2CH_2COO)_2Cu$.
Properties: Blue triboluminescent crystals; m.p. 130°C. Slightly soluble in water and alcohol; insoluble in hydrocarbons, ethers and ketones.
Grades: Anhydrous; hydrate.
Uses: Catalyst for rapid biochemical assimilation of iron; electroplating baths.

copper hemioxide. See copper oxide, red.

copper hydrate. See copper hydroxide.

copper hydroxide (cupric hydroxide; copper oxide; hydrated; copper hydrate) $Cu(OH)_2$.
Properties: Blue powder. Soluble in acids and ammonium hydroxide; insoluble in water. Sp. gr. 3.368; m.p., decomposes.
Derivation: Interaction of a solution of a copper salt with an alkali.
Grade: Technical.
Hazard: Moderately toxic.
Uses: Copper salts; mordant; cuprammonium rayon; pigment; staining paper; feed additive; pesticide and fungicide.

copper 8-hydroxyquinoline. See copper 8-quinolinolate.

"Copper Inhibitor 50."[28] Trademark for 50% disalicylalpropylenediamine, and 50% aromatic solvent. Used to prevent catalytic action of copper on oxidation of natural and synthetic rubbers.
Hazard: Flammable, moderate fire risk.

copper iodide. See cuprous iodide.

"Copperized CZC Chromated Zinc Chloride."[28] Trademark for a wood preservative and flame retardant containing cupric chloride, zinc chloride, and sodium dichromate.

copper lactate (cupric lactate) $Cu(C_3H_5O_3)_2 \cdot 2H_2O$.
Properties: Greenish-blue crystals or granular powder; soluble in water and ammonium hydroxide.
Hazard: Toxic by ingestion.
Uses: Source of copper in copper plating; fungicides.

copper mercury iodide. See mercuric cuprous iodide.

copper metaborate (copper borate; cupric borate) $Cu(BO_2)_2$.
Properties: Bluish-green, crystalline powder; insoluble in water; soluble in acids. Sp. gr. 3.859.
Derivation: Interaction of copper hydroxide and boric acid.
Grades: Technical; C.P.
Hazard: Toxic by ingestion.
Uses: Oil pigments; painting on porcelain; insecticides (especially wheat-rust compounds).

copper methane arsenate CH_3AsO_3Cu.
Properties: Greenish solid.
Derivation: Reaction of disodium methyl arsenate with copper salts.
Hazard: Highly toxic by ingestion.
Use: Algicide.
Shipping regulations: (Rail) Arsenical compounds, n.o.s., Poison label. (Air) Arsenates, n.o.s., Poison label.

copper molybdate $CuMoO_4$. Crystals; density 3.4 g/cc; m.p. ca. 500°C. Insoluble in water.
Grade: 99.98% pure. Electronic and optical equipment.
Uses: Paint pigment and protective coatings; corrosion inhibitor.

copper monoxide. See copper oxide, black.

copper naphthenate.
Properties: Green-blue solid. High germicidal power. Soluble in gasoline, benzene, and mineral oil distillates.
Derivation: Addition of solution of cupric sulfate to aqueous solution of sodium naphthenate.
Grades: 6, 8, 11½% copper.
Containers: 1- to 55-gal drums.
Hazard: Moderately toxic by ingestion. Flammable, moderate fire risk (solution).
Uses: Wood, canvas, and rope preservative; insecticide; fungicide; antifouling paints.
See also soap (2).

copper nitrate (cupric nitrate) (a) $Cu(NO_3)_2 \cdot 3H_2O$; (b) $Cu(NO_3)_2 \cdot 6H_2O$.
Properties: Blue, deliquescent crystals. Soluble in water and alcohol. Sp. gr. (a) 2.32 (25°C); (b) 2.074. M.p. (a) 114.5°C; (b) loses $3H_2O$ at 26.4°C; (a) decomposes 170°C.
Derivation: By treating copper or copper oxide with nitric acid. The solution is evaporated and product recovered by crystallization.
Grades: Technical; C.P.
Hazard: Moderately toxic. Oxidizing material; in contact with organic materials it may cause violent combustion or explosion.
Uses: Medicine; preparation of light-sensitive papers; analytical reagent; dyes; insecticide for vines; coloring copper black; electroplating; production of burnished effect on iron; paints, varnishes, enamels; pharmaceutical preparations; textiles; catalyst.
Shipping regulations: Nitrates, n.o.s. (Rail) Yellow label. (Air) Oxidizer label.

copper nitrite. (copper nitrite, basic; cupric nitrite) $Cu(NO_2)_2 \cdot 3Cu(OH)_2$; variable.
Properties: Green powder. Decomposes at 120°C. Soluble (with decomposition) in dilute acids, ammonium hydroxide; slightly soluble in water.

copper nitrite, basic. See copper nitrite.

copper octoate. See soap (2).

copper oleate (cupric oleate) $Cu(C_{18}H_{33}O_2)_2$.
Properties: Brown powder or greenish-blue mass. Soluble in ether; insoluble in water. Combustible.
Derivation: Interaction of copper sulfate and sodium oleate.
Hazard: Moderately toxic.
Uses: Medicine; preserving fish nets and marine lines; fungicide; insecticide; ore flotation agent.
See also soap (2).

copper oxide, black (cupric oxide; copper monoxide) CuO. For native black copper oxide see tenorite.
Properties: Brownish-black powder. Soluble in acids; difficultly soluble in water. Sp. gr. 6.32; decomposes at 1026°C.
Derivation: Ignition of copper carbonate or copper nitrate.
Grades: Technical; C.P.
Hazard: Toxic by ingestion.
Uses: Ceramic colorant; reagent in analytical chemistry; insecticide for potato plant; catalyst; purification of hydrogen; batteries and electrodes; aromatic acids from cresols; electroplating; solvent for chromic iron ores; desulfurizing oils; rayon; medicine; paints.

copper oxide, hydrated. See copper hydroxide.

copper oxide, red. (cuprous oxide; copper protoxide; copper hemioxide; copper suboxide) Cu_2O. For the native ore see cuprite.
Properties: Reddish-brown octahedral crystals. Soluble in acids and ammonium hydroxide; insoluble in water. Sp. gr. 5.75–6.09; m.p. 1235°C; b.p. 1800°C.
Derivation: (a) Oxidation of finely divided copper. (b) Addition of bases to cuprous chloride. (c) Action of glucose on cupric hydroxide.
Grades: Technical; C.P.; 97% min (for pigments); also USN Type I (97%); USN Type II (90%).
Hazard: Toxic by ingestion.
Uses: Copper salts; ceramics; porcelain red glaze; red glass; electroplating; antifouling paints; fungicide.

copper oxinate. See copper 8-quinolinolate.

copper oxychloride (cupric oxychloride) composition variable, possibly $3CuO \cdot CuCl_2 \cdot 3\frac{1}{2}H_2O$.
Properties: Bluish-green powder. Soluble in acids, ammonia; insoluble in water.
Hazard: Toxic by ingestion.
Use: Pigment; pesticide; fungus control in grapevines.

copper-2,4-pentanedione. See copper acetylacetonate.

copper phenolsulfonate (copper sulfocarbolate) $[C_6H_4(OH)SO_3]_2Cu \cdot 6H_2O$.
Properties: Green prismatic crystals. Soluble in water and alcohol. Combustible.
Derivation: Interaction of barium phenolsulfonate and copper sulfate.
Use: Medicine.

copper phosphate (copper orthophosphate) $Cu_3(PO_4)_2 \cdot 3H_2O$.
Properties: Light-blue powder. Soluble in acids, ammonium hydroxide; insoluble in water.
Hazard: Toxic by ingestion.
Uses: Analysis; medicine; fungicide.

copper phosphide Cu_3P_2.
Properties: Grayish-black, metallic powder. Insoluble in water; soluble in nitric acid; insoluble in hydrochloric acid. Sp. gr. 6.67.
Derivation: By heating copper and phosphorus.
Hazard: Dangerous; spontaneously flammable and toxic phosphine is evolved on reaction with water.
Use: Manufacturing phosphor-bronze.

copper phthalate $C_8H_4O_4Cu$.
Properties: Fine blue powder; assay, minimum 95%; insoluble or very slightly soluble in common organic solvents or water. Combustible.
Hazard: Toxic by ingestion.
Use: Fungicide.

copper phthalocyanine blue. See Pigment Blue 15.

copper phthalocyanine green. See Pigment Green 7.

copper potassium ferrocyanide (potassium copper ferrocyanide) $K_2CuFe(CN)_6 \cdot H_2O$.
Properties: Brownish-red powder. Insoluble in water. Used for pigment. Low toxicity.

copper protoxide. See copper oxide, red.

copper pyrites. See chalcopyrite.

copper 8-quinolinolate (copper-8; copper oxinate; copper 8-hydroxyquinoline) $Cu(C_9H_6ON)_2$.
Properties: Yellow-green nonhygroscopic, odorless powder. Insoluble in water. Somewhat soluble in weak acids, soluble in strong acids. Insoluble in most organic solvents, but somewhat soluble in pyridine and quinoline. Solubilized copper-8 refers to the product formed by heating copper-8-quinolinolate with certain organic acids (naphthenic, lactic, stearic, etc.) or their salts. In such products the copper-8-quinolinolate does not settle out on standing, even after dilution with various solvents.
Derivation: From 8-quinolinol and copper salt such as copper acetate.
Grade: 10% active salt (1.8% Cu), solubilized.
Containers: Drums.
Hazard: May be toxic.
Uses: Fungicide and mildew-proofing of fabrics; analysis for copper.

copper resinate (cupric resinate).
Properties: Green powder. Soluble in ether and oils; insoluble in water. Combustible.
Derivation: By heating copper sulfate and rosin oil and filtering and drying the precipitate.
Hazard: Moderately toxic.
Uses: Antifouling paints; insecticide.
See also soaps (2).

copper ricinoleate $Cu(C_{17}H_{32}OHCOO)_2$. Green plastic solid; soluble in water and aliphatic hydrocarbons; partially soluble in alcohols and glycols; soluble in ketones and aromatic hydrocarbons. Softening point 64°C. Combustible.
Hazard: Toxic by ingestion.
Uses: Fungicides; insecticides.

copper scale. A coating formed on copper after heating, composed of cupric and cuprous oxides.

copper selenate (cupric selenate) $CuSeO_4 \cdot 5H_2O$.
Properties: Light-blue crystals. Soluble in acids, ammonium hydroxide, water; insoluble in alcohol. Sp. gr. 2.559. Loses $4H_2O$ at 50–100°C.

copper silicate. A complex mixture precipitated by solutions of copper salts from sodium silicate solutions. Used in pigments, catalysts and insecticides.

copper silicide. See silicon-copper.

copper silicofluoride. See copper fluosilicate.

copper sodium chloride (sodium copper chloride) $CuCl_2 \cdot 2NaCl \cdot 2H_2O$.
Properties: Light-green crystals. Soluble in water.

copper sodium cyanide. See sodium-copper cyanide.

copper stearate (cupric stearate) $Cu(C_{18}H_{35}O_2)_2$.
Properties: Light blue powder. Soluble in ether, chloroform, benzene and turpentine; insoluble in water. M.p. 125°C. Combustible.
Derivation: By the interaction of copper sulfate and sodium stearate.
Hazard: Moderately toxic.
Uses: Bronzing plaster statues; paint; see also soaps (2).

copper subacetate. See copper acetate, basic.

copper suboxide. See copper oxide, red.

copper sulfate (cupric sulfate; blue vitriol; blue stone; blue copperas) $CuSO_4 \cdot 5H_2O$.
Properties: Blue crystals or blue crystalline granules or powder, slowly efflorescing in air; white when dehydrated; nauseous metallic taste. Soluble in water, methanol; slightly soluble in alcohol and glycerol. Sp. gr. 2.284.
Derivation: Action of dilute sulfuric acid on copper or copper oxide (often as oxide ores) in large quantities, with evaporation and crystallization.
Method of purification: Recrystallization.
Grades: Technical; C.P.; N.F.; also sold as monohydrate. Available as crystals or powder.
Containers: Multiwall paper sacks; drums.
Hazard: Toxic by ingestion.
Uses: Agriculture (soil additive, pesticides, Bordeaux mixture); feed additive; germicides; textile mordant; leather industry; pigments; electric batteries; electroplated coatings; copper salts; reagent in analytical chemistry; medicine; wood preservative; preservation of pulp wood and ground pulp; process engraving and lithography; ore flotation; petroleum industry; synthetic rubber; steel manufacture; treatment of natural asphalts. The anhydrous salt is used as a dehydrating agent.

copper sulfate, ammoniated (cupric ammonia sulfate; ammonio-cupric sulfate; copper aminosulfate) $Cu(NH_3)_4SO_4 \cdot H_2O$.
Properties: Dark blue, crystalline powder; decomposes in air; soluble in water; insoluble in alcohol.
Derivation: By dissolving copper sulfate in ammonium hydroxide and precipitating with alcohol.
Hazard: Toxic by ingestion.
Uses: Calico printing; manufacturing copper arsenate; insecticide; treating fiber products.

copper sulfate, tribasic $CuSO_4 \cdot 3Cu(OH)_2 \cdot H_2O$.
Properties: Aqua colored powder of extremely fine particle size; water-insoluble; stable in storage; forms essentially neutral water dispersion.
Hazard: Toxic by ingestion.
Uses: A fixed copper fungicide. Also micronutrient for plants. Compatible with DDT, arsenicals, organic insecticides, sulfur and cryolite. Used as spray or dust. Does not inhibit photosynthesis.

copper sulfide (cupric sulfide) CuS.
Properties: Black powder or lumps. Soluble in nitric acid; insoluble in water. Occurs as the mineral covellite. Sp. gr. 3.9–4.6; decomposes 220°C.
Derivation: By passing hydrogen sulfide gas into a solution of a copper salt.
Hazard: Toxic by ingestion.
Uses: Antifouling paints; dyeing with aniline black.
See also cuprous sulfide.

copper sulfocarbolate. See copper phenolsulfonate.

copper sulfocyanide. See cuprous thiocyanate.

copper tallate. See soaps (2).

copper trifluoroacetylacetonate $Cu[OC(CH_3):CHCO(CF_3)]_2$.
Properties: Solid; m.p. 188–190°C.
Hazard: May be toxic.
Uses: Metal analysis standards, vapor phase deposition of metals, laser studies.

copper tungstate (cupric tungstate) $CuWO_4 \cdot 2H_2O$.
Properties: Light-green powder; soluble in ammonium hydroxide; slightly soluble in acetic acid; insoluble in alcohol and water.

copper, yellow. See chalcopyrite.

copper zinc chromate. Variable in composition. Used as a fungicide.

"Coppralyte."[28] Trademark for a group of products for electroplating copper.

"Coprantine."[443] Trademark for dyes requiring after-treatment with copper compounds.

copra oil. The name applied to lower grades of coconut oil (q.v.).

"Co-Ral."[181] Trademark for O,O-diethyl-O-3-chloro-4-methyl-2-oxo-2H-1-benzopyran-7-yl phosphorothioate.

"Coral."[278] Trademark for a stereospecific polyisoprene rubber consisting essentially of cis-1, 4-polyisoprene.
Properties: Similar to those of natural rubber, both unvulcanized and vulcanized.
Use: Replacement for natural rubber.

coral. Skeletons of the coral polyps found in the warmer oceans and consisting mainly of calcium carbonate colored with ferric oxide. Coral rock is porous. It occurs in the form of reefs east of Australia, and at other locations in the southwest Pacific. It has been found to be the habitat of organisms that are rich in prostaglandins (q.v.).
Hazard: Toxic; will infect open wounds.
Uses: Building stone; cement; road and air field construction; jewelry, etc.

"Coramine."[305] Trademark for nikethamide.

"Coray."[51] Trademark for a series of general-purpose naphthenic base oils that serve as lubricants, in process applications, and as components in proprietary formulations.

"Corcel" 46[19a] Trademark for cellular glass microspheres in the form of fine white powder, obtained by beneficiation of a natural siliceous earth; has high water absorption value, low thermal conductivity, bulk density of 3 to 4.5 lb/cubic foot. Used as filler for plastics, thermal insulating material, admixture for concrete, surface coatings and catalyst carrier.

"Cordex."[74] Trademark for a cordage oil solution of copper naphthenate containing 8% copper metal.
Use: Fungicide.

cordite. A smokeless powder which is a mixture of nitrocellulose and nitroglycerin with about 5% petrolatum added to thicken and stabilize the mixture. Materials are dissolved in acetone and mixed. Evaporation of the excess acetone leaves a gelatinous mass which is extruded into cords.
Shipping regulations: (Rail, Air) Not listed. Consult regulations.

"Corephen 10."[470] Trademark for phenol formaldehyde condensation product; used as vehicle in highly acid-resistant paints.

"Coresinblak No. 3."[133] Trademark for a paste consisting of a jet black impingement carbon thoroughly dispersed in an alkyd resin. Composition: 18% carbon black, 32% alkyd resin, and 50% solvent. Compatible with most medium and short oil alkyds.
Containers: 40- and 400-lb metal drums.
Use: For making high grade enamels, both air dry and baking.

"Corexit."[29] Trademark for a nonionic surfactant designed to disperse oil slicks on sea water. Can be applied in spray form. Claimed to be nontoxic to marine organisms.

"CoRezyn."[463] Trademark for a variety of polymer resins tailored for specific uses, and for materials used with the polymers in making finished products.

"Corgard."[20] Trademark for a borosilicate glass armored with an opaque, filament-wound laminate of glass fiber and a modified polyester resin.

"Corial Bottom."[307] Trademark for a series of leather-finishing agents consisting of plasticized acrylic resin emulsion; 40–41% solids.

coriander oil.
Properties: An essential oil. Colorless or slightly yellowish liquid; aromatic odor; warm, spicy taste. Soluble in alcohol, ether, and chloroform. Sp. gr. 0.863–0.878; refractive index 1.4665 (20°C); optical rotation +8 to +15°. May be irritating when ingested. Combustible.
Chief known constituents: Linalool; pinene.
Derivation: Distilled from the fruit of Coriandrum sativum.
Grades: Technical; U.S.P.; F.C.C.
Containers: Tins; glass bottles.
Use: Flavoring (gin).

coriandrol. See *d*-linalool.

Cori ester. See glucose 1-phosphate.

cork. A form of cellulose comprising the light outer bark of the oak tree known as Quercus suber. It grows naturally in southern Europe and northern Africa and has been cultivated in southwestern U.S. Its special properties are extreme lightness, relative imperviousness to water, resilient structure, and low rate of heat transfer. These account for its usefulness as bottle stoppers, insulation, wallboard, life preservers, gaskets, and sound-deadening insertions. Specific gravity about 0.4. Combustible; nontoxic.

corkboard. A mixture of ground cork and paper pulp formed into thick sheets for insulating purposes.

"Corlar."[28] Trademark for chemical resistant finishes having a base of polyamide catalyzed epoxy resins.

corn oil (maize oil).
Properties: Pale yellow liquid; characteristic taste

and odor. Insoluble in water; soluble in ether, chloroform, amyl acetate, benzene, and carbon disulfide and slightly soluble in alcohol. Sp. gr. 0.914–0.921; saponification value 188–193; iodine value 102–128. Flash point 490°F. Combustible. Nontoxic and nondrying. Moderate tendency to spontaneous heating.
Chief constituents: linoleic and oleic acids (unsaturated), palmitic and stearic (saturated).
Derivation: The germ of common corn (Indian corn, Zea mays) is removed from the grain and pressed.
Grades: Crude; refined; U.S.P.; technical.
Containers: Drums; tank cars.
Uses: Foodstuffs; soap; lubricants; leather dressing; factice; margarine; salad oil; hair dressing; solvent.

cornstarch. A carbohydrate polymer derived from corn of various types; composed of 25% amylose and 75% amylopectin; a white powder which swells in water. It is the most widely used starch in the U.S. The so-called waxy variety (made from waxy corn) contains only branched amylopectin molecules. Its chief uses are as a source of glucose; in the food industry as a filler in baking powder and thickening agent in various food products and in adhesives and coatings. It has been proposed as an additive to plastics to promote rapid degradation in such products as bottles and waste containers. See also starch.

corn steep liquor. The dilute aqueous solution obtained by soaking corn kernels in warm 0.2% sulfur dioxide solution for 48 hours as the first step in the recovery of corn starch, corn oil, and gluten from corn. The solution contains mineral matter as well as soluble organic material extracted from the corn. It is used as a growth medium for penicillin and other antibiotics, and it is also concentrated and used as an ingredient of cattle feeds.

corn sugar. See dextrose.

corn syrup. See glucose syrup.

"Corobex."[159] Trademark for a series of organotin salt compounds, phenylmercuric salt compounds and quaternary compounds used as bacteriostatic and fungistatic finish in the textile, plastics, and rubber industries.
Hazard: Probably toxic by ingestion.

corona. An electrical discharge effect which causes ionization of oxygen and the formation of ozone. It is particularly evident near high-tension wires and in spark-ignited automotive engines. The ozone formed can have a drastic oxidizing effect on wire insulation, cable covers and hose connections; for this reason such accessories are made of oxidation-resistant materials such as nylon, neoprene, and other synthetics.

"Corragel."[99] Trademark for a colloidal clay that functions as a unique starch adhesive additive.

corresponding states (reduced states). Two substances are in corresponding states when their pressures, volumes (or densities) and temperatures are proportional, respectively, to their critical pressures, volumes (or densities) and temperatures. If any two of these ratios are equal, the third must also be equal. This principle has been useful in the development of physical and thermal properties of substances.

corrosion. (1) The electrochemical degradation of metals or alloys due to reaction with their environment; it is accelerated by the presence of acids or bases. In general, the corrodability of a metal or alloy depends upon its position in the activity series (q.v.). Corrosion products often take the form of metallic oxides; in the case of aluminum and stainless steel, this is actually beneficial, for the oxide forms a strongly adherent coating which effectively prevents further degradation. Hence these metals are widely used for structural purposes. The rusting of iron is a familiar example of corrosion, which is catalyzed by moisture. Acidic soils are highly corrosive; sulfur is a corrosive agent in automotive fuels and in the atmosphere (as SO_2). Sodium chloride in the air at locations near the sea is also strongly corrosive, especially at temperatures above 70°F. Copper, nickel, chromium, and zinc are among the more corrosion-resistant metals, and are widely used as protective coatings for other metals. (See also tarnish). Excellent corrosion-resistant alloys are stainless steel (18 Ni-8Cr), monel metal (66 Ni-34Cu) and duralumin. See also protective coatings; paints.

(2) The destruction of body tissues by strong acids and bases. See corrosive material.

"Corrosion Inhibitor CS."[108] Trademark for a synergistic combination of sodium nitrate-borax and organic inhibitors; used to prevent corrosion of ferrous and non-ferrous metal and alloy surfaces in low-makeup closed cooling and heating systems.

"Corrosion Inhibitor NPA."[219] Trademark for nonylphenoxyacetic acid.

corrosive sublimate. Obsolete term for mercuric chloride.

corrosive material. Any solid, liquid, or gaseous substance that burns, irritates, or destructively attacks organic tissues, most notably the skin and, when taken internally, the lungs and stomach. Among the more widely used chemicals that have corrosive properties are the following:

acetic acid, glacial	hydrofluoric acid
acetic anhydride	nitric acid
bromine	potassium hydroxide
chlorine	sodium hydroxide
fluorine	sulfuric acid
hydrochloric acid	

See also toxic materials

"Corrosol."[526] Trademark for phosphoric acid-type metal conditioners and rust removers used to remove rust from steel and prepare it for further processing or prepare normally nonreceptive zinc and aluminum surfaces for paint.

"Cortate."[321] Trademark for deoxycorticosterone acetate.

"Cortef."[327] Trademark for hydrocortisone.

corticoid hormone. A hormone produced or isolated from the cortex (external layer) of the adrenal gland. Corticoid hormones now used in medicine include cortisone, hydrocortisone, deoxycorticosterone, fludrocortisone, prednisone, prednisolone, methyl prednisolone, triamcinolone, dexamethasone, corticotropin (ACTH), and aldosterone. Some occur naturally in adrenal extract; others are modifications of the natural hormones. All are now made synthetically. They are derivatives of cyclopentanophenanthrene.
See also cortisone, ACTH.

corticosterone $C_{21}H_{30}O_4$. One of the less active adrenal cortical steroid hormones.
Properties: Crystalline plates; m.p. 180–182°C. Soluble in organic solvents; insoluble in water.
Derivation: Isolation from adrenal cortex extract; synthesis from deoxycholic acid.
Use: Biochemical research.

corticotropin. See ACTH.

"Cortifan."[321] Trademark for hydrocortisone (q.v.).

cortisol. See hydrocortisone.

cortisone (11-dehydro-17-hydroxycorticosterone) $C_{21}H_{28}O_5$. An adrenal cortical steroid hormone. It affects carbohydrate and protein metabolism.
Properties: White crystalline solid; m.p. 220–224° (dec). Dextrorotatory in solutions. Slightly soluble in water; sparingly soluble in ether, benzene, and chloroform; fairly soluble in methanol, ethanol, and acetone.
Derivation: From adrenal gland extract (usually from cattle) (historical method); synthetically, from bile acids, from other steroids or sapogenins.
Hazard: May have damaging side effects, e.g., sodium retention.
Use: Medicine (usually as acetate salt) in treatment of acute arthritis, Addison's disease, inflammatory diseases of eyes and skin, and others.

"Cortogen" Acetate.[321] Trademark for cortisone acetate.

"Cortone."[123] Trademark for cortisone (q.v.).

"Cortril."[299] Trademark for hydrocortisone (q.v.).

corundum (emery) Al_2O_3. Natural aluminum oxide, sometimes with small amounts of iron, magnesium, silica, etc.
Occurrence: New York; Greece; Asia Minor.
Hazard: Slight, by inhalation. A nuisance dust.
Use: Various polishing and abrasive operations; grinding wheels.
See also aluminum oxide; diaspore; sapphire.

corynine. See yohimbine.

cosmetic. Any preparation in the form of a liquid, semi-liquid, paste, or powder applied to the skin to improve its appearance, and for cleaning, softening or protecting the skin or its adjuncts, but without specific medicinal or curative effects; they include hairsprays, shampoos, nail polish, deodorants, shaving creams, facial creams, dusting powders, rouge, etc. Detergents such as common soap and bactericidal agents are not themselves classed as cosmetics, though they may be components of cosmetic mixtures.
A partial list of cosmetic ingredients is the following:

- animal fats (lanolin)
- vegetable oils, waxes
- alcohols (glycerol, glycols)
- surfactants (alkyl sulfonates)
- UV blocking agents (PABA)
- pnehylene diamine
- aluminum chlorohydrate
- FDC organic dyes
- talc (magnesium silicate)
- essential oils
- inorganic pigments
- chlorophyllins
- nitrocellulose lacquers
- steroid hormones

Knowledge of the structure and function of the skin is essential for proper cosmetic formulation (cosmeticology). All ingredients must be tested for possible toxic effects, since the skin is an important means of access for poisons. Addition of proteins to cosmetic preparations is of questionable value.

"Cosmo."[528] Trademark for imitation flavors made from essential oils and synthetic chemicals with alcohol and propylene glycol as solvents.

cosmochemistry. See astrochemistry; chemical planetology.

"Cosol."[21] Trademark for high-boiling coal-tar solvents for use in alkyd resin enamels and synthetic lacquers.

"Costyreneblak."[133] Trademark for a series of carbon black-polystyrene dispersions. Used for coloring polystyrene scrap. Available in chip form in several grades.

cotton. Staple fibers, surrounding the seeds of various species of Gossypium. Both Egyption and Sea Island cotton have unusually long staple (about 2 inches). Cotton is the major textile fiber and an important source of cellulose, which constitutes 88–96% of the fiber. So-called "absorbent cotton" is almost pure cellulose.
Properties: Tenacity, 3 to 6 g per denier (dry), 4 to 8 g per denier (wet); elongation 3–7%; sp. gr. 1.54; moisture regain 7% (70°F, 65% relative humidity); yellows slowly at 250°F, decomposes about 300°F; low permanent set; decomposed by acids; swells in caustic but is undamaged. Soluble in cuprammonium hydroxide. Subject to mildew. May be dyed by direct, vat, azoic, sulfur, and basic dyes. Combustible.
Sources: United States; Brazil; Egypt; India.
Hazard: Moderately toxic and flammable in the form of dust or linters; fiber ignites readily. Tolerance (dust), 0.2 mg per cubic meter of air.
Uses: Apparel; industrial and household fabrics; automobile tires; upholstery; medicine.
Shipping regulations: (as linters, hulls, or oily waste) (Rail) Yellow label. Not acceptable by express. (Air) Not acceptable. See also cellulose.

cotton, acetylated. Cotton fibers, threads, or fabrics treated with acetic anhydride, acetic acid, and perchloric and catalyst to improve the heat, rot, and mildew resistance by forming a surface coating of cellulose acetae.
Hazard: Ignites readily; not self-extinguishing.

cotton, aminized. A cotton fabric produced by reacting 2-aminoethylsulfuric acid with the cellulose of the fabric in a strongly alkaline solution. The treated cotton can take acid wool dyes and can be made rot-resistant and water-repellent.

cotton, cyanoethylated. Cotton treated with acrylonitrile; it is passed through a caustic bath, which induces mild swelling of the fiber and catalyzes the subsequent reaction with acrylonitrile. The fabric is then neutralized with acetic acid, washed and dried. The treatment leaves 3–5% nitrogen attached to the cellulose polymer. The cyanoethylated fiber is claimed to have permanent rot- and mildew-resistance, greater retention of strength after exposure to heat, improved receptiveness to dyes, and higher abrasion- and stretch-resistance.

cotton linters. Short fleecy fibers which adhere to cottonseed after it has been passed once through a cotton-gin. They are removed from the seed by a second ginning.

Hazard: Flammable, dangerous fire risk.
Uses: Rayon manufacture; cellulosic plastics; nitrocellulose lacquers; soil-cement binder in road construction; explosives.
Shipping regulations: (Rail, Air) Consult regulations.

cotton, mercerized. Cotton which has been strengthened by passing through a 25 to 30% solution of sodium hydroxide under tension, and then washed with water while under tension. This causes the fibers to shrink, increases their strength and attraction for colors, as well as imparting luster. A process using liquid ammonia for this purpose has been introduced in the U.K.

cotton oil (cotton spraying oil). A compounded oil sprayed (in the form of a fine-mist) onto cotton to condition the fibers for yarn-making operations. Used to lubricate the fibers, to reduce static, "fly," and dust and generally improve the suppleness and strength of the fibers.

cottonseed. The seed of the cotton plant, Gossypium hirsutum. It contains about 22% crude fiber, 20% protein, 20% oil, 10% moisture, and 24% nitrogen-free extract; also contains from 1 to 2% of toxic pigment gossypium (q.v.); specially processed kernels are free of this principle.
Hazard: Generates dangerous amounts of heat if piled or stored wet or hot. Powerful allergen; may cause asthma and other respiratory difficulty on inhalation.
Uses: Source of cottonseed oil and meal; cotton linters; source of nutritional protein after removal of gossypol by centrifugation.
Shipping regulations: (Air) Consult regulations.

cottonseed meal (cottonseed cake). The pulverized cottonseed press cake. Depending on the extractive process, varying percentages of protein will remain in the meal and it is normally sold with 36 to 45% protein content. The 42% product contains approximately 42% crude protein; 6% crude fiber; 25% nitrogen-free extract; 10% ether extract (fat) and 7% ash. The total digestible nutrient averages 79%. The ash is high in potash and phosphate; some types contain the toxic principle gossypium (q.v.).
Containers: Bulk or bags.
Uses: Animal feeds; fertilizer ingredient; filler for plastics.

cottonseed oil.
Properties: Pale yellow or yellowish-brown to dark ruby-red or black-red, fixed, semi-drying oil depending on the nature and condition of the seed. The pure oil is odorless and has a bland taste. Soluble in ether, benzene, chloroform and carbon disulfide; slightly soluble in alcohol. Sp. gr. 0.915–0.921; solidification range 31–35°C; saponification value 190–198; iodine value 109–116. Flash point 486°F. Combustible.
Chief constituents: Glycerides of palmitic, oleic and linoleic acids.
Derivation: From cotton seeds by hot-pressing or solvent extraction.
Grades: Crude; refined; prime summer yellow; bleachable; U.S.P.
Containers: Drums; tank cars.
Uses: Leather dressing; soap stock; lubricant; glycerol; base for cosmetic creams; hydrogenated to semisolid for use in food products; waterproofing compositions.

Cottrell, Frederick G. (1877–1948). A native of California, Cottrell obtained his doctorate from Liebig in 1902. His major contribution to industrial chemistry was his discovery of a practical method of dust elimination by electrical precipitation. Used in factory stacks and other large units, this process has contributed greatly to purifying the atmosphere of industrial areas. The principle involves charging a suspended wire with electricity. This creates a field which ionizes the surrounding air, the particles assuming the charge on contact and then moving to the wall of the stack where they are electrically discharged and precipitated.

coumaphos. Generic name for O,O-diethyl-O-(3-chloro-4-methyl-2-oxo-2H-1-benzopyran-7-yl)phosphorothioate; $C_9H_3O_2(CH_3)ClOPS(OC_2H_5)_2$.
Properties: Crystals; m.p. 95°C. Insoluble in water; soluble in aromatic solvents. Cholinesterase inhibitor.
Hazard: Toxic. Use may be restricted.
Uses: Insecticide; anthelmintic.

coumarin (cumarin; benzopyrone; tonka bean camphor) $C_9H_6O_2$. A lactone.
Properties: Colorless crystals, flakes or powder; fragrant odor similar to vanilla; bitter, aromatic burning taste; m.p. 69°C; b.p. 290°C. Soluble in 10 vols of 95% alcohol, in ether, chloroform, and fixed volatile oils; slightly soluble in water. Combustible.
Derivation: (a) By heating salicylic aldehyde, sodium lactate and acetic anhydride; (b) fine grades are isolated from Tonka beans.
Hazard: Toxic. Use in food products prohibited (FDA).
Uses: Deodorizing and odor-enhancing agent; pharmaceutical preparations.

coumarone (benzofuran). C_8H_6O. A bicyclic ring compound derived from coal-tar naphtha; the parent substance of coumarone-indene resins (q.v.).

coumarone-indene resin. A thermosetting resin derived by heating a mixture of coumarone and indene (q.v.) with sulfuric acid, which induces polymerization. At room temperature it is soft and sticky; on heating it hardens to a resinous solid. Soluble in hydrocarbon solvents, pyridine, acetone, carbon disulfide and carbon tetrachloride; insoluble in water and alcohol. Combustible; nontoxic. Said to have been the first synthetic polymer.
Uses: Adhesives, printing inks, floor tile binder, friction tape. See also "Cumar," "Nevidene," "Paradene."

countercurrent. Descriptive of a process in which a liquid and a vapor stream, or two streams of immiscible liquids or a liquid and a solid, are caused to flow in opposite directions and past or through one another with more or less intimate contact, so that the individual substances present are more or less completely transferred to that stream in which they are more soluble or stable under the conditions existing. The streams leaving such a process are usually of higher purity that can be attained otherwise at equal cost (see extraction, liquid-liquid). Distillation with a fractionating column is also a typical countercurrent process, in which rising vapor is purified by contact with descending liquid (reflux). Leaching, washing and chemical reaction are frequently carried out in a countercurrent manner.

count. (1) The external indication given by a radiation detector such as a Geiger counter of the amount of radioactivity to which the detector is exposed. The background counts are those which come from a source external to that being measured.

(2) The number of warp and filler threads in a linear inch of a textile fabric, e.g., the count of a sheeting may be 80 × 60.

coupling. (1) The combination of an amine or phenol with a diazonium compound to give an azo compound, the reaction of which azo dyes are prepared; thus meta-phenylenediamine $C_6H_4(NH_2)_2$ couples with benzene diazonium chloride $C_6H_5N_2Cl$ to produce the dye chrysoidine $C_6H_5N_2C_6H_3(NH_2)_2$.

(2) Oxidative coupling (q.v.).

(3) An agent, e.g., a vinyl silane, used to protect fiberglass laminates from effects of water absorption.

(4) A condensation polymerization of amino acids to form proteins; it can be done synthetically only by suppressing certain active sites on the amino acid molecules.

covalent bond. See bond, chemical.

"Covarnishblak."[133] Trademark for a series of dry powders consisting of carbon black dispersed in a hydrocarbon fossil resin. Used for pigmenting alkyd and oleoresinous enamels and coloring rubber cements. Can be dissolved in the vehicle by simple stirring.

covering power. See opacity.

Cox chart. A special semilogarithmic plot of vapor pressure vs. temperature especially useful for the petroleum hydrocarbons. The graph corresponding to each separate hydrocarbon is a straight line. All the lines appear to intersect at a point outside the chart.

C.P. Abbreviation for chemically pure, an accepted grade of drugs and fine chemicals that contain no detectable impurities.

"CP-40."[62] Trademark for chlorination derivatives of paraffin wax.

Uses: Rendering fabrics waterproof and fire retardant; fire-retardant paints; plasticizer and extender for certain plastic materials; etc.

"CPA-1800."[244] Trademark for a compound containing fluorine compounds and trivalent chromium.

Uses: In the electroplating industry as an additive to chromium plating solutions, acts as catalyst to improve performance.

"C-P-B."[248] Trademark for dibutyl xanthogen disulfide.

Properties: Amber-colored, free-flowing liquid; sp. gr. 1.15; soluble in acetone, benzene, gasoline, and ethylene dichloride; insoluble in water.

Use: Accelerator for pure gum hand-made druggists sundries and medical supplies, bathing shoes, bathing caps, novelties, and cold-cure cements.

"CP" Bond.[30] Trademark for custom-tailored dry polyvinyl alcohol-based adhesive blends.

"CPDA."[505] Trademark for the dianhydride of 1,2,3,4-cyclopentanetetracarboxylic acid. Available in both granular and fine ground forms.

CPR. Abbreviation for cyclonene-pyrethrin-rotenone. Applied to various insecticide formulations containing as active ingredients approximately 10 parts piperonyl cyclonene, 5 parts rotenone, and 1 part pyrethrin.

"CPTA."[505] Trademark for 1,2,3,4-cyclopentanecarboxylic acid.

Cr Symbol for chromium.

"CR-39."[177] Trademark for allyl diglycol carbonate, an optical plastic.

Properties: Clear optical plastic highly resistant to impact and abrasion. Furnished as a clear liquid. Thermosetting.

Uses: Ophthalmic lenses, shields, instrument panels, marine and aircraft glazing.

cracking. The decomposition by heat, with or without a catalyst, of petroleum or heavy petroleum fractions, with production of gasoline, fuel oil, or other products. If no catalysts are used the process is called thermal cracking. Steam cracking of naphtha to ethylene and other petrochemicals is widely practiced. The term is also applied to other thermal decomposition processes; thus ammonia (NH_3) may be cracked to nitrogen and hydrogen, and natural gas hydrocarbons such as methane (CH_4) are cracked into carbon and hydrogen, or into other hydrocarbons.

See also hydrocracking, reforming, catalysis.

"Crag."[214] Trademark for agricultural chemicals including:

(1) Fly Repellent (active ingredient, butoxy polypropylene glycol). Colorless liquid, 100% active material.

(2) Fungicide 974 (active ingredient, 3,5-dimethyltetrahydro-1,3,5-2H-thiadiazine-2-thione). Wettable powder, 85% active material. Toxic and irritant.

(3) Glyodin Solution (active ingredient, 2-heptadecyl glyoxaldine acetate) 34% active solution.

(4) Herbicide-1 (SES) (active ingredient, sodium 2,4-dichlorophenoxyethyl sulfate). Water-soluble powder, 90% active material. Tolerance, 10 mg per cubic meter of air.

crambe seed oil. A vegetable oil obtained from the seeds of the crambe, a plant related to mustard and rape. Growth and processing techniques have been studied by the Agricultural Research Service of USDA. The plant can be grown on marginal and strip-mined land; its high content of erucic acid and the high protein content of its meal make it economically attractive; the oil itself is useful as an industrial lubricant, especially for molds for continuous steel casting.

cream of tartar. See potassium bitartrate.

creatine (N-methyl-N-guanylglycine; (alpha-methylguanido)acetic acid) $HN{:}C(NH_2)N(CH_3)CH_2COOH$. A nitrogenous acid widely distributed in the muscular tissue of the body.

Properties: (monohydrate) Prisms from water; anhydrous at 100°C; decomposes 303°C; slightly soluble in water; insoluble in ether.

Source: Commercially isolated from meat extracts.

Grades: Technical; C.P.

Use: Biochemical research.

creatinine $C_4H_7N_3O$. The anhydride of creatine (q.v.); a metabolic waste product.

Properties: Leaflets from water; decomposes about 270°C; soluble in water; slightly soluble in alcohol; nearly insoluble in ether, acetone, and chloroform.

Use: Biochemical research.

creep. See cold flow.

creosol (2-methoxy-4-methylphenol; 4-methylguaiacol) $CH_3O(CH_3)C_6H_3OH$. One of the active constituents of creosote, derived from beechwood tar.

Properties: Colorless to yellow liquid. Sp. gr. 1.092 (25°C); m.p. 5.5°C; b.p. 220°C; slightly soluble in water; soluble in alcohol, benzene, chloroform, ether, acetic acid.

creosote carbonate.
Properties: Clear, colorless or yellowish, viscous liquid; slight creosote odor and taste. Soluble in alcohol; insoluble in water.
Derivation: Mixture of carbonates of various constituents of creosote.
Use: Medicine.

creosote, coal-tar (creosote oil; liquid pitch oil; tar oil).
Properties: Yellowish to dark green-brown, oily liquid; clear at 38°C or higher; naphthenic odor; frequently contains substantial amounts of naphthalene and anthracene; distilling range 200–400°C; flash point 165°F (closed cup); soluble in alcohol, benzene, and toluene; immiscible with water. Combustible.
Derivation: Fractional distillation of coal-tar.
Method of purification: Rectification.
Grades: Technical; crude; refined.
Containers: Drums; tank cars.
Hazard: Toxic by inhalation of fumes; skin and eye irritant. Use may be restricted.
Uses: Wood preservative (ties, telephone poles, marine piling, etc.); disinfectants.

para-**cresidine.** See 5-methyl-ortho-anisidine.

"Creslan."[57] Trademark for an acrylic fiber.

cresol (methyl phenol; hydroxymethylbenzene) $CH_3C_6H_4OH$. A mixture of isomers obtained from coal tar or petroleum.
Properties: Colorless, yellowish, or pinkish liquid; phenolic odor; sp. gr. 1.030–1.047; wt/gal 8.66–8.68 lb; flash point approx. 180°F; m.p. 11–35°C; b.p. 191-203°C. Soluble in alcohol, glycol, and dilute alkalies. Combustible.
Derivation: Coal tar (from coke and gas works); also synthetic.
Grades: Various, depending on phenol content, or other properties. N.F. grade contains not more than 5% phenol.
Containers: Drums; tank cars; tank trucks.
Hazard: Toxic and irritant; corrosive to skin and mucous membranes; absorbed through skin. Tolerance, 5 ppm in air. Safety data sheet available from Manufacturing Chemists Assn., Washington, D.C.
Uses: Disinfectant; phenolic resins; tricresyl phosphate; ore flotation; textile scouring agent; organic intermediate, mfg. of salicylaldehyde, coumarin, and herbicides; surfactant; synthetic food flavors (para isomer only).
Shipping regulations: (Air) Poison label.
See also cresylic acids.

meta-**cresol** (meta-cresylic acid; 3-methylphenol) $CH_3C_6H_4OH$.
Properties: Colorless to yellowish liquid; phenol-like odor. Soluble in alcohol, ether, and chloroform; soluble in water. Sp. gr. 1.034; m.p. 12°C; b.p. 203°C; wt/gal 8.66 lb flash point 187°F. Combustible. Autoignition temp. 1038°F.
Derivation: By fractional distillation of crude cresol (from coal tar); also synthetically.
Method of purification: Rectification.
Grade: Technical (95–98%).
Toxicity, uses, see cresol.

ortho-**cresol** (ortho-cresylic acid; 2-methylphenol) $CH_3C_6H_4OH$.
Properties: White crystals; phenol-like odor. Soluble in alcohol, ether, chloroform, and hot water. Sp. gr. 1.047; m.p. 30.9°C; flash point 178°F. Autoignition temp. 1110°F; b.p. 191°C; lb/gal 8.68. Combustible.
Derivation: (a) By fractional distillation of crude cresol from coal tar. (b) Interaction of methanol and phenol.
Method of purification: Crystallization.
Grades: According to freezing point: 25°, 29°, 30°, 30.5°C, etc.
Toxicity, uses, see cresol.

para-**cresol** (para-cresylic acid; 4-methylphenol) $CH_3C_6H_4OH$.
Properties: Crystalline mass; phenol-like odor. Soluble in alcohol, ether, chloroform, and hot water; wt/gal 8.67 lb. Sp. gr. 1.039; b.p. 202°C; m.p. 35.26°C; flash point 187°F. Combustible. Autoignition temp. 1038°F.
Derivation: (a) By fractional distillation of crude cresol. (b) From benzene by the cumene process (see phenol).
Method of purification: Crystallization.
Grades: Technical; 98%; 99.0% min purity or 34°C min F.P.
Toxicity, uses, see cresol.

cresolphthalein $C_6H_4COOC(C_6H_3(OH)CH_3)_2$. An acid-base indicator, changes from colorless to red between pH 8.2 and 9.8. See also indicator.

cresol purple $C_6H_4SO_2OC(C_6H_3(OH)CH_3)_2$. Meta-cresolsulfonphthalein, an acid-base indicator, showing color change from red to yellow over the range pH 1.2 to 2.8 and from yellow to purple over the range pH 7.4 to 9.0. See also indicator.

cresol red $C_6H_4SO_2OC(C_6H_3(OH)CH_3)_2$. Ortho-cresolsulfonphthalein, an acid-base indicator, changes from red to yellow between pH 0.2 and 1.8 and from yellow to red between pH 7.0 and 8.8. See also indicator.

cresotic acid (cresotinic acid; hydroxytoluic acid) $CH_3C_6H_3(OH)COOH$. Ten possible isomers; most common is 2-hydroxy-3-methylbenzoic acid, also known as ortho-cresotic acid or ortho-homosalicylic acid. The description which follows is of this isomer.
Properties: White crystals or powder; m.p. 166°C; b.p. about 250°C; insoluble in water; soluble in alcohol and ether. Combustible.
Derivation: Treatment of ortho-cresol with caustic and carbon dioxide under pressure.
Containers: Fiber cans; drums.
Uses: Dye intermediate; research on plant growth inhibition.

"Crestalkyd."[263] Trademark for a group of oil-modified alkyd resins. Oil length from 30% to 80%, A.V. < 15.
Uses: High-quality paints, lacquers, enamels, and varnishes. Plasticizing resin for nitrocellulose and similar finishes.

"Crestapol."[263] Trademark for a series of plasticizers for polyvinyl chloride; also a designation for specialty polyesters used in dispersing media.

meta-cresyl acetate (meta-tolyl acetate) $CH_3C_6H_4OCOCH_3$.
Properties: Colorless oily liquid; odor similar to phenol; b.p. near 112°C; distils with steam; insoluble in water; soluble in common organic solvents. Combustible.
Use: Medicine.

ortho-**cresyl acetate** (ortho-tolyl acetate) $CH_3COOC_6H_4CH_3$.
Properties: Liquid; b.p. 83–85°C (10 mm); 208°C (760 mm). Nearly insoluble in cold, soluble in hot water; soluble in organic solvents. Combustible.
Use: Flavoring.

para-**cresyl acetate** (para-tolyl acetate) $CH_3C_6H_4OCOCH_3$.
Properties: Colorless liquid; floral odor. Soluble in 2.5 vols of 70% alcohol; in most fixed oils; insoluble in glycerol. Sp. gr. 1.0532 (15°C); optical rotation (100 mm) 0°; refractive index 1.500 to 1.504; acid value 0.7; ester value 341.6. Combustible.
Grades: Technical; F.C.C.
Uses: Perfumery; flavoring.

cresyl diglycol carbonate (diethylene glycol bis(cresylcarbonate)) $C_{20}H_{22}O_7$.
Properties: Colorless liquid of low volatility. Sp. gr. 1.19 (20/4°C); boiling point approximately 250°C (2 mm); flash point 475°C; refractive index 1.523 (20°C). Insoluble in water (very stable to hydrolysis). Widely soluble in organic solvents. Compatible with many resins and plastics. Combustible. Low toxicity.
Uses: Plasticizer.

cresyl diphenyl phosphate (cresyl phenyl phosphate) $(CH_3C_6H_4)(C_6H_5)_2PO_4$. Probably seldom a pure compound, but a mixture of ortho, meta, and para-cresyl and phenyl phosphates.
Properties: Clear transparent liquid; very slight odor; insoluble in water; soluble in most organic solvents except glycerol. Sp. gr. 1.20 (20/20°C); f.p. -38°C; boiling range 235–255°C (4 mm). Flash point 450°F. Combustible.
Uses: Plasticizer; extreme-pressure lubricant; hydraulic fluids; gasoline additive; food packaging.

ortho-**cresyl**-alpha-**glyceryl ether.** See mephenesin.

cresylic acids. Commercial mixtures of phenolic materials boiling above the cresol range. An arbitrary standard in use for cresylic acids is that 50% must boil above 204°C. If the boiling point is below 204°C, the material is called cresol (q.v.). Cresylic acid varies widely according to its source and boiling range.

A typical commercial cut, b.p. 220–250°C, has the composition meta, para-cresols, 0–1%; 2,4- and 2,5-xylenols, 0–3%; 2,3- and 3,5-xylenols, 10–20%; 3,4-xylenol, 20–30%, and C_9 phenols, 50–60%. Excellent electrical insulators.
Derivation: Petroleum; coal tar. Imported cresylic acid is derived from coal tar (gas works); also made synthetically.
Containers: 5-, 55-gal drums; tank trucks; tank cars.
Hazard: Corrosive to skin; absorbed by skin.
Uses: Phosphate esters, phenolic resins, wire enamel solvent, plasticizers, gasoline additives, laminates, coatings for magnet wire for small electric motors. Disinfectants, metal-cleaning compounds, phenolic resins, flotation agents, surfactants, chemical intermediates, oil additives; solvent refining of lubricating oils, scouring compounds, pesticides.
Shipping regulations: (Air) Poison label.

para-**cresyl isobutyrate** (para-tolyl isobutyrate) $CH_3H_6H_4OCOCH(CH_3)_2$. Low toxicity.
Use: Flavoring.

cresyl phenyl phosphate. See cresyl diphenyl phosphate.

cresyl silicate $(CH_3C_6H_4O)_4Si$.
Properties: Colorless liquid; b.p. 450°C.
Derivation: Reaction of cresol and silicon tetrachloride.
Use: Heat-transfer fluid.

cresyl-para-**toluene sulfonate** (tolyl-para-toluene sulfonate) $CH_3C_6H_4SO_3C_6H_4CH_3$.
Properties: Brown, oily liquid; faint odor; sp. gr. 1.207; flash point 365°F; m.p. 68.70°C. Combustible.
Derivation: From reaction of para-toluenesulfonyl chloride with para-cresol.
Hazard: May be toxic.
Use: Plasticizer.

"C.R.I."[239] Trademark for a concentrated rust-inhibiting germicide containing 12.5% (by weight) alkyl (mostly C_{12}, C_{14}, C_{16} but ranging from C_8 to C_{18}) dimethylbenzylammonium chloride as active ingredient. Inert ingredients are 40% 2-hydroxypropylamine nitrite, water (47.5%), and Fastusol Turquoise Blue.

critical constant. A maximum or minimum value for a physical constant which is characteristic of a substance, e.g., the critical temperature of a gas is the temperature above which it cannot be liquefied by an increase in pressure.

criticality The state of a nuclear reactor when it is sustaining a chain reaction. (Atomic Energy Commission).

critical mass. The minimum mass of fissionable material (U-235 or plutonium) that will support a self-sustaining chain reaction. A critical assembly (nuclear reactor) is a system of fissionable material and moderator sufficient to sustain a chain reaction at a low and controllable power level. See also fission, nuclear.

critical solution temperature. The temperature above or below which two liquids are miscible in all proportions. Some pairs of liquids have both an upper and a lower critical solution temperature, that is, they can exist in two phases only in a medium temperature range.

crocein acid (croceic acid; Bayer's acid; 2-naphthol-8-sulfonic acid) $C_{10}H_6OHSO_3H$.
Derivation: Sulfonation of beta-naphthol at a low temperature and recrystallization from a salt.
Containers: 100-lb fiber drums (sodium salt).
Use: Dye intermediate.

Crocein Scarlet MOO. See Brilliant Crocein.

crocetin $C_{20}H_{24}O_4$. A dicarboxylic carotenoid derived from saffron.
Properties: Red rhomboid crystals; soluble in pyridine and dilute sodium hydroxide; slightly soluble in water and organic solvents; m.p. 285°C. Combustible; nontoxic.
Uses: Experimental treatment of arteriosclerosis by increasing oxygen diffusion through arterial walls, thus decreasing build-up of cholesterol.

"Crolyn." Trademark for chromium dioxide (q.v.).

"Cromophthal."[443] Trademark for pigments of the azo and vat classes.

"Cronar."[28] Trademark for a series of polyester photographic film.

cross-linking. Attachment of two chains of polymer molecules by bridges composed of either an element, a group, or a compound which join certain carbon atoms of the chains by primary chemical bonds, as indicated in the schematic diagram:

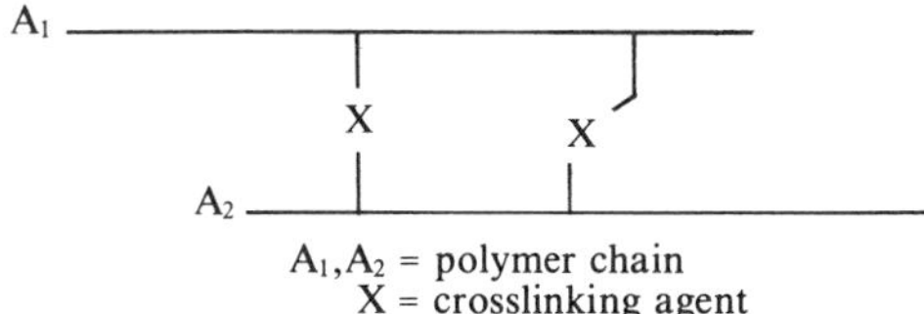

A_1,A_2 = polymer chain
X = crosslinking agent

Cross-linking occurs in nature in substances made up of polypeptide chains, which are joined by the disulfide bonds of the cystine residue, as in keratins, insulin, and other proteins. Polysaccharide molecules can also cross-link to form stable gel structures (dextran). This can also be effected artificially, either by addition of a chemical substance (cross-linking agent) and exposing the mixture to heat, or by subjecting the polymer to high-energy radiation. Examples are: (1) vulcanization of rubber with sulfur or organic peroxides; (2) cross-linking of polystyrene with divinylbenzene; (3) cross-linking of polyethylene by means of high-energy radiation or with an organic peroxide; (4) cross-linking of cellulose with dimethylol carbamate (10% solution) in durable-press cotton textiles. Cross-linking has the effect of changing a plastic from thermoplastic to thermosetting. Thus it also increases strength, heat- and electrical resistance, and especially resistance to solvents and other chemicals. See also vulcanization; polyethylene (cross-linked); keratin.

cross-section. (1) A measure of the probability that a nuclear reaction will occur. Usually measured in barns, it is the apparent (or effective) area presented by a target nucleus (or particle) to an oncoming particle or other nuclear radiation, such as a photon or gamma radiation. (Atomic Energy Commission). Also called capture cross-section.

(2) A section made by a plane cutting through a solid, usually at right angles. Tissue cross-sections are widely used for microscopic observation.

crotonaldehyde (2-butenal; crotonic aldehyde; beta-methyl acrolein) $CH_3CH{:}CHCHO$. Commercial crotonaldehyde is the trans isomer.
Properties: Water-white, mobile liquid; pungent, suffocating odor; turns to a pale yellow color in contact with light and air. A lachrymator. Very soluble in water; miscible in all proportions with alcohol, ether, benzene, toluene, kerosine, gasoline, solvent naphtha. Sp. gr. 0.8531 (20/20°C); b.p. 102°C; flash point 55°F; f.p. -69°C; vapor pressure 30 mm (20°C).
Derivation: Aldol condensation of two molecules of acetaldehyde.
Grades: Technical; 87% water-wet form.
Hazard: Flammable, dangerous fire risk. Toxic; especially irritating to the eyes and skin. Tolerance, 2 ppm in air.
Uses: Intermediate for n-butyl alcohol and 2-ethylhexyl alcohol; solvent; preparation of rubber accelerators; purification of lubricating oils; insecticides; tear gas; fuel-gas warning agent; organic synthesis; leather tanning; alcohol denaturant.
Shipping regulations: (Rail) Red label. (Air) Flammable Liquid label. Not acceptable on passenger planes.

crotonic acid (2-butenoic acid; beta-methacrylic acid) $CH_3CH{:}CHCOOH$. Exists in cis and trans isomeric forms, the latter being the stable isomer used commercially. The cis form melts at 15°C and is sometimes called isocrotonic acid.
Properties: White crystalline solid; sp. gr. 0.9730; m.p. 72°C; b.p. 185°C; soluble in water, ligroin, ethanol, toluene, acetone. Flash point (190°F (C.O.C.). Combustible.
Derivation: Oxidation of crotonaldehyde.
Grade: 97%.
Containers: Glass bottles; fiber drums.
Hazard: Strong irritant to tissue.
Uses: Synthesis of resins, polymers, plasticizers, drugs.
Shipping regulations: (Air) Corrosive label.

crotonic aldehyde. See crotonaldehyde.

croton oil (tiglium oil).
Properties: Brownish-yellow liquid. Sp. gr. 0.935–0.950 (25°C); refractive index (n 40/D) 1.470–1.473. Soluble in ether, chloroform and fixed or volatile oils; slightly soluble in alcohol.
Chief constituents: Glycerides of stearic, palmitic, myristic, lauric and oleic acids and croton resin, a vesicant.
Derivation: By expression from the seeds of Croton tiglium.
Hazard: Toxic; strong skin irritant; ingestion of small amounts may be fatal.
Use: Medicine.

crotonolic acid. See tiglic acid.

crotonylene (2-butyne; dimethylacetylene) $CH_3C \cdot CCH_3$.
Properties: Liquid; b.p. 27°C; flash point 64°F.
Hazard: Flammable, dangerous fire risk. Moderate explosion hazard.
Shipping regulations: (Rail) Flammable liquid, n.o.s., Red label. (Air) Flammable Liquid label.

crotyl alcohol (2-buten-1-ol, 3-methylallyl alcohol) $CH_3CH{:}CHCH_2OH$.
Properties: Clear, stable liquid; sp. gr. 0.8550 (20/20°C); boiling range 121–126°C; flash point (TOC) 113°F. Partially soluble in water (17%), wholly soluble in alcohol and ether. Combustible.
Hazard: Toxic by ingestion. Strong eye and skin irritant. Moderate fire risk.
Uses: Chemical intermediate; source of monomers; herbicide and soil fumigant.

crown filler. A mineral filler, usually calcium sulfate or carbonate or a mixture thereof used in paper manufacture.

crown glass. See glass, optical.

cryochemistry. That branch of chemistry devoted to the study of reactions occurring at extremely low temperatures (-200°C and lower). It permits synthesis of compounds that are too unstable or too reactive to exist at normal temperatures.

cryogenics. Study of the behavior of matter at temperatures below -200°C. The use of the liquefied gases, oxygen, nitrogen and hydrogen at about -260°C is standard industrial practice. Examples: Use of liquid nitrogen for quick-freezing of foods, and of liquid oxygen in steel production. Some electronic devices and specialized instruments, such as the cryogenic gyro, operate at liquid helium temperatures (about 4°K). Many lasers and computer circuits require low temperatures. See also superconductivity.

cryolite (Greenland spar, icetone) Na_3AlF_6. A natural fluoride of sodium and aluminum, or made synthetically from fluorspar, sulfuric acid, hydrated alumina and sodium carbonate.
Properties: Colorless to white; sometimes red, brown or black; luster vitreous to greasy; hardness 2.5; sp. gr. 2.95–3.0. Refractive index 1.338; m.p. 1000°C; soluble in fused aluminum and ferric salts.
Occurrence: Greenland (only commercial source); Colorado; U.S.S.R.
Derivation: Synthetic product is made by fusing sodium fluoride and aluminum fluoride.
Uses: Electrolyte in the reduction of alumina to aluminum; ceramics; insecticide; binder for abrasives; electric insulation; explosives; polishes.

"Cryovac."[311] Trademark for a light, shrink-film, transparent packaging material based on polyvinylidene chloride. Used especially for meats and other perishables.

cryptocyanine (1,1′-diethyl-4,4′-carbocyanine iodide) $C_{25}H_{25}N_2I$.
Properties: Solid; m.p. 250.5–253°C.
Use: Organic dye solution used as a chemical shutter in laser operation.
See also cyanine dye.

cryptopine $C_{21}H_{23}NO_5$.
Properties: White crystalline alkaloid. Soluble in chloroform and boiling alcohol; insoluble in water and ether. M.p. 217–221°C.
Derivation: From opium, by extraction and crystallization.
Method of purification: Recrystallization.
Hazard: Highly toxic. A narcotic drug.
Use: Medicine.

cryptoxanthin (provitamin A; hydroxy-beta-carotene) $C_{40}H_{56}O$. A carotenoid pigment with vitamin A activity.
Properties: Garnet-red prisms with metallic luster; m.p. 170°C; soluble in chloroform, benzene, and pyridine; slightly soluble in hydrocarbon solvents, alcohol and methanol.
Occurrence: In many plants, egg yolk, butter, blood serum.
Uses: Nutrition; medicine.

crystal. The normal form of the solid state of matter. Crystals have characteristic shapes and cleavage planes due to the arrangement of their atoms, ions, or molecules, which comprise a definite pattern called a lattice. Crystals may be face-centered, body-centered, cubic, orthorhombic, monoclinic, prismatic, etc.; they have flat surfaces, sharp edges, and a definite angle between a given pair of surfaces. The form of a crystal is called its "habit." One of the most important features of a crystal is its optical properties, chief of which is its index of refraction, i.e., the extent to which a beam of light is deflected on passing through the crystal. Depending on the manner of light transmission, a crystal may be isotropic or anisotropic (q.v.). Anisotropic crystals can polarize light (see also optical isomerism, optical rotation). Crystals also have electrical and magnetic properties now being used in computers and other electronic devices. Crystals are almost always imperfect and contain impurities (atoms of other elements). These are utilized in semiconductors (q.v.). For methods of growing crystals see nucleation.
Single crystals are used in masers, lasers, semiconductors, miniaturized components, computer memory systems, and as "whiskers" (q.v.). Many metals are now available in large single crystal form, and such natural crystals as ruby, garnet, sapphire, etc. are used in the foregoing applications. See also crystallization; nucleation; liquid crystal.

crystal, liquid. See liquid crystal.

crystallite. That portion of a crystal whose constituent atoms, ions, or molecules form a perfect lattice, without strains or other imperfections. Single crystals may be quite large, but crystallites are usually in the microscopic range. See also crystal.

crystallization. The phenomenon of crystal formation by nucleation (q.v.) and accretion. The freezing of water into ice is one of the commonest examples of crystallization in nature. Industrially, it is used as a means of purifying materials by evaporation and solidification. The sugar of commerce is made in this way. Similarly, salt cake is derived from crystallization of natural brines (Searles Lake). Nucleated crystallization is also used to form polycrystalline ceramic structures. See also crystal.

crystallography. The study of the crystal formation of solids, including x-ray determination of lattice structures, crystal habit, and the shape, form, and defects of crystals. When applied to metals, this science is called metallography (q.v.).

crystal violet. See methyl violet.

"Crystamet."[428] Trademark for sodium metasilicate pentahydrate.

"Crystarose"[188] Trademark for a highly purified grade of trichloromethylphenylcarbinyl acetate.

"Crystex."[1] Trademark for a rubber-insoluble sulfur used as vulcanizing agent in natural and synthetic rubbers; 85% of the sulfur is insoluble at the usual milling temperatures. This metastable form is converted to the stable soluble sulfur at usual vulcanizing temperatures.

"Crysticillin."[412] Trademark for procaine penicillin G.

"Crystic."[263] Trademark for a series of unsaturated polyester resins. Available in several grades as well as self-extinguishing and thixotropic types.
Use: Low-pressure laminating for glass fiber reinforced plastics, potting, casting, embedding, and coating.

"Crystodigin."[100] Trademark for digitoxin. U.S.P.

"Crystolon."[249] Trademark for silicon carbide.

cs. Abbreviation for centistoke.

Cs Symbol for cesium.

"CS-137."[304] Trademark for a barium-sodium organic-complex vinyl stabilizer.
Properties: Creamy-white paste, sp. gr. 1.54; refractive index 1.48.
Containers: Supplied as a 70% solids paste in DOP. Metal drums containing 60 lb net.
Uses: Stabilizer for transparent organosols and solution coatings to impart superior light and weathering resistance.

C_2S. Abbreviation for dicalcium silicate, as used in cement. See cement, Portland.

C_3S. Abbreviation for tricalcium silicate, as used in cement. See cement, Portland.

"CSC."[319] Trademark for choline chloride aqueous solutions and choline chloride feed supplements.

C-stage resin (resite). The fully cross-linked phenol-formaldehyde type resin which is infusible and insoluble in all solvents. See A-stage resin; phenol-formaldehyde resin.

CTC. Abbreviation for chlortetracycline (q.v.).

CTFE. See chlorotrifluoroethylene.

CTP. Abbreviation for cytidine triphosphate (q.v.).

Cu Symbol for copper.

cuam. Abbreviation for cuprammonium, the copper ammonium radical.

cube root. A powdered insecticidal preparation containing 5% rotenone (q.v.).

"Cubex."[308] Trademark for an oriented silicon-iron alloy in rolled sheet form for use as cores for transformers and other inductive devices. The alloy sheet comprises cube grains with faces parallel to the sheet surface. In one form the sheet is doubly oriented with two directions of easy magnetization parallel to the surface of the sheet. One direction of easy magnetization is parallel to the rolling direction and the second is perpendicular to the rolling direction.

cubic centimeter. See milliliter.

"Cubidow."[233] Trademark for compacted salt comprising either or both calcium and sodium chloride.

"Cubond."[296] Trademark for a copper brazing paste. Consists of metallic copper powder on cuprous oxide pigments of high purity in organic or petroleum vehicles which impart satisfactory suspension properties.

"Cubor Dusts."[147] Trademark for insecticides containing 0.75 and 1.0% pure rotenone.

cue-lure. Generic name for 4-(para-hydroxphenyl)-2-butanone acetate, $CH_3COCH_2CH_2C_6H_4OOCCH_3$. Liquid; boiling range 117–124°C. Insoluble in water; soluble in most organic solvents.
Use: Insect attractant.

cuen. Abbreviation for cupriethylenediamine.

cullet. Broken or waste glass added to the batch in glass manufacture.

"Culofix."[300] Trademark for dye fixatives of fatty or resin cationic type for application to dyed textiles.

cumaldehyde. See cuminic aldehyde.

"Cumar."[175] Trademark for a series of neutral, stable, synthetic resins of the coumarone-indene type, manufactured from selected distillates of tar. See also coumarone-indene resin.
Uses: Softener and tackifier in varnishes, floor tile, rubber products, printing ink, adhesives, and waterproofing materials; leather, electrical, radio, paper, and other industries.

"Cumate."[69] Trademark for preparation of copper dimethyldithiocarbamate $[(CH_3)_2NC(S)S]_2Cu$.
Properties: Dark brown powder, sp. gr. 1.75; melts above 325°C; moderately soluble in acetone, benzene, chloroform; insoluble in water, alcohol, gasoline.
Uses: In SBR, primary accelerator; secondary accelerator with thiazoles. In butyl rubber, primary accelerator. For molded and extruded goods.

cumene (isopropylbenzene) $C_6H_5CH(CH_3)_2$.
Properties: Colorless liquid. Soluble in alcohol, carbon tetrachloride, ether and benzene; insoluble in water. Sp. gr. 0.8620; b.p. 152.7°C; wt/gal 7.19 lb (25°C); freezing point -96°C; refractive index 1.489 (25°C); flash point 115°F. Autoignition temp. 795°F. Combustible.
Derivation: (a) Alkylation of benzene with propylene (phosphoric acid catalyst); (b) distillation from coal-tar naphtha fractions or from petroleum.
Grades: Technical; research; pure.
Containers: Drums; tank cars; tank trucks.
Hazard: Toxic by ingestion, inhalation, and skin absorption. Moderate fire risk. Tolerance, 50 ppm in air. Narcotic in high concentrations.
Uses: Production of phenol, acetone, and alpha-methylstyrene; solvent.

cumene hydroperoxide (alpha, alpha-dimethylbenzyl hydroperoxide) $C_6H_5C(CH_3)_2OOH$.
Properties: Colorless to pale yellow liquid, slightly soluble in water; readily soluble in alcohol, acetone, esters, hydrocarbons. Chlorinated hydrocarbons. Flash point 175°F. Combustible.
Derivation: A solution or emulsion of cumene is oxidized with air at about 130°C.
Hazard: Toxic by inhalation and skin absorption. Strong oxidizing agent; may ignite organic materials.
Uses: Production of acetone and phenol; polymerization catalyst, particularly in redox systems, used for rapid polymerization.
Shipping regulations: (Rail) Yellow label. (Air) (solution exceeding 96%) Not acceptable; (solution not exceeding 96% in nonvolatile solvent) Organic Peroxide label.

cumic alcohol. See cuminic alcohol.

cumic aldehyde. See cuminic aldehyde.

cumidine (para-isopropylaniline) $(CH_3)_2CHC_6H_4NH_2$.
Properties: Colorless liquid; sp. gr. 0.957 (20/4°C); f.p. -63°C; b.p. 225°C. Insoluble in water; soluble in alcohol and ether. See also pseudocumidine.

cuminic alcohol (para-isopropylbenzyl alcohol; cuminyl alcohol; cumic alcohol) $CH_2OH(C_6H_4)CH(CH_3)_2$. Found in caraway seed.
Properties: Colorless liquid; caraway-like odor; aromatic taste; sp. gr. 0.981 (15°C); b.p. 248°C; refractive index (n 24/D) 1.522. Insoluble in water; miscible with alcohol, ether. Low toxicity. Combustible.
Hazard: Moderate fire risk.
Use: Flavoring.

cuminic aldehyde (cumic aldehyde; cumaldehyde; para-isopropylbenzaldehyde) $(CH_3)_2CHC_6H_4CHO$.
Properties: Colorless to yellow liquid with a cumin odor; sp. gr. 0.986 (22°C); b.p. 235°C. Insoluble in water; soluble in alcohol and ether. Combustible.
Hazard: Moderate fire risk.
Use: Perfumery; flavoring.

cumin oil.
Properties: Colorless or yellowish, limpid liquid; characteristic odor of cumin; sharp, spicy taste. Soluble in alcohol, ether and chloroform. Sp. gr. 0.900–0.930; optical rotation +4° to +8°; refractive index 1.497–1.509 (20°C). Combustible; low toxicity.

Chief known constituents: Cumene; cumic aldehyde.
Derivation: Distilled from the fruit of Cuminum cyminum.
Grades: Technical; F.C.C.
Uses: Medicine; flavoring; perfumery.

cumulene. A chain of up to 6 double-bonded carbon atoms derived from acetylene.

cumyl phenol $C_6H_5C(CH_3)_2C_6H_4OH$.
Properties: White to tan crystals with characteristic phenol odor. Solidifying point 72.0°C; density 1.115 g/ml (25°C); distillation range 188.9–190.9°C (10 mm); flash point 320°F. Combustible.
Uses: Intermediate for resins, insecticides, lubricants.

"Cunife."[166] Trademark for a ductile permanent magnet material composed of 60% copper, 20% nickel and 20% iron. Forms available: wire, strip, finished magnets.

"Cunilate."[8] Trademark for "solubilized copper-8-quinolinolate." One type contains 10% copper-8-quinolinolate with 2-ethylhexoic acid and an aromatic solvent as carrier, with 40% total solids. Combustible.
Properties: Sp. gr. 0.9542–0.9545 at 77°F; flash point 110°F; color, greenish yellow; pH 5.5–6.0.
Containers: Steel drums.
Hazard: Moderate fire risk. Moderately toxic.
Uses: Fungicidal treatment of fabrics, wood (by immersion or pressure), cotton rope and other materials as well as in combination with water repellents and auxiliary chemicals.

"Cunimene."[8] Trademark for a series of metal complexes of dehydroabietyl amine 8-hydroxyquinolinium 2-ethylhexoate.
Uses: Fungicide and bactericide used in vinyl plasticizers, rubber, cordage fibers, textiles, etc.

"Cunisil-647."[324] Trademark for an alloy containing 97.50% copper, 1.90% nickel, and 0.60% silicon. This high-strength, corrosion-resistant alloy is available in round rod, with or without the final precipitation-hardening heat treatment. Used in electrical equipment.

cupellation process. A process for freeing silver, gold or other nonoxidizing metals from base metals which can be oxidized. The metallic mixture is placed in a cupel, which is a shallow, porous cup, and roasted in a blast of air. The base metal oxides are absorbed in the cupel, leaving the pure metal to be decanted.

cupferron (ammonium nitroso-beta-phenylhydroxylamine) $C_6H_5N(NO)ONH_4$.
Properties: Creamy-white crystals; m.p. 163–164°C. Soluble in water, alcohol and ether.
Derivation: By treating an ethereal solution of beta-phenylhydroxylamine with dry ammonia gas and amyl nitrite.
Use: Reagent in analytical chemistry.

cuprammonium process. A minor process for making rayon by dissolving cellulose in an ammoniacal copper solution and reconverting it to cellulose by treatment with acid. Relatively pure alpha-cellulose (from cotton linters or treated wood pulp) is dissolved in a solution of ammonia and copper hydroxide or basic copper sulfate until the cellulose content reaches 7–10%. During an aging period, absorption of atmospheric oxygen causes degradation of the cellulose and reduces the viscosity of the solution. When the proper viscosity has been attained, the solution is filtered, deaerated, and extruded into a bath of dilute sulfuric acid. The fibers are closely similar to those made by the viscose process, which is the major method of rayon production.

cupreine (hydroxycinchonine) $C_{19}H_{22}O_2N_2 \cdot 2H_2O$. One of the cinchona alkaloids.
Properties: Colorless crystals; m.p. (anhydrous) 198°C. Soluble in alcohol; slightly soluble in water, chloroform, ether, benzene.
Derivation: From cuprea bark Remijia pedunculata.
Use: Medicine.

cupric. Form of the word copper used in naming copper compounds in which the copper has a valence of 2. See the corresponding compound under copper.

cupric chromate, acid. See "Celcure."

cupric chromate, basic. See copper chromate.

cupriethylene diamine. Purple liquid; ammoniacal odor. Dissolves cellulose products.
Hazard: Strong irritant to tissue. Toxic.
Shipping regulations: (solution) (Rail) White label. (Air) Corrosive label.

cuprite (copper ore, ruby; red oxide of copper) Cu_2O. Crimson, scarlet, vermilion, deep or brownish-red, secondary mineral; adamantine or dull luster; brownish-red streak. Soluble in nitric and concentrated hydrochloric acids. Sp. gr. 5.85–6.15; Mohs hardness 3.5–4.
Occurrence: United States; England; Germany; France; Siberia; Australia; China; Peru; Bolivia.
Use: Source of copper.

"Cuprochrome."[155] Trademark for an electronic tube alloy of 98.2% copper and 1.8% chromium.

cupronickel. An alloy of copper and nickel used in coinage, condenser and heat-exchanger tubes. Most types contain from 10–30% nickel. Strongly corrosion-resistant, especially to seawater.

"Cupron."[158] Trademark for an alloy of 55% copper and 45% nickel (constantan).
Forms: Wire, ribbon; strip.
Uses: Rheostats and controls; resistors for electrical instruments.

cuprotungsten. An alloy of copper and tungsten.

cuprous. Form of the word copper used in naming copper compounds in which the copper has a valence of 1.

cuprous acetate, ammoniacal. See copper ammonium acetate.

cuprous acetylide Cu_2C_2.
Properties: Amorphous red powder. A salt of acetylene.
Derivation: Reaction of acetylene with aqueous solution of cuprous salts.
Hazard: Severe explosion risk when shocked or heated.
Uses: Detonators and other explosive devices.
Shipping regulations: (Rail, Air) Explosives, n.o.s., Not acceptable.

cuprous chloride $CuCl$ or Cu_2Cl_2.
Properties: White cubical crystals; sp. gr. 4.14; m.p. 430°C; b.p. 1490°C; becomes greenish on exposure to air, and brown on exposure to light. Slightly soluble in water; soluble in acids, ammonia, ether.
Derivation: Copper and cupric chloride solution or copper and hydrochloric acid in air.
Grades: Technical; reagent; single crystals.
Hazard: Toxic by ingestion.
Uses: Catalyst; preservative and fungicide; desulfuriz-

ing agent; decolorizing agent in petroleum industry; absorbent for carbon monoxide.

cuprous cyanide $Cu_2(CN)_2$ or CuCN.
Properties: Cream-colored powder; insoluble in water; soluble in sodium cyanide and potassium cyanide, in hydrochloric acid and ammonium hydroxide; sp. gr. 2.9; m.p. 475°C.
Containers: 100-lb drums.
Hazard: Highly toxic.
Use: Copper and brass cyanide electroplating solutions.
Shipping regulations: (Rail) Cyanides of copper, etc., Poison label. (Air) Cyanides, dry, Poison label.

cuprous iodide CuI.
Properties: White to brownish-yellow powder. Soluble in ammonia and potassium iodide solutions; insoluble in water. Sp. gr. 5.653 at 15°C; b.p. 1290°C; m.p. 606°C.
Derivation: Interaction of solutions of potassium iodide and copper sulfate.
Hazard: Moderately toxic in greater than trace amounts.
Use: Feed additive; in table salt as source of dietary iodine (up to 0.01%).

cuprous oxide. See copper oxide, red.

cuprous sulfide. Cu_2S.
Properties: Black powder or lumps; soluble in nitric acid and ammonium hydroxide; insoluble in water. Occurs as the mineral chalcocite. Sp. gr. 5.52–5.82; m.p. about 1100°C.
Derivation: By heating cupric sulfide in a stream of hydrogen.
Grades: Technical; single crystals.
Hazard: Moderately toxic.
Uses: Antifouling paints; solar cells.

cuprous sulfite $Cu_2SO_3 \cdot H_2O$.
Properties: White crystalline powder; soluble in ammonium hydroxide, hydrochloric acid (with decomposition); insoluble in water. Sp. gr. 3.83.

cuprous thiocyanate (copper sulfocyanide) CuSCN.
Properties: Yellow-white powder. Insoluble in water; soluble in ammonia; sp. gr. 2.843: m.p. 1084°C.
Hazard: Moderately toxic.
Uses: Manufacture of organic chemicals, antifouling paints; printing textiles.

"Curafos."[108] Trademark for food grade sodium hexametaphosphate and sodium tripolyphosphate used for curing, preventing undesirable color change, and loss of moisture in meat.

"Curalon."[248] Trademark for a series of urethane curatives.
"Curalon L." A mixture of hindered aromatic primary diamines.
"Curalon M." p,p′-Methylene bis-(ortho-chloroaniline).

curare. A highly toxic mixture of about 40 alkaloids occurring in several species of South American trees. Its action is on the central nervous system, and derivatives are used to some extent in medicine for treatment of cardiac disease. See also snake venom.

curative. (1) Any substance or agent that effects a fundamental and desirable change in a material. See curing.
(2) Any substance that combats disease by killing bacteria or that restores health by chemical means.

"Curavis."[108] Trademark for a dry, pulverized, food grade, polymeric phosphate composition exhibiting high viscosity in water solution. For use in the meat packing industry.

curcumin (tumeric yellow; 1,7-bis(4-hydroxymethoxyphenyl)-1,6-heptadiene-3,5-dione) $[CH_3OC_6H_3(OH)CH{:}CHCO]_2CH_2$.
Properties: Orange-yellow needles, m.p. 183°C. Soluble in water and ether; soluble in alcohol. C.I. 75300.
Derivation: The coloring principle from curcuma.
Uses: Dye; analytical reagent; food dye; biological stain. As an acid-base indicator it is brownish-red with alkalies, yellow with acids (pH range 7.4–8.6); also an indicator for boron.

"Cure-Set."[65] Trademark for a two-component adhesive (special latex and a catalyst vulcanizer) for bonding rubber and vinyl tile in on-grade installation where moisture may be a problem.

curie (abbreviation ci.). The official unit of radioactivity, defined as exactly 3.70×10^{10} disintegratoins per second. This decay rate is nearly equivalent to that exhibited by one gram of radium in equilibrium with its disintegration products. A millicurie (mc) is 0.001 curie. A microcurie (μc) is one millionth curie.

Curie, Marie S. (1867–1934). Born in Warsaw, Poland, she and her husband Pierre made an intensive study of the radioactive properties of uranium. They isolated polonium in 1898 from pitchblende ore. By devising a tedious and painstaking separation method they obtained a salt of radium in 1912, receiving the Nobel prize in physics for this achievement in 1903, jointly with Becquerel. In 1911, Mme. Curie alone received the Nobel prize in chemistry. Her work laid the foundation of radioactive elements which culminated in control of nuclear fission. (See also Rutherford).

curing. Conversion of a raw product to a finished and useful condition usually by application of heat and/or chemicals which induce physicochemical changes. Many food products require aging under specified temperature conditions. The more common types of curing are as follows:

(a) Meats: use of sodium chloride, sugars, sodium nitrite, sodium nitrate, ascorbic acid; these not only act as preservatives but also aid in color retention. Some types are smoked subsequently. Conversion of collagen to gelatin occurs as a result of "hanging" meat for several days.

(b) Leather: treatment of hides and skins with tanning agents of vegetable or mineral origin; this converts the protein structure into a firm and durable product as a result of complexing reactions. See also tanning.

(c) Tobacco: Exposure for several days to temperatures from 100 to 150°F to discharge the chlorophyll and reduce water content.

(d) Cheese: Precipitated casein of many varieties which is aged for many weeks at moderate temperature (40 to 50°F). The process is also called ripening.

(e) Rubber: addition of sulfur and accelerator, followed by exposure to heat which effects cross-linking. This converts the material from a thermoplastic to a thermosetting product. High-energy radiation can also be used. See also vulcanization.

curium Cm Synthetic radioactive element of atomic number 96; atomic weight 244. See also actinide elements. Valences 3, 4. Isotopes available: 244, and 242 (gram quantities).
Properties: Silvery white metal; sp. gr. 13.5; m.p. 1340°C; chemically reactive, more electropositive than aluminum. An alpha emitter. Biologically it is a bone-seeking element. Forms compounds such as CmO_2, Cm_2O_3, CmF_3, CmF_4, $CmOH_3$, $CmCl_3$, $CmBr_3$, CmI_3, $Cm_2(C_2O_4)_3$.
Uses: Thermoelectric power generation for instrument operation in remote locations on earth or in space vehicles.

"Curona."[173] Trademark for sodium isoascorbate as a curing aid in cured and comminuted meat products. It is used as an anti-oxidant in the meat packing industry for preserving the natural color of meat products, to shorten processing time and reduce shrinkage.

current density. In an electroplating bath or solution, the electric current per unit area of the object or surface being plated. Expressed in amperes per square centimeter, or more usually, amperes per square decimeter.

cutting fluid. A liquid applied to a cutting tool to assist in the machining operation by washing away the chips or serving as a lubricant or coolant. Commonly used cutting fluids are: water; water solutions or emulsions of detergents and oils; mineral oils; fatty oils; chlorinated mineral oils; sulfurized mineral oils; mixtures of the foregoing oils. Transparent grades are available.

"CWS."[58] Trademark for fumaric acid, food and technical grades.

"Cyamon."[57] Trademark for an ammonium nitrate blasting agent designed for safe handling in the field. It is not sensitive to a blasting cap, rifle slug, primacord, flame or impact of heavy steel weights.

"Cyan."[57] Trademark for a series of blue and green phthalocyanine pigments.

"Cyana."[57] Trademark used in connection with the textile finishes obtained by applying "Aerotex" Resins and similar products.

"Cyanamer."[57] Trademark for an acrylic polymer.

cyanamide.
(1) (cyanogenamide; carbodiimide) HN:C:NH or $N\vdots CNH_2$.
Properties: Deliquescent crystals; m.p. 42°C; sp. gr. 1.08. Very soluble in water, alcohol, ether. Flash point 285°F; combustible.
Derivation: From calcium cyanamide and sulfuric acid.
Hazard: Moderately toxic.
(2) Calcium cyanamide (q.v.)

cyanic acid. See isocyanic acid.

cyanide pulp. The mixture obtained by grinding crude gold and silver ore and dissolving the precious metal content in sodium cyanide solution.

cyanine dye. One of a series of dyes consisting of two heterocyclic groups (usually quinoline nuclei) connected by a chain of conjugated double bonds containing an odd number of carbon atoms. Example: cyanine blue $C_2H_5NC_9H_6{:}CHC_9H_6NC_2H_5$. They include the isocyanines, merocyanines, cryptocyanines, and dicyanines.
Use: Sensitizers for photographic emulsions.

cyanoacetamide (malonamide nitrile; propionamide nitrile) $CNCH_2CONH_2$.
Properties: White crystals; b.p., decomposes; m.p. 119°C; soluble in water and alcohol. Flash point 419°F. Combustible.
Derivation: Ammonolysis of cyanoacetic ester or dehydration of ammonium cyanoacetate.
Uses: Organic pharmaceutical synthesis; plastics.

cyanoacetic acid (malonic nitrile) $CNCH_2COOH$.
Properties: White crystals; hygroscopic. Soluble in water, alcohol and ether. M.p. 66.1–66.4°C; decomposed at 160°C.
Derivation: Interaction of sodium chloroacetate and potassium cyanide solution.
Use: Organic synthesis.

cyanoacrylate adhesive. An adhesive based on the alkyl 2-cyanoacrylates. (See, for example, methyl 2-cyanoacrylate.) The latter are prepared by pyrolyzing the poly(alkyl)-2-cyanoacrylates produced when formaldehyde is condensed with the corresponding alkyl cyanoacetates. These adhesives have excellent polymerizing and bonding properties. To prevent premature polymerization, inhibitors are added. Supplied commercially as "Eastman 910."

"Cyanobrik."[28] Trademark for 98% sodium cyanide in 1-oz, pillow-shaped, briquette form.
Hazard: Highly toxic.

cyanocarbon. Any of a class of compounds in which the cyanide radical (—CN) replaces hydrogen in organic compounds, as in tetracyanoethylene, $(CN)_2C{:}C(CN)_2$. The compounds are quite reactive and form colored complexes with aromatic hydrocarbons. See also nitrile.

"Cyanocel."[57] Trademark for chemically modified (cyanoethylated) cellulose.

cyanocobalamine $C_{63}H_{88}CoN_{14}O_{14}P$. One of the active forms of vitamin B_{12} (q.v.). A grade using radioactive cobalt is also available for use as diagnostic aid.
Properties: Dark red crystals or red powder. Very hygroscopic; odorless and tasteless. Slightly soluble in water; soluble in alcohol; insoluble in acetone and ether.
Grade: U.S.P.
Containers: Vials; tins; drums.
Uses: Vitamin; food supplement; medicine.

2-cyanoethyl acrylate $CH_2{:}CHCOOCH_2CH_2CN$.
Properties: Liquid. Sp. gr. 1.0690; b.p., polymerizes when heated; f.p. −16.9°C; lb/gal 8.9; flash point 255°F (COC); soluble in water; combustible.
Uses: Forms polymers, copolymers for viscosity index improvers; adhesives; textile finishes and sizes.
See also cyanoacrylate adhesive.

cyanoethylation. Process for introducing the group, $—OCH_2CH_2CN$, into an organic molecule by reaction of acrylonitrile with a reactive hydrogen, such as that on a hydroxyl or amino group.

cyanoethyl sucrose (CES)
$C_{12}H_{14}O_3(OH)_{0.7}(OC_2H_4CN)_{7.3}$.
Properties: Clear, very viscous, pale yellow liquid. Sp. gr. 1.20 (20/20°C); f.p. sets to a glass at −10°C; b.p. greater than 300°C; flash point over 375°F; combustible; refractive index (n 25°C/D) 1.615. Viscosity decreases very rapidly when heated. Has high volume resistivity, low power factor, unusually high dielectric constant.
Uses: Capacitor impregnation, phosphor binding in electroluminescent panels, modification of electrical properties in coatings.

cyanoformic chloride CNCOCl.
Properties: Oily liquid. B.p. 126–128°C (750 mm).
Derivation: Reaction between phthaloyl chloride and the amide of ethyl oxalate.

"Cyanogas."[57] Trademark for a pesticide containing not less than 42% calcium cyanide (q.v.); evolves hydrocyanic acid gas on exposure to atmospheric moisture.
Hazard: Highy toxic.

cyanogen (dicyan; oxalonitrile) C_2N_2.
Properties: Colorless gas; pungent penetrating odor; burns with a purple-tinged flame. Soluble in water, alcohol and ether. Sp. gr. 1.8064 (air = 1); f.p. -28°C; b.p. -20.7°C.
Derivation: (a) Potassium cyanide solution is slowly dropped into copper sulfate solution; (b) mercury cyanide is heated.
Grades: Technical; pure.
Containers: (Liquefied) cylinders.
Hazard: Highly toxic. Flammability limits in air 6 to 32%. Store away from light and heat. Tolerance, 10 ppm in air.
Uses: Organic synthesis; welding and cutting metals; fumigant; rocket propellant.
Shipping regulations: (Rail) Poison gas label. Not accepted by express. (Air) Not acceptable.

cyanogenamide. See cyanamide (1).

cyanogen bromide (bromine cyanide) CNBr.
Properties: Crystals; penetrating odor; slowly decomposed by cold water. Corrodes most metals. Soluble in alcohol, benzene, ether; sparingly soluble in water. Sp. gr. 2.02; b.p. 61.3°C (750 mm); m.p. 52°C; vapor density 3.6.
Derivation: (a) Action of bromine on potassium cyanide. (b) Interaction of sodium bromide, sodium cyanide, sodium chlorate, and sulfuric acid.
Hazard: Highly toxic. Strong irritant to skin and eyes.
Uses: Organic synthesis; parasiticide; fumigating compositions; rat exterminants; cyaniding reagent in gold extraction processes.
Shipping regulations: (Rail, Air) Poison label.

cyanogen chloride CNCl.
Properties: Colorless gas or liquid. Soluble in water, alcohol and ether. Sp. gr. 1.2; b.p. 12.5°C; f.p. -65°C; vapor density 2.1. Min. purity 97 mole %.
Derivation: Action of chlorine on moist sodium cyanide suspended in carbon tetrachloride and kept cooled to -3°C, followed by distillation.
Containers: Cylinders (as liquefied gas).
Hazard: Highly toxic by inhalation or ingestion; strong irritant to eyes and skin.
Uses: Organic synthesis; tear gas; warning agent in fumigant gases.
Shipping regulations: (Rail) Containing less than 0.9% water. Poison gas label. Not accepted by express. (Air) Not acceptable.

cyanogen fluoride (fluorine cyanide) CNF.
Properties: Colorless gas. Forms a white, pulverulent mass if cooled strongly and sublimes at -72°C. Insoluble in water.
Derivation: Interaction of silver fluoride and cyanogen iodide.
Containers: Glass containers (eventually, under the influence of light, it attacks the glass).
Hazard: Toxic by inhalation or ingestion; strong irritant to eyes and skin. Tolerance (as F), 2.5 mg per cubic meter of air.
Uses: Organic synthesis; tear gas.

cyanogen iodide (iodine cyanide) CNI.
Properties: Colorless needles; very pungent odor; acrid taste. Soluble in water, alcohol, and ether; m.p. 146.5°C; sp. gr. 2.84.
Derivation: By heating a metal cyanide with iodine.
Containers: Glass bottles.
Hazard: Toxic by inhalation or ingestion; strong irritant to eyes and skin.
Use: Taxidermists' preservatives.

"Cyanogran."[28] Trademark for a 98% sodium cyanide in granular form. White crystalline solid crushed to pass 100% through 10 mesh, retained on 50 mesh.
Hazard: Highly toxic by inhalation or ingestion; strong irritant to eyes and skin.

cyanoguanidine. See dicyandiamide.

cyano(methylmercuri)guanidine (methylmercury dicyandiamide) $CH_3HgNHC(:NH)NHCN$. Crystals; m.p. 156°C; soluble in water.
Hazard: Highly toxic by inhalation or ingestion. Strong skin irritant.
Uses: Seed fungicide and disinfectant.
Shipping regulations: (Rail, Air) Mercury compounds, solid, Poison label.

3-cyanopyridine C_5H_4NCN.
Properties: Colorless liquid. B.p. 206.2°C; m.p. 49.6°C; soluble in water.
Hazard: Probably toxic.
Containers: Drums; tank cars.
Uses: Organic synthesis.

4-cyanopyridine C_5H_4NCN.
Properties: Colorless liquid. B.p. 195.4°C; m.p. 78.5°C; partially soluble in water; soluble in most organic solvents.
Hazard: Probably toxic.
Containers: Drums; tank cars.
Uses: Organic synthesis.

cyanuramide. See ammelide.

cyanurdiamide. See ammeline.

cyanuric acid (tricarbimide; tricyanic acid) $HOCNC(OH)NC(OH)N \cdot 2H_2O$ (ring). See also isocyanuric acid, the ketone isomer.
Properties: White crystals; odorless; slight bitter taste. Soluble in water and hot mineral acids. Insoluble in alcohol and acetone; solubility in water, 0.2 g/100 ml. Sp. gr. 1.768; decomp. to cyanic acid at 320°C. Low toxicity.
Uses: Intermediate for chlorinated bleaches; selective herbicide; whitening agents.

cyanuric chloride (2,4,6-trichloro-1,3,5-triazine) $C_3N_3Cl_3$ (cyclic).
Properties: Crystals with pungent odor; sp. gr. 1.32; m.p. 146°C; b.p. 194°C (764 mm). Soluble in chloroform, carbon tetrachloride, hot ether, dioxane, ketones. Very slightly soluble in water (hydrolyzes in cold water). Low toxicity.
Use: Chemical synthesis; dyestuffs; herbicides; optical brighteners.

cyanurtriamide. See melamine.

"Cyclaine."[123] Trademark for an anesthetic consisting of hexylcaine hydrochloride.

cyclamate. Group name for synthetic nonnutritive sweetening agents derived from cyclohexylamine or cyclamic acid; the series includes sodium, potassium, and calcium cyclamates. As a result of a study made on laboratory animals in 1970, which indicated that these compounds cause incidence of genetic damage in chick embryos and cancer in rats from high dosage of cyclamates, their use in beverages and food products was banned in the U.S. More recent research has failed to confirm the carcinogenicity of these compounds in laboratory animals, even at levels up to 240 times human intake. Notwithstanding these results, FDA has not withdrawn its ban on use of cyclamates as food additives or as table-top sweeteners, in view of the continuing uncertainty about its safety (1976). The subject is still controversial (1976). See also sweetener, nonnutritive.

cyclamen alcohol. The alcohol corresponding to cyclamen aldehyde (q.v.), used as a stabiizer of cyclamen aldehyde.

cyclamen aldehyde (methyl para-isopropylphenylpropyl aldehyde) $(CH_3)_2CHC_6H_4CH(CH_3)CH_2CHO$.
Properties: Colorless liquid; floral odor. Sp. gr. 0.949–0.959; refractive index 1.507–1.520. Soluble in 1 volume of 80% alcohol; in most oils.
Grades: F.C.C.
Use: Perfumery; soap perfumes; flavoring.

cyclamic acid. USAN name for cyclohexanesulfamic acid (cyclohexylsulfamic acid) $C_6H_{11}NHSO_3H$.
Properties: Odorless, white crystalline solid with a sweet-sour taste; m.p. 178–181°C. Strong, stable acid, soluble in water and alcohol; insoluble in oils.
Hazard: May be carcinogenic.
Uses: Nonnutritive sweetener; acidulant.
See also cyclamate.

cyclethrin 3-(2-cyclopentenyl)-2-methyl-4-oxo-2-cyclopentenyl ester of chrysanthemummonocarboxylic acid. See also furethrin, barthrin, ethythrin.
Properties: Viscous brown liquid, soluble in petroleum solvents and other common organic solvents. Formulated principally as liquid for spray applications corresponding to natural pyrethrins.
Hazard: Moderately toxic by inhalation and ingestion.
Uses: Insecticide with applications similar to allethrin and other analogs.

cyclic compound. An organic compound whose structure is characterized by one or more closed rings; it may be mono-, bi-, tri-, or polycyclic depending on the number of rings present. There are three major groups of cyclic compounds: (1) alicyclic, (2) aromatic (also called arene), and (3) heterocyclic. For more detailed information, consult specific entries, and see also organic compound.

cyclizine hydrochloride (1-diphenylmethyl-4-methylpiperazine hydrochloride) $(C_6H_5)_2CHC_4H_8N_2CH_3 \cdot HCl$.
Properties: White crystalline powder or small colorless crystals. Odorless or nearly so; bitter taste; m.p. 285°C with decomposition; slightly soluble in water, alcohol, chloroform; insoluble in ether; pH (2% solution) 4.5–5.5.
Grade: U.S.P.
Use: Medicine (antiemetic).

cycloaliphatic epoxy resin (cycloalkenyl epoxides). A polymer prepared by epoxidation of multicycloalkenyls (polycyclic aliphatic compounds containing carbon-carbon double bonds) with organic peracids such as peracetic acid. Resistant to high temperatures.
Uses: Space vehicles; outdoor electrical installations in polluted and humid atmospheres; high-temperature adhesives.

cyclobarbital [5-(1-cyclohexenyl)-5-ethylbarbituric acid; tetrahydrophenobarbital] $C_{12}H_{16}N_2O_3$.
Properties: White crystals or crystalline powder; odorless; bitter taste; m.p. 170–174°C; soluble in alcohol or ether; very slightly soluble in cold water or benzene.
Derivation: Hydrogenation of phenobarbital with colloidal palladium in alcohol as a catalyst.
Hazard: See barbiturate.
Use: Medicine.

cyclobutane (tetramethylene) C_4H_8.
Properties: Colorless gas. Sp. gr. 0.7083 (11°C); b.p. 13°C; f.p. −80°C. Insoluble in water; soluble in alcohol and acetone.
Derivation: From petroleum.
Hazard: Flammable. Dangerous fire risk.

cyclobutene (cyclobutylene) C_4H_6.
Properties: Gas; sp. gr. 0.733; b.p. 2.0°C.
Derivation: From petroleum.
Hazrd: Flammable. Dangerous fire risk.

"Cyclocel."[99] Trademark for a reforming catalyst for gasolines and naphthas. Developed primarily for the cyclo-version process which incorporates catalyst reactivation facilities. Mesh grades 4/8, 4/10.

cyclocitrylideneacetone. See ionone.

cyclocumarol $C_{20}H_{18}O_4$. A synthetic blood anticoagulant.
Properties: White, crystalline powder with slight odor. M.p. 164–168°. Insoluble in water; slightly soluble in alcohol.
Use: Medicine.

"Cyclodex."[74] Trademark for water-dispersible driers with certified metal content of cobalt, lead, or manganese.

cycloheptane (heptamethylene; suberane) C_7H_{14}.
Properties: Colorless liquid. Soluble in alcohol; insoluble in water. Sp. gr. 0.809; b.p. 117°C; f.p. −12°C; aniline equivalent −6; flash point below 100°F.
Grades: Technical.
Hazard: Moderately toxic; narcotic action by inhalation. Flammable, dangerous fire risk.
Use: Organic synthesis.

cycloheptanone (suberone) $C_7H_{12}O$.
Properties: Colorless liquid; peppermint odor; b.p. 179°C; sp. gr. 0.95; insoluble in water; soluble in alcohol. Combustible.
Hazard: Moderately toxic.
Uses: Research; intermediate.

5-[1-cyclohepten-1-yl]-5-ethylbarbituric acid. See heptabarbital.

cyclohexane (hexamethylene; hexanaphthene; hexahydrobenzene) C_6H_{12}.
Structure: A typical alicyclic hydrocarbon. It may exist in two modifications, called the "boat" and the

"chair," as shown. This is due to slight distortion of the bond angles, in accordance with the modified version of Baeyer's strain theory.

Properties: Colorless, mobile liquid. Pungent odor. Sp. gr. 0.779 (20/4°C); b.p. 80.7°C; f.p. 6.3°C; refractive index 1.4263; aniline equiv. 7. Insoluble in water. Soluble in alcohol, acetone, benzene. Flash point (98% grade) -1°F (closed cup). Autoignition temp. 500°C. Flammable limits in air, 1.3 to 8.4%.

Derivation: (a) Catalytic hydrogenation of benzene. (b) Constituent of crude petroleum.

Grades: 85, 98, 99.86%; spectrophotometric.

Containers: Special metal cans and drums; tank cars; pipelines.

Hazard: Flammable, dangerous fire risk. Moderately toxic by inhalation and skin contact. Tolerance, 300 ppm in air. Safety data sheet available from Manufacturing Chemists Assn., Washington, D.C.

Uses: Manufacture of nylon; solvent for cellulose ethers, fats, oils, waxes, bitumens, resins, crude rubber; extracting essential oils; chemical (organic synthesis, recrystallizing medium); paint and varnish remover; glass substitutes; vapor has been used as lubricant for steel (experimental).

Shipping regulations: (Rail) Red label. (Air) Flammable Liquid label.

1,4-cyclohexanebis(methylamine) $C_6H_{10}(CH_2NH_2)_2$. The commercial product is about 40% cis and 60% trans.

Properties: Clear liquid; sp. gr. 0.9419 (20/4°C); b.p. 239–244°C. Soluble in all proportions in water, alcohol, and most other organic solvents. Combustible; low toxicity.

Uses: Intermediate; resins.

cyclohexanecarboxylic acid. See hexahydrobenzoic acid.

1,2-cyclohexanedicarboxylic anhydride. See hexahydrophthalic anhydride.

1,4-cyclohexanedimethanol (CHDM) $C_6H_{10}(CH_2OH)_2$. Cis and trans isomers are known and are present in the commercial product in about 30 to 70%.

Properties: Liquid; b.p. 286.0°C (735 mm, cis-isomer), 283°C (735 mm, trans-isomer); m.p. 41–61°C; sp. gr. (super-cooled) 1.0381 (25/4°C); flash point 330°F (COC); refractive index (n 20/D) 1.4893. Soluble in water; ethyl alcohol. Combustible.

Uses: Polyester films and protective coatings; reduction of reaction time in esterification.

cyclohexanesulfamic acid. See cyclamic acid.

cyclohexanol (hexahydrophenol) $C_6H_{11}OH$.

Properties: Colorless, oily liquid, camphor-like odor; hygroscopic. Sparingly soluble in water; miscible with most organic solvents and oils. Sp. gr. 0.937 (37/4°C); m.p. 23°C; b.p. 160.9°C; wt/gal approximately 8 lb; flash point 154°F; refractive index 1.465 (22°C). Combustible; autoignition temp. 572°F.

Derivation: Phenol is reduced with hydrogen over active nickel at 160 to 170°F. The cyclohexanone is removed by condensing with benzaldehyde in the presence of alkali.

Grades: Technical (contains freezing inhibitor).

Containers: Drums; tank cars.

Hazard: Toxic by skin absorption and inhalation. Tolerance, 50 ppm in air. Narcotic in high concentrations.

Uses: Soap making, to incorporate solvents and phenolic insecticides; source of adipic acid for nylon; textile finishing; solvent; blending agent; lacquers; paints and varnishes; finish removers; dry cleaning; emulsified products; leather degreasing; polishes; plasticizers; plastics; germicides.

cyclohexanol acetate (cyclohexanyl acetate) $CH_3COOC_6H_{11}$.

Properties: Colorless liquid. Odor resembling that of amyl acetate. Miscible with most lacquer solvents and diluents, and with halogenated and hydrogenated hydrocarbons. Soluble in alcohol; insoluble in water; sp. gr. 0.966; b.p. 177°C. Combustible.

Hazard: Narcotic in high concentrations.

Uses: Solvent for nitrocellulose, cellulose ether; bitumens, metallic soaps, basic dyes, blown oils, crude rubber, many natural and synthetic resins and gums; lacquers.

cyclohexanone (pimelic ketone; ketohexamethylene) $C_6H_{10}O$.

Properties: Water-white to pale yellow liquid with acetone- and peppermint-like odor. Slightly soluble in water. Miscible with most solvents. B.p. 156.7°C; f.p. -47°C; sp. gr. 0.948; flash point 111°F; refractive index (n 20/D) 1.4507; vapor pressure (212°F) 136 mm. Autoignition temp. 788°F. Combustible.

Derivation: By passing cyclohexanol over copper with air at 280°F; also by oxidation of cyclohexanol with chromic acid or oxide.

Containers: Cans; drums; tank cars.

Hazard: Moderate fire risk. Toxic by inhalation and skin contact. Tolerance, 50 ppm in air.

Uses: Organic synthesis, particularly of adipic acid and caprolactam (about 95%); polyvinyl chloride and its copolymers, and methacrylate ester polymers; wood stains; paint and varnish removers; spot removers; degreasing of metals; polishes; leveling agent in dyeing and delustering silk; lube oil additive; general solvent.

cyclohexanone peroxide (1-hydroperoxycyclohexyl 1-hydroxycyclohexyl peroxide) $C_6H_{10}(OOH)OOC_6H_{10}OH$.

Properties: Grayish paste; insoluble in water; soluble in most organic solvents.

Hazard: Dangerous fire risk in contact with organic materials. Strong oxidizing agent.

Shipping regulations: (Rail) (up to 85%) Yellow label. (Air) (not over 50%) Organic Peroxide label; (50–85%) Organic Peroxide label. Not acceptable on passenger planes. (Over 85%) Not acceptable.

cyclohexanyl acetate. See cyclohexanol acetate.

cyclohexene (1,2,3,4-tetrahydrobenzene) C_6H_{10}.

Properties: Colorless liquid. Sp. gr. 0.811 (20/4°C); b.p. 83°C; f.p. -103.7°C; refractive index 1.445 (25°C); flash point 11°F; aniline equivalent 10; wt/gal 6.7 lb (25°C). Soluble in alcohol; insoluble in water.

Grades: Technical 95%; 99%; research 99.9 mole %.

Containers: Tank cars; barges.

Hazard: Moderately toxic by inhalation. Tolerance, 330 ppm in air. Flammable, dangerous fire risk.

Uses: Organic synthesis; catalyst solvent; oil extraction.

Shipping regulations: Flammable liquid, n.o.s. (Rail) Red label. (Air) Flammable Liquid label.

3-cyclohexene-1-carboxaldehyde (1,2,3,6-tetrahydrobenzaldehyde $CH_2CH{:}CHCH_2CH_2CHCHO$.
Properties: Liquid; sp. gr. 0.9721; b.p. 164.2°C; f.p. -100°C; lb/gal 8.1 (20°C); flash point 135°F; slightly soluble in water. Combustible.
Containers: Drums.
Hazard: Strong irritant to tissue.
Uses: Intermediates; improves water resistance of textiles.
Shipping regulations: (Air) Corrosive label.

cyclohexene oxide $C_6H_{10}O$.
Properties: Colorless liquid with a strong odor; b.p. 129–130°C; sp. gr. 0.967 (25/4°C); refractive index (n 25/D) 1.4503; flash point 81°F. Soluble in alcohol, ether, acetone; insoluble in water.
Grade: 98% pure.
Hazard: Flammable, dangerous fire risk. Moderately toxic.
Use: Chemical intermediate.

cyclohexenylethylbarbituric acid. See cyclobarbital.

cyclohexenylethylene. See vinylcyclohexene.

cyclohexenyltrichlorosilane. $C_6H_9SiCl_3$.
Properties: Colorless liquid. B.p. 202°C; sp. gr. 1.263 (25/25°C); refractive index (n 25/D) 1.488; flash point (COC) 200°F. Readily hydrolyzed by moisture, with liberation by hydrochloric acid. Combustible.
Grades: Technical.
Hazard: Strong irritant; corrosive to skin.
Use: Intermediate for silicones.
Shipping reuglations: (Rail) White label. (Air) Corrosive label. Not accepted on passenger planes.

cycloheximide. Generic name for 3-[2-(3,5-dimethyl-2-oxocyclohexyl)-2-hydroxyethyl]glutarimide. $CH_2CH(CH_3)CH_2CH(CH_3)COCHCH(OHCH_2$-

$CHCH_2CONHCOCH_2$.

By-product in the manufacture of streptomycin.
Properties: Crystals; m.p. 115.5–117°C. Slightly soluble in water; soluble in acetone, alcohol, and chlorinated solvents.
Hazard: Moderately irritant.
Uses: Fungicide; antibiotic; abscission of citrus fruit in harvesting.

cyclohexylamine (hexahydroaniline; aminocyclohexane) $C_6H_{11}NH_2$.
Properties: Colorless liquid, with unpleasant odor. B.p. 134.5°C; sp. gr. 0.8647 (25/25°C); f.p. -18°C. Strong organic base; pH of 0.01% aqueous solution 10.5; forms an azeotrope with water, b.p. 96.4°C. Miscible with most solvents. Flash point 90°F (open cup). Autoignition temp. 560°F.
Grades: Technical (98%).
Containers: Drums; tank cars.
Hazard: Flammable; moderate fire risk; toxic by ingestion, inhalation, skin absorption. Tolerance, 10 ppm in air.
Uses: Boiler-water treatment; rubber accelerator; intermediate.

cyclohexylbenzene. See phenylcyclohexane.

N-cyclohexyl-2-benzothiazolesulfenamide (benzothiazyl 2-cyclohexylsulfenamide) $C_6H_4SNCSNHC_6H_{11}$.
Properties: Cream-colored powder; sp. gr. 1.27; melting range 93–100°C. Insoluble in water; soluble in benzene.
Use: Rubber accelerator.

cyclohexyl bromide $C_6H_{11}Br$.
Properties: A liquid, not more than faintly yellow, having a penetrating odor. Sp. gr. 1.32–1.34 (25/25°C); refractive index 1.4926–1.4936 (25°C).

cyclohexyl chloride $C_6H_{11}Cl$.
Properties: Colorless liquid. F.p. -43°C; b.p. 142°C; flash point 89°F; sp. gr. 0.992.
Hazard: Flammable, moderate fire risk.

2-cyclohexylcyclohexanol $C_6H_{11}C_6H_{10}OH$.
Properties: Colorless liquid; freezing point 29°C; b.p. 271–277°C; sp. gr. 0.977 (25/25°C); lb/gal 8.13; refractive index 1.495 (25°C); flash point 255°F. Soluble in methanol and ether. Slightly soluble in water. Combustible.

2-cyclohexyl-4,6-dinitrophenol. See dinitrocyclohexylphenol.

cyclohexyl isocyanate $C_6H_{11}NCO$. Used to form cyclohexyl carbamates or ureas for agricultural chemicals or pharmaceutical use.
Hazard: Toxic.
Shipping regulations: (Air) Poison label. Not accepted on passenger planes.

cyclohexyl methacrylate $H_2C{:}C(CH_3)COOC_6H_{11}$.
Properties: Colorless monomeric liquid with pleasant odor. B.p. 210°C; refractive index (20°C) 1.4578; sp. gr. (20/20°C) 0.9626; viscosity (25°C) 5.0 centipoises; insoluble in water. Combustible.
Uses: Optical lens systems; dental resins; encapsulation of electronic assemblies.

1-cyclohexyl-2-methylaminopropane. See propylhexedrine.

para-**cyclohexylphenol** $C_6H_{11}C_6H_4OH$.
Properties: Crystals; m.p. 120°C (min). Combustible.
Grades: Technical.
Hazard: May be toxic.
Use: Intermediate for resins and organic synthesis.

cyclohexylphenyl-1-piperidinepropanol hydrochloride. See trihexyphenidyl hydrochloride.

N-cyclohexylpiperidine $C_6H_{11}NC_5H_{10}$. Yellow liquid; refractive index (n 20/D) 1.4856. Combustible.
Use: Intermediate.

cyclohexyl stearate $C_6H_{11}OOCC_{17}H_{35}$.
Properties: Pale yellow powder; sp. gr. 0.882 at 30/15.5°C; m.p. 26–28°C. Soluble in benzene, toluene and acetone; insoluble in water.
Uses: Plasticizer for natural and synthetic resins.

cyclohexylsulfamic acid. See cyclamic acid.

N-cyclohexyl para-**toluenesulfonamide** $C_6H_{11}NHSO_2C_6H_4CH_3$.
Properties: Yellow-brown fused mass; relatively light-stable; m.p. 86°C; b.p. 350°C. Combustible. Low toxicity. Soluble in alcohol, esters, ketones, aromatic hydrocarbons and vegetable oils; insoluble in water. Combustible with a wide variety of resins including most of the cellulosic and vinyl resins.
Use: Resin plasticizer.

cyclohexyl trichlorosilane $C_6H_{11}SiCl_3$.
Properties: Colorless to pale yellow liquid. B.p. 206°C; sp. gr. 1.226 (25/25°C); refractive index (n 25/D) 1.4759; flash point (COC) 185°F. Readily hydrolyzed by moisture, with liberation of hydrochloric acid. Combustible.

Derivation: By Grignard reaction of silicon tetrachloride and cyclohexylmagnesium chloride.
Grade: Technical.
Containers: 1-, 10-lb bottles; 100-lb drums.
Hazard: Toxic by ingestion and inhalation; strong irritant to tissue.
Use: Intermediate for silicones.
Shipping regulations: (Rail) White label. (Air) Corrosive label. Not acceptable on passenger planes.

"Cyclolube."[500] Trademark for a series of oils composed principally of cycloparaffins. Used as plasticizers for nonpolar polymers, lubricants for polar polymers and extenders for relatively saturated polymers.

cyclomethycaine (3-(2-methylpiperidine)propyl-para-cyclohexyloxybenzoate) $C_{22}H_{33}NO_3$.
Properties: White, odorless, crystalline powder; sparingly soluble in water, alcohol, and chloroform and very slightly soluble in acetone, ether and dilute acids.
Use: Medicine.

cyclomethycaine sulfate $(C_{22}H_{33}NO_3)_2 \cdot H_2SO_4$.
Properties: Crystals; somewhat soluble in water.
Use: Medicine (topical anesthetic).

cyclonite (sym-trimethylene trinitramine; hexahydro-1,3,5-trinitro-sym-triazine; trinitrotrimethylenetriamine; cyclotrimethylenetrinitramine; RDX)
$N(NO_2)CH_2N(NO_2)CH_2N(NO_2)CH_2$.
Properties: White crystalline solid; sp. gr. 1.82; m.p. 203.5°C. Soluble in acetone; insoluble in water, alcohol, carbon tetrachloride, and carbon disulfide; slightly soluble in methanol and ether.
Derivation: Reaction of hexamethylenetetramine with concentrated nitric acid.
Containers: Special lined drums.
Hazard: High explosive, easily initiated by mercury fulminate. Toxic by inhalation and skin contact. Tolerance, 1.5 mg per cubic meter of air.
Use: Explosive, 1.5 times as powerful as TNT.
Shipping regulations: (Rail) Consult regulations. (Air) Not acceptable. Legal label name, cyclotrimethylenetrinitramine.

3-cycloöctadiene C_8H_{12}. Intermediate for such compounds as suberic acid.

1,5-cyclooctadiene $HC{:}CH(CH_2)_2CH{:}CHCH_2CH_2$. A butadiene dimer.
Properties: Liquid; freezing point −56.39°C; distillation range 301–303°F (technical); b.p. 149.34°C (pure); sp. gr. 0.88328 (20/4°C); lb/gal 7.38; vapor pressure 0.50 psia (100°F); refractive index 1.4933 (20/D); flash point 100°F. Combustible.
Derivation: Catalytic dimerization of butadiene.
Grades: Technical 95%; 99%; 99.8 mole %.
Hazard: Moderate fire risk.
Uses: Resin intermediate; third monomer in EPT rubber.

cycloöctane C_8H_{16}.
Properties: Colorless liquid. Sp. gr. 0.835; b.p. 148°C; m.p. 14°C. Combustible.

cycloöctatetraene C_8H_8.
Properties: Colorless liquid; f.p. −7°C; b.p. 140°C; sp. gr. 0.943 (0/4°C); refractive index 1.5394 (n 20/D). It behaves like an aliphatic hydrocarbon, is

```
     CH=CH
   /       \
 HC         CH
 ||         ||
 HC         CH
   \       /
     CH=CH
```

relatively reactive and resinifies on standing in air. Combustible.
Derivation: Polymerization of acetylene by Reppe process developed about 1940 in Germany.
Uses: Organic research.

cycloölefin. An alicyclic hydrocarbon having two or more double bonds, e.g., the very reactive and widely used cyclopentadiene derived from coal tar, as well as cyclohexadiene and cyclooctatetraene, containing six and eight carbon atoms, respectively. The latter has four double bonds and is a polymer of acetylene.

cycloparaffin. An alicyclic hydrocarbon in which three or more of the carbon atoms in each molecule are united in a ring structure, and each of these ring carbon atoms is joined to two hydrogen atoms, or alkyl groups. The simplest members are cyclopropane (C_3H_6), cyclobutane (C_4H_8), cyclopentane (C_5H_{10}), cyclohexane (C_6H_{12}), and derivatives of these such as methylcyclohexane ($C_6H_{11}CH_3$).
Hazard: All members of the cycloparaffin series are narcotic, and may cause death through respiratory paralysis. For most of the members there appears to be a narrow range between the concentrations causing deep narcosis and those causing death.

1,3-cyclopentadiene.

```
      CH=CH
     /     |
  H2C      |
     \     |
      CH=CH
```

Colorless liquid with sp. gr. 0.805 and b.p. 42.5°C; insoluble in water; soluble in alcohol, ether and benzene.
Derivation: From coal tar and cracked petroleum oils.
Hazard: Decomposes violently at high temperatures. Moderately toxic. Tolerance, 75 ppm in air.
Uses: Chemical intermediate; organic synthesis; starting material for synthetic prostaglandin; chlorinated insecticides; formation of sandwich compounds by chelation; e.g., cyclopentadienyl iron dicarbonyl dimer $[C_5H_5Fe(CO)_2]_2$.

cyclopentamine hydrochloride (1-cyclopentyl-2-methylaminopropane hydrochloride)
$C_5H_9CH_2CH(CH_3)NHCH_3 \cdot HCl$.
Properties: White, crystalline powder. Mild characteristic odor and bitter taste. M.p. 113.0–116.0°C. Freely soluble in water, alcohol and chloroform; soluble in benzene; slightly soluble in ether. pH (1% solution) about 6.2.
Grade: N.F.
Use: Medicine.

cyclopentane (pentamethylene) C_5H_{10}.

```
      CH2
     /   \
  CH2     CH2
   |       |
  CH2-----CH2
```

Properties: Colorless liquid; soluble in alcohol; insoluble in water. Sp. gr. 0.7445 (20/4°C); b.p. 49.27°C; f.p. -94°C; refractive index 1.406 (20/D); flash point -35°F.
Derivaton: From petroleum.
Grades: Technical; 95%, 99%; research.
Hazard: Moderately toxic by ingestion and inhalation. Flammable, dangerous fire risk.
Use: Solvent for cellulose ethers; motor fuel; azeotropic distillation agent.
Shipping regulations: (Rail) Red label. (Air) Flammable Liquid label.

1,2,3,4-cyclopentanetetracarboxylic acid $C_5H_6(COOH)_4$.
Properties: Crystalline powder; m.p. 195–196°C; soluble in water but insoluble in most organic solvents.
Uses: Curing agent for resins; imparts thermal stability and high-temperature properties.

1,2,3,4-cyclopentanetetracarboxylic acid, dianhydride $C_5H_6(C_2O_3)_2$.
Properties: White solid; m.p. 220–221°C; moderately soluble in dimethyl formamide, butyrolactone, and dimethylsulfoxide, but only slightly soluble in acetone. Insoluble in hydrocarbons.
Uses: Curing agent for epoxy resins.

cyclopentanol (cyclopentyl alcohol)
$\overline{CH_2CH_2CH_2CH_2C}HOH$.
Properties: Colorless, viscous liquid; pleasant odor; sp. gr. 0.946 (20/4°C); refractive index 1.4575 (20°C); b.p. 139–140°C; flash point 124°F. Slightly soluble in water; soluble in alcohol. Combustible.
Hazard: Moderate fire risk.
Uses: Perfume and pharmaceutical solvent; intermediate for dyes, pharmaceuticals and other organics.

cyclopentanone C_5H_8O.
Properties: Water-white, mobile liquid; distinctive ethereal odor, somewhat like peppermint; b.p. 125–126°C (630 mm); sp. gr. 0.943; refractive index 1.437; flash point (closed cup) 87°F. Insoluble in water; soluble in alcohol and ether.
Hazard: Flammable, moderate fire risk. Narcotic in high concentrations.
Use: Intermediate for pharmaceuticals, biologicals, insecticides, and rubber chemicals.

cyclopentanone oxime C_5H_8NOH. Nearly colorless and odorless crystalline solid; m.p. 56°C; b.p. 196°C; soluble in water, alcohol. Used as intermediate in synthesis of amino acids, proline and ornithine.

cyclopentene $\overline{CH:CHCH_2CH_2C}H_2$.
Properties: Colorless liquid; sp. gr. 0.772; b.p. 44°C; f.p. -135.21°C; refractive index (20/D) 1.4225; flash point below 0°F.
Grades: Technical; research (99.89 mole %).
Hazard: Moderately toxic; narcotic action. Flammable, dangerous fire risk.
Use: Organic synthesis; polyolefins; epoxies; cross-linking agent.

cyclopentenyl acetone [1-(1-cyclopentenyl)-2-propanone] $C_5H_7CH_2COCH_3$.
Properties: Clear, colorless liquid; ketone odor; b.p. 170°C; refractive index 1.4545–1.4550 at 25°C. Combustible.

3-(2-cyclopentenyl)-2-methyl-4-oxo-2-cyclopentenyl ester of chrysanthemummonocarboxylic acid. See cyclethrin.

1-(1-cyclopentenyl)-2-propanone. See cyclopentenyl acetone.

cyclopentylacetone (1-cyclopentyl-2-propanone) $C_5H_9CH_2COCH_3$.
Properties: Liquid; sp. gr. 0.893 (25/25°C); b.p. 180–184°C; refractive index 1.4420 (25°C). Combustible.

cyclopentyl alcohol. See cyclopentanol.

cyclopentyl bromide (bromocyclopentane) C_5H_9Br.
Properties: Clear, mobile liquid with sweet aromatic odor. B.p. 137–138°C; sp. gr. 1.3866 (20/4°C); wt/gal 11.6 lb (20°C); refractive index (n 20/D) 1.4885; flash point (closed cup) 108°F. Insoluble in water. Combustible.
Hazard: Moderate fire risk. Moderately toxic.
Uses: Organic synthesis (pharmaceuticals).

1-cyclopentyl-2-methylaminopropane hydrochloride. See cyclopentamine hydrochloride.

cyclopentyl phenyl ketone $C_5H_9COC_6H_5$.
Properties: Colorless to light yellow liquid; b.p. 145–146°C (15 mm). Soluble in most common organic solvents; insoluble in water. Combustible.
Use: Pharmaceutical intermediate.

1-cyclopentyl-2-propanone. See cyclopentyl acetone.

cyclopentylpropionic acid $C_5H_9CH_2CH_2COOH$.
Properties: Liquid; b.p. 130–132°C (12 mm); flash point 116°F; insoluble in water. Combustible.
Hazard: Moderate fire risk. Moderately toxic.
Use: Intermediate; wood preservatives.

cyclopentylpropionic acid chloride. See cyclopentylpropionyl chloride.

cyclopentylpropionyl chloride (cyclopentylpropionic acid chloride) $C_5H_9CH_2CH_2COCl$.
Properties: Liquid; b.p. 81–82°C (10 mm); flash point 104°F; soluble in water; combustible.
Hazard: Moderate fire risk. Moderately toxic.
Use: Intermediate.

"Cyclophos."[1] Trademark for colorless crystalline, hydrated sodium polyphosphate. Cyclic compound. F.C.C. grade. Coagulates albumin and certain other proteins. Soluble in water.
Use: Precipitation of proteins.

cyclopropane (trimethylene) C_3H_6.
Properties: Colorless gas of characteristic odor resembling that of solvent naphtha and having a pungent taste. Sp. gr. 0.72–0.79; b.p. -32.9°C; f.p. -126.6°C. Soluble in alcohol and water. Autoignition temp. 928°F.
Derivation: Reaction of zinc dust with methylene chloride.
Grades: Technical; U.S.P.; 99.5% min.
Containers: Steel cylinders.
Hazard: Highly flammable. Forms flammable and explosive mixtures with air or oxygen. Explosive limits in air, 2.4 to 10.3 % by volume. Moderately toxic by inhalation; narcotic in high concentration.
Uses: Organic synthesis; anesthetic.
Shipping regulations: (Rail) Red Gas label. (Air) Flammable Gas label. Not acceptable on passenger planes.

cyclopropanespirocyclopropane. See spiropentane.

cyclosilane. See silane.

"Cyclo Sol."[125] Trademark for a series of hydrocarbon solvents composed of 50 to 99% aromatic hydrocarbons.
Hazard: Flammable, dangerous fire risk.

"Cyclotene."[233] Trademark for a synthetic aromatic chemical, methylcyclopentenolone; used in soap, cosmetic perfumery, and flavor compounding.

cyclotrimethylenetrinitramine. Legal label name for cyclonite (q.v.).

cycloversion. A process using bauxite as a catalyst for (1) desulfurization, (2) reforming, and (3) cracking of petroleum to form high-octane gasoline.

cyclyd. A coined term referring to the cyclic alkyd coatings prepared from "Polycyclol 1222" (q.v.); used in baking metal primers and air-drying maintenance paints.

"Cycocel"[57] Trademark for a plant growth regulator (2-chloroethyltrimethyl ammonium chloride), said to be effective for cereal grains, tomatoes and peppers.

"Cycolac."[525] Trademark for a series of acrylonitrile-butadiene-styrene polymers. See ABS resin. Especially suitable for metal plating. Also used for stiletto heels and telephone handsets.

"Cycolon."[525] Trademark for medium-impact material possessing most of the properties of ABS resins (q.v.).
Uses: Automotive parts, housewares, toys, refrigeration parts.

"Cycoloy"[525] Trademark for a family of alloys of ABS and other thermoplastic resins, providing a high-performance balance of properties.
Uses: Automotive, sporting goods, electrical/electronics markets.

"Cycopol"[11] Trademark for a group of copolymer resins including styrenated alkyds, acrylonitrile-modified styrenated alkyd, vinyl toluene type, special acrylic type resins.
Uses: Vehicle for solvents, hammer finishes, coil coatings, metal decorating.

"Cycovin"[525] Trademark for alloys of ABS and PVC, providing a family of materials with flame-retardant properties in addition to the balanced features of ABS.
Uses: Instructional and electrical applications.

"Cyfor."[57] Trademark for rosin size for use in paper manufacture.

"Cygon."[57] Trademark for a systemic insecticide whose active ingredient is O,O-dimethyl S-(N-methylcarbamolymethyl)phosphorodithioate.
See dimethoate.

"Cykelin."[64] Trademark for dicyclopentadiene copolymers of linseed oil that produce vehicles of quick hard dry and excellent water resistance.
Uses: Varnishes, enamels, aluminum paints, cold-cut varnishes, reinforcement oil.

"Cyklosit."[470] Trademark for cyclized rubber; used primarily in printing inks.

"Cymel."[57] Trademark for a series of synthetic resins based on melamine-formaldehyde filled with alpha-cellulose, cellulose, chopped fabric, glass fiber, asbestos fiber.

cymene (cymol; isopropyltoluene; methylpropylbenzene) $CH_3C_6H_4CH(CH_3)_2$. The ortho-, meta, and paraisomers are known.
Properties: Colorless, transparent liquids; aromatic odor. Combustible. Sp. gr.: Ortho 0.8748, meta 0.862, para 0.8551; f.p.: Ortho −182°C, meta −25°C, para −73.5°C; b.p.: Ortho 177°C, meta 175.6°C, para 176.5°C; refractive index, para 1.489 (20°C). Soluble in alcohol, ether, and chloroform; insoluble in water. Flash point (para) 117°F (closed cup).
Derivation: Mixed cymenes are produced from toluene by alkylation. Para-cymene occurs in several essential oils, and is made from monocyclic terpenes by dehydrogenation. These terpenes can be made from turpentine, or obtained as a by-product from the sulfite digestion of spruce pulp in paper manufacture.
Method of purification: Washing with sulfuric acid, water, and alkali.
Grade: Technical.
Containers: Drums; glass bottles.
Hazard: Moderate fire risk. Moderately toxic.
Uses: Solvents; synthetic resin manufacture; metal polishes; organic synthesis (oxidation to hydroperoxides used as catalysts for synthetic rubber manufacture. Cymene alcohols are made by hydrogenating the hydroperoxides.) Pure para-cresol and carvacrol are made from para-cymene.

"Cynol."[57] Trademark for a series of rewetting, softening and defoaming agents used in the manufacture of paper. Available in liquid form.

"Cynorex."[134] Trademark for bright cyanide copper plating additive.

"Cypan."[57] Trademark for a synthetic organic chemical used to modify and control the properties of oil well drilling fluids.

"Cypip."[57] Trademark for diethylcarbamazine medicated premix; used in veterinary medicine.

"Cyprex."[57] Trademark for dodine.

"Cyrea."[57] Trademark for a urea feed compound.

"Cyron."[57] Trademark for a synthetic sizing material for the paper industry. "Cyron" permits sizing of alkaline as well as acid paper and may be added either to the pulp or as a coating.

Cys Abbreviation for cysteine.

cysteamine. See 2-aminoethanethiol.

cysteine (alpha-amino-beta-thiolpropionic acid; beta-mercaptoalanine) $HSCH_2CH(NH_2)COOH$. A non-essential amino acid derived from cystine, occurring naturally in the L(+) form.
Properties: Colorless crystals; soluble in water, ammonium hydroxide, and acetic acid; insoluble in ether, acetone, benzene, carbon disulfide, and carbon tetrachloride.
Derivation: Hydrolysis of protein; degradation of cystine. Found in urinary calculi.
Uses: Biochemical and nutrition research; reducing agent in bread doughs (up to 90 ppm). Available commercially as L(+)-cysteine hydrochloride.

cystine (beta, beta'-dithiobisalanine; di[alpha-amino-beta-thiolpropionic acid])
$HOOCCH(NH_2)CH_2SSCH_2CH(NH_2)COOH$. A non-essential amino acid.
Properties: White crystalline plates; soluble in water; insoluble in alcohol. Optically active.
DL-cystine, m.p. 260°C.
D(+)-cystine, m.p. 247–249°C.
L(−)-cystine, m.p. 258–261°C with decomposition; the naturally occurring form.
Derivation: Hydrolysis of protein (keratin); organic synthesis. Occurs as small hexagonal crystals in urine.
Grade: F.C.C.

Use: Biochemical and nutrition research; nutrient and dietary supplement.

"Cystokon."[329] Trademark for a 30% solution of sodium acetrizoate.

cytidine $C_9H_{13}N_3O_5$. The nucleoside consisting of D-ribose and cytosine.
Properties: White, crystalline powder; soluble in water, acid, alkali; insoluble in alcohol.
Derivation: From yeast ribonucleic acid. Also available as the hemisulfate, $(C_9H_{13}N_3O_5)_2 \cdot H_2SO_4$.

cytidine phosphates. Nucleotides used by the body in growth processes; important in biochemical and physiological research. Those isolated and commercially available (as sodium salts) are the monophosphate (CMP; see cytidylic acid), the diphosphate (CDP) and the triphosphate (CTP).

cytidylic acid (cytosylic acid; cytidinephosphoric acid; cytidine monophosphate; CMP) $C_9H_{14}N_3O_8P$. The monophosphoric ester of cytosine; i.e., the nucleotide containing cytosine, D-ribose and phosphoric acid. The phosphate may be esterified to either the 2,3, or 5 carbon of ribose yielding cytidine-2′-phosphate, cytidine-3′-phosphate, and cytidine-5′-phosphate, respectively.
Properties: (cytidine-3′-phosphate): White, crystalline powder; odorless; mild sour taste; m.p.: crystals from 50% alcohol, 230–233°C (with decomposition); crystals from water, 227°C (with decomposition). Soluble in water and dilute alkalies; slightly soluble in 50% alcohol; insoluble in alcohol and other organic solvents.
Derivation of commercial product: From yeast nucleic acid by hydrolysis. The 5′-monophosphate is made synthetically by phosphorylation and hydrolysis of isopropylidene cytidine.
Use: Biochemical research.

cytochemistry. The branch of biochemistry devoted to study of the chemical composition of cells and cell membranes, including chromosomes, genes, and the complex reactions involved in cell growth and replication, as well as the mechanism of enzyme activity. See also molecular biology.

cytochrome. A class of iron-porphyrin proteins (see porphyrin) of great importance in cell metabolism. They are pigments occurring in the cells of nearly all animals and plants. Several types have been identified; cytochrome C is the most abundant and has been obtained in pure forms. The cytochromes and cytochrome oxidase have important functions in cell respiration. The latter is an iron-porphyrin-containing protein which is an important enzyme in cell respiration. It catalyzes the oxidation of cytochrome C and is reduced itself in the reaction; it is then reoxidized by oxygen.

cytokinins. See kinins.

"Cytomel."[71] Trademark for a brand of liothyronine.

cytoplasm. The extra-nuclear components of the living cell, containing mitochondria, plastids, spherosomes, etc. This, together with the nucleus, constitutes the protoplasm (q.v.). The chemical constituents are chiefly proteins, plus a high percentage of water.

"Cytosar."[327] Trademark for cytosine arabinoside (q.v.).

cytosine $C_4H_5N_3O$. 2-Oxo-4-aminopyrimidine. A pyrimidine found in both ribonucleic and deoxyribonucleic acids, and certain coenzymes.
Properties: (monohydrate): Lustrous platelets; decomposes at 320–325°C. Slightly soluble in water and alcohol; insoluble in ether.
Derivation: Isolation following hydrolysis of nucleic acids; organic synthesis.
Use: Biochemical research.

cytosine arabinoside (Ara-C). A drug synthesized in 1969 at Salk Institute; it is useful in combating myelocytic leukemia in adults, and has been approved by FDA as a prescription drug.

cytosine monophosphate. See cytidylic acid.

cytosylic acid. See cytidylic acid.

"Cytrel"[352] Trademark for a tobacco substitute derived from cellulose. It is free from nicotine, and has from 15 to 30% of the tar content of tobacco.

"Cytrol" Amitrole-T.[57] Trademark for 3-amino-1,2,4-triazole liquid formulation herbicide.

"Cyuram."[57] Trademark for tetramethylthiuramdisulfide pellets. Rubber accelerator.

"Cyzine."[57] Trademark for 10% 2-acetylamino-5-nitrothiazole feed supplement.

D

d- Prefix indicating that a substance is dextrorotatory. A plus sign (+) is now preferred.

D Symbol for deuterium.

D- Prefix indicating the right-handed enantiomer of an optical isomer.
See also glyceraldehyde; asymmetry; enantiomer.

2,4-D (2,4-dichlorophenoxyacetic acid) $Cl_2C_6H_3OCH_2COOH$.
Properties: White to yellow crystalline powder; difficultly soluble in water or oils; soluble in alcohols. Stable; m.p. 138° C; b.p. 160° C (0.4 mm).
Derivation: Reaction of 2,4-dichlorophenol and chloroacetic acid in aqueous sodium hydroxide.
Forms available: Sodium salt (60–85% acid), amine salts (10–60% acid); esters (10–45% acid). These forms are dispersible in water or oils (esters) and can be applied as sprays.
Grade: Technical.
Containers: Bags; 100- and 250-lb drums; (butyl, isopropyl esters) tank cars.
Hazard: Moderately toxic and irritant. Tolerance, 10 mg per cubic meter of air. Use may be restricted.
Uses: Selective weed killer and defoliant.

DAA. Abbreviation for diacetone acrylamide (q.v.).

"DABCO."[472] Trademark for triethylenediamine (q.v.).

"Dacamine."[244] Trademark for a combination of N-oleyl 1,3-propylene diamine salt or 2,4-dichlorophenoxyacetic acid. Herbicide.

"Dacolyte."[28] Trademark for an addition agent for acid copper plating solution for heavy deposits. Dark brown, amorphous, fine powder; sharp odor.

"Daconate."[244] Trademark for a monosodium methylarsonate and inert organic compound. Used in post-emergence weed control.
Hazard: Toxic by ingestion.

"Dacote."[244] Trademark for a precipitated calcium carbonate with high brightness and opacifying power. Used especially in and on paper.

"Dacovin."[244] Trademark for a compound of PVC resins. Used in production of rigid pipe, blow-molding bottles, rigid film and sheet and a variety of rigid molded products.

"Dacron."[28] Trademark for a polyester fiber made from polyethylene terephthalate. Available as filament yarn, staple, tow, and fiberfill. See also dimethyl terephthalate, terephthalic acid.
Properties: Sp. gr. 1.38; tensile strength (psi) 4 to 5 grams/denier (about 50 to 60 thousand psi); break elongation 10–36%; moisture regain 0.4%; high elastic recovery; good insect resistance; difficult to ignite; self-extinguishing; soluble in meta-cresol (hot) trifluoroacetic acid, and ortho-chlorophenol; m.p. 250° C.
Derivaton: Reaction of dimethyl terephthalate and ethylene glycol. The resulting polymer is melt-extruded through a spinneret and stretched.
Uses: Textile fabrics and suitings, often combined with wool and other fibers; cordage; fire hose, etc.

"Dacthal."[244] Trademark for a composition of dimethyl ester 2,3,5,6-tetra-chloroterephthalic acid. Commonly known as DCPA. Used in pre-emergence weed control.

dactinomycin. USAN for actiomycin D, $C_{62}H_{86}N_{12}O_{16}$, an antibiotic produced from Streptomyces. Used in medicine.

"dag" Dispersions.[46] Trademark for a series of dispersions useful as lubricants. Most are colloidal. The base types are: graphite-water, graphite-oil, graphite-solvent, molybdenum disulfide resin-bonded solid film (with alkyd, epoxy or phenolic resin solutions), and some special types.

DAHQ. See 2,5-di-tert-amylhydroquinone.

Dakin's solution. An aqueous solution containing 0.5% sodium hypochlorite, used as an antiseptic, especially for wound treatment.

"Dalamar."[28] Trademark for an azo yellow pigment.

dalapon. Generic name for 2,2-dichloropropionic acid, CH_3CCl_2COOH. The sodium salt (sodium 2,2-dichloropropionate), is commonly used.
Properties: (free acid): Liquid; b.p. 185–190° C (760 mm); 90–92° C (14 mm); sp. gr. 1.389 (22.8/4° C). Very soluble in water and alcohol; soluble in ether. (sodium salt): Crystals; decompose 174–176° C; salty taste. Corrosive to iron. Soluble in water; aqueous solutions hydrolyze above 70° C.
Hazard: Moderately toxic; strong irritant to eyes and skin.
Use: Herbicide.

"Dalar."[28] Trademark for a polyester resin of the polyethylene terephthalate type, used for plastic bottles for carbonated beverages.

"Dalbon."[244] Trademark for dispersions formulated for industrial coatings from "Dalvor" polyvinyl fluoride resin. Used to protect surfaces from attack by chemicals and various weather and severe use conditions.

"Dalpac."[266] Trademark for several grades of butylated hydroxytoluene (di-tert-butyl-para-cresol). Supplied as flakes, powder, or liquid.
Uses: Antioxidant in food, animal feed, and nonfood industrial uses.

"Dalpad."[233] Trademark for a coalescing agent, a stable, low-odor, low-temperature film-forming aid for polyvinyl acetate and acrylic latex paints.

dalton. A unit of mass introduced in comparatively recent years. It designates 1/16th of the mass of oxygen-16, the lightest and most abundant isotope of oxygen. Since this is 15.9949, the dalton is equivalent to 0.9997 mass unit.

Dalton, John (1766–1844). The first theorist since the Greek philosopher Democritus to conceive of matter in terms of small particles. The founder of the atomic theory on which all succeeding chemical investigation has been based (1807). His essential concept of the indivisibility of the atom was not called into question until 1910, when radioactive decay was established by Rutherford. Dalton's theories relating to pressures of gases and atomic combinations led to the basic generalizations stated in the law of multiple proportions, the law of constant composition, and the law of conservation of matter. Dalton's law of partial pressures states that in any mixture of gases each constituent exerts its pressure independently, as if the other constituents were absent, and that the solubility of mixed gases in a liquid is proportional to the partial pressure of each. (See Priestley; chemical laws).

"Dalvor."[244] Trademark for polyvinyl fluoride, a thermoplastic fluorocarbon polymer.

"Dalyde."[348] Trademark for dibromosalicylaldehyde, (q.v.), used in medicine.

dammar. A group of tree-derived resins soluble in hydrocarbon and chlorinated solvents; partially soluble in alcohols; insoluble in water. Used in colorless and overprint varnishes, cellulosic lacquers, alkyd baking enamels, and paper and textile coatings.

danthron. See 1,8-dihydroxyanthraquinone.

"Dantoin."[73] Trademark for a series of hydantoin compounds (q.v.).

DAP. (1) Abbreviaton for diallyl phthalate. (2) Abbreviation for diammonium phosphate. See ammonium phosphate, dibasic.

"Dapon."[55] Trademark for a series of diallyl phthalate resins. Used for molding compounds, prepregs, and coatings.

dapsone. USAN and U.S.P. name for 4,4′-sulfonyldianiline (q.v.).

"Daran."[311] Trademark for polyvinylidene chloride latexes; used as barrier coatings for packaging papers, paperboards, plastic films, and specialty saturants.

"Daratak."[311] Trademark for polyvinyl acetate homopolymer emulsions; used in adhesives and specialty coatings.

"Darco."[89] Trademark for activated carbon. Available in various grades for use in sugar refining; removal of impurities from electroplating solutions; purification of drycleaning solvents; drug and chemical purification; and purification and decolorization of animal and vegetable oils, fats, and waxes.

"Darex."[311] Trademark for styrene-butadiene latexes and related vehicles.
Uses: Textile coatings, rug backings, saturants, shoe products, coatings, paints, and adhesives.

"Dariloid."[322] Trademark for a series of milk-soluble algin compositions.

"Darvan."[311] Trademark for a series of anionic surfactants used as dispersants, emulsion stabilizers and latex stabilizers.

"Darvon"[100]. Trademark for *d*-propoxyphene hydrochloride, an analgesic drug.

DAS. Abbreviation for 4,4′-diamino-2,2′-stilbenedisulfonic acid.

dating, chemical. See chemical dating.

daughter element. The element formed when another element undergoes radioactive decay. The latter is called the parent. The daughter may or may not be radioactive.

daunorubicin. See anticancer agent.

"Davco."[241] Trademark for nitrogen solutions used as fertilizers.

Davy, Sir Humphry (1778–1829). Born in Cornwall, Davy was the first to isolate the alkali metals and recognize the identity of chemical and electrical energy. A pioneer in the science of electrochemistry, he carried out basic studies of electrolysis of salts and water, and his application of electricity to the decomposition of molten caustic potash led to the isolation of metallic potassium.

"Dawsterol."[502] Trademark for a vegetable oil solution of vitamin D_3 for food fortification.

"Daxad."[311] Trademark for anionic, polymer-type dispersing agents. Supplied as light-colored powders or aqueous solutions. Effective dispersant for aqueous suspensions of insoluble dyestuffs, polymers, clays, tanning agents and pigments.
Uses: Manufacture of dyestuff pastes; textile backings, latex paints and paper coatings, retanning and bleaching of leather; dye resist in leather dyeing; dispersion of pitch in paper manufacture; pre-floc prevention in the manufacture of synthetic rubber.

"Daxan."[244] Trademark for a concentrated (minimum 95% available chlorine), dry form of chlorine specially formulated for disinfection of swimming pools.

"Daxtron."[233] Trademark for herbicides containing salt formulations of 2,3,5,6-tetrachloro-4-pyridinol.

2,4-DB. Abbreviation for 2,4-dichlorophenoxybutyric acid.

DBCP. Abbreviation for 1,2-dibromo-3-chloro-propane.

DBM. Abbreviation for dibutyl maleate.

DBMC. Abbreviation for 4,6-di-tert-butyl-meta-cresol.

"DB" Oil.[202] Trademark for a castor oil specially refined to minimum acidity and moisture content for dielectric and sonar applications, and for urethane polymers.

DBP. Abbreviation for dibutyl phthalate.

"dbpc."[11] Trademark for 2,6-di-tert-butyl-para-cresol, in technical, food and feed grades.

DBS. Abbreviation for dibutyl sebacate.

DCA. Abbreviation for deoxycorticosterone acetate.

DCB. Abbreviation for 1,4-dichlorobutane.

"DC Filtrol."[217] Trademark for an acid-activated clay used as a purifying agent in the recovery of drycleaning solvents.

DCHP. Abbreviation for dicyclohexyl phthalate.

DCO. Abbreviation for dehydrated castor oil. See castor oil, dehydrated.

DCP. Abbreviation for dicapryl phthalate.

DCPA. Abbreviation for dimethyl 2,3,5,6-tetrachloroterephthalate.

DCPC. Abbreviation for dichlorophenyl methyl carbinol. See di(para-chlorophenyl)ethanol.

DDB. Abbreviation for dodecylbenzene.

DDBSA. Abbreviation for dodecylbenzenesulfonic acid.

DDD. Abbreviation for dichlorodiphenyldichloroethane. See TDE.

DDDM. Abbreviation for 2,2′-dihydroxy-5,5′-dichlorodiphenylmethane. See dichlorophene.

DDE. Abbreviation for dichlorodiphenyldichloroethylene, $(CiC_6H_4)_2C{:}CCl_2$. It is a degradation product of DDT, found as an impurity in DDT residues.

DDH. Abbreviation for dichlorodimethylhydantoin.

"DDI Diisocyanate."[259] Trademark for an intermediate made from a 36-carbon dimer aliphatic dibasic acid; exact structure is a complex mixture of isomers. May be used for reaction with diamines to give polyureas with highly desirable properties; also a source of polyurethanes.

DDM. (1) Abbreviation for diaminodiphenylmethane (q.v.).
(2) Abbreviation for n-dodecyl mercaptan.

DDNP. Abbreviation for diazodinitrophenol.

DDP. Abbreviation for dodecyl phthalate.

DDQ. See 2,3-dichloro-5,6-dicyanobenzoquinone.

DDS. Abbreviation for diaminodiphenylsulfone. See sulfonyldianiline.

DDT (dichlorodiphenyltrichloroethane; dicophane; chlorophenothane; 1,1,1-trichloro-2,2-bis(chlorophenyl)ethane) $(ClC_6H_4)_2CHCCl_3$.
Properties: Colorless crystals or white to slightly off-white powder. Odorless or with slight aromatic odor. Insoluble in water; soluble in acetone, ether, benzene, carbon tetrachloride, kerosine, dioxane, and pyridine. Not compatible with alkaline materials.
Derivation: Condensing chloral or chloral hydrate with chlorobenzene in presence of sulfuric acid.
Grades: Technical; purified; aerosol; U.S.P., as chlorophenothane.
Containers: Bottles; tins; bags; fiber drums.
Hazard: Though not particularly toxic to humans under ordinary conditions, DDT is not biodegradable and is ecologically damaging. For the latter reasons agricultural use of DDT was prohibited in the U.S. in 1973, though its manufacture for export is permitted. Claims of human carcinogenicity have not been proved. DDT can be used for a few specialized purposes, e.g., to combat the tussock moth. Tolerance, 1 mg per cubic meter of air; 5 ppm in foods.
Uses: Insecticide, especially for tobacco and cotton; pesticide (tussock moth).

DDVP. Abbreviation for dimethyl dichlorovinyl phosphate. See dichlorovos.

D.E. Abbreviation for dextrose equivalent (q.v.).

DEA. Abbreviation for diethanolamine; also abbreviation for Drug Enforcement Administration, a government agency replacing the Bureau of Narcotics.

DEAC. Abbreviation for diethylaluminum chloride.

"Deacidite."[184] Trademark for a commercial grade anion exchanger consisting of a porous aliphatic polyamine weak base. Used in streptomycin conversion and strong acid adsorption.

Deacon process. A method of converting hydrogen chloride (HCl) to chlorine by oxidation of HCl with oxygen at 400 to 500°C over a copper salt catalyst: $2HCl + O_2 \longrightarrow Cl_2 + H_2O$. It is a means of producing chlorine without caustic and of utilizing the large amounts of by-product HCl from the chlorination of organic compounds. When conducted in the presence of an organic compound which reacts with the chlorine formed it is known as oxychlorination, e.g., $CH_2{=}CH_2 + 2HCl + \frac{1}{2}O_2 \longrightarrow CH_2ClCH_2Cl + H_2O$.

DEAE. Abbreviation for diethylaminoethyl-.

DEAE-cellulose (diethylaminoethyl cellulose). A cellulose ether containing the group $(C_2H_5)_2NCH_2CH_2$- bound to the cellulose in an ether linkage. An anionic ion-exchange material.
Use: Chromatography.

DEAE-dextran. A diethylaminoethyl ether of dextran; an electropositively charged polymer.

deanol. See 2-dimethylaminoethanol.

"Dearborn Red."[141] Trademark for azo color pigments, light, medium, and deep red shades. Used in paints and enamels.

deblooming agent. A substance added to mineral oils to mask fluorescence. Nitronaphthalene and yellow coal-tar dyes are among the products so used.

Debye, Peter J. M. (1884–1966). A Dutch chemist and physicist who received the Nobel Prize in 1936 for his pioneer studies of molecular structure by x-ray diffraction methods. The interference patterns are still called Debye-Sherrer rings. He also made outstanding contributions to knowledge of polar molecules and to fundamental electrochemical theory. (See also Debye-Huckel theory).

Debye-Hückel theory. A theory advanced in 1923 for quantitatively predicting the deviations from ideality of dilute electrolytic solutions. It involves the assumption that every ion in a solution is surrounded by an ion atmosphere of opposite charge. Results deduced from this theory have been verified for dilute solutions of strong electrolytes, and it provides a means of extrapolating the thermodynamic properties of electrolytic solutions to infinite dilution.

DEC. Abbreviation for beta-diethylaminoethyl chloride hydrochloride.

decaborane $B_{10}H_{14}$.
Properties: Colorless crystals; stable indefinitely at room temperatures; decomposes slowly into boron and hydrogen at 300°C; density (25/4°C) 0.94; m.p. 99.7°C; sp. gr. 0.78 (100°C); b.p. 213°C. Autoignition temp., 147°C. Slightly soluble in cold water; hydrolyzes in hot water; soluble in benzene, hexane, toluene.
Derivation: By-product of the pyrolysis of diborane.
Grades: Technical 95%; high purity 99%.
Hazard: Highly toxic. Tolerance, 0.05 ppm in air. Absorbed by skin. May explode in contact with heat or flame, or with oxygenated and halogenated solvents. Safety data sheet available from Manufacturing Chemists Assn., Washington, D.C.
Uses: Polymer synthesis; corrosion inhibitor; fuel addi-

tive; stabilizer; rayon delustrant; mothproofing agent; dye stripping agent; reducing agent; fluxing agent; oxygen scavenger; propellant.
Shipping regulatons: (Rail) Yellow label. (Air) Flammable Solid label. Not acceptable on passenger planes.

decachloro-1,1'-bis-2,4-cyclopentadienyl. See bis(pentachloro-2,4-cyclopenta dien-l-yl).

decachloro-octahydro-1,2,4-metheno-2H-cyclobuta-[cd]-pentalen-2-one $C_{10}C_{10}O$.
Properties: Solid; m.p. 359°C (dec.). Soluble in oxygenated solvents such as acetone; fairly soluble in organic solvents.
Uses: Insecticide; fungicide.

decaglycerol. See polyglycerol.

decahydronaphthalene $C_{10}H_{18}$. Cis- and trans- forms are known.
Properties: Colorless liquid; aromatic odor. Insoluble in water; soluble in alcohol and ether. Cis: Sp. gr. (20/4°C 0.8927; f.p. -43.2°C; b.p. 194.6°C; refractive index (n 20/D) 1.48113. Trans: Sp. gr. (20/4°C) 0.8700; f.p. -31.5°C; b.p. 185.5°C; refractive index (n 20/D) 1.46968. Flash point 136°F (closed cup). Autoignition temp. 482°F. Combustible.
Derivation: By treatment of naphthalene in a fused state (above 100°C) with hydrogen in the presence of a copper or nickel catalyst.
Grade: Technical.
Containers: 1-, 2-, 5-, 10-, 50-gal drums.
Hazard: Moderate fire risk. Moderately toxic; irritant to eyes and skin.
Uses: Solvent for oils, fats, waxes, resins, rubber, etc.; substitute for turpentine; cleaning machinery; stain-remover; shoe creams; floor waxes, etc.; cleaning fluids; lubricants.

Δ-decalactone. Artificial flavoring for margarine. Approved by FDA.

"Decalin."[28] Trademark for decahydronaphthalene.

"Decalso."[184] Trademark for a commercial grade cation exchanger consisting of a synthetic aluminosilicate gel (inorganic).
Uses: Water softening, separation of amino acids; radioactive waste treatment; milk treatment.

decamethonium bromide (decamethylenebis(trimethylammonium bromide) $(CH_3)_3NBr(CH_2)_{10}NBr(CH_3)_3$.
Properties: Crystals; decomposes 255–267°C. Soluble in water and alcohol; very slightly soluble in chloroform; insoluble in ether.
Use: Medicine (anesthetic).

decamethylene-bis(4-aminoquinaldinium acetate). See dequalinium.

decamethylenebis(trimethylammonium bromide). See decamethonium bromide.

n-**decanal** (capraldehyde; capric aldehyde; n-decyl aldehyde; aldehyde C-10) $CH_3(CH_2)_8CHO$.
Properties: Colorless to light yellow liquid; floral-fatty odor; sp. gr. 0.831–0.838 (15°C); refractive index (n 20/D) 1.427–1.431. Soluble in 80% alcohol, fixed oils, volatile oils, mineral oil; insoluble in water and glycerol. Combustible. Low toxicity.
Derivation: Occurs in lemongrass, citronella, orange, and many other oils. Synthetically by oxidation of the corresponding alcohol or reduction of the acid.
Grades: Technical; F.C.C.
Containers: Cans; drums.
Use: Perfumery; flavoring.

n-**decane** (decyl hydride) $CH_3(CH_2)_8CH_3$.
Properties: Colorless liquid; sp. gr. 0.7298; b.p. 174°C; f.p. -30°C; refractive index (n 20/D) 1.4114; flash point 111°F (closed cup). Autoignition temp. 482°F. Soluble in alcohol; insoluble in water. Combustible.
Grades: Technical; 95%; 99%; research.
Containers: Bottles; drums.
Hazard: Moderate fire risk. Narcotic in high concentrations.
Uses: Organic synthesis; solvent; standardized hydrocarbon; jet fuel research.

decanedioic acid. See sebacic acid.

decanoic acid. See capric acid.

1-decanol (n-decyl alcohol; alcohol C-10) $CH_3(CH_2)_8CH_2OH$.
Properties: Colorless, water-white liquid. Sweet odor; sp. gr. 0.829; b.p. 232.9°C; m.p. 6°C; flash point (open cup) 180°F; refractive index (n 20/D) 1.4372. Insoluble in water (25°C); soluble in alcohol and ether. Combustible. Low toxicity.
Derivation: Reduction of coconut oil fatty acids; from C_9 olefin and synthesis gas, by the Oxo process.
Grades: Technical; high purity.
Containers: Drums; tank cars.
Uses: Plasticizers; detergents; synthetic lubricants; solvents; perfumes; flavorings.

decanoyl chloride (sometimes called caproyl chloride) $CH_3(CH_2)_8COCl$. Available in bottles; carboys and drums. Intermediate; polymerization initiator.

decanoyl peroxide $CH_3(CH_2)_8C(O)OOC(O)(CH_2)_8CH_3$.
Properties: Soft white granules; m.p. 38–42°C (dec.); insoluble in water and alcohol; soluble in ether and benzene.
Hazard: Strong oxidizing agent; fire risk in contact with organic materials.
Use: Polymerization catalyst.
Shipping regulations: (Rail) Oxidizing material, n.o.s., Yellow label. (Air) Peroxides, organic, solid, n.o.s., Organic Peroxide label.

"Decanox."[154] Trademark for decanoyl peroxide (q.v.).

decarboxylase One of a group of enzymes in the living cell that removes carbon dioxide from various carboxylic acids without oxidation.

decay. Spontaneous disintegration of an unstable atomic nucleus, e.g., uranium, radium, with emission of alpha, beta, and gamma radiation and eventual formation of another element of lower atomic weight. See radioactivity; half-life.

1-decene. See decylene.

"Dechlorane."[62] Trademark for perchloropentacyclodecane.

"Decholin."[272] Trademark for dehydrocholic acid (q.v.).

"Declomycin."[315] Trademark for demethylchlortetracycline hydrochloride (q.v.).

decoction. Pharmaceutical term for a liquid produced by boiling one or more drugs in water and filtering.

decoic acid. See capric acid.

decolorizing agent. Any material that removes color by a physical or chemical reaction. Charcoals, blacks, clays, earths or other materials of highly adsorbent character used to remove undesirable color, as from sugar, vegetable and animal fats and oils, etc. Also refers to bleaches involving a chemical reaction for removing color.

decomposition. (1) A fundamental type of chemical change. In simple decomposition, one substance breaks down into two simpler substances, e.g., water yields hydrogen and oxygen. In double decomposition, two compounds break down and recombine to form two different compounds, e.g.,
$2HCl + CaCO_3 \longrightarrow CaCl_2 + H_2CO_3$.
In some cases heat is absorbed, and in others it is evolved. Decomposition may also be induced by radiation (sunlight), as in the breakdown of chlorine-containing fluorocarbons in the upper atmosphere and of biodegradable polymer structures (also called photodecomposition or photolysis); and by bacterial action (degradation, q.v.).

(2) Conversion of a carbonaceous raw material into carbon and volatile organic compounds by exposure to high temperature in the absence of air, i.e., without combustion. Thermal decomposition products of coal are coke, coal tar and coal gas; similarly hydrocarbon mixtures are obtained from shale oil. The term thermal decomposition is virtually synonymous with pyrolysis and destructive distillation.

decontamination. Removal of radioactive poisons from skin, clothing, equipment, etc. Skin can often be decontaminated by washing with soap and water; application of titanium dioxide paste or a saturated solution of potassium permanganate followed by a rinse of 5% sodium bisulfite is approved procedure. Contaminated clothing should not be sent to commercial laundries nor burned in open incinerators. Water, steam, and detergents are effective on painted or metal surfaces.

"Decroline."[307] Trademark for a series of stripping agents consisting of zinc sulfoxylate formaldehyde.

decyl acetate (acetate C-10) $CH_3(CH_2)_9OOCCH_3$.
Properties: Liquid with floral orange-rose odor; b.p. 187–190°C; sp. gr. 0.862–0.864; refractive index 1.426. Soluble in 80% alcohol, ether, benzene, glacial acetic acid; insoluble in water. Combustible.
Grades: Technical.
Use: Perfumery.

n-decyl alcohol. See 1-decanol.

n-decyl aldehyde. See n-decanal.

n-decylamine $CH_3(CH_2)_9NH_2$.
Properties: Water-white liquid; amine odor; boiling range 215–221°C; sp. gr. 0.797 (20/20°C); refractive index 1.437 (20°C); flash point 210°F. Combustible.
Hazard: May be toxic and irritant.

decyl carbinol. See 1-undecanol.

decylene (1-decene) $C_{10}H_{20}$ or $H_2C{:}CH(CH_2)_7CH_3$.
Properties: Colorless liquid; sp. gr. 0.7396 (20/4°C); b.p. 172°C; f.p. −66.3°C; refractive index (n 20/D) 1.4220. Soluble in alcohol; insoluble in water. Combustible.
Grades: Technical; high purity.
Use: Organic synthesis of flavors, perfumes, pharmaceuticals, dyes, oils, resins.

decyl hydride. See n-decane.

decylic acid. See capric acid.

decyl mercaptan $C_{10}H_{21}SH$.
Properties: Liquid, f.p. −26°C; b.p. 114°C (13 mm); sp. gr. 0.8410 (20/4°C); refractive index 1.4536 (n 20/D). Combustible; strong odor.
Grade: 95% (min) purity.
Hazard: Toxic by ingestion and inhalation.
Uses: Intermediate; synthetic rubber processing.

decyl-octyl methacrylate
$H_2C{:}C(CH_3)COO(CH_2)CH_3$.
Containers: Drums.
Uses: Monomer for plastics, molding powders, solvent coatings, adhesives, oil additives; emulsions for textile, leather, and paper finishing.

"DeeGee".[212] See "Dee O."

"Deenax."[29] Trademark for di-tert-butyl-para-cresol (DBPC), an oxidation inhibitor used especially in waxes and natural fats.

"Dee O."[212] Trademark for a glucose oxidase enzyme system catalyzing the reaction of glucose to gluconic acid with the uptake of oxygen.
Uses: Beverages; salad dressing and other oxygen sensitive foods; analytical; stabilizing egg solids by desugaring whites, yolks or whole eggs.

deet. See N,N-diethyl-meta-toluamide.

"DEF."[181] Trademark for S,S,S-tributyl phosphorotrithioate (q.v.).

defecation. Purification; used specifically of the industrial clarification of sugar solutions.

deflagration. Very rapid autocombustion of particles of explosive as a surface phenomenon. Initiated by contact of a flame or spark, but may be caused by impact or friction. Deflagration is a characteristic of low explosives. See also detonation.

defoaming agent. A substance used to reduce foaming due to proteins, gases, or nitrogenous materials, which may interfere with processing. Examples are 2-octanol, sulfonated oils, organic phosphates, silicone fluids, dimethylpolysiloxane, etc. For restrictions on their use in foods, see FDA regulations.

defoliant. An herbicide (q.v.) that removes leaves from trees and growing plants. They may be either organic or inorganic. Some examples: (organic) phenoxyacetic acids, trichloropicolinic acid, carbamates, and nitro compounds; (inorganic) arsenic compounds, cyanides, thiocyanates, and chlorates. Several of the more persistent types have been used in military operations.

DEG. (1) Abbreviation for diethylene glycol. (2) Abbreviation for diethanolglycine.

DEGN. Abbreviation for diethylene glycol dinitrate.

degradation. A type of decomposition characteristic of high molecular weight substances such as proteins, polymers, branched-chain sulfonates, etc. It may result from oxidation, heat, solvents, bacterial action, or in the case of body proteins from infectious microorganisms. See also biodegradability; decomposition.

degras. Crude wool grease obtained by solvent-washing of wool. It is a dark brown semisolid with strong, unpleasant odor and high water-absorbing capacity. A type known as moellen degras is a by-product of tanning chamois leather with various fish oils. The chief use of degras is as the source of lanolin; minor uses are in leather dressing and printing inks. Available in several grades (neutral, common, and technical).

"deGreen."[1] Trademark for a cotton defoliant of which the active ingredient is S,S,S-tributyl phosphorotrithioate (q.v.).

degree of polymerization (D.P.). The number of monomer units in an average polymer molecule in a given sample; for natural cellulose it is about 3000, but in most polymers it is still higher. It can be controlled by appropriate processing techniques. D.P. is an important factor in plastics technology, as it directly affects the viscosity of solutions and properties of the end product.
See also polymerization; shortstopping agent.

"D.E.H."[233] Trademark for a variety of polyamines and polyamides suitable for curing epoxy resins.

dehumidification. The removal of moisture (water vapor) from air. Also sometimes extended to analogous processes of removing a vapor from a gas mixture.

"Dehybor."[441] Trademark for anhydrous sodium tetraborate (q.v.).

"Dehydratine."[205] Trademark for bituminous water barrier coatings.

dehydration. Removal of 95% or more of the water from a material, usually a foodstuff, by exposure to high temperature by various means. Its primary purpose is to reduce the volume of the product, increase its shelf-life, and lower transportation costs. Special equipment for dehydration includes tunnel dryers, vacuum (shelf) dryers, drum dryers, etc., in which the bulk product is exposed to a hot-air environment. Another method is spray-drying, in which a liquid product is ejected from a nozzle into hot air; dried milk and egg-white are prepared in this way. The term dehydration is not applied to loss of water by evaporation or sun-drying. See also drying.

"Dehydrite."[16] Trademark for anhydrous granular magnesium perchlorate (q.v.).

dehydroabietic acid $C_{20}H_{28}O_2$. Solid; used as a basis for thermoplastic resins.

dehydroabietylamine $C_{19}H_{27}NH_2$ (having a phenanthrene ring system). A rosin-derived compound. See abietic acid.

dehydroacetic acid (DHA; methylacetopyranone)

$CH_3C{:}CHC(O)CH(COCH_3)C(O)O$.

Properties: Colorless, odorless, tasteless crystals. M.p. 108.5°C. Soluble in acetone, alcohol, and ether; insoluble in water; highly reactive; combustible.
Derivation: (a) By action of N-bromosuccinimide on ketene dimer; (b) by strong heating of acetoacetic ester.
Grades: Technical; F.C.C.
Hazard: Moderately toxic by ingestion.
Uses: Fungicide and bactericide; plasticizer; chemical intermediate.

dehydroascorbic acid

$OCOCOCOCHCHOHCH_2OH$. The oxidized form of ascorbic acid, with the same vitamin activity.
Properties: Needles; m.p. 225° (dec); soluble in water at 60°C.
Derivation: Synthesized from ascorbic acid.
Uses: Nutrition; medicine.

7-dehydrocholesterol (provitamin D_3) $C_{27}H_{44}O \cdot H_2O$. A sterol found in the skin of man and animals which forms vitamin D_3 upon ultraviolet irradiation.

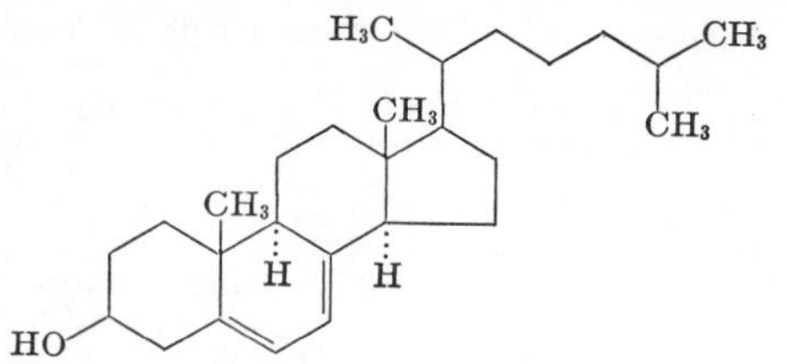

Properties: Slender platelets from ether-methanol; m.p. 150°C; insoluble in H_2O; soluble in organic solvents.
Uses: Nutrition; medicine; biochemical research.
See also cholecalciferol.

dehydrocholic acid $C_{24}H_{34}O_5$. A polycyclic compound.
Properties: White, fluffy, odorless powder with bitter taste; m.p. 231–240°C. Almost insoluble in water; slightly soluble in ether and alcohol; soluble in chloroform, glacial acetic acid, and solutions of alkali hydroxides and carbonates.
Derivation: Oxidation of cholic acid.
Grade: N.F.
Hazard: May have toxic reactions.
Uses: Medicine; pharmaceutical intermediate.

dehydrocyclodimerization. A method of converting paraffin (straight-chain) hydrocarbons containing from 3 to 5 carbon atoms into aromatic (ring-type) hydrocarbons. Its main steps are: (a) removal of hydrogen from the paraffins; (b) dimerization of the resulting olefins; (c) aromatization of the dimerized olefins and diolefins; (d) isomerization or transalkylation to C_8 to C_{10} alkyl benzene isomers. Metallic catalysts are essential in some or all of these steps. The process is not in large-scale use.

dehydroepiandrosterone. See dehydroisoandrosterone.

dehydrogenase. An enzyme which catalyzes oxidation by the removal of hydrogen. See oxidase.

dehydrogenation. The process whereby hydrogen is removed from compounds by chemical means. Dehydrogenation of primary alcohols yields the group of compounds called aldehydes (q.v.). It is considered to be a form of oxidation, as two hydrogen atoms, each of which contains an electron, have been removed, as in the reaction $CH_3CH_2OH \longrightarrow CH_3CH = O + H_2$.

11-dehydro-17-hydroxycorticosterone. See cortisone.

dehydroisoandrosterone (dehydroepiandrosterone) $C_{19}H_{28}O_2$. An androgenic steroid; a metabolic product of the adrenal steroid hormones, with about one-third of the androgenic activity of androsterone (q.v.).
Properties: Dimorphous: Needles with m.p. 140–141°C; leaflets with m.p. 152–153°C; precipitated by digitonin; soluble in benzene, alcohol, and ether. Sparingly soluble in chloroform and petroleum ether. Also available as the acetate salt.
Derivation: Isolated form male urine; synthesis from cholesterol or sitosterol.
Uses: Medicine; biochemical research.

"Dehydrol."[141] Trademark for dehydrated castor oil used as a drying oil in the manufacture of varnishes and alkyd resins.

3-dehydroretinol (vitamin A_2) $C_{20}H_{27}OH$.
Properties: Golden yellow oil (may be a mixture of stereoisomers). Readily affected by oxygen.
Derivation: From pike liver oils.

Use: Nutrition.
See also vitamin A.

dehydrothio-para-toluidine $CH_3C_6H_3SC(C_6H_4NH_2)N$.
Properties: Long, yellowish iridescent needles. Solutions have a violet-blue fluorescence. M.p. 191°C; b.p. 434°C. Soluble in alcohol; very slightly soluble in water.
Derivation: By heating para-toluidine and primuline base with sulfur and separation from the primuline base by distillation in vacuo.
Uses: Dyestuffs; intermediate.

deicing compound. See calcium chloride; sodium chloride; alcohol.

de-inking. The removal of printing inks from paper by use of strong alkaline solutions such as soda-ash liquor, caustic soda or lime which dissolve varnish and free the ink carbon. Removal of the carbon is accomplished by use of colloidal agents such as talc or bentonite and by mechanical agitation with water.

"Dekatyl."[28] Trademark for a series of dyes for dyeing and printing 65% "Dacron" polyester fiber and 35% cotton.

deKhotinsky cement. A thermoplastic adhesive mixture of shellac and pine tar. It is not attacked by water, sulfuric acid, nitric acid, hydrochloric acid, carbon disulfide, benzene, gasoline, or turpentine; very little affected by ether, chloroform, alkalies, but readily dissolved by ethyl alcohol.

"Delac."[248] Trademark for a series of delayed action rubber accelerators.

"Delactol."[503] Trademark for a vegetable oil solution of vitamin D_2; used in dairy products.

"Delamin."[266] Trademark for a series of fatty amines.
P. Technical grade, high molecular weight, primary amine derived from 18-carbon fatty acids of tall oil origin.
80. A low-cost flotation reagent designed especially for beneficiation of phosphate and other nonmetallic ores; consists principally of stearyl and oleyl amines.
100. A high molecular primary amine designed for flotation of silica and silicate minerals; derived from 18-carbon fatty acids.

delhi hard. A ferrous alloy (sp. gr. 7.75; m.p. 1500°C) containing in addition to iron 16.5 to 18% chromium, 1 to 1.1% carbon, 0.75 to 1% silicon, 0.35 to 0.5% manganese. It is resistant to cold ammonium hydroxide in all concentrations, and to mine and sea waters and moist sulfurous atmospheres.

deliquescent. Tending to absorb atmospheric water vapor and become liquid. The term refers specifically to water-soluble chemical salts in the form of powders, which dissolve in the water absorbed from the air. Such salts should be kept closely stoppered or otherwise enclosed. See also hygroscopic.

"Delnav."[266] Trademark for technical grade of 2,3-para-dioxanedithiol S,S-bis(O,O-diethylphosphorodithioate).

"Delrin."[28] Trademark for a type of acetal resin. White and colors available. Also supplied as pipe and fittings. Thermoplastic.
Containers: 50-lb bags; pipe in 20-ft lengths, or coils of 500 ft.
Uses: Injection-molded and extruded parts, door handles, bushings, other mechanical items; underground pipe; automotive parts.

"Delsan."[28] Trademark for fungicide-insecticide seed treatment containing 60% thiram and 15% dieldrin.
Hazard: Toxic by ingestion and inhalation.

delta acid. See 2-naphthylamine-7-sulfonic acid.

"Deltyl."[227] Trademark for a mixture of isopropyl esters of lauric, myristic and palmitic acids. "Deltyl Extra" is predominantly isopropyl myristate; "Deltyl Prime," isopropyl palmitate.
Uses: Replaces vegetable or mineral oils in cosmetics; emollient and auxiliary emulsifying agent.

delustrant. A substance used to produce dull surfaces on a textile fabric; chiefly used are barium sulfate, clays, chalk, etc. They are applied in the finishing coat. See also weighting agent.

"Delvex."[100] Trademark for dithiazanine iodide (q.v.).

demecarium bromide $C_{32}H_{52}Br_2N_4O_4$.
Properties: White, slightly hygroscopic powder; decomposes 162–167°C. Freely soluble in water, alcohol; sparingly soluble in acetone; insoluble in ether. Aqueous solutions are neutral, stable and may be sterilized by heat.
Use: Medicine.

"Demerol" Hydrochloride.[162] Trademark for meperidine hydrochloride (q.v.).

11-demethoxyreserpine. See deserpidine.

demethylchlortetracycline hydrochloride
$C_{21}H_{21}ClN_2O_8 \cdot HCl$.
Properties: Yellow crystalline powder; odorless and has a bitter taste. Partially soluble in water and slightly soluble in alcohol.
Grade: N.F.
Use: Medicine (antibiotic).

demeton. Generic name for a mixture of O,O-diethyl O-2-(ethylthio)-ethyl phosphorothioate (demeton-O), and O,O-diethyl S-2-(ethylthio)ethyl phosphorothioate (demeton-S). See also "Systox." A cholinesterase inhibitor.
Properties (of mixture): Pale yellow liquid; b.p. 134°C; (2 mm); sp. gr. 1.118. Slightly soluble in water; soluble in most organic solvents.
Hazard: Highly toxic; absorbed by skin. Tolerance, 0.1 mg per cubic meter of air. Use may be restricted.
Use: Systemic insecticide (absorbed by plant, which then becomes toxic to sucking and chewing insects).
Shipping regulations: Organic phosphate, liquid, n.o.s., (Rail) Poison label. (Air) Poison label. Not acceptable on passenger planes.
Note: Approved by EPA as substitute for certain uses of DDT.

demineralization. Removal from water of mineral contaminants, usually present in ionized form. The methods used include ion-exchange techniques, flash distillation, or electrodialysis (q.v.). Acid mine wastes may be purified in this way, thus aiding the pollution problem. See also desalination; water conditioning.

Democritus (about 465 B.C.). A Greek philosopher, the first thinker of record to conceive of matter as existing in the form of small, indivisible particles, which he called atoms. However, this concept was overshadowed by Aristotle's theories and it was not until some 2000 years later that it was developed by John Dalton in England—an astonishing length of dormancy for one of the most creative ideas in the history of science. (See also Dalton).

demulsification. The process of destroying or "breaking" an unwanted emulsion, especially water-in-oil types occurring in crude petroleum. Both chemical and physical means are used: addition of polyvalent ions to neutralize electrical charges, or of a strong acid; physical means include heating, centrifuging or use of high-potential alternating current. See also emulsion; nonylphenol.

demurrage. A fee imposed on shippers of chemicals and other products by the railroads for retaining freight cars at loading docks for more than 24 hours. Until recently the retention time allowed was 48 hours.

"D.E.N."[233] Trademark for a series of epoxy novolacs for multi functional resins for all uses where maximum chemical or heat resistance is required.

denatonium benzoate. USAN for benzyldiethyl-[(2,6-xylylcarbamoyl)methyl]ammonium benzoate ("Bitrex"); a bitter-tasting compound approved as a denaturant for alcohol.

denaturant. See alcohol, denatured.

denaturation. A change in the molecular structure of globular proteins that may be induced by bringing a protein solution to its boilng point, or by exposing it to acids or alkalies or to various detergents. Denaturation reduces the solubility of proteins and prevents crystallization. It involves rupture of hydrogen bonds, so that the highly ordered structure of the native protein is replaced by a looser and more random structure. It is usually irreversible, but in some cases is reversible, depending on the protein and the treatment involved. See also degradation.

denatured alcohol. See alcohol, denatured.

denier. A unit used in the textile industry to indicate the fineness of a filament. If 9000 meters of a filament weighs 1 gram, the filament is 1 denier; if 10,000 meters weighs 1 gram, the filament is 1 grex. Sheer women's hosiery usually runs from 15 to 20 denier.

"Densitol" Brominated Sesame Oil.[3] Trademark for an oil used to adjust specific gravity of citrus oils to provide cloud in beverages.

density. Weight per unit volume, usually expressed in grams per cubic centimeter or in pounds per cubic foot or gallon. Apparent density is the weight of a unit volume of powder, usually expressed as grams per cubic centimeter, determined by a specified method. (M.P.A. definition, M.P.A. Standard 9-50T). Bulk density is an alternate term for apparent density. See also specific gravity; current density.

"Denzox."[468] Trademark for lead-free zinc oxide. Used in ceramics.

"Deo-Base."[45] Trademark for light petroleum distillate; superfine grade of kerosine without its objectionable odor.

deodorant. A substance used to remove or mask an unpleasant odor. It may or may not have a distinctive odor of its own. Deodorants act (1) by adsorption (activated carbon, charcoal, chlorophyllin); (2) by replacement (pine oil or other perfume); (3) by neutralization (aluminum chlorohydrate); and (4) by oxidation or hydrogenation e.g., of fish oils. The cosmetic industry supplies a wide variety of deodorants and antiperspirants chiefly based on neutralization. Mouth washes and breath "sweeteners" often contain calcium iodate, thymol, peppermint, or similar substance to mask or replace odors. See also odor; cosmetic.

deoxidizer. An agent which removes oxygen from a compound or from a molten metal.

deoxy-. Preferred prefix indicating replacement of hydroxyl by hydrogen in the parent compound. The meaning is the same as that of desoxy- and the two prefixes are used interchangeably.

deoxyanisoin (4′-methoxy-2-(para-methoxyphenyl)acetophenone) $CH_3OC_6H_4COCH_2C_6H_4OCH_3$.

Properties: Off-white to buff, crystalline powder with a sweet, faint cinnamon-like odor; m.p. 110–112°C.

Use: Intermediate.

deoxybenzoin (alpha-phenylacetophenone; benzyl phenyl ketone) $C_6H_5CH_2COC_6H_5$.

Properties: Colorless crystals; m.p. 53–60°C; slightly soluble in hot water; soluble in alcohols and ketones.

Uses: Intermediate.

deoxycholic acid (desoxycholic acid) $C_{24}H_{40}O_4$. A bile acid; contains one less hydroxyl group than cholic acid.

Properties: Crystals; m.p. 172–173°C. Not precipitated by digitonin. Practically insoluble in water and benzene; slightly soluble in chloroform and ether; soluble in acetone and solutions of alkali hydroxides and carbonates; freely soluble in alcohol. Also available as sodium salt.

Derivation: Isolation from bile; organic synthesis.

Grades: Technical; F.C.C. (as desoxycholic acid).

Uses: Medicine; precursor for organic synthesis of cortisone; emulsifying agent in foods (up to 0.1%).

deoxycorticosterone (4-pregnen-21-ol-3,20-dione; 11-deoxycorticosteroid) $C_{21}H_{30}O_3$. An adrenal cortical steroid hormone. Active in causing the retention of salt and water by the kidney.

Properties: Crystalline plates; m.p. 141–142°C. Freely soluble in alcohol and acetone.

Derivation: From adrenal cortex extract; synthesis from other steroids.

Use: Medicine (usually as acetate or pivalate).

deoxyribonuclease. One of a group of enzymes which cause the splitting of deoxyribonucleic acids. Pancreatic deoxyribonuclease, the most widely studied, cleaves the acid at the 3′-phosphate bond. Other deoxyribonucleases cleave the 5′-phosphate bond.

deoxyribonucleic acid (DNA). A complex sugar-protein polymer or nucleoprotein (q.v.) which contains the complete genetic code for every enzyme in the cell. It occurs as a major component of the genes, which are located on the chromosomes in the cell nucleus. The DNA molecule is a unique and vastly intricate structure; it is comprised of from 3000 to several million nucleotide units arranged in a double helix containing phosphoric acid, 2-deoxyribose, and the nitrogenous bases adenine, guanine, cytosine, and thymine. The spiral consists of two chains of alternating phosphate and deoxyribose units in continuous linkages. The nitrogenous bases project toward the axis of the spiral and are joined to the chains by hydrogen bonds. Adenine always unites with thymine, and cytosine with guanine. The complementarity of the bases on the joined chains allows each chain to act as a template for replication of the other when the chains are separated, thus producing two new strands of DNA. The sequence of the bases on the chains varies with the individual, and it is this sequence that governs the genetic code. DNA works in conjunction with ribonucleic acid (RNA). Synthesis of self-replicating DNA was reported late in 1967. See also ribonucleic acid; gene; nucleic acid; genetic code; replication; recombinant DNA.

D-**deoxyribose** $CH_2OHCHOHCHOHCH_2CHO$. A five-carbon-atom sugar that is unusual because there is no oxygen atom attached to the second carbon atom. It is a constituent of deoxyribonucleic acid (q.v.).

"Deoxy-Sol."[552] Trademark for an aqueous solution of hydrazine; used for boiler feed treatment.

2,4-DEP Tris-(2,4-dichlorophenoxy) ethyl phosphite. An herbicide.

DEP. Abbreviation for diethyl phthalate.

DEPC. Abbreviation for diethylpyrocarbonate (q.v.).

Department of Transportation (DOT). The Federal agency which has been responsible since 1967 for the regulation and control of transportation of hazardous materials.

"Depban."[233] Trademark for paraffin inhibitors for use in oil-well equipment.

DEPC. Abbreviation for gamma-diethylaminopropyl chloride hydrochloride.

"Dependip."[200] Trademark for a petroleum solvent prepared by straight-run overhead distillation.
Properties: Water-white; sp. gr. 0.758 (60°F); wt/gal 6.31 lb (60°F); flash point (TOC) 52°F; considered non-toxic.
Hazard: Flammable, dangerous fire risk.
Use: Solvent in rubber-dipping cement.

dephlegmation. Partial condensation of vapor from a distillation operation to produce a liquid richer in higher-boiling constituents than the original vapor. The residual vapor is richer in the lower-boiling constituents.

depilatory. A substance used to remove hair from skin. Sulfides are largely used for this purpose. The leather industry uses large amounts of sodium sulfide for unhairing hides. The cosmetic industry also offers various sulfide preparations for removing unsightly body hair.

"Depilin."[57] Trademark for a series of organic dehairing agents for dehairing hides in the leather industry.

dequalinium. Short for decamethylene-bis(4-aminoquinaldinium acetate), an antibiotic oral antiseptic.

"D.E.R."[233] Trademark for a series of epoxies, including liquid resins, solid resins and solutions, flexible resins, and flame-retardant resins.

"Deraspan."[233] Trademark for a group of epoxy resins and curing agents.

"Dergon."[300] Trademark for a series of liquid detergents used for textile scouring.

"Dergopal."[300] Trademark for a series of fluorescent whitening agents for natural and synthetic textile fibers. Different members applied to specific fibers. Includes both coumarin and stilbene types.

"Deriphat."[259] Trademark for a series of amphoteric surfactants.
Uses: Cosmetic and detergent formulations.

"Deronil."[321] Trademark for a brand of dexamethasone.

derris root. The root of the shrubs Derris elliptica and D. malaccensis. Chief active constituent is rotenone (q.v.). Used as an insecticide.

DES. Abbreviation for diethylstilbestrol.

desalination (desalting). Any of several processes for removing dissolved mineral salts from ocean water and other brines. The most important are: (1) Distillation (q.v.), with reuse of vapors by compressive distillation or multiple-effect evaporation. Solar distillation has been in use on the Greek islands for some years. (2) Electrodialysis (q.v.), an ion-exchange method more efficient for purification of brackish water than sea-water (see dimineralizaton). (3) Reverse osmosis, which uses pressure applied to the surface of a saline solution which is separated from pure pure water by a semipermeable membrane which ions cannot easily penetrate. (See also osmosis). The pressure forces the water component of the solution through the membrane, thus effectively separating the components of the solution. Membranes used are cellulose acetate or graphitic oxide. This method is planned for use in a desalination plant proposed for the brackish waters of the lower Colorado River, and is said to be the world's largest. It is also used in a Potomac River installation. (4) Flash distillation (q.v.) appears to be the most effective method so far developed for sea water desalination, accounting for about 90% of world production capacity.

There are about 350 desalination plants in the U.S. producing over 65 million gallons of fresh water a day. Development is under control of the Office of Saline Water, Dept. of the Interior.

"Desiccite."[217] Trademark of adsorbent used for static dehumidification in protective packaging of metal equipment, food, and pharmaceuticals.

desiccant. A hygroscopic substance such as activated alumina, calcium chloride, silica gel, or zinc chloride. Such substances adsorb water vapor from the air and are used to maintain a dry atmosphere in containers for food packaging, chemical reagents, etc. See also molecular sieve.

"Desicote."[274] Trademark for a mixture of hydrophobic monomers stabilized in chlorinated hydrocarbon and aromatic solvents. Rapidly decomposes on contact with sorbed water on glass surfaces, leaving surface water-repellent.
Derivation: Mixed silanes.
Grades: Green label (approved for U.S. Parcel Post); yellow label (approved for Railway Express); red label (special precautions required for shipment).
Note: The Green label is the commercial stock in general use.
Containers: Two-ounce glass bottles with heavy molded cap and special cap liners packed in friction top steel can with absorbent filler to conform to safety requirements, approved by U.S. Post Office.
Hazard: Moderately toxic; skin irritant.
Uses: May be used safely on the most delicate glassware such as absorption cells of "Pyrex," "Vycor,"

and fused silica and which will not give any interference in the visible or ultraviolet range; also for pH-sensitive glass electrodes and the like.
Shipping regulations: See above, Grades.

desiodothyroxine. See thyronine.

"Desmocoll."[470] Trademark for a series of isocyanate-modified polyesters for adhesive application.

"Desmodur."[470] Trademark for a group of isocyanates and isocyanate prepolymers for urethane coatings, foams, adhesives, etc.

"Desmophen."[470] Trademark for a group of polyesters and polyethers for crosslinking with isocyanates.

desorption. The process of removing an adsorbed material from the solid on which it is adsorbed. See adsorption. Desorption may be accomplished by heating, by reduction of pressure, by the presence of another more strongly adsorbed substance, or by a combination of these means.

desoxy-. See deoxy-.

desoxycholic acid. F.C.C. name for deoxycholic acid.

"Desoxyn".[3] Trademark for methamphetamine hydrochloride.

destructive distillation. An operation in which a highly carbonaceous material, such as coal, oil shale, or tar sands is subjected to high temperature in the absence of air or oxygen, resulting in decomposition to solids, liquids and gases. As the solid end product is carbon, the term carbonization is often used. Other terms with the same general meaning as destructive distillation are pyrolysis and thermal decomposition. Destructive distillation of coal is carried out in the temperature range of 350 to 1000°C, yielding coal tar, coal gas, and char (coke, carbon). The coal tar is then distilled into various fractions; it may also be hydrogenated to produce synthetic petroleum. Fuel gas can be obtained from shale oil by destructive distillation. See also pyrolysis; thermal decomposition.

"Destun".[158] Trademark for perfluidone (q.v.).

"Detamide" 95.[428] Trademark for N,N-diethyl-meta-toluamide.

detergent. Any substance that reduces the surface tension of water; specifically, a surface-active agent which concentrates at oil-water interfaces, exerts emulsifying action, and thus aids in removing soils. The older and still widely used types are the common sodium soaps of fatty acids, which are relatively weak. The much stronger synthetic detergents are classed as anionic, cationic, or nonionic, depending on their mode of chemical action. The latter functions by a hydrogen bonding mechanism. The most widely used group comprises linear alkyl sulfonates (LAS), often aided by "builders" (q.v.). LAS are preferable to alkyl benzene sulfonates (ABS), because they are readily decomposed by microorganisms (biodegradable). LAS are straight-chain compounds having ten or more carbon atoms in the chain; the branched-chains characteristic of ABS resist decomposition; these have been largely replaced by LAS because of water pollution. Replacement of LAS with linear C_{12} to C_{18} alcohols may be expected. See surface tension; emulsion; wetting agent; soap (1); alkylate (3); biodegradability; eutrophication; builder.

"Dethdiet."[342] Trademark for red squill rodenticide concentrates.
Hazard: Toxic by ingestion.

"Dethmor"[342] Trademark for warfarin rodenticide concentrates (q.v.).
Hazard: Highly toxic by ingestion.

detonation. The extremely rapid, self-propagating decomposition of an explosive accompanied by a high pressure-temperature wave that moves at from 1000 to 9000 meters per second. May be initiated by mechanical impact, friction, or heat. Detonation is a characteristic of high explosives, which vary considerably in their sensitivity to shock, nitroglycerin being one of the most dangerous in this respect. See also explosive, high; deflagration.

deuterium (heavy hydrogen). Symbol D. An isotope of hydrogen whose nucleus contains one neutron and one proton, and is therefore twice as heavy (at. wt. 2.014) as the nucleus of normal hydrogen. The ratio in nature is 1 part deuterium to 6500 parts normal hydrogen. See deuteron.
Properties: Almost identical with hydrogen. Sp. gr. (H = 1) 2.0; f.p. -254.5°C (121 mm); b.p. -249.5°C. Ignition point 1085°F. Noncorrosive, nontoxic.
Derivation: Electrolysis of high purity heavy water.
Grades: 98, 99.5 atom %.
Containers: High-pressure steel cylinders.
Hazard: Highly flammable and explosive. Explosive range 4 to 74%.
Uses: Bombardment of atomic nuclei; tracer element. See also deutero-; heavy water.
Shipping regulations: (Rail) Red Gas label. (Air) Flammable Gas label. Not accepted on passenger planes.

deuterium oxide. See heavy water.

deutero- (deuterated). Prefix indicating that one or more of the hydrogens in a compound is the deuterium isotope. Example: deuteroborane solution, used for labeling olefinic unsaturation. The adjective form, deuterated, has the same meaning. Deuterated ethylene, sometimes written ethylene-1,1-d_2, has the formula $CH_2{:}CD_2$.

deuteron (deuton). A nuclear particle having mass 2 and a positive charge of 1; identical with the nucleus of the deuterium atom.

Devarda's metal (Devarda's alloy).
Properties: Gray powder. Contains copper, aluminum, and zinc in the proportion of 50:45:5. Slightly soluble in hydrochloric acid.
Grade: Reagent (20-mesh and finer).
Use: Analysis (testing for nitrogen).

developer. (1) A term applied in the dyeing industry to certain organic compounds which, in combination with some other organic compound already deposited upon the fiber, will develop a colored compound, or if united with a dye already upon the fiber, will form a new coloring matter possessing a more desirable or a faster color.

(2) A substance used in photography to convert a latent image to a visible one by chemical reduction of a silver compound to metallic silver more rapidly in the portions exposed to light than in those not exposed. Such reducing agents as hydroquinone, pyrogallol, and para-phenylenediamine are used. See also photographic chemistry.

"Devlex."[233] Trademark for oxazolidinone polymers.

devitrification. Unwanted crystallization of silica on heating or cooling. The term is used largely in the glass industry. The tendency to devitrify results from the unstable nature of glasses. It usually occurs if the melt is cooled too slowly.

devulcanization. Technically a misnomer, since vulcanization is irreversible. The term is used to describe the softening of a vulcanizate caused by heat and chemical additives during reclaiming.

dew point. Temperature at which air is saturated with moisture, or in general the temperature at which a gas is saturated with respect to a condensable component.

dexamethasone (9-alpha-fluoro-16-alpha-methylprednisolone) $C_{22}H_{29}FO_5$. A corticosteroid.
Properties: Crystals; m.p. 262–264°C. Insoluble in water; somewhat soluble in organic solvents.
Grade: N.F.
Use: Medicine and veterinary medicine.

"Dexedrine."[71] Trademark for dextroamphetamine sulfate.

"Dexet."[233] Trademark for aluminous cement, and related ingredients for use chiefly in oil wells.

"Dexon."[181] Trademark for para-dimethylaminobenzenediazo sodium sulfonate (q.v.).

dextran (macrose). Certain polymers of glucose which have chain-like structures and molecular weights up to 200,000. Produced from sucrose by Leuconostoc bacteria; occurs as slimes in sugar refineries, on fermenting vegetables or in dairy products. Clinical dextran is standardized to a low molecular weight (75,000); is made by partial hydrolysis and fractional precipitation of the hgh molecular weight particles.
Properties: Stable to heat and storage. Soluble in water, making very viscous solutions. Solutions can be sterilized. Combustible; nontoxic.
Uses: Blood plasma substitute or expander; confections; lacquers; oil-well drilling muds; filtration gel; food additive.

dextranase. An enzyme reported to be effective in reducing dental caries.

dextran sulfate. See sodium dextran sulfate.

"Dextrid."[236] Trademark for an organic polymer used for control of filtration, mud rheology and solids in drilling muds. Stabilized against microbiological degradation. A low percentage controls filtration without appreciable viscosity increase.

dextrin (starch gum). A group of colloidal products formed by the hydrolysis of starches. Industrially it is made by treatment of various starches with dilute acids or by heating dry starch. The yellow or white powder or granules are soluble in water and insoluble in alcohol and ether.
Uses: Adhesives; thickening agent; sizing paper and textiles; substitute for natural gums; food industry; glass-silvering compositions, printing inks; felt manufacture; substitute for lactose in penicillin manufacture.

"Dextrinase" A.[212] Trademark for a fungal amylase which converts starches and dextrins to maltose and dextrose.
Uses: Syrups and other products high in reducing sugars.

dextromethorphan hydrobromide. (*d*-3-methoxy-N-methylmorphinan hydrobromide)
$C_{18}H_{25}NO \cdot HBr \cdot H_2O$.
Properties: Practically white crystals or crystalline powder possessing a faint odor; slightly soluble in water; freely soluble in alcohol and chloroform; insoluble in ether. Specific rotation 200 mg/10 ml solution +26 to +28°; pH (1 in 50 solution) 5.2–6.5.
Grade: N.F.
Use: Medicine.

dextrorotatory. Having the property when in solution of rotating the plane of polarized light to the right or clockwise. Dextrorotatory compounds are given the prefix *d* or (+) to distinguish them from their levorotatory, *l* or (−) isomers. The plus sign (+) is now preferred.
See optical rotation.

dextrose. See glucose, which is the preferred term.

dextrose equivalent (D.E.). The total amount of reducing sugars expressed as dextrose that is present in a corn syrup, calculated as a percentage of the total dry substance. The usual technique for determining D.E. in the corn products industry is the volumetric alkaline copper method. See also glucose syrup.

DFDT (difluorodiphenyltrichloroethane)
$(FC_6H_4)_2CHCCl_3$. Fluorine analog of DDT.
Properties: A low-melting white solid; m.p. 45.5°C. Odor resembling ripe apples; does not have broad killing power of DDT toward all insects but is more effective against flying insects, especially house flies.
Derivation: By condensing chloral and fluorobenzene in the presence of sulfuric acid or chlorosulfonic acid.
Uses: Insecticide.

"DFL No. 3."[28] Trademark for a solution of buffered phosphate esters, used as a lubricant release agent and corrosion inhibitor for synthetic rubber driers.

DFP. Abbreviation for diisopropyl fluorophosphate.

DHA. Abbreviation for dihydroxyacetone.

"DHA."[233] Trademark for fungicides comprised of dehydroacetic acid and its salts.

"DHP-MP."[203] Trademark for 1,4-bis(2-hydroxypropyl) 2-methylpiperazine.

DHS. Abbreviation for dihydrostreptomycin.

Di Symbol for didymium.

di-. Prefix meaning two. See also bi-.

"Diabestos."[218] Trademark for a filteraid of "Dicalite" diatomite and selected asbestos fibers.

"Diabinese."[299] Trademark for chloropropamide.

"Diablo."[244] Trademark for a group of chemically inert liquids containing 70% chlorine. Used in flammable compounds to impart flame retardance.

diacetic acid. See acetoacetic acid.

diacetin (glyceryl diacetate)
$CH_2O(OCCH_3)CHOHCH_2O(OCCH_3)$.
Properties: Hygroscopic liquid. It is a mixture of isomers. Sp. gr. 1.18; b.p. 259°C (approx.); refractive index 1.44. Miscible with water, benzene, and alcohol; the commercial mixture gels about −30°C. Combustible. Low toxicity.
Derivation: Heating one mole of glycerin with two moles of glacial acetic acid.
Grades: Technical.
Uses: Plasticizer and softening agent; solvent for cellulose derivatives, "Glyptal" resins, shellac.

diacetone acrylamide (DAA). A vinyl monomer.
Properties: White, crystalline solid; purity 99+%; highly soluble in water and most organic solvents; polymerizes readily. (The DAA homopolymer is insoluble in water).
Containers: Fiber drums; car lots.
Uses: Imparts water tolerance and vapor permeability to copolymer films; latex and water-based coating compositions; adhesion improver for cellulosics, concrete, glass; cross-linking agent in polyester resins; color photography.

diacetone alcohol (diacetone; 4-hydroxy-4-methylpentanone-2; 4-hydroxy-2-keto-4-methylpentane) $CH_3COCH_2C(CH_3)_2OH$.
Properties: Colorless liquid, pleasant odor; sp. gr. 0.9406 at 20/20°C; b.p. 169.1°C; flash point (commercial grades) 55°F (O.C.); wt/gal 7.8 lb (20°C); viscosity 0.032 poise (20°C); f.p. −42.8°C; refractive index 1.42416 (20°C). Autoignition temp. 1118°F.
Miscible with alcohols, aromatic and halogenated hydrocarbons, esters, and water. A constant-boiling mixture with water has b.p. 99.6°C and contains approx. 13% diacetone alcohol.
Derivation: Condensation of acetone.
Grades: Technical; acetone-free; reagent.
Hazard: Flammable. Dangerous fire risk. Toxic and irritant. Tolerance, 50 ppm in air. Flammable limits in air, 1.8 to 6.9%.
Containers: Drums; tank cars.
Uses: Solvent for nitrocellulose, cellulose acetate, various oils, resins, waxes, fats, dyes, tars; lacquers, dopes, coating compositions; wood preservatives; stains; rayon and artificial leather; imitation gold leaf; dyeing mixtures; antifreeze mixtures; extraction of resins and waxes; preservative for animal tissue; metal-cleaning compounds; hydraulic compression fluids; stripping agent (textiles); laboratory reagent. The technical grade, containing acetone, has greater solvent power.
Shipping regulations: (Rail) Flammable liquid, n.o.s., Red label. (Air) Flammable Liquid label.

diacetonyl sulfide $(CH_3COCH_2)_2S$.
Properties: Crystals; b.p. 136–137°C (15 mm); m.p. 47°C.
Derivation: Interaction of chloroacetone and hydrogen sulfide gas.

diacetyl (biacetyl; butanedione; diketobutane; dimethyl diketone; dimethylglyoxal) $CH_3COCOCH_3$.
Properties: Yellow liquid; strong odor. Soluble in water, alcohol, and ether. Sp. gr. 0.990 (15/15°C); m.p. ⩾3 to ⩾4°C; b.p. 88–91°C; refractive index (n 18/D) 1.3933; flash point below 80°F.
Derivation: Special fermentation of glucose; synthesis from methyl ethyl ketone.
Grades: Technical; flavor grade; F.C.C.
Hazard: Flammable, dangerous fire risk.
Use: Aroma carrier in food products.
Shipping regulations: (Rail) Flammable liquid, n.o.s., Red label. (Air) Flammable Liquid label.

diacetylaminoazotoluene (4-ortho-tolylazo-ortho-diacetotoluide) $[CH_3C_6H_4N{:}NC_6H_3(CH_3)N(CH_3CO)_2]$.
Properties: Crystalline powder. Color varies from yellowish-red, through rose to red. Acted upon by atmospheric water vapor. Soluble in alcohol, chloroform, and ether; also in fats, oils, and greases; insoluble in water.
Constants: M.p. 74–76°C.
Use: Medicine (external).

diacetyldihydroxydiphenylisatin $C_{24}H_{19}O_5N$.
Properties: White, odorless, tasteless, crystalline powder; m.p. 241–242°C. Soluble in alcohol, ether, and benzene.
Use: Medicine.

diacetylene. An unsaturated hydrocarbon containing two triple bonds, with the type formula C_nH_{2n-6}. The simplest is butadiyne or biacetylene HC⋮CC⋮CH, a gas which boils at 10°C. Combustible. Low toxicity.
Hazard: Ignites spontaneously in contact with moist silver salts.

1,2-diacetylethane. See acetonylacetone.

1,1′-diacetyl ferrocene $(C_5H_4COCH_3)_2Fe$. Red crystalline solid; m.p. 122–124°C. Used as an intermediate. See ferrocene.

diacetylmethane. See acetylacetone.

diacetylmorphine (diamorphine; heroin) $C_{17}H_{17}NO(C_2H_3O_2)_2$.
Properties: White, odorless, bitter crystals or crystalline powder. Soluble in alcohol. M.p. 173°C.
Derivation: By acetylization of morphine.
Hazard: Highly toxic, habit-forming narcotic; 1/6th grain may be fatal. Cannot be legally sold in U.S.

diacetyl peroxide. See acetyl peroxide.

"Diadem Chrome."[232] Trademark for a series of chrome dyestuffs suitable for application by the after-chrome method.

"Di-Ademil."[412] Trademark for hydroflumethiazide (q.v.).

"Diagnex Blue."[412] Trademark for azuresin.

"Diak."[28] Trademark for a series of rubber accelerators used to vulcanize "Viton" fluoroelastomer and polyacrylate elastomers.

dialdehyde starch. See starch dialdehyde.

dialkylchloroalkylamine hydrochloride. A group of amine salts having the formula RCl · HCl, when R represents such groups as $(CH_3)_2NCH_2CH_2$— (beta dimethylaminoethyl chloride hydrochloride); $(CH_3)_2NCH_2CH(CH_3)$— (beta-dimethylaminoisopropyl chloride hydrochloride), etc. Used in organic synthesis.

"Diall."[175] Trademark for a series of diallyl phthalate thermosetting molding compounds.
Properties: Excellent electrical resistance; high physical strength; flame resistance; good dimensional stability; resistance to virtually all solvents and most chemicals; outstanding colorfastness in sunlight and heat; fungus resistant.
Grades: Mineral-filled, synthetic fiber-filled and glass fiber-filled.

di-allate. See 2,3-dichloroallyl diisopropylthiocarbamate.

diallyl adipate $C_3H_5OOC(CH_2)_4COOC_3H_5$.
Properties: Liquid; color-maximum #100 Pt-Co, characteristic odor; sp. gr. (20°C) 1.025. Combustible.
Use: Monomer.

diallylamine (di-2-propenylamine) $(CH_2{:}CHCH_2)_2NH$.
Properties: Liquid; sp. gr. (20°C) 0.7889; b.p. 112°C; f.p. −100°C; refractive index (n 20/D) 1.4404. Soluble in water. Combustible. Low to moderate toxicity.
Derivation: From allylamine.
Containers: Drums; tank cars.
Use: Intermediate.

diallylbarbituric acid (5,5-diallylbarbituric acid) $C_{10}H_{12}N_2O_3$.
Properties: White, odorless, crystals or crystalline

powder; slightly bitter taste; soluble in alcohol or ether; slightly soluble in water; m.p. 171–173°C.
Hazard: Moderately toxic. See barbiturate.
Use: Medicine (sedative).

diallyl chlorendate $(C_3H_5OOC)_2C_7H_2Cl_6$. Solid; f.p. 29.5°C; viscosity (20°C) 4.0 cp; sp. gr. (20°C) 1.47. Used as a monomer for allyl resins, especially in flame-retardant compositions.

diallyl cyanamide $(H_2C{:}CHCH_2)_2NCN$.
Properties: Liquid; f.p. less than −70°C; b.p. 222°C; sp. gr. 0.90.
Hazard: Yields cyanide fumes on heating.
Use: Organic intermediate; polymers.

diallyl diglycollate $(C_3H_5OOCCH_2)_2O$.
Properties: Liquid; color-maximum #100 Pt-Co; characteristic odor; sp. gr. (20°C) 1.1113.
Use: Monomer.

diallyldimethylammonium chloride. See "Cat-floc."

diallyl isophthalate $C_6H_4(COOH_2C{:}CHCH_2)_2$.
Properties: Monomer is liquid; color-maximum #175 Pt-Co; mild characteristic odor; sp. gr. (20°C) 1.124. Prepolymer is solid; sp. gr. (25°C) 1.256.
Uses: Molding and laminating; cross-linker for polyesters.

diallyl maleate $C_3H_5OOCCH{:}CHCOOC_3H_5$.
Properties: Colorless or straw-colored liquid. B.p. 109–110°C (3 mm); sp. gr. 1.077 (20°C); refractive index (n 20/D) 1.4699. Polymerizes readily when exposed to light or temperatures above about 50°C. Combustible.
Hazard: Moderately toxic by ingestion; irritant to skin.
Uses: Polymers and copolymers; insecticide formulations.

diallylmelamine $(C_3H_5)_2N\underline{CNC(NH_2)NC(NH_2)N}$.

Properties: White crystalline solid; m.p. 142°C; sp. gr. 1.24 (30°C). Combustible.
Hazard: Moderately toxic by ingestion; irritant to skin. Evolves cyanide on heating.
Use: Monomer for resins.

diallyl phosphite $(CH_2{:}CHCH_2O)_2PHO$. Water-white liquid; f.p. 0°C; b.p. 62°C (1 mm); refractive index (n 25/D) 1.444; sp. gr. 1.080 (25/15°C). Combustible.
Hazard: May be toxic.
Use: Synthesis of organophosphorus compounds.

diallyl phthalate (DAP) $C_6H_4(COOCH_2CH{:}CH_2)_2$. The name is also used for the polymer.
Properties: Nearly colorless oily liquid; limited solubility in gasoline, mineral oil, glycerin, glycols and certain amines. Soluble in most other organic liquids. Insoluble in water. Sp. gr. 1.120 (20/20°C); f.p. −70°C (viscous liquid); boiling range 158–165°C (4 mm); odor mild, lachrymatory; flash point 330°F; viscosity 13 cp (20°C). Combustible. Low toxicity.
Containers: Drums; tank cars.
Uses: Primary plasticizer which will polymerize if not inhibited; a monomer which will polymerize with heat and catalyst. Forms low-pressure laminates with various fillers such as glass cloth, paper, etc. for electrical insulation.

diallyl sulfide. See allyl sulfide.

dialysis. The separation of small molecules from macromolecules in a solution by means of a semipermeable membrane, such as parchment or collodion. The rates of diffusion of the small and the large molecules are so widely different that the former will readily pass through the membrane, whereas the latter will penetrate with extreme difficulty. For example, the diffusion rates are about 2.3 for sodium chloride, 7 for cane sugar, and from 50 to 100 for proteins and other macromolecules. This differential led Thomas Graham to define substances that would pass through the membrane easily as crystalloids, and those having a tendency to be retained by the membrane as colloids. See also colloid chemistry; electrodialysis; Graham.

"Diam."[250] Trademark for a series of fatty diamines, $RNH(CH_2)_3NH_2$.
Uses: Corrosion inhibitors, petroleum additives, asphalt emulsifiers and chemical intermediates.

diamantane. See adamantane.

diamide hydrate. See hydrazine hydrate.

diamine. See hydrazine.

3,6-diaminoacridine. See acriflavine.

3,6-diaminoacridinium hydrogen sulfate. See proflavine sulfate.

meta-**diaminoazobenzene hydrochloride.** See chrysoidine.

diaminoazoxytoluene (azoxytoluidine) $C_6H_3(CH_3)(NH_2)N_2OC_6H_3(NH_2)(CH_3)$.
Properties: Yellow or orange crystals. M.p. 168°C; soluble in alcohol; insoluble in water. Combustible.
Derivation: By alkaline reduction of para-nitro-orthotoluidine.
Use: Dye intermediate.

diaminobenzene. See phenylenediamine.

3,3′-diaminobenzidine (3,3′,4,4′-biphenyltetramine) $(H_2N)_2C_6H_3C_6H_3(NH_2)_2$. Solid; m.p. 178–180°C.
Hazard: Probably toxic.
Use: Copolymerized with diphenyl isophthalate to make high-temperature-resistant polybenzimidazoles (q.v.).

1,3-diaminobutane $NH_2CH_2CH_2CHNH_2CH_3$.
Properties: Water-white liquid; amino odor; boiling range 143–150°C; sp. gr. 0.858 (20/20°C); refractive index 1.450 (20°C); flash point 125°F. Combustible.
Hazard: Moderate fire risk.

alpha, epsilon-**diaminocaproic acid.** See lysine.

diaminochrysazin $(NH_2)_2(OH)_2C_{14}H_4O_2$. 1,8-diamino-4,5-dihydroxyanthraquinone.
Use: Colorimetric determination of boron.

trans-**1,2-diaminocyclohexanetetraacetic acid monohydrate** (CDTA) $C_6H_{10}[N(CH_2COOH)_2]_2 \cdot H_2O$.
Properties: White, crystalline solid; m.p. 200–220°C. Very slightly soluble in water and insoluble in most common organic solvents. Partially soluble in dimethylformamide and dimethylsulfoxide upon heating. Forms stable complexes.
Use: Chelating agent similar to ethylenediaminetetraacetic acid.

diaminodiethyl sulfide $S(CH_2CH_2NH_2)_2$.
Properties: Mobile, colorless liquid with amine-like odor. Miscible with water and benzene; insoluble in aliphatic hydrocarbons. B.p. 230–240°C; sp. gr. 1.054 (25°C). Combustible.

1,8-diamino-4,5-dihydroxyanthraquinone. See diaminochrysazin.

diaminodihydroxyarsenobenzene dihydrochloride. See arsphenamine.

di-para-**aminodimethoxydiphenyl.** See dianisidine.

diaminodiphenic acid (benzidinedicarboxylic acid) $C_6H_3(CO_2H)NH_2C_6H_3(CO_2H)NH_2$.
Properties: White crystals; soluble in alcohol and ether; insoluble in water.
Derivation: By boiling meta-nitrobenzaldehyde with caustic soda, reducing with zinc dust and acidifying.
Hazard: Probably toxic. See benzidine.
Use: Dyestuff.

para-**diaminodiphenyl.** See benzidine.

diaminodiphenylamine $HN(C_6H_4NH_2)_2$.
Properties: Yellowish crystals; soluble in alcohol and ether; insoluble in water. M.p. 158°C.
Uses: Dye intermediate; detection of hydrogen cyanide.

diaminodiphenylethylene. See para-diaminostilbene.

para,para′-**diaminodiphenylmethane** (4,4′-methylenedianiline; MDA) $H_2NC_6H_4CH_2C_6H_4NH_2$.
Properties: Crystals from water or benzene; m.p. 92–93°C; b.p. 398–399°C (78 mm). Slightly soluble in cold water. Very soluble in alcohol, benzene, ether. Flash point 440°F. Combustible.
Hazard: Reported to cause toxic hepatitis.
Uses: Determination of tungsten and sulfates; polymer and dye intermediate; corrosion inhibitor; epoxy resin hardening agent; isocyanate resins; polyamides.

4,4′-diaminodiphenyl sulfone. See sulfonyldianiline.

diaminodiphenylthiourea (diaminothiocarbanilide) $(NH_2C_6H_4NH)_2CS$.
Properties: Colorless plates or crystalline solid; soluble in alcohol and ether; sparingly soluble in water. M.p. 195°C.
Derivation: By boiling para-phenylenediamine with carbon disulfide.

diaminodiphenylureadisulfonic acid $CO(NHC_6H_3NH_2SO_3H)_2$.
Properties: Colorless, needle-like crystals; slightly soluble in water.
Derivation: Action of phosgene upon either para-phenylenediaminesulfonic acid or 4-nitroaniline-3-sulfonic acid.
Use: Dye manufacture.

3,3′-diaminodipropylamine. See 3,3′-imino-bispropylamine.

diaminoditolyl. See tolidine.

para,para′-**diaminoditolylmethane.** $NH_2C_7H_6CH_2C_7H_6NH_2$.
Properties: Glistening, crystalline plates; soluble in alcohol and ether; m.p. 149°C.
Derivation: By heating formaldehyde and orthotoluidine.

1,2-diaminoethane. See ethylenediamine.

6,9-diamino-2-ethoxyacridine lactate monohydrate. See ethodin.

di-para-**aminoethoxydiphenyl.** See ethoxybenzidine.

diaminoethyl ether tetraacetic acid $(HOOCCH_2)_2NCH_2CH_2OCH_2CH_2N(CH_2COOH)_2$. Slightly soluble in water; purity, 98% min. Used as a chelating agent.

1,6-diaminohexane. See hexamethylenediamine.

3,6-diamino-10-methylarcridinium chloride. See acriflavine.

diaminonaphthalene. See naphthylenediamine.

1,5-diaminopentane. See cadaverine.

2,4-diaminophenol $C_6H_3OH(NH_2)_2$.
Properties: Colorless crystals; m.p. 78–80°C with decomposition. Soluble in alcohol and ether.
Derivation: By reduction of 2,4-dinitrophenol.
Hazard: May be toxic or skin irritant.
Uses: Photographic developer; organic synthesis.

2,5-diaminophenol $C_6H_3OH(NH_2)_2$.
Properties: Colorless crystals; m.p. 68°C; soluble in water.
Derivation: By reduction of 2,5-dinitrophenol.
Hazard: May be toxic or skin irritant.
Use: Organic synthesis.

2,4-diaminophenol hydrochloride (amidol) $C_6H_3(NH_2)_2OH \cdot 2HCl$.
Properties: Grayish-white crystals; soluble in water; slightly soluble in alcohol.
Derivation: By interaction of dinitrophenol with iron and hydrochloric acid.
Hazard: May be toxic or skin irritant.
Uses: Photographic developer; dyeing furs and hair; analytical reagent.

2,4-diamino-6-phenyl-s-triazine. See benzoguanamine.

1,2-diaminopropane (propylenediamine) $NH_2CH_2CH(NH_2)CH_3$.
Properties: Colorless, very hygroscopic, strongly alkaline liquid. Very soluble in water. Ammoniacal odor. Sp. gr. 0.8732 (20/20°C); refractive index 1.4460 (n 20/D); flash point 160°F; b.p. 117°C. Combustible.
Grades: Technical; 75%; 90%; 98% solution.
Hazard: Strong irritant to skin and tissue. Moderate fire risk.
Uses: Synthesis of medicinals, dyes, rubber accelerators; electroplating; analytical reagent.
Shipping regulations: (Air) Corrosive label. Label name propylene diamine.

1,3-diaminopropane (1,3-propanediamine) $NH_2CH_2CH_2CH_2NH_2$.
Properties: Water-white mobile liquid; amine odor. Sp. gr. 0.8881 (20/20°C); b.p. 139.7°C; f.p. −12°C; refractive index 1.459 (20°C); flash point 120°F (TOC). Completely soluble in water, methanol, and ether. Combustible.
Hazard: Moderate fire risk. Strong irritant to eyes and skin.
Use: Intermediate.

2,6-diaminopyridine $NC_5H_3(NH_2)_2$.
Properties: Crystals; m.p. 120.8°C; b.p. 285°C (760 mm); soluble in water. Combustible.
Derivation: From 2-aminopyridine.

para-**diaminostilbene** (diaminodiphenylethylene) $C_6H_4(NH_2)CHCHC_6H_4(NH_2)$.
Properties: Colorless needles or plates; soluble in alcohol and ether; insoluble in water. M.p. 227°C. Combustible.
Derivation: Reduction of dinitrostilbene.

4,4′-diamino-2,2′-stilbenedisulfonic acid (DAS) $C_6H_3(NH_2)(SO_3H)CHCHC_6H_3(SO_3H)(NH_2)$.
Properties: Yellowish microscopic needles; soluble in alcohol and ether; insoluble in water.
Derivation: Boiling sodium salt of para-nitrotoluene-ortho-sulfonate in water with caustic soda and reduction with zinc dust.

Hazard: Toxic by ingestion.
Use: Dyestuffs.

diaminothiocarbanilide. See diaminodiphenylthiourea.

di-alpha-amino-beta-thiolpropionic acid. See cystine.

diaminotoluene. See toluene-2,4-diamine.

4,6-diamino-meta-toluenesulfonic acid. See meta-tolylenediaminesulfonic acid.

4,6-diamino-s-triazine-2-ol. See ammeline.

2,5-diaminovaleric acid. See ornithine.

diammonium ethylenebisdithiocarbamate $NH_4S_2CNH(CH_2)_2NHCS_2NH_4$.
Properties: M.p. 72.5°C; very soluble in water.
Grades: 42% solution in water.
Hazard: May be toxic.
Uses: Fungicide; intermediate; corrosion inhibitor. See also nabam.

diammonium hydrogen phosphate. See ammonium phosphate, dibasic.

diammonium phosphate. See ammonium phosphate, dibasic.

diamond. An allotropic form of carbon (q.v.), crystallized isometrically; consists of carbon atoms covalently bound by single bonds only in a predominantly octahedral structure. This accounts for its extreme hardness (Mohs 10) and great stability. It has a high refractive index (2.42); sp. gr. 3.50; coefficient of friction 0.05. M.p. 3700°C; b.p. 4200°C. The purest diamonds used for gems are mined in South Africa; lower grades in Brazil, Venezuela, India, Borneo, Arkansas; also made synthetically by heating carbon and a metal catalyst in an electric furnace at about 3000°F under high pressure. See also diamond, industrial.

diamond, industrial. Low-grade diamonds (bort and carbonado), as well as those made synthetically in an electric furnace (3000°F, 1.3 million psi).
Uses: Oil-well drill bits; primary grinding of steel; wire-drawing dies; glass and metal cutting. As dust, in abrasive wheels.

diamond pyramid hardness. See hardness.

"Diamox."[315] Trademark for acetazolamide (q.v.).

diamthazole dihydrochloride $C_{15}H_{23}N_3OS \cdot 2HCl$. 6-(2-Diethylaminoethoxy)-2-dimethylaminobenzothiazole dihydrochloride.
Properties: Crystals; decomposes 269°C. Soluble in water, ethanol and methanol.
Use: Topical therapy (medicine); antifungal agent.

di-n-amylamine (di-n-pentylamine) $(C_5H_{11})_2NH$.
Properties: Colorless liquid; b.p. 202–3°C (745 mm); very slightly soluble in water, soluble in alcohol and ether. Sp. gr. 0.77–0.78 (20°C); refractive index 1.430 (20°C). Flash point 124°F. Combustible.
Derivation: From reaction of amyl chloride and ammonia.
Containers: Drums; tank cars.
Hazard: Moderate fire risk. May be toxic.
Uses: Rubber accelerators; flotation reagents; dyestuffs and corrosion inhibitors; solvent for oils, resins, and some cellulose esters.

N,N-diamylaniline (mixed isomers) $C_6H_5N(C_5H_{11})_2$.
Properties: Dark amber liquid. Sp. gr. 0.898 (20°C); boiling range 276–292°C; faint aniline odor; flash point 260°F. Combustible.
Hazard: May be toxic.

di-tert-amyl disulfide $CH_3CH_2C(CH_3)_2SSC(CH_3)_2CH_2CH_3$.
Properties: Liquid; sp. gr. 0.931 (15.5/15.5°C; vacuum distillation range 86–102°C; refractive index 1.495 (20°C); flash point 220°F. Combustible.

diamylene $C_{10}H_{20}$.
Properties: Colorless liquid; f.p. below −50°C; b.p. 150°C; flash point 118°F (open cup); sp. gr. 0.77; low toxicity. Combustible.
Hazard: Moderate fire risk.
Uses: Solvent; organic synthesis.

2,5-di(tert-amyl)hydroquinone (DAHQ; 2,5-di(tert-pentyl)hydroquinone) $(C_5H_{11})_2C_6H_2(OH)_2$.
Properties: Buff powder; m.p. 176°C; sp. gr. 1.05 (25°C). Slightly soluble in water; soluble in alcohol and benzene.
Uses: Antioxidant for uncured rubber and for unsaturated resins and oils; food packaging; polymerization inhibitor.

diamyl maleate $(CHCOOC_5H_{11})_2$.
Properties: Water-white liquid. Sp. gr. 0.981 (20°C); boiling range 263–300°C; odor faintly alcoholic; flash point 270°F. Combustuble.

diamyl phenol (1-hydroxy-2,4-diamylbenzene) $(C_5H_{11})_2C_6H_3OH$. Commercial form is a mixture of isomers including both secondary amyl and tertiary amyl groups mainly in 2,4 positions.
Properties: Light straw-colored liquid with mild phenolic odor; miscible with both aliphatic and aromatic hydrocarbons; insoluble in water and 10% aqueous alkalies. Boiling range (ASTM 5–95%) 280–295°C; sp. gr. 0.930 (20°C); wt/gal 7.8 lb (20°C); flash point (TOC) 260°F. Combustible.
Containers: Drums; tank cars; tank trucks.
Hazard: May be toxic and irritating to skin.
Uses: Synthetic resins; lubricating oil additives; rust preventives; plasticizers; synthetic detergents; antioxidants and antiskinning agents; rubber chemicals; rodenticide; fungicide.

diamyl phthalate $C_6H_4(COOC_5H_{11})_2$.
Properties: Colorless, nearly odorless oily liquid. Sp. gr. (20°C) 1.022; wt/gal 8.52 lb (20°C); refractive index (25°C) 1.488; b.p. 342°C; f.p. less than −55°C; flash point (closed cup) 357°F. Combustible. Low toxicity.
Derivation: By esterification of phthalic anhydride with amyl alcohol in the presence of approximately 1% concentrated sulfuric acid as catalyst.
Use: Plasticizer.

diamyl sulfide (amyl sulfide) $(C_5H_{11})_2S$. A mixture of isomers.
Properties: Yellow liquid; sp. gr. 0.85–0.91 (20/20°C); distillation range 170–180°C; flash point (open cup) 185°F; refractive index 1.477 (19°C). Combustible. Obnoxious odor.
Hazard: Moderately toxic and irritant by inhalation, ingestion and skin absorption.
Uses: Organic sulfur compounds by addition reactions; flotation agent in metallurgical processes; odorant.

"Dianabol."[305] Trademark for methandrostenolone. (q.v.).

dianhydrosorbitol. See sorbide.

1,4,3,6-dianhydrosorbitol. See isosorbide.

dianisidine (di-para-aminodi-meta-methoxydiphenyl; 3,3′-dimethoxybenzidine) $[C_6H_3(OCH_3)NH_2]_2$.
Properties: Colorless crystals; soluble in alcohol and ether; insoluble in water. M.p. 137° C. Flash point 403° F. Combustible.
Derivation: The methyl ether of ortho-nitrophenol is reduced by zinc dust and caustic soda to the hydrazo compound, which is then rearranged with hydrochloric acid.
Hazard: May be toxic. See anisidine, benzidine.
Use: Dye intermediate.

dianisidine diisocyanate (3,3′-dimethoxybenzidine 4,4′-diisocyanate) $[OCN(CH_3O)C_6H_3]_2$.
Properties: Grey to brown powder; m.p. 112°C min; soluble in ketones and esters.
Hazard: Toxic by inhalation and ingestion at high temperatures. Skin irritant.
Uses: Polymers and adhesive systems; high-strength backbone or cross-linking intermediate.

di-para-**anisyl-**para-**phenetylguanidine hydrochloride** (acoin; guanicaine)
$C_2H_5OC_6H_4NC(NHC_6H_4OCH_3)_2 \cdot HCl$.
Properties: White crystalline powder; incompatible with iodine and alkaline iodides. Soluble in water and alcohol. M.p. 176° C.
Use: Medicine (local).

diaphenadione. An anticoagulant drug that has been found to reduce blood cholesterol in experimental animals.

diaphragm cell. A type of electrolytic cell for the production of caustic soda and chlorine from sodium chloride brine. The cell contains anode and cathode compartments separated by a porous diaphragm or membrane to prevent mixing of the solutions. Asbestos fibers are usually used for this diaphragm, though a recent development is a plastic material made from perfluorosulfonic acid (see "Nafion"). The brine is fed continuously to the anode compartment, where chlorine is released at the graphite anode, and flows through the diaphragm to the steel cathode, where hydrogen is liberated. Caustic soda accumulates in the liquid and is continuously drained from the cathode compartment. The Hooker cell and the Dow cell are two widely used types of diaphragm cell.

"Di-Aqua."[244] Trademark for sodium alkyl aryl sulfonate in which the alkyl aryl portion is substantially all dodecylbenzene. Available in 40% and 80% active strengths, both in powder and flake form, also 40% beads.
Uses: Wetting agent; detergent; emulsifier.

"Diaron."[36] Trademark for powdered melamine adhesives.

"Diasan."[570] Trademark for a quaternary ammonium bacteriostate for textile sanitizing.

diaspore $Al_2O_3 \cdot H_2O$. A natural hydrous aluminum oxide, occurring in bauxite (q.v.), and with corundum and dolomite.
Properties: White, gray, yellowish, and greenish; luster vitreous to pearly; sp. gr. 3.35–3.45; Mohs hardness 6.5–7.
Occurrence: Arkansas, Missouri, Pennsylvania; Switzerland; U.S.S.R.; Czechoslovakia.
Uses: Refractory; abrasive.

diastase, malt. A commercial mixture containing amylolytic enzymes.
Properties: Yellowish-white amorphous powder, or syrupy liquid. Soluble in water; almost insoluble in alcohol. Nontoxic.
Derivation: The filtrate from the mash of malted grain is concentrated at low temperatures in vacuum. The sugar acts as preservative. The diastase hydrolyzes starch to malt sugar.
Uses: Desizing of textiles; calico printing; finishing of textiles; medicine; bread-making; malted foods.
See also amylase.

"Diastazyme."[309] Trademark for liquid enzyme of malt origin. Used for sizing starch-sized fabrics.

diastereoisomer. In any group of four optical isomers occurring in compounds containing two asymmetric carbon atoms, such as 4-carbon sugars, there are two pairs of enantiomers (structures which are mirror images of each other), indicated by the letters D and L. The two D and the two L isomers are not mirror images, and these are called diastereoisomers. For example, in the structures below, (1) and (2) are enantiomers, and so are (3) and (4); (1) and (3) and (2) and (4) are disastereoisomers:

```
        CHO                 CHO
         |                   |
      H—C—OH            HO—C—H
         |                   |
(1)  HO—C—H              H—COH     (2)
         |                   |
        CH2OH               CH2OH
         L                   D

        CHO                 CHO
         |                   |
     HO—C—H              H—C—OH
         |                   |
(3)  HO—C—H              H—C—OH    (4)
         |                   |
        CH2OH               CH2OH
                             D
```

See also optical isomer; enantiomer; epimer; anomer.

diatomaceous earth (diatomite; kieselguhr).
Properties: Soft bulky solid material (88% silica) composed of skeletons of small prehistoric aquatic plants related to algae (diatoms). They have intricate geometric forms. Available as light-colored blocks, bricks, powder, etc. True sp. gr. 1.9 to 2.35; bulk density from 5 to 15 lb per cubic foot. Insoluble in acids except hydrofluoric; soluble in strong alkalies. Absorbs 1.5 to 4 times its weight of water; also has high oil absorption capacity. Poor conductor of sound, heat, and electricity. Noncombustible.
Occurrence: Western U.S., Europe; Algeria; U.S.S.R.
Grades: Natural; chemical; airfloated.
Containers: Bulk; bags; multiwall paper sacks.
Hazard: Moderate risk from inhalation of dust. Otherwise nontoxic.
Uses: Filtration, clarifying, and decolorizing; insulation; absorbent; mild abrasive; drilling mud thickener; extender in paints, rubber and plastic products; ceramics; paper coating; anticaking agent in fertilizers; asphalt compositions; chromatography.

diatomic. Descriptive of a gas whose molecules are composed of two atoms, e.g., oxygen, nitrogen, chlorine, hydrogen. Gases in which the element is present as single atoms are called monatomic, e.g., argon, neon.

diatrizoate sodium. See sodium diatrizoate.

1,4-diazabicyclo[2.2.2]octane

$\overline{CH_2CH_2NCH_2CH_2NCH_2CH_2}$.

Properties: Crystals, hygroscopic; m.p. 158°C; b.p. 174°C; forms crystalline hydrate; sublimes easily; soluble in water and organic solvents.
Uses: Possible catalyst for urethane foams and coatings; chemical intermediate.

"Diazald."[541] Trademark for N-methyl-N-nitrosopara-toluenesulfonamide.

diazepam USAN name for 7-chloro-1,3-dihydro-1-methyl-5-phenyl-2H-1,4-benzodiazepin-2-one, $C_{16}H_{13}ClN_2O$. Listed in N.D.
Properties: Slightly yellowish crystalline powder; practically no odor. M.p. 131.5–134.5°C. One gram of diazepam dissolves in about 350 ml of water, in about 15 ml of 95% ethanol or in about 2 ml of chloroform.
Hazard: Central nervous system depressant. Toxic in high concentrations. Manufacture and dosage restricted.
Use: Medicine (tranquilizer).

"Diazine."[243] Trademark for a group of direct dyes, applied to cotton, diazotized and then coupled onto phenols or amines.

1,3-diazine (pyrimidine, miazine)
$\overline{CHN(CH)_3N}$.
Properties: Liquid or crystalline mass with a penetrating odor; melting point 20–22°C; b.p. 123–124°C; soluble in water, alcohol, and ether.
See also pyrimidine.

diazinon. Generic name of an insecticide, O,O-diethyl O-(2-isopropyl-4-methyl-6-pyrimidinyl) phosphorothioate $[(CH_3)_2CHC_4N_2H(CH_3)O]PS(OC_2H_5)_2$.
Properties: Colorless liquid; b.p. 83–84°C (0.002 mm); slightly soluble in water, freely soluble in petroleum solvents, alcohol and ketones. More stable in alkaline than neutral or acid solutions.
Hazard: Toxic by ingestion, inhalation, and skin absorption. Cholinesterase inhibitor. Use may be restricted.
Use: Insecticide.
Shipping regulations: (Rail) Organic phosphate compound, liquid, n.o.s., Poison label. (Air) No label required.

diazoaminobenzene (diazobenzeneanilide; 1,3-diphenyltriazene; benzeneazoanilide) $C_6H_5NNNHC_6H_5$.
Properties: Golden-yellow scales. Soluble in alcohol, ether, and benzene; insoluble in water. M.p. 96°C; explodes when heated to 150°C.
Derivation: Interaction of nitrous acid and an alcoholic solution of aniline.
Hazard: Explodes on heating to 150°C. Dangerous!
Uses: Organic synthesis; dyes; insecticide.
Shipping regulations: Consult authorities. Not listed.

4,4′-diazoaminodibenzamidine
$NH_2C(NH)C_6H_4NNNHC_6H_4C(NH)NH_2$.
Properties: (trihydrate) Yellow powder; m.p. 203°C (with decomposition); soluble in water at pH 6.1; solutions are unstable.
Use: Veterinary medicine.

diazobenzeneanilide. See diazoaminobenzene.

para-**diazobenzenesulfonic acid** $C_6H_4SO_3N_2$.
Properties: White or slightly red crystals, or white paste. Soluble in water and ether; insoluble in alcohol.
Derivation: From sulfanilic acid, sodium nitrite and sulfuric acid.
Hazard: Explodes when shocked or heated.
Uses: Dyestuffs; reagent.
Shipping regulations: Consolut authorities. Not listed.

para-**diazodimethylaniline zinc chloride double salt** (para-dimethylaminobenzene diazonium chloride, zinc chloride double salt; para-diazotized aminodimethylaniline zinc chloride double salt)
$(CH_3)_2NC_6H_4N_2Cl \cdot ZnCl_2$.
Properties: Yellow to orange (light-sensitive) crystals.
Specifications: Moisture content 5–20%; zinc 17–23%; chloride 31–35%.
Containers: Bottles; fiber drums.
Hazard: May be toxic and irritant.
Uses: Rapid diazotype coupler, used in coatings for light-sensitive paper.

2-diazo-4,6-dinitrobenzene-1-oxide. See diazodinitrophenol.

diazodinitrophenol (2-diazo-4,6-dinitrobenzene-1-oxide; 5,7-dinitro-1,2,3-benzoxadiazole; DDNP)
$(NO_2)_2C_6H_2ON_2$ (bicyclic).
Properties: Yellow, crystalline compound; darkens rapidly by exposure to sunlight. Specific gravity 1.6. Soluble in nitrobenzene, acetone, acetic acid, and nitroglycerin. Desensitized by water.
Derivation: Diazotization of picramic acid in aqueous solution with sodium nitrite and hydrochloric acid.
Hazard: Explodes when shocked or heated. Dangerous! An initiating explosive.
Use: Primary charge in blasting caps.
Shipping regulations: (Rail) Consult regulations. Not accepted by express. (Air) Not accepted.

para-**diazodiphenylamine sulfate.**
$(C_6H_5NHC_6H_4N_2)_2SO_4$. Yellow-green solid with unpleasant odor. Sensitive to light; soluble in water. Used as a light sensitive diazo compound for coating on reproduction paper, giving direct positive prints of various colors with different developers or coupling agents.

1,2-diazole. See pyrazole.

diazomethane (azimethylene). $H_2C{=}N^+{=}N^-$
Properties: Yellow gas at room temp.; f.p. −145°C; b.p. −23°C; sp. gr. 1.45.
Hazard: Highly toxic by inhalation; strong irritant. Tolerance, 0.2 ppm in air. Severe explosion risk when shocked; may explode on contact with alkali metals, rough surfaces, or heat (100°C).
Shipping regulations: (Rail, Air) Not listed. Consult authorities.

1-diazo-2-naphthol-4-sulfonic acid $C_{10}H_5N_2OSO_3H$.
Properties: Yellow needles in paste or dry form. Slightly soluble in water. Also available as sodium salt, m.p. 168°C. Combustible.
Derivation: Diazotization of 1-amino-2-naphthol-4-sulfonic acid and filtering of the diazo compound.
Hazard: May explode if heated above 100°C.
Uses: Azo dyes; valuable chrome dyestuff component.
Shipping regulations: (Rail, Air) Not listed. Consult authorities.

"Diazopon AN."[307] Trademark for a nonionic dispersing and stabilizing agent consisting of polyoxyethylated fatty alcohol.

Properties: Clear, oily, yellow liquid; sp. gr. 1.03–1.04; soluble in water; stable to strong acids; alkalies and metallic ions.
Uses: Dispersing and stabilizing agent in the naphthol dyeing process to improve fastness to rubbing. Used in dissolving fast color salts, it has a stabilizing effect on the diazonium compound.

diazotization. The reaction of a primary aromatic amine with nitrous acid in the presence of excess mineral acid to produce a diazo (—N=N—) compound. Widely used in organic synthesis, especially production of dyes.

diazotizing salts. See sodium nitrite.

"Diazyme."[212] Trademark for an amyloglucosidase which splits starch almost completely to glucose.

DIBA. See diisobutyl adipate.

DIBAC. Abbreviation for diisobutylaluminum chloride.

"Dibactol."[430] Trademark for myristyldimethylbenzylammonium chloride (q.v.).

DIBAL-H. Abbreviation for diisobutylaluminum hydride (q.v.).

dibasic. See monobasic.

dibenzanthrone (violanthrone) $C_{34}H_{16}O_2$. Violet-blue vat dye.
Properties: Bluish-black powder. Soluble in nitrobenzene, conc. sulfuric acid.
Derivation: From benzanthrone.
Use: Intermediate.

3,3′-dibenzanthronyl (also called the 13,13′-compound) $C_{34}H_{18}O_2$.
Properties: Dark yellow needles. Soluble in conc. sulfuric acid. M.p. 412°C.
Use: Intermediate.

4,4′-dibenzanthronyl (also called the 2,2′-compound) $C_{34}H_{18}O_2$.
Properties: Yellow needles. Soluble in nitrobenzene; slightly soluble in benzene, alcohol, ether. M.p. 320°C.
Use: Intermediate.

2,3,6,7-dibenzoanthracene. See pentacene.

dibenzocycloheptadienone $C_{15}H_{12}O$. A tricyclic compound; light yellow to amber solid. M.p. 28.5°C; b.p. 203–204°C (7 mm); sp. gr. 1.1635 (20°C); refractive index (n 20/D) 1.6324. Soluble in alcohol and most organic solvents. Insoluble in water. Combustible.
Use: Intermediate.

dibenzofuran. See diphenylene oxide.

"Dibenzo G-M-F."[248] Trademark for dibenzoyl-para-quinonedioxime, $(C_6H_5COON)_2C_6H_4$.
Properties: Brownish gray powder; sp. gr. 1.37; starts to decompose above 200°C; good storage stability. Insoluble in acetone, benzene, gasoline, ethylene dichloride and water.
Grades: Available in superdispersing grades.
Uses: Non-sulfur vulcanizing agent; in tire-curing bags, gaskets and wire insulation to impart heat resistance.
See also benzoyl peroxide.

dibenzopyran. See xanthene.

dibenzopyrone. See xanthone.

dibenzopyrrole. See carbazole.

dibenzothiophene $\overline{C_6H_4C_6H_4S}$.
Properties: Colorless crystals; m.p. 97–98°C. Combustible.
Uses: Cosmetics and pharmaceuticals; intermediate.

dibenzoyl. See benzil.

trans-**1,2-dibenzoylethylene** $C_6H_5COCHCHCOC_6H_5$.
Properties: Yellow orange crystals; m.p. 111°C; soluble in glacial acetic acid, ethyl acetate, benzene and chloroform; sparingly soluble in alcohol; insoluble in water and petroleum ether. Combustible.
Containers: Polyethylene-lined fiber drums.
Uses: Enzyme inhibitor, bactericide and intermediate.

dibenzoylmethane (1,3-diphenyl-1,3-propanedione) $C_6H_5COCH_2COC_6H_5$.
Properties: Crystals; m.p. 80°C; b.p. 219–221°C (18 mm). Partially soluble in alcohol; soluble in ether, chloroform, and aqueous sodium hydroxide. Combustible.
Use: Colorimetric determination of uranium.

dibenzoyl peroxide. See benzoyl peroxide.

dibenzoyl-para-**quinonedioxime.** See "Dibenzo G-M-F."

2,4-dibenzoylresorcinol $C_6H_5COC_6H_2(OH)_2COC_6H_5$.
Properties: Light yellow crystals; nearly odorless; m.p. 125–128°C; soluble in alcohol, ethyl acetate, methyl ethyl ketone; insoluble in water.
Use: Light absorber, best at 280–370μ.

dibenzyl. See sym-diphenylethane.

N,N-**dibenzylamine** $HN(CH_2C_6H_5)_2$.
Properties: Colorless to light yellow liquid; b.p. 300°C (partial decomp.); m.p. -26°C; ammonia-like odor; combustible; sp. gr. 1.017 (20°C); refractive index 1.5730–1.5740 (n 25/D); distilling range 168–172°C (10 mm); partially soluble in water, acetone, benzene and gasoline.
Hazard: Toxic by ingestion, inhalation.
Uses: Intermediate; rubber activator.

N,N-**dibenzyl**-para-**aminophenol** $(C_6H_5CH_2)_2NC_6H_4OH$.
Properties: Brown powder. Soluble in acetone, benzene, anhydrous methanol. M.p. not lower than 110°C.

N,N-**dibenzylaniline** $C_6H_5N(CH_2C_6H_5)_2$.
Properties: Yellowish-white crystals. Soluble in alcohol and ether; insoluble in water. M.p. 70°C; b.p. above 300°C.
Hazard: Probably toxic. See aniline.

dibenzyl disulfide (benzyl disulfide) $C_6H_5CH_2SSCH_2C_6H_5$.
Properties: Pink solid; odor similar to benzaldehyde. M.p. 70–72°C. Dissolves in most organic solvents; very slightly soluble in water. Combustible.
Use: Antioxidant and antisludging agent for petroleum oils; extreme-pressure lube oils and greases.

dibenzyl ether. $C_6H_5CH_2OCH_2C_6H_5$.
Properties: Colorless, unstable liquid. Faint, almond odor. Insoluble in water; soluble in most organic solvents. Sp. gr. 1.035; b.p. 298–300°C; flash point 275°F. Combustible.
Grade: Technical.
Hazard: Irritant and narcotic in high concentrations.
Uses: Plasticizer for various resins; perfumery (solvent for nitro-musks); flavoring.

"Dibenzyline."[71] Trademark for phenoxybenzamine hydrochloride (q.v.).

N,N-**dibenzylmethylamine** $CH_3N(CH_2C_6H_5)_2$.
Properties: Colorless to light yellow liquid; sp. gr. 0.99 (25°C); refractive index 1.5560–1.5590 (25°C); distilling range: 152–158°C (11 mm). Partially soluble in water; soluble in organic solvents. Combustible.
Containers: 200-lb, 400-lb steel drums.
Uses: Intermediate; oil-soluble rust inhibitor; cutting oils; hydraulic fluids; lubricants.

dibenzyl sebacate $C_6H_5CH_2OOC(CH_2)_8COOCH_2C_6H_5$.
Properties: Light straw-colored liquid; m.p. 28°C; b.p. 265°C (4 mm); sp. gr. 1.055 (30/20°C); complete nonvolatility; gives excellent low-temperature flexibility. Flash point 450°F. Combustible.
Containers: Drums; tank cars.
Use: Plasticizer, especially for plastic linings for containers.

dibenyl succinate. See benzyl succinate.

2,5-dibiphenylyloxazole (BBO) $C_{27}H_{21}NO$.
Properties: Crystalline solid, m.p. 237–239°C.
Grade: Purified.
Use: Scintillation counter or as wavelength shifter in solution scintillators.

diborane (diboron hexahydride; boroethane) B_2H_6.
Properties: Colorless gas with repulsive odor. B.p. -92.5°C; f.p. -165°C; density 0.18 g/ml (17°C. Soluble in carbon disulfide; decomposes in water. Highly reactive; flash point -130°F. Autoignition temp. 100–125°F.
Derivation: (a) From boron trichloride or bromide and hydrogen; (b) by reaction of lithium aluminum hydride and boron trichloride in ether solution.
Grades: Technical 95%; high purity 99+%.
Containers: Steel cylinders.
Hazard: Highly flammable; dangerous fire risk. Highly toxic by inhalation; strong irritant. Tolerance, 0.1 ppm in air. Reacts violently with oxidizing materials. Safety data sheet available from Manufacturing Chemists Assn., Washington, D.C.
Uses: Synthesis of organic boron compounds and metal borohydrides; polymerizaton catalyst for ethylene; fuel for air-breathing engines, and rockets; reducing agent; doping agent in electronics.
Shipping regulations: (Air) Not acceptable. (Rail) Not listed. Consult authorities.

"Dibromantin."[109] Trademark for 1,3-dibromo-5,5-dimethylhydantoin (q.v.).

"Dibrom" 8 Emulsive.[253] Trademark for an insecticide, the active ingredient of which is 60% naled. (q.v.).

dibromoacetylene (dibromomethyne) BrC⋮CBr.
Properties: Heavy, colorless liquid. Disagreeable odor. Soluble in most organic solvents. Sp. gr. (approx.) 2; m.p. 76–76.5°C.
Derivation: (a) Interaction of magnesium dibromoacetylene and in ethereal solution of cyanogen bromide. (b) Interaction of tribromoethylene and alcoholic potash.
Hazard: Explodes on contact with oxygen or on heating! Dangerous! Highly toxic by inhalation and injection; strong irritant.
Use: Organic synthesis (halogenated ethylene).
Shipping regulations: (Rail) Not listed. Consult regulations. (Air) Explosives, n.o.s. Not acceptable.

9,10-dibromoanthracene $C_6H_4C_2Br_2C_6H_4$ (tricyclic).
Properties: Yellow crystals. Soluble in chloroform; slightly soluble in alcohol and ether; insoluble in water. M.p. 221°C; b.p., sublimes.
Derivation: Bromination of anthracene.
Use: Organic synthesis.

ortho-**dibromobenzene** (benzene dibromide) $C_6H_4Br_2$.
Properties: Heavy liquid with pleasant, aromatic odor. B.p. 225.5°C; freezing point 7.13°C; sp. gr. 1.9767 (25/4°C); refractive index (n 20/D) 1.6155. Combustible. Miscible with alcohol, acetone, ether, benzene, carbon tetrachloride, and n-heptane; insoluble in water.
Derivation: Interaction of benzene with an excess of bromine in presence of iron.
Uses: Solvent for oils; motor fuels; top-cylinder compounds; organic synthesis; ore flotations.

para-**dibromobenzene** (benzene dibromide) $C_6H_4Br_2$.
Properties: Colorless crystals. Soluble in alcohol and ether. M.p. 89°C; b.p. 219°C; sp. gr. 2.261; refractive index (n 99/D) 1.5743.
Hazard: Moderately toxic by inhalation.
Uses: Organic synthesis of dyestuffs and drugs; manufacture of intermediates; fumigant.

N,N-**dibromobenzenesulfonamide** $C_6H_5SO_2NBr_2$.
Properties: Solid; m.p. 109–111°C. Active bromine 50.4%.
Use: Halogenating agent.

dibromochloromethane $CHBr_2Cl$.
Properties: Clear, colorless heavy liquid. Sp. gr. 2.38; b.p. 116°C.
Hazard: Probably irritant and narcotic.
Use: Organic synthesis.

1,2-dibromo-3-chloropropane $CH_2BrCHBrCH_2Cl$.
Properties: Colorless (when pure) liquid; sp. gr. 2.05 (20°C); b.p. 195.5°C; f.p. 6.7°C; refractive index 1.5530 (20/D); flash point (TOC) 170°F. Combustible. Slightly soluble in water; miscible with oils.
Hazard: Strong irritant to skin, eyes, and throat. Absorbed by skin. Use may be restricted.
Use: Soil fumigant for nematodes.

dibromodiethyl sulfide $(CH_2CH_2Br)_2S$. The bromine analog of mustard gas.
Properties: White crystals. Soluble in alcohol, benzene, ether; insoluble in water. Sp. gr. 2.05 (15°C); b.p. 240°C (decomposes); m.p. 31–34°C.
Derivation: Action of hydrobromic acid on an aqueous solution of thiodiglycol.
Hazard: Highly toxic by inhalation; strongly irritant poison gas.
Use: Organic synthesis.
Shipping regulations: (Rail, Air) Not listed. Consult authorities.

dibromodiethyl sulfone $(BrCH_2CH_2)_2SO_2$.
Properties: Plates. Soluble in alcohol, benzene, ether. M.p. 111–112°C.
Derivation: Interaction of dibromodiethyl sulfide, chromic anhydride, and dilute sulfuric acid.

dibromodiethyl sulfoxide $(BrCH_2CH_2)_2SO$.
Properties: Glittering crystals. Soluble in alcohol, benzene, ether. M.p. 100–101°C.
Derivation: Interaction of benzoyl peroxide and a hot solution of dibromodiethyl sulfide in chloroform.

dibromodifluoromethane CF_2Br_2.
Properties: Colorless heavy liquid; f.p. -141°C; b.p. 24.5°C; sp. gr. 2.288 (15/4°C); refractive index 1.399 (12°C). Insoluble in water; soluble in methanol and ether. Nonflammable.

Derivation: Vapor phase bromination of difluoromethane.
Grades: Pure (95.0% min).
Containers: Cylinders.
Hazard: Moderately toxic and irritant.
Uses: Synthesis of dyes, pharmaceuticals, quaternary ammonium compounds, fire-extinguishing agent.

1,3-dibromo-5,5-dimethylhydantoin $BrNCONBrCOC(CH_3)_2$.
Properties: Free-flowing cream-colored powder with slight bromine odor. M.p. 187–191°C (decomposes); quite stable at 75°C. Soluble in benzene, chloroform, glacial acetic acid; slightly soluble in water and carbon tetrachloride; insoluble in hexane. Contains 55% active bromine, which is slowly released in aqueous solution.
Derivation: Bromination of dimethylhydantoin.
Uses: Controlled bromination and oxidation of organic compounds; water treatment; polymerization catalyst; potential germicide and sanitizer.

1,2-dibromoethane. See ethylene dibromide.

4,5′-dibromofluorescein $C_{20}H_{10}Br_2O_5$.
Properties: Yellow powder (clean shade); m.p. 265–267°C.
Grades: 99+%.
Use: Lipstick dye.

2,4-dibromofluorobenzene $C_6H_3Br_2F$.
Properties: Colorless liquid; sp. gr. (20°C) 2.047; b.p. 214°C; refractive index (n 25/D) 1.5790. Insoluble in water; soluble in alcohol, acetone, ether, benzene, chloroform, ethyl acetate, and glacial acetic acid.
Hazard: Irritating to skin and eyes.
Uses: Intermediate for agricultural and pharmaceutical chemicals.

dibromoformoxime CBr_2NOH.
Properties: Crystals. M.p. 70–71°C. Distills between 75 and 85°C (3 mm).
Hazard: Evolves highly toxic fumes on heating. A military poison.
Shipping regulations: (Rail, Air) Not listed. Consult authorities.

dibromoiodoethylene Br_2CCHI.
Properties: Liquid; sp. gr. 2.952 (24°C); b.p. 91°C (15mm).
Derivation: Reaction of iodine and dibromoacetylene.

dibromomalonic acid $HOOC \cdot CBr_2COOH$.
Properties: Light yellow needles or prisms; m.p. 147°C (decomposes).
Use: Intermediate for drugs and fine chemicals.

dibromomalonyl chloride $ClOCCBr_2COCl$.
Properties: Yellowish oily liquid; b.p. 75–77°C (15 mm).
Use: Chemical intermediate.

dibromomethane. See methylene bromide.

dibromomethyl ether $(CH_2Br)_2O$.
Properties: Colorless liquid. Decomposed by water. Soluble in acetone, benzene, ether; insoluble in water. Sp. gr. 2.2; b.p. 154–155°C; f.p. −34°C.
Derivation: (a) The reaction product of paraformaldehyde and sulfuric acid is treated with ammonium bromide. (b) Interaction of hydrobromic acid and paraformaldehyde.
Hazard: Evolves highly toxic fumes on heating, strong irritant to eyes; fire risk in contact with oxidizing materials.
Uses: Military poison (lachrymator).
Shipping regulations: (Rail, Air) Not listed.

9,10-dibromooctadecanoic acid (9,10-dibromostearic acid) $CH_3(CH_2)_7CHBrCHBr(CH_2)_7CO_2H$.
Properties: Yellow solid or liquid; sp. gr. 1.2458 (30/4°C); m.p. 29–30°C; refractive index 1.4893 (42°C). Insoluble in water; soluble in alcohols, ketones, aromatic and chlorinated hydrocarbons. Also available as methyl ester.
Grade: Technical (amber liquid).
Use: Chemical intermediate.

1,5-dibromopentane. See pentamethylene dibromide.

1,3-dibromopropane. See trimethylene bromide.

dibromopropanol (2,3-dibromo-1-propanol) $CH_2BrCHBrCH_2OH$.
Properties: Colorless liquid. Sp. gr. (20/4°C) 2.120; b.p. 219°C. Soluble in acetone, alcohol, ether and benzene.
Uses: Intermediate in preparation of flame retardants, insecticides and pharmaceuticals.

dibromoquinonechlorimide. A reagent used for spot visualizations in chromatographic systems.
Hazard: Highly sensitive to heat. Explodes at 120°C; decomposes with rapid heat evolution at 60°C.
Shipping regulations: (Rail, Air) Not listed. Consult authorities.

3,5-dibromosalicylaldehyde (3,5-dibromo-2-hydroxybenzaldehyde) $Br_2(OH)C_6H_2CHO$.
Properties: Pale yellow crystals; m.p. 86°C; slightly soluble in water; soluble in ether, benzene, chloroform, alcohol and acetic acid.
Use: Medicine (external); fungicide.

4′,5-dibromosalicylanilide. See dibromsalan.

9,10-dibromostearic acid. See 9,10-dibromoöctadecanoic acid.

2,5-dibromoterephthalic acid $C_6H_2Br_2(COOH)_2$. A flame-retardant monomer for production of polyester fibers, which are made by reacting this acid with dimethyl terephthalate and ethylene glycol. Permanent lowering of flammabilty is said to be gained by this method of incorporating bromine in molecular combination.

sym-**dibromotetrafluoroethane** $CBrF_2CBrF_2$.
Properties: Liquid; b.p. 47.3°C; density 2.18 g/cc (21.1°C). Nonflammable.
Uses: Refrigerant; fire-extinguishing agent; control fluid.

dibromsalan. USAN for 4′,5-dibromosalicylanilide $BrC_6H_3(OH)CONHC_6H_4Br$.
Use: Disinfectant.

dibucaine $C_{20}H_{29}N_3O_2$. 2-n-Butoxy-N-(2-diethylaminoethyl)cinchoninamide.
Properties: Colorless or almost colorless powder; m.p. 62–65°C. Odorless; somewhat hygroscopic; affected by light. Soluble in alcohol and acetone; slightly soluble in water. Also available as the hydrochloride.
Grade: N.F.
Hazard: May have serious allergic side effects.
Use: Medicine (anesthetic).

dibutoline sulfate $(C_{15}H_{33}N_2O_2)_2SO_4$. Bisdibutylcarbamate of ethyl(2-hydroxyethyl)dimethylammonium sulfate.
Properties: Hygroscopic powder; decomposes 166°C; soluble in water and benzene.
Use: Surface-active agent in medicine.

2,5-dibutoxyaniline $C_6H_3(OC_4H_9)_2NH_2$.
Properties: M.p. 18°C; insoluble in water; soluble in organic solvents.

Hazard: May be toxic. See aniline.
Uses: Dyes; synthesis.

1,4-dibutoxy benzene. See hydroquinone di-n-butyl ether.

dibutoxyethyl adipate $(C_2H_4COOC_2H_4OC_4H_9)_2$.
Properties: Colorless, oily liquid. Sp. gr. (20/20°C) 0.997; f.p. -34°C; boiling range 205–215°C (4 mm); mild, butyl type odor; flash point 370°F; refractive index, 1.442 (25°C); wt/gal 8 lb. Insoluble or only slightly soluble in mineral oil, glycerine, glycols and some amines; soluble in most other organic liquids. Combustible. Low toxicity.
Uses: Primary plasticizer for most resins, imparting flexibility at very low temperature, as well as stability to ultraviolet light.

dibutoxyethyl phthalate (n-butyl glycol phthalate) $C_6H_4(COOC_2H_4OC_4H_9)_2$.
Properties: Colorless liquid; f.p. -55°C; sp. gr. 1.06 (20°C); b.p. 270°C; wt/gal 8.86 lb; fast to light, water-resistant. Flash point 407°F. Combustible; low toxicity. Soluble in organic solvents.
Containers: Drums; tank cars.
Uses: Plasticizer for polyvinyl chloride, polyvinyl acetate, and other resins.

dibutoxymethane $CH_2(OC_4H_9)_2$.
Properties: Colorless liquid. Wt/gal 6.97 lb (20°C); refractive index 1.40615 (20°C); sp. gr. 0.838 (20/20°C); flash point 140°F (closed cup); boiling range 164–186°C. Insoluble in water. Low toxicity. Combustible.
Hazard: Moderate fire risk.

dibutoxytetraglycol (tetraethylene glycol dibutyl ether) $(C_4H_9OC_2H_4OC_2H_4)_2O$.
Properties: Pratically colorless liquid with characteristic odor. Slightly soluble in water (1.3% by wt); sp. gr. 0.9436 (20/20°C); lb/gal 7.85 (20°C); b.p. 237°C (50 mm), 330°C; vapor pressure less than 0.01 mm (20°C); freezing point -20°C; flash point 355°F; solubility of water in product 4.8% by wt (20°C); refractive index 1.4357. Combustible. Low toxicity.
Containers: 5-, 55-gal drums.
Use: Solvent, especially for DDT.

N,N-**di-n-butyl acetamide** $CH_3CON)C_4H_9)_2$.
Properties: Colorless liquid. Sp. gr. 0.890 (20°C); boiling range 245–250°C; odor faint. Flash point 225°F. Combustible. Low toxicity.

di-n-butylamine $(C_4H_9)_2NH$.
Properties: Colorless liquid with amine odor. B.p. 159.6°C; freezing point -62°C; sp. gr. (20/20°C) 0.7613; wt/gal (20°C) 6.33 lb; refractive index (n 20/D) 1.4175; flash point (open cup) 125°F. Partially soluble in water; soluble in alcohol and ether; miscible with hydrocarbons. Combustible.
Derivation: By reaction of butanol or butyl chloride with ammonia.
Grade: Technical.
Containers: Drums; tank cars.
Hazard: Moderate fire risk.
Uses: Corrosion inhibitor; intermediate for emulsifiers, rubber accelerators, dyes, insecticides, and flotation agents; inhibitor for butadiene.

di-sec-butylamine $(CH_3CHCH_2CH_3)_2NH$.
Properties: Water-white liquid; amine odor; boiling range 132–135°C; sp. gr. 0.754 (20/20°C); refractive index 1.412 (20°C); flash point 75°F.
Hazard: Flammable, dangerous fire risk.

dibutylamine pyrophosphate.
Grade: Available as 40% solution in ethyl alcohol-acetone.
Hazard: Flammable, dangerous fire risk.
Uses: Anticorrosion agent in lacquers and cotton solutions; for light and heat stabilization of vinyl chloride and vinyl copolymer resins.

N,N-**di-n-butylaminoethanol** $(C_4H_9)_2NCH_2CH_2OH$.
Properties: Colorless liquid. Sp. gr. 0.859 (20°C); boilng range 224–232°C; odor faint, amine-like. Flash point 200°F. Combustible.
Containers: Drums; tank cars.
Uses: Synthesis.

dibutylammonium oleate. See "Barak."

di-n-butylammonium tetrafluoroborate $(C_4H_9)_2NH_2BF_4$. Solid; m.p. 266°C. Used for lubricating, surface treating, fluxing of aluminum.

N,N-**di-n-butylaniline** $C_6H_5N(C_4H_9)_2$.
Properties: Amber liquid; faint aniline odor. Sp. gr. 0.904 (20°C); boiling range 267–275°C; refractive index 1.519 (20°C). Soluble in alcohol and ether; insoluble in water. Flash point 230°F. Combustible.
Hazard: May be toxic. See aniline.

2,5-di-tert-butyl benzoquinone $[C(CH_3)_3]_2C_6H_2O_2$.
Properties: Yellow crystals; insoluble in water; soluble in ethyl acetate, acetone, benzene; slightly soluble in ethyl alcohol. M.p. 149–151°C.
Hazard: May be toxic. See quinone.
Uses: Oxidant; polymerization catalyst.

dibutyl butyl phosphonate $C_4H_9P(O)(OC_4H_9)_2$.
Properties: Colorless liquid, with mild odor. Stable. Insoluble in water, miscible with most common organic solvents; sp. gr. 0.948 (20/4°C); b.p. 127–128°C (2.5 mm); flash point 310°F (COC). Combustible.
Uses: Heavy-metal extraction and solvent separation; gasoline additives; antifoam agent; plasticizer; textile conditioner and antistatic agent.

dibutyl "Carbitol."[214] Trademark for diethylene glycol dibutyl ether (q.v.).

dibutyl "Cellosolve."[214] Trademark for ethylene glycol dibutyl ether (q.v.).

dibutyl chlorophosphate $(C_4H_9O)_2P(O)Cl$.
Properties: Water-white liquid; b.p. 103–106°C (1.5 mm); sp. gr. 1.0742 (25°C); refractive index 1.4289 (n 25/D). Soluble in alcohols; hydrolyzes slowly.
Use: Intermediate in organic synthesis.

4,6-di-tert-butyl-meta-cresol (DBMC; 4,6-tert-butyl-3-methylphenol) $[C(CH_3)_3]_2CH_3C_6H_2OH$.
Properties: Crystalline solid; m.p. 62.1°C; b.p. 282°C; sp. gr. 0.912 (80/4°C); viscosity 9.9 centistokes (80°C), 1.42 centistokes (160°C); flash point 262°F (open cup); combustible. Low toxicity; very soluble in ethanol, benzene, carbon tetrachloride, ethyl ether, and aceotne; essentially insoluble in water, ethylene glycol, and 10% aqueous sodium hydroxide.
Uses: Rubber reclaiming; surface-active agents; resins and plasticizers; antioxidants and perfumes.

2,6-di-tert-butyl-para-cresol (DBPC; 2,6-di-tert-butyl- $[C(CH_3)_3]_2CH_3C_6H_2OH$.
Properties: White, crystalline solid; freezing point 70°C; b.p. 265°C; sp. gr. 1.048 (20/4°C); viscosity 3.47 centistokes (80°C), 1.54 centistokes (120°C); refractive index (n 75/D) 1.4859; soluble in methanol,

ethanol, isopropanol, "Cellosolve" (12°C), naphtha, benzene, methyl ethyl ketone and linseed oil; insoluble in water and 10% sodium hydroxide. Flash point (COC) 275°F. Combustible. Low toxicity.
Grades: Technical; feed; F.C.C.
Containers: Drums; tank cars.
Hazard: Use in foods restricted. Consult FDA regulations for details.
Uses: Antioxidant for petroleum products, jet fuels, rubber, plastics and food products; food packaging; animal feeds. Satisfies ASTM D910-64T for use in aviation gasoline.

2,6-di-tert-**butyl**-alpha-**dimethylamino**-para-**cresol** ("Ethyl" Antioxidant 703) $(C_4H_9)_2C_6H_2OH[CH_2N(CH_3)_2]$.
Properties: Light yellow crystalline solid; m.p. 201°F; flash point 280°F (open cup). Insoluble in water and 10% sodium hydroxide; soluble in organic solvents. Combustible.
Use: Antioxidant in gasoline and oils, including jet-engine oils.

di-tert-**butyl diperphthalate** $(CH_3)_3COOCOC_6H_4COOOC(CH_3)_3$. Available as a 50% solution in dibutyl phthalate.
Properties: Clear liquid, insoluble in water; soluble in most organic solvents. Flash point 145°F (open cup); combustible. Sp. gr. 1.056.
Hazard: May ignite organic materials. Strong oxidizing agent. May explode when shocked or in contact with reducing materials.
Use: High-temperature polymerization catalyst for vinyl and polyester resins.
Shipping regulations: Peroxides, organic, liquid, n.o.s., (Rail) Yellow label. (Air) Not acceptable.

dibutyl diphenyl tin $(CH_3CH_2CH_2CH_2)_2Sn(C_6H_5)_2$.
Properties: Clear, slightly greenish liquid; contains 30.7% Sn; b.p. 175°C (2 mm); refractive index 1.563 (17.5°C); sp. gr. 1.19. Combustible.
Hazard: Toxic. Tolerance (as Sn) 0.1 mg per cubic meter of air.

di-tert-**butyl disulfide** $C(CH_3)_3SSC(CH_3)_3$.
Properties: Liquid; sp. gr. 0.9291 (60/60°F); boiling range 375–405°F; refractive index 1.491 (20°C); flash point 170°F (approx). Combustible.
Use: Intermediate; diesel and jet fuel additive; lubricant additive.

n-**dibutyl ether.** See butyl ether.

dibutyl fumarate $(C_4H_9)OOCCH{:}CHCOO(C_4H_9)$.
Properties: Colorless liquid; sp. gr. (20°C) 0.9873; b.p. 285.2°C; f.p. −15.6°C; refractive index n (20°C) 1.4466; insoluble in water; flash point 300°F; combustible; nontoxic.
Containers: Drums; tank trucks; tank cars.
Uses: Monomeric plasticizers; copolymers; intermediate.

dibutyl hexahydrophthalate $C_6H_{10}(COOC_4H_9)_2$.
Properties: Liquid; sp. gr. 1.005; b.p. 185–190°C; flash point 305°F. Combustible.
Use: Plasticizer for nitrocellulose.

2,5-di-tert-**butyl hydroquinone** $[C(CH_3)_3]_2C_6H_2(OH)_2$.
Properties: White powder; soluble in acetone, alcohol, benzene; insoluble in water; aqueous alkali. M.p. 210–212°C.
Uses: Polymerization inhibitor; antioxidant; stabilizer against ultraviolet deterioration of rubber.

di-n-butyl itaconate $CH_2{:}C(COOC_4H_9)CH_2(COOC_4H_9)$.
Properties: Clear, colorless liquid with slight odor. B.p. 145°C (10 mm); sp. gr. 0.9833 (22°C); refractive index (n 25/D) 1.442. Insoluble in water. Combustible.
Uses: Resins; lube oil additives; plasticizers.

N,N-**di-n-butyl lauramide** $C_{11}H_{23}CON(C_4H_9)_2$.
Properties: Straw-colored liquid. Sp. gr. 0.861 (20°C); boiling range 200–230°C (3 mm); odor lauric acid. Flash point 375°F. Combustible.

dibutyl maleate (DBM) $C_4H_9OOCCH{:}CHCOOC_4H_9$.
Properties: Colorless oily liquid. B.p. 280.6°C; sets to a glass below −85°C; sp. gr. 0.9964 (20/20°C); wt/gal 8.3 lb (20°C); flash point (open cup) 285°F. Insoluble in water. Combustible. Low toxicity.
Containers: Drums; tank cars.
Uses: Copolymers; plasticizers; intermediate.

2,6-di-tert-**butyl-4-methylphenol.** See 2,6-di-tert-butyl-para-cresol.

4,6-di-tert-**butyl-3-methylphenol.** See 4,6-di-tert-butyl-meta-cresol.

dibutyl oxalate $(COOC_4H_9)_2$.
Properties: Water-white, high-boilng liquid. Mild odor. B.p. 240–250°C; refractive index 1.425; f.p. −30°C; wt/gal 8.24 lb (approx) (20°C); flash point 220°F (closed cup). Miscible with most alcohols, ketones, esters, oils, hydrocarbons. Combustible.
Derivation: By the standard esterification process using normal butyl alcohol and oxalic acid.
Grades: According to ester content 90%; 95%; 99–100%.
Containers: Drums; tank cars.
Hazard: Toxic by ingestion and inhalation; strong skin irritant.
Uses: Organic synthesis; solvent.

di-tert-**butyl peroxide** (tert-butyl peroxide; DTBP) $(CH_3)_3COOC(CH_3)_3$.
Properties: Clear, water-white liquid. Sp. gr. 0.791 (25/25°C); f.p. −40°C; b.p. 111°C; refractive index 1.389 (20°C); flash point 65°F (closed cup). Soluble in styrene, ketones, most aliphatic and aromatic hydrocarbons; insoluble in water.
Hazard: Moderately toxic. Flammable, dangerous fire hazard. Strong oxidizing agent; may ignite organic materials or explode when shocked or in contact with reducing materials.
Uses: Polymerization catalyst for resins, including olefins, styrene, styrenated alkyds, and silicones; ignition accelerator for diesel fuel; organic synthesis; intermediate.
Shipping regulations: Peroxides, organic, liquid, n.o.s. (Rail) Yellow label. (Air) Not acceptable.

2,4-di-tert-**butylphenol** $[(CH_3)_3C]_2C_6H_3OH$.
Properties: Tan crystalline solid; m.p. 52°C; b.p. 152–157°C (25 mm); sp. gr. 0.907 (60/4°C); lb/gal 7.57 (60°C); flash point 265°F. Soluble in methanol, ether, very slightly soluble in water. Combustible.
Hazard: May be toxic. See phenol.
Uses: Intermediate; antioxidant; stabilizer; germicide.

2,6-di-tert-**butylphenol** $[(CH_3)_3C]_2C_6H_3OH$.
Properties: Light straw crystalline solid; m.p. 37°C; sp. gr. 0.914 (20°C); b.p. 253°C; flash point 245°F. Combustible. Soluble in alcohol and benzene; insoluble in water.
Hazard: May be toxic. See phenol.
Uses: Intermediate; antioxidant; satisfies ASTM D910-64T for use as antioxidant in aviation gasoline.

N,N′-**di**-sec-**butyl**-para-**phenylenediamine** $CH_3CH_2CH(CH_3)NHC_6H_4NHCH(CH_3)CH_2CH_3$.
Properties: Amber to red liquid (normally supercooled below 18°C). Sp. gr. 0.94 (15.5/15.5°C); f.p.

20° C; flash point 290° F (COC). Soluble in gasoline, absolute alcohol, and benzene; insoluble in water or caustic solutions. Combustible.
Hazard: Toxic by ingestion, inhalation, and skin absorption. Strong irritant to tissue; causes skin burns.
Uses: Oxidation inhibitor and stabilizer in gasoline (satisfies ASTM D910-64T as antioxidant in aviation gasoline); prevents decomposition of tetraethyl lead in gasoline.

dibutyl phosphite $(C_4H_9O)_2PHO$.
Properties: Water-white liquid; b.p. 95°C (1 mm); sp. gr. 0.9860 (25°C); refractive index 1.4228 (n 25/D), flash point 120°F; soluble in common organic solvents. Combustible.
Hazard: Moderate fire risk.
Uses: Solvent; antioxidant; intermediate.

dibutyl phthalate (DBP) $C_6H_4(COOC_4H_9)_2$.
Properties: Colorless, odorless, stable, oily liquid. Sp. gr. 1.0484 (20/20° C); f.p. -35° C; viscosity 0.203 poise (20° C); distillation range 227–235° C (37 mm); flash point (COC) 340°F; wt/gal 8.72 lb (68°F); refractive index 1.4915 (25° C); b.p. 340.0°C; vapor pressure 1.1 mm (150°C). Miscible with common organic solvents; insoluble in water. Combustible. Autoignition temp., about 750° F.
Derivation: By treating n-butyl alcohol with phthalic anhydride followed by purification, which results in a product unusually free from odor and color.
Grades: Technical; 99–100% dibutyl phthalate.
Containers: Drums; tank cars and trucks.
Hazard: Toxic. Tolerance, 5 mg per cubic meter of air.
Uses: Plasticizer in nitrocellulose lacquers, elastomers, explosives, nail polish and solid rocket propellants; solvent for perfume oils; perfume fixative; textile lubricating agent; safety glass; insecticides; printing inks; resin solvent; paper coatings; adhesives.

2,5-di-tert-butylquinone $[C(CH_3)_3]_2C_6H_2O_2$. Yellow powder; m.p. 149–151° C; insoluble in water; soluble in alcohol, acetone, ethyl acetate, and benzene.
Hazard: Fire risk in contact with organic materials.
Use: Oxidizing agent.

dibutyl sebacate (DBS) $C_4H_9OCO(CH_2)_8OCOC_4H_9$.
Properties: Clear, colorless, odorless liquid. B.p. 349°C (760 mm), 180°C (3 mm); f.p. -11° C; sp. gr. 0.936 (20/20° C); wt/gal 7.81 lb (20° C); refractive index 1.4395 (25° C); flash point 350° F. Insoluble in water. Combustible. Low toxicity.
Grade: Technical.
Containers: Drums; tank cars.
Uses: Plasticizer; rubber softener; dielectric liquid; cosmetics and perfumes; sealing food containers; flavoring.

N,N-**dibutylstearamide** $C_{17}H_{35}CON(C_4H_9)_2$.
Properties: Yellow liquid; sp. gr. 0.860 (20/20° C); boiling range 173–175°C (0.4 mm); flash point 420° F; fatty-acid odor. Combustible. Low toxicity.

di-tert-butyl sulfide (butyl sulfide) $[(CH_3)_3C]_2S$.
Properties: Liquid; f.p. 12.3° F; boiling range 297–303° F; sp. gr. 0.8316 (60/60° F); wt/gal 6.93 lb; refractive index 1.451 (20° C); flash point 125° F. Combustible.
Hazard: Moderate fire risk.
Uses: Intermediate; flavoring.

dibutyl tartrate $C_4H_9OOCCHOHCHOHCOOC_4H_9$.
Properties: Light tan liquid. M.p. 21° C; b.p. approx. 204° C (26 mm); refractive index 1.4463 (20° C); flash point 195° F (closed cup); combustible; low toxicity; autoignition temp. 544° F; wt/gal 9.07 lb (68° F). Miscible with the common organic solvents, oils, hydrocarbons.
Uses: Solvent and plasticizer for cellulose esters and ethers, elastomers; lubricant; rubberized fabrics; lacquers; dopes; transfer inks.

dibutylthiourea $C_4H_9NHCSNHC_4H_9$.
Properties: White to light tan powder; m.p. 59–69° C; slightly soluble in water; soluble in methanol, ether, acetone, benzene, ethyl acetate; insoluble in gasoline; low toxicity.
Uses: Corrosion inhibitor; for pickling cast iron or carbon steel; reducing corrosion of ferrous metals and aluminum alloys in brine; intermediate.

dibutyltin bis(lauryl) mercaptide $(C_4H_9)_2Sn(SC_{12}H_{25})_2$. Yellow liquid; tin content 18.5%; soluble in toluene and heptane.
Hazard: Toxic; tolerance, 0.1 mg per cubic meter of air.
Uses: Antioxidant and metal cleaning (or protective) agent.

dibutyltin diacetate $(C_4H_9)_2Sn(C_2H_3O_2)_2$.
Properties: Clear yellow liquid. B.p. 130° C (2 mm); f.p. below 12° C. Soluble in water and most organic solvents. Flash point 290° F. Combustible.
Derivation: Reaction of acetic acid with dibutyltin oxide.
Hazard: Toxic. Tolerance, 0.1 mg per cubic meter of air.
Uses: Stabilizer for chlorinated organics; catalyst for condensation reactions.

dibutyltin dichloride $(C_4H_9)_2SnCl_2$.
Properties: White crystalline solid; m.p. 43°C; b.p. 135°C (10 mm); sp. gr. 1.36 (24/4°C); refractive index 1.4991 (51°C). Insoluble in cold water; hydrolyzed by hot water. Soluble in many organic solvents. Flash point 335° F; combustible.
Derivation: Reacton of butylmagnesium chloride with tin tetrachloride.
Hazard: Toxic. Tolerance, 0.1 mg per cubic meter of air.
Use: Organotin intermediate.

dibutyltin di-2-ethylhexoate $(C_4H_9)_2Sn(O_2CC_7H_{15})_2$.
Properties: Waxy white solid. M.p. 52–54° C. Insoluble in water; soluble in most organic solvents.
Derivation: Reaction of dibutyltin oxide with 2-ethylhexoic acid.
Hazard: Toxic. Tolerance, 0.1 mg per cubic meter of air.
Uses: Catalyst for silicone curing; polyether foams.

dibutyltin dilaurate $(C_4H_9)_2Sn(OCOC_{10}H_{20}CH_3)_2$.
Properties: Clear, pale yellow liquid. Sp. gr. 1.066 (20/20°C); f.p. 8.0°C; lb/gal 8.84 (20° C); flash point 440° F; soluble in acetone and benzene; insoluble in water. Combustible.
Hazard: Toxic. Tolerance, 0.1 mg per cubic meter of air.
Uses: Stabilizer for vinyl resins, lacquers, elastomers; catalyst for urethane.

dibutyltin maleate $[(C_4H_9)_2Sn(OOCCH)_2]_x$.
Properties: White amorphous powder; m.p. 110° C. Insoluble in water; soluble in benzene and organic esters. Flash point 400° F; combustible.
Derivation: Reaction of maleic acid with di-butyltin oxide.

Hazard: Toxic. Tolerance, 0.1 mg per cubic meter of air.
Uses: Stabilizer for polyvinyl chloride resins; condensation catalyst.

dibutyltin oxide $[(C_4H_9)_2SnO\text{-}]_x$.
Properties: White powder; m.p. (decomposes). Insoluble in water. Combustible.
Derivation: Hydrolysis of dibutyltin dichloride with caustic.
Hazard: Toxic. Tolerance, 0.1 mg per cubic meter of air.
Uses: Condensation catalysts; intermediate for other organotins.

dibutyltin sulfide $[(C_4H_9)_2SnS]_3$.
Properties: Colorless oily liquid. Combustible.
Derivation: Reaction of dibutyltin oxide with hydrogen sulfide.
Hazard: Toxic. Tolerance, 0.1 mg per cubic meter of air.
Uses: Vinyl stabilizer; antioxidant; lubricating additive.

2,6-di-tert-**butyl-para-tolyl N-methylcarbamate** $[(CH_3)_3C]_2CH_3C_6H_2OOCNHCH_3$. White solid; m.p. 200°C; insoluble in water; soluble in alcohol.
Use: Herbicide.

1,1-dibutylurea (N,N-dibutylurea) $NH_2CON(C_4H_9)_2$. Liquid. M.p. 22–25°C; boiling range 118–119°C (2–3 mm); soluble in alcohol and ether. Flash point 279°F. Combustible.
Use: Thermoplastic resins, by copolymerization with simple urea with formaldehyde catalyst.

dibutyl xanthogen disulfide. See "C-P-B."

DIC. Abbreviation for beta-diisopropylaminoethyl chloride hydrochloride (q.v.).

dicalcium magnesium aconitate (calcium magnesium aconitate) $[C_3H_3(COO)_3]_2Ca_2Mg$.
Properties: White crystalline powder or lumps.
Derivation: Precipitation from molasses with lime.
Uses: Conversion to aconitic acid, tributyl aconitate and similar ester plasticizers.

dicalcium orthophosphate. See calcium phosphate, dibasic.

dicalcium orthophosphite. See calcium phosphite.

dicalcium phosphate dihydrate. See calcium phosphate, dibasic.

dicalcium silicate $2CaO \cdot SiO_2$. One of the components of cement. See cement, Portland. Also obtained as a by-product in electric furnace operation; used to neutralize acid soils. Noncombustible.

"Dicalite."[218] Trademark for a group of products made from either diatomite or perlite; used in filters and filteraids. See also diatomaceous earth.

dicamba. Generic name for 3-6-dichloro-ortho-anisic acid (2-methoxy-3,6-dichlorobenzoic acid) $HOOC(Cl)C_6H_2Cl(OCH_3)$.
Properties: Crystals; m.p. 114–116°C. Slightly soluble in water; moderately soluble in xylene; very soluble in alcohol.
Uses: Herbicide; pest control.

dicapryl adipate $C_8H_{17}OOC(CH_2)_4COOC_8H_{17}$. See also also capryl compounds.
Properties: Almost water-white liquid; b.p. (4 mm) 213–216.5°C. Flash point 352°F; combustible. Low toxicity.
Use: Plasticizer for vinyl resins and cellulose esters.

dicapryl phthalate (DCP; di-(2-octyl) phthalate) $(C_8H_{17}OOC)_2C_6H_4$. See also capryl compounds.
Properties: Colorless, viscous liquid; b.p. (4.5 mm) 227–234°C; f.p. -60°C; refractive index (20°C) 1.480; sp. gr. (25°C) 0.965; flash point 395°F. Insoluble in water; compatible with vinyl chloride resins and some cellulosic resins. Combustible. Low toxicity.
Containers: Drums; tank cars.
Use: Monomeric plasticizer for vinyl and cellulosic resins.

dicapryl sebacate $C_8H_{17}OOC(CH_2)_8COOC_8H_{17}$. See also capryl compounds.
Properties: Light straw-colored liquid; b.p. (4 mm) 231.5–239°C; nonvolatile; gives excellent low-temperature flexibility. Flash point 445°F. Combustible. Low toxicity.
Containers: Drums; tank cars.
Use: Plasticizer for vinyl resins and acrylonitrile rubber.

dicapthon. Generic name for O-(2-chloro-4-nitrophenyl) O,O-dimethyl phosphorothioate $(CH_3O)_2P(:S)OC_6H_3(Cl)NO_2$.
Properties: White solid; m.p. 51–52°C. Insoluble in water; soluble in acetone, cyclohexanone, ethyl acetate, toluene, and xylene. Cholinesterase inhibitor.
Hazard: Highly toxic. See parathion. Use may be restricted.
Use: Insecticide.
Shipping regulations: (Rail, Air) Organic phosphate, dry, n.o.s., Poison label. Not accepted on passenger planes.

dicarboxylic acid. A carboxylic acid (q.v.) containing two —COOH groups, e.g., adipic, oxalic, phthalic, sebacic and maleic acids.

dicetyl. See dotriacontane.

dicetyl ether (dihexadecyl ether)
Properties: Crystals; f.p. 54°C; b.p., decomposes at 300°C; sp. gr. 0.8117 (54/4°C).
Uses: Electrical insulators; water repellents; lubricants in plastic molding and processing; antistatic substances; chemical intermediates.

dicetyl sulfide (dihexadecyl thioether; dihexadecyl sulfide) $(C_{16}H_{33})_2S$.
Properties: Solid; m.p. 57–58°C; b.p. decomposes; sp. gr. 0.8253 (60/4°C).
Uses: Organic synthesis (formation of sulfonium compounds).

dichlobenil. Generic name for 2,6-dichlorobenzonitrile $Cl(Cl)C_6H_3CN$.
Properties: White solid; m.p. 144°C. Almost insoluble in water; soluble in organic solvents.
Hazard: Toxic by ingestion and inhalation.
Use: Herbicide.

dichlone. Generic name for 2,3-dichloro-1,4-naphthoquinone, $C_{10}H_4Cl_2O_2$.
Properties: Yellow needles; m.p. 193°C; soluble in xylene and ortho-dichlorobenzene; slightly soluble in ethyl alcohol, glacial acetic acid and carbon tetrachloride; almost insoluble in water.
Hazard: Toxic by ingestion and inhalation; irritating to skin and eyes.
Uses: Seed disinfectant; fungicide for foilage and textiles; insecticide; organic catalyst.
Shipping regulations: (Rail, Air) Not listed.

dichlor-. See dichloro-.

dichloramine-T (para-toluenesulfondichloramide) $CH_3C_6H_4SO_2NCl_2$. See also chloramine-T.
Properties: Pale yellow crystals, containing not less

than 28% nor more than 30% active chlorine; chlorine odor; stable when pure; decomposed slowly by air, rapidly by impurities, petrolatums, kerosine, olive oil, and alcohol. Soluble in glacial acetic acid, chlorinated paraffin hydrocarbons, eucalyptol, benzene, chloroform and carbon tetrachloride; almost insoluble in water. M.p. 80°C. Low toxicity.
Derivation: Reaction between toluene-para-sulfonamide and calcium hypochlorite solution is acidified with acetic acid and extracted with chloroform. The chloroform solution is dried chemically, filtered and evaporated.
Use: Medicine (external).

"Dichloran."[430] Trademark for a high-alkyl dimethyl dichlorobenzyl ammonium chloride. Used as an antiseptic.

dichlorethylene. Legal label name for dichloroethylene (q.v.).

dichlorisone acetate
Properties: A chlorinated steroid. Available as liquid and cream.
Containers: Aerosol containers; bottles.
Use: Topical therapy in medicine.

dichloroacetaldehyde $CHCl_2CHO$.
Properties: Colorless liquid with a penetrating pungent odor; sp. gr. (25°C) 1.436; wt. per gal 12.1 lb.
Containers: 55-gal drums; tank cars. Flash point 140°F (closed cup). Combustible.
Hazard: Moderate fire hazard. Toxic by ingestion and inhalation; strong skin irritant.
Use: Manufacture of insecticides.

dichloroacetic acid CHC_2COOH.
Properties: Colorless liquid; sp. gr. 1.5724 (13°C); f.p. -4°C; b.p. 193–194°C. Soluble in water, alcohol, and ether. Crystalline form, m.p. +9.3°C.
Derivation: Chlorination of acetic acid in presence of iodine.
Hazard: Toxic by ingestion and inhalation. Strong irritant to eyes and skin.
Uses: Intermediate; pharmaceuticals.
Shipping reuglations: (Rail) Corrosive liquid, n.o.s., White label. (Air) Corrosive label.

sym-**dichloroacetic anhydride.** See chloroacetic anhydride.

2,5-dichloroacetoacetanilide
$CH_3COCH_2CONHC_6H_3Cl_2$. White, crystalline solid. M.p. 96°C.

alpha,alpha-**dichloroacetophenone** $C_6H_5COCHCl_2$.
Crystals; sp. gr. 1.34 (15°C); b.p. 247°C (dec); m.p. 20–21.5°C.
Hazard: Toxic by ingestion and inhalation; strong irritant to eyes and skin.

dichloroacetyl chloride $Cl_2CHCOCl$.
Properties: Fuming liquid; acrid odor; sp. gr. 1.5315 (16/4°C); b.p. 107–108°C; refractive index 1.4638 (16°C). Soluble in ether; decomposes in water and alcohol.
Hazard: Toxic; strong irritant to skin and eyes.
Use: Intermediate.
Shipping regulations: (Rail) Corrosive liquid, n.o.s., White label. (Air) Corrosive label.

2,3-dichloroallyl diisopropylthiolcarbamate (di-allate) $[(CH_3)_2CH]_2NCOSCH_2C(Cl){:}CHCl$.
Properties: Brown liquid; b.p. 159°C (9 mm). Almost insoluble in water; soluble in acetone, benzene, chloroform, kerosine, and xylene.
Hazard: Toxic. Use may be restricted. Absorbed by skin.
Use: Herbicide.

dichloroaniline, liquid. Shipping regulations: (Air) Poison label.

2,5-dichloroaniline $C_6H_3NH_2Cl_2$.
Properties: Light brown or amber-colored crystalline mass. Slightly soluble in water; soluble in alcohol, benzene, and dilute hydrochloric acid. M.p. 47–50°C; b.p. 251–252°C. Combustible.
Derivation: Nitration of para-dichlorobenzene with subsequent reduction.
Containers: Up to tank cars.
Hazard: Toxic. See aniline.
Use: Dye intermediate.

3,4-dichloroaniline $C_6H_3NH_2Cl_2$.
Properties: Crystals; m.p. 68–72°C; b.p. 272°C. Slightly soluble in water; soluble in most organic solvents. Flash point 331°F. Combustible.
Derivation: Nitration of ortho-dichlorobenzene, with subsequent reduction.
Containers: Up to tank cars.
Hazard: Toxic. See aniline.
Uses: Intermediate for manufacture of dyes and pesticides.

dichlorobenzaldehyde $C_6H_3CHOCl_2$. (2,4-, 2,5-, and 3,4-isomers).
Properties: White crystalline solid; b.p. 233°C; m.p. 65–67°C. (2,4-); 35°C (3,4-). 2,4-isomer soluble in methanol, ethanol, ether, and acetone; insoluble in water. 2,5-isomer soluble in ethanol and ether. 3,4-isomer soluble in ethanol, ether, and acetone, slightly soluble in methanol and amyl ether; insoluble in water.
Containers: Fiber drums.
Use: Intermediate in the manufacture of pharmaceuticals, dyes, and other organic chemicals.

meta-**dichlorobenzene** (1,3-dichlorobenzene) $C_6H_4Cl_2$.
Properties: Colorless liquid. Sp. gr. 1.288 (20/4°C); b.p. 172°C; f.p. -24°C; refractive index (20.9°C) 1.547. Soluble in alcohol and ether; insoluble in water. Combustible.
Derivation: Chlorination of monochlorobenzene.
Uses: Fumigant and insecticide.

ortho-**dichlorobenzene** (1,2-dichlorobenzene) $C_6H_4Cl_2$.
Properties: Colorless liquid. Pleasant odor; a mixture of isomers containing at least 85% ortho- and varying percentages of para- and meta-. Flash point 150°F; combustible. Miscible with most organic solvents; insoluble in water. Autoignition temp. 1198°F. Sp. gr. 1.284; wt. per gal 10.7 lb; b.p. 172–179°C; frz. pt. below -20°C.
Derivation: Chlorination of monochlorobenzene.
Method of purification: Rectification.
Grades: Purified; technical.
Containers: Drums; tank cars; tank trucks.
Hazard: Irritant; moderately toxic by inhalation and ingestion. Tolerance, 50 ppm in air. Safety data sheet available from Manufacturing Chemists Assn., Washington, D.C.
Uses: Mfg. of 3,4-dichloroaniline; solvent for a wide range of organic materials and for oxides of nonferrous metals; solvent carrier in production of toluene diisocyanate; dye mfg; fumigant and insecticide; de-

greasing hides and wool; metal polishes; industrial odor control.

para-**dichlorobenzene** (1,4-dichlorobenzene; PDB) $C_6H_4Cl_2$.
Properties: White crystals; volatile (sublimes readily); penetrating odor. Sp. gr. 1.458; b.p. 173.7°C; m.p. 53°C; flash point 150°F. (closed cup). Soluble in alcohol, benzene, and ether; insoluble in water. Combustible.
Derivation: Chlorination of monochlorobenzene.
Grade: Technical.
Containers: Semibulk cartons; fiber drums.
Hazard: Moderately toxic; irritant to eyes. Tolerance, 75 ppm in air.
Uses: Moth repellent; general insecticide; germicide; space odorant; manufacture of 2,5-dichloroaniline; dyes; intermediates; pharmacy; agriculture (fumigating soil).

N,N-dichlorobenzenesulfonamide $C_6H_5SO_2NCl_2$.
Properties: Solid fine crystals; white color; m.p. 68–71°C.
Use: A source of positive chlorine.

3,3′-dichlorobenzidine $C_6H_3ClNH_2C_6H_3ClNH_2$.
Properties: Gray to purple crystalline solid; insoluble in water; soluble in alcohol and ether; m.p. 133°C.
Hazard: Causes tumors in experimental animals; worker exposure should be minimized. Absorbed by skin. May be a carcinogen.
Uses: Intermediate for dyes and pigments; curing agent for isocyanate-terminated resins for urethane plastics.

2,4-dichlorobenzoic acid $Cl_2C_6H_3COOH$.
Properties: White to slightly yellowish powder; m.p. 158–162°C. Soluble in alcohol, ether, acetone, 5% caustic. Insoluble in water and heptane.
Uses: Intermediate for antimalarials, dyes, fungicides, pharmaceuticals, etc. other organic chemicals.

3,4-dichlorobenzoic acid. $Cl_2C_6H_3COOH$.
Properties: White to slightly yellowish powder; m.p. 202–204°C; soluble in alkali, alcohol, ether, and acetone; slightly soluble in diacetone; insoluble in water, ethylene dichloride, and toluene.
Uses: Intermediate for pharmaceuticals, dyes, etc.

2,6-dichlorobenzonitrile. See dichlobenil.

3,4-dichlorobenzotrichloride $Cl_2C_6H_3CCl_3$.
Properties: Water-white liquid; boiling range 276–285°C; f.p. (approx.) 24.0°C; sp. gr. 1.585–1.590 (25/15°C). Soluble in alcohol, ether, and acetone; insoluble in water. Combustible.
Use: Intermediate for pharmaceuticals, dyes, etc.

dichlorobenzoyl chloride $Cl_2C_6H_3COCl$.
Properties: Colorless liquid; boiling range 250–260°C; f.p. 15–16°C; sp. gr. 1.500–1.510)25/15°C). Soluble in alcohol, ether, and acetone; slightly soluble in heptane; insoluble in water. Combustible.
Use: Intermediate in the manufacture of pharmaceuticals, dyes, and other organic chemicals.

dichlorobenzyl chloride $Cl_2C_6H_3CH_2Cl$.
Properties: Colorless liquid; boiling range 245–252°C; sp. gr. 1.415–1.420 (25/15°C). Soluble in alcohol, ether, and acetone; insoluble in water. Combustible.
Hazard: Probably toxic.
Use: Intermediate for organic chemicals, pharmaceuticals, and dyes; insecticide.

1,1-dichloro-2,2-bis (para-**chlorophenyl**) **ethane**. See TDE.

1,1-dichloro-2,2-bis(para-**ethylphenyl**)**ethane** (diethyldiphenyldichloroethane; "Perthane"). $CHCl_2CH(C_6H_4C_2H_5)_2$.
Properties: Crystals; m.p. 56–57°C. Soluble in acetone, kerosine, diesel fuel.
Hazard: Moderately toxic; absorbed by skin.
Use: Insecticide, formulated as emulsifiable concentrate or wettable powder. Used especially in aerosols against insects, including moths.

1,4-dichlorobutane (tetramethylene dichloride; DCB) $ClCH_2(CH_2)_2CH_2Cl$.
Properties: Colorless, mobile liquid; pleasant odor. Sp. gr. 1.141 (20/4°C); boiling point 155°C; flash point 104°F (TOC); refractive index (n 20/D) 1.4542. Insoluble in water; soluble in most common organic solvents. Combustible.
Containers: 9-lb (1-gal) containers; 45-, 500-lb drums; 76,000-lb tank cars.
Hazard: Moderate fire risk.
Use: Organic synthesis, including adiponitrile.

1,3-dichlorobutene-2 $ClH_2CCH:CClCH_3$.
Properties: Clear to straw-colored liquid; b.p. 125–130°C. Insoluble in water; soluble in organic solvents. Flash point 80°F (COC).
Hazard: Flammable, moderate fire risk.

1,4-dichlorobutene-2 $ClH_2CCH:CHCH_2Cl$.
Properties: Colorless liquid, distinct odor. Miscible with benzene, alcohol, carbon tetrachloride. Immiscible with ethylene glycol, glycerine, and water. B.p. 158°C, 60°C (20 mm); m.p. 3.5°C; sp. gr. 1.1858 (25/4°C); refractive index (n 25/D) 1.4863. Combustible.
Grades: Available as 95–98% trans-isomer, 2–5% cis-isomer. Above constants are for the pure transisomer.
Containers: Up to tank cars.
Hazard: Toxic. Irritant to skin and eyes; causes blisters. Moderate fire hazard.
Use: Intermediate.

dichlorocarbene CCl_2. Exists only at low temperatures and pressures. F.p. –114°C; b.p. –20°C; decomposes on distillation at normal pressure to hexachloroethane and hexachlorobenzene.
Derivation: Reaction of carbon tetrachloride vapor with carbon at 1300°C and 10^{-3} mm Hg.
Hazard: Explosive reaction with carbon; forms phosgene on reaction with oxygen.
Use: Research.

1,1′-dichlorocarboyl ferrocene (ferrocenoyl dichloride) $[C_5H_4(COCl)]_2Fe$.
Properties: Red crystalline solid; m.p. 93–95°C.
Use: Intermediate.

4,6-dichloro-N-(2-chlorophenyl)-1,3,5-triazin-2-amine ("Dyrene") $ClC_6H_4NHC_3N_3Cl_2$.
Properties: Tan crystalline solid; m.p. 159–160°C; insoluble in water.
Hazard: Moderately toxic. See aniline.
Uses: Foliage fungicide.

dichloro(2-chlorovinyl)arsine. See chlorovinyldichloroarsine.

3,3′-dichloro-4,4′-diaminodiphenylmethane. See 4,4′-methylenebis(2-chloroaniline).

2,3-dichloro-5,6-dicyanobenzoquinone (DDQ) OCC(Cl):C(Cl)COC(CN):C(CN) (ring closure from first C to last C).
Properties: Bright yellow-orange solid; m.p. 213–215°C.
Use: Highly selective oxidizing agent for organic compounds.

2,2′-dichlorodiethyl ether. See dichloroethyl ether.

dichlorodiethyl formal. See dichloroethyl formal.

dichlorodiethyl sulfide (mustard gas; dichloroethyl sulfide) $S(CH_2CH_2Cl)_2$.
Hazard: Highly toxic. Vesicant war gas; causes conjunctivitis and blindness! Can be decontaminated by chloroamines or bleaching powder. Vapor is extremely poisonous and can be absorbed through skin.
Derivation: Bubbling ethylene through sulfur chloride; also from thiodiglycol and hydrogen chloride.
Grades: Pure; technical (containing excess sulfur as a polysulfide).
Uses: Organic synthesis; poison gas; medicine.
Shipping regulations: (Rail) Poison gas label. Not accepted by express. Legal label name mustard gas. (Air) Not acceptable.

dichlorodiethyl sulfone $(ClCH_2CH_2)_2SO_2$.
Properties: Colorless crystals. B.p. 179–181°C (14–15 mm); m.p. 52°C. Soluble in alcohol, chloroform, and ether; slightly soluble in water.
Hazard: Toxic; strong irritant to eyes and skin.

2,2-dichloro-1,1-difluoroethyl methyl ether (methoxyflurane) $HCCl_2CF_2OCH_3$.
Properties: Clear, colorless liquid; fruity odor; b.p. 104.65°C; f.p. -35°C; sp. gr. 1.4223 (25°C); completely stable in the presence of alkali, air, light, or moisture. Slightly soluble in water. Combustible.
Grade: N.D.
Use: Anesthetic.

dichlorodifluoromethane (difluorodichloromethane; fluorocarbon-12). CCl_2F_2.
Properties: Colorless, odorless, noncorrosive gas. B.p. −29.8°C; f.p. −158°C; sp. gr. 1.486 (−30); critical pressure 43.2 atm. Soluble in water; soluble in most organic solvents. Nonflammable.
Derivation: (a) Reaction of carbon tetrachloride and anhydrous hydrogen fluoride, in the presence of an antimony halide catalyst; (b) high temperature chlorination of vinylidene fluoride (vinylidene fluorides made by addition of hydrogen fluoride to acetylene).
Grade: 99.9% min. purity.
Containers: Cylinders.
Hazard: Narcotic in high concentrations. Tolerance, 1000 ppm in air.
Uses: Refrigerant and air conditioner; aerosol propellant (?); plastics; blowing agent; low-temperature solvent; leak-detecting agent; freezing of foods by direct contact; chilling cocktail glasses.
Shipping regulations: (Rail) Green label. (Air) Nonflammable Gas label.
See also chlorofluorocarbon.

1,3-dichloro-5,5-dimethylhydantoin (DDH) $ClNCONClOC(CH_3)_2$.
Properties: White powder with mild chlorine odor. M.p. approximately 130°C; sublimes about 100°C without decomposition. Contains approximately 36% active chlorine. Slightly soluble in water with gradual liberation of hypochlorous acid; soluble in benzene, chloroform, ethylene dichloride, alcohol. Combustible, with evolution of chlorine at 210°C.
Derivation: Chlorination of dimethylhydantoin.
Grades: Technical.
Hazard: Highly toxic by inhalation. Tolerance, 0.2 mg per cubic meter of air. Skin irritant.
Uses: Household laundry bleach; water treatment; mild chlorinating agent.

dichlorodimethylsilane. See dimethyldichlorosilane.

dichlorodiphenyldichloroethane. See TDE.

dichlorodiphenyldichloroethylene. See DDE.

dichlorodiphenyltrichloroethane. See DDT.

1,1-dichloroethane. See ethylidene chloride.

1,2-dichloroethane. See ethylene dichloride.

dichloroether. See dichloroethyl ether.

dichloroethoxymethane. See dichloroethylformal.

1,2-dichloroethyl acetate $CH_3COOCHClCH_2Cl$.
Properties: Water-white liquid. Sp. gr. 1.296 (20°C); boiling range: 58–65°C (13 mm); f.p. < -32°C; refractive index 1.444 (20°C); b.p., dec. Flash point 307°F. Combustible. Low toxicity. Miscible with alcohol and ethyl ether. Immiscible with water.
Use: Organic synthesis.

para-**di(2-chloroethyl)aminophenylalamine.** See melphalan.

dichloroethylarsine. See ethyldichloroarsine.

dichloroethyl carbonate $(ClH_2CCH_2O)_2CO$.
Properties: Colorless liquid. Slowly hydrolyzed by alkalies. Volatile in steam. Sp. gr. 1.3506 (20°C); b.p. 240°C (partial decomposition). Insoluble in water.
Derivation: By heating ethylene chlorohydrin and trichloromethylchloroformate together (under reflux).

sym-**dichloroethylene** (1,2-dichloroethylene; acetylene dichloride). $ClHC{:}CHCl$. Exists as cis and trans isomers.
Properties: Colorless, low-boiling liquid. Pleasant odor. It decomposes slowly on exposure to air, light and moisture. Soluble in most organic solvents; slightly soluble in water. Trans-isomer; sp. gr. 1.257; b.p. 47–49°C. Cis-isomer: sp. gr. 1.282; b.p. 58–60°C. Flash point 39°F; f.p. −80°C.
Derivation: Two stereoisomeric compounds made by the partial chlorination of acetylene.
Grades: Technical; as cis, trans, and mixture of both.
Containers: 300-, 550-lb drums.
Hazard: Toxic by ingestion, inhalation and skin contact; irritant and narcotic in high concentrations. Tolerance, 200 ppm in air. Flammable, dangerous fire hazard.
Uses: General solvent for organic materials; dye extraction; perfumes; lacquers; thermoplastics; organic synthesis; medicine.
Shipping regulations: (Rail) Red label. (Air) Flammable Liquid label.

dichloroethyl ether (dichloroether; dichloroethyl oxide; 2,2′-dichlorodiethyl ether, bis(2-chloroethyl) ether; sym-dichloroethyl ether) $ClCH_2CH_2OCH_2CH_2Cl$.
Properties: Colorless liquid. Odor like that of ethylene dichloride. B.p. 178.5°C; sp. gr. 1.2220 (20/20°C); wt/gal 10.2 lb (20°C); refractive index 1.457 (20°C); flash point (closed cup) 131°F; f.p. -51.8°C. Autoignition temp. 696°F. Miscible with most organic solvents; insoluble in water. Combustible.
Derivation: Chlorination of ethyl ether.
Grades: Technical.
Containers: Glass bottles; iron drums; tank cars.
Hazard: Toxic by inhalation and ingestion; absorbed by skin; strong irritant. Tolerance, 5 ppm in air. Moderate fire hazard.
Uses: General solvent; selective solvent for produc-

tion of high-grade lubricating oils; textile scouring and cleansing; fulling compounds; wetting and penetrating compounds; organic synthesis; paints; varnishes, lacquers; finish removers; spotting and dry cleaning; soil fumigant.

sym-**dichloroethyl ether.** See dichloroethyl ether.

dichloroethyl formal (dichlorodiethyl formal) $CH_2(OCH_2CH_2Cl)_2$.
Properties: Colorless liquid; b.p. 218.1°C; f.p. -32.8°C; sp. gr. 1.2339 (20/20°C); wt/gal 10.3 lb (20°C); flash point (open cup) 230°F. Slightly soluble in water; decomposed by mineral acids. Combustible.
Hazard: Toxic by inhalation and ingestion; strong irritant.
Uses: Solvent; intermediate for polysulfide rubber.

dichloroethyl oxide. See dichloroethyl ether.

dichloroethyl sulfide. See dichlorodiethyl sulfide.

dichlorofluoromethane (fluorodichloromethane; fluorocarbon 21) $CHCl_2F$.
Properties: Colorless, nearly odorless, heavy gas. B.p. 8.9°C; f.p. -135°C; sp. gr. 1.426 (0°C); critical pressure 51.0 atm. Soluble in alcohol and ether; insoluble in water. Nonflammable.
Derivation: Reaction of chloroform and hydrogen fluoride.
Grade: Technical.
Containers: Drums; cylinders; tanks.
Hazard: Narcotic in high concentrations. Tolerance, 1000 ppm in air.
Use: Fire extinguishers; solvent; refrigerant; aerosol propellant (?).
See also chlorofluorocarbon.

dichloroformoxime CCl_2NOH.
Properties: Colorless, prismatic crystals. Disagreeable, penetrating odor. High vapor pressure. Slowly decomposes at normal temperatures, the rate depending on temperature and humidity. B.p. 53–54°C (28 mm); m.p. 39–40°C. Soluble in water, alcohol, ether, and benzene.
Derivation: (a) Action of chlorine on fulminic acid, HONC. (b) Reduction of trichloronitrosomethane with either aluminum amalgam or hydrogen sulfide.
Hazard: Highly toxic. Strong irritant to skin and eyes.
Use: A military poison.

alpha-**dichlorohydrin** (alpha-propenyldichlorohydrin; glycerol dichlorohydrin; GDCH) $CH_2ClCHOHCH_2Cl$.
Properties: Colorless, slightly viscous, unstable liquid; faint chloroform-like odor. The commercial product is a mixture of two isomers. Miscible with most organic solvents, vegetable oils; slightly soluble in water. Sp. gr. 1.36–1.39; f.p. −4°C; b.p. 174°C; refractive index 1.47–1.48; flash point 165°F; combustible. Vapor pressure 7 mm.
Derivation: Interaction of glycerol and dry hydrogen chloride gas and subsequent distillation.
Hazard: Moderately toxic by inhalation and ingestion; moderate fire risk.
Uses: General solvent; intermediate in organic synthesis; paints, varnishes, lacquers; celluloid cements; water colors' binder; photographic lacquers.

5,7-dichloro-8-hydroxyquinaldine. See chlorquinaldol.

dichloroisocyanuric acid (dichloro-s-triazine-2,4,6-trione)
OCNClCONClCONH.
Properties: White, slightly hygroscopic, crystalline powder, granules. Density: (loose bulk, approx.) powder 34 lb/cu ft; (granular) 53 lb/cu ft. Active ingredient approximately 70% available chlorine; decomp. 225°C.
Containers: 200-lb fiber drums.
Hazard: Oxidizing material. May ignite organic materials with which it is in contact. Moderately toxic.
Uses: Household dry bleaches, dishwashing compounds, scouring powders, and detergent sanitizers; replacement for calcium hypochlorite.
Shipping regulations: Dry, containing more than 39% available chlorine, (Rail) Yellow label. (Air) Oxidizer label.

sym-**dichloroisopropyl alcohol.** See alpha-dichlorohydrin.

dichloroisopropyl ether $[ClCH_2C(CH_3)H]_2O$.
Properties: Colorless liquid; sp. gr. 1.1135 (20/20°C); b.p. 187.4°C (760 mm); vapor pressure 0.10 mm (20°C); flash point 185°F; wt/gal 9.3 lb (20°C); coefficient of expansion 0.00096 (20°C); viscosity 0.0230 poise (20°C). Miscible with most oils and organic solvents; immiscible with water. Combustible.
Containers: 1-gal cans; 5- and 55-gal drums.
Hazard: Strong irritant to tissue.
Uses: Solvent for fats, waxes, greases; extractant; paint and varnish removers; spotting agents and cleaning solutions.
Shipping regulations: (Air) Corrosive label.

dichloromethane. Legal label name for methylene chloride (q.v.).

3′,4′-dichloro-2-methylacrylanilide. See dicryl.

dichloromethylchloroformate $ClCOOCHCl_2$.
Properties: Colorless liquid. Decomposed by water and alkalies. Soluble in alcohol, benzene, and ether. Sp. gr. 1.56 (15°C); b.p. 110–111°C; vapor density 5.7 (air = 1).
Derivation: (a) By chlorinating methyl formate; (b) by chlorinating methylchloroformate. In both methods the mixture of chloro-derivatives is then separated by fractionation.
Hazard: Toxic; strong irritant to eyes and skin.

sym-**dichloromethyl ether** $O(CH_2Cl)_2$.
Properties: Colorless, volatile liquid; suffocating odor. Decomposed by heat and water; soluble in acetone, benzene, ethyl alcohol, and methyl alcohol; insoluble in water. Sp. gr. 1.315 (20°C); b.p. 105°C.
Derivation: (a) Action of chlorine on methyl ether; (b) interaction of hydrochloric acid and formaldehyde, with subsequent dehydration of the chloromethyl alcohol formed.
Hazard: Strong irritant to eyes and mucous membranes.

5,7-dichloro-2-methyl-8-quinolinol. See chlorquinaldol.

dichloromethylsilane CH_3SiHCl_2. Colorless liquid; b.p. 41°C; sp. gr. 1.113 (25/25°C); refractive index (n 20/D) 1.3983. Combustible.
Hazard: Corrosive to tissue.
Use: Intermediate.
Shipping regulations: (Rail, Air) Not listed. Consult authorities.

dichloromethyl sulfate $(ClCH_2O)_2SO_2$.
Properties: Colorless, odorless liquid; soluble in alcohol, benzene, and ether. Sp. gr. 1.60 (20°C); b.p. 96–97°C (14 mm). Combustible.
Derivation: (a) By bubbling sulfur trioxide through (cooled) dichloromethyl ether; (b) by heating chlorosulfonic acid with formaldehyde.

dichloronaphthalene. See chlorinated naphthalenes.

2,3-dichloro-1,4-naphthoquinone. See dichlone.

1,2-dichloro-4-nitrobenzene $Cl_2C_6H_3NO_2$. Solid; m.p. 43°C; b.p. 255–256°C; sp. gr. 1.4266 (100/4°C). Insoluble in water; soluble in hot alcohol, ether.
Hazard: Probably toxic. Fire risk by spontaneous reaction.
Use: Intermediate.

2,5-dichloronitrobenzene $Cl_2C_6H_3NO_2$.
Properties: Pale yellow crystals; sp. gr. 1.669 (22°C); m.p. 55°C; b.p. 266°C. Insoluble in water; soluble in chloroform and hot alcohol.
Hazard: Probably toxic; fire risk by spontaneous reaction.
Use: Intermediate.

1,1-dichloro-1-nitroethane $H_3CC(Cl)_2NO_2$.
Properties: B.p. 124°C; sp. gr. 1.4153 (20/20°C); flash point 168°F (open cup); combustible.
Hazard: Toxic, strong irritant. Tolerance, 10 ppm in air.
Uses: Grain fumigant; solvent.

dichloropentane $C_6H_{10}Cl_2$. A mixture of the dichloroderivatives of both normal and isopentane. About 40% are amylene dichlorides having two chlorine atoms attached to adjacent carbon atoms.
Properties: Light-yellow liquid; sp. gr. 1.06–1.08 (20°C); acidity as HCl not over 0.025%; distillation 95% between 130–200°C; flash point 97–105°F; combustible; wt/gal 8.94 lb. Water solubility negligible; water azeotrope at 80–97°C, 66% $C_6H_{10}Cl_2$ (approx.).
Containers: Cans; drums; tank cars.
Hazard: Moderate fire risk. Moderately toxic by inhalation.
Uses: Solvent for oils, greases, rubber, resins and bituminous materials; removal of tar; reclaiming rubber; paint and varnish removers; degreasing of metals; insecticide; soil fumigant; removal of wax deposits on oil-well equipment.

dichlorophenarsine hydrochloride (3-amino-4-hydroxyphenyldichloroarsine hydrochloride) $C_6H_3(AsCl_2)(OH)NH_2 \cdot HCl$.
Properties: White, odorless powder; m.p. 200°C; soluble in water, solutions of alkali hydroxides and carbonates, and in dilute mineral acids.
Hazard: Toxic by ingestion.
Use: Medicine (syphilis treatment).
Shipping regulations: (Rail, Air) Arsenical compounds, Poison label.

dichlorophene (2,2′-dihydroxy-5,5′-dichlorodiphenylmethane; DDDM) $(C_6H_3ClOH)_2CH_2$.
Properties: Light tan, free-flowing powder; weakly phenolic odor; m.p. 177°C; vapor pressure 10^{-4} mm (100°C) and about 10^{-10} mm (25°C) (extrapolated value). Soluble in acetone and alcohols; slightly soluble in benzene, toluene, carbon tetrachloride; insoluble in water.
Derivation: Condensation of para-chlorophenol with formaldehyde in the presence of sulfuric acid.
Grades: Pure and technical.
Hazard: Moderately toxic.
Uses: Fungicide and bactericide; textile preservative; some dermatological and cosmetic applications; veterinary medicine.

2,4-dichlorophenol $Cl_2C_6H_3OH$.
Properties: White, low melting solid; b.p. 210°C; m.p. 45°C; flash point 237°F; combustible. Soluble in alcohol and carbon tetrachloride; slightly soluble in water.
Derivation: By chlorination of phenol.
Hazard: Strong irritant to tissue; toxic by ingestion.
Use: Organic synthesis.

2,4-dichlorophenoxyacetic acid. See 2,4-D.

2,4-dichlorophenoxybutyric acid (2,4-DB) $Cl_2C_6H_3O(CH_2)_3COOH$. M.p. 118–120°C. Used as an herbicide.
Hazard: May be toxic.

di-(4-chlorophenoxy)methane $(ClC_6H_4O)_2CH_2$.
Properties: Solid; m.p. 65°C; insoluble in water and oils; soluble in ether and acetone.
Hazard: May be toxic.
Use: Acaricide.

2-(2,4-dichlorophenoxy)propionic acid (dichloroprop) $Cl_2C_6H_3OCH(CH_3)COOH$.
Properties: Solid; m.p. 117–118°C. Soluble in acetone, alcohol, and ether; insoluble in water.
Hazard: Moderately toxic.
Use: Herbicide.

2,4-dichlorophenylbenzenesulfonate $Cl_2C_6H_3OSO_2C_6H_5$.
Properties: Waxy solid; vapor pressure 2.7×10^{-4} mm at 30°C. Insoluble in water; soluble in most organic solvents. Low toxicity.
Use: Acaricide; insecticide.

O-(2,4-dichlorophenyl)-O,O,diethyl phosphorothioate $Cl_2C_6H_3OP(S)(OC_2H_5)_2$.
Properties (pure compound): Liquid; b.p. 120–123°C (0.2 mm); slightly soluble in water; miscible with most organic solvents.
Hazard: Highly toxic; absorbed by skin. Cholinesterase inhibitor. Use may be restricted.
Use: Nematocide; insecticide.
Shipping regulations: Organic phosphorus compound, liquid, n.o.s., (Rail, Air) Poison label. Not accepted on passenger planes.

3-(3,4-dichlorophenyl-1,1-dimethylurea. See diuron.

di(para-chlorophenyl)ethanol (1,1-bis(para-chlorophenyl)ethanol; di(para-chlorophenyl)methyl carbinol; DMC; DCPC) $CH_3C(C_6H_4Cl)_2OH$.
Properties: Colorless crystals; m.p. 70°C; insoluble in water; soluble in common organic solvents.
Derivation: Reaction of 4,4′-dichlorobenzophenone with methyl magnesium bromide, followed by treatment with water.
Hazard: Moderately toxic.
Use: Insecticide.

3,4-dichlorophenyl isocyanate $Cl_2C_6H_3NCO$. White to yellow crystalline solid.
Hazard: Strong irritant to tissue, especially eyes and mucous membranes.
Uses: Chemical intermediate; organic synthesis.

3-(3,4-dichlorophenyl)-1-methoxy-1-methylurea. See linuron.

di(para-chlorophenyl)methyl carbinol. See di(para-chlorophenyl)ethanol.

O-(2,4-dichlorophenyl) O-methyl isopropylphosphoramidothioate (DMPA) $Cl_2C_6H_3OP(S)(OCH_3)NHCH(CH_3)_2$.
Properties: Solid; m.p. 51.4°C. Vapor pressure 2 mm at 150°C. Slightly soluble in water (5 ppm); freely soluble in acetone, benzene, and carbon tetrachloride.

Hazard: Highly toxic. Cholinesterase inhibitor. Use may be restricted.
Use: Herbicide; insecticide.
Shipping regulations: (Rail) Poison label. (Air) Poison label. Not acceptable on passenger planes.

2,4-dichlorophenyl 4-nitrophenyl ether $Cl_2C_6H_3OC_6H_4NO_2$.
Properties: Dark brown solid; setting point 62.5°C. Soluble in acetone, methanol, and xylene.
Use: Herbicide.

dichlorophenyltrichlorosilane $Cl_2C_6H_3SiCl_3$. A mixture of isomers.
Properties: Straw-colored liquid; sp. gr. 1.562; b.p. 260°C. Refractive index (n 20/D) 1.5638; flash point (COC) 286°F. Readily hydrolyzed, with liberation of hydrochloric acid. Soluble in benzene, perchloroethylene. Combustible.
Derivation: Chlorination of phenyltrichlorosilane.
Grades: Technical.
Hazard: Strong irritant to skin and eyes.
Use: Intermediate for silicones.
Shipping regulations: (Rail) Corrosive liquid, n.o.s., White label. (Air) Corrosive label. Not acceptable on passenger planes.

3,6-dichlorophthalic acid $C_6H_2Cl_2(COOH)_2$.
Properties: Colorless, thick crystals; soluble in hot water.
Derivation: By oxidizing dichloronaphthalene tetrachloride (see chloronaphthalene) with nitric acid.

dichloroprop. See 2-(2,4-dichlorophenoxy)-propionic acid.

1,2-dichloropropane. See propylene dichloride.

1,2-dichloro-3-propanol. See alpha-dichlorohydrin.

1,3-dichloro-2-propanol. See alpha-dichlorohydrin.

1,3-dichloropropene (1,3-dichloropropylene). $CHCl{:}CHCH_2Cl$.
Properties: Exists is cis and trans isomeric forms, both colorless liquids. Sp. gr. 1.225 (20/4°C); flash point (open cup) 95°F; insoluble in water; soluble in acetone, toluene, octane. Cis isomer b.p. 104°C, trans 112°C; refractive index (n 20/D) cis 1.469, trans 1.475.
Derivation: Chlorination of propylene.
Hazard: Strong irritant; flammable, moderate fire risk.
Uses: Organic synthesis; soil fumigants.

dichloropropene-dichloropropane mixture.
Hazard: Highly toxic. May be fatal if swallowed, inhaled, or absorbed through skin. Corrosive to tissue, skin and eyes.
Uses: Pesticide; insecticide.
Shipping regulations: (Air) Corrosive label.

3,4-dichloropropionanilide. See propanil.

2,2-dichloropropionic acid. See dalapon.

2,6-dichlorostyrene $C_6H_3(CH{:}CH_2)Cl_2$.
Properties: Colorless liquid. B.p. 92–94°C (5 mm). Flash point 225°F (O.C.) insoluble in water; soluble in most organic solvents. Polymerizes slowly on standing, unless inhibited. Combustible.
Uses: Monomer and co-monomer in plastic research.

para-N,N-**dichlorosulfamylbenzoic acid.** See halazone.

dichlorosulfonphthalein. See chlorphenol red.

2,6-dichloroquinonechlorimide. Reagent used for spot visualizations in chromatographic systems.
Hazard: Explodes readily on slight heating.

dichlorotetrafluoroacetone $CClF_2COCClF_2$.
Properties: Colorless liquid; b.p. 45.2°C; f.p. < -100°C. Soluble in water and most organic solvents. Stable to acids but not alkalies.
Uses: Solvent in acidic media; complexing agent for active hydrogen compound separation.

sym-**dichlorotetrafluorethane** (fluorocarbon-114; tetrafluorodichloroethane); $CClF_2CClF_2$.
Properties: Colorless, nearly odorless, nonflammable gas. B.p. 3.55°C; f.p. −94°C; critical pressure 32.3 atm. Insoluble in water.
Derivation: By treating perchloroethylene with hydrogen fluoride.
Grades: Technical 95%.
Containers: Cylinders up to 1 ton.
Hazard: Moderately toxic by inhalation. Tolerance, 1000 ppm in air.
Uses: Solvent; fire extinguishers; refrigerant and air conditioner; aerosol propellants (?); blowing agent, dielectric fluid.
See also chlorofluorocarbon.

2,5-dichlorothiophene $C_4H_2Cl_2S$ (cyclic).
Properties: Colorless to light-yellow liquid; b.p. 161°C. Combustible.
Use: Intermediate.

alpha, alpha-**dichlorotoluene.** See benzyl dichloride.

dichlorotoluene (chlorobenzyl chloride) $C_6H_3CH_3Cl_2$.
Properties: Colorless liquid; boiling range 200–200°C; f.p. (approx.) -13°C; sp. gr. 1.245–1.247 (25/15°C); refractive index 1.5480 (22°C). Soluble in alcohol, ether, and acetone; insoluble in water. Exists as ortho and para isomers. Combustible.
Containers: Glass carboys or drums.
Hazard: May be toxic and irritant. See benzyl chloride.
Uses: High-boiling solvent; intermediate for organic synthesis.

dichloro-sym-**triazine-2,4,6-trione.** See dichloroisocyanuric acid.

beta,beta'-**dichlorovinylchloroarsine** $(ClCH{:}CH)_2AsCl$.
Properties: Yellowish liquid; b.p. 230°C (decomp.); sp. gr. 1.70.
Hazard: Highly toxic by inhalation and ingestion. Strong irritant to skin and mucous membranes.
Use: Poison gas.
Shipping regulations: (Rail) Arsenical compounds, n.o.s., Poison label. (Air) Not acceptable.

2,2-dichlorovinyl dimethyl phosphate. See dichlorovos.

beta,beta'-**dichlorovinylmethylarsine** $(ClCH{:}CH)_2AsCH_3$.
Properties: Liquid. B.p. 140–145°C (10 mm).
Derivation: Interaction of acetylene and methyldichloroarsine in the presence of aluminum chloride.
Hazard: Highly toxic by inhalation and ingestion. Strong irritant to skin and mucous membranes.
Use: Poison gas.
Shipping regulations: Poisonous liquid or gas, n.o.s., (Rail) Poison Gas label; not accepted by express. (Air) Not acceptable.

dichlorovos. Generic name for 2,2-dichlorovinyl dimethyl phosphate (DDVP) $(CH_3O)_2P(O)OCH{:}CCl_2$.
Properties: Liquid; b.p. 120°C (14 mm). Slightly soluble in water and glycerine; miscible with aromatic and chlorinated hydrocarbon solvents and alcohols.
Hazard: Toxic. Tolerance, 1 mg per cubic meter of air.
Uses: Insecticide; fumigant.
Shipping regulations: (Rail) Not listed. (Air) Poison label.

dichroic. A term used in crystallography to denote crystals which refract incident light in two directions, thus displaying two colors when observed from different angles, for example, calcite. See also anisotropic; birefringent.

dichromatic. Characterizing certain dyes and indicators for which different colors may be seen depending on the thickness of the solution viewed.

"Di-Clad."[281] Trademark for a series of copper-clad laminated plastics consisting of either electrolytic or rolled copper laminated on either one or both sides of a supporting medium made from various base materials and resins. Available in sheets for use primarily in printed circuit applications.

dicobalt octacarbonyl. See cobalt tetracarbonyl.

"Dicodid."[9] Trademark for dihydrocodeinone; used in medicine as bitartrate and hydrochloride salts.

dicofol. Generic name for 4,4′-dichloro-alpha-trichloromethyl)benzhydrol. See 1,1′-bis(chlorophenyl)-2,2,2-trichloroethanol.

"Dicom."[79] Trademark for commercial steam-distilled dipentene (q.v.).

dicoumarol. See bishydroxycoumarin.

dicresyl glyceryl ether (glyceryl ditolyl ether). This may be a mixture of ortho-, meta-, and para-isomers. $CH_3C_6H_4OCH_2CHOHCH_2OC_6H_4CH_3$.
Properties: Similar to cresyl glyceryl ether (see mephenesin) in properties. Sp. gr. 1.136; refractive index 1.549; boiling range 328–340°C.

dicresyl glyceryl ether acetate
$CH_3C_6H_4OCH_2CHOOCCH_3CH_2OC_6H_4CH_3$.
Properties: Fairly stable liquid. Sp. gr. 1.115; b.p. 360°C; refractive index 1.53. Combustible.

dicryl. Generic name for 3′,4′-dichloro-2-methylacrylanilide $Cl_2C_6H_3NHCOC(CH_3){:}CH_2$.
Properties: Solid; m.p. 128°C. Insoluble in water; soluble in acetone, alcohol, isophorone, and dimethyl sulfoxide. Probably low toxicity.
Uses: Herbicide; pest control.

"Dicrylan."[443] Trademark for acrylic resin coating and bonding agent.

"Dicumarol."[355] Trademark for bishydroxycoumarin (q.v.).

dicumyl peroxide $[C_6H_5C(CH_3)_2O]_2$.
Hazard: Strong oxidizing material; may ignite organic materials on contact.
Use: Polymerization catalyst and vulcanizing agent.
Shipping regulations: (Rail) (solid and 50% solution), Yellow label. (Air) (solid and solution) Organic Peroxide label.

"Di-cup."[266] Trademark for a series of vulcanizing, and polymerization agents containing dicumyl peroxide (q.v.).

dicyan. See cyanogen.

dicyandiamide (cyanoguanidine) $NH_2C(NH)(NHCN)$.
Properties: Pure white crystals; sp. gr. 1.400 (25°C). Stable when dry. Melting range 207–209°C. Soluble in water and alcohol; sparingly soluble in ether. Nonflammable.
Derivation: Polymerization of cyanamide in the presence of bases.
Grades: 99% pure; technical.
Containers: 100-lb multiwall paper bags.
Uses: Fertilizers; nitrocellulose stabilizer; organic synthesis, especially of melamine, barbituric acid and guanidine salts; pharmaceutical products; dyestuffs; explosives; retarding rancidity in fats and oils; fire-proofing compounds; case-hardening preparations; cleaning compounds; soldering compounds; accelerator; thinner for oil-well drilling muds; stabilizer in detergent compositions; modifier for starch products; catalyst for epoxy resins.

dicyanine. See cyanine dye.

ortho-**dicyanobenzene.** See phthalonitrile.

1,4-dicyanobutane. See adiponitrile.

2,4-dicyanobutene-1. See methylene glutaronitrile.

dicyclohexyl. See bicyclohexyl.

dicyclohexyl adipate $(\text{—}CH_2CH_2COOC_6H_{11})_2$.
Properties: White, crystalline solid. Odorless. Compatible with most natural and synthetic resins. Soluble in most organic solvents; insoluble in water. Sp. gr. 0.913–0.919 (15/15°C); b.p. 256°C; f.p. −0.1°C; density 45 lb/cu ft; acidity (as adipic acid) less than 0.05%.
Use: Plasticizer.

dicyclohexylamine $(C_6H_{11})_2NH$.
Properties: Colorless liquid with faint amine odor; sp. gr. 0.913–0.919 (15/15°C); b.p. 256°C; f.p. -0.1°C; refractive index 1.4823 (n 25/D); flash point 210°F. Combustible. Slightly soluble in water; miscible with organic solvents. Strongly basic.
Containers: Drums; tank cars.
Hazard: Strong irritant to skin and mucous membranes.
Uses: Intermediate; insecticides; plasticizer; corrosion inhibitors; antioxidants in rubber, lubricating oils, fuels; catalysts for paint, varnishes and inks; detergents; extractant.

dicyclohexyl carbodiimide $C_6H_{11}NCNC_6H_{11}$.
Properties: White crystalline solid with a heavy, sweet odor. Set point 29–30°C; b.p. 138–140°C (2 mm); soluble in organic solvents.
Hazard: Vapor may be toxic.
Use: Chemical intermediate.

dicyclohexyl phthalate (DCHP). $C_6H_4(COOC_6H_{11})_2$.
Properties: White, granular solid; nonvolatile; mildly aromatic odor; sp. gr. (25/25°C) 1.20; m.p. 62–65°C. Soluble in most organic solvents; insoluble in water; compatible with a large number of polymers. Flash point 405°F. Combustible.
Uses: Plasticizer for nitrocellulose, ethyl cellulose, chlorinated rubber, polyvinyl acetate, polyvinyl chloride, and other polymers.

dicyclomine hydrochloride
$(C_6H_{11})(C_6H_{10})CO_2(CH_2)_2N(C_2H_5)_2 \cdot HCl$. 2-Diethylaminoethyl bicyclohexyl-1-carboxylate hydrochloride.
Properties: Crystals; m.p. 164–166°C. Soluble in water.
Use: Medicine.

dicyclopentadiene $C_{10}H_{12}$.
Properties: Liquid; sp. gr. 0.979 (20/20°C); b.p. 172°C; m.p. 33.6°C; 8.2 lb/gal (60°F); refractive index 1.5073 (n 31/D); flash point 90°F. Soluble in alcohol; insoluble in water.

Containers: Drums; tank cars.
Derivation: Olefin manufacture.
Hazard: Flammable, moderate fire risk. Toxic by ingestion.
Uses: Chemical intermediate for insecticides; EPDM elastomers; metallocenes; paints and varnishes; flame retardants for plastics.

dicyclopentadiene dioxide $C_{10}H_{12}O_2$.
Properties: White crystalline powder; m.p. 180–184°C; sp. gr. 1.331 (25°C); slightly soluble in water; soluble in acetone and benzene.
Containers: Up to tank car lots.
Use: Intermediate for epoxy resins, plasticizers, protective coatings.

dicyclopentadienylcobalt. See cobaltocene.

dicyclopentadienyliron (bis-cyclopentadienyliron). The first organometallic "sandwich" compound (synthesized in 1951), which served as a prototype for metallocenes (q.v.). Such compounds are based on cyclic unsaturates combined with a transition metal or its halide. See also ferrocene.

dicyclopentadienylnickel. See nickelocene.

dicyclopentadienyl osmium. See osmocene.

dicyclopentadienyltitanium dichloride. See titanocene dichloride.

dicyclopentadienylzirconium dichloride. See zirconocene dichloride.

DIDA. See diisodecyl adipate.

didecyl adipate $(-CH_2CH_2COOC_{10}H_{21})_2$.
Properties: Light-colored liquid; sp. gr. 0.9181 (20/20°C); 7.7 lb/gal (20°C); b.p. 245°C (5 mm); vapor pressure 0.58 mm (200°C); insoluble in water; viscosity 26.3 cp (20°C). Flash point 425°F. Combustible. Low toxicity.
Use: Plasticizer.

didecylamine $[CH_3(CH_2)_9]_2NH$.
Properties: Light straw-colored liquid; faint amine odor; boiling range 195–215°C (12 mm); sp. gr. 0.840 (20/20°C) (solid). Combustible.

didecyl ether $(C_{10}H_{21})_2O$.
Properties: Liquid; m.p. 16°C; b.p. 170–180°C (6 mm); sp. gr. 0.819 (20/4°C); refractive index 1.4418 (n 20/D). Combustible.
Grades: 95% (min) purity.
Uses: Electrical insulators; water repellent; lubricant in plastic molding and processing; antistatic agent; intermediate.

didecyl phthalate (DDP) $C_6H_4(COOC_{10}H_{21})_2$.
Properties: Light-colored liquid; sp. gr. 0.9675 (20/20°C); 8.05 lb/gal (20°C); b.p. 261°C (5 mm); insoluble in water; vapor pressure 0.3 mm Hg (200°C); viscosity 113.2 cp (20°C); flash point 445°F. Combustible. Low toxicity.
Containers: Drums, tank cars; tank trucks.
Use: Plasticizer, especially for vinyl resins.

didecyl sulfide (didecyl thioether) $(C_{10}H_{21})_2S$.
Properties: Liquid; solidifies at 72°F; b.p. 205–206°C (4 mm); sp. gr. 0.831 (20/4°C); refractive index 1.4569 (n 33.5°C/D). Combustible.
Grades: 95% (min.) purity.
Use: Organic synthesis (formation of sulfonium compounds).

didecyl thioether. See didecyl sulfide.

didodecylamine. See dilaurylamine.

didodecyl ether. See dilauryl ether.

didodecyl 3,3′-thiodipropionate. See dilauryl thiodipropionate.

didodecyl thioether. See dilauryl sulfide.

DIDP. See diisodecyl phthalate.

didymium Di
Commercial mixture of rare earth elements obtained from monazite sand by extraction followed by the elimination of cerium and thorium. The name is used like that of an element in naming mixed oxides and salts. The approximate composition of didymium from monazite, expressed as rare earth oxides is 46% lanthana, La_2O_3; 10% praseodymia, Pr_6O_{11}; 32% neodymia, Nd_2O_3; 5% samaria, Sm_2O_3; 0.4% yttrium earth oxides; 1% ceria, CeO_2; 3% gadolinia Gd_2O_3; 2% others. Commercially used didymium salts are acetate, carbonate, chloride, fluoride, nitrate, etc. Didymium oxide; Di_2O_3; brown powder; insoluble in water; soluble in acids.
Containers: Glass bottles; fiber and steel drums.
Uses (as salts): Coloring and decolorizing glass; in temperature-compensating capacitors for radio, television, and radar; in carbon arc cores (fluoride); in stainless steel (oxide); metallurgical research; textile treatment.
Shipping regulations: (didymium nitrate) (Rail) Nitrates, n.o.s., Yellow label. (Air) Oxidizer label.

die casting. Shaping of metal products by forcing a molten metal or alloy under high pressure into a negative die cavity by means of a plunger or ram. The die is usually made of an alloy steel. Metals commonly used for die casting are zinc, aluminum, copper, lead and their alloys, some of which also include silicon. Die castings can be held to tolerances as low as 0.001 to 0.0015 in., and sizes from 75 to 100 lb (aluminum) are possible. The largest end-use area for die castings is automobile and airplane parts; they are also used in washing and drying machines, electrical equipment and appliances, and for various military applications.

dieldrin (HEOD) $C_{12}H_{10}OCl_6$. Generic name for an insecticidal product containing not less than 85% of 1,2,3,4,10,10-hexachloro-6,7-epoxy-1,4,4a,5,6,7,8,8a-octahydro-1,4-endo,exo-5,8-dimethanonaphthalene, and not less than 15% active related compounds. See also endrin, a stereoisomer of dieldrin.
Properties: Light-tan, flaked solid, m.p. 175°C. Insoluble in water, methanol, aliphatic hydrocarbons; soluble in acetone, benzene. Compatible with most fertilizers, herbicides, fungicides and insecticides.
Derivation: By oxidation of aldrin (q.v.) with peracids.
Containers: Fiber drums.
Hazard: Highly toxic by ingestion, inhalation and skin absorption. Penetrates intact skin. Tolerance, 0.25 milligram per cubic meter of air. Use restricted to nonagricultural applications. Said to be carcinogenic.
Use: Insecticide.

dielectric. A substance that has very low electrical conductivity, i.e., an insulator. Such substances have electrical conductivity of less than 1,000,000th mho per cm. Those with a somewhat higher conductivity (10^{-6} to 10^{-3} mho per cm) are called semiconductors (q.v.). Among the more common solid dielectrics are glass, rubber and similar elastomers, wood and other cellulosics. Liquid dielectrics are hydrocarbon oils, askarel (q.v.) and silicone oils. See also transformer oil.

dielectric constant. A value that serves as an index of the ability of a substance to resist the transmission of an electrostatic force from one charged body to another, as in a condenser. The lower the value, the greater the resistance. The standard apparatus utilizes a vacuum, whose dielectric constant is 1; in reference to this various materials interposed between the charged terminals have the following values at 20°C: air, 1.00058; glass, 3; benzene, 2.3; acetic acid, 6.2; ammonia, 15.5; ethyl alcohol, 25; glycerol, 56; and water, 81. The exceptionally high value for water accounts for its unique behavior as a solvent and in electrolytic solutions. Most hydrocarbons have high resistance (low conductivity). Dielectric constant values decrease as the temperature rises.

dielectric strength. The maximum electric field that an insulator or dielectric can withstand without breakdown, usually measured in kilovolts per centimeter. At breakdown a considerable current passes as an arc, usually with more or less decomposition of the material along the path of the current.

Diels-Alder reaction. An important organic reaction for the synthesis of six-membered rings, discovered in 1928. It involves the addition of an ethylenic double bond to a conjugated diene, i.e., a compound containing two double bonds separated by one single bond, as in 1,3-butadiene ($CH_2=CH-CH=CH_2$) or cyclopentadiene. The ease of addition of the ethylenic compound is greatly enhanced by adjacent carbonyl groups; hence, maleic anhydride reacts quantitatively with hexachlorocyclopentadiene to form chlorendic anhydride.

dien. Abbreviation for diethylenetriamine, as used in formulas for coordination compounds. See also en; pn; py.

diene. See diolefin.

"Diene."[278] Trademark for a commercial stereospecific polybutadiene rubber.
Properties: Unvulcanized: Narrow molecular weight distribution; soluble in aliphatic or aromatic hydrocarbons; free from gel; low "tackiness"; no elastic memory ("nerve").
Vulcanized: High resilience; excellent hysteretic properties; excellent resistance to abrasion and cold (brittle point $< -125°F$).
Containers: Boxes (75 lb in polyethylene bags).
Uses: To enhance the low temperature and hysteretic properties of other elastomers.

dienestrol $HOC_6H_4C(CHCH_3)C(CHCH_3)C_6H_4OH$. 3,4-Bis(para-hydroxyphenyl)-2,4-hexadiene; a synthetic with estrogenic activity.
Properties: Colorless, odorless needles or powder; m.p. 227°C. Soluble in alcohol; practically insoluble in water. Sensitive to light.
Grade: N.F.
Use: Medicine.

diesel ignition improver. A substance such as amyl nitrate, which is added to diesel fuels to improve fuel ignition and to raise the cetane number.

diesel oil (Fuel oil No. 2). Fuel for diesel engines obtained from distillation of petroleum. Its efficiency is measured by the cetane number (q.v.). It is composed chiefly of unbranched paraffins. Its volatility is similar to that of gas oil. Flash point 110–190°F; sp. gr. less than 1. Combustible.
Containers: Tank trucks; tank cars; pipeline.
Hazard: Moderate fire risk. Environmental hazard.
Uses: Fuel for trucks, ships, and other heavy automotive equipment; drilling muds; mosquito control (coating on breeding waters).
See also fuel oil.

"Diesel-Treat."[108] Trademark for dry, granular, orange sodium dichromate, used as a corrosion inhibitor. Sold in 50-lb drums.
Uses: Closed cooling systems, particularly diesel engines; cooling tower systems.

dietary food supplement. Any food product to which enough vitamins and minerals have been added to furnish more than 50% of the recommended daily allowance in a single serving (FDA). Such foods must have added ingredients identified on labels.

diethanolamine (DEA; di(2-hydroxyethyl)-amine) $(HOCH_2CH_2)_2NH$.
Properties: Colorless crystals or liquid; active base; m.p. 28.0°C; b.p. 217°C (150 mm); sp. gr. 1.092 (30/20°C); flash point 306°F (o.c.); very soluble in water and alcohol; insoluble in ether, benzene. Combustible. Low toxicity.
Derivation: Ethylene oxide and ammonia.
Containers: Drums; tank cars.
Uses: Liquid detergents for emulsion paints, cutting oils, shampoos, cleaners and polishes; textile specialties; absorbent for acid gases; chemical intermediate for resins, plasticizers, etc.; solubilizing 2,4-D.

N,N-**diethanolglycine** (DEG) $(HOCH_2CH_2)_2NCH_2COOH$. Used as a chelating agent. Also available as the sodium salt, $(HOCH_2CH_2)_2NCH_2COONa$.

2,5-diethoxyaniline $NH_2C_6H_3(OC_2H_5)_2$.
Properties: White to gray powder; m.p. 83–85°C; insoluble in water; soluble in organic solvents.
Hazard: May be toxic. See aniline.
Use: Intermediate.

1,4-diethoxybenzene. See hydroquinone diethyl ether.

1,1-diethoxyethane. See acetal.

diethoxyethyl phthalate. See diethyl glycol phthalate.

diethylacetal. See acetal.

diethyl acetaldehyde. See 2-ethylbutyraldehyde.

N,N-**diethylacetamide** $CH_3CON(C_2H_5)_2$.
Properties: Colorless liquid. Sp. gr. 0.920 (20°C); boiling range 182–186°C; faint odor. Flash point 170°F. Combustible.

diethylacetic acid. See 2-ethylbutyric acid.

N,N-**diethylacetoacetamide** $CH_3COCH_2CON(C_2H_5)_2$.
Properties: Liquid; sp. gr. 0.9950 (20/20°C); b.p., decomposes; f.p. -70°C (sets to glass below this temperature). Completely soluble in water; flash point 250°F. Combustible.
Use: Intermediate for pigments.

diethyl adipate $C_2H_5OCO(CH_2)_4OCOC_2H_5$.
Properties: Colorless liquid. Sp. gr. (25°C) 1.002; refractive index (25°C) 1.426; b.p. 245°C; f.p. −21°C. Insoluble in water. Combustible.
Use: Plasticizer.

diethylaluminum chloride (aluminum diethyl monochloride; DEAC) $(C_2H_5)_2AlCl$.
Properties: Colorless pyrophoric liquid. B.p. 208°C; f.p. −50°C.
Derivation: Reaction of triethylaluminum with ethylaluminum sesquichloride.

Hazard: Highly flammable in contact with air; reacts violently on contact with water. Dangerous fire and explosion hazard.
Uses: Polyolefin catalyst; intermediate in production of organometallics.
Shipping regulations: (Rail) Pyrophoric liquid, n.o.s., Red label. (Air) Not acceptable.

diethylaluminum hydride $(C_2H_5)_2AlH$. A pyrophoric mixture with triethylaluminum.
Derivation: Action of ethylene and hydrogen on aluminum.
Hazard: Highly flammable in contact with air; reacts violently on contact with water. Dangerous fire and explosion hazard.
Use: Catalyst reducing agent.
Shipping reuglations: Pyrophoric liquid, n.o.s., (Rail) Red label. (Air) Not acceptable.

diethylamine $(C_2H_5)_2NH$.
Properties: Colorless liquid; ammoniacal odor; alkaline reaction. B.p. 55.5°C; freezing point -49.8°C; sp. gr. (20/20°C) 0.7062; wt/gal (20°C) 5.91 lb; autoignition temp. 594°F; flash point less than -15°F. Miscible with water, alcohol, most organic solvents.
Derivation: From ethyl chloride and ammonia under heat and pressure.
Grade: Technical.
Containers: Drums; tank cars.
Hazard: Highly flammable, dangerous fire risk. Moderately toxic by ingestion; strong irritant. Tolerance, 25 ppm in air. Flammable limits in air 1.8–10.1%. Safety Data Sheet available from Mfg. Chemists Assn., Washington, D.C.
Uses: Rubber chemicals; textile specialties; selective solvent; dyes; flotation agents; resins; pesticides; polymerization inhibitors; pharmaceuticals; petroleum chemicals; electroplating; corrosion inhibitors.
Shipping regulations: (Rail) Red label. (Air) Flammable Liquid label.

alpha-**diethylaminoaceto-2,6-xylidide.** See lidocaine.

1-diethylamino-4-aminopentane. See 5-diethylamino-2-aminopentane.

5-diethylamino-2-aminopentane (1-diethylamino-4-aminopentane) $CH_3CH(NH_2)(CH_2)_2CH_2N(C_2H_5)_2$.
Properties: Liquid with an amine odor; sp. gr. 0.82; b.p. 142–144°C; soluble in water, alcohol and ether. Combustible.
Use: Pharmaceuticals.

diethylaminoaniline. See para-aminodiethylaniline.

diethylaminocellulose (DEAE cotton). A cellulose derivative containing a tertiary amine group which acts as a catalyst for epoxide reactions. It is also used in ion-exchange fractionations. It is made by adding beta-chloroethyldiethylamine hydrochloride to cellulose in sodium hydroxide. Repeated treatments increase the nitrogen content of the cotton (cellulose) to over 1 per cent with beneficial effect on crease resistance. Combustible. See also epoxide.

2-diethylaminoethanethiol hydrochloride $(C_2H_5)_2NCH_2CH_2SH \cdot HCl$. Solid; m.p. 170°C; soluble in water and alcohol; insoluble in benzene. Used as a thio intermediate.

diethylaminoethanol (N,N-diethylethanolamine) $(C_2H_5)_2NCH_2CH_2OH$.
Properties: Colorless, hygroscopic liquid base combining the properties of amines and alcohols; b.p. 161°C; sp. gr. 0.88–0.89 (20/20°C); vapor pressure 21 mm (20°C); flash point 140°F (open cup); wt/gal 7.14 lb (20°C); hygroscopic; freezing point -70°C; combustible; soluble in water, alcohol, benzene. Low toxicity.
Grade: Technical.
Containers: Cans; drums; tank cars.
Hazard: Moderate fire risk.
Uses: Water-soluble salts; fatty acid derivatives; textile softeners; pharmaceuticals; antirust compositions; emulsifying agents in acid media; derivatives containing tertiary amine groups; curing agent for resins.

diethylaminoethoxyethanol $(C_2H_5)_2NC_2H_4OC_2H_4OH$.
Properties: Colorless liquid. Sp. gr. 0.930–0.950 (20/20°C); boiling range 95% distills between 215.0–228.0°C. Combustible.
Use: Intermediate.

beta-**diethylaminoethyl chloride hydrochloride** (DEC) $(C_2H_5)_2NCH_2CH_2Cl \cdot HCl$. Intermediate in the manufacture of pharmaceuticals, and as an organic intermediate for attaching the diethylaminoethyl radical.

N,N′-**diethyl-3-amino-4-methoxy-benzene-sulfonamide** $NH_2(CH_3O)C_6H_3SO_2N(C_2H_5)_2$. White to pink crystals; m.p. 100–103°C; insoluble in water and ether, partially soluble in alcohol.
Use: Intermediate.

1-diethylamino-2-methylbenzene. See N,N-diethylorthotoluidine.

7-diethylamino-4-methylcoumarin (MDAC; 4-methyl-7-(diethylamino)coumarin) $CH_3C_9H_4O(O)N(C_2H_5)_2$.
Properties: Granular; light-tan color; m.p. 68–72°C; gives a bright blue-white fluorescence in very dilute solutions; soluble in aqueous acid solutions, resins, varnishes, vinyls and nearly all common organic solvents; slightly soluble in aliphatic hydrocarbons.
Uses: Optical bleach in textile industry; in coatings for paper, labels, book covers, etc.; to lighten plastics, resins, varnishes and lacquers; invisible marking agent.

1-diethylamino-4-pentanone. See 5-diethylamino-2-pentanone.

5-diethylamino-2-pentanone (1-diethylamino-4-pentanone) $CH_3CO(CH_2)_3N(C_2H_5)_2$. Liquid with an amine odor. Combustible.
Use: Pharmaceuticals.

meta-**diethylaminophenol** $C_6H_4OHN(C_2H_5)_2$.
Properties: White, crystalline solid. M.p. 78°C; b.p. 276–280°C. Soluble in alcohol, caustic soda, ether.
Derivation: Diethylaniline is sulfonated with oleum, and resulting diethylaniline-meta-sulfonic acid fused with caustic soda.
Hazard: May be toxic; see phenol.
Use: Dyes.

3-diethylaminopropylamine $(C_2H_5)_2NCH_2CH_2CH_2NH_2$.
Properties: Water-white liquid; amine odor; b.p. 169°C; f.p. -100°C (sets to a glass); sp. gr. 0.82 (20/20°C); refractive index 1.442 (10°C); flash point 138°F (O.C.). Combustible.
Hazard: Moderate fire risk. Irritant to skin.
Uses: Curing agent for epoxy resins; intermediate.
Shipping regulations: (Air) Corrosive label.

gamma-**diethylaminopropyl chloride hydrochloride** (DEPC) $(C_2H_5)_2NCH_2CH_2CH_2Cl \cdot HCl$. Used in manufacture of pharmaceuticals; intermediate for attaching the diethylaminopropyl radical.

N,N-**diethylaniline** $(C_2H_5)_2NC_6H_5$.
Properties: Colorless to yellow liquid. Sp. gr. 0.9351; f.p. −38 to −39°C; b.p. 215–216°C; flash point

185°F. Soluble in alcohol and ether; slightly soluble in water. Combustible.
Derivation: By heating aniline hydrochloride with alcohol at 180°C under pressure.
Grade: Technical.
Containers: Drums; tank cars.
Hazard: May be toxic. See aniline.
Uses: Organic synthesis; dyestuff intermediate.

diethylbarbituric acid. See barbital.

diethylbenzene $C_6H_4(C_2H_5)_2$. The commercial product is either a mixture of isomers or the para isomer, the latter being available in both pure and technical grades.
Properties: Colorless liquid. Boiling range 179.8–184.8°C; sp. gr. 0.865 (25/25°C); flash point 132°F; soluble in alcohol, benzene, carbon tetrachloride, ether; insoluble in water; wt/gal 7.22 lb; refractive index 1.49. Autoignition temp. 806°F. Combustible.
Containers: 55-gal drums; tank cars.
Hazard: Moderate fire risk. Moderately toxic.
Uses: Intermediate; solvent.

diethylbromoacetamide $C(C_2H_5)_2BrCONH_2$.
Properties: Crystalline powder; bitter, cooling taste; camphor-like odor. Decomposed by hot water.
Constants: M.p. 66–67°C. Slightly soluble in cold water; soluble in alcohol, benzene, ether, oils.
Use: Medicine.

di(2-ethylbutyl) azelate $C_6H_{13}OOC(CH_2)_7COOC_6H_{13}$. (2-Ethylbutyl is $(C_2H_5)_2CHCH_2$- or C_6H_{13}-.)
Properties: Pale-yellow to water-white liquid; sp. gr. 0.9340 (20/20°C); viscosity 56 sec Saybolt (100°F); flash point 385°F; freezing point below -40°C; acid number below 1.0; faint odor; stable to heat, light, and hydrolysis. Combustible.
Uses: As plasticizer for polyvinyl chloride and its copolymers and for cellulose esters.

di(2-ethylbutyl) phthalate (dihexyl phthalate) $C_6H_4(COOC_6H_{13})_2$.
Properties: Oily, slightly aromatic liquid; sp. gr. 1.010–1.016 (20/20°C); b.p. 350°C (735 mm); acidity (as phthalic acid) 0.01% max; f.p. −50°C; ester content 98% max. flash point 381°F; combustible. Low toxicity.
Grade: Technical.
Containers: Steel drums.
Use: Plasticizer for cellulose ester and vinyl plastics.

diethyl cadmium $(C_2H_5)_2Cd$.
Properties: Colorless pyrophoric oily liquid; b.p. 64°C (19 mm); f.p. −21°C.
Derivation: Reaction of cadmium acetate with triethyl aluminum.
Hazard: Ignites spontaneously in air; dangerous fire hazard. Toxic. See cadmium.
Uses: TEL production; synthesis of ketones from acid chlorides.
Shipping regulations: Pyrophoric liquid, n.o.s., (Rail) Red label. (Air) Not acceptable.

diethylcarbamazine citrate $C_{10}H_{21}N_3O \cdot C_6H_8O_7$. 1-Diethylcarbamyl-4-methylpiperazine dihydrogen citrate.
Properties: White, crystalline powder. Odorless or has slight odor; slightly hygroscopic. Very soluble in water; sparingly soluble in alcohol; practically insoluble in acetone, chloroform, and ether. M.p. 135–138°C.
Grade: U.S.P.
Use: Medicine.

N,N'-**diethylcarbanilide.** See sym-diethyldiphenylurea.

diethyl carbinol. See 3-pentanol.

diethyl "Carbitol."[214] Trademark for diethylene glycol diethyl ether (q.v.).

1,1'-diethyl-4,4'-carbocyanine iodide. See cryptocyanine.

diethyl carbonate (ethyl carbonate) $(C_2H_5)_2CO_3$.
Properties: Colorless liquid. Mild odor; stable. Sp. gr. 0.975 (20/4°C); b.p. 126°C; f.p. −43°C; flash point (open cup) 115°F. Miscible with alcohols, ketones, esters, aromatic hydrocarbons, some aliphatic solvents; insoluble in water. Combustible.
Derivation: The steps in its manufacture are: (a) reacting chlorine and carbon monoxide to produce phosgene ($COCl_2$); (b) reacting phosgene with ethyl alcohol to make ethyl chlorocarbonate ($ClCO_2C_2H_5$); (c) reacting ethyl chlorocarbonate with anhydrous ethyl alcohol to produce diethyl carbonate.
Grade: Technical.
Containers: Steel drums; tank cars.
Hazard: Moderate fire risk. Moderately toxic by ingestion and inhalation; strong irritant.
Uses: Solvent for nitrocellulose, cellulose ethers, many synthetic and natural resins; organic synthesis.

diethyl chlorophosphate $(C_2H_5O)_2P(O)Cl$.
Properties: Water-white liquid; b.p. 60°C (2 mm); sp. gr. 1.1915 (25°C); refractive index 1.4153 (n 25/D). Soluble in alcohols. Combustible.
Grade: Technical.
Hazard: Vapor is strong irritant to eyes and lungs; liquid is corrosive to skin and tissue. Reacts violently with water; absorbed by skin.
Use: Intermediate for organic synthesis.

diethylcyclohexane $(C_2H_5)_2C_6H_{10}$.
Properties: Liquid; sp. gr. 0.8037 (20/20°C); b.p. 174°C; f.p. -100°C; insoluble in water; flash point 125°F (open cup). Autoignition temp. 465°F. Combustible.
Hazard: Moderate fire risk.

N,N-**diethylcyclohexylamine** $C_6H_{11}N(C_2H_5)_2$.
Properties: Colorless liquid; b.p. 194.5°C. Soluble in ether and benzene; slightly soluble in water. Low toxicity. Combustible.
Grade: Technical.
Uses: Solvent; intermediate.

diethyl-1-(2,4-dichlorophenyl)2-chlorovinyl phosphate $(C_2H_5O)_2PO \cdot OC(C_6H_3Cl_2){:}CHCl$.
Properties: Liquid; b.p. 167–170°C (5 mm); insoluble in water. Miscible with acetone, ethanol, kerosine, and xylene.
Hazard: Highly toxic; cholinesterase inhibitor.
Use: Insecticide.
Shipping regulations: Organic phosphate, liquid, n.o.s., (Rail, Air) Poison label. Not accepted on passenger planes.

diethyldichlorosilane $(C_2H_5)_2SiCl_2$.
Properties: Colorless liquid; b.p. 130.4°C; sp. gr. 1.053 (25/25°C); refractive index (n 25/D) 1.4309; flash point (COC) 77°F. Readily hydrolyzed, with liberation of HCl.
Derivation: Reaction of powdered silicon and ethyl chloride at 300°C, in presence of copper powder.
Grade: Technical.
Containers: Bottles; 85-lb drums.

Hazard: Flammable and corrosive to tissue; dangerous fire risk.
Use: Intermediate for silicones.
Shipping regulations: (Rail) White label. (Air) Corrosive label. Not acceptable on passenger planes.

O,O-**diethyl S-2-diethylaminoethyl phosphorothioate hydrogen oxalate.** See tetram.

diethyl diethylmalonate $(C_2H_5)_2C(COOC_2H_5)_2$. Colorless liquid; sweet odor; sp. gr. 0.984 (25/25°C). Used as an intermediate.

3,3-diethyl-2,4-dioxopiperidine. See dihyprylone.

diethyldiphenyldichloroethane. See 1,1-dichloro-2,2-bis (para-ethylphenyl)ethane.

sym-**diethyldiphenylurea** (N,N′-diethylcarbanilide; ethyl centralite; carbamite)
$C_2H_5(C_6H_5)NCON(C_6H_5)C_2H_5$.
Properties: White crystalline solid; peppery odor; m.p. 79°C; b.p. 325–330°C; sp. gr. 1.12 (20°C); insoluble in water; soluble in organic solvents. Flash point 302°F.
Hazard: Severe explosion hazard when shocked or heated.
Uses: Stabilizer for nitrocellulose-based smokeless powder and in solid rocket propellants.
Shipping regulations: Not listed. Consult authorities.

diethylenediamine. See piperazine.

1,4-diethylene dioxide. See 1,4-dioxane.

diethylene disulfide. See 1,4-dithiane.

diethylene ether. See 1,4-dioxane.

diethylene glycol (dihydroxydiethyl ether; diglycol, DEG) $CH_2OHCH_2OCH_2CH_2OH$.
Properties: Colorless, practically odorless, syrupy liquid; sweetish taste; extremely hygroscopic; noncorrosive; lowers freezing point of water. Miscible with water, ethyl alcohol, acetone, ether, ethylene glycol; immiscible with benzene, toluene, carbon tetrachloride. B.p. 245.0°C; freezing point -8.0°C; sp. gr. 1.1184 (20/20°C); wt/gal 9.35 (15°C); refractive index 1.446 (25°C); surface tension 48.5 dynes/cm (25°C); viscosity 0.30 poise (25°C); vapor pressure 0.01 mm (30°C); flash point 255°F. Combustible. Autoignition temp. 444°F.
Derivation: By-product of manufacture of ethylene glycol.
Grade: Technical.
Containers: Drums, tank cars.
Hazard: Toxic by ingestion. FDA regards it as hazardous for household use in concentrations of 10% or more.
Uses: Polyurethane and unsaturated polyester resins; triethylene glycol; textile softener; petroleum solvent extraction; dehydration of natural gas; plasticizers and surfactants; solvent for nitrocellulose, and many dyes and oils; humectant for tobacco, casein, synthetic sponges, paper products; cork compositions; book-binding adhesives; dyeing assistant; cosmetics.

diethylene glycol acetate. See diethylene glycol monoacetate.

diethylene glycol bis(allyl carbonate) (allyl diglycol carbonate $O[CH_2CH_2OCOO(C_3H_5)]_2$. Liquid; f.p. -4°C; b.p. 160°C (4 mm); sp. gr. (20°C) 1.143; viscosity (20°C) 9 cp. Monomer for allyl resins, particularly in optically clear castings.

diethylene glycol bis(n-butylcarbonate). See butyl diglycol carbonate.

diethylene glycol bis(chloroformate). See diglycol chloroformate.

diethylene glycol bis(cresyl carbonate). See cresyl diglycol carbonate.

diethylene glycol bis(2,2-dichloropropionate). An herbicide; see "Garlon."

diethylene glycol bis(phenyl carbonate). See phenyl diglycol carbonate.

diethylene glycol diacetate (diglycol acetate) $(CH_3COOCH_2CH_2)_2O$.
Properties: Colorless liquid. Miscible with water. Sp. gr. 1.1159; b.p. 250°C; m.p. 19.1°C; flash point 275°F; vapor pressure 0.02 mm. Combustible. Nontoxic.
Grades: Technical.
Uses: Solvent for cellulose esters, printing inks, lacquers.

diethylene glycol dibenzoate.
$C_6H_5COO(CH_2)_2O(CH_2)_2OOCC_6H_5$.
Properties: Liquid; b.p. 225–227°C (3 mm). Flash point 450°F. Combustible. Nontoxic.
Use: Plasticizer.

diethylene glycol dibutyl ether (dibutyl "Carbitol"). $C_4H_9O(C_2H_4O)_2C_4H_9$.
Properties: Practically colorless liquid. Slightly soluble in water; sp. gr. 0.8853 (20/20°C); 7.36 lb/gal (20°C); b.p. 256°C; vapor pressure 0.02 mm (20°C); freezing point -60.2°C; viscosity 2.39 cp (20°C); flash point 245°F. Combustible. Low toxicity.
Containers: Drums.
Uses: High-boiling, inert solvent with application in extraction processes and in coatings and inks; diluent in vinyl chloride dispersions; extractant for uranium ores.

diethyleneglycol dicarbamate. See diglycol carbamate.

diethylene glycol diethyl ether (diethyl "Carbitol"). $C_2H_5O(C_2H_4O)_2C_2H_5$.
Properties: Colorless liquid; extremely stable; sp. gr. 0.9082 (20/20°C); b.p. 189°C; flash point 180°F; wt/gal 7.56 lb (20°C); freezing point -44.3°C. Soluble in hydrocarbons and water. Combustible. Nontoxic.
Containers: Cans; drums.
Uses: Solvent for nitrocellulose, resins, lacquers; high-boiling medium and solvent for organic synthesis.

diethylene glycol dimethyl ether ("diglyme"; diglycol methyl ether) $CH_3(OCH_2CH_2)_2OCH_3$.
Properties: Colorless liquid with mild odor. B.p. 162.0°C; f.p. −68.0°C; sp. gr. 0.9451 (20/20°C); flash point 153°F; viscosity 1.089 cp (20°C). Miscible with water and hydrocarbons. Combustible. Nontoxic.
Grade: Technical.
Uses: Solvent; anhydrous reaction medium for organometallic syntheses.

diethylene glycol dinitrate (DEGN; diglycol nitrate; dinitroglycol) $(O_2NOCH_2CH_2)_2O$.
Properties: Liquid; sp. gr. 1.377 (25/4°C); f.p. -11.3°C; b.p. 161°C; slightly soluble in water and alcohol; soluble in ether.
Hazard: Moderately toxic by ingestion. Severe explosion hazard when shocked or heated.
Uses: Rocket propellant; plasticizer in solid rocket propellants.
Shipping regulations: (Rail) Consult regulations. (Air) Not accepted.

diethylene glycol dipelargonate $(C_8H_{17}COOCH_2CH_2)_2O$. A simple ester of pelargonic acid. Acid number 2.0; b.p. 229°C (5 mm); pour point 10°F, viscosity (S.U.V. at 110°C) 36 seconds; flash point 410°F. Combustible. Low toxicity.
Use: Secondary plasticizer for vinyl resins.

diethylene glycol disterate. See diglycol stearate.

diethylene glycol monoacetate (diethylene glycol acetate) $HO(CH_2)_2O(CH_2)_2OOCCH_3$. Miscible with water and aromatic hydrocarbons. Solvent for nitrocellulose, cellulose acetate, camphor and rosin.

diethylene glycol monobutyl ether (butyl "Carbitol"). $C_4H_9OCH_2CH_2OCH_2CH_2OH$.
Properties: Colorless liquid; faint, butyl odor; b.p. 230.6°C; sp. gr. 0.9536 (20/20°C); wt/gal 7.94 lb (20°C); refractive index 1.4316 (20°C); viscosity 0.0649 poise (20°C); vapor pressure 0.01 mm (20°C); specific heat 0.546 cal/g (20–25°C); flash point 172°F; autoignition temp. 442°F. Coefficient of expansion 0.00088 (per °C) to 20°C; f.p. −68.1°C. Soluble in oils and water. Combustible. Nontoxic.
Grade: Technical.
Containers: Cans; drums; tank cars; tank trucks.
Uses: Solvent for nitrocellulose, oils, dyes, gums, soaps, polymers; plasticizer intermediate.

diethylene glycol monobutyl ether acetate (butyl "Carbitol" acetate). $CH_3CO(OC_2H_4)_2OC_4H_9$.
Properties: Colorless liquid, miscible with most organic liquids. Sp. gr. 0.9810 (20/20°C) b.p. 246.7°C; vapor pressure < 0.01 mm (20°C); flash point 240°F; wt/gal 8.16 lb (20°C); nitrocellulose-xylene dilution ratio 1.8; coefficient of expansion 0.0010 (20°C); f.p. −32.3°C; viscosity 0.0356 poise (20°C). Autoignition temp. 570°F. Combustible; nontoxic.
Grade: Technical.
Containers: Cans; drums; tank cars; tank trucks.
Uses: Solvent for oils, resins, gums, also for cellulose nitrate and polymeric coatings; plasticizer in lacquers and coatings.

diethylene glycol monoethyl ether ("Carbitol" solvent). $CH_2OHCH_2OCH_2CH_2OC_2H_5$.
Properties: Colorless, hygroscopic liquid; mild, pleasant odor; slightly viscous; stable; b.p. 195–202°C; sp. gr. 1.0273 (20/20°C); refractive index 1.425 (n 25/D); flash point 205°F; wt/gal 8.55 lb (20°C); miscible with water and the common organic solvents. Combustible.
Grade: Technical.
Containers: 1-gal cans; 5- and 55-gal drums; tank cars.
Uses: Solvent for dyes, nitrocellulose, and resins; mutual solvent for mineral oil-soap and mineral oil-sulfonated oil mixtures; nonaqueous stains for wood; for setting the twist and conditioning yarns and cloth; textile printing; textile soaps; lacquers; organic synthesis; brake fluid diluent.

diethylene glycol monoethyl ether acetate ("Carbitol" acetate) $CH_3COOCH_2CH_2OCH_2CH_2OC_2H_5$.
Properties: Colorless liquid; sp. gr. 1.0114 (20/20°C); b.p. 217.4°C; vapor pressure 0.05 mm (20°C); flash point 230°F; wt/gal 8.4 lb (20°C); coefficient of expansion 0.00105 (20°C); f.p. −25°C; refractive index (30/D) 1.418; viscosity 0.0279 poise (20°C). Soluble in water; miscible with most organic solvents. Combustible. Nontoxic.
Grade: Technical.
Containers: 1-gal cans; 5-, 55-gal drums; tank cars.
Uses: Solvent for cellulose esters, gums, resins; coatings and lacquers; printing inks.

diethylene glycol monohexyl ether (n-hexyl "Carbitol") $C_6H_{13}OC_2H_4OC_2H_4OH$.
Properties: Water-white liquid; sp. gr. 0.9346 (20/20°C); 7.8 lb/gal (20°C); b.p. 259.1°C; vapor pressure < 0.01 mm (20°C); f.p. −33°C; viscosity 8.6 cp (20°C). Flash point 285°F; combustible; low toxicity.
Containers: 1-gal cans; 5-, 55-gal drums.
Use: High-boiling solvent.

diethylene glycol monolaurate. See diglycol laurate.

diethylene glycol monomethyl ether (2-beta-methyl "Carbitol"); methoxyethoxy ethanol $CH_3OCH_2CH_2OCH_2CH_2OH$.
Properties: Colorless liquid; refractive index (n 27/D) 1.4264; sp. gr. 1.0211 (20/4°C); b.p. 194°C; soluble in water; flash point 200°F; wt/gal 8.51 lb (20°C). Combustible.
Containers: Drums; tank cars.
Uses: Solvent; brake fluid component; intermediate.

diethylene glycol monomethyl ether acetate (methyl "Carbitol" acetate). $CH_3COOC_2H_4OC_2H_4OCH_3$.
Properties: Colorless liquid; b.p. 209.1°C; flash point 180°F (open cup); sp. gr. 1.04 (20/20°C); vapor pressure 0.12 mm (20°C). Combustible.
Use: Solvent.

diethylene glycol monooleate. See diglycol oleate.

diethylene glycol monoricinoleate. See diglycol ricinoleate.

diethylene glycol monostearate. See diglycol monostearate.

diethylene glycol phthalate. See diglycol phthalate.

diethylene glycol stearate. See diglycol stearate.

diethylene oxide. See 1,4-dioxane.

diethylenetriamine $NH_2C_2H_4NHC_2H_4NH_2$.
Properties: Yellow liquid; ammoniacal odor. Strongly alkaline, hygroscopic, somewhat viscous liquid. Soluble in water, hydrocarbons. Corrosive to copper and its alloys. B.p. 206.7°C; f.p. −39°C; sp. gr. 0.9542 (20/20°C); vapor pressure 0.37 mm (20°C); flash point 215°F; wt/gal 7.9 lb (20°C); viscosity 0.0714 poise (20°C); coefficient of expansion 0.00088.
Grade: Technical.
Containers: Drums; tank cars.
Hazard: Toxic by ingestion and inhalation; strong irritant to eyes and skin. Safety Data Sheet available from Mfg. Chemists Assn., Washington, D.C.
Uses: Solvent for sulfur, acid gases, various resins, dyes; saponification agent for acidic materials; fuel component (see hydyne).
Shipping regulations: (Air) Corrosive label.

diethylenetriamine pentaacetic acid $HOOCCH_2N[CH_2CH_2N(CH_2COOH)_2]_2$.
Properties: White, crystalline solid; m.p. 230°C (dec); slightly soluble in cold water; soluble in hot water.
Grade: Technical.
Hazard: May be toxic.
Use: Chelating agent.

N,N-**diethylethanolamine.** See diethylaminoethanol.

diethyl ether. See ethyl ether.

diethyl ethoxymethylenemalonate $C_2H_5OCH{:}C(COOC_2H_5)_2$.
Properties: Liquid; sp. gr. 1.0855 (15/15°C); refractive index (n 20/D) 1.4625; b.p. 279–281°C with decomposition; flash point 190°F; insoluble in water.
Grade: 98% min (purity). Combustible.
Containers: 55-gal drums.
Use: Synthesis.

uns-**diethylethylene.** See 2-ethyl-1-butene.

N,N-**diethylethylenediamine** $(C_2H_5)_2NC_2H_4NH_2$.
Properties: Colorless liquid; b.p. 145.2°C; sets to a glass below -100°C; sp. gr. 0.8211 (20/20°C); wt/gal 6.8 lb (20°C); flash point (open cup) 115°F. Miscible with water. Combustible.
Hazard: Moderate fire risk.
Use: Intermediate.

para, para'-**(1,2-diethylethylene)diphenol.** See hexestrol.

diethyl ethylmalonate $C_2H_5CH(COOC_2H_5)_2$.
Properties: Colorless liquid; ester odor; sp. gr. 0.9994 (25/25°C). Combustible.
Use: Intermediate.

diethyl ethylphosphonate $C_2H_5P(O)(OC_2H_5)_2$.
Properties: Colorless liquid with mild odor. Stable. Miscible with most common organic solvents. Slightly soluble in water; soluble in alcohol. Sp. gr. 1.025 (20/4°C); b.p. 82–83°C (11 mm); flash point 220°F (COC). Combustible.
Containers: Steel drums.
Uses: Heavy metal extraction and solvent separation; gasoline additives; antifoam agent; plasticizer; textile conditioner and antistatic agent.

1,1'-diethyl ferrocenoate (1,1'-ferrocene dicarboxylic acid diethyl ester) $(C_2H_5OOCC_5H_4)_2Fe$. Orange crystalline solid; m.p. 38–40°C.
Uses: Intermediate; high-temperature plasticizer.

diethylgermanium dichloride $(C_2H_5)_2GeCl_2$.
Properties: Colorless liquid; f.p. −38°C; b.p. 175°C. Decomposed by water.
Uses: Biocide; intermediate.

diethylglycol phthalate (diethoxyethyl phthalate) $(C_2H_5OCH_2CH_2OOC)_2C_6H_4$.
Properties: Water-white to pale straw liquid; sp. gr. 1.115–1.120 (20/20°C); wt/gal 9.31 lb. Flash point 343°F. Combustible.
Uses: Plasticizer.

di(2-ethylhexyl) adipate (DOA; dioctyl adipate) $[CH_2CH_2COOCH_2CH(C_2H_5)C_4H_9]_2$.
Properties: Light-colored oily liquid; sp. gr. 0.9268 (20/20°C); refractive index 1.4472; flash point 385°F; pour point -75°C; b.p. 417°C; 214°C (5 mm); vapor pressure 2.60 mm (200°C); insoluble in water; viscosity 13.7 cps (20°C); 7.7 lb/gal (20°C). Combustible. Low toxicity.
Assay: 99% min.
Containers: 55-gal drums; tank cars; tank trucks.
Uses: Plasticizer, commonly blended with general purpose plasticizers, such as DOP and DIOP in processing polyvinyl and other polymers; solvent; aircraft lubes.

di(2-ethylhexyl)amine (dioctylamine) $[C_4H_9CH(C_2H_5)CH_2]_2NH$.
Properties: Water-white liquid with slightly ammoniacal odor; sp. gr. 0.8062 (20/20°C); 6.7 lb/gal (20°C); b.p. 281.1°C; vapor pressure 0.01 mm (20°C); viscosity 3.70 cp (20°C); flash point 270°F; high solubility in hydrocarbons and low solubility in water; solubility of water in, 0.17% by wt; refractive index (n 20/D) 1.4420. Combustible; low toxicity.
Uses: Synthesis of dyestuffs, insecticides, emulsifying agents, etc.

di(2-ethylhexyl)aminoethanol. See di(2-ethylhexyl) ethanolamine.

di(2-ethylhexyl) azelate (DOZ; dioctyl azelate) $(CH_2)_7[COOCH_2CH(C_2H_5)C_4H_9]_2$.
Properties: Colorless, odorless liquid; sp. gr. 0.919 (20/20°C); refractive index 1.4472; b.p. 376°C; flash point 430°F. Combustible.
Grade: 99% pure.
Uses: Plasticizer for vinyls; especially used as low-temperature plasticizer; base for synthetic lubricants.

di(2-ethylhexyl)ethanolamine (di(2-ethylhexyl)aminoethanol; dioctylaminoethanol) $[C_4H_9CH(C_2H_5)CH_2]_2N(CH_2)_2OH$.
Properties: Colorless liquid; insoluble in water; wt/gal 7.2 lb. Flash point 280°F. Combustible.
Uses: Emulsifier; acid-stable wetting agent.

di(2-ethylhexyl) ether $[C_4H_9CH(C_2H_5)CH_2]_2O$.
Properties: Colorless, stable liquid; mild odor. Almost insoluble in water; sp. gr. 0.8121 (20/20°C); 6.6 lb/gal (20°C); b.p. 269.4°C; vapor pressure $<$ 0.01 mm (20°C); sets to glass below -95°C; viscosity 2.89 cp (20°C); refractive index (n 20/D) 1.4525. Combustible.
Uses: High-boiling, inert reaction medium; component of certain foam breakers.

di(2-ethylhexyl) 2-ethylhexylphosphonate (bis(2-ethylhexyl) 2-ethylhexylphosphonate) $C_8H_{17}PO(OC_8H_{17})_2$.
Properties: Colorless liquid with a mild odor. Sp. gr. 0.908 (20/4°C); b.p. 160–161°C (0.26 mm); flash point 420°F. Insoluble in water; miscible with most common organic solvents. Combustible. Probably low toxicity.
Containers: 5-, 55-gal drums.
Uses: Heavy-metal extraction; solvent separation; gasoline additive; anti-foam agent; plasticizer; stabilizer; textile conditioner and antistatic agent.

di(2-ethylhexyl) fumarate (dioctyl fumarate; DOF) $C_8H_{17}OOCCH{:}CHCOOC_8H_{17}$.
Properties: Clear, mobile liquid; sp. gr. 0.937–0.940 (25/25°C); b.p. 211–220°C; flash point 365°F; combustible; low toxicity.
Uses: Monomer for polymerization and co-polymerization.

di(2-ethylhexyl) hexahydrophthalate (dioctyl hexahydrophthalate) $C_6H_{10}[COOCH_2CH(C_2H_5)C_4H_9]_2$.
Properties: Light-colored liquid; sp. gr. 0.9586 (20/20°C); 8.0 lb/gal (20°C); b.p. 216°C (5 mm); vapor pressure 2.2 mm (200°C); insoluble in water; viscosity 42.1 cp (20°C); flash point 425°F; combustible. Low toxicity.
Use: Plasticizer.

di(2-ethylhexyl) hydrogen phosphate (bis(2-ethylhexyl hydrogen phosphate) $(C_8H_{17})_2HPO_4$.
Properties: Solid; sp. gr. 0.972 (20/4°C); flash point (COC) 340°F. Insoluble in water. Combustible.
Use: Heavy-metal extraction.

di(2-ethylhexyl) isophthalate (dioctyl isophthalate) $C_6H_4[COOCH_2CH(C_2H_5)C_4H_9]_2$.
Properties: Colorless liquid. B.p. 258°C at 10 mm; sp. gr. 0.984 (20/20°C); 8.2 lb/gal; pour point +46°C; insoluble in water; viscosity 86.5 cp (20°C). Combustible.
Use: Plasticizer.

di(2-ethylhexyl) maleate (dioctyl maleate; DOM) $C_8H_{17}OOCCH{:}CHCOOC_8H_{17}$.
Properties: Liquid; b.p. 209°C (10 mm); f.p. sets to glass below -60°C; sp. gr. 0.9436 (20/20°C); wt/gal 7.9 lb (20°C); flash point (open cup) 365°F. Insoluble in water. Combustible. Low toxicity.
Containers: 1-gal cans; 5-, 55-gal drums.
Uses: Copolymers; intermediate.

di(2-ethylhexyl) phosphite (bis(2-ethylhexyl) phosphite $(C_8H_{17}O)_2PHO$.
Properties: Mobile, colorless liquid with mild odor and a high degree of thermal stability. Insoluble in water (hydrolyzes very slowly); miscible with most common organic solvents. Sp. gr. 0.937 (20/4°C); b.p. 163-164°C (3 mm); refractive index n 25/D 1.444; flash point 330°F. Combustible.
Containers: 5-gal, 55-gal drums.
Uses: Lubricant additive; intermediate; adhesive.

di(2-ethylhexyl)phosphoric acid (dioctyl phosphoric acid) $[C_4H_9CH(C_2H_5)CH_2]_2HPO_4$.
Properties: Liquid having strong acid properties; sp. gr. 0.973 (25/25°C); f.p. -60°C; refractive index 1.4420 (n 25/D); flash point 385°F; wt/gal 8.2 lb. Insoluble in water; soluble in organic solvents. Combustible.
Hazard: Toxic and irritant.
Uses: Metal extraction and separation; intermediate for wetting agents and detergents.
Shipping regulations: (Air) Corrosive label.

di(2-ethylhexyl) phthalate (dioctyl phthalate; DOP) $C_6H_4[COOCH_2CH(C_2H_5)C_4H_9]_2$.
Properties: Light-colored odorless liquid; sp. gr. 0.9861 (20/20°C); pour point -46°C; refractive index 1.4836; flash point 425°F; 8.20 lb/gal (20°C); b.p. 231°C (5 mm); vapor pressure 1.32 mm Hg (200°C); viscosity 81.4 cp (20°C); insoluble in water; miscible with mineral oil. Combustible.
Derivation: Reaction of 2-ethylhexyl alcohol and phthalic anhydride.
Containers: Drums; tank cars; tank trucks.
Uses: Plasticizer for many resins and elastomers.

di(2-ethylhexyl) sebacate (dioctyl sebacate) $(\text{-}C_4H_8COOC_8H_{17})_2$.
Properties: Pale straw-colored liquid; sp. gr. 0.91 (25°C); refractive index 1.447 (28°C); b.p. 248°C (4 mm); f.p. −55°C; flash point 410°F (COC); insoluble in water; partially compatible with cellulose acetate and cellulose acetate butyrate; compatible with ethyl cellulose, polystyrene, polyethylene, vinyl chloride, and vinyl chloride acetate. Combustible. Low toxicity.
Containers: Drums; tank cars.
Use: Plasticizer.

di(2-ethylhexyl) sodium sulfosuccinate. See dioctyl sodium sulfosuccinate.

di(2-ethylhexyl) succinate (dioctyl succinate) $C_8H_{17}OCOCH_2CH_2COOC_8H_{17}$.
Properties: Liquid; b.p. 257°C (50 mm); f.p., sets to glass below -60°C; sp. gr. 0.9346 (20/20°C); wt/gal 7.8 lb (20°C); flash point (open cup) 315°F; vapor pressure <0.01 mm (20°C); solubility in water $<0.01\%$ by wt (20°C). Combustible. Low toxicity.
Containers: 1-gal cans; 5-, 55-gal drums.
Uses: Plasticizer; intermediate.

diethylhydroxylamine $(C_2H_5)_2NOH$.
Properties: Liquid; refractive index (n 20/D) 1.4238. Combustible.
Grade: 85%.
Uses: Photographic developer; antioxidant; corrosion inhibitor.

diethyl isoamylethylmalonate
$(C_2H_5)(C_5H_{11})C(COOC_2H_5)_2$. Colorless liquid; sweet odor; sp. gr. 0.950 (25/25°C). Combustible. Used as an intermediate.

diethylketone (metacetone; propione; 3-pentanone; ethyl propionyl) $C_2H_5COC_2H_5$.
Properties: Colorless, mobile liquid; acetone-like odor; soluble in alcohol and ether; slightly soluble in water. Autoignition temp. 846°F. B.p. 101°C; sp. gr. 0.816; f.p. −42°C; flash point (open cup) 55°F. Low toxicity.
Derivation: By distilling sugar with an excess of lime.
Method of purification: Rectification.
Grade: Technical.
Containers: Iron drums; glass bottles.
Hazard: Flammable, dangerous fire hazard.
Uses: Medicine; organic synthesis.
Shipping regulations: (Rail) Flammable liquid, n.o.s., Red label. (Air) Flammable Liquid label.

diethyl maleate $C_2H_5OOCCH{:}CHCOOC_2H_5$.
Properties: Water-white liquid; sp. gr. 1.0687; 8.92 lb/gal (20°C); refractive index (n 20/D) 1.4400; b.p. 225°C; f.p. −11.5°C (approx.); viscosity 3.567 cp (20°C); flash point 200°F (COC); dielectric constant 2.18 (calc) (25°C); surface tension 37.0 dynes/cm (20°C). Readily soluble in alcohol, diethyl ether, paraffinic hydrocarbons and common organic solvents; soluble in water; readily hydrolyzed by alkaline solutions. Combustible.
Derivation: Reaction of maleic anhydride with ethyl alcohol in the presence of a catalyst.
Containers: 1-gal cans; 5-, 55-gal drums; tank cars.
Hazard: Irritant to eyes and skin.
Uses; Organic synthesis; flavoring.

diethyl malonate. See ethyl malonate.

diethylmalonylurea. See barbital.

diethyl (1-methylbutyl)malonate
$[C_3H_7CH(CH_3)]CH(COOC_2H_5)_2$. Colorless liquid; ester odor; sp. gr. 0.969 (25/25°C).
Uses: Intermediate; organic synthesis.

diethylmethylmethane. See 3-methylpentane.

O,O-**diethyl** O-para-**nitrophenyl phosphorothioate.** See parathion.

diethyl oxalate. See ethyl oxalate.

diethyl oxide. See ether.

di(para-**ethylphenyl)dichloroethane.** See 1,1-dichloro-2,2-bis(para-ethylphenyl)ethane.

N,N-**diethyl**-para-**phenylenediamine.** See para-aminodiethylaniline.

diethyl phosphite $(C_2H_5O)_2HPO$.
Properties: Water-white liquid; b.p. 138°C; sp. gr. 1.069 (25°C); refractive index 1.4061 (n 25/D); flash point 195°F (COC). Soluble in water, common organic solvents. Combustible.
Uses: Paint solvent; lubricant additive; antioxidant; reducing agent; intermediate for flame retardants and insecticides; phosphorylating agent.

O,O-**diethyl phosphorochloridothioate** (ethyl PCT) $(C_2H_5O)_2P(S)Cl$.
Properties: Colorless to light amber liquid; sp. gr.

1.196 (25/25°C); f.p. below −75°C; b.p. 49°C (< 1 mm) refractive index 1.4705 (n 25/D); insoluble in water; soluble in most organic solvents. Stable at room temperature; slowly isomerizes at 100°C.
Hazard: Probable cholinesterase inhibitor. May be highly toxic; irritant to eyes and lungs.
Uses: Intermediate for pesticides; oil and gasoline additives; flame retardants; flotation agents.
Shipping regulations: Organic phosphate, liquid, n.o.s., (Rail) Poison label, (Air) Poison label. Not acceptable on passenger planes.

diethyl phthalate (ethyl phthalate; DEP) $C_6H_4(CO_2C_2H_5)_2$.
Properties: Water-white, stable, odorless, liquid; bitter taste; f.p. −40.5°C; refractive index 1.5002 (25°C); surface tension 37.5 dynes/cm (20°C); viscosity 31.3 centistokes (0°C); vapor pressure 14 mm (163°C), 30 mm (182°C), 734 mm (295°C); b.p. 298°C; flash point 325°F (O.C.); wt/gal 9.31 lb (approx.) (20°C); sp. gr. 1.120 (25/25°C). Miscible with alcohols, ketones, esters, aromatic hydrocarbons; partly miscible with aliphatic solvents; insoluble in water. Combustible.
Derivation: By reacting phthalic anhydride with ethyl alcohol, followed by careful purification.
Grade: Technical.
Containers: Drums; tank cars.
Hazard: Moderately toxic by ingestion and inhalation; strong irritant to eyes and mucous membranes. Narcotic in high concentrations.
Uses: Solvent for nitrocellulose, cellulose acetate; plasticizer; wetting agent; insecticidal sprays; camphor substitute; plastics; perfumery as fixative and solvent; alcohol denaturant; mosquito repellents; plasticizer in solid rocket propellants.

2,2-diethyl-1,3-propanediol $HOCH_2C(C_2H_5)_2CH_2OH$.
Properties: Colorless liquid; m.p. 61.3°C; b.p. 160°C (50 mm); sp. gr. (at m.p.) 0.949; wt/gal 8.2 lb (60°C); flash point (open cup) 215°F. Soluble in water. Combustible. Low toxicity.
Grades: Technical; pharmaceutical.
Uses: Emulsifying agent; intermediate; medicine.

O,O-diethyl-O-2-pyrazinyl phosphorothioate. See thionazin.

diethylpyrocarbonate (DEPC). $C_2H_5OC(O)OC(O)OC_2H_5$.
Properties: Colorless liquid; sweet, ester-like odor. Miscible with ethanol and methanol. Refractive index 1.395–1.398 (25°C).
Grade: F.C.C.
Hazard: Use in food products prohibited (FDA). Irritant to eyes and skin.
Use: Fermentation inhibitor.

diethylstilbestrol (stilbestrol; DES). $HOC_6H_4C(C_2H_5):C(C_2H_5)C_6H_4OH$. 3,4-Bis(parahydroxyphenyl)-3-hexene. A non-steroid, synthetic estrogen, always in the trans form. It is the most active of the commonly used stilbene compounds.
Properties: White, odorless crystalline powder; m.p. 169–172°C; almost insoluble in water; soluble in alcohol, chloroform, ether, fatty oils, and dilute alkali hydroxide.
Derivation: From anethole hydrobromide; from anisole; from anisoin.
Grade: U.S.P.
Hazard: Probable carcinogen. Under USDA regulations, no residues are permitted in tissues of slaughtered animals. Not permitted in cattle feeds (FDA).
Uses: Biochemical research; emergency contraceptive.

diethyl succinate $(\text{-}CH_2COOC_2H_5)_2$.
Properties: Colorless liquid with faint pleasant odor. B.p. 216.2°C; f.p. −21°C; sp. gr. 1.0418 (20/20°C); wt/gal 8.7 lb (20°C); refractive index (n 20/D) 1.4201; flash point (open cup) 230°F. Miscible with alcohol and ether; slightly soluble in water. Combustible. Low toxicity.
Uses: Plasticizer; intermediate; flavoring.

diethyl sulfate (ethyl sulfate) $(C_2H_5)_2SO_4$.
Properties: Colorless liquid; faint, ethereal odor; irritating after-effect. Noncorrosive; soluble in alcohol and ether; insoluble in water. Sp. gr. 1.1803; b.p. 208°C (dec); vapor pressure 0.19 mm (20°C); flash point 220°F; autoignition temp. 817°F; wt/gal 9.8 lb (20°C) f.p. -24.4°C; viscosity 1.79 cp (20°C). Combustible.
Derivation: Action of fuming sulfuric acid on ethyl alcohol.
Method of purification: Rectification in vacuo.
Grade: Technical.
Containers: Drums; tank cars.
Hazard: Highly toxic by ingestion and inhalation; strong irritant.
Use: Ethylating agent in organic synthesis.
Shipping reuglations: (Air) Poison label.

diethyl sulfide. See ethyl sulfide.

diethyl tartrate $C_4H_4O_6(C_2H_5)_2$.
Properties: Colorless thick, oily liquid; b.p. 280°C; m.p. 17°C; soluble in water and alcohol; sp. gr. 1.204 (20/4°C). Combustible.
Uses: Plasticizer for automobile lacquers; solvent for nitrocellulose, gums, and resins.

diethylthioglycol $(C_2H_5OCH_2CH_2)_2S$.
Properties: Liquid; volatile in steam. Soluble in alcohol, benzene, and ether; slightly soluble in water. Sp. gr. 0.9672 (20°C); b.p. 225°C (746 mm). Combustible.

1,3-diethylthiourea $C_2H_5NHCSNHC_2H_5$.
Properties: Buff solid; m.p. 68–71°C; slightly soluble in water; soluble in methanol, ether, acetone, benzene, and ethyl acetate; insoluble in gasoline.
Uses: Inhibitor of corrosion in metal pickling solutions; accelerator activator in elastomers.

N,N-**diethyl**-meta-**toluamide** (deet) $CH_3C_6H_4CON(C_2H_5)_2$.
Properties: Colorless liquid; mild bland odor; b.p. 160°C (19 mm); sp. gr. 0.996–1.002 (25/25°C); refractive index 1.5200–1.5235 (25°C); slightly soluble in water; soluble in alcohol, ether and benzene. Combustible.
Grade: U.S.P.
Hazard: Irritant to eyes and mucous membranes.
Uses: Insect repellents; resin solvent; film formers.

N,N-**diethyl**-meta-**toluidine** $CH_3C_6H_4N(C_2H_5)_2$.
Properties: Light amber oil; b.p. 231°C; refractive index (n 20/D) 1.5361.
Hazard: Probably toxic. See toluidine.
Use: Dye intermediate.

N,N-**diethyl**-ortho-**toluidine** (1-diethylamino-2-methylbenzene) $CH_3C_6H_4N(C_2H_5)_2$.
Properties: Prisms from water; m.p. 72.3°C; b.p. 209°C; soluble in water, alcohol, and ether.
Derivation: From ortho-toluidine.
Hazard: Probably toxic. See toluidine.
Use: Dye intermediate.

O,O-diethyl O-3,5,6-trichloro-2-pyridyl phosphorothioate $Cl_3C_5NHOP(S)(OC_2H_5)_2$.
Properties: Solid; m.p. 41.5–43°C; soluble in acetone, benzene, ether; almost insoluble in water.

Hazard: Toxic. Cholinesterase inhibitor.
Use: Insecticide.
Shipping regulations: (Rail, Air) Poison label. Not accepted on passenger planes.

3,9-diethyl-6-tridecanol (heptadecanol) $C_4H_9CH(C_2H_5)C_2H_4CH(OH)C_2H_4CH(C_2H_5)_2$.
Properties: White solid; sp. gr. 0.8475; b.p. 309° C; flash point 310° F (open cup); refractive index 1.4531 (20° C). Insoluble in water. Combustible.
Uses: Intermediate for synthetic lubricants, defoamers and surfactants.

1,1-diethylurea $NH_2CON(C_2H_5)_2$. White solid; m.p. 75°C; soluble in water, alcohol and benzene. When copolymerized with simple urea by the use of formaldehyde, it yields modified resins that differ in nature from those made from monosubstituted ureas. These resins tend to be permanently thermoplastic.

diethylzinc (ethylzinc; zinc ethyl; inc diethyl) $Zn(C_2H_5)_2$. The first known organometallic compound.
Properties: Colorless, pyrophoric liquid; sp. gr. 1.207 (20°C); f.p. −28°C; b.p. 118°C. Soluble in most saturated hydrocarbons.
Derivation: Action of ethyl iodide on zinc and sodium-zinc; or by interaction of zinc chloride with triethyl aluminum.
Grade: Technical.
Containers: Sealed tubes; steel cylinders.
Hazard: Ignites spontaneously on contact with air. Dangerous fire hazard. Decomposes violently in water.
Uses: Organic synthesis; catalyst for polymerization of olefins; high energy aircraft and missile fuel; production of ethyl mercuric chloride.
Shipping regulations: (Rail) Pyrophoric liquids, n.o.s., Red label. (Air) Not acceptable.

differential gravimetric analysis (DGA). A variation of differential thermal analysis in which additional information is obtained by determining the rate of change in weight during the heating process.

differential thermal analysis (DTA). The method of precisely measuring the temperature, and the rate of temperature change as heat is added to or abstracted from a sample of material that is in a controlled constant environment. The method determines whether the sample is a pure substance or a mixture, and yields information about its composition and thermal properties.

diffraction, x-ray. A method of spectroscopic analysis involving the reflection or scattering of x-radiation by the atoms of a substance (lattice) as the rays pass through it. The rays are reflected by the atoms at an angle that is characteristic of the substance, yielding a spectrum that indicates its atomic or molecular structure. The spectra thus obtained are well-defined and specific; from them the properties of elements, and the structure of both crystalline and amorphous materials can be obtained. For example, unvulcanized rubber gives an amorphous pattern, while vulcanized rubber is crystalline; the cellulose macromolecule has been found to have alternating crystalline and amorphous areas. X-ray diffraction was one of the earliest and most successful methods of instrumental analysis; developed by Bragg and von Laue early in this century, it was used with dramatic effect by Moseley (1912) in establishing the location of several elements in the Periodic System.
See also lattice; crystal; x-radiation.

diffusion. The spontaneous mixing of one substance with another when in contact or separated by a permeable membrane or microporous barrier. The rate of diffusion is proportional to the concentrations of the substances and increases with temperature. Diffusion occurs most readily in gases, less so in liquids, and least in solids. The theoretical principles are stated in Fick's laws. In gases, diffusion takes place counter to gravity, and the rate at which different gases diffuse into a particular gas (e.g. air) is inversely proportional to the square roots of the densities. Carbon dioxide and chlorine vapor will diffuse in air until a uniform mixture results. Diffusion occurs in the cell walls of plants and animals (see osmosis). Many substances diffuse through a parchment membrane. See also dialysis.

diffusion, gaseous. A technique used for separating the light isotope of uranium (U-235) from the heavy isotope (U-238). The uranium is allowed to diffuse through a series of microporous barriers, whose apertures are of molecular dimensions, in the form of the gas uranium hexafluoride; this is a mixture of $U^{228}F_6$ and $U^{235}F_6$ in a ratio of 140 to 1. Because of the vastly greater number of the heavier molecules and the extremely small difference in their masses, the mixture must pass through the barrier a great many times to obtain a high concentration of the 235 isotope. Assuming that the diffusion rate of two gases through a porous barrier is inversely proportional to the square root of their molecular weights, the ideal separation factor is $\sqrt{M_1 M_2}$, where M_2 is the molecular weight of $U^{238}F_6$ and M_1 that of $U^{235}F_6$. This method is still in use for uranium enrichment for nuclear power plant fuel.

diffusion length. A property of materials used in reactors for moderators or reflectors. It is a measure of the distance a thermal neutron diffuses after it is thermalized until it is captured. It is related to the density of the material and to the scattering and absorption cross sections.

difluophosphoric acid. See difluorophosphoric acid.

1,1,1-difluorochloroethane. See 1,1,1-chlorodifluoroethane.

difluorochloromethane. See chlorodifluoromethane.

difluorodiazine FN:NF.
Properties: Gas; can exist as cis and trans isomers.
Grade: All trans isomer; 95–99.8%.
Hazard: May be toxic. See fluorine.
Use: Trans form—preparation of ionic fluorine compounds; cis form—polymerization initiator.

difluorodichloromethane. See dichlorodifluoromethane.

4,4′-difluorodiphenyl $FC_6H_4C_6H_4F$.
Properties: White, crystalline powder; aromatic odor. Soluble in alcohol, ether, chloroform, and oils; insoluble in water. Sp. gr. 1.04; m.p. 92–95°C; b.p. 254–255°C.
Use: Medicine.

difluorodiphenyltrichloroethane. See DFDT.

1,1-difluoroethane (ethylidene fluoride) CH_3CHF_2.
Properties: Colorless, odorless gas. B.p. −24.7° C;

f.p. −117°C; sp. gr. 1.004 (−25°C); index of refraction 1.255 (20°C). Insoluble in water.
Derivation: By adding hydrogen fluoride to acetylene.
Grades: Technical, 98%.
Containers: Cylinders.
Hazard: Flammable, dangerous fire risk. Moderately toxic by inhalation. Narcotic in high concentrations; lung irritant. Flammable limits in air, 3.7–18%.
Uses: Intermediate.
Shipping regulations: (Rail) Red Gas label. (Air) Flammable Gas label. Not acceptable on passenger planes.

1,1-difluoroethylene. (Air) legal label name for vinylidene fluoride (q.v.).

difluoromethane CH_2F_2.
Properties: Gas; b.p. −51.6°C. Soluble in alcohol; insoluble in water. High thermal stability.
Uses: Refrigeration; organic synthesis.

difluoromonochloroethane. Legal label name (Rail) for 1,1,1-chlorodifluoroethane (q.v.).

difluoromonochloromethane. See chlorodifluoromethane.

difluorophosphoric acid (difluophosphoric acid) HPO_2F_2.
Properties: Mobile, strongly fuming, colorless liquid; sp. gr. 1.583 (25/4°C); f.p. −75°C; b.p. 116°C. Corrosive to glass and fabric. Noncombustible.
Hazard: Highly toxic when heated; corrosive to tissue.
Uses: Chemical polishing agent; protective coatings for metal surfaces; catalyst.
Shipping regulations: (Rail) White label. (Air) Corrosive label. Not acceptable on passenger planes.

"Difolatan."[253] Trademark for cis-N-[(1,1,2,2,-tetrachloroethyl)thio]4-cyclohexene-1,2-dicarboximide (q.v.).

"Digesta."[1] Trademark for a series of fine calcium phosphates with varying proportions of calcium and phosphate.
Grade: F.C.C.
Use: Feed supplement for livestock and poultry.

digester. A cylindrical metal vessel used chiefly in the preparation of wood pulp for papermaking, in which lignin is separated from cellulose by chemical means. It operates at about 150 psi and 170°C. The wood is fed to the digester in the form of chips, to which the cooking liquor is added. Standard digesters are 12 ft in diameter and 45 ft high, with a wall thickness of 2 inches. These hold about 20 cords of wood, and some are even larger. The cooking cycle varies from 2.5 hr for board stock to 5 hr for bleached paper. Heat supply is by circulating steam and heat exchanger, though some types have direct steam injection. Digesters are designed for both batch and continuous operation. They are also used in reclaiming fabricated rubber products (tires, boot and shoe, etc.). See also pulp, paper; digestion (2).

digestion. (1) The physiological processes involved in the assimilation of nutrients from ingested foods by the animal organism. Hydrochloric acid in the gastric juice plays a prominent part, aided by the saliva, which initiates carbohydrate breakdown, and by bile and pancreatic secretions in the intestine. Numerous types of enzymes catalyze these processes. See also metabolism; nutrition.

(2) In chemical engineering the term refers to several processes: (a) the preferential dissolution of certain mineral constituents in some ore concentrations; (b) the liquefaction of organic waste materials by microbiological action, as in activated sludge; (c) removal of lignin from wood by hot chemical solutions in the manufacture of chemical cellulose and paper pulp; (d) separation of fabric from scrap tires by hot sodium hydroxide solution in the reclaiming of rubber. The equipment for (c) and (d) is called a digester (q.v.).

digitalis. A drug obtained from dried leaves of the purple foxglove; native to southern Europe but grown in the U.S. Used in treatment of cardiac diseases, both human and animal. Allergic reactions are infrequent. Contains both digitonin and digitoxin. Toxic symptoms result from overdose.

digitonin $C_{56}H_{92}O_{29}$. A saponin derived from digitalis seeds; used for the determination of cholesterol (an insoluble addition compound is formed). Also used as an analytical reagent.

digitoxin $C_4H_{64}O_{13}$. Most active glycoside of Digitalis purpurea.
Properties: White, odorless, bitter leaflets or powder; m.p. 255–256°C; slightly soluble in water or ether; soluble in alcohol.
Derivation: From digitalis leaves, usually digitalis purpurea (foxglove).
Grade: U.S.P.
Hazard: Highly toxic by ingestion. Overdose can be fatal.
Use: Medicine (cardiac treatment).

diglycerol. See polyglycerols.

1,3-diglycidyloxybenzene. See resorcinol diglycidyl ether.

diglycol. See diethylene glycol.

diglycol acetate. See diethylene glycol diacetate.

"Diglycolamine."[137] Trademark for 2-(2-aminoethoxy)-ethanol.

diglycol carbamate (diethylene glycol dicarbamate) $O(CH_2CH_2OCONH_2)_2$.
Properties: White crystalline solid; relatively stable to acid hydrolysis, but less stable to basic conditions.
Use: Manufacture of resins.

diglycol chloroformate [diethylene glycol bis(chloroformate)] $O(CH_2CH_2OCOCl)_2$.
Properties: Liquid; b.p. 125–127°C (5 mm); flash point 295°F; combustible; nontoxic; soluble in acetone, alcohol, ether, chloroform, and benzene.
Uses: Preparation of nonvolatile plasticizers or modifying agent.

diglycol chlorohydrin. $ClCH_2CH_2OCH_2CH_2OH$.
Properties: Colorless liquid; miscible with water; sp. gr. 1.1698; b.p. 196.8°C; flash point 225°F; combustible; vapor pressure 0.17 mm.
Hazard: Moderately toxic by ingestion and inhalation; an irritant.

diglycolic acid $O(CH_2COOH)_2$.
Properties: White crystalline solid; m.p. 148°C; soluble in water and alcohol; pH of 10% aqueous solution 1.4. Forms a nonhygroscopic monohydrate at relative humidities above 72% at 25°C. Low toxicity.
Containers: Multiwall paper bags.
Uses: Manufacture of resins and plasticizers; organic synthesis; sequestering agent; emulsion breaker in petroleum.

diglycol laurate (diethylene glycol monolaurate) $C_{11}H_{23}COOC_2H_4OC_2H_4OH$.
Properties: Light straw-colored, oily liquid practically odorless, nontoxic, and edible. Sp. gr. 0.96. Dis-

persible in water; soluble in methanol, ethanol, toluene, naphtha, and mineral oil. Miscible in certain proportions in cottonseed oil, acetone, and ethyl acetate. Flash point 290°F; low toxicity; combustible.
Derivation: Lauric acid ester of diethylene glycol.
Uses: Emulsifying agent for oils and hydrocarbon solvents; emulsions for lubrication, sizing, finishing, of textiles, paper, and leather; fluid emulsions of oils for hand lotions, hair dressings, etc.; cutting and spraying oils; dry-cleaning soap base; antifoaming agent.

diglycol methyl ether. See diethylene glycol dimethyl ether.

diglycol monostearate (diethylene glycol monostearate) $C_{17}H_{35}COOC_2H_4OC_2H_4OH$.
Properties: Small white flakes, available in regular or water-dispersible types. Nontoxic.
Uses: Emulsifier and thickener in cosmetics; mold release lubricant for die casting; temporary binder for ceramics and grinding wheels.

diglycol nitrate. See diethylene glycol dinitrate.

diglycol oleate (diethylene glycol monoöleate) $C_{17}H_{33}COOC_2H_4OC_2H_4OH$.
Properties: Light red, oily liquid; fatty odor. Soluble in ethanol, naphtha, ethyl acetate, methanol; partly soluble in cottonseed oil; insoluble in water. Sp. gr. 0.93; iodine value 65–75; titer below 0°C; pH (25°C) 7.7–8.2 (5% aqueous dispersion). Combustible; nontoxic.
Derivation: Oleic acid ester of diethylene glycol.
Uses: Emulsifying agent for fluid water-in-oil emulsions for the manufacture of furniture and automobile polish; water-emulsion paints; agricultural sprays.

diglycol phthalate (diethylene glycol phthalate) $C_6H_4(COOC_2H_4OH)_2$.
Properties: Pale yellow, liquid resin. Soluble in methanol, ethanol, acetone, ethyl acetate; partly soluble in toluene, naphtha, mineral oil, cottonseed oil; insoluble in water. Sp. gr. 1.29; saponification value 430–450; acid value 170–175. Combustible; nontoxic.
Use: Plasticizer; emulsifier.

diglycol ricinoleate (diethylene glycol monoricinoleate) $C_{17}H_{32}(OH)COOC_2H_4OC_2H_4OH$.
Properties: Light yellow liquid; f.p. below −60°C; sp. gr. 0.980 (25°C). Soluble in alcohol, acetone, and ethyl acetate; insoluble in water. Combustible; nontoxic.
Use: Plasticizer for high polymers and elastomers.

diglycol stearate (diethylene glycol distearate) $(C_{17}H_{35}COOC_2H_4)_2O$.
Properties: White, wax-like solid; faint fatty odor. Disperses in hot water; soluble (hot) in alcohol, oils, and hydrocarbons. M.p. 54–55°C; sp. gr. 0.9333 (20/4°C). Combustible; nontoxic.
Derivation: Stearic acid ester of diethylene glycol.
Grades: Technical; cosmetic.
Containers: Drums.
Uses: Emulsifying agent for oils, solvents, and waxes; lubricating agent for paper and cardboard; suspending medium for powders in the manufacture of polishes, cleaners, and textile delusterants; temporary binder for abrasive powders and clays for ceramic insulation; protective coating for hygroscopic powders; thickening agent; pharmaceuticals.

diglyme. Abbreviation for diglycol methyl ether. See diethylene glycol dimethyl ether.

digoxin. A cardiotonic digitalis glycoside, $C_{41}H_{64}O_{14}$.
Properties: Colorless to white crystals or white crystalline powder; odorless. Melts at about 235°C (dec). Insoluble in water, chloroform, and ether; soluble in pyridine and dilute alcohol.
Derivation: From the leaves of Digitalis lanata.
Grade: U.S.P.
Use: Medicine (cardiac diseases).

"Di-Halo."[225] Trademark for bromochlorodimethylhydantoin (q.v.).

diheptyl-para-phenylenediamine. See N,N′-bis(1,4-dimethylpentyl)-para-phenylenediamine.

dihexadecylamine. See dipalmitylamine.

dihexadecyl ether. See dicetyl ether.

dihexadecyl sulfide. See dicetyl sulfide.

dihexadecyl thioether. See dicetyl sulfide.

dihexyl. See n-dodecane.

di-n-hexyl adipate $(\text{-}CH_2CH_2COOC_6H_{13})_2$.
Properties: Liquid; color, water-white to maximum 100 Pt-Co; sp. gr. (20°C) 0.939; refractive index (25°C) 1.438; surface tension (20°C) 32.7 dynes/cm; viscosity (20°) 8.8 cp; b.p. (4 mm) 183–192°C (midpoint 191°C); water solubility (25°C) 0.1%; gasoline and oil solubility, complete. Flash point 325°F; combustible; low toxicity.
Use: Low-temperature plasticizer for SBR elastomers.

di-n-hexylamine $[CH_3(CH_2)_5]_2NH$.
Properties: Water-white liquid; b.p. 233–243°C; sp. gr. 0.788 (20/20°C); refractive index 1.434 (20°C); flash point 220°F. Combustible; low toxicity.

di-n-hexyl maleate $C_6H_{13}OOCCH{:}CHCOOC_6H_{13}$.
Properties: Liquid; sp. gr. 0.9602 (20/20°C); b.p. 179°C (10 mm); refractive index 1.449 (20°C); vapor pressure, less than 0.01 mm (20°C); f.p. −70°C; viscosity 10.2 cp (20°C); solubility in water, less than 0.01% by wt (20°C). Combustible. Low toxicity.
Use: Preparation of resins.

dihexyl phthalate. See di(2-ethylbutyl) phthalate.

dihexyl sebacate $(\text{-}CH_2CH_2CH_2CH_2COOC_6H_{13})_2$.
Properties: Light straw-colored liquid; b.p. (4 mm) 203°C. Flash point 415°F; combustible; low toxicity.
Derivation: By reacting dodecyl alcohol with sebacic acid.
Containers: Drums; tank cars.
Use: Plasticizer for vinyl resins.

dihydrazine sulfate $(N_2H_4)_2 \cdot H_2SO_4$.
Properties: White, crystalline flakes; m.p. approx. 104°C; decomposes at about 180°C; soluble in water; insoluble in most organic solvents.
Grades: 95% grade is available commercially.
Containers: Drums.
Hazard: Moderately toxic and irritant.
Use: Reducing agent.

dihydric. Containing two hydroxyl groups connected to different carbon atoms, e.g., a dihydric alcohol. Dihydric alcohols are collectively called glycols (diols). See also ethylene glycol.

dihydroabietyl alcohol (hydroabietyl alcohol) $C_{19}H_{31}CH_2OH$.
Properties: Solid; sp. gr. 1.007–1.008; refractive index 1.5280; vapor pressure 1.5×10^{-5} mm (25°C); m.p. 32–33°C; flash point 190°C; insoluble in water. Combustible.
Containers: Drums.
Use: Plasticizer.

1,8-dihydroacenaphthylene. See acenaphthene.

dihydrochalcone. Any of a group of disaccharide sugars having a sweet taste; they are derived from such flavanones as naringin and hesperidin. Basic research on the conversion of these compounds into acceptable synthetic sweeteners has been in progress for some years; may be introduced in certain pharmaceuticals, toothpastes, etc., with the possibility of future use in beverages and foods. See also flavanone.

dihydrocholesterol (beta-cholestanol; 3-beta-hydroxycholestane) $C_{27}H_{47}OH$. A sterol found in the feces. Unlike cholesterol, it has no double bond.
Properties: White crystals; m.p. (monohydrate) 142°C; optical rotation α (25/D) +23°. Soluble in fat solvents; insoluble in water.
Derivation: By a series of oxidation and reduction reactions from cholesterol.
Use: Biochemical research; pharmaceutical preparations.

dihydrocodeinone bitartrate $C_{18}H_{21}NO_3 \cdot C_4H_6O_6 \cdot 2\frac{1}{2}H_2O$.
Properties: White, odorless, crystalline powder. An alkaloid derivative. Fairly soluble in water; slightly soluble in alcohol; insoluble in ether or chloroform. Affected by light.
Grade: N.F.
Hazard: May cause addiction.
Use: Medicine (narcotic)

dihydrocoumarin. See benzodihydropyrone.

"Dihydrocyclol."[19] Brand name for norbornane-2-methanol (q.v.).

6,7-dihydrodipyrido(1,2-a:2′,1′-c)pyrazidinium salt. See diquat.

dihydrofolic reductase. An enzyme acting as an essential catalyst of DNA synthesis.

dihydrogen ferrous EDTA. See ethylenediaminetetraacetic acid (note).

dihydro-2(3)-imidazolone. See ethylene urea.

2,3-dihydroindene. See indan.

9,20-dihydro-9-oxoanthracene. See anthrone.

2,3-dihydropyran C_5H_8O.
Properties: Colorless, mobile liquid; ether-like odor; b.p. 84.3°C; freezing point −70°C; sp. gr. 0.927 (20/4°C); refractive index 1.4180 (25/D); flash point: 0°F. Soluble in water and alcohol.
Hazard: Highly flammable; dangerous fire risk. Moderately toxic by ingestion and inhalation.
Uses: Intermediate.
Shipping regulations: (Rail) Flammable liquid, n.o.s., Red label. (Air) Flammable Liquid label.

1,2-dihydro-3,6-pyridazinedione. See maleic hydrazide.

dihydrostreptomycin (DHS) $C_{21}H_{41}N_7O_{12}$. A derivative of streptomycin (q.v.) in which the carbonyl group of the streptose (q.v.) portion has been reduced by the addition of two hydrogen atoms. It has antibiotic properties similar to streptomycin and is mainly used in the treatment of tuberculosis.
Derivation: By hydrogenation of streptomycin.
Uses: Medicine (usually as sulfate salt).

dihydrotachysterol $C_{28}H_{45}OH$.
Properties: Colorless or white crystals. Odorless. Practically insoluble in water. Soluble in alcohol, ether, chloroform; sparingly soluble in vegetable oils; m. p. 123.5–129°C.
Grade: U.S.P.
Use: Medicine.

2,5-dihydrothiophene-1,1-dioxide (sulfolene) $O_2SCH_2CH{:}CHCH_2$ (ring).
Properties: Solid; b.p. decomposes; sp. gr. 1.314 (70°C); flash point 235°F. Partially soluble in water, acetone and toluene. Combustible.

1,2-dihydro-2,2,4-trimethylquinoline ("Agerite" Resin D). $C_{12}H_{15}N$.
Properties: Brown pellets or flakes; sp. gr. 1.03 (25°C); melting range, initial melt 75–100°C.
Use: Rubber antioxidant.

dihydroxyacetone (DHA; dihydroxypropanone) $HOCH_2COCH_2OH$.
Properties: Colorless, crystalline solid; m.p. 80°C; hygroscopic; soluble in water and alcohol; nearly insoluble in ether; insoluble in naphtha, characteristic odor; sweet cooling taste.
Derivation: Action of sorbose bacterium on glycerol.
Uses: Medicine; intermediate; emulsifier; humectant; plasticizers; fungicides; cosmetics (creates synthetic suntan).

2,4-dihydroxyacetophenone. See 4-acetylresorcinol.

dihydroxyaluminum sodium carbonate $Al(OH)_2OOCONa$. White powder.
Derivation: Reaction of aluminum isopropoxide and an aqueous solution of sodium bicarbonate.
Use: Medicine (antacid).

1,3-dihydroxy-2-amino-4-octadecene. See sphingosine.

1,8-dihydroxyanthranol. See anthralin.

1,2-dihydroxyanthraquinone. See alizarin.

1,4-dihydroxyanthraquinone. See quinizarin.

1,5-dihydroxyanthraquinone (anthrarufin) $C_{14}H_6O_2(OH)_2$.
Properties: Yellow crystals; soluble in alcohol; sparingly soluble in water. M.p. 280°C.
Derivation: By heating anthraquinone with boric acid and sulfuric anhydride.
Use: Dyes.

1,8-dihydroxyanthraquinone (chrysazin; danthron) $C_{14}H_6O_2(OH)_2$.
Properties: Orange-colored powder or reddish-brown needles; m.p. 191°C; soluble in alcohol; sparingly soluble in water.
Derivation: From 1,8-anthraquinone potassium disulfonate.
Grades: Technical; N.F.
Uses: Dyes; medicine.

meta-**dihydroxybenzene.** See resorcinol.

ortho-**dihydroxybenzene.** See pyrocatechol.

para-**dihydroxybenzene.** See hydroquinone.

2,4-dihydroxybenzenecarboxylic acid. See beta-resorcyclic acid.

2,4-dihydroxybenzoic acid. See beta-resorcylic acid.

2,5-dihydroxybenzoic acid. See gentisic acid.

3,5-dihydroxybenzoic acid. See alpha-resorcylic acid.

2,4-dihydroxybenzophenone $C_6H_5COC_6H_3(OH)_2$.
Properties: Light-yellow, crystalline solid; m.p. 142°C; b.p. 194°C (1 mm); density 5.8 lb/gal (20°C). Insoluble in water; soluble in ethyl alcohol, methyl alcohol, methyl ethyl ketone, and ethyl acetate.
Use: Ultraviolet absorber in polymers.

2,5-dihydroxybenzoquinone $C_6H_2(OH)_2O_2$.
Properties: Yellow-orange solid, m.p. 216°C (dec.); soluble in concentrated sulfuric acid; slightly soluble in ethyl alcohol, acetone, water, benzene. Insoluble in naphtha.
Derivation: From hydroquinone.
Hazard: Irritant to eyes and skin.
Uses: Metal chelating; insecticides; polymerization inhibitor; tanning agent; dyestuff manufacture.

2,3-dihydroxybutane. See 2,3-butylene glycol.

2,5-dihydroxychlorobenzene. See chlorohydroquinone.

3-(3,4-dihydroxycinnamoyl) quinic acid. See chlorogenic acid.

dihydroxydiaminomercurobenzene $OHNH_2C_6H_3HgC_6H_3OHNH_2$. A mercury compound analogous to arsphenamine, used in medicine as a source of mercury.
Hazard: Highly toxic. See mercury.
Shipping regulations: (Rail, Air) Mercury compounds, n.o.s. Poison label.

2,2′-dihydroxy-5,5′-dichlorodiphenylmethane. See dichlorophene.

dihydroxydiethyl ether. See diethylene glycol.

2,2′-dihydroxy-5,5′-difluorodiphenyl sulfide $FC_6H_3(OH)S(OH)C_6H_3F$.
Properties: White amorphous solid; m.p. 119–121°C. Soluble in acetone, ether, chloroform, ethanol, ethyl acetate and glacial acetic acid; moderately soluble in benzene; insoluble in water.
Hazard: May be toxic.
Uses: Fungicide (textile); agricultural chemical.

2,4-dihydroxy-3,3-dimethylbutyric acid, gamma **lactone.** See pantolactone.

N-**(2,4-dihydroxy-3,3-dimethylbutyryl)**-beta-**alanine.** See pantothenic acid.

5,7-dihydroxydimethylcoumarin $C_{11}H_{10}O_4$. Properties and uses closely resemble those of 5,7-dihydroxy-4-methylcoumarin (q.v.)

dihydroxydiphenyl sulfone (sulfonyl bisphenol) $(C_6H_4OH)_2SO_2$. The commercial product is a mixture of isomers.
Properties: White, free-flowing, odorless crystals. M.p. 215–240°C. Soluble in alcohol and acetone; insoluble in water.
Grade: Technical.
Containers: 150-lb fiber drums.
Uses: Electroplating; phenolic resins; polyvinyl chloride; intermediate.

5,5′-dihydroxy-7,7′-disulfonic-2,2′-dinaphthylurea (6,6′-ureylenebis-1-naphthol-3-sulfonic acid) $HOC_{10}H_5(SO_3H)NHCONHC_{10}H_5(SO_3H)OH$.
Properties: (Crude) light gray paste. Soluble in water, very soluble in alkaline solution.
Derivation: Phosgenation of J acid.
Hazard: May be toxic.
Use: Dye intermediate.

para-**di-(2-hydroxyethoxy)benzene.** See hydroquinone, di(beta-hydroxyethyl)ether.

di(2-hydroxyethyl)amine. See diethanolamine.

N,N-**dihydroxyethyl ethylenediamine** $(CH_2NHC_2H_4OH)_2$.
Properties: Solid crystals; m.p. 98°C; b.p. 196°C (10 mm); flash point 355°F. Combustible. Low toxicity.
Use: Manufacture of textile-finishing assistants.

dihydroxyethyl sulfide. See thiodiglycol.

N,N-**dihydroxyethyl**-meta-**toluidine** $CH_3C_6H_4N(C_2H_4OH)_2$.
Properties: Light-gray solid; m.p. 62°C; dist. range 175–185°C (2 mm).

2,2′-dihydroxy-3,5,6,3′,5′,6′-hexachlorodiphenyl methane. See hexachlorophene.

1,3-dihydroxy-4-hexylbenzene. See hexyl resorcinol.

3′,4′-dihydroxy-2-isopropylaminoacetophenone hydrochloride $C_6H_3(OH)_2COCH_2NH(C_3H_7)\cdot HCl$. Light-colored crystalline powder with a faint odor. Used as an intermediate.

3,4-dihydroxy-alpha-**(methylaminomethyl)benzyl alcohol.** See epinephrine.

1,8-dihydroxy-3-methylanthraquinone. See chrysophanic acid.

5,7-dihydroxy-4-methylcoumarin $C_{10}H_8O_4\cdot H_2O$.
Properties: Yellow to white solid; fluoresces blue, absorbs ultraviolet light; melting range 270–285°C; insoluble in water, benzene, ether; soluble in alcohol and sodium hydroxide.
Derivation: From phloroglucinol.
Uses: In suntan oils as a sun screen; in wall paints as a whitening agent.

1,1′-dihydroxymethylferrocene $[C_5H_4(CH_2OH)]_2Fe$. Yellow crystalline solid; m.p. 85–86°C. Used as an intermediate.
See also ferrocene.

1,3-dihydroxynaphthalene (naphthoresorcinol) $C_{10}H_6(OH)_2$.
Properties: Transparent, crystalline plates; m.p. 124–125°C; soluble in alcohol, ether and water. Combustible.
Derivation: By heating naphthalene-1,3-disulfonic acid with alkali at 230°C under pressure.
Grades: Technical; reagent.
Uses: Dyes; pharmaceuticals; analytical reagent for sugars, oils, glucuronic acid.
Note: There are several other isomeric forms of dihydroxynaphthalene (1,5-; 1,6-; 1,7-; 1,8-; 2,3-; 2,6-; 2,7-). They are derived by heating a naphthalene disulfonic acid isomer with caustic soda; are soluble in alcohol and ether and sparingly soluble in water. M.p. ranges from 136°C (1,6-) to 260°C (2,6-). Their chief use is in organic dyes.

4,5-dihydroxy-2,7-naphthalenedisulfonic acid (chromotropic acid) $C_{10}H_4(OH)_2(SO_3H)_2$.
Properties: White needles; very soluble in water; insoluble in alcohol or ether.
Hazard: May be toxic.
Uses: Dyestuff intermediate; analytical reagent.

2,3-dihydroxynaphthalene-6-sulfonic acid, sodium salt $C_{10}H_5(OH)_2SO_3Na$. Coupler in diazo copying.

2,8-dihydroxy-3-naphthoic acid ($C_{10}H_5(OH)_2COOH$).
Properties: Light-green powder; slightly soluble in hot water; soluble in alcohol and acetone. M.p. 235–240°C.
Use: Intermediate.

L-**dihydroxyphenylalanine** (L-dopa). An amino acid used in treating Parkinson's disease, manganese poisoning and muscular dystonia. Derived from several types of beans, including vanilla; also made synthetically. Classified by FDA as experimental drug.

1,2-dihydroxypropane. See 1,2-propylene glycol.

dihydroxypropanone. See dihydroxyacetone.

8,9-dihydroxystearic acid $C_{17}H_{33}(OH)_2COOH$.
Properties: White crystals, odorless, tasteless. Soluble in alcohol and ether; insoluble in water. M.p. 135°C.
Derivation: By heating dibromide of isooleic acid with silver oxide.

dihydroxysuccinic acid. See tartaric acid.

3,5-dihydroxytoluene. See orcin.

DII. Abbreviation for diesel ignition improver.

diiodoacetylene (diiodoethyne) $IC \vdots CI$.
Properties: White crystals. Unpleasant odor. Light acts upon it, causing a gradual change in color to red and a separation of iodine. Highly volatile! Soluble in alcohol, ether, benzene; insoluble in water. M.p. 78.5°C (decomp.).
Derivation: By dissolving iodine in liquid ammonia and passing acetylene into the solution.
Hazard: Highly toxic by inhalation. Vapors irritating to eyes and mucous membranes.
Use: Organic synthesis; military poison.
Shipping regulations: (Rail, Air) Not listed. Consult authorities.

diiodoaniline $C_6H_3I_2NH_2$.
Properties: Shining, brown crystals. Soluble in alcohol, ether, chloroform, ethyl acetate, and carbon disulfide; insoluble in water. M.p. 96°C; sp. gr. 2.75.
Derivation: By the action of iodine chloride on acetanilide, followed by saponification and distillation with steam.
Hazard: May be toxic. See aniline.
Use: Medicine.

diiodobrassidinic acid ethyl ester. See ethyl diiodobrassidate.

sym-**diiododibromoethylene** BrIC:CIBr.
Properties: Crystals; m.p. 95–96°C.
Derivation: Reaction of iodine and dibromoacetylene.
Use: Organic synthesis.

diiododiethyl sulfide $(ICH_2CH_2)_2S$.
Properties: Bright-yellow prisms. Slowly decomposes, the rate being accelerated by light and by heat. Hydrolyzed by alkali solutions. Soluble in alcohol, benzene, ether; insoluble in water. M.p. 62°C.
Derivation: Interaction of dichlorodiethyl sulfide with an acetic acid solution of sodium iodide.
Hazard: Highly toxic by ingestion and inhalation. Strong irritant.
Shipping regulations: (Rail, Air) Not listed. Consult authorities.

diiodomethane. See methylene iodide.

3,5-diiodosalicylic acid $I_2C_6H_2(OH)COOH$.
Properties: White to pale pink crystalline powder; slightly soluble in water.
Uses: Source of iodine for animal nutrition.

diiodothyronine $HOC_6H_4OC_6H_2I_2CH_2CH(NH_2)COOH$. 3,5-Diiodothyronine. A thyronine derivative which is an intermediate obtained in the manufacture of synthetic thyroxine; also, probably an intermediate in the synthesis of thyroxine by the thyroid gland.

diisobutyl adipate (DIBA) $[C_2H_4COOCH_2CH(CH_3)_2]_2$.
Properties: Colorless liquid. Odorless. Compatible with most natural and synthetic polymers. Soluble in most organic solvents; insoluble in water. Sp. gr. 0.950 (25°C); b.p. 278–280°C; f.p. −20°C; wt/gal 7.95 lb; acidity (as adipic acid) less than 0.05%. Combustible.
Containers: 1-, 5-gal cans; 55-gal drums; tank cars.
Use: Plasticizer.

diisobutyl aluminum chloride (DIBAC) $[(CH_3)_2CHCH_2]_2AlCl$.
Properties: Colorless liquid; density 0.905 g/ml, f.p. -39.5°C.
Derivation: Reaction of isobutylene and hydrogen on aluminum.
Hazard: May be irritant to tissue.
Use: Polyolefin catalyst.

diisobutyl aluminum hydride (DIBAL-H) $[(CH_3)_2CHCH_2]_2AlH$.
Properties: Colorless pyrophoric liquid; f.p. -80°C; density 0.798 g/ml; b.p. 105°C (0.2 mm); miscible in hydrocarbon solvents; dilute solutions nonpyrophoric.
Derivation: Reaction of isobutylene and hydrogen with aluminum.
Hazard: Ignites spontaneously in air; dangerous fire risk.
Use: Reducing agent in pharmaceuticals.
Shipping regulations: (Rail) Pyrophoric liquid, n.o.s., Red label. (IATA) Not acceptable.

diisobutylamine $[(CH_3)_2CHCH_2]_2NH$.
Properties: Water-white liquid; sp. gr. 0.745 (20°C); boiling range 136–140°C; amine odor. Flash point 85°F. Low toxicity.
Hazard: Flammable, moderate fire risk.
Use: Intermediate.

diisobutylcarbinol. See 2,6-dimethyl-4-heptanol.

diisobutyl carbinyl acetate. See nonyl acetate.

diisobutylene. A group of isomers of the formula C_8H_{16}, of which 2,4,4-trimethylpentene-1 and 2,4,4-trimethylpentene-2 are the most important since they are formed in appreciable amounts when isobutene (isobutylene) is polymerized.
Properties: Colorless liquids. Sp. gr. 0.7227 (60°F); boiling range 214–220°F.
Hazard: Fire risk. Narcotic in high concentrations.
Uses: Alkylation; intermediates; antioxidants; surfactants; lube additives; plasticizers; rubber chemicals.

alpha-**diisobutylene.** See 2,4,4-trimethylpentene-1.

beta-**diisobutylene.** See 2,4,4-trimethylpentene-2.

diisobutyl ketone (2,6-dimethyl-4-heptanone) $(CH_3)_2CHCH_2COCH_2CH(CH_3)_2$.
Properties: Colorless liquid. Stable; mild odor. Miscible with most organic liquids; immiscible with water. Sp. gr. 0.8089 (20/20°C); b.p. 168.1°C; vapor pressure 1.7 mm (20°C); flash point 140°F; wt/gal 6.7 lb (20°C); freezing point -41.5°C; coefficient of expansion 0.00101 (20°C). Combustible.
Grade: Technical.
Containers: 1-gal cans; 5-, 55-gal drums; tank cars.
Hazard: Toxic. Tolerance 25 ppm in air.
Uses: Solvent for nitrocellulose, rubber, synthetic

resins; lacquers; coating compositions; organic synthesis; roll-coating inks; stains.

diisobutyl phenol. See octyl phenol.

diisobutyl phthalate $C_6H_4[COOCH_2CH(CH_3)_2]_2$.
Properties: Liquid; refractive index 1.4900 (n 25/D); sp. gr. 1.040 (20/20°C); flash point 385°F; b.p. 327°C. Combustible. Low toxicity.
Containers: 55-gal drums, tank trucks, tank cars.
Use: Plasticizer.

diisocyanate. An organic compound with two isocyanate groups (-NCO), formed by treating diamines (e.g., toluene-2,4-diamine, hexa-methylenediamine, para, para'-diaminodiphenylmethane) with phosgene. Combustible.
Uses: Production of polyurethane foams, and elastomers; in phenol-formaldehyde resins to improve water and alkali resistance; bonding rubber to rayon or nylon.
See also polyurethane.

diisodecyl adipate (DIDA) $C_{10}H_{21}OOC(CH_2)_4COOC_{10}H_{21}$.
Properties: Light-colored, oily liquid; mild odor. Sp. gr. 0.918 (20/20°C); f.p. -71°C; boiling range 239–246°C (4 mm); refractive index 1.450 (25°C); wt/gal 7.5 lb. Flash point 225°F. Combustible. Low toxicity. Insoluble in glycerol, glycols and some amines. Soluble in most other organics.
Containers: 1-, 5-gal cans; 55-gal drums; tank cars; tank trucks.
Uses: Primary plasticizer for polymers.

diisodecyl-4,5-epoxytetrahydrophthalate ("Flexol" PEP) $OC_6H_8(COOC_{10}H_{21})_2$.
Properties: Liquid; sp. gr. 0.9867 (20/20°C); 8.2 lb/gal; pour point 38°C; oxirane oxygen 3%; combustible.
Use: Plasticizer-stabilizer resistant to fungi; in vinyl plastics for outdoor use.

diisodecyl phthalate (DIDP) $C_6H_4(COOC_{10}H_{21})_2$.
Properties: Clear liquid with a mild odor; sp. gr. 0.966 (20/20°C); f.p. -50°C; b.p. 250–257°C (4 mm); refractive index (n 25°C) 1.483; viscosity 108 cp (20°C); wt/gal 8 lb. Flash point 450°F; combustible. Insoluble in glycerol, glycols and some amines; soluble in most other organics.
Grade: Technical.
Containers: 55-gal drums; tank cars; tank trucks.
Hazard: Moderately toxic and irritant.
Use: Plasticizer.

diisononyl adipate $[C_9H_{19}OOC(CH_2)_4COOC_9H_{19}]$. A low-volatility plasticizer based on isononyl alcohol. Combustible.

diisooctyl acid phosphate $(C_8H_{17})_2HPO_4$.
Hazard: Strong irritant to skin and eyes. Toxic.
Shipping regulations: (Rail) White label. (Air) Corrosive label.

diisooctyl adipate (DIOA) $C_8H_{17}OOC(CH_2)_4COOC_8H_{17}$.
Properties: A light straw-colored liquid; mild odor; sp. gr. 0.924 (25°C); b.p. 214–226° (4 mm); f.p. $\geqslant$75°C. Flash point 370°F; combustible. Low toxicity.
Containers: 1-, 5-gal cans; 55-gal drums; tank cars.
Use: Plasticizer, especially at low temperatures.

diisooctyl azelate (DIOZ) $C_8H_{17}OOC(CH_2)_7COOC_8H_{17}$. Liquid diester of azelaic acid. B.p. 237°C (5 mm); pour point -85°F, acid number 1.0; viscosity (min) 3.2 cs (210°F). Sp. gr. 0.92 at 20°C. Flash point 415°F; combustible. Low toxicity.
Containers: Drums; tanks.
Use: Plasticizer for vinyl resins; base for synthetic lubricants.
See also di-2-ethylhexyl azelate.

diisooctyl phthalate (DIOP) $(C_8H_{17}COO)_2C_6H_4$. Isomeric esters obtained from phthalic anhydride and the mixed octyl alcohols made by the Oxo process (see isooctyl alcohol).
Properties: Nearly colorless, viscous liquid; mild odor; b.p. 370°C; sp. gr. (20/20°C) 0.980–0.983; wt/gal (20°C) 8.20 lb; flash point 450°F. Combustible. Low toxicity. Insoluble in water; compatible with vinyl chloride resins and some cellulosic resins.
Grade: Technical.
Containers: Drums; tank cars; tank trucks.
Use: Plasticizer for vinyl, cellulosic, and acrylate resins and synthetic rubber.

diisooctyl sebacate (DIOS) $C_8H_{17}OOC(CH_2)_8COOC_8H_{17}$.
Properties: Liquid; sp. gr. 0.915 (25/25°C); flash point 440°F; pour point -40°C; viscosity 24 cp (20°C); 7.65 lb/gal. Combustible. Low toxicity.
Containers: Drums; tank cars.
Use: Plasticizer.

diisopropanolamine (DIPA) $(CH_3CHOHCH_2)_2NH$.
Properties: White crystalline solid; sp. gr. 0.9890 (45/20°C); b.p. 248.7°C; 8.2 lb/gal (45°C); vapor pressure 0.02 mm (42°C); melting point 42°C; viscosity 1.98 poise (45°C); miscible with water. Flash point 260°F; combustible; low toxicity.
Containers: Drums; tank cars.
Uses: Emulsifying agents for polishes, textile specialties, leather compounds, insecticides, cutting oils, and water paints.

diisopropyl. See 2,3-dimethylbutane.

diisopropylamine $[(CH_3)_2CH]_2NH$.
Properties: Colorless, volatile liquid with amine odor. B.p. 84.1°C; freezing point -96.3°C; sp. gr. (20/20°C) 0.7178; wt/gal (20°C) 6.0 lb; refractive index 1.3924 (n 20/D); flash point (open cup) 30°F. Slightly soluble in water; soluble in most organic solvents.
Derivation: From isopropyl chloride and ammonia.
Grade: Technical.
Containers: 5-, 55-gal drums; tank cars.
Hazard: Flammable; dangerous fire risk. Toxic by inhalation and ingestion; strong irritant.
Uses: Intermediate; catalyst.
Shipping regulations: (Rail) Red label. (Air) Flammable Liquid label.

diisopropylaminoethanol. See N,N-diisopropylethanolamine.

beta-**diisopropylaminoethyl chloride hydrochloride** (DIC) $[(CH_3)_2CH]_2NCH_2CH_2Cl \cdot HCl$.
Use: Organic synthesis, especially for introduction of the beta-diisopropylaminoethyl radical.

meta-**diisopropylbenzene** $C_6H_4(C_3H_7)_2$.
Properties: Colorless liquid; sp. gr. 0.8559 (20/4°C); f.p. -63°C; b.p. 203°C; refractive index 1.4883 (20°C); flash point 170°F. Insoluble in water; miscible with alcohol, ether, acetone, benzene, and carbon tetrachloride. Combustible. Autoignition temp. 840°F.
Uses: Solvent; intermediate.

para-**diisopropylbenzene** $C_6H_4(C_3H_7)_2$.
Properties: Colorless liquid; sp. gr. 0.8568 (20/4°C); f.p. −17°C; b.p. 210°C; refractive index 1.4898 (20°C). Insoluble in water; miscible with alcohol, ether, acetone, benzene, and carbon tetrachloride. Combustible.
Uses: Solvent; intermediate.

diisopropylbenzene hydroperoxide. Available as a 52% solution in a nonvolatile solvent. Colorless to pale yellow liquid. Strong oxidizing agent.
Hazard: Dangerous fire risk in contact with organic materials.
Shipping regulations: (Rail) Yellow label. (Air) (less than 60% in nonvolatile solvent) Organic Peroxide label; (over 60%) Not acceptable.

N,N-**diisopropylbenzothiazyl-2-sulfenamide** $C_6H_4NC[SN(C_3H_7)_2]S$. Rubber accelerator.

diisopropyl carbinol (2,4-dimethylpentanol-3) $[(CH_3)_2CH]_2CHOH$.
Properties: Colorless liquid. B.p. 140°C; f.p. below −70°C; wt/gal 6.9 lb; flash point 120°F. Low toxicity. Combustible.
Containers: 55-gal drums.
Hazard: Moderate fire risk.
Uses: Solvent; organic synthesis (intermediate); denaturant.

diisopropyl cresol. Antioxidant or stabilizer in medicine (external). See also isopropyl cresol.

N,N′-**diisopropyldiamidophosphoryl fluoride** (mipafox) $(CH_3)_2CHNHPO(F)NHCH(CH_3)_2$.
Hazard: Toxic. Tolerance (as F), 2.5 mg per cubic meter of air.
Use: Insecticide.

diisopropyl dixanthogen $(C_3H_7OCS_2)_2$.
Properties: Yellow to greenish pellets; sp. gr. 1.28; m.p. 52°C (min); purity 98% (min.). Insoluble in water; soluble in ethyl alcohol, acetone, benzene and gasoline.
Hazard: Toxic by ingestion and inhalation; strong irritant.
Uses: Modifier in polymerization reactions; additive for lubricants; flotation reagent; fungicide; weed killer.

N,N-**diisopropylethanolamine** (diisopropylaminoethanol) $[(CH_3)_2CH]_2NCH_2CH_2OH$.
Properties: Colorless liquid; sp. gr. 0.8742 (20°C); vapor pressure 0.08 mm (20°C); freezing point −39.3°C; b.p. 191°C, flash point 175°F; slightly soluble in water. Combustible.
Hazard: Strong irritant to tissue.
Uses: Organic synthesis.
Shipping regulations: (Air) Corrosive label.

diisopropyl ether. Legal label name for isopropyl ether (q.v.).

diisopropyl fluorophosphate (DFP; isoflurophate) $[(CH_3)_2CHO]_2POF$. Oily liquid; in the presence of moisture forms hydrogen fluoride. One member of a series of compounds, the fluorophosphate alkyl esters, characterized by extremely high toxicity, marked miotic effects noted even in concentrations that are chemically undetectable.
Hazard: Highly toxic. Cholinesterase inhibitor.
Uses: Medicine.

diisopropyl ketone (2,4-dimethylpentanone-3) $[(CH_3)_2CH]_2CO$.
Properties: Colorless liquid. B.p. 123.7°C; wt/gal 6.9 lb.
Use: Solvent.

diisopropyl oxide. See isopropyl ether.

diisopropyl peroxydicarbonate (isopropyl percarbonate; isopropyl peroxydicarbonate; IPP) $(CH_3)_2CHOC(O)OOC(O)OCH(CH_3)_2$.
Properties: Colorless, crystalline solid; sp. gr. 1.080 (15.5/4°C); m.p. 8–10°C; refractive index (n 20/D) 1.4034. Almost insoluble in water; miscible with aliphatic and aromatic hydrocarbons, esters, ethers, chlorinated hydrocarbons.
Derivation: By reaction of sodium peroxide with isopropyl chloroformate.
Grades: Stabilized and unstabilized.
Hazard: Spontaneous decomposition at room temperature releases flammable and corrosive products. Dangerous fire hazard. Store in open containers at low temperature, with adequate ventilation. Explodes on heating.
Use: Low temperature polymerization catalyst.
Shipping regulations: (Rail) (freight) Stabilized, White label; Unstabilized, Yellow label; (express) Not accepted. (Air) Not acceptable. Legal label name: isopropyl percarbonate.

2,6-diisopropylphenol $C_6H_3OH[CH(CH_3)_2]_2$.
Properties: Light straw-colored liquid; f.p. 18°C; sp. gr. 0.955 (20°C); b.p. 242°C; flash point 240°F. Soluble in toluene and alcohol; insoluble in water. Combustible.
Use: Intermediate for synthetic, polymers, plasticizers, surface-active agents.

diisopropyl-para-**phenylenediamine** $H_7C_3NHC_6H_4NHC_3H_7$.
Properties: Dark red liquid; sp. gr. 0.88. Soluble in alcohol.
Uses: Gasoline antioxidant and sweetener (permissible for aviation gasoline, ASTM D910-64T).

N,N′-**diisopropylthiourea** $(CH_3)_2CHNHCSNHCH(CH_3)_2$.
Properties: Grayish white solid; m.p. 138.5–142.5°C; slightly soluble in water; soluble in methanol, acetone, and ethyl acetate; insoluble in ether, benzene and gasoline.
Uses: Corrosion inhibitor; metal pickling with hydrochloric or sulfuric acid; for reducing corrosion of ferrous metals and aluminum alloys in brine; intermediate.

diketene (acetyl ketene) $CH_2{:}CCH_2C(O)O$.
Properties: Colorless, non-hygroscopic liquid; pungent odor; readily polymerizes on standing; sp. gr. 1.096 (20/20°C); f.p. −7.5°C; b.p. 127.4°C. Soluble in common organic solvents; soluble in water. Flash point (TOC) 93°F.
Derivation: By spontaneous polymerization of ketene obtained by thermal decomposition of acetone, or from bromoacetylbromide and zinc.
Containers: Steel drums; insulated tank trucks, cooled to 40–50°F.
Hazard: Flammable, moderate fire risk. Moderately toxic and irritant.
Uses: Production of acetoarylamides; pigments and toners; pesticides; food preservatives; pharmaceutical intermediates.

diketobutane. See diacetyl.

2,5-diketohexane. See acetonylacetone.

2,5-diketopyrrolidone. See succinimide.

2,5-diketotetrahydrofurane. See succinic anhydride.

"Dilantin."[330] Trademark for diphenylhydantoin and diphenylhydantoin sodium.

dilatancy. A system is said to be dilatant if its rate of increase of strain decreases with increased shear. Among the better known systems exhibiting dilatant behavior are pastry doughs, highly pigmented paints and many other industrially important materials. Dilatancy is usually associated with suspensions, especially those containing high concentrations of suspended matter, which is often of colloidal dimensions.

"Dilaudid."[9] Trademark for dihydromorphinone; used in medicine as the sulfate and hydrochloride.

dilaurylamine (didodecylamine) $(C_{12}H_{25})_2NH$.
Properties: Liquid. M.p. 45°C; sp. gr. 0.89; almost insoluble in water.
Use: Chemical intermediate.

dilauryl ether (didodecyl ether) $(C_{12}H_{25})_2O$.
Properties: Liquid. M.p. 33°C; b.p. 190–195°C (1 mm); sp. gr. 0.8147 (33/4°C). Combustible.
Grade: 95% (min.) purity.
Uses: Electrical insulators; water repellents; lubricants for plastic molding and processing; antistatic substances; chemical intermediates.

dilauryl phosphite $(C_{12}H_{25}O)_2PHO$. Water-white liquid. Combustible.
Uses: Synthesis of organophosphorus compounds for extreme pressure lubricants, adhesives, textile finishing agents, pesticides; catalyst in polymerization of unsaturated compounds.

dilauryl sulfide (didodecyl thioether) $(C_{12}H_{25})_2S$.
Properties: Liquid. M.p. 40–40.5°C; b.p. 260–263°C (4 mm); sp. gr. 0.8275 (40/4°C). Combustible.
Grade: 95% (min.) purity.
Use: Organic synthesis (formation of sulfonium compounds).

dilauryl thiodipropionate (didodecyl 3,3′-thiodipropionate; thiodipropionic acid, dilauryl ester) $(C_{12}H_{25}OOCCH_2CH_2)_2S$.
Properties: White flakes having sweetish odor; m.p. 40°C; sp. gr. (solid, 25°C) 0.975. Insoluble in water; soluble in most organic solvents. Extremely resistant to heat and hydrolysis. Nontoxic. Combustible.
Grade: F.C.C.
Containers: Drums; trucks.
Uses: Antioxidant; additive for high-pressure lubricants and greases; plasticizer and softening agent; antioxidant for edible fats and oils (up to 0.02% of oil content).

"Dilecto."[281] Trademark for a series of laminated plastics consisting of a variety of base materials and resins, available in sheets, tubes, or rods, of numerous grades for a variety of special mechanical and electrical insulation applications.

dilinoleic acid $C_{34}H_{62}(COOH)_2$.
Properties: Light yellow viscous liquid with slight odor; sp. gr. 0.921 (100°C); refractive index 1.4851 (40°C); iodine value 80; combustible.
Uses: Modifier in alkyd and polyamide resins; polyester or metallic soap for petroleum additive; emulsifying agent; adhesives; shellac substitute; to upgrade drying oils.

dilithium sodium phosphate Li_2NaPO_4. A commercial source of lithium found in Searles Lake brine.

dilituric acid. See 5-nitrobarbituric acid.

"Dillydap."[458] Trademark for dilauryl thiodipropionate (q.v.).

"Diloderm" Acetate.[321] Trademark for dichlorisone acetate (q.v.).

diluent. (1) An ingredient used to reduce the concentration of an active material to achieve a desirable and beneficial effect. Examples are combination of diatomaceous earth with nitroglycerine to form the much less shock-sensitive dynamite; addition of sand to cement mixes to improve workability with no serious loss of strength; addition of an organic liquid having no solvent power to a paint or lacquer to reduce viscosity and achieve suitable application properties (see also thinner). (2) Low-gravity materials used primarily to reduce cost, i.e., blown asphalt, wood flock, whiting, etc. in rubber and plastic mixes. In this sense there is no clear distinction between a diluent and an extender (q.v.). (3) An ingredient of rocket fuels such as helium, hydrazine, or hydrogen.

"Diluex."[98] Trademark for a fine-mesh attapulgite clay. "Diluex" A is employed under conditions that require higher sorptivity and fineness.
Use: Granular pesticide carrier.

dilution ratio (hydrocarbon tolerance). The maximum number of unit volumes of hydrocarbon that can be added per unit volume of active solvent to cause the first trace of gelation to occur when the concentration of nitrocellulose in the solution is 8 grams per 100 milliliters. This may be used to evaluate the solvent power of active solvents by comparing them with a standard hydrocarbon, or to evaluate hydrocarbon solvents by comparing them with an active solvent. See solvent; lacquer.

dimagnesium orthophosphate. See magnesium phosphate, dibasic.

dimagnesium phosphate. See magnesium phosphate, dibasic.

"Dimazine."[55] Trademark for uns-dimethylhydrazine (q.v.).

dimedone (1,1-dimethyl-3,5-diketocyclohexane) $(CH_3)_2C_6H_6O_2$.
Properties: Greenish-yellow needles, or prisms; m.p. 148–149°C; slightly soluble in cold water and naphtha; soluble in alcohol, chloroform, benzene.
Use: Reagent for the detection of ethyl alcohol and the identification of aldehydes.

dimefox. Generic name for bis(dimethylamino)fluorophosphate; tetramethyldiamidophosphoric fluoride (BFPO). $[(CH_3)_2N]_2POF$.
Properties: Liquid; fishy odor; sp. gr. 1.1151 (20/4°C); b.p. 67°C (4.0 mm), 86°C (15 mm); refractive index 1.4267 (n 20/D). Soluble in water, ether, benzene; aqueous solutions are stable.
Hazard: Highly toxic; use may be restricted. Tolerance (as F), 2.5 mg per cubic meter of air.
Use: A systemic pesticide, primarily for ornamental and non-food plants.
Shipping regulations: (Rail) Not listed. (Air) Poison label. Not accepted on passenger planes.

dimenhydrinate $C_{17}H_{22}NO \cdot C_7H_6ClN_4O_2$. 2-(Benzohydryloxy)-N,N-dimethylethylamine-8-chlorotheophyllinate.
Properties: Crystalline, white, odorless powder. Freely soluble in alcohol and chloroform; soluble in benzene; sparingly soluble in ether; slightly soluble in water. M.p. 102–107°; pH (saturated solution) 6.8–7.3.

Grade: U.S.P.
Use: Medicine (antihistamine; antiemetic).

dimer. An oligomer whose molecule is composed of two molecules of the same chemical composition. For example, 1,4-dioxane is a dimer of ethylene oxide. See also polymer.

A dimer acid is a high molecular weight dibasic acid, which is liquid (viscous), stable, resistant to high temperatures, and which polymerizes with alcohols and polyols to yield a variety of products, such as plasticizers, lube oils, hydraulic fluids. It is produced by dimerization of unsaturated fatty acids at mid-molecule and usually contains 36 carbons. Trimer acid, which contains three carboxyl groups and 54 carbons, is similar.

dimercaprol. U.S.P. name for 2,3-dimercaptopropanol (q.v.).

2,3-dimercaptopropanol (BAL; British Anti-Lewisite; dimercaprol; 1,2-dithioglycerol) $CH_2(SH)CH(SH)CH_2OH$.
Properties: Colorless, oily, viscous liquid with strong, offensive odor of mercaptans. B.p. 80° C (1.9 mm), 140° C (40 mm); m.p. 77° C; sp. gr. 1.2385 (25/4° C); refractive index (n 25/D) 1.5720. Soluble in vegetable oils; moderately soluble in water with decomposition; soluble in alcohol.
Derivation: Bromination of allyl alcohol followed by reaction with sodium hydrosulfide.
Grade: U.S.P., as dimercaprol.
Hazard: Intensely irritating to eyes and mucous membranes.
Uses: Medicine; antidote to Lewisite, organic arsenicals and heavy metals.

dimetan ($C_{11}H_{17}NO_3$). Generic name for 5,5-dimethyldihydroresorcinol dimethylcarbamate.
Properties: Yellow crystals; m.p. 43–45°C; slightly soluble in water and oils but readily soluble in organic solvents.
Hazard: Toxic by ingestion and inhalation.
Use: Insecticide.

dimethallyl. See 2,5-dimethylhexadiene-1,5.

dimethicone $CH_3[Si(CH_3)_2O]Si(CH_3)_3$.
Properties: Colorless silicone oil consisting of dimethylsiloxane polymers (range in viscosities from 0.65 to 1,000,000 centistokes at room temperature). Viscosity grades above 50 centistokes are immiscible in water. Miscible with chloroform, ether. Low toxicity.
Uses: Ointments and topical drug ingredient; skin protectant.

dimethisoquin hydrochloride
$CH_3(CH_2)_2CH_2C_9H_5NOCH_2CH_2N(CH_3)_2 \cdot HCl$. 3-Butyl-1-(2-dimethylaminoethoxy)isoquinoline hydrochloride.
Properties: White powder, odorless, with bitter, numbing taste. M.p. 144–147°C. Freely soluble in alcohol; very slightly soluble in ether; soluble in water; pH (1% solution) 3.5–5.0.
Grade: N.F.
Use: Medicine.

dimethoate. Generic name for O,O-dimethyl S-(N-methylcarbamoylmethyl) phosphorodithioate. $(CH_3O)_2PSSCH_2CONHCH_3$.
Properties: White solid; m.p. 51–52° C. Moderately soluble in water; soluble in most organic solvents except hydrocarbons.
Hazard: Highly toxic. A cholinesterase inhibitor. Use may be restricted.
Use: Insecticide.
Shipping regulations: (Rail, Air) Organic phosphate, dry, Poison label. Not accepted on passenger planes.

dimethoxane (6-acetoxy-2,4-dimethyl-meta-dioxane) $CH_3COOC_4H_5O_2(CH_3)_2$.
Properties: Clear yellow to light-amber liquid; sp. gr. 1.069–1.076 (25/25°C); refractive index 1.431–1.438 (20°C); b.p. 66–68°C (3 mm); f.p. less than -25°C. Soluble in or miscible with water and organic solvents. Combustible.
Use: Antimicrobial agent.

1,2-dimethoxy-4-allylbenzene. See methyl eugenol.

2,5-dimethoxyaniline $NH_2C_6H_3(OCH_3)_2$.
Properties: Gray flakes; m.p. 69–73° C. Insoluble in cold water; soluble in organic solvents and hot water.
Hazard: May be toxic. See aniline.
Uses: Intermediate for dyes, pharmaceuticals and insecticides; antioxidant.

2,5-dimethoxybenzaldehyde $(CH_3O)_2C_6H_3CHO$.
Properties: Flaked solid; m.p. 46–49°C; soluble in organic solvents; insoluble in water. Combustible.
Use: Organic synthesis.

1,2-dimethoxybenzene. See veratrole.

1,3-dimethoxybenzene. See resorcinol dimethyl ether.

1,4-dimethoxybenzene. See hydroquinone dimethyl ether.

dimethoxybenzidine. See dianisidine.

3,3′-dimethoxybenzidine 4,4′-diisocyanate. See dianisidine diisocyanate.

3,4-dimethoxybenzyl alcohol $C_6H_3(OCH_3)_2CH_2OH$.
Properties: Viscous, brown liquid or low-melting solid. Combustible.
Use: Organic synthesis.

2,4-dimethoxy-5-chloroaniline $NH_2(Cl)C_6H_2(OCH_3)_2$.
Properties: Violet-gray crystals; m.p. 70–71°C. Insoluble in water; soluble in alcohol, benzene, and other organic solvents.
Grade: 98.5%.
Hazard: Toxic. See aniline.
Use: Intermediate for dyes and other organics

para, para′-**dimethoxydiphenylamine** $(CH_3OC_6H_4)_2NH$. A rubber antioxidant.

1,2-dimethoxyethane. See ethylene glycol dimethyl ether.

(2-dimethoxyethyl) adipate
$CH_3OC_2H_4OOC(CH_2)_4COOC_2H_4OCH_3$.
Properties: Liquid; sp. gr. 1.075 (25°C); refractive index 1.439 (25°C); b.p. 185–190°C (11 mm); f.p. -16°C; slightly soluble in water. Combustible. Low toxicity.
Use: Plasticizer.

di(2-methoxyethyl) phthalate
$C_6H_4(COOCH_2CH_2OCH_3)_2$.
Properties: Oily liquid with mild odor; sp. gr. 1.172 (20/20°C); b.p. 340°C; f.p. -45°C; flash point 381°F (open cup). Combustible. Low toxicity.
Containers: 55-gal drums; tank cars; tank trucks.
Use: Plasticizer, especially for cellulose acetate; solvent.

dimethoxymethane. See methylal.

2,5-dimethoxy-4-methylamphetamine (STP, DOM). A hallucinogenic, habit-forming drug; used in medicine; mfg. and use controlled by law in U.S. See also amphetamine.

3,4-dimethoxyphenethylamine. See homoveratrylamine.

3,4-dimethoxyphenylacetic acid. See homoveratric acid.

2,6-dimethoxyphenyllithium $(CH_3O)_2C_6H_3Li$.
Properties: White to tan, free-flowing pyrophoric powder. Moderately soluble in toluene and benzene; slightly soluble in ethyl ether. Stable indefinitely in sealed containers.
Hazard: Ignites spontaneously in air.
Shipping regulations: Not specified. Consult authorities.

1-(3,4-dimethoxyphenyl)-2-nitro-1-propene $(CH_3O)_2C_6H_3CH{:}C(NO_2)CH_3$. Yellow crystals; m.p. 68–75°C.
Use: Intermediate.

3-(dimethoxyphosphinyloxy)-N,N-dimethyl-cis-crotonamide $(CH_3O)_2P(O)OC(CH_3){:}CHC(O)N(CH_3)_2$.
Properties: Brown liquid; b.p. 400°C (760 mm); miscible with water, ethanol, xylene; very slightly soluble in kerosine.
Hazard: Toxic. Cholinesterase inhibitor. Use may be restricted.
Use: Insecticide.
Shipping regulations: Organic phosphate, liquid, n.o.s., (Rail) Poison label. (Air) Poison label. Not accepted on passenger planes.

dimethoxystrychnine. See brucine.

dimethoxytetraglycol (tetraethylene glycol dimethyl ether; bis(2-methoxyethoxyethyl ether) $CH_3(OCH_2CH_2)_4OCH_3$.
Properties: Water-white, practically odorless liquid. Stable; soluble in hydrocarbons, water. Sp. gr. 1.0132 (20/20°C); b.p. 275.8°C, 189°C (100 mm); vapor pressure < 0.01 mm (20°C); flash point 285°F; wt/gal 8.4 lb (20°C); freezing point −29.7°C; viscosity 0.0405 poise (20°C); coefficient of expansion 0.00091 (20°C). Combustible. Low toxicity.
Grade: Technical.
Containers: 1-gal cans; 5-, 55-gal drums.
Use: Solvent.

dimethrin. Generic name for 2,4-dimethylbenzyl-2,2-dimethyl-3-(2-methylpropenyl) cyclopropane carboxylate (2,4-dimethylbenzyl chrysanthemumate) $C_{19}H_{26}O_2$.
Properties: Amber liquid; sp. gr. 0.986 (20°C); b.p. 175°C (3.8 mm). Insoluble in water; soluble in petroleum hydrocarbons, aromatic petroleum derivatives, alcohols, and methylene chloride. Decomposed by strong alkali.
Hazard: May be toxic.
Use: Insecticide.

dimethyl. See ethane.

dimethylacetal (ethylidenedimethyl ether) $CH_3CH(OCH_3)_2$.
Properties: Colorless liquid; strongly aromatic odor. Soluble in water, alcohol, ether, and chloroform. Sp. gr. 0.848 (25°C); b.p. 62–63°C. Flash point below 80°F.
Derivation: By heating acetaldehyde with methyl alcohol and glacial acetic acid, and distilling.
Hazard: Flammable, dangerous fire risk. Moderately toxic by ingestion and inhalation.
Uses: Medicine; organic synthesis.
Shipping regulations: (Rail, Air) Not listed.

N,N-dimethyl acetamide (DMAC) $CH_3CON(CH_3)_2$.
Properties: Colorless liquid; b.p. 166°C; sp. gr. 0.9366 (25°C); refractive index 1.4351 (25°C); miscible with water, aromatics, esters, ketones, and ethers. Flash point 171°F (T.O.C.). Combustible.
Derivation: From dimethylamine.
Grades: Technical; high purity certified.
Containers: 55-gal drums.
Hazard: Toxic by inhalation; absorbed by skin; strong irritant. Tolerance, 10 ppm in air.
Uses: Solvent for plastics, resins, gums, and electrolytes; intermediate; catalyst; paint remover; high purity solvent for crystallization and purification.

N,N-dimethylacetoacetamide $CH_3COCH_2CON(CH_3)_2$.
Properties: Liquid; b.p. 220°C; sp. gr. 1.048–1.053 (20/20°C); refractive index (n 20/D) 1.4379; miscible in water and organic solvents. Flash point 252°F (COC). Combustible.
Hazard: May be toxic.
Use: Chemical intermediate.

2,4-dimethyl acetophenone $CH_3COC_6H_3(CH_3)_2$.
Properties: Colorless liquid, odor suggesting mimosa; sp. gr. 0.994–0.997; refractive index 1.532–1.534; soluble in four volumes of 60% alcohol. Combustible.
Uses: Perfumery; flavoring.

dimethylacetylene. See crotonylene.

dimethylamine (DMA) $(CH_3)_2NH$.
Properties: (anhydrous). Gas with a strong ammoniacal odor; sp. gr. 0.6865 at −6°C; b.p. 6.88°C; f.p. −92.2°C. (25% water solution): flash point 0°F; wt/gal approx 7.8 lb (68°F); soluble in alcohol, ether, and water. Autoignition temp. 806°F.
Derivation: Interaction of methanol and ammonia over a catalyst at high temperatures. The mono-, di-, and trimethylamines are all produced.
Method of separation: Azeotropic or extractive distillation.
Grades: Technical (anhydrous; 25% and 40% aqueous solutions); 99%.
Containers: Solution: 1-gal bottles; 5-, 55-gal drums; tank cars and trucks. Anhydrous: 25-, 50-, 100-, 1400 lb cylinders.
Hazard: Flammable, dangerous fire risk. Explosive limits in air 2.8 to 14%. Moderately toxic and irritant. (Anhydrous) Safety Data Sheet available from Mfg. Chemists Assn., Washington, D.C.
Uses: Acid gas absorbent; solvent; antioxidants; mfg. of dimethylformamide and dimethylacetamide; dyes; flotation agent; gasoline stabilizers; pharmaceuticals; textile chemicals; rubber accelerators; electroplating; dehairing agent; missile fuels; pesticide propellant; rocket propellants; surfactants.
Shipping regulations: (anhydrous) (Rail) Red Gas label; (Air) Flammable Gas label. Not acceptable on passenger planes. (aqueous solution) (Rail) Red label; (Air) Flammable Liquid label.

dimethylaminoaniline. See para-aminodimethylaniline.

dimethylaminoantipyrine. See aminopyrine.

dimethylaminoazobenzene (methyl yellow; butter yellow) $C_6H_5NNC_6H_4N(CH_3)_2$.
Properties: Yellow crystalline leaflets; m.p. 116°C; soluble in alcohol, ether, strong mineral acids, and oils; insoluble in water.
Derivation: Action of benzenediazonium chloride on dimethyl aniline.
Hazard: Causes cancer in experimental animals; worker exposure should be minimized. May not be used in foods or beverages (FDA).
Uses: Organic research.

para-**dimethylaminobenzaldehyde** $C_6H_4[N(CH_3)_2]CHO$.
Properties: Colorless crystalline plates. Soluble in hot water, alcohol and ether. M.p. 73°C; b.p. 176–177°C (17 mm). Low toxicity. Combustible.
Derivation: By mixing dimethylaniline, anhydrous chloral and phenol and allowing the mixture to stand. The phenol is removed by shaking with dilute caustic soda and the residue dissolved in water and hydrochloric acid and crystallized.
Grades: Technical; reagent.
Uses: Dyes; medicine; reagent.

para-**dimethylaminobenzene diazonium chloride, zinc chloride double salt.** See para-diazodimethylaniline, zinc chloride double salt.

para-**dimethylaminobenzenediazo sodium sulfonate** ("Dexon") $(CH_3)_2NC_6H_4NNSO_3Na$.
Properties: Solid; melts with decomposition above 200°C. Soluble in water.
Hazard: Toxic by ingestion.
Use: Fungicide for protection of germinating seed and seedlings.

3-dimethylaminobenzoic acid $(CH_3)_2NC_6H_4COOH$.
Properties: Pale yellow crystals; m.p. 147–153°C.
Use: Intermediate.

dl-**dimethylamino-4,4-diphenyl-3-heptanone hydrochloride.** See methadone hydrochloride.

2-dimethylaminoethanol (deanol; dimethylethanolamine) $(CH_3)_2NCH_2CH_2OH$.
Properties: Colorless liquid with amine odor. B.p. 134.6°C; f.p. -59.0°C; sp. gr. (20/20°C) 0.8879; wt/gal (20°C) 7.4 lb; refractive index (20°C) 1.4300; flash point (open cup) 105°F; low toxicity. Combustible. Miscible with water, acetone, ether, and benzene.
Preparation: From ethylene oxide and dimethylamine.
Grades: Anhydrous and 70% aqueous soln.
Containers: Drums; tank cars.
Hazard: Moderate fire risk.
Uses: Intermediate in the synthesis of dyestuffs, textile auxiliaries, pharmaceuticals and corrosion inhibitors; medicine; curing of epoxy, amine and polyamide resins; emulsifier.

beta-**dimethylaminoethyl chloride hydrochloride** (DMC) $(CH_3)_2NCH_2CH_2Cl \cdot HCl$. Used in manufacture of antihistamics and other pharmaceuticals. Organic intermediate for introduction of beta-dimethylaminoethyl radical.

dimethylaminoethyl methacrylate
$CH_2{:}C(CH_3)COOCH_2CH_2N(CH_3)_3$.
Properties: Liquid; sp. gr. 0.933 (25°C); b.p. 182–190°C; refractive index 1.4376 (25°C); flash point 165°F (TOC). Combustible.
Hazard: Irritant to skin, eyes, and mucous membranes. Strong lachrymator.
Uses: Binders for coatings, textile chemicals, dispersing agents for nonaqueous systems, antistatic agents, stabilizers for chlorinated polymers, ion exchange resins, emulsifying agents, cationic precipitating agents.

dimethylaminomethyl phenol $C_6H_4OHCH_2N(CH_3)_2$. Exists as ortho-, meta- and para- isomers; the commercial material is a mixture of ortho- and para-.
Properties: Dark red liquid; odor phenolic, free of methylamine; sp. gr. 1.020 (25/25°C); refractive index 1.530 (25°C); distillation range 80–130°C (2 mm); water content (Karl Fischer) 0.5%. Soluble in organic solvents; moderately soluble in water.
Hazard: May be toxic. See phenol.
Uses: Antioxidants; stabilizers; catalysts; intermediates.

1-dimethylamino-2-propanol $(CH_3)_2NCH_2CHOHCH_3$.
Properties: Water-white liquid; amine odor; b.p. 125.6°C; sp. gr. 0.850 (20/20°C); refractive index 1.421 (20°C); flash point 90°F; soluble in water and most organic solvents.
Hazard: Flammable, moderate fire risk.
Use: Organic synthesis.

3-dimethylaminopropylamine
$(CH_3)_2NCH_2CH_2CH_2NH_2$.
Properties: Colorless liquid; b.p. 123°C; f.p. -70°C, sets to a glass below this temperature; sp. gr. 0.8100 (30°C); refractive index (n 25/D) 1.4328; flash point 95°F (TCC). Soluble in water and organic solvents.
Hazard: Toxic by ingestion and inhalation; strong irritant. Flammable, moderate fire risk.
Uses: Curing agent for epoxy resins; organic intermediate.

1-dimethylamino-2-propyl chloride (beta-dimethylaminoisopropyl chloride) $(CH_3)_2NCH_2CHClCH_3$. Yellow liquid which darkens with age; distillation range 113–120°C; refractive index 1.422–1.423 (25°C). Used as an intermediate.

1-dimethylamino-3-propyl chloride hydrochloride (DMPC) $(CH_3)_2NCH_2CH_2CH_2Cl \cdot HCl$.
Use: Intermediate for pharmaceutical and organic synthesis.

4-dimethylamino-3-methylphenolmethyl carbamate (ester). $(CH_3)_2NC_6H_3(CH_3)OOCNHCH_3$.
Properties: Tan crystalline solid; m.p. 93–94°C. Slightly soluble in water; moderately soluble in aromatic solvents. Soluble in most polar organic solvents; unstable in highly alkaline media.
Hazard: Toxic by ingestion, inhalation and skin absorption.
Use: Insecticide.

4-dimethylamino-3,5-xylyl N-methylcarbamate
$(CH_3)_2N(CH_3)_2C_6H_2OOCNHCH_3$. White crystalline solid; m.p. 85°C. Almost insoluble in water; soluble in benzene, alcohol and chloroform.
Hazard: Toxic by ingestion.
Use: Systemic insecticide.

N,N-**dimethylaniline** (aniline N,N-dimethyl)
$C_6H_5N(CH_3)_2$.
Properties: Yellowish to brownish oily liquid. Soluble in alcohol and ether; slightly soluble in water. Sp. gr. 0.954; m.p. 2.5°C; b.p. 192.5–193.5°C; flash point 145°F (C.C.); refractive index 1.5582. Combustible. Autoignition temp. 700°F.
Derivation: By heating a mixture of aniline, aniline hydrochloride and methyl alcohol (free from acetone) in an autoclave and distilling.
Grades: Technical; reagent; 99.9% pure.
Containers: Drums; tank cars.
Hazard: Toxic. Tolerance, 5 ppm in air. Absorbed by skin.
Uses: Dyes; intermediates; solvent; manufacture of vanillin; stabilizer (acid acceptor).

dimethyl anthranilate (N-methyl methyl anthranilate) $CH_3OOCC_6H_4NHCH_3$.
Properties: Colorless or pale yellow liquid, grape-like odor; sp. gr. 1.132–1.138 (15°C); refractive index 1.578–1.581 (20°C); soluble in 3 volumes or more of 80% alcohol, in benzyl benzoate, diethyl phthalate, fixed oils, mineral oils and volatile oils; insoluble in glycerin; slightly soluble in propylene glycol; congealing point 18°C (4% impurity) to 10°C (20% impurity). Combustible.

Derivation: Methylation of methyl anthranilate or esterification of N-methyl anthranilic acid.
Containers: Glass; aluminum or tin-lined cans.
Uses: Perfumes; flavorings and drugs.

dimethylarsinic acid. See cacodylic acid.

1,2-dimethylbenzene. See ortho-xylene.

1,3-dimethylbenzene. See meta-xylene.

1,4-dimethylbenzene. See para-xylene.

3,3-dimethylbenzidine. See ortho-tolidine.

dimethylbenzylcarbinyl acetate. See alpha, alpha-dimethylphenethyl acetate.

2,5-dimethylbenzyl chloride (2,5-dimethyl-alpha-chlorotoluene; alpha-chloro-para-xylene) $C_6H_3(CH_3)_2CH_2Cl$.
Properties: Colorless to pale-yellow liquid; sharp, pungent odor. A lachrymator. Sp. gr. 1.035–1.045 (25/25°C); b.p. 221–226°C; refractive index (n 25/D) 1.5350–1.5360. Soluble in alcohols, ethers; insoluble in water. Combustible.
Hazard: Irritant to eyes and mucous membranes.
Uses: Intermediate for pharmaceuticals, dyes, perfumes, plasticizers, resins, wetting agents, germicides, etc.

2,4-dimethylbenzyl chrysanthemumate. See dimethrin.

alpha, alpha-**dimethylbenzyl hydroperoxide.** See cumene hydroperoxide.

dimethyl benzylphosphonate $(CH_3O)_2P(O)CH_2C_6H_5$. Water-white liquid; b.p. 114°C (0.75 mm). Used for extraction of mineral salts from special solutions; fire retardant; low-temperature plasticizer.

3,3′-dimethyl-4,4′-biphenylene diisocyanate (4,4′-bi-orthotolylene diisocyanate). Flaked white crystalline solid; assay 98% min; melting point 69°C min.
Uses: High-strength elastomers, coatings, rigid plastics.

1,1′-dimethyl-4,4′-bipyridinium salt. See paraquat.

dimethyl brassylate $CH_3OOC(CH_2)_{11}COOCH_3$. An ester of a 13-carbon saturated aliphatic dibasic acid. Waxy, low-melting solid; made by an ozone oxidation process. Used in preparation of ethylene brassylate, a synthetic musk; chemical intermediate.

2,2-dimethylbutane. See neohexane.

2,3-dimethylbutane (diisopropyl) $(CH_3)_2CHCH(CH_3)_2$.
Properties: Colorless liquid; b.p. 57.9°C; sp. gr. 0.66164 (20°C); f.p. −128.41°C; refractive index 1.37495 (20°C); flash point −20°F. Autoignition temp. 788°F.
Derivation: Alkylation of ethylene with isobutane using aluminum chloride catalyst.
Grades: Technical 95%; 99%; 99.8 mole %.
Containers: Bottles; drums.
Hazard: Flammable, dangerous fire and explosion risk.
Uses: High-octane fuel; organic synthesis.
Shipping regulations: (Rail) Flammable liquid, n.o.s., Red label. (Air) Flammable Liquid label.

2,2-dimethyl-1,3-butanediol
$CH_3CH(OH)C(CH_3)_2CH_2OH$.
Properties: Liquid; sp. gr. 0.9700; b.p. 202.4°C; f.p. −12.8°C; wt/gal 8.1 lb; very soluble in water. Combustible.
Hazard: Moderately toxic.

2,3-dimethylbutene-1 $CH_3CH(CH_3)C(CH_3){:}CH_2$.
Properties: Liquid; f.p. −157.27°C; b.p. 55.6°C; sp. gr. 0.678 (20/4°C); weight 5.68 lb/gal (60°F); refractive index 1.390 (20°C); flash point −20°F.
Hazard: Flammable; dangerous fire and explosion risk.
Uses: Perfume synthesis; isomerization reactions.
Shipping regulations: Flammable liquid, n.o.s., (Rail) Red label. (Air) Flammable Liquid label.

2,3-dimethylbutene-2 $CH_3C(CH_3){:}C(CH_3)CH_3$.
Properties: Liquid; f.p. −74.3°C; b.p. 73.2°C; sp. gr. 0.708 (20/4°C); weight 5.94 lb/gal (60°F); refractive index 1.412 (20°C); flash point 0°F.
Hazard: Flammable, dangerous fire and explosion risk.
Use: Isomerizatoin reactions.
Shipping regulations: Flammable liquid, n.o.s., (Rail) (Air) Flammable Liquid label.

2,4-dimethyl-6-tert-**butyl phenol**
$(CH_3)_2C_6H_2OH[C(CH_3)_3]$.
Properties: A low-viscosity, straw-colored liquid, readily soluble in gasoline and oils; sp. gr. 0.961. Combustible.
Containers: 55-gal drums, tank cars, tank trucks.
Uses: Antiskinning agent; antioxidant, especially for gasoline. (ASTM D910 64T).

dimethyl carbate $C_{11}H_{14}O_4$. Generic name for bicyclo (2,2,1)-5-heptene-2,3-dicarboxylic acid dimethyl ester.
Properties: Clear, oily liquid or crystalline solid; sp. gr. 1.165 (35/4°C); insoluble in water. Low toxicity.
Derivation: Esterification of Diels-Alder condensation product of maleic anhydride and cyclopentadiene.
Use: Insect repellent.

dimethylcarbinol. See isopropyl alcohol.

dimethyl carbonate. Legal label name (Air) for methyl carbonate (q.v.).

dimethyl chloroacetal (chloroacetaldehyde dimethyl acetal) $ClCH_2CH(OCH_3)_2$.
Properties: Colorless liquid with a pleasant odor; boiling range 126–132°C; flash point 110°F; sp. gr. 1.082–1.092 (25/4°C); refractive index (n 25/D) 1.4110–1.4130; purity 97% (min.); wt/gal 9.07 lb. Combustible.
Grade: Technical.
Containers: Drums; tank cars.
Hazard: Moderate fire risk.
Uses: Organic synthesis; pharmaceuticals; solvent.

2,5-dimethyl-alpha-**chlorotoluene.** See 2,5-dimethylbenzyl chloride.

dimethyl cyanamide $(CH_3)_2NCN$.
Properties: Colorless, mobile liquid; f.p. −41°C; sp. gr. 0.876; b.p. 160°C; flash point 160°F (TCC). Combustible.
Hazard: May be toxic by ingestion.
Uses: Chemical intermediate; solvent.

dimethylcyclohexane (hexahydroxylene). Mixture of ortho-, meta-, and para-isomers. $C_6H_{10}(CH_3)_2$.
Properties: Water-white liquid of mild odor; sp. gr. 0.776 (15/15°C); boiling range 120 to 129°C; f.p. $<$ −65°C; flash point about 50°F (C.C.); soluble in most common solvents; almost insoluble in water. Low toxicity.
Hazard: Flammable, dangerous fire risk.
Uses: Synthesis; special solvent.

Shipping regulations: (1,4 isomer) (Air) Flammable Liquid label.
Note: There are several isomers of dimethylcyclohexane, i.e., 1,3-; 1,4-; cis-1,2-; trans-1,2-. They have properties and uses closely similar to those given above. All are flammable, and of low toxicity.

dimethyl 1,4-cyclohexanedicarboxylate $CH_3OOCC_6H_{10}COOCH_3$.
Properties: Partially crystallized solid; sp. gr. 1.102 (35/4°C). Consists of about 60% cis and 40% trans isomers; b.p. (mixed isomer) 265°C; weight 9.18 lb/gal (20°C). Soluble in all proportions in most organic solvents.
Uses: Plasticizers; polymers.

N-**dimethylcyclohexaneethylamine.** See propylhexedrine.

N,N-**dimethylcyclohexylamine** $(CH_3)_2NC_6H_{11}$.
Properties: Water-white liquid; distilling range 157–160°C; sp. gr. 0.8490 (20/20°C); f.p. below −77°C; flash point 110°F (COC). Partly soluble in water; miscible with alcohol, benzene, acetone. Combustible.
Hazard: Moderate fire risk.
Uses: Catalyst for polyurethane foams; intermediate for rubber accelerators; treatment of textiles.

dimethylcyclohexylamine dibutyldithiocarbamate. See "RZ-50-A."

1,2-dimethylcyclopentane $C_5H_8(CH_3)_2$.
Properties: Colorless liquid. Cis: b.p. 99.5°C; sp. gr. 0.772 (20°C). Trans: b.p. 91.8°C; sp. gr. 0.751 (20°C). Combustible.
Grade: Technical.
Use: Organic synthesis.

2,2′-dimethyl-1,1′-dianthraquinone. $C_{30}H_{18}O_4$.
Properties: Yellow crystals; soluble in hot nitrobenzene, aniline and chlorobenzene. M.p. 365–367°C.
Use: Dye intermediate.

2,5-dimethyl-2,5-di(tert-**butylperoxy)hexane** $C_4H_9OOC(CH_3)_2CH_2CH_2C(CH_3)_2OOC_4H_9$.
Properties: Stable colorless liquid; b.p. 50–52°C (0.1 mm Hg). Active oxygen 10.5% min. F.p. 8°C; flash point 185°F. Soluble in alcohol; insoluble in water. Combustible.
Hazard: Irritating to eyes and skin. Strong oxidizer. May ignite organic materials.
Uses: Catalyst in polyethylene cross-linking, styrene polymerization; polyester resins.
Shipping regulations: (Rail) Peroxides, organic liquid, Yellow label. (Air) Peroxides, organic liquid, n.o.s., oxidizing material. Not acceptable.

dimethyldichlorosilane (dichlorodimethylsilane) $(CH_3)_2SiCl_2$.
Properties: Colorless liquid; b.p. 70°C; f.p. −86°C; sp. gr. 1.062 (20°C); refractive index (n 25/D) 1.4023; flash point (COC) 16°F. Reacts with water to form complex mixture of dimethylsiloxanes, and liberates hydrochloric acid. Soluble in benzene and ether.
Derivation: Action of silicon on methyl chloride in presence of a copper catalyst, or by Grignard reaction from methyl chloride and silicon tetrachloride.
Grade: Technical.
Containers: Steel drums.
Hazard: Probably toxic. Flammable, dangerous fire and explosion risk.
Use: Intermediate for silicone products.
Shipping regulations: (Rail) Red label. (Air) Flammable Liquid label. Not acceptable on passenger planes.

dimethyl dichlorovinyl phosphate. See DDVP.

5,5-dimethyldihydroresorcinol dimethylcarbamate. See dimetan.

1,1-dimethyl-3,5-diketocyclohexane. See dimedone.

dimethyldiketone. See diacetyl.

N,N′-**dimethyl**-N,N′-**di(1-methylpropyl)**-para-**phenylene diamine.** Volatile, reddish-brown liquid. Forms a continuous protective film.
Use: Antiozonant in rubber.

N,N-**dimethyl**-N,N-**dinitrosoterephthalamide** $C_6H_4[CON(CH_3)NO]_2$. Used as a blowing agent, liberating nitrogen at 100°C with a residue of dimethyl terephthalate.

dimethyl dioxane $\overline{OCH(CH_3)CH_2OCH_2}CH(CH_3)$.
Properties: Water-white liquid. Soluble in water. Sp. gr. 0.9268; b.p. 117.5°C; flash point 75°F; vapor pressure 15.4 mm at 20°C.
Hazard: Flammable, dangerous fire risk. Moderately toxic irritant to skin.

N,N-**dimethyl-2,2-diphenylacetamide.** See diphenamid.

dimethylenemethane. See allene.

dimethylethanolamine. See dimethylaminoethanol.

dimethyl ether (methyl ether; methyl oxide; wood ether) CH_3OCH_3.
Properties: Colorless, compressed gas, or liquid; soluble in water and alcohol. Sp. gr. 0.661; b.p. −24.5°C; f.p. −141.5°C; flash point −42°F. Autoignition temp. 662°F. Soluble in water and organic solvents.
Derivation: By dehydration of methanol.
Grades: Technical; 99.5%.
Containers: 25-, 50-, 100-, and 150-lb pressure cylinders.
Hazard: Highly flammable. Dangerous fire and explosion hazard.
Uses: Refrigerant; solvent; extraction agent; propellant for sprays; chemical (reaction medium); catalyst and stabilizer in polymerization.
Shipping regulations: (Rail) Red gas label. (Air) Flammable Gas label. Not acceptable on passenger planes.

dimethyl ethyl carbinol. See tert-amyl alcohol.

dimethylethylene. See butene-2.

sym-**dimethylethylene glycol.** See 2,3-butylene glycol.

O,O-**dimethyl** S-**2-(ethylsulfinyl)ethyl phosphorothioate** (oxydemetonmethyl) $(CH_3O)_2P(O)SC_2H_4SOC_2H_5$.
Properties: Amber liquid; b.p. 106°C (0.01 mm); sp. gr. 1.28 (20/4°C); soluble in water in all proportions.
Hazard: Toxic. Cholinesterase inhibitor. Use may be restricted.
Use: Systemic insecticide.
Shipping regulations: Organic phosphate, liquid, n.o.s., (Rail) Poison label. (Air) Poison label. Not accepted on passenger planes.

dimethyl ferrocenoate (1,1′-ferrocene dicarboxylic acid dimethyl ester) $(C_5H_4COOCH_3)_2Fe$.
Properties: Orange crystalline solid; m.p. 114–115°C. See also ferrocene.

N,N-**dimethylformamide** (DMF) $HCON(CH_3)_2$.
Properties: Water-white liquid; a dipolar aprotic solvent; b.p. 152.8°C; f.p. −61°C; refractive index (n 25/D) 1.4269; sp. gr. 0.953–0.954 (15.6/15.6°C). Flash point 136°F; autoignition temp. 833°F. Miscible with water and most organic solvents (except halogenated hydrocarbons). Combustible.

Derivation: Reaction of methyl formate with dimethylamine.
Containers: 55-gal drums; tank cars; tank trucks.
Hazard: Strong irritant to skin and tissue. Tolerance, 10 ppm in air. Moderate fire risk.
Uses: Solvent for vinyl resins and acetylene, butadiene, acid gases; catalyst in carboxylation reactions; organic synthesis.

dimethylfuran $OC(CH_3)CHCHC(CH_3)$.
Properties: Colorless liquid; insoluble in water; sp. gr. 0.8900; b.p. 94°C; flash point 45°F; low toxicity.
Hazard: Flammable, dangerous fire risk.
Shipping regulations: Flammable liquid, n.o.s. (Rail) Red label. (Air) Flammable Liquid label.

dimethyl glycol phthalate $C_6H_4(COOCH_2CH_2OCH_3)_2$.
Properties: Colorless liquid; sp. gr. 1.17; b.p. 230°C. Combustible.
Uses: Solvent mixtures for cellulose esters; plasticizing mixtures for cellulose esters.

dimethylglyoxal. See diacetyl.

dimethylglyoxime (butane dioxime) $CH_3C(NOH)C(NOH)CH_3$.
Properties: White crystals or powder; m.p. 204–242°C; soluble in alcohol and ether; very slightly soluble in water.
Uses: Analytical chemistry, especially as a reagent for nickel; biochemical research.

2,6-dimethyl-4-heptanol (diisobutylcarbinol) $[(CH_3)_2CHCH_2]_2CHOH$.
Properties: Colorless liquid; refractive index 1.423 (21°C); sp. gr. 0.8121 (20°C); f.p., sets to glass below −65°C; b.p. 178°C (750 mm); insoluble in water; soluble in alcohol and ether. Flash point (TOC) 162°F. Combustible. Low toxicity.
Containers: Drums; tank cars.
Uses: Surface-active agents; lubricant additives; rubber chemicals; flotation agents; antifoam agent.

2,6-dimethyl-4-heptanone. See diisobutyl ketone.

2,6-dimethyl-5-hepten-1-al $(CH_3)_2C{:}CH(CH_2)_2CH(CH_3)CHO$.
Properties: Yellow liquid; moderately stable, but not likely to cause discoloration. Sp. gr. 0.845–0.855 (25/25°C); refractive index 1.441–1.447 (20°C); flash point (TCC) 144°F. Soluble in 2 parts of 70% alcohol. Combustible.
Use: Perfumery.

2,6-dimethylheptene-3 $(CH_3)_2CHCH{:}CHCH_2CH(CH_3)_2$. Mixed cis and trans isomers.
Properties: Liquid; distillation range 128 to 129°C; sp. gr. (60/60°F) 0.722; refractive index (20/D) 1.412; flash point 70°F (TOC).
Grade: 95%.
Containers: Bottles; 5-gal drums.
Hazard: Flammable, dangerous fire risk.
Shipping regulations: (Rail) Flammable liquid, n.o.s., Red label. (Air) Flammable Liquid label.

2,5-dimethylhexadiene-1,5 (dimethallyl) $CH_2{:}C(CH_3)CH_2CH_2C(CH_3){:}CH_2$.
Properties: Water-white liquid with hydrocarbon odor; sp. gr. 0.740–0.760 (25/25°C); refractive index 1.426–1.429 (25°C); ASTM distillation, 90% between 114–123°C; soluble in hydrocarbons; insoluble in water. Flash point 56°F.
Hazard: Flammable, dangerous fire risk.
Use: Solvent.
Shipping regulations: (Rail) Flammable liquid, n.o.s., Red label. (Air) Flammable Liquid label.

2,5-dimethylhexane 2,5-dihydroperoxide $(CH_3)_2C(OOH)CH_2CH_2C(OOH)(CH_3)_2$.
Properties: Fine powder; 90% peroxide. M.p. 102–104°C; insoluble in hydrocarbons; slightly soluble in water; soluble in alcohols.
Hazard: Dangerous fire risk; strong oxidizing agent; store away from organic materials.
Use: High-temperature catalyst for polyester premix compounds and silicone resins.
Shipping regulations: (Rail) (dry) Not accepted; (wet) Yellow label. (Air) (dry or wet with less than 30% water by weight), Not acceptable; (wet with not less than 30% water by weight), Organic Peroxide label. Not acceptable on passenger planes.

dimethylhexanediol (2,5-dimethylhexane-2,5-diol) $(CH_3)_2COH(CH_2)_2COH(CH_3)_2$.
Properties: White crystals; m.p. 88.5–89°C; b.p. 214–215°C; sp. gr. (20/20°C) 0.898. Soluble in water, acetone, and alcohol; insoluble in benzene, carbon tetrachloride, and kerosine. Combustible.
Containers: Fiber drums.
Uses: Chemical intermediate.

2,5-dimethylhexane-2,5-diperoxybenzoate $C_6H_5C(O)OOC(CH_3)_2CH_2CH_2C(CH_3)_2OOC(O)C_6H_5$.
Properties: Fine white granules; m.p. 114°C; insoluble in alcohols and hydrocarbons; soluble in acetone and chlorinated hydrocarbons.
Hazard: Strong oxidant; fire risk in contact with organic materials.
Uses: Oxidizing agent; polymerization agent.
Shipping regulations: (Rail) Oxidizing material, n.o.s., Yellow label. (Air) Peroxides, organic, solid, n.o.s., Organic Peroxide label.

dimethylhexynediol (2,5-dimethyl-3-hexyne-2,5-diol) $(CH_3)_2COHC{\equiv}CCOH(CH_3)_2$.
Properties: White crystals; m.p. 94–95°C; b.p. 205–206°C; sp. gr. (20/20°C) 0.949. Soluble in water; slightly soluble in benzene, carbon tetrachloride, naphtha; very soluble in acetone, alcohol, and ethyl acetate.
Uses: Wire-drawing lubricant; antifoaming agent; coupling agent in resin coatings; chemical intermediate.

dimethyl hexynol $HC{\vdots}CCOH(CH_3)CH_2CH(CH_3)_2$ (3,5-dimethyl-1-hexyne-3-ol). A tertiary acetylenic alcohol.
Properties: Colorless liquid with camphor-like odor; b.p. 150–151°C; sets to a glass below −68°C; sp. gr. (20/20°C) 0.8545. Slightly soluble in water. Flash point 134°F (TOC). Combustible.
Containers: Drums.
Hazard: Moderate fire risk.
Uses: Stabilizer for chlorinated organic compounds; surface active agent; intermediate; solvent; lubricant.

5,5-dimethylhydantoin (DMH) $HNCONHCOC(CH_3)_2$.
Properties: White, crystalline solid; m.p. 178°C; soluble in water, alcohol, and ether.
Derivation: (a) From acetone, urea, and ammonium carbonate; (b) from acetone, potassium cyanate and hydrocyanic acid.
Hazard: Central nervous system depressant.
Uses: Synthesis; preparation of water-soluble polymers.

dimethylhydantoin-formaldehyde polymer.
Properties: Light-colored brittle resin containing 0.3% max. of formaldehyde; sp. gr. 1.30; softening point 59–80°C; dissolves readily in cold and hot water, methanol, ethyl acetate, methyl ethyl ketone, chloroform, methylene chloride, and hot glycerol; insoluble in benzene, xylene, diethyl ether, trichloroethylene, and carbon tetrachloride. Low toxicity.
Uses: Sizing; adhesives; blending agent; aerosol hair sprays.

1,1-dimethylhydrazine $(CH_3)_2NNH_2$. An unsymmetrical compound.
Properties: Colorless liquid with ammonia-like odor; f.p. −58°C; b.p. 63°C; sp. gr. 0.782 (25°C); soluble in water and alcohol. Flash point about 5°F; autoignition temp. 480°F.
Derivation: (a) Reaction of dimethylamine and chloramine; (b) catalytic oxidation of dimethylamine and ammonia.
Containers: Drums; tank cars.
Hazard: Highly toxic. Tolerance, 0.5 ppm in air. Absorbed by skin. Flammable, dangerous fire risk.
Uses: Component of jet and rocket fuels; chemical synthesis; stabilizer for organic peroxide fuel additives; absorbent for acid gases; photography; plant growth control agent.
Shipping regulations: (Rail) Red label. (Air) Flammable Liquid label. Not acceptable on passenger planes.

dimethyl hydroquinone. See hydroquinone dimethyl ether.

dimethylhydroxybenzene. See xylenol.

dimethyl 3-hydroxyglutaconate dimethyl phosphate. See "Bomyl."

dimethylisopropanolamine $(CH_3)_2NCH_2CH(OH)CH_3$.
Properties: Colorless liquid. Sp. gr. 0.8645 (25/20°C); 7.4 lb/gal (20°C); b.p. 125.8°C; completely soluble in water; viscosity 1.51 cp (20°C); vapor pressure 9 mm (20°C); f.p. sets to glass below −85°C; refractive index 1.4189 (n 20/D); flash point 95°F (O.C.); solubility of water in compound, complete at 20°C. Low toxicity.
Hazard: Flammable, moderate fire risk.
Uses: Synthesis of methadone; other chemical syntheses. Combines the properties of tertiary amine and secondary alcohol.

dimethyl itaconate $CH_2{:}C(COOCH_3)CH_2(COOCH_3)$.
Properties: White crystals with slight odor; m.p. 36°C; b.p. 91.5°C (10 mm); sp. gr. 1.27 (24°C); refractive index (n 20/D) 1.441. Slightly soluble in water. Combustible.
Uses: Polymers and copolymers; plasticizers; intermediate.

dimethylketol. See acetylmethylcarbinol.

dimethylketone. See acetone.

dimethyl maleate $CH_3OOCCH{:}CHCOOCH_3$.
Properties: Colorless liquid; sp. gr. 1.153; 9.62 lb/gal; b.p. 200.4°C; flash point 235°F. Combustible. Low toxicity.

dimethyl malonate $CH_2(COOCH_3)_2$.
Properties: Colorless liquid; b.p. 180–181°C;. f.p. −62°C; refractive index 1.4140; flash point 194°F. Very slightly soluble in water; soluble in alcohol and ether. Combustible.
Containers: 55-gal steel drums.
Use: Chemical intermediate.

dimethylmethane. See propane.

2,6-dimethylmorpholine
$OCH(CH_3(CH_2NHCH_2CH(CH_3)$.
Properties: Liquid; sp. gr. 0.99346; b.p. 146.6°C; f.p. −85°C; soluble in water; wt/gal 7.8 lb; flash point 112°F. Combustible.
Hazard: Moderate fire risk. Moderately toxic.
Uses: Corrosion inhibitors; stabilizers for chlorinated solvents; rubless floor polishes, rubber accelerators, germicides, and textile finishing agents.

2-(2,6-dimethyl-4-morpholinothio)benzothiazole $C_{13}H_{16}N_2OS_2$.
Properties: Cream to light-yellow powder; m.p. 88°C; sp. gr. 1.26 (25°C).
Use: Delayed-action vulcanization accelerator.

dimethyl-alpha-naphthylamine $C_{10}H_7N(CH_3)_2$.
Properties: Colorless liquid; soluble in alcohol and ether; insoluble in water. Sp. gr. 1.045; b.p. 275°C.
Derivation: Action of methylsulfate on alpha-naphthylamine.
Grades: C.P.; analytical.
Containers: Glass bottles.
Use: Determination of nitrites.

dimethyl-beta-**naphthylamine** $C_{10}H_7N(CH_3)_2$.
Properties: Crystalline solid; soluble in alcohol and ether; insoluble in water; sp. gr. 1.039 (70/70°C); m.p. 46°C; b.p. 305°C.
Derivation: Interaction of dimethylamine and beta-naphthol.

dimethylnitrobenzene. See nitroxylene.

O,O-**dimethyl** O-para-**nitrophenyl phosphorothioate.** See methyl parathion.

dimethyl nitrosamine. See N-nitrosodimethylamine.

N,N-**dimethyl**-para-**nitrosoaniline** (para-nitrosodimethylaniline) $(CH_3)_2NC_6H_4NO$.
Properties: Green leaflets. Soluble in alcohol and ether; insoluble in water; m.p. 93°C. Flammable.
Derivation: Nitrous acid and N-dimethylaniline.
Uses: Production of methylene blue; vulcanization accelerator.
Shipping reguations: (Rail) Flammable solid, n.o.s., Yellow label. (Air) Flammable solid label.

3,6-dimethyl-3,6-octanediol
$C_2H_5(CH_3)COH(CH_2)_2COH(CH_3)C_2H_5$.
Properties: White, waxy solid; m.p. 44°C; b.p. 241–242°C; sp. gr. (20/20°C) 0.919. Soluble in water, acetone, alcohol, benzene, carbon tetrachloride.
Uses: Non-foaming surface-active agent; chemical intermediate.

3,7-dimethyl-1,7-octanediol. See hydroxycitronellol.

dimethyloctanoc acid. See isodecanoic acid.

3,6-dimethyl-3-octanol
$(C_2H_5)CHCH_3(CH_2)_2COHCH_3(C_2H_5)$.
Properties: Colorless liquid; sweet odor; sp. gr. 0.8366 (20/20°C); refractive index 1.4370 (n 20/D); b.p. 202–203°C; freezing point −67.5°C. Combustible.
Uses: Perfumery (floral odors); flavoring.

3,7-dimethyl-1-octanol (tetrahydrogeraniol).
$(CH_3)_2CH(CH_2)_3CH(CH_3)CH_2CH_2OH$.
Properties: Colorless liquid, sweet odor. Soluble in mineral oil and in propylene glycol. Insoluble in glycerin. Refractive index 1.4350–1.4450 (20°C); sp. gr. 0.826–0.842 (25°C). Combustible.
Grade: F.C.C.
Uses: Flavoring agent; perfumery.

3,7-dimethyl-3-octanol. See tetrahydrolinalool.

2,6-dimethyl-1,5,7-octatriene. See ocimene.

2,6-dimethyl-2,4,6-octatriene. See alloocimene.

3,7-dimethyl-6-octenal. See citronellal.

3,7-dimethyl-6(or 7)**-octen-1-ol.** See citronellol.

dimethyloctynediol
$C_2H_5(CH_3)COHC \cdot CCOH(CH_3)C_2H_5$ (3,6-dimethyl-4-octyne-3,6-diol).
Properties: White crystals; m.p. 55–56°C; b.p. 222°C; sp. gr. (solid, 20°C) 0.923, (liquid, 60°C) 0.908. Moderately soluble in water; slightly soluble in kerosine; very soluble in acetone, alcohol, benzene, and carbon tetrachloride. Combustible.
Containers: Fiber drums.
Hazard: May be toxic.
Uses: Surface-active agent; intermediate.

dimethylol ethylene urea
$OCN(CH_2OH)CH_2CH_2N(CH_2OH)$. A cyclic urea, used in wrinkle-resistant textile finishes.

dimethylolethyltriazone
$OCN(CH_2OH)CH_2N(C_2H_5)CH_2N(CH_2OH)$. Used in wrinkle-resistant textile finishes.

dimethylolpropionic acid (DMPA: 2,2-bis(hydroxymethyl)-propionic acid) $CH_3C(CH_2OH)_2COOH$.
Properties: Off-white crystalline solid; m.p. 192–194°C. Soluble in water and methanol; slightly soluble in acetone and insoluble in benzene.
Uses: Water-soluble alkyd resins; textile finishing; cosmetics; plasticizers.

dimethylolurea (DMU; 1,3-bishydroxymethylurea) $CO(NHCH_2OH)_2$.
Properties: Colorless crystals; m.p. 126°C (tech. 85–90°C); sp. gr. (20°C) 1.34; slight formaldehyde odor; soluble in water and methanol; insoluble in ether; capable of polymerization.
Derivation: Combination of urea and formaldehyde in the presence of salts or alkaline catalysts.
Containers: 50-, 100-lb multiwall paper bags.
Hazard: Irritant to skin.
Uses: First stage of ureaformaldehyde resins; impregnating wood to increase hardness and fire resistance and to form self-binding laminations for plywood manufacture; permanent-press fabrics.

2,3-dimethylpentaldehyde
$CH_3CH_2CH(CH_3)CH(CH_3)CHO$. A branched-chain heptanal.
Properties: Liquid; sp. gr. 0.8293; b.p. 140.5°C (760 mm); f.p. −110°C; 6.91 lb/gal (20°C); flash point 94°F. Slightly soluble in water.
Hazard: Flammable, moderate fire risk.
Use: Intermediate.

2,4-dimethylpentane $(CH_3)_2CHCH_2CH(CH_3)_2$.
Properties: Colorless liquid; sp. gr. 0.6684 (25°C); b.p. 80.5°C; refractive index 1.382 (n 20/D); flash point 10°F; f.p. −119°C; soluble in alcohol; insoluble in water.
Grades: 95%; 99%; research, 99.7 mole %.
Hazard: Flammable, dangerous fire hazard.
Use: Organic synthesis.
Shipping regulations: Flammable liquid, n.o.s. (Rail) Red label (Air) Flammable Liquid label.
Note: Other isomers (2,2-; 2,3-; 3,3-) of closely similar properties are available.

2,4-dimethylpentanol-3. See diisopropyl carbinol.

2,4-dimethylpentanone-3. See diisopropyl ketone.

alpha, alpha-**dimethylphenethyl acetate** (dimethylbenzylcarbinyl acetate) $C_6H_5CH_2C(CH_3)_2OOCCH_3$.
Properties: Colorless liquid; floral, fruity odor. Solidifies at room temperature. Soluble in mineral oil; insoluble in glycerin. Refractive index 1.4910–1.4950 (20°C) in supercooled liquid form; sp. gr. 0.995–1.002 in supercooled liquid form; m.p. 29–30°C. Combustible. Nontoxic.
Grade: F.C.C.
Use: Flavoring agent.

dimethylphenol. See xylenol.

dimethyl-para-**phenylenediamine.** See para-aminodimethylaniline.

N,beta-**dimethylphenylethylamine.** See phenylpropylmethylamine.

2,3-dimethyl-1-phenyl-3-pyrazolin-5-one. See antipyrine.

dimethyl phosphite $(CH_3O)_2P(O)H$.
Properties: Mobile, colorless liquid; mild odor; sp. gr. 1.200 (20/4°C); refractive index 1.400 (25/D); b.p. 72–73°C (25 mm); flash point 205°F. Soluble in water, and miscible with most common organic solvents. Combustible.
Containers: 55-gal drums.
Uses: Lubricant additives; intermediate; adhesive.

O,O-**dimethyl phosphorochloridothioate** (methyl PCT) $(CH_3O)_2P(S)Cl$.
Properties: Colorless to light amber liquid; b.p. 66–67°C (16 mm); sp. gr. 1.320 (25°C); refractive index 1.4795 (n 25/D). Soluble in alcohol, benzene, acetone, carbon tetrachloride, chloroform, ethyl acetate; slightly soluble in hexane; insoluble in water.
Grades: 96–100% purity.
Hazard: Toxic; strong irritant to eyes, skin, and mucous membranes. Cholinesterase inhibitor. Use may be restricted.
Uses: Intermediate for insecticides, pesticides, fungicides; oil and gasoline additives; plasticizers; corrosion inhibitors; flame retardants; flotation agents.
Shipping regulations: Organic phosphate, liquid, n.o.s., (Rail, Air) Poison label. Not accepted on passenger planes.

dimethyl phthalate $C_6H_4(COOCH_3)_2$.
Properties: Colorless, odorless liquid; refractive index 1.5138 (25°C); heat of combustion 5769 cal/g; sp. gr. 1.189 (25/25°C); b.p. 282°C; flash point 300°F; wt/gal 9.93 lb (68°F); vapor pressure <0.1 mm (20°C). Miscible with common organic solvents; insoluble in water. Combustible; autoignition temp. 1032°F.
Grade: Technical.
Containers: 55-gal drums; tank cars; tank trucks.
Hazard: Toxic by ingestion and inhalation; irritant. Tolerance, 5 mg per cubic meter of air.
Uses: Plasticizer for nitrocellulose and cellulose acetate, resins, rubber and in solid rocket propellants; lacquers; plastics; rubber; coating agents; safety glass; molding powders; insect repellent; perfumes.

N,N′-**dimethylpiperazine** (1,4-dimethylpiperazine) $(CH_3)_2C_4H_8N_2$.
Properties: Colorless, mobile liquid. Sp. gr. (20/4°C) 0.8565; b.p., 131°C; f.p. −1°C; flash point (TOC) 176°F. Combustible.

Uses: **Curing agent for polyurethane foams;** intermediate for cationic surface-active agents.
Note: Two other isomers are available: *cis*-2,5- and *trans*-2,5-, neither is flammable, but both are combustible. Toxicity presumed low.

2,6-dimethylpiperidine (2,6-lupetidine) $(CH_3)_2C_5H_8NH$.
Properties: Liquid. B.p. 127.9°C; sp. gr. (20/20°C) 0.8199; refractive index 1.4383 (n 20/D); soluble in water at 20°C in all proportions. Combustible.
Use: Intermediate.

dimethylpolysiloxane. A liquid defoaming agent; refractive index 1.40; viscosity 300 centistokes. Use in foods limited to 10 ppm (0 in milk).

dimethyl-POPOP. See 1,4-bis[2-(4-methyl-5-phenyloxazolyl)]benzene.

2,2-dimethylpropane. See neopentane.

2,2-dimethyl-1,3-propanediol See neopentyl glycol.

dimethylpyridine. See lutidine.

2,7-dimethylquinoline $(CH_3)_2C_9H_5N$.
Properties: Liquid. F.p. −40°C (approx.); distillation range 140–150°C (20 mm); soluble in benzene and diethyl ether. Combustible.
Uses: Organic synthesis; dye intermediate.

dimethyl resorcinol. See resorcinol dimethyl ether.

dimethyl sebacate $[(CH_2)_4COOCH_3]_2$.
Properties: Liquid, water-white; sp. gr. 0.9896 (25/20°C); m.p. 24.5°C; flash point 293°F; b.p. approx. 294°C; refractive index 1.4376 (20°C). Combustible.
Grade: Technical.
Containers: Cans; tank cars.
Uses: Solvent or plasticizer for nitrocellulose, vinyl resins; intermediate.

dimethyl silicone. General term for a family of silicones of composition $[(CH_3)_2SiO]_x$, the more volatile materials formed on hydrolysis of dimethyldichlorosilane. Colorless oils with b.p. ranging from 134°C (760 mm) (for x = 3) to 188°C (20 mm) (for x = 9).

dimethyl sulfate (methyl sulfate) $(CH_3)_2SO_4$.
Properties: Colorless liquid. Soluble in alcohol and ether; soluble in water. Sp. gr. 1.3516; f.p. −26.8°C; b.p. 188°C (dec). Flash point 182°F (closed cup); combustible.
Derivation: By adding fuming sulfuric acid to methyl alcohol and distilling in vacuo.
Grade: Technical.
Containers: Returnable drums; tank cars.
Hazard: Strong irritant; absorbed by skin. May be carcinogenic. Induces tumors in animals. Protective clothing required. Safety Data Sheet available from Mfg. Chemists Assn., Washington, D.C.
Use: Methylating agent for amines and phenols.
Shipping regulations: (Rail) White label. (Air) Corrosive label. Not acceptable on passenger planes.

dimethyl sulfide (methyl sulfide) $(CH_3)_2S$.
Properties: Colorless volatile liquid; disagreeable odor. Soluble in alcohol and ether; insoluble in water. Sp. gr. 0.845 (20°C); f.p. −83°C; b.p. 37.5°C. Evolves sulfur dioxide when heated. Autoignition temp. 403°F. Flash point 0°F.
Derivation: (a) From kraft pulping black liquor, by heating with inorganic sulfur compounds; (b) by interaction of a solution of potassium sulfide and methyl chloride in methanol.
Containers: Drums.
Hazard: Flammable. Dangerous fire risk; moderate explosion risk. Probably highly toxic. Flammable limits in air 2.2 to 19.7%.
Uses: Gas odorant; solvent for many inorganic substances; catalyst impregnator.
Shipping regulations: (Rail) Red label. (Air) Flammable Liquid label. Legal label name, methyl sulfide.

dimethylsulfonyloxybutane. See busulfan.

dimethyl sulfoxide (methyl sulfoxide; DMSO) $(CH_3)_2SO$.
Properties: Colorless hygroscopic liquid; b.p. 189°C; m.p. 18.5°C; sp. gr. 1.01 (20/20°C); specific heat 0.7; flash point 203°F (open cup); dielectric constant 48.9 (20°C). Nearly odorless; slightly bitter taste; miscible with water. Readily penetrates skin and other tissues. Combustible. Extremely powerful aprotic solvent.
Derivation: Oxidation of dimethyl sulfide with nitrogen tetroxide under anhydrous conditions; sulfide waste liquors.
Containers: Drums; tank cars.
Hazard: Harmful if ingested; must be applied to surface of skin only. Must comply with FDA regulations.
Uses: Solvent for polymerization and cyanide reactions; analytical reagent; spinning polyacrylonitrile and other synthetic fibers; industrial cleaners, pesticides, paint stripping; hydraulic fluids; preservation of cells at low temperatures; diffusion of drugs, etc., into blood stream by topical application; medicine; plant pathology and nutrition; pharmaceutical products.

dimethyl terephthalate (DMT) $C_6H_4(COOCH_3)_2$.
Properties: Colorless crystals; m.p. 140°C; sublimes above 300°C; insoluble in water; soluble in ether and hot alcohol.
Derivation: Oxidation of para-xylene or mixed xylene isomers, followed by esterification.
Uses: Polyester resins for film and fiber production, especially polyethylene terephthalate; intermediate.

dimethyl 2,3,5,6-tetrachloroterephthalate. $C_6Cl_4(COOCH_3)_2$.
Properties: Crystals; m.p. 156°C. Insoluble in water; slightly soluble in acetone and benzene.
Use: Herbicide.

3,5-dimethyltetrahydro-1,3,5(2H)-thiadiazine-2-thione $C_5H_{10}N_2S_2$.
Properties: Crystals; m.p. 100°C; sp. gr. 1.30 (68°F). Slightly soluble in water and alcohol; soluble in acetone.
Hazard: Toxic by ingestion and inhalation. Tolerance, 15 mg per cubic meter of air.
Uses: Herbicide; nematocide; preservative for adhesives and proteinaceous additives.
See also "Crag."

dimethyltin dichloride $(CH_3)_2SnCl_2$.
Properties: White crystals; m.p. 106–108°C; soluble in water, alcohol, hydrocarbons.
Hazard: Highly toxic. Tolerance, 0.1 mg per cubic meter of air.
Uses: Electroluminescence; PVC stabilizer; catalyst.

dimethyltin oxide $(CH_3)_2SnO$.
Properties: White powder, 98.5% min. purity.
Hazard: Highly toxic. Tolerance, 0.1 mg per cubic meter of air.
Uses: Intermediate; PVC stabilizer.

N,N′**-dimethylurea** (DMU; sym-dimethylurea; 1,3-dimethylurea) $(CH_3NH)_2CO$.
Properties: Colorless prisms; sp. gr. 1.14; m.p. 106°C;

b.p. 270°C; soluble in water and alcohol; insoluble in ether.
Use: Intermediate in synthesis of drugs.

1,3-dimethylxanthine. See theophylline.

3,7-dimethylxanthine. See theobromine.

dimetilan. Generic name for 1-dimethylcarbamoyl-5-methyl-3-pyrazolyl dimethylcarbamate.
Properties: Solid; m.p. 68–71°C. Soluble in water and most organic solvents.
Hazard: Toxic by ingestion and inhalation.
Use: Insecticide.

"Dimocillin-RT."[412] Trademark for buffered sodium methicillin.

dimolybdenum trioxide. See molybdenum sesquioxide.

"Dī-MōN."[93] Trademark for modified diammonium phosphate. Used as a low cost source of phosphorus and nitrogen as nutrients in biological waste treatment.

dimyristyl amine (ditetradecylamine) $(C_{14}H_{29})_2NH$. Solid; m.p. 52°C; sp. gr. 0.89. Almost insoluble in water.
Use: Intermediate.

dimyristyl ether (ditetradecyl ether) $(C_{14}H_{29})_2O$.
Properties: Liquid. M.p. 38–40°C; b.p. 238–248°C (4 mm); sp. gr. 0.8127 (45/4°C). Combustible. Low toxicity.
Grade: 95% (min.) purity.
Uses: Electrical insulators; water repellents; lubricants in plastic molding; antistatic substances; chemical intermediates.

dimyristyl sulfide (ditetradecyl sulfide; dimyristyl thioether) $(C_{14}H_{29})_2S$.
Properties: Solid; m.p. 49–50°C; b.p. decomposes; sp. gr. 0.8258 (50/4°C).
Grade: 95% (min.) purity.
Use: Organic synthesis (formation of sulfonium compounds).

N,N′-**di**-beta-**2-naphthyl**-meta-**phenylenediamine** $C_6H_4(NHC_{10}H_7)_2$.
Properties: Colorless needles; m.p. 191°C; sparingly soluble in alcohol; insoluble in water and ether.
Derivation: By heating meta-phenylenediamine with beta-naphthol and extraction with alcohol.
Use: Organic synthesis.

N,N′-**di**-beta-**naphthyl**-para-**phenylenediamine** (sym-di-beta-naphthyl-para-phenylenediamine; DNPD) $C_6H_4(NHC_{10}H_7)_2$.
Properties: Gray powder; sp. gr. 1.25; set point 225°C (min); purity 98% (min); insoluble in water, slightly soluble in acetone and chlorobenzene.
Uses: Antioxidant; stabilizer; polymerization inhibitor; intermediate in organic synthesis.

dinitraniline orange (Permanent Orange). A pigment made from dinitroaniline and beta-naphthol. It is a reddish shade orange that has excellent light-fastness.
Hazard: Probably toxic by ingestion.

dinitroaminophenol. See picramic acid.

2,4-dinitroaniline (2,4-dinitraniline) $C_6H_3NH_2(NO_2)_2$.
Properties: Yellow crystals; slightly soluble in alcohol; insoluble in water. S. gr. 1.615; m.p. 188°C. Flash point 435°F. Combustible.
Derivation: Nitration of para-nitroaniline with hot mixed acid.
Grades: Technical; pure.
Hazard: Toxic by ingestion and inhalation; strong irritant.
Use: Intermediate for azo pigments; toner pigment in printing inks; corrosion inhibitor.
Shipping regulations: (Liquid) (Air) Poison label.

2,4-dinitroanisole (2,4-dinitrophenyl methyl ether) $CH_3OC_6H_3(NO)_2$.
Properties: Colorless to yellow monoclinic needles from water or alcohol; m.p. 88°C; sp. gr. 1.341 (20/4°C), sublimes; slightly soluble in hot water; soluble in alcohol and ether.
Hazard: Probably toxic by ingestion.
Use: Ovicide, effective against moths, furniture and carpet beetles, cockroaches, and body lice.

dinitrobenzene $C_6H_4(NO_2)_2$. Meta, ortho and para isomers.
Properties: Yellow crystals; soluble in benzene; slightly soluble in water. Sp. gr.: meta 1.546, ortho 1.565, para 1.6; m.p. meta 89.9°C, ortho 117.9°C, para 172–173°C; b.p. meta 302.8°C, ortho 319°C, para 299°C.
Derivation: Nitration of nitrobenzene with hot mixed acid.
Method of purification: Crystallization.
Grades: Technical.
Hazard: Toxic by inhalation and ingestion; strong irritant. Tolerance, 1 mg per cubic meter of air. Absorbed by skin.
Uses: Organic synthesis; dyes; camphor substitute in celluloid production.
Shipping regulations: (Rail, Air) Poison label.

5,7-dinitro-1,2,3-benzoxadiazole. See diazodinitrophenol.

3,5-dinitrobenzoyl chloride $(NO_2)_2C_6H_3COCl$.
Properties: Yellow crystals; m.p. 66–68°C; b.p. 196°C (12 mm); decomposed by water and alcohol.
Hazard: May be toxic and irritant.
Use: Reagent.

2-(2,4-dinitrobenzyl)pyridine $C_5H_4NCH_2C_6H_3(NO_2)_2$. Behaves photochromically. Used to make plastics sensitive to light. See photochromism.

2,4-dinitro-6-sec-butylphenol. **(2-sec-butyl-4,6-dinitrophenol; dinoseb; DNBP).** $CH_3(C_2H_5)CHC_6H_2(NO_2)_2OH$.
Properties: Reddish brown liquid, slightly soluble in water, soluble in alcohol and other organic solvents. Forms salts with metals and organic bases.
Hazard: Highly toxic; absorbed by skin. Possible fire risk. Strong irritant.
Uses: Insecticide and ovicide, but must be used in the dormant growth season or as a salt form to reduce toxicity. Also as herbicide for preemergence treatment; increases yield of corn 5 to 10%.
Shipping regulations: (Air) Poison label. Legal label name dinoseb.

dinitrochlorobenzol. Label name for dinitrochlorobenzene.

dinitrochlorobenzene. See 1-chloro-2,4-dinitrobenzene.

4,6-dinitro-ortho-**cresol (4,6-dinitro-2-methylphenol)** $CH_3C_6H_2(NO_2)_2OH$.
Properties: Yellow solid; m.p. 85.8°C; slightly soluble in water.
Hazard: Highly toxic; absorbed by skin. Tolerance, 0.2 mg per cubic meter of air. Use may be restricted.

Use: Dormant ovicidal spray for fruit trees (highly phytotoxic and cannot be used successfully on actively growing plants).
Shipping regulations: (Air) Poison label.

2,6-dinitro-para-**cresol** (DNPC) $(NO_2)_2CH_3C_6H_2OH$. Light yellow crystalline solid.
Grade: Presscake (36–43% active 2,6-DNCP)
Hazard: Probably toxic.
Uses: Parent compound for intermediates, dyes and pharmaceuticals.

dinitrocyclohexylphenol. (2-cyclohexyl-4,6-dinitrophenol; dinitro-ortho-cyclohexylphenol; DNOCHP) $C_6H_{11}C_6H_2(NO_2)_2OH$.
Properties: Crystalline solid.
Hazard: Probably toxic.
Use: Control of mites on citrus fruits.

3,5-dinitro-2,6-dimethyl-4-tert-**butylacetophenone.** See musk ketone.

2,4-dinitrofluorobenzene (1-fluoro-2,4-dinitrobenzene) $C_6H_3F(NO_2)_2$.
Properties: Crystals; soluble in ether, benzene, propylene glycol; m.p. 26°C; b.p. 137°C.
Use: Reagent in protein analysis.

dinitrogen tetroxide. See nitrogen dioxide.

dinitrogen trioxide. See nitrogen trioxide.

dinitroglycol. See diethylene glycol dinitrate.

2,4-dinitro-4-hydroxydiphenylamine $(NO_2)_2C_6H_3NHC_6H_4OH$.
Properties: Yellow solid; m.p. 190°C; insoluble in water.
Derivation: Condensation of 2,4-dinitro-1-chlorobenzene and para-aminophenol.

2,6-dinitro-3-methoxy-4-tert-**butyltoluene.** See musk ambrette.

dinitro(1-methylheptyl)phenyl crotonate. See dinocap.

4,6-dinitro-2-methylphenol. See 4,6-dinitro-ortho-cresol.

dinitronaphthalene $C_{10}H_6(NO_2)_2$. Isomers: (a) 1,5-; (b) 1,8-.
Properties: (a) Yellowish-white needles, (b) yellowish-white, thick, crystalline tablets. M.p. (a) 217°C; (b) 172°C. (a) Sparingly soluble in pyridine; (b) soluble in pyridine. B.p. (a) sublimes; (b) decomposes.
Derivation: By dissolving alpha-nitronaphthalene in sulfuric acid and adding nitric acid.
Method of purification: Crystallization.
Hazard: Probably toxic. Moderate fire and explosion risk.
Uses: Dyes, especially sulfur colors; intermediates.

2,4-dinitro-1-naphthol-7-sulfonic acid (flavianic acid) $C_{10}H_6O_8N_2S$.
Properties: Yellow needles; m.p. 151°C; very soluble in water.
Hazard: Probably toxic.
Uses: Intermediate; precipitant for organic bases; reagent for amino acids.

dinitrophenol $C_6H_3OH(NO_2)_2$. Commercial product is usually mixture of 2,3-, 2,4-, and 2,6-isomers.
Properties: Yellow crystals. (2,3) sp. gr. 1.681; m.p. 144°C. (2,4) sp. gr. 1.683; m.p. 114–115°C. (2,6) m.p. 63°C. Soluble in alcohol and ether, also benzene and chloroform; slightly soluble in water.
Derivation: (a) By heating phenol with dilute sulfuric acid, cooling the product, and then nitrating, keeping the temperature below 50°C. (b) By nitration with mixed acid with careful temperature control.
Method of purification: Crystallization.
Hazard: Highly toxic; absorbed by skin. Dust inhalation may be fatal. Severe explosion hazard when dry.
Uses: Dyes, especially sulfur colors; picric acid; picramic acid; preservation of lumber; manufacture of the photographic developer diaminophenol hydrochloride; explosives manufacture.
Shipping regulations: (Solutions) (Rail, Air) Poison label. (Dry, or wet with less than 15% water) (Air) Not acceptable; (Wet with not less than 15% water) (Air) Flammable Solid label.

2,4-dinitrophenylhydrazine $(NO_2)_2C_6H_3NHNH_2$.
Properties: Red, crystalline powder; m.p. about 200°C. Slightly soluble in water and alcohol; soluble in moderately dilute inorganic acids; readily soluble in diglyme.
Hazard: Severe explosion and fire risk.
Uses: Explosive; reagent.
Shipping regulations: (Air) (dry or wet with less than 10% water), Not acceptable; (wet with not less than 10% water), Flammable Solid label. (Rail) Consult authorities.

2,4-dinitrophenyl methyl ether. See 2,4-dinitroanisole.

3,5-dinitrosalicylic acid $C_6H_2(OH)(NO_2)_2COOH$.
Properties: Yellow crystals; slightly soluble in water; soluble in alcohol and benzene; m.p. 174°C.
Derivation: Nitration of salicylic acid.
Use: Determination of glucose.

dinitrosopentamethylenetetramine (DNPT) $(NO)_2C_5H_{10}N_4$. A bicyclic compound.
Properties: Light cream-colored powder; decomposes in air at 190–200°C. Soluble in dimethyl formamide; somewhat soluble in pyridine, methyl ethyl ketone, and acetonitrile. Combustible.
Hazard: Moderate fire risk.
Use: Blowing agent for rubber and plastics.

2,4-dinitrosoresorcinol $C_6H_2(OH)_2(NO)_2 \cdot H_2O$.
Properties: Light-brown powder; m.p. 162–163°C. Decomposes, sometimes violently. Soluble in water and most organic solvents.
Grade: Technical (13.7%N).
Hazard: Severe explosion risk when shocked or heated. Toxic and irritant.
Uses: Chelation of heavy metals; cross-linking agent; blasting caps and explosive primers.
Shipping regulations: (Rail) Consult regulations. (Air) Explosives, n.o.s., Not acceptable.

3,5-dinitro-ortho-**toluamide.** See zoalene.

dinitrotoluene (DNT) $C_6H_3CH_3(NO_2)_2$ (a) 2,4-; (b) 3,4-; (c) 3,5-.
Properties: Yellow crystals; soluble in alcohol and ether; very slightly soluble in water. Sp. gr. (a) 1.3208, (b) 1.32, (c) 1.277; m.p. (a) 70.5°C, (b) 61°C, (c) 92.3°C. A commercial grade consisting of a mixture of the three isomers in an oily liquid. Combustible.
Derivation: Nitration of nitrotoluene with mixed acid.
Method of purification: Crystallization.
Containers: Drums; liquid, in heated tanks.
Hazard: Highly toxic; absorbed by skin. Tolerance, 1.5 mg per cubic meter of air. Moderate fire and explosion hazard. Safety Data Sheet available from Mfg. Chemists Assn., Washington, D.C.
Uses: Organic synthesis; toluidines; dyes; explosives.
Shipping regulations: (Liquid) (Air) Poison label.

dinocap. Generic name for 2-(2-methylheptyl)-4,6-dinitrophenyl crotonate.
$CH_3CH{:}CHCOOC_6H_2(NO_2)_2CH(CH_3)C_6H_{13}$.
Properties: Brown liquid; b.p. 138–140°C (0.05 mm). Insoluble in water; soluble in most organic solvents.

Hazard: Probably toxic by ingestion and inhalation.
Uses: Acaricide; fungicide.

dinonyl adipate Ester of nonyl alcohol.
Properties: Colorless liquid; b.p. 201–210°C (1 mm); sp. gr. 0.926 (25°C); refractive index (n 20/D) 1.4523; viscosity 14.9 centistokes (100°F). Combustible. Low toxicity.
Use: Plasticizer where special low-temperature properties are desired.

dinonyl carbonate $(C_9H_{19})_2CO_3$. Ester of nonyl alcohol; colorless liquid; b.p. 135–140°C (0.3 mm); sp. gr. 0.894 (25°C); refractive index (n 20/D) 1.4427. Combustible. Low toxicity.

dinonyl ether $C_9H_{19}OC_9H_{19}$
Properties: Colorless liquid; b.p. 148–153°C (5 mm); sp. gr. 0.817 (25°C); refractive index (n 20/D) 1.4405. Can be made from nonyl alcohol plus nonyl halide by the Williamson synthesis. Combustible. Low toxicity.

dinonyl maleate $C_9H_{19}OOCCH{:}CHCOOC_9H_{19}$. Ester of nonyl alcohol; colorless liquid; b.p. 157–167°C (0.1 mm); sp. gr. 0.941 (25°C); refractive index (n 20/D) 1.4586; viscosity 6900 centistokes (−40°F), 17.47 centistokes (100°F), 3.50 centistokes (210°F). Combustible. Low toxicity.

dinonyl phenol $(C_9H_{19})_2C_6H_3OH$.
Properties: Colorless liquid; insoluble in water; soluble in common organic solvents. Combustible.
Hazard: May be toxic. See phenol.
Use: Solvent.

dinonyl phthalate (DNP) $C_6H_4(COOC_9H_{19})$. Ester of nonyl alcohol.
Properties: Colorless liquid. B.p. 205–220°C (1 mm); sp. gr. 0.979 (25°C); refractive index (n 20/D) 1.4871; viscosity 55.3 centistokes (100°F); flash point 420°F. Combustible. Low toxicity.
Uses: General-purpose low-volatile plasticizer for vinyl resins; pure grade as stationary liquid phase in chromatography.

dinoseb. Legal label name for 2,4-dinitro-6-sec-butylphenol (q.v.).

DIOA. Abbreviation for diisooctyl adipate.

dioctadecylamine. See distearylamine.

2,6-dioctadecyl-para-cresol. See 2,6-distearyl-para-cresol.

diotadecyl ether. See distearyl ether.

dioctadecyl sulfide. See distearyl sulfide.

3,3′-dioctadecyl thiodipropionate. See distearyl thiodipropionate.

dioctyl adipate. Now more exactly named di(2-ethylhexyl) adipate (q.v.).

dioctylamine. See di-2-ethylhexylamine.

dioctylaminoethanol. See di(2-ethylhexyl)ethanolamine.

dioctyl azelate. See di(2-ethylhexyl) azelate.

dioctyl chlorophosphate (dioctyl phosphorochloridate) $(C_8H_{17}O)_2P(O)Cl$.
Properties: Water-white liquid; sp. gr. 0.991 (25°C); refractive index 1.445 (n 25/D); decomposes on distillation; soluble in common inert organic solvents; insoluble in water. Combustible.
Hazard: Probably toxic.
Use: Intermediate; insecticide.

di(n-octyl, n-decyl)adipate (DNODA).
Properties: Clear, oily liquid; sp. gr. 0.912–0.920 (25/25°C); refractive index 1.443–1.447 (25°C). Combustible. Low toxicity.
Containers: Drums; tank cars.
Use: Low-temperature plasticizer.

di(n-octyl, n-decyl) phthalate (DNODP).
Properties: Clear, oily liquid; odor slight; acidity (as phthalic acid) 0.01% max; sp. gr. 0.968–0.977 (25/25°C); crystallizing point −30°C; b.p. 232–267°C (4 mm); flash point 426°F. Combustible. Low toxicity.
Containers: 5-, 55-gal drums; tank cars.
Uses: Plasticizer for polyvinyl chloride and other vinyls.

di-n-octyl diphenylamine $C_8H_{17}C_6H_4NHC_6H_4C_8H_{17}$.
Properties: Light tan powder; sp. gr. 0.99; m.p. 80–90°C. Soluble in benzene, gasoline, acetone, and ethylene dichloride; insoluble in water.
Use: Antioxidant for petroleum-based and synthetic lubricants and plastics.

dioctyl ether $(C_8H_{17})_2O$.
Properties: Liquid; f.p. −7°C; b.p. 291.7°C; sp. gr. 0.805 (17/4°C); refractive index 1.4329 (n 24/D). Combustible. Low toxicity.
Grades: 95% (min) purity.
Uses: Electrical insulator; water repellent; mold lubricant; antistatic agent; intermediate.

dioctyl fumarate. See di(2-ethylhexyl) fumarate.

dioctyl hexahydrophthalate. See di(2-ethylhexyl) hexahydrophthalate.

N,N′-**di-n-octyl**-para-**phenylenediamine** $C_6H_4(NHC_8H_{17})_2$.
Properties: Colorless liquid; b.p. about 390°C; sp. gr. 0.912 (15°C); pour point −4°C; flash point (Pensky-Martin) 395°F; refractive index (n 20/D) 1.5129. Completely miscible in methanol, pentane and benzene; vapor pressure (absolute) 0.33 mm (150°C). Combustible. Low toxicity.
Containers: Steel drums, tank cars, tank trucks.
Uses: Antioxidant, antiozonant for gasoline, mercaptans, synthetic rubber.

dioctyl phosphite (dioctyl phosphonate) $(C_8H_{17}O)_2P(O)H$.
Properties: Water-white liquid; b.p. 150–155°C (2–3 mm); sp. gr. 0.929 (25°C); refractive index 1.4418 (n 25/D); soluble in common organic solvents. Combustible.
Uses: Solvent; antioxidant; intermediate.

dioctyl phosphoric acid. See (di(2-ethylhexyl) phosphoric acid.

dioctyl phosphorochloridate. See dioctyl chlorophosphate.

dioctyl phthalate. See di(2-ethylhexyl) phthalate.

di(2-octyl) phthalate. See dicapryl phthalate.

dioctyl sebacate. See di(2-ethylhexyl) sebacate.

dioctyl sodium sulfosuccinate (di(2-ethylhexyl) sodium sulfosuccinate; sodium dioctyl sulfosuccinate) $C_8H_{17}OOCCH_2CH(SO_3Na)COOC_8H_{17}$. An anionic surface-active agent.
Properties: White, wax-like solid with characteristic odor. Slowly soluble in water; freely soluble in alco-

hol and glycerin; very soluble in petroleum ether. Saponification value 240–253; stable in acid and neutral solutions; hydrolyzes in alkaline solutions. Low toxicity.
Derivation: By esterification of maleic anhydride with 2-ethylhexyl alcohol followed by addition of sodium bisulfite.
Hazard: Use in food products restricted.
Grade: N.F.; F.C.C.
Uses: Food additive (processing aid in sugar industry, stabilizer for hydrophilic colloids); wetting agent; dispersant; emulsifier.

dioctyl succinate. See di(2-ethyl hexyl) succinate.

dioctyl sulfide (dioctyl thioether) $(C_8H_{17})_2S$.
Properties: Liquid; m.p. 0.5°C; b.p. 180°C (10 mm); sp. gr. 0.8419 (17/17°C); refractive index 1.4606 (n 20/D). Combustible.
Grades: 95% (min) purity.
Uses: Organic synthesis (formation of sulfonium compounds).

dioctyl thioether. See dioctyl sulfide.

dioctyl thiopropionate. See 3,3′-(2-ethylhexyl) thiodipropionate.

di(n-octyl)tin S,S′ bis-isooctylmercaptoacetate. A heat stabilizer for PVC food packaging materials, especially for clear plastic bottles. Approved by FDA for all foods except malt beverages, carbonated soft drinks, milk and other dairy products.

"Diodoquin."[70] Trademark for diiodohydroquin (q.v.).

"Diodrast."[162] Trademark for iodopyracet (q.v.).

"Diofan."[440] Trademark for dispersions of vinylidene copolymers for paper coating.

diol. Synonym for glycol or dihydric alcohol.

diolefin. (diene; alkadiene). An aliphatic compound (olefin) containing two double bonds, e.g., butadiene.

"Dion."[244] Trademark for a series of polyester resins.
Uses: Translucent panel resins; surfacing, laminating, molding, and casting resins.

"Dionin."[123] Trademark for ethylmorphine hydrochloride.
Hazard: Highly toxic by ingestion; narcotic and habit-forming.

"Dion" Polymercaptan Resins.[244] Resins in which the mercaptan-terminated liquid polymers are capable of being crosslinked to form elastomers at ambient temperatures by the addition of organic or inorganic oxidizing agents.
Uses: Sealants, caulking compounds, adhesives, molding compounds and coatings.

DIOP. Abbreviation for diisooctyl phthalate.

DIOS. Abbreviation for diisooctyl sebacate (q.v.).

1,4-dioxane (diethylene ether; 1,4-diethylene dioxide; diethylene oxide; dioxyethylene ether)
$OCH_2CH_2OCH_2CH_2$ (ring).

Properties: Colorless liquid; ethereal odor; stable; miscible with water and most organic solvents. B.p. 101.3°C; f.p. 10–12°C; sp. gr. 1.0356 (20/20°C); wt/gal 8.61 lb (20°C); refractive index 1.4221 (20°C); flash point 65°F (ASTM open cup). Autoignition temp. 356°F.
Derivation: (a) Ethylene glycol by treatment with acid; (b) from beta, beta-dichloroethyl ether by treatment with alkali.
Grades: Reagent; technical; spectrophotometric; scintillation.
Containers: Drums; tank cars.
Hazard: Highly toxic by inhalation; absorbed by skin. Tolerance, 50 ppm in air. Flammable, dangerous fire risk. May form explosive peroxides (MCA).
Uses: Solvent for cellulosics and wide range of organic products; lacquers; paints; varnishes; paint and varnish removers; wetting and dispersing agent in textile processing, dye baths, stain and printing compositions; cleaning and detergent preparations; cements; cosmetics; deodorants; fumigants; emulsions; polishing compositions; stabilizer for chlorinated solvents; scintillation counter.
Shipping regulations: (Rail) Flammable liquid, n.o.s., Red label. (Air) Flammable Liquid label.

dioxathion. Generic name for para-dioxane-2,3-diyl ethyl phosphorodithioate $C_4H_6O_2[SPS(OC_2H_5)_2]_2$.
Properties: Viscous brown liquid; f.p. −20°C; density 1.257 g/ml (26°C); insoluble in water; soluble in organic solvents. The technical material is a mixture of cis and trans isomers.
Hazard: Toxic by inhalation and ingestion. Cholinesterase inhibitor. Use may be restricted.
Uses: Insecticide; miticide.
Shipping regulations: (Rail, Air) Organic phosphate, liquid, n.o.s. Poison label. Not accepted on passenger planes.

dioxin (2,3,7,8-tetrachlorodibenzo-*p*-dioxin (TCDD). A toxic chlorinated hydrocarbon which occurs as an impurity in the herbicide 2,4,5-T. It can be removed by extraction with coconut charcoal. Its half-life in soil is about one year.

"Dioxitol."[125] Trademark for diethylene glycol monoethyl ether, b.p. 202°C; sp. gr. 0.990 (20/20°C).

3,5-dioxo-1,2-diphenyl-4-n-butylpyrazolidine. See phenylbutazone.

dioxolane $OCH_2CH_2OCH_2$ (ring). A cyclic acetal.

Properties: Water-white liquid; soluble in water; stable under neutral or slightly alkaline conditions. Sp. gr. 1.065; b.p. 74°C; flash point 35°F (open cup); vapor pressure 70 mm (20°C); wt/gal 8.2 lb (20°C).
Derivation: Reaction of formaldehyde with ethylene glycol.
Hazard: Flammable, dangerous fire risk. Moderately toxic by inhalation and ingestion.
Use: Low-boiling solvent and extractant for oils, fats, waxes, dyes, and cellulose derivatives.
Shipping regulations: (Rail) Flammable liquid, n.o.s., Red label. (Air) Flammable Liquid label.

dioxolone-2. See ethylene carbonate.

dioxopurine. See xanthine.

dioxyanthraquinones. See dihydroxyanthraquinones.

dioxybenzenes. See dihydroxybenzenes.

dioxyethylene ether. See 1,4-dioxane.

dioxyline phosphate $C_{22}H_{25}NO_4 \cdot H_3PO_4$.
Properties: White crystalline, odorless solid with a bitter taste; soluble in water; melts (with decomposition) 197–199°C.
Use: Medicine.

dioxynaphthalenes. See dihydroxynaphthalenes.

DIOZ. Abbreviation for diisooctyl azelate.

DIPA. Abbreviation for diisopropanolamine.

"Dipac."[204] Trademark for diisopropylbenzothiazyl-2-sulfenamide (q.v.).

dipalmitylamine (dihexadecylamine) $(C_{16}H_{33})_2NH$. Solid; m.p. 65°C; sp. gr. 0.83; slightly soluble in water.
Use: Intermediate.

"Dipan."[530] Trademark for diphenylacetonitrile (q.v.).

"Di-Paralene" Hydrochloride.[3] Trademark for chlorcyclizine hydrochloride.

"Dipaxin."[327] Trademark for diphenadione (q.v.).

dipentaerythritol $(CH_2OH)_3CCH_2OCH_2C(CH_2OH)_3$. Occurs in technical pentaerythritol. An off-white, freeflowing powder. The molecule contains six primary hydroxyl groups, all esterifiable. M.p. 212–220°C; sp. gr. 1.33 (25/4°C). Combustible. Low toxicity.
Use: Paints and coatings.

dipentamethylenethiuram tetrasulfide $[CH_2(CH_2)_4NCS]_2S_4$.
Properties: Light gray powder; sp. gr. 1.53; m.p. 110°C min. Soluble in chloroform, benzene, acetone; insoluble in water.
Use: Ultra-accelerator for rubber.

"Dipentek."[138] Trademark for dipentaerythritol, technical.

dipentene (cinene; limonene, inactive; *dl*-para-mentha-1,8-diene; cajeputene) $C_{10}H_{16}$. Commercial form is high in dipentene content, but also contains other terpenes and related compounds in varying amounts.
Properties: Colorless liquid; lemon-like odor; sp. gr. (15.5/15.5°C) 0.847; b.p. 175–176°C; refractive index 1.473 (20°C); flash point 113°F (closed cup); f.p. −97°C; wt/gal 7.15 lb (15.5°C). Autoignition temp. 458°F. Miscible with alcohol; insoluble in water. Combustible.
Derivation: (a) From various essential oils. (b) By close fractionation of wood turpentine. (c) By-product in making synthetic camphor. Generally low toxicity.
Grades: Steam-distilled; destructively distilled.
Containers: 30-, 55-gal drums; tank cars.
Hazard: Moderate fire risk. Moderately toxic.
Uses: Solvent for oleoresinous products, rosin, ester gum, alkyd resins, waxes, metallic soap driers, rubber, etc.; rubber compounding and reclaiming; dispersing agent for oils, resins, resin-oil combinations, pigments and driers; paints, enamels, lacquers, and varnishes; general wetting and dispersing agent; printing inks; perfumes; floor waxes and furniture polishes; synthetic resins, polyterpenes, chemicals.

"Dipentene 200."[79] Trademark for commercial steam-distilled dipentene (q.v.).

dipentene dioxide (limonene dioxide) $C_{10}H_{16}O_2$.
Properties: Liquid; sp. gr. 1.0287 (20°C); b.p. 242°C; f.p. −100°C; soluble in water.
Uses: Intermediate for plasticizers, epoxy resins; pharmaceuticals.

dipentene glycol. See terpin hydrate.

dipentene monoxide (limonene monoxide) $C_{10}H_{14}O$.
Properties: Liquid; sp. gr. 0.929 (20°C); f.p. −6°C; b.p. 75°C; flash point 152°F. Combustible.
Uses: Organic intermediate; epoxy resins.

"Dipentite."[62] Trademark for diphenylpentaerythritol diphosphite (q.v.).

di-n-pentylamine. See di-n-amylamine.

2,5-di(tert-pentyl)hydroquinone. See 2,5-di(tert-amyl)-hydroquinone.

"Diphacin."[536] Trademark for 2-diphenylacetyl-1,3-indandione. See diphenadione.
Uses: Rodenticide; anticoagulant.

diphacinone. Generic name for 2-diphenylacetyl-1,3-indandione; used as a rodenticide. See diphenadione.
Hazard: Harmful by ingestion.

diphemanil methyl sulfate (4-diphenylmethylene-1,1-dimethylpiperidinium methyl sulfate)
$(C_6H_5)_2CCCH_2CH_2N(CH_3)_2CH_2CH_2 \cdot CH_3SO_4$.
Properties: White bitter crystalline solid with faint characteristic odor. M.p. 189–196°C. Very slightly soluble in ether; slightly soluble in alcohol, chloroform, and water. Stable to heat and light; somewhat hygroscopic; pH (1% solution) 4.0–6.0.
Grade: N.F.
Use: Medicine.

diphenadione (2-diphenylacetyl-1,3-indanedione; diphacinone) $C_{23}H_{16}O_3$.
Properties: Yellow, odorless crystals or crystalline powder; m.p. 145–147°C. Practically insoluble in water, slightly soluble in acetone and in alcohol; soluble in benzene, ether, and in glacial acetic acid. Melting range: 144–150°C.
Grade: N.F.
Hazard: Harmful by ingestion. Inhibits blood clotting.
Uses: Medicine (anticoagulant); rodenticide.

diphenamid. Generic name for N,N-dimethyl-2,2-diphenylacetamide $(C_6H_5)_2CHCON(CH_3)_2$.
Properties: White solid; m.p. 134.5–135.5°C. Very slightly soluble in water; moderately soluble in acetone, dimethyl formamide, and phenyl "Cellosolve."
Hazard: Moderately toxic by ingestion.
Uses: Herbicide; plant growth regulator.

diphenatrile. See diphenylacetonitrile.

diphenhydramine hydrochloride (2-benzhydryloxy)-N,N-dimethylethylamine hydrochloride
$(C_6H_5)_2CHOCH_2CH_2N(CH_3)_2 \cdot HCl$.
Properties: White, odorless, crystalline powder. Darkens slowly on exposure to light; m.p. 166–170°C. Solutions practically neutral to litmus paper. Freely soluble in acetone; very slightly soluble in benzene and ether.
Grades: U.S.P.
Hazard: Prescription only. Do not use with alcohol or other CNS depressants.
Use: Medicine (antihistamine).

diphenic acid (2,2′-biphenyldicarboxylic acid) $HOOCC_6H_4C_6H_4COOH$.
Properties: White needles; m.p. 228–229°C; soluble in hot water.
Uses: Synthesis of dyes, detergents, pharmaceuticals.

diphenolic acid. See 4,4-bis(4-hydroxyphenyl)pentanoic acid.

diphenyl (biphenyl) $C_6H_5C_6H_5$. Several crystalline forms are known.

Properties: White scales; pleasant odor. Soluble in alcohol and ether; insoluble in water. Sp. gr. approx.

l; m.p. 70°C; b.p. 256°C. Flash point 235°F; combustible.
Derivation: (a) By slowly passing benzene through a red hot iron tube. (b) By heating bromobenzene and sodium, with subsequent distillation.
Hazard: Toxic. Tolerance, 0.2 ppm in air.
Uses: Organic synthesis; heat-transfer agent; fungistat in packaging of citrus fruit; plant disease control; mfg. of benzidine; dyeing assistant for polyesters.
See also chloridiphenyl.

diphenylacetic acid $(C_6H_5)_2CHCOOH$.
Properties: Colorless, odorless crystals; b.p. sublimes; m.p. 147.8–148.2°C; soluble in hot water, alcohol, ether, chloroform.

diphenylacetonitrile (diphenatrile) $(C_6H_5)_2CHCN$.
Properties: Yellow crystalline powder; m.p. 73–73.5°C; insoluble in water and very soluble in alcohol.
Uses: Preparation of diphenylacetic acid; synthesis of antispasmodics; herbicide.

diphenylacetylene. See tolan.

2-diphenylacetyl-1,3-indianedione. See diphenadione.

diphenylamine (DPA; N-phenylaniline) $(C_6H_5)_2NH$.
Properties: Colorless to grayish crystals. Soluble in carbon disulfide, benzene, alcohol, and ether; insoluble in water. Sp. gr. 1.159; m.p. 52.85°C; b.p. 302°C; flash point 307°F; autoignition temp. 1173°F; combustible.
Derivation: By heating equal formula weights of aniline and aniline hydrochloride in an autoclave. The product is boiled with dilute hydrochloric acid to remove the unaltered aniline, and the residue is distilled.
Grades: Technical; refined, flake and fused.
Containers: 50-lb polyethylene-lined paper bags; fiber drums; tank trucks; tank cars.
Hazard: Toxic, absorbed by skin. Tolerance, 10 mg per cubic meter of air.
Uses: Rubber antioxidants and accelerators; solid rocket propellants; pesticides; dyes; pharmaceuticals; veterinary medicine; storage preservation of apples; stabilizer for nitrocellulose; analytical chemistry.

diphenylamine chloroarsine (adamsite; phenarsazine chloride;DM) $C_6H_4(AsCl)(NH)C_6H_4$.
Properties: Canary-yellow crystals. Sublimes readily. Sp. gr. 1.65; m.p. 195°C; b.p. 410°C (dec); insoluble in water; soluble in benzene, xylene, carbon tetrachloride.
Derivation: By heating diphenylamine with arsenic trichloride.
Hazard: Highly toxic by inhalation and ingestion; strong irritant.
Uses: Military poison gas; wood treating.
Shipping regulations: (Air) Poison label. Not accepted on passenger planes. (Rail) Tear gas label.

9,10-diphenylanthracene $C_{14}H_8(C_6H_5)_2$.
Properties: Crystals; m.p. 248–250°C; insoluble in water and alcohol; slightly soluble in toluene.
Grade: Purified.
Use: Primary fluor or wavelength shifter in solution scintillators.

1,4-diphenylbenzene. See terphenyl.

diphenylbenzidine $C_6H_5HNC_6H_4C_6H_4NHC_6H_5$.
Properties: White powder. Insoluble in water; slightly soluble in alcohol and aromatic hydrocarbons. Sensitive to light. M.p. 242°C.
Derivation: Diphenylamine and fuming sulfuric acid.
Containers: Glass bottles.
Hazard: May be carcinogenic. See benzidine.
Use: Determination of zinc and nitrites.

2,5-diphenyl-para-benzoquinone $C_6H_5C_6H_2O_2C_6H_5$.
Properties: Greenish-yellow solid; m.p. 210–214°C. Slightly soluble in styrene, benzene, acetone, and ethyl acetate.
Use: Polymerization inhibitor.

diphenylbromoarsine $(C_6H_5)_2AsBr$.
Properties: White crystals. M.p. 54–56°C.
Derivation: (a) Hydrobromic acid and diphenylarsenious oxide are heated together for about 4 hours at 115–120°C; (b) by action of arsenic tribromide on triphenyl arsine at 300–350°C.
Hazard: Highly toxic; strong irritant.
Shipping regulations: (Rail, Air) Arsenical compounds, solids, n.o.s., Poison label.

1,3-diphenyl-2-buten-1-one. See dypnone.

diphenylcarbazide $(C_6H_5NHNH)_2CO$.
Properties: White crystals or flakes. Insoluble in water; soluble in alcohol and benzene. M.p. 173°C. Decomposes in light.
Derivation: Phenylhydrazine and urea.
Use: Determination of copper and other metals.

diphenylcarbinol. See benzhydrol.

diphenyl carbonate $(C_6H_5O)_2CO$.
Properties: White, crystalline solid. Can be halogenated and nitrated in characteristic manner. Readily undergoes hydrolysis and ammonolysis. Soluble in acetone, hot alcohol, benzene, carbon tetrachloride, ether, glacial acetic acid and other organic solvents; insoluble in water. B.p. 302°C; m.p. 78°C; sp. gr. 1.1215 (87/4°C).
Grade: Technical.
Uses: Plasticizer and solvent; synthesis of polycarbonate resins.

diphenylchloroarsine $(C_6H_5)_2AsCl$.
Properties: Colorless crystals or dark-brown liquid, which slowly becomes semi-solid. Decomposed by water (slowly). Soluble in carbon tetrachloride, chloropicrin, phenyldichloroarsine; practically insoluble in water. Sp. gr. 1.363 (40°C) (solid), or 1.358 (45°C) (liquid); b.p. 333°C (in CO_2 atmosphere); m.p. 41°C.
Derivation: Benzene and arsenic trichloride are heated in presence of aluminum chloride.
Grade: Technical.
Hazard: Highly toxic by inhalation; strong irritant to tissue.
Use: Military poison gas.
Shipping regulations: (Rail) Tear Gas label. (Air) Poison label. Not accepted on passenger planes.

diphenyldecyl phosphite $(C_6H_5O)_2POC_{10}H_{21}$.
Properties: Nearly water-white liquid; sp. gr. 1.023 (25/15.5°C); m.p. 18°C; refractive index 1.5160 (n 25/D). Combustible.
Uses: Chemical intermediate; stabilizer for polyvinyl and polyolefin resins.

diphenyldichlorosilane $(C_6H_5)_2SiCl_2$.
Properties: Colorless liquid; b.p. 305°C; f.p. −22°C; sp. gr. 1.19 (20°C); refractive index (n 25/D) 1.5773; flash point (COC) 288°F. Readily hydrolyzed by moisture, with liberation of hydrochloric acid. Combustible.
Derivation: (a) Reaction of powdered silicon and chlorobenzene in the presence of copper powder as catalyst; (b) reaction of phenylmagnesium chloride with silicon tetrachloride.
Grade: Technical.
Hazard: Highly toxic; strong irritant to tissue.
Use: Intermediate for silicone lubricants.

Shipping regulations: (Rail) White label. (Air) Corrosive label. Not acceptable on passenger planes.

diphenyldi-n-dodecylsilane $(C_6H_5)_2Si(C_{12}H_{25})_2$.
Properties: Colorless oil.
Derivation: Reaction of didodecyldichlorosilane with phenyl lithium.
Hazard: Probably toxic.
Use: High-temperature lubricant.

diphenyldimide. See azobenzene.

diphenyldimethoxysilane $(C_6H_5)_2Si(OCH_3)_2$.
Properties: Liquid; sp. gr. 1.080 (25°C); b.p. 191°C (53 mm); refractive index 1.5404 (25°C). Soluble in acetone, benzene, methyl alcohol; combustible.
Hazard: Probably toxic.
Uses: Treatment of powders, glass, paper, and fabrics.

diphenyleneimine. See carbazole.

alpha-**diphenylenemethane.** See fluorene.

diphenylene oxide (dibenzofuran) $C_{12}H_8O$ (tricyclic).
Properties: Crystalline solid; m.p. 87°C; b.p. 288°C; insoluble in water; slightly soluble in alcohol, ether and benzene.
Derived from coal-tar.
Hazard: Probably toxic.
Use: Insecticide.

1,1-diphenylethane. See uns-diphenylethane.

1,2-diphenylethane. See sym-diphenylethane.

uns-**diphenylethane** (1,1-diphenylethane) $(C_6H_5)_2CHCH_3$.
Properties: Colorless liquid. Soluble in chloroform, ether, carbon disulfide. B.p. 286°C; sp. gr. 1.004 (20°C); f.p. −21.5°C. Flash point 264°F; combustible.
Derivation: Action of acetaldehyde on benzene in presence of concentrated sulfuric acid.
Uses: Solvent for nitrocellulose; organic synthesis.

sym-**diphenylethane** (bibenzyl; dibenzyl; 1,2-diphenylethane) $C_6H_5CH_2CH_2C_6H_5$.
Properties: White, crystalline needles or small plates. Soluble in alcohol, chloroform, ether, carbon disulfide; insoluble in water. Sp. gr. 0.9782; b.p. 284°C; m.p. 52°C.
Derivation: (a) By treating benzyl chloride with metallic sodium. (b) Action of benzyl chloride on benzylmagnesium chloride.
Use: Organic synthesis.

diphenyl ether. See diphenyl oxide.

diphenylethylene. See stilbene.

N,N-**diphenylethylenediamine** (ethyl diphenyldiamine) $C_6H_5NHCH_2CH_2NHC_6H_5$.
Properties: Cream-colored solid; sp. gr. 1.14; softening point 54°C; insoluble in water; soluble in acetone, ethylene dichloride, benzene, and gasoline.
Use: Antioxidant in rubber compounding.

diphenylglycolic acid. See benzilic acid.

diphenylguanidine (DPG; melaniline) $HN{:}C(NHC_6H_5)_2$.
Properties: White powder; bitter taste; slight odor; sp. gr. 1.13; m.p. 147°C; decomposes above 170°C; soluble in ethyl alcohol, carbon tetrachloride, chloroform, hot benzene and toluene; slightly soluble in water.
Derivation: Treatment of aniline with cyanogen chloride.
Uses: Basic rubber accelerator; primary standard for acids.

1,6-diphenylhexatriene (DPH) $C_6H_5HC{:}CHCH{:}CHCCH{:}CHC_6H_5$.
Use: Wavelength shifter in solution scintillation counting.

diphenyl isophthalate (DPIP) $C_6H_5OOCC_6H_4COOC_6H_5$.
Properties: White solid; m.p. 138–139°C. Combustible.
Use: Manufacture of polybenzimidazoles, high temperature-resistant polymers.

diphenylketone. See benzophenone.

diphenylmethane (benzylbenzene) $(C_6H_5)_2CH_2$.
Properties: Long colorless needles. Soluble in alcohol and ether; insoluble in water. Sp. gr. 1.0056; m.p. 26.5°C; b.p. 264.7°C. Flash point 266°F; combustible.
Derivation: Condensation of benzyl chloride and benzene in presence of aluminum chloride.
Hazard: Probably irritant and narcotic in high concentrations.
Uses: Organic synthesis; dyes; perfumery.

diphenylmethane-4,4′-diisocyanate (MDI; methylene-diparaphenylene isocyanate; methylenebis(phenyl isocyanate) $CH_2(C_6H_4NCO)_2$.
Properties: Light-yellow, fused solid; solidification point 37°C; sp. gr. (70°C) 1.197; soluble in acetone, benzene, kerosine and nitrobenzene; combustible.
Derivation: para, para′-Diaminodiphenylmethane and phosgene.
Hazard: Highly toxic by inhalation of fumes. Strong irritant. Tolerance, 0.02 ppm in air.
Uses: Preparation of polyurethane resin and spandex fibers; bonding rubber to rayon and nylon.

diphenylmethanol. See benzhydrol.

diphenylmethyl bromide (benzhydryl bromide) $BrCH(C_6H_5)_2$.
Properties: Solid; m.p. 45°C; b.p. 193°C (26 mm). Decomposes in hot water; soluble in alcohol; very soluble in benzene.
Hazard: Strong irritant to eyes and skin.
Use: Organic synthesis.
Shipping reguations: (Air) Corrosive label. (Rail) Corrosive material, n.o.s., White label.

diphenylmethylchlorosilane $(C_6H_5)_2(CH_3)SiCl$.
Properties: Colorless liquid; sp. gr. 1.107 (25°C); b.p. 295°C; flash point 135°F. Combustible.
Derivation: Grignard reactoin of diphenyldichlorosilane with methylmagnesium chloride.
Hazard: Moderate fire risk. Probably toxic.
Use: Intermediate; end stopper for silicone oils.

diphenylnaphthylenediamine $C_{10}H_6(NHC_6H_5)_2$.
Properties: Silvery, crystalline plates. Slightly soluble in alcohol; insoluble in water. M.p. 164°C.
Derivation: By heating 2,7-dihydroxynaphthalene with aniline and aniline hydrochloride.
Use: Organic synthesis.

diphenylnitrosamine. See N-nitrosodiphenylamine.

2,5-diphenyloxazole (DPO) $OOC_6H_5{:}CC_6H_5$.
Properties: White, fluffy solid; m.p. 70–72°C.
Grade: Scintillation.

Containers: Glass bottles.
Use: Primary fluor used as scintillation counter or wavelength shifter.

diphenyl oxide (phenyl ether; diphenyl ether) $(C_6H_5)_2O$.
Properties: Colorless crystals or liquid; geranium-like odor. Soluble in alcohol and ether; insoluble in water. Sp. gr. 1.072–1.075; m.p. 27°C; b.p. 259°C. Flash point 239°F. Combustible; low toxicity. Autoignition temp. 1144°F.
Derivation: Reaction of bromobenzene and sodium phenate heated under pressure.
Grades: Technical; perfume; industrial.
Containers: Glass bottles; aluminum drums.
Uses: Perfumery, particularly soaps; heat-transfer medium; resins for laminated electrical insulation; chemical intermediate for such reactions as halogenation, acylation, alkylation, etc.

diphenylpentaerythritol diphosphite
$C_6H_5OP(OCH_2)_2C(CH_2O)_2POC_6H_5$(spiro).
Properties: White powder; b.p. 190–200°C (0.1 mm).
Containers: 12-gal pails, 25-lb net.
Use: Stabilizer for resins.

N,N′-**diphenyl**-meta-**phenylenediamine**
$C_6H_4(NHC_6H_5)_2$.
Properties: Crystalline needles. Soluble in hot alcohol; insoluble in water. M.p. 95°C. Combustible.
Derivation: By heating resorcinol with aniline in presence of calcium chloride and zinc chloride.
Use: Organic synthesis.

N,N′-**diphenyl**-para-**phenylenediamine** (DPPD)
$(C_6H_5NH)_2C_6H_4$.
Properties: Gray powder; sp. gr. 1.28; m.p. 145–152°C; purity 9.2% (min). Insoluble in water; soluble in acetone, benzene, monochlorobenzene and isopropyl acetate. Combustible; low toxicity.
Uses: Flex-resistant antioxidant in rubbers; stabilizer; polymerization inhibitor; retards copper degradation; intermediate for dyes, drugs, plastics, and detergents.

diphenyl phosphite $(C_6H_5O)_2PHO$.
Properties: Clear straw-colored liquid; m.p. 12°C; refractive index (n 25/D) 1.557; sp. gr. 1.221 (25/15°C); flash point 350°F. Combustible; low toxicity.
Containers: Specially lined 55-gal drums.
Use: Synthesis of organophosphorus compounds.

diphenyl phthalate $C_6H_4(COOC_6H_5)_2$.
Properties: Yellow-white powder; m.p. 68–70°C; sp. gr. 1.28 (20°C); flash point 435°F; b.p. 405°C; wt/gal 10.68 lb; refractive index 1.572 (74°C). Combustible.
Hazard: Moderately toxic.
Uses: Plasticizer for ethylcellulose, nitrocellulose, and various polymers.

diphenyl-4-piperidimemethanol hydrochloride. See azacyclonol hydrochloride.

diphenyl-4-piperidylmethane $(C_6H_5)_2(C_5H_{10}H)CH$.
Properties: White solid; difficultly soluble in water but readily soluble in dilute acids; moderately soluble in organic solvents; f.p. 99.7°C min.
Use: Intermediate.

1,3-diphenyl-1,3-propanedione. See dibenzoylmethane.

N,N′-**diphenylpropylenediamine**
$C_6H_5NHCH_2CH(CH_3)NHC_6H_5$.
Properties: Clear, deep reddish-brown, thick liquid; sp. gr. 1.07; stable in storage; insoluble in water; soluble in acetone, ethylene dichloride, benzene, and gasoline; readily disperses. Combustible; low toxicity.
Uses: Antioxidant for rubber latexes.

diphenyl-4-pyridylcarbinol $(C_6H_5)_2(C_5H_4N)COH$.
Properties: White solid; very weak base; slightly soluble in most organic solvents; soluble in hot glacial acetic acid; m.p. 236–241°C.
Use: Intermediate.

diphenyl-4-pyridylmethane $(C_6H_5)_2(C_5H_4N)CH$.
Properties: White to pale-yellow crystalline solid; moderately soluble in common organic solvents; b.p. 234°C (20 min); f.p. 123°C min.
Use: Intermediate.

diphenylsilanediol $(C_6H_5)_2Si(OH)_2$.
Properties: White solid; m.p. 130–150°. Derived from hydrolysis of diphenyldichlorosilane. Used as a silicone chemical.

para,para′-**diphenylstilbene**
$C_6H_5C_6H_4CH{:}CHC_6H_4C_6H_5$.
Properties: Crystals; m.p. 308–310°C.
Uses: In purified form as fluor in plastic scintillators.

N,N′-**diphenylthiourea.** See thiocarbanilide.

1,3-diphenyltriazene. See diazoaminobenzene.

diphenylurea (carbanilide) $(NHC_6H_5)CO(NHC_6H_5)$.
Properties: Colorless prisms. Soluble in alcohol and ether; very slightly soluble in water. Sp. gr. 1.239; m.p. 235°C; b.p. 260°C.
Use: Organic synthesis.

diphenyl-ortho-**xenyl phosphate**
$(C_6H_5O)_2(C_6H_5C_6H_4O)PO$.
Properties: Sp. gr. 1.20 (20°C); refractive index 1.582–1.590 (60°C); boiling range 250–285°C (5 mm); flash point 225°F; insoluble in water. Combustible. Low toxicity.
Use: Plasticizer.

diphosgene. Legal label name (Air) for trichloromethylchloroformate (q.v.).

diphosphopyridine nucleotide. See nicotinamide adenine dinucleotide.

diphosphoric acid. See pyrophosphoric acid.

dipicrylamine. See hexanitrodiphenylamine.

dipicryl sulfide. Legal label name (Air) for hexanitrodiphenyl sulfide (q.v.).

1,3-di-4-piperidylpropane (4-di-pip)
$(HNC_5H_9)CH_2CH_2CH_2C_5H_9NH$.
Properties: Solid; f.p. 67°C; b.p. 329°C; soluble in water; a stable, high-boiling strong organic base.
Use: Intermediate.

dipole. An assemblage of atoms or subatomic particles having equal electric charges of opposite sign separated by a finite distance; for instance the nucleus and orbital electron of a hydrogen atom, or the hydrogen and chlorine atoms of an HCl molecule. See also polar.

dipole moment. Molecules in which the atoms and their electrons and nuclei are so arranged that one part of the molecule has a positive electrical charge while the other part is negatively charged. The molecule therefore becomes a small magnet or dipole. Changing electrical or magnetic fields, causes the molecule to turn or rotate in one direction or another, depending on the charge of the field. The dipole moment (μ) is the distance in centimeters between the charges multiplied by the quantity of charge in electrostatic units. See also polar.

dipotassium orthophosphate. See potassium phosphate, dibasic.

dipropargyl (bipropargyl; 1,5-hexadiyne) $HC{:}CCH_2CH_2C\ CH$.
Properties: Colorless liquid. Soluble in alcohol; insoluble in water; sp. gr. 0.805; b.p. 85°C; f.p. −6.0°C.
Hazard: Moderate fire and explosion risk when exposed to heat.

dipropenyl. See 2,4-hexadiene.

di-2-propenylamine. See diallylamine.

di-n-propylamine $(C_3H_7)_2NH$.
Properties: Liquid; f.p. -63°C; sp. gr. 0.741 (20°C); b.p. 109°C; color water-white; odor amine; wt/gal 6.2 lb; flash point 63°F (TOC); slightly soluble in water. Low toxicity.
Containers: 5-gal cans; 55-gal drums; tank cars.
Hazard: Flammable, dangerous fire risk.
Use: Intermediate.
Shipping regulations: Flammable liquid, n.o.s. (Rail) Red label. (Air) Flammable Liquid label.

dipropylene. See 2,4-hexadiene.

dipropylene glycol $(CH_3CHOHCH_2)_2O$.
Properties: Colorless, slightly viscous liquid. Soluble in toluene, water. Sp. gr. 1.0252 (20/20°C); b.p. 233°C; vapor pressure 0.01 mm (20°C); flash point 280°F (o.c.) wt/gal 8.5 lb (20°C); coefficient of expansion 0.00073 (20°C); viscosity 1.07 poise (20°C). Combustible.
Grade: Technical.
Containers: Drums; tank cars.
Hazard: Moderately toxic by ingestion.
Uses: Polyester and alkyd resins; reinforced plastics; plasticizers; solvent.

dipropylene glycol dibenzoate $C_6H_5COOCH(CH_3)CH_2OCH_2CH(CH_3)OOCC_6H_5$.
Properties: Light-colored liquid; sp. gr. 1.1271 (20/20°C); 9.4 lb.gal (20°C); b.p. 250°C (10 mm); m.p. (200°C); insoluble in water; viscosity 227 cps (20°C). Combustible.
Use: Plasticizer.

dipropylene glycol dipelargonate $[CH_3(CH_2)_7COOCH(CH_3)CH_2]_2O$. A synthetic lubricant.

dipropylene glycol monomethyl ether $CH_3OC_3H_6OC_3H_6OH$.
Properties: Colorless liquid. Sp. gr. 0.950 (25/4°C); b.p. 189°C (760 mm), 74.5°C (10 mm); f.p. -80°C; viscosity 3.5 cp (25°C); refractive index 1.419 (25°C); flash point 185°F (o.c.); completely miscible with water, VM & P naphtha, acetone, ethanol, benzene, carbon tetrachloride, ether, methanol, monochlorobenzene and petroleum ether. Combustible.
Containers: Drums; tank cars.
Hazard: Moderately toxic. Absorbed by skin. Tolerance, 100 ppm in air.
Uses: Solvent; hydraulic brake fluids.

dipropylene glycol monosalicylate (dipropylene glycol monoester) $C_3H_6(OOCC_6H_4OH)OC_3H_6OH$.
Properties: Light-colored oil; fragrant odor; sp. gr. 1.16 (40°C); refractive index, about 1.52; soluble in alcohol; insoluble in water.
Uses: Ultraviolet light-screening agents; protective coatings; plasticizers.

dipropylene triamine. See 3,3′-iminobispropylamine.

dipropyl ketone (butyrone; 4-heptanone) $(CH_3CH_2CH_2)_2CO$.
Properties: Stable, colorless liquid. Pleasant odor. Miscible with many organic solvents. B.p. 143.7°C; f.p. −32.1°C; sp. gr. 0.8162 (20/20°C); wt/gal 6.79 lb (20°C); refractive index. 1.4068 (20°C); surface tension 25.2 dynes/cm (25°C) viscosity 0.0074 poise (20°C); vapor pressure 5.2 mm (20°C); flash point (c.c.) 120°F.
Grade: Technical.
Hazard: Moderate fire risk. Toxic by inhalation; skin irritant.
Uses: Solvent for nitrocellulose, raw and blown oils, resins and polymers; lacquers; flavoring.

dipropylmethane. See heptane.

dipropyl phthalate $C_6H_4(COOC_3H_7)_2$.
Properties: Colorless liquid. Sp. gr. 1.071 (25°C); refractive index 1.494 (25°C); b.p. 129–132°C (1 mm); solubility in water 0.015% by weight. Combustible; low toxicity.
Use: Plasticizer.

"Dipterex."[181] Trademark for trichlorfon (q.v.).

alpha, alpha-**dipyridyl** (2,2′-bipyridine) $(C_5H_4N)_2$.
Properties: White crystals; m.p. 69–70°C; b.p. 272–273°C; slightly soluble in water; insoluble in alcohol, ether, benzene, chloroform, and petroleum ether.
Grade: Reagent.
Use: Reagent for iron determination.

2,2′-dipyridylamine $(C_5H_4N)_2NH$.
Properties: Solid; f.p. 92.3°C (min); b.p. 222°C (50 mm); very slightly soluble in water. Combustible.
Derivation: From 2-aminopyridine.
Use: Intermediate.

dipyridylethyl sulfide $[C_5H_4N(CH_2)_2]_2S$.
Properties: Liquid. Sp. gr. 1.113 (25°C); refractive index 1.5841 (n 20/D); m.p. 1.5°C; soluble in water and common organic solvents. Combustible.
Grade: Technical (95% purity).
Hazard: May be toxic.
Uses: Synthesis of pharmaceuticals, dyestuffs, rubber chemicals, flotation agents, insecticides, fungicides, plasticizers, textile assistants, herbicides, oil additives, rust preventives, and pickling inhibitors.

diquat. Generic name for 6,7-dihydrodipyrido(1,2-a:2′,1′-c)pyrazidinium salt (1,1′-ethylene 2,2′-dipyridinium dibromide) $(C_5H_4NCH_2\text{-})_2Br_2$. Yellow crystals; m.p. 335°C; soluble in water.
Hazard: Tolerance, 0.5 mg per cubic meter of air.
Use: Herbicide and plant growth regulator.

direct dye. A soluble dye taken up directly by fibers, presumably due to selective adsorption. Usually applied to cotton or union goods (cotton-wool mixtures). Dyeing assistants such as sodium chloride or sodium sulfate are used to obtain a higher concentration of dye on the fibers. See also dye, synthetic.

diresorcinol (tetrahydroxydiphenyl) $(OH)_2C_6H_3C_6H_3(OH)_2$.
Properties: White to slight yellowish crystalline powder. Soluble in hot water and alcohol. M.p. 310°C.
Derivation: By fusing resorcinol and phenol with caustic soda.
Use: Organic synthesis.

diresorcinolphthalein. See fluorescein.

N,N′-**disalicylidene-1,2-diaminopropane** (N,N′-disalicylidene propylenediamine; disalicylalaminopropane; disalicylalpropylenediimine. $HOC_6H_4CH{:}NCH_2CH(CH_3)N{:}CHC_6H_4OH$.

Properties (80% active compound): Liquid; sp. gr. 1.08 (60/60°F); density 9.0 lb/gal; pour point 0°F; flash point about 70°F (TOC); viscosity 25 cs (100°F). Insoluble in water; miscible with benzene and xylene.
Hazard: Flammable, dangerous fire risk.
Use: Metal deactivator in motor fuels.

disc (disk). A small, thin, circular section or platelet of a material, especially of a biological specimen. "Disc" is the spelling preferred by scientists.

"Discaloy."[308] Trademark for an austenitic iron-base alloy containing nickel, chromium and relatively small proportions of molybdenum, titanium, silicon, and manganese. This alloy is precipitation-hardening and was developed primarily to meet the need for improved gas turbine discs, one of the most critical components of jet engines.

discharging agent. A substance capable of destroying a dye or mordant present within the fibers of a fabric. There are various methods of utilizing this property so that it is possible to produce a colorless figure upon a colored ground or a colored figure upon a different-colored ground. Some examples are titanous sulfate, sodium hydrosulfide, zinc formaldehyde sulfoxylate. See also stripping (2).

"Discolite."[159] Trademark for sodium sulfoxylate formaldehyde.
Uses: Dye discharge, vat printing, stripping of fibers; accelerator in emulsion polymerization systems.

disilanyl. See silane.

disiloxane. See siloxane.

disinfectant. A substance used on inanimate objects which destroys harmful microorganisms or inhibits their activity. Disinfectants are either complete or incomplete. Complete disinfectants destroy spores as well as vegetative forms of microorganisms; incomplete disinfectants destroy vegetative forms of the organisms, but do not injure spores.

Some representative disinfectants are: (1) mercury compounds (mercuric chloride, phenylmercuric borate); (2) halogens and halogen compounds (chlorine, iodine, fluorine, bromine, calcium and sodium hypochlorite); (3) phenols, including cresol from coal tar, ortho-phenylphenol; (4) synthetic detergents (anionic, such as sodium alkyl benzene sulfonates and cationic, such as quaternary ammonium compounds); (5) alcohols (of low molecular weight except methanol); (6) natural products (pine oil); (7) gases (sulfur dioxide, formaldehyde, ethylene oxide). Heat and electromagnetic waves are used as disinfectants.

A number of compounds (mercurous and mercuric chlorides, copper sulfate or carbonate, and a mixture of zinc oxide and zinc hydroxide) have been employed as seed disinfectants.

Effectiveness of disinfectants is rated by the phenol coefficient (q.v.). See also antiseptic.

dislocation. Any variation from perfect order and symmetry in a crystal lattice. In some cases the imperfection is due to one or more missing atoms, resulting in "holes" in the lattice; or one or more atoms of another element may be present, effecting important changes in the conductivity, hardness, and other properties of the crystal. Disordered arrangement of the constituent atoms may also occur. See also impurity; semiconductor; hole.

disodium acetarsenate (aricyl) $NaOOCCH_2As(OH)O(ONa) \cdot 2H_2O$.
Properties: White, crystalline powder. Soluble in water.
Derivation: By reacting sodium arsenite with sodium monochloracetate.
Hazard: Highly toxic.
Shipping regulations: Arsenical compounds, n.o.s. (Rail, Air) Poison label.

2,7-disodium dibromo-4-hydroxymercurifluorescein. See merbromin.

disodium dibutyl-*ortho*-phenylphenoldisulfonate.
Properties: Light-brown paste; soluble in alcohol, acetone, dibutyl tartrate, ethylene glycol.
Uses: Wetting, penetrating and spreading agent used in kier-boiling; scouring and dyeing textiles; industrial cleaners; deodorant preparations; insecticidal formulations; metal cleaning; stabilizer and wetting agent in latex used to treat cord or other fabrics.

disodium dihydrogen pyrophosphate. See sodium pyrophosphate acid.

disodium 1,2-dihydroxybenzene-3,5-disulfonate (4,5-dihydroxy-meta-benzenedisulfonic acid disodium salt; sodium catechol disulfonate) $C_6H_2(OH)_2(SO_3Na)_2$.
Properties: Non-hygroscopic crystals. Freely soluble in water; produces water-soluble, colored compounds with metal salts.
Use: Colorimetric reagent for iron, manganese, titanium, molybdenum.

disodium diphosphate. See sodium pyrophosphate, acid.

disodium edetate (ethylenediaminetetraacetic acid, disodium salt; disodium EDTA) $C_{10}H_{14}N_2Na_2O_8 \cdot 2H_2O$.
Properties: White, crystalline powder; freely soluble in water; 6.5 lb/gal; pH (5% solution) between 4 and 6. Low toxicity.
Grades: U.S.P., F.C.C.
Uses: Medicine; food preservative; sequestrant in vinegar.

disodium endothall. See endothall.

disodium ethylenebisdithiocarbamate. See nabam.

disodium guanylate. See sodum guanylate.

disodium hydrogen phosphate. See sodium phosphate, dibasic.

disodium inosinate. See sodium inosinate.

disodium methylarsonate (DMA; disodium methanearsonate; methanearsonic acid, disodium salt) $CH_3AsO(ONa)_2$, sometimes with $6H_2O$.
Properties: Colorless crystalline solid; hygroscopic; m.p. >355°C; (hexahydrate m.p. 132–139°C); soluble in water and methanol.
Derivation: Reaction of methyl chloride with sodium arsenate.
Grades: 55–65% powder concentrate; 31.5% blend.
Hazard: Highly toxic by ingestion and inhalation.
Uses: Pharmaceuticals; herbicide (crabgrass killer).
Shipping regulations: (Rail, Air) Arsenical compounds, n.o.s., Poison label.

disodium orthophosphate. See sodium phosphate, dibasic.

disodium phosphate. See sodium phosphate, dibasic.

disodium pyrophosphate. See sodium pyrophosphate, acid.

disodium tartrate. See sodium tartrate.

"Dispersall."[159] Trademark for ethylene oxide condensate. Used for after soaping of vat and naphthol colors to prevent bleeding.

"Dispersed Sulfur 70."[1] Trademark for an aqueous disperson of sulfur in the form of a flowable paste. Sulfur particles are virtually colloidal, averaging less than 2 microns in diameter.
Uses: Manufacture of water-base paints, compounding of latexes, sulfonation reactions in aqueous media; in general, for chemical processes requiring sulfur in the presence of water.

disperse phase. See phase (2); colloid chemistry.

dispersing agent. A surface-active agent (q.v.) added to a suspending medium to promote uniform and maximum separation of extremely fine solid particles, often of colloidal size. Applications include wet-grinding of pigments and sulfur; preparation of ceramic glazes; oil well drilling muds; insecticidal mixtures; carbon black in rubber; and water-insoluble dyes. True dispersing agents are polymeric electrolytes (condensed sodium silicates, polyphosphates, lignin derivatives); in nonaqueous media sterols, lecithin, and fatty acids are effective. See also emulsion; detergent.

dispersion. (1) A two-phase system of which one phase consists of finely divided particles (often in the colloidal size range) distributed throughout a bulk substance, the particles being the disperse or internal phase and the bulk substance the continuous or external phase. Under natural conditions the distribution is seldom uniform, but under controlled conditions the uniformity can be increased by addition of wetting or dispersing agents (surfactants) such as a fatty acid. The various possible systems are: gas/liquid (foam), solid/gas (aerosol), gas/solid (foamed plastic), liquid/gas (fog), liquid/liquid (emulsions, solid/liquid paint), and solid/solid (carbon black in rubber). Some types, such as milk and rubber latex, are stabilized by a protective colloid which prevents agglomeration of the dispersed particles by an abherent coating. Solid-in-liquid colloidal dispersions (loosely called solutions) can be precipitated by adding electrolytes which neutralize the electrical charges on the particles. Larger particles will either gradually coalesce and rise to the top or settle out, depending upon their specific gravity. See also suspension; colloid chemistry.

(2) In the field of optics, dispersion denotes the retardation of a light ray, usually resulting in a change of direction, as it passes into or through a substance, the extent of this effect depending on the frequency. Dispersion is a critically important property of optical glass. See also refractive index.

"Dispersite."[248] Trademark for water dispersions of natural, synthetic, and reclaimed rubbers and resins.
Uses: Adhesives for textiles, paper, shoes, leather, tapes; coatings for metal, paper, fabrics, carpets; protective (strippable) for saturating paper, felt, book covers, tape, jute pads; for dipping tire cords. Can be applied by spraying, spreading, impregnation, saturation.

"Disperso."[104] Trademark for wettable grades of zinc, calcium and other metallic stearates. Used where easy dispersion in water is desired.

"Dispersol OS."[206] Trademark for an oil-soluble emulsifying agent comprised of an 8% solution of a polyethenoxy compound in isopropanol. Designed especially for dispersion of oil spills in sea water. Claimed to be biodegradable and to have low toxicity for fish and other marine organisms. Amount needed said to be from 20 to 25% of the oil volume.

"Dispersol" VL.[325] Trademark for an ethylene oxide condensate. Nonionic dispersant and retardant for vat dyes and dispersant for acetate dyes. 20% aqueous solution.

displacement series. See activity series.

disproportionation. A chemical reaction in which a single compound serves as both oxidizing and reducing agent, and is thereby converted into a more oxidized and a more reduced derivative. Thus a hypochlorite upon appropriate heating yields a chlorate and a chloride, and an ethyl radical formed as an intermediate is converted into ethane and ethylene. See also transalkylation.

dissociation. The process by which a chemical combination breaks up into simpler constituents as a result of either (1) added energy, as in the case of gaseous molecules dissociated by heat, or (2) the effect of a solvent on a dissolved substance, e.g., water on hydrogen chloride. It may occur in the gaseous, solid, or liquid state, or in solution. All electrolytes dissociate to a greater or less extent in polar solvents. The degree of dissociation can be used to determine the equilibrium constant for dissociation, an important factor in ascertaining the extent of a chemical process. See also ionization.

dissolved oxygen (D.O.). One of the most important indicators of the condition of a water supply for biological, chemical and sanitary investigations. Adequate dissolved oxygen is necessary for the life of fish and other aquatic organisms and is an indicator of corrosivity of water, photosynthetic activity, septicity, etc. See also biochemical oxygen demand.

distearylamine (dioctadecylamine) $(C_{18}H_{37})_2NH$.
Properties: Solid; sp. gr. 0.85; m.p. 69°C. Almost insoluble in water.
Use: Intermediate.

2,6-distearyl-para-cresol (2,6-dioctadecyl-para-cresol) $(C_{18}H_{37})_2CH_3C_6H_2OH$. A viscous pale yellow liquid; soluble in most nonpolar solvents; refractive index 1.4825–1.4855 (25°C). Combustible.
Uses: Antioxidant; heat stabilizer for polypropylene.

distearyl ether (dioctadecyl ether) $(C_{18}H_{37})_2O$.
Properties: Solid; m.p. 58–60°C; b.p.; decomposes.
Grade: 95% (min) purity.
Uses: Electrical insulators; water repellents; lubricants in plastic molding and processing; antistatic agent; intermediates.

distearyl sulfide (dioctadecyl sulfide; distearyl thioether) $(C_{18}H_{37})_2S$.
Properties: Solid; m.p. 68–69°C; b.p., decomposes; sp. gr. 0.8148 (70/4°C).
Grades: 95% (min) purity.
Uses: Organic synthesis (formation of sulfonium compounds).

distearyl thiodipropionate (3,3′-dioctadecyl thiodipropionate; thiodipropionic acid, distearyl ester) $(C_{18}H_{37}OOCCH_2CH_2)_2S$.
Properties: White flakes, m.p. 58–62°C; m.p. 55°C. Insoluble in water; very soluble in benzene and olefin polymers. Resistant to heat and hydrolysis. Low toxicity.
Uses: Antioxidant; plasticizer; softening agent.

distearyl thioether. See distearyl sulfide.

"Disterdap."[458] Trademark for distearyl thiodipropionate.

distillate. Any product of distillation, especially of petroleum.

distillation. A separation process in which a liquid is converted to vapor and the vapor then condensed to a liquid. The latter is referred to as the distillate, while the liquid material being vaporized is the charge or distilland. Distillation is thus a combination of evaporation, or vaporization, and condensation. Simple examples are the natural cycle of evaporation and condensation that produces rain, and the vaporization and condensation of steam from a tea kettle on a cold surface.

The usual purpose of distillation is purification, or separation of the components of a mixture. This is possible because the composition of the vapor is usually different from that of the liquid mixture from which it is obtained. Alcohol has been so purified for generations to separate it from water, fusel oil, and aldehydes produced in the fermentation process. Gasoline, kerosine, fuel oil and lubricating oil are produced from petroleum by distillation. It is the key operation in removing salt from sea water. Distillation is extensively used in chemical analysis, in laboratory research, and for manufacture of many chemical products. See also destructive distillation, batch distillation, extractive distillation, rectification, dephlegmation, flash distillation, continuous distillation, simple distillation, reflux, fractional distillation, azeotropic distillation, vacuum distillation, molecular distillation, and hydrodistillation.

disulfiram. See tetraethylthiuram disulfide.

3,5-disulfobenzoic acid $C_6H_3(HSO_3)_2COOH$.
Properties: White powder. Soluble in water.
Grade: C.P.
Uses: Intermediate for detergents, dyes and pharmaceuticals.

disulfoton. Generic name for O,O-diethyl S-[2-(ethylthio)ethyl]phosphorodithioate. $(C_2H_5O)_2P(S)SCH_2CH_2SCH_2CH_3$.
Properties: Pure compound, yellow liquid; technical compound, brown liquid; b.p. 62°C (0.01 mm); density 1.144 (20°C). Insoluble in most organic solvents.
Hazard: Highly toxic by ingestion and inhalation. Cholinesterase inhibitor. Absorbed by skin. Use may be restricted. Tolerance, 0.1 mg per cubic meter of air.
Uses: Systemic insecticide; acaricide.
Shipping regulations: Organic phosphate, liquid, n.o.s. (Rail) (Air) Poison label. Not accepted on passenger planes.

disulfuryl chloride. See pyrosulfuryl chloride.

"Di-Syston."[181] Trademark for disulfoton (q.v.).

ditetradecylamine. See dimyristylamine.

ditetradecyl ether. See dimyristyl ether.

ditetradecyl sulfide. See dimyristyl sulfide.

"Dithane."[23] Trademark for agricultural fungicides based on salts of ethylene bisdithiocarbamate. Supplied in zinc, manganese, and sodium forms as wettable powder or liquid concentrate.

1,4-dithiane (diethylene disulfide) $\overline{SCH_2CH_2SCH_2CH_2}$.
Properties: White crystals. Volatile in steam. Soluble in alcohol, ether; slightly soluble in water. B.p. 115.6°C (60 mm); m.p. 108°C.
Derivation: Interaction of dichloroethyl sulfide with sodium or potassium sulfide.
Use: Organic synthesis.

dithiane methiodide $C_4H_8S_2 \cdot CH_3I$.
Properties: Crystals. Soluble in hot water; slightly soluble in alcohol; insoluble in ether. M.p. 174°C.
Derivation: Interaction of dichloroethyl sulfide and methyl iodide.

dithiazanine iodide $C_{23}H_{23}IN_2S_2$. 3-Ethyl-2-[5-(3-ethyl-2-benzothiazolinylidene)-1,3-pentadienyl]benzothiazolium iodide.
Properties: Dark greenish, crystalline powder; decomposes at 248°C. Slightly soluble in methylene chloride and in dimethylformamide; insoluble in ether.
Grade: U.S.P.
Uses: Medicine; photography.

beta,beta'-**dithiobisalanine.** See cystine.

2,2'-dithiobisbenzothiazole (benzothiazolyl disulfide; benzothiazyl disulfide; 2-mercaptobenzothiazyl disulfide; MBTS) $(C_6H_4SCN)_2S_2$.
Properties: Pale-yellow, free-flowing, odorless powder; sp. gr. 1.54; m.p. 168°C.
Uses: Vulcanization accelerator; retarder in neoprene.

dithiocarbamic acid (aminodithioformic acid) NH_2CS_2H.
Properties: Colorless needles, soluble in alcohol. Low toxicity.
Uses: The metal salts of the acid are important as strong (ultra) rubber accelerators, as are the thiuram disulfide derivatives. See thiuram, selenium diethyldithiocarbamate, zinc dibutyldithiocarbamate, zinc diethyldithiocarbamate; ziram. Also used as seed disinfectant.

2,2'-dithiodibenzoic acid $(C_6H_4COOH)_2S_2$.
Properties: Tan to gray powder; m.p. 280°C (min).
Use: Intermediate for pharmaceuticals.

4,4'-dithiodimorpholine $C_4H_8ONSSNOC_4H_8$.
Properties: Gray to tan powder; m.p. 122°C min; sp. gr. 1.36 (25°C).
Hazard: Moderately toxic by ingestion, inhalation, and skin absorption.
Use: Vulcanizing agent.

1,2-dithioglycerol. See 2,3-dimercaptopropanol.

6,8-dithiooctanoic acid. See *dl*-alpha-lipoic acid.

dithioöxamide (rubeanic acid) $SC(NH_2)C(NH_2)S$.
Properties: Stable orange-red powder; decomposes at 140°C; insoluble in water; soluble in acetone and chloroform. Forms highly colored stable complexes with many metal ions; can form a series of N,N'-derivatives.
Use: Chemical intermediate.

dithiothreitol (Cleland's reagent).
Properties: Solid; m.p. 42–43.5°C. 99% pure. Available in 1-, 5-, 25-gram quantities.
Uses: Reducing agent for proteins and enzymes; biochemical research.

dithymol diiodide. See thymol iodide.

1,4-di-para-**toluidinoanthraquinone** (FD & C Green No. 6) $C_{14}H_6O_2(NHC_6H_4CH_3)_2$.
Properties: Solid; m.p. 213°C.
Use: Dye, approved for restricted use in drugs and cosmetics.

di-ortho-**tolyl carbodiimide** $(CH_3)C_6H_4N{:}C{:}NC_6H_4(CH_3)$.
Properties: Yellow to red-brown liquid with faint acrid odor. Sp. gr. 1.063 (25/4°C); b.p. 140°C (0.9 mm); refractive index (n 20/D) 1.6248. Soluble in organic solvents such as chloroform, carbon tetrachloride, benzene, and dioxane. Combustible.
Containers: 1-, 5-, and 55-gal.
Hazard: Toxic by inhalation and ingestion. Irritant to skin and eyes.
Uses: Surface coatings; textile processing; stabilizers of polyesters and urethanes; scavenger compounds for materials sensitive to active hydrogen.

N,N-**di-**ortho-**tolylethylenediamine** $CH_3C_6H_4NHCH_2CH_2NHC_6H_4CH_3$.
Properties: Light-brown to purple granular solid; sp. gr. 1.13; softens at about 57°C; insoluble in water; soluble in acetone, ethylene dichloride, benzene and gasoline.
Use: Antioxidant for light-colored rubber goods.

di-ortho-**tolylguanidine** (DOTG) $(CH_3C_6H_4NH)_2CNH$.
Properties: White powder; non-hygroscopic. Very slightly soluble in water; soluble in warm alcohol, from which it crystallizes on cooling. Sp. gr. 1.10; m.p. 179°C. Low toxicity.
Derivation: Desulfurization of di-ortho-tolylthiourea with a lead compound in the presence of ammonia.
Containers: Drums.
Use: Basic rubber accelerator.

2,7-di-para-**tolylnaphthylenediamine** $C_{10}H_6(NHC_6H_4CH_3)_2$.
Properties: Fine needles. Sparingly soluble in alcohol; insoluble in water. M.p. 237°C.
Derivation: By heating 2,7-dihydroxynaphthalene with para-toluidine and para-toluidine hydrochloride.

1,3-di-para-**tolylphenylenediamine** $C_6H_4(NHC_6H_4CH_3)_2$.
Properties: Long needles. Soluble in alcohol and ether; insoluble in water. M.p. 137°C.
Derivation: By heating resorcinol and para-toluidine in presence of zinc chloride.

di-ortho-**tolylthiourea** (DOTT) $SC(NHC_6H_4CH_3)_2$.
Properties: Colorless, crystalline leaflets; pungent odor; not hygroscopic. Soluble in alcohol, ether and benzene; insoluble in water. M.p. 144–148°C. Low toxicity.
Derivation: By the interaction of ortho-toluidine and carbon disulfide.
Containers: Drums.
Use: Metal pickling inhibitor.

ditridecyl phthalate (DTDP) $C_6H_4(COOC_{13}H_{27})_2$.
Properties: Colorless liquid. Sp. gr. 0.951 (20/20°C); b.p. >285°C at 5 mm Hg; pour point −37°C; viscosity 190 cps (25°C). Flash point 460°F. Combustible. Low toxicity.
Containers: Drums; tank trucks.
Use: Plasticizer.

ditridecyl thiodipropionate (3,3′-tetramethylnonyl thiodipropionate; thiodipropionic acid, ditridecyl ester) $(C_{13}H_{27}OOCCH_2CH_2)_2S$.
Properties: Colorless liquid. Sp. gr. (25°C) 0.932. Insoluble in water; soluble in most organic solvents. Combustible.
Uses: Stabilizer; plasticizer and softening agent for plastics; lubricant additive.

diuron. Generic name for 3-(3,4-dichlorophenyl)-1,1-dimethylurea, $C_6H_3Cl_2NHCON(CH_3)_2$.
Properties: White crystalline solid; m.p. 153–155°C; vapor pressure (30°C) 2×10^{7} mm. Very low solubility in hydrocarbon solvents; approx. 42 ppm at 25°C in distilled water. Stable toward oxidation and moisture. Decomposes at 180°C.
Hazard: Irritant to skin and mucous membranes.
Use: Pre-emergence herbicide.

divanadyl tetrachloride. See vanadyl chloride.

divinyl acetylene $H_2C{:}CHC{:}CCH{:}CH_2$. Trimer of acetylene, formed by passing it into a hydrochloric acid solution containing metallic catalyst. Intermediate in manufacture of neoprene.

divinylbenzene (DVB; vinylstyrene) $C_6H_4(CH{:}CH_2)_2$, existing as ortho-, meta- and para-isomers. The commercial form contains the 3 isomeric forms together with ethylvinylbenzene and diethylbenzene.
Properties: (pure meta- isomer) Water-white liquid easily polymerized. B.p. 199.5°C; f.p. −66.90°C; sp. gr. 0.9289 (20°C); viscosity 1.09 centipoise (20°C); refractive index (n 20/D) 1.5772. Flash point 165°F. (Divinylbenzene, 55%) Pale straw-colored liquid; f.p. −87°C; b.p. 195°C; sp. gr. 0.918 (25/25°C); insoluble in water; soluble in methanol and ether. Combustible.
Grades: 50–60%; 20–25%.
Containers: Drums; tank cars.
Hazard: Moderately toxic by inhalation. Velocity of polymerization involves an explosion risk. If uninhibited, store at less than 90°F.
Uses: Polymerization monomer for special synthetic rubbers, drying oils, ion-exchange resins, casting resins, and polyesters.
Note: Should contain inhibitor when stored or shipped.
Shipping regulations: (Rail, Air) Not listed. Consult authorities.

divinyl ether. Legal label name (Air) for vinyl ether (q.v.).

divinyl oxide. See vinyl ether.

3,9-divinylspirobi-meta-**dioxane** (3,9-divinyl-2,4,8,10-tetraoxaspiroundecane.) $[CH_2{:}CHCH(OCH_2)_2]_2C$. Bicyclic.
Properties: Liquid; m.p. 42°C; 120°C (2 mm); sp. gr. 1.251 (20/20°C); flash point 290°F (COC); slightly soluble in water. Combustible.
Uses: Intermediate and monomer.

divinyl sulfide $(CH_2{:}CH)_2S$.
Properties: Mobile liquid. Unpleasant odor. Polymerizes readily. Sp. gr. 0.9174 (15°C); b.p. 85–86°C. Combustible.
Derivation: Interaction of dichlorodiethyl sulfide and an alcoholic solution of potassium hydroxide.
Grades: Technical.

divinyl sulfone $Ch_2{:}CHSO_2CH{:}CH_2$.
Properties: Liquid; sp. gr. 1.1788 (20/20°C); b.p. 234°C; f.p. −26°C; soluble in water; flash point 255°F. Combustible.
Uses: Monomer used in manufacture of polymers with diols, urea, and malonic esters; shrinkage control agent (textiles).

di-ortho-**xenyl phenyl phosphate** $(C_6H_5C_6H_4O)_2(C_6H_5O)PO$.
Properties: Liquid. Sp. gr. 1.20 (60°C); refractive index 1.603–5 (60°C); boiling range 285–330°C (5

mm); flash point 250°C; insoluble in water. Combustible.
Use: Plasticizer.

"Dixie" Clay.[69] Trademark for a hard kaolin.
Properties: White to cream; sp. gr. 2.62 ± .03; fineness (through 325 mesh) 99.8%.
Use: Filler and reinforcing agent for rubber and plastics; paper coating.

"Dixsol."[319] Trademark for nitrogen solutions containing 37 to 50.6% total nitrogen by weight, used as, or in production of, soil fertilizers. Sp. gr. at 60/60°F varies from 0.984 to 1.188; lb/gal at 60°F, 8.20 to 9.89. Are composed of free ammonia, ammonium nitrate, and water (5.5–18.2% by weight).

di-para-**xylylene** $(\text{-}CH_2C_6H_4CH_2\text{-})_2$. Stable, white crystals; m.p. 280°C.
Derivation: Pyrolysis of para-xylene at 750°F in presence of steam. An organic quench (benzene or toluene) gives the dimer in yields of 10–15%.
Use: Parylene resins.

DKP. Abbreviation for dipotassium phosphate. See potassium phosphate, dibasic.

DL-. A prefix indicating that a compound contains equal parts of D and L stereoisomers. See also racemic; meso-.

dl-. A prefix denoting a crystal that rotates the plane of polarized light equally to both left and right, resulting in optical inactivity. The symbol ± is preferably used.

DM. See diphenylamine chloroarsine.

DMA. Abbreviation for dimethylamine or disodium methyl arsonate (q.v.).

DMAC. Abbreviation for dimethylacetamide (q.v.).

DMB. Abbreviation for dimethoxybenzene. See hydroquinone dimethyl ether.

DMC.
(1) Abbreviation for dichlorophenyl methyl carbinol. See di(para-chlorophenyl) ethanol (q.v.).
(2) Abbreviation for beta-dimethylaminoethyl chloride hydrochloride (q.v.).

DMDT. Abbreviation for dimethoxydiphenyltrichloroethane. See methoxychlor.

DMF. Abbreviation for dimethyl formamide (q.v.).

DMH. Abbreviation for dimethyl hydantoin (q.v.).

DMHF. Abbreviation for dimethylhydantoin formaldehyde. See methylol dimethylhydantoin.

"DMP."[23] Trademark for dimethylaminomethyl-substituted phenols.
Uses: Chemical intermediates, curing agents for epoxy resins, antioxidants.

DMPA.
1. See O-(2,4-dichlorophenyl) O-methyl isopropylphosphoramidothioate.
2. See dimethylol propionic acid.

DMSO. Abbreviation for dimethyl sulfoxide (q.v.).

DMT. Abbreviation for dimethyl terephthalate (q.v.).

DMU. Abbreviation for dimethylurea and dimethylolurea (q.v.).

DNA. Abbreviation for deoxyribonucleic acid (q.v.).

DNBP. Abbreviation for dinitro-ortho-sec-butylphenol (q.v.).

DNC. Abbreviation for dinitrocresol, (q.v.).

DNOC. Abbreviation for 4,6-dinitro-ortho-cresol (q.v.).

DNOCHP. Abbreviation for dinitro-ortho-cyclohexylphenol (q.v.).

DNODA. Abbreviation for di(n-octyl,n-decyl)adipate (q.v.).

DNODP. Abbreviation for di(n-octyl,n-decyl)phthalate (q.v.).

DNP. Abbreviation for dinonyl phthalate (q.v.).

DNPC. See 2,6-dinitro-para-cresol.

DNPD. Abbreviation for N,N′-di-beta-naphthyl-para-phenylenediamine (q.v.).

DNPT. See dinitrosopentamethylenetetramine (q.v.).

DNT. Abbreviation for dinitrotoluene (q.v.).

D.O. Abbreviation for dissolved oxygen (q.v.).

DOA. Abbreviation for dioctyl adipate. See the more exact name, di(2-ethylhexyl) adipate.

D.O.C. Abbreviation for dichromate oxygen consumed. See oxygen consumed.

documentation. The techniques involved in recording, coding, storing, and retrieving technical data; it is also called information processing and data processing. It can be performed mechanically by means of a punched card system, of which there are many variations; electronically by means of input to a computer; or by a combination of these methods. One of the most sophisticated techniques for storage and retrieval of chemical information has been developed by Chemical Abstracts Service (q.v.), which includes structural formulas. Pioneer development work in documentation was done in the 1950's by Dyson, Perry, Crane, and others. A journal serving this field is published by ACS.

n-**docosane** $C_{22}H_{46}$.
Properties: Solid; m.p. 45.7°C; b.p. 230°C (15 mm); sp. gr. 0.778 (45/4°C); refractive index 1.4400 (n 45/D). Combustible.
Grades: 95% (min) purity.
Uses: Organic synthesis; calibration; temperature-sensing devices.

docosanoic acid. See behenic acid.

1-docosanol. See behenyl alcohol.

cis-**13-docosenoic acid.** See erucic acid.

doctor treatment. A method of improving or "sweetening" the odor of gasoline, petroleum solvents, or kerosine. A doctor solution of sodium plumbite, Na_2PbO_2, is made by dissolving litharge in caustic soda solution; the feed to be sweetened is passed through the doctor solution. The action of the sodium plumbite and the lead sulfide formed from it, in conjunction with free sulfur (either naturally present, or added) converts the malodorous mercaptans to the pleasanter disulfides.

dodecachlorooctahydro-1,3,4-metheno-2H-**cyclobuta**-[c,d]**pentalene.** See mirex.

dodecahydrosqualane. See squalane.

dodecanal. See lauryl aldehyde.

n-**dodecane** (dihexyl) $CH_3(CH_2)_{10}CH_3$.
Properties: Colorless liquid; sp. gr. 0.749 (20/4°C); f.p. −10°C; b.p. 213°C; refractive index 1.4221 (20/D); flash point 160°F. Soluble in alcohol, acetone,

ether; insoluble in water. Combustible. Autoignition temp. 400°F.
Grades: 95, 99%, 99.7 mole %.
Containers: 1-, 5-, and 10-gal cans; drums.
Uses: Solvent; organic synthesis; distillation chaser; jet fuel research.

1,12-dodecanedioic acid $HOOCC_{10}H_{20}COOH$. A twelve-carbon straight-chain saturated aliphatic dibasic acid.
Properties: White, crystalline powder; m.p. 130–132°C; soluble in hot toluene, alcohols, hot acetic acid; slightly soluble in hot water.
Use: Intermediate.

dodecanoic acid. See lauric acid.

n-**didecanol.** See lauryl alcohol.

dodecanoyl peroxide. See lauryl peroxide.

dodecene $C_{12}H_{24}$. Many possible isomers. See 1-dodecene; tetrapropylene; sodium dodecylbenzenesulfonate.

1-dodecene (alpha-dodecylene) $H_2C{:}CH(CH_2)_9CH_3$.
Properties: Colorless liquid; sp. gr. 0.7600 (20/4°C); f.p. −33.6°C; b.p. 213°C; refractive index (n 20/D) 1.4327; soluble in alcohol, acetone, ether, petroleum, coal tar solvents; insoluble in water. Combustible.
Containers: 1-, 5-, 10-gal cans.
Hazard: Probably irritant and narcotic in high concentration.
Uses: Flavors; perfumes; medicine; oils; dyes; resins.

dodecenylsuccinic acid $HOOCCH(C_{12}H_{23})CH_2COOH$.
Properties: Extremely viscous liquid; practically insoluble in water; completely soluble in oil. Combustible.
Uses: Synthesis; corrosion inhibitor in oils; waterproofing.

dodecenylsuccinic anhydride $C_{12}H_{23}\overline{CHCOOOCC}H_2$. The normal and at least two branched-chain dodecenyls are used commercially. The following properties are those of a branched chain compound.
Properties: Light-yellow, clear, viscous oil; b.p. 180–182°C (5 mm); sp. gr. (25°C) 1.002; flash point 352°F (COC); viscosity (20°C) 400 centipoises, (70°C) 15.5 centipoises. Combustible. Low toxicity.
Containers: Drums.
Uses: Alkyd, epoxy, and other resins, anticorrosion agents, plasticizers, wetting agents for bituminous compounds, and vulcanizable products.

dodecyl acetate (acetate C-12; lauryl acetate) $C_{12}H_{25}OOCCH_3$.
Properties: Colorless liquid; fruity odor; sp. gr. 0.860–0.862; refractive index 1.430–1.433; b.p. 150.5–151.5°C; soluble in 3 volumes of 80% alcohol. Combustible.
Use: Perfumery; flavoring.

n-**dodecyl alcohol.** See lauryl alcohol.

dodecyl aldehyde. See lauryl aldehyde.

dodecylaniline $C_{12}H_{25}C_6H_4NH_2$. (Probably the para-isomer).
Hazard: May be toxic. See aniline.
Properties: Colorless liquid. Sp. gr. (25/25°C) 0.907–0.912; b.p. 340–350°C. Oil-soluble aromatic amine. Insoluble in water; soluble in most organic solvents. Combustible.
Use: Intermediate.

dodecylbenzene (detergent alkylate).
$C_{12}H_{25}C_6H_5$. A commercial blend of isomeric, predominantly monoalkyl benzenes. The side chains are saturated, averaging twelve carbon atoms. Combustible; flash point 285°F.
Derivation: Alkylation of benzene with isomeric dodecenes, obtained usually by polymerization of propylene. See tetrapropylene. For linear (normal) dodecylbenzene, see sodium dodecylbenzenesulfonate.
Hazard: Moderately toxic by ingestion.
Containers: Drums; tank cars.
Uses: Detergents of ABS or LAS type.

dodecylbenzenesulfonic acid (DDBSA). See sodium dodecylbenzenesulfonate.

dodecylbenzyl mercaptan $C_{12}H_{25}C_6H_4CH_2SH$. Offered as branched-chain isomers.
Properties: Light-yellow oil; unpleasant odor; b.p. approx. 150°C (0.5 mm); soluble in acetone, benzene and heptane; insoluble in water and alcohol. Combustible.
Hazard: Toxic and irritant.
Uses: Polymerization modifier; intermediate; metal-cleaning and polishing compounds.
See also thiol.

n-**dodecyl bromide.** See lauryl bromide.

6-dodecyl-1,2-dihydro-2,2,4-trimethylquinoline $C_{12}H_{25}C_9H_5N(CH_3)_3$.
Properties: Dark, viscous liquid; sp. gr. 0.93 (45°C). Combustible.
Uses: Rubber antioxidant (flex-cracking).

dodecyldimethyl(2-phenoxyethyl)ammonium bromide. See domiphen bromide.

N-**dodecylguanidine acetate.** See dodine.

n-**dodecyl mercaptan** (DDM) $C_{12}H_{25}SH$. Many isomers of dodecyl mercaptan are possible and a variety of these occur in the technical material known as tert-dodecyl mercaptan or lauryl mercaptan or simply dodecyl mercaptan. See lauryl mercaptan.
Properties of n-isomer: Colorless liquid, boiling point 143°C; refractive index 1.4589; soluble in ether and alcohol; insoluble in water; flash point 262°F; combustible.
See also thiol.

4-dodecyloxy-2-hydroxybenzophenone $C_{12}H_{25}OC_6H_3(OH)C(O)C_6H_5$.
Properties: Pale-yellow flakes; setting point 43°C; soluble in polar and nonpolar organic solvents.
Use: Ultraviolet light inhibitor in plastics.

dodecylphenol $C_{12}H_{25}C_6H_4OH$. A mixture of isomers.
Properties: Straw-colored liquid; sp. gr. 0.94 (20/20°C); flash point 325°F; boiling range 310–335°F; phenolic odor; soluble in organic solvents; insoluble in water. Combustible.
Containers: Drums; tank cars.
Hazard: Probably toxic. See phenol.
Uses: Solvent; intermediate for surface-active agents; oil additives; resins; fungicides; bactericides; dyes; pharmaceuticals; adhesives; rubber chemicals.

dodecyltrichlorosilane $C_{12}H_{25}SiCl_3$.
Properties: Colorless to yellow liquid; b.p. 288°C; sp. gr. 1.026 (25/25°C); refractive index (n 25/D) 1.4521. Readily hydrolyzed with liberation of hydrochloric acid.

Derivation: By Grignard reaction of silicon tetrachloride and dodecylmagnesium chloride.
Hazard: Toxic. Strong irritant to tissue.
Use: Intermediate for silicones.
Shipping regulations: (Rail) White label. (Air) Corrosive label. Not acceptable on passenger planes.

dodecyltrimethylammonium chloride $C_{12}H_{25}N(CH_3)_3Cl$.
Properties: White liquid; surface tension, 0.1% in water, 33 dynes/cm. (25°C); soluble in water and alcohol.
Uses: Germicides; moldicides; fungicides; textile fiber softeners; cationic emulsifiers; flotation reagents.

dodine. Generic name for N-dodecylguanidine acetate, $C_{12}H_{25}NHC(:NH)NH_2 \cdot CH_3COOH$.
Properties: Crystals; m.p. 136°C; soluble in hot water and alcohol.
Hazard: Strong irritant to eyes and skin over 50% concentration.
Use: Fungicide.

DOF. See di(2-ethylhexyl) fumarate.

"Dolcoseal."[528] Trademark for spray-dried flavoring products such as essential oils, aromatic chemicals, flavor and perfume compounds.

"Dolocotone."[528] Trademark for purified benzodihydropyrone (q.v.).

dolomite $CaMg(CO_3)_2$. A carbonate of calcium and magnesium.
Properties: Color gray, pink, white, etc.; vitreous luster; sp. gr. 2.85; Mohs hardness 3.5–4; good cleavage in three directions; similar to calcite, but less soluble in acids (reacts with acid when powdered or with hot acid). Noncombustible, nontoxic.
Uses: Refractory for furnaces; manufacture of magnesium compounds and magnesium metal; as building material; in fertilizers; stock feeds; paper making; ceramics; mineral wool; removal of sulfur dioxide from stack gases.

"Dolophine" Hydrochloride.[100] Trademark for methadone hydrochloride (q.v.).

"Doloxide."[139] Trademark for a calcined dolomite, powdered; 91% through 200 mesh; used in chemical processes where a high magnesium burned lime is required.

dolphin oil. See porpoise oil.

DOM. (1) See di(2-ethylhexyl) maleate. (2) A hallucinogenic drug. Use restricted by FDA.

domiphen bromide
(dodecyldimethyl(2-phenoxyethyl)-ammonium bromide) $C_6H_5OC_2H_4N(C_{12}H_{25})(CH_3)_2Br$. A quaternary ammonium salt.
Properties: Crystals; m.p. 112°C. Soluble in water and organic solvents.
Use: Medicine.

donor. An atom which furnishes a pair of electrons to form a covalent bond or linkage with another atom, called the acceptor. See also bond, chemical; Lewis electron theory.

DOP. Abbreviation for dioctyl phthalate. See the more exact name, di(2-ethylhexyl) phthalate.

L-dopa. See L-dihydroxyphenylalanine.

dope (1) Sizing formulation consisting of solutions of nitrocellulose, cellulose acetate, or other cellulose derivations applied to crepe yarn to set the twist and assist creping, to leather to form a high-gloss finish. (2) A combustible, such as wood pulp, starch, sulfur, etc., used in "straight" dynamites. (3) A trace impurity introduced into ultrapure crystals to obtain desired physical properties, especially electrical properties. Examples: erbium oxide doped with thulium for use as laser crystals, germanium or silicon doped with boron or arsenic for use as semiconductors.

"Doriden."[305] Trademark for glutethimide. (2-ethyl-2-phenylglutarimide) (q.v.).

"Dorisyl."[233] Trademark for an aromatic chemical useful as a perfume base.

"Dorlone."[233] Trademark for fumigants and insecticides containing ethylene dibromide and 1,3-dichloropropene and related chlorinated hydrocarbons.

"Dormison."[321] Trademark for meparfynol (3-methyl-1-pentyn-3-ol) (q.v.).

"Dortan."[51] Trademark for light-colored, sulfurized cutting oils which permit work visibility.

"Dorvon" FR.[233] Trademark for molded polystyrene, flame-retardant foam. Used as a thermal insulation in building.

dosimetry (radiation). Measurement of the amount of radiation delivered to the body of an individual. The permissible dose is the quantity of radiation which may be received by an individual over a given period with no detectable harmful effects. For x- or gamma ray exposure the permissible dose is 0.3 roentgen per week, measured in air. All working with radioactive materials are expected to wear some device for detecting incident radiation.

DOT. Abbreviation for Department of Transportation, the agency responsible for shipping regulations for hazardous products in the U.S.

DOTG. Abbreviation for di-ortho-tolylguanidine (q.v.).

dotriacontane (dicetyl) $CH_3(CH_2)_{30}CH_3$ or $C_{32}H_{66}$.
Properties: Crystals; sp. gr. 0.823; b.p. 310°C; m.p. 70°C.
Use: Research.

DOTT. Abbreviation for di-ortho-tolylthiourea (q.v.).

double bond. See unsaturation.

double salt. A hydrated compound resulting from crystallization of a mixture of ions in aqueous solution. Common examples are the alums, made by crystallizing from solution either potassium or ammonium sulfate and aluminum sulfate; Rochelle salt (potassium sodium tartrate), made from a water solution of potassium acid tartrate treated with sodium carbonate; and Mohr's salt (ferrous ammonium sulfate), crystallized from mixed solutions of ferrous sulfate and ammonium sulfate. See also nickel ammonium sulfate.

"Dowanol."[233] Trademark for a series of glycol monoethers; used as solvents; intermediates for plasticizers; bactericidal agents; and fixatives for soap and perfumes.

"Dowcarb."[233] Trademark for calcium carbonate slurry for use in the paper industry.

"Dowclene."[233] Trademark for a series of solvents for specialized cleaning.
10. A stabilized emulsion of caustic soda, a detergent, and a sequestering agent.
EC. A colorless liquid; f.p. –56.6; b.p. 77–122°C; sp. gr. 1.381.
WR. Inhibited 1,1,1-trichloroethane.

"Dow Corning."[149] Trademark for a wide range of silicone and polysiloxane products, including emulsions, lubricants, greases, mold-release agents, laminating polymers, electrical varnishes, and heat-resistant coatings.

"Dowetch."[233] Trademark for magnesium photoengraving sheet, plate, and extruded tube. Also applies to chemicals used in the one-step engraving process.

"Dowex."[233] Trademark for a series of synthetic ion-exchange resins made from styrene-divinylbenzene copolymers, having a large number of ionizable or functional groups attached to this hydrocarbon matrix. These functional groups determine the chemical behavior and type of ion-exchange resin. The strong acid cation resins are capable of exchanging cations—for example, sodium for calcium and magnesium as in softening water. The strong base anion resins are capable of exchanging anions.

"Dowfax" 9N.[233] Trademark for a series of nonylphenolethylene oxide adducts; used as surfactants.
Uses: Textile, pulp and paper, leather, latex paint.

"Dowflake."[233] Trademark for calcium chloride, 77–80%, in a special flake form.

"Dowfrost."[233] Trademark for a heat-transfer medium consisting of inhibited propylene glycol; used for the immersion freezing of poultry and other foods.

"Dowfume."[233] Trademark for a series of proprietary products used as fumigants and pesticides.

"Dowicil" 100.[233] Trademark for 1-(3-chloroallyl)-3,5,7-triaza-1-azoniaadamantane chloride. Water-soluble bactericide with fungicidal properties. Used as preservative in latexes, polishes, adhesives, inks, starches, joint cements, etc.

"Dowmetal."[233] Trademark for a series of magnesium alloys containing more than 85% of magnesium.

"Dow-Per."[233] Trademark for perchloroethylene of a special drycleaning grade.

"Dowpon."[233] Trademark for a grass-killer, based on dalapon.

"Dowsol."[233] Trademark for alkyl phosphates for use in metal extraction.

"Dowtherm."[233] Trademark for a series of liquid heat-transfer media.

"Dowzene."[233] Trademark for formulations of piperazines; used in veterinary medicine.

doxorubicin. See adriamycin.

DOZ. Abbreviation for dioctyl azelate. See the more exact name, di(2-ethylhexyl) azelate.

D.P. Abbreviation for degree of polymerization (q.v.).

DPA. Abbreviation for diphenylamine; also for diphenolic acid.

DPG. Abbreviation for diphenylguanidine (q.v.).

DPH. Abbreviation for 1,6-diphenylhexatriene.

DPIP. See diphenyl isophthalate.

DPN. Abbreviation for diphosphopyridine dinucleotide. See nicotinamide adenine dinucleotide.

DPO. Abbreviation for 2,5-diphenyloxazole (q.v.).

DPPD. Abbreviation for N,N′-diphenyl-para-phenylenediamine (q.v.).

"DPS."[342] Trademark for bacitracin methylene disalicylate (q.v.).

"D.P. Solution."[58] Trademark for dibutylamine pyrophosphate (q.v.).

"Drakeol."[25] Trademark for a series of colorless, odorless, tasteless, white mineral oils.

"Dramamine."[70] Trademark for dimenhydrinate (q.v.).

"Drawcote."[204] Trademark for a class of compounds used as dry film lubricants in the cold working of metals.

"Draw-ex."[51] Trademark for a group of drawing compounds suitable for hot and cold metal working. Oil- and water-soluble grades are available, some pigmented.

"Dresinate."[266] Trademark for liquid, paste, and powder forms of sodium and potassium soaps of rosins and modified rosins used as emulsifiers, detergents, and dispersants in soluble oils, cleaning compounds, and other compositions.

"Dresinol."[266] Trademark for 40 to 45% solids dispersions of rosins, modified rosins, or ester resins using aqueous ammonia as a dispersant plus a protective colloid stabilizing agent.

"Drewmulse."[555] Trademark for a series of glycerol and glycol esters and sorbitan and polyoxyethylene sorbitan esters of fatty acids. Used as emulsifiers and opacifiers.

"Driacin."[416] Trademark for an ash-free oganic salt of a hydrophobic, film-forming corrosion inhibitor.
Uses: Sludge-dispersant additive for extending the storage life of cracked fuel oil and preventing filter or burner tip clogging; ingredient in rust-preventive formulations such as slushing oils.

"Dri-Clor."[204] Trademark for a powdered laundry bleach with not less than 38% available chlorine.

"Dricoid."[204] Trademark for a series of algin-emulsifier compositions.
"Dricoid." Sodium alginate.
K. Propylene glycol alginate-carrageenan.
KB. Propylene glycol alginate-vegetable gums.
Uses: Stabilizer-emulsifier compositions for ice cream and pressurized whipped cream.

"Drierite."[345] Trademark for a special form of anhydrous calcium sulfate having a highly porous granular structure and a high affinity for water. Absorbs water vapor both by hydration and capillary action.
Grades: Regular; Indicating (turns blue to red in use); "Du-Cal" (for drying air and gases.)
Uses: Drying of solids, liquids, gases.

drier. A substance used to accelerate the drying of paints, varnishes, printing inks and the like by catalyzing the oxidation of drying oils or synthetic resin varnishes, such as alkyds. The usual driers are salts of metals with a valence of two or greater and unsaturated organic acids. The approximate order of effectiveness of the more common metals is cobalt, manganese, cerium, lead, chromium, iron, nickel, uranium and zinc. These are usually prepared as the linoleates, naphthenates and resinates of the metals. Paste driers are commonly the metal salts as acetates, borates, or oxalates dispersed in a dry oil. See also soap (2).

Note: The spelling "dryer" refers to equipment used for drying (of paper, textiles, food products, etc.).

"Dri-Film."[245] Trademark for a group of silicone resins designed to impact durable moisture and weather resistance to surfaces.

Uses: Dri-Film 88 is used as a protective coating for electric motors, transformers, field coils, etc.; Dri-Film 144, a masonry water repellent; Dri-Film 1040 and 1042 imparts durable water repellency and other properties including water-borne spot and stain resistance, improved hand and drape, increased flex abrasion resistance, and improved tear strength and wrinkle recovery.

drilling fluid (drilling mud). A suspension of barytes and bentonite or attapulgite clay in either water or a petroleum oil. When circulated through oil-well drilling pipes, it acts as a coolant and lubricant and keeps the hole free from bore cuttings. To be effective it must have a specific gravity of at least 2.0 and should be thixotropic, with appreciable gel strength. Lignosulfonates are used as thinners in the water-based type. Oil-based muds require thickening additives such as blown asphalt and metallic soaps of tall oil and rosin acids.

"Driltreat."[236] Trademark for a phosphatide liquid used as a stabilizer in oil emulsion drilling muds to offset effects of water contamination. Also effectively controls filtration.

"Drimix."[267] Trademark for a form of dry dispersion of liquid or solid materials in a siliceous carrier. Also called dry liquid concentrates. Used to dry up liquids to convert them to free-flowing powders for easier handling in rubber and plastics industries.

"Drinalfa."[412] Trademark for methamphetamine hydrochloride (q.v.).

"Drinox."[401] Trademark for a series of insecticides, used in seed treatment.

H-34. A liquid containing 24.5% heptachlor.

PX. A planter-box seed treatment containing 25% heptachlor.

Hazard: Highly toxic by ingestion, inhalation and skin absorption.

"Driocel."[99] Trademark for a solid, granular desiccant for drying process liquids and gases. Manufactured from a selected grade of natural bauxite, reduced to the required particle size, and thermally activated to its maximum absorbing activity. Mesh grades 4/8, 4/10, 8/14.

"Driocel" S.[99] Trademark for a drying (dehydrating) medium for light liquid hydrocarbons. Developed specifically to overcome souring of light hydrocarbon liquids, such as LPG products in the final drying stage. Mesh grades 4/8, 8/14, tamped bulk density 55–57 lb/cu ft.

"Dri-Pax."[241] Trademark for a group of silica gel products used in packaging pharmaceuticals and similar packaged products. Extends shelf life and deodorizes products having unpleasant odors.

"Drisdol."[162] Trademark for crystalline vitamin D_2.

"Dri-Sol."[319] Trademark for low moisture nitrogen solutions used as, or in production of, soil fertilizers. Toral nitrogen, % by wt, 46.2 to 58.5; approx. sp. gr. at 60/60°F, 0.930 to 1.162; lb/gal at 60°F, 7.75 to 9.68. Are composed of free ammonia, ammonium nitrate, and less than 0.5% by wt. of water.

"Drisoy."[64] Trademark for a group of chemically treated soybean oils, for replacement of linseed oil.

"Dritomic."[50] Trademark for a micron-fine wettable sulfur agricultural fungicide.

"Dri-Tri."[1] Trademark for sodium phosphate, tribasic, Na_3PO_4.

"Driwall."[448] Trademark for transparent coatings (silicone); prevents water penetrating exterior walls of brick, stone, masonry.

dromostanolone propionate (USAN). 17beta-Hydroxy-2alpha-methylandrostan-3-one propionate; 2alpha-methylandrostan-17beta-ol-3-one propionate. $C_{23}H_{36}O_3$.

Properties: White crystals; m.p. 127–133°C. Soluble in alcohol, acetone, and other common organic solvents.

Grade: N.D.

Use: Medicine.

"Drucomine" 9650.[555] Trademark for a cationic fatty amino amide. Used as a softener and finisher.

drug. (1) A substance that acts on the central nervous system, e.g., a narcotic, hallucinogen, barbiturate, or psychotropic drug. (2) A substance that kills or inactivates disease-causing infectious organisms. (3) A substance that affects the activity of a specific bodily organ or function. (4) According to the FDA, a drug is a substance "intended for use in the diagnosis, cure, mitigation, treatment, or prevention of disease, or to affect the structure or function of the body." Under this definition, high-potency preparations of vitamins A and D are classified as drugs, though other vitamins are not; antiperspirants and suntan lotions are also considered to be drugs.

The Drug Enforcement Administration is the federal agency responsible for supervision and control of the manufacture and use of drugs, especially those in class (1). The term "ethical drug" is equivalent to "prescription drug", i.e., a drug sold only on a physician's prescription. See also pharmaceutical.

"Drupene."[555] Trademark for nonionic ethoxylated fatty alcohols; available as liquid or paste. Used as detergents and penetrants.

"Drusil."[555] Trademark for series of silicone emulsions used as water repellents.

"Drustat."[555] Trademark for a series of cationic fatty amino amide condensates; alkyl sulfates. Used as antistatic agents.

"Drutergent E."[555] Trademark for a fatty acid amine condensate.

dry cell (Lechanché cell). A primary battery having a zinc anode, a carbon or graphite cathode surrounded by manganese dioxide, and a paste containing ammonium chloride as electrolyte. Such batteries are not reversible and therefore have a limited operating life. Their chief use is in flashlights and similar devices requiring low voltage. See also battery.

dry chemical. A mixture of inorganic substances containing sodium bicarbonate (or frequently potassium bicarbonate) with small percentages of added ingredients to render it free-flowing and water repellent. Used as a fire extinguisher on fires in electric equipment, oils, greases, gasoline, paints and flammable gases.

"Dry-Flo."[53] Trademark for starch ester derivative containing hydrophobic groups. Especially free-flowing in dry state. Cannot be wetted with water, yet when moistened with a water-miscible solvent, it can be gelatinized and used to produce films with water-repellent properties.

Uses: Dusting powder; no-offset spray for printing; approved by FDA for use in foods.

"Dry Gas."[580] A proprietary preparation used as a gasoline line antifreeze. See antifreeze (2).

dry ice. See carbon dioxide (solid).

drying. (1) Polymerizaton of the glycerides of unsaturated vegetable oils induced by exposure to air or oxygen. See drying oil; drier.

(2) Removal of from 90 to 95% of the water from a material, usually by exposure to heat. Industrial drying is performed by both continuous and batch methods. The type of equipment and the temperatures used depend on the physical state of the material, i.e., whether liquid (solution or slurry), semiliquid (paste), solid units, or sheet. *Continuous drying:* the rotating-drum dryer is used for flaked or powdered products (soap flakes); a heated metal drum revolves slowly in contact with a solution of the material, the dry product being removed with a doctor knife. In paper manufacturing drying is performed by a battery

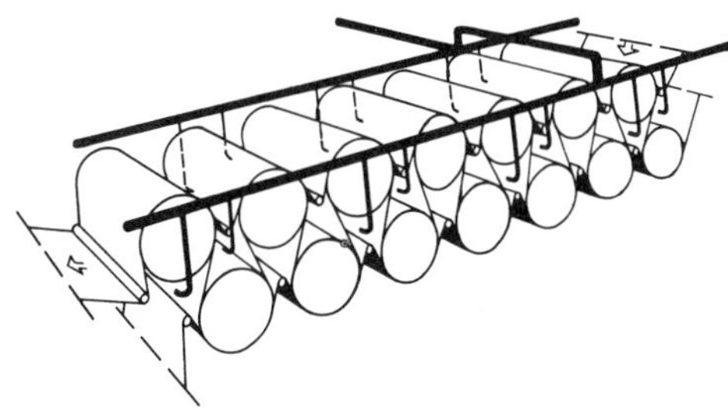

of staggered steam-heated revolving drums located at the dry end of the fourdrinier machine, the paper passing around the drums at high speed; the moisture content is thus reduced from 60% to about 5%. In spray drying, milk, egg-white and other liquid food products are passed through an atomizing device into a stream of hot air. In tunnel drying the product travels on a conveyor belt through a heated chamber of considerable length. *Batch drying:* steam-jacketed pans are used if the material is in paste or slurry form, or in removable trays placed in an oven, if in solid units (fruits, vegetables, meats, etc.). The revolving-tube dryer, used for granular solids and coarse powders, is a long, horizontal cylinder in which a current of warm air runs countercurrent to the movement of the material. Freeze-drying is a specialized technique utilizing high vacuum and low temperatures. See also dehydration; evaporation; freeze-drying.

drying oil. An organic liquid which, when applied as a thin film, readily absorbs oxygen from the air and polymerizes to form a relatively tough, elastic film. The oxidation is catalyzed by such metals as cobalt and manganese. See drier. Drying oils are usually natural products such as linseed, tung, perilla, soybean, fish, and dehydrated castor oils, but are also prepared by combination of natural oils or their fatty acids with various synthetic resins. The drying ability is due to the presence of unsaturated fatty acids, especialy linoleic and linolenic, usually in the form of glycerides. The degree of unsaturation of an oil, and hence its drying ability, is expressed by its iodine number (q.v.). The drying oils have the greatest capacity for iodine and the nondrying oils the least.

For further information refer to American Oil Chemists Assn., 35 E. Wacker Drive, Chicago, Ill.

"Drymet."[428] Trademark for sodum metasilicate, anhydrous.

"Dryolene."[200] Trademark for a petroleum solvent prepared by straightrun overhead distillation.

Properties: Water-white; boiling range 205–287°F; sp. gr. 0.747 (60°F); flash point (Tag closed cup) 25°F; wt/gal 6.22 lb (60°F).

Hazard: Flammable; dangerous fire risk.

Use: Paint and rubber cement for fast setting and relatively slow final drying.

"Dryorth."[428] Trademark for anhydrous sodium orthosilicate.

"Dryseq."[428] Trademark for sodium sesquisilicate, anhydrous (q.v.).

"Dryspersion."[267] Trademark for dry dispersion of rubber compounding chemicals in powder form. Deagglomerated and treated with non-staining oil.

"DS-207."[304] Trademark for a mixture of dibasic lead salts of C_{16}-C_{18} fatty acids.

Properties: Soft, white, unctuous powder; sp. gr. 2.0; refractive index, 1.6; non-melting at high temperatures; apparent lubricity of normal lead stearate.

Uses: Stabilizer-lubricants for vinyls.

DSP. Abbreviation for disodium phosphate. See sodium phosphate, dibasic.

DTA. See differential thermal analysis.

DTBP. Abbreviation for di-tert-butyl peroxide (q.v.).

DTDP. Abbreviation for ditridecyl phthalate (q.v.).

"Duclean."[28] Trademark for acids containing pickling inhibitors. Used in pickling iron and steel.

No. 1. Sulfuric acid, 60° Bé.; sp. gr. 1.706; f.p. below -10.8°C.

No. 2. Hydrochloric acid, technical; sp. gr. 1.142; f.p. below -40°C.

Containers: Tank cars and trucks.

Hazard: Toxic and corrosive liquids.

Dühring's rule. Relates the vapor pressures of similar substances at different temperatures. A straight or nearly straight line results if the temperature at which a liquid exerts a particular pressure is plotted on a graph against the temperature at which some similar reference liquid exerts the same vapor pressure. Water is most frequently used as a reference liquid since its vapor pressure at various temperatures is well known.

dulcin (para-phenetolecarbamide) $H_2NOCNHC_6H_4OC_2H_5$. This substance should not be confused with dulcitol (q.v.) which is sometimes called dulcin.

Properties: White, needle crystals or powder with taste about two hundred times as sweet as sugar. M.p. 173–174°C; soluble in alcohol and ether; moderately soluble in hot water.

Derivation: From para-aminophenol.

Uses: Artificial sweetening agent. Prohibited by FDA for use in foods.

dulcitol (dulcite; dulcin; dulcose). $C_6H_8(OH)_6$. A sugar.

Properties: White, crystalline powder; slightly sweet taste. Soluble in hot water; slightly soluble in cold water; very slightly soluble in alcohol. Sp. gr. 1.466; m.p. 188.5°C.

Derivation: By hydrogenation of lactose; occurs naturally in Melampyrum nemorosum.
Grades: Technical; reagent.
Containers: Glass bottles; fiber containers.
Uses: Bacteriology; medicine.

Dulong and Petit's law. The atomic heat capacity (atomic weight times specific heat) of elementary substances is a constant whose average value at room temperature is 6.2. A few elements, notably boron, carbon, and silicon, obey the law only at high temperatures.

dumortierite. Perhaps $8Al_2O_3 \cdot 6SiO_2 \cdot B_2O_3 \cdot H_2O$. A natural basic aluminum silicate. Bright, smalt-blue to greenish-blue in color. Vitreous luster. Sp. gr. 3.26–3.36; Mohs hardness 7.
Occurrence: United States (New York, Arizona, Nevada); France; Silesia; Norway.
Uses: Spark-plug porcelain; special refractories.

"Duobel."[28] Trademark for high-velocity permissible explosives furnished in five grades based upon velocity and cartridge count.
Hazard: Explosion risk.
Use: For mining coal where lump coal is not a factor.

"Duocarb."[155] Trademark for carbonized nickel from which all excess carbon has been removed and both surfaces polished.

"Duoferm."[319] Trademark for antibiotic feed supplement containing vitamin B_{12} and bacitracin.

"Duolite."[244] Trademark for anion and cation exchange resins, electron exchanger and resinous adsorbent.

"Duomac."[15] Trademark for N-alkyl trimethylenediamine diacetate salts made from long-chain amines.
Containers: 55-gal open head drums.
Uses: Algicides; corrosion inhibitors; flushing agents; fungicides.

"Duomeen."[15] Trademark for a series of N-alkyl trimethylenediamines. The alkyl group ranges from C_8 to C_{18}. Available in mixtures derived from coco, soya or tallow fatty acids.
Containers: 55-gal drums; tank cars; tank trucks.
Uses: Anti-icers; biocides; bonding agents; chemical intermediates; corrosion inhibitors; dispersants; emulsifiers; mineral flotation and flocculation; wetting agents.

"Duo-Sol" process. A proprietary process for refining lubricating oils by extraction with a solvent consisting of liquid propane and a cresol base.

"Duponol."[28] Trademark for a series of surface-active agents based on lauryl sulfate. These have detergent emulsifying, dispersing, and wetting properties, and are used in the textile, paper, leather, cosmetic, electroplating industries; in dental and medical preparations, and agricultural products.

"Duracillin."[100] Trademark for procaine penicillin G, U.S.P.

"Duradene."[278] Trademark for a stereoregular copolymer of butadiene and styrene with exceptional purity. Soluble in aromatics and alipahtics.
Uses: Rubber products having superior abrasion resistance, good wet-skid resistance, flex-cracking resistance and good resilience. Automotive tire treads, belt conveyors, shoe soles and resilient products.

duralumin. A high-strength aluminum alloy. It contains 4% copper, 0.5% magnesium, 0.25–1.0% manganese, and low percentages of iron and silicon. Resistant to corrosion by acids and sea water.
Uses: Aircraft parts, railroad cars, boats, machinery.

"Duramir."[244] Trademark for a chromic acid formulation that has been significantly modified in many plating properties by the addition of special catalysts. Used in plating steel, copper, brass and bright nickel.

"Duran."[573] Trademark for a series of polyvinylidine chloride packaging materials.

"Durana."[197] Trademark for a nitrogen fertilizer solution used in the manufacture of solid fertilizer compositions.

"Duranickel" 301.[283] Trademark for an alloy of 94% nickel and 4.5% aluminum.

"Duraphos."[55] Trademark for a complex sodium-calcium-aluminum polyphosphate containing 67% phosphate. Used as a slow-dissolving glassy phosphate.

"Duraplex."[23] Trademark for drying and non-drying oil-modified alkyd resins derived from phthalic anhydride, polyhydric alcohols, and vegetable oils. Air drying, baking, and non-drying grades, in solvent solution or viscous 100% resins. Produce tough, glossy, light-colored coatings with excellent durability.
Uses: Primers, lacquers, and enamels; metal decorating; automotive coatings; furniture finishes; architectural enamels; inks.

"Duratone."[236] Trademark for a synthetic organophilic colloid used to control filtration and aid in suspending drilled solid particles in oil-well muds. It is effective at high temperature, and is affected little by dissolved salts in the emulsified water phase.

"Durax."[69,265] Trademark for N-cyclohexyl-2-benzothiazolesulfenamide.
Uses: Primary accelerator for SBR.

"Durco" D-10.[47] Trademark for a nickel-base alloy. Typical analysis: nickel, 65%; chromium, 23%; iron, 5%; copper, 3.5%; molybdenum, 2%; manganese, 1%.
Properties: Wt. 0.310 lb/in.^{3}; tensile strength, 60,000 psi; hardness (Brinell) 180; corrosion-resistant; machinable.

"Durcon" 6.[47] Trademark for an epoxy resin; corrosion resistant to a variety of materials while retaining high impact resistance and low moisture absorption.
Properties: Wt. 0.069 lb/in.^{3}; tensile strength, 13,000 psi; hardness (Rockwell M) 114.

durene (durol; sym-1,2,4,5-tetramethylbenzene) $C_6H_2(CH_3)_4$.
Properties: Colorless crystals; camphorlike odor. Soluble in alcohol, ether, and benzene; insoluble in water. Sublimes and is volatile with steam. Sp. gr. 0.838; m.p. 77°C; b.p. 196–197°C; flash point 165°F (COC). Combustible.
Derivation: By heating ortho-xylene and methyl chloride in presence of aluminum chloride. Occurs in coal tar.
Use: Organic synthesis; plasticizers; polymers; fibers.

"Durez."[62] Trademark for synthetic resins, plastics, molding compositions, phenolformaldehyde resins, furfuryl alcohol resins, polyester resins, and polyurethane resins.
Uses: Industrial chemicals; curing agents, rubber accelerators, plasticizers and binders.

"Durichlor" 51.[47] Trademark for a high silicon iron alloy. Typical analysis: silicon, 14.5%; chromium, 4.5; carbon 0.9%; manganese 0.65%.
Properties: Wt. 0.255 lb/in.^{3}; hardness (Brinell) 520; tensile strength, 16,000 psi; resistant to moist chlorine, ferric chloride, chlorinated solutions, and hydrochloric acid.

"Durimet 20."[47] Trademark for an austenitic stainless steel. Typical analysis; nickel, 29%; chromium, 20%; copper, 3%; molybdenum, 2%; silicon, 1%; carbon 0.07% max.
Properties: Wt., 0.287 lb/in.^{3}; tensile strength, 62,500 psi; hardness (Brinell) 120–150; machinable; sulfuric acid-resistant; weldable.

"Duriron."[47] Trademark for a high-silicon iron alloy. Typical analysis: silicon, 14.5%; carbon, 0.9%; manganese, 0.65%.
Properties: Wt. 0.255 lb/in.^{3}; hardness (Brinell) 520; tensile strength, 16,000 psi; resistant to abrasion, erosion, and corrosion; machined by grinding.

"Durisite."[326] Trademark for a furan acid- and alkali-resistant resin cement mortar.

"Durite."[65] Trademark for a series of phenolformaldehyde resins used in the manufacture of grinding wheels, brake linings, clutch facings, lamp-basing cements.

"Duro."[446] Trademark for an acid-proof brick made from materials of very low flux content. When fired stable minerals are formed which are insoluble in various acids and corrosive acidic solutions and resist alkaline solutions. Insoluble in mineral acids, except hydrofluoric.
Uses: Domes and dished bottoms of cylindrical tanks; capping tile; separators for electrolytic tanks; girders and supports for acid tower packing.

"Durobrite."[28] Trademark for zinc cyanide plating brightening agents. Amber liquids.
Hazard: Highly toxic.

"Durol."[243] Trademark for milling blacks for wool and nylon.

Durometer hardness. See hardness.

"Duron."[116] Trademark for a malleable cast iron.

"Durpon" Compounds.[309] Trademark for drawing lubricants and compounds for ferrous metals, steels and alloys, ultra high strength metals and refractories. Available in both oil soluble and water soluble series which include EP additives.

"Dursban."[233] Trademark for insecticides containing O,O-diethyl O-3,5,6-trichloro-2-pyridyl phosphorothioate.
Hazard: Highly toxic. Use may be restricted.
Uses: Control of chinch bugs in Gulf Coast states; tick control on cattle and sheep (Australia).

dust, industrial. Finely divided solid particles that may have damaging effects on personnel by inhalation, or that constitute a fire hazard. They are air suspensions of particles 10 microns or less in diameter, though sizes up to 50 microns may be present. Such dusts include (1) metallic particles of all types, some being more harmful than others; (2) silica, mica, talc, quartz, graphite, clays, calcium carbonate, asbestos; (3) organic materials such as chemicals, pesticides, flour or other cereals, cellulose, coal, etc. The size range of mineral dust most damaging to the lungs is between 0.5 and 2 microns. Silicosis is caused by chronic exposure to uncombined silica (quartz, cristobalite) in mines and quarries. Dust suspensions in enclosed industrial areas are an imminent fire hazard regardless of the chemical nature of the dust; an explosion may be initiated by static sparks or any open flame. Baghouses, dust collectors, or electrostatic precipitators can be used for dust control. TLV tolerances for industrial dusts are given by ACGIH. See threshold limit value.

dusting agent. A powdery solid used as an abherent and mold-release agent in the plastics and rubber industries. Typical materials in general use are talc (soapstone), mica, slate flour, and clay. Graphite and mica have a flat crystal structure which causes them to act as lubricants, and thus are especially effective in preventing sheets or slabs of hot solid mixtures from sticking together when stacked.

Dutch oil. See ethylene dichloride.

Dutch process. Process for making white lead. See lead carbonate, basic.

"Dutrex."[125] Trademark for a series of aromatic hydrocarbon concentrates derived from petroleum.
Properties: Colors ranging from light amber to black; viscosities from 3 cp to 10,000 cp at 210°F; sp. gr. approx. 1.0; odor slight to none; very low volatilities.
Hazard: Probably flammable.
Uses: Rubber processing and extending oils; plasticizer extenders for polyvinyl chloride; resin solvents; tackifier for nitrile-butadiene rubber.

DVB. Abbreviation for divinylbenzene.

Dy Symbol for dysprosium.

"Dyal."[141] Trademark for a series of resins and resin solutions of the alkyd type.

"Dyasist."[300] Trademark for a series of dyeing assistants.

"Dybar."[28] Trademark for pellets containing 25% fenuron (q.v.).

dyclonine hydrochloride (4′-butoxy-3-piperidinopropiophenone hydrochloride) $C_{18}H_{27}NO_2 \cdot HCl$.
Properties: Crystals; m.p. 175–176°C; soluble in water, alcohol, acetone. Phenol coefficient 3.6.
Grade: N.D.
Use: Medicine.

"Dycril."[28] Trademark for a photosensitive plastic bonded to steel, aluminum or "Cronar" base supports. Used in printing plates. Exposure to an ultraviolet light source renders exposed areas insoluble to a subsequent mild alkaline washout solution so that a relief image of the exposed pattern remains. Depth of image is dependent on the thickness of the photosensitive plastic layer. There are eight types available.

dye, certified. See food color; FD & C colors.

dye, fiber-reactive. A synthetic dye containing reactive groups capable of forming covalent linkages with certain portions of the molecules of natural or synthetic fibers, i.e., covalently bound azo dye on cellulose. This problem was not solved till 1953, when chlorotriazinyl dyes were introduced. Since then, hundreds of fiber-reactive dyes for cellulose have been patented.

dyeing assistant. Any material added to a dye bath to promote or control dyeing. The action of assistants differs with the classes of dyes, but in most cases they aid in level deposition of the dye, either by delaying its absorption, increasing its solubility, or assisting the dye solution to penetrate the material. Examples: sodium sulfate decahydrate; pine oils; alkylaryl sulfonates.

dyeing, solvent. The use of organic solvents instead of water in dyeing synthetic fibers and fabrics. The solvents used are the chlorinated hydrocarbons: 1,1,1-trichloroethane, trichloroethylene, and perchloroethylene. They are applied most successfully to nylon, polyester and acrylic fibers; the process is especially applicable to carpet dyeing.

dye intermediate. See also intermediate. Specifically, most dye intermediates are ring compounds, usually containing amino, hydroxy, sulfonic acid or quinone groups or their combinations. An archaic nomenclature is still common, including proper names (Cleve's acid) and alphabetical names (J acid; 2R acid). Some are toxic and flammable.

dye, metal. An organic dye suitable for use with a metal such as aluminum or steel. Alizarin Cyanin RR, Alizarin Green S, Nigrosine 2Y and Naphthalene Blue RS are used for this purpose.

dye, natural. An organic colorant obtained from an animal or plant source. Few of these are now in major use. Among the best known are madder, cochineal, logwood, and indigo. The distinction between natural dyes and natural pigments is often arbitrary. (See colorant; pigment).

dye retarding agent. An additive to dye baths to prevent, by decreasing absorption of the dyes, the rapid exhaustion of the bath. Examples: sodium lauryl sulfate; sulfonated oils.

dye, synthetic. An organic colorant, derived from coal tar- and petroleum-based intermediates and applied by a variety of methods, to impart bright, permanent colors to textile fibers, leather, plastics, rubber, aluminum, and other industrial products. They were first synthesized by Perkin (q.v.) in England (1856) and were later developed by Hofmann (q.v.) in Germany. Some, termed "fugitive," are unstable to sunlight, heat, and acids or bases; others, called "fast," are not. Direct (or substantive) dyes can be used effectively without "assistants"; indirect dyes require either chemical reduction (vat type) or a third substance (mordant), usually a metal salt or tannic acid, to bind the dye to the fiber. A noteworthy development (1953) is the fiber-reactive group (q.v.), wherein the dye reacts chemically with cellulose. Dyes may be either acidic or basic, and their effectiveness on a given fiber depends on this factor. Some types are soluble in water; others are not, but can be made so by specific chemical treatment.

The central problem is the affinity of a dye for a fiber, and this involves both the chemical nature and the physical state of the dye, i.e., whether acidic or basic, and whether colloidal, molecular, or ionic. Neither colloidal particles nor dissociated ions are accepted by a fiber: the dye must be in molecular dispersion to be effective. Further details will be found in the following entries: acetate dye, anthraquinone dye, acid dye, azo dye, alizarin, aniline, fiber-reactive dye, eosin, intermediate, resist, stilbene dye, sulfide dye. The chemical classes of coloring matters and their arrangement according to chemical structures have been designated numerically in the "Colour Index," a British publication. See also colorant; pigment.

dye, disperse. A dye that may be in any of three clearly defined chemical classes: (a) nitroarylamine; (b) azo and (3) anthraquinone, and almost all contain amino or substituted amino groups but no solubilizing sulfonic acid groups. They are water-insoluble dyes introduced as a dispersion or colloidal suspension in water and are absorbed by the fiber after which they may remain untreated or be after-treated (diazotized) to produce the final color. Their use is primarily for cellulose acetate and nylon, polyester and other synthetic fibers and for thermoplastics.

"Dykon."[553] Trademark for sodum diacetate (q.v.).

"Dylan."[11] Trademark for polyethylene made by the high pressure process.

Uses: Film packaging, flexible pipe, squeeze bottles, wire insulation, filaments, paper and cloth laminates, housewares. See "Super Dylan."

"Dylene."[11] Trademark for polystyrene; available in regular, medium, and high-impact grades.

"Dylex" Latices.[11] Trademark for acrylic, styrene-butadiene latices. Used in exterior and interior paints.

"Dylite."[11] Trademark for expandable polystyrene beads and pellets. They expand on application of heat to a foam with completely closed cell structure. Used as thermal insulator, protective packaging, flotation equipment, and containers.

"Dylopon."[300] Trademark for a low-temperature dyeing auxiliary for wool.

"Dylox."[181] Trademark for trichlorfon (q.v.).

"Dylux."[28] Trademark for a photosensitive paper, free from silver compounds, and coated with organic dyes that are activated by UV radiation. Gives good continuous tones and high resolution. Used in lithography, photoproofing, and information-handling systems. Permits instantaneous direct recording without processing.

"Dymal."[141] Trademark for resins of the maleic anhydride or fumaric acid alkyd type.

"Dymerex" Resin.[266] Trademark for a light-colored, hard, thermoplastic resin; dimerized resin acids; softening point 152°C; acid number 135; USDA color N. Available in solid and flake forms.

"Dymetracin."[342] Trademark for sodium bacitracin methylene disalicylate (q.v.).

"Dymid."[530] Trademark for diphenamid (q.v.).

dynamite. An industrial explosive detonated by blasting caps. The principal explosive ingredient is nitroglycerin (straight dynamite) or especially sensitized ammonium nitrate (ammonia gelatin dynamite) dispersed in carbonaceous materials. Diethyleneglycol dinitrate, which is also explosive, is often added as a freezing point depressant. A dope such as wood pulp, and an antacid, as calcium carbonate, are also essential.

Hazard: Of moderate sensitivity to shock or heat, dynamite is a rather serious explosion hazard; also may ignite or explode as a result of contact with powerful oxidizing agents.

Shipping regulations: (Rail) Consult regulations. (Air) Not acceptable.

See also explosive.

Note: Use of dynamite is decreasing; it may be entirely replaced by safer and more efficient explosives within a few years.

"Dyne."[284] Trademark for "Tamed Iodine" complex used as a non-foam pipeline detergent-germicide for dairy industry.

"Dynel."[214] Trademark for copolymer of vinyl chloride and acrylonitrile used as a textile fiber. A modacrylic fiber. Manufacture discontinued in 1975.

"Dyphene."[141] Trademark for pure phenolic resins characterized by extreme hardness, excellent chemical resistance and fast-drying properties.

"Dyphenite."[141] Trademark for rosin-modified phenolic resins.

"Dyphos."[304] Trademark for lead phosphite, dibasic. Used as a stabilizer for vinyls and other chlorinated resins.

dyphylline [7-(2,3-dihydroxypropyl)theophylline; hyphylline] $C_{10}H_{14}N_4O_4$.
Properties: Crystals; bitter taste; m.p. 158°C; soluble in water, alcohol, and chloroform.
Use: Medicine.

dypnone (phenyl alpha-methylstyryl ketone; 1,3-diphenyl-2-butene-1-one) $C_6H_5COCHC(CH_3)C_6H_5$.
Properties: Stable, light-colored liquid with a mild fruity odor. B.p. (50 mm) 246°C; sp. gr. 1.093 (20/20°C); f.p., sets to a glass below -30°C; almost insoluble in water; flash point 350°F. Combustible. Low toxicity.
Uses: Softening agent, platicizer, and perfume base. High absorption of ultraviolet light, low water solubility, low evaporation rate, and good solvent action make it useful in light-stable coatings.

"Dyrene."[181] Trademark for 2,4-dichloro-6-ortho-chloroaniline-s-triazine (q.v.)

dysprosia. See dysprosium oxide.

dysprosium Dy Rare earth or lanthanide element having atomic number 66; atomic weight 162.5; valence 3. Six stable isotopes.
Properties: Non-corrosive metal; does not react with moist air to form hydroxide; m.p. 1407°C; b.p. 2330°C; sp. gr. 8.54. Reacts slowly with water and halogen gases; soluble in dilute acids. High cross-section for thermal neutrons. Low toxicity.
Derivation: Reduction of the fluoride with calcium.
Forms: High-purity lumps; ingots; sponge; powder.
Uses: Measurement of neutron flux; reactor fuels; fluorescence activator in phosphors.
See also rare earth metal.

dysprosium nitrate $Dy(NO_3)_3 \cdot 5H_2O$.
Properties: Yellow crystals; m.p. 88.6°C (in its water of crystallization). Soluble in water.
Derivation: Treatment of oxides, carbonates or hydroxide with nitric acid.
Grades: Up to 99.9% pure.
Containers: Glass bottles; fiber drums.
Hazard: Strong oxidizing agent; may ignite organic materials.
Shipping regulations: (Rail) Nitrates, n.o.s., Yellow label. (Air) Oxidizer label.

dysprosium oxide (dysprosia) Dy_2O_3.
Properties: White powder; much more magnetic than ferric oxide; slightly hygroscopic, absorbing moisture and carbon dioxide form the air. Sp. gr. 7.81 (27/4°C); m.p. 2340 ± 10°C. Soluble in acids and alcohol.
Derivation: Ignition of hydroxides and oxyacids (carbonates, oxalates, sulfates, etc.).
Containers: Glass bottles; fiber drums.
Uses: With nickel, in cermets used as nuclear reactor control rods that do not require water-cooling.

dysprosium salts. Salts available commercially are the chloride, $DyCl_3 \cdot xH_2O$; fluoride, $DyF_3 \cdot 2H_2O$; arsenide, antimonide and phosphide. The last three are used as high-purity binary semiconductors.

dysprosium sulfate $Dy_2(SO_4)_3 \cdot 8H_2O$.
Properties: Brilliant yellow crystals; stable at 110°C and completely dehydrated at 360°C. Soluble in water.
Derivation: Dissolving the hydroxide, carbonate or oxide in dilute sulfuric acid.
Use: Atomic weight determination.

"Dyterge."[300] Trademark for a one-bath scouring and dyeing assistant for wool and wool blends.

"Dythal."[304] Trademark for lead phthalate, dibasic (q.v.). Used as a stabilizer for vinyl resins.

"Dytol."[23] Trademark for aliphatic primary alcohols derived from natural fats and oils. (C_{10} to C_{18}).
Uses: Additives for cosmetic creams, polymerization regulators for elastomers and plastics, detergents and viscosity index improvers for lubricating oils, finishing and softening agents for textiles, preparation of quaternary ammonium compounds, surfactants, water evaporation control and anti-foam.

E

EADC. Abbreviation for ethylaluminum dichloride, q.v.

EAK. Abbreviation for ethyl amyl ketone, q.v.

earth. (1) Any siliceous or clay-like compound or mixture, e.g., fullers earth, diatomaceous earth (q.v.). (2) A natural metallic oxide, sometimes used as a pigment, e.g., red and yellow iron oxide, ocher, or umber. (3) A series of chemically related metals that are difficult to separate from their oxides or other combined forms; specifically, rare earths, alkaline earths. See also specific entry.

earth wax. General name for ozocerite, ceresin and montan waxes. See also wax.

EASC. Abbreviation for ethylaluminum sesquicloride (q.v.).

"Eastacryl."[256] Trademark for dyes for acrylic fibers.

East India. A type of fossil or semirecent resin similar to dammar (q.v.). Varieties are batu, black, and pale. Used in spirit and oleoresinous varnishes.

"Eastobond."[256] Trademark for a series of hot-melt adhesives used in the packaging industry for bonding paper, board, film, foil and glassine.

"Eastone.'[256] Trademark for disperse dyestuffs for use with acetate and nylon fibers.

"Eastozone."[256] Trademark for rubber antiozonants based on the para-phenylenediamine.

"EB-5."[233] Trademark for fumigant containing approximately 5% ethylene bromide (q.v.). Concentrations other than 5% are indicated by appropriate number.
Hazard: May be toxic. See ethylene dibromide.

ebonite. See rubber, hard.

"Ebonol."[142] Trademark for various blackening agents for metals. Materials are supplied as powders that are added to water at various temperatures to accomplish blackening.

"Eclipse."[492] Trademark for a complete series of thin-boiling starches developed for the textile industry. They are based on corn starch which has been partially hydrolyzed by sulfuric acid, weakening the cross-linkages and breaking some of the chains.
Uses: Warp sizing, especially in the sizing of spun synthetics, particularly viscose and viscose-acetate blends. Lettered alphabetically in order of increasing fluidity.

"Ecolid" Chloride.[305] Trademark for chlorisondamine chloride (q.v.).

"Econo-chrome."[244] Trademark for a wide spectrum chromic acid compound containing a special catalyst. Used in both decorative and hard chromium plating.

economic poison. See pesticide.

economics, chemical. See chemical economics.

"Econo-Sour."[244] Trademark for a product consisting chiefly of fluorine compounds. White powder; sparingly soluble in water.
Use: Laundry sour.

"Econovat."[496] Trademark for a high molecular weight carbohydrate natural gum blend; used as a base for printing emulsion.

ECTEOLA-cellulose (epichlorohydrin triethanolamine cellulose). A dry powdered celleulose derivative containing tertiary amine groups which is used as an anion exchanger in chromatography. It is less basic than DEAE-cellulose (q.v.) and serves to separate viruses, nucleic acids, and nucleoproteins.

EDB. Abbreviation for ethylene dibromide (q.v.).

Edeleanu process. A solvent extraction process using liquid sulfur dioxide for the removal of undesirable aromatics from heavy lubrication oils.

edestin. A protein having a molecular weight of 310,000; source of leucine.

edible oil. As commonly used, the term refers to any fatty oil obtained from the flesh or seeds of plants that is used primarily in foodstuffs (margarine, salad dressing, shortening, etc.). Among these are olive, safflower, cottonseed, coconut, peanut, soybean and corn oils, some of which may be hydrogenated to solid form. They vary in degree of unsaturation, ranging from 78% for safflower to about 10% for coconut. Castor oil, though technically edible, is not usually considered in this classification, nor are medicinal oils derived from animal sources (cod liver, mineral oil, etc.).

"Edifas."[206a] Trademark for methyl ethyl cellulose (q.v.).

EDTA. Abbreviation for ethylenediaminetetraacetic acid (q.v.).

EDTAN. Abbreviation for ethylenediaminetetraacetonitrile (q.v.).

EDTA Na_4. Abbreviation for ethylenediaminetetraacetic acid tetrasodium salt. See tetrasodium EDTA.

edetate. See ethylenediaminetetraacetic acid (note).

effect. An evaporation-condensation unit. See evaporation.

efflorescence. Loss of combined water molecules by a hydrate when exposed to air, resulting in partial decomposition indicated by presence of a powdery coating on the material. This commonly occurs with washing soda ($Na_2CO_3 \cdot 10H_2O$) which loses almost all its water constituent spontaneously.

effluent. Any gas or liquid emerging from a pipe or similar outlet; usually refers to waste products from chemical or industrial plants as stack gases or liquid mixtures.

egg yolk
Properties: Yellow semi-solid mass; sp. gr. 0.95; m.p. 22°C; high cholesterol content.

Grades: Technical; edible.
Uses: Baking; dairy products; mayonnaise; pharmaceuticals; soap; perfumery. See also albumin, egg.

EHEC. Abbreviation for ethyl hydroxyethyl cellulose (q.v.).

Ehrlich, Paul (1854–1915). A native of Silesia, Ehrlich is considered the founder of the science of chemotherapy, or the treatment of diseases by chemical agents. He did fundamental work on immunity which gave him the Nobel prize in medicine in 1908, and also developed the famous neoarsphenamine (salvarsan or 606) treatment for syphilis (1910) which was not improved upon until the discovery of penicillin.

eicosane $C_{20}H_{42}$. Most technical eicosane is a mixture of predominantly straight-chain hydrocarbons averaging 20 carbon atoms to the molecule.
Properties (pure n-eicosane): White crystalline solid; f.p. 36.7°C; b.p. 205°C (15 mm); refractive index 1.4348 (n 20/D); sp. gr. 0.778 (at melting point). Insoluble in water; soluble in ether. Can be readily chlorinated. Combustible.
Grades: Pure normal (99 + %); technical.
Uses: Cosmetics; lubricants; plasticizers.

eicosanoic acid. See arachidic acid.

1-eicosanol. See arachidyl alcohol.

5,8,11,14-eicosatetraenoic acid. See arachidonic acid.

einsteinium Es A synthetic radioactive element with atomic number 99 first discoverd in the debris from the 1952 hydrogen bomb explosion. Eisteinium has since been prepared in a cyclotron by bombarding uranium with accelerated nitrogen ions, in a nuclear reactor by irradiating plutonium or californium with neutrons, and by other nuclear reactions. The element is named for Albert Einstein. It has chemical properties similar to those of the rare earth holmium. Isotopes are known with mass numbers ranging from 246–253.
See also actinide element.

ekahafnium. One of the recently discovered transuranic elements with atomic number 104. It has two alpha-emitting isotopes (257 and 259) and possibly a third (258). The former, made by bombardment of californium-249 with carbon-12 nuclei, has a half-life of 5 seconds and decays into nobelium-253. The 259 isotope, made by merging a carbon-13 nucleus with californium-249, has a half-life of 3 seconds and decays to nobelium-255.

"Ekonol."[280] Trademark for an engineering plastic composed of poly-p-oxybenzoate. Resistant to temperatures above 600°C; self-lubricating surface. Possible uses are in pumps handling corrosive liquids, protective coating for titanium skins on supersonic transports, disk brakes, etc.

"ELA."[28] Brand name for elastomer lubricating agent, a mixture of phosphate esters. Light amber liquid. Used for unvulcanized rubbers.

elaboration. A term used in biochemistry to describe chemical transformations within an organism resulting in formation of specific types of substances; for example, plants elaborate proteins and fats, and poisonous snakes elaborate their venom. It also refers to formation of metabolic end products such as purines and uric acid.

elaidic acid (trans-9-octadecenoic acid) $CH_3[CH_2]_7CH{:}CH[CH_2]_7COOH$. The trans form of an unsaturated fatty acid of which the cis-form is oleic acid (q.v.).
Properties: White solid; sp. gr. 0.8505 (79/4°C); m.p. 43.7°C; b.p. 288°C (100 mm); 234°C (15 mm); refractive index 1.4358 (79°C). Insoluble in water; soluble in alcohol, ether, benzene and chloroform. Combustible.
Derivation: Synthesized from oleic acid by elaidinization (q.v.)..
Grade: Purified, 99+%.
Uses: Medical research; reference standard in chromatography.

elaidinization. Originally, the reaction by which oleic acid is converted into elaidic acid but now used in a more general sense to indicate the conversion of any unsaturated fatty acid or related compound from the geometric cis to the corresponding trans form. Nitrous acid and selenium compounds are commonly used as catalysts for this reaction. The resulting trans acids are more stable to oxidation.

"Elastex."[175] Trademark for a series of plasticizers.

elastic modulus. See modulus of elasticity.

elasticity Recovery, partial or complete, of original shape after deforming forces are removed. One measure of deformation or strain is the Poisson ratio (the ratio of the decrease in thickness or diameter to the elongation that occurs when a deforming force is applied). In elastic bodies the deformation or strain always results in a stress, or tendency to resume the original shape. Stress is expressed as force per unit area. Young's modulus is the ratio of stress to strain in elongation or retraction. Fundamental particles of matter are perfectly elastic, i.e., atomic collisions result in no deformation or energy loss. Glass is 100% elastic, but has extremely low elongation; vulcanized rubber is over 90% elastic after extension to break, with exceedingly high elongation. See also modulus of elasticity; elastomer.

elastin. A scleroprotein which occurs in elastic tissue.
Properties: Yellow fibrous mass; insoluble in water, in dilute acids, alkalies and salt solutions, and in alcohol. Is partially digested by pepsin solution and wholly by trypsin.

elastomer. A general term introduced about 1940 to embrace all high polymers having the property of extensibility and elastic recovery, i.e., the ability to be stretched to at least twice their original length and to retract very rapidly to approximately their original length when released. Before the advent of synthetic rubber-like polymers, vulcanized natural rubber was the only material that displayed this property. See also rubber; rubber, synthetic; rubber, natural; elasticity.

"Elastopar."[58] Trademark for N-methyl-N,4-dinitrosoaniline (33 1/3%) plus diluent (66 2/3%). Used for chemically modifying butyl rubber.

"Elastothane."[27] Trademark for urethane rubber.
Uses: Conveyor rollers and belting; hose liners and covers; seals, packings and gaskets; other molded and extruded goods.

"Elbelan."[232] Trademark for a series of neutral-dyeing pre-metallized dyestuffs of outstanding fastness properties.

"Elchem."[28] Trademark for a series of addition agents used for electroplating brighteners in copper cyanide plating baths.

"Elcide."[530] Trademark for thimerosal (q.v.).

electric steel. Steel made in an electric furnace.

electrochemical equivalent. The number of grams of an element or group of elements liberated by the passage of one coulomb of electricity (one ampere for one second).

Electrochemical Society. Established in 1902, this society was organized to promote the advance of the science of electrochemistry and related fields. It is comprised of eleven divisions, each devoted to a special branch of electrochemistry, e.g., corrosion, batteries, rare metals, electrodeposition, etc. It publishes a journal and sponsors books relating to its major interests. Its office is at 30 E. 42nd Street, New York.

electrochemistry. A branch of chemistry which deals with the chemical changes produced by an electric current and with the production of electricity from the energy released in a chemical reaction. Applications are in battery science, electroplating, purification of copper, production of aluminum, fuel cells, and corrosion of metals. It also has certain applications in biochemistry, e.g. nerve reactions and specialized electric organs in fish.

electrocoating. A process of applying primer paint to household appliances, automobiles, etc., in which the metal piece to be coated becomes the anode in a tank of water-based paint. The coating deposited on the metal is uniform regardless of the shape of the article. Large-scale use of electrocoating is on automobile bodies. See also electrostatic coating.

electroconductive polymer. See polymer, electroconductive.

electrocratic. Descriptive of a liquid colloidal dispersion of insoluble solid particles whose stability is maintained by either positive or negative charges on the particles. As the like charges are mutually repellent, they offset the attraction of gravity and prevent the particles from sinking or coalescing. A colloidal gold suspension is a well-known example. Electrocratic dispersions can be readily precipitated by addition of an oppositely charged electrolyte, as in the purification of water with Al^{+++} ions from aluminum sulfate.

"Electrocure" Trademark for a process for hardening or curing paint films on plastics by use of low-voltage electron beams. Paints so treated cure in four seconds or less at room temperature; the finishes produced are said to be superior in resistance to abrasion and chemicals. The process can also be applied to other substrates, such as wood, glass, metals, etc.

electrode. Either of two substances having different electromotive activity, which enables an electric current to flow in the presence of an electrolyte. Electrodes are sometimes called plates or terminals. Commercial electrodes are made of a number of materials which vary widely in electrical conductivity, i.e., lead, lead dioxide, zinc, aluminum, copper, iron, manganese dioxide, nickel, cadmium, mercury, titanium, and graphite; research electrodes may be calomel (mercurous chloride), platinum, glass, or hydrogen. Electrodes are essential components of both batteries and electrolytic cells; in batteries the negative plate is the anode and the positive plate the cathode, whereas in electrolytic cells the reverse is the case. Electrodes are also used in welding devices. See also anode; cathode; battery; electrolytic cell.

electrode, glass. A thin glass membrane which, when immersed in a suitable liquid medium, develops a measurable electrical potential that can be readily related to the activity of ionic species present in the solution. By appropriate manipulation of the glass composition, careful pretreatment of the glass surface, and reproducible experimental conditions, electrodes can be devised which not only yield information about the concentration of ions in solution but also have the ability to discriminate, in terms of a selective response, between a number of different ions of similar chemical characteristics. Because of their ability to give both qualitative and quantitative information about ions in solution, glass electrodes are widely used for purposes of chemical measurement, especially in electrochemical research.

electrode, hydrogen. A platinum surface coated with platinum black, immersed in a solution and bathed with a stream of pure hydrogen gas. The potential developed depends on the equilibrium between the hydrogen gas and the hydrogen ions in solution. Used as a standard reference electrode.

electrodeposition. The precipitation of a material at an electrode as the result of the passage of an electric current through a solution or suspension of the material, for example, alkaline-earth carbonates; rubber from latex; paint films on metal. The electrode is in the shape of the desired article. See also electrophoresis.

electrodialysis. A form of dialysis (q.v.) in which an electric current aids the separation of substances that ionize in solution. Seawater can be desalted by this method on a large scale by placing it in the center chamber of a 3-compartment container having two semipermeable membranes and a positive electrode in one end chamber and a negative electrode in the other; the ions migrate to their respective electrodes under a difference of potential, leaving the water salt-free. See also dialysis; desalination; demineralization.

electroforming. An electrolytic plating process for manufacturing metal parts. A mold of the object to be reproduced is made in a soft metal or in wax (by impression). The mold surface is made conducting by coating with graphite. Some suitable metal is then deposited electrolytically on the mold surface. This mold is then (in most cases) a negative of the object to be produced. Other industrial applications are phonograph records, plastic tile, ducting, tubing, etc.

"electroless" coating. A protective coating of copper, cobalt, nickel, gold, or palladium deposited in a bath without application of an electric current, i.e., by chemical reduction.

electroluminescence. Luminescence generated in crystals by electric fields or currents in the absence of bombardment or other means of excitation. It is a solid-state phenomenon involving p and n-type semiconductors, and is observed in many crystalline substances, especially silicon carbide, zinc sulfide, gallium arsenide, as well as in silicon, germanium and diamond. See also phosphor.

electrolysis. Dissociation of an ionizable compound by passing an electric current through it; the compound forms positive and negative ions which are carried by the current to the oppositely charged electrodes, where they are collected (if wanted) or released (if unwanted). A simple electrolysis is the separation of water into oxygen and hydrogen; this is one method of producing hydrogen. Somewhat more complicated is electrolysis of brine to chlorine and sodium hydroxide; this is carried out in electrolytic cells of the diaphragm or mercury type, with water taking part in the reaction. In electroplating, metal salts dissociate into their constituent ions, the positively charged metal ions coating the cathode. There are a number of variations of this process (electrodeposition, electrocoating, electroforming). See also electrolytic cell.

electrolyte. A substance that will provide ionic conductivity when dissolved in water or when in contact with it; such compounds may be either solid or liquid. Familiar types are sulfuric acid and sodum chloride, which ionize in solution. A recently developed solid electrolyte, used originally in fuel cells, is a polymer of perfluorinated sulfonic acid used as the core of a water electrolysis cell for production of hydrogen and oxygen. When saturated with water it has high conductivity. The most common application of electrolytes is in electroplating of metals, in which dissolved (ionized) metal salts are the electrolytes. See also electrolysis; electroplating.

electrolytic cell. An electrochemical device in which electrolysis occurs when an electric current is passed through it. Ionizable compounds dissociate in the aqueous solution with which the electrodes are in contact. Such cells are of two types: (1) the diaphragm cell, which has two compartments separted by a porous membrane, and (2) the mercury cell, in which mercury is the cathode. The anodes of both types have long been made of graphite; because this decomposes rapidly as electrolysis progresses, they are being replaced with dimensionally stable types consisting of titanium coated with oxides of ruthenium and other rare metals, which are also much more efficient. In electrolysis of sodium chloride, the current causes chloride ion to migrate to the anode, where it is collected as chlorine gas; sodium hydroxide and hydrogen are also formed by decomposition of the water, the hydrogen being discharged. This principle is applied in the electroplating of metals, electrodeposition of colloids, and similar processes. See also diaphragm cell; mercury cell; electroplating.

electromagnetic separation. Separation of isotopes, especially those of uranium, by first accelerating them by means of an electrostatic field and then passing them through a magnetic field. The effect of this is to cause all the particles to take a curved path; the heavier ones, having higher kinetic energy, describe a wider arc than the lighter ones. Thus two isotopes of closely similar masses can be separated and collected. See also mass spectrometry.

electromagnetic spectrum. See radiation.

electromotive series. See activity series.

electron. Discovered by J. J. Thomson in 1896, the electron is a fundamental particle of matter that can exist either as a constituent of an atom or in the free state. It has a negative electric charge (4.8×10^{-10} e.s.u.) and a mass 1/1837th that of a proton (q.v.), equivalent to 9.1×10^{-28} gram. The number of electrons in an atom of any element is the same as the number of protons in the nucleus, i.e., the atomic number. Thus the range is from 1 electron in hydrogen to 103 in lawrencium. As the negative charge of the electrons equals the positive charge of the protons, all atoms are electrically neutral. Electrons are arranged in from 1 to 7 shells around the nucleus; the maximum number of electrons in each shell is strictly limited by the laws of physics. The tendency of electrons to form complete outer shells accounts for the valence of an element, and they play an essential part in chemical bonding (q.v.). The outer shells are not always filled: sodium has 2 electrons in the first shell, 8 in the second, and only one in the third. A single electron in the outer shell may be attracted into an incomplete shell of another element, leaving the original atom with a net positive charge. The latter then is called an ion (q.v.). Valence electrons are those which can be captured by or shared with another atom.

Electrons can be removed from the atoms of metals and some other elements by heat, light, electric energy, or bombardment with high-energy particles (see radiation, ionizing). In such cases they are totally free from the atomic orbit, and their energy can be utilized by means of a conductor (electricity) or a vacuum tube or semiconductor. Current is generated by detaching the electrons of a metallic conductor (silver, copper) by means of an electric or magnetic field; the electrons then flow along the conductor to a positively charged terminal. The entire science of electronics is made possible by the tendency of a heated metal cathode to emit a continuous stream of electrons in a vacuum tube.

Free electrons, called beta particles are spontaneously emitted by decaying radioactive nuclei; they have comparatively low energy, but can be accelerated to velocities approaching that of light. The basic nature of the electron has been the subject of much research of the highest order of mathematical rigor. In simplest terms, the electron has the properties of both a particle and a wave, i.e., a standing wave is associated with an electron moving in its orbit. The energy state of any electron is described by four quantum numbers (q.v.).

See also shell; atom; orbital theory.

electronegativity. All atoms (except those of helium) that have fewer than 8 electrons in their highest principal quantum level have low energy orbital vacancies capable of accommodating electrons from outside the atom. The existence of these vacancies is evidence that within these regions the nuclear charge can exert a significant attraction for such electrons, even though as a whole the atom is electrically neutral. This attraction is called "electronegativity." To the extent that the initially neutral atom may be able to acquire electrons from outside, it will acquire also their negative charge. The word "electronegativity" means "tendency to become negatively charged." The concept of electronegativity is an extremely useful one in chemistry, because all chemical bonding originates with attractions of nuclei for electrons.

electron-volt (eV). An extremely small unit used in measuring the energy of electrons and other nuclear constituents. It is the energy developed by an electron in falling through a potential difference of 1 volt, equivalent to 1.6×10^{-19} joule. The rupture of a carbon-to-carbon bond has been calculated to yield about 5 eV.

electrophile. See nucleophile; Lewis acid.

electrophoresis. Migration of suspended or colloidal particles in a liquid such as rubber latex, due to the effect of potential difference across immersed electrodes. The migration is toward electrodes of charge opposite to that of the particles. Most solids, being negatively charged, migrate to the anode, the exception being basic dyes, hydroxide sols, and colloids which have adsorbed positive ions, all of which are positively charged and migrate to the cathode. Migrating particles lose their charge at the electrode, and generally agglomerate around it. Clay suspensions can be filtered by means of forced flow electrophoresis.

Electrophoresis is important in the study of proteins because the molecules of such materials act like colloidal particles and their charge is positive or negative according to whether the surrounding solution is acidic or basic. Thus, the acidity of the solution can be used to control the direction in which a protein moves upon electrophoresis. See also electrodeposition.

electroplating. The deposition of a thin layer or coating of metal (e.g., chromium, nickel, copper, silver, etc.) on an object by passing an electric current through an aqueous solution of a salt containing ions of the element being deposited, for example Cu^{++}. The material being plated (usually a metal, but often a plastic) constitutes the cathode. The anode is often composed of the metal being deposited; ideally, it dissolves as the process proceeds. The thin layer deposited is sometimes composed of two or more metals, in which case it is an alloy. The solution, or plating bath, contains dissolved salts of all the metals being deposited. Electrolytic cells (q.v.) are used for this process.

The anode must be an electrical conductor, but may or may not be of the same chemical composition as the material being deposited, and may or may not dissolve during the process. The purpose of electroplating is usually protection of the base metal from corrosion. Silver is electroplated on copper for economy reasons; plastics may be electroplated for decorative effects.

See also electrophoresis; protective coating; electroless coating; throwing power; current density.

electropolishing. A nonmechanical method of polishing metal surfaces by a method that is actually the reverse of electroplating. This is achieved by making the object to be polished the anode in an electrolytic circuit, the cathode usually being carbon. The electrolytes used are phosphoric, hydrofluoric, nitric, or sulfuric acids (sometimes called polishing acids).

electrostatic coating. A metal painting technique in which electrostatically charged pigment particles are sprayed onto a substrate metal, followed by baking. The electric charge attracts the particles to the metal and holds them in place until heat treatment is applied. Maintenance of the charge is thus essential; factors affecting this are relative humidity (the lower the better) and the chemical nature of the pigment, e.g., phthalocyanine blue retains the charge much longer than titanium dioxide.

electrowinning. The technique of extracting a metal from its soluble salt by an electrolytic cell. It is used in recovery of zinc, cobalt, chromium, and manganese, and has recently been applied to copper when in the form of a silicate ore. For any specific metal, the salt in solution is subjected to electrolysis and is electrodeposited on a cathode made of the metal being extracted.

"Electrunite."[251] Trademark for steel tubing, including both stainless and carbon steels.

element. One of the 106 presently known kinds of substance that comprise all matter at and above the atomic level. According to one theory that has gained some acceptance, the elements originated in less than half an hour from a primordial complex called ylem, a mixture of neutrons and electromagnetic radiation. The smallest unit of any element is the atom (q.v.). All the atoms of a given element are identical in nuclear charge and number of electrons and protons, but they may differ in mass, e.g., hydrogen has mass numbers of 1,2, and 3, called hydrogen, deuterium, and tritium, respectively. These are the isotopes of hydrogen; most elements have isotopic forms, which are due to the presence of one or more extra neutrons in the nucleus. The atomic number (q.v.) of an element indicates its position in the Periodic Table (q.v.) and represents the number of protons present, which is the same as the number of electrons.

All elements above lead are unstable and radioactive. About 90% of the earth's crust is made up of elements with even numbers of protons and neutrons. No stable elements heavier than nitrogen have an odd number of both protons and neutrons. Elements of even atomic number normally have several isotopes, while those of odd atomic number never have more than two stable isotopes. All elements beyond uranium (transuranic) were nonexistent in 1940. They are artificially created by bombardment of other elements with neutrons or other heavy particles. The Lawrence Berkeley Laboratory pioneered research on new elements; it announced discovery of Element 106 in 1974.

See also Periodic Table; isotope; radioactivity; abundance.

Element 104. See rutherfordium.

Element 105. See hahnium.

elemi. A soft, balsam-like resin obtained from a tree in the Philippines; soluble in coal-tar hydrocarbons, but not in petroleum solvents, alcohols, and ketones.

Uses: Plasticizer; adhesion of lacquers to metals; cements and adhesives; wax compositions; printing inks; textile and paper coatings; perfumery; waterproofing; engraving.

"Elorine" Chloride.[100] Trademark for tricyclamol chloride (q.v.).

"Elprene."[41] Trademark for a self-curing synthetic-rubber coating of the neoprene type used as a general maintenance coating.

"El-Sixty."[58] Trademark for 1,3-bis-(2-benzothiazolylmercaptomethyl)urea (q.v.), a rubber accelerator.

elutriation. A process of washing, decantation and settling which separates a suspension of a finely divided solid into parts according to their weight. It is especially useful for very fine particles below the usual screen sizes and is used for pigments, clay dressing, and ore flotation.

"Elvacet."[28] Trademark for a non-molding polyvinyl acetate resin available as an emulsion or solution.

"Elvanol."[28] Trademark for various grades of polyvinyl alcohol (q.v.).

"EMA." Resins.[58] Trademark for ethylene-maleic anhydride copolymers Water-soluble resins which serve as dispersing agents in cold-water detergents, thickeners, binders, stabilizers and emulsifiers.

embrittlement. Hardening of a metal (especially steel) or of an ABS resin, resulting in loss of strength and impairment of other physical properties. In metals the primary cause is exposure to hydrogen, though other factors such as corrosion also are involved. In copolymer plastics such as ABS resins embrittlement is due to formation of a vitreous matrix as well as to oxidation of the butadiene particles in the matrix.

"Emcol."[104] Trademark for a series of emulsifiers, germicides and detergents.

"Emeen."[242] Trademark for fatty imidazolines.

emerald green. A pigment consisting of copper acetoarsenite.
Hazard: Highly toxic by ingestion.

emeralds, synthetic. Artificial crystal produced by a high pressure-high temperature process from beryllium aluminum silicate containing a small amount of chromium.
Uses: Lasers; masers; semiconductors.

"Emerest.'[242] Trademark for a series of fatty esters. Used as wetting agents, detergents, softeners, antistats, defoamers, lubricants, stabilizers, emulsifiers, corrosion inhibitors and other applications not requiring food-grade products.

"Emerez."[242] Trademark for a series of dimer acid-based polyamide resins. Used for the manufacture of printing inks, hot melt adhesives, and thixotropic paints.

"Emerox."[242] Trademark for azelaic acid. Supplied in two grades.

"Emersol."[242] Trademark for a series of fatty acids including linoleic, oleic, palmitic, and stearic acid of improved stabilty and color.

emery. See corundum; abrasive.

"Emete-con."[299] Trademark for an antiemetic agent (q.v.).

emetine $C_{29}H_{40}O_4N_2$. An alkaloid from ipecac.
Properties: White powder; m.p. 74°C; very bitter taste; darkens on exposure to light. Soluble in alcohol and ether; slightly soluble in water.
Derivation: By extraction from root of Cephalis ipecacuanha (ipecac), or synthetically.
Hazard: Toxic by ingestion.
Use: Medicine (usually as hydrochloride).

"Emfac."[242] Trademark for pelargonic acid (q.v.).

"EMI-24."[472] Trademark for 2-ethyl-4-methyl imidazole.

"Emid."[242] Trademark for a series of fatty amides.

"Emisaloy."[155] Trademark for an alloy composed of nickel, 80% and cobalt, 20%.
Properties: M.p. 1500°C; sp. gr. 8.84; resistivity at 20°C 11.0 microohm cm; tensile strength at 20°C, 85,000 lb/sq. in.
Use: Electron tubes.

emission spectroscopy. Study of the composition of substances and identification of elements by observation of the wavelengths of radiation they emit as they return to a normal state after excitation by an external energy source. When atoms or molecules are excited by energy input from an arc, spark, or flame, they respond in a characteristic manner; their identity and composition are signaled by the wavelengths of incident light they emit. The spectra of elements are in the form of lines of distinctive color, such as the yellow sodium D line of sodium; those of molecules are groups of lines called bands. The number of lines present in an emission spectrum depends on the number and position of the outermost electrons and the degree of excitation of the atoms. The first application of emission spectra was identification of sodium in the solar spectrum (1814). See also spectroscopy.

emodin (frangula emodin; frangulic acid) $C_{14}H_4O_2(OH)_3CH_3$. 1,3,8-Trihydroxy-6-methylanthraquinone. Occurs either free or combined with a sugar in a glucoside, in rhubarb, cascara sagrada, and other plants. A synthetic product is also available. Orange crystals; m.p. 256°C; soluble in alcohol; insoluble in water. Used in medicine.

"Emolein."[242] Trademark for synthetic lubricant esters. Used for the compounding of synthetic jet engine lubricants to meet both military and civilian specifications. They include diisooctyl azelate; di-2-ethylhexyl azelate; dipropylene glycol dipelargonate; isodecyl pelargonate.

empirical formula. See formula, chemical.

"Empol."[242] Trademark for dimer acids; 36-carbon aliphatic dibasic acids; used as modifiers of synthetic polymers, soaps, corrosion inhibitors, etc. Six grades available.

"Emralon."[46] Trademark for a series of resin-bonded tetrafluoroethylene solid film lubricants. Colloidal TFE resin is dispersed in phenolic resin, acrylic resin latex, epoxy resin, or thermoplastic resin solutons. Flash point varies from 40 to 58°F.
Hazard: Flammable, dangerous fire hazard.

"Emsal."[242] Trademark for sulfated fatty alcohols.

"Emsorb.'[242] Trademark for sorbitan and ethoxylated sorbitan esters.

"Emtal."[242] Trademark for a line of fractionated tall-oil fatty acids and distilled tall oils.

EMTS. Abbreviation for ethylmercury-para-toluenesulfonanilide.

"Emulphogene BC."[307] Trademark for a series of tridecyloxopoly(ethyleneoxy)ethanols.

"Emulphor."[307] Trademark for a series of nonionic emulsifying agents and dispersants. Some are polyozyethylated vegetable oils, alcohols, and fatty acids.

emulsifier. A surface-active agent. See emulsion.

"Emulsifier STH."[307] Trademark for a corrosion inhibitor, sodium salt of an N-(alkylsulfonyl)glycine; 88% active (min).

emulsifying oil. See soluble oil.

emulsin (synaptase; amygdalase; beta-glucosidase). An enzyme catalyzing the production of glucose from beta-glucosides.
Properties: White powder; odorless and tasteless; capable of hydrolyzing glucosides such as amygdalin

to glucose and the other component substances. Soluble in water; insoluble in ether and alcohol.
Source: Sweet almonds.
Derivation: By extracting an emulsion of almonds with ether filtering the clear solution and precipitating the emulsion with alcohol.

emulsion. A stable mixture of two or more immiscible liquids held in suspension by small percentages of substances called emulsifiers. These are of two types: (1) Proteins or carbohydrate polymers, which act by coating the surfaces of the dispersed fat or oil particles, thus preventing them from coalescing; these are sometimes called protective colloids.(2) Long-chain alcohols and fatty acids, which are able to reduce the surface tension at the interface of the suspended particles because of the solubility properties of their molecules. Soaps behave in this manner; they exert cleaning action by emulsifying the oily components of soils. All such substances, both natural and synthetic, are known collectively as detergents (q.v.).

Polymerization reactions are often carried out in emulsion form; a wide variety of food and industrial products are emulsions of one kind or another, e.g., floor and glass waxes, drugs, paints, shortenings, textile and leather dressings, etc.

All emulsions are comprised of a continuous phase and a disperse phase; in an oil-in-water (o/w) emulsion (milk), water is the continuous phase and butter fat (oil) the disperse phase; in a water-in-oil (w/o) emulsion (butter) free fat (from crushed fat globules) is the continuous phase, and unbroken fat globules plus water droplets are the disperse phase. See also colloid, protective; phase (2); detergent; surface-active agent; wetting agent.

emulsion breaker. See demulsification.

emulsion paint. See paint, emulsion.

emulsion polymerization. Polymerizatoin (q.v.) reaction carried out with the reactants in emulsified form. Performed at normal pressure at −20 to +60°C. Many copolymers (synthetic rubbers) are made in this way.

"Emulsol."[244] Trademark for a series of raw fish oils emulsified with soap.
Uses: Leather processing.

"Emulvin."[470] Trademark for non-ionic emulsifying, stabilizing and wetting agent for latex processing.

en. Abbreviation for ethylenediamine used in formulas for coordination compounds, e.g., the cobalt complex $Co[en]_3(NO_3)_3$. See also dien; pn; py.

enamel. (1) A type of paint consisting of an intimate dispersion of pigments in a varnish or resin vehicle. The vehicle may be an oil-resin mixture, or entirely synthetic resin. Those containing drying oils are converted to films by oxidation; those comprised wholly of synthetic resins may be converted by either heat or oxygen, or both. See also baking finish. (2) Porcelain enamel (q.v.).

enamine. A group of amino olefins; the name refers especially to unsaturated tertiary amines of the general formula $R_2N—\overset{|}{C}=\overset{|}{C}—$, where R is any alkyl group. Though of little use as end products, enamines are valuable intermediates for many organic syntheses.

enanthaldehyde. See heptanal.

enanthic acid. See n-heptanoic acid.

enanthyl alcohol. See 1-heptanol.

enantiomer (enantiomorph). One of a pair of optical isomers containing one or more asymmetric carbon atoms C* whose molecular configurations have left- and right-hand (chiral) forms. These forms are conventionally designated dextro (D) and levo (L) because they compare to each other structurally as do the right and left hands when the carbon atoms are lined up vertically. This is apparent in the enantiomorphic forms of glyceraldehyde; the two structures are mirror images of each other and cannot be made to coincide:

```
      CHO                    CHO
       |                      |
 HO—C*—H     (L)        H—C*—OH    (D)
       |                      |
     CH2OH                  CH2OH
```

Several pairs of enantiomers are possible, depending on the number of asymmetric carbon atoms in the molecule. Compounds in which an asymmetric carbon is present display optical rotation. See also asymmetry; optical isomer; optical rotation.

enantiomorph. See enantiomer.

"Enbond."[142] Trademark for alkaline compounds for the cleaning of metal surfaces prior to electroplating, painting, or other processes.

encapsulation. The process in which a material, or an assembly of small, discrete units, is coated with, or imbedded in, a molten film, sheath, or foam, usually of an elastomer. A foam-forming plastic may be used to fill the spaces between various electrical or electronic components so that they are imbedded in and supported by the foam. Plastics and other materials used for this purpose are often called potting compounds. A specialized use of this technique is in growing crystals for semiconductors, in which a coating of liquid boric oxide is the encapsulating agent.
See also microencapsulation.

"Endic" Anhydride.[316] Trademark for endo-cis-bicyclo-(2.21)-5-heptene-2,3-dicarboxylic anhydride ($C_9H_8O_3$).
Properties: White crystals; m.p. 163°C; soluble in aromatic hydrocarbons, acetone, ethyl alcohol.
Uses: Elastomers, plasticizers, fire retardant chemicals, resins, and epoxy curing systems.

endo-. A prefix used in chemical names to indicate an inner position, specifically (a) in a ring rather than in a side chain, or (b) attached as a bridge within a ring. See also exo-.

endophenolphthalein. See diacetyldihydroxydiphenylisatin.

"Endor."[28] Trademark for a rubber peptizing agent containing activated zinc salt of pentachlorothiophenol $(C_6Cl_5S)_2Zn$, and 80% inert filler. Grayish green powder; sp. gr. 2.39.

endosulfan. Generic name for 6,7,8,9,10,10-hexachloro-1,5,5a,6,9,9a-hexahydro-6,9-methano-2,4,3-benzodioxathiepin 3-oxide $C_9H_6Cl_6O_3S$.
Properties: (commercial product): Brown crystals; m.p. 70–100°C (pure m.p. 106°C); mixture of two isomers, m.p. 108–109°C and 206–208°C.
Hazard: Toxic by ingestion, inhalation, and skin absorption. Tolerance, 0.1 mg per cubic meter of air. Use may be restricted.
Use: Insecticide for vegetable crops.

endothal. Generic name for 7-oxalobicyclo-[2.2.1]-heptane-2,3-dicarboxylic acid $C_8H_{10}O_5$.
Hazard: Toxic; strong irritant to eyes and skin.

Uses: Plant growth regulator.
Note: Various derivatives of endothal are used as defoliants, herbicides, etc.; the mono-N,N-dimethylalkylamine salt in aqueous solution shows promise of increasing yields from sugarcane.

endothermic. A process or change that takes place with absorption of heat and requires high temperature for initiation and maintenance. An example is production of water gas by passing steam over hot coal: $H_2O + C + heat \longrightarrow CO + H_2 - heat$.

endothion. Generic name for S-[(5-methoxy-4-oxo-4H-pyran-2-yl)-methyl] O,O-dimethyl phosphorothioate
$(CH_3O)_2P(O)SCH_2C{:}CHC(O)C(OCH_3){:}CHO$.
Properties: Crystals; m.p. 90–91°C. Very soluble in water; soluble in chloroform and ethanol.
Hazard: Highly toxic by ingestion. Cholinesterase inhibitor. Use may be restricted.
Use: Insecticide.
Shipping regulations: (Rail, Air) Organic phosphate, solid, n.o.s., Poison label. Not acceptable on passenger planes.

"Endox."[142] Trademark for alkaline rust-removal and descaling products supplied in powder form. These products are added to water and the solutions are used both electrolytically and non-electrolytically for removal of rust, etc. from iron and steel.

endoxan. See cyclophosphamide.

end point. (1) In chemical analysis, the point during a titration at which a marked color change is observed, indicating that no more titrating solution is to be added. See indicator.
(2) The highest temperature reached during an assay distillation of hydrocarbon liquids, indicating the overall volatility of the liquid (ASTM).

endrin ($C_{12}H_8OCl_6$). Generic name for 1,2,3,4,10,10-hexachloro-6,7-epoxy-1,4,4a,5,6,7,8,8a-octahydro-1,4-endo-endo-5,8-dimethanonaphthalene. A stereoisomer of dieldrin which is the endo-exo isomer. See also aldrin.
Properties: White crystalline powder; m.p. approx. 200°C (rearranges above this point); insoluble in water, methanol. Moderately soluble in other common organic solvents. Compatible with non-acidic fertilizers, herbicides, fungicides and insecticides.
Hazard: Highly toxic by inhalation and skin absorption. Tolerance, 0.1 milligram per cubic meter in air. Use may be restricted. A suspected carcinogen.
Use: Insecticide.
Shipping regulations: (Rail) Organic phosphate, solid, n.o.s., Poison label. (Air) Poison label.

"Enduro."[251] Trademark for a series of stainless and heat-resisting steels: low nickel group (200 Series); chromium-nickel group (300 Series); straight-chromium group (400 Series).

enercology. A coined word defined as the balanced relationship between energy and ecology. A foundation devoted to practical applications of this relationship has been established at Alma College, Michigan.

energy. The fundamental active entity in the universe, defined by physicists as either $E = mc^2$ (Einstein) or $E = h\nu$ (Planck). More simply, it is considered as the capacity for doing work. These equations show that energy cannot be completely divorced from mass, as the two are to some extent interconvertible. The law of conservation of energy, simply stated, is that the sum total of energy in the universe is constant; therefore, energy cannot be either created or destroyed, but only converted from one form to another.

Radiant energy (light) comprises the electromagnetic spectrum; all wavelengths of light are composed of photons, or packets of energy, traveling at the speed of light. Theoretically, they have no mass except that associated with their speed. Protons, electrons, and neutrons are forms of highly condensed energy which possess determinable mass but move at lower speeds.

Energy is directly related to chemical phenomena in the formation and decomposition of compounds, the many important reactions that occur in electrochemistry, and in the release of energy in nuclear fission and fusion. Free energy is a thermodynamic function; in chemical reactions it is a measure of the extent to which a substance can react. Kinetic energy (the energy of motion) is most clearly exhibited in gases, in which molecules have much greater freedom of motion than in liquids and solids. See also radiation; matter; thermodynamics; free energy.

Note: "One of the most difficult challenges we face is to find ways to ensure that all peoples of the world share more equitably the vast human benefits that energy can bring.

"The foundation of worldwide energy policy must be based on energy conservation and the development of additional sources through a judicious application of science and technology. . . ."

Glenn T. Seaborg, ACS meeting, April, 1976.

energy converter. Any element or compound having the ability to convert the radiant energy of sunlight into electrical, thermal or chemical energy. Promiment among them are silicon, selenium and tellurium, as well as the chlorophyll of plants in photosynthesis. See also solar cell; magnetohydrodynamics.

Energy Research and Development Administration. A Federal agency established in 1974 to plan and execute a research and development program for energy sources of all types. It replaces the former Atomic Energy Commission.

energy sources. See coal; geothermal energy; nuclear energy; breeder; fission; fusion; oil sands; ocean water (note); shale oil; methane; solar cell; waste treatment; natural gas; gasoline; petroleum; gasification; biogas; cellulose; fuel oil; synthetic natural gas; thermoelectricity; fuel cell; battery; radiation; hydrogen.

Mechanical energy sources to which considerable engineering research is being devoted include water power facilities, harnessing of tides, and use of windmills. Commercial production of energy from alternative sources is not expected to attain significant volume before 1985. Of those listed above, coal and nuclear energy appear to be the most immediate possibilities, though there is growing interest in solar energy.

In 1973 sources of energy for U.S. requirements were approximately as follows: petroleum and natural gas liquids, 46%; natural gas, 31%; coal, 18%; hydroelectric, 4%; nuclear 1% (Bureau of Mines). A new scientific journal entitled "Energy Sources: an International Interdisciplinary Journal of Science and Technology" made its appearance in 1974.

enfleurage. Extraction of odoriferous components of flowers by means of fats or mixtures of fat and tallow, the process being carried out at room temperature to avoid decomposition of the desired perfumes. The latter are separated from the fat by washing with alcohol. See also essential oil; perfume.

engineering plastic. See plastic.

enhancer. A food additive that brings out the taste of a food product without contributing any taste of its own. Sodium glutamate (q.v.) is the most widely used substance of this class; its effective concentration is in parts per thousand. See also potentiator.

"Enjay."[29] Trademark for a series of elastomeric materials, including copolymers and terpolymers, in both solid, semi-liquid and liquid states.

enol. A chemical grouping containing both a double bond (ene) and a hydroxyl group (OH), forming an intermediate and reversible product. Enols are characteristic of racemic compounds (q.v.).

enolase. An enzyme active in glycolysis which catalyzes the conversion of 2-phosphoglyceric acid to the phosphorylated enol form of pyruvic acid.

"Enovid."[70] Trademark for norethynodrel with mestranol. Oral contraceptive approved by FDA.

enrichment. (1) In food technology, the addition to a foodstuff of various nutrient substances during manufacture to increase the dietary value of the food, e.g., addition to wheat flour of vitamins B_1, B_2, niacin, and iron. In this way the food is brought up to a specific nutritional standard.

(2) Increase in the abundance of certain isotopes of an element by any of several methods. (a) By a chemical reaction accompanied by irradiation from a laser beam; enrichment of boron, chlorine and sulfur isotopes has been achieved in this way in a number of reasearch laboratories. Uranium enrichment is also possible, either by adding previously prepared U-235 to natural uranium or by the laster technique. (b) Uranium can also be enriched by gas centrifugation and gaseous diffusion. The latter is the usual procedure.

(3) Addition of oxygen to air to increase its combustion-supporting ability.

enterokinase. An enzyme found in the small intestine which converts trypsinogen into trypsin.

"Enth-Acid."[142] Trademark for a blend of acid salts, activators and surfactants which can be used as a replacement for liquid acids. Used for acid dipping of iron, steel, brass, copper, or zinc die castings prior to plating.

"Enthol."[142] Trademark for phosphoric acid-solvent mixtures designed for degreasing and oxide removal for such metals as steel, aluminum, and zinc. Materials are supplied in the liquid form.

"Enthonite."[142] Trademark for a polyvinyl platisol coating used for coating metals; primarily for imparting resistance to acid, alkaline, and electroplating solutions. Contains 100% solids and is applied by dipping.

"Enthox."[142] Trademark for salts that are added to water for producing chromate coatings on zinc and cadmium to withstand 100 or more hours in 20% salt spray.

entrainer. An additive for liquid mixtures that are difficult to separate by ordinary distillation. The entrainer usually forms an azeotrope with one of the compounds of the mixture and thereby aids in the separation of such a compound from the remainder of the mixture.

environmental chemistry. That aspect of chemistry associated with water pollution and purification, pesticides, air pollution, and waste disposal of various types (solid, liquid, gaseous, and radioactive). See also eutrophication; biodegradability; smog; waste, industrial; ecology; chemodynamics.

Environmental Protection Agency (EPA). A Federal agency established in 1970 for the purpose of controlling all aspects of air and water pollution and the manufacture and transportation of any toxic materials that may enter into or affect the environment. It includes an Office of Hazardous Materials Control which administers Congressional legislation pertaining to this field. The EPA may require manufacturers to conduct tests on materials or products which may adversely affect the environment or public health and safety. One of the most important aspects of its activities is the establishment and supervision of automotive emission standards. It is also concerned with pesticides, fungicides and other potentially detrimental materials, as well as industrial waste disposal. It operates in close conjunction with the U.S. Dept. of Agriculture and the Food and Drug Administration. The construction of new plants for manufacturing a wide range of chemicals and chemically treated products must conform to EPA standards, especially as regards effluents that contribute to water pollution. See also toxic material.

"Enzose."[30] Trademark for a liquid carbohydrate in two grades with dextrose equivalent (D.E.) of 72 or 87 and pH of 3.8–4.4 or 4.0–5.0. Shipped in tank cars.
Uses: Chemicals, colorants; drugs, pharmaceuticals.

enzyme. A complex organic substance formed in the living cells of plants and animals; they are catalysts for the chemical reactions of biological processes, such as digestion. The first synthesis of an enzyme was reported in 1969 (ribonuclease). Enzymes are highly specific in their catalytic behavior; a given enzyme is effective for only one particular reaction. Enzymes are often classified by the kind of substance (substrate) consumed in the reactions catalyzed. Enzymes that break down proteins by hydrolytic cleavage are called proteolytic.

A partial list of enzymes follows; a number of these are used in industrial processes, especially those involving fermentation.

amylase	starch hydrolysis
carboxylase	decomposes pyruvic acid
cellulase	converts cellulose to glucose
invertase	converts sucrose to glucose and fructose
lipase	hydrolysis of fats
maltase	converts maltose to glucose
pepsin	hydrolysis of proteins
protease	hydrolysis of peptide linkages
rennin	hydrolysis of proteins
ribonuclease	decomposes RNA
trypsin	splits proteins to amino acids
urease	decomposes urea to NH_4 and CO_2
zymase	converts sugars to alcohol and CO_2

eosin (bromeosin; tetrabromofluorescein) $C_{20}H_8Br_4O_5$. CI No. 45380.
Properties: Red, crystalline powder; soluble in alcohol and acetic acid; insoluble in water; the potassium and sodium salts are soluble in water.
Derivation: Bromination of fluorescein.
Uses: Dyeing silk, cotton, and wool; red writing ink; coloring motor fuel.

EP. (1) Abbreviation for extreme pressure, as applied to lubricants. (2) Abbreviation for ethylene-propylene.

EPA. Abbreviation for Environmental Protection Agency (q.v.).

EPC black. Abbreviation for easy processing channel black. See carbon black.

EPDM. Abbreviation for a terpolymer elastomer made from ethylene-propylene diene monomer. See ethylene-propylene terpolymer.

"Epex."[267] Trademark for a series of epoxy resin extenders.
Uses: Extension and dilution of epoxy resin systems for all applications.

ephedrine (1-phenyl-2-methylaminopropanol) $C_6H_5CH(OH)CH(NHCH_3)CH_3$. Optically active (levorotatory) form. See racephedrine for the inactive mixture of isomers.
Properties: White to colorless granules, pieces or crystals; unctuous to touch; hygroscopic; gradually decomposes on exposure to light. Soluble in water, alcohol, ether, chloroform, and oils. M.p. 33–40°C; b.p. 255°C (decomposes).
Derivation: Isolation from stems or leaves of Ephedra, especially Ma huang (China and India).
Grades: Technical; N.F.
Use: Medicine (usually as hydrochloride).

epi. (1) A prefix denoting a bridge or intramolecular connection, e.g., epoxide.
(2) An abbreviation for epichlorohydrin.

"Epiall."[175] Trademark for a series of epoxy molding compounds.
Properties: Excellent electrical resistivity; high physical strength; flame resistance; good dimensional stability; outstanding resistance to high temperature (500°F). Resistant to virtually all solvents and most chemicals; good colorfastness in sunlight and heat; fungus resistance; water absorption, 0.17% in 48 hours.
Available in mineral glass-filled and long glass fiber-filled grades.

epichlorohydrin (chloropropylene oxide; epi) An epoxide. CH_2OCHCH_2Cl.
Properties: Highly volatile, unstable, liquid. Chloroform-like odor; miscible with most organic solvents; slightly soluble in water. Sp. gr. 1.1761 (20/20°C); b.p. 115.2°C; wt/gal 9.78 lb; vapor pressure 12.5 mm (20°C); f.p. -58.1°C; viscosity 1.12 cp (20°C); refractive index (n 25/D) 1.4358; flash point 93°F (TOC).
Derivation: By removing hydrogen chloride from dichlorohydrin.
Hazard: Toxic by inhalation, ingestion, and skin absorption. Strong irritant. Tolerance, 5 ppm in air. Flammable, moderate fire risk.
Uses: Major raw material for epoxy and phenoxy resins; mfg. of glycerol; curing propylene-based rubbers; solvent for cellulose esters and ethers; high wet-strength resins for paper industry.
See also "Hydrin," "Herdor."

epichlorohydrin triethanolamine cellulose. See ECTEOLA-cellulose.

"Epi-Cure."[474] Trademark for curing agents for epoxy resins.

epimer. An isomer which differs from the compound with which it is being compared only in the relative positions of an attached hydrogen and hydroxyl. The isomerism may be represented as —HCOH— and —HOCH—. It is common in sugars.
See also diastereoisomer.

epinephrine (*l*-methylaminoethanolcatechol; "Adrenalin.") $C_6H_3(OH)_2CHOHCH_2NHCH_3$. A hormone of the adrenal glands.
Properties: Light brown or nearly white, odorless crystalline powder; affected by light; m.p. 211–212°C; specific rotation (25°C) -50° to -53.5°; sparingly soluble in water; insoluble in alcohol, chloroform, ether, acetone, oils. Readily soluble in aqueous solutions of mineral acids, sodium hydroxide and potassium hydroxide.
Derivation: From the adrenal glands of sheep and cattle or synthetically from pyrocatechol.
Grade: U.S.P.
Use: Medicine.

"Epiphen."[65] Trademark for an epoxy resin in liquid form. "Epiphen" ER-823 is used in adhesives for rubber, steel, aluminum or glass. Catalyst is supplied for specific end uses.

epitaxial growth. An oriented crystal growth between two crystalline solid surfaces in which the surface of one crystal offers suitable positions for deposition of a second crystal. This behavior is characteristic of some types of high polymers.

"E P Mudlube."[236] Trademark for a solution of modified higher organic acids which imparts extreme pressure lubricating properties to drilling muds.

EPN (O-ethyl O-para-nitrophenyl phenylphosphonothioate) $C_6H_5P(C_2H_5O)(S)OC_6H_4NO_2$.
Properties: Light-yellow crystals; m.p. 36°C; sp. gr. 1.5978 (30°C). Insoluble in water; soluble in most organic solvents. Decomposes in alkaline solutions.
Grades: Wettable powders and dusts.
Hazard: Toxic; a cholinesterase inhibitor. Tolerance, 0.5 mg per cubic meter of air. Absorbed by skin. Use may be restricted.
Uses: Insecticide; acaricide.
Shipping regulations: (Rail, Air) Organic phosphate, solid, n.o.s. Poison label. Not accepted passenger planes.

"Epocryl" Resins.[125] Trademark for a group of thermosetting resins, best described as epoxy acrylates, designed to combine the performance properties of epoxy resins and the application features of unsaturated polyesters. Used in reinforced structures.

"Epolene." Trademark for a series of low-molecular-weight polyethylene resins. Available in both emulsifiable and nonemulsifiable types.

"Epon" Curing Agent.[125] Trademark for curing agents by which "Epon" resins can be hardened to form clear tough polymers with high physical strength, excellent chemical resistance, and good electrical resistance.

"Eponite" 100.[125] Trademark for a water-dispersible liquid epoxy resin. Used in the textile industry alone or in combination with hydrolyzed or partially hydrolyzed polyvinyl acetate, starches, gums, cellulose ethers, selected resins, or other chemical finishing agents to impart crease resistance, shrinkage control, embossing, stiffness, softness or other fabric properties.

"Eponol" Resins.[125] Trademark for high molecular weight linear copolymers of bisphenol A and epichlorohydrin; produce outstanding surface coatings by solvent evaporation alone.

"Epon" Resins.[125] Trademark for a series of condensation products of epichlorohydrin and bisphenol-A having excellent adhesion, strength, chemical resistance and electrical properties when formulated into protective coatings, adhesives and structural plastics.

"Eporal."[418] Trademark for sulfonyldianiline.

"Epotuf."[36] Trademark for epoxy resins (q.v.).

epoxide. An organic compound containing a reactive group resulting from the union of an oxygen atom with two other atoms (usually carbon) that are joined in some other way, as indicated:

```
     O
    / \
 —C———C—
```

This group, commonly called "epoxy," characterizes the epoxy resins (q.v.). Epichlorohydrin and ethylene oxide are well-known epoxides. The compounds are also being used in recently developed types of cellulose derivatives and fluorocarbons.

"Epoxols."[152] Trademark for epoxidized oils and esters.
7-4. High purity epoxidized soybean oil with a minimum of 7% oxirane oxygen.
9-5. A high oxirane (9% minimum) epoxidized triglyceride.
5-2E. An epoxidized, higher alkyl ester.
7-4 and 9-5 approved for use in food packaging materials.
Uses: Vinyl films, sheetings, extrusions and coatings, polyvinylidene chloride, chlorinated rubber, chlorinated paraffins, nitrocellulose, and other compunds.

'Epoxybond."[41] Trademark for an epoxy adhesive putty in stick form.

1,2-epoxybutane. See 1,2-butylene oxide.

3,4-epoxycyclohexane carbonitrile $O(C_6H_9)CN$.
Properties: Liquid; sp. gr. 1.0929 (20/20°C); b.p. 244.5°C; f.p. -33°C; soluble in water.
Hazard: Toxic by skin absorption; moderately by ingestion and inhalation.
Uses: Intermediate; stabilizer.

epoxyethane. See ethylene oxide.

2,3-epoxy-2-ethylhexanol $C_3H_7\underline{CHOC}(C_2H_5)CH_2OH$.
Properties: Liquid; sp. gr. 0.9517 (20/20°C); b.p., decomposes; f.p. -65°C; slightly soluble in water. Low toxicity. Combustible.
Hazard: Moderately irritant to skin.
Uses: Stabilizer; intermediate.

epoxy novolak. Epoxy resin made by the reaction of epichlorohydrin with a novolak resin (phenol-formaldehyde; see novolak). These have a repeating epoxide structure which offers better resistance to high temperatures than epichlorohydrin-bisphenol A type, and are especially useful as adhesives.

2,3-epoxy-1-propanol. See glycidol.

epoxy resin. A thermosetting resin based on the reactivity of the epoxide group (q.v.). One type is made from epichlorohydrin and bisphenol A. Aliphatic polyols such as glycerol may be used instead of the aromatic bisphenol A. Molecules of this type have glycidyl ether structures, $—OCH_2\overline{CHOCH_2}$, in the terminal positions, have many hydroxyl groups, and cure readily with amines.

Another type is made from polyolefins oxidized with peracetic acid. These have more epoxide groups, within the molecule as well as in terminal positions, and can be cured with anhydrides, but require high temperatures. Many modifications of both types are made commercially. Halogenated bisphenols can be used to add flame-retardant properties. See also epoxy novolak. The reactive epoxies form a tight cross-linked polymer network, and are characterized by toughness, good adhesion, corrosion, chemical resistance, and good dielectric properties.
Hazard: Strong irritant to skin in uncured state.
Uses: Surface coatings, as on household appliances and gas storage vessels; adhesives for composites and for metals, glass, and ceramics; casting metal-forming tools and dies; encapsulation of electrical parts; filament-wound pipe and pressure vessels; floor surfacing and wall panels; neutron-shielding materials; cements and mortars; nonskid road surfacing; rigid foams; oil wells (to solidify sandy formations); matrix for stained glass windows.

EPR. Abbreviation for ethylene propylene rubber; also for electron paramagnetic resonance.

epsilon acid. See 1-naphthylamine-3,8-disulfonic acid.

Epsom salts. See magnesium sulfate.

EPT. Abbreviation for ethylene-propylene terpolymer (q.v.).

"Eptac No. 1."[28] Trademark for zinc dimethyldithiocarbamate, an ultra-accelerator for rubber.

"Eptam."[1] Trademark for a selective herbicide containing ethyl-N,N-di-n-propylthiolcarbamate.

EPTC (S-ethyl di-N,N-propylthiocarbamate) $C_2H_5SC(O)N(C_3H_7)_2$. A pre-emergence herbicide. Available in liquid and granular formulations.

eq. Abbreviation for gram equivalent weight, i.e., the equivalent weight in grams. Recommended as an international unit.

equilibrium. (1) Chemical equilibrium is a condition in which a reaction and its opposite or reverse reaction occur at the same rate, resulting in a constant concentration of reactants; for example, ammonia synthesis is at equilibrium when ammonia molecules form and decompose at equal velocities ($N_2 + 3H_2 \rightleftharpoons 2NH_3$).

(2) Physical equilibrium is exhibited when two or more phases of a system are changing at the same rate, so that the net change in the system is zero. An example is the liquid-to-vapor/vapor-to-liquid interchange in an enclosed system, which reaches equilibrium when the number of molecules leaving the liquid is equal to the number entering it.

equilibrium constant. A number that relates the concentrations of starting materials and products of a reversible chemical reaction to one another. For example, for a chemical reaction represented by the equation $aAB + bCD \rightleftharpoons cAD + dBC$, the equilibrium constant would be $K = [(AD)^c(BD)^d]/[(AB)^aCD)^b]$ where a,b,c, and d are the numbers of molecules of AB, CD, AD and BC that occur in the balanced equation and (AD), (BC), (AB) and (CD) are the molecular concentrations of AD, BC, AB and CD in any mixture that is at equilibrium. At any one temperature, K is usually at least approximately constant regardless of the relative quantities of the several substances, so that when K is known, it is often possible to predict concentrations of products when those of the starting materials are known. The constant changes markedly with temperature. The constant can often be calculated from the relations of thermodynamics if the free energy for the chemical reaction is known, or can be calculated by measuring all concentrations in one or more carefully conducted experiments.

equivalent weight (combining weight). The weight of an element that combines chemically with 8 grams of oxygen or its equivalent. Since 8 grams of oxygen combines with 1.008 grams of hydrogen, the latter is considered equivalent to 8 grams of oxygen. When 8 grams is selected for the combining weight of oxygen, no element has a combining weight value of less than 1. The equivalent weight of an acid is the weight that contains one atomic weight of acidic hydrogen, i.e., the hydrogen that reacts during neutralization of acid with base. The equivalent weight of a base or hydroxide is the weight that will react with an equivalent weight of acid. Equivalent weights of other substances are defined in a similar manner.

Er Symbol for erbium.

"Erasol."[232] Trademark for a series of sulfoxylate reducing agents for use in the textile and paper industries.

"Eratrope."[232] Trademark for a series of discharge agents for removing vat dyes from cellulosic material.

erbia. See erbium oxide.

erbium Er Element with atomic number 68; atomic weight 167.26; valence 3. One of the rare earth elements of the yttrium subgroup. See rare earth metals.
Properties: Soft, malleable solid with metallic luster; insoluble in water; soluble in acids; salts are pink to red; sp. gr. 9.16 (15°C); m.p. 1522°C; b.p. 2500°C (approx). Toxicity low. High electrical resistivity.
Derivation: Reduction of the fluoride with calcium; electrolysis of the fused chloride.
Source: Rare earth minerals.
Forms: Lumps; ingots of high purity; sponge; powder.
Hazard: Flammable in finely divided form.
Uses: Nuclear control; special alloys; room-temperature laser.

erbium nitrate $Er(NO_3)_3 \cdot 5H_2O$.
Properties: Large reddish crystals; soluble in water, alcohol, ether, and acetone; loses $4H_2O$ at 130°C.
Derivation: Treatment of oxides, carbonates or hydroxides with nitric acid.
Grade: 99.9%.
Hazard: May explode if shocked or heated.
Shipping regulations: (Rail) Nitrates, n.o.s., Yellow label. (Air) Oxidizer label.

erbium oxalate $Er_2(C_2O_4)_3 \cdot 10H_2O$.
Properties: Reddish microcrystalline powder; decomposes at 575°C; sp. gr. 2.64. Soluble in water and dilute acids.
Hazard: Toxic and corrosive.
Use: Oxalates of the rare earth metals are used to separate the latter from common metals.

erbium oxide (erbia) Er_2O_3.
Properties: Pink powder which readily absorbs moisture and carbon dioxide from the atmosphere. Sp. gr. 8.64; specific heat 0.065; infusible; insoluble in water; slightly soluble in mineral acids.
Derivation: By heating the oxalate or other oxy-acid salts.
Grades: 98–99%.
Uses: Phosphor activator; infrared-absorbing glass. See also rare earth.

erbium sulfate $Er_2(SO_4)_3 \cdot 8H_2O$.
Properties: Pink monoclinic crystals; soluble in water; sp. gr. 3.217; dehydrated at 400°C.
Derivation: Dissolving hydroxides, carbonates, or oxides in dilute sulfuric acid.
Grade: 99.9%.
Use: To determine atomic weight of the rare-earth element.

erbon. Generic name for 2-(2,4,5-trichlorophenoxy)-ethyl 2,2-dichloropropionate $(Cl)_3C_6H_2OC_2H_4OC(O)C(Cl)_2CH_3$.
Properties: Solid; m.p. 49–50°C (technical). Insoluble in water; soluble in most organic solvents.
Hazard: Moderately toxic by inhalation and ingestion.
Use: Herbicide.

"Ercusol."[470] Trademark for an aqueous dispersion of an acrylonitrile-based copolymer and other monomers.

ERDA. Abbreviation for Energy Research and Development Administration (q.v.).

erepsin. A mixture of peptidase enzymes, formerly thought to be a single enzyme, which catalyzes the hydrolysis of peptides in the small intestine.

ergocalciferol (calciferol; vitamin D_2) $C_{28}H_{44}O$.
Properties: White, odorless crystals; affected by air and light. Insoluble in water; soluble in alcohol, chloroform, ether and fatty oils; m.p. 115–118°C; specific rotation +103° to +106°.
Derivation: From ergosterol by irradiation with ultraviolet light.
Grades: U.S.P.; F.C.C., as vitamin D_2.
Uses: Medicine; nutrition; dietary supplement.

ergosterol (provitamin D_2) $C_{28}H_{44}O$. A plant sterol widely distributed in nature.

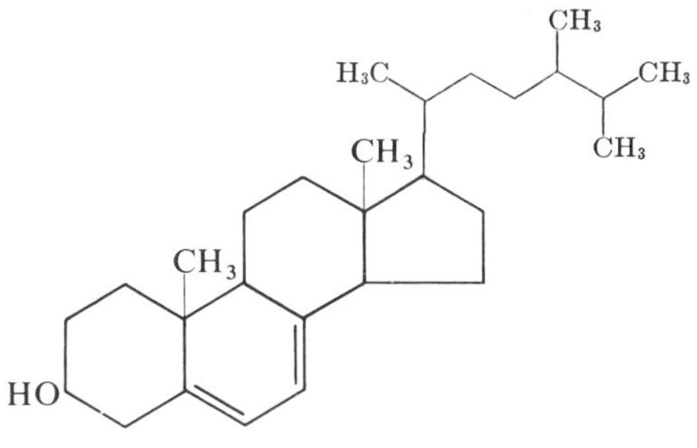

Properties: Colorless crystals; m.p. 166°C (with 1½ H_2O); b.p. 250°C (0.01 mm); sp. gr. 1.04; specific rotation −135° (in chloroform); insoluble in water; soluble in alcohol, benzene, ether.
Derivation: Synthesized by yeast from simple sugars; obtained from fungus ergot.
Hazard: Due to its ability to catalyze calcium deposition in the bony structure (thus preventing rickets), overdosage of vitamin D may be harmful.
Use: Medicine; when irradiated with ultraviolet light, it has vitamin D activity; source of estradiol (q.v.).

ergot (secale cornutum; rye ergot). A fungus growth, Claviceps purpurea, on rye.

Habitat: Europe; cultivated in Spain and Russia.
Grades: Spanish; Russian.
Containers: Bats of variable size; tin-lined drums.
Hazard: Toxic by ingestion.
Uses: Source of many alkaloids; medicine.

erucamide (erucylamide) $C_{21}H_{41}CONH_2$.
Properties: Solid; sp. gr. 0.888; m.p. 75–80°C; iodine value 70–80; soluble in isopropanol; slightly soluble in alcohol and acetone. Combustible.
Uses: Foam stabilizer; solvent for waxes and resins; emulsions; antiblock agent for polyethylene.

erucic acid (cis-13-docosenoic acid) $C_8H_{17}CH{:}CH(CH_2)_{11}COOH$. A C_{22} (solid) fatty acid with one double bond. A homolog of oleic acid with four more carbons.
Properties: M.p. 33–34°C; b.p. 264°C (15 mm); iodine value 75; combustible.
Derivation: Fats and oils from mustard seed, rapeseed and crambe seed.
Uses: Preparation of dibasic acids and other chemicals; polyethylene film additive; water-resistant nylon.

erucyl alcohol $C_{22}H_{43}OH$. A C_{22} (solid) fatty alcohol having one double bond. White, soft solid almost odorless. Sp. gr. 0.8486; cloud point 81°F; boiling range 334–376°C; iodine value 82; flash point 395°F; soluble in alcohol and most organic solvents. Combustible.
Derivation: Sodium reduction of erucic acid.
Uses: Lubricants; surfactants; petrochemicals; plastics; textiles; rubber.

"Erusticator."[204] Trademark for a rust remover which dissolves rust rapidly from fabrics.

"Erusto."[204] Trademark for a series of laundry and dry-cleaning products.

"Ervol."[45] Trademark for white mineral oil.

erythorbic acid (trivial name adopted officially for *d*-erythro-ascorbic acid, formerly called isoascorbic acid) $C_6H_8O_6$.
Properties: Shiny, granular crystals; decomp. 164–169°C. Soluble in water, alcohol, pyridine; moderately soluble in acetone; slightly soluble in glycerol.
Grade: F.C.C.
Uses: Antioxidant (industrial and food) especially in brewing industry; reducing agent in photography.

erythrite.
(1) Synonym for erythritol.
(2) (cobalt bloom) $Co_3(AsO_4)_2 \cdot 8H_2O$. A natural hydrated cobalt arsenate. Crimson, peach, red, pink or pearl gray. Contains 37.5% cobalt oxide. Soluble in hydrochloric acid. Sp. gr. 2.91–2.95; Mohs hardness 1.5–2.5; m.p. decomposes.
Occurrence: United States (California, Colorado, Idaho, Nevada); Ontario.
Hazard: Highly toxic by ingestion and inhalation.
Uses: Coloring glass and ceramics.
Shipping regulations: Arsenical compounds n.o.s., (Rail, Air) Poison label.

erythritol (tetrahydroxybutane; erythrite) $CH_2OHCHOHCHOHCH_2OH$. A tetrahydric alcohol found in Protococcus vulgaris and other lichens of Rocella species. Can be made synthetically. Combustible.
Properties: White, sweet crystals; m.p. 121–122°C; b.p. 329–331°C; sp. gr. 1.45; soluble in water; slightly soluble in alcohol; insoluble in ether.
Use: Mfg. of erythrityl tetranitrate.

erythrityl tetranitrate (erythrol tetranitrate) $CH_2ONO_2(CHONO_2)_2CH_2ONO_2$.
Properties: Crystals; m.p. 61°C. Soluble in alcohol, ether, and glycerol; insoluble in water.
Derivation: By nitration of erythritol.
Hazard: Severe explosion risk when shocked or heated.
Use: Medicine (diluted with lactose in non-explosive tablets).
Shipping regulations: (Rail) Consult regulations; (Air) Not acceptable.

erythro-ascorbic acid. See erythorbic acid.

"Erythrocin."[3] Trademark for erythromycin (q.v.).

erythrocyte. A red blood cell, containing hemoglobin, the iron-carrying protein of the blood. See also leucocyte.

erythrogenic acid. See isanolic acid.

erythrol tetranitrate. See erythrityl tetranitrate.

erythromycin. $C_{37}H_{67}NO_{13}$. An antibiotic produced by growth of Streptomyces erythreus Waksman. It is effective against infections caused by Gram-positive bacteria, including some beta-hemolytic streptococci, pneumococci, and staphylococci.
Properties: White or slightly yellow, odorless, bitter crystalline powder; m.p. 133–138°C. Freely soluble in alcohol, chloroform, and ether; very slightly soluble in water. Slightly hygroscopic. pH (saturated solution) 8–10.5. pH less than 4 is destructive. Alcoholic solution is levorotatory.
Grade: U.S.P.
Use: Medicine (antibiotic); various salts are available.

erythrosine $C_{20}H_6I_4Na_2O_5$. Sodium (or potassium) salt of iodeosin. CI No. 45430.
Properties: Brown powder; forms cherry red solution in water; soluble in alcohol.
Use: Colorant.

Es Symbol for einsteinium.

"Escalol 106."[10] Trademark for glyceryl para-aminobenzoate. $CH_2OHCH(OH)CH_2OOCC_6H_4NH_2$.
Use: Sun-screening compound.

"Escalol 506."[10] Trademark for amyl para-dimethylaminobenzoate, $C_5H_{11}OOCC_6H_4N(CH_3)_2$. Used as an ultraviolet screen.

"Escon."[29] Trademark for a polypropylene thermoplastic.

eserine. See physostigmine.

"Eskabarb."[71] Trademark for phenobarbital (q.v.).

"Eskaserp."[71] Trademark for reserpine (q.v.).

"Es-min-el."[93] Trademark for a granular mixture of compounds of manganese, copper, zinc, iron, magnesium and boron. Used in horticulture to correct trace element deficiencies.

"Espantone."[505] Trademark for a ketone derived from petrochemical starting materials.
Properties: Colorless to straw oily liquid; sp. gr. 1.069; refractive index (n 20/D) 1.513; carbonyl value 340; acid value 1.
Use: Perfumery (for lavender and spike fragrance).

esparto. The leaves of a desert plant (Stipa tenacissima and Lygeum spartum) of the Mediterranean area. Relatively high content of cellulose and alphacellulose make it usable for high-quality paper for intaglio color printing. A wax is also derived from this plant.

essential (1) Containing the characteristic odor or flavor (i.e., the essence) of the original flower or fruit; an essential oil (q.v.), usually obtained by steam distillation of the flowers or leaves or cold-pressing of the skin.

(2) As applied to certain amino acids, fatty acids, and vitamins, this term is used by biochemists to mean that the compound in question is a necessary nutritional factor that is not synthesized within the body of the animal and thus must be obtained from external sources. Eight amino acids are classified as essential on this basis. See also amino acid.

essential oil. A volatile oil derived from the leaves, stem, flower or twigs of plants, and usually carrying the odor or flavor of the plant. Chemically, they are often principally terpenes (hydrocarbons), but many other types also occur. Essential oils (except for those containing esters) are unsaponifiable. Some are nearly pure single compounds, as oil of wintergreen, which is methyl salicylate. Others are mixtures, as turpentine oil (pinene, dipentene), and oil of bitter almond (benzaldehyde, hydrocyanic acid). Some contain resins in solution and are called oleoresins or balsams (q.v.).

Properties: Pungent taste, and odor, usually nearly colorless when fresh, but becoming darker and thick on exposure to the air; optically active; sp. gr. 0.850–1.100. Soluble in alcohol, carbon disulfide, carbon tetrachloride, chloroform, petroleum ether and fatty oils; insoluble in water, except for individual constituents of some oils which may be partially water-soluble, resulting in a loss of these constituents during steam distillation.

Derivation: (a) By steam distillation; (b) by pressing (fruit rinds); (b) by solvent extraction; (c) by maceration of the flowers and leaves in fat and treating the fat with a solvent; (d) by enfleurage (q.v.).

Uses: Perfumery; flavors; thinning precious metal preparations used in decorating ceramic ware.

See also terpeneless oil, and specific entries. Further information can be obtained from the Essential Oil Association of U.S.

Note: Many essential oils are now made synthetically for a wide variety of fragrances and flavoring agents. Use of these synthetics is increasing because of a shortage of natural products.

"Esskol."[64] Trademark for modified linseed oils used in redwood finishes, varnishes, and enamels.

"Essotane."[51] Trademark for liquefied petroleum gases for domestic and industrial uses.

"Essowax."[51] Trademark for fully refined paraffin wax available in slabs and in liquid form in wide range of melting points and hardness.

"Estan."[51] Trademark for light colored, general purpose, lime-base greases. Available in wide range of consistencies and suitable for all methods of application. Made with an oil having a minimum of internal friction and bearing drag.

"Estane"[119] Trademark for thermoplastic polyester and polyether urethane elastomers which provide good physical and chemical properties without curing. Extremely tough and abrasion resistant with high tensile strength at high ultimate elongation; good solvent resistance, particularly to gasoline; low air permeability, and exceptional low-temperature flexibility.

Uses: Wire and cable jacketing, fuel hose and tanks, belting, shoe heels, coated fabrics, free film, adhesives.

"Estar."[115] Trademark for polyester film base.

ester. An organic compound corresponding in structure to a salt in inorganic chemistry. Esters are considered as derived from acids by the exchange of the replaceable hydrogen of the latter for an organic radical. The usual reaction is that of an acid (organic or inorganic) with an alcohol or other organic compound rich in OH groups. Esters of acetic acid are called acetates and esters of carbonic acid carbonates.

See also fatty ester.

ester gum. Hard, semisynthetic resin produced by esterification of natural resins (especially rosin) with polyhydric alcohols (principally glycerol, but also pentaerythritol). Flash point 375°F. Combustible.

Grades: By color; also as gum rosin or wood rosin.

Containers: 300-lb barrels and drums; multiwall paper sacks.

Uses: Paints, varnishes, and cellulosic lacquers.

"Esteron."[233] Trademark for a series of weed and brush control products; they are formulated esters of 2,4-D and 2,4,5-T.

Hazard: Moderately toxic.

"Estinyl."[321] Trademark for ethinyl estradiol.

"Estonmite."[88] Trademark for para-chlorophenyl parachlorobenzene sulfonate; miticide; available as a dust base, wettable powder and emulsifiable solution; used as an ovicide, specific against the eggs of spider mites.

"Estonox."[88] Trademark for toxaphene in a dust base, wettable powder and in a stabilized emulsifiable carrier; used for control of insects on cotton, seed alfalfa, sugar beets, beans and potatoes.

Hazard: Probably toxic. See toxaphene.

estradiol $C_{18}H_{24}O_2$. A female sex hormone. It occurs in two isomeric forms, alpha and beta. Beta-estradiol has the greatest physiological activity of any naturally occurring estrogen. The alpha form is relatively inactive. Commonly used preparations, are the benzoate, dipropionate, and valerate, as well as ethinylestradiol (q.v.).

Properties of beta form: White or slightly yellow; small crystals or crystalline powder; odorless; m.p. 173–179°C; stable in air; almost insoluble in water; soluble in alcohol, acetone, dioxane, and in solutions of alkali hydroxides; sparingly soluble in vegetable oils.

Derivation: Isolation from human and mare pregnancy urine; commercial synthesis from cholesterol or ergosterol.

Grade: N.F. (beta form).

Use: Medicine.

estragole (chavicol methyl ether; methyl chavicol) $C_6H_4(C_3H_5)(OCH_3)$.

Properties: Colorless liquid; anise odor; sp. gr. 0.965–0.975 (20/4°C); (n 17.5/D) 1.5230; b.p. 216°C. Soluble in alcohol and chloroform. Low toxicity.

Occurrence: In estragon oil; basil oils; anise bark oil, and others.

Uses: Perfumes; flavors.

estragon oil (tarragon oil).

Properties: Colorless to yellowish-green essential oil; anise-like odor; aromatic but not sweet taste. Keep well stoppered. Solubility in alcohol: in 6 to 11 vols and more of 80% alcohol; in 1 vol and more of 90% alcohol. Sp. gr. 0.914–0.956 (25°C); optical rotation +2° to +9°; refractive index 1.502 to 1.514; acid value up to 1; ester value 1 to 9, after acetylation 15. Low toxicity. Combustible.

Derivation: Steam distillation of the flowering herb artemisia dracunculus L.
Containers: Glass bottles; copper flasks.
Grades: Technical; F.C.C. (as tarragon oil).
Use: Flavoring; cosmetics; fragrances; medicine.

"Estrex."[152] Trademark for a series of methyl, butyl, and propyl, glyceryl, and polyethylene glycol esters of fatty acids.
Uses: Adhesives, agricultural sprays, aluminum rolling, antioxidants, cosmetics, cutting oils, detergents, leather tanning, lubricants and lubricating additives, plasticizers, rubber and textile finishes.

estriol $C_{18}H_{24}O_3$.
Properties: White, odorless, microcrystalline powder; m.p. 282°C. Exhibits reddish fluorescence under filtered ultraviolet light. Undergoes phase change at 270–275°C. Practically insoluble in water; soluble in alcohol, dioxane, and oils.
Derivation: Isolation from human pregnancy urine; isolation from human placenta; organic synthesis.
Use: Medicine.

estrogen. A general term for female sex hormones. They are responsible for the development of the female secondary sex characteristics such as the deposition of fat and the development of the breasts. The naturally occurring estrogens, such as estradiol, estrone, and estriol, are steroids. Estrogens are produced by the ovary, and to a lesser degree, by the adrenal cortex and testis. Some synthetic nonsteroid compounds, such as diethylstilbestrol and hexestrol, have estrogenic activity.
Hazard: May be carcinogenic; may have damaging side effects (thromboembolism) as used in oral contraceptives.
Use: Medicine; biochemical reseach; oral contraceptives.
See also antifertility agent.

"Estron."[256] Trademark for synthetic yarn and staple fiber, acetate tow for use in cigarette filter tips, tobacco smoke filters, and tobacco smoke filter tip rods.

estrone $C_{18}H_{22}O_2$. A steroid with some estrogenic activity.
Properties: Small, white crystals or white crystalline powder; m.p. 258–262°C; odorless; stable in air; insoluble in water; soluble in alcohol, acetone, dioxane, and in solutions of fixed alkali hydroxides.
Derivation: Isolation from human pregnancy urine; synthesis from ergosterol.
Grades: N.F.
Use: Medicine.

"Estynox."[202] Trademark for epoxidized esters of castor or other fatty acids produced by reacting the double bonds in these oils to produce oxirane or epoxy groups.
Uses: Stabilizing plasticizers for polyvinyl chloride and nitrocellulose resins. Acid scavengers in cutting fluids, lubricants, textile processing, alkyd manufacture; corrosion inhibitors.

Et Informal abbreviation for ethyl. Example: EtOH, ethyl alcohol.

"Etamon."[330] Trademark for tetraethylammonium chloride (q.v.).

"Ethafoam."[233] Trademark for a lightweight, low-density polyethylene foam.

ethanal. See acetaldehyde.

ethanamide. See acetamide.

ethane (dimethyl; methylmethane) C_2H_6.
Properties: Colorless, odorless gas; insoluble in water; soluble in alcohol; relatively inactive chemically; b.p. −88.63°C; f.p. −183.23°C (triple point); sp. gr. of liquid 0.446 (0°C); sp. gr. of vapor (air = 1) 1.04 (0°C); critical temperature 32.1°C; critical pressure (absolute) 718 psi; specific heat at constant pressure 0.897; specific heat at constant volume 0.325; ratio of specific heats (cp/cv) 1.224; heat of combustion (approx.) 22,300 Btu/lb or 1800 Btu/cu ft; flash point −211°F; autoignition temp. 959°F. An asphyxiant gas.
Derivation: Fractionation of natural gas.
Grades: 95%; 99%; research, 99.98%.
Containers: Steel cylinders.
Hazard: Severe fire risk if exposed to sparks or open flame. Flammable limits in air, 3 to 12.5%.
Uses: Petrochemicals (source of ethylene, halogenated ethanes); refrigerant fuel.
Shipping regulations: (Rail) Red Gas label. (Air) Flammable Gas label. Not accepted on passenger planes.

ethanedioyl chloride. See oxalyl chloride.

ethane hydrate. See gas hydrate.

ethanethiol (ethyl sulfhydrate; ethyl mercaptan). C_2H_5SH.
Properties: Colorless liquid. Has one of the most penetrating and persistent odors known (skunk). Slightly soluble in water; soluble in alcohol, ether, petroleum naphtha. Sp. gr. 0.83907 (20/4°C); b.p. 36°C; f.p. −121°C; refractive index (n 20/D) 1.4305; flash point below 80°F (C.C.); autoignition temp. 570°F.
Derivation: By saturating potassium hydroxide solution with hydrogen sulfide, mixing with calcium ethylsulfate solution and distilling on a water bath.
Containers: 55-gal drums; 25-lb cylinders; tank cars.
Hazard: Flammable; dangerous fire risk. Highly toxic by ingestion and inhalation. Tolerance, 0.5 ppm in air.
Uses: LPG odorant; adhesive stabilizer; chemical intermediate.
Shipping regulations: (Rail) Red label (Air) Flammable Liquid label. Legal label name, ethyl mercaptan.
Note: Tomato juice is reported to deodorize materials contaminated with this compound.

ethanethiolic acid. See thioacetic acid.

ethanoic acid. See acetic acid.

ethanol. See ethyl alcohol.

ethanolamine (MEA; monoethanolamine; colamine; 2-aminoethanol; 2-hydroxyethylamine) $HOCH_2CH_2NH_2$.
Properties: Colorless, moderately viscous liquid. Ammoniacal odor. Strong base. Miscible with water; soluble in carbon tetrachloride, alcohol, chloroform. Sp. gr. 1.0179 (20/20°C); b.p. 170.5°C; melting point 10.5°C; vapor pressure 0.48 mm (20°C); flash point (open cup) 200°F; wt/gal 8.5 lb (20°C). Combustible.
Derivation: Reaction of ethylene oxide and ammonia gives a mixture of mono-, di-, and triethanolamines.
Grades: Technical; N.F.
Containers: 1-, 5-gal cans; 55-gal drums; tank cars.
Hazard: Toxic and irritant. Tolerance, 3 ppm in air.
Uses: Scrubbing acid gases (H_2S, CO_2), especially in synthesis of ammonia, from gas streams; nonionic detergents used in drycleaning, wool treatment, emulsion paints, polishes, agricultural sprays; chem-

ical intermediates; pharmaceuticals; corrosion inhibitor; rubber accelerator.
Shipping regulations: (Air) Corrosive label; (Solutions) Flammable Liquid label.

ethanol formamide $HOCH_2CH_2NHOCH$.
Properties: Somewhat viscous liquid completely miscible with water, alcohol and glycerol; compatible with polyvinyl alcohol, many cellulosic and natural resins; b.p. 143°C (2.5 mm); freezing point below -72°C; sp. gr. (25/4°C) 1.170; flash point 347°F; combustible. Low toxicity.

ethanol hydrazine. See beta-hydroxyethylhydrazine.

ethanolurea $NH_2CONHCH_2CH_2OH$. Liquid. Solidification point 71–74°C; formaldehyde condensation products are permanently thermoplastic and water-soluble. As increasing amounts of simple urea are mixed with ethanolurea, the condensation products gradually change from pliable film-forming resins into the brittle types. Thus almost any degree of water-solubility and flexibility may be obtained in the final resin. The modified resins formed with ethanolurea are compatible with polyvinyl alcohol, methyl cellulose, cooked starch and other water-dispersible materials.

"Ethasan."[58] Trademark for zinc diethyldithiocarbamate, a rubber accelerator.

"Ethavan."[58] Trademark for ethyl vanillin.

"Ethazate."[248] Trademark for zinc diethyldithiocarbamate, a rubber accelerator.

ethchlorvynol (1-chloro-3-ethyl-1-penten-4-yn-3-ol; beta-chlorovinyl ethyl ethynyl carbinol) $HC\vdots CCOH(C_2H_5)CH{:}CHCl$.
Properties: Colorless to yellow liquid; pungent aromatic odor; darkens on exposure to light and to air; sp. gr. 1.068–1.071; refractive index, 1.4765–1.4800 (n 25/D). Immiscible with water; miscible with most organic solvents.
Grade: N.F.
Use: Medicine (sedative).

ethene. See ethylene.

ethenol. See vinyl alcohol.

ether. A class of organic compounds in which an oxygen atom is interposed between two carbon atoms (organic groups) in the molecular structure, giving the generic formula ROR. They may be derived from alcohols by elimination of water, but the major method is catalytic hydration of olefins. Only the lowest member of the series, methyl ether, is gaseous; most are liquid, and the highest members are solid (cellulose ethers). The term "ether" is often used synonymously with "ethyl ether", and is the legal label name for it.
Hazard: The lower m.w. ethers are dangerous fire and explosion hazards; when containing peroxides, they can detonate on heating.
Uses: See ethyl ether; polymer, water-soluble; ethylene oxide; propylene oxide; diethylene glycol; ethyl cellulose; polyether.
Note: An illogical and archaic use of the term "ether" survives in such names as petroleum ether, referring to its volatility and "ethereal" nature.

ethical drug. A prescription drug. See also drug.

ethinylestradiol (ethynylestradiol) $C_{20}H_{24}O_2$. An estrogen (female sex hormone), 19-nor-17alpha-pregna-1,3,5(10)-trien-20-yne-3,17-diol.
Properties: Fine, white to creamy white, odorless, crystalline powder; sensitive to light. M.p. 142–146°C. May also exist in a polymorphic modification, m.p. 180–186°C. Soluble in acetone, alcohol, chloroform, dioxane, ether and vegetable oils; practically insoluble in water; soluble in solutions of sodium or potassium hydroxide. Slightly dextrorotatory in dioxane solution.
Derivation: Preparation from estrone.
Grade: U.S.P.
Use: Medicine.

ethion Generic name for O,O,O′,O′-tetraethyl-S,S′-methylenediphosphorodithioate $[(C_2H_5O)_2P(S)S]_2CH_2$.
Properties: Liquid; sp. gr. 1.220 (20°C); f.p. -13°C. Slightly soluble in water; soluble in lactones, xylene, chloroform and methylated naphthalene.
Hazard: Highly toxic; cholinesterase inhibitor. Use may be restricted.
Use: Insecticide and miticide.
Shipping regulations: Organic phosphate, liquid, n.o.s., (Rail, Air) Poison label. Not accepted on passenger planes.

ethisterone (pregneninolone; anhydrohydroxyprogesterone; ethynyltestosterone) $C_2H_{28}O_2$. A female sex hormone; a derivative of progesterone with similar activity. Effective when given by mouth.
Properties: White or slightly yellow crystals or as crystalline powder. Odorless; stable in air. Affected by light. M.p. 267–275°C (dec.). Practically insoluble in water; slightly soluble in alcohol, chloroform, ether, and vegetable oils.
Derivation: From progesterone and other steroids.
Grade: N.F.
Use: Medicine.

"Ethocel."[233] Trademark for ethylcellulose resins which are able to withstand shock and maintain toughness over a temperature range of 200° to -40°F. Available in transparent, translucent, and opaque colors. Insoluble in water; soluble in most organic solvents.
Uses: Household articles; automotive parts; tools for aircraft industry.

ethodin (6,9-diamino-2-ethoxyacridine lactate monohydrate) $C_{15}H_{15}N_3O \cdot C_3H_6O_3 \cdot H_2O$.
Properties: Pale yellow crystals; darken at 200°C. M.p. 235°C; slowly soluble in 15 parts water, soluble in 9 parts boiling water; soluble in 110 parts alcohol (22°C). Solutions are yellow, fluorescent and stable to boiling.
Purity: 97% (dry basis).
Containers: 1-oz, ¼ lb, 1 lb.
Uses: Bactericide; surgical antisepsis; preparation of pure gamma globulin.

"Ethoduomeen."[15] Trademark for a series of polyethoxylated high-molecular-weight aliphatic diamines.
Containers: 55-gal open head drums; tank cars; tank trucks.
Uses: Cationic emulsifiers; wetting agents.

"Ethofat."[15] Trademark for a series of polyethoxylated fatty acids. Nonionic.

Containers: 55-gal open head or bung-type drums; tank cars; tank trucks.
Uses: Detergents; emulsifiers.

"Ethofil."[539] Trademark for glass fiber-reinforced linear polyethylene.

ethoheptazine (ethyl heptazine) $C_{16}H_{23}NO_2$. 1-Methyl-4-carbethoxy-4-phenylhexamethyleneimine.
Properties: Liquid; sp. gr. 1.038 (26/4°C); b.p. 133–134°C (1.0 mm); refractive index 1.5210 (26°C).
Use: Medicine.

ethohexadiol. U.S.P. name for 2-ethylhexanediol-1,3.

"Ethomeen."[15] Trademark for a series of polyethoxylated aliphatic amines with alkyl groups ranging from C_8 to C_{18}, also for naturally occurring mixtures of these alkyl groups. Can be obtained with varying degrees of cationic strength from quite strong to almost nonionic.
Containers: 55-gal drums; tank cars; tank trucks.
Uses: Emulsifiers; wetting agents.

"Ethomid."[15] Trademark for polyethoxylated high-molecular-weight amides.
Containers: 55-gal drums; tank cars; tank trucks.
Uses: Emulsifiers; wetting agents.

"Ethone."[227] Trademark for alpha-methylanisalacetone (q.v.).

"Ethoquad."[15] Trademark for a series of polyethoxylated quaternary ammonium chlorides derived from long-chain amines.
Containers: 55-gal drums; tank cars; tank trucks.
Uses: Antistatic agents; dye levelers; electroplating bath additives.

"Ethosperse."[73] Trademark for a series of surface-active compounds, the reaction products of fatty alcohols, glycols, glycerol, and sorbitol and ethylene oxide.
Uses: Food, cosmetics, pharmaceuticals and industrial applications.

ethosuximide (SAN) (alpha-ethyl-alpha-methylsuccinimide) $C_7H_{11}NO_2$.
Properties: White to off-white waxy powder or solid having a characteristic odor; soluble in water, alcohol and ether.
Grade: N.D.
Use: Medicine.

para-**ethoxyacetanilide.** See acetophenetidin.

ethoxybenzidine (di-para-aminoethoxydiphenyl) $C_6H_4NH_2C_6H_3(OC_2H_5)NH_2$.
Properties: Glistening, flat needles. Soluble in alcohol; sparingly soluble in water. M.p. 135°C.
Derivation: By heating ethoxybenzidine sulfonic acid, obtained from benzeneazophenetolesulfonic acid, with water in an autoclave.
Hazard: May be carcinogenic. See benzidine.
Use: Medicine.

2-ethoxy-3,4-dihydro-2H-pyran $OCH{:}CHCH_2CH_2CHOC_2H_5$ (ring closed between O and CH).
Properties: Liquid; sp. gr. 0.970 (20/20°C); b.p. 143°C; f.p. −100°C; sets to glass below this temperature; flash point 111°F (O.C.). Very slightly soluble in water. Combustible.
Hazard: Moderate fire risk. Moderately toxic.
Uses: Stabilizer; intermediate.

6-ethoxy-1,2-dihydro-2,2,4-trimethylquinoline (ethoxyquin) $C_{14}H_{19}NO$.
Properties: Yellow liquid; b.p. 125°C (2 mm); m.p. below 0°C; refractive index 1.569–1.572 (25°C); sp. gr. 1.029–1.031 (25°C). Discolors and stains badly.
Containers: 55- and 5-gal drums.
Uses: Insecticide; antioxidant; flex-cracking inhibitor; post-harvest preservation of apples (scald inhibitor).

2-ethoxyethyl-para-methoxycinnamate $CH_3OC_6H_4CH{:}CHCOOC_2H_4OC_2H_5$.
Properties: Slightly yellow viscous liquid; practically odorless; sp. gr. 1.1000–1.1035 (25/25°C); refractive index (1.5650–1.5675 (20°C); flash point above 212°F (TCC). Miscible with alcohol and isopropyl alcohol; almost insoluble in water. Combustible.
Use: Ultraviolet absorber in suntan preparations.

3-ethoxy-4-hydroxybenzaldehyde. See ethyl vanillin.

1-ethoxy-2-hydroxy-4-propenylbenzene. See propenyl guaethol.

4-ethoxy-3-methoxybenzaldehyde $C_6H_3(OC_2H_5)(OCH_3)CHO$. White to light brown crystals having a slight vanillin odor; m.p. 62–64°C.
Use: Intermediate. Combustible.

4-ethoxy-3-methoxyphenylacetic acid $C_6H_3(OC_2H_5)(OCH_3)CH_2COOH$. An off-white powder; m.p. 119–122°C.
Use: Intermediate.

1-ethoxy-2-methoxy-4-propenylbenzene. See isoeugenol ethyl ether.

ethoxymethylenemalononitrile $C_2H_5OCH{:}C(CN)_2$.
Properties: Colorless liquid. B.p. 160°C; m.p. 64°C; flash point 311°F. Combustible.
Hazard: May be toxic.
Uses: Chemical intermediate.

4-ethoxyphenol. See hydroquinone monoethyl ether.

ethoxyquin. Coined name for 6-ethoxy-1,2-dihydro-2,2,4-trimethylquinoline (q.v.).

ethoxytriglycol $C_2H_5O(C_2H_4O)_3H$.
Properties: Colorless liquid. Sp. gr. 1.0208 (20/20°C); 8.5 lb/gal (20°C); b.p. 255.4°C; vapor pressure less than 0.01 (20°C); freezing point −18.7°C; viscosity 7.80 cp (20°C); completely soluble in water. Flash point 275°F (closed cup); combustible; low toxicity.
Use: Chemical intermediate.

"Ethrel."[574] Trademark for a plant-growth regulator that promotes release of ethylene when applied to plants thus hastening ripening and making possible higher product yield. Claimed useful as stimulant for rubber latex formation and pineapple growth.

"Ethron."[233] Trademark for polyethylene resins.

"Ethycol."[233] Trademark for pigments dispersed in ethylcellulose and plasticizer.
Uses: Protective coatings and inks in which ethylcellulose is a major ingredient.

ethyl abietate $C_{19}H_{29}COOC_2H_5$.
Properties: Amber-colored, viscous liquid; hardens upon oxidation. Soluble in ether, most varnish solvents; insoluble in water. B.p. 350°C; flash point 352°F; m.p. 45°C; refractive index 1.4980; sp. gr. 1.02. Combustible.
Derivation: (a) By heating together ethyl chloride and an alcoholic solution of rosin and caustic soda. (b) By reacting ethyl iodide with silver abietate.
Hazard: Moderately toxic; irritant.
Uses: Varnishes, lacquers, and coating compositions.

"Ethylac."[204] Trademark for 2-benzothiazyl-N,N-diethylthiocarbamylsulfide (q.v.).

N-ethylacetamide $CH_3CONHC_2H_5$.
Properties: Colorless liquid; sp. gr. 0.920 (20/20°C); boiling range 206–208.5°C; flash point 230°F; faint odor. Combustible.

ethyl acetamidocyanoacetate (acetamidocyanoacetic ester; ethyl N-acetyl-alpha-cyanoglycine) $NCCH(NHCOCH_3)COOC_2H_5$. Solid; m.p. 129°C.
Use: Synthesis of amino acids and related compounds.

N-ethylacetanilide (ethylphenylacetamide) $C_6H_5NC_2H_5COCH_3$.
Properties: White, crystalline solid. Faint odor. Soluble in most organic solvents. Insoluble in water. Sp. gr. 0.994; b.p. 258°C; flash point 126°F; m.p. 54°C. Low toxicity. Combustible.
Grade: Technical.
Hazard: Moderate fire risk.
Use: Substitute for camphor in nitrocellulose.

ethyl acetate (acetic ether; acetic ester; vinegar naphtha) $CH_3COOC_2H_5$.
Properties: Colorless, fragrant liquid. Soluble in chloroform, alcohol and ether; slightly soluble in water. B.p. 77°C; vapor pressure (20°C) 73 mm; f.p. -83.6°C; density 0.8945 g/ml (25°C); flash point 24°F. Autoignition temp. 800°F.
Derivation: By heating acetic acid and ethyl alcohol in presence of sulfuric acid, and distilling.
Grades: Commercial, 85–88%; 95–98%; 99%; N.F. (99%); F.C.C.
Containers: Drums; tank cars.
Hazard: Flammable, dangerous fire and explosion risk. Flammable limits in air, 2.2 to 9%. Moderately toxic by inhalation and skin absorption. Irritating to eyes and skin. Tolerance, 400 ppm in air. Safety data sheet available from Mfg. Chemists Assn., Washington, D.C.
Uses: General solvent in coatings and plastics; organic synthesis; smokeless powders; pharmaceuticals.
Shipping regulations: (Rail) Red label. (Air) Flammable Liquid label.

ethyl acetate, anhydrous. Ethyl acetate, grade 99%.

ethylacetic acid. See butyric acid.

ethyl acetoacetate (diacetic ester; acetoacetic ester) $CH_3COCH_2COOC_2H_5$ (keto form). This compound is a tautomer, at room temperature consisting of about 93% keto form and 7% enol form, $CH_3C(OH){:}CHCOOC_2H_5$.
Properties: Colorless liquid; fruity odor. Soluble in alcohol; soluble in water. Sp. gr. 1.0250 (20/4°C); f.p. (enol) -80°C, (keto) -39°C; b.p. 180–181°C; wt/gal 8.5 lb; vapor pressure 0.8 mm (20°C); flash point 185°F (COC); coefficient of expansion 0.00101/°C. Combustible.
Derivation: Action of metallic sodium on ethyl acetate, with subsequent distillation.
Grades: Technical; 98%.
Containers: Drums; tank cars.
Hazard: Moderately toxic by ingestion and inhalation; irritant to skin and eyes.
Uses: Organic synthesis; antipyrine; lacquers; dopes; plastics; manufacture of dyes, pharmaceuticals, antimalarials, vitamin B; flavoring.

ethyl acetone. See methyl propyl ketone.

ethylacetylene (1-butyne) $C_2H_5C{\vdots}CH$.
Properties: Available as liquefied gas; b.p. 8.3°C; sp. gr. 0.669 (0/0°C); f.p. −130°C; flash point (TOC) <20°F; specific volume 7.2 cu ft/lb (70°F); insoluble in water.
Hazard: Flammable, dangerous fire risk.
Uses: Specialty fuel; chemical intermediate.
Shipping regulations: (Rail) Red Gas label. (Air) Flammable Gas label. Not acceptable on passenger planes.

ethyl N-acetyl-alpha-cyanoglycine. See ethyl acetamidocyanoacetate.

ethyl acetylsalicylate $C_2H_5OOCC_6H_4OOCCH_3$.
Properties: Colorless liquid. Insoluble in water; soluble in many organic solvents. Sp. gr. 1.153; b.p. 272°C.
Use: Medicine.

ethyl acrylate $CH_2{:}CHCOOC_2H_5$.
Properties: Colorless liquid; b.p. 99.4°C; f.p. -72.0°C; sp. gr. 0.9230 (20/20°C); refractive index 1.4037 (25/D); wt/gal 7.6 lb (20°C); soluble in water and alcohol; readily polymerized. Flash point 60°F (open cup).
Derivation: (a) Ethylene cyanohydrin, ethyl alcohol, and dilute sulfuric acid; (b) Oxo reacton of acetylene, carbon monoxide, and ethyl alcohol in the presence of nickel or cobalt catalyst.
Grades: Technical (inhibited, usually with hydroquinone or its monomethyl ether); pure uninhibited.
Containers: Drums; tank cars.
Hazard: Toxic by ingestion, inhalation, skin absorption; extremely irritating to skin and eyes. Tolerance, 25 ppm in air. Flammable, dangerous fire and explosion hazard. Safety data sheet available from Mfg. Chemists Assn., Washington, D.C.
Uses: Polymers; acrylic paints; intermediates.
Shipping regulations: (Rail) Flammable liquid, n.o.s., Red label (Air) (inhibited) Flammable Liquid label; (uninhibited) Not acceptable.
See also acrylate and acrylic resin.

ethyl alcohol (alcohol; grain alcohol; ethanol; EtOH) C_2H_5OH. 47th highest-volume chemical produced in U.S. (1975).
Properties of pure 100% absolute alcohol (dehydrated): Colorless, limpid, volatile liquid; b.p. 78.3°C; f.p. -117.3°C; ethereal, vinous odor; pungent taste. Miscible with water, methyl alcohol, ether, chloroform, and acetone.
Properties: (95%) Refractive index 1.3651 (15°C); surface tension 22.3 dynes/cm (20°C); viscosity 0.0141 poise (20°C); vapor pressure 43 mm (20°C); specific heat 0.618 cal/g (23°C); flash point 55°F; sp. gr. 915.56°C) 0.816; b.p. 78°C; f.p. -114°C. Autoignition temp. 793°F.
Derivation: (a) From ethylene by direct catalytic hydration or with ethyl sulfate as intermediate. (b) Fermentation of blackstrap molasses. (c) Fermentation of lactose from whey. (d) Enzymatic hydrolysis of cellulose (see also cellulase). Methods (c) and (d) are experimental but may be expected to move into production before 1980.
Grades: U.S.P. (95% by volume); absolute; pure; completely denatured; specially denatured; industrial; various proofs (one half the proof number is the percent of alcohol by volume).
Containers: Drums, tank trucks, tank cars.
Hazard: Flammable, dangerous fire risk; flammable limits in air, 3.3 to 19%. Tolerance, 1000 ppm in air. Ethyl alcohol is classified as a depressant drug. Though it is rapidly oxidized in the body and is

therefore noncumulative, ingestion of even moderate amounts causes lowering of inhibitions, often succeeded by dizziness, headache, or nausea. Larger intake causes loss of motor nerve control, shallow respiration, and in extreme cases unconsciousness and even death. Degree of intoxication is determined by concentration of alcohol in the brain. Of primary importance is the fact that intake of even moderate amounts together with barbiturates or similar drugs is extremely dangerous, and may even be fatal.
Uses: Solvent for resins, fats, oils, fatty acids, hydrocarbons, alkali hydroxides; extractive medium; manufacture of intermediates, organic derivatives (especially acetaldehyde), dyes, synthetic drugs, esters, elastomers, detergents, cleaning solutions, surface coatings, cosmetics, pharmaceuticals, explosives, antifreeze; beverages; antisepsis; medicine; gasoline additive; yeast-growth medium.
Shipping regulations: (Rail) Red label. (Air) Flammable Liquid label.
See also: alcohol, denatured; industrial alcohol.

ethyl aldehyde. See acetaldehyde.

ethyl alpha-**allylacetoacetate**
$CH_3COCH(CH_2CH:CH_2)COOC_2H_5$.
Properties: Water-white liquid; sp. gr. (20°C) 0.989; wt/gal (20°C) 8.24 lb. Combustible. Low toxicity.
Uses: Intermediate for pharmaceuticals, perfumes, fungicides, insecticides, fine chemicals.

ethyl aluminum dichloride (EADC) $C_2H_5AlCl_2$.
Properties: Clear, yellow pyrophoric liquid. B.p. (extrapolated) 194°C; f.p. 32°C; density 1.222 g/ml; weight 10.28 lb/gal (25°C).
Derivation: Reaction of aluminum chloride with ethyl aluminum sesquichloride.
Hazard: Ignites on contact with air; dangerous fire risk; reacts violently with water.
Uses: Catalyst for olefin polymerization, aromatic hydrogenation; intermediate.
Shipping regulations: (Rail) Red label. (Air) Not acceptable.

ethyl aluminum sesquichloride (EASC) $(C_2H_5)_3Al_2Cl_3$.
Properties: Clear yellow pyrophoric liquid; b.p. 204°C; f.p. -50°C. Density 1.08 g/ml.
Derivation: Reaction of ethyl chloride and aluminum.
Grades: Commercial.
Containers: Tank cars and trucks.
Hazard: Ignites on contact with air; dangerous fire risk; reacts violently with water.
Uses: Catalyst for olefin polymerization, aromatic hydrogenation; intermediate.
Shipping regulations: (Rail) Red label. (Air) Not acceptable.

ethylamine (monoethylamine; aminoethane)
$CH_3CH_2NH_2$.
Properties: Colorless, volatile liquid (or gas); ammonia odor; strong alkaline reaction. B.p. 16.6°C; freezing point -81.2°C; sp. gr. (liquid, 15/15°C) 0.689; wt/gal (20°C) 5.7 lb; flash point (open cup) below 0°F. Autoignition temp. 723°F. Miscible with water, alcohol, and ether.
Derivation: From ethyl chloride and alcoholic ammonia under heat and pressure.
Grades: Technical (anhydrous and 70% aqueous solution); pure, 98.5% min.
Containers: 1-gal cans; 5-, 55-gal drums; tank cars; tank trucks.
Hazard: Flammable, dangerous fire risk.Toxic; strong irritant. Tolerance, 10 ppm in air. Flammable limits in air 3.5 to 14%.
Use: Dye intermediates; solvent extraction; petroleum refining; stabilizer for rubber latex; detergents; organic synthesis.
Shipping regulations: (Rail) Red label. (Air) Flammable Liquid label. Legal label name monoethylamine.

ethylamine hydrobromide $C_2H_5NH_2 \cdot HBr$.
Properties: White, practically odorless granules; m.p. 158–161°C; very soluble in water.
Hazard: Probably toxic.
Use: Intermediate (where liquid ethylamine or liquid hydrobromic acid cannot be used).

ethyl-ortho-**aminobenzoate.** See ethyl anthranilate.

ethyl-para-**aminobenzoate** (benzocaine, anesthesin). $C_6H_4NH_2CO_2C_2H_5$.
Properties: White, crystalline, odorless, tasteless powder; stable in air; m.p. 88–92°C. Soluble in dilute acids, less so in chloroform, ether, and alcohol, very slightly soluble in water. Low toxicity.
Derivation: Ethylation of para-nitrobenzoic acid, followed by reduction.
Grades: Technical; pure; N.F. (as benzocaine).
Containers: Glass bottles; drums.
Use: Medicine (local anesthetic); suntan preparations.

ethylaminoethanol. See ethylethanolamine. Mixed ethylaminoethanols (sold in up to tank car lots) may also contain diethylaminoethanol (q.v.).

2-ethylamino-4-isopropylamino-6-methylthio-s-triazine $C_2H_5HNC_3N_3(SCH_3)NHCH(CH_3)_2$.
Properties: White crystalline powder; m.p. 84–85°C; slightly soluble in water; soluble in organic solvents.
Hazard: Toxic by ingestion.
Use: Weed-killing agent in pineapple and sugar cane.

ethyl-1-(para-**aminophenyl)-4-phenylisonipecotate**
See anileridine.

ethyl amyl ketone (EAK; 5-methyl-3-heptanone)
$CH_3CH_2CO(CH_2)_4CH_3$.
Properties: Colorless liquid, pungent odor. Insoluble in water. Soluble in 4 vols of 60% alcohol. Sp. gr. 0.819–0.824; refractive index 1.416; flash point 138°F. Combustible.
Containers: Drums; tank cars.
Hazard: Narcotic in high concentrations. Moderate fire risk. Tolerance, 25 ppm in air.
Uses: Perfumery; solvent.

N-**ethylaniline** $C_2H_5NHC_6H_5$.
Properties: Colorless liquid, becoming brown on exposure to light. Soluble in alcohol; insoluble in water. Sp. gr. 0.9631; f.p. -63.5°C; b.p. 206°C; refractive index (n 20/D) 1.5559. Flash point 185°F (O.C.). Combustible.
Derivation: By heating aniline and ethyl alcohol in presence of sulfuric acid, with subsequent distillation.
Containers: Iron drums; tank cars.
Hazard: Toxic by ingestion, inhalation, skin absorption.
Use: Organic synthesis.

ortho-**ethylaniline** $C_6H_4(NH_2)C_2H_5$.
Properties: Brown liquid; f.p. -44°C; sp. gr. (20°C) 0.982; b.p. 214°C; flash point (open cup) 185°F. Soluble in alcohol and toluene; insoluble in water. Combustible.
Hazard: Toxic by ingestion, inhalation, skin absorption.
Uses: Intermediate for pharmaceuticals, dyestuffs, pesticides, and other products.

ethyl anthranilate (ethyl-ortho-aminobenzoate) $C_6H_4(NH_2)COOCH_2CH_3$.
Properties: Colorless liquid, fruity odor. Sp. gr. 1.117; refractive index 1.564; b.p. 260°C. Soluble in alcohol and propylene glycol. Combustible.
Grades: Technical; F.C.C.
Uses: Perfumery and flavors, similar to methyl anthranilate (q.v.).

2-ethylanthraquinone $C_{14}H_8O_2C_2H_5$.
Properties: Buff to light yellow paste. M.p. 108°C.
Use: Synthesis, especially of hydrogen peroxide.

"Ethyl" Antiknock Compounds.[313] Trademark for a series of fuel additives containing various percentages of tetraethyl lead, ethylene dibromide, ethylene dichloride, dye, kerosine, and antioxidant. All are used to improve the octane rating of motor fuels.

"Ethyl" Antioxidants.[313] Trademark for a series of gasoline antioxidants based on phenols (chiefly di-tert-butyl phenol). They inhibit formation of gum and peroxides in gasoline, and the formation of decomposition products of jet fuels in storage. Also used for steam-turbine and industrial oils, and to retard decomposition of antiknock compunds in gasoline.

ethylarsenious oxide C_2H_5AsO.
Properties: Colorless oil. Garlic-like, nauseating odor. Oxidizes in air and forms colorless crystals. Soluble in acetone, benzene, ether. Sp. gr. 1.802 (11°C); b.p. 158°C (10 mm).
Hazard: Highly toxic.
Use: Organic synthesis.
Shipping regulations: (Rail, Air) Arsenical compounds, n.o.s., Poison label.

"Ethyl" Automate Liquid Dyes.[313] Trademark for single-phase organic dyes in aromatic solvent. Full range of liquid dyes in various concentrations, colors, and combinations.

2-ethylaziridine. See ethylethyleneimine.

ethylbenzene (phenylethane) $C_6H_5C_2H_5$. 20th highest-volume chemical produced in U.S. (1975).
Properties: Colorless liquid; aromatic odor; vapor heavier than air; boiling point 136.187°C; refractive index 1.49594 (20°C); sp. gr. 0.867 (20°C); f.p. -94.-975°C; wt/gal 7.21 lb (25°C); flash point 59°F; autoignition temp. 810°F; specific heat 0.41/cal/gal/°C; viscosity 0.64 centipoise (25°C). Soluble in alcohol, benzene, carbon tetrachloride, and ether; almost insoluble in water.
Derivation: (a) By heating benzene and ethylene in presence of aluminum chloride, with subsequent distillation; (b) by fractionation directly from the mixed xylene stream in petroleum refining.
Grades: Technical; pure; research.
Containers: Drums; tank cars.
Hazard: Flammable, dangerous fire risk. Moderately toxic by ingestion, inhalation, and skin absorption. Irritant to skin and eyes. Tolerance, 100 ppm in air.
Uses: Intermediate in production of styrene; solvent.
Shipping regulations: (Rail) Flammable Liquid, n.o.s., Red label. (Air) Flammable Liquid label.

ethyl benzoate $C_6H_5CO_2C_2H_5$.
Properties: Colorless, aromatic liquid; soluble in alcohol and ether; insoluble in water. Sp. gr. 1.043–1.046; f.p. -32.7°C; b.p. 212.9°C; refractive index 1.505. Flash point 200°F; combustible; low toxicity.
Derivation: By heating ethyl alcohol and benzoic acid in presence of sulfuric acid.
Grade: Technical; F.C.C.
Containers: Drums; glass bottles.
Uses: Flavoring; perfumery; solvent mixtures; lacquers; solvent for many cellulose derivatives and natural and synthetic resins.

ethyl benzoylacetate $C_6H_5COCH_2COOC_2H_5$.
Properties: Light yellow oil; boiling range 144–148°C (10 mm); sp. gr. 1.111–1.117 (20°C); flash point 275°F. Soluble in most organic solvents; insoluble in water. Combustible. Low toxicity.
Derivation: Reaction of ethyl acetate and ethyl benzoate with metallic sodium.
Method of purification: Vacuum distillation.
Grade: 95% pure.
Use: Dye and pharmaceutical intermediate.

ethyl ortho-**benzoylbenzoate** $C_6H_5COC_6H_4COOC_2H_5$.
Properties: Yellowish white solid; odorless; insoluble in water; soluble in alcohol, acetone, ethyl acetate, and benzene; m.p. 56–58°C; b.p. 325°C. Combustible.
Uses: Plasticizer for nitrocellulose and synthetic resins.

ethylbenzylaniline $C_6H_5N(C_2H_5)CH_2C_6H_5$.
Properties: Colorless oil; soluble in alcohol and ether; insoluble in water. Sp. gr. 1.034; b.p. 286°C. Combustible.
Derivation: By heating ethyl aniline, benzyl chloride, and aqueous caustic soda, with subsequent distillation.
Hazard: May be toxic.
Uses: Dyestuffs; organic synthesis.

ethylbenzyl chloride (1-chloromethylethylbenzene) $ClCH_2C_6H_4C_2H_5$. Consists of 70% para- and 30% ortho-ethylbenzyl chloride.
Properties: Colorless liquid; sp. gr. 1.0460–1.0475 (25/25°C); refractive index 1.5290–1.5305 (n 25/D); soluble in alcohols; insoluble in water.
Hazard: Irritant to eyes; a lachrymator.
Use: Intermediate.

ethyl biscoumacetate (ethyl bis(4-hydroxycoumarinyl)-acetate) $C_{22}H_{16}O_8$. A synthetic derivative of bishydroxycoumarin.
Properties: White, odorless, bitter, crystalline solid. M.p. 177–182°C. Another form melts 154–157°C. Soluble in acetone and benzene; slightly soluble in alcohol and ether; insoluble in water.
Grade: N.F.
Use: Medicine.

ethyl bis(4-hydroxycoumarinyl) acetate. See ethyl biscoumacetate.

ethyl borate. Legal label name for triethyl borate (q.v.).

ethyl bromide (bromoethane) C_2H_5Br.
Properties: Colorless liquid; soluble in alcohol and ether; sparingly soluble in water. Sp. gr. 1.431 (20/4°C); b.p. 38.4°C; wt/gal 12–12.1 lb; vapor pressure 386 mm (20°C); autoignition temp. 952°F; f.p. -119°C; constant minimum boiling mixture of ethyl alcohol 7 mole% and ethyl bromide 93 mole% 37.6°C.
Derivation: From ethyl alcohol or ethylene and hydrogen bromide. One process uses gamma radiation to initiate the combination.
Grade: Technical (98%).
Containers: Drums; tank cars.

Hazard: Toxic by ingestion, inhalation, and skin absorption. Strong irritant. Tolerance, 200 ppm in air; explosion limits in air, 6 to 11%. Flammable, dangerous fire hazard.
Uses: Organic synthesis; medicine (anesthetic); refrigerant; solvent; grain and fruit fumigant.

ethyl bromoacetate $CH_2BrCOOC_2H_5$.
Properties: Clear, colorless liquid. Partially decomposed by water. Soluble in alcohol, benzene, ether; insoluble in water. Sp. gr. 1.53 (4°C); b.p. 168°C; f.p. -13.8°C; vapor density 5.8.
Derivation: Interaction of bromine and acetic acid in the presence of red phosphorus.
Grade: Technical.
Hazard: Highly toxic by ingestion, inhalation, and skin absorption. Strong irritant.
Shipping regulations: (Air) Not acceptable.

ethyl butanoate. See ethyl butyrate.

2-ethylbutanol. See 2-ethylbutyl alcohol.

2-ethyl-1-butene (uns-diethylethylene) $CH_3CH_2(C_2H_5)C{:}CH_2$.
Properties: Colorless liquid; sp. gr. 0.6894 (20/4°C); b.p. 64.95°C; refractive index (n 20/D) 1.3969; soluble in alcohol, acetone, ether, petroleum, and coal-tar solvents; insoluble in water. Combustible.
Grade: 95% pure.
Uses: Organic synthesis of flavors, perfumes, medicines, dyes, resins.

3-(2-ethylbutoxy)propionic acid $CH_3CH_2CH(C_2H_5)CH_2OCH_2CH_2COOH$.
Properties: Water-white liquid; sp. gr. (20/20°C) 0.9600; b.p. (100 mm) 200°C; vapor pressure (20°C) $<$0.1 mm; f.p., glass below -90°C; insoluble in water. Flash point 280°F. Combustible. Low toxicity.
Uses: Preparation of metallic salts for paint driers and gelling agents.

2-ethylbutyl acetate $C_2H_5CH(C_2H_5)CH_2OOCCH_3$.
Properties: Colorless liquid; mild odor. Sp. gr. 0.875–0.881 (20/20°C); boiling range, below 155°C none, above 164°C none; purity not less than 90% ethylbutyl acetate; average wt/gal 7.33 lb (20°C). Flash point 130°F (open cup).
Hazard: Moderate fire risk.
Uses: Solvent for nitrocellulose lacquers; flavoring.

2-ethylbutyl alcohol (2-ethylbutanol; hexyl alcohol, pseudo-) $CH_3CH_2CH(C_2H_5)CH_2OH$.
Properties: Colorless liquid; stable. Miscible with most organic solvents; slightly soluble in water. B.p. 148.9°C; f.p. -114°C; sp. gr. 0.8328 (20/20°C); wt/gal 6.93 lb (20°C); refractive index 1.4229 (20°C); flash point (ASTM open cup) 137°F; low toxicity; vapor pressure 0.9 mm (20°C). Combustible.
Containers: Drums; tank cars.
Hazard: Moderate fire risk.
Use: Solvent for oils, resins, waxes, dyes; diluent; synthesis of perfumes, drugs; flavoring.

N-ethylbutylamine $C_2H_5NHCH_2CH_2CH_2CH_3$.
Properties: Water-white; amine odor; b.p. 108°C; f.p. -78°C; sp. gr. 0.7401 (20/20°C); refractive index 1.407 (20°C); flash point 65°F. Partly soluble in water.
Hazard: Flammable; dangerous fire risk.
Use: Intermediate.

ethylbutyl carbonate $C_2H_5CO_3C_4H_9$.
Properties: Colorless liquid used as solvent for many natural and synthetic resins; in mixtures for nitrocellulose. Sp. gr. 0.92 to 0.93 (20°C); b.p. 135–175°C; flash point 122°F (closed cup). Combustible.
Hazard: Moderate fire risk.

ethyl n-butyl ether (n-butyl ethyl ether) $C_2H_5OC_4H_9$.
Properties: Liquid; sp. gr. (20°C) 0.7528; f.p. -103°C; b.p. 92.2°C; flash point 40°F; vapor pressure (20°C) 43 mm; slightly soluble in water.
Containers: Cans; drums.
Hazard: Flammable, dangerous fire risk.
Uses: Extraction solvent; inert reaction medium.
Shipping regulations: (Rail) Flammable liquid, n.o.s., Red label. (Air) Flammable Liquid label.

ethyl butyl ketone (3-heptanone) $CH_3CH_2CH_2CH_2COCH_2CH_3$.
Properties: Clear liquid; sp. gr. 0.8191 (20/20°C); f.p. -39°C; boiling range 142.8 to 147.8°C; 95% purity; wt/gal 6.8 lb. Flash point 115°F (open cup). Insoluble in water; soluble in alcohol. Combustible.
Containers: Drums; tank cars.
Hazard: Moderate fire risk. Moderately toxic; irritant. Tolerance, 50 ppm in air.
Uses: Solvent mixtures for air-dried and baked finishes; for polyvinyl and nitrocellulose resins.

2-ethyl-2-butylpropanediol-1,3 (2-butyl-2-ethylpropanediol-1,3) $HOCH_2C(C_2H_5)(C_4H_9)CH_2OH$.
Properties: White crystals or liquid; sp. gr. (50/20°C) 0.931; b.p. (50 mm) 178°C; f.p. 41.4°C; solubility in water (20°C) 0.8% by wt. Flash point 280°F; combustible.
Uses: Synthesis of lubricants, emulsifying agents; insect repellents; plastics.

2-ethylbutyl silicate $[CH_3CH_2CH(C_2H_5)CH_2O]_4Si$.
Properties: Colorless liquid; b.p. (1 mm) 164°C.
Derivation: Reaction of silicon tetrachloride with 2-ethylbutanol.
Uses: Hydraulic fluid; heat-transfer liquid.

2-ethylbutyraldehyde (diethylacetaldehyde) $(C_2H_5)_2CHCHO$.
Properties: Colorless liquid; insoluble in water. Sp. gr. 0.8164 (20/20°C); b.p. 116.8°C; vapor pressure 13.7 mm (20°C); flash point 70°F (open cup); wt/gal 6.8 lb (20°C); f.p. -89°C.
Grade: Technical.
Containers: 1-gal cans; 5-, 55-gal drums.
Hazard: Flammable, dangerous fire risk. Irritant to eyes and skin.
Uses: Organic synthesis; pharmaceuticals; rubber accelerators; synthetic resins.
Shipping regulations: (Rail) Flammable liquid. n.o.s., Red label. (Air) Flammable Liquid label.

ethyl butyrate (ethyl butanoate) $C_3H_7CO_2C_2H_5$.
Properties: Colorless liquid; pineapple-like odor. Soluble in alcohol and ether; almost insoluble in water and glycerin. Sp. gr. 0.8788; f.p. -93.3°C; b.p. 120.6°C; refractive index (n 20/D) 1.400. Flash point 78°F (closed cup). Autoignition temp. 865°F.
Derivation: Ethyl alcohol and butyric acid heated in presence of sulfuric acid, with subsequent distillation.
Grades: Technical; F.C.C.
Containers: Iron drums; glass bottles.
Hazard: Flammable, dangerous fire risk. Irritating to eyes and mucous membranes; narcotic in high concentrations.
Uses: Flavoring extracts; perfumery; solvent mixture for cellulose esters and ethers; many natural and synthetic resins; lacquers; safety glass.
Shipping regulations: (Air) Flammable Liquid label.

2-ethylbutyric acid (diethyl acetic acid) $(C_2H_5)_2CHCOOH$.
Properties: Water-white liquid. Resembles butyric acid in most properties except that its odor is less pronounced and its water solubility limited. Sp. gr. 0.9225 (20/20°C); b.p. 190°C; vapor pressure 0.08 mm (20°C); flash point 210°F; wt/gal 7.7 lb (20°C); f.p. -15°C. Combustible; low toxicity.
Uses: Ester formation; intermediate for drugs, dyestuffs, chemicals; flavoring.

ethyl caffeate (ethyl 3,4-dihydroxycinnamate) $C_6H_3(OH)_2CH{:}CHCOOC_2H_5$.
Properties: Yellow to tan crystals; characteristic, aromatic odor; insoluble in water; very soluble in alcohol.
Grade: C.P.
Containers: Bottles; fiber drums.
Use: Food antioxidant.

ethyl caprate (ethyl decanoate) $C_9H_{19}COOC_2H_5$.
Properties: Colorless liquid; fragrant odor. Soluble in alcohol and ether; insoluble in water; sp. gr. 0.862; b.p. 243°C. Low toxicity. Combustible.
Derivation: By heating capric acid, absolute alcohol and sulfuric acid, with subsequent distillation.
Uses: Organic synthesis; manufacturing wine-bouquet and cognac essence.

ethyl caproate (ethyl hexoate; ethyl hexanoate) $C_5H_{11}COOC_2H_5$.
Properties: Colorless to yellowish liquid; pleasant odor. Soluble in alcohol and ether; insoluble in water and glycerin. Sp. gr. 0.873; b.p. 167°C. Low toxicity. Combustible.
Derivation: Heating absolute alcohol and n-caproic acid in presence of sulfuric acid, with subsequent distillation.
Grades: Technical; F.C.C.
Uses: Organic synthesis; artificial fruit essences.

ethyl caprylate (ethyl octoate; ethyl octanoate) $CH_3(CH_2)_6COOC_2H_5$.
Properties: Colorless liquid; pineapple odor. Soluble in alcohol and ether; insoluble in water and glycerin. Sp. gr. 0.865–0.869 (25°C); f.p. -48°C; b.p. 207–209°C. Low toxicity. Combustible.
Derivation: Heating caprylic acid, alcohol, and sulfuric acid with subsequent distillation.
Grades: Technical; F.C.C.
Uses: Flavoring; fruit essences.

ethyl carbamate. See urethane.

N-ethylcarbazole (9-ethylcarbazole) $(C_6H_4)_2NC_2H_5$ (tricyclic).
Properties: Leaflets; soluble in ether and hot alcohol. M.p. 69–70°C; b.p. 175°C (5 mm).
Derivation: Action of ethyl chloride on the potassium salt of carbazole.
Uses: Intermediate for dyes, pharmaceuticals; agricultural chemicals.

ethyl carbonate. See diethyl carbonate.

ethylcellulose. An ethyl ether of cellulose.
Properties: White, granular, thermoplastic solid; lowest density of the commercial cellulose plastics; properties vary with extent to which hydroxyl radicals of cellulose have been replaced by ethoxy groups. Standard commercial product has: 47–48% ethoxy content; sp. gr. 1.07–1.18; refractive index 1.47; high dielectric strength; softening point 100–130°C. Soluble in most organic liquids, and compatible with resins, waxes, oils, and plasticizers; inert to alkalies and dilute acids. Insoluble in water and glycerol. Nontoxic. Combustible.
Derivation: From alkali cellulose and ethyl chloride or sulfate; from cellulose and ethyl alcohol in presence of dehydrating agents.
Grades: Technical; N.F.; F.C.C.
Uses: Hot-melt adhesives and coatings for cables, paper, textiles, etc.; extrusion wire insulation; protective coatings; pigment-grinding bases; toughening agent for plastics; printing inks; molding powders; proximity fuses; vitamin preparations; casing for rocket propellants; food and feed additive.

ethyl centralite. See sym-diethyldiphenylurea.

ethyl chloride (chloroethane) C_2H_5Cl.
Properties: Gas at room temperature; when compressed, a colorless, volatile liquid. Ether-like odor, burning taste. Stable and noncorrosive when dry but will hydrolyze in the presence of water or alkalies. Miscible with most of the commonly used solvents; slightly soluble in water. Sp. gr. 0.9214; f.p. -140.85°C; b.p. 12.5°C; critical point 187.2°C (52 atm; sp. gr. 0.33); vapor pressure 1000 mm (20°C); flash point (closed cup) -58°F. Autoignition temp. 966°F.
Derivation: (a) From ethylene and hydrogen chloride; (b) by passing hydrogen chloride into a solution of zinc chloride and ethyl alcohol.
Grades: Technical; U.S.P.
Containers: Cylinders; drums; tank cars.
Hazard: Highly flammable; severe fire and explosion risk. Flammable limits in air 3.8–15.4%. Moderately toxic; irritant to eyes. Tolerance, 1000 ppm in air. Safety data sheet available from Mfg. Chemists Assn., Washington, D.C.
Uses: Manufacture of tetraethyl lead and ethylcellulose; anesthetic; organic synthesis, alkylating agent; refrigeration; analytical reagent; solvent for phosphorus, sulfur, fats, oils, resins and waxes; insecticides.
Shipping regulations: (Rail) Red label. (Air) Flammable Liquid label. Not acceptable on passenger planes.

ethyl chloroacetal $ClCH_2CH(OC_2H_5)_2$.
Properties: Water-white liquid with pleasant odor; sp. gr. 1.022 (20°C); boiling range; 54–61°C (20 mm); 149–153°C; f.p. -32°C); flash point 117°F; refractive index 1.418 (20°C); soluble in alcohol and ethyl ether; insoluble in water. Combustible.
Hazard: Moderate fire risk. May be toxic.

ethyl chloroacetate $CCH_2ClCO_2C_2H_5$.
Properties: Water-white, mobile liquid; pungent, fruity odor. Decomposed by hot water and alkalies. Soluble in alcohol, benzene, and ether; insoluble in water. Sp. gr. 1.1585 (20°C); b.p. 144.2°C; vapor density 4.23–4.46; flash point 179°F; combustible; refractive index (n 20/D) 1.4227.
Derivation: (a) Action of chloroacetyl chloride on alcohol; (b) by treating chloroacetic acid with alcohol and sulfuric acid.
Containers: Bottles; carboys; 55-gal drums.
Hazard: Highly toxic by ingestion and inhalation.
Uses: Solvent; organic synthesis; military poison gas; vat dyestuffs.
Shipping regulations: (Air) Poison label.

ethylchlorocarbonate (ethyl chloroformate) $ClCOOC_2H_5$.

Properties: Water-white liquid with irritating odor; sp. gr. 1.135–1.139 (20/20°C); b.p. 93–95°C; refractive index (n 20/D) 1.3974. Flash point 61°F (closed cup). Decomposes in water and alcohol; soluble in benzene, chloroform, and ether.
Derivation: Reaction of carbon monoxide with gaseous chlorine, producing phosgene ($COCl_2$) which is then treated with anhydrous ethyl alcohol giving ethyl chlorocarbonate and splitting off hydrochloric acid.
Grades: Technical.
Hazard: Highly toxic; strong irritant to eyes and skin. Flammable, dangerous fire risk.
Use: Organic synthesis; intermediate in making diethyl carbonate, flotation agents, polymers, isocyanates.
Shipping regulations: (Rail) White label. (Air) Corrosive label. Not acceptable on passenger planes. Legal label name: ethyl chloroformate.

ethyl chloroformate. Legal label name for ethyl chlorocarbonate.

ethylchlorosulfonate $C_2H_5OClSO_2$.
Properties: Colorless, oily liquid. Pungent odor; fumes in moist air. Decomposed by water. Attacks lead and tin, but copper mildly. Iron and steel not affected. Soluble in chloroform and ether, insoluble in water. Sp. gr. 1.379 (0°C); b.p. 153–153°C; vapor density 5 (air = 1); volatility 18,000 mg/cu m (20°C).
Derivation: (a) Action of fuming sulfuric acid on ethylchloroformate; (b) interaction of ethylene and chlorosulfonic acid.
Grades: Technical.
Hazard: Strong irritant to eyes and skin. Evolves phosgene when heated.
Uses: Organic synthesis; military poison gas.

ethyl cinnamate (ethyl phenylacrylate; cinnamylic ether) $C_6H_5CH{:}CHCOOC_2H_5$.
Properties: Limpid, oily liquid; strawberry-like odor. Soluble in alcohol and ether; insoluble in water. Sp. gr. 1.045–1.048; refractive index (n 20/D) 1.560; congealing point 7°C (min); b.p. 271°C. Combustible; low toxicity.
Derivation: Heating ethyl alcohol and cinnamic acid in presence of sulfuric acid.
Uses: Perfumery; flavoring extracts.

ethyl citrate. See triethyl citrate.

ethyl crotonate. $CH_3CH{:}CHCOOC_2H_5$.
Properties: Water-white solid or liquid. Characteristic pungent persistent odor. Soluble in alcohol and ether; insoluble in water. Sp. gr. 0.9207 (20/20°C); m.p. (solid) 45°C; b.p. (solid) 209°C, liquid 139°C; flash point 36°F; refractive index 1.4242 (20°C); wt/gal 7.65 lb (20°).
Hazard: Flammable, dangerous fire risk. Strong irritant.
Uses: Solvent and softening agent; lacquers; organic synthesis.
Shipping regulations: (liquid): (Rail) Flammable liquid, n.o.s., Red label. (Air) Flammable Liquid label.

ethyl cyanide (propionitrile; propanenitrile) C_2H_5CN.
Properties: Mobile, colorless liquid; ethereal odor. Soluble in alcohol and water. Sp. gr. 0.7829 (20/20°C); refractive index (n 20/D) 1.3664; b.p. 97.4°C; f.p. −92.9°C; flash point 61°F (open cup).
Derivation: Heating barium ethyl sulfate and potassium cyanide, with subsequent distillation.
Containers: Iron drums; glass bottles.
Hazard: Highly toxic by ingestion and inhalation. Flammable, dangerous fire risk.
Uses: Solvent; dielectric fluid; intermediate.
Shipping regulations: (Rail) Not listed. (Air) Cyanide solutions, n.o.s., Poison label.

ethyl cyanoacetate (malonic ethyl ester nitrile) $CNCH_2COOC_2H_5$.
Properties: Colorless liquid; b.p. 206–208°C; f.p. −22.5°C; refractive index 1.41751 (20°C/D). Soluble in alcohol and ether; slightly soluble in water and alkaline solutions. Flash point 230°F. Combustible.
Derivation: Esterification of cyanoacetic acid with ethanol; reaction of an alkali cyanide and chloroacetic ethyl ester.
Method of purification: Vacuum distillation.
Grades: Reagent; technical.
Containers: Tin-lined steel drums.
Hazard: Toxic by ingestion and inhalation.
Uses: Organic synthesis; pharmaceuticals; dyes.

5-ethyl-5-cycloheptenylbarbituric acid. See heptabarbital.

ethylcyclohexane $C_2H_5C_6H_{11}$.
Properties: Colorless liquid; sp. gr. 0.787; boiling point 131.8°C; refractive index (n 20/D) 1.4330. Flash point 95°F; autoignition temp. 504°F.
Hazard: Flammable, moderate fire risk. Flammable limits in air, 0.9 to 6.6%.
Use: Organic synthesis.

ethylcyclopentane $C_2H_5C_5H_9$.
Properties: Colorless liquid. Sp. gr. 0.766; b.p. 103.5°C; refractive index (n 20/D) 1.4198. Autoignition temp. 504°F.
Hazard: Flammable, moderate fire risk. Flammable limits in air, 1.1 to 6.7%.

ethyl cyclopentanone-2-carboxylate. See 2-carbethoxycyclopentanone.

ethyl decanoate. See ethyl caprate.

ethyldichloroarsine (dichloroethylarsine) $C_2H_5AsCl_2$.
Properties: Colorless, mobile liquid. Becomes yellowish under the action of light and air. Fruit-like odor (high dilution). Decomposed by water. Attacks brass, but not iron (dry). Soluble in alcohol, benzene, ether, and water. Sp. gr. 1.742 (14°C); b.p. 156°C (decomposes); f.p. −65°C; coefficient of thermal expansion 0.0011; vapor density 6 (air = 1); volatility 20,000 mg/cu m (20°C); vapor pressure 2.29 mm (21.5°C).
Derivation: Chlorination of ethyl arsenious oxide.
Hazard: Highly toxic by ingestion, inhalation, skin absorption. Strong irritant.
Use: Miltary poison gas.
Shipping regulations: (Rail) Poison gas label. Not accepted by express; (Air) Not acceptable.

ethyl 4,4′-dichlorobenzilate. See chlorobenzilate.

ethyl dichlorophenoxyacetate. See 2,4-D.

ethyldichlorosilane $C_2H_5SiHCl_2$.
Properties: Colorless liquid; b.p. 75.5°C; sp. gr. 1.088 (25/25°C); flash point (COC) 30°F. Readily hydrolyzed by moisture, with the liberation of hydrogen and hydrochloric acid.
Derivation: By Grignard reaction of trichlorosilane and ethylmagnesium chloride.
Hazard: Flammable, dangerous fire risk. Strong irritant to eyes and skin.
Use: Intermediate for silicones.
Shipping regulations: (Rail) Red label. (Air) Flammable Liquid label. Not acceptable on passenger planes.

ethyldiethanolamine $C_2H_5N(CH_2CH_2OH)_2$.
Properties: Water-white liquid; sp. gr. 1.015 (20°C);

boiling range 246–252°C; odor amine. Flash point 255°F. Combustible. Low toxicity.
Uses: Solvent; detergents.

ethyldimethylmethane. See isopentane.

ethyldipropylmethane. See 4-ethylheptane.

S-ethyl di-N,N-propylthiocarbamate. See EPTC.

ethyl enanthate (ethyl oenanthate; ethyl heptanoate; cognac oil) $CH_3(CH_2)_5COOC_2H_5$.
Properties: Clear, colorless oil with fruity odor and taste; sp. gr. 0.87; b.p. 187°C; soluble in alcohol, chloroform and ether; insoluble in water. Low toxicity. Combustible.
Derivation: By heating oenanthic acid and ethyl alcohol in presence of sulfuric acid, and subsequent recovery by distillation.
Grade: Technical.
Containers: Iron drums; glass bottles.
Uses: Artificial cognac flavor; flavor for liqueurs and fruity-type soft drinks.

ethylene (bicarburetted hydrogen; ethene) $H_2C:CH_2$. 5th highest-volume chemical produced in U.S. (1975).
Properties: Colorless gas with sweet odor and taste; freezing point −169°C; boiling point −103.9°C; flash point −213°F; sp. gr. of liquid at 0°C 0.610; vapor density (0°C) (air = 1) 0.975; critical temperature 9.5°C; autoignition temp. 842°F; critical pressure (absolute) 744 psi. Purity not less than 96% ethylene by gas volume, not more than 0.5% acetylene, not more than 4% methane and ethane; 13.4 cu ft/lb (15.6°C), slightly soluble in water, alcohol, and ethyl ether. An asphyxiant gas.
Derivation: Thermal cracking of propane, ethane, butane, naphtha, and refinery off-gases. Occurs naturally in plant organs and tissues.
Hazard: Extremely flammable; dangerous fire and explosion risk. Explosive limits in air: upper, 3% by vol., lower 32% by vol.
Grades: Technical (95% min.); 99.5% min.; 99.9 mole %; N.F.
Containers: Steel cylinders; tube trailers; tank cars (as liquid); pipeline; ocean tankers.
Uses: Polyethylene, polypropylene, ethylene oxide, ethylene dichloride, ethylene glycols, aluminum alkyls, vinyl chloride, vinyl acetate, ethyl chloride, ethylene chlorohydrin, acetaldehyde, linear alcohols, polystyrene, styrene, polyvinyl chloride, SBR, polyester resins, trichloroethylene, etc; refrigerant; welding and cutting of metals; anesthetic; in orchard sprays to accelerate fruit ripening.
Shipping regulations: (Rail) Red Gas label. (Air) Flammable Gas label. Not accepted on passenger planes.

ethylenebis(iminodiacetic acid). See ethylenediaminetetraacetic acid.

ethylene bis(oxyethylenenitrilo)tetraacetic acid. See ethylene glycol-bis(beta-aminoethyl ether)-N,N-tetraacetic acid.

ethylene bromide. See ethylene dibromide.

ethylene bromohydrin (glycol bromohydrin; 2-bromoethyl alcohol) $BrCH_2CH_2OH$.
Properties: Hygroscopic liquid; sp. gr. 1.7629 (20°C); b.p. 149–150°C (750 mm); refractive index 1.4915 (20°C). Soluble in most organic solvents and completely miscible with water, alcohol, or ether. Aqueous solutions have a sweet, burning taste. Hydrolysis of aqueous solutions is accelerated by heat, acids, and alkalies. Combustible.
Derivative: Action of hydrobromic acid on ethylene oxide.
Hazard: Irritant to eyes and mucous membranes.
Use: Organic synthesis.

ethylene carbonate (glycol carbonate; dioxolone-2) $(—CH_2O)_2CO$.
Properties: Colorless, odorless solid or liquid. M.p. 36.4°C; b.p. 248°C; sp. gr. (39/4°C) 1.3218; refractive index (n 50/D) 1.4158; flash point 290°F (O.C.). Miscible (40°) with water, alcohol, ethyl acetate, benzene, and chloroform. Soluble in ether, n-butanol, and carbon tetrachloride. Combustible. Low toxicity.
Derivation: Interaction of ethylene glycol and phosgene.
Uses: Solvent for many polymers and resins; solvent extraction; synthesis of pharmaceuticals, rubber chemicals, textile finishing agents.

ethylene chloride. See ethylene dichloride.

ethylene chlorobromide. See sym-bromochloroethane.

ethylene chlorohydrin (2-chloroethyl alcohol; glycol chlorohydrin) $ClCH_2CH_2OH$.
Properties: Colorless liquid; faint ethereal odor. Soluble in most organic liquids and completely miscible with water. Sp. gr. 1.2045 (20/20°C); b.p. 128.7°C, refractive index (n 20/D) 1.4419; vapor pressure 4.9 mm (20°C); flash point 140°F (open cup); wt/gal 10.0 lb (20°C); coefficient of expansion 0.00089 (20°C); f.p. −62.6°C; viscosity 0.0343 poise (20°C). Autoignition temp. 797°F. Combustible.
Derivation: Action of hypochlorous acid on ethylene.
Grades: Anhydrous; 38%.
Containers: (36 to 40% grade) bottles; jugs; carboys; (anhydrous) tank cars.
Hazard: Toxic by ingestion and inhalation; skin absorption may be fatal. Strong irritant. Tolerance, 5 ppm in air. Moderate fire hazard. Penetrates ordinary rubber gloves and protective clothing (MCA).
Uses: Solvent for cellulose acetate, ethylcellulose; introduction of hydroxyethyl group in organic synthesis; to activate sprouting of dormant potatoes; mfg. of ethylene oxide and ethylene glycol.
Shipping regulations: (Air) Poison label.

ethylene cyanide (ethylene dicyanide; succinonitrile) $C_2H_4(CN)_2$.
Properties: Colorless, waxy solid; soluble in alcohol, water, and chloroform. M.p. 57–57.5°C; b.p. 265.7°C; flash point 270°F. Combustible.
Derivation: Interaction of ethylene dibromide and potassium cyanide in presence of alcohol.
Hazard: Highly toxic. See cyanide.
Use: Organic synthesis.
Shipping regulations: (Rail, Air) Cyanides, dry, Poison label.

ethylene cyanohydrin (beta-hydroxypropionitrile) $HOCH_2CH_2CN$.
Properties: Straw-colored liquid; f.p. −46°C; b.p. 227–228°C (dec); sp. gr. 1.0404 (25/4°C); vapor pressure 0.08 mm (25°C); 20 mm (117°C). Miscible in all proportions with water, acetone, methyl ethyl ketone, ethanol, chloroform and diethyl ether. Insoluble in benzene, carbon tetrachloride, and naphtha. Combustible.

Derivation: Ethylene oxide and hydrocyanic acid.
Containers: Steel drums; tank cars.
Hazard: Toxic by ingestion.
Uses: Solvent for certain cellulose esters and inorganic salts. Organic intermediate for acrylates.

ethylenediamine (1,2-diaminoethane) $NH_2CH_2CH_2NH_2$.
Properties: Colorless, alkaline liquid; ammonia odor. Strong base. Soluble in water, alcohol; slightly soluble in ether; insoluble in benzene. Readily absorbs carbon dioxide from air. Sp. gr. 0.8995 (20/20°C); wt/gal 7.50 lb (20°C); b.p. 116–117°C; vapor pressure 10.7 mm (20°C); m.p. 8.5°C; viscosity 0.0154 poise (25°C); flash point 93°F (closed cup); refractive index 1.4540 (26°C); pH of 25% solution 11.9 (25°C).
Derivation: Heating ethylene dichloride and ammonia, with subsequent distillation.
Method of purification: Redistillation.
Grades: Technical; U.S.P., 97%; solutions of various strengths.
Containers: 55-gal tin-lined drums; tank cars.
Hazard: Strong irritant to skin and eyes; toxic by inhalation and skin absorption. Tolerance 10 ppm in air. Flammable, moderate fire risk.
Uses: Fungicides; manufacture of chelating agents (EDTA); dimethylolethylene-urea resins; chemical intermediate.

ethylenediamine carbamate. See "Diak."

ethylenediamine dihydroiodide $(—CH_2NH_2)_2 \cdot 2HI$.
Use: Feed additive.

ethylenediamine di-*ortho*-hydroxyphenylacetic acid $[—CH_2NHCH(COOH)C_6H_4OH]_2$. A phenolic analog of ethylenediaminetetraacetic acid; m.p. 218°C (dec.). Insoluble in water and most organic solvents; soluble in mineral acid; also forms water-soluble alkali metal and ammonium salts. Used for chelating iron in mildly alkaline solutions.

ethylenediamine tartrate. Used to make piezo-electric crystals for control of electric frequencies, etc., as in television.

ethylenediaminetetraacetic acid (EDTA; ethylenebisiminodiacetic acid; ethylenedinitrilotetraacetic acid) $(HOOCCH_2)_2NCH_2CH_2N(CH_2COOH)_2$. An organic chelating agent.
Properties: Colorless crystals, decomposing at 240°C. Slightly soluble in water; insoluble in common organic solvents; neutralized by alkali metal hydroxides to form a series of water-soluble salts containing from one to four alkali metal cations. Low toxicity.
Derivation: (a) Addition of sodium cyanide and formaldehyde to a basic solution of ethylenediamine (forms the tetrasodium salt); (b) heating tetrahydroxyethylethylenediamine with sodium or potassium hydroxide with cadmium oxide catalyst.
Uses: Detergents, liquid soaps, shampoos, agricultural chemical sprays; metal cleaning and plating; metal chelating agent; treatment of chlorosis; decontamination of radioactive surfaces; metal deactivator in vegtable oils, oil emulsions, pharmaceutical products, etc.; anticoagulant of blood; eluting agent in ion exchange; to remove insoluble deposits of calcium and magnesium soaps; in textiles to improve dyeing, scouring, and detergent operations; antioxidant; clarification of liquids; analytical chemistry; spectrophotometric titration; aid in reducing blood cholesterol; in medicine to treat lead poisoning and calcinosis; food additive (preservative).
Note: A number of salts of EDTA are available with uses identical or similar to the acid. The U.S.P. salts are called edetates (calcium disodium, disodium edetates); others are usually abbreviated to EDTA (tetrasodium, trisodium EDTA). Other salts, known chiefly under trademarked names, are the sodium ferric, dihydrogen ferrous, and a range of disodium salts with magnesium, divalent cobalt, manganese, copper, zinc and nickel.

ethylenediaminetetraacetonitrile (EDTAN) $[-CH_2N(CH_2CN_2]_2$.
Properties: White crystalline solid; melting range 126–132°C; bulk density 48.4 lb/cu ft. Slightly soluble in water; soluble in acetone.
Hazard: May be toxic.
Uses: Chelating agent and intermediate.

ethylene dibromide (EDB; 1,2-dibromoethane; ethylene bromide) $BrCH_2CH_2Br$.
Properties: Colorless, nonflammable liquid. Sweetish odor. Emulsifiable. Miscible with most solvents and thinners; slightly soluble in water. Sp. gr. 2.17–2.18 (20°C); wt/gal wt/gal 18.1 lb; b.p. 131°C; vapor pressure 17.4 mm (30°C); f.p. 9.10°C; refractive index 1.5337 (25°C); flash point, none.
Derivation: Action of bromine on ethylene.
Hazard: Toxic by inhalation, ingestion, and skin absorption. May be carcinogenic. Strong irritant to eyes and skin. Tolerance, 20 ppm in air.
Uses: Scavenger for lead in gasoline; grain fumigant; general solvent; waterproofing preparations; organic synthesis; fumigant for tree crops.
Shipping regulations: (Air) Poison label.
Note: May poison platinum catalysts.

ethylene dichloride (sym-dichloroethane; 1,2-dichloroethane; ethylene chloride; Dutch oil). 16th highest-volume chemical produced in U.S. (1975) $ClCH_2CH_2Cl$.
Properties: Colorless, oily liquid; chloroform-like odor; sweet taste. Stable to water, alkalies, acids, or active chemicals. Resistant to oxidation. Will not corrode metals. Miscible with most common solvents; slightly soluble in water. B.p. 83.5°C; f.p. -35.5°C; sp. gr. 1.2554 (20/4°C); wt/gal 10.4 lb; refractive index 1.444; flash point 56°F.
Derivation: Action of chlorine on ethylene with subsequent distillation, with metallic catalyst.
Grades: Technical; spectrophotometric.
Containers: Drums; tank cars.
Hazard: Toxic by ingestion, inhalation, skin absorption. Strong irritant to eyes and skin. Tolerance, 50 ppm in air. Flammable, dangerous fire risk; explosive limits in air 6 to 16%. Safety data sheet available from Mfg. Chemists Assn., Washington, D.C.
Uses: Vinyl chloride (Wulff process); solvent; lead scavenger in antiknock gasoline; paint, varnish and finish removers; metal degreasing; soaps and scouring compounds; wetting and penetrating agents; organic synthesis; ore flotation.
Shipping regulations: (Rail) Red label. (Air) Flammable Liquid label.

ethylene dicyanide. See ethylene cyanide.

ethylenedinitrilotetraacetic acid. See ethylenediaminetetraacetic acid.

ethylenedinitrilotetra-2-propanol. See N,N,N′,N′-tetrakis(2-hydroxypropyl) ethylenediamine.

ethylene diphenyldiamine. See N,N-diphenylethylenediamine.

1,1′-ethylene-2,2′-dipyridinium dibromide. See diquat.

ethylene glycol (ethylene alcohol; glycol; 1,2-ethanediol). CH_2OHCH_2OH. The simplest glycol.
Properties: Clear, colorless, syrupy liquid; sweet taste; hygroscopic; lowers freezing point of water. Relatively non-volatile. Odorless. Soluble in water, alcohol, and ether. Sp. gr. 1.1155 (20°C); b.p. 197.2°C; f.p. -13.5°C; wt/gal 9.31 lb (15/15°C); refractive index 1.430 (25°C); flash point 240.8°F; combustible; autoignition temp. 775°F.
Derivation: (1) Air oxidation of ethylene followed by hydration of the ethylene oxide formed; (2) heating ethylene chlorohydrin with a solution of an alkali carbonate or bicarbonate; (3) from formaldehyde, water and carbon monoxide, with hydrogenation of the resulting glycolic acid; (4) acetoxylation (q.v.); (5) from carbon monoxide and hydrogen (synthesis gas) from coal gasification. Methods (2) and (3) are obsolescent; methods (4) and (5) are recent innovations.
Grade: Technical.
Containers: Drums; tank cars.
Hazard: Toxic by ingestion and inhalation; lethal dose reported to be 100 cc. Tolerance (vapor), 100 ppm in air.
Uses: Coolant and antifreeze; asphalt-emulsion paints; heat-transfer agent in refrigeration and electron tubes; low-pressure laminates; brake fluids; glycol diacetate; polyester fibers and films; low-freezing dynamite; solvent; extractant for various purposes; solvent mixtures for cellulose esters and ethers, especially cellophane; cosmetics (up to 5%); lacquers; alkyd resins; printing inks; wood stains; adhesives; leather dyeing; textile processing; tobacco; ingredient of deicing fluid for airport runways.

ethylene glycol-bis(beta-**aminoethyl ether**)-N,N-**tetraacetic acid** (ethylene bis(oxyethylenenitrilo)tetraacetic acid) $[-CH_2OC_2H_4N(CH_2COOH)_2]_4$. Crystals; m.p. 241° (dec.). Soluble in water.
Use: Chelating agent.

ethylene glycol bis(mercaptopropionate). See glycol dimercaptopropionate.

ethylene glycol bisthioglycolate. See glycol dimercaptoacetate.

ethylene glycol diacetate (glycol diacetate) $CH_3COOCH_2CH_2OOCCH_3$.
Properties: Colorless liquid; faint odor. Soluble in alcohol, ether, benzene; slightly soluble in water (10%). Sp. gr. 1.1063 (20/20°C); b.p. 190.5°C; vapor pressure 0.3 mm (20°C); flash point 205°F. (o.c.); wt/gal 9.2 lb (20°C); f.p. -41.5°C; refractive index (n 20/D) 1.415. Combustible. Low toxicity.
Derivation: (a) Ethylene glycol and acetic acid; (b) ethylene dichloride and sodium acetate.
Uses: Solvent for cellulose esters and ethers; resins; lacquers; printing inks; perfume fixative; non-discoloring plasticizer for ethyl and benzyl cellulose.

ethylene glycol dibutyl ether $C_4H_9OC_2H_4OC_4H_9$.
Properties: Practically colorless liquid; slight odor. Slightly soluble in water; sp. gr. 0.8374 (20/20°C); 7.0 lb/gal (20°C); b.p. 203.1°C; vapor pressure 0.09 mm (20°C); freezing point -69.1°C; flash point 185°F. Combustible.
Containers: 1-gal cans; 5-, 55-gal drums.
Uses: High-boiling inert solvent; specialized solvent and extraction applications.

ethylene glycol dibutyrate (glycol dibutyrate) $(-CH_2OCOC_3H_7)_2$.
Properties: Colorless liquid; sp. gr. (0°C) 1.024; refractive index (25°C), 1.424; b.p. 240°C; f.p. less than -80°C; solubility in water, 0.050% by weight. Combustible.
Use: Plasticizer.

ethylene glycol diethyl ether $C_2H_5OCH_2CH_2OC_2H_5$.
Properties: Colorless liquid; slight odor; stable. Sp. gr. 0.8417 (20/20°C); b.p. 121.4°C; vapor pressure 9.4 mm (20°C); flash point 95°F; wt/gal 7 lb (20°C); f.p. -74°C. Immiscible with water.
Grade: Technical.
Containers: 1-gal cans; 5-, 55-gal drums.
Hazard: Flammable, moderate fire risk.
Uses: Organic synthesis (reaction medium); solvent and diluent for detergents.

ethylene glycol diformate (glycol diformate) $HCOOCH_2CH_2OOCH$.
Properties: Water-white liquid, soluble in water, alcohol and ether. Sp. gr. 1.2277 (20/20°C); 10.2 lb/gal (20°C); b.p. 177.1°C; flash point 200°F; combustible. Vapor pressure 0.5 mm (20°C); f.p. -10°C. Hydrolyzes slowly, liberating formic acid.
Hazard: Moderately toxic.
Use: Embalming fluids.

ethylene glycol dimethyl ether (GDME; glycol dimethyl ether; glyme; 1,2-dimethoxyethane) $CH_3OCH_2CH_2OCH_3$.
Properties: Water-white liquid with a mild odor. Sp. gr. 0.8683 (20°C); b.p. 85.2°C; f.p. -69°C; refractive index 1.3792 (20/D); flash point 34°F (open cup); soluble in water and hydrocarbons; pH 8.2.
Hazard: Flammable, dangerous fire risk.
Use: Solvent.
Shipping regulations: Flammable liquid, n.o.s., (Rail) Red label. (Air) Flammable Liquid label.

ethylene glycol dinitrate. A freezing-point depressant for nitroglycerine, used in low-freezing dynamites.
Hazard: Toxic; can penetrate the skin. Tolerance, 0.2 ppm in air.

ethylene glycol dipropionate (glycol propionate; glycol dipropionate) $(-CH_2OCOC_2H_5)_2$. Liquid; sp. gr. (15°C) 1.054; refractive index (25°C) 1.419; b.p. 211°C; f.p. less than -80°C; solubility in water, 0.16% by weight. Combustible.
Use: Plasticizer.

ethylene glycol monoacetate (glycol monoacetate) $HOCH_2CH_2OOCCH_3$.
Properties: Colorless liquid; almost odorless; soluble in alcohol, ether, benzene, and toluene; partially soluble in water. B.p. 181–182°C; sp. gr. 1.108. Flash point 215°F. Combustible. Low toxicity.
Derivation: (a) Heating ethylene glycol with acetic acid (glacial) or acetic anhydride; (b) passing ethylene oxide into hot acetic acid containing sodium acetate or sulfuric acid.
Use: Solvent for nitrocellulose, cellulose acetate, camphor.

ethylene glycol monobenzyl ether (benzyl "Cellosolve"). $C_6H_5CH_2OC_2H_4OH$.
Properties: Water-white liquid; faint rose-like odor; sp. gr. 1.070 (20/20°C); b.p. 255.9°C; vapor pressure 0.02 mm (20°C); flash point 265°F; wt 8.9 lb/gal (20°C). Combustible; autoignition temp. 665°F.

Containers: Glass jugs; glass-stoppered carboys.
Hazard: Moderately toxic by ingestion.
Uses: Solvent for cellulose acetate, dyes, inks, resins, perfume fixative; organic synthesis (selective hydroxyethylating agent); coating compositions for leather, paper, and cloth; lacquers.

ethylene glycol monobutyl ether (2-butoxyethanol, butyl "Cellosolve"). $HOCH_2CH_2OC_4H_9$.
Properties: Colorless liquid; mild odor; high dilution ratio with petroleum hydrocarbons; soluble in alcohol and water. B.p. 171.2°C; sp. gr. 0.9019 (20/20°C); wt/gal 7.51 lb (20°C); refractive index 1.4190 (25°C); vapor pressure 0.76 mm (20°C); flash point 142°F. Autoignition temp. 472°F. Combustible. Low toxicity.
Grade: Technical.
Containers: 1-gal cans; 5- and 55-gal drums; tank cars and trucks.
Hazard: Moderately toxic. Tolerance, 50 ppm in air.
Uses: Solvent for nitrocellulose resins; spray lacquers; quick-drying lacquers; varnishes; enamels; drycleaning compounds; varnish removers; textile (preventing spotting in printing or dyeing); mutual solvent for "soluble" mineral oils to hold soap in solution and to improve the emulsifying properties.

ethylene glycol monobutyl ether acetate (butyl "Cellosolve" acetate). $C_4H_9OCH_2CH_2OOCCH_3$.
Properties: Colorless liquid; fruity odor. Soluble in hydrocarbons and organic solvents; insoluble in water. B.p. 192.3°C; sp. gr. 0.9424 (20/20°C); f.p. -63.5°C; flash point 190°F. Combustible. Low toxicity.
Grade: Technical.
Containers: 1-, 5-, 55-gal drums; tank cars; tank trucks.
Uses: High-boiling solvent for nitrocellulose lacquers, epoxy resins, multicolor lacquers; film coalescing aid for polyvinyl acetate latex.

ethylene glycol monobutyl ether laurate (butoxyethyl laurate) $C_{11}H_{23}COO(CH_2)_2OC_4H_9$.
Properties: Liquid; sp. gr. (25°C) 0.985; f.p. -10 to -15°C; insoluble in water. Combustible. Probably low toxicity.
Use: Plasticizer.

ethylene glycol monobutyl ether oleate (butoxyethyl oleate) $C_{17}H_{33}COOCH_2CH_2OC_4H_9$.
Properties: Liquid; sp. gr. (25°C) 0.892; f.p. less than -45°C; insoluble in water. Combustible. Probably low toxicity.
Use: Plasticizer.

ethylene glycol monobutyl ether stearate (butoxyethyl stearate) $C_{17}H_{35}COOC_2H_4OC_4H_9$.
Properties: Colorless liquid. Sp. gr. 0.882 (20°C); (25°C) 1.446; vapor pressure $<$ 0.01 mm (20°C); b.p. 210–233°C (4 mm); m.p. 16.5°C; insoluble in water. Combustible.
Uses: Plasticizer and solvent.

ethylene glycol monoethyl ether (2-ethoxyethanol; "Cellosolve" solvent). $HOCH_2CH_2OC_2H_5$.
Properties: Colorless liquid, practically odorless; b.p. 135.6°C; sp. gr. 0.9311 (20/20°C); wt/gal 7.74 lb (20°C); refractive index 1.4060 (25°C); flash point 120°F; pour point $<$ 100°F. Autoignition temp. 460°F. Miscible with hydrocarbons and water. Combustible.
Grade: Technical.
Containers: 1-gal cans; 5- and 55-gal drums; tank cars.
Hazard: Moderate fire risk. Moderately toxic; absorbed by skin. Tolerance, 100 ppm in air.
Uses: Solvent for nitrocellulose, natural and synthetic resins; mutual solvent for formulation of soluble oils; lacquers and lacquer thinners; dyeing and printing textiles; varnish removers; cleaning solutions; leather; anti-icing additive for aviation fuels.

ethylene glycol monoethyl ether acetate ("Cellosolve" acetate). $CH_3COOCH_2CH_2OC_2H_5$.
Properties: Colorless liquid; pleasant, ester-like odor; b.p. 156.3°C; sp. gr. 0.9748 (20/20°C); wt/gal 8.1 lb (20°C); refractive index 1.4030 (25°C); viscosity 1.32 cp. (20°C); flash point 120°F; f.p. -61.7°C; vapor pressure 2 mm (20°C). Miscible with aromatic hydrocarbons; slightly miscible with water. Combustible.
Grade: Technical.
Containers: 1-gal cans; 5- and 55-gal drums; tank cars up to 10,000 gals.
Hazard: Moderately toxic by ingestion; moderate fire risk.
Uses: Solvent for nitrocellulose, oils and resins; retards "blushing" in lacquers; varnish removers; wood stains; textiles; leather.

ethylene glycol monoethyl ether laurate $C_{11}H_{23}COO(CH_2)_2OC_2H_5$.
Properties: Liquid; sp. gr. (25°C) 0.89; f.p. -7 to -11°C; insoluble in water. Combustible.
Uses: Plasticizer.

ethylene glycol monoethyl ether ricinoleate $C_{17}H_{32}(OH)COO(CH_2)_2OC_2H_5$.
Properties: Liquid; sp. gr. (25°C) 0.929; f.p. less than -10°C; insoluble in water. Combustible.
Use: Plasticizer.

ethylene glycol monohexyl ether (n-hexyl "Cellosolve"). $C_6H_{13}OCH_2CH_2OH$.
Properties: Water-white liquid; sp. gr. 0.8887 (20/20°); 7.4 lb/gal (20°C); b.p. 208.1°C; vapor pressure 0.05 mm (20°C); f.p. -50.1°C; flash point 195°F. Combustible. Low toxicity.
Containers: 1-gal can; 5-, 55-gal drums.
Use: High-boiling solvent.

ethylene glycol monomethyl ether (2-methoxyethanol, methyl "Cellosolve"). $CH_3OCH_2CH_2OH$.
Properties: Colorless liquid; mild, agreeable odor; stable; miscible with hydrocarbons, alcohols, ketones, glycols, water. B.p. 124.5°C; sp. gr. 0.9663 (20/20°C); wt/gal 8.0 lb (20°C); refractive index 1.4021 (20°C); flash point 110°F; f.p. -85.1°C. Autoignition temp. 551°F. Combustible.
Derivation: From ethylene oxide.
Grade: Technical.
Containers: 1-, 5-gal cans; 55-gal drums; tank cars; up to 10,000 gal.
Hazard: Moderately toxic by ingestion and inhalation; moderate fire risk. Tolerance, 25 ppm in air.
Uses: Solvent for nitrocellulose, cellulose acetate, alcohol-soluble dyes, natural and synthetic resins; solvent mixtures; lacquers; enamels; varnishes; leather; perfume fixative; wood stains; sealing moistureproof cellophane; jet fuel deicing additive.

ethylene glycol monomethyl ether acetate (methyl "Cellosolve" acetate). $CH_3COOCH_2CH_2OCH_3$.
Properties: Colorless liquid; pleasant odor; stable; miscible with common organic solvents; soluble in water. Sp. gr.1.0067 (20/20°C); b.p. 145°C; vapor pressure 2 mm (20°C); flash point 120°F; wt/gal 8.4 lb (20°C); f.p. -65.1°C. Combustible.
Containers: 1-, 5-gal cans; 55-gal drums; tank cars.
Hazard: Moderately toxic by ingestion and inhalation; moderate fire risk. Tolerance, 25 ppm in air.
Uses: Solvent for nitrocellulose, cellulose acetate,

various gums, resins, waxes, oils; textile printing; photographic film; lacquers; dopes.

ethylene glycol monomethyl ether acetyl ricinoleate $C_{17}H_{32}(OCOCH_3)COOCH_2CH_2OCH_3$.
Properties: Liquid; sp. gr. 0.966; refractive index 1.460; boiling range 220–260°C; f.p. less than -60°C; flash point 425°F; insoluble in water. Combustible. Low toxicity.
Use: Plasticizer.

ethylene glycol monomethyl ether ricinoleate $C_{17}H_{32}(OH)COOCH_2CH_2OCH_3$.
Properties: Liquid; sp. gr. (25°C) 0.935; f.p. less than -60°C; insoluble in water. Combustible. Low toxicity.
Use: Plasticizer.

ethylene glycol monomethyl ether stearate $C_{17}H_{35}COOCH_2CH_2OCH_3$.
Properties: Liquid; sp. gr. 0.890; m.p. 21°C; insoluble in water. Combustible. Low toxicity.
Use: Plasticizer.

ethylene glycol monooctyl ether $C_4H_9CHC_2H_5CH_{23}OCH_2CH_2OH$.
Properties: Colorless, odorless liquid; b.p. 228.3°C; sp. gr. 0.8859; flash point 230°F; vapor pressure 0.02 mm (20°C). Combustible. Low toxicity.
Uses: Solvent for cellulose esters; plasticizers.

ethylene glycol monophenyl ether (2-phenoxyethanol, phenyl "Cellosolve"). $C_6H_5OCH_2CH_2OH$.
Properties: Colorless liquid; faint aromatic odor; stable in presence of acids and alkalies; partially soluble in water. Sp. gr. 1.1094 (20/20°C); b.p. 244.9°C; f.p. 14°C; vapor pressure $<$0.01 mm (20°C); flash point 250°F; phenol 0.3% (max.); 9.2 lb/gal. Combustible; low toxicity.
Grade: Technical.
Containers: 1-, 5-gal cans; 55-gal drums.
Uses: Solvent for cellulose acetate, dyes; inks, resins; perfume and soap fixative; bactericidal agent; organic synthesis of plasticizers, germicides, perfume materials and pharmaceuticals.

ethylene glycol monoricinoleate $C_{17}H_{32}(OH)COO(CH_2)_2OH$.
Properties: Clear, moderately viscous, pale yellow liquid; mild odor; miscible with most organic solvents. Sp. gr. 0.965 (25/25°C); saponification value 170; hydroxyl value 270; solidifies at -20°C; insoluble in water. Combustible. Low toxicity.
Derivation: Castor oil and ethylene glycol.
Grade: Technical.
Containers: 5-gal cans; 55-gal drums.
Uses: Plasticizer; greases; urethane polymers.

ethylene glycol monostearate (glycol stearate) $C_{17}H_{35}COO(CH_2)_2OH$.
Properties: Yellow waxy solid; m.p. 57–60°C; sp. gr. 0.96 (25°C); soluble in alcohol, hot ether, and acetone; insoluble in water. Combustible.

ethylene glycol silicate $(HOCH_2CH_2)_4SiO_4$.
Properties: Colorless liquid; slowly hydrolyzed by acids; miscible with water.
Uses: Nonvolatile bonding agent for pigments; weather-proofing paints for protecting concrete, stone, brick, and plastic surfaces.

ethylene-hexene-1-copolymer. Product developed especially for use in blow-molded items; other possible uses are coatings for pipe, wire, and cables, sheeting, and monofilament.

ethylene hydrate. See gas hydrate.

ethyleneimine (aziridine; ethylenimine) $H_2\overline{CNHCH}_2$.
Properties: Clear, colorless liquid with amine odor; b.p. 57°C; sp. gr. 0.832 (20/4°C); refractive index 1.4123 (25°C); f.p. -78°C; flash point 12°F. Miscible with water and most organic solvents. Autoignition temp. 612°F.
Containers: Steel cylinders; plastic-lined drums.
Derivation: From ethylene dichloride and ammonia by use of an acid acceptor.
Hazard: Highly toxic and corrosive. Absorbed by skin. Tolerance, 0.5 ppm in air. Dangerous fire and explosion hazard. Flammable limits in air 3.6 to 46%. Causes tumors in experimental animals; worker exposure should be minimized. A known carcinogen.
Uses: Intermediate and monomer for fuel oil and lubricant refining, ion exchange, protective coatings, including paper and textiles, pharmaceuticals, adhesives, polymer stabilizers, surfactants. Alkyl substituted forms, called alkyl aziranes, are used as intermediates and for microbial control; aziridinyl compounds are also used as polymers and intermediates.
Shipping regulations: (inhibited) (Rail) Red label. (Air) Flammable Liquid label. Not accepted on passenger planes. (uninhibited) (Rail) Not listed. (Air) Not acceptable.

ethylene-maleic anhydride copolymer.
Properties: Fine, water-soluble powder available as both straight-chain and crosslinked polymers in a variety of molecular weights. May be in form of free acid, sodium or amide-ammonium salts. Reacts readily with alcohols and amines.
Uses: Oil-well drilling muds; stabilizers and thickeners in liquid detergents, cosmetics, paints; textile sizes; printing inks; suspending agents; ceramic binders.

ethylenenaphthalene. See acenaphthene.

ethylene oxide (epoxyethane; oxirane) $\overline{CH_2CH_2O}$. 26th highest-volume chemical produced in U.S. (1975).
Properties: Colorless gas at room temperature; soluble in organic solvents; miscible with water in all proportions; f.p. -111.3°C; b.p. 10.73°C; sp. gr. (20/20°C) 0.8711; wt/gal (20°C) 7.25 lb; viscosity (0°C) 0.32 cp; flash point (TOC) below 0°F; autoignition temp. 804°F.
Derivation: (a) oxidation of ethylene in air or oxygen with silver catalyst; (b) action of an alkali on ethylene chlorohydrin.
Grades: Technical; pure (99.7%).
Containers: Cylinders; tank cars.
Hazard: Highly flammable; dangerous fire and explosion risk. Flammable limits in air 3 to 100%. Toxic; irritrating to eyes and skin. Tolerance, 50 ppm in air. Safety data sheet available from Mfg. Chemists Assn.,Washington, D.C.
Uses: Manufacture of ethylene glycol and higher glycols; surfactants; acrylonitrile; ethanolamines; petroleum demulsifier; fumigant; rocket propellant; industrial sterilant, e.g., medical plastic tubing.
Shipping regulations: (Rail) Red label. (Air) Flammable Liquid label. Not acceptable on passenger planes.

ethylene-propylene-diene monomer. See ethylene-propylene terpolymer.

ethylene-propylene rubber (EPR). An elastomer made by the stereospecific copolymerization of ethylene and propylene. Has no unsaturation; cannot be vulcanized with sulfur, but can be cured with peroxides.

ethylene-propylene terpolymer (EPT, EPDM). An elastomer based on stereospecific linear terpolymers of ethylene, propylene and small amounts of a nonconjugated diene; e.g., a cyclic or aliphatic diene (hexadiene, dicyclopentadiene, or ethylidene norbornene). The unsaturated part of the polymer molecule is pendant from the main chain, which is completely saturated. Can be vulcanized with sulfur.
Properties of vulcanizate: Light cream to white; excellent resistance to ozone, to high and low temperatures (from -60 to +300°F), and to acids and alkalies; good electrical resistance. Susceptible to attack by oils.
Uses: Automotive parts; gaskets; cable coating; mechanical rubber products; cover strips for tire sidewalls; tire tubes; safety bumpers; coated fabrics; footwear; wire and cable coating. See also "Nordel."

ethylene thiourea $\overline{NHCH_2CH_2NHC}S$. (2-imidazolidinethione).
Properties: White to pale green crystals, faint amine odor; m.p. 199–204°C; slightly soluble in cold water; very soluble in hot water; slightly soluble at room temperature in methanol, ethanol, acetic acid, naphtha.
Containers: Fiber drums (150 lb).
Uses: Electroplating baths; intermediate for antioxidants, insecticides; fungicides, vulcanization accelerators, dyes, pharmaceuticals, synthetic resins.

ethylene urea (2-imidazolidinone; dihydro-2(3)-imidazolone; 2-imidazolidone) $\overline{CH_2CH_2NHCONH}$.
Properties: White lumpy powder; odorless; m.p. 125–128°C; soluble in water.
Use: Drip-dry fabrics and textiles. See dimethylolethylene urea.

ethylene-vinyl acetate copolymer (EVA). An elastomer used to improve adhesion properties of hot-melt and pressure-sensitive adhesives; also for conversion coatings and thermoplastics. See also "Ultrathene."

ethylenimine. See ethyleneimine.

N-ethylethanolamine (ethylaminoethanol) $C_2H_5NHCH_2CH_2OH$.
Properties: Colorless liquid. Sp. gr. 0.914 (20°C); boiling range 167–169°C; odor amine. Soluble in water, alcohol, and ether. Flash point 160°F (open cup). Combustible. Low toxicity.
Containers: 5-gal cans; 55-gal drums; tank cars.
Uses: Solvent; intermediate.

ethyl ether (ether, diethyl ether; sulfuric ether; ethyl oxide; diethyl oxide) $(C_2H_5)_2O$.
Properties: Colorless, volatile, mobile liquid; hygroscopic; aromatic odor; burning and sweet taste. B.p. 34.5°C; freezing point -116.2°C; sp. gr. 0.7147 (20/20°C); surface tension 17.0 dynes/cm (20°C); refractive index (n 20/D) 1.3526; viscosity 0.00233 poise (20°C); vapor pressure 442 mm (20°C); specific heat 0.5476 cal/g (30°C); flash point -49°F; autoignition temp. 356°F; latent heat of evaporation 83.96 cal/g at b.p.; electric conductivity 4×10^{-13} recip. ohms (25°C); wt/gal 6 lb (20°C). Soluble in alcohol, chloroform, benzene, solvent naphtha, and oils; slightly soluble in water.
Derivation: By the action of sulfuric acid on ethyl alcohol or ethylene, followed by distillation.
Method of purification: Rectification, dehydration, treatment with alkali and charcoal.
Grades: U.S.P. (for anesthesia); A.C.S. Reagent; A.C.S. Absolute; C.P.; concentrated; U.S.P. 1880; washed; motor; electronic.
Containers: 30-, 100-lb drums; tank trucks.
Hazard: Extremely flammable, severe fire and explosion hazard when exposed to heat or flame. Forms explosive peroxides. Explosive limits in air 1.85 to 48%. CNS depressant by inhalation and skin absorption. Tolerance, 400 ppm in air. Safety data sheet available from Mfg. Chemists Assn., Washington, D.C.
Uses: Manufacture of ethylene and other chemical synthesis; industrial solvent (smokeless powder); analytical chemistry; anesthetic; perfumery; extractant.
Shipping regulations: (Rail) Red label. (Air) Flammable Liquid label.

ethyl 3-ethoxypropionate. $C_2H_5OOCCH_2CH_2OC_2H_5$.
Properties: Liquid; sp. gr. (20°C) 0.9496; b.p. 170.1°C; vapor pressure (20°C) 0.9 mm; sets to glass at -100°C; slightly soluble in water. Flash point 180°F (O.C.). Combustible.
Uses: Intermediate for vitamin B_1; other chemicals.

ethylethylene. See butene-1.

ethylethyleneimine (2-ethylaziridine) $C_2H_5H\overline{CNHC}H_2$.
Properties: Water-white liquid; b.p. 87–90°C; sp. gr. (25/25°C) 0.812–0.816.
Hazard: Toxic.
Uses: Organic intermediate whose derivatives are used in the textile, paper, rubber and pharmaceutical industries.

ethyl ferrocenoate (ferrocenecarboxylic acid ethyl ester) $(C_5H_4COOC_2H_5)_2Fe$. Orange crystalline solid; m.p. 63–64°C. Used as an intermediate.

ethylfluoroformate $FCOOC_2H_5$.
Properties: Liquid. Sp. gr. 1.11 (33°C); b.p. 57°C.
Derivation: Interaction of ethylchloroformate and a reactive fluoride.
Hazard: Highly toxic; strong irritant.
Shipping regulations: (Rail, Air) Not listed.

ethylfluorosulfonate $C_2H_5OFSO_2$.
Properties: Liquid. Ethereal odor.
Hazard: Highly toxic; strong irritant.

ethyl formate $HCOOC_2H_5$.
Properties: Water-white, unstable liquid. Pleasant, aromatic odor. Miscible with benzene, ether, alcohol; slightly soluble in water; gradual decomposition in water. Flash point -4°F (closed cup); explosive limits in air, 2.8–13.5%. Autoignition temp. 851°F. Sp. gr. 0.9236 (20/20°C); f.p. -80.5°C; b.p. 54.3°C; vapor pressure 200 mm (20.6°C), 300 mm (30.2°C); wt/gal 7.61 lb (68°F); refractive index 1.35975 (20°C).
Derivation: Heating ethyl alcohol with formic acid in presence of sulfuric acid.
Grades: Technical; F.C.C.
Hazard: Highly flammable; dangerous fire and explosion risk. Narcotic and irritating to skin and eyes. Tolerance, 100 ppm in air. Use in foods restricted to 0.0015%.
Uses: Solvent for cellulose nitrate and acetate; acetone substitute; fumigant; larvicide; synthetic flavors; synthetic resins; medicine.
Shipping regulations: (Rail) Red label. (Air) Flammable Liquid label.

ethyl 3-formylpropionate $C_2H_5OOCC_2H_4C(O)H$.
Properties: Liquid; sp. gr. 1.0625 (20/20°C); b.p. 190.9°C; f.p. less than -80°C; wt/gal 8.9 lb; flash point 200°F. Somewhat soluble in water. Combustible.
Uses: Solvent for lacquers; antibiotic extraction; acetic acid separation; coalescing acids for emulsion paints.

ethyl furoate $C_4H_3OCO_2C_2H_5$.
Properties: White leaflets or prisms. Insoluble in water; soluble in alcohol and ether. Sp. gr. 1.1174 (20.8/4°C); m.p. 34°C.

4-ethylheptane (ethyldipropylmethane) $CH_3(CH_2)_2CHC_2H_5(CH_2)_2CH_3$.
Properties: Colorless liquid. Sp. gr. 0.730; b.p. 141.2°C; refractive index (n 20/D) 1.4109. Combustible.
Grade: Technical.
Use: Organic synthesis.

ethyl heptanoate. See ethyl enanthate.

ethyl heptazine. See ethoheptazine.

2-ethylhexaldehyde (butylethylacetaldehyde; octyl aldehyde; 2-ethylhexanal) $C_4H_9CH(C_2H_5)CHO$.
Properties: Colorless, high-boiling liquid. Mild odor. Miscible with most organic solvents; slightly soluble in water. Sp. gr. 0.8205 (20°C); b.p. 163.4°C; vapor pressure 1.8 mm (20°C); flash point 125°F (open cup); wt/gal 6.8 lb. Combustible.
Grades: Technical.
Hazard: Moderate fire risk.
Uses: Organic synthesis; perfumes.

2,2′-(2-ethylhexamido)diethyl di(2-ethylhexoate) ("Flexol" 8N8) $(C_7H_{15}OCOC_2H_4)_2NCOC_7H_{15}$.
Properties: Light-colored liquid; sp. gr. 0.9564 (20/20°C); 8.0 lb/gal (20°C); b.p. 256°C (5 mm); flash point 420°F; vapor pressure 0.60 mm (200°C); insoluble in water; viscosity 139.2 cp (20°C). Combustible.
Use: Plasticizer.

2-ethylhexanal. See 2-ethylhexaldehyde.

2-ethylhexanediol-1,3 (ethohexadiol; 2-ethyl-3-propyl-1,3-propanediol) $C_3H_7CH(OH)CH(C_2H_5)CH_2OH$.
Properties: Colorless, slightly viscous, odorless liquid; hygroscopic. Sp. gr. 0.9422 (20/20°C); 7.8 lb/gal (20°C); b.p. 244°C; vapor pressure less than 0.01 mm (20°C); flash point 260°F; freezing point below -40°C; refractive index 1.4465–1.4515; viscosity 323 cp (20°C); soluble in alcohol and ether; partially soluble in water. Low toxicity; combustible.
Grades: U.S.P. (as ethohexadiol); industrial.
Uses: Insect repellent; cosmetics; vehicle and solvent in printing inks; medicine; chelating agent for boric acid.

ethyl hexanoate. See ethyl caproate.

2-ethylhexanol. See 2-ethylhexyl alcohol.

2-ethylhexenal. See 2-ethyl-3-propylacrolein.

2-ethyl-1-hexene $CH_3(CH_2)_3(C_2H_5)C{:}CH_2$.
Properties: Colorless liquid; sp. gr. 0.7270 (20/4°C); b.p. 120°C; refractive index (n 20/D) 1.4157; soluble in alcohol, acetone, ether, petroleum, coal-tar solvents; insoluble in water. Combustible.
Grade: 95% min purity.
Hazard: Moderately toxic by ingestion and inhalation.
Use: Organic synthesis of flavors, perfumes, medicines, dyes, resins.

ethyl hexoate. See ethyl caproate.

2-ethylhexoic acid (butylethylacetic acid) $C_4H_9CH(C_2H_5)COOH$.
Properties: Mild-odored liquid; slightly soluble in water; sp. gr. 0.9077 (20/20°C); 7.6 lb/gal (20°C); b.p. 226.9°C; vapor pressure 0.03 mm (20°C); freezing point -83°C; viscosity 7.73 cp (20°C); acid number 370; flash point 260°F; combustible.
Grade: 99%.
Containers: Drums; tank cars and trucks.
Uses: Paint and varnish driers (metallic salts). Ethylhexoates of light metals are used to convert some mineral oils to greases. Its esters are used as plasticizers.

2-ethylhexyl. An eight-carbon radical of the formula $CH_3(CH_2)_3CH(C_2H_5)CH_2$—. Many of its compounds were formerly called octyl (q.v.).

2-ethylhexyl acetate $CH_3COOCH_2CHC_2H_5C_4H_9$.
Properties: Water-white, stable liquid; very slightly soluble in water; miscible with alcohol.Sp. gr. 0.8733 (20°C); b.p. 198.6°C; f.p. -93°C; vapor pressure 0.4 mm (20°C); flash point 180°F; wt/gal 7.3 lb (20°C). Combustible; low toxicity.
Grade: Technical (about 95%).
Containers: 1-, 5-gal cans; 55-gal drums.
Use: Solvent for nitrocellulose, resins, lacquers, baking finishes.

2-ethylhexyl acrylate $CH_2{:}CHCOOCH_2CH(C_2H_5)C_4H_9$.
Properties: Liquid, pleasant odor; sp. gr. 0.8869; b.p. 214–218°C; vapor pressure at 20°C 0.1 mm; sets to glass at -90°C; flash point 180°F (O.C.); insoluble in water. Combustible; low toxicity.
Containers: 400-lb drums; tank cars.
Uses: Monomer for plastics, protective coatings, paper treatment; water-based paints.

2-ethylhexyl alcohol (2-ethylhexanol; octyl alcohol) $CH_3(CH_2)_3CHC_2H_5CH_2OH$.
Properties: Colorless liquid. Miscible with most organic solvents; insoluble in water. Sp. gr. 0.83 (20°C); b.p. 183.5°C; f.p. -76°C; vapor pressure 0.36 mm (20°C); refractive index 1.4300 (20°C); wt/gal 6.9 lb (20°C); flash point 178°F. Combustible; low toxicity.
Derivation: (a) Oxo process, from propylene and synthesis gas. (b) Aldolization of acetaldehyde or butyraldehyde, followed by hydrogenation.
Grade: Technical.
Containers: 1-, 5-gal cans; 55-gal drums; tank cars.
Uses: Plasticizer for PVC resins; defoaming agent; wetting agent; organic synthesis; solvent mixtures for nitrocellulose, paints, lacquers, baking finishes; penetrant for mercerizing cotton; textile finishing compounds; plasticizers; inks; rubber; paper; lubricants; photography; dry cleaning.

2-ethylhexylamine $C_4H_9CH(C_2H_5)CH_2NH_2$.
Properties: Colorless liquid; sp. gr. 0.7894 (20/20°C); 6.56 lb/gal (20°C); b.p. 169.2°C; vapor pressure 1.2 (20°C); viscosity 1.11 cp (20°C); flash point 140°F (open cup). Soluble in water; solubility of water in (20°C), 25.3%. Combustible.
Hazard: Moderate fire risk. Moderately toxic.
Uses: Synthesis of detergents, rubber chemicals, oil additives and insecticides.

N-2-ethylhexylaniline $C_6H_5NHCH_2CH(C_2H_5)C_4H_9$.
Properties: Light yellow liquid with mild odor. Sp.

gr. (20/20°C) 0.9119; b.p. (50 mm) 194°C; vapor pressure (20°C) <0.01 mm; freezing point, sets to a glass below -70°C; viscosity (20°C) 7.4 cp; solubility in water (20°C)<0.01%; flash point 325°F (COC). Combustible.
Hazard: Toxic on chronic exposure.
Uses: Solvent; organic synthesis.

2-ethylhexyl bromide $C_4H_9CH(C_2H_5)CH_2Br$.
Properties: Water-white liquid, b.p. 56–58°C (6 mm); insoluble in water.
Uses: Introduction of the 2-ethylhexyl group in organic synthesis. Preparation of disinfectants, pharmaceuticals.

2-ethylhexyl chloride $C_4H_9CH(C_2H_5)CH_2Cl$.
Properties: Colorless liquid. Sp. gr. 0.8833 (20°C); b.p. 172.9°C; refractive index 1.4310; 7.33 lb/gal; flash point 140°F (open cup); freezing point -135°C. Insoluble in water. Combustible.
Hazard: Moderate fire risk. Moderately toxic.
Use: Synthesis of cellulose derivatives, dyestuffs, pharmaceuticals, textile auxiliaries, insecticides, resins.

2-ethylhexyl cyanoacetate
$CNCH_2COOCH_2CH(C_2H_5)C_4H_9$. Liquid; b.p. 150°C; refractive index (n 20/D) 1.4389; insoluble in water; soluble in alcohol, benzene, and ether. Combustible.
Containers: 55-gal steel drums.
Use: Organic intermediate.

2-ethylhexyl isodecyl phthalate
$C_4H_9CH(C_2H_5)CH_2OOCC_6H_5COOC_7H_{14}CH(CH_3)_2$.
Properties: Colorless, high-boiling liquid; good dielectric properties. Combustible. Refractive index 1.478–1.488 (25°C); sp. gr. 0.969–0.977 (25/25°C). Miscible with most common solvents, thinners and oils.
Containers: Tank cars and trucks.
Use: Plasticizer.

2-ethylhexylmagnesium chloride
$C_4H_9CH(C_2H_5)CH_2MgCl$. Grignard reagent available commercially in tetrahydrofuran solution.
Containers: 55-gal drums.

2-ethylhexyl octylphenyl phosphite
$(C_8H_{17}O)_2(C_8H_{17}C_6H_{17}C_6H_4O)P$.
Properties: Colorless to light-yellow liquid with characteristic odor. Sp. gr. (20/4°C) 0.935–0.950; flash point 385°F; insoluble in water. Combustible.
Containers: 5, 55-gal drums.
Uses: Antioxidant; plasticizer; flame retardant; lubricating oil additive.

3,3′-(2-ethylhexyl) thiodipropionate (dioctyl thiopropionate) $(C_8H_{17}OOCCH_2Ch_2)_2S$.
Properties: Colorless liquid; sp. gr. (25°C) 0.952. Insoluble in water; soluble in most organic solvents. Combustible.
Uses: Antioxidant; stabilizer and lubricant.

ethyl 2-hydroxy-2,2-bis(4-chlorophenyl) acetate. See chlorobenzilate.

ethyl hydroxyethyl cellulose (EHEC). A cellulose ether.
Properties: White granular solid; available in extra low, and high-viscosity types. Soluble in mixtures of aliphatic hydrocarbons containing alcohol; soluble in water. Combustible.
Uses: Stabilizer; thickener; binder; film former in silk screen and gravure printing inks; protective coatings; aqueous, aqueous-organic, and organic solvent systems.

ethyl alpha-hydroxyisobutyrate $(CH_3)_2COHCOOC_2H_5$.
Properties: Water-white liquid; sp. gr. 0.978–0.986 (20°C); b.p. 149–150°C; soluble in water, alcohol, and ether. Combustible.
Uses: Solvent for nitrocellulose and cellulose acetate; solvent mixtures for cellulose ethers; organic synthesis; pharmaceuticals.

ethylidene chloride (1,1-dichloroethane) CH_3CHCl_2.
Properties: Colorless, neutral, mobile liquid; aromatic ethereal odor; saccharin taste. Soluble in alcohol, ether, fixed and volatile oils; very sparingly in water. Sp. gr. 1.174 (17°C); b.p. 57–59°C; freezing point -98°C; refractive index (n 20/D) 1.4166. Combustible.
Hazard: Moderately toxic. Tolerance, 200 ppm in air.
Use: Medicine; extraction solvent; fumigant.

ethylidenediethyl ether. See acetal.

ethylidenedimethyl ether. See dimethylacetal.

ethylidene fluoride. See 1,1-difluoroethane.

5-ethylidene-2-norbornene (ENB). A diene used as the third monomer in EPDM elastomers.
Hazard: Toxic; tolerance, 0.2 ppm in air.

ethyl iodide (iodoethane) C_2H_5I.
Properties: Colorless liquid; turns brown on exposure to light. Soluble in alcohol and ether; slightly soluble in water. Sp. gr. 1.90–1.93 (25/25°C); f.p. -108°C; b.p. 72°C; refractive index (n 15/D) 1.5168. Combustible.
Derivation: By digesting red phosphorus with absolute ethyl alcohol, after which iodine is added. The mixture is heated under a reflux condenser and finally distilled.
Hazard: Moderately toxic by inhalation and skin absorption. Narcotic in high concentrations.
Uses: Medicine; organic synthesis.

ethyl iodoacetate $CH_2ICOOC_2H_5$.
Properties: Dense, colorless liquid. Decomposed by light and air; also (very slowly) by alkaline solutions and water. Sp. gr. 1.8; b.p. 179°C; vapor density 7.4; vapor pressure 0.54 mm (20°C).
Derivation: Interaction of potassium iodide with ethyl bromo- or chloroacetate.
Hazard: Strong irritant to eyes and skin.

ethyl iodophenylundecylate. See iophendylate.

5-ethyl-5-isoamylbarbituric acid. See amobarbital.

ethylisobutylmethane. See 2-methylhexane.

ethyl isobutyrate $(CH_3)_2CHCOOC_2H_5$.
Properties: Colorless, volatile liquid. Soluble in alcohol and ether; slightly soluble in water. Sp. gr. 0.870; b.p. 110–111°C; f.p. -88°C; refractive index (n 20/D) 1.3903. Combustible.
Derivation: Heating isobutyric acid and ethyl alcohol, with subsequent distillation.
Hazard: Moderately toxic.
Uses: Organic synthesis; flavoring extracts.

ethyl isocyanate C_2H_5NCO. Liquid; sp. gr. 0.898; b.p. 60°C; soluble in chlorinated and aromatic hydrocarbons.
Hazard: Highly toxic.
Use: Pharmaceutical and pesticide intermediate.
Shipping regulations: (Air) Poison label. Not acceptable on passenger planes.

2-ethylisohexanol $(CH_3)_2CHCH_2CH(C_2H_5)CH_2OH$.
Properties: Liquid; sp. gr. 0.825–0.835 (20/20°C); boiling range 173.0–181.0°C; refractive index (n

20/D) 1.4235; density 6.89 lb/gal (20°C); flash point 170°F (COC). Combustible. Low toxicity.
Containers: Tank cars and trucks.
Use: Chemical intermediate.

ethyl isothiocyanate. See ethyl thiocarbimide.

ethyl isovalerate (ethyl valerate; ethyl 2-methylbutyrate) $(CH_3)_2CHCH_2COOC_2H_5$.
Properties: Colorless oily liquid with fruity odor. B.p. 135°C; f.p. -99°C; sp. gr. 0.864 (20/20°C); refractive index (n 20/D) 1.3950–1.3990. Slightly soluble in water; miscible with alcohol, ether, and benzene. Low toxicity. Combustible.
Derivation: Heating sodium valerate and ethyl alcohol in presence of sulfuric or hydrochloric acid, with subsequent distillation.
Grades: Technical; F.C.C.
Uses: Essential oils; perfumery; artificial fruit essences; flavoring.

ethyl lactate $CH_3CHOHCOOC_2H_5$.
Properties: Colorless liquid. Mild odor. Miscible with water, alcohols, ketones, esters, hydrocarbons, oils. Sp. gr. 1.020–1.036 (20/20°C); b.p. 154°C; flash point 115°F (closed cup); wt/gal 8.55 lb (approx.) (20°C). Autoignition temp. 752°F. Low toxicity. Combustible.
Derivation: (a) By the esterification of lactic acid with ethyl alcohol; (b) by combining acetaldehyde with hydrocyanic acid to form aceltadehyde cyanohydrin, which is converted into ethyl lactate by treatment with ethyl alcohol and an inorganic acid.
Grades: Technical (96%). Combustible.
Containers: 5-, 55-gal steel drums.
Hazard: Moderate fire risk.
Uses: Solvent for nitrocellulose, cellulose acetate, many cellulose ethers, resins; lacquers; paints; enamels; varnishes; stencil sheets; safety glass; flavoring.

ethyl levulinate $CH_3CO(CH_2)_2COOC_2H_5$.
Properties: Colorless liquid; sp. gr. 1.012; b.p. 205–206°C; soluble in water; miscible with alcohol; refractive index (n 20/D) 1.4229. Combustible.
Uses: Solvent for cellulose acetate and starch ethers; flavoring.

ethyllithium (lithium ethyl).
Properties: Transparent crystals, decomposed by water. The commercial product is a 2*M* suspension of C_2H_5Li in benzene.
Hazard: Flammable; reacts with oxidizing materials.
Uses: Grignard-type reactions.

ethylmagnesium bromide C_2H_5MgBr, dissolved in ether.
Properties: Liquid; sp. gr. 1.01.
Containers: Glass bottles; 5-, 55-gal drums.
Hazard: Flammable; dangerous fire risk.
Uses: Grignard-type reactions.

ethylmagnesium chloride C_2H_5MgCl, dissolved in ether. (Is also offered commercially in tetrahydrofuran).
Properties: Liquid; sp. gr. 0.85.
Containers: Glass bottles; 5-, 55-gal drums.
Hazard: Flammable; dangerous fire risk.
Uses: Grignard-type reactions.

ethyl malonate (malonic ester; diethyl malonate) $CH_2(COOC_2H_5)_2$.
Properties: Colorless liquid; sweet ester odor. Insoluble in water; soluble in alcohol, ether, chloroform, and benzene. B.p. 198°C; f.p. -50°C; sp. gr. 1.055 (25/25°C); flash point 200°F. Combustible. Low toxicity.
Derivation: By passing hydrogen chloride into cyanoacetic acid dissolved in absolute alcohol, with subsequent distillation.
Containers: Carboys; bottles.
Uses: Intermediate for barbiturates and certain pigments; flavoring.

ethyl mercaptan. Legal label name for ethanethiol, q.v.

ethylmercuric acetate $C_2H_5HgOOCCH_3$.
Properties: White crystalline powder; m.p. 178°C; slightly soluble in water; soluble in many organic solvents; may be steam-distilled.
Hazard: Highly toxic; strong irritant. See mercury.
Use: Seed fungicide as dust or slurry with water.
Shipping regulations: (Rail, Air) Mercury compounds, solid, n.o.s., Poison label.

ethylmercuric chloride C_2H_5HgCl.
Properties: Crystals; sp. gr. 3.482; m.p. 193°C. Insoluble in water; slightly soluble in ether; soluble in hot alcohol. Sublimes readily.
Derivation: Reaction of zinc diethyl and mercuric chloride.
Hazard: Highly toxic; strong irritant. See mercury.
Uses: Fungicide for seed or bulb treatment either alone or with other organic mercury compounds.
Shipping regulations: (Rail, Air) Mercury compounds, solid, n.o.s., Poison label.

ethylmercuric phosphate $(C_2H_5HgO)_3PO$.
Properties: White powder, soluble in water; garlic-like odor.
Derivation: Reaction of ethylmercuric acetate with phosphoric acid.
Hazard: Highly toxic; strong irritant. See mercury.
Uses: Seed fungicide; timber preservative.
Shipping regulations: (Rail, Air) Mercury compounds, solid, n.o.s., Poison label.

ethylmercurithiosalicylic acid, sodium salt. See thimerosal.

ethylmercury 2,3-dihydroxypropyl mercaptide $C_2H_5HgSCH_2CHOHCH_2OH$. Organic mercurial compound used as a fungicidal dust or in slurry treatment for control of seedborne diseases and to reduce losses from seed decay and damping-off of wheat, oats, rye, etc.
Hazard: Highly toxic. See mercury.
Shipping regulations: (Rail, Air) Mercury compounds, solid, n.o.s., Poison label.

ethylmercury-*para*-toluenesulfonanilide (EMTS: "Ceresan" M) $C_6H_5N(HgC_2H_5)SO_2C_6H_4CH_3$.
Properties: Crystals; pungent odor; m.p. 154–157°C; nearly insoluble in water; soluble in acetone and chloroform.
Hazard: Highly toxic. See mercury.
Uses: Dust or slurry for control of seed-borne diseases and of fungi by treatment of seeds or bulbs.
Shipping regulations: (Rail, Air) Mercury compounds, solid, n.o.s., Poison label.

"Ethyl" Metal Deactivator.[313] Trademark. Contains 80% N,N′-disalicylidene-1,2-diaminopropane and 20% toluene solvent. Amber liquid; density 1.0672 g/ml at 68°F; flash point (open cup) 84°F. Soluble

in gasoline, insoluble in water. Used to neutralize the catalytic effect of copper in promoting fuel oxidation.
Hazard: Flammable; moderate fire risk. Irritant to eyes and skin.

ethyl methacrylate $H_2C{:}CCH_3COOC_2H_5$.
Properties: Colorless liquid; b.p. 119°C; f.p. below −75°C; sp. gr. 0.911; refractive index (n 25/D) 1.4116; flash point (open cup) 70°F (O.C.); insoluble in water; readily polymerized.
Derivation: Reaction of methacrylic acid or methyl methacrylate with ethyl alcohol.
Grade: Technical (inhibited).
Containers: Drums; tank cars.
Hazard: Moderately toxic and irritant. Flammable; dangerous fire and explosion hazard.
Uses: Polymers; chemical intermediates.
Shipping regulations: (Air) (inhibited) Flammable Liquid label; (uninhibited) Not acceptable.
See also acrylic resin.

N-ethyl-3-methylaniline. See N-ethyl-meta-toluidine.

5-ethyl-5(1-methyl-1-butenyl)barbituric acid. See vinbarbital.

ethyl 2-methylbutyrate. See ethyl isovalerate.

ethyl methylcellulose. A water-soluble cellulose ether used for thickening, sizing, emulsifying and dispersing. See also polymer, water-soluble.

ethyl methyl ether $C_2H_5OCH_3$.
Properties: Colorless liquid; sp. gr. 0.725; b.p. 10.8°C. Soluble in water, miscible with alcohol and ether; flash point −35°F; autoignition temp. 374°F.
Hazard: Highly flammable; dangerous fire and explosion risk.
Use: Medicine.
Shipping regulations: (Rail) Red label. (Air) Flammable Liquid label.

2-ethyl-4-methylimidazole $C_2H_5C_3N_2H_2CH_3$. A supercooled amber liquid which if crystallized is a low-melting solid; m.p. 45°C; b.p. 154°C (10 mm).
Use: Curing epoxy resin systems.

ethyl methyl ketone. Legal label name for methyl ethyl ketone (q.v.).

ethyl methylphenylglycidate (so-called aldehyde C-16; "strawberry aldehyde")
$CH_3(C_6H_5)\underline{COC}HCOOC_2H_5$.
Properties: Colorless to yellowish liquid, having a strong odor, suggestive of strawberry. Soluble in 3 vols of 60% alcohol. Sp. gr. 1.104–1.123; refractive index 1.509–1.511. Low toxicity. Combustible.
Grade: Technical; F.C.C. (as aldehyde C-16).
Uses: Perfumery; flavors.

1-ethyl-1-methylpropyl carbamate. See emylcamate.

alpha-**ethyl**-alpha-**methylsuccinimide.** See ethosuximide.

7-ethyl-2-methyl-4-undecanol (tetradecanol)
$C_4H_9CH(C_2H_5)C_2H_4CH(OH)CH_2CH(CH_3)_2$.
Properties: Liquid; sp. gr. 0.8355 (20/20°C); b.p. 264°C; flash point 285°C; insoluble in water. Combustible.
Use: Intermediate for synthetic lubricants, defoamers and surfactants.

ethylmorphine hydrochloride $C_{19}H_{23}NO_3 \cdot HCl \cdot 2H_2O$.
Properties: White crystalline powder; odorless; soluble in water and alcohol; slightly soluble in ether and chloroform. M.p., about 123°C (dec).
Derivation: Action of hydrochloric acid on ethylmorphine which is made by action of ethyl iodide on morphine in alkaline solution.
Grades: Technical; N.F.
Hazard: Highly toxic by ingestion. Habit-forming narcotic.
Use: Medicine.

N-ethylmorpholine $\underline{CH_2CH_2OCH_2CH_2N}CH_2CH_3$.
Properties: Colorless liquid; ammoniacal odor. Miscible with water. Sp. gr. 0.916 (20/20°C); f.p. −63°C; b.p. 138°C; wt/gal 7.6 lb (20°C); flash point 90°F (O.C.).
Grade: Technical.
Containers: Drums; tank cars.
Hazard: Flammable; moderate fire risk. Toxic; irritant to skin and eyes; absorbed by skin. Tolerance, 20 ppm in air.
Uses: Intermediate for dyestuffs, pharmaceuticals, rubber accelerators and emulsifying agents. Solvent for dyes, resins, oils. Catalyst in making polyurethane forms.

"Ethyl" Multi-Purpose Additive.[313] Trademark. Contains 52% mixed substituted oleamides, 37% isopropyl alcohol, 7% aromatic solvent, and 4% water. Clear amber liquid; density 0.888 g/ml at 68°F; flash point (closed cup) 75°F; pour point 50°F. Used to remove and prevent deposits on the throat walls of carburetors, to prevent carburetor icing, and as a corrosion preventive.
Hazard: Flammable; dangerous fire risk.

ethyl mustard oil. See ethyl thiocarbimide.

ethyl myristate (ethyl tetradecanoate)
$CH_3(CH_2)_{12}COOC_2H_5$.
Properties: Liquid; sp. gr. 0.856; m.p. 12°C; b.p. 295°C. Insoluble in water; soluble in alcohol; slightly soluble in ether. Combustible. Low toxicity.
Use: Flavoring.

N-ethyl-alpha-**naphthylamine** (N-ethyl-1-naphthylamine) $C_{10}H_7NCH_2H_5$.
Properties: Colorless liquid; refractive index 1.6475; b.p. 305°C. Insoluble in water; soluble in alcohol and ether. Combustible.
Use: Intermediate.

ethyl nitrate $C_2H_5NO_3$.
Properties: Colorless liquid; pleasant odor; sweet taste. Soluble in alcohol and ether; insoluble in water. Sp. gr. 1.116; f.p. −112°C; b.p. 87.6°C; vapor three times heavier than air. Flash point 50°F (closed cup); lower explosive limit 3.8%.
Derivation: By heating alcohol, urea nitrate and nitric acid, with subsequent distillation.
Hazard: Flammable; dangerous fire and explosion risk.
Uses: Organic synthesis; drugs; perfumes; dyes; rocket propellant.
Shipping regulations: (Rail) Red label. (Air) Not acceptable.

ethyl nitrite $C_2H_5NO_2$.
Properties: Yellowish, volatile liquid. Soluble in alcohol and ether; decomposes in water. Sp. gr. 0.90; b.p. 16.4°C. Flash point −31°F. Decomposes spontaneously at 194°F.
Hazard: Highly flammable; dangerous. Explodes at 194°F. Explosive limits in air 3 to 50%. Narcotic in high conc.
Uses: Organic reactions; medicine; synthetic flavoring.
Shipping regulations: (Rail) Red label. (Air) Flammable Liquid label.

ethyl nonanoate. See ethyl pelargonate.

ethyl octanoate. See ethyl caprylate.

ethyl octoate. See ethyl caprylate.

ethyl oenanthate. See ethyl enanthate.

ethyl oleate $C_{17}H_{33}COOC_2H_5$.
Properties: Light-colored, yellowish oleaginous liquid. Insoluble in water; soluble in alcohol and ether. Solubility of water in product 1.0 cc/100 cc (202°C). Wt/gal 7.27 lb (20°C); refractive index 1.45189 (20°C); f.p. −32°C (approx); sp. gr. 0.867; flash point 348°F. Combustible; low toxicity.
Uses: Solvent; plasticizer; lubricant; water-resisting agent; flavoring.

ethyl orthoacetate $CH_3C(OC_2H_5)_3$.
Properties: Colorless liquid. B.p. 144–148°C; refractive index (n 25/D) 1.395; insoluble in water; soluble in alcohol and ether; flash point 131°F. Combustible.
Containers: 55-gal steel drums.
Hazard: Moderate fire risk.
Use: Intermediate.

ethyl orthoformate $CH(OC_2H_5)_3$.
Properties: Colorless liquid. B.p. 145.9°C; f.p. −76°C; refractive index (n 20/D) 1.392; slightly soluble in water; soluble in alcohol and ethers; flash point 85°F.
Containers: 55-gal steel drums.
Hazard: Flammable; fire risk.
Use: Intermediate.

ethyl orthopropionate $CH_3CH_2C(OC_2H_5)_3$.
Properties: Colorless liquid. B.p. 155–160°C; refractive index (n 20/D) 1.401; insoluble in water; soluble in alcohol and ether; flash point 140°F. Combustible.
Containers: 55-gal steel drums.
Hazard: Moderate fire risk.
Use: Intermediate.

ethyl oxalate (diethyl oxalate) $(COOC_2H_5)_2$.
Properties: Colorless, unstable, oily, aromatic liquid. Miscible in all proportions with alcohol, ether, ethyl acetate, and other common organic solvents; slightly soluble in water and gradually decomposed by it. Sp. gr. 1.09 (20/20°C); b.p. 186°C; f.p. −40.6°C; wt/gal 8.96 lb (20°C) (approx); flash point 168°F (closed cup). Combustible.
Derivation: By standard esterification procedure using ethyl alcohol and oxalic acid. The final purification calls for unusual technique and equipment. The last traces of water are most difficult to remove and this is accomplished by a special step in the rectification.
Containers: Drums; tank cars.
Hazard: Strong irritant to skin and mucous membranes, especially by ingestion.
Uses: Solvent for cellulose esters and ethers, many natural and synthetic resins; radio tube cathode fixing lacquers; dye intermediate; pharmaceuticals; perfume preparations; organic synthesis.

ethyl 2-oxocyclopentanecarboxylate. See 2-carbethoxycyclopentanone.

ethyl parathion. See parathion.

ethyl PCT. See O,O-diethylphosphorochloridothioate.

ethyl pelargonate (ethyl nonanoate; wine-ether) $CH_3(CH_2)_7COOC_2H_5$.
Properties: Colorless liquid; fruity odor; refractive index 1.4220 (20°C); sp. gr. 0.866 (18/4°C); b.p. about 220°C; f.p. −44°C; insoluble in water; soluble in alcohol and ether. Low toxicity. Combustible.
Grades: Technical; F.C.C.
Uses: Flavoring alcoholic beverages; perfumes; chemical intermediate.

3-ethylpentane (triethylmethane) $(C_2H_5)_3CH$.
Properties: Colorless liquid; b.p. 93.5°C; freezing point −118.6°C; sp. gr. 0.69818 (20°C); refractive index (n 20/D) 1.3934; soluble in alcohol; insoluble in water. Combustible.
Grades: Technical.
Use: Organic synthesis.

meta-**ethylphenol** (3-ethylphenol) $HOC_6H_4C_2H_5$.
Properties: Colorless liquid; f.p. −4°C; b.p. 214°C; sp. gr. 1.001; very slightly soluble in water; miscible with alcohol and ether. Combustible.
Hazard: May be toxic. See phenol.

para-**ethylphenol** (4-ethylphenol) $HOC_6H_4C_2H_5$.
Properties: Colorless needles; m.p. 46°C; b.p. 219°C; flash point 219°F; soluble in alcohol or ether; slightly soluble in water. Combustible.
Hazard: May be toxic. See phenol.

ethylphenylacetamide. See ethylacetanilide.

ethyl phenylacetate $C_6H_5CH_2COOC_2H_5$.
Properties: Colorless liquid, honey-like odor. Sp. gr. 1.027–1.032; refractive index, 1.498; b.p. 276°C. Soluble in 8 parts of 60% alcohol; insoluble in water. Combustible.
Grades: Technical; F.C.C.
Uses: Perfumery; flavors; intermediate, especially for phenobarbital; herbicide.

ethyl phenylacrylate. See ethyl cinnamate.

5-ethyl-5-phenylbarbituric acid. See phenobarbital.

ethyl phenylcarbamate (phenylurethane; ethyl phenylurethane) $C_6H_5NHCOOC_2H_5$.
Properties: White, crystalline solid; aromatic odor, clove-like taste. M.p. 51°C. Soluble in alcohol, ether and boiling water; insoluble in cold water.
Derivation: Action of ethyl alcohol on phenyl isocyanate.

ethyl phenyl dichlorosilane $C_2H_5(C_6H_5)SiCl_2$.
Properties: Colorless liquid which fumes strongly in moist air.
Hazard: Highly toxic by inhalaton and ingestion. Strong irritant to eyes and skin.
Shipping regulations: (Rail) White label. (Air) Corrosive label. Not acceptable on passenger planes.

N,N-**ethylphenylethanolamine** $C_6H_5NC_2H_5CH_2CH_2OH$.
Properties: Liquid; sp. gr. 1.04 (20/20°C); b.p. (740 mm) 268°C; wt/gal 8.7 lb (20°C). Combustible; low toxicity.
Containers: 1-gal cans; 5-, 55-gal drums.
Uses: Organic synthesis; dyestuffs.

2-ethyl-2-phenylglutarimide (glutethimide) $C_{13}H_{15}NO_2$.
Properties: White, crystalline powder. A saturated solution is slightly acid. Freely soluble in acetone, ethyl acetate, and chloroform; soluble in ethanol and methanol; practically insoluble in water. Melting range; 85–87°C.
Grade: N.F. (as glutethimide).
Hazard: Manufacture and use controlled by law.
Use: Medicine (sedative).

ethyl phenyl ketone. See propiophenone.

ethyl phenylurethane. See ethyl phenylcarbamate.

ethylphosphoric acid $C_2H_5H_2PO_4$.
Properties: Pale straw-colored liquid; sp. gr. 1.33 (25°C); can be neutralized with alkalies or amines to give water-soluble salts. Combustible.
Purity: 97%, remainder being orthophosphoric acid and ethyl alcohol.
Hazard: May be irritant. See phosphoric acid.
Uses: Catalyst; rust remover; soldering flux; intermediate.

ethyl phthalate. See diethyl phthalate.

ethyl phthalyl ethyl glycolate
$C_2H_5OCOC_6H_4COOCH_2COOC_2H_5$.
Properties: Liquid with slight odor. Sp. gr. (25°C) 1.180; refractive index (25°C) 1.498; b.p. (5 mm) 190°C; flash point 193°C; solubility in water, 0.018% by wt. Miscible with most organic solvents. Combustible. Low toxicity.
Use: Plasticizer.

5-ethyl-2-picoline. See 2-methyl-5-ethylpyridine.

1-ethyl-1-propanol. See 3-pentanol.

ethyl propionate $C_2H_5COOC_2H_5$.
Properties: Water-white liquid. Odor resembling pineapples; soluble in alcohol and ether; in water 2.5% at 15°C; sp. gr. 0.888 (25°C); b.p. 99°C; f.p. −73°C; flash point 54°F (closed cup); autoignition temp. 890°F; refractive index (n 20/D) 1.3844. Toxicity probably low.
Derivation: Treating ethyl alcohol with propionic acid.
Grades: Commercial, 85–90% ester content; F.C.C.
Containers: 50-, 100-gal drums.
Hazard: Flammable; dangerous fire risk.
Uses: Solvent for cellulose ethers and esters, various natural and synthetic resins; flavoring agent; fruit syrups; cutting agent for pyroxylin.
Shipping regulations: (Rail) Flammable liquid, n.o.s., Red label. (Air) Flammable Liquid label.

ethylpropionyl. See diethyl ketone.

2-ethyl-3-propylacrolein (2-ethylhexenal)
$C_3H_7CH{:}C(C_2H_5)CHO$.
Properties: Yellow liquid. Powerful odor. Sp. gr. 0.8518 (20/20°C); b.p. 175.0°C; vapor pressure 1.0 mm (20°C); flash point 155°F (open cup); wt/gal 7.1 lb (20°C); viscosity 0.113 poise (20°C). Combustible.
Containers: 1-gal cans; 5-, 55-gal drums.
Hazard: Moderately toxic by inhalation and ingestion; strong irritant.
Uses: Insecticide; organic synthesis (intermediate); warning agents and leak detectors.

alpha-**ethyl**-beta-**propylacrylanilide**
$C_6H_5NHCOC(C_2H_5){:}CH(CH_2)_2CH_3$.
Use: Rubber accelerator.

2-ethyl-3-propylacrylic acid $C_3H_7CH{:}C(C_2H_5)COOH$.
Properties: Liquid; sp. gr. (20°C) 0.9484; f.p. −7.8°C; b.p. 232.1°C; vapor pressure (20°C) less than 0.01 mm; flash point 330°F; insoluble in water. Combustible.
Uses: Pharmaceuticals; resins and plastics; lubricants.

2-ethyl-3-propylglycidamide. See oxanamide.

ethyl propyl ketone. See 3-hexanone.

3-ethyl-3-propyl-1,3-propanediol. See 2-ethylhexanediol-1,3.

2-(5-ethyl-2-pyridyl)ethyl acrylate
$CH_2{:}CHCOOC_2H_4C_5H_3NC_2H_5$.
Properties: Liquid; sp. gr. 1.0458 (20°C); b.p. 181°C (50 mm); f.p. −75°C; very slightly soluble in water. Combustible.
Hazard: Moderately toxic.
Uses: Manufacture of plastics and fibers; adhesives, textile finishes and sizes.

ethyl pyrophosphate. See tepp.

ethyl salicylate $C_6H_4(OH)COOC_2H_5$.
Properties: Colorless liquid, with a faint odor of methyl salicylate. Soluble in ether and alcohol; insoluble in water. Sp. gr. 1.127–1.130; refractive index 1.523; b.p. 231–234°C. Combustible.
Uses: Perfumery; flavors.

ethyl silicate (tetraethyl orthosilicate) $(C_2H_5)_4SiO_4$.
Properties: Colorless liquid; faint odor; slightly soluble in benzene; hydrolyzed to an adhesive form of silica. Sp. gr. 0.9356 (20/20°C); b.p. 168.1°C; m.p. 110°C (sublimes); vapor pressure 1.0 mm (20°C); flash point 125°F (open cup); wt/gal 7.8 lb (20°C); f.p. −77°C; viscosity 0.0179 poise (20°C). Combustible.
Grades: 29% SiO_2, 40% SiO_2.
Containers: 5-, 55-gal (tin-lined) drums; tank cars.
Hazard: Moderate fire risk. Strong irritant to eyes, nose and throat. Tolerance, 100 ppm in air.
Uses: Weatherproof and acidproof mortar and cements; refractory bricks; other molded objects; heat-resistant paints; chemical-resistant paints; protective coatings for industrial buildings and castings; lacquers; bonding agent; intermediate.

ethyl silicate, condensed. Light-yellow liquid with mild odor consisting of 85% by wt tetraethyl orthosilicate and 15% polyethoxysiloxanes. On hydrolysis or ignition, yields high purity, refractory silica. Used as an intermediate for siloxane compounds; for precision casting of high-melting alloys; pigments binder for paints; surface hardener for sandstones.
Hazard: See ethyl silicate.

ethyl sodium oxalacetate
$C_2H_5OOCC(ONa){:}CHCOOC_2H_5$. Light yellow powder; 92% pure.
Derivation: Reaction of pure ethyl acetate and diethyl oxalate with metallic sodium.
Containers: 175-lb fiber drums.
Hazard: May be toxic.
Uses: Dyes; synthesis.

ethyl sulfate. See diethyl sulfate.

ethyl sulfhydrate. See ethanethiol.

ethyl sulfide (diethyl sulfide) $(C_2H_5)_2S$.
Properties: Colorless, oily liquid; garlic-like odor; soluble in alcohol and ether; slightly soluble in water; sp. gr. 0.837; f.p. −102°C; b.p. 92–93°C; refractive index (n 20/D) 1.4423. Combustible.
Derivation: Heating potassium ethyl sulfate and potassium sulfide, with subsequent distillation.
Hazard: Probably toxic.
Uses: Organic synthesis; special solvent.

ethylsulfuric acid (acid ethylsulfate) $C_2H_5HSO_4$.
Properties: Colorless, oily liquid; soluble in water, alcohol and ether; sp. gr. 1.316; b.p. 280°C (decomposes). Combustible.
Derivation: Action of sulfuric acid on ethyl alcohol.
Hazard: Strong irritant to eyes and skin.
Uses: Medicine; precipitant for casein; organic synthesis.

4-ethyl-1,4-thiazane $C_4H_8NSC_2H_5$.
Properties: Colorless, mobile liquid; soluble in water;

sp. gr. 0.9929 (15°C); b.p. 184°C (763 mm). Combustible.
Derivation: Interaction of dichlorodiethyl sulfide and an aliphatic amine in the presence of alcohol and sodium carbonate.

ethyl thiocarbimide (ethyl mustard oil; ethy isothiocyanate) C_2H_5NCS.
Properties: Colorless liquid; pungent odor; soluble in alcohol; insoluble in water. Sp. gr. 1.004 (15/4°C); f.p. -5.9°C; b.p. 131–132°C.
Hazard: Highly toxic by inhalation; strong irritant to skin and mucous membranes.
Use: Military poison gas.
Shipping regulations: (Rail, Air) Not listed. Consult authorities.

ethyl thioethanol $C_2H_5SC_2H_4OH$.
Properties: Pale straw liquid; sp. gr. 1.015–1.025 (20/20°C); distillation range; 180–184°C; refractive index 1.486 (20°C). Combustible.
Grade: 95% min.
Containers: 5-, 55-gal drums; tank cars.
Uses: Synthesis; intermediate.

ethylthiopyrophosphate. See sulfotepp.

N-ethyl para-**toluenesulfonamide** $C_2H_5NHSO_2C_6H_4CH_3$.
Properties: Colorless crystals; m.p. 64°C. Soluble in alcohol. Flash point 260°F; combustible.
Grades: A mixture of the ortho- and para-isomers is available commercially.
Hazard: May be toxic.
Use: Plasticizer.

ethyl para-**toluenesulfonate** $CH_3C_6H_4SO_3C_2H_5$.
Properties: Unstable solid; m.p. 33°C; b.p. 221.3°C; sp. gr. 1.17; soluble in many organic solvents; insoluble in water. Flash point 316°F; combustible.
Hazard: Probably toxic.
Uses: Plasticizer for cellulose acetate; ethylating agent.

N-ethyl-meta-**toluidine** (N-ethyl-3-methylaniline) $CH_3C_6H_4NHC_2H_5$. Light amber liquid; refractive index (n 20/D) 1.5451; b.p. 221°C. Combustible.
Containers: Drums; tanks.
Hazard: Probably toxic.
Use: Chemical intermediate.

N-ethyl-ortho-**toluidine** $CH_3C_6H_4NHC_2H_5$.
Properties: Colorless to yellowish oil; soluble in alcohol, ether, and hydrochloric acid; insoluble in water. Sp. gr. 0.9534; b.p. 214°C. Combustible.
Derivation: Heating ethyl alcohol with ortho-toluidine and hydrochloric acid.
Hazard: Probably toxic.

ethyl triacetylgallate $C_2H_5OOCC_6H_2(OOCCH_3)_3$.
Properties: Colorless crystals, or white crystalline powder. Insipid taste; odorless; soluble in warm alcohol and acetone; slightly soluble in ether and alcohol; insoluble in water. M.p. 134–136°C.

ethyltrichlorosilane $C_2H_5SiCl_3$.
Properties: Colorless liquid. B.p. 99.5°C; sp. gr. 1.236 (25/25°C); refractive index (n 25/D) 1.4257; flash point 72°F (open cup). Soluble in benzene, ether, heptane, perchloroethylene. Readily hydrolyzed with liberation of hydrochloric acid.
Derivation: By reaction of ethylene and trichlorosilane in the presence of a peroxide catalyst.
Hazard: Highly toxic; strong irritant. Flammable, dangerous fire risk. May form explosive mixtures with air.
Uses: Intermediate for silicones.
Shipping regulations: (Rail) Red label. (Air) Flammable Liquid label. Not acceptable on passenger planes.

ethyl urethane. See urethane.

ethyl valerate. See ethyl isovalerate.

ethyl vanillate $(CH_3O)HOC_6H_3COOC_2H_5$.
Properties: Solid; m.p. 44°C; b.p. 291–293°C. Insoluble in water; soluble in alcohol and ether.
Uses: Food preservative; medicine; sunburn preventive.

ethyl vanillin (3-ethoxy-4-hydroxybenzaldehyde) $OHC_6H_3(OC_2H_5)CHO$.
Properties: Fine, white crystals; intense odor of vanillin; affected by light; m.p. 76.5°C; soluble in alcohol, chloroform, and ether; slightly soluble in water. Combustible; low toxicity.
Grades: N.F.; F.C.C.
Containers: 100-lb fiber drums.
Uses: Flavors; replacement or fortifier of vanillin.

ethyl vinyl ether. See vinyl ethyl ether.

ethylxanthic acid. See xanthic acids.

ethylzinc. See diethylzinc.

ethyne. See acetylene.

ethynylation. Condensation of acetylene with a reagent such as an aldehyde to yield an acetylenic derivative. The best example is the union of formaldehyde and acetylene to produce butynediol.

1-ethynylcyclohexanol $HC{\vdots}CC_6H_{10}OH$.
Properties: Colorless, low-melting solid with sweet odor. M.p. 30–31°C; b.p. 180°C; sp. gr. (20/20°C) 0.967. Slightly soluble in water. Combustible.
Containers: Cans; drums.
Hazard: May be highly toxic and irritant to eyes and skin.
Uses: Stabilization of chlorinated organic compounds; intermediate; corrosion inhibitor for mineral acids.

1-ethynylcyclohexyl carbamate. See ethinamate.

ethynylestradiol. See ethinylestradiol.

ethythrin. Ethyl analog of allethrin, used as insecticide with applicatoins similar to allethrin (q.v.). Moderately toxic. See also barthrin, cyclethrin and furethrin.

Eu Symbol for europium.

eucalyptol (cineol; cajeputol) $C_{10}H_{18}O$.
Properties: Colorless essential oil; a terpene ether having a camphor-like odor and pungent, cooling, spicy taste. Slightly soluble in water; miscible with alcohol, chloroform, ether, glacial acetic acid and fixed or volatile oils. Sp. gr. 0.921–0.923 (25°C); b.p. 174–177°C; congealing point not below 0°C; refractive index 1.4550–1.4600 (20°C). Combustible.
Derivation: By fractionally distilling eucalyptus oil, followed by freezing. The oil is imported from Spain, Portugal, and Australia.
Grades: Technical; N.F.
Containers: Drums.
Uses: Pharmaceuticals (cough syrups, expectorants); flavoring; perfumery.

eucryptite. A lithium silicate mineral ($LiAlSiO_4$) containing up to 4.8% lithia. Used in glass manufacture and as a source of lithium.

eugenic acid. See eugenol.

eugenol (4-allyl-2-methoxyphenol; caryophyllic acid; eugenic acid) $C_3H_5C_6H_3(OH)OCH_3$.
Properties: Colorless or yellowish liquid; oily; becomes brown in air; spicy odor and taste. Soluble in alcohol, chloroform, ether, and volatile oils; very slightly soluble in water. Sp. gr. 1.064–0.070; b.p. 253.5°C; refractive index 1.5400–1.5420 (20°C); optically inactive. Low toxicity; combustible.
Derivation: By extraction of clove oil with aqueous potash, liberation with acid and rectification in a stream of carbon dioxide.
Grades: Technical; U.S.P.; F.C.C.
Uses: Perfumes; essential oils; medicine (active germicide); production of isoeugenol for the manufacture of vanillin; flavoring.

eugenyl mthyl ether. See methyl eugenol.

"Eulan" CN.[307] Trademark for a permanent anionic moth-proofing agent, consisting of sodium pentachlorodihydroxytriphenylmethanesulfonate.
Properties: Fine powder. Compatible with acid or chrome dyes as well as with anionic and non-ionic surfactants. Incompatible with basic dyes, cationic surfactants and leveling agents based on albumin decomposition products.

"Eulysine" A.[307] Trademark for an organic amine used as a dyeing assistant.

"Eunaphtol" AS.[307] Trademark for a naphthol dyeing assistant, composed of a modified lignin-sulfonate.

"Eureka."[28] Trademark for a soldering flux crystal composition based on zinc chloride and ammonium chloride.

"Euresol."[9] Trademark for resorcinol acetate (q.v.).

europia. See europium oxide.

europium Eu Atomic number 63; one of the lanthanide or rare-earth elements of the cerium subgroup. Atomic weight 151.96. Valences 2,3. Two stable isotopes.
Properties: Steel-gray metal, difficult to prepare. Quite soft and malleable (DPH = 20); m.p. 826°C; b.p. 1489°C (approx); sp. gr. 5.24. Oxidizes rapidly in air; may burn spontaneously. Most reactive of the rare earth metals; liberates hydrogen from water. Toxicity unknown.
Derivation: Reduction of the oxide with lanthanum or misch metal.
Source: See rare-earth minerals.
Grade: High purity (ingots; lumps).
Hazard: Highly reactive; may ignite spontaneously in powder form.
Uses: Neutron absorber in nuclear control; doping agent in plastic lasers. Color TV phosphors to activate yttrium; phosphors in postage stamp glues to permit electronic recognition of first-class mail.

europium chloride $EuCl_36H_2O$. Yellow needles, soluble in water; sp. gr. 4.89 (20°C); m.p. 850°C. Obtained by treating the oxide with hydrochloric acid.

europium fluoride EuF_33H_2O. Crystals; m.p. 1390°C; b.p. 2280°C. Insoluble in water and dilute acids.

europium nitrate $Eu(NO_3)_3 \cdot 6H_2O$. Colorless to pale pink crystals; m.p. 85°C (sealed tube); soluble in water. Obtained by treating the oxide with nitric acid.

europium oxalate $Eu_2(C_2O_4)_3 \cdot 10H_2O$. White powder; insoluble in water; slightly soluble in acids. Grades: 25–50% and 99.8% europium salt. Impure grades may be colored.

europium oxide (europia) Eu_2O_3.
Properties: Pale rose powder; sp. gr. 7.42; insoluble in water, but soluble in acids to give the corresponding salt.
Derivation: Calcination of the oxalate; solvent extraction or liquid ion-exchange processes.
Grades: 25–50%; 99.9%.
Uses: Nuclear-reactor control rods; especially in red- and infrared-sensitive phosphors.

europium sulfate $Eu_2(SO_4)_3$, anhydrous and $8H_2O$. Colorless to pale pink crystals; sp. gr. (anhydrous) 4.99 (20°C). Hydrate loses $8H_2O$ at 375°C, and is slightly soluble in water.

eutectic. The lowest melting point of an alloy or solution of two or more substances (usually metals) that is obtainable by varying the percentage of the components. Eutectic alloys are relatively few; they are the particular alloys that have definite and minimum melting points compared with other combinations of the same metals.
See also alloy, fusible.

eutrophication. The unintentional enrichment or "fertilization" of either fresh or salt water by chemical elements or compounds present in various types of industrial wastes. Phosphates and nitrogenous compounds in detergent and chemical processing wastes are particularly effective eutrophying agents, but they supply nutrients to algae, which proliferate so abundantly that a large proportion die for lack of light; their decomposition depletes the water of its dissolved oxygen and thus causes the death of fish and other marine life. One process for removing phosphates involves addition of a metal ion source to waste effluents to insolubilize dissolved phosphates, after which the particles are agglomerated by anionic polymers.
See also: algae; nitrilotriacetic acid (note).

eV. Abbreviation for electron-volt.

EVA. Abbreviation for ethylene-vinyl acetate copolymer.

"Evanacid 3CS."[312] Trademark for carboxymethylmercaptosuccinic acid (q.v.).

"Evanohm."[155] Trademark for an alloy of 75% nickel, 20% chromium, 2.5% aluminum and 2.5% copper.
Forms: Wire; insulated wire.
Use: Precision-wound resistors.

Evans blue $C_{34}H_{24}N_6Na_4O_{14}S_4$. A diazo dye used in medicine to measure blood plasma volume.
Properties: Bluish green or brown iridescent powder. Odorless and hygroscopic. Very soluble in water; very slightly soluble in alcohol; practically insoluble in benzene, carbon tetrachloride, chloroform, and ether. pH (0.5% solution) 5.5–7.5.

evaporation. The change of a substance from the solid or liquid phase to the gaseous or vapor phase. Evaporation of solids (ice, snow, dry ice, and a few others) occurs at or below room temperature; the substance does not go through a liquid phase and the phenomenon is called sublimation (q.v.). The rate of evaporation of liquids varies with their chemical nature, and with the temperature; in general, organic liquids (benzene, gasoline) evaporate at lower temperatures and higher rates than water. The thermal energy required to vaporize a given volume of a liquid is known as its latent heat of vaporization; it remains in the vapor (steam, in the case of water heated to its boiling point) and is released when the vapor condenses. For water, this latent heat value is 540 cal/g. In chemical processing installations requiring a series of evaporations and condensations, the units are set up in series, and the latent heat of vaporization from one unit is utilized to supply energy for the next. Such units are called "effects" in engineering parlance, as, e.g., a triple-effect evaporator. See also distillation.

EVE. Abbreviation for ethyl vinyl ether. See vinyl ethyl ether.

"Evenglo."[11] Trademark for a durable lightweight polystyrene, varying in its properties of light transmission, degree of diffusion and color. Used in lighting applications.

"Everdur."[324] Trademark for a group of five copper-silicon alloys, with compositions adjusted to hot and cold working, hot forging, welding, free machining, and for ingots for remelting and casting.

"Everflex."[311] Brand name for polyvinyl acetate copolymer emulsions; used in paints, paper coating, and acoustic tile coatings.

"Evipal."[162] Trademark for hexobarbital (q.v.).

exchange reaction. A chemical reaction in which atoms of the same element in two different molecules or in two different positions in the same molecule transfer places. Exchange reactions are usually studied with the aid of a tracer or tagged atom.

excipient. A natural inert and somewhat tacky material used in the pharmaceutical industry as a binder in tablets, etc. Commonly used materials are gum arabic, honey, and beet pulp.

excited state. A higher than normal energy level (vibrational frequency) of the electrons of an atom, radical, or molecule resulting from absorption of photons (quanta) from a radiation source (arc, flame, spark, etc.) in any wavelength of the electromagnetic spectrum. X-ray, ultraviolet, visible, infrared, microwave, and radiofrequencies are used for excitation in various types of spectroscopy. When the energizing source is removed or discontinued the atom or molecule returns to its normal or stable state either by emitting the absorbed photons or by transferring the energy to other atoms or molecules. The increased vibrational activity of the atom or molecule yields line or band spectra characteristic of its structure, thus permitting identification. Photochemical reactions are induced by excited chemical entities, which are also responsible for the phenomena of luminescence (phosphorescence and fluorescence). See also: spectroscopy; photochemistry; absorption (2).

exhaust emission control. See air pollution.

"Exkin 1,2,3"[74] Trademark for a series of anti-skinning agents of the volatile oxime type.
Use: Paints.

exo-. A prefix used in chemical names to indicate attachment to a side chain rather than to a ring. See also endo-.

"Exobond."[244] Trademark for a compound of sodium silicate-ferrosilicon. Used as a binder in a self-setting exothermic process, in moldings.

"Exolvent."[141] Trademark for a solvent made from aliphatic hydrocarbons.
Hazard: Probably flammable.

"Exon."[35] Trademark for a series of polyvinyl resins, compounds and latexes composed of polymers and copolymers based on vinyl chloride.
Containers: Up to tank cars and trucks.
Uses: Molding, sheeting film, strip coatings, protective coatings, extrusions, paints, ink, adhesives, plastisols, flooring, phonograph records.

exothermic. A process or chemical reaction which is accompanied by evolution of heat, for example, combustion reactions.

exotic. A term applied to materials of various functional types that have extra power, and that are often derived from unusual sources. Examples are rocket propellants derived from boron hydrides and certain special-purpose solvents.

expander. (1) A mixture of lampblack, barium sulfate, and an organic material usually derived from lignin that increases the capacity of storage batteries, presumably by coating the anode and thus preventing the deposit of lead sulfate on the underlying lead metal.

(2) A substance used in medicine as a substitute for blood plasma, e.g., dextran, gelatin, or polyvinylpyrollidone.

explosive, high. A chemical compound, usually containing nitrogen, that detonates as a result of shock or heat (see detonation). Dynamite was the most widely used explosive for blasting and other industrial purposes until 1955, when it was largely replaced by prills-and-oil and slurry types. The former consists of 94% ammonium nitrate prills and 6% fuel oil, (ANFO). Slurry blasting agents (SBA) are based on thickened or gelatinized ammonium nitrate slurries sensitized with TNT, other solid explosives, or aluminum. An unusual type of explosive is represented by acetylides of copper and silver, which are examples of commercially used explosives that contain neither oxygen nor nitrogen.

High explosives vary greatly in their shock sensitivity; most sensitive are mercury fulminate and nitroglycerin, while TNT and ammonium nitrate are comparatively difficult to detonate, requiring the use of blasting caps or similar activating device. For further information, refer to Institute of Makers of Explosives, 420 Lexington Ave., New York.

explosive, initiating. An explosive composition used as a component of blasting caps, detonators, and primers. They are highly sensitive to flame, heat, impact, or friction.Examples are lead azide; silver acetylide, mercury fulminate, diazodinitrophenol, nitrosoguanidine, lead styphnate and pentaerythritol tetranitrate.

explosive limits. The range of concentration of a flammable gas or vapor (% by volume in air) in which explosion can occur upon ignition in a confined area. Explosive limits for some common substances are:

	Lower (%)	Upper (%)
carbon disulfide	1	50
benzene	1.5	8
methane	5	15
butadiene	2	11.5
butane	1.9	8.5
propane	2.4	9.5
natural gas	3.8	17
hydrogen	4	75
acetylene	2.5	80

explosive, low. An explosive which deflagrates rather than detonates, such as black powder. See deflagration.

explosive, permissible. Explosives approved for use in coal mines. Usually they are modified dynamites.

expression. Removal of a liquid from a solid by hydraulic pressure, as in manufacture of vegetable oils from seeds, rinds, or meal cake. Worm devices similar to extrusion machines are also used for this purpose; they are called expellers.

extender. A low-gravity material used in paint, ink, plastic and rubber formulations chiefly to reduce cost. Extenders include diatomaceous earth, wood flock, mineral rubber, liquid asphalt, etc. Microscopic droplets of water fixed permanently in a plastic matrix have been introduced as an efficient extender for polyester resins. In the food industry the term refers to certain extruded proteins, especially those derived from soybeans, which are used in meat products to provide equivalent nutrient values at lower cost. Made from defatted soy flour, they are often called textured proteins.
See also diluent (2); filler.

extinguishing agent. See fire extinguishment.

extraction, solvent. See solvent extraction.

extractive distillation. A variety of distillation that always involves the use of a fractionating column, and is characterized by use of a purposely added substance which modifies the vaporization characteristics of the materials undergoing separation, to make them easier to separate. The additive is often called a solvent, and is usually chosen to be much less volatile than any of the substances being separated. It is added to the downflowing liquid reflux stream near the top of the column, and is removed from the still pot or reboiler at the base. The addition of furfural to mixtures of butadiene and butene hydrocarbons to separate the butadiene more easily is an example of extractive distillation.

extractive metallurgy. That portion of metallurgical science devoted to the technology of mining and processing of metals and their ores.

extreme-pressure additive (EP additive)
(1) Material added to cutting fluids to impart high film strength. They are mainly sulfur, chlorine, and occasionally phosphorus compounds. Actual conditions, amounts, etc. are proprietary.
(2) Lubricating oil and grease additives that prevent metal-to-metal contact in highly loaded gears. (See "Aroclor"). Some react with the metal gears to form a protective coating. Saponified lead salts are often used.

extrusion. A fundamental processing operation in many industries, in which a material is forced through a metal forming die, followed by cooling or chemical hardening. The material may be liquid (molten glass or a polymer dispersion); a viscous polymer, as in injection molding; a semisolid mass such as a rubber or plastic mixture; or a hot metal billet. High-viscosity materials are fed into a rotating screw of variable pitch, which forces it through the die with considerable pressure; a ram is used for metals at temperatures from 1000 to 2000°F. Film is made by passing a low-viscosity mixture through a narrow slit. Molten glass and polymer suspensions are forced through a nozzle having a tiny orifice (spinneret); the latter are hardened after extrusion by immersion in a bath of formaldehyde or similar agent. Food items (spaghetti, etc.) are also extruded. Extrusion involves rheological principles of some complexity, critical factors being viscosity, temperature, flow rate, and die design. See also injection molding.

F Symbol for fluorine; the molecular formula is F_2.

"FA."[224] Trademark for furfuryl alcohol (q.v.).

"Fabrez."[36] Trademark for cyclic ethylene urea and urea-formaldehyde resins.

fabric. A textile structure composed of mechanically interlocked fibers or filaments; it may be randomly integrated (nonwoven) or closely oriented by warp and filler strands at right angles to each other (woven). While the word usually refers to wool, cotton, or synthetic fibers, fabrics can also be made of glass fiber and other inorganic materials, such as graphite.

fabrication. The molding, forming, machining, assembly and finishing of metals, rubber, and plastics into enduse products. In the paper industry the term "converting" is used in this sense.

"Fabrikoid."[28] Trademark for pyroxylin-coated fabrics which are water-resistant, soap- and water-washable.
Uses: Book binding; luggage.

"Fabrilite."[28] Trademark for vinyl coated fabrics and selected vinyl compounds without fabric backing.
Uses: Pocketbooks, bags, upholstery, etc.

F acid. See 2-naphthol-7-sulfonic acid and 2-naphthylamine-7-sulfonic acid.

factice (vulcanized oil). A soft mealy material made by reaction of sulfur or sulfur chloride with a vegetable oil.
Uses: Erasers; rubber goods (bath spray tubing, etc.) to give soft "hand."

FAD. Abbreviation for flavin adenine dinucleotide (q.v.).

Fahrenheit. The scale of temperature in which 212° is the boiling point of water at 760 mm Hg, and 32° is the freezing point of water. See centigrade for method of converting from Fahrenheit to centigrade. See absolute temperature for converting Fahrenheit to absolute Rankine.

"Fairprene."[28] Trademark for a variety of products including:
Industrial cements made from synthetic rubber compounds for cementing rubber sheets and coated fabrics to leather, fabric, paper, metal and wood.
Synthetic rubber coated fabrics and sheet stock including silicone rubber and "Viton" A fluoroelastomer. Used for industrial applications in the automotive, aviation, gas, chemical, petroleum and marine fields.

fallout. Deposition upon the earth of the radioactive particles resulting from a nuclear explosion. The most dangerous component is strontium-90 (q.v.).

"Falone."[248] Trademark for tris(2,4-dichlorophenoxyethyl) phosphite.
Properties: Viscous amber liquid; sp. gr. 1.434; m.p. 70–72°C; soluble in benzene, xylene and aromatic hydrocarbons; insoluble in water. Available as an emulsifiable concentrate and a granular solid.
Use: A pre-emergence herbicide.

famphur. Generic name for O,O-dimethyl O-[para-(dimethylsulfamoyl)phenyl]phosphorothioate. $(CH_3O)_2P(S)OC_6H_4SO_2N(CH_3)_2$.
Properties: Crystalline powder; m.p. 55°C; very soluble in chloroform and carbon tetrachloride; slightly soluble in water.
Hazard: Highly toxic. Cholinesterase inhibitor. Use may be restricted.
Use: Insecticide.
Shipping regulations: (Rail, Air) Organic phosphate, solid, n.o.s. Poison label. Not accepted on passenger planes.

"Fanal."[307] Trademark for phosphotungstic lakes. Used for printing inks. Characterized by brilliancy of shade and good fastness to light.

"Fanox."[51] Trademark for a line of low-sulfur nonstaining cutting fluids for ferrous and nonferrous metals.

Faraday, Michael (1791–1867). A native of England, Faraday did more to advance the science of electrochemistry than any other scientist. A profound thinker and accurate experimentalist and observer, he was the first to propound correct ideas as to the nature of electrical phenomena, not only in chemistry but in other fields. His contributions to chemistry include the basic laws of electrolysis, electrochemical decomposition (the basis of corrosion of metals) as well as of battery science and electrometallurgy. His work in physics led to the invention of the dynamo. Faraday was in many respects the exemplar of a true scientist, combining meticulous effort and interpretive genius.

faraday. The quantity of electricity that can deposit (or dissolve) one gram-equivalent weight of a substance during electrolysis (about 96,500 coulombs).

farnesol. Generic name for 3,7,11-trimethyl-2,6,10-dodecatrienol, $C_{15}H_{25}OH$.
Properties: Colorless liquid; delicate floral odor. Soluble in 3 vols of 70% alcohol. Sp. gr. 0.885 (15°C); b.p. 145–146°C (3 mm). Combustible. Low toxicity.
Derivation: Found in nature in many flowers and essential oils, such as cassie, neroli, cananga, rose, balsams, ambrette seed.
Uses: Perfumery; flavoring; insect hormone.

fast. (1) Descriptive of a dye or pigment whose color is not impaired by prolonged exposure to light, steam, high temperature or other environmental conditions. Inorganic pigments are normally superior in this respect to organic dyes.

(2) In nuclear technology, the term refers to neu-

trons moving at the speed at which they emerge from a ruptured nucleus, as opposed to "slow" or thermal neutrons whose speed has been reduced by impinging on a neutral substance called a moderator (q.v.).

fat. A glyceryl ester of higher fatty acids such as stearic and palmitic. Such esters and their mixtures are solids at room temperatures and exhibit crystalline structure. Lard and tallow are examples. There is no chemical difference between a fat and an oil, the only distinction being that fats are solid at room temperature and oils are liquid. The term "fat" usually refers to triglycerides specifically, whereas "lipid" is all-inclusive. See also lipid.

fat dyes. Oil-soluble dyes for candles, wax, etc.

fatliquoring agent. An oil-in-water emulsion usually made from raw oils such as neatsfoot, cod, etc., made soluble by dispersing agents such as sulfonated oils.
Uses: Leather processing, to replace natural oils removed from hides by tanning operations.
See also neatsfoot oil; emulsion.

fat splitting. See hydrolysis.

fatty acid. A carboxylic acid derived from or contained in an animal or vegetable fat or oil. All fatty acids are composed of a chain of alkyl groups containing from 4 to 22 carbon atoms (usually even-numbered) and characterized by a terminal carboxyl radical -COOH. The generic formula for all above acetic is $CH_3(CH_2)COOH$ (the carbon atom count includes the radical). Fatty acids may be saturated or unsaturated (olefinic), and either solid, semisolid, or liquid. They are classed among the lipids (q.v.), together with soap and waxes.

Saturated. The most important of these are butyric (C_4), lauric (C_{12}), palmitic (C_{16}) and stearic (C_{18}). They have a variety of special uses (see specific entry). Stearic acid leads all other fatty acids in industrial use, primarily as a dispersing agent and accelerator activator in rubber products, and in soaps. *Note:* There is still no conclusive proof that increase in body cholesterol as a result of high dietary intake of animal-derived saturated fats or fatty acids is causatively related to atherosclerosis (see also cholesterol).

Unsaturated. These are usually vegetable-derived and consist of alkyl chains containing 18 or 22 carbon atoms with the characteristic end group -COOH. Most vegetable oils are mixtures of several fatty acids or their glycerides; the unsaturation accounts for the broad chemical utilty of these substances, especially of drying oils (q.v.). The most common unsaturated acids are oleic, linoleic, and linolenic (all C_{18}). Safflower oil is high in linoleic acid; peanut oil contains 21%. Olive oil is 83% oleic acid. Palmitoleic acid is abundant in fish oils. Aromatic fatty acids are now available (see phenylstearic acid). *Note:* Linoleic, linolenic and arachidonic acids are called essential fatty acids by biochemists, because they are necessary nutrients that are not synthesized in the animal body.
Uses: Special soaps; heavy-metal soaps; lubricants; paints and lacquers (drying oils); candles; salad oil; shortening; synthetic detergents; cosmetics; emulsifiers.
For further details refer to Fatty Acid Producers' Council, 485 Madison Ave., New York.

fatty acid enol ester. A fatty acid reacted with the enolic form of acetone for the purpose of increasing the chemical reactivity of the acid. Stearic acid (18-carbon) combined with acetone (3-carbon) gives isopropenyl stearate (21-carbon). This is effective in making the fatty stearoyl group available for synthesis of polymers, medicinals, and the like. See also fatty ester.

fatty acid pitch. A by-product residue from (a) soap stock and candle stock manufacture; (b) refining of vegetable oils; (c) refining of refuse greases; (d) refining of wool grease.
Properties: Dark brown to black; properties analogous to complex hydrocarbons; contains fixed carbon (5% to 35%). Soluble in naphtha and carbon disulfide.
Uses: Manufacture of black paints and varnishes; tarred papers; printers' rolls; rubber filling agent; impregnating agent; electrical insulations; marine caulking; waterproofing; sealant.

fatty alcohol. A primary alcohol (from C_8 to C_{20}), usually straight-chain. High molecular weight alcohols are produced synthetically, by the Oxo and Ziegler processes. Those from C_8 to C_{11} are oily liquids; above C_{11} they are solids. Other methods of production are (a) reduction of vegetable seed oils and their fatty acids with sodium, (b) catalytic hydrogenation at elevated temperatures and pressures and (c) hydrolysis of spermaceti and sperm oil by saponification and vacuum fractional distillation. The more important commercial saturated alcohols are octyl, decyl, lauryl, myristyl, cetyl and stearyl. The commercially important unsaturated alcohols such as oleyl, linoleyl and linolenyl are also normally included in this group. The odor tends to disappear as the chain length increases.
Uses: Solvent for fats, waxes, gums and resins; pharmaceutical salves and lotions; lube oil additives; detergents and emulsifiers; textile antistatic and finishing agents.

fatty amine. A normal aliphatic amine derived from fats and oils. May be saturated or unsaturated, primary, secondary or tertiary, but the alkyl groups are straight-chain and have an even number of carbons in each. The length varies from 8 to 22 carbon atoms.
Derivation: Fatty acids are treated with ammonia and heated to form fatty acid amides which are converted to nitriles and reduced to the amine.
Uses: Organic bases; soaps; plasticizers; tire cords; fabric softeners; water-resistant asphalt; hair conditioners; cosmetics; medicinals.

fatty ester. A fatty acid with the active hydrogen replaced by the alkyl group of a monohydric alcohol. The esterification of a fatty acid, RCOOH, by an alcohol, R'OH, yields the fatty ester RCOOR'. The most common alcohol used is methanol, yielding the methyl ester, $RCOOCH_3$. The methyl esters of fatty acids have higher vapor pressures than the corresponding acids and are distilled more easily.

fatty nitrile (RCN). An organic cyanide derived from a fatty acid.
Derivation: Fatty acids are treated with ammonia and heated to form fatty acid amides, which are converted to nitriles.
Hazard: May be toxic.
Uses: Intermediates for fatty amines; lube oil additives; plasticizers.

"Faxam."[51] Trademark for a series of general-purpose, paraffin-base oils that serve as lubricants, in process applications, and as components in proprietary formulations.

FCC. (1) Abbreviation for Food Chemicals Codex, a publication giving specifications and test methods for chemicals used in foods.

(2) Abbreviation for fluid-cracking catalyst, as used in the petroleum refining industry. Examples are powdered silica-alumina, in which alumina is impregnated with dry synthetic silica gel, and various natural clays impregnated with alumina.

FDA. Abbreviation for Food and Drug Administration (q.v.).

FD&C colors. A series of colorants permitted in food products, marking inks, etc., certified by the FDA. Among the more important are the following:

Blue No. 1: Disodum salt of 4-((4-(N-ethyl-*p*-sulfobenzylamino)-phenyl)-(2-sulfoniumphenyl)-methylene)-(1-(N-ethyl-N-*p*-sulfobenzyl)-$\Delta^{2,5}$-cyclohexadienimine).

Blue No. 2: Disodium salt of 5,5′-indigotindisulfonic acid.

Green No. 3: Disodium salt of 4-((4-(N-ethyl-*p*-sulfobenzylamino)-phenyl-(4-hydroxy-2-sulfoniumphenyl)-methylene)-(1-(N-ethyl-N-*p*-sulfobenzyl)-$\Delta^{2,5}$-cyclohexadienimine).

Red No. 2: Trisodium salt of 1-(4-sulfo-1-naphthylazo)-2-naphthol-3,6-disulfonic acid. The largest-volume food color in commercial use. A suspected carcinogen. Use prohibited by FDA in 1975. Red No. 40 is permissible substitute.

Red No. 3: Disodium salt of 9-*o*-carboxyphenyl-6-hydroxy-2,4,5,7-tetraiodo-3-isoxanthone.

Red No. 4: Disodium salt of 2-(5-sulfo-2,4-xylylazo)-naphthyl-4-sulfonic acid. Use in foods prohibited by FDA.

Violet No. 1: Monosodium salt of 4-((4-(N-ethyl-*p*-sulfobenzylamino)-phenyl)-(4(N-ethyl-*p*-sulfoiumbenzylamino)-phenyl)-methylene)-(N,N,dimethyl-$\Delta^{2,5}$-cyclohexadienimine). Use prohibited by FDA in 1973 and by USDA in 1976.

Yellow No. 5: Trisodium salt of 3-carboxy-5-hydroxy-1-*p*-sulfophenyl-4-sulfophenylazopyrazole.

Yellow No. 6: Disodium salt of 1-*p*-sulfophenylaze-2-naphthol-6-sulfonic acid.

See also food color; food additive.

Fe Symbol for iron.

Federal Trade Commission (FTC). A consumer protection agency.

feedstock. Gaseous or liquid petroleum-derived hydrocarbons or mixtures of hydrocarbons from which gasoline, fuel oil, and petrochemicals are produced by thermal or catalytic cracking. It is also called charging stock. Feedstocks commonly used include ethane, propane, butane, butene, benzene, toluene, xylene, naphtha and gas oils.

"FE Fatty Esters."[487] Trademark for a series of esters derived from straight chain (normal) even-carbon alcohols ranging from C_8 (octyl) to C_{18} (stearyl) and straight chain (normal) fatty acids ranging from C_8 to C_{18} and including mixtures of these esters. Example: FE 12:18 is dodecyl octadecanoate.

Uses: Cosmetics, lubricants, softeners, water repellents, paper coatings, protective coatings and wax substitutes.

Fehling's solution. A reagent used as a test for sugars aldehydes, etc. It consists of two solutions, one of copper sulfate, the other of alkaline tartrate, which are mixed just before use. Benedict's modification is a one-solution preparation. For details, see Book of Methods, Association of Official Analytical Chemists.

feldspar (potassium aluminosilicate). General name for a group of sodium, potassium, calcium, and barium aluminum-silicates. Commercially, feldspar usually refers to the potassium feldspars with the formula $KAlSi_3O_8$, usually with a little sodium. Noncombustible; nontoxic, except as fine-ground powder.

Grades: Usually based on silicon dioxide content, potassium-sodium ratio, iron content, and fineness of grinding.

Occurrence: North Carolina, Colorado, New Hampshire, South Dakota, California, Arizona, Wyoming, Virginia, Texas.

Uses: Pottery, enamel, and ceramic ware; glass; soaps; abrasive; bond for abrasive wheels; cements and concretes; insulating compositions; fertilizer; poultry grit; tarred roofing materials.

"Feltmetal."[551] Trademark for randomly interlocked structure of metal fibers which are sintered together to provide strong, permeable materials with precisely controlled properties.

Uses: Acoustical applications where high noise levels, high velocity gas flows, elevated temperatures, corrosive atmospheres or other situations exist which preclude the use of conventional material.

FEMA. Abbreviation of Flavor Extract Manufacturers Association. It makes recommendations to FDA on safety aspects of flavoring materials.

fenchone $C_{10}H_{16}O$.

Properties: Oil with camphor-like odor. Sp. gr. 0.9465 (19°C); b.p. 193°C. Soluble in ether; insoluble in water. Combustible; nontoxic.

Derivation: A ketone found (a) as dextrofenchone in oil of fennel; (b) as levofenchone in oil of thuja.

Use: Flavoring.

fenchyl alcohol (fenchol; 2-fenchanol; 1-hydroxyfenchane) $C_{10}H_{17}OH$; structural formula:

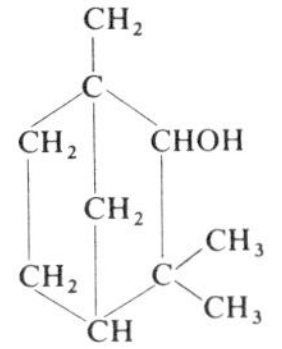

CH_2 — C — CH_2, CHOH — CH_2 — CH_2, C(CH_3)(CH_3) — CH

Properties: Colorless, oily liquid (*d* and *l* forms) or solid (*dl* form). Sp. gr. approx. 0.96; b.p. 201°C; m.p. 39°C; refractive index 1.473.

Derivation: Pine oil; fennel oil; also made synthetically.

Uses: Solvent; organic intermediate; odorant; flavoring.

fenoprop. See silvex.

fenson. Generic name for para-chlorophenyl benzenesulfonate, $ClC_6H_4OSO_2C_6H_5$.

Properties: Colorless crystals; m.p. 61–62°C. Soluble in organic solvents; insoluble in water.

Hazard: May be toxic.

Use: Acaricide.

fenthion. Generic name for O,O-dimethyl O-[4-(methylthio)-meta-tolyl] phosphorothioate $(CH_3O)_2P(S)OC_6H_3(CH_3)SCH_3$.

Properties: Brown liquid; b.p. 105°C (0.01 mm); insoluble in water; soluble in most organic solvents.

Hazard: Highly toxic by ingestion, inhalation, skin absorption. Use may be restricted. Cholinesterase inhibitor.

Use: Insecticide; acaricide.

Shipping regulations: Organic phosphate, liquid,

n.o.s., (Rail, Air) Poison label. Not acceptable on passenger planes.

fenuron. Generic name for 3-phenyl-1,1-dimethylurea, $C_6H_5NHCON(CH_3)_2$.
Properties: White crystalline solid; almost insoluble in water (3850 ppm at 25°C); sparingly soluble in hydrocarbon solvents. Stable towards oxidation and moisture; m.p. 127–129°C.
Use: Weed and brush killer.

"Feosol."[71] Trademark for medicines containing ferrous sulfate.

FEP resin. Abbreviation for fluorinated ethylene-propylene resin (q.v.).

"Feran."[197] Trademark for a nitrogen fertilizer solution containing ammonium nitrate and water. Used in manufacturing solid fertilizers and for direct application.

ferbam. Generic name for ferric dimethyldithiocarbamate, $[(CH_3)_2NCSS]_3Fe$.
Properties: Black or dark colored, fluffy powder; decomposes above 180°C; usually readily dispersible but very slightly soluble in water; pH of saturated solution 5.0.
Derivation: By addition of carbon disulfide to an alcoholic solution of dimethylamine and precipitation with a ferric salt.
Grades: 76% wettable powder; 87% technical powder.
Hazard: Irritant to eyes and mucous membranes. Tolerance, 15 mg per cubic meter of air.
Use: Fungicide.

"Ferberk."[81] Trademark for ferbam (q.v.).

"Fergon."[508] Trademark for ferrous gluconate (q.v.).

"Fermate."[28] Trademark for a wettable powder containing 76% ferbam (q.v.).

"Fermentase."[114] Trademark for a diastatic malt syrup; used in baking.

fermentation. A chemical change induced by a living organism or enzyme (q.v.), specifically bacteria or the microorganisms occurring in unicellular plants such as yeast, molds, or fungi. The reaction usually involves the decomposition of sugars and starches to ethyl alcohol and carbon dioxide, the acidulation of milk, or the oxidation of nitrogenous organic compounds. The basic reaction is $C_6H_{12}O_6 \xrightarrow[\text{alyst}]{\text{cat-}} 2C_2H_5OH + 2CO_2$. Enzymes are usually involved in such reactions; with yeast, the effective enzyme is zymase. Fermentation is essential in the preparation of breads and other food products, in the manufacture of beer, wine, and other alcoholic beverages, as well as of citric acid, gluconic acid, sodium gluconate, and synthetic biopolymers (q.v.). Much of the industrial alcohol used in the U.S. is made by fermentation of blackstrap molasses, a by-product of sugar manufacture. Antibiotics are produced by various forms of microorganisms active in molds, especially bacteria and Actinomycetes. The activated sludge process for sewage digestion is a form of fermentation. A continuous fermentation process for deriving edible protein from petroleum has been introduced. Fermentation is also used in making synthetic amino acids. Research in this field is being directed toward conversion of agricultural, urban and animal wastes to fuel oil by fermentation processes. See also yeast, sewage sludge, antibiotics, bacteria.

fermentation alcohol. See ethyl alcohol.

"Fermex."[173] Trademark for a diastatic (protease-amylase) enzyme supplement for baking fermentation used to improve quality and uniformity of bread and other yeast-raised bakery products.

Fermi, Enrico (1901–1954). Italian physicist who later became a U.S. citizen. He developed a statistical approach to fundamental problems of physical chemistry based on Pauli's exclusion principle. He discovered induced or artificial radioactivity resulting from neutron impingement, as well as slow or thermal neutrons. He was Professor of Physics at Columbia (1939) and was awarded the Nobel Prize in Physics in 1938. He was the first to achieve a controlled nuclear chain reaction, directed the construction of the first nuclear reactor at University of Chicago (1942), and worked on the atomic bomb at Los Alamos. He also carried on fundamental research on subatomic particles using sophisticated statistical techniques. Element 100 (fermium) is named after him.

fermium (element 100) Fm atomic weight; 254; valence +3. Half-life 3 hours. A synthetic radioactive element with atomic number 100 discovered in 1952. Fermium has since been prepared in a nuclear reactor by irradiating californium, plutonium, or einsteinium with neutrons, in a cyclotron by bombarding uranium with accelerated oxygen ions, and by other nuclear reactions. The element is named for Enrico Fermi. It has chemical properties similar to those of the rare earth erbium. Isotopes are known with mass numbers 254, 255, and 256. Used in tracer studies. See also actinide element.

ferrate. See ferrite (2).

ferredoxin. An iron-containing protein thought to be involved in photosynthesis as an acceptor of energy-rich electrons from chlorophyll. It occurs in green plants and in bacteria that metabolize elemental hydrogen.

ferric acetate, basic (iron acetate, basic) $Fe(C_2H_3O_2)_2OH$.
Properties: Red powder. Soluble in alcohol and acids; insoluble in water. Combustible.
Derivation: Action of acetic acid on ferric hydroxide with subsequent crystallization.
Uses: Medicine; textile industries.

ferric acetylacetonate $Fe[OC(CH_3){:}CHC(O)CH_3]_3$.
Properties: Crystalline powder; m.p. 179–182°C. Slightly soluble in water, soluble in most organic solvents. Resistant to hydrolysis. A chelating non-ionizing compound. Combustible.
Uses: Moderating and combustion catalyst; solid fuel catalyst; bonding agent; curing accelerator; intermediate.

ferric ammonium alum. See ferric-ammonium sulfate.

ferric ammonium citrate (iron ammonium citrate).
Properties: Thin, transparent, garnet red scales or granules, or as a brownish yellow powder; odorless (or slight ammonia odor); saline, mildly ferruginous taste; deliquescent; affected by light. Soluble in water; insoluble in alcohol. Combustible. Low toxicity.
Derivation: Addition of citric acid to ferric hydroxide, then adding ammonium hydroxide, followed by filtration.
Grade: Technical.
Uses: Medicine; blueprint photography; feed additive.

ferric ammonium oxalate (iron ammonium oxalate; ammonioferric oxalate) $(NH_4)_3Fe(C_2O_4)_3 \cdot 3H_2O$.

Properties: Green crystals. Soluble in water and alcohol; sensitive to light.
Derivation: Interaction of ammonium binoxalate and ferric hydroxide.
Hazard: Moderately toxic and irritant to skin and mucous membranes.
Use: Blueprint photography.

ferric ammonium sulfate (iron ammonium sulfate; ferric ammonium alum; ammonio-ferric sulfate) $FeNH_4(SO_4)_2 \cdot 12H_2O$.
Properties: Lilac to violet efflorescent crystals. Sp. gr. 1.71; m.p. 39–41°C; b.p. ($-12H_2O$) 230°C. Soluble in water; insoluble in alcohol.
Derivation: By mixing solutions of ferric sulfate and ammonium sulfate, followed by evaporation and crystallization.
Uses: Medicine; analytical chemistry; textile dyeing (mordant).

ferric ammonium tartrate (ammonium iron tartrate; iron ammonium tartrate).
Properties: Reddish-brown scales. Transparent. Sweet taste. Soluble in water; insoluble in alcohol.
Use: Medicine.

ferric arsenate $FeAsO_4 \cdot 2H_2O$.
Properties: A green or brown powder; sp. gr. 3.18; decomposes on heating. Insoluble in water; soluble in dilute mineral acids. Nonflammable.
Hazard: Highly toxic by ingestion and inhalation; strong irritant.
Use: Insecticide.
Shipping regulations: (Rail, Air) Poison label.

ferric arsenite. $2FeAsO_3 \cdot Fe_2O_3 \cdot 5H_2O$. A basic salt of variable composition.
Properties: Brownish-yellow powder. Soluble in acids; insoluble in water. Nonflammable.
Hazard: Highly toxic by ingestion and inhalation; strong irritant.
Use: Combined with ammonium citrate (ferric ammonium citrate) and used in medicine.
Shipping regulations: (Rail, Air) Poison label.

ferric benzoate (iron benzoate) $Fe_2(OOCC_6H_5)_6$.
Properties: Brown powder. Slightly soluble in water, alcohol, and ether. Combustible.
Derivation: Interaction of ferric hydroxide and benzoic acid.
Use: Medicine.

ferric bromide (ferric tribromide; iron bromide) $FeBr_3$.
Properties: Dark-red, deliquescent crystals. Soluble in water, alcohol, and ether. M.p., sublimes. Low toxicity.
Derivation: By the action of bromine on iron filings.
Containers: Boxes; glass bottles. Keep cool, well closed, and protected from light.
Uses: Medicine; analytical chemistry; bromine salts.

ferric cacodylate $Fe[(CH_3)_2AsO_2]_3$.
Properties: Yellowish-brown powder; odorless. Moderately soluble in cold water, more so in hot water, less so in alcohol.
Hazard: Highly toxic.
Use: Medicine.
Shipping regulations: (Rail, Air) Arsenical compounds, n.o.s., Poison label.

ferric chloride, anhydrous (ferric trichloride; ferric perchloride; iron chloride; iron trichloride; iron perchloride) $FeCl_3$.
Properties: Black-brown solid; sp. gr. 2.898 (25°C); m.p. 306°C (decomp.) b.p. 319°C; soluble in water, alcohol, glycerol, methanol and ether. Noncombustible.
Derivation: Action of chlorine on ferrous sulfate or chloride.
Grades: Anhydrous 96%; 42° Bé solution, photographic and sewage grades.
Containers: (Solid) barrels; (solution) carboys; tank cars and trucks.
Hazard: Strong irritant to skin and tissue.
Uses: Treatment of sewage and industrial wastes; etching agent for engraving, photography, and printed circuitry; catalyst; mordant; oxidizing, chlorinating and condensing agent, disinfectant, pigment, medicine; feed additive; water purification.
Shipping regulations: (Air) Corrosive label.

ferric chloride hydrate $FeCl_3 \cdot 6H_2O$.
Properties: Orange-yellow very deliquescent crystals. M.p. 37°C; b.p. 280°C; decomposes to yield hydrochloric acid if exposed to moist air or light.
Uses: See ferric chloride, anhydrous.

ferric chromate (iron chromate) $Fe_2(CrO_4)_3$.
Properties: Yellow powder; soluble in acids; insoluble in water and alcohol; soluble in HCl.
Derivation: By adding sodium chromate to a solution of a ferric salt.
Hazard: Toxic by inhalation. May be carcinogenic. Strong irritant. Moderate fire risk by reaction with reducing agents. Tolerance, 0.1 mg per cubic meter of air.
Uses: Metallurgy; ceramics (color); paint pigment.

ferric citrate (iron citrate) $FeC_6H_5O_7$ *2·* $5H_2O$.
Properties: Reddish-brown scales. Keep away from light. Soluble in water; insoluble in alcohol.
Derivation: By the action of citric acid on ferric hydroxide, and crystallization.
Uses: Medicine; blueprint paper.

ferric dichromate (iron dichromate; ferric bichromate) $Fe_2(Cr_2O_7)_3$.
Properties: Reddish-brown granules. Soluble in water and acids.
Derivation: By heating aqueous chromic acid and moist ferric hydroxide.
Hazard: Toxic by inhalation; strong irritant. Moderate fire risk by reaction with reducing agents.
Use: Preparation of pigments.

ferric dimethyldithiocarbamate. See ferbam.

ferric ferrocyanide (iron ferrocyanide; Prussian Blue). Blue pigment described under iron blue (q.v.). Low toxicity.

ferric fluoride (iron fluoride) FeF_3.
Properties: Green crystals; sp. gr. 3.52; soluble in acids and in water; insoluble in alcohol and ether.
Hazard: Toxic; strong irritant. Tolerance (as F), 2.5 mg per cubic meter of air.
Use: Ceramics (porcelain, pottery).

ferric glycerophosphate (iron glycerophosphate) $Fe_2[C_3H_5(OH)_2PO_4]_3 \cdot xH_2O$.
Properties: Yellowish scales, odorless and nearly tasteless; soluble in water; insoluble in alcohol.
Use: Pharmaceutical.

ferric hydroxide (ferric hydrate; iron hydroxide; iron hydrate; iron oxide, hydrated; ferric oxide, hydrated) $Fe(OH)_3$.

Properties: Brown flocculant precipitate which dries to the oxide; sp. gr. 3.4–3.9; m.p., loses water below 500°C; soluble in acids; insoluble in water, alcohol and ether. Noncombustible; low toxicity.
Derivation: Addition of ferrous sulfate solution to ammonia solution.
Uses: Pharmaceutical preparations; water purification; manufacturing pigments; rubber pigment.

ferric hypophosphite (iron hypophosphite) $Fe(H_2PO_2)_3$.
Properties: White or grayish-white powder; odorless; tasteless. Slightly soluble in water; more so in boiling water.
Hazard: Explosion may occur if triturated or heated with nitrates, chlorates or other oxidizing agents.
Use: Medicine.

"Ferriclear."[1] Trademark for a grade of ferric sulfate largely used as a conditioner of alkaline soils.

ferric malate (iron malate) $Fe_2(C_4H_4O_5)_2$.
Properties: Brown, hygroscopic crystals. Soluble in water and alcohol.
Derivation: By the interaction of ferric hydroxide and malic acid.
Use: Medicine.

ferric naphthenate
Properties: A heavy-metal soap. See also soaps (2). Combustible.
Derivation: Fusion method, by heating naphthenic acids with the metallic oxide.
Containers: Drums.
Uses: Conditioning and waterproofing agent, sludge preventive, fungicide and paint drier.

ferric nitrate (iron nitrate) $FeNO_3)_3 \cdot 9H_2O$.
Properties: Violet crystals; sp. gr. 1.684; m.p. 47.2°C; dec. 125°C; soluble in water, and alcohol.
Derivation: Action of concentrated nitric acid on scrap iron or iron oxide, and crystallizing.
Hazard: Dangerous fire risk in contact with organic materials; strong oxidant and irritant.
Uses: Dyeing (mordant for buffs and blacks); medicine; tanning; analytical chemistry.
Shipping regulations: (Rail) Nitrates, n.o.s., Yellow label. (Air) Oxidizer label.

ferric octoate. See soaps (2).

ferric oleate (iron oleate) $Fe(C_{18}H_{33}O_2)_3$.
Properties: Brownish-red lumps. Soluble in alcohol, ether, and acids; insoluble in water. Combustible. See also soap (2).

ferric oxalate $Fe_2(C_2O_4)_3$.
Properties: Pale-yellow amorphous scale or powder; odorless; decomposes on heating to 100°C. Soluble in water and acids; insoluble in alkali. Combustible.
Hazard: Moderately toxic by ingeston.
Uses: Catalyst in making oxygen; silvertone photographic printing papers.

ferric oxide (ferric oxide, red; iron oxide; red iron trioxide; ferric trioxide) Fe_2O_3. See also iron oxide reds.
Properties: Dense, dark-red powder or lumps; sp. gr. 5.12–5.24; m.p. 1565°C; soluble in acids; insoluble in water.
Grades: Technical; 99.5% pure; electronic.
Uses: Metallurgy; gas purification; paint and rubber pigment; component of thermite; polishing compounds; mordant; laboratory reagent; catalyst (parahydrogen); medicine; feed additive; electronic pigments for TV; permanent magnets; memory cores for computers; magnetic tapes.

ferric oxide, yellow. Impure ferric oxide (water and calcium sulfate are the usual impurities). See iron oxide yellows. Natural hydrated yellow iron oxide (limonite) is sometimes referred to as yellow ferric oxide.

ferric perchloride. See ferric chloride.

ferric phosphate (iron phosphate) $FePO_4 \cdot 2H_2O$.
Properties: Yellowish-white powder. Insoluble in water; soluble in acids; sp. gr. 2.87. Low toxicity.
Derivation: by adding a solution of sodium phosphate to a solution of ferric chloride. The product is filtered and then dried.
Uses: Medicine; fertilizers; feed and food additive.

ferric potassium citrate (iron potassium citrate).
Properties: Brown scales. Hygroscopic. Odorless. Contains 16% (approx) iron. Soluble in water; insoluble in alcohol.
Use: Medicine.

ferric pyrophosphate (iron pyrophosphate) $Fe_4(P_2O_7)_3 \cdot xH_2O$.
Properties: Yellowish-white powder; insoluble in water, soluble in dilute acid. 24% Fe minimum. Not to be confused with ferric pyrophosphate, soluble (q.v.).
Uses: Source of nutritional iron and for enrichment of foods not subject to rancidity.

ferric pyrophosphate, soluble. A combination of ferric pyrophosphate and sodium citrate.
Properties: Apple-green crystals; very soluble in water; insoluble in alcohol. Protect from light. 11% Fe.
Uses: Medicine; feed additive.

ferric resinate (iron resinate).
Properties: Reddish-brown powder; soluble in ligroin, carbon disulfide, ether, oil of turpentine; slightly soluble in alcohol; insoluble in water.
Use: Drier (paints, varnish). See also soap (2).

ferric sodium oxalate (iron-sodium oxalate) $Na_3Fe(C_2O_4)_3 \cdot 5\frac{1}{2}H_2O$.
Properties: Emerald-green crystals; decomposed by heat or light; soluble in water and alcohol. Sp. gr. 1.973 (18°C); dec. 300°C. Protect from light.
Derivation: By the interaction of sodium acid oxalate and ferric hydroxide.
Hazard: Moderately toxic.
Uses: Photography; blueprinting.

ferric stearate (iron stearate) $Fe(C_{18}H_{35}O_2)_3$.
Properties: Light-brown powder; soluble in alcohol and ether; insoluble in water. Combustible.
Derivation: Interaction of solutions of ferric sulfate and sodium stearate.
Use: Varnish driers; photocopying. See also soap (2).

ferric succinate (iron succinate) $(C_4H_4O_4)_2Fe_2(OH)_2$.
Properties: Reddish-brown powder; insoluble in cold water; partly decomposed by hot water forming a more basic salt; soluble in dilute acids. Protect from light. Combustible.
Derivation: By addition of ferric chloride to a succinate solution.
Use: Medicine.

ferric sulfate (iron sulfate; ferric trisulfate; iron tersulfate; iron persulfate) (a) $Fe_2(SO_4)_3$; (b) $Fe_2(SO_4)_3 \cdot 9H_2O$.
Properties: Yellow crystals or grayish-white powder. Sp. gr. (a) 3.097, (b) 2.0–2.1; m.p., decomposes (480°C). (a) Slightly soluble in water. (b) Very soluble in water. Keep well closed and protected from light. Noncombustible. Low toxicity.
Derivation: By adding sulfuric acid to ferric hydroxide.
Grades: Technical; C.P.; partly hydrated.
Uses: Pigments; medicine; reagent; iron alum manufacture; etching aluminum; disinfectant; textiles

(dyeing and calico printing); flocculant in water purification; soil conditioner.

ferric tallate. See soap (2).

ferric tannate (iron tannate; iron gallotannate) $Fe_2(C_{14}H_7O_9)(OH)_3$.
Properties: Dark-brown or bluish black powder, variable compositoin. Soluble in alkalies; insoluble in water, alcohol and ether; soluble in dilute acids. Combustible. Low toxicity.
Derivation: Interaction of ferric acetate and tannic acid solutions.
Use: Medicine.

ferric tribromide. See ferric bromide.

ferric trichloride. See ferric chloride.

ferric trioxide. See ferric oxide.

ferric trisulfate. See ferric sulfate.

ferric vanadate (iron metavanadate) $Fe(VO_3)_3$.
Properties: Grayish-brown powder; soluble in acids; insoluble in water and alcohol. Noncombustible; low toxicity.
Derivation: By adding a solution of a ferric salt to the liquor obtained by leaching vanadium ores with caustic soda solution or by lixiviating the slags obtained when vanadium ores are fused with soda ash, etc.
Grade: Technical.
Use: Metallurgy.

"Ferri-Floc."[93] Trademark for ferric sulfate. Used as a coagulant in water, sewage and industrial waste treatment; in manufacture of fungicides, chelated iron products, chemical intermediate, and in pickling metals.

ferrite.
(1) Iron in the body-centered cubic form; commonly occurs in steels, cast iron and pig iron below 910°C. Alpha and beta iron are the common varieties of ferrite, and the name is also applied to delta iron.
(2) A compound, a multiple oxide, of ferric oxide with another oxide, as sodium ferrite, $NaFeO_2$, but more commonly, a multiple oxide crystal.

Ferrites are made by dissolving hydrated ferric oxide in concentrated alkali solution, by fusing ferric oxide with alkali metal chloride, carbonate or hydroxide, or by simply heating metal oxides with ferric oxide. Ceramic ferrites are made by press-forming powdered ingredients (with a binder) into a sheet, then sintering, or firing.

The oxide ferrites are used as rectifiers, on memory and record tapes, for permanent magnets, semiconductors, insulating materials, dielectrics, high frequency components, and various related uses in radio, television, radar, computers, and automatic control systems.

ferro-alloy. An alloy of iron with some element other than carbon used as a vehicle for introducing such an element into steel during its manufacture. The element may alloy with the steel by solution or as the carbide, neutralize the harmful impurities by combining with them and separating from the steel as flux or slag before solidification.

ferroboron. A ferro-alloy used as hardening agent in special steels. It also is an efficient deoxidizer. Boron steel is used in controlling the operating rate of nuclear reactors. Two grades available, 10 and 17% boron.

"Ferrocarbo."[280] Trademark for briquetted or granular silicon carbide (q.v.).
Uses: Cupola addition in the production of gray iron, or as a ladle addition to steel. It decomposes into its component elements and acts as a powerful deoxidizer and graphitizer. Machinability and strength of the iron or steel are increased with no loss of hardness.

ferrocene (dicyclopentadienyliron) $(C_5H_5)_2Fe$. A coordination compound of ferrous iron and 2 molecules of cyclopentadiene in which the organic portions have typically aromatic chemical properties. Its activity is intermediate between phenol and anisole. The first compound shown to have the "sandwich" structure found in certain types of metallocene molecules.
Properties: Orange, crystalline solid; camphor-like odor; m.p. 173–174°C; resists pyrolysis at 400°C. Insoluble in water, slightly soluble in benzene, ether and petroleum ether. Iron content 29.4–30.6%.
Derivation: From ferrous chloride and cyclopentadiene sodium.
Containers: Bottles; fiber drums.
Hazard: Evolves toxic products on decomposition and heating. Moderate fire risk.
Uses: Additive to fuel oils to improve efficiency of combustion and eliminate smoke; antiknock agent; catalyst; coating for missiles and satellites; high-temperature lubricant; intermediate for high temperature polymers; ultraviolet absorber.
See also metallocene.

ferrocenecarboxylic acid ethyl ester. See ethyl ferrocenoate.

1,1′-ferrocenedicarboxylic acid diethyl ester. See 1,1′-diethyl ferrocenoate.

1,1′-ferrocenedicarboxylic acid dimethyl ester. See dimethyl ferrocenoate.

ferrocenoyl chloride. See chlorocarbonyl ferrocene.

ferrocenoyl dichloride. See 1,1′-dichlorocarbonyl ferrocene.

ferrocenylborane polymer.
Properties: Long-term heat resistance at 600°F; short-term stability about 1500°F. Good resistance to oxidation and hydrolysis. Contains up to 30% iron directly combined.
Uses: Specialty plastics, coatings, fibers; ablative material for space vehicles.

ferrocenyl methyl ketone. See acetylferrocene.

ferrocerium. A pyrophoric alloy (q.v.) of iron and misch metal.
Shipping regulations: (Rail) Flammable solid, n.o.s., Yellow label. (Air) Flammable solid label.

ferrocholinate (iron choline citrate chelate) $C_{11}H_{20}FeNO_9$.
Properties: Greenish-brown to reddish-brown amorphous solid; soluble in water, yielding a stable solution; soluble in acid and alkaline media.
Derivation: Reaction of choline dihydrogen citrate with ferric hydroxide.
Grade: N.D.
Use: Medicine.

ferrochromium (ferrochrome). An alloy composed principally of iron and chromium used as a means of adding chromium to steels (low-, medium-, and high-carbon) and to cast iron. Available in several classifications and grades, generally containing between 60–70% chromium; in crushed sizes and lumps up to 75 pounds, which readily dissolve in molten steel.

ferroconcrete. See concrete.

ferroelectric. A crystalline material such as barium titanate, monobasic potassium phosphate or potassium-sodium tartrate (Rochelle salts) that over certain limited temperature ranges has a natural or inherent deformation (polarization) of the electrical fields or electrons associated with the atoms and groups in the crystal lattice. This results in the development of positive and negative poles and a consequent "direction" of polarization, which can be reversed when the crystal is exposed to an external electric field. Ferroelectric crystals are internally strained and as a consequence show unusual piezoelectric and elastic properties.
Uses: Capacitors; transducers; computer technology.
See also ceramic, ferroelectric.

ferroin chelation group. A functional group characteristic of heterocyclic ring nitrogen compounds:

=N—C(‖)—C(‖)—N=

Among such compounds are 2,2′-bipyridine, 1,10-phenanthroline, and the 2-pyridyl triazines. These provide a large number of terminal (=C—H) groups in which the hydrogen can be replaced by many chemical groupings (carboxyl, hydroxyl, halogen, etc.). Thus synthesis of an almost endless number of substituted ferroin reactants is possible. About 200 such chelation reagents have been synthesized. Ferroin chelation chemicals in general form complex undissociated cations with divalent metal ions, e.g., $[(C_{12}H_8N_2)_3Fe]^{+2}$.

"Ferrolon."[232] Brand name for a series of organic sequestering agents usedin textile applications.

ferromagnesite. An iron-bearing variety of magnesite (q.v.) used for refractories owing to its ability to bond under heat.

ferromagnetic oxide. See ferrite (2).

ferromanganese. An alloy consisting of manganese (about 48%), plus iron and carbon; used as a vehicle for adding manganese to steel. Available in standard, low-carbon, and medium carbon grades in ground, crushed and lump sizes ranging from 80 mesh to 75-lb lumps, suitable for ladle or furnace addition.
See also manganese steel.

ferromolybdenum. An alloy composed largely of iron and molybdenum used as a means of adding molybdenum to steel. Engineering steels rarely contain more than 1% molybdenum, stainless steels may contain 3% and tool steels as much as 10%. Ferromolybdenum is available in several grades in which molybdenum ranges from 55% to 75% and the maximum carbon content is either 0.10%, 0.60%, or 2.50%. It is generally added in the furnace, since it does not oxidize under steel-making conditions. M.p. 2965°F (approx). Available in crushed sizes up to one inch.

ferron. See 7-iodo-8-hydroxyquinoline-5-sulfonic acid.

ferrophosphorus. An alloy of iron and phosphorus used in the steel industry for adjustment of phosphorus content of special steels; particularly useful in preventing thin sheets from sticking together when rolled and annealed in bundles. Produced in two grades: (a) 18% phosphorus; (b) 25% phosphorus.

ferrosilicon. An alloy of iron and silicon used to add silicon to steel and iron. Sp. gr. 5.4. Insoluble in water. Small quantities of silicon deoxidize the iron and larger amounts impart special properties. It is also used in Pidgeon process for producing metallic magnesium. Available in six grades containing from 15 to 95% silicon.
Hazard: Ferrosilicon containing from 30–90% silicon is flammable and evolves toxic gases in presence of moisture.
Shipping regulations: (Air) (30 to 90% silicon) Flammable Solid label. (Rail) Not listed.

ferrosoferric oxide. See iron oxide, black.

ferrotitanium. An alloy composed principally of iron and titanium used to add titanium to steel. It is often made from titanium scrap. Three classifications are available: low, high, and medium carbon content. Furnished in various lump, crushed, and ground sizes.

ferrotungsten. An alloy of iron and tungsten used as a means of adding tungsten to steel. See tungsten steels. Contains 70–80% tungsten and no more than 0.6% carbon. Melting range 3000–5000°F; dissolves readily in molten steel. Furnished in ground and crushed sizes up to one inch.

ferrous acetate (iron acetate) $Fe(C_2H_3O_2)_2 \cdot 4H_2O$.
Properties: Greenish crystals when pure and unexposed to air; usually partly brown from action of air; soluble in water and alcohol; oxidizes to basic ferric acetate in air. Combustible.
Derivation: Action of acetic acid or pyroligneous acid on iron, with subsequent crystallization.
Uses: Textile dyeing; medicine; dyeing leather; wood preservative.

ferrous ammonium sulfate (Mohr's salt; iron ammonium sulfate) $Fe(SO_4) \cdot (NH_4)_2SO_4 \cdot 6H_2O$.
Properties: Light-green crystals. Soluble in water; insoluble in alcohol. Sp. gr. 1.865; dec. 100–110°C. Deliquescent; affected by light.
Derivation: By mixing solutions of ferrous sulfate and ammonium sulfate, followed by evaporation and subsequent crystallization.
Uses: Medicine; analytical chemistry; metallurgy.

ferrous arsenate (iron arsenate) $Fe(AsO_4)_2 \cdot 6H_2O$.
Properties: Green, amorphous powder. Insoluble in water; soluble in acids.
Derivation: Interaction of solutions of sodium arsenate and ferrous sulfate.
Uses: Medicine; insecticide.
Hazard: Highly toxic.
Shipping regulations: (Rail, Air) Poison label.

ferrous bromide (iron bromide) $FeBr_2 \cdot 6H_2O$.
Properties: Green crystalline powder; very deliquescent. Affected by light. Soluble in water and alcohol; sp. gr. 4.636; m.p. 27°C. Low toxicity.
Derivation: Action of bromine on iron filings.
Use: Medicine.

ferrous chloride (iron chloride; iron dichloride; iron protochloride). (a) $FeCl_2$; (b) $FeCl_2 \cdot 4H_2O$.
Properties: Greenish-white crystals. Soluble in alcohol and water; sp. gr. (a) 3.16 (25°C); (b) 1.93. M.p. (a) 670–674°C. Deliquescent; affected by light. Low toxicity.
Derivation: Action of hydrochloric acid on an excess of iron, with subsequent crystallization.
Uses: Mordant in dyeing; metallurgy; pharmaceutical preparations; manufacture of ferric chloride; sewage treatment.

ferrous 2-ethylhexoate. A paint drier. See soap (2).

ferrous fluoride (iron fluoride) $FeF_2 \cdot 8H_2O$. (also known: FeF_2 and $FeF_2 \cdot 4H_2O$.)
Properties: Green crystals. Soluble in acids; slightly soluble in water; insoluble in alcohol and ether; sp. gr. 4.09 (anhydrous).
Hazard: Toxic; strong irritant. Tolerance (as F), 2.5 mg per cubic meter of air.
Use: Ceramics.

ferrous fumarate $FeC_4H_2O_4$. Anhydrous salt of a combination of ferrous iron and fumaric acid. Stable, odorless, substantially tasteless, reddish-brown anhydrous power. Contains 33% iron by weight. Does not melt at temperatures up to 280°C. Insoluble in alcohol; very slightly soluble in water. Combustible. Low toxicity.
Grade: U.S.P.
Uses: Medicine; dietary supplement.

ferrous gluconate (iron gluconate) $Fe(C_6H_{11}O_7)_2 \cdot 2H_2O$.
Properties: Yellowish gray or pale greenish yellow, fine powder or granules with slight odor. Solution (1 in 20) is acid to litmus. Soluble in water and glycerin; insoluble in alcohol. Combustible. Low toxicity.
Grades: Pharmaceutical; N.F.
Uses: Medicine; feed and food additive; vitamin tablets.

ferrous iodide (iron iodide; iron protoiodide) $FeI_2 \cdot 4H_2O$.
Properties: Crystalline, grayish-black. Soluble in water and alcohol. Sp. gr. 2.873; dec. 90–98°C; m.p. (anhydrous) 177°C. Deliquescent; affected by light. Low toxicity.
Derivation: By the action of iodine on iron filings.
Uses: Manufacture of alkali metal iodides; pharmaceutical preparations.

ferrous lactate (iron lactate) $Fe(C_3H_5O_3) \cdot 3H_2O$.
Properties: Greenish-white crystals, slight peculiar odor. Moderately soluble in water; slightly soluble in alcohol. Deliquescent; affected by light. Combustible. Nontoxic.
Derivation: By interaction of calcium lactate with ferrous sulfate or direct action of lactic acid on iron filings.
Uses: Food additive and dietary supplement.

ferrous naphthenate. A soap based on mixed naphthenic acids. Available commercially as a liquid containing 6% iron. See soap (2).

ferrous octoate. A paint drier. See soap (2).

ferrous oxalate (iron oxalate) $FeC_2O_4 \cdot 2H_2O$.
Properties: Pale yellow, odorless crystalline powder. Soluble in acids; insoluble in water. Sp. gr. 2.28; decomposes 160°C, releasing carbon monoxide.
Derivation: By the interaction of solutions of ferrous sulfate and sodium oxalate.
Hazard: Evolves carbon monoxide on heating.
Uses: Medicine; photographic developer.

ferrous oxide (iron monoxide) FeO.
Properties: Black powder; sp. gr. 5.7; m.p. 1420°C. Insoluble in water; soluble in acid.
Derivation: Prepared from the oxalate by heating but the product contains some ferric oxide.
Hazard: Tolerance (as fume): 10 mg per cubic meter of air.

ferrous phosphide (iron phosphide) Fe_2P. A ferrophosphorus (q.v.).
Properties: Bluish-gray powder. Sp. gr. 6.56; m.p. 1290°C.
Grade: 24–26% phosphorus.
Containers: Barrels; bulk.
Hazard: Evolves toxic and flammable products on exposure to moisture or acids.
Uses: Iron and steel manufacture.

ferrous sulfate (iron sulfate; iron vitriol; copperas; green vitriol; sal chalybis) $FeSO_4 \cdot 7H_2O$.
Properties: Greenish or yellow-brown crystals or granules. Odorless; slightly soluble in water, with saline taste; insoluble in alcohol. Sp. gr. 1.89; m.p. 64°C; loses $7H_2O$ by 300°C; pH (10% solution) 3.7. Low toxicity. Hygroscopic.
Derivation: (a) By-product from the pickling of steel and many chemical operations. (b) By action of dilute sulfuric acid on iron. (c) Oxidation of pyrites in air followed by leaching and treatment with scrap iron.
Method of purification: Recrystallization.
Grades: Technical; anhydrous; C.P.; U.S.P.
Uses: Iron oxide pigment; other iron salts; ferrites; water and sewage treatment; catalyst, especially for synthetic ammonia; fertilizer; feed additive; flour enrichment.

ferrous sulfide (iron sulfide; iron protosulfide) FeS.
Properties: Dark-brown or black metallic pieces, sticks or granules. Soluble in acids; insoluble in water. Sp. gr. 4.75; m.p. 1195°C; b.p., decomposes.
Derivation: By fusing iron and sulfur.
Uses: Generating hydrogen sulfide; ceramics; other sulfides.
See also pyrite.

ferrovanadium. An iron-vanadium alloy used to add vanadium to steel. Vanadium is used in engineering steels to the extent of 0.1–0.25%, and in high speed steels to the extent of 1–2.5% or higher. Several grades are available containing from 50 to 80% vanadium. Produced by reduction of the oxide with aluminum or silicon in presence of iron in an electric furnace. Melting range 2700–2770°F. Furnished in a variety of lump, crushed, and ground sizes.
Hazard (dust): Tolerance, 1 mg per cubic meter of air. Moderate fire risk.

ferrozirconium. Alloys used in the manufacture of steel. 12 to 15 percent zirconium alloy: Approximate analysis: zirconium 12–15%, silicon 39–43%; iron 40–45%. Application: Steel of high silicon content. 35–40 percent zirconium alloy: Approximate analysis: zirconium 35–40%, silicon 47–52%, iron 8–12%. Application: Steel of low silicon content.

ferrum. Latin name for iron; hence the symbol Fe.

fertile material. In nuclear technology, any substance not capable of sustaining a chain reaction but which can be converted into a fissionable material in a nuclear reactor. Uranium 238 (converted to plutonium 239) and thorium 232 (converted to uranium 233) are the most important fertile materials.

fertilizer. A substance or mixture that contains one or more of the primary plant nutrients and sometimes also secondary and/or trace nutrients. The primary nutrients are nitrogen (supplied as anhydrous ammonia, or solutions containing nitrogen derived from ammonia, ammonium nitrate or urea); phosphorus (as superphosphates derived from phosphate rock); and potassium (in the form of KCl from sylvite ore or natural brines). Secondary nutrients are calcium, magnesium, and sulfur. Trace elements (iron, copper, boron, manganese, zinc and molybdenum) are also among the twelve elements considered essential for plant growth. Nitrogen solutions and anhydrous ammonia are used both in fertilizer manufacture and for direct application to the soil. Substantial amounts of both separate materials and mixtures are used in liquid form. Controlled-release fertilizers are those whose particles are coated with polymeric sulfur by a proprietary process. Their advantages include more uniform supply of nutrient, lower labor costs, and reduced leaching losses in areas of irrigation and high rainfall. See also superphosphate; nutrient solution. For further information refer to National Fertilizer Solutions Assn., Peoria, Ill.

Shipping regulations: (Rail) Green label. (Air) Fertilizer ammoniating solution containing free ammonia (pressurized): Nonflammable Compressed Gas label. Not acceptable on passenger planes.

F.F.A. Abbreviation for free fatty acid. Used in describing specifications for fatty esters, glycerides, oils, etc.

FFPA. Abbreviation for free from prussic acid.

FGAN. Fertilizer grade ammonium nitrate (q.v.). Used in blasting agents as well as fertilizers, because its coating of kieselguhr and its prilled form make it safer to handle than the usual grades.

fiber. A fundamental form of solid (usually crystalline), characterized by relatively high tenacity and an extremely high-ratio of length to diameter (several hundred to one). Natural fibers are animal, such as wool and silk (proteins); vegetable, such as cotton (cellulose); and mineral (asbestos). Cotton fiber is called staple, and rarely exceeds 2″ in length.

Semisynthetic fibers include rayon and inorganic substances extruded in fibrous form, such as glass, boron, boron carbide, boron nitride, carbon (graphite), aluminum silicate, fused silica and some metals (steel). Synthetic fibers are made from high polymers (polyamides, polyesters, acrylics, and polyolefins) by extruding from spinnerets (nylon, "Orlon," etc.). Some are being used in specialty papers, though the primary use is in textile fabrics. For ceramic fibers, see "Fiberfrax."

Metals are used in the form of fibers in several forms: (1) As "whiskers," which are single-crystal fibers up to 2 in. long having extremely high tensile strength; they are made from tungsten, cobalt, tantalum and other metals, and are used largely in composite structures for specialized functions. (2) As metal fibers, which are alloys drawn through diamond dies to diameters as small as 0.002 cm; steel for tire cord and antistatic devices has been developed for such applications. (3) As metallic fibers, which are biconstituent structures composed of a metal and a polymeric material, for example, aluminum filament covered with cellulose acetate butyrate.

Hollow fibers of cellulose acetate and nylon are used as membranes in the reverse osmosis method of water purification. See also filament, denier, whiskers, glass fiber.

fiber, biconstituent. A composite fiber comprising a dispersion of fibrils of one synthetic material within, and parallel to the axis of another; also a fiber made up of a polymeric material and a metal or alloy filament.

fiberfill. A fiber designed specifically for use as a filling material in such products as pillows, comforters, quilted linings, furniture battings, e.g., sisal, jute.

"Fiberfrax."[280] Trademark for ceramic fiber made from alumina and silica. Available in bulk as blown, chopped and washed; long staple; paper; rope; roving; blocks.

Properties: Retains properties to 2300°F and under some conditions used to 3000°F; light weight; inert to most acids and unaffected by hydrogen atmosphere; resilient.

Uses: High-temperature insulation of kilns and furnaces; packing expansion joints; heating elements, burner blocks; rolls for roller hearth furnaces and piping; fine filtration; insulating electrical wire and motors; insulating jet motors; sound deadening.

"Fiberglas."[191] Trademark for a variety of products made of or with glass fibers or glass flakes including insulating wools, mats and rovings, coarse fibers, acoustical products, yarns, electrical insulation, and reinforced plastics.

See also glass fiber; reinforced plastics.

fiber, optical. An extremely fine-drawn glass fiber of exceptional purity that will transmit laser light impulses with high fidelity. Such fibers, which are in the experimental stage, are made from high-purity quartz coated with germanium-doped silica by vapor deposition. After drawing to a microscopically small diameter, the strands are assembled into cables containing 144 filaments; these are intended for use in telephone systems as replacements for copper cables. See also glass fiber.

fiber-reactive dye. See dye, fiber-reactive.

"Fibertex."[236] Trademark for a fibrous material used as an additive to rotary drilling mud to prevent or restore lost circulation in gravelly, fractured, or creviced formations.

"Fibra Flo."[247] Trademark for a series of filter aids comprised of physical mixtures in different proportions of either asbestos fiber and diatomite, asbestos fiber and cellulose fiber, or asbestos fiber alone.

fibrid. Generic name for a fibrous form of synthetic polymeric material, used for example as a binder material in the manufacture of textryl (q.v.).

fibrinogen. A sterile fraction of normal human blood plasma, dried from the frozen state. In solution it has the property of being converted into insoluble fibrin when thrombin is added. It is an essential factor in blood-clotting mechanism.

Properties: White or grayish amorphous substance.

Grade: U.S.P.

Use: Medicine.

fibrinolysin (plasmin). A proteolytic enzyme which dissolves fibrin and hastens the solution of clots that may form in the bloodstream. It is prepared by activating a fraction of normal human plasma with highly purified streptokinase. Listed in N.D.

fibroin. The fibrous material in silk, a scleroprotein containing glycine and alanine; light yellow silk-like mass. Insoluble in water; soluble in concentrated alkalies and concentrated acids.

fibrolite. See sillimanite.

ficin. A proteolytic enzyme hydrolyzing casein, meat, hide powder, edestin, fibrin, liver, and other protein-like material.

Properties: Buff to cream-colored powder with an acrid odor; very hygroscopic; partially soluble in water; insoluble in the usual organic solvents.

Source: Fig latex or sap. Commercially prepared by filtering and drying the latex.
Use: Brewing, cheese, meat, leather, and textile industries; diagnostic aid in medicine.

"Ficoll."[485] Trademark for an inert non-ionized synthetic, high polymer made by the crosslinking reaction of epichlorohydrin and sucrose. Contains no ionized groups and its extreme solubility is due to the high content of hydroxyl groups (23%).
Uses: Density gradients for centrifugation, electrophoresis and specific gravity determinations; component in the isotonic solutions for preparation of cells and cell fragments; concentrating solutions of sensitive materials by means of dialysis.

fictile. Descriptive of certain molecules which have no permanent structure but are constantly changing their shapes and arrangements. An example is the metal carbonyl $Fe_3(CO)_{12}$ in which, according to Dr. F. Albert Cotton, originator of the term, "carbonyl groups readily move from one iron atom to another through the rapid formation and dissolution of carbonyl bridges between iron atoms."

filament. A continuous fiber, usually made by extrusion from a spinneret (nylon, rayon, glass, polyethylene). It also may be a drawn metal (tungsten, gold) or a metal carbide. (See fiber).

filament winding. The process of winding fibers under tension on to a prepared core. Before or during the winding operation, the assembly is impregnated with a thermosetting resin. Structures of considerable size and strength can be made in this way. The fibers used are chiefly glass, boron, or silicon carbide. (See filament).

filler. (1) An inert mineral powder of rather high specific gravity (2.00–4.50) used in plastic products and rubber mixtures to provide a certain degree of stiffness and hardness, and to decrease cost. Examples are calcium carbonate (whiting), barytes, blanc fixe, silicates, glass spheres and bubbles, slate flour, soft clays, etc. Fillers have neither reinforcing nor coloring properties, and the term should not be applied to materials that do, i.e., reinforcing agents or pigments. Fillers are similar to extenders and diluents in their cost-reducing function; exact lines of distinction between these terms are difficult, if not impossible, to draw. Use of fillers and extenders in plastics has increased in recent years due to shortages of basic materials.

(2) The cross or transverse thread in a fabric or other textile structure.

(3) A metal or alloy used in brazing and soldering to effect union of the metals being joined.

See also diluent; extender; reinforcing agent.

film. An extremely thin, continuous sheet of a substance, which may or may not be in contact with a substrate. There is no precise upper limit of thickness, but a reasonable assumption is 0.010 in. The protective value of any film depends on its being 100% continuous, i.e., without holes or cracks, since it must form an efficient barrier to molecules of atmospheric water vapor, oxygen, etc. A long-chain fatty acid or alcohol on water produces a film whose "thickness" is the length of one molecule (about 200 A.) The fatty acid molecules are oriented with the radical end in the water. Such films are good evaporation barriers, and have been successfully imposed on glass. Soap bubbles are elastic films about one micron thick, and have considerable strength.

Film-forming agents (drying oils) are essential in paints and lacquers. Oxide films formed automatically on the surface of aluminum protect it from corrosion. Thin metallic oxide films are widely used in electronic and semiconductor devices. Electrodeposited metals (chromium, copper, nickel) are conventionally (and perhaps illogically) called coatings.

The term film is also applied to sheets of cellophane, polyethylene, polyvinylidene chloride, etc., used for wrapping and packaging of food products, meats, and poultry (especially shrink films, which are stretched before application). These function as a moisture vapor barrier. Plastic films are also used as slip surfaces in concrete structures such as air strips, ice rinks, and highways. Photographic film is made from cellulose acetate.

"Filmfast."[50] Trademark for a spray compound for improving distribution of insecticidal and fungicidal sprays.

"Filmcol."[125] Trademark for an authorized ethyl alcohol. Specially denatured alcohol. No. 1 (190 proof) 100 parts by volume; isopropyl alcohol, 10 parts by volume; methyl isobutyl ketone, 1 part by volume.
Hazard: Flammable; dangerous fire risk.
Uses: Solvent, purifying and recrystallizing operations in chemical industry; photographic, textile, rubber latex, printing industries.

filter aid. See filtration.

filter alum. See aluminum sulfate.

filter sand. Sand used to separate sediment and suspended matter from water.

"Filtrasorb Carbon."[108] Trademark for specially engineered activated carbon granules having mechanical strength great enough to withstand repeated filter backwashings and regeneration at high temperature. Used in conjunction with polymeric water-treatment chemicals for coagulating solids in raw sewage, with subsequent filtration over the carbon granules.

filtration. The process of separating suspended solids from a liquid by forcing the mixture through a porous barrier. This may be a fabric of heavy weave, a wire screen, or (on micro scale) a thin plastic membrane of high porosity (several million pores per square inch). These retain bacteria and are used in industrial sterilization. Clay suspensions are filtered by forced flow electrophoresis. Many fibrous or particulate materials (often called filter aids) can effect filtration; examples are special papers (Whatman), glass fibers, diatomaceous earth, fly ash, sand, etc.

The construction and operation of the many kinds of filtration equipment are too detailed to permit description. The most widely used types may be classified as follows: (1) gravity filters, used largely for water purification and consisting of thick beds of sand and gravel, which retain the flocculated impurities as the water passes through; (2) pressure filters of plate-and-frame or shell-and-leaf construction, which utilize filter cloths of coarse fabric as a separating medium; (3) vacuum or suction filters of the rotating drum or disk type, used on thick sludges and slurries; (4) edge filters; and (5) clarification filters. Gel filtration (q.v.) is a chromatographic technique involving separation at the molecular level.

"Filtrol."[217] Trademark for acid-activated clays used as decolorizing adsorbents and catalysts.

fine chemical. A chemical produced in comparatively small quantities and in a relatively pure state. Examples are pharmaceutical and biological products, perfumes, photographic chemicals, and reagent chemicals.

fines. The portion of a powder composed of particles which are smaller than a specified size (M.P.A. definition, M.P.A. Standard 9-50T).

"Finetex."[1] Trademark for a phosphate leavening acid.

finishing compound. A substance used in the final or finishing stages of manufacture of a product, usually textiles and leather, to make them suitable for specific purposes. Such compounds contain materials that impart softness, flexibility, stiffness, color, water and fire resistance, etc.

fireclay. See refractory.

fire extinguishment. Fires are divided into four classes, each requiring special treatment. The essential point in extinguishing all types is exclusion of air from the fire by any effective means.

Class A includes fires in combustible materials, such as wood, paper, and cloth, where the quenching and cooling effect of quantities of water, or of solutions containing a high percentage of water, is of first importance. Fire extinguishers utilizing the pressure of carbon dioxide to throw a stream of water onto the fire are the most widely used for this class. In the soda-acid extinguisher the carbon dioxide is generated within the cylinder at the time of use. In another type, carbon dioxide gas is stored in the cylinder under pressure and is released by opening a valve.

Class B includes fires in flammable liquids, where a blanketing or smothering effect is essential. Carbon dioxide gas, dry chemical (q.v.), or foam are suitable. Water should not be used. Class C includes fires in electrical equipment. The use of carbon dioxide gas or dry chemical extinguishers is recommended. Water should not be used. Class D fires are burning metals. A powder formulation such as "Met-L-X" (q.v.), powdered graphite or trimethoxyboroxine will extinguish a metal fire; water should not be used.

In general, for small fires salt (sodium chloride) and sodium bicarbonate are effective, either dry or in concentrated solution. Carbon tetrachloride and methyl bromide are to be avoided as extinguishing agents because of the toxicity of their decomposition products, for example, phosgene.

See also foam, fire-extinguishing.

"Firefrax."[280] Trademark for a group of refractory cements made from kaolin or fireclay base materials for applications where aluminum silicate cements are best suited.

Containers: 100-lb bags.

Uses: Laying and repairing fireclay and silica brick work; bond for crushed firebrick or ganister for patching furnace linings and for making rammed-up or monolithic linings; patching material for by-product coke ovens; and as a wash for small pouring ladles in non-ferrous foundry.

"Firemaster" BP4A.[426] Trademark for tetrabromobisphenol-A (q.v.).

"Firemaster" PHT4.[426] Trademark for tetrabromophthalic anhydride (q.v.).

"Firemaster" T23P.[426] Trademark for tris(2,3-dibromopropyl) phosphate (q.v.).

fire point. The lowest temperature at which a liquid evolves vapors fast enough to support continuous combustion. It is usually close to the flash point. See also autoignition point.

fire-retarding agent. See flameretarding agent.

fire sand. See furnace sand.

"Fi-Retard."[300] Trademark for a group of flame retardants for cotton, rayon, nylon and paper.

fir needle oil (fir oil). An essential oil obtained by the steam distillation of needles and twigs of several varieties of coniferous trees (Abies) native to both Canada and Siberia.

Properties: Colorless to faintly yellow liquid, pleasant odor. Soluble in most fixed oils; insoluble in glycerin. Sp. gr. 0.87–0.91 (25°C); refractive index 1.473 (20°C); optically active.

Constituents: Esters calculated as bornyl acetate: 8–16% (Canadian), 32–44% (Siberian).

Grades: Technical; F.C.C.

Containers: Cans.

Uses: Odorant in perfumery; flavoring agent; medicine.

Fischer, Emil (1852–1919). German organic chemist, recipient of Nobel Prize in chemistry (1902) for his original research in the chemistry of purines and sugars. He was Professor of Chemistry at University of Berlin (1882), succeeding Hoffmann. He synthesized fructose and glucose and elucidated their stereochemical configurations; also established the nature of uric acid and its derivatives. Additional work included enzyme chemistry, proteins, synthetic nitric acid and ammonia production.

Fischer, Hans (1881–1945). German biochemist who studied under Emil Fischer. He was awarded the Nobel Prize in chemistry in 1930 for his synthesis of the blood pigment hemin. He also did important fundamental research on chlorophyll, the porphyrins and carotene.

Fischer's reagent. A reagent used as a test for sugars.

Preparation: 3 parts of sodium acetate and 2 parts of phenylhydrazine hydrochloride in 20 parts of water.

Note: Do not confuse with Karl Fischer reagent (q.v.).

Fischer's salt. See cobalt potassium nitrite.

Fischer-Tropsch process. Any of several catalytic processes originating in Germany and using synthesis gas mixtures of carbon monoxide and hydrogen to produce aliphatic straight-chain hydrocarbons and oxygenated derivatives. It is also called the synthine process. It is one of a number of processes that may be applied to production of liquid fuels from coal. See also gasification.

fisetin $C_{15}H_{10}O_6$. 3,7,3′-Tetrahydroxyflavone. See flavanol.

Fisher's solution. See physiological salt solution.

fish glue. An adhesive derived from the skins of commercial fish (chiefly cod). A ton of skins yields about 50 gallons of liquid glue. Bond strength on wood is about 2500 psi; pH about 6.5–7.2. Compatible with animal glues, some dextrins, some polyvinyl acetate emulsions, and with rubber latex. Chief applications are in gummed tape, cartons, blueprint paper, and letterpress printing plates. Fish glue can be made light-sensitive by adding ammonium bichromate, and water-insoluble by ultraviolet radiation; hence its usefulness in the photoengraving process. See also adhesive.

fish-liver oil. An oil containing a high percentage of vitamin A; high-potency livers, as from cod, shark and halibut, contain from 100,000 to 1,500,000 A units per gram. The oil is extracted by cooking the livers under low-pressure steam and removing the oil, which floats on the condensate. Livers of low oil content are processed with a weak solution of sodium hydroxide or sodium carbonate, which extracts the oil in emulsified form. Uses are chiefly as medicine and dietary supplement. See also fish oil.

fish meal. A fishery by-product consisting essentially of processed scrap from the filleting operation or from whole fish. In the dry process the waste from cod, halibut and haddock heads is disintegrated and dried. The oil and proteins are largely retained. In the wet process the whole fish (chiefly menhaden and pilchard) are used. These are steam-cooked and run through a screw press to remove the oil; the resulting meal is then dried and packed. Its chief use is now for animal feeds, and as a raw material for fish protein concentrate (q.v.).

Hazard: Strong tendency to spontaneous heating; flammable.

Shipping regulations: (containing less than 6% or more than 12% moisture) (Rail) Yellow label. Not accepted by rail express. (Air) Not listed.

fish oil. A drying oil obtained chiefly from menhaden, pilchard, sardine, and herring. Extracted from the entire body of the fish by cooking and compressing. Should not be confused with fish-liver oil (q.v.). It contains about 20% polyunsaturated fatty acids, which enables it to lower cholesterol content of the human diet. Chemically modified fish oil is used in soaps, detergents, protective coatings, and alkyd resins. The hydrogenated product is used as a base for margarines and shortenings, and as an industrial dispersing agent.

Hazard: Subject to spontaneous heating.

fish protein concentrate (FPC). A flour or paste-like product prepared from whole fish, including bones and viscera, of a size and type not acceptable for market. Both biological (enzymatic) and chemical (solvent extraction) methods are used to extract proteins. Fish meal is sometimes used as a raw material. FPC contains a high percentage of proteins at very low cost, and is considered a valuable dietary supplement. Approved by FDA.

fissiochemistry. The process by which a chemical change or reaction is brought about by nuclear energy; for example, the production of anhydrous hydrazine from liquid ammonia in a nuclear reactor.

fission, nuclear. The splitting of an atomic nucleus induced by bombardment with neutrons from an external source and propagated by the neutrons so released. When a fissionable (unstable) nucleus, such as uranium-235 or plutonium, is struck by a neutron in a critical area, the following events occur: (1) the nucleus disintegrates to form several other elements, called fission products or fragments, all of which are radioactive and have high kinetic energy; (2) the disrupted nucleus emits an average of 2.5 neutrons, which in turn split other nuclei of the fissionable material in a chain reaction which is self-perpetuating; (3) it also emits the energy equivalent of the mass defect (q.v.) of the nucleus, usually about 200 MeV per nucleus, some of which is in the form of gamma rays (q.v.).

A nuclear explosion will not occur until a critical mass (q.v.) of fissionable material is attained, that is, the smallest amount capable of sustaining a chain reaction. Similarly, a nuclear reactor will not produce power until the assembly achieves a critical activity. This occurs as follows. The neutrons introduced to the system are continually escaping; some are lost through the walls, some are captured by structural materials, and some are absorbed by the fissionable atoms themselves without fission taking place. When the conditions are so balanced that the number of neutrons entering the system is exactly equal to those escaping from it, the assembly is said to be critical. When the neutrons entering the system are very slightly in excess of those lost to it, the reactor is supercritical, and measurable power generation takes place. The ratio is carefully controlled, the rate of energy production rising exponentially. Control rods made of cadmium absorb neutrons so readily that precise levels of power-generating activity are possible.

See also breeder; nuclear reaction.

Note: Fission-derived energy is very gradually being adapted to electric power production; expansion is being delayed by safety and environmental considerations. It has been officially stated that the risk of accidental release of radioactive material from a nuclear power plant is extremely slight, and that it is impossible for it to explode like an atomic bomb, since the amount of fissionable material involved is far below the critical mass required. Nonetheless, some instances of leakage have occurred, necessitating long and costly shutdowns. Use of nuclear fission for a major percentage of power production in the U.S. appears to be many years away. In 1975 it was estimated to be 9% of total power production. Constant inspection for plant defects and observation of the strictest safety controls will always be essential.

Fittig's synthesis. The preparation of aromatic hydrocarbons by condensation of aryl halides with alkyl halides in the presence of metallic sodium.

"Fixanol" PA.[325] Trademark for a resinous condensation product; a water-colored syrupy liquid used as a fixation product for direct dyes.

fix (1) To cause an unreactive element to combine into a chemical compound, as in ammonia synthesis. See also nitrogen fixation.

(2) To hold a dye permanently on a fiber or fabric by chemical or mechanical action, or a combination of both.

(3) To retard the evaporation rate of the volatile components of essential oils, as in perfumes.

See also following entries.

fixation. See nitrogen fixation.

fixative. (1) See fixing agent, perfume.

(2) A substance applied as a spray or solution to harden and preserve objects for microscopic examination, or to pencil and ink drawings to prevent blurring, e.g., a sodium silicate solution.

fixed oil. A nonvolatile, fatty oil characteristic of vegetables, as opposed to the volatile essential oils of flowers.

fixing agent, chemical.
(1) A substance which aids fixation of mordants upon textiles by uniting chemically with them and holding them on the fiber until the dyes can react with them.
(2) A substance which causes actual precipitation of mordant on the fiber by double decomposition.

fixing agent, mechanical.
(1) A substance (e.g., albumin) capable of holding pigments permanently upon textile fibers.
(2) Certain gums and starches which hold dyes and other substances upon textile fibers long enough to permit a desirable reaction to take place.

fixing agent, perfume (fixative). A substance which prevents too rapid volatilization of the components of a perfume, and tends to equalize their rates of volatilization. It thus increases the odor life of a perfume and keeps the odor unchanged. For many years the chief fixatives were animal products (ambergris, civet, musk, castoreum), but these have been largely replaced by synthetics. See also perfume.

flake lead. See lead carbonate, basic.

"Flamenol."[245] Trademark for electrical conductors insulated with a vinyl halide resin such as plasticized polyvinyl chloride.

flame-retarding agent. A material used as a coating on or a component of a combustible product to raise its ignition point. The protection provided is only partial, and most materials so treated will burn when exposed to high temperatures. There are three types: (1) nondurable, consisting of water-soluble inorganic salts, which are easily removed by washing or accidental exposure to water; (2) semi-durable (removed by repeated laundering); and (3) durable (not affected by laundering or dry-cleaning). The latter type includes organic compounds of bromine and chlorine, and insoluble metal salts; antimony trioxide, tricresyl phosphate and other phosphate esters, chlorendic acid, etc. are effective, as well as cellulose-reactive agents. Zinc carbonate in high-volume concentration will render a rubber or plastic compound self-extinguishing.

flammability. The ease with which a material (gas, liquid, or solid) will ignite, either spontaneously (pyrophoric), from exposure to a high-temperature environment (autoignition), or to a spark or open flame. It also involves the rate of spreading of a flame once it has started. The more readily ignition occurs, the more flammable the material; less easily ignited materials are said to be combustible, but the line of demarcation is often indefinite, and depends on the state of subdivision of the material as well as on its chemical nature. The Flammable Fabrics Act (1967) established standards of flammability to which all textile manufacturers must conform.
See also flammable material; combustible material.

flammable material. Any solid, liquid, vapor, or gas that will ignite easily and burn rapidly. Flammable solids are of several types: (1) dusts or fine powders (metals or organic substances such as cellulose, flour, etc.); (2) those that ignite spontaneously at low temperatures (white phosphorus); (3) those in which internal heat is buit up by microbial or other degradation activity (fish meal, wet cellulosic materials); (4) films, fibers, and fabrics of low-ignition point materials.
Flammable liquids are defined by the National Fire Protection Association as those having a flash point below 100°F (Closed Cup) and a vapor pressure of not over 40 psia at 100°F. The shipping regulatory authorities use 80°F (Open Cup) and 73°F (Closed Cup), and Manufacturing Chemists Association uses 80°F (Open Cup) as the flash point below which liquids are considered flammable.
Flammable gases are ignited very easily; the flame and heat propagation rate is so great as to resemble an explosion, especially if the gas is confined. The most common flammable gases are hydrogen, carbon monoxide, acetylene and other hydrocarbon gases. Oxygen, through essential for the occurrence of combustion, is not itself either flammable or combustible; neither are the halogen gases, sulfur dioxide or nitrogen. Flammable gases are extremely dangerous fire hazards, and require precisely regulated storage conditions.
Note: The terms "flammable," "nonflammable," and "combustible" are difficult to delimit. Since any material that will burn at any temperature is combustible by definition, it follows that this word covers all such materials, irrespective of their ease of ignition. Thus the term "flammable" actually applies to a special group of combustible materials that ignite easily and burn rapidly. Some materials (usually gases) classified in shipping and safety regulations as nonflammable are actually noncombustible. The distinction between these terms should not be overlooked. For example, sodium chloride, carbon tetrachloride, and carbon dioxide are noncombustible; sugar, cellulose, and ammonia are combustible but nonflammable.
See also combustible material.

flash. The overflow of cured rubber or plastic from a mold. Correct design will minimize flash, which is an economic loss.

flash distillation. Distillation in which an appreciable proportion of a liquid is quickly converted to vapor in such a way that the final vapor is in equilibrium with the final liquid. This method is now widely used for desalination of sea water.

flash photolysis. A method of investigating the mechanism of extremely rapid photochemical reactions involving the formation of free radicals (both inorganic and organic) by irradiating a given reaction mixture with a flash of high-intensity light, thus producing the short-lived radicals which activate photochemical reactions. These products are instantaneously analyzed spectroscopically, which permits identification of the radical species from the spectra obtained. The time lapses involved in this technique are about 1/100,000th second. It has also been applied to study of the exceedingly fast reactions occurring in flames and explosions. See also photochemistry; free radical; photolysis.

flash point. The temperature at which a liquid or volatile solid gives off a vapor sufficient to form an ignitable mixture with the air near the surface of the liquid or within the test vessel (NFPA). For the purposes of the official shipping regulations, the flash point is determined by the Tagliabue open-cup method (ASTM D1310-63), usually abbreviated TOC. (IATA also permits the Abel or Abel-Pensky closed-cup tester.) Other methods used, generally for the higher flash points, are the Tag closed cup (TCC) and Cleveland open cup (COC). The open cup method more nearly approximates actual conditions.
See also flammable material.

flatting agent. A substance ground into minute particles of irregular shape and used in paints and varnishes to disperse incident light rays, so that a dull or "flat" effect is produced. Standard flatting agents are heavy-metal soaps, finely divided silica, and diatomaceous earth.

flavanol (3-hydroxyflavone). A derivative of flavanone (q.v.); crystallizes in yellow needles melting at 169°C. Has violet fluorescence in conc. sulfuric acid. It is a flavonoid pigment (q.v.). Dyes cotton a bright yellow when mordanted with aluminum hydroxide. Other hydroxyflavones are chrysin, fisetin and quercitin. Eleven different flavonols are known. Not identical with flavonol.

flavanone (2,3-dihydroflavone). A group of colorless derivatives of flavone (q.v.) distributed in higher plants either in free form or as glucosides. About 25 different types have been isolated. It comprises one of the major groups of flavonoids (q.v.). Examples are hesperidin and naringen.

flavanthrene (indanthrene yellow; chloranthrene yellow) $C_{28}H_{12}O_2N_2$.
Properties: Brownish-yellow needles. Soluble in dilute alkaline solutions.
Derivation: Action of antimony pentachloride on beta-aminoanthraquinone in boiling nitrobenzene.
Use: Vat dye for textiles, etc.

"Flavaxin."[162] Trademark for riboflavin (q.v.).

flavianic acid. See 2,4-dinitro-1-naphthol-7-sulfonic acid.

flavin.
(1) Isoalloxazine, $C_{10}H_6N_4O_2$, the nucleus of various natural yellow pigments. See riboflavin and flavin enzymes.
(2) Tetrahydroxyflavanol, $C_{15}H_{10}O_7 \cdot 2H_2O$, a yellow dye derived from oak bark.

flavine. Acriflavine hydrochloride, a bacteriostatic agent.

flavin enzyme (flavoprotein). An enzyme composed of protein linked to coenzymes which are mono- or dinucleotides containing riboflavin. Because of their distinctive color they are also called "yellow enzymes." The flavin enzymes function in tissue respiration as dehydrogenases, the hydrogen atoms being taken up by the riboflavin group.

flavin mononucleotide. See riboflavin phosphate.

flavone (2-phenylchromone). One of a group of flavonoid plant pigments, existing as colorless needles, insoluble in water and melting at 100°C. It fluoresces violet in conc. sulfuric acid. It can be synthesized. Treatment with alcoholic alkali yields flavanone (q.v.). The flavones produce ivory and yellow colors in plants and flowers. See also flavonoid.

flavonoid. A group of aromatic, oxygen-containing heterocyclic pigments widely distributed among higher plants. They constitute most of the yellow, red, and blue colors in flowers and fruits. Exceptions are the carotenoids (q.v.). The flavonoids include the following subgroups: (1) catechins, (2) leucoanthocyanidins and flavonones, (3) flavanols, flavones, and anthocyanins; and (4) flavonols. For details, consult specific entries.

flavonol (flavon-3-ol). A flavonoid plant pigment giving ivory and yellow colors to flowers. Not identical with flavanol (q.v.).

flavoprotein. See flavin enzyme.

flavor. (1) The simultaneous physiological and psychological response obtained from a substance in the mouth that includes the senses of taste (salty, sour, bitter, sweet), smell (fruity, pungent), and feel. The sense of feel as related to flavor encompasses only the effect of chemical action on the mouth membranes, such as heat from pepper, coolness from peppermint, and the like. (Institute of Food Technologists). No reliable correlation of taste with chemical structure has yet been possible. Flavor is a critical factor in the acceptability of foods, medicines, confectionery, and beverages. Flavors are used in insect and animal baits to induce ingestion of the bait; also to prevent rodent attack on organic materials, e.g., tributyl tin in cable covers. Substances that affect flavor often have a synergistic effect (for example, monosodium glutamate and certain nucleotides). Sodium chloride is classed as a seasoning agent. See also potentiator; enhancer.
(2) Any substance or mixture of substances that contributes a positive taste to a food product, such as vanillin, cacao, and fruit extracts among natural products, together with numerous synthetic compounds that imitate or duplicate these tastes. Undesirable or "off-flavors" occur in milk, meat and other food products as a result of improper preparation, oxidation, and incipient rancidity. There are over 1500 flavoring materials listed as food additives under provisions of the Food, Drug and Cosmetics Act. See also odor.

flax. Bast fibers, approximately 20 in. long, obtained from the stems of the linseed plant, linum usitatissimum. Stronger and more durable than cotton. Combustible.
Uses: Apparel fabrics (linens); thread; rope; twine; cigarette paper; duplicating papers.

"Flaxedil."[57] Trademark for gallamine triethiodide (q.v.).

flaxseed oil. See linseed oil.

flea seed. See psyllium.

"Flectol" H.[58] Trademark for polymerized 1,2-dihydro-2,2,4-trimethylquinoline.
Use: Rubber antioxidant, especially for belting and tire carcasses.

Fleming, Sir Alexander (1881–1955). A Scottish biochemist and bacteriologist who discovered (1928) the bactericidal properties of molds produced from the plant Penicillium notatum. The broad spectrum of antibiotics developed from this discovery. See also antibiotic.

"Flexalyn" 80M.[266] Trademark for a pale, balsamic resin; the diethylene glycol ester of rosin.

"Flexamine G."[248] Trademark for a mixture of N,N'-diphenyl-para-phenylenediamine and a complex diarylamine-ketone reaction product.
Properties: Brownish gray granules; sp. gr. 1.20; melting range 75–90°C; soluble in acetone, benzol and ethylene dichloride; insoluble in water and gasoline.
Uses: Antioxidant used in tires, camelback, wire insulation, neoprene belting and soles.

"Flexane."[445] Trademark for room-temperature-curing urethane.

"Flexbond."[144] Trademark for polyvinyl acetate copolymer emulsions.
Properties: 48% to 57% solids; 60 to 3800 cp; 4.0 to 6.5 pH; 0.14μ to 1.5μ average particle size; nonionic; borax stable except for 803 grade; mechanically stable.
Containers: Bulk, 55-gal. lined steel or fiber drums.
Uses: Adhesive bases for paper, cloth, plastic, and leather; vehicles for house paints or industrial finishes; binders for pigments in paper coatings.

"Flexclad."[265] Trademark for polyester resin powders for use in coating applications.

"Flexichem."[152] Trademark for a series of soaps composed of sodium stearate, sodium tallowate, calcium stearate, zinc stearate, potassium stearate, and sodium oleate. Food grade materials available.
Uses: Gelling and thickening agents, lubricant, concrete water absorption reducing agents, plasticizer, and processing aids for rubber and metal products.

"Fleximet."[152] Trademark for a series of soaps used as rod and wire processing lubricants.

"Flexol."[214] Trademark for a series of plasticizers and stabilizers including phthalates, adipates, polyalkylene glycol derivatives, polymeric epoxies, decanoates, octoates, hexoates, tri(2-ethylhexyl)phosphate and dibutyltin dilaurate.
Uses: Film and sheeting; flooring; coated fabrics; wire and cable and other extrusions; organosols, plastisols, and plastigels; lacquers and rubbers.

"Flexon."[51] Trademark for a series of petroleum process oils mainly used as compounding and extender oils in synthetic rubber and resins.

"Flexo Wax C."[73] Trademark for a synthetic, noncrystalline hydrocarbon wax; m.p. 66–74°C; has cold flow of 63–65°C; high adhesive and waterproofing properties.
Uses: Adhesives; coatings for paper; greaseproofing; lacquers and enamels; nickel alloy stamping, plating; water-repellent coatings.

"Flexricin."[202] Trademark for a group of castor oil derivatives including (1) alkyl ricinoleates, compatible with cellulosic resins, polyvinyl butyral, polyamide, rosin and shellac and used as plasticizers; (2) acetyl ricinoleates, the alkyl esters of acetylated ricinoleic acid used as plasticizers for nitrocellulose, ethylcellulose, polyvinyl chloride resins and copolymers, natural and synthetic rubbers, and other polymers and as textile lubricants; and (3) polyol ricinoleates used for synthesis, cosmetics, brake fluids, in waxes and greases, and as anti-foam agents.

"Flexzone."[248] Trademark for a series of antiozonants, antioxidants and stabilizers based on paraphenylenediamine.

flint. A crystalline form of native silica or quartz.
Properties: Color, smoky gray, brownish, blackish, dull yellowish; luster waxy to greasy; Mohs hardness 6.5–7; sp. gr. 2.60–2.65. More easily soluble in hot caustic alkali than crystallized quartz.
Occurrence: Europe; U.S.
Uses: Abrasive; balls for ball mills; paint extender; filler for fertilizer, insecticides, rubber, plastics and road asphalt; ceramics; chemical tower packing.

"Flintflex."[28] Trademark for air- or force-dry organic coating system used for linings in interiors of containers that haul dry bulk ladings of edibles or chemicals. Complies with FDA Regulations. Accepted by the Meat Inspection Division of the U.S. D.A. for interiors of freight cars, motor trucks and trailers and in federally inspected meat-processing plants.
Containers: Up to 55-gal drums.

flint glass. See glass, optical.

floatation. Purification and/or classification of finely divided solids, e.g. clays, by passing them through an airblast. Do not confuse with flotation (q.v.).

flocculant. A substance that induces flocculation (q.v.). Flocculants are used in water purification, liquid waste treatment and other special applications. Inorganic flocculants are lime, alum, and ferric chloride; polyelectrolytes (q.v.) are examples of organic flocculants.

flocculation. The combination or aggregation of suspended solid particles in such a way that they form small clumps or tufts resembling wool. The word is derived from this appearnce. Carbon black displays a tendency to flocculate in rubber when improperly dispersed, and some colloidal clays have the same property. Oil-well drilling muds are made alkaline to prevent flocculation of their components. Flocculation can often be reversed by agitation, as the cohesive forces are relatively weak. This is not true of other forms of aggregation (coalescence and coagulation), which are irreversible.

flock. A light powder comprised of ground wood or cotton fibers used as an extender or filler in plastics, low-grade rubber and flooring compositions.

"Floform."[319] Trademark for carbon black used for various industrial purposes and especially for compounding with rubber.

"Flo-Gard."[177] Trademark for amorphous calcium polysilicate used as an anticaking agent for salt.

"Flogel."[266] Trademark for regular, high-density, and aluminized grades of water-resistant, nonnitroglycerin slurry explosives so formulated that a strong booster is required for their detonation.

"Flomet-Z."[329] Trademark for a fine, white, grit-free powder containing 12.5 to 14.0% zinc oxide. Used as a lubricant in powdered iron metallurgy.

"Florco."[98] Trademark for a granular mineral absorbent processed from attapulgite. Used to absorb liquids from concrete, terrazzo and other surfaces.

"Florex."[98] Trademark for attapulgite which has been substantialaly improved in adsorption efficiency by high-pressure extrusion and other special processes. Recommended where good adsorption capacity is essential. Available in various grades.

"Florigel H-Y."[98] Trademark for a high-yield salt water drilling clay.

"Florisil."[98] Trademark for a highly selective adsorbent of extremely white, hard granular or powdered magnesia-silica gel (magnesium silicate).
"Florisil" TLC. Special grade for thin layer chromatography.
Uses: Preparative and analytical chromatography.

"Florite" Activated Bauxite.[98] Trademark for products produced from carefully selected domestic and imported bauxite ores. Used in the refining of petroleum oils, distillates, petrolatums, waxes and many other organic solvents; also as desiccant for gases and liquids.

"Floropryl."[123] Trademark for diisopropyl fluorophosphate.

flotation. A process for separating minerals from waste rock, or solids of different kinds from one another by agitating the pulverized mixture of solids with water, oil, and special chemicals which cause preferential wetting of solid particles of certain types by the oil, while other kinds are not wet. The unwetted particles are carried to the surface by the air bubbles and thus separated from the wetted particles. A frothing agent is also used to stabilize the bubbles in the form of a froth which can be easily separated from the body of the liquid (froth flotation). Do not confuse with floatation (q.v.).

"Flotox."[253] Trademark for flotation sulfur products.

"Flotronics."[535] Trademark for metal membrane filters of uniformly porous silver, available in distinct porosity grades from 0.2 to 5 micron maximum pore diameter. No bonding agent or fiber is used; the filters are of uniform particles of pure silver only. Used for filtration, clarification, and cold sterilization of all types of solutions.

"Flovis."[73] Trademark for a modified polyoxyethylene fatty-acid ester.
Properties: Cream to tan solid; m.p. 39–42°C; sp. gr. 1.02 (25°C); pH 5% aqueous dispersion 3–5 (25°C).
Uses: Textile and adhesive industries for stabilizing starch solutions, fluid or heavy paste, against "setting-up."

"Flowbrite."[142] Trademark for a formulation of oils used at elevated temperature for the bright flowing of electroplated tin.

"Flowco."[329] Trademark for a group of stearates. Used as finishes and additives in paper and pulp.

flow diagram (flow sheet). A chart or line drawing used by chemical engineers to indicate successive steps in the production of a chemical, materials input and output, by-products, waste, and other relevant data.

flowers. A fine powder usually resulting from sublimation of a substance, e.g., flowers of sulfur. The term is now obsolescent.

flox. A mixture of liquid fluorine (30%) and liquid oxygen (70%), designed for use as a space vehicle propellant.
Containers: Stainless steel tanks.
Hazard: Highly flammable.
Shipping reuglations: Consult authorities.

"Flozan."[244] Trademark for an anhydrous soda ash product. Contains a bicarbonate of soda and a monohydrate of soda ash. White, free-flowing and will absorb approximately 30% of its weight.
Uses: Cleansing and detergent preparations.

"Flozene."[45] Trademark for a series of lubricant white mineral oils; sp. gr. 0.860–0.870 (60°F).
Use: Industrial lubricants.

fludrocortisone acetate (9alpha-fluorohydrocortisone acetate) $C_{23}H_{31}FO_6$. 9Alpha-17alpha-hydroxycorticosterone-21 acetate.
Properties: White, odorless polymorphic crystals; m.p. 233–234°C. Sparingly soluble in alcohol and very slightly soluble in water.
Uses: Medicine; veterinary medicine.

"Fludrocortone."[123] Trademark for fludrocortisone.

fluid. Any material or substance that changes shape or direction uniformly in response to an external force imposed upon it. The term applies not only to liquids, but to gases and to finely divided solids. Fluids are broadly classified as Newtonian and non-Newtonian depending on their obedience to the laws of classical mechanics. See also liquid, Newtonian; rheology; fluidization; hydraulic fluid.

"Fluid Ball."[84] Trademark for a propellant casting powder consisting of colloidal nitrocellulose having an average particle diameter of 50 microns or less. Composition can include liquid or solid modifiers. Used as binder constituent of modified double base solid rocket propellants.
Containers: Fiber drums, 100-lb net weight.
Hazard: Highly flammable; dangerous fire risk.

fluidization. A process in which a finely divided solid is caused to behave like a fluid by suspending it in a moving gas or liquid. The solids so treated are frequently catalysts; hence, the term fluid catalysis. The fluidized catalyst, e.g., alumina-silica gel, is brought into intimate contact with the suspending liquid or gas mixture, usually a petroleum fraction. Local overheating of the catalyst is greatly reduced, and portions of catalyst can be easily removed for regeneration without shutting down the unit. There are also non-catalytic applications of fluidization, e.g., reduction of iron ore. Important uses of the fluidized bed process are (1) cracking of petroleum fractions; (2) gasification of coal; and (3) application of organic coatings to metals (fusion bond method).

fluoboric acid (fluoroboric acid) HBF_4.
Properties: Colorless, strongly acid liquid; sp. gr. approx 1.84; b.p. 130° with decomposition. Miscible with water, alcohol.
Derivation: Action of boric and sulfuric acids on fluorspar.
Grades: Technical (about 48%); pure.
Hazard: Highly toxic; a corrosive irritant.
Uses: Production of fluoborates; specially purified solution used for electrolytic brightening of aluminum. Also to form stabilized diazo salts.
Shipping regulations: (Rail) Corrosive liquid, n.o.s., White label. (Air) Corrosive label.

"Fluokarb."[184] Trademark for an activated bonechar used for fluoride removal.

fluor. See phosphor.

fluophosphate alkyl ester. See diisopropyl fluorophosphate.

fluoranthene (idryl) $C_{16}H_{10}$. A tetracyclic hydrocarbon.
Properties: Colored needles; f.p. 107°C; b.p. 250°C (60 mm); insoluble in water; soluble in ether and benzene. Combustible.
Derivation: From coal tar.
Hazard: Moderately toxic.

fluorapatite. See apatite.

fluorbenside. Generic name accepted for para-chlorobenzyl para-fluorophenyl sulfide, $ClC_6H_4CH_2SC_6H_4F$. Crystals; m.p. 36°C. Insoluble in water; soluble in acetone and oils. Used as an acaricide.

"Fluorel"[158] Trademark for a fully-saturated fluorinated polymer containing more than 60 per cent fluorine by weight; non-flammable.
Uses: O-rings, gaskets; hoses; wire and fabric coatings; diaphragms; fuel cells; expellent bladders; sealants; insulation; containers.

fluorene (alpha-diphenylenemethane) $\overline{C_6H_4CH_2C_6H_4}$. A tricylic hydrocarbon.
Properties: Small, white, crystalline plates; fluorescent when impure. Soluble in alcohol, ether, benzene and carbon disulfide; insoluble in water. M.p. 116°C; b.p. 295°C (with decomposition). Combustible.
Derivation: By reduction of diphenylene ketone with zinc; from coal tar.
Grades: Technical; 98% pure.
Uses: Resinous products; insecticides; dyestuffs.

fluorescein (resorcinolphthalein; diresorcinolphthalein; see also uranine) $C_{20}H_{12}O_5$.
Properties: Orange-red, crystalline powder; very dilute alkaline solutions exhibit intense, greenish-yellow fluorescence by reflected light, while the solution is reddish-orange by transmitted light; m.p., decomposes at 290°C; soluble in dilute alkalies, boiling alcohol, ether and dilute acids, glacial acetic acid; insoluble in water, benzene, chloroform. Combustible; low toxicity.
Derivation: By heating phthalic anhydride and resorcinol.
Grades: The sodium salt (uranine) and potassium salt are marketed.
Uses: Dyeing sea water for spotting purposes; tracer to locate impurities in wells; dyeing silk and wool; medicine; indicator and reagent for bromine.

fluorescence. A type of luminescence (q.v.) in which an atom or molecule emits visible radiation in passing from a higher to a lower electronic state. The term is restricted to phenomena in which the time interval between absorption and emission of energy is extremely short (10^{-8} to 10^{-3} second). This distinguishes fluorescence from phosphorescence, in which the time interval may extend to several hours. Fluorescent materials may be liquid or solid, organic or inorganic. Fluorescent crystals such as zinc or cadmium sulfide are used in lamp tubes, television screens, scintillation counters, and similar devices. Fluorescent dyes are used for labeling molecules in biochemical research. See also phosphorescence; phosphor; resonance (2).

fluoridation. Addition to public drinking water supplies of about 1 ppm of a fluorine salt for the purpose of reducing the incidence of dental caries. The chemicals most commonly used for city fluoridation programs are fluosilicic acid, sodium silicofluoride, and sodium fluoride. The concentration used has been established to be far below the permissible level of toxicity of fluorine-containing compounds in the human body. The program was successfully tested for over twenty years on local populations, and since then has been widely adopted in large cities in the U.S. The protection is especially effective for children, whose teeth are usually more susceptible to caries than those of adults. Fluorine is a bone-seeking element; tooth protection is due to the ability of fluoride ion to replace other ions in hydroxyapatite (q.v.), the chief mineral component of bones and teeth. Fluorides are used in toothpastes and other dentrifices.

fluorinated ethylene-propylene resin (FEP resin). A copolymer of tetrafluoroethylene and hexafluoropropylene with properties similar to polytetrafluoroethylene resin. The repeating structure of the molecule is $[-CF_2-CF_2-CF_2-CF(CF_3)-]_n$. See also "Teflon."
Properties: Similar to polytetrafluoroethylene (q.v.). Capable of continuous service from -450° to +400°F. Has higher coefficient of friction.
Available as extrusion and molding powder, aqueous dispersion, film, monofilament fiber, and nonsticking finish.
Uses: Wire and cable insulation; pipe linings; lining for processing equipment. Fibers are used for filtration screening and mist separators.

fluorine F Nonmetallic halogen element in group VIIA of the periodic classification. Atomic number 9. Atomic weight 18.9984; valence 1. No stable isotopes. The most electronegative and most reactive element. The molecular formula is F_2.
Properties: Pale yellow diatomic gas or liquid; pungent odor; b.p. -188°C; f.p. -219.6°C; density of gas 1.695 (air = 1); sp. gr. of liquid (-188°C) 1.108; sp. vol. 10.2 cu ft/lb (70°F). Reacts vigorously with most oxidizable substances at room temperature, frequently with ignition. Forms fluorides with all elements except helium, neon, and argon.
Occurrence: Widely distributed to the extent of 0.03% of the earth's crust. The chief minerals are fluorapatite, cryolite, and fluorspar.
Derivation: Electrolysis of molten anhydrous hydrofluoric acid—potassium fluoride melts with special copper-bearing carbon anodes, steel cathodes, containers and monel screens.
Containers: Special steel cylinders; tank trailers (as liquid). Available as a liquid and a compressed gas.
Hazard: Highly toxic; strong corrosive irritant. Flammable, dangerous fire risk. Tolerance 1 ppm in air. Powerful oxidizing agent.
Uses: Oxidizer in rocket fuels; production of metallic and other fluorides; production of fluorocarbons (q.v.).
Shipping regulations: (Rail) Red gas label. (Air) Not acceptable.

fluorine cyanide. See cyanogen fluoride.

fluorite. See fluorspar.

fluoroacetic acid CH_2FCOOH.
Properties: Colorless crystals; m.p. 33°C; b.p. 165°C; soluble in water, alcohol. Nonflammable.
Hazard: Toxic and irritant.

fluoroacetone CH_2FCOCH_3.
Properties: Yellow liquid. Pungent odor; sp. gr. 0.967 (20°C); b.p. 72.5°C. Nonflammable.

fluoroacetophenone (phenacyl fluoride; phenyl fluoromethylketone) $C_6H_5COCH_2F$.
Properties: Brown liquid. Pungent odor. B.p. 98°C (8 mm). Nonflammable.
Derivation: By Friedel-Crafts synthesis.
Hazard: Toxic and irritant. Tolerance (as F), 2.5 mg per cubic meter of air.

para-**fluoroaniline** $FC_6H_4NH_2$.
Properties: Liquid; sp. gr. (25°C) 1.1524; b.p. 187.4°C; m.p. 0.82°C; refractive index (n 20/D) 1.5395. Nonflammable.
Hazard: Probably toxic. See aniline.
Uses: Dye intermediate; preparation of para-fluorophenol.

fluorobenzene (phenyl fluoride) C_6H_5F.
Properties: Colorless liquid with benzene odor; sp. gr. (20°C) 1.0252; refractive index (n 25/D) 1.4636; b.p. 84.9°C; freezing point 41.9°C. Insoluble in water; miscible with alcohol, ether. Nonflammable.
Hazard: Toxic and irritant. Tolerance (as F), 2.5 mg per cubic meter of air.
Uses: Insecticide and larvicide intermediate; identification reagent for plastic or resin polymers.

fluoroboric acid. See fluoboric acid.

fluorocarbon. Any of a number of organic compounds analogous to hydrocarbons, in which the hydrogen atoms have been replaced by fluorine. The term is loosely used to include fluorocarbons that contain chlorine; these should properly be called chlorofluorocarbons or fluorocarbon chlorides, since it is these which are thought to deplete the ozone layer of the upper atmosphere. Studies of this possibility are in progress; meanwhile the use of chlorofluorocarbons as aerosol propellants has been reduced.

Properties: Fluorocarbons are chemically inert, nonflammable, and stable to heat up to 500–600°F. They are denser and more volatile than the corresponding hydrocarbons, and have low refractive indices, low dielectric constants, low solubilities, low surface tensions, and viscosities comparable to hydrocarbons. Some are compressed gases; others are liquids.

Hazard: React violently with reactive substances, e.g., barium, sodium, and potassium.

Uses: Aerosol propellants, refrigerants, solvents, blowing agents, fire extinguishment; lubricants and hydraulic fluids; flotation and damping fluids; dielectrics; plastics; electrical insulation; wax coatings for alkali cleaning tanks; air conditioning.

Note: The rather considerable number of these compounds is designated by a number system, preceded by the word "refrigerant," "propellant," "fluorocarbon" or by a trademark ("Freon," "Ucon," "Genetron.") They are cross-referenced in this book as follows:

11. See trichlorofluoromethane.
12. See dichlorodifluoromethane.
13. See chlorotrifluoromethane.
14. See tetrafluoromethane.
21. See dichlorofluoromethane.
22. See chlorofluoromethane.
23. See fluoroform.
113. See 1,1,2-trichloro-1,2,2-trifluoroethane.
114. See 1,2-dichloro-1,1,2,2-tetrafluoroethane.
115. See chloropentafluoroethane.
116. See hexafluoroethane.

fluorocarbon polymer. This term includes polytetrafluoroethylene, polymers of chlorotrifluoroethylene, fluorinated ethylene-propylene polymers, polyvinylidene fluoride, hexafluoropropylene, etc.

Properties: Thermoplastic; resistant to chemicals and oxidation; noncombustible; broad useful temperature range (up to 550°F); high dielectric constant; resistant to moisture, weathering, ozone, and ultraviolet radiation. Their structure comprises a straight backbone of carbon atoms symmetrically surrounded by fluorine atoms.

Forms: Powders, dispersions, film, sheet, tubes, rods, tapes, and fibers.

Uses: High-temperature wire and cable insulation; electrical equipment; food, drug and chemical equipment; coating of cooking utensils; piping; gaskets; continuous sheet.

See also fluoroelastomer.

fluorochemical. Organic compounds, not necessarily hydrocarbons, in which a large percentage of the hydrogen directly attached to carbon has been replaced by fluorine. The presence of two or more fluorine atoms on a carbon atom usually imparts stability and inertness to the compound and fluorine usually increases the acidity of organic acids.

Derivation: (a) Electrolysis of solutions in hydrogen fluoride (Simons process); (b) replacement of chlorine or bromine by fluorine with hydrogen fluoride in the presence of a catalyst (antimony trifluoride or pentafluoride); (c) addition of hydrogen fluoride to olefins or acetylene.

Uses: Dielectric and heat-transfer liquids; pump sealants; surfactants; metering devices; special solvents.

See also fluorocarbon; fluoroelastomer.

fluorodichloromethane. See dichlorofluoromethane.

fluoroelastomer. Any elastomeric high polymer containing fluorine; they may be homopolymers or copolymers. Fluorocarbon polymers (q.v.) include a large group of fluoroelastomers. Besides these, a relatively recent development is a copolymer in which the molecular skeleton is a —P=N— chain containing approximately equal numbers of tri- and heptafluoroethoxy side groups. Such polymers are amorphous, thermally stable, noncombustible, have low glass transition temperature (-77°C), and are generally resistant to attack by solvents and chemicals.

fluoroethylene. See vinyl fluoride.

fluoroform (trifluoromethane; propellant 23; refrigerant 23) CHF_3.

Properties: Colorless, nonflammable gas; b.p. -84°C (1 atm); f.p. -160°C (1 atm); sp. vol. 5.5 cu ft/lb (70°F, 1 atm).

Grade: 98% min. purity.

Containers: Up to 70 lb steel cylinders.

Hazard: Moderately toxic and irritant. Narcotic in high concentrations.

Uses: Refrigerant; intermediate in organic synthesis; direct coolant for infrared detector cells; blowing agent for urethane foams.

Shipping reuglations: (Air) Nonflammable Gas label.

fluoroformyl fluoride. See carbonyl fluoride.

"Fluoroinert."[158] Trademark for a series of perfluorinated liquids used for cleaning electronic components after testing. B.p. range 88–345°F; high dielectric strength; colorless; nonflammable.

"Fluorolubes."[62] Trademark for polymers of trifluorovinyl chloride $(\text{—}CF_2\text{—}CFCl\text{—})_x$ containing 49% fluorine and 31% chlorine. Products are light oils, heavy oils, and grease-like materials.

Uses: Lubricant and sealant for plug cocks, valves, and vacuum pumps; impregnant for gaskets and packings; fluid for hydraulic equipment, heat exchange and instrument damping.

fluoromethane (methyl fluoride) CH_3F.

Properties: Colorless gas; b.p. −78.2°C; f.p. −142°C; density 0.877.

Hazard: Toxic and flammable. Tolerance (as F), 2.5 mg per cubic meter of air.

Shipping regulations: (Rail) Not listed. (Air) Flammable Gas label. Not accepted on passenger planes. Legal label name methyl fluoride.

fluorometholone. A synthetic derivative of prednisolone, 21-desoxy-9α-fluoro-6-methylprednisolone. Listed in N.D. Used in medicine.

para-**fluorophenol** FC_6H_4OH.

Properties: White crystalline solid; sp. gr. (56°C) 1.1889; m.p. 48.2°C (stable form), 28.5°C (unstable form); b.p. 185.6°C (760 mm), 78°C (15 mm). Soluble in water.

Hazard: Toxic and irritant.

Use: Fungicide; intermediate for pharmaceuticals and fungicides.

fluorophosphoric acid H_2PO_3F.
Properties: Colorless viscous liquid; sp. gr. 1.1818 (25°C); miscible with water.
Hazard: Highly toxic. Strong irritant to tissue.
Uses: Metal cleaners; electrolytic or chemical polishing agents, formation of protective coatings for metal surfaces; catalyst.
Shipping regulations: (anhydrous) (Rail) Corrosive liquids, n.o.s., White label. (Air) Corrosive label. Not acceptable on passenger planes.
See also difluorophosphoric acid and hexafluorophosphoric acid.

fluorosilicic acid. Legal label name (Air) for fluosilicic acid.

fluorosulfonic acid. Legal label name (Air) for fluosulfonic acid (q.v.).

fluorosulfuric acid. See fluosulfonic acid.

fluorothene. See chlorotrifluoroethylene polymer.

fluorotrichloromethane. See trichlorofluoromethane.

fluorspar (fluorite, florspar) CaF_2. Natural calcium fluoride. Color yellow, green, or purple crystals; Mohs hardness 4; sp. gr. 3.2; m.p. 1350°C.
Grades: Metallurgical, ceramic, and acid, containing more than 85, and 98% CaF_2, respectively.
Occurrence: U.S.; Canada; Europe; Mexico.
Containers: Bulk in railroad cars or barges; 125-lb bags; 500-lb barrels.
Use: Principal source of fluorine and its compounds, by way of hydrogen fluoride; flux in open hearth steel furnaces and in metal smelting; in ceramics; for synthetic cryolite; in carbon electrodes; emery wheels; electric arc welders; certain cements; dentifrices; phosphor; paint pigment; catalyst in wood preservatives; optical equipment.

fluosilicate. A salt of fluosilicic acid H_2SiF_6. For possible synonyms, see fluosilicic acid.

fluosilicic acid (hydrofluosilicic acid; fluorosilicic acid; hexafluorosilicic acid; hydrogen hexafluorosilicate; hydrosilicofluoric acid; hydrofluorosilicic acid; sand acid) H_2SiF_6.
Properties: Colorless, fuming liquid. Attacks glass and stoneware. Soluble in water.
Derivation: By-product of the action of sulfuric acid on phosphate rock containing fluorides and silica or silicates. The hydrofluoric acid acts on the silica to produce silicon tetrafluoride, SiF_4, which reacts with water to form fluosilicic acid, H_2SiF_6.
Grades: Technical; C.P.
Containers: Special plastic bottles; 50-gal barrels; tank cars.
Hazard: Highly toxic; extremely corrosive by skin contact and inhalation.
Uses: Water fluoridation; ceramics (to increase hardness); disinfecting copper and brass vessels; hardening cement, etc.; technical paints; wood preservative and impregnating compounds; electroplating; manufacture of aluminum fluoride, synthetic cryolite (q.v.), and hydrogen fluoride.
Shipping reuglations: (Rail) White label. (Air) Corrosive label. Legal label name (Rail) hydrofluosilicic acid, (Air) fluorosilicic acid.

fluosulfonic acid (fluorosulfuric acid; fluorosulfonic acid) HSO_3F.
Properties: Colorless, fuming liquid; sp. gr. 1.745 (15°C); f.p. −87°C; b.p. 165°C; soluble in nitrobenzene, soluble in water with partial decomposition.
Derivation: Reaction of anhydrous hydrogen fluoride with sulfuric acid or sulfuric acid anhydride.
Hazard: Highly toxic. Extremely irritating to eyes and tissue.
Uses: Catalyst in organic synthesis; electropolishing.
Shipping regulations: (Rail) White label. (Air) Corrosive label. Not accepted on passenger planes. Legal label name (Air) fluorosulfonic acid.

flushed color. A pigment dispersed in oil, varnish, etc., the transfer from the water phase to the oil phase having been effected without the usual drying and subsequent grinding of the dry pigment. It is claimed that flushed colors are ready for use without grinding.

flux.
(1) A substance that promotes the fusing of minerals or metals or prevents the formation of oxides. For example, in metal refining lime is added to the furnace charge to absorb mineral impurities in the metal. A slag (q.v.) is formed which floats on the bath and is run off.
(2) A substance applied to metals that are to be united. On application of heat, it aids the flow of solder and prevents formation of oxides.
(3) Any readily fusible glass or enamel used as a base or ground in ceramic processing.
(4) The rate of flow or transfer of electricity, magnetism, water, heat, energy, etc., the term being used to denote the quantity that crosses a unit area of a given surface in a unit of time.
(5) The intensity of neutron radiation, expressed as the number of neutrons passing through 1 square centimeter in 1 second.
(6) A mixture of sodium nitrate and sodium nitrite; oxidizing agent used as a low explosive.

fluxing lime. See calcium oxide.

fly ash. The very fine ash produced by combustion of powdered coal with forced draft, and often carried off with the flue gases from such processes. Special equipment is necessary for effective recovery, e.g., electrostatic precipitators. Considerable percentages of CaO, MgO, silica and alumina are present. Fly ash is used as a cement additive for oil-field well casings; as an absorbent for oil spills (silicone-coated); as a filler in plastics; and (in England) as a source of germanium. Experimental work has been carried out on its use for separation of oil-sand tailings; it may also be useful in removal of heavy metals from industrial waste waters.
Note: A new cement, utilizing from 50 to 70% fly ash and said to be superior to portland cement in several respects, has been developed in Belgium.

Fm Symbol for fermium.

FMN. Abbreviation for flavin mononucleotide. See riboflavin phosphate.

foam. A dispersion of a gas in a liquid or solid. The gas globules may be of any size, from colloidal to macroscopic, as in soap bubbles. Bakers' bread and sponge rubber are examples of solid foams. Typical liquid foams are those used in fire-fighting, shaving creams, etc. In such foams the liquid must have sufficient cohesion to form an elastic film, e.g., soap, oil, protein, fatty acids, etc. Surfactant-induced foams have been developed to increase the efficiency of fuel cells.

Foams made by mechanical incorporation of air are widely used in the food industry, e.g., whipped cream, eggwhite, ice cream, etc. Useful foams for auto seats, mattresses and similar uses are made from natural and synthetic latexes, e.g., polystyrene, polyurethane.

A glass foam is based on sodium silicate and rock wool, and vitreous ceramic foams are also available. Metals can be caused to foam. Concrete foams are also in general use. See also foam, metal; foam, fire-extinguishing; foam, plastic.

foam, fire-extinguishing. An agglomeration of small bubbles of gas, produced by two methods: (1) by chemical reaction between aluminum sulfate and sodium bicarbonate to generate carbon dioxide (chemical foams), and (2) by mixing or agitation of air with water containing the foaming ingredients (mechanical foams). The two types are equally efficient in extinguishing ability.

Besides the foaming ingredients, the foams contain stabilizing agents to assure permanence; there are many of these, for example, soaps, proteins, extract of licorice root, fatty acids and sulfite liquors. The ingredients of chemical foams are assembled in two separate units, which generate the foam on blending.

Fire foams are used primarily on fires in hydrocarbon liquids (Class B fires). There are many special types tailored for specific uses. Recently developed fire protection systems for aircraft include an instant-generating foam for cabin interiors, using a 2.5% aqueous solution of alkyl sulfonate, and rigid polyurethane foams for use in fuel tanks. See also fire extinguishment.

"Foam-Cel."[160] Trademark for a foam-polyethylene insulation used on solid copper conductor with impervious aluminum sheath.

Uses: Signal transmission in all high-frequency applications including community antennae, closed circuit TV, electronic circuits of all types.

"Foamex."[73] Trademark for a mixture of aliphatic esters.

Properties: Very pale yellow liquid. Insoluble in water; sp. gr. 0.96–0.97.

Uses: Foam retardation and prevention in water solutions of glue, casein, shellac, gelatin, etc.

foam, metal. A cellular metallic structure, usually of aluminum or zinc alloys, made by incorporating titanium or zirconium hydride in the base metal. This subsequently evolves hydrogen to produce a uniform, foam-like material. Its specific gravity is about that of sea-water, so that it is weightless when submerged. The principal use of foamed metals is in absorption of shock impact without elastic rebound. Fiber-reinforced light-metal foams have potential application in reducing weight of automobile bodies.

foam, plastic. A cellular plastic (q.v.) which may be either flexible or rigid. Flexible foams may be polyurethane, rubber latex, polyethylene or vinyl polymers; rigid foams are chiefly polystyrene, polyurethane, epoxy, and polyvinyl chloride. The blowing agents used are sodium bicarbonate, halocarbons such as CCl_3F, and hydrazine. Flexible polystyrene foam is available in extruded sheets and also in the form of "beads" made by treating a polystyrene suspension with pentane; these expand from 30 to 50 times on heating, and are used as automobile radiator sealants. Rigid foams are widely used for building insulation, boat construction, filtration, fillers in packing cases, absorption of oil spills, etc.

Hazard: The most widely used types of organic foam plastics (polystyrene, polyurethane, polyisocyanurate) are combustible; even when fire-retardant agents are incorporated, such foams will burn. The extent of burning or fire severity will vary with surface treatment, end-use location, recipe, and degree of protection. Thin coatings of fire-retardant paint, metal, or automatic sprinkler systems may not adequately protect against rapid fire spread. Organic foamed plastic surfaces should not be left exposed. Multiple adjacent surfaces like walls and ceiling create a most severe hazard because of the chemical kinetics associated with radiative, conductive and convective currents developed during a building fire.

New methods of making such plastic foams as polyurethane that are reported to reduce their combustibility have been developed, for example, use of trichlorobutylene oxide instead of propylene oxide.

f.o.b. Abbreviation for "freight on board," a designation used in shipping a material to indicate that freight charges are to be paid by the purchaser. This is in contrast to "freight prepaid and allowed," indicating that freight charges are paid by the manufacturer.

fog. A suspension of liquid droplets in air; an aerosol. The size of the droplets ranges from colloidal to macroscopic. "Synthetic" fogs can be produced on a laboratory scale by ultrasonic vibrations, and natural fogs can be precipitated by the same means. Mists of fogs comprised of atomized particles of oil are used as military concealment screens and for insecticidal purposes in orchards and truck gardens. See also smog; chemical smoke; aerosol.

folacin. See folic acid.

"Folex."[40] Trademark for a 75% emulsifiable concentrate of tributylphosphorotrithioite for use as a cotton defoliant. See also "Merphos."

"Foliafume."[342] Trademark for pyrethrin-rotenone plant spray concentrates.

folic acid (pteroylglutamic acid; folacin; PGA) $C_{19}H_{19}N_7O_6$. Considered a member of the vitamin B complex. At least three substances with folic acid activity (see also folinic acid) occur in nature, one of which, pteroylglutamic acid, is made synthetically.

Properties (pteroylglutamic acid): Orange-yellow needles or platelets; tasteless; odorless; slightly soluble in water; insoluble in lipid solvents; soluble in dilute alkali hydroxide and carbonate solutions; stable to heat in neutral and alkaline solution; destroyed by heating with acid; inactivated by light.

Sources: Green plant tissue, fresh fruit, liver, and yeast. Synthetic pteroylglutamic acid made by the reaction of 2,3-di-bromopropanol, 2,4,5-triamino-6-hydroxypyrimidine and para-aminobenzoyl glutamic acid.

Grade: 10% feed grade; U.S.P.

Containers: Fiber drums.

Uses: Medicine; nutrition; feed additive.

folinic acid (5-formyl-5,6,7,8-tetrahydropteroyl-L-glutamic acid; citrovorum factor; leucoverin) $C_{20}H_{23}N_7O_7$. A member of the folic acid group of vitamins and a growth factor for the bacterium Leuconostoc citrovorum. Folinic acid is an important

metabolite of folic acid and may be the active form in cellular metabolism. It is an effective hematopoietic factor. Ascorbic acid and vitamin B_{12} are essential for the conversion of folic acid to folinic acid.
Properties: *dl*-L-form: Crystals; decompose 240–250°C; sparingly soluble in water.
Derivation: (a) Prepared by catalytic reduction of folic acid; (b) produced microbially.
Uses: Medicine; nutrition; biochemical research.

folpet. Generic name for N-(trichloromethylthio)-phthalimide, $C_6H_4(CO)_2NSCCl_3$.
Properties: Light colored powder; insoluble in water; slightly soluble in organic solvents.
Hazard: May be toxic.
Uses: Fungicide-bactericide for vinyls, paints and enamels.

"Folvite."[315] Trademark for folic acid (q.v.).

"Fomade."[69] Trademark for a combination of foam former, foam stabilizer and vinyl stabilizer.
Properties: Light tan, semi-fluid paste; sp. gr. 1.02.
Uses: Foaming agent for polyvinyl chloride.

"Fomrez."[104] Trademark for a series of polyester and polyether resins, stannous octoate catalysts, and coupling agents used in the manufacture of urethane foams.

"Fonoline."[45] Trademark for petrolatum, of soft consistency and low melting point, with color range of white to yellow and meeting U.S.P. or N.F. purity requirements for petrolatum.

food. Any substance or mixture which, when ingested by man or animals, contributes to the maintenance of vital processes. With the exception of sodium chloride and water, all foods are derived from plants, either by direct consumption or by ingestion of animal tissue or such animal products as eggs, milk, etc., which are derived metabolically from vegetable sources. Basic foods are composed of proteins, fats (lipids), and carbohydrates, together with vitamins and minerals. Ancillary items that are associated with foods, though with little or no nutritive value, are collectively called food additives, e.g., flavorings, spices, preservatives and colorants. Many of these are also plant-derived, though some are now made synthetically. See also plant (1); nutrient; and following entries.

food additive.
(a) Intentional: The Food Protection Committee of the National Research Council states that a food additive is "a substance or mixture other than a basic foodstuff that is present in food as a result of any aspect of production, processing, storage, or packaging." See also additive (note).
(b) Unintentional: Substances that may become part of a food product as a result of chance contamination, such as insecticide residues, fertilizers, and the like. The permissible content of insecticide residues has been established by the FDA.
The Food Additives Amendment to the Food, Drug, and Cosmetic Act empowers the FDA to disapprove or discontinue any food additive that it determines to be unsafe at the level of intended use, based on data supplied by the manufacturer.

food color (certified color). A colorant which may be either a dye (soluble) or a lake (insoluble) permissible for use in foods, drugs, or cosmetics by FDA. The dyes color by solution and the lakes by dispersion. All must satisfy strict regulations as to toxicity. See also FD&C colors.

Food and Drug Administration (FDA). The Federal agency responsible for administering and enforcing the Food, Drug, and Cosmetic Act, including the Food Additives Amendment which went into effect in March, 1960. It has the authority to require proof of the efficacy and safety of drugs, foods, and pharmaceuticals, to conduct and evaluate screening tests, and to compel withdrawal from the market of any such product that it finds ineffective or hazardous. It establishes tolerances on food and animal feed additives of all types, including pesticides, as well as on cosmetic products, flammable fabrics, and packaging and labeling materials. It can also require specific statement on labels of the components or ingredients of a product, as well as precautionary warnings. See also Environmental Protection Agency; food additive.

food engineering. Application of engineering principles to the design of equipment for large-scale food processing, e.g., automatic harvesting devices; dryers of various types; crystallizers; ovens and heat-exchangers; comminuting and mixing equipment; distillation units; packaging machines. Food engineering requires an understanding of thermodynamics, conditions of state and equilibrium, rate processes, and transport phenomena. The unit operations of chemical engineering and basic physics and mechanics are also involved.

food technology. Practice of the techniques used in the preparation of foods for large-scale human use. Among others, these include harvesting; post-harvest treatment; all forms of cooking; tenderizing; preservation by chemicals, heating, dehydration, drying, and freezing; distillation and solvent extraction; milling; refining; hydrogenation; emulsification; packaging materials; and storage, labeling and transportation. Other aspects of food technology are bacteriology, sanitation, quality control, and formulation of ingredients for a wide variety of end products. A recent development of importance is the growth of convenience and quick-service foods.

foots (soapstock). The mixture of soap, oil, and impurities that precipitates when natural fatty oils are refined by treatment with caustic soda or soda ash. Usually contains 30 to 50% of free and combined fatty acids. Used for manufacture of relatively low-grade soaps, as a source of free fatty acids. A related meaning is the suspended solid matter in crude oils.

"Foramine."[36] Trademark for urea-formaldehyde adhesives.

"Forasite."[36] Trademark for a phenolic water-soluble adhesive used in bonding comminuted wood products, applicable to both wet and dry processes.

"Foray."[548] Trademark for a monoammonium phosphate-based formulation used to extinguish fires in flammable liquids (Class B fires) and in combustible materials such as wood and paper (Class A fires).

"Foremul."[309] Trademark for a series of polyethylene esters of fatty acids and polyglycol fatty acid condensate.
Uses: Emulsifier in the fatliquoring of leather; formulation of insecticides and agricultural sprays; emulsifier for vegetable, animal and mineral oils, waxes, chlorinated hydrocarbons.

forensic chemistry. See legal chemistry.

formal. See methylal.

"Formaldafil."[539] Trademark for glass fiber reinforced acetal.

formaldehyde (oxymethylene; formic aldehyde; methanal) HCHO. A readily polymerizable gas. It is commercially offered as a 37 to 50% aqueous solution, which may contain up to 15% methanol to inhibit polymerization. These commercial grades are called formalin. It is one of the few organic compounds known to exist in outer space. 22nd highest-volume chemical produced in U.S. (1975)
Properties (gas): Strong, pungent odor. Sp. gr. 0.82; vapor density. 1.08; b.p. −19°C; f.p. −118°C; autoignition point 806°F; soluble in water and alcohol. (Aqueous 37% solution with 15% methanol): B.p. 101°C; flash point 122°F; (methanol-free): b.p. 101°C; flash point 185°F. Sp. gr. 0.82.
Derivation: Oxidation of synthetic methanol or low-boiling petroleum gases such as propane and butane. Silver, copper, or an iron-molybdenum oxide are the most common catalysts.
Grades: Aqueous solutions: 37%, 44%, 50% inhibited (with varying percentages of methanol) or stabilized or unstabilized (methanol-free); also available in solution in n-butanol, methanol, or urea; U.S.P. (37% aqueous solution containing methanol). See also paraformaldehyde, the polymerized, solid form.
Containers: Drums, bottles, carboys, tank cars, tank trucks.
Hazard (gas): Highly toxic by inhalation; strong irritant. Moderate fire risk. Tolerance, 2 ppm in air. Explosive limits in air 7 to 73%. (Solution): Avoid breathing vapor and skin contact. Safety data sheet available from Manufacturing Chemists Assn., Washington, D.C.
Uses: Urea and melamine resins; polyacetal resins; phenolic resins; ethylene glycol; pentaerythritol; hexamethylenetetramine; fertilizer; dyes, medicine (disinfectant, germicide); embalming fluids; preservative; hardening agent; reducing agent, as in recovery of gold and silver; corrosion inhibitor in oil wells; durable-press treatment of textile fabrics; possible condensation to sugars and other carbohydrates for food use (experimental); industrial sterilant; treatment of grain smut.

formaldehyde aniline (formaniline) $C_6H_5NCH_2$.
Properties: Colorless to yellowish crystals; initial m.p. 133°C; b.p. 271°C; sp. gr. 1.14, but these vary somewhat from sample to sample. Soluble in water, ether, alcohol.
Derivation: Condensation of formaldehyde and aniline.
Hazard: Toxic by ingestion.
Uses: Rubber accelerator; intermediate.

formaldehyde cyanohydrin. See glycolonitrile.

formaldehyde-para-**toluidine** (methylene-para-toluidine) $(CH_3C_6H_4NCH_2)_x$.
Properties: White powder with grayish-yellow cast. Aromatic odor. Not toxic to skin; soluble in acetone; sp. gr. 1.11.
Derivation: Reaction between formaldehyde and para-toluidine.
Uses: Rubber accelerator; dyes.

formalin. An aqueous 37 to 50% solution of formaldehyde which may contain 15% methyl alcohol. See formaldehyde.

formamide (methanamide) $HCONH_2$.
Properties: Colorless, hygroscopic oily liquid; sp. gr. 1.146; b.p. 200–212°C with partial decomposition beginning about 180°C; m.p. 2.5°C. Soluble in water and alcohol. Flash point 310°F; combustible.
Derivation: Interaction of ethyl formate and ammonia, with subsequent distillation.
Containers: Drums, tank cars.
Hazard: Moderately toxic.
Uses: Solvent; softener; intermediate in organic synthesis.

formanilide (phenylformamide) C_6H_5NHCHO.
Properties: Colorless to yellowish crystals; soluble in alcohol and water; m.p. 48–50°C; b.p. 271°C.
Derivation: Reaction of aniline and formic acid.
Use: Medicine.

formaniline. See formaldehyde aniline.

"Formaset" LC-1.[42] Trademark for a syrup type urea formaldehyde resin. Used as a stabilizing and bodying agent for textiles.

"Formcel."[352] Trademark for a series of water-free formaldehyde solutions in alcohols.
Uses: Alcoholated urea and melamine resins, coatings, laminating and textile-treating resins; embalming fluids.

"Formica."[13] Trademark for high-pressure laminated sheets of melamine and phenolic plastics for decorative applications as surfacing; insulating material; adhesives for bonding laminated plastic to other surfaces; flakeboard for use as corestock with laminated plastic; industrial plastics using various thermosetting resins, combined with various base materials for electrical, mechanical and chemical applications.

formic acid (hydrogen carboxylic acid; methanoic acid) HCOOH.
Properties: Colorless, fuming liquid; pungent penetrating odor. Soluble in water, alcohol and ether; sp. gr. 1.2201 (20/4°C); m.p. 8.3°C; b.p. 100.8°C; flash point 156°F (open cup); lb/gal (20°C) 10.16; refractive index 1.3719 (20°C). Combustible. Autoignition temp. 1114°F. Strong reducing agent.
Derivation: (a) By treatment of sodium formate and sodium acid formate with sulfuric acid at low temperatures, and distilling in vacuo; (b) by acid hydrolysis of methyl formate; (c) as a by-product in the manufacture of acetaldehyde and formaldehyde.
Method of purification: Rectification.
Grades: Technical; 85%; 90%; C.P.; F.C.C.
Containers: Carboys; stainless steel tanks and drums.
Hazard: Corrosive to skin and tissue. Tolerance, 5 ppm in air. Avoid skin contact!
Uses: Dyeing and finishing of textiles and paper, leather treatment; chemicals (formates, oxalic acid, organic esters); manufacture of fumigants, insecticides, refrigerants, solvents for perfumes, lacquers; electroplating; medicine; brewing (antiseptic); silvering glass; cellulose formate; natural latex coagulant; ore flotation; vinyl resin plasticizers; animal feed additive.
Shipping regulations: (Rail) White label. (Air) Corrosive label.

formic aldehyde. See formaldehyde.

formonitrile. See hydrocyanic acid.

"Formopon."[23] Trademark for sodium formaldehyde hydrosulfite. "Formopon" Extra is the basic zinc salt.

"Formrez."[104] Trademark for a series of resins and prepolymers used to produce urethane elastomers (solid cast urethanes).

"Formula 40."[233] Trademark for alkanolamine salts of 2,4-D weed killers.

formula, chemical. A written representation, using symbols, of a chemical entity or relationship. There are several kinds of formulas, as follows:

(1) Empirical: Expresses in simplest form the *relative* number and the kind of atoms in a molecule of one or more compounds; it indicates composition only, not structure. Example: CH is the empirical formula for both acetylene and benzene.

(2) Molecular: shows the actual number and kind of atoms in a chemical entity (i.e., a molecule, group, or ion). Examples: H_2 (1 molecule of hydrogen), $2(H_2SO_4)$ (2 molecules of sulfuric acid); CH_3 (a methyl group); $Co(NH_3)_6^{++}$ (an ion).

(3) Structural: Indicates the location of the atoms, groups, or ions relative to one another in a molecule, as well as the number and location of chemical bonds: Examples:

```
           CH3
           |
    CH2=C—CH=CH2 (isoprene)
```

```
        H   H
        |   |
        C=C
       /     \
  H—C         C—H
      \\     //
        C—C
        |   |
        H   H
```

benzene

Since all molecules are 3-dimensional they cannot properly be shown in the plane of the paper. This is sometimes indicated by extra-heavy lines or 3-dimensional artwork (configurational formula) as in this representation of an ethanol molecule:

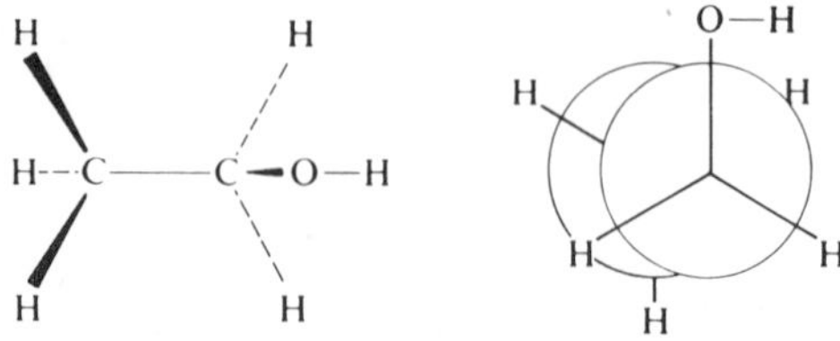

(4) Generic: Expresses a generalized type of organic compound, where the variables stand for the number of atoms or for the kind of radical in a homologous series (q.v.). Examples:

C_nH_{2n+2}	C_nH_{2n}
a paraffin	an olefin
ROR	ROH
an ether	an alcohol

(R = a hydrocarbon radical)

(5) Electronic: A structural formula in which the bonds are replaced by dots indicating electron pairs, a single bond being equivalent to one pair of electrons shared by two atoms. Example: the electronic formula for methane is:

```
   H
   ..
 H:C:H
   ..
   H
```

formula, product. A list of the ingredients and their amounts or percentages required in an industrial product. Such formulas (or recipes) are mixtures, not compounds; they are generally used in such industries as adhesives, food, paint, rubber and plastics. See also formulation.

formulation. Selection of components of a product formula or mixture to provide optimum specific properties for the end use desired. Formulation by experienced technologists is essential for products intended to meet specifications or special service conditions.

formula weight. The sum of the atomic weights represented in a chemical formula. Thus, since the atomic weight of hydrogen is 1, and oxygen is 16, the formula weight of water (H_2O) is 18. (Approximate atomic weights used).

"Formvar."[58] Trademark for polyvinyl formal resins.
Uses: Wire enamels; electrical insulation; coatings; adhesives; films and molded materials; electron microscopy.

2-formyl-3,4-dihydro-2H-pyran. See acrolein dimer.

formyl fluoride HCOF.
Properties: Colorless, mobile liquid. Decomposes slowly with formation of hydrofluoric acid and carbon monoxide. Soluble in water (dec). B.p. -26°C.
Derivation: Interaction of benzoyl chloride and a formic acid solution of potassium fluoride.
Grades: Technical.
Hazard: Toxic by ingestion and inhalation; strong irritant to tissue. Tolerance (as F), 2.5 mg per cubic meter of air.
Uses: Organic synthesis.

1-formylpiperidine $\overline{C_5H_5N}CH{=}O$.
Properties: Colorless liquid; liquid from −30 to 222°C; aprotic; low-volatility. Miscible with alcohols, esters, ketones, amines, amides, inorganic acids, organometallics. Soluble in water and hexane.
Uses: Solvent for polar and nonpolar compounds, as well as many high polymers; gas absorption; plastics modifiers.
See also N,N-dimethylformamide.

"Forthane."[100] Trademark for methylhexaneamine (q.v.).

"Forticel."[352] Trademark for a cellulosic thermoplastic for use in injection molding, extrusion, rotational casting, and blow molding.
Properties: Pellets (crystals, translucent, metallic and opaque colors). Sp. gr. 1.20. Highest use temp. 175°F. Combustible. Soluble in organic solvents such as ketones and esters. Insoluble in mineral oils.
Containers: Fiber drums.
Uses: Pen and pencil barrels, telephone bases, spectacle frames, tool handles, sheeting, steering wheels, etc.

fortification. In food technology, addition to a food ingredient or product of nutrients that are not normally present, for example, addition of vitamin D to milk or of vitamin C to cake-fillings. Nutritionists apply this term to foods especially designed for school children and elderly persons. See also nutrification; enrichment.

"Fortiflex."[352] Trademark for a high-density polyethylene consisting mainly of long molecules with occasional short side branches. Thermoplastic.
Properties: Milk-white translucent pellets. Colors are also available. Density 0.95; melt index 0.2–8; tensile strength 3,100 to 3,700 psi; highest use temperature 225° F; combustible.
Uses: See polyethylene.

"Fortisan"[352] Trademark for a cellulosic fiber manufactured by partial saponification of stretched cellulose acetate. A semisynthetic product (q.v.). It resembles cellulose (cotton) in many respects. The high-tenacity product has a dry strength of 5 to 7 pounds per denier (100,000 to 130,000 psi); wet strength is 85% of dry strength. It has relatively low elongation under stress; elastic recovery is about 70% after extension to break; immediate elastic recovery is 46%, delayed recovery 30%, at 5% strain. Young's modulus 1650. Can be dyed in the same way as cotton. The monofilament can be produced to a fineness of 1 denier. Specific gravity 1.50. Resists stretching both dry and wet. Combustible.

"Fortracin."[342] Trademark for animal feed supplements of bacitracin methylene disalicylate.

"Fortrel."[206,352] Trademark for a polyester type synthetic fiber.

"Fosbond."[204] Trademark for a group of chemicals used to provide a corrosion-resistant bond between zinc or ferrous metals and a paint film.

"Foscoat."[204] Trademark for a class of chemicals designed to provide a phosphate coating prior to cold working.

"Fosfodril."[55] Trademark for a glassy phosphate of high molecular weight (sodium hexametaphosphate).
Uses: Thickener for drilling muds; water treatment in oil-well flooding operations.

"Fosfo" Rosin.[79] Trademark for limed FF wood rosin.
Uses: Box toes; matches; printing ink; smoking molds; pipe bending.

"Foslube."[204] Trademark for a class of lubricants used to impregnate a phosphate coating prior to cold working.

"Fosrinse."[204] Trademark for a class of chemicals used to render insoluble the acid salts remaining after phosphate treatment of ferrous metals.

fossil. Any material that results from an animal or vegetable source in past geologic ages and has been buried (compressed) in the earth. Examples are fossil fuels (petroleum, natural gas, coal, lignite), fossil waxes (ozocerite, montan), fossil resins (amber) and fossil woods partially preserved (petrified) by the action of silica.

"Fostalite."[568] Trademark for light-stable polystyrenes, specially formulated for fluorescent light fixtures.

"Fostarene."[568] Trademark for molding and extrusion grade polystyrenes.

"Fosterite."[308] Trademark for a family of resins. Largest application is as "solventless" varnishes for electric insulation, also as a photoelastic resin and as a bond for impregnating and laminating asbestos sheets. Rods made of this plastic will carry a beam of light without the dispersion which occurs in air, making it possible to bend the beam.

"Fotoceram."[20] Trademark for a photosensitive, crystalline ceramic; used in electronics and industrial arts.

"Fotocol."[319] Trademark for special industrial alcohol solvents. Available in regular and anhydrous formulations.
Containers: Up to 55-gal drums.
Uses: Printing, photography, and rubber industries.

"Fotoform."[20] Trademark for photosensitive glass; used in electronics.

"Foundrez."[36] Trademark for a group of water-soluble phenol-formaldehyde and urea-formaldehyde resins for foundry applications.

foundry sand (greensand; molding sand). Sand mixed with suitable binders used in making molds for casting metals. The sand/binder mixture is either rammed into place around the mold or baked into a core at 400–500° F (dry-sand molding). The binders used are resins of various types, casein, etc.

Fourdrinier. The machine most widely used for paper-making, named for its English inventors, who introduced it in the mid-19th century. It produces a wide range of papers from heavy board to light tissue, and is of impressive size and complexity. Its unique feature is the traveling wire mesh belt onto which the slurry of fiber and water is run from the headbox. The sheet is formed on this wire almost instantaneously, most of the water draining through the interstices of the wire. After leaving the wire the web passes through the press section of the machine, where a number of rollers express enough of the remaining water to enable the sheet to hold together. It then moves into the multi-roller drying section (described under drying). The dried sheet (4 to 6% moisture content) is then fed to a high-speed calender for compaction and finishing. The entire process is continuous and rapid, the machine often operating for several days without shutdown. See also calendering; drying (2); paper.

"FP" Acids.[484] Trademark for a series of fluorophosphoric acids.

FPC. (1) Abbreviation for fish protein concentrate (q.v.). (2) Abbreviation for Federal Power Commission.

Fr Symbol for francium.

fraction. Any portion of a mixture characterized by closely similar properties. The most important fractions of petroleum are naphtha, gasoline, fuel oil, kerosene, and tarry or waxy residues. These are obtained by fractional distillation (q.v.). See also separation.

fractional distillation. Distillation in which rectification is used to obtain product as nearly pure as possible. A part of the vapor is condensed and the resulting liquid contacted with more vapor, usually in a column with plates or packing. The term is also applied to any distillation in which the product is collected in a series of separate components of similar boiling range. See also reflux.

fractionation. In general, the separation or isolation of components of a mixture or a micromolecular complex. In distillation this is done by means of a tower or column in which rising vapor and descending liquid are brought into contact (counter-current flow). Macromolecular components (proteins and other high polymers) can be separated by a number of methods, including electrophoresis, gel filtration, chromatography, centrifugation, foam fractionation, and partition. See also reflux.

fracturing, hydraulic. A method of inducing flow in oil wells by injecting water or brine under pressure into the oil-bearing strata. The viscosity of the water is often increased by adding gelling agents such as guar gum, cellulose derivatives, or polyacrylamides.

francium Fr Element of atomic number 87, group IA of the periodic system. Atomic weight 223; valence +1. It appears to exist only as radioactive isotopes. One isotope is actinium K (Fr^{223}). Other isotopes have been made artificially. Francium 223 is the longest-lived isotope, having a half-life of 21 minutes and is the only natural isotope. Francium is the heaviest of the alkali-metal family.

fragrance. An odorant used to impart a pleasant smell to shaving lotions, toothpastes, men's accessories, etc. Balsamic and piny odors are typical.

franklinite (Fe,Mn,Zn)$(FeMn)_2O_4$. Black mineral resembling magnetite (q.v.).

Frary metal. A lead-base bearing metal containing 97–98% lead alloyed with 1–2% each of barium and calcium. Excellent for low-pressure bearings at moderate temperatures.

Frasch process. A process by which much of the world's sulfur is obtained. Developed about 1900 by Herman Frasch, the process involves melting sulfur underground by introducing superheated water through a pipe under pressure and forcing the molten sulfur to the surface by compressed air.

Fraunhofer lines. See spectroscopy.

free energy. An exact thermodynamic quantity used to predict the maximum work obtainable from the spontaneous transformation of a given system. It also provides a criterion for the spontaneity of a transformation or reaction, and predicts the greatest extent to which the reaction can occur, i.e., its maximum yield. Transformation of a system can be brought about by either heat or mechanical work. Free energy is derived from the internal energy and entropy of a system in accordance with the laws of thermodynamics.

free radical. A molecular fragment having one or more unpaired electrons, usually short-lived and highly reactive. In formulas a free radical is conventionally indicated by a dot, as $Cl^{\cdot}$,$(C_2H_5)^{\cdot}$. In spite of their transitory existence, they are capable of initiating many kinds of chemical reactions by means of a chain mechanism. Free radicals are formed only by the splitting of a molecular bond. A chain can result only if (1) radicals attack the substrate, and (2) the radicals lost by this reaction are regenerated. Chain mechanisms for the thermal decomposition of many substances have been established. Free radicals are known to be formed by ionizing radiation, and thus play a part in the deleterious degradation effects that occur in irradiated tissue. They also act as initiators or intermediates in such basic phenomena as oxidation, combustion, photolysis, and polymerization. See also carbonium ion.

free sulfur. Sulfur which is left chemically uncombined after vulcanization of a rubber compound. When this exceeds 1%, the upper limit of solubility of sulfur in rubber, blooming will occur. Most rubber products are vulcanized with as low a sulfur content as possible, so that the free sulfur content of the product is seldom over 0.5%. See also bloom; vulcanization.

freeze-drying (lyophilization). A type of dehydration for separating water from biological materials. The material is first frozen and then placed in a high vacuum so that the water (ice) vaporizes in the vacuum (sublimes) without melting, and the non-water components are left behind in an undamaged state. Used for blood plasma, certain antibiotics, vaccines, hormone preparations, food products such as coffee and vegetables. One technique prepares freeze-dried ceramic pellets from water solutions of metal salts.

"Freezene."[45] Trademark for a series of refrigeraton white mineral oils.
Use: Low-temperature lubrication.

freezing point. See melting point.

"Freon."[28] Trademark for a series of fluorocarbon products used in refrigeration and air-conditioning equipment, as aerosol propellants, blowing agents, fire extinguishing agents, and cleaning fluids and solvents.
Properties: Clear, water-white liquids. Vapors have a mild, somewhat ethereal odor and are not irritating; low toxicity; nonflammable, nonexplosive; noncorrosive and essentiallay stable and inert. For listing of specific types, see fluorocarbon.
Note: Many types contain chlorine as well as fluorine, and should be called chlorofluorocarbons.

"Freon" E.[28] Trademark for a series of hydrogen end-capped tetrafluoroethylene epoxide polymers having a D.P. up to 10. Boiling range 39–490°C. High dielectric constant. Used as coolants in electronic devices.

"Freon" C-51-12. See perfluorodimethylcyclobutane.

"Frianite."[118] Trademark for a processed anhydrous potassium aluminum silicate. Typical chemical analysis: Silicon dioxide, 74.7%; aluminum oxide, 13.7%; potassium oxide, 5.6%; other oxides, 4.3%.
Properties: Fine, pinkish, inert powder containing no soluble salts. Sp. gr. 2.37; pH 5.4 to 6.5; density 47 lb/cu ft. Free flowing liquid holding capacity is about 1½% max.
Use: Diluent for dry blending or formulating dusting pesticides.

Friedel-Crafts reaction. Originally defined as the condensation of alkyl or aryl halides with benzene and its homologs in the presence of anhydrous aluminum chloride. This definition has been widened to include analogous processes.
Uses: Alkylation and acylation in general. Some examples are the production of high-octane gasoline; production of cumene, ethylbenzene, detergent alkylate, various plastics and elastomers.

frit. A ground glass used in making glazes and enamels and also for making so-called frit seals. Finely powdered glass may be called a frit. The term is also used for finely ground inorganic minerals, mixed with fluxes and coloring agents, which turn into a glass or enamel on heating.

"FR-N."[278] Trademark for a series of butadiene-acrylonitrile elastomeric polymers and latices.
Properties: Oil and solvent resistance combined with good flexibility and resistance to low temperatures, water absorption and permanent set.

Uses: Oil-resistant seals, shoe soles, gasoline hose, belt conveyors, plasticizers, paper saturation, adhesives, leather finishes and carpet backing.

froth flotation. See flotation.

FRP. Abbreviation for glass fiber-reinforced plastic. See reinforced plastic.

"FR-S."[278] Trademark for general-purpose rubbers and latexes, composed of copolymers of butadiene and styrene.
Uses: Rubber: Tires, hose, belting and packing; molded and extruded automotive and industrial products; soles and heels; hard rubber. Latex: Adhesives; foamed rubber; textile and rug backing; paper coating and impregnation; modification of plastics to produce high impact strength; asphalt additive.

"FRTP."[539] Trademark for glass fiber-reinforced thermoplastic.

fructose (fruit sugar; D(-)-fructose; levulose) $C_6H_{12}O_6$. A sugar occurring naturally in a large number of fruits and in honey. It is the sweetest of the common sugars.
Properties: White crystals; soluble in water, alcohol and ether; m.p. 103–105°C (dec); specific rotation -89 to -91°. Combustible; nontoxic.
Derivation: Hydrolysis of inulin; hydrolysis of beet sugar followed by lime separation; from cornstarch by enzymic or microbial action.
Grades: Technical; N.F.; food; parenteral.
Containers: Wooden barrels; tins; fiber drums.
Uses: Foodstuffs; medicine; preservative.

fructose-1,6-diphosphate (FDP; fructosediphosphoric acid; Harden-Young ester) $H_2PO_4(C_6H_{10}O_4)H_2PO_4$. Can be prepared from fructose and certain other sugars by the use of yeasts. It is known to take part in cell metabolism; an intermediate in carbohydrate metabolism. Usually handled in the form of its barium or calcium salts, white amorphous powders, soluble in ice water and dilute acid solutions; insoluble in hot water and alcohol.
Uses: Organic synthesis; research in cell metabolism.

fruit sugar. See fructose.

F salt. The sodium salt of F acid; a dye intermediate. See 2-naphthol-7-sulfonic acid.

FT black. Abbreviation for fine thermal black. See thermal black.

FTC. Abbreviation for Federal Trade Commission (q.v.).

"Fuadin."[162] Trademark for stibophen (q.v.).

fuchsin (basic fuchsin; magenta). A synthetic rosaniline dyestuff, a mixture of rosaniline and pararosaniline hydrochlorides.
Properties: Dark green powder or greenish crystals with a bronze luster; faint odor. Soluble in water and alcohol.
Grades: N.F.
Uses: Textile and leather industries; as a red dye; pharmaceutical.

fuel. Any substance that evolves energy in a controlled chemical or nuclear reaction. The most common type of chemical reaction is combustion, the type of oxidation occurring with petroleum products, natural gas, coal, and wood; more rapid oxidation takes place in rocket fuels (hydrogen, hydrogen peroxide, hydrazine) which approaches the rate of an explosion. The nuclear fuels used for power generation release their energy by fission of the atomic nucleus (uranium, plutonium, thorium). See also combustion; fission.

fuel cell. (1) An electrochemical device for continuously converting chemicals—a fuel and an oxidant—into direct-current electricity. It consists of two electronic-conductor electrodes separated by an ionic-conducting electrolyte with provision for the continuous movement of fuel, oxidant and reaction product into and out of the cell. The fuel can be gaseous, liquid or solid, the electrolyte liquid or solid, the oxidant gaseous or liquid. The electrodes are solid, but may be porous and may contain a catalyst. Fuel cells differ from batteries in that electricity is produced from chemical fuels fed to them as needed, so that their operating life is theoretically unlimited. The cell products can be regenerated externally into fuel for return to the cell, e.g., CO_2 from the cell can be reacted with coal to form CO for feed to the cell. Fuel is oxidized at the anode (negative electrode) giving electrons to an external circuit; the oxidant accepts electrons from the anode and is reduced at the cathode. Simultaneously with the electron transfer, an ionic current in the electrolyte completes the circuit. One type of electrolyte is a solid polymer of perfluorinated sulfonic acid. The fuels range from hydrogen, carbon monoxide, and carbonaceous materials to redox compounds, alkali metals, and biochemical materials. Fuel cells based on hydrogen and oxygen have a significant future as a primary energy source. Large-scale development of fuel cells for on-site power generation for housing units is well advanced, and research is under way for construction of a 26-megawatt cell capable of serving the needs of a small community. C. A. Hampel

(2) An aircraft fuel tank or container made of or lined with an oil-resistant synthetic rubber.

fuel element. A fabricated rod, form, or other shape which consists of or contains the fissionable fuel for a nuclear reactor. The term does not refer to a chemical element, but rather to a device from which power is derived.

fuel oil. Any liquid petroleum product that is burned in a furnace for the generation of heat, or used in an engine for the generation of power, except oils having a flash point below 100°F and oils burned in cotton or woolwick burners. The oil may be a distillated fraction of petroleum, a residuum from refinery operations, a crude petroleum, or a blend of two or more of these.

Because fuel oils are used with burners of various types and capacities, different grades are required. A.S.T.M. has developed specifications for six grades of fuel oil. No. 1 is a straight-run distillate, a little heavier than kerosine, used almost exclusively for domestic heating. No. 2 (diesel oil) is a straight-run or cracked distillate used as a general purpose domestic or commercial fuel in atomizing-type burners. No. 4 is made up of heavier straight-run or cracked distillates and is used in commercial or industrial burner installations not equipped with preheating facilities. The viscous residuum fuel oils, Nos. 5 and 6, sometimes referred to as bunker fuels, usually must be preheated before being burned. A.S.T.M. specifications list two grades of No. 5 oil, one of which is lighter and under some climatic conditions may be

handled and burned without preheating. These fuels are used in furnaces and boilers of utility power plants, ships, locomotives, metallurgical operations, and industrial power plants. See also diesel oil; fuel oil, synthetic.
Containers: Tank cars and trucks; barges; pipelines.
Uses: Domestic and industrial heating; power for heavy units (ships, trucks, trains); source of synthesis gas.

fuel oil, synthetic. Any type of oil that can be used for fuel energy that is derived from sources other than petroleum. The most promising raw materials are organic wastes of various kinds, including urban, agricultural, and animal wastes. The two methods best adapted for this purpose are flash pyrolysis and fermentation, the latter involving use of carbon monoxide as a reactant. Small-scale production is already under way by pyrolysis, and the fermentation processes are approaching pilot-plant scale.

fuller's earth. A porous colloidal aluminum silicate (clay) which has high natural adsorptive power. Gray to yellow color. Nontoxic; noncombustible. See also bentonite; diatomite.
Occurrence: Florida; England; Canada.
Containers: Bulk; burlap bags; paper bags.
Uses: Decolorizing of oils and other liquids; oil-well drilling muds; insecticide carrier; floor-sweeping compounds; cosmetics; rubber filler; carrier for catalysts; filtering medium.

fulminate of mercury. Legal label name for mercury fulminate (q.v.).

"Fulvicin."[321] Trademark for a brand of griseofulvin (q.v.).

fumagillin $C_{27}H_{36}O_7$. An antibiotic substance produced by Aspergillus fumigatus.
Properties: Light yellow crystals from dilute methanol; m.p. 189–194°C. Insoluble in water, dilute acids, saturated hydrocarbons. Soluble in most other organic solvents.
Use: Medicine.

fumaric acid (boletic acid; lichenic acid; allomaleic acid; trans-butenedioic acid) HOOCCH:CHCOOH. The trans-isomer of maleic acid.
Properties: Colorless, odorless crystals; fruit acid taste; stable in air; sp. gr. 1.635; sublimes at 290°C; m.p. 287°C (sealed tube); solubility in water (25°C) 0.63 g/100 g; solubility in alcohol (30°C) 5.76 g/100 g; insoluble in chloroform and benzene. Low toxicity. Combustible.
Derivation: (a) isomerization of maleic acid; (b) catalytic oxidation of benzene.
Grades: Technical; crystals; F.C.C.
Containers: 250-lb drums; bags.
Uses: Modifier for polyester, alkyd, and phenolic resins; paper-size resins; plasticizers; rosin esters and adducts; alkyd resin coatings; upgrading natural drying oils (especially tall oil) to improve drying characteristics; in foods, to replace citric and tartaric acids and acidulant and flavoring agent (FDA approved); mordant; organic synthesis; printing inks.

fumaryl chloride ClCOCH:CHCOCl.
Properties: Clear, straw-colored liquid; b.p. 158–160°C (760 mm), 62–64°C (13 mm); sp. gr. (20°C) 1.408.
Hazard: Toxic. Corrosive to eyes and skin.
Uses: Chemical intermediate for pharmaceuticals, dyestuffs, and insecticides.
Shipping regulations: (Rail) Corrosive liquid, n.o.s., White label. (Air) Corrosive label.

"Fumazone."[233] Trademark for fumigants and nematocides containing 1,2-dibromo-3-chloropropane and related halogenated C_3 hydrocarbons.

fume. The particulate, smoke-like emanation from the surface of heated metals. Also, the vapor evolved from concentrated acids (sulfuric, nitric); from evaporating solvents; or as a result of combustion or other decomposition reactions (exhaust fume). Many of these fumes are toxic.

fumigant. A toxic agent in vapor form that destroys rodents, insects, and infectious organisms; a type of pesticide. The most effective temperature for their use is about 70°F. They are used chiefly in enclosed or limited areas (barns, greenhouses, ships' holds, and the like); also are applied locally to soils, grains, fruits and garments. Some commonly used fumigants are formaldehyde, sulfur dioxide and other sulfur compounds, hydrogen cyanide, methyl bromide, carbon tetrachloride, p-dichlorobenzene, and ethylene oxide. Care must be used when handling and applying fumigants in view of their toxicity. See also repellent; pesticide.

fuming. A characteristic of some highly active liquids which evolve visible smoke-like emanations on contact with air. Most familiar are the forms of nitric and sulfuric acids designated as "fuming." These are not pure, concentrated acids; low percentages of nitrogen dioxide and water are present in fuming nitric acid, and fuming sulfuric acid contains sulfur trioxide. Hydrofluoric acid (a mixture of hydrogen fluoride and water) also fumes. Pure compounds in which fuming occurs are fluosilicic acid and hydrazine.

fundamental particle (elementary particle). One of the many constituents of atoms in either the normal or the excited state. Since only a few such particles are involved in strictly chemical reactions, the definitions in this dictionary are limited to these, i.e., protons, neutrons, electrons. Other subatomic species are mesons, positrons, neutrinos, hyperons, etc. Photons are light quanta, i.e., "particles" of energy, which have no rest mass. See also particle.

fundamental research (basic research). Scientific investigations undertaken primarily to increase knowledge of a given field on a long-range basis. It is free from the time factor usually present in applied research (q.v.), and is comparatively unlimited by economic restrictions. In general, it seeks basic causes for phenomena rather than immediate results. It has no predetermined goal or purpose. None the less, tremendous achievements in chemistry and other sciences have resulted, and it will always remain the essential cornerstone of science. An example is the fundamental research of Carothers (q.v.) on polymerization that led to the discovery of nylon. More recent examples are the study of rocks and soil samples obtained from the moon and of the surface of the planet Mars made by direct contact of space vehicles.

fungal protease enzyme. See acid fungal protease.

fungicide. Any substance which kills or inhibits the growth of fungi. Older types include mixtures of lime and sulfur, copper oxychloride, and Bordeaux mixture. Copper naphthenate has been used to impregnate textile fabrics such as tenting and military clothing. Dithiocarbamate and quinone types were introduced about 1940. Mercury compounds are also effective, but were discontinued in 1971 because of their toxicity to humans. Ferbam, zineb and 8-hydroxyquinoline are now in wide agricultural use, as well as hypochlorite solutions in swimming pools and water-cooled heat exchangers. Some types of fungi that infect the human body are extremely hard to eradicate and require highly specific medical treatment.

fungus. Any of a plant-like group of organisms that does not produce chlorophyll; they derive their food either by decomposing organic matter from dead plants and animals or by parasitic attachment to living organisms, thus often causing infections and disease. Examples of fungi are molds, mildews, mushrooms, and the rusts and smuts that infect grain and other plants. They grow best in a moist environment at temperatures of about 25°C, little or no light being required.

Funk, Casimir (1884–1948). Born in Poland and later becoming an American citizen, Funk in 1911 isolated a food factor extracted from rice hulls which he found to be a cure for a disease due to malnutrition (beriberi). Believing this to be an amine compound essential to life, he coined the name "vitamine," from which the final "e" was later dropped. The various types and functions of vitamins were not differentiated till some years later as a result of the work of McCollum, Szent-Gyorgi, R. J. Williams, and others.

furacrylic acid. See furylacrylic acid.

"Furadan" (2,3-dihydro-2,2-dimethyl-7-benzofuranyl-methylcarbamate). Trademark for a pesticide designed to combat corn rootworm and rice water weevil. Approved by USDA. Also effective on alfalfa, sugar cane, rice, peanuts and potatoes.

"Furafil."[224] Trademark for lignocellulose produced by the pressure digestion of the acidified residue remaining after extraction of furfural from agricultural raw materials.

Uses: Additive for bulk, absorbency or conditioning action; extender for phenolic glues for plywood; foundry facings; oil well drilling muds; fertilizer.

"Fur-Ag."[224] Trademark for a sterilized grade of "Furafil" lignocellulose. Used as an anti-caking agent and organic conditioner in mixed fertilizers.

furamide. See furoamide.

furan (furfuran; tetrol) HC:CHCH:CHO. A heterocyclic compound. Its basic structure is:

```
C——C
‖    ‖
C    C
 \  /
  O
```

Properties: Colorless liquid turning brown on standing. This color change is retarded if a small amount of water is added. Sp. gr. 0.938 (20/4°C); f.p. -86°C; b.p. 31.4°C; flash point less than 32°F (TOC); refractive index 1.4216 (n 20/D); insoluble in water; soluble in alcohol and ether.

Derivation: Dry distillation of furoic acid from furfural.

Containers: Drums; tank cars.

Hazard: Toxic; absorbed by skin. Flammable, dangerous fire risk; flammable limits 2 to 14%. Forms peroxides on exposure to air.

Uses: Organic synthesis, especially for pyrrole, tetrahydrofuran, thiophene.

Shipping regulations: Flammable liquid, n.o.s., (Rail) Red label. (Air) Flammable Liquid label.

See also furan polymer.

furancarboxylic acid. See furoic acid.

2,5-furandione. See maleic anhydride.

2-furanmethanethiol. OCH:CHCH:CCH$_2$SH.

Properties: Liquid; sp. gr. 1.1319 (20/4°C); b.p. 155°C; refractive index 1.5324 (20°C). Insoluble in water. Low toxicity.

Uses: Ingredient for synthetic coffee compositions and fortifier for natural coffee blends and flavor adjunct. Also inhibits the corrosive power of nitric acid.

furan polymer. A plastic derived (1) from furfuryl alcohol or (2) from furfural or reaction products of furfural and a ketone. The materials are dark-colored and are resistant to solvents, most non-oxidizing acids and alkalies and to specific corrosives such as dinitrogen pentoxide. Physical properties of a typical asbestos-reinforced furan polymer are specific gravity 1.7, tensile strength 5,000 psi, flexural strength 7,500 psi, impact strength 0.5 ft lb/in. notch, water absorption 0.2 percent and coefficient of thermal expansion $3.0 -10^{-5}$ in/in/°F. Furan polymers are used for coating asphaltic pavements, foundry sand cores, shell molding and for corrosion resistant materials of construction. Since furfural is readily obtainable by heating pentosan-containing products such as corn cobs with mineral acid, these resins are inexpensive and have great potential use where products with their characteristics are required.

furazolidone $C_8H_7N_3O_5$ (N-(5-nitro-2-furfurylidene-3-amino)-2-oxazolidinone).

Properties: Yellow powder; odorless; m.p. 255°C. Slightly soluble in polyethylene glycol; insoluble in water, alcohol; and peanut oil.

Derivation: Synthetically from furfural, hydroxyethylhydrazine, and diethyl carbonate.

Grade: N.F.

Hazard: A suspected carcinogen; use suspended by FDA. See also nitrofuran.

Uses: Veterinary medicine; ingredient of poultry feeds.

furcellaran.

Properties: Seaweed gum (a natural phycocolloid) available as an odorless white powder; soluble in warm water. Form gels at low concentrations reputed to be more stable to heat and acids than those of other vegetable gums. Nontoxic. Available in the form of salts.

Derivation: From the sea-weed Furcellaria fastigiata.

Use: Gelling agent; viscosity control agent; puddings, jams, tooth pastes; bacterial cultural media.

furethrin. Generic name for 3-furfuryl-2-methyl-4-oxo-2-cyclopenten-1-yl chrysanthemumate, $C_{21}H_{27}O_4$. A synthetic analog of allethrin substituting the 2-furfuryl for allyl in the side chain. Low toxicity.

Properties: Yellow liquid; b.p. 187–188°C (0.4 mm). Insoluble in water; soluble in light oils.

Use: Insecticide, use like allethrin.

furfural (ant oil, artificial; pyromucic aldehyde; furfuraldehyde; bran oil) C_4H_3OCHO or OCH:CHCH:CCHO.

Properties: Colorless liquid when very pure; becomes reddish brown upon exposure to light and air. Odor

somewhat similar to benzaldehyde. Forms condensation products with many types of compounds, phenol, amines, urea, etc. Soluble in alcohol, ether, and benzene; 8.3% soluble in water at 20°C. Sp. gr. 1.1598 (20/4°C); f.p. −36.5°C; b.p. 161.7°C; heat of vaporization 107.5 cal; refractive index 1.5260 (n 20/D); flash point 140°F (closed cup). Combustible; autoignition temp. 600°F.
Derivation: From oat hulls, rice hulls, corn cobs, bagasse and other cellulosic waste materials by steam-acid digestion.
Grades: Technical; refined.
Containers: Drums; tank cars.
Hazard: Toxic; absorbed by skin. Tolerance, 5 ppm in air. Irritant to eyes, skin, and mucous membranes.
Uses: Solvent refining of lubricating oils, butadiene, rosin and other organic materials; solvent for nitrocellulose, cellulose acetate, shoe dyes; intermediate for tetrahydrofuran and furfuryl alcohol; phenolic and furan polymers; wetting agent in manufacture of abrasive wheels and brake linings; weed killer; fungicide; adipic acid and adiponitrile; road construction; production of lysine; refining of rare earths and metals; flavoring.

furfuralacetic acid. See furylacrylic acid.

furfuraldehyde. See furfural.

furfuramide. See hydrofuramide.

furfuran. See furan.

furfuryl acetate $C_4H_3OCH_2OOCCH_3$.
Properties: Colorless liquid turning brown upon exposure to light and air; pungent odor. Insoluble in water; soluble in alcohol and ether. Sp. gr. 1.1175 (20/4°C); b.p. 175–177°C; refractive index 1.4627 (D). Nontoxic. Combustible.
Derivation: By treatment of furfuryl alcohol with acetic anhydride.
Grade: Refined.
Use: Flavor.

furfuryl alcohol (furyl carbinol) $C_4H_3OCH_2OH$ or OCH:CHCH:CCH₂OH.

Properties: Colorless, mobile liquid becoming brown to dark-red upon exposure to light and air. With acid catalysts it autopolymerizes, often with explosive violence, to form a thermosetting resin that cures to an insoluble, infusible solid, highly resistant to chemical attack. Soluble in alcohol, chloroform, benzene, and water; insoluble in paraffin hydrocarbons. See also furan polymer. Sp. gr. 1.1285 (20/4°C); b.p. 170°C; refractive index 1.4850 (25°C/D); flash point (open cup) 167°F. Combustible. Autoignition temp. 915°F.
Derivation: Continuous vapor-phase hydrogenation of furfural.
Grades: Technical; refined.
Containers: Drums; tank cars.
Hazard: Approved for food products, but toxic by inhalation. May react explosively with mineral and some organic acids. Tolerance, 5 ppm in air.
Uses: Wetting agent; furan polymers; corrosion-resistant sealants and cements, foundry cores; modified urea-formaldehyde polymers; penetrant; solvent for dyes and resins; flavoring.

alpha-**furfuryl amine** $C_4H_3OCH_2NH_2$.
Properties: Colorless liquid; soluble in water, alcohol, and ether. Sp. gr. 1.0550 (17°C); b.p. 145°C (757 mm); refractive index 1.4900 (17°C); flash point 99°F (open cup).
Derivation: Furfural and ammonia.
Hazard: Flammable, moderate fire risk. Moderately toxic.
Uses: Corrosion inhibitor; component of soldering flux; chemical intermediate.

6-furfurylaminopurine. Kinetin. See kinins.

furfuryl mercaptan. See 2-furanmethanethiol.

furnace black. A carbon black made by burning natural gas or vaporized aromatic hydrocarbon oil in a closed furnace with about 50% of the air required for complete combustion. The combustion products are cooled by a water spray and finely divided carbon is separated from the gases. Furnace black produced from oil can be made in a wide range of closely controlled particle sizes.
Grades: Conducting furnace black (CF); fine (FF); high modulus (HMF); high elongation (HEF); reinforcing (RF); semi-reinforcing (SRF); high abrasion (HAF); super abrasion (SAF); fast extruding (FEF); general purpose (GPF); intermediate super abrasion (ISAF); channel replacement (CRF); easy processing furnace black (EPF).
Uses: Rubber reinforcing agent; plastics.
See also thermal black, carbon black.

furnace oil Usually No. 1 fuel oil. See fuel oil.

furnace sand (fire sand). Sand used to line furnace bottoms or walls, particularly in open hearth steel furnaces.

"Furnex."[133] Trademark for a series of semi-reinforcing furnace carbon black (SRF).

furnish. Term used by paper makers for mixtures containing the constituents of paper as supplied to the fourdrinier wire on which the sheet is formed.

furoamide (pyromucamide; furamide) C_4H_3OCON.
Properties: Crystals; sublimes partly at 100°C. M.p. 142°C.
Derivation: Treatment of furoyl chloride with ammonia.

furoic acid (pyromucic acid; furancarboxylic acid) C_4H_3OCOOH or OCH:CHCH:CCOOH.

Properties: Colorless crystals; m.p. 133–134°C; sublimes at 130°C (50–60 mm). Slightly soluble in cold water; very soluble in hot water, alcohol and ether; insoluble in paraffin hydrocarbons. Combustible; low toxicity.
Derivation: Cannizzaro reaction from furfural; oxidation of furfural.
Purification: Sublimation; fractional crystallization from hot water.
Grade: Technical.
Uses: Preservative; bactericide; furoates for perfume and flavoring; fumigant; textile processing; chemical intermediate.

Furol viscosity. The efflux time in seconds (SFS) of 60 ml of sample flowing through a calibrated Furol orifice in a Saybolt viscometer under specified conditions. Furol viscosity is about one-tenth of Saybolt Universal viscosity (q.v.) and is used for fuel oil and residual materials of relatively high viscosity. Furol is derived from the words "fuel and road oils."

furoyl chloride C_4H_3OCOCl.
Properties: Colorless liquid; powerful lachrymator. Soluble in ether; decomposes in water. M.p. −2°C; b.p. 176°C. Combustible.
Derivation: Treatment of furoic acid with phosphorus pentachloride.
Hazard: Strong irritant to eyes and skin.

Use: Substitute for chloropicrin in disinfecting grain elevators.

furylacrylic acid (furfural acetic acid; furacrylic acid) $C_4H_3OCH{:}CHCOOH$.

Properties: White powder; slightly soluble in cold water; easily soluble in hot water; soluble in alcohol, ether, and glacial acetic acid. M.p. 141°C; b.p. 117°C (8 mm), 286°C.

Derivation: From furfural.

Uses: Derivatives used in perfumes.

furyl carbinol. See furfuryl alcohl.

fused ring. A ring having one or more of its sides in common with another ring, as shown:

fused salt. A salt (i.e., ionic compund) in the molten state. Halides and nitrates are the salts most commonly used. High temperatures, (500–1000°C) for the alkali halides, are usually required. Most fused salts are liquids with viscosities, diffusion coefficients, thermal conductivities, and surface tensions in the same range as water. They conduct electricity exceptionally well.

Uses: Production of sodium by electrolysis; heat-transfer agents; reaction medium in chemical synthesis; heat-treatment of metals (from 350–2400°F); solvents for the metals corresponding to their cations; nuclear power reactors.

See also salt bath.

fused silica. See silica.

fusel oil (amyl alcohol, fermentation; grain oil; potato oil). A volatile, oily mixture consisting largely of amyl alcohols. Isoamyl alcohol (isobutyl carbinol) and active amyl alcohol (2-methyl-1-butanol) are chief constituents. Ethyl, propyl, butyl, hexyl and heptyl alcohols as well as other alcohols have been separated. Acids, esters, and aldehydes are also present. Normal primary amyl alcohol (1-pentanol) is not found in fusel oil. Combustible.

Containers: 10-gal cans; drums; tank cars.

Hazard: Moderate fire risk. Toxic by ingestion and inhalation.

Uses: Chemicals (amyl ether, amyl acetate, pure amyl alcohols, nitrous ether, various esters); explosives (gelatinizing agent); solvent for fats and oils; intermediate; pharmaceuticals; nitrocellulose plastics; synthetic rubber; varnishes; lacquers; solvent for resins and waxes; perfumery.

fusible alloy. See alloy, fusible.

fusion. (1) A synonym for melting of a crystalline substance. Since melted substances tend to mix readily, the word has assumed the meaning of "melt and blend." See also fused salt; melting point.

(2) A nuclear reaction in which the nuclei of light atoms unite at temperatures of 10 to 15 million degrees Kelvin to form heavier nuclei, thus releasing vast amounts of energy. Such reactions occur in the sun by the union of hydrogen nuclei to form helium.

Uncontrolled release of fusion energy has been achieved in the hydrogen bomb, in which a fission reaction supplies the necessary high temperature, which endures only for a time span measured in nanoseconds. A fusion reaction has been induced on a micro scale with a laser beam impinging on deuterium, causing emission of high-energy neutrons.

Controlled fusion reactions are the ultimate means of solving the energy problem. Maintenance of the high temperatures necessary for continuous energy yield at controlled rates is one of the central problems, as is also the selecton of suitable materials. Use of lasers will undoubtedly be an important feature of future development of thermonuclear power. At least 20 years of basic research will be necessary. See also thermonuclear reaction.

"Fybrene."[45] Trademark for petrolatum, U.S.P., of medium melting point and medium consistency. Used in paper industry.

"Fybrol."[498] Trademark for a series of wool lubricants based on petroleum sulfonates, fatty esters and mineral oil.

"Fyrel"[312a] Trademark for a flexible multifilament fiber made of a nickel-chromium alloy. Withstands heat up to 4000°F. Used for reentry parachute yarns and other space applications.

"Fyrex."[1] Trademark for a substantially neutral ammonium phosphate; fine crystals; soluble in water.

Uses: Flameproofing textiles, wood, and fibers; manufacture of matches to prevent afterglow.

"Fyrol."[1] Trademark for a series of phosphate flame-retardant plasticizers.

Used in polyesters, rigid urethane foam, polyurethane, phenolic, epoxy, PVC, PVA, nitrocellulose, cellulose acetae, and ethyl cellulose resins.

G

g (1) Abbreviation of gram. (2) Acceleration due to gravity.

μg Abbreviation of microgram (q.v.).

"G-4."[12] Trademark for a brand of dichlorophene (q.v.).

"G-942."[28] Trademark for an aqueous solution of the partial sodium salt of a polymeric carboxylic acid, used as a tanning agent.
Containers: 500-lb drums.

Ga Symbol for gallium.

G acid. 2-Naphthol-6,8-disulfonic acid.

gadolinium Gd A rare-earth element of the lanthanide series; atomic number 64; group III B of the periodic table; atomic weight 157.25; valence 3. Seven natural isotopes.
Properties: Lustrous metal; sp. gr. 7.87; m.p. 1312°C; b.p. 3000°C. Reacts slowly with water; soluble in dilute acid; insoluble in water. Exhibits a high degree of magnetism, especially at low temperatures. Salts are colorless. Has highest neutron absorption cross-section of any known element; has superconductive properties. Combustible; burns in air to form the oxide.
Derivation: (a) Reduction of the fluoride with calcium; (b) electrolysis of the chloride with NaCl or KCl in iron pot, which serves as anode, and graphite cathode.
Grades: Ingots, lumps, turnings, powder, up to 99.9+% pure.
Uses: Neutron shielding; garnets in microwave filters; phosphor activator; catalyst; scavenger for oxygen in titanium production.

gadolinium chloride $GdCl_3 \cdot xH_2O$. Colorless crystals, soluble in water. Purities up to 99.9%.
Containers: Glass bottles, fiber drums.
Uses: Gadolinium sponge metal, by contact with a reducing metal vapor.

gadolinium fluoride $GdF_3 \cdot 2H_2O$. Available up to 99.9% purity.
Containers: Glass bottles, fiber drums.
Hazard: Toxic. Tolerance (as F), 2.5 mg per cubic meter of air.

gadolinium nitrate $Gd(NO_3)_3 \cdot xH_2O$. Colorless crystals, soluble in water; decomposes 110°C. Purities up to 99.9% gadolinium salt.
Containers: Glass bottles, fiber drums.
Hazard: Fire risk in contact with organic materials.
Shipping regulations: Nitrates, n.o.s. (Rail) Yellow label. (Air) Oxidizer label.

gadolinium oxalate $Gd_2(C_2O_4)_3 \cdot 10H_2O$. White powder, insoluble in water, slightly soluble in acids. Loses $6H_2O$ at 110°C. Purities up to 99.9% gadolinlium salt.
Containers: Glass bottles, fiber drums.

gadolinium oxide Gd_2O_3. White to cream-colored powder; sp. gr. 7.41 m.p. 2330°C; insoluble in water soluble in acids to form the corresponding salts. Hygroscopic; absorbs carbon dioxide from the air. Purities up to 99.8% gadolinium oxide.
Derivation: Calcination of gadolinium salts; liquid/-liquid ion-exchange separation.
Containers: Glass bottles, fiber drums.
Uses: Neutron shields; catalysts; dielectric ceramics; filament coatings; special glasses; TV phosphor activator; lasers, masers, and telecommunication; laboratory reagent.

gadolinium sulfate $Gd_2(SO_4)_3 \cdot 8H_2O$. Colorless crystals, slightly soluble in hot water, more soluble in cold; sp. gr. 3.01 (15°C). Purities up to 99.9% gadolinium salt.
Uses: Cryogenic research; the selenide is used in thermoelectric generating devices.

"Gafac."[307] Trademark for a series of anionic, complex organic phosphate esters with good emulsifying and detergent properties.
Uses: Surfactants in emulsion polymerization (FDA approved).

"Gafamide."[307] Trademark for a series of fatty acid ethanolamides used as foam boosters, stabilizers, thickeners, detergents, and emulsifiers.

"Gafanol."[307] Trademark for a series of polyethylene glycol polymers ranging from free-flowing liquids to waxes. Used for surfactant intermediates, solvents for pharmaceuticals, plasticizers, etc.

"GAF" Carbonyl Iron Powders.[307] Trademark for microscopic spheres of extremely pure iron. Produced in eleven carefully controlled grades, ranging in particle size from 3 to 20 microns in diameter. The iron content of some types is as high as 99.5%.
Uses: High-frequency cores for radio, telephone, television, short wave transmitters, radar receivers, direction finders; alloying agents, catalysts; powder metallurgy; magnetic fluids.

"Gafcol."[307] Trademark for a series of mild-odored glycol ethers used as solvents for nitrocellulose lacquers, dyes, insecticides, and intermediates in the manufacture of plasticizers. Miscible with most liquids.

"Gafstat"[307] Trademark for a series of free acids of complex phosphate esters. Antistatic agents for plastic packaging films (FDA approved).

Gal Abbreviation for galactose.

galactose $C_6H_{12}O_6$. A monosaccharide commonly occurring in milk sugar or lactose.
Properties: White crystals; soluble in water and alcohol; slightly soluble in glycerol; m.p. 165–168°C.
Derivation: By acid hydrolysis of lactose.
Uses: Organic synthesis; medicine.

D(+)-galacturonic acid $COOH(CHOH)_4CHO$. A major constituent of plant pectins (q.v.). It exhibits mutarotation, having both an alpha and a beta form.
Properties: The alpha form melts with decomposition at 159–160°C. Soluble in water; slightly soluble in hot alcohol; insoluble in ether.
Derivation: Hydrolysis of pectins.
Use: Biochemical research.

"Galecron."[219] Trademark for chlordimeform (q.v.).

galena (galenite; lead glance) PbS. Natural lead sulfide.
Properties: Color lead gray; streak lead gray; luster

metallic; good cubic cleavage; sp. gr. 7.4–7.6; Mohs hardness 2.5. Soluble in strong nitric acid; also in excess of hot hydrochloric acid.
Occurrence: Western U.S.; Canada; Africa; South America.
Use: Chief ore of lead. Frequently recovered for the silver it sometimes contains.

"Galex."[122,235] Trademark for a stable non-oxidizing rosin, principally dehydroabietic acid.
Properties: Light amber to red brown solid. Sp. gr. 1.05; B & R softening point 65–70°C. Nontoxic.
Grades: W-100, W-100-D, NXD, G-75.
Uses: Tackifier.

gallic acid (3,4,5-trihydroxybenzoic acid) $C_6H_2(OH)_3CO_2H \cdot H_2O$.
Properties: Colorless or slightly yellow, crystalline needles or prisms. Soluble in alcohol and glycerol; sparingly soluble in water and ether. Sp. gr. 1.694; m.p. 222–240°C.
Derivation: Action of mold on solutions of tannin or by boiling the latter with strong acid or caustic soda.
Uses: Photography; writing ink; dyeing; manufacture of pyrogallol; tanning agent and manufacture of tannins; paper manufacture; pharmaceuticals; engraving and lithography; analytical reagent.

gallimycin 36. A liquid form of erythromycin used for veterinary treatment of mastitis. FDA approved.

gallium Ga Metallic element of atomic number 31, of group III A of the periodic system. Atomic weight 69.72; valences 2, 3. Two stable isotopes.
Properties: Silvery-white liquid at room temperature; m.p. 29.7°C; b.p. 2403°C; may be under-cooled to almost 0°C without solidifying; sp. gr. 5.9 (25°C), more dense as a liquid than as a solid; soluble in acid, alkali and slightly soluble in mercury. Reacts with most metals at high temperatures. Nontoxic. Grades up to 99.9999% purity are available.
Occurrence: Prepared commercially from zinc ores and bauxite.
Derivation: Extraction of gallium as gallium chloride by ethyl ether or isopropyl ether and subsequent electrodeposition from an alkaline gallium oxide solution.
Uses: The metal has no significant commercial uses. Its compounds are used as semiconductors.

gallium antimonide GaSb. Available in an electronic grade.
Uses: Semiconducting devices.

gallium arsenide GaAs.
Properties: Crystals; m.p. 1238°C. Electroluminescent in infrared light.
Grades: Ingots; polycrystalline form in high purity electronic grade; single crystals. Often alloyed with gallium phosphide or indium arsenide.
Hazard: Highly toxic. See arsenic.
Uses: Semiconductor in light-emitting diodes for telephone dials; injection lasers; solar cells; magnetoresistance devices; thermistors; microwave generation.
Shipping regulations: (Rail, Air) Arsenical compounds, n.o.s., Poison label.

gallium oxides. The sesquioxide, Ga_2O_3, and suboxide, Ga_2O, are known. Both are stable at ordinary temperatures. See gallium sesquioxide.

gallium phosphide GaP.
Properties: Pale orange, transparent crystals or whiskers up to 2 cm long; made by vapor phase reaction, at relatively low temperatures, between phosphorus and gallium suboxide. These crystals are intermediate between normal semiconductors and insulators or phosphors. They operate over a temperature range of −55 to 500°C. Gallium phosphide is electroluminescent in visible light.
Grades: Polycrystalline form in high-purity electronic grade; single crystals; whiskers.
Uses: Semiconducting devices.

gallium sesquioxide Ga_2O_3.
Properties: (alpha and beta forms) Alpha form: White crystals; sp. gr. 6.44; m.p. 1900°C. Changes to beta form at 600°C. Beta form: White crystals; sp. gr. 5.88. Both forms insoluble in hot acid.
Grade: High-purity electronic.
Uses: Spectroscopic analysis.

gallocyanine. $C_{15}H_{12}N_2O_5$. A dye made from gallic acid. Used as a biological stain.

gallotannic acid. See tannic acid.

galls (nutgalls). Excrescences on various kinds of oak trees resulting from the deposition of insect eggs.
Grades: The best grades (55–60% tannic acid) come from Iran, Syria, Turkey, and Tripoli.
Uses: Source of gallic and gallotanic acids; leather tanning; writing inks; medicine; textile printing; pharmaceuticals.

"Galvan."[275] Trademark for battery and electrically conductive carbon black.

galvanizing. Coating of a ferrous metal by passing it through a bath of molten zinc or by electrodeposition of zinc. In the former process, the iron and zinc combine to form an intermetallic compound at the interface, the outer surface being relatively pure zinc, which crystallizes as it cools, to form the characteristic "spangle." The electrodeposition method gives a uniform surface which may be either dull or bright. Duration of corrosion protection is directly related to the thickness of the zinc coating. See also sacrificial protection.

"Galvaseal."[526] Trademark for a light chromate-type conversion coating for all types of zinc or galvanized steel surfaces, producing salt spray resistance and white rust prevention as well as base for paint. Application by spray, immersion or roller coat.

"Galvoline."[233] Trademark for a cored magnesium ribbon used as a continuous anode for the cathodic protection of buried pipe lines and other metal structures. Combustible.

"Galvomag."[233] Trademark for magnesium alloy composition used in anodes in cathodic protection.

"Galvopak."[233] Trademark for a product consisting of a magnesium anode packed with back-fill material. Used in the cathodic protection of buried pipe lines and other metal structures. Combustible.

"Galvorod."[233] Trademark for a cored magnesium rod used as an anode in the cathodic protection of water-heater tanks. Combustible.

gametocide. A substance which can control pollination of plants by selectively killing plant sex cells (gametes). Some suggested gametocides are maleic hydrazide and sodium alpha, beta-dichloroisobutyrate.

gamma (γ). A prefix having meanings analogous to those of alpha (q.v.), namely, to designate locations of substituents in a compound or a particular form or modification of an organic substance (gamma-globulin) or a metal crystal. It also identifies the most intense form of short-wave radiation. See gamma ray.

gamma acid. See 7-amino-1-naphthol-3-sulfonic acid.

"Gammacorten."[305] Trademark for dexamethasone (q.v.).

"Gamma-Dent."[1] Trademark for a phosphate cleaning agent used in dentrifrices.

gamma-globulin. See globulin.

gamma ray. Electromagnetic radiation of extremely short wavelength and intensely high energy. Gamma rays originate in the atomic nucleus; they usually accompany alpha and beta emission, as in the decay of radium, and always accompany fission. Gamma rays are extremely penetrating and are best absorbed by dense materials like lead and depleted uranium.

Hazard: Exposure to gamma radiation may be lethal; complete protection is essential.

Uses: To initiate chemical syntheses (ethyl bromide) and crosslinking of polyethylene and other polymers; biochemical research; food preservation.

See also xray; radiation.

"Gamtox."[253] Trademark for a pesticide formulation containing benzene hexachloride.

Hazard: See benzene hexachloride.

"Ganex."[307] Trademark for a series of modified polyvinylpyrrolidone-based products. Available in a wide range of physical forms from liquid to waxy solid to granular solid. Used as emollients and lubricity additives, dispersing and suspending agents, pour point depressants, sizing additives.

gangue. The minerals and rock mined with a metallic ore but valueless in themselves or used only as a by-product. They are separated from the ore in the milling and extraction processes, often as slag. Common gangue materials are quartz, calcite, limonite, feldspar, pyrite, etc.

"Gantrez" AN.[307] Trademark for an interpolymer of methyl vinyl ether and maleic anhydride.

Properties: Soluble in water and many organic solvents. Compatible with a wide range of gums, resins, and plasticizers, and with most metallic salts. Stable in acid and alkaline solutions. Available in a range of molecular weights.

Uses: Ammonium nitrate slurry explosives, for its suspending action and crosslinking ability; rust-preventive films; antistatic and finishing agent in natural and synthetic textiles and glass fibers; adhesives and coatings; thickening agent and protective colloid; flocculant and foam stabilizer in papermaking; photoreproduction, pharmaceutical preparations; cosmetics, etc.

"Gantrez" M.[307] Trademark for polyvinyl methyl ether (q.v.).

"Gardenol."[227] Trademark for methyl phenyl carbinyl acetate.

"Garlon."[233] Trademark for herbicides containing diethylene glycol bis(2,2-dichloropropionate) and esters of 2-(2,4,5-trichlorophenoxy)propionic acid.

garnet, natural. A group of silicate minerals. Garnets in nature are usually composed of mixtures of various garnet subspecies.

Uses: Abrasive; blast cleaning of buildings.

garnet, synthetic. Single-crystal garnets designed, by the introduction of bismuth, calcium, and vanadium, for microwave devices such as low-frequency ferrimagnetic resonators. Special yttrium, iron and aluminum garnets are used in lasers and electronic devices.

garnierite $(Ni,Mg)_6(OH)_6Si_4O_{11} \cdot H_2O$. A natural nickel-magnesium silicate, occuring as a natural alteration of magnesium silicate rocks. A nickel ore.

gas. One of the three states of matter, characterized by very low density and viscosity (relative to liquids and solids); comparatively great expansion and contraction with changes in pressure and temperature; ability to diffuse readily into other gases; and ability to occupy with almost complete uniformity the whole of any container. Gases may be either elements (argon) or compounds (carbon dioxide); elemental gases may be monatomic (helium), diatomic (chlorine) or triatomic (ozone). All exist in the gaseous state at standard temperature and pressure, but can be liquefied by pressure. The most abundant gases are oxygen, hydrogen, nitrogen (diatomic), and carbon dioxide. Gases are used for fundamental research on the behavior of matter largely because the low concentration permits isolation of the phenomena far better than is possible in liquids or solids.

A "perfect" or ideal gas is one which closely conforms to the simple gas laws for expansion and contraction (Boyle's Law, Charles' Law).

Note: Use of the word "gas" for gasoline, natural gas, or the anesthetic nitrous oxide, is acceptable only in informal communication.

See also kinetic theory; compressed gas; noxious gas; noble gas; vapor.

gas, asphyxiant. See asphyxiant gas.

gas black. See carbon black.

"Gas-Chrom."[425] Trademark for a series of diatomaceous earths designed for use as supports in gas chromatography. They have been treated in various ways, such as flux-calcined and screened, washed with acids and bases, and/or silanized for inertness. "Gas-Chrom" Q, which has been treated with dichlorodimethylsilane, is specially recommended for the analytical separation of steroids, pesticides, and other high molecular weight materials.

gas chromatography (GC; gas-liquid chromatography; GLC; vapor-phase chromatography; VPC). The process in which the components of a mixture are separated from one another by volatilizing the sample into a carrier gas stream which is passing through and over a bed of packing consisting of a 20 to 200 mesh solid support. The surface of the latter is usually coated with a relatively nonvolatile liquid (the stationary phase). This gives rise to the term gas-liquid chromatography. If the liquid is not present, the process is gas-solid chromatography, which is also widely useful for analysis. Different components move through the bed of packing at different rates, and so appear one after another at the effluent end, where they are detected and measured by thermal conductivity changes, density differences, or ionization detectors.

Gas chromatography is advantageous as a means of analysis of minute quantities of complex mixtures from industrial, biological, and chemical sources, and is also of potential value in actually preparing moderate quantities of highly purified compounds otherwise difficult to separate from the mixtures in which they occur.

gas, compressed. See compressed gas; liquefied petroleum gas.

gaseous diffusion. See diffusion, gaseous.

gas hydrate. A clathrate compound (q.v.) formed by a gas (both noble and reactive) and water. The compounds are crystalline solids, and are insoluble in water. They usually form (only at relatively low temperatures and high pressures) directly by contact of gas and liquid water. From 6 to 18 molecules of water may combine with each molecule of gas, depending upon the nature of the gas.

The best known gas hydrates are those of ethane, ethylene, propane and isobutane. Others include: methane and 1-butene, most of the fluorocarbon refrigerant gases, nitrous oxide, acetylene, vinyl chloride, carbon dioxide, methyl and ethyl chloride, methyl and ethyl bromide, cyclopropane, hydrogen sulfide, methyl mercaptan and sulfur dioxide.

Interest in the gas hydrates originated mainly because of the nuisance of such compound formation in gas pipelines. In recent years, propane has been successfully used to precipitate water from salt solution (or sea water), thus yielding potable water.

gasification. Production of gaseous or liquid fuels from coal. The term may mean either (1) catalytic addition of hydrogen from an external source to coal, or (2) production of a synthesis gas mixture of hydrogen, carbon monoxide, carbon dioxide and some methane from coal by reaction with steam in the presence of air. In the latter case catalytic methanation is required to increase the methane content and Btu value. An example of the first method is the Hydrane process developed by the Bureau of Mines; an instance of the second method is the Lurgi process, originally developed in Scotland. High temperatures and pressures are required. A number of related processes are under active research, some of which are as follows:

Process	Developer
Hygas	Institute of Gas Technology
Bi-Gas	Bituminous Coal Research
Koppers-Totzek	Heinrich Koppers
Synthane	Bureau of Mines
Atgas	Applied Technology, Inc.
Coalcon	Union Carbide Corp.
Co_2 Acceptor	Consolidated Coal Co.
Coed	FMC Corp.
Cogas	Consortium
Fischer-Tropsch	M. W. Kellogg Co.
H-Coal	Hydrocarbon Research, Inc.
Winkler	Davy Powergas, Inc.
Synthoil	Bureau of Mines

Also under experimentation is an underground gasification process in which air is pumped down to strata of unmined coal that has been ignited and the product gases collected. Substantial contribution of coal gasification to the energy needs of the U.S. before 1985 is unlikely. See also coal; hydrogenation; synthetic natural gas.

gas, inert. See noble gas; inert.

gas laws. See Boyle's law; Charles' law.

gas, liquefied petroleum. See liquefied petroleum gas.

gas liquid. See light hydrocarbon.

gas, natural. See natural gas.

gas, noble. See noble.

gas oil. A liquid petroleum distillate with viscosity and boiling range between kerosine and lubricating oil. Boiling range between 450 and 800°F. Flash point 150°F. Autoignition temp. 640°F. Combustible.

Uses: Absorption oil; manufacture of ethylene.

gasoline (gasolene). A mixture of volatile hydrocarbons suitable for use in a spark-ignited internal combustion engine and having an octane number of at least 60. The major components are branched-chain paraffins, cycloparaffins, and aromatics. There are several methods of production: distillation or fractionation, which yields straight-run product of relatively low octane number, used primarily for blending; thermal and catalytic cracking; reforming; polymerization; isomerization; and dehydrocyclodimerization. All but the first are various means of converting hydrocarbon gases into motor fuels by modifications of chemical structure, usually involving catalysis. The present source of gasoline is petroleum, but it may also be produced from shale oil and Athabasca tar sands, as well as by hydrogenation or gasification of coal.

antiknock gasoline. A gasoline to which a low percentage of tetraethyllead or similar compound, has been added to increase octane number and eliminate knocking. Such gasolines have an octane number of 100 or more and are now used chiefly as aviation fuel. Since 1972 gasolines having a research octane number of about 90 have been in general automotive use; the percentage of lead compound added has been gradually reduced. The maximum permitted by EPA is now (1976) 1.4 grams per gallon, but this will gradually be lowered to 0.5 gram per gallon in 1979. See also antiknock agent; octane number.

casinghead gasoline. See natural gasoline (below).

cracked gasoline. Gasolines produced by the catalytic decomposition of high-boiling components of petroleum. In general such gasolines have higher octane ratings (from 80 to 100) than gasoline produced by fractional distillation. The difference is due to the prevalence of unsaturated, aromatic and branched-chain hydrocarbons in the cracked gasoline. The actual properties vary widely with the nature of the starting material, and the temperature, time, pressure and catalyst used in cracking (q.v.).

high-octane gasoline. A gasoline with an octane number of about 100. See antiknock gasoline; octane number.

lead-free gasoline. An automotive fuel containing no more than 0.05 gram of lead per gallon, designed for use in engines equipped with catalytic converters.

natural gasoline. A gasoline obtained by recovering the butane, pentane, and hexane hydrocarbons present in small proportion in certain natural gases. Used in blending to produce a finished gasoline with adjusted volatility, but low octane number. Do not confuse with natural gas (q.v.).

polymer gasoline. A gasoline produced by polymerization of low molecular weight hydrocarbons such as ethylene, propene, and butenes. Used in small amounts for blending with other gasolines to improve their octane number.

pyrolysis gasoline. Gasoline produced by thermal cracking as a by-product of ethylene manufacture. It is used as a source of benzene by the hydrodealkylation process.

reformed gasoline. A high octane gasoline obtained from low octane gasoline by heating the vapors to a high temperature or by passing the vapors through a suitable catalyst.

straight-run gasoline. Gasoline produced from petroleum by distillation, without use of cracking or other chemical conversion processes. Its octane number is low.

white gasoline. An unleaded gasoline expecially designed for use in motorboats; it is uncracked and strongly inhibited against oxidation to avoid gum formation, and is usually not colored to distinguish it from other grades. It also serves as a fuel for camp lanterns and portable stoves.

Containers: Tank cars and tank trucks; barges; pipeline.

Hazard: Highly flammable; dangerous fire and explosion risk.

Shipping regulations: (Rail) Red label (Air) Flammable Liquid label.

gas, perfect. See ideal gas.

"Gastex."[275] Trademark for semi-reinforcing gas furnace carbon black for rubber.

gastric juice. A mixture of hydrochloric acid and pepsin secreted by glands in the stomach in response to a conditioned nerve reflex. Its pH is about 2.0. Its action in the metabolic breakdown of food components is essential to the digestive process, though carbohydrate decompositoin is initiated by the saliva. See also digestion (1).

Gay-Lussac's law. A modification of Charles' Law to state the following: At constant pressure, the volume of a confined gas is proportional to its absolute temperature. The volumes of gases involved in a chemical change can always be represented by the ratio of small whole numbers.

GC. Abbreviation for gas chromatography.

Gd Symbol for gadolinium.

GDCH. Abbreviation for glycerol dichlorohydrin. See alpha-dichlorohydrin.

GDME. Abbreviation for glycol dimethyl ether. See ethylene glycol dimethyl ether.

GDP. Abbreviation for guanosine diphosphate. See guanosine phosphates.

Ge Symbol for germanium.

gel. A colloid in which the disperse phase has combined with the continuous phase to produce a viscous, jelly-like product. Only 2% gelatin in water forms a stiff gel. A gel is made by cooling a solution, whereupon certain kinds of solutes (gelatin) form submicroscopic crystalline particle groups which retain much solvent in the interstices (so-called "brush-heap" structure). Gels are usually transparent but may become opalescent. See also pectin.

gelatin. A mixture of proteins obtained by hydrolysis of collagen by boiling skin, ligaments, tendons, etc. Its production differs from that of animal glue in that the raw materials are selected, cleaned and treated with special care so that the product is cleaner and purer than glue. Type A gelatin is obtained from acid-treated raw materials, and Type B from alkali-treated raw materials. Gelatin is strongly hydrophilic, absorbing up to ten times its weight of water and forming reversible gels of high strength and viscosity. It can be chemically modified to make it insoluble in water for such special applications as microencapsulation of fish nutrients for fish culture.

Grades: Edible; photographic; technical; U.S.P.

Containers: Cans; fiber drums.

Uses: Photographic film; lithography; sizing; plastic compounds; textile and paper adhesives; cements; capsules for medicinals; matches; light filters; clarifying agent; desserts, jellies, etc.; culture meduim for bacteria; blood plasma volume expander; microencapsulation.

gelatin dynamite. A high explosive which contains about 1% of nitrocellulose in addition to nitroglycerin. The product is a gelatinized mass, less sensitive to shock and friction than straight dynamite.

"Gelcarin."[124] Trademark for carrageenan extractives from Irish moss. Dry, free-flowing powder packed in fiber drums.

Grade: Food grade, meeting FDA requirements.

Uses: Gelling, suspending, binding and viscosity-building.

gel filtration. A type of fractionation procedure in which molecules are separated from each other according to differences in size and shape; the action is similar to that of molecular sieves. Dextran gels (three-dimensional networks of polysaccharide chains) are usually used in this method, known as gel filtration chromatography. See also fractionation; molecular sieve.

"Gelgard."[233] Trademark for a synthetic polymeric water gelling material; used in fire control.

"Gel-Kote."[448] Trademark for pigmented polyester resin coatings for polyester products.

"Gelloid."[309] Trademark for a series of purified extracts of various types of Irish moss seaweed, rich in mucilagous content. Manufactured into dry, odorless, edible powders used in food and pharmaceutical applications.

gel paint (thixotropic paint). A paint formulation which has a semi-solid or gel consistency when undisturbed but which flows readily under the brush or when stirred or shaken. After removal of the stress, it becomes stiff again and has little tendency to spill, drip, or run. The thixotropic quality is obtained by the carefully controlled reaction of a relatively small proportion of a polyamide resin with an alkyd resin vehicle. See thixotropy.

"Geltone."[236] Trademark for a synthetic organophilic colloid that imparts gel to any oil/mud system.

"Gelva."[58] Trademark for vinyl acetate polymers.

Uses: Adhesives; binders; chewing gum bases; coatings; hot melt adhesives; paints; paper treatment; permanent starches; slush molding; textile sizes and finishes; and thickeners.

"Gelvatex."[58] Trademark for aqueous emulsions of vinyl acetate polymers and compounded compositions therof.

"Gelvatol."[58] Trademark for polyvinyl alcohol resins (q.v.).

Uses: Adhesive; emulsifier; hydraulic cement additive; textile sizes; paper coating.

"Gelwhite."[471] Trademark for a series of white grades of water-refined montmorillonite clays.

Grades: GP and L, high sodium content in exchange position and low sodium content.

gem-. Prefix; abbreviation of geminate, meaning two identical groups attached to the same carbon atom.

"Gemon-3010."[245] Trademark for a glass-reinforced thermosetting polyimide molding compound with excellent high-temperature properties, mechanical performance and processing ease. Used for space applications, electric and electronic industries.

"Genacron."[307] Trademark for a series of disperse dyes for polyester fibers.

"Genacryl."[307] Trademark for a series of basic dyes for acrylic fibers.

"Genamid."[259] Trademark for resinous amine-based coreactants for epoxy resins.

Uses: Coatings, castings, potting adhesives, laminates, tooling.

gene. A complex of nucleoproteins (chiefly DNA) that is the active transmitter of genetic information. Genes occur on the chromosomes of every living cell, where they are arranged in a linear order. Genetic mechanism is the same in all organisms, ranging from the lowest forms of life, both plant and animal, to man. Every organism has a large number of different genes; there are over 10,000 in each cell of the human body. These control the intricate and well-balanced system of biochemical reactions in the cell. The first synthesis of a gene was reported in 1970. In 1976 it was announced that a synthetic gene was successfully introduced into a microorganism in which it functioned in the same manner as would a normal gene.

Because of the ability of DNA to store and transfer "coded" genetic instructions, genes determine the sequence of amino acids in specific polypeptides (see proteins); thus they prescribe the structure of the proteins synthesized. In viruses the genes consist of ribonucleic acid (RNA). Mutation of genes is a change in the basic sequence of amino acids in DNA, and may be induced by ionizing radiation. Mutations can occur in any living cell. See also genetic code; radiation, ionizing; recombinant DNA.

"GenEpoxy."[259] Trademark for liquid and solid resins having terminal epoxide groups.

Uses: Coatings, structural adhesives, laminates, casting, encapsulation, sealants, concrete topping.

"Genesolv."[50] Trademark for ultrapure solvents of the halogenated hydrocarbons of the methane and ethane series.

Properties: Nonflammable, nonexplosive, thermally stable, and relatively nonhydrolyzable. Have high densities and low viscosities. Combined residue (soluble plus insoluble matter) is less than 1 part per million.

Grades: standard and electronic; Genesolv A, B, C, D.

Uses: Removal of greases and oils from metals, plastic, elastomer and paint or varnish surfaces. Used with all cleaning techniques on assembled motors and parts, electronic devices, precision components, motion picture film, refrigeration systems, etc. Also used for isolation of viruses, for fire-extinguishing, and as dielectric coolants.

genetic code. Information stored in the genes (q.v.) which "program" the linear sequence of amino acids within the protein polypeptide chain synthesized during cell develoment. The agencies involved are DNA (deoxyribonucleic acid, q.v.) and RNA (ribonucleic acid, q.v.). It is this information code (analogous to a computer language) that determines (1) the nature and type of an organism, i.e., its heredity, and (2) the kind of cell structure to be formed (muscle, bone, organ, etc.). See also protein; amino acid; nucleotide.

"Genetron."[50] Trademark for a group of fluorinated hydrocarbons.

Properties: Low moisture content, noncorrosive, nonflammable, stable, low power requirement per unit of refrigeration, high dielectric strength. For specific identification, see fluorocarbon.

Geneva System*. A system of nomenclature for organic compounds recommended in 1892. It is based on compounds derived from hydrocarbons as a starting point, the names corresponding to the longest straight carbon chain present. The position is indicated by numbers applied to the carbons of the straight chain, beginning with the end carbon nearest the substituent element or group.

Normal compounds, in which the carbons are linked in a straight chain, are named ethane, propane, pentane, etc. Normal propane is:

```
    H  H  H
    |  |  |
H—C—C—C—H
    |  |  |
    H  H  H
```

Branched compounds are so named as to show the proper deviation, e.g., 2-methyl propane

```
  1    2     3
H3C—CH—CH3
        |
       CH3
```

is a derivation of propane, not of methane. Similarly, isopentane

```
  1    2     3     4
H3C—CH—CH—CH
        |
       CH3
```

is named 2-methyl butane. The four carbons in a straight line give the name butane, and the side group is attached to the second carbon. Another pentane has the structure:

```
       CH3
        |
H3C—C—CH3
        |
       CH3
```

and was once called tetramethyl pentane. Under the Geneva system it is named in terms of two methyl groups joined to carbon number 2 in a propane chain—2,2-dimethyl propane.

Some special recommendations of the Geneva system define the use of certain suffixes: open-chain hydrocarbons with one double bond end in -ene; those with two double bonds end in -diene (alkenes and alkadienes rather than olefins and diolefins). Triple-bonded compounds end in -yne.

Alcohols are named by use of the hydrocarbon name with *-ol* as the characteristic suffix, i.e., methanol, ethanol, etc. In order to extend this plan to polyhydric alcohols, a syllable such as *di*, *tri*, and *tetra* is inserted between the name of the parent hydrocarbon and the suffix *ol*.

CH_2OHCH_2OH 1,2-ethane*di*ol

1 2 3

$CH_2OHCHOHCH_2OH$ 1,2,3-propane*tri*ol

Sulfides, disulfides, sulfoxides and sulfones are named like ethers, the oxy- term being replaced with thio-, dithio-, sulfinyl and sulfonyl. The acids are also named in accordance with the Geneva System.

*Based on article "Chemical Nomenclature" by F. A. Griffitts in "Encyclopedia of Chemistry", (Hampel and Hawley, 3rd Ed., Van Nostrand Reinhold, New York, 1973).

See also chemical nomenclature.

"Gen-Flo Latices."[179] Trademark for styrene-butadiene emulsion polymers with varying amounts of styrene and butadiene.
Uses: Latex paints, paper coatings, adhesives, textile printing inks, sizes.

"Genicop."[50] Trademark for a DDT-copper formulation which acts as an insecticide and fungicide.

genin. The steroid portion linked to a sugar residue in certain glycosides. Important genins are found in the digitalis glycosides, used as heart stimulants.

"Genite" 923.[50] Trademark for 2,4-dichlorophenyl ester benzenesulfonic acid; available as 50% emulsifiable or 50% wettable powder. Miticide specific for European red mite and clover mite.

"Gensol No. 6."[79] Brand name for terpene solvent.
Uses: Odorant for masking other odors; solvent and softener in rubber reclaiming; solvent in printing ink manufacture.

"Gen-Tac Latex."[179] Trademark for a latex containing vinyl pyridine, butadiene, and styrene.
Uses: Adhesive for rubber-to-fabric constructions, as in tires and mechanical products.

"Genthane-S."[179] Trademark for a polyurethane elastomer.
Uses: Mechanical goods, grommets, packings, and extrusions.

gentian violet. U.S.P. name for methyl violet.

gentisic acid (2,5-dihydroxybenzoic acid) $C_6H_3(OH)_2COOH$.
Properties: Crystals; m.p. 199–200°C; soluble in water, alcohol, and ether; insoluble in carbon disulfide, chloroform, and benzene.
Use: Medicine, as sodium gentisate.

"Gentro."[179] Trademark for a series of vulcanizable polymers containing butadiene and styrene manufactured by the cold process.
Properties: Contains approximately 23.5% styrene and 76.5% butadiene. Polymers mixed with reinforcing pigments give a variety of properties, abrasion resistance, flexibility, color, etc.
Uses: Tires, mechanical goods, proofed goods, extruded goods, shoe soles, heels, sponge, etc.

"Gentro-Jet."[179] Trademark for a series of carbon blacks co-precipitated with SBR polymer.
Uses: Tires, tread rubber, mechanical goods, conveyor belts, V-belts, etc.

geochemistry. The study of the chemical composition of the earth in terms of the physicochemical and geological processes and principles that produce and modify minerals and rocks. Of practical importance in discovering and establishing the limits of ore deposits, petroleum, tar sands, salt, sulfur, and other valuable resources.

geometric isomer. A type of stereoisomer in which a chemical group or atom occupies different spatial positions in relation to the double bond. If the double bond is between two carbon atoms the isomers are called *cis* and *trans*, as in crotonic acid, where the H and COOH reverse locations:

```
CH3         COOH        CH3          H
   \       /               \       /
    C=C                     C=C
   /       \               /       \
  H         H             H         COOH
      cis                     trans
```

If the double bond is between a nitrogen and a carbon atom the isomers are named *anti* and *syn*, as in benzaldoxime, where the OH group shifts location:

```
C6H5—C—H        C6H5—C—H
     ‖               ‖
  HO—N               N—OH
   anti              syn
```

This phenomenon also occurs in saturated ring compounds having three or more members in the ring, as well as in certain coordination compounds. See also stereochemistry.

"Geon"[119] Trademark for a group of polyvinyl chloride polymers, available as general-purpose, rigid, insulation, compounded, latex, paste, polyblend, or solution types. Also a high-temperature type (polyvinyl dichloride) that will withstand temperatures 60°F higher than other vinyls, and can be processed in standard equipment. See also polyvinyl chloride.

geothermal energy. Superheated steam at high prespure and temperatures ranging from 120 to 350°C, either trapped in rock strata or emanating from vents called fumaroles, located in areas characterized by volcanic activity or by geologic intrusions of igneous rock (magma). The former is the cause of such "wells" in New Zealand, Iceland, and southern Italy, whereas the latter is responsible for the geysers, hot springs and fumaroles in the Yellowstone area. The presence of subterranean steam pockets is indicated by a steeply rising geothermal gradient. This energy source was utilized for power production in Italy as long ago as 1935, and is now being used for this purpose in New Zealand, as well as for interior heating, papermaking, etc. It is also in limited use in Iceland and in western U.S. Studies have shown that geothermal steam plants are much less efficient than those fired by coal or oil, and that they emit a high level of atmospheric contaminants. These adverse effects can undoubtedly be mitigated by further engineering study.

geranial. See citral.

geranialdehyde. See citral.

geraniol (trans-3,7-dimethyl-2,6-octadien-1-ol) $(CH_3)_2C{:}CH(CH_2)_2C(CH_3){:}CHCH_2OH$. A terpene alcohol.
Properties: Colorless to pale yellow, liquid oil; pleasant geranium-like odor; sp. gr. 0.870–0.890 (15°C); f.p. −15°C; b.p. 230°C; refractive index (n 20/D) 1.4710–1.4780; optical rotation −2° to +2°; soluble in alcohol and ether, mineral oil, fixed oils; insoluble in water, and glycerol. Combustible. Nontoxic.
Derivation: From citronella oil (Java); citronellol-free grades from palmerosa oil and (synthetically) from pinene. These are of higher quality.
Grades: Standard; soap; synthetic; FCC; EOA.
Uses: Perfumery; flavoring; source of linalool.

geranium oil.
Properties: Pale yellow or greenish liquid; rose-like odor. Slightly soluble in water; soluble in alcohol and ether and most oils; sp. gr. 0.886–0.898; optical rotation −7° to −12°; refractive index (n 20/D) 1.4650–1.4700. Combustible; nontoxic.

Chief known constituents: Geraniol; citronellol.
Derivation: Distilled from the herb of several species of Pelargonium.
Grades: Algerian; Bourbon; F.C.C. (Algerian).
Uses: Perfumery; flavoring agent.
See also palmerosa oil.

geranyl acetate (geraniol acetate) $CH_3COOC_{10}H_{17}$.
Properties: Clear, colorless liquid; odor of lavender; sp. gr. 0.907–0.918 (15°C); b.p. 128–129°C (16 mm); optical rotation −2° to +2°; refractive index (n 20/D) 1.4580–1.4640; soluble in alcohol and ether; insoluble in water and glycerol. Combustible; nontoxic.
Derivation: (a) Constituent of several essential oils. (b) By heating geraniol and sodium acetate with acetic anhydride.
Grade: Technical; F.C.C.
Use: Perfumery; flavoring.

geranyl butyrate (geraniol butyrate) $C_3H_7COOC_{10}H_{17}$.
Properties: Colorless liquid; b.p. 151°C (18 mm); rose-like odor; sp. gr. 0.9008 (17/4°C); insoluble in water and glycerin; soluble in alcohol, ether. Occurs in several essential oils. Combustible. Nontoxic.
Grade: F.C.C.
Uses: Perfumes and soaps; flavoring.

geranyl formate (geraniol formate) $HCOOC_{10}H_{17}$.
Properties: Colorless liquid; b.p. 113°C (15 mm); sp. gr. 0.927 (20/4°C); rose-like odor; insoluble in water and glycerin; soluble in alcohol and ether. Occurs in several essential oils. Combustible; nontoxic.
Grade: F.C.C.
Uses: Perfumes and soaps; flavoring.

geranyl propionate (geraniol propionate) $C_2H_5COOC_{10}H_{17}$.
Properties: Colorless liquid; rose-like odor; sp. gr. 0.896–0.913 (25°C); refractive index 1.4570–1.4650 (20°C); soluble in most oils; insoluble in glycerol. Combustible; nontoxic.
Grades: Technical; F.C.C.
Uses: Perfumery; flavoring.

germane. A germanium hydride of the general formula Ge_nH_{2n+2}. See germanium tetrahydride.

germanium Ge Nonmetallic element of atomic number 32, atomic weight 72.59; valences 2, 4; group IVA of the periodic system.
Properties: Grayish-white metal; a p-type semiconductor; conductivity depends largely on added impurities. Sp. gr. 5.323; m.p 937.4°C; b.p. 2830°C; oxidizes readily at 600–700°C; does not volatilize below 1350°C; hardness 6 on Mohs scale. Attacked by nitric acid and aqua regia; stable to water, acids, and alkalies in absence of dissolved oxygen. Nontoxic.
Derivation: Recovered from residues from refining of zinc and other sources, by heating in the presence of air and chlorine. It is also present in some coals and can be recoverd from their combustion.
Occurrence: Missouri, Kansas, Oklahoma.
Method of purification: The chloride is distilled, and then hydrolyzed to the oxide, which is reduced by hydrogen to the metal. Zone-melting is used for final purification; single crystals are made by vaporization of germanium diiodide. The impurities in germanium are of controlling importance in its use in transistors. These are added to high-purity germanium in trace amounts during growth of single crystals.
Grades: Transistor, i.e., impurities 1 part in 10^{10}.
Forms: Ingots; single crystals, pure or doped; powder.
Uses: Solid state electronic devices (transistors, diodes); semiconducting applications; brazing alloys; phosphors; gold and beryllium alloys.

germanium dichloride $GeCl_2$. White powder; m.p. decomposes; decomposes in water; soluble in germanium tetrachloride; insoluble in alcohol, chloroform. Low toxicity.

germanium dioxide (germanium oxide) GeO_2.
Properties: White powder; hexagonal, tetragonal, and amorphous.
Grades: Technical; semiconductor 99.999% pure.
Uses: Special glass mixtures; phosphors; transistors and diodes.

germanium monoxide GeO. Black solid; m.p. 710°C (sublimes); insoluble in water; soluble in oxidizing agents. Available commercially 99.999% pure. Low toxicity.

germanium potassium fluoride (potassium germanium fluoride) K_2GeF_6.
Properties: White crystals; soluble in water (hot); insoluble in alcohol.
Grade: Technical.
Hazard: Toxic. Tolerance (as F), 2.5 mg per cubic meter of air.

germanium telluride GeTe. M.p. 725°C. An efficient semiconductor.
Hazard: May be toxic.

germanium tetrachloride $GeCl_4$.
Properties: Colorless liquid; sp. gr. 1.874 (25/25°C); f.p. -49.5°C (1 atm); b.p. 83.1°C; refractive index 1.464. Decomposes in water; insoluble in conc. HCl; soluble in carbon disulfide, chloroform, benzene, alcohol, and ether.
Derivation: Reaction of chlorine with elemental Ge.
Containers: Ampoules of Pyrex or quartz.
Use: Research.

germanium tetrahydride GeH_4. Colorless gas; sp. gr. 3.43 g/l; f.p. −165°C; b.p. −88°C; decomp. at 350°C; insoluble in water; soluble in liquid ammonia; slightly soluble in hot hydrochloric acid.
Hazard: Highly toxic. Tolerance, 0.2 ppm in air.

germicide. See bactericide.

"Germ-I-Tol."[430] Trademark for higher alkyl dimethyl benzyl ammonium chloride. Meets the requirements for benzalkonium chloride, U.S.P.

getter. See scavenger.

ghatti gum. Exudation from the stem of Anogeissus latifolia.
Properties: Colorless to pale yellow tears, rounded or vermiform. Almost tasteless and odorless; partially soluble in water. Can be solubilized by autoclaving. Nontoxic.
Uses: Thickener and protective colloid.

gibberellic acid $C_{19}H_{22}O_6$. A plant-growth-promoting hormone. It is a tetracyclic dihydroxylactonic acid.
Properties: Crystals; m.p. 233–235°C; slightly soluble in water; soluble in methanol, ethanol, acetone. Soluble in aqueous solutions of sodium bicarbonate and sodium acetate.

Uses: Agriculture and horticulture; malting of barley with improved enzymatic characteristics.

gibberellin. A group of hormones, widely distributed in flowering plants, which promotes elongation of shoots and coleoptiles. They differ from auxins (q.v.) in not stimulating the growth of roots, and in various other properties. However, the presence of auxin appears to be necessary for gibberellins to function. All gibberellins have closely related structures, being weak acids and having a ring system containing double bonds.
See also gibberellic acid; plant growth regulator.

Gibbs, Josiah Willard (1839–1903). The father of modern thermodynamics. During his lifelong post as professor of mathematical physics at Yale, he stated the fundamental concepts embraced by the three laws of thermodynamics, especially the nature of entropy. A theorist rather than an experimenter, Gibbs was the first to expound with mathematical rigor the "relation between chemical, electrical and thermal energy and capacity for work." It has been said that through out his adult life, Gibbs did nothing but think. The results have established him as one of the foremost creative scientists in history. See also thermodynamics.

Gibbs-Duhem equation (GDE). An exact thermodynamic relation that permits computation of the changes of chemical potential for one component of a uniform mixture over a range of compositions, provided the changes of potential for each of the other components have been measured over the same range. For binary systems, a useful form of the equation in terms of mole fraction is

$$\left(\frac{\partial \mu_1}{\partial x_1}\right)_{P,T} = -\frac{x_2}{x_1}\left(\frac{\partial \mu_2}{\partial x_1}\right)_{P,T}$$

"Gibrel."[123] Trademark for compound for promoting plant growth; the potassium salt of gibberellic acid.

"Gib-Sol."[530] Trademark for gibberellic acid.

"Gib-TAbs."[530] Trademark for gibberellic acid.

giga-. Prefix meaning 10^9 units (symbol = G). 1 Gg = 1 gigagram = 10^9 grams.

gilsonite An asphaltic material, or solidified hydrocarbon, found only in Utah and Colorado. One of the purest (9.9%) natural bitumens (q.v.). Said to be the first solid hydrocarbon to be converted to gasoline.
Hazard: Moderately toxic; irritant; skin sensitizer.
Uses: Acid, alkali and waterproof coatings; black varnishes, lacquers, baking enamels and japans; wire-insulation compounds; linoleum and floor tile; paving; insulation; diluent in low-grade rubber compounds; a possible source of gasoline, fuel oil, and metallurgical coke.
See also asphalt; bitumen.

gin. An alcoholic beverage made by distilling alcohol through a mixture of herbs and berries (juniper, coriander, etc.) and adjusting to 85 to 100 proof. Flash point 90°F.
Hazard: Flammable; moderate fire risk. Slight irritant; intoxicant.

Girard's reagent.
"P:" Carboxymethylpyridinium chloride hydrazide; acethydrazidepyridinium chloride.
$C_5H_5NClCH_2CONHNH_2$.
"T:" Carboxymethyltrimethyl ammonium chloride hydrazide; trimethylacethydrazide ammonium chloride. $(CH_3)_3NClCH_2CONHNH_2$.
Properties: White to faintly pinkish crystals with little or no odor; m.p. 190–200°C; soluble in water; insoluble in oils. "T" is hygroscopic.
Uses: Separation of aldehydes and ketones from natural oily or fatty materials; extraction of hormones.

Girbotol absorption (amine absorption). A process for the removal of hydrogen sulfide or carbon dioxide from a gaseous mixture. An organic amine (ethanolamine or diethanolamine, which are basic) is allowed to flow down a tortuous path through a tower where it is contacted by and absorbs (acidic) hydrogen sulfide or carbon dioxide from the gas to be purified as it moves up the tower. The amine, contaminated with these products, is then sent from the bottom of the tower to a steam stripper, where it flows countercurrent to steam, which strips the hydrogen sulfide or carbon dioxide from it. The amine is then returned to the top of the tower. The process is widely used in the petroleum industry for purifying refinery and natural gases and for recovery of hydrogen sulfide for sulfur manufacture. Removal of carbon dioxide from gases is usually done with monoethanolamine.

"Giv-Tan F."[12] Trademark for 2-ethoxyethyl paramethoxycinnamate.

glacial. A term applied to a number of acids, e.g., acetic and phosphoric, which have a freezing point slightly below room temperature when in a highly pure state. For example, glacial acetic acid is 99.8% pure and crystallizes at 62°F.

glance. A mineralogical term meaning brilliant or lustrous, used to describe hard, brittle materials that exhibit a bright, reflecting surface when fractured. Examples of such materials are hard asphalts (glance pitch) and ores of certain metals such as lead glance (galena).

glass. A ceramic material consisting of a uniformly dispersed mixture of silica (sand) (75%), soda ash (20%), and lime (5%), often combined with such metallic oxides as those of calcium, lead, lithium, cerium, etc., depending on the specific properties desired. The blend (or "melt") is heated to fusion temperature (about 2000°F) and then gradually cooled (annealed) to a rigid, friable state, often referred to as vitreous. Technically, glass is an amorphous, undercooled liquid of extremely high viscosity which has all the appearances of a solid. It has almost 100% elastic recovery. See also glass, optical.
Properties: (soda-lime glass) Lowest electrical conductivity of any common material (less than 10^{-6} mho/cm). Low thermal conductivity. High tensile and structural strength. Relatively impermeable to gases. Inert to all chemicals except hydrofluoric, fluosilicic, and phosphoric acids and hot, strong alkaline solutions. Continuous upper use temperature about 250°F, but may be higher, depending on composition. Noncombustible and nontoxic. Good thermal insulator in fibrous form. Molten glass is extrudable into extremely fine filaments. Glass is almost opaque to ultraviolet radiation; in the absence of added colorant it transmits 95 to 98% of light to which it is exposed.
Occurrence: Natural glass is rare, but exists in the form of obsidian (q.v.) in areas of volcanic activity and meteor strikes. Excellent sand for glass-making occurs in Virginia (James River), Pennsylvania, Massachusetts, New Jersey, West Virginia, Illinois

and Maryland; also in southern Germany and Czechoslovakia.

Forms available: Plate; sheet; fiber; filament; fabric; rods; tubing; pipe; powder; beads; flakes; hollow spheres. See also sodium silicate.

Uses: Windows, structural building blocks, chemical reaction equipment, pumps and piping, vacuum tubes, light bulbs, glass fibers (q.v.), yarns and fabrics, containers, optical equipment. Minute glass spheres with partial vacuum interior and treated exterior are available for compounding with resins for use in deep-sea floats, potting compounds and other composites. See also following entries.

glass, borosilicate. See glass, heat-resistant.

glass-ceramic. A devitrified or crystallized form of glass whose properties can be made to vary over a wide range.

Properties: Sp. gr. 2.5; rupture modulus up to 50,000 psi; thermal shock resistance 900°C; upper continuous use temperature 700°C. Glass-ceramics lie between borosilicate glasses and fused silica in high-temperature capability.

Derivation: A standard glass formula, to which a nucleating agent such as titania has been added, is melted, rolled into sheet, and cooled. It is then heated to a temperature at which nucleation occurs, causing formation of crystals. See also nucleation.

Uses: Range and stove tops; laboratory bench tops; architectural panels; restaurant heating and warming equipment; telescope mirrors.

See also "Pyroceram"; "Cer-vit."

glass electrode. See electrode, glass.

glass enamel. A finely ground flux, basically lead borosilicate, intimately blended with colored ceramic pigments. Different grades give characteristics of acid resistance, alkali resistance, sulfide resistance, or low lead release to meet requirements for various uses. Firing range 1000–1400°F (540–760°C).

Uses: For fired-on labels and decorations on glassware, tumblers, milk bottles, beverage bottles, glass containers, illuminating ware, architectural glass, and signs.

See also porcelain enamel.

glass fiber. Generic name for a manufactured fiber in which the fiber-forming substance is glass (Federal Trade Commission).

Properties: Tensile strength 15 g per denier; elongation 3 to 4%; sp. gr. 2.54; moisture regain, none; loses strength above 600°F, softens about 1500°F. Noncombustible. See also under glass.

Derivation: Molten glass is extruded at high speed through extremely small orifices.

Uses: Thermal, acoustic, and electrical insulation (coarse fibers in bats or sheets); decorative and utility fabrics such as drapes, curtains, table linen, carpet backing, tenting, etc.; tire cord, as belt between tread and carcass; filter medium; reinforced plastics (q.v.); light transmission for communication signals; reinforcement of cement products for construction use.

glass, heat-resistant. (1) A soda-lime glass containing about 5% boric oxide, which lowers the viscosity of the silica without increasing its thermal expansion. Such glasses (known as borosilicates) have a very low expansion coefficient and high softening point (about 1100°F). Tensile strength is about 10,000 psi. Continuous use temperature 900°F. Transmits ultraviolet light in higher wavelengths, and is used in "sunlight" lamps and similar equipment for this reason. See also "Pyrex."

(2) A pure silica glass trademarked "Vycor" (q.v.) which softens at about 2700°F. See also silica, fused.

glassine. A thin, transparent, and very flexible paper obtained by excessive beating of the pulp; it may contain an admixture of urea-formaldehyde to improve strength. Use is for packaging, dust covers for books, general household purposes.

glass, optical. Glasses intended for vision-correcting and such applications as lenses for cameras, microscopes, and other instruments; must be of extremely high quality and uniformity to meet requirements for refractive index and light dispersion. Optical glass may be either crown (lime) or flint (lead). Lead oxide is a major ingredient of flint glass, imparting high refractive index and dispersion, as well as surface brilliance. Flint glasses are also used in vacuum tubes and electrical equipment. Many special ingredients are used in both crown and flint glasses for specific refractivity and dispersion properties.

glass, photochromic. A glass that changes color on exposure to light and returns to its original color when the light has been removed. One type is a silicate glass containing dispersed crystals of colloidal silver halide which is precipitated within the melt during cooling. Alkali borosilicates are the most suitable types of glasses for this purpose. Photochromic glasses are used in variable-tint prescription lenses, which darken in sunlight and return to original clearness indoors (85% light transmission when clear; 45% in sunlight).

glass, photosensitive. A glass containing a small amount of a photosensitive substance, such as a gold, silver, or copper compound. When ultraviolet light is passed through a photographic negative onto the surface of this glass, a latent image is formed within the glass, which is converted to a visible image made up of tiny metal particles when the glass is heated. In a special type of photosensitive glass (photosensitive opal), the metal particles of the photographic image within the glass serve as nuclei for the growth of nonmetallic crystals; crystal growth is confined to the area of the image. These crystalline areas are dissolved much more rapidly than the adjacent glass by hydrofluoric acid. Thus the glass can be formed into intricate shapes without the use of mechanical tools.

glass, plate. Plate glass has the same composition as window glass (soda-lime-silica), differing from it only in method of manufacture. These differences are primarily (1) the longer time of annealing (q.v.), (three or four days) which eliminates the distortion and strain effects of rapid cooling, and (2) intensive grinding and polishing, which removes local imperfections and produces a bright, highly reflecting finish.

glass, ruby. A deep-red glass made by incorporating colloidal gold in the silicate mixture. It is used chiefly in the decorative arts.

glass, safety. See safety glass.

glass transition temperature (T_g). The temperature at which an amorphous material (such as glass or a high polymer) changes from a brittle, vitreous state to a plastic state. Many high polymers, such as acrylics and their derivatives, have this transition point, which is related to the number of carbon atoms in the ester group. The T_g of glass depends on its composition and extent of annealing.

glass, water. See sodium silicate.

"Glasteel."[522] Trademark for an engineering laminate formed by spraying a slurry of powdered glass onto a base of mild steel, followed by high-temperature firing. A chemical reaction takes place during the firing between the glass and steel, forming a chemical bond which locks the pieces firmly together.
Uses: Corrosion-resistant process equipment; piping; heat exchangers; pumps and valves.

Glauber's salt. See sodium sulfate decahydrate.

glaucarubin $C_{25}H_{36}O_{10}$. Crystals; decomposing at 250–255°C; slightly soluble in water. Obtained from the meal from Simaruba glauca seeds. Listed in N.D. Used in medicine.

glauconite $K_2(Mg, Fe)_2Al_6(Si_4O_{10})_3(OH)_{12}$. A natural silicate of potassium, aluminum, iron, and magnesium, found in greensands, and other sedimentary rocks. Color green, luster earthy; sp. gr. 2. 3.
Occurrence: New Jersey, Virginia.
Uses: Water softener; foundry molds; fertilizer.

glaze. A mixture similar to porcelain enamel, applied to a ceramic substrate. It may refer to (1) a vitreous coating on pottery or enamelware; (2) the mixed dry powders of the batch to be used for the coating; or (3) a water suspension of these materials (wet glaze). Glazes must be low in sodium, and are usually mixtures of silicates and flint, lead compounds, boric acid, calcium carbonate, etc. See also frit; porcelain enamel.

glaze stain. Finely ground calcined oxide of cobalt, copper, iron and manganese used for coloring ceramic glazes.

GLC. Abbreviation for gas-liquid chromatography. See gas chromatography.

gliadin (prolamin). A group of simple vegetable proteins or globulins found in gluten, the protein of wheat, rye and other grains. Wheat gliadin has the composition: 52.7% C, 17.7% N, 21.7% O, 6.9% H, 1.0% S. It is composed of 18 amino acids, 40% being glutamic acid. Insoluble in water, soluble in 70–90% alcohol, soluble in dilute acid and in alkali.
Uses: Chemical synthesis of spinal anesthetics; pharmaceutical preparations.

"Glidco" Cordage Tars.[296] Trademark for a series of tars derived from the distillation and decomposition of oleoresinous southern pine.
Use: Coating and impregnation of paper, twine, fibers, oakum, rope and nets.

"Glidco" Dipentene-300.[296] Trademark for commercial cut of terpene hydrocarbons with range similar to that of *dl*-limonene.
Uses: Paints; rubber reclaiming agent; resin manufacturing.

"Glidco" Terpene-S.[296] Trademark for a terpene cut similar to pine oil containing terpineol, plus other monocyclic terpene alcohols and terpene hydrocarbons. Typical purity of 66% as terpene alcohols.

"Glidco" Terpineol-350.[296] Trademark for a water-white viscous liquid with a characteristic woody lilac odor. Contains a minimum alcohol content as terpineol of 97% consisting principally of the alpha isomer. Used in the preparation of essential oils; industrial and soap perfumes.

"Globar."[280] Trademark for silicon carbide heating elements and resistors and accessories.
Properties: Elements have a working temperature up to 2750°F whch can be extended to 3000°F for short periods; low coefficient of expansion; structure not affected by rapid heating and quick cooling; resistance remains practically constant above 900°F.
Uses: Electric resistors and heating elements; terminals and other accessories for electric heating elements; electric heating appliances; electric furnaces.

globulin. (1) Any of a group of proteins synthesized by the body when invaded by infective organisms. They are coagulated by heat, insoluble in water, soluble in dilute solutions of salts, strong acids and strong alkalies. Enzymes and acids cause hydrolysis to amino acids as the only products. Examples are immune serum or gamma globulins in blood, myosin in muscle. The blood globulins are used in immunizing against specific diseases and in medical research. The Gamma globulin molecule is reported to consist of 19,996 atoms associated in 1320 amino acid units.
(2) A protein occurring naturally in wheat and other cereal grains.

"Gloria."[45] Trademark for white mineral oil, U.S.P. Sp. gr. 0.875–0.885 (60°F).
Uses: Pharmaceutical and cosmetic formulations; plastics; tobacco; paper; animal husbandry.

"Gloval."[309] Trademark for a sperm and mineral oil blend; used in tanning industry.

Glu Abbreviation for glutamic acid.

glucagon (hyperglycemic-glycogenolytic factor; HG-factor; HGF). Produced by the alpha cells of the islands of Langerhans and also by the gastric mucosa. It is opposite in effect to insulin. It appears to be a straight-chain polypeptide with a molecular weight of about 3500. Small amounts have been detected in commercial insulin preparations.
Grade: U.S.P.
Uses: Medicine; biochemical research.

"Glucarine B."[73] Trademark for glycol carbohydrate complex.
Properties: Water-white, syrupy, clear fluid. Soluble in water, alcohols, glycols and glycerol.
Uses: Replaces glycerin where a cheaper, colorless product is desired, especially for cosmetic and technical purposes.

glucase. See maltase.

gluconic acid (glyconic acid; glycogenic acid) $CH_2OH(CHOH)_4COOH$.
Properties: Light brown syrupy liquid with mild taste. Soluble in water and alcohol. Nontoxic.
Derivation: Bacterial, chemical, or electrochemical oxidation of dextrose. It is the chief acid in honey.
Grades: Technical, 50% solution.
Containers: Barrels, kegs, carboys, tank cars.
Uses: Pharmaceutical and food products; cleaning and pickling metals; sequestrant; cleansers for bottle washing; paint strippers; alkaline derusters.

glucono-delta-**lactone** $CH_2OH\overline{CH(CHOH)_3C(O)O}$.
Properties: White crystals. M.p. 155°C; b.p. (decomposes); readily soluble in water, slightly soluble in alcohol. Nontoxic.
Derivation: From gluconic acid.

Grade: F.C.C.
Uses: Acid; leavening agent; sequestrant.

D(+)-**glucosamine** $CH_2OH(CHOH)_3CHNH_2CHO$.
Properties: Colorless needles; m.p. 110°C (dec); very soluble in water; slightly soluble in methanol and ethanol; insoluble in ether and chloroform.
Use: Biochemical research.

glucose (dextrose; grape sugar; corn sugar) $C_6H_{12}O_6 \cdot H_2O$.
Properties: Colorless crystals or white granular powder; odorless; sweet taste. Sp. gr. 1.544; m.p. 146°C; soluble in water; slightly soluble in alcohol. It has the D (right-handed) configuration and is dextrorotatory. Combustible. Nontoxic.
Occurrence: Formed in plants by photosynthesis; also in the blood.
Derivation: Hydrolysis of corn starch with acids or enzymes; a component of invert sugar and glucose syrup. In a recently (1974) discovered process, glucose has been obtained from cellulose by enzyme hydrolysis.
Grades: Technical; U.S.P.; anhydrous; hydrated.
Uses: Confectionery; infant foods; medicine; brewing and wine-making; intermediate; caramel coloring; baking and canning.
See also glucose syrup; invert sugar.

glucose oxidase. An enzyme commercially available in different grades for the removal of excess glucose or oxygen.
Uses: Food preservative; analytical reagent for glucose.

glucose 1-phosphate (glucose 1-phosphoric acid; Cori ester) $C_6H_{11}O_5 \cdot H_2PO_4$. An intermediate in carbohydrate metabolism.

glucose 6-phosphate (glucose 6-phosphoric acid; Robison ester) $C_6H_{11}O_5 \cdot H_2PO_4$. An intermediate in carbohydrate metabolism, usually as the barium or dipotassium salt, which are water-soluble.

glucose 1-phosphoric acid. See glucose 1-phosphate.

glucose syrup (corn syrup). A mixture of D-glucose, maltose and maltodextrins made by hydrolysis of cornstarch by the action of acids or enzymes. The degree of conversion of the starch varies, with consequent effect on the dextrose equivalent (D.E.) or reducing power of the syrup. Its primary uses are in the food industry as a sweetener (high D.E.) or as thickeners or bodying agents in soft drinks (low D.E.). See also glucose.

alpha-**glucosidase.** See maltase.

beta-**glucosidase.** See emulsin.

glucosides. See glycosides.

glucosulfone sodium $C_{24}H_{34}N_2O_{18}S_3Na_2$. para, para'-Diaminodiphenylsulfone-N,N'-di(dextrose sodium sulfonate). Available only in a mixture containing 11.5% of dextrose.
Properties: White to faintly yellow, odorless, sweet, amorphous solid. Freely soluble in water.
Use: Medicine.

D(+)-**glucuronic acid** $COOH(CHOH)_4CHO$. A widely distributed substance in both plants and animals; usually occurs as part of a larger molecule as in various gums, or combined with phenols or alcohols.
Properties: Exhibits mutarotation. The beta form has m.p. 165°C. Soluble in water and alcohol. Nontoxic.
Derivation: From gum acacia.
Uses: Biochemical research; medicine.

glucuronolactone $C_6H_8O_6$. The gamma lactone of glucuronic acid. Found in plant gums and animal connective tissues.
Properties: Colorless, odorless, white powder. Sp. gr. 1.76 (30/4°C); m.p. 172–178°C; soluble in water. Nontoxic.
Derivation: From glucuronic acid.
Uses: Growth factor; medicine; pharmaceutical intermediate.

glue. A colloidal suspension of various proteinaceous materials in water. Most familiar are those derived by boiling animal hides, tendons, or bones which are high in collagen. Chief sources are slaughterhouse wastes and fish scraps. Other animal-derived glues are made from casein (milk) and blood. The most important vegetable glue is made from soybean protein. Combustible in solid form. Nontoxic, but inedible. See also adhesive, fish glue, collagen, gelatin.
Containers: Bags (40–500 jelly grams); carlots.

"Gluflex."[157] Trademark for a clear, noncrystallizing product consisting of sucrose, levulose, dextrose, sodium bisulfite and water, and having a solids content of 181.5%.
Uses: Animal glues; gummed tapes.

"Glueglis." Trademark for a composition containing chiefly glue and glycerol, used in printing press rollers. Has soft, rubbery consistency but is easily decomposed by heat.

gluside. See saccharin.

glutamic acid (alpha-aminoglutaric acid; 2-amino pentanedioc acid) $COOH(CH_2)_2CH(NH_2)COOH$. A nonessential amino acid. The naturally occurring form is L(+)-glutamic acid. Nontoxic.
Properties:
DL-glutamic acid; (synthetic racemic mixture) crystals; m.p. 225–227°C (dec); slightly soluble in ether, alcohol, and petroleum ether; sp. gr. 1.4601 (20/4°C).
D(−)-glutamic acid: leaflets from water; m.p. 247–249°C (dec); sp. gr. 1.538 (20/4°C).
L(+)-glutamic acid: crystals; sublimes at 200°C; decomposes at 247–249°C; nearly soluble in ether, acetone, cold glacial acetic acid; insoluble in ethyl alcohol and methanol; sp. gr. 1.538 (20/4°C). Specific rotation (25°C) +37 to +38.9°. Available commercially.
Derivation: Hydrolysis of vegetable protein (e.g., beet sugar waste, wheat gluten); organic synthesis based on acrylonitrile. It comprises 40% of the gliadin in wheat gluten.
Grade: F.C.C. (L-form).
Containers: Fiber drums.
Uses: Medicine; biochemical research; salt substitute; flavor enhancer; synthetic leather.
See also sodium glutamate.

glutamic acid hydrochloride
$COOH(CH_2)_2CH(NH_2)COOH \cdot HCl$.
Properties: White crystalline powder; sp. gr. 1.525; m.p. 202–213°C (dec); specific rotation (25°C) +23.5° to +25.5°. Very soluble in water, liberating hydrochloric acid; almost insoluble in alcohol and ether.
Derivation: Hydrolysis of gluten; organic synthesis.
Grades: N.F.; F.C.C.
Use: Medicine; salt substitute; flavor enhancer.

glutamine $H_2NC(O)(CH_2)_2CH(NH_2)COOH$. (2-Amino-4-carbamoylbutanoic acid). A nonessential amino acid. Both the L- and DL- forms are available.
Properties: White, crystalline powder; soluble in water; insoluble in most organic solvents. Should be kept dry and refrigerated. M.p. (L-form) 184–185°C (dec); (DL-form) 176°C. Nontoxic. Its presence in wheat gluten contributes to the elastic properties of flour by hydrogen bonding and disulfide cross-linking.
Derivation: Action of enzymes on gluten; from beet roots; constituent of many proteins.
Uses: Medicine; culture media; biochemical research; feed additive.

gamma-**glutamylcysteinylglycine.** See glutathione.

"Glutan H-C-L."[315] Trademark for glutamic acid hydrochloride (q.v.).

glutaraldehyde $OHC(CH_2)_3CHO$.
Properties: Liquid; sp. gr. 0.72; b.p. 188°C (decomposes); f.p. -14°C; vapor pressure 17 mm (20°C); soluble in water and alcohol. No flash point. Nonflammable.
Grades: 99%; 50% biological solution; 25% solution.
Hazard: Moderately toxic and irritant.
Uses: Intermediate; fixative for tissues; for cross-linking protein and polyhydroxy materials; tanning of soft leathers.

glutaric acid (n-pyrotartaric acid; pentanedioic acid) $COOH(CH_2)_3COOH$.
Properties: Colorless crystals; m.p. 97°C; refractive index (n 106/D) 1.419; b.p. 302–304°C; soluble in water, alcohol, and ether. Nontoxic.
Derivation: From cyclopentanone.
Use: Organic synthesis.

glutaric anhydride (pentanedioic acid anhydride) $CH_2(CH_2CO)_2O$.
Properties: Solid. M.p. 56.5°C; b.p. 303°C. Soluble in benzene and toluene; soluble in water on complete hydrolysis.
Hazard: Moderately toxic and irritant.
Uses: Plasticizers; resin; lubricant; adhesive synthesis; dyes and pharmaceuticals.

glutaronitrile (trimethylenedicyanide; pentanedinitrile) $NC(CH_2)_3CN$.
Properties: Colorless to straw-colored viscous liquid; b.p. 144–146°C (13 mm); sp. gr. 0.989; soluble in water and alcohol, insoluble in ether and carbon disulfide. Combustible.
Hazard: Moderately toxic.
Use: Chemical intermediate.

glutathione (gamma-glutamylcysteinylglycine) $C_{10}H_{17}O_6N_3S$. A universal component of the living cell. Contains glutamic acid, cysteine, and glycine. These are chemically bound but can be obtained by hydrolysis.
Properties: White crystalline powder; odorless; m.p. 190–192°C; mild sour taste; soluble in water; insoluble in alcohol. Nontoxic.
Use: Nutritional and metabolic research.

gluten. A mixture of manay proteins in which gliadin, glutenin, globulin and albumin predominate; it occurs in highest percentage in wheat (Manitoba wheat contains about 12%) and also to some extent in other cereal grains, usually associated with starch. It comprises 18 amino acids. Gluten is insoluble in water and is hydrophilic. Its specific adaptability to bread making is due to its elastic, cohesive nature that enables it to retain the bubbles of carbon dioxide evolved by leavening agents; this also imparts to doughs their characteristic dilatant properties. This behavior is due to disulfide crosslinks and hydrogen bonding between the proteins or their constituent amino acids.
Containers: Drums and cans.
Uses: Special breakfast foods and other cereals and foods; cattle food; adhesives; production of certain amino acids.

glutenin. One of the proteins present in wheat flour in substantial percentages. It is composed of 18 amino acids.

glutethimide. N.F. name for 2-ethyl-2-phenylglutarimide.

Gly Abbreviation for glycine.

"Glycamide."[123] Trademark for glycarbylamide (q.v.).

glycarbylamide (4,5-imidazoledicarboxamide) $C_3H_2N_2(CONH_2)_2$. White powder, melting above 360°C; insoluble in water. A coccidiostat for chickens.

glyceraldehyde (glyceric aldehyde) $HOCH_2CHOHCHO$. Isomeric with dihydroxyacetone. It is produced by the oxidation of sugars in the body. As the simplest aldose, the conformation of D and L-glyceraldehydes has been designated the reference standard for D- and L- carbohydrates and derivatives.

$$\begin{array}{ccc} \text{L} & \qquad & \text{D} \\ CHO & & CHO \\ | & & | \\ HO—C^*—H & & H—C^*—OH \\ | & & | \\ CH_2OH & & CH_2OH \end{array}$$

In these isomers the central carbon atom (C*) is asymmetric.
Properties: (DL-glyceraldehyde) Tasteless crystals from alcohol-ether mixture; m.p. 145°C; insoluble in benzene, petroleum ether, pentane. Nontoxic.
Grades: 40% aqueous solution.
Uses: Biochemical research; intermediate; nutrition; preparation of polyesters, adhesives; cellulose modifier; leather tanning.
See also asymmetry; optical isomer.

glyceride. An ester of glycerol and fatty acids in which one or more of the hydroxyl groups of the glycerol have been replaced by acid radicals. The latter may be identical or different, so that the glyceride may contain up to three different acid groups. Glycerides can be made synthetically. The most common are based on fatty acids which occur naturally in oils and fats. See also mono-, di-, and triglyceride.

glycerin. See glycerol.

glycerin carbonate (hydroxymethylethylene carbonate) $CH_2O(CO)OCHCH_2OH$.
Properties: Pale yellow, odorless, hygroscopic liquid. Boiling range 125–130°C (0.1–0.2 mm); freezing point, supercools to a glass; sp. gr. 1.4000 (20/4°C); refractive index (n 20/D) 1.4580; flashpoint 415°F. Miscible with water, alcohol, ether; soluble in ethylene dichloride; insoluble in carbon tetrachloride, benzene, and alihatic hydrocarbons. Combustible; nontoxic.
Grades: Technical.
Uses: Solvent; intermediate.

glycerol (glycerin; glycyl alcohol; 1, 2, 3-propanetriol) $C_3H_5(OH)_3$. A trihydric (polyhydric) alcohol.

Properties: Clear, colorless, odorless, syrupy liquid; sweet taste; hygroscopic; sp. gr. (anhydrous 1.2653; (U. S. P.) greater than 1.249 (25/25°C); (dynamite) 1.2620; m.p. 18°C; b.p. 290°C; soluble in water and alcohol (aqueous solutions are neutral); insoluble in ether; benzene and chloroform and in fixed and volatile oils. Flash point 320°F. Combustible; low toxicity. Autoignition temp. 739°F.
Derivation: (1) By-product of soap manufacture; (2) from propylene and chlorine to form allyl chloride, which is converted to the dichlorohydrin with hypochlorous acid; this is then saponified to glycerol with caustic solution; (3) isomerization of propylene oxide to allyl alcohol, which is reacted with peracetic acid; the resulting glycidol is hydrolyzed to glycerol; (4) hydrogenation of carbohydrates with nickel catalyst; (5) from acrolein and hydrogen peroxide.
Method of purification: Redistillation; ion-exchange techniques.
Grades: U.S.P., C. P. (pharmaceutical and commercial, where highest grade is required); saponification, soap lye, crude yellow distilled (for commercial purposes where color and extreme purity are not factors); high gravity or dynamite (dehydrated to 99.8–99.9% purity); natural; synthetic; F.C.C.
Containers: Drums, tank cars.
Uses: Alkyd resins; explosives; ester gums; pharmaceuticals; perfumery; plasticizer for regenerated cellulose; cosmetics; foodstuffs, conditioning tobacco; liqueurs; solvent; printer's ink rolls; polyurethane polyols; emulsifying agent; rubber stamp and copying inks; binder for cements and mixes; paper coatings and finishes; special soaps; lubricant and softener; bacteriostat; penetrant; hydraulic fluid; humectant.

glycerol boriborate. Pale yellow liquid obtained by heating glycerin, sodium borate and boric acid. Composition varies. Soluble in cold water, absolute alcohol, other alcohols, glycerin. Used as adhesive, binder, fabric softener, fire retardant on fabrics.

glycerol dichlorohydrin. See alpha-dichlorohydrin.

glycerol 1,3-distearate (glyceryl 1,3-distearate) $CH_3(CH_2)_{16}COOCH_2CH(OH)CH_2OOC(CH_2)_{16}CH_3$.
Properties: Solid; m.p. 29.1°C. Very slightly soluble in cold alcohol and ether; soluble in hot organic solvents.

glycerol monolaurate (glyceryl monolaurate) $C_{11}H_{23}COOCH_2CHOHCH_2OH$.
Properties: Cream-colored paste; faint odor. Dispersible in water; soluble in methyl and ethyl alcohols, toluene, naphtha, mineral oil, cottonseed oil, ethyl acetate M.p. 23–27°C; sp. gr. 0.98; F. F. A. less than 2.5%; iodine value 6–8; pH 8.0–8.6 (25°C) (5% aqueous dispersion). Combustible; nontoxic.
Derivation: See monoglyceride.
Grades: Edible; technical.
Uses: Emulsifying and dispersing agent for food products, oils, waxes and solvents; antifoaming agent; dry-cleaning soap base.

glycerol monooleate (glyceryl monooleate) $C_{17}H_{33}COOCH_2CHOHCH_2OH$.
Properties: Yellow oil or soft solid; sp. gr. 0.95; m.p. 14–19°C, depending on purity; iodine value 65–80; insoluble in water; somewhat soluble in alcohol and most organic solvents. Combustible; nontoxic.
Derivation: See monoglyceride.
Grades: Edible; technical.
Uses: Foods, pharmaceuticals and cosmetics; rust-preventive oils, textile finishing; vinyl light stabilizers; odorless base paints; flavoring.

glycerol monoricinoleate (glyceryl monoricinoleate) $C_6H_{13}CHOHC_{10}H_{18}COOCH_2CHOHCH_2OH$.
Properties: Yellow liquid; sp. gr. 1.01; m.p. <-5°C; iodine value 65–70; dispersible in water; soluble in most organic solvents. Combustible.
Derivation: See monoglyceride.
Uses: Non-drying emulsifying agent; solvent; plasticizer; in polishes, in cosmetics, in textile, paper and leather processing; low-temperature lubricant. Stabilizes latex paints against breakdown due to repeated freeze-thaws.

glycerol monostearate (GMS; glyceryl monostearate; monostearin) $(C_{17}H_{35})COOCH_2CHOHCH_2OH$.
Properties: Pure white or cream-colored, wax-like solid with faint odor, and fatty, agreeable taste. Affected by light. M.p. 58–59°C (capillary tube); sp. gr. 0.97; F.F.A. less than 5%; iodine value 3–4; dispersible in hot water. Soluble (hot) in alcohol, oils and hydrocarbons. Combustible; nontoxic.
Derivation: See monoglyceride.
Grades: Edible; cosmetic; N. F.
Uses: Thickening and emulsifying agent for margarine, shortenings and other food products; flavoring; emulsifying agent for oils, waxes and solvents; protective coating for hygroscopic powders; cosmetics; opacifier; detackifier; resin lubricant.

glycerol phthalate. See glyceryl phthalate.

glycerol tributyrate. See glyceryl tributyrate.

glycerol tripropionate. See glyceryl tripropionate.

glycerol tristearate. See stearin.

glycerophosphoric acid $C_3H_5(OH)_2H_2PO_4$.
Properties: Colorless, odorless liquid. Soluble in water and alcohol. Sp. gr. 1.60; f.p. −20°C. Combustible.
Derivation: Interaction of glycerol and phosphoric acid.
Uses: Medicine; manufacture of glycerophosphates.

glyceryl abietate. An ester gum used as additive in citrus-flavored beverages.

glyceryl para-**aminobenzoate.** See "Escalol 106."

glyceryl alpha-**chlorohydrin.** See chlorohydrin.

glyceryl diacetate. See diacetin.

glyceryl 1,3-distearate. See glycerol 1,3-distearate.

glyceryl ditolyl ether. See dicresyl glyceryl ether.

glyceryl monoacetate. See acetin.

glyceryl monolaurate. See glycerol monolaurate.

glyceryl monooleate. See glycerol monooleate.

glyceryl monoricinoleate. See glycerol monoricinoleate.

glyceryl monostearate. See glycerol monostearate.

glyceryl phthalate (glycerol phthalate).
Properties: Water-white, solid resin. Insoluble in water. Soluble (hot) in methyl and ethyl alcohols, acetone, ethyl acetae. Partly soluble in toluene, naphtha. Sp. gr. 1.29; saponification value 605–615; acid value 300–315; softening point about 67°C.
Grade: Technical.
Uses: Varnishes, lacquers, etc. See also alkyd resin.

glyceryl ricinoleate. See glyceryl triricinoleate.

glyceryl triacetate. See triacetin.

glyceryl tri-(12-acetoxystearate) (castor oil, acetylated and hydrogenated) $C_3H_5(OOCC_{17}H_{34}OCOCH_3)_3$.
Properties: Clear, pale yellow, oily liquid; mild odor; soluble in most organic solvents; insoluble in water. Sp. gr. 0.955 (25/25°C); saponification value 298; iodine value 3; solidifies at 4°C. Combustible; low toxicity.
Derivation: Hydrogenation of acetylated castor oil.
Grade: Technical.
Containers: 5-gal cans; 55-gal drums; tank cars.
Uses: Plasticizer for nitrocellulose, ethylcellulose and polyvinyl chloride; lubricants; protective coatings.

glyceryl tri-(12-acetylricinoleate) (castor il, acetylated) $C_3H_5(OOCC_{17}H_{32}OCOCH_3)_3$.
Properties: Clear, pale yellow, oily liquid; mild odor; soluble in most organic liquids; insoluble in water. Sp. gr. 0.967 (25/25°C); saponification value 300; iodine value 76; solidifies at −40°C. Combustible; low toxicity.
Grade: Technical.
Derivation: Acetylation of castor oil.
Containers: 5-gal cans; 55-gal drums; tank cars.
Uses: Plasticizer for nitrocellulose, ethylcellulose and polyvinyl chloride; lubricants; protective coatings.

glyceryl tributyrate (tributyrin; butyrin; glycerol tributyrate) $C_3H_5(OCOC_3H_7)_3$.
Properties: Colorless, oily liquid; sp. gr. 1.035 (20°C); refractive index (20°C) 1.4359; b.p. 315°C; solubility in water 0.010%; soluble in alcohol and ether. Combustible; low toxicity.
Containers: Drums.
Grades: Technical; F.C.C.
Uses: Plasticizer; flavoring.

glyceryl tri(12-hydroxystearate) (castor oil, hydrogenated) $C_3H_5(OOCC_{17}H_{34}OH)_3$. Glyceryl triricinoleate in which hydrogen has saturated the ricinoleic groups.
Properties: Hard, brittle wax-like solid, yellowish to milk-white in color. M.p. 86–88°C; sp. gr. (100/25°C) 0.899.
Uses: Lubricants; heavy-metal soaps; waxes; plasticizers, cosmetics; chemical intermediate. The lithium compound is used in high-temperature greases.

glyceryl trinitrate. See nitroglycerin.

glyceryl trioleate. See olein.

glyceryl tripalmitate. See tripalmitin.

glyceryl tripropionate (glycerol tripropionate; tripropionin ($C_3H_5(OCOC_2H_5)_3$.
Properties: Solid; sp. gr. (20°C) 1.078; refractive index (20°C) 1.431; b.p. (20 mm) 177–182°C; f.p. less than −50°C; solubility in water, 0.313% of weight; soluble in alcohol.
Use: Plasticizer.

glyceryl triricinoleate (glyceryl ricinoleate) $C_3H_5(OOCC_{17}H_{32}OH)_3$. The triglyceride of ricinoleic acid. It constitutes about 80% of castor oil.
Properties: Light amber oily liquid.
Use: Emulsifying agent.

glyceryl tristearate. See stearin.

glycidol (2,3-epoxy-1-propanol) $CH_2OHCHCH_2$ (with O bridging the last two carbons). An epoxide.
Properties: Colorless liquid; b.p. 162°C; soluble in water, alcohol and ether. Combustible. Sp. gr. 1.12.
Derivation: Treatment of monochlorohydrin with bases; reaction product of allyl alcohol and peracetic acid.
Hazard: Moderately toxic. Tolerance, 50 ppm in air.
Uses: Stabilizer for natural oils; demulsifier; dye-leveling agent; stabilizer for vinyl polymers.

gama-**glycidoxypropyltrimethoxysilane** $\underline{OCH_2C}HCH_2O(CH_2)_3Si(OCH_3)_3$.
Properties: Liquid; sp. gr. 1.070 (25°C); b.p. approx. 120°C (2 mm); refractive index 1.4280 (25°C). Soluble in acetone, benzene, ether. Reacts with water.
Derivation: Addition of hypochlorite to allyl alcohol and reaction with soda lime.
Containers: Up to 55-gal drums.
Uses: Coupling agent for glass- and mineral-filled plastics.

glycidyl acrylate $H_2C{:}CHCOOCH_2\underline{CHCH_2O}$.
Properties: Liquid; sp. gr. 1.1074 (20/20°C); b.p. 57°C (2 mm) with polymerization; f.p. −41.5°C; flash point 141°F (TOC). Insoluble in water. Combustible.
Containers: 55-gal drums.
Hazard: May be irritant to skin and eyes.
Use: See acrylate.

glycidyl ether $OCH_2\overline{CHOCH_2}$. Liquid. Appears as the terminal group of epoxy resin structures resulting from reaction of epichlorohydrin and bisphenol A with alkaline catalyst. Also reacts with novolac resins (q.v.). Combustible.

glycine (1) Aminoacetic acid. A nonessential amino acid. NH_2CH_2COOH.
Properties: White, sweet odorless crystals; m.p. 232–236°C with decomposition; sp. gr. 1.1607; combines with hydrochloric acid to form the hydrochloric; soluble in water; insoluble in alcohol and ether.
Derivation: Action of ammonia on chloroacetic acid; occurs in many proteins, and is especially abundant in silk fibroin, gelatin, and sugar cane.
Grades: Technical; N.F.; F.C.C.
Containers: Drums; carlots.
Hazard: Use in food products restricted to 0.01%.
Uses: Organic synthesis; medicine; biochemical research; buffering agent; chicken-feed additive; reduces bitter taste of saccharin; retards rancidity in animal and vegetable fats.
(2) The extreme dilution of methylacetophenone gives a perfume resembling the odor of the climbing plant glycine (Wistaria sinensis), native to China and cultivated elsewhere. The name is also given to bouquets made from violet, lilac and jasmin ottos.
(3) Para-hydroxyphenylglycine (q.v.). A photographic developer.

glycocholic acid (cholylglycine) $C_{26}H_{43}NO_6$. The sodium salt occurs in bile, where it is formed by the combination of glycine with cholic acid (q.v.). It aids in the digestion and absorption of fats.
Properties: Crystallizes from water with 1.5 moles H_2O. Becomes anhydrous at 100°C. Anhydrous form decomposes at 154–155°C. Practically insoluble in water. The sodium salt is soluble in water and alcohol. Nontoxic.
Derivation: Precipitation from bile.
Uses: Biochemical research; food emulsifying agent (up to 0.1%).

glycocoll. See glycine (1).

glycocoll-para-phenetidine hydrochloride. See phenocoll hydrochloride.

glycogen (animal starch; liver starch) $(C_6H_{10}O_5)_n$. A glycose polysaccharide; the storage carbohydrate of the animal organism, found especially in the liver and rested muscle.
Properties: White powder; forms a dextrorotatory colloidal solution; partially soluble in water; sweet taste; nontoxic.
Derivation: Isolated from liver by treatment with 30% sodium hydroxide solution.
Use: Biochemical research.

glycogenic acid. See gluconic acid.

glycol. See ethylene glycol; it is also a general term for dihydric alcohols, which are physically and chemically similar to glycerol (q.v.).

glycol bromohydrin. See ethylene bromohydrin.

glycol carbonate. See ethylene carbonate.

glycol chlorohydrin. See ethylene chlorohydrin.

glycol diacetate. See ethylene glycol diacetate.

glycol dibutyrate. See ethylene glycol dibutyrate.

glycol diformate. See ethylene glycol diformate.

glycol dimercaptoacetate (ethylene glycol bisthioglycolate) $HSCH_2COOCH_2CH_2OOCCH_2SH$.
Properties: Liquid; sp. gr. 1.313; b.p. 137–139°C (2 mm); refractive index 1.519 (25/D); insoluble in water; soluble in alcohol, acetone, and benzene. Combustible.
Uses: Crosslinking agent for rubbers; accelerator in curing epoxy resins.

glycol dimercaptopropionate [ethylene glycol bis(mercaptopropionate)] $(HSCH_2CH_2COOCH_2)_2$.
Properties: Liquid; sp. gr. 1.219 (25°C); b.p. 175–195°C; refractive index 1.5150 (25/D); insoluble in water and hexane; soluble in alcohol, acetone and benzene. Combustible.
Uses: Crosslinking agent for polymers, especially epoxy resins; chemical intermediate.

glycol dimethyl ether. See ethylene glycol dimethyl ether.

glycol dipropionate. See ethylene glycol dipropionate.

glycolic acid. See hydroxyacetic acid.

glycol monoacetate. See ethylene glycol monoacetate.

glycolonitrile (glyconitrile; formaldehyde cyanohydrin) $HOCH_2CN$.
Properties: Mobile, colorless, odorless oil. Supplied commercially as a 70% aqueous solution stabilized with phosphoric acid. B.p. 183°C (slight decomposition); m.p., does not solidify when cooled to −72°C. Sp. gr. 1.1039 (19°C); refractive index (n 25/D) 1.4090; electrolytic dissociation constant $K = 0.843 \times 10^{-5}$ (25°C).
Derivation: Formaldehyde and hydrocyanic acid.
Containers: (70% solution) Drums; tank cars.
Hazard: Toxic by skin absorption.
Uses: Solvent and organic intermediates.

glycol propionate. See ethylene glycol dipropionate.

glycol stearate. See ethylene glycol monostearate.

glycothiourea. See 2-thiohydantoin.

glycolylurea. See hydantoin.

glycolysis. Enzymatic (anaerobic) decompositon of sugars, starches and other carbohydrates, with release of energy, a type of reaction occurring in yeast fermentation and in certain metabolic processes. Lactic acid is one of the products formed.

"Glycomuls."[73] Trademark for a series of sorbitol fatty acid esters, ranging from liquids to relatively high-melting wax-like solids and with varying surface-active characteristics. Used in foods, cosmetics, pharmaceuticals, chemical specialties.

glyconic acid. See gluconic acid.

glyconitrile. See glycolonitrile.

glycoside. One of a group of organic compounds, of abundant occurrence in plants, which can be resolved by hydrolysis into sugars and other organic substances, known as aglycones. Specifically glycosides are acetals which are derived from a combination of various hydroxy compounds with various sugars. They are designated individually as glucosides, mannosides, galactosides, etc. Glycosides were formerly called glucosides, but the latter term now refers to any glycoside having glucose as its sugar constituent.

glycyl alcohol. See glycerol.

glycyrrhizin. A glycoside of the triterpene group, the active principle of licorice root, from which it is extracted. It has an intensely sweet taste and is used as a humectant in tobacco and a flavoring in confectionery and pharmaceutical products. The ammoniated derivative, which is 50 times as sweet as sucrose, is used as a foaming agent in root beer and mouthwashes, as a sweetener in chocolate, cocoa, and chewing gum, and as a taste-masking agent in pharmaceuticals such as aspirin. Its ability to exert strong synergistic action with sucrose makes it useful in low-calorie foods (from 30 to 100 ppm are effective). See also sweetener, nonnutritive.

"Glydag" B.[46] Trademark for a dispersion of colloidal graphite in 1,3-butylene glycol. Used as a rubber lubricant and soluble-oil additive.

"Glyecine" A.[307] Trademark for a dyeing assistant, comprising thiodiethylene glycol; 100% active.
Uses: Hygroscopic agent in textile printing; solvent for basic colors.

glyme. Abbreivation for glycol dımethyl ether. See ethylene glycol dimethyl ether.

glyodin. Generic name for 2-heptadecyl-2-imidazoline acetate (2-heptadecylglyoxalidine acetate) $C_{17}H_{35}C_3H_5N_2 \cdot CH_3COOH$.
Properties: Light orange crystals; m.p. 94°C; insoluble in water. Toxicity probably low; high concentrations are irritating to skin and eyes.
Uses: Fungicide (fruits).

glyoxal OHCCHO.
Properties: Yellow crystals or light yellow liquid; mild odor; m.p. 15°C; b.p. 51°C; sp. gr. 1.26 (20/20°C); 10.0 lb/gal (20°C). Vapor has a green color and burns with a violet flame; refractive index (n 20/D) 1.3826; polymerizes on standing or in presence of a trace of water. An aqueous solution contains monomolecular glyoxal, and reacts weakly to acid. Undergoes many addition and condensation reactions with amines, amides, aldehydes, and hydroxyl-containing materials. Glyoxal VP resists discoloration.
Derivation: Oxidation of acetaldehyde.

Grades: 40% solution; pure, solid; VP.
Containers: (solid) Tins or fiber drums; (liquid) drums; tank cars.
Hazard: Moderately toxic by ingestion and inhalation; irritant.
Uses: Permanent-press fabrics; dimensional stabilization of rayon and other fibers. Insolubilizing agent for compounds containing polyhydroxyl groups (polyvinyl alcohol, starch, and celulosic materials); insolubilizing of proteins (casein, gelatin and animal glue); embalming fluids; leather tanning; paper coatings with hydroxyethylcellulose; reducing agent in dyeing textiles.

glyoxaline. See imidazole.

glyoxyldiureide. See allantoin.

glyoxylic acid HC(O)COOH. Supplied as a 50% solution, sp. gr. 1.42 (20/4°C); miscible with water and alcohol; insoluble in ether and hydrocarbons.
Uses: Intermediate for flavorings, perfumes, pharmaceuticals, dyes, plastics, agricultural chemicals.

"Glyptal."[245] Trademark for a group of alkyd-type polymers and plasticizers.

"G-M-F."[248] Trademark for para-quinonedioxime, $HONC_6H_4NOH$, a rubber accelerator.

GMP. Abbreviation for guanosine monophosphate. See guanosine phosphates; guanylic acid; sodium guanylate.

GMS. Abbreviation for glycerol monostearate.

"G.N.S. No. 5."[79] Trademark for a pine oil. Used in ore flotation.

goa powder. Araroba. See chrysarobin.

gold Au Metallic element of atomic number 79, Group IB of the Periodic Table. Atomic weight 196.9665; valence 1, 3. No stable isotopes.
Properties: Yellow, ductile metal; relatively soft. Does not corrode in air, but is tarnished by sulfur. Chemically nonreactive and nontoxic; attacked by chlorine and cyanide solutions in presence of oxygen. Soluble in aqua regia; insoluble in acids. Sp. gr. 19.3; m.p. 1063°C; b.p. 2808°C. Excellent reflector or infrared and heat; electrical resistivity (0°C) 2.06 microhm-cm. Extremely high light reflectivity. Nontoxic.
Occurrence: South Africa, USSR, Northwest Canada, U.S. (South Dakota, Nevada, Utah, Alaska, California) Australia. Oceans are estimated to contain 70 million tons in solution, with ten billion additional tons on the ocean floor. There is no present method of exploiting these resources.
Derivation: Ore is treated with cyanide solution; the dissolved gold cyanate is recovered by precipitation with zinc dust, aluminum, or by hydrolysis. Placer methods are also used.
Forms available: Ingots, sheet, wire, tubing, powder, leaf; alloys with copper or other metals, the gold content being expressed in carats (the number of parts of gold in 24 parts of alloy). Single crystals and aqueous colloidal suspensions. Leaf may be made in near-colloidal thickness; 1 troy ounce covers 68 sq ft (0.0001 inch).
Uses: Infrared reflectors; electrical contact alloys; brazing alloys; polarographic electrodes; spinnerets; laboratory ware; decorative arts (ceramics); in electronics, for bonding transistors and diodes to wires, for metallizing ceramic and mica capacitors, and for printed circuits; space vehicle instruments; dental alloys; jewelry. Colloidal dispersions are used in coloring glass, as nucleating agent in photosensitive glasses, and for specialized medical treatments; gold leaf is used in surgery.

gold 198. Radioactive gold of mass number 198.
Properties: Half-life, 2.7 days.
Derivation: Neutron irradiation of the element.
Forms available: Gold metal, colloidal gold (see radiogold) and gold sodium thiosulfate. The N.F. solution is colloidal Au-198.
Hazard: Moderate, from beta and gamma radiation.
Uses: Internal radiation therapy; to detect leaks in bacterial filters; to locate solidification boundary in continuously cast aluminum; to determine metallic silver in photographic materials.
Shipping regulations: Consult regulations.

gold, artificial. See stannic sulfide.

gold bromide (aurous bromide) AuBr.
Properties: Yellowish-gray mass. Decomposes at 165°C (approx). Insoluble in water.
Grades: Technical.
See also gold tribromide.

gold bronze powder. See aluminum bronze powder.

gold chloride. See gold trichloride.

gold-cobalt alloys. See cobalt-gold alloys.

gold cyanide (auric cyanide; cyanoauric acid) $Au(CN)_3\ 3H_2O$ or $HAu(CN)_4 \cdot 3H_2O$.
Properties: Colorless, hygroscopic crystals. M.p. 50°C (dec); very soluble in water; soluble in alcohol and ether.
Hazard: Highly toxic.
Use: Electrolyte in the electroplating industry.
Shipping regulations: (Rail, Air) Cyanides, dry, Poison label.

gold, filled. A thin gold alloy bonded to a base-metal; also called clad stock. Used in inexpensive jewelry.

gold hydroxide (auric hydroxide; gold hydrate) $Au(OH)_3$.
Properties: Brown powder. Sensitive to light. Keep in amber bottle! Probably a hydrated gold oxide (Au_2O_3), and loses water easily. Soluble in hydrochloric acid, solutions of sodium cyanide and alkali hydroxides; insoluble in water.
Uses: Gilding liquids; medicine; porcelain; gold plating.

gold, liquid bright. See "Liquid Bright Gold."

gold oxide (auric oxide; auric trioxide; gold trioxide) Au_2O_3.
Properties: Brownish-black powder; decomposed by heat. Keep in dark bottle. Soluble in hydrochloric acid; insoluble in water.
Uses: Gilding liquids; medicine; porcelain; gold plating.

gold potassium chloride. See potassium gold chloride.

gold potassium cyanide. See potassium gold cyanide.

gold potassium iodide. See potassium gold iodide.

gold salts. See sodium gold chloride.

Goldschmidt, Hans. See thermite.

gold-silicon alloy (silicon-gold alloy). Formed in amorphous foils 10 microns thick by cooling molten gold and silicon almost instantaneously by spreading on a moving wheel. The atoms are "frozen" before crystals can form. Used in electronics.

gold sodium chloride. See sodium gold chloride.

gold sodium cyanide. See sodium gold cyanide.

gold sodium thiomalate
$NaOOCCH(SAu)CH_2COONa \cdot H_2O$.
Properties: White to yellowish-white, odorless powder with metallic taste; affected by light. Very soluble in water; practically insoluble in alcohol and ether. Aqueous solutions are colorless to pale yellow; pH (5% solution) 5.8–6.5.
Derivation: Reaction of sodium thiomalate with a gold halide.
Grade: U.S.P.
Use: Medicine.

gold sodium thiosulfate. See sodium gold thiosulfate.

gold solder. A solder usually composed of gold, silver, copper, zinc, or brass; used principally by jewelers.

gold thioglucose. See aurothioglucose.

gold tin precipitate. See gold tin purple.

gold tin purple (purple of Cassius; gold tin precipitate).
Properties: Brown powder. Insoluble in water; soluble in ammonia.
Derivation: By the reaction of a neutral solution of gold trichloride with stannous and stannic chlorides, yielding a mixture of colloidal gold and tin oxide in varying proportions.
Grade: Technical.
Containers: Tins; glass bottles.
Uses: Manufacture of ruby glass; coloring enamels; painting porcelain.

gold tribromide (auric bromide; gold bromide) $AuBr_3$.
Properties: Brownish-black powder; m.p. 160°C, with decomposition. Soluble in alcohol, ether; slightly soluble in water.
Uses: Analysis (testing for alkaloids, spermatic fluid); medicine.

gold tribromide, acid. See bromoauric acid.

gold trichloride (a) $AuCl_3$ (auric chloride; gold chloride); (b) $AuCl_3 \cdot 2H_2O$; (c) $AuCl_2 \cdot HCl \cdot 4H_2O$ or $HAuCl_4 \cdot 4H_2$ (chlorauric acid; chloroauric acid; gold trichloride, acid).
Properties: Yellow to red crystals; decomposed by heat; soluble in water, alcohol and ether.
Derivation: Action of aqua regia on gold.
Grades: Technical; C.P., usually as chlorauric acid.
Uses: Photography; gold plating; special inks; medicine; ceramics (enamels, gilding and painting porcelain); glass (gilding, ruby glass); manufacture of finely divided gold and purple of Cassius.

gold trioxide. See gold oxide.

gold, white. A jeweler's alloy consisting of about 58% gold, 17% nickel, 7% zinc and 17% copper.

Goodyear, Charles (1800–1860). Born in Woburn, Mass., Goodyear was the first to realize the potentialities of natural rubber Frustrated by its lack of stability to temperature and other weaknesses in the uncured state, he experimented with additives such as magnesium and sulfur. The discovery of vulcanization was not accidental, as is often stated, but the result of intelligent trials and correct evaluation of their results. Though Goodyear's patents were contested by Hancock in England, he well merits the credit for making rubber usable in countless ways, and helping to make the automobile possible.

gossypol. $C_{30}H_{30}O_8$. A natural polyphenol.
Properties: Yellow, crystalline pigment, having three modifications. Insoluble in water; soluble in alcohol.
Occurrence: Cottonseed kernels.
Hazard: Toxic by ingestion; but is inactivated by heat; 0.04% max. allowed in foods.
Uses: Stabilizer for vinyl polymers; has possibilities as a biodegradable insecticide.

GPF black. Abbreviation for general purpose furnace black. See carbon black.

grade. Any of a number of purity standards for chemicals and chemical products established by various specifications. Some of these grades are as follows:
ACS (American Chemical Society specifications)
reagent (analytical reagent quality)
C.P. (chemically pure)
USP (conforms to U.S. Pharmacopeia specifications)
N.F. (conforms to National Formulary specifications)
purified
FCC (Food Chemicals Codex specifications)
technical (industrial chemicals)

food	spectro
feed	commercial
semiconductor	chemical
radio	injectable
research	nitration

"Grafoil."[521] Trademark for pure flexible graphite tape with highly directional properties similar to pyrolytic graphite. Thermal insulating properties up to 6600°F.

grafting. A deposition technique whereby organic polymers can be bonded to a wide variety of other materials, both organic and inorganic, in the form of fibers, films, chips, particles, or other shapes. Grafting occurs at specific catalyst sites on the "host" materials, which must have some capacity for ion exchange, methathesis, or complex formation. Ionizable groups may be added artificially.

One proprietary application is polymerization of acrylonitrile with wood pulp fibers to make synthetic soil blocks; the polymer imparts high water-holding capacity to the pulp. Plant nutrient materials are added, and the mixture pressed into blocks to be used for starting seedlings.

graft polymer. A copolymer molecule comprised of a main backbone chain, to which side chains containing different atomic constituents are attached at various points. The main chain may be either a homopolymer or a copolymer. This process may be applied to the union of cellulosic molecules (cotton, rayon) with synthetic polymers (except polyesters, acrylics, and polypropylene) to form modified fibers having improved flame resistance, dimensional stability, resilience, and bacterial resistance. An intermediate called cellulose thiocarbonate (q.v.) is formed in this proprietary process.
See also polyorganosilicate graft polymer.

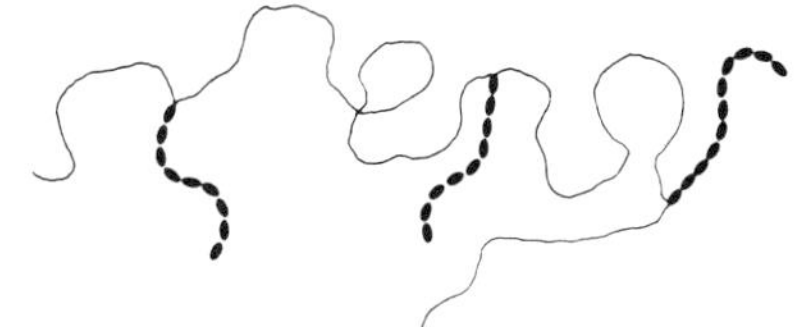

Graham, Thomas (1805–1869). Born in Scotland, Graham is famous for his basic studies in diffusion which led to the development of colloid chemistry. He was the first to observe a marked differnce in the rate of passage of certain types of substances through a parchment membrane. Those which readily crystallize, like sugar, pass rapidly through the membrane, but gelatinous types are "slow in the extreme." The latter, which comprise albumin, starch, gums, etc., he designated as colloids, and their solutions as colloidal solutions. The former, which he called crystalloids, form "true" or molecularly dispersed solutions. See also colloid chemistry.

Graham's salt. See sodium metaphosphate.

grain (1) The smallest unit of weight in the avoirdupois system; one ounce contains 437.5 grains.
(2) Any cereal plant, as wheat, corn, barley, etc.

grain alcohol. See ethyl alcohol.

grain oil. See fusel oil.

gram. One one-thousandth kilogram (q.v.). It is the weight of one milliliter of water at 4°C. One pound contains 453.5 grams, or 15.4 grains.

gram atomic weight. The atomic weight of an element in grams, i.e., the gram atomic weight of carbon is 12.01115 grams. See mole.

gramicidin. An antibiotic produced by the metabolic processes of the bacteria Bacillus brevis. It is a polypeptide which is active against most Gram-positive pathogenic bacteria. It is one of the two antibiotic components of tyrothricin (q.v.) but has been isolated and used alone.
Properties: White crystalline platelets; m.p. 229–230°C; soluble in lower alcohols, acetic acid and pyridine; moderately soluble in dry acetone and dioxane; almost insoluble in water, ether, and hydrocarbons. Depresses surface tension; forms a fairly stable colloidal emulsion in distilled water.
Derivation: From tyrothricin by extraction with a mixture of equal volumes of acetone and ether, followed by concentration in vacuo and dissolving in hot acetone.
Grade: N.F.
Use: Medicine.

gram molecular weight. The molecular weight of a compound in grams, i.e., the gram molecular weight of carbon dioxide is 44.01 grams. See mole.

Gram-positive, -negative. A characteristic property of bacteria in reacting to a staining method developed by Gram about 1880. The bacteria are stained with crystal violet, treated with Gram's solution, and again stained with safranine. If the dye is retained the bacteria are called Gram-positive; and vice versa. See also bacteria.

grapefruit oil. See citrus peel oil.

grape sugar. See glucose.

"Graphallast."[82] Trademark for a group of graphite and hydrocarbon oilless materials. Used for low friction bushings, bearings, and seals in submerged applications in many corrosive chemicals at normal temperatures below 150°F. Resistant to scuffing or abrasion.

"Graphalloy."[82] Trademark for a series of oilless, self-lubricating, long life, low-friction materials consisting of graphite and a metal or alloy such as Babbitt, bronze, cadmium, copper, gold, iron, nickel, or silver. Widely used for bearings, bushings, seals, electric brushes, brush assemblies, and non-freezing electric contacts. Many grades are available to meet most chemical applications. Cryogenic applications to −450°F, or to +750°F in air.

"Graphic Red."[141] Trademark for lithol red pigments composed of sodium, barium, or calcium salts of diazotized Tobias acid coupled with beta-naphthol.
Uses: Printing inks, paints, rubber, floor coverings, crayons.

"Graphilm."[82] Trademark for a superior liquid lubricant containing graphite and a carrying agent which dries to a film approximately 0.001″ thick. The resulting film is insoluble in water and adheres tenaciously to the surface upon which it has been baked. Suitable for high temperature applications to 1000°F and may be applied by brushing, dipping, or spraying.
Uses: Provides lubrication for hard vacuum, high temperatures, and cryogenic applications.

graphite (black lead, plumbago). The crystalline allotropic form of carbon. Occurs naturally in Madagascar, Ceylon, Mexico, Korea, Austria, USSR, and China. Also produced synthetically by heating petroleum coke to about 3000°C in an electric resistance furnace. About 70% used in U.S. is synthetic.
Properties: Relatively soft greasy feel; steel gray to black color with a metallic sheen; sp. gr. 2.0–2.25 depending upon origin; apparent specific gravity artificial graphite 1.5–1.8. High electrical and thermal conductivity. Specific heat 0.17 at room temperature. 0.48 at 1500°C; tensile strength 400 to 2000 psi; compressive strength usually about 2000–8000 psi. Coefficient of friction 0.1μ. Resistant to oxidation and thermal shock. Sublimes at 6600°F. Low toxicity.
Grades: Powdered; flake; crystals; rods; plates; fibers (q.v.).
Containers: Bags; fiber drums.
Hazard: (powder, natural) Fire risk. Tolerance, 15 million particles per cubic foot of air.
Uses: Granular or flake forms: pencils; crucibles, retorts, foundry facings, molds, lubricants, paints and coatings, boiler compounds, powder glazing, electrotyping; monochromator in x-ray diffraction analysis
Fabricated forms: Molds, crucibles, electrodes, bricks, chemical equipment, motor and generator brushes, seal rings, rocket nozzles, moderator in nuclear reactors; cathodes in electrolytic cells.
See also carbon; graphite fiber; carbon, industrial.

graphite fiber. High-tensile fibers or whiskers (q.v.) made from (1) rayon or (2) polyacrylonitrile.
Properties: (1) Amorphous structure; oxidizes readily, requiring silicon carbide coating. Resistant to acids and bases, including hydrofluoric acid. Tensile strength 50,000 to 150,000 psi; elastic modulus 4 to 9 million psi. Upper temperature limit in oxidizing atmosphere 500°C. Self-lubricating. Resistant to electricity. Lightweight. (2) Polycrystalline structure. Tensile strength up to 350,000 psi; elastic modulus up to 70 million psi. Smooth surface. Lightweight.
Derivation: (1) Heating rayon in air at 1400–1700°C. (2) Heating polyacrylonitrile in air at 220°C (20 hrs) to oxygenate, then in hydrogen to 1000°C (24 hrs) to carbonize, finally in argon at 2500°C (2 hrs). (3) Heating pitch materials (petroleum residues, asphalt, etc.) to carbonization temperature.
Forms: Filament, yarn, fabric, whiskers. Fibers may be 7 to 8 microns in diameter.
Uses: (1) Heating pads (combined with glass fiber); protective clothing for electrical workers. (2) Polyes-

ter and epoxy composites for jet engine components, space craft, compressor blades, airframe structures; electrodes for spark-hardening metals.
See also fiber; whiskers; composite; carbon, industrial.

graphite, pyrolytic (pyrographite). A dense, nonporous graphite, stronger and more resistant to heat and corrosion than ordinary graphite, intended for rocket nozzles, missiles in general, and nuclear reactors; exhibits tensile strengths 5–10 times higher than commercial graphite and maintains its strength above 6000°F. It is ultrapure, impermeable to all fluids, and is an anisotropic thermal and electrical conductor.

graphitic oxide (GO). $C_7O_2(H)_2$.
Properties: Light-yellow flakes or plates which deposit in layers to form a membrane, usually supported on a cellulose ester.
Derivation: Slow oxidation of graphite with potassium nitrate in nitric and sulfuric acids.
Uses: Reverse-osmosis membrane for desalination of sea water.

"Graphlon."[82] Trademark for a group of graphite and resin materials that exhibit extremely high chemical inertness for submerged applications. "Graphlon" bearings, bushings, and seals withstand highly corrosive chemicals. Operating range from −450° to +450°F in air.

"Graphmetex."[82] Trademark for a group of special insulating resins employed as insulating spacers and holders for electrical components.

GRAS. Abbreviation for "Generally recognized as safe," applied to food additives approved by FDA.

gravimetric analysis. A type of quantitative analysis involving precipitation of a compound which can be weighed and analyzed after drying. It is also used in determining specific gravity.

"Gredag."[46] Trademark for specialized dispersions based on graphite, aluminum, and molybdenum disulfide. Used in die casting compounds, release agents, plunger lubricants, die pretreatments, toggle assembly and machine and ejector pin lubricants.

green. Used in the chemical industries to mean uncured or untreated material, e.g., greensand.

greenhouse effect. Absorption or trapping by the carbon dioxide and water vapor in the atmosphere of radiation wavelengths above the infrared as they are radiated away from the earth. This effect is important in controlling the earth's surface temperature.

Greenland spar. See cryolite, natural.

green liquor. See liquor.

greenockite CdS. A native cadmium sulfide containing 77.7% cadmium. Ore of cadmium.

greensalt. A wood preservative containing chromated copper arsenate.
Hazard: Highly toxic.

greensand. See foundry sand.

green soap. A liquid soap made with potassium hydroxide and a vegetable oil (except coconut and palm kernel oil). See also soap.

green verditer. A paint pigment consisting of the hydroxycarbonate of copper. See copper carbonate; malachite.

"Greenz."[48] Trademark for an ammonium lignin sulfonate containing 4.5% iron, for use as an agricultural spray or soil additive in the treatment of iron chlorosis.

grex. See denier.

"Griffco."[309] Trademark for a polyvinyl acetate emulsion useful as a base for adhesives and paint.

Grignard reagent. An important class of reagents used in synthetic organic chemistry, made by union of metallic magnesium with an organic chloride, bromide, or iodide, usually in the presence of an ether, and in the complete absence of water. General formula RMgX, where R is an alkyl or aryl or other organic group, and X a halogen. The value of the reagents lies in their ease of reaction with water, carbon dioxide, alcohols, aldehydes, ketones, amines, etc., to produce a great variety of organic compounds, usually with good yields. Examples are ethyl magnesium chloride (C_2H_5MgCl), methyl magnesium bromide (CH_3MgBr), etc.
Hazard: Since all Grignard reagents react rapidly with both water and oxygen, contact must be avoided. Ordinary materials of construction are satisfactory for use with Grignard reagents. Because the heat of decomposition of Grignard reagents with water is great and the ether they are dissolved in is highly volatile and flammable, they must be handled with extreme care. Some, especially the solution of MeMgBr in ethyl ether, may ignite spontaneously on contact with water or even damp floors; nearly all will ignite on a wet rag or similar material.

griseofulvin $C_{17}H_{17}ClO_6$.
Properties: White to creamy white, odorless powder. Very slightly soluble in water; soluble in acetone, and chloroform; sparingly soluble in alcohol.
Derivation: By growth of Penicillium griseofulvum or other means.
Grade: U.S.P.
Use: Medicine (antifungal); feed additive.

"Groco."[410] Trademark for a series of fatty acids which include stearic acid, tallow and hydrogenated tallow, oleic acid, white oleine, coconut, soya bean, corn, cottonseed and palm oils.

grog. Crushed refractory materials added to ceramic mixes to reduce lamination in plastic clays and also to reduce shrinkage on drying. Materials crushed for this purpose are pottery, fire brick, quartz, quartzite, burned ware, saggers.

ground-nut oil. See peanut oil.

groundwood. A type of wood pulp produced by direct friction of a rotating stone on the end of a log. In the paper trade this is called mechanical pulp. It is used largely in newsprint and other low-grade papers. See also pulp, paper.

group. (1) One of the major classes or divisions into which elements are arranged in the Periodic Table (vertical columns). The classification is made according to the properties of the elements, those whose properties are similar occupying one group. Groups I through VII are divided into subgroups (A and B), but Group VIII is not so divided. The noble gas group has no number, though it was formerly referred to as Group 0. See also Periodic Table.
Note: Though often loosely so used, the word "group" should not be applied to a number of elements of similar properties that are not actual groups in the

Periodic Table; the proper term for these is "series," i.e., lanthanide series, rare-earth series, etc.

(2) A combination of two or more closely associated elements which tend to remain together in reactions, usually behaving chemically as if they were individual entities, i.e., in respect to valence, ionization, and related properties. Among the more familiar are OH (hydroxyl), COOH (carboxyl), CO_3 (carbonate), NH_4 (ammonium), SO_4 (sulfate), CH_3 and homologs (methyl, etc.), SH (sulfhydryl), and CO (carbonyl). When a group acquires an electric charge it is called a radical.

(3) Any combination of elements that has a specific functional property, for example, a chromophore group in dyes.

growth. (1) In biochemistry, the continuous process of cell division and reproduction characteristic of all living organisms. The basic phenomenon is considered to be osmosis, by which nutrients are transferred through cell walls and tissue structures; it is thus essential to the metabolic functioning of the organism.

(2) In crystallography, the process of crystal formation and development by nucleation and accretion. Crystals of many kinds are artificially produced for a variety of uses (e.g., lasers) by vapor condensation, electrodeposition, or rapid cooling of a saturated solution.

growth hormone. See somatotropic hormone.

growth regulator. See plant growth regulator.

G salt. The sodium or potassium salt of 2-naphthol-6,8-disulfonic acid (G acid). Used as a dye intermediate.

GTP. Abbreviation for guanosine triphosphate. See guanosine phosphates.

guaiac (guaiac gum; guaiac resin). A resin from certain Mexican and West Indian trees, especially Guaiacum santum and G. officinale.
Soluble in alcohol, ether, acetone, chloroform and caustic soda.
Uses: Flavoring agent in foods.

guaiacol (methylcatechol; pyrocatechol methyl ether, pyrocatechol methyl ester; ortho-methoxyphenol; ortho-hydroxyanisole) $OHC_6H_4OCH_3$.
Properties: Faintly yellowish, limpid, oily liquid or yellow crystals; aromatic odor. Guaiacol constitutes 60–90% of beechwood creosote. Soluble in alcohol, ether, chloroform and glacial acetic acid; moderately soluble in water. Sp. gr. 1.1395; m.p. 27.9°C; b.p. 205°C. Flash point 180°F (open cup); combustible.
Derivation: (a) By extracting beechwood creosote with alcoholic potash, washing with ether, crystallizing the potash compound from alcohol and decomposing it with dilute sulfuric acid. (b) Also from ortho-anisidine by diazotization and subsequent action of dilute sulfuric acid.
Hazard: Moderately toxic by ingestion and skin absorption.
Containers: (Crystals). Drums and tins; (liquid) drums and carboys.
Use: Food flavoring agent.

guaiacol benzoate $C_6H_5COOC_6H_4OCH_3$.
Properties: White odorless, almost tasteless powder; m.p. 57–58°C; slightly soluble in water, soluble in hot alcohol, ether and chloroform.
Use: Medicine.

guaiacol carbonate (neutral guaiacol carbonate) $(C_7H_7O)_2CO_3$.
Properties: Small colorless crystals or white crystalline powder. Either slight aromatic odor and taste or odorless and tasteless. Soluble in chloroform, ether; less soluble in alcohol; insoluble in water; m.p. 86–88°C.
Derivation: (a) Reaction between sodium guaiacolate and carbonyl chloride. (b) Reaction between guaiacol and methyl chloroformate.
Use: Medicine.

guaiacwood oil.
Properties: Yellow to amber semisolid mass with floral odor. Soluble in alcohol, ether, and chloroform; insoluble in water. Sp. gr. 0.965–0.975; optical rotation −6 to −7°.
Derivation: Steam distillation of guaiacwood (Paraguay).
Chief constituent: Guaiol.
Containers: Cans.
Uses: Perfume fixative and modifier; soap odorant; fragrances; production of guaiacwood acetate.

"Guai-A-Phene."[296] Trademark for a phenolic type antiskinning agent used for the prevention of gelling, skinning, and oxidation in paint, varnishes, printing inks, linoleum, etc.

guaiol $C_{15}H_{26}O$. A bicyclic sesquiterpene alcohol found in guaiac wood oil. Crystalline solid; m.p. 91°C. Insoluble in water; soluble in alcohol.

guanethidine sulfate $(C_{10}H_{22}N_4)_2 \cdot H_2SO_4$. [2-(Hexahydro-1(2H)-azocinyl)ethyl]guanidine sulfate.
Properties: White, crystalline powder; strong odor. Very soluble in water; slightly soluble in alcohol; practically insoluble in chloroform.
Grade: U.S.P.
Use: Medicine (antihypertensive agent).

guanicaine. See di-para-anisyl-para-phenetylguanidine hydrochloride.

guanidine (carbamidine; iminourea) $NHC(NH_2)_2$.
Properties: Colorless crystals; m.p. 50°C; decomposes at 160°C. Soluble in water and alcohol.
Derivation: (a) By heating calcium cyanamide with ammonium iodide. (b) By treating urea with ammonia under pressure. (c) From guanidine carbonate.
Combustible; low toxicity.
Use: Organic synthesis.

guanidine-aminovaleric acid. See arginine.

guanidine carbonate $(H_2NCNHNH_2)_2 \cdot H_2CO_3$.
Properties: White granules. Soluble in water, slightly soluble in alcohol and acetone. Decomposes at 197–199°C without melting; sp. gr. 1.25. Combustible; low toxicity.
Derivation: From dicyandiamide.
Grade: Technical, over 95% pure.
Hazard: Moderately toxic by ingestion.
Containers: Fiber drums.
Uses: As a strong organic alkali; organic intermediate; soap and cosmetic products.

guanidine hydrochloride $NHC(NH_2)_2 \cdot HCl$.
Properties: White powder; m.p. about 183°C. Soluble in water and alcohol; pH of aqueous solution 6.2 for 10% solution.
Grades: 88% and 95% pure.
Containers: Paper bags or fiber drums.
Hazard: Moderately toxic by ingestion; evolves hydrochloric acid fumes on heating.
Uses: Exceptionally water-soluble source of guanidine for organic syntheses.

guanidine nitrate $H_2NC(NH)NH_2 \cdot HNO_3$.
Properties: White granules. Soluble in water and al-

cohol; slightly soluble in acetone. Melting range 206–212°C.
Derivation: From cyanamide or dicyandiamide.
Grade: Technical, over 95% pure.
Containers: Fiber drums, multiwall paper sacks.
Hazard: Strong oxidant; may ignite organic materials on contact; may explode by shock or heat.
Uses: Manufacture of explosives; disinfectants; photographic chemicals.
Shipping regulations: (Rail) Yellow label. (Air) Oxidizer label.

guanine $C_5H_5N_5O$ (2-amino-6-oxypurine). A purine constituent of ribonucleic acid and deoxyribonucleic acid. Usual sources are guano, sugar beets, yeast, clover seed, and fish scales.
Properties: Colorless rhombic crystals. M.p. 360°C (dec). Insoluble in water; sparingly soluble in alcohol and ether; freely soluble in ammonium hydroxide, alkali hydroxides, and dilute acids. Available as hydrochloride or hemisulfate.
Derivation: Isolation following hydrolysis of nucleic acids (usually from yeast); organic synthesis.
Uses: Biochemical research; cosmetics.
See also pearl essence.

guanosine (guanine riboside) $C_{10}H_{13}N_5O_5$. The nucleoside containing guanine and D-ribose.
Properties: White, crystalline, odorless powder with mild, saline taste. M.p. 237–240°C (dec). Very slightly soluble in cold water; soluble in boiling water, dilute mineral acids, hot acetic acid, and dilute bases; insoluble in alcohol, ether, chloroform and benzene.
Derivatoin: Found in pancreas, clover, coffee plant, and pollen of pines; prepared from yeast nucleic acid.
Use: Biochemical research.

guanosine monophosphate. See guanylic acid.

guanosine phosphates. Nucleotides used by the body in growth processes; important in biochemical and physiological research. Those isolated are the monophosphate (GMP), the diphosphate (GDP) and the triphosphate (GTP).

guanosine phosphoric acid. See guanylic acid.

"Guantal."[58] Trademark for diphenylguanidine phthalate (q.v.), a rubber accelerator.

guanylic acid (GMP; guanosine monophosphate; guanosine phosphoric acid) $C_{10}H_{14}N_5O_8P$. The monophosphoric ester of guanine; i.e., the nucleotide containing guanine, D-ribose and phosphoric acid. The phosphate may be esterified to either the 2,3, or 5 carbon of ribose yielding guanosine 2′-phosphate, guanosine-3′-phosphate, and guanosine-5′-phosphate, respectively. It is important in growth processes of the body.
Derivation of commercial product: Isolation from nucleic acid of yeast or pancreas; also made synthetically.
Use: Biochemical research.

guanyl nitrosaminoguanylidene hydrazine. A high explosive.
Shipping regulations: (Rail) Not accepted by express. Consult regulations: (Air) Not acceptable.

guanyl nitrosaminoguanyl tetrazene. See tetrazene.

guanylurea sulfate. (carbamylguanidine sulfate) $(C_2H_6ON_4)_2 \cdot H_2SO_4 \cdot 2H_2O$.
Properties: White powder, over 97% pure. Soluble in water and alcohol.
Derivation: From cyanamide or dicyandiamide.
Uses: Analytical reagent for nickel; manufacture of dyes; organic synthesis.

"Guardkote."[125] Trademark for fast-setting two-component liquid systems based on "Epon" resins and designed specifically for highway resurfacing and repair. They are used in combination with sharp aggregates to waterproof and deslick portland cement or bituminous concrete pavements.

guar gum. A water-soluble plant mucilage obtained from the ground endosperms of Cyanopsis tetragonoloba, cultivated in Pakistan as livestock feed. The water-soluble portion of the flour (85%) is called guaran, and consists of 35% galactose, 63% mannose, probably combined in a polysaccharide; 5–7% protein.
Properties: Yellowish-white powder. Dispersible in hot or cold water. has 5–8 times the thickening power of starch; reduces friction drag of water on metals.
Grades: Industrial; technical; F.C.C.
Uses: Paper manufacture; cosmetics; pharmaceuticals; interior coating of fire-hose nozzles; fracturing aid in oil wells; textiles; printing; polishing; thickener and emulsifier in food products, e.g., cheese spreads, ice cream and frozen desserts.

Guggenheim process. A process for the manufacture of sodium nitrate from the Chilean nitrate ore, caliche, in which heat is efficiently utilized and handling costs are kept to a minimum.

"Guidon."[446] Trademark for a magnesia-chrome refractory made from electrically fused grains. Has high temperature strength about 10 times that of conventional magnesia-chrome refractories which makes it well suited to abrasive environments.
Uses: Electric furnace sidewalls; copper converter tuyere lines; open hearth front walls and roof.

Guignet's green. A chrome green pigment made by fusing potassium chromate and boric acid.

gum arabic. See arabic, gum.

gum camphor. See camphor.

gum, gasoline. A viscous oxidation product occurring after long standing in gasoline that is not stabilized with an antioxidant.

gum, natural. A carbohydrate high polymer that is insoluble in alcohol and other organic solvents, but generally soluble or dispersible in water. Natural gums are hydrophilic polysaccharides composed of monosaccharide units joined by glycosidic bonds. They occur as exudations from various trees and shrubs in tropical areas or as phycocolloids (algae), and differ from natural resins in both chemical composition and solubility properties. Some contain acidic components and others are neutral. Their chief use is as protective colloids and emulsifying agents in food products and pharmaceuticals; as sizing for textiles; and in electrolytic deposition of metals. For details, see arabic, tragacanth, guar, karaya.

Note: The terminology of natural gums and resins is inconsistent and often confusing. The word "gum," often used as an adjective, seems to acquire a different meaning from the noun. For example, the resinous products obtained from pine pitch (produced by the parenchyma cells of softwoods) are conventionally called "gum turpentine" and "gum rosin." There are also such "gum resins" as gum benzoin, gum camphor, and others. The so-called ester gum is a semisynthetic reaction product of rosin and a polyhydric alcohol. All these are actually resinous products having properties quite different from those of natural gums. Furthermore, resins are complex *mixtures*, whereas gums are chemical *compounds* that can be represented by a formula. Still further complicating the matter is the common application of the word "gum" to such plant latices as chicle and natural rubber, which are different from both carbohydrate gums and resins. It is probable that the confusion originated in the casual use of "gum" to refer to any soft, sticky product derived from trees. In view of this situation, any specific definition of these terms is likely to be controversial. See also resin, natural; polymer, water-soluble.

gum rosin. See rosin; gum, natural (note).

gum sugar. See arabinose.

gum, synthetic. See ester gum.

gum turpentine. See turpentine; gum, natural (note).

guncotton. See nitrocellulose.

gun metal. An alloy of copper with 10% tin.

gunpowder. See black powder.

"Guthion."[181] Trademark for O,O-dimethyl S-4-oxo-1,2,3-benzotriazin-3(4H)-ylmethyl phosphorodithioate (q.v.). See methyl azinphos.
Hazard: A cholinesterase inhibitor.

gutta percha. See rubber, natural (note).

glyocardia oil. See chaulmoogra oil.

gyplure. Generic name for cis-9-octadecen-1, 12-diol-12-acetate. It is a synthetic product, used as a sex attractant for the male gypsy moth. The natural product, found in female moths, is said to be *d*-10-acetoxy-1-hydroxy-cis-7-hexadecene.

gypsum $CaSO_4 \cdot 2H_2O$. Natural hydrated calcium sulfate. See calcium sulfate.

gypsum cement (plaster of Paris; Keene's cement; Parian cement; Martin's cement; Mack's cement). A group of cements which consist essentially of calcium sulfate and are produced by the partial dehydration of gypsum to the hemihydrate, $CaSO_4 \cdot \frac{1}{2}H_2O$. They usually contain additions of various sorts. For example, Keene's cement contains alum or aluminum sulfate, Mack's cement contains sodium or potassium sulfate, Martin's cement contains potassium carbonate and Parian cement contains borax.

H

H Symbol for hydrogen; the molecular formula is H_2.

Ha Symbol for hahnium.

Haber, Fritz (1868–1934). Born in Breslau, Germany, Haber's great contribution of chemistry, for which he was given the Nobel prize in 1918, was his development (with Bosch) of a workable method of synthesizing ammonia from nitrogen and hydrogen. It was the first successful attempt to "fix" atmospheric nitrogen in an industrial process. This discovery was developed to production scale about 1912; it enabled Germany to manufacture an independent supply of explosives for World War I. Haber left Germany to escape persecution in 1933. See also ammonia; explosives.

habit. The type of geometric structure that a given crystalline material invariably forms, e.g., cubic, orthorhombic, monoclinic, tetragonal, hexagonal, etc. Each of these types has several subclasses. Thus crystals may have the form of thin sheets or plates, cubes, rhomboids, and even more complicated geometric structures. For example, the crystal habit of mica is monoclinic, with formation of extremely thin sheets. See also crystal.

H acid. See 1-amino-8-naphthol-3,6-disulfonic acid.

hafnia. See hafnium oxide.

hafnium Hf Metallic element of atomic number 72, Group IVB of the Periodic Table. Atomic weight 178.49; valences 2, 3, 4. There are 6 stable isotopes.
Properties: Generally similar to zirconium. Gray crystals; sp. gr. 13.1; m.p. about 2150°C; b.p. over 5400°C; high thermal neutron cross-section (115 barns). Good corrosion resistance and high strength. Nontoxic. See also zirconium.
Occurrence: Present in most zirconium ores.
Derivation: Extremely difficult to separate from zirconium. Most important methods are: (1) solvent extraction of the thiocyanates by hexone; (2) solvent extraction of the nitrates by tributyl phosphate; (3) fractional crystallization of the double fluorides.
Forms available: Powder; rods; single crystals.
Hazard: Explosive in powder form, either dry or wet with less than 25% water. Tolerance, 0.5 mg per cubic meter of air. Safety data sheet available from Manufacturing Chemists Assn., Washington, D.C.
Uses: Control rods in water-cooled nuclear reactors; light-bulb filaments; electrodes; special glasses; getter in vacuum tubes.
Shipping regulations: Powder or sponge form, wet or dry: (Rail) Yellow label. (Air) Flammable Solid label. Not accepted on passenger planes. Consult rail and air regulations for details.

hafnium boride HfB_2. Crystalline solid; m.p. 3100°C.
Use: Refractory metal.

hafnium carbide HfC.
Properties: High thermal neutron absorption cross section; very high melting point, 3890°C (7030°F). Most refractory binary substance known.
Use: Control rods in nuclear reactors.

hafnium disulfide HfS_2. Available in a particle size of 40 microns as a solid lubricant. The diselenide and ditelluride are also available.

hafnium nitride HfN. Yellow-brown crystals; m.p. 3305°C. The most refractory of known metal nitrides.

hafnium oxide (hafnia) HfO_2.
Properties: White solid; sp. gr. 9.68 (20°C); m.p. 2812°C; b.p. about 5400°C. Insoluble in water.
Use: Refractory metal oxide.

hahnium. Ha Artificial element of atomic number 105 and atomic weight 260; discovered at Lawrence Radiation Laboratory and named for the German physicist Otto Hahn.

"Halane."[203] Trademark for a dry chlorine-liberating compound, dichlorodimethylhydantoin (q.v.).

halazone (para-N,N-dichloro-sulfamylbenzoic acid; para-sulfondichloraminobenzoic acid) $HOOCC_6H_4SO_2NCl_2$.
Properties: White crystalline powder; strong chlorine odor; affected by light. Soluble in glacial acetic acid, benzene; slightly soluble in water, chloroform; insoluble in petroleum ether. M.p. 195° with decomposition.
Grade: N.F.
Use: Water disinfectant.

half-life. The time required for an unstable element or nuclide to lose one-half of its radioactive intensity in the form of alpha beta and gamma radiation; it is a constant for each unstable element or nuclide. Half-lives vary from fractions of a second for some artificially produced radioactive elements to millions of years.

halibut liver oil (haliver oil).
Properties: Pale yellow to dark red liquid; fishy odor and taste; soluble in alcohol, ether, chloroform, and carbon disulfide; insoluble in water. Sp. gr. 0.920–0.930; saponification number 160–180; iodine number 120–136; refractive index about 1.47. Nontoxic.
Derivation: By expressing and boiling halibut livers.
Grades: Crude, refined.
Uses: Medicine; source of vitamins A and D; leather dressing.

halides. Binary compounds of the halogens (q.v.).

halite. See sodium chloride.

Hall, Charles Martin (1863–1914). A native of Ohio, Hall invented a method of reducing aluminum oxide in molten cryolite by electrochemical means. This discovery made possible the large-scale production of metallic aluminum and resulted in formation of the Aluminum Co. of America. The process requires high electric power input. Hall is generally considered as the founder of the aluminum industry. See also Hall process.

"Hallcomid."[94] Trademark for a series of N,N′-dimethyl amides of fatty acids.

halloysite $Al_2O_3 \cdot 3SiO_2 \cdot 2H_2O$. A clay used to some extent in refractories and as a cracking catalyst.

Hall process. The electrolytic recovery of aluminum from bauxite or more specifically, from the alumina extracted from it (see Bayer process). A typical cell for this process consists of a rectangular steel shell, lined with insulating brick and block carbon. The cell holds a molten cryolite-alumina electrolyte, commonly called the "bath." The carbon bottom is covered by a pad of molten aluminum, and serves as the cathode. The anodes are prebaked carbon blocks, suspended in the electrolyte. The cathodic current is collected from the carbon bottom by imbedded steel bars that protrude through the shell to connect with the cathode bus. During electrolysis, aluminum is deposited on the metal pad and the oxygen, liberated at the anode, reacts with the carbon to CO_2, some of which is reduced to CO by secondary reactions. At 24–48 hour intervals, aluminum is tapped from the cell by a siphon. The process requires large amounts of electric power (from 4 to 5% of total U.S. production). Disposition of the toxic fluoride waste is a problem.

hallucinogen. Any of a number of drugs acting on the central nervous system in such a way as to cause memtal disturbance, imaginary experiences, coma, and even death. Many of these are narcotics (q.v.) and/or alkaloids; some are derived from plants and others made synthetically. They differ in degree of addiction and hallucinatory effect. Their sale and possession (other than by physicians) is illegal in the U.S. Most common hallucinogens are cannabis (marijuana; hashish), lysergic acid (LSD), amphetamine, and numerous morphine derivatives.

halocarbon. A compound containing carbon, one or more halogens, and sometimes hydrogen. The lower members of the various homologous series are used as refrigerants, propellant gases, fire-extinguishing agents, and blowing agents for urethane foams. When polymerized they yield plastics characterized by extreme chemical resistance, high electrical resistivity, and good heat resistance. See also fluorocarbon.

halogenation. Incorporation of one of the halogen elements, usually chlorine or bromine, into a chemical compound. Thus benzene (C_6H_6) is treated with chlorine to form chlorobenzene (C_6H_5Cl), and ethylene (C_2H_4) is treated with bromine to form ethylene dibromide ($C_2H_4Br_2$). Compounds of chlorine and bromine are sometimes used as the source of the halogen, phosphorus pentachloride.

halogen. One of the electronegative elements of Group VII A of the Periodic Table (fluorine, chlorine, bromine, iodine, and astatine). Listed in order of their activity, fluorine being the most active of all chemical elements.

"Halon."[175] Trademark for tetrafluoroethylene polymer.
Properties: Inertness to almost all chemicals; resistance to high and low temperatures; zero moisture absorption; high impact strength; excellent dielectric properties; nonstick surface with low coefficient of friction; self-extinguishing (ASTM D-635).
Grades: Vary with particle size as, 600, 300, 20 micron average size.
See also polytetrafluoroethylene; fluorocarbon polymer.

"Halopont."[28] Trademark for pigments used for tinting white paper.

"Halotestin."[327] Trademark for fluoxymesterone; used in medicine.

halothane (2-bromo-2-chloro-1, 1, 1-trifluoroethane) $CF_3CHBrCl$.
Properties: Noncombustible, colorless, volatile liquid; sweetish odor. Sp. gr. (20/4°C) 1.872–1.877; b.p. 50.2°C, 20°C (243 mm). Light-sensitive; may be stabilized with 0.01% thymol. Slightly soluble in water; miscible with many organic solvents.
Grade: U.S.P.
Use: Medicine (anesthetic).

"Halowax."[11] Trademark for chlorinated hydrocarbons of varying chlorine content, including chlorinated naphthalenes, ranging from a low viscosity oil to a hard microcyrstalline wax.
Hazard: May be toxic by ingestion.
Uses: Electrical insulating and fire-resisting materials; impregnants; sealing compounds; crankcase additive; ingredient in penetrating oils; plasticizer; protective coatings.

"Hamp-ene."[517] Trademark for a series of chelating agents including ethylenediaminetetraacetic acid; the disodium salt, dihydrate; the trisodium salt, monohydrate; the tetrasodium salt, monohydrate; the tetrasodium salt; and the tetrasodium salt, dihydrate.

"Hamp-ex."[517] Trademark for diethylenetriaminepentaacetic acid pentasodium salt. (40.2% as DETPA Na_5). Used when slightly higher metal chelate stability is required at a sacrifice in stoichiometric chelating activity.

"Hamp-ol" 120[517] Trademark for N-hydroxyethylethylenediaminetriacetic trisodium salt (41.3% as HEEDTA Na_3). Used as a general purpose chelating agent for the control of iron in the pH range from 6.5 to 9.5 as well as calcium and magnesium.

"Hamposyl."[517] Trademark for a series of fatty acid sarcosinates.

"Han."[51] Trademark for a high-aromatic solvent, boiling range 355–535°F, used as a vehicle for chemical insecticides and herbicides. Combustible.

hand (handle). A term used chiefly in the textile industry to describe a fabric in a qualitative manner, as determined by the sensory perception of its feeling. It is also used to some extent in the leather industry. Such pragmatic physical-perception tests are of great practical value, although the properties being ascertained are often not objectively definable.

handedness. See asymmetry; enantiomer; chiral.

"Hansa."[307] Trademark for a group of yellow to orange insoluble azo pigments based on toluidine and beta-naphthol. Have good lightfastness in deep shades but tend to fade in pastels. Notably poor resistance to bleeding; good weather resistance and are relatively unaffected by acids and alkalies. Nontoxic. Used chiefly in emulsion paints, toy enamels, and other applications.

"Harbide."[446] Trademark for a silicon carbide brick, formed by impact pressing; low permeability, dense impervious surfaces, high resistance to oxidation.
Uses: Refractory for ceramic kilns, furnace linings,

recuperator tubes, radiant tubes, retorts, and in applications subject to mechanical abrasion.

"Harchemex."[189] Trademark for a compound of mainly C_{14} and C_{16} straight-chain primary alcohols in the approximate ratio of 2 to 1.
Uses: Wetting agents; germicidal quaternary ammonium compounds; lubricating oil additive.

"Harchem 2-SL."[189] Trademark for inorganic ester of sebacic acid suitable for use as a base for synthetic lubricants.

Harden-Young ester. See fructose 1,6-diphosphate.

hardness. (1) The resistance of a material to deformation of an indenter of specific size and shape under a known load. This definition applies to all types of hardness scales except the Mohs scale (q.v.), which is based on the concept of scratch hardness and is used chiefly for minerals. The most generally used hardness scales are Brinell (for cast iron), Rockwell (for sheet metal and heat-treated steel), diamond pyramid, Knoop and scleroscope (for metals). Durometer hardness is used for softer materials like rubber and plastics.

(2) The proportion of calcium carbonate or calcium sulfate contained in a given sample of water.

(3) The resistance of a pesticide or detergent to decomposition or biodegradation; also called persistence.

(4) Extremely short-wave radiation, e.g., hard x-rays.

(5) Loosely, having the potential for affecting the nervous system, as hard liquor or hard drugs.

"Hardnox Alkali."[244] Trademark for a fused, flake combination of caustic soda and phosphates.

hard rubber. See rubber, hard.

hardwood. In papermaking terminology, the wood from deciduous trees (maple, oak, birth), regardless of whether it is actually hard or soft. Hardwoods are used as components of paper pulp, but in much less volume than softwoods (q.v.).

"Harflex."[189] Trademark for a series of polymeric plasticizers for vinyl resins, synthetic rubbers and cellulose esters. Characterized by permanence and resistance to extraction.

Hargreaves process. The manufacture of sodium sulfate (salt cake) from sodium chloride and sulfur dioxide. A mixture of sulfur dioxide and air is passed over briquettes of sodium chloride in a countercurrent manner to produce sodium sulfate and hydrochloric acid. This process accounts for only a small amount of the salt cake produced in the U.S.

hartshorn. See ammonium carbonate.

hartshorn oil. See bone oil.

HAS. Abbreviation for hydroxylamine acid sulfate.

hashish. Extract of cannabis (marijuana), more concentrated and thus more powerful than the base drug. See also cannabis.

"Hastelloy."[214] Trademark for a series of high-strength, nickel-base, corrosion-resistant alloys. With the exception of "Hastelloy alloy D," which is cast only, all are produced in the forms of sheet, plate, bars, rods, welding electrodes, and wire and can be fabricated into all types of process equipment.

"Haveg."[349] Trademark for a series of corrosion-resistant mixtures fabricated into chemical process equipment. They are composites of various synthetic polymers with asbestos, graphite, glass fibers, etc. Especially useful for drums and containers for shipment of chemicals.

"Havelast."[349] Trademark for an elastomeric binder or impregnant for various reinforcing materials such as "Sil-Temp," as fabric or rovings, asbestos, glass or graphite. Used in the rocket and missile industry when resiliency is desired.

"Haystellite."[214] Trademark for tungsten carbide products, principally in the form of hard-surfacing rods for protecting parts from severe abrasion.

hazardous material. Any material or substance which, if improperly handled, can be damaging to the health and well-being of man. Such materials cover a broad range of types which may be classified as follows:

(1) Poisons or toxic agents (see poison; toxicity), including drugs, chemicals, and natural or synthetic products that are in any way harmful, ranging from those that cause death to skin irritants and allergens.

(2) Corrosive chemicals, such as sodium hydroxide or sulfuric acid, that burn or otherwise damage the skin and mucous membranes on external contact or inhalation.

(3) Flammable materials (q.v.) including (a) organic solvents, (b) finely divided metals or powders, (c) some classes of fibers, textiles, or plastics, and (d) chemicals that either evolve or absorb oxygen during storage, thus constituting a fire risk in contact with organic materials.

(4) Explosives (q.v.) and strong oxidizing agents such as peroxides and nitrates.

(5) Materials in which dangerous heat build-up occurs on storage, either by oxidation or microbiological action. Examples are fish meal, wet cellulosics and other organic waste materials.

(6) Radioactive chemicals that emit ionizing radiation (q.v.).

Packaging, labeling, and shipping of hazadous materials by rail and truck is regulated by the Hazardous Materials Regulation Board of the Dept. of Transportation (DOT) (1920 L St. N.W., Washington, D.C.); by the Coast Guard (a division of DOT) for shipment by coastwise boats and inland waterway barges; and by the Federal Aviation Administration for air shipment. The latter regulations follow those established by the International Air Transportation Authority (IATA) whose central office is in Geneva, Switzerland.

The Manufacturing Chemists Association (1825 Connecticut Ave., Washington, D.C.) issues a series of authoritative data sheets on the properties of a number of hazardous materials, as well as instructions for preparation of warning labels. The American Conference of Governmental Industrial Hygienists (Box 1937, Cincinnati, Ohio) issues a periodically revised list of Threshold Limit Values for substances in workroom air. The *Journal of Hazardous Materials* (Elsevier) began publication in 1975.

An approach is being made to the problem of formulating an international code of regulations for hazardous materials; legislation pertaining to toxic materials testing and control is imminent.

See also flammable material; toxic material; poison (1).

Hb Symbol for hemoglobin.

"HB."[58] Trademark for a series of plasticizers for vinyl compounds.

"HBPA."[58] Trademark for hydrogenated bisphenol A.
Use: Unsaturated polyester raw material.

"HCA."[50] Trademark for a highly potent herbicide whose active ingredient is hexachloroacetone. Used in the control of all weeds in non-crop areas. It increases the initial and residual effectiveness of oil sprays. Available in mixtures or concentrate.
Hazard: Moderately toxic and irritant.

HCCH. Abbreviation for hexachlorocyclohexane (q.v.).

HCG. Abbreviation for human chorionic gonadotropin.

HCH. Abbreviation for hexachlorocyclohexane (q.v.).

"HCR."[182] Trademark applied to ion exchange resin used in water treating and chemical process applications; strong acid cation exchange resin, 8% divinylbenzene crosslinked.

He Symbol for helium.

HEA. Abbreviation for 2-hydroxyethyl acrylate (q.v.).

heat. A mode of energy associated with and proportional to molecular motion. It can be transferred from one body to another by radiation, conduction or convection; sensible heat is accompanied by a change in temperature, but latent heat is not.
—*of combustion:* The heat evolved when a definite quantity of a substance is completely oxodized (burned).
—*of crystallization:* The heat evolved or absorbed when a crystal forms from a saturated solution of a substance.
—*of dilution:* The heat evolved per mole of solute when a solution is diluted from one specific concentration to another.
—*of formation:* The heat evolved or absorbed when a compound is formed in its standard state from elements in their standard states at a specified temperature and pressure.
—*of fusion:* The heat required to convert a substance from the solid to the liquid state with no temperature change (also called latent heat of fusion or melting).
—*of hydration:* The heat associated with the hydration or solvation of ions in solution; also the heat evolved or absorbed when a hydrate of a compound is formed.
—*of reaction:* The heat evolved or absorbed when a chemical reaction occurs, in which the final state of the system is brought to the same temperature and pressure as that of the initial state of the reacting system.
—*of solution:* The heat evolved or absorbed when a substance is dissolved in a solvent.
—*of sublimation:* The heat required to convert a unit mass of a substance from the solid to the vapor state (sublimation) at a specified temperature and pressure without the appearance of the liquid state.
—*of transition:* The heat evolved or absorbed when a unit mass of a given substance is converted from one crystalline form to another.
—*of vaporization:* The heat required to convert a substance from the liquid to the gaseous state with no temperature change (also called latent heat of vaporization).

heat exchanger. A vessel in which an outgoing hot liquid or vapor transfers a large part of its heat to an incoming cool liquid; in the case of vapors, the latent heat of condensation is thus utilized to heat the entering liquid. The shell-and-tube type is widely used, in which the hot liquid or vapor is contained in the shell while the cool liquid passes through the tubes, which are usually arranged in coils for maximum contact with the heat source. Heat exchangers are used in many chemical operations, e.g., evaporation and pulp manufacture, as well as to produce steam from the heat developed in nuclear reactors for power generation. See also evaporation; heat transfer.

heat transfer. Transmission of thermal energy from one location to another by means of a temperature gradient existing between the two locations. It may take place by conduction, convection or radiation. Heat transfer is involved in many types of industrial operations, including distillation, evaporation, canning of foods, baking, curing, etc. In some cases a heat exchanger is utilized. Fluids of high heat capacity are widely used to remove unwanted heat, or to transfer it from one place to another within a system; examples are air, water, ethylene glycol. See also coolant.

heavy. A nontechnical word used in a number of scientific senses: (1) referring to atomic weight (heavy water, heavy metal); (2) referring to production volume (heavy chemical); (3) referring to physical weight (heavy spar); (4) referring to thickness (heavy-gauge wire); referring to distillation range (heavy oil).

heavy chemical. A chemical produced in tonnage quantities, often in a relatively impure state. Examples are sodium chloride, sulfuric acid, soda ash, salt cake, sodium hydroxide, etc. See also fine chemical.

heavy hydrogen. See deuterium.

heavy metal. A metal of atomic weight greater than sodium (22.9) that forms soaps on reaction with fatty acids, e.g., aluminum, lead, cobalt. See soap (2).

heavy oil. An oil distilled from coal-tar between 230 and 330°C, the exact range not at all definite. See coal tar.

heavy oxygen. See oxygen 18.

heavy spar. See barite.

heavy water (deuterium oxide; D_2O) Water composed of 2 atoms of heavy hydrogen (deuterium) and 1 atom of oxygen. In lower states of purity the proportion of heavy water molecules decreases. Deuterium oxide is present to the extent of about 1 part of 6,500 of ordinary water. Deuterium oxide freezes at 3.8°C, boils at 101.4°C and has sp. gr. of 1.1056 at 25°C.
Derivation: There are several methods of separating or concentrating D_2O: (a) Fractional distillation; (b) Girdler-Spevack process; (c) hydrogen-ammonia exchange process; (d) electrolysis; (e) cryogenic methane distillation.
Commercially available: D_2O 99.75% pure, in up to 5000 g lots.
Use: Moderator in some types of nuclear reactors.
See also tritium (hydrogen of atomic weight 3) which combines with oxygen to give another variety of heavy water.

hecto-. Prefix meaning 10^2 units (symbol h); e.g., 1 hg = 1 hectogram = 100 grams.

hectorite (hector clay $Mg_{2,67}Li_{0,33}Si_4O_{10}(OH)_2$. One of the montmorillonite group of minerals that are principal constituents of bentonite (q.v.).

hedeoma oil (American pennyroyal oil; pulegium oil).
Properties: Yellowish limpid essential oil; mint-like odor and taste. Sensitive to light. Soluble in two or more parts of 70% alcohol, ether and chloroform; slightly soluble in water; sp. gr. 0.925–0.940; refractive index 1.482. Combustible; low toxicity.
Chief constituent: Pulegone.
Derivation: Distilled from the leaves and tops of Hedeoma pulegioides.
Uses: Medicine; insectifuge; perfumery; manufacture of pulegone and its derivatives.

hedonal (methylpropylcarbinolurethane) $CH_3CH_2CH(CH_3)OCONH_2$.
Properties: White crystalline powder; faint aromatic odor and taste; soluble in alcohol, ether, organic solvents; sparingly soluble in cold water; more soluble in hot water. Fusing point 76°C. B.p. 215°C.
Use: Medicine (sedative).

HEDTA. See hydroxyethylethylenediaminetriacetic acid.

Hehner number. The percent by weight of water-insoluble fatty acids in oils and fats.

Heisenberg, Werner K. (1901–1976). A native of Germany, Heisenberg received his doctorate from the University of Munich in 1923, after which he was closely associated for several years with Niels Bohr in Copenhagen. He was awarded the Nobel prize in physics in 1932 for his brilliant work in quantum mechanics. In 1946 he became Director of the Max Planck Institute. His notable contributions to theoretical physics, best known of which was the Uncertainty Principle, imparted new impetus to nuclear physics and made possible a better understanding of atomic structure and chemical bonding. See also uncertainty principle.

helenine. A nucleoprotein derived from the mold Penicillium funiculosum and used with some success as an antiviral drug.

Helianthine B. See methyl orange.

"Helindon."[307] Trademark for vat dyestuffs used for dyeing wool.

"Helio."[307] Trademark for organic dyestuffs, used for paints, lacquers, printing inks, wallpaper, coated paper, ruber and organic plastics.

"Heliogen."[307] Trademark for phthalocyanine dyestuffs used for paint, lacquers, printing inks, wallpaper, coated paper, rubber, and plastics.

heliotropyl acetate. See piperonyl acetate.

"He-Li-Ox" 500 Series.[329] Trademark for a series of PVC heat stabilizers for extruding, injection molding, calendering, and blow molding. All ingredients conform to F.D.A. requirements for food packaging.

"Heliozone."[28] Trademark for a blend of waxy material used to retard sun checking and cracking of rubbers.

helium He Inert element of atomic number 2; first element in the noble gas group of the periodic table. Atomic weight 4.0026; valence 0 (not known to combine with any other element). Helium nuclei are alpha particles (q.v.). Most important isotope is helium-3.
Properties: Colorless, noncombustible gas; odorless and tasteless; liquefies at 4.2°K to form helium I; at 2.2°K there is a transition (lambda) point at which helium II is formed (see below). B.p. −268.9°C (1 atm); f.p. −272.2°C (25 atm) lowest of any substance. Density 0.1785 g/l, or 0.138 (air = 1) at 0°C. Very slightly soluble in water; insoluble in alcohol. Rate of diffusion through solids is 3 times that of air. Helium is an asphyxiant gas. See also noble gas.
Occurrence: Texas, Oklahoma, Kansas, New Mexico, Arizona, Canada. Originally discovered in sun's atmosphere and recently confirmed in atmosphere of Jupiter.
Derivation: From natural gas, by liquefaction of all other components, followed by purification over activated charcoal.
Grades: U.S.P.; technical; 99.995% pure.
Containers: (compressed gas) Cylinders; tank cars; semitrailers; (liquid) vacuum-insulated semitrailers; pipeline.
Uses: To pressurize rocket fuels; welding; inert atmospheres (growing germanium and silicon crystals); inflation of weather and research balloons; heat-transfer medium; leak detection; chromatography; cryogenic research; magnetohydrodynamics; luminous signs; geological dating; aerodynamic research; diving and space vehicle breathing equipment. Possible future uses include coolant for nuclear fusion power plants and in superconducting electric systems.
Shipping regulations: (gaseous) (Rail) Green label; (Air) Nonflammable Gas label; (liquid, nonpressurized) (Air) Not acceptable; (liquid, pressurized) (Air) Nonflammable Gas label. Not acceptable on passenger planes; (liquid, low-pressure) (Air) No label required. Helium-oxygen mixtures Green label; (Air) Nonflamamble Gas label.
Note: Liquid helium has unique thermodynamic properties too complex to be adequately described here. Liquid helium I has refractive index 1.026, sp. gr. 0.125, and is called a "quantum fluid" because it exhibits atomic properties on a macroscopic scale. Its b.p. is near absolute zero and vicosity is 25 micropoises (water = 10,000). See also superconductivity. Helium II, formed on cooling helium I below its transition point, has the unusual property of "superfluidity," extremely high thermal conductivity, and viscosity approaching zero.

"Helix" Rosin.[79] Trademark for a specially processed pale limed wood rosin; contains very little free lime.
Uses: Linoleum print paint; varnish-air drying; varnish-baking; varnish-gloss oil; printing ink.

"Heloplex."[85] Trademark for the sodium bisulfite complex of heliotropin $[(CH_2O_2)C_6H_3CH(OH)SO_3Na$ (bicyclic)].
Properties: Dry, white to gray powder; slightly soluble in water.
Containers: 100-lb fiber drums.
Use: Zinc plating industry for compounding of brighteners.

HEMA. See hydroxyethyl methacrylate.

hematein $C_{16}H_{12}O_6$. A tetracyclic compound. An oxidation product of hematoxylin, the coloring principle of logwood. Not to be confused with hematin.
Properties: Dark purple solid; m.p. >320°C; insoluble in water; slightly soluble in alcohol and ether; soluble in dilute sodium hydroxide giving a bright red color; soluble in ammonia with brownish-violet color.

Derivation: By adding ammonia to logwood extract and exposing to air.
Use: Indicator; biological stain.

hematin $C_{34}H_{32}N_4 \cdot FeOH$. The hydroxide of heme (q.v.). Not to be confused with hematein.
Properties: Blue to brown-black powder; decomposes at 200°C without melting; soluble in alkalies; hot alcohol or ammonia; slightly soluble in hot pyridine; insoluble in water, ether and chloroform.
Derivation: By dissolving hemin in dilute potassium hydroxide, precipating with acetic acid and recrystallizing from pyridine.
Use: Biochemical research.

hematite, red (red iron ore; bloodstone). Iron oxide (Fe_2O_3), with impurities.
Properties: Brilliant black to blackish red or brick red mineral with brown to cherry red streak and metallic to dull luster. Sp. gr. 4.9–5.3; Mohs hardness about 6. Noncombustible, except as powder.
Uses: The most important ore of iron. Also certain varieties are used as paint pigments and for rouge.
See also iron oxide reds and ferric oxide.

hematoporphyrin $C_{34}H_{38}O_6N_4$. Deep red crystals; soluble in alcohol; sparingly soluble in ether; insoluble in water. Obtained from hemin or hematin by the action of strong acids. It is nontoxic and is reported to be preferentially absorbed by cancerous tissues, making them fluoresce under ultraviolet light.
Use: Medical and biochemical research.

heme (hem) $C_{34}H_{32}FeN_4O_4$. The nonprotein portion of hemoglobin and myoglobin, consisting of reduced (ferrous) iron bound to protoporphyrin (see porphyrin, hemin).
Use: Medical and biochemical research.

hemel. Generic name for hexamethylmelamine (HMM) $C_3N_3[N(CH_3)_2]_3$(cyclic).
Properties: Solid; insoluble in water; soluble in acetone.
Use: Chemosterilant for insects.

hemicellulose. Cellulose (q.v.) having a degree of polymerization of 150 or less. A collective term for beta and gamma cellulose. The pure form is obtained from corn grain hulls. It is not an important component of cellulosic products and is of chiefly theoretical interest. Hemicellulose obtained by treating a mixture of hard- and softwoods with steam has been used as an animal feed supplement.

hemimellitene (1,2,3-trimethylbenzene) $C_6H_3(CH_3)_3$. Liquid; sp. gr. 0.8944 (20/4°C); f.p. −25.5°C; b.p. 176°C. Insoluble in water; soluble in alcohol. Occurs in some petroleums. Combustible.

hemin (Teichmann's crystals) $C_{34}H_{32}N_4O_4FeCl$. The chloride of heme.
Properties: Crystals which are brown by transmitted light and steel blue by reflected light. Sinters at 240°C. Freely soluble in ammonia water; soluble in strong organic bases; insoluble in carbonate solutions, dilute acid solutions; insoluble but stable in water.
Derivation: By heating hemoglobin with acetic acid and sodium chloride.
Use: Identification of blood stains; biochemical research.

hemisulfur mustard (2-chloroethyl methyl sulfide) $ClCH_2CH_2SCH_3$.
Properties: Liquid; sp. gr. 1.1155 (20/4°C); b.p. 138°C (744 mm); refractive index (n 20/D) 1.4908. Soluble in fat solvents. Characteristic mustard odor.

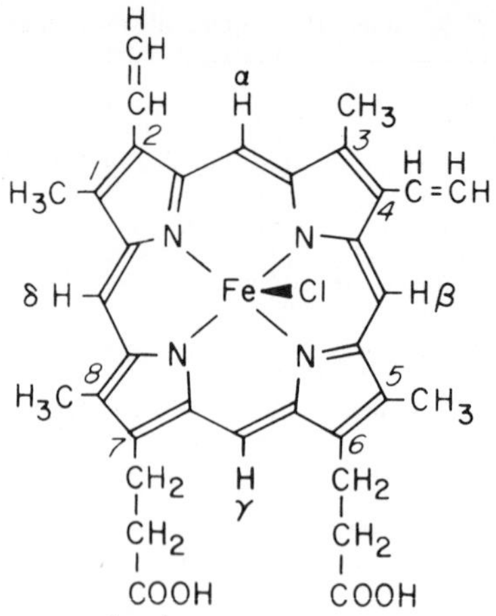

hemin
(Teichmann's crystals)

Hazard: Highly toxic. irritating to skin, eyes, and mucous membranes.
Use: Medicine.

hemoglobin (Hb). Approximate formula (molecular weight 65,000): $(C_{738}H_{1166}FeN_{208}S_2)_4$. The respiratory protein of the red blood cells; it transfers oxygen from the lungs to the tissues and carbon dioxide from the tissues to the lungs. Its affinity for carbon monoxide is over 200 times that for oxygen.

Hemoglobin is a conjugated protein consisting of approximately 94% globin (protein portion) and 6% heme (q.v.). Each molecule can combine with one molecule of oxygen to form oxyhemoglobin (HbO_2). The iron (in the heme portion) must be in the reduced (ferrous) state to enable the hemoglobin to combine with oxygen.

Oxyhemoglobin is available commercially as a brownish red powder or crystals; soluble in water. Used in medicine; usually called hemoglobin.

hemp. Soft, white fibers, 3–6 ft long, obtained from the stems of Cannabis sativa. It is coarser than flax but stronger, more glossy, and more durable than cotton. Combustible, not self-extinguishing. See also abaca.
Sources: Central Asia, Italy, U.S.S.R., India, U.S.
Hazard: May ignite spontaneously when wet.
Use: Blended with cotton or flax in toweling and heavy fabrics; twine; cordage; packing.
See also cannabis.

hempa. Generic name for hexamethylphosphoric triamide (hexamethylphosphoramide; HMPA) $[(N(CH_3)_2]_3PO$.
Properties: Water-white liquid, mild amine odor; soluble in water and polar and nonpolar solvents; b.p. 230–232°C (739.4 mm); sp. gr. 1.021 (60/60°F). Combustible.
Hazard: Toxic; a carcinogen.
Uses: Ultraviolet inhibitor in polyvinyl chloride; chemosterilant for insects; promoting stereospecific reactions; specialty solvent.

hempseed oil. A drying oil similar in properties and uses to linseed. Nontoxic and edible. Iodine value about 160; sp. gr. 0.923; refractive index 1.470–1.472. Contains about 10% saturated fatty acids (palmitic and stearic); unsaturated acids present are linoleic, linolenic, and oleic. Saponification value

190–193. Produced chiefly in U.S.S.R., southern Europe, Japan, Chile; little in U.S. Combustible.

henda-compounds. See corresponding unde-compound.

n-**heneicosanoic acid** $CH_3(CH_2)_{19}COOH$. A saturated fatty acid not normally found in natural fats or waxes. White crystalline solid; m.p. 74.3°C. Synthetic product available for organic synthesis; 99% purity. Combustible.

henequen. Hard, strong, reddish fibers obtained from the leaves of Agave fourcroydes. It is similar to sisal but coarser and stiffer. Combustible. Denier ranges from 300 to 500.
Source: Mexico, Cuba.
Use: Binder twine; cordage.

"Hennig Purifier."[177] Trademark for a preparation having a soda-ash base and other materials. Produced as walnut-sized briquettes. Packed in 100-lb paper bags. Used as a ladle addition to produce cleaner steel by aiding in removal of dissolved oxides and silicates and fluxing non-metallic inclusions to slag.

Henry's law. When a liquid and a gas remain in contact, the weight of the gas that dissolves in a given quantity of liquid is proportional to the pressure of the gas above the liquid. Thus if air is kept in contact with water at standard atmospheric pressure, each kilogram of water dissolves 0.017 gram of oxygen at 20°C; if this pressure is halved (by doing the experiment at high altitude where the pressure is only half an atmosphere) the water dissolves only 0.0085 gram of oxygen. The law holds true only for equilibrium conditions, i.e., when enough time has elapsed so that the quantity of gas dissolved is no longer increasing or decreasing.

hentriacontane. $C_{31}H_{64}$ or $CH_3(CH_2)_{29}CH_3$.
Properties: Crystals; sp. gr. 0.781 (68°C); b.p. 302°C (15 mm); m.p. 68°C. Combustible.

HEOD. See dieldrin.

heparin. A complex organic acid (mucopolysaccharide) present in mammalian tissues; a strong inhibitor of blood coagulation. A dextrorotatory polysaccharide built up from hexosamine and hexuronic acid units containing sulfuric acid ester groups. Precise chemical formula and structure uncertain; a formula of $(C_{12}H_{16}NS_2Na_3)_{20}$ and molecular weight of 12000 have been suggested for sodium heparinate.
Properties: White or pale-colored, amorphous powder; nearly odorless, hygroscopic; soluble in water; insoluble in alcohol, benzene, acetone, chloroform, and ether; pH in 17% solution between 5.0 and 7.5.
Derivation: Animal livers or lungs.
Grade: U.S.P.
Use: Medicine; biochemical research; rodenticides.

heptabarbital (5-[1-cyclohepten-1-yl]-5-ethylbarbituric acid; 5-ethyl-5-cycloheptenylbarbituric acid) $C_{13}H_{18}N_2O_3$.
Properties: White, odorless crystalline powder; slightly bitter taste. M.p. 174°C. Very sparingly soluble in water; slightly soluble in alcohol; soluble in alkaline solutions. Forms water-soluble sodium, magnesium and calcium salts.
Use: Medicine. See also barbiturate.

heptachlor $C_{10}H_7Cl_7$. Generic name for 1,4,5,6,7,8,8-heptachloro-3a, 4, 7, 7a-tetrahydro-4, 7-methanoindene.
Properties: White to light tan, waxy solid; m.p. 95–96°C; sp. gr. 1.57–1.59. Insoluble in water; soluble in xylene, and alcohol.
Containers: Drums.
Hazard: Toxic by ingestion, inhalation and skin absorption. Tolerance, 0.5 mg per cubic meter of air. Use has been restricted. May be carcinogenic.
Use: Insecticide.

heptachlorepoxide $C_{10}H_9Cl_7O$. A degradation product of heptachlor which also acts as an insecticide.

heptachlorotetrahydromethanoindene. See heptachlor.

n-**heptacosane** $C_{27}H_{56}$ or $CH_3(CH_2)_{25}CH_3$.
Properties: Crystals; soluble in alcohol; insoluble in water; sp. gr. 0.804; b.p. 270°C (15 mm); m.p. 59.5°C. Combustible.

n-**heptadecane** $C_{17}H_{36}$ or $CH_3(CH_2)_{15}CH_3$.
Properties: Leaflets; soluble in alcohol; insoluble in water; sp. gr. 0.778; b.p. 303°C; m.p. 22.5°C. Combustible.

n-**heptadecanoic acid** (margaric acid) $CH_3(CH_2)_{15}COOH$. A saturated fatty acid not normally found in natural fats or waxes.
Properties: Colorless crystals; m.p. 61°C; sp. gr. 0.8355 (90.6/4°C); b.p 363.8°C, 230.7°C (16 mm); refractive index 1.4324 (70°C). Soluble in alcohol and ether; insoluble in water. Available as a 99% pure synthetic product.
Uses: Organic synthesis.

heptadecanol. Any saturated C_{17} alcohol. See, for example, n-heptadecanol and 3,9-diethyl-6-tridecanol. Combustible; low toxicity.

n-**heptadecanol** $C_{17}H_{35}OH$.
Properties: Colorless liquid. Slightly soluble in water; sp. gr. 0.8475 (20/20°C); b.p. 308.5°C; vapor pressure <0.01 mm (20°C); flash point 310°F; wt/gal 7.1 lb (20°C). Combustible. Low toxicity.
Grade: Technical.
Containers: 1-gal cans; 5-, 55-gal drums.
Uses: Organic synthesis; plasticizer; intermediates; perfume fixatives; soaps and cosmetics; manufacture of wetting agents and detergents.

2-(8-heptadecenyl)-2-imidazoline-1-ethanol. See amine 220.

2-heptadecylglyoxalidine (2-heptadecylimidazoline) $C_{20}H_{40}N_2$ or $C_{17}H_{35}C_3H_5N_2$.
Properties: Waxy solid; m.p. 85°C; b.p. 200°C (2 mm); slightly soluble in water; soluble in alcohol, benzene; hydrolyzes on standing to form N-2 aminoethyl stearamide. Combustible.
Derivation: By reacting stearic acid with ethylene diamine.
Use: Fungicide.

2-heptadecylglyoxalidine acetate. See glyodin.

2-heptadecylimidazoline. See 2-heptadecylglyoxalidine.

2-heptadecyl-2-imidazoline acetate. See glyodin.

heptafluorobutyric acid (perfluorobutyric acid) C_3F_7COOH.
Properties: Colorless hygroscopic liquid with sharp odor. B.p. 120°C (735 mm); f.p. −17.5°C; sp. gr. 1.641 (25°C); refractive index (n 25/D) 1.290; surface tension 15.8 dynes/cm (30°C). Miscible with water, acetone, ether, and petroleum ether; soluble

in benzene and carbon tetrachloride; insoluble in carbon disulfide and mineral oil.
Derivation: By electrolysis of a solution of butyric acid in hydrogen fluoride.
Hazard: Irritant to tissue.
Uses: Intermediate; surfactant; acidulant.

"Heptagran."[147] Brand name for a granular insecticide containing 2½, 10, or 25% heptachlor.
Hazard: Toxic. See heptachlor.

heptaldehyde. See heptanal.

heptalin acetate. See methylcyclohexanol acetate.

heptamethylene. See cycloheptane.

heptamethylnonane $C_{16}H_{34}$. Isomer of cetane (hexadecane); ignition value 15 on cetane-alpha-methylnaphthalene scale. In 1964 it replaced alpha-methylnaphthalene as ignition standard for diesel fuels.
Hazard: Flammable, moderate fire risk.

heptanal (heptaldehyde; enanthaldehyde; aldehyde C-7) $C_6H_{13}CHO$.
Properties: Oily, colorless liquid; penetrating fruity odor. Hygroscopic. Soluble in 3 volumes of 60% alcohol; slightly soluble in water, soluble in ether. Sp. gr. 0.814–0.819; refractive index 1.42; m.p. 43°C; b.p. 153°C. Combustible.
Derivation: Castor oil, from decomposition of the ricinoleic acid glyceride.
Containers: Up to 55-gal drums.
Uses: Manufacture of 1-heptanol; organic synthesis; perfumery; pharmaceuticals; flavoring.

heptane (dipropylmethane) $CH_3(CH_2)_5CH_3$.
Properties: Volatile, colorless liquid; freezing point −90.595°C; b.p. 98.428°C; refractive index (n 20/D) 1.38764; sp. gr. 0.68368 (20°C); flash point 25°F (closed cup). Soluble in alcohol, ether, chloroform; insoluble in water. Distillation range 200–210°F; vapor pressure 2.0 psi absolute (100°F) (max); color, Saybolt +30 (min); maximum sulfur content 0.01 wt %; corrosion, passes ASTM D 130-30 test; autoignition temp. 433°F.
Derivation: Fractional distillation of petroleuem. Purified by rectification.
Grades: Commercial; 99%; spectro; ASTM reference fuel; research, 99.92 mole %.
Containers: Bottles; drums; special tank cars.
Hazard: Flammable, dangerous fire risk; moderately toxic by inhalation. Tolerance, 500 ppm in air.
Uses: Standard for octane rating determinations (pure normal heptane has zero octane number); anesthetic; solvent; organic synthesis; preparation of laboratory reagents.
See also octane number.
Shipping regulations: (Rail) Red label (Air) Flammable Liquid label.

1,7-heptanedicarboxylic acid. See azelaic acid.

1,7-heptanedioic acid. See pimelic acid.

n-**heptanoic acid** (enanthic acid; n-heptylic acid; heptoic acid) $CH_3(CH_2)_5COOH$.
Properties: Clear, oily liquid; unpleasant odor. Soluble in alcohol and ether; insoluble in water. Sp. gr. 0.9181 (20/4°C); f.p. −7.0°C; b.p. 221.9°C; refractive index 1.4229. Combustible; low toxicity.
Derivation: By oxidizing heptanal with potassium dichromate and sulfuric acid.
Uses: Organic synthesis; production of special lubricants for aircraft and brake fluids.

1-heptanol. See heptyl alcohol.

2-heptanol. See methyl amyl carbinol.

3-heptanol $CH_3CH_2CH(OH)C_4H_9$.
Properties: Liquid; sp. gr. (20°C) 0.8224; f.p. −70°C; b.p. 156.2°C; flash point 140°F (open cup); slightly soluble in water.
Containers: Cans, drums, tank cars.
Hazard: Moderately toxic by inhalation and ingestion. Moderate fire risk.
Uses: Flotation frother; solvent and diluent in organic coatings; intermediates.

2-heptanone. See methyl n-amyl ketone.

3-heptanone. See ethyl butyl ketone.

4-heptanone. See dipropyl ketone.

"Hepteen Base."[248] Trademark for a heptaldehydeaniline reaction product.
Properties: Dark-brown, free-flowing liquid; sp. gr. 0.93; soluble in acetone, benzene, and ethylene dichloride; moderately soluble in gasoline; insoluble in water.
Hazard: Probably toxic by ingestion.
Uses: Rubber accelerator.

1-heptene (1-heptylene) C_7H_{14} or $CH_3(CH_2)_4CH{:}CH_2$.
Properties: Colorless liquid; sp. gr. 0.6968 (20/4°C); b.p. 93.3°C; f.p. −119.2°C; refractive index (n 20/D) 1.3994; soluble in alcohol, acetone, ether, petroleum, coal-tar solvents; insoluble in water.
Hazard: Probably flammable.
Grade: 95% purity.
Uses: Organic synthesis.
Shipping regulations: Flammable liquid, n.o.s., (Rail) Red label; (Air) Flammable Liquid label.

2-heptene (2-heptylene) $CH_3(CH_2)_3CH{:}CHCH_3$. (cis and trans isomers).
Properties: Colorless liquid; sp. gr. (cis) 0.708, (trans) 0.704, (commercial) 0.7010–0.7050 (20/4°C); b.p. (trans) 98.0°C, (cis) 98.5°C, (commercial) 97–99°C; refractive index (n 20/D), 1.406; flash point (commercial) 28°F; soluble in alcohol, acetone, ether, petroleume, coal-tar solvents; insoluble in water. Low toxicity.
Hazard: Flammable, dangerous fire risk.
Use: Organic synthesis.
Shipping regulations: Flammable liquid, n.o.s., (Rail) Red label; (Air) Flammable Liquid label.

3-heptene (3-heptylene) C_7H_{14} or $C_3H_7CH{:}CHC_2H_5$.
Properties: (mixed cis or trans isomers): Colorless liquid; b.p. 95°C; sp. gr. (60/60°F) 0.705; refractive index (20/D) 1.405; flash point 21°F.
Hazard: Flammable, dangerous fire risk.
Use: Suggested as plant growth regardant.
Shipping regulations: Flammable liquid, n.o.s., (Rail) Red label; (Air) Flammable Liquid label.

heptene C_7H_{14}. A mixture of isomers.
Properties: Liquid. Sp. gr. (15.56/15.56°C) 0.711; b.p. 189.5°C.
Derivation: Olefin fraction produced by catalytic polymerization of propylene and butylene.
Containers: Up to tank cars.
Hazard: Flammable, dangerous fire risk.
Uses: Lubricant additive; catalyst; surfactants.
Shipping regulations: Flammable liquid, n.o.s., (Rail) Red label; (Air) Flammable Liquid label.

heptoic acid. See heptanoic acid.

heptyl acetate $C_7H_{15}OOCCH_3$. Liquid with fruity odor.
Hazard: Experiments on animals indicate irritant properties.
Use: Artificial fruit essences.

heptyl alcohol (1-heptanol; enanthyl alcohol). $C_7H_{15}OH$.
Properties: Colorless fragrant liquid; f.p. −34.6°C; b.p. 175°C; sp. gr. 0.824 (20/4°C); refractive index (n 20/D) 1.4233. Flash point 170°F. Combustible. Low toxicity.
Derivation: From heptaldehyde by reduction.
Containers: Up to 55-gal lined drums.
Uses: Organic intermediate, solvent, cosmetic formulations.

heptylamine $C_7H_{15}NH_2$.
Properties: Colorless liquid; sp. gr. 0.777 (20/4°C); f.p. −23°C; b.p. 155°C; flash point 140°F (open cup); slightly soluble in water; soluble in alcohol or ether. Combustible.

heptyl formate $HCOOC_7H_{15}$.
Properties: Colorless liquid with fruity odor; b.p. 176.7°C; sp. gr. 0.894 (0°C). Combustible.
Use: Artificial fruit essences.

heptyl heptoate. $C_7H_5OOCC_6H_{13}$.
Properties: Colorless liquid with fruty odor; sp. gr. 0.865 (19°C); b.p. 273–274°C (754 mm). Combustible.
Use: Artificial fruit essences.

n-**heptylic acid.** See heptanoic acid.

heptyl pelargonate $C_7H_{15}OOCC_8H_{17}$.
Properties: Liquid, with pleasant odor; sp. gr. 0.866 (15.5/15.5°C); b.p. 300°C; refractive index 1.4360. Combustible.
Uses: Flavors and perfumes.

"Hepzide."[123] Trademark for nithiazide for treatment of blackhead in turkeys and chickens.

"Herban."[266] Trademark for norea; a preemergence herbicide.

"Herbandin."[505] Trademark for an ester derived from petrochemicals.
Properties: Colorless oily liquid; sp. gr. 1.046; refractive index (n 20/D) 1.498; saponification value 245; acid value 0.7.
Use: Extension and support of lavandin oil.

herbicide (weed killer). A pesticide (q.v.), either organic or inorganic, used to destroy unwanted vegetation, especially various types of weeds, grasses, and woody plants. Until 1924 inorganics such as sodium chlorate, sodium chloride, ammonium sulfamate, arsenic, and boron compounds were used. At that time the more specific organics were introduced, typified by 2,4-dichlorophenoxyacetic acid (2,4-D). Herbicides may be of two major types: (1) selective, such as 2,4-D; 2,4,5-T; phenols; carbamates, and urea derivatives, permitting elimination of weeds without injury to the crop; and (2) nonselective, comprising soil sterilants (sodium compounds) and silvicides (ammonium sulfamate). The latter kill wood plants and trees. Some types act as overstimulating growth hormones. Many herbicides are highly toxic and should be handled and applied with care; use of chlorinated types may be restricted. See also defoliant.

"Herclor."[266] Trademark for a group of specialty elastomers based on epichlorohydrin, claimed to have unique service performance properties. Uses include automotive and aircraft parts, wire and cable coating, seals and gaskets, packings, hose, belting, and coated fabrics. "Herclor H" is a homopolymer, and "Herclor C" a copolymer with ethylene oxide, both having high resistance to ozone, heat, solvents, and chemical attack.

"Hercoflat."[266] Trademark for a polypropylene texturing agent.

"Hercoflex."[266] Trademark for a series of plasticizers.
150. Di(n-octyl, n-decyl) phthalate.
290. Di(n-octyl, n-decyl) adipate.
600. High-boiling ester of pentaerythritol and a saturated aliphatic acid.
707. High molecular weight polyol ester. Used in high-temperature vinyl electrical insulation.
900. High molecular weight polyester plasticizer for polyvinyl acetate.
J15. Saturated aliphatic ester of pentaerythritol for plasticizing vinylidene chloride.

"Hercogel" A.[266] Trademark for a water-resistant, plastic, ammonia gelatin. A permissible dynamite used under the most severe water conditions.
Hazard: High explosive.

"Hercolube."[266] Trademark for synthetic lubricant base stocks derived from pentaerythritol esters of saturated fatty acids.

"Hercolyn" D.[266] Trademark for a pale, viscous liquid, the hydrogenated methyl ester of rosin. Used as a plasticizing resin.

"Hercomix" 1.[266] Trademark for premixed, prill- and fuel-oil nitrocarbonitrate blasting agents.Used where water conditions are not too severe.
Hazard: High explosive.

"Herco."[266] Trademark for high-quality pine oil; predominantly secondary and tertiary cyclic terpene alcohols.
Use: Odorant.

"Herculoid."[266] Trademark for nitrocellulose containing 10.9 to 11.2% nitrogen. Used for pyroxylin plastics.
Hazard: See nitrocellulose.

"Herculon."[266] Trademark for polypropylene olefin fibers. Available in bulked continuous and continuous multifilament yarns, staple, and uncut tow.
Uses: Apparel, home furnishings, and industrial applications.

"Heresite."[17] Trademark for a series of pure phenol-formaldehyde resinous coatings and related products of the thermosetting type. Applied by spraying, dipping, or roller-coating, followed by curing at temperatures of approximately 400°F.
Uses: Anticorrosive lining for shipping containers, machinery and equipment for chemicals, food, drugs, and petroleum industries. Lining of tank cars for sulfuric acid and other corrosive chemials.

heroin. See diacetylmorphine.

"Herox."[28] Trademark for nylon filaments, used in brushes.

herring oil. See fish oil.

hesperidin $C_{28}H_{34}O_{15}$. A natural bioflavonoid of the flavanone grup.
Properties: Fine needles. M.p. 258–262°C; soluble in dilute alkalies, pyridine; very slightly soluble in water, acetone, benzene, and chloroform. Tasteless.
Derivation: Extraction from citrus fruit peel.
Uses: Synthetic sweetener research.

Hess's law. The heat evolved or absorbed in a chemical process is the same whether the process takes place in one or in several steps; also known as the law of constant heat summation.

"Het" Acids.[62] Trademark for chlorendic acid (q.v.).

heteroaromatic. See heterocyclic.

heterocyclic. Designating a closed-ring structure, usually of either 5 or 6 members, in which one or more of the atoms in the ring is an element other than carbon, e.g., sulfur, nitrogen, etc. Examples are pyridine, pyrrole, furan, thiophene, and purine.

```
CH—CH
‖    ‖
CH  CH
  \S/
```

thiophene   pyridine

heterogeneous (Latin, "different kinds"). Any mixture or solution comprised of two or more substances, whether or not they are uniformly dispersed. Common examples are such diverse materials as air (a mixture of 1/5th oxygen and 4/5ths nitrogen), milk, marble, paint, gasoline, blood, mayonnaise. In all such cases the mixtures can be separated mechanically into their components. "Homogenized" milk is technically as heterogeneous as regular milk, and the term is, strictly speaking, a misnomer. See also homogeneous; mixture.

heterogeneous catalysis. See catalysis, heterogeneous.

heteromolybdates (heteropolymolybdates). A large group of complex molybdenum salts and acids in which the anion contains oxygen atoms and from 2 to 18 hexavalent molybdenum atoms, as well as one or more other metal or nonmetal atoms (phosphorus, arsenic, iron and tellurium). The latter are referred to as hetero atoms and any of about 35 elements may be present in this manner. Example: $Na_3PMo_{12}O_{40}$, sodium phospho-12-molybdate. The molecular weights of these compounds range up to 3000. The acids and most of the salts are very soluble in water, and the acids and some salts are soluble in organic solvents.
Uses: Phosphomolybdates and phosphotungstates are used as precipitants for basic dyes to form lakes and toners. The phospho- and silicomolybdate groups are of key importance in the functioning of certain enzymes. There are many uses in analytical chemistry.

HETP.
(1) Abbreviation for hexaethyl tetraphosphate.
(2) Abbreviation for height equivalent to a theoretical plate. It is the height of a distillation or fractionating column which gives a separation equivalent to that of a theoretical plate in the physical separation process involved. A theoretical plate is one that produces the same difference in composition as exists at equilibrium between two phases.

"Hetron."[62] Trademark for a series of polyester resins.

hexabromoethane C_2Br_6.
Properties: Yellowish-white, rhombic needles. Slightly soluble in water, alcohol. M.p. 149°C (decomposes with separation of bromine).
Derivation: Action of bromine on diiodoacetylene.
Use: Organic synthesis.

hexacalcium phytate. See calcium phytate.

hexachloroacetone (hexachloro-2-propanone) $Cl_3CCOCCl_3$.
Properties: Yellow liquid; b.p. 204°C; f.p. −3°C; sp. gr. 1.744 (12/12°C); slightly soluble in water; soluble in acetone. Combustible.
Hazard: Moderately toxic by ingestion and inhalation; strong irritant. Evolves phosgene when heated.
Uses: Desiccant; herbicide.

hexachlorobenzene (perchlorobenzene) C_6Cl_6.
Properties: White needles. Soluble in benzene and boiling alcohol; insoluble in water. M.p. 229°C; b.p. 326°C. Combustible. Flash point 468°F.
Hazard: EPA recommends upper limit of 0.5 ppm in fatty tissues of cattle, pigs, and sheep, subject to change after further toxicological research. Previous work on laboratory animals had indicated a low level of toxicity. Do not confuse with benzene hexachloride.
Uses: Organic synthesis; fungicide for seeds; wood preservative.

hexachlorobutadiene C_4Cl_6, or $Cl_2C{:}CClCCl{:}CCl_2$.
Properties: Clear, colorless liquid with mild odor; freezing range −19 to −22°C; boiling range 210–220°C; refractive index (n 20/D) 1.552 (± .001); flash point, none; sp. gr. 1.675 (15.5/15.5°C); 13.97 lb/gal (15.5°C); purity 98% (min); vapor pressure 22 mm (100°C); 500 mm (200°C); viscosity (100°F) 2.447 cp; 1.479 centistokes; (210°F) 1.131 cp; 0.724 centistokes; insoluble in water; compatible with numerous resins; soluble in alcohol and ether. Nonflammable.
Containers: 55-gal steel drums.
Uses: Solvent for elastomers; heat-transfer liquid; transformer and hydraulic fluid; wash liquor for removing C_4 and higher hydrocarbons.

1,2,3,4,5,6-hexachlorocyclohexane (BHC; HCCH: HCH: TBH: benzene hexachloride) $C_6H_6Cl_6$. A systemic insecticide. The gamma isomer is lindane.
Properties: White or yellowish powder or flakes; musty odor; color, odor, melting point vary with isomeric compositoin. Sp. gr. 1.87. Vapor pressure about 0.5 mm (60°C); stable toward moderate heat but decomposed by alkaline substances. Melting points of the pure isomers are: (alpha-trans) 157–158°C; (beta-cis) 297°C (sublimes); (gamma) 112.5°C; (delta) 138–139°C; (epsilon) 217–219°C. Insoluble in water; soluble 100% alcohol, chloroform, and ether.
Derivation: Chlorination of benzene in actinic light.
Method of purification: Fractional crystallization. The technical grade may run 10–15% gamma isomer, but can be brought up to 99% (lindane).
Grades: Technical (mixture of isomers); 25% gamma isomer; 99% gamma isomer (lindane).
Containers: Bags, fiber drums.
Hazard: Highly toxic by ingestion, moderately by inhalation. Absorbed by skin; strong irritant to skin and eyes. CNS depressant. Tolerance (lindane), 0.5 mg per cubic meter of air. Use may be restricted.
Uses: Component of insecticides toxic to flies, cockroaches, aphids, grasshoppers, wire worms, and boll weevils.

hexachlorocyclopentadiene (perchlorocyclopentadiene) C_5Cl_6.
Properties: Pale yellow liquid; pungent odor. B.p. 239°C; freezing point 9.6°C; sp. gr. 1.717 (15/15°C); wt/gal 14.30 lb (15.5°C); refractive index (n 25/D) 1.563; flash point, none. Nonflammable.
Grade: Technical.
Containers: 55-gal drums; tank cars.
Hazard: Highly toxic by ingestion, inhalation, and skin absorption. Tolerance, 0.01 ppm in air.
Uses: Intermediate for resins, dyes, pesticides, fungicides, pharmaceuticals.

hexachlorodiphenyl oxide $C_{12}H_4Cl_6O$.
Properties: Light yellow, very viscous liquid. B.p. 230–260°C (8 mm); sp. gr. 1.60 (20/20°C); lb/gal 13.12 at 25°C; refractive index 1.621 (25°C); flash point, none. Soluble in methanol, ether. Very slightly soluble in water. Nonflammable.
Hazard: Probably toxic.
Uses: Solvent; intermediate.

1,1,1,4,4,4-hexachloro-1,4-disilabutane. See bis(trichlorosilyl)ethane.

hexachloroendomethylenetetrahydrophthalic acid. See chlorendic acid.

hexachloroendomethylenetetrahydrophthalic anhydride. See chlorendic anhydride.

hexachloroethane (perchloroethane; carbon trichloride; carbon hexachloride) Cl_3CCCl_3.
Properties: Colorless crystals; camphor-like odor; sp. gr. 2.091; m.p. 185°C; b.p. sublimes at 185°C. Soluble in alcohol and ether; insoluble in water.
Hazard: Toxic by ingestion and inhalation. Strong irritant. Absorbed by skin. Tolerance, 1 ppm in air.
Uses: Organic synthesis; retarding agent in fermentation; camphor substitute in nitrocellulose; rubber accelerator; pyrotechnics and smoke devices; solvent; explosives.

hexachloromethylcarbonate (triphosgene) $(OCCl_3)_2CO$.
Properties: White crystals. Odor similar to that of phosgene. Decomposed by hot water and alkali hydroxides. Only slowly acted upon by cold water. Soluble in alcohol, benzene, ether. Sp. gr. 2 (approx); b.p. 205–206°C (partial decomposition); m.p. 78–79°C.
Derivation: Chlorination of dimethyl carbonate exposed to direct sunlight.
Hazard: Toxic; strong irritant to eyes and skin.
Use: Tear gas.
Shipping regulations: (Rail, Air) Tear gas materials, Poison label or Tear Gas label. Not acceptable on passenger planes.

hexachloromethyl ether $O(CCl_3)_2$.
Properties: Liquid. Phosgene-like odor. Sp. gr. 1.538 (18°C); b.p. 98°C (partial decomposition).
Derivation: Chlorination of dichloromethyl ether.
Hazard: Toxic; strong irritant to eyes and skin.

hexachloronaphthalene $C_{10}H_2Cl_6$.
Properties: White solid.
Hazard: Highly toxic by inhalation; strong irritant. Absorbed by skin. Tolerance, 0.2 mg per cubic meter of air.

hexachlorophene (2,2′-methylene bis-(3,4,6-trichlorophenol); bis-(3,5,6-trichloro-2-hydroxyphenyl)methane. $(C_6HCl_3OH)_2CH_2$.
Properties: White, free-flowing powder; odorless; m.p. 161–167°C. Soluble in acetone, alcohol, ether; chloroform; insoluble in water.
Derivation: Condensation of 3,4,5-trichlorophenol with formaldehyde in the presence of sulfuric acid.
Containers: Cans; drums.
Hazard: Toxic properties have caused FDA to prohibit its use unless prescribed by a physician.
Uses: Formerly as bactericide and bacteriostat.

hexachloro-2-propane. See hexachloroacetone.

hexachloropropylene (hexachloropropene; perchloropropylene) $CCl_3CCl{:}CCl_2$.
Properties: Water-white liquid; b.p. 210°C; insoluble in water; miscible with alcohol, ether, chlorinated compounds.
Hazard: May be toxic.
Uses: Solvent; plasticizer; hydraulic fluid.

hexacontane $C_{60}H_{122}$. High molecular weight hydrocarbon.
Properties: Waxy solid. M.p. 101°C. Combustible; nontoxic.

hexacosanoic acid. See cerotic acid.

n-**hexadecane** (cetane) $C_{16}H_{34}$.
Properties: Colorless liquid; sp. gr. 0.77335 (20/4°C); b.p. 286.5°C; m.p. 18.14°C; refractive index (n 20/D) 1.43435. Soluble in alcohol, acetone, ether; insoluble in water. Combustible. Autoignition temp. 401°F.
Grades: Technical; ASTM.
Uses: Solvent; organic intermediate; ignition standard for diesel fuels. See also cetane number.

hexadecanoic acid. See palmitic acid.

1-hexadecanol. See cetyl alcohol.

hexadecanoyl chloride. See palmitoyl chloride.

1-hexadecene (cetene; alpha-hexadecylene) $CH_3(CH_2)_{13}CH{:}CH_2$.
Properties: Colorless liquid; m.p. 4°C; b.p. 274°C; sp. gr. 0.784 (15/4°C); refractive index (n 20/D) 1.4441; insoluble in water; soluble in alcohol, ether, petroleum and coal-tar solvents. Combustible.
Derivaton: Treatment of cetyl alcohol with phosphorus pentoxide.
Grade: 95% purity.
Uses: Organic synthesis.

cis-**9-hexadecenoic acid.** See palmitoleic acid.

6-hexadecenolide. See ambrettolide.

hexadecyl mercaptan. See cetyl mercaptan.

tert-**hexadecyl mercaptan** $C_{16}H_{33}SH$.
Properties: Colorless liquid; unpleasant odor; boiling range (5 mm) 121 to 149°C; sp. gr. (60/60°F) 0.874; refractive index (n 20/D) 1.474; flash point 265°F. Combustible.
Containers: Drums and tank cars.
Hazard: Probably toxic.
Use: Polymer modification.
See also thiol.

1-hexadecylpyridinium chloride. See cetylpyridinium chloride.

hexadecyltrichlorosilane $C_{16}H_{33}SiCl_3$.
Properties: Colorless to yellow liquid. B.p. 269°C; sp. gr. 0.996 (25/25°C); refractive index (n 25/D) 1.4568; flash point 295°F. Combustible.
Derivation: By Grignard reaction of silicon tetrachloride and hexadecylmagnesium chloride.
Grade: Technical.
Hazard: Highly toxic; strong irritant; evolves hydrochloric acid in presence of moisture.
Use: Intermediate for silicones.
Shipping reuglations: (Rail) White label (Air) Corrosive label. Not acceptable on passenger planes.

hexadecyltrimethylammonium bromide. See cetyl trimethylammonium bromide.

1,4-hexadiene C_6H_{10} or $CH_2{:}CHCH_2CH{:}CHCH_3$.
Properties: Colorless liquid; sp. gr. 0.6996 (20/4°C);

b.p. 64°C (745 mm); refractive index 1.4162 (20/D); flash point −6°F; insoluble in water.
Derivation: Reaction between ethylene and butadiene with a special catalyst.
Hazard: Highly flammable. Explosive limits in air 2 to 6.1%.
Use: As third monomer in EPDM synthetic elastomers.
Shipping regulations: (Rail) Flammable liquid, n.o.s., Red label. (Air) Flammable Liquid label.

2,4-hexadienoic acid. See sorbic acid.

1,5-hexadiyne. See dipropargyl.

"Hexadow."[233] Trademark for insecticide formulations containing benzene hexachloride.
Hazard: May be toxic.

hexa-2-ethylbutoxydisiloxane $[(CH_3CH(C_2H_5)CH_2CH_2O)_3Si]_2O$.
Properties: Colorless oil; b.p. 195°C (0.2 mm). Combustible.
Derivation: Reaction of silicon tetrachloride, 2-ethylbutanol and water.
Uses: Aircraft hydraulic fluid.

hexaethyl tetraphosphate (HETP). A mixture of ethyl phosphates and ethyl pyrophosphate (TEPP).
Properties: Yellow liquid. Sp. gr. 1.26–1.28 (25/4°C); f.p. −90°C; refractive index 1.427; decomposes at high temperatures; soluble or miscible in water and many organic solvents except kerosine; hydrolyzes in low concentrations; hygroscopic.
Containers: 55-gal drums.
Hazard: Highly toxic by ingestion, inhalation, and skin absorption. Cholinesterase inhibitor.
Use: Contact insecticide.
Shipping regulations: (Rail, Air) Poison label. Not acceptable on passenger planes. *Note:* Hexethyl tetraphosphate and compressed gas mixtures not accepted by air or by rail express. For details consult regulations.

hexafluoroacetone CF_3COCF_3.
Properties: Colorless, hygroscpic, highly reactive gas. B.p. −27°C; f.p. −122°C; liquid density 1.33 g/ml (25°C). Reacts vigorously with water and other substances, releasing considerable heat. Nonflammable. Minimum purity 95%.
Containers: Compressed gas in steel cylinders.
Hazard: Highly toxic by inhalation. Tolerance, 0.1 ppm in air.
Use: Intermediate in organic synthesis.
Shipping reuglations: (Rail) Poisonous gas, n.o.s., Poison Gas label. Not accepted by express. (Air) Nonflammable Gas label. Not acceptable on passenger planes.

hexafluorobenzene C_6F_6. Liquid; b.p. 80.26°C; m.p. 5.2°C; sp. gr. 1.613. Combustible.
Use: Chemical intermediate; solvent in NMR spectroscopy.

hexafluoroethane (fluorocarbon 116) CF_3CH_3.
Properties: A gas; b.p. −78.2°C. Sp. gr. 1.59. One of the most stable of all organic compounds. Insoluble in water; slightly soluble in alcohol. Nonflammable; nontoxic.
Grade: 99.6% pure.
Containers: Steel cylinders.
Uses: Dielectric and coolant; aerosol propellant; refrigerant.
Shipping regulations: (Air) Nonflammable Gas label.

hexafluorophosphoric acid HPF_6.
Properties: (65% solution): Colorless, fuming liquid; sp. gr. 1.81; m.p. 31°C ($6H_2O$). Stable in neutral and alkaline solutions.
Hazard: Highly toxic; strong irritant to tissue.
Uses: Metal cleaners, electrolytic or chemical polishing agents, for the formation of protective coatings for metal surfaces, and as a catalyst.
Shipping regulations: (Rail) White label. (Air) Corrosive label. Not acceptable on passenger planes.

hexafluoropropylene (perfluoropropene) $CF_3CF{:}CF_2$.
Properties: Gas; f.p. −156°C; b.p. −29°C; sp. gr. 1.583 (−40/4°C). Nonflammable.
Shipping regulations: (Rail) Green label. (Air) Nonflammable Gas label.

hexafluoropropylene epoxide (HFPO) $CF_2CF_2CF_2$.
Derivation: Oxidation of hexafluoropropylene with alkaline hydrogen peroxide at about −30°C.
Use: Monomer for HFPO polymers that are heat-resistant to 410°C, noncombustible and of low toxicity. See also "Freon E" and "Krytox."

hexafluorosilicic acid. See fluosilicic acid.

hexaglycerol. See trimethylolpropane; polyglycerol.

"Hexagon."[319] Trademark for grain alcohol, pure and specially denatured.

hexahydric alcohol. See mannitol, sorbitol, and dulcitol.

hexahydroaniline. See cyclohexylamine.

hexahydrobenzene. See cyclohexane.

hexahydrobenzoic acid (cyclohexanecarboxylic acid, a naphthenic acid) $C_6H_{11}COOH$.
Properties: Colorless monoclinic prisms; m.p. 31°C; b.p. 233°C; sp. gr. 1.048 (15/4°C); refractive index 1.4561 (33.8°C); slightly soluble in water, soluble in alcohol and ether.
Uses: Paint and varnish driers, drycleaning soaps, lubricating oils; stabilizer for rubber.

hexahydrocresol. See methylcyclohexanol.

hexahydromethylphenol. See methylcyclohexanol.

hexahydrophenol. See cyclohexanol.

hexahydrophthalic anhydride (1,2-cyclohexanedicarboxylic anhydride) $C_6H_{10}(CO)_2O$.
Properties: Clear, colorless, viscous liquid which becomes a glassy solid at 35–36°C; b.p. 158°C (17 mm); sp. gr. (40°C) 1.19; miscible with benzene, toluene, acetone, carbon tetrachloride, chloroform, ethanol and ethyl acetate; slightly soluble in petroleum ether.
Containers: 5-gal tins; 55-gal drums.
Hazard: Strong irritant to eyes and skin; toxic by inhalation.
Uses: Intermediate for alkyds, plasticizers, insect repellents and rust inhibitors; hardener in epoxy resins.

hexahydropyridine. See piperidine.

hexahydrotoluene. See methylcyclohexane.

hexahydro-1,3,5-trinitro-sym-**triazine.** See cyclonite.

hexahydroxycyclohexane. See inositol.

hexahydroxylene. See dimethylcyclohexane.

hexakis(methoxymethyl)melamine
NC(NR)NC(NR)NC(NR) where $R = (CH_2OCH_3)_2$.
Uses: Crosslinking agent for alkyds, epoxies, cellulosics and vinyls.

n-**hexaldehyde** (caproic aldehyde) $CH_3(CH_2)_4CHO$.
Properties: Colorless liquid. Sharp, aldehyde odor. Sp. gr. 0.8156 (20/20°C); b.p. 128.6°C; vapor pres-

sure 10.5 mm (20°C); flash point 90°F (open cup); wt/gal 6.9 lb (20°C); f.p. −56.3°C. Immiscible with water.
Grades: Technical.
Containers: 5-, 55-gal drums.
Hazard: Flammable, moderate fire risk.
Uses: Organic synthesis of plasticizers, rubber chemicals, dyes, synthetic resins, insecticides.

"Hexalin."[28] Trademark for cyclohexanol (usually shipped with 2.25% methanol as antifreeze).
Hazard: Toxic by ingestion.

hexamethonium chloride $(CH_3)_3NCl(CH_2)_6NCl(CH_3)_3$. Hexamethylenebis(trimethylammonium chloride).
Properties: White, crystalline, hygroscopic powder with faint odor. M.p. 289–292°C (dec); very soluble in water; soluble in alcohol, methanol and n-propanol; insoluble in chloroform and ether. Available commercially as unhydrated form or as dihydrate.
Use: Medicine.

hexamethylbenzene $C_{12}H_{18}$ or $C_6(CH_3)_6$.
Properties: Colorless plates; soluble in alcohol; insoluble in water. B.p. 265°C; m.p. 165.5°C. Combustible.

hexamethyldiaminoisopropanol diiodide. See propiodal.

hexamethyldisilazane (HMDS) $(CH_3)_3SiNHSi(CH_3)_3$.
Properties: Liquid; sp. gr. 0.77 (25°C); refractive index 1.4057 (25°C); b.p. 125°C; flash point 77°F. Soluble in acetone, benzene, ethyl ether, heptane, perchloroethylene. Reactive with methyl alcohol and water.
Purity: 99% min.
Hazard: Flammable, moderate fire risk.
Uses: Chemical intermediate; chromatographic packings.

hexamethylene. See cyclohexane.

hexamethylenediamine (1,6-diaminohexane) $H_2N(CH_2)_6NH_2$.
Properties: Colorless leaflets; m.p. 39–42°C; b.p. 205°C; somewhat soluble in water, ethyl alcohol, and ether. Combustible.
Derivation: (1) Reaction of adipic acid and ammonia (catalytic vapor-phase) to yield adiponitrile, followed by liquid-phase catalytic hydrogenation.
(2) Chlorination of butadiene followed by reaction with sodium cyanide (cuprous chloride catalyst) to 1,4-dicyanobutylene, and hydrogenation.
Hazard: Toxic by ingestion; strong irritant to tissue.
Use: Formation of high polymers, e.g., nylon 66.
Shipping regulations: (Rail) (solution) White label. (Air) (solid or solution) Corrosive label.

hexamethylenediamine carbamate. See "Diak."

hexamethylene diisocyanate $OCN(CH_2)_6NCO$.
Properties: Liquid; sp. gr. 1.04 (25/15.5°C); flash point 284°F. Combustible.
Hazard: Probably toxic.
Use: Chemical intermediate.

1,6-hexamethylene glycol. See 1,6-hexanediol.

hexamethyleneimine $C_6H_{12}NH$ (cyclic).
Properties: Clear, colorless liquid with an ammonia-like odor; b.p. 138°C; f.p. −37°C; sp. gr. 0.8799 (20/4°C).
Use: Intermediate for pharmaceutical, agricultural, and rubber chemicals.
Shipping regulations: (Air) Corrosive label.

hexamethylenetetramine (methenamine; HMTA; aminoform; hexamine; erroneously "Hexamethyleneamine") $(CH_2)_6N_4$. A heterocyclic fused ring structure.

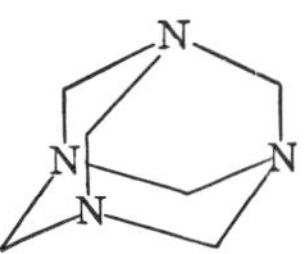

Properties: White crystalline powder, or colorless, lustrous crystals; practically odorless; sp. gr. 1.27 (25°C); soluble in water, alcohol, and chloroform; insoluble in ether; sublimes about 200°C, partly decomposing.
Derivation: Action of ammonia on formaldehyde.
Grades: Technical; N.F. (as methenamine).
Hazard: Moderately toxic; skin irritant. Flammable, dangerous fire risk.
Uses: Catalyst in phenolformaldehyde and resorcinol-formaldehyde resins; ingredient in rubber-to-textile adhesives; protein modifier; organic synthesis; pharmaceuticals; ingredient of high explosive cyclonite (q.v.); fuel tablets.
Shipping regulations: (Rail) Flammable solid, n.o.s., Yellow label. (Air) Flammable Solid label.

hexamethylenetetramine mandelate (methenamine mandelate) $C_6H_{12}N_4 \cdot C_6H_5CHOHCOOH$.
Properties: White crystalline powder with sour taste; practically no odor; m.p. 127–130°C; very soluble in water; pH solutions 4; soluble in alcohol.
Grade: U.S.P. (as methenamine mandelate).
Use: Medicine.

hexamethylenetetramine salicylate (methenamine salicylate) $(CH_2)_6N_4 \cdot C_6H_4OHCOOH$.
Properties: White crystalline powder; pleasant acidulous taste; soluble in alcohol and water.
Use: Medicine.

1,1,2,3,3,5-hexamethylindan methyl ketone $C_{17}H_{24}O$. See "Phantolid."

hexamethylmelamine. See hemel.

hexamethylpararosaniline chloride. See methyl violet.

hexamethylphosphoramide. See hempa.

hexamethylphosphoric triamide. See hempa.

hexamethyltetracosahexaene. See squalene.

hexamethyltetracosane. See squalane.

hexamine. See hexamethylenetetramine.

hexanaphthene. See cyclohexane.

n-**hexane** $CH_3(CH_2)_4CH_3$.
Properties: Colorless, volatile liquid; faint odor; sp. gr. 0.65937 (20/4°C); b.p. 68.742°C; f.p. −95°C; refractive index (n 20/D) 1.37486; flash point −9°F. Autoignition temp. 500°F. Soluble in alcohol, acetone and ether; insoluble in water.
Derivation: By fractional distillation from petroleum (molecular sieve process).
Containers: Bottles; drums; special tank cars.
Grades: 85%, 95%, 99%, spectro, research, and nanograde.
Hazard: Flammable, dangerous fire risk. Moderately toxic by inhalation and ingestion. Tolerance, 500 ppm in air.

Uses: Solvent, especially for vegetable oils; low temperature thermometers; calibrations; polymerization reaction medium; paint diluent; alcohol denaturant.
Shipping regulations: (Rail) Red label. (Air) Flammable Liquid label.

hexanedioic acid. See adipic acid.

1,6-hexanediol (1,6-hexamethylene glycol) $CH_2OH(CH_2)_4CH_2OH$.
Properties: Water-white liquid; m.p. 41–42°C; b.p. 244°C; density 0.9531 g/cc (50°C); miscible with water and isopropyl alcohol; flash point 266°F. Combustible. Low toxicity.
Uses: Solvent; resin intermediate; coupling agent.

hexanedione-2,5. See acetonyl acetone.

1,2,6-hexanetriol $HOCH_2CH(OH)CH_2CH_2CH_2CH_2OH$.
Properties: Water-white liquid; sp. gr. 1.1063; sets to glass below −20°C (f.p. under controlled conditions 32.8°C); b.p. at 5 mm 178°C; flash point 380°F; miscible with water. Combustible. Low toxicity.
Containers: Cans.
Uses: Alkyd and polyester resin intermediate; softener, moistening agent, and solvent.

hexanitrodiphenyl amine (hexil; hexyl; hexite; dipicrylamine) $(NO_2)_3C_6H_2NHC_6H_2(NO_2)_3$.
Properties: Yellow solid; m.p. 238–244°C; decomposes violently at higher temperatures; insoluble in water, and alcohol; soluble in alkalies and warm acetic or nitric acid.
Derivation: Nitration of diphenylamine; also from dinitrochlorobenzene.
Hazard: Explodes on shock or exposure to heat. Dangerous.
Uses: Booster explosive; analysis for potassium.
Shipping regulations: (Rail) Consult regulations. (Air) Dry or wet with less than 10% water. Not acceptable; wet with not less than 10% water, Flammable Solid label.

hexanitrodiphenyl sulfide (dipicryl sulfide) $[(NO_2)_3C_6H_2]_2S$.
Properties: Golden-yellow leaflets; m.p. 234°C; nontoxic. Sparingly soluble in alcohol and ether; more soluble in glacial acetic acid and acetone.
Derivation: Interaction of picryl chloride and sodium thiosulfate in alcohol solution in the presence of magnesium carbonate.
Hazard: Explodes on shock or exposure to heat. Dangerous.
Use: High explosive.
Shipping regulations: (Rail) Not accepted by express. Consult regulations. (Air) Explosives, n.o.s., Not acceptable.

hexanitromannite. See mannitol hexanitrate.

hexanoic acid. Legal label name for caproic acid (q.v.).

1-hexanol. See hexyl alcohol.

2-hexanone. See methyl n-butyl ketone.

3-hexanone (ethyl propyl ketone) $C_2H_5CO(CH_2)_2CH_3$.
Properites: Colorless liquid; b.p. 124°C; sp. gr. 0.813 at 22°C. Combustile.
Hazard: Moderately toxic by ingestion and inhalation; strong irritant. Moderate fire risk.
Uses: Solvent.

hexanoyl chloride $CH_3(CH_2)_4COCl$.
Properties: Colorless liquid, b.p. 151–153°C; refractive index 1.4867 (n 20/D); decomposed by water and alcohol; soluble in ether and chloroform. Combustible.
Use: Chemical intermediate.

hexaphenyldisilane $[(C_6H_5)_3Si]_2$.
Properties: White powder; m.p. 352 °C.
Derivation: Sodium condensation of triphenylchlorosilane.
Use: High-temperature applications.

"Hexaphos."[55] Trademark for a glassy phosphate of high molecular weight having superior water-softening properties.
Uses: Water-softening; boiler-scale control; component of cleansers; laundry mixes, dishwashing compounds, pitch control in pulp industry, prevention of lime soap deposits in textile operations.

hexatriacontane $C_{36}H_{74}$.
Properties: Waxy solid. Combustible. Nontoxic. Sp. gr. 0.797; m.p. 75°C. See also paraffin wax.

"Hex-Cem."[480] Trademark for liquids and pastes of octoates of calcium, cobalt, iron, lead, manganese, nickel and zinc.
Uses: Driers for paint and printing inks.

1-hexene (hexylene) $CH_3CH_2CH_2CH_2CH{:}CH_2$.
Properties: Colorless liquid; sp. gr. 0.6734 (20/4°C); b.p. 63.55°C; f.p. −139.8°C; refractive index (in 20/D) 1.3876; flash point − 15°F. Insoluble in water; soluble in alcohol.
Grades: 95%, 99%; research.
Hazard: Highly flammable, dangerous fire risk; moderately toxic and irritant.
Uses: Synthesis of flavors, perfumes, dyes, resins; polymer modifier.
Shipping regulations. (Rail) Flammable liquid, n.o.s., Red label. (Air) Flammable liquid, n.o.s., Flammable Liquid label.

2-hexene $CH_3CH_2CH_2CH{:}CHCH_3$.
Properties (of mixed cis and trans isomers): Colorless liquid; b.p. 68°C; f.p. −146°C; refractive index (20/D) 1.3948; sp. gr. (60/60°F) 0.686; flash point −5°F. Low toxicity. Insoluble in water; soluble in alcohol.
Grades: 95%; 99%.
Hazard: Highly flammable; dangerous fire risk.
Uses: Chemical intermediate.
Shipping regulations: See 1-hexene.

5-hexene-2-one. See allylacetone.

hexenol (leaf alcohol) $C_6H_{11}OH$.
Properties: Liquid; odor of green leaves. B.p. 156°C; refractive index 1.438; sp. gr. 0.85. Combustible.
Occurrence: Grasses, leaves, herbs, tea, etc.
Uses: Odorant in perfumery.

hexestrol $HOC_6H_4CH(C_2H_5)CH(C_2H_5)C_6H_4OH$. para,-para'-(1,2-Diethylethylene)diphenol. A nonsteroid synthetic estrogen.
Properties: Odorless, white crystalline powder; m.p. 185–188°C. Soluble in ether, acetone, alcohol, and methanol; practically insoluble in water. Sensitive to light.
Derivation: From anethole; by reaction of diacetyl peroxide on para-methoxy-n-propylbenzene.
Use: Medicine.

hexetidine $C_{21}H_{45}N_3$. 5-Amino-1,3-bis(beta-ethylhexyl)-5-methylhexahydropyrimidine.
Properties: Liquid; sp. gr. 0.860–0.875 (25/25°C); b.p. 172–176°C (1 mm); refractive index (n 25/D) 1.460–1.466. Soluble in methanol, benzene, petroleum ether; insoluble in water. Combustible.
Grades: Technical; N.D.
Uses: Fungicide, bactericide, algicide; antistatic agent for synthetics; insect repellent; medicine.

hexil. See hexanitrodiphenyl amine.

hexite. See hexanitrodipheyl amine.

hexobarbital (N-methyl-5-cyclohexenyl-5-methylbarbituric acid) $C_{12}H_{16}N_2O_3$.
Properties: White crystals; m.p. 145–147°C.
Use: Medicine (sedative)
See also barbiturate.

"Hexogen."[230] Trademark for a series of paint driers made with odorless solvents, essentially solutions of metallic salts of 2-ethylhexoic acid. Supplied in a variety of high metal concentrations including calcium 4%, calcium 5%, cobalt 6%, lead 24%, manganese 6%, iron 6%, and zinc 8%.

hexoic acid. See caproic acid.

hexokinase. An enzyme which catalyzes the formation of adenosine diphosphate and hexose-6-phosphate from adenosine triphosphate, and glucose or fructose.
Use: Biochemical research.

hexone. See methyl isobutyl ketone.

hexyl. (1) The straight-chain group C_6H_{13}. (2) Hexanitrodiphenyl amine (q.v.)

hexyl acetate. $CH_3COOC_6H_{13}$.
Properties: Colorless liquid; sweet ester odor; insoluble in water; very soluble in alcohol or ether. B.p. 169.2°C; sp. gr. 0.890. Combustible.
Derivation: From primary and sec-hexyl alcohols.
Uses: Solvent for cellulose esters and other resins; spray base.

sec.-**hexyl acetate.** See methyl amyl acetate.

hexyl alcohol (1-hexanol; amyl carbinol) $CH_3(CH_2)_4CH_2OH$.
Properties: Colorless liquid; sp. gr. 0.8186; f.p. −51.6°C; b.p. 157.2°C; wt/gal 6.8 lb (20°C); refractive index 1.1469 (25°C); flash point 149°F (TOC); autoignition temp. 559°F. Slightly soluble in water; soluble in alcohol and ether. Combustible; low toxicity.
Derivation: (a) By reduction of ethyl caproate; (b) from olefins by the Oxo process.
Grades: Technical (90–99%); purified (99.8%).
Containers: 1-, 5-gal cans; 55-gal drums; tank cars.
Uses: Pharmaceuticals (introduction of hexyl group into hypnotics, antiseptics, perfume esters, etc.); solvent; plasticizer; intermediate for textile and leather finishing agents.

n-**hexylamine** $CH_3(CH_2)_5CH_{24}$.
Properties: Water-white; amine odor; boiling range 126–132°C; f.p. −21°C; sp. gr. 0.767 (20/20°C); refractive index 1.419 (20°C); flash point 85°F (open cup). Slightly soluble in water. Low toxicity.
Hazard: Flammable, moderate fire risk.

n-**hexyl bromide** (1-bromohexane) $CH_3(CH_2)_5Br$.
Properties: Colorless to slightly yellow liquid; sp. gr. 1.165 (20/20°C); b.p. 155.5°C; soluble in alcohol; esters, ethers; insoluble in water.
Grade: 96–98% pure.
Use: Intermediate, for introduction of hexyl group.

n-**hexyl "Carbitol."**[214] Trademark for diethylene glycol monohexyl ether (q.v.).

n-**hexyl "Cellosolve."**[214] Trademark for ethylene glycol monohexyl ether (q.v.).

hexyl cinnamaldehyde $C_6H_{13}C(CHO):CHC_6H_5$.
Properties: Pale yellow liquid; jasmin-like odor, particularly on dilution; sp. gr. 0.953–0.959 (25°C); refractive index 1.5480–1.5520 (20°C). Soluble in most fixed oils and in mineral oil; insoluble in glycerin and in propylene glycol. Nontoxic; combustible.
Grade: F.C.C.
Use: Flavoring agent.

hexylene. See 1-hexene.

hexylene glycol (4-methyl-2,4-pentanediol) $(CH_3)_2COHCH_2CHOHCH_3$. Colorless, nearly odorless liquid; sp. gr. 0.9216 (20/4°C); b.p. 198.3°C; refractive index (n 20/D) 1.4276; flash point (open cup) 215°F; wt/gal 7.69 lb. Miscible with water, hydrocarbons, and fatty acids. Combustible.
Containers: 5-, 55-gal drums; tank cars.
Hazard: Irritating to skin, eyes, and mucous membranes.
Uses: Hydraulic brake fluids; printing inks; coupling agent and penetrant for textiles; fuel and lubricant additive; emulsifying agent; inhibitor of ice formation in carburetor.
See also 1,6-hexanediol.

n-**hexyl ether** $C_6H_{13}OC_6H_{13}$.
Properties: Colorless liquid with characteristic odor. Very slightly soluble in water; sp. gr. 0.7942 (20/20°C); 6.6 lb/gal (20°C); f.p. −43.0°C; viscosity 1.68 cp (20°C); flash point 170°F. Combustible. Autoignition temp. 369°F.
Uses: Extraction processes; manufacture of collodion, photographic film, and smokeless powder.

hexylic acid. See caproic acid.

hexyl mercaptan $C_6H_{13}SH$.
Properties: Colorless liquid; b.p. 149–150°C (768 mm); sp. gr. 0.8450 (20/4°C); refractive index 1.4492 (n 20/D); unpleasant odor. Combustible.
Grades: 95% min purity.
Hazard: Moderately toxic.
Uses: Intermediates; synthetic rubber processing. See also thiol.

hexyl methacrylate $C_6H_{13}OOCC(CH_3):CH_2$. Liquid; sp. gr. 0.88; b.p. 67–85°C (8 mm). Combustible.
Containers: Drums.
Uses: Monomer for plastics, molding powder, solvent coatings, adhesives, oil additives; emulsions for textile, leather, and paper finishing.

para-tert-**hexylphenol** $C_6H_{13}C_6H_4OH$.
Properties: Water-white liquid, faint phenol odor; sp. gr. 0.986 (20/20°C); boiling range 155–165°C; refractive index 1.520 (20°C); flash point 285°F. Combustible.
Uses: Organic synthesis; preparation of resinous condensation products.

hexylresorcinol (1,3-dihydroxy-4-hexylbenzene) $C_6H_{13}C_6H_3(OH)_2$.
Properties: White to yellowish-white needle-shaped crystals with a faint, fatty odor and sharp, astringent taste which produces a sensation of numbness when placed on the tongue; slightly soluble in water; freely soluble in alcohol, glycerin, and vegetable oils; m.p. 62–67°C; b.p. 178–180°C (8 mm).
Grade: N.F.
Hazard: Irritant to respiratory tract and skin. Alcohol solutions are vesicant.
Containers: Bottles, fiber drums.
Use: Medicine.

hexyltrichlorosilane $C_6H_{13}SiCl_3$.
Properties: Colorless liquid with a sharp penetrating odor; fumes strongly in moist air. Combustible.

Hazard: Toxic by ingestion and inhalation. Strong irritant.
Use: Chemical intermediate.
Shipping regulations: (Rail) White label. (Air) Corrosive label. Not acceptable on passenger planes.

1-hexyne (butyl acetylene) $C_4H_9C{:}CH$.
Properties: Water-white liquid with characteristic odor. Sp. gr. 0.7152 (20/4°C); refractive index 1.3990 (20°C); b.p. 71.4°C; f.p. −132°C.
Hazard: Probably flammable.

hexynol (1-hexyn-3-ol) $CH_3(CH_2)_2CHOHC{:}CH$.
Properties: Light yellow liquid with strong odor; b.p. 142°C; sp. gr. (20/20°C) 0.882; slightly soluble in water, and miscible with most hydrocarbons, chlorinated solvents, ketones, alcohols and glycols.
Containers: 7-, 35- and 380-lb drums.
Hazard: Probably flammable. Toxic by ingestion and inhalation; absorbed by skin.
Uses: Corrosion inhibitor against mineral acids; high-temperature oil well-acidizing inhibitor.

Hf Symbol for hafnium.

HFPO. See hexafluoropropylene epoxide.

Hg Symbol for mercury (Latin: hydrargyrum).

HGF. See glucagon.

"HGI."[62] Trademark for lindane (q.v.)

"HGR."[182] Trademark for an ion exchange resin used in water treating and chemical process applications; strong acid cation exchange resin, 10% divinylbenzene crosslinked.

HHDN. Abbreviation for hexachlorohexahydrodimethanonaphthalene. See aldrin.

"Hiblak."[133] Trademark for a series of aqueous carbon black dispersions designed for darkening concrete and mortar.

"Hi Calcium Phosphate."[1] Trademark for a crystalline monocalcium phosphate with high calcium content.
Use: Ingredient of calcium-enriched flour.

"Hi-Cap."[28] Trademark for a series of low cost ammonia dynamites used in mining and quarrying softer ores and minerals.

"Hi-Carb."[275] Trademark for agglomerates of carbon black and carbon of high purity for carburizing metals; used for conducting electricity or for any use for which hard carbon may be used.

"Hi-D."[319] Trademark for ammonium nitrate fertilizer. White, roughly cube-like high-density granular material screened to optimum particle size, with special moisture-proof coating.

hiding power. See opacity.

"Hi-Eff."[425] Trademark for a series of polyester stationary phases for use in gas chromatography. Examples are "Hi-Eff" 4B, butane-1, 4-diol succinate; "Hi-Eff" 1AP, a pretested diethylene glycol adipate.

"Hi-fax."[266] Trademark for plastic molding powders of high-density polyethylene. Offered in three densities, natural or colored.
Uses: Blown containers and industrial items, wire insulation, cable jacketing, conduit, monofilament, sheet, injection-molded items, and shape extrusions.

hi-flash naphtha. See naphtha (2a).

"Hi-Flo."[241] Trademark for triple superphosphates containing at least 46% available phosphorus pentoxide (P_2O_5). Various forms available.

"Hi-Flow."[275] Trademark for carbon blacks used for printing inks.

"Hifos"[79] Trademark for a limed FF wood rosin.
Uses: Linoleum cements; printing inks; dark gloss oils; adhesives; paint and varnish.

high explosive. See explosive, high.

"High Speed."[28] Trademark for tinning flux based on zinc chloride and ammonium chloride.

high vacuum distillation. See molecular distillation.

"Hi-Jell."[471] Trademark for a Wyoming sodium bentonite; recommended where extra high gelling and bonding strength are required.

hindered isocyanate. See isocyanate generator.

hindrance. See steric hindrance.

"Hi-N-Dri."[319] Trademark for low moisture nitrogen solution used as, or in production of, soil fertilizers.

"Hio-Dine."[536] Trademark for a source of hypohalous iodine for swimming pool water disinfection in dry powder form.

"Hiperco."[308] Trademark for an alloy of cobalt and iron with varying percentages of cobalt, 27%, 35%, 50%, with the highest magnetic saturation of any known commercial alloy. Permeability above 100 oersteds is far superior to iron and the electrical sheet steels.
Uses: Core material for motors, generators, and transformers where minimum weight and size are prime requisities.

"Hipernik."[308] Trademark for an alloy of approximately equal proportions of iron and nickel.
Uses: Audio and instrument transformers, relays, magnetic bridges, shields for electronic tubes; saturable reactors; magnetic amplifiers.

"Hipernom."[308] Trademark for a nickel-base alloy containing molybdenum and iron which exhibits extremely high initial and maximum permeability at low magnetizing forces, with minimum hysteresis loss. Chief application is for magnetic shielding.

"Hipersil."[308] Trademark for an oriented silicon-iron alloy in rolled sheet form for use as cores for transformers and other inductive devices. The alloy sheet has a preferred grain orientation with the direction of the easiest magnetization in the rolling direction.

"Hippuran."[329] Trademark for brand of iodohippurate sodium, a water-soluble x-ray contrast medium.

hippuric acid. (benzaminoacetic acid; benzoylaminoacetic acid; benzoylglycocoll; benzoylglycin) $C_6H_5CONHCH_2COOH$.
Properties: Colorless crystals; sp. gr. 1.371 (20°C); m.p. 188°C; decomposes on further heating; soluble in hot water, alcohol or ether.
Uses: Organic synthesis and medicine.

"Hi-Ratio Silicate."[244] Trademark for an anhydrous homogeneous combination of caustic alkali and sodium silicate. Used for metal cleaning, laundering, and other applications.

His Abbreviation for histidine.

"Hi-Sil."[177] Trademark for a group of hydrated, amorphous silicas used as reinforcing pigments in elastomers as fillers and brightening agents in paper and paints, and as flow-conditioners.

"Hi-Sol."[457] Trademark for a series of aromatic solvents. Flash points vary from 31 to 250°F. Used as lacquer diluents, in synthetic polymer and enamel

formulations, in printing inks, as solvents for insecticides, and as general industrial solvents.
Hazard: Flammable, dangerous fire risk if flash point is less than 100° F.

histaminase. An enzyme occurring in the animal digestive system; it converts histidine to histamine.

histamine (4-aminoethylglyoxaline; 4-(2-aminoethyl) imidazole; 4-imidazole ethylamine $NH_2CH_2CH_2C_3H_3N_2$.

```
HC════CCH₂CH₂NH₂
|      |
N      NH
 \\   /
   C
   |
   H
```

Properties: White crystals; m.p. 83–84°C; b.p. 209–210°C (18 mm); soluble in water; slightly soluble in alcohol. A product of the degradation of hisitidine, histamine occurs in animal and human body tissues and is liberted by injury to the tissue, or whenever a protein is decomposed by putrefactive bacteria. See also antihistamine.
Uses (as hydrochloride or phosphate): Medicine.
Hazard: A strong allergen.

histidine (alpha-amino-beta-imidazolepropionic acid) $HOOCCH(NH_2)CH_2C_3H_3N_2$. An amino acid essential for rats. It is found naturally in the L(−) form.

```
H—C════CCH₂CH(NH₂)COOH
  |     |
  N     NH
   \\  /
    CH
```

Properties: colorless crystals; soluble in water; insoluble in alcohol and ether; shows optical activity.
DL-histidine, m.p. 285–6°C with decomposition.
D(+)-histidine, m.p. 287–8°C.
L(−)-histidine, m.p. 277°C with decomposition.
Available commercially as L(+)-histidine hydrochloride, and as the free base.
Derivation: Hydrolysis of protein; organic synthesis.
Uses: Medicine; feed additive; biochemical research; dietary supplement.

histochemistry. A branch of biochemistry devoted to the study of the chemical composition and structure of animal and plant tissues. It involves the use of microscopic, x-ray diffraction, and radioactive tracer techniques in examining the cellular composition and structure of bones, blood, muscle, and other animal and vegatable tissues. It is also applied to study of the action of herbicides, defoliants, etc. See also cytochemistry.

"Hitec."[28] Trademark for a eutectic mixture comprised of sodium nitrite, sodium nitrate, and potassium nitrate; used as a heat transfer medium for both heating and cooling operations in the range of 300–1,000° F, such as maintaining reactor temperature, high temperature distillation, and preheating of reactants.
Containers: 150-, 400-lb fiber drums.

"Hi-Tri."[233] Trademark for a grade of trichloroethylene for specific industrial applications.

"Hi-White"[285] Trademark for a group of hydrous aluminum silicates (sedimentary kaolins) from Georgia.

HMDS. Abbreviation for hexamethyldisilazane.

HMF black. Abbreviation for high-modulus furnace black. See carbon black.

HMM. Abbreviation for hexamethylmelamine. See hemel.

HMPA. Abbreviation for hexamethylphosphoramide. See hempa.

HMRB. Abbreviation for Hazardous Materials Regulation Board.

HMTA. Abbreviation for hexamethylenetetramine.

HNM. Abbreviation for hexanitromannite. See mannitol hexanitrate.

Ho Symbol for holmium.

"Hodag."[512] Trademark for a series of nonionic, surface-active chemicals, including sorbitan and polyoxyethylene sorbitan esters; polyhydric alcohol, glycol, polyglycol and glycerol esters; polyoxyethylene ethers; alkanolamine condensates. Some are FDA-cleared for food applications.
Uses: Emulsifiers, antifoam agents, solubilizers, lubricants.

hog. A large power-driven blade, either straight or serrated, with a short stroke, used for cutting and disintegrating bulky solids of irregular shape, such as pine stumps, bales of crude rubber, and the like.

Hofmann, August Wilhelm (1818–1892). German organic chemist who studied under Liebig. While professor of chemistry at the Royal College of Chemistry in London he did original research on coal-tar derivatives which later led him into a study of organic dyes. Perkin, who first synthesized the dye mauveine in England, was a student of Hofmann. When the latter returned to Germany he continued his work in the field of dyes, which became the basis of German leadership in synthetic dye manufacture which continued until World War I.

Hofmann's reaction. Reaction used for preparation of a primary amine from an amide by treatment with a halogen (bromine, usually) and caustic soda. The resulting amine has one less carbon atom than the amide used.

hole. In semiconductor terminology, a hole is an energy deficit in a crystal lattice due to (1) electrons ejected from unsatisfied covalent bonds at sites where an atom is missing, i.e., a vacancy, or (2) to electrons supplied by atoms of impurities in the crystal, e.g., arsenic or boron. The free electrons from these sources move through the crystal, leaving positively charged energy deficits, which are considered to move as they become alternately filled and vacated by electrons, creating a flow of positive electricity. See also semiconductor.

holmium Ho Metallic element of atomic number 67; group IIIB of the periodic table; one of the rare-earth elements of the yttrium subgroup. Atomic weight 164.9303; valence 3. No stable isotopes. See rare-earth metals.
Properties: Crystalline solid with metallic luster; sp. gr. 8.803; m.p. 1470°C; b.p. 2720°C. Reacts slowly with water; soluble in dilute acids. Has one of the highest nuclear moments of any rare earth. Important magnetic and electrical properties. Low toxicity.
Occurrence: In gadolinite and monazite.

Derivation: Reduction of the fluoride by calcium.
Grades: Lumps; ingots; bulk sponge; powder. Highest purity is nuclear grade, 99.9+%.
Uses: Getter in vacuum tubes; research in electrochemistry; spectroscopy.

holmium chloride $HoCl_3$. Light yellow solid; m.p. 718°C; b.p. 1500°C. Soluble in water.
Hazard: Probably toxic.

holmium fluoride HoF_3. Light yellow solid; m.p. 1143°C; b.p. greater than 2200°C. Insoluble in water.
Hazard: Probably toxic.

holmium oxide (holmia) Ho_2O_3.
Properties: Light yellow solid; slightly hygroscopic.
Grades: 98–99%.
Containers: Glass bottles; fiber drums.
Uses: Refractories; special catalyst.

"Holocaine" Hydrochloride.[162] Trademark for phenacaine hydrochloride (q.v.).

holocellulose. The entire water-insoluble carbohydrate fraction of wood (60 to 80%). The balance is lignin.

holopulping. A method for making paper pulp without use of sulfur compounds which may eventually replace the kraft process (sodium sulfate). More selective delignification of the wood fibers is obtained by alkaline oxidation of extremely thin (.03 inch) wood chips at low temperatures and pressures, followed by solubilization of the lignin fraction. Holopulping has the following advantage over the kraft process:

(a) A 65 to 80% carbohydrate yield compared to 45 to 50% for kraft.

(b) Holopulp may be used for a dense paper such as glassine or for a bulky board. Its use in tissue and printing grades offers improved strength.

(c) Low temperatures and atmospheric pressure. (Kraft pulping is carried out at 170°C and under pressure.) Readily adaptable to continuous operation and automatic control.

(d) Air pollution is greatly reduced because the organic materials are burned and few odorous compounds are formed. Stream pollution is minimized by countercurrent washing. The remaining calcium sludge waste is easily disposed of without harmful effects.

(e) Comparable costs.

holothurin. Steroid glucoside (saponin) having antibiotic properties; extracted from the sea cucumber. It is reported to have suppressed growth of tumors in mice.

"Holzon."[205] Trademark for a chlorinated rubber coating.

homatropine $C_{16}H_{21}NO_3$. An alkaloid.
Properties: White crystals. Slightly soluble in water; m.p. 95.5°C.
Derivation: Condensation of tropine and mandelic acid.
Hazard: Toxic by ingestion and inhalation.
Use: Medicine (usually in the form of its salts).

homo-. A prefix meaning the same or similar; usually designating a homolog of a compound, differing in formula from the latter by an increase of CH_2. See homologous series.

homocyclic. A ring compound containing only one kind of atom in the ring structure, e.g., benzene. See also heterocyclic.

homogeneous (Latin, "the same kind"). This term, in its strict sense, describes the chemical constitution of a compound or element. A compound (q.v.) is homogeneous, since it is composed of one and only one group of atoms represented by a formula (q.v.). For example, pure water is homogeneous, as it contains no other substance than is indicated by its formula H_2O. Homogeneity is a characteristic property of compounds and elements (collectively called substances (q.v.), as opposed to mixtures. The term is often loosely used to describe a mixture or solution comprised of two or more compounds or elements that are uniformly dispersed in each other. Actually, no solution or mixture can be homogeneous; the situation is more accurately described by the phrase "uniformly dispersed." Thus so-called "homogenized" milk is not truly homogeneous; it is a mixture in which the fat particles have been mechanically reduced to a size that permits uniform dispersion and consequent stability. See also mixture; compound; heterogeneous; substance.

homogeneous catalysis. See catalysis, homogeneous.

homogeneous reaction. A chemical reaction in which the reacting substances are in the same phase of matter, i.e., solid, liquid, or gaseous. See also catalysis, homogeneous.

homogenization. A mechanical process for reducing the size of the fat particles of an emulsion (usually milk) to uniform size, thus creating a colloidal system that is unaffected by gravity. The original diameter of the fat particles (from 6 to 10 microns) is reduced to from 1 to 2 microns, with an increase in total surface area of from four to six times. This is done by passing the milk through a homogenizer (or colloid mill), a machine having small channels, under a pressure of from 2000 to 2500 psi at a speed of about 700 feet a second. This operation not only brings about a permanently stable system, but also changes the properties of the milk in respect to taste, color, and the chemical nature of the protective coating on the fat particles. It also increases its sensitivity to light and its tendency to foam. The forces involved are shear, impingement, distention, and cavitation. See also homogeneous.

homologous series. A series of organic compounds in which each successive member has one more CH_2 group in its molecule than the next preceding member. For instance CH_3OH (methanol), C_2H_5OH (ethanol), C_3H_7OH (propanol), C_4H_9OH (butanol), etc., form a homologous series.

homomenthyl salicylate (3,3,5-trimethylcyclohexyl salicylate) $(CH_3)_3C_6H_8OOCC_6H_4OH$. A homolog of menthyl salicylate.
Properties: Light yellow almost odorless oil, neutral and nonirritating to the skin. Absorbs ultratiolet radiation in sunlight, (about 2940 to 3200 Å). Insoluble in water; soluble in alcohol, chloroform and ether.
Uses: Ultraviolet filter for antisunburn creams.

homomorphs. Molecules similar in size and shape. They need have no other characteristics in common. Many properties of several homomorphs can be predicted by knowing properties of one.

homophthalic acid $C_6H_4(CH_2COOH)COOH$. Light tan powder, used as an intermediate.

homopolymer. A natural or synthetic high polymer derived from a single monomer; an example of a natural homopolymer is rubber hydrocarbon, whose monomer is isoprene; a synthetic homopolymer is typified by polychloroprene or polystyrene, whose monomers are, respectively, chloroprene and styrene. See also polyblend.

ortho-**homosalicylic acid.** See cresotic acid.

4-homosulfanilamide hydrochloride. See mafenide hydrochloride.

homoveratric acid (3,4-dimethoxyphenylacetic acid) $(CH_3O)_2C_6H_3CH_2COOH$. Crystals; m.p. 94–101°C; very slightly soluble in water; soluble in most organic solvents.

homoveratrylamine (3,4-dimethoxyphenylethylamine) $(CH_3O)_2C_6H_3(CH_2)_2NH_2$. Colorless to pale yellow liquid with slight vanilla odor; sp. gr. 1.09 (25/25°C); solidifies 15°C; b.p. 295°C (with decomposition); refractive index 1.5442–1.5452 (25°C).

honey. A unique mixture of a number of low molecular weight sugars (except sucrose) but including invert sugar. It is considerably sweeter than glucose. It has been used as a food and sweetener since the beginning of civilization. It also has applications in medicine and tobacco curing.

Hooke's law. When a load is applied to any elastic body so that the body is deformed or strained, then the resulting stress (the tendency of the body to resume its normal condition) is proportional to the strain. Stress is measured in units of force per unit area, strain is the extent of the deformation. For example when a bar of metal is subjected to a stretching load, the extent of the increase in length of the bar is directly proportional to the force per unit area, i.e., to the stretching load or stress. In general Hooke's law applies only up to a certain stress called the yield strength.

hopcalite. A mixture of oxides of copper, cobalt, manganese and silver, used in gas masks as a catalyst converting carbon monoxide to carbon dioxide.
Hazard: Not safe when nitroparaffin vapors are present.

horizon. See soil.

"Hormodin."[123] Trademark for a formulation of indolebutyric acid.

hormone, animal. An organic compound secreted by an endocrine (ductless) gland, whose products are released into the circulating fluid. Hormones regulate such physiological processes as metabolism, growth, reproduction, molting, pigmentation, and osmotic balance. They are sometimes called "chemical messengers." Hormones produced by one species usually show similar action in other species. They vary widely in chemical nature; some are steroids (q.v.)—estrogen, progesterone, cortisone—while others are amino acids (thyroxine), polypeptides (vasopressin), low molecular weight proteins (insulin) and conjugated proteins. Amino acid and steroid hormones have been isolated, and many (including insulin) have been synthesized and are manufactured for medical purposes. Other types are made directly from the endocrine organs of animals.

hormone, plant. See plant growth regulator.

"Horne's Dry Lead."[50] Trademark for a laboratory chemical used for clarification in sugar analysis. Adopted by the Association of Official Agricultural Chemists. Available in 1-lb, 5-lb, and 10-lb bottles and 25-lb cartons.

"Hornstone."[205] Trademark for zinc fluosilicate concrete hardener.

hot. Slang for highly radioactive, e.g., hot laboratory.

hot-melt composition. See adhesive, hot-melt; sealant; asphalt; bitumen.

Houdry process. (1) Catalytic cracking. The vaporized charging stock is passed through a fixed bed of catalyst (activated hydrosilicate of alumina). Cracking and regeneration of catalyst are carried out alternately in two reaction chambers; when the catalyst is fouled, the vapor is diverted to a second chamber while the first is being cleaned by burning off the deposited coke with air. The cracked products are separated in conventional equipment. The process requires complicated timing mechanisms to control the alternate cracking and regeneration periods.

(2) Butadiene production. A one-step process the essential feature of which is adiabatic dehydrogenation of butane, in which the heat of reaction during the on-stream cycle is supplied by burning off coke during the regeneration cycle. The butane is preheated and passed through a fixed bed pelleted alumina-chromia catalyst. Several reactors are used in a series of cycles allowing for cleaning and regeneration of catalyst.

HPA. Abbreviation for hydroxypropyl acrylate (q.v.).

HPC black. Abbreviation for hard-processing channel black. See carbon black.

"HPC."[266] Trademark for solutions of ammonium nitrate, ammonia, and urea; available in various concentrations and proportions from 20 to 49% nitrogen.

HS. Abbreviation for hydroxylamine sulfate.

"HSM."[65] Trademark for hemisulfur mustard seed in medicine.

"H.T."[58] Trademark for monocalcium phosphate. Free-flowing beads.

"HT-44."[212] Trademark for an extremely heat-stable liquefying enzyme (an amylase) from a bacterial source high in alpha-amylase activity.
Uses: Textiles, starch adhesives, paper, brewing, industrial grain alcohol.

"HTH."[84] Trademark for a high-test calcium hypochlorite product commercially available as a stable, water-soluble material in both granular and tablet form, containing a minimum of 70% available chlorine as calcium hypochlorite.
Uses: Bleaching; sterilizing; oxidizing.

"HTH-15."[84] Trademark for an all-purpose germicide, disinfectant and stain remover. Contains 15% of available chlorine and yields sodium hypochlorite solutions directly when added to water.
Uses: Dairy and poultry farm sanitation; for sterilizing glasses and food utensils, and general sanitation.

HTST. Abbreviation for high-temperature short-time; refers to processes such as pasteurization, sterilization, etc.

HTU. Abbreviation for height of a transfer unit: the height of a distillation column or fractionating tower in which unit separation is achieved by transfer from liquid to vapor or vice versa, of the materials being separated. Unit separation is defined by the differential equation that takes into account the varying concentrations along the column. HTU is also applied to extraction and other countercurrent separation processes.

Huber's reagent. An aqueous solution of ammonium molybdate and potassium ferrocyanide used for detecting free mineral acid. With the exception of boric acid and arsenic trioxide, free mineral acids produce a reddish-brown precipitate, or a turbidity with the reagent.

Hubl's reagent.
(a) 50 grams iodine dissolved in 1 liter of 95% alcohol.
(b) 60 grams mercuric chloride dissolved in 1 liter of alcohol.
(c) Make up an iodine monochloride solution from (a) and (b). Add in excess to a known weight of the fat or oil dissolved in chloroform. The excess of iodine chloride can be estimated by the potassium iodide and thiosulfate method. By running a blank test, the amount of iodide absorbed can be estimated.
Use: Determination of iodine values of oils and fats.

humectant. A substance having affinity for water, with stabilizing action on the water content of a material. A humectant keeps within a narrow range the moisture content caused by humidity fluctuations. Example: glycerol.
Uses: Tobacco; baked products; dentifrices.

humic acid. The brown, polymeric constituent of soil humus; it is soluble in bases, but insoluble in mineral acids and alcohols. It is not a well-defined compound, but a mixture of polymers containing aromatic and heterocyclic structures, carboxyl groups, and nitrogen. An excellent chelating agent, important in the exchange of cations in soils. It is a natural stream pollutant and is thought to be capable of triggering the "red tide" phenomenon due to microorganisms in sea water. Detectable to 0.1 ppm in water.

humidity, absolute. The pounds of water vapor per pound of dry air in an air-water vapor mixture.

humidity indicator. A cobalt salt (e.g., cobaltous chloride) that changes color as the humidity of the environment changes. Cobaltous compounds are pink when hydrated and greenish-blue when anhydrous.

humidity, relative. The percentage relation between the actual amount of water vapor in a given volume of air at a definite temperature and the maximum amount of water vapor that would be present if the air were saturated with water vapor at that temperature.

humus. The organic component of soils, containing humic acid (q.v.), fulvic acid, and humin. It is formed by the decay of leaves, wood, and other vegetable matter. It is used as a top soil additive in horticulture, golf courses and truck gardens, and is available in 100-lb paper bags and bulk shipments. See soil (1).

hyaluronic acid. A polymer of acetylglycosamine, $C_8H_{15}NO_6$, and glucuronic acid, occurring as alternate units, with a high molecular weight. Found in vitreous humor (of the eye), synovial fluid, pathologic joints, group A and C hemolytic streptococci and skin. It appears to bind water in the interstitial spaces, forming a gel-like substance which holds the cells together. Its solutions are highly viscous. The polymeric structure is broken down by the enzyme hyaluronidase.

"Hyamine."[23] Trademark for quaternary-ammonium-type bactericides, algaecides, and fungicides, supplied as water-soluble crystals or aqueous solutions. Non-irritating; low toxicity.

Hyatt, John Wesley (1837–1920). Hyatt is generally credited as being the father of the plastics industry. In 1869 he and his brother patented a mixture of cellulose nitrate and camphor which could be molded and hardened. Its first commercial use was for billard balls. The trade-mark "Celluloid" was the first ever applied to a synthetic plastic product; its flammability hazard limits its use

"Hybase."[544] Trademark for overbased petroleum sulfonates used as additives in motor oils to reduce corrosion; overbased sulfonates have base numbers ranging from 65 to 400.

"Hycar."[119] Trademark for various types of synthetic elastomers.
Nitrile Rubbers. Copolymers of butadiene and acrylonitrile.
Polyacrylic Rubbers. Polymers of an acrylic acid ester. Useful in applications where oil and solvent-resistance at high temperatures (to 425°F) is required.
Uses: Oil and gasoline hose, packings, automotive transmission gaskets, and conveyor belts.
Styrene Rubber, Copolymer of styrene and butadiene. Oil soluble.
Latexes. Nitrile and acrylic polymers, available as water emulsions with total solids ranging from 40 to 52%.
Uses: Adhesives, abrasion-resistant coatings.

"Hycryl."[22] Trademark for a series of thickening agents. Available as
A-1000—modified ammonium polyacrylate (water solution).
A-2000—Latex dispersion which can be converted to ammonium polyacrylate solution through the addition of ammonia.
Uses: Antisettling stabilizer for water dispersions of pigments; compounding rubber latices; thickener for alcoholic nylon coating solutions.

"Hydan."[28] Trademark for methionine hydroxy analogue calcium 90%.
Use: A source of methionine (an essential amino acid) for poultry, dog and livestock feeds.

hydantoin (glycolylurea) $NHCONHCOCH_2$.
Properties: White, odorless solid, crystallizing in needles. Slightly soluble in water, ether; soluble in alcohols and solutions of alkali hydroxides. M.p. 220°C.
Uses: Intermediate in the synthesis of pharmaceuticals, textile lubricants, and certain high polymers, including epoxy resins.

hydrabamine penicillin V (hydrabamine phenoxymethylpenicillin) $2C_{16}H_{18}N_2O_5S \cdot C_{42}H_{64}N_2$.
Properties: A water-insoluble mixture of crystalline phenoxymethylpenicillin salts consisting chiefly of the salt of N,N′-bis(dehydroabietyl)ethylenediamine, with smaller amounts of the salts of the dihydro and tetrahydro derivatives.
Use: Medicine.

"Hydraid."[108] Trademark for a family of water soluble organic polymers, some cationic, anionic and nonionic of various molecular weight and coagulation properties.
Uses: Paper and pulp mill retention, drainage and clarification aids.

"Hydralase."[309] Trademark for a series of fungal enzymes used principally in the food industries for conversion of starches or dextrins to dextrose and of proteins to amino acids.

hydralazine hydrochloride (1-hydrazinophthalazine hydrochloride) $C_8H_5N_2NHNH_2 \cdot HCl$.
Properties: White, odorless, crystalline powder. M.p. 270–280°C (dec); very slightly soluble in ether and alcohol; soluble in water; pH (2% solution) 3.5–4.5.
Grade: N.F.
Use: Medicine (antihypertensive agent).

"Hydral" 700 Series.[226] Trademark for several grades of hydrated aluminum oxide, $Al_2O_3 \cdot 3H_2O$ or $Al(OH)_3$, of extremely fine, uniform particle size. Fluffy, snow-white powders used as fillers in rubber, paper, plastics, adhesives, polishes, inks, paints, cosmetics.

Hydrane process. A method for direct hydrogenation of coal. See hydrogenolysis; gasification.

"Hydraphthal."[28] Trademark for a combination solvent and detergent for textile scouring.

"Hydrar" Process.[416] Proprietary process for the catalytic hydrogenation of benzene to cyclohexane, or higher aromatics to their corresponding cycloparaffins. The hydrogenation of benzene is virtually stoichiometric. The purity of the cyclohexane is a function of the purity of the benzene feed.

hydrase. See hydrolase.

"Hydrasperse."[285] Trademark for chemically treated hydrous aluminum silicate (kaolin).

hydrate. See hydration.

hydrated aluminum oxide. See alumina trihydrate.

hydrated silica. See silicic acid.

"Hydratex."[285] Trademark for a group of hydrous aluminum silicates (kaolin).

hydration. (1) The reaction of molecules of water with a substance, in which the H—OH bond is not split. The products of hydration are called hydrates, i.e.,

$$CuSO_4 + 5H_2O \longrightarrow CuSO_4 \cdot 5H_2O$$

A given compound often forms more than one hydrate; the hydration of sodium sulfate can give $Na_2SO_4 \cdot 10H_2O$ (decahydrate), $Na_2SO_4 \cdot 7H_2O$ (heptahydrate), and $Na_2SO_4 \cdot H_2O$ (monohydrate). In formulas of hydrates, the addition of the water molecules is conventionally indicated by a centered dot. The water is usually split off by heat, yielding the anhydrous compound. See also water of crystallization; gas hydrate.
(2) Less specifically, the term hydration is also applied to the sorption of water molecules by other substances, of which cellulose is a good example. In the paper industry it is used to describe the combination of water with wood pulp in the beater, as a result of which fiber-to-fiber adhesion is increased by hydrogen bonding, thus increasing the strength of the finished sheet.

hydraulic cement. See cement, hydraulic.

hydraulic fluid. A liquid or mixture of liquids designed to transfer pressure from one point to another in a system on the basis of Pascal's Law, i.e., pressure on a confined liquid is transmitted equally in all directions. For industrial use, such fluids are based on paraffinic and cycloparaffinic petroleum fractions, usually with added antioxidant and viscosity index improver. Flame-resistant types include additives such as phosphate esters or emulsions of water and ethylene glycol. The brake fluids used in autos are composed of (1) a lubricant (polypropylene glycol of 1000–2000 molecular weight, a castor oil derivative, or a synthetic polymeric mixture of monobutyl ethers of oxyethylene and oxypropylene glycols); (2) a solvent blend (mixtures of glycol ethers); and (3) additives for corrosion resistance, buffering, etc. Boiling points range from 375 to 550°F. The composition and performance characteristics are specified by the Society of Automotive Engineers.

hydraulic lime. See lime, hydraulic.

hydrazine (hydrazine base; hydrazine, anhydrous; diamine) H_2NNH_2.
Properties: Colorless, fuming, hygroscopic liquid; m.p. 2.0°C; b.p. 113.5°C: m.p. 1.4°C; sp. gr. 1.004 (25/4 °C); wt/gal 8.38 lbs; flash point (open cup) 126°F. Miscible with water and alcohol; insoluble in chloroform and ether. Strong reducing agent and diacidic, but weak base. Autoignition temp. 518°F. Combustion of hydrazine is highly exothermic, yielding 148.6 kcal/mole; nitrogen and water are products.
Derivation: (a) Sodium hydroxide, chlorine, and ammonia react in aqueous solution to form dilute solution of hydrazine (Raschig process). This is concentrated by distillation to the monohydrate, which is converted to anhydrous hydrazine by azeotropic distillation with aniline. (b) Reaction of ammonia, hydrogen peroxide and methyl ethyl ketone in presence of an amide and a phosphate forms an intermediate that hydrolyzes to yield hydrazine (75% yield reported). Not yet in commercial use.
Grades: To 99% pure.
Containers: Carboys; steel or aluminum drums.
Hazard: Highly toxic by ingestion, inhalation, and skin absorption. Strong irritant to skin and eyes. Severe explosion hazard when exposed to heat or by reaction with oxidizing materials. Tolerance, 1 ppm in air.
Uses: Rocket propellant; agricultural chemicals (maleic hydrazide); drugs (antibacterials, antihypertension): polymerization catalyst; blowing agent; short-stopping agent; Spandex fibers; antioxidants (petroleum, detergents); plating metals on glass and plastics; fuel cells; solder fluxes; scavenger for gases; explosives; photographic developers; corrosion inhibitors; oil-well drilling in soils containing kaolinite; buoyancy agent for undersea salvage; diving equipment.
Shipping regulations: (Rail) White label. (Air) Corrosive label. Anhydrous, hydrate, or water solutions with 50% or less of water not accepted on passenger planes. Solutions with more than 50% water, Corrosive label.

hydrazine acid tartrate (hydrazine tartrate) $N_2H_4 \cdot C_4H_6O_6$.
Properties: Colorless crystals; m.p. 182–183°C; soluble in water.
Hazard: See hydrazine.

hydrazine dihydrochloride $N_2H_4 \cdot 2HCl$.
Properties: Colorless crystals; sp. gr. 1.42; m.p. 198°C (loses HCl); b.p. 200°C (dec). Soluble in water; slightly soluble in alcohol.

hydrazine hydrate (diamide hydrate). $H_2NNH_2 \cdot H_2O$.
Properties: Colorless, fuming liquid; f.p. −51.7°C; b.p. 119.4°C; sp. gr. 1.032; wt/gal 8.61 lb; flash point (open cup) 163°F. Miscible with water and alcohol; insoluble in chloroform and ether. Strong reducing agent; weak base. Combustible.
Derivation: See hydrazine.
Hazard: See hydrazine.
Uses: Chemical intermediate; azodicarbonamide.
Shipping regulations: See hydrazine.

hydrazine monobromide $N_2H_4 \cdot HBr$.
Properties: White, crystalline flakes; m.p. 81–87°C; decomposes at about 190°C; soluble in water and lower alcohols; insoluble in most organic solvents.
Grade: 95%.
Use: Soldering flux.

hydrazine monochloride $N_2H_4 \cdot HCl$.
Properties: White crystalline flakes; m.p. 87–92°C, decomposes at about 240°C; soluble in water (37g/100g H_2O at 20°C); somewhat soluble in lower alcohols; insoluble in most organic solvents.

hydrazine nitrate $N_2H_4NO_3$.
Hazard: Severe explosion risk.
Shipping regulations: (Rail) Consult regulations. (Air) Not acceptable.

hydrazine perchlorate $N_2H_4 \cdot HClO_4 \cdot \frac{1}{2}H_2O$.
Properties: Solid; sp. gr. 1.939; m.p. 137°C; b.p. 145°C (dev). Decomposes in water; soluble in alcohol; insoluble in ether, benzene, chloroform, and carbon disulfide.
Hazard: Severe explosion risk.
Use: Rocket propellant.
Shipping regulations: (Rail) Consult regulations. (Air) Not acceptable.

hydrazine sulfate (diamine sulfate; diamidogen sulfate) $NH_2NH_2 \cdot H_2SO_4$.
Properties: White crystalline powder; very soluble in hot water; soluble in 1 part in 33 cold water; insoluble in alcohol; stable in storage; strong reducing agent. Sp. gr. 1.37; m.p. 85°C.
Uses: Manufacture of chemicals; condensation reactions; catalyst in making acetate fibers. Analysis of minerals, slags and fluxes; determination of arsenic in metals; separation of polonium from tellurium; fungicide; germicidie; reported useful in cancer treatment.

hydrazinophthalazine hydrochloride. See hydralazine hydrochloride.

hydrazobenzene (N,N′-diphenylhydrazine) $C_6H_5NHNHC_6H_5$.
Properties: M.p. (min) 126°C; soluble in alcohol; nearly insoluble in water.
Grade: 95%.
Use: Synthesis.

hydrazoic acid (hydrogen azide) HN_3.
Properties: Colorless, volatile liquid; soluble in water. Obnoxious odor. F.p. −80°C; b.p. 37°C.
Derivation: Reaction of hydrazine and nitrous acid, or of nitrous oxide and sodium amide (with heat).
Hazard: Toxic; strong irritant to eyes and mucous membranes. Dangerous explosion risk when shocked or heated.
Shipping regulations: (Rail) Consult regulations. (Air) Explosives, n.o.s., Not acceptable.

hydrazone. An exotic fuel formed by the action of hydrazine or one of its derivatives on a compound containing the carbonyl group CO.
Hazard and Shipping regulations: See hydrazine.

"Hydrea."[412] Trademark for hydroxy urea (q.v.), used in cancer therapy.

"Hydrex."[189] Trademark for hydrogenated fatty acids and glycerides.
Uses: Buffering compounds; rubber compounding; candles; lubricating grease; crayons; emulsifiers; paper coatings; plasticizers; cosmetics; metallic stearates; textile finishes, polishes.

hydride. An inorganic compound of hydrogen with another element. Some are ionic and others are covalent. Hydrides may be either binary or complex; the latter are transition-metal complexes, e.g., carbonyl hydrides and cyclopentadienyl hydrides. Most common are hydrides of sodium, lithium, aluminum, boron, etc.
Hazard: Highly toxic and irritant; flammable, dangerous fire risks; react violently with water and oxidizing agents.
See lithium aluminum hydride; sodium borohydride.

"Hydrholac."[23] Trademark for plasticized nitrocellulose lacquer emulsions, including clear finishes, binders and colors. Produce flexible, lacquer-type, cleanable leather finishes from aqueous systems.
Uses: Finishes on glove, garment, handbag and shoe leather.

"Hydricin."[202] Trademark for a castor-derived base oil for industrial functional fluids and SAE hydraulic brake fluids.

hydridene. See indan.

"Hydrin."[119] Trademark for a series of synthetic rubbers based on epichlorohydrin.
Uses: Industrial and automotive molded goods.

hydriodic acid HI
Properties: Colorless or pale yellow liquid (an aqueous solution of hydrogen iodide, which is a gas at room temperature). A constant-boiling solution is formed of sp. gr. 1.7 containing 57% hydrogen iodide. Hydriodic acid is a strong acid and an active reducing agent. For anhydrous hydrogen iodide: sp. gr. 4.3737; f.p. −51.3°C; b.p. −35.6°C.
Derivation: (a) By passing hydrogen with iodine vapor over warm platinum sponge which acts as a catalyzer, and absorption in water. (b) By the action of iodine on a solution of hydrogen sulfide.
Grades: Technical, 47%; N.F., diluted, 10%
Hazard: Toxic; strong irritant to eyes and skin.
Uses: Preparation of iodine salts; organic preparations; analytical reagent; disinfectant; pharmaceuticals.
Shipping regulations: (Rail) White label. (Air) Corrosive label.

hydroabietyl alcohol. See dihydroabietyl alcohol.

"Hydro-Benzo #200-K."[457] Trademark for an aliphatic petroleum naphtha used as a rubber solvent. Distillation range (ASTM D850) 140–195°F.
Hazard: Flammable, dangerous fire risk.

hydrobiotite. A natural ore of magnesium, iron, and aluminum silicate occurring in Montana. Source of verxite (q.v.).

hydroboration. The reaction of diboranes with alkenes (olefins) to form trialkylboron compounds, used in organic synthesis. Prostaglandins have been synthesized by this method.

hydrobromic acid. Hydrogen bromide in aqueous solution. See also hydrogen bromide, anhydrous.
Properties: Colorless or faintly yellow liquid consisting of an aqueous solution of hydrogen bromide, which is a gas at room temperature. F.p. −87°C; b.p. −67°C. Soluble in water and alcohol. A constant-boiling solution is formed, of sp. gr. 1.49, containing 48% hydrogen bromide. Hydrobromic acid is a strong acid and sensitive to light. Sp. gr. (48% solution) 1.488 (20/4C). Noncombustible.
Derivation: By dissolving the gas in water, or by distilling from a mixture of sodium bromide and 50% sulfuric acid.
Grades: Technical 40%; medicinal 48%; 62%.
Containers: Glass bottles; carboys.
Hazard: Highly toxic; strong irritant to eyes and skin.
Uses: Medicine; analytical chemistry; solvent for ore

minerals; manufacture of inorganic and some alkyl bromides; alkylation catalyst.
Shipping regulations: (Rail) White label. (Air) (exceeding 49% strength), Not acceptable; (not exceeding 49% strength), Corrosive label.

hydrocarbon. An organic compound consisting exclusively of the elements carbon and hydrogen. Derived principally from petroleum, coal tar, and vegetable sources. Following is a resume of the principal types.
I. Aliphatic (straight-chain)
(1) Paraffins (alkanes): generic formula C_nH_{2n+2}. Saturated, single bonds only.
(2) Olefins: generic formula C_nH_{2n}.
(a) alkenes: unsaturated (one double bond)
(b) alkadienes: unsaturated (two double bonds) (butadiene)
(3) Acetylenes: generic formula C_nH_{2n-2}. Unsaturated (triple bond).
(4) Acyclic terpenes. Unsaturated. (Polymers of isoprene, C_5H_8).
Note: Some aliphatic compounds have branched chains in which the subchain also contains carbon atoms (isobutane); both chains are essentially straight.
II. Cyclic (closed ring).
(1) Alicyclic: three or more carbon atoms in a ring structure, with properties similar to those of aliphatics.
(a) Cycloparaffins (naphthenes): saturated compounds often having a boat or chair structure, e.g., cyclohexane, cyclopentane.
(b) Cycloolefins: unsaturated, two or more double bonds, e.g., cyclopentadiene (2), cyclooctatetraene (4).
(c) Cycloacetylenes (cyclynes): unsaturated (triple bond).
(2) Aromatic: unsaturated; hexagonal ring structure (three double bonds); single rings and double or triple fused rings.
(a) benzene group (1 ring)
(b) naphthalene group (2 rings)
(c) anthracene group (3 rings)
(3) Cyclic terpenes: monocyclic (dipentene) dicyclic (pinene).

hydrocarbon, halogenated. A hydrocarbon in which one or more of the hydrogen atoms has been replaced by fluorine, chlorine, bromine, or iodine. Examples: carbon tetrachloride, chlorobenzene, chloroform, trifluoromethane. This greatly increases the anesthetic and narcotic action of aliphatic hydrocarbons. Many halogenated hydrocarbons are highly toxic; some may detonate on contact with barium. A number of the chlorinated types are used as insecticides. See also fluorocarbon; chlorofluorocarbon.

hydrocellulose. See cellulose, hydrated.

hydrochloric acid. Hydrogen chloride in aqueous solution. See also hydrogen chloride.
Properties: Colorless or slightly yellow, fuming, pungent liquid. F.p. −115°C; b.p. −85°C; flash point, none. A constant-boiling acid containing 20% hydrogen chloride is formed. Hydrochloric acid is a strong, highly corrosive acid. The commercial "concentrated" or fuming acid contains 38% of hydrogen chloride and has a sp. gr. 1.19. Soluble in water, alcohol and benzene. Noncombustible.
Derivation: (a) By-product of the chlorination of benzene and other hydrocarbons; by furnace combustion of chlorine and hydrogen.
Grades: U.S.P. (35–38%); N.F. diluted (10%); technical (usually 18, 20, 22, 23° Bé, corresponding to approx. 28, 31, 35, 37% HCl); F.C.C.
Containers: Glass bottles; carboys; rubber-lined steel drums; rubber-linked tank cars.
Hazard: Highly toxic by ingestion and inhalation; strong irritant to eyes and skin. Safety data sheet available from Manufacturing Chemists Assn., Washington, D.C.
Uses: Acidizing (activation) of petroleum wells; chemical intermediate; ore reduction; food processing (corn syrup, sodium glutamate); pickling and metal cleaning; industrial acidizing; general cleaning, e.g., of membranes in desalination plants; alcohol denaturant.
Shiping regulations: (Rail) White label. (Air) Corrosive label.

hydrocinnamic acid (3-phenylpropionic acid) $C_6H_5CH_2CH_2COOH$.
Properties: Crystals with hyacinth-rose odor. M.p. 46°C.
Derivation: Reduction of cinnamic acid with sodium amalgam.
Uses: Fixative for perfumes; flavoring.

hydrocinnamic alcohol. See phenylpropyl alcohol.

hydrocinnamic aldehyde. See phenylpropyl aldehyde.

hydrocinnamyl acetate. See phenylpropyl acetate.

hydrocodone bitartrate (USAN). See dihydrocodeinone bitartrate.

hydrocolloid. A hydrophilic colloidal material used largely in food products as emulsifying, thickening and gelling agents. They readily absorb water, thus increasing viscosity and imparting smoothness and body texture to the product, even in concentrations of less than 1%. Natural types are plant exudates (gum arabic), seaweed extracts (agar), plant seed gums or mucilages (guar gum), cereal gums (starches), fermentation gums (dextran), and animal products (gelatin). Semisynthetic types are modified celluloses and modified starches. Completely synthetic types are also available, e.g., polyvinylpyrrolidone. Most are carbohydrate polymers, but a few, such as gelatin and casein, are proteins.

hydrocortisone (17-hydroxycorticosterone; cortisol; hydrocortisone alcohol) $C_{21}H_{30}O_5$. An adrenal cortical steroid hormone.
Properties: White, odorless, crystalline powder; sensitive to light; bitter taste; m.p. 212–220°C with some decomposition. Freely soluble in dioxane and methanol; insoluble in ether and water; soluble in alcohol and acetone.
Derivation: Isolation from extracts of adrenal glands; synthesis from other steroids.
Grade: U.S.P.
Use: Medicine (anti-inflammatory agent). Also used as the acetate and sodium succinate salts.
See also cortisone.

"Hydrocortone."[123] Trademark for hydrocortisone (q.v.).

hydrocracking. The cracking of petroleum or its products in the presence of hydrogen. Special catalysts are used, for example, platinum on a solid base of mixed silica and alumina, or zinc chloride. See also hydrogenation; hydrogenolysis; hydroforming.

hydrocyanic acid (prussic acid; hydrogen cyanide; formonitrile) HCN.
Properties: Water-white liquid at temperatures below

26.5°C; faint odor of bitter almonds; usual commercial material is 96–99.5% pure; sp. gr. (liquid) 0.688 (20/4°C), (gas) 0.938 g/l; b.p. 25.6°C; freezing point −13.3°C; flash point 0°F; soluble in water. The solution is weakly acidic; sensitive to light. When not absolutely pure or stabilized, hydrogen cyanide polymerizes spontaneously with explosive violence. Miscible in all proportions with water, alcohol; soluble in ether. Autoignition temp. 1000°F.

Derivation: (a) By catalytically reacting ammonia and air with methane or natural gas. (b) By recovery from coke oven gases. (c) From bituminous coal and ammonia at 1250 C. HCN occurs naturally in some plants (almond, oleander).

Grades: Technical (96–98%); 2, 5 and 10% solutions. All grades usually contain a stabilizer, usually 0.05% phosphoric acid.

Containers: Bottles; steel cylinders; tank cars.

Hazard: Highly toxic by ingestion, inhalation, and skin absorption. Flammable, dangerous fire risk. Explosive limits in air, 6 to 41%. Tolerance, 10 ppm in air; absorbed by skin. Safety data sheet available from Manufacturing Chemists Assn., Washington, D.C.

Uses: Manufacture of acrylonitrile, acrylates, adiponitrile, cyanide salts, dyes; chelates.

Shipping regulations: (Rail) (liquid or liquefied) Poison Gas label. Not accepted by express; (unstabilized) Not accepted; (solutions) Poison label. (Air) (solutions not exceeding 5%) Poison label. Not acceptable on passenger planes; all other forms, Not acceptable.

"Hydrodarco."[89] Trademark for activated carbons used in municipal and industrial water purification.

hydrodealkylation (HDA). A type of hydrogenation used in petroleum refining in which heat and pressure in the presence of hydrogen are used to remove methyl or larger alkyl groups from hydrocarbon molecules, or to change the position of such groups. The process is used to upgrade products of low value, such as heavy reformate fractions, naphthenic crudes, or recycle stocks from catalytic cracking. Also toluene and pyrolysis gasoline (q.v.) are converted to benzene, and methyl naphthalenes to naphthalene by this process. See also transalkylation.

hydrodistillation (steam distillation). Removal of essential oils from plant components (flowers, leaves, bark, etc.) by the use of high-temperature steam. The process is used chiefly in the perfume and fragrance industry.

hydrofining. A petroleum refining process in which a limited amount of hydrogenation converts the sulfur and nitrogen in a petroleum fraction to forms in which they can be easily removed. Hydrofining is generally a separate treatment prior to more extensive hydrogenation. The usual catalysts are oxides of cobalt and molybdenum. Desulfurization, ultrafining, and catfining have a similar meaning.

hydroflumethiazide (trifluoromethylhydrothiazide) $C_8H_8F_3N_3O_4S_2$.

Properties: White, crystalline, odorless solid; m.p. 260–275°C. Insoluble in water and acid; soluble in dilute alkali but unstable in alkaline solutions.

Grade: N.D.

Use: Medicine.

hydrofluoric acid (hydrogen fluoride in aqueous solution). See also hydrogen fluoride, anhydrous.

Properties: Colorless, fuming, mobile, liquid. Will attack glass and any silica-containing material.

Derivation: Hydrogen fluoride gas is distilled from a mixture of calicum fluoride (fluorspar) and sulfuric acid. The gas is absorbed in water.

Grades: C.P.; technical; various strengths to 70%.

Containers: Polyethylene bottles; polyethylene-lined steel drums; steel tank cars.

Hazard: Highly corrosive to skin and mucous membranes; highly toxic by ingestion and inhalation.

Uses: Aluminum production; fluorocarbons; pickling stainless steel; etching glass; acidizing oil wells; fluorides; gasoline production (alkylation); processing uranium.

Shipping regulations: (Rail) White label. (Air) Corrosive label.

hydrofluorosilicic acid. See fluosilicic acid.

hydrofluosilicic acid. See fluosilicic acid.

"Hydrofol."[221] Trademark for a line of fatty acids and hydrogenated glycerides, including some in food grades. Available up to ton lots.

hydroforming. The use of hydrogen in the presence of heat, pressure and catalysts (usually platinum) to convert petroleum hydrocarbons to molecular structures giving high-octane gasoline for automobiles and airplanes. Dehydrogenation of naphthenic hydrocarbons to form aromatics, and isomerization to more highly branched molecules are typical of hydroforming reactions. Catforming and other words ending in "-forming" are often used with similar meaning.

hydrofuramide (furfuramide) $OC_4H_3CH(NCHC_4H_3O)_2$.

Properties: Light brown to white powder. M.p. 117°C; boils about 250°C with decomposition. Insoluble in cold water; soluble in alcohol and ether.

Derivation: Treatment of furfural with ammonia.

Uses: Rubber accelerator; hardening agent for resins; fungicides.

hydrogasification. Production of gaseous or liquid fuels by catalytic addition of hydrogen to coal. See also gasification.

hydrogen H. Element of atomic number 1. Group IA of Periodic Table. Atomic weight 1.008; valence 1. Isotopes: deuterium (H^2), tritium (H^3). Discovered by Cavendish in 1766; named by Lavoisier in 1783 (water-maker).

Properties: A diatomic gas; density 0.0899 g/l; sp. gr. 0.0694 (air = 1.0); specific volume 193 cu ft/lb (70°F); frz. pt. −259°C; b.p. −252°C. Autoignition temp. 1075°F. Very slightly soluble in water, alcohol and ether. Noncorrosive. Can exist in crystalline state at from 4 to 1 degree Kelvin. Classed as an asphyxiant gas.

Occurrence: Chiefly in combined form (water, hydrocarbons, and other organic compounds); traces in earth's atmosphere. Unlimited quantities in sun and stars. It is the most abundant element in the universe.

Derivation: (1) Reaction of steam with natural gas (steam reforming) and subsequent purification; (2) partial oxidation of hydrocarbons to carbon monoxide, and interaction of carbon monoxide and steam; (4) dissociation of ammonia; (5) thermal or catalytic decomposition of hydrocarbon gases (see "Hypro" process); (6) catalytic reforming of petroleum; (7) reaction of iron and steam; (8) catalytic reaction of methanol and steam; (9) electrolysis of water; (10) Thermochemical decomposition of water is under consideration as a large-scale source of hydrogen; it is said to involve addition to water of calcium, mercury, and bromine and temperatures between 500 and 800°C (Euratom plan). The chief industrial method is (1); but (7) is being developed in connection with hydrogasification, and (9) has long-range future possibilities by use of a special plastic electrolyte originally developed for fuel cells.

Method of purification: By scrubbing with various solutions (see, especially, the Girbitol absorption process). For very pure hydrogen, by diffusion through palladium.

Grades: Technical; pure, from an electrolytic grade of 99.8% to ultra-pure, with less than 10 ppm impurities. See also para-hydrogen.

Containers: Steel cylinders; tank cars (in cylinders); pipeline.

Hazard: Highly flammable and explosive; dangerous when exposed to heat or flame. Explosive limits in air 4 to 75% by volume.

Uses (Gas): Production of ammonia and methanol; hydrocracking, hydroforming and hydrofining of petroleum; hydrogenation of vegetable oils; hydrogenolysis of coal; reducing agent for organic synthesis and metallic ores; reducing atmospheres to prevent oxidation; as oxyhydrogen flame for high temperatures; atomic-hydrogen welding; instrument-carrying balloons; making hydrochloric and hydrobromic acids; production of high-purity metals.

(Liquid): Coolant and propellant; fuel for nuclear rocket engines for hypersonic transport (mach 6); missile fuel; cryogenic research.

Shipping regulations: (gas) (Rail) Red Gas label; (Air) Flammable Gas label. Not accepted on passenger planes. (liquid) (Rail) Red Gas label; not accepted by express. (Air) Not accepted. For other carriers, consult authorities.

Note: In view of the energy-releasing capacity of hydrogen, especially its isotope tritium (H^3), which occurs in thermonuclear reactions, hydrogen is regarded by some authorities as the ultimate energy source. Splitting of the water molecule by thermochemical or nuclear reactor technology is considered to be an active future possibility for producing hydrogen in high volume for use as an energy source. The electrolytic method is too inefficient for this purpose. Research on development of a controlled hydrogen fusion reaction is in progress, but positive results are unlikely in this century. The use of hydrogen as a transportation fuel for naval ships and aircraft is under active investigation.

ortho-**hydrogen.** See para-hydrogen.

para-**hydrogen**. Type of molecular hydrogen preferred for rocket fuels. Molecular hydrogen (H_2) exists in two varieties, ortho and para, named according to their nuclear spin types. Ortho-hydrogen molecules have a parallel spin; para- an antiparallel spin. By cooling to liquid air temperature and use of a ferric oxide gel catalyst, the normal equilibrium of 3 ortho- to 1 para- is displaced and para-hydrogen may be isolated. It is being produced with less than 5 ppm impurities.

hydrogenation. Any reaction of hydrogen with an organic compound. It may occur either as direct addition of hydrogen to the double bonds of unsaturated molecules, resulting in a saturated product, or it may cause rupture of the bonds of organic compounds, with subsequent reaction of hydrogen with the molecular fragments. Examples of the first type (called addition hydrogenation) are the conversion of aromatics to cycloparaffins and the hydrogenation of unsaturated vegetable oils to solid fats by addition of hydrogen to their double bonds. Examples of the second type (called hydrogenolysis) are hydrocracking of petroleum and hydrogenolysis of coal to hydrocarbon fuels. See also hydrogenolysis; hydrocracking; hydroforming.

hydrogen azide NH_3. See hydrazoic acid.

hydrogen bond. An attractive force, or bridge (q.v.) occurring in polar compounds such as water, in which a hydrogen atom of one molecule is attracted to two unshared electrons of another. The hydrogen atom is the positive end of one polar molecule and forms a linkage with the electronegative end of another such molecule. In the formula below, the hydrogen atom in the center is the "bridge."

$$\begin{array}{c} H:\ddot{\underset{..}{O}}:H:\ddot{\underset{..}{O}}:H:\ddot{\underset{..}{O}}: \\ \quad H \quad\; H \quad\; H \end{array}$$

Hydrogen bonds are only one-tenth to one-thirtieth as strong as covalent bonds (q.v.), but they have pronounced effects on the properties of substances in which they occur, especially as regards melting and boiling points and crystal structure. They are found in compounds containing such strongly electronegative atoms as nitrogen, oxygen and fluorine. They play an important part in the bonding of cellulosic compounds, e.g., in the paper industry, and occur also in many complex structures of biochemical importance, e.g., adenine-uracil linkage in DNA.

hydrogen bromide. See hydrobromic acid; hydrogen bromide, anhydrous.

hydrogen bromide, anhydrous HBr.

Properties: Colorless gas; sp. gr. 2.71 (referred to air); f.p. −86°C; b.p. −66.4°C; specific volume 4.8 cu ft/lb (70°F, 1 atm). Soluble in water and alcohol. Nonflammable.

Derivation: (a) By passing hydrogen with bromine vapor over warm platinum sponge, which acts as a catalyst. (b) As a byproduct in the bromination of organic compounds. The gas is liquefied under a pressure of 350 psi at 25°C.

Grades: Up to 99.8% min purity.

Containers: Cylinders.

Hazard: Toxic by inhalation; strong irritant to eyes and skin. Tolerance 3 ppm in air.

Uses: Organic synthesis; makes bromides by direct reaction with alcohols; intermediates for barbiturate manufacture; acts as intermediate in the manufacture of synthetic hormones; alkylation catalyst.

Shipping regulations: (Rail) Green label. (Air) Nonflammable Gas label. Not acceptable on passenger planes.

See also hydrobromic acid.

hydrogen chloride HCl.

Properties: Colorless, fuming gas with a suffocating odor; sp. gr. 1.268 (referred to air); f.p. −114°C; b.p. −85°C; specific volume 10.9 cu ft/lb (70°F, 1 atm). Very soluble in water; soluble in alcohol and ether. Nonflammable.

Derivation: The large-scale commercial processes are designed to produce hydrochloric acid (q.v.). Anhydrous hydrogen chloride may be made by (a) a fractional distillation process under pressure, or (b) independent of the acid processes, by passing hot burner gases (source of hydrogen) over anhydrous calcium chloride.

Containers: Cylinders.

Hazard: Toxic by inhalation; strong irritant to eyes and skin. Tolerance, 5 ppm in air. Safety data sheet available from Manufacturing Chemists Assn., Washington, D.C.
Uses: Production of vinyl chloride from acetylene and alkyl chlorides from olefins; hydrochlorination (see rubber hydrochloride), polymerization, isomerization, alkylation, and nitration reactions.
Shipping regulations: (Rail) Green label. (Air) Nonflammable Gas label. Not accepted on passenger planes.
See also hydrochloric acid.

hydrogen cyanide. See hydrocyanic acid.

hydrogen dioxide. See hydrogen peroxide.

hydrogen electrode. See electrode, hydrogen.

hydrogen fluoride. See hydrofluoric acid; hydrogen fluoride, anhydrous.

hydrogen fluoride, anhydrous HF.
Properties: Colorless, fuming, mobile, liquid, or colorless gas, very soluble in water. The liquid and gas consist of associated molecules; the vapor density corresponds to HF only at high temperatures. F.p. −83°C; b.p. 19.5°C; sp. gr. liquid 0.988 (14°C); sp. vol. 17 cu ft/lb (70°F, 1 atm). Nonflammable.
Derivation: Distillation from the reaction product of calcium fluoride and sulfuric acid; also from fluosilicic acid.
Grade: To 99.9% min purity.
Containers: Cylinders; tank cars.
Hazard: Highly toxic by ingestion and inhalation; strong irritant to eyes, skin and mucous membranes. Tolerance, 3 ppm in air.
Uses: Catalyst in alkylation, isomerization, condensation, dehydration, and polymerization reactions; fluorinating agent in organic and inorganic reactions; production of fluorine and aluminum fluoride; additive in liquid rocket propellants; refining of uranium.
Shipping regulations: (Rail) White label. (Air) Corrosive label. Not acceptable on passenger planes.
See also hydrofluoric acid.

hydrogen hexafluorosilicate. See fluosilicic acid.

hydrogen iodide, anhydrous HI. See also hydriodic acid.
Hazard: Toxic; strong irritant.
Shipping regulations: (Air) Nonflammable Gas label. Not accepted on passenger planes.

hydrogen ion concentration. See pH.

hydrogenolysis (destructive hydrogenation). A type of hydrogenation reaction in which molecular cleavage of an organic compound occurs, with addition of hydrogen to each portion. An important application is hydrocracking (hydrogenative splitting) of large organic molecules, with formation of fragments that react with hydrogen by use of catalysts and high temperatures. Hydrogenolysis of coal to gaseous and liquid fuels was used in Germany in the 1940's; a similar method (Hydrane process) is under development in the U.S. The German process used pulverized coal made into a paste with heavy oil and a metallic catalyst; the mixture plus the necessary hydrogen was subjected to from 300 to 700 atm at about 500°C. The coal was converted into heavy oil, distillable oil, gasoline, and hydrocarbon gases. Large quantities of hydrogen are necessary. See also gasification; hydrogenation.

hydrogen peroxide H_2O_2 (molecular formula); H—O—O—H (structural formula).
Properties (pure anhydrous): Density, solid, 1.71 grams per cc; density, liquid, 1.450 grams per cc at 20°C; viscosity, liquid, 1.245 centipoises; surface tension, 80.4 dynes per cm at 20°C; f.p., −0.41°C; b.p. 150.2°C. Soluble in water and alcohol. (Solutions): Pure hydrogen peroxide solutions, completely free from contamination, are highly stable; a low percentage of an inhibitor such as acetanilide or sodium stannate, is usually added to counteract the catalytic effect of traces of impurities, such as iron, copper, and other heavy metals. A relatively stable sample of hydrogen peroxide typically decomposes at the rate of about 0.5% per year at room temperature.
Derivation: (a) autoxidation of an alkyl anthrahydroquinone, such as the 2-ethyl derivative, in a cyclic continuous process in which the quinone formed in the oxidation step is reduced to the starting material by hydrogen in the presence of a supported palladium catalyst; (b) by electrolytic processes in which aqueous sulfuric acid or acidic ammonium bisulfate is converted electrolytically to the peroxydisulfate which is then hydrolyzed to form H_2O_2; (c) by autoxidation of isopropyl alcohol. Method (a) is most widely used.
Grades: U.S.P. (3%); technical (3, 6, 27.5, 30, 35, 50 and 90%); F.C.C. Most common commercial strengths are 27.5, 35, 50, and 70%.
Containers: Amber glass bottles; carboys; aluminum drums; non-returnable polyethylene drums with fiber and steel overpacks; tank trucks and tank cars.
Hazard: Concentrated solutions are highly toxic and strong irritants. Dangerous fire and explosion risk. Strong oxidizing agent. Tolerance, 1 ppm in air. Safety data sheet available from Manufacturing Chemists Assn., Washington, D.C. for conc. over 27.5%.
Uses: Bleaching and deodorizing of textiles, wood pulp, hair, fur, etc; source of organic and inorganic peroxides; pulp and paper industry; plasticizers; rocket fuel; foam rubber; manufacture of glycerol; antichlor; dyeing; electroplating; antiseptic; laboratory reagent; epoxidation; hydroxylation; oxidation and reduction; viscosity control for starch and cellulose derivatives; refining and cleaning metals; bleaching and oxidizing agent in foods; neutralizing agent in wine distillation; seed disinfectant; substitute for chlorine in water and sewage treatment.
Shipping regulations: (Rail) (water solutions containing over 8% H_2O_2 by weight) White label. (Air) solutions containing not more than 8% H_2O_2 by weight, not restricted; solutions containing more than 8% but not more than 40% H_2O_2 by weight, Corrosive label; solutions containing more than 40% H_2O_2 by weight, not acceptable. (Solid) Organic peroxide label.

hydrogen phosphide. See phosphine.

hydrogen, phosphoretted. See phosphine.

hydrogen selenide H_2Se.
Properties: Colorless gas; soluble in water, carbon disulfide, phosgene; b.p. −42°C; f.p. −64°C; sp. gr. 2.00.
Grade: 98% pure.
Containers: Cylinders.
Hazard: Highly toxic by inhalation. Strong irritant to skin; damaging to lungs and liver. Dangerous fire and explosion risk; reacts violently with oxidizing materials. Tolerance, 0.05 ppm in air.
Uses: Preparation of metallic selenides and organoselenium compounds; in doping gas mixtures for preparation of semiconductor materials containing controlled amounts of significant impurities.
Shipping regulations: (Air) Not acceptable.

hydrogen slush. A mixture of solid and liquid hydrogen at the hydrogen triple point, 13.8°K and 1.02 psia. It is denser and less hazardous than liquid hydrogen.

hydrogen sulfide (sulfuretted hydrogen) H_2S.
Properties: Colorless gas; offensive odor; sweetish taste. Soluble in water and alcohol; sp. gr. 1.1895 referred to air; f.p. −83.8°C; b.p. −60.2°C; sp. vol. 11.23 cu ft/lb (70°F, 1 atm); autoignition temp. 500°F.
Derivation: (a) By the action of dilute sulfuric acid on a sulfide, usually iron sulfide; (b) by direct union of hydrogen and sulfur vapor at a definite temperature and pressure; (c) as a by-product of petroleum refining.
Containers: Steel cylinders; as liquid at 350–400 psi in 60-ton tank cars or 13-ton trailers (specially designed equipment is required for bulk shipment).
Grades: Technical 98.5%; purified 99.5% min.; C.P.
Hazard: Toxic by inhalation; strong irritant to eyes and mucous membranes. Highly flammable, dangerous fire risk. Explosive limits in air 4.3 to 46%. Tolerance, 10 ppm in air. Safety data sheet available from Manufacturing Chemists Assn., Washington, D.C.
Uses: Purificatoin of hydrochloric and sulfuric acid; precipitating sulfides of metals; reagent; manufacture of elementary sulfur.
Shipping regulations: (Rail) Red Gas label. (Air) Flammable Gas label. Not acceptable on passenger planes.

hydrogen tellurate. See telluric acid.

hydrogen telluride H_2Te. Colorless gas or yellow needles; sp. gr. (liq.) 2.57 (−20°C); f.p. −49°C; b.p. −2°C. Soluble in water but unstable; soluble in alcohol and alkalies.
Hazard: See hydrogen selenide.
Shipping regulations: Not listed.

hydrol. See tetramethyldiaminobenzhydrol.

hydrolase (hydrase). An enzyme which catalyzes the removal of water from the substrate. See enzyme.

"Hydrolin."[84] Trademark for an ammonium nitrate base blasting agent which requires specially constructed primers for detonation.
Use: For seismic prospecting at sea.
Hazard: See ammonium nitrate.

hydrolube. A water-glycol base noncombustible hydraulic fluid.

"Hydrolux" D.[42] Trademark for a clear straw-colored liquid, completely miscible in water. A complex organic, stabilized reducing agent for vat color dyeing and printing.

hydrolysis. A chemical reaction in which water reacts with another substance to form two or more new substances. This involves ionization of the water molecule as well as splitting of the compound hydrolyzed, e.g., $CH_3COOC_2H_5 + H \cdot OH \longrightarrow CH_3COOH + C_2H_5OH$. Examples are conversion of starch to glucose by water in the presence of suitable catalysts; conversion of sucrose (cane sugar) to glucose and fructose by reaction with water, in the presence of an enzyme or acid catalyst; conversion of natural fats into fatty acids and glycerin by reaction with water in one process of soap manufacture; and reaction of the ions of a dissolved salt to form various products, such as acids, complex ions, etc.

"Hydromagma."[123] Trademark for magnesium hydroxide (q.v.).

"Hydron."[307] Trademark for sulfide blues derived from carbazole.

hydronium ion. An ion (H_3O^+) formed by the transfer of a proton (hydrogen nucleus) from one molecule of H_2O to another; a companion ion (OH^-) is also formed; the reaction is $2H_2O \longrightarrow H_3O^+ + OH^-$. Formation of such ions is statistically rare, resulting from the interaction of water molecules in a ratio of 1 to 556 million.

hydroperoxide. An organic peroxide having the generalized formula ROOH. An example is ethyl hydroperoxide (C_2H_5OOH). Methyl and ethyl hydroperoxides are unstable, and thus are strong oxidizing agents and explosion hazards; those of higher molecular weight are more stable. Hydroperoxides can be derived by oxidation of saturated hydrocarbons, or by alkylating hydrogen peroxide in a strongly acidic environment; they are used as polymerization initiators.

hydrophilic. Having a strong tendency to bind or absorb water, which results in swelling and formation of reversible gels. This property is characteristic of carbohydrates, such as algin, vegetable gums, pectins, and starches, and of complex proteins such as gelatin and collagen.

hydrophobic. Antagonistic to water; incapable of dissolving in water. This property is characteristic of all oils, fats, waxes and many resins, as well as of finely divided powders like carbon black and magnesium carbonate.

hydroponics. See nutrient solution.

"Hydro-Pruf."[300] Trademark for a silicone water repellent for fabrics. Applied with a catalyst at high curing temperatures.

hydroquinol. See hydroquinone.

hydroquinone (quinol; hydroquinol; para-dihydroxybenzene) $C_6H_4(OH)_2$.
Properties: White crystals; soluble in water, alcohol and ether. Sp. gr. 1.330; m.p. 170°C; b.p. 285°C. Flash point 329°F; combustible; autoignition temp. 960°F.
Derivation: Aniline is oxidized to quinone by manganese dioxide and is then reduced to hydroquinone.
Grades: Technical; photographic.
Containers: Glass bottles; multiwall paper sacks; fiber drums.
Hazard: Toxic by ingestion and inhalation; irritant. Tolerance, 2 mg per cubic meter of air.
Uses: Photographic developer (except color film); dye intermediate; medicine; antioxidant; inhibitor; stabilizer in paints and varnishes, motor fuels and oils; antioxidant for fats and oils; inhibitor of polymerization.

hydroquinone benzyl ether. See para-benzyloxyphenol.

hydroquinone dibenzyl ether
$C_6H_5CH_2OC_6H_4OCH_2C_6H_5$.
Properties: Tan powder; m.p. 119°C (min); purity 90% (min). Insoluble in water; soluble in acetone, benzene, and chlorobenzene. Combustible. Low toxicity.
Uses: Solvent; perfumes, soap, plastics and pharmaceuticals.

hydroquinone di-n-butyl ether (1,4-dibutoxybenzene) $C_6H_4[O(CH_2)_3CH_3]_2$.
Properties: White flakes with no appreciable odor; melting point 45–46°C; b.p. 124°C (1.3 mm), 158°C (15.0 mm); insoluible in water; soluble in benzene, acetone, ethyl acetate, and alcohol. Combustible. Low toxicity.

hydroquinone diethyl ether (1,4-diethoxybenzene) $C_6H_4(OC_2H_5)_2$.
Properties: White, granular solid with anise-like odor; m.p. 71–72°C; b.p. 246°C; neither boiling caustic nor acid solution cause any hydrolysis; absorbs ultraviolet light; insoluble in water; soluble in benzene, acetone, ethyl acetate, and alcohol. Combustible. Low toxicity.

hydroquinone, di(beta-**hydroxyethyl) ether** (para-di-[2-hydroxyethoxy]benzene) $C_6H_4(OC_2H_4OH)_2$.
Properties: White solid. M.p. 99°C; b.p. 185–200°C (0.3 mm). Slightly soluble in water and most organic solvents; miscible with water at 80°C. Combustible. Low toxicity.
Containers: Lined steel drums.
Uses: Preparation of polyester, polyolefins, polyurethanes and hard waxy resins; organic synthesis.

hydroquinone dimethyl ether (1,4-dimethoxybenzene; DMB; dimethyl hydroquinone) $C_6H_4(OCH_3)_2$.
Properties: White flakes with sweet clover odor; b.p. 213°C; m.p. 56°C; density 1.0293 g/ml (65°C); viscosity 1.04 cp (65°C); dielectric constant 2.8; absorbs ultraviolet light in range 2800–3000 Å. Soluble in benzene and alcohol; insoluble in water. Combustible. Low toxicity.
Containers: Glass bottles; fiber drums.
Uses: Weathering agent in paints and plastics; fixative in perfumes; dyes; resin intermediate; cosmetics, especially suntan preparations; flavoring.

hydroquinine mono-n-butyl ether $CH_3(CH_2)_3OC_6H_4OH$.
Properties: White flakes; m.p. 64–65°C; b.p. 115°C (1.4 mm); insoluble in water; soluble in benzene, acetone, ethyl acetate, and alcohol. Combustible. Low toxicity.

hydroquinone monoethyl ether (4-ethoxyphenol) $C_2H_5OC_6H_4OH$. White solid; m.p. 63–65°C; b.p. 246–247°C; slightly soluble in water; soluble in benzene, acetone, ethyl acetate, and alcohol. Combustible. Low toxicity.
Uses: See hydroquinone monomethyl ether.

hydroquinone monomethyl ether (4-methoxyphenol; para-hydroxyanisole) $CH_3OC_6H_4OH$.
Properties: White waxy solid; m.p. 52.5°C; b.p. 243°C; sp. gr (20/20°C) 1.55. Slightly soluble in water; readily soluble in benzene, acetone, ethyl alcohol, and ethyl acetate. Combustible.
Containers: Fiber drums.
Hazard: Moderately toxic.
Uses: Manufacture of antioxidants, pharmaceuticals, plasticizers, dyestuffs; stabilizer for chlorinated hydrocarbons and ethyl cellulose; inhibitor for acrylic monomers and acrylonitriles; ultraviolet inhibitor.

hydrosilicofluoric acid. See fluosilicic acid.

hydrosolvation. Solvent extraction of coal (containing up to 5% sulfur) under hydrogen pressure with use of a catalyst such as zinc chloride; pressures from 1000 to 2000 psi are necessary for suitable conversion. This process offers a means of deriving fuel oil and petrochemical feedstocks directly from coal.

hydrosulfite. Sodium hydrosulfite (q.v.).

hydrosulfite-formaldehyde. One of several mixtures of sodium formaldehyde hydrosulfite and sodium formaldehyde bisulfite, used as discharges and stripping or reducing agents in dyeing and other textile operations. In some cases the zinc derivatives are used. Derivation is by the action of formaldehyde on aqueous sodium hydrosulfite, or from zinc, formaldehyde, sulfur dioxide and sodium hydroxide.
Hazard: Probably toxic by ingestion.

hydrotrope. A chemical which has the property of increasing the aqueous solubility of various slightly soluble organic chemicals. Used especially in the formulation of liquid detergents.

hydroxocobalamin (USAN) $C_{62}H_{89}CoN_{13}O_{15}P$. alpha-(5,6-Dimethylbenzimidazolyl)hydroxocobamide. A form of vitamin B_{12}; used in medicine. See vitamin B_{12}.

hydroxyacetal. See hydroxycitronella dimethyl acetal.

para-**hydroxyacetanilide.** See para-acetylaminophenol.

hydroxyacetic acid (glycolic acid). $CH_2OHCOOH$.
Properties: Colorless crystals, deliquescent; m.p. 78–79°C; b.p., decomposes; soluble in water, alcohol and ether. Available commercially as a 70% solution; light straw-colored liquid; odor like burnt sugar; sp. gr. 1.27; m.p. 10°C; combustible.
Derivation: From chloroacetic acid by boiling with water or aqueous alkali; by oxidation of glycol. Occurs naturally in sugar cane syrup.
Grades: Technical, 70% solution; pure crystals.
Containers: Drums; tank cars.
Hazard: Moderately toxic by ingestion.
Uses: Leather dyeing and tanning; textile dyeing; cleaning, polishing, and soldering compounds; copper pickling; adhesives; electroplating; breaking of petroleum emulsions; chelating agent for iron.

hydroxyacetone. See hydroxy-2-propanone.

ortho-**hydroxyacetophenone** $C_6H_4(OH)COCH_3$.
Properties: Greenish-yellow liquid with minty odor; sp. gr. (20.8°C) 1.1307; b.p. (717 min) 213°C; refractive index (n 20/D) 1.5580; slightly soluble in water Combustible.

2-hydroxyadipaldehyde $OHCCH_2CH_2CH_2CHOHCHO$.
Properties: (25% aqueous solution): Sp. gr. (20°C) 1.066; b.p. 37°C (50 mm); vapor pressure (20°C) 17 mm; f.p. −3.5°C; pH approx 3.0. Combustible.
Containers: 55-gal drums.
Hazard: May be toxic.
Uses: Intermediate; insolubilizing agent for proteins and polyhydroxy materials; crosslinking agent for polyvinyl compounds; shrinkage control agent (textiles).

beta-**hydroxyalanine.** See serine.

5-hydroxy-3-(beta-**aminoethyl)indole.** See serotonin.

hydroxyamphetamine [para-(2-aminopropyl)phenol] $HOC_6H_4CH_2CHNH_2CH_3$.
Properties: Crystals with m.p. 125–126°C. Soluble in water, alcohol, chloroform, and ethyl acetate.
Derivation: From para-nitrobenzyl chloride and a salt of nitroethane or from anisaldehyde and nitroethane.
Hazard: May be toxic.
Use: Medicine (also as hydrobromide).

hydroxyaniline. See aminophenol.

ortho-**hydroxyanisole.** See guaiacol.

para-**hydroxyanisole.** See hydroquinone monomethyl ether.

9-hydroxyanthracene. See anthranol.

hydroxyapatite $Ca_{10}(PO_4)_6(OH)_2$. The major constituent of bone and tooth mineral. It is a finely divided, crystalline, non-stoichiometric material rich in surface ions (carbonate, magnesium, citrate) which are readily replaced by fluoride ion, thus affording protection to the teeth. See also fluoridation; calcium phosphate, tribasic.

para-**hydroxyazobenzene**-para-**sulfonic acid** $HOC_6N_4NNC_6H_4SO_3H$.
Properties: Orange-red crystals; very soluble in water.
Uses: Analytical reagent; precipitant for numerous organic bases.

meta-**hydroxybenzaldehyde** HOC_6H_4CHO.
Properties: Orange-pink crystals; m.p. 101.5°C; slightly soluble in cold water; very soluble in hot water and aromatic hydrocarbons.
Hazard: May be toxic by ingestion.
Uses: Intermediate for dyes, plastics, pharmaceuticals and bactericides; color reagent for Schiff's reagent; sensitizing agent in photographic emulsions.

ortho-**hydroxybenzaldehyde.** See salicylaldehyde.

para-**hydroxybenzaldehyde** HOC_6H_4CHO.
Properties: Colorless needles; soluble in alcohol, ether, or water; sp. gr. 1.129; m.p. 116°C (sublimes).
Uses: Pharmaceuticals.

ortho-**hydroxybenzamide.** See salicylamide.

hydroxybenzene. See phenol.

2-hydroxy-1′,2′-benzocarbazole-3-carboxylic acid $C_{17}H_{11}NO_3$. A four-ring structure. Light green powder; m.p. 315–320°C; soluble in ethyl alcohol and acetone; insoluble in water. Used in the manufacture of dye intermediates and other organic chemicals.

meta-**hydroxybenzoic acid** $C_6H_4(OH)COOH$.
Properties: White powder; m.p. 200°C. Soluble in water and hot alcohol. Nontoxic.
Uses: Intermediate for plasticizers; resins; light stabilizers; petroleum additives; pharmaceuticals.

ortho-**hydroxybenzoic acid.** See salicylic acid.

para-**hydroxybenzoic acid** $C_6H_4(OH)COOH \cdot H_2O$.
Properties: Colorless crystals; soluble in alcohol, water, and in ether. M.p. 210°C. Nontoxic.
Derivation: Interaction of para-aminobenzoic acid and nitrous acid.
Uses: Intermediate; synthetic drugs; food preservative (up to 0.1%) (approved by FDA). Its methyl, propyl and butyl esters are preservatives for cosmetics and pharmaceuticals.

2-hydroxybenzophenone $C_6H_5COC_6H_4OH$. A solid; m.p. 41°C; b.p. 210°C (27 mm). Insoluble in water; soluble in alcohol. An ultraviolet absorber used in plastics.

ortho-**hydroxybenzyl alcohol.** See salicyl alcohol.

beta-**hydroxybutyraldehyde.** See aldol.

beta-**hydroxybutyric acid** $CH_3CH(OH)CH_2COOH$.
Properties: Viscid, yellow mass; m.p. 48–50°C; b.p. 130°C (12 mm); very soluble in water, alcohol, and ether.
Derivation: Interaction of acetoacetic acid and sodium amalgam.
Grades: Technical; reagent.
Use: Intermediates.

2-hydroxycamphane. See borneol.

hydroxy-beta-**carotene.** See cryptoxanthin.

hydroxychloroquine sulfate $C_{18}H_{26}ClN_3O \cdot H_2SO_4$.
Properties: White, crystalline, odorless powder. Bitter taste; pH of solutions 4.5. Freely soluble in water. Insoluble in alcohol, chloroform, ether. Exists in two forms: m.p. usual form 240°C; m.p. other form 198°C.
Grade: U.S.P.
Use: Medicine.

3-β-hydroxycholestane. See dihydrocholesterol.

hydroxycitronellal (citronellal hydrate; 3,7-dimethyl-7-hydroxyoctanal) $(CH_3)_2C(OH)(CH_2)_3CH(CH_3)CH_2CHO$.
Properties: Viscous, colorless or faintly yellow liquid; sweet lily-type odor; sp. gr. 0.925–0.930 (15°C); refractive index (n 20/D) 1.448–1.450; optical rotation (Java type) +9 to +10.5°; boiling range 94–96°C (1 mm). Soluble in alcohol (50%), fixed oils; slightly soluble in water, glycerol, and mineral oil. Combustible; nontoxic.
Derivation: From citronellal (Java citronella or Eucalyptus citriodora).
Grades: Perfume; F.C.C.
Containers: Cans.
Uses: Perfumery (fixative, muguet odor); flavoring; soap and cosmetic fragrances.

hydroxycitronellal dimethyl acetal (hydroxyacetal) $(CH_3)_2C(OH)(CH_2)_3CH(CH_3)CH_2CH(OCH_3)_2$.
Properties: Colorless liquid; light floral odor; sp. gr. 0.925–0.930 (25/25°C); refractive index 1.4410–1.4440 (20°C). Soluble in most fixed oils, mineral oil, propylene glycol. Insoluble in glycerin. Combustible; nontoxic.
Grades: Perfume; F.C.C.
Uses: Flavoring agent in foods; perfumery.

hydroxycitronellal-methyl anthranilate Schiff base $(CH_3)_2C(OH)(CH_2)_3CH(CH_3)CH_2CH{:}N\text{-}C_6H_4COOCH_3$.
Properties: Linden-orange flower odor; yellow honey-like viscous liquid; stable; refractive index (n 20/D) 1.5350–1.5460; flash point (TCC) 206°F. Soluble in 2 parts of 70% alcohol, 1 part of 80% alcohol. Combustible; nontoxic.
Use: Perfumery.

17-hydroxycorticosterone. See hydrocortisone.

2-hydroxy-para-**cymene.** See carvacrol.

3-hydroxy-para-**cymene.** See thymol.

1-hydroxy-2,4-diamylbenzene. See diamyl phenol.

para-**hydroxydiphenylamine** (anilinophenol) $C_6H_5NHC_6H_4OH$. Gray solid leaflets; m.p. 70°C (approx); purity 98% (min); b.p. 330°C; insoluble in water; soluble in alcohol, ether, acetone, chloroform, alkali, and benzene.
Hazard: Probably toxic and irritant.

hydroxydiphenyl. See phenylphenol.

2-hydroxyethanesulfonic acid. See isethionic acid.

hydroxyethylacetamide. See N-acetylethanolamine.

2-hydroxyethyl acrylate (HEA). A functional monomer for the manufacture of thermosetting acrylic resins (q.v.).

2-hydroxyethylamine. See ethanolamine.

2-hydroxyethyl carbamate $H_2NCOOCH_2CH_2OH$.
Properties: Crystalline deliquescent solid; m.p. 43°C; b.p. 130–135°C (1 mm); density 1.2852 g/cc (20°C); flash point 370°F. Miscible with water; soluble in alcohol and acetone; insoluble in benzene and chloroform. Combustible.
Use: Chemical intermediate, especially for wash-and-wear cotton finishing agents.

hydroxyethylcellulose ("Cellosize").
Properties: Nonionic, water-soluble ether of cellulose. White, free-flowing powder insoluble in organic solvents; soluble in hot or cold water. Stable in concentrated salt solutions; grease and oil resistant. Nontoxic. Combustible.
Containers: Fiber drums.
Uses: Thickening and suspending agent; stabilizer for vinyl polymerization; retards evaporation of water in cements, mortars, and concrete; binder in ceramic glazes; films and sheeting.

hydroxyethylethylenediamine (aminoethylethanol amine) $NH_2CH_2CH_2NHCH_2CH_2OH$.
Properties: Hygroscopic liquid. Mild, ammoniacal odor. Soluble in water. Sp. gr. 1.0304 (20/20°C); b.p. 243.7°C; vapor pressure 0.01 mm (20°C); flash point 275°F; wt/gal 8.6 lb (20°). Combustible. Low toxicity.
Uses: Textile finshing compounds (antifuming agents, dyestuffs, cationic surfactants); resins, rubber products, insecticides, and certain medicinals.

hydroxyethylethylenediaminetriacetic acid (HEDTA) $HOOCCH_2N(CH_2CH_2OH)CH_2CH_2N(CH_2COOH)_2$.
Properties: Solid; m.p. 159°C; soluble in water and methanol.
Use: Chelating compound.

N-**(2-hydroxyethyl)ethyleneimine.** See 1-aziridineethanol.

1-(2-hydroxyethyl)-2-n**-heptadecenyl-2-imidazoline** $C_{22}H_{42}N_2O$. A liquid cationic detergent.
Uses: Corrosion inhibitor; acid-stable emulsifier.

beta-**hydroxyethylhydrazine** (ethanolhydrazine) $HOCH_2CH_2NHNH_2$.
Properties: Colorless, slightly viscous liquid; sp. gr. (20°C) 1.11; f.p. −70°C; boiling range (25 mm) 145–153°C; flash point 224°F; completely miscible with water; soluble in lower alcohols; slightly soluble in ether. Combustible.
Grade: 70%.
Hazard: May be toxic and irritant. Moderate fire risk.
Use: Intermediate.

2-hydroxyethyl methacrylate (HEMA).
Properties: Clear mobile liquid; sp. gr. 1.064 (77/60°F); f.p. −12°C; refractive index 1.4505 (25/D). Miscible with water and soluble in common organic solvents. An inhibitor is usually added to solutions to prolong shelf life. The 30% solution is made with xylene.
Hazard: 30% grade (with xylene) is flammable; moderate fire risk.
Grades: 30%; 96%.
Containers: 55-gal drums.
Uses: Acrylic resins; binder for nonwoven fabrics; enamels.

N-**hydroxyethylmorpholine.** See N-morpholine ethanol.

N-**hydroxyethyl piperazine** $HOCH_2CH_2NCH_2CH_2NHCH_2CH_2$ (ring).
Properties: Liquid; sp. gr. 1.0614 (20/20°C); b.p. 246.3°C; f.p. −10°C; flash point 255°F. Miscible with water. Combustible. Low toxicity.
Uses: Intermediate for pharmaceuticals, anthelmintics, surface-active agents, and synthetic fibers.

N-**2-hydroxyethylpiperidine.** See 2-piperidinoethanol.

hydroxyethyltrimethylammonium bicarbonate (choline bicarbonate) $(CH_3)_3NCH_2CH_2OH \cdot HCO_3$.
Properties: Colorless liquid; sp. gr. 1.0965 (25/4°C); f.p. −21.3°C; refractive index 1.3967 (25/D); miscible with water. Low toxicity.
Grade: Industrial grade is a 44–46% aqueous solution of a mixture of carbonate and bicarbonate.
Uses: Alkaline catalyst; intermediate for choline salts and surfactants.

beta-**hydroxyethyltrimethylammonium hydroxide.** See choline.

1-hydroxyfenchane. See fenchyl alcohol.

4-hydroxy-2-keto-4-methylpentane. See diacetone alcohol.

hydroxyl group. The chemical group —OH, occurring in many inorganic compounds that ionize in solution to yield OH^- radicals. Also the characteristic group of alcohols. See also -ol.

hydroxylamine (oxammonium) NH_2OH.
Properties: Colorless crystals. Soluble in alcohol, acids and cold water. Sp. gr. 1.227; m.p. 33°C. The free base is unstable.
Derivation: By decomposing hydroxylamine hydrochloride or sulfate with a base and distilling in vacuo.
Hazard: Decomposes rapidly at room temperature, violently when heated. Strong irritant to tissue.
Uses: Reducing agent; organic synthesis.

hydroxylamine acid sulfate (hydroxylammonium acid sulfate; hydroxylamine hemisulfate; HAS) $NH_2OH \cdot H_2SO_4$.
Properties: White to brown crystalline solid; very hygroscopic; soluble in water and methanol; slightly soluble in alcohol. Wt/gal 15–16 lb (20°C); m.p. indefinite (decomposes); pH of 0.1 M aqueous solution 1.6.
Uses: Reducing agent; photographic developer; purification agent for aldehydes and ketones; synthesis of dyes, pharmaceuticals; rubber chemicals; reagent.

hydroxylamine hydrochloride (hydroxylammonium chloride) $NH_2OH \cdot HCl$.
Properties: Colorless hygroscopic crystals; sp. gr. 1.67 (17°C); soluble in water, glycerol and alcohol; insoluble in ether. M.p. 152°C (decomposes); pH of 0.1 M aqueous solution 3.4.
Derivation: Reduction of ammonium chloride, frequently by electrolysis.
Uses: Organic synthesis; photographic developer; medicine; controlled reduction reactions; nondiscoloring short-stopper for synthetic rubbers; antioxidant for fatty acids.

hydroxylamine sulfate (HS; hydroxylammonium sulfate) $(NH_2OH)_2 \cdot H_2SO_4$.
Properties: Colorless crystals; solution has a corrosive action on the skin; m.p. 177°C (dec); soluble in water; slightly soluble in alcohol.
Uses: Reducing agent; photographic developer; purification agent for aldehydes and ketones; chemical synthesis; textile chemical; oxidation inhibitor for fatty acids; catalyst; biological and biochemical research; making oximes for paints and varnishes; rustproofing; nondiscoloring short-stopper for synthetic rubbers.

hydroxylammonium acid sulfate. See hydroxylamine acid sulfate.

hydroxylammonium chloride. See hydroxylamine hydrochloride.

hydroxylammonium sulfate. See hydroxylamine sulfate.

hydroxymercurichlorophenol (2-chloro-4-(hydroxymercuri)phenol) $C_6H_3Cl(HgOH)(OH)$. Insoluble in water and common organic solvents. Soluble in solutions of acids and alkalies with the formation of salts.
Hazard: Highly toxic by ingestion and inhalation; strong irritant to eyes and skin. Use may be restricted. Tolerance, 0.05 mg per cubic meter of air.
Uses: Seed disinfectant and fungicide.
Shipping regulations: (Rail, Air) Mercury compounds, n.o.s., Poison label.

hydroxymercuricresol $C_6H_3CH_3(HgOH)(OH)$. A pesticide.
Hazard: Highly toxic by ingestion and inhalation; strong irritant to eyes and skin. Use may be restricted. Tolerance, 0.05 mg per cubic meter of air.
Shipping regulations: (Rail, Air) Mercury compounds, n.o.s., Poison label.

hydroxymercurinitrophenol $C_6H_3NO_2(HgOH)(OH)$. A pesticide.
Hazard: Highly toxic by ingestion and inhalation; strong irritant to eyes and skin. Use may be restricted. Tolerance, 0.05 mg per cubic meter of air.
Shipping regulations: (Rail, Air) Mercury compounds, n.o.s., Poison label.

2-hydroxy-3-methylbenzoic acid. See cresotic acid.

3-hydroxy-3-methylbutanone-2 $CH_3COH(CH_3)C(O)CH_3$. Colorless liquid; sp. gr. 0.95 (25/25°C). Used as an intermediate. Combustible.

7-hydroxy-4-methylcoumarin. See beta-methyl umbelliferone.

4-hydroxymethyl-2,6-di-tert-**butylphenol** $[(CH_3)_3C]_2C_6H_2(OH)(CH_2OH)$.
Properties: White crystalline powder; m.p. 140–141°C; b.p. 162°C (2.6 mm); flash point 375°F; vapor pressure 0.03 (100°C); bulk density 26 lb/cu ft; no odor. Partially soluble in methanol, ethanol, and acetone; insoluble in water. Combustible.
Hazard: May be toxic by ingestion.
Uses: Antioxidant for gasoline and other hydrocarbons.

hydroxymethylethylene carbonate. See glycerin carbonate.

DL-alpha-**hydroxy**-gamma-**methylmercaptobutyric acid, calcium salt.** See methionine hydroxy-analog calcium.

2-hydroxymethyl-5-norbornene. See 5-norbornene-2-methanol.

4-hydroxy-4-methylpentanone-2. See diacetone alcohol.

2-(2′-hydroxy-5′-methylphenyl)benzotriazole $HOC_6H_3(CH_3)N_3C_6H_4$.
Properties: Off-white, crystalline powder; m.p. 129–130°C; b.p. 225°C (10 mm); insoluble in water; soluble in methyl ethyl ketone, methyl methacrylate.
Uses: Protector against ultraviolet radiation; thermal stabilizer, antioxidant; chelation of metals.

3-hydroxy-2-methyl-1,4-pyrone. See maltol.

hydroxynaphthalene. See naphthol.

3-hydroxy-2-naphthoic acid (beta-hydroxynaphthoic acid; 3-naphthol-2-carboxylic acid; beta-oxynaphthoic acid) $C_{10}H_6OHCOOH$.
Properties: Yellow rhombic leaflets; soluble in alcohol and ether. M.p. 217.5–219°C. Insoluble in water.
Derivation: By treating sodium 2-naphtholate with carbon dioxide under pressure.
Hazard: Moderately toxic and irritant.
Uses: Dyes and pigments.

beta-**hydroxynaphthoic anilide** (naphthol AS; azoic coupling component 2) $C_{10}H_6OHCONHC_6H_5$.
Properties: Cream-colored crystals. Sodium salt is soluble in water; m.p. 246.0°C.
Derivation: Condensation of beta-hydroxynaphthoic acid and aniline.
Hazard: May be toxic.
Use: Dyes.

2-hydroxy-1,4-naphthoquinone $C_{10}H_5O_2(OH)$.
Properties: Yellow to orange-yellow needles or powder; m.p. 192–195°C (dec.). Redox potential 0.362 volt; soluble in cold water, benzene, carbon tetrachloride and petroleum ether.
Uses: Intermediate for pharmaceuticals, henna hair and wool dye, bactericides; seed disinfectant.

5-hydroxy-1,4-naphthoquinone (juglone). This compound, derived from walnuts, is the starting point in the 16-step synthesis of oxytetracycline.

N-(7-hydroxyl-1-naphthyl)acetamide (1-acetylamino-7-naphthol) $C_{10}H_6(OH)NHCOOH_3$. A granular paste used as a chemical intermediate.

4-hydroxy-3-nitrobenzenearsonic acid (3-nitro-4-hydroxyphenylarsonic acid) $HOC_6H_3(NO_2)AsO(OH)_2$.
Properties: Pale-yellow crystals.
Derivation: Heating phenol with arsenic and treating the para-hydroxyphenyl arsonate with nitric and sulfuric acids.
Hazard: Toxic by ingestion.
Shipping regulations: (Rail, Air) Arsenical compounds, n.o.s., Poison label.

4-hydroxynonanoic acid, gamma-lactone. See gamma-nonyl lactone.

cis-**12-hydroxyoctadec-9-enoic acid.** See ricinoleic acid.

12-hydroxyoleic acid. See ricinoleic acid.

15-hydroxypentadecanoic acid lactone. See pentadecanolide.

3-hydroxyphenol. See resorcinol.

2-hydroxy-2-phenylacetophenone. See benzoin.

beta-para-**hydroxyphenylalanine.** See tyrosine.

2-(2′-hydroxyphenyl)benzotriazole $HOC_6H_4N_3C_6H_4$. An ultraviolet light absorber used in plastics.

para-**hydroxyphenyl benzyl ether.** See para-benzyloxyphenol.

4-(para-**hydroxyphenyl)-2-butanone acetate.** See cuelure.

para-**hydroxyphenylglycine** (glycine[photographic]; photo-glycin) $HOC_6H_4NHCH_2COOH$.
Properties: White to buff crystals or powder; m.p.

240°C (with decomposition); slightly soluble in water; soluble in alkaline solutions.
Derivation: By condensation of para-aminophenol with chloroacetic acid.
Uses: Photographic developer; cellulose and nitrocellulose acetate lacquers and varnishes.

2-hydroxyphenylmercuric chloride (chloromercuriphenol) HOC_6H_4HgCl.
Properties: White to faint pink, fine crystals; 0.1 parts in 100 soluble in water (25°C); soluble in hot water, alkali, and alcohol. M.p. 152°C.
Hazard: Highly toxic. Skin irritant. Tolerance, 0.05 mg per cubic meter of air.
Uses: Antiseptic; fungicide.
Shipping regulations: (Rail, Air) Mercury compounds, n.o.s., Poison label.

1-(para-hydroxyphenyl)-2-methylaminoethanol tartrate. See synephrine tartrate.

hydroxyphenylstearic acid. A derivative of phenylstearic acid (q.v.) potentially useful as oxidation and corrosion inhibitor.

11-alpha-hydroxyprogesterone $C_{21}H_{30}O_3$.
Properties: White crystalline powder; m.p. 163°C approx; specific rotation +179°; insoluble in water; soluble in alcohol.
Derivation: From progesterone by microbiological oxidation.
Use: A steroid intermediate; biochemical research.

hydroxyproline (Hyp) HOC_4H_7NCOOH. (4-hydroxy-2-pyrrolidinecarboxylic acid.)
Properties: Colorless crystals; very soluble in water; slightly soluble in alcohol; insoluble in ether; optically active.
DL-hydroxyproline, m.p. 261–262°C with decomposition.
L-hydroxyproline, m.p. 270°C (natural).
D-hydroxyproline, m.p. 274°C.
Derivation: Hydrolysis of protein (gelatin); organic synthesis.
Use: Biochemical research. Available commercially as L-hydroxyproline.

2-hydroxy-1,2,3-propanetricarboxylic acid. See citric acid.

hydroxy-2-propanone (acetol; acetonyl alcohol; acetylcarbinol; hydroxyacetone; pyruvic alcohol) CH_3COCH_2OH.
Properties: Colorless liquid; sp. gr. 1.0824 at 20/20°C; b.p. 146°C; f.p. −17°C. Soluble in water, alcohol, and ether, Combustible. Low toxicity.
Derivation: (a) By action of potassium acetate or potassium formate on a solution of bromo- or chloroacetone in dry methanol; (b) by bacterial fermentation of propylene glycol.
Grade: Technical
Use: Solvent for nitrocellulose.

alpha-**hydroxypropionic acid.** See lactic acid.

3-hydroxypropionitrile. See ethylene cyanohydrin.

alpha-**hydroxypropionitrile.** See lactonitrile.

beta-**hydroxypropionitrile.** See ethylene cyanohydrin.

hydroxypropyl acrylate (HPA). A functional monomer used in manufacture of thermosetting acrylic resins for surface coatings.

2-hydroxypropylamine. See isopropanolamine.

hydroxypropyl cellulose. A cellulose ether with hydroxypropyl substitution.
Properties: White powder; soluble in water, methyl and ethyl alcohols and other organic solvents. Combustible. Nontoxic. Thermoplastic; can be extruded and molded. Insoluble in water warmer than 100°F.
Grades: F.C.C.
Uses: Emulsifier; film former; protective colloid; stabilizer; suspending agent; thickener; food additive.

hydroxypropylglycerin.
Properties: Pale straw-colored liquid; sp. gr. 1.084 (25/25°C); refractive index 1.459 (25°C); flash point 380°F; pour point −23°C; soluble in water and methanol. Combustible; low toxicity.
Uses: Intermediate for alkyd resins and polyesters; plasticizer for cellulosics, glue, starch, etc.

hydroxypropyl methacrylate
$CH_3CHOHCH_2OOCC(CH_3){:}CH_2$.
Properties: Clear, mobile liquid; sp. gr. 1.066 (25/16°C); refractive index (n 25/D) 1.446; flash point 206°F. Limited solubility in water; soluble in common organic solvents. Combustible. Low toxicity.
Uses: Monomer for acrylic resins; nonwoven fabric binders; detergent lube oil additives.

hydroxypropyl methylcellulose (methylcellulose, propylene glycol ether).
Properties: White powder. Swells in water producing clear to opalescent, viscous, colloidal solution. Insoluble in anhydrous alcohol, in ether, and in chloroform. Combustible. Nontoxic.
Grades: N.F.; F.C.C.
Uses: Medicine (suspending agent); in food products as thickening agent, stabilizer, emulsifier; thickener in paint-stripping preparations.

2-(alpha-**hydroxypropyl)piperidine.** See conhydrine.

N-beta-**hydroxypropyl**-ortho-**toluidine**
$CH_3C_6H_4NHCH_2CH(OH)CH_3$.
Properties: Color, amber; distillation range, 170-180°C (20 mm); sp. gr. 1.035–1.045 (20/20°C); refractive index 1.540–1.550 (20°C).
Hazard: May be toxic.
Use: Dye intermediate.

4-hydroxy-2H-pyran-3,3,5,5(4H,6H)tetramethanol. See anhydroenneaheptitol.

2-hydroxypyridine-N-oxide. Bactericidal agent related to aspergillic acid; made from pyridine-N-oxide.

1-hydroxy-2-pyridine thione (2-pyridinethiol-1-oxide) $C_5H_4NOH(S)$. Apparently exists in equilibrium with the -SH form. Forms chelates with iron, manganese, zinc, etc.
Hazard: May be toxic.
Uses: Fungicide; bactericide.

4-hydroxy-2-pyrrolidinecarboxylic acid. See hydroxyproline.

8-hydroxyquinoline (8-quinolinol; oxyquinoline; oxine) C_9H_6NOH.
Properties: White crystals or powder; darkens when exposed to light; technical grade usually tan; almost insoluble in water; soluble in alcohol, acetone, chloroform, benzene, also in formic, acetic, hydrochloric and sulfuric acids, and in alkalies; phenolic odor; m.p. 73–75°C; b.p. 267°C.
Grades: C.P.; technical.
Hazard: Moderately toxic.
Uses: Precipitating and separating metals; preparation of fungicides.

8-hydroxyquinoline benzoate $C_9H_6NOH \cdot C_6H_5COOH$.
Properties: Yellowish-white crystals with a saffron odor; m.p. 56–61°C; almost insoluble in water; soluble in alcohol and glycerol.

Hazard: Moderately toxic.
Uses: Antiseptics; fungicide, recommended for Dutch elm disease.

8-hydroxyquinoline sulfate $(C_6H_7NO)_2 \cdot H_2SO_4$.
Properties: Pale yellow powder; slight saffron odor; burning taste. Melting range 167–182°C. Soluble in water; slightly soluble in alcohol; insoluble in ether.
Hazard: Moderately toxic.
Uses: Antiseptic, antiperspirant, deodorant; fungicide.

8-hydroxyquinoline-5-sulfonic acid $C_9H_5N(OH)SO_3H$.
Properties: Pale yellow, needlelike crystals or powder; soluble in water; slightly soluble in organic solvents; m.p. 213°C with decomposition.
Hazard: Moderately toxic.

4-hydroxysalicylic acid. See beta-resorcylic acid.

12-hydroxystearic acid
$CH_3(CH_2)_5(CHOH)(CH_2)_{10}COOH$. A C_{18} straight chain fatty acid with an —OH group attached to the carbon chain; m.p. 79–82°C. It is produced by hydrogenation of ricinoleic acid. Combustible. Low toxicity.
Uses: Lithium greases; chemical intermediates.

1,12-hydroxystearyl alcohol (1,12-octadecanediol). A long-chain fatty alcohol made by reduction of 12-hydroxystearic acid by replacing the —COOH group with a —CH_2OH. Combustible. Low toxicity.
Properties: M.p. 69°C; boiling range 315–335°C.
Uses: Chemical intermediate; synthetic fibers; organic synthesis; pharmaceuticals; surface-active agents; plastics and resins; protective coatings.

hydroxysuccinic acid. See malic acid.

alpha-**hydroxytoluene.** See benzyl alcohol.

hydroxytoluene. See cresol.

hydroxytoluic acid. See cresotic acid.

5-hydroxytryptamine. See serotonin.

4-hydroxyundecanoic acid, gamma-**lactone.** See gamma-undecalactone.

hydroxyurea ("Hydrea"). A drug used in cancer therapy, said to inhibit biosynthesis of DNA; especially useful for melanoma.

hydroxyzine hydrochloride. $C_{21}H_{27}ClN_2O_2 \cdot 2HCl$.
Properties: White, odorless powder. Very soluble in water; fully soluble in chloroform, and practically insoluble in ether; melting range 196–204°C (dec.).
Grade: N.F.
Use: Medicine.

"Hydrozin."[309] Trademark for normal zinc formaldehyde sulfoxylate used for discharge printing on acetate grounds and for stripping wool, acetates and nylon. Also used as a catalyst for polymerization of vinyl monomers.

hydrozincite (zinc bloom) $Zn_5(OH)_6(CO_3)_2$. A natural basic carbonate of zinc, found in the upper zones of zinc deposits.
Properties: Color white to gray or yellowish; luster dull to sikly; fluorescent in ultraviolet light; sp. gr. 3.5–4.0; hardness 2.0–2.5.
Occurrence: Missouri, Pennsylvania, Utah, California, Nevada; Europe.
Use: An ore of zinc.

hydyne. Mixture of 60% (by weight) of uns-dimethylhydrazine and 40% diethylenetriamine; used as a high-energy fuel.

"Hyfac."[242] Trademark for saturated or hydrogenated fatty acids and glycerides and castor oil derivatives.

"Hyform."[57] Trademark for water emulsions of pure paraffin wax, microcrystalline wax, or a modification of one of these waxes.
Uses: Binders for pressed ceramic pieces, lubricants for die or mold release, and plasticizers during mold forming.

Hygas process. A method of coal gasification in which hydrogen from an external source is reacted with low-grade coal such as lignite, producing a gas with a Btu value of about 1000. It requires high-pressure fluidization techniques and subsequent methanation to obtain the heating values necessary. The sources of hydrogen may be natural gas reforming (steam-oxygen process) or the steam-iron process. A 75-ton per day pilot plant is in operation, and a full-scale commercial plant is under construction (1975). See also gasification.

"Hygromix."[530] Trademark for hygromycin B, a feed and food additive.

hygromycin $C_{23}H_{29}NO_{12}$.
Properties: White powder; weakly acidic; freely soluble in water and alcohol. Produced by Streptomyces hygroscopicus.
Use: Veterinary medicine.

hygroscopic. Descriptive of a substance that has the property of adsorbing moisture from the air, such as silica gel, calcium chloride or zinc chloride. The water-vapor molecules are held by the molecules of the agent, which is called a desiccant (q.v.) when used primarily for this purpose. Paper and cotton fabrics are hygroscopic, normally containing from 5 to 8% water after standing in an atmosphere of normal humidity; they are usually kept in constant-humidity rooms before use. Many dry chemicals are hygroscopic, and should be kept in well-stoppered bottles or tightly closed containers. See also deliquescent.

"Hylene."[28] Trademark for a series of organic isocyanates.
M-50. 50% methylenebis(4-phenyl isocyanate) $(C_6H_4NCO)_2$, in monochlorobenzene. Used in adhesives. See diphenylmethane diisocyanate.
MP. Bisphenol adduct of methylenebis(4-phenyl isocyanate). Used as bonding agent for adhering "Dacron" fiber to rubber compositions.
T. Toluene-2,4-diisocyanate; used for urethane products.
TM. 80% toluene-2,4-diisocyanate and 20% toluene-2,6-diisocyanate.
TM-65. 65% 2,4-compound and 35% 2,6-compound.

"Hylox."[214] Trademark for a polysulfone matrix used as the base for stereotype printing plates. The plate is produced by extruding polypropylene onto the matrix and pressure-forming.

"Hyonic."[309] Trademark for a series of products used in liquid detergents, shampoos, textile scouring, metal cleaners, dairy detergents, rug and upholstery shampoos, hard surface cleaners.

Hyp Abbreviation for hydroxyproline.

"Hypalon."[28] Trademark for chlorosulfonated polyethylene, a synthetic rubber.

Properties: White chips; sp. gr. 1.10–1.28. Resistant to ozone, as well as the weather, oil, solvents, chemicals, head and abrasion.
Uses: Insulation for wire and cable; shoe soles and heels; automotive components; building products; coatings; flexible tubes and hoses, seals, gaskets, diaphragms. "Hypalon" 45 can accept large amounts of filler and is used as a binder for powdered metal to produce magnetic gaskets for doors, and sheet goods for x-ray barriers.

"Hypaque" Sodium.[162] Trademark for sodium diatrizoate (q.v.).

hyperglycemic-glycogenolytic factor. See glucagon.

hypergolic fuel. A liquid rocket fuel or propellant which consists of combinations of fuels and oxidizers, which ignite spontaneously on contact.

hypersorption. Process in which activated carbon selectively adsorbs less volatile components from a gaseous mixture, while the more volatile compoents pass on unaffected. Particularly applicable to separations of low-boiling mixtures such as hydrogen and methane, ethane from natural gas, ethylene from refinery gas, etc.

hypertensin. See angiotensin.

"Hypertensin-CIBA."[305] Trademark for angiotensin amide.

"Hy-Phos."[62] Trademark for a form of anhydrous glassy sodium phosphate. Used in detergent formulations.

hypnone. See acetophenone.

hypo-. (1) A prefix used in chemical terminology to indicate a compound (usually an acid) in its lowest oxidation state, or containing the lowest proportion of oxygen in a series of compounds E.g., nitric acid (HNO_3), nitrous acid (HNO_2), hyponitrous acid ($H_2N_2O_2$).
(2) Common term for a photographic chemical. See sodium thiosulfate.

hypoallergenic. A term recently introduced in the cosmetics industry, and defined by FDA as describing a cosmetic product that is less likely to cause adverse allergenic reactions than competing products. Claims of hypoallergenicity must be substantiated by specific dermatological tests made by the manufacturer of the product.

hypochlorite solution. An aqueous solution of a metallic salt of hypochlorous acid (q.v.). Strong oxidizing agent.
Hazard: Moderately toxic; irritant to skin and eyes.
Uses: Bleaching of textiles; antiseptic agent.
Shipping regulations: (Rail) (Containing more than 7% available chlorine by weight). White label. (Air) Corrosive label.

hypochlorous acid. HOCl
Properties: Greenish-yellow aqueous solution; highly unstable, weak acid, decomposing to hydrochloric acid and oxygen. Can exist only in dilute solutions.
Derivation: Water solution of chloride of lime (bleaching powder).
Hazard: Moderately toxic; irritant to skin and eyes.
Uses: Textile and fiber bleaching; water purification; antiseptic.
See also hypochlorite solution; calcium hypochlorite; sodium hypochlorite.

alpha-**hypophamine.** See oxytocin.

beta-**hypophamine.** See vasopressin.

hypophosphorous acid H_3PO_2.
Properties: Colorless, oily liquid or deliquescent crystals; sour odor. Soluble in water. Sp. gr. 1.439; m.p. 26.5°C. A strong monobasic acid and reducing agent; sold in solution.
Derivation: Heating concentrated baryta water with white phosphorus and decomposing the barium hypophosphite with sulfuric acid, filtering the liquid and concentrating under reduced pressure.
Grades: Technical; N.F. (30–32% solution, sp. gr. 1.13); 50% purified.
Containers: Bottles; carboys; drums.
Hazard: Fire and explosion risk in contact with oxidizing agents.
Use: Preparation of hypophosphites; electroplating baths.

hypoxanthine $C_5H_4N_4O$. An intermediate in the metabolism of animal purines; also widely distributed in the vegetable kingdom.
Properties: White to cream powder; decomposes at 150°C; almost insoluble in cold water; slightly soluble in boiling water; soluble in dilute acids and alkalies.
Derivation: Deamination of adenine; reduction of uric acid.
Uses: Biochemical research; biological media.

hypoxanthine riboside. See inosine.

hypoxanthine riboside-5-phosphoric acid. See inosinic acid.

"Hyprin" GP30.[233] Trademark for hydroxypropylglycerin (q.v.).

"Hypro" Process.[416] Propietary method for the production of hydrogen from hydrocarbon gas streams. A hydrocarbon, such as methane, is contacted with a catalyst under moderate conditions of temperature and pressure and decomposed into carbon, which remains on the catalyst, and hydrogen, which is mechanically removed. The hydrogen produced is about 94% pure when the charge stock is methane.

"Hyprose" SP80.[233] Trademark for octakis-(2-hydroxypropyl)sucrose (q.v.).

"Hyros."[79] Trademark for a special "FF" heat-treated wood rosin. M.p. (capillary tube) 64°C, (ball and ring) 85°C; sp. gr. 1.08 (25/25°C); color "FF."
Uses: Adhesive tape; battery wax; belt dressings; box toes; branding paint; brewers' pitch; core oil; dry core binders; fireworks; linoleum cement; matches; printing ink; rock wool; roofing cement; rubber cement; smoking molds; synthetic rosin oil; tree banding; Venice turpentine; wire-coating compounds.

hysteresis. Hysteresis (derived from the Greek word meaning "to lag behind") is a retardation of the effect, as if from viscosity, when the forces acting upon the body are changing. A common illustration is the retentivity of induction in ferromagnetic materials, such as iron and its alloys, when the magnetizing force is changed. When such a substance is placed in a magnetizing coil and the magnetizing field is gradually increased to a given value, and then decreased, the magnetic induction in decreasing does not follow the same relation to the magnetizing field that it did when the field was increasing, but lags behind the decreasing field.

Hysteresis is analogous to mechanical inertia, and the energy lost is analogous to that lost in mechanical friction. It presents a major problem in the design of electrical machines with iron cores, such as transformers and rotating armatures. In instruments designed for very high frequencies the retardation and losses are so great as to render iron cores useless.

The stress-stain curves of vulcanized rubber, also display hysteresis, in that strain (elongation, crystallization) persists when the deforming stress is removed, thus producing a hysteresis *loop* instead of a reversible pathway of the curves. This loop indicates a *loss* of resilient energy. Norman E. Gilbert.

"Hystrene."[473] Trademark for high-purity saturated fatty acids including caprylic, capric, lauric, myristic, palmitic, triple pressed stearic acid, etc.

"Hytakerol."[162] Trademark for dihydrotachysterol (q.v.).

"Hytrol" O.[28] Trademark for cyclohexanone (q.v.).

"Hyvar X."[28] Trademark for a wettable powder containing 80% bromacil.

"Hyvar" X-WS is a water-soluble powder containing 50% Bromacil.

I

I Symbol for iodine.

I-131. See iodine 131.

IAA. Abbreviation for 3-indoleacetic acid (q.v.).

IBIB. Abbreviation for isobutyl isobutyrate.

IBP. Abbreviation for initial boiling point; used when a range of boiling temperatures is given. See also DP.

-ic. A suffix used in naming inorganic compounds which indicates that the central element is present in its highest oxidation state; thus in ferric chloride ($FeCl_3$), the iron has an oxidation number of +3, equivalent to its valence; in an ionized state it would have three positive charges (Fe^{+++}). (A recommended change in this system of nomenclature is to use the common name of the element (iron) together with a Roman numeral showing the oxidation number; thus ferric chloride would be iron (III) chloride).

ICC. Abbreviation for ignition control compound.

ice (H_2O). An allotropic, crystalline form of water; melting point 0°C (32°F); latent heat of melting: 80 calories per gram; sp. gr. 0.91; its property of melting under pressure accounts for "slipperiness." Occurs in nature as Ice I, but several other forms are known.

Uses: Refrigeration; preservation of fish at sea.

See also water.

Iceland spar. A form of calcite having unique optical properties. Used in polarizing light (nicol prism).

ichthammol (ammonium ichthosulfonate).

Properties: Brownish-black, syrupy liquid, burning taste, tarry odor. Incompatible with acids, alkaloids, carbonates, hydroxides, mercuric chloride. Soluble in water, alcohol-ether or alcohol-ether-water mixtures; partially soluble in alcohol and ether. Miscible with glycerol. Low toxicity.

Derivation: Aqueous solution of sulfonated ammonium compounds derived from the action of sulfuric acid upon distillates from bituminous shales.

Grade: N.F.

Use: Pharmaceitical products such as skin ointments, cosmetic preparations, special dermatological soaps.

"Ichthymall."[329] Trademark for ichthammol.

"Ichthyol." Trademark for ichthammol.

-ide. A suffix used in naming compounds comprised of two elements; in such names, the first element, being electropositive, retains its name without change, but the second, being electronegative, utilizes the suffix -ide as a modification of the elemental name. Examples are: sodum hydroxide, magnesium chloride, hydrogen sulfide, etc. Similarly, oxygen is modified to oxide, fluorine to fluoride, phosphorus to phosphide, and carbon to carbide.

ideal gas (perfect gas). A gas in which there is complete absence of cohesive forces between the component molecules; the behavior of such a gas can be predicted accurately by Boyle's Law, Charles' Law or the ideal gas equation through all ranges of temperature and pressure. The concept is theoretical, as no actual gas meets the ideal requirement; carbon dioxide especially lacks conformity.

ideal solution. A solution which exhibits no change of internal energy on mixing and complete uniformity of cohesive forces. Its behavior is described by Raoult's Law (q.v.) over all ranges of temperature and concentrations.

identity period. The repeating unit or monomer which occurs *n* times in a natural or synthetic polymer molecule; for example, the anhydroglucose unit in cellulose is enclosed in brackets:

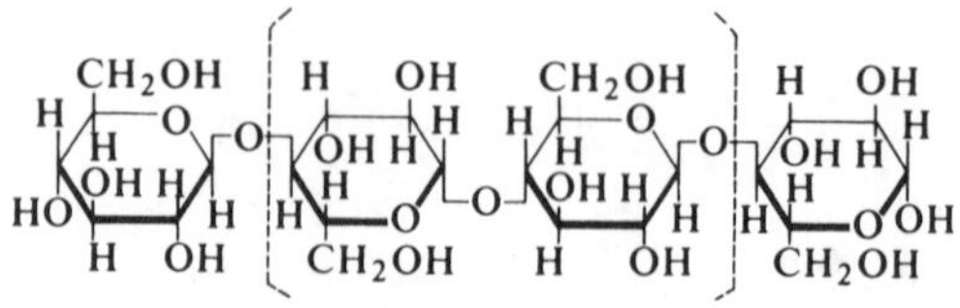

"Igenal."[307] Trademark for a series of dyestuffs for chrome-tanned leather. Characterized by unusual tinctorial power.

"Igepal."[307] Trademark for a series of biodegradable nonionic surfactants used as detergents, dispersants, emulsifiers, and wetting agents.

CA, CO, DM, RC: A series of alkylphenoxypoly(oxyethylene)ethanols, resulting from the combination of an alkylphenol with ethylene oxide. The general formula is $RC_6H_4O(CH_2CH_2O)_nCH_2CH_2OH$ in which R may be C_8H_{17} or a higher homolog.

LO and A: A series of linear alkylphenol ethoxylates (LO series) and a series of linear aliphatic ethoxylates (A series).

"Igepon."[307] Trademark for a series of anionic surfactants used as detergents, wetting agents, emulsifiers, dispersants and foaming agents. T and C types are sulfo-amides derived from N-methyltaurine or N-cyclohexyltaurine and fatty acids and have the general formula: $RCON(R')CH_2CH_2SO_3Na$. A types are sulfo-esters derived from isethionic acid and a fatty acid and have the general formula: $RCOOCH_2CH_2SO_3Na$. (R and R′ are alkyl groups).

ignition control compound. A substance such as methyldiphenyl phosphate, or trimethyl phosphate which is added to gasoline to control spark plug fouling, surface ignition, and motor rumble.

ignition point. See autoignition point.

ignotine. See carnosine.

"Illium."[491] Trademark for a series of superstainless steel alloys with high corrosion resistance.

ilmenite (titanic iron ore) $FeO \cdot TiO_2$. Iron-black mineral; black to brownish-red streak; submetallic luster. Resembles magnetite in appearance but is readily distinguished by feeble magnetic character. Sp. gr. 4.5–5; Mohs hardness 5–6.

Occurrence: Widely in U.S.; Canada; Sweden; U.S.S.R.; India. Also made synthetically.
Uses: Titanium paints and enamel; source of titanium metal; welding rods; titanium alloys; ceramics.

"Ilosone."[100] Trademark for propionyl erythromycin ester lauryl sulfate.

"Ilotycin."[100] Trademark for erythromycin, U.S.P.

Imhoff slude. See sewage sludge.

imidazole (glyoxalin) HNCHNCHCH. A dinitrogen ring compound. An antimetabolite (q.v.) and inhibitor of histamine. Colorless crystals, m.p. 90°C; b.p. 257°C. Soluble in water, alcohol, and ether. Used in biological control of pests, especially fabric-feeding insects, often in combination with *dl-p*-fluorophenylalanine, an amino-acid inhibitor; also as a contact insecticide in an oil spray. The mechanism is that of structural antagonism rather than active toxicity. See also antihistamine; antagonist, structural.

4,5-imidazoledicarboxamide. See glycarbylamide.

4-imidazole ethylamine. See histamine.

2-imidazolidinone. See ethylene urea.

2-imidazolidone. See ethylene urea.

imidazo(4,5-d)pyrimidine. See purine.

imide. A nitrogen-containing acid having two double bonds. See succinimide; phthalimide.

imine. A nitrogen-containing organic substance having a carbon-to-nitrogen double bond R—CH(=NH). Such compounds are highly reactive, even more so than the carbon-nitrogen triple bond characteristic of nitriles.

3,3′-iminobispropylamine (dipropylene triamine; 3,3′-diaminodipropylamine) $H_2NC_3H_6NHC_3H_6NH_2$.
Properties: Colorless liquid; sp. gr. 0.9307 (20/20°C); b.p. 240.6°C; f.p. −6.1°C; flash point 175°F (closed cup); soluble in water and polar organic solvents. Combustible.
Hazard: Moderately toxic by ingestion and inhalation; irritant.
Uses: Intermediate for soaps, dyestuffs, rubber chemicals, emulsifying agents, petroleum specialties, insecticides, and pharmaceuticals.

iminodiacetonitrile $HN(CH_2CN)_2$.
Properties: Light tan, crystalline solid; m.p. 77–78°C; soluble in water and acetone.
Hazard: May be toxic.
Use: Chemical intermediate.

iminourea. See guanidine.

"IML-1."[28] Trademark for a creamy-white nondusting powder of sodium alkyl sulfates, used as an internal lubricant in elastomer compounds. Improves flow characteristics but does not appreciably reduce tensile strength.

"Imlar."[28] Trademark for vinyl resin-base finish used where extreme resistance to abnormal chemical exposure is required.

"Immedial."[307] Trademark for a series of sulfur dyestuffs. used for the dyeing of cotton and rayon. Characterized by very good fastness to light and good fastness to washing and perspiration.

immiscible. Descriptive of substances of the same phase or state of matter that cannot be uniformly mixed or blended. Though usually applied to liquids such as oil and water, the term also may refer to powders that differ widely in some physical property, e.g., specific gravity, such as magnesium carbonate and barium sulfate. See also miscibility.

immune serum globulin. A sterile solution of globulins which contains those antibodies normally present in adult blood. Not less than 90% of the total protein is globulin. It is a transparent, nearly colorless, nearly odorless liquid. Must be kept refrigerated.
Derivation: From a plasma or serum pool of venous or placental blood from 1000 or more individuals.
Grade: U.S.P.
Use: Medicine.
See also antigen.

immunochemistry. That branch of chemistry concerned with the various defense mechanisms of the animal organism against infective agents, particularly the response between the body and foreign macromolecules (antigens), and the interaction between the products of the response (antibodies) and the agents that have elicited them. This involves study of the many proteins (serum globulins, enzymes, bacteria, and viruses) involved in these responses. It developed from the original work of Jenner (1775) and Pasteur (1880). See also antigen-antibody; chemotherapy.

"Imol."[51] Trademark for a group of fire-resistant hydraulic fluids.

IMP. (1) Abbreviation for inosine monophosphate. See inosinic acid or sodium inosinate.
(2) Abbreviation for insoluble metaphosphate. See Maddrell's salts under sodium metaphosphate.

impalpable. Descriptive of a state of subdivision of particles so fine that the individual particles cannot be distinguished as such by pressing a powder between the thumb and index finger.

"Impedex."[28] Trademark for sodium propionate (q.v.).

imperial green. See copper acetoarsenite.

"Impermex."[236] Trademark for a water-dispersible organic colloid, developed for the purpose of decreasing the water loss of drilling muds, even in those contaminated with salt, salt water, cement, or other water-soluble electrolyte.

"Implex."[23] Trademark for thermoplastic, high-impact acrylic modling powder, supplied in natural and colored forms. Maximum toughness, gloss, stain- and heat-resistant grades.
Uses: Shoe heels, business-machine and musical instrument keys, housings, automotive parts, knobs, metallized parts, etc.

"Impranil."[422] Trademark for isocyanate-based adhesives used for textile finishes.

"Impregnole."[42] Trademark for a series of water-repellents.
FH. A zirconium-synthetic wax complex.
367. A solvent solution of a mixture of silicone polymers and other ingredients.

"Impruvol."[11] Trademark for a solution of di-tert-butyl-para-cresol in a high quality transformer oil.

impurity. The presence of one substance in another in such low concentration that it cannot be measured quantitatively by ordinary analytical methods. It is impossible to prepare an ideally pure substance. In certain metal crystal lattices foreign substances can exist in as low a concentration as one millionth of an atomic percent. For example, arsenic atoms are present in germanium crystals in this percentage; this fact is largely responsible for the semiconducting properties of germanium. Here the impurity is beneficial, but often it is detrimental, for example, in graphite used as a moderator in nuclear reactors, and in many metallic catalysts. In the air, trace amounts of sulfur dioxide and carbon monoxide are potentially dangerous impurities in concentrations of 5 ppm of SO_2 and 50 ppm for CO. See also purity, chemical; trace element; air pollution; semiconductor. See also purification.

"Imron."[28] Trademark for a polyurethane caulking compound and floor finish.

Imuran. See azothiaprine.

In Symbol for indium.

INAH. Abbreviation for isonicotinic acid hydrazide. See isoniazid.

"Inceloid."[176] Trademark for plastic sheets, rods, tubes, moulded products, adhesives, lacquers and dyes.

incendiary gel.

(1) Mixtures of thermite (q.v.) suspended in oil set to a jelly with a small amount of soap, which undergo spontaneous ignition on contact with air. Another type may contain magnesium in jellied oil.

(2) Jellied gasoline combined with thickening agents such as napalm (q.v.) or finely divided magnesium.

inclusion complex (adduct). An unbonded association of two molecules, in which a molecule of one component is either wholly or partly locked within the crystal lattice of the other. There are several types of such complexes, the most familiar being the so-called clathrates (from Latin, crossbars of a grating). The clathrate compound $3C_6H_4(OH)_2 \cdot SO_2$ may be depicted as:

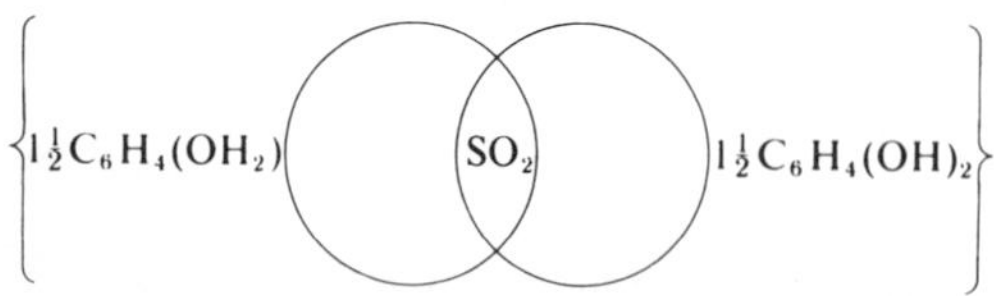

where the interlocked rings denote mutual enclosure of two identical cages. The formula for any clathrate compound is determined by the ratio of available cavities to the amount of cage material. Inclusion compounds can be used to separate molecules of different shapes, e.g., straight-chain hydrocarbons from those containing side chains, as well as structural isomers. They can also be used as templates for directing chemical reactions. See also clathrate compound; gas hydrate.

"Incoloy."[283] Trademark for a group of corrosion-resistant alloys of nickel, iron, and chromium.

"Inconel."[283] Trademark for a group of corrosion-resistant alloys of nickel and chromium.

"Indalone."[55] Trademark for n-butyl mesityl oxide oxalate (see butopyronoxyl). Used in insect repellents which can be applied directly on the skin. Low toxicity.

indan (hydrindene; 2,3-dihydroindene)

$CHCHCHCHCCCH_2CH_2CH_2$. Bicyclic.

Properties: Colorless liquid; b.p. 176.5°C; f.p. −51.4°C; refractive index 1.5388 (16.4°C); sp. gr. 0.965 (20/4°C); insoluble in water; soluble in alcohol and ether. Combustible.
Derivation: From coal tar.
Hazard: May be irritant to skin and eyes.
Use: Organic synthesis.

indanthrene. See indanthrone.

"Indanthrene."[307] Trademark for a series of vat dyestuffs used for dyeing and printing cotton, rayon, and silk. Characterized by excellent fastness to light, washing, chlorine, etc.

indanthrene yellow. See flavanthrene.

indanthrone (indanthrene) $C_{28}H_{14}N_2O_4$. A blue vat dye or pigment, also called Indanthrene Blue RS, (6,15-dihydro-5,9,14,18-anthrazinetetrone). The molecule consists of two anthraquinone nuclei linked through two NH groups.
Properties: Excellent durability and light-fastness, not decomposed by heating to 250°C. Soluble in dilute alkaline solutions.
Derivation: By fusion of beta-aminoanthraquinone with caustic potash in the presence of potassium nitrate.
Uses: Dyeing unmordanted cotton; pigment in quality paints and enamels.

indene $CHCHCHCHCCCHCHCH_2$.

Properties: Colorless liquid, sp. gr. 1.006 (20/4°C), f.p. −3.5°C; b.p. 182°C; refractive index 1.5726 (n 25/D); flash point 173°F. Insoluble in water; soluble in most organic solvents; oxidizes readily in air; forms polymers on exposure to air and sunlight. Combustible. Low toxicity.
Derivation: Contained in the fraction of crude coal tar distillates which boils from 176 to 182°C.
Uses: Preparation of coumarone-indene resins; intermediate.
See also coumarone.

Indian red (iron saffron). A red (maroon) pigment made by calcining copperas to obtain red ferric oxide. Fine particle size.
Uses: Pigment in paint, rubber, and plastics; polishing agent.
See also iron oxide red; rouge.

Indian yellow.

(1) (aureolin). A yellow pigment distinguished by being unaffected by hydrogen sulfide. It is durable, without action upon other pigments, and is permanent in oils and water color. It consists of a double nitrite of cobalt and potassium and is prepared by adding excess of potassium nitrite solution to a solution of cobalt nitrate acidified with acetic acid. See cobalt potassium nitrite.

(2) Also sometimes used for the yellow synthetic dye primuline.

indicator. An organic substance (usually a dye or intermediate) which indicates by a change in its color the presence or absence or concentration of some other substance, or the degree of reaction between two or more other substances. The most common example is the use of acid-base indicators such as litmus, phenolphthalein, and methyl orange to indicate the presence or absence of acids and bases, or the approximate concentration of hydrogen ion in a solution. Their chief use is in analytical chemistry. The pH ranges of several typical indicators are as follows:

		color in acid	color in base
methyl orange	3.1–4.4	red	yellow
phenolphthalein	8.3–10.0	colorless	red
phenol red	6.8–8.4	yellow	red
litmus	4.4–8.3	red	blue
Congo red	3.0–5.2	blue	red
bromthymol blue	6.0–7.6	yellow	blue
chlorphenol red	5.2–6.8	yellow	red
cresol purple	7.4–9.0	yellow	purple

See also titration; pH.

indigo (indigotin; synthetic indigo blue) $C_{16}H_{10}N_2O_2$. C.I. No. 73000. A double indole derivative.
Properties: Dark blue, crystalline powder; bronze luster; sp. gr. 1.35; sublimes at 300°C (decomposes); soluble in aniline, nitrobenzene, chloroform, glacial acetic acid and concentrated sulfuric acid; insoluble in water, ether, and alcohol.
Derivation: From aniline and chloroacetic acid, and fusing the resulting phenylglycine with alkali and sodium amide. Formerly from plants of genus Indigofera.
Grades: Technical; pure.
Containers: Tins; fiber drums.
Uses: Textile dyeing and printing inks; manufacture of indigo derivatives; paints; analytical reagents.

"Indigolite." Trademark for a combination of sodium formaldehyde sulfoxylate and a sulfonated quaternary base used to give a discharge on indigo-dyed grounds and discharge printing of vat dye-stuffs.

indirect dye. A mordant dye (q.v.).

indium In Metallic element of atomic number 49, of Group IIIA of the periodic system. Atomic weight 114.82; valences 1,3; two stable isotopes.
Properties: Ductile, shiny, silver-white metal; softer than lead. Soluble in acids. Insoluble in alkalies. Nontoxic. Sp. gr. 7.362; m.p. 156°C. b.p. 1450°C. Corrosion-resistant at room temperature, but oxidizes readily at higher temperatures.
Occurrence: Not found native, but in a variety of zinc and other ores. The indium content is generally very low, rarely exceeding 0.001%. Indium-bearing ores occur in western U.S., Canada, Peru, Japan, Europe, U.S.S.R.
Forms available: Small ingots or bars, shot, pencils, wire, sheets, powder, single crystals.
Purity: Technical; high purity (less than 10 ppm impurities).
Uses: Automobile bearings; electronic and semiconductor devices; low-melting brazing and soldering alloys; reactor control rods; electroplated coatings on silver-plated steel aircraft bearings which are tarnish-resistant; radiation detector.

indium acetylacetonate $In(C_5H_7O_2)_3$. M.p. 186°C. Used as a catalyst.

indium antimonide InSb. M.p. 535°C. Electronic grade is used for semiconductor devices.
Hazard: May be toxic.

indium arsenide InAs. Crystals; m.p. 943°C. Insoluble in acids. Electronic grade is used for semiconductor devices; in injection lasers.
Hazard: Highly toxic by ingestion.
Shipping regulations: (Rail, Air) Arsenical compound, n.o.s., Poison label.

indium chloride (indium trichloride) $InCl_3$.
Properties: White powder; deliquescent; soluble in alcohol and water. Sp. gr. 3.46 (25°C); m.p. 586°C; sublimes at 300°C.
Derivation: Direct union of the elements or by the action of hydrochloric acid on the metal.

indium oxide (indium sesquioxide; indium trioxide) In_2O_3.
Properties: White to light-yellow powder in both amorphous and crystalline forms, depending on temperature. Soluble in hot acid (amorphous), insoluble (crystalline); sp. gr. 7.179.
Derivation: By burning the metal in air or heating the hydroxide, nitrate, or carbonate.

indium phosphide InP. A brittle metallic mass; m.p. 1070°C. Slightly soluble in mineral acids. Electronic grade is used in semiconductor devices and injection lasers.

indium sesquioxide. See indium oxide.

indium sulfate $In_2(SO_4)_3$.
Properties: Grayish powder; deliquescent; soluble in water; sp. gr. 3.438; decomposed by heat.

indium trichloride. See indium chloride.

indium trioxide. See indium oxide.

"Indo Carbon."[307] Trademark for sulfur dyestuffs. Used for the dyeing and printing of cotton and rayon.

"Indocin." Trademark for the drug indomethacin (q.v.), used in treatment of arthritis.

"Indofast."[438] Trademark for vat dyestuff pigments, including carbazole dioxazine violet. Used in paints, printing inks, and plastics.

indole (2,3-benzopyrrole) CHCHCHCHCCCHCHNH.
Properties: White to yellowish scales, turning red on exposure to light and air; unpleasant odor in high concentrations, but pleasant in dilute solutions. Soluble in alcohol, ether, hot water, and fixed oils; insoluble in mineral oil and glycerol. M.p. 52°C; b.p. 254°C.
Derivation: From indigo, and by numerous syntheses. Also can be produced from 220 to 260° fraction from coal tar.
Grades: Technical; C.P.; F.C.C.
Hazard: A carcinogen.
Uses: Chemical reagent; perfumery.

3-indoleacetic acid (IA; IAA; beta-indoleacetic acid; beta-indolylacetic acid) $C_8H_6NCH_2COOH$. A plant hormone.
Properties: Crystals, m.p. 168–170°C. The natural material is levorotatory; specific rotation 20/D is −3.8° in alcohol. Insoluble in water and chloroform; soluble in alcohol and ether. Low toxicity.
Use: Agriculture and horticulture; growth-promoting hormone.
See also auxin; plant growth regulator.

indole-alpha-**aminopropionic acid.** See tryptophan.

3-indolebutyric acid (hormodin) $C_8H_6N(CH_2)_3COOH$.
Properties: White or off-white powder, essentially odorless. M.p. 123°C. Insoluble in water; soluble in alcohols and ketones. Low toxicity.
Containers: Glass bottles; fiber drums.
Uses: Plant hormone, especially used in rooting plants.
See also auxin; plant growth regulator.

indole-2,3-dione. See isatin.

indomethacin. USAN name for 1-(para-chlorobenzoyl)-5-methoxy-2-methylindole-3-acetic acid $C_{19}H_{16}ClNO_4$. Used in medicine (for treating arthritis).

"Indopol."[216] Trademark for synthetic mono-olefin polymers of relatively high molecular weight used in caulking compounds, industrial sealants, adhesives, electrical insulation surgical tapes, and also as chemical intermediates.

"Indulin."[228] Trademark for a series of lignin products obtained from wood pulp.
Containers: 50-lb multiwall paper bags.
Uses: Latex extension; protein removal and recovery; secondary oil recovery; in storage batteries as a negative plate expander; retarding agent for vat and sulfur dyes.

"Indusoil."[228] Trademark for distilled or fractionated tall oils and tall oil products.
Uses: For coatings and resins; drying oils; heavy-metal soaps; detergents; cleaners; polishes; core oils; lubricants; floor coverings and flotation.

industrial alcohol. See alcohol, industrial.

industrial carbon. See carbon, industrial.

industrial diamonds. See diamonds, industrial.

industrial dust. See dust, industrial

industrial waste. See waste treatment.

inert. A term used in chemistry to indicate chemical inactivity of an element or compound. Helium, neon and argon are inert gaseous elements; carbon dioxide is an inert gaseous compound. Ingredients added to mixtures chiefly for bulk and weight purposes are said to be inert. See also noble; extender.

inflammable. This term is no longer on approved use.

infrared. The region of the electromagnetic spectrum including wave lengths from 0.78 micron to about 300 microns (i.e., longer than visible light and shorter than microwave). (See also radiation).
Use: Spectroscopic analysis; medicine; baking of enamels; drying; photography.

infrared spectroscopy. An analytical technique which may measure either (1) the range of wavelengths in the infrared that are absorbed by a specimen, which characterize its molecular constitution (absorption spectroscopy), or (2) the infrared waves emitted by excited atoms or molecules (emission spectroscopy). Extremely hot bodies (stars) emit spectra in which the atomic composition can be determined by characteristic lines such as the sodium D line in the sun's spectrum. Infrared absorption bands identify molecular components and structures, some of which are:

absorption band (μ)	structure indicated
2.3–3.2	OH and NH groups; H_2
3.2–3.33	aromatics, olefins
3.33–3.55	aliphatics
5.7–6.1	aldehydes, ketones, acids, amides

See also microwave spectroscopy; absorption (2)

"Infrax."[280] Trademark for a refractory insulation used as primary linings of fuel-fired and electric furnaces only when protected by a cement facing. Available in brick form.

infusion. An aqueous solution obtained by treating drugs with hot or cold water, without boiling. Generally prepared by pouring boiling water upon the vegetable substance and macerating the mixture in a tightly closed vessel until the liquid cools. When not otherwise specified, they are of 5% strength by weight.

ingot iron. Highly refined steel with a maximum of 0.15% impurity. Due to high purity it has excellent ductility and resistance to rusting.

ingrain dye. An insoluble dye developed by impregnating a fabric with one or more intermediates and then producing the dye by reaction with a different intermediate.

inhibitor. (1) A compound (usually organic) that retards or stops an undesired chemical reaction, such as corrosion, oxidation, or polymerization. Examples are acetanilide which retards decomposition of hydrogen peroxide, and salicylic acid, used to prevent prevulcanization of rubber. Such substances are sometimes called negative catalysts. See also antioxidant. (2) A biological antagonist used to retard growth of pests and insects, and in medicine. See antagonist, structural.

"Inhibitor NPH."[329] Trademark for a synthetic organic chemical which provides an effective means of controlling hard polymer formation in synthetic rubber production.
Properties: Fine, white to yellow-white platelets; density 5 lb/gal; m.p. 160–164° (with decomposition); ammoniacal odor.

initiating explosive. See explosive, initiating.

initiator. An agent used to start the polymerization of a monomer. Its action is similar to that of a catalyst, except that is usually consumed in the reaction. Organic peroxides and similar compounds are often used, and short-wave radiation has a similar initiating effect. Free radicals usually play a part. See also activator; free radical.

injection molding. A plastic-molding operation, introduced about 1935, performed in a single machine capable of producing both small articles of complex geometry (combs) and large units that could not be made economically in any other way (auto body parts, tubs, etc.). Though used primarily for thermoplastics, the injection method can be modified to handle thermosets. A simplified description is as follows: (1) A molding powder is fed into the heating chamber of the machine, which holds several times as much material as is necessary to fill the mold. The powder is heated to a viscous liquid. (2) An amount of molding powder that is just sufficient to fill the mold cavity is then forced into the rear of the heating chamber by a plunger, thus injecting an equal amount of liquid plastic from the front of the heating chamber into the mold. (3) The material remains in the mold under high pressure until it cools, and is then ejected. The rheological properties of the fluid plastic are of critical importance, as it must flow readily through the sprue and mold gate and fill the mold uniformly. The amount of material injected into a mold can range from less than an ounce to 25 pounds or more. Modern injection-molding machines have many specialized mechanical features, and are of impressive dimensions.

ink. See printing ink; writing ink.

inorganic. Any chemical compound that does not contain the element carbon, with the exception of the oxides of carbon; compounds containing a carbonate group (CO_3) such as calcium carbonate; carbon disulfide; phosgene; carbonyl sulfide; and metallic carbonyls. Inorganic compounds comprise by far the greater part of the earth's crust, and range from those that are almost wholly inert (sand, limestone, clay) to highly active and corrosive materials (hydrofluoric acid). In recent years the most rewarding progress in inorganic chemistry has been in the field of coordination compounds and transition metals. See also organometallic.

inosine (hypoxanthine riboside) $C_{10}H_{12}N_4O_5$. An important intermediate in animal purine metabolism. Also available as its barium salt.
Properties: (dihydrate) Crystallizes in needles from water; m.p. 90°C; (anhydrous) 218°C (dec). Levorotatory in solution. Slightly soluble in water; soluble in alcohol.
Derivation: By deamination of adenosine.
Use: Biochemical research.

inosinic acid (IMP; hypoxanthine riboside-5-phosphoric acid). $C_{10}H_{13}N_4O_8P$. An important intermediate in the synthesis and metabolism of animal purines.
Properties: Syrup with agreeable sour taste. Freely soluble in water and formic acid; very sparingly soluble in alcohol and ether.
Derivation: From meat extract or by enzymatic deamination of muscle adenylic acid.
Uses: Biochemical research; flavor enhancer (see sodium inosinate).

inositol (hexahydroxycyclohexane) $C_6H_6(OH)_6 \cdot 2H_2O$. A constituent of body tissue. There are 9 isomeric forms of inositol (myo-inositol or meso-inositol or specifically the cis-1,2,3,5-trans-4,6-hexahydroxycyclohexane) is the one having vitamin activity.
Properties (i-inositol): White crystals; odorless; m.p. 224–227°C; sweet taste; soluble in water; insoluble in absolute alcohol and ether; stable to heat, strong acid and alkali; sp. gr. 1.524 (dihydrate); 1.752 when anhydrous. Dihydrate melts at 215–216°C. Nontoxic.
Source: Food source: Vegetables, citrus fruits, cereal grains, liver, kidney, heart and other meats. Commercial source: From corn steep liquor by precipitation and hydrolysis of crude phytate.
Units: Amounts are expressed in milligrams of inositol.
Grades: N.F.; F.C.C.
Containers: Bottles, cans; drums.
Uses: Medicine; nutrition; intermediate.

inositol hexaphosphoric acid. See phytic acid.

inositol hexaphosphoric acid ester, sodium salt. See sodium phytate.

INPC. Abbreviation for isopropyl N-phenylcarbamate. See IPC.

insecticide. A type of pesticide designed to control insect life that is harmful to man, either directly as disease vectors, or indirectly as destroyers of crops, food products, or textile fabrics. General types are as follows: (1) Inorganic; arsenic, lead, and copper (inorganic compounds and mixtures); the use of these has diminished sharply in recent years because of the development of more effective types that are less toxic to man. (2) Natural organic compounds, such as rotenone and pyrethrins (relatively harmless to man, since they quickly decompose to nontoxic compounds), nicotine, copper naphthenate, and petroleum derivatives. (3) Synthetic organic compounds: (a) chlorinated hydrocarbons, such as DDT, dieldrin, endrin, chlordane, lindane, p-dichlorobenzene; (b) the organic esters of phosphorus (the parathions and related substances). Another group of organics, of which imidazole is typical, act on the principle of metabolic inhibition (see antimetabolite). Besides direct application, these can be fed to growing plants, with the result that the plants can no longer serve as nutrients for the specific insect.
Hazard: Insecticides are toxic to man in varying degrees. Among the safest are the pyrethrins, rotenone, and methoxychlor. Most of the organophosphorus types (parathion and related compounds) are highly toxic, but are reasonably biodegradable. The chlorinated hydrocarbons resist biodegradation; their use has been restricted, and in some cases (DDT) banned, for agricultural application, due to their harmful ecological effects. The EPA has carefully evaluated substitutes for DDT and has approved several types for some uses of DDT; among those approved, are demeton, parathion, phorate and methyl parathion.
See also pesticide; fumigant; herbicide; rodenticide; repellent.

"Insecti-sol."[25] Trademark for a highly purified odorless solvent exceeding CSMA minimum requirements for an insecticide base. The deodorized hydrocarbon distillate is water white and has a 170°F (min.) flash point, a 465–480°F distillation end point, and a 98% unsulfonatable residue. Also useful in food processing and cosmetic manufacture. Combustible.

insect wax. See Chinese wax.

"Insidol."[300] Trademark for a wetting agent used in textile processing. It is a sulfodicarboxylic acid ester composition.

"Instant Acid."[233] Trademark for a dry acid in granular form. Water-soluble and easily stored. Used in oil-well production.

"Instant-Dri."[108] Trademark for a solid rinsing agent. Reduces surface tension of water.

instantizing. A method of increasing the solubility of dehydrated food products in powder form by moistening the surface of the powders enough to cause them to form aggregates or clumps, after which the material is redried. The resulting product will have slightly darker color and about half the density of the original dehydrated material, but it will be more readily reconstituted.

"Instantreat."[108] Trademark for a high purity specially processed powdered complex phosphate. Contains enough chlorinated ingredient to disinfect feed solution and feed equipment.
Uses: Controls corrosion, lime scale and red water trouble in homes, hotels, restaurants and small industrial water systems.

instrumental analysis. See analytical chemistry; spectroscopy.

"Insular."[52] Trademark for polyvinyl chloride resins identified by type numbers corresponding roughly to their molecular weights PVC 250, PVC 230, PVC 200, PVC 185, and PVC 155.

insulating oil. See transformer oil.

"Insulgrease."[245] Trademark for a series of grease-like silicone dielectric compounds. Used as electrical insulation in connectors, electrical and electronic equipment, as heat transfer media, moisture and corrosion resistant coatings, anti-seize compounds, and insulator coatings.

insulin. A polypeptide hormone having a molecular weight of 5733. It is formed in the isles of Langerhans located in the pancreas, and was so named for this reason. Insulin is composed of 16 amino acids arranged in a coiled chain and cross-linked in several places by the disulfide bonds of cystine residues. The sequence of amino acids has been elucidated. The insulin molecule was synthesized in 1963, but commercial production has not yet been attained. Insulin regulates carbohydrate metabolism in the body by decreasing the blood glucose level. A systemic deficiency leads to diabetes.

Properties: White powder or hexagonal crystals; readily soluble in dilute acids; soluble in water.

Derivation: Extraction of minced pancreas with acidified dilute alcohol, followed by precipitation with absolute alcohol.

Grades: U.S.P., in various solutions of suspensions, which include insulin injection; isophane insulin suspension; protamine zinc insulin suspension; N.F., as globin zinc insulin injection.

Containers: Glass bottles, ampules, phials.

Hazard: Overdosage can be fatal.

Uses: Medicine (for diabetes control).

See also Banting.

"Insuloxide."[337] Trademark for a mixture containing a minimum of 94% zirconium oxide and a maximum of 5% silicon dioxide. M.p. 4500°F. Used for thermal insulation; super-refractories.

"Insurok."[63] Trademark for a series of laminated and molding plastics characterized by durability, lightweight, resistance to many chemicals, and high dielectric strength.

Use: Replacement for cast aluminum in airplane parts; switches, distributors, commutators, etc.

"Intalox."[326] Trademark for a particular shape of tower filling materials available in procelain, chemical stoneware and carbon.

interface. The area of contact between two immiscible phases of a dispersion, which may involve either the same or different states of matter. Five types are possible: (1) solid/solid (carbon black/rubber); (2) liquid/liquid (water/oil); (3) solid/gas (smoke/air); (4) solid/liquid (clay/water); (5) liquid/gas (water/air). At a fresh surface of either liquid or solid the molecular attraction exerts a net inward pull. Hence the characteristic property of a liquid is surface tension (q.v.) and that of a solid surface is adsorption (q.v.). Both have the same cause, namely, the inward cohesive forces acting on the molecules at the surface. These phenomena provide to some degree the fundamental mechanism for many industrially important processes (catalysis, emulsification, mixing, alloying) and products (detergents, adhesives, lubricants, paints). Such properties as wettability of solid powders, spreading coefficients of liquids, and protective action of colloidal substances are intimately associated with interfacial behavior. See also surface, surface tension, catalysis, emulsion, detergent, wetting agent.

interferon. An antiviral protein produced by vertebrate cells in response to virus infection. Discovered in 1957, it is a product of the infected cell, rather than of the disease-inducing virus. It can be formed by any type of infected cell, and is non-specific in its protection, i.e., it is effective against many viruses, but only in the cells of the organism that produced it. It is quite different from an antibody, which is produced only by specialized cells, and acts by combining directly with a specific virus. Interferon does not inactivate viruses directly, but reacts with susceptible cells, which then resist virus multiplication. The molecular weight of interferon is only 30,000 compared with 100,000 or more for antibody. The mechanism of its formation is controlled by RNA, but only when the latter is double-stranded and protein-free. A plastic copolymer called "Pyran" (q.v.), which has a structure resembling RNA, has been found to increase the formation of interferon in the body. See also antigen-antibody; virus.

intermediate. An organic compound, either cyclic (derived from coal tar or petroleum products such as benzene, toluene, naphthalene, etc.) or acyclic (e.g., ethyl and methyl alcohol). These compounds may be considered as chemical stepping stones between the parent substance and the final product. The cyclic type (e.g., aniline, beta-naphthol, and benzoylbenzoic acid) still predominate as intermediates for synthetic dyes and have few other uses; the acyclic type in general have many independent uses. Exceptions are hexamethylenetetramine, an acyclic intermediate for phenolformaldehyde resin, and butadiene for synthetic elastomers. Intermediates are the foundation of the modern approach to organic technology. The distinction between an intermediate and an end product is not always precise.

intermetallic compound. (1) A compound or alloy formed by two metals that have been placed in intimate contact during the process of brazing or coating, the compound occurring at the interface between the metal surfaces. In some cases, as in galvanizing, the metals form bimetallic compounds; in others they form alloys or solid solutions of varying composition. (2) A two- or three-component metal system prepared with special metals having semiconducting properties, e.g., gallium, for use in lasers, diodes, transistors, etc. See also galvanizing; semiconductor

International Union of Pure and Applied Chemistry (IUPAC). A voluntary, nonprofit association of national organizations representing chemists in 45 member countries. It was formed in 1919 with the object of facilitating international agreement and uniform practice in both academic and industrial aspects of chemistry. Examples are nomenclature, atomic weights, symbols and terminology, physico-chemical constants, and certain methods of analysis and assay. Reports on such subjects are presented at the biennial conferences of the Union by the numerous Commissions dealing with them, discussed, and, after approval, are published. Its offices are in Basle, Switzerland.

interstitial. (1) Descriptive of a nonstoichiometric compound of a metal and a nonmetal whose structure conforms to a simple chemical formula, but exists over a limited range of chemical composition. Interstitial compounds are represented by borides, nitrides and carbides of the transition metals. (2) Descriptive of an atom of an impurity that causes a defect or dislocation in a crystal lattice, e.g., an atom of carbon or nitrogen in an iron crystal, or of arsenic in a semiconductor. (3) In a biological sense, the term describes cells located between or within layers of tissue.

intumescence. The foaming and swelling of a plastic or other material when exposed to high surface temperatures or flames. Used with respect to polyurethane base coating materials for rocket reentry.

"Invar." Trademark for an iron-nickel alloy containing 40–50% nickel and characterized by an extremely low coefficient of thermal expansion.
Uses: Precision instruments, measuring tapes, weights, etc.

invention. The chief requirement of an invention is that it be unobvious to a person having ordinary skill in the art to which the claim pertains and knowing everything that has gone before, as shown by publication anywhere or public use in the U.S. When reliance for patentability (q.v.) is placed on a new mixture of components that have been used separately, it must be shown that there is some unexpected coaction between the ingredients and not just the additive effects of the several materials. The inventor is the one who contributes the inventive concept in workable detail, not the one who demonstrates it or tests it.

"Invermul."[236] Trademark for a basic emulsifier in the preparation of water in oil-type emulsion drilling muds for oil wells.

invertase (sucrase; invertin). Enzyme produced by yeast and by the lining of the intestines. It is a white powder, soluble in water. It catalyzes the conversion of sucrose (ordinary sugar) to glucose and levulose (fructose) during fermentation of sugars.
Uses: Production of invert sugar (q.v.) for syrups and candy; analytical reagent for sucrose.

invert sugar. A mixture of 50% glucose and 50% fructose obtained by the hydrolysis of sucrose. It absorbs water readily, and is usually only handled as a syrup. Because of its fructose content, invert sugar is levorotatory in solution, and sweeter than sucrose. Invert sugar is often incorporated in products where loss of water must be avoided. Commercially it is obtained from the inversion of a 96% cane sugar solution.
Use: Food industry; brewing industry; medicine; confectionery; tobacco curing.

investment casting. See lost wax process.

in vitro. A condition in which a reaction is carried out in a laboratory experiment (i.e., a glass container, test tube, or beaker), as opposed to a reaction occurring in a living organism (in vivo).

"Iobac."[284] Trademark for a "Tamed Iodine" compound combining the germicidal values of iodine with adequate detergent and wetting ingredients. Used for santizing dairy equipment.

"Ioclide."[239] Trademark for a water-soluble form of iodine, containing a nonionic detergent. Contains 15.8% polyethoxopolypropoxypolyethoxyethanol-iodine complex, 15.2% nonylphenoxypolyethoxyethanol-iodine complex (provides 3.1% available iodine), and 0.2% hydrogen chloride, remainder water and other inert ingredients. Used as a germicide.

iodeosin (tetraiodofluorescein) $C_{20}H_8I_4O_5$.
Properties: Red powder; soluble in dilute alkalies; slightly soluble in alcohol and ether; insoluble in water.
Derivation: Interaction of fluorescein and iodine in presence of iodic acid.
Use: Indicator in analytical chemistry.

iodic acid HIO_3.
Properties: Colorless, rhombic crystals or white, crystalline powder. A moderately strong acid. Soluble in cold and hot water. Sp. gr. 4.629; m.p. 110°C (decomposes).
Derivation: By adding sulfuric acid to a solution of barium iodate and subsequent filtration and crystallization.
Hazard: Toxic by ingestion; strong irritant to eyes and skin.
Uses: Analytical chemistry; medicine (1 to 3% sol'n).

iodic acid anhydride (iodine pentoxide) I_2O_5.
Properties: White, crystalline powder. Soluble in water and nitric acid; insoluble in absolute alcohol, chloroform, ether, carbon disulfide. Sp. gr. 4.799; m.p. 300°C (decomposes).
Hazard: Toxic by ingestion; strong irritant to eyes and skin. Strong oxidizing agent.
Uses: Oxidizing agent; organic synthesis.

iodine I Nonmetallic, halogen element of atomic number 53, group VIIA of the periodic table; the least reactive of the halogens. Atomic weight 126.9045; Valences 1, 3, 5, 7; no stable isotopes but many artificial radioactive isotopes.
Properties: Heavy, grayish-black plates or granules, having a metallic luster and characteristic odor. Readily sublimed, having a violet vapor. Sp. gr. 4.98; m.p. 113.5°C; b.p. 184°C. Soluble in alcohol, carbon disulfide, chloroform, ether, carbon tetrachloride, glycerol, and alkaline iodide solutions, insoluble in water. Noncombustible.
Derivation: (a) from brine wells in Michigan, Oklahoma, Japan, Indonesia; (b) from mother liquors of Chilean saltpeter. It can also be extracted from kelp.
Method of purification: Sublimation.
Grades: Crude; C.P.; U.S.P.
Containers: Bottles; 100- and 200-lb kegs; drums.
Hazard: Highly toxic by ingestion and inhalation; strong irritant to eyes and skin. Tolerance, 0.1 ppm in air.
Uses: Dyes (aniline dyes, phthalein dyes); catalyst; iodides; iodates; pharmaceuticals; process engraving and lithography; special soaps; analytical reagent; iodized salt; medicine; in special lubricants for titanium and stainless steel parts. See also iodine tincture.

iodine 131 (I-131). Radioactive iodine of mass number 131.
Properties: Half-life 8 days; radiation, beta and gamma.
Derivation: By pile irradiation of tellurium and from the fission products of nuclear reactor fuels.
Forms available: As elemental iodine and in a weakly basic solution of sodium iodide in sodium sulfite; iodine 131 is also available in tagged compounds such as dithymol diiodide, potassium iodate, diiodofluorescein, insulin, ACTH, etc. See also Grades.
Grades: U.S.P. lists iodinated I-131 serum albumin, rose bengal sodium I-131 injection, sodium iodide I-131 capsules and solution, sodium iodohippurate I-131 injection.
Hazard: Moderately toxic.
Uses: Diagnosis and treatment of goiter hyperthyroidism, and other thyroid disorders; internal radiation therapy; in film gauges to measure film thicknesses of the order of one micron; for detecting

leaks in water lines; as a source of radiation in oil field tests; as a tracer in chemical analysis; as a tracer in studying diet iodine for cattle, the functions of the thyroid gland, the efficiency of mixing pulp fibers, the thermal stability of potassium iodate in bread dough, chemical reaction mechanisms, etc.
Shipping regulations: Consult regulations.

iodine bisulfide. See sulfur iodine.

iodine bromide. See iodine monobromide.

iodine chloride. See iodine monochloride and iodine trichloride.

iodine cyanide. See cyanogen iodide.

iodine disulfide. See sulfur iodine.

iodine monobromide (bromine iodide) IBr.
Properties: Crystalline, purplish-black mass. Soluble in water with decomposition, alcohol, and ether. M.p. 42°C; b.p. 116°C (dec); sp. gr. 4.41.
Derivation: By the interaction of iodine and bromine.
Hazard: Highly toxic by ingestion and inhalation; vapors corrosive to tissue.
Use: Organic synthesis.

iodine monochloride ICl.
Properties: Reddish-brown, oily liquid; two solid forms, alpha and beta. Soluble in alcohol, water (with decomposition), and dilute hydrochloric acid. M.p. (alpha) 27°C, (beta) 14°C; b.p. 101°C (dec); sp. gr. (alpha) 3.18 (0°C), (beta), 3.24 (liquid, at 34°C).
Derivation: By the action of dry chlorine on iodine.
Hazard: Highly toxic by ingestion and inhalation; strong irritant to eyes and skin.
Uses: Analytical chemistry; organic synthesis.
Shipping regulations: (Rail) White label. (Air) Corrosive label. Not acceptable on passenger planes.

iodine number (iodine value). The percentage of iodine that will be absorbed by a chemically unsaturated substance (vegetable oils, rubber, etc.) in a given time under arbitrary conditions. A measure of unsaturation.

iodine pentafluoride IF_5.
Properties: Fuming liquid; b.p. 98°C; m.p. 9.4°C; sp. gr. (liquid) 3.189 (25°C); attacks glass.
Derivation: By igniting iodine in a stream of fluorine. Available in cylinders at 98.0% min purity.
Hazard: Highly toxic by ingestion and inhalation. Corrosive to skin and mucous membranes. Reacts violently with water. Dangerous fire risk. Tolerance (as F), 2.5 mg per cubic meter of air.
Uses: Fluorinating and incendiary agent.
Shipping regulations: (Air) Corrosive label. Not acceptable on passenger planes.

iodine pentoxide. See iodic acid, anhydride.

iodine tincture. A solution of iodine and potassium iodide or sodium iodide in alcohol; a reddish-brown liquid having the odors of iodine and alcohol; contains from 44–50% by volume of alcohol, 2 g of iodine and 2.4 g sodium iodide per 100 cc.
Grade: U.S.P.
Hazard: Toxic by ingestion. Avoid using on open cuts.
Use: Antiseptic (use on skin surface only).

iodine trichloride ICl_3.
Properties: Orange-yellow, deliquescent, crystalline powder; pungent, irritating odor. Soluble in water (with decomposition), alcohol and benzene. M.p. 33°C; sp. gr. 3.11.
Derivation: By interaction of iodine and chlorine.
Hazard: Toxic by ingestion and inhalation; corrosive to tissue.
Uses: Medicine; agent for introducing iodine and chlorine in organic synthesis.

iodine value. See iodine number.

iodipamide $(CH_2)_4(CONHC_6HI_3COOH)_2$.
Properties: White, nearly odorless, crystalline powder, very slightly soluble in alcohol, chloroform, ether, and water; pH of saturated solution is between 3.5 and 3.9.
Hazard: Probably toxic.
Use: Medicine (x-ray contrast medium).

iodisan. See propiodal.

iodized oil. An iodine addition product of vegetable oil or oils containing 38–42% organically combined iodine.
Properties: Thick, viscous, oily liquid. Alliaceous odor. Oleaginous taste. Affected by air and light. Soluble in solvent naphtha.
Hazard: Probably toxic by ingestion.
Use: Medicine (radiopaque medium).

iodoethane. See ethyl iodide.

iodoethylene. See tetraiodoethylene.

iodoform (triiodomethane) CHI_3.
Properties: Small, greenish yellow or lustrous crystals or powder; penetrating odor. Soluble in alcohol, glycerol, chloroform, carbon disulfide and ether; insoluble in water. Sp. gr. 4.08; m.p. 115°C.
Derivation: (a) By heating acetone or methyl alcohol with iodine in presence of an alkali or alkaline carbonate. (b) Electrolytically, by passing a current through a solution containing potassium iodide, alcohol and sodium carbonate.
Grades: Technical; N.F.
Hazard: Moderately toxic and irritant. Decomposes violently at 400°F.
Use: Medicine (antiseptic for external use only).

7-iodo-8-hydroquinoline-5-sulfonic acid (ferron; loretin; 8-quinolinol-7-iodo-5 sulfonic acid) $C_9H_4N(I)(OH)(SO_3H)$.
Properties: Sulfur-yellow, almost odorless, tasteless, crystalline powder; m.p. 260–270°C, with some decomposition. Slightly soluble in water and alcohol; insoluble in ether.
Derivation: Obtained from the potassium salt of 8-hydroxyquinoline-5-sulfonic acid by the action of potassium iodide, bleaching powder and hydrochloric acid.
Hazard: May be toxic.
Use: Medicine; colorimetric reagent for ferric iron.

iodomethane. See methyl iodide.

iodopanoic acid. See iopanoic acid.

iodophor.
(1) A complex of iodine with certain types of surface-active agents that have detergent properties.
(2) More generally, any carrier of iodine.

iodophosphonium. See phosphonium iodide.

2-iodopropane. See isopropyl iodide.

iodopyracet $C_5H_2I_2NOCH_2COONH_2(CH_2CH_2OH)_2$. A sterile water solution is listed in NF.
Hazard: May be toxic by ingestion.
Use: Medicine (radio-opaque iodine compound).

N-**iodosuccinimide** (succiniodimide) $(—CH_2CO)_2NI$.
Properties: Colorless crystals; m.p. 200–201°C; solu-

ble in acetone, methanol; insoluble in carbon tetrachloride and ether. Decomposes in water.
Hazard: May be toxic by ingestion.
Use: Iodinizing agent in synthetic organic chemistry.

"Iokel."[244] Trademark for a phosphoric acid solution of a polyoxyethanol aklyl phenol condensate complex of elemental iodine.
Properties: Viscous, dark liquid with acidic reaction. Miscible in all proportions with water.
Uses: Detergent sanitizer; milk stone remover.

"Iomag."[329] Trademark for a potassium iodide mixture containing 90% potassium iodide and made free-flowing with 8% magnesium carbonate and 2% potassium hydroxide.

ion. An atom or radical that has lost or gained one or more electrons, and has thus acquired an electric charge. Positively charged ions are cations, and those having a negative charge are anions. An ion often has entirely different properties from the element (atom) from which it was formed. Examples: In sodium chloride solutions, the sodium exists as sodium ion (Na^+), i.e., sodium atoms that have lost one electron. The chlorine is present as chloride ion (Cl^-), i.e., chlorine atoms that have gained one electron. Copper sulfate solution contains copper ion (Cu^{++}), i.e., copper atoms that have lost two electrons, and sulfate ion (SO_4^{--}), i.e., sulfate radicals that have gained two electrons.

Ions occur in water solution or in the fused state (except in the case of gases). Compounds that form ions are called electrolytes, because they enable the solution to conduct electricity. Ion formation causes an abnormal increase in the boiling point of water and also lowers the freezing point, the extent depending on the concentration of the solution. Ions are also formed in gases as a result of electrical discharge. See also ionizatoin; electrolysis; ion exchange.

ion exchange. A reversible chemical reaction between a solid (ion exchanger) and a fluid (usually a water solution) by means of which ions may be interchanged from one substance to another The superficial physical structure of the solid is not affected. The customary procedure is to pass the fluid through a bed of the solid, which is granular and porous, and has only a limited capacity for exchange. The process is essentiallay a batch type in which the ion exchanger, upon nearing depletion, is regenerated by inexpensive brines, carbonate solutions, etc. Ion exchange occurs extensively in soils.

Ion exchange resins are synthetic resins containing active groups (usually sulfonic, carboxylic, phenol, or substituted amino groups) that give the resin the property of combining with or exchanging ions between the resin and a solution. Thus a resin with active sulfonic groups can be converted to the sodium form and will then exchange its sodium ions with the calcium ions present in hard water. "Amberlite" resins are of this type.

Some specific applications of ion exchange: Water softening; milk softening (substitution of sodium ions for calcium ions in milk); removal of iron from wine (substitution of hydrogen ions); recovery of chromate from plating solutions; uranium from acid solutions; streptomycin from broths; removal of formic acid form formaldehyde solutions; demineralization of sugar solutions; recovery of valuable metals from wastes; recovery of nicotine from tobacco-dryer gases; catalysis of reaction between butyl alcohol and fatty acids; recovery and separation of radioactive isotopes from atomic fission; chromatography; establishment of mass micro standards; in cigarette filters to remove polonium from smoke.
See also zeolite.

ion-exchange chromatography. A chromatographic method based on the ability of polymers to sorb ionized solutes reversibly, e.g., cross-linked resins with exchangeable hydrogen or hydroxyl ions. It can be carried out both in columns and on sheets.

ion-exchange resin. See ion exchange.

ion exclusion. The process in which a synthetic resin of the ion exchange type absorbs nonionized solutes such as glycerine or sugar while it does not absorb ionized solutes that are also present in a solution in contact with the resin. Thus sodium chloride and glycerine can be separated by passage of their aqueous solution through a bed of particles of an ion exclusion resin.

ionic atmosphere. A diffuse population of ions surrounding a charged particle in a solution; the particle may be an ion, a molecule, or a colloidal micelle. Most of the ions in the "atmosphere" have a charge opposite from that of the particle they surround, and the net strength of the charge diminishes with their distance from the central particle.

ionic bond. See bond, chemical.

ionic detergent. See detergent.

"Ionite."[271] Trademark for a lignite-type material used for oil-well drilling muds, as a low-density filler in dark-colored rubber, as an organic base filler for fertilizers, and as a source of humic acid.

ionization. The separation or dissociation of a molecule into ions of opposite electrical charge. This occurs spontaneously in many salts when dissolved in water, or melted. Thus sodium chloride yields positive sodium ions and negative chloride ions. Molecules or atoms of gases are ionized by passage of an electric current through the gas. In this case electrons are removed, leaving a positive charge. See also ion; electrochemistry; electromotive series.

ionizing radiation. See radiation, ionizing.

"Ionol."[125] Trademark for 2,6-di-tert-butyl-4-methylphenol.

ionomer resin. A copolymer of ethylene and a vinyl monomer with an acid group such as methacrylic acid. They are crosslinked polymers in which the linkages are ionic as well as covalent bonds (see bond, chemical). There are positively and negatively charged groups which are not associated with each other, and this polar character makes these resins unique.
Properties: Transparent, electrically conductive, resilient and thermoplastic. Cannot be completely dissolved in any commercial solvents. High resistance to abrasion, cracking, corona attack; high tensile strength. Working temperature about −160 to 150°F.
Uses: Break-resistant transparent bottles; packaging films; mercury flasks; protective equipment; pipe and tubing; electric distribution elements. As foam, insulation of fresh concrete.

ionone (alpha- or beta-cyclocitrylidenacetone) $C_{13}H_{20}O$.
Properties: Light yellow to colorless liquid; violet odor; sp. gr. 0.927–0.933 (25°C); b.p. 126–128°C (12 mm); refractive index (n 20/D) 1.4970–1.5020. Soluble in alcohol, ether, mineral oil, and propylene glycol; insoluble in water and glycerin.
Grade: 95%; 99%; mixed isomers; F.C.C.
Derivation: (1) Condensing citral with acetone, followed by ring closure with an acid; (2) reaction of acetylene with acetone, followed by hydrogenation, condensation with diketone, and further reaction with acetylene.
Uses: Perfumery; chemical synthesis; flavoring; vitamin A production (beta isomer).

ion retardation. A process based on amphoteric (bifunctional) ion-exchange resins containing both anion and cation adsorption sites. These sites will associate with mobile anions and cations in solution and thus remove both kinds of ions from solutions. These ions may be eluted by rinsing with water. Process can make clean separations of ionic-nonionic mixtures. Has also been suggested for demineralization of salt solutions. See also ion.

iopanoic acid (iodopanoic acid; 3-amino-alpha-ethyl-2,4,6-triiodohydrocinnamic acid) $C_6HI_3NH_2CH_2CH(COOH)C_2H_5$.
Properties: Cream-colored, tasteless powder with faintly aromatic odor; m.p. 152–158°C (dec); darkens on exposure to light; soluble in acetone, ether, alcohol, chloroform and dilute alkalies; insoluble in water.
Grade: U.S.P.
Hazard: May be toxic.
Use: Medicine (radiopaque medium).

IPA. (1) Abbreviation for isophthalic acid. (2) Abbreviation for isopropyl alcohol.

IPAE. See isopropylaminoethanols.

IPC (INPC; isopropyl N-phenylcarbamate) $C_6H_5NHCOOCH(CH_3)_2$.
Properties: White to gray crystalline needles, odorless when pure; m.p. 84°C (technical grade); soluble in alcohol, acetone, isopropyl alcohol; slightly soluble in water.
Hazard: Toxic by ingestion and inhalation.
Use: Pre-emergence herbicide.

IPC, chloro-. See chloro-IPC.

"IPCF."[177] Trademark for isopropyl chloroformate (q.v.).

ipecac. Dried root of Cephaelis ipecacuanha.
Habitat: Brazil and Bolivia; cultivated in India.
Grades: Technical; U.S.P.
Hazard: May have adverse side effects.
Use: Medicine (emetic); production of emetine.

"IPP."[177] Brand name for isopropyl percarbonate. See diisopropyl peroxydicarbonate.

Ir Symbol for iridium.

"Irco" Aluminum Coatings.[526] Trademark for chromate conversion coatings for aluminum which provide corrosion resistance by themselves and also increase the corrosion resistance and adhesion of subsequent paint films. Application of spray, dip or roller coat.

"Irco" Bond.[526] Trademark for crystalline zinc phosphates used as a base for oil and paint, as an aid to drawing and forming of steel, as a nonembrittling etchant for steel, and as a wearing surface on certain types of steel parts. Application by dip, automatic or barrel methods.

"Iroc" Iron Phosphates.[526] Trademark for specially blended compounds of acidic salts which produce an iridescent iron phosphate coating on cold-rolled steel surfaces, providing paint base by increasing adhesion and corrosion resistance. Application by spray or immersion.

"Ircolene."[526] Trademark for special polar-type rust-proofing oils for indoor-outdoor protection of ferrous or nonferrous metals. Application by spray, dip or dip-spin centrifugal methods.

"Irco Lube.[526] Trademark for oil-absorptive manganese phosphate coatings for mating moving steel surfaces. Application by dip, automatic or barrel methods.

"Irganox."[219] Trademark for a series of complex, high-molecular-weight stabilizers that inhibit oxidation and thermal degradation of many organic materials. They contain multifunctional chemical groupings. Several are hindered polyphenols. All are white, crystalline, free-flowing powders; non-staining, non-volatile, and odorless.

iridic chloride (iridium chloride; iridium tetrachloride) $IrCl_4$.
Properties: Brownish-black mass. Hygroscopic. Soluble in water, alcohol, and dilute hydrochloric acid.
Derivation: Action of chlorine or aqua regia on the ammonium salt, $(NH_4)_2IrCl_6$.
Uses: Analysis (testing for nitric acid in the presence of nitrous acid); microscopy; plating solution.

iridium Ir Metallic element of atomic number 77, one of the platinum metals, group VIII of the periodic system. Atomic weight 192.22; valences 1,2,3,4,6; two stable isotopes.
Properties: Silver-white. Low ductility. Does not tarnish in air. On heating strongly a slightly volatile oxide is formed. Insoluble in acids, slowly soluble in aqua regia and in fused alkalies. Density 22.65 g/cc (20°C) (calculated) (the heaviest element known); m.p. 2443°C; b.p. about 4500°C. Most corrosion-resistant element; modulus of elasticity is one of highest: 75,000,000 psi. Brinell hardness (cast) 218. Highly resistant to chemical attack. Toxicity probably low.
Occurrence: Canada, So. Africa, U.S.S.R., Alaska.
Derivation: Occurs with platinum; remains insoluble when the crude platinum is treated with aqua regia; occurs as iridosmine. The powder is obtained by hydrogen reduction of ammonium chloroiridate.
Forms: Powder; single crystals.
Use: Alloy with platinum for ammonia fuel-cell catalyst; electric contacts and thermocouples; commercial electrodes, and resistance wires; laboratory ware; extrusion dies for glass fibers; jewelry. Primary standards of weight and length.

iridium 192. Radioactive iridium of mass number 192.
Properties: Half-life, 74 days; radiation, beta and gamma.
Derivation: Pile irradiation of iridium.
Forms available: Iridium metal, or potassium or sodium chloroiridate in hydrochloric acid solution.
Uses: Radiography of light castings; treatment of cancer.
Shipping regulations: Consult regulations.

iridium bromide. See iridium tribromide.

iridium chloride. See iridic chloride; iridium trichloride.

iridium potassium chloride (potassium chloroiridate; potassium iridium chloride) K_2IrCl_6.
Properties: Dark-red crystals. Soluble in water (hot). Sp. gr. 3.549.
Use: Black pigment (porcelain decoration).
The Ir (III) salt is also known: K_3IrCl_6, greenish yellow.

iridium sesquioxide Ir_2O_3.
Properties: Black powder. Slightly soluble in hydrochloric acid (conc); insoluble in water. Dec. at 400°C.
Derivation: By heating the chloroiridate K_2IrCl_6 with sodium carbonate.
Use: Ceramics (porcelain decoration).

iridium tetrachloride. See iridic chloride.

iridium tribromide (iridium bromide) $IrB_3 \cdot 4H_2O$. Olive-green, brown or black crystals; soluble in water; insoluble in alcohol. Prepared by action of hydrobromic acid on iridium trihydroxide.

iridium trichloride (iridium chloride) $IrCl_3$. Dull green to blue-black particles; m.p. 763°C. (dec); insoluble in water and alcohol. Prepared by action of chlorine on iridium powder at 600°C.

iridosmine (osmiridium). A natural alloy of iridium and osmium containing some platinum, rhodium, ruthenium, iron, copper, palladium. Tin-white to light steel gray in color; streak, same; metallic luster. Composition is variable ranging from 10.0–77.2% iridium, 17.2–80.0% osmium, 0–10.1% platinum, 0–17.2% rhodium, 0–8.9% ruthenium, 0–1.5% iron, 0–0.9% copper, trace, palladium, Unattacked by aqua regia. Sp. gr. 18.8–21.12; Mohs hardness 6–7.
Occurrence: Alaska; South Africa.
Uses: Fountain-pen point tips; surgical needles; watch pivots; compass bearings; hardening platinum (standard weights, jewelry); source of iridium and osmium.

Irish moss See carrageenan.

"Irisol."[307] Trademark for a fast alizarine direct violet.

"Irisone."[227] Trademark for a series of compounds consisting of alpha- and beta-ionones.

"IRN."[299] Trademark for a group of magnetic iron oxides, black ferroso-ferric oxides (Fe_3O_4) and brown gamma ferric oxides (Fe_2O_3), with application in ferric cores and electronic parts, magnetic inks, data-handling accounting systems, recording and instrumentation tapes.

iron Fe Metallic element of atomic number 26; Group VIII of the Periodic Table. Atomic weight 55.847; valences 2, 3. Four stable isotopes; four artificially radioactive isotopes.
Properties: Silver-white, malleable metal; tensile strength 30,000 psi; Brinell hardness 60; m.p. 1536°C; b.p. 3000°C; specific gravity 7.87 (20°C); magnetic permeability 88,400 gauss (25°C). The only metal that can be tempered (q.v.). Mechanical properties are altered by impurities, especially carbon.

Iron is highly reactive chemically; a strong reducing agent; oxidizes readily in moist air; reacts with steam when hot to yield hydrogen and iron oxides. Dissolves in nonoxidizing acids (sulfuric and hydrochloric) and in cold dilute nitric acid. For biochemical properties, see below.
Major ores: Hematite, limonite, magnetite, siderite; also taconite (low-grade, 25–30% iron).
Occurrence: Minnesota (Mesabi), Alabama, Labrador, Yukon, Europe, South America.
Major types:

(1) Molten (or pig) iron. Derived (a) by smelting ore with limestone and coke in blast furnaces (purity 91–92%); (b) by continuous direct reduction in which iron ore and limestone are preheated in a fluidized bed, followed by heating to 1700°F, by melting at 3500°F, and reduction to iron at 3000°F with powdered coal (purity 99%) (proprietary process).
Uses: Steels of various types; other alloys (cast and wrought iron); source of hydrogen by reaction with steam.
See also steel.

(2) Powdered iron. Derived (a) by treatment of ore or scrap with hydrochloric acid to give ferrous chloride solution, which is then purified by filtration, vacuum-crystallized, and dehydrated to ferrous chloride dehydrate powder; this is reduced at 800°C to metallic iron (briquettes or powder) of 99.5% purity. (b) By thermal decomposition of iron carbonyl [$Fe(CO_5)$] at 250°C (99.6–99.9% pure); (c) By hydrogen reduction of high-purity ferric oxide or oxalate. (d) By electrolytic deposition from solutions of a ferrous salt (99.9% pure)
Uses: Powder metallurgy products; magnets; high-frequency cores; auto parts; catalyst in ammonia synthesis.

(3) Cast iron and wrought iron are mixtures of iron and other materials; see specific entry.

(4) Single crystals and whiskers are also available.

(5) Iron sponge (q.v.)

Hazard: Tolerance (as oxide fume): 10 mg per cubic meter of air. Dust and fine particles suspended in air are flammable and an explosion risk.
Biochemistry: Iron is a constituent of hemoglobin (q.v.) and is essential to plant and animal life; an important factor in cellular oxidation mechanisms. Used in medicine and dietary supplements. Essentially nontoxic.
For further information refer to the American Iron and Steel Institute, 150 East 42 St. New York, N.Y.

iron 55. Radioactive iron of mass number 55.
Properties: Half-life 2.91 years; decays through K capture. See iron 59.
Hazard: Highly toxic.
Shipping regulations: Consult regulations.

iron 59. Radioactive iron of mass number 59.
Properties: Half-life 46.3 days; radiation, beta and gamma.
Derivation: Pile irradiation of iron metal, giving a product which contains iron 55 impurity. Both iron 55 and iron 59 are produced pure in cyclotron. (See iron 55). Enriched samples of each are also available.
Hazard: Highly toxic.
Uses: Medicine; tracer element in biochemical and metallurgical research.
Shipping regulations: Consult regulations.

iron acetate liquor (iron liquor; black liquor; black mordant; iron pyrolignite).
Properties: Intensely black liquor, sometimes containing copperas or tannin. Absorbs oxygen from the air. Sp. gr. 1.09–1.115, containing 5–5.5% iron.
Derivation: (a) By the action of pyroligneous acid on

iron filings; (b) double decomposition of ferrous sulfate with calcium pyrolignite (calcium acetate).
Uses: Mordant, especially for alizarine and nitroso dyes.

iron alum. See ferric potassium sulfate.

iron black (contains no iron).
Properties: Fine black powder.
Derivation: Action of zinc upon an acid solution of an antimony salt, a black antimony being precipitated as a fine powder.
Uses: Imparting the appearance of polished steel to papier maché and plaster of Paris.

iron blue. A pigment (of which there are several varieties) prepared by precipitating a ferrous ferrocyanide from a soluble ferrocyanide and ferrous sulfate. Subsequent oxidation produces a complex ferriferrocyanide whose shade and pigment properties are dependent upon the oxidizing agent, reactant concentrations, pH, temperature, size of batch and other conditions of manufacture. Common oxidants are nitric acid, sulfuric acid, potassium dichromate with sulfuric acid, perchlorates, and peroxides. Low toxicity.
Properties: Insoluble in water, oils, alcohol, hot paraffin, organic solvents, and unaffected by dilute acids. Unstable to alkalies of all concentrations or reducing media. Resistant to light and ordinary baking temperatures.
Containers: Barrels, carloads.
Uses: Paints, printing inks, plastics; cosmetics (eye shadow); artist colors, laundry blue, paper dyeing, fertilizer ingredient, baked enamel finishes for autos and appliances; industrial finishes.

iron carbonate, precipitated. See iron oxide, brown.

iron carbonyl. See iron pentacarbonyl.

iron, cast. See cast iron.

iron compounds. See corresponding ferric or ferrous compounds.

iron, electrolytic. See iron.

iron mass. See iron sponge (2).

iron octoate. Same as ferric octoate. Used as a catalyst for curing silicone resins and rubbers.

iron-ore cement. Cements in which ferric oxide replaces a large part of the alumina. There must be some alumina present, however. Iron-ore cement is rather slow setting and hardening, but is more resistant to sea water than is Portland cement. It is light to chocolate brown in color and has a specific gravity about 3.31, higher than Portland cement.

iron ore, chrome. See chromite.

iron ore, magnetic. See magnetite.

iron ore, red. See hematite, red.

iron oxide, black (ferrosoferric oxide; ferroferric oxide; iron oxide, magnetic; black rouge)
$FeO \cdot Fe_2O_3$ or Fe_3O_4. See also the mineral form, magnetite.
Properties: Reddish or bluish black amorphous powder; sp. gr. 5.18; m.p. 1538°C (dec); soluble in acids; insoluble in water, alcohol and ether.
Derivation: (a) Action of air, steam or carbon dioxide on iron. (b) Specially pure grade by precipitating hydrated ferric oxide from a solution of iron salts, dehydrating and reducing with hydrogen. (c) Occurs in nature as the mineral magnetite.
Grades: Technical; pure (96% min).
Containers: Bags; carloads.
Uses: Pigment; polishing compound; metallurgy; medicine; magnetic inks and in ferrites for electronic industry; coatings for magnetic tape.

iron oxide, brown (iron subcarbonate; iron carbonate, precipitated)
Properties: Reddish-brown powder, containing ferric carbonate with ferric hydroxide $Fe(OH)_3$, and ferrous hydroxide $Fe(OH)_2$ in varying quantities. Not a true oxide. Soluble in acids; insoluble in water and alcohol.
Derivation: By the interaction of solution of ferrous sulfate and sodium carbonate.
Grade: Technical.
Containers: Bags; carloads.
Use: Paint pigment.

iron oxide, hydrated. See ferric hydroxide.

iron oxide, metallic brown. A naturally occurring earth, principally ferric oxide, to which extenders have been added. Used as a paint pigment.

iron oxide process. A process for the removal of sulfides, from a gas by passing the gas through a mixture of iron oxide, Fe_2O_3, (called iron sponge or iron mass) and wood shavings. The iron oxide is converted to iron sulfide and can be regenerated by allowing the iron sulfide to contact air. See iron sponge, spent.

iron oxide red (burnt sienna, Indian red; red iron oxide; red oxide; rouge (1); Turkey red). Pigments composed mainly of ferric oxide, Fe_2O_3. See also ferric oxide for a description of the pure material.
Containers: Multiwall paper bags; fiber drums.
Uses: Marine paints, metal primers; polishing compounds; pigment in rubber and plastic products; theatrical rouge; grease paints.

iron oxide yellow. Hydrated ferric oxide, $Fe_2O_3 \cdot H_2O$. A precipitated pigment of finer particle size and greater tinctorial strength than the naturally occurring oxides such as ocher (q.v.); excellent lightfstness and resistance to alkali.
Containers: Fiber drums, multiwall paper bags.
Uses: Paints; rubber products; plastics.

iron pentacarbonyl (iron carbonyl) $Fe(CO)_5$.
Properties: Mobile yellow liquid. Evolves carbon monoxide on exposure to air or to light. Soluble in nickel tetracarbonyl and most organic solvents; soluble with decomposition in acids and alkalies; insoluble in water. Sp. gr. 1.466 (18°C); b.p. 102.8°C (749 mm); decomposes at 200°C; m.p. −21°C. Flash point 5°F.
Derivation: Finely divided iron is treated with carbon monoxide, in the presence of a catalyst such as ammonia.
Hazard: Toxic by ingestion, inhalation, and skin absorption. Flammable, dangerous fire risk.
Uses: Organic synthesis; antiknock agent; source of carbonyl iron powder.
Shipping regulations: Flammable liquid, n.o.s. (Rail) Red label. (Air) Flammable Liquid label.

iron powder. See iron.

iron protochloride. See ferrous chloride.

iron protoiodide. See ferrous iodide.

iron protosulfide. See ferrous sulfide.

iron pyrite. See pyrite.

iron pyrolignite. See iron acetate liquor.

iron pyrophosphate. See ferric pyrophosphate.

iron red. Red varieties of ferric oxide that are used as pigments. See iron oxide red.

iron sponge.
(1) Sponge iron, a finely divided porous form of iron made by reducing an iron oxide at such low temperatures that melting does not occur, usually by mixing iron oxide and coke and applying limited increase in temperature.
Uses: For precipitating copper or lead from solutions of their salts, and in electric furnace steel operations.
(2) Iron mass (iron oxide). Finely divided iron oxide, distributed on a support so as to give a large surface area. One form is a mixture of wood shavings covered with a hydrated iron oxide. This may be made by mixing wet wood shavings with iron borings or similar material and allowing rusting to occur to produce finely divided iron oxide. In another method wood shavings are mixed with a slurry of the hydrated ferric oxide produced in purifying alum, and then dried. The iron sponge or iron mass is used for removing sulfur from coal gas or similar materials. See iron oxide process. See also iron sponge, spent.
Shipping regulations: (not properly oxidized) (Rail) Yellow lable; not accepted by express. (Air) Not acceptable. Not restricted if properly oxidized.

iron sponge, spent (iron mass, spent; spent oxide). Iron sponge after saturation with sulfur. It is liable to spontaneous heating. See iron oxide process.
Shipping regulations: (Rail) Yellow label; not accepted by express. (Air) Not accepted.

iron, stainless. Alloys containing 3 to 38% chromium, with or without traces of nickel; essentially magnetic and ferritic in character. High chromium irons are brittle after welding. Most popular composition for fabrication is 15–18% chromium, 0.1% carbon (max). See ferritic stainless steel, under steel, stainless.

irradiation. Exposure to radiation of wavelengths shorter than those of visible light (gamma, xray or ultraviolet) either for medical purposes (cancer therapy, removal of skin blemishes), for destruction of bacteria in milk and other foodstuffs, or for inducing polymerization of monomers or vulcanization of rubber. UV irradiation was formerly used to induce activation of vitamin D in milk, and has been used for some time to sterilize the air in operating rooms, etc.

"Irrathene."[245] Trademark for a thermosetting form of polyethylene (q.v.) formed by irradiation of polyethylene with high energy cathode rays (electrons). Does not melt (up to 250°C) but oxidizes rapidly at elevated temperatures unless protected by an inhibitor. Its resistance to acids, alkalies, and solvents is superior to that of polyethylene and it has excellent electrical resistance even at 200°C. It is available in films and tapes used for packaging and electrical insulation.

irreversible. A permanent chemical or physicochemical change. Examples are: (1) A chemical reaction that can proceed only to the right, giving a product that is stable and that cannot revert to the original constituents; most reactions are of this type. (2) A colloidal system that cannot be restored to its original form after coagulation or precipitation, for example, the hardening of egg white or milk protein by heat, the formation of butter from milk (mechanical action), and the coagulation of rubber latex by acid (chemical action).
See also: reversible.

"Irron."[169] Trademark for 8-quinolinol-7-iodo-5-sulfonic acid used in the colorimetric determination of iron.

"IR-S Latices."[52] Trademark for a series of emulsion copolymers of butadiene and styrene.
Uses: Paper and fabric impregnations; adhesives for laminating; textile and carpet backing; mastics; tire cord dipping; asphalt emulsions; caulking compounds; chewing gum bases.

"Irtran" 4.[115] Trademark for an optical material made from zinc selenide (q.v.)

ISAF black. Abbreviation for intermediate super abrasion furnace black.

isanolic acid (erythrogenic acid). An 18-carbon fatty acid with acetylene triple bonds at the 9 and 11 positions and having also an ethylene bond but no hydroxyl substituent.

isatin (ortho-aminophenylglyoxalic lactim; ortho-aminobenzoylformic anhydride; isatic acid anhydride; indole-2,3-dione) $C_6H_4COC(OH)N$(bicyclic).
Properties: Yellowish-red or orange cyrstals; bitter taste. Soluble in water, alcohol, and ether. M.p. 200–203°C.
Derivation: From indigo by oxidation.
Uses: Dyestuffs; pharmaceuticals; analytical reagent.

isatoic anhydride $C_8NO_3H_5$. A bicylic molecule composed of a benzene ring attached at the ortho and meta positions to a heterocyclic ring. It forms useful anthranilic acid derivatives by reaction with hydrogen.
Properties: Tan powder.
Grade: Technical, 96% min.
Containers: Bags; drums; bulk.
Uses: Intermediate for polymer curing agents, anthranilic esters, heterocyclic compounds, and benzyne.

isethionic acid (2-hydroxyethanesulfonic acid) $HOCH_2CH_2SO_3H$.
Properties: Liquid; b.p. 100°C (dec); very soluble in water; insoluble in alcohol.
Uses: Detergents; surfactants; synthesis.

"Ismelin" Sulfate.[305] Trademark for guanethidine sulfate; used in medicine.

iso-. A prefix meaning "the same," as in such terms as isomer (the same part), isotope (the same place) isometric (the same measure), isobar (the same pressure), etc. In organic chemistry it denotes an isomer of a compound, especially one having a subordinate chain of one or more carbon atoms attached to a carbon of the straight chain. See isobutane; branched chain.

isoalloxazine (flavin) $C_{10}H_6N_4O_2$. A derivative of isoalloxazine widely distributed in plants and animals, usually as yellow pigments. See also riboflavin.

isoamyl acetate $CH_3COOCH_2CH_2CH(CH_3)_2$.
Properties: Colorless liquid; b.p. 142°C; f.p. −78.5°C; sp. gr. (15/4°C) 0.876; wt/gal (15°C) 7.30 lb; flash point 80°F. Slightly soluble in water; miscible with alcohol and ether. Banana-like odor.
Derivation: Rectification of commercial amyl acetate.
Grades: Reagent; technical.
Hazard: Flammable, moderate fire risk. Moderately toxic, and irritant. Tolerance, 100 ppm in air. Explosive limits in air 1 to 7.5%.
Uses: Flavoring; perfumes; solvent.
See also amyl acetate.

isoamyl alcohol, primary (3-methyl-1-butanol; isobutyl carbinol) $(CH_3)_2CHCH_2CH_2OH$.
Properties: Colorless liquid; pungent taste; disagreeable

odor. B.p. 132.0°C; f.p. −117.2°C; sp. gr. 0.813 (15/4°C); wt/gal 6.79 lb; refractive index (n 20/D) 1.4075; flash point (closed cup) 109°F; autoignition temp. 657°F. Slightly soluble in water; miscible with alcohol and ether. Combustible.
Derivation: Distillation of fusel oil or the mixed alcohols resulting from the chlorination and hydrolysis of pentane.
Grade: Technical.
Containers: Drums, tank cars.
Hazard: Moderate fire risk. Moderately toxic and irritant. Tolerance, 100 ppm in air. Explosive limits in air 1.2 to 9%.
Uses: Photographic chemicals; organic synthesis; pharmaceutical products; medicine; solvent; determination of fat in milk; microscopy; flavoring.
See also fusel oil.

isoamyl alcohol secondary. Similar to the primary form except b.p. is 113°C; flash point 103°F, density 0.819. combustible.

isoamyl benzoate (amyl benzoate) $C_6H_5COOC_5H_{11}$.
Properties: Colorless liquid, with fruity odor. Sp. gr. 0.986 to 0.989; refractive index 1.493; b.p. 260°C. Soluble in water. Combustible.
Uses: Perfumery; flavors.

isoamyl benzyl ether (benzyl isoamyl ether) $C_5H_{11}OCH_2C_6H_5$.
Properties: Colorless liquid; fruity odor; sp. gr. 0.904–0.908; refractive index 1.481–1.485; soluble in 4 parts of 80% alcohol. Combustible.
Grade: Technical.
Use: Soap perfumes.

isoamyl butyrate $C_5H_{11}OOCC_3H_7$.
Properties: Practically water-white. Sp. gr. 0.866 to 0.868 (15.5°C); boiling range 150 to 180°C. Soluble in alcohol and ether; very slightly soluble in water. Combustible. Low toxicity.
Derivation: By treating isoamyl alcohol with butyric acid.
Method of purification: Distillation.
Grades: Commercial, 95 to 100% ester content; F.C.C. (as amyl butyrate).
Containers: Carboys; tin-lined drums.
Uses: Flavoring extracts; solvent and plasticizer for cellulose acetate.

isoamyl chloride $C_5H_{11}Cl$. Any of several compounds or mixtures thereof may be referred to by this name, since numerous isomers are possible, the most common of which is 1-chloro-3-methylbutane $(CH_3)_2CH(CH_2)_2Cl$. Combustible. Low toxicity.
Properties: Colorless or slightly yellow liquid; b.p. 99.7°C (758 mm); sp. gr. 0.893; refractive index 1.410; insoluble in water; soluble in alcohol and ether.
Derivation: Isoamyl alcohol and hydrogen chloride, or chlorination of isopentane.
Uses (Mixtures, usually also containing normal amyl chloride):solvent (nitrocellulose, varnishes, lacquers, neoprene); rotogravure inks; soil fumigation; organic compounds.

isoamyldichloroarsine $C_5H_{11}AsCl_2$.
Properties: Oily liquid. Sweetish odor. Decomposed by water. B.p. 88.5 to 91.5°C (15 mm). Combustible.
Derivation: Interaction of phosphorus trichloride and isoamylarsenic acid.
Hazard: Highly toxic; strong irritant.
Shipping regulations: (Rail, Air) Arsenic compounds, n.o.s., Poison label.

alpha-**isoamylene.** See 3-methyl-1-butene.

beta-**isoamylene.** See 3-methyl-2-butene.

isoamyl furoate $C_4H_3OCO_2C_5H_{11}$.
Properties: Colorless liquid, becoming brown in light. Insoluble in water; soluble in alcohol and ether. Sp. gr. 1.0335 (20/4°C); b.p. 232 to 234°C, 135–137°C (25 mm); refractive index 1.4720.

isoamyl isovalerate. See isamyl valerate.

sec-**isoamyl mercaptan** (2-methylbutyl 3-thiol) $(CH_3)_2CHCH(SH)CH_3$.
Properties: Colorless liquid; distillation range 101–127°C; sp. gr. 0.841 (20/4°C); density 7.04 lb/gal (60°F); refractive index 1.445 (20°C); flash point 46°F.
Containers: Up to 54-gal.
Hazard: Probably toxic. Flammable, dangerous fire risk.
Uses: Polymerization modifier; insecticide intermediate; vulcanization accelerator intermediate; nonionic surface-active agent.
Shipping regulations: Flammable liquid, n.o.s., (Rail) Red label. (Air) Flammable Liquid label.

isoamyl nitrite. See amyl nitrite.

isoamyl pelargonate $(CH_3)_2CHCH_2CH_2OOCC_8H_{17}$.
Properties: Liquid with fruity odor; sp. gr. 0.860 (15.5/15.5°C); b.p. 260°C; refractive index 1.4300 (20°C). Combustible. Low toxicity.
Uses: Flavors and perfumes; chemical intermediate.

isoamyl propionate. See amyl propionate.

isoamyl salicylate (amyl salicylate) $C_6H_4OHCOOC_5H_{11}$.
Properties: Water-white liquid sometimes having a faint yellow tinge which should not be pink or red. Orchid-like odor. Sp. gr. 1.053–1.059 (15°C); refractive index (n 20/D) 1.5050–1.5080; optical rotation 0 to +2.30°; b.p. 280°C. Soluble in alcohol, ether; insoluble in water and glycerol. Combustible. Low toxicity.
Derivation: By esterifying salicylic acid with amyl alcohol. The usual article of commerce is the isoamyl ester.
Method of purification: Distillation.
Grades: A pure grade of at least 99% ester content which should not exceed 100% on analysis (indicating lower esters).
Containers: Carboys; drums.
Uses: Soap perfumes.

isoamyl valerate ("apple essence"; "apple oil"; isoamyl isovalerate; amyl valerianate; amyl valerate) $C_4H_9CO_2C_5H_{11}$.
Properties: Clear liquid; odor of apples when diluted with alcohol; sp. gr. 0.8812; b.p. 203.7°C; soluble in alcohol and ether; slightly soluble in water. Combustible. Low toxicity.
Derivation: By adding sulfuric acid to a mixture of amyl alcohol and valeric acid. Subsequent recovery by distillation.
Grades: Technical; F.C.C. (as isoamyl isovalerate).
Uses: Medicine; fruit essences; flavoring agent.

isoascorbic acid. See erythorbic acid.

isobars. Nuclides having the same mass number but different atomic numbers, in contrast to isotopes, which have the same atomic number but different mass numbers. C-14 and N-14 are isobars.

isobenzan. Generic name for 1,3,4,5,6,7,8,8-octachloro-1,3,3a,4,7,7a-hexahydro-4,7-methanoisobenzofuran, $C_9H_6OCl_8$.
Properties: Solid; m.p. 248–257°C. Soluble in acetone and ether; slightly soluble in alcohols and kerosine; insoluble in water.

Hazard: Probably toxic.
Use: Insecticide.

isoborneol $C_{10}H_{17}OH$. A geometrical isomer of borneol.
Properties: White solid with camphor odor; m.p. 216°C (sublimes). More soluble in most solvents than borneol. Low toxicity.
Derivation: By reduction of camphor.
Uses: Perfumery; chemical esters.

isobornyl acetate $C_{10}H_{17}OOCCH_3$.
Properties: Colorless liquid. Pine-needle odor. Sp. gr. 0.978 ± 0.001 (20°C); b.p. 220 to 224°C. Soluble in most fixed oils and in mineral oil. Insoluble in glycerol and water. Combustible. Low toxicity.
Derivation: By heating camphene (50 to 60°C) with glacial acetic acid and sulfuric acid and separating by adding water.
Grades: Technical; F.C.C.
Uses: Compounding pine-needle odors; toilet waters; bath preparations; antiseptics; theater sprays; soaps; making synthetic camphor; flavoring agent.

isobornyl salicylate.
Properties: Viscous, colorless oil. Sweet odor. Ester content 96%. Combustible. Low toxicity.
Grade: Technical.
Uses: Perfumery (fixative); cosmetics (filter for suntan preparations).

isobornyl thiocyanoacetate $C_{10}H_{17}OOCCH_2SCN$. The technical grade contains 82% or more of isobornyl thiocyanoacetate, also other terpenes and derivatives.
Properties: Yellow, oily liquid; terpene-like odor. Sp. gr. (25/4°C) 1.1465; refractive index (25/D) 1.512; acid number 1.19. Very soluble in alcohol, benzene, chloroform, and ether; practically insoluble in water. Relatively low toxicity. Combustible.
Derivation: By treating isoborneol with chloroacetyl chloride and potassium thiocyanate.
Uses: Insecticide, cattle spray; medicine.

isobornyl valerate $(CH_3)_2CHCH_2COOC_{10}H_{17}$.
Properties: Colorless, neutral liquid, oily taste, peculiar aromatic odor. Does not irritate the stomach. Soluble in alcohol, ether; sparingly soluble in water. B.p. 132–138°C (12 mm); sp. gr. 0.954.

isobutane (2-methylpropane; trimethylmethane). Molecular formula $(CH_3)_2CHCH_3$; Structural formula (branched-chain):

```
      H         H         H
      |         |         |
  H—C————C————C—H
      |         |         |
      H     H—C—H      H
                |
                H
```

A liquefied petroleum gas.
Properties: Colorless gas; slight odor; stable, does not react with water; has no corrosive action on metals; b.p. −11.73°C; f.p. −159.6°C; sp. gr. 0.5572 (20°C, at saturation pressure); sp. gr. (air = 1) 2.01; soluble in water; slightly soluble in alcohol; soluble in ether; flash point −117°F; autoignition point 864°F. Nontoxic.
Derivation: An important component of natural gasoline, refinery gases, wet natural gas; also obtained by isomerization of butane.
Grades: Technical; 99 mole % (pure grade), 99.96 mole % (research grade), and other high-purity grades.
Containers: Pressure cylinders; tank cars; tank trucks.
Hazard: Highly flammable; dangerous fire and explosion risk. Explosive limits in air 1.9 to 8.5%.
Uses: Organic synthesis; refrigerant; fuel; aerosol propellant; high-octane gasoline (aviation fuel); synthetic rubber; instrument calibration fluid.
Shipping regulations: (Rail) Red Gas label. (Air) Flammable Gas label. Not acceptable on passenger planes.

isobutane hydrate. See gas hydrates.

isobutanol. See isobutyl alcohol.

isobutanolamine. See 2-amino-2-methyl-1-propanol.

isobutene (2-methylpropene; isobutylene) $(CH_3)_2C:CH_2$. A liquefied petroleum gas.
Properties: Colorless volatile liquid or easily liquefied gas; coal gas odor; b.p. −6.9°C; f.p. −139°C; flash point −105°F; sp. gr. 0.6 (20°C); soluble in organic solvents. Polymerizes easily and also reacts easily with numerous materials. Nontoxic. Autoignition point 869°F.
Derivation: Gas mixtures containing considerable isobutene are obtained by fractionation of refinery gases resulting from cracking of petroleum.
Containers: Tank cars; cylinders.
Hazard: Highly flammable; dangerous fire and explosion risk. Explosive limits in air 1.8 to 8.8%.
Uses: Production of isooctane, high-octane aviation gasoline; butyl rubber, polyisobutene resins, tert-butyl chloride, tert-butanol methacrylates and other derivatives; copolymer resins with butadiene, acrylonitrile, etc.
Shipping regulations: (Rail) Red Gas label. (Air) Flammable Gas label. Not acceptable on passenger planes.

isobutyl acetate $C_4H_9OOCCH_3$.
Properties: Colorless, neutral liquid; fruit-like odor. Soluble in alcohols, ether, and hydrocarbons; immiscible with water. B.p. 116–117°C; flash point 64°F (c.c.) sp. gr. 0.8685 (15°C); refractive index 1.392 (approx.); wt/gal 7.23 lb; f.p. −99°C. Autoignition temp. 793°F.
Derivation: Treating isobutyl alcohol with acetic acid in the presence of catalysts.
Grades: Technical; solvent; perfume; F.C.C.
Containers: 55-gal drums; tank cars.
Hazard: Flammable, dangerous fire risk. Tolerance, 150 ppm in air.
Uses: Solvent for nitrocellulose; in thinners, sealants, and topcoat lacquers; perfumery; flavoring agent.
Shipping regulations: (Air) Flammable Liquid label.

isobutyl acrylate $(CH_3)_2CHCH_2OOCCH:CH_2$.
Properties: Liquid; b.p. 61–63°C (51 mm); sp. gr. 0.884 (25°C); refractive index 1.4124 (25/D); flash point 86°F (TOC). Contains 100 ppm monomethyl ether hydroquinone as inhibitor.
Containers: Tank cars and trucks.
Hazard: Flammable, moderate fire risk.
Use: Monomer for acrylate resins.

isobutyl alcohol (isopropylcarbinol; 2-methyl-1-propanol) $(CH_3)_2CHCH_2OH$.
Properties: Colorless liquid. Soluble in water, alcohol, and ether. Sp. gr. 0.806 (15°C); b.p. 107°C; flash point 100°F (open cup); f.p. −108°C; refractive index 1.397 (15°C). Autoignition point 800°F.
Derivation: Byproduct of synthetic methanol production, purified by rectification.
Containers: 55-gal drums, tank cars and trucks.

Hazard: Flammable, moderate fire risk. Moderately toxic; strong irritant. Tolerance, 100 ppm in air.
Uses: Organic synthesis; latent solvent in paints and lacquers; intermediate for amino coating resins; substitute for *n*-butyl alcohol. Paint removers; fluorometric determinations; liquid chromatography.

isobutyl aldehyde. See isobutyraldehyde.

isobutylamine $(CH_3)_2CHCH_2NH_2$.
Properties: Colorless liquid; amine odor; strongly caustic. Soluble in water, alcohol, ether, and hydrocarbons. Sp. gr. 0.731 (20°C); boiling range 66–69°C: flash point 15°F. Autoignition temp. 712°F.
Containers: 55-gal drums; tank cars.
Hazard: Flammable; dangerous fire risk. Toxic; strong irritant to skin and mucous membranes.
Uses: Organic synthesis; insecticides.
Shipping regulations: (Rail) Red label. (Air) Flammable Liquid label.

isobutyl-para-aminobenzoate $NH_2C_6H_4COOCH_2CH(CH_3)_2$.
Properties: White crystalline scales; m.p. 64–65°C. Almost insoluble in water; soluble in alcohol and vegetable oils.
Use: Medicine.

isobutylbenzene $(CH_3)_2CHCH_2C_6H_5$.
Properties: Liquid; sp. gr. 0.8532 (20/4°C); f.p. −51.6°C; b.p. 171.1°C; refractive index 1.486 (20°C); flash point 140°F. Autoignition temp. 802°F. Combustible.
Hazard: Moderate fire risk. May be toxic in high concentrations.
Uses: Pharmaceutical intermediate; perfume synthesis; flavoring.

isobutyl benzoate (eglantine) $C_6H_5CO_2CH_2CH(CH_3)_2$.
Properties: Colorless liquid; characteristic odor; sp. gr. 1.002; b.p. 237°C; insoluble in water; miscible with alcohol and ether. Combustible; low toxicity.
Uses: Perfumes; flavors.

isobutyl carbinol. See isoamyl alcohol, primary.

isobutyl cinnamate $C_4H_9OOCCH{:}CHC_6H_5$. Colorless oil, amber fragrance. Sp. gr. 1.001 to 1.004; refractive index 1.541. Soluble in 2 vols. of 70% alcohol. Combustible; low toxicity.
Use: Perfumery.

isobutyl cyanoacrylate. A tissue adhesive effective in surgery and medicine to retard bleeding from internal organs. Also applicable to the mounting of pearls and other jewels. Available as an aerosol spray.

isobutylene. See isobutene.

isobutylene-isoprene rubber. See butyl rubber.

isobutyl furoate $C_4H_3OCO_3C_4H_9$.
Properties: Colorless liquid becoming brown in light. Insoluble in water; soluble in alcohol and ether. Sp. gr. 1.0383 (26.5/4°C); b.p. 221–223°C (corr.); refractive index 1.4676 (26.5°C). Combustible; low toxicity.

N-isobutylhendecenamide. See N-isobutylundecylenamide.

isobutyl heptyl ketone. See 2,6,8-trimethyl-4-nonanone.

isobutyl isobutyrate (IBIB) $(CH_3)_2CHCOOCH_2CH(CH_3)_2$.
Properties: Colorless liquid; fruity odor; sp. gr. 0.853–0.857 (20/20°C); f.p. −80.7°C; b.p. 148.7°C; refractive index (n 20/D) 1.3999; insoluble in water; soluble in alcohol and ether. Combustible.
Containers: Drums; tank cars and trucks.
Use: Flavoring; insect repellent; nitrocellulose lacquers and thinners; substitute for methyl amyl acetate.

isobutyl mercaptan. See 2-methyl-1-propanethiol.

isobutyl methacrylate $(CH_3)_2CHCH_2OOCC(CH_3){:}CH_2$.
Properties: Liquid; density 0.882 g/ml (25/15°C); boiling range 155°C; refractive index 1.4172 (25°C); flash point 120°F (TOC). Contains 25 ppm hydroquinone monomethyl ether as inhibitor. Combustible.
Containers: Tanks.
Hazard: Moderate fire risk.
Use: Monomer for acrylic resins.

isobutyl phenylacetate $(CH_3)_2CHCH_2OOCCH_2C_6H_5$.
Properties: Colorless liquid; honey-like odor. Soluble in most fixed oils; insoluble in glycerin, mineral oil, and propylene glycol. Sp. gr. 0.984–0.988 (25°C); refractive index 1.4860–1.4880 (20°C). Combustible.
Grades: Technical; F.C.C.
Uses: Flavoring agent; perfumes.

isobutyl propionate $CH_3CH_2COOCH_2CH(CH_3)_2$.
Properties: Water-white liquid; sp. gr. 0.86–0.8635 (20/20°C); b.p. 138°C; f.p. −71.4°C; refractive index 1.3975 (20°C); insoluble in water; very soluble in alcohol and ether. Combustible; low toxicity.
Containers: 55-gal drums; tank cars.
Use: Paint, varnish, and lacquer solvent.

isobutyl salicylate $HOC_6H_4COOCH_2CH(CH_3)_2$.
Properties: Colorless liquid. Sp. gr. 1.064–1.065 (25°C); may have slightly yellowish tinge; b.p. 259°C. Soluble in alcohol and mineral oil; insoluble in water and glycerol. Combustible; low toxicity.
Grades: Technical; F.C.C.
Uses: Perfumery; flavoring.

N-isobutylundecylenamide (N-isobutylhendecenamide) $CH_3(CH_2)_7CH{:}CHCONHC_4H_9$. A synergist for pyrethrum, used in insecticides.

isobutyl vinyl ether. See vinyl isobutyl ether.

isobutyraldehyde (isobutyl aldehyde) $(CH_3)_2CHCHO$.
Properties: Transparent, colorless, highly refractive liquid; pungent odor. Sp. gr. 0.794 (20/4°C); b.p. 65°C; f.p. −66°C; refractive index (n 20/D) 1.3730; flash point (closed cup) −40°F, (open cup) −11°F, explosive limits in air, 1.6 to 10.6. Autoigition point 490°F. Soluble in alcohol; insoluble in water.
Derivation: (a) Oxo process reaction of propylene with carbon monoxide and hydrogen; (b) dehydrogenation of isobutyl alcohol.
Containers: 55-gal drums; tank cars.
Hazard: Highly flammable; dangerous fire and explosion risk. Irritant to skin and eyes.
Uses: Intermediate for rubber antioxidants and accelerators, for neopentyl glycol; organic synthesis.
Shipping regulations: Flammable liquid, n.o.s., (Rail) Red label. (Air) Flammable Liquid label.

isobutyric acid $(CH_3)_2CHCOOH$ (2-methylpropionic acid).
Properties: Colorless liquid; soluble in water, alcohol, and ether. Sp. gr. 0.946 to 0.950 (20/20°C); b.p. 154.4°C; f.p. −47°C; refractive index (n 20/D) 1.3930. Flash point 170°F (TOC); low toxicity. Autoignition temp. 935°F. Combustible.
Grade: Technical.
Containers: Carboys; steel and aluminum drums; tank cars and trucks.
Hazard: Strong irritant to tissue.
Uses: Manufacture of esters for solvents, flavors and perfume bases; disinfecting agent; varnish; deliming hides; tanning agent.

Shipping regulations: (Air) Corrosive label. (Rail) Corrosive liquid, n.o.s., White label.

isobutyric anhydride $[(CH_3)_2CHCO]_2O$. Liquid with boiling range 180 to 187°C;. sp. gr. 0.951–0.956 (20/20°C). Flash point 139°F; autoignition temp. 665°F. Combustible.
Containers: 55-gal drums, tank trucks, tank cars.
Hazard: Strong irritant to tissue.
Uses: Chemical intermediate.
Shipping regulations: (Air) Corrosive label. (Rail) Corrosive liquid, n.o.s., White label.

isobutyronitrile (2-methylpropanenitrile; isopropyl cyanide) $(CH_3)_2CHCN$.
Properties: Colorless liquid; sp. gr. 0.773 (20/20°C); b.p. 107°C; f.p. −75°C; slightly soluble in water; very soluble in alcohol and ether. Combustible.
Containers: Tank cars.
Hazard: May be toxic.
Uses: Intermediate for insecticides, etc.

isobutyroyl chloride (isobutyryl chloride; 2-methylpropanoyl chloride) $(CH_3)_2CHCOCl$. Colorless liquid; refractive index 1.4079; sp. gr. 1.017 (20/4°C); f.p. −90°C; b.p. 92°C; soluble in ether; reacts with water and alcohol.
Hazard: May be toxic and irritant.
Use: Chemical intermediate.

isocetyl laurate $C_{11}H_{23}COOC_{16}H_{33}$.
Properties: Oily liquid; almost odorless; sp. gr. 0.858; f.p. below −65°C; viscosity 19.6 cp. at 25°C; insoluble in water; soluble in most organic solvents. Combustible; low toxicity.
Uses: Cosmetics and pharmaceuticals (lubricant, fixative and solvent); plasticizer; mold release agent; textile softener.

isocetyl myristate $C_{13}H_{27}COOC_{16}H_{33}$.
Properties: Oily liquid with practically no odor; sp. gr. 0.857; f.p. −39°C; viscosity 25.6 cp at 25°C; insoluble in water; soluble in most organic solvents. Combustible; low toxicity.
Uses: See isocetyl laurate.

isocetyl oleate $C_{17}H_{33}COOC_{16}H_{33}$.
Properties: Oily liquid with practically no odor; sp. gr. 0.862; f.p. −57°C; viscosity 29.0 cp at 25°C; insoluble in water; soluble in most organic solvents. Combustible; low toxicity.
Uses: See isocetyl laurate.

isocetyl stearate $C_{17}H_{35}COOC_{16}H_{33}$.
Properties: Oily liquid with practically no odor; sp. gr. 0.857; f.p. 0°C; viscosity 32.0 cp at 25°C; insoluble in water; soluble in most organic solvents. Combustible; low toxicity.
Uses: See isocetyl laurate.

isocil. Generic name for 5-bromo-3-isopropyl-6-methyluracil. $OCCBrC(CH_3)NHC(O)NCH(CH_3)_2$. Crystals; m.p. 158°C; soluble in absolute alcohol.
Hazard: May be toxic.
Use: Herbicide.

isocinchomeronic acid (2,5-pyridinedicarboxylic acid) $HOOC(C_5H_3N)COOH$.
Properties: Light-tan powder, leaflets or prisms; no odor; m.p. 254°C, sublimes as nicotinic acid above this temperature; insoluble in cold water, alcohol, ether, benzene; slightly soluble in boiling water, boiling alcohol; soluble in hot dilute mineral acids.
Use: Intermediate for nicotinic acid, insecticides, polymers, dyestuffs.

isocrotonic acid. See crotonic acid.

isocyanate. A compound containing the isocyanate radical, —NCO. Monoisocyanates are in use, as in the treatment of cellulose to obtain a cellulose tricarbamate, but the term isocyanate usually refers to a diisocyanate (q.v.).

isocyanate generator (hindered isocyanate). An isocyanate derivative that decomposes to an isocyanate upon heating. In one type phenol is combined with an isocyanate, and the resulting urethane is stable at room temperature, but dissociates at 160°C to the original phenol and isocyanate. These generators are used commercially in a mixture with a polyester which can be stored indefinitely, but which upon heating produces a polyurethane resin.

isocyanate resin. See polyurethane resin.

isocyanic acid (cyanic acid) HN=C=O. A gas resulting from depolymerization of cyanuric acid at 300–400°C in a stream of carbon dioxide. An intermediate product in the formation of urethanes and allophanates.
Hazard: Strong irritant to eyes, skin, and mucous membranes; severe explosion risk.

isocyanine. See cyanine dye.

isocyanurate. A compound closely related to isocyanate, but containing three NCO groups. Its products are similar to polyurethane resins and are particularly useful as rigid foams for insulation in the building and construction industry. For combustibility, see foam, plastic. See also isocyanuric acid.

isocyanuric acid (s-triazine-2,4,6-trione) OCNHCONHCONH. White crystalline powder; the ketone isomer of cyanuric acid (q.v.). Derivatives of isocyanuric acid, as dichloro- and trichloroisocyanuric acid, and potassium and sodium dichloroisocyanurate, are used as bleaches and sanitizers.

isodecaldehyde $C_9H_{19}CHO$ (mixed isomers).
Properties: Liquid. Sp. gr. 0.8290; b.p. 197.0°C; f.p. −80°C; insoluble in water; weight 6.9 lb/gal; flash point 185°F. Combustible.
Hazard: Probably toxic; strong irritant.
Uses: Intermediate for pharmaceuticals, dyes, resins.

isodecane. See 2-methylnonane.

isodecanoic acid $C_{10}H_{20}O_2$ (mixture of branched chain acids, primarily trimethylheptanoic and dimethyloctanoic).
Properties: Liquid; b.p. 254°C, 137°C (10 mm); sp. gr. 0.9019 (20/20°C); f.p., glass below −60°C, very slightly soluble in water; viscosity 12.9 cp. at 20°C; refractive index 1.4358 (n 20/D). Combustible.
Uses: Intermediate for metal salts, ester type lubricants, plasticizers.

isodecanol $C_{10}H_{21}OH$ (mixed isomers).
Properties: Colorless liquid. Sp. gr. 0.8395; insoluble in water; flash point 220°F; b.p. 220°C. Combustible.
Use: Antifoaming agent in textile processing.

isodecyl chloride $C_{10}H_{21}Cl$ (mixed isomers).
Properties: Colorless liquid. Sp. gr. 0.8767; b.p. 210.6°C; f.p. −180°C (sets to a glass); insoluble in water; flash point 200°F. Combustible. Low toxicity.
Uses: Solvent for oils, fats, greases, resins, gums; ex-

tractants, cleaning compounds; intermediate for insecticides, pharmaceuticals, plasticizers, polysulfide rubbers, resins, and cationic surfactants.

isodecyl octyl adipate.
Properties: Light-colored, oily liquid; sp. gr. 0.924 (20/20°C); mid-b.p. 227°C (4 mm); refractive index 1.448 (25°C); viscosity 20 cp (20°C); flash point 400°F: combustible. Low toxicity.
Use: Plasticizer.

isodecyl pelargonate. A synthetic lubricant. $(CH_3)_2CH(CH_2)_6CH_2OOC(CH_2)CH_3$.

isodrin. Generic name for 1,2,3,4,10,10-hexachloro-1,4,4a,5,8,8a-hexahydro-1,4-endo, endo-5,8-dimethanonaphthalene $C_{12}H_8Cl_6$. An isomer of aldrin.
Properties: Crystals; decomposes above 100°C.
Hazard: Highly toxic. Use may be restricted.
Use: Insecticide.
Shipping regulations: (Air) Poison label.

isodurene (1,2,3,5-tetramethylbenzene) $(CH_3)_4C_6H_2$.
Properties: Liquid. Soluble in alcohol and ether; insoluble in water. Sp. gr. 0.896; b.p. 197°C; f.p. −24°C. Combustible.
Derivation: From coal tar.
Grade: Technical.
Use: Organic synthesis.

isoelectric point. The pH at which the net charge on a molecule in solution is zero. At this pH amino acids exist almost entirely in the zwitterion state, that is, the positive and negative groups are equally ionized. A solution of proteins or amino acids at the isoelectric point exhibits minimum conductivity, osmotic pressure, and viscosity. Proteins coagulate best at this point. Typical isoelectric points (pH) are: glycine 6.6; gelatin 4.7; serum albumin 5.4.

isoeugenol $(CH_3CHCH)C_6H_3OHOCH_3$. 1-Hydroxy-2-methoxy-4-propenylbenzene.
Properties: Pale yellow oil, spice-clove type odor. Soluble in alcohol, ether, and other organic solvents; slightly soluble in water. Sp. gr. 1.081–1.084; m.p. 19°C; b.p. 268°C; refractive index (n 19/D) 1.5739. Combustible; low toxicity.
Derivation: From eugenol by isomerization with caustic potash.
Grades: Perfumer's grade; F.C.C.
Uses: Perfumes; vanillin; flavoring agent.

isoeugenol acetate. See acetylisoeugenol.

isoeugenol ethyl ether (1-ethoxy-2-methoxy-4-propenylbenzene) $C_3H_5(CH_3O)C_6H_3OC_2H_5$.
Properties: Synthetic white crystalline powder; m.p. 64°C; insoluble in water; soluble in alcohol, ether, benzene. Combustible; low toxicity.
Uses: Sweetening agent and odorant fixative.

isoflurophate. See diisopropyl fluorophosphate.

"Isoforming." Proprietary process for fixed-bed hydroisomerization, requiring a non-noble metal catalyst. Claimed to give high yields of C_8 (xylene) isomers with low hydrogen consumption and minimal catalyst regeneration.

isoheptane. See 2-methylhexane.

isohexane C_6H_{14}. A mixture of branched-chain isomers.
Properties: Colorless liquid. Boiling range 54 to 61°C; sp. gr. 0.671 (60/60°F); flash point −26°F. Low toxicity.
Grade: Commercial.
Containers: 5-, 54-gal drums.
Hazard: Highly flammable; dangerous fire and explosion risk. Explosive limits in air, 1 to 7%.
Uses: Solvent; freezing-point depressant.
Shipping regulations: Flammable liquid, n.o.s. (Rail) Red label. (Air) Flammable Liquid label.

isolan. See 1-isopropyl-3-methyl-5-pyrazolyl dimethylcarbamate.

isoleucine (2-amino-3-methylpentanoic acid; Ile). $CH_3CH_2CH(CH_3)CH(NH_2)COOH$. An essential amino acid, found naturally in the L(+) form.
Properties: Crystalline; slightly soluble in water; nearly insoluble in alcohol; insoluble in ether.
Derivation: Hydrolysis of protein (zein, edestin); amination of the alpha-bromo-beta-methylvaleric acid.
Uses: Medicine; nutrition; biochemical research.

"Isoline."[550] Trademark for a dehydrated castor oil, a synthetic drying oil used in lithographic inks, linoleum, paints and varnishes. Supplied in various grades.

"Isomax" Process.[416,564] Proprietary hydrocracking process to convert hydrocarbon distillates (including vacuum gas oils) and residua into gasolines and/or high quality middle distillates and also to manufacture liquid petroleum gas. The C_5—C_6 light naphtha has a leaded (3ml. TEL/gal) research octane number of about 100. Leaded research octane ratings for the heavy naphtha vary from about 78–95. Premium quality low freeze point (less than −76°F) jet fuels and low pour point, high cetane number diesel fuels can also be manufactured. The process requires catalysts trademarked "Isocracking."

"Isome."[342] Trademark for di-n-propyl 6,7-methylene-dioxy-3-methyl-1,2,3,4-tetrahydronaphthalene-1,2-dicarboxylate. Used in insecticides. See propyl isome.

isomer. (1) One of two or more molecules having the same number and kind of atoms and hence the same molecular weight, but differing in respect to the arrangement or configuration of the atoms. Butyl alcohol (C_4H_9OH or $C_4H_{10}O$) and ethyl ether ($C_2H_5OC_2H_5$ or $C_4H_{10}O$) have the same molecular formulas but are entirely different kinds of substances; normal butyl alcohol ($CH_3CH_2CH_2CH_2OH$) and isobutyl alcohol ($[CH_3]_2CHCH_2OH$) are the same kinds of substances, differing chiefly in the shape of the molecules; sec-butyl alcohol ($CH_3CHOHCH_2CH_3$) exists in two forms, one being a mirror image of the other (enantiomer). Isomers often result from location of an atom or group of a compound at various positions on a benzene ring, e.g., xylene, dichlorobenzene. See also geometric isomer; optical isomer. (2) Nuclides (i.e., kinds of atomic nuclei) having the same atomic and mass numbers, but existing in different energy states. One is always unstable with respect to the other, or both may be unstable with respect to a third. In the latter instance the energy of transformation in the two cases will differ.

isomerase. A class of enzymes. See enzyme.

isomerism. See isomer.

isomerization. Process for converting a hydrocarbon or other organic compound into an isomer (q.v.). An important example is the conversion of normal butane into branched-chain isobutane, in connection with the production of isooctane and other high-grade motor fuels.

isometheptene (2-methylamino-6-methyl-5-heptene) $(CH_3)_2C:CHCH_2CH_2CH(NHCH_3)CH_3$.
Properties: Colorless or slightly yellow, oily liquid with an amine-like odor. B.p. 175–177°C; sp. gr. 0.794–0.798 (25/25°C); refractive index 1.4428–1.4438 (n 25/D). Miscible with the dilute mineral acids and

most organic solvents; volatile with steam. Insoluble in water.
Hazard: Moderately toxic by ingestion.
Use: Medicine.
Available also as the isometheptene tartrate, the hydrochloride, and the mucate.

alpha-**isomethylionone** (gamma-methylionone) $C_{14}H_{22}O$.
Properties: Slightly yellow liquid; sp. gr. 0.925–0.929 (25/25°C); refractive index (20°C) 1.5000–1.5010; flash point 217°F (TCC). Soluble in 5 parts of 70% alcohol. Combustible; low toxicity. A synthetic product.
Uses: Floral perfumes, particularly of a violet character; flavoring.

isomorphism. The state in which two or more compounds that form crystals of similar shape have similar chemical properties and can usually be represented by analogous formulas, e.g., Ag_2S and Cu_2S.

"Isomate."[520] Trademark for isocyanate foam systems. Available as non-burning, pour-in-place froth, or spray foams.

isoniazid (N-isonicotinyl hydrazine; INAH; isonicotinic acid hydrazide) $C_5H_4NCONHNH_2$.
Properties: Colorless or white crystals; odorless; affected by air and light; m.p. 170–173°C; sparingly soluble in alcohol, slightly soluble in benzene and ether; freely soluble in water. Solutions practically neutral to litmus.
Derivation: From gamma-picoline.
Grade: U.S.P.
Use: Medicine.

isonicotinic acid CHCHNCHCHCCOOH (ring). Pyridine-4-carboxylic acid.
Properties: White, practically odorless powder; m.p. 314–317°C (sealed capillary); slightly soluble in water; pH of saturated aqueous solution at 20° 3.6.
Containers: Fiber drums.
Use: Synthesis of isoniazid and similar substances.

isonipecaine hydrochloride. See meperidine hydrochloride.

"Isonol C100."[520] An aromatic reinforcing polyol. $C_6H_5N[CH_2CH(CH_3)OH]_2$
Properties: Amber liquid. Viscosity (50°C) 1000 cps (max.); sp. gr. (23°C) 1.055. Water content 0.05%. Combustible.
Uses: Ingredient of polyurethane foams, coatings, sealants, and elastomers; intermediate in organic synthesis.

isononyl alcohol $C_8H_{17}CH_2OH$. A higher alcohol developed in early 1968; used as a basis of plasticizers such as diisononyl adipate. Combustible.

isooctane (2,2,4-trimethylpentane). Molecular formula $(CH_3)_3CCH_2CH(CH_3)_2$; structural formula:

```
      CH3
      |
CH3—C—CH2—CH—CH3
      |         |
      CH3       CH3
```

A branched-chain hydrocarbon.
Properties: Colorless liquid; sp. gr. 0.6919 (20/4/°C); f.p. −107.4°C; b.p. 99.2°C; refractive index (n 20/D) 1.3914; flash point 10°F. Insoluble in water, slightly soluble in alcohol and ether. Autoignition temp. 784°F.
Grades: Technical; pure; research; spectrophotometric.
Containers: 5-, 54-gal drums; tanks.
Hazard: Flammable, dangerous fire risk. Moderately toxic by ingestion and inhalation. Explosive limits in air 1.1 to 6%.
Uses: Organic synthesis; solvent; motor fuel; used with normal heptane to prepare standard mixtures to determine anti-knock property of gasoline. See octane number.
Shipping regulations: (Rail) Red label. (Air) Flammable Liquid label.

isooctene C_8H_{16}. Mixture of isomers.
Properties: Colorless liquid; boiling range 190–200°F; bromine number 137; sp. gr. 0.726 (60/60°F). Flash point below 20°F.
Hazard: Flammable, dangerous fire risk.
Shipping regulations: (Rail) Red label. (Air) Flammable Liquid label.

isooctyl adipate $(C_8H_{17}OOCCH_2CH_2-)_2$. Plasticizer providing low-temperature stability. Used in calendering film, sheeting, vinyl dispersions, extrusions.

isooctyl alcohol (isooctanol). General term applied to any isomer of the formula $C_7H_{15}CH_2OH$ in which the eight carbon atoms form a branched chain. Usually refers to a mixture of isomers made by the Oxo process. A selected C_7 hydrocarbon fraction is reacted with hydrogen and carbon monoxide in the presence of a catalyst at pressures up to 3000 psi. The crude alcohol is recovered and purified.
Properties: Clear liquid. Distillation range 182–195°C; wt/gal 6.95 lb; sp. gr. (20/20°C) 0.832; flash point (Tag open cup) 180°F. Combustible. Low toxicity.
Containers: 55-gal drums; tank cars.
Uses: Ingredient of plasticizers; intermediate for nonionic detergents and surfactants; synthetic drying oils, cutting and lubricating oils, hydraulic fluids; resin solvent; emulsifier; antifoaming agent; intermediate for introducing the isooctyl group into other compounds.

isooctyl isodecyl phthalate $C_8H_{17}OOCC_6H_4COOC_{10}H_{21}$.
Properties: Clear liquid; sp. gr. (20/20°C), 0.976; flash point 445°F; combustible; low toxicity; mild odor.
Grade: Technical.
Containers: Drums; tank trucks; tank cars.
Use: Plasticizer.

isooctyl palmitate $C_8H_{17}OOCC_{15}H_{31}$.
Properties: Clear liquid; sp. gr. 0.863 (20°); acidity 0.2% max (palmitic); moisture 0.05% max; m.p. 6–9°C; b.p. 228°C (5 mm). Soluble in most organic solvents. Combustible; low toxicity.
Uses: Secondary plasticizer for synthetic resins; extrusion aid and plasticizer.

isooctylphenoxypolyoxyethylene ethanol isooctylphenylpolyethylene glycol ether) $(CH_3)_3CCH_2C(CH_3)_2C_6H_4O(CH_2)_2O(C_2H_4O)_7C_2H_4OH$.
Properties: Slightly viscous pale amber-colored liquid; oily musty odor; m.p. 2–5°C; b.p. 150°C (initial) at 1 micron; density 1.06 g/ml (20°C); flash point 227°C. Combustible; low toxicity.
Use: Surface-active agent.

isooctylphenylpolyethylene glycol ether. See isooctylphenoxypolyoxyethylene ethanol.

isooctyl thioglycolate $HSCH_2COOCH_2C_7H_{15}$.
Properties: Water-white liquid with faint fruity odor; b.p. 125°C (17 mm); sp. gr. 0.9736 (25°C); refractive index 1.4606 (21°C); acid no., less than 1. Combustible.
Grade: 99% (minimum purity).
Uses: Antioxidants, fungicides, oil additives, plasticizers, insecticides, stabilizers, polymerization modifiers.

"Isopar."[51] Trademark for a group of high-purity isoparaffinic materials. Used as odorless solvents, reaction diluents, in proprietary formulations, etc. Have low levels of skin irritation; little tendency to migrate through polyethylene containers, and are permitted under a number of FDA regulations for food-related applications.

isopentaldehyde C_4H_9CHO. A mixture of isomeric 5-carbon aldehydes.
Properties: Water-white liquid with sharp odor; sp. gr. 0.8089 (20/20°); b.p. 98.6°C; f.p. −95.4°C; water dissolves 0.85% aldehyde at 20°C; water-soluble to 2.2% in the aldehyde. Combustible.
Hazard: Probably toxic.
Uses: Possible intermediate for bis-phenols, epoxy and polycarbonate resins, and modified formaldehyde resins.

isopentane (2-methylbutane; ethyldimethylmethane $(CH_3)_2CHCH_2CH_3$.
Properties: Colorless, liquid; pleasant odor; f.p. −159.890°C; b.p. 27.854°C; sp. gr. 0.61967 (20°C); flash point −70°F; soluble in hydrocarbons, oils, ether; very slightly soluble in alcohol; insoluble in water. Low toxicity. Autoignition temp. 788°F.
Derivation: Fractional distillation from petroleum; purified by rectification.
Grades: Research (99.99%), pure (99%), technical (95%), commercial.
Containers: Pure: 55-gal drums. Technical and commercial: drums: tank cars.
Uses: Solvent; manufacture of chlorinated derivatives; blowing agent for polystyrene.
Hazard: Highly flammable, dangerous fire risk.
Shipping regulations: (Rail) Red label. (Air) Flammable Liquid label.

isopentanoic acid C_4H_9COOH. A mixture of isomeric 5-carbon acids.
Properties: Water-white liquid; penetrating odor; sp. gr. 0.9388 (20/20°C); b.p. 183.2°C; vapor pressure 0.14 mm at 20°C; f.p. −44°C; water dissolves 3.24 wt% of acid at 20°C; acid dissolves 10.4% water at 20°C. Combustible.
Hazard: Strong irritant to tissue.
Uses: Intermediate for plasticizers, synthetic lubricants, pharmaceuticals, metallic salts, vinyl stabilizers; extractant for mercaptans from hydrocarbons.
Shipping regulations: (Air) Corrosive label.

isophane insulin suspension. See insulin.

isophorone (3,5,5-trimethyl-2-cyclohexen-1-one) $C(O)CHC(CH_3)CH_2C(CH_3)_2CH_2$.
Properties: Water-white liquid; sp. gr. 0.9229 (20/20°C); 7.7 lb/gal (20°C); b.p. 215.2°C; vapor pressure 0.2 mm (20°C); f.p. −8.1°C; viscosity 2.62 cp (20°C); flash point 205°F. Combustible. Autoignition temp. 864°F. Has high solvent power for vinyl resins, cellulose esters and ether, and many substances soluble with difficulty in other solvents; slightly soluble in water.
Containers: Drums; tank cars.
Hazard: Toxic; irritant to skin and eyes. Tolerance, 10 ppm in air.
Uses: In solvent mixtures for finishes; for polyvinyl and nitrocellulose resins; pesticides; stoving lacquers.

isophthalic acid (meta-phthalic acid; IPA) $C_6H_4(COOH)_2$.
Properties: Colorless crystals; m.p. 345–348°C; sublimes. Slightly soluble in water; soluble in alcohol acetic acid; insoluble in benzene and petroleum ether. Combustible; low toxicity.
Derivation: (a) Oxidation of meta-xylene; (b) liquid phase oxidation of mixed xylenes; (c) direct oxidation of mixed alkyl aromatics with heavy metal salts and bromine as catalysts.
Containers: Drums; tank cars.
Grade: Technical.
Uses: Polyester, alkyd, polyurethane, and other high polymers; plasticizers.

isophthaloyl chloride (meta-phthalyl dichloride) $C_6H_4(COCl)_2$.
Properties: Crystalline solid; m.p. 41°C; b.p. 276°C; flash point 356°F (COC); combustible; soluble in ether and other organic solvents; reactive with water and alcohol.
Hazard: Irritant to eyes and skin.
Uses: Intermediate; dyes; synthetic fibers; resins; films; protective coatings; laboratory reagent.

isopolyester. A polyester resin based on isophthalic acid.

isoprene (3-methyl-1,3-butadiene; 2-methyl-1,3-butadiene) $CH_2{=}C(CH_3)CH{=}CH_2$. The molecular unit of natural rubber.
Properties: Colorless, volatile liquid; f.p. −146°C; b.p. 34.08°C; refractive index 1.4216 (n 20/D); sp. gr. 0.6808 (20/4°C); flash point −55°F; insoluble in water; soluble in alcohol and ether. Autoignition temp. 802°F.
Derivation: (1) From cracked products of heavy petroleum oils; (2) dehydrogenation of isopentene; (3) pyrolysis of methyl pentene or of isobutylene-formaldehyde condensation products; (4) dehydration of methyl butenol.
Grades: Polymerization (min. purity 99%); research (99.99%).
Containers: Drums; tank cars; tank trucks.
Hazard: Highly flammable, dangerous fire and explosion risk; moderately toxic and irritant.
Uses: Monomer for manufacture of polyisoprene; chemical intermediate.
Shipping regulations: (Rail) Red label (Air) Flammable Liquid label.
See also polyisoprene; rubber, natural.

isoprenoid. A compound based on the isoprene structure. These include many naturally occurring materials such as terpenes, rubber, cholesterol and other steroids.

isopropanol. Legal label name for isopropyl alcohol (q.v.).

isopropanolamine (MIPA; 2-hydroxypropylamine; 1-amino-2-propanol) $CH_3CH(OH)CH_2NH_2$.
Properties: Liquid; slight ammonia odor; sp. gr. 0.9619; m.p. 1.4°C; b.p. 159.9°C; refractive index 1.4462 (20/D); flash point 170°F. Combustible; low toxicity; soluble in water.
Containers: Cans; drums.
Uses: Emulsifying agent; drycleaning soaps; soluble textile oils; wax removers; metal cutting oils; cosmetics; emulsion paints; plasticizers; insecticides.

isopropenyl acetate $CH_3COOC(CH_3){:}CH_2$.
Properties: Water white liquid; sp. gr. 0.9226; b.p. 97.4°C; f.p. −92.9°C; solubility in water 3.25% by

weight (20°C); refractive index 1.4020 (20/D); flash point 60°F (o.c.). Low toxicity.
Hazard: Flammable; dangerous fire risk.

isopropenylacetylene (2-methyl-1-buten-3-yne) $H_2C{:}C(CH_3)C{\vdots}CH$.
Properties: Colorless liquid; b.p. 33–34°C; freezing point −113°C: sp. gr. (20/20°C) 0.695; refractive index (n 20/D) 1.4168; flash point (TOC) <20°F; very slightly soluble in water and miscible with acetone, alcohol, benzene, carbon tetrachloride and kerosine.
Hazard: Flammable; dangerous fire risk.
Use: Specialty fuel; chemical intermediate.
Shipping regulations: Flammable liquid, n.o.s. (Rail) Red label. (Air) Flammable Liquid label.

isopropenyl chloride. See chloropropene.

isopropenylchloroformate $ClCOOC(CH_3){:}CH_2$.
Properties: Liquid. Sp. gr. 1.103 (20°C); b.p. 93°C (746 mm).
Derivation: Distillation of the reaction products of acetone and phosgene.
Hazard: Toxic; strong irritant to eyes and skin.

para-**isopropoxydiphenylamine** $C_6H_5NHC_6H_4OCH(CH_3)_2$.
Properties: Dark gray flakes; sp. gr. 1.10; set point 80–86°C; purity 92% (min); ash 0.10% (max); insoluble in water; soluble in ethyl alcohol (2B), acetone, benzene and gasoline. Low toxicity.
Use: Rubber antioxidant.

beta-**isopropoxypropionitrile** $(CH_3)_2CHO(CH_2)_2CN$.
Properties: Colorless to straw-colored liquid. Combines the chemical and physical properties of ethers and nitriles. F.p. −67°C; b.p. 82–86°C (25 mm) 65–65.5°C (10 mm). Sp. gr. 0.9058 (25°C). Slightly soluble in water; soluble in organic solvents. Flash point 155°F; combustible.
Hazard: May be toxic.

isopropyl acetate $CH_3COOCH(CH_3)_2$.
Properties: Colorless liquid; aromatic odor; b.p. 89.4°C; sp. gr. 0.8690 (25/4°C); refractive index 1.378 (20°C); sp. ht. 0.46 cal/g; f.p. −73.4°C; heat of vaporization 135 Btu/lb; viscosity 0.49 cp (25°C); solubility in water 2.9 wt %; flash point 40°F; 7.17 lb/gal (20°C). Miscible with most organic solvents. Autoignition temp. 860°F.
Derivation: By reacting isopropyl alcohol with acetic acid in the presence of catalysts.
Grades: 95%; 85 to 88%.
Containers: Drums; tank cars.
Hazard: Flammable, dangerous fire risk. Moderately toxic.
Uses: Solvent for nitrocellulose, resin gums, etc.; paints, lacquers, and printing inks, organic synthesis.
Shipping regulations: (Rail) Red label. (Air) Flammable Liquid label.

N-isopropylacrylamide (NIPAM). Crystalline solid. Homopolymers and copolymers prepared with this material show inverse solubility in water. They are used as binders in textiles, paper, adhesives, detergents, cosmetics. See also acrylic resin.

isopropyl alcohol (IPA; dimethylcarbinol; sec-propyl alcohol; isopropanol; 2-propanol) $(CH_3)_2CHOH$.
Properties: Colorless liquid. Pleasant odor; b.p. 82.4°C; sp. gr. 0.7863 (20/20°C); refractive index 1.3756 (20°C); sp. ht. 0.65 cal/g; f.p. −86°C; critical temperature 235°C; critical pressure 53 atmospheres; vapor pressure 33 mm at 20°C; flash point 59°F (TOC); heat of combustion 14,346 Btu/lb; heat of vaporization 288 Btu/lb; viscosity 2.1 cp (25°C). Autoignition temp. 750°F. Soluble in water, alcohol and ether.
Derivation: By treatment of propylene with sulfuric acid and hydrolyzing.
Method of purification: Rectification.
Grades: 91%; 95%; 99%; N.F. (99%); nanograde.
Containers: Drums; tank trucks and cars.
Hazard: Flammable, dangerous fire risk. Moderately toxic by ingestion and inhalation. Tolerance, 400 ppm in air. Explosive limits in air 2 to 12%.
Uses: Manufacture of acetone and its derivatives; manufacture of glycerol and isopropyl acetate; solvent for essential and other oils, alkaloids, gums, resins, etc.; latent solvent for cellulose derivatives; coatings solvent; deicing agent for liquid fuels; pharmaceuticals; perfumes; lacquers; extraction processes; dehydrating agent; preservative.
Shipping regulations: (Rail) Red label. (Air) Flammable Liquid label. Legal label name isopropanol.

isopropylamine (2-aminopropane) $(CH_3)_2CHNH_2$.
Properties: Colorless volatile liquid; amine odor; strong alkaline reaction. B.p. 32.4°C; f.p. −95.2°C; sp. gr. (20/20°C) 0.6881; wt/gal (20°C) 5.7 lb, refractive index (n 15/D) 1.3770; flash point (open cup) −35°F; autoignition temp. 756°F. Miscible with water, alcohol, and ether.
Derivation: From isopropyl chloride and ammonia under pressure.
Containers: Drums; tank cars.
Hazard: Toxic by ingestion and inhalation; strong irritant to tissue. Tolerance, 5 ppm in air. Highly flammable, dangerous fire risk. Safety data sheet available from Manufacturing Chemists Assn., Washington, D.C.
Uses: Solvent; intermediate in synthesis of rubber accelerators, pharmaceuticals, dyes, insecticides, bactericides, textile specialities, and surface-active agents; dehairing agent; solubilizer for 2,4-D acid.
Shipping regulations: (Rail) Flammable liquid, n.o.s., Red label. (Air) Flammable Liquid label.

para-**isopropylaminodiphenylamine.** See N-isopropyl-N′phenyl-para-phenylenediamine.

isopropylaminoethanol (IPAE). A commercial mixture of approximately 60% isopropylethanolamine, $(CH_3)_2CHNHCH_2CH_2OH$, and 40% of isopropyldiethanolamine, $(CH_3)_2CHN(CH_2CH_2OH)_2$.
Properties: Amber to straw-colored liquid, distillation range 110–265°C; f.p. below −50°C; sp. gr. 0.91–0.94 (20/20°C); flash point 145–155°F (open cup). Combustible.
Use: Synthesis of emulsifiers.

N-isopropylaniline $C_6H_5NHCH(CH_3)_2$.
Properties: Yellowish liquid; b.p. 206°C; pour point less than −67°C; refractive index 1.5365 (20°C); flash point 190°F (COC). Combustible.
Hazard: May be toxic.
Uses: Dyeing acrylic fibers; chemical intermediate.

para-**isopropylaniline.** See cumidine.

isopropyl antimonite $[(CH_3)_2CHO]_3Sb$.
Properties: Colorless liquid; b.p. 82°C at 7 mm Hg pressure.

Derivation: Reaction of antimony trichloride with isopropanol.
Hazard: May be toxic.
Uses: Cross-linking agent; flameproofing agent.

para-**isopropylbenzaldehyde.** See cuminic aldehyde.

isopropylbenzene. See cumene.

para-**isopropylbenzyl alcohol.** See cuminic alcohol.

isopropyl bromide $CH_3CHBrCH_3$.
Properties: Colorless liquid; sp. gr. 1.304 (25/25°C); b.p. 58.5–60.5°C; f.p. −90°C; refractive index 1.422 (n 25/D); flash point none; slightly soluble in water; soluble in methanol, ether. Probably low toxicity. Nonflammable.
Uses: Synthesis of pharmaceuticals, dyes, other organics.

isopropyl butyrate $(CH_3)_2CHOOCC_3H_7$.
Properties: Colorless liquid; sp. gr. 0.8652 (13°C) b.p. 128°C.
Uses: Solvent for cellulose ethers; flavoring.

isopropylcarbinol. See isobutyl alcohol.

isopropyl chloride $CH_3CHClCH_3$.
Properties: Colorless liquid; sp. gr. 0.858 (25/25°C); b.p. 34.8°C; f.p. −117.6°C; refractive index 1.374 (n 25/D); flash point −26°F; autoignition point 1100°F; slightly soluble in water; soluble in methanol, ether.
Hazard: Highly flammable; fire and explosion risk. Explosive limits in air 2.8 to 10.7%.
Uses: Solvent; intermediate; isopropylamine.
Shipping regulations: Flammable liquid, n.o.s. (Rail) Red label. (Air) Flammable Liquid label.

N-isopropyl-alpha-**chloroacetanilide.** See 2-chloro-N-isopropylacetanilide.

isopropyl 3-chlorocarbanilate. See chloro-IPC.

isopropyl chloroformate $(CH_3)_2CHOOCCl$. Colorless liquid; a phosgene derivative.
Hazard: Toxic by inhalation.
Uses: Chemical intermediate for free-radical polymerization initiators; organic synthesis.

isopropyl N-(3-chlorophenyl)carbamate. See chloro-IPC.

isopropyl-meta-**cresol.** See thymol.

isopropyl-ortho-**cresol.** See carvacrol.

isopropyl cresol. A mixture of di- and monoisopropyl cresols used as an antioxidant. See thymol and carvacrol.

isopropyl cyanide. See isobutyronitrile.

isopropyl dichlorophenoxyacetate. See 2,4-D.

isopropyldiethanolamine. See isopropylaminoethanol.

2-isopropyl-4-dimethylamino-5-methylphenyl 1-piperidinecarboxylate methyl chloride
$C_3H_7C_6H_2(CH_3)N(CH_3)_2OOCNC_5H_{10} \cdot CH_3Cl$.
Properties: White solid; m.p. 151–152°C; insoluble in ether; soluble in methanol.
Uses: A plant tranquilizer or antigibberellin, which causes some plants to become dwarfs without otherwise affecting their growth or health.

isopropylethanolamine. See isopropylaminoethanol.

isopropyl ether (diisopropyl ether; diisopropyl oxide) $(CH_3)_2CHOCH(CH_3)_2$.
Properties: Colorless, volatile liquid. Ethereal odor. Somewhat similar to ethyl ether in properties but does tend to form peroxides more readily than ethyl ether. (See ether). Consequently the presence or absence of peroxides should be determined and if present should be destroyed with sodium sulfite before distillation. B.p. 67.5°C; sp. gr. 0.723 (15.5/4°C); refractive index 1.368; f.p. −88°C; solubility in water 0.65% wt (25°C); flash point approx. 0°F; autoignition temp. 830°F; 6.05 lb/gal (60°F). Miscible with most organic solvents and water.
Grade: Technical.
Containers: Drums, tank cars.
Hazard: Flammable, dangerous fire risk. Toxic by ingestion and inhalation. Strong irritant. Explosive limits in air 1.4 to 21%. Tolerance, 250 ppm in air.
Uses: Solvent for animal, vegetable, mineral oils, waxes and resins; extraction of acetic acid from aqueous solutions; solvent for dyes in presence of small amount of alcohol; paint and varnish removers; spotting compositions; rubber cements.
Shipping regulations: (Rail) Flammable liquid, n.o.s., Red label. (Air) Flammable Liquid label. Legal label name diisopropyl ether.

isopropylethylene. See 3-methyl-1-butene.

isopropyl furoate $C_4H_3OCO_2C_3H_7$.
Properties: Colorless liquid, becoming brown in light. Insoluble in water; soluble in alcohol and ether; sp. gr. 1.0655 (23.7/4°C); b.p. 198.6°C (corr); refractive index 1.4682 (23.7°C).

isopropylideneacetone. See mesityl oxide.

para, para′-**isopropylidenediphenol.** See bisphenol A.

isopropyl iodide (2-iodopropane) CH_3CHICH_3.
Properties: Colorless liquid that discolors in air and light; miscible with chloroform, ether, alcohol, and benzene; slightly soluble in water; sp. gr. 1.703; f.p. −90°C; b.p. about 90°C; refractive index (n 20/D) 1.5026.
Uses: Organic synthesis; pharmaceuticals.

isopropyl meprobamate. See N-isopropyl-2-methyl-2-propyl-1,3-propanediol dicarbamate.

isopropyl mercaptan $(CH_3)_2CH(HS)$.
Properties: Liquid; extremely unpleasant odor; sp. gr. 0.814 (60/60°F); boiling range 51–55°C; flash point −30°F.
Derivation: Propylene and hydrogen sulfide.
Hazard: Highly flammable, dangerous fire hazard. Toxic by inhalation; strong irritant.
Uses: Standard for petroleum analysis; intermediate.
Shipping regulations: (Rail) Red label (Air) Flammable Liquid label.

2-isopropyl-5-methylbenzoquinone. See para-thymoquinone.

5-isopropyl-2-methyl-1,3-cyclohexadiene. See alpha-phellandrene.

3-isopropyl-6-methylene-1-cyclohexene. See beta-phellandrene.

1-isopropyl-2-methylethylene. See 4-methyl-2-pentene.

N-isopropyl-2-methyl-2-propyl-1,3-propanediol dicarbamate (isopropyl meprobamate; carisoprodol) $(CH_3)_2CHNHCOOCH_2C(CH_3)(C_3H_7)CH_2COONH_2$.
Properties: Crystals; m.p. 92–93°C. Sparingly soluble in water; insoluble in vegetable oils; soluble in many common organic solvents. Stable in dilute acids and alkalies.
Use: Medicine.

1-isopropyl-3-methyl-5-pyrazolyl dimethylcarbamate (isolan) $C_{10}H_{17}N_3O_2$.
Properties: Liquid; sp. gr. 1.07 (20°C); b.p. 103°C (0.7 mm). Miscible with water.

Derivation: By treating 1-isopropyl-3-methyl-5-pyrazolone with dimethylcarbamoyl chloride.
Hazard: May be toxic.
Use: Insecticide.

isopropyl myristate $CH_3(CH_2)_{12}CO_2CH(CH_3)_2$.
Properties: Colorless oil; practically odorless; sp. gr. 0.850–0.860; freezing point 3°C; refractive index (20°C) 1.435–1.438. Soluble in most organic solvents; insoluble in water. Combustible; nontoxic.
Grade: Double-distilled from coconut oil.
Use: Cosmetics.

isopropyl nitrate (2-propanol nitrate) $(CH_3)_2CHNO_3$
Properties: Colorless liquid; b.p. 102°C.
Hazard: Oxidizing material; fire risk in contact with organic materials.
Shipping regulations: (Rail) Nitrates, n.o.s., Yellow label. (Air) Oxidizer label.

isopropyl palmitate $(CH_3)_2CHOOCC_{15}H_{31}$
Properties: Colorless liquid; sp. gr. 0.850–0.855 (25/25°C) refractive index 1.4350–1.4390 (20°C); m.p. 14°C. Soluble in 4 parts 90% alcohol, soluble in mineral and fixed oils; insoluble in water. Combustible; nontoxic.
Uses: Emollient and emulsifier in lotions, creams, and similar cosmetic products.

isopropyl percarbonate. Legal label name for diisopropyl peroxydicarbonate (q.v.).

isopropyl peroxydicarbonate. See diisopropyl peroxydicarbonate.

meta, para-**isopropylphenol** $(CH_3)_2CHC_6H_4OH$.
Properties: A solid mixture of the meta and para isomers, completely soluble in 10% sodium hydroxide; f.p. (meta) 25.9°C, (para) 63.2°C; b.p. (meta) 228.6°C, (para) 228.5°C. Combustible.
Hazard: May be toxic.

ortho-**isopropylphenol** $(CH_3)_2CHC_6H_4OH$.
Properties: Light yellow liquid, b.p. 214°C; f.p. 17°C; sp. gr. 0.995 at 20°C; flash point (open cup) 220°F; insoluble in water; soluble in isopentane, toluene, ethyl alcohol, 10% sodium hydroxide. Combustible.
Hazard: May be toxic.
Uses: Intermediate for synthetic resins, plasticizers, surface active agents, perfumes.

isopropyl N-phenylcarbamate. See IPC.

N-isopropyl-N′-phenyl-para-**phenylenediamine** (para-isopropylaminodiphenylamine) $C_3H_7NHC_6H_4NHC_6H_5$.
Properties: Dark gray to black flakes; f.p. range 72–76°C; sp. gr. 1.04 (25°C); soluble in benzene, gasoline; insoluble in water.
Use: Protection of rubbers against oxidation, ozone, flexcracking, and poisoning by copper and manganese.

4-isopropylpyridine $C_5NH_4C_3H_7$.
Properties: Liquid. B.p. 182.2°C; sp. gr. 0.9282 at 20°C; refractive index 1.4960 (n 20/D); solubility in 100 g of water at 20°C, 1.17 g; solubility of water in 100 g at 20°C, 19.4 g.
Hazard: May be toxic.

isopropyl titanate. See tetraisopropyl titanate.

isopropyltoluene. See cymene.

isopropyl 2,4,5-trichlorophenoxyacetate. See 2,4,5-trichlorophenoxyacetic acid.

isopropyltrimethylmethane. See 2,2,3-trimethylbutane.

isopulegol (1-methyl-4-isopropenylcyclohexan-3-ol) $C_{10}H_{17}OH$. A terpene derivative.
Properties: Water-white liquid. Mint-like odor. Sp. gr. 0.904–0.911; refractive index 1.471–1.474. Combustible. Low toxicity.
Uses: Perfumery (geranium and rose compounds); flavoring.
Also available as the acetate.

isopurpurin. See anthrapurpurin.

isoquinoline CHCHCHCHCCCHCHNCH.
Properties: Colorless plates or liquid; sp. gr. 1.099 (20°C); m.p. 23°C; b.p. 243°C; insoluble in water, soluble in dilute mineral acids and most organic solvents; refractive index (n 25/D) 1.6223. Combustible.
Derivation: From coal-tar; also synthetic.
Containers: Technical (95% min).
Hazard: Probably toxic.
Uses: Manufacture of pharmaceuticals (such as nicotinic acid), dyes, insecticides, rubber accelerators, and in organic synthesis.

1,3-isoquinolinediol $C_9H_5N(OH)_2$.
Properties: Cream colored paste; solids, approx. 80%.
Hazard: Probably toxic.
Use: Intermediate.

isosafrole $C_{10}H_{10}O_2$. Bicyclic.
Properties: Colorless, fragrant liquid; odor of anise; sp. gr. 1.117–1.120; refractive index 1.576; b.p. 253°C. Soluble in alcohol, ether, and benzene, Combustible. Probably low toxicity.
Derivation: Treatment of safrole with alcoholic potash.
Uses: Manufacture of heliotropin; perfumes; flavors; pesticide synergists.

isosafrole, n-octyl sulfoxide. See sulfoxide.

isosorbide (1,4,3,6-dianhydrosorbitol)
$OCH_2CHOHCHCHCHOHCH_2O$. A polyol with a hydroxyl group attached to each of two cis-oriented saturated furan rings. Intermediate for pharmaceuticals. Combustible.

isostearic acid. A coined name for a C_{18} saturated fatty acid of the formula $C_{17}H_{35}COOH$. It is a complex mixture of isomers, primarily of the methyl-branched series, that are mutually soluble and virtually inseparable.
Use: Similar to stearic or oleic acids.

isosterism. Similarity in physical properties of elements, ions, or compounds, due to similar or identical outer shell electron arrangements.

isostilbene. See stilbene.

isotactic polymer. See polymer, isotactic.

"Isothan."[328] Trademark for a series of liquid cationic detergents.
Uses: Emulsifying; mothproofing; bacteriostat; fungicide and germicide.

isotherm. Constant temperature line used on climatic maps or in graphs of thermodynamic relations, particularly the graph of pressure-volume relations at constant temperature.

isotone. A nuclide which has the same excess of neutrons over protons as another nuclide.

isotonic. A solution having the same osmotic pressure as another solution, for example, human blood and physiological salt solution (q.v.).

isotope. One of two or more forms or species of an element that have the same atomic number, i.e., the same position in the Periodic Table, but different masses. The difference in mass is due to the presence of one or more extra neutrons in the nucleus. Thus "regular" hydrogen, with atomic number 1 and a mass of 1 (proton), is one of the three isotopes of hydrogen. The oher two are the naturally occurring deuterium (q.v.), which has a neutron in its nucleus as well as a proton, giving a mass of 2; and the artifically produced tritium (1 proton and 2 neutrons) with a mass of 3 (approx). The atomic weight of an element is the average weight percent of all its natural isotopes. The heavier isotopes usually occur very rarely in the atomic population (1 part in 4500 for hydrogen-2, and 1 part in 140 for U-235; in the exceptional case of chlorine, the ratio of isotopes 35 and 37 is about 3 to 1).

The occurrence of isotopes among the 83 most abundant elements is widespread, but separation methods are complicated and costly. 21 elements have no isotopes, each consisting of only one kind of atom. (See note below). The remaining 62 natural elements have from 2 to 10 isotopes each. There are 287 different isotopic species in nature; noteworthy among them are oxygen-17, carbon-14, uranium-235, cobalt-60, and strontium-90, all but the first being radioactive.

There are three kinds of isotopes: (1) natural non-radioactive, (2) natural radioactive, and (3) artificial radioactive (made by neutron bombardment).

Uses: (Nonradioactive) Preparation of heavy water to moderate nuclear reactors. (Radioactive): Tracers in biochemical, metallurgical, and medical research; in geochemical and archeological research (C-14); irradiation source for polymerization, sterilization, etc.; therapeutic agents in various diseases (iodine, sodium, gold. etc.); electric power generation.

See also Aston; chemical dating; radioactivity; heavy water; decay; tracer; nuclide.

Note: According to this definition, it is strictly improper to refer to elements that exist in only one atomic form as having "one isotope"; actually such elements as beryllium, aluminum, arsenic, iodine, and others have *no* isotopes, that is, they have no other atomic form that is like them in all respects except mass. The term "isotope" requires the existence of at least *two* elemental forms, in the same sense that the word "twin" requires the existence of a pair.

"Isotox."[253] Trademark for a type of insecticide containing lindane (q.v.).

"Isotron."[204] Trademark for fluorinated hydrocarbons. Used as refrigerants, propellants, blowing agents, and solvents.

11. Trichlorofluoromethane.
12. Dichlorodifluoromethane.
13. Chlorotrifluoromethane.
113. Trichlorotrifluoroethane.
114. Dichlorotetrafluoroethane.

M Solvent. Trichlorofluoromethane.
T Solvent. Trichlorotrifluoroethane.
Precision Solvent Cleaner. Trichlorotrifluoroethane. An ultrapure solvent.

isotropic. Descriptive of the property of transmitting light equally in all direction; cubic (isometric) crystals have this property, as well as liquids, gases, and most glasses. See also anisotropic.

isovaleral. See isovaleraldehyde.

isovaleraldehyde (isovaleral; isovaleric aldehyde; 3-methylbutyraldehyde) $(CH_3)_2CHCH_2CHO$. Occurs in orange, lemon, peppermint and other essential oils.

Properties: Colorless liquid; apple-like odor; sp. gr. 0.785; f.p. −51°C; b.p. 92°C; refractive index (n 20/D) 1.390. Soluble in alcohol and ether; slightly soluble in water. Combustible; low toxicity.

Derivation: Oxidation of isoamyl alcohol; also by Oxo process from petroleum.

Uses: Flavoring; perfumes; pharmaceuticals; synthetic resins; rubber accelerators.

isovaleric acid (isopropylacetic acid) $(CH_3)_2CHCH_2COOH$. Occurs in valerian, hop oil, tobacco and other plants.

Properties: Colorless liquid; disagreeable taste and odor; sp. gr. 0.931 (20/20°C); b.p. 176°C; refractive index (n 20/D) 1.4043; f.p. −29°C; slightly soluble in water; soluble in alcohol and ether. Combustible; low toxicity.

Derivation: With other valeric acids, by distillation from valerian; by oxidation of isoamyl alcohol.

Uses: Medicine; flavors; perfumes.

isovaleric aldehyde. See isovaleraldehyde.

isovaleroyl chloride (3-methylbutanoyl chloride) $(CH_3)_2CHCH_2COCl$.

Properties: Colorless liquid; refractive index (n 24/D) 1.4136; sp. gr. 0.9854 (24/4°C); b.p. 113°C; soluble in ether; reacts with water and alcohols.

Use: Intermediate in synthesis.

isovaleryl-para-phenetidine $C_2H_5OC_6H_4NHCOCH_2CH(CH_3)_2$.

Properties: White, glistening needles. Almost insoluble in water and ether; soluble in alcohol and chloroform.

Derivation: By heating isovaleric acid with para-phenetidine.

"Isuprel" Hydrochloride.[162] Trademark for isoproterenol hydrochloride.

itaconic acid (methylene succinic acid) $CH_2:C(COOH)CH_2COOH$.

Properties: White, odorless crystals; m.p. 167–168°C; soluble in water, alcohols and acetone; sparingly soluble in other organic solvents. Low toxicity.

Derivation: Submerged fermentation by mold or various carbohydrates.

Uses: Copolymerizations; resins; plasticizers; lube oil additives; intermediate.

-ite. A suffix indicating an intermediate oxidation state of a metallic salt, analogous to -ous for acids, e.g., sodium sulfite ($NaSO_3$), containing one less oxygen atom than the sulfate.

"ITP- 43, 63A, 67."[19] Trademark for a series of imine-terminated polymers derived from poly(alkylene ether) glycols and polyester backbones.

Uses: Solid rocket propellant binders and liners, insulation materials, polymeric plasticizers, sealants, adhesives, coatings and potting compounds.

IUPAC. Abbreviation for International Union of Pure and Applied Chemistry (q.v.).

IVE. Abbreviation for isobutyl vinyl ether. SEe vinyl isobutyl ether.

ivory black. An animal black produced from ivory. The term is sometimes erroneously applied to other animal blacks. Chief use is as a pigment, especially as artists' color in oil.

J

J acid. See 2-amino-5-naphthol-7-sulfonic acid.

J acid urea. See 5,5′-dihydroxy-7,7′-disulfonic-2,2′-disulfonic-2,2′-dinaphthylurea.

japan. A varnish yielding a hard, glossy, dark-colored film. Japans are usually dried by baking at relatively high temperatures. (ASTM D16-52). True Japan varnishes contain a strongly irritant chemical; more recent types contain kauri or copal resin, linseed oil, lead oxide, pigments and solvents such as kerosene or turpentine.
Hazard: Flammable; irritant to eyes and skin.
Uses: Coatings for miscellaneous wood and metal products.

Japan wax. (Japan tallow; sumac wax).
Properties: Pale yellow solid; tallow-like rancid odor. Contains 10–15% palmatin and other glycerides. Soluble in benzene and naphtha. Insoluble in water and in cold alcohol. Sp. gr. 0.970–0.980; m.p. 53°C. Combustible; nontoxic.
Derivation: From a species of Rhus by boiling the fruit in water.
Uses: Candles; wax matches; special soaps; substitute for beeswax; food packaging.

jasmine aldehyde. See alpha-amylcinnamic aldehyde.

jasmine oil.
Properties: Light yellow essential oil; characteristic odor. Soluble in alcohol, ether and chloroform. Sp. gr. 1.007–1.018; optical rotation +2.5° to +3.5°. Combustible; low toxicity.
Chief known constituents: Benzyl acetate, linalyl acetate, linalool, indole, and jasmone.
Derivation: Flowers of Spanish jasmine, Jasmium grandiflorum.
Uses: Perfumery; flavoring.

jasmone $C_{11}H_{16}O$. 3-Methyl-2-(2-pentenyl)-2-cyclopenten-1-one. A ketone found in jasmine oil and other flower oils. Odor of jasmine; sp. gr. 0.944 (22/°0°C).

"Jasmonyl."[227] Trademark for isomeric mixture of nonane-1,3-diol monoacetates.

jatrophone. A diterpenoid growth inhibitor isolated from an alcohol extract of the plant *Jatropha gossypiifolia*. Its unique structure includes a 12-membered ring. It is readily attacked by nucleophiles. Useful in study of tumor growth inhibition and other biochemical research.

"Javollal."[188] Trademark for an aromatic concentrate used as a substitute for oil of citronella.

"Jayflex."[29] Trademark for a group of resin plasticizers, including dihexyl phthalate, diisooctyl phthalate, di-2-ethylhexyl phthalate, diisodecyl phthalate, ditridecyl phthalate and hydrocarbon plasticizers.

"Jaysol."[29] Trademark for an ethyl alcohol composition containing an aliphatic solvent, methyl isobutyl ketone, and ethyl acetate.

"Jaysolve."[29] Trademark for a group of glycol ether solvents which possess unique solubility characteristics because they have an alcohol and ether group in a single molecule. They dissolve many synthetic and natural resins and are used in a wide variety of surface coating applications.

"Jeffersol."[137] Trademark for a group of glycol ether solvents.

"Jefron."[233] Trademark for hematinic iron products.

"Jellitic."[103] Trademark for a prepared dry wheat paste which is pre-gelatinized over hot rolls to make the starch water-soluble.

jelly. A modified form of the word "gel" (q.v.) widely used in popular language, but also used in chemical literature to refer to the mechanical strength of the gel structures occurring with pectins, gelatin and various natural gums. "Jelly strength" is frequently specified in the food industry. Other uses of the word are found in "petroleum jelly" obtained as a distillation product of petroleum residues (petrolatum) and in the so-called "royal jelly," a natural nutrient mixture of proteins and carbohydrates produced by bees as food for the queen bee. See also gel.

"Jet II."[1] Trademark for a series of synthetic lubricants for aviation engines.

jet fuel. A fuel for jet (turbine) engines, usually a petroleum product similar to kerosine. A number of types with somewhat different compositions and properties have been used. The important military jet fuels are as follows:
JP-1: F.p. below −46°C; autoignition temp. 442°F; flash point 95°F.
JP-4: 65% gasoline and 35% light petroleum distillate, with rigidly specified properties. Flash point −10°F; autoignition temp. 468°F.
JP-5: A specially refined kerosine having a flash point of 95°F and a freezing point of −40°C. Used by carrier-based aircraft because it can be stored aboard. Autoignition temp. 475°F.
JP-6: A higher kerosine cut than JP-4, with fewer impurities, used in advanced engines. Flash point 100°F; autoignition temp. 435°F.
Commercial jet planes use ASTM Type A, A-1, or B. A and A-1 are kerosine types. The A-1 has lower freezing point and is used for long range flights; type A is the large volume fuel for short and medium range flights. Type B is a gasoline-kerosine type similar to JP-4.
Hazard: Flammable; dangerous fire risk.
Shipping regulations: (Rail) Flammable liquid, n.o.s., Red label. (Air) Flammable Liquid label.

"Jet Jel."[553] Trademark for guar gum drilling mud flocculant.

"Jetron."[179] Trademark for a series of carbon blacks coprecipitated with synthetic SBR polymer.

Jetset. A fast-setting cement developed by the Portland Cement Assn. Reported to harden in 20 minutes after pouring. Accelerating agent has not been disclosed.

"Jet-X"[270] Trademark for a protective fire foam for use in passenger aircraft. The aqueous foaming agent is a 2.5% solution of detergent-type alkyl sulfonates reinforced with protein hydrolyzates. The foam acts as both thermal insulator and barrier against toxic fumes; said to fill an aircraft cabin within 30 seconds; does not irritate skin or damage textiles.

JH (methyl-*cis*-10,11-epoxy-7-ethyl-3,11,dimethyl-*trans*, *trans*-2,6-tridecadienoate). A synthetic hormone containing a 13-carbon chain; said to have possibilities as an insecticide. It acts by preventing insects from maturing. Its future depends on the possibility of large-scale production.

"JHR Compound."[554] Trademark for a thermoplastic compound impervious to mineral acid; does not decompose hydrogen peroxide. Used to coat interiors of tanks and containers for shipment of hydrogen peroxide, acids, etc.

"Jiffix."[329] Trademark for an acid-hardening, ammonium thiosulfate fixing bath. Ready-mixed and rapid-acting.

jojoba oil
Properties: Colorless, odorless waxy liquid. Chemically similar to sperm oil.
Derivation: By crushing seeds of an evergreen desert shrub found in southwestern U.S. and northern Mexico. Experimental cultivation in California and Israel. Yield of oil from seeds approaches 50%.
Uses: Substitute for sperm oil, especially in transmission lubricants; high-pressure lubricant; antifoam agent (antibiotic fermentation); substitute for carnauba wax and beeswax; cosmetic preparations.

Joule-Thomson coefficient. The change in temperature per atmosphere change of pressure on a gas or other fluid when the enthalpy remains constant. It is found by measuring the temperature change from T' to T when the pressure P' of a gas on one side of a porous plug changes to P on the other side. The change of temperature $(T' - T)$ and of $(P' - P)$ are measured under conditions such that no heat is gained or lost, and the pressure of the plug is great enough to insure a nearly constant pressure in the incoming and outgoing gas. The ratio of $(T' - T)/(P' - P)$ at several pressure ranges is extrapolated to the limiting case as $(P' - P)$ approaches zero. This limiting value is the Joule-Thomson coefficient, μ. Thus

$$\mu = \left(\frac{\alpha T}{\alpha P}\right)_H$$

The subscript H indicates that the enthalpy, H, remains constant during the expansion $(\Delta H = 0)$.

juglone. See 5-hydroxy-1,4-naphthoquinone.

jute. Bast fibers, 4–10 ft long, obtained from the stems of several species of Corchorus, especially C. capularis. Contains a higher proportion of lignin and less cellulose than any other commercial vegetable fiber and has relatively poor strength and durability. The fibers are soft and lustrous but lose strength when wet. Combustible; not self-extinguishing.
Sources: Bengal, Pakistan.
Containers: Bales.
Hazard: May ignite spontaneously when wet. Flammable in form of dust.
Uses: Burlap; sacking; linoleum; twine; carpet backing; packing; coarse paper.

"JZF."[248] Trademark for N,N'-diphenyl-para-phenylene diamine (q.v.).

K

k Abbreviation for kilo-, as in kcal.

K Symbol for potassium (from Latin kalium); symbol for Kelvin scale.

"K4 and **K6."**[471] Trademark for Western bentonites for bonding foundry sands and steel castings.

"KA-101."[475] Trademark for active alumina. Manufactured by carefully controlled calcination of beta [alumina] trihydrate. It is spherical, extremely absorbent, chemically inert. Used extensively to dehydrate fluids and gases in a variety of commercial processes.

K acid. See 1-amino-8-naphthol-4,6-disulfonic acid.

"Kadox."[268] Trademark for lead-free zinc oxides manufactured by the French process; produced from zinc metal.
Uses: Rubber, paints and lacquers; sealants for food industry; fast-curing latex.

kainite $MgSO_4 \cdot KCl \cdot 3H_2O$. A natural hydrated double salt of potassium and magnesium. Color, white, gray, reddish or colorless; streak, colorless; vitreous luster. Contains 30% potassium chloride. Sp. gr. 2.05–2.13; hardness 2.5–3.
Occurrence: Germany. One of the Stassfurt minerals. See potash.
Uses: Chemicals (potassium salts); fertilizer (as such).

"Kaleidoscope" Diphenolic Acid.[424] Trademark for 4,4-bis(4-hydroxypheny) pentanoic acid (q.v.).

"Kalite."[244] Trademark for calcium carbonate, surface-coated. Used in rubber, plastics, and drawing compounds.

kalsomine. See calcimine.

kanamycin sulfate $C_{18}H_{36}N_4O_{11} \cdot H_2SO_4$. A wide-spectrum antibiotic.
Properties: White odorless crystalline powder; decomposes over a wide range above 250°C; soluble in water; practically insoluble in methanol and ethanol.
Grade: U.S.P.
Use: Medicine.

"Kaocast."[455] Trademark for an alumina-silica refractory which can withstand temperatures up to 3000°F.

"Kaolex."[285] Trademark for a series of hydrous aluminum silicates (sedimentary kaolins) from Georgia and South Carolina.
Properties: P.C.E. 33–35; sp. gr. 2.60; DMR to 350 psi for casting, jiggering, pressing and extruding. Air-floated or water-washed (lump or pulverized).
Containers: 50-lb multiwall bags or bulk.
Uses: Ceramics; refractories; rubber and plastic products.

kaolin (China clay). A white-burning aluminum silicate which, due to its great purity, has a high fusion point and is the most refractory of all clays.
Composition: Mainly kaolinite (40% alumina, 55% silica, plus impurities and water).
Properties: White to yellowish or grayish fine powder; sp. gr. 1.8–2.6. Darkens and develops clay odor when moistened. Insoluble in water, dilute acids, and alkali hydroxides. Has high lubricity (slipperiness). Nontoxic; noncombustible.
Occurrence: Southeastern U.S., England, France.
Grades: Technical; N.F.; also graded on basis of color and particle size.
Containers: Cartons; paper bags; drums; bulk.
Uses: Filler and coatings for paper; rubber; refractories; ceramics; cements; fertilizers; chemicals (especially aluminum sulfate); catalyst carrier; anticaking preparations; cosmetics; insecticides; paint; source of alumina (see Toth process).
See also kaolinite; aluminum silicate; clay.

kaolinite $Al_2O_3 \cdot 2SiO_2 \cdot 2H_2O$. A clay mineral, rarely found pure; the main constituent of kaolin and some other clays.

"Kaomul."[455] Trademark for a mullite firebrick that combines the properties of alumina-diaspore and mullite refractories; m.p. 3280°F; chemical analysis 61% alumina and 35% silica.

"Kao-Tab."[455] Trademark for a tabular alumina, high-strength castable refractory for service up to 3300°F.

"Kaowool."[455] Trademark for a stable, high-temperature alumina-silica ceramic fiber. Can be used up to 2300°F; m.p. 3200°F; diameter of fibers 2.8 microns; length up to 10 inches. Used as insulating material; high temperature filter; pipe and joint protection; for sound absorption. Available in bulk, strip or blanket forms. Noncombustible.

kapok. Cotton-like fibers obtained from the seed pods of various species of Ceiba and Bombax. Extremely light and resilient but too brittle for spinning. Combustible; not self-extinguishing.
Sources: Indonesia, Philippines, Ecuador, West Africa.
Uses: Life jackets; insulation; pillows; upholstery.

"Kapton."[28] Trademark for a polyimide film (1 to 5 mils thick).

"Karathane."[23] Trademark for an agricultural fungicide-miticide based on dinitro(1-methylheptyl) phenyl crotonate (see dinocap) and supplied as a wettable powder or liquid concentrate. May be combined with most other insecticides and fungicides, except oil-based products.
Use: Controls powdery mildew and various species of mites on plants.

karaya gum (sterculia gum; India tragacanth, kadaya gum). A hydrophilic gum which exudes from certain Indian trees of the genus Sterculia. Color varies from white to dark brown or black.
Properties: A carbohydrate polymer of varying chemical composition. Properties depend on freshness and time of storage. Viscosity greatly decreases

Superior numbers refer to Manufacturers of Trade Mark Products. For page number see Contents.

over six months storage. Forms a translucent colloidal gel in water. Nontoxic.
Grades: Technical; F.C.C.
Uses: Pharmaceuticals; textile coatings; ice cream and other food products; adhesives; protective colloid; stabilizer; thickener; emulsifier.
See also gum, natural.

"Karbate."[214] Trademark for carbon and graphite made impervious to fluids under pressure by impregnation with chemically resistant materials. Strength is increased but thermal conductivity is not lowered by this impregnation nor are the other properties modified to any extent. Resistant to thermal shock and to attack by most nonoxidizing chemicals.
Forms: Supplied as complete equipment items; also in blocks, cylinders, tubes.
Uses: Pipe; fittings; valves; pumps; heat exchangers; towers and absorbers for chemical process equipment.
See also carbon, industrial.

"Kardel."[214] Trademark for biaxially oriented polystyrene film. Used for window envelopes and packaging produce.

Karl Fischer reagent. A solution of iodine, sulfur dioxide and pyridine in methanol or methyl "Cellosolve." It is used in the determination of water.
Note: Do not confuse with Fischer's reagent (q.v.).

"Karmex."[28] Trademark for a wettable powder containing 80% diuron, a selective or nonselective weed control. "Karmex" DL is a suspension containing 28% diuron for pre-emergence control in cotton.
Containers: 4-lb bags; 50-lb fiber drums.
Hazard: May be toxic by inhalation.

"Kasil."[201] Trademark for a group of soluble silicates of potassium which vary from a clear, thin syrupy solution to colorless plates or lumps.
Uses: Detergents, deflocculating agents, adhesives, cements, films and coatings, sols and gels, and paints.

"Katadyne" process. A proprietary method of sterilizing water and other potable liquids with a specially prepared form of silver.

kauri. A fossil (hard) copal resin (q.v.), derived from the kauri pine (Agathis australis) of New Zealand. Soluble in alcohols and ketones; acid value 60–80; must be heat-treated (cracked) before use in varnishes. Combustible.
Uses: Varnishes and lacquers; paints; organic cements; to evaluate the solvent power of hydrocarbons.

kauri-butanol value. A measure of the aromatic content, and hence the solvent power, of a hydrocarbon liquid. Kauri gum is readily soluble in butanol but insoluble in hydrocarbons. The k.b. value is the measure of the volume of solvent required to produce turbidity in a standard solution containing kauri gum dissolved in butanol. Naphtha fractions have a k.b. value of about 30, and toluene about 105.

kcal Abbreviation for kilogram calorie; Cal has the same meaning.

K-capture (K-radiation). A type of radioactive decay in which an electron is captured by an atomic nucleus and immediately combines with a proton to form a neutron. The product of this radioactivity has the same mass number as the parent but the atomic number is one unit less. Thus Fe-55 with atomic number 26 decays by K-capture to form Mn-55, with atomic number 25. Terms synonymous with K-capture are K-electron capture and orbital electron capture.

Keene's cement. See gypsum cement.

"Keflin.'[100] Trademark for cephalothin (sodium salt).

"Kelacid."[322] Trademark for alginic acid (q.v.).
Uses: Tablet disintegrant; hemostatic agent.

Kekulé, August (1829–1896). Born in Darmstadt, Germany, Kekulé laid the basis for the ensuing development of aromatic chemistry. His idea of a hexagonal structure for benzene in 1865 was a monumental contribution to theoretical organic chemistry. "This had been preceded in 1858 by the remarkable notion that carbon was tetravalent and that carbon atoms could be joined to each other in molecules. The theory of the benzene ring has been called the 'most brilliant piece of scientific prediction to be found in the whole field of organic chemistry,' for besides promulgating the idea, he had predicted the number and types of isomers which might be expected in various substitutions on the ring." (L. B. Clapp)

"Kelco-gel."[322] Trademark for refined sodium alginate. Available as HV and LV grades, which vary in viscosity. A hydrophilic colloid.
Uses: Thickening, suspending, stabilizing, binding and gelling agent for foods, pharmaceuticals, cosmetics, welding rods, ceramics, latex paints, industrial gels, pastes, coatings, films.

"Kelcoloid."[322] Trademark for propylene glycol alginate products, available in a wide variety of forms.
Uses: A hydrophilic colloid. Emulsifier, thickener, stabilizer, suspending, foam stabilizing and whipping agent in aqueous media below pH 7.0. Suitable for food, pharmaceutical, cosmetic and industrial uses.

"Kelecin."[64] Trademark for a group of soybean lecithins.
Uses: Emulsifying and dispersing agents for protective coatings, mastics, animal feeds, automotive specialty chemicals, textiles, rubber processing, plastics, cosmetics, and food products.

"Kel-F."[158] Trademark for a series of fluorocarbon products, including polymers of chlorotrifluoroethylene and certain copolymers available as extrusion and molding powders, resins, dispersions, gums, oils, waxes and greases, that are characterized by high thermal stability, resistance to chemical corrosion, high dielectric strength, high impact, tensile and compressive strength.
Uses: Corrosion control, contamination prevention, insulation, electrical equipment, molded and fabricated industrial equipment, lubricants, gyro and damping fluids. Especially useful under extreme conditions, including jet and space technology.
See also fluorocarbon polymer.

"Kellin."[64] Trademark for a chemically treated linseed oil which polymerizes rapidly, dries fast, has excellent water resistance. Used in printing inks.

"Kelo-Form."[19] Trademark for esters of para-aminobenzoic acid, such as butyl, ethyl, isobutyl and propyl; used as a local anesthetic in ointments, powders, lozenges and other medicinals; also in preparation of sun tan lotions.

kelp. A large, coarse seaweed occurring chiefly off the coast of California. It is a type of algae (q.v.) and is mechanically harvested by specially equipped barges. Dried kelp contains from 2 to 4% ammonia, 1 to 2% phosphoric acid, 15 to 20% potash and traces of iodine. Kelp can be used for fertilizers, plastics, and conversion to methane by mictoorganisms. Large-scale growth off California is in process.

"Kelprint."[322] Trademark for an exceptionally high viscosity sodium alginate used in textile printing.

"Kelthane."[23] Trademark for an agricultural miticide based on 1,1-bis(chlorophenyl)-2,2,2-trichloroethanol and supplied as a wettable powder or emulsifiable concentrate.

"Keltrol."[64] Trademark for vinyltoluene copolymers. Used in fast air dry, forced dry or baking finishes for clear sealers, sanding sealers, primers, aerosol enamels, paper coatings, tub caulks, tile adhesives, hardboard and plywood sealers and primers.

"Kelzan."[322] Trademark for a polysaccharide known as xanthan gum produced by fermentation. Functions as a hydrophilic colloid to thicken, suspend, and stabilize systems where other colloids are unsuitable. See xanthan gum.

"Kemamides."[473] Trademark for a group of fatty amides including behenamide, erucamide, oleamide, stearamide.

"Kemamines."[473] Trademark for a group of fatty nitrogen derivatives including primary and secondary amines, 1,3-propylene diamines, and quaternary ammonium chlorides.

"Kemesters."[473] Trademark for a group of methyl esters of fatty acids.

"Kempore."[511] Trademark for a series of azodicarbonamide foaming agents, for vinyls, rubbers, and thermoplastic resins in general.

"Kemprint."[141] Trademark for a textile pigment emulsion used for printing on fabrics.

"Kenacort."[412] Trademark for triamcinolone (q.v.).

kenaf. Bast fibers obtained from the stems of Hibiscus sabdariffa, grown in Brazil and the West Indies. Used as a substitute for jute in burlap coffee bags.

"Kenalog."[412] Trademark for triamcinolone acetonide in cream, paste, suspension, lotion, ointment and spray forms.

"Kenamine."[473] Trademark for a series of straight-chain amines from primary through quaternary; available in chain lengths from C_{12} through C_{22}. They are strongly hydrophobic and are biodegradable.
Uses: Cationic intermediates; removal of moisture from surfaces.

"Kenflex."[267] Trademark for a synthetic polymer of aromatic hydrocarbons.
Uses: Processing and compounding aid for neoprene, "Hypalon," SBR, vinyl compounds and other plastics; potting compounds; protective coatings, paper and textile coatings; insecticides; inks; chemical synthesis.

"Kennametal."[347] Trademark for a group of hard cemented carbides whose principal ingredient is tungsten carbide plus cobalt with additions of carbides of tantalum, titanium, niobium. Available in varied degrees of hardness from 82.0 to 94.0 Rockwell A and of shock resistance by variation of composition, manufacturing techniques, and structure. There are six major series.

"Kennertium."[347] Trademark for a group of high-density tungsten alloys. Grades W-2 and W-10 are 18.5 and 17.0 g/cc, respectively. Easily machinable; used for high inertial applications, balancing and radioactive shielding.

"Kenplast."[267] Trademark for a series of hydrocarbon oils, both aromatic and naphthenic, compatible with all elastomers and polymers.
Uses: PVC plasticizers, reclaim oil, SBR, neoprene, butyl, "Thiokol."

"Kensol."[167] Trademark for a group of products of close-cut petroleum fractions.
Uses: Ink oils, metal rolling oils, petroleum solvents, and solvent naphtha high flash.

"Kentanium."[347] Trademark for a group of hard cemented carbides whose principal ingredient is titanium carbide.

"Keranol."[300] Trademark for a modified cationic softener for fabrics. Also used on synthetic fibers as antistatic agent.

keratin. A class of natural fibrous proteins occurring in vertebrate animals and man; they are characterized by their high content of several amino acids, especially cystine, arginine, and serine. They are generally harder than the fibrous collagen group of proteins; the softer keratins are components of the external layers of skin, wool, hair, and feathers, while the harder types predominate in such structures as nails, claws, and hoofs. The hardness is largely due to the extent of cross-linking by the disulfide bonds of cystine by the mechanism shown below:

$$\begin{array}{c}\cdots R_1CH\text{—}CO\text{—}HN\cdot CH\cdot CO\text{—}NH\text{—}CHR_2\cdots\\ |\\ CH_2\\ |\\ S\\ |\\ S\\ |\\ CH_2\\ |\\ \cdots R_3CH\cdot OC\text{—}HN\cdot CH\cdot CO\text{—}NH\text{—}CHR_4\cdots\end{array}$$

Keratins are insoluble in organic solvents but do absorb and hold water. The molecules contain both acidic and basic groups, and are thus amphoteric. Other than wool, keratins have limited industrial application. A foam based on keratin extracted from chicken feathers is reported useful in protecting growing vegetables from frost damage. See also cross-linking; protein; wool.

"Keripon."[300] Trademark for a water soluble synthetic fatty ester with wetting and rewetting properties for textile, leather and paper.

kernite $Na_2B_4O_7 \cdot 4H_2O$. A natural sodium borate, found in Kern County, California. Colorless to white; two good cleavages; luster vitreous to pearly; Mohs hardness 3; sp. gr. 1.95. Noncombustible.
Use: Major source of borax and boron compounds.

kerogen. The organic component of oil shale, ranging from 30 to 60% of total volume. See shale oil.

kerosene (kerosine).
Properties: Water-white, oily liquid; strong odor; sp. gr. 0.81; boiling range 180 to 300°C. Flash point 100 to 150°F; autoignition point 444°F. Combustion properties can be greatly improved by a proprietary hydrotreating process involving a selective catalyst.
Derivation: Distilled from petroleum.
Containers: Drums; trucks; tank cars.
Hazard: Moderate fire risk. Moderately toxic by in-

gestion and inhalation. Explosive limits in air 0.7 to 5.0%.
Uses: Rocket and jet engine fuel; domestic heating; solvent; insecticidal sprays; diesel and tractor fuels.

"Kessco."[15] Trademark for a series of fatty acid esters used as plasticizers, softeners, wetting agents, emulsifiers, etc.

ketene $H_2C{:}CO$.
Properties: Colorless gas; disagreeable taste. Readily polymerizes; cannot be shipped or stored in a gaseous state. M.p. −151°C; b.p. −56°C.
Derivation: Pyrolysis of acetone or acetic acid by passing its vapor through a tube at 500–600°C.
Containers: Steel bottles for intraplant transfer only.
Hazard: Highly toxic by ingestion and inhalation; strong irritant to skin and mucous membranes. Tolerance, 0.5 ppm in air.
Uses: Acetylating agent, generally reacting with compounds having an active hydrogen atom; reacts with ammonia to give acetamide. Starting point for making various commercially important products, especially acetic anhydride and acetate esters.

ketimine. A type or class of curing agent for epoxy resins which makes it possible to use very high solids content-coatings in spray equipment. Reacts with epoxies very slowly, and thus delays curing time, which prevents setting up of the resin during spraying operation. In presence of water or water vapor, ketimine breaks down to a polyamine and a ketone. Epoxy coatings cured with ketimine should not exceed thickness of ten mils.

4-ketobenzotriazine (benzazimide; 4-keto-(3H)-1,2,3-benzotriazine) $C_7H_5N_3O$ (bicyclic).
Properties: Tan powder; m.p. 210°C (dec.); soluble in alkaline solutions and organic bases.
Uses: Organic synthesis.

alpha-**ketoglutaric acid** (2-oxopentanedioic acid) $HOOCCH_2CH_2COCOOH$. M.p. 113.5°C; soluble in water and alcohol. Important in cell metabolism.

beta-**ketoglutaric acid** (ADA; acetonedicarboxylic acid) $HO_2CCH_2COCH_2CO_2H$.
Properties: Colorless needles; m.p. 135°C (dec); soluble in water and alcohol; insoluble in benzene and chloroform.
Derivation: By heating dehydrated citric acid and concentrated sulfuric acid together.
Use: Organic synthesis.

ketohexamethylene. See cyclohexanone.

ketone. A class of liquid organic compounds in which the carbonyl group, C=O, is attached to two alkyl groups; they are derived by oxidation of secondary alcohols. The simplest member of the series is acetone, $CH_3C(O)CH_3$, but many more complex ketones are known. Ketones are used primarily as solvents, especially for cellulose derivatives, in lacquers, paints, explosives, and textile processing.
See also acetone, diethyl ketone, methyl ethyl ketone.

"Ketone BD 9."[505] Trademark for a bicyclic ketone.
Properties: Pale straw-colored liquid; sp. gr. 0.952; refractive index (n 20/D) 1.488; b.p. 290°C. Used in woody perfumes.

ketone, Michler's. See tetramethyldiaminobenzophenone.

ketonimine dye. A dye whose molecules contain the —NH=C= chromophore group. There are only two members in the class: auramine, and a closely related homolog, methyl aurin in which a methyl group replaces one of the hydrogen atoms of aurin. These are basic dyes used on cotton with tannin or tartar emetic as mordant.

alpha-**etopropionic acid.** See pyruvic acid.

gamma-**ketovaleric acid.** See levulinic acid.

"Kevlar."[28] Trade mark for an aromatic polyamide fiber of extremely high tensile strength and greater resistance to elongation than steel. Its high energy-absorption property makes it particularly suitable for use in belting radial tires, for which it was specifically developed; it is also used as a reinforcing material for plastic composites, and in cordage products. See also aramid

Keyes process. A distillation process involving the addition of benzene to the constant-boiling 95% alcohol-water solution to obtain absolute (100%) alcohol. On distillation a ternary azeotrope mixture containing all three components leaves the top of the column while anhydrous alcohol leaves the bottom. The azeotrope (which separates into two layers) is redistilled separately for recovery and reuse of the benzene and alcohol.

"Keyval" TN.[496] Trademark for a modified EDTA complex; used as a high-capacity iron sequestrant.

kg Abbreviation for kilogram = 1000 grams.

"KH-31."[475] Trademark for hydrated alumina. Used in the manufacture of cracking catalysts for the petroleum industry; in the production of many types of glass to increase resistance to chemical attack, thermal and mechanical shock; to improve the surface luster and smoothness of vitreous enamel and whiteware.

khurchatovium. Artificial element No. 104 discovered by U.S.S.R. physicists in 1967.

Kick's law. The amount of energy required to crush a given quantity of material to a specified fraction of its original size is the same, regardless of the original size.

kieselguhr. See diatomaceous earth.

kieserite $MgSO_4 \cdot H_2O$. A natural magnesium sulfate occurring in enormous quantities in the Stassfurt salt beds, Germany, in Austria and India. See also epsomite and magnesium sulfate.

"killed" steel. Steel deoxidized by the addition of aluminum, ferrosilicon, etc., while the mixture is maintained at melting temperature until all bubbling ceases. The steel is quiet and begins to solidify at once without any evolution of gas when poured into the ingot molds.

"Kilmag."[50] Trademark for a formulation of calcium arsenate for the control of certain fly maggots under poultry cages.
Hazard: Highly toxic.

kilo-. Prefix meaning 10^3 units (symbol k). E.g., 1 kg = 1 kilogram = 1,000 grams.

kilogram.
(1) A mass identical with that of the international kilogram, at the International Bureau of Weights and Measures in France. It is the mass of a liter of water at 4°C.
(2) A force equal to the weight of one kilogram mass, measured at sea level.

kinematic viscosity. See viscosity.

kinetics, chemical. Chemical phenomena can be studied from two fundamental approaches: (1) *thermodynamics*, a rigorous and exact method concerned with equilibrium conditions of initial and final states of chemical changes (see thermodynamics, chemical); and (2) *kinetics*, which is less rigorous and deals with the rate of change from initial to final states under nonequilibrium conditions. The two methods are related. Thermodynamics, which yields the driving potential—a measure of the tendency of a system to change from one state to another—is the foundation upon which kinetics is built. The rate at which a change will occur depends upon two factors: (a) directly with driving force or potential, and (b) inversely with a resistance. A measure of the tendency of a system to resist chemical change is the so-called activation energy, which is independent of the driving force or so-called free energy of the reaction.

The diagram is a mechanical analogy illustrating the difference between activation energy and driving potential. The chemical system is represented by a sphere resting in a valley. The initial equilibrium state A is at a higher elevation than the final state B. The difference in elevation between A and B is a measure of the free energy change of the reaction, that is, the driving force which will take the system from A to B. This quantity ΔG is determined by the classical methods of thermodynamics. Now both A and B are equilibrium states represented by the valleys. For the system to go from A to B it must first overcome the hill separating the valleys. The elevation of this hill from the valley of the initial state is a measure of the resistance to change in the system in going from A to B. The quantity ΔG^*, known as free energy of activation, is determined by the methods of kinetics.

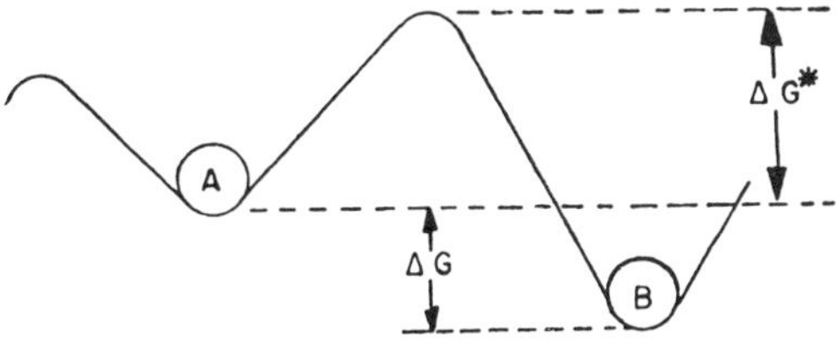

The system of molecules which is undergoing reaction consists of these molecules in different energy states. If the temperature of a gas is raised, there is an increase in the energy of these states and hence an increase in the collisions of molecules which have the necessary activation energy; as a result, the rate of the reaction will increase. Also, if by means of a catalyst, the activation energy is decreased, more colliding molecules will react and again the rate of reaction will increase.

Note: Adapted from article by Roger Gilmont in "Encyclopedia of Chemistry."

See also thermodynamics.

kinetic theory. A theory of matter based on the mathematical description of the relationship between the volumes, temperatures, and pressures of gases (P-V-T phenomena). This relationship is summarized in the so-called gas laws, as follows: (1) Boyle's law (at constant temperature, the volume of a gas is inversely proportional to its pressure); (2) Charles' law (at constant volume, the pressure exerted by a gas is proportional to its absolute temperature); (3) Avogadro's law (equal volumes of the same or different gases under the same conditions of temperature and pressure contain the same number of molecules).

The theory involves the basic concept of matter as comprised of atoms and/or molecules which move more rapidly (gases) or vibrate more energetically (solids) as temperature increases. Thus crystals melt at a point where the heat or energy input exceeds the bond energy of the solid state. See also kinetics, chemical; gas; thermodynamics.

king's green. See copper acetoarsenite.

kinin (cytokinin). One of a group of plant growth regulators (q.v.) which promote cell division and differentiation. Best known is kinetin, $C_{10}H_9N_5O$ (6-furfurylaminopurine), derived from deoxyribonucleic acid.

Kjeldahl test. An analytical method for determination of nitrogen in certain organic compounds. It involves addition of a small amount of anhydrous potassium sulfate to the test compound, followed by heating the mixture with concentrated sulfuric acid, often with a catalyst such as copper sulfate. As a result ammonia is formed. After alkalyzing the mixture with sodium hydroxide the ammonia is separated by distillation, collected in standard acid and the nitrogen determined by back-titration.

"Kleanrol."[28] Trademark for a soldering flux crystal based on zinc chloride and ammonium chloride.

"Klearol."[45] Trademark for a white mineral oil, technical grade. Sp. gr. 0.828–0.838.
Uses: Cosmetic preparations; shell egg preservation.

"Kleenup."[253] Trademark for a type of insecticide and fungicide containing petroleum oils.

Klein's reagent. A saturated solution of cadmium borotungstate, formula variously given, possibly $2CdO \cdot B_2O_3 \cdot 9WO_3 \cdot 18H_2O$. Sp. gr. 3.28.
Hazard: Probably toxic.
Use: Separation of minerals by specific gravity.

"Klucel."[266] Trademark for a hydroxypropyl cellulose (q.v.).
Properties: Water- and organo-soluble; thermoplastic; nontoxic; nonionic; surface-active.
Uses: Thickener, stabilizer, film-forming agent; binder.

knock. Ignition of a portion of the gasoline in the cylinder head due to spontaneous oxidation reactions rather than to the spark. It causes serious power loss, especially in high-compression engines. See also octane number.

Knoop hardness. See hardness.

Koch's acid. See 1-naphthylamine-3,6,8-trisulfonic acid.

"Kodaflex AD-2."[256] Trademark for adhesion-promoting plasticizer for vinyl coatings.

"Kodar."[115] Trademark for a thermoplastic polyester resin (1,4-cyclohexylenedimethylene terephthalate/isophthalate copolymer). It is made from terephthalic acid, isophthalic acid, and the glycol cyclohexylenedimethanol. Its main field of use is in so-called blister packaging, for which it is claimed to be superior to polyvinylchloride and polyethylene terephthalate. Available in clear-colored amorphous pellets. Complies with food packaging safety regulations, including meats and poultry.

"Kodel."[256] Trademark for a polyester-type fiber.

kojic acid [5-hydroxy-2-(hydroxymethyl)-4-pyrone]$C_6H_6O_4$. An antibiotic.
Properties: Crystals; m.p. 152–154°C; soluble in water, acetone, alcohol; slightly soluble in ether; insoluble in benzene; mildly antibiotic.
Derivation: Fermentation of starches and sugars by certain molds.
Uses: Chemical intermediate; metal chelates; insecticide; antifungal and antimicrobial agent.

kola. See cola.

Kolbe-Schmidt reaction. The preparation of salicyclic acid or its derivatives from carbon dioxide and sodium or potassium phenolate.

"Kolene" DGS Salt.[557] Trademark for an anhydrous molten oxidizing salt bath using a sodium hydroxide base with additives necessary to provide controlled chemical oxiding and dissolving properties.
Uses: Descaling of heat treat and hot work oxides and scales; deglassing (removal of glass drawing lubricants), investment, and silica removal; removal of burned-in carbon deposits; cleaning of oils, greases and organic materials from the surface of metals.

"Kolineum."[11] Trademark for a highly refined creosote obtained from coal tar by fractional distillation followed by refrigeration and filtration.

"Kollidon."[440] Trademark for polyvinylpyrrolidone for the production of pharmaceuticals.

"Koloc."[248] Trademark for a resin emulsion for application to cotton and wool. It stops the shrinking and felting of wool, increases tensile strength, improves wearing quality, reduces fiber loss in processing and service.

korax. Generic name for 1-chloro-1-nitropropane $ClCH_2CH(CH_3)NO_2$.
Properties: Liquid; b.p. 170.6°C (745 mm). Miscible with most organic solvents; slightly soluble in water. Flash point 144°F. Combustible.
Hazard: Toxic; strong irritant. Moderate fire risk. Tolerance, 20 ppm in air.
Use: Fungicide.

"Koreon."[292] Trademark for a group of one-bath chrome-tanning compounds, usually of the basic chromic sulfate type. Range of chromic oxide equivalence 23.5–25.5%; basicity range 33.0–59.0%. Soluble in water.
Containers: Dry: 75-lb paper bags; 231-lb fiber drums. Solution: 700-lb steel drums; tank trucks.

"Koresin."[307] Trademark for a rubber tackifier; para-tert-butylphenol-acetylene resin; 100% active.

"Koresin."[440] Trademark for condensation product of a substituted phenol and acetylene for improving the tackiness and processing properties of synthetic rubber mixes and blends of these with natural rubber.

"Korez."[41] Trademark for a silica-filled, synthetic-resin, acid-proof cement of the phenol-formaldehyde type especially good as a mortar cement for electrorefining work. Can be used up to 360°F.

"Korlan."[233] Trademark for O,O-dimethyl O-2,4,5-trichlorophenyl phosphorothioate. See ronnel.
Hazard: Toxic. See ronnel.
Use: Insecticide.

"Korundal XD."[446] Trademark for a 90% alumina brick with high density. Resistant to corrosion and penetration by molten slags and fluxes, thermal shock, and can withstand load at temperatures up to 3000°F and higher.
Uses: Top checker courses of glass tank furnaces; aluminum alloy furnaces; electric furnaces producing alloys.

"Kovar."[308] Trademark for an iron-nickel-cobalt alloy (29% nickel, 17% cobalt, 53% iron, 1% minor ingredients). It has thermal expansion characteristics matching those of hard glass. Used for matched-expansion joints between metals and glass or ceramics. Sp. gr. 8.36; m.p. 1450°C.

Kr Symbol for krypton.

kraft paper. A strong and relatively cheap paper made chiefly from pine by digestion with a mixture of caustic soda, sodium sulfate, sodium carbonate and sodium sulfide. It is by far the largest volume paper made in U.S.
See also holopulping process.

"Kralac A-EP."[248] Trademark for a high styrene-butadiene copolymer.
Use: With natural and chemical rubbers, especially recommended for high-grade soles, tiling and molded parts. Main function is that of greatly increasing the hardness of elastomeric compounds.

"Kralastic."[248] Trademark for a series of ABS resins.
Properties: Granular rubber-plasticized resins; rigid and tough; dimensionally stable, light in weight; chemically resistant; good electrical resistance.
Uses: Injection and extrusion applications; chemical pipe; cathode edge strips; cams, gears, cable floats, wheels, etc.

"Kralon."[248] Trademark for a series of rigid thermoplastic resin-rubber blends used for conduit and irrigation pipe, tool handles, automotive air ducts and other applications.

"Kraton" 101.[125] Trademark for a styrene-butadiene elastomer which requires no vulcanization while displaying most of the properties of conventional vulcanized polymers. White, free-flowing crumb; readily soluble in a large number of commercially used solvents. Used in pressure sensitive, wet lay-up and hot-melt adhesives, dip coating, spraying, and spreading applications.

Krebs cycle. See TCA cycle.

"Krene."[214] Trademark for flexible vinyl film and sheeting supplied in continuous roll form for packaging, laminations, seat covers, and shower curtains.

"Krenite" 10.[28] Trademark for a solution of urea and ammonium nitrate in aqueous ammonia; contains 43.5% nitrogen. Used in the manufacture of mixed fertilizers.
Containers: tank cars.

Kroll process. A widely used process for obtaining titanium metal. Titanium tetrachloride is reduced with magnesium metal at red heat and atmospheric pressure, in the presence of an inert gas blanket of helium or argon. Magnesium chloride and titanium metal are produced. The reacton is $TiCl_4 + 2Mg \rightarrow Ti + 2MgCl_2$. Essentially the same process is also used for obtaining zirconium.

"Kroma-Clor."[329] Trademark for a turf fungicide with turf color whose active ingredients are mercuric dimethyldithiocarbamate, potassium chromate, and cadmium succinate.
Hazard: Toxic by ingestion and inhalation.

"Kromarc."[308] Trademark for a chromium-nickel steel containing manganese and molybdenum and small amounts of other additives which is readily formed and exhibits high strength and oxidation resistance at temperatures up to 1200°F. Chief applications are steam turbines and cryogenic equipment.

"Kromatherm."[299] Trademark for high-temperature pigments designed for silicone- and fluorocarbon resin-based paint vehicles.

"Kromfax" Solvent.[214] Trademark for thiodiethylene glycol (q.v.).

"Kromik."[141] Trademark for a multiple pigment rust-inhibiting primer for shop coatings or for first field coat on structural steel.

"Kromosperse."[74] Trademark for a pigment dispersant to produce tinting pastes compatible with many aqueous and oil solutions.

"Kroniflex."[55] Trademark for a series of flame retartant phosphate plasticizers and autiform agents.

"Kronisol."[55] Trademark for a series of plasticizers.

"Kronitex."[55] Trademark for a series of synthetic phosphate esters to replace such natural products as tricresyl and cresyl diphenyl phosphates.
Uses: Flame-retardant plasticizers for vinyls; dust filter medium; gas additives; wood-treating chemical; foam control.

"Kryocide."[214] Trademark for a natural cryolite insecticide.

"Kryolith."[204] Trademark for sodium fluoaluminate. See cryolite, synthetic.

krypton Kr Element of atomic number 36, noble gas group of the periodic system. Atomic weight 83.80; valence 2 (possibly others); has six stable isotopes and a number of artificially radioactive forms.
Properties: Colorless, odorless gas; b.p. −152.9°C (1 atm); f.p. −157.1°C; sp. gr. 2.818 (air = 1); sp. vol. 4.61 cu ft/lb (70°F, 1 atm); only slightly soluble in water. Known to combine with fluorine, at liquid nitrogen temperature, by means of electric discharges or ionizing radiation, to make KrF_2 and KrF_4. These compounds decompose at room temperature. Nontoxic; noncombustible. See noble gas.
Derivation: By fractional distillation of liquid air. Air contains 0.000108% of krypton, by volume.
Containers: Hermetically sealed glass flasks; steel containers.
Uses: Incandescent bulbs and fluorescent light tubes; lasers; high-speed photography.
Shipping regulations: (Air) Nonflammable gas label.
Note: Solid krypton exists, at cryogenic temperatures, as a white, crystalline substance, m.p. 116°K.

krypton 85. Radioactive krypton of mass number 85.
Properties: Half-life 10.3 years; radiations, beta, with a small component of gamma. Low radiotoxicity.
Derivation: A fission product extracted from irradiated nuclear fuel.
Forms available: Gas of high chemical purity, but mixed with other isotopes of krypton, in sealed glass flasks. See also kryptonates.
Uses: Activation of phosphors for self-luminous markers; detecting leaks; medicine, to trace blood flow.

krypton 86. Isotope of krypton used in measurement of standard meter (q.v.).

kryptonates. Materials impregnated with krypton-85 in such a way that the radioactive atoms are held within the crystal lattice structure. Elements, alloys, glasses, inorganic compounds, rubbers and plastics have been so impregnated with tracer atoms.

"Krytox."[28] Trademark for a series of hexafluoropropylene epoxide polymers of medium molecular weight, used as lubricating oils and greases. For use in high temperature or corrosive conditions; good chemical inertness, even with boiling sulfuric acid; low solubility in most solvents; good lubricity under load; nonflammable; have thermal stability up to 500°F.

"K-Stay."[69,466] Trademark for mixture of oil-soluble sulfonic acid of high molecular weight in a petroleum base oil.
Properties: Amber liquid, sp. gr. 0.88–0.90; acid no. 1.0–1.1
Uses: Processing aid for elastomers.

"KT."[280] Trademark for self-bonded silicon carbide shapes of very high density. Used for various purposes where resistance to high temperature and/or corrosive or erosive conditions is desired.

"KT-1060" and **"KT-1061."**[475] Trademarks for tabular aluminas. A high purity aluminum oxide that has been converted to the corundum state. It will not fuse below 3600°F. Used in the manufacture of high temperature refractory bricks, castables and ramming mixes, in electrical insulators, and in many other high grade ceramic bodies.

KTPP. Abbreviation for potassium tripolyphosphate.

"Kuron."[233] Trademark for a hormone-type weed and brush killer.

Kurroll's salt $NaPO_3$ (IV). A high temperature crystalline form of sodium metaphosphate (q.v.).

"Kutwell."[51] Trademark for emulsifiable or soluble cutting oils consisting of good quality base mineral oils to which emulsifiers are added. Water emulsions are used as coolants in metal-cutting operations.

"Kydex."[23] Trademark for a thermoforming acrylic-polyvinyl chloride alloy plastic sheet.
Uses: Housings, trays, covers, containers, protective guards and decorative parts exposed to severe service conditions.

"Kymene."[266] Trademark for a series of wet-strength resins.
557 and 709. Cationic, polyamide-epichlorohydrin resins with wet-strength development independent of papermaking pH.
882 and 917. Cationic, acid-curing, urea-formaldehyde resins.

"Kynol."[280] Trademark for a flame-resistant fiber available as 1.5 inch staple, 1.7 denier. It is a cross-linked amorphous phenolic polymer, inert to all solvents and with fair resistance to oxidizing acids and strong alkalies. Will not ignite up to 2500°C, but will char slowly at 500°F. Potential uses are as ablative agent in spacecraft, flameproof apparel, protective clothing, etc.

"Kyro AC."[487] Trademark for a modified amine condensate surfactant.

Properties: An amber colored, free-flowing liquid with a cloud point of approximately 50° F.

Uses: As a detergent, emulsifier and dispersing agent in textile applications.

"Kyro EOB."[487] Trademark for a neutral nonionic surfactant of the 100% ethoxylated alcohol type.

Properties: Water-white liquid readily soluble in hot or cool water. Liquid in all concentrations except about 50%, at which level it forms a gel at room temperature and lower.

Uses: A moderate sudsing nonionic for use where excessive foaming may be objectionable.

L

l Abbreviation for liter.

l- Prefix indicating that a compound is levorotatory. A minus sign (−) is now preferred.

L Prefix indicating the left-handed enantiomer of an optical isomer. See also D.

"L-26."[304] Trademark for the normal lead salt of 2-ethylhexoic acid. Straw-colored viscous liquid; soluble in many common organic solvents; stable at comparatively high temperatures. Used as a curing agent for silicone paints and insulating varnishes.

La Symbol for lanthanum.

labile. Descriptive of a substance that is inactivated by high temperature or radiation, e.g., a heat-labile vitamin; unstable.

laboratory conditions. An ideal set of conditions in which all variant factors except the one under test can be held constant, as for example, rooms provided with constant temperature and humidity control, clean rooms, and the like. Less specifically, the term refers to experimental or small batch conditions, as opposed to large-scale production.

label. (1) A printed notice attached to or stamped on an article held for sale or delivery or offered for transportation. Special labels are legally required for toxic and flammable products and certain food products by the Dept. of Agriculture, the Federal Trade Commission, and the Food and Drug Administration. Transportation warning labels are prescribed by the Hazardous Materials Regulation Board of the Dept. of Transportatoin (DOT) for rail, water, and truck shipment (the Coast Guard is an agency of DOT), and by the International Air Transport Associatoin for airplane shipment (applied in the U.S. by the Federal Aviation Administration). The transportation labels used are:

	DOT	IATA
Flammable solids	Yellow	Flammable solid
Flammable liquids	Red	Flammable liquid
Oxidizing material	Yellow	Oxidizer
Corrosive material	White	Corrosive
Flammable gas	Red gas	Flammable gas
Nonflammable compressed gas	Green	Nonflammable compressed gas
Poisonous solids and liquids	Poison	Poison
Explosive materials	Explosive	Explosive
Radioactive material	Radioactive	Radioactive

Many hazardous materials are required to carry warning labels for safe handling, storage, and in-plant movement.

The Manufacturing Chemists Association issues instructions for preparation of warning labels in its publication "Guide to Precautionary Labeling of Hazardous Chemicals." See also hazardous material; toxic material; flammable material.

(2) A radioactive isotope or fluorescent dye added to a chemical compound to trace its course and behavior through a series of reactions, usually in living organisms. This technique has also been used in frictional wear of moving parts in automotive engines. See also tracer; tagged compound.

laccase. An enzyme that oxidizes phenols to ortho- and para-quinones. It occurs in the latex of the lac tree, in potatoes, sugar beets, apples, cabbages and other plants.

lac dye. A brilliant red dye obtained by maceration of crude lac. See shellac.

lachrymator (lacrimator). A gas that is strongly irritant to the eyes; tear gas.

lacmoid (resorcinol blue) $(HO)_2C_6H_3N[C_6H_2(OH)_3]_2$.

Properties: Lustrous, dark-violet, crystalline scales. Soluble in alcohol, ether, acetone, phenol, and glacial acetic acid; slightly soluble in water; pH 4.4–6.2.

Derivation: From resorcinol by treatment with sodium nitrite.

Use: Indicator in analytical chemistry.

"Lacolene."[457] Trademark for aliphatic petroleum naphthas; used in nitrocellulose lacquer formulations, in printing and roto types of inks. Distillation range (ASTM D86-59) 195–275°F.

lacquer. A protective or decorative coating that dries primarily by evaporation of solvent, rather than by oxidation or polymerization. Lacquers were originally comprised of high-viscosity nitrocellulose, a plasticizer (dibutyl phthalate or blown castor oil), and a solvent. Later, low-viscosity nitrocellulose became available; this was frequently modified with resins such as ester gum or rosin. The solvents used are ethyl alcohol, toluene, xylene, and butyl acetate. Modern automotive lacquers are characterized by their exceptionally high drying rate, a fact which did much to make possible the large-scale production of automobiles in the twenties. Together with nitrocellulose, alkyd resins are used to improve durability. The nitrocellulose (q.v.) used for lacquers has a nitrogen content of 11 to 13.5%, and is available in a wide range of viscosities, compatibilities, and solvencies. Chief uses of nitrocellulose-alkyd lacquers is for coatings for metal, paper products, textiles, plastics, furniture, and nail polish. Various types of modified cellulose are also used as lacquer bases, combined with resins and plasticizers. Many non-cellulosic materials, such as vinyl and acrylic resins, have been introduced in recent years. Bitumens are also used in lacquers, with or without drying oils, resins, etc. See also nitrocellulose.

Hazard: Flammable, dangerous fire and explosion risk.

Shipping regulations: (Lacquer base or chips, dry) (Rail) Yellow label. (Air) Flammable Solid label. (Lacquer base or chips, plastic, wet with alcohol or solvent) (Rail) Red label. (Air) Flammable Liquid label.

Superior numbers refer to Manufacturers of Trade Mark Products. For page number see Contents.

lactalbumin. A component of skimmilk protein (2 to 5%). Can be crystallized. Exact function is not known, but probably aids in stabilization of the fat particles. See also milk.

lactam. A cyclic amide produced from amino acids by the removal of one molecule of water. An example is caprolactam (q.v.), $\overline{CH_2(CH_2)_4CONH}$, derived from epsilon-aminocaproic acid, $NH_2(CH_2)_5COOH$.

lactase. An enzyme present is intestinal juices and mucosa which catalyzes the production of glucose and galactose from lactose.
Use: Biochemical research.

lactic acid (alpha-hydroxypropionic acid; milk acid) $CH_3CHOHCOOH$. (For isomeric structure, see under asymmetry).
Properties: Colorless or yellowish, odorless, hygroscopic syrupy liquid. B.p. (15 mm) 122°C; m.p. 18°C; sp. gr. 1.2. Miscible with water, alcohol, glycerin; soluble in ether; insoluble in chloroform, petroleum ether, carbon disulfide. Cannot be distilled at atmospheric pressure without decomposition; when concentrated above 50%, it is partialy converted to lactic anhydride. It has one asymmetric carbon and two enantiomorphic isomers. The commercial form is a racemic mixture.
Derivation: (a) By fermenting starch, milk whey, molasses, potatoes, etc. and neutralizing the acid as soon as formed with calcium or zinc carbonate. The solution of lactates is concentrated and decomposed with sulfuric acid; (b) synthetically by hydrolysis of lactonitrile (q.v.).
Grades: Technical, 22% and 44%; food, 50 to 80%; plastic, 50 to 80%; U.S.P. (85–90%); C.P.; F.C.C.
Containers: Barrels; carboys; drums; tank trucks.
Uses: Cultured dairy products; as acidulant; flavoring; preservative; chemicals (salts, plasticizers, adhesives, pharmaceuticals).

lactic acid dehydrogenase. An enzyme found in animal tissues and yeast which takes part in controlling carbohydrate metabolism in the cell.
Use: Biochemical research.

lactogenic hormone. See luteotropin.

lactoglobulin. A protein occurring in milk. It comprises from 7 to 12% of the skimmilk protein, and is closely associated with casein.

"Lactol."[106] Trademark for a solvent naphtha. An aliphatic naphtha in the toluene evaporation range.

lactone. An inner ester of a carboxylic acid, formed by intramolecular reaction of hydroxylated or halogenated carboxylic acids, with elimination of water. They occur in nature as odor-bearing components of various plant products; also made synthetically. See butyrolactone; propiolactone.

lactonitrile (alpha-hydroxypropionitrile; acetaldehyde cyanohydrin) $CH_3CHOHCN$.
Properties: Straw-colored liquid, acid to methyl red; f.p. −40°C; b.p. 182–184°C (slight decomposition); sp. gr. 0.9919 (18.4°C); refractive index (18.4/D) 1.4058. Soluble in water and alcohol. Insoluble in petroleum ether and carbon disulfide. Flash point 170°F. Combustible.
Derivation: Acetaldehyde and hydrocyanic acid.
Grades: Technical; 95–97% purity.
Containers: Carboys.
Hazard: Evolves hydrocyanic acid in presence of alkali.
Uses: Solvent; intermediate in production of ethyl lactate and lactic acid.

lactose (milk sugar; saccharum lactis) $C_{12}H_{22}O_{11} \cdot H_2O$.
Properties: White, hard, crystalline mass or white powder; sweet taste, odorless. Stable in air. Soluble in water; insoluble in ether and chloroform; very slightly soluble in alcohol. Sp. gr. 1.525; m.p. decomposes at 203.5°C. Nontoxic.
Derivation: From whey, by concentration and crystallization. Cows' milk contains about 5% lactose.
Grades: Crude; fermentation; spray-dried; edible; U.S.P.
Uses: Pharmacy; infant foods; bacteriology; baking and confectionery; margarine and butter manufacture; manufacture of penicillin, yeast, edible protein, and riboflavin.

lac wax. A wax obtained from lac consisting of myricyl and ceryl alcohols, free and combined with various fatty acids. Combustible; nontoxic.

LAD. Abbreviation for lithium aluminum deuteride (q.v.).

ladder polymer. An ordered molecular network of double-stranded chains connected by hydrogen or chemical bonds located at regular intervals along the chains. Many complex proteins, including DNA, are of this nature.

LAH. Abbreviation for lithium aluminum hydride.

lake. An organic pigment produced by the interaction of an oil-soluble organic dye, a precipitant, and an absorptive inorganic substrate. Insoluble in water. Poor light-fastness makes them unsuitable for use in exterior paints.
Uses: Printing inks; wallpaper inks; metal decorative coatings; coated fabrics; rubber; plastics; food colorants.
See also toner; alizarin; madder lake.

Lake Red C. Red pigments made by coupling 2-chloro-5-aminotoluene-4-sulfonic acid with beta-naphthol and forming various metal salts.
Properties: Good resistance to bleeding; reasonable light resistance, good transparency; produces inks with good flow.
Grades: Resinated and non-resinated.
Uses: General purpose color for letterpress, gravure, flexographic, moisture set, heat set inks; specially for offset printing inks.

Lake Red C Amine. See 5-amino-2-chlorotoluene-4-sulfonic acid.

"Laktane."[51] Trademark for a solvent especially prepared for use as lacquer diluent and in rotogravure printing inks. Its boiling range is typically 218–228°F.

"Lambast."[58] Trademark for a contact herbicide containing 25.5% 2,4-bis(3-methoxypropylamino)-6-methylthio-s-triazine $[CH_3O(CH_2)_3NH]_2C_3N_3SCH_3$.
Properties: White solid; m.p. 55°C. Insoluble in water; slightly soluble in ethanol; very soluble in acetone and benzene.

lamepon. An acetylated peptide used as a surface-active agent.

"Laminac."[57] Trademark for a proprietary grade of polyester resin used mainly in the manufacture of reinforced plastics. Most widely used reinforcement is glass fiber. Typical products fabricated include speedboats, radomes, tanks for water and chemical storage, and sports car bodies.

laminate. A composite (q.v.) made of any one of several types of thermosetting plastic (phenolic, polyester, epoxy, or silicone) bonded to paper, cloth, asbestos, wood, or glass fiber. High tensile and dielectric strength, and low moisture absorption are characteristic of these products. Available as sheet, rod, or tubing in mechanical, electrical and general-purpose grades (National Electrical Mfrs. Assn.). Plywood is composed of a veneer, with grain oriented at a 90° angle on successive layers, and bonded with a thermosetting adhesive of the urea or phenol-formaldehyde type to give a high strength, dimensionally stable, weather-resistant construction material. It can be made non-flammable by treatment with salt solution. Polyvinyl butyral sheet is used in safety glass. (See also reinforced plastic).

lampblack. A black or gray pigment made by burning low-grade heavy oils or similar carbonaceous materials with insufficient air, and in a closed system such that the soot can be collected in settling chambers. Properties are markedly different from carbon black. Strongly hydrophobic. Nonflammable.
Uses: Black pigment for cements, ceramic ware, mortar, inks, linoleum, surface coatings, crayons, polishes, carbon paper, soap, etc.; ingredient of insulating compositions, liquid-air explosives, matches, fertilizers, furnace lutes, lubricating compositions; carbon brushes; reagent in cementation of steel.
See also carbon black.

"Lanacid" C.[460] Trademark for crude lanolin fatty acids consisting of normal branched and hydroxy acids. A light tan, plastic waxy solid.
Uses: Polishes, cleaners; paints; inks; floor waxes; intermediate for heavy-metal soaps; greases; leather specialties.

"Lanacron."[443] Trademark for metal-complex dyes for wool and polyamide fibers.

"Lanamid."[243] Trademark for neutral-dyeing, premetallized dyes.

"Lanaset."[57] Trademark for a melamine-formaldehyde resin applied to woven and knitted wool fabrics to control wool shrinkage and felting.

langbeinite $K_2Mg_2(SO_4)_3$. A natural sulfate of potassium and magnesium, found in salt deposits.
Properties: Colorless, yellowish, reddish, greenish; luster vitreous; Mohs hardness 3.5–4; sp. gr. 2.83.
Occurrence: New Mexico, Germany, India.
Use: Source of potash.

Langmuir, I. (1881–1957). A brilliant American physical chemist who was awarded the Nobel Prize in 1932 for his fundamental research in surface chemistry, especially monomolecular films. This led to development of modern knowledge of emulsification and detergency. Langmuir also investigated electrical discharges in gases and did pioneer work on cloud-seeding techniques.

lanital. A regenerated protein fiber developed in Italy. The protein used most successfully is casein. It is subject to attack by microorganisms to about the same extent as wool, for which it was designed as a substitute.

"Lanitol."[300] Trademark for a group of alkylarylsulfonate type detergents.
A: Liquid biodegradable type.
F: Flake, sodium salt.
CW: Powder; same as flake, plus alkaline builders.

"Lanole B."[42] Trademark for a blend of sodium oleate and solvents. Amber colored liquid which disperses in water at all temperatures. Used as a scouring agent for the removal of tar, grease, and paint from woolen textile fabrics.

lanolin.
Properties: (Hydrous): Yellowish to gray semisolid containing from 25 to 30% water; slight odor. (Anhydrous): Brownish-yellow semisolid containing no more than 0.25% water, but can be mixed with about twice its weight of water without separation.
Derivation: Purification of degras, a crude grease obtained by solvent-treatment of wool. Contains cholesterol esters of higher fatty acids. Hydrogenated, ethoxylated and acetylated derivatives are available.
Grades: Technical; cosmetic; U.S.P.
Containers: Drums; tubes.
Uses: Ointments; leather finishing; soaps; face creams; facial tissues; hair-set and suntan preparations.
See also degras.

lanosterol (isocholesterol) $C_{30}H_{50}O$. An unsaturated sterol closely related to cholesterol; m.p. 139-140°C; optically active crystals.

"Lanoxin."[301] Trademark for preparations of digoxin (q.v.).

lanthana. See lanthanum oxide.

lanthanide series (lanthanoid series). The rare earth series of elements, atomic numbers 58 through 71. See also rare earth metal.

lanthanum La Element of atomic number 57; group IIIB of the periodic table; a rare earth of the cerium group. Atomic weight 138.905; valence 3. Two stable isotopes.
Properties: White, malleable, ductile metal; oxidizes rapidly in air. Sp. gr. 6.18–6.19; m.p. 920°C; b.p. 3454°C; corrodes in moist air; soluble in acids; decomposes water to lanthanum hydroxide and hydrogen. Superconducting below 6°K.
Derivation: By cracking of monazite or bastnasite ores with conc. sulfuric acid and subsequent separation.
Forms available: Ingots; rods; 20-mil sheets; powdered 99.9% pure.
Hazard: Ignites spontaneously in powdered form.
Uses: Lanthanum salts; electronic devices; pyrophoric alloys; rocket propellants; reducing agent catalyst for conversion of nitrogen oxides to nitrogen in exhaust gases (usually in combination with cobalt, lead, or other metals); phosphors in x-ray screens.
See also rare-earth metal.

lanthanum acetate $La(C_2H_3O_2)_3 \cdot xH_2O$.
Properties: White powder, soluble in water. Purities up up to 99.9+%. Soluble in acids.

lanthanum ammonium nitrate $La(NO_3)_3 \times 2NH_4NO_3 \times 4H_2O$.
Properties: Colorless crystals; soluble in water.
Grade: Purities to 99.9+%.
Hazard: Explosion and fire risk.
Shipping regulations: Nitrates, n.o.s., (Rail) Yellow label. (Air) Oxidizer label.

lanthanum antimonide LaSb. A binary semiconductor.

lanthanum arsenide LaAs. Made in high purity for use as a binary semiconductor.
Hazard: Highly toxic.
Shipping reuglations: (Rail, Air) Arsenical compounds, n.o.s., Poison label.

lanthanum carbonate $La_2(CO_3)_3 \cdot 8H_2O$.
Properties: White powder; insoluble in water; soluble in acids. Sp. gr. 2.6.
Grades: Up to 99.9+% La salts.
Containers: Glass bottles, fiber drums.

lanthanum chloranilate $La_2(O{:}C_6Cl_2O_2{:}O)_3 \cdot nH_2O$. Used as a reagent for fluoride determination.

lanthanum chloride $LaCl_3 \cdot 7H_2O$.
Properties: White crystals; transparent; hygroscopic; m.p. (hydrate) 91°C (dec); (for anhydrous) sp. gr. 3.842 (25°C); m.p. 872°C. Soluble in alcohol, water; acids.
Derivation: Treatment of lanthanum carbonates or oxides with hydrochloric acid in an atmosphere of dry hydrogen chloride.
Grades: Purities to 99.9+%. Single crystals available.
Containers: Glass bottles, fiber drums.
Hazard: Moderately toxic.
Use: Anhydrous trichloride of rare-earth metal is often used to prepare the metal.

lanthanum fluoride LaF_3.
Properties: White powder, insoluble in water, acids.
Grades: Purities up to 99.9+%. Single crystals available.
Containers: Glass bottles, fiber drums.
Hazard: Toxic by ingestion.
Uses: Phosphor lamp coating (gallium arsenide solid-state lamp); carbon arc electrodes; lasers.

lanthanum nitrate $La(NO_3)_3 \cdot 6H_2O$.
Properties: White crystals; hygroscopic. B.p. 126°C; m.p. 40°C. Soluble in alcohol, water, acids.
Grades: Purities to 99.9+%.
Containers: Glass bottles, fiber drums.
Hazard: Explosion and fire risk; moderately toxic.
Uses: Antiseptic; gas mantles.
Shipping regulations: Nitrates, n.os., (Rail) Yellow label. (Air) Oxidizer label.

lanthanum oxalate $La_2(C_2O_4)_3 \cdot 9H_2O$.
Properties: White powder, insoluble in water; soluble in acids.
Grades: Purities to 99.9+%.
Containers: Glass bottles, fiber drums.

lanthanum oxide (lanthana; lanthanum trioxide; lanthanum sesquioxide) La_2O_3.
Properties: White or buff amorphous powder; sp. gr. 6.51 (15°C); m.p. 2315°C; b.p. 4200°C. Soluble in acids; insoluble in water.
Derivation: Ignition of hydroxide or oxyacid (oxalate, sulfate, nitrate, etc.); direct combustion of free metal (burns with brilliant, white light).
Grades: Purities to 99.9+%.
Containers: Boxes, glass bottles, fiber drums.
Uses: Calcium lights; optical glass; technical ceramics; cores for carbon-arc electrodes; fluorescent phosphors; refractories.

lanthanum phosphide LaP. Made in high purity for use as a binary semiconductor.

lanthanum sesquioxide. See lanthanum oxide.

lanthanum sulfate $La_2(SO_4)_3 \cdot 9H_2O$.
Properties: White crystals. Sp. gr. 2.821; refractive index (n 20/D) 1.564; insoluble in alcohol; slightly soluble in water, acids.
Derivation: By dissolving hydroxide, carbonate or oxide in dilute sulfuric acid.
Grades: Purities to 99.9+%.
Containers: Glass bottles, fiber drums.
Use: The sulfates of the rare-earth elements are often used for atomic weight determination of the element.

lanthanum trioxide. See lanthanum oxide.

lanthionine $S(CH_2CHNH_2COOH)_2$. A non-essential amino acid first obtained from deaminated wool.

"Lanum."[123] Trademark for purified wool fat prepared for medicinal and pharmaceutical use.

lard. Purified internal fat of the hog.
Properties: Soft, white unctuous mass, faint odor, bland taste. Soluble in ether, chloroform, light petroleum hydrocarbons, carbon disulfide, insoluble in water. M.p. 36–42°C. High in saturated fats. Combustible; nontoxic.
Chief constituents: Stearin, palmitin, olein.
Uses: Cooking; pharmacy (ointments, cerates); perfumery (pomades); source of lard oil.

lard oil.
Properties: Colorless or yellowish liquid; peculiar odor and bland taste. Soluble in benzene, ether, chloroform, and carbon disulfide; slightly soluble in alcohol. M.p. −2°C; refractive index (20/D) 1.470; sp. gr. 0.915; saponification value 195–196; iodine value 56–74. Subject to spontaneous heating. Combustible. Nontoxic. Flash point 420°F. Autoignition temp. 883°F.
Chief constituents: Olein, with a small percentage of the glycerides of solid fatty acids.
Derivation: By cold-pressing lard.
Grades: Prime winter edible; prime winter inedible; antibiotic; off prime; extra no. 1; no. 1; no. 2.
Containers: Drums; tank cars; tank wagons.
Uses: Lubricant; metal cutting compounds; oiling wool; soap manufacture; antibiotic fermentation.

"Larex."[152] Trademark for a series of nine grades of lard oil.
Uses: Sulfonating, sulfurizing, cutting oils, textile lubricant, nutrient and defoamer for antibiotic fermentations.

"Laromin."[440] Trademark for polyamines for epoxy resins.

"Larvacide."[401] Trademark for various products containing chloropicrin.
15. A liquid fumigant containing chloropicrin (15%), carbon tetrachloride, and carbon disulfide.
100. Commercially pure chloropicrin.
70 Aerosol. An aerosol fumigant containing 70% chloropicrin.
85 Aerosol. A fumigant aerosol containing chloropicrin (85%) and methyl chloride (15%).
Soil Fumigant. A pre-plant soil fumigant containing chloropicrin (93.5%) and ethylene dibromide (6.5%).

LAS. See alkyl sulfonate, linear.

laser. A device which produces a beam of coherent or monochromatic light as a result of photon-stimulated emission. Such beams have extremely high energy, as they consist of a single wavelength and frequency. Materials capable of producing this effect are certain high-purity crystals (ruby, yttrium garnet, and metallic tungstates or molybdates doped with rare-earth ions); semiconductors such as gallium arsenide; neodymium-doped glass; various gases, including carbon dioxide, helium, argon, and neon; and plasmas. A chemical laser is one in which the excitation energy is furnished by a chemical reaction, e.g., $H + Cl_2 \longrightarrow HCl(\text{active}) + Cl$; or combustion of carbon monoxide to form excited carbon dioxide.

Hazard: Laser radiation can irreparably damage the eyes. Proper shielding is essential at all times.
Uses: Laser beams are used in industry for cutting diamonds for wire-drawing dies, in flash photolysis, spectroscopy and photography. They also have developing applications in medicine and surgery. In research they are being studied as a possible initiator for controlled fusion reactions and for biomedical investigations. It is also possible to increase the abundance of certain isotopes of such elements as uranium, chlorine and boron by use of laser irradiation. Research on uranium enrichment by this method has been under way for several years. See also fusion (2); enrichment.

LATB. See lithium aluminum tri-tert-butoxyhydride.

latent heat. The quantity of energy in calories per gram absorbed or given off as a substance undergoes a change of state, that is, as it changes from liquid to solid (freezes), from solid to liquid (melts), from liquid to vapor (boils), or from vapor to liquid (condenses). No change in temperature occurs. Water has unusually high latent heat values: the latent heat of fusion (melting) of ice is 80 cal per gram, and the latent heat of condensation of steam (latent heat of vaporization of water) is 540 cal. per gram. The considerable energy delivered by steam condensation is utilized for power generation and for heating a variety of chemical plant equipment (dryers, evaporators, reactors and distillation columns). See also evaporation; heat.

latent solvent. See solvent, latent.

laterite. A low-grade ore similar to bauxite, but containing only half as much aluminum oxide. Possible substitute for bauxite.

latex. A white, free-flowing liquid, obtained from some species of shrubs or trees, in which microscopically small particles or globules of natural rubber are suspended in a watery serum. Natural rubber latex, obtained from the tree Hevea Braziliensis, contains about 60% water, 35% rubber hydrocarbon, and 5% proteins and other substances. Coagulation is prevented by protective colloids, but can be induced by addition of acetic or formic acid. Synthetic latexes include polystyrene, SBR rubber, neoprene, polyvinyl cyloride, etc. Both natural and synthetic latexes are available in vulcanized form (see "Vultex"). The function of latex in the plant is not known. Nonflammable.
Containers: Steel drums; tank cars.
Grades: Regular; concentrated.
Hazard: Dangerous to ingest, as coagulation will occur internally.
Uses: Thin rubber products (surgeons' gloves, drug sundries); girdles, pillows, etc.; emulsion paints; adhesives; tire cord coating.

latex paint. See paint, emulsion.

"Lanthanol" LAL.[243] Trademark for a highly refined sodium "lauryl" sulfoacetate; a biodegradable organic detergent possessing wetting, scouring, emulsifying and dispersing properties; a foaming agent.
Properties: White, dry powder; pH 6.9–7.1 in 0.25% water solution; stable to hard water; stable to acid and alkali in a pH range of 5.0–8.5; solubility in water solution 1% at 25°C, 25% at 100°C; hygroscopic; sp. gr. 0.55; pleasant odor. Nontoxic. Tasteless.
Uses: Tooth pastes; tooth powders; liquid dentrifices; foaming bath salts; shampoos; synthetic detergents.

"Laticrete."[248] Trademark for a latex-based surfacing compound, a "flexible concrete" combining the resilient and long-wearing properties of rubber with the structural characteristics of concrete repairing; highways; playgrounds, tennis courts, etc.

lattice. (1) The structural arrangement of atoms in a crystal. Accurate information is obtained by x-rays, which are diffracted by the lattice at various angles. As the atoms are from 1.5 to 3 angstrom units apart in most crystals, the lattice acts as a diffraction grating. See also crystal; dislocation.
(2) The array of nuclear fuel elements and moderator in a nuclear reactor.

"Latyl."[28] Trademark for a group of disperse dyes developed particularly for coloration of "Dacron" polyester fiber, on which they have exceptionally good light and wet fastness properties.

laudanidine (levo-laudanine; tritopine) $C_{20}H_{25}O_4N$. An alkaloid.
Properties: White crystals; m.p. 182–185°C; insoluble in water; soluble in alcohol, benzene, chloroform and slightly soluble in ether.
Derivation: From opium.
Use: Medicine.

laudanum. A tincture of opium.
Hazard: Highly toxic by ingestion and inhalation.
Use: Medicine.

laughing gas. See nitrous oxide.

laundry sour. See sour.

lauraldehyde. See lauryl aldehyde.

Laurent's acid. See 1-naphthylamine-5-sulfonic acid.

"Laurex."[248] Trademark for the zinc salts of a mixture of fatty acids in which lauric acid predominates.
Properties: Yellowish white granulated waxy powder; sp. gr. 1.15; m.p. 95–105°C; soluble in benzene; insoluble in acetone, gasoline, ethylene dichloride, and water. Combustible; nontoxic.
Uses: Accelerator activator and plasticizer for rubber.

lauric acid (dodecanoic acid) $CH_3(CH_2)_{10}COOH$. A fatty acid occurring in many vegetable fats as the glyceride, especially in coconut oil and laurel oil. Combustible; low toxicity.
Properties: Colorless needles; sp. gr. 0.833; m.p. 44°C; b.p. 225°C (100 mm); refractive index 1.4323 (n 45/D); insoluble in water, soluble in alcohol and ether.
Derivation: Fractional distillation of mixed coconut or other acids.
Grades: 99.8% pure; technical; F.C.C.
Containers: 55-gal drums, tank cars.
Uses: Alkyd resins; wetting agents; soaps; detergents; cosmetics; insecticides; food additives.

lauric aldehyde. See lauryl aldehyde.

"Laurine."[227] Trademark for hydroxycitronellal (q.v.).

laurone. An aliphatic ketone; insoluble in water; stable to high temperatures, acids, alkalies. Compatible with high-melting vegetable waxes, fatty acids, paraffins, etc. Incompatible with resins, polymers, and organic solvents at room temperature, but compatible with them at high temperature. Used as antiblock agent.

N-lauroyl-para-aminophenol $HO(C_6H_4)NHCOCH_2(CH_2)_9CH_3$.
Properties: White to off-white powder; m.p. 123–126°C. Insoluble in water; soluble in polar organic solvents (especially when heated) including alcohol, acetone, and dimethylformamide.
Use: Rubber antioxidant.

lauroyl chloride $C_{11}H_{23}COCl$.
Properties: Water-white liquid; refractive index 1.445 (20°C); f.p. −17°C; b.p. 145°C (18 mm); decomposes in water and alcohol; soluble in ether.
Containers: 13-gal carboys; drums.
Uses: Surfactant, polymerization initiator, antienzyme agent, foamer; synthesis of lauroyl peroxide, sodium N-lauroyl sarcosinate and other sarcosinates.

lauroyl peroxide (dodecanoyl peroxide) $(C_{11}H_{23}CO)O_2$.
Properties: White, coarse powder; tasteless; faint odor; soluble in oils and in most organic solvents; slightly soluble in alcohols; insoluble in water; m.p. 49°C.
Grades: Technical (about 95%).
Hazard: Toxic by ingestion and inhalation; strong irritant to skin. Dangerous fire and explosion risk; will ignite organic materials. Strong oxidizing material.
Uses: Bleaching agent, intermediate and drying agent for fats, oils, and waxes; polymerization catalyst.
Shipping regulations: (Rail) Yellow label. (Air) Organic Peroxide label.

N-lauroylsarcosine $CH_3(CH_2)_{10}CON(CH_3)CH_2COOH$.
Properties: White solid; m.p. 31–35°C; sp. gr. 0.970.
Containers: Drums; tank cars and trucks.
Uses: Surfactant, antienzyme, in cosmetics and pharmaceuticals. Also used in form of sodium N-lauroylsarcosinate.

lauryl acetate. See dodecyl acetate.

lauryl alcohol (alcohol C-12; n-dodecanol; dodecyl alcohol) $CH_3(CH_2)_{10}CH_2OH$.
Properties: Colorless solid; floral odor; sp. gr. 0.830–0.836; refractive index 1.428; m.p. 24°C; b.p. 259°C. Insoluble in water. Soluble in 2 parts of 70% alcohol. Flash point (C.C.) >212°F. Combustible; low toxicity.
Derivation: Reduction of coconut oil fatty acids.
Grades: Technical; F.C.C.
Containers: 55-gal drums; 8000-gal tank cars.
Uses: Synthetic detergents; lube additives; pharmaceuticals; rubber; textiles; perfumes; flavoring agent.

lauryl aldehyde (lauric aldehyde; dodecyl aldehyde; aldehyde C-12 lauric; dodecanal; lauraldehyde) $CH_3(CH_2)_{10}CHO$.
Properties: Colorless solid or liquid; strong fatty floral odor; sp. gr. 0.828–0.836; refractive index 1.433–1.440; m.p. 44°C. Soluble in 90% alcohol; insoluble in water. Combustible; low toxicity.
Grades: Technical; F.C.C.
Uses: Perfumery; flavoring agent.

lauryl bromide (n-dodecyl bromide; 1-bromododecane) $C_{12}H_{25}Br$.
Properties: Amber liquid with coconut odor and low volatility. Sp. gr. 1.026 (25/25°C); boiling range (5–95% at 45 mm) 151–208°C; f.p. −15.5°C. Flash point 291°F. Combustible.
Grade: Technical, approx. 60% pure.
Derivation: Coconut oil.
Use: Intermediate for quaternary ammonium compounds, organometallics and vinyl stabilizers.

lauryl chloride. Commercially, a mixture of n-alkyl chlorides, with $C_{12}H_{25}Cl$ dominant. A clear, water-white, oily liquid, with a faint fatty odor. Completely miscible with most organic solvents; slightly miscible with alcohol; immiscible with water.
Properties: Sp. gr. 0.863 (15.5/15.5°C); crystallization point −19°C; distillation range 112–160°C (5 mm); flash point 235°F; combustible. Low toxicity.
Grades: Refined; technical.
Containers: Drums; tank cars.
Uses: Synthesis of esters, sulfides, lauryl mercaptan (used in styrene-butadiene polymerization), other organics.

lauryldimethylamine $CH_3(CH_2)_{11}N(CH_3)_2$. A liquid cationic detergent.
Uses: Corrosion inhibitor; acid-stable emulsifier.

lauryldimethylamine oxide $CH_3(CH_2)_{11}N(CH_3)_2 \cdot H_2O$. A nonionic detergent, used as a foam stabilizer; stable at high concentrations of electrolytes and over a wide pH range.

lauryl lactate. See "Ceraphyl" 31.

lauryl mercaptan (n-dodecyl mercaptan) $C_{12}H_{25}SH$ (approx.).
Properties (technical material, mixture of isomers): Water-white or pale-yellow liquid; mild odor; sp. gr. 0.85 (20/20°C); f.p. −7.5°C; distillation range 200–235°C; refractive index 1.45–1.47; insoluble in water; soluble in methanol, ether, acetone, benzene, gasoline, and ethyl acetate. Flash point 210°F. Combustible.
Grades: 95% min.
Containers: Steel drums; carboys; tank cars.
Hazard: May be injurious to eyes.
Use: Manufacture of synthetic rubber and plastics, also in the synthesis of pharmaceuticals, and in insecticides and fungicides; nonionic detergent.
See also thiol.

lauryl methacrylate $CH_2{:}C(CH_3)COO(CH_2)_{11}CH_3$. The commercial material is a mixture, containing also lower and higher fatty derivatives. Boiling range 272–344°C; density 0.868 g/ml; flash point 270°F (COC). Combustible. Probably low toxicity.
Containers: Drums.
Uses: Polymerizable monomer for plastics, molding powders, solvent coatings, adhesives, oil additives; emulsions for textile, leather, and paper finishing. See also acrylic resin.

lauryl pyridinium chloride $C_5H_5NClC_{12}H_{25}$.
Properties: Mottled tan semisolid. Soluble in water and organic solvents. Flash point 347°F; combustible.
Grade: Technical, contains higher and lower fatty acid derivatives.
Uses: Cationic detergent; dispersing and wetting agent; ingredient of fungicides and bactericides.

lauryl pyridinium 5-chloro-2-benzothiazyl sulfide. See "Vancide 26EC."

lautal. A hard aluminum alloy containing 4–5% copper, 1.5–2% silicon and fractional percentages of other metals such as iron, manganese, or magnesium.

"Lauxein."[58] Trademark for casein and soybean adhesives, dry powders good for low-temperature applications and glue bonding where water-resistance is desired.
Containers: Multiwall bags and fiber drums.
Uses: Bonding and cold-setting glues used in the manufacture of plywood furniture.

"Lauxite."[58] Trademark for a series of urea, phenolic, melamine and resorcinol resins. Available in dry powders or liquids. Used for bonding, cold-setting and impregnating adhesives and glues for furniture, plywood and aircraft; hot and cold pressing; radio frequency equipment; molding of diversified components from granulated wood.

"Lava." Trademark for synthetic block talc or soapstone used in electron tubes and other devices as an insulator.

lavandin oil, abrial. Essential oil obtained by steam distillation of plant material of a hybrid of true lavender and spike lavender, lavandula abrialis.
Properties: Yellow liquid with camphoraceous odor which is suggestive of lavender. Sp. gr. 0.885–0.893 (25°C); refractive index 1.4605–1.4640 (20°C). Soluble in most fixed oils, propylene glycol and mineral oil; relatively insoluble in glycerol.
Grades: Sold on basis of ester content. F.C.C. requires 28–35% ester calculated as linalyl acetate.
Uses: Perfumery; flavoring agent.

lavender oil.
Properties: Essential oil; colorless, yellowish or greenish-yellow; sweet odor; slightly bitter taste. Sp. gr. 0.875–0.888 (25°C); refractive index 1.4590–1.4700 (20°C); angular rotation −3 to −10°; one volume oil dissolves in 4 volumes 70% alcohol. Combustible; nontoxic.
Chief constituents: Linalool; linalyl acetate; geraniol; pinene; limonene; cineol.
Derivation: Steam-distilled from the fresh flowering tops of Landula officinalis.
Grades: N.F.; F.C.C. Both require not less than 35% of esters, calculated as linalyl acetate.
Uses: Perfumery; flavoring agent.

lavender oil, terpeneless
Concentration: About 1.75–2 times that of natural lavender oil. Sp. gr. 0.893–0.898; optical rotation, about −5°. Solubility in alcohol: 15 parts per 100 parts of 60% alcohol; 55 parts per 100 parts of 70% alcohol.

lavender-spike oil. See spike oil.

"Lavenol."[188] Trademark for a series of synthetic lavender oil substitutes of various types.

Lavoisier, Antoine Laurent (1743–1794). French chemist generally regarded as the "father" of chemistry. His "Traité Elementaire de Chimie" (1789) listed 30 elements, clarified the nomenclature of acids, bases and salts, and described the composition of numerous organic substances. He erroneously believed that oxygen is the characteristic element of acids. However, his fundamental work on combustion, as a result of which he identified and named nitrogen (azote), and on the separation of hydrogen from water by a unique reduction experiment carried out in a heated gun barrel, earned him a leading position among early chemists. (See also Mendeleéf).

"Lavol."[188] Trademark for a substitute for linalyl acetate.

lawrencium Lr A synthetic radioactive element with atomic number 103, discovered in 1961. Atomic weight 257. Only one other isotope is known (256). The 257 isotope has a half-life of 8 seconds. It has been made by bombarding californium with boron ions. It exhibits alpha radiation. See actinide series.

LC_{50} (lethal concentration, 50%). That quantity of a substance administered by inhalation that is necessary to kill 50% of test animals exposed to it within a specified time. This test applies not only to gases and vapors but to fume, dusts and other particulates suspended in air.

LCL. Abbreviation for "less than carload lot"; used by shippers, traffic managers, railroads, etc.

LD_{50} (lethal dose, 50%). That quantity of a substance administered either orally or by skin contact necessary to kill 50% of exposed animals in laboratory tests within a specified time.

leaching. See solvent extraction.

lead Pb (from Latin plumbum). Metallic element of atomic number 82, Group IVA of the periodic table. Atomic weight 207.2; valences 2, 4; 4 stable isotopes. The isotopes are the end products of the three series of natural radioactive elements uranium (206), thorium (208), and actinium (207).
Properties: Heavy, ductile, soft gray solid. Sp. gr. 11.35; m.p. 327.4°C; b.p. 1755°C; soluble in dilute nitric acid; insoluble in water but dissolves slowly in water containing a weak acid; resists corrosion; relatively impenetrable to radiation. Poor electrical conductor; good sound and vibration absorber. Noncombustible.
Occurrence: U.S., Mexico, Canada, S. America, Australia, Africa, Europe.
Derivation: Roasting and reduction of galena (lead sufide), anglesite (lead sulfate), and cerussite (lead carbonate). Also from scrap.
Purification method: Desilvering (Parkes process); electrolytic refining (Betts process); pyrometallurgical refining (Harris process). Bismuth is removed by Betterton-Kroll process.
Grades: High purity (less than 10 ppm impurity); pure (99.9+); powdered (99% pure); pig lead; paste.
Forms available: Ingots, sheet, pipe, shot, buckles or straps, grids, rod, wire, etc.; paste; powder; single crystals.
Hazard: Toxic by ingestion and inhalation of dust or fume. Tolerance (lead, fumes and dusts, and inorganic compounds) 0.15 mg per cubic meter of air. A cumulative poison. FDA regulations require zero lead content in foods and 0.05% in house paints.
Uses: Storage batteries; tetraethyllead (gasoline additive); radiation shielding; cable covering; ammunition; sheet and pipe; solder and fusible alloys; type metal; vibration damping in heavy construction; foil; numerous alloys.
For further information refer to Lead Industries Association, 292 Madison Ave., New York.

lead acetate (sugar of lead) $Pb(C_2H_3O_2)_2 \cdot 3H_2O$.
Properties: White crystals or flakes (commercial grades are frequently brown or gray lumps). Sweetish taste. Absorbs carbon dioxide when exposed to air, becoming insoluble in water. Soluble in water; slightly soluble in alcohol; freely soluble in glycerol. Sp. gr. 2.50; m.p. loses H_2O at 75°C; at 200°C decomposes; b.p. (anhydrous) 280°C. Combustible.
Derivation: By the action of acetic acid on litharge or thin lead plates.
Grades: Powdered; granular; crystals; flakes; C.P.
Containers: Multiwall paper sacks; drums; carloads.
Uses: Medicine; lead salts; dyeing of textiles; waterproofing; varnishes; lead driers; chrome pigments; gold cyanidation process; insecticide; antifouling paints; analytical reagent.
Hazard: Highly toxic by ingestion, inhalation, skin absorption. Absorbed by skin. Use may be restricted.
Shipping regulations: (Air) Poison label.

lead alkyl, mixed. A mixture containing various methyl and ethyl derivatives of tetraethyl lead and tetramethyl lead. Thus methyl triethyl lead, dimethyl diethyl lead and ethyl trimethyl lead may all be present with or without tetraethyl and tetramethyl lead.
Hazard: Toxic by ingestion and skin absorption.
Uses: Antiknock agents in aviation gasoline.

lead antimonate (Naples yellow; antimony yellow) $Pb_3(SbO_4)_2$.
Properties: Orange-yellow powder. Insoluble in water. Sp. gr. 6.58 (20°C). Noncombustible.
Derivation: Interaction of solutions of lead nitrate and potassium antimonate, concentratoin and crystallization.
Hazard: Toxic by inhalation. Tolerance, 0.15 mg per cubic meter of air.
Uses: Staining glass, crockery and porcelain.

lead arsenate (lead orthoarsenate) $Pb_3(AsO_4)_2$.
Properties: White crystals. Soluble in nitric acid; insoluble in water; sp. gr. 7.80; m.p. 1042°C (decomposes).
Derivation: By the action of a soluble lead salt on a solution of sodium arsenate, concentration and crystallization.
Uses: Insecticide; herbicide.
Hazard: Highly toxic. Tolerance, 0.15 mg per cubic meter of air.
Shipping regulations: (Rail, Air) Poison label.

lead arsenite $Pb(AsO_2)_2$.
Properties: White powder; soluble in nitric acid; insoluble in water. Sp. gr. 5.85.
Hazard: Highly toxic.
Use: Insecticide.
Shipping regulations: (Rail, Air) Poison label.

lead azide $Pb(N_3)_2$.
Properties: Colorless needles; an initiating explosive. Should always be handled submerged in water.
Derivation: Reaction of sodium azide with a lead salt.
Hazard: Severe explosion risk; detonates at 350°C (660°F). Highly toxic. Tolerance, 0.15 mg per cubic meter of air.
Use: Primary detonating compouind for high explosives.
Shipping regulations: (Air) Not accepted. (Rail) Consult regulations. Not accepted by express.
Note: Explosions have occurred in cases where azide compounds have reacted with the lead in plumbing after being washed down sinks.

lead-base Babbitt. See Babbitt metal.

lead biorthophosphate. See lead phosphate, dibasic.

lead, blue. A term applied to galena to distinguish it from white lead ore. It is also applied to blue basic lead sulfate.

lead borate $Pb(BO_2)_2 \cdot H_2O$.
Properties: White powder. Soluble in dilute nitric acid; insoluble in water. Sp. gr. 5.598; m.p. 160°C (loses water). Noncombustible.
Derivation: Interaction of solutions of lead hydroxide and boric acid, with subsequent crystallization.
Hazard: Highly toxic by inhalation. Tolerance, 0.15 mg per cubic meter of air.
Uses: Varnish and paint drier; waterproofing paints; lead glass; galvanoplastic products.

lead borosilicate. A constituent of optical glass, composed of a mixture of the borate and silicate of lead.

lead bromate $Pb(BrO_3)_2 \cdot H_2O$.
Properties: Colorless crystals. Soluble in hot water; sp. gr. 5.53; decomposes at about 180°C.

lead bromide $PbBr_2$.
Properties: White powder. Slightly soluble in hot water; insoluble in alcohol. Sp. gr. 6.66; b.p. 916°C; m.p. 373°C.
Hazard: Highly toxic by inhalation. Tolerance, 0.15 mg per cubic meter of air.

lead carbolate. See lead phenate.

lead carbonate $PbCO_3$. See lead carbonate, basic.
Properties: White, powdery crystals. Soluble in acids; insoluble in water and alcohol. Sp. gr. 6.6; decomposed by heat 315°C. Noncombustible.
Derivaton: By adding a solution of sodium bicarbonate to a solution of lead nitrate. Occurs in nature as cerussite.
Hazard: Toxic by inhalation. Tolerance, 0.15 mg per cubic meter of air.
Impurities: Basic lead carbonate.
Use: Industrial paint pigment.

lead carbonate, basic (lead subcarbonate; white lead; BCWL; lead flake) $2PbCO_3 \cdot Pb(OH)_2$.
Properties: White, amorphous powder. Soluble in acids; insoluble in water; decomposes at 400°C; sp. gr. 6.86. Noncombustible.
Derivation: (a) Dutch process. By the corrosion of lead buckles in pots by acetic acid and carbon dioxide generated by the fermentation of waste tanbark. (b) Carter process. By treating very finely divided lead in revolving wooden cylinders with dilute acetic acid and carbon dioxide.
Hazard: Toxic by inhalation. Tolerance, 0.15 mg per cubic meter of air.
Uses: Industrial paint pigment; putty; ceramic glazes.

lead chloride $PbCl_2$.
Properties: White crystals. Slightly soluble in hot water; insoluble in alcohol and cold water. Sp. gr. 5.88; m.p. 498°C; b.p. 950°C. Noncombustible.
Derivation: By the addition of hydrochloric acid or sodium chloride to a solution of a lead salt, with subsequent crystallization.
Hazard: Toxic by inhalation. Tolerance, 0.15 mg per cubic meter of air.
Uses: Preparation of lead salts; lead chromate pigments; analytical reagent.

lead chlorosilicate complex. See "Lectro" 60.

lead chromate. $PbCrO_4$.
Properties: Yellow crystals. Soluble in strong acids and alkalies; insoluble in water. Sp. gr. 6.123; m.p. 844°C.
Derivation: Reaction of sodium chromate and lead nitrate in solution.
Hazard: Toxic by ingestion and inhalation. May be a carcinogen. Tolerance, 0.1 mg per cubic meter of air.
Uses: Pigment in industrial paints, rubber, plastics, ceramic coatings.
See also chrome pigment.

lead coating. Coatings of lead or lead-rich alloys are (1) deposited by dipping into the molten metal, after applying a layer of tin to secure good adhesion of the lead coating; (2) by electroplating from a fluosilicate or fluoborate bath, or (3) by spraying.

lead cyanide $Pb(CN)_2$.
Properties: White to yellowish powder. Slightly soluble in water; decomposes in acid.
Derivation: Interaction of solutions of potassium cyanide and lead acetate.

Hazard: Highly toxic by ingestion, inhalation, and skin absorption. Tolerance (as CN), 5 mg per cubic meter of air.
Use: Metallurgy.
Shipping regulations: (Rail, Air) Poison label.

lead dimethyldithiocarbamate $Pb[SCSN(CH_3)_2]_2$.
Properties: White powder; sp. g. 2.43 ± .03; melting range 310°C; insoluble in all common organic solvents; slightly soluble in cyclohexanone.
Use: Vulcanization accelerator with litharge.

lead dioxide (lead oxide, brown; plumbic acid, anhydrous; lead peroxide; lead superoxide) PbO_2.
Properties: Brown, hexagonal crystals. Soluble in glacial acetic acid; insoluble in water and alcohol; sp. gr. 9.375; m.p. 290°C (dec). An oxidizing agent.
Derivation: By adding bleaching powder to an alkaline solution of lead hydroxide.
Hazard: Toxic. Dangerous fire risk in contact with organic materials. Tolerance, 0.15 mg per cubic meter of air.
Uses: Oxidizing agent; electrodes; lead-acid storage batteries; curing agent for polysulfide elastomers; textiles (mordant, discharge in dyeing with indigo); matches; explosives; analytical reagent.
Shipping regulations: (Rail) Oxidizing material, n.o.s., Yellow label. (Air) Oxidizer label.

lead dross (lead scrap). Consists of the scrap, dross, or waste from sulfuric acid tanks; a mixture of metallic lead, lead sulfate, and free sulfuric acid.

lead, electrolytic. Pure lead obtained by electrolytic deposition. See Betts process.

"Lead/Epox."[504] Trademark for a two-component shielding material designed to be formed after mixing. It is then a putty-like high-density material which can be used as a mortar. No. 250 consists of 94.5 wt.% lead, 5.5 wt.% epoxy resin, for protection against gamma rays. No. 251 consists of 93.8 wt.% lead, 5.7 wt.% epoxy resin and 0.5 wt.% boron.

lead ethylhexoate. See "L-26."

lead flake. See lead carbonate, basic.

lead fluoride PbF_2.
Properties: Crystalline solid; density 8.2 g/cc; m.p. (approx) 824°C; very slightly soluble in water. Noncombustible.
Grade: Crystal, 99.93%.
Hazard: Highly toxic; strong irritant. Tolerance, 0.15 mg per cubic meter of air.
Uses: Electronic and optical applications; starting materials for growing single crystal solid-state lasers; high temperature dry film lubricants in the form of ceramic-bonded coatings.

lead fluosilicate (lead silicofluoride) $PbSiF_6 \cdot 2H_2O$.
Properties: Colorless crystals; soluble in water; decomposes when heated.
Hazard: Highly toxic; strong irritant. Tolerance, 0.15 mg per cubic meter of air.
Use: Solution for electrorefining lead.

lead formate $Pb(CHO_2)_2$.
Properties: Brownish-white, lustrous, very finely divided, crystalline substance; sp. gr. 4.63; soluble to water. Decomposes at 190°C. Noncombustible.
Hazard: Highly toxic.
Use: Reagent in analytical determinations.

lead fumarate, tetrabasic. See "Lectro" 78.

lead glass. See glass.

lead hydroxide (lead hydrate; hydrated lead oxide) $Pb(OH)_2$ or $Pb_2O(OH)_2$.
Properties: White, bulky powder. Soluble in alkalies; slightly soluble in water; soluble in nitric and acetic acid; sp. gr. 7.592; m.p., decomposes at 145°C; absorbs carbon dioxide from air. Noncombustible.
Derivation: By the addition of sodium or ammonium hydroxide to a solution of a lead salt with subsequent filtration and drying.
Hazard: Toxic. Tolerance, 0.15 mg per cubic meter of air.
Use: Lead salts; lead dioxide.

lead hyposulfite. See lead thiosulfate.

lead iodide PbI_2.
Properties: Golden-yellow crystals or powder; odorless. Soluble in potassium iodide and concentrated sodium acetate solutions; insoluble in water and alcohol; sp. gr. 6.16; m.p. 402°C; b.p. 954°C. Noncombustible.
Derivation: Interaction of lead acetate and potassium iodide.
Hazard: Tolerance, 0.15 mg per cubic meter of air.
Uses: Bronzing; mosaic gold; printing; photography; cloud seeding.

lead linoleate (lead plaster) $Pb(C_{18}H_{31}O_2)_2$.
Properties: Yellowish-white paste. Soluble in oils; insoluble in water. Combustible.
Derivation: By heating a solution of lead nitrate with sodium linoleate.
Grades: Technical; fused (contains 26.5% Pb).
Hazard: Toxic; absorbed by skin.
Uses: Medicine; drier in paints and varnishes.

lead maleate, tribasic.
Properties: Soft, yellowish white crystalline powder; sp. gr. 6.3; refractive index 2.08.
Hazard: Toxic; absorbed by skin.
Uses: Vulcanizing agent for chlorosulfonated polyethylene. Highly basic stabilizer with high heat stability in vinyls.

lead metavanadate. See lead vanadate.

lead molybdate $PbMoO_4$.
Properties: Yellow powder. Soluble in acids; insoluble in water and alcohol. Absolute density 5.9 g/cc; m.p. 1060–1070°C. Noncombustible.
Derivation: By adding a solution of lead nitrate to a solution of ammonium molybdate, concentration and crystallization.
Hazard: Toxic. Tolerance, 0.15 mg per cubic meter of air.
Uses: Analytical chemistry; pigments (see molybdate oranges). Single crystals available for electronic and optical uses.

lead monohydrogen phosphate. See lead phosphate, dibasic.

lead mononitroresorcinate $PbO_2C_6H_3NO_2$.
Hazard: An initiating explosive; dangerous.
Shipping regulations: (Rail, Air) Not accepted.

lead monoxide. See litharge; massicot.

lead beta-**naphthalenesulfonate** $Pb(C_{10}H_7SO_3)_2$.
Properties: White crystalline powder. Soluble in alcohol; insoluble in water. Combustible.
Derivation: By the action of lead acetate on beta-naphthalenesulfonic acid.

Hazard: Toxic. Absorbed by skin.
Use: Organic preparations.

lead naphthenate.
Properties: Soft, yellow, resinous semi-transparent. Gives deposits in highly acid oils, but not when mixed with suitable quantities of cobalt or manganese. Soluble in alcohol; m.p. approx. 100°C. Combustible.
Derivation: Addition of lead salt to aqueous sodium naphthenate solution.
Grades: Liquid; 16%, 24% Pb; solid: 37% Pb.
Containers: Steel drums; fiber drums.
Hazard: Toxic. Absorbed by skin.
Uses: Paint and varnish drier; wood preservative; insecticide; catalyst for reaction between unsaturated fatty acids and sulfates in the presence of air; lube oil additive.
See also soap (2).

lead nitrate $Pb(NO_3)_2$.
Properties: White crystals. Soluble in water and alcohol; sp. gr. 4.53; decomposes at 470°C.
Derivation: By the action of nitric acid on lead.
Hazard: Toxic. Strong oxidizing material. Dangerous fire risk in contact with organic materials. Tolerance, 0.15 mg per cubic meter of air.
Uses: Lead salts; mordant in dyeing and printing calico; matches; paint pigment; mordant for staining mother-of-pearl; oxidizer in the dye industry; sensitizer in photography; explosives; tanning; process engraving and lithography.
Shipping regulations: (Rail) Yellow label. (Air) Oxidizer label.

lead nitrite (basic lead nitrite; lead subnitrite).
Properties: Light-yellow powder; variable composition, essentially $3PbO \cdot N_2O_3 \cdot H_2O$. Soluble in dilute nitric acid. Easily decomposed.
Hazard: Toxic. Tolerance, 0.15 mg per cubic meter of air.

lead ocher. See massicot (1).

lead octoate. See soap (2).

lead oleate $[CH_3(CH_2)_7CH{:}CH(CH_2)_7COO]_2Pb$.
Properties: White powder or ointment-like granules or mass. Soluble in alcohol, ether, turpentine, and benzene; insoluble in water. Combustible.
Derivation: Reaction of oleic acid with lead hydrate or carbonate, or of lead acetate and sodium oleate.
Hazard: Toxic; absorbed by skin.
Uses: Varnishes; lacquers; paint drier; high-pressure lubricants.
See also soap (2).

lead orthoarsenate. See lead arsenate.

lead orthophosphate, normal. See lead phosphate.

lead orthosilicate. See lead silicate.

lead oxide, black. See lead suboxide.

lead oxide, brown. See lead dioxide.

lead oxide, hydrated. See lead hydroxide.

lead oxide, red (red lead; minium; plumboplumbic oxide) Pb_3O_4.
Properties: Bright red powder; partly soluble in acids; insoluble in water. Sp. gr. reported variously 8.32–9.16; decomposes between 500 and 530°C. An oxidizing agent; may react with reducing agents.
Derivation: By carefully heating litharge in a furnace in a current of air.
Grades: Technical; 95%, 97%, 98%.
Hazard: Toxic as dust. Tolerance, 0.15 mg per cubic meter of air. Safety data sheet available from Manufacturing Chemists Assn., Washington, D.C.
Uses: Storage batteries; glass; metal-protective paints; pottery and enameling; varnish; purification of alcohol; packing pipe joints.

lead oxide, yellow. See litharge.

lead perchlorate $Pb(ClO_4)_2 \cdot 3H_2O$. White crystals; sp. gr. 2.6; m.p. 100°C (dec). Very soluble in cold water; soluble in alcohol.
Hazard: Toxic; dangerous in contact with organic materials. Strong oxidizing agent.
Shipping regulations: (Rail) Perchlorates n.o.s., Yellow label. (Air) Oxidizer label.

lead peroxide. See lead dioxide.

lead phenate (lead phenolate; lead carbolate) $Pb(OH)OC_6H_5$.
Properties: Yellowish to grayish-white powder. Soluble in nitric acid; insoluble in water and alcohol.
Derivation: By boiling phenol with litharge.
Hazard: Toxic; absorbed by skin.

lead phenolsulfonate (lead sulfocarbolate) $Pb(C_6H_4OHSO_3)_2 \cdot 5H_2O$.
Properties: White crystals or powder. Soluble in water and alcohol.
Hazard: Toxic; absorbed by skin.

lead phosphate (normal lead orthophosphate) $Pb_3(PO_4)_2$.
Properties: White powder; sp. gr. 6.9–7.3; m.p. 1014°C. Insoluble in water; soluble in acids and alkalies.
Hazard: Toxic. Tolerance, 0.15 mg per cubic meter of air.

lead phosphate, dibasic (lead monohydrogen phosphate; lead biorthophosphate) $PbHPO_4$.
Properties: Soft white powder of fine plate-like crystals; sp. gr. 5.66 (15°C); m.p., decomposes.
Hazard: Toxic. Tolerance, 0.15 mg per cubic meter of air.
Uses: Imparting heat resistance and pearlescence to polystyrene and casein plastics.

lead phosphite, dibasic $2PbO \cdot PbHPO_3 \cdot \frac{1}{2}H_2O$.
Properties: Fine white acicular crystals; sp. gr. 6.94; refractive index 2.25. Insoluble in water.
Containers: 40-, 350-lb fiber drums.
Uses: Heat and light stabilizer for vinyl plastics and chlorinated paraffins. As an ultraviolet screening and antioxidizing stabilizer for vinyl and other chlorinated resins in paints and plastics.
Hazard: Toxic. Tolerance, 0.15 mg per cubic meter of air. Store in closed containers, away from open flame or sparks and at temperatures not to exceed 400°F.
Shipping regulations: (Rail, Air) Not listed. Consult authorities.

lead phthalate, dibasic $C_6H_4(COO)_2Pb \cdot PbO$.
Properties: Fluffy, white crystalline powder; insoluble in water; sp. gr. 4.5; refractive index 1.99 (av.).
Derivation: By boiling litharge with phthalic acid.
Hazard: Toxic. Absorbed by skin.
Use: Heat and light stabilizer for general vinyl use.

lead plaster. See lead linoleate.

lead protoxide. See litharge.

lead, red. See lead oxide, red.

lead resinate $Pb(C_{20}H_{29}O_2)_2$.
Properties: Brown lustrous translucent lumps or yellow-white powder, or yellowish-white paste. Insoluble in most solvents. Combustible.
Derivation: By heating a solution of lead acetate and rosin oil.
Grades: Precipitated, 23% Pb.
Hazartd: Toxic; absorbed by skin.
Uses: Paint and varnish drier; textile water-proofing agent.

lead salicylate $Pb(OOCC_6H_4OH)_2 \cdot H_2O$.
Properties: Soft, creamy white crystalline powder; sp. gr. 2.3; refractive index 1.78; soluble in hot water and alcohol. Combustible.
Hazard: Probably toxic.
Uses: Stabilizer or costabilizer for flooring and other vinyl compounds requiring good light stability.

lead selenide PbSe. Gray crystals; sp. gr. 8.10 (15°C); m.p. 1065°C. Insoluble in water; soluble in nitric acid. A semiconductor used in infrared detectors and thermoelectric devices.
Hazard: Toxic. Moderate fire hazard as dust or in presence of moisture. Tolerance, 0.15 mg per cubic meter of air.

lead sesquioxide Pb_2O_3.
Properties: Reddish-yellow powder. Soluble in alkalies and acids; insoluble in water. Decomposes at 370°C.
Derivation: By gently heating metallic lead.
Hazard: Toxic. Tolerance, 0.15 mg per cubic meter of air.
Uses: Medicine; ceramics; ceramic cements; metallurgy; varnishes.

lead silicate (lead metasilicate) $PbSiO_3$. (?)
Properties: White, crystalline powder; insoluble in most solvents. Noncombustible.
Derivations: Interaction of lead acetate and sodium silicate.
Hazard: Toxic. Tolerance, 0.15 mg per cubic meter of air.
Uses: Ceramics; fireproofing fabrics.

lead silicate, basic (white lead silicate; lead silicate sulfate). A pigment made up of an adherent surface layer of basic lead silicate and basic lead sulfate cemented to silica.
Properties: Excellent film-forming properties with drying oils combined with low specific gravity.
Derivation: Fine silica is mixed with litharge and sulfuric acid. The mixture is then furnaced in a rotary kiln and ground to break up agglomerates.
Hazard: Toxic. Tolerance, 0.15 mg per cubic meter of air.
Use: Pigment in industrial paints.

lead silicochromate. Any of a group of yellow pigments. Normal lead silicon chromate is used as a yellow prime pigment for traffic marking paints. Basic lead silicon chromate is used as a corrosion inhibitive pigment for metal protective coatings, primers and finishes. Also for industrial enamels requiring a high gloss.
Hazard: Toxic. Tolerance, 0.15 mg per cubic meter of air.

lead silicofluoride. See lead fluosilicate.

lead-silver Babbitt. See Babbitt metal.

lead-soap lubricants. Lead salts saponified with fats; hard at low temperatures, viscous at medium temperatures, but become somewhat fluid on heating by friction. Used as "extreme-pressure lubricants," but are not suited for high-speeds.
See lead naphthenate; lead oleate; lead stearate; also soap (2).

lead sodium hyposulfite. See lead sodium thiosulfate.

lead sodium thiosulfate (lead sodium hyposulfite; sodium lead hyposulfite; sodium lead thiosulfate) $PbS_2O_3 \cdot 2Na_2S_2O_3$.
Properties: Heavy, small, white crystals. Soluble in solutions of thiosulfates.
Hazard: Toxic. Tolerance, 0.15 mg per cubic meter of air.
Use: Matches.

lead stannate $PbSnO_3 \cdot 2H_2O$.
Properties: Light-colored powder. Insoluble in water. Approximate temperature of dehydration 170°C.
Hazard: Toxic. Tolerance, 0.15 mg per cubic meter of air.
Uses: Additive in ceramic capacitors; pyrotechnics.

"Leadstar."[304] Trademark for normal lead stearate.

lead stearate $Pb(C_{18}H_{35}O_2)_2$.
Properties: White powder; m.p. 100–115°C; sp. gr. 1.4; soluble in ether; insoluble in alcohol; slightly soluble in water. Combustible.
Derivation: By heating a solution of lead acetate with sodium stearate.
Hazard: Toxic; absorbed by skin.
Uses: Varnish and lacquer drier; high-pressure lubricants; lubricant in extrusion processes; stabilizer for vinyl polymers; corrosion inhibitor for petroleum; component of greases, waxes and paints.

lead styphnate. Legal label name for lead trinitroresorcinate (q.v.).

lead subcarbonate. See lead carbonate, basic.

lead subnitrite. See lead nitrite.

lead suboxide (lead oxide, black; litharge, leaded) Pb_2O.
Properties: Black, amorphous material; sp. gr. 8.342; decomposes on heating. Insoluble in water; soluble in acids and bases.
Hazard: Toxic. Tolerance, 0.15 mg per cubic meter of air.
Use: In storage batteries.

lead, sugar of. See lead acetate.

lead sulfate $PbSO_4$.
Properties: White, rhombic crystals. Slightly soluble in hot water; insoluble in alcohol. Sp. gr. 6.12–6.39; m.p. 1170°C. Noncombustible.
Derivation: Interaction of solutions of lead nitrate and sodium sulfate.
Hazard: Toxic. Strong irritant to tissue. Tolerance, 0.15 mg per cubic meter of air.
Uses: Storage batteries; paint pigments.
Shipping regulations: (Air) Containing more than 3% free acid: Corrosive label.

lead sulfate, basic (white lead, sublimed; white lead sulfate). Approximate formula $PbSO_4 \cdot PbO$.
Properties: White monoclinic crystals; sp. gr. 6.92; m.p. 977°C; only slightly soluble in hot water or acids. Noncombustible.

Grades: Vary form 72 to 85% lead sulfate and remainder lead oxide. Sold dry or ground in oil.
Derivation: Three methods are used: (a) Lead sulfide ore (galena) is subjected to high temperatures in an oxidizing atmosphere. (b) Molten lead is sprayed into a jet of ignited fuel gas and air in the presence of sulfur dioxide gas. (c) Atomized metallic lead is mixed with water and sulfuric acid is added under controlled conditions.
Containers: Barrels; multiwall paper sacks.
Hazard: Toxic. Tolerance, 0.15 mg per cubic meter of air.
Uses: Paints; ceramics; pigments.

lead sulfate, blue basic (sublimed blue lead; blue lead).
Composition: Lead sulfate (min) 45%, lead oxide (min) 30%, lead sulfide (max) 12%, lead sulfite (max) 5%, zinc oxide 5%, carbon and undetermined matter (max) 5%.
Properties: Blue-gray corrosion-inhibiting pigment; insoluble in water or alcohol. Sp. gr. 6.2. Noncombustible.
Derivation: By heating lead ores in special furnaces.
Hazard: Toxic. Tolerance, 0.15 mg per cubic meter of air.
Uses: Component of structural-metal priming coat paints; rust-inhibitor in paints; lubricants; vinyl plastics, and rubber products.

lead sulfate, tribasic $3PbO \cdot PbSO_4 \cdot H_2O$.
Properties: Fine, white powder; sp. gr. 6.4; refractive index 2.1. Used for electrical and other vinyl compounds requiring high heat stability.
Hazard: Toxic. Tolerance, 0.15 mg per cubic meter of air.

lead sulfide (plumbous sulfide) PbS.
Properties: Silvery, metallic crystals or black powder. Soluble in acids; insoluble in water and alkalies. Sp. gr. 7.13–7.7; m.p., 1114°C; sublimes at 1281°C.
Derivation: (a) Found in nature as the mineral galena (q.v.); (b) by passing hydrogen sulfide gas into an acid solution of lead nitrate.
Grades: Technical; C.P.; electronic.
Hazard: Toxic by ingestion and inhalation. Tolerance, 0.15 mg per cubic meter of air.
Uses: Ceramics; metallic lead; infrared radiation detector; semiconductor.

lead sulfite $PbSO_3$.
Properties: White powder; decomposes on heating. Insoluble in water; soluble in nitric acid.
Hazard: Toxic by ingestion and inhalation. Tolerance, 0.15 mg per cubic meter of air.

lead sulfocarbolate. See lead phenolsulfonate.

lead sulfocyanide. See lead thiocyanate.

lead superoxide. See lead dioxide.

lead tallate. A lead derivative of tall oil. Combustible.
Grades: Liquid, 16% Pb; 24% Pb; solid, 30% Pb.
Derivation: By the fusion process.
Hazard: Toxic. Absorbed by skin.
Uses: See soap (2).

lead telluride PbTe.
Properties: Crystalline solid; sp. gr. 8.2; m.p. 905°C. Insoluble in water and most acids.
Hazard: Toxic by ingestion and inhalation. Tolerance, 0.15 mg per cubic meter of air.
Uses: Single crystals used as photoconductor and semiconductor in thermocouples.

lead tetraacetate $Pb(CH_3COO)_4$.
Properties: Colorless or faintly pink crystals, sometimes moist with glacial acetic acid. M.p. 175°C; sp. gr. 2.228 (17°C); soluble in benzene, chloroform, nitrobenzene, hot glacial acetic acid. Combustible.
Derivation: From red lead (Pb_3O_4) and glacial acetic acid in the presence of acetic anhydride.
Containers: Glass bottles; fiber drums.
Hazard: Toxic. Absorbed by skin.
Uses: Selective oxidizing agent in organic synthesis; laboratory reagent.

lead tetraethyl. See tetraethyl lead.

lead thiocyanate (lead sulfocyanate) $Pb(SCN)_2$.
Properties: White or light-yellow, crystalline powder; soluble in potassium thiocyanate, nitric acid and slightly soluble in cold water; decomposes in hot water. Sp. gr. about 3.8.
Containers: Drums (100 lbs net).
Hazard: Toxic by ingestion and inhalation. Tolerance, 0.15 mg per cubic meter of air.
Uses: Ingredient of priming mixture for small-arms cartridges; safety matches; dyeing.

lead thiosulfate (lead hyposulfite) PbS_2O_3.
Properties: White crystals. Sp. gr. 5.18. Soluble in acids and sodium thiosulfate solution; insoluble in water. M.p., decomposes.
Derivation: By the interaction of solutions of lead nitrate and sodium thiosulfate, concentration and crystallization.
Hazard: Toxic by ingestion and inhalation. Tolerance, 0.15 mg per cubic meter of air.

lead titanate $PbTiO_3$.
Properties: Pale-yellow solid; insoluble in water. Sp. gr. 7.52.
Derivation: Interaction of oxides of lead and titanium at a high temperature. Contains lead sulfate and lead oxide as impurities.
Hazard: Toxic by ingestion and inhalation. Tolerance, 0.15 mg per cubic meter of air.
Use: Industrial paint pigment.

lead trinitroresorcinate (lead styphnate) $C_6H(NO_2)_3(O_2Pb)$.
Properties: Sp. gr. 3.1 for monohydrate and 2.9 for anhydrous. Monohydrate is monoclinic orange-yellow crystals; practically insoluble in water.
Derivation: Prepared by adding a solution of magnesium styphnate (from magnesium oxide and styphnic acid) to a lead salt solution.
Hazard: Detonates at 500°F; dangerous explosion risk. An initiating explosive.
Shipping regulations: (Air) Not accepted. (Rail) Consult regulations. Not accepted by express.

lead tungstate (lead wolframate) $PbWO_4$.
Properties: Yellowish powder. Soluble in acid; insoluble in water. Sp. gr. 8.235; m.p. 1130°C.
Derivation: By mixing solutions of lead nitrate and sodium tungstate, concentrating and crystallizing.
Hazard: Toxic. Tolerance, 0.15 mg per cubic meter of air.
Use: Pigment.

lead vanadate (lead metavanadate) $Pb(VO_3)_2$.
Properties: Yellow powder; insoluble in water; decomposes in nitric acid.
Hazard: Toxic. Tolerance, 0.15 mg per cubic meter of air.
Uses: Preparation of other vanadium compounds; pigment.

"Lead V-1 (Dibasic)."[74] Precipitated dibasic lead stearate. Uniform white powder, 50% lead content. Used in plastics.

lead water. A 1% solution of basic lead acetate. Toxic. See lead acetate.

lead, white. See lead carbonate, basic; also lead silicate, basic; and lead sulfate, basic.

lead wolframate. See lead tungstate.

lead wool. Fine filaments or threads of metallic lead, prepared and used for packing pipe joints.

lead zirconate titanate (LZT) $PbTiZrO_3$. Forms piezoelectric crystals. Used as an element in hi-fi sets and as a transducer for ultrasonic cleaners; ferroelectric material in computer memory units.

"Leafseal."[563] Trademark for a formulation of decenylsuccinic acid and its esters. Used for direct application to plants to enable them to resist frost and drought.

leather. An animal skin or hide that has been permanently combined with a tanning agent, which causes a physicochemical change in the protein components of the skin. This change renders it resistant to putrefactive bacteria, enzymes, and hot water, increases its strength and abrasion resistance, and makes it serviceable for long periods of time. Tanning agents are either vegetable, mineral, or synthetic. (See tanning). Hides from cows or steers are chiefly used for men's shoes, transmission belting, and other heavy-duty service. These are usually vegetable-tanned. Lighter grades made from the skins of sheep, calves, or reptiles are used for shoe uppers, luggage, gloves, and similar end products (chrome-tanned).

Leather is a naturally poromeric material which retains the microporosity of the original skin; this property makes it uniquely applicable to footwear; to a limited extent it is able to conform to the contour of the individual foot. Leather is made in many colors, weights and finishes. However, it is being replaced to an increasing extent by plastics for many minor uses and by synthetics for shoe uppers and soling. For further information refer to American Leather Chemists' Assn., University of Cincinnati, Cincinnati, Ohio. See also poromeric.

leavening agent. See yeast; baking powder.

Le Blanc (1742–1806). French inventor of the first successful process for making soda ash. His patent was confiscated by the Revolutionist government and the process was used widely for years without either acknowledgment or remuneratoin. His original formula was 100 parts salt cake, 100 parts limestone and 50 parts coal.

Le Chatelier (1850–1936). French physical chemist, famous chiefly for his statement of the equilibrium principle (often known as Le Chatelier's Law). His work included investigations of cements, alloys, and gaseous combustion. The principle may be stated: every system in equilibrium is conservative, and tends to resist the changes upon it by reacting in such a way as to help nullify the imposed change.

Lechanché cell. See dry cell.

lecithin. Pure lecithin is a phosphatidyl choline.
$CH_2(R)CH(R')CH_2OPO(OH)O(CH_2)_2N(OH)(CH_2)_3$, R and R′ being fatty acid groups. The lecithins are mixtures of diglycerides of fatty acids linked to the choline ester of phosphoric acid. The lecithins are classed as phosphoglycerides or phosphatides (phospholipids). Commercial lecithin is a mixture of acetone-insoluble phosphatides. F.C.C. specifies not less than 50% acetone-insoluble matter (phosphatides).

Properties: Light brown to brown, viscous semiliquid with a characteristic odor; partly soluble in water and acetone; soluble in chloroform and benzene. Nontoxic.

Derivation: Usually from soybean oil; also from corn, other vegetable seeds, egg yolk and other animal sources.

Grades: Technical, unbleached, bleached; fluid; plastic; edible; F.C.C.; 96+% (for biochemical or chromatographic standards).

Uses: Emulsifying, dispersing, wetting, penetrating agent and antioxidant; in margarine, mayonnaise, chocolate and candies, baked goods, animal feeds; paints; petroleum industry (drilling, leaded gasoline); printing inks; soaps and cosmetics; mold release for plastics; blending agent in oils and resins; rubber processing; lubricant for textile fibers.

"Lecton."[28] Trademark for acrylic resin-coated glass fabrics and laminates used as electric insulating material because of thermal stability up to 130°C. Resistant to fluorinated hydrocarbons.

"Lectro."[304] Trademark for a series of lead vinyl stabilizers.

60. A lead chlorosilicate complex. A fine white powder; sp. gr. 3.9; refractive index, 2.1. Used for vinyl electrical insulation and tapes.

78. Tetrabasic lead fumarate. Creamy white powder; sp. gr. 6.5; refractive index, 2.1. Used for heat stabilizer for electrical grade plastisols, phonograph records and electrical insulation.

Containers: Up to 325-lb fiberboard drums.

"Ledate."[69] Trademark for lead dimethyldithiocarbamate (q.v.).

Leeuwenhoek, A. van (1632–1723). A self-educated hobbyist and experimenter of Delft, Holland. He was a member of several scientific societies. His chief contribution was invention of the simple microscope about 1700.

legal chemistry (forensic chemistry). The application of chemical knowledge and procedures to matters involving civil or criminal law, and to all questions where control of chemical compounds, products, or processes is vested in agencies of Federal or state governments. Leal chemistry applies to the following areas:

(1) Crime detection: primarily identification of poisons (see toxicology), of bloodstains, writing and typewriter inks, and a host of miscellaneous materials such as textile fibers from clothing, hair, skin, etc. A variety of analytical methods are used in police laboratories, including microscopes, spot tests, color reactions, and spectrophotometry.

(2) Food, drugs, and cosmetics are under the control of the U.S. Food and Drug Administration. New products and proposed additives must be submitted by the manufacturer and approved before being placed on public sale. Control of the manufacture of illicit drugs is an important phase of legal chemistry.

(3) Pesticides are subject to Federal regulation. New products must be registered; labeling must be specific as to chemical composition, active and inert ingredients, and directions for use.

(4) Marketing and competitive pricing of chemical products' fair trade agreements and discriminatory

practices are also under Federal supervision (Robinson-Patman Act). This includes mergers, tie-in sales, and other merchandising practices.

(5) Interstate shipment and labeling of hazardous chemicals is regulated by the Department of Transportation, and the Federal Aviation Agency, as well as by state and local laws (see labeling).

(6) Patent law comprises a vast body of legal practice and court decisions. The patent system is designed to protect inventions and new discoveries, and most chemical companies retain legal counsel in this field.

(7) Water pollution is subject to Federal regulations (Federal Water Pollution Control Act, 1956). This covers the discharge of contaminating industrial waste, sewage, oil, etc., into navigable streams and their tributaries as well as into coastal waters.

(8) Flammability of fabrics.

(9) Use of volatile, toxic solvents on an industrial scale.

(10) Air pollution, including gases and particulates from industrial stacks and auto exhaust emissions.

"Leguval."[470] Brand name for a series of polyester resins for plastic applications.

"Lehigh."[268] Brand name covering a range of leaded zinc oxides.
Uses: Exterior paints and primers.

"Lekutherm."[470] Trademark for epoxy-based coating resins.

"Lemac."[65] Trademark for a series of polyvinyl acetates as water-white, odorless, tasteless, uniform free flowing beads or alkali-soluble, copolymer beads. Excellent heat, light, and aging stability; specific heat, 0.39 cal/g/C°; tensile strength, up to 6000 psi; water absorption about 2% at 20°C; sp. gr. varies from 1.19 to 1.20; softening point, from 66°C to 155°C. Used in hot melt adhesives; seam sealers; adhesives for plastics; paper coatings.

"Lemasize."[65] Trademark for alkali-soluble resin beads, which are dissolved in ammonia or soda ash solution, and applied in warpsizing of acetate and synthetic fibers.

"Lemoflex."[65] Trademark for a series of internally-plasticized powdered polyvinyl alcohols having excellent flexibility and cold-water solubility. Sp. gr. varies between 1.2–1.3.
Uses: Water-soluble films; gas barrier coatings and films; textile sizes, grease resistant coatings, extruded tubing, protective colloid; mold release agent.

"Lemol."[65] Trademark for a series of polyvinyl alcohols in partially and fully hydrolyzed form at various molecular weights. Used in adhesives, emulsions, polymerization, film-coatings, polyester release agents, textile printing, finishing and sizing. Supplied as nondusting white granules with sp. gr. 1.2–1.3.

lemon chrome. See barium chromate.

lemongrass oil (verbena oil, Indian). An essential oil obtained by steam distillation of a grass (Cymbopogon).
Properties: Dark yellow to light brown-red; pronounced heavy lemon-like odor; sp. gr. 0.900–0.910 (15/15°C); optical rotaton −3° to +1°; refractive index (n 20/D) 1.4830–1.4890; soluble in alcohol; slightly soluble in glycerol. Combustible; nontoxic.
Source: India; East and West Indies; Guatamala.
Chief constituents: Citral (75–85%), geraniol, methylheptenone.
Containers: Cans; drums.
Uses: Perfumes; flavoring; isolates and ionones; source of citral.

lemon oil. See citrus peel oils.

"Lenigallol."[9] Trademark for pyrogallol triacetate.

lepidine (gamma-methylquinoline; cincholepidine) $C_9H_6NCH_3$. An alkaloid.
Properties: Oily liquid; quinoline-like odor; turns red-brown on exposure to light. Sp. gr. 1.086; b.p. 266°C; solidifies about 0°C. Soluble in alcohol, ether, and benzene; slightly soluble in water.
Derivation: From cinchonine.
Hazard: Probably toxic.
Use: Organic preparations.

lepidolite (lithia mica) $K_2Li_3Al_4Si_7O_{21}(OH,F)_3$. A flusilicate of potassium, lithium, and aluminum, found in pegmatites. Rubidium occurs as an impurity. A variety of mica (q.v.).
Properties: Color pink and lilac to gray; luster pearly; perfect micaceous cleavage; Mohs hardness 2.5–4; sp. gr. 2.8–3.0.
Occurrence: California, South Dakota, New Mexico, South Africa.
Uses: Source of lithium and rubidium; flux in glass and ceramics production.

"Lethane."[23] Trademark for a group of thiocyanate insecticides.
60. 2-Thiocyanoethyl laurate.
384. Beta-butoxy-beta′-thiocyanodiethyl ether.
A-70. Diethylene glycol dithiocyanate.

Leu Symbol for leucine.

leucine (alpha-amino-gamma-methylvaleric acid; alpha-aminoisocaproic acid)
$(CH_3)_2CHCH_2CH(NH_2)COOH$. An essential amino acid. Found naturally in the L(−) form.
Properties: White crystals; soluble in water; slightly soluble in alcohol; insoluble in ether; optically active (natural form). DL-leucine m.p. 332°C with decomposition. L(−)-leucine m.p. 295°C; sp. gr. 1.239 (18/4°C).
Derivatoin: Hydrolysis of protein (edestin, hemoglobin, zein); organic synthesis from the alpha-bromo acid.
Grades: Commercial (DL-); F.C.C. (L-).
Containers: Drums.
Uses: Nutrient and dietary supplement; biochemical research.

leuco-compounds. See vat dyes.

leucocyte. A white blood cell.

"Leucosol."[28] Trademark for a series of vat colors especially prepared for textile printing.

leucovorin. A preferred name for folinic acid (q.v.).

"Leukanol."[23] Trademark for synthetic tanning assistants of the sulfonic-type, supplied in liquid and solid grades. Powerful dispersants for vegetable tannins and bleaches for chrome-tanned leather.

"Leukeran."[301] Trademark for chlorambucil (q.v.).

"Levair."[1] Trademark for a phosphate leavening agent.

levallorphan tartrate (*l*-N-allyl-3-hydroxymorphinan bitartrate) $C_{19}H_{25}NO \cdot C_4H_6O_6$.
Properties: White, odorless crystalline powder. M.p. 174–177°C. Soluble in water; sparingly soluble in alcohol; insoluble in ether and chloroform.
Grade: N.F.
Use: Medicine.

"Levapren 450."[470] Trademark for ethylene-vinyl acetate copolymer containing approximately 45% vinyl acetate.

levarterenol (*l*-norepinephrine; *l*-arterenol; *l*-alpha-(aminomethyl)-3,4-dihydroxybenzyl alcohol) $C_6H_3(OH)_2CH(OH)CH_2NH_2$. A peripheral vasoconstrictor.

Properties: Microcrystals; occurs in adrenal glands. Decomposes at 217°C.

Use: Medicine. Available also as bitartrate (U.S.P.).

leveling. (1) A term used in the paint industry to describe the application properties of a paint, i.e., its ability to cover a dry surface easily and to hold its level without sagging or running. (2) The ability of a nickel-plated coating to cover surface irregularities of the substrate, achieved by the incorporation of one or more brighteners (q.v.) in the plating formulation. (3) Aiding the uniform dispersion of a dye in a dye bath or solution by addition of a suitable material, e.g., lignin.

"Levelume."[288] Trademark for bright, high-leveling nickel process. Prepared from nickel sulfate, nickel chloride, boric acid and organic addition agents. Applications are in electrical appliance, automotive trim, plumbing fixtures.

levorotatory. Having the property when in solution of rotating the plane of polarized light to the left or counterclockwise. Levorotatory compounds may have the prefix *l*- to distinguish them from their dextrorotatory or *d*- isomers, but the minus sign (−) is now in more general use.

levulinic acid (gamma-ketovaleric acid; acetylpropionic acid; 4-oxopentanoic acid; levulic acid) $CH_3CO(CH_2)_2COOH$.

Properties: Crystals. B.p. 245–246°C; m.p. 33–35°C; sp. gr. 1.1447 (25/4°C); refractive index 1.442 (16/D). Completely soluble in water, alcohol, esters, ethers, ketones, aromatic hydrocarbons. Insoluble in aliphatic hydrocarbons. Combustible; low toxicity.

Containers: 5-, 55-gal containers; carloads.

Uses: Intermediate for plasticizers, solvents, resins, flavors, pharmaceuticals; acidulant and preservative; chrome plating; solder flux; stabilizer for calcium greases; control of lime deposits.

levulose. See fructose.

Lewis, Gilbert N. (1875–1946). American chemist, native of Massachusetts; Professor of chemistry at M.I.T. from 1905 to 1912, after which he became dean of chemistry at University of California at Berkeley. His most creative contribution was the electron-pair theory of acids and bases which laid the groundwork for coordination chemistry. He was also a leading authority on thermodynamics, and his textbook on this subject, written with Merle Randall and published in 1923, became world-famous. See also following entries.

Lewis acid. Any molecule or ion (also called an electrophile) that can combine with another molecule or ion by forming a covalent bond with two electrons from the second molecule or ion. An acid is thus an electron acceptor. Hydrogen ion (proton) is the simplest substance that will do this, but many compounds, such as boron trifluoride, BF_3, and aluminum chloride, $AlCl_3$, exhibit the same behavior and are therefore properly called acids. Such substances show acid effects on indicator colors and when dissolved in the proper solvents.

Lewis base. A substance that forms a covalent bond by donating a pair of electrons, neutralizaton resulting from reaction between the base and the acid with formation of a coordinate covalent bond. It is also called a nucleophile. See also Lewis electron theory.

Lewis electron theory. A theory involving acid and base formation, neutralization, and related phenomena on the basis of exchange of electrons between substances and the formation of coordinate bonds. It represented an important advance in chemical theory, largely replacing earlier concepts. Advanced in 1923 by Gilbert N. Lewis, it contributed much to the development of coordination chemistry, in which the base is represented by the ligand and the acid by the metal ion.

A comparatively recent concept involves the terms "hard" and "soft" acids and bases, which refer to the ease with which the electron orbitals can be disturbed or distorted. Hard acids have a high positive oxidation state, and their valence electrons are not readily excited; soft acids and bases have little or no positive charge and easily excited valence electrons. Hard acids combine preferentially with hard bases, and soft acids with soft bases. Soft acids tend to accept electrons and form covalent bonds more readily than hard acids. For example, the halogen acids arranged in a series by increasing atomic weight (and decreasing chemical activity) show a progression from hard (HF) to soft (HI).

lewisite. Legal label name for beta-chlorovinyl-dichloroarsine (q.v.).

"Lewisol" 28.[266] Trademark for a pale, hard resin; a maleic-modified glycerol ester of rosin. Acid number 36, softening point 141°C; USDA color WG.

Lewis, Warren K. (1882–1974) Born in Laurel, Maryland; graduated from MIT in 1905; Ph.D. from University of Breslau, Germany in 1908. Became professor of chemical engineering at MIT in 1910. He is often regarded as the father of chemical engineering in the U.S., as his outstanding books and other publications did much to establish the fundamental theory and practice of unit operations. He was coauthor of the famous textbook "Principles of Chemical Engineering" (1923).

"Lexan."[245] Trademark for thermoplastic carbonate-linked polymers produced by reacting bisphenol A and phosgene. Used in molding applications and other industrial arts.

See also polycarbonate resin.

"L-310 Fatty Acid."[487] Trademark for a fatty acid derived from linseed oil. The major component acids are oleic, linoleic and linolenic. Light yellow liquid at ambient temperature; obtained from naturally occurring triglycerides; nontoxic, nonirritating.

Containers: Tank cars and trucks.

Uses: Chemical intermediate; paint and varnishes; alkyd resins and soaps.

Li Symbol for lithium.

"Librium" Hydrochloride.[190] Trademark for chlordiazepoxide hydrochloride (q.v.). Manufacture and use restricted.

licanic acid (4-keto-9,11,13-octadecatrienoic acid) $CH_3(CH_2)_3(CH:CH)_3(CH_2)_4COCH_2COOH$.
Properties: White crystals, alpha-Licanic acid (naturally occurring isomer) melts at 74–75°C. Readily isomerizes to the beta-form, m.p. 99.5°C. Soluble in organic solvents.
Derivation: Occurs in oiticica and other oils as glycerides.

lichenic acid. See fumaric acid.

licorice. See glycyrrhizin.

lidocaine (alpha-diethylaminoaceto-2,6-xylidide) $C_6H_3(CH_3)_2NHCOCH_2N(C_2H_5)_2$.
Properties: White or slightly yellow crystalline powder; characteristic odor; m.p. 66–69°C; b.p. 180–182°C (at 4 mm); soluble in alcohol, ether or chloroform; insoluble in water.
Derivation: By action of diethylamine on chloroacetylxylidide.
Grade: U.S.P.
Use: Medicine (also available as the hydrochloride).

Liebig, Justus Von (1803–1873). German chemist who founded the *Annalen*, a world-famous chemical journal. He was a great teacher of chemistry, training such men as Hofmann, who did basic work on organic dyes. Liebig contributed original research in the fields of human physiology, plant life, soil chemistry, and was the discoverer of chloroform, chloral and cyanogen compounds. He was the first to recommend addition of nutrients to soils and thus may be considered the originator of the fertilizer industry.

ligand. A molecule, ion, or atom that is attached to the central atom of a coordination compound, a chelate, or other complex. Thus the ammonia molecules in $[Co(NH_3)_6]^{+++}$, and the chlorine atoms in $[PtCl_6]^{--}$ are ligands. Ligands are also called complexing agents, as for example, EDTA, ammonia, etc. See also chelate; coordination compound.

ligase. A class of enzymes. See enzyme.

light hydrocarbon (gas liquid). One of a group of hydrocarbon products derived from natural gas or petroleum (ethane, propane, iso- and normal butane, and natural gasoline (C_5 and heavier). Produced largely in southwest Texas and Louisiana, these are used as feedstocks for a wide variety of organics. See also liquefied petroleum gas.

light metal. In engineering terminology, a metal of specific gravity less than 3 that is strong enough for construction use (aluminum, magnesium, beryllium).

light oil (coal tar light oil). A fractional distillate from coal-tar, with b.p. range from 110–210°C, consisting of a mixture of benzene, pyridine, toluene, phenol and cresols. The term is also sometimes used for oils of about the same b.p. range, but from other sources.
Grade: Technical.
Containers: Tank cars; iron drums.
Hazard: Highly flammable, dangerous fire risk.
Uses: Source of benzene, solvent naphthas, toluene, phenol and cresols.
Shipping regulations: (Rail) Red label. (Air) Flammable Liquid label. Legal label name: coal tar light oil or distillate.

light water. (1) A fire-fighting agent consisting of a water solution of perfluorocarbon compounds mixed with a water-soluble thickener of the polyoxyethylene compound type. It can be used simultaneously with dry chemical to smother gasoline or similar fires.
(2) Ordinary water (as distinct from heavy water) used to cool and moderate nuclear reactors.

lignin. The major noncarbohydrate constituent of wood (about one fourth). It functions as a natural plastic binder for the cellulose fibers. Its exact formula is unknown. Quinone methides are the most reactive intermediates in its formation. Lignin is removed from wood by both the sulfite and sulfate pulp processes, and limited amounts have been recovered from these sources and other wood waste.
Containers: 70-lb bags; 250-lb drums.
Uses: Components of groundwood paper (newsprint); stabilization of asphalt emulsions; ceramic binder and deflocculant; dye leveler and dispersant; drilling fluid additive; precipitation of proteins; extender for phenolic plastics; special molded products; vanillin (see lignin sulfonate); source of a component of battery expanders.

lignin sulfonate (lignosulfonate). A metallic sulfonate salt made from the lignin of sulfite pulp-mill liquors. Molecular weights range from 1000 to 20,000.
Properties: Light-tan to dark-brown powder; no pronounced odor; stable in dry form and relatively stable in aqueous solution; nonhygroscopic; no definite m.p.; decompose above 200°C; sp. gr. about 1.5. Forms colloidal solutions or dispersions in water; practically insoluble in all organic solvents.
Uses: Dispersing agent in concrete and carbon black-rubber mixes; extender for tanning agents; oil-well drilling mud additives; ore flotation agents; production of vanillin, industrial cleaners, gypsum slurries, dyestuffs, pesticide formulatons.
Commercially available as the salts of most metals and of ammonium.
See also "Lignosol."

lignite. A low rank of coal between peat and sub-bituminous. The distinction of lignite from these materials is not sharp, as the transition from one to the other is gradual. Brown coal is a form of lignite closely related to peat. Though it has low Btu value, it is being studied as a possible source of pipeline fuel gas, hydrogen, and electric power. It occurs abundantly in East Germany, Poland and the Netherlands. See also gasification.
Hazard: Flammable in form of dust.

lignoceric acid (n-tetracosanoic acid) $CH_3(CH_2)_{22}COOH$. A long-chain saturated fatty acid found in minor quantities in most natural fats.
Properties: Crystals; m.p. 84.2°C; b.p. 272°C (10 mm); sp. gr. 0.8207 (100/4°C); refractive index 1.4287 (100°C); nearly insoluble in ethanol. Nontoxic.
Source: Lignite and beechwood tar; peanut oil; sphingomyelin.
Grades: Technical; 99%.
Use: Biochemical research.

"Lignosol."[476] Trademark for a series of calcium, sodium, and ammonium lignosulfonates. These compounds are mixtures of lignosulfonates and wood sugars. Special grades are available which have low wood sugar content.
Uses: Binders, dispersing agents and tanning agents.

"Lignosol AA3."[476] Trademark for a mixture of sodium lignosulfonate and aldonic acids. Used as a substitute for gluconic acid in alkaline paint stripping.

lignosulfonate. See lignin sulfonate.

"Lignox."[236] Trademark for a soluble calcium lignosulfonate in dry powder form. Used for treatment of drilling mud containing calcium ions and in brine emulsion muds.

ligroin A saturated, volatile fraction of petroleum boiling in the range 20–135°C (58–275°F) (based

on A.S.T.M. definition). There is a special grade of ligroin known as petroleum benzin.
Hazard: Highly flammable; dangerous fire risk. Toxic by ingestoin and inhalation.
Shipping regulatoins: (Rail) Red label. (Air) Flammable Liquid label. Legal label name: benzine or petroleum ether.

"Lilial."[227] $C_{14}H_{20}O$. Trademark for para-tert-butyl-alpha-methylhydrocinnamaldehyde.

"Lily."[84] Trademark for sublimed flowers of sulfur.
Uses: For high purity powdered sulfur requirements in pharmaceuticals, stock feeds, and chemicals.

lime. Specifically, calcium oxide (CaO); more generally, any of the various chemical and physical forms of quicklime, hydrated lime, and hydraulic lime. (Adapted from ASTM definition C41-47.) Noncombustible (see Hazard). For further information see National Lime Assn., 925 16th St. N.W., Washington, D.C.
Hazard: Unslaked lime (quicklime) yields heat on mixture with water, and is a caustic irritant.
Uses: See calcium oxide; calcium hydroxide.

lime acetate. See calcium acetate.

lime, agricultural. Lime slaked with a minimum amount of water to form calcium hydroxide.

lime, air-slaked. Lime which has absorbed carbon dioxide and moisture from the atmosphere. It consists of a powder composed of calcium carbonate and calcium hydroxide.

lime-ammonium nitrogen. Ammonium nitrate with dolomite.

lime citrate. See calcium citrate.

lime, chlorinated (chloride of lime; bleaching powder); approximately $CaCl(ClO) \cdot 4H_2O$.
Properties: White powder; chlorine odor; m.p. (dec); decomposes in water, acids.
Derivation: By conducting chlorine into a box-like structure containing slaked lime spread upon perforated shelves.
Grades: 35–37% active chlorine; technical.
Hazard: Evolves chlorine and at higher temperatures oxygen. With acids or moisture evolves chlorine freely at ordinary temperatures.
Uses: Textile and other bleaching applications; organic synthesis; deodorizer; disinfectant.
Shipping regulations: See calcium hypochlorite.
See also calcium hypochlorite; bleach.

"Limed Topit."[79] Trademark for a limed tall oil pitch used in adhesives and asphalt tile manufacture.

lime, fat. A pure lime which combines readily with water to form a fine white powder, free from grit and makes a smooth stiff paste with excess of water. Must not be loaded hot.
See also lime, lean.

lime, hydrated. See calcium hydroxide.

lime, hydraulic. A variety of calcined limestone which, when pulverized, absorbs water without swelling or heating and gives a cement that hardens under water. The limestone burned for this purpose usually contains from 10–17% silica, alumina and iron and from 40–45% lime, magnesia sometimes replacing lime. Must not be loaded hot.

lime hypophosphite. See calcium hypophosphite.

lime, lean. A lime which does not slake freely with water because it has been prepared from limestone containing a high percentage of impurities, e.g., silica, iron, alumina, etc. Must not be loaded hot.

lime-nitrogen. See calcium cyanamide.

lime oil, distilled. Colorless to greenish yellow volatile oil obtained by distillatoin from the juice or whole crushed fruit of Citrus aurantifolia Swingle.
Properties: Sp. gr. 0.855–0.863 (25°C); refractive index 1.4745–1.4770 (20°C); angular rotation +34° to +47°. Soluble in most fixed oils and mineral oil. Insoluble in glycerin and propylene glycol. Combustible; nontoxic.
Chief constituents: Terpineol; citral.
Grade: F.C.C. (contains between 0.5 and 2.5% of aldehydes, calculated as citral).
Uses: Extracts; flavoring; perfumery; toilet soaps; cosmetics.

lime oil, expressed. See citrus peel oil.

lime saltpeter. See calcium nitrate.

lime, slaked. See calcium hydroxide.

limestone $CaCO_3$. A noncombustible solid characteristic of sedimentary rocks and composed mainly of calcium carbonate in the form of the mineral calcite (q.v.). Mohs hardness about 3. Limestones are sometimes classed according to the impurities contained, e.g.
Dolomitic limestone: Usually a limestone containing more than 5% magnesium carbonate. See dolomite.
Magnesium limestone: Dolomitic limestone. Used as a solid diluent and carrier in pesticides.
Argillaceous limestone: Contains clays; used in cement manufacture, as "cement rock."
Siliceous limestone: A limestone containing sand or quartz.
Limestones are also named according to the formation in which they occur. See also marble.
Uses: Building stone; metallurgy (flux); manufacture of lime; source of carbon dioxide; agriculture; road ballast; cement (Portland and natural); alkali manufacture; removal of SO_2 from stack gases.

lime, sulfurated (calcium sulfide, crude) A mixture of calcium sulfide and calcium sulfate.
Properties: Yellowish-gray or grayish-white powder; odor of hydrogen sulfide. Soluble in acids; insoluble in water and alcohol. Noncombustible.
Derivation: By roasting calcium sulfate with coke.
Uses: Medicine; depilatory; luminous paint.

lime-sulfur solution. A solution made by boiling together lime (50 lbs), sulfur 100 (100 lbs) and water (100 gals) and diluting to one-tenth strength. Contains calcium polysulfide, free sulfur, and calcium thiosulfate. Used extensively as a fungicidal spray on fruit trees and as a sheep dip.

lime, unslaked. See calcium oxide.

lime water (calcium hydroxide solution).
Properties: Clear, colorless, odorless, alkaline aqueous solution of calcium hydroxide containing not less than 0.14 g of $Ca(OH)_2$ in each 100 ml at 25°C. (Note: The strength varies with the temperature at which the solution is stored.) Sp. gr. about 1.00 (25°C). Absorbs carbon dioxide from the air.
Grade: U.S.P. (as calcium hydroxide solution).
Use: Medicine (external).

limonene $C_{10}H_{16}$. A widely distributed optically active terpene, closely related to isoprene. It occurs naturally in both D- and L-forms. The racemic mixture of the two isomers is known as dipentene (q.v.).
Properties: Colorless liquid. (a) Sp. gr. 0.8411 (20°C); b.p. 176–176.4°C. (b) Sp. gr. 0.8422 (20°C); b.p. 176–176.4°C. Oxidizes to film in air; oxidation behavior similar to that of rubber or drying oils.
Derivation: (a) Lemon, bergamot, caraway, orange and other oils. (b) Peppermint and spearmint oils.
Uses: Flavoring, fragrance and perfume materials; solvent; wetting agent; resin manufacture.

limonene dioxide. See dipentene dioxide.

limonene, inactive (or racemic or *dl*). See dipentene.

limonene monoxide. See dipentene monoxide.

limonite. See hematite, brown.

"Lin-All."[480] Trademark for a liquid, paste or flake form of tallates of calcium, cobalt, copper, iron, lead, manganese and zinc.
Uses: Drier for paint and printing inks.

linalool (linalol) $(CH_3)_2C{:}CHCH_2CH_2C(CH_3)OHCH{:}CH_2$. 3,7-Dimethyl-1,6-octadien-3-ol. Linalool is the *l*-isomer; coridandrol is the *d*-isomer.
Properties: Colorless liquid; odor similar to that of bergamot oil and French lavender. Soluble in alcohol and ether. Sp. gr. 0.858–0.868 (25°C); b.p. 195–199°C; angular rotation −2° to +2°. Soluble in fixed oils. Combustible; nontoxic.
Derivation: Derived from many essential oils, particularly from bois de rose, petigrain, linaloe wood, bergamot, and others. Made synthetically from geraniol.
Method of purification: Rectification.
Grades: Ex bois de rose oil; synthetic; F.C.C.
Use: Perfumery; flavoring agent.

linalool oxide (tetrahydro-alpha, alpha, 5-trimethyl-5-vinylfurfuryl alcohol) $C_{10}H_{17}O_2$.
Properties: Liquid; refractive index (n 20/D) 1.4523.
Derivation: Synthetically from acetone.
Containers: 55-gal drums.
Uses: Perfuming and flavoring agent.

linalyl acetate $C_{10}H_{17}C_2H_3O_2$.
Properties: Clear, colorless, oily liquid; odor of bergamot; b.p. 108–110°C; sp. gr. 0.908–0.920; refractive index (n 20/D) 1.450–1.458; angular rotation −1 to +1°; soluble in alcohol, ether, diethyl phthalate, benzyl benzoate, mineral oil, fixed oils, alcohol; slightly soluble in propylene glycol; insoluble in water, glycerin. Combustible; low toxicity.
Derivation: Action of acetic anhydride on linalool in presence of sulfuric acid. May also be obtained from bergamot and other oils.
Method of purification: Rectification.
Grades: Ex bois de rose oil 92%, 96–98%; F.C.C. (natural and synthetic).
Uses: Extracts; perfumery; flavoring agent; substitute for petigrain oil.

linalyl formate $C_{10}H_{17}OOCH$.
Properties: Sp. gr. 0.915 (25/4°C); b.p. 100–103°C (10°C); refractive index 1.456–1.457 (20°C). Insoluble in water; soluble in alcohol. Combustible; low toxicity.
Derivation: Synthetically from acetone.
Containers: 55-gal drums.
Uses: Perfume; flavoring.

linalyl isobutyrate $C_{10}H_{17}OOCCH(CH_3)_2$.
Properties: Liquid; sp. gr. 0.890; refractive index 1.4490. Combustible; low toxicity.
Derivation: Synthetically from acetone.
Containers: 55-gal drums.
Uses: Perfume; flavoring.

linalyl propionate $C_{10}H_{17}OOCC_2H_5$.
Properties: Colorless liquid; floral odor similar to bergamot oil; sp. gr. 0.895–0.902 (25°C); refractive index 1.4500 to 1.4550 (20°C). Soluble in most fixed oils and in mineral oil; slightly soluble in propylene glycol; insoluble in glycerin. Combustible; low toxicity.
Derivation: Synthetically, starting with acetone.
Grade: F.C.C.
Uses: Perfumery; flavoring agent.

"Linaqua."[64] Trademark for 85% solids, water-soluble linseed oil vehicle for formulating gloss or flat exterior house paints and primers.

"Lincocin."[327] Trademark for lincomycin hydrochloride monohydrate.

lincomycin $C_{18}H_{34}N_2O_6S$. An antibacterial substance produced by streptomyces lincolnensis var. lincolnensis. The hydrochloride is stable in the dry state and in aqueous solution for at least 24 months. Soluble in water at room temperature in concentrations up to 500 mg/ml.
Use: Medicine.

lindane (gamma-benzene hexachloride) $C_6H_6Cl_6$. Legal label name for gamma isomer of 1,2,3,4,5,6-hexachlorocyclohexane (q.v.).
Hazard: Highly toxic. Use may be restricted. Tolerance, 0.5 mg per cubic meter of air.

"Lindol"[1] Trademark for a product consisting of tricresyl phosphate (q.v.).

"Lindsite."[88] Trademark for a rare earth oxide glass polishing agent. Contains cerium and other rare earth oxides in the proportion in which they occur naturally.

linear alkyl sulfonate. See alkyl sulfonate, linear.

linoleic acid (linolic acid) $CH_3(CH_2)_4HC{:}CHCH_2CH{:}CH(CH_2)_7COOH$. A polyunsaturated fatty acid (2 double bonds) existing in both conjugated and unconjugated forms. A plant glyceride essential to human diet.
Properties: Colorless to straw-colored liquid; sp. gr. 0.905 (15/4°C); f.p. −5°C; b.p. 228°C (14 mm); refractive index 1.4710 (15°C). Insoluble in water; soluble in most organic solvents. Combustible; nontoxic.
Commercial sources: Linseed oil; safflower oil; tall oil.
Grades: Technical, purified (99+%); edible.
Uses: Soaps; special driers for protective coatings; emulsifying agents; medicine; feeds; biochemical research; dietary supplement; margarine.
Note: Do not confuse with linolenic acid.

linolein. A glyceride of linoleic acid. It is one of the constituents of linseed oil which induces drying.

linolenic acid (9,12,15-octadecatrienoic acid) $CH_3CH_2CH{:}CHCH_2CH{:}CHCH_2CH{:}CH(CH_2)_7COOH$. A polyunsaturated fatty acid (3 double bonds), which occurs as the glyceride in many seed fats. It is an essential fatty acid in the diet. It is also a constituent of drying oils.
Properties: Colorless liquid; soluble in most organic solvents, insoluble in water; sp. gr. 0.916 (20/4°C);

f.p. −11°C; b.p. 230°C (17 mm). Combustible; nontoxic.
Grade: Purified 99+%.
Uses: Medicine; biochemical research; drying oils.
Note: Do not confuse with linoleic acid.

linolenin. A glyceride of linolenic acid. Like linolein it is a constituent of drying oils. See linseed oil.

linoleum. A product composed of a coarse fiber backing coated with a mixture of linseed oil, cork filler, and rosin or other binder, plus desired colorants. Now almost completely replaced by plastics.

linolenyl alcohol (octadecatrienol) $CH_3CH_2CH{:}CHCH_2CH{:}CHCH_2CH{:}CH(CH_2)_7CH_2OH$. The fatty alcohol derived from linolenic acid. Available commercially as 50% pure.
Properties: Colorless solid; iodine value 190; cloud point 50.0°F; sp. gr. 0.864; combustible; low toxicity.
Derivation: Reduction of acid made from linseed oil.
Impurities: Oleyl and linoleyl alcohols with some saturated alcohols.
Uses: Paints; flotation; lubricants; surface-active agents; resins; synthetic fibers.

linoleyl alcohol $CH_3(CH_2)_4CH{:}CHCH_2CH{:}CH(CH_2)_7CH_2OH$. The fatty alcohol derived from linoleic acid. Available commercially as 50–60% pure alcohol.
Properties: Colorless solid; iodine value 137; sp. gr. 0.855; cloud point 59°F. Combustible; low toxicity.
Derivation: Reduction of linoleic acid.
Impurities: Mostly oleyl alcohol with some linolenyl, and saturated alcohols.
Uses: Paints; flotation; paper; surface-active agents; resins; leather.

linoleyltrimethylammonium bromide. An amorphous solid from yellow to very light brown in color. Soluble in water, alcohol.
Uses: Germicide; deodorant; algicide; slime control.

linolic acid. See linoleic acid.

linseed cake. The press cake formed when the seeds are crushed and the oil is extracted. See linseed oil meal.

linseed oil (flaxseed oil).
Properties: Golden-yellow, amber or brown drying oil with peculiar odor and bland taste. Sp. gr. 0.931–0.936; iodine value 177; saponification value 189–195; acid no. (max) 4 (ASTM D 234–48); boiling point, none; flash point 432°F; autoignition point about 650°F. Polymerizes on exposure to air. Soluble in ether, chloroform, carbon disulfide, and turpentine; slightly soluble in alcohol. Heats spontaneously; combustible; nontoxic.
Chief constituents: Glycerides of linolenic, oleic, linoleic and saturated fatty acids. The drying property is due to the linoleic and linolenic groups.
Derivation: From seeds of the flax plant Linum usitatissimum by expression or solvent extraction. Various refining and bleaching methods are used.
Grades: Raw; boiled; double-boiled; blown; varnish makers; refined.
Containers: Cans; drums; tank cars.
Uses: Paints; varnishes; oilcloth; putty; printing inks; core oils; linings and packings; alkyd resins; soap; pharmaceuticals.
Note: Use of linseed oil in paints has decreased sharply since the introduction of emulsion paints.

linseed oil, blown. Linseed oil whose viscosity is increased by air bubbled through it at 200°F. The reaction is mainly oxidation followed by polymerization. The resulting product dries to a harder film than heat-bodied oils and is used in interior paints and enamels.

linseed oil, boiled. The term is a misnomer since the oil does not boil. Small amounts of driers (e.g., oxides of manganese, lead, or cobalt, or their naphthenates, resinates or linoleates) are added to hot linseed oil to accelerate drying. The "boiled oil" becomes thicker and darker.

linseed oil, heat-bodied. Linseed oil which has been polymerized by heating at 550–600°F. This increases viscosity and acid content and reduces iodine value. Bodied oil dries much faster than unprocessed oil.

linseed oil meal (linseed cake). The crushed and extracted residue from flexseed (linseed), generally prepared by crushing the seeds, cooking with steam, and hydraulic expression of the oil. The resulting cake is sold by the protein content.

linters, cotton. See cotton linters.

linuron. Generic name for 3-(3,4-dichlorophenyl)-1-methoxy-1-methylurea. $C_6H_3Cl_2NHC(O)N(OCH_3)CH_3$.
Properties: Solid; m.p. 93–94°C; slightly soluble in water; partially soluble in acetone and alcohol.
Hazard: May be toxic.
Use: Herbicide.

"Lipal."[555] Trademark for a series of polyoxyethylene esters and ethers of fatty acids. Used as detergents, emulsifiers, solubilizers, wetting agents, and coupling agents.

lipase. Any of a class of enzymes that hydrolyze fats to glycerol and fatty acids. Lipase is abundant in the pancreas, but also occurs in gastric mucosa, in the small intestine, and in fatty tissue. It is found in milk, wheat germ, and various fungi. Commercial pancreatin and most trypsin preparations contain lipase.
Derivation: From pork pancreas or calf glands.
Uses: Manufacture of cheese and similar foods; for removal of fat spots in dry cleaning or grease accumulations; in analytical chemistry of fats because it selectively hydrolyzes only fatty acids on the ends of triglycerides.

lipid (lipide). An inclusive term for fats and fat-derived materials. It includes all substances which are (1) relatively insoluble in water but soluble in organic solvents (benzene, chloroform, acetone, ether, etc.); (2) related either actually or potentialaly to fatty acid esters, fatty alcohols, sterols, waxes, etc.; and (3) utilizable by the animal organism. One of the chief structural components of living cells. See also phospholipid; fat; fatty acid.

dl-alpha-**lipoic acid** (6,8-dithiooctanoic acid; thioctic acid; POF) $\underline{SSCH_2CH_2CH}(CH_2)_4COOH$. A pyruvate oxidation factor. Pyruvate is a normal intermediate in carbohydrate metabolism.
Properties: Crystals; m.p. 60–61°C; b.p. 160–165°C. Practically insoluble in water; soluble in fat solvents; forms a water-soluble sodium salt.
Sources: Food sources; yeast and liver.
Uses: Nutrition; biochemical research.

"Lipoiodine."[305] Trademark for ethyl diiodobrassidate.

"Lipopure."[425] Brand name for a group of solvents purified to remove contaminants which might interfere with lipid procedures.

lipotropic agent. An agent which, because of its affinity for fats and oils, helps to regulate the metabolism of fat and cholesterol in the animal body. Inositol in an example.

Lipowitz's metal. A fusible alloy.

lipoxidase. An enzyme which catalyzes the addition of oxygen to the double bonds of unsaturated fatty acids of plant origin.
Uses: Biochemical research; whitening bread.

liquidation. The separation of two or more components of a mixture by heating to a temperature at which one component melts, leaving the others as solids. Used in the separation of alloy components.

liquefied natural gas (LNG). See natural gas.

liquefied petroleum gas (compressed petroleum gas; liquefied hydrocarbon gas; LPG). A compressed or liquefied gas obtained as a by-product in petroleum refining or natural gasoline manufacture, e.g., butane, isobutane, propane, propylene, butylenes, and their mixtures. Colorless, noncorrosive, nontoxic. Flash point −100°F. Autoignition temp. 800–1000°F.
Containers: Cylinders and drums; tank cars and tank trucks; pipelines; ocean tankers.
Hazard: Highly flammable; dangerous fire and explosion risk. Tolerance, 1000 ppm in air.
Uses: Domestic and industrial fuel; automotive fuel; welding, brazing, and metal-cutting.
Shipping regulations: (Rail) Red Gas label. (Air) Flammable Gas label. Not accepted on passenger planes.
See also compressed gas; natural gas.

"Liquibor."[441] Trademark for alkali borate condensation products.

liquid. An amorphous (noncrystalline) form of matter intermediate between gases and solids, in which the molecules are much more highly concentrated than in gases, but much less concentrated than in solids. The molecules of liquids are free to move within the limits set by intermolecular attractive forces. At the air/liquid interface the vibration of the molecules causes some to be ejected from the liquid at a rate depending on the surface tension. (See kinetic theory). The tendency of molecules to escape from a liquid surface is called fugacity, and is largely responsible for evaporation, which occurs when the air space above the liquid is unrestricted. In a closed system, where the air space is restricted, the escaping molecules eventually saturate the air and thus the number of molecules leaving the liquid will be equal to those returning to it as a result of molecular attraction. In these circumstances the liquid/air system is said to be in equilibrium.

Liquids vary greatly in viscosity, melting point, vapor pressure, and surface tension. Mercury has a specific gravity of 13.6 and the highest surface tension of all liquids. Glass has the highest viscosity. Polar liquids are those whose molecules have opposite electrical charges on their terminal atoms or groups, which impart a force called dipole moment (q.v.). Water is a polar liquid with high dielectric constant. Pure hydrocarbon liquids are generally nonpolar.
See also liquid, Newtonion; glass; amorphous; solid; liquid crystal.

liquid chromatography. An analytical method based on separation of the components of a mixture in solution by selective adsorption. All systems include a moving solvent; a means of producing solvent motion, such as gravity or (in more recently developed equipment) a pump; a means of sample introduction; a fractionating column; and a detector. Innovations in functional systems provide the analytical capability for operating in three separation modes: (1) Liquid/liquid: partition in which separations depend on relative solubilities of sample components in two immiscible solvents (one of which is usually water). (2) Liquid/solid: adsorption where the differnces in polarities of sample components and their relative adsorption on an active surface determine the degree of separation. (3) Molecular size separations which depend on the effective molecular size of sample components in solution.

Solvents, often referred to as carriers, include isooctane, methyl ethyl ketone, acetone/chloroform, tetrahydrofuran, hexane, and toluene.

Packing materials in columns of varying lengths, include silica gel, alumina, glass beads, polystyrene gel and ion exchange resins.
See also gas chromatography; paper chromatography; thin-layer chromatography.

"Liquid Bright Gold."[28] Trademark for a formulation of gold compounds designed to be painted or sprayed on ceramic or glass products, plastics, etc., and subjected to firing or curing temperatures. Used for decoration, printed circuits, etc.

liquid crystal. An organic compound in a transition state between solid and liquid forms. They are viscous, jelly-like materials that resemble liquids in certain respects (viscosity) and crystals in other properties (light scattering and reflection). The various esters of cholesterol are the best-known examples. They have the ability to indicate small temperature differences by changing color, for example, when applied in a thin film to the skin. They are available in microencapsulated form and are applied in water slurry by brush or spray. There are three types: nematic, smectic, and cholesteric. Chief uses are in medicine, color TV and electronic display tubes; temperature-sensitive tape; electronic drives for clocks and calculators; integrated circuit inspection and other devices dependent on temperature determination. They can align with dichroic dye molecules in a thin-layer cell to produce color changes in transmitted light.

liquid dioxide. See nitrogen dioxide.

liquid, Newtonian. Characteristic of liquids is their ability to flow, a property depending largely on their viscosity, and sometimes also on the rate of shear. A Newtonian liquid is one that flows immediately on application of force, and for which the rate of flow is directly proportional to the force applied. Water, gasoline and motor oils at high temperatures are examples. Some liquids have abnormal flow response when force is applied, that is, their viscosity is dependent on the rate of shear. Such liquids are said to exhibit non-Newtonian flow properties. Some will not flow until the force exerted is greater than a definite value called the yield point. [W. A. Gruse].

"Liquidow."[233] Trademark for dust-laying products based on calcium chloride.

liquid pitch oil. See creosote, coal tar.

liquid rosin. See tall oil.

liquid rubber. See rubber, liquid.

"Liquid Scintillator HF."[109] Trademark for a solution of 5g diphenyloxazole and 0.1g dimethyl-POPOP per liter in a high-flash-point solvent. Maximum fluorescence 4300A.

"Liquid Stainless Steel."[289] Trademark for a protective and decorative coating for steel and other exposed materials consisting of polished stainless steel flake pigment and vinyl, epoxy and other systems.

"Liqui-Moly."[289] Trademark for a series of extreme-condition lubricants compounded with submicronized molybdenum disulfide (MoS_2) in a wide range of petroleum and synthetic oils and greases, volatile hydrocarbons and various resin vehicles.

liquor. In chemical technology, any aqueous solution of one or more chemical compounds. In sugar manufacturing it refers to the sirups obtained from various refining steps (mother liquors). The paper industry uses this term extensively as follows: (a) black liquor is liquid digester waste (also called spent sulfate liquor) containing sulfonated lignin, rosin acids, and other waste wood components, from which tall oil is made; (b) green liquor is a solution made by dissolving chemicals recovered in the alkaline pulping process in water; (c) white liquor is made by adding caustic soda to sodium sulfide solution. In dyeing technology, red liquor is an alternate name for mordant rouge.

"Lissolamine" 100[325] Trademark for a water-soluble polymer. Stripping assistant for vat dyed fibers.

liter. The volume of one kilogram of water at its temperature of maximum density (4°C) at standard atmospheric pressure. A liter is 1.05 quarts, or 0.26 gallon.

"Lithaflux."[250] Trademark for lepidolite (q.v.). Used for batch material for glass, porcelain enamel ground coats; ceramic body flux and glaze constituents.

"Lithafrax."[280] Trademark for a ceramic material made from beta-spodumene.

litharge (lead oxide, yellow; plumbous oxide; lead monoxide) PbO. An oxide of lead made by controlled heating of metallic lead.

Properties: Yellow crystals; sp. gr. 9.53; m.p. 888°C. Insoluble in water; soluble in acids and alkalies. A strong base. Commercial grades are yellow to reddish, depending on treatment and purity. See also massicot.

Containers: Cans; drums.

Derivation: By oxidizing metallic lead in air. Various forms of lead and various temperatures from 500 to 1000°C are used.

Grades: C.P.; fused; powdered.

Hazard: Toxic by ingestion and inhalation. Tolerance, 0.15 mg per cubic meter of air. Safety data sheet available from Manufacturing Chemists Assn., Washington, D.C.

Uses: Storage batteries; ceramic cements and fluxes, pottery and glazes; glass; chromium pigments; oil refining; varnishes, paints, enamels, ink, linoleum; insecticides; cement (with glycerin); acid-resisting compositions; match-head compositions; other lead compounds; rubber accelerator (dry heat only).

litharge-glycerin cement. Made by mixing glycerin with one sixth to one half portion of water and mixing with enough litharge to give a paste of desired consistency. Must be used as soon as mixed. Fillers retard the setting and avoid cracking. The product is somewhat resistant to acids.

Hazard: Toxic by ingestion.

litharge, leaded. See lead suboxide.

"Lithcote."[145] Trademark for a series of protective coatings available as LC-19, LC-24, LC-25, LC-34, LC-73 baked phenolic or modified epoxy linings, or as LC-82, LC-610 catalyzed epoxy coatings.

Uses: Tank cars; storage tanks; pipe, etc., to prevent corrosion and product contamination.

lithia. See lithium oxide.

lithic acid. See uric acid.

lithium Li Metallic element of atomic number 3, group IA of Periodic Table; atomic weight 6.941; valence 1; two isotopes. It is the lightest and least reactive of the alkali metals and the lightest solid element.

Properties: Very soft silvery metal. Sp. gr. 0.534 (20°C); m.p. 179°C; b.p. 1317°C; Mohs hardness 0.6. Viscosity of liquid lithium less than water; heat capacity about the same as water. Reacts exothermally with nitrogen in moist air at room temperature. High electrical conductivity. Soluble in liquid ammonia. Generally low toxicity. Combustible.

Sources: Spodumene, lepidolite, ambylgonite occurring in U.S., Canada, central Africa, Brazil, Argentina, Australia, Europe. Also desert lake brines.

Derivation: By electrolysis of fused mixture of lithium chloride and potassium chloride; high-temperature extracton from spodumene by sodium carbonate; solar evaporation of lake brines.

Grades: 99.86% to 99.9999%. Available as ingots, rods, wire, ribbon, and pellets.

Containers: Glass bottles containing kerosine; hermetically sealed copper or aluminum cartridges; special steel drums.

Hazard: Ignites in air near melting point. Dangerous fire and explosion risk when exposed to water, nitrogen, acids or oxidizing agents. Extinguish fires with chemical only (See "Lith-X").

Uses: Production of tritium (q.v.); reducing and hydrogenating agents; alloy hardeners; pharmaceuticals; Grignard reagents. Scavenger and degasifier for stainless and mild steels in molten state; modular iron; lithium soaps; deoxidizer in copper and copper alloys; catalyst; heat-transfer liquid; storage batteries (with sulfur, selenium, tellurium and chlorine). Rocket propellants; vitamin A synthesis; silver solders; under-water buoyancy devices; nuclear reactor coolant.

Shipping regulations: (Rail) Yellow label. (Air) Flammable Solid label. Not acceptable on passenger planes unless in cartridges.

lithium aluminate $LiAlO_2$.

Properties: White powder; m.p. 1900–2000°C; sp. gr. (25°C) 2.55. Insoluble in water. Low toxicity.

Grade: Ceramic.

Containers: Fiber drums.

Use: As a flux in high-refractory porcelain enamels.

lithium aluminum deuteride (LAD) $LiAlD_4$.

Properties: White to gray crystals; sp. gr. 1.02. Stable in dry air at room temperature, but very sensitive to moisture. Decomposes above 140°C liberating deuterium. Soluble in diethyl ether, tetrahydrofuran. Slightly soluble in other low-molecular-weight ethers.

Preparation: By acting aluminum chloride with lithium deuteride.
Containers: Glass bottles; vials.
Hazard: Flammable; dangerous fire risk. Requires special handling.
Uses: Introduction of deuterium atom into molecule by reduction of same groups attacked by lithium aluminum hydride.
Shipping regulations: Not listed. Consult authorities.

lithium aluminum hydride (LAH) $LiAlH_4$.
Properties: White powder. Sp. gr. 0.917. Sometimes turns gray on standing. Stable in dry air at room temperature, but highly sensitive to moisture, including atmospheric. Decomposes to lithium hydride, aluminum metal and hydrogen above 125°C without melting. Soluble in diethyl ether, tetrahydrofuran, dimethyl "Cellosolve." Slightly soluble in dibutyl ether. Insoluble or very slightly soluble in hydrocarbons and dioxane.
Preparation: Reaction of aluminum chloride with lithium hydride.
Containers: Glass bottles; polyethylene bags placed in metal cans, up to 6-gallon capacity; steel drums; fiber cans. Obtain precautionary instructions before opening containers.
Hazard: Flammable, dangerous fire risk. May ignite spontaneously on grinding or rubbing, or from static sparks. Fires must be extinguished by smothering with powdered limestone.
Uses: Reducing agent for over 60 different functional groups, especially for pharmaceutical, perfume, and fine organic chemicals; converts esters, aldehydes, and ketones to alcohols, and nitriles to amines; source of hydrogen; propellant; catalyst in polymerizations.
Shipping regulations: (Rail) Yellow label. (Air) Flammable Solid label. Not acceptable on passenger planes.

lithium aluminum hydride, ethereal. $LiAlH_4$, plus ether.
Properties: Colorless solution in ether; very reactive to water.
Derivation: From lithium hydride and ether solution of aluminum chloride.
Hazard: Flammable; dangerous fire risk.
Use: See lithium aluminum hydride.
Shipping regulations: (Rail) Red label. (Air) Flammable Liquid label. Not acceptable on passenger planes.

lithium aluminum tri-tert-**butoxyhydride** (LATB; lithium tri-tert-butoxyaluminohydride) $LiAl[OC((CH_3)_3]_3H$.
Properties: White powder; sp. gr. 1.03. Stable in dry air but sensitive to moisture. Soluble in the dimethyl ether of diethylene glycol, tetrahydrofuran, diethyl ether; slightly soluble in other ethers.
Containers: Glass bottles.
Hazard: Evolves flammable hydrogen at 400°C. Dangerous.
Uses: For stereospecific reductions of steroid ketones and for reductoins of acid chlorides to aldehydes.
Shipping regulations: Not listed. Consult authorities.

lithium amide $LiNH_2$.
Properties: White crystalline solid; ammonia-like odor; sp. gr. 1.18; melts 380–400°C; decomposes in water, to form ammonia.
Derivation: Reaction of lithium hydride with ammonia.
Grades: 92–95% lithium amide.
Containers: 25-, 100-, 300-lb fiber drums.
Hazard: Flammable; dangerous fire risk. Do not use water to extinguish. (See "Lith-X").
Uses: Organic synthesis, including antihistamines and other pharmaceuticals.
Shipping regulations: (powdered) (Rail) Yellow label. (Air) Flammable Solid label.

lithium arsenate Li_3AsO_4.
Properties: White powder; sp. gr. 3.07 (15°C). Slightly soluble in water; soluble in dilute acetic acid.
Hazard: Highly toxic.
Shipping regulations: (Rail, Air) Arsenical compounds, n.o.s., Poison label.

lithium benzoate $LiC_7H_5O_2$.
Properties: White crystals or powder; soluble in water and alcohol. Low toxicity.
Derivation: Reaction of benzoic acid with lithium carbonate.

lithium bicarbonate $LiHCO_3$. A lithium salt formed by dissolving lithium carbonate in water with excess carbon dioxide. The solution, called lithia water, is used in medicine and prepared mineral waters.

lithium borate. See lithium metaborate, lithium tetraborate.

lithium borohydride $LiBH_4$.
Properties: White to gray crystalline powder; decomposes in vacuum above 200°C; in air at 275°C; soluble in lower primary amines and ethers; sp. gr. 0.66; extremely hygroscopic.
Derivation: Reaction of sodium borohydride and lithium chloride.
Containers: Glass bottles.
Hazard: Flammable; dangerous fire and explosion risk.
Uses: Source of hydrogen and other borohydrides; reducing agent for aldehydes, ketones and esters.
Shipping regulatons: (Rail) Flammable solid, n.o.s., Yellow label. (Air) Flammable Solid label. Not acceptable on passenger planes.

lithium bromide LiBr.
Properties: White cubic deliquescent crystals, or as a white to pinkish white granular powder; odorless; sharp, bitter taste. Sp. gr. 3.464; m.p. 547°C; b.p. 1265°C; very soluble in water, alcohol, and ether. Slightly soluble in pyridine; soluble in methanol, acetone, glycol. A hot concentrated solution dissolves cellulose. Forms addition compounds with ammonia and amines. Forms double salts with $CuBr_2$, $HgBr_2$, HgI_2, $Hg(CN)_2$, and $SrBr_2$. Greatly depresses vapor pressure over its solutions. Low toxicity.
Derivation: Reaction of hydrobromic acid with lithium carbonate.
Grades: 53% (min) LiBr brine; solid; single crystals.
Containers: Glass jars; bags; brine in steel drums.
Uses: Pharmaceuticals; air conditioning; humectant; drying agent; batteries; low temperature heat-exchange medium.

lithium butyl. See butyllithium.

lithium carbide Li_2C_2. Crystalline white powder; sp. gr. 1.65 (18°C); decomposes in water; soluble in acid, with evolution of acetylene.
Hazard: Fire risk in contact with acids.

lithium carbonate Li_2CO_3.
Properties: White powder; sp. gr. 2.111; m.p. 735°C; b.p., decomposes at 1200°C. Sparingly soluble in water. Low toxicity.
Derivation: (a) Finely ground ore is roasted with sulfuric acid at 250°C. Lithium sulfate is leached from the mass and converted to the carbonate by precipitation with soda ash. (b) Reaction of lithium oxide with carbon dioxide or ammonium carbonate solution.

Grades: Technical; C.P.
Containers: Fiber drums; barrels; bags.
Hazard: Water solution is strong irritant.
Uses: Ceramics and porcelain glazes; pharmaceuticals; catalyst; other lithium compounds; coating of arc-welding electrodes; nucleonics; luminescent paints, varnishes and dyes; glass ceramics; aluminum production.

lithium chlorate $LiClCO_3$.
Properties: Needlelike crystals, deliquescent; sp. gr. 1.119 (18°C); m.p. 128°C; decomposes at 270°C; more soluble in water than any other inorganic salt (313 g per 100 ml water at 18°C); very soluble in alcohol.
Hazard: Dangerous explosion hazard when shocked or combined with organic materials. Strong oxidant. (See "Lith-X").
Uses: Air conditioning; inorganic and organic chemicals; propellants.
Shipping regulatons: Chlorates, n.o.s. (Rail) Yellow label. (Air) Oxidizer label.

lithium chloride LiCl.
Properties: White deliquescent crystals; sp. gr. 2.068; m.p. 614°C; b.p. 1360°C; very soluble in water, alcohols, ether, pyridine, nitrobenzene. One of the most hygroscopic salts known. Low toxicity, but should not be used as dietary salt substitute.
Derivation: Reaction of lithium ores with chlorides; natural brines.
Grades: Technical, 99% (min) assay; 35–40% brine, inhibited; single crystals.
Containers: Crystals: 25-, 50-, 100-lb paper-lined drums; brine: 75-, 275-, and 500-lb steel drums.
Uses: Air conditioning; welding and soldering flux; dry batteries; heat-exchange media; salt baths; desiccant; humectant; production of lithium metal; soft drinks and mineral water to reduce escape of carbon dioxide.

lithium chromate $Li_2CrO_4 \cdot 2H_2O$.
Properties: Yellow, crystalline deliquescent powder; soluble in water and forms a eutectic at −60°C. Soluble in alcohols.
Uses: Corrosion inhibitor in alcohol-base antifreezes and water-cooled reactors. Oxidizing agent for organic material, especially in the presence of light.

lithium citrate $Li_3C_6H_5O_7 \cdot 4H_2O$.
Properties: White powder or granules; loses $4H_2O$ at 105°C; m.p., decomposes; soluble in water; slightly soluble in alcohol. Low toxicity.
Derivation: Reaction of citric acid with lithium carbonate.
Uses: Beverages; pharmaceuticals; dispersion stabilizer (clay deflocculant).

lithium cobaltite $LiCoO_2$. Dark blue powder; insoluble in water. The compound exhibits both the fluxing property of lithium oxide and the adherence-promoting property of cobalt oxide. Low toxicity.
Use: Ceramics.

lithium deuteride LiD.
Properties: Gray crystals; sp. gr. 0.906. Reacts slowly with moist air. Thermally stable to its melting point of 680°C.
Containers: Glass bottles.
Use: Source of deuterium.

lithium dichromate $Li_2Cr_2O_7 \cdot 2H_2O$.
Properties: Yellowish-red, crystalline powder; sp. gr. 2.34 (30°C); m.p. 130°C; deliquescent; soluble in water; forms eutectic at −70°C.
Uses: Dehumidifying; refrigeration.

lithium ferrosilicon.
Properties: Dark crystalline brittle metallic lumps or powder; evolves flammable gas in contact with moisture. Must be kept cool and dry.
Hazard: Flammable, dangerous fire risk.
Shipping regulations: (Rail) Yellow label. (Air) Flammable Solid label. Not acceptable on passenger planes.

lithium fluoride LiF.
Properties: Fine white powder; sp. gr. 2.635 (20°C); m.p. 842°C; b.p. 1670°C; slightly soluble in water; does not react with water at red heat; soluble in acids and hydrofluoric acid; insoluble in alcohol.
Derivatoin: Reaction of hydrofluoric acid with lithium carbonate.
Grades: Guaranteed 98% (min) LiF; C.P.; single pure crystals.
Hazard: Toxic. Strong irritant to eyes and skin. Tolerance: 2.5 mg per cubic meter of air (as F).
Uses: Welding and soldering flux; ceramics; heat-exchange media; synthetic crystals in infrared and ultraviolet instruments; rocket fuel component; radiation dosimetry. Component of fuel for molten salt reactors; x-ray diffraction.

lithium fluorophosphate $LiF \cdot Li_3PO_4 \cdot H_2O$.
Properties: White crystals.
Derivation: Interaction of lithium fluoride and lithium phosphate.
Use: Ceramics.

lithium gallate. A material having strong piezoelectric properties; used in electronic devices.

lithium grease. A grease using a lithium soap of the higher fatty acids as a base. Water-resistant; stable when heated above melting point and cooled again. Used in aircraft and low temperature service. See lithium stearate; lubricating grease. Lithium hydroxystearate from hydrogenated castor oil is also widely used.

lithium hydride LiH.
Properties: White, translucent, crystalline mass or powder. Commercial product is light bluish-gray due to minute amount of colloidally dispersed lithium. Sp. gr. (20°C) 0.82; m.p. 680°C; decomposition pressure nil at 25°C, 0.7 mm at 500°C, 760 mm at approx. 850°C. Decomposed by water, forming hydrogen and lithium hydroxide; insoluble in benzene and toluene; soluble in ether.
Derivation: Reaction of molten lithium with hydrogen.
Grades: 93–95%, based on hydrogen evolution.
Containers: Cans; cases; drums.
Hazard: Flammable, dangerous fire risk; ignites spontaneously in moist air. Use dry chemical to extinguish. Tolerance, 0.025 mg per cubic meter of air. (See "Lith-X").
Uses: Desiccant; source of hydrogen; condensing agent in organic synthesis; preparation of lithium amide and double hydrides; nuclear shielding material.
Shipping regulations: (Rail) Yellow label. (Air) Flammable Solid label. Not acceptable on passenger planes.

lithium hydroxide LiOH.
Properties: Colorless crystals; sp. gr. 1.46; m.p.

462°C; b.p. 924°C, decomposes; slightly soluble in alcohol and water.
Derivation: Causticizing of lithium carbonate; action of water on metallic lithium; or by addition of Li_2O to water.
Hazard: Water solutions are strong irritants.
Uses: Storage battery electrolyte; carbon dioxide absorbent in space vehicles; lubricating greases; ceramics.

lithium hydroxystearate $LiOOC(CH_2)_{10}CHOH(CH_2)_5CH_3$.
Properties: White powder; m.p. 205°C. Dissolves in hot petroleum oil to form greases. Combustible; low toxicity.
Derivation: From hydrogenated castor oil.
Use: Greasemaking.

lithium hypochlorite LiOCl.
Hazard: Strong oxidant; may ignite organic materials.
Uses: Bleach; sanitizing agent.
Shipping regulations: (dry, containing more than 39% available chlorine) (Rail) Yellow label. (Air) Oxidizer label.

lithium iodate $LiIO_3$.
Properties: White powder, sp. gr. 4.487 (25°C); m.p. 50–60°C (transition point from alpha to beta form); soluble in water; insoluble in alcohol. An oxidizing material.

lithium iodide trihydrate $LiI \cdot 3H_2O$.
Properties: White crystals; soluble in water and in alcohol. Sp. gr. 3.48 (25°C); m.p. 72°C, loses water; (anhydrous) 450°; b.p. 1171°C. Extremely hygroscopic.
Derivation: Action of hydriodic acid on lithium hydroxide, with subsequent crystallization.
Uses: Air conditioning; catalyst in acetal formation; solubilizes lithium in propylamine; nucleonics.

lithium lactate $LiC_3H_5O_3$.
Properties: White odorless powder; very soluble in water with practically neutral reaction. Lithium lactate is nonhygroscopic and stable, whereas sodium lactate can be prepared in solution only.
Use: Wherever a dry alkali lactate is required.

lithium manganite Li_2MnO_3.
Properties: Reddish-brown powder. Insoluble in water. Extremely stable.
Hazard: Moderately toxic.
Uses: Smelter addition in the manufacture of frit; as a mill addition; ceramic-bonded grinding wheels.

lithium metaborate dihydrate $LiBO_2 \cdot 2H_2O$.
Properties: White crystalline powder; soluble in water; m.p. 840°C (anhydrous).
Uses: Ceramics (flux in enamel cover coats; increases resistance to torsion); welding and brazing (anhydrous form); nucleonics.

lithium metasilicate (lithium silicate) Li_2SiO_3.
Properties: White powder; m.p. 1201°C; sp. gr. (25°C) 2.52; insoluble in water.
Uses: Flux in glazes and ceramic enamels; welding rod coating.

lithium methoxide (lithium methylate) $LiOCH_3$.
Properties: White, free-flowing powder; soluble in methanol. Must be protected from moisture.
Grade: Commercially available as a 10% solution in methanol.
Hazard: Solution highly toxic by ingestion.
Uses: Strong base and chemical intermediate where water is undesirable.

lithium methylate. See lithium methoxide.

lithium molybdate Li_2MoO_4. A white crystalline compound soluble in water. M.p. 705°C; sp. gr. 2.66.
Uses: Steel coating; petroleum cracking catalyst.

lithium niobate (lithium metaniobate) $LiNbO_3$. Single crystals available. M.p. 1250°C; sizes up to 6″ long and 1/2″ diam.
Uses: Infrared detectors; transducer in laser technology.

lithium nitrate $LiNO_3$.
Properties: Colorless powder; sp. gr. 2.38; m.p. 261°C; soluble in water and alcohol.
Derivatoin: Reaction of nitric acid with lithium carbonate.
Grades: Technical; commercially pure; reagent.
Hazard: Dangerous explosion risk when shocked or heated. Strong oxidizing agent.
Uses: Ceramics; pyrotechnics; salt baths; heat-exchange media; refrigeration systems; rocket propellant.
Shipping regulations: Nitrates, n.o.s., (Rail) Yellow label. (Air) Oxidizer label.

lithium nitride Li_3N.
Properties: Brownish-red crystals of hexagonal structure, or fine, free-flowing red powder. Water vapor in moist air causes slow decomposition. Reacts with water giving lithium hydroxide and ammonia. Insoluble in most organic solvents, but sparingly soluble in polyethers. Density approx. 1.3g/cc (25°C); m.p. 845°C; decomposition pressure not measurable below 1250°C.
Hazard: Ignites at high temperatures. Use dry chemicals to extinguish (see "Lith-X").
Uses: Nitriding agent in metallurgy; reducing and nucleophilic reagent in organic reactions; source of bound nitrogen in inorganic reactions.
Shipping regulatoins: (Rail) Flammable solid, n.o.s., Yellow label. (Air) Flammable Solid label. Not acceptable on passenger planes.

lithium orthophosphate. See lithium phosphate.

lithium oxide (lithia) Li_2O. White powder; m.p. 1427°C; density 2.023 g/cc (25°C). A strong alkali.
Hazard: Water solutions of high concentratoin are strong irritants.
Uses: Ceramics and special glass formulations; carbon dioxide absorbent (mineral water).

lithium perchlorate $LiClO_4$.
Properties: Colorless deliquescent crystals; sp. gr. 2.429; m.p. 236°C; decomposes at 430°C; has more available oxygen than liquid oxygen (on a volume basis). Reacts with $4NH_3$ to form an ammoniate. Forms $LiClO_4 \cdot 3H_2O$, colorless crystals with sp. gr. 1.84; m.p. 75°C. Soluble in water and alcohol.
Hazard: Oxidizing agent; dangerous fire and explosion risk in contact with organic materials. Irritating to skin and mucous membranes.
Use: Solid rocket propellants.
Shipping regulatoins: Perchlorates, n.o.s., (Rail) Yellow label. (Air) Oxidizer label.

lithium peroxide Li_2O_2.
Properties: Fine white powder; m.p., decomposes; density 2.14 g/cc (20°C). In closed container no detectable loss of available oxygen. Soluble in water, 8% (20°C); anhydrous acetic acid, 5.6% (20°C); insoluble in absolute alcohol (20°C).
Derivatoin: Addition of hydrogen peroxide to lithium hydroxide.
Hazard: Dangerous fire and explosion risk in contact with organic materials. Strong oxidizing agent.
Uses: As supplier of active oxygen; commercial samples have 32.5% to 34% available oxygen content.

Shipping regulations: (Rail) Yellow label. (Air) Oxidizer label.

lithium prosphate (lithium orthophosphate) $2Li_3PO_4 \cdot H_2O$.
Properties: White, crystalline powder; soluble in acids; slightly soluble in water; sp. gr. 2.41.

lithium ricinoleate $LiOOCC_{17}H_{32}OH$.
Properties: White powder. M.p. 174°C. Insoluble or limited solubility in most organic solvents. Combustible.
Derivation: Castor oil.
Uses: Alcoholysis and ester interchange catalyst.

lithium silicate. See lithium metasilicate.

lithium silicon. An alloy of lithium and silicon.
Properties: Black, shiny lumps or powder with sharp irritating odor. Keep cool and dry.
Hazard: Dangerous fire risk; reacts with water to form flammable gases.
Shipping regulations: (Rail) Yellow label. (Air) Flammable Solid label. Not acceptable on passenger planes.

lithium stearate $LiC_{18}H_{35}O_2$.
Properties: White crystals; sp. gr. 1.025; m.p. 220°C; insoluble in cold and hot water, alcohol, and ethyl acetate; forms gels with mineral oils.
Derivation: Reaction of stearic acid with lithium carbonate.
Grades: Grease; cosmetic.
Uses: Cosmetics; plastics; waxes; greases; lubricant in powder metallurgy; corrosion inhibitor in petroleum; flatting agent in varnishes and lacquers; high-temperature lubricant.

lithium sulfate $Li_2SO_4 \cdot H_2O$.
Properties: Colorless crystals; sp. gr. 2.06; m.p. 130°C; soluble in water, insoluble in 80% alcohol. Does not form alums.
Derivation: Reaction of sulfuric acid with lithium carbonate or with spodumene ore.
Grades: Technical and pharmaceutical.
Uses: Pharmaceutical products; ceramics.

lithium tetraborate $Li_2B_4O_7 \cdot 5H_2O$.
Properties: White crystalline powder; m.p., loses water 200°C; very soluble in water, insoluble in alcohol.
Derivation: Reaction of boric acid with lithium carbonate.
Uses: Ceramics; vacuum spectroscopy; metal refining and degassing.

lithium titanate Li_2TiO_3. White powder. Insoluble in water. Has strong fluxing properties in small percentages in titanium-bearing enamels. The insolubility permits its use in vitreous and semi-vitreous glazes.

lithium tri-tert-butoxyaluminohydride. See lithium aluminum tri-tert-butoxyhydride.

lithium tungstate Li_2WO_4. White crystals; sp. gr. 3.71; soluble in water.

lithium vanadate (lithium metavanadate) $LiVO_3 \cdot 2H_2O$.
Properties: Yellowish powder; soluble in water.

lithium zirconate Li_2ZrO_3. White powder; insoluble in water. Efficient flux in glasses containing zirconium dioxide; recommended as a flux in zirconium-opacified enamels, glazes and porcelains.

lithium-zirconium silicate $Li_2O \cdot ZrO_2 \cdot SiO_2$. White powder. A strong flux in enamels, glazes and porcelains. It can be used in place of lithium zirconate.

lithocholic acid $C_{24}H_{39}O_2OH$. A bile acid.
Properties: Crystallizes in leaflets from alcohol; m.p. 184–186°C. Not precipitated by digitonin. Freely soluble in hot alcohol; soluble in ethyl acetate; slightly soluble in glacial acetic acid; insoluble in water, ligroin.
Derivatoin: From bile and gallstones; from deoxycholic acid or cholic acid.
Use: Biochemical research.

lithol red. One of a group of pigments made by combining Tobias acid and beta-naphthol.
Available as sodium, barium, and calcium lithols. Poor resistance to sunlight and weathering; generally good resistance to bleeding and to chemicals.
Uses: Industrial enamels; toys and dipping enamels; rubber, plastics, etc.

lithol rubine $OOCC_{10}H_5(OH)N{:}NC_6H_3(CH_3)SO_2OCa$.
The calcium salt of an azo pigment, made by diazotizing para-toluidine-meta-sulfonic acid and coupling with 3-hydroxy-2-naphtholic acid. Has poor hiding power; good resistance to bleeding and baking; fair lightfastness and alkali resistance.
Uses: Paints, plastics, printing inks, cosmetics.

lithopone. A white pigment consisting of 28% zinc sulfide and 72% barium sulfate, formerly used widely in paints, white rubber goods, paper, white leather, etc. It has been largely replaced by titanium dioxide pigments.

"Lithosol."[28] Trademark for a group of pigments, specialty dyes, acids and bases. The acids and bases are specially prepared for the manufacture of lakes and organic pigments.

"Lith-X."[548] Trademark for a graphite-base dry chemical extinguishing agent suitable for fires of lithium and of metals such as titanium and zirconium.

litmus (lichen blue).
Properties: A blue, amorphous powder (frequently compressed into small cakes or stocks). Soluble in water; changes color with acidity of solution; red at pH 4.5, blue at pH 8.3.
Derivation: By treating lichens (particularly variolaria lecanora and V. rocella) with ammonia and potash and then fermenting the mass.
Use: Indicator in analytical themistry where precision is not required; soil testing.

Little, Arthur D. (1863–1935). Born in Boston, Little was a pioneer in the field of industrial research and chemical consulting. Originally an authority on paper technology, he established a consulting industrial chemical laboratory in 1886, which has since become a large institution of world-wide reputation, located in Cambridge, Mass. It has served as a prototype of many industrially oriented consulting firms that have become a significant factor in the growth of research in the last half century. It has made significant contributions in such fields as flavors, food chemistry and acceptability, paper chemistry and rubber chemistry, as well as in corporate management.

liver. A term formerly used to characterize compounds containing sulfides or sulfates, possibly because of the dark brown color, e.g., liver of sulfur, liver of lime, etc. The word is outmoded and should be avoided.

livering. A sharp or pronounced increase in the viscosity of a paint or printing ink, as a result of which it becomes a coagulated and unusable mass. This may be due to oxidation or to partial polymerization of the components.

lixiviation. See leaching.

"LNA."[65] Trademark for leucyl aminopeptidase substrate used as diagnostic agent in medicine.

LNG. Abbreviation for liquefied natural gas. See natural gas.

loam. A soil comprised of 25–50% sand, 30–50% silt, and 5–25% clay. See also soil.

"Lo-Bax."[84] Trademark for a chlorine sterilizer designed especially for handlers of milk and other dairy and food products who require clear, fast-killing bactericidal solutions. It is a stable, quick-dissolving powder containing 50% available chlorine.

"Loctite"[477] Trademark for anaerobic polymers which retain liquid while exposed to air and which automatically harden without heat or catalysts when confined between closely fitted metal parts. Applied to mating surfaces before, during, and after assembly. Available in range of strengths and viscosities for a wide variety of appliations.

locust-bean gum. See carob-seed gum.

"Lodex."[245] Trademark for fine particle permanent magnet material and permanent magnets.

logwood (hematoxylin) $C_{10}H_{14}O_6 \cdot 3H_2O$.
See hematein; dye, natural.

"Loloss."[553] Trademark for guar gum drilling fluid additive.

"Lomar."[309] Trademark for condensed naphthalene sulfonate. Used as dipersing agents in paper-making, cement and concrete, gypsum wallboard, emulsion polymerization, dyeing, etc.

"Lomax."[416] Trademark for the catalyst used in the "Isomax" process (q.v.).

London purple, solid. An insecticide containing arsenic trioxide, aniline, lime, and ferrous oxide. Insoluble in water.
Hazard: Highly toxic by ingestion and inhalation.
Shipping regulations: (Rail, Air) Poison label.

loretin (ferron). See 7-iodo-8-hydroxyquinoline-5-sulfonic acid.

"Lorfan" Tartrate.[190] Trademark for a brand of levallorphan tartrate (q.v.).

"Lorite."[148] Trademark for a natural extender composed of predominantly acicular particles containing approximately 80% calcium carbonate and 20% diatomaceous earth. Has a flatting and bodying effect in paints and is recommended for sealers, flats, semi-glosses, undercoaters, and emulsion paints.

"Lorol."[28] Trademark for a group of straight-chain (normal) even-numbered-carbon alcohols ranging from C-8 (octyl) to C-18 (stearyl), and including mixtures of these.
Properties: Colorless liquids or white flakes.
Grades: Technical; N.F. ("Lorol" 24, cetyl alcohol); U.S.P. ("Lorol" 28, stearyl alcohol).
Containers: Various up to tank trucks and tank cars.
Uses: Intermediates in manufacture of detergents, lubricating oil additives quaternary ammonium compounds, mercaptans, textile finishing agents, plasticizers, and perfumes; antifoam agents; emulsifying and softening agents in cosmetics, salves, and ointments.

"Lorothidol."[162] Trademark for bithionol (q.v.).

"Lorox."[28]. Trademark for either a powder or solution containing linuron (q.v.).

lost wax process. A ceramic or metal casting method used to make reproductions of sculptured pieces, and adapted to industry for manufacture of precision metal parts. It is generally known as precision investment casting. The sequence of operations is (1) a wax prototype is made in a metal mold; (2) several of these are attached to a central member, also made of wax to form a "tree"; (3) the tree is dipped and dried eight times in a ceramic glaze, thus building up a coating or "investment"; (4) the assembly is baked in an oven, thus melting out the wax, leaving a cavity which is then used as a mold for liquid metal. Great accuracy can be obtained with this method. The exact nature of the waxes and coatings are not disclosed.

"Lotol."[248] Trademark for a series of compounded latices, based on all synthetic and natural types.

"Loupole."[309] Trademark for a detergent used for boiling-off rayon piece goods at low and elevated temperatures; particularly recommended for graphite streak removal; has good compatibility with soap and other detergents.

"Lovibond."[581] Trademark for a color analysis technique for petroleum products used chiefly in England, but never adopted by ASTM.

low-alloy steel. See alloy steel.

low explosive. See explosive, low.

low-melting alloy. See alloy, fusible.

low-pressure resin. See contact resin.

low-soda alumina. Aluminum oxide (Al_2O_3) with less than 0.15% sodium oxide (Na_2O) content.
Use: High-grade electric insulators and other ceramic bodies.

LOX. An abbreviation for liquid oxygen, especially when used as a rocket fuel.

"Loxite."[277] Trademark for rubber cements, dispersions, and latices, consisting of natural and synthetic rubbers and reclaimed rubbers in latex form, dispersed in water or dissolved in solvents.
Containers: 55-gal drums.
Uses: Impregnating and spreading paper and textiles; adhesives for rubber to metal, rubber to rubber, etc.

LPG. Abbreviation for liquefied petroleum gas (q.v.).

Lr Symbol for element lawrencium.

LSD. Abbreviation for lysergic acid diethylamide (q.v.).

"L-70 Series Surfactants."[214] Trademark for a class of organo-siliones.
Uses: Wetting agents, emulsifiers and coemulsifiers, penetrants, foamers and foam stabilizers, defoamers, lubricants, leveling agents, antistats, and textile softeners.

"LSW Dyes"[256] Trademark for a series of disperse dyes for polyester fibers.

LTH. Abbreviation for luteotropic hormone. See luteotropin.

Lu Symbol for lutetium.

lube oil additive. A chemical added in small amounts to lubricating oils to impart special qualities, such as low pour point when chlorinated hydrocarbons are added. Other special properties are:

low viscosity index	butene polymers
detergent and suspensoid properties	metallic stearate soaps
oxidation stability	calcium stearate
reduced foaming tendency	silicone compounds
resistance to high operating temperatures	phosphorus pentasulfide, zinc dithiophosphate

Note: The effect of automotive antipollution devices, which increase the operating temperatures of engines, upon lubricating oils and their additives is being thoroughly researched by lube oil manufacturers. Results may increase the use of some types and reduce or eliminate others.

lubricant, solid. A material having a characteristic crystal habit which causes it to shear into thin, flat plates, which readily slide over one another and thus produce an antifriction or lubricating effect. See mica, graphite, molybdenum disulfide, talc, boron nitride.

lubricant, synthetic. Any of a number of organic fluids having specialized and effective properties that are required in cases where petroleum-derived lubricants are inadequate. Each type has at least one property not found in conventional lubricants. Though their cost is much higher, they can be used over a wide range of temperatures and are stable to heat and oxidation. The major types are polyglycols (hydraulic and brake fluids), phosphate esters (fire-resistant), dibasic acid esters (aircraft turbine engines), chlorofluorocarbons (aerospace), silicone oils and greases (electric motors, antifriction bearings), silicate esters (heat-transfer agents and hydraulic fluids), neopentyl polyol esters (turbine engines), and polyphenyl ethers (excellent heat and oxidation resistance, but poor performance at low temperatures). An unusual property of synthetic lubricants is their exceptional resistance to ionizing radiation.

lubricating grease. A mixture of a mineral oil or oils with one or more soaps. The most common soaps are those of sodium, calcium, barium, aluminum, lead, lithium, potassium and zinc. Oils thickened with residuum, petrolatum or wax may be called greases. Some form of graphite may be added. Greases range in consistency from thin liquids to solid blocks and in color from transparent to black. The specifications for a grease are determined by the speed, load, temperature, environment, and metals in the desired application. Texture of grease may be smooth, buttery, ropy or stringy, fibrous, spongy, or rubbery. The texture does not necessarily indicate the viscosity of the grease, but is related to the formulation and methods of manufacture.

See also lubricating oil.

lubricating oil (lube oil). A selected fraction of refined mineral oil used for lubrication of moving surfaces, usually metallic, and ranging from small precision machinery (watches) to the heaviest equipment. Lubricating oils usually have small amounts of additives to impart special properties such as viscosity index and detergency. They range in consistency from thin liquids to grease-like substances. In contrast to lubricating greases, lube oils do not contain solid or fibrous materials. See also porpoise oil; transformer oil; lubricant, synthetic; extreme-pressure additive (2); lube oil additive.

lubrication. The introduction of a substance of low viscosity between two adjacent solid surfaces, one of which is in motion (bearing). From an engineering point of view, the chemical nature of the substance is not of critical importance. Thus materials as diverse as air, water, and molasses could theoretically be used as lubricants under appropriate conditions. Air and water have been used, as well as some solids such as graphite; but in general oils, fats, and waxes are utilized. The ability of a substance to act as a lubricant is sometimes called lubricity. See also lubricant, solid.

"Lubricin" N-1.[202] Trademark for a low viscosity ricinoleate derivative that increases the oiliness and wetting power of mineral oils; decreases corrosion, and exhibits detergent action on tar, varnish, and carbon deposits.

Used as an additive for lubricating oils, motor fuels, and cutting oils.

"Lubricin" V-1.[202] Trademark for a lubricant which facilitates the processing of rigid vinyl plastics.

lubricity. See lubrication.

"Lubriseal."[16] Trademark for a lubricating, sealing, and rust-inhibiting stopcock grease or spray.

"Lubrol" 90.[325] Trademark for an oleic acid-ethylene oxide condensate. Emulsifier for mineral oil, kerosine, fatty acid esters.

"Lucalox."[245] Trademark for a transparent, polycrystalline alumina; used in special lamp bulbs to produce a golden white light considered superior for industrial, commercial, outdoor and consumer lighting applications.

"Lucidol."[154] Trademark for benzoyl peroxide (q.v.).

luciferase. An enzyme occurring only in fireflies. See luciferin.

luciferin. An albumin present in some animals (notably fireflies) which, under the influence of the enzyme luciferase, exhibits bioluminescence. When luciferin (along with luciferase) comes in contact with adenosine triphosphate, present in all living cells, a chemical reaction occurs which causes luminescence. By correlating its intensity with the amount of adenosine triphosphate present, a means of measuring bacterial growth is obtained. Applications in biomedical technology are envisaged for this so-called "luminescence biometer."

"Lucite."[28] Trademark for acrylic resins consisting of a series of polymeric esters of methacrylic acid, $CH_2C(CH_3)COOR$, in which R is methyl, ethyl, n-butyl, isobutyl, or combinations of these alkyl groups. These are water-white, transparent, thermoplastic resins in granular or solution forms; used in lacquers, coatings, adhesives, modifiers for other resins. "Lucite" is also a trademark for an acrylic monomer, methyl methacrylate; for acrylic resins, in the form of injection molding and extrusion powders; for acrylic resin dental materials; and for acrylic syrup, a liquid methyl methacrylate for use in translucent and decorative panels. See acrylic resin.

"Ludox."[28] Trademark for series of aqueous colloidal silica sols.
Uses: Floor polishes; paints; adhesives; paper coatings; catalyst supports; latex products; textile cleaning treatments; photosensitized paper, binder for inorganic fibrous materials; reactant for synthetic silicates.

"Lugatol."[440] Trademark for anionic dyes having good tinctorial and fastness properties for all kinds of leather.

Lugol's solution. An aqueous solution containing 5 g iodine and 10 g potassium iodide per 100 ml water.
Grade: U.S.P.
Uses: Medicine and pharmaceuticals.

"Lumatex."[440] Trademark for a series of mineral and organic pigment dyestuffs which can be fixed on all types of textile fibers with suitable binders.

"Lumigraphic" Pigments.[266] Trademark for fluorescent dyestuffs on a resin base. Used in poster paints, inks, plastics, etc., where extreme brightness and daylight fluorescing properties are desirable.

"Luminal."[162] Trademark for phenobarbital (q.v.).
See also barbiturate.

luminescence. The emission of visible or invisible radiation unaccompanied by high temperature by any substance as a result of absorption of exciting energy in the form of photons, charged particles, or chemical change. It is a general term which includes both fluorescence and phosphorescence. Special types are chemiluminescence, bioluminescence, electroluminescence, photoluminescence, and triboluminescence. Common examples are light from the firefly, fluorescent lamp tubes and television screens. See also fluorescence; phosphorescence; phosphor.

lupanine $C_{15}H_{24}N_2O$. An alkaloid. Both dextro and levo forms are known. The latter is a colorless oil. The material has also been described as a yellow syrupy liquid with green fluorescence.
Properties: (*d*-form) White needles; m.p. 40°C; distils undecomposed at 220°C (10 mm); slightly soluble in water, with an alkaline reacton; soluble in alcohol, ether, and chloroform; refractive index (n 24/D) 1.5444.
Derivation: From the seeds of Lupinus albus and Lupinus angustifolius.
Use: Medicine.

"Luperco."[154] Trademark for a series of organic peroxides including aromatic diacyl, ketone, and alkyl peroxides.
Hazard: See peroxide (1).
Uses: Crosslinking polyethylene; curing synthetic and natural rubber; vinyl polymerization initiator; catalyst in polyester and vinyl monomers; mild organic oxidizing agents.

"Luperox."[154] Trademark for a group of organic peroxides.
2,5-2,5. 2,5-Dimethylhexane-2,5-dihydroperoxide.
6. Technical bis(1-hydroxycyclohexyl) peroxide.
118. 2,5-Dimethylhexane-2,5-diperoxybenzoate.
Hazard: See peroxide (1).
Uses: Catalyst for polymerization of polyesters and vinyls; elastomer vulcanization.

"Lupersol."[154] Trademark for a group of organic compounds including aliphatic diacyl peroxides, ketone peroxides, alkyl peroxyesters, and alkyl peroxides.
Hazard: See peroxide (1).
Uses: Polyethylene crosslinking, elastomer vulcanization, vinyl and polyester polymerization; replacement for MEK peroxide; pharmaceutical and fumigant applications.

2,6-lupetidine. See 2,6-dimethylpiperidine.

"Lupolen."[440] Trademark for series of high- and low-density polyethylenes.

"Luposec."[309] Trademark for a stable cationic emulsion of waxes with aluminum salts used in one-bath method for showerproofing or water-repelling all textile fabrics.

"Luposol."[309] Trademark for a prepared solution of locust bean gum used for finishing or as a printing paste.

"Luran S" (ASA polymer).[440] Trademark for acrylic ester-modified styrene-acrylonitrile terpolymer. A thermoplastic material claimed to be excellent for outdoor use. Highly resistant to sunlight and weathering.

"Lurex."[565] Trademark for metallic yarns produced in two forms: (1) lamination of clear plastic films and metallic foil or metallized film; (2) plastic film metallized and coated on both sides with a protective plastic lacquer.

Lurgi process. A coal gasification method in which synthesis gas is obtained from coal by reaction with steam and oxygen. The gas mixture is then catalytically methanated in a fixed-bed reactor at high temperature. A methane content of 90 to 95% and a Btu value of about 1000 is possible. The Lurgi process can also be adapted to productoin of methyl alcohol. Some 15 plants using this process are in production throughout the world. See also gasification.

"LuSoil."[99] Trademark for a mineral soil conditioner derived from attapulgite.

"Lustan."[51] Trademark for a pair of glass mold release agents; one grade is an emulsion-type product, the other is graphite-in-oil type.

luster. The appearance of the surface of a substance in reflected light. The term is used particularly in describing minerals. Types of luster are: (a) metallic, like metals, or the mineral pyrite; (b) vitreous, like glass or quartz; (c) adamantine, exceedingly brilliant, like diamond; (d) resinous, like resin, or sphalerite; (e) dull, not bright or shiny; like chalk.
See also delustrant.

"Lustran."[58] Trademark for ABS and SAN molding and extruding compounds.

"Lustrasol."[36] Trademark for acrylic and acrylic-modified alkyd solutions.

"Lustre-Phos."[58] Trademark for dentifrice grade of calcium orthophosphate.

"Lustrex."[58] Trademark for styrene molding compound, and extruding and calendering material.

"Lustrone"[23] Trademark for plasticized nitrocellulose lacquers tinted with dyes.
Use: Transparent color effects on leather.

lutein. A yellow pigment isolated from the corpus luteum, and found in body fats and egg yolks. It is a carotenoid and is similar to, or identical with xanthophyll (q.v.).

luteotropin (adenohypophyseal luteotropin; prolactin; lactogenic hormone; LTH). One of the hormones

secreted by the anterior lobe of the pituitary gland. It aids in causing growth of the mammary gland and initiates milk secretion by the mammary gland; it also influences the activity of the corpus luteum, including the secretion of progesterone.
Properties: Crystalline protein; molecular weight about 33,300; almost insoluble in water; soluble in dilute acids and acidified methanol and ethanol.
Use: Medicine.

lutetia. See lutetium oxide. See also rare earths.

lutetium Lu Metallic element. Atomic number 71; group IIIB of the periodic table; a lanthanide rare-earth. Atomic weight 174.97; valence 3. It has one natural radioactive isotope (Lu-176) with half-life of 2.2×10^{10} years. See rare-earth metals.
Properties: Metallic luster. Soft and ductile. Sp. gr. 0.849; m.p. 1652°C; b.p. 3327°C; reacts slowly with water; soluble in dilute acids. Difficult to isolate. Generally low toxicity.
Occurrence: Monazite (about 0.003%) (India, Brazil, Africa, Australia, U.S.).
Derivation: Reduction of the fluoride or chloride with calcium.
Grades: Regular, high purity (ingots, lumps).
Use: Nuclear technology.

lutetium chloride $LuCl_3 \cdot xH_2O$. (Anhydrous) sp. gr. 3.98; m.p. 905°C. Water-soluble.
Purity: Up to 99.9% lutetium salts.
Containers: Glass bottles; fiber drums.

lutetium fluoride $LuF_3 \cdot 2H_2O$. (Anhydrous) m.p. 1182°C; b.p. 2200°C. Insoluble in water.
Purity: Up to 99.9% lutetium salts.
Containers: Glass bottles; fiber drums.
Hazard: Toxic and irritant. Tolerance (as F): 2.5 mg per cubic meter of air.

lutetium nitrate $Lu(NO_3)_3 \cdot 6H_2O$.
Purity: Up to 99.9% lutetium salts.
Containers: Glass bottles; fiber drums.
Hazard: Fire and explosion risk.
Shipping reguations: Nitrates, n.o.s., (Rail) Yellow label. (Air) Oxidizer label.

lutetium oxide (lutetia) Lu_2O_3.
Properties: White solid; slightly hygroscopic; m.p. about 2500°C; sp. gr. 9.42; absorbs water and carbon dioxide.
Derivation: Monazite sand.
Grades: Up to 99.9% purity.
Containers: Glass bottles; fiber drums.

lutetium sulfate $Lu_2(SO_4)_3 \cdot 8H_2O$. Soluble in water.
Purity: Up to 99.9% lutetium salts.
Containers: Glass bottles; fiber drums.

2,6-lutidine (2,6-dimethylpyridine)
$NC(CH_3)CHCHCHCCH_3$.
Properties: Colorless oily liquid; peppermint odor; sp. gr. 0.932; b.p. 143°C; f.p. −6.6°C; refractive index 1.4973 (n 20/D). Derived from coal tar.
Hazard: May be toxic and flammable.
Uses: Pharmaceuticals; resins; dyestuffs; rubber accelerators; insecticides.
Also available as 2,3-, 2,4-, 2,5-, 3,4-, 3,5- isomers.

"Lutonal."[440] Trademark for vinyl ether polymers, solid and in solution.

"Luxapole" W-1054-B.[309] Trademark for a specially prepared sulfonated oil finish; used in textile industry.

"Luxol."[28] Trademark for a series of spirit and lacquer-soluble dyes. Used in lacquers, wood stains, stamp inks, ball point pen inks.

"Luxolene."[555] Trademark for glycerol and sulfated glycerol trioleate.

"LX-685."[21] Trademark for a heat-reactive resin used in the manufacture of ready-mixed aluminum paints, grease- and gasoline-resistant coatings, floor and deck enamels and concrete curing compounds.

"Lyamine."[123] Trademark for lysine for use as an animal feed supplement.

"Lycedan."[91] Trademark for a brand of adenosine-5-phosphoric acid for medicinal use.

lycopene $C_{40}H_{56}$. A long-chain hydrocarbon with thirteen double bonds, of which eleven are conjugated. Related to carotene (q.v.). Combustible; nontoxic.
Properties: Red crystals; m.p. 175°C; soluble in chloroform and carbon disulfide; slightly soluble in alcohol; insoluble in water.
Source: The main pigment of tomato, paprika, rose hips, etc.

"Lycra."[28] Trademark for a spandex fiber (q.v.) in the form of continuous monofilaments. Available in yarn with denier from 40 to 2240.
Properties: Sp. gr. 1.21 tensile strength 0.8 gpd; break elongation 550%; moisture regain 1.3%; m.p. 230°C; tensile recovery 95% from 50% elongation; soluble in dimethyl formamide (hot). Combustible, but self-extinguishing.
Uses: Foundation garments, swim wear, surgical hose, and other elastic products.

lye. See sodium hydroxide and potassium hydroxide.

"Lykopon."[23] Trademark for sodium hydrosulfite (q.v.).

lyocratic. See lyophilic.

"Lyofix."[443] Trademark for thermosetting resins and aftertreating agents used in textile processing.

lyophilic. Characterizing a material which readily goes into colloidal suspension in a liquid; if into water, it is called hydrophilic. The colloid is stabilized by the formaton of an adsorbed layer of molecules of the dispersing medium about the suspended particles. Systems of this type are said to be lyocratic. Examples: glue, gelatin; milk fat particles.

lyophilizaton. See freeze-drying.

lyophobic. Characterizing a material which exists in the colloidal state but with a tendency to repel liquids; if the liquid is water, the material is called hydrophobic. Such colloids are generally stabilized by the adsorption of ions, and coagulate when the charge is neutralized. Examples: colloidal gold, colloidal arsenic sulfide.

Lys Symbol for lysine.

lysergic acid $C_{16}H_{16}N_2O_2$. Structure based on a condensed 4-ring nucleus.
Properties: Crystallized (with 1-2 molecules of water) in plates from water. M.p. 240°C (dec). It is amphoteric; moderately soluble in pyridine; slightly soluble in water and neutral organic solvents; soluble in alkaline and acid solutions.
Derivation: Alkaline hydrolysis of ergot alkaloids; organic synthesis.
Uses: Medical research; synthesis of ergonovine.

D-lysergic acid diethylamide (LSD) $C_{20}H_{25}N_3O$. Synthetic derivative of lysergic acid. Only the D-form is active. Tasteless, colorless, odorless.
Hazard: A strong hallucinatory and habit-forming drug, with possible mutagenic effects. Manufacture and sale are under legal restraint.

lysidine (methylglyoxalidine) $CH_3\overline{CNCH_2CH_2N}H$.
Properties: Colorless, hygroscopic crystals; mousy odor; m.p. 105°C; b.p. 195°C. Soluble in water, alcohol, and ether.
Derivatin: From ethylenediamine hydrochloride and sodium acetate by dry distillation, decomposing the hydrochloride of the new base with concentrated potassium hydroxide and crystallizing.
Method of purification: Recrystallization.
Grades: Technical; 50% solution.
Containers: Glass bottles.
Use: Medicine.

lysine (alpha, epsilon-diaminocaproic acid; Lys.) $NH_2(CH_2)_4CH(NH_2)COOH$. An essential amino acid.
Properties: Colorless crystals; soluble in water; slightly soluble in alcohol; insoluble in ether; optically active. D(−)-lysine, m.p. 224°C with decomposition. L(+)-lysine, m.p. 224°C with decomposition. Nontoxic.
Derivation: Extraction of natural proteins; synthetically by fermentatoin of glucose or other carbohydrates, and by synthesis from caprolactam.
Uses: Biochemical and nutritional research; pharmaceuticals; culture media; fortification of foods and feeds (wheat flour); nutrient and dietary supplement; animal feed additive. Commercially available as DL- and L-lysine monohydrochloride.

lysozyme. An enzyme found in egg white. By hydrolyzing certain sugar linkages in glycoproteins, it can dissolve the mucopolysaccharides found in the walls of certain bacteria, and hence acts as a mild antiseptic. Its three-dimensional structure has been determined by x-ray crystallography. It is a molecule containing about 2200 atoms composing 129 amino acid units strung together and intricately folded (molecular weight 14,500).

M

m Abbreviation for meter.

M Abbreviation for molar, used to characterize the concentration of a solution. A molar solution contains one mole of a substance in one liter of solution.

m- Abbreviation for meta- (q.v.).

μm Abbreviation for micrometer (q.v.).

mμ Abbreviation for millimicron (q.v.).

MAC. Abbreviation for methyl allyl chloride (q.v.).

"Macaloid."[236] Trademark for a creamy white beneficiated clay (a magnesium montmorillonite) used as a thickening agent, adsorbent, flocculating agent, and stabilizer for emulsions and suspensions in aqueous systems. Used primarily in paints, cosmetics, and pharmaceuticals.

"Mace" (also called "Chemical Mace"). Trademark for a riot control gas dispersed as an aerosol. See chloroacetophenone.

M acid. See 1-amino-5-naphthol-7-sulfonic acid.

macromolecule. A molecule, usually organic, comprised of an aggregation of hundreds or thousands of atoms. Such giant molecules are in general of two types. (1) Individual entities (compounds) that cannot be subdivided without losing their chemical identity; typical of these are proteins, many of which have molecular weights running into the millions. (2) Combinations of repeating chemical units (monomers) linked together into chain or network structures called polymers; each monomer has the same chemical constitution as the polymer, e.g., isoprene (C_5H_8) and polyisoprene ($(C_5H_8)_x$). Synthetic elastomers (plastics) are typical of this kind of macromolecule; cellulose is the most common example found in nature. Most macromolecules are in the colloidal size range. See also polymer, high; protein; colloid chemistry.

macrose. See dextran.

madder. A natural dyestuff. See alizarin; lake; dye, natural.

Maddrell's salts (IMP). Insoluble sodium metaphosphate, $NaPO_3$-II and $NaPO_3$-III. See sodium metaphosphate.

mafenide hydrochloride (para-aminomethylbenzenesulfonamide hydrochloride; marfanil) $(SO_2NH_2)C_6H_4CH_2NH_2 \cdot HCl$.

Properties: White crystals; freely soluble in water; m.p. 260–265°C.

Use: Medicine (ear infection; treatment of burns).

"MagAmp."[241] Trademark for magnesium ammonium phosphate (q.v.).

"Magcarb."[123] Trademark for magnesium carbonate (q.v.).

magenta. See fuchsin.

"Mag-Li-Kote."[139] Trademark for a burned dolomite, ground; used for coating pig casting machine molds, cinder ladles, stools and ingot molds.

"Maglite."[123] Trademark for light magnesium oxide.

mag-lith. A magnesium-lithium alloy suggested as a structural metal in space vehicles.

magma.

(1) In medicine, a class of preparations in which finely divided, freshly precipitated, insoluble, inorganic hydroxides are suspended in water to form a viscous, opaque mixture which may settle out on standing. Magmas of bismuth, magnesium and iron are used, commonly called milk of bismuth, milk of magnesia, etc.

(2) In geology, a molten mass within the earth's crust (e.g., lava). The source of igneous rock.

magnalium. An alloy of aluminum and magnesium.

magnesia. Magnesium oxide that has been specially processed. See magnesium oxide.

magnesia alba. See magnesium carbonate; magnesium carbonate, basic.

magnesia-alumina $MgO \cdot Al_2O_3$. A synthetic spinel.

magnesia, burnt. See magnesite, dead-burned.

magnesia, calcined. See magnesite, caustic-calcined.

magnesia, caustic-calcined. See magnesite, caustic-calcined.

magnesia-chromia $MgO \cdot Cr_2O_3$. A synthetic spinel.

magnesia, dead-burned. See magnesite, dead-burned.

magnesia, fused. Used as a refractory and to handle electricity at high temperatures. See "Magnorite."

magnesia, lightburned. A special high-purity magnesium oxide.

magnesite. Natural magnesium carbonate ($MgCO_3$). The term magnesite is loosely used as a synonym for magnesia, as are also the terms caustic-calcined magnesite, dead-burned magnesite, and synthetic magnesite.

Properties: White, yellowish, grayish-white or brown crystalline solid. Sp. gr. 3–3.12; Mohs hardness 3.5–4.5.

Occurrence: United States (California, Washington, Nevada); Austria; Greece.

Uses: To make the various grades of magnesium oxide; to produce carbon dioxide.

See also magnesium carbonate.

magnesite, burnt. See magnesite, dead-burned.

magnesite, caustic-calcined (caustic-calcined magnesia; calcined magnesite; calcined magnesia). Principally magnesia (magnesium oxide) MgO. The product obtained by firing magnesite, or other substances convertible to magnesia upon heating, at temperatures

below 1450°C so that some carbon dioxide is retained (2–10%) and the magnesia displays adsorptive capacity or activity.
Grades: Technical; chemical; synthetic rubber; U.S.P. (light, medium light, heavy).
Containers: Bags; cartons; car lots.
Uses: Magnesium oxychloride and oxysulfate cements; 85% magnesium oxide insulation; rubber (reinforcing agent, accelerator); uranium processing; chemical processing; rayon; refractories; paper pulp; acid-neutralizing fertilizers; welding rod coatings, fillers, glass constituents, abrasives.
See also magnesium oxide.

magnesite, dead-burned (burnt magnesia; dead-burned magnesia; refractory magnesia; burnt magnesite) Magnesium oxide, MgO. The granular product obtained by burning (firing) magnesite, or other substances convertible to magnesia upon heating, above 1450°C long enough to form ... granules suitable for use as a refractory ... (ASTM). Synthetic magnesium hydroxide or chloride is sometimes used instead of magnesite as a source.
Grades: 85–87% (from magnesite ores); 97–99% (from sea water and brines).
Containers: Bulk; car lots.
Uses: Refractories, as grains or basic brick, the latter especially in open hearth furnaces for steel, furnaces for nonferrous metal smelting, and in cement and other kilns.
See also magnesium oxide.

"Magnesite H-W."[446] Trademark for an over 90% burned magnesia brick with minor added components to control properties. Resistant to molten metal and basic slags. Used in open hearth furnaces, electric steel furnaces, copper and nickel converters, refining industries, and other metallurgical industries.

magnesite, synthetic. Magnesium oxide, MgO, as obtained from sea water, sea water bitterns, or well brines. The preliminary product is usually magnesium hydroxide or chloride, which is then heated, or sometimes treated with steam and heated in the case of the chloride, to obtain the oxide. Synthetic magnesite constitutes the purer grades of dead-burned magnesite.

magnesium Mg Metallic element of atomic number 12, Group IIA of the Periodic Table. Atomic weight 24.305; valence 2. There are three isotopes.
Properties: Silvery, moderately hard metal, readily fabricated by all standard methods. Lightest of the structural metals. Strong reducing agent. Electrical conductivity similar to aluminum. Sp. gr. 1.74; m.p. 650°C; b.p. 1107°C. Soluble in acids; insoluble in water. Magnesium is the central element of the chlorophyll molecule; it is also an important component of red blood corpuscles.
Sources: Magnesite and dolomite; sea water and brines.
Derivation: (a) Electrolysis of fused magnesium chloride (Dow sea-water process); (b) reduction of magnesium oxide with ferrosilicon (Pidgeon process).
Forms available: Ingots; bars; fine powder (up to 99.6% pure); sheet and plate; rods; tubing; ribbon; flakes.
Hazard: (solid metal) Combustible at 650°C; (powder, flakes, etc.). Flammable, dangerous fire hazard. Use dry sand or talc to extinguish.
Uses: Alloys for structural parts, die-cast auto parts, missiles, space vehicles; powder for pyrotechnics and flash photography; production of iron, nickel, zinc, etc.; magnesium compounds and Grignard syntheses; cathodic protection; reducing agent; precision instruments; optical mirrors; dry and wet batteries.
Shipping regulations: (Rail) (powder, ribbon, etc.) Yellow label; (dross) Consult regulations. (Air) (pellets, powder, ribbon, etc.) Flammable Solid label; (dross, wet) Not acceptable; (dross, dry) Not restricted.

magnesium acetate (a) $Mg(OOCCH_3)_2$ or (b) $Mg(OOCCH_3)_2 \cdot 4H_2O$.
Properties: Colorless crystalline aggregate or monoclinic crystals. Acetic acid odor. (a) M.p. 323°C, sp. gr. 1.42, (b) m.p. 80°C, sp. gr. 1.45. Soluble in water and (dilute) alcohol.
Derivation: Interaction of magnesium carbonate and acetic acid.
Uses: Dye fixative in textile printing; medicine; deodorant, disinfectant and antiseptic.

magnesium acetylacetonate $MgC_5H_7O_2)_2$. Crystalline powder. Slightly soluble in water. Resistant to hydrolysis. A chelating nonionizing compound.

magnesium ammonium orthophosphate. See magnesium ammonium phosphate.

magnesium ammonium phosphate (magnesium ammonium orthophosphate) $MgNH_4PO_4 \cdot 6H_2O$.
Properties: White powder; sp. gr. 1.71; m.p. decomposes to magnesium pyrophosphate, $Mg_2P_2O_7$; soluble in acids; insoluble in alcohol and water.
Derivation: By the interaction of solutions of a magnesium salt and ammonium phosphate.
Uses: Medicine; fire retardant for fabrics; fertilizer.

magnesium arsenate $Mg_3(AsO_4)_2 \cdot xH_2O$. White powder when pure, insoluble in water. Technical material is highly hydrated and made from magnesium carbonate and arsenic acid.
Containers: Fiber cans; drums; multiwall paper sacks.
Hazard: Highly toxic by ingestion and inhalation.
Use: Insecticide.
Shipping regulations: (Rail, Air) Poison label.

magnesium benzoate $Mg(C_7H_5O_2)_2 \cdot 3H_2O$.
Properties: White crystalline powder; loses $3H_2O$ at 110°C; m.p. about 200°C; soluble in water and alcohol.
Use: Medicine.

magnesium biphosphate. See magnesium phosphate, monobasic.

magnesium borate $3MgO \cdot B_2O_3$ (orthoborate) or $Mg(BO_2)_2 \cdot 8H_2O$ (metaborate).
Properties: Transparent, colorless crystals or white powder. Soluble in alcohol, acetic acid, and inorganic acids; slightly soluble in water. Low toxicity.
Derivation: By heating magnesium oxide, boric anhydride and potassium hydrogen fluoride.
Containers: Glass bottles; fiber cans.
Uses: Preservative; antiseptic; fungicide.

magnesium borocitrate
$Mg(BO_2)_2 \cdot Mg_3(C_6H_5O_7)_2 \cdot 14H_2O$.
Properties: White powder or small, white, lustrous scales. Soluble in water. Low toxicity.
Derivation: By mixing magnesium borate and magnesium citrate.
Use: Medicine.

magnesium-boron fluoride
Grade: Technical.
Containers: 145-lb steel drums.
Hazard: Toxic. Strong irritant. Tolerance (as F): 2.5 mg per cubic meter of air.
Use: Metal flux.

magnesium bromate $Mg(BrO_3)_2 \cdot 6H_2O$.
Properties: White crystals or crystalline powder. Soluble in water; insoluble in alcohol. Sp. gr. 2.29; m.p. loses $6H_2O$ (200°C); b.p. decomposes.
Derivation: By adding magnesium sulfate to a solution of barium bromate.
Hazard: Moderately toxic. Dangerous fire risk in contact with organic materials.
Use: Analytical reagent; oxidizing agent.
Shipping regulation: (Air) Oxidizer label.

magnesium bromide $MgBr_2 \cdot 6H_2O$.
Properties: Colorless, very deliquescent crystals; bitter taste. Soluble in water; slightly soluble in alcohol. Sp. gr. 2.00; m.p. 172°C; anhydrous m.p. 700°C.
Derivation: Reaction of hydrobromic acid with magnesium oxide with subsequent crystallization.
Hazard: Moderately toxic.
Uses: Medicine; organic syntheses.

magnesium-calcium chloride. See calcium-magnesium chloride.

magnesium carbonate $MgCO_3$. The term magnesium carbonate is generally reserved for the synthetic, pure variety.
Properties: Light, bulky white powder; bulk density about 4 lb per cubic foot; actual sp. gr. about 3.0; decomposes 350°C. Refractive index about 1.52. Soluble in acids; insoluble in water and alcohol. Nontoxic. Noncombustible.
Derivation: By mixing solutions of magnesium sulfate and sodium carbonate, filtering and drying.
Containers: Bags; car lots.
Grades: Technical; N.F.; F.C.C.
Uses: Magnesium salts; heat insulation and refractory; rubber reinforcing agent; inks; glass; pharmaceuticals, dentrifices and cosmetics; free-running table salts; mineral waters; filtering medium. Used in foods as an alkali, drying agent, color retention agent, anticaking agent, carrier.

magnesium carbonate, basic (magnesia alba). Various formulas are given and may all be possible, because of the method of derivatoin (see magnesium carbonate). A typical formula is $Mg(OH)_2 \cdot 3MgCO_3 \cdot 3H_2O$. Properties and uses are almost identical with those listed above.
Derivation: Precipitation from magnesium salt solution.

magnesium chlorate $Mg(ClO_3)_2 \cdot 6H_2O$.
Properties: White powder. Bitter taste. Very hygroscopic. Soluble in water; slightly soluble in alcohol. Sp. gr. 1.8; m.p. 35°C; (decomposes at 120°C).
Hazard: Dangerous fire risk in contact with organic materials; strong oxidizing agent. Moderately toxic.
Uses: Medicine; defoliant; desiccant.
Shipping regulations: Chlorates, n.o.s. (Rail) Yellow label. (Air) Oxidizer label.

magnesium chloride (a) $MgCl_2$; (b) $MgCl_2 \cdot 6H_2O$.
Properties: Colorless or white crystals; deliquescent. Sp. gr. (a) 2.32, (b) 1.56; m.p. (a) 708°C, (b) loses $2H_2O$ at 100°C; if heated rapidly, melts at 116–118°C; b.p. (a) 1412°C, (b) decomposes to oxychloride. Soluble in water and alcohol. Combustible.
Derivation: Action of hydrochloric acid on magnesium oxide or hydroxide, especially the latter when precipitated from sea water or brines (Great Salt Lake).
Method of purification: Recrystallization.
Grades: Technical (crystals, fused, flakes, granulated); C.P.
Containers: Bags; drums.
Hazard: Moderately toxic.
Uses: Source of magnesium metal; disinfectants; fire extinguishers; fireproofing wood; magnesium oxychloride cement; refrigerating brines; ceramics; cooling drilling tools; textiles (size, dressing and filling of cotton and woolen fabrics, thread lubricant; carbonization of wool); paper manufacture; road dust-laying compounds; floor sweeping compounds; flocculating agent; catalyst.

magnesium chromate $MgCrO_4 \cdot 5H_2O$, is in the form of small, readily soluble, yellow crystals. Since it does not produce a fusible alkaline residue when thermally decomposed, it is used as a corrosion inhibitor in the water coolant of gas turbine engines. Insoluble basic magnesium chromates also are available. Their potential applications are in the treatment of light metal surfaces.

magnesium citrate, dibasic (acid magnesium cirtate) $MgHC_6H_5O_7 \cdot 5H_2O$.
Properties: White or slightly yellow, odorless granules or powder. Soluble in water, insoluble in alcohol. Nontoxic. Combustible.
Derivation: Reaction of citric acid and magnesium hydroxide or carbonate.
Use: Medicine; dietary supplement.

magnesium dichromate $MgCr_2 \cdot 6H_2O$, is characterized by high solubility in water. It is an orange-red, deliquescent, crystalaline hydrate. Potential applications are in formulations for corrosion prevention and metal treatment. Noncombustible.

magnesium dioxide. See magnesium peroxide.

magnesium fluoride (magnesium flux) MgF_2.
Properties: White crystals; exhibits fluorescence by electric light. Soluble in nitric acid; insoluble in alcohol and water. Sp. gr. 3.0; m.p. 1263°; b.p. 2239°C. Noncombustible.
Derivation: By adding sodium fluoride or hydrofluoric acid to a solution of magnesium salt.
Grades: Technical; C.P.; single crystals.
Hazard: Toxic. Strong irritant. Tolerance (as F), 2.5 mg per cubic meter of air.
Uses: Ceramics; glass; single crystals for polarizing prisms, lenses, and windows.

magnesium fluosilicate (magnesium silicofluoride) $MgSiF_6 \cdot 6H_2O$.
Properties: White, efflorescent crystalline powder; sp. gr. 1.788; decomposes 120°C. Soluble in water.
Derivation: By treating magnesium hydroxide or carbonate with hydrofluosilicic acid.
Grades: Technical (crystals; solution).
Containers: Crystals: 400-lb barrels; drums. Solution: 400-lb barrels.
Hazard: Toxic. Strong irritant. Tolerance (as F), 2.5 mg per cubic meter of air.
Uses: Ceramics; concrete hardeners; water-proofing; mothproofing; laundry sour; magnesium casting.

magnesium flux. See magnesium fluoride.

magnesium formate. $Mg(CHO_2)_2 \cdot 2H_2O$.
Properties: Colorless crystals; soluble in water; insoluble in alcohol and ether. Combustible.
Derivation: Action of formic acid on magnesium oxide.
Uses: Analytical chemistry; medicine.

magnesium gluconate $Mg(C_6H_{11}O_7)_2 \cdot 2H_2O$.
Properties: Odorless, almost tasteless, white powder

or fine needles. Soluble in water. Nontoxic. Combustible.
Derivation: Magnesia or magnesium carbonate dissolved in gluconic acid.
Grade: Pharmaceutical.
Uses: Medicine; vitamin tablets.

magnesium glycerinophosphate. See magnesium glycerophosphate.

magnesium glycerophosphate (magnesium glycerinophosphate) $CH_2(OH)CH(OH)CH_2OP(O)O_2Mg$.
Properties: Colorless powder; soluble in water; insoluble in alcohol. Nontoxic. Combustible.
Derivation: Action of glycerophosphoric acid on magnesium hydroxide.
Uses: Medicine; stabilizer for plastics.

magnesium hydrate. See magnesium hydroxide.

magnesium hydrogen phosphate. See magnesium phosphate, dibasic.

magnesium hydroxide (magnesium hydrate; in aqueous suspension: milk of magnesia (q.v.); magnesia magma) $Mg(OH)_2$.
Properties: White powder; odorless. Soluble in solutions of ammonium salts and dilute acids; almost insoluble in water and alcohol; sp. gr. 2.36; m.p. decomposes at 350°C. Nontoxic. Noncombustible.
Derivation: Precipitation from a solution of a magnesium salt by sodium hydroxide; precipitation from sea water with lime. It occurs naturally as brucite (q.v.).
Grades: Technical; N.F.; F.C.C.
Uses: Intermediate for obtaining magnesium metal; sugar refining; medicine; residual fuel oil additive; sulfite pulp; uranium processing; dentifrices; in foods as alkali, drying agent, color retention agent; frozen desserts.

magnesium hypophosphite $Mg(H_2PO_2)_2 \cdot 6H_2O$.
Properties: White efflorescent crystals; sp. gr. 1.59; decomposes at 100°C to evolve phosphine. Soluble in water; insoluble in alcohol. Nontoxic.
Derivation: Action of hypophosphorous acid on magnesium oxide.
Hazard: Evolves toxic gas when heated to 100°C.
Use: Medicine.

magnesium hyposulfite. See magnesium thiosulfate.

magnesium iodide (a) MgI_2; (b) $MgI_2 \cdot 8H_2O$.
Properties: White, deliquescent, crystalline powder; discolors in air; soluble in water, alcohol, and ether. M.p. 632°C; decomposes above 700°C; sp. gr. (a) 4.48. (b) Decomposes at 41°C.
Derivation: Heating magnesium in iodine vapors.
Use: Medicine.

magnesium lactate $Mg(C_3H_5O_3)_2 \cdot 3H_2O$.
Properties: White crystals; very bitter taste; soluble in water; slightly soluble in alcohol. Nontoxic. Combustible.
Derivation: Action of lactic acid on magnesium oxide, with subsequent crystallization.
Use: Medicine.

magnesium lauryl sulfate $Mg(OSO_3C_{12}H_{25})_2$.
Properties: Pale yellow liquid with a mild odor; soluble in methanol, acetone and water; insoluble in kerosine. Low toxicity. Combustible.
Derivation: Sulfonation of lauryl alcohol, and interaction with a magnesium salt.
Containers: Drums; tanks.
Uses: Surfactant and anionic detergent; foaming, wetting and emulsifying agent.

magnesium lime. Same as magnesium limestone. See limestone.

magnesium limestone. See limestone.

magnesium methoxide (magnesium methylate) $(CH_3O)_2Mg$.
Properties: Colorless crystalline solid; decomposes on warming.
Derivation: Reaction of magnesium and methanol.
Uses: Dielectric coating; cross-linking agent to form stable gels; catalyst.

magnesium methylate. See magnesium methoxide.

magnesium molybdate $MgMoO_4$. Crystalline powder; absolute density 2.8 g/cc; m.p. 1060°C (approx); soluble in water.
Uses: Electronic and optical applications.

magnesium nitrate $Mg(NO_3)_2 \cdot 2H_2O$.
Properties: White crystals; soluble in water and alcohol. Deliquescent. Sp. gr. 2.03 (25°C); m.p. 129°C; decomposes at 330°C.
Derivation: Action of nitric acid on magnesium oxide, with subsequent crystallization.
Hazard: Dangerous fire and explosion risk in contact with organic materials. Strong oxidizing agent.
Use: Pyrotechnics.
Shipping regulations: (Rail) Yellow label. (Air) Oxidizer label.

magnesium oleate $Mg(C_{18}H_{33}O_2)_2$.
Properties: Yellowish mass; soluble in linseed oil, hydrocarbons, alcohol, and ether; insoluble in water. Combustible. Low toxicity. Combustible.
Derivation: Interaction of magnesium chloride and sodium oleate.
Uses: Varnish driers; in dry-cleaning solvents (to prevent spontaneous ignition); emulsifying agent; lubricant for plasticizers.

magnesium orthophosphate. See magnesium phosphate.

magnesium oxide (magnesia) MgO. Two forms are produced, one a light fluffy material prepared by a relatively low-temperature dehydration of the hydroxide, the other a dense material made by high-temperature furnacing of the oxide after it has been formed from the carbonate or hydroxide. See also periclase.
Properties: White powder, either light or heavy depending upon whether it is prepared by heating magnesium carbonate or the basic magnesium carbonate; sp. gr. about 3.6 (varies); m.p. 2800°C; b.p. 3600°C; slightly soluble in water; soluble in acids and ammonium salt solutions. Noncombustible.
Derivation: (a) By calcining magnesium carbonate or magnesium hydroxide; (b) by treating magnesium chloride with lime and heating, or by heating it in air; (c) from sea water via the hydroxide.
Grades: Technical; C.P.; U.S.S.; F.C.C.; 99.5%; fused; low boron; rubber; semiconductor; single crystals.
Containers: Fiber drums; multiwall paper sacks; tonnage lots.
Hazard: Toxic by inhalation of fume. Tolerance, 10 mg per cubic meter of air.
Uses: Refractories, especially for steel furnace linings; polycrystalline ceramic for aircraft windshields; electrical insulation; pharmaceuticals and cosmetics; inorganic rubber accelerator; oxychloride and oxysulfate cements; paper manufacture; fertilizers; removal of sulfur dioxide from stack gases; adsorption and catalysis; semiconductors; pharmaceuticals; food and feed additive.

magnesium oxychloride cement (Sorel cement). A mixture of magnesium chloride and magnesium oxide that reacts with water to form a solid mass, presumed to be magnesium oxychloride. Fillers such as wood flour, sawdust, sand, powdered stone, talc, etc. are usually present. A variety of proprietary mixtures are available. Strength ranges from 7000 to 10,000 psi. Copper powder minimizes water solubility.

magnesium palmitate $Mg(C_{16}H_{31}O_2)_2$.
Properties: Crystalline needles or white lumps; m.p. 121.5°C; insoluble in water and alcohol. Low toxicity. Combustible.
Uses: Varnish drier; lubricant for plastics.

magnesium pemoline. A combination of 2-imino-5-phenyl-4-oxazolidinone and magnesium hydroxide. Said to be a stimulant of the nervous system because of its action on the enzyme, RNA polymerase.

magnesium perborate $Mg(BO_3)_2 \cdot 7H_2O$.
Properties: White powder; sparingly soluble in water; decomposes with evolution of oxygen.
Derivation: Action of peroxide or electrolytic oxidation of borate solutions.
Hazard: Moderate fire risk in contact with organic materials.
Uses: Driers; bleaching; antiseptic.

magnesium perchlorate (a) $Mg(ClO_4)_2$; (b) $Mg(ClO_4)_2 \cdot 6H_2O$.
Properties: White, crystalline, deliquescent; very soluble in water and alcohol; (a) sp. gr. 2.21 (18°C); decomposes at 251°C; (b) sp. gr. 1.98; decomposes at 185–190°C. Strong oxidizing material.
Derivatoin: Reaction of magnesium hydroxide and perchloric acid.
Hazard: Moderately toxic. Dangerous fire and explosion risk in contact with organic materials.
Uses: (a) As a regenerable drying agent for gases, and (b) Oxidizing agent.
Shipping regulations: (Rail) Yellow label. (Air) Oxidizer label.

magnesium permanganate $Mg(MnO_4)_2 \cdot 6H_2O$.
Properties: Bluish-black, friable, deliquescent crystals. Sp. gr. 2.18; m.p., decomposes. Soluble in water. Strong oxidizing material.
Hazard: Moderately toxic; fire hazard in contact with organic materials.
Use: Medicine (strong antiseptic).

magnesium peroxide (magnesium dioxide) MgO_2.
Properties: White, tasteless, odorless powder. Decomposes above 100°C. Insoluble in water; soluble in dilute acids with formation of hydrogen peroxide. Available oxygen 28.4%. Keep cool and dry. A powerful oxidizing material.
Derivation: From sodium or barium peroxide with magnesium sulfate in a concentrated solution.
Grades: Technical; 15, 25, and 50%.
Hazard: Dangerous fire risk; reacts with acidic materials and moisture.
Uses: Bleaching and oxidizing agent; medicine.
Shipping regulations: (Rail) Yellow label. (Air) Oxidizer label.

magnesium phosphate (magnesium orthophosphate). See magnesium phosphate, dibasic; magnesium phosphate, monobasic; or magnesium phosphate, tribasic.

magnesium phosphate, dibasic (dimagnesium orthophosphate; dimagnesium phosphate; magnesium phosphate, secondary; magnesium hydrogen phosphate) $MgHPO_4 \cdot 3H_2O$.
Properties: White, crystalline powder; decomposes to pyrophosphate on heating. Soluble in dilute acids; slightly soluble in water. Sp. gr. 2.13; loses one H_2O at 205°C; dec. 550–650°C. Nontoxic. Nonflammable.
Derivation: Action of orthophosphoric acid on magnesium oxide.
Grades: Technical; F.C.C.
Use: Medicine; stabilizer for plastics; food additive.

magnesium phosphate, monobasic (magnesium biphosphate; acid magnesium phosphate; magnesium tetrahydrogen phosphate) $MgH_4(PO_4)_2 \cdot 2H_2O$.
Properties: White, hygroscopic, crystalline powder; decomposes to metaphosphate on heating. Soluble in water and acids; insoluble in alcohol. Nontoxic. Nonflammable.
Derivation: Action of orthophosphoric acid on magnesium hydroxide.
Uses: Medicine; fireproofing wood; stabilizer for plastics.

magnesium phosphate, neutral. See magnesium phosphate, tribasic.

magnesium phosphate, secondary. See magnesium phosphate, dibasic.

magnesium phosphate, tribasic (magnesium phosphate, neutral; trimagnesium phosphate) $Mg_3(PO_4)_2 \cdot 8H_2O$ or $5H_2O$.
Properties: Soft, bulky white powder; odorless and tasteless; loses all water at 400°C. Soluble in acids; insoluble in water. Nontoxic. Nonflammable.
Derivation: Reaction of magnesium oxide and phosphoric acid at high temperatures.
Grades: Technical; reagent; N.F. ($5H_2O$ variety); F.C.C. (4, 5, or $8H_2O$).
Containers: Bags; barrels; drums.
Uses: Dentifrice polishing agent; pharmaceutical antacid, adsorbent, stabilizer for plastics, food additive and dietary supplement.

magnesium pyrophosphate $Mg_2P_2O_7 \cdot 3H_2O$.
Properties: White powder; soluble in acids; insoluble in alcohol and water. Sp. gr. 2.56; loses water at 100°C; m.p. (anhydrous) 1383°C.

magnesium ricinoleate $Mg(OOCC_{17}H_{32}OH)_2$.
Properties: Coarse, yellow granules with faint fatty acid odor; m.p. 98°C; sp. gr. 1.03 (25/25°C). Used in cosmetics. Combustible.

magnesium salicylate $Mg(C_7H_5O_3)_2 \cdot 4H_2O$.
Properties: Colorless, efflorescent crystalline powder. Soluble in water and alcohol.
Derivation: Action of salicylic acid on magnesium hydroxide.
Use: Medicine.

magnesium silicate $3MgSiO_3 \cdot 5H_2O$ (variable). The F.C.C. specifies a ratio of $2MgO{:}5SiO_2$. See also magnesium trisilicate, serpentine, and talc.
Properties: Fine, white powder; insoluble in water or alcohol. An absorbent. Noncombustible; may be toxic by inhalation.
Derivation: Interaction of a magnesium salt and a soluble silicate.
Grades: Technical; F.C.C.
Hazard: Use in foods restricted to 2%.
Uses: Rubber filler; ceramics; glass; refractories; absorbent for crude oil spills; manufacture of permanently dry resins and resinous compositions; paints,

varnishes, and paper (filler); animal and vegetable oils (bleaching agent); odor absorbent; filter medium; catalyst and catalyst carrier; anticaking agent in foods. See also asbestos.

magnesium silicofluoride. See magnesium fluosilicate.

magnesium stannate $MgSnO_3 \cdot 3H_2O$.
Properties: White crystalline powder. Soluble in water. Approximate temperature of decomposition 340°C.
Hazard: Toxic by inhalation. Tolerance, 2 mg per cubic meter of air.
Use: Additive in ceramic capacitors.

magnesium stearate $Mg(C_{18}H_{35}O_2)_2$, or with one H_2O. Technical grade contains small amounts of the oleate and 7% magnesium oxide MgO.
Properties: Soft white light powder; sp. gr. 1.028; m.p. 88.5°C (pure), 132°C (technical); tasteless; odorless. Insoluble in water and alcohol. Nontoxic. Nonflammable.
Grades: Technical; U.S.P.; F.C.C.
Containers: Fiber cans; multiwall paper sacks.
Uses: Dusting powder; lubricant in making tablets; drier in paints and varnishes; flatting agent; in medicines; stabilizer and lubricant for plastics; emulsifying agent in cosmetics; in foods as anticaking agent, binder, emulsifier.

magnesium sulfate (a) $MgSO_4$; (b) (epsom salts) $MgSO_4$ $7H_2O$.
Properties: Colorless crystals; saline, bitter taste; neutral to litmus; sp. gr. (a) 2.65; (b) 1.678; (a) decomposes 1124°C; (b) loses $6H_2O$ at 150°C; $7H_2O$ at 200°C; soluble in glycerol; very soluble in water; sparingly soluble in alcohol. Low toxicity. Noncombustible.
Derivation: (a, b) Action of sulfuric acid on magnesium oxide, hydroxide or carbonate; (b) mined in a high degree of purity.
Grades: Technical; C.P.; U.S.P.; F.C.C.
Uses: Fireproofing; textiles (warp-sizing and loading cotton goods, weighting silk, dyeing and calico printing); mineral waters; catalyst carrier; ceramics; fertilizers; paper (sizing); cosmetic lotions; dietary supplement; medicine (antidote).

magnesium sulfide MgS.
Properties: Red brown crystalline solid; sp. gr. 2.84; decomposes above 2000°C. Decomposes in water. Low toxicity.
Uses: Source of hydrogen sulfide; laboratory reagent.

magnesium sulfite $MgSO_3 \cdot 6H_2O$.
Properties: White, crystalline powder; slightly soluble in water; insoluble in alcohol. Sp. gr. 1.725; m.p., loses $6H_2O$ at 200°C; b.p., decomposes. Low toxicity.
Derivation: Action of sulfurous acid on magnesium hydroxide.
Uses: Medicine; paper pulp.

magnesium tetrahydrogen phosphate. See magnesium phosphate, monobasic.

magnesium thiosulfate (magnesium hyposulfite) $MgS_2O_3 \cdot 6H_2O$.
Properties: Colorless crystals; soluble in water; insoluble in alcohol. Sp. gr. 1.818; loses $3H_2O$ at 170°C.
Use: Medicine.

magnesium titanate Mg_2TiO_4. Used in electronics.

magnesium trisilicate. U.S.P. specifies not less than 20% MgO and 45% SiO_2; similar to the F.C.C. requirements under magnesium silicate. See also talc.
Properties: Fine, white, odorless, tasteless powder; free from grittiness. Insoluble in water and alcohol; readily decomposed by mineral acids. Noncombustible.
Derivation: By reaction of soluble magnesium salts with soluble silicates.
Grades: Technical; U.S.P.
Uses: Industrial odor absorbent; decolorizing agent; antioxidant; medicine.

magnesium tungstate (magnesium wolframate) $MgWoO_4$.
Properties: White crystals; sp. gr. 5.66; soluble in acids; insoluble in water and alcohol. Low toxicity. Noncombustible.
Derivation: Interaction of solutions of magnesium sulfate and ammonium tungstate.
Uses: Fluorescent screens for x-rays; luminescent paint.

magnesium zirconate MgO,ZrO_2.
Properties: Powder; sp. gr. 4.23; m.p. 2060°C.
Use: Electronics.

magnesuim zirconium silicate $MgZrSiO_5$, or $MgO \cdot ZrO_2 \cdot SiO_2$.
Properties: White solid; m.p. 1760°C; density 80 lb/cu ft; insoluble in water; alkalies; slightly soluble in acids. Noncombustible.
Containers: 80-lb paper bags; 500-lb drums.
Uses: Electrical resistor ceramics; glaze opacifier.

"Magnesol."[55] Trademark for a synthetic adsorptive magnesium silicate.
Uses: Solvent purification, clarification and recovery; oil refining; deodorizing and decolorizing of oils and fats.

magnetic separation. Removal of bits of iron and other tramp metal from a material as it passes to a screen or classifying device by means of a magnet placed close to the stream of particles.

magnetite (lodestone; iron ore, magnetic) Fe_3O_4, often with titanium or magnesium. A component of taconite (q.v.).
Properties: Black mineral; black streak; submetallic, or dull to metallic luster. Contains 72.4% iron. Readily recognized by strong attraction by magnet. Soluble in powder form in hydrochloric acid. Decomposes at 1538°C to ferric oxide Fe_2O_3. Sp. gr. 4.9–5.2; hardness 5.5–6.5.
See also iron oxide, black.

magnetochemistry. A subdivision of chemistry concerned with the effect of magnetic fields on chemical compounds; analysis and measurement of these effects (e.g., magnetic moment and magnetic susceptibility) are important tools in crystallographic research and determination of molecular structures. Substances that are repelled by a magnetic field are diamagnetic (water, benzene); those that are attracted are paramagnetic (oxygen, transition element compounds). Diamagnetic materials have only induced magnetic moment; paramagnetic materials have permanent magnetic moment. Magnetochemistry has been useful in detection of free radicals, elucidation of molecular configurations of highly complex compounds, and in its application to catalytic and chemisorption phenomena. See also nuclear magnetic resonance.

magnetohydrodynamics (MHD). The behavior of high-temperature ionized gases passed through a magnetic field. A power-generating method using MHD involves an open cycle in which hot combustion gases seeded with cerium or potassium to increase electrical conductivity constitute the working fluid. These are sent through a nozzle surrounded by a magnet; the electricity induced by movement of the ionized gas through the magnetic field is passed to electrodes, and the gas sent to a steam generator. Efficiency is rated at 50–60%, compared with 40% for conventional fossil fuel plants and 33% for plants using nuclear fuels. Two-phase liquid-metal systems are being studied as auxiliary units for a number of energy converters. MHD is an important field of expansion of research activity on new sources of energy; its high efficiency and low pollution factor indicate that it may have a significant future in electric power supply. See also plasma (2).

"Magnorite."[249] Trademark for fused magnesium oxide refractory products; fusion point up to 4750°F (2620°C) depending on purity. Used for cements; lining metal melting furnaces and chemical reactors; electrical insulating medium.

"Magon."[169] Trademark for 1-azo-2-hydroxy-3-(2,4-dimethylcarboxanilido)naphthalene-1′-(2-hydroxybenzene), used in the colorimetric determination of magnesium.

"Magron."[233] Trademark for vegetation maturant containing magnesium chlorate.

Maillard reaction. See browning reaction.

"Makrofol."[470] Trademark for electric insulating films based on polycarbonates.

"Makrolon."[470] Trademark for a series of polycarbonate resins.

malachite green (benzaldehyde green; victoria green). A triphenylmethane dye (q.v.), the zinc double chloride, oxalate, or ferric double chloride of tetramethyl-para-aminotriphenylcarbinol (see triphenylcarbinol, bis(4-dimethylamino)-. The term is sometimes applied to the mineral malachite.

malachite green toner. See phosphotungstic pigment.

"Malaphos."[88] Trademark for antidusting, wettable powder and emulsifiable solution containing malathion. Used as insecticide.
Hazard: See malathion.

malathion. Generic name for S-[1,2-bis(ethoxycarbonyl)ethyl] O,O-dimethyl phosphorodithioate, $(CH_3O)_2P(S)SCH(COOC_2H_5)CH_2COOC_2H_5$.
Properties: Yellow, high-boiling liquid; (b.p. 156–157°C, under 0.7 mm with slight decomposition); m.p. 2.85°C; refractive index (n 25/D) 1.4985; sp. gr. 1.2315 (25°C); vapor pressure (20°C) approximately 0.00004 mm. Miscible with most polar organic solvents. Slightly soluble in water. Combustible.
Purity: Technical grade is 95+ % pure.
Derivation: From diethyl maleate and dimethyldithiophosphoric acid.
Hazard: Toxic by ingestion and inhalation. Absorbed by skin. Tolerance, 10 mg per cubic meter of air. Cholinesterase inhibitor.
Use: Insecticide.
Note: Approved by EPA as substitute for certain uses of DDT.

maleic acid (maleinic acid; cis-butenedioic acid) HOOCCH:CHCOOH (cis isomer).
Properties: Colorless crystals; repulsive, astringent taste; faint odor. Soluble in water, alcohol, and acetone; very slightly soluble in benzene; sp. gr. 1.59; m.p. 130–131°C; at temperatures slightly above the m.p., is converted partly to fumaric acid. Combustible.
Derivation: Same as for maleic anhydride, with special recovery conditions.
Grades: Technical; reagent.
Hazard: Moderately toxic; strong irritant to tissue.
Uses: Organic synthesis (malic, succinic, aspartic, tartaric, propionic, lactic, malonic, acrylic, hydracrylic acids); dyeing and finishing of cotton, wool and silk; preservative for oils and fats.

maleic anhydride (2,5-furandione) HC:CHC(O)OC(O).
Properties: Colorless needles; sp. gr. 0.934 (20/4°C); m.p. 53°C; b.p. 200°C; flash point 218°F; soluble in acetone, hydrocarbons, ether, chloroform, petroleum ether. Hydrolyzes slowly in water. Combustible; autoignition temp. 890°F.
Derivation: (a) By passing a mixture of benzene vapor and air over a vanadium oxide catalyst at about 450°C; (b) byproduct from manufacture of phthalic anhydride from naphthalene; (c) catalytic oxidation of butylenes.
Grades: Technical; rods, flakes, lumps, briquettes, and molten.
Containers: Fiber drums (rod, flake, and lump); iron drums (fused); tank cars (molten).
Hazard: Strong irritant to tissue. Tolerance, 0.25 ppm in air. Safety data sheet available from Manufacturing Chemists Assn., Washington, D.C.
Uses: Polyester resins; alkyd coating resins; fumaric and tartaric acid manufacture; pesticides; preservative for oils and fats; paper; permanent-press resins (textiles).

maleic hydrazide (1,2-dihydro-3,6-pyridazinedione) HC:CHC(O)NHNHC(O). A plant growth regulator.
Properties: Crystals; m.p. 297°C; slightly soluble in hot alcohol; more soluble in hot water.
Derivation: By treating maleic anhydride with hydrazine hydrate.
Uses: Systemic herbicide; treatment of tobacco plants; post-harvest sprouting inhibitor; weed control; sugar content stabilizer in beets.

maleinic acid. See maleic acid.

maleo-pimaric acid. Reaction product of maleic anhydride and *l*-pimaric acid; derived from pine gum.
Properties: Crystalline solid; m.p. about 225°C; soluble in most organic solvents; insoluble in water or aliphatic hydrocarbons.
Use: Resins.

malic acid (hydroxysuccinic acid; apple acid) $COOHCH_2CH(OH)COOH$. (Do not confuse with maleic acid).
Properties: Colorless crystals; sour taste; sp. gr. (*dl*-form) 1.601, (*d* or *l* form) 1.595 (20/4°C); m.p. (*dl*) 128°C, (*d* or *l*) 100°C; b.p. (*dl*) 150°C (dec), (*d* or *l*) 140°C (dec). Very soluble in water and alcohol; slightly soluble in ether. Combustible; low toxicity.
Derivation: Occurs naturally in unripe apples, and other fruits. Made synthetically by catalytic oxidation of benzene to maleic acid, which is converted to malic acid by heating with steam under pressure.
Grades: Technical, active and inactive; F.C.C. The natural material is levorotatory, but the synthetic material is inactive.
Containers: Glass jars; fiber cans; drums.

Uses: Medicine; manufacture of various esters and salts; wine manufacture; chelating agent; food acidulant; flavoring.

"Malix 138."[79] Trademark for a limed M wood rosin used in linoleum print paint; printing inks; gloss oils and varnishes.

"Mallcosorb" A.R.[329] Trademark for a high-capacity, granular absorbent for CO_2, HCl, SO_2, H_2S, and other acidic gases. Changes from pink to yellowish tan as it loses absorptive capacity.

malonamide nitrile. See cyanoacetamide.

malonic acid (methanedicarbonic acid) $CH_2(COOH)_2$.
Properties: White crystals; soluble in water, alcohol and ether. M.p. 132–134°C; b.p. decomposes; sp. gr. 1.63.
Derivation: From monochloroacetic acid by reaction with potassium cyanide, followed by hydrolysis.
Use: Intermediate for barbiturates and pharmaceuticals.

malonic dinitrile (malononitrile) $CH_2(CN)_2$.
Properties: Colored crystals. M.p. 32.1°C; b.p. 220°C.
Use: Organic synthesis; leaching agent for gold.

malonic ester. See ethyl malonate.

malonic ethyl ester nitrile. See ethyl cyanoacetate.

malonic methyl ester nitrile. See methyl cyanoacetate.

malonic mononitrile. See cyanoacetic acid.

malononitrile. See malonic dinitrile.

malonylurea. See barbituric acid.

malt. Yellowish or amber-colored grains of barley which have been partially germinated by artificial means. It contains dextrin, maltose and amylase; characteristic odor and taste. Black malt is grain which has been scorched in the drying process.
Uses: Brewing; malted milk and similar food products; extract of malt (with 10% glycerol).

maltase (glucase; alpha-glucosidase). An enzyme that hydrolyzes maltose to glucose. Occurs in the small intestine, in yeast, molds and malt; usually associated with the enzyme amylase.

malt extract (maltine).
Properties: Light brown, sweet, viscous liquid; contains dextrin, maltose, a little glucose, and an amylolytic enzyme. It is capable of converting not less than five times its weight of starch into water-soluble sugars. Soluble in cold water, but more readily soluble in warm water. Sp. gr. not less than 1.350 and not more than 1.430 (25°C).
Derivation: By infusing malt with water at 60°C, concentrating the expressed liquid at a temperature not exceeding 60°C, and adding 10% by weight of glycerol.
Use: Nutrient; emulsifying agent.

maltha. A black, viscous natural bitumen, a component of asphalt, consisting of a complex mixture of waxy hydrocarbons.

maltol (3-hydroxy-2-methyl-4-pyrone)
$CH_3C_5H_2O(O)(OH)$. White crystalline powder having a characteristic caramel-butterscotch odor, and suggestive of a fruity-strawberry aroma in dilute solution. Slightly soluble in water; more soluble in alcohol and propylene glycol. Melting range 160–164°C. Nontoxic.
Grade: F.C.C.
Use: Flavoring agent.

maltose (malt sugar; maltobiose) $C_{12}H_{22}O_{11} \cdot H_2O$. The most common reducing disaccharide; composed of two molecules of glucose. Found in starch and glycogen.
Properties: Colorless crystals; m.p. 102–103°C; soluble in water; soluble in alcohol; insoluble in ether. Combustible; nontoxic.
Derivation: By the enzymatic action of diastase (usually obtained from malt extract) on starch.
Containers: Glass bottles; fiber cans; drums.
Uses: Nutrient; sweetener; culture media; stabilizer for polysulfides.

"Mam."[342] Trademark for dimethyl anthranilate (q.v.). Used in perfumery.

Man Symbol for mannose; also abbreviation for methacrylonitrile (q.v.).

mandarin oil. See citrus peel oils.

"Mandelamine."[546] Trademark for hexamethylenetetramine mandelate (q.v.).

mandelic acid (phenylglycolic acid; alpha-phenylhydroxyacetic acid; benzoglycolic acid; known also as amygdalic acid) $C_6H_5CHOHCOOH$. Exists in stereoisomeric forms. The properties are those of the *dl*-form.
Properties: Large white crystals or powder with a faint odor; sp. gr. 1.30; m.p. 117–119°C. Darkens on exposure to light. Soluble in ether; slightly soluble in water and alcohol.
Derivation: Hydrolysis of the cyanohydrin formed from benzaldehyde, sodium bisulfite, and sodium cyanide. Can be obtained from amygdalin (q.v.).
Hazard: May be toxic.
Use: Medicine; organic synthesis.

mandelonitrile (benzaldehyde cyanohydrin)
$C_6H_5CH(OH)CN$.
Properties: Oily yellow liquid; sp. gr. 1.1165 (20/4°C); f.p. −10°C; b.p. 170°C (dec). Soluble in alcohol, chloroform, ether; nearly insoluble in water.

maneb. Generic name for manganese ethylenebisdithiocarbamate $(SSCNCH_2CH_2NHCSS)Mn$.
Properties: Brown powder; decomposes on heating. See also zineb.
Derivation: Reaction of disodium ethylenebisdithiocarbamate and a manganese salt.
Hazard: Toxic by ingestion and inhalation. Irritant to skin and eyes.
Uses: Fungicide for foliage.

manganese Mn Metallic element of atomic number 25, Group VIIB of Periodic Table. Atomic weight 54.938; valences 2, 3, 4, 6, 7. No stable isotopes; 4 artificial radioactive isotopes.
Properties: There are four allotropic forms, of which alpha is most important. Brittle, silvery metal; sp. gr. 7.44; Mohs hardness 5; m.p. 1245°C; b.p. 2097°C. Decomposes water; readily dissolves in dilute mineral acids. Pure Mn cannot be fabricated. Manganese is considered essential for plant and animal life.
Occurrence: Usually associated with iron ores in submarginal concentrations. Important ores of Mn are pyrolusite, manganite, psilomelane, rhodochrosite. An important source of Mn is open-hearth slags. Ores occur chiefly in Brazil, India, South Africa, Australia, Cuba, Montana. 90% of U.S. consumption is imported. So-called nodules rich in manganese and containing also cobalt, nickel and copper have been found in huge quantities (estimated at 1.5 trillion tons) on the floor of the Pacific south of Hawaii. Equipment and techniques for salvaging

these is under development. Such nodules have been located in other areas, as well as in Lake Michigan.
Derivation: Reduction of the oxide with aluminum or carbon. Pure Mn is obtained electrolytically from sulfate or chloride solution.
Grades: Technical; pure or electrolytic; powdered.
Containers: Drums; car lots.
Hazard: Prolonged inhalation of fume or dust is damaging to central nervous system. Tolerance, 5 mg per cubic meter of air applies to the metal and all its compounds (as Mn). Dust or powder is flammable. Use dry chemical to extinguish.
Uses: Ferroalloys (steel manufacture); nonferrous alloys (improved corrosion resistance and hardness); high-purity salt for various chemical uses; purifying and scavenging agent in metal production; manufacture of aluminum by Toth process (q.v.).

manganese acetate $Mn(C_2H_3O_2)_2 \cdot 4H_2O$.
Properties: Pale red crystals; very soluble in water and alcohol; sp. gr. 1.59; m.p. 80°C. Combustible.
Derivation: Action of acetic acid on manganese hydroxide.
Containers: Drums.
Hazard: Moderately toxic.
Uses: Textile dyeing; oxidation catalyst; paint and varnish (drier, boiled oil manufacture); fertilizers; food packaging; feed additive.

manganese ammonium sulfate (manganous ammonium sulfate) $MnSO_4 \cdot (NH_4)_2SO_4 \cdot 6H_2O$.
Properties: Light-red crystals; sp. gr. 1.83; soluble in water.

manganese arsenate. See manganous arsenate.

manganese binoxide. See manganese dioxide.

manganese black. See manganese dioxide.

manganese borate MnB_4O_7.
Properties: Reddish-white powder; insoluble in water.
Derivation: By the action of boric acid on manganese hydroxide.
Grades: Technical.
Containers: Fiber drums.
Uses: Varnish and oil drier.

manganese-boron. An alloy of manganese and boron used in making brass, bronze, and other alloys.

manganese bromide. See manganous bromide.

manganese bronze. Alloy of 55 to 60% copper, 38 to 42% zinc, up to 3.5% manganese, with or without small amounts of iron, aluminum, tin or lead.

manganese carbonate (manganous carbonate) $MnCO_3$ (found as rhodocrosite, q.v.).
Properties: Rose-colored crystals; almost white when precipitated. Soluble in dilute inorganic acids; almost insoluble in common organic acids, both concentrated and dilute; insoluble in water. Sp. gr. 3.125; m.p., decomposes.
Derivation: (1) A precipitate from the addition of sodium carbonate to a solution of a manganese salt. (2) Hydrometallurgical treatment of manganiferous iron ore.
Grades: Chemical (46% Mn).
Containers: Bags; carloads.
Uses: Manufacture of manganese salts; medicine.

manganese chloride. See manganous chloride.

manganese chromate. See manganous chromate.

manganese citrate (manganous citrate) $Mn_3(C_6H_5O_7)_2$.
Properties: White powder; soluble in water in presence of sodium citrate. Combustible.
Derivation: Action of citric acid on manganese hydroxide.
Uses: Medicine; feed additive; food additive and dietary supplement.

manganese cyclopentadienyl tricarbonyl.
Hazard: Highly toxic; absorbed by skin. Tolerance (as Mn), 0.1 mg per cubic meter of air.

manganese dioxide (manganese binoxide; manganese black; battery manganese; manganese peroxide) MnO_2.
Properties: Black crystals or powder; soluble in hydrochloric acid; insoluble in water. Sp. gr. 5.026; m.p., decomposes to Mn_2O_3 and oxygen at 535°C.
Derivation: (a) Natural as pyrolusite (q.v.) and as a special African ore of different atomic structure used exclusively for the battery grade; (b) By electrolysis; (c) By heating manganese oxide in presence of oxygen; (d) By decomposition of manganese nitrate.
Grades: Technical; C.P.; Battery.
Containers: Bottles; multiwall paper bags; drums; carloads, African grade: paper-lined burlap bags; paper bags.
Hazard: Oxidizing agent; may ignite organic materials; moderately toxic.
Uses: Oxidizing agent; depolarizer in dry cell batteries (African and synthetic types only); pyrotechnics, matches, etc.; catalyst; laboratory reagent; scavenger and decolorizer; textile dyeing.

manganese dithionate MnS_2O_6. Crystals; sp. gr. 1.76; soluble in water.

manganese ethylenebisdithiocarbamate. See maneb.

manganese ethylhexoate. See manganese octoate.

manganese fluoride. See manganous fluoride.

manganese gluconate $Mn(C_6H_{11}O_7)_2 \cdot 2H_2O$.
Properties: Light pinkish powder or coarse pink granules. Soluble in water; insoluble in alcohol and benzene. Combustible. Nontoxic.
Grades: Pharmaceutical; F.C.C.
Containers: Drums.
Uses: Feed additive; food additive and dietary supplement; vitamin tablets.

manganese glycerophosphate $CH_2OHCHOHCH_2OP(O)O_2Mn$.
Properties: Yellow-white or pinkish powder; odorless; nearly tasteless. Soluble in water in presence of citric acid; slightly soluble in water; insoluble in alcohol.
Derivation: Action of glycerophosphoric acid on manganese hydroxide.
Grades: Technical; F.C.C.
Containers: Glass bottles; boxes.
Uses: Medicine (for its glycerophosphate content); food additive and dietary supplement.

manganese green. See barium manganate.

manganese hydrate. See manganic hydroxide; manganous hydroxide.

manganese hydrogen phosphate. See manganous phosphate, acid.

manganese hydroxide. See manganic hydroxide; manganous hydroxide.

manganese hypophosphite $Mn(H_2PO_2)_2 \cdot H_2O$.
Properties: Pink crystals or powder; odorless; tasteless. Soluble in water; insoluble in alcohol. Nontoxic.
Derivation: Interaction of manganese sulfate and calcium hypophosphite.
Grades: Technical; F.C.C.
Containers: Drums.
Hazard: Dangerous fire and explosion risk when heated or in presence of strong oxidizing agents.
Use: Food additive and dietary supplement.

manganese iodide. See manganous iodide.

manganese lactate $Mn(C_3H_5O_3)_2 \cdot 3H_2O$.
Properties: Pale red crystals; soluble in water and alcohol. Combustible.
Derivation: Action of lactic acid on manganese hydroxide.
Use: Medicine.

manganese linoleate $Mn(C_{18}H_{31}O_2)_2$.
Properties: Dark brown, plaster-like mass; soluble in linseed oil. Combustible. Low toxicity.
Derivation: By boiling a manganese salt, sodium linoleate and water.
Uses: Paint and varnish drier; pharmaceutical preparations.

manganese monoxide. See manganous oxide.

manganese naphthenate.
Properties: Hard, brown, resinous mass. Combustible. When precipitated in the cold it is a pale buff in color, but darkens immediately on solution. Soluble in mineral spirits, hardens on exposure to air. M.p. (approx) 130–140°C. Commercial solution contains 6% Mn.
Derivation: Precipitation from mixture of soluble manganese salts and aqueous sodium naphthenate solution.
Containers: 50-lb steel drums.
Hazard: Solution is flammable.
Uses: Paint and varnish drier.

manganese nitrate. See manganous nitrate.

manganese octoate (manganese ethylhexoate) $Mn(OOCC_6H_{13}[C_2H_5])_2$. Commercially formed from 2-ethylhexoic acid and manganous hydroxide. Sold as a clear brown solution in a petroleum solvent containing 6% manganese.
Hazard: Solution is flammable.
Uses: Primarily as a drier for paints, enamels, varnishes, and printing inks.

manganese oleate $Mn(C_{18}H_{33}O_2)_2$.
Properties: Brown, granular mass; soluble in oleic acid and ether; insoluble in water.
Derivation: By boiling manganese chloride, sodium oleate and water.
Hazard: Solution is flammable.
Uses: Paint and varnish drier; medicine.

manganese oxalate $MnC_2O_4 \cdot 2H_2O$.
Properties: White crystalline powder; soluble in dilute acids; very slightly soluble in water. Sp. gr. 2.453; loses $2H_2O$ at 100°C. Combustible.
Derivation: By adding sodium oxalate to manganese chloride.
Hazard: May be toxic.
Uses: Paint and varnish drier.

manganese peroxide. See manganese dioxide.

manganese phosphate. See manganous phosphate. See also manganous phosphate, acid.

manganese phosphate, dibasic. See manganous phosphate, acid.

manganese protoxide. See manganous oxide.

manganese pyrophosphate. See manganous pyrophosphate.

manganese resinate $Mn\ (C_{20}H_{29}O_2)_2$.
Properties: Dark, brownish-black mass or flesh-colored powder. Soluble in hot linseed oil; insoluble in water.
Derivation: By boiling manganese hydroxide, resin oil and water.
Containers: Drums.
Hazard: Flammable. Dangerous fire risk.
Uses: Varnish and oil drier.
Shipping regulations: (Air) Flammable Solid label.

manganese sesquioxide. See manganic oxide.

manganese silicate. See manganous silicate.

manganese sulfate. See manganous sulfate.

manganese sulfide. See manganous sulfide.

manganese sulfite. See manganous sulfite.

manganese tallate. Manganese salts of tall oil fatty acids. Marketed as a solution containing 6% manganese. Used as a drier. Combustible

manganese-titanium. An alloy containing manganese, titanium, aluminum, iron, silicon.
Properties: (regular) M.p. 2650°F; (special) m.p. 2430°F.
Uses: (regular) Deoxidizer in high grade steel; (special) non-ferrous alloys deoxidizer.

manganic acetylacetonate
$Mn[OC(CH_3){:}CHCO(CH_3)]_3$.
Properties: Brown crystalline solid; m.p. 172°C. Combustible.
Derivation: Reaction of a manganese salt with acetylacetone and sodium carbonate.

manganic fluoride MnF_3.
Properties: Red crystalline solid; sp. gr. 3.54; decomposed by water and by heat; soluble in acids.
Hazard: Toxic. Tolerance (as F), 2.5 mg per cubic meter of air.
Use: Fluorinating agent.

manganic hydroxide (manganese hydroxide; hydrated manganic oxide) $Mn(OH)_3$. Rapidly loses water to form $MnO(OH)$.
Properties: Brown powder; sp. gr. 4.2–4.4; m.p., decomposes; decomposes in acids; insoluble in water.
Derivation: By the action of oxygen on precipitated manganous hydroxide.
Containers: Drums.
Uses: Pigment for fabrics; ceramics.

manganic oxide (manganese oxide; manganese sesquioxide) Mn_2O_3. In nature as manganite; a manganese ore.
Properties: Black, lustrous powder, sometimes tinged brown; very hard; sp. gr. 4.5; decomposes at 1080°C; soluble in cold hydrochloric acid, hot nitric acid (decomposes), hot sulfuric acid; insoluble in water. Noncombustible.
Hazard: Toxic by inhalation of dust. Flammable in finely divided form.

manganic oxide, hydrated. See manganic hydroxide.

"Manganin."[155] Trademark for a resistance alloy of copper, 87% and manganese, 13%.
Uses: Thermocouples and electrical instruments.

manganous ammonium sulfate. See manganese ammonium sulfate.

manganous arsenate (manganese arsenate; manganous arsenate, acid) $MnHAsO_4$.
Properties: Reddish-white powder. Hygroscopic. Soluble in acids; slightly soluble in water.
Hazard: Highly toxic.
Shipping regulations: (Rail, Air) Arsenates, n.o.s., Poison label.

manganous bromide (manganese bromide) $MnBr_2 \cdot 4H_2O$.
Properties: Red crystals; loses H_2O at 64°C; very soluble in water; deliquescent. Noncombustible.
Derivation: Action of hydrobromic acid with manganous carbonate, or manganous hydroxide.
Hazard: Moderately toxic and irritant.

manganous carbonate. See manganese carbonate.

manganous chloride (manganese chloride) (a) $MnCl_2$; (b) $MnCl_2 \cdot 4H_2O$.
Properties: Rose-colored crystals; deliquescent. Sp. gr. (a) 2.98; (b) 1.913; m.p. (a) 650°C; (b) 87.5°C; b.p. (a) 1190°C. Very soluble in water; slightly soluble in alcohol; insoluble in ether. Noncombustible.
Grades: C.P.; anhydrous.
Containers: Drums.
Uses: Catalyst in the chlorination of organic compounds; paint drier; dyeing; pharmaceutical preparations; fertilizer compositions, feed additive; dietary supplement.

manganous chromate (manganese chromate; manganous chromate, basic) $2MnO \cdot CrO_3 \cdot 2H_2O$.
Properties: Brown powder; slightly soluble in water with hydrolysis.
Hazard: Toxic by inhalation. Tolerance (as CrO_3): 0.1 mg per cubic meter of air.

manganous citrate. See manganese citrate.

manganous fluoride (manganese fluoride) MnF_2.
Properties: Reddish powder; soluble in acids; insoluble in water, alcohol, and ether. Sp. gr. 3.98; m.p. 856°C. Noncombustible.
Derivation: Action of hydrofluoric acid on manganous hydroxide.
Grade: Technical.
Hazard: Toxic. Tolerance (as F), 2.5 mg per cubic meter of air.

manganous hydroxide (manganese hydroxide) $Mn(OH)_2$. Occurs naturally as pyrochroite.
Properties: White to pink crystals; sp. gr. 3.258; Mohs hardness 2.5; decomposes with heat; insoluble in water and alkali; soluble in acids and ammonium salts.

manganous iodide (manganese iodide) (a) MnI_2; (b) $MnI_2 \cdot 4H_2O$.
Properties: (a) White deliquescent, crystalline mass; (b) rose crystals; sp. gr. (a) 5.01; m.p. 9a) 638°C (in vacuum); b.p. (a) 1061°C; soluble in water with gradual decomposition.
Derivation: Action of hydriodic acid on manganous hydroxide.
Hazard: May be toxic.

manganous nitrate (magnesium nitrate) $Mn(NO_3)_2 \cdot 6H_2O$.
Properties: Pink crystals; very soluble in water; deliquescent; soluble in alcohol. Sp. gr. 1.82; b.p. 129°C; m.p. 26°C.
Hazard: Fire and explosion risk in contact with organic materials.
Uses: Ceramics; intermediates; catalyst; manganese dioxide.
Shipping regulations: Nitrates, n.o.s., (Rail) Yellow label. (Air) Oxidizer label.

manganous orthophosphate. See manganous phosphate.

manganous oxide (manganese protoxide; manganese monoxide; manganese oxide) MnO.
Properties: Grass-green powder; soluble in acids; insoluble in water. Sp. gr. 5.45; m.p. 1650°C, but converted to Mn_3O_4 if heated in air. Nontoxic. Noncombustible.
Derivation: (a) By reduction of the dioxide in hydrogen. (b) By heating the carbonate with exclusion of air.
Grades: Technical.
Uses: Medicine; textile printing; analytical chemistry; catalyst in manufacture of allyl alcohol; ceramics; paints; colored glass; bleaching tallow; animal feeds; fertilizers; food additive and dietary supplement.

manganous phosphate (manganese phosphate; manganous orthophosphate) $Mn_3(PO_4)_2 \cdot 7H_2O$.
Properties: Reddish-white powder; soluble in mineral acids; insoluble in water. Low toxicity.
Derivation: By the action of orthophosphoric acid on manganous hydroxide.
Use: Conversion coating of steels, aluminum and other metals.

manganous phosphate, acid (manganese hydrogen phosphate; manganous phosphate, secondary; manganese phosphate, dibasic) $MnHPO_4 \cdot 3H_2O$.
Properties: Pink powder. Contains some tribasic phosphate. Soluble in acids; slightly soluble in water.
Use: Feed additive.

manganous pyrophosphate (manganese pyrophosphate) (a) $Mn_2P_2O_7$; (b) $Mn_2P_2O_7 \cdot 3H_2O$.
Properties: White powder; sp. gr. (a) 3.71; m.p. 1196°C. Soluble in solutions of potassium or sodium pyrophosphate; insoluble in water.

manganous silicate (manganese silicate) $MnSiO_3$. Occurs naturally as rhodonite.
Properties: Red crystals or yellowish-red powder. Insoluble in water. Sp. gr. 3.72; m.p. 1323°C. Noncombustible.
Derivation: Interaction of manganous salts with sodium silicate.

manganous sulfate (manganese sulfate) $MnSO_4 \cdot 4H_2O$.
Properties: Translucent, pale rose-red, efflorescent prisms. Soluble in water; insoluble in alcohol. Sp. gr. 2.107; m.p. 30°C; anhydrous, m.p. 700°C; decomposes at 850°C.
Derivation: Byproduct of production of hydroquinone; or by the action of sulfuric acid on manganous hydroxide or carbonate.
Grades: Technical; C.P.; fertilizer; feed.
Containers: Bags; car lots; trucks.
Uses: Fertilizers; feed additive; paints and varnishes; ceramics; textile dyes; medicines; fungicides; ore flotation; catalyst in viscose process; synthetic manganese dioxide.

manganous sulfide (manganese sulfide) MnS. Green crystals; sp. gr. 3.99. Decomposes on melting. Almost insoluble in water. Used as an additive in steel.

manganous sulfite (manganese sulfite; manganous sulfite, normal) $MnSO_3$.
Properties: Grayish-black or brownish-red powder. Soluble in solution of sulfur dioxide; insoluble in water.

"Manganox."[250] Trademark for brown manganese oxide.
Uses: Welding electrode coatings; oxidizer and slag former; ceramic colorant.

"Mangrid."[155] Trademark for an electronic tube alloy of 95% nickel and 5% manganese.

Manila fiber. See abaca.

Manila resin. A type of copal resin (q.v.) similar to Congo and kauri.
Properties: Soluble in ether, methyl alcohol and ethyl alcohol; partially soluble in amyl alcohol; insoluble in water. Sp. gr. 1.062; m.p. 230–250°C. Combustible; nontoxic.
Occurrence: Philippine Islands and East Indies.
Uses: Varnishes; paints; lacquers; printing inks.

"Manilyl."[188] Trademark for a replacement for natural ylang ylang oil.

mannitol (manna sugar; mannite) $C_6H_8(OH)_6$. A straight-chain hexahydric alcohol.
Properties: White, crystalline powder or granules; odorless; faint, sweet taste. Soluble in water; slightly soluble in lower alcohols and amines; almost insoluble in other organic solvents. Sp. gr. 1.52; m.p. 165–167°C; specific rotation (20°C) between +23° and +24°; b.p. 290–295°C (at 3–3.5 mm). Nontoxic; combustible.
Derivation: By hydrogenation of corn sugar or glucose.
Grades: Reagent; commercial; N.F.; F.C.C.
Containers: Fiber drums.
Uses: Base for tableting; ingredient in electrolytic condensers; basis of dietetic foods; sweetener, thickener, and stabilizer in food products.

mannitol hexanitrate (hexanitromannite; HNM; nitromannite; nitromannitol) $C_6H_8(ONO_2)_6$.
Properties: Colorless crystals, m.p. 112–113°C. Soluble in alcohol, acetone, ether, insoluble in water.
Derivation: By nitrating mannitol with mixed acid, purifying by precipitation from organic solvents, and stabilizing.
Grades: Technical; pharmaceutical.
Containers: Water-tight wooden barrels for wet shipment.
Hazard: Dangerous fire and explosion risk. An initiating explosive.
Uses: Explosive cap ingredient; medicine (admixed with a large proportion of carbohydrate).
Shipping regulations: (Rail) Consult regulations. Not accepted by express. (Air) Not accepted. Legal label name nitromannite.

D(+)-mannose $C_6H_{12}O_6$. A carbohydrate occurring in some plant polysaccharides.
Properties: Crystals from alcohol or acetic acid; sweet taste with bitter after-taste; m.p. 132°C (dec).
Use: Biochemical research.

"Mansulox."[250] Trademark for a high oxygen-bearing manganese sulfide for use on rimming steels to offset the "killing" action of heavy sulfur additions.

Manufacturing Chemists Association, Inc. (MCA). Founded in 1862, this nonprofit trade association of chemical manufacturers represents more than 90% of the chemical industry of U.S. and Canada. It is particularly active in the fields of packaging, transportation, safe handling of hazardous chemicals, and wording of precautionary labels, as well as in progressive chemical education and all aspects of pollution control. It issues a continuing series of safety data sheets on numerous toxic and flammable materials. Its publications in this field include "Guide to Precautionary Labeling of Hazardous Chemicals" and "Chem-Card Manual." It also has instituted an emergency telephone information service called "Chem-trec" to provide instant information for safety precautions in accidents involving chemicals. Its offices are located at 1825 Connecticut Ave. N.W., Washington, D.C. 20009.

"Mapharsen."[330] Trademark for oxophenarsin hydrochloride.
Hazard: Toxic by ingestion.

"Mapico."[133] Trademark for a series of oxides of iron used in paints, cement products, asbestos shingles, stucco, leather finishes, rubber and electronic ferrites. Colors: Black, brown, red, tan, and yellow. See iron oxide.

"MAPO."[19] Trademark for tris[1-(2-methyl)-aziridinyl]-phosphine oxide (q.v.).

"Mapp."[233] Trademark for a stabilized gas or liquid mixture of methylacetylene, propadiene and hydrocarbon gases, used as an industrial cutting fuel. See also methylene-propadiene, stabilized.

"Maprofix."[328] Trademark for a series of sodium, ammonium, and magnesium lauryl sulfates and lauryl ether sulfates, sodium myristyl sulfate, and several di- and triethanolamine lauryl sulfates. These are anionic detergents, used for foaming, wetting, and emulsifying in cosmetic, household and industrial uses.

"Maprosyl."[328] Trademark for sodium N-lauroyl sarcosinate.

"MAPS."[19] Trademark for tris[1-(2-methyl)-aziridinyl]-phosphine sulfide, $C_9H_{18}N_3PS$. Properties similar to "MAPO" but has lower water solubility.

"Marabond."[121] Trademark for a partially purified calcium lignosulfonate.
Uses: Oil well cement retarders, foundry supplies, ceramic products.

"Maracarb."[121] Trademark for a series of chelating humectant and dispersing agents which are complex mixtures of the salts of lower molecular weight lignosulfonic acids and the salts of the alkaline reversion products of hexoses and pentoses which are produced from wood in the sulfite pulping process. Available in liquid and powder form.
Uses: Fertilizers, agricultural chemicals, dyestuff pastes.

"Maracell-E."[121] Trademark for a partially desulfonated sodium lignosulfate, developed for use as an organic agent for internal boiler treatments over a wide range of temperatures and pressure. Claimed to prevent or inhibit scale formation in boiler tubes, injectors, feed lines and economizers.

"Maracon."[121] Trademark for water-reducing admixtures, based on medium to high molecular weight lignosulfonate fractions.
Containers: 50-lb bags.
Uses: Cement and concrete.

maraging steel. An alloy containing iron, nickel, cobalt and titanium.
Uses: Solid rocket cases.

"Maraglas."[478] Trademark for a clear epoxy resin available in liquid, rods, or sheets.
Uses: Models and encapsulation.

"Maraset."[478] Trademark for an epoxy resin available in liquid or powder foam.
Uses: Adhesive, coating, electrical insulation, and plastic tooling.

"Marasperse."[121] Trademark for a series of lignosulfonates used as dispersants or emulsion-stabilizing agents. The basic lignin monomer unit is a substiuted phenyl propane. Available in various types for specific uses.
Properties: Brown powder; completely soluble in water, insoluble in oils and most organic solvents.
Uses: Dyestuffs; oil-well drilling fluids; gypsum board; agricultural chemical formulations; industrial cleaners; carbon black dispersions; ceramics.

"Maratan."[121] Trademark for highly purified lignosulfonate tanning agents used in conjunction with chrome or vegetable tans for high-quality leather. Available in powder (50-lb paper bags) or liquid form (drums, tank cars).

Marathon-Howard process. A treatment of waste sulfite liquor from sulfite pulp manufacture to recover chemicals and reduce stream pollution. The waste sulfite is treated with lime and precipitates (1) calcium sulfite for use in preparing fresh cooking acid for the sulfite pulp process and (2) a basic calcium salt of lignin sulfonic acid (see lignin sulfonates) which can be pressed and used as a fuel or used as raw material for vanillin, lignin plastics, and other chemicals. The remaining liquor, with its biological oxygen demand reduced 80 percent, is the effluent.

marble. A metamorphic form of calcium carbonate, usually containing admixtures of iron and other minerals, which impart variegated color patterns. Marble chips are often used as source of CO_2 in laboratory experiments.

"Marblette."[478] Trademark for a phenol-formaldehyde resin.
Uses: Imitation jewelry, knife handles, miscellaneous molded articles.

"Marbon."[525] Trademark for a series of high-styrene rubber-reinforcing resins.
800A. An easy-processing resin used with natural and synthetic rubbers for smoother calendering and extrusion, reduced shrinkage, faster mixing with lowered power consumption. Light color; sp. gr. 1.04.
8000AE. Electrical grade resin.

"Marcol."[51] Trademark for low viscosity N.F. white mineral oils used in pharmaceutical and cosmetic products. Also used on food, cigar and cigarette, candy and baking machinery where a nontoxic lubricant is necessary.

"Marex."[322] Trademark for low viscosity ammonium alginate. Used as a stabilizer, moisture-controlling, suspending and plasticizing agent.

"Marezine."[301] Trademark for cyclizine lactate. Used in medicine.

marfanil. See mafenide hydrochloride.

margaric acid. See n-heptadecanoic acid.

marijuana. See cannabis.

"Marincate."[123] Trademark for magnesium trisilicate.

"Marinco."[123] Trademark for magnesium hydroxides, oxides and carbonates.

marjoram oil (calamintha oil).
Properties: Colorless, yellowish or greenish-yellow essential liquid; strong, penetrating odor. Soluble in alcohol, ether, and chloroform. Sp. gr. 0.890–0.910; optical rotation +5° to +18°. Low toxicity. Combustible.
Chief known constituents: Terpineol; terpenes.
Derivation: Distilled from the flowering herb of Origanum marjorana L.
Uses: Perfuming soaps; toilet preparations; flavoring.

marjoram oil, Spanish.
Properties: Volatile, slightly yellow essential oil; odor of camphor. Soluble in most fixed oils; insoluble in glycerin, propylene glycol and mineral oil. Sp. gr. 0.907–0.915 (25°C); refractive index 1.4630–1.4660 (20°C); angular rotation −5° to +10°. Low toxicity. Combustible.
Derivation: Steam distillation of the flowering plant Thymus mastichina L.
Grade: F.C.C.
Use: Flavoring.

"Mark II."[134] Trademark for semibright, sulfur-free nickel plating additive.

"Marlate."[28] Trademark for technical grade methoxychlor insecticides, supplied in 50% wettable powder and an emulsifiable formulation containing 2 lb/gal.

"Marlex."[303] Trademark for a complete family of polyolefin plastics including polyethylenes, low, medium, and high density (ASTM Type I, II, and III); ethylene copolymers, high-density (ASTM Type III); and high-density tailored resins (ASTM Type III). Densities range from 0.914 to 0.970.
Uses (high-density): Heavy-duty automotive greases that are stable to temperature change; blow molding; injection molding, film and sheets.
(Low-density): Coatings; films; wire and cable coatings.

"Marmix."[525] Trademark for a series of multipolymer latices made from monomers including styrene, butadiene, acrylonitrile. Range from soft, rubbery products to the high-styrene reinforcing resinous polymers.

"Marplan."[190] Trademark for isocarboxazid (q.v.).

marsh gas. See methane.

martensite. The chief constituent of hardened carbon tool steels. It is a solution of C or Fe_3C in beta-iron, or an exceedingly fine-grained alpha-iron with C or Fe_3C in atomic or molecular dispersion. Carbon content up to 1%. Easily obtained by quenching small bodies of hypereutectoid steel in cold water. More difficult to obtain in low-carbon steels.

"Marvinol."[248] Trademark for polyvinyl chloride thermoplastic resins.
Properties: White powder or pellets. Actual properties vary with selection of stabilizer, plasticizer, pigment, and filler.
Grades: Thirteen types for calendering, molding, extrusion, and coating including high, medium, and low molecular weight resins, plastisol resins and copolymer resins. Also includes rigid, electrical, flexible extrusion and injection molding compounds.

mash. Mixture of malted barley (or other grain) and water used for preparing wort in brewing operations.

Also mixture of grain, etc., prepared for fermentation in distilling, e.g., "sour mash whiskey."

"Masonex."[92] Trademark for a hemicellulose extract.
Properties: Liquid containing 65% solids, 55% carbohydrates. pH 5.5; water-soluble. The spray-dried product is also water-soluble, with 97% solids and 84% carbohydrates (pH 3.7).
Derivation: From wood fibers after mild acid hydrolysis in steam.
Uses: (liquid) Intermediate in furfuryl production; animal feeds; (dried) binder in refractory bricks; tackifier in adhesives.

"Masonite."[92] Trademark for composition hardboard made by treating wood chips with steam at high pressure, and compressing the resulting fibers into mats from which rigid panels are made by hot-pressing. The fiber is waterproofed with an emulsion having a paraffin base.

masonry cement. A group of special cements more workable and plastic than Portland cement. Some are similar to waterproofed Portland cement while others are Portland cement mixed with hydrated lime, crushed limestone, diatomaceous earth, or granulated slag.

mass. The quantity of matter contained in a particle or body, regardless of its location in the universe. Mass is constant, whereas weight is affected by the distance of a body from the center of the earth, (or of any other planet or satellite, e.g., the moon). At extremely high temperatures, for example, the sun's interior, mass is converted into energy. According to the Einstein equation $E = mc^2$, all forms of energy, such as radiant energy and energy of motion, possess a mass equivalent, even though they have no independent rest mass (photons); thus there is no absolute distinction between mass and energy. See also energy; matter.

mass action, law. See chemical laws (1).

mass conservation law. See chemical laws (4).

mass defect. The difference between the total mass of the constituents of an atomic nucleus (protons and neutrons) taken independently, and the actual mass of the nucleus as a whole. The latter is always slightly less than the sum of the masses of the constituents, the difference being the mass equivalent of the energy of formation (binding energy) of the nucleus. This accounts for the high energy release obtained from nuclear fission. See also fission, nuclear; mass number.

massecuite. A term used in the sugar industry for the mixture of sugar and molasses prior to removal of the molasses.

massicot.

(1) (lead ocher) Natural lead monoxide, PbO. Contains 92.8% lead. Found in United States (Colorado, Idaho, Nevada, and Virginia).

(2) An oxide of lead corresponding to the same formula as litharge (PbO) but having a different physical state. It is a yellow powder formed by the oxidation of a bath of metallic lead at a temperature of about 345°C so that the oxide formed is not melted; sp. gr. 9.3, m.p. 600°C. If the oxide is melted, it is converted into litharge.

mass number. The number of neutrons and protons in the nucleus of an atom. Thus the mass number of normal helium is 4, of carbon 12, of oxygen 16, and of uranium 238. A given nuclide is characterized by its atomic number, equivalent to the number of protons, which give it its charge and thus determine the kind of element, and the number of neutrons, which make up the remainder of its mass. Helium has two protons and two neutrons, (mass number 4 and atomic number 2). Protons and neutrons each have very close to unit mass, and since the mass change associated with binding the particles together into the nucleus is very small, the mass number is always within one-tenth unit of the atomic weight of the nuclide. See also mass; mass defect; atomic weight.

mass spectrometry. A method of chemical analysis in which ions are passed in a vacuum first through an accelerating electric field and then through a strong magnetic field. This has the effect of separating the ions according to their mass, as they traverse the magnetic field at different velocities (electromagnetic separation). Because of their greater kinetic energy, the heavier ions describe a wider arc than the lighter ones. The ions are collected in appropriate devices and are identified on the basis of their mass.

mastic. (1) A solid, resinous exudation of the tree Pistachia lentiscus, of recent geologic origin. Found in the Mediterranean area. Soluble in alcohol and ether; balsamic odor; turpentine taste. Used in chewing gum, varnishes, and to some extent in adhesives and dentistry. (2) A soft, putty-like sealant, often packed in cartridges and applied by a gun with a nozzle of appropriate size, or in bulk for application by a knife or other spreading device. Such mastics often contain bituminous ingredients or polymerized rosin acids, used for laying floor and wall tiles and similar applications.

"Matacil."[181] Trademark for 4-dimethylamino metatolyl methylcarbamate.

material. A nonspecific term used with various shades of meaning in the technical literature. It should not be used as a synonym for substance, but is generally used in the collective expressions "raw materials" and "materials handling." The term "material balance" in chemical engineering denotes the sum of all the substances entering a reaction and all those that leave it in a given time period. "Material" also loosely refers to closely associated mixtures, either of natural origin (petroleum, wood, ores) or man-made (glass, cement, composites). See also substance.

material balance. See material.

materials handling. A general term that includes the methods used for in-plant transportation, distribution, and storage of raw materials and semi-processed products, as by fork-lift trucks, elevators, conveyor systems, pipelines, etc., as well as safe practices for storage and movement of toxic and flammable substances.

matte. A product containing a metal sulfide, as obtained after roasting and fusion of sulfide minerals. Oxides or metals may also be present. Common examples are copper matte and nickel matte.

matter. Anything that has mass or occupies space. (See also mass).

(a) ***states.*** There are three generally accepted states (phases) in which substances can exist, namely, solid, liquid and gas (vapor). From time to time it has been proposed that specialized forms of matter be regarded as states, such as the vitreous (glassy) state, the colloidal state, and the plasma state; but none of these suggestions has gained substantial acceptance. See phase (1); solid; liquid; gas.

(b) *levels*. Matter is basically composed of particles in the following levels of size and complexity: (1) subatomic (protons, neutrons, electrons); (2) atomic and molecular (less than 10 angstroms); (3) colloidal (from 10 angstroms to 1 micron); (4) microscopic; (5) macroscopic; (6) space or celestial. Levels (1) and (2) are invisible by any means; (3) can be resolved in an electron microscope; (4) in an optical microscope; (5) is visible to the naked eye; (6) requires telescopes for detailed observation. See also particle.

Note. The scanning electron microscope is reported to be capable of resolving particles at level (2). See atom.

mayonnaise. A semiliquid salad dressing composed of vegetable oil, vinegar and egg yolk. It is a typical colloidal emulsion in which the mutually repellent base ingredients are brought together and held in a stable form by the lecithin in egg yolk. See also emulsion.

"Maxad."[123] Trademark for magnesium oxide for adsorption application.

"Maxipen."[299] Trademark for potassium phenethicillin.

Mayer's reagent. See mercuric potassium iodide.

"M-B-C."[88] Trademark for a methyl bromide fumigant with chloropicrin added.
Hazard: Toxic by inhalation.

MBMC. Abbreviation for monobutyl-meta-cresol. See tert-butyl-meta-cresol.

MBT. Abbreviation for mercaptobenzothiazole (q.v.).

MBTS. Abbreviation for 2-mercaptobenzothiazyl disulfide. See 2,2′-dithiobisbenzothiazole.

"MBX."[11] Trademark for 2,4-dimethyl-6-tert-butylphenol liquid. Used as an antioxidant.

mc. Abbreviation for millicurie. See curie.

"MC-3."[204] Trademark for a special blend of mild alkalies, complex phosphates, and wetting agents. Especially designed for dairy plant and farm equipment. Sequesters hard water.

MCA. (1) Abbreviation for Manufacturing Chemists' Association, Inc. (q.v.). (2) Abbreviation for monochloroacetic acid. See chloroacetic acid.

"MCC."[84] Trademark for caustic soda flake. Used in cleaning and sterilizing milk, beer, and beverage bottles.

MCPA (2-methyl-4-chlorophenoxyacetic acid; 4-chloro-2-methylphenoxyacetic acid)
$CH_3ClC_6H_3OCH_2COOH$.
Properties: White cyrstalline solid; m.p. 118–119°C. Free acid insoluble in water but sodium and amine salts are soluble.
Grades: Emulsifiable concentrates.
Use: Selective herbicide.

MCPB. 4-(2-methyl-4-chlorophenoxy) butyric acid.
Hazard: Toxic by ingestion.
Use: Herbicide.

MCPP. Abbreviation for 2-(2-methyl-4-chlorophenoxy)-propionic acid. See mecoprop.

Md Symbol for mendelevium.

MDA. (1) Abbreviation for metal deactivator (q.v.). (2) Abbreviation for para,para′-methylenedianiline. See para,para′-diaminodiphenylmethane.

"M.D.A."[138] Trademark for methylene disalicylic acid (q.v.).

MDAC. Abreviation for 4-methyl-7-diethylaminocoumarin. See 7-diethylamino-4-methylcoumarin.

MDI. Abbreviation for methylene di-para-phenylene isocyanate. See diphenylmethane 4,4′-diisocyanate.

Me Symbol for methyl.

MEA. Abbreviation for monoethanolamine. See ethanolamine.

mecamylamine hydrochloride (methylaminoisocamphane hydrochloride) $C_{11}H_{21}N \cdot HCl$.
Properties: White crystalline powder; almost odorless; m.p. 245°C with some decomposition; soluble in water, alcohol, chloroform; somewhat soluble in benzene, isopropyl alcohol; insoluble in ether.
Grade: U.S.P.
Use: Medicine.

mecarbam. Generic name for S-[(ethoxycarbonyl) methylcarbamoyl]methyl O,O-diethyl phosphorodithioate.
$(C_2H_5O)_2P(S)SCH_2CON(CH_3)COOC_2H_5$.
Properties: Yellow oil; b.p. 144°C (0.02 mm).
Hazard: Highly toxic. A cholinesterase inhibitor. Use may be restricted.
Use: Insecticide.
Shipping regulations: Organic phosphate, liquid, n.o.s., (Rail) Poison label (Air) Poison label. Not accepted on passenger planes.

mechlorethamine hydrochloride [methyl-bis-(2-chloroethyl)amine hydrochloride] $CH_3N(CH_2CH_2Cl)_2 \cdot HCl$. A nitrogen mustard.
Properties: White, crystalline, hygroscopic powder; nasal irritant and a vesicant; soluble in water; m.p. 108–111°C.
Grade: U.S.P.
Hazard: Highly toxic; vesicant and strongly irritating to mucous membranes.
Use: Medicine.

meclizine hydrochloride $C_{25}H_{27}ClN_2 \cdot 2HCl \cdot H_2O$. (1-para-Chlorobenzhydryl-4-methylbenzylpiperazine dihydrochloride).
Properties: White or yellowish powder or crystals; slight odor; insoluble in water and ether; very soluble in chloroform, pyridine, and acid-alcohol-water mixtures; slightly soluble in dilute acids and alcohol. Low toxicity.
Grade: U.S.P.
Use: Medicine (antihistamine).

"Mecopex."[401] Trademark for a broad-leaf herbicide containing potassium salt of mecoprop (31.5%).
Hazard: See mecoprop.

mecoprop (2-(4-chloro-2-methylphenoxy)propionic acid; MCPP; CMPP) $ClC_6H_3(CH_3)OCH(CH_3)COOH$.
Properties: Solid; m.p. 93–94°C. Insoluble in water; soluble in alcohol, acetone, and ether.
Hazard: Toxic by ingestion and inhalation; irritant to skin and eyes.
Use: Herbicide.

"Mecostrin."[412] Trademark for dimethyl tubocurarine chloride (q.v.).

"Medi-Calgon."[108] Trademark for a pure white, powdered, sodium hexametaphosphate which has the ability to form a firm coagulum with tissue exudates.
Use: Medicine (topical applications).

medicinal chemistry. A subdivision of chemistry which deals with the effects of drugs and pharmaceuticals on the human body and on various infective organisms, and with the synthesis of compounds specifically for certain diseases, such as antimalarials and antihypertensive agents. It also is concerned with immunology, hormone activity. etc.
See also: clinical chemistry:

medlure. Generic name for sec-butyl-4-(or 5)-chloro-2-methylcyclohexanecarboxylate $C_{12}H_{21}ClO_2$.
Properties: Liquid; b.p. 78–79° (0.25 mm). Insoluble in water; soluble in most organic solvents.
Use: Insect attractant.

"Medomin."[219] Trademark for heptabarbital. See barbiturate.

"Medrol."[327] Trademark for the 6-methyl derivative of prednisolone (q.v.).

medroxyprogesterone acetate (17-hydroxy-6*a*-methylpreg-4-ene-3,20-dione 17 acetate) $C_{24}H_{34}O_4$. A hormone derivative.
Properties: White to off-white, odorless crystalline powder; stable in air. Insoluble in water; freely soluble in chloroform; sparingly soluble in alcohol. Melting range 200 to 208°C.
Use: Medicine (injectable contraceptive).

mega-. Prefix meaning 10^6 units (symbol M). E.g., 1 megaton = 1,000,000 tons.

megatomoic acid (*trans*-3, *cis*-5-tetradecadienoic acid). The active ingredient of the sex attractant in the female carpet beetle. Research is under way on synthesis of this substance to be used as a lure in control of this pest.

megaton. One million tons; usually used in defining the blast effect of a nuclear explosion. A 1-megaton bomb is equivalent in destructive potential to 1 million tons of TNT.

meglumine diatrizoate (methylglucamine diatrizoate; diatrizoate methylglucamine)
$(CH_3CONH)_2C_6I_3COOH \cdot CH_3NHCH_2(CHOH)_4 \cdot CH_2OH$.
Properties: Available in solution for injection; pH between 6.0 and 7.6.
Grade: U.S.P. (as injection).
Use: Medicine (radiopaque medium).
Note: The iodipamide and iothalamate are also available.

MEK. Abbreviation for methyl ethyl ketone.

"Mekomask."[188] Trademark for an industrial odorant especially effective with methyl ethyl ketone.

"Melafix."[443] Trademark for chlorine buffer used in chlorination of wool.

"Melaglaze."[175] Trademark for a synthetic modified resin used primarily as a coating to impart superior gloss and enhanced stain resistance to melamine-formaldehyde products.

melamine (cyanurtriamide; 2,4,6-triamino sym-triazine) $H_2NCNC(NH_2)NC(NH_2)N$. A cyclic trimer of cyanamide.
Properties: White, monoclinic crystals. Sparingly soluble in water, glycol, glycerol, pyridine; very slightly soluble in ethanol; insoluble in ether, benzene, carbon tetrachloride. Sp. gr. 1.573 (14°C); m.p. 354°C; b.p. sublimes. Low toxicity. Nonflammable.
Derivation: (1) From urea. (2) From cyanamide, dicyandiamide, or cyanuric chloride. (3) From ammonia and carbon dioxide (Dutch process).
Method of purification: Recrystallization from water.
Grade: 99% minimum.
Containers: 80-lb bags; fiber drums.
Uses: Melamine resins; organic syntheses; leather tanning.

melamine resin. A type of amino resin (q.v.) made from melamine and formaldehyde. The first step in resin formation is the production of trimethylol melamine, $C_3N_3(NHCH_2OH)_3$, the molecules of which contain a ring with 3 carbon and 3 nitrogen atoms, the $-NHCH_2OH$ groups being attached to the carbon atoms. This molecule can combine further with others of the same kind by a condensation reaction. Excess formaldehyde or melamine can also react with trimethylol melamine or its polymers, providing many possibilities of chain growth and cross-linking. The nature and degree of polymerization depend upon pH, but heat is always needed for curing. Melamine resins are more water-and heat-resistant than urea resins. They may be water-soluble syrups (low m.w.) or insoluble powders (high m.w.) dispersible in water. Widely used as molding compounds with alpha-cellulose, wood flour or mineral powders as fillers, and with coloring materials; also for laminating, boilproof adhesives, increasing wet strength of paper, textile treatment, leather processing, and for dinnerware and decorative plastic items.

Butylated melamine resins are formed by incorporating butyl or other alcohols during resin formation, whereupon the $-NHCH_2OH$ groups convert to $-NHCH_2OC_4H_9$. These resins are soluble in paint and enamel solvents and in surface coatings, often in combination with alkyds. They give exceptional curing speed, hardness, wear resistance, and resistance to solvents, soaps, and foods.

Melamine-acrylic resins are water-soluble and are used for formation of water-base industrial and automotive finishes.
See also urea-formaldehyde resin.

melaniline. See diphenylguanidine.

melanin. A brownish-black pigment that occurs normally in the retina, skin, and hair of higher animals with the exception of albinos. Formed from tyrosine by the action of tyrosinase.

"Melilotine."[227] Trademark for benzodihydropyrone (q.v.).

melissic acid (triacontanoic acid) $CH_3(CH_2)_{28}COOH$. A long-chain fatty acid.
Properties: Crystalline solid; m.p. 94°C; soluble in benzene and hot alcohol; insoluble in water. Combustible; low toxicity.
Derivation: By oxidation of 1-triacontanol; occurs in minor amounts in many plant and insect waxes and in montan wax.
Use: Biochemical research.

melissyl alcohol. See 1-triacontanol and 1-hentriacontanol.

melittin. A polypeptide derived from bee venom that has strong antibacterial activity, especially against Staphylococcus aureus 80, which is resistant to penicillin. It inhibits growth of many Gram-positive and Gram-negative bacteria.

"Mellotone."[188] Trademark for 3-hydroxy-2-methyl-1,4-pyrone. Caramel-butterscotch flavor; used in perfumes and flavors for a soft, malt-like effect.

"Melmac."[57] Trademark for products molded from melamine-formaldehyde resins.

"Melonal."[227] Trademark for 2,6-dimethyl-5-hepten-1-al (q.v.).

"Melostrength."[57] Trademark for a melamine-formaldehyde paper resin designed as a pulp additive to improve wet and dry strength properties of paper.

"Meloxine."[327] Trademark for methoxsalen (q.v.).

melphalan (para-di(2-chloroethyl)aminophenylalanine; formerly called sarcolysin) $(ClCH_2CH_2)_2NC_6H_4CH_2CH(NH_2)COOH$. Melphalan is both the USAN name for the acid, and the generic name for the hydrochloride.
Properties: A nitrogen mustard (q.v.). Crystals; m.p. 180°C.
Grade: N.D. (in medicine, for the acid).
Hazard: Strong irritant to eyes and mucous membranes.
Uses: Medicine; insect chemosterilant.

melting point. The melting point or freezing point of a pure substance is the temperature at which its crystals are in equilibrium with the liquid phase at atmospheric pressure. The term "melting point" is used when the equilibrium temperature is approached by heating the solid. Ordinarily melting points refer to temperatures above 0°C, the melting point of ice. The terms melting point and freezing point are often used interchangeably. The number of calories required to convert one mole of pure crystals to the liquid state is called the molar heat of fusion.

"Meltopax."[337] Trademark for a double silicate, containing 55.5–57.5% zirconium oxide, 27–29% silicon dioxide, and 13.5–15.5% sodium oxide. M.p. 2600°F. Used as an ingredient for enamel frits and special glasses.

"Melurac.'[57] Trademark for urea-melamine-formaldehyde condensation products used mainly as adhesives for bonding of veneers for the production of exterior grade plywood.

"Mema."[147] Trademark for a liquid seed disinfectant which contains 11.4% methoxyethylmercury acetate.
Hazard: Toxic by ingestion.

membrane, semipermeable. A microporous structure, either natural or synthetic, which acts as a highly efficient filter in the range of molecular dimensions, allowing passage of ions, water and other solvents, and very small molecules, but almost impermeable to macromolecules (proteins) and colloidal particles. The thickness is about 100 Angstrom; the pore diameter is from 8 Angstrom for the walls of tissue cells to 100 Angstrom or more for manufactured membranes. Plant cell wall membranes are proteinaceous substances which function in natural osmosis. Membranes of cellophane, collodion, asbestos fiber, etc. are used in such industrial operations as waste liquor recovery, desalination, and electrolysis. See also osmosis; dialysis.

memtetrahydrophthalic anhydride (methyl norbornene dicarboxylic anhydride).
Properties: Clear, transparent, slightly viscous liquid. Colorless to light yellow.
Hazard: Highly toxic; strong irritant to eyes and skin.
Uses: Curing epoxy resins; electrical laminating and filament winding; intermediate for polyesters, alkyd resins and plasticizers.
Shipping regulations: (Rail) White label. (Air) Corrosive label. Not acceptable on passenger planes.

MENA. Abbreviation for the methyl ester of naphthaleneacetic acid. See alpha-naphthaleneacetic acid, methyl ester.

menadione (2-methyl-1,4-naphthoquinone; menaphthone; vitamin K_3) $C_{10}H_5CH_3O_2$.
Properties: Yellow, crystalline powder; nearly odorless; m.p. 105–107°C; affected by sunlight. Soluble in alcohol, benzene, and vegetable oils; insoluble in water.
Derivation: Oxidation of beta-methylnaphthalene.
Grade: U.S.P.
Hazard: Irritant to skin and mucous membranes, especially alcoholic solution.
Uses: Medicine; fungicides; animal feed additive.

menazon. Generic name for S-[(4,6-diamino-s-triazin-2-yl)methyl]O,O-dimethyl phosphorodithioate $(CH_3O)_2P(S)SCH_2\underline{C:NC(NH_2):NC(NH_2):N}$.
Properties: Off-white solid; m.p. 160–162°C. Slightly soluble in water and organic solvents.
Hazard: Highly toxic. A cholinesterase inhibitor.
Uses: Acaricide; insecticide.
Shipping regulations: (Rail, Air) Organic phosphate, solid, n.o.s., Poison label. Not accepted on passenger planes.

Mendéleef, D. I. (1834–1907). Born in Siberia, Mendéleef in 1869 made one of the greatest contributions to chemistry of any scientist by establishing the principle of periodicity of the elements. His first Periodic Table was compiled on the basis of arranging the elements in ascending order of atomic weight and grouping them by similarity of properties. So accurate was Mendeleef's thinking that he predicted the existence and atomic weights of several elements that were not actually discovered till years later. The original table has been modified and corrected several times, notably by Moseley, but it has accommodated the discovery of isotopes, rare gases, etc. Its importance in the development of chemical theory can hardly be overestimated. (See Becquerel, Moseley, periodic table).

mendelevium Md Synthetic radioactive element produced in a cyclotron by bombarding einsteinum with alpha particles; atomic number 101. Atomic weight 256; four isotopes; valence 3. Mendelevium decays by spontaneous fission with a half-life of 1½ hours. The heaviest isotope, Md-258, has half-life of 60 days. Md is thought to have chemical properties similar to those of the rare earth thulium. It is made in research quantities only and no uses are reported. See actinide series.

menhaden oil
Properties: Yellowish-brown or reddish-brown drying oil; characteristic odor. Soluble in ether, benzene, naphtha, and carbon disulfide. Sp. gr. 0.927–0.933; saponification value 191–196; iodine value 139–180; refractive index 1.480. Combustible. Nontoxic.
Derivation: By cooking or pressing the body of the menhaden fish.
Method of purification: Filtration and bleaching with fuller's earth.
Grades; Prine crude; brown strained; strained; bleached; winter oil; bleached winter white oil.
Containers: Drums; tank cars.
Hazard: Subject to spontaneous heating.

Uses: Hydrogenated fats for cooking and industrial use (soap, rubber compounding); printing inks; animal feed; margarine; leather dressing; lubricant; paint drier; cleansers; lipstick.

meniscus. The concave curve of a liquid surface in a graduate or narrow tube caused by surface tension. In reading the value (e.g., 5 cc) it is conventional to ignore the higher liquid around the perimeter. In the case of mercury, which does not wet the tube, the meniscus is convex.

menthanediamine (para-menthane-1,8-diamine) $(CH_3)_2C(NH_2)CHCH_2CH_2C(CH_3)(NH_2)CH_2CH_2$. A primary alicyclic diamine; also a tert-alkylamine.
Properties: Clear liquid; terpene odor; boiling range 107–126°C (10 mm); f.p. −45°C; refractive index 1.4794 (25/D); miscible with water and most organic solvents.
Containers: Drums; tank cars.
Hazard: Strong irritant to eyes and skin. Use eye protection.
Uses: Curing agent for epoxy resins; chemical intermediate.

para-**menthane hydroperoxide.**
Properties: Clear, pale yellow liquid; sp. gr. 0.910–0.925 (15.5/4°C); refractive index 1.460–1.475 (20°C).
Hazard: Strong irritant to skin and eyes; strong oxidizing agent; dangerous in contact with organic materials.
Uses: Catalyst for rubber and polymerization reactions; coatings.
Shipping regulations: (Air) Solution over 60%, not acceptable; not over 60% in non-volatile solvent, Organic Peroxide label; (Rail) Oxidizing material, n.o.s., Yellow label.

para-**menhtan-3-one.** See menthone.

menthol (hexahydrothymol; methylhydroxyisopropylcyclohexane; para-menthan-3-ol; peppermint camphor) $CH_3C_6H_9(C_3H_7)OH$. It may be *l*- (from natural sources) or *dl*- (natural or synthetic).
Properties: White crystals with cooling odor and taste; m.p. 41–43°C (*l*-form); congealing temperature 27–28°C (*dl*-form); specific rotation (25/D) (*l*-menthol) −45 to −51°, (*dl*-menthol) −2 to +2°. Soluble in alcohol, light petroleum solvents, glacial acetic acid, and fixed or volatile oils; slightly soluble in water. Low toxicity. Combustible.
Occurrence: Brazil (natural product); Japan.
Derivation: By freezing from peppermint oil; by synthesis from citronellal, thymol, or turpentine oil.
Grades: Technical; U.S.P.; F.C.C.
Hazard: Moderate irritant to mucous membranes on inhalation.
Uses: Perfumery; cigarettes; liqueurs; flavoring agent; chewing gum; chest rubs; cough drops.

menthol acetic ester. See menthyl acetate.

menthol valerate (menthyl isovalerate) $(CH_3)_2CHCH_2COOC_{10}H_{19}$.
Properties: Colorless liquid; mild odor; cooling, faintly bitter taste; sp. gr. 0.907 (15.4°C); insoluble in water; soluble in alcohol, chloroform, ether, and oils.
Derivation: By action of valeric acid on menthol.
Uses: Medicine; flavoring.

menthone (para-menthan-3-one) $C_{10}H_{18}O$.
Properties: Colorless, oily, mobile liquid; slight peppermint odor; slightly soluble in water; soluble in organic solvents. Sp. gr. 0.897 (15°C); b.p. 207°C. Low toxicity. Combustible.
Derivation: A ketone found in oil of peppermint.
Use: Flavoring.

menthyl acetate (menthol acetic ester) $C_{10}H_{19}OOCCH_3$.
Properties: Colorless liquid. Menthol-like odor. Slightly soluble in water; miscible with alcohol and ether. B.p. 227 to 228°C; sp. gr. 0.922–0.927; optical rotation −72° 47′ to −73° 18′; refractive index 1.447. Low toxicity. Combustible.
Derivation: (a) By Boiling menthol with acetic anhydride in the presence of sodium acetate. (b) Peppermint oil.
Uses: Perfumery; flavoring.

menthyl isovalerate. See menthol valerate.

menthyl salicylate $C_6H_4(OH)COOC_{10}H_{19}$.
Properties: Colorless liquid; miscible with alcohol, ether, chloroform, and fatty oils in all proportions. Insoluble in water; soluble in organic solvents. Combustible. Low toxicity.
Uses: Medicine; sunscreen preparations.
See also homomenthyl salicylate.

"Mentor."[51] Trademark for nonviscous oils of the mineral seal type, used mainly as absorption media in vapor recovery systems, such as those recovering coal-tar solvents and casinghead gasoline.

MEP. Abbreviation for methyl ethyl pyridine.

meperidine hydrochloride $C_{15}H_{21}NO_2HCl$. An addictive drug. Use by prescription only.

meprobamate (2-methyl-2-n-propyl-1,3-propanediol dicarbamate) $H_2NCOOCH_2C(CH_3)(C_3H_7)CH_2OOCNH_2$.
Properties: White powder; m.p. 103–107°C; characteristic odor and bitter taste; slightly soluble in water and ether; soluble in alcohol and acetone.
Grade: N.F.
Hazard: Central nervous system depressant. Not addictive. Use restricted by law.
Use: Medicine (antianxiety agent).

meq Abbreviation for milliequivalent. See eq.

"Merac."[204] Trademark for a liquid dithiocarbamate rubber accelerator.

meralluride. Methoxyhydroxymercuripropylsuccinylurea ($C_9H_{16}HgN_2O_6$) and theophylline ($C_7H_8N_4O_2 \cdot H_2O$) in approximately molecular proportions.
Properties: White to slightly yellow powder; affected slowly by light; soluble in glacial acetic acid and solutions of alkali hydroxides; slightly soluble in water; moderately soluble in hot water. Saturated solution is acid to litmus.
Grade: U.S.P.
Hazard: Toxic by ingestion.
Use: Medicine.
Shipping regulations: (Rail, Air) Mercury compounds, n.o.s., Poison label.

merbromin (dibromohydroxymercurifluorescein disodium salt; 2,7-disodiumdibromo-4-hydroxymercurifluorescein) $C_{20}H_8Br_2HgNa_2O_6$.
Properties: Iridescent, green scales or granules; odorless; soluble in water; insoluble in alcohol, acetone, chloroform, or ether. Stable in air.
Derivation: From dibromofluorescein and mercuric acetate.
Grades: Technical; N.F.
Hazard: Toxic by ingestion. Tolerance, 0.05 mg per cubic meter of air.
Use: Medicine (antiseptic).

Shipping regulations: (Rail, Air) Mercury compounds, n.o.s., Poison label.

"Mercadium"[266] Trademark for calcined coprecipitations of cadmium sulfide and mercury sulfide, with or without barium sulfate; available in concentrated and lithopone forms; colors: orange, red, maroon.
Hazard: Toxic by ingestion.
Uses: Artists' colors; plastics colorants.

mercaptamine. See 2-aminoethanethiol.

mercaptan. See thiol.

"Mercaptate."[458] Trademark for mercaptocarboxylate esters of polyols. High boiling, viscous liquids with a sharp, sulfur-like odor. Useful for chemically and physically modifying polymeric systems.

mercaptoacetic acid. See thioglycolic acid.

2-mercaptobenzoic acid. See thiosalicylic acid.

2-mercaptobenzothiazole (MBT)

CHCHCHCHCCSC(SN)N.

Properties: Yellowish powder. Distinctive odor (depends on degree of purification). Soluble in dilute caustic, alcohol, acetone, benzene, chloroform; insoluble in water and gasoline. Sp. gr. 1.52; melting range 164–175°C. Low toxicity. Combustible.
Grades: Technical; 97%.
Containers: 1-, 5- and 10-lb fiber cans; 55-lb drums; multiwall paper sacks.
Uses: Vulcanizatoin accelerator for rubber (requires use of stearic acid for full activation); tire treads and carcasses, mechanical specialties, etc.; fungicide; corrosion inhibitor in cutting oils and petroleum products; extreme-pressure additive in greases.

2-mercaptobenzothiazyl disulfide. See 2,2′-dithiobisbenzothiazole.

mercaptoethanol $HSCH_2CH_2OH$.
Properties: Water-white, mobile liquid with disagreeable odor; sp. gr. 1.1168 (20/20°C); 9.29 lb/gal (20°C); b.p. 157.1°C; vapor pressure 1.0 mm (20°C); viscosity 3.43 cp. (20°C; completely soluble in water, and most organic solvents; flash point 165°F; f.p., sets to a glass below −100°C; refractive index (n 20/D) 1.5011. Combustible.
Containers: Fiber cans; drums; multiwall paper sacks.
Hazard: Moderately toxic by inhalation and ingestion.
Uses: Solvent for dyestuffs; intermediate for producing dyestuffs, pharmaceuticals, rubber chemicals, flotation agents, insecticides, plasticizers, textile assistants and other compounds; water-soluble reducing agent; biochemical reagent.

beta-**mercaptoethylamine hydrochloride** $HS(CH_2)_2NH_2 \cdot HCl$.
Properties: Hygroscopic white powder; m.p. 71°C; very soluble in water and ethyl alcohol.
Hazard: Moderately toxic by inhalation and ingestion.

N(**2-mercaptoethyl)benzenesulfonamide** S-(O,O-**diisopropyl phosphorodithioate**) (N-(beta-diisopropyldithiophosphorylethyl)benzenesulfonamide) $C_6H_5SO_2NH(CH_2)_2SP(S)[OCH(CH_3)_2]_2$.
Properties: Colorless liquid, slight odor; sp. gr. 1.25 (22°C). Insoluble in water; soluble in kerosine and xylene.
Hazard: Highly toxic. Cholinesterase inhibitor. Use may be restricted.
Use: Herbicide.
Shipping regulations: Organic phosphate, liquid, n.o.s. (Rail) Poison label. (Air) Poison label. Not accepted on passenger planes.

2-mercapto-4-hydroxypyrimidine. See thiouracil.

2-mercaptoimidazoline. See "NA-22."

beta-**mercaptopropionic acid** $HSCH_2CH_2COOH$.
Properties: Clear liquid; sp. gr. 1.218 (21°C); m.p. 16.8°C; b.p. 111°C (15 mm); soluble in water and alcohol.
Uses: Stabilizer; antioxidant; catalyst; chemical intermediate.

6-mercaptopurine (6-MP; 6-purinethiol) $C_5H_4N_4S \cdot H_2O$. A sulfur-containing purine base not found in animal nucleoproteins.
Properties: Yellow, crystalline powder; m.p. 308°C (decomposes); nearly odorless; insoluble in water; soluble in hot alcohol and dilute alkali solutions; slightly soluble in dilute H_2SO_4.
Grade: U.S.P.
Use: Medicine (to prevent rejection of kidney transplants).

mercaptosuccinic acid. See thiomalic acid.

2-mercaptothiazoline C_2H_4NSSH. White crystals.
Use: Synthesis of pharmaceuticals.

D-**3-mercaptovaline.** See penicillamine.

mercerized cotton. See cotton, mercerized.

mercerizing assistant. A compound used to increase the penetration of mercerizing baths. Cresylic acid derivatives, special sulfonated oils, and other wetting agents are typical.

mercuric acetate (mercury acetate) $Hg(C_2H_3O_2)_2$.
Properties: White, crystalline powder. Soluble in alcohol and water; sensitive to light. Sp. gr. 3.2544.
Hazard: Highly toxic by ingestion, inhalation, and absorption; strong irritant. Tolerance, 0.01 mg per cubic meter of air.
Uses: Medicine; catalyst in organic synthesis; pharmaceuticals.
Shipping regulations: (Rail, Air) Poison label.

mercuric ammonium chloride (mercury ammonium chloride) $HgCl_2 \cdot 2NH_4Cl \cdot 2H_2O$.
Properties: White poweder; soluble in water; slightly soluble in alcohol.
Hazard: Highly toxic by ingestion, inhalation, and skin absorption; strong irritant. Tolerance, 0.05 mg per cubic meter of air.
Uses: Medicine; veterinary medicine.
Shipping regulations: (Rail, Air) Poison label.
See also mercury, ammoniated.

mercuric arsanilate. See mercury atoxylate.

mercuric arsenate (mercury arsenate) $HgHAsO_4$.
Properties: Yellow powder. Soluble in hydrochloric acid; slightly soluble in nitric acid; insoluble in water.
Hazard: Highly toxic by ingestion, inhalation, and skin absorption; strong irritant. Tolerance, 0.05 mg per cubic meter of air.
Uses: Medicine; waterproof paints; antifouling paints.
Shipping regulations: (Rail, Air) Poison label.

mercuric barium bromide (barium mercury bromide; mercury barium bromide) $HgBr_2 \cdot BaBr_2$.
Properties: Colorless, crystalline mass. Very hygroscopic. Soluble in water.
Hazard: Highly toxic by ingestion, inhalation, and

skin absorption; strong irritant. Tolerance, 0.05 mg per cubic meter of air.
Shipping regulations: (Rail, Air) Poison label.

mercuric barium iodide (barium mercury iodide; mercury barium iodide) $HgI_2 \cdot BaI_2 \cdot 5H_2O$.
Properties: Reddish or yellow, crystalline mass; unstable; deliquescent. Soluble in alcohol and water.
Hazard: Highly toxic by ingestion, inhalation, and skin absorption; strong irrirant. Tolerance, 0.05 mg per cubic meter of air.
Uses: Microanalysis (testing for alkaloids); preparing Rohrbach solution.
Shipping regulations: (Rail, Air) Poison label.

mercuric benzoate (mercury benzoate) $Hg(C_7H_5O_2)_2 \cdot H_2O$.
Properties: White crystals; m.p. 165°C; sensitive to light. Soluble in solutions of sodium chloride and ammonium benzoate; slightly soluble in alcohol and water.
Derivation: By the interaction of a mercuric salt and sodium benzoate.
Hazad: Highly toxic by ingestion, inhalation, and skin absorption; strong irritant. Tolerance, 0.05 mg per cubic meter of air.
Use: Medicine.
Shipping regulations: (Rail, Air) Poison label.

mercuric bromide (mercury bromide) $HgBr_2$.
Properties: White, rhombic crystals. Sensitive to light. Soluble in alcohol and ether; sparingly soluble in water. Sp. gr. 6.11; m.p. 235°C; b.p. 322°C.
Derivation: By adding potassium bromide to a solution of a mercuric salt and crystallizing.
Hazard: Highly toxic by inhalation, ingestion, and skin absorption. Strong irritant. Tolerance, 0.05 mg per cubic meter of air.
Use: Medicine.
Shipping regulations: (Rail, Air) Poison label. Not acceptable on passenger planes.

mercuric chloride (mercury bichloride; mercury chloride, corrosive) $HgCl_2$.
Properties: White crystals or powder; odorless. Soluble in water, alcohol, ether, pyridine, glycerine and acetic acid ester. Sp. gr. 5.44 (25°C); 276°C; b.p. 303°C.
Derivation: (a) Direct combination of chlorine with mercury heated to volatilizing point; (b) by subliming mercuric sulfate with common salt.
Method of purification: Recrystallization and sublimation.
Impurities: Mercurous chloride.
Grades: Technical; crystals; granular; powder; C.P.; N.F.
Containers: 100-lb drums.
Hazard: Highly toxic by ingestion, inhalation, and skin absorption. May be fatal! Tolerance, 0.05 mg per cubic meter of air.
Uses: Manufacture of calomel and other mercury compounds; disinfectant; organic synthesis; analytical reagent; metallurgy; tanning; catalyst for vinyl chloride; sterilant for seed potatoes; fungicide, insecticide and wood preservative; embalming fluids; textile printing; dry batteries; photography; process engraving and lithography.
Shipping regulatoins: (Rail, Air) Poison label. Not acceptable on passenger planes.

mercuric chloride, ammoniated. See mercury, ammoniated.

mercuric cuprous iodide (copper mercury iodide; mercury copper iodide) $HgI_2 \cdot 2CuI$.
Properties: Dark red, crystalline powder; sp. gr. 6.12. Insoluble in alcohol and water.
Hazard: Highly toxic. Tolerance, 0.05 mg per cubic meter of air.
Use: Thermoscopy (detecting overheating of machine bearings) by reversible color change.
Shipping regulations: (Rail, Air) Mercury compounds, Poison label.

mercuric cyanate. See mercury fulminate.

mercuric cyanide (mercury cyanide) $Hg(CN)_2$.
Properties: Colorless, transparent prisms, darkened by light. Soluble in water and alcohol. Sp. gr. 4.018; m.p., decomposes.
Derivation: Interaction of mercuric oxide and an aqueous solution of hydrocyanic acid.
Grades: Technical; C.P.
Hazard: Highly toxic by ingestion, inhalation, and skin absorption. Tolerance,0.05 mg per cubic meter of air.
Uses: Medicine; germicidal soaps; manufacturing cyanogen gas; photography.
Shipping regulations: (Rail, Air) Poison label.

mercuric dichromate. See mercury dichromate.

mercuric dimethyldithiocarbamate $C_6H_{12}N_2S_4Hg$. Used in turf fungicides.
Hazard: Highly toxic by ingestion, inhalation, and skin absorption. Tolerance, 0.05 mg per cubic meter of air.
Shipping regulations: (Rail, Air) Mercury compounds, n.o.s. Poison label.

mercuric dioxysulfate. See mercuric subsulfate.

mercuric fluoride (mercury fluoride) HgF_2.
Properties: Transparent cyrstals; sp. gr. 8.95; m.p. 645°C, decomposes; moderately soluble in water and alcohol.
Derivation: Mercuric oxide and hydrofluoric acid.
Hazard: Highly toxic by ingestion, inhalation, and skin absorption. Strong irritant. Tolerance, 0.05 mg per cubic meter of air.
Use: Synthesis of organic fluorine compounds.
Shipping regulations: (Rail, Air) Mercury compounds, n.o.s., Poison label.

mercuric iodate (mercury iodate) $HgIO_2)_2$.
Properties: White powder. Soluble in hydrochloric acid, hydrobromic acid, hydriodic acid, water (containing sodium chloride or potassium iodide); insoluble in alcohol and water.
Hazard: Highly toxic by ingestion, inhalation, and skin absorption. Strong irritant. Tolerance, 0.05 mg per cubic meter of air.
Use: Medicine.
Shipping regulations: (Rail, Air) Mercury compounds, n.o.s., Poison label.

mercuric iodide (mercury iodide, red; mercury iodide, yellow) HgI_2.
Properties: (a) Red, tetragonal crystals; turn yellow when heated to 150°C, returning to red on cooling; (b) Yellow, rhombic crystals. Soluble in boiling alcohol, and in solutions of sodium thiosulfate or potassium iodide or other hot alkali chloride solutions; almost completely insoluble in water. Sp. gr. (a) 6.2–6.35, (b) 5.91–6.06; m.p. (b) 259°C, b.p. 349°C.
Derivation: (a) By the direct union of mercury and iodine. (b) As a precipitate by adding potassium iodide to a solution of a mercuric salt. Yellow form precipitates from alcoholic solutions.
Grades: Technical; reagent.
Containers: Bottles, drums.

Hazard: Highly toxic by ingestion, inhalation, and skin absorption. Strong irritant. Tolerance, 0.05 mg per cubic meter of air.
Uses: Medicine; analytical reagents (Nessler's reagent; Mayer's reagent).
Shipping regulations: (Rail, Air) Poison label.
See also mercuric potassium iodide.

mercuric lactate $Hg(C_3H_5O_3)_2$.
Properties: White crystalline powder; soluble in water; decomposed by heat.
Hazard: Highly toxic by ingestion, inhalation, and skin absorption. Tolerance, 0.01 mg per cubic meter of air.
Use: Fungicide.
Shipping regulations: (Rail, Air) Mercury compounds, n.o.s., Poison label.

mercuric naphthenate (mercury naphthenate)
Properties: Dark amber liquid; soluble in lubricating oils and mineral spirits. Wt/gal 10.4 lbs.
Grades: 25% mercury.
Containers: 80-lb drums.
Hazard: Highly toxic by ingestion, inhalation, and skin absorption. Tolerance, 0.05 mg per cubic meter of air.
Uses: Mildew-resistance promoter in paints; anti-knock compound.
Shipping regulations: (Rail, Air) Mercury compounds, n.o.s., Poison label.

mercuric nitrate (mercury nitrate; mercury pernitrate) $Hg(NO_2)_2 \cdot H_2O$.
Properties: Colorless crystals or white deliquescent powder. Sp. gr. 4.3; m.p. 79°C; decomposed by heat. Soluble in water and nitric acid; insoluble in alcohol.
Derivation: By action of hot nitric acid on mercury.
Hazard: Highly toxic. Dangerous fire risk in contact with organic materials. Tolerance, 0.05 mg per cubic meter of air.
Uses: Nitration of aromatic organic compounds; medicine; felt manufacture; mercury fulminate.
Shipping regulations: Nitrates, n.os., (Rail) Yellow label. (Air) Oxidizer label.

mercuric oleate (mercury oleate)
Properties: Yellowish to red liquid, semi-solid or solid mass. Soluble in fixed oils; insoluble in water; slightly soluble in alcohol or ether.
Derivation: By mixing yellow mercuric oxide with oleic acid.
Grades: Technical; pharmaceutical.
Uses: Medicine; antiseptic; antifouling paints.
Shipping regulations: (Rail, Air) Poison label.

mercuric oxide, red (red precipitate; mercury oxide, red) HgO.
Properties: Heavy, bright, orange-red powder. Soluble in dilute hydrochloric and nitric acids; slightly soluble in water, more so after boiling; insoluble in alcohol and ether. Sp. gr. 11.00–11.29; m.p., decomposes.
Derivation: By heating mercurous nitrate.
Grades: Technical; paint; A.C.S. reagent.
Containers: 100-lb drums.
Hazard: Highly toxic. Fire risk in contact with organic materials. Tolerance, 0.05 mg per cubic meter of air.
Uses: Chemicals; paint pigment; perfumery and cosmetics; pharmaceuticals for topical disinfection; ceramics (pigment); batteries, especially for miniaturized equipment; polishing compounds; analytical reagent; antifouling paints; fungicide.
Shipping regulations: (Rail, Air) Poison label.

mercuric oxide, yellow (mercury oxide, yellow; yellow precipitate) HgO.
Properties: Light, orange-yellow powder; odorless; stable in air but turns dark on exposure to light; finer powder than the red form; sp. gr. 11.03 (27.5°C); m.p., decomposes. Slightly soluble in cold water, more so after boiling; soluble in dilute hydrochloric and nitric acids, potassium iodide solution, concentrated solutions of alkaline-earth chloride, magnesium chloride; insoluble in alcohol.
Derivation: (a) By the action of either potassium hydroxide or sodium hydroxide on mercuric chloride. (b) By the action of sodium carbonate upon mercuric nitrate solution.
Grades: C.P.; technical; N.F.; paint.
Containers: 100-lb drums.
Hazard: Highly toxic. Fire risk in contact with organic materials. Tolerance, 0.05 mg per cubic meter of air.
Uses: Medicine; mercury compounds; see also mercuric oxide, red.
Shipping regulations: (Rail, Air) Poison label.

mercuric oxycyanide $HgO \cdot Hg(CN)_2$.
Properties: White crystalline powder; sp. gr. 4.44. Moderately soluble in water.
Hazard: Highly toxic. Detonates on heating; dangerous explosion risk. Tolerance, 0.05 mg per cubic meter of air.
Use: Medicine.
Shipping regulations: (Rail, Air) Consult authorities.

mercuric phosphate (normal mercuric phosphate; neutral mercuric phosphate; trimercuric orthophosphate; mercuric phosphate, tertiary; mercury phosphate) $Hg_3(PO_4)_2$.
Properties: Heavy, white or yellowish powder. Soluble in acids; insoluble in alcohol, water.
Hazard: Highly toxic. Tolerance, 0.05 mg per cubic meter of air.
Shipping regulations: (Rail, Air) Mercury compounds, n.o.s., Poison label.

mercuric potassium cyanide (mercury potassium cyanide) $Hg(CN)_2 \cdot 2KCN$.
Properties: Colorless crystals. Soluble in water and alcohol.
Derivation: By mixing mercuric and potassium cyanides and crystallizing.
Hazard: Highly toxic by ingestion, inhalation, and skin absorption. Tolerance, 0.05 mg per cubic meter of air.
Use: Silvering glass in mirror manufacture.
Shipping regulations: (Rail, Air) Poison label.

mercuric-potassium iodide (Mayer's reagent; potassium mercuric iodide). See also Nessler's reagent. K_2HgI_4 or $2KI \cdot HgI_2$.
Properties: Odorless, yellow crystals, deliquescent in air. Crystallizes with either 1, 2, or 3 molecules of water. The commercial product is the anhydrous form containing about 25.5% mercury. Sp. gr. 4.29. Neutral or alkaline to litmus. Very soluble in water; soluble in alcohol, ether and acetone.
Derivation: (a) Action of hydrochloric acid and potassium iodide on mercuric cyanide or mercuric chloride. (b) Action of potassium iodide on mercuric oxide.

Containers: 1-lb and 5-lb bottles.
Hazard: Highly toxic by ingestion, inhalation, and skin absorption. Tolerance, 0.05 mg per cubic meter of air.
Uses: Medicine; chemical analysis.
Shipping regulations: (Rail, Air) Poison label.

mercuric salicylate (salicylated mercury).
Properties: White powder, yellow or pink tinge. A compound of mercury and salicylic acid of somewhat varying composition, mercury replacing both phenolic and carboxylic hydrogen. Contains not less than 54% nor more than 59.5% mercury. Odorless; tasteless. Soluble in solutions of the fixed alkalies or their carbonates, and in warm solutions of the alkali halides; almost insoluble in water and alcohol.
Derivation: By gently heating freshly precipitated yellow mercuric oxide and salicylic acid in presence of water.
Hazard: Highly toxic by ingestion, inhalation, and skin absorption. Tolerance, 0.05 mg per cubic meter of air.
Use: Medicine.
Shipping regulations: (Rail, Air) Poison label.

mercuric silver iodide (mercury silver iodide; silver mercury iodide) $HgI_2 \cdot 2AgI$.
Properties: Yellow powder; becomes red at 40–50°C. Sp. gr. 6.08; soluble in solutions of potassium cyanide or potassium iodide; insoluble in acids (dilute), and water.
Hazard: Highly toxic by ingestion, inhalation, and skin absorption. Tolerance, 0.05 mg per cubic meter of air.
Use: Thermoscopy (detecting overheating in journal bearings).
Shipping regulations: (Rail, Air) Mercury compounds, n.o.s., Poison label.

mercuric stearate (mercury stearate) $(C_{17}H_{35}CO_2)_2Hg$.
Properties: Yellow, granular powder; soluble in fatty oils; slightly soluble in alcohol.
Hazard: Highly toxic by ingestion, inhalation, and skin absorption. Tolerance, 0.05 mg per cubic meter of air.
Uses: Medicine; germicide.
Shipping regulations: (Rail, Air) Mercury compounds, n.o.s., Poison label.

mercuric subsulfate (basic mercuric sulfate; mercuric dioxysulfate) $Hg(HgO)_2SO_4$.
Properties: Heavy, lemon-yellow powder or bright-yellow scales. Soluble in dilute acids; very slightly soluble in water, more so in hot water. Sp. gr. 6.444.
Derivation: Addition of water to normal mercuric sulfate.
Hazard: Highly toxic by ingestion, inhalation, and skin absorption.
Use: Medicine.
Shipping regulations: (Rail, Air) Poison label.

mercuric sulfate (mercury persulfate; mercury sulfate) $HgSO_4$.
Properties: White, crystalline powder. Soluble in acids; insoluble in alcohol; decomposes in water. Sp. gr. 6.466; m.p., decomposes at red heat.
Derivation: By the action of sulfuric acid on mercury, with subsequent crystallization.
Hazard: Highly toxic by ingestion, inhalation, and skin absorption. Tolerance, 0.05 mg per cubic meter of air.
Uses: Medicine; producing calomel and corrosive sublimate; catalyst in the conversion of acetylene to acetaldehyde; extracting gold and silver from roasted pyrites; galvanic batteries.
Shipping reguations: (Rail, Air) Poison label.

mercuric sulfate, basic. See mercuric subsulfate.

mercuric sulfide, black (mercury sulfide, black) HgS.
Properties: Black powder. Soluble in sodium sulfide solution; insoluble in water, alcohol, and nitric acid. Sp. gr. 7.55–7.70; m.p., sublimes at 446°C; m.p. 583°C.
Derivation: By passing hydrogen sulfide gas into a solution of a mercury salt or the reaction of mercury with sulfur.
Hazard: Highly toxic by ingestion, inhalation, and skin absorption. Tolerance, 0.05 mg per cubic meter of air.
Use: Pigment.
Shipping regulations: (Rail, Air) Mercury compounds, n.o.s., Poison label.

mercuric sulfide, red (vermilion; quicksilver vermilion; Chinese vermilion; red mercury sulfide; artificial cinnabar; red mercury sulfuret) HgS.
Properties: Fine, bright scarlet powder. Insoluble in water and alcohol. Sp. gr. 8.06–8.12; m.p., sublimes at 583°C.
Derivation: By heating mercury and sulfur, with subsequent recovery by sublimation. A precipitated form is known as English vermilion.
Hazard: Highly toxic by ingestion, inhalation, and skin absorption. Tolerance, 0.05 mg per cubic meter of air.
Uses: Medicine; pigment.
Shipping regulations: (Rail, Air) Mercury compounds, n.o.s., Poison label.

mercuric thiocyanate (mercuric sulfocyanate; mercuric sulfocyanide; mercury sulfocyanate; mercury thiocyanate) $Hg(SCN)_2$.
Properties: White powder. Slightly soluble in alcohol; insoluble in water. M.p., decomposes.
Derivation: By precipitation of mercuric nitrate with ammonium sulfocyanate and subsequent solution in a large amount of hot water and crystallizing.
Hazard: Highly toxic by ingestion, inhalation, and skin absorption. Tolerance, 0.05 mg per cubic meter of air.
Uses: Photography; producing "Pharoah's serpents."
Shipping regulations: (Rail, Air) Poison label.

"Mercurochrome."[348] Trademark for merbromin (q.v.).
Hazard: Highly toxic by ingestion; but less irritating than iodine to skin and tissue.
Use: General antiseptic, commonly in the form of a 2% aqueous solution.

mercurol (mercury nucleate).
Composition: Contains 20% mercury.
Properties: Brown powder. Soluble in water; insoluble in alcohol.
Hazard: Highly toxic by ingestion, inhalation, and skin absorption.
Use: Medicine.
Shipping regulations: (Rail, Air) Poison label.

mercurous acetate (mercury proto-acetate; mercury acetate) $HgC_2H_3O_2$.
Properties: Colorless scales or plates. Decomposed by boiling water and by light into mercury and mercuric acetate. Slightly soluble in water; insoluble in alcohol and ether; soluble in dilute nitric acid.
Derivation: Reaction of sodium acetate with mercurous nitrate solution acidified with nitric acid.
Grade: Technical.

Hazard: Highly toxic by ingestion, inhalation, and skin absorption. Tolerance, 0.01 mg per cubic meter of air.
Shipping regulations: (Rail, Air) Poison label.

mercurous acetylide Hg_2C_2.
Properties: White powder. A salt of acetylene.
Derivation: Reaction of acetylene with aqueous solution of mercurous salts.
Hazard: Severe explosion risk when shocked or heated. Highly toxic.
Shipping regulations: (Rail, Air) Explosive. Not acceptable.

mercurous bromide (mercury bromide) HgBr or Hg_2Br_2.
Properties: White powder or colorless crystals. Odorless, tasteless. Becomes yellow on heating, returning to white on cooling. Darkens on exposure to light. Soluble in fuming nitric acid (prolonged heating), hot concentrated sulfuric acid, hot ammonium carbonate or ammonium succinate solutions; sparingly soluble in water; insoluble in alcohol and ether. Sp. gr. 7.307; sublimes at 340–350°C; m.p. 405°C.
Derivation: (a) Action of potassium bromide on solution of mercurous nitrate in dilute nitric acid. (b) Sublimation from mixture of mercury and mercuric bromide.
Hazard: Highly toxic by ingestion, inhalation, and skin absorption. Tolerance, 0.05 mg per cubic meter of air.
Use: Medicine.
Shipping regulations: (Rail, Air) Poison label.

mercurous chlorate (mercury chlorate) $Hg_2(ClO_3)_2$.
Properties: White crystals. Sp. gr. 6.409; m.p. 250°C (decomposes). Soluble in alcohol, water, and acetic acid.
Hazard: Highly toxic. Explodes in contact with organic or combustible materials. Keep away from light.
Shipping regulations: Chlorates, n.o.s. (Rail) Yellow label. (Air) Oxidizer label.

mercurous chloride (mercury monochloride; mercury protochloride; mercury chloride, mild; calomel) Hg_2Cl_2.
Properties: White, rhombic crystals or crystalline powder. Odorless, stable in air, but darkens on exposure to light. Insoluble in water, ether, alcohol and cold dilute acids. Sp. gr. 6.993; m.p. 302°C; b.p. 384°C. Decomposed by alkalies.
Derivation: By heating mercuric chloride and mercury, with subsequent sublimation.
Grades: Technical; C.P.; N.F.
Hazard: Toxic in large doses; must be removed from body by cathartics.
Uses: Fungicide; electrodes; pharmaceuticals.
Shipping regulations: Not restricted.

mercurous chromate (mercury chromate) Hg_2CrO_4.
Properties: Brick-red powder. Variable composition. Decomposes on heating. Soluble in nitric acid (conc.); insoluble in alcohol and water.
Hazard: Highly toxic. Moderate fire hazard in contact with organic materials. Tolerance, 0.05 mg per cubic meter of air.
Use: Ceramics (coloring green).
Shipping regulations: (Rail, Air) Mercury compounds, n.o.s., Poison label.

mercurous iodide (mercury protoiodide) HgI or Hg_2I_2.
Properties: Bright yellow powder, becoming greenish on exposure to light due to decomposition into metallic mercury and mercuric iodide. Becomes dark yellow, orange and orange-red on heating. Undergoes same color change in opposite order on cooling. Odorless and tasteless. Soluble in castor oil, liquid ammonia, aqua ammonia; insoluble in water, alcohol, and ether. Sp. gr. 7.6445–7.75; sublimes at 140°C; m.p. 290°C (with partial decomposition).
Derivation: (a) Action of potassium iodide on a mercurous salt. (b) Boiling a solution of mercurous nitrate containing nitric acid with excess of iodine.
Grade: Technical.
Hazard: Highly toxic by ingestion, inhalation, and skin absorption. Tolerance, 0.05 mg per cubic meter of air.
Use: Medicine (external).
Shipping regulations: (Rail, Air) Poison label.

mercurous nitrate, hydrated $HgNO_3 \cdot 2H_2O$.
Properties: Short prismatic crystals; effloresces and becomes anhydrous in dry air. Sensitive to light. Soluble in small quantities of warm water (hydrolyzes in larger quantities), water acidified with nitric acid. Sp. gr. 4.785 (3.9°C); m.p. 70°C, decomposes.
Derivation: Action of cold dilute nitric acid upon an excess of mercury and warming slightly.
Hazard: Highly toxic. May be explosive if shocked or heated. Tolerance, 0.05 mg per cubic meter of air.
Uses: Medicine; cosmetics; analytical agent.
Shipping regulations: (Rail, Air) Poison label.

mercurous oxide Hg_2O.
Properties: Black powder; sp. gr. 9.8; decomposes at 100°C; soluble in acids; insoluble in water.
Derivation: Action of sodium hydroxide on mercurous nitrate.
Hazard: Highly toxic. Tolerance, 0.05 mg per cubic meter of air.
Shipping regulations: (Rail, Air) Poison label.

mercurous sulfate Hg_2SO_4.
Properties: White to yellow crystalline powder; soluble in hot sulfuric acid, dilute nitric acid; almost insoluble in water. Sp. gr. 7.56. Decomposes on heating.
Derivation: (a) Dissolving mercury in sulfuric acid (2.3) and heating gently. (b) Adding sulfuric acid to mercurous nitrate solution.
Hazard: Highly toxic. Tolerance, 0.05 mg per cubic meter of air.
Uses: Chemical (admixed with sulfuric acid as a catalyst, in oxidation of naphthalene to phthalic acid); laboratory batteries (Clark cell, Weston cell).
Shipping regulations: (Rail, Air). Poison label.

mercury (quicksilver; hydrargyrum) Hg Metallic element of atomic number 80, group IIb of the periodic table. Atomic weight 200.59; valences 1, 2. There are 4 stable isotopes and 12 artificially radioactive isotopes.
Properties: Silvery, extremely heavy liquid, sometimes found native. Insoluble in hydrochloric acid; soluble in sulfuric acid upon boiling; readily soluble in nitric acid. Insoluble in water, alcohol, and ether. Soluble in lipids. Extremely high surface tension (480 dynes per cm), giving it unique rheological behavior. High electric conductivity. Sp. gr. 13.59; f.p. −38.85°C; b.p. 356.6°C. Noncombustible.
Chief ore: Cinnabar (q.v.).
Occurrence: Spain, Yugoslavia, Mexico, Italy, Canada, U.S.S.R.
Derivation: By heating cinnabar in air, or with lime, and condensing the vapor.

Method of purification: Distillation. An important proportion of used mercury is recovered by redistillation.
Grades: Technical; virgin; redistilled; A.C.S.
Containers: 1-, 5-, 10-lb bottles; 76-lb flasks.
Hazard: (1) Mercury, metallic: Highly toxic by skin absorption and inhalation of fume or vapor. Tolerance, 0.05 mg per cubic meter of air. Absorbed by respiratory and intestinal tract; accidental intake of small amounts is stated to be harmless (Merck Index). FDA permits zero addition to the 20 micrograms of Hg contained in average daily diet.
(2) All inorganic compounds of Hg are highly toxic by ingestion, inhalation, and skin absorption. Tolerance, 0.05 mg per cubic meter of air.
(3) Most organic compounds of Hg are highly toxic. Inorganic Hg can be converted to methylmercury by bacteria in water. Tolerance (alkyl compounds), 0.01 mg per cubic meter of air; all others, 0.05 mg.
Note: Spillage may be a toxic hazard due to droplet proliferation. Clean-up requires special care.
Uses: Fungicides and bactericides; amalgams; catalyst; electrical apparatus; cells for production of chlorine and caustic soda; instruments (thermometers, barometers, etc.); antifouling paints; mildew-proofing preparations; pulp and paper manufacture; pharmaceuticals; mercury vapor lamps.

mercury, ammoniated (mercuric chloride, ammoniated; ammonobasic mercuric chloride; ammoniated mercury chloride; white precipitate; white precipitate, fusible; aminomercuric chloride; mercury cosmetic) $HgNH_2Cl$.
Properties: White, pulverulent lumps or powder; earthy, metallic taste; odorless; stable in air; darkens on exposure to light. Soluble in ammonium carbonate and sodium thiosulfate solutions and in warm acids; insoluble in water and alcohol.
Derivation: By precipitating mercuric chloride with ammonium hydroxide in excess.
Grades: U.S.P.; technical.
Containers: 100-lb drums.
Hazard: Highly toxic. Tolerance, 0.05 mg per cubic meter of air.
Use: Medicine (local anti-infective); pharmaceuticals.
Shipping regulations: (Rail, Air) Poison label. See also mercuric ammonium chloride.

mercury atoxylate (mercury para-aminophenylarsenate; mercuric arsanilate) $Hg[OOAs(OH)C_6H_4NH_2]_2$.
Properties: White powder containing 31.8% mercury; very slightly soluble in water. Used as a 10% suspension in olive oil, or as a 5% ointment.
Hazard: Highly toxic. Tolerance, 0.05 mg per cubic meter of air.
Use: Medicine.
Shipping regulations: (Rail, Air) Poison label.

mercury bichloride. Legal label name (Rail) for mercuric chloride.

mercury cell. An electrolytic cell for the production of caustic soda and chlorine from sodium chloride brine. Continuously fed brine is decomposed in one compartment between graphite anodes, where chlorine is liberated, and a mercury cathode, where a sodium amalgam is formed. The amalgam flows continuously or intermittently to a second compartment where it is decomposed with water, forming a caustic solution. The decomposition is usually performed electrolytically by making the amalgam anodic with respect to an iron or graphite cathode. Pure water is supplied to the decomposition compartment at such a rate as to maintain a constant concentration of caustic in the product. With respect to the diaphragm cell, the mercury cathode cell has the advantages of producing a very pure caustic and, generally, a more concentrated solution (50–70%); it has the disadvantages of a higher operating voltage and lower efficiency (52–55%), and a high capital investment in mercury. Examples of mercury-cathode cells are Castner cell and DeNora cell.

mercury compounds. See corresponding mercurous or mercuric compound, e.g., mercury chlorate, see mercurous chlorate; mercury oxide, see mercuric oxide.

mercury dichromate (mercuric dichromate; mercury bichromate) $HgCr_2O_7$.
Properties: Heavy, red, crystalline powder; soluble in acids; insoluble in water.
Hazard: Highly toxic. Tolerance, 0.05 mg per cubic meter of air.
Shipping regulations: (Rail, Air) Mercury compounds, n.o.s., Poison label.

mercury fulminate (mercuric cyanate) $Hg(CNO)_2$.
Properties: Gray crystalline powder. Soluble in alcohol, ammonium hydroxide and hot water; slightly soluble in cold water. Sp. gr. 4.42; m.p., explodes.
Derivation: By treating mercury with strong nitric acid and alcohol.
Grade: Technical.
Containers: Canvas bags in stone crocks filled with water; in lots of five pounds in glass bottles.
Hazard: Explodes readily when dry; keep moist till used. An initiating explosive. Highly toxic. Tolerance, 0.05 mg per cubic meter of air.
Uses: Manufacture of caps and detonators for producing explosions of military, industrial and sporting purposes.
Shipping regulations: (Air) Not accepted. Legal label name: fulminate of mercury. (Rail) Consult regulations. Not accepted by express.

mercury selenide HgSe. Sp. gr. 8.266. Sublimes in a vacuum. Insoluble in water. Used as a semiconductor in solar cells, thin-film transistors, infrared detectors, ultrasonic amplifiers.
Hazard: Highly toxic. Tolerance, 0.05 mg per cubic meter of air.
Shipping regulations: (Rail, Air) Mercury compounds, n.o.s., Poison label.

mercury telluride HgTe. Grade 99.99%. Used as a semiconductor in solar cells, thin-film transistors, infrared detectors, ultrasonic amplifiers.
Shipping regulations: (Rail, Air) Mercury compounds, n.o.s., Poison label.

"Merginamide."[451] Trademark for a polyamide resin formed by reacting aliphatic polyamides with an alkyl-aryl polycarboxylic acid. It is claimed to impart to epoxy coatings chemical resistance, controlled curing, low viscosity, high gloss retention, and good wetting characteristics.

myricyl alcohol. See 1-triacontanol.

"Merlon."[567] Trademark for a polycarbonate resin. Transparent, slightly straw-colored resin; sp. gr. 1.2; refractive index 1.587. Rods, tubes, pipes, sheets, film, extrusions and special shapes may be produced by extrusion, thermoforming, injection and blow moldings.
Uses: Electrical and electronic component industries, protective equipment, graphic arts and photographic

film, cable wrapping and protective overlays for other thermoplastic sheeting; missile applications.

"Mermeth."[123] Trademark for veterinary sulfamethazine.

merocyanine. See cyanine dye.

"Merox"[416] Proprietary process for extracting easily removed mercaptans and converting the remaining mercaptans in naphthas to disulfides, thereby yielding a product of low mercaptan content. The combination of extraction and sweetening is applicable to gasoline and lighter hydrocarbons, and the sweetening step to many jet fuels and kerosines. It employs caustic soda which is regenerated by blowing with air in the presence of a special catalyst to oxidize the mercaptans to disulfides. Disulfides are decanted from the caustic in the extraction application, while in the sweetening application they remain in the hydrocarbon.

"Merpentine."[28] Trademark for a surface-active, compounded anionic wetting, penetrating, and leveling agent used in leather and textile fields.

"Merphos."[40] Trademark for tributylphosphorotrithioite (q.v.).

"Merpol."[28] Trademark for a series of surface-active agents used as dyeing assistants.

"Mersize."[58] Trademark for paper-sizing agents. "Mersize" TFL 70%, 77% and "Mersize" RM Dry are forms of chemically fortified rosin-based sizes. "Mersize" CD-2 and CD-2 Dry are used in conjunction with typical rosin sizes.

"Mersolite."[81] Trademark for a series of phenyl mercuric compounds. Bactericide and fungicide used in paint, adhesive and agriculture industries.
Hazard: Highly toxic.

"Mer-Sorb."[123] Trademark for solid sorbitol, industrial grade.

"Mertaste."[123] Trademark for straight ribonucleotides used as food flavoring.

"Merthiolate"[100] Trademark for a 0.1% solution of thiomersal (q.v.) in alcohol. Widely used as general antiseptic.
Hazard: Highly toxic by ingestoin; for external use only.

"Mertone" WB-4.[58] Trademark for a 30% colloidal dispersion of silica in water; used as a precoat for reproduction papers.

"Mesamoll."[470] Trademark for an alkyl sulfonic ester of phenol and cresol. Used as plasticizer for vinyls and as lubricating agent in urethane foaming machines.

mescaline (3,4,5-trimethoxyphenethylamine) $(CH_3O)_3C_6H_2CH_2CH_2NH_2$. An alkaloid derived from a Mexican cactus.
Properties: Crystals; m.p. 35–36°C; b.p. 180°C; soluble in water, alcohol, chloroform, benzene; nearly insoluble in ether.
Hazard: Toxic by ingestion and inhalation.
Uses: Biochemical and medical research; hallucinogenic drug.

mesitylene (1,3,5-trimethylbenzene; sym-trimethylbenzene) $C_6H_3(CH_3)_3$. Liquid; sp. gr. 0.863; f.p. −52.7°C; b.p. 164.6°C; insoluble in water; soluble in alcohol and ether. Derived from coal tar. Combustible.
Containers: Drums; tank trucks.
Hazard: Toxic by inhalation. Moderate fire hazard. Tolerance, 25 ppm in air.
Uses: Intermediate, including anthraquinone vat dyes; ultraviolet oxidation stabilizers for plastics.

mesityl oxide (isopropylideneacetone; methyl isobutenyl ketone; 4-methyl-3-penten-2-one) $(CH_3)_2C{:}CHCOCH_3$.
Properties: Oily, colorless liquid. Honey-like odor. Soluble in water, alcohols, and ethers. Sp. gr. 0.8569 (20/20°C); b.p. 130–131°C; vapor pressure 8.7 mm (20°C); flash point 90°F; wt/gal 7.1 lb (20°C); freezing point −46.4°C; viscosity 0.0060 poise (20°C). Autoignition temp. 652°F.
Derivation: Dehydration of acetone or diacetone alcohol.
Grade: Technical.
Containers: 1-gal cans; 5-, 55-gal drums; tank cars.
Hazard: Flammable, moderate fire risk. Toxic by ingestion, inhalation, and skin absorption. Tolerance, 25 ppm in air.
Uses: Solvent for cellulose esters and ethers, oils, gums, resins, lacquers, roll-coating inks, stains, ore flotation; paint and varnish-removers; insect repellent.

meso-. A prefix meaning middle or intermediate; specifically:
(1) An optically inactive stereoisomeric form resulting from the presence of an even number of dextro- and levorotatory isomers in a natural substance. This cancels the optical activity and causes formation of an intermediate structure which is its own mirror image. This effect is called internal compensation. Tartaric acid is an example. See also racemization; tartaric acid.
(2) An intermediate hydrated form of an inorganic acid.
(3) Designating a middle position in certain cyclic organic compounds, or
(4) A ring system characterized by a middle position of certain rings.

mesomerism. See resonance. (1).

messenger RNA. See ribonucleic acid.

mesyl chloride. See methanesulfonyl chloride.

Met Symbol for methionine.

meta-. A prefix. For definition of meta-compounds see under ortho-.

metabolism. The chemical transformations occurring in an organism from the time a nutrient substance enters it until it has been utilized and the waste products eliminated. In animals and man, digestion and absorption are primary steps, followed by a complicated series of degradations, syntheses, hydrolyses and oxidations, in which agents such as enzymes (q.v.), bile acids, and hydrochloric acid take part. These transformations are often localized with respect to organs, tissues, and types of cells involved.

Basal metabolism is the rate of total heat production of an individual who is awake, but in complete mental and physical respose, at comfortable temperature, and without having had food for at least 12 hours. Under these conditions, oxidation of stored nutrients provides the sole source of energy expended, and heat is measurable by calorimetry. See also digestion (1).

metabolite. An intermediate material produced and used in the processes of a living cell or organism. Metabolites are used for replacement and growth in living tissue, and are also broken down to be a source of energy in the body. Examples are nucleic acids, enzymes, glucose, cholesterol, and many similar substances.

"Meta Bond."[526] Trademark for microcrystalline zinc phosphates. Used in the preparation of steel, galvanized steel and aluminum for decorative and protective finishing. Application by spray or immersion.

metaboric acid HBO_2. White crystals; sp. gr. 2.486; m.p. 236 ± 1°C; slightly soluble in water.

metacetone. See diethylketone.

"Metadelphene."[266] Trademark for a commercial grade of N,N-diethyl-meta-toluamide; an insect repellent.

metaformaldehyde. See sym-trioxane.

"Metafos."[164] Trademark for a sodium phosphate glass commonly called sodium hexametaphosphate. A linear polymer containng a minimum of 67% P_2O_5. See sodium polyphosphate.

metal. An element that forms positive ions when its compounds are in solution and whose oxides form hydroxides rather than acids with water. About 75% of the elements are metals, which occur in every group of the Periodic Table except VIIA and the noble gas group. Most are crystalline solids with metallic luster, conductors of electricity, and have rather high chemical reactivity; many are quite hard and have high physical strength. They also readily form solutions (alloys) with other metals. The presence of very low percentages of other elements (not necessarily metals) profoundly affects the properties of many metals, e.g., carbon in iron. Mercury and cesium are liquid at room temperature. Geologically, metals usually occur in the form of compounds that must be physically or chemically processed to yield the pure metal; common methods are application of heat (smelting), carbon reduction, electrolysis, and reduction with aluminum or magnesium. Metals fall into the following classifications, which are not mutually exclusive (see specific entry for further information):

alkali metals	rare metals
alkaline-earth metals	rare-earth metals
transition metals	actinide metals
noble metals	light metals
platinum metals	heavy metals

"Metalam."[233] Trademark for packaging films based on aluminum foil.

"Metalate."[244] Trademark for a white crystalline material containing 29.8% total sodium oxide; used in laundry industry.

metal deactivator. A compound added to gasoline to neutralize the catalytic effect of copper in promoting fuel oxidation.

metaldehyde $(CH_3CHO)_n$. A polymer of acetaldehyde, in which n usually is 4 to 6.

Properties: White prisms; decomposes with partial regeneration of acetaldehyde, when heated above 80°C. Soluble in benzene, chloroform; slightly soluble in alcohol, ether; insoluble in water. Sublimes 112–115°C; m.p. 246°C in sealed tube.

Hazard: Flammable, dangerous fire risk. Toxic; strong irritant to skin and mucous membranes.

Uses: Fuel to replace alcohol; to destroy snails and slugs.

Shipping regulations: (Rail) Flammable solid, n.o.s., Yellow label. (Air) Flammable Solid label.

metal dye. See dye, metal.

metal fiber. See fiber.

metal, formed. A light metal such as aluminum or zinc, to which titanium hydride has been added. The latter involves hydrogen to give a blown structure with about the same specific gravity as water.

metallic bond. See bond, chemical.

metallic soap. See soap (2).

"Metalliding."[245] A proprietary surface-alloying process in which ions are electrolytically dissolved from a metal anode in a bath of molten fluoride salts and diffuse into the surface of the cathode. The latter may be any of a number of transition metals or rare earths. The alloys are intended for use in space vehicles, and also have potential application in the automotive field.

metallized dye. A soluble dye including any one of a variety of metals chemically combined, applied to wool in an acid bath, by use of sodium chloride to salt out the dye onto the fiber.

metallizing. Coating a plastic, or similar material, with a deposit of metal (usually aluminum) by means of vacuum deposition (q.v.). The thickness of such films may vary from .01 to as much as 3 mils. Metallized plastic films are used for yarns, packaging, stamping foil, labels, etc.

metallocene. An organometallic coordination compound obtained as a cyclopentadienyl derivative of a transition metal or metal halide. The metal is bonded to the cyclopentadienyl ring by electrons moving in orbitals extending above and below the plane of the ring (pi bond). There are three types of metallocenes: (a) dicyclopentadienyl-metals with the general formula $(C_5H_5)_2M$; (b) dicyclopentadienyl-metal halides with the general formula $(C_5H_5)_2MX_{1-3}$; (c) monocyclopentadienyl-metal compounds with the general formula $C_5H_5MR_{1-3}$, where R is CO, NO, halide group, alkyl group, etc. Types (a) and (b) are known as molecular sandwiches since the two cyclopentadiene rings lie above and below the plane on which the metal atom is situated.

Most metallocenes are crystalline and those belonging to the first transition series of the periodic table, from vanadium to nickel, have the same melting point, 173°C. The metal complexes are soluble in many organic solvents while the halides are soluble primarily in polar solvents. When the central metal atom is in a stable oxidation state the metallocene is not decomposed by high temperatures, air, water, dilute acids or bases.

Some metallocenes, such as ferrocene, undergo a wide variety of aromatic ring substitution reactions, including Friedel-Crafts acylation, arylation, and sulfonation; a few, such as nickelocene and cobaltocene, are too unstable to be directly substituted.

Derivation: The most important industrial method is the reaction of a cyclopentadienide (for example the sodium salt) with a transition metal halide in an organic solvent.

Uses: Catalysts, polymers, ultraviolet absorbers, reducing agents, free radical scavengers, antiknock agents.

See also ferrocene, cobaltocene, nickelocene, titanocene dichloride, zirconocene dichloride, uranocene.

metalloid. See nonmetal; semiconductor.

metallurgical coke. See coke.

metal, powdered. Metals are produced in powdered form for a variety of uses in several industries. In this form they are the raw materials for powder metallurgy, in which the powders are pressed in molds and heated (sintered) at high temperature. Metal powders range in size from −325 mesh (0.045–0.060 mm "diameter") to +100 mesh, and are available in practically all industrial metals. They are produced by machining, milling, shotting, granulation, atomizing, condensation, reduction, chemical precipitation or electrodeposition. Their properties and purities vary with the method of preparation.
Hazard: Dangerous fire risk, especially as dust. Toxic by inhalation.
Uses: Electric, automotive, machinery, tool and refractory-metal industries; paint pigments; flares and incendiary bombs; brazing materials; calorizing, metal-spraying, metallurgical agents; heat-generating agents; catalysts, etc.

"Metalyn."[266] Trademark for a series of distilled methyl esters of tall oil fatty acids.

"Metandren."[305] Trademark for methyltestosterone (q.v.).

metanilic acid (meta-sulfanilic acid; meta-aminobenzenesulfonic acid) $C_6H_4(NH_2)SO_3H$.
Properties: Small colorless needles. Soluble in water, alcohol, and ether.
Derivation: By the reduction of metal-nitrobenzenesulfonic acid. Nitrobenzene is sulfonated until the product is soluble in water. The mixture is then poured into water and reduced with iron, made alkaline with lime and the lime salt dissociated with sodium carbonate.
Hazard: Toxic; see aniline.
Uses: Dyes; medicine.

metaphosphoric acid. See phosphoric acid, meta-.

metapon A narcotic drug, methyldihydromorphine. See also narcotic.

metaraminol bitartrate (*l*-meta-hydroxynorephedrine bitartrate; *l*-alpha-(1-aminoethyl)-meta-hydroxybenzyl alcohol bitartrate)
$HOC_6H_4CH(OH)CH(CH_3)NH_2 \cdot C_4H_6O_6$.
Properties: White, practically odorless, crystalline powder; melting range 170–176°C; soluble in water; somewhat soluble in alcohol; insoluble in ether and in chloroform; pH of 5% solution 3.2 and 3.5.
Grade: U.S.P.
Use: Medicine.

"Metasap"[309] Trademark for a series of heavy-metal soaps. Palmitates, stearates and hydroxystearates are used in compounding them.
Uses: Paints, inks, greases, waxes, cosmetics, water repellents, lubricants, stabilizers, pharmaceutical tablets.

"Metasol."[123] Trademark for a series of chemicals used as fungicides, mildewicides, bactericides, slime control agents for pulp and paper mill systems and as preservatives for all types of systems. A recently developed type (J-26) is claimed to be a broad-spectrum antimicrobial agent with a minimum ecological hazard.

"Meta-Systox R."[181] Trademark for O,O-dimethyl-S-2-(ethylsulfinyl)ethyl phosphorothioate (q.v.).

metepa. Generic name for tris(methyl-1-aziridinyl) phosphine oxide $(C_3H_6N)_3PO$.
Properties: Amber liquid, amine odor. B.p. 118°C at 1 mm; sp. gr. 1.079. Miscible with water and organic solvents.
Hazard: Toxic by ingestion and skin absorption. Strong irritant to skin.
Uses: Insect chemosterilant; addition products for textile treatments, adhesives, paper and rubber processing; crosslinking agent in polymer systems which contain active hydrogens; monomer.
Shipping regulations: (Solution) (Air) Corrosive label.

meter. The basic unit of length of the metric system. (39.37 inches). Originally defined as one ten-millionth of the distance from the equator to the North Pole. Now defined as 1,650,763.73 wave lengths of the orange-red line of the isotope krypton 86.

"Methac."[65] Trademark for a series of blends of methyl acetate with methanol in varying proportions.
Hazard: Toxic by ingestion.
Uses: Lacquer solvents, paint removers, organic synthesis.

"Methacrol."[28] Trademark for a series of acrylic resin dispersions used as antisnag and bodying agents for nylon hosiery.

methacrolein (methacrylaldehyde) $CH_2{:}C(CH_3)CHO$.
Properties: Liquid; sp. gr. 0.8474 (20/20°C); b.p. 68.0°C; flash point 5°F (O.C.); solubility in water at 20°C 5.9% by weight. Shipped with 0.1% hydroquinone as polymerization inhibitor.
Hazard: Flammable, dangerous fire risk. Toxic; strong irritant.
Uses: Copolymers; resins.
Shipping regulations: Flammable liquid, n.o.s., (Rail) Red label. (Air) Flammable Liquid label.

methacrylamide $CH_2{:}C(CH_3)CONH_2$. Solid; m.p. 110°C. A monomer for acrylic resins.

methacrylate ester. Ester of methacrylic acid having the formula $CH_2{:}C(CH_3)COOR$, where R is usually methyl, ethyl, isobutyl, or n-butyl-isobutyl (50–50). They are supplied commercially as the polymers. See acrylic resin.

methacrylate resin. See acrylic resin.

methacrylatochromic chloride.
$H_2C{:}C(CH_3)\underline{C{:}OCrCl_2OHCrCl_2O}$.
Properties: Water-soluble solid.
Derivation: Reaction of methacrylic acid with basic chromic chloride.
Hazard: Probably toxic.
Uses: Water repellent; nonadhesive; insolubilizer for vinyl polymer.

methacrylic acid (monomer) (alpha-methacrylic acid) $CH_2{:}C(CH_3)COOH$.
Properties: Colorless liquid; m.p. 15–16°C; b.p. 161–162°C; sp. gr. 1.015 (20°C). Flash point 170°F; combustible. Soluble in water, alcohol, ether, most organic solvents. Polymerizes readily to give water-soluble polymers.
Derivation: Reaction of acetone cyanohydrin and dilute sulfuric acid; oxidation of isobutylene.
Grades: 40% Aqueous solution, b.p. 76–78°C (25 mm); crude monomer 85% pure; glacial (98%).

Containers: Carboys; drums; tank cars.
Hazard: Strong irritant to skin.
Uses: Monomer for large-volume resins and polymers; organic synthesis. Many of the polymers are based on esters of the acid, as the methyl, butyl, or isobutyl esters. See acrylic resin.

alpha-**methacrylic acid.** See methacrylic acid.

beta-**methacrylic acid.** See crotonic acid.

methacrylonitrile (MAN; "Vistron")
Properties: Clear, colorless liquid; b.p. 90.3°C; f.p. −35.8°C; flash point (Tag O.C.) 55°F; sp. gr. 0.789. Thermoplastic; resistant to solvents, alkalies, acids.
Hazard: Flammable, dangerous fire risk. Highly toxic by ingestion, inhalation, and skin absorption.
Uses: Vinyl nitrile monomer; copolymer with styrene, butadiene, etc.; elastomers, coatings, plastics.
Shipping regulations: Flammable liquid, n.o.s., (Rail) Red label. (Air) Flammable Liquid label.

gamma-**methacryloxypropyltrimethoxysilane** $CH_2{:}C(CH_3)COOCH_2CH_2CH_2Si(OCH_3)_3$.
Properties: Liquid; sp. gr. 1.045 (25°C); b.p. approx. 80°C (1 mm); refractive index 1.4285 (25°C); flash point 135°F. Soluble in acetone, benzene, ether, methanol, hydrocarbons. Combustible.
Grade: 97% min. purity.
Containers: Up to 55-gal.
Hazard: Moderate fire risk. May be toxic.
Uses: Coupling agent for promotion of resin-glass, resin-metal, and resin-resin bonds; for formulation of adhesives having "built in" primer systems.

methadone hydrochloride (*dl*-6-dimethylamino-4,4-diphenyl-3-heptanone hydrochloride). A synthetic narcotic. $(C_6H_5)_2C(COC_2H_5)CH_2CH(CH_3)N(CH_3)_2 \cdot HCl$.
Properties: Crystalline substance with bitter taste; no odor; m.p. 232–235°C; soluble in water, alcohol, and chloroform; practically insoluble in ether and glycerol. pH (1% aqueous solution) 4.5–6.5.
Grade: U.S.P.
Hazard: Toxic. Addictive narcotic. Use restricted.
Use: Medicine (sedative, treating heroin addiction).

methallenestril (beta-ethyl-6-methoxy-alpha, alpha-dimethyl-2-naphthalenepropionic acid) $CH_3OC_{10}H_6CH(C_2H_5)C(CH_3)_2COOH$.
Properties: Crystals; m.p. 132.5°C. Soluble in ether, vegetable oils.
Use: Medicine.

methallyl acetate. See methylallyl acetate.

methallyl alcohol. See methylallyl alcohol.

beta-**methallyl chloride.** See beta-methylallyl chloride.

methallylidene diacetate $CH_2{:}C(CH_3)CH(OCOCH_3)_2$.
Properties: Liquid; sp. gr. 1.510 (20/20°C); b.p. 191.0°C; f.p. −15.4°C; flash point 215°F (COC). Slightly soluble in water. Combustible.
Uses: Chemical intermediate; can provide controlled release of methacrolein in acid solution.

methamphetamine hydrochloride. See amphetamine.

methanal. See formaldehyde.

methanamide. See formamide.

methanation. A reaction by which methane is formed from the hydrogen, carbon monoxide, and carbon dioxide derived from coal gasification. It requires a catalyst, e.g., nickel, and temperatures in the range of 500°C. In one process the reaction is performed in an adiabatic fixed-bed reactor. See also gasification.

methandrostenolone (17*a*-methyl-17*b*-hydroxyandrosta-1,4-dien-3-one). An androgenic hormone; anabolic synthetic steroid.
Use: Medicine.

methane (marsh gas; firedamp) CH_4. The first member of the paraffin (alkane) hydrocarbon series.
Properties: Colorless, odorless, tasteless gas; lighter than air, practically inert toward sulfuric acid, nitric acid, alkalies, and salts but reacts with chlorine and bromine in light (explosively in direct sunlight); flash point −306°F; b.p. −161.6°C; f.p. −182.5°C; autoignition temp. 1000°F; density of vapor 0.554 (0°C) (760 mm, air = 1); critical temperature −82.1°C; critical pressure 672 psia; heating value 1009 Btu/cu ft. Soluble in alcohol, ether; slightly soluble in water. Methane is an asphyxiant gas.
Occurrence: Natural gas and coal gas; from decaying vegetation and other organic matter in swamps and marshes.
Derivation: (1) From natural gas by absorption or adsorption; (2) from coal mines for use as fuel gas; (3) from a mixture of carbon monoxide and hydrogen (synthesis gas) obtained by reaction of hot coal with steam; the mixed gas is passed over a nickel-based catalyst at high temperature. See methanation. Methane can also be obtained by a nickel-catalyzed reaction of carbon dioxide and hydrogen; (4) anaerobic decomposition of manures and other agricultural wastes.
Containers: High-pressure pipe lines; high-pressure cylinders; as liquid at −260°F; in insulated tank ships.
Grades: Research, 99.99%; C.P., 99%; technical, 95%; Btu grade, must have heating value of 1000 ±3 Btu/cu ft at 60°F and 30 inches Hg pressure.
Hazard: Severe fire and explosion hazard; forms explosive mixtures with air (5–15% by volume).
Uses: Source of petrochemicals by conversion to hydrogen and carbon monoxide by steam cracking or partial oxidation. Important products are methanol, acetylene, hydrogen cyanide. Chlorination gives carbon tetrachloride, chloroform, methylene chloride, and methyl chloride. In the form of natural gas, methane is used as a fuel, as a source of carbon black, and as the starting material for manufacture of synthetic proteins.
Shipping regulations: (Rail) Red gas label. (Air) Flammable gas label. Not acceptable on passenger planes.
See also natural gas; synthetic natural gas; biogas.

methanecarboxylic acid. See acetic acid.

methanedicarbonic acid. See malonic acid.

methanesulfonic acid CH_3SO_3H.
Properties: Liquid at room temperature; sp. gr. 1.4812 (18/4°C); m.p. 17 to 20°C; b.p. 167°C (10 mm); refractive index 1.4317 (16°C). Soluble in water, alcohol, ether. Flash point, none.
Grade: 70%.
Hazard: Corrosive to tissue (eyes, skin and mucous membranes).
Uses: Catalyst in esterification, alkylation, olefin polymerization, peroxidation reactions.

methanesulfonyl chloride (mesyl chloride) CH_3SO_2Cl.
Properties: Pale yellow liquid; sp. gr. 1.485 (20/20°C); b.p. 164°C; f.p. −32°C; soluble in most organic solvents; insoluble in water (slowly hydrolyzes).
Containers: 5-gal, 55-gal drums.
Grades: 98%; 99+%.
Hazard: Probably toxic.

Use: Intermediate; flame-resistant products; stabilizer for liquid sulfur trioxide; biological chemicals.

methanethiol (methyl mercaptan) CH_3SH.
Properties: Water-white liquid when below boiling point, or colorless gas; powerful unpleasant odor. F.p. −121°C; sp. gr. 0.87 (20°C); flash point below 0°C; b.p. 5.96°C; slightly soluble in water; soluble in alcohol, ether, petroleum naphtha.
Derivation: Methanol and hydrogen sulfide.
Grade: 98.0% purity.
Containers: 180-lb cylinders; tank cars.
Hazard: Flammable, dangerous fire risk. Highly toxic; strong irritant. Tolerance, 0.5 ppm in air. Explosive limits in air 3.9 to 21.8%.
Uses: Synthesis, especially of methionine, jet fuel additives, fungicides; also as catalyst.
Shipping regulations: (Rail) Red Gas label. (Air) Flammable Gas label. Not acceptable on passenger planes. Legal label name methyl mercaptan.

methanethiomethane. See dimethyl sulfide.

methanoic acid. See formic acid.

methanol. See methyl alcohol.

"Methasan."[58] Trademark for zinc dimethyldithiocarbamate (q.v.).

"Methazate."[248] Trademark for zinc dimethyldithiocarbamate. White powder or pellets.
Uses: Accelerator for rubber products.

"Methedrine."[301] Trademark for methamphetamine hydrochloride. See amphetamine.

methenamine. See hexamethylenetetramine.

"Methendic" Anhydride.[316] Trademark for a mixture of bicyclic unsaturated dibasic anhydrides as a relatively nonvolatile liquid at room temperature.
Properties: Pale amber liquid; color, Gardner, 3–6; sp. gr. at 25°C, 1.2–1.3; (n 27/D) 1.5052; flash point (COC), 275–285°F; miscible with acetone, aromatic and aliphatic hydrocarbons in all proportions at room temperature. Combustible. Low toxicity.
Uses: Cross-linking or curing agent for epoxy type resin systems.

methenyl tribromide. See bromoform.

methenyl trichloride. See chloroform.

methiodal sodium. See sodium methiodal.

methionine (2-amino-4-(methylthio)butyric acid) $CH_3SCH_2CH_2CH(NH_2)COOH$. An optically active essential sulfur-containing amino acid important in biological trans-methylation processes. The levo form is biologically active.
Properties (of DL racemic mixtures): White, crystalline platelets or powder. Faint odor. Soluble in water, dilute acids, and alkalies; very slightly soluble in alcohol; practically insoluble in ether. pH (1% aqueous solution) 5.6–6.1.
Derivation: Hydrolysis of protein; synthesized from hydrogen cyanide, acrolein, and methyl mercaptan.
Containers: Glass vials, to fiber drums.
Grades: N.F.; feed, 98%.
Uses: Pharmaceuticals; feed additive; vegetable oil enrichment; single-cell protein.

methionine hydroxy-analog calcium (DL-alpha-hydroxygamma-methylmercaptobutyric acid, calcium salt; 2-hydroxy-4-methylthiobutyric acid, calcium salt) $(CH_3SCH_2CH_2CHOHCOO)_2Ca$. Free methionine hydroxy analog is a metabolite in methionine utilization.
Properties: Free-flowing light tan powder; soluble in water; insoluble in common organic solvents.
Uses: Animal feed; synthesis of pharmaceuticals.

methiotepa. Generic name for tris(2-methyl-1-aziridinyl)-phosphine sulfide.
Hazard: Highly toxic. See metepa.

methocarbamol. (3-(ortho-methoxyphenoxy)-1,2-propanediol 1-carbamate) $CH_3OC_6H_4OCH_2CH(OH)CH_2OOCNH_2$.
Properties: M.p. 92–94°C; soluble in water, alcohol and propylene glycol.
Grade: N.D.
Use: Medicine.

"Methocel."[233] Trademark for methylcellulose (q.v.).

"Metholene."[242] Trademark for a series of methyl esters of fatty acids, from methyl caproate through methyl stearate.

methotrexate (amethopterin; 4-amino-10-methylfolic acid).
Properties: Orange-brown crystalline powder. Insoluble in water, alcohol, chloroform, ether. Slightly soluble in dilute hydrochloric acid; soluble in dilute solutions of alkali hydroxides and carbonates. Folic acid antagonist.
Grade: U.S.P.
Hazard: Highly toxic in massive dosage.
Use: Medicine; insect chemosterilant; cancer treatment.

"Methoxone."[147] Trademark for MCP acid; available as the sodium salt, amine or ester. Used as a selective weed killer.
Hazard: Moderately toxic.

methoxsalen (8-methoxypsoralen) $C_{12}H_8O_4$, tricyclic.
Properties: White to cream-colored, odorless, crystalline solid. Slightly soluble in alcohol, practically insoluble in water. Combustible.
Use: Suntan accelerator, sunburn protector.

methoxyacetaldehyde. A highly reactive aldehyde derivative available in 77% aqueous solution. Clear, colorless liquid, with odor characteristic of lower aldehydes. Miscible with water and many organic solvents. Resembles butyraldehyde in structure and some properties. Claimed as possible antimicrobial agent, preservative and polymer modifier.

methoxyacetic acid CH_3OCH_2COOH.
Properties: Liquid; m.p. (min) 7.7°C; boiling range (733 mm) (min) 197–198°C; sp. gr. (25/4°C) 1.1738; refractive index 1.415 (25°C); flash point 260°F; acid number (min) 612. Combustible.
Use: Synthesis.

para-**methoxyacetophenone** (para-acetoanisole; acetanisole; par-acetylanisole) $CH_3OC_6H_4COCH_3$.
Properties: Crystalline solid with pleasant odor; b.p. 258°C; congealing point 36.5°C. Soluble in alcohol, ether, fixed oils. Combustible.
Derivation: Interaction of anisole and acetyl chloride in the presence of aluminum chloride and carbon disufide.
Grades: Technical; F.C.C.
Uses: Perfumery, for floral odors; flavoring.

methoxyacetyl-para-**phenetidine** $CH_3OCH_2CONHC_6H_4OC_2H_5$.
Properties: White needles; tasteless but becoming bitter on chewing. Soluble in alcohol, ether, chloro-

form, volatile oils, and boiling water; much less so in cold water. M.p. 98°C.
Derivation: By heating methoxyacetic acid and para-phenetidine.
Use: Medicine.

ortho-**methoxyaniline.** See ortho-anisidine.

para-**methoxyaniline.** See para-anisidine.

methoxybenzaldehyde. See anisaldehyde.

methoxybenzene. See anisole.

para-**methoxybenzoic acid.** See anisic acid.

para-**methoxybenzyl acetate.** See anisyl acetate.

para-**methoxybenzyl alcohol.** See anisic alcohol.

para-**methoxybenzyl formate.** See anisyl formate.

2-methoxy-4,6-bis(isopropylamino)-s-triazine (2,4-bis-(isopropylamino)-6-methoxy-s-triazine) $CH_3OC_3N_3[NHCH(CH_3)_2]_2$. White solid; almost insoluble in water. Used as an herbicide.
Hazard: Probably toxic.

3-methoxybutanol $CH_3CH(OCH_3)CH_2CH_2OH$.
Properties: Liquid; sp. gr. 0.9229; b.p. 161.1°C; flash point 165°F; vapor pressure (20°C) 0.9 mm; sets to glass at −85°C; soluble in water. Combustible; low toxicity.
Uses: High-boiling lacquer solvent; coupling agent for brake fluids; intermediate for plasticizers, herbicides; film-forming additive in PVA emulsions; solvent for pharmaceuticals.

1-methoxycarbonyl-1-propen-2-yl dimethylphosphate. See mevinphos.

methoxychlor (methoxy DDT; DMDT) Generic name for 2,2-Bis (para-methoxyphenyl)-1, 1, 1-trichloroethane; $Cl_3CCH(C_6H_4OCH_3)_2$.
Properties: White, crystalline solid; m.p. 89°C; insoluble in water. Not compatible with alkaline materials.
Derivation: Reaction of methyl phenyl ether and chloral hydrate.
Containers: Drums.
Hazard: Less toxic than DDT. Tolerance, 10 mg per cubic meter in air.
Use: Insecticide effective against mosquito larvae and house flies. Especially recommended for use in dairy barns.

2-methoxy-3,6-dichlorobenzoic acid. See dicamba.

2-methoxy-3,6-dichlorobenzoic acid, dimethylamine salt. See "Banvel" D.

2-(beta-**methoxyethoxy)ethanol.** See diethylene glycol monomethyl ether.

methoxyethylene. See vinyl methyl ether.

2-methoxyethylmercury acetate
$CH_3OCH_2CH_2HgOOCCH_3$. A fungicide and disinfectant used in treating seeds.
Hazard: Highly toxic. Tolerance, 0.01 mg per cubic meter of air.
Shipping regulations: (Rail, Air) Mercury compounds, n.o.s., Poison label.

methoxyethyl oleate $CH_3OCH_2CH_2OOCC_{17}H_{33}$.
Properties: Oily liquid, mild odor. F.p. below −18°C; sp. gr. 0.898 (25°C); boiling range 180–206°C (4 mm); flash point (open cup) 385°C; viscosity, 8 cp (25°C). Combustible.
Uses: Plasticizer and solvent.

methoxyethyl stearate $CH_3OCH_2CH_2OOCC_{17}H_{35}$.
Properties: Oily liquid; mild odor; f.p. 19 to 24°C; boiling range 186–205°C (4 mm); flash point (open cup) 378°C; viscosity 9 cp at 25°C. Combustible.
Uses: Plasticizer and solvent.

methoxyflurane. See 2,2-dichloro-1,1-difluoroethyl methyl ether.

3-methoxy-4-hydroxybenzaldehyde. See vanillin.

methoxyhydroxymercuripropylsuccinyl urea
$C_9H_{16}HgN_2O_6$. 3-Hydroxymercuri-2-methoxypropylcarbamoylsuccinamic acid.
Properties: Bitter crystals, m.p. 198.5°C.
Derivation: Made by the mercuration of allylsuccinylurea. See meralluride.
Hazard: Highly toxic.
Shipping regulations: (Rail, Air) Mercury compounds, n.o.s., Poison label.

4′-methoxy-2-(para-**methoxyphenyl)acetophenone.** See deoxyanisoin.

2-methoxy-5-methylaniline. See 5-methyl-ortho-anisidine.

4-methoxy-4-methylpentanol-2 ("Pent-Oxol") $CH_3C(CH_3)(OCH_3)CH_2CHOHCH_3$.
Properties: Liquid. Boiling range 163.8–167°C. Combustible.
Use: Solvent for resin-coating formulation.

4-methoxy-4-methylpentanone-2 ("Pent-Oxone") $CH_3C(CH_3)(OCH_3)CH_2COCH_3$.
Properties: Water-white liquid; boiling range 147–163°C; flash point 141°F. Combustible.
Derivation: Diacetone alcohol.
Hazard: Moderate fire risk. Irritant to skin and eyes.
Use: Solvent for a variety of resin coatings.

2-methoxy-4-methylphenol. See creosol.

2-methoxynaphthalene. See beta-naphthyl methyl ether.

1-methoxy-4-nitrobenzene. See para-nitroanisole.

methoxyphenamine hydrochloride
$CH_3OC_6H_4CH_2CH(CH_3)NHCH_3 \cdot HCl$. 2-(ortho-Methoxyphenyl)isopropyl methylamine hydrochloride.
Properties: Crystalline, white powder; odorless and bitter; m.p. 124–128°C; freely soluble in alcohol, chloroform, and water; slightly soluble in ether and benzene; pH (5% solution) 5.3–5.7.
Use: Medicine.

4-methoxyphenol (para-methoxyphenol). See hydroquinone monomethyl ether.

ortho-**methoxyphenol.** See guaiacol.

3-(ortho-**methoxyphenoxy)-1,2-propanediol 1-carbamate.** See methocarbamol.

para-**methoxyphenylacetic acid** (para-methoxy-alpha-toluic acid) $CH_3OC_6H_4CH_2COOH$. Off-white to pale yellow flakes; m.p. 85°C.
Uses: Preparation of pharmaceuticals; other organic compounds.

para-**methoxyphenylbutanone.** See anisylacetone.

methoxypolyethylene glycol. One of a series of compounds with properties similar to the polyethylene glycols of comparable molecular weight. Slightly viscous liquids to soft wax-like solids. Used for manufacture of detergents and emulsifying and dispersing agents through the preparation of the mono-fatty-acid derivatives.

methoxy propanol (monopropylene glycol methyl ether). $CH_3OCH_2CH(OH)CH_3$.
Properties: Colorless liquid. B.p. 120°C; flash point 102°F; specific gravity 0.92 (25°C). Low toxicity. Combustible.
Hazard: Moderate fire risk.
Uses: Antifreeze and coolant for diesel engines.

para-**methoxypropenylbenzene.** See anethole.

para-**methoxypropiophenone** $C_2H_5COC_6H_4OCH_3$.
Properties: Clear, colorless liquid; distillation range 110–140°C (3 mm); refractive index, 1.543 to 1.545 (25°C).

3-methoxypropylamine (3-MPA) $CH_3OCH_2CH_2CH_2NH_2$.
Properties: Colorless liquid; b.p. 119°C; sp. gr. 0.873 (20/20°C); refractive index (n 25/D) 1.4153; flash point 80°F (Tag Closed Cup); miscible with water, ethanol, toluene, acetone, carbon tetrachloride, hexane, and ether.
Hazard: Flammable, moderate fire risk. Moderately toxic by ingestion and inhalation.
Uses: Organic intermediate; emulsifier in anionic coatings and wax formulations.

8-methoxypsoralen. See methoxsalen.

para-**methoxytoluene.** See para-methylanisole.

para-**methoxy**-alpha-**toluic acid.** See para-methoxyphenylacetic acid.

methoxytriethylene glycol acetate. See methoxytriglycol acetate.

methoxytriglycol $CH_3O[C_2H_4O]_3H$.
Properties: Colorless liquid. Sp. gr. 1.0494; b.p. 249°C; f.p. −44°C; flash point 245°F; soluble in water. Combustible; low toxicity.
Use: Plasticizer intermediate.

methoxytriglycol acetate (methoxytriethyleneglycol acetate) $CH_3COO(C_2H_4O)_3CH_3$.
Properties: Colorless liquid with fruity odor; sp. gr. 1.0940 (20/20°C); 9.2 lb/gal (20°C); b.p. 244.0°C; flash point 260°F; combustible; low toxicity. Water-soluble; low volatilty.
Uses: Antidusting agent for finely powdered materials, especially as an "antisneeze" for certain dyestuffs.

methyl abietate $C_{19}H_{29}COOCH_3$.
Properties: Colorless to yellow liquid; sp. gr. 1.033–1.043 (20°C); refractive index 1.525–1.535; flash point 360°F; b.p. 365°C. Miscible with most organic solvents. Combustible; low toxicity.
Containers: Drums; carloads.
Uses: Solvent and plasticizer; lacquers; varnishes; coating compositions.

N-methylacetanilide $C_6H_5N(CH_3)COCH_3$.
Properties: Needles or long tablet-like crystals; soluble in hot water and dilute alcohol. B.p. 240–250°C; m.p. 101°C.
Derivation: By heating acetylchloride and methylaniline.
Use: Medicine.

methyl acetate $CH_3CO_2CH_3$.
Properties: Colorless, volatile, liquid; fragrant odor. Miscible with the common hydrocarbon solvents; soluble in water. Sp. gr. 0.92438; f.p. −98.05°C; b.p. 54.05°C; flash point 15°F; refractive index 1.3619 (20°C); wt/gal 7.76 lb (20°C). Autoignition temp., 935°F.
Derivation: By heating methyl alcohol and acetic acid in presence of sulfuric acid and distilling.
Grades: Technical; C.P.
Containers: 5-gal cans; 55-gal drums; tank cars.
Hazard: Highly flammable; dangerous fire and explosion risk. Explosive limits in air 3 to 16%. Tolerance, 200 ppm in air. Moderately toxic by inhalation.
Uses: Paint remover compounds; lacquer solvent; intermediate; synthetic flavoring.
Shipping reguations: (Rail) Red label. (Air) Flammable Liquid label.

methylacetic acid. See propionic acid.

methyl acetoacetae $CH_3COCH_2CO_2CH_3$.
Properties: Colorless liquid; soluble in alcohol; slightly soluble in water; sp. gr. 1.0785 (20/20°C); b.p. 171.7°C; vapor pressure 0.7 mm (20°C); flash point 158°F (TOC); wt/gal 9.0 lb (20°C); f.p. −31.9°C. Combustible. Low toxicity.
Containers: Carboys; drums; tank cars; trucks.
Uses: Solvent for cellulose ethers; ingredient of solvent mixtures for cellulose esters; organic synthesis.

methyl acetone.
Properties: Water-white, anhydrous liquid, consisting of various mixtures of acetone, methyl acetate, and methyl alcohol. Miscible with hydrocarbons, oils, and water. Flash point near 0°F.
Derivation: A byproduct in the wood-distillation industry; also synthetic.
Grades: Technical (natural and synthetic).
Containers: Drums; tank cars.
Hazard: Highly flammable, dangerous fire risk. Toxic by ingestion.
Uses: Solvent for nitrocellulose, cellulose acetate, rubber, gum, resins; lacquers; paint and varnish removers; extracts; extracting perfumes; dewaxing natural gums.
Shipping regulations: (Rail) Red label. (Air) Flammable Liquid label.

methylacetophenone (methyl para-tolyl ketone) $CH_3C_6H_4COCH_3$.
Properties: Colorless or pale-yellow liquid; fragrant, coumarin odor. Soluble in 7 parts of 50% alcohol and in most fixed oils. Sp. gr. 1.001–1.004; refractive index 1.533–1.535. Combustible; low toxicity.
Derivation: Action of acetic anhydride on toluene.
Grade: Technical; F.C.C.
Uses: Perfumery; flavoring. See also glycine, (2).

methylacetopyranone. See dehydroacetic acid.

methylacetylene (allylene; propyne) $CH_2C⫶CH$.
Properties: Colorless liquefied gas; b.p. −23.1°C; f.p. −101.5°C; sp. vol. 9.7 cu ft/lb (70°F).
Uses: Specialty fuel; chemical intermediate.
Hazard: Flammable, dangerous fire risk; moderately toxic by inhalation. Tolerance, 1000 ppm in air.
Shipping regulations: (Rail) Red Gas label. (Air) Flammable Gas label. Not accepted on passenger planes.

methylacetylene-propadiene, stabilized
Properties: Colorless liquefied gas; sp. gr. (liquid) 0.576 (15/15°C; boiling range −39 to −20°C; flame temperature in oxygen 5300°F. A mixture containing 60–66.5% methylacetylene and propadiene, balance propane and butane.
Hazard: Flammable, dangerous fire risk; moderately toxic by inhalation. Tolerance, 1000 ppm in air.
Uses: Industrial fuel gas for cutting, welding, brazing, heat treating, metallizing.

Shipping regulations: (Rail) Red Gas label. (Air) Flammable Gas label. Not acceptable on passenger planes.
See also "Mapp."

methyl acetylricinoleate $C_{17}H_{32}(OCOCH_3)COOCH_3$.
Properties: Pale yellow, low viscosity, oily liquid; mild odor; sp. gr. 0.938 (25/25°C); solidifies at −26°C. Soluble in most organic liquids; insoluble in water. Combustible; low toxicity.
Derivation: Castor oil, methyl alcohol and acetic anhydride.
Containers: Drums; tank cars.
Uses: Plasticizer; lubricant; protective coatings; syntetic rubbers; vinyl compounds.

methyl acetylsalicylate $CH_3COOC_6H_4COOCH_3$.
Properties: White crystals; m.p. 52°C; b.p. 134–136°C (9 mm).
Derivation: By heating methyl salicylate with a slight excess of acetic anhydride, adding alcohol, then water, and separting the precipitate.
Uses: Medicine; synthesis.

methyl acid phosphate. See methylphosphoric acid.

beta-**methylacrolein.** See crotonaldehyde.

methyl acrylate $CH_2:CHCOOCH_3$.
Properties: Colorless, volatile liquid; b.p. 80.5°C; f.p. −76.5°C; vapor pressure (20°C) 65 mm; sp. gr. (20/20°C) 0.9574; wt/gal 8.0 lb; slightly soluble in water; readily polymerized. Flash point 25°F (TOC).
Derivation: (a) Ethylene cyanohydrin, methanol, and dilute sulfuric acid; (b) Oxo reaction of acetylene, carbon monoxide, and methanol in the presence of nickel or cobalt catalyst; (c) from beta-propiolactone.
Grades: Technical (inhibited).
Containers: Drums; tank cars.
Hazard: Flammable, dangerous fire and explosion risk. Highly toxic by inhalation, ingestion and skin absorption. Tolerance, 10 ppm in air. Irritant to skin and eyes. Safety data sheet available from Manufacturing Chemists Assn., Washington, D.C.
Uses: Polymers; amphoteric surfactants; vitamin B_1; chemical intermediate.
Shipping regulations: (Air) (inhibited) Flammable Liquid label; (uninhibited) Not acceptable. (Rail) Not listed.
See also acrylate.

beta-**methylacrylic acid.** See crotonic acid.

methylal (dimethoxymethane; formal) $CH_3OCH_2OCH_3$.
Properties: Colorless, volatile, liquid; chloroform-like odor; pungent taste; m.p. 104.8°C; sp. gr. 0.86 (20/4°C); b.p. 42.3°C; soluble in water at 20°C to extent of 32 wt %; completely soluble in alcohol, ether, and hydrocarbons; flash point (open cup) 0°F (approx.); autoignition temp. 459°F.
Containers: Glass bottles; steel drums.
Hazard: Flammable, dangerous fire and explosion risk. Moderately toxic by ingestion and inhalation. Tolerance, 1000 ppm in air.
Uses: Solvent; starting material for organic synthesis; perfumes, adhesives and protective coatings; special fuel.
Shipping regulations: (Rail) Flammable liquid, n.o.s., Red label. (Air) Flammable Liquid label.

2-methylalanine. See aminoisobutyric acid.

methyl alcohol (methanol; wood alcohol) CH_3OH. Eighteenth highest volume chemical produced in U.S. (synthetic) (1975).
Properties: Clear, colorless, mobile, highly polar liquid. Miscible with water, alcohol, and ether; sp. gr. 0.7924; f.p. −97.8°C; b.p. 64.5°C; wt/gal 6.59 lb (20°C); refractive index 1.329 (20°C); surface tension 22.6 dynes/dm (20°C); viscosity 0.00593 poise (20°C); vapor pressure 92 mm (20°C); flash point (open cup) 54°F; autoignition temp. 867°F.
Derivation: (a) By high pressure catalytic synthesis from carbon monoxide and hydrogen. (b) Partial oxidation of natural gas hydrocarbons.
Method of purification: Rectification.
Containers: 4000-gal tank cars; tank trucks and barges.
Grades: Technical; C.P. (99.85%); electronic (used to cleanse and dry components); fuel.
Hazard: Flammable, dangerous fire risk. Toxic by ingestion (causes blindness). Tolerance, 200 ppm in air. Explosive limits in air 6–36.5% by volume. Safety data sheet available from Manufacturing Chemists Assn., Washington, D.C.
Uses: Manufacture of formaldehyde and dimethyl terephthalate; chemical synthesis (methyl amines, methyl chloride, methyl methacrylate, etc.); aviation fuel (for water injection); automotive antifreeze; solvent for nitrocellulose, ethylcellulose, polyvinyl butyral, shellac, rosin, manila resin, dyes; denaturant for ethyl alcohol; dehydrator for natural gas; fuel for utility plants (methyl fuel); feedstock for manufacture of synthetic proteins by continuous fermentation.
Shipping regulations: (Rail) Red label. (Air) Flammable Liquid label. Legal label name (Rail) wood alcohol.

methylallyl acetate (methallyl acetate) $CH_2:C(CH_3)CH_2OOCCH_3$.
Properties: Colorless liquid; sp. gr. 0.9162 (20°C); wt/gal 7.6 lb.
Hazard: Probably flammable and moderately toxic.
Uses: Monomer; preparation of methallyl derivatives.

methylallyl alcohol (methallyl alcohol; 2-methyl-2-propen-1-ol) $H_2C:C(CH_3)CH_2OH$.
Properties: Colorless liquid with pungent odor; sp. gr. 0.8515 (20/4°C); b.p. 115°C; refractive index 1.4255 (25/D); flash point 92°F. Soluble in water, alcohols, esters, ketones, and hydrocarbons.
Grade: 98.5% min purity.
Hazard: Flammable, moderate fire risk. Toxic and irritant to eyes and skin.
Use: Intermediate.

beta-**methylallyl chloride** (beta-methallyl chloride; 3-chloro-2-methyl-1-propene; MAC) $CH_2:C(CH_3)CH_2Cl$.
Properties: Colorless to straw-colored volatile liquid with a sharp, penetrating odor; sp. gr. 0.925 (20°C); b.p. 73°C; refractive index (n 25/D) 1.427; flash point −3°F (TOC).
Hazard: Flammable, dangerous fire risk. Moderately toxic by ingestoin; irritant to eyes and skin. Explosive limits in air, 2.3 to 9.3%.
Uses: Intermediate for production of insecticides, plastics, pharmaceuticals, other organic chemicals; fumigant for grains, tobacco, and soil.
Shipping regulations: Flammable liquid, n.o.s., (Rail) Red label. (Air) Flammable Liquid label.

methylaluminum sesquibromide $(CH_3)_3Al_2Br_3$.
Properties: Cloudy yellow liquid at 25°C; f.p. −4°C; b.p. (extrapolated) 166°C; density 1.514 g/ml (25°C).
Hazard: Ignites spontaneously in air; reacts violently with water. Keep out of contact with air and moisture.

Uses: Catalyst for polymerization of olefins and hydrogenation of aromatics; chemical intermediate.
Shipping reguations: (Rail) Red label. (Air) Not acceptable.

methylaluminum sesquichloride $(CH_3)_3Al_2Cl_3$.
Properties: Colorless liquid at 25°C; f.p. 22.8°C; b.p. (extrapolated) 143.7°C; density 1.1629 g/ml (25°C), 9.705 lb/gal (25°C).
Hazard: Ignites spontaneously in air; reacts violently with water. Keep out of contact with air and moisture.
Uses: Catalyst for polymerization of olefins; catalyst for hydrogenation of aromatics.
Shipping reuglatoins: (Rail), Red label. (Air) Not acceptable.

methylamine (monomethylamine; aminomethane) CH_3NH_2.
Properties: Colorless gas. Strong, ammoniacal odor; b.p. −6.79°C; f.p. −92.5°C; flash point (gas) 14°F, (30% solution) 34°F (TOC); soluble in water, alcohol, ether. Autoignition temp. 806°F.
Derivation: Interaction of methanol and ammonia over a catalyst at high temperature. The mono-, di-, and trimethylamines are all produced, and yields are regulated by conditions. They are separated by azeotropic or extractive distillation.
Grade: Technical (anhydrous; 30–40% solutions).
Containers: Steel drums; tank cars; cylinders (anhydrous).
Hazard: (gas and liquid) Flammable, dangerous fire risk. Toxic; strong irritant to tissue. Explosive limits in air, 5 to 21%. Tolerance, 10 ppm in air. Safety data sheet available from Manufacturing Assn., Washington, D.C.
Uses: Intermediate for accelerators, dyes, pharmaceutials, insecticides, fungicides, surface active agents; tanning; dyeing of acetate textiles; fuel additive; polymerization inhibitor; component of paint removers; solvent; photographic developer; rocket propellant.
Shipping regulations: (Rail) (gas) Red gas label; (solution) Red label. Legal label name monomethylamine. (Air) (gas) Flammable Gas label. Not accepted on passenger planes; (solution) Flamamble Liquid label.

methyl ortho-**aminobenzoate.** See methyl anthranilate.

methylaminoacetic acid. See sarcosine.

methylaminodimethylacetal $(CH_3O)_2CHCH_2NHCH_3$.
Properties: Water-white to slightly yellow, clear liquid having a sharp ammoniacal odor; refractive index 1.406–1.409 (n 20/D); sp. gr. 0.924 (25/25°C). Combustible.

l-**methylaminoethanolcatechol.** See epinephrine.

para-**methylaminoethanolphenol tartrate.** See synephrine tartrate.

2-(methylamino)glucose. See N-methylglucosamine.

2-methylaminoheptane $CH_3(CH_2)_4CH(NHCH_3)CH_3$. Oily liquid with a slight amine odor; b.p. 155°C. Somewhat soluble in water. Used in medicine, usually as the hydrochloride.

methyl meta-**amino**-para-**hydroxybenzoate** (orthoform) $CH_3OOCC_6H_3(NH_2)(OH)$.
Properties: White powder, odorless, tasteless; soluble in alcohol; almost insoluble in water; m.p. 141–143°C.
Use: Medicine.

3-methylaminoisocamphane hydrochloride. See mecamylamine hydrochloride.

2-methylamino-6-methyl-5-heptene. See isometheptene.

N-**methyl**-para-**aminophenol** $CH_3NHC_6H_4OH$.
Properties: Colorless needles. Soluble in water, alcohol and ether. M.p. 87°C. Combustible.
Derivation: (a) Interaction of hydroquinone and methylamine. (b) Methylation of para-aminophenol hydrochloride.
Containers: Glass bottles; fiber drums; multiwall paper sacks.
Hazard: Moderately toxic; eye and skin irritant.
Uses: Organic synthesis; photographic developer.

N-**methyl**-para-**aminophenol sulfate** $HOC_6H_4NHCH_3 \cdot \frac{1}{2}H_2SO_4$.
Properties: Colorless needles; m.p. 250–260°C with decomposition; soluble in water and alcohol; insoluble in ether. Discolors in air. Combustible.
Derivation: By methylation of para-aminophenol and conversion of the resulting methylated base by neutralization with sulfuric acid.
Grades: C.P.; photographic.
Containers: Kegs; bottles; barrels; fiber cans.
Hazard: Moderately toxic; eye and skin irritant.
Use: Photographic developer.

methylamyl acetate (methylisobutyl carbinol acetate; sec.-hexyl acetate; 4-methyl-2-pentanol acetate) $CH_3COOCH(CH_3)CH_2CH(CH_3)_2$.
Properties: Colorless liquid. Mild odor. Sp. gr. 0.8598 (20/20°C); b.p. 146.3°C; f.p. −64°C; vapor pressure 3 mm (20°C); flash point 110°F (O.C.); wt/gal 7.1 lb (20°C). Insoluble in water; soluble in alcohol. Combustible.
Grade: Technical.
Containers: Drums; tank cars.
Hazard: Moderate fire risk.
Uses: Solvent for nitrocellulose and other lacquers.

methylamyl alcohol (methylisobutyl carbinol; MIBC; 4-methylpentanol-2) $(CH_3)_2CHCH_2CH(CH_3)OH$.
Properties: Colorless, stable liquid. Miscible with most common organic solvents; slightly soluble in water. B.p. 131.8°C; sp. gr. 0.8079 (20/20°C); wt/gal 6.72 lb (20°C); sets to a glass below −90°C; refractive index 1.4089 (25°C); vapor pressure 2.8 mm (20°C); flash point 105°F (O.C.). Combustible.
Derivation: From methyl isobutyl ketone.
Grade: Technical.
Containers: Drums; tank cars.
Hazard: Moderate fire risk. Explosive limits in air, 1 to 5.5%. Tolerance, 25 ppm in air.
Uses: Solvent for dyestuffs, oils, gums, resins, waxes, nitrocellulose and ethylcellulose; organic synthesis; froth flotation; brake fluids.

methyl-n-amyl carbinol (heptanol-2; 2-methyl-1-pentanol). $CH_3(CH_2)_4CHOHCH_3$.
Properties: Stable colorless liquid; mild odor; miscible with common organic liquids. Sp. gr. 0.8187 (20/20°C); b.p. 160.4°C; vapor pressure 1.0 mm (20°C); flash point 130°F (O.C.). Low toxicity; wt/gal 6.8 lb (20°C). Combustible.
Grade: Technical.
Containers: Drums; tank cars.
Hazard: Moderate fire risk.
Uses: Solvent for synthetic resins; frothing agent in ore flotation.

methyl-n-amylketone (2-heptanone) $CH_3CH_2CH_2CH_2CH_2COCH_3$.

Properties: Water-white liquid. Soluble in water; miscible with organic lacquer solvents. Sp. gr. 0.8166 (20/20°C); b.p. 150.6°C; vapor pressure 2.6 mm (20°C); flash point 120°F; autoignition temp. 991°F; refractive index 1.4110 (20°C); wt/gal 6.8 lb (20°C) nitrocellulose-toluene dilution ratio 3.9; f.p. −35°C. Combustible.
Grade: Technical.
Containers: Drums; carloads.
Hazard: Moderate fire risk. Skin irritant; moderately toxic by inhalation; narcotic in high concentrations. Tolerance, 100 ppm in air.
Uses: Solvent for nitrocellulose lacquers; inert reaction medium; synthetic flavoring.

N-**methylaniline** $C_6H_5NH(CH_3)$.
Properties: Colorless to reddish-brown oily liquid; discolors on standing. Soluble in alcohol, ether, water and chloroform. Sp. gr. 0.991; f.p. −57°C; b.p. 190–191°C. Combustible.
Derivation: By heating methyl iodide with aniline and subsequent distillation.
Grade: Technical.
Containers: Iron drums; tank cars.
Hazard: Toxic by ingestion, inhalation and skin absorption.
Uses: Organic synthesis; solvent; acid acceptor.

alpha-**methylanisalacetone.** (1-[para-methoxyphenyl]-1-penten-3-one) $CH_3OC_6H_4CH{:}CHCOCH_2CH_3$.
Properties: White to pale yellow solid; sharp odor; stable; m.p. 60°C min. One gram is soluble in 5 ml 95% alcohol. Combustible; low toxicity.
Grade: 99% pure min.
Use: Flavoring.

5-methyl-ortho-**anisidine** (meta-amino-para-cresol, methyl ether; 2-methoxy-5-methylaniline) $CH_3C_6H_3(NH_2)OCH_3$.
Properties: White crystals; m.p. 51.5°C; b.p. 235°C. Soluble in alcohol and ether; sparingly soluble in hot water; volatile with steam.
Derivation: 2-Nitro-para-cresol, obtained by the action of nitrous and excess nitric acids upon para-toluidine, is methylated and reduced.
Grade: Technical.
Containers: Drums.
Hazard: Probably toxic. See anisidine.
Use: Dyes.

para-**methylanisole** (para-cresyl methyl ether; para-methoxytoluene; methyl-para-cresol) $CH_3C_6H_4OCH_3$.
Properties: Colorless liquid; strong floral odor; sp. gr. 0.966–0.970; refractive index 1.5100–1.5130 (20°C). One part dissolves in 3 parts of 80% alcohol. Combustible.
Grades: F.C.C.; technical.
Uses: Perfumery; flavoring.

1-methylanthracene. See alpha-methylanthracene.

alpha-**methylanthracene** (1-methylanthracene) $C_{15}H_{12}$ or $C_6H_4(CH)_2C_6H_3CH_3$ (a tricyclic aromatic).
Properties: Colorless leaflets. Soluble in alcohol; insoluble in water. Sp. gr. 1.101; b.p. 200°C; m.p. 86°C. Combustible.
Grade: Technical.
Use: Organic synthesis.

methyl anthranilate (methyl ortho-aminobenzoate; neroli oil, artificial) $H_2NC_6H_4CO_2CH_3$.
Properties: Colorless to pale-yellow liquid with bluish fluorescence; grape-type odor; sp. gr. 1.167–1.175 (15°C); refractive index (n 20/D) 1.5820–1.5840; b.p. 135°C; congealing point 23.8°C (min); soluble in 5 volumes or more of 60% alcohol; soluble in fixed oils, propylene glycol, volatile oils; slightly soluble in water, mineral oil; insoluble in glycerin. Nontoxic. Combustible.
Derivation: By heating anthranilic acid and methyl alcohol in presence of sulfuric acid, with subsequent distillation. Occurs in many flower oils.
Grades: Technical; F.C.C.
Uses: Flavoring; perfume (cosmetics and pomades).

methylanthraquinone $CH_3C_6H_3(CO)_2C_6H_4$ (tricyclic).
Properties: White needles; soluble in ether and benzene; very slightly soluble in alcohol; insoluble in water. M.p. 177°C; b.p. sublimes. Low toxicity; combustible.
Derivation: By heating anthraquinone and methyl alcohol in presence of sulfuric acid.
Use: Organic synthesis.

methyl apholate. Generic name for 2,2,4,4,6,6-hexahydro-2,2,4,4,6,6-hexakis(2-methyl-1-aziridinyl)-1,3,5,2,4,6-triazatriphosphorine.
$N_3P_3(\overline{NCH_2CH}CH_3)_6$.
Hazard: Probably toxic.
Use: Insect chemosterilant.

methyl arachidate (methyl eicosanoate) $CH_3(CH_2)_{18}COOCH_3$. The methyl ester of arachidic acid.
Properties: Waxlike solid; m.p. 45.8°C; b.p. 284°C (100 mm), 216°C (10 mm); refractive index 1.4352 (50°C). Insoluble in water, soluble in alcohol and ether. Low toxicity.
Derivation: Esterification of arachidic acid with methanol and vacuum distillation.
Grade: Purified (99.8%+).
Uses: Special synthesis; intermediate for pure arachidic acid; reference standard for gas chromatography; medical research.

methyl arecaidinate. See arecoline.

methyl azinphos (O,O-dimethylS-4-oxo-1,2,3-benzotriazin-3(4H)-yl methyl phosphorodithioate; "Guthion"). $C_{10}H_{12}N_3O_3PS_2$.
Properties: Brown, waxy solid; m.p. 73°C. Slightly soluble in water; soluble in most organic solvents.
Hazard: Highly toxic; cholinesterase inhibitor. Tolerance, 0.2 milligram per cubic meter in air.
Use: Insecticide for fruit. Use may be restricted.
Shipping regulations: (Rail, Air) Organic phosphate, n.o.s., Poison label. Not acceptable on passenger planes.

methyl behenate (methyl docosanoate) $CH_3(CH_2)_{20}COOCH_3$. The methyl ester of behenic acid.
Properties: Waxlike solid; m.p. 53.2°C; b.p. 215.5°C (3.75 mm); refractive index 1.4262 (80°C). Insoluble in water, soluble in alcohol and ether. Combustible; low toxicity.
Derivation: Esterification of behenic acid with methanol followed by fractional distillation.
Grade: Purified (99.8%+).
Uses: Special synthesis; intermediate for pure behenic acid; biochemical and medical research; reference standard in gas chromatography.

methylbenzaldehydes. See tolyl aldehydes.

methylbenzene. See toluene.

methylbenzethonium chloride $(CH_3)_3CCH_2C(CH_3)_2C_6H_3(CH_3O)(CH_2)_2O(CH_2)_2$-$N(CH_3)_2(CH_2C_6H_5)Cl \cdot H_2O$.

Benzyldimethyl(2-[2-(para-1,1,3,3-tetramethylbutyl-cresoxy) ethoxy] ethyl) ammonium chloride. A quaternary ammonium compound.
Properties: Colorless, odorless crystals with bitter taste; m.p. 161–163°C; readily soluble in alcohol, hot benzene, "Cellosolve," chloroform, and water; insoluble in carbon tetrachloride and ether.
Grade: N.F.
Use: Medicine (bactericide).

methyl benzoate (niobe oil) $C_6H_5CO_2CH_3$.
Properties: Liquid of fragrant odor; colorless; oily; sp. gr. 1.085–1.088; refractive index 1.514; f.p. −12.3°C; b.p. 198.6°C. Flash point 181°F. Soluble in 3 parts of 60% alcohol; in most fixed oils; in ether; insoluble in water. Combustible.
Derivation: (a) By heating methyl alcohol and benzoic acid in presence of sulfuric acid. (b) Passing dry hydrogen chloride through a solution of benzoic acid in methanol. (c) Occurs naturally in oils of clove, ylang ylang, tuberose.
Grades: Technical; F.C.C
Hazard: May be toxic by ingestion.
Uses: Perfumery; solvent for cellulose esters and ethers, resins, rubber; flavoring.

methylbenzoic acid. See ortho-, meta-, and para-toluic acid.

methyl ortho-**benzoylbenzoate** $C_6H_5COC_6H_4COOCH_3$.
Properties: Colorless liquid. Sp. gr. 1.190 (25°C); refractive index 1.587 (25°C); vapor pressure 4.0 mm (175°C); b.p. 351°C; m.p. 40°C; flash point, 350°F; very slightly soluble in water. Combustible; low toxicity.
Use: Plasticizer.

alpha-**methylbenzyl acetate** (methylphenylcarbinyl acetate; styralyl acetate; sec-phenylethyl acetate; phenylmethylcarbinyl acetate) $C_6H_5CH(CH_3)OOCCH_3$.
Properties: Colorless liquid; strong floral odor; sp. gr. 1.023–1.026 (25/25°C); refractive index 1.4935–1.4960 (20°C); flash point 178°F (TCC). Soluble in 70% alcohol, glycerin, and mineral oil; insoluble in water. Combustible; low toxicity.
Grades: 98% min; F.C.C.
Uses: Perfumery; flavoring.

alpha-**methylbenzyl alcohol** (styralyl alcohol; phenylmethylcarbinol; 1-phenylethan-1-ol-; sec.-phenethyl alcohol; methylphenylcarbinol) $C_6H_5CH(CH_3)OH$.
Properties: Colorless liquid; mild floral odor; congeals below room temperature. Sp. gr. 1.009-1.014 (25°C); m.p. 20.7°C; b.p. 204°C; refractive index 1.525–1.529 (20°C); flash point 205°F (COC). Soluble in alcohol, glycerin, mineral oil; slightly soluble in water. Combustible; low toxicity.
Grade: F.C.C.
Use: Perfumery; flavoring; dyes; laboratory reagent.

alpha-**methylbenzylamine** $C_6H_5CH(CH_3)NH_2$.
Properties: Water-white liquid, mild ammoniacal odor; sp. gr. 0.9535 (20/20°C); refractive index 1.5366 (20°C); b.p. 188.5°C; vapor pressure 0.5 mm (20°C); f.p., sets to a glass below −65°C; flash point, 175°F (COC); soluble in most organic solvents and hydrocarbons; somewhat soluble in water. Combustible; low toxicity.
Uses: Synthesis; emulsifying agent.

alpha-**methylbenzyldiethanolamine** $C_6H_5CH(CH_3)N(C_2H_4OH)_2$.
Properties: Dark amber liquid, ammonia-like odor; sp. gr. 1.0812 (20°C); b.p. 244°C (50 mm); flash point 370°F (O.C.) vapor pressure less than 0.01 mm (20°C); sets to glass at −7°C; moderately soluble in water. Combustible; low toxicity.
Uses: Emulsifying agents, textile specialties, quaternaries.

alpha-**methylbenzyldimethylamine** $C_6H_5CH(CH_3)N(CH_3)_2$.
Properties: Colorless liquid. Sp. gr. 0.9044 (20/20°C); b.p. 195.6°C; vapor pressure 0.6 mm (20°C); f.p., sets to a glass below −70°C; refractive index 1.5024 (20°C); flash point 175°F (O.C.) viscosity 1.85 cp (20°C); slightly soluble in water. Combustible; low toxicity.
Use: Polymerization catalyst.

alpha-**methylbenzyl ether** $C_6H_5CH(CH_3)OCH(CH_3)C_6H_5$.
Properties: Straw yellow, mobile liquid with faint odor; sp. gr. 1.0017 (20/20°C); b.p. 286.3°C (760 mm); vapor pressure less than 0.01 mm (20°C); f.p., sets to a glass below −30°C; very slightly soluble in water; flash point 275°F (COC); soluble in most organic solvents. Combustible.
Uses: Solvent; styrenating agent; softener for synthetic rubbers.

methyl-bis(2-chlorethyl)amine hydrochloride. See mechlorethamine hydrochloride.

methyl blue $C_{37}H_{27}N_3O_3S \cdot 2NaSO_3$. Sodium triphenyl-para-rosaniline sulfonate, a dark blue powder or dye used in medicine as an antiseptic and in biological and bacteriological stains.

methyl borate. See trimethyl borate.

methyl bromide (bromomethane) CH_3Br.
Properties: Colorless, transparent, easily liquefied gas, or volatile liquid; burning taste; chloroform-like odor. Miscible with most organic solvents; forms a voluminous crystalline hydrate with cold water. Sp. gr. 1.732 (0°C); b.p. 3.46°C; vapor pressure 1250 mm (20°C); f.p. −94°C; flash point, none; autoignition temp. 998°F in range 10 to 16%. Nonflammable.
Derivation: Action of bromine on methyl alcohol in presence of phosphorus, with subsequent distillation.
Grades: Technical; pure (99.5% min).
Containers: Steel cylinders.
Hazard: Toxic by ingestion, inhalation and skin absorption; strong irritant to skin. Tolerance, 15 ppm in air. Safety data sheet available from Manufacturing Chemists Assn., Washington, D.C.
Uses: Soil and space fumigant; disinfestation of potatoes, tomatoes, and other crops; organic synthesis.
Shipping regulations: (liquid) (Rail, Air) Poison label. Not acceptable on passenger planes; (methyl bromide and ethylene dibromide mixture, liquid), Poison label. Not acceptable on passenger planes; (methyl bromide and nonflammable, nonliquefied compressed gas mixtures, liquid), Poison label. Not acceptable on passenger planes. (Rail) (methyl bromide and chlorpicrin mixture, liquid) Poison label.

methyl bromoacetate $BrCH_2COOCH_3$.
Properties: Colorless to straw-colored liquid; f.p. below −50°C; b.p. 145.0–146.7°C; sp. gr. 1.655 (25/25°C); refractive index 1.456 (25°C); very slightly soluble in water; soluble in methanol, ether.
Hazard: Vapor is strong irritant to eyes.
Uses: Synthesis of weed killers, dyes, vitamins, pharmaceuticals; lachrymator.

2-methyl-1,3-butadiene. See isoprene.

2-methylbutanal. See 2-methylbutyraldehyde.

2-methylbutane. See isopentane.

2-methyl-2-butanethiol (tert-butyl mercaptan). $(CH_3)_2CSH(C_2H)_5$.
Properties: Boiling range 95–119°C; sp. gr. 0.828 (60/60°F); refractive index 1.438 (20/D); flash point 30°F. Strong, offensive odor.
Grades: 95%.
Containers: Bottles and drums.
Hazard: Flammable, dangerous fire risk.
Use: Odorant; intermediate; bacterial nutrient.
Shipping regulations: (Rail) Red label. (Air) Flammable Liquid label.

2-methyl-1-butanol (amyl alcohol, primary, active; sec-butyl carbinol) $CH_3CH_2CH(CH_3)CH_2OH$. The active alcohol from fusel oil. The synthetic product is a racemic mixture of both dextro- and levorotatory compounds and, therefore, not optically active.
Properties: Colorless liquid; sp. gr. 0.81–0.82 (20°C); f.p., less than −70°C; b.p. 128°C; refractive index (20°C) 1.41. Slightly soluble in water; miscible with alcohol and ether. Flash point 115°F (O.C.). Combustible.
Derivation: Occurs in fusel oil (q.v.); is made synthetically by fractional distillation of the mixed alcohols resulting from the chlorination and alkaline hydrolysis of pentane.
Containers: 1- and 5-gal cans; 55-gal drums.
Hazard: Moderate fire and explosion risk. Moderately toxic by ingestion, inhalation and skin absorption.
Uses: Solvent; organic synthesis (introduction of active amyl group); lubricants; plasticizers, additives for oils and paints.

2-methyl-2-butanol. See tert-amyl alcohol.

3-methyl-1-butanol. See isoamyl alcohol, primary.

3-methyl-2-butanone. See methyl isopropyl ketone.

methylbutanoyl chloride. See isovaleroyl chloride.

2-methyl-1-butene C_5H_{10} or $H_2C{:}C(CH_3)CH_2CH_3$.
Properties: Colorless, volatile liquid; disagreeable odor; b.p. 31.11°C; refractive index 1.378 (n 20/D); sp. gr. 0.650 (20/20°C); f.p. −137.52°C; flash point below −20°F; soluble in alcohol; insoluble in water.
Derivation: Refinery gas.
Grades: 95%, 99%, and research.
Containers: Bottles, drums.
Hazard: Highly flammable, dangerous fire and explosion risk.
Use: Organic synthesis.
Shipping regulations: (Rail) Flammable liquid, n.o.s., Red label. (Air) Flammable Liquid label.

2-methyl-2-butene. See 3-methyl-2-butene.

3-methyl-1-butene (isopropylethylene; alpha-isoamylene) C_5H_{10} or $H_2C{:}CHCH(CH_3)_2$.
Properties: Colorless extremely volatile liquid or gas; disagreeable odor; b.p. 20.1°C; refractive index 1.3643 (n 20/D); sp. gr. 0.6272 (20°C); f.p. −168.5°C; flash point −70°F; soluble in alcohol; insoluble in water.
Derivation: Cracking of petroleum; a component of refinery gas.
Grades: Research; 99% min; technical 95% min.
Containers: Cylinders under pressure.
Hazard: Highly flammable; dangerous fire and explosion risk. Explosive limits 1.6 to 7.7%.
Uses: Organic synthesis; high-octane fuel manufacture.
Shipping regulations: Flammable liquid, n.o.s., (Rail) Red label. (Air) Flammable Liquid label.

3-methyl-2-butene (2-methyl-2-butene; trimethylethylene; beta-isoamylene) C_5H_{10} or $H_3CCH{:}C(CH_3)_2$.
Properties: Colorless volatile liquid, disagreeable odor; b.p. 38.51°C; refractive index 1.387 (n 20/D); sp. gr. 0.6623 (20/4°C); f.p. −133.83°C; flash point −50°F; soluble in alcohol; insoluble in water.
Derivation: Cracking of petroleum; a component of refinery gas.
Grades: 90%, 95% (technical), 99% (pure), and research.
Containers: Bottles; drums; tank cars.
Hazard: Highly flammable; dangerous fire and explosion risk.
Uses: Organic synthesis; dental and surgical anesthetic; high octane fuel manufacture.
Shipping regulations: Flammable liquid, n.o.s., (Rail) Red label. (Air) Flammable Liquid label.

cis-**2-methyl-2-butenoic acid.** See angelic acid.

trans-**2-methyl-2-butenoic acid.** See tiglic acid.

2-methyl-1-buten-3-yne. See isopropenylacetylene.

1-methylbutyl alcohol. See 2-pentanol.

N-**methylbutylamine** $CH_3CH_2CH_2CH_2NHCH_3$.
Properties: Liquid; sp. gr. 0.7335; b.p. 91.1°C (760 mm); f.p. −75.0°C. Completely soluble in water; flash point 35°F (TOC).
Hazard: Flammable, dangerous fire risk. Moderate toxicity.
Use: Intermediate.
Shipping regulations: Flammable liquid, n.o.s., (Rail) Red label. (Air) Flammable Liquid label.

methyl butyl ketone (propylacetone; 2-hexanone). $CH_3COC_4H_9$.
Properties: Colorless liquid. B.p. 127.2°C; sp. gr. 0.830 (20/20°C); refractive index 1.4024 (20°C); vapor pressure 10 mm (20°C); soluble in water, alcohol, ether. Flash point 95°F (O.C.).
Grade: Technical.
Hazard: Flammable, moderate fire risk. Explosive limits 1.2 to 8% in air. Irritant to eyes and mucous membranes; narcotic in high concentrations. Tolerance, 100 ppm in air.
Use: Solvent.

2-methyl-6-tert-**butylphenol** $C_6H_3(OH)(CH_3)$(tert-C_4H_9).
Properties: Crystalline solid; light straw color; m.p. 28°C; sp. gr. 0.9618 (30°C); b.p. 230°C; flash point (open cup) 220°F, soluble in methyl ethyl ketone, ethyl alcohol, benzene, isooctane; insoluble in water. Combustible.
Use: Chemical intermediate.

2-methylbutyl 3-thiol. See sec-isoamyl mercaptan.

2-methyl-4-tert-**butylthiophenol** (4-tert-butyl-ortho-thiocresol) $(CH_3)_3CC_6H_3(CH_3)SH$.
Properties: Water-white liquid; no mercaptan odor; sp. gr. 0.983 (25°C); refractive index (n 25/D) 1.546;

2-methylbutyl 3-thiol. See sec-isoamyl mercaptan.

2-methyl-4-tert-**butylthiophenol** (4-tert-butyl-ortho-thiocresol) $(CH_3)_3CC_6H_3(CH_3)SH$.
Properties: Water-white liquid; no mercaptan odor; sp. gr. 0.983 (25°C); refractive index (n 25/D) 1.546;
f.p. −4°C; b.p. 177°C. (100 mm). Soluble in aliphatic and aromatic hydrocarbons; insoluble in water. Combustible.
Use: Chemical intermediate.

methylbutynol HC:CCOH$(CH_3)_2$. 2-Methyl-3-butyn-2-ol.
Properties: Colorless liquid with fragrant odor; b.p. 104–105°C; m.p. 2.6°C; sp. gr. (20/20°C) 0.8672; refractive index (n 20/D) 1.4211; flash point (TOC) 77°F. Miscible with water; soluble in most organic solvents. Low toxicity.
Grade: Technical, 95% min.
Containers: Drums; tank trucks and tank cars.
Hazard: Flammable, dangerous fire risk.
Uses: Stabilizer in chlorinated solvents; viscosity reducer and stabilizer; electroplating brightener; intermediate.
Shipping regulations: Flammable liquid, n.o.s., (Rail) Red label. (Air) Flammable Liquid label.

2-methylbutyraldehyde (2-methylbutanal) $CH_3CH_2CH(CH_3)CHO$.
Properties: Liquid; sp. gr. 0.8029 (20/4°C); b.p. 92–93°C; refractive index 1.3869 (20°C). Soluble in alcohol and ether; insoluble in water. Combustible.
Use: Flavoring.

3-methylbutyraldehyde. See isovaleraldehyde.

methyl butyrate $CH_3CH_2CH_2COOCH_3$.
Properties: Colorless liquid. Slightly soluble in water; soluble in alcohol. Sp. gr. 0.898 (20°C); b.p. 102°C; refractive index (n 20/D) 1.3875. Flash point 57°F (C.C.).
Grade: Technical.
Hazard: Flammable, dangerous fire risk.
Uses: Solvent for ethylcellulose; solvent mixture for nitrocellulose; flavoring.
Shipping regulations: (Rail) Flammable liquids, n.o.s., Red label. (Air) Flammable Liquid label.

2-methylbutyric acid. See isopentanoic acid.

3-methylbutyric acid. See isopentanoic acid.

methyl caprate (methyl decanoate) $CH_3(CH_2)_8COOCH_3$.
Properties: Colorless liquid; sp. gr. 0.8733 (20/4°C); f.p. −13.3°C; b.p. 224°C; 130.6°C (30 mm); refractive index 1.4237 (25°C). Insoluble in water, soluble in alcohol and ether. Combustible; low toxicity.
Derivation: Esterification of capric acid with methanol or alcoholysis of coconut oil; purified by fractional vacuum distillation.
Grades: Technical; 99.8% pure.
Uses: Intermediate for detergents, emulsifiers, wetting agents, stabilizers, resins, lubricants, plasticizers.

methyl caproate (methyl hexanoate) $CH_3(CH_2)_4COOCH_3$. The methyl ester of caproic acid.
Properties: Colorless liquid; sp. gr. 0.8850 (20/4°C); f.p. −71°C; b.p. 151.2°C, 63.0°C (30 mm); refractive index 1.4054 (n 20/D); insoluble in water; soluble in alcohol and ether. Low toxicity; combustible.
Derivation: Esterification of caproic acid with methanol or alcoholysis of coconut oil.
Grades: Technical; 99.8+%.
Uses: Intermediate for caproic acid detergents, emulsifiers, wetting agents, stabilizers, resins, lubricants, plasticizers; flavoring.

methyl caprylate (methyl octanoate) $CH_3(CH_2)_6COOCH_3$. The methyl ester of caprylic acid.
Properties: Colorless liquid; sp. gr. 0.8784 (20/4°C); f.p. −37.3°C; b.p. 192°C (759 mm), 98.3 (30 mm); refractive index 1.4152 (25°C). Insoluble in water; soluble in alcohol and ether. Low toxicity; combustible.
Derivation: (a) Esterification of caprylic acid with methanol, (b) alcoholysis of coconut oil.
Grades: Technical; 99.8%.
Containers: 55-gal drums.
Uses: Intermediate for caprylic acid detergents, emulsifiers, wetting agents, stabilizers, resins, lubricants, plasticizers; flavoring.

methyl "Carbitol."[214] Trademark for diethylene glycol monomethyl ether (q.v.).

methyl "Carbitol" acetate. Trademark for diethylene glycol monomethyl ether acetate (q.v.).

methyl carbonate (dimethyl carbonate) $CO(OCH_3)_2$.
Properties: Colorless liquid. Pleasant odor. Miscible with acids and alkalies. Stable in the presence of water. Soluble in most organic solvents; insoluble in water. Sp. gr. 1.0718 (20°C); b.p. 90.6°C; m.p. −0.5°C.
Derivation: Interaction of phosgene and methyl alcohol.
Grade: Technical.
Hazard: Flammable, dangerous fire risk. Toxic by inhalation; strong irritant.
Uses: Organic synthesis; specialty solvent.
Shipping regulations: (Rail) Flammable Liquid, n.o.s., Red label. (Air) Flammable Liquid label. Legal label name, dimethyl carbonate.

methyl "Cellosolve."[214] Trademark for ethylene glycol monomethyl ether (q.v.).

methyl "Cellosolve" acetate.[214] Trademark for ethylene glycol monomethyl ether acetate (q.v.).

methylcellulose (cellulose methyl ether; "Methocel").
Properties: Grayish white, fibrous powder; aqueous suspensions neutral to litmus. Swells in water to a viscous, colloidal solution. Insoluble in alcohol, ether, chloroform, and in water warmer than 123°F. Soluble in glacial acetic acid; unaffected by oils and greases; stable up to about 300°C; stable to light. Combustible; nontoxic.
Molecular weights may vary from 40,000 to 180,000. Specifications call for methoxy group content of narrow or wide ranges within 25–33%.
Derivation: From cellulose by conversion to alkali cellulose and then reacting this with methyl chloride, dimethyl sulfate, or methyl alcohol, and dehydrating agents. The proportions of the reacting materials are varied to control the properties of the product, such as water solubility and viscosity of the water solutions.
Grades: U.S.P.; technical; F.C.C.
Containers: Cartons, bags, carloads.
Uses: Protective colloid in water-based paints to prevent flocculation of pigment; film and sheeting; binder in ceramic glazes; leather tanning; dispersing, thickening and sizing agent; adhesive; food additive.
See also cellulose, modified; carboxymethylcellulose; hydroxyethylcellulose.

methylcellulose, propylene glycol ether. See hydroxypropyl methylcellulose.

methyl cerotate (methyl hexacosanoate) $CH_3(CH_2)_{24}COOCH_3$. The methyl ester of cerotic acid.
Properties: Waxlike solid. Insoluble in water, soluble

in alcohol and ether; m.p 62.9°C; b.p. 237°C (1.95 mm); refractive index 1.4301 (80°C). Combustible.
Derivation: Esterification of cerotic acid with methanol.
Grade: Purified (99+%).
Uses: Intermediate in special synthesis; medical research; reference standard for gas chromatography.

methyl chloride (chloromethane) CH_3Cl.
Properties: Colorless, compressed gas or liquid. Faintly sweet, ethereal odor; sp. gr. 0.92 (20°C); b.p. −23.7°C; f.p. −97.6°C; flash point below 32°F (O.C.) refractive index 1.3712 (−23.7°C); critical temperature, 143°C; critical pressure 970 psi absolute; autoignition temp. 1170°F; wt/gal 7.68 lb (20°C); slightly soluble in water, by which it is decomposed; soluble in alcohol, chloroform, benzene, carbon tetrachloride, glacial acetic acid. Attacks aluminum, magnesium and zinc.
Derivation: (a) Chlorination of methane; (b) action of hydrochloric acid on methanol, either in vapor or liquid phase.
Grades: Pure (99.5% min), technical, and 2 refrigerator grades.
Containers: 60-, 100-, 300-, 1300-lb cylinders, and in tank cars and tank trucks.
Hazard: Flammable, dangerous fire risk; explosive limits in air 10.7 to 17%. Toxic in high concentrations. Tolerance, 100 ppm in air. Safety data sheet available from Manufacturing Chemists Assn., Washington, D.C.
Uses: Catalyst carrier in low-temperature polymerization (butyl rubber); tetramethyl lead; silicones; refrigerant; medicine; fluid for thermometric and thermostatic equipment; methylating agent in organic synthesis; such as methylcellulose; extractant and low-temperature solvent; propellant in high-pressure aerosols; herbicide.
Shipping regulations: (Rail) Red gas label. (Air) Flammable Gas label. Not acceptable on passenger planes.

methyl chloroacetate $ClCH_2COOCH_3$.
Properties: Colorless liquid with sweet pungent odor; sp. gr. 1.236 (20/4°C); f.p. −32.7°C; b.p. 131°C; refractive index 1.419–1.420 (25/D); slightly soluble in water; miscible with alcohol and ether. Combustible.
Hazard: Toxic by ingestion and inhalation.
Uses: Solvent; intermediate.

methylchloroform. See 1,1,1-trichloroethane.

methyl chloroformate (methyl chlorocarbonate). $ClCOOCH_3$.
Properties: Colorless liquid. Decomposed by hot water. Stable to cold water. Soluble in methyl alcohol, ether, and benzene. Sp. gr. 1.23 (15°C); b.p. 71.4°C; vapor density 3.9 (air = 1). Flash point 54°F.
Derivation: Reaction between methyl alcohol and carbonyl chloride.
Grade: Technical (95% min).
Hazard: Highly toxic; corrosive and irritant to skin and eyes. Flammable, dangerous fire risk.
Uses: Military poison gas (lachrymator); organic synthesis; insecticides.
Shipping regulations: (Rail) White label. (Air) Corrosive label. Not acceptable on passenger planes.

methylchloromethyl ether $ClCH_2OCH_3$.
Properties: Colorless liquid; sp. gr. 1.0625 (10/4°C); f.p. −103.5°C; b.p. 59.5°C; decomposes in water; soluble in alcohol and ether.
Hazard: Flammable; dangerous fire and explosion risk. A known carcinogen.
Shipping regulations: (anhydrous) (Rail) Red label. Not accepted by express. (Air) Not acceptable.

2-methyl-4-chlorophenoxyacetic acid. See MCPA.

4-(2-methyl-4-chlorophenoxy)butyric acid. See 4-MCPB.

2-(2-methyl-4-chlorophenoxy-propionic acid. See mecoprop.

methyl chlorosilane. One of several intermediates in the formation of silicones or siloxanes; they react with hydroxyl groups on many types of surfaces to produce a permanent thin surface film of silicone that imparts water-repellency. Examples are methyltrichlorosilane, dimethyldichlorosilane, and trimethylchlorosilane (q.v.).
Hazard: Highly toxic by ingestion and inhalation; strong irritant to skin and eyes.

methyl chlorosulfonate CH_3OSO_2Cl.
Properties: Colorless liquid. Pungent odor. Decomposed by water. Soluble in alcohol, carbon tetrachloride, chloroform; insoluble in water. Sp. gr. 1.492 (10°C); b.p. 133–135°C (dec); f.p. −70°C; vapor density 4.5 (air = 1).
Derivation: Interaction of sulfuryl chloride and methyl alcohol.
Containers: Steel bottles.
Grade: Technical.
Hazard: Highly toxic by ingestion and inhalation; strong irritant to skin and eyes.
Uses: Organic synthesis; military poison gas.
Shipping regulations: (Rail, Air) Not accepted.

methylcholanthrene. A polynuclear hydrocarbon with formula $C_{21}H_{16}$.

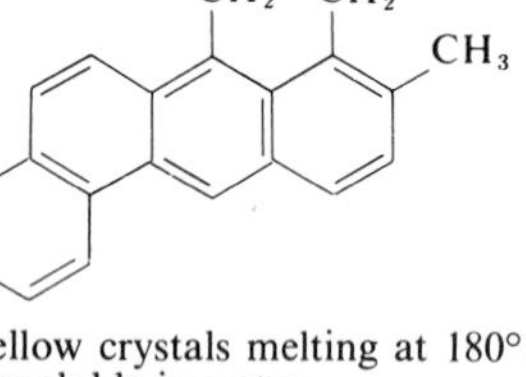

CH_2 CH_2 CH_3

Properties: Yellow crystals melting at 180°C. Soluble in benzene; insoluble in water.
Derivation: From bile acids via 1,2-benzanthracene.
Hazard: One of the most powerful carcinogenic agents known.
Uses: Biochemical research.

methyl cinnamate $C_6H_5CH{:}CHCOOCH_3$.
Properties: White crystals; strawberry-like odor. Soluble in alcohol and ether, in glycerol, most fixed oils, and mineral oil; insoluble in water. Sp. gr. 1.0415; m.p. 34°C; b.p. 259.6°C. Combustible; low toxicity.
Derivation: By heating methyl alcohol, cinnamic acid and sulfuric acid, with subsequent distillation.
Grades: Technical; F.C.C.
Containers: 1-, 5-, 10-lb bottles; 25-lb tins.
Use: Perfumes; flavoring.

methylcoumarin $C_{10}H_8O_2$.
Properties: White crystals with vanilla flavor; exists as alpha and beta forms; m.p. (alpha) 90°C, (beta) 82°C; both forms are soluble in alcohol. Combustible; low toxicity.
Uses: Perfumes; flavoring.

methyl-para-**cresol.** See para-methylanisole.

cis-alpha-**methylcrotonic acid.** See angelic acid.

trans-alpha-**methylcrotonic acid.** See tiglic acid.

methyl cyanide. See acetonitrile.

methyl cyanoacetate (malonic methyl ester nitrile) $CNCH_2COOCH_3$.
Properties: Colorless liquid; b.p. 203°C (115°C at 16 mm); f.p. −22.5°C; sp. gr. 1.1225 (15/4°C) refractive index 1.419–1.420 (n 20/D); soluble in water, alcohol and ether. Combustible.
Derivation: Esterification of cyanoacetic acid with methanol; reaction of an alkali cyanide and chloroacetic methyl ester.
Hazard: May be toxic.
Uses: Organic synthesis; pharmaceuticals; dyes.

methyl-2-cyanoacrylate $CH_2{:}C(CN)COOCH_3$.
Properties: Colorless liquid; b.p. 48–49°C (2.5–2.7 mm); sp. gr. 1.1044 (27°/4°C); viscosity 2.2 cp (25°C).
Uses: Adhesive; dentistry. See also cyanoacrylate adhesives.

methyl cyanoformate $CNCOOCH_3$.
Properties: Colorless liquid. Ethereal odor. Decomposed by alkalies and water. Soluble in alcohol, benzene, ether. Sp. gr. (approx) 1.00 (20°C); b.p. 100°C.
Derivation: Methylchloroformate is dissolved in methanol and subjected to the action of (hot) sodium or potassium cyanide.
Hazard: May be toxic.
Use: Organic synthesis.

methylcyclohexane (hexahydrotoluene) $CH_3C_6H_{11}$.
Properties: Colorless liquid; sp. gr. 0.769; b.p. 100.8°C; f.p. −126.9°C; refractive index 1.42312; flash point 25°F (C.C.); autoignition temp. 545°F.
Source: Petroleum.
Grades: Technical (95%); 99%, and research.
Containers: Glass bottles; drums.
Hazard: Flammable, dangerous fire risk. Moderately toxic; tolerance 500 ppm. Lower explosive limit 1.2% in air.
Uses: Solvent for cellulose ethers; organic synthesis.
Shipping regulations: (Rail) Flammable liquids, n.o.s., Red label. (Air) Flammable Liquid label.

methylcyclohexanol (hexahydromethyl phenol; hexahydrocresol) $CH_3C_6H_{10}OH$. Ortho-, meta-, and para-forms.
Properties: Colorless, viscous liquid; aromatic, menthol-like odor; b.p. 155–180°C; sp. gr. 0.924; flash point 154°F (C.C.). Combustible.
Derivation: (a) A mixture of three isomeric (ortho, meta and para) cyclic secondary alcohols made by the hydrogenation of cresol; (b) catalytic oxidation of methylcyclohexane.
Grade: Technical.
Containers: Cans; drums.
Hazard: Toxic by ingestion, inhalation, and skin absorption. Tolerance, 50 ppm in air.
Uses: Solvent for cellulose esters and ethers for lacquers; antioxidant for lubricants; blending agent for special textile soaps and detergents.

methylcyclohexanol acetate (heptalin acetate; methyl cyclohexyl acetate). $C_7H_{13}OOCCH_3$.
Properties: Colorless, liquid. Ester-like odor; slower rate of evaporation than amyl acetate. B.p. 176–193°C; sp. gr. 0.941; flash point 147°F (C.C.); toluene dilution ratio 2.5. Combustible.
Derivation: Catalytic hydrogenation and esterification of cresols by means of acetic acid.
Use: Solvent.

methylcyclohexanone $CH_3C_5H_9CO$.
Properties: Water-white to pale-yellow liquid. Acetone-like odor. A mixture of cyclic ketones. Closely resembles cyclohexanone in physical properties, miscibility, tolerance for nonsolvents, and solvent action. B.p. 160–170°C; sp. gr. 0.925; flash point 138°F. Combustible.
Derivation: By high-temperature, catalytic hydrogenation of cresols or by the dehydrogenation of methylcyclohexanol.
Containers: Cans; drums.
Hazard: Toxic by ingestion, inhalation, and skin absorption. Tolerance (ortho-isomer) 50 ppm in air. Moderate fire risk.
Uses: Solvent; lacquers.

methylcyclohexanone glyceryl acetal $CH_3C_6H_9O_2$-C_3H_5OH(spiro rings).
Properties: Colorless liquid. Sp. gr. 1.074 (20°C); refractive index 1.474 (20°C); b.p. 130–140°C (20 mm); flash point 200°F. Combustible. Insoluble in water.
Use: Plasticizer.

methylcyclohexanyl oxalate $(CH_3C_6H_{10}OOC)_2$.
Properties: Colorless, odorless, neutral, stable liquid comprising a mixture of isomers. Miscible with most lacquer solvents and diluents.
Hazard: May be toxic.

4-methylcyclohexene-1 $CH_3C_6H_9$.
Properties: Colorless liquid; f.p. −121.1°C; distillation range 110°C to 117°C; sp. gr. 0.818 (60/60°F); refractive index 1.450 (20°C); flash point 30°F. Insoluble in water; soluble in alcohol.
Grades: Pure, 99.0 mole%; technical, 95.0 mole%.
Containers: Up to 54-gal drums.
Hazard: Flammable, dangerous fire risk. May be irritant to skin and eyes.
Use: Intermediate.
Shipping regulations: Flammable liquid, n.o.s., (Rail) Red label. (Air) Flammable Liquid label.

6-methyl-3-cyclohexene carboxaldehyde
$CH_3CHCH_2CH{:}CHCH_2CHCHO$.

Properties: Colorless liquid. Sp. r. 0.9484; b.p. 176.4°C; f.p. −39.0°C; solubility in water, 0.3% by weight at 20°C. Probably low toxicity.
Use: Intermediate.

N-methyl-5-cyclohexenyl-5-methylbarbituric acid. See hexobarbital.

N-methylcyclohexylamine $C_6H_{11}NHCH_3$.
Properties: Water-white liquid; sp. gr. 0.86 (20°C); soluble in alcohol and ether; slightly soluble in water. Purity 99%; distillation range, 5–95 cc within 2°C, including 149°C, corrected to 760 mm. Combustible. Probably low toxicity.
Uses: Intermediate; solvent; acid acceptor.

methylcyclopentadiene dimer (methyl-1,3-cyclopentadiene) $C_{12}H_{16}$.
Properties: Colorless liquid. Sp. gr. 0.9341 (20/4°C); b.p. 78–183°C; flash point 140°F (TOC). Insoluble in water; very soluble in alcohol, benzene and ether. Combustible.
Hazard: Moderate fire risk.
Uses: High-energy fuels; curing agents; plasticizers; resins; surface coatings; pharmaceuticals; dyes.

methylcyclopentadienyl manganese tricarbonyl $CH_3C_5H_4Mn(CO)_3$.

Derivation: Methylcyclopentadiene with manganese carbonyl.
Hazard: Probably toxic and irritant.
Use: Antiknock agent, either alone or admixed with TEL.

methylcyclopentane $C_5H_9CH_3$.
Properties: Colorless liquid. Sp. gr. 0.750 (20/4°C); f.p. −142.5°C; refractive index 1.40983 (20°C); b.p. 72°C (742 mm); flash point below 20°F. Immiscible with water.
Grades: Technical (95%), 99%, and research.
Containers: Glass bottles; cans; steel drums.
Hazard: Flammable, dangerous fire and explosion risk. May be irritant and narcotic.
Uses: Organic synthesis; extractive solvent; azeotropic distillation agent.
Shipping regulations: (Rail) Red label. (Air) Flammable Liquid label. Legal label name (Rail) cyclopentane, methyl.

methyl cyclopentenolone. See 2-hydroxy-3-methyl-2-cyclopentene-1-one.

methylcytosine $C_5H_7N_3O$. 5-Methyl-2-oxy-4-aminopyrimidine. A pyrimidine found in deoxyribonucleic acids, nucleotides, and nucleosides. Crystallizes in prisms from water; m.p. 270° (dec).

methyl decanoate. See methyl caprate.

methyl demeton. Generic name for a mixture of O,O-dimethyl O(and S)-(2-ethylthio)ethyl phosphorothioates. Slightly soluble in water; soluble in most organic solvents.
Hazard: Highly toxic. Cholinesterase inhibitor. Use may be restricted. Tolerance, 0.5 mg per cubic meter of air.
Use: Systemic insecticide.
Shipping regulations: Organic phosphate mixture: (Rail, Air) Poison label. Not acceptable on passenger planes.

methyl dichloroacetate $Cl_2CHCOOCH_3$.
Properties: Liquid; sp. gr. 1.3759–1.3839 (20/20°C); refractive index (n 20/D) 1.4374–1.4474; b.p. 143°C. Slightly soluble in water; soluble in ether and alcohol. Flash point 176°F. Combustible.
Grades: 99.0% pure.
Containers: Carboys; drums.
Hazard: Forms corrosive products on hydrolysis. Keep dry.
Use: Organic intermediate.
Shipping regulations: (Air) Corrosive label.

methyldichloroarsine CH_3AsCl_2.
Properties: A colorless, mobile liquid having an agreeable odor. Decomposed by water. B.p. 136°C; m.p., −59°C; sp. gr. 1.838.
Hazard: Highly toxic and irritant.
Uses: Military poison gas; intermediate.
Shipping regulations: (Rail) Poison gas label. (Air) Not acceptable.

methyl 3,4-dichlorocarbanilate. See swep.

methyl N-3,4-dichlorophenylcarbamate. See swep.

methyldichlorosilane CH_3SiHCl_2.
Properties: Colorless liquid; b.p. 41°C; sp. gr. 1.10 (27°C); flash point −26°F. Soluble in benzene, ether, heptane.
Derivation: Reaction of methyl chloride with silicon and copper.
Hazard: Highly toxic and flammable; dangerous fire risk.
Use: Manufacture of siloxanes.
Shipping regulations: (Rail) Red label. (Air) Flammable Liquid label. No accepted on passenger planes.

methyl dichlorostearate $C_{17}H_{33}Cl_2COOCH_3$ (approx).
Properties: Light yellow, oily liquid with a slight odor. Completely miscible with most organic solvents; freezing range +7 to −5°C; b.p. decomposes 250°C; sp. gr. 0.997 (15.5/15.5°C); refractive index 1.4599 (n 25/D); flash point 358°F. Combustible; low toxicity.
Containers: 55-gal steel drums.
Uses: Intermediate; plasticizer extender.

methyldiethanolamine $CH_3N(C_2H_4OH)_2$.
Properties: Colorless liquid with amine-like odor. Miscible with benzene, water. Sp. gr. 1.0418 (20°C); b.p. 247.2°C; wt/gal 8.7 lb; vapor pressure $<$0.01 mm (20°C); f.p. −21.0°C; refractive index 1.4699; flash point 260°F (COC). Combustible; low toxicity.
Grade: Technical.
Containers: Glass bottles; cans; drums.
Uses: Intermediate; absorption of acidic gases; catalyst for polyurethane foams; pH control agent.

methyldiethylamine $CH_3N(C_2H_5)_2$.
Properties: Water-white to straw-colored liquid; b.p. 62.5°C; sp. gr. 0.724 (20°C); has inverse water solubility. Combustible.
Uses: Desalination of brackish water; chemical intermediate; acid neutralizer.

4-methyl-7-(diethylamino)coumarin. See 7-diethylamino-4-methylcoumarin.

6-methyldihydromorphinone hydrochloride. See metopon hydrochloride.

3-methyl-2,5-dihydrothiophene-1,1-dioxide (3-methylsulfolene) $CH_3C_4H_5SO_2$ (cyclic).
Properties: Solid; m.p. 63°C; b.p., decomposes. Slightly soluble in water, acetone, and toluene. Combustible.
Uses: Chemical intermediate; catalyst.

methyl 3-(dimethoxyphosphinyloxy)crotonate. See mevinphos.

methyl N,3,7-dimethyl-7-hydroxyoctyliden anthranilate. See hydroxycitronellal-methyl anthranilate Schiff base.

Methyl "Dioxitol."[125] Trademark for diethylene glycol monomethyl ether (q.v.) having an A.S.T.M. distillation range 192–196°C.

methyldioxolane (2-methyl-1,3-dioxolane)
Properties: Water-white liquid. Soluble in water. $\overline{OCH(CH_3)OCH_2CH_2}$. Sp. gr. 0.982 (20/20°C); b.p. 81°C. Combustible.
Hazard: Moderately toxic; irritant.
Uses: Extractant and solvent for oils, fats, waxes, dyestuffs and cellulose derivatives.

methyl diphenyl phosphate (methyl phenyl phosphate) $CH_3OP(O)(OC_6H_5)_2$.
Properties: Clear, oily liquid; sp. gr. 1.225–1.235 (25°C); refractive index (n 20/D) 1.5370; pour point −85°F. Nonflammable.
Containers: Drums; tank cars and trucks.
Use: Ignition control compound.

methyldipropylmethane. See 4-methylheptane.

2-methyl-1,2-di-3-pyridyl-1-propanone. See metyrapone.

methyl docosanoate. See methyl behenate.

methyl dodecanoate. See methyl laurate.

methyl eicosanoate. See methyl arachidate.

methyl elaidate $CH_3(CH_2)_7CH{:}CH(CH_2)_7COOCH_3$. The methyl ester of elaidic acid (trans-octadec-9-enoic acid).
Properties: Colorless liquid, insoluble in water, soluble in most organic solvents. Sp. gr. 0.8702 (25°C); m.p. < 15°C; b.p. 213.5°C (15 mm); refractive index 1.4462 (25°C). Combustible. Low toxicity.
Derivation: Prepared from oleic acid by elaidinization and esterification.
Use: Pure grade (99+%) used in biochemical research.

methylene. See carbene.

N,N′-**methylenebisacrylamide** $CH_2(NHCOCH{:}CH_2)_2$.
Properties: Colorless, crystalline powder; m.p. 185°C;
Hazard: Moderately toxic by ingestion.
Uses: Organic intermediate, cross-linking agent.

4,4′-methylenebis(2-chloroaniline) (3,3′-dichloro-4,4′-diaminodiphenylmethane; para, para′-methylene-bis-(ortho-chloroaniline) $CH_2(C_6H_4ClNH_2)_2$.
Properties: Tan-colored pellets; sp. gr. 1.44; melting range 99–107°C; soluble in hot methyl ethyl ketone, acetone, esters and aromatic hydrocarbons.
Hazard: Highly toxic. Causes cancer in animals. Tolerance, 0.02 ppm in air. Absorbed by skin.
Use: Curing agent for polyurethanes and epoxy resins.

para,para′-**methylenebis**(ortho-**chloroaniline).** See 4,4′-methylenebis(2-chloroaniline).

2,2′-methylenebis(4-chlorophenol). See dichlorophene.

4,4′-methylenebis(2,6-di-tert-**butylphenol)** $[(C_4H_9)_2C_6H_2(OH)]_2CH_2$. A sterically hindered bisphenol.
Properties: Light yellow powder; m.p. 155°C; b.p. 217°C (1 mm); sp. gr. 0.99 (20°C). Insoluble in water and 1.0 *N* sodium hydroxide. Combustible.
Uses: Oxidation inhibitor and antiwear agent for motor oils, aviation piston engine oils, industrial oils; antioxidant for rubbers, resins, adhesives.

3,3′-methylenebis(4-hydroxycoumarin). See bishydroxycoumarin.

methylenebis(phenylene isocyanate). See diphenylmethane diisocyanate.

2,2′-methylenebis(3,4,6-trichlorophenol). See hexachlorophene.

methylene blue (methylthionine chloride) $C_{16}H_{18}N_3SCl \cdot 3H_2O$ (medicinal); $(C_{16}H_{18}N_3SCl)_2 \cdot ZnCl_2 \cdot H_2O$ (dye).
Properties: Dark green crystals or powder with bronze-like luster; odorless or slight odor; stable in air. Soluble in water, alcohol, chloroform. Water solutions are deep blue.
Derivation: By oxidation of para-aminodimethylaniline with ferric chloride in the presence of hydrogen sulfide. The dye is the zinc chloride double salt of the chloride; the medicinal product is the hydrochloride of the base.
Grades: U.S.P.; technical.
Containers: Bottles; drums.
Hazard: Moderately toxic by ingestion.
Uses: Dyeing cotton and wool; biological and bacteriological stains; reagent in oxidation-reduction titrations in volumetric analysis; indicator.

methylene bromide (dibromomethane) CH_2Br_2.
Properties: Clear, colorless liquid; sp. gr. 2.47; solidifies −52°C; b.p. 97°C; slightly soluble in water; miscible with alcohol, ether, chloroform and acetone. Low toxicity. Nonflammable.
Containers: 5-gal carboys.
Uses: Organic synthesis; solvent.

methylene chloride (methylene dichloride; dichloromethane) CH_2Cl_2.
Properties: Colorless, volatile liquid; penetrating ether-like odor. Soluble in alcohol and ether; slightly soluble in water. Sp. gr. 1.335 (15/4°C); f.p. −97°C; b.p. 40.1°C; 11.07 lb/gal (20°C); refractive index (n 20/D) 1.4244; viscosity (20°C) 0.430 cp. Nonflammable and nonexplosive in air. Autoignition temp. 1224°F.
Derivation: Chlorination of methyl chloride and subsequent distillation.
Containers: Glass bottles; 55-gal drums; tank trucks; tank cars.
Hazard: Moderately toxic by inhalation, ingestion and skin absorption. Irritant to eyes. Tolerance, 250 ppm in air.
Uses: Paint removers; propellant for aerosol sprays; solvent degreasing; plastics processing; blowing agent in foams; solvent extraction.

methylene chlorobromide. See bromochloromethane.

4,4′-methylenedianiline. See para,para′-diaminodiphenylmethane.

methylene dichloride. See methylene chloride.

3,4-methylenedioxybenzaldehyde. See piperonal.

1,2-(methylenedioxy-4-[2-(octylsulfinyl)propyl]benzene. See sulfoxide.

3,4-methylenedioxypropylbenzene (dihydrosafrole) $C_3H_7C_6H_3O_2CH_2$.
Properties: Colorless liquid. Sp. gr. 1.065 (25/25°C); somewhat soluble in alcohol. Combustible.
Uses: Essential oil compositions.

methylene-di-para-**phenylene isocyanate.** See diphenylmethane-4,4′-diisocyanate.

5,5′-methylenedisalicylic acid $CH_2[C_6H_3(OH)COOH]_2$.
Properties: Nonhygroscopic, light tan, coarse powder, stable in air (darkens in light); tends to decarboxylate at very high temperatures. Combustible; low toxicity.
Uses: Alkyd resins and modified phenolic compositions for paints and varnishes; intermediate for dyestuffs and printing ink.

methylene glutaronitrile (acrylonitrile dimer; 2,4-dicyanobutene-1) $NCC_2H_4C({:}CH_2)CN$.
Properties: Colorless liquid; b.p. 103°C (5 mm); f.p. −9.6°C; refractive index (n 25/D) 1.4505; sp. gr. 0.9756 (25/4°C). Slightly soluble in water; insoluble in aliphatic and alicyclic hydrocarbons; soluble in aromatic hydrocarbons and polar organic solvents.
Derivation: From acrylonitrile by catalytic dimerization.
Hazard: Probably toxic.
Uses: Vinyl monomer for polymerization and copolymerization; intermediate in making fibers and pharmaceuticals.

methylene iodide (diiodomethane) CH_2I_2.
Properties: Yellow liquid. Soluble in alcohol and ether; insoluble in water. Sp. gr. 3.33; m.p. 6°C; b.p. 180°C (decomp.).
Containers: Steel drums; glass bottles.
Hazard: May be irritant and narcotic in high concentrations.

Uses: Separating mixtures of minerals; organic synthesis.

5-methylene-2-norbornene. A copolymer in EPDM elastomers.

methylene succinic acid. See itaconic acid.

methylene-para-**toluidine.** See formaldehyde-para-toluidine.

methyl ester. Any of a group of fatty esters derived from coconut and other vegetable oils, tallow, etc.; alkyl groups range from C_8 to C_{18}, in varying percentages. Used as lubricants for metal-cutting fluids, high-temperature grinding, cold-rolling of steel.

N-**methylethanolamine** $CH_3NHC_2H_4OH$.
Properties: Liquid; sp. gr. 0.9414; b.p. 159.5°C; vapor pressure 0.7 mm (20°C); f.p. −4.5°C; flash point 165°F; soluble in water. Combustible; low toxicity.
Containers: Cans; drums.
Uses: Textile chemicals; pharmaceuticals.

methyl ether. See dimethyl ether.

4-methyl-7-ethoxycoumarin
$C_2H_5OC_6H_3C(CH_3):CHC(O)O$.
Properties: White solid with a walnut odor; m.p. 113–114°C. Slightly soluble in alcohol.
Uses: Perfumery; flavoring.

2-(1-methylethoxy)phenol methylcarbamate
$(CH_3)_2CHOC_6H_4OOCNHCH_3$.
Properties: White, odorless crystalline powder; m.p. 91°C; soluble in most polar solvents; very slightly soluble in water. Unstable in highly alkaline media; stable under normal conditions.
Hazard: Toxic by ingestion and inhalation. Tolerance, 0.5 mg per cubic meter of air.
Use: Insecticide.

methylethylcarbinol. See sec-butyl alcohol.

methylethylcellulose. The methyl ether of ethylcellulose in which both methyl and ethyl groups are attached to anhydroglucose units by ether linkages.
Properties: White to pale cream-colored fibrous solid or powder. Practically odorless; disperses in cold water to form aqueous solutions which undergo a reversible transformation from sol to gel upon heating and cooling, respectively. Combustible; nontoxic.
Grades: Technical; F.C.C.
Uses: Emulsifier; stabilizer; foaming agent.
See also cellulose, modified.

methyl ethyl diketone. See acetyl propionyl.

2-methyl-2-ethyl-1,3-dioxolane
$(CH_3)(C_2H_5)COCH_2CH_2O$.
Properties: Colorless liquid. Sp. gr. 0.9392; b.p. 117.6°C; f.p. −81.96°C; flash point 74°F (O.C.); solubility in water 2.2% by weight.
Hazard: Flammable, dangerous fire risk.
Use: Solvent.
Shipping regulations: Flammable liquid, n.o.s. (Rail) Red label. (Air) Flammable Liquid label.

sym-**methylethylethylene.** See 2-pentene.

methylethylglyoxal. See acetyl propionyl.

methylethylhydantoin formaldehyde resin. A reaction product of 5,5-methylethylhydantoin, $NHCONHCOC(CH_3)C_2H_5$, and formaldehyde. Clear, pale solid resin; softening point (B & R) 85°C; sp. gr. 1.30; pH (10% aqueous solution) 6.5–7.5. See "Dantoin" 7.

methyl ethyl ketone (ethyl methyl ketone; 2-butanone; MEK) $CH_3COC_2H_5$.
Properties: Colorless liquid; acetone-like odor; b.p. 79.6°C; sp. gr. 0.8255 (0/4°C), 0.805 (20/4°C), and 0.7997 (25/4°C); refractive index 1.379 (20°C); sp. heat 0.549 cal/g; f.p. −86.4°C; viscosity 0.40 cp (25°C); solubility in water 22.6 wt %; solubility of water 9.9 wt %; flash point (TOC) 24°F; 6.71 lb/gal (20°C). Autoignition temp. 960°F. Soluble in benzene, alcohol, and ether; miscible with oils.
Derivation: (a) From mixed n-butylenes and sulfuric acid to cause hydrolysis, followed by distillation to separate sec-butyl alcohol, which is dehydrogenated; (b) by controlled oxidation of butane.
Grade: Technical.
Containers: Drums; tank cars.
Uses: Solvent in nitrocellulose coatings and vinyl films; "Glyptal" resins; paint removers; cements and adhesives; organic synthesis; manufacture of smokeless powder; cleaning fluids; printing; catalyst carrier; lube oil dewaxing; acrylic coatings. *Note:* Does not dissolve celleulose acetate and moxt waxes.
Hazard: Flammable, dangerous fire risk. Explosive limits in air 2 to 10%. Tolerance, 200 ppm in air. Narcotic by inhalation. Safety data sheet available from Manufacturing Chemists Assn., Washington, D.C.
Shipping regulations: (Rail) Red label. (Air) Flammable Liquid label. Legal label name, ethyl methyl ketone.

methyl ethyl ketone peroxide (ethyl methyl ketone peroxide) $C_8H_{16}O_4$.
Properties: Colorless liquid; strong oxidizing agent.
Hazard: Strong irritant to skin and tissue. Fire risk in contact with organic materials. Tolerance, 1 mg per cubic meter of air.
Uses: Manufacture of acrylic resins; hardening agent for fiber glass reinforced plastics.
Shipping regulations: (Rail) Peroxides, organic, liquid, Yellow label. (Air) (with not more than 60% by weight of peroxide in dimethyl phthalate). Organic Peroxide label; (with more than 60% by weight of peroxide), Not acceptable.

2-methyl-5-ethylpyridine (MEP; aldehydine; aldehyde collidine; 5-ethyl-2-picoline) $CH_3C_5H_3NC_2H_5$.
Properties: Colorless liquid; sharp penetrating odor; sp. gr. 0.921 (20/20°); b.p. 178.3°C; f.p. −70.3°C; flash point 165°F (COC); refractive index (n 20/D) 1.4970; slightly soluble in water. Combustible.
Derivation: Paraldehyde is treated with ammonia under high pressure and in the presence of ammonium acetate as a catalyst. Picolines and other substituted pyridines are byproducts.
Grade: Technical.
Containers: Drums; tank cars.
Hazard: Strong irritant to tissue.
Uses: Nicotinic acid and nicotinamide; vinyl pyridines for copolymers; intermediates for germicides and textile finishes; corrosion inhibitor for chlorinated solvents.
Shipping regulations: (Air) Corrosive label.

methyl eugenol. Generic name for 4-allyl-1,2-dimethoxybenzene (4-allyl veratrole; 1,2-dimethoxy-4-allylbenzene; eugenyl methyl ether)
$(CH_3O)_2C_6H_3CH_2CH:CH_2$.
Properties: Colorless to pale yellow liquid; b.p. 91–95°C (0.3 mm); sp. gr. 1.032–1.036 (25°C). Insoluble

in water; soluble in most organic solvents. Combustible; low toxicity.
Grade: Technical; F.C.C.
Containers: 25-lb cans.
Uses: Insect attractant; flavoring.

methyl fluoride. Legal label name (Air) for fluoromethane (q.v.).

N-methylformanilide $C_6H_5H(CH_3)CHO$.
Properties: Colorless to light yellow liquid; refractive index 1.5570–1.5600 (n 25/D); distillation range 127–131°C (16 mm).
Grade: 95% min.
Containers: 30-, 55-gal steel drums.
Hazard: May be toxic.
Use: Organic synthesis.

methyl formate $HCOOCH_3$.
Properties: Colorless liquid; agreeable odor. Saponified by water or alkaline solutions. Soluble in water, alcohol and ether. Sp. gr. 0.950–0.980 (20/20°C); f.p. −99.8°C; b.p. 31.8°C; flash point −2°F; wt/gal 8.03 lbs (68°F); refractive index 1.3431 (20°C); autoignition temp. 853°F.
Derivation: By heating methyl alcohol with sodium formate and hydrochloric acid, with subsequent distillation.
Grades: Technical; refined; F.C.C.
Containers: Drums; tank cars.
Hazard: Flammable, dangerous fire and explosion risk. Explosive limits in air 5.9 to 20%. Tolerance 100 ppm in air. Moderately toxic and irritant.
Uses: Organic synthesis; cellulose acetate solvent; military poison gases; fumigant; larvicides.
Shipping regulations: (Rail) Red label. (Air) Flammable Liquid label.

methyl fuel. A proposed auxiliary fuel for automotive equipment, electric power production, fuel cells, etc. A mixture of methyl alcohol and alcohols of up to four carbon atoms, it is made by catalytic treatment of synthesis gas derived from coal. It is stated to have a research octane number of 130 and can be blended with gasoline up to 20% by volume. Its combustion products contain less polluting components than No. 5 fuel oil. See also methyl alcohol.

2-methylfuran $C_4H_3OCH_3$.
Properties: Colorless liquid; ether-like odor; f.p. −88.68°C; b.p. 63.2–65.6°C; sp. gr. 0.913 (20/4°C); refractive index: 1.4320 (20/D); flash point −22°F. Insoluble in water, 0.3 g/100 g water. Infinitely miscible with most organic solvents. Forms a binary azeotrope with methanol, a ternary azeotrope with methanol-water.
Containers: 1-, 5-, 55-gal containers; shipped with 0.01% hydroquinone as oxidation inhibitor.
Hazard: Highly flammable, dangerous fire and explosion risk. Moderately toxic and irritant.
Use: Chemical intermediate.
Shipping regulations: (Rail) Flammable liquids, n.o.s., Red label. (Air) Flammable Liquid label.

N-methylfurfurylamine $C_4H_3OCH_2NHCH_3$. Colorless to light yellow; refractive index 1.4700–1.4720 (25°C); distilling range 144–153°C. Combustible.
Use: Intermediate.

methyl 2-furoate (methyl pyromucate) $C_4H_3OCO_2CH_3$.
Properties: Colorless liquid turning yellow in light. Pleasant odor. Insoluble in water; soluble in alcohol and ether. Sp. gr. 1.1739 (15/15°C); b.p. 181.3°C (corr); refractive index 1.4860 (20°C/D). Combustible.
Derivation: By esterification of furoic acid.
Uses: Solvent; organic synthesis.

methyl gallate. (methyl 3,4,5-trihydroxybenzoate). $C_6H_2CO_2CH_3(CH)_3$. White crystalline power. Used as industrial antioxidant. Shipped in polyethylene-lined fiber drums.

N-methylglucamine $CH_2OH(CHOH)_4CH_2NHCH_3$.
Properties: White crystals; m.p. 128°C. Soluble in water; slightly soluble in alcohol.
Preparation: From glucose and methylamine.
Uses: Detergents, pharmaceuticals, dyes.

methyl-alpha-**D-glucopyranoside.** See methyl glucoside.

N-methyl-L-glucosamine (2-methylaminoglucose). A hexosamine found in streptomycin.
Properties (of hydrochloride): Crystals with m.p. 160–163°C. Freely soluble in water. Exhibits mutarotation in solution.
Derivation: Hydrolysis of streptomycin.
Use: Medicine.

methyl glucoside (methyl alpha-d-glycopyranoside
$CH_2OHCH(CHOH)_3CHOOCH_3$.

Properties: Odorless, white crystals. M.p. 168°C; b.p. 200°C (0.2 mm); specific optical rotation (aqueous solution) + 158.9° (20°C); sp. gr. (30/4°C) 1.46. Soluble in water and 80% alcohol; slightly soluble in methanol; insoluble in ether. Combustible, low toxicity.
Derivation: (a) By treating dextrose with methanol in the presence of hydrochloric acid or cation exchange resin; (b) enzymatic synthesis from yeast.
Grade: Technical.
Containers: 100-lb multiwall paper bags.
Uses: Plasticizer for phenolic, amine, and alkyd resins; nonionic surfactants.

methyl glycocoll. See sarcosine.

methyl glycol. See propylene glycol.

methylglyoxalidine. See lysidine.

methyl group. The simplest alkyl group, CH_3, formed by dropping a hydrogen atom from methane (CH_4). It occurs at both ends of paraffinic molecules having two or more carbon atoms in the chain, as well as in many other organic compounds. See also alkyl.

4-methylguaiacol. See creosol.

N-methyl-N-gyanylglycine. See creatine.

methyl heneicosanoate $CH_3(CH_2)_{19}COOCH_3$. The methyl ester of heneicosanoic acid.
Properties: White waxlike solid. Insoluble in water, soluble in alcohol and ether. M.p. 48–9°C; b.p. 207°C (3.75 mm). Combustible.
Grades: Purified 96%, 99.5%.
Uses: Intermediate in organic synthesis; medical research.

methyl heptadecanoate (methyl margarate) $CH_3(CH_2)_{15}COOCH_3$. The methyl ester of heptadecanoic acid (margaric acid).
Properties: White waxlike solid. Insoluble in water, soluble in alcohol and ether. M.p. 29°C; b.p. 184–7°C; 130°C (1 mm). Combustible.
Grades: Purified 96%, 99.5%.
Uses: Intermediate in organic synthesis; medical research.

2-methylheptane $(CH_3)_2CH(CH_2)_4CH_3$. See isooctane.

3-methylheptane $C_2H_5CH(CH_3)(CH_2)_3CH_3$.
Properties: Colorless liquid; f.p. −120.5°C; b.p. 118.9°C; sp. gr. 0.70582 (20/4°C); refractive index 1.39849 (n 20/D).
Grades: 99%, 95%.
Hazard: Flammable, dangerous fire risk.
Uses: Calibration; organic synthesis.
Shipping regulations: (Rail, Air) Flammable liquids, n.o.s., Flammable Liquid label.

4-methylheptane (methyldipropylmethane) C_8H_{18} or $CH_3(CH_2)_2CHCH_3(CH_2)_2CH_3$.
Properties: Colorless liquid. Soluble in alcohol and ether; insoluble in water. Sp. gr. 0.7161; b.p. 122.2°C.
Grades: Technical.
Hazard: Flammable, dangerous fire risk.
Use: Organic synthesis.
Shipping regulations: (Rail, Air) Flammable liquids, n.o.s., Flammable Liquid label.

3-methyl-1,4,6-heptatriene. See butadiene dimers.

methylheptenone (6-methyl-5-heptene-2-one) $(CH_3)_2C:CH(CH_2)_2COCH_3$. Constituent of lemongrass oil and many other essential oils.
Properties: Colorless liquid; insoluble in water but miscible with alcohol or ether. Sp. gr. 0.860 (20°C); f.p. −67.1°C; b.p. 173–174°C. Combustible; low toxicity.
Derivation: From oil of lemon grass or by controlled oxidation of corresponding secondary alcohol.
Containers: 25-lb cans.
Uses: Organic synthesis; inexpensive perfumes; flavoring.

methyl heptine carbonate (methyl 2-octynoate) $CH_3(CH_2)_4C{\equiv}CCOOCH_3$.
Properties: Colorless liquid; strong violet-type odor. Sp. gr. 0.919–0.923; refractive index 1.446–1.450 (20°C). Soluble in most fixed oils and mineral oil. Soluble in 5 parts of 70% alcohol. Combustible; low toxicity.
Derivation: From heptaldehyde.
Grades: Technical; F.C.C.
Containers: Bottles; cans.
Uses: Perfumery; flavoring.

2-(1-methylheptyl)-4,6-dinitrophenyl crotonate. See dinocap.

methyl hexacosanoate. See methyl cerotate.

methyl hexadecanoate. See methyl palmitate.

methylhexamine. See 4-methyl-2-hexylamine.

2-methylhexane (ethylisobutylmethane; isoheptane) $(CH_3)_2CH(CH_2)_2CH_2CH_3$.
Properties: Colorless liquid. Soluble in alcohol; insoluble in water. Sp. gr. 0.6789; b.p. 90.0°C; f.p. −118.5°C; refractive index 1.38498 (20°C). Flash point below 0°F.
Grade: Technical.
Containers: Glass bottles; tins; drums.
Hazard: Flammable, dangerous fire risk. Explosive limits in air 1 to 6%.
Use: Organic synthesis.
Shipping regulations: Flammable liquid, n.o.s., (Rail) Red label. (Air) Flammable Liquid label.

3-methylhexane $H_3CCH_2CH(CH_3)CH_2CH_2CH_3$.
Properties: Colorless liquid; b.p. 92°C; sp. gr. 0.692 (60/60°F); refractive index 1.388 (n 20/D); flash point 25°F.
Grades: Technical 95%.
Containers: Bottles and 5-gal drums.
Hazard: Flammable, dangerous fire risk. Explosive limits in air 1 to 6%.
Uses: Organic synthesis; oil extender solvent.
Shipping regulations: Flammable liquid, n.o.s., (Rail) Red label. (Air) Flammable Liquid label.

methylhexaneamine. See 4-methyl-2-hexylamine.

methyl hexanoate. See methyl caproate.

5-methyl-2-hexanone. See methyl isoamyl ketone.

1-methyl-2-hexylamine. See 2-aminoheptane.

methyl hexyl ketone (2-octanone) $CH_3COC_6H_{13}$.
Properties: Colorless liquid with pleasant odor; camphor taste; sp. gr. 0.82 (20/4°C) m.p. 20.9°C; b.p. 173.5°C; distillation range 166–173°C; flash point 160°F; refractive index 1.416 (20°C); insoluble in water; soluble in alcohol, hydrocarbons, ether, esters, etc. Combustible; low toxicity.
Preparation: By distilling sodium ricinoleate with caustic soda.
Containers: Glass bottles; tins; drums.
Uses: Perfumes; high-boiling solvent, especially for epoxy resin coatings; leather finishes; flavoring; odorant; antiblushing agent for nitrocellulose lacquers.

methylhydrazine (monomethylhydrazine; MMH) CH_3NHNH_2.
Properties: Colorless hygroscopic liquid; ammonia-like odor; sp. gr. 0.874 (25°); f.p. −52.4°C; b.p. 87.5°C. Soluble in water; soluble in alcohol and ether. Flash point below 80°F.
Hazard: Vapors explode at high temperatures. Flammable, dangerous fire risk.
Uses: Missile propellant; intermediate; solvent.
Shipping regulations: (Rail) Red label. (Air) Flammable Liquid label. Not acceptable on passenger planes.

methyl hydrogen sulfate. See methylsulfuric acid.

methyl para-**hydroxybenzoate.** See methyl paraben.

methylhydroxybutanone (3-methyl-3-hydroxy-2-butanone) $(CH_3)_2COHCOCH_3$.
Properties: Clear, colorless liquid with sweet, camphor-like odor; b.p. 140.3°C; freezing point −86.5°C; sp. gr. (20/20°C) 0.9553; refractive index (n 20/D) 1.4153; miscible with water, acetone, benzene, mineral spirits. Combustible; low toxicity.
Uses: Specialty solvent; chemical intermediate; flavor formulations.

methyl 3-hydroxy-alpha-**crotonate dimethyl phosphate**. See mevinphos.

methylhydroxyisopropylcyclohexane. See menthol.

methyl 12-hydroxystearate $C_{17}H_{34}OHCOOCH_3$.
Properties: White, waxy solid in the form of short flat rods; m.p. 48°C; acid value 4; saponification value 177; iodine value 5; insoluble in water, limited solubility in organic solvents. Combustible; low toxicity.
Uses: Adhesives; inks; cosmetics; greases.

2-methylimidazole (2MZ) $\overline{CHCHNC(CH_3)NH}$. Solid; m.p. 142–143°C.
Uses: Dyeing auxiliary for acrylic fibers; plastic foams.

3-methylindole. See skatole.

methyl iodide (iodomethane) CH_3I.
Properties: Colorless liquid; turns brown on exposure to light; sp. gr. 2.24–2.27 (25/25°C); f.p. −66.1°C; b.p. 42°C; refractive index 1.526–1.527 (25°C). Sol-

uble in alcohol and ether; slightly soluble in water. Nonflammable.
Derivation: Interaction of methyl alcohol, sodium iodide and sulfuric acid, with subsequent distillation.
Hazard: Toxic by ingestion, inhalation, and skin absorption. Narcotic; irritant to skin. Tolerance, 5 ppm in air.
Uses: Medicine; organic synthesis; microscopy; testing for pyridine.

methylionone (irone) $C_{14}H_{22}O$.
Properties: Colorless to amber-yellow liquid; floral odor.
Grades: Several isomers are available, as alpha, beta, delta, gamma, and mixtures. The constants are approximately: Sp. gr. 0.926–0.939; refractive index 1.501–1.504; b.p. 144°C (16 mm); soluble in alcohol; insoluble in water. Nontoxic.
Derivation: Oil of orris.
Containers: Cans; drums.
Uses: Perfumery; flavoring.
See also alpha-isomethylionone.

methyl isoamyl ketone (5-methyl-2-hexanone; MIAK) $CH_3COC_2H_4CH(CH_3)_2$.
Properties: Colorless, stable liquid; pleasant odor. Sp. gr. 0.8132 (20/20°C); refractive index 1.4062 (n 20/D); b.p. 144°C; f.p. −73.9°C; wt/gal 6.77 lb; flash pt. 110°F (open cup). Slightly soluble in water; miscible with most organic solvents. Combustible.
Grade: 97.5%.
Containers: Drums; tank cars.
Hazard: Moderate fire risk. Tolerance, 100 ppm in air.
Uses: Solvent for nitrocellulose, cellulose acetate butyrate, acrylics, and vinyl copolymers.

methyl isobutenyl ketone. See mesityl oxide.

methylisobutyl carbinol. (MIBC) See methylamyl alcohol.

methylisobutyl carbinol acetate. See methylamyl acetate.

methyl isobutyl ketone (hexone; 4-methyl-2-pentanone; MIKB) $(CH_3)_2CHCH_2COCH_3$.
Properties: Colorless, stable liquid. Pleasant odor. Slightly soluble in water; miscible with most organic solvents. Sp. gr. 0.8042 (20/20°C); b.p. 115.8°C; f.p. −80.4°C; wt/gal 6.68 lbs (20°C); vapor pressure 15.7 mm (20°C); refractive index 1.3959 (20°C). Flash point 73°F. Autoignition temp. 860°F.
Derivation: Mild hydrogenation of mesityl oxide.
Grades: Technical, 98.5%.
Containers: Cans; drums; tank cars.
Hazard: Avoid ingestion and inhalation. Flammable, dangerous fire risk. Explosive limits in air 1.4 to 7.5%. Tolerance, 100 ppm in air.
Uses: Solvent for paints, varnishes, nitrocellulose lacquers; manufacture of methyl amyl alcohol; extraction processes, including extraction of uranium from fission products; organic synthesis; denaturant for alcohol.
Shipping regulations: (Air) Flammable Liquid label.
Note: Use restricted in some states due to possible toxicity.

methyl isocyanate CH_3NCO. Colorless liquid; sp. gr. 0.9599 (20/20°C); b.p. 39.1°C; reacts with water; flash point less than 20°F.
Hazard: Flammable, dangerous fire risk. Highly toxic. Tolerance, 0.02 ppm in air.
Use: Intermediate.
Shipping regulations: (Air) Poison label. Not acceptable on passenger planes.

methylisoeugenol (propenyl guaiacol) $CH_3CH{:}CHC_6H_3(OCH_3)_2$.
Properties: Colorless to light-yellowish liquid, with spicy odor. Sp. gr. 1.050–1.053; b.p. 262–264°C; refractive index 1.566–1.569. Soluble in 2 parts of 70% alcohol; almost insoluble in mineral oil; insoluble in glycerol. Combustible; low toxicity.
Grades: Technical; F.C.C.
Containers: Glass bottles; drums.
Uses: Perfumery, flavoring agent.

methyl isonicotinate $C_5NH_4COOCH_3$.
Properties: Clear amber to red liquid; mild odor; sp. gr. 1.15 (20/20°C).
Hazard: May be toxic.
Use: Intermediate for synthesis of isonicotinic acid hydrazide.

1-methyl-4-isopropenylcyclohexan-3-ol. See isopulegol.

methyl isopropenyl ketone $CH_3COC(CH_3){:}CH_2$.
Properties: Colorless liquid; sp. gr. 0.854 (20°C); pleasant odor and sweet taste; polymerizes readily.
Hazard: Flammable, dangerous fire risk.
Use: Plastics.
Shipping regulations: (inhibited) (Rail) Red label. (Air) Flammable Liquid label. (uninhibited). (Air) Not acceptable.

methyl isopropyl ketone (3-methyl-2-butanone) $CH_3COCH(CH_3)_2$.
Properties: Colorless liquid; b.p. 93°C; f.p. −92°C; refractive index (n 16/D) 1.38788; sp. gr. 0.815 (15/4°C); very slightly soluble in water; soluble in alcohol, ether.
Derivation: Synthetic; also by fermentation.
Hazard: May be flammable.
Use: Solvent for nitrocellulose lacquers. See also ketones.

2-methyl-5-isopropylphenol. See carvacrol.

5-methyl-2-isopropylphenol. See thymol.

methyl-para**-isopropylphenyl propyl aldehyde.** See cyclamen aldehyde.

methyl lactate $CH_3CHOHCOOCH_3$.
Properties: Colorless liquid. Miscible with most organic liquid, water. B.p. 144.8°C; f.p. (approx) −66°C; refractive index 1.4156 (20°C); flash point 125°F (closed cup); autoignition temp. 725°F; wt/gal 9 lb (68°F). Combustible.
Grade: Technical.
Containers: Cans; steel drums.
Hazard: Moderate fire risk.
Uses: Solvent for cellulose acetate, nitrocellulose, cellulose acetobutyrate, cellulose acetopropionate; lacquers; stains.

methyllactonitrile. See acetone cyanohydrin.

methyl laurate (methyl dodecanoate) $CH_3(CH_2)_{10}COOCH_3$. The methyl ester of lauric acid.
Properties: Water-white liquid; sp. gr. 0.8702 (20/4°C); m.p. 4.8°C; b.p. 262°C (766 mm), 160°C (30 mm); refractive index 1.4301 (25°C). Insoluble in water; non-corrosive. Combustible; low toxicity.
Derivation: From coconut oil.
Method of purification: Vacuum fractional distillation.
Grades: 69, 74, 90, 96, 99.8%.
Containers: Cans; drums; tank cars.

Uses: Intermediate for detergents, emulsifiers, wetting agents, stabilizers, resins, lubricants, plasticizers, textiles; flavoring.

methyl lauroleate $CH_3CH_2CH{:}CH(CH_2)_7COOCH_3$. The methyl ester of lauroleic acid.
Properties: Colorless liquid; insoluble in water; soluble in common organic solvents. Combustible; low toxicity.
Grades: Purified product, 99.5%.
Uses: In medical research, organic synthesis and as a reference standard for gas chromatography.

methyl lignocerate (methyl tetracosanoate) $CH_3(CH_2)_{22}COOCH_3$. The methyl ester of lignoceric acid.
Properties: Waxlike solid. Insoluble in water, soluble in alcohol and ether; m.p. 57.8°C; b.p. 232°C (3.75 mm); refractive index 1.4283 (80°C). Combustible; low toxicity.
Derivation: Esterification of lignoceric acid with methanol followed by vacuum distillation.
Grades: Purified (99.8%+).
Uses: Intermediate in special synthesis; medical research; reference standard in gas chromatography.

methyl linoleate
$CH_3(CH_2)_4CH{:}CHCH_2CH{:}CH(CH_2)_7COOCH_3$. The methyl ester of linoleic acid (cis, cis-octadec-9,12-dienoic acid).
Properties: Colorless oil, insoluble in water; soluble in alcohol and ether. Sp. gr. 0.8886 (18/4°C); f.p. −35°C; b.p. 212°C (16 mm); refractive index 1.4593 (25°C). Combustible. Low toxicity.
Derivation: Urea fractionation and vacuum distillation of methyl esters of safflower oil.
Grades: Technical; purified (99+%).
Uses: Intermediate for detergents, emulsifiers, wetting agents, stabilizers, resins, lubricants, plasticizers, textiles; reference standard in gas chromatography; biochemical research.

methyl linolenate
$CH_3CH_2CH{:}CHCH_2CH{:}CHCH_2CH{:}CH(CH_2)_7$-$COOCH_3$. The methyl ester of linolenic acid (cis, cis, cis-octadec-9,12,15-trienoic acid).
Properties: Colorless liquid; insoluble in water; soluble in alcohol and ether. Sp. gr. 0.892 (20/4°C); m.p. less than 15°C; b.p. 207°C (14 mm); refractive index 1.4709 (n 20/D). Combustible. Low toxicity.
Derivation: Esterification and vacuum fractional distillation.
Uses: Organic synthesis; biochemical research.

methyllithium CH_3Li. Commercially available as a 5% solution in ether. Used in Grignard-type reactions.
Hazard: Flammable, dangerous fire and explosion risk.
Shipping regulations: Not listed. Consult authorities.

methylmagnesium bromide CH_3MgBr. Available in solution in ether.
Derivation: Reaction of magnesium and methylbromide.
Containers: Glass bottles; 5- and 55-gal drums.
Hazard: Flammable, dangerous fire and explosion risk.
Uses: Alkylating agent in organic synthesis; Grignard reagent.
Shipping regulations: (Rail) (conc. not over 40%) Red label. (Air) (all conc.) Not acceptable.

methylmagnesium chloride CH_3MgCl. Available as a solution in tetrahydrofuran.
Containers: Up to 55-gallon drums.
Hazard: Flammable, dangerous fire and explosion risk.
Uses: Alkylating agent in organic synthesis; Grignard reagent.
Shipping regulations: Not listed, but see methylmagnesium bromide.

methylmagnesium iodide CH_3MgI. Available as solution in ether.
Derivation: Reaction of magnesium and methyl iodide.
Containers: Glass bottles; 5- and 55-gal drums.
Hazard: Flammable, dangerous fire and explosion risk.
Uses: Alkylating agent in organic synthesis; Grignard reagent.
Shipping regulations: Not listed, but see methylmagnesium bromide.

methylmaleic acid. See citraconic acid.

methylmaleic anhydride. See citraconic anhydride.

methyl margarate. See methyl heptadecanoate.

methyl mercaptan. Legal label name for methanethiol (q.v.).

methylmercury acetate $CH_3HgOOCCH_3$.
Hazard: Highly toxic by ingestion; absorbed by skin. Tolerance, 0.01 mg per cubic meter of air.
Use: Seed disinfectant.
Shipping regulations: (Rail, Air) Mercury compounds, n.o.s., Poison label.

methylmercury cyanide (methylmercury nitrile) CH_3HgCN.
Properties: Crystals; m.p. 95°C. Soluble in water and organic solvents.
Hazard: Highly toxic by ingestion; absorbed by skin. Tolerance, 0.01 mg per cubic meter of air.
Uses: Fungicide; seed disinfectant.
Shipping regulations: (Rail, Air) Mercury compounds, n.o.s., Poison label.

methylmercury dicyandiamide. See cyano (methylmercuri) guanidine.

methylmercury 2,3-dihydroxypropylmercaptide $CH_3HgSCH_2CHOHCH_2OH$.
Hazard: Highly toxic by ingestion; absorbed by skin. Tolerance, 0.01 mg per cubic meter of air.
Uses: Seed disinfectant.
Shipping regulations: (Rail, Air) Mercury compounds, n.o.s., Poison label.

methylmercury 8-hydroxyquinolate. See methylmercury quinolinolate.

methylmercury quinolinolate (methylmercury 8-hydroxyquinolate; methylmercury oxyquinolinate) $C_9H_6NOHgCH_3$.
Properties: Yellow crystals; m.p. 133–137°C.
Hazard: Highly toxic. Tolerance, 0.01 mg per cubic meter of air.
Use: Seed disinfectant.
Shipping regulations: (Rail, Air) Mercury compounds, n.o.s., Poison label.

methyl methacrylate $CH_2{:}C(CH_3)COOCH_3$. Acrylic resin monomer.
Properties: Colorless, volatile liquid; b.p. 101°C; f.p. −48.2°C; sp. gr. (25/25°C) 0.940; flash point (open cup) 50°F; autoignition temp. 790°F, slightly soluble in water; soluble in most organic solvents. Readily polymerized by light, heat, ionizing radiation, and catalysts. Can be copolymerized with other methacrylate esters and many other monomers. Low toxicity.
Derivation: Acetone cyanohydrin, methanol, and dilute sulfuric acid.

Grade: Technical (inhibited).
Containers: 440-lb drums; tank cars and trucks.
Hazard: Flammable, dangerous fire risk. Explosive limits in air, 2.1 to 12.5%. Tolerance, 100 ppm in air.
Uses: Monomer for polymethacrylate resins; impregnation of concrete. See also acrylic resin.
Shipping regulations: (inhibited) (Rail) Red label; (Air) Flammable Liquid label. (uninhibited) (Air) Not acceptable.

methylmethane. See ethane.

N-**methyl methyl anthranilate.** See dimethyl anthranilate.

7-methyl-3-methylene-1,6-octadiene. See myrcene.

methylmorphine. See codeine.

N-**methyl morpholine** $CH_2CH_2OCH_2CH_2NCH_3$.

Properties: Water-white liquid with ammonia odor. Forms constant-boiling mixture with 25% water and boiling at 97°C. Miscible with benzene, water. Sp. gr. 0.921 (20/20°C); b.p. 115.4°C; f.p. −66°C; flash point 75°F (TOC).
Grade: Technical.
Containers: Glass bottles; tins; drums.
Hazard: Flammable, dangerous fire risk. Moderately toxic; skin irritant.
Uses: Catalyst in polyurethane foams; extraction solvent; stabilizing agent for chlorinated hydrocarbons; self-polishing waxes; oil emulsions; corrosion inhibitors; pharmaceuticals.
Shipping regulations: Flammable liquid, n.o.s., (Rail) Red label. (Air) Flammable Liquid label.

methyl myristate (methyl tetradecanoate) $CH_3(CH_2)_{12}COOCH_3$. The methyl ester of myristic acid.
Properties: Colorless liquid; m.p. 17.8°C; b.p. 186.8°C (30 mm), 157.5°C (1 mm); refractive index 1.4351 (25°C). Insoluble in water. Combustible; low toxicity.
Derivation: (a) Esterification of myristic acid with methanol, (b) alcoholysis of coconut oil with methanol.
Method of purification: Vacuum fractional distillation.
Grades: Technical (93%); purified (99.8+%).
Containers: 55-gal drums.
Uses: Intermediate for myristic acid detergents, emulsifiers, wetting agents, stabilizers, resins, lubricants, plasticizers, textiles, animal feeds; standard for gas chromatography; flavoring.

methyl myristoleate $CH_3(CH_2)_3CH{:}CH(CH_2)_7COOCH_3$. The methyl ester of myristoleic acid (cis-tetradec-9-enoic acid).
Properties: Colorless liquid; insoluble in water; soluble in alcohol and ether. B.p. 108.9°C (1 mm). Combustible; low toxicity.
Uses: Purified product used in medical research and organic synthesis.

alpha-**methylnaphthalene** $C_{10}H_7CH_3$.
Properties: Colorless liquid. Sp. gr. 1.025; f.p. −32°C; b.p. 240–243°C; refractive index 1.6140 (25°C); insoluble in water; soluble in alcohol and ether. Autoignition temp. 984°F. Combustible.
Derivation: From coal tar.
Hazard: Moderate fire risk. Probably toxic.
Uses: Organic synthesis.

beta-**methylnaphthalene** $C_{10}H_7CH_3$.
Properties: Solid; sp. gr. 0.994 (40/4°C); b.p. 241–242°C; m.p. 34°C; refractive index 1.6015 (25°C); insoluble in water; soluble in alcohol and ether. Combustible.
Derivation: From coal tar.
Grades: Technical, 95% min.
Containers: 1-, 5-, 55-gal drums.
Uses: Organic synthesis; insecticides.

2-methylnaphthioquinone. See menadione.

methylnaphthyldodecyldimethylammonium chloride $C_{10}H_7HC_2C_{12}H_{25}(CH_3)_2NCl$. Quaternary ammonium salt.
Properties: White to slightly yellow crystalline powder with mild odor and taste. M.p. 159–160°C; density 5 lb/gal; soluble in cold water, lower alcohols, glycerin, and acetone; pH of 5% sol. 8.9.
Grade: Min. purity 97%.
Use: Germicide.

methyl naphthyl ether. See beta-naphthyl methyl ether.

methyl nitrate CH_3NO_3.
Properties: Colorless liquid. B.p. 66°C (explodes); sp. gr. 1.217 (15°C); slightly soluble in water; soluble in alcohol and ether.
Derivation: By reaction of nitric acid and methanol in the presence of urea.
Hazard: Explodes when heated; severe hazard when exposed to heat or shock. Narcotic in high concentrations.
Use: Rocket propellant.
Shipping reglations: Not listed. Consult authorities.

methyl nitrite CH_3NO_2. A gas; b.p. −12°C; f.p. −17°C. Sp. gr. 0.991 at 15°C.
Hazard: Severe explosion risk when shocked or heated; narcotic in high concentrations.
Uses: Synthesis of nitrile and nitroso esters.
Shipping regulatoins: (Rail) Not listed. (Air) Flammable Gas label. Not acceptable on passenger planes.

methylnitrobenzene. See nitrotoluene.

4-methyl-2-nitrophenol. See 2-nitro-para-cresol.

methyl-4-nitrosoperfluorobutyrate. A monomer for a nitroso ester terpolymer; prepared by reacting methyl nitrite with perfluorosuccinic and perfluoroglutaric anhydrides. The reaction is exothermic and gives high conversoin to liquid nitrile esters, which yield nitroso esters of pyrolysis or photolysis.

N-**methyl**-N-**nitroso**-para-**toluenesulfonamide** $CH_3C_6H_4SO_2N(NO)CH_3$.
Properties: Fine yellow crystals; m.p. 56–59°C soluble in ether and most organic solvents; insoluble in water.
Use: Reagent for the preparation of diazomethane.

methyl nonadecanoate $CH_3(CH_2)_{17}COOCH_3$. The methyl ester of nonadecanoic acid.
Properties: White waxy solid; m.p. 39.5°C; b.p. 190.5°C (3.75 mm); insoluble in water; soluble in alcohol and ether. Combustible. Low toxicity.
Grades: Purified 96%, 99.5%.
Uses: Intermediate in organic synthesis; medical research.

2-methylnonane (isodecane) $(CH_3)_2CH(CH_2)_6CH_3$.
Properties: Colorless liquid; sp. gr. 0.728; f.p. −74.7°C; b.p. 167°C. Combustible. Low toxicity.

methyl nonanoate (methyl pelargonate) $CH_3(CH_2)_7COOCH_3$. The methyl ester of pelargonic acid.
Properties: Colorless liquid with fruity odor; f.p. −35°C; b.p. 213.5°C; sp. gr. 0.877 (18°C); refractive index 1.4302 (25°C). Insoluble in water; soluble in alcohol and ether. Combustible. Low toxicity.
Derivation: Esterification of nonanoic (pelargonic) acid with methanol followed by fractional distillation.
Grade: Purified (96+%).
Uses: Perfumes; flavors; reference standard for gas chromatography; intermediate in organic synthesis; medical research.

methyl 2-nonenoate. $CH_3(CH_2)_5CH:CHCOOCH_3$. Colorless to slightly yellow liquid with strong violet-leaf odor; sp. gr. 0.893–0.898 (25/25°C); refractive index 1.4400–1.4440. Stable; soluble in alcohol. Combustible. Low toxicity.
Use: Perfumes.

methylnonylacetaldehyde (aldehyde C-12 MNA; methylundecanal) $CH_3(CH_2)_8CH(CH_3)CHO$.
Properties: Colorless liquid; fruity odor. Soluble in 3 volumes of 80% alcohol; in fixed oils, mineral oil. Sp. gr. 0.824–0.828; refractive index 1.432–1.435 (20°C). Combustible. Low toxicity.
Grades: Technical; F.C.C. (as methylundecanal).
Uses: Perfumery; flavoring.

methyl nonyl ketone (2-undecanone) $CH_3COC_9H_{19}$.
Properties: Oily liquid, strong odor. Soluble in 2 parts of 70% alcohol. Sp. gr. 0.822–0.826; b.p. 225°C; refractive index 1.429–1.433. Flash point 192°F. Combustible. Low toxicity.
Derivation: Oil of rue; also made synthetically.
Uses: Perfumery; flavoring.

methyl norbornene dicarboxylic anhydride. Legal label name for memtetrahydrophthalic anhydride (q.v.).

methyl octadecanoate. See methyl stearate.

methyl 2-octynoate. See methyl heptine carbonate.

methylolacrylamide. Available in 60% aqueous solution.
Uses: Intermediate for copolymerization of vinyl acetate and acrylic acid; polymers for coatings, varnishes, adhesives; crease-proof and wrinkle-resistant fabrics; permanent-press textiles by irradiation bonding.

methylol dimethylhydantoin (dimethylhydantoin formaldehyde; DMHF) $(CH_3)_2\underline{CN(CH_2OH)CONHCO}$.
Properties: White, odorless crystalline solid; m.p. 110–117°C; soluble in water, methanol, acetone; insoluble in hydrocarbons.
Uses: Textile and paper finishing; source of formaldehyde. See also dimethylhydantoinformaldehyde resin.

methyl oleate $CH_3(CH_2)_7CH:CH(CH_2)_7COOCH_3$. The methyl ester of oleic acid (cis-octadec-9-enoic acid).
Properties: Clear to amber liquid. Faint fatty odor. Soluble in alcohols and most organic solvents; insoluble in water; sp. gr. 0.8739 (20°C); f.p. −19.9°C; b.p. 218.5°C (20 mm); refractive index 1.4521 (20°C). Combustible; low toxicity.
Derivation: Esterification of oleic acid; vacuum fractional distillation; solvent crystallization.
Grades: Technical; purified 99+%.
Uses: Intermediate for detergents, emulsifiers, wetting agents, stabilizers, textile treatment; plasticizers for duplicating inks, rubbers, waxes, etc.; biochemical research; chromatographic reference standard.

methylol imidazolidone $C_3H_4N_2(O)CH_2OH$.
Properties: Straw-colored to water white clear liquid; mild odor. Density 10 lb/gal (60°F).
Use: Wash and wear fabrics.

methylol riboflavin. A mixture of methylol derivatives of riboflavin (q.v.) exhibiting the same activity.
Properties: Orange to yellow hygroscopic powder; nearly odorless; soluble in water; nearly insoluble in alcohol, benzene, chloroform, and ether. Dextrorotatory. Dry powder is unstable and on standing loses biological activity by liberation of formaldehyde.
Derivation: Action of formaldehyde on riboflavin in weakly alkaline solutions.
Uses: Nutrition; medicine.

methylol urea $H_2NCONHCH_2OH$.
Properties: Colorless crystals; m.p. 11°C; soluble in water and methanol; insoluble in ether; capable of polymerization to synthetic resin.
Derivation: Combination of urea and formaldehyde, in the presence of salts or alkaline catalysts.
Containers: Fiber cans and drums.
Uses: Urea-formaldehyde resins; molding adhesives; treating textiles and wood. See A-stage resin; urea-formaldehyde resin.

"Methylon."[245] Trademark for resinous compositions made from condensation products of substituted aromatic hydrocarbons and aldehydes used in the surface coating and protective arts.

methyl orange (para-(para-dimethylamino phenylazo)-benzene sulfonate of sodium; Helianthine B; orange III, gold orange; tropeolin D) $(CH_3)_2NC_6H_4NNC_6H_4SO_3Na$.
Properties: Orange-yellow powder; soluble in water, insoluble in alcohol. C.I. 13025.
Use: Acid-base indicator, red in acid, yellow-orange in alkaline, pH range 3.1–4.4
See indicator.

methyl-orthophosphoric acid. See methylphosphoric acid.

methyl oxide. See dimethyl ether.

Methyl "Oxitol."[125] Trademark for ethylene glycol monomethyl ether (q.v.) whose A.S.T.M. distillation range is 123.5–125.5°C.

methyl palmitate (methyl hexadecanoate) $CH_3(CH_2)_{14}COOCH_3$. The methyl ester of palmitic acid.
Properties: Colorless liquid; m.p. 29.5°C; b.p. 211.5°C (30 mm), 180.5°C (10 mm); refractive index 1.4310 (45°C). Insoluble in water, soluble in alcohol and ether. Combustible. Low toxicity.
Derivation: Esterification of palmitic acid with methanol or alcoholysis of palm oil. Vacuum distillation.
Grades: 80%; pure (99.8%).
Uses: Intermediate for detergents, emulsifiers, wetting agents, stabilizers, resins, lubricants, plasticizers; animal feeds; medical research.

methylparaben (methyl para-hydroxybenzoate) $CH_3OOCC_6H_4OH$.
Properties: Colorless crystals, or white crystalline powder; m.p. 125–128°C; odorless or faint characteristic odor; slight burning taste; almost insoluble in water; soluble in alcohol, ether; slightly soluble in benzene and in carbon tetrachloride. Low toxicity.
Grades: U.S.P.; F.C.C.
Hazard: Use in foods restricted to 0.1%.
Uses: Medicine; food additive (preservative); antimicrobial agent.

methylparafynol. See 3-methyl-1-pentyn-3-ol.

methyl parathion (O,O-dimethyl O-para-nitrophenyl-phosphorothioate) $(CH_3O)_2P(SO)OC_6H_4NO_2$. The methyl homolog of parathion (q.v.).
Properties: White crystalline solid; m.p. 35–36°C; sp. gr. 1.358 (20/4°C); refractive index 1.5515 (35°C). Slightly soluble in water; miscible in all proportions with acids and alcohols, esters and ketones. Slowly decomposed by acid solutions, rapidly in dilute alkalies. The commercial product is a tan liquid (xylene solution) with pungent odor; decomposes violently at 248°F. It is not classed as a "hard" insecticide, but is relatively biodegradable.
Forms available: Emulsifiable concentrates; wettable powders; dusts; technical (80%); liquid solution; encapsulated.
Hazard: Highly toxic by skin absorption, inhalation and ingestion. Cholinesterase inhibitor. Explosion risk when heated. Use may be restricted. Tolerance, 0.2 mg per cubic meter of air. Approved by EPA as substitute for some uses of DDT.
Uses: Insecticide, especially for cotton; also permissible for some food crops.
Shipping regulations: (liquid) (Rail, Air) Poison label. Not acceptable on passenger planes; content of mixtures must be less than 2%. For details, consult regulations.
Note: An encapsulated form (water suspension of polyamide microcapsules containing methyl parathion) is approved by EPA for cotton, corn, alfalfa. It provides slow, uniform release of the pesticide by diffusion through the pores of the capsule.

methyl PCT. See O,O-dimethyl phosphorochloridothioate.

methyl pelargonate. See methyl nonanoate.

methyl pentadecanoate $CH_3(CH_2)_{13}COOCH_3$. The methyl ester of pentadecanoic acid.
Properties: Colorless liquid. Insoluble in water; soluble in alcohol and ether. Sp. gr. 0.8618 (25/4°C); m.p. 18.5°C; b.p. 199°C (30 mm); index of refraction 1.4374 (25°C). Combustible.
Grades: Reagent; 96% and 99.5%.
Uses: Intermediate in organic synthesis; reagent; medical research.

methylpentadiene C_6H_{10}. Numerous isomers are possible. Commercially available mixture contains 2- and 4-methyl-1,3-pentadiene.
Properties: Colorless liquid. Sp. gr. 0.7184 (20/4°C); b.p. 75–77°C; flash point −30°F; reactive with halogens, hydrohalogens, sulfur dioxide and maleic anhydride.
Hazard: Highly flammable; dangerous fire risk.
Uses: Organic synthesis; alkyd and other polymers.
Shipping regulations: (Rail) Flammable liquid, n.o.s., Red label. (Air) Flammable Liquid label.

2-methylpentaldehyde $C_3H_7CH(CH_3)CHO$.
Properties: Colorless liquid. Sp. gr. 0.8092; b.p. 118.3°C; f.p. −100°C; solubility in water 0.42% by weight; flash point 68°F. (O.C.).
Hazard: Flammable dangerous fire risk. Strong irritant to skin and mucous membranes.
Uses: Intermediates for dyes, resins, pharmaceuticals.
Shipping regulations: Flammable liquid, n.o.s., (Rail) Red label. (Air) Flammable Liquid label.

2-methylpentane (dimethylpropylmethane) $CH_3(CH_2)_2CH(CH_3)_2$.
Properties: Colorless liquid; f.p. −153°C; b.p. 60°C; refractive index 1.372 (20°C); sp. gr. 0.658 (60/60°F); flash point −10°F; autoignition temp. 583°F.
Grades: 95%; 99%; research.
Containers: 1-gal bottles; 5-gal drums.
Hazard: Flammable, dangerous fire risk; reacts vigorously with oxidizing materials.
Uses: Organic synthesis; solvent.
Shipping regulations: (Rail) Flammable liquid, n.o.s., Red label. (Air) Flammable Liquid label.

3-methylpentane (diethylmethylmethane) C_6H_{14} or $CH_3CH_2CH(CH_3)CH_2CH_3$.
Properties: Colorless liquid. Soluble in alcohol; insoluble in water; slightly soluble in ether. Sp. gr. 0.6645 (20/4°C); b.p. 64.0°C; refractive index 1.37662 (20°C); flash point below 20°F.
Grades: Technical (95%); 99%; research.
Hazard: Flammable, dangerous fire risk.
Uses: Organic synthesis; solvent.
Shipping regulations: See 2-methylpentane.

2-methyl-1,3-pentanediol
$C_2H_5CH(OH)CH(CH_3)CH_2OH$.
Properties: Colorless liquid. Sp. gr. 0.9745; b.p. 220.3°C; f.p. −30°C; infinitely soluble in water. Low toxicity. Combustible.
Uses: Solvent; coupling agent.
See also hexylene glycol, a mixture of isomers.

2-methyl-2,4-pentanediol
$CH_3CHOHCH_2COH(CH_3)CH_3$.
Properties: Colorless liquid. Completely miscible with water and most organic solvents including lower aliphatic hydrocarbons. Sp. gr. 0.9235 (20/20°C); b.p. 197.1°C; viscosity 34 cp (20°C); vapor pressure 10.8 mm (95.2°C); 334 mm (169.7°C); flash point 201°F (open cup); wt/gal 7.59 lb (20°C). Low toxicity. Combustible.
Uses: Coupling agent; chemical synthesis.
See also hexylene glycol, a mixture of isomers.

4-methyl-2,4-pentanediol. See 2-methyl-2,4-pentanediol.

2-methylpentanoic acid. $CH_3CH_2CH_2CH(CH_3)COOH$.
Properties: Water-white liquid; sp. gr. 0.9242 (20/20°C); b.p. 196.4°C; vapor pressure 0.02 mm (20°C); f.p. sets to glass below −85°C. Flash point 225°F. Solubility in water 1.3% by wt (20°C); solubility of water in, 2.9% by wt (20°C). Combustible.
Uses: Plasticizers, vinyl stabilizers; metallic salts; alkyd resins.

4-methylpentanoic acid $(CH_3)_2CH(CH_2)_2COOH$.
Properties: Colorless liquid; b.p. 197°C; sp. gr. 0.921 (20/4°C); density 7.66 lb/gal (20°C). Soluble in all proportions in alcohol, benzene, and acetone. Low solubility in water. Combustible.
Use: Intermediate for plasticizers, pharmaceuticals, and perfumes.

2-methyl-1-pentanol $C_3H_7CH(CH_3)CH_2OH$.
Properties: Colorless liquid. Sp. gr. 0.8252; b.p. 148.0°C; vapor pressure 1.1 mm (20°C); solubility in water 0.31% by weight; flash point 135°F. (O.C.). Combustible.
Hazard: Moderate fire risk.
Uses: Solvent; intermediate.

4-methyl-2-pentanol. See methylamyl alcohol.

4-methyl-2-pentanol acetate. See methylamyl acetate.

3-methyl-3-pentanol carbamate. See emylcamate.

4-methyl-2-pentanone. See methyl isobutyl ketone.

2-methyl-1-pentene (1-methyl-1-propylethylene) $CH_2{:}C(CH_3)CH_2CH_2CH_3$.
Properties: Colorless liquid; sp. gr. 0.6820 (20/4°C); b.p. 62.2°C; f.p. −135°C; refractive index 1.3925 (n 20/D); flash point −15°F; soluble in alcohol, acetone, ether, petroleum, coal-tar solvents; insoluble in water. Low toxicity.
Grades: 95%; 99%; research.
Hazard: Flammable, dangerous fire risk.
Uses: Organic synthesis; flavors; perfumes; medicines; dyes; oils; resins.
Shipping regulations: Flammable liquid, n.o.s., (Rail) Red label. (Air) Flammable Liquid label.

4-methyl-1-pentene $H_2C{:}CHCH_2CH(CH_3)CH_3$.
Properties: Colorless liquid; f.p. −244.7°F; b.p. 128.4°F; sp. gr. 0.6640 (20/4°C); refractive index 1.3826 (n 20/D); flash point −25°F. Low toxicity.
Grades: 95%, 99%; research.
Containers: Gallon bottles; drums.
Hazard: Same as for 2-methyl-1-pentene.
Uses: Organic synthesis; monomer for plastics used in automobiles, electronic components, and laboratory ware.
Shipping regulations: See 2-methyl-1-pentene.

4-methyl-2-pentene cis, trans, mixture (1-isopropyl-2-methylethylene) $CH_3CH{:}CHCH(CH_3)_2$.
Properties: Colorless liquid; sp. gr. 0.670 (20/4°C); b.p. (mixture) 131°F; (cis) 133.2°F); (trans) 137.1°F; refractive index 1.388 (n 20/D); flash point −20°F; soluble in alcohol, acetone, ether, petroleum, coal-tar solvents; insoluble in water.
Containers: Up to 5-gal bottles.
Hazard: Highly flammable; dangerous fire risk.
Use: Organic synthesis.
Shipping regulations: See 2-methyl-1-pentene.

4-methyl-3-penten-2-one. See mesityl oxide.

3-methyl-1-pentyn-3-ol (meparfynol; methylparafynol; methyl pentynol). $HC{\vdots}CCOH(CH_3)CH_2CH_3$.
Properties: Colorless liquid; b.p. 121.4°C; f.p. −30.6°C; sp. gr. 0.8721 (20/20°C); refractive index (n 20/D) 1.4318; flash point 101°F (TOC). Moderately soluble in water; miscible with acetone, benzene, carbon tetrachloride, ethyl acetate. Combustible.
Grades: High purity, 98.5%; technical, 95.0% min; pharmaceutical.
Hazard: Toxic in high concentrations. Moderate fire risk.
Uses: Stabilizer in chlorinated solvents; viscosity reducer; electroplating brightener; intermediate in syntheses of hypnotics and isoprenoid chemicals; solvent for polyamide resins; acid inhibitor; prevention of hydrogen embrittlement; medicine (soporific and anesthetic).

methylphenethylamine. See amphetamine.

methyl phenylacetate $C_6H_5CH_2COOCH_3$.
Properties: Colorless liquid; honey-like odor. Soluble in 5 parts of 60% alcohol, in fixed oils; insoluble in water. Sp. gr. 1.062–1.066 (25°C); refractive index 1.506–1.509 (20°C); b.p. 218°. Low toxicity. Combustible.
Containers: 25-lb cans.
Grades: Technical; F.C.C.
Uses: Perfumery; flavors for tobacco; flavoring.

methylphenylcarbinol. See alpha-methylbenzyl alcohol.

methylphenylcarbinyl acetate. See alpha-methylbenzyl acetate.

methylphenyldichlorosilane $CH_3(C_6H_5)SiCl_2$.
Properties: Colorless liquid; b.p. 82°C (13 mm), 205°C; refractive index 1.5199 (25°C); sp. gr. 1.19; soluble in benzene, ether, methanol. Flash point 83°F.
Derivation: From chlorobenzene Grignard reagent and methyltrichlorosilane, or from benzene and methyldichlorosilane.
Hazard: Toxic and irritant. Flammable, moderate fire risk; reacts strongly with oxidizing materials.
Use: Manufacture of silicones.

methyl phenyl ether. See anisole.

4-methyl-1-phenyl-2-pentanone (benzyl isobutyl ketone) $C_6H_5CH_2C(O)CH_2CH(CH_3)_2$. Combustible; low toxicity. Used in flavoring.

methyl phenyl phosphate. See methyl diphenyl phosphate.

2-methyl-2-phenylpropane. See tert-butyl-benzene.

3-methyl-1-phenyl-2-pyrazolin-5-one (3-methyl-1-phenyl-5-pyrazolone; 1-phenyl-3-methyl-5-pyrazolone) $C_6H_5NN{:}C(CH_3)CH_2CO$.
Properties: White powder or crystals; soluble in water; slightly soluble in alcohol or benzene; insoluble in ether; b.p. 287°C (205 mm); m.p. 127°C; vapor pressure $<$ 0.01 mm (20°C).
Derivation: By condensation of phenylhydrazine with ethylacetoacetate.
Uses: Intermediate for dyes and drugs; sensitive reagent for detection of cyanide.

methylphloroglucinol (2,4,6-trihydroxytoluene) $C_6H_2(OH)_3CH_3$.
Properties: Cream to light tan; fine crystals; odorless; m.p. 210–214°C; soluble in water, alcohol, and ether; insoluble in benzene. Combustible.
Uses: Reactive coupling agent; dye and plastic intermediate.

methylphosphonic acid $CH_3PO(OH)_2$. White solid; m.p. 103–104°C.
Containers: 55-gal drums; tank trucks.
Use: Organic synthesis.

methylphosphoric acid (methyl orthophosphoric acid; methyl acid phosphate) $CH_3H_2PO_4$.
Properties: Pale straw-colored liquid; sp. gr. 1.42 (25°C); can be neutralized with alkalies or amines to give water-soluble salts. Combustible.
Purity: 97% (remainder orthophosphoric acid and methyl alcohol).
Containers: Glass bottles; 1-, 5- and 10-gal tins; 55-gal steel drums.
Uses: Textile and paper processing compounds; catalysts in urea-resin formation; polymerizing agent for resin and oils; rust remover; soldering flux; chemical intermediate.

methyl phthalyl ethyl glycolate $C_2H_5OOCCH_2OOC\text{-}C_6H_4COOCH_3$.
Properties: Colorless liquid with slight characteristic odor; sp. gr. 1.217–1.227. Flash point 375°F. Combustible. Low toxicity. Miscible with most organic solvents; very slightly soluble in water.
Uses: Plasticizer; solvent.

methyl picrate. See trinitroanisole.

N-methylpiperazine $CH_3NCH_2CH_2NHCH_2CH_2$.
Properties: Colorless liquid; sp. gr. 0.9038; b.p. 138.0°C; f.p. −6.4°C; hygroscopic; amine-like odor. Flash point 108°F. Combustible.
Hazard: Moderate fire risk.

Uses: Intermediate for pharmaceuticals, surface agents, synthetic fibers.

2-methylpiperidine (2-pipecoline) $C_5H_{10}NCH_3$.
Properties: Colorless liquid; b.p. 118.2°C; f.p. −4.2°C; sp. gr. 0.8401 (20/20°C); refractive index 1.4457 (n 20/D); soluble in water in all proportions at 20°C. Combustible.

1-methyl-4-piperidinol $CH_3C_5H_9NOH$.
Properties: Colorless to light amber, oily liquid with amine-like odor. Crystals may be present, or material may be solid, but will melt on warming; refractive index (n 25/D) 1.4757–1.4777; congealing temperature 27.5°C min. Combustible.
Use: Organic synthesis.

3(2-methylpiperidino)propyl 3,4-dichlorobenzoate. See piperalin.

methylprednisolone $C_{22}H_{30}O_5$. A steroid.
Properties: White, odorless, crystalline powder; m.p. 240°C with decomposition; sparingly soluble in alcohol; slightly soluble in acetone; insoluble in water.
Grade: N.F.
Use: Medicine (hormone).

2-methylpropane. See isobutane.

2-methylpropanenitrile. See isobutyronitrile.

2-methyl-1-propanethiol (isobutyl mercaptan). $(CH_3)_2CHCH_2SH$.
Properties: Liquid, boiling 85–95°C; unpleasant odor; sp. gr. 0.8363 (60/60°F); flash point 15°F.
Grade: 95%.
Containers: Tank cars.
Hazard: Flammable, dangerous fire risk; toxic.
Shipping regulations: Flammable liquid, n.o.s., (Rail) Red label. (Air) Flammable Liquid label.

2-methyl-2-propanethiol (tert-butyl mercaptan). $(CH_3)_3CSH$.
Properties: Colorless liquid with strong skunk odor; sp. gr. 0.79–0.82 (60/60°F); distillation range 62–67°C; refractive index (20/D) 1.422; flash point −15°F; wt/gal 6.71 lb.
Containers: 1-qt cans; 1-, 5-, 54-gal drums; tank cars.
Hazard: Flammable, dangerous fire risk. Highly toxic. Tolerance, 0.5 ppm in air.
Use: Intermediate; gas odorant for detecting leaks.
Shipping regulations: (Rail) Red label. (Air) Flammable Liquid label. Legal label name butyl mercaptan.

2-methyl-1-propanol. See isobutyl alcohol.

2-methyl-2-propanol. See tert-butyl alcohol.

2-methylpropanoyl chloride. See isobutyrol chloride.

2-methylpropene. See isobutene.

2-methyl-2-propen-1-ol. See methylallyl alcohol.

methyl propionate $CH_3CH_2COOCH_3$.
Properties: Colorless liquid. Soluble in most organic solvents; somewhat soluble in water. Sp. gr. 0.937 (4°C); refractive index 1.3769 (n 20/D); boiling range 78.0–79.5°C; autoignition temp. 876°F; flash point 28°F (C.C.); wt/gal 7.58 lb. Low toxicity.
Grade: Technical.
Containers: 55-gal drums; tank cars.
Hazard: Flammable, dangerous fire risk. Explosive limits in air 2.5 to 13%.
Uses: Solvent for cellulose nitrate; solvent mixtures for cellulose derivatives; lacquers, paints, varnishes; coating compositions; flavoring.
Shipping regulatoins: (Rail) Flammable liquid, n.o.s., Red label. (Air) Flammable Liquid label.

2-methylpropionic acid. See isobutyric acid.

methylpropylbenzene. See cymene.

methyl propyl carbinol. See 2-pentanol.

methylpropylcarbinolurethane. See hedonal.

methyl propyl ketone (ethyl acetone, 2-pentanone; MPK) $CH_3COC_3H_7$.
Properties: Water-white liquid. The commercial material consists of a mixture of methyl propyl and diethyl ketones in the approximate ratio of 3 to 1 and contains at least 97% of these ketones, the balance being sec-amyl alcohol. Soluble in alcohol and ether; slightly soluble in water. Sp. gr. 0.809 (20/20°C); f.p. −77.5°C; b.p. 101.7°C; refractive index 1.3895 (20°C); viscosity 0.473 centipoise (25°C); flash point 45°F. Autoignition temp. 941°F.
Grade: Technical.
Containers: Glass bottles; tins; steel drums.
Hazard: Flammable, dangerous fire risk. Moderately toxic. Tolerance, 200 ppm in air. Explosive limits in air 1.6 to 8.2%.
Uses: Solvent; substitute for diethyl ketone; flavoring.
Shipping regulations: (Rail) Flammable liquid, n.o.s., Red label. (Air) Flammable Liquid label.

2-methyl-2-n-propyl-1,3-propanediol dicarbamate. See meprobamate.

methylpyridine. See picoline.

methyl pyromucate. See methyl 2-furoate.

N-methylpyrrole $C_4H_4NCH_3$.
Properties: Colorless liquid; b.p. 112°C; f.p. −57°C; density 0.914 (20°C); refractive index 1.4898 (n 17/D); flash point 61°F; insoluble in water; soluble in alcohol.
Grade: 98% min. purity.
Containers: Steel drums; cans.
Hazard: Flammable, dangerous fire risk.
Use: Organic synthesis.
Shipping regulations: Flammable liquid, n.o.s., (Rail) Red label. (Air) Flammable Liquid label.

N-methylpyrrolidine $CH_3NCH_2CH_2CH_2CH_2$.
Properties: Colorless liquid, ammonia-like odor; refractive index 1.4200 to 1.4230 (25°C); b.p. 80.5°C; f.p. −90°C; sp. gr. 0.805; flash point 7°F.
Hazard: Flammable, dangerous fire risk. Irritant to skin and eyes.
Shipping regulations: Flammable liquid, n.o.s., (Rail) Red label. (Air) Flammable Liquid label.

N-methyl-2-pyrrolidone $CH_3NCH_2CH_2CH_2CO$.
Properties: Colorless liquid with mild amine odor; sp. gr. 1.027; f.p. −24°C; b.p. 202°C; flash point 204°F; miscible in all proportions with water; various organic solvents, castor oil. Combustible.
Derivation: High-pressure synthesis from acetylene and formaldehyde.
Uses: Solvent for resins, acetylene, etc; pigment dispersant; petroleum processing; spinning agent for polyvinyl chloride; intermediate.
Note: A proprietary adaptation of this solvent to clean-up of vinyl chloride reaction vessels is available under trademark of "M-Pyrol."

alpha-**methylquinoline.** See quinaldine.

gamma-**methylquinoline.** See lepidine.

2-methylquinone. See toluquinone.

6-methyl-2,3-quinoxalinedithiol cyclic carbonate (6-methyl-2-oxo-1,3-dithio(4,5-b)quinoxaline) $C_{10}H_6N_2S_2O$.
Properties: Yellow crystalline powder; m.p. 172°C; insoluble in water; slightly soluble in acetone and alcohol; soluble in organic solvents.
Hazard: Toxic by ingestion and inhalation.
Uses: Acaricide; insecticide; fungicide.

methyl red $(CH_3)_2NC_6H_4NNC_6H_4COOH$. para-Dimethylaminoazobenzenecarboxylic acid. C. I. 13020.
Properties: Dark-red powder or violet crystals; m.p. 180°C; insoluble in water; soluble in alcohol, ether, glacial acetic acid.
Use: Acid-base indicator in the range pH 4.2–6.2 (red to yellow).
See indicator.

methylresorcinol. See orcin.

methyl ricinoleate
$CH_3(CH_2)_5CH(OH)CH_2CH{:}CH(CH_2)_7COOCH_3$.
The methyl ester of ricinoleic acid.
Properties: Colorless liquid, insoluble in water; soluble in alcohol and ether. Sp. gr. 0.9236 (22/4°C); f.p. −4.5°C; b.p. 245°C (10 mm); refractive index 1.4628. Combustible; low toxicity.
Derivation: Esterification of ricinoleic acid or alcoholysis of castor oil; purification by vacuum distillation.
Grades: Technical; purified (99+%).
Containers: 5-gal cans; 55-gal drums.
Uses: Platicizer; lubricant; cutting oil additive; wetting agent.

methylrosaniline chloride. See methyl violet.

methyl salicylate (gaultheria oil; wintergreen oil; betula oil; sweet-birch oil) $C_6H_4OHCOOCH_3$.
Properties: Yellow to red liquid; odor of wintergreen; refractive index 1.535–1.538; sp. gr. 1.180–1.185; f.p. −8.3°C; flash point 214°F; autoignition temp. 850°F; b.p. 222.2°C; natural oil optically inactive; synthetic oil, angular rotation not more than −1.5°; soluble in 7 parts of 70% alcohol; soluble in ether and in glacial acetic acid; sparingly soluble in water. Combustible.
Derivation: By heating methanol and salicylic acid in presence of sulfuric acid, or by distillation from leaves of Gaultheria procumbens or bark of Betula lenta.
Method of purification: Rectification.
Grades: Technical; U.S.P.; F.C.C.
Containers: Bottles; drums.
Hazard: Toxic by ingestion. Use in foods restricted by FDA.
Uses: Flavor in foods, beverages, pharmaceuticals; odorant; perfumery; UV-absorber in sunburn lotions.

"Methyl Selenac."[69] Trademark for selenium dimethyldithiocarbamate $[(CH_3)_2NC(S)S]_4Se$.
Properties: Yellow powder, sp. gr. 1.58 ± .03; melting range 140–172°C; slightly soluble in carbon disulfide, benzene, chloroform; insoluble in water, dilute caustic, gasoline.
Uses: Vulcanizing agent.

methyl silicone. See dimethyl silicone. See also silicone and siloxane.

methyl stearate (methyl octadecanoate)
$CH_3(CH_2)_{16}COOCH_3$. The methyl ester of stearic acid.
Properties: Semisolid or liquid; m.p. 37.8°C; b.p. 234.5°C (30 mm), 204.5°C (10 mm); flash point 307°F; refractive index 1.4328 (50°C). Insoluble in water; soluble in ether and alcohol. Combustible.
Derivation: Esterification of stearic acid with methanol or alcoholysis of stearin with methanol.
Method of purification: Vacuum fraction distillation.
Impurities: Most technical methyl stearate is 55% stearate and 45% methyl palmitate.
Grades: Distilled; pressed; technical; pure (99.8+%).
Containers: Cans; drums.
Uses: Intermediate for stearic acid detergents, emulsifiers, wetting agents, stabilizers, resins, lubricants, plasticizers and textiles; biochemical and medical research.

alpha-**methylstyrene** $C_6H_5C(CH_3){:}CH_2$.
Properties: Colorless liquid, subject to polymerization by heat or catalysts; b.p. 165.38°C; f.p. −23.21°C; sp. gr. 0.9062 (25/25°C); viscosity 0.940 cp (20°C); flash point 129°F; autoignition temp. 1065°F; refractive index 1.5359 (25/25°C); insoluble in water. A polymerization inhibitor such as tert-butyl catechol is usually present in commercial quantities. Combustible.
Derivation: From benzene and propylene by use of aluminum chloride and hydrogen chloride to yield cumene which is then dehydrogenated.
Containers: Glass bottles; steel drums.
Hazard: Moderately toxic. Avoid inhalation and skin contact. Moderate fire risk. Explosive limits in air 1.9 to 6.1%. Tolerance, 100 ppm in air.
Use: Polymerization monomer, especially for polyesters.

methyl styryl ketone. See benzylidene acetone.

methylsuccinic acid. See pyrotartaric acid.

methyl sulfate. See dimethyl sulfate.

methyl sulfide. Legal label name (Rail) for dimethyl sulfide, q.v.

3-methylsulfolane. See 3-methyltetrahydrothiophene-1,1-dioxide.

3-methylsulfolene. See 3-methyl-2,5-dihydrothiophene-1,1-dioxide.

methyl sulfoxide. See dimethyl sulfoxide.

methylsulfuric acid (acid methyl sulfate; methyl hydrogen sulfate) CH_3OSO_2OH or CH_3HSO_4.
Properties: Oily liquid. Soluble in anhydrous ether; slightly soluble in alcohol, water. B.p. 188°C; sp. gr. 1.352; f.p. −27°C.
Derivation: Interaction of methyl alcohol and chlorosulfonic acid.
Uses: Sulfonating agent; specialty solvent.

N-methyltaurine. Available in commercial quantities as an aqueous solution of the sodium salt, sodium N-methyltaurate, or $CH_3NHCH_2CH_2SO_3Na$.
Properties (of solution): Clear, light-colored liquid, about 34–36% sodium salt, sp. gr. 1.21 (25/4°C). At freezing point (−28°C average) becomes a suspension of white crystals.
Uses: Intermediate for detergents, dyestuffs, pharmaceuticals, and other organics.

methyltestosterone $C_{20}H_3O_2$. A synthetic androgenic steroid.
Properties: White or creamy white crystals or crystalline powder; odorless; stable in air; slightly hygroscopic; affected by light; m.p. 163–168°C. Soluble in alcohol, methanol, ether, and other organic solvents; insoluble in water.
Derivation: By organic synthesis.

Grade: N.F.
Use: Medicine (hormone).

methyl tetracosanoate. See methyl lignocerate.

methyl tetradecanoate. See methyl myristate.

2-methyltetrahydrofuran $C_4H_7CH_3$.
Properties: Colorless liquid; ether-like odor; b.p. 80.2°C; f.p. −136°C; sp. gr. 0.854 (20/4°C); refractive index 1.4025 (25/D); flash point 12°F (TOC). Solubility in water 15.1 g/100 g water (25°C). Solubility in water increases with a decrease in temperature. Infinitely soluble in most organic solvents.
Containers: 1-, 5-, and 55-gal drums.
Hazard: Flammable, dangerous fire risk. May be toxic.
Uses: Chemical intermediate; reaction solvent.
Shipping regulations: Flammable liquid, n.o.s., (Rail) Red label. (Air) Flammable Liquid label.

3-methyltetrahydrothiophene-1,1-dioxide (3-methylsulfolane) $CH_3C_4H_7SO_2$.
Properties: Colorless liquid; sp. gr. 1.194 (20/4°C); m.p. 0°C; b.p. 276°C; flash point 325°F. Completely soluble in water, acetone and toluene. Combustible; low toxicity.
Containers: Up to 54-gal drums.
Uses: Solvent and extractive medium.

methyl 2-thienyl ketone. See 2-acetylthiophene.

meta-**(methylthio)aniline** $H_2NC_6H_4SCH_3$.
Properties: Pale yellow oil. Sp. gr. 1.140 (25°C); b.p. 163–165°C (16 mm); f.p. −3.0°C. Insoluble in water; soluble in alcohol, benzene, acetic acid. Combustible.
Hazard: Probably toxic. See aniline.
Use: Pharmaceutical intermediate.

4-(methylthio)-3,5-dimethylphenyl N-methylcarbamate [(4-methylthio)-3,5-xylyl methylcarbamate] $CH_3S(CH_3)_2C_6H_2OOCNHCH_3$. Solid; m.p. 121°C. Insoluble in water; soluble in alcohol and acetone.
Hazard: May be toxic.
Use: Pesticide.

methylthionine chloride. See methylene blue.

methyltin. See organotin compound.

methyl para-**toluate** $CH_3C_6H_4COOCH_3$.
Properties: White crystalline solid; f.p. 34°C; sp. gr. 1.058 (25/15.6°C).
Use: Organic synthesis.

methyl para-**toluenesulfonate** $CH_3C_6H_4SO_3CH_3$.
Properties: White damp crystals; solidification point 24°C; b.p. 157°C (8 mm); decomposes 262°C. A vesicant material. Insoluble in water; soluble in alcohol and benzene.
Grade: 97% min.
Hazard: Highly toxic by inhalation; strong irritant to skin and eyes.
Uses: Accelerator; methylating agent; catalyst for alkyd resins.

methyl para-**tolyl ketone.** See methyl acetophenone.

methyltrichlorosilane CH_3SiCl_3.
Properties: Colorless liquid. B.p. 66.4°C; sp. gr. 1.270 (25/25°C); refractive index (n 25/D) 1.4085; flash point 8°F. Readily hydrolyzed by moisture, with the liberation of hydrochloric acid.
Derivation: By Grignard reaction of silicon tetrachloride and methylmagnesium chloride.
Hazard: Highly toxic; strong irritant. Flammable, dangerous fire risk. May form explosive mixtures with air.
Use: Intermediate for silicones.
Shipping regulations: (Rail) Red label. (Air) Flammable Liquid label. Not acceptable on passenger planes.

methyl tricosanoate $CH_3(CH_2)_{21}COOCH_3$. The methyl ester of triosanoic acid.
Properties: White waxlike solid. Insoluble in water, soluble in alcohol and ether. M.p. 55–56°C. Combustible.
Grades: Purified 96% and 99.5%.
Uses: Intermediate in organic synthesis; medical research.

methyl tridecanoate $CH_3(CH_2)_{11}COOCH_3$. The methyl ester of tridecanoic acid.
Properties: Colorless liquid. Insoluble in water, soluble in alcohol and ether; m.p. 5.5°C; b.p. 130–132°C (4 mm); refractive index 1.4327 (25°C). Combustible; low toxicity.
Derivation: Esterification of tridecanoic acid with methanol followed by fractional distillation.
Grades: Purified 96% and 99.5%.
Uses: Intermediate in organic synthesis; medical research; reference standard in gas chromatography.

methyl trimethylolmethane. See trimethylol ethane.

"Methyl Trithion."[1] Trademark for an insecticide-acaricide containing various percentages of S-(para-chlorophenylthio)methyl O,O-dimethyl phosphorodithioate. A vailable as liquid or powder.
Hazard: Probably toxic.

"Methyl Tuads."[69] Trademark for tetramethylthiuram disulfide. $[(CH_3)_2NC(S)S]_2$.
Uses: Vulcanizing agent and primary accelerator in natural, nitrile and butyl rubbers and in SBR. As secondary accelerator in natural and nitrile rubbers and SBR. Used in coated fabrics, extruded and molded goods, steam hose. Increases heat resistance.

beta-**methylumbelliferone** (7-hydroxy-4-methylcoumarin; BMU) $C_{10}H_8O_3$.
Properties: White to light tan powder; m.p. 186–188°C; soluble in concentrated sulfuric acid; partly soluble in ethanol, isopropanol, 5% aqueous sodium carbonate solution; very slightly soluble in water; very dilute aqueous alkaline solutions give a bright blue-white fluorescence in daylight or ultraviolet light.
Grade: Technical.
Containers: Fiber drums.
Uses: Optical bleach in soaps, starches, and laundry products; suntan lotions.

2-methylundecanal. See methylnonylacetaldehyde.

methyl undecanoate $CH_3(CH_2)_9COOCH_3$. The methyl ester of undecanoic acid.
Properties: Colorless liquid. Insoluble in water, soluble in alcohol and ether; b.p. 123°C (10 mm); refractive index 1.4270 (25/4°C). Combustible; low toxicity.
Derivation: Esterification of undecanoic acid with methanol followed by fractional distillation.
Grades: Purified 96%; 99.5%.
Uses: Organic intermediate for synthesis; flavoring; medical research.

5-methyluracil. See thymine.

methylvinyldichlorosilane $(CH_3)(C_2H_3)SiCl_2$.
Properties: Colorless liquid; b.p. 92°C; sp. gr. 1.08 (25°C); refractive index 1.4270 (25°C); flash point

40°F. Soluble in benzene, ether; reacts with methanol and water.
Derivation: From methyldichlorosilane and acetylene or vinyl chloride.
Hazard: Highly toxic and irritant. Flammable, dangerous fire risk.
Uses: Manufacture of silicones, as coupling agents.
Shipping regulations: Flammable liquid, n.o.s., (Rail) Red label. (Air) Flammable Liquid label.

methyl vinyl ether. See vinyl methyl ether.

methyl vinyl ketone. Legal label name for vinyl methyl ketone (q.v.).

2-methyl-5-vinylpyridine $CH_3C_5H_3NCH{:}CH_2$.
Properties: Clear to faintly opalescent liquid; sp. gr. 0.978–0.982 (20/20°C); b.p. 181°C; refractive index 1.5400–1.5454 (20°C); f.p. (anhydrous) −14.3°C; flash point 165°F (TOC). Combustible.
Containers: Drum, tank trucks; tank cars.
Uses: Monomer for resins; oil additive; ore flotation agent; dye acceptor.

methyl violet (crystal violet; gentian violet; methylrosaniline chloride). A confusing series of names for closely related and overlapping derivatives of pararosaniline, a triphenylmethane type of dye. Gentian violet is the U.S.P. name for the medicinal product.
Uses: Medicine (topical use only as antiallergen and specific bactericide); acid-base indicator; denaturant for alcohol; biological stain; dye for textiles; pigments.

methyl yellow. See dimethylaminoazobenzene.

"Methyl Zimate."[69] Trademark for zinc dimethyldithiocarbamate, $[(CH_3)_2NC(S)S]_2Zn$.
Uses: Ultra-accelerator for natural and synthetic rubbers.

methy prylon (3,3-diethyl-5-methyl-2,4-piperidinedione) $C_5NH_4(O)_2(C_2H_5)_2(CH_3)$.
Properties: Nearly white, crystalline powder; slight characteristic odor; bitter taste; melting range 74–77°C; soluble in water; very soluble in alcohol, in chloroform, in ether, and benzene.
Grade: N.F.
Use: Medicine.

"Met-L-KYL."[548] Trademark for a dry chemical which extinguishes fires caused by pyrophoric liquids such as triethylaluminum and adsorbs the spilled metal alkyl to prevent reignition.

"Met-L-X."[548] Trademark for a dry chemical based on sodium chloride, approved for use on sodium, potassium, sodium potassium alloy, and magnesium fires.

"Metol."[134] Trademark for methyl-para-aminophenol sulfate (q.v.).

"Metopirone."[305] Trademark for metyrapone (q.v.).

metopon hydrochloride (6-methyldihydromorphinone hydrochloride) $C_{18}H_{21}O_3N \cdot HCl$. A morphine derivative.
Properties: White, odorless, crystalline powder; very soluble in water; sparingly soluble in alcohol; insoluble in benzene.
Hazard: Highly toxic. See alkaloid; narcotic. Prescription only.
Use: Medicine (sedative).

"Metso."[201] Trademark for a series of detergents which include the pentahydrate of sodium metasilicate, hydrated sodium sesquisilicate; and technical anhydrous sodium orthosilicate. Available in lump and powder forms.

"Metubine Iodide."[100] Trademark for dimethyl-tubocurarine iodide.

"Metycaine."[100] Trademark for piperocaine hydrochloride, U.S.P.

metyrapone. (USAN) (2-methyl-1,2-di-3-pyridyl-1-propanone) $C_5H_4NC(O)C(CH_3)_2C_5H_4N$.
Properties: White to light amber, fine crystalline powder, having a characteristic odor. Darkens on exposure to light. Soluble in methanol and chloroform. Forms water-soluble salts with acids.
Grade: U.S.P.
Use: Medicine (also as ditartrate).

MeV. Abbreviation for million electron-volts.

mevinphos (2-carbomethoxy-1-methylvinyl dimethyl phosphate) $(CH_3O)_2P(O)OC(CH_3){:}CHCOOCH_3$. Generic name for methyl 3-hydroxy-alpha-crotonate dimethyl phosphate.
Properties: Liquid; b.p. 99–103°C (0.03 mm). Slightly soluble in oils; miscible with water and benzene.
Hazard: Highly toxic by ingestion, inhalation, and skin absorption. Use may be restricted.
Uses: Systemic insecticide and acaricide.
Shipping regulations: Organic phosphate, liquid, (Rail, Air) Poison label. Not acceptable on passenger planes.

MFC. Abbreviation of miracle fruit concentrate. See sweetener, nonnutritive.

Mg Symbol for magnesium.

mg Abbreviation of milligram (q.v.).

"MGN."[192] Trademark for methylene glutaronitrile.

MH. See maleic hydrazide.

"MH-30."[248] Trademark for a 30% solution of maleic hydrazide (q.v.).

"MHA."[58] Trademark for methionine hydroxy analog, calcium salt (q.v.).

MHD. See magnetohydrodynamics.

MIAK. See methyl isoamyl ketone.

miazine. See pyrimidine.

MIBC. Abbreviation for methylisobutyl carbinol. See methylamyl alcohol.

MIBK. Abbreviation for methyl isobutyl ketone (q.v.).

mica. Any of several silicates of varying chemical composition but with similar physical properties and crystal structure. All characteristically cleave into thin sheets, which are flexible and elastic. Synthetic mica is available. It has electrical and mechanial properties superior to those of natural mica; it is also water-free.
Properties: Soft, translucent solid. Sp. gr. 2.6–3.2; Mohs hardness 2.8–3.2 refractive index 1.56–1.60; dielectric constant 6.5–8.7; noncombustible; heat-resistant to 600°C; colorless to slight red (ruby); brown to greenish yellow (amber).
Derivation: From muscovite (ruby mica), phlogopite (amber mica), and pegmatite. Synthetic single crystals are "grown" electrothermally.
Occurrence: U.S., Canada, Madagascar, India, So. Africa, So. America, U.S.S.R.
Forms: Block, sheet, powder; single crystals.
Containers: Bags; carloads.

Grades: Dry-ground; wet-ground.
Hazard (dust): Irritant by inhalation; may be damaging to lungs. Tolerance, 20 million particles per cubic meter in air.
Uses: Electrical equipment; vacuum tubes; incandescent lamps; dusting agent; lubricant; windows in high-temperature equipment; filler in exterior paints; cosmetics; glass and ceramic flux; roofing; rubber; mold-release agent; specialty paper for insulation and filtration; wallpaper and wallboard joint cement; oil-well drilling muds.

"Micabond."[281] Trademark for an electrical insulation material consisting primarily of mica with electrical insulating binders.
Forms: Tape; tubing; segments; plate; fabricated parts.
Uses: Motors; insulation against heat.

"Mica-Flex.'[116] Trademark for a flexible electrical insulation containing a large percentage of mica; used mainly on high voltage motor and generator coils.

"Micarta."[308] Trademark for a group of laminated plastics composed of paper or fabric made from cellulose, glass, asbestos, or synthetic fibers bonded with phenolic or melamine resins and cured at elevated temperature and pressure.
Uses: Plating barrels; rayon-manufacturing equipment; pickling tanks; electrical and thermal insulation; oil-handling equipment; steel rolling-mill bearings; chemical-handling valve bodies; paper-mill suction box covers and equipment.

"Micatex."[236] Trademark for mica prepared for addition to drilling fluids to reduce water loss to the formation and for overcoming mild losses of circulation. An effective seal is formed over mildly permeable formations when the mud in which it is entrained forces the material against the formation. Will not disintegrate appreciably, nor will it corrode or abrade slush-pump liners or other metal or moving parts of the mud system.

micelle. An electrically charged colloidal particle, usually organic in nature, composed of aggregates of large molecules, e.g., in soaps and surfactants. The term is also applied to the casein complex in milk. See also colloid chemistry.

Michler's hydrol. See tetramethyldiaminobenzhydrol.

Michler's ketone. See tetramethyldiaminobenzophenone.

micro-. Prefix meaning 10^{-6} unit (symbol μ). E.g., 1 microgram = 0.000001 gram. See also micron.

microanalysis. See microchemistry.

"Microballoon."[572] Trademark for hollow, finely divided, hole-free, low-density particles of synthetic resins or similar film-forming materials. Glass is one of the materials used.
Uses: To form a protecting layer of the tiny spheres over liquid surfaces, such as oils in big tanks, to reduce evaporation; to separate helium from natural gas because of the wide difference in relative rates of diffusion through the spheres; as an extender in plastics to achieve low density.

"Micro-Cel."[247] Trademark for a group of finely divided hydrated synthetic calcium silicates.
Properties: White to light gray color range dependent on grade. Density (apparent) 5-10 lb/cu ft; pH 7–10; absorption (water) 300–600%. Noncombustible.
Uses: Inert extenders; absorbents; bulking agents; pesticide-carrier.

microchemistry. A branch of analytical chemistry that involves procedures that require handling of very small quantities of materials. Specifically, it refers to carrying out various chemical operations (weighting, purification, quantitative and qualitative analysis) on samples ranging from 0.1 to 10 milligrams; this often involves use of a microscope, and still more often of chromatography (q.v.). See also microscopy, chemical.

"Micro-Cote."[98] Trademark for a micromesh grade of attapulgite; used as an anticaking agent, especially for ammonium nitrate, ammonium sulfate and other fertilizers.

microcrystalline structure. A form in which a number of high-polymeric substances have been prepared and on which active research is continuing. They include cellulose (see "Avicel"), chrysotile asbestos, amylose (starch), collagen, and nylon. On the microscopic level, these substances are composed of colloidal microcrystals connected by molecular chains. The process involves breaking up the network of microcrystals (by acid hydrolysis in the case of cellulose) and separating them by mechanical agitation. The size range of the microcrystals is from 2.5 to 500 nanometers (millimicrons). The products form extremely stable gels which have a number of commercial use possibilities. Petroleum-derived waxes of high molecular weight have been available in microcrystalline form for many years. Chlorophyll has a naturally microcystalline structure. See also cellulose; wax, microcrystalline.

microcurie. See curie.

microencapsulation. Enclosure of a material in capsules ranging from 20 to 150 microns in diameter and composed of polymeric substances such as polyamides, or in some cases of gelatin, which act as semipermeable membranes. Enzymes are immobilized in this way until they reach the desired site in an enzymatic process. A growing list of applications of this slow-release technique includes fertilizers, nutrients for fish culture, pesticides, pharmaceuticals, insect sex attractants, deodorants, etc.

"Microfined."[443] Trademark for finely ground and dispersed anthraquinone vat pastes.

"Microgel."[93] Trademark for a neutral agricultural fungicide. The blue powder assays 52% copper.

microgram (μg). One millionth (10^{-6}) gram.

"Microlith."[443] Trademark for organic pigment stir-in dispersions compatible with a broad range of organic solvents and polymers.

"Micro-Mag."[481] Trademark for refined lime and lime admixtures.

"Micromet."[108] Trademark for a specially formulated phosphate glass, slowly soluble in water, and used to inhibit scale, corrosion, and red water in water systems and air conditioning systems.

micrometer (μm). One millionth (10^{-6}) meter, or 1 micron (10,000 Angstrom units).

micron (μ). See micrometer.

"Micronex."[133] Trademark for series of pelleted impingement carbon blacks produced from natural gas.

Grades available are Standard Micronex (medium processing channel); Micronex W-6 (easy processing channel). Available in 25- and 50-lb bags and hopper cars. See also carbon blacks.

"Micronized" Stearates.[74] Trademark for stearates that have been specially processed to fine particle size providing a greatly increased surface area. Available as calcium-precipitated, calcium-densified and zinc U.S.P.

micronutrient. See trace element.

microorganism. A living organism of microscopic size generally considered to include bacteria, molds (actimonyces), and fungi, but excluding viruses. See also bacteria.

"Micro-Pen."[530] Trademark for special coated procaine penicillin G.

microscopy, chemical. Use of a microscope primarily for study of physical structure and identification of materials. This is especially useful in forensic chemistry (q.v.) and police laboratories. Many types of microscopes are used in industry; most important are the optical, ultra-, polarizing, stereoscopic, electron, and x-ray microscopes. Organic dyes of various types are used to stain samples for precise identification.

"Microsol."[443] Trademark for aqueous pigment dispersions for spin-coloring of regenerated cellulose fibers.

"Micro-Sorb."[98] Trademark for a microfine attapulgite clay. Available in two grades: RVM and LVM.
Uses: Pharmaceutical preparations, pressure-sensitive copy paper, cosmetics, and polishing compounds.

"Microthene."[192] Trademark for a group of coating materials.
Grades: Regular, powdered polyethylene resins and to coat large, irregular shaped objects; F, microfine polyolefins for coating specialty papers.

microwave spectroscopy. A type of absorption spectroscopy used in instrumental chemical analysis which involves use of that portion of the electromagnetic spectrum having wavelengths in the range between the far infrared and the radiofrequencies, i.e., between 1 millimeter and 30 centimeters. Substances to be analyzed are usually in the gaseous state. Klystron tubes are used as microwave source.

middle oil. A fraction distilled from coal tar. See coal tar.

middlings. The granular part of the interior of the wheat berry obtained in the process of milling. This product, when reduced by grinding to the desired fineness, produces the finest quality of flour.

Midgley, Thomas, Jr. (1889–1944). American chemist and inventor. One of the most creative and brilliant chemists of his era, Midgley's early work was in the field of rubber chemistry and technology, especially in the development of synthetic and substitute rubbers, which were being introduced in the 1930's. He worked with Kettering at General Electric and then became vice-president of Ethyl Corporation, as well as of the Ohio State University Research Foundation. His innovative genius was responsible for the develoment of organic lead compounds for antiknock gasoline and later for the discovery of fluorocarbon refrigerants, on which he did the basic research. He was recipient of many of chemistry's highest honors, including the Nichols Medal, the Perkin Medal and the Priestley Medal.

migration area. A term used in nuclear technology as a measure of the moderation or slowing down of neutrons. It is one sixth of the mean square distance a neutron travels before thermal capture.

mike. Term adopted by the American Standards Association for a microinch, or 10^{-6} inch.

"Mikrovan."[51] Trademark for microcrystalline waxes in a broad selection of melting points and hardnesses.

mil One thousandth (0.001) inch. Used chiefly in reference to surface coatings, metal sheet, films, cable covers, friction tape, etc.

mildew preventive. A compound used to prevent the growth of parasitic fungi, usually stain-producing, on such organic materials as textiles, leather, paper, farinaceous products, etc. Compounds used widely employed include cresols, phenols, benzoic acid, formaldehyde and organic derivatives or salts of copper, zinc and mercury.

"Milezyme" P.[212] Trademark for an enzyme system standardized for protease activity for use in feeding all classes of livestock.

milk. A heterogeneous system composed (for cows' milk) of about 87% water, 3.8% emulsified particles of fat and fatty acids, 3% casein, 5% sugar (lactose), serum proteins, calcium, phosphorus, potassium, iron, magnesium, copper, and several vitamins. The fat particles are from 6 to 10 micrometers in diameter, much larger than colloidal size; they are coated with an adsorbed layer of protein (protective colloid), which maintains the emulsion. The casein is closely associated with the calcium, forming micelles of calcium caseinate which are of colloidal dimensions. The white color of milk is largely due to light scattering by these particles, and to some extent by the fat particles, rather than to the presence of a pigment. The casein complex coagulates (a) when high temperature or bacteria convert the lactose to lactic acid, as in souring, or (b) when acid or certain enzymes (rennet) are intentionally added. The serum proteins lactalbumin and lactoglobulin are also colloidally dispersed. The lactose and mineral salts are in true molecular solution. Thus milk is a complex system exhibiting several levels of dispersion, from the molecular through colloidal and into the microscopic size range. Important processes applied to milk on an industrial scale include pasteurization, homogenization, coagulation, dehydration, and condensation (see specific entries). The milks of animals other than cows show considerable variations in composition, especially fat and protein content.
See also colloid chemistry; emulsion; casein.

milk of lime (lime water). Calcium hydroxide suspended in water.

milk of magnesia (magnesia magma). A white, opaque, more or less viscous suspension of magnesium hydroxide in water from which varying proportions of water usually separate on standing.
Grade: U.S.P.
Use: Medicine (laxative).

milk sugar. See lactose.

milli-. Prefix meaning 10^{-3} unit, or 1/1000th part.

millicurie. See curie.

milligram (mg). One-thousandth gram (10^{-3} gram).

milliequivalent (meq). One-thousandth of the equivalent weight of a substance.

milliliter (ml). One thousandth liter; the volume occupied by one gram of pure water at 4°C and 760 mm pressure. One milliliter (ml) equals 1.000027 cubic centimeters (cc).

millimeter (mm). One-thousandth meter; about 0.03937 inch.

millimicron ($m\mu$). One-thousandth micron; 10 angstrom units; 1 billionth meter; 1 nanometer.

"Milltrox."[288] Trademark for wet-milled zircon (q.v.).

"Milorganite." Trademark for an activated sludge marketed in dry granular form by the Milwaukee sewage disposal plant. Contains 5–10% moisture, 6.5–7.5% ammonia, 2.5–3.5% available phosphoric acid, 3–4% total phosphoric acid.
Use: Fertilizer.

Milori blue. Any of a number of the varieties of iron blue pigments. See iron blues.

"Miltown." Trademark for meprobamate (q.v.).

"Milvex."[259] Trademark for a group of specialty nylon resins said to differ from conventional nylons in the following ways: (1) Highly water-resistant (dimensionally stable); (2) clarity, as opposed to the opacity of existing types; (3) adhesion to metals, glass, and other smooth surfaces; (4) tensile strength (up to 7500 psi) and elongation (up to 600%); (5) high flexibility and internal plasticization.

mineral. A loose and inexact term that may be defined chemically as any element, inorganic compound or mixture occurring or originating in the earth's crust and atmosphere, including all metals and nonmetals (except carbon), their compounds and ores. Inorganic gases (H_2, O_2, N_2, CO_2, Cl_2) are regarded as minerals. The term has long been used by biologists to indicate the nonliving "kingdom" of nature. Carbon-rich substances and mixtures, such as coal, petroleum, asphalt and non-vegetable waxes, often loosely called minerals, are organic materials that are not original components of the earth's crust, and thus are not true minerals. The term was illogically used by early chemists to describe a variety of substances; many of these uses are obsolescent, but a few persist, e.g.:
mineral acid: any of the major inorganic acids (sulfuric, nitric, hydrochloric)
mineral black: inorganic black pigments
mineral blue: varieties of blue pigments
mineral dust: industrial dust; nuisance dust
mineral green: copper carbonate
mineral oil: a liquid petroleum derivative
mineral pitch: asphalt
mineral red: iron oxide red
mineral rubber: blown asphalt
mineral spirits: a grade of naphtha
mineral water: natural spring water containing sulfur, iron, etc.
mineral wax: a wax found in the earth (ozocerite), or derived from petroleum
mineral wool: fibers made by blowing air or steam through slag.
(2) As used by nutritionists the term refers to such components of foods as iron, copper, phosphorus, calcium, iodine, selenium, fluorine and trace micronutrients.

minium Pb_3O_4. Natural red oxide of lead. Found in Colorado, Idaho, Utah, Wisconsin. See also lead oxide, red.

"Minol."[79] Trademark for series of steam-distilled pine oil used in disinfectants; cleaning compounds; insecticides; paint and varnish; textile processing.

"Min-U-Gel."[98] Trademark for colloidal attapulgite clay. Available in various grades. Forms stable gel solutions in presence of electrolytes, such as saturated salt solutions.
Uses: Adhesives, cosmetics, paints, pharmaceuticals, suspension fertilizers and pesticide formulations.

"Min-U-Sil."[436] Trademark for high-purity, crystalline silica in uniform micron sizes, 5, 10, 15 and 30.
Uses: Filler in epoxy resin formulations, silicone rubber, solid polyurethane and other elastomers; industrial and latex paints; ceramic applications; high-temperature insulation products.

"Miokon Sodium."[329] Trademark for sodium diprotrizoate (q.v.), a water-soluble x-ray contrast medium.

MIPA. Abbreviation for monoisopropanolamine. See isopropanolamine.

mipafox. See N,N'-diisopropyldiamidophosphoryl fluoride.

miracle fruit concentrate (MFC). See sweetener, non-nutritive.

"Mira-Cleer."[492] Trademark for a group of starches used as a nongelling food thickener.

"Miradon."[321] Trademark for a brand of anisindione (q.v.).

"Mira-Film."[492] Trademark for a series of acetate esters of corn starch used as warp size for synthetic yarns.

"Mirasol."[223] Trademark for a series of alkyd type resins. Epoxy resin esters are also marketed under this name. Available in all modifications including drying oils, semi-drying oils, non-drying oils, natural and phenolic resins.
Uses: Air-drying and baking finishes including architectural, lacquer, wrinkle, hammer and other industrial enamels; also printing inks and textile finishes.

"Miravar."[223] Trademark for a group of oleoresinous varnishes.
Uses: Industrial finishes and wrinkle enamels.

mirbane oil. See nitrobenzene.

"Mirrex."[482] Trademark for a calendered, unplasticized PVC film. Available in film or sheeting for a wide range of packaging applications.

mirror-image molecules. See optical isomerism; enantiomorph; chiral.

misch metal. The primary commercial form of mixed rare-earth metals (95%), prepared by the electrolysis of fused rare earth chloride mixtures. Sp. gr. about 6.67; m.p. about 1200°F. Form: Waffle-like plates weighing 40 to 60 lb packed in oiled paper, immersed in oil, or coated with vinyl paint.
Hazard: Flammable, dangerous fire risk.
Uses: Lighter flints; ferrous and non-ferrous alloys; cast iron; aluminum, nickel, magnesium and copper alloys; getter in vacuum tubes.

Shipping regulations: (Rail) Flammable solids, n.o.s., Yellow label. (Air) Flammable Solid label.

miscibility. The ability or tendency of one liquid to mix or blend uniformly with another. Alcohol is miscible with water; gasoline and water are immiscible. See also solubility, often used in the same sense in reference to liquids.

"Mist."[84] Trademark for a wettable rubbermakers' grade of sulfur.

"Mistron."[38] Trademark for ultra-fine particle size magnesium silicates available in a variety of grades. See also talc.

miticide. A substance which kills mites, small animals of the spider class, among them the European red mite and the common red spider which infest fruit trees.

mitochondria. Particles of cytoplasm (q.v.) found in most respiring cells. They synthesize most of the cell's adenosine triphosphate and are the chief energy sources of living cells. They are highly plastic, mobile structures which may fragment or fuse together at random. Many enzymes, especially those involved in converting food-derived energy into a form usable by the cell, are located in the mitochondria, and DNA molecules have also been found there. Yeast is a particularly rich source of mitochondria for research purposes.

mitosis. The division of a cell nucleus to produce two new cells, each having the same chemical and genetic constitution as the parent cell. The deoxyribose (nucleic acid) component of the chromosomes is present in duplicate in the original nucleus. The amount of nucleic acid is doubled just before cell division begins; subsequent events (called phases) permit separation of the products of replication to form the new nuclei. Each half-chromosome carries the identical nucleic acids of the original chromosome. See also cell (1).

"Mitox."[253] Trademark for para-chlorobenzyl para-chlorophenyl sulfide. Used as a specific miticide or acaracide.

mixed acid (nitrating acid). A mixture of sulfuric and nitric acids used for nitrating, e.g., in the manufacture of explosives, plastics, etc. Consists of 36% nitric acid and 61% sulfuric acid.

Hazard: Causes severe burns; highly toxic and irritant by ingestion and inhalation. May cause nitrous gas poisoning; spillage may cause fire or liberate dangerous gas. Safety data sheet available from Manufacturing Chemists Assn., Washington, D.C.

Shipping reuglations: (Rail) White label. (Air) Corrosive label. Not acceptable on passenger planes. Legal label name, Nitrating acid.

mixing. Effecting a uniform dispersion of liquid, semi-solid, or solid ingredients of a mixture by means of mechanical agitation. Thin liquids and slurries are blended by means of impellers, i.e., rotating rods at the end of which blades of various shapes and sizes are attached. Also called dispersers, these are widely used in the paint industry, though roller mills and ball mills are also employed. Viscous liquids require the use of revolving paddles of various shapes, as in the food and candy industries; thick mixtures involving volatile solvents are blended in closed containers provided with a rotating shaft equipped with fin-like projections (sometimes called churns), used in preparing organic cements. Mixing of thick plastic masses containing no solvent is done in kneaders, either open or enclosed. The open type is comprised of a metal trough in which an agitating device of various designs revolves slowly. The enclosed type is typified by the Banbury mixer, in which oppositely turning involuted rotors provide the blending action. These can handle batches up to 1000 pounds of material and are used in the rubber and plastics industries. Particulate solids can be effectively blended in tumbling barrels or similar devices.

mixture. A heterogeneous blend of elements or compounds, which may or may not be uniformly dispersed. The components or phases (q.v.) of a mixture are not chemically combined and can be separated by mechanical means, i.e., by gravity, by distillation, by freezing, by melting, etc. All solutions are uniformly dispersed mixtures. Examples are air, milk, gasoline, petroleum, glass, cement, alloys. See also compound; solution; homogeneous; heterogeneous.

MKP. Abbreviation for monopotassium phosphate. See potassium phosphate, monobasic.

ml Abbreviation for milliliter (q.v.).

MLA. Abbreviation for mixed lead alkyls (q.v.).

mm Abbreviation for millimeter (q.v.).

MMH. Abbreviation for monomethylhydrazine. See methylhydrazine.

Mn Symbol for manganese.

Mo Symbol for molybdenum.

"Mobilcer."[331] Trademark for a group of wax emulsions.

Uses: Sizes for paper, paperboard, particle board, textiles and cordage; binders in ceramics; carriers for pesticides; coating of fruits and vegetables; plasticizing of laundry starch; treatment of nursery stock; end-checking treatments for timber and lumber; in the manufacture of foam rubber and polymeric latex products; curing of concrete.

mobility. The ease with which a liquid moves or flows. Hydrocarbon liquids (nonpolar), which have low viscosity, surface tension and specific gravity, respond more readily to an applied force than does water (a polar liquid). For this reason fires involving hydrocarbon liquids should be extinguished with foam rather than with a direct stream of water.

"Mobil-Kote."[331] Trademark for a group of wax based blends with resins, polymers and copolymers.

Uses: Improved gloss and barrier coatings for paper packaging, folding cartons, and corrugated cartons.

"Mobilpar."[331] Trademark for a group of emulsified, or emulsifiable petroleum products.

Uses: Lubrication of textile fibers; plasticizers for starch formulas in the textile industry; softeners and wetting-out agents in the paper and textile industries; foam control agents in aqueous processes.

"Moca."[28] Trademark for 4,4′-methylenebis(2-chloroaniline) (q.v.).

modacrylic fiber. Generic name for a manufactured fiber in which the fiber-forming substance is any long chain synthetic polymer composed of less than 85% but at least 35% by weight of acrylonitrile units, —$CH_2CH(CN)$— (Federal Trade Commission). Other chemicals such as vinyl chloride are incorporated as modifiers. Characterized by moderate tenacity, low water absorption, and resistance to combustion. Self-extinguishing.

Uses: Deep pile and fleece fabrics; industrial filters; carpets; underwear; blends with other fibers.
See also fiber; acrylonitrile.

"Mod-Epox."[58] Trademark for modifier for epoxy resins. Reduces viscosity of liquid epoxy resins, accelerates cure; improves strength, electrical and adhesive characteristics. Used in tool and die manufacture; electric potting compounds; encapsulations; adhesives; surface coatings and body solders.

moderator. A substance of low atomic weight such as beryllium, carbon (graphite), deuterium (in heavy water) or ordinary water, which is capable of reducing the speed of neutrons but which has little tendency toward neutron absorption. The neutrons lose speed when they collide with the atomic nuclei of the moderator. Moderators are used in nuclear reactors, since slow neutrons are most likely to produce fission. A typical graphite-moderated reactor may contain 50 tons of uranium for 472 tons of graphite. Reactors in the U.S. are cooled and moderated with light water.

"Moderil."[299] Trademark for rescinnamine (q.v.).

"Modicol" VD.[309] Trademark for a modified sodium polyacrylate solution; used as a thickener.
See acrylates.

modification. A chemical reaction in which some or all of the substituent radicals of a high polymer are replaced by other chemical entities, resulting in a marked change in one or more properties of the polymer without destroying its structural identity. Cellulose, for example, can be modified by substitution of its hydroxyl groups by carboxyl or alkyl radicals along the carbon chain. These reactions are usually not stoichiometric. Their products have many properties foreign to the original cellulose, e.g., water solubility, high viscosity, gel and film-forming ability. Other polymeric substances that can undergo modification are rubber, starches, polyacrylonitrile, and some other synthetic resins. See also cellulose, modified.

"Modulan."[493] Trademark for a specially prepared acetylated lanolin. An anhydrous, practically odorless, pale yellow unctuous solid; soluble in mineral oil.
Uses: Hypoallergenic cosmetics.

modulus of elasticity (elastic modulus). A coefficient of elasticity representing the ratio of strain to stress as a material is deformed under dynamic load. It is a measure of the softness or stiffness of the material (Young's modulus). Typical values of Young's modulus for a stress of 100,000 lb/sq in. are carbon steels 300, copper 170, soda-lime glass 100, polystyrene 5, graphite 1.

"Modx."[94] Trademark for a blend of an inorganic acetate with N,N′-diphenylethylenediamine. A rubber accelerator.

moellon degras. See degras.

"Mogul."[275] Trademark for channel carbon blacks; used for inks, paints, paper, and plastics.

mohair. A natural fiber similar to wool obtained from angora goats. Used in fabrics for outer clothing, draperies, upholstery. Combustible. Tenacity 14 grams per denier.

Mohr's salt. See ferrous-ammonium sulfate.

Mohs scale. A scale of hardness of minerals, running from 1 to 10, with talc the softest and diamond the hardest. Each mineral in the scale with scratch all those below it. The steps are not of equal value; the difference in hardness between 9 and 10 is much greater than between 1 and 2.

Diamond	10	Apatite	5
Corundum	9	Fluorite	4
Topaz	8	Calcite	3
Quartz	7	Gypsum	2
Orthoclase (Feldspar)	6	Talc	1

A recent addition is 11 (fused zirconia).

moiety. An indefinite portion of a sample.

molal. A concentration in which the amount of solute is stated in moles and the amount of solvent in kilograms. The unit of molality is moles of solute per kilogram of solvent and is designated by a small m; 1 mole of NaCl in 1 kg of water is a 1 molal concentration.
Note: Do not confuse with molar (q.v.).

molar. A concentration in which one molecular weight in grams (one mole) of a substance is dissolved in one liter of solution. Molarity is indicated by an italic capital M. Molar quantities are proportional to the molecular weights of the substances.

molasses. The thick liquid left after sucrose has been removed from the mother liquor in sugar manufacture. Blackstrap molasses is the syrup from which no more sugar can be obtained economically. It contains about sucrose 20%, reducing sugars 20%, ash 10%, organic nonsugars 20%, water 20%. Combustible. Nontoxic.
Uses: Feed; food; raw material for various alcohols, acetone, citric acid, and yeast propagation. Sodium glutamate is made from Steffens molasses, a waste liquor from beet sugar manufacture.
See also fermentation.

molding. Forming a plastic or rubber article in a desired shape by application of heat and pressure, either in a negative cavity, usually of metal, or in contact with a contoured metal or phenolic resin surface. For details see injection molding; blow molding; compression molding.

molding powder. A mixture in granular or pelleted form of a plastic base material together with necessary modifying ingredients (filler, plasticizer, pigment, etc.). Such mixtures are normally prepared by resin manufacturers and sold as such to processors ready for use in injection molding or extrusion operations.

molding sand. See foundry sand.

mole. The amount of pure substance containing the same number of chemical units as there are atoms in exactly 12 grams of carbon-12 [i.e., 6.023×10^{23}]. This involves the acceptance of two dictates—the scale of atomic masses and the magnitude of the gram. Both have been established by international agreement. Formerly, the connotation of "mole" was "gram molecular weight." Current usage tends to apply the term "mole" to an amount containing Avogadro's number of whatever units are being considered. Thus it is possible to have a mole of atoms, ions, radicals, electrons, or quanta. This usage makes unnecessary such terms as "gram-atom," "gram-formula weight," etc. All stoichiometry essentially is based on the evaluation of the number of moles of substance. The most common involves the measurement of mass. Thus 25.000 grams of H_2O will contain 25.000/18.015 moles of H_2O; 25.000 grams of sodium will contain 25.000/22.990 moles of sodium. The convenient measurements on gases are pressure, volume, and temperature. Use of the ideal gas law constant R allows direct calculation of the number of moles: $n = (P \times V)/(R \times T)$. T is the absolute temperature; R must be chosen in units appropriate for P, V, and T. The acceptance of Avogadro's law is inherent in this calculation; so too are approximations of the ideal gas. See also Avogadro's law. William F. Kieffer.

molecular biology. A subdivision of biology that approaches the subject of life at the level of molecular size. This applies to phenomena occurring within the cell nucleus, where the chromosomes and genes are located. These structures, which determine heredity, are in turn composed of nucleic acids (q.v.), which direct the selection and assembly of amino acids in the dividing chromosomes. Much of the essential mechanism of life can be understood by study of specific protein molecules (DNA and RNA) and their effect on the amino acid composition of the genes. See also genetic code; deoxyribonucleic acid; recombinant DNA.

molecular distillation (high vacuum distillation). Distillation at low pressures of the order of 0.001 mm. A molecular distillation is distinguished by the fact that the distance from the surface of the liquid being vaporized to the condenser is less than the mean free path (the average distance traveled by a molecule between collisions) of the vapor at the operating pressure and temperature. This distance is usually of the order of magnitude of a few inches. This process is useful in separation of extremely high boiling and heat-sensitive materials such as glycerides and some vitamins.

molecular formula. See formula, chemical.

molecular rearrangement. See rearrangement.

molecular sandwich (sandwich molecule). See metallocene.

molecular sieve. A group of adsorptive desiccants which are crystalline aluminosilicate materials, chemically similar to clays and feldspars and belonging to a class of minerals known as zeolites (q.v.). The outstanding characteristic of these materials is their ability to undergo dehydration with little or no change in crystal structure. The dehydrated crystals are interlaced with regularly spaced channels of molecular dimensions. This network of uniform pores comprises almost 50 per cent of the total volume of crystals.

The empty cavities in activated "molecular sieve" crystals have a strong tendency to recapture the water molecules that have been driven off. This tendency is so strong that if no water is present they will accept any material that can get into the cavity. However, only those molecules that are small enough to pass through the pores of the crystal can enter the cavities and be adsorbed on the interior surface. This sieving or screening action, which makes it possible to separate smaller molecules from larger ones, is the most unusual characteristic of molecular sieves. They are used in many fields of technology: to dry gases and liquids; for selective molecular separations based on size and polar properties; as ion-exchangers; as catalysts; as chemical carriers; in gas chromatography; and in the petroleum industry to remove normal paraffins from distillates. See also zeolite; gel filtration.

molecular weight. The sum of the atomic weights of the atoms in a molecule. That of methane (CH_4) is 16.032, the atomic weights being C = 12, H = 1.008. The chemical formula used in such a calculation must be the true molecular formula of the substance designated. For example, the molecular formula of oxygen is O_2, and its molecular weight is 32 (atomic weight of O = 16). For ozone the molecular formula is O_3, and the molecular weight is 48. The true molecular weight of a gas or vapor is found by measuring the volume of a given weight and then calculating the weight of 22.4 liters at 0°C and 760 mm. The molecular weight of many complex organic molecules runs as high as a million or more (proteins and high polymers). See also Avogadro; atomic weight; mole.

molecule. Molecules are chemical units composed of one or more atoms. The simplest molecules contain one atom each; for example, helium atoms (one atom per molecule) are identical with helium molecules. Oxygen molecules (O_2) are composed of two atoms, and ozone (O_3) of three. Molecules may contain several different sorts of atoms. Water (H_2O) contains two different kinds, hydrogen and oxygen, and dimethyl amine [$(CH_3)_2NH$] has three kinds. Molecules of many common gases [hydrogen (H_2), oxygen (O_2), nitrogen (N_2), and chlorine (Cl_2)] consist of two atoms each. The atoms of a molecule are held together by chemical bonds. Molecules vary in size from less than 1 to more than 500 millimicrons. The latter are called macromolecules (q.v.). See also bond, chemical; atom.

"Molex"[416] Proprietary process employing molecular sieves to separate normal paraffins from mixtures with isoparaffins and other types of hydrocarbons. Products consist of a normal paraffin stream of high purity and a second stream containing the remaining hydrocarbons in the original mixture. The normal paraffins are excellent components of jet fuels or as raw materials for further chemical synthesis. The denormalized stream, if within the gasoline boiling range, will have higher antiknock quality than the original mixture containing the normal paraffins. The present commercial application is in producing normal paraffins for conversion into "soft" detergent alkylates.

"Moli-Spray."[456] Trademark for a suspension of finely divided molybdenum disulfide powder. Forms a fast drying, adhesive film on all surfaces.

Containers: 16-oz aerosol can.

molten salt. See fused salt.

"Molytropren."[470] Trademark for urethane polyester and polyether foams.

"Moly."[67] Trademark for a series of molybdenum-containing compounds used for seed treatment as a foliar spray, fertilizer additive and for similar related uses.

molybdate chrome orange. See molybdate orange.

molybdate orange (molybdenum orange; molybdate chrome orange). A solid solution of lead chromate, lead molybdate, and lead sulfate.

Properties: Fine dark orange or light red powder.

Derivation: By adding solutions of sodium chromate, sodium molybdate and sodium sulfate to a lead nitrate solution under carefully controlled conditions and filtering off the precipitate.

Containers: Barrels.
Hazard: Toxic by ingestion.
Uses: Pigment in printing inks, paints, plastics.

molybdenite (molybdenum glance) MoS_2. Natural molybdenum sulfide found in igneous rocks and metallic veins.
Properties: Color, bluish-lead gray; stream gray-black; luster metallic; one perfect cleavage; greasy feel; Mohs hardness 1–1.5; sp. gr. 4.6–4.8. Similar in appearance to graphite. Soluble in sulfuric and strong nitric acids.
Occurrence: Colorado, Utah, New Mexico; Chile.
Use: Principal ore of molybdenum.

molybdenite concentrate. Commercial molybdenite ore after the first processing operations. Contains about 90% molybdenum disulfide along with quartz, feldspar, water, and processing oil.

molybdenum Mo Metallic element of atomic number 42, in group VIB of the periodic table. Atomic weight 95.94; valences 2,3,4,5,6. Seven stable isotopes.
Properties: Gray metal or black powder. Does not occur free in nature (see molybdenite). A necessary trace element in plant nutrition. Insoluble in hydrochloric or hydrofluoric acid, ammonia, sodium hydroxide, or dilute sulfuric acid; soluble in hot concentrated sulfuric or nitric acids; insoluble in water. Sp. gr. 10.2; m.p. 2610°C; b.p. 5560°C; high strength at very high temperatures. Nontoxic. Oxidizes rapidly above 1000°F in air at sea level, but is stable in upper atmosphere.
Derivation: By aluminothermic, hydrogen, or electric furnace reduction of molybdenum trioxide or ammonium molybdate.
Forms available: Ingots, rods, wire, powder; ingots (from powder); high ductility sheets; concentrates; also as large single crystals.
Purity: Rods and wire 99.9%; powder 99.9%.
Containers: (powder) Drums.
Hazard: Flammable in form of dust or powder. Tolerance of Mo compounds: insoluble, 10 mg per cubic meter of air; soluble, 5 mg per cubic meter of air.
Uses: Alloying agent in steels and cast iron; high-temperature alloys; tool steels; pigments for printing inks, paints, and ceramics; catalyst; solid lubricants; missile and aircraft parts; reactor vessels; cermets; die-casting copperbase alloys; special batteries.
See also ferromolybdenum; heteromolybdates.

molybdenum acetylacetonate $Mo(C_5H_7O_2)$. Crystalline powder. Slightly soluble in water. Resistant to hydrolysis. A chelating nonionizing compound.
Uses: Catalyst for polymerization of ethylene and formation of polyurethane foam.

molybdenum aluminide. A cermet which can be flame-sprayed.

molybdenum anhydride. See molybdenum trioxide.

molybdenum anhydride. See molybdenum trioxide.

molybdenum boride. Several borides are known: Mo_2B, m.p. 2000°C; Mo_3B_2, m.p. 2070°C; MoB (ordinary and β-forms) m.p. 2180°C; Mo_2B_5, m.p. 1600°C (transforms to MoB_2).
Derivation: By heating molybdenum powder and boron to 1500–1600°C in hydrogen.
Uses: Brazes to join molybdenum, tungsten, tantalum and niobium parts, especially electronic components; corrosion and abrasion-resistant parts; cutting tools; refractory cermets.

molybdenum carbonyl. See molybdenum hexacarbonyl.

molybdenum dioxide MoO_2.
Properties: Lead-gray, nonvolatile powder; insoluble in hydrochloric and hydrofluoric acids and alkalies; sparingly soluble in sulfuric acid; sp. gr. about 6.4.
Derivation: Reduction of molybdenum trioxide or molybdates by hydrogen; partial oxidation of metallic molybdenum.
Hazard: Tolerance, 10 mg per cubic meter of air.

molybdenum diselenide $MoSe_2$. Available as a 40-micron powder. Used as a solid lubricant.

molybdenum disilicide $MoSi_2$. A cermet.
Properties: Dark gray, crystalline powder; sp. gr. 6.31 (20°C); m.p. 1870–2030°C. Not affected by air up to 3000°F; not attacked by most inorganic acids, including aqua regia, but very soluble in hydrofluoric and nitric acids. Has high stress-rupture strength.
Derivation: By fusion of hydrogen-reduced molybdenum with silicon.
Forms: Cylinders: lumps; granules; powder; whiskers. May be coated on materials by vapor deposition and by flame spraying.
Grades: 98%; 99.5%; mesh size-200 and -325.
Hazard: Tolerance, 5 mg per cubic meter of air.
Uses: Electrical resistors; protective coatings at high temperatures; engine parts in space vehicles (molybdenum coated with molybdenum disilicide).

molybdenum disulfide (molybdic sulfide, molybdenum sulfide) MoS_2. See also molybdenite.
Properties: Black, crystalline powder; sp. gr. 4.80; m.p. 1185°C; soluble in aqua regia, sulfuric acid (conc); insoluble in water. Mohs hardness 1; coefficient of friction .02–.06.
Hazard: Tolerance, 5 mg per cubic meter of air.
Derivation: Purification of molybdenite; reaction of sulfur or hydrogen sulfide on molybdenum trioxide.
Uses: Lubricants in greases, oil dispersions, resin-bonded films, dry powders, etc., especially at extreme pressures and high vacua; hydrogenation catalyst.

molybdenum ditelluride $MoTe_2$. Available as a 40-micron powder. Used as a solid lubricant.

molybdenum glance. See molybdenite.

molybdenum hexacarbonyl (molybdenum carbonyl) $Mo(CO)_6$.
Properties: White shiny crystals; decomposes at 150°C without melting; sp. gr. 1.96; b.p. about 155°C; vapor pressure at 20°C about 0.1 mm at 101°C about 43 mm; insoluble in water; soluble in ceresin, paraffin oil, benzene, aminoanthraquinone; slightly soluble in ether and other organic solvents.
Derivation: From molybdenum pentachloride by reaction with zinc dust and carbon monoxide in ether at high pressures.
Hazard: Toxic. Above 150°C, decomposes to evolve carbon monoxide. Tolerance, 5 mg per cubic meter of air.
Uses: Plating molybdenum, i.e., molybdenum mirrors; intermediate.

molybdenum hexafluoride MoF_6.
Properties: White crystalline compound; m.p. 17.5°C; b.p. 35°C; sp. gr. (liq.) about 2.5. Reacts readily with water.
Derivation: Direct action of fluorine on molybdenum metal.

Hazard: Toxic; strong irritant. Tolerance, 5 mg per cubic meter of air.
Uses: Separation of molybdenum isotopes.

molybdenum lake. See phosphomolybdic pigment.

molybdenum metaphosphate $Mo(PO_3)_6$. Yellow powder; sp. gr. 3.28 (0°C); insoluble in water, in most acids. Slightly soluble in hot aqua regia.

molybdenum naphthalene. Dark purple, viscous liquid; soluble in most hydrocarbons; catalyst for commercial production of propylene oxide using hydroperoxides.
Hazard: Tolerance, 5 mg per cubic meter of air.

molybdenum orange. See molybdate orange.

molybdenum III oxide. See molybdenum sesquioxide.

molybdenum oxides. See molybdenum sesquioxide; molybdenum dioxide; molybdenum trioxide.

molybdenum pentachloride $MoCl_5$.
Properties: Green-black solid; dark red as liquid or vapor; m.p. 194°C; b.p. 268°C; sp. gr. 2.9. Hygroscopic, reacting with water and air; soluble in dry ether, dry alcohol, and other organic solvents.
Derivation: Direct action of chlorine on finely divided molybdenum metal.
Containers: Inert gas-filled glass bottles.
Hazard: Toxic and irritant. Tolerance, 5 mg per cubic meter in air.
Uses: Chlorination catalyst; vapor-deposited molybdenum coatings; component of fire-retardant resins; brazing and soldering flux; intermediate for organometallic compounds, e.g., molybdenum hexacarbonyl.

molybdenum sesquioxide (dimolybdenum trioxide, molybdenum III oxide) Mo_2O_3. Known only in the hydrated form, $Mo(OH)_3$, although commonly assigned the formula Mo_2O_3. A compound formed by a dry reaction of molybdenum and oxygen which approximates the composition of the sesquioxide is probably a mixture of molybdenum and molybdenum dioxide.
Properties: Gray-black powder; slightly soluble in acids; insoluble in alkalies and water.
Derivation: Zinc reduction of acid solutions of molybdic acids and molybdates; electrolytic deposition from acid solutions of molybdates.
Hazard: Tolerance, 10 mg per cubic meter of air.
Uses: Catalyst in organic synthesis; decoration and protection for metal articles; feed additive.

molybdenum silicide. Alloy of 60% molybdenum, 30% silicon, and 10% iron, used as means of introducing molybdenum into steel.

molybdenum sulfide. See molybdenum disulfide.

molybdenum trioxide (molybdenum anhydride; molybdic oxide; molybdic acid anhydride) MoO_3.
Properties: White powder at room temp.; yellow at elevated temperatures; sp. gr. 4.69; m.p. 795°C; sublimes starting at 700°C. Sparingly soluble in water; very soluble in excess alkali with formation of molybdates; soluble in concentrated mixtures of nitric and hydrochloric acids or nitric and sulfuric acids. Two hydrates are known: $MoO_3 \cdot H_2O$ and $MoO_3 \cdot 2H_2O$. Readily combines with acids and bases to form a series of polymeric compounds.
Derivation: Roasting of molybdenite; by ignition of the metal, the sulfides, the lower oxides, and of molybdic acids.
Purification: Sublimation.
Grades: Technical; pure; reagent, A.C.S.
Containers: Drums; carload lots.
Hazard: Tolerance, 5 mg per cubic meter of air.
Uses: Source of molybdenum compounds; agriculture; analytical chemistry; manufacture of metallic molybdenum; introduction of molybdenum in alloys; corrosion inhibitor; ceramic glazes; enamels; pigments; catalyst in petroleum industry; medicine.

molybdic acid. Molybdic acid of commerce is either an ammonium molybdate (molybdic acid 85%) or molybdenum trioxide. The use of the term interchangeably for these compounds has caused confusion. Solutions of molybdic acid are very complex chemically since they show a great tendency to polymerize.

molybdic acid, anhydride. See molybdenum trioxide.

molybdic oxide. See molybdenum trioxide.

molybdic sulfide. See molybdenum disulfide.

molybdophosphates. See heteromolybdates.

12-molybdophosphoric acid. See phosphomolybdic acid.

molybdosilicates. See heteromolybdates.

12-molybdosilicic acid (silicomolybdic acid) $H_4SiMo_{12}O_{40} \cdot xH_2O$ where x is usually 6–8.
Properties: Yellow crystalline powder; sp. gr. 2.82. Soluble in water, ethyl alcohol, acetone; insoluble in benzene and cyclohexane. Decomposes in strongly basic solutions. Thermally stable.
Grade: Reagent.
Hazard: Strong oxidizing agent in aqueous solution. Tolerance, 5 mg per cubic meter in air.
Uses: Catalysts; reagents; photography; precipitants and ion exchangers in atomic energy; additives in plating processes; imparting water resistance to plastics, adhesives, and cement.
See also heteromolybdates.

"Moly-Gro."[433] Trademark for molybdenum compounds used to correct molybdenum deficiency in soils and crops.

"Molykote."[149] Trademark for lubricants used in friction problems which involve galling, welding and siezing. Also used for similar mating materials; freeting, high starting friction; and rubber to metal. Available as powders, greases, dispersions, sticks and bonded lubricant coatings.

"Molynamel."[289] Trademark for resin-bonded molybdenum disulfide lubricating enamels.

"Molysulfide."[67] Trademark for molybdenum disulfide (q.v.).

"Monacide."[405] Trademark for a series of insecticides containing 5% DDVP.

"Monamids."[405] Trademark for a group of dialkylolamides which include various grades of coconut fatty acid monoethanolamide, coconut fatty acid monopropanolamide, lauric acid monoethanolamide, lauric acid monoisopropanolamide, and stearic acid monoethanolamide.

"Monamines."[405] Trademark for a group of dialkylolamides used as detergent, detergent additives, foam boosters, wetters, emulsifiers, dispersing agents, thickeners and conditioners.

"Monamulse."[405] Trademark for a series of compounded emulsifiers for paints, plastics, graphic arts, agricultural sprays and other fields.

"Monaquats."[405] Trademark for quaternary ammonium compounds based on 1-ethyl hydroxy-2-alkyl imidazolines.

"Monaquests."[405] Trademark for a series of ethylene diaminetetraacetic acid and (hydroxyethyl)ethylenediaminetriacetic acid sequestering agents.

"Monastral."[28] Trademark for a group of quinacridone colorants, chiefly red, maroon, blue, and violet. Excellent covering power, lightfastness and chemical resistance. Used in lacquers, paints, plastics, rubber. Stable at high temperatures.

"Monastrip."[405] Trademark for a solvent-stripper for uncured and cured epoxy, polyester, and silicone rubber casting and encapsulating compounds.

"Monaterics."[405] Trademark for a special group of substituted imidazolines classified as amphoteric surfactants. Excellent wetting, emulsifying, penetrating and spreading properties in systems requiring broad pH ranges.

monatomic. See diatomic.

"Monawets."[405] Trademark for a group of surfactants of di-octyl, di-hexyl, di-isobutyl and di-tridecyl sulfosuccinates known for their wetting, spreading, penetrating and emulsifying power.

monazite. A natural phosphate of the rare earth metals, principally the cerium and lanthanide metals, usually with some thorium. Yttrium, calcium, iron, and silica are frequently present. Monazite sand is the crude natural material and is usually purified from other minerals before entering commerce.

Properties: Color yellowish to reddish brown; luster vitreous to resinous; streak white; Mohs hardness 5–5.5; sp. gr. 4.9–5.3.

Occurrence: North Carolina, South Carolina, Idaho, Colorado, Montana, Florida; Brazil; India; Australia; Canada.

Uses: Source of thorium, cerium, and other rare-earth metals and compounds.

"Monazolines."[405] Trademark for a series of cationic imidazolines useful as emulsifiers, antistatic agents, water displacers and corrosion inhibitors in agricultural sprays, acid and solvent cleaners, cosmetic and water-oil systems.

Mond process. Mixed ores obtained from roasting crude ores are heated from 50–80°C in a stream of producer gas. Oxides other than nickel are reduced to the metallic state while nickel forms nickel carbonyl [$Ni(CO)_4$] which passes off as a vapor. The vapor is subsequently resolved into carbon monoxide and free nickel.

"Mondur."[567] Trademark for a series of isocyanates.

CB-60 and CB-75. Polyisocyanates.

HX. Hexamethylene diisocyanate.

O. Octadecyl isocyanate.

P. Phenyl isocyanate.

S and SH. Stabilized polyisocyanate adduct.

TM. Triphenylmethane triisocyanate.

Uses: Surface coatings, adhesives, chemical intermediate, hydrophobic agent to increase water repellency of textiles, leather and paper products.

"Monel."[283] Trademark for a large group of corrosion-resistant alloys of predominantly nickel and copper and very small percentages of carbon, manganese, iron, sulfur, and silicon. Some also contain a small percentage of aluminum, titanium, and cobalt. Have good corrosion-resisting properties. Used in electronics industry.

"Monex."[248] Trademark for tetramethylthiuram monosulfide (q.v.).

mono-. Prefix denoting one; see under specific compound; e.g., monochloroacetic acid, see chloroacetic acid. Only a few of the many possible cross references from mono-compounds are given here. (See below).

mono acid F. See 2-naphthol-7-sulfonic acid.

monoanhydrosorbitol. See sorbitan.

monoazo dye. See azo dye.

monobasic. Descriptive of acids having one displaceable hydrogen atom per molecule; acids having two, three or more displaceable H atoms are called dibasic, tribasic, and polybasic, respectively. See also acid.

"Monobed."[23] Trademark for intimate mixtures of "Amberlite" cation and anion exchange resins.

Use: Complete removal of ionizable impurities from water and other solutions in a one-step treatment.

"Monobel."[28] Trademark for a low-velocity permissible dynamite furnished in six grades based upon velocity and cartridge count. For mining coal where lump coal is a factor.

monobenzone (para-benzyloxyphenol; monobenzyl ether of hydroquinone) $C_6H_5CH_2OC_6H_4OH$. See also hydroquinone benzyl ethers.

Properties: White, odorless, crystalline powder; little taste; melting range 115–118°C; soluble in alcohol and acetone; insoluble in water.

Grade: N.F.

Use: Medicine.

monocalcium phosphate. See calcium phosphate, monobasic.

monochloroacetic acid. Legal label name (Rail) for chloroacetic acid (q.v.).

monochloroacetone. Legal label name (Rail) for chloroacetone (q.v.).

monochlorobenzene. See chlorobenzene.

monochlorodifluoromethane. Legal label name (Rail) for chlorodifluoromethane (q.v.).

monochloropentafluoroethane. Legal label name (Rail) for chloropentafluoroethane (q.v.).

monochlorophenol. See chlorophenol.

monochlorotetrafluoroethane. Legal label name (Rail) for chlorotetrafluoroethane (q.v.).

monochlorotriazinyl dye. A fiber-reactive dye for cellulose fibers. See dye, fiber-reactive.

monochlorotrifluoromethane. Legal label name (Rail) for chlorotrifluoromethane (q.v.).

"Monochrome."[307] (U.S.A.). Trademark for a series of mordant dyestuffs. Used for the dyeing of wool. Characterized by good fastness properties.

monoethanolamine. See ethanolamine.

monoethylamine. Legal label name (Rail) for ethylamine (q.v.).

monofilament. A single, continuous strand of glass or synthetic fiber as extruded from a spinneret. See filament.

"Monofrax."[442] Trademark for a line of fused-cast refractories.

monoglyceride. A glycerol ester of fatty acids in which only one acid group is attached to the glycerol group. A typical formula is $RCOOCH_2CHOHCH_2OH$. Small amounts of monoglycerides occur naturally.
Derivation: Produced synthetically by the alcoholysis of fats with glycerol, yielding a mixture of mono-, di-, and triglycerides which is predominantly monoglycerides.
Uses: Emulsifiers; cosmetics; lubricants.
See glycerol monostearate, glycerol monolaurate, etc.

monohydric alcohol. An alcohol in which a hydroxyl group (—OH) has replaced one of the hydrogen atoms of a hydrocarbon; for example:

C_2H_5H	C_2H_5OH
ethane	*ethanol*
RH	ROH
alkane	*alkyl alcohol*

There are a number of classifications analogous to those of hydrocarbons: (1) paraffinic or simple alcohols, whose formula may be represented as C_nH_{2n+1}; (2) olefinic or fatty alcohols, which contain one or more double bonds; (3) alicyclic alcohols, closed ring structures which may or may not contain a double bond, e.g., cyclohexanol; (4) aromatic alcohols, in which the hydroxyl group is attached to a benzene nucleus, as in phenol; (5) heterocyclic alcohols, based on the pentagonal furan ring; and (6) polycyclic alcohols of high molecular weight, known collectively as sterols. Any of these types that contain 12 or more carbon atoms are semisolid to solid, and have a wax-like consistency; the others are colorless liquids.

Monohydric alcohols are also classified as primary, secondary, or tertiary, on the basis of the number of alkyl (methyl) groups substituted for the hydrogen atoms on the central or methanol carbon atom. See also primary.

monomer (momer). A molecule or compound usually containing carbon and of relatively low molecular weight and simple structure, which is capable of conversion to polymers, synthetic resins or elastomers by combination with itself or other similar molecules or compounds. Thus, styrene is the monomer from which polystyrene resins are produced; vinyl chloride is the monomer of polyvinyl chloride. Other common monomers are methyl methacrylate; adipic acid and hexamethylenediamine.

monomethylamine. Legal label name (Rail) for methylamine (q.v.).

monomolecular fillm. See film.

"Monopentek."[138] Trademark for monopentaerythritol. See pentaerythritol.

"Monoplas."[263] Trademark for a group of monomeric phthalates used as plasticizers.

"Monoplex."[23] Trademark for monomeric liquid plasticizers for polyvinyl chloride and other high polymers. Primarily esters, but also some epoxides which impart heat and light stability.
Uses: Plasticizers, stabilzers, processing aids.

"Monopole oil."[309] Trademark for a double sulfonated castor oil used as a textile auxiliary.

monopropellant. A propellant which combines fuel and oxidizer in one compund or mixture. Gunpowder is an example of a solid monopropellant. Liquid monopropellants, for rockets, include: methyl nitrate; nitromethane; a mixture of hydrocarbons with tetranitromethane; a mixture of methyl nitrate and methanol. See also rocket fuel.

monosaccharide. Any of several simple sugars having the formula $C_6H_{12}O_6$; the best-known are glucose, fructose and galactose. Monosaccharides combine to form more complex sugars known as oligo- and polysaccharides.

monosodium glutamate. See sodium glutamate.

monostearin. See glycerol monostearate.

monosulfonic acid F. See 2-naphthol-7-sulfonic acid.

"Monosulph.'[309] Trademark for straight sulfated castor oil for defoaming and leveling. Used to improve flow characteristics of coating formulations.

"Monotan."[309] Trademark for a series of synthetic tanning compounds of the aminoplast resin type.

"Mono-Thiurad."[58] Trademark for bis(dimethylthiocarbamyl)sulfide (q.v.).

montan wax (lignite wax).
Properties: White, hard earth wax; crude product, dark brown; m.p. 80–90°C. Soluble in carbon tetrachloride, benzene, and chloroform; insoluble in water. Combustible; nontoxic.
Derivation: By countercurrent extraction of lignite. American and German lignite are usual sources.
Method of purification: Distillation with superheated steam.
Grades: Crude; refined.
Uses: Substitute for carnauba and beeswax; shoe and furniture polishes; phonograph records; roofing paints; rendering paints waterproof; adhesive pastes; electric insulating compositions; paper-sizing compositions; carbon papers; wire coating; sun-crack preventive in rubber products.

"Montar."[58] Trademark for a series of related brown to black synthetic resins; softening point, 120–250°C. Used to impart fire resistance.

montmorillonite. A type of clay whose composition is approximately $Al_2O_3 \cdot 4SiO_2 \cdot H_2O$. One of the major components of bentonite (q.v.).

monuron. Generic name for 3-(para-chlorophenyl)-1,1-dimethylurea (CMU) $ClC_6H_4NHCON(CH_3)_2$.
Properties: White, crystalline, odorless solid; m.p. 175°C. Very low solubility in water and hydrocarbon solvents. Slightly soluble in oils. Stable toward oxidation and moisture.
Hazard: Moderately toxic by ingestion.
Use: Herbicide.

"Mo-Permalloy."[155] Trademark for a magnetic alloy.
Properties: Density, 8.72 gm/cm^3; tensile strength, 85,000 psi.
Used in laminated cores for high quality communication inductors, transformers and magnetic field detectors.

mordant. A substance capable of binding a dye to a textile fiber. The mordant forms an insoluble lake (q.v.) in the fiber, the color depending on the metal of the mordant. The most important mordants are trivalent chromium complexes, metallic hydroxides, tannic acid, etc. Mordants are used with acid dyes, basic dyes, direct dyes, and sulfur dyes. Premetallized dyes contain chromium in the dye molecule. A mordant dye is a dye requiring use of a mordant to be effective. See also dye, fiber-reactive.

"Mordantine."[165] Trademark for liquid antimony lactate containing 11% available antimony oxide. Recommended as a replacement for technical tartar emetic. See "Antilac."

mordanting assistant. A chemical such as lactic, oxalic and sulfuric acids, tartrates, etc., used in conjunction with mordants to bring about a gradual decomposition of the latter, and to assist in producing a uniform deposition of the actual mordant upon and within textile materials.

mordant rouge (red liquor; red acetate). A solution used in dyeing and calico printing. It is made by dissolving aluminum hydroxide in acetic acid, giving aluminum acetate, or by decomposing lead or calcium acetates with aluminum sulfate or alum.

"Morecrop."[139] Trademark for a dolomitic, hydrated lime having a neutralizing value reported as 104% in terms of calcium carbonate; used for adjusting soil pH and furnishing the elements magnesium and calcium.

"Morestan."[181] Trademark for 6-methyl-2,3-quinoxalinedithiol cyclic carbonate (q.v.).

"Morflex."[299] Trademark for specialty plasticizers (including phthalates, adipates, azelates, sebacates, trimellitates, triacetin and polymerics) for vinyl compounds.

morin (2′,3,4′,5,7-pentahydroxyflavone) $C_{15}H_{10}O_7 \cdot 2H_2O$. One of the two coloring principles of yellow Brazil-wood. (See brasilin.)

Properties: Colorless needles; m.p. 299°C (dec). Soluble in boiling alcohol, alkaline solutions; slightly soluble in water. Low toxicity. Combustible.

Uses: Mordant dye; analytical reagent; indicator.

"Morlex."[214] Trademark for corrosion inhibitors used for steam boilers and steam heating systems. For example: Corrosion Inhibitor A. A mixture of 91% morpholine in water.

"Moroc."[244] Trademark for a silicate binder, containing $Na_2O \cdot 4SiO_2 \cdot 2H_2O$ plus organic compounds.

Uses: Sand core and mold binder for foundries, cured with carbon dioxide.

"Morpasol."[170] Trademark for a nonaqueous dispersion of polyvinyl chloride resins in plasticizers plus heat-and-light stabilizers, fillers, colors and other special-purpose chemicals. Resin articles are solvated by the plasticizer when heated to 350°F and fuse into a homogeneous solid plastic without loss of weight. The cured plastisol is abrasion resistant, fire-proof, grease-proof, water-proof, freeze resistant and electrically insulating. Viscosity is adjusted for use in rotary casting and slush molding, reverse roller coating, knife coating, cold dipping, flow coating, hot dipping, etc.

morphine $C_{17}H_{19}NO_3 \cdot H_2O$.

Properties: White crystalline alkaloid. Slightly soluble in water, alcohol and ether. M.p. 254°C; b.p. (vacuum) 191°C; sp. gr. 1.31.

Derivation: From opium by extraction and crystallization. Opium contains about 10% morphine.

Containers: Cans (100 oz.).

Hazard: Narcotic, habit-forming drug; sale restricted by law in U.S.

Uses: Medicine (in form of acetate, hydrochloride, tartrate, and other soluble salts).

para-**morphine.** See thebaine.

morphine benzyl ether hydrochloride. See peronine.

morpholine (tetrahydro-1,4-oxazine)

$\overline{OCH_2CH_2NHCH_2CH_2}$, or C_4H_8ONH.

Properties: Colorless, hygroscopic liquid, amine-like odor. A mild base. Soluble in water and organic solvents. B.p. 128.9°C; f.p. −4.9°C; sp. gr. (20/20°C) 1.002; wt/gal (20°C) 8.34 lb; vapor pressure (20°C) 6.6 mm; viscosity (20°C) 2.23 cp; flash point (open cup) 100°F; autoignition temp. 590°F.

Derivation: By reaction of ethylene oxide and ammonia.

Grades: Technical; 98%.

Containers: Drums; tank cars.

Hazard: Moderately toxic by ingestion and inhalation; irritant to skin; absorbed by skin. Tolerance, 20 ppm in air. Flammable, moderate fire risk.

Uses: Rubber accelerator; solvent; organic synthesis; additive to boiler water; waxes and polishes; optical brightener for detergents; corrosion inhibitor, preservation of book paper.

morpholine borane $C_4H_8ONH \cdot BH_3$. White needle-shaped crystalline compound that melts at 93°C. Soluble in hot water, alcohol; insoluble in carbon tetrachloride.

Uses: Reducing agent for aldehydes and ketones. Useful in acid media where sodium borohydride is ineffectual because of its instability in acid.

morpholine ethanol (N-hydroxyethylmorpholine) $C_4H_8ONCH_2CH_2OH$.

Properties: Colorless liquid; miscible with water; sp. gr. 1.0724; b.p. 225.5°C; flash point 210°F. Combustible; low toxicity.

6-(N-morpholino)-4,4-diphenyl-3-heptanone hydrochloride $C_4H_8ONCH(CH_3)CH_2C(C_6H_5)_2COC_2H_5 \cdot HCl$.

Properties: M.p. 225°C (dec); (free base melts at 76°C); soluble in water, alcohol, methanol, chloroform; insoluble in acetone, benzene, ethyl acetate.

Use: Medicine.

2-(morpholinothio)benzothiazole (N-oxydiethylene-2-benzothiazolesulfenamide) $C_{11}H_{12}N_2OS_2$.

Properties: Buff to brown flakes with sweet odor; m.p.. 80°C min; sp. gr. 1.34 (25°C). Insoluble in water; soluble in benzene, acetone, methanol.

Use: Delayed action vulcanization accelerator.

morphothion. Generic name for O,O-dimethyl S-(morpholinocarbonylmethyl) phosphorodithioate $(CH_3O)_2P(S)SCH_2C(O)\underline{NCH_2CH_2OCH_2CH_2}$.

Properties: Colorless solid; m.p. 65°C. Soluble in acetone, dioxane, and acetonitrile.

Hazard: Highly toxic. Cholinesterase inhibitor. Use may be restricted.

Use: Insecticide.

Shipping regulations: (Rail, Air) Organic phosphate, solid, n.o.s. Poison label. Not acceptable on passenger planes.

"Morosodren."[401] Trademark for a fungicide concentrate containing methylmercury dicyandiamide (2.2%).

Hazard: May be toxic by ingestion.

mortar. (1) A mixture of cement, lime, sand and water. See cement. (2) A hollow ceramic receptacle used by pharmacists for preparing mixtures.

mortar, metallic. A mixture of powdered metal and other ingredients that have been mixed with water. Lead, tungsten and depleted uranium have been used as the metal component. These mortars resist weathering, mild acids and alkalies, intense radiation and extreme temperature variation. Used in space technology.

morzid. Generic name for bis(1-aziridinyl)-morpholinophosphine sulfide.
$CH_2CH_2OCH_2CH_2NP(NCH_2CH_2)_2{:}S$. Used as an insect chemosterilant.

Moseley, Henry (1887–1915). A British chemist, Moseley studied under Rutherford and brilliantly developed the application of X-ray spectra to study of atomic structure; his discoveries resulted in a more accurate positioning of elements in the Periodic Table by closer determination of atomic numbers. Tragically for the development of science, Moseley was killed in action at Gallipoli in 1915.

"Moskene."[227] Trademark for 1,1,3,3,5-pentamethyl-4,6-dinitroindane (q.v.).

Mössbauer effect. A nuclear phenomenon discovered in 1957. Defined as the elastic (recoil-free) emission of a gamma particle by the nucleus of a radioactive isotope and the subsequent absorption (resonance scattering) of the particle by another atomic nucleus. Occurs in crystalline solids and glasses, but not in liquids. Examples of gamma-emitting isotopes are: iron 57, nickel 61, zinc 67, tin 119. The Mössbauer effect is used to obtain information on isomer shift, on vibrational properties and atomic motions in a solid, and on location of atoms within a complex molecule.

mother. (1) A mold or bacterial complex containing enzymes which promote fermentation, as in manufacture of vinegar from cider or of cultured dairy products from milk.
(2) A substance secreted by epithelial cells of the oyster. See nacre.
(3) A mother liquor is a concentrated solution from which the product is obtained by evaporation and/or crystallizaton, e.g., in sugar manufacture.

motor octane. See octane number.

mountain blue (copper blue).
Derivation: The mineral azurite, in ground form.
Use: Paint pigment.

6-MP. Abbreviation for 6-mercaptopurine.

MPA. Abbreviation for multipurpose additive.

"M-P-A.'[202] Trademark for a colloidal thixotropic agent used in paints, inks and lacquers to prevent pigment settling and excess film flow. Dispersed paste available in mineral spirits, xylene or toluene.

3-MPA. Abbreviation for 3-methoxypropylamine.

MPC black. Abbreviation for medium processing channel black. See carbon black.

MPK. Abbreviation for methyl propyl ketone (q.v.)

"MPS-500."[62] Trademark for a stabilized chlorinated ester of a fatty acid. A viscous, light yellow liquid recommended as a low-cost plasticizer for polyvinyl chloride formulations.

"MpT."[266] Trademark for a technical grade (96%) of methyl para-toluate (q.v.).

"M-Pyrol."[307] Trademark for a solvent, N-methyl-2-pyrrolidone (q.v.). Claimed to be effective cleaner of vinyl chloride reaction equipment.

"M-S."[241] Trademark for microspherical silica alumina; used as a cracking catalyst.

MSG. Abbreviation for monosodium glutamate. See sodium glutamate.

MSP. Abbreviation for monosodium phosphate. See sodium phosphate, monobasic.

MT black. Abbreviation for medium thermal black.

"M & T Catalysts."[288] Trademark for a series of one-shot catalysts for urethane foams.

MTD. Abbreviation for meta-tolylenediamine. See toluene-2,4-diamine.

mucic acid (saccharolactic acid; galactaric acid; tetrahydroxyadipic acid) $HOOC(CHOH)_4COOH$.
Properties: White crystalline powder; m.p. about 210°C (dec); soluble in water; insoluble in alcohol. Combustible; low toxicity.
Derivation: Oxidation of lactose or similar carbohydrates with nitric acid.
Uses: Substitute for tartaric acid; sequesterant for metal ions (calcium, iron); retards hardening of concrete; intermediate for synthesis of heterocyclic compounds (pyrroles).

mucilage. A plant product obtained from seeds, roots or other parts of plants by extraction with either hot or cold water. Mucilages give slippery or gelatinous solutions, e.g., those from guar bean, linseed, locust bean and related leguminous plant seeds. Generally plant mucilages are insoluble in alcohol, but some are partly soluble in water and partly soluble in alcohol. From various types of salt-water algae the so-called seaweed mucilages, such as agar, algin and carrageenin (sometimes referred to as algal polysaccharides) may be obtained by extraction with hot water. Mucilages are closely related to gums, and the distinction between them is not always clear. See also adhesives; gum, natural.

mucopolysaccharide. A polysaccharide composed of alternate units of uronic acids (q.v.) and amino sugars (in which a hydroxyl group is replaced with an amino group, which in turn may be N-substituted by other groups). The mucopolysaccharides act as structural support for connective tissue and mucous membranes of animal organisms.

mud, drilling. See drilling fluid.

"Mullfrax."[280] Trademark for refractory products made from mullite produced in electric furnaces.
Use: Construction materials for furnaces and kilns.

mullite $3Al_2O_3 \cdot 2SiO_2$. A stable form of aluminum silicate formed by heating other aluminum silicates (such as cyanite, sillimanite, and andalusite) to high temperatures; also found in nature.
Properties: Colorless crystals; sp. gr. 3.15; m.p. 1810°C; insoluble in water.
Containers: 50-lb bags.
Uses: Refractories; glass. See also aluminum silicate.

"Mulram."[455] Trademark for a ramming mix with a fused mullite base, and a use limit up to 3200°F. Available in several grain sizes. Used as a refractory where resistance to metal or slag penetration is needed.

"Multifex MM."[244] Trademark for ultrafine calcium carbonate. Used in paints, plastics, and inks.

"Multimet."[214] Trademark for a high-temperature alloy composed of approximately equal parts of iron, nickel, cobalt and chromium. Used for applications involving high stress up to 1500°F and moderate stress to 2000°F.

multiple proportions law. See chemical laws (3).

"Multisorb" A.R.[329] Trademark for a grade of manganese dioxide used as a solid absorbent for SO_2 and NO_2 in analytical chemistry.

"Multi-Sperse."[141] Trademark for stir-in pulps of yellows, oranges, reds, blues, greens, violets and black compatible with all types of latex paints, including acrylic, butadiene-styrene types and polyvinyl acetate.

"Multiwax."[45] Trademark for refined microcrystalline wax obtained from crude petroleum. Composed primarily of alkylated naphthenes and isoparaffins with small amounts of normal paraffins.

"Multranil 176."[567] Trademark for a polymer which in the solid form is similar in appearance to SBR rubber. A base resin of a two-component system which when mixed with the proper curing agents forms a versatile adhesive.

"Multrathane."[567] Trademark for series of compounds used mainly in formulations for solid urethane elastomers. Some are also used in the formulations of spandex fibers, urethane coatings and adhesives.

"Multron."[567] Trademark for a series of polyesters used in the formulation of solderable urethane wire coatings, surface coatings, urethane foams, and adhesives.

municipal waste. See sewage sludge; urban waste.

Muntz metal. An alloy containing about 60% copper and 40% zinc; a low percentage of lead is sometimes added for free-cutting. It is classified as a brass, and is used primarily for condenser tube plates and other electrical applications. It is formed by hot-working, but is not amenable to cold-working. See also brass.

"Murano."[270] Trademark for a synthetic nacreous pigment which has twin inherent colors, one being observed by reflected light and the second by transmitted light, the two colors being complementary. The pigment particles consist of a lead compound in the thin, plate-like crystals. The twin colors are usually seen simultaneously, since part of the light reaches the eye by direct reflection from the surface of the crystals and part of the light is reflected from within the crystal layer and is transmitted through the intervening crystals. (See also nacreous pigment.)
Use: To produce a play of colors when coated on surfaces or incorporated in transparent plastics or plastic film.

muriatic acid. Obsolete name for hydrochloric acid. The related term muriate, indicating presence of chlorine in an inorganic compound, is also obsolete.

muscone. See musk.

muscovite (white mica; potassium mica; isinglass $KAl_2(AlSi_3O_{10})(OH)_2$. A natural hydrous potassium aluminum silicate of the mica group (q.v.).

musk
Properties: An unctuous brownish semi-liquid when fresh; dried, in grains or lumps with color resembling dried blood. Strong characteristic odor providing an animal note to products in which it is used, which has become increasingly popular. The odor-bearing constituent is muscone or muskone, $CH_3C_{15}H_{27}O$, a 15-carbon ring with ketone oxygen.
Derivation: Natural: Secretion from preputial follicles of the musk deer. Synthetic: (a) ketones and lactones with 15- or 16-carbon rings, structurally resembling the odoriferous principles of natural musk, civet, and musky-type plants. Among these are ambrettolide, civetone, muskone, exaltolide. (b) Nitrated compounds, usually nitrated tert-butyl-toluenes or xylenes or related compounds. The three most commonly used in perfumery are musk ambrette, musk ketone, and musk xylene.
Uses: Cosmetics and perfumery (fixative); fragrances; mothproofing agent.

mustard gas. Legal label name for dichlorodiethyl sulfide (q.v.).

mustard oil, artificial. See allyl isothiocyanate.

mustard oil. Any of several organic compounds having the formula R—N=C=S, in which R is an alkyl or aryl radical, —NCS an isothiocyanate group. Its best known member is allyl isothiocyanate (q.v.), the characteristic ingredient of mustard oils. See also nitrogen mustard.
Hazard: Highly toxic and irritant to mucous membranes.

mutagenic agent. (1) Any of a number of chemical compounds able to induce mutations in DNA and in living cells. The alkyl mustards, as well as dimethyl sulfate, diethyl sulfate, and ethylmethane sulfonate, comprise a group of so-called alkylating agents, reacting with the nitrogen atoms of guanine, a constituent of both RNA and DNA. This reaction affects the guanine molecule in such a way as ultimately to induce a mutation in DNA by depurination. Nitrous oxide can deaminate both guanine and cytosine. If DNA having transforming activity is exposed to such deamination conditions, it is slowly deactivated. Nitrous oxide also produces mutants in whole cells, whole bacteriophage, some viruses, and in DNA having transforming ability.
(2) Ionizing radiation.

"MV33."[507] Trademark for a vitreous refractory mullite; used in ceramics. Impervious to gas; compressive strength, 150,000 psi; max. service temperature 3200°F; hardness (Mohs) 9. Fabricated by casting.

MVE. Abbreviation for methyl vinyl ether. See vinyl methyl ether.

m.w. Abbreviation for molecular weight.

"MX."[280] Trademark for fiber-bonded abrasives.
Properties: High tensile strength and resistance to impact and heat shock; unusually resilient.
Uses: For finishing and polishing flutes of taps, drill end mills, reamers, etc.; removing burrs from milling and drilling operations; breaking edges of cast aluminum parts, etc.; cleaning cast iron molds; removing flash from molded plastics.

"Mycifradin" Sulfate.[327] Trademark for neomycin sulfate (q.v.).

"Mycoban."[299] Trademark for sodium and calcium propionates. These salts inhibit the growth of many fungi and of some microorganisms, particularly Bacillus mesentericus, for commercially significant periods of time. Used in many industries, particularly to inhibit mold and rope in bakery products.

"Mycostatin."[412] Trademark for nystatin (q.v.).

"Mydel" 550.[233] Trademark for a water-soluble acrylamide-type paper resin; used to improve the strength properties of chemical and groundwood pulps.

myelin. A unique, sheathlike structure which encloses major nerve trunks, somewhat like insulation around a wire. It is comprised of about 80% lipid, the balance being madeup of proteins, polysaccharides, salts, and water. The lipid fraction is composed of sphingolipids and glycerophosphates, which in turn contain long-chain fatty acids. It has a low concentration of polyunsaturated lipids and high concentrations of long-chain sphingolipids. Its composition is essentially constant in different species of animals, and also as between adults and infants. The breakdown of the lipid structure of myelin is a characteristic of multiple sclerosis.

"Mylar."[28] Trademark for a polyester film. Seven available types used for electrical, industrial, and packaging purposes.
Forms: Roll and sheet.

"Mylase."[173] Trademark for a fungal amylase enzyme.

"Myleran."[301] Trademark for busulfan (q.v.).

myoglobin. A protein-iron-porphyrin molecule similar to hemoglobin. The chief difference is that myoglobin complexes one heme group per molecule, whereas hemoglobin complexes 4 heme groups. See also heme.

myo-inositol. See inositol.

myokinase. An enzyme found in muscle and other tissues that catalyzes the reacton 2 ADP $\rightleftharpoons$ ATP + AMP.

myrcene (7-methyl-3-methylene-1,6-octadiene) $C_{10}H_{16}$. A triply unsaturated aliphatic hydrocarbon found in oil of bay, verbena, hops, and others.
Properties: Yellow, oily liquid; pleasant odor; b.p. 167°C; sp. gr. of 81% myrcene 0.806 (15.5/15.5°C); refractive index of 81% myrcene 1.471 (20°C). Insoluble in water; soluble in alcohol, chloroform, ether, glacial acetic acid. Combustible; low toxicity.
Uses: Preparation of perfume chemicals; flavoring.

"Myrcene 85."[296] Trademark for a special grade of the triply unsaturated aliphatic hydrocarbon, $C_{10}H_{16}$, 7-methyl-3-methylene-1,6-octadiene. Minimum purity 75%. Balance mainly *l*-limonene.

myrcia oil (bay oil; bayleaf oil).
Properties: Essential oil; yellow color, becoming brown on exposure to air; pleasant clove-like odor; pungent, spicy taste; phenol content 50–65%. Sp. gr. 0.950–0.990 (25/25°C), going as low as 0.939 for poor-quality oils; optical rotation 0 to −3°; refractive index 1.507–1.516 (20°C). Soluble in alcohol and glacial acetic acid. Combustible; nontoxic.
Derivation: By distillation of the leaves of Pimenta racemosa (Pimenta acris). Found in West Indies, Virgin Islands, and Puerto Rico.
Containers: Drums.
Uses: Bay rum; perfumes; flavors.

myricyl alcohol. See 1-triacontanol and 1-hentriacontanol.

myricyl palmitate $C_{30}H_{61} \cdot C_{16}H_{31}O_2$ (approx). A wax ester found in beeswax.

myristic acid (tetradecanoic acid) $CH_3(CH_2)_{12}COOH$.
Properties: Oily, white crystalline solid. Soluble in alcohol and ether; insoluble in water. Sp. gr. 0.8739 (80°C); b.p. 326.2°C; 204.3°C (20 mm); m.p. 54.4°C; refractive index 1.4310 (n 60/D). Combustible; nontoxic.
Derivation: Fractional distillation of coconut acid.
Containers: Bags; tank cars.
Grades: Technical; 99.8%; F.C.C.
Uses: Soaps, cosmetics; synthesis of esters for flavors and perfumes; component of food-grade additives.

myristica oil. N.F. name for nutmeg oil.

myristin (glyceryl trimyristate) $C_3H_5(OOCC_{13}H_{27})_3$. A triglyceride occurring, usually to a small extent, in natural fatty oils.

myristoleic acid (cis-tetradec-9-enoic acid) $CH_3(CH_2)_3CH{:}CH(CH_2)_7COOH$. Colorless liquid; m.p. −4°C. Found in fat of some seeds and in fish oil.

myristoyl peroxide $(C_{13}H_{27}CO)_2O_2$.
Properties: Soft granules; 90% peroxide.
Hazard: Oxidizing material, dangerous fire and explosion risk.
Use: Catalyst for vinyl type monomers.
Shipping reuglations: (Air) Peroxides, organic, solid, n.o.s., Organic Peroxide label.

myristyl alcohol (1-tetradecanol) $C_{14}H_{29}OH$.
Properties: White solid; sp. gr. 0.8355 at 20/20°C; b.p. 264.1°C, (20 mm) 171.5°C; m.p. 38°C; flash point 285°F; wt 7.0 lb/gal (20°C). Insoluble in water; soluble in ether; partially soluble in ethyl alcohol. Combustible. Low toxicity.
Grade: Technical.
Containers: Cans; drums; tanks.
Uses: Organic synthesis; plasticizers; anti-foam agent; intermediate; perfume fixative for soaps and cosmetics; wetting agents and detergents; ointments and suppositories; shampoos, toothpaste; specialty cleaning preparations.

myristyl chloride. See tetradecyl chloride.

myristyldimethylamine $CH_3(CH_2)_{13}N(CH_3)_2$. A liquid cationic detergent.
Uses: Corrosion inhibitor, acid-stable.

myristyldimethylbenzylammonium chloride $C_{14}H_{29}(CH_3)_2C_6H_5CH_2NCl$. A quaternary ammonia compound. Free flowing powder; used as a surfactant and detergent.

myristyl lactate. See "Ceraphyl" 50.

myristyl mercaptan. See tetradecyl thiol.

myrrh. The resin of Commiphora myrrha or other species of myrrh. Partly soluble in water and partly in alcohol and ether. Combustible; low toxicity.
Habitat: East Africa and Arabia.
Uses: Dentrifices; perfumery; pharmaceuticals; odorant.

N

N (1) Symbol for nitrogen (q.v.). The names of certain compounds (such as N,N-dibutyl urea) contain this symbol as an indication that the group or groups appearing next in the name (i.e., the butyl groups in the example cited) are joined to the nitrogen atoms in the molecule. The molecular formula is N_2. (2) Mathematical symbol for Avogadro's Number.

N Abbreviation for normal solution. See normal (2).

n Symbol for refractive index; nD^{20} is refractive index under standard conditions of temperature and wavelength (sodium D line).

n- Abbreviation for normal. See normal (1).

Na Symbol for sodium.

"NA-22."[28] Trademark for 2-mercaptoimidazoline $CH_2CH_2NC(SH)NH$ (ring).

Properties: A white powder; sp. gr. 1.42; m.p. not lower than 195°C.
Use: To accelerate vulcanization of neoprene.

nabam (disodium ethylenebisdithiocarbamate) $NaSSCNHCH_2CH_2NHCSSNa$.
Properties: Colorless crystals when pure; easily soluble in water.
Derivation: Addition of carbon disulfide to an alcoholic solution of ethylenediamine followed by neutralization with sodium hydroxide; or by reaction of ethylenediamine with carbon disulfide in aqueous sodium hydroxide.
Grades: 19% aqueous solution.
Hazard: Toxic and irritant to skin and mucous membranes; narcotic in high concentrations. Use may be restricted.
Uses: Plant fungicide; starting material for derivatives that are also pesticides.

"Nabor."[243] Trademark for basic dyes intended especially for dyeing acrylic fibers.

NAC. Abbreviation for National Agricultural Chemicals Association (q.v.).

"Nacan."[243] Trademark for meta-nitrobenzenesulfonic acid, sodium salt.

"Naccogene."[243] Trademark for azoic compositions.

"Nacconate."[243] Trademark for a group of diisocyanates.
Containers: 5- and 55-gal special lined drums.
Uses: In preparation of flexible, semi-rigid, and rigid urethane foams; elastomers; protective coatings, adhesives; spandex-type fibers, and textile finishes.

"Nacconol."[243] Trademark for a group of technical biodegradable linear alkyl sulfonate detergents. See alkyl sulfonate, linear.
Uses: Industrial cleaners, specialty compounding; air entrainment; emulsion polymerization; textile scouring agent; detergent intermediate.

"Naccosol" A.[243] Trademark for a sulfonated alkylated aromatic compound, primarily a wetting agent.

"Naccotan" A.[243] Trademark for an alkyl aryl sodium sulfonate. Used in the retanning of chrome-tanned leather; useful in the dispersion of thick slurries.

"Naccufix."[243] Trademark for a copper organocomplex for fixing direct dyes on cellulose fibers.

"Nacelan."[243] Trademark for water-insoluble dyes dispersed in water and forming solid solutions in synthetic fibers.

nacre (mother of pearl). A form of calcium carbonate secreted by the epithelial cells in the mantle of the oyster. The crystals are bonded by conchiolin ($C_{32}H_{98}N_2O_{11}$); the layers built up by excretion form pearls.

nacreous pigment. A pigment containing guanine crystals obtained from fish scales or skin, which gives a pearly, lustrous effect. Produces a pearly luster. May be applied as surface coatings, as in simulated pearls, or incorporated in plastics. The pigment particle is generally a very thin platelet of high index of refraction. The crystals are readily oriented into parallel layers because of their shape. Being transparent, each crystal reflects only part of the incident light reaching it, and transmits the remainder to the crystals below. The nacreous effect is obtained from the simultaneous reflection of light from the many parallel microscopic layers.

"Nacromer."[270] Trademark for synthetic nacreous pigments, consisting of lead salts crystallized in the form of extremely thin plates and suspended in a suitable liquid vehicle.

NAD. See nicotinamide adenine dinucleotide.

"Nadic Methyl Anhydride."[175] Trademark for methylbicyclo[2.2.1]heptene-2,3-dicarboxylic anhydride isomers. $C_{10}H_{10}O_3$.
Properties: Colorless to light-yellow viscous liquid; viscosity 175–225 cp (25°C); refractive index 1.500–1.506 (n 20/D); sp. gr. 1.200–1.250 (20/20°C). Miscible in all proportions with acetone, benzene, naphtha, and xylene.
Uses: Curing agent for epoxy resins; intermediate for polyesters, alkyd resins, and plasticizers; electrical laminating and filament-winding.

"Nadone."[243] Trademark for cyclohexanone (q.v.).

NADP. See nicotinamide adenine dinucleotide phosphate.

"Nafion."[28] Trademark for a perfluorosulfonic acid membrane used in manufacture of chlorine and caustic soda. It is a chemically stable ion-exchange resin.

NaK (sodium-potassium alloy).
Properties: Soft silvery solid or liquid; must be kept away from air and moisture. The liquid forms come

Superior numbers refer to Manufacturers of Trade Mark Products. For page number see Contents.

under the class name potassium (or sodium), metallic liquid alloy.
Grades: (a) 78% K, 22% Na; m.p. −11°C; b.p. 784°C; density 0.847 g/cc (100°C); (b) 56% K, 44% Na; m.p. 19°C; b.p. 825°C; density 0.886 g/cc (100°C).
Containers: Stainless steel cans (3, 10, 25, 200 lbs); small glass ampules.
Hazard: Ignites in air; explodes in presence of moisture, oxygen, halogens, acids. Store under kerosene. Use *dry* salt or soda ash to extinguish, *not* water, foam or carbontetrachloride.
Uses: Heat exchange fluid; electric conductor; for organic synthesis and catalysis.
Shipping regulations: (Rail) Yellow label. (Air) Flammable Solid label. Not acceptable on passenger planes. Legal label name, sodium-potassium alloy.

"Nalan."[28] Trademark for durable water repellents used in textile industry.

"Nalcamine."[182] Trademark for a group of organic amines.

"Nalco."[182] Trademark for a broad class of chemicals, organic or otherwise, employed in the treatment of water and hydrocarbons; also paper-making chemicals, cleaning compounds, combustion aids, weed and brush controls, lubricating and antilubricating compositions, apparatus, pumps and mechanisms for proportioning chemicals.

"Nalcolyte" 671.[182] Trademark for a synthetic high polymer used for clarifying industrial plant water and municipal water supplies. A coagulant behaving as a polyelectrolyte (q.v.). Effective at concentration of less than 1 ppm; also used in still lower concentrations as a filter aid.

naled. Generic name for 1,2-dibromo-2,2-dichloroethyl dimethyl phosphate.
$(CH_3O)_2P(O)OC(Br)\ HCBr(Cl)_2$.
Properties: Pure compound is a solid; m.p. 26°C. Technical compound is a moderately volatile liquid; b.p. 110°C (0.5 mm). Insoluble in water; slightly soluble in aliphatic solvents; very soluble in aromatic solvents. Hydrolyzes in water.
Hazard: Cholinesterase inhibitor. Use may be restricted.
Use: Insecticide.
Shipping regulations: Organic phosphate, liquid, n.o.s. (Rail) Poison label. (Air) Poison label. Not acceptable on passenger planes.

"Nalgene."[559] Trademark for plastic laboratory ware for industry, research, and education. Made of polypropylene, polyethylene, "Teflon" FEP, polyallomer and polycarbonate.

nalidixic acid (USAN). 1-Ethyl-7-methyl-1,8-naphthyridin-4-one-3-carboxylic acid. Listed in N.D. An antibacterial compound used in medicine.

"Nalkylene."[544] Trademark for biodegradable detergent alkylates made from straight-chain hydrocarbons. Available in molecular-weight ranges to replace branch-chain dodecylbenzene or tridecylbenzene; the linear alkylbenzenes sulfonate readily with SO_3 or oleum by batch or continuous processes to form low-viscosity "soft" sulfonic acid; neutralization gives soft LAS slurries.

"Nalline."[123] Trademark for nalorphine (q.v.).

nalorphine (N-allylnormorphine). The allyl ($-CH_2-CH{=}CH_2$) derivative of morphine. It is able to "antagonize" or neutralize most of the effects of narcotic drugs (morphine, codeine), but not those of other types of depressants. Nalorphine is used as a biochemical research tool for studying the mechanism of narcotic action; also as an antidote for acute morphine poisoning. See also narcotic.

"Nalquat."[182] Trademark for a group of organic quaternary salts.

NaMBT. See sodium MBT.

"Nankor."[233] Trademark for a group of organophosphorus compounds.

nano-. Prefix meaning 10^{-9} unit (symbol n). 1 ng = 1 nanogram = 0.000000001 gram; 1 nanometer = 1 millimicron.

"Nanograde."[329] Trademark for a grade of chemical purity. Impurities guaranteed to be no more than ten parts per trillion.

nanometer (nm). One billionth (10^{-9}) meter, equal to 1 millimicron.

"Nap-All."[480] Trademark for liquid naphthenates of calcium, cobalt, copper, iron, lead, manganese, and zinc.
Uses: Drier for paint and printing inks.

napalm. An aluminum soap of a mixture of oleic, naphthenic and coconut fatty acids.
Properties: Granular powder; mixed with gasoline, it forms a sticky gel that is stable from −40° to 100°C.
Hazard: Flammable, dangerous fire risk.
Uses: Flame throwers and fire bombs; gasoline storage.
Shipping regulations: (Rail, Air) Not listed. Consult authorities.

naphtha
(1) *petroleum* (petroleum ether). A generic term applied to refined, partly refined, or unrefined, petroleum products and liquid products of natural gas not less than 10 percent of which distil below 347°F (175°C), and not less than 95 percent of which distil below 464°F (240°C) when subjected to distillation in accordance with the Standard Method of Test for Distillation of Gasoline, Naphtha, Kerosine, and Similar Petroleum Products (ASTM D 86). F.p. −73°C; b.p. 30–60°C; flash point −57°F; autoignition temp. 550°F; sp. gr. 0.6.
Hazard: Flammable, dangerous fire risk. Explosive limits in air 1 to 6%.
Uses: Source (by various cracking processes) of gasoline, special naphthas, petroleum chemicals, especially ethylene. Cracking for ethylene also produces propylene, butadiene, pyrolysis gasoline and fuel oil; source of synthetic natural gas (q.v.).
Shipping regulations: (Rail) Red label. (Air) Flammable Liquid label.
(a) *V. M. & P* (Varnish makers and Painters); (petroleum spirits; petroleum thinner.
Any of a number of narrow-boiling-range fractions of petroleum with boiling points of about 200 to 400°F, according to the specific use; distillation range 246 to 290°F; sp. gr. 0.7543; density 6.280 lb/gal; pour point below −70°F; flash point 57°F (TCC).
Hazard: Flammable, dangerous fire risk.
Uses: Thinners in paints and varnish.
Shipping regulations: (Rail) Red label. (Air) Flammable Liquid label.
(b) *blending*. A petroleum fraction with volatility similar to the higher boiling fractions of gasoline. It is used primarily in blending with natural gasoline to produce a finished gasoline of specified volatility.

(c) *cleaners'*. A dry-cleaning fluid derived from petroleum and similar to Stoddard solvent (q.v.), but not necessarily meeting all its specifications. Flash point 100°F.

(2) *coal-tar*

(a) *heavy* (high-flash naphtha)

Properties: Deep amber to dark red liquid; a mixture of xylene and higher homologs; sp. gr. 0.885–0.970; b.p. 160–220°C (about 90% at 200°C); flash point 100°F; evaporation 303 minutes.

Derivation: From coal-tar by fractional distillation.

Containers: Drums; tank cars.

Hazard: Toxic by ingestion, inhalation, and skin absorption. Moderate fire risk. Tolerance, 100 ppm in air.

Uses: Coumarone resins; solvent for asphalts, road tars, pitches, etc.; cleansing compositions; process engraving and lithography; rubber cements (solvent); naphtha soaps; mfg. of ethylene and acetic acid.

(b) *solvent* (160° benzol).

Properties: A mixture of small percentages of benzene and toluene with xylene and higher homologs from coal-tar. (a) Crude: dark straw-colored liquid. (b) Refined: water-white liquid. Sp. gr. (a) 0.862–0.892, (b) 0.862–0.872; b.p. (a) about 160°C (80%), (b) about 160°C (90%); flash point (a) and (b) about 78°F.

Derivation: From coal-tar by fractional distillation.

Grades: Dark straw; water-white.

Containers: Drums; tank cars.

Hazard: Flammable, dangerous fire risk.

Uses: Solvent; xylene; cumene; nitrated, for incorporation in dynamite.

Shipping regulations: (Rail) Red label. (Air) Flammable Liquid label.

naphthacene (tetracene; rubene) $C_{18}H_{12}$. The molecule consists of four fused benzene rings.

Properties: Orange solid; sp. gr. 1.35; m.p. about 350°C; not easily soluble; slight green fluorescence in daylight.

Occurrence: In commercial anthracene and coal tar.

Hazard: Explodes when shocked; reacts with oxidizing materials.

Uses: Organic synthesis.

naphthalene (tar camphor) $C_{10}H_8$.

Properties: White crystalline, volatile flakes; strong coal-tar odor. Soluble in benzene, absolute alcohol and ether; insoluble in water. Sp. gr. 1.145 (20/4°C); m.p. 80.2°C; b.p. 217.96°C; flash point 176°F. Sublimes at room temperature. Autoignition temp. 979°F. Combustible.

Derivation: (a) From coal-tar oils boiling between 200 to 250°C (middle oil) by crystallization and distillation. (b) From petroleum fractions after various catalytic processing operations.

Grades: By melting point, 74°C min (crude) to above 79°C (refined); scintillation (80–81°C).

Forms: Flakes, cubes, spheres, powder.

Containers: Cans; drums; tank cars.

Hazard: Toxic by inhalation. Safety data sheet available from Manufacturing Chemists Assn., Washington, D.C. Tolerance, 10 ppm in air.

Uses: Intermediate (phthalic anhydride, naphthol, "Tertralin," "Decalin," chlorinated naphthalenes, naphthyl and naphthol derivatives, dyes); moth repellent; fungicide; explosives; cutting fluid; lubricant; synthetic resins; synthetic tanning; preservative; solvent; textile chemicals; emulsion breakers; scintillation counters.

Shipping regulations: (Air) Crude or refined: Flammable Solid label.

alpha-**naphthaleneacetic acid** (1-naphthylacetic acid) $C_{10}H_7CH_2COOH$.

Properties: White crystals, odorless; m.p. 132–135°C. Soluble in acetone, ether, and chloroform; slightly soluble in water and alcohol.

Grades: Usually supplied in dilute form, either as a powder or liquid solution ready for use.

Containers: Powder; fiber cans or multiwall paper sacks; solution: glass bottles and carboys.

Hazard: Moderately toxic; skin irritant.

Uses: Inducing rooting of plant cuttings; spraying apple trees to prevent early drop.

alpha-**naphthaleneacetic acid, methyl ester** (MENA). $C_{10}H_7CH_2COOCH_3$. A plant growth regulator or hormone, used for delaying sprouting of potatoes, weed control, thinning of peaches, etc.

naphthalene, chlorinated. See chlorinated naphthalene.

naphthalenediamine. See naphthylenediamine.

1,5-naphthalene diisocyanate $C_{10}H_6(NCO)_2$. White to light yellow crystalline solid; m.p. 127–131°C.

Hazard: Toxic and irritant.

Use: Manufacture of polyurethane solid elastomers.

naphthalene-1,5-disulfonic acid (Armstrong's acid) $C_{10}H_6(SO_3H)_2$.

Properties: White crystalline solid; soluble in water.

Derivation: Sulfonation of naphthalene with fuming sulfuric acid at low temperature and separation from the 1,6 isomer. Combustible.

Hazard: May be toxic.

Uses: Intermediate for dyes.

naphthalene-2,7-disulfonic acid $C_{10}H_6(SO_3H)_2$.

Properties: White crystalline solid; soluble in water. Combustible.

Derivation: Sulfonation of naphthalene at high temperature and separation from 2,6-isomer.

Hazard: May be toxic.

Use: Intermediate for dyes.

alpha-**naphthalenesulfonic acid** $C_{10}H_7SO_3H \cdot H_2O$.

Properties: Deliquescent crystals; soluble in water, alcohol, and ether. M.p. 90°C. Combustible.

Derivation: Interaction of naphthalene and sulfuric acid.

Hazard: May be toxic.

Uses: Starting point in the manufacture of alpha-naphthol, alpha-naphtholsulfonic acid, alpha-naphthylaminesulfonic acid; solvent (sodium salt) for phenol in the manufacture of disinfectant soaps.

beta-**naphthalenesulfonic acid** $C_{10}H_7SO_3H$ or $C_{10}H_7SO_3H \cdot H_2O$.

Properties: Non-deliquescent, white plates; m.p. 124–125°C. Soluble in water, alcohol, and ether. Combustible.

Derivation: Sulfonation of naphthalene.

Hazard: May be toxic.

Uses: Starting point in the manufacture of beta-naphthol, beta-naphtholsulfonic acid, beta-naphthylaminesulfonic acid; etc.

2-naphthalenethiol. See "RPA."

1,3,6-naphthalenetrisulfonic acid, trisodium salt $C_{10}H_5(SO_3Na)_3$. Fine buff crystals. Used as a diazo type stabilizer.

1,8-naphthalic acid anhydride $C_{12}H_6O_3$. Light tan powder; m.p. 268–270°C; used for dyestuffs; organic synthesis in general.

"Naphthanil."[28] Trademark for a series of dye bases. Prior to coupling the bases must first be diazotized to form the diazo salt. Widelv used on cotton and rayon textiles. Also represents a series of diazo pigments.

naphtha, petroleum. See naphtha (1).

naphtha, solvent. See naphtha (2b).

naphtha, V.M.&P. See naphtha (1a).

naphthene. See cycloparaffin. The term naphthene is misleading and obsolescent.

naphthenic acid. Any of a group of saturated higher fatty acids derived from the gas-oil fraction of petroleum by extraction with caustic soda solution and subsequent acidification. Gulf and West-coast crudes are relatively high in these acids. The commercial grade is a mixture, usually of dark color and unpleasant odor; corrosive to metals. The chief use of naphthenic acids is in the production of metallic naphthenates for paint driers and cellulose preservatives. Other uses are as solvents, detergents, rubber reclaiming agent, etc.

naphthionic acid (1-aminonaphthalene-4-sulfonic acid; 1-naphthylamine-4-sulfonic acid; 4-amino-1-naphthalenesulfonic acid) $C_{10}H_6(NH_2)SO_3H$.
Properties: White crystals or powder. Soluble in water; slightly soluble in alcohol and ether.
Derivation: Heating equimolar amounts of alpha-naphthylamine and concentrated sulfuric acid.

alpha-**naphthol** (1-naphthol; 1-hydroxynaphthalene) $C_{10}H_7OH$.
Properties: Colorless or yellow prisms or powder; disagreeable taste. Soluble in benzene, alcohol and ether; insoluble in water. Sp. gr. 1.224 (4°C), 1.0954 (95/4°C); m.p. 96°C; b.p. 278°C; volatile in steam; sublimes; refractive index 1.6206 (98.7°C). Combustible.
Derivation: By fusing sodium alpha-naphthalene sulfonate and caustic soda. The melt is decomposed with hydrochloric acid and distilled.
Uses: Dyes; organic synthesis; synthetic perfumes.

beta-**naphthol** (2-naphthol; 2-hydroxynaphthalene) $C_{10}H_7OH$.
Properties: White, lustrous, bulky leaflets, or white powder; darkens with age; faint phenol-like odor; stable in air but darkens on exposure to sunlight; sp. gr. 1.217; m.p. 121.6°C; b.p. 285°C; flash point 307°F. Soluble in alcohol, ether, chloroform, glycerol, oils and alkaline solutions; insoluble in water. Combustible.
Derivation: By fusing sodium beta-naphthalene sulfonate with caustic soda. The product is distilled in vacuo and then sublimed.
Grades: Technical; sublimed; resublimed.
Uses: Dyes; pigmets, antioxidants for rubber, fats, oils; insecticides; synthesis of fungicides, pharmaceuticals, perfumes.

naphthol AS. See beta-hydroxynaphthoic anilide.

beta-**naphthol benzoate.** See benzonaphthol.

3-naphthol-2-carboxylic acid. See 3-hydroxy-2-naphthoic acid.

1-naphthol-3,6-disulfonic acid $C_{10}H_5(SO_3H)_2OH$.
Derivation: Fusion of sodium naphthalene-1,3,6-trisulfonate with caustic soda, or by diazotization of 1-naphthylamine-3,6-disulfonic acid and treatment with sulfuric acid.
Hazard: May be toxic.
Use: Dye intermediate.

1-naphthol-4,8-disulfonic acid (Schoelkopf's acid) $C_{10}H_5OH(SO_3H)_2$.
Properties: Colorless crystals.
Derivation: Decomposition of 1-naphthylamine-4,8-disulfonic acid by diazotization and acidifying with heat.
Hazard: May be toxic.
Use: Dye intermediate.

2-naphthol-3,6-disulfonic acid (R acid; beta-naphtholdisulfonic acid) $C_{10}H_5(OH)(SO_3H)_2$.
Properties: Deliquescent, colorless, silky needles; soluble in water, alchol and ether.
Derivation: By heating beta-naphthol with sulfuric acid (98%), dissolving the melt in water and adding salt.
Grades: Technical, with varying amounts of 2-naphthol-6-sulfonic acid. The common article of trade is the disodium salt, an odorless, gray powder.
Hazard: Probably toxic.
Uses: Manufacturing azo dyes for textiles. The disodium salt is used as a reagent in the detection of nitrogen dioxide in the air.

2-naphthol-3,6-disulfonic acid, disodium salt (R salt). See 2-naphthol-3,6-disulfonic acid.

2-naphthol-6,8-disulfonic acid, potassium salt (G salt) $C_{10}H_5OH(SO_3K)_2$.
Properties: White needles, dry or in paste. Very soluble in water.
Derivation: Sulfonating beta-naphthol in oleum, diluting in water and salting out with potassium sulfate.
Hazard: Probably toxic.
Use: Azo dyes for textiles and lakes.

Naphthol Green B. A dye used in crystallizing solar salt; it increases the evaporation rate by added absorption of energy; 5 ppm is said to increase salt production from 15 to 20%.

beta-**naphthol methyl ether.** See beta-naphthyl methyl ether.

beta-**naphthol sodium.** See sodium beta-naphtholate.

1-naphthol-4-sulfonic acid (NW acid; Neville and Winther's acid; alpha-naphtholsulfonic acid) $C_{10}H_6(OH)SO_3H$.
Properties: Transparent plates. Soluble in water. M.p. 170°C.
Derivation: From the sodium salt of naphthionic acid by hydrolyzing the amino group.
Grades: Technical. Also commonly handled as the sodium salt, an odorless cream to light tan, dry-ground powder.
Hazard: May be toxic.
Use: Dyes.

1-naphthol-5-sulfonic acid (L acid; alpha-naphtholsulfonic acid) $C_{10}H_6(OH)SO_3H$.
Properties: White solid. Soluble in water.
Derivation: (a) From naphthalene-1,5-disulfonic acid by fusion with caustic soda. (b) From 1-naphthyl-

amine-5-sulfonic acid by diazotizing and boiling the diazo solution with dilute sulfuric acid.
Grades: Technical. See also 1-naphthol-5-sulfonic acid, sodium salt.
Hazard: May be toxic.
Use: Dyes.

2-naphthol-1-sulfonic acid (beta-naphtholsulfonic acid) $C_{10}H_6(OH)SO_3H$.
Properties: White crystalline solid; soluble in water.
Derivation: By sulfonating beta-naphthol with 2 to 2.5 parts of 90 to 92% sulfuric acid at about 40°C.
Hazard: May be toxic.
Use: Intermediate for Tobias acid.

2-naphthol-6-sulfonic acid (Schaeffer's acid; beta-naphtholsulfonic acid) $C_{10}H_6(OH)SO_3H$.
Properties: White leaflets; soluble in water and alcohol. M.p. 122°C.
Derivation: By sulfonation of beta-naphthol and separation from the crocein acid formed simultaneously.
Grades: Technical. Also commonly sold as the sodium salt, a cream to light tan, dry ground powder with faint naphthol odor.
Hazard: May be toxic.
Use: Dyes.

2-naphthol-7-sulfonic acid (Cassella's acid; mono-sulfonic acid F; Facid; mono acid F; beta-naphtholsulfonic acid) $C_{10}H_6(OH)(SO_3H)$.
Properties: White crystals; soluble in water and alcohol; m.p. 89°C.
Derivation: By fusion of naphthalene-2,7-disulfonic acid with caustic soda or by heating the acid with an aqueous solution of caustic soda in an autoclave.
Grades: Technical. Also sold as the sodium salt, F salt.
Containers: Fiber drums.
Hazard: May be toxic.
Use: Intermediate for dyes and pigments.

2-naphthol-8-sulfonic acid. See crocein acid.

alpha-**naphtholsulfonic acids.** See 1-naphthol-4-sulfonic acid and 1-naphthol-5-sulfonic acid.

beta-**naphtholsulfonic acids.** See 2-naphthol-1-sulfonic acid, 2-naphthol-6-sulfonic acid, 2-naphthol-7-sulfonic acid, and crocein acid.

1-naphthol-4-sulfonic acid, sodium salt. See 1-naphthol-4-sulfonic acid.

1-naphthol-5-sulfonic acid, sodium salt (L-acid salt) $C_{10}H_6(OH)(SO_3Na)$.
Properties: Gray to light tan dry powder. Odorless.
Containers: 175 lb fiber drums.

2-naphthol-6-sulfonic acid, sodium salt. See 2-naphthol-6-sulfonic acid.

2-naphthol-7-sulfonic acid, sodium salt. See 2-naphthol-7-sulfonic acid.

1,2-naphthoquinone (beta-naphthoquinone) $C_{10}H_6O_2$.
Properties: Orange-red crystals; m.p. 120°C (dec). Soluble in water, ether, benzene, alcohol.
Hazard: Moderately toxic and irritant.
Uses: Chemical reagent and intermediate.

1,4-naphthoquinone (alpha-naphthoquinone) $C_{10}H_6O_2$.
Properties: Greenish yellow powder; odor like benzoquinone; m.p. 123–126°C; sublimes at 100°C; very slightly soluble in water; soluble in ethyl alcohol, ethyl ether, chloroform, benzene, and acetic acid. Combustible.
Hazard: Moderately toxic and irritant.
Uses: Polymerization regulator for rubber and polyester resins; synthesis of dyes and pharmaceuticals; fungicide; algicide.

naphthoquinone oxime. See 1-nitroso-2-naphthol; 2-nitroso-1-naphthol.

1,2-naphthoquinone-4-sulfonic acid (beta-naphthoquinone-4-sulfonic acid) $C_{10}H_5(O)_2SO_3H$.
Derivation: Oxidation with nitric acid of 2-amino-1-naphthol-4-sulfonic acid or 1-amino-2-naphthol-4-sulfonic acid.
Hazard: May be toxic.
Uses: Dye intermediate; identification of sulfonamide derivatives.

naphthoresorcinol. See 1,3-dihydroxynaphthalene.

1,8-naphthosultam-2,4-disulfonic acid (sultam acid) $C_{10}H_4(SO_3H)_2NHSO_2$. (The $-NHSO_2-$ group forms a third ring by attachment at the 1,8 positions.)
Properties: White solid; soluble in water; slightly soluble in alcohol.
Derivation: Oleum sulfonation of 1-naphthylamine-8-sulfonic acid or 1-naphthylamine-4,8-disulfonic acid.
Hazard: May be toxic.
Use: Intermediate for 8-amino-1-naphthol-5,7-disulfonic acid (Chicago acid).

beta-**naphthoxyacetic acid** (2-naphthoxyacetic acid) $C_{10}H_7OCH_2COOH$. A plant hormone.
Properties: Crystals; m.p. 156°C. Soluble in water, alcohol, acetic acid.

1-naphthylacetic acid. See naphthaleneacetic acid.

alpha-**naphthylamine** $C_{10}H_7NH_2$.
Properties: White crystals becoming red on exposure to air. Soluble in alcohol and ether; slightly soluble in water. Sp. gr. 1.13; m.p. 50°C; b.p. 301°C. Combustible. Flash point 315°C.
Derivation: Reduction of alpha-nitro-naphthalene with iron and hydrochloric acid. The mass is then mixed with milk of lime and distilled.
Method of purification: Crystallization.
Hazard: Toxic, especially if containing the beta isomer. Safety data sheet available from Manufacturing Chemists Assn., Washington, D.C.
Uses: Dyes and dye intermediates; agricultural chemicals.
Shipping regulations: (Air) Poison label.

beta-**naphthylamine** $C_{10}H_7NH_2$.
Properties: White to reddish lustrous leaflets. Soluble in hot water, alcohol, ether, and benzene. Commercial: m.p. 109.5°C; sp. gr. 1.061 (98/4°C); b.p. 306°C. Combustible.
Derivation: From beta-naphthol by heating in an autoclave with ammonium sulfite and ammonia.
Method of purification: Distillation.
Hazard: Toxic by ingestion, inhalation, and skin absorption. A carcinogen; must conform to strict safety regulations. Safety data sheet available from Manufacturing Chemists Assn., Washington, D.C.

1-naphthylamine-3,8-disulfonic acid (epsilon acid) $C_{10}H_5(NH_2)(SO_3H)_2$.
Properties: White crystalline scales; soluble in hot water.
Derivation: Naphthalene-1,5- and 1,6-sulfonic acids are nitrated and reduced, resulting in 1-naphthylamine-3,8- and 4,8-disulfonic acids. The separation is

effected by crystallizing out the acid sodium salts of 1-naphthylamine-3,8-disulfonic acid.
Use: Dyes.

1-naphthylamine-4,8-disulfonic acid. See derivation under 1-naphthylamine-3,8-disulfonic acid.

2-naphthylamine-3,6-disulfonic acid (amino R acid) $C_{10}H_5(NH_2)(SO_3H)_2$.

2-naphthylamine-4,8-disulfonic acid (C acid; Cassella acid; 3-amino-1,5-naphthalenedisulfonic acid) $C_{10}H_5(NH_2)(SO_3H)_2$.
Properties: White crystalline solid; soluble in water (slightly).
Derivation: Reduction of 2-nitronaphthalene-4,8-disulfonic acid.
Method of purification: Recrystallization of sodium salt from water.
Hazard: May be toxic.
Use: Dyestuffs.

2-naphthylamine-5,7-disulfonic acid (amono-J acid; 6-amino-1,3-naphthalenedisulfonic acid) $C_{10}H_5(NH_2)(SO_3H)_2$.
Properties: Crystallizes in white lustrous leaflets from water and in long needles from hydrochloric acid solutions.
Derivation: By sulfonation of either 2-naphthylamine-5-sulfonic acid or 2-naphthylamine-7-sulfonic acid.
Hazard: May be toxic.
Use: Dyes.

2-naphthylamine-6,8-disulfonic acid (amino-G acid; 7-amino-1,3-naphthalenedisulfonic acid) $C_{10}H_5(NH_2)(SO_3H)_2$.
Properties: White crystalline solid; soluble in water.
Derivation: (a) From G acid by heating sodium salt with ammonia and sodium bisulfite solution in an autoclave under pressure. (b) Sulfonation by beta-naphthylamine.
Hazard: May be toxic.
Use: Dyes.

alpha-**naphthylamine hydrochloride** $C_{10}H_7NH_2 \cdot HCl$.
Properties: White to gray, crystalline powder; soluble in water, alcohol, and ether.
Derivation: By the action of hydrochloric acid on alpha-naphthylamine.
Hazard: May be toxic.
Uses: Dyes; organic synthesis.

1-naphthylamine-4-sulfonic acid. See naphthionic acid.

1-naphthylamine-5-sulfonic acid (Laurent's acid; L acid; 5-amino-1-naphthalenesulfonic acid) $C_{10}H_6NH_2(SO_3H)$.
Properties: Anhydrous, white or pinkish crystalline needles; greenish fluorescence in dilute aqueous solution. Soluble in water.
Derivation: (a) From alpha-naphthylamine by sulfonation with oleum. (b) From alpha-naphthalenesulfonic acid by nitration, reduction and separation from 1-naphthylamine-8-sulfonic acid also formed.
Hazard: May be toxic.
Use: Azo dyes.

1-naphthylamine-6 and 7-sulfonic acid (Cleve's acid; 5- and 8-amino-2-naphthalenesulfonic acids) $C_{10}H_6NH_2SO_3H$.
Properties: Colorless needles; not very soluble in water. M.p. (1,6-) over 330°C.
Derivation: Sulfonation, nitration, reduction and either separation of the combined 1,6 and 1,7 acids or separating by different concentrations and salting out.
Method of purification: Recrystallization.
Grades: Technical; either mixture of 1,6 plus 1,7 acids or each separate.
Hazard: May be toxic.
Uses: Azo and diazo dyes.

1-naphthylamine-8-sulfonic acid (peri acid; S acid; Schoelkopf's acid; 8-amino-1-naphthalenesulfonic acid) $C_{10}H_6NH_2SO_3H$.
Properties: White needles; slightly soluble in water.
Derivation: Together with 1-naphthylamine-5-sulfonic acid by sulfonating naphthalene, nitrating, reducing and separating the combined precipitated acids with soda ash. The insoluble 1,8 sodium salt is filtered off from the 1,5 sodium salt solution and transposed into free acid with hydrochloric acid.
Grade: Technical, with less than 1% 1,5 acid.
Hazard: May be toxic.
Use: Mostly as phenyl-1-naphthylamine-8-sulfonic acid for azo dyes.

2-naphthylamine-1-sulfonic acid (Tobias acid; 2-amino-1-naphthalenesulfonic acid) $C_{10}H_6(NH_2)(SO_3H)$.
Properties: Crystallizes in white needles. Soluble in hot water.
Derivation: Sodium 2-naphthol-1-sulfonate (from beta-naphthol and sulfuric acid at 40°C) is heated with ammonium hydrogen sulfite and ammonia in an autoclave at from 100 to 150°C.
Method of purification: Precipitation from dilute solution of sodium salt.
Hazard: May be toxic.
Use: Intermediate for dyes, pigments, J-acid and optical brighteners.

2-naphthylamine-6-sulfonic acid (Broenner's acid; 6-amino-2-naphthalenesulfonic acid) $C_{10}H_6(NH_2)SO_3H$.
Properties: Colorless needles; soluble in boiling water.
Derivation: By heating sodium 2-naphthol-6-sulfonate with concentrated ammonia in an autoclave at 180°C.
Grades: Technical. Also commonly sold in the form of the sodium salt, an odorless, light gray to pink powder.
Containers: Fiber drums.
Hazard: May be toxic.
Use: Intermediate for dyes and pigments.

2-naphthylamine-7-sulfonic acid (Cassella's acid F; Bayer's acid; F acid; delta acid) $C_{10}H_6(NH_2)SO_3H$.
Properties: Colorless crystals; soluble in water, alcohol, and ether.
Derivation: (Cassella's acid F.) By heating sodium 2-naphthol-7-sulfonate with aqueous ammonia and ammonium acid sulfite in an autoclave.
Hazard: May be toxic.
Use: Dyes.

2-naphthylamine-8-sulfonic acid (Badische acid). Similar to other naphthylaminesulfonic acids.

1-naphthylamine-3,6,8-trisulfonic acid (Koch's acid; 8-amino-1,3,6-naphthalenetrisulfonic acid) $C_{10}H_4(NH_2)(SO_3H)_3$.
Properties: White solid; soluble in water (slightly).
Derivation: Naphthalene is sulfonated to naphthalene-1,3,6-trisulfonic acid using oleum; and this trisulfonic acid is nitrated cold and then reduced with iron.
Hazard: May be toxic.
Use: Dyes.

1-naphthylamine-4,6,8-trisulfonic acid. See "Osmon."

beta-**naphthyl benzoate.** See benzonaphthol.

ortho-**2-naphthyl** meta, **N-dimethylthiocarbanilate.** See tolnaftate.

naphthylenediamine (diaminonapthalene; naphthalenediamine) $C_{10}H_6(NH_2)_2$. There are eight isomers. The following properties are those of the 1,5 isomer.
Properties: Colorless crystals; m.p. 190°C; b.p., sublimes. Soluble in alcohol and hot water; very sparingly soluble in cold water. Combustible.
Derivation: (a) By the reduction of alpha-dinitronaphthalene. (b) By heating dihydroxynaphthalene with aqueous ammonia.
Use: Organic synthesis.

N-alpha-**naphthylethylenediamine dihydrochloride** $C_{10}N_7NHCH_2CH_2NH_2 \cdot 2HCl$. Colorless crystals; soluble in water. Used as reagent for the quantitative determination of sulfa drugs; for the detection of nitrogen dioxide in air.

beta-**naphthyl ethyl ether** (nerolin) $C_{10}H_7OC_2H_5$.
Properties: White crystals; orange-blossom odor; congealing pt. 35°C; soluble in 5 parts of 95% alcohol. Combustible.
Derivation: Interaction of beta-naphthol and ethyl alcohol in presence of sulfuric acid.
Uses: Perfumes; soaps; flavoring.

1-naphthyl N-methylcarbamate. See carbaryl.

alpha-**naphthylmethyl chloride.** See 1-chloromethylnaphthylene.

beta-**naphthyl methyl ether** (beta-naphthol methyl ether; 2-methoxynaphthalene; methyl naphthyl ether) $C_{10}H_7OCH_3$.
Properties: White, crystalline scales; soluble in alcohol and ether; insoluble in water. M.p. 72°C; b.p. 274°C. Combustible.
Derivation: (a) By heating beta-naphthol and methyl alcohol in presence of sulfuric acid. (b) By methylating beta-naphthol with dimethyl sulfate.
Use: Perfumery (soaps).

alpha-**naphthylphenyloxazole** (NPO; ANPO; 2-(1-naphthyl)-5-phenyloxazole) $C_{19}H_{13}NO$.
Properties: Fluorescent yellow needles; m.p. 104–106°C.
Grade: Scintillation.
Containers: 100-g bottles.
Use: Scintillation counter or wave length shifter in solution scintillators.

N-1-naphthylphthalamic acid $C_{10}H_7NHCOC_6H_4COOH$.
Properties: Crystalline solid; m.p. 185°C; almost insoluble in water; slightly soluble in acetone, benzene, and ethanol; not stable in solutions above pH 9.5 nor at temperatures in excess of 180°C; noncorrosive; combustible; low toxicity. Do not store near seeds or fertilizers.
Use: Selective, pre-emergence herbicide.

alpha-**naphthylthiourea** (ANTU) $C_{10}H_7NHCSNH_2$.
Properties: Odorless gray powder; m.p. 198°C. Insoluble in water and only very slightly soluble in most organic solvents.
Derivation: From alpha-naphthylthiocarbamide and alkali or ammonium thiocyanate.
Containers: Fiber cans.
Hazard: Toxic by ingestion.
Use: Rodenticide.
Shipping regulations: (Air) Poison label. Legal label name, antu.

naphite. See trinitronaphthalene.

Naples yellow. See lead antimonate.

"Napon."[53] Trademark for a starch derivative containing both hydrophilic and hydrophobic groups.
Uses: Emulsifying agent.

narceine $C_{23}H_{27}O_8N \cdot 3H_2O$.
Properties: White, silky crystals; bitter taste; odorless. Soluble in alcohol and boiling water; less soluble in cold water; insoluble in ether. M.p. 150–160°C (commercial), 170–171°C (pure base).
Derivation: Occurs in opium. May be obtained from narcotine.
Use: Medicine.

narcosine. See noscapine.

narcotic. (1) A natural, semisynthetic, or synthetic nitrogen-containing heterocyclic drug that characteristically effects sleep (coma) and relief of pain, but also may result in addiction, i.e., a biochemical situation in which the body tissues become so adapted to the drug that they can no longer function normally without it. Natural narcotics are the plant products morphine and codeine (constituents of opium), both of which are alkaloids (q.v.). Opium is obtained from the seed of the oriental poppy, Papaver somniferens. Semisynthetic narcotics are modifications of the morphine molecule, e.g., diacetylmorphine (heroin), ethylmorphine ("Dionin"), and methyldihydromorphine (metopon).
Synthetic narcotics are meperidine, methadone, and phenazocine. (There are other addictive agents that are not narcotics). The sale of narcotics is strictly controlled by law in the United States.
(2) Inducing sleep or coma. Many chemicals that are not narcotics in sense (1) have this property (chloroform, barbiturates, benzene, etc.).

l-alpha-**narcotine.** See noscapine.

naringin (naringenin-7-rhamnoglucoside; naringenin-7-rutinoside; aurantiin) $C_{27}H_{32}O_{14}$.
Properties: A flavanone glycoside (bioflavonoid). Crystals; m.p. 171°C; bitter taste. Soluble in acetone, alcohol, warm acetic acid.
Source: Extracted from flowers and rind of grapefruit and immature fruit.
Containers: Fiber drums.
Uses: Beverages; sweetener research.

nascent. Descriptive of the abnormally active condition of an element, for example, the atomic oxygen released from hydrogen peroxide and sulfur atoms evolved from thiuramsulfide accelerators. The term is now obsolesent.

"Natac."[285] Trademark for a resin acid; density 1.075–1.085, odorless, used as tackifier, bloom retarder, and processing acid. Neutral cure rate.

National Agricultural Chemicals Association (NAC). An organization of chemical manufacturers concerned with the safety aspects of pesticides. Its Pesticide Safety Team Network can be called upon to decontaminate accidental spills of the more toxic pesticides.

"National" Boronated Graphites.[214] Trademark for a series of nuclear grades of graphite with unique characteristics of the combination of neutron absorption with heat dissipation, accomplished by dispersing boron as boron carbide particles in graphite. Safe to handle before and after exposure. See also graphite.

Uses: Moderators, reflectors, thermal columns, shielding, control rods, fuel elements, crucibles, and molds.

"Natox."[1] Trademark for sodium oxalate. $Na_2C_2O_4$.
Uses: Wallboard cement; insulating materials; tanning of kid skin; fireworks.

natrium. The Latin name for sodium; hence the symbol Na in chemical nomenclature.

natrolite $Na_2Al_2Si_3O_{10} \cdot 2H_2O$. A mineral of the zeolite group. See zeolite.
Properties: Colorless or white to gray, yellow, greenish or red; sp. gr. 2.2–2.25; Mohs hardness 5–5.5.

natron. A complex salt found in dry lake beds of Egypt. Originally an ingredient of ceramic glazes. It has the percentage composition: sodium carbonate, 4.9; sodium bicarbonate, 12.6; sodium chloride, 30.6; sodium sulfate, 20.6; silica, 10; calcium carbonate, 2; magnesium carbonate, 1.9; alumina, 0.7; iron oxide, 0.3; water, 4.7, and organic matter, 11.7.

"Natron 86."[53] Trademark for a cationic polyelectrolyte.
Uses: Water purification; clarification of paper mill effluents; removal of colored waste and other solids from water.

"Natrosol."[266] Trademark for hydroxyethylcellulose.

"Natsyn."[265] Trademark for a series of cis-1,4-polyisoprene synthetic rubbers essentially duplicating the chemical structure of natural rubber.

Natta catalyst. A stereospecific catalyst (q.v.) made from titanium chloride and aluminum alkyl or similar materials by a special process which includes grinding the materials together to produce an active catalyst surface. See also Ziegler catalyst.

natural. Descriptive of a substance or mixture which occurs in nature; the opposite of synthetic or man-made. Elements from 1 to 92 are natural substances that may occur in either the free or combined state. The transuranium elements and all artificial isotopes of other elements are synthetic. Many mixtures occur naturally, e.g., petroleum, wood, metallic ores, natural gas; others are man-made combinations of natural compounds, e.g., glass, cement, paper. Such materials may be considered to be semisynthetic (q.v.). The term "synthetic natural" is often applied to synthesized compounds that are identical with the natural substance; for example, synthetic natural gas and synthetic natural rubber. The term "natural product" is defined as any organic compound formed by living organisms; "natural gasoline" has a specialized meaning (see under gasoline). See also synthesis; natural gas.

natural gas. A mixture of low-molecular weight hydrocarbons obtained in petroleum-bearing regions throughout the world. Its composition is 85% methane, 10% ethane, the balance being made up of propane, butane, and nitrogen. In the U.S. it occurs chiefly in the southwestern states and Alaska. Natural gas is classed as a simple asphyxiant. It should not be confused with natural gasoline. About 3% of the natural gas consumed in the U.S. is used as feedstocks by the chemical industries.
Properties: Colorless flammable gas or liquid; almost odorless; autoignition temperature 900–1100°F. Heating value 1000 Btu per cu ft. A warning odor is added to household fuel gas as a safety precaution.
Containers: As gas, in pipelines; as liquid (LNG) in ocean tankers, barges, tank trucks, pipelines.
Hazard: Flammable, dangerous fire and explosion risk. Explosive limits in air, 3.8 to 17%.
Uses: Fuel and cooking gas; ammonia synthesis; petrochemical feedstocks; carbon black manufacture (channel process); experimental auto fuel.
Shipping regulations: (Rail) Red Gas label. (Air) Flammable Gas label. Not acceptable on passenger planes. Legal label name Liquefied petroleum gas.
See also liquefied petroleum gas; synthetic natural gas.

natural gasoline. See under gasoline.

"Naturetin."[412] Trademark for bendroflumethiazide.

"Naugapol."[248] Trademark for a series of butadiene-styrene copolymers which have received special processing for minimum water-soluble salts and low ash content. Included are "Naugapol K" series which are masterbatches of high styrene resin ("Kralac" A—EP) and "Naugapol" elastomers. Uses include wire and cable insulation, mechanical goods, adhesives, and cements. The "K" series are used for shoe soles, floor tile and wire and cable insulation.

"Naugatex."[248] Trademark for a series of synthetic latices.

"Naugawhite."[248] Trademark for an alkylated phenol antioxidant.
Uses: General-purpose antioxidant for rubber and latex in foam sponge, tire carcass, refrigerator gaskets, footwear, proofing, wire insulation and sundries.

naval stores. Historically, the pitch and rosin used on wooden ships. The term now includes all products derived from pine wood and stumps, including rosin, turpentine, pine oils, tall oil and its derivatives.

"Navane."[299] Trademark for the *cis* isomer of *N,N*-dimethyl-9 [3-(4-methyl-1-piperazinyl) propylidene]-thioxanthene-2-sulfonamide. An antipsychotic drug (thiothixene) said to be as effective as most of the phenothiazines. Approved by FDA.

"Naxol."[243] Trademark for cyclohexanol (q.v.).

"Naxonate" Hydrotropes.[536] Trademark for products used to obtain uniform, clear, and mobile synthetic detergent formulatons. Some examples of the chemicals used in solutions and powders are: sodium, potassium, and ammonium salts of xylenesulfonate, toluenesulfonate, and benzenesulfonate.

Nb Symbol for niobium.

NBA. Abbreviation for N-bromoacetamide.

"NBC."[28] Designation for nickel dibutyldithiocarbamate (q.v.).

NBR. Abbreviation for nitrile-butadiene rubber.

NBS. Abbreviation for N-bromosuccinimide.

NC. Abbreviation for nitrocellulose.

NCS. Abbreviation for N-chlorosuccinimide.

Nd Symbol for neodymium.

NDGA. Abbreviation for nordihydroguaiaretic acid.

"N-Dure."[197] Trademark for a heavy liquid solution containing urea and formaldehyde; used to manufacture solid fertilizer compositions in which the nitrogen is slowly available to growing plants.

Ne Symbol for neon.

"Neantine."[227] Trademark for diethyl phthalate (q.v.).

"Neatex."[152] Trademark for various grades of neatsfoot oil (q.v.).

neatsfoot oil
Properties: A fixed, pale yellow oil with a peculiar odor. Soluble in alcohol, ether, chloroform, and kerosine. Sp. gr. 0.916; saponification value 194–199; flash point 470°F; autoignition temp. 828°F; iodine value 70. Combustible. Low toxicity.
Derivation: By boiling in water the shinbones and feet (deprived of hoofs) of cattle and separating the oil from the fat obtained.
Grades: 15°; 20°; 30°; 40°F cold test (the temperature in °F at which stearin separates).
Containers: Tins; drums.
Hazard: Subject to spontaneous heating.
Uses: Leather industry for fat-liquoring and softening; lubricant; oiling wool.

neburon. Generic name for 1-n-butyl-3-(3,4-dichlorophenyl)-1-methylurea (q.v.).

"Nectadon."[123] Trademark for noscapine (q.v.).

"Neelium."[41] Trademark for a synthetic rubber coating of the neoprene type which can be applied in films up to 20 mils. It is supplied in a two-package system of liquid and accelerator.

"Neetol."[244] Trademark for a series of alkaline oils, based primarily on neatsfoot oil.
Use: Leather industry (referred to as mayonnaise type fat-liquor).

"Negatan."[100] Trademark for negatol (q.v.).

negatol. A condensation product of meta-cresolsulfonic acid with formaldehyde. A polymerized dihydroxydimethyldiphenylmethanedisulfonic acid. It is dispersible in water forming colloidal solutions which are very acidic. The pH of a 5% dispersion is about 1.0.
Use: Medicine.

"NegGram."[162] Trademark for nalidixic acid (q.v.).

"Neguvon."[181] Trademark for trichlorfon (q.v.).

"Nekal."[307] Trademark for a series of wetting and dispersing agents.

"Nelio."[420] Trademark for a series of pine products (rosin, pinene, turpentine, etc.) characterized by extreme purity and uniformity.

"Nemafume."[88] Trademark for 1,2-dibromo-3-chloropropane; 99% purity.

"Nemagon."[125] Trademark for a soil fumigant containing 95% 1,2-dibromo-3-chloropropane.
Hazard: Toxic by inhalation and skin absorption. Irritant to skin and eyes. Moderate fire risk.

nematocide. An agent which is destructive to nematodes (roundworms or threadworms).

"Nembutal."[3] Trademark for pentobarbital, calcium or sodium. See barbiturate.

neo-.
(1) A prefix meaning new and designating a compound related in some way to an older one.
(2) A prefix indicating a hydrocarbon in which at least one carbon atom is connected directly to four other carbon atoms; as, neopentane, neohexane.

"Neobee."[555] Trademark for series of cosmetic and pharmaceutical grades of synthetic oils derived from natural fatty products.

"Neobon."[41] Trademark for a synthetic rubber coating of the neoprene type for protection against corrosion. It is supplied in a two-package system of liquid and accelerator.

"Neochel."[288] Trademark for a wetting agent used in copper plating. A liquid replacement for Rochelle salt with proprietary additives.

"NeoCoat."[204] Trademark for a self-curling liquid neoprene maintenance coating for steel, concrete, and wood surfaces. Resistant to corrosive atmospheric conditions, splash and spill of chemicals, and to abrasion.

"NeoCryl."[516] Trademark for acrylic polymers and copolymers.
Uses: Protective and decorative coatings in the leather, floor polish, paper, paint, textile and adhesive fields.

neodecanoic acid $C_9H_{19}COOH$. Clear, colorless liquid in 97% purity; available commercially. Its derivatives are especially effective as paint driers and are being widely used. Applications as plasticizers and lubricants are also possible.

"Neodol 25."[125] Trademark for a C_{12}—C_{15} linear, primary alcohol.
Uses: Manufacture of biodegradable surfactants, dispersants, solvents, emulsifiers, and chemical intermediates.

neodymia. See neodymium oxide.

neodymium Nd Metallic element having atomic number 60; group IIIB of the periodic table; atomic weight 144.24; valence 3. A rare-earth element of the lanthanide (cerium) group. There are 7 isotopes.
Properties: Soft, malleable, yellowish metal; tarnishes easily; sp. gr. 7.0; m.p. 1024°C; b.p. about 3030°C; ignites to oxide (200–400°C); liberates hydrogen from water; soluble in dilute acids. High electrical resistivity; paramagnetic. Readily cut and machined. Store under mineral oil or inert gas to prevent tarnish and corrosion.
Derivation: Monazite, bastnasite, allanite. Ores are cracked by heating with sulfuric acid.
Forms available: Ingots, rods, sheet, powder to 99.9+% purity.
Hazard: (Salts) Irritant to eyes and abraded skin.
Uses: Neodymium salts; electronics; alloys; colored glass, especially in astronomical lenses and lasers; to increase heat resistance of magnesium; metallurgical research; yttrium-garnet laser dope.
See also didymium.

neodymium acetate $Nd(C_2H_3O_2)_3 \cdot xH_2O$. Pink powder, soluble in water.

neodymium ammonium nitrate
$Nd(NO_3)_3 \cdot 2NH_4NO_3 \cdot 4H_2O$. Pink crystals, soluble in water. Technical grade contains 75% neodymium salt. Principal impurities praseodymium and samarium.

neodymium carbonate $Nd(CO_3)_3 \cdot xH_2O$. Pink powder, insoluble in water; soluble in acids. Grades 75%, 95% and 99% neodymium salt.

neodymium chloride (a) $NdCl_3$; (b) $NdCl_3 \cdot 6H_2O$.
Properties: (a) Violet crystals; sp. gr. 4.134 (25°C); m.p. 784°C; b.p. 1600°C. Very soluble in water, soluble in alcohol; insoluble in ether and chloroform. (b) Red crystals; m.p. 124°C; loses $6H_2O$ at 160°C. Very soluble in water and alcohol.
Grades: 75%, 95%, 99% and 99.9%.

neodymium fluoride NdF_3.
Properties: Pink powder; m.p. 1410°C; b.p. 2300°C. Insoluble in water.
Grades: 65%, 75%, 99% and 99.9%.
Hazard: Toxic and irritant.

neodymium nitrate $Nd(NO_3)_3 \cdot 6H_2O$.
Properties: Pink crystals; very soluble in water; soluble in alcohol and in acetone.
Grades: 75%, 95%, 99% and 99.9%.
Hazard: Toxic; possible explosion risk.

neodymium oxalate $Nd_2(C_2O_4)_3 \cdot 10H_2O$.
Properties: Pink powder; insoluble in water; slightly soluble in acids.
Grades: 75%, 95% and 99%.

neodymium oxide (neodymia) Nd_2O_3.
Properties: Pure product a blue-gray powder; technical grade a brown powder; sp. gr. 7.24; m.p. approx. 1900°C. Insoluble in water; soluble in acids. Hygroscopic; absorbs carbon dioxide from the air.
Grades: 65%, 75%, 85%, 95%, 99% and 99.9% oxide.
Uses (65%): To counteract color of iron in glass. (Purified grades): Ceramic capacitors; coloring glass; catalysts; refractories; carbon arc-light electrodes.

neodymium sulfate $Nd(SO_4)_3 \cdot 8H_2O$.
Properties: Pink crystals; sp. gr. 2.85; m.p. 1176°C; soluble in cold water; sparingly soluble in hot water.
Grades: 75%, 99% and 99.9%.
Uses: Decolorizing glass; coloring glass used in glass blowers and welders' goggles, tableware, etc.

"Neo-Fat."[15] Trademark for a series of fatty acids produced by fractional crystallization and distillation processes. See phenylstearic acid.

"Neoflex."[125] Trademark for a series of linear primary alcohols, from C_6 through C_{11}, including blends.
Properties: (C_7): Colorless liquid; sp. gr. 0.8217 (25/25°C); distillation range 174–182°C. Low toxicity.
Hazard: Combustible, moderate fire risk.
Uses: Solvent; reaction intermediate; esterifying agent.

"Neofinish."[159] Trademark for a fatty-based ester derivative used as non-ionic softener for all fibers.

"Neofolione."[227] Trademark for methyl 2-nonenoate (q.v.).

neohexane (2,2-dimethylbutane) C_6H_{14} or $C_2H_5C(CH_3)_3$.
Properties: Colorless volatile liquid; b.p. 49.7°C; refractive index 1.3659 (25°C); sp. gr. 0.6570 (25°C); freezing point −99.7°C; flash point −54°F; octane rating 100+; autoignition temp. 797°F. Low toxicity.
Derivation: By the thermal or catalytic union (alkylation) of ethylene and isobutane, both recovered from refinery gases.
Grades: 95%, 99%, research.
Hazard: Highly flammable; dangerous fire and explosion risk; lower explosion limit 3%.
Uses: Component of high-octane motor and aviation fuels; intermediate for agricultural chemicals.
Shipping regulations: (Rail) Red label. (Air) Flammable Liquid label.

"Neolan."[443] Trademark for metal-complex dyes for wool and polyamide fibers.

"NeoLine."[204] Trademark for a liquid neoprene, externally catalyzed, for coating immersed surfaces of steel and concrete. Applied by brush, spray or roller for heat curing or air curing.

"Neolite."[265] Trademark for a vinyl shoe-soling material.

"Neolith."[434] Trademark for pellet-type; free-flowing dustfree litharge; designed for ease of handling. Used in the ceramic industry.

"Neolyn."[266] Trademark for a series of soft-to-medium hard modifying rosin-derived alkyd resins. Used in adhesives, lacquers, organosols, plastisols, and floor tile. Available in solution and/or solid forms of various grades.

neomycin sulfate. The sulfate salt of neomycin.
Properties: White to slightly yellow crystals or powder; odorless or practically odorless; hygroscopic. Solutions are dextrorotatory. Very soluble in water, very slightly soluble in alcohol and insoluble in acetone, chloroform and ether.
Grades: U.S.P.; commercial.
Use: Medicine (antibiotic); cosmetic, textile, paper industries; feed supplement; food additive.

neon Ne Inert element of atomic number 10, noble gas group of the periodic system. Atomic weight 20.179. Three stable isotopes.
Properties: Colorless, odorless, tasteless gas; does not form compounds, but ionizes in electric discharge tubes. Liquefies at −245.92°C; f.p. −248.6°C; sp. gr. 0.6964 (air = 1); sp. vol. 11.96 cu ft/lb (70°F, 1 atm). Noncombustible; nontoxic. Slightly soluble in water.
Derivation: By fractional distillation of liquid air. It constitutes 0.0012% of normal air.
Grades: Technical; highest purity.
Containers: Technical, steel cylinders; H.P., hermetically sealed glass flasks.
Uses: (Gas) Luminescent electric tubes and photoelectric bulbs; electronic industry; in lasers. (Liquid) cryogenic research.
Shipping regulations: (Gas) (Rail) Green label. (Air) Nonflammable Gas label. (Liquid, nonpressurized) (Air) Not acceptable. (Liquid, low-pressure) (Air) No label required; not acceptable on passenger planes. (Liquid, pressurized) (Air) Nonflammable Gas label; not acceptable on passenger planes.

neonicotine. See anabasine.

"Neonyl." Trademark for neutral dyeing dyes for polyamide fibers.

neopentane (2,2-dimethylpropane; tetramethylmethane) C_5H_{12} or $C(CH_3)_4$. Present in small amounts in natural gas.
Properties: Colorless gas or very volatile liquid; b.p. 9.5°C; sp. gr. 0.591 (20/4°C); f.p. $\geqslant$17°C; soluble in alcohol; insoluble in water. Flash point −85°F. Autoignition temp. 842°F; explosive limits in air, 1.4 to 7.5%.
Grades: Technical 95%, pure 99%, research 99.9%.
Containers: Cylinders under low pressure.
Hazard: Highly flammable; dangerous fire and explosion risk.
Uses: Research; butyl rubber.

neopentanoic acid. See trimethylacetic acid.

neopentyl glycol (2,2-dimethyl-1,3-propanediol) $HOCH_2C(CH_3)_2CH_2OH$.
Properties: White, crystalline solid; boiling range, 95% between 204–208°C; m.p. 120–130°C; sp. gr. 1.066 (25/4°C).
Containers: Drums; carloads.
Uses: Polyester foams; insect repellent.

neoprene (polychloroprene). $(C_4H_5Cl)_n$. A synthetic elastomer available in solid form, as a latex, or as a flexible foam. Vulcanized with metallic oxides rather than sulfur. Sp. gr. 1.23; resistant to oils, oxygen,

ozone, corona discharge, and electric current. Combustible, but less so than natural rubber. An isocyanate-modified form has high flame resistance.
Uses: (Solid) mechanical rubber products; lining oil-loading hose and reaction equipment; adhesive cement; binder for rocket fuels; coatings for electric wiring; gaskets and seals. (Liquid) Specialty items made by dipping or electrophoresis from the latex. (Foam) Adhesive tape to replace metal fasteners for automotive accessories; seat cushions; carpet backing; sealant.

"NeoPrime"[204] Trademark for chlorinated rubber base primers; bonding agent for neoprene coatings.

"Neoprontosil."[162] Trademark for azosulfamide (q.v.).

neopyrithiamine. See pyrithiamine.

"NeoRez."[516] Trademark for polystyrene copolymer emulsions and urethane prepolymers.
Uses: Floor polish, textile and paper fields, leather finishes, adhesives and protective coatings.

"Neo-Silvol."[330] Trademark for colloidal silver iodide (q.v.).

"Neosol"[125] Trademark for ethyl alcohol based on a formulation approved by the Bureau of Internal Revenue.

"Neo Spectra."[133] Trademark for a series of impingement carbon blacks for automotive enamels.

neostigmine bromide (3-dimethylcarbamoxyphenyltrimethylammonium bromide)
$(CH_3)_2NCOOC_6H_4N(CH_3)_3Br$.
Properties: White, crystalline powder; odorless; bitter taste; m.p. 167°C (dec); very soluble in water; soluble in alcohol; insoluble in ether.
Grade: U.S.P.
Use: Medicine.

"Neo-Synephrine" Hydrochloride.[162] Trademark for phenylephrine hydrochloride (q.v.).

"Neotex."[133] Trademark for furnace carbon blacks used in rubber, printing inks and protective coatings. Characterized by low structure and low oil absorption.
Containers: 25- and 50-lb bags; hopper cars.

"Neo-Tone Sour."[244] Trademark for a laundry sour of the iron-removing type, consisting chiefly of fluorine compounds and complex phosphates.
Hazard: Probably toxic.

"Neotran."[233] Trademark for insecticidal preparations with di-(4-chlorophenoxy)methane as active ingredient.
Hazard: Probably toxic.

neotridecanoic acid $C_{12}H_{25}COOH$. Colorless liquid of 97% purity. Suggested for plasticizers, lubricants, paint driers, fungicides, cosmetics, alkyd resins.

"NeoVac."[516] Trademark for a series of polymer and copolymer emulsions used in textiles, adhesives, papercoatings and plastic starches.

"Neowet."[159] Trademark for a complex polyethylene ether non-ionic wetting agent.
Uses: General wetting purposes, particularly useful with enzymatic desizing agents; scouring all types of textile fabrics; good for use with resin finishes.

"Neozone."[28] Trademark for a group of rubber antioxidants.
A. N-Phenyl-alpha-naphthylamine.
D. N-Phenyl-beta-naphthylamine.
C. Fusion of "Neozone" A and toluene-2,4-diamine.

"Neozymes."[159] Trademark for a series of pancreatic and bacterial enzymes, proteolytic and amylolytic activity for low and high temperatures, continuous or batch desizing.
Grades: Regular, HT, 3-LC and 5-LC, available in powder and liquid form.
Uses: Desizing fabrics coated with starch or gelatin.

nephelite (nepheline) $(Na,K)(Al,Si)_2O_4$. Essentially a silicate of sodium, found in silica-poor igneous rocks.
Properties: Colorless, white, yellowish; luster vitreous to greasy; hardness 5.5–6; sp. gr. 2.55–2.65.
Occurrence: U.S.S.R.; Ontario; Norway; South Africa; Maine, Arkansas, New Jersey.
Uses: Ceramic and glass manufacture; enamels; source of potash and aluminum (U.S.S.R.).

nephelometry. Photometric analytical techniques for measuring the light scatterd by finely divided particles of a substance in suspension. It is used to estimate the extent of turbidity in such products as beer and wine, in which colloidally dispersed particles are present.

"Nepinol."[420] Trademark for a commercial grade of pine oil.

"Nepoxide."[41] Trademark for a coating of the epoxy type which exhibits excellent adhesive properties and resistance to general chemicals and solvents. Can be deposited in high film thickness.

neptunium Np A radioactive transuranic element, having atomic number 93, first formed by bombarding uranium with high-speed deuterons. Atomic weight 237.0482. Valences 3,4,5,6. Sp. gr. 20.45. Neptunium 237, the longest-lived of the 11 isotopes, has been found naturally in extremely small amounts in uranium ores. It is produced in weighable amounts as a by-product in the production of plutonium.

Metallic neptunium is obtained by first preparing neptunium trifluoride, which is reduced with barium vapor at 1200°C. It is a silvery white metal; m.p. 640°C. Neptunium is similar chemically to uranium; it forms such intermetallic compounds as $NpAl_2$ and $NpBe_{13}$, as well as NpC, $NpSi_2$, NpN, NpF_3, NpF_6, NpF_4, NpO_2, Np_3O_8, etc. Soluble in HCl; strong reducing agent.
Neptunium 237 is used in neutron detection instruments.

neptunium dioxide NpO_2.
Properties: Dark olive, free-flowing powder.
Derivation: Neptunium oxalate is precipitated from solutions containing nitric acid and neptunium. The neptunium oxalate is calcined at 500–550°C, producing neptunium dioxide.
Uses: Fabrication by powder metallurgy into target elements to be irradiated to produce plutonium 238.

nerol (cis-3,7-dimethyl-2,6-octadien-1-ol)
$(CH_3)_2C{:}CH(CH_2)_2C(CH_3){:}CHCH_2OH$. The cis isomer of geraniol.
Properties: Colorless liquid; rose-neroli odor; combustible.
Derivation: Iodization of geraniol with hydriodic acid, followed by treatment with alcoholic soda.
Containers: Glass bottles; drums.
Uses: Perfumery; flavoring.

nerolidol (3,7,11-trimethyl-1,6,10-dodecatrien-3-ol) $(CH_3)_2C{:}CH(CH_2)_2C(CH_3){:}CH(CH_2)_2C(CH_3)\cdot(OH)CH{:}CH_2$.
A sesquiterpene alcohol.
Properties: Straw-colored liquid with an odor similar to rose and apple. Sp. gr. 0.878; refractive index 1.480–1.482; angular rotation (natural) +11 to +14°; (synthetic) optically inactive; stable in air; soluble in alcohol and most fixed oils; insoluble in glycerol. Combustible; nontoxic.
Derivation: Occurs naturally in Peru balsam and oils of orange flower, neroli, sweet orange, and ylang ylang. Also made synthetically.
Grade: F.C.C.
Containers: Up to 55-gal drums.
Uses: Perfumery; flavoring.

nerolin. See beta-naphthyl ethyl ether.

neroli oil (orange flower oil).
Properties: Essential oil; pale yellow, fluorescent liquid becoming brownish red on exposure to light; pleasant odor; bitter aromatic taste; sp. gr. 0.863–0.880 (25/25°C); optical rotation +1.5° to +9.1° (25°C). Soluble in an equal volume of alcohol. Combustible; nontoxic.
Chief known constituent: Limonene, linalool, methyl anthranilate, geraniol, linalyl acetate.
Derivation: Distilled from the flower of Citrus aurantium L. Also synthetically. See methyl anthranilate.
Grade: N.F.
Uses: Perfumery; flavoring.

"Nerosol."[342] Trademark for a blend of sesquiterpenic alcohols used in perfumery.

nerve gas (nerve poison). One of several toxic chemical warfare agents developed in Germany during World War II. They are organic derivatives of phosphoric acid (principally, alkyl phosphates, fluorophosphates, and thiophosphates). They inhibit the enzyme cholinesterase and cause acetylcholine poisoning and cessation of nerve transmission. They are colorless, odorless, tasteless liquids of low volatility, and are absorbed rapidly through the eyes, lungs, or skin; they are lethally toxic to higher animals and man. Atropine sulfate is used in the treatment of nerve gas poisoning. The principal German nerve gases were sarin, soman, tabun (q.v.). Many modern pesticides have the same general structure. See also insecticide; parathion.
Note: Manufacture of nerve gases for military use has been discontinued in U.S.

"Nesol."[420] Trademark for a commercial dipentene consisting essentially of monocyclic terpene hydrocarbons.

Nessler's reagent. Solution of mercuric iodide in potassium iodide, used in detecting the presence of ammonia, particularly in very small amounts.
Hazard: Highly toxic.

"Neto."[492] Trademark for a dual acid-enzyme converted product with a dextrose equivalent of approximately 42. Has a maltose content three times that of an acid-converted corn syrup of the same degree of converson. Used in confections.

Neuberg blue. A mixture of copper blue (powdered azurite) and an iron blue (Prussian blue). It can be more easily ground in oil than pure copper blue.

neurine $CH_2{:}CHN(CH_3)_3OH$ (trimethylvinylammonium hydroxide). A poisonous ptomaine formed during putrefaction by the dehydration of choline.
Properties: Syrupy liquid; fishy odor; absorbs carbon dioxide from the air; soluble in water and alcohol.
Hazard: Highly toxic.
Use: Biochemical research.

"Neustrene."[473] Trademark for a group of refined glycerides.

"Neutralite."[184] Trademark for calcite granules of approximately 98% $CaCO_3$ (16-50 mesh) used to neutralize low pH waters.

neutralization. The reaction between hydrogen ion from an acid and hydroxyl ion from a base to produce water, or in nonaqueous solvents, the reaction between the positive and negative ions of the solvent to produce solvent and another salt-like compound.

neutral oil. A lubricating oil of medium or low viscosity obtained by distillation and dewaxing of crude petroleum or its cracking products.

neutral red (toluylene red)
$(CH_3)_2NC_6H_3N_2C_6H_2CH_3NH_2\cdot HCl$ (tricyclic). 3-Amino-7-(dimethylamino)-2-methylphenazine monohydrochloride. C.I. No. 50040.
Properties: Green powder; dissolves in water or alcohol to give red color.
Use: Acid-base indicator in the range pH 6.8–8.0 (red in acid, yellow brown in alkali). See indicator.

"Neu-Tri."[233] Trademark for a grade of trichloroethylene for specific industrial applications.

"Neutrol."[217] Trademark for an acid-activated clay used as decolorizing adsorbent for vegetable and animal fats and oils. See also "Filtrol."

"Neutroleum."[188] Trademark for an all-purpose deodorizing agent.

"Neutrolox."[204] Trademark for a high-grade ammonium chloride used in textile finishing plants.
Uses: To neutralize textiles containing caustic soda from mercerizing, scouring, or bleaching operations.

neutron. Discovered by Chadwick in 1932, the neutron is a fundamental particle of matter having a mass of 1.009 but no electric charge. It is a constituent of the nucleus of all elements except hydrogen, the number of neutrons present being the difference between the mass number and the atomic number of the element. Neutrons may be liberated from the nucleus by fission (q.v.) of uranium-235, plutonium, and a few other elements, each nucleus yielding an average of 2.5 neutrons; they can also be produced by bombardment of other elements, e.g. beryllium, with positively charged particles.

Free neutrons are radioactive and have tremendrous penetrating power as a result of their electrical neutrality; hence they have a highly damaging effect on living tissue, requiring the use of shielding (q.v.) of all equipment in which they are produced. Neutrons directly emitted from atomic nuclei are termed "fast"; it is these that bring about the chain reaction in the atomic bomb. In a nuclear power reactor, where a less rapid reaction is desired, the energy of fast neutrons is partially absorbed by the moderator (q.v.), and the neutrons so retarded are called "slow" or thermal. See also electron; proton; fission.
Uses: Nuclear fission; manufacture of plutonium and radioactive isotopes; cancer therapy; activation analysis.

neutron activation analysis. A type of nondestructive testing which utilizes a stream of neutrons produced by a nuclear reactor or other source. The sample to be analyzed is bombarded by the neutrons, which activate the elements composing it; the measurement of their radioactivity and its various energy levels provides an index of chemical composition. This method is used successfully for on-site analysis of concrete, and may be applied also to other crystalline materials.

"Neutronyx."[328] Trademark for a group of nonionic detergents composed of alkylphenol polyglycol ethers containing ethylene oxide, or of polyethylene glycol fatty acid esters. "Neutronyx" S-60 and S-30 are anionics, the ammonium and sodium salts of a sulfated alkylphenol polyglycol ether.
Uses: Detergents; wetting, emulsifying, dispersing agents.

"Neutroscents."[188] Trademark for a series of perfumes designed particularly to cover objectionable odors; available in water-soluble form for sprays, air-conditioning apparatus, and other dispersion devices. Also available in a highly concentrated form.

"Nevillac."[21] Trademark for a series of (alkyl) hydroxy resins. Used in adhesives, lacquers, paper coatings, special inks and varnishes.

Neville and Winther's acid. See 1-naphthol-4-sulfonic acid.

"Nevindene."[21] Trademark for high-melting coumarone-indene resins of extreme hardness used for dental compounds, fast-drying varnishes, rotogravure inks, aluminum paints and insulating compounds.

"Nevinol."[21] Trademark for a plasticizing and solvent oil used as a stable plasticizer for resins and gums; also used in fly paper, adhesives, inks, aluminum pastes, water-proofing compounds, and rubber-resin finishes.

Newtonian flow. See liquid, Newtonian.

"NFB."[58] Trademark for nonfuming aluminum bright dip bath based on phosphorus.

Ni Symbol for nickel.

"Niacet."[214] Trademark for various metallic acetate salts, including aluminum formoacetate, copper acetate, potassium acetate, sodium acetate, sodium diacetate, zinc acetate and "Niaproof" aluminum acetate (q.v.).

"Niacide."[55] Trademark for fungicidal products containing dimethyl dithiocarbamates used mainly for scab control.

niacin (nicotinic acid; pyridine-3-carboxylic acid) C_5H_4NCOOH. The antipellagra vitamin, essential to many animals for growth and health. In man, niacin is believed necessary along with other vitamins for the prevention and cure of pellagra. It functions in protein and carbohydrate metabolism. As a component of two important enzymes, coenzyme I and coenzyme II, it functions in glycolysis and tissue respiration.
Properties: Colorless needles; odorless; m.p. 236°C; sublimes above melting point; sour taste; soluble in water and alcohol; insoluble in most lipid solvents; quite stable to heat and oxidation. Sp. gr. 1.473. A vasodilator in high concentrations.
Units: Amounts of niacin are expressed in milligrams.
Sources: Food sources; meat, fish, milk, whole grains, yeast. Commercial sources: synthetic niacin is made by oxidation of nicotine, quinoline, or 2-methyl-5-ethylpyridine (from ammonia and formaldehyde or acetaldehyde).
Containers: Glass vials, bottles, fiber cans and drums.
Grades: N.F.; F.C.C.; blended with soy flour (animal feeds).
Uses: Medicine (cholesterol-lowering agent); nutrition; feeds; enriched flours; dietary supplement.
See also niacinamide.

niacinamide (nicotinamide; nicotinic acid amide) $C_5N_4NCONH_2$. Same biological function as niacin.
Properties: Colorless needles; m.p. 129°C; sp. gr. 1.40. Very soluble in water, ethyl alcohol, and glycerol; bitter taste.
Sources: Synthetic made by conversion of niacin to the amide.
Grades: U.S.P.; F.C.C.
Containers: 50-kilogram drums.
Uses: Medicine; dietary supplement.
Also commercially available as the hydrochloride.

niacinamide ascorbate. A complex of ascorbic acid and niacinamide.
Properties: Lemon-yellow powder; odorless or with a very slight odor. May gradually darken upon exposure to air. Soluble in water and alcohol; sparingly soluble in glycerin; practically insoluble in benzene.
Grade: F.C.C.
Use: Dietary supplement.

"Niagathal."[62] Trademark for tetrachlorophthalic anhydride.

"Niagra."[433] Trademark for iron powders.

"Nial."[155] Trademark for a nickel-base thermocouple alloy, wherein the negative thermal E.M.F. with reference to pure platinum is obtained by adding manganese, aluminum, and silicon. Low percentages of iron, cobalt, zirconium and magnesium are added to control both the E.F.M. and type of oxide.
Properties: Magnetic, m.p. 2550°F; sp. gr. 8.47 g/cc; sp. heat at 20°C, 0.12 cal/gm; tensile strength, 83,000 psi.

nialamide $C_5H_4NCO(NH)_2(CH_2)_2CONHCH_2C_6H_5$. Designated chemically as 1-(2-benzylcarbamyl)ethyl-2-isonicotinoylhydrazine. It is a white, crystalline powder of low solubility in water and good solubility in slightly acid solutions. It is stable in crystalline form, suspension and solution. Used in medicine as an amine oxidase inhibitor.

"Nialk."[62] Trademark for chlorine, caustic soda, caustic potash, carbonate of potash, paradichlorobenzene and trichloroethylene.

"Niaproof."[214] Trademark for a water-repellent compound. Substantially a soluble basic aluminum acetate salt.
Uses: Source of aluminum ion for water-repellent finishes for textile, paper, and leather products, particularly in processes using wax or soap emulsions.

"Niax."[214] Trademark for a polyurethane foam preparation packaged in two units: (1) the resin base and (2) the expander or catalyst (see "Niax" ES).

"Nibrite."[134] Trademark for barrel nickel-plating additive.

nicarbazin. Equimolar complex of 4,4′-dinitrocarbanilide and 2-hydroxy-4,6-dimethylpyrimidine. Forms crystals, decomposes at 265–275°C; insoluble in water. used as a coccidiostat.

niccolite (arsenical nickel) NiAs. An ore of nickel.
Properties: Pale copper-red mineral with dark tarnish, metallic luster. Contains 43.9% nickel; soluble in concentrated nitric acid. Sp. gr. 7.3–7.67; hardness 5–5.5.
Hazard: Highly toxic by inhalation of dust.

"Nichrome."[350] Trademark for an alloy containing 60% nickel, 24% iron, 16% chromium, 0.1% carbon. It is used principally for electric resistance purposes. It also offers good resistance to mine and sea waters and moist sulfurous atmospheres.

nickel Ni Metallic element of atomic number 28, group VIII of the Periodic Table. Atomic weight 58.71; valences 2, 4. Five stable isotopes.
Properties: Malleable, silvery metal; readily fabricated by hot- and cold-working; takes high polish; excellent resistance to corrosion. Sp. gr. 8.908; m.p. 1455°C; b.p. 2900°C; electrical resistivity (20°C) 6.844 microhm-cm. Attacked slightly by hydrochloric and sulfuric acids, somewhat more by nitric; highly resistant to strong alkalies. Low toxicity.
Occurrence: As pentlandite in Ontario and garnierite in New Caledonia; Cuba; Norway; U.S.S.R.
Derivation: Nickel ores are of two types: sulfide and oxide, the former accounting for 2/3 of the world's consumption. Sulfide ores are refined by flotation and roasting to sintered nickel oxide, and either sold as such or reduced to metal, which is cast into anodes and refined electrolytically or by the carbonyl process (see Mond process). Oxide ores are treated by hydrometallurgical refining, e.g., leaching with ammonia. Much secondary nickel is recovered from scrap.
Grades: Electrolytic; ingot; pellets; shot; sponge; powder; high-purity strip; single crystals (wire 2 × .05–.005 in.).
Hazard: Flammable and toxic as dust or powder. Tolerance, 1 mg per cubic meter of air (metal and soluble compounds).
Uses: Alloys (low-alloy steels, stainless steel, copper and brass, permanent magnets, electrical resistance alloys); electroplated protective coatings; electroformed coatings; alkaline storage battery; fuel cell electrodes; ceramics; catalyst for methanation of fuel gases and hydrogenation of vegetable oils.
Shipping regulations: (Rail) Nickel catalyst, finely divided, spent or activated, Yellow label. (Air) dry or wet with less than 40% water, Not acceptable; wet with not less than 40% water, Flammable Solid label; not acceptable on passenger planes.
See also Raney's nickel.

nickel acetate $Ni(OOCCH_3)_2 \cdot 4H_2O$.
Properties: Green, monoclinic crystals. Effloresces somewhat in air. Sp. gr. 1.74; decomposes on heating to 250°C. Soluble in water and alcohol.
Derivation: By heating nickel hydroxide with acetic acid in the presence of metallic nickel.
Containers: Drums; truckloads.
Uses: Textiles (mordant); catalyst.

nickel aluminide. A cermet that can be flame-sprayed.

nickel ammonium chloride (ammonium nickel chloride) (a) $NiCl_2 \cdot NH_4Cl$; (b) $NiCl_2 \cdot NH_4Cl \cdot 6H_2O$.
Properties: (a) Yellow powder; (b) green crystals; sp. gr. 1.65. Soluble in water; deliquescent.
Hazard: See under nickel.
Uses: Electroplating; dyeing (mordant).

nickel ammonium sulfate. (nickel salts, double; ammonium nickel sulfate) $NiSO_4 \cdot (NH_4)_2SO_4 \cdot 6H_2O$.
Properties: Green crystals; decomposed by heat. Soluble in water; less in ammonium sulfate solution; insoluble in alcohol. Sp. gr. 1.929.
Derivation: An aqueous solution of nickel sulfate is acidified with sulfuric acid; then an aqueous solution of ammonium sulfate is added. On concentrating, crystals of the double sulfate separate out.
Hazard: See under nickel.
Use: Nickel electrolyte for electroplating.

nickel arsenate (nickelous arsenate) $Ni_3(AsO_4)_2 \cdot 8H_2O$.
Properties: Yellow-green powder; soluble in acids; insoluble in water. Sp. gr. 4.98.
Hazard: Highly toxic.
Use: Catalyst (hardening fats used in soap).
Shipping reguations: Arsenates, n.o.s. (Rail, Air) Poison label.

nickel bromide (nickelous bromide) (a) $NiBr_2$; (b) $NiBr_2 \cdot 3H_2O$.
Properties: (a) Brownish-yellow solid or yellow, lustrous scales; sp. gr. 5.098; m.p. 963°C. (b) Deliquescent, greenish scales; loses $3H_2O$ at 300°C. Soluble in water, alcohol, ether, and ammonium hydroxide.
Derivation: Bromination of nickel powder or nickel carbonyl.
Hazard: See under nickel.

nickel carbonate, basic. Uncertain composition, approximately $NiCO_3 \cdot 2Ni(OH)_2 \cdot 4H_2O$.
Properties: Light green crystals or brown powder. Sp. gr. 2.6. Insoluble in water, soluble in ammonia and dilute acids.
Derivation: (1) In nature as the mineral zaratite. (2) Synthetically, by addition of soda ash to a solution of nickel sulfate.
Containers: Drums; bags; truckloads.
Hazard: See under nickel.
Uses: Electroplating; preparation of nickel catalysts; ceramic colors and glazes.

nickel carbonyl (nickel tetracarbonyl) $Ni(CO)_4$. A zerovalent compound. The four carbonyl groups form a tetrahedral arrangement and are linked covalently to the metal through the carbons.
Properties: Colorless liquid. Soluble in alcohol and many organic solvents; soluble in concentrated nitric acid; insoluble in water. Sp. gr. 1.3185; f.p. −25°C; b.p. 43°C; vapor pressure, 400 mm at 25.8°C.
Derivation: By passing carbon monoxide gas over finely divided nickel.
Grade: Technical.
Containers: Iron cylinders pressurized with carbon monoxide.
Hazard: Highly toxic by inhalation. Flammable, dangerous fire risk. Tolerance, 0.001 ppm in air. A carcinogenic agent.
Uses: Production of high-purity nickel powder by Mond process (q.v.); continuous nickel coatings on steel and other metals.
Shipping regulations: (Rail) Red label. Not accepted by express. (Air) Not acceptable.

nickel chloride (nickelous chloride) (a) $NiCl_2$ (b) $NiCl_2 \cdot 6H_2O$.
Properties: (a) Brown scales; deliquescent. (b) Green scales; deliquescent. Soluble in water, alcohol and ammonium hydroxide. (a) Sp. gr. 3.55; m.p. 1001°C. Nonflammable.
Derivation: Action of hydrochloric acid on nickel.
Uses: Electroplated nickel coatings; reagent chemical.

nickel cyanide $Ni(CN)_2 \cdot 4H_2O$.
Properties: Apple-green plates or powder. Soluble in ammonium hydroxide and potassium cyanide solu-

tion; insoluble in water and acids. M.p., loses $4H_2O$ at 200°C; b.p. decomposes.
Derivation: By adding potassium cyanide to a solution of a nickel salt.
Hazard: Highly toxic.
Uses: Metallurgy, electroplating.
Shipping regulations: (Rail, Air) Poison label.

nickel dibutyldithiocarbamate $Ni[SC(S)N(C_4H_9)_2]_2$.
Properties: Dark green flakes; sp. gr. 1.26; m.p. 86°C min.
Uses: Antioxidant for synthetic rubbers.

nickel formate $(HCOO)_2Ni \cdot 2H_2O$.
Properties: Green crystals; sp. gr. 2.15; soluble in water.
Derivation: (a) Reaction of sodium formate and nickel sulfate; (b) dissolving nickel hydroxide in formic acid.
Containers: Bags; truck loads.
Hazard: See under nickel.
Use: Production of nickel catalysts.

nickel hydroxide. See nickelous hydroxide; nickelic hydroxide.

nickelic hydroxide $Ni(OH)_3$.
Properties: Black powder; m.p. decomposes.
Derivation: By adding a hypochlorite to a solution of a nickel salt.
Use: Nickel salts.

nickelic oxide (nickel peroxide; nickel sesquioxide; black nickel oxide) Ni_2O_3.
Properties: Gray-black powder; soluble in acids; insoluble in water. Sp. gr. 4.84; m.p., is reduced to NiO at 600°C.
Derivation: By gentle heating of the nitrate or chlorate.
Containers: Barrels.
Hazard: See under nickel.
Use: Storage batteries.

nickel iodide (nickelous iodide) NiI_2 or $NiI_2 \cdot 6H_2O$ (loses water at 43°C).
Properties: Black, crystalline powder or blue-green crystals. Hygroscopic. Sublimes at 797°C without melting. Soluble in alcohol, water; sp. gr. 5.834.
Derivation: Direct combination of nickel and I_2.
Hazard: Toxic by ingestion. See under nickel.

nickel-iron alloy. See iron-nickel alloy.

"Nickel-Lume."[288] Trademark for a bright nickel electroplating process. Materials used are nickel sulfate, nickel chloride, boric acid and organic addition agents.

nickel matte. See matte.

nickel nitrate (nickelous nitrate) $Ni(NO_3)_2 \cdot 6H_2O$.
Properties: Green, deliquescent crystals. Soluble in water, ammonium hydroxide and alcohol. Sp. gr. 2.065; m.p. 55°C; b.p. 136.7°C.
Derivation: By the action of nitric acid on nickel, or on nickel oxide.
Grades: Technical; reagent.
Containers: Drums; bags; truck loads.
Hazard: Dangerous fire risk; strong oxidizing agent. Tolerance (as Ni), 1 mg per cubic meter of air.
Uses: Nickel plating; preparation of nickel catalysts; manufacture of brown ceramic colors.
Shipping regulations: (Rail) Nitrates, n.o.s., Yellow label. (Air) Nitrates, n.o.s., Oxidizer label.

nickel nitrate, ammoniated (nickel nitrate tetrammine) $Ni(NO_3)_2 \cdot 4NH_3 \cdot 2H_2O$.
Properties: Green crystals. Soluble in water; insoluble in alcohol; decomposes in air.
Derivation: By adding ammonium hydroxide to a nitric acid solution of nickel oxide, with subsequent crystallization.
Hazard: See nickel nitrate.
Use: Nickel plating.
Shipping regulatoins: See nickel nitrate.

nickelocene (dicyclopentadienylnickel) $(C_5H_5)_2Ni$.
Properties: Dark green crystals; m.p. 171–173°C; soluble in most organic solvents; insoluble in water; decomposes in acetone, alcohol and ether; highly reactive compound which decomposes rapidly in air.
Containers: Supplied in an inert atmosphere in bottles and steel containers.
Hazard: Toxic by inhalation or skin contact.
Uses: Catalyst; antiknock agent for fuels; complexing agent.
See also metallocene.

nickelous arsenate. See nickel arsenate.

nickelous bromide. See nickel bromide.

nickelous chloride. See nickel chloride.

nickelous hydroxide (nickel hydroxide) $Ni(OH)_2$.
Properties: Fine green powder; sp. gr. 4.15; m.p. (dec) 230°C. Very slightly soluble in water; soluble in acids and ammonium hydroxide.
Derivation: By adding caustic soda to a solution of nickelous salt.
Hazard: See under nickel.
Use: Nickel salts.

nickelous iodide. See nickel iodide.

nickelous nitrate. See nickel nitrate.

nickelous phosphate. See nickel phosphate.

nickel oxide (nickelous oxide; nickel protoxide; green nickel oxide) NiO.
Properties: Green powder, becoming yellow; soluble in acids and ammonium hydroxide; insoluble in water and caustic solutions. Sp. gr. 6.6–6.8; absorbs oxygen at 400°C forming Ni_2O_3 which is reduced to NiO at 600°C.
Derivation: By heating nickel above 400°C in presence of oxygen.
Hazard: See under nickel.
Uses: Nickel salts; porcelain painting; fuel cell electrodes.

nickel oxide, black. See nickelic oxide.

nickel oxide, green. See nickel oxide.

nickel peroxide. See nickelic oxide.

nickel phosphate (nickelous phosphate; trinickelous orthophosphate) $Ni_3(PO_4)_2 \cdot 7H_2O$.
Properties: Light-green powder. Soluble in acids, ammonium hydroxide; insoluble in water.
Use: Electroplating.

nickel potassium sulfate (potassium nickel sulfate) $NiSO_4 \cdot K_2SO_4 \cdot 6H_2O$.
Properties: Blue-green crystals. Soluble in water. Sp. gr. 2.124.

nickel protoxide. See nickel oxide.

nickel-rhodium. Alloys containing nickel and from 25–80% of rhodium; but sometimes also some platinum, iridium, palladium, molybdenum, tungsten, copper, iron, or cobalt.
Uses: Electrodes; chemical apparatus; reflectors.

nickel salt, double. See nickel ammonium sulfate.

nickel salt, single. See nickel sulfate.

nickel sesquioxide. See nickelic oxide.

nickel silver. Nonferrous alloy of nickel, copper and zinc having a silver appearance.
Uses: Etching, enameling, silver-plating and chromium-plating.

nickel stannate $NiSnO_3 \cdot 2H_2O$.
Properties: Light colored crystalline powder; approx. temperature of dehydration 120°C.
Hazard: Highly toxic by ingestion and inhalation. Tolerance, 1 mg per cubic meter of air.
Use: Additive in ceramic capacitors.

nickel sulfate (nickel salts, single; blue salt) (a) $NiSO_4$; (b) $NiSO_4 \cdot 6H_2O$; (c) $NiSO_4 \cdot 7H_2O$.
Properties: (a) Yellow-green crystals; (b) blue or emerald green crystals; (c) green crystals. All the sulfates are soluble in water; (b) and (c) are soluble in alcohol; (a) is insoluble in alcohol and ether. Sp. gr. (a) 3.68, (b) 2.031, (c) 1.98; m.p. (a) 840°C (loses SO_3); (b) and (c) lose $6H_2O$ at 103°C.
Derivation: Action of sulfuric acid on nickel.
Grades: Technical; C.P.; single crystals.
Containers: Bags; carloads.
Hazard: Tolerance, 1 mg per cubic meter of air.
Uses: Manufacture of nickel ammonium sulfate; nickel catalysts; nickel plating; mordant in dyeing and printing textiles; coatings; ceramics.

nickel tetracarbonyl. See nickel carbonyl.

nickel titanate. See "Solfast Titan Yellow."

nicol. An optical material (Iceland spar) that functions as a prism, separating light rays that pass through it into two portions, one of which is reflected away, and the other transmitted. The transmitted portion is called plane-polarized light. See also calcite; optical isomerism.

"Nicomo-12."[241] Trademark for a hydrodesulfurization catalyst of nickel, cobalt, and molybdenum on alumina.

"Nicon."[169] Brand name for diethyldithiocarbamate used in the colorimetric determination of nickel.

nicotinamide. See niacinamide.

nicotinamide adenine dinucleotide. Name recommended by the International Union of Biochemistry and IUPAC. (NAD; diphosphopyridine nucleotide; DPN; coenzyme I; Co I; codehydrogenase I). $C_6H_6N_2O \cdot C_5H_8O_3 \cdot PO_3 \cdot O \cdot HPO_3 \cdot C_5H_8O_3 \cdot C_5H_4N_5$. A coenzyme necessary for the alcoholic fermentation of glucose, and the oxidative dehydrogenation of other substrates. Isolated from yeast. The phosphate is also available.
Uses: Biochemical research; chromatography.

nicotine (beta-pyridyl-alpha-N-methylpyrrolidine) $C_{10}H_{14}N_2$ or $C_5H_4NC_4H_7NCH_3$.
Properties: Alkaloid from tobacco; thick water-white levorotatory oil, turning brown on exposure to air. Hygroscopic; soluble in alcohol, chloroform, ether, kerosine, water, and oils. B.p. 247°C (dec); sp. gr. 1.00924. Also in form of dust or powder. Autoignition temp. 471°F. Combustible.
Derivation: By distilling tobacco with milk of lime and extracting with ether.
Hazard: Highly toxic by ingestion, inhalation, and skin absorption. Tolerance, 0.5 mg per cubic meter of air.
Uses: Medicine; insecticide (horticultural); tanning. Use as insecticide may be restricted. Available as the dihydrochloride, salicylate, sulfate, and bitartrate.
Shipping regulations: (Rail, Air) Poison label.

nicotine salts.
(a) Hydrochloride: $C_{10}H_{14}N_2 \cdot HCl$.
(b) Salicylate: $C_{10}H_{14}N_2 \cdot C_7H_6O_3$.
(c) Sulfate: $(C_{10}H_{14}N_2)_2 \cdot H_2SO_4$.
(d) Tartrate: $C_{10}H_{14}N_2 \cdot 2C_4O_6H_6 \cdot H_2O$.
Properties: (a) Colorless oil; (b) White crystals. M.p. 117.5°C. (c) White crystals; (d) White plates. M.p. 89°C. All the salts are soluble in water, alcohol and ether.
Derivation: By the action of the respective acid on the alkaloid.
Hazard: All highly toxic.
Shipping regulations: (Rail, Air) Poison label.

nicotinic acid. See niacin.

nicotinic acid amide. See niacinamide.

nicotinyl alcohol (USAN) (3-pyridinemethanol) $C_5H_4NCH_2OH$.
Properties: Hygroscopic liquid; b.p. 144–145°C (16 mm); refractive index 1.5425 (n 20/D). Soluble in water and ether. Combustible.
Hazard: Toxic.
Use: Medicine.

nigrosine. A class of dark blue or black dyes, some soluble in water, some in alcohol and some in oil, used in manufacture of ink and shoe-polish and in dyeing leather, wood, textiles, etc. It is also used as shark-repellent.

nigre. The dark-colored layer, containing some soap as well as salts and impurities, formed in soap manufacture between the layers of soap proper and caustic solution.

"Nilite."[28] Trademark for a series of nitrocarbonitrate blasting agents.

"Nilofoam."[300] Trademark for a silicone defoaming agent.

"Nilox U."[79] Trademark for a disproportionated rosin.
Uses: Adhesives; greases; stitching waxes; rubber reclaiming; rubber compounding, manufacture of toners; wax paper coatings.

"Nilstain."[155] Trademark for a group of stainless steel alloys.

ninhydrin (triketohydrindene hydrate) $C_9H_4O_3 \cdot H_2O$.
Properties: White crystals or powder; m.p. 240–245°C; becomes red when heated above 100°C. Soluble in water and alcohol; slightly soluble in ether and chloroform.
Hazard: Toxic and irritant.
Uses: Chemical intermediate; reagent for determination of amines, amino acids, ascorbic acid.

niobe oil. See methyl benzoate.

niobic acid. Any hydrated form of Nb_2O_5. It forms as a white insoluble precipitate when a potassium hydrogen sulfate fusion of a niobium compound is leached with hot water or when niobium fluoride solutions are treated with ammonium hydroxide. Soluble in concentrated sulfuric acid, concentrated hydrochloric acid, hydrofluoric acid, and in bases. Important in analytical determination of niobium.

niobite. See columbite.

niobium Nb (also columbium, Cb). The name niobium is officially approved by chemical authorities, but columbium is still used, chiefly by metallurgists. Metallic element, atomic number 41; Group VB of the Periodic Table. Atomic weight 92.9064; valences 2, 3, 4, 5. No stable isotopes.

Properties: Gray or silvery, ductile metal; does not tarnish or oxidize at room temperature; sp. gr. 8.57; m.p. 2468°C. Reacts with oxygen and halogens only when heated. Less corrosion-resistant than tantalum at high temperature. Not attacked by nitric acid up to 100°C, but vigorously attacked by mixture of nitric and hydrofluoric acids. Hot, conc. hydrochloric, sulfuric, and phosphoric acids attack it, but hot, conc. nitric does not. Unaffected at room temperature by most acids and by aqua regia. It is attacked by alkaline solutions to some extent at all temperatures.

Occurrence: Found in two major ores, columbite (q.v.) and pyrochlore (a carbonate-silicate rock). Chief sources are Brazil, Nigeria, Canada.

Derivation: Niobium is so closely associated with tantalum that they msut be separated by fractional crystallization or by solvent extraction, with subsequent purification.

Grades: Plates; rods; powder; single crystals.

Uses: Alloy steels; getter in vacuum tubes; nuclear reactors; cermets; missiles and rockets; cryogenic equipment.

niobium carbide NbC.

Properties: Lavender-gray powder; insoluble in water and in all acids except a mixture of nitric and hydrofluoric. M.p. about 3500°C; hardness 2400 kg/sq mm; sp. gr. 7.82.

Derivation: By direct combination of niobium with carbon or by the reduction of niobium oxide with lampblack.

Uses: Cemented carbide tipped tools; special steels; preparation of niobium metal.

niobium chloride (niobium pentachloride) $NbCl_5$.

Properties: Yellow crystalline solid; soluble in alcohol, ether, carbon tetrachloride, hydrochloric acid, conc. sulfuric acid. M.p. 205°C; b.p. 254°C; sp. gr. 2.75. Deliquescent; decomposes in moist air with evolution of hydrogen chloride fumes. Available in commercial quantities.

Derivation: Direct combination of niobium and chlorine; chlorination of niobium oxide in the presence of carbon.

Hazard: May evolve toxic fumes of HCl. Keep dry.

Uses: Preparation of pure niobium; intermediate.

niobium diselenide $NbSe_2$.

Properties: Gray-black solid; m.p. above 1316°C; vacuum-stable from −430 to 2400°F. Has higher electrical conductivity than graphite.

Uses: Lubricant and conductor at high temperatures and high vacuum.

niobium oxalate $NbO(HC_2O_4)_3 \cdot 4H_2O$. White crystals; 99.99% pure. Very soluble in water. Used as an intermediate and in special catalysts.

niobium oxide (niobium pentoxide) Nb_2O_5.

Properties: White powder; insoluble in acids except hydrofluoric; soluble in fused potassium hydrogen sulfate, or carbonates or hydroxides of the alkali metals. Sp. gr. 4.5–5.0; m.p. 1460°C.

Derivation: Strong ignition of niobic acid.

Uses: Intermediate; electronics.

niobium pentachloride. See niobium chloride.

niobium pentoxide. See niobium oxide.

niobium potassium oxyfluoride (potassium niobium oxyfluoride; potassium oxyfluoniobate) $K_2NbOF_5 \cdot H_2O$.

Properties: White lustrous leaflets. Greasy to touch. Soluble in water.

Hazard: Toxic by ingestion; strong irritant. Tolerance (as F), 2.5 mg per cubic meter of air.

Uses: Separation of niobium from tantalum; electrolytic preparation of niobium metal.

niobium silicide $NbSi_2$. Crystalline solid; m.p. 1950°C. Used as a refractory material.

niobium-tin Nb_3Sn. Alloy used for special wire for super-conducting magnets to obtain high magnetic fields for use in communication, and containment of thermonuclear fusion plasmas.

niobium-titanium. Alloy used for magnetic devices of fields up to 100,000 gauss.

niobium-uranium. Niobium alloyed with 20% uranium yields a nuclear fuel which maintains tensile strength and hardness at 1600°F.

NIOSH. Abbreviation for National Institute for Occupational Safety and Health. It establishes work exposure standards in toxic environments. It is located in Rockville, Md.

"Niox."[28] Trademark for U.S.P. grade of ammonium nitrate. Used in manufacture of nitrous oxide gas.

"Ni-Plex."[288] Trademark for a process for and materials used in the stripping of nickel from all basic metals without attack on the basic metals.

"Nirez."[79] Trademark for a group of resins including polyterpenes, terpene phenols, zinc resinates, high melting limed resins, and medium-colored, high-melting metal resinates.

Uses: Adhesives, rubber compounding, paper coatings, textile coatings, printing inks, paint and varnish.

"Niomet" 46.[155] Trademark for a magnetic alloy composed of nickel, 45%; iron, 55%.

Properties: Density, 8.17 gm/cm^3; tensile strength, 70,000 psi. Used in cores and armatures for relays, motors, inductors, and transformers.

"Niron" 52.[155] Trademark for a magnetic alloy composed of nickel, 50% and iron, 50%.

Properties: Density 8.46 gm/cm^3; tensile strength, 70,000 psi. Used in leads and armatures for glass-sealed relays, cores for magnetic amplifiers.

niter (nitre; saltpeter) KNO_3. A natural potassium nitrate (q.v.).

niter cake. See sodium bisulfate.

niter, Chile. See sodium nitrate.

nithiazide $C_6H_8N_4O_3S$. 1-Ethyl-3-(5-nitro-2-thiazolyl)-urea.

Properties: Crystals; practically insoluble in water; decomposes 228°C.

Use: Veterinary medicine.

nitralloy. See nitriding.

"Nitramex."[28] Trademark for a blasting agent which is not detonated by impacts from rifle bullets, sledge hammers or heat, but rather by specially constructed primers. Used for blasting in extremely hard shooting quarries.

nitralin. An herbicide used largely to control weeds in cotton and soybeans. Research indicates possible use in plant breeding; treatment of corn roots induces abnormal chromosome formation and cell wall deterioration.

nitramine. See tetryl.

"Nitramite."[28] Trademark for a series of nitrocarbonitrate blasting agents.

"Nitrana."[197] Trademark for any of several nitrogen fertilizer solutions containing water, ammonia, and ammonium nitrate. It is used in manufacturing solid fertilizer and for direct application to the soil.

nitranilic acid (2,5-dihydroxy-3,6-dinitroquinone) $C_6O_2(NO_2)_2(OH)_2$.

Properties: Flat yellow crystals; loses water at 100°C; decomposes explosively at 170°C; soluble in water and alcohol; insoluble in ether.

Hazard: Explosion risk when heated; evolves highly toxic fumes on decomposition.

nitraniline. See nitroaniline.

nitrating acid. Legal label name for mixed acid (q.v.).

nitration. A reaction in which a nitro group ($-NO_2$) replaces a hydrogen on a carbon atom by the use of nitric acid or mixed acid. An example is the nitration of cellulose to nitrocellulose. It is widely used in aromatic reactions to form such compounds as nitrobenzene, trinitrotoluene, nitroglycerin and other explosives. Aromatic nitrations are usually effected with mixed acid, a mixture of nitric and sulfuric acids, at 0 to 120°C. Aliphatic nitration is less common than aromatic, but propane can be nitrated under pressure to yield nitroparaffins.

"Nitrelmang."[250] Trademark for a high purity nitrided manganese for use in the production of free-machining steels; high nitrogen tinplate; and special high temperature alloy steels.

"Nitrex."[248] Trademark for synthetic rubber latices of the butadiene-acrylonitrile type; used for paper saturation, leather finishing, and as a plasticizer for resin latices.

nitric acid (aqua fortis; engraver's acid; azotic acid) HNO_3. Eleventh highest-volume chemical produced in U.S. (1975).

Properties: Transparent, colorless or yellowish, fuming, suffocating, caustic and corrosive liquid. Will attack almost all metals. The yellow color is due to release of nitrogen dioxide on exposure to light. Strong oxidizing agent. Miscible with water; decomposes in alcohol. B.p. (decomposes) 86°C; f.p. −41.59°C; sp. gr. 1.504 (25/4°C); vapor pressure 62 mm (25°C); refractive index 1.3970 (n 24/D); viscosity 0.761 cp (25°C).

Derivation: (1) Oxidation of ammonia by air or oxygen with platinum catalyst. (*Note:* A pelleted catalyst not containing platinum or other noble metals is available.) Air oxidation yields 60% acid. Concentration is achieved by (a) distillation with sulfuric acid, (b) extractive distillation with magnesium nitrate, or (c) by neutralizing the weak acid with soda ash, evaporating to dryness, and treating with sulfuric acid. Method (c) yields synthetic niter cake ($NaHSO_4$) as a byproduct. (2) High-pressure oxidation of nitrogen tetroxide (yields 98% acid).

Strength of solutions: 36, 38, 40, 42°Bé; 58–63.5%; 95%.

Containers: Bottles; carboys; tank cars and trucks.

Hazard: Highly toxic by inhalation; corrosive to skin and mucous membranes; strong oxidizing agent. Dangerous fire risk in contact with organic materials. Tolerance, 2 ppm in air. Safety data sheet available from Manufacturing Chemists Assn., Washington, D.C.

Uses: Manufacture of ammonium nitrate for fertilizer and explosives; organic synthesis (dyes, drugs, explosives, cellulose nitrate, nitrate salts); metallurgy, photoengraving; etching steel; ore flotation; medicine; reagent.

Shipping regulations: (Rail) White label. (Air) Oxidizer label *and* Corrosive label. Not acceptable on passenger planes.

nitric acid, fuming.

(1) White fuming nitric acid (WFNA) contains more than 97.5% nitric acid, less than 2% water, and less than 0.5% oxides of nitrogen. It is a colorless or pale yellow liquid which fumes strongly. It is decomposed by light or elevated temperatures, becoming red in color from nitrogen dioxide.

(2) Red fuming nitric acid (RFNA) contains more than 86% nitric acid, approximately 6–15% oxides of nitrogen (as nitrogen dioxide) and less than 5% water.

Derivation: From dilute nitric acid, nitrogen dioxide, and oxygen.

Containers: Stainless steel or aluminum drums.

Hazard: Highly toxic by inhalation; corrosive to skin and mucous membranes. Strong oxidizing agent; may explode in contact with strong reducing agents. Dangerous fire risk in contact with organic materials.

Uses: Preparation of nitro-compounds; rocket fuels; laboratory reagent.

Shipping regulations: See nitric acid.

nitric oxide NO. See also nitrogen dioxide.

Properties: Colorless gas (readily reacts with oxygen at room temperature to form nitrogen dioxide, NO_2, a reddish brown gas). B.p. −152°C; f.p. −164°C; sp. gr. at b.p. 1.27; slightly soluble in water. Noncombustible.

Derivation: Oxidation of ammonia above 500°C; decomposition of nitrous acid (aqueous solution). Also from atmospheric oxygen and nitrogen in the electric-arc process for fixation of nitrogen.

Grade: Pure (99%).

Containers: Cylinders.

Hazard: Highly toxic by inhalation; strong irritant to skin and mucous membranes; supports combustion. Tolerance, 25 ppm in air.

Uses: To prepare nitrosyl carbonyls. Very dry, high purity gas is proposed for filling incandescent lamps and as in inert gas in electrical systems.

Shipping regulations: (Rail) Poison Gas label. Not accepted by express. (Air) Not acceptable.

nitriding. A process of case hardening in which a ferrous alloy, usually of special composition, is heated in an atmosphere of ammonia or in contact with nitrogenous material to produce surface hardening by absorption of the nitrogen without quenching. The alloys used for nitriding are known as nitroalloys. Several types are available with ranges of composition as follows: aluminum 0.85–11.2%; carbon 0.20–0.45%; chromium none to 1.8%; molybdenum 0.15–1.00%; manganese 0.4–0.7%; silicon 0.2–0.4%.

nitrile. An organic compound containing the —CN grouping; for example, acrylonitrile $CH_2{:}CHCN$; it may also contain a triple bond as in acetonitrile ($CH_3C{\equiv}N$).

Hazard: Some organic cyanide compounds are toxic and flammable (acetonitrile, acrylonitrile).

nitrile rubber (acrylonitrile rubber; acrylonitrile-butadiene rubber; nitrile-butadiene rubber; NBR). A synthetic rubber made by random polymerization of

acrylonitrile with butadiene by free radical catalysis. Alternating copolymers using Natta-Ziegler catalyst have been developed. About 20% of the total is used as latex; also available in powder form. Its repeating structure may be represented as
—$CH_2CH{=}CHCH_2CH_2CH(CN)$—.

Typical properties: (Medium acrylonitrile): Sp. gr. (polymer) 0.98; tensile strength (psi) 1000 to 3000; elongation (%) 100 to 700; maximum service temperature (°F) 250 to 300. Combustible.

Uses: (High acrylonitrile): Oil well parts; fuel tank liners; fuel hose; gaskets; packing oils seals; hydraulic equipment.

(Medium acrylonitrile): General-purpose oil-resistant applications; shoe soles; kitchen mats, sink topping, and printing rolls.

(Low acrylonitrile): Gaskets, grommets, and O-rings (flexible at a very low temperature); adhesives. Binder-fuel in solid rocket propellants.

nitrile-silicone rubber (NSR). Combines the characteristic properties of silicones with the oil resistance of nitrile rubber. Resistant to jet fuels, solvents and hot oils.

nitrilotriacetic acid (NTA; triglycine; TGA; triglycollamic acid) $N(CH_2COOH)_3$.

Properties: White crystalline powder; m.p. 240°C (with decomposition); insoluble in water and most organic solvents; forms mono-, di-, and tribasic salts which are water-soluble. Combustible; 70% biodegradable.

Hazard: A confirmed carcinogen in rats and mice.

Uses: Synthesis; chelating agent; eluting agent in purification of rare-earth elements.

Also available as the di- and trisodium salts.

Note: This substance and its sodium salt, sodium nitrilotriacetate, have been recommended as substitutes for phosphates in detergent builders, but their use for this purpose has not been officially approved because of possible toxic effects. A scientific study published in 1973 indicated that the material is sufficiently biodegradable, even in combination with toxic metals, to preclude environmental damage.

nitrilotriacetonitrile (NTAN) $N(CH_2CH)_3$. White crystalline solid; m.p. 130–134°C; insoluble in water; soluble in acetone. Used as an intermediate and chelating agent.

Hazard: May be toxic. See nitrile.

para-**nitroacetanilide** $NO_2C_6H_4NHCOCH_3$.

Properties: White crystals; soluble in alcohol and ether; very slightly soluble in cold water. Soluble in hot water, in potassium hydroxide solution. M.p. 214–216°C. Combustible.

Derivation: By acetylating aniline, then nitrating.

Use: Manufacture of para-nitraniline.

para-**nitro**-ortho-**aminophenol** $C_6H_3OHNH_2NO_2$.

Properties: Yellow-brown leaflets containing water of crystallization melting at 80 to 90°C; anhydrous melts at 154°C. Soluble in acid.

Derivation: From dinitrophenol.

Hazard: Probably toxic.

Use: Dyes.

meta-**nitroaniline** (meta-nitraniline) $NO_2C_6H_4NH_2$.

Properties: Yellow needles; sp. gr. 1.43; m.p. 111.8°C; b.p. 306°C; soluble in alcohol and ether; slightly soluble in water.

Derivation: From aniline by nitration after acetylation, with subsequent removal of the acetyl group by hydrolysis.

Hazard: Highly toxic; absorbed by skin. Moderate fire risk; Safety data sheet available from Manufacturing Chemists Assn., Washington, D.C.

Uses: Color test for pine wood; dye intermediate.

Shipping regulations: (Air) Poison label.

ortho-**nitroaniline** (ortho-nitraniline) $NO_2C_6H_4NH_2$.

Properties: Orange-red needles; sp. gr. 1.443; m.p. 69.7°C. Soluble in alcohol and ether; slightly soluble in water. Not light-fast. Flash point 335°F. Autoignition temp. 970°F.

Derivation: From aniline by nitration after acetylation, with subsequent removal of the acetyl group by hydrolysis.

Containers: Drums; kegs.

Hazard: Highly toxic; absorbed by skin. Explosion risk. Safety data sheet available from Manufacturing Chemists Assn., Washington, D.C.

Uses: Dye intermediate; synthesis of photographic antifogging agent, ortho-phenylenediamine, coccidiostats; interior paint pigment.

Shipping regulations: (Rail, Air) Poison label. (Rail) Legal label name, under Ortho-.

para-**nitroaniline** (para-nitraniline) $NO_2C_6H_4NH_2$.

Properties: Yellow needles; sp. gr. 1.437; m.p. 148°C; flash point 390°F. Combustible. Soluble in alcohol and ether; insoluble in water.

Derivation: (a) From p-chloronitrobenzene; (b) from aniline by nitration after acetylation.

Containers: Drums; carloads.

Hazard: Highly toxic, absorbed by skin. Tolerance, 1 ppm in air. Explosion risk. Safety data sheet available from Manufacturing Chemists Assn., Washington, D.C.

Uses: Dye intermediate, especially para-nitraniline red; intermediate for antioxidants, gasoline gum inhibitors; medicinals for poultry; corrosion inhibitor.

Shipping regulations: (Rail, Air) Poison label. (Rail) Legal label name under Para-.

ortho-**nitroanisole** $C_6H_4OCH_3NO_2$.

Properties: Light reddish or amber liquid. Soluble in alcohol and ether; insoluble in water; sp. gr. 1.255 (20/20°C); crystallizing point 9.6°C; boiling range 268–271°C; refractive index 1.5602 (n 20/D). Combustible.

Derivation: From ortho-nitrophenol by methylation or from ortho-nitrochlorobenzene by action of methanol and caustic soda.

Grades: Technical.

Containers: Tank cars.

Uses: Organic synthesis; manufacture of intermediates for dyes and pharmaceuticals.

para-**nitroanisole** (1-methoxy-4-nitrobenzene) $NO_2C_6H_4OCH_3$.

Properties: Colored monoclinic crystals; m.p. 54°C; b.p. 260°C. Insoluble in water; soluble in alcohol, ether. Combustible.

Grades: Technical.

Containers: Drums.

Use: Intermediate.

5-nitrobarbituric acid (dilituric acid) $O_2NHCCONHCONHCO$.

Properties: Prisms and leaflets from water; m.p. 176°C (dec); slightly soluble in water; soluble in al-

cohol and sodium hydroxide solution; insoluble in ether.
Use: Microreagent for potassium.

3-nitrobenzaldehyde (meta-nitrobenzaldehyde) $NO_2C_6H_4CHO$.
Properties: Yellowish, crystalline powder; m.p. 58°C; b.p. (23 mm) 164°C; almost insoluble in water; soluble in alcohol, chloroform, ether.
Uses: Synthesis of dyes, pharmaceuticals, surface active agents; vapor phase corrosion inhibitor; antioxidant for chlorophyll; mosquito repellent.

nitrobenzene (oil of mirbane) $C_6H_5NO_2$.
Properties: Greenish-yellow crystals or yellow, oily liquid. Soluble in alcohol, benzene, and ether; slightly soluble in water. Sp. gr. 1.19867; m.p. 5.70°C; b.p. 210.85°C. Flash point 190°F. Combustible. Autoignition temp. 900°F.
Derivation: From benzene by nitrating with nitric acid-sulfuric acid mixture.
Method of purification: By washing and distilling with steam, then redistilling.
Grades: Technical; redistilled; 97%.
Containers: Glass bottles; drums; tank cars.
Hazard: Highly toxic by ingestion, inhalation, and skin absorption. Tolerance, 1 ppm in air. Safety data sheet available from Manufacturing Chemists Assn., Washington, D.C.
Uses: Manufacture of aniline; solvent for cellulose ethers; modifying esterification of cellulose acetate; ingredient of metal polishes and shoe polishes; manufacture of benzidine, quinoline, azobenzene, etc.
Shipping regulations: (Rail, Air) Poison label. (Rail) Legal label name, nitrobenzol.

para-**nitrobenzeneazoresorcinol.** $NO_2C_6H_4N_2C_6H_3(OH)_2$.
Properties: Red crystals. Slightly soluble in water; soluble in nitrobenzene. M.p. 198°C. Combustible.
Derivation: Diazotized para-nitroaniline is coupled with resorcinol.
Grade: Analytical.
Hazard: Probably toxic.
Use: Determination of magnesium.

meta-**nitrobenzenesulfonic acid** (3-nitrobenzenesulfonic acid) $C_6H_4NO_2SO_3H$. Crystals; m.p. 70°C. Soluble in water and alcohol. Used in organic synthesis. The sodium salt is a protective antireduction agent.

6-nitrobenzimidazole $O_2NC_6H_4NHCH{:}N$ (ring closure between C_6H_4 and N).
Properties: Solid; m.p. 203°C. Very soluble in alcohol; slightly soluble in water, ether, and benzene.
Use: Antifogging agent in photographic developers. See photographic chemistry.

nitrobenzoic acid $C_6H_4(NO_2)COOH$. (a) meta-, (b) ortho-, (c) para- (nitrodracylic acid).
Properties: Yellowish-white crystals. (a) Soluble in alcohol and ether; slightly soluble in water; (b) Soluble in water, alcohol and ether; (c) Soluble in alcohol; sparingly soluble in water. 9a) Sp. gr. 1.494; m.p. 140–141°C; (b) Sp. gr. 1.575; m.p. 147.7°C; (c) Sp. gr. 1.5497; m.p. 238°C. Combustible; low to moderate toxicity.
Derivation: (a and b) Nitration of benzoic acid; (c) Oxidation of para-nitrotoluene by hot chromic acid mixture.
Containers: Fiber drums; car loads.
Uses: (a and b) Organic synthesis; (c) preparations of anesthetics and as intermediate in the manufacture of dyes and sun-screening agents.

nitrobenzol. Legal label name (Rail) for nitrobenzene.

meta-**nitrobenzotrifluoride** (3-nitrobenzotrifluoride; meta-nitrotrifluoromethylbenzene) $NO_2C_6H_4CF_3$.
Properties: Pale straw, thin oily liquid, aromatic odor; distillation range 200.5–208.5°C; f.p. −5.0°C; sp. gr. (15.5°C) 1.437; b.p. 203°C; flash point (open cup) 214°F; wt/gal 11.98 lb (15.5°C); viscosity 2.35 cp (100°F); soluble in organic solvents; insoluble in water. Combustible.
Hazard: Toxic. Tolerance (as F), 2.5 mg per cubic meter of air.

meta-**nitrobenzoyl chloride** $NO_2C_6H_4COCl$.
Properties: Yellow to brown liquid; partially crystallized at room temperature; m.p. 34°C (approx); b.p. 278°C; soluble in ether; decomposes in water and alcohol. Combustible.
Uses: Manufacture of dyes for fabrics and color photography; intermediate in preparation of pharmaceuticals.

para-**nitrobenzoyl chloride** $NO_2C_6H_4COCl$.
Properties: Yellow crystalline solid; m.p. 72°C; b.p. 154°C (15 mm); decomposes in water and alcohol; soluble in ether. Combustible.
Uses: Intermediate for procaine hydrochloride; dyestuffs.

para-**nitrobenzyl cyanide** (para-nitro-alpha-tolunitrile) $NO_2C_6H_4CH_2CN$.
Properties: Crystals; m.p. 116–118°C. Insoluble in water; soluble in alcohol and ether.
Derivation: Action of concentrated nitric acid on benzyl cyanide.
Uses: Intermediate for dyestuffs and pharmaceuticals; preparation of para-nitrophenylacetic acid.

ortho-**nitrobiphenyl** (ONB; 4-nitrodiphenyl) $C_6H_5C_6H_4NO_2$.
Properties: Light-yellow to reddish solid or liquid; sp. gr. 1.203 (25/25°C); 10 lb/gal; crystallizing pt. 34.5°C (min); refractive index 1.613 approx (25°C); b.p. 330°C approx. Flash point 290°F; autoignition temp. 356°F. Soluble in carbon tetrachloride, mineral spirits, pine oil, turpentine, benzene, acetone, glacial acetic acid, and perchloroethylene. Combustible. Insoluble in water.
Derivation: By controlled nitration of biphenyl.
Hazard: Toxic by ingestion. Carcinogenic.
Uses: Organic research.

nitrobromoform. See bromopicrin.

"Nitro BT."[64] Trademark for tetrazolium salt (q.v.).

2-nitro-1-butanol $CH_3CH_2CHNO_2CH_2OH$.
Properties: Colorless liquid; solubility in water 20 g/100 cc (20°C); sp. gr. 1.133 (20/20°C); b.p. 105°C (10 mm); f.p. −48 to −47°C; wt/gal 9.44 lb (20°C); refractive index 1.4390 (20°C); pH of 0.1 M solution 4.51. Combustible.
Containers: 5- and 55-gal drums; 1-gal cans.
Use: Organic synthesis.

nitro carbo nitrate. A blasting agent consisting of ammonium nitrate sensitized with diesel oil. Will burn with explosive violence.
Hazard: Dangerous explosion risk. Strong oxidizing agent.
Shipping regulations: (Rail) Yellow label. (Air) Oxidizer label.

nitrocellulose (cellulose nitrate; nitrocotton; guncotton; pyroxylin). Formula approximately $C_6H_7O_2(ONO_2)_3$.
Properties: Pulpy, cotton-like, amorphous solid (dry); colorless liquid to semisolid (solution). Contains from 10 to 14% nitrogen. High-nitrogen form (ex-

plosives) is soluble in acetone, insoluble in ether-alcohol mixtures. Low-nitrogen form (pyroxylin) is soluble in ether-alcohol mixtures and acetone (collodion and lacquers). Sp. gr. 1.66; flash point 55°F; autoignition point 338°F. Low toxicity.
Derivation: Treatment of cellulose (as cotton linters, wood pulp) with mixtures of nitric and sulfuric acids. By varying strength of acids, temperature and time of reaction, and acid/cellulose ratio, widely different products are obtained.
Forms: Colloided, block; colloided, flake or granular; flakes; powder; solutions of several viscosities (from 1/4 to 1000 seconds). May be dry or wet with alcohol or water.
Containers: Barrels; carlots.
Hazard: Highly flammable; dangerous fire and explosion risk. Somewhat less flammable when wet.
Uses: Fast-drying automobile lacquers; high explosives; collodion; rocket propellant; medicine; printing ink base; flashless propellant powder; coating book-binding cloth; leather finishing.
Shipping regulations: Consult regulations. Explosive types not acceptable for air freight.

nitrocellulose laquer. See lacquer.

nitrochlorobenzene. Legal label name (Air) for chloronitrobenzene.

nitrochloroform. See chloropicrin.

para-**nitro**-ortho-**chlorophenyl dimethyl thionophosphate.** See dicapthon.

nitrocobalamin $C_{62}H_{90}N_{14}O_{16}PCo$. One of the active forms of vitamin B_{12} (q.v.) in which a nitro group is attached to the central cobalt atom.

"Nitrocols."[223] Trademark for products consisting of pigments (carbon black or titanium dioxide) dispersed in various proportions of nitrocellulose and dibutyl phthalate. Used in high-gloss white, industrial-type, and high-grade jet black lacquer finishes. Available in two forms, chip and paste.

nitrocotton. See nitrocellulose.

2-nitro-para-**cresol** (4-methyl-2-nitrophenol) $NO_2(CH_3)C_6H_3OH$.
Properties: Yellow crystals; density 1.24 g/ml (38/4°C); m.p. about 35°C; b.p. 234°C; slightly soluble in water; soluble in alcohol, ether. Combustible.
Containers: Steel drums.
Hazard: Toxic by ingestion, inhalation, and skin absorption.
Use: Intermediate.

nitrodichloro derivative. See the corresponding dichloronitro derivative.

ortho-**nitrodiphenyl.** See ortho-nitrobiphenyl.

ortho-**nitrodiphenylamine** $C_6H_5NHC_6H_4NO_2$.
Properties: Red-brown crystalline powder; m.p. 75–76°C.
Use: Intermediate.

nitrodracylic acid. See para-nitrobenzoic acid.

nitro dye. A dye whose molecules contain the NO_2 chromophore group.

nitroethane $CH_3CH_2NO_2$. A nitroparaffin.
Properties: Colorless liquid. Solubility in water 4.5 cc/100 cc (20°C); solubility of water in nitroethane 0.9 cc/100 cc (20°C). Sp. gr. 1.052 (20/20°C); freezing point −90°C; b.p. 114°C; vapor pressure (15.6 mm) (20°C); flash point 82°F; autoignition temp. 779°F; wt/gal 8.75 lb (68°F); refractive index 1.3917 (20°C).
Derivation: By reaction of propane with nitric acid under pressure.
Containers: 5- and 55-gal drums; tank cars.
Hazard: Flammable, moderate fire risk. Moderately toxic. Tolerance, 100 ppm in air.
Uses: Propellant; solvent for nitrocellulose, cellulose acetate, cellulose acetopropionate, cellulose acetobutyrate, vinyl, alkyd, and many other resins, waxes, fats and dyestuffs; chemical synthesis.

2-nitro-2-ethyl-1,3-propanediol $CH_2OHC(C_2H_5)NO_2CH_2OH$.
Properties: White, crystalline solid; m.p. 56–75°C; b.p. decomposes (10 mm); pH 0.1 *M* aqueous solution 5.48; soluble in organic solvents; very soluble in water.
Containers: Fiberpak boxes; drums.
Use: Organic synthesis.

"Nitroform."[266] Trademark for a long-lasting, slow-release, high-nitrogen fertilizer. Nitrogen (total 38+%) released from urea-formaldehyde polymers by soil bacteria. Used for fertilizer manufacture and direct application.

nitrofuran. Any of several synthetic antibacterial drugs used to treat mammary gland infections in cows and to inhibit disease in swine, chickens, etc. Among them are nitrofurazone, furazolidone, nihydrazone, and furaltadone. All of the latter have been found to cause cancer in laboratory animals, and their use has been discontinued. See furazolidone.

nitrofurantoin $C_8H_6N_4O_5$. N-(5-nitro-2-furfurylidene)-1-aminohydantoin.
Properties: Yellow, bitter powder with slight odor. M.p. (dec) 270–272°C. Very slightly soluble in alcohol and practically insoluble in ether and water.
Grade: U.S.P.
Use: Medicine (antibacterial agent).

N-**(5-nitro-2-furfurylidene-3-amino)-2-oxazolidinone.** See furazolidone.

"Nitrogation."[125] Trademark for anhydrous ammonia specifically intended for direct injection into the irrigation stream for purpose of soil fertilization.

nitrogen N. Gaseous element of atomic number 7, of group VA of the periodic system. Atomic weight 14.0067; valences 1, 2, 3, 4, 5. There are two stable and 4 radioactive isotopes. The molecular formula is N_2. Eighth highest-volume chemical produced in U.S. (1975).
Properties: Colorless, odorless, tasteless diatomic gas constituting about four-fifths of the air; colorless liquid, chemically unreactive; sp. gr. (gas) 0.96737, referred to air; (liquid) 0.804; (solid) 1.0265; f.p. −210°C; b.p. −195.5°C. Slightly soluble in water; slightly soluble in alcohol. Noncombustible; an asphyxiant gas.
Derivation: From liquid air by fractional distillation.
Impurities: Argon and other "rare gases"; oxygen.
Grades: U.S.P.; prepurified 99.966% min; extra dry 99.7% min; water pumped 99.6% min.
Containers: Steel cylinders; pipelines.
Uses: Production of ammonia, acrylonitrile, cyanamide, cyanides, nitrides; inert gas for purging, blanketing, and exerting pressure; electric and electronic industries; in-transit food refrigeration and

freeze drying, pressurizing liquid propellants; quick-freezing foods; chilling in aluminum foundries; bright annealing of steel; cryogenic preservation; inert pressuring and blanketing gas in missiles; food antioxidant; source of pressure in oil wells; inflating tires.
Shipping regulations: (Gas) (Rail) Green label. (Air) Nonflammable Gas label; (liquid, pressurized) Nonflammable Gas label. Not accepted on passenger planes; (liquid nonpressurized), No label required; (liquid, low-pressure). No label required.

nitrogen 15. A stable isotope with an atomic mass of 15.00011 which is present in naturally occurring nitrogen to the extent of 0.37%. Many nitrogen 15 compounds are commercially available.

nitrogenase. Enzyme which fixes nitrogen and can be isolated from soil bacteria. It is possible to synthesize ammonia from nitrogen and hydrogen without high temperatures and pressures by means of nitrogenase. Pyruvic acid is an adjunct of the reaction.

nitrogen chloride. See nitrogen trichloride.

nitrogen dioxide NO_2
Properties: A red-brown gas or yellow liquid; becomes colorless solid at −11.2°C, which exists in varying equilibrium with other oxides of nitrogen as the temperature is varied. A component of automotive exhaust fumes. M.p. (liquid) −9.3°C; b.p. (gas) 21°C. Noncombustible.
Derivation: By oxidation of nitric oxide, an intermediate stage in the oxidation of ammonia to nitric acid.
Containers: 125-, 150-, 2000-lb cylinders; tank cars.
Grades: Pure, 99.5% min.
Hazard: Highly toxic; inhalation may be fatal. Tolerance, 5 ppm in air. Can react strongly with reducing materials.
Uses: Production of nitric acid; nitrating agent; oxidizing agent; catalyst; oxidizer for rocket fuels; polymerization inhibitor for acrylates.
Shipping regulations: (Rail) (liquid) Poison Gas label. Not accepted by express. (Air) Not acceptable.

nitrogen fixation. Utilization of atmospheric nitrogen to form chemical compounds. In nature, this function is performed by bacteria located on the root hairs of plants; as a result, plants are able to synthesize proteins. Industrial nitrogen fixation is exemplified by the synthesis of ammonia (q.v.) from a gaseous mixture containing carbon monoxide, hydrogen, and nitrogen. See also synthesis gas. Nitrogen can also be fixed by the electric arc process and the cyanamide process.

nitrogen monoxide. See nitrous oxide.

nitrogen mustard. A class of compounds with fishy odor and lachrymatory properties. They are named from their similarity in structure to mustard gas (dichlorodiethyl sulfide). The sulfur of the mustard gas is replaced by an amino nitrogen. Typical nitrogen mustards are halogenated alkylamines, such as methyl bis(2-chloroethyl)amine, $(ClCH_2CH_2)_2NCH_3$ (See mechloroethamine hydrochloride). Other examples are triethylene melamine, and triethylene thiophosphoramide (q.v.), and triethylene phosphoramide.
Hazard: Highly toxic; strong irritant to sissues; lachrymatory.
Use: Medicine; military poison gas.
Shipping regulations: See specific compound.

nitrogen oxides. The nitrogen oxides provide an example of the Law of Multiple Proportions and the +1 to +5 oxidation states of nitrogen. They are as follows: N_2O (nitrous oxide); NO (nitric oxide); N_2O_3 (nitrogen trioxide or nitrogen sesquioxide); N_2O_4 (dinitrogen tetroxide or nitrogen peroxide); NO_2 (nitrogen dioxide); N_2O_5 (dinitrogen pentoxide); N_3O_4 (trinitrogen tetroxide); and NO_3, which is unstable. These oxides, especially NO_2, are a factor in air pollution from automotive emissions.

nitrogen solutions. Aqueous solutions of ammonium nitrate, ammonia, and/or urea. They are graded according to total nitrogen content and composition.
Containers: Tank cars and trucks.
Uses: Direct application to soil as fertilizer; neutralizing superphosphate fertilizers.

nitrogen trichloride (nitrogen chloride) NCl_3.
Properties: Yellow oil or rhombic crystals; sp. gr. 1.653; b.p. less than 71°C; f.p. less than −40°C; insoluble in cold water, decomposes in hot water; soluble in chloroform, phosphorus trichloride, and carbon disulfide.
Hazard: May explode when heated to about 200°F or when exposed to direct sunlight. Moderately toxic by ingestion and inhalation. Strong irritant.
Shipping regulations: Consult authorities.

nitrogen triiodide NI_3.
Properties: Black crystals.
Hazard: Explodes at slightest touch when dry; when handled it should be kept wet with ether. Too sensitive to be used as explosive as it cannot be stored, handled, or transported safely.

nitrogen trifluoride NF_3.
Properties: Colorless, stable gas having a moldy odor; sp. gr. 1.537 (−129°C); gas density 0.1864 lb/cu ft (70°F); f.p. −206.6°C; b.p. −128.8°C. Very slightly soluble in water. Nonflammable.
Containers: Steel cylinders.
Hazard: Toxic and corrosive to tissue. Tolerance, 10 ppm in air. Severe explosion hazard!
Uses: Oxidizer for high-energy fuels; chemical synthesis.
Shipping regulations: (Rail) Not listed. (Air) Nonflammable Gas label. Not acceptable on passenger planes.

nitrogen trioxide (dinitrogen trioxide; nitrogen sesquioxide) N_2O_3.
Properties: Blue liquid; sp. gr. 1.447 (2°C); b.p. 3.5°C (1 atm); f.p. −102°C (1 atm). Partially dissociates into nitric oxide (NO) and nitrogen dioxide (NO_2).
Derivation: Prepared by passing nitric oxide into nitrogen tetroxide at about 0°C, until the stoichiometric amount has been absorbed.
Hazard: Highly toxic by inhalation; strong irritant.
Uses: Oxidant in special fuel systms; identification of terpenes; preparation of pure alkali nitrites.
Shipping regulations: (Rail) Poisonous liquids, n.o.s., Poison label. (Air) Not acceptable.

nitroglycerin (glyceryl trinitrate; trinitroglycerin) $CH_2NO_3CHNO_3CH_2NO_3$.
Properties: Pale yellow, viscous liquid. Soluble in alcohol and ether; slightly soluble in water. Sp. gr. 1.6009; freezing point 13.1°C; explosion point 260°C. Autoignition temp. 518°F.
Derivation: By dropping glycerol through cooled, mixed acid and stirring, followed by repeated washing with water.
Grade: Technical.
Hazard: Severe explosion risk; highly sensitive to shock and heat. Highly toxic by ingestion, inhala-

tion and skin absorption. Tolerance 0.2 ppm in air. Flammable, dangerous fire risk.
Uses: High explosive; production of dynamite and other explosives; medicine; combating fires in oil wells; rocket propellants.
Shipping regulations: (Rail) Consult regulations. Not accepted by express. (Air) Not acceptable. See regulations for spirits of nitroglycerin.

nitroguanidine $H_2NC(NH)NHNO_2$. Exists in two forms, alpha and beta.
Properties: (a) Long thin flat, flexible, lustrous needles; (b) small, thin elongated plates. Both melting ranges are from 220 to 250°C. Beta-form appears to be more soluble in water.
Derivation: (a) Results when guanidine nitrate is dissolved in concentrated sulfuric acid and the solution is poured into water. (b) Nitration of a mixture of guanidine sulfate and ammonium sulfate which results from the hydrolysis of dicyandiamide by sulfuric acid.
Hazard: Dangerous fire and explosion risk.
Uses: High explosives, especially flashless propellant powder (with nitrocellulose); chemical intermediate.
Shipping regulations: (Rail) (dry) Consult regulations. Not accepted by express. (Wet with 20% water) Yellow label. (Air) (dry or wet with less than 20% water, Not acceptable; (wet with 20% water) Flammable Solid label.

nitrohydrochloric acid. See aqua regia.

3-nitro-2-hydroxybenzoic acid. See meta-nitrosalicylic acid.

4-nitro-3-hydroxymercuri-ortho-**cresol anhydride.** See nitromersol.

3-nitro-4-hydroxyphenylarsonic acid. See 4-hydroxy-3-nitrobenzenearsonic acid.

"Nitrojection Ammonia."[125] Trademark for anhydrous ammonia specifically intended for direct injection into the soil for fertilization.

nitromannite. See mannitol hexanitrate.

nitromersol (4-nitro-3-hydroxymercuri-ortho-cresol anhydride) $C_6H_2(CH_3)(NO_2)(OHg)$.
Properties: Brownish yellow or yellow granules or powder. Odorless and tasteless. Insoluble in water; almost insoluble in alcohol, acetone, ether; soluble in solutions of alkalies, ammonia, by opening the anhydride ring and salt formation.
Grade: N.F.
Hazard: Toxic by ingestion and inhalation.
Use: Disinfectant for skin in extremely dilute solution (1 part in 1000 is maximum strength).

nitromethane CH_3NO_2. A nitroparaffin.
Properties: Colorless liquid. Soluble in water and alcohol; solubility of water in nitromethane 2.2 cc/100 cc (20°C). Sp. gr. 1.139 (20/20°C); b.p. 101°C; vapor pressure 27.8 mm (20°C); flash point 95°F; wt/gal 9.5 lb; refractive index 1.3817 (20°C); freezing point −29°C. Autoignition temp. 785°F.
Derivation: By reaction of methane or propane with nitric acid under pressure.
Containers: 5-, and 55-gal drums and 1-gal cans.
Hazard: Toxic by ingestion and inhalation; lower explosion limit 7.3% in air. Dangerous fire and explosion risk. Tolerance, 100 ppm in air.
Uses: Solvent for cellulosic compounds, polymers, waxes, fats, etc; chemical synthesis; rocket fuel; gasoline additive.
Shipping regulations: (Air) Flammable Liquid label.

2-nitro-4-methoxyaniline $NO_2C_6H_3(OCH_3)NH_2$. Orange-red powder; m.p. 118–120°C; sparingly soluble in cold water and alcohol; soluble in hot water and dioxane. Used as a chemical intermediate.
Hazard: Probably toxic.

2-nitro-2-methyl-1,3-propanediol $CH_2OHC(CH_3)NO_2CH_2OH$.
Properties: White, crystalline solid. Solubility in water 80 g/100 cc (20°C). M.p. 147–149°C; b.p. decomposes (10 mm); pH 0.1 M solution 5.42.
Hazard: Probably toxic.
Use: Intermediate.

nitromuriatic acid. See aqua regia.

nitron (1,4-diphenyl-3,5-endo-anilino-4,5-dihydro-1,2,4-triazole) $C_{20}H_{16}N_4$.
Properties: Lemon-yellow, fine crystalline needles. Soluble in chloroform, acetone, and acetic acid ester; slightly soluble in ether and alcohol.
Use: Reagent for detection of the nitrate ion $(NO_3)^-$ in very dilute solutions.

alpha-**nitronaphthalene** (1-nitronaphthalene) $C_{10}H_7NO_2$.
Properties: Yellow crystals. Soluble in alcohol and ether; insoluble in water. Sp. gr. 1.331; m.p. 55–56°C; b.p. 304°C; flash point 327°F; combustible.
Derivation: Action of a mixture of nitric and sulfuric acids on finely ground naphthalene.
Hazard: Moderate irritant.
Uses: Dyes; naphthylamine; added to mineral oils to mask fluorescence.
Note: The beta isomer is highly toxic by ingestion.

1-nitronaphthalene-5-sulfonic acid. (Laurent's alpha acid) $C_{10}H_6(NO_2)(SO_3H)$.
Properties: Pale yellow needles. Soluble in water, alcohol, and ether. Combustible.
Derivation: By sulfonating nitronaphthalene with a mixture of chlorohydrin and sulfuric acid.
Hazard: May be toxic.
Use: Dyes.

nitronium perchlorate NO_2ClO_4.
Properties: White crystalline solid; m.p. 120–140°C; hygroscopic; noncorrosive; soluble in water to form nitric and perchloric acids; highly reactive.
Derivation: From ozone, nitrogen dioxide, chlorine dioxide.
Hazard: Highly irritant to skin and mucous membranes. Strong oxidizing agent; may explode in contact with organic materials.
Use: Suggested as propellant oxidizer.
Shipping regulations: Perchlorates, n.o.s., (Rail) Yellow label. (Air) Oxidizer label.

para-**nitrophenetole** $NO_2C_6H_4OC_2H_5$.
Properties: Crystallizes in prisms. Soluble in alcohol and ether. M.p. 58°C; b.p. 283°C.
Derivation: By ethylation of para-nitrophenol with ethyl chloride.
Uses: Dyes and other intermediates.

nitrophenide [bis(3-nitrophenyl) disulfide]$(NO_2C_6H_4)_2S$.
Properties: Yellow, rhomboid crystals; m.p. 83°C; insoluble in water; soluble in ether; slightly soluble in alcohol.
Derivation: Reduction of meta-nitrobenzenesulfonyl chloride with hydriodic acid.

Uses: Veterinary medicine and pharmaceutical intermediate.

meta-**nitrophenol** $NO_2C_6H_4OH$.
Properties: Pale yellow crystals; sp. gr. 1.485 (20°C); m.p. 96–97°C; b.p. 149°C (70 mm). Combustible. Slightly soluble in water; soluble in alcohol.
Derivation: Diazotized meta-nitroaniline is boiled with water and sulfuric acid.
Use: Indicator.

ortho-**nitrophenol** $NO_2C_6H_4OH$.
Properties: Yellow crystals; sp. gr. 1.295 (45°C), 1.657 (20°C); m.p. 44–45°C; b.p. 214°C; soluble in hot water, alcohol, ether.
Derivation: Action of dilute nitric acid on phenol at low temperature; para-nitrophenol formed at same time. They are separated by steam distillation.
Containers: Glass bottles; fiber cans; drums.
Uses: Intermediate in organic synthesis; indicator.

para-**nitrophenol** $NO_2C_6H_4OH$.
Properties: Yellowish monoclinic prismatic crystals; sp. gr. 1.479–1.495 (20°C); m.p. 111.4–114°C (sublimes); b.p. 279°C (dec); soluble in hot water, alcohol, ether.
Derivation: (a) From p-chloronitrobenzene; (b) See orthonitrophenol.
Containers: Glass bottles; fiber cans; drums.
Use: Intermediate in organic synthesis; production of parathion; fungicide for leather.

para-**nitrophenol, sodium salt** (sodium para-nitrophenolate; para-nitro sodium phenolate) $NO_2C_6H_4ONa$. Yellow crystalline solid; soluble in water. Used as an intermediate.

para-**nitrophenylacetic acid** (para-nitro-alpha-toluic acid) $NO_2C_6H_4CH_2COOH$.
Properties: Colored needles. M.p. 152.3°C. Slightly soluble in cold water; soluble in alcohol and chloroform. Combustible.
Derivation: Hydrolysis of para-nitrobenzyl cyanide with 50% sulfuric acid.
Hazard: May be toxic.
Uses: Intermediate for dyestuffs, pharmaceuticals, penicillin precursors, local anesthetics.

4-nitrophenylarsonic acid $NO_2C_6H_4AsO(OH)_2$.
Properties: Crystalline solid.
Derivation: Nitration of phenylarsonic acid.
Hazard: Toxic by ingestion. An arsenic compound.
Use: Veterinary medicine.
Shipping regulatoins: (Rail, Air) Arsenical compounds, n.o.s., Poison label.

para-**nitrophenylazosalicylate sodium.** See alizarin yellow R.

"Nitrophoska."[440] Trademark for a group of complete (N-P-K) fertiizers for agricultural and horticultural crops.

nitrophosphate. A nitrogen-phosphorus fertilizer produced by the action of nitric acid or a mixture of nitric and sulfuric or phosphoric acids on phosphate rock. Potassium salts usually are added to produce complete fertilizers. Typical analysis: Available nitrogen 15%, available P_2O_5 15%, available K_2O 15%.
Containers: Bags, bulk, carloads.
Use: Fertilizer.
See also superphosphate; triple superphosphate.

1-nitropropane $CH_3CH_2CH_2NO_2$. A nitroparaffin.
Properties: Colorless liquid. Solubility in water 1.4 ml/100 ml (20°C); solubility of water in 1-nitropropane 0.5 cc/100 cc (20°C). Sp. gr. 1.003 (20/20°C); b.p. 132°C; vapor pressure 7.5 mm (20°C); flash point 120°F (T.O.C.); wt/gal 8.4 lb (68°F); refractive index 1.4015 (20°C); freezing point −108°C; autoignition temp. 789°F. Combustible.
Derivation: By reaction of propane with nitric acid under pressure.
Containers: Drums, tank cars.
Hazard: Moderately toxic; moderate fire risk. Tolerance, 25 ppm in air. Moderate explosion hazard when shocked or heated.
Uses: Solvent; chemical synthesis; rocket propellant; gasoline additive.

2-nitropropane $CH_3CHNO_2CH_3$. A nitroparaffin.
Properties: Colorless liquid. Solubility in water 1.7 ml/100 ml (20°C); solubility of water in 2-nitropropane 0.6 cc/100 cc (20°C). Sp. gr. 0.992 (20/20°C); b.p. 120°C; vapor pressure 12.9 mm (20°C); flash point 103°F (T.O.C.); wt/gal 8.3 lb (68°F); refractive index 1.3941 (20°C); freezing point −93°C; autoignition temp. 802°F. Combustible.
Derivation: By reaction of propane with nitric acid under pressure.
Containers: Drums; tank cars.
Hazard: Moderately toxic; moderate fire risk. Tolerance, 25 ppm in air. Moderate explosion hazard when shocked or heated.
Uses: Solvent especially for vinyl and epoxy coatings; chemical synthesis; rocket propellant; gasoline additive.

meta-**nitrosalicylic acid** (3-nitro-2-hydroxybenzoic acid) $C_6H_3COOH(OH)NO_2$.
Properties: Yellowish crystals. Soluble in water and in alcohol. M.p. 144°C.
Derivation: Nitration of salicylic acid.
Uses: Intermediate; azo dyes.

nitrosamine. Any of a group of organic compounds in which NNO is attached to an alkyl or aryl group, e.g., $(C_6H_5)_2NNO$. Some members of this group have given evidence of carcinogenicity.

"Nitrosan."[28] Trademark for a yellow powdered blowing agent of 70% N,N′-dimethyl-N,N′-dinitrosoterephthalamide and 30% white mineral oil. Controlled thermal decomposition liberates terepthalate. Nonstaining, nondiscoloring gas generant. Used for formation of plastic foams.
Containers: 5-lb fiber drums.

N-nitrosodimethylamine (dimethylnitrosamine; DMN) $(CH_3)_2N_2O$.
Properties: Yellow liquid; soluble in water, alcohol, and ether. B.p. 152°C; sp. gr. 1.006. Combustible.
Hazard: Causes cancer in experimental animals.
Uses: Rocket fuels; solvent; organic research.

para-**nitrosodimethylaniline.** See N,N-dimethyl-para-nitrosoaniline.

nitroparaffin (nitroalkane). Any of a homologous series of compounds whose generic formula is $C_nH_{2n+1}NO_2$, the nitro groups being attached to a carbon atom through the nitrogen.
Properties: Colorless, mobile liquids; pleasant odor; b.p. range, 101–131°C; f.p. range, −28 to −104°C; sp. gr. range, 0.983–1.131; dielectric constant range, 23–35; slightly soluble in water. Flash point range, 82–120°F. Combustible. Moderately toxic by inhalation.
Derivation: Nitration of propane and other paraffin hydrocarbons under pressure.
Uses: Solvents for dyes, waxes, resins, gums and various polymeric substances; gravure and flexographic

inks; propellants and fuel additives; chemical intermediates.
See also nitroethane, nitromethane, nitropropane.

N-nitrosodiphenylamine (diphenylnitrosamine; nitrous diphenylamide) $(C_6H_5)_2NNO$.
Properties: Yellow to brown or orange powder or flakes; m.p. 64–66°C; sp. gr. 1.23; insoluble in water; soluble in alcohol, acetone, benzene, and ethylene dichloride. Somewhat soluble in gasoline.
Containers: Fiber drums.
Hazard: May be carcinogenic.
Uses: Retarder of vulcanization of rubber; pesticide.

para-**nitro sodium phenolate.** See para-nitrophenol, sodium salt.

nitroso dyes (quinone oxime dyes). Dyes whose molecules contain the —NO or ═NOH chromophore group.

nitroso ester terpolymer. See nitroso polymer.

nitrosoguanidine ONNHC(NH)NH_2. Pale yellow crystalline powder.
Derivation: Cooling the reaction products of nitroguanidine, ammonium chloride and zinc dust.
Hazard: Severe explosion risk.
Use: An initiating explosive.
Shipping regulations: (Rail) Consult regulations. Not accepted by express. (Air) Not acceptable.

1-nitroso-2-naphthol (naphthoquinone oxime, one of several isomers. Alpha-nitroso-beta-naphthol).
Properties: Yellow needles. Soluble in alcohol and ether; insoluble in water. M.p. 110°C. Combustible.
Derivation: Action of nitrous acid on beta-naphthol.
Uses: Organic synthesis; prevention of gum formation in gasoline; analytical reagent.

para-**nitrosophenol** C_6H_4OHNO.
Properties: Crystallizes in light brown leaflets which decompose at 140°C. Soluble in alcohol, ether and acetone; moderately soluble in waer. M.p. 140°C.
Derivation: From phenol by action of nitrous acid in the cold.
Hazard: Dangerous fire and explosion risk; reacts violently with acids and alkalies; may explode spontaneously. Intraplant transport must be in tightly covered steel barrels.
Use: Dyes.
Shipping regulations: Not listed.

nitroso polymer (nitroso rubber). A flame- and heat-resistant copolymer derived from trifluoronitrosomethane (CF_3NO) and tetrafluoroethylene ($CF_2 = CF_2$). A third monomer (methyl-4-nitrosoperfluorobutyrate) provides more easily cross-linked functional groups than heretofore available. This terpolymer, called nitroso ester terpolymer (NET), is a notable improvement on the earlier copolymer, and is especially effective for fireproof interiors of aerospace vehicles and airplanes. Cross-linking is by a peroxide mechanism. Properties of the terpolymer are a glass transition temperature of about −50°C, decomposition temperature of 275°C, tensile strength 750 psi, elongation 500%, Shore hardness 78. Nonflammable.

nitrostarch (starch nitrate) $C_{12}H_{12}(NO_2)_8O_{10}$.
Properties: Orange-colored powder; contains 16.5% nitrogen; soluble in ether-alcohol.
Containers: Steel drums or earthenware pots containing water.
Hazard: Severe explosion risk when dry. Flammable, dangerous fire risk.
Use: Explosives.
Shipping regulations: (Rail) Wet with 20% water, Yellow label. Wet with 30% alcohol, Red label. Dry, not accepted by express. (Air) Dry or wet with less than 20% water or 30% solvent. Not acceptable. Wet with not less than 30% solvent, Flammable Liquid label. Wet with not less than 20% water, Flammable Solid label.

beta-**nitrostyrene** $C_6H_5CH{:}CHNO_2$.
Properties: Liquid. M.p. 58°C. Available as a 30% solution in styrene.
Hazard: Moderate fire and explosion risk.
Use: Chain stopper in styrene type polymerization.

para-**nitrosulfathiazole** $NO_2C_6H_4SO_2NHC_3H_2NS$.
Properties: Pale yellow powder; odorless; slightly bitter taste; m.p. 258–266°C. Slightly soluble in alcohol, very slightly soluble in chloroform, ether and water and practically insoluble in benzene.
Use: Medicine.

nitrosyl chloride NOCl. One of the oxidizing agents in aqua regia.
Properties: Yellow-red liquid or yellow gas; b.p. −5.5°C; f.p. −61.5°C; sp. gr. (gas) 2.3 (0°C, air = 1); sp. gr. (liquid) 1.273 (20°C); decomposed by water; dissociates into nitric oxide and chlorine on heating; soluble in fuming sulfuric acid. Nonflammable.
Derivation: By action of chlorine on sodium nitrate.
Grades: Pure, 93% min.
Containers: Nickel cylinders.
Hazard: Highly toxic; strong irritant, especially to lungs and mucous membranes.
Uses: Synthetic detergents; catalyst; intermediate.
Shipping regulations: (Rail) Green label. (Air) Nonflammable Gas label. Not acceptable on passenger planes.

nitrosylsulfuric acid $ONOSO_3H$ with H_2SO_4.
Properties: Clear, straw-colored, oily liquid. Approximately 40% nitrosylsulfuric acid and 54% sulfuric acid. Pure $ONOSO_3H$: crystalline, m.p. 73°C (dec).
Containers: Carboys; drums.
Hazard: Strong irritant to skin and mucous membranes.
Uses: Manufacture of dyes and intermediates.
Shipping regulations: Not listed.

meta-**nitrotoluene** (meta-methylnitrobenzene) $NO_2C_6H_4CH_3$.
Properties: Yellow crystals; sp. gr. 1.1571 (20/4°C); m.p. 15°C; b.p. 232.6°C; refractive index 1.5466 (20°C). Insoluble in water; soluble in alcohol; miscible with ether. Flash point 223°F; combustible.
Derivation: From meta-nitro-para-toluidine.
Hazard: Toxic by inhalation, ingestion, skin absorption. Tolerance, 5 ppm in air.
Use: Organic synthesis.
Shipping regulations: (Air) Poison label.

ortho-**nitrotoluene** (ortho-methylnitrobenzene) $NO_2C_6H_4CH_3$.
Properties: Yellow liquid; sp. gr. 1.1629 (20°C); f.p. −9.3°C; b.p. 220.4°C; refractive index 1.544 (25°C). Insoluble in water; miscible with alcohol and ether. Flash point 223°F; combustible.
Derivation: From toluene by nitration and separation by fractional distillation.
Hazard: Toxic by inhalation, ingestion, skin absorption. Tolerance, 5 ppm in air.

Uses: For production of toluidine, tolidine, fuchsin, and various synthetic dyes.
Shipping regulations: (Air) Poison label.

para-**nitrotoluene** (para-methylnitrobenzene) $NO_2C_6H_4CH_3$.
Properties: Yellow crystals; sp. gr. 1.299 (0/4°C); m.p. 51.7°C; b.p. 238.3°C; refractive index 1.5382 (15°C). Insoluble in water; very soluble in alcohol, ether, and benzene. Flash point 223°F; combustible.
Derivation: From toluene by nitration and separation by fractional distillation.
Hazard: Toxic by inhalation, ingestion, skin absorption. Tolerance, 5 ppm in air.
Uses: For production of toluidine, fuchsin, tolidine and various synthetic dyes.
Shipping regulations: (Air) Poison label.

para-**nitro**-alpha-**toluic acid.** See para-nitrophenylacetic acid.

meta-**nitro**-para-**toluidine** (3-nitro-4-toluidine) $NO_2C_6H_3(CH_3)NH_2$.
Properties: Orange-red crystals; soluble in alcohol and concentrated sulfuric acid. M.p. 117°C. Flash point 315°F; combustible.
Derivation: From acetyl-para-toluidine by nitration.
Containers: Fiber drums; multiwall paper sacks.
Use: Intermediate for dyes and pigments.

para-**nitro**-ortho-**toluidine** (4-nitro-2-toluidine) $NO_2C_6H_3(CH_3)NH_2$.
Properties: Yellow crystalline solid. Soluble in alcohol and ether. M.p. 104°C. Flash point 315°F; combustible.
Derivation: From ortho-toluidine by nitration.
Containers: Barrels; fiber drums; multiwall paper sacks.
Use: Intermediate for dyes and pigments.

para-**nitro**-alpha-**tolunitrile.** See para-nitrobenzyl cyanide.

nitrotrichloromethane. See chloropicrin.

meta-**nitrotrifluoromethylbenzene.** See meta-nitrobenzotrifluoride.

2-nitro-4-trifluoromethylbenzonitrile $NO_2(CF_3)C_6H_3CN$.
Properties: Liquid. B.p. 156–158°C (18–19 mm); m.p. 47–48°C.
Derivation: From a cyclic halogen compound by heating with copper cyanide in the presence of amines.
Hazard: May be toxic.
Use: Dyes.

2-nitro-4-trifluoromethyl chlorobenzene $NO_2C_6H_3(CF_3)Cl$.
Properties: Pale yellow oil; f.p. −6°F; soluble in acetone, alcohol and other solvents; insoluble in water.
Hazard: May be toxic.
Uses: Intermediate for dyes and organic chemicals.

nitrourea $NH_2CONHNO_2$.
Properties: White crystalline powder; slightly soluble in water; soluble in alcohol or ether; m.p. 158–159°C.
Derivation: Urea nitrate and concentrated sufuric acid.
Containers: Steel drums.
Hazard: Severe explosion risk.
Use: Explosives.
Shipping regulations: (Rail) Consult regulations. Not accepted by express. (Air) Not acceptable.

nitrous acid HNO_2
Properties: A weak acid occurring only in the form of a light blue solution.
Derivation: Reaction of strong inorganic acids with nitrites, e.g., HCl + $NaNO_2$.
Uses: Formation of diazotizing compounds by reaction with primary aromatic amines; source of nitric oxide.

nitrous diphenylamide. See N-nitrosodiphenylamine.

nitrous oxide (nitrogen monoxide) N_2O.
Properties: Colorless, sweet-tasting gas; noncombustible; sp. gr. gas, 1.52 referred to air; liquid, 1.22 (−89°C); f.p. −90.8°C; b.p. −88.5°C; soluble in alcohol, ether, and concentrated sulfuric acid; slightly soluble in water. An asphyxiant gas.
Derivation: Thermal decomposition of ammonium nitrate; controlled reduction of nitriles or nitrates.
Grades: Pure, 98.0% min; U.S.P. (97% min).
Containers: Steel cylinders.
Hazard: Moderate fire risk; supports combustion; can form explosive mixtures with air.
Uses: Anesthetic in dentistry and surgery; propellant gas in food aerosols; leak detection.
Shipping regulations: (Rail) Green label. (Air) Nonflammable Gas label.

"Nitrox."[292] Trademark for a product designed to provide in a single operation the cleansing and neutralizing action of caustic soda and the corrosion protection properties of sodium nitrite. Used in cleaning, protection and neutralizing of iron and steel during processing and storage as, for example, tanks of petroleum tankers carrying light oils.
Containers: 400-lb steel drums.

nitroxanthic acid. See picric acid.

nitroxylene (dimethylnitrobenzene) $C_6H_3(CH_3)_2NO_2$. There are three isomers (2,4-; 3.4-; and 2,5-).
Properties: Yellow liquid (2,4- and 2,5-); crystalline needles (3,4-). Sp. gr. 1.135; m.p. 2°C; b.p. 246°C. Soluble in alcohol and ether; insoluble in water.
Derivation: By nitrating xylene and separating from the resulting mixture by rectification.
Hazard: Highly toxic by ingestion, inhalation and skin absorption.
Uses: Organic synthesis; gelatinizing accelerators for pyroxylin.
Shipping regulations: (Rail, Air) Poison label. Legal label name: nitroxylol.

"Nivco."[308] Trademark for a precipitation-hardening cobalt-base alloy with good strength-to-weight ratio for use at temperatures up to 1200°F. It contains somewhat more than 20% nickel and small amounts of titanium, zirconium, and iron. Used as a damping material for steam turbine blade applications.

nm Abbreviation for nanometer (q.v.).

NMR Abbreviation for nuclear magnetic resonance (q.v.).

No Symbol for the element nobelium (q.v.).

Nobel, Alfred B. (1833–1896). A native of Sweden, Nobel devoted most of his life to a study of explosives, and was the inventor of a mixture of nitroglycerin and diatomaceous earth which he called dynamite. He also invented blasting gelatin and smokeless powder. With the fortune he accumulated from his work, Nobel established the foundation that bears his name, which annually recognizes outstanding work in physics, chemistry, medicine, literature, and human relations. The Nobel prize is still the world's most valued scientific award.

nobelium No Synthetic radioactive element 102; one of the actinide series of elements; its discovery has been claimed by research groups in Sweden, U.S.S.R., and California. It can be produced in a cyclotron by bombarding curium with nuclei of carbon 13 accelerated to high energies. The name nobelium has been accepted by the IUPAC Commission on Nomenclature of Inorganic Chemistry. Its isotopes are so short-lived that its chemical properties have not been determined. It has no known uses or compounds.

noble. In chemical terminology this term describes an element which either is completely unreactive or reacts only to a limited extent with other elements. Six noble gases constitute a group in the Periodic Table that is variously called the zero group (as the first three of its members have a valence of zero), the inert gas group, and the noble gas group. The last is preferable, as three of the gases, though unreactive, are not inert. The gases of this group are helium, neon, argon, krypton, xenon, and radon. The noble metals are generally considered to be gold, silver, platinum, palladium, iridium, rhodium, mercury, ruthenium, and osmium. The term has no reference to the commercial value of these metals. See also inert.

"Noctec."[412] Trademark for chloral hydrate (q.v.).

NODA. Abbreviation for n-octyl, n-decyl adipate.

nodular iron. A ductile form of cast iron.

"Nogas."[28] Trademark for a metal pickling agent.
Properties: Amorphous brown powder, soluble in water and acids. Will give foam-producing solutions when dissolved in water or acids.
Use: Addition agent in acid pickling of iron and steel products to produce a foam blanket and to improve the atmosphere around pickling operations by elimination of acid mist particles usually given off by the pickling operation.

"Nolane."[342] Trademark for 2,6-allyl-dichlorosilane-resorcinol, used for priming glass fibers, i.e., prior to laminating.

"Noludar."[190] Trademark for methyprylon (q.v.).

nomenclature, chemical. See chemical nomenclature.

nomograph. A chart in which a straight line, either drawn or indicated by a ruler, intersects three scales at points that represent values that satisfy an equation or a given set of physical conditions. It thus enables one to determine the value of a dependent variable when that of the independent variable is known. Used by chemical engineers to determine relationships between the properties of materials.

"Nomex."[28] Trademark for a heat-resistant aromatic polyamide fiber. It retains useful properties after long-term exposure to temperatures up to 220°C; does not support combustion.
Uses: Military flight suits, electrical insulation, heat-resistant paper, coated fabrics; upholstery.
See also aramid.

nonadecane $C_{19}H_{40}$ or $CH_3(CH_2)_{17}CH_3$.
Properties: Leaflets; soluble in alcohol; insoluble in water; soluble in ether; sp. gr. 0.777; m.p. 32°C; b.p. 330°C. Combustible; low toxicity.
Use: Organic synthesis.

n-**nonadecanoic acid** $CH_3(CH_2)_{17}COOH$. A saturated fatty acid normally not found in natural vegetable fats or waxes. Combustible; low toxicity.
Properties: Colorless crystals; m.p. 68.7°C; b.p. 297°C (100 mm); soluble in alcohol and ether; insoluble in water. Synthetic product available 99% pure for organic synthesis.

gamma-**nonalactone.** See gamma-nonyl lactone.

nonanal (pelargonic aldehyde; n-nonyl aldehyde; aldehyde C-9) $C_8H_{17}CHO$.
Properties: Colorless liquid with an orange-rose odor; sp. gr. 0.822–0.830; refractive index 1.424–1.429. Soluble in 3 volumes of 70% alcohol, in mineral oil; insoluble in glycerol. Combustible.
Grades: Technical; F.C.C.
Uses: Perfumery; flavoring agent.

nonane (nonyl hydride) C_9H_{20} or $CH_3(CH_2)_7CH_3$.
Properties: Colorless liquid. Soluble in alcohol; insoluble in water. Sp. gr. 0.722; b.p. 150.7°C; f.p. −54°C; refractive index 1.40561 (20°C); flash point 86°F. (Closed cup); autoignition temp. 403°F.
Grades: Technical (95%); 99%; research.
Containers: Glass bottles; drums.
Hazard: Flammable, moderate fire risk. Irritant; narcotic in high concentrations.
Uses: Organic synthesis; biodegradable detergents; distillation chaser.

nonanedioic acid. See azelaic acid.

nonane-1,3-diol monoacetate $C_9H_{18}OOCCH_3$.
Properties: A mixture of isomers. Colorless to slightly yellow liquid; stable; soluble in alcohol. Combustible; low toxicity.
Uses: Perfumery; flavoring.

n-**nonanoic acid.** See pelargonic acid.

nonanol. See nonyl alcohol.

nonallyl chloride. See pelargonyl chloride.

noncombustible material. A solid, liquid or gas that will not ignite or burn, regardless of how high a temperature it is exposed to.

nondestructive testing. A test that does not involve destruction of the sample or material tested. Such tests are usually carried out by radiographic methods on large, finished metal products (castings and machine components) to determine the presence of internal defects likely to cause operational failure. An infrared camera scanning device for detection of internal weaknesses in tires is a recent development in nondestructive testing. X-radiation is used to determine the authenticity of paintings and other objects of art.

nondrying oil. See drying oil.

nonelectrolyte. A compound that resists passage of electricity (nonconductor) both in liquid form and in solution. Included in this classification are most organic compounds, with the exception of acids and amides, and such inorganic compounds as nonmetal halides. Nonelectrolytes are covalently bonded; some inorganic compounds having covalent bonds form electrolytic solutions in aqueous solution. See also electrolyte.

nonene. See nonylene.

nonflammable material. A gas or liquid whose flash point is above 100°F, or a solid whose ignition point is above 500°F. See combustible material and note under flammable material.

"Nonic."[204] Trademark for a group of non-ionic surface-active agents based on polyethylene glycol tert-dodecylthioether or related compounds.

Uses: Detergents; paper manufacture; grease removal; wetting agents; emulsifier; clouding agent.

"Nonisol."[219] Trademark for a series of nonionic surface-active fatty acid esters of higher polyglycols (polyoxyalkylenes).
Uses: Cosmetics; solvent emulsions; polishes; rust-preventive oils; insecticide and agricultural sprays.

nonmetal. (1) Any of a number of elements whose electronic structure, bonding characteristics, and consequent physical and chemical properties differ markedly from those of metals, particularly in respect to electronegativity and thermal and electrical conductivity. In general, nonmetals have very low to moderate conductivity and high electronegativity. The 25 elements classified as nonmetals may be considered in two groups: (a) those having moderate electrical conductivity (semiconductors), all of which are solids; and (b) those having very low conductivity, many of which are gases. The semiconductors of group (a) were formerly called metalloids since they more nearly resemble metals than those of group (b), but this term is no longer used by chemists. The nonmetals are given below based on this subgrouping, though any such list is open to challenge:

(a)		(b)
arsenic	polonium	halogens
antimony	phosphorus	hydrogen
boron	selenium	nitrogen
carbon	silicon	oxygen
germanium	tellurium	noble gases
		sulfur

(2) Loosely, any material that is not a metal, e.g., petroleum, plastics, waxes, etc. The term is widely used in this sense by engineers and specification writers.

n-**nonoic acid.** See pelargonic acid.

nonoic acid. An acid of the formula $C_8H_{17}COOH$, of which there are many possible isomers. Pelargonic acid (q.v.) is the normal or straight-chain acid. Various mixtures of branched chain nonoic acids are recovered from products of the Oxo process.

nonviscous neutral oil. A neutral oil of viscosity lower than 135 S.U.S. at 100°F.

nonwoven fabric. A fabric made from staple lengths of cotton, rayon, glass or thermoplastic synthetic fiber mechanically positioned in a random manner, and bonded with a synthetic adhesive or rubber latex. The sheets thus formed can be pressed together to form porous mats of high absorptivity and good elastic recovery on deformation. Permanent bonds are formed where the fibers touch each other as a result of heat treatment when the fibers are thermoplastic, or by use of a high-polymer binder.

A specialty nonwoven (so-called "melded") fabric trademarked "Cambrelle" (q.v.) is a composite of two different polymers; heating the exterior layer to its melting point causes the fibers to fuse into a fabric.

Applications are as a backing for plastic film; padding for surgical dressings, diapers, sanitary napkins; drapes and other decorative textile products; filtration; shoe liners; industrial wiping and polishing fabrics; disposable clothing; carpet backing.

n-**nonyl acetate** (acetate C-9) $CH_3COO(CH_2)_8CH_3$.
Properties: Colorless liquid; strong, pungent odor; sp. gr. 0.864–0.868; refractive index 1.422–1.426. Soluble in 4 volumes of 70% alcohol. Flash point 155°F. Combustible. Low toxicity. Several isomers exist.
Uses: Perfumery; flavoring.

n-**nonyl alcohol** (nonanol; alcohol C-9; octyl carbinol; pelargonic alcohol) $CH_3(CH_2)_7CH_2OH$.
Properties: Colorless liquid, with floral odor; sp. gr. 0.826–0.829; refractive index 1.431–1.435; f.p. −5°C; b.p. 215°C. Flash point 165°F. Combustible; low toxicity. Soluble in 7 volumes of 50% alcohol; insoluble in water. Several isomers exist.
Uses: Perfumery; flavoring.

n-**nonylamine** $C_9H_{19}NH$.
Properties: Colorless liquid. B.p. 75–85°C (20 mm); sp. gr. 0.798 (25°C); refractive index 1.4366 (n 20/D); may be prepared from nonyl halides by conventional techniques. Combustible.
Hazard: Moderate fire risk; moderately toxic and skin irritant.

tert-**nonylamine.** Principally tert-$C_9H_{19}NH_2$ and tert-$C_{10}H_{21}NH_2$.
Properties: Colorless liquid. Boiling range 160–174°C; sp. gr. 0.789 (25°C); refractive index 1.428 (25°C); flash point 120°F; insoluble in water; soluble in common organic solvents; especially in petroleum hydrocarbons. Combustible.
Containers: Drums; tank cars.
Hazard: Moderate fire risk; moderately toxic and skin irritant.
Uses: Intermediate for rubber accelerators, insecticides, fungicides, dyestuffs, pharmaceuticals.

nonylbenzene (1-phenylnonane) $C_6H_{19}C_6H_5$.
Properties: Light-straw liquid; aromatic odor; boiling range 245–252°C; sp. gr. 0.864 (20/20°C); refractive index 1.488 (20°C); viscosity 41.9 cp (20°C); flash point 210°F (closed cup). Combustible.
Hazard: Probably irritant and narcotic in high concentrations.
Use: Manufacture of surface-active agents.

nonyl bromide $C_9H_{19}Br$.
Properties: Liquid. B.p. 81–85°C (10 mm); sp. gr. 1.101 (25°C); refractive index 1.4583 (n 20/D). Nonflammable.
Derivation: High yields of nonyl bromide are obtained by passing hydrogen bromide into the alcohol while heating or by refluxing with aqueous hydrobromic acid in the presence of an acid catalyst.

nonyl chloride $C_9H_{19}Cl$.
Properties: Liquid. B.p. 58–63°C (8 mm); sp. gr. 0.878 (25°C); refractive index 1.4379 (n 20/D). Nonflammable.
Derivation: Nonyl alcohol reacts with hydrogen chloride at elevated temperatures and pressure to give nonyl chloride. It can also be made by refluxing a mixture of concentrated hydrochloric acid with the alcohol in presence of zinc chloride.

1-nonylene (1-nonene) C_9H_{18} or $CH_3(CH_2)_6CH{:}CH_2$.
Properties: Colorless liquid; soluble in alcohol; insoluble in water. Sp. gr. 0.7433; b.p. 149.9°C. Combustible.
Derivation: From propylene.
Uses: Organic synthesis; wetting agent; lube oil additive; polymer gasoline.

nonyl hydride. See nonane.

n-**nonylic acid.** See pelargonic acid.

gamma-**nonyl lactone** (aldehyde C-18; 4-hydroxynonanoic acid, gamma lactone; gamma-nonalactone)

$CH_3(CH_2)_4CHCH_2CH_2C(O)O$.
Properties: Yellowish to almost colorless liquid; coconut-like odor; sp. gr. 0.956–0.963; refractive index 1.447; soluble in 5 volumes of 50% alcohol; soluble in most fixed oils and mineral oil; practically insoluble in glycerol. Combustible; low toxicity.
Grades: Technical; F.C.C.
Uses: Perfumery; flavors.

nonyl nonanoate (nonyl pelargonate) $C_9H_{19}OOCC_8H_{17}$.
Properties: Liquid; sp. gr. 0.863 (25°C); b.p. 315°C; refractive index 1.4419 (n 20/D); floral odor; combustible; low toxicity.
Uses: Flavors; perfumes; organic synthesis.

nonylphenol $C_9H_{19}C_6H_4OH$. A mixture of isomeric monoalkyl phenols, predominantly para-substituted.
Properties: Pale yellow, viscous liquid with a slight phenolic odor. Insoluble in water; soluble in most organic solvents. Sp. gr. (20/20°C) 0.950; b.p. 315°C; f.p. −10°C (sets to glass below this temperature); viscosity 563 cp (20°C); flash point 285°F. Combustible.
Grade: Technical.
Containers: Drums; tank cars.
Uses: Non-ionic surfactant (nonbiodegradable); lube oil additives; stabilizers, petroleum demulsifiers, fungicides; antioxidants for plastics and rubber.

nonylphenoxyacetic acid $C_9H_{19}C_6H_4OCH_2COOH$.
Properties: Light amber liquid; miscible with organic solvents; insoluble in water; soluble in alkali. Flow point 0°C; viscosity (25°C) 6500 cp. Combustible.
Uses: Corrosion inhibitor for turbine oils, lubricants, fuels, greases; hydraulic fluids, cutting oils and metal coolants.

nonyl thiocyanate $C_9H_{19}SCN$.
Properties: B.p. 84–86.5°C (1 mm); sp. gr. 0.919 (25°C); refractive index 1.4696 (n 20/D). Nonyl thiocyanate can be made from nonyl chloride by refluxing with alcoholic sodium thiocyanate solution using conventional techniques.

nonyltrichlorosilane (trichlorononylsilane) $C_9H_{19}SiCl_3$.
Properties: Water-white liquid with a pungent irritating odor. Fumes readily in moist air.
Hazard: Highly toxic; strong irritant to skin and mucous membranes.
Shipping regulations: (Rail) White label. (Air) Corrosive label. Not acceptable on passenger planes.

"Nopalcol."[309] Trademark for a series of diethylene glycol and polyethylene glycol fatty esters.
Uses: General-purpose emulsifying agents for oils, solvents, essential oils, opacifying agents for cosmetics.

"Nop-cap."[309] Trademark for calcium pantothenate-containing products for the fortification of foods and feeds.

"Nopcocell."[309] Trademark for polyurethane foamed plastic for general uses.

"Nopcodraw" Compounds.[309] Trademark for a group of die box lubricants. Powdered dry compounds for drawing and heading ferrous, alloy, stainless steel and refractory metals. Full range of formulations, based on soda, potash, aluminum, zinc, calcium, and barium soaps combined with inorganic and organic lubricants plus anti-wear agents.

"Nopcogen."[309] Trademark for a series of fatty acid derivatives, including alkylolamides, polyamine condensates, alkylolamine condensates, and sulfonated amine condensates, used as softeners, detergents, wetting, or emulsifying agents.

"Nopcosant."[309] Trademark for sodium salt of condensed naphthalene sulfonic acid.
Use: Dispersing agent for pigments.

"Nopcoset."[309] Trademark for a series of polyvinyl acetate emulsions for use in the paper, textile, paint, leather and wood-working industries.

"Nopcosize."[309] Trademark for polyacrylic sizes for nylon filament.

"Nopcosulf."[309] Trademark for a series of sulfated oils, fatty acids and fatty esters.
Uses: Mineral oil emulsifiers, dyestuff dispersants, latex paint additives, cosmetic emulsifiers, textile wetting and re-wetting agents, paper wetting agents and softeners.

"Nopcote" C-104.[309] Trademark for a fluid, stable, 50% calcium stearate dispersion for lubricating protein, starch or latex coating colors for friction control.

nopinene. See beta-pinene.

NOPON. See para-bis[2-(5-alpha-naphthyloxazolyl)]-benzene.

"Norad."[121] Trademark for a water-soluble adhesive used to prevent slippage of palletized multiwall bags. Shipped in 55-gal steel drums.

"Norane."[42] Trademark for a group of water repellents used in the textile industry for a wide variety of fabrics. Include formulations of anionic wax emulsions, silicone resins, thermosetting resins, fluorophilic compounds, and fatty acid amide condensates.

"Norbide."[249] Trademark for boron carbide (q.v.).

norbormide. Generic name for 5-(alpha-hydroxy-alpha-2-pyridylbenzyl)-1-(alpha-2-pyridylbenzylidene)-5-norbornene-2,3-dicarboximide.
Properties: White solid; m.p. 180–190°C. Insoluble in water; soluble in dilute acids.
Hazard: Highly toxic by ingestion.
Use: Rodenticide.

5-norbornene-2-methanol (2-hydroxymethyl-5-norbornene; "Cyclol") $C_7H_9CH_2OH$.
Properties: Stable, colorless liquid; sp. gr. 1.022–1.024 (25/25°C); refractive index 1.4985–1.4995 (25/D); a high-boiling solvent, b.p. 192–198°C; miscible with most common organic solvents; slightly soluble in hot water. Flash point 183°F (TOC). Combustible.
Containers: 55-gal drums.
Hazard: May be toxic.
Uses: Monomer for the modification of condensation and addition polymers for coatings.

5-norbornene-2-methyl acrylate.
Properties: Colorless liquid; b.p. 103–105°C; sp. gr. (25/25°C) 1.027–1.031; soluble in most organic solvents; inhibitor 0.02% MEHQ.
Use: A difunctional monomer particularly suited for in situ crosslinking of vinyl acetate emulsions polymer systems.

"Nordel."[28] Trademark for an elastomer based on an ethylene-propylene-hexadiene terpolymer; sulfur-curable.

Uses: Automotive and appliance components; wire insulation; electrical accessories, belts, hose, mechanical products.

nordhausen acid. Fuming sulfuric acid of sp. gr. 1.86–1.90.
Hazard: Corrosive to skin and tissue.
Shipping regulations: (Air) Corrosive label. Not acceptable on passenger planes.
See sulfuric acid, fuming.

nordihydroguaiaretic acid (NDGA) $[C_6H_3(OH)_2CH_2CH(CH_3)]_2$ 4,4′-(2,3-Dimethyltetramethylene)dipyrocatechol).
Properties: Crystals from acetic acid, m.p. 184–185°C. Soluble in methanol, ethyl alcohol, ether; slightly soluble in hot water, chloroform; nearly insoluble in benzene, petroleum ether.
Grades: Technical; F.C.C.
Derivation: Extraction from guaiac; also synthetically.
Hazard: Use in foods prohibited by FDA.
Use: Antioxidant to retard rancidity of fats and oils.

norea. Generic name for 3-(hexahydro-4,7-methanoindan-5-yl)-1,1-dimethylurea, $C_{10}H_{14}NHCON(CH_3)_2$.
Properties: White solid; m.p. 168–169°C. Very slightly soluble in water; soluble in polar solvents such as cyclohexanone, acetone, alcohol; less soluble in nonpolar solvents such as benzene, zylene, hexane, and kerosine.
Hazard: May be toxic.
Use: Herbicide.

***dl*-norephedrine hydrochloride.** See phenylpropanolamine hydrochloride.

***l*-norepinephrine.** See levarterenol.

norgestrel. A synthetic steroid hormone used as ingredient of oral contraceptives. Approved by FDA. See "Ovral."

"Norit."[107] Trademark for activated adsorption carbons of vegetable origin.
Containers: 40-, 50-, 60-lb bags (powdered and pelletized); 100 to 150 lb drums (granular).
Uses: Purification and decolorization of chemical solutions; bleaching agent for glycerin, vegetable oils, and fats; dechlorination, taste and odor control in water treatment; catalyst and carrier for catalysts; recovery of volatile solvents; air purification; gas treatment and separation.

"Norlestrine."[330] Trademark for an oral contraceptive.

norleucine (alpha-aminocaproic acid) $CH_3(CH_2)_3CH(NH_2)COOH$. A nonessential amino acid found naturally in the L(+) form.
Properties: Leaflets, crystallized from water. DL-norleucine. Soluble in water; slightly soluble in alcohol; soluble in acids; decomposes 327°C. Available commercially.
L(+)-norleucine: Slightly sweet; sublimes 275–280°C; m.p. 301°C (dec).
D(−)-norleucine: Bitter; partially sublimes 275–280°C; m.p. 301°C (dec).
Derivation: Found in traces in proteins; also synthetically.
Uses: Biochemical research; medicine.

"Norlig."[121] Trademark for a line of unmodified or partially modified lignosulfonates derived from spent sulfite liquor. Available in both powder and liquid forms. Liquids are darker brown in color and more sticky as solids concentration increases.
Uses: Road binder; linoleum paste; foundry products; leather tanning; gypsum board; soil stabilization; feed pellets.

normal. (1) Of hydrocarbon molecules: containing a single unbranched chain of carbon atoms, usually indicated by the prefix *n*-, as *n*-butane, *n*-propane, etc.
(2) Of solutions: containing one equivalent weight (q.v.) of a dissolved substance per liter of solution; a standard measure of concentration, indicated by *N*, e.g., 2*N*, 0.5*N*, etc. (See also molar).
(3) Perpendicular or at right angles to, as for example, incident light rays normal to a surface.

normalize. To temper steels and other alloys by heating to a predetermined temperature and cooling at a controlled rate to relieve internal stresses and improve strength (analogous to annealing of glass).

"Normasal."[304] Trademark for normal lead salicylate vinyl stabilizer.

"Norox."[310] Trademark for tert.-butyl peroxy benzoate (q.v.).

norphytane. See pristane.

"Nortron." Trademark for an herbicide and weed-killing agent (made by Fisons, Ltd, U.K.) Active ingredient reported to be 2-ethoxy-2,3-dihydro-3,3-dimethyl-5-benzofuranil methane sulfonate.

"Norva."[51] Trademark for nonsoap grease for specialized application as, for instance, where high temperature prevails. Has water resistance and good adhesiveness.

Norway saltpeter. See ammonium nitrate.

Norwegian saltpeter. See calcium nitrate.

n.o.s. Abbreviation for "not otherwise specified." Used in shipping regulations for classes of substances to which a restriction applies, individual members of which are not listed in the regulations.

noscapine (*l*-alpha-narcotine; narcosine). An isoquinoline alkaloid of opium. $C_{22}H_{23}NO_7$.
Properties: Fine white crystalline powder. M.p. 176°C; sublimes at 150–160°C. Insoluble in water; practically insoluble in vegetable oils; slightly soluble in hot solutions of potassium and sodium hydroxide; soluble in most organic solvents. Salts formed with acids are dextrorotatory and unstable in water.
Derivation: Prepared from the seed capsules of Papaver somniferum.
Grade: U.S.P.
Containers: Cans.
Hazard: Toxic and narcotic. Use legally restricted.
Use: Medicine (also as hydrochloride).

"Novadelox."[282] Trademark for a mixture of benzoyl peroxide and an inert filler used to bleach flour.

"Novege."[233] Trademark for compositions containing 2,4,5-T (q.v.).

novobiocin $C_{31}H_{36}H_2O_{11}$.
Properties: A light yellow to white antibiotic produced by Streptomyces niveus. Available as calcium and sodium salts.
Hazard: May have damaging side effects.

"Novocain."[162] Trademark for a brand of procaine hydrochloride. Used as local anesthetic in dentistry and minor surgery.

novolak (novolac). A thermoplastic phenol-formaldehyde type resin obtained primarily by the use of acid catalysts and excess phenol. Generally alcohol-soluble and require reaction with hexamethylenetetramine, paraformaldehyde, etc. for conversion to cured, cross-linked structures by heating to 200–400°F.

Uses: Molding materials; bonding materials; bonding agent in brake linings, abrasive grinding wheels; electrical insulation; clutch facings; air-drying varnishes; reinforcing agent and modifier for nitrile rubber.

"Novoviol."[188] Trademark for a series of ionones.

noxious gas. Any natural or by-product gas or vapor that has specific toxic effects on humans or animals. (Military poison gases are not included in this group). Examples of noxious gases are ammonia, carbon monoxide, nitrogen oxides, hydrogen sulfide, sulfur dioxide, ozone, fluorine, and vapors evolved by benzene, carbon tetrachloride, and a number of chlorinated hydrocarbons. Gases which act as simple asphyxiants are not classified as noxious. See also air pollution.

Np Symbol for neptunium.

"NP-10."[256] Trademark for a polymeric plasticizer; used for vinyl formulations.

N-P-K mixtures. Fertilizers containing nitrogen, phosphorus and potassium. These are usually characterized by numbers such as 5-10-10, meaning 5% nitrogen, 10% phosphorus as P_2O_5, and 10% potassium as K_2O. The percentages refer to the amount of N, P, or K present in available form rather than to the amount of the compounds.

NPN. Abbreviation for n-propyl nitrate.

NPO. See alpha-naphthylphenyloxazole.

NQR spectroscopy. The initials stand for nuclear quadrupole resonance, an analytical technique which requires no magnets, as does NMR. It utilizes instead the electric fields naturally present in the crystals under analysis. Chlorine has been studied extensively by this method. See also analytical chemistry; nuclear magnetic resonance.

NR. Abbreviation for natural rubber.

NRC. Abbreviation of Nuclear Regulatory Commission (q.v.).

"500 NS."[545] A high potency compound whose active ingredients are aryl disulfides. Used to reclaim natural and synthetic rubber.

"NSAE" Powder.[328] Trademark for a sodum alkylnaphthalenesulfonate; an agricultural dispersing agent for sulfur.

"N-Sol."[201] Trademark for a series of patented processes used in water treatment. Dilute alkali silicate is partially neutralized with an acid reactant to develop an activated silica for use as a coagulant aid in clarifying raw and waste waters.

NSR. Abbreviation for nitrile silicone rubber.

NTA. Abbreviation for nitrilotriacetic acid (q.v.).

NTAN. Abbreviation for nitrilotriacetonitrile (q.v.).

"Nuactant."[300] Trademark for a fiber reactant to improve crush resistance and dimensional control in cellulosic fabrics.

"Nuact."[74] Trademark for lead hydroxynaphthenate paste (see lead naphthenate).

"Nuade."[74] Trademark for a rubber-based tackifying agent used as a mastication aid.

"Nubrite."[134] Trademark for barrel nickel-plating additive.

"Nucerite."[522] Trademark for a material of construction for high-temperature corrosive environments. A family of composites consisting of a *crystalline* glass ceramic bonded to steel, e.g., "Nucerite" 7000 ceramic. Used for nozzles, flange faces, reaction equipment, etc. See also "Glasteel"; glass ceramic.

"Nuchar."[228] Trademark for activated carbon of vegetable origin.

Grades: Industrial and water purification.

Containers: Up to hopper cars and trucks.

Uses: Taste and odor control of municipal and potable water.

nuclear chemistry. The division of chemistry dealing with changes in or transformations of the atomic nucleus (see nucleus) (1). It includes spontaneous and induced radioactivity, the fission or splitting of nuclei, and their fusion, or union; also the properties and behavior of the reaction products and their separation and analysis. The reactions involving nuclei are usually accompanied by large energy changes, millions of times greater than those of chemical reactions; they are carried out in nuclear reactors for military purposes and for electric power production; also (in research work) in cyclotrons. See also nuclear energy, fission, fusion (2).

nuclear energy. The energy liberated by (1) the splitting or fission (q.v.) of an atomic nucleus; (2) the union or fusion (q.v.) of two atomic nuclei; and (3) the radioactive decay of a nucleus (transmutation). The energy emitted in (1) is the equivalent of the slight difference in mass between the nucleus as a whole and its constituents (see mass defect). The transformation of mass to energy is stated by Einstein's equivalence equation $E = mc^2$. In the fission process the release of energy is of the order of nanoseconds; it may involve a vast number of nuclei (atomic bomb) or comparatively few (nuclear reactor), where the rate of release is under strict control. In the fusion process energy release is much greater than in fission; no means of controlling it has as yet been found, though research on this immensely useful possibility is continuing. In radioactive decay, the rate of release is extremely slow, extending in some cases for millions of years. See also radioactivity, radiation synthesis, ionizing radiation; fission; fusion (2).

Note: In 1975 nuclear power provided about 9% of the electricity generated in the U.S. and about 2.3% of the total energy consumption. After balancing the pros and cons of safety, environmental factors, costs, etc. it may be regarded as an acceptable risk in future energy production, always provided that the very highest standards of expertise and plant maintenance are observed.

nuclear fuel. Any material which is the source of energy in a nuclear reactor. At the present state of technology the term usually refers to uranium, thorium, or plutonium, either as natural materials, or enriched or synthesized isotopes. See also nuclear reaction; breeder; plutonium; fission.

nuclear fuel element. A rod, tube, plate, or other mechanical shape or form into which a nuclear fuel is fabricated for use in a reactor. The word "element" is used here in its electrical engineering rather than its chemical sense.

nuclear magnetic resonance (NMR). A type of radiofrequency spectroscopy, based on the magnetic field generated by the spinning of the electrically charged nucleus of certain atoms. This nuclear magnetic field is caused to interact with a very large (10,000–50,000 gauss) magnetic field of the instrument magnet. Each nuclear species of atom requires a different instrumental setting to obtain resonance of the magnetic field frequency of the instrument with that of the nucleus. The setting also varies with the electron distribution around each atom, so that it is possible for instance to distinguish between the three kinds of hydrogen atoms in ethyl alcohol. Frequently used nuclear species are the elements: hydrogen, deuterium, fluorine, boron, phosphorus, carbon-13, oxygen-17, nitrogen-15, and silicon-29.

The direct relation of NMR spectral bank height to concentration of a chemical compound allows NMR to be used for quantitative analysis and the following of chemical rates of reaction and equilibria of chemicals in solution. The splitting of bands due to interactions between adjacent nuclei with spin in a compound is used to aid in establishing structural formulas. This technique is a useful method of determining whether hydrogen atoms in a molecule are part of a CH_3— group, a —CH_2— group, an —OH group or other characteristic grouping. It also indicates the number of hydrogen atoms in each category. See also magnetochemistry; nucleus (note).

nuclear reaction. A reaction that involves a change in the nucleus of an atom, as opposed to a chemical reaction, in which only the electrons take part. Nuclear reactions usually result in release of tremendous amounts of energy; the energy obtained from chemical reactions is slight, e.g., rupture of a chemical bond evolves about 5 eV compared with 200 million eV resulting from the fission of an atomic nucleus. There are three types of nuclear reaction, namely, (1) transmutation (radioactive disintegration), (2) fission, and (3) fusion (thermonuclear). See these entries for details.

Nuclear Regulatory Commission. A Federal agency established in 1975 to regulate all commercial uses of atomic energy, including construction and operation of nuclear power plants, nuclear fuel reprocessing plants, and research applications of radioactive materials. It is also responsible for safety and environmental protection.

nucleation. The process by which crystals are formed from liquids, supersaturated solutions (gels), or saturated vapors (clouds). Crystals originate on a minute trace of a foreign substance acting as a nucleus. These are provided by impurities or by container walls in laboratory apparatus. Crystals form initially in tiny regions of the parent phase and then propagate into it by accretion. Rain and snow are formed in this way in moisture-laden air; the moisture condenses on minute particles of dust or ice. Cloud seeding with silver iodide or carbon dioxide is based on this principle. The former has a crystal structure similar to that of ice, and the latter causes rapid formation of ice nuclei by intense local lowering of temperature. See also crystal; whiskers.

There are various methods for growing crystals, including: (a) evaporation of a solution; (b) cooling of a saturated solution, or melt; (c) condensation of a vapor (plasmas are sometimes used to generate the vapor); (d) electrodeposition; (e) growth in gel media (in order to slow growth). Methods (b) through (e) are particularly suited for the growth of large crystals, which are widely used in modern technology.

nucleic acid. Any of several complex compounds occurring in living cells, usually chemically bound to proteins to form nucleoproteins. Nucleic acids are of high molecular weight and are easily changed by many mild chemical reagents. They contain carbon, hydrogen, oxygen, nitrogen (15–16%), and phosphorus (9–10%).

The fundamental units of nucleic acid are nucleotides (q.v.); nucleic acids are polynucleotides in which the nucleotides are linked by phosphate bridges. Upon extensive heating in the presence of water (hydrolysis), nucleic acids yield a mixture of purines and pyrimidines, D-ribose or D-deoxyribose, and phosphoric acid. Nucleic acids are subdivided into two types: ribonucleic acid (RNA), containing the sugar D-ribose; and deoxyribonucleic acid (DNA) containing the sugar D-deoxyribose. See deoxyribonucleic acid; ribonucleic acid; nucleoside; nucleotide; nucleoprotein.

Good sources of nucleic acids are salmon, thymus, yeast, and wheat kernel embryo.

Uses: Biochemical and medical research in genetics, virus diseases and cancer.

nucleon. General name applied to neutrons and protons, the essential constituents of atomic nuclei, and also used as a class name for fundamental particles of that mass. The study of subatomic particles is often called nucleonics.

nucleophile. An ion or molecule that donates a pair of electrons to an atomic nucleus to form a covalent bond; the nucleus that accepts the electrons is called an electrophile. This occurs, for example, in the formation of acids and bases according to the Lewis concept, as well as in covalent carbon bonding in organic compounds. See also Lewis base; donor.

nucleoprotein. A type of protein universally present in the nuclei and the surrounding cytoplasm of living cells. A nucleoprotein is composed of a protein, which is rich in basic amino acids, and a nucleic acid (q.v.). The nucleic acid portions can be isolated and used in medical and biochemical research. See also deoxyribonucleic acid; chromatin; virus.

nucleoside. A compound of importance in physiological and medical research, obtained during partial decomposition (hydrolysis) of nucleic acids (q.v.), and containing a purine or pyrimidine base linked to either D-ribose, forming ribosides, or D-deoxyribose, forming deoxyribosides. They are nucleotides minus the phosphorus group. For specific nucleosides see adenosine, cytidine, guanosine and uridine.

nucleotide. A fundamental unit of nucleic acids (q.v.); some are important coenzymes. The four nucleotides found in nucleic acids are phosphate mono-esters of nucleosides (q.v.): adenylic acid, guanylic acid, uridylic acid, and cytidylic acid. Great progress has been made in determining the nucleotide sequence in fundamental materials, such as yeast genes.

The term is also applied to compounds not found in nucleic acids and which contain substances other than the usual purines and pyrimidines. Such compounds are modified vitamins, and function as coenzymes; examples are riboflavin phosphate (flavin mononucleotide), flavin adenine nucleotide, nicotinamide adenine dinucleotide, nicotine adenine dinucleotide phosphate, and coenzyme A. The nucleotides inosine-5′-monophosphate and guanosine-5′-monophosphate are used as flavor potentiators.

nucleus. (1) The positively charged central mass of an atom; it contains essentially the total mass in the form of protons and neutrons. The nucleus of the hydrogen atom consists of one proton, while that of uranium is comprised of 92 protons and 146 neutrons. The atomic nuclei of certain unstable (radioactive) elements, such as uranium, can be split by bombardment with protons or neutrons from an external source, yielding 200 MeV of energy and an average of 2.5 free neutrons for every nucleus split; gamma radiation is also present.

(2) The central portion of a living cell, consisting primarily of nucleoplasm in which chromatin (q.v.) is dispersed. It is enclosed by a membrane that separates it from the surrounding cytoplasm. All the most important functions of the cell, including the mechanics of division (mitosis) and the programming of the genetic code, take place in the nucleus. (See also gene, chromosome).

(3) The characteristic structure of a group of chemical compounds, e.g., the benzene nucleus.

(4) Any small particle which can serve as the basis for crystal growth (see nucleation).

Note: The multiple meanings of "nucleus" and "resonance" can be a source of confusion, especially when these terms are closely associated, as in nuclear magnetic resonance and resonance of a molecular nucleus. In the first of these expressions, nucleus is used in sense (1) under nucleus, and resonance in sense (2) under resonance. In the second expression, nucleus is used in sense (3) under nucleus and resonance in sense (1) under resonance.

nuclide. A particular species of atom, characterized by the mass, the charge (number of protons), and the energy content of its nucleus. A radionuclide is a radioactive nuclide. Example: carbon 14 is a radionuclide of carbon. See also isotope.

"Nuclon."[177] Trademark for a polycarbonate thermoplastic resin.
Properties: High impact resistance not appreciably reduced by temperature fluctuations; good dimensional stability; good electrical resistance. Naturally transparent, of light straw color.
Uses: Engineering plastic for "hard service" parts and components.

"Nucon."[446] Trademark for a magnesia-chrome refractory made by a high-fired direct bonding process.
Uses: Copper and lead furnace roofs; linings of rotary kilns used to burn lime, dolomite and magnesia; steel furnace sidewalls and roofs.

"Nuflo."[285] Trademark for a high quality, extra fine kaolin treated with a surfactant by an intimate mixing process. A number of grades are obtained by applying various percentages of additive.
Uses: Conditioning prilled ammonium nitrate and other high analysis granular fertilizers.

"Nugreen."[28] Trademark for a urea fertilizer containing 45% nitrogen. Can be used as solid or spray.

nuisance particulate. Fine particles (dusts) that are relatively harmless in low concentrations. Among them are clay, calcium carbonate, emery, glass fiber, silicon carbide, gypsum, starch, Portland cement, marble, and titanium dioxide. See also dust, industrial.

"Nu-Iron."[93] Trademark for an iron chelate. A fine yellow powder used as a spray, dust or soil application to correct iron deficiencies in agriculture and horticulture. It reduces the hazard of arsenical sprays.

"Nulocrystal."[157] Trademark for a partial invert sugar, proprietary pure-food noncrystallizing product, consisting of sucrose, levulose, and dextrose.
Uses: Confectionery; baking; beverages; syrups; pharmaceuticals; animal and vegetable adhesives; paper (as a plasticizer); and varied other products where glycerin can be used.

"Nulomoline."[157] Trademark for invert sugar; pure-food product consisting of levulose and dextrose.
Uses: See "Nulocrystal."

"Nu-Manese."[93] Trademark for manganous oxide. Available in a variety of grades for use in mixed fertilizers, foliar sprays or dusts, animal nutrition and certain industrial applications.

numerals. For their use and meaning in chemical names see Geneva System; benzene; chemical nomenclature.

"Numet."[439] Trademark for depleted uranium.
Properties: Uranium metal wherein U-235 isotopic content has been reduced below 0.7% as found in normal uranium. Has high structural strength coupled with high density of 18.9 g/cc.
Derivation: Processed from uranium hexafluoride as tailing from gaseous diffusion plants.
Uses: Balance weights for aircraft, high speed rotors in gyro-compasses, gamma radiation shielding, radio-isotope transportation casks and fuel element transfer casks; in general as structural material applicable to radiation shielding.

"Nuocure 28."[74] Trademark for a curative containing 28% tin octoate for silicone resins, one-shot polyurethane foams, and rubber, offering a wide range of curing time.

"Nuodex 100 VT."[74] Trademark for quaternary ammonium naphthenate.
Use: Fungicide for vinyl formations.

"Nuogel" AO.[74] Trademark for aluminum octoate. Used in protective coatings and printing inks.
718 & 753. Aluminum soap. Used as gelling agent.

"Nuolates."[74] Trademark for driers based on metallic salts of tall oil acids.

"Nuophene."[74] Trademark for a techical grade of dihydroxydichlorodiphenylmethane.
Use: Industrial fungicide.

"Nuoplaz 1046."[74] Trademark for a non-phthalate ester used as plasticizer for stain-resistant vinyl compounds.

"Nuostabe."[74] Trademark for a series of vinyl stabilizers and fungicides. Many of them are metallic soaps or metal-organic complexes.

"Nupercainal."[305] Trademark for dibucaine (q.v.).

nuplex. An acronym of nuclear-powered agroindustrial complex. The concept is based on use of nuclear reactors to provide heat for desalination of sea water and generation of electric power.

"Nu-Pon."[448] Trademark for epoxy resin primers and enamels for household appliances, metal products, and corrosion-resistant applications.

"Nuroz."[79] Trademark for a polymerized wood rosin.
Uses: Adhesives; gloss oils; paper label coatings; oleoresinous varnishes; solder flux; spirit varnishes; waxed paper and hot melt compounds; synthetic resins.

"Nusat."[134] Trademark for proprietary satin finish nickel-plating additive.

"Nuso."[51] Trademark for highly aromatic oils used as resin plasticizers.

Nusselt number. A value used in heat transfer studies and calculations to compare heat losses by conduction from various shaped objects under various conditions. It combined into a single number the actual heat loss (Q), the temperature difference (ΔT) between the body and its surroundings, the size (d) and shape of the body and the thermal conductivity (k) of the fluid surrounding the object, in the equation $Nu = Qd/\Delta Tk$.

"Nuto."[51] Trademark for lubricating oils of good color and high resistance to oxidation; recommended for circulating and hydraulic systems.

"Nutralac."[244] Trademark for a hydrated compound consisting of sodium carbonate and sodium bicarbonate.
Uses: Dairy and food industry for neutralizing acidity in cream and related foods; dishwashing preparations; leather tanning; and textile processing.

nutrient. Any element or compound that is essential to the life and growth of plants or animals, either as such or as transformed by chemical or enzymatic reactions. In plants, nutrients include numerous mineral elements as well as nitrogen, carbon dioxide and water. In animals and man the primary nutrients are the proteins, carbohydrates and fats obtained from plants, either directly or indirectly, supplemented by vitamins and minerals. Water and oxygen are included in this definition. All told, there are 43 basic nutrients. See also food.

nutrient solution. A water solution of minerals and their salts necessary for plant growth which is used instead of soil, the plants being supported by mechanical means. Such solutions contain combined nitrogen, potassium, phosphorus, calcium, sulfur, and magnesium, together with traces of iron, boron, zinc, and copper. They are extensively used for commercial growing of flowers and vegetables particularly on islands, and also to some extent for house plants.

nutrification. Addition of nutrients to a food either to replace those lost in processing (restoration), to provide nutrients that are not normally present in the food (fortification), or to bring the food into conformity with a specific standard for that food.

nutrition. The effects of nutrients on living organisms and the biochemical mechanisms involved in bringing them about; also, that subdivision of biochemistry which deals specifically with these effects. In plant nutrition the essential requirements are carbon dioxide and water, from which the plant forms carbohydrates by photosynthesis (q.v.); nitrogen, which is essential for the synthesis of proteins by the plant, with the aid of nitrogen-fixing bacteria; as well as phosphorus, calcium, potassium and a number of trace elements (micronutrients). Besides proteins and carbohydrates, plants also synthesize vitamins and various fats and oils. Thus they provide a basis for human nutrition, both directly (grain and other vegetables) and indirectly (meats and dairy products), though the conversion to protein values for human nutrition is only about 10% for meats.

Human diet requires proteins (milk, eggs, fish and some vegetables); carbohydrates (plants); fats (oils) from both plants and animals; minerals from milk and meats; salt (chloride); vitamins from green vegetables and citrus fruits; and water. Micronutrients are furnished by sea food, cereals, vegetables, and fruit.

Human digestive processes involve primarily the hydrolysis of complex carbohydrates to simple sugars, of proteins to a mixture of amino acids, and of fats to glycerol and higher fatty acids. Hydrolysis is catlyzed by various enzymes (q.v.) in the saliva and digestive tract. The end products of digestion are absorbed across a semipermeable membrane in the intestine and thus enter the blood stream, unusable products being eliminated. The efficiency of digestion plus absorption is about 92% for protein, 95% for fat and 98% for carbohydrates. See also metabolism; digestion (1); plant (1); nutrient; RDA.

nut shells. In a fine-ground state the shells of coconuts and other nuts are a source of decolorizing carbon; the pits of peaches and similar fruits have been used for gas-adsorbent carbon.

nux vomica. See strychnine.

"Nu-Z."[93] Trademark for a fine cream-white powder assaying 52% zinc. Used as a foliar application to correct zinc deficiencies in plants, and in animal nutrition.

NW acid. Abbreviation for Neville and Winther's acid. See 1-naphthol-4-sulfonic acid.

"Nydrazid."[412] Trademark for isoniazid (q.v.).

"NyeBar."[483] Trademark for a liquid applied around oiled areas to retard oil spreading. A solvent evaporates leaving a polymer film across which lubricants do not spread or creep.

"NyeFret."[483] Trademark for a dolphin oil lubricant which reduces fretting corrosion for small springs and other instrument components. Viscosity 6 centistokes at 100°F.

"NyeSolve."[483] Trademark for a nonflammable, low-toxicity solvent for cleaning oils and oil-held soils from fine instruments.

"Nylafil."[539] Trademark for glass fiber-reinforced nylon.

nylidrin hydrochloride $HOC_6H_4CH(OH)CH(CH_3) \cdot NHCH(CH_3)(CH_2)_2C_6H_5HCl$. para-Hydroxy-alpha-[1-(1-methyl-3-phenyl-propylamino)-ethyl] benzyl alcohol hydrochloride.
Properties: White, odorless, tasteless, crystals or powder; slightly soluble inwater, alcohol; very slightly soluble in chloroform, ether; pH of 1% solution is between 4.5 and 6.5.
Grade: N.F.
Use: Medicine (treatment of heart disease).

nylon. Generic name for a family of polyamide polymers characterized by the presence of the amide group —CONH. By far the most important are nylon

66 (75% of U.S. consumption) and nylon 6 (25% of U.S. consumption). Except for slight difference in melting point, the properties of the two forms are almost indentical, though their chemical derivations are quite different. Other types are nylons 4, 9, 11, and 12 (see Grades).

Properties: Crystalline, thermoplastic polymers. May be extruded as monofilament over a wide dimensional range. Filaments are oriented by cold-drawing. Tensile strength (high-tenacity) up to 8 grams per denier (about 100,000 psi). Sp. gr. 1.14. Melting point (66) 264°C; (6) 223°C. Low water absorption. Good electrical resistance, but accumulates static charges. Highly elastic, with rather high percentage of delayed recovery at low strain values; low permanent elongation. Moisture absorption 4% at 65% R. H. Wet strength about 90% of dry strength. Can be dyed with ionic and nonionic dyestuffs. Attacked by mineral acids, but resistant to alkalies and cold abrasion. Soluble in hot phenols, cresols and formic acids; insoluble in most organic solvents. Difficult to ignite; self-extinguishing; melts, forming beads. Resistant to attack by moths, carpet beetles, etc. Compatible with wool and cotton; increases wear and crease resistance in 30% blends with natural fibers. Nontoxic. Rods and blanks are machinable.

Forms: Monofilaments, yarns, bristles, molding powders, rods, bars, sheets. Microcrystalline nylon is now available.

Grades: Nylon 66 is a condensation product of adipic acid and hexamethylenediamine developed by Carothers (q.v.) in 1935. Adipic acid is obtained by catalytic oxidation of cyclohexane. Nylon 6 is a polymer of caprolactam (q.v.), originated by I. G. Farbenindustrie in 1940. Nylon 4 is based on butyrolactam (2-pyrrolidone); its tenacity, abrasion resistance, and melting point are said to be about the same as for the 6 and 66 grades. It has excellent dyeability. See "Tajmir."

Nylon 610 (trademarked "Tynex") is obtained by condensation of sebacic acid and hexamethylenediamine, and nylon 11 (trademarked "Rilsan") from castor bean oil (developed in France). Nylon 12 (also called "Rilsan" 12) is made from butadiene, also by a French process involving photonitrosation of cyclododecane by actinic light from mercury lamps. Its properties are similar to those of nylon 11. Nylon 9 can be made from 9-aminononanoic acid, present in soybean oil. It has properties specifically desired in metal coatings and electrical parts; higher electrical resistance than 6 and 66; absorbs less moisture; and has better distortion resistance.

Uses: Tire cord; hosiery; wearing apparel component; bristles for toothbrushes, hairbrushes, paint brushes (nylon 610); cordage and towlines for gliders; fish nets and lines; tennis rackets; rugs and carpets; molded products; turf for athletic fields; parachutes; composites; sails; automotive upholstery; film; gears and bearings; wire insulaton; surgical sutures; artificial blood vessels; metal coating; pen tips; osmotic membranes; fuel tanks for automobiles.

See also polyamide; aramid.

Note: Not all nylons are polyamide resins, nor are all polyamide resins nylons, e.g. "Versamide." One class of polyamide resins distinct from nylons is derived from ethylenediamine; they may be liquids or low-melting solids and have lower molecular weight than nylons. Another class, called aramids, is aromatic in nature.

"NyoGel."[483] Trademark for a series of low shear thixotropic greases and semifluid instrument lubricants for use where nonspreading properties are critical.

"NyoSil."[483] Trademark for a wide temperature silicone instrument oil halogenated for improved wear properties. Viscosity 55 centistokes at 100°F.

nystatin (fungicidin) $C_{46}H_{77}NO_{19}$. An antifungal agent.

Properties: Yellow to light tan powder; odor suggestive of cereals; hygroscopic; affected by light, heat, air and moisture. Sparingly soluble in methanol, ethanol; very slightly soluble in water; insoluble in chloroform, ether and benzene. In solution is rapidly inactivated by acids and bases. Nontoxic.

Derivation: Produced by fermentatoin with Streptomyces noursei and aureus.

Grades: U.S.P.

Use: Medicine; feed additive.

"Nytal."[69] Trademark for talc or magnesium-calcium silicate.

Uses: Dusting uncured rubber; filler in specialized applications.

nytril. Generic name for a manufactured fiber containing at least 85% of a long-chain polymer of vinylidene dinitrile. $—CH_2C(CH)_2—$, where the vinylidene dinitrile content is no less than every other unit in the polymer chain (Federal Trade Commission).

Properties: Soft, resilient fabric is obtained; is easy to clean; does not pill; resists wrinkling, and retains shape after pressing.

Uses: Fur-like pile fabrics; sweaters; yarns; blended fabrics for coats and suits.

O

O Symbol for oxygen; the molecular formula is O_2.

o-. Abbreviation for ortho- (q.v.).

O.C. Abbreviation for oxygen consumed (q.v.).

Occupational Safety and Health Administration (OSHA). A Federal agency responsible for establishing and enforcing standards for exposure of workers to harmful materials in industrial atmospheres, and other matters affecting the health and well-being of industrial personnel. Its present standards (1975) are based on the threshold limit values (TLV) recommended by the American Conference of Governmental Industrial Hygienists for 1968, which were adopted by OSHA in 1972. New and stricter limitations on a number of toxic substances including asbestos, lead, and toluene, have been proposed. See also threshold limit value.

ocean water (sea water). A uniform solution of essentially constant composition containing 96.5% water and 3.5% ionized salts, associated compounds, elements and ionic complexes. Na^+ and Cl^- are completely ionized, but $MgSO_4$ and AuCl remain in bound form. Dissolved gases, e.g., oxygen and nitrogen, are also present.
Composition: Over 60 elements, by far the most important of which are (in tons per cubic mile): chlorine 89,500,000; sodium 49,500,000; magnesium 6,400,000; sulfur 4,200,000; calcium 1,900,000; potassium 1,800,000; bromine 306,000; carbon 132,000. Other elements present include (tons per cubic mile): copper 14; zinc 47; silver 1; lead 0.1; gold 0.02. Total solids content per cubic mile is about 166 million tons.
Properties: Colorless liquid; sp. gr. 1.03; pH 7.8–8.2; f.p. 27°F; b.p. 214°F; average temperature about 5°C; bitter taste; faint odor (depending on organic impurities).
Hazard: Ingestion of substantial amounts will create bodily chloride imbalance with harmful effects.
Uses: Source of magnesium, bromine, sodium chloride, and fresh water; possible energy source.
Note: Dr. Donald F. Othmer has stated: "It has been estimated that the heat in the warm water of the Gulf Stream could generate over 75 times the entire power production of the U.S. Thus the trillions of BTU's in a cubic mile of warm seawater which may be passed to colder seawater may develop mechanical energy during the process to make electric power. The total amount of water to be handled may be less than that required to produce the same amount of power in a hydroelectric plant...." Feasibility studies of this possibility are in process.

"Ocenol."[28] Trademark for technical grades of oleyl alcohol rich in cetyl and unsaturated alcohols.

ocher. Any of various colored earthy powders consisting essentially of hydrated ferric oxides mixed with clay, sand, etc. Some grades are calcined (burnt ocher). The colors are yellow, brown, or red. Noncombustible, nontoxic.
Uses: Paint pigments; cosmetics; theatrical make-up. See also umber; sienna.

ocimene (2,6-dimethyl-1,5,7-octatriene) $CH_2{:}C(CH_3)(CH_2)_2CH{:}C(CH_3)CH{:}CH_2$. A terpene obtained from sweet basil oil; b.p. 81°C (30 mm); sp. gr. 0.8031 (15°C). Used in flavors and perfumes. Combustible.

ocotea oil. A volatile oil derived from Ocotea cymbarum, and used for its safrole content for the manufacture of heliotropin.
Containers: Drums.

"OCPN."[233] Trademark for 2-chloro-4-nitroaniline (q.v.).

octadecadienoic acid. See linoleic acid.

n-**octadecane** $C_{18}H_{38}$ or $CH_3(CH_2)_{16}CH_3$.
Properties: Colorless liquid; sp. gr. 0.7767 (28/4°C); b.p. 318°C; m.p. 28.0°C; refractive index 1.4369 (n 28/D). Soluble in alcohol, acetone, ether, petroleum, coal-tar hydrocarbons; insoluble in water. Combustible; low toxicity.
Uses: Solvents; organic synthesis; calibration.

1,12-octadecanediol. See 1,12-hydroxystearyl alcohol.

n-**octadecanoic acid.** See stearic acid.

1-octadecanol. See stearyl alcohol.

n-**octadecanoyl chloride.** See stearoyl chloride.

octadecatrienol. See linolenyl alcohol.

9-octadecen-1,12-diol. See ricinoleyl alcohol.

1-octadecene $C_{18}H_{36}$ or $CH_3(CH_2)_{15}CH{:}CH_2$.
Properties: Colorless liquid; sp. gr. 0.7884 (20/4°C); refractive index (n 20/D) 1.4456; b.p. 180°C (15 mm); soluble in alcohol, acetone, ether, petroleum and coal-tar solvents; insoluble in water. Combustible.
Grade: 95% min. purity.
Containers: Glass bottles.
Uses: Organic synthesis; surfactants.

octadecene-octadecadieneamine. See oleyl-linoleylamine.

cis-**9-octadecenoic acid.** See oleic acid.

trans-**9-octadecenoic acid.** See elaidic acid.

octadecenol. See oleyl alcohol.

cis-**9-octadecenoyl chloride.** See oleoyl chloride.

octadecenyl aldehyde (oleyl aldehyde) $C_{17}H_{35}CHO$.
Properties: Liquid. B.p. 167°C (20 mm); refractive index 1.4620 (25°C); sp. gr. 0.847 (25°C).
Hazard: Flammable, moderate fire risk. Irritant to skin and mucous membranes.
Uses: Intermediate for vulcanization accelerators, rubber antioxidants, synthetic drying oils and pesticides.

octadecyl alcohol. See stearyl alcohol.

octadecyldimethylbenzylammonium chloride $C_{18}H_{37}(CH_3)_2(C_6H_5CH_2)NCl$. White, crystalline powder. Soluble in water, chloroform, benzene, acetone, xylene. A quaternary ammonium salt.

octadecyl isocyanate. A straight-chain, saturated monoisocyanate consisting principally of a mixture of C_{18} and C_{16} alkyls, $CH_3(CH_2)_{16}CH_2NCO$ and $CH_3(CH_2)_{14}CH_2NCO$.
Properties: Colorless, slightly cloudy liquid at ordinary temperatures; f.p. 10–20°C; b.p. 170°C (2 mm); flash point (open cup) 355°F. Combustible.
Containers: 55-gal steel drums.
Hazard: Moderately toxic by inhalation; irritant to skin.
Uses: Intermediate in synthesis; water-repellent textiles, paper, and other surfaces.

octadecyl mercaptan. See stearyl mercaptan; thiol.

octadecyltrichlorosilane (trichlorooctadecylsilane) $C_{18}H_{37}SiCl_3$.
Properties: Water-white liquid; pungent odor; sp. gr. 0.984 (25°C); refractive index 1.4580 (25°C); b.p. 380°C. Flash point 193°F. Combustible. Soluble in benzene, ethyl ether, heptane, and perchloroethylene.
Hazard: Highly toxic; strong irritant to skin and mucous membranes.
Use: Intermediate for silicones.
Shipping regulations: (Rail) White label. (Air) Corrosive label. Not acceptable on passenger planes.

octafluoro-2-butene (perfluoro-2-butene) $CF_3CF{:}CFCF_3$.
Properties: Colorless nonflammable gas or liquid; density of liquid (at b.p.) 1.5297 g/cc; b.p. 1.2°C; f.p. −136 to −134°C; specific volume 2.7 cu ft/lb (70°F, 1 atm). Nontoxic.
Grade: 95%.
Uses: Organic intermediate; fluorocarbon polymers.

octafluoropropane (perfluoropropane) C_3F_8.
Properties: Colorless, nonflammable gas; sp. gr. 1.29; b.p. −36.7°C; f.p. approx. −160°C; specific volume 2.02 cu ft/lb (70°F, 1 atm). Nontoxic.
Grade: 98% min purity.
Derivation: Electrofluorination of various organic compounds.
Uses: Refrigerant (when combined with chlorofluorohydrocarbons); gaseous insulator, especially for radar wave guides.

octakis(2-hydroxypropyl)sucrose.
Properties: Viscous straw-colored liquid; sp. gr. 1.170 (70/20°C); refractive index 1.485 (25°C); pour point 38°C; flash point 485°F; soluble in water, methanol and ether. Combustible.
Uses: Crosslinking agent for urethane foams; plasticizer for cellulosics, glue, starch and many resins.

octamethyl pyrophosphoramide. See schradan.

"Octamine."[248] Trademark for a reaction product of diphenylamine and diisobutylene.
Properties: Light brown granular waxy solid; sp. gr. 0.99; m.p. 75–85°C; good storage stability. Soluble in gasoline, benzol, ethylene dichloride and acetone. Insoluble in water.
Use: Rubber antioxidant.

octanal (n-octyl aldehyde; aldehyde C-8; caprylic aldehyde) $CH_3(CH_2)_6CHO$.
Properties: Colorless liquid with strong fruity odor; sp. gr. 0.820–0.830; refractive index 1.418–1.425; b.p. 163°C. Soluble in 70% alcohol and mineral oil; insoluble in glycerol. Low toxicity. Flash point 125°F (c.c.). Combustible.
Grades: Technical; F.C.C.
Hazard: Moderate fire risk.
Uses: Perfumery; flavors.

n-octane C_8H_{18} or $CH_3(CH_2)_6CH_3$.
Properties: Colorless liquid; sp. gr. 0.7026 (20/4°C); refractive index (n 20/D) 1.39745; b.p. 125.667°C; f.p. −56.798°C; flash point 56°F; autoignition temp. 428°F. Soluble in alcohol, acetone; insoluble in water.
Grades: 95%; 99% research.
Containers: Bottles; 5-gal drums; tank cars.
Hazard: Flammable, dangerous fire risk. Tolerance, 400 ppm in air.
Uses: Solvent; organic synthesis; calibrations; azeotropic distillations.
Shipping regulations: (Rail) Flammable liquid, n.o.s., Red label. (Air) Flammable Liquid label.

1,8-octanedicarboxylic acid. See sebacic acid.

octanedioic acid. See suberic acid.

octane number. A number indicating the antiknock properties of an automotive fuel mixture under standard test conditions. Pure normal heptane (a very high-knocking fuel) is arbitrarily assigned an octane number of zero, while isooctane (a branched-chain paraffin) is assigned 100. Thus a rating of 80 for a given fuel indicates that its degree of knocking in a standard test engine is that of a mixture of 80 parts isooctane and 20 parts n-heptane.

Octane ratings as high as 115 have been obtained by addition of tetraethyllead to isooctane. Premium leaded gasolines have a Research octane rating of about 100, but this value drops to from 85 to 90 for unleaded gasolines. The octane rating scale ends at 125, and any higher figure is meaningless.

Research octane numbers are those obtained under test or "laboratory" conditions; they generally run about ten points higher than the so-called Motor octane numbers, which represent actual road operating conditions. Thus the 90 Research o.n. gasoline (unleaded), necessitated by air pollution requirements, has a Motor o.n. of about 80. See also under gasoline.

octanoic acid (octylic acid; octoic acid; caprylic acid). $CH_3(CH_2)_6COOH$.
Properties: Colorless oily liquid; slight odor; rancid taste; sp. gr. 0.9105 (20/4°C); m.p. 16°C; b.p. 237.9°C, 147.9°C (30 mm); refractive index 1.4278 (20°C). Flash point 270°F. Slightly soluble in water; soluble in alcohol and ether. Combustible; low toxicity.
Derivation: By saponification and subsequent distillation of coconut oil.
Method of purification: Crystallization or rectification.
Grades: Techical; 99%; F.C.C.
Containers: Drums; tanks; carload lots.
Uses: Synthesis of various dyes, drugs, perfumes, flavors; antiseptics and fungicides; ore separations.

octanol. See octyl alcohol.

2-octanone. See methyl hexyl ketone.

octanoyl chloride (capryloyl chloride; sometimes called caprylyl chloride) $CH_3(CH_2)_6COCl$.
Properties: Water-white to straw-colored liquid with pungent odor. Miscible with most common solvents;

reacts with alcohol and water. Freezing point below −70°C; distillation range 183–212°C; sp. gr. 0.9576 (15.5/15.5°C); refractive index 1.4357 (n 20/D); flash point 180°F; combustible.
Containers: 5-, 13-gal carboys.
Hazard: Irritant to eyes and skin.
Use: Synthesis.

"Octasols."[134] Trademark for metal salts of 2-ethylhexoic acid, in liquid form. Soluble in most hydrocarbons, vegetable oils, paint vehicles, etc. Used as paint driers, wetting agents, catalysts, etc.

1-octene (1-octylene; 1-caprylene) C_8H_{16} or $CH_3(CH_2)_5CH{:}CH_2$.
Properties: Colorless liquid; sp. gr. 0.7160 (20/4°C); b.p. 121.27°C; f.p. −102.4°C; refractive index (n 20/D) 1.4088; flash point 70°F (TOC); soluble in alcohol, acetone, ether, petroleum and coal-tar solvents; insoluble in water.
Grades: 95%, 99%, research.
Containers: Glass bottles; drums.
Hazard: Flammable, dangerous fire risk.
Uses: Organic synthesis; plasticizer; surfactants.
Shipping regulations: Flammable liquid, n.o.s., (Rail) Red label. (Air) Flammable Liquid label.

2-octene C_8H_{16} or $CH_3(CH_2)_4CH{:}CHCH_3$. Cis and trans forms exist.
Properties: Colorless liquid; sp. gr. cis 0.7243, trans 0.7199, commercial 0.7185–0.7200 (20/4°C); b.p. cis 125.6°C, trans 125.0°C, commercial 124.5-127°C; f.p. −94.04°C; refractive index cis 1.4150, trans 1.4132, commercial 1.4120–1.4145 (n 20/D); flash point (mixed isomers) 70°F (TOC); soluble in alcohol, acetone, ether, petroleum and coal-tar solvents; insoluble in water.
Containers: Glass bottles; drums.
Grades: 95; 99 mole %.
Hazard: Flammable, dangerous fire risk.
Uses: Organic synthesis; lubricants.
Shipping regulations: Same as 1-octene.

"Octin."[9] Trademark for isometheptene (q.v.).

octoic acid. An eight-carbon acid; usually designates caprylic acid.

octyl. The general name describing all eight-carbon radicals having the formula C_8H_{17}; often used interchangeably for the 2-ethylhexyl isomer (q.v.).

n-**octyl acetate** (acetate C-8; caprylyl acetate) $CH_3COO(CH_2)_7CH_3$.
Properties: Colorless liquid; floral-fruity odor. Slightly soluble in water; soluble in alcohol and most other organic liquids. Sp. gr. 0.865–0.869; flash point 180°F (O.C.); refractive index 1.419–1.422; b.p. 199°C. Combustible; low toxicity.
Containers: Glass bottles.
Uses: Perfumery; flavors.

n-octyl alcohol, primary (1-n-octanol; alcohol C-8; heptyl carbinol) $CH_3(CH_2)_6CH_2OH$. In industrial practice, the term octyl alcohol has been used for both 1-octanol and 2-ethylhexanol. The latter is also sometimes called isooctanol. The term capryl alcohol has been used for both 1-octanol and 2-octanol. It therefore seems preferable to designate the normal primary alcohol as 1-n-octanol.
Properties: Colorless liquid with penetrating, aromatic odor; sp. gr. 0.826 (20°C); b.p. 194–195°C, 108.7°C (30 mm); refractive index 1.430 (20°C); f.p. −16°C. Flash point 178°F; combustible. Low toxicity. Miscible with alcohol, chloroform, mineral oil; immiscible with water and glycerol.
Derivation: By reduction of caprylic acid.
Grades: Technical; C. P.; pure; perfume; F.C.C.
Containers: Drums; tank cars.
Uses: Perfumery; cosmetics; organic synthesis; solvent; manufacture of high-boiling esters; antifoaming agent; flavoring agent.

sec.-n-**octyl alcohol** (2-n-octanol; methyl hexyl carbinol) $CH_3(CH_2)_5CHOHCH_3$. Frequently called capryl alcohol (q.v.).
Properties: Colorless, oily liquid; refractive; aromatic odor. Miscible with alcohol, ether; slightly soluble in water. Sp. gr. 0.825 (15°C); b.p. 178–179°C; f.p. −38°C; refractive index 1.437 (20°C). Flash point 190°F. Combustible; low toxicity.
Derivation: By distilling sodium ricinoleate with an excess of sodium hydroxide.
Grades: Technical 92–99%; pure.
Containers: Bottles; drums; tank cars.
Uses: Solvent; manufacture of plasticizers, wetting agents, foam control agents, hydraulic oils, petroleum additives, perfume intermediates; masking of industrial odors.

octyl aldehyde. See octanal and 2-ethylhexaldehyde.

n-**octalamine.** $CH_3(CH_2)_7NH_2$.
Properties: Water-white liquid; amine odor; boiling range 170–179°C; sp. gr. 0.779 (20/20°C); refractive index 1.431 (20°C); flash point 140°F. Low toxicity. Combustible.

tert-**octylamine** $(CH_3)_3CCH_2C(CH_3)_2NH_2$.
Properties: Colorless liquid; amine odor; b.p. 137–143°C; sp. gr. 0.771 (25°C); refractive index 1.423 (25°C); flash point 92°F (o.c.); insoluble in water; soluble in common organic solvents, especially petroleum hydrocarbons.
Containers: Drums; tank cars.
Hazard: Flammable, moderate fire risk. Skin irritant.
Uses: Intermediate for rubber accelerators, insecticides, fungicides, dyestuffs; pharmaceuticals.

N-**octylbicycloheptene dicarboximide** (N-(2-ethylhexyl)-bicyclo(2,2,1)-5-heptene-2,3-dicarboximide) $C_8H_{17}NC_9H_8O_2$.
Properties: Liquid; b.p. 158°C (2 mm); sp. gr. 1.05 (18°C); refractive index 1.505 (n 20/D). Miscible with most organic solvents and oils. Combustible.
Derivation: From maleic anhydride, cyclopentadiene and 2-ethylhexylamine.
Use: Insecticide and pesticide synergist.

n-**octyl bromide** (capryl bromide; caprylic bromide) $CH_3(CH_2)_6CH_2Br$.
Properties: Colorless liquid. Miscible with alcohol, ether; immiscible with water. Sp. gr. 1.118 (15°C); b.p. 202°C; f.p. −55°C; refractive index 1.4503 (25°C). Low toxicity.
Grades: Technical.
Use: Synthesis of quaternary ammonium compounds, organometallics, vinyl stabilizers.

octyl carbinol. See nonyl alcohol.

n-**octyl chloride** $CH_3(CH_2)_6CH_2Cl$.
Properties: Colorless liquid; soluble in most organic solvents. Sp. gr. 0.8697 (25/15.5°C); refractive index (n 25/D) 1.4288; f.p. −62°C; b.p. 181.6°C; flash point 158°F. Combustible.
Containers: 55-gal steel drums; tank cars.
Uses: Chemical intermediate; manufacture of organometallics.

n-octyl n-decyl adipate (NODA).
Properties: Liquid, mild odor; sp. gr. 0.92–0.98 (20/20°C); f.p. −50°C; boiling range 220–254°C (4 mm); refractive index 1.447. Combustible.
Containers: 5-gal cans; 55-gal drums; tank cars.
Use: Low-temperature plasticizer.

n-octyl-decyl alcohol. A blend of alcohols. Combustible. Available in tank cars and trucks. Used as intermediate for plasticizers.

n-octyl n-decyl phthalate.
Properties: Clear liquid; mild characteristic odor. Sp. gr. 0.972–0.976 (20/20°C); f.p. −40°C; boiling range 232–267°C (4 mm); refractive index 1.482 (25°C). Flash point 455°F. Combustible.
Containers: 55-gal drums; tank cars and trucks.
Uses: Plasticizer for vinyl resins.

octylene. See octene.

octylene glycol titanate.
Properties: Light-yellow solid.
Derivation: Reaction of butyl titanate with octylene glycol.
Uses: Cross-linking agent; surface-active agent.
See also titanium chelate.

oxtylene oxide Mixed $CH_3(CH_2)_5\overline{CHCH_2O}$ and $CH_3(CH_2)_4\overline{CHCH(CH_3)O}$. Sp. gr. of liq. 0.830 (25°C). Combustible.
Uses: Organic intermediate; epoxy resins.

octyl formate $C_8H_{17}OOCH$.
Properties: Colorless liquid; fruity odor. Soluble in mineral oil; practically insoluble in glycerin. One ml dissolves in 5 ml of 70% alcohol; sp. gr. 0.869–0.872 (25°C); refractive index 1.4180–1.4200 (20°C). Combustible.
Grade: F.C.C.
Use: Flavoring agent.

octylic acid. See caprylic acid.

2-octyl iodide (caprylic iodide; secondary normal capryl iodide) $CH_3(CH_2)_5)CHICH_3$.
Properties: Oily liquid. Sp. gr. 1.318 (18°C); b.p. (approx) 210°C (dec).
Hazard: Moderately toxic by ingestion and inhalation. Keep away from light and air.
Use: Organic synthesis.

octylmagnesium chloride $C_8H_{17}MgCl$. A Grignard reagent available in tetrahydrofuran solution.
Containers: Up to 55-gal drums.

n-octyl mercaptan $C_8H_{17}SH$.
Properties: Water-white liquid with mild odor; b.p. 199°C; sp. gr. 0.8395 (25/4°C); refractive index 1.4497 (25°C); flash point 115°F (o.c.). Combustible.
Hazard: Moderate fire risk. Probably toxic.
Uses: Polymerization conditioner; synthesis.

tert-octyl mercaptan $C_8H_{17}SH$.
Properties: Colorless liquid. Boiling range 154–166°C; sp. gr. 0.848 (60/60°F); refractive index 1.454 (n 20/D); flash point 105°F. Combustible.
Grade: 95%.
Containers: Drums; tank cars.
Hazard: Moderate fire risk. Probably toxic.
Uses: Polymer modification; lubricant additive.

n-octyl methacrylate $H_2C{:}C(CH_3)COOC_8H_{17}$.
Properties: Water-insoluble colorless liquid; polymerizes to a resin if unstabilized.

octyl peroxide (caprylyl peroxide).
Properties: Straw-colored liquid with sharp odor. Immiscible with water.
Hazard: Dangerous fire risk. Strong oxidizing agent.
Shipping regulations: (Solution) (Rail) Yellow label. (Air) Organic Peroxide label. Legal label name: caprylyl peroxide.

octyl phenol (diisobutyl phenol) $C_8H_{17}C_6H_4OH$. Probably a mixture of isomers.
Properties: White flakes; congeals 72–74°C; sp. gr. 0.89 (90°C); b.p. 280–302°C; hydroxyl coefficient 259–275. Insoluble in hot and cold water. Limited solubility in alkalies. Soluble in 1:1 mixture of methanol and 50% aqueous potassium hydroxide, also in alcohol, acetone, fixed oils. Available commercially as a liquid. Combustible.
Derivation: (p-tert-isomer): Catalytic alkylation of phenol with olefins.
Containers: Bags; drums; carloads; tank cars.
Use: Nonionic surfactants; plasticizers; antioxidants; fuel oil stabilizer; intermediate for resins, fungicides, bactericides, dyestuffs, adhesives, rubber chemicals.

para-tert-octylphenoxy polyethoxyethanol $(CH_3)_3CCH_2C(CH_3)_2C_6H_4O(CH_2CH_2O)_xH$. Anhydrous liquid mixture of mono-para-(1,1,3,3-tetramethylbutyl)phenyl esters of polyethylene glycols in which x varies from 5 to 15.
Properties: Yellow, viscous liquid; faint odor; bitter taste, sp. gr. 1.060; refractive index 1.489 (25°C); soluble in water, alcohol, acetone, benzene, toluene; insoluble in hexane; pH is between 7 and 9. Combustible.
Grade: N.F.
Uses: Food packaging, probably as a plasticizer for films.

para-octylphenyl salicylate $C_6H_4OHCOOC_6H_4C_8H_{17}$.
Properties: White solid; m.p. 72–74°C; density 5.6 lb/gal (20°C). Soluble in hexane, benzene, acetone and ethanol. Insoluble in water.
Use: Prevents photoxidation in polyethylene and polypropylene.

octyl phosphate. See triioctyl phosphate.

n-octyl sulfoxide isosafrole. See sulfoxide.

octyl trichlorosilane (trichlorooctylsilane) $C_8H_{17}SiCl_3$.
Properties: Water-white liquid with pungent irritating odor. Fumes readily in moist air to evolve corrosive vapors.
Hazard: Highly toxic by ingestion and inhalation; strong irritant to skin and mucous membranes. Moderate fire risk in contact with oxidizing materials.
Use: Intermediate silicones.
Shipping regulations: (Rail) White label. (Air) Corrosive label. Not accepted on passenger planes.

"Odamask."[342] Trademark for deodorizing and reodorizing aromatics for industrial use.

ODPN. Abbreviation for beta, beta′-oxydipropionitrile.

odor. An important property of many substances, manifested by a physiological sensatoin due to contact of their molecules with the olfactory nervous system. Odor and flavor are closely related, and both are profoundly affected by submicrogram amounts of volatile compounds. Attempts to correlate odor with chemical structure have produced no definitive results. Objective measurement techniques involving chromatography are under development. Even potent odors must be present in a concentration of 1.7×10^7 molecules per cc to be detected. It has been authentically stated that the nose is 100 times as sensitive in detection of threshold odor values as the best analytical apparatus.

Many compounds have a characteristic odor that is an effective means of identification. Toxic and noxious gases have distinctive odors often utilized for warning purposes. An important exception is carbon monoxide, which is almost odorless. The penetrating banana-like odor of amyl acetate has been used in mine rescue work. Among the most powerful unpleasant odors are those of organic sulfur compounds, especially ethyl mercaptan (skunk). Organic substances having a pleasant odor are broadly designated as aromatic, regardless of chemical nature. The cyclic aromatic (benzene) series of hydrocarbons was so named for this reason. Most essential oils have a pleasant odor and are the basis of perfumes and fragrances. Odor research, including evaluation by test panels, is conducted at the Olfactronics and Odor Sciences Center at Illinois Institute of Technology, Chicago.

odorant. A substance having a distinctive, sometimes unpleasant, odor which is deliberately added to essentiallay odorless materials to provide warning of their presence. For example, mercaptan derivatives may be added to natural gas for this purpose. In a broad sense, perfumes are odorants which are added to cosmetics, toilet goods, etc. largely for consumer appeal. See also odor; deodorant; perfume; fragrance.

"Oenethyl."[9] Trademark for 2-methyl-aminoheptane; employed as hydrochloride salt.

oenology. The study of wines.

"OFHC."[433] Trademark for high-conductivity oxygen-free copper in cast or wrought form.

oil. The word "oil" has little specific meaning, as it is applied to a wide range of substances that are quite different in chemical nature. Oils derived from animals or from plant seeds or nuts are chemically identical with fats, the only difference being one of consistency at room temperature. They are composed largely of glycerides of the fatty acids, chiefly oleic, palmitic, stearic and linolenic. As a rule the more hydrogen the molecule contains, the thicker the oil becomes. Petroleum (rock oil) is quite different from animal and vegetable oils, being composed of hydrocarbon mixtures comprising hundreds of chemical compounds. It is thought to be derived from the remains of tiny sea animals laid down in past geologic ages. Following is a classification of oils by type and function.

I. *Mineral*
  1. Petroleum
    (a) Aliphatic or wax-base (Pennsylvania)
    (b) Aromatic or asphalt-base (California)
    (c) Mixed-base (Midcontinent)
  2. Petroleum-derived
    (a) Lubricants: engine oil, machine oil, cutting oil
    (b) Medicinal: refined paraffin oil

II. *Vegetable* (chiefly from seeds or nuts)
  1. Drying (linseed, tung, oiticica)
  2. Semidrying (soybean, cottonseed)
  3. Nondrying (castor, coconut)
  4. Inedible soap stocks (palm, coconut)

III. *Animal*
  These usually occur as *fats* (tallow, lard, stearic acid). The liquid types include fish oils, fish-liver oils, oleic acid, sperm oil, etc. They usually have a high fatty acid content.

IV. *Essential*
  Complex volatile liquids derived from flowers, stems and leaves, and often the entire plant. They contain terpenes (pinene, dipentene, etc.) and are used chiefly for perfumery and flavorings. Usually resinous products are admixed with them. Turpentine is a highly resinous essential oil.

V. *Edible*
  Derived from fruits or seeds of plants and used chiefly in foodstuffs (margarine, etc.). Most common are corn, coconut, soybean, olive, cottonseed.

oil black. A carbon black made from oil, usually an aromatic-type petroleum oil. See furnace black.

"Oil Blue A."[28] Trademark for hydrocarbon-soluble dye for imparting distinctive color to gasoline and other petroleum products. Essentially 1,4-di(isopropylamino)anthraquinone. Available in liquid and flake form.

oil bois de rose (rosewood oil). An essential oil derived from linaloe wood; formerly an important perfume component, but has been largely replaced by synthetics in recent years. Its main constituent is linalool (q.v.).

"Oil Bronze."[28] Trademark for blend of Oil Red and Oil Orange, petroleum dyes used to color gasoline, etc.

oil cake. The residue obtained after the expression of vegetable oils from oil-bearing seeds, used as cattle feed and fertilizer. When ground they are known as meal. See also cottonseed cake and meal; peanut oil meal, etc.

oil, chloronaphthalene.

Properties: Almost colorless, thin, mobile liquid. Sp. gr. 1.20–1.25 (68°F); liquid down to −25°F; congealing point −30°F; flash point (approx) 350°F; volatile at 212°F; b.p. 480–550°F. Combustible. Soluble in practically all organic solvent liquids and oils.

Derivation: By chlorinating naphthalene.

Containers: 55-gal steel drums; tank cars.

Hazard: Toxic by inhalation; strong skin irritant.

Uses: Plasticizer; carbon softener and remover; heat-transfer medium; solvent for rubber, aniline and other dyes, mineral and vegetable oils, varnish gums and resins, waxes.

"Oildag."[46] Trademark for a dispersion of colloidal graphite in petroleum oil for general industrial and automotive lubricants.

"Oilfos."[58] Trademark for glassy sodium phosphate, a dry powder used exclusively for controlling viscosity of oil-well drilling muds.

oil gas. A gas made by the reaction of steam at high temperatures on gas oil or similar fractions of petroleum, or by high-temperature cracking of gas oil. One typical analysis is: heating value about 554 Btu/cu ft, illuminants 4.2%, carbon monoxide 10.4%,

hydrogen 47.6%, methane 27.0%, carbon dioxide 4.6%, oxygen 0.4%, nitrogen 5.8%. Autoignition temp. 637°F.
Hazard: Highly toxic by inhalation. Flammable; dangerous fire and explosion risk.
Shipping regulations: (Air) Flammable gas label. Not acceptable on passenger planes.

oiliness. That property of a lubricant that causes a difference in coefficient of friction when all the known factors except the lubricant itself are the same. This concept is also expressed by the term lubricity.

oil of bitter almond. The volatile essential oil distilled from ground kernels of bitter almonds, which contain amygdalin (q.v.), imported from Spain, Portugal, and France. It consists of about 90% benzaldehyde and 10% hydrocyanic acid, which must be removed before use. Sp. gr. 1.04–1.06. Used in cosmetic creams, perfumes, and pharmacy. See also benzaldehyde; benzoin.

oil of mirbane. See nitrobenzene.

oil of sweet almond. The fatty nondrying oil comprised chiefly of unsaturated fatty acids expressed from the kernels of either bitter or sweet almonds. As sweet almonds are too valuable as such to be used for oil, which is almost identical in its properties with that from bitter almond, the term "sweet almond oil" is actually a misnomer. Most of the commercial product is derived from bitter almonds. Its sp. gr. is 0.91 and iodine no. 95–105. Used as emollient in cosmetics, and in perfumes.

oil of vitriol. See sulfuric acid.

oil of wintergreen. See methyl salicylate.

"Oil Orange."[28] Trademark for a petroleum dye used to color gasoline, etc. Essentially phenylazo-2-naphthol.

"Oil Red."[28] Trademark for a petroleum dye used to color gasoline, etc. Essentially methyl derivatives of azobenzene 4-azo-2-naphthol.

oil sands (tar sands). Porous geological structures located in the Athabaska region of northern Alberta which bear a high ratio of petroleum-to-solids content. These structures lie not only on the surface but extend to considerable depth beneath it. Only a small fraction of the oil values can be extracted by strip mining. Though these deposits represent a potential energy source of considerable magnitude, the high cost of extraction has retarded its development. Experimental investigations have been carried on for some time without significant results.

oil shale. Extensive sedimentary rock deposits in Western U.S. (chiefly in Wyoming, Colorado and Utah) having a relatively high (30 to 60%) content of bituminous matter (kerogen) from which shale oil can be derived by heating in absence of air. Some shales yield up to 65 gallons of oil per ton. See also shale oil.

oil spill treatment. Several methods have been tried, none of which has proved more than partially satisfactory: (1) Mechanical containment by log booms or similar devices; (2) combustion; (3) absorption by straw, colloidal silica, or polyurethane foam; (4) chemical detergents or emulsifiers. Of these the absorption method shows the most promise, not only for its direct effectiveness but because it does less ecological damage than chemical dispersants, which expose marine life to toxic components of petroleum hydrocarbons. Combustion is ineffective, as the heat applied is too rapidly dissipated to ignite the oil. Research is being conducted on development of active bacteria that can decompose petroleum by enzymatic means. An international system for monitoring petroleum in the oceans was instituted in 1975.

oil varnish. See varnish.

oil, vulcanized. See factice.

oil white. One of several mixtures of lithopone and white lead or zinc white (q.v.). It may also contain gypsum, magnesia, whiting, or silica. Used as a white-lead substitute.

"Oil Yellow NB."[28] Trademark for a petroleum dye used to color gasoline, etc. Essentially para-diethylaminoazobenzene.

oiticica oil.
Properties: Light yellow drying oil which slowly solidifies unless first heated for about 30 minutes at 210–220°C. Properties vary according to whether the oil is raw or heat-treated (semi-polymerized). Combustible; nontoxic.
Derivation: By expression from the seeds of the Brazilian oiticica tree, Licania rigida.
Chief constituents: Glycerides of alpha-licanic acid (4-keto-9,11,13-octadecatrienoic acid).
Containers: Steel drums; tank trucks.
Uses: Drying oil in paints, varnishes, etc.

"O K Liquid."[487] Trademark for a combination of anionic and nonionic surfactants plus suitable solvents for use as a foaming agent in air and gas drilling of oil wells.

"OKO."[64] Trademark for a series of polymerized linseed oils produced by vacuum polymerization to remove all traces of decomposed matter. Low acid, light, color, minimum after-yellowing.
Uses: Paints, varnishes, enamels, flats, printing inks, brake linings.

-ol. A suffix indicating that one or more hydroxyl groups (OH) are present in an organic compound, e.g., alcohol, phenol, menthol. Thiol is an exception, the oxygen of the OH group being replaced by sulfur. There are a few other exceptions among the essential oils, e.g., eucalyptol.

oleamide cis-$CH_3(CH_2)_7CH{:}CH(CH_2)_7CONH_2$.
Properties: Ivory-colored powder; approximate melting point 72°C; sp. gr. 0.94. Combustible; low toxicity.
Grade: Refined.
Uses: Slip-agent for extrusion of polyethylene; wax additive; ink additive.

"Oleandomycin."[299] Trademark for oleandomycin phosphate.

oleandromycin phosphate $C_{35}H_{61}NO_{12} \cdot H_3PO_4$.
Properties: White, odorless, crystals or powder; soluble in water, alcohol; slightly soluble in ether.
Derivation: Streptomyces antibioticus.
Grade: N.F.
Use: Medicine.

"Olefane."[429] Trademark for polypropylene film.

olefin (alkene). A class of unsaturated aliphatic hydrocarbons having one or more double bonds, obtained by cracking naphtha or other petroleum

fractions at high temperatures (1500 to 1700°F). Those containing one double bond are called alkenes, and those with two alkadienes, or diolefins. They are named after the corresponding paraffins by adding "-ene" or "-ylene" to the stem. Alpha-olefins are particularly reactive because the double bond is on the first carbon. Examples are 1-octene and 1-octadecene, which are used as the starting point for medium-biodegradable surfactants. Other olefins (ethylene, propylene, etc.) are starting points for manufactured fibers, of which the polymer comprises at least 85% by weight.
See also diolefin.

oleic acid (cis-9-octadecenoic acid; red oil) $CH_3(CH_2)_7CH:CH(CH_2)_7COOH$. A monounsaturated fatty acid; a component of almost all natural fats as well as tall oil. Most oleic acid is derived from animal tallow or vegetable oils.
Properties: Commercial grades: Yellow to red oily liquid; lardlike odor; darkens on exposure to air. Insoluble in water; soluble in alcohol, ether and most organic solvents, fixed and volatile oils. Solvent for other oils, fatty acids and oil-soluble materials.
Purified grades: Water-white liquid; sp. gr. 0.895 (20/4°C); m.p. 13.2°C; b.p. 286°C (100 mm), 225°C (10 mm); refractive index 1.4599 (20°C); acid value 196–204; iodine value 83–103; saponification value 196–206. Flash point 372°F; combustible; nontoxic.
Derivation: The free fatty acid is obtained from the glyceride by hydrolysis, steam distillation and separation by crystallization or solvent extraction. Filtration from the press cake results in the oleic acid of commerce (red oil) which is purified and bleached for specific uses.
Grades: Variety of technical grades; grade free from chick edema factor; U.S.P.; F.C.C.; 99+%. A purified technical oleic acid containing 90% or more oleic, 4% maximum linoleic and 6% maximum saturated acids is available.
Containers: Drums; tank trucks; tank cars.
Uses: Soap base; manufacture of oleates; ointments; cosmetics; polishing compounds; lubricants; ore flotation; intermediate; surface coatings; food-grade additives.

olein (triolein; glyceryl trioleate) $(C_{17}H_{33}COO)_3C_3H_5$. The triglyceride of oleic acid, occurring in most fats and oils. It constitutes about 70–80% of olive oil.
Properties: Yellow, oily liquid; sp. gr. 0.915; m.p. −4 to −5°C; soluble in chloroform, ether, carbon tetrachloride; slightly soluble in alcohol. Combustible; nontoxic.
Impurities: Stearin, linolein.
Derivation: Refined natural oils.
Use: Textile lubricants.

oleoresin. Any of a number of mixtures of essential oils and resins characteristic of the tree or plant from which they are derived. Most types are semisolid and tacky at room temperature, becoming soft and sticky at higher temperatures. They have various distinctive odors. See also balsam; rosin.

oleoyl chloride (cis-9-octadecenoyl chloride) $CH_3(CH_2)_7CH:CH(CH_2)_7COCl$.
Properties: Liquid; b.p. 175–180°C (3 mm); soluble in hydrocarbons and ethers; reacts slowly with water. Combustible; low toxicity.
Containers: Bottles; carboys; 55-gal drums.
Use: Chemical intermediate.

N-oleoylsarcosine $C_{17}H_{33}C(O)N(CH_3)CH_2COOH$.
Properties: Amber liquid; sp. gr. 0.955 (20/20°C); refractive index 1.4703 (n 20/D). About 95% pure. Combustible.
Containers: Drums; tanks.
Use: Surfactants.

oleum. The Latin name for oil. Also applied to fuming sulfuric acid (q.v.).

oleyl alcohol (octadecenol) $CH_3(CH_2)_7CH:CH(CH_2)_7CH_2OH$. The unsaturated alcohol derived from oleic acid. Clear viscous liquid at room temperature. Iodine value 88, cloud point 20°F, boiling point 333°C; f.p. −7.5°C. Sp. gr. 0.84. Combustible; low toxicity.
Impurities: Linoleyl, myristyl and cetyl alcohols.
Derivation: Reduction of oleic acid; occurs in fish and marine mammal oils.
Grades: Technical; commercial (80–90% pure).
Uses: Chemical synthesis, petroleum additives, surfactants; polymers; plasticizer; antifoaming agent.

oleyl aldehyde. See octadecenyl aldehyde.

oleylhydroxamic acid $C_{17}H_{33}CONHOH$.
Properties: Waxy solid; off-white color; sp. gr. 0.897 (70/25°C); insoluble in water; soluble in aqueous potassium hydroxide and organic solvents.

oleyl-linoleylamine (octadecene-octadecadieneamine).
Properties: Highly unsaturated primary amine; soluble in many organic solvents; insoluble in water. Sp. gr. 0.83; m.p. 19°C; b.p. 198–209°C; amine no. 200–210; iodine value 90 min.
Use: Organic intermediate.

oleyl methyl tauride. See sodium N-methyl-N-oleoyl taurate.

oligo- A prefix meaning "a few" or "very little." See following entries.

oligodynamic. Literally, active in small amounts. In technical literature, the term describes the sterilizing or purifying action of a substance, e.g., silver.

oligomer. A polymer molecule consisting of only a few monomer units (dimer, trimer, tetramer).

oligopeptide. A peptide made up of not more than ten amino acids.

oligosaccharide. A carbohydrate containing from two up to ten simple sugars linked together (e.g., sucrose, composed of dextrose and fructose). Beyond ten they are called polysaccharides.

olive oil.
Properties: Pale yellow or greenish-yellow liquid; a nondrying oil; slight odor and taste. Soluble in ether, chloroform and carbon disulfide; sparingly soluble in alcohol. Sp. gr. 0.910–0.918; saponification value 188–196; iodine value 77–88. Flash point 437°F; combustible; nontoxic. Cloud point 20–30°F.
Derivation: By expressing the pulp of the fruit of the olive tree, Olea europea, cultivated in Spain, Greece, Tunisia and Turkey.
Chief constituents: oleic acid, palmitic acid, linoleic acid.
Grades: U.S.P.; edible; commercial; sulfur oil (olive oil foots). The edible and commercial oils are obtained by expression, and the last grade by extraction, usually with carbon disulfide.
Containers: Drums.
Uses: Salad dressings and other foods; ointments; liniments, etc.; Castile soap; special textile soaps; lubricant; sulfonated oils; cosmetics.

olivine (chrysolite) $(Mg, Fe)_2SiO_4$. Natural magnesium-iron silicate, found in igneous and metamorphic

rocks, meteorites, and blast furnace slags. A complete series exists from Fe_2SiO_4 to Mg_2SiO.
Grades: Crude, 20 mesh, 100 mesh.
Uses: Refractories; cements.

"OLOA."[151] Trademark for mixtures of metal-organic and/or organic compounds in a lubricating oil carrier. Used to fortify well-refined base stocks to yield motor oils that minimize engine deposits, engine wear, bearing corrosion, engine rusting, and friction between rubbing surfaces.

"OM-4."[299] Trademark for an antibiotic preparation for addition to animal feeds containing oleandomycin.

"Omadine."[84] Trademark for a series of derivatives of pyridinethione [such as 1-hydroxy-2-pyridinethione, $C_5H_4NOH(S)$] having bactericide-fungicide properties. Used in cosmetics, textiles, cutting oils and coolant systems, vinyl films and rubber products.

"Omazene."[84] Trademark for copper dihydrazinium sulfate (q.v.). Available as 50% wettable powder.
Use: Foliage fungicide.

OMC. Abbreviation of oxidized microcystalline waxes.

"Omnadin."[162] Trademark for prolipin (q.v.).

OMPA. Abbreviation for octamethyl pyrophosphoramide. See schradan.

"Onamine."[328] Trademark for a series of liquid cationic detergents.
Uses: Intermediate; acid-stable emulsifier; corrosion inhibitor.

ONB. Abreviation for ortho-nitrobiphenyl.

oncogenic. Designating any substance that will cause tumors in test animals, either benign or malignant. EPA pesticide regulations use this term instead of "tumorogenic" and "carcinogenic."

"Ondal" A.[28] Trademark for an oxidizing agent for one-bath soaping and oxidizing of vat colors.

one-step resin. See A-stage resin.

"Onyxcide 75%."[328] Trademark for a 75% concentration of alkenyl dimethyl ethyl ammonium bromide. Used as an algicide.

"Onyx-ol."[328] Trademark for a series of fatty acid ethanolamine and isopropanolamine condensates.
Uses: Foam stabilizers; thickeners; nonionic detergents.

"OP-100,101,102."[210] Trademark for a series of specialty chemicals used in paper wet strengthening; selective oxidatoin of carbohydrates, amino acids, etc.; cotton grafting; photographic paper, etc.

opacity. The optical density of a material, usually a pigment; the opposite of transparency. A colorant or paint of high opacity is said to have good hiding power or covering power, by which is meant its ability to conceal another tint or shade over which it is applied. Apparatus for measuring opacity is available.

"Opalon."[58] Trademark for vinyl chloride extruding and calendering material and molding compounds.

"Opalwax."[202] Trademark for hydrogenated castor oil. See "Castorwax" and glyceryl tri-12-hydroxystearate.

"Opax."[337] Trademark for a brand of zirconium oxide containing 88% minimum of zirconium oxide and 7% maximum silicon dioxide. M.p. 4500°F. Used in ceramic enamels and glazes. See also zirconium oxide.

OPDN. Abbreviation for beta, beta-oxydipropionitrile (q.v.).

"Opex."[141] Nitrocellulose lacquer for automotive and industrial finishes.

OPG. Abbreviation for oxypolygelatin (q.v.).

opium. A plant alkaloid.
Derivation: The air-dried, milky exudate obtained from the unripe capsules of Papaver somniferum.
Containers: Tins.
Hazard: Highly toxic. A habit-forming narcotic. Importation and sale restricted by law in U.S.
Use: Medicine; source of morphine.

"Oppanol."[440] Trademark for a series of polyisobutylenes, varying from oily liquids through highly viscous materials to rubberlike solids according to degree of polymerization.

optical activity. See optical rotation.

optical brightener (optical bleach; colorless dye; fluorescent brightener). A colorless fluorescent organic compound which absorbs ultraviolet light and emits it as visible blue light. The blue light masks the undesirable yellow of textiles, paper, detergents and plastics. Some examples are: derivatives of 4,4′-diaminostilbene-2,2′-disulfonic acid; coumarin derivatives such as 4-methyl-7-diethylaminocoumarin.

optical crystal. A comparatively large crystal, either natural or synthetic, used for infrared and ultraviolet optics, piezoelectric effects, and short wave radiation detection. Examples are sodium chloride, potassium iodide, silver chloride, calcium fluoride, and (for scintillation counters) such organic materials as anthracene, naphthalene, stilbene, and terphenyl.

optical glass. See glass.

optical isomer. Either of two kinds of optically active three-dimensional isomers (stereoisomers). One kind is represented by mirror-image structures called enantiomers, which result from the presence of one or more asymmetric carbon atoms in the compound (glyceraldehyde, lactic acid, sugars, tartaric acid, amino acids). The other kind is exemplified by diastereoisomers, which are not mirror images. These occur in compounds having two or more asymmetric carbon atoms; thus such compounds have 2^n optical isomers, where n is the number of asymmetric carbons. See also enantiomer; diastereoisomer; optical rotation; asymmetry.

optical microscopy. Microscopy which utilizes light in the visible wavelengths of the spectrum, as opposed to ultraviolet microscopy and electron microscopy, which utilize shorter wavelengths.

optical rotation. The change of direction of a ray of plane-polarized light to either the right or the left as it passes through a molecule containing one or more asymmetric carbon atoms, e.g., sugars. The direction of rotation, if to the right, is indicated by either a plus sign (+) or an italic *d*; if to the left, by a minus sign (−) or an italic *l*. Molecules having a right-handed configuration (D) usually are dextrorotatory, D(+), though they may be levorotatory, D(−); those having a left-handed configuration (L) are usually levorotatory, L(−), but may be dextrorotatory, L(+). Compounds having this property are said to be optically active, and are isomeric. The amount of rotation varies with the compound, but is the same for any two isomers, though in opposite directions. See also optical isomer; nicol.

optical spectroscopy. Absorption or emission spectroscopy utilizing radiation in the visible portion of the electromagnetic spectrum.

"Oragrafin."[412] Trademark for calcium ipodate (q.v.).

oral contraceptive. See antifertility agent.

orange III. See methyl orange.

orange IV. See torpeolin OO.

orange cadmium. See cadmium sulfide.

orange flower oil. See neroli oil.

orange mineral. A red lead oxide pigment made in a furnace by roasting lead carbonate or sublimed litharge; it is a very bright orange, but has low tinting strength.
Properties: Fine powder; −325 mesh; sp. gr. 9.0; contains 95.5% red lead (Pb_3O_4).
Use: Pigment in printing inks and primers.

orange peel oil. See citrus peel oils.

orange seed oil. See citrus seed oils.

"Orange Crystals.'[227] Trademark for methyl betanaphthyl ketone.
"Orange Liquid." Liquid mixture of alpha- and betanaphthylmethyl ketone.

"Orasol."[443] Trademark for solvent dyes for transparent coatings, inks, and plastics.

"Oratol."[309] Trademark for a neutralized sulfonated ester of a higher alkanolamide, containing added solvents for increased scouring efficiency.

orbital theory. The quantum theory of matter applied to the nature and behavior of the electron, either in single atoms (atomic orbital) or in united atoms (molecular orbital). A combination of Schrödinger's wave mechanics and Heisenberg's uncertainty principle, the orbital theory was formulated in 1926. It has yielded a better understanding of the electron and its critical part in chemical bonding than is possible with Newtonian mechanics. In simple language, the orbital theory considers the electron not as a particle, but as a three-dimensional wave that can exist at several energy levels; its exact location and position in the "shell" (which in most elements is a group of orbitals) cannot be precisely determined, but only predicted by the laws of mathematical probability. The orbital levels and the movement of electrons within them are expressed by wave functions and quantum numbers (q.v.). The probability that an electron will be found in a given volume (i.e., the square of the one-electron wave function) is called the orbital of that electron, and the shape of the orbital is defined by surfaces of constant probability (i.e., spheres and doughnut-shaped ellipses). The electron orbital, described in terms of probability, is like a cloud, with indefinite boundaries. The energy state of each electron is given by four quantum numbers which describe its principal level, its angular momentum, its magnetic moment, and its spin. This concept has exerted a profound effect on modern ideas about chemical bonding, transition metal complexes, semiconductors, and solid state physics.

orcin (dihydroxytoluene; methylresorcinol; orcinol) $CH_3C_6H_3(OH)_2 \cdot H_2O$. 1-Methyl-3,5-dihydroxybenzene.
Properties: White, crystalline prisms, becoming red in air; intensely sweet, unpleasant taste. Soluble in water, alcohol, and ether. Sp. gr. 1.2895; m.p. (anhydrous) 107°C, (hydrated) 56°C; b.p. 287–290°C.
Derivation: By fermentation of various species of lichens (rocella), and extraction.
Uses: Medicine; reagent for certain carbohydrates.

order of magnitude. A range of values applied to numbers, distances, dimensions, etc. which begins at any value and extends to ten times that value; e.g., 2 is of the same order of magnitude as any number between itself and 20; 5 miles is of the same order of magnitude as any distance between 5 and 50 miles. The expression usually applies to extremely large or extremely small units, i.e., the size ranges of atoms, molecules, colloidal particles, etc., or astronomical distances.

"Ordess."[510] Trademark for a series of ortho esters (including tridecyl orthoformate). Useful for removing small quantities of water from ethers or other solvents where acid catalysis can be employed.

ore. An aggregate of valuable minerals and gangue (q.v.) from which one or more metals can be extracted at a profit.

ore flotation. See flotation.

"Oreton."[321] Trademark for a brand of testosterone (q.v.) and its derivatives.

organelle. A portion of a cell having specific functions, distinctive chemical constituents and characteristic morphology; it is a unit subsystem of a cell. Examples are mitochondria and chromosomes (q.v.). Organelles are often closely associated with enzymes. The lysosome (an enzyme-bearing organelle) has been synthesized.

organic compound. Any substance that contains the element carbon, with the exception of carbon oxides, and various carbonates. Some 700,000 organic substances have been identified. A brief classification follows:

I. Aliphatic (straight-chain)
  1. Hydrocarbons (petroleum and coal-derived)
    (a) paraffins or alkanes (saturated) (C_nH_{2n+2}) methane and homologs
    (b) Olefins (unsaturated)
      (1) alkenes (one double bond) (C_nH_{2n}) ethylene and homologs
      (2) alkadienes (two double bonds) (C_nH_{2n-2}) butadiene, allene
    (c) acetylenes or alkynes (triple bond)
  2. Alcohols (ROH): methanol and homologs
  3. Ethers (ROR): methyl ether and homologs
  4. Aldehydes (RCHO): formaldehyde, acetaldehyde
  5. Ketones: (RCOR): acetone, methylethylketone
  6. Carboxylic acids (RCOOH)
  7. Carbohydrates ($C_nH_{2x}O_x$)
    (a) Sugars: glucose, fructose, sucrose
    (b) Starches: wheat, corn, potato
    (c) Cellulose: cotton, plant fibers
II. Cyclic (closed ring)
  1. Alicyclic hydrocarbons (properties similar to aliphatics):

(a) cycloparaffins (naphthenes) (saturated): cyclohexane, cyclopentane, etc.
(b) cycloolefins (unsaturated): cyclopentadiene, cyclooctatetraene
(c) cycloacetylenes (triple bond)
2. Aromatic hydrocarbons (arenes): unsaturated compounds; hexagonal ring structure with three double bonds; single and multiple fused rings
(a) benzene group (1 ring)
(b) naphthalene group (2 rings)
(c) anthracene group (3 rings)
3. Heterocyclic: unsaturated; usually pentagonal rings, containing at least one other element besides carbon.
(a) pyrroles
(b) furans
(c) thiazoles
(d) porphyrins
III. Combinations of aliphatic and cyclic structures
1. terpene hydrocarbons
2. amino acids (some are aliphatic and others combinations)
3. proteins and nucleic acids (coiled or helical formations)

organoclay (organopolysilicate). A clay such as kaolin or montmorillonite, to which organic structures have been chemically bonded; since the surfaces of the clay particles, which have a lattice-like arrangement, are negatively charged, they are capable of binding organic radicals. When this type of structure is in turn reacted with a monomer such as styrene, a complex results that is known as a polyorganosilicate graft polymer.

organoleptic. A term widely used to describe consumer testing procedures for food products, perfumes, wines, and the like, in which samples of various products, flavors, etc., are submitted to groups or panels. Such tests are a valuable aid in determining the acceptance of the products, and thus may be viewed as a marketing technique. They also serve psychological purposes, and are an important means of evaluating the subjective aspects of taste, odor, color, and related factors.

organopolysilicate. See organoclay.

organometallic compound. A compound comprised of a metal attached directly to carbon (RM); such compounds have been prepared of practically all the metals. Metallic salts (soaps) of organic acids are excluded from this group. Examples are diethylzinc (the first known organometallic); Grignard compounds such as methyl magnesium iodide (CH_3MgI); metallic alkyls such as butyllithium (C_4H_9Li); tetraethyllead; triethyl aluminum; tetrabutyl titanate; sodium methylate; copper phthalocyanine; metal carbonyls; and metallocenes (q.v.). Some are highly toxic or flammable; others are coordination compounds. See specific compound for details.

Reactive and moderately reactive organometallic compounds will react with all functional groups; two major types of reaction in which they are involved are oxidation and cleavage by acids. Probably the most important organometallic reactions are those involving addition to an unsaturated linkage.

organophosphorus compound. Any organic compound containing phosphorus as a constituent. These fall into several groups, chief of which are the following: (1) phospholipids, or phosphatides, which are widely distributed in nature in the form of lecithin, certain proteins, and nucleic acids; (2) esters of phosphinic and phosphonic acids, used as plasticizers, insecticides, resin modifiers, and flame retardants; (3) pyrophosphates, for example, tetraethyl pyrophosphate, which are the basis for a broad group of cholinesterase inhibitors used as insecticides; (4) phosphoric esters of glycerol, glycol, sorbitol, etc., which are components of fertilizers. While many of these compounds play an important part in animal metabolism, those in group (3) are toxic and should be handled with extreme care.

"Organ-O-Sil."[275] Trademark for treated fumed silicon dioxide for use as a free-flowing or anti-caking agent.

organosol. Colloidal dispersion of any insoluble material in an organic liquid; specifically the finely divided or colloidal dispersion of a synthetic resin in plasticizer in which dispersion the volatile content exceeds 5% of the total. See platisol.

organotin compound. A family of alkyl tin compounds widely used as stabilizers for plastics, especially rigid vinyl polymers used as piping, construction aids and cellular structures. Some have catalytic properties. They include butyl tin trichloride, dibutyltin oxide, etc., and various methyltin compounds. They are both liquids and solids, and all are highly toxic, with a tolerance of 0.1 mg per cubic meter of air. For specific data, see dibutyltin entries.

"Oriex.'[482] Trademark for a bi-axially oriented vinyl. Available in roll and sheet form for shrink packaging.

origanum oil, Spanish.
Properties: Light yellow essential oil; spicy odor; sp. gr. 0.935–0.960 (25°C); refractive index 1.5020–1.5080 (20°C); angular rotation −2 to +3°. Soluble in 2 volumes 70% alcohol; soluble in fixed oils and mineral oil; insoluble in glycerol. Combustible.
Components: (F.C.C.) 60–75% by volume of phenols.
Derivation: By steam distillation from the flowering herb. Thymus capitatus, and various species of Origanum.
Grade: F.C.C.
Hazard: May be toxic by ingestion.
Uses: Pharmacy; flavoring.

"Orlon."[28] Trademark for a copolymer containing at least 85% acrylonitrile. Available in various types of staple and tow.
Properties: Sp. gr. 1.14–1.17; tensile strength (psi) 32,000–39,000; break elongation 20–28%; moisture regain 1.5% (70°F, 65% R.H.); softens at 455°F; soluble in butyrolactone (hot), dimethyl formamide (hot), ethylene carbonate (hot). Resistant to mineral acids; fair to good resistance to weak alkalies. Insoluble in alcohol, acetone, benzene, carbon tetrachloride, and petroleum ether; soluble in dimethyl sulfoxide; maleic anhydride, ethylene carbonate, nitriles, and nitrophenols.
Hazard: Combustible; burns freely and rapidly.
Containers: Bales and cartons.
Uses: In apparel, usually blended with wool or other fibers.

Orn Abbreviation for ornithine.

ornithine (2,5-diaminovaleric acid) $NH_2(CH_2)_3CH(NH_2)COOH$. A nonessential amino acid produced by the body and important in protein metabolism.

Properties: L(+)-ornithine: Crystals from alcohol-ether; m.p. 140°C; soluble in water and alcohol.

DL-ornithine: Crystals from water; slightly soluble in alcohol.

Derivation: Isolated from proteins after hydrolysis with alkali.

Use: Biochemical research.

"Oronite."[151] Trademark for a series of petroleum products used as detergents, chelating agents, hydraulic fluids, odorants, lube-oil additives, sealants, etc.

"Oropon."[23] Trademark for proteolytic enzyme concentrates, formulated with or without deliming salts. Various grades include enzymes of pancreatic, fungal or bacterial origin, to remove undesirable protein matter and excess lime from dehaired animal skins.

Use: Bating of skins in preparation for tanning.

"Orotan" TV.[23] Trademark for a synthetic tanning agent with attributes of vegetable tannins. Dark-red viscous solution, 31% tannin. Imparts high degree of tannage, strength, fullness and solidity to leather. Solubilizing, penetrating and bleaching agent.

orotic acid (uracil-6-carboxylic acid; 6-carboxyuracil) $C_4N_2H_3(O)_2COOH$. Occurs in cow's milk and has also been isolated from certain strains of molds (Neurospora). A growth factor for certain microorganisms.

Properties: Crystals with m.p. 345–346°C.

Use: Biochemical research, especially the biosynthesis of nucleic acids.

orpiment. Obsolete name for arsenic trisulfide.

orthamine. See ortho-phenylenediamine.

ortho-. A prefix meaning "straight ahead"; *meta-* means "beyond"; *para-* means "opposite." These prefixes are used in organic chemistry in naming disubstitution products derived from benzene in which the substituent atoms or radicals are located in certain definite positions on the benzene ring. This is illustrated in the diagram, where A and B represent the substituent atoms or groups. When attached to adjoining carbon atoms, B is in the ortho position in respect to A (also called the 1,2-position). If B is located on the third carbon atom in respect to A, it is in the meta position (also called 1,3-); when B is attached to the opposite carbon atom, it is the para position (1,4).

ortho meta para

In organic compounds these prefixes conventionally appear in italics (often abbreviated *o-*, *m-*, and *p-*) and are ignored in alphabetizing.

In inorganic chemistry the prefix ortho- designates the most highly hydrated acid, or its salt, to contrast with the meta- or less hydrated acid or salt. For example, $H_3PO_4(P_2O_5 \cdot 3H_2O)$ is orthophosphoric acid and $HPO_3(P_2O_5 \cdot H_2O)$ is metaphosphoric acid.

orthoarsenic acid. See arsenic acid.

orthoboric acid. See boric acid.

"Orthochrom."[23] Trademark for pigmented plasticized nitrocellulose lacquers and thinners. Produce durable, washable, flexible, colored lacquer finishes of good light fastness.

Use: Finishing of belt, garment, upholstery and other leathers.

"Orthocide."[253] Trademark for a type of fungicide products containing captan.

"Orthoclear."[23] Trademark for permanently plasticized nitrocellulose binders and lacquers in various solvents. Produce clear, durable, flexible finishes.

Use: Topcoat finishes for glazing or high gloss leather coatings.

orthoform. See methyl meta-amino-para-hydroxybenzoate.

"Ortho-Klor."[253] Trademark for insecticide products containing chlordan.

Hazard: Probably toxic; absorbed by skin.

"Ortholeum."[28] Trademark for a series of lubricant additives.

"Ortholite."[23] Trademark for clear and pigmented vinyl lacquers, binders, and solvents. Produce finishes of outstanding abrasion resistance and low-temperature flexibility.

Use: Finishes on upholstery, automotive, luggage, and case leathers.

"Orthol-K."[253] Trademark for pesticide products containing phytonomic or similar type petroleum ols.

"Orthophen" 278.[204] Trademark for a special blend of amyl phenol isomers.

Properties: Straw colored liquid; sp. gr. 0.94–0.95; distillation range, 75–270°C.

Use: Antiskin agent for paint industry.

"Orthophos."[253] Trademark for a type of insecticide containing parathion.

Hazard: Highly toxic; absorbed by skin.

"Orthorix."[253] Trademark for a type of parasiticide containing lime sulfur.

"Orthosil."[204] Trademark for a practically anhydrous, water-soluble, sodium orthosilicate in granular form. Quick-acting detergent used in heavy-duty metal cleaning. Special types made for still tank cleaning have suitable wetting and water softening additions.

"Orthoxine" Hydrochloride.[327] Trademark for methoxyphenamine hydrochloride.

"Ortolan."[440] Trademark for series of dyes for dyeing and printing textiles of animal fibers and wool-nylon mixtures.

"Orvus."[487] Trademark for a series of surfactants and detergents.

"Orzan."[48] Trademark for a group of dark brown lignosulfonates (spent sulfite liquor products).

Containers: Free-flowing powder, up to 50-lb bags; 50% solid solution, up to tank cars.

Uses: Dispersant; flotation reagent; chelating agent; emulsion stabilizer; binder; tannin extender; adhesive ingredient.

Os Symbol for osmium.

-ose. A suffix indicating a carbohydrate compound or polymer, usually a simple or complex sugar, for example, sucrose, fructose, glucose, maltose, etc., and also cellulose, cellobiose, amylose (starch).

OSHA. Abbreviation for Occupational Safety and Health Administration (q.v.).

osmic acid anhydride (osmium tetroxide; perosmic acid anhydride; perosmic oxide) OsO_4.
Properties: A dimorphic compound with both crystalline and amorphous forms. Colorless; pungent, disagreeable odor. Soluble in water, alcohol, and ether. Sp. gr. 4.90; m.p. 40°C; b.p. 130°C.
Derivation: By heating powdered osmium in air, or by treating it with nitric acid, aqua regia, or chlorine.
Hazard: Highly toxic by inhalation; strong irritant to eyes and mucous membranes. Tolerance, 0.002 mg per cubic meter of air.
Uses: Microscopic staining; photography; catalyst in organic synthesis.
Shipping regulations: (Air) Poison label.

osmiridium. See iridosmine.

osmium Os Metallic element having atomic number 76; in group VIII of the periodic system. Atomic weight 190.2; valences 2,3,4,6,8. Has 7 stable isotopes.
Properties: Hard white metal of the platinum group. On heating in air gives off poisonous fume of osmium tetroxide. Insoluble in acids and aqua regia; attacked by fused alkalies. Sp. gr. 22.5; m.p. 3000°C; b.p. 5500°C. It has the highest specific gravity and melting point of the platinum metals. Metallurgically unworkable.
Occurrence: Tasmania, So. Africa, U.S.S.R., Canada.
Derivation: Occurs with platinum from which it is recovered during the purification process. Also occurs with iridium as a natural alloy, iridosmine.
Hazard: Irritant to skin.
Uses: Alloys; hardener for platinum; phonograph needles; instrument pivots; laboratory catalyst.

osmium ammonium chloride. See ammonium hexachloroosmate.

osmium chloride (osmium dichloride; osmous chloride) $OsCl_2$.
Properties: Dark green needles. Hygroscopic. Keep away from air! Soluble in alcohol, ether; insoluble in water.
Hazard: May be toxic.

osmium dichloride. See osmium chloride.

osmium potassium chloride. See potassium hexachloroosmate.

osmium sodium chloride. See sodium hexachloroosmate.

osmium tetroxide. See osmic acid anhydride.

osmocene (dicyclopentadienylosmium) $(C_5H_5)_2Os$.
Properties: Stable white solid; m.p. 229–230°C.
Uses: Intermediate; high-temperature applications; derivatives used as an ultraviolet radiation absorber. See metallocene.

"Osmon."[169] Trademark for 1-naphthylamine-4,6,8-trisulfonic acid used in the colorimetric determination of osmium.

osmosis. Passage of a pure liquid (usually water) into a solution (e.g., of sugar and water) through a membrane that is permeable to the pure water but not to the sugar in the solution. This passage can also occur when the two phases consist of solutions of different concentrations. The membrane is called semipermeable when the molecules of the solvent, but not those of the solute, can penetrate it. This pushing of water through a membrane into a solution results from the greater tendency of water molecules to escape from water than from a solution. The term osmosis is usually restricted to movement through a solid or liquid barrier that prevents the phases from mixing rapidly. In test apparatus parchment or collodion membranes are used; in plants and animals the cell wall acts as a diffusion barrier. The pressure exerted by osmosis is substantial, and accounts for the elevation of sap from root systems to the tops of trees. Osmosis is considered an essential characteristic of growth. See also diffusion.

Reverse osmosis is utilized as a method of desalting sea water (see desalination), recovering waste water from paper mill operations, pollution control, industrial water treatment, chemical separations, and food processing. This method involves application of pressure to the surface of a saline solution, thus forcing pure water to pass from the solution through a membrane that is too dense to permit passage of sodium and chloride ion. Hollow fibers of cellulose acetate or nylon are used as membranes, since their large surface area offers more efficient separation.
See also dialysis; membrane.

osmous chloride. See osmium chloride.

otto. See attar.

ouabain (g-strophanthin; strophanthin thoms) $C_{29}H_{44}O_{12} \cdot 8H_2O$. A glucoside.
Properties: White odorless crystals; stable in air but affected by light; soluble in water and alcohol; specific rotation (alpha) −31 to −32.5° 20/D (anhydrous). M.p. 190° (dec).
Grades: U.S.P.
Use: Medicine.

ouricury wax. A vegetable wax exuded by the leaves of Cocos coronapa (South America).
Properties: Brown; acid value 10; saponification value 80; sp. gr. 0.970 (15°C); m.p. 83°C; foreign matter (dirt, etc.) sometimes 18%. Combustible; nontoxic.
Grades: Crude; refined.
Containers: Bags.
Use: Substitute for carnauba wax.

-ous. A suffix used in naming inorganic compounds which indicates that the central element has an oxidation number of +2, or a valence of 2. For example, in ferrous chloride ($FeCl_2$), the iron atom is in an intermediate oxidation state. It can also be expressed as iron (II) chloride.

outgassing. The removal of gas from a metal by heating at a temperature somewhat below melting, while maintaining a vacuum in the space around the metal. Usually done before melting but sometimes afterward. Used in manufacture of tubes and other vacuum devices.

ovalbumin. See albumin, egg.

ovex. Generic name for para-chlorophenyl para-chlorobenzenesulfonate, $ClC_6H_4OSO_2C_6H_4Cl$.
Properties: White crystalline solid; m.p. 86.5°C; insoluble in water; soluble in acetone and aromatic solvents.

Hazard: Toxic by ingestion. Strong irritant to skin.
Uses: Insecticide and acaricide.

"Ovocylin" Dipropionate.[305] Trademark for estradiol dipropionate.

"Ovral."[24] Trademark for an oral contraceptive containing norgestrel, a synthetic progestogen. Contains 0.5 mg norgestrel and 0.05 mg ethinyl estradiol.

oxa-. Prefix indicating the presence of oxygen in a heterocyclic ring.

"OXAF."[248] Trademark for the zinc salt of 2-mercaptobenzothiazole. $Zn(SCNSC_6H_4)_2$.
Properties: White to pale yellow powder. Sp. gr. 1.63; melting range, decomposes without melting when heated to 200°C or over. Excellent storage stability. Slightly soluble in ethylene dichloride and actone. Insoluble in water, benzol, and gasoline. Available in pelletized form.
Use: Rubber accelerator; especially latex, foam sponge, wire insulation, air-cured footwear, druggist sundries and specialties.

oxalic acid $HOOCCOOH \cdot 2H_2O$.
Properties: Transparent, colorless crystals; m.p. 187°C of anhydrous form, 101.5°C for dihydrate.
Derivation: Occurs naturally in many plants (wood sorrel, rhubarb, spinach) and can be made by alkali extraction of sawdust. Now manufactured by reaction of sodium formate with sodium hydroxide, followed by distillation of the resulting dihydrate crystals.
Grades: Technical (crystals and powder); C.P.
Containers: Bags; drums; carlots.
Hazard: Toxic by inhalation and ingestion; strong irritant. Tolerance, 1 mg per cubic meter of air.
Uses: Automobile radiator cleanser; general metal and equipment cleaning; purifying agent and intermediate for many compounds; leather tanning; catalyst; laboratory reagent; stripping agent for permanent press resins; bleaching of textiles; rare-earth processing.
Shipping regulations: (Rail) Poisonous solids, n.o.s., Poison label. (Air) Poison label.

7-oxalobicyclo-[2.2.1]-heptane-2,3-dicarboxylic acid. See endothall.

oxalonitrile. See cyanogen.

oxalyl chloride (ethanedioyl chloride) $(COCl)_2$.
Properties: Colorless liquid. If cooled to −12°C, solidifies to a white, crystalline mass. Gives off carbon monoxide on heating. Decomposed by water and alkaline solutions. Soluble in ether, benzene, chloroform. B.p. 64°C; f.p. −12°C; sp. gr. 1.43.
Derivation: Interaction of oxalic acid and phosphorus pentachloride.
Containers: Steel drums.
Hazard: Toxic by inhalation and ingestion.
Uses: Military poison gas; chlorinating agent in organic synthesis.
Shipping regulations: (Rail) Poisonous solids, n.o.s., Poison label. (Air) Poison label.

oxamide $NH_2COCONH_2$.
Properties: White, odorless powder; m.p. 419°C (dec) (probably the highest melting organic compound); slightly soluble in water; very slightly soluble in alcohol and ether; not hygroscopic; decomp. to ammonia and carbonic acid.
Use: Stabilizer for nitrocellulose preparations; possible substitute for urea as fertilizer.

oxammonium. See hydroxylamine.

oxamycin. See amino-3-isoxazolidone.

"Oxanal."[443] Trademark for dyes for coloring anodized aluminum.

oxazole $\overline{OCH{:}NCH{:}CH}$. A five-membered heterocyclic compound valuable for its derivatives. The dihydro forms, 2- and 4-oxazoline, are the parents of increasingly useful commercial compounds, e.g., surface-active agents, detergents, etc.; 2-oxazoline is $\overline{OCH_2NHCH{:}CH}$.

oxazoline wax. A series of synthetic waxes having the oxazoline structure. See oxazole. They can be made to fairly exact specifications; are miscible with most natural and synthetic waxes, and can be applied to the same uses.

oxetane (trimethylene oxide). $\overline{CH_2OCH_2CH_2}$. An oxetane group $(=\underline{COCH_2C}=)$ is one kind of epoxy group. See "Penton."

oxidase. An enzyme whose activity results in the transfer of electrons on the substrate; an oxidizing enzyme.

oxidation. The term "oxidation" originally meant a reaction in which oxygen combines chemically with another substance, but its usage has long been broadened to include any reaction in which electrons are transferred. Oxidation and reduction always occur simultaneously (redox reactions), and the substance which gains electrons is termed the oxidizing agent. For example, cupric ion is the oxidizing agent in the reaction: Fe (metal) + $Cu^{++} \longrightarrow Fe^{++}$ + Cu (metal); here, 2 electrons (negative charges) are transferred from the Fe atom to the Cu atom; thus the Fe becomes positively charged (is oxidized) by loss of 2 electrons while the Cu receives the 2 electrons and becomes neutral (is reduced). Electrons may also be displaced within the molecule without being completely transferred away from it. Such partial loss of electrons likewise constitutes oxidation in its broader sense and leads to the application of the term to a large number of processes which at first sight might not be considered to be oxidations. Reaction of a hydrocarbon with a halogen, for example, $CH_4 + Cl_2 \longrightarrow CH_3Cl + HCl$, involves partial oxidation of the methane; halogen addition to a double bond is regarded as an oxidation.

Dehydrogenation is also a form of oxidation, when two hydrogen atoms, each having one electron, are removed from a hydrogen-containing organic compound by a catalytic reaction with air or oxygen, as in oxidation of alcohols to aldehydes. See dehydrogenation.

oxidation number. The number of electrons that must be added to or subtracted from an atom in a combined state to convert it to the elemental form; i.e., in barium chloride ($BaCl_2$) the oxidation number of barium is +2 and of chlorine is −1. Many elements can exist in more than one oxidation state. See also valence.

oxidation-reduction indicator. A substance that has a color in the oxidized form different from that of the reduced form, and can be reversibly oxidized and reduced. Thus if diphenylamine is present in a ferrous sulfate solution to which potassium dichromate is being added, a violet color appears with the first drop of excess dichromate. See also indicator.

oxidative coupling. A polymerization technique for certain types of linear high polymers. Oxidation of 2,6-dimethylphenol with an amine complex of a copper salt as catalyst forms a polyether, with splitting off of water. The product is soluble in aromatic and chlorinated hydrocarbons, insoluble in alcohols, ketones, and aliphatics. It is thermoplastic and unaffected by acids, bases, and detergents. It has a very broad useful temperature range (from −275 to +375°F). It is also dimensionally stable and has good electrical resistance. Oxidative coupling of diacetylenes and dithiols also yields promising polymers. See also "PPO."

"Oxidex."[188] Trademark for an antioxidant for soap fats and oils. It can be added to the oils before saponification or it can be added by milling into the finished soap in the same manner that perfume is incorporated. The correct proportion for solid soaps is 0.1%.

oxidizing material. Any compound that spontaneously evolves oxygen either at room temperature or under slight heating. The term includes such chemicals as peroxides, chlorates, perchlorates, nitrates, and permanganates. These can react vigorously at ambient temperatures when stored near or in contact with reducing materials such as cellulosic and other organic compounds. Storage areas should be well ventilated and kept as cool as possible.

oxine. See 8-hydroxyquinoline.

oxirane. A synonym for ethylene oxide, $H_2\underline{COCH}_2$.

An oxirane group is one having the structure $=\overline{COC}=$, and is one kind of epoxy group.

oxirene (oxacyclopropene). An organic intermediate containing four pi electrons, reported to result from oxidation of acetylene.

"Oxitol."[125] Trademark for ethylene glycol monoethyl ether having an A.S.T.M. distillation range 134–136°C.

2-oxohexamethylenimine. See caprolactam.

"Oxone."[28] Trademark for an acidic, white granular, free-flowing solid containing the active ingredient potassium peroxymonosulfate; readily soluble in water; 1% solution has pH of 2–3. Minimum active oxygen content 4.5%. Strong oxidizing agent.
Containers: 50-lb bags.
Hazard: Fire risk in contact with organic materials.
Uses: Manufacture of dry laundry bleaches, detergent-bleach washing compound, scouring powders, plastic dishware cleaners, and metal cleaners; hair wave neutralizers and pharmaceuticals; general oxidizing reactions.

oxonium ion. Formerly called hydronium ion; it results from ionization of water: $2H_2O \rightarrow H_3O^+ + OH^-$.

2-oxopentanedioic acid. See alpha-ketoglutaric acid.

4-oxopentanoic acid. See levulinic acid.

Oxo process. Production of alcohols, aldehydes and other oxygenated organic compounds by pasage of olefin hydrocarbon vapors over cobalt catalysts in the presence of carbon monoxide and hydrogen. Aldehydes are formed as products, but in most cases these are hydrogenated at once to the corresponding alcohol. Propylene produces normal and isobutyraldehyde; higher olefins produce a mixture of aldehydes containing one more carbon atom than the olefins. n-Butyl, isobutyl, amyl, isooctyl, decyl and tridecyl alcohols are produced in large quantities.

oxosilane. See siloxane.

oxybenzoic acid. See hydroxybenzoic acid.

para,para′-**oxybis(benzenesulfonylhydrazide)** $H_2NNHSO_2C_6H_4OC_6H_4SO_2NHNH_2$.
Properties: Fine white crystalline powder; odorless; sp. gr. 1.52; mp. decomposes at 150–160°C; soluble in acetone; moderately soluble in ethanol and polyethylene glycols; insoluble in gasoline and water. Combustible.
Use: Blowing agent for sponge rubber and expanded plastics.

oxyconiine. See conhydrine.

oxydemetonmethyl. Generic name for S-[2-(ethylsulfinyl)ethyl] O,O-dimethyl phosphorothioate. See O,O-dimethyl S-2-(ethylsulfinyl)ethyl phosphorothioate.

N-**oxydiethylene-2-benzothiazolesulfenamide.** See 2-(morpholinothio)benzothiazole.

beta,beta′-**oxydipropionitrile** (ODPN). $O(CH_2CH_2CN)_2$.
Properties: Colorless liquid; f.p. −26.3°C; b.p. 120°C (1 mm); b.p. 155°C (5 mm); sp. gr. 1.0405 (30°C); viscosity (30°C) 8.00 cp; refractive index (n 25/D) 1.4392; flash point 180°F (TOC); combustible; soluble in water. It is thermally unstable, yielding acrylonitrile and water above 175°C. Hydrolyzed by strong acids and bases; quite immiscible with paraffin hydrocarbons, but dissolves aromatics. Combustible.
Derivation: From acrylonitrile.
Hazard: Toxic by inhalation and ingestion.
Use: Solvent in fractional extraction.

"Oxyfume."[214] Trademark for a sterilant mixture of ethylene oxide and either carbon dioxide or dichlorodifluoromethane. Colorless and tasteless gas; excellent penetrating ability and ether odor. Available as: "Oxyfume" Sterilant-20, 20% ethylene oxide and 80% carbon dioxide by weight and gas volume; "Oxyfume" Sterilant-12, a mixture of 12% ethylene oxide and 88% dichlorodifluoromethane by weight.

oxygen O. Gaseous element of atomic number 8; Group VIA of the Periodic Table; atomic weight 15.9994; Valence 2; isotopes 16, 17, 18. Molecular oxygen is O_2 and ozone O_3. Oxygen was discovered by Priestley in England in 1774. Fourth highest-volume chemical produced in U.S. (1975).
Properties: Colorless, odorless, tasteless diatomic gas. Liquefiable at −183°C to slightly bluish liquid; solidifiable at −218°C. It constitutes about 20% by volume of air at sea level. Sp. gr. (gas) 1.10535, referred to air; (liquid) 1.14 (−183°C). Soluble in water and alcohol. Oxygen is noncombustible, but actively supports combustion (q.v.).
Derivation: Commercial method for tonnage oxygen is fractionation of liquid air; electrolysis of water is used for small amounts and for laboratory demonstration. Note: Atmospheric oxygen originated from photosynthesis (q.v.).
Grades: Low purity; high purity; U.S.P.
Containers: (Comporessed gas) steel cylinders, tank cars, tube trailers; (Liquid) vacuum-jacketed cylinders, truck cargo tanks, tank cars. (Gas) Pipelines for distances of about 50 miles.

Hazard: (Gaseous) Moderate fire risk as oxidizing agent. Therapeutic overdoses can cause convulsions. (Liquid): May explode on contact with heat or oxidizable materials; irritant to skin and tissue.
Uses: Blast furnaces; copper smelting; steel production (basic oxygen converter process); manufacture of synthesis gas (q.v.) for production of ammonia, methyl alcohol, acetylene, etc; oxidizer for liquid rocket propellants; resuscitation; heart stimulant; decompression chambers; spacecraft; chemical intermediate; to replace air in oxidation of municipal and industrial organic wastes; to counteract effect of eutrophication in lakes and reservoirs; coal gasification.
Shipping regulations: (Rail) (gas and pressurized liquid) Green label. (Air) (gas) Nonflammable gas label; (liquid, pressurized and nonpressurized) Not acceptable.
See also ozone.

oxygen 18 (heavy oxygen). Oxygen isotope of atomic weight 18. Occurs in proportion of 8 parts to 10,000 of ordinary oxygen in water, air, rocks, etc. The proportion may be increased by passing carbon dioxide gas repeatedly through a packed column down which water is passed. The carbon dioxide leaving the top of tower is enriched in heavy oxygen and the water leaving the bottom is depleted. Used in tracer experimentation. See also heavy water.

"Oxygenated Hydrocarbons."[569] Trademark for a series of petroleum-derived oxidates composed primarily of organic acids and esters. Designated by TC or TX followed by a four digit number e.g., TC-5416, TC-6664.
Uses: Corrosion inhibitors; surface-active components in wax emulsions; emulsifiable lubricants; plasticizer; intermediate; leather and cordage oils; lubricity agents and solubilizers.

oxygen consumed (C.O.D.) (O.C.) (D.O.C.). A measure of the quantity of oxidizable components present in water. Since the carbon and hydrogen, and not the nitrogen, in organic matter are oxidized by chemical oxidants, the oxygen consumed is a measure only of the chemically oxidizable components and is dependent upon the oxidant, structure of the organic compound and manipulative procedure. Since this value does not differentiate stable from unstable organic matter, it does not necessarily correlate with the biochemical oxygen demand value. It is also known as chemical oxygen demand (C.O.D.) and dichromate oxygen consumed (D.O.C.). See also biochemical oxygen demand; dissolved oxygen.

oxygen fluoride (oxygen difluoride, fluorine monoxide). OF_2. An unstable, colorless gas; f.p. −224°C; b.p. −145°C; slightly soluble in water and alcohol. Suggested as oxidizer for rocket propellants (q.v.).
Hazard: Highly toxic; explodes on contact with water, air, and reducing agents. Tolerance, 0.05 ppm in air.
Shipping regulations: (Rail, Air) Not listed. Consult authorities.

"Oxygen Scavenger K-91."[108] Trademark for a technical grade of sodium sulfite in a dry, tan, water-soluble powder form.
Containers: 100-lb bags.
Uses: To chemically remove dissolved oxygen left in the feed-water after mechanical deaeration. Specially catalyzed for quick reaction. Usable in cold water.

oxyhemoglobin. See hemoglobin.

"Oxylone."[327] Trademark for fluorometholone (q.v.).

oxyluminescence. See chemiluminescence.

oxymethylene. See formaldehyde.

beta-**oxynaphthoic acid.** See 3-hydroxy-2-naphthoic acid.

oxyneurine. See betaine.

oxyphosphorane. One of a class of compounds derived from trialkyl phosphites and orthoquinones. Their molecules have a five atom ring, $\overline{\text{OCCOP}}(OR)_3$ in which the two carbon atoms are part of an aromatic ring. They react by liberating a phosphate ester.

oxypolygelatin (OPG). A purified gelatin treated with glyoxal, followed by oxidation with hydrogen peroxide. A possible plasma substitute.

oxyquinoline. See 8-hydroxyquinoline.

oxytetracycline $C_{22}H_{24}N_2O_9 \cdot 2H_2O$. An antibiotic obtained by fermentation from Streptomyces rimosus, an actinomycete. Its chemical structue is that of a modified naphthacene molecule having six asymmetric centers; also recently synthesized. Nontoxic.
Properties: Dull yellow, odorless, slightly bitter crystalline powder. M.p. 179–182°C (dec). Soluble in acids and alkalies; very slightly soluble in acetone, alcohol, chloroform and water; practically insoluble in ether. Stable in air; affected by sunlight. Deteriorates in solution with pH below 2; destroyed rapidly by alkali hydroxide solutions; pH (saturated solution) about 6115.
Grade: N.F.
Use: Medicine (antibiotic). Inhibitor of lethal yellowing in coconut palm trees. The hydrochloride is trademarked "Terramycin."

oxytocin (alpha-hypophamine) $C_{43}H_{66}N_{12}O_{12}S_2$. A hormone secreted by the posterior lobe of the pituitary gland. Its chief action is stimulation of the contraction of the smooth muscle of the uterus. It contains eight different amino acids. In 1955 duVigneaud elucidated its amino acid sequence, the first such determination ever made; it may be represented:

```
Cys—Tyr—Ileu
 |        |
Cys—Asp—Glu
 |        |
Pro—Leu—Gly
```

It is available as a solution for injection (oxytocin injection, U.S.P.).

"Oxytrol."[492] Trademark for a chemically modified, cold-water-swelling starch that produces noncongealing pastes. Free-flowing, light cream, small granules that form straw-colored sols that are mildly alkaline in water. Used for sizing cotton yarns and other fabrics. A low B.O.D. in sizing wastes is claimed.

oyster shells. Shells of Ostrea virginica, taken from the Gulf of Mexico coast in Texas and Louisiana and from Chesapeake Bay. Average analysis: $CaCO_3$ 93–97%; $MgCO_3$ 1%; silica 0.5–2.0%; SO_4 (as $CaSO_4$) 0.3–0.4%; also miscellaneous substances.
Uses: Source of lime; drilling muds, road beds; poultry and cattle feeds.

"Ozene."[292] Trademark for an emulsifiable ortho-dichlorobenzene (q.v.).

"Ozlo."[141] Trademark for leaded zinc oxides produced by the co-fumed process. Composition varies as follows: zinc oxide 50–88% and basic lead sulfate, $PbSO_4 \cdot PbO$, 12–55% depending upon grade. Used in paint pigments.

ozocerite (mineral eax; fossil wax; ozokerite).
Properties: Waxlike hydrocarbon mixture, yellow-brown to black or green, translucent when pure and having a greasy feel. Soluble in light petroleum hydrocarbons, benzene, turpentine, kerosine, ether, carbon disulfide; slightly soluble in alcohol; insoluble in water. Sp. gr. 0.85–0.95; m.p. 55–110°C, usually about 70°C. Combustible; nontoxic.
Occurrence: Utah; Australia; near the Caspian Sea.
Method of purification: Filtration.
Grade: Technical.
Uses: Electric insulation; rubber products; paints; leather polish; lithographic and printing inks; electrotypers' wax; carbon paper; source of ceresin; floor polishes; impregnating furniture and parquet floor lumber; lubricating compositions; grease crayons; sizing and glossing paper; waxed paper; cosmetics; ointments; matrices for galvanoplastic work; textile sizings; waxed cloth; substitute for carnauba and beeswax.

ozone O_3 An allotropic form of oxygen.
Properties: Unstable blue gas with pungent odor; liquefiable at −12°C; more active oxidizing agent than oxygen. Contributes to formation of photochemical smog. Deterioration of rubber is accelerated by traces of ozone. B.p. −112°C; f.p. −192°C; sp. gr. (liquid) about 1.6. More soluble in water than oxygen.
Occurrence: Formed locally in air from lightning, in stratosphere by UV radiation; inhibits penetration of UV. Also occurs in automobile engines, and by electrolysis of alkaline perchlorate solutions. Commercial mixtures containing up to 2% ozone are produced by electronic irradiation of air. It is usually manufactured on the spot, as it is too expensive to ship. Tonnage quantities are used.
Hazard: Toxic by inhalation; strong irritant. Dangerous fire and explosion risk in contact with organic materials. Tolerance, 0.1 ppm in air (industrial workrooms); U.S. standard 0.080 ppm.
Uses: Purification of drinking water; industrial waste treatment; deodorization of air and sewage gases; bleaching waxes, oils, wet paper and textiles; production of peroxides; bactericide. Oxidizing agent in several chemical processes (acids, aldehydes, ketones from unsaturated fatty acids); steroid hormones; removal of chlorine from nitric acid; oxidation of phenols and cyanides.
Note: Depletion of the ozone layer in the stratosphere, which acts as a shield against penetration of UV light in the sun's rays, has become a matter of concern. This is thought to be due to light-induced chlorofluorocarbon decomposition resulting from increased use of halocarbon aerosol propellants. Research on the extent of this effect is in progress; meanwhile use of halocarbon aerosols has been reduced or discontinued by some manufacturers.

ozonide. A product of ozonolysis (q.v.).

ozonolysis. (1) Oxidation of an organic material by means of ozone, i.e., tall oil, oleic acid, safflower oil, cyclic olefins; carbon treatment; peracetic acid production. See also ozone. (2) The use of ozone as a tool in analytical chemistry to locate double bonds in organic compounds, and a similar use in synthetic organic chemistry for preparing new compounds. Under proper conditions, ozone attaches itself at the double bond of an unsaturated compound to form an ozonide. Since many ozonides are explosive, it is customary to decompose them in solution and deal with the final product.

P

P Symbol for phosphorus.

p-. Abbreviation for para-.

Pa Symbol for protactinium.

PA. Abbreviation for phthalic anhydride and for polyamide.

PABA. Abbreviation for para-aminobenzoic acid.

PABA sodium. See sodium para-aminobenzoate.

packaging. The operation of placing materials in suitable containers or protective covering for purposes of storage, distribution, and sale. Some packages act merely as containers, but others protect perishable materials (especially foodstuffs) from environmental damage, contamination, and biological deterioration; in this respect the critical factor is exclusion of moisture vapor, bacteria, and oxygen. Some packages perform both functions simultaneously. Common packaging materials are:

(for nonperishable products)	(for perishable products)
wooden boxes, kegs, barrels	glass bottles
fiber drums	"tin" cans
glass bottles (perfumes, pharmaceuticals)	plastic film
polyvinyl chloride bottles (detergents)	cellophane (tobacco)
aluminum tubes (toothpaste)	polypropylene
paperboard cartons	polyvinylidene chloride
polyethylene film (textiles)	polyethylene
ceramic jars (cosmetic creams)	paraffin-coated paper and board
steel cylinders (gases)	aluminum cans (beer, soft drinks)

packing. (1) A collar or gasket used to seal mechanical devices to prevent leakage of oil or water; often made of specially compounded rubber or of a flexible plastic. (2) The operation of placing solid materials or objects in shipping containers in such a way as to secure maximum space economy and freedom from damage by vibration or impact. Barriers of paperboard, foamed plastic or glass fiber are widely used.

PAHA. See para-aminohippuric acid.

paint. A uniformly dispersed mixture having a viscosity ranging from a thin liquid to a semisolid paste and consisting of (1) a drying oil, synthetic resin, or other film-forming component, called the binder; (2) a solvent or thinner; and (3) an organic or inorganic pigment. The binder and the solvent are collectively called the vehicle (q.v.). Paints are used (1) to protect a surface from corrosion, oxidation, or other type of deterioration, and (2) to provide decorative effects.

Hazard: Flammable, dangerous fire risk (except water-based). May be toxic if vapors are inhaled over a long period. The lead content of household paints is limited to 0.5%.

Shipping regulations: (except water-based) (Rail) Red label. (Air) Flammable Liquid label.

For further information refer to National Paint, Varnish and Lacquer Association, 1500 R. I. Ave., Washington, D.C.

See also paint, emulsion; vehicle; protective coating; antifouling paint.

paint, emulsion (latex paint). A paint composed of two dispersions: (1) dry powders (colorants, fillers, extenders) and (2) a resin dispersion. The former is obtained by milling the dry ingredients into water. The resin dispersion is either a latex formed by emulsion polymerization or a resin in emulsion form. The two dispersions are blended to produce an emulsion paint. Surfactants and protective colloids are necessary to stabilize the product. Emulsion paints are characterized by the fact that the binder is in a water-dispersed form, whereas in a solvent paint it is in solution form. The principal latex paints are styrene-butadiene, polyvinyl acetate, and acrylic resins. Percentage composition may be 25 to 30% dry ingredients, 40% latex, and 20–30% water, plus stabilizers. The unique properties of emulsion paints are ease of application, absence of disagreeable odor, and nonflammability. They can be used on both interior and exterior surfaces.

paint, inorganic. A potassium silicate-based corrosion-resistant coating designed for use on bridges and other metal work subject to marine environments.

paint, metallic. A paint in which the primary pigment is a finely divided metal dispersed in the vehicle. Most common is aluminum paint, but other metals are also used.

paint remover (varnish remover). A mixture in liquid or paste form containing volatile solvents and nonvolatile components which retard evaporation of the solvent, thereby prolonging its action. Typical solvents are: methanol, denatured ethyl alcohol, methylene chloride, toluene, benzene, and ethyl acetate. Paraffin is often used as the retarder. Caustic removers contain sodium phosphate, sodium silicate, caustic soda, or the like.

paint, water-based. See paint, emulsion.

"Palacet."[440] Trademark for a series of organic pigments used for dyeing and printing on acetae, nylon and polyester fibers.

"Palanil."[440] Trademark for a series of dyes for polyester fibers.

"Palanthrene."[440] Trademark for a series of vat dyestuffs, mainly anthraquinonoid types.

"Palatine."[307] Trademark for metallized acid dyestuffs approaching the fastness of chrome colors.

"Palatone."[233] Trademark for a synthetic aromatic chemical; used in soap, cosmetic perfumery, and flavor compounding.

"Palconate."[126] Trademark for the sodium salt of complex organic acids extracted from redwood bark.
Uses: Dispersing agent, phenol and tannin replacement, ceramic binder, ore flotation, battery expander, leather tanning, and water treatment.

"Paliofast."[440] Trademark for high-grade organic pigments for dyeing lacquers, varnishes, printing colors, and plastics.

palladium Pd Metallic element of atomic number 46; atomic weight 106.4; valences 2,3,4; Group VIII of the Periodic Table. There are six stable isotopes.
Properties: Silver-white, ductile metal which does not tarnish in air. It is the least noble (most reactive) of the platinum group. Absorbs up to 800 times its own volume of hydrogen. Attacked by hot, concentrated nitric acid and boiling sulfuric acid; soluble in aqua regia and fused alkalies. Insoluble in organic acids. Good electrical conductor. Sp. gr. 12.0; m.p. 1554°C; b.p. 2927°C; Brinell hardness (annealed) 46. Nontoxic; noncombustible, except as dust.
Occurrence: Siberia, Ural Mts (U.S.S.R.); Ontario; So. Africa.
Derivation: In ores with platinum, gold, copper, etc. Concentrated ores are dissolved in aqua regia; after gold and platinum are removed by chemical treatment, palladium is precipitated by ammonia, followed by hydrochloric acid. After further purification treatment, ignition yields palladium metal.
Forms: Wire; leaf; powder; single crystals.
Grades: C.P. (99.99%); technical (99.0%).
Uses: Alloys for electrical relays and switching systems in telecommunication equipment; catalyst for reforming cracked petroleum fractions and hydrogenation; metallizing ceramics; "white gold" in jewelry; resistance wires; hydrogen valves (in hydrogen separation equipment); aircraft spark plugs; protective coatings.

palladium chloride (palladous chloride; palladium dichloride) (a) $PdCl_2$ (b) $PdCl_2 \cdot 2H_2O$.
Properties: Dark brown, deliquescent powder or crystals; soluble in water, hydrochloric acid, alcohol and acetone. (a) Sp. gr. 4.0 (18°C); m.p. 501°C (decomposes).
Derivation: By solution of palaldium in aqua regia and evaporation.
Grades: Technical; reagent.
Uses: Analytical chemistry; "electroless" coatings for metals; manufacture of stearin, porcelains, germicides.

palladium dichloride. See palladium chloride.

palladium iodide (palladous iodide PdI_2.
Properties: Black powder; sp. gr. 6.003 (18°C); soluble in a solution of potassium iodide; insoluble in alcohol, water, and ether. Decomposes at 350°C.

palladium monoxide. See palladium oxide.

palladium nitrate (palladous nitrate) $Pd(NO_3)_2$.
Properties: Brown salt, deliquescent; decomposed by heat; soluble in water with turbidity; soluble in dilute nitric acid.
Hazard: Oxidizing agent. May react with organic materials.
Use: Analytical reagent.
Shipping regulations: (Rail) Nitrates, n.o.s., Yellow label. (Air) Nitrates, n.o.s., Oxidizer label.

palladium oxide (palladium monoxide) PdO.
Properties: Black-green or amber solid; sp. gr. 8.70 (20°C); m.p. (dec) 750°C; soluble in dilute acids.
Derivation: Careful ignition of the nitrate or prolonged heating of the finely divided metal at about 800°C.
Use: Reduction catalyst in organic synthesis.

palladium potassium chloride (palladous potassium chloride; potassium palladium chloride) $PdCl_2 \cdot 2KCl$.
Properties: Reddish-brown crystals; soluble in water; slightly soluble in hot alcohol. Sp. gr. 2.67; m.p. 524°C.
Use: Reagent for carbon monoxide determination.

palladium sodium chloride (palladous sodium chloride; sodium palladium chloride) $PdCl_2 \cdot 3H_2O$.
Properties: Brown salt; hygroscopic; soluble in alcohol and water.
Use: Analysis (testing for carbon monoxide, ethylene, illuminating gas, iodine).

"Palldon."[169] Trademark for para-nitrosodimethylaniline used in the colorimetric determination of palladium and platinum.

palladous chloride. See palladium chloride.

palladous iodide. See palladium iodide.

palladous nitrate. See palladium nitrate.

palladous potassium chloride. See palladium potassium chloride.

palladous sodium chloride. See palladium sodium chloride.

palmarosa oil (geranium oil, Turkish).
Properties: Colorless or light yellow, essential oil, occasionally colored green by copper. Rose-like odor; sp. gr. 0.885–0.897 (15°C); optical rotation -2° to +3°C; refractive index (n 20/D) 1.4730–1.4775; soluble in 2 volumes of 70% alcohol, benzyl benzoate, fixed oils, propylene glycol, mineral oil; insoluble in glycerin. Combustible; nontoxic.
Chief known constituent: Geraniol.
Derivation: By distilling Cymbopogon martini, var. motia, found in the East Indies and Java.
Grades: Technical; F.C.C.
Uses: Perfumery; manufacture of geraniol; flavoring.
See also geranium oil.

palm butter. See palm oil.

"Palmetto."[84] Trademark for agricultural dusting sulfur; used as an insecticide and fungicide.

"Palmex."[152] Trademark for a series of metal processing oils available in several grades for hot dip thinning, pickling, and rolling of sheet, strip and thin tinplate.

palmitic acid (hexadecanoic acid; cetylic acid) $CH_3(CH_2)_{14}COOH$. A saturated fatty acid. It occurs in natural fats and oils and in tall oil, and in most commercial-grade stearic acid.
Properties: White crystals; soluble in alcohol and ether; insoluble in water; sp. gr. 0.8414 (80/4°C); m.p. 62.9°C; b.p. 351.5°C, 271.5°C (100 mm), 139.0°C (1 mm); refractive index 1.4309 (70°C). Combustible; nontoxic.
Derivation: From spermaceti by saponification; hydrolysis of natural fats.
Method of purification: Crystallization.
Grades: Technical; 99.8%; F.C.C.
Uses: Manufacture of metallic palmitates; soaps; lube oils; waterproofing; food-grade additives.

palmitic acid cetyl ester. See cetin.

palmitin. See tripalmitin.

palmitoleic acid (cis-9-hexadecenoic acid) $CH_3(CH_2)_5CH{:}CH(CH_2)_7COOH$. An unsaturated fatty acid found in nearly every fat, especially in marine oils (15–20%).
Properties: Colorless liquid; m.p. 1.0°C; b.p. 140–141°C (5 mm). Insoluble in water; soluble in alcohol and ether. Combustible; nontoxic.
Grade: Purified product 99%.
Uses: Organic synthesis; chromatographic standard.

palmitoyl chloride (hexadecanoyl chloride; palmityl chloride, so-called) $CH_3(CH_2)_{14}COCl$.
Properties: Colorless liquid; soluble in ether, decomposes in water or alcohol; m.p. 11–12°C; b.p. 194.5°C (17 mm).

palmityl alcohol. See cetyl alcohol.

palm nut cake (palm cake). The cakes formed in the press when the palm nut kernels are expressed to obtain the oil. Contains various useful constituents, such as unexpressed oil, carbohydrates, proteins, and salts. Typical analysis: proteins 30.4%; fats 8.4%; fiber 41.0%; water 9.5%; ash 10.6%.
Containers: Bags; bulk.
Uses: Cattle-food; fertilizer ingredient.

palm oil.
Properties: Yellow-brown, buttery, edible solid at room temperature. Sp. gr. 0.952; m.p. about 30°C; iodine number 13.5; saponification number 247.6. Soluble in alcohol, ether, chloroform, carbon disulfide. Combustible. Nontoxic.
Chief constituents: Triglycerides of oleic, palmitic, and lauric acids, the percentage depending on whether kernel or fruit pulp is used as source.
Derivation: Separation of fat from palm fruits by expression or centrifugation. Palms grow in West Africa, Nigeria, Malaya, Indonesia; chief commercial source is Malaysia.
Grades: Crude; refined.
Containers: Drums; tank wagons.
Uses: Soap manufacture; pharmacy; food shortening; cutting-tool lubricant; hot-dipped tin coating; terne plating; cosmetics; softener in rubber processing; cotton goods finishing; substitute for tallow.

PAM. Abbreviation of 2-pyridine aldoxine methiodide (q.v.).

"Pamak."[266] Trademark for various tall oil products including a series of tall oil fatty acids and distilled tall oils containing varying percentages of rosin acids. "Pamak" TP and WTP are residues from fractionation of crude tall oil in the manufacture of tall oil fatty acids.

pamaquine naphthoate $C_{42}H_{45}N_3O_7$.
Properties: Yellow to orange yellow odorless almost tasteless powder. Insoluble in water; soluble in alcohol and acetone.
Use: Medicine (antimalarial).

"Pamisyl."[330] Trademark for para-aminosalicylic acid. "Pamisyl" Sodium: sodium para-aminosalicylate.

"Panaflex."[216] Trademark for liquid hydrocarbon plasticizers; used as secondary plasticizers in compounding polyvinyl chloride; in vinyl plastisols; and in molded and extruded products.

"Panapol."[216] Trademark for liquid hydrocarbon polymers produced by polymerization of olefins and diolefins; available in a wide color range with various physical properties.
Containers: Tank cars or 55-gal drums.
Use: Extender for drying oils, or where a low-cost resinous product is desirable.

"Panasol."[216] Trademark for petroleum aromatic solvents available in a variety of boiling ranges. Used in paint and varnish applications, and in the formulation of insecticides.

pancreatin. A mixture of enzymes, principally pancreatic amylase, trypsin, and pancreatic lipase. Obtained from the pancreas of hog or ox.
Properties: Cream-colored amorphous powder; characteristic odor; acts upon starch and proteins. Soluble in water; insoluble in alcohol. It changes protein into proteoses and derived substances, and starch into dextrins and sugars. Its greatest activity is in neutral or slightly alkaline media.
Derivation: Pancreas gland is extracted by macerating with chloroform, water, dilute boric acid, glycerol, or alcohol, filtered and evaporated.
Grades: N.F.
Containers: Glass bottles; fiber cans.
Uses: Medicine, as an emulsifying agent, and as a ferment; in bating compounds for leather; to remove starch and protein sizings from textiles.

"Pan Glaze."[149] Trademark for a resinous silicone coating developed to replace pan grease in the baking industry. It provides easy release for 100 or more bakes.

"Panmycin" Phosphate.[327] Trademark for tetracycline phosphate complex.

"Pano-drench."[401] Trademark for a liquid soil treatment concentrate containing 0.6% cyano(methylmercuri)-guanidine.
Hazard: May be toxic by ingestion.

"Panogen."[401] Trademark for a series of fungicides containing methylmercury dicyandiamide. Toxic by ingestion.

pantethine $C_{22}H_{42}N_4O_8S_2$. The disulfide form of N-pantothenylthioethanolamine. Lactobacilus bulgaricus growth factor (LBF). A fragment of coenzyme A, a pantothenic acid derivative.
Use: Biochemical research.

panthenol. USAN name for pantothenol.

"Pantholin."[100] Trademark for racemic calcium pantothenate, U.S.P.

pantolactone (2,4-dihydroxy-3,3-dimethylbutyric acid, gamma lactone) $HC(OH)C(CH_3)_2CH_2OCO$.
Properties: Crystals; sp. gr. (20/20°C), 1.180; m.p. 79.2°C; soluble in water.
Grade: 80% aqueous solution.
Use: Preparation of pantothenic acid.

pantothenic acid [N-(2,4-dihydroxy-3,3-dimethylbutyryl)-beta-alanine]
$HOCH_2C(CH_3)_2CHOHCONH(CH_2)_2COOH$. A member of the vitamin B complex; it is a component of coenzyme A (q.v.), and may be considered a beta-alanine derivative with a peptide linkage. As a component of coenzyme A, it is involved in the release of energy from carbohydrate utilization, and is necessary for synthesis and degradation of fatty acids, sterols and steroid hormones; it also functions in the formation of porphyrins. It occurs in all living cells and tissues. Nontoxic. The natural product is dex-

trorotatory (D+), and is the only form having vitamin activity.

Sources: Food sources: liver, kidney, yeast, crude molasses, milk, whole grain cereals, rice. Commercial sources: produced synthetically from 2,4-dihydroxy-3,3-dimethylbutyric acid and beta-alanine. See calcium pantothenate.

pantothenol (D[+]-pantothenyl alcohol; panthenol, USAN name)
$HOCH_2C(CH_3)_2CHOHCONH(CH_2)_2CH_2OH$. The alcohol corresponding to pantothenic acid, with vitamin activity.

Properties: Viscous liquid; soluble in water, ethanol, methanol; specific rotation +28.36° to 30.7° in water (c = 5); refractive index 1.497 (20°C).

Uses: Biochemical research; food additive and dietary supplement.

papain (papayotin)

Properties: White or gray slightly hygroscopic powder; soluble in water and glycerine; insoluble in other common organic solvents. The most thermostable enzyme known. Digests proteins.

Derivation: Obtained as dried and purified latex of Carica papaya.

Grades: Technical; purified. Technical grade is susceptible to decompositoin in storage.

Containers: Glass bottles; lined fiber drums.

Uses: Meat tenderizer; other food industries (mainly to prevent protein haze on chilling beer); tobacco, pharmaceutical, cosmetic, leather, textiles.

papaverine $(CH_3O)_2C_6H_3CH_2C_9H_4N(OCH_3)_2$. 6,7-Dimethoxy-1-veratrylisoquinoline. An alkaloid.

Properties: White crystalline powder. Soluble in chloroform, hot benzene, aniline, glacial acetic acid, and acetone; slightly soluble in alcohol and ether; insoluble in water. M.p. 147°C.

Derivatoin: From opium.

Hazard: Moderately toxic narcotic.

Use: A vasodilator used for treatment of hypertension (also as the hydrochloride which is soluble in water.)

paper. A semisynthetic product made by chemically processing cellulosic fibers. Wide varieties of sources have been used for specialty papers (flax, bagasse, esparto, straw, papyrus, bamboo, jute, and others), but by far the largest quantity is made from softwoods (coniferous trees), such as spruce, hemlock, pine, etc.; some is also made from such hardwoods as poplar, oak, etc., as well as from synthetic fibers. Papermaking technology involves the following basic steps: (1) chipping or other subdivision of the logs (see groundwood); (2) manufacture of chemical or semichemical pulp by digestion in acid or alkaline solutions, which separates the cellulose from the lignin (see pulp, paper); (3) beating the pulp to break down the fibers and permit proper bonding when the sheet is formed; (4) addition of starches, resins, clays, and pigments to the liquid stock (or "furnish"); (5) formation of the sheet continuously on a fourdrinier machine, where the water is screened out and the sheet dried by passing over a series of heated drums; (6) high-speed calendering for brightness and finish; (7) coating, either by machine application or (for heavy finishes) by brushes.

Note: Wet paper stock and waste are flammable, and are considered as dangerous fire risks.

Further information can be obtained from the Technical Association of the Pulp and Paper Industry, 360 Lexington Ave., N.Y. (q.v.), or from the Institute of Paper Chemistry, Appleton, Wisconsin.

"Paperad."[489] Trademark for a finely precipitated trihydrated alumina pigment filler and extender for use in the paper industry. Characterized by high brightness and blue-white color.

paper chromatography (PC). A micro type of chromatography (q.v.). A drop of the liquid to be investigated is placed near one end of a strip of paper. This end is immersed in solvent, which travels down the paper and distributes the materials present in the original drop selectively. Comparison with known substances makes identification possible.

paper, coated. A paper that is covered on one or both sides with a suspension of clays, starches, casein, rosin, wax, or combinations of these, to serve special purposes. Machine-coated paper is required for standard book printing; the rather light coating is applied by a roll-coating device. Heavier coatings are applied as a final step by means of brushes or spreading devices. These are required for high-grade printing of magazines, art books, etc., where excellent photographic reproduction is essential. Special-purpose coatings, as for packaging, are applied in a separate operation.

paper, synthetic. Paper or paper-like material made from a polyolefin; polypropylene is usually selected. This product reached pilot-plant production stage in 1974. A paper made from styrene copolymer fibers has been developed to production stage in Japan. Plastic-coated cellulosic papers are available for children's books, posters, and similar applications.

"Papi."[520] Trademark for polymethylene polyphenylisocyanate. A dark non-volatile liquid; average viscosity 250 cp at 25°C. Used in one-shot rigid urethane foams.

"Papi" 50. 50% solution in monochlorobenzene. Used in adhesives (rubber to metals and synthetic fabrics); coating intermediate.

"Papricol."[342] Trademark for an oleoresin of paprika for food coloring and flavoring.

para- (p-). A prefix. For definition of para-compounds, see ortho-. For para-compounds, see specific compound; para-cresol is listed under cresol and para-dichlorobenzene under dichlorobenzene.

"Parabens."[583] Trademark for the methyl, propyl, butyl, and ethyl esters of para-hydroxybenzoic acid. Antimicrobial agents for foods and pharmaceuticals. Nontoxic. Approved by FDA.

paracasein. See casein.

paracetaldehyde. See paraldehyde.

"Paracol."[266] Trademark for a series of wax and wax-rosin emulsions produced from paraffin waxes, microcrystalline waxes, or combinations of these waxes with rosin. Used to impart water resistance to paper and allied materials.

"Paracort."[330] Trademark for prednisone (q.v.).

"Paracril."[248] Trademark for a group of synthetic rubbers of the Buna-N or nitrile type, produced by the copolymerization of butadiene and acrylonitrile. Resist deterioration by aliphatic hydrocarbon, mineral and vegetable oils, and animal fats and oils and are particularly resistant to petroleum products.

"Paracril" is also used as a plasticizer for vinyls and other thermoplastic and thermosetting resins.

"Paradene."[21] Trademark for low-priced, dark, thermoplastic, coal-tar resins (coumarone-indene) available in low to high softening point ranges; used in rubber compounding.

"Paradi."[62] Trademark for para-dichlorobenzene (q.v.).

"Paradors."[12] Trademark for products for use in the neutralization of objectionable odors in rubber.

"Paradow."[233] Trademark for para-dichlorobenzene (q.v.).

paraffin (1) Also called alkane. A class of aliphatic hydrocarbons characterized by a straight or branched carbon chain; generic formula C_nH_{2n+2}. Their physical form varies with increasing molecular weight from gases (methane) to waxy solids. They occur principally in Pennsylvania and Midcontinent petroleum.
(2) Paraffin wax (q.v.).

paraffin distillate. A distilled petroleum fraction which, when cooled, consists of a mixture of crystalline wax and oil.

paraffin, chlorinated. A paraffin oil or wax in which some of the hydrogen atoms have been replaced by chlorine. Nonflammable; low toxicity. Used in high-pressure lubricants; as flame retardant in plastics and textiles; as plasticizer for polyvinyl chloride in polyethylene sealants; and in detergents.

paraffin oil. An oil either pressed or dry-distilled from paraffin distillate. Liquid petrolatum is also known as paraffin oil. Combustible.
Grades: By viscosity and color.
Containers: Up to tank cars.
Uses: Floor treatment; lubricant; when purified, medicine.

paraffin wax (paraffin scale; paraffin).
Properties: White, translucent, tasteless, odorless solid; consisting of a mixture of solid hydrocarbons of high molecular weight, e.g., $C_{36}H_{74}$. Soluble in benzene, ligroin, warm alcohol, chloroform, turpentine, carbon disulfide and olive oil; insoluble in water and acids; sp. gr. 0.880–0.915; m.p. 47–65°C. Combustible; essentially nontoxic. Flash point 390°F; autoignition temp. 473°F.
Grades: Yellow crude scale; white scale; refined wax; ASTM; N.F. Also graded by melting point, in °F and color. The higher-melting grades are the more expensive.
Containers: Kegs; cases (slabs); tank cars.
Hazard: (fume): Tolerance, 1 mg per cubic meter of air.
Uses: Candles; paper coating; protective sealant for food products, beverages, etc.; glass-cleaning preparations; hot-melt carpet backing; biodegradable mulch (hot melt-coated paper); impregnating matches; lubricants; crayons; surgery; stoppers for acid bottles; electrical insulation; floor polishes; cosmetics; photography; anti-frothing agent in sugar refining; packing tobacco products; protecting rubber products from sun-cracking; chewing gum base (to ASTM specifications).

"Paraflint." Trademark for a polymethylene wax. It is white, odorless, and has congealing point of 205°F. Available in flaked form. Approved by F.D.A.

"Paraflow."[29] Trademark for a mixture of synthetic organic compounds of high molecular weight, which are added to lubricating oils to reduce their pour points.

"Para-Flux."[94] Trademark for a series of petroleum-based plasticizers for natural and synthetic rubbers.

paraformaldehyde (paraform) $HO(CH_2O)_nH$. A polymer of formaldehyde in which n equals 8 to 100. Not to be confused with the trimer, symtrioxane (q.v.).
Properties: White solid with slight odor of formaldehyde; insoluble in alcohol and ether. Soluble in strong alkali solution. The higher polymers are insoluble in water. Melting range 120–170°C; flash point 160°F (closed cup). Combustible. Autoignition temp. 572°F.
Derivation: By evaporating an aqueous solution of formaldehyde.
Forms: Flake; powder.
Grades: Bags; carlots.
Hazard: Toxic by ingestion; irritant to tissue. Safety data sheet available from Manufacturing Chemists Assn., Washington, D.C.
Uses: Fungicides, bactericides, and disinfectants; adhesives; hardener and waterproofing agent for gelatin.

paraldehyde (2,4,6-trimethyl-1,3,5-trioxane) $C_6H_{12}O_3$. A cyclic polymer (trimer) of acetaldehyde. A depressant drug.
Properties: Colorless liquid. Disagreeable taste; agreeable odor. Decomposes on standing. Stable toward alkalies but slowly decomposed to acetaldehyde when treated with a trace of mineral acid. Miscible with most organic solvents, volatile oils; soluble in water. Sp. gr. 0.9960 at 20°C/20°C; b.p. 124.5°C; m.p. 12.6°C; vapor pressure 25.3 mm (20°C); flash point 96°F (TOC); specific heat 0.434; refractive index 1.40 to 1.42 (20°C); wt 8.27 lb/gal (20°C). Autoignition temp. 460°F.
Derivation: Action of hydrochloric or sulfuric acid upon acetaldehyde.
Grades: Technical; U.S.P.
Containers: Drums; tank cars.
Hazard: Flammable, moderate fire risk; toxic by ingestion and inhalation.
Uses: Substitute for acetaldehyde; rubber accelerators; rubber antioxidants; synthetic organic chemicals; dyestuff intermediates; medicine; solvent for fats, oils, waxes, gums, resins; leather; solvent mixtures for cellulose derivatives; sedative.

"Paramine."[300] Trademark for a cationic finishing agent for textiles; an aminofatty condensation product solubilized with a volatile acid.

paranitraniline. See nitroaniline, para-.

paranitraniline red. See para red.

"Paranox."[29] Trademark for a series of additives compounded for use in lubricating oils. Imparts high temperature detergency, low temperature sludge inhibition and anti-wear characteristics to oils compounded for lubrication of automobile and diesel engines and for automatic transmission fluids.

para-oxon (diethyl para-nitrophenyl phosphate) $(C_2H_5O)_2P(O)OC_6H_4NO_2$. Generic name for the oxygen analog of parathion.
Properties: Odorless, reddish-yellow oil; b.p. 148–151°C (1 mm); sp. gr. 1.269 (25/25°C); refractive index 1.5060 (25°C). Slightly soluble in water, soluble in most organic solvents. Decomposes rapidly in alkaline solutions.
Hazard: Highly toxic by ingestion, inhalation, and skin absorption. Cholinesterase inhibitor. Use may be restricted.

Uses: Insecticide; nerve poison.
Shipping regulatoins: (Rail, Air) Organic phosphate, liquid, n.o.s., Poison label. Not accepted on passenger planes.

"Parapastels."[188] Trademark for a group of perfume and color combinations designed to impart agreeable fragrance and pleasing color to products made of para-dichlorobenzene or naphthalene.

"Paraplex."[23] Trademark for polymeric plasticizers for resinous coatings. Primarily polyester, but some grades also epoxides which impart heat and light stability. Supplied as viscous liquids, 100% solids or solutions in petroleum hydrocarbons. Most grades compatible with nitrocellulose, ethyl cellulose, polyvinyl butyral, and other high polymer film formers.
Use: Coatings for wood, metal, fabrics, paper, and rubber. Also in formulation of free films, caulking and sealing compounds.

"Parapol."[29] Trademark for a copolymer of isobutylene and styrene. This material is useful as a free film, as a supported film, and as a modifier for paraffin.

"Parapon."[300] Trademark for a highly sulfated fatty ester dyeing assistant and leveling agent for dyed fabrics.

paraquat. Generic name for 1,1′-dimethyl-4,4′-bipyridinium salt, $[CH_3(C_5H_4N)_2CH_3] \cdot 2CH_3SO_4$.
Properties: Yellow solid; soluble in water.
Hazard: Highly toxic by ingestion, inhalation, and skin absorption. Tolerance, 0.5 mg per cubic meter of air. Use may be restricted.
Use: Herbicide.
Shipping regulations: Not listed.

para red (paranitraniline red) $C_{10}H_6(OH)NNC_6H_4NO_2$. A pigment formed by coupling diazotized paranitroaniline with beta-naphthol. Ther term is also used to refer to a group of lakes based on this dye. See para toner.

"Para Resin" 2457.[94] Trademark for a petroleum base resin. Used as a plasticizer and softener for rubbers.

pararosaniline $HOC(C_6H_4NH_2)_3$. A triphenylmethane dye. C.I. No. 42500.
Properties: Colorless to red crystals; m.p. 205°C (approx.). Soluble in alcohol; very slightly soluble in water and ether. Combustible.
Hazard: May be toxic.
Use: Dye (usually as the hydrochloride). Component of fuchsin (q.v.).

pararosolic acid. See aurin.

"Parasepts."[138] Trademark for a group of neutral esters of para-hydroxybenzoic acid; includes the methyl, ethyl, propyl, benzyl, and butyl esters. White powder.
Use: Antimicrobial agents.

"Paratac."[29] Trademark for a high molecular weight isobutylene polymer used as an additive for both lubricating oils and greases, particularly in producing so-called "nondrip," "nonspatter" oils.

"Paratex."[300] Trademark for a water softener used in textile processing.

parathion. Generic name for O,O-diethyl O-para-nitrophenyl phosphorothioate (ethyl parathion; O,O-diethyl para-nitrophenyl thiophosphate; AATP) $(C_2H_5O)_2P(S)OC_6H_4NO_2$.
Properties: Deep brown to yellow liquid; usually has faint odor. Refractive index (n 25/D) 1.5367; sp. gr. 1.26 (25/4°C); b.p. 157–162°C (0.6 mm); 375°C; vapor pressure 0.003 mm (24°C); very slightly soluble in water (20 ppm); completely soluble in esters, alcohols, ketones, ethers, aromatic hydrocarbons, animal and vegetable oils; insoluble in petroleum ether, kerosine, spray oils. Stable in distilled water and in acid solution. Hydrolyzed in the presence of alkaline materials; slowly decomposes in air.
Purity: Technical grade is about 95% pure. Also supplied diluted with inert carriers of various types, and in various proportions.
Derivaton: From sodium ethylate, thiophosphoryl chloride and sodium para-nitrophenate.
Containers: Tins; steel drums.
Hazard: Highly toxic by skin contact, inhalation or swallowing. Cholinesterase inhibitor. Repeated exposure may, without symptoms, be increasingly hazardous. Fatalities have resulted from its accidental use. Tolerance, 0.1 mg per cubic meter of air. Approved by EPA.
Uses: Insecticides and acaricide.
Shipping regulatoins: (Rail, Air) (liquid) Poison label. Not acceptable on passenger planes. (Rail) (dry and liquid mixtures) Poison label. (Air) Consult regulations for varous mixtures, all of which require Poison label. Parathion and compressed gas mixture, Not acceptable.
See also methyl parathion.

"Paratone."[29] Trademark for organic compounds used as additives designed to improve viscosity index of lubricating oils.

para toner. An insoluble red pigment derived from beta-naphthol and para-nitroaniline. The former is sometimes partly replaced by mono-acid F, 2-naphthol-7-sulfonic acid. Through varying the conditions of temperature and acid concentration, different shades may be obtained.
Containers: Barrels.
Uses: Paint and printing ink pigments; making para lakes.

"Parco."[62] Trademark for phosphoric acid and phosphate compounds for dissolving rust from the surface of metal.

"Parcolene."[62] Trademark for chemicals for treating metal surfaces to remove extraneous matter and/or condition the surface prior to other finishing operations.

paregoric. A derivative of opium that is habit-forming on continued use. Its use is restricted by FDA. Used in medicine, especially for digestive disorders.

parethoxycaine hydrochloride. See beta-diethylaminoethyl-para-ethoxybenzoate hydrochloride.

"Parex" process.[505] A proprietary continuous method for extracting para-xylene from mixed xylene feedstocks. See para-xylene.

"Parez."[57] Trademark for a series of melamine-formaldehyde and urea-formaldehyde resins, designed for use in papermaking.

"Paricin."[202] Trademark for various alkyl hydroxystearates (soft, low melting point waxes useful as firming agents in cosmetics and specialty inks and as coupling agents for incompatible mixtures of polar and non-polar materials in hot melt applications) and acetoxystearates (plasticizers for nitrocellulose, cellulose acetate butyrate, ethylcellulose, and vinyl resins.)

"Pariflux."[250] Trademark for fluorspar. Used in welding electrode coatings for arc stabilization, fluxing weld metal and forming protective slag.

Paris green. See copper acetoarsenite.

Paris white. See whiting.

"Parkerex."[62] Trademark for resins to be used for ion exchange purposes.

"Parkerizing."[343] Trademark for a process for the inhibition of rust formation on iron or steel by coating surface with a phosphate layer, achieved by dipping in acid phosphate solution.

Parkes process. A standard process for the separation of silver from lead. From 1–2% molten zinc is added to the lead-silver mix, heated to above the melting point of zinc. A scum containing most of the silver and zinc forms on the surface. This is separated and the silver recovered. The separation of silver is not complete and the process is repeated several times.

"Parlon."[266] Trademark for chlorinated rubber. White, odorless, nonflammable, granular powder. Available in five viscosity grades.
Uses: Film-former in paints, inks, adhesives, and concrete-treating compounds; traffic and marine paints; corrosion-resistant coatings.

"Parmo."[51] Trademark for petrolatums meeting U.S.P. or N.F. requirements and having melting points, consistencies and colors suitable for pharmaceuticals and cosmetics.

"Parnol" 40.[309] Trademark for sodium dodecylbenzene sulfonate in the form of a fine flake, containing approximately 40% of active matter.
Uses: Textile scouring and dyeing operations; synthetic detergents, and a household detergent additive.

"Parolite."[159] Trademark for normal zinc sulfoxylate formaldehyde. Used for the stripping of all forms of wool and synthetics, (nylon, acetates, Dacron, etc.).

paromomycin sulfate. Antibiotic from a strain of Streptomyces.
Properties: Creamy white, odorless, hygroscopic powder. Soluble in water; insoluble in chloroform and ether.
Grade: N.D.
Use: Medicine (antibiotic).

partial pressure. The pressure due to one of the several components of a gaseous or vapor mixture. In general this pressure cannot be measured directly but is obtained by analysis of the gas or vapor and calculation by use of Dalton's law. See also Raoult's law.

particle. Any discrete unit of material structure; the particulate basis of matter is a fundamental concept of science. The size ranges of particles may be summarized as follows: (1) Subatomic: protons, neutrons, electrons, deuterons, etc. These are collectively called fundamental particles. They are invisible by any means. (2) Molecular: includes atoms and molecules with size ranging from a few angstroms to about half a micron. (3) Colloidal: includes macromolecules, micelles, and ultrafine particles such as carbon black; resolved in electron microscope; size ranges from 1 millimicron up to lower limit of the optical microscope (1 micron). (4) Microscopic: units that can be resolved by an optical microscope (includes bacteria). (5) Macroscopic: all particles that can be resolved by the naked eye. See also matter; particle size.

particle accelerator. A device in which the speed of charged subatomic particles (protons, electrons) and heavier particles (deuterons, alpha particles) can be greatly increased by applicaton of electric fields of varying intensity, often in conjunction with magnetic fields. It is possible to accelerate electrons and protons to speeds approaching the speed of light if sufficiently high voltage is used. Straight-line (linear) accelerators are used for protons, and doughnut-shaped betatrons for electrons; other types are the Van de Graaf electrostatic generator, the synchrotron, and the cyclotron. Before the development of nuclear reactors, the cyclotron was used to accelerate deuterons for use in bombarding stable nuclei to produce neutrons for inducing artificial radioactivity, fission, and formation of synthetic (transuranic) elements.

particle size. This term refers chiefly to the solid particles of which industrial materials are composed (carbon black, zinc oxide, clays, pigments, and the like). The smaller the particle, the greater will be the total exposed surface area of a given volume. Activity is a direct function of surface area; that is, the finer a substance is, the more efficiently it will react, both chemically and physically. A colloidal pigment is a more effective colorant than a coarse one because of the greater surface area of its particles. A pound of channel carbon black has a surface area of about eighteen acres, which largely accounts for its powerful reinforcing effect in rubber. Thus ultrafine grinding of powders is of utmost importance in such products as paints, cement, plastics, rubber, dyes, pharmaceuticals, printing inks, and numerous others. See also particle; surface chemistry; colloid chemistry; sedimentation.

partition chromatography. See liquid chromatography.

"Parvan."[51] Trademark for fully refined paraffin wax available in slabs and in liquid form in wide range of melting points and hardness.

"Parvo."[57] Trademark for folic acid feed supplement.

parylene. Generic name for thermoplastic film polymers based on para-xylylene and made of vapor-phase polymerization.
Derivation: para-Xylene, $CH_3C_6H_4CH_3$, is heated with steam at 950°C to produce the cyclic dimer di-para-xylylene, a solid which can be separated in pure form. The dimer is then pyrolyzed at 550°C to produce monomer vapor of para-xylene $CH_2{:}C_6H_4{:}CH_2$ which is then cooled below 50°C and condenses on the desired object as a polymer having the repeating structure—$(CH_2C_6H_4CH_2—)_n$, with n about 5000 and molecular weights of about 500,000. The polymer is used as a protective coating. Films as thin as 500 A to 5 mils are obtained.
Uses: Thin coatings of high purity and uniformity on almost any substrate which will resist a high vacuum, as paper, fabric, polyethylene and polystyrene film, ceramics, metals, many solid chemicals; electronic miniaturization systems; capacitors; thin film circuits.

"Parzate."[28] Trademark for a series of fungicides.
"Parzate" Liquid is a solution containing 22% nabam to be combined with zinc sulfate in the spray tank.
"Parzate" C is a wettable powder containing 75% zineb.
"Parzate" D is a finely divided powder containing 85% zineb.
Hazard: Irritant to skin and mucous membranes.

PAS. Abbreviation for para-aminosalicylic acid. See 4-aminosalicylic acid.

PAS sodium. See sodium para-aminosalicylate.

passivity. A property shown by iron, chromium and related metals, involving loss of their normal chemical activity in an electrochemical system or in a corrosive environment after treatment with strong oxidizing agents like nitric acid, and when oxygen is evolved upon them during electrolysis, forming an oxide coating.

paste. (1) An adhesive composition of semisolid consistency, usually water-dispersible. The common pastes are based upon starch, dextrin, or latex, often with the addition of gums, glue, and antioxidants. They are widely employed for the adhesion of paper and paperboard. (2) More generally, a soft, viscous mass of solids dispersed in a liquid. For example, paste resins are finely divided resins mixed with plasticizer to form fluid or semifluid mixtures, without use of low boiling solvents or water emulsions.

paste solder. A paste (2) containing flux, cleaner, tinning agent and powdered metallic solder.

Pasteur, Louis (1822–1895). French chemist and bacteriologist who made three notable contributions to science: (1) As a result of extensive study of fermentation, which led him to conclude that it is caused by infective bacteria, he extended the work of Jenner on smallpox serum made from cowpox (1775) to development of the concept of immunizing serums and the antibody-antigen relationship (1880). Pasteur was the first to inoculate for rabies and anthrax, and suggested the term vaccination (from Latin vaccus = cow) in recognition of Jenner's achievement. (2) initiation of the practice of heat-treating wine, and later milk and other food products, to kill or inactivate toxic microorganisms, especially the tuberculosis bacillus; (3) discovery of the optical properties of tartaric acid, present in wine residues, which laid the basis for modern knowledge of optical isomers (right- and left-handed molecular structure), a phenomenon now often called chirality. See also pasteurizatoin; optical isomer.

pasteurization. Heat treatment of milk, fruit juices, canned meats, egg products, etc., for the purpose of killing or inactivating disease-causing organisms. For milk, the minimum exposure is 62°C for 30 minutes, or 72°C for 15 seconds, the latter being called flash pasteurization. Though this treatment kills all pathogenic bacteria and also inactivates enzymes which cause deterioration of the milk, the shelf-life is limited; to prolong storage life temperatures of 80 to 88°C for 20 to 40 seconds must be used. Complete sterilization (q.v.) requires ultrahigh pasteurization at from 94°C for 3 seconds to 150°C for 1 second; in-can heating at 116°C for 12 minutes and 130°C for 3 minutes is also employed for maximum stability and long storage life. Some meat products are pasteurized by alpha-radiation.

PAT. Abbreviation for polyaminotriazole (q.v.).

patchouli oil.
Properties: Yellow-greenish brown or brown essential oil; penetrating camphoraceous odor; sp. gr. (15°C) 0.950–0.995; refractive index (n 20/D) 1.5070 to 1.5200; optical rotation −48° to −68°; soluble in 10 volumes of 90% alcohol. Soluble in ether, chloroform, benzyl benzoate, fixed oils, mineral oil; insoluble in glycerol. Combustible; nontoxic.
Derivation: Direct steam distillation of the dried leaves of Pogostemon potchouly.
Chief constituents: Patchouly alcohol (a tricyclic sesquiterpene alcohol), eugenol, cinnamic aldehyde, cadinene.
Uses: Perfumery; flavoring.

patentability. The qualifications for obtaining a patent on an invention (q.v.) or chemical process. These are: (1) the invention must not have been published in any country or in public use in the U.S., in either case for more than one year prior to date of filing the application; (2) it must not have been known in the U.S. before date of invention by the applicant; (3) it must not be obvious to an expert in the art; (4) it must be useful for a purpose not immoral and not injurious to the public welfare; (5) it must fall within the five statutory classes on which only patents may be granted, namely, (a) composition of matter, (b) process of manufacture or treatment, (c) machine, (d) design (ornamental appearance), or (e) a plant produced asexually. Special regulations relate to atomic energy developments and subjects directly affecting national security.

Robert Calvert

patent alum. See aluminum sulfate.

patent leather. Fashion leather used chiefly for formal shoes, bags, etc., characterized by a high, glossy finish applied as the final step in processing. The finish greatly reduces the poromeric nature of the leather.

pathfinder element. An element present in small proportions (less than 1%), generally metallic in nature, associated with ore deposits at the time of formation. Mapping of the concentration variation of the selected element serves to locate the main ore deposit. Examples are zinc as the "pathfinder" for lead, copper and silver ores and molybdenum associated with copper deposits.

"Pathilon"[315] Trademark for tridihexethyl chloride (q.v.).

pathway. A term used largely by biochemists for the series of chemical changes that occur during the transformation of one biochemical entity to another.

patina. Variously used to refer to an ornamental and/or corrosion-resisting film on the surface of copper, copper alloys, including bronzes, and also sometimes of iron and other metals. Such a film is formed by exposure to the air, or by a suitable chemical treatment.

patronite. A mixture of vanadium-bearing substances with the approximate formula VS_4, found in Peru.

Pattinson process. Process for the removal of silver from lead. The silver-lead mixture is melted in one of a series of pots and allowed to cool slowly. The lead which is free from silver or poorer in silver separates out as crystals which are removed, leaving the silver-rich lead in the molten state. From a number of such operations in series a lead rich in silver is obtained, collected, and the silver recovered. See also Parkes process.

Pauli exclusion principle. A fundamental generalization concerning the energy relationships of electrons within the atom, namely, that no two electrons in the same atom have the same value for all four quantum numbers; corollary to this is the fact that only two electrons can occupy the same orbital, in which case they have opposite spin, i.e., $+\frac{1}{2}$ and $-\frac{1}{2}$. This principle has an important bearing on the sequence of elements in the Periodic Table and on the limiting numbers of electrons in the shells (2 in the first, 8 in the second, 18 in the third, 32 in the fourth, etc.). See also quantum number; shell; orbital theory.

"Paveril Phosphate."[100] Trademark for dioxyline phosphate (q.v.).

Pb Symbol for lead.

PBAA. Abbreviation for polybutadieneacrylic acid copolymer.

PBD. See 1,3,4-phenylbiphenylyloxadiazole.

PBI. See polybenzimidazole.

PBPB. Abbreviation for pyridinium bromide perbromide.

PCB. Abbreviation for polychlorinated biphenyl (q.v.).

"PC-1244" Defoamer.[58] Trademark for a clear light yellow liquid; sp. gr. at 60/60°F, 0.86; 7.72 lb/gal. Used as a foam control in non-aqueous systems and also imparts a leveling action in paraffin wax formulations.

P.C.E. Abbreviation for pyrometric cone equivalent, a scale of melting or fusion points of refractory materials, based on comparison with the temperature at which pyrometric cones (q.v.) melt.

P-cellulose. See phosphorylation.

PCNB. Abbreviation for pentachloronitrobenzene (q.v.).

"PCON."[233] Trademark for 4-chloro-2-nitroaniline (q.v.).

PCP. Abbreviation for pentachlorophenol (q.v.).

PCTFE. Abbreviation for polychlorotrifluoroethylene. See chlorotrifluoroethylene resin.

pcu. Abbreviation for pound centigrade unit, the amount of heat needed to raise one pound of water from 15 to 16°C. See also chu.

Pd Symbol for palladium.

PDB. Abbreviation for para-dichlorobenzene (q.v.).

PE. Abbreviation for pentaerythritol and for polyethylene.

"PE-100."[210] Trademark for a cationic polyelectrolyte containing a quaternary ammonium group. Used as a conductive coating, coagulant aid, and metal complexing agent.

peach-kernel oil. See persic oil.

peacock blue. A blue organic pigment used especially in inks for multicolor printing. It is a lake of acid glaucine blue dye on alumina hydrate. Structurally, the dye is alpha, alpha-bis[N-ethyl-N-(4-sulfobenzyl)aminophenyl]-alpha-hydroxy-ortho-toluenesulfonic acid sodium salt,
HSO $C_6H_4COH[C_6H_4N(C_2H_5)CH_2C_6H_4SO_3Na]_2$,
and is prepared from aniline, ethanol, benzyl chloride, ortho-chlorobenzaldehyde, sulfuric acid, and sodium bisulfite.
Containers: 250-lb barrels.
Note: The term peacock blue is sometimes applied to other pigments of similar color, such as Prussian blue which has been treated with phosphotungstic acid.

peanut cake. The press cake resulting from the extraction of oil from the peanut. See peanut oil meal.

peanut oil (arachis oil; groundnut oil). A fixed nondrying oil.
Properties: Yellow to greenish yellow. Soluble in ether, petroleum ether, carbon disulfide and chloroform; insoluble in alkalies, but saponified by alkali hydroxides with formation of soaps; insoluble in water; slightly soluble in alcohol. Sp. gr. 0.912–0.920 (25°C); solidifying point −5 to +3°C; saponification value 186–194; iodine number 88–98; refractive index 1.4625–1.4645 (40°C). Flash point 540°F. Combustible. Nontoxic. Autoignition temp. 883°F.
Chief known constituents: Principally glycerides of oleic and linoleic acids, with lesser amounts of the glycerides of palmitic, stearic, arachidic, behenic, and lignoceric acids.
Derivation: By pressing ground peanut meats or by extraction with hot or cold solvents.
Method of purification: Bleaching with fuller's earth or carbon. Hot-pressed oil is frequently allowed to stand to deposit stearin (which it will do even at ordinary temperatures) and then filtered.
Grades: U.S.P., crude; refined; edible.
Containers: 5-gal tins; 50-gal drums; tank cars.
Uses: Substitute for olive oil and other edible oils, both hydrogenated and unhydrogenated; soaps; vehicle for medicines; salad oil; mayonnaise; margarine.

peanut oil meal. The crushed form of peanut cake resulting from the extraction of oil from the seed. Prepared with or without the shells; the oil meal of commerce contains between 39–45% crude protein and is sold on that basis. Typical analysis of 39% protein meal: 39.1% crude protein; 5.3% crude fiber; 34.3% nitrogen-free extract; 6.2% ether-soluble (fats); 5.3% ash; total digestible nutrient approximately 80%.
Containers: Bulk or bags.
Uses: Animal feeds; fertilizer ingredient.

pearl alum. See aluminum sulfate.

pearl ash. See potassium carbonate.

"Pearlescent Pigment."[304] Trademark for a lead monohydrogen phosphate.

pearl essence. See nacreous pigment.

pearl pigment. See nacreous pigment.

pearl white. See bismuth oxychloride; bismuth subnitrate.

pear oil. See amyl acetate.

peat. Partly decayed vegetable matter which has accumulated in marshes. Essentially the first stage in the development of lignite.

pectic acid. An acid derived from pectin by treating it with sodium hydroxide solution, washing with isopropyl alcohol, adding alcoholic hydrogen chloride and finally washing again with isopropyl alcohol and drying. Acidulant in pharmaceuticals.

pectin. A high molecular weight hydrocolloidal substance (polyuronide) related to carbohydrates and found in varying proportions in fruits and plants. Pectin consists chiefly of partially methoxylated galacturonic acids joined in long chains.
Properties: White powder or syrupy concentrates. Commonest characteristic of pectins is their property of jelling at room temperature, after addition of sugar to fruit juices in the preparation of jams or jellies. Soluble in water; insoluble in organic solvents. Nontoxic.
Derivation: By dilute-acid extraction of the inner portion of the rind of citrus fruits, or of fruit pomaces, usually apple.
Method of purification: Following decolorization, the

extracts are concentrated by evaporation or the pectins precipitated with alcohol or acetone.
Grades: Pure (N.F.) containing not less than 6.7% methoxy groups and not less than 74% galacturonic acid; 150-, 200-, 250-jelly grades, containing various diluents.
Uses: Jellies, foods, cosmetics, drugs; protective colloids; emulsifying agents; dehydrating agents.

pectinase. An enzyme present in most plants. It catalyzes the hydrolysis of pectin to sugar and galacturonic acid.
Use: Biochemical research; juice and jelly industry.

"Pectinol."[23] Trademark for formulated enzyme concentrates, of fungal origin, with varying degrees of pectinase activity which hydrolyze pectic substances.
Uses: Clarification of wines and fruit juices and processing of jellies.

pectin sugar. See L-arabinose.

PEG. Abbreviation for polyethylene glycol (q.v.).

"Pegosperse."[73] Trademark for a series of polyglycol esters of fatty acids.
Uses: Plasticizers; softeners; wetting agents; detergents; lubricants; emulsifying agents.

"Peladow."[233] Trademark for calcium chloride, 94–97%, available as white, deliquescent pellets.

"Pelargol."[188] Trademark for a perfume base for soap perfumes of the geranium and rose types.

pelargonic acid (n-nonoic acid; n-nonanoic acid; n-nonylic acid) $CH_3(CH_2)_7COOH$.
Properties: Colorless or yellowish oil with slight odor; sp. gr. 0.9052 (20/4°C); m.p. 12.5°C; b.p. 255.6°C, 162.4°C (32 mm); refractive index 1.4322 (20°C); soluble in alcohol, ether, and organic solvents; almost insoluble in water. Combustible. Low toxicity.
Derivation: By the oxidation of nonyl alcohol or nonyl aldehyde; the oxidation of oleic acid, especially by ozone.
Grades: Technical; 99%.
Containers: Drums; tank cars.
Hazard: May be skin irritant.
Uses: Organic synthesis; lacquers; plastics; production of hydrotropic salts; pharmaceuticals; synthetic flavors and odors; flotation agent; esters for turbojet lubricants; vinyl plasticizer; gasoline additive.
See also nonoic acid.

pelargonic alcohol. See nonyl alcohol.

pelargonic aldehyde. See nonanal.

pelargonyl chloride (n-nonanoyl chloride) $CH_3(CH_2)_7COCl$.
Properties: B.p. 80–85°C (5 mm); min assay 97%; soluble in hydrocarbons and ethers; decomposes in water.
Containers: Bottles, carboys, drums.
Hazard: May be skin irritant.
Use: Intermediate in organic synthesis.

pelargonyl peroxide $(C_8H_{17}COO—)_2$.
Properties: Water-white liquid with a faint odor; sp. gr. 0.926 min (25/25°C); m.p. 10°C; refractive index 1.443 min (25°C). Insoluble in water and glycerol; soluble in alcohol and hydrocarbons.
Hazard: Strong skin irritant and oxidizing agent. Dangerous fire risk in contact with organic materials.
Use: Initiator of polymerization reactions.
Shipping regulations: (Rail) Peroxides, liquid, n.o.s., Yellow label. (Air) Not acceptable.

"Pelaspan."[233] Trademark for a series of expandable polystyrenes in bead or pellet form. Each bead contains its own expanding agent, which is activated by heat.

Peligot's salt. See potassium chlorochromate.

"Pelletex."[275] Trademark for pelleted gas furnace carbon black for rubber.

pelletierine $C_5H_{10}N(CH_2)_2CHO$. beta-2(-Piperidyl) propionaldehyde.
Properties: Liquid alkaloid from the root of the pomegranate. Soluble in water, alcohol, ether, chloroform, benzene. Sp. gr. 20/4°C 0.988; b.p. 195°C.
Hazard: Probably toxic.
Use: Medicine (in form of its salts, sulfate, tannate, valerate).

"Peltex."[476] Trademark for a ferrochrome complex of sodium lignosulfonate; used as a specialized dispersant in oil well drilling fluids.

"Pemco."[296] Trademark for ceramic frits and coloring oxides.

"Penacolite"[11] Trademark for resorcinolformaldehyde and resorcinol-phenol-formaldehyde adhesives and thermosetting resins in a variety of formulations.
Uses: Bonding of wood and cellulosic products, some thermosetting and thermoplastic resins and plastics, natural and some synthetic rubbers.

"Penbro.'[79] Trademark for B wood resin.
Uses: Adhesives; animal depilatories; aslphalt emulsions; paper size; cleaning compounds (electroplating).

"Penchlor."[204] Trademark for cold-setting silicate-type acid-resistant cements. Completely stable in dry storage.
Grades: Acid-Proof, S-25, Fireproof, and FCC.
Containers: Cement powder available in 100-lb bags, solution in 50-lb and 600-lb steel drums.
Use: Quick setting cement mortar used with acid-proof brick to resist acids and strong oxidizing agents, except hydrofluoric acid, up to 2000°F.

"Penco."[204] Trademark for a group of agricultural chemical products, including many insecticides, herbicides, defoliants, and fungicides.

"Pencogel."[342] Trademark for extract of Irish gums.

"Pendane."[342] Trademark for lindane insecticide concentrates.
Hazard: See lindane.

penetrant. Any agent used to increase the speed and ease with which a bath or liquid permeates a material being processed by effectively reducing the interfacial tension between the solid and liquid. Penetrants are widely used in the textile, tanning and paper industries for improving dyeing, finishing, etc., operations. Sulfonated oils, soluble pine oils and soaps are popular among the older penetrants, and the salts of sulfated higher alcohols are typical of the synthetic organics developed for this purpose. See also wetting agent.

"Penex" Process.[416] Proprietary process for isomerization of normal pentane and normal hexane to corresponding branched-chain hydrocarbons in an environment of hydrogen over a specially prepared platinum-containing catalyst of high activity. Preferably, the charge should be either predominantly normal pentane or normal hexane, but mixtures can be processed. Used in preparing high octane gasoline blending stock from the lower quality normal paraffin stocks.

"Pen-Gleam."[204] Trademark for a mildly alkaline powdered cleaning compound used for manual washing of painted surfaces.

"Pen-Glo."[204] Trademark for a powdered, inhibited acidic material used for the removal of soil and oxides from railroad car bodies without damaging paint; 400-lb fiber drums with polyethylene liners.

"Penglo 65."[79] Trademark for a pale maleic modified pentaerythritol ester of a special tall oil in mineral spirits.
Uses: Paint and varnish.

penicillamine (USAN) (D-3-mercaptovaline) $(CH_3)_2C(SH)CH(NH_2)COOH$. Crystals, dec. 178°C. Used in medicine as a chelate for copper.

penicillin $(CH_3)_2C_5H_3NSO(COOH)NHCOR$ (bicyclic). A group of isomeric and closely related anitibiotic compounds with outstanding antibacterial activity, obtained from the liquid filtrate of the molds, Penicillium notatum and Penicillium chrysogenum, or by a synthetic process which includes fermentation.
Derivation: The mold is grown in a nutrient solution such as corn steep liquor, lactose, or dextrose. After several days of cultivation the mold excretes penicillin into its liquid culture medium. This liquid is then filtered off and the penicillin extracted and purified by countercurrent extraction with amyl acetate, adsorption on carbon, or other methods. Different varieties of penicillin are produced biosynthetically by adding the proper precursors to the nutrient solution.
Grades: Aluminum penicillin G; benzathine penicillin G; chloroprocaine penicillin O; hydrabamine penicillin V; phenoxymethylpenicillin; potassium penicillin G; potassium alpha-phenoxyethylpenicillin; potassium phenoxymethylpenicillin; procaine penicillin G; sodium methicillin; sodium penicillin G.
Hazard: Strong allergen; reaction may be severe in susceptible people.
Uses: Medicine (antibiotic); food and feed additive.

penicillinase. An enzyme which antagonizes the antibacterial action of penicillin. Such enzymes are found in many bacteria.
Uses: Phamaceutical; biological research.
See also antagonist, structural.

pennyroyal oil, European.
Properties: Yellowish to reddish-yellow essential oil, sometimes with bluish or greenish fluorescence; aromatic mint-like odor. Soluble in 1.5 to 2.5 vols. of 70% alcohol; soluble in fixed oils; insoluble in glycerol. Sp. gr. 0.928–0.940 (25°C); optical rotation +15° to +25°; refractive index 1.483–1.4875 (20°C). Combustible; low toxicity.
Chief constituents: Pulegone, limonene, dipentene, menthol, menthone.
Derivation: By distillation of Mentha pulegium.
Grades: Technical; F.C.C.
Uses: Medicine; insectifuge; perfumery; manufacture of pulegone and its derivatives; flavoring.

"Penros."[79] Trademark for polymerized wood rosin.
Uses: Adhesives; gloss oils; paper label coatings; oleoresinous varnishes; solder flux; spirit varnishes; waxed paper and hot melt compounds; synthetic resins.

pentaborane B_5H_9.
Properties: Colorless liquid with pungent odor; f.p. −46.6°C; b.p. 58°C; sp. gr. 0.61; vapor pressure (0°C) 66 mm; decomposes at 150°C; ignites spontaneously in air if impure. Flash point (pure) 86°F. Hydrolyzes slowly in water.
Derivation: Hydrogenation of diborane.
Grades: Technical 95%; high purity 99%.
Hazard: Highly toxic by ingestion and inhalation; strong irritant. Tolerance, 0.005 ppm in air. Highly flammable; dangerous fire and explosion risk. Safety data sheet available from Manufacturing Chemists Assn., Washington, D.C.
Uses: Fuel for air-breathing engines; propellant.
Shipping regulations: (Rail) Red label. Not accepted by express. (Air) Not acceptable.

"Pentac."[62] Trademark for miticide whose active ingredient is bis(pentachloro-2,4-cyclopentadien-1-yl).

pentacene (2,3,6,7-dibenzoanthracene) $C_{22}H_{14}$. Highly reactive aromatic compound consisting of five fused benzene rings.
Properties: Deep blue-violet solid; sublimes 290–300°C; decomposes in air above 300°C. Insoluble in water; slightly soluble in organic solvents.
Uses: Suggested as organic photoconductor (instead of selenium) in copying systems.

"Pentacetate."[204] Trademark for synthetic amyl acetate. used as an extractant for antibiotics; lacquer solvent.

pentachloroethane (pentalin) $CHCl_2CCl_3$.
Properties: Dense, high-boiling colorless liquid; sp. gr. 1.685 (15/4°C); b.p. 159.1°C; f.p. −22°C); refractive index 1.503 (24°C). Insoluble in water.
Derivation: By chlorination of trichloroethylene, obtained by a two-step process involving chlorination of acetylene to obtain tetrachloroethane, and removal of hydrochloric acid by action of alkali.
Containers: Drums; tank cars.
Hazard: Toxic by inhalation and ingestion. Moderate fire and explosion risk.
Uses: As solvent for oil and grease in metal cleaning. See tetrachloroethane for other uses. Also used for separation of coal from impurities by density difference.
Shipping regulations: (Rail) Poisonous liquid, n.o.s., Poison label. (Air) Poison label.

pentachloronaphthalene $C_{10}H_3Cl_5$. White powder.
Hazard: Toxic. Tolerance, 0.5 mg per cubic meter of air. Toxic action similar to chlorinated naphthalenes and chlorinated diphenyls.

pentachloronitrobenzene (PCNB) $C_6Cl_5NO_2$.
Properties: Cream crystals; musty odor; sp. gr. 1.718 (25/4°C); m.p. 142–145°C; b.p. 328°C (some decomposition). Practically insoluble in water; slightly soluble in alcohols; somewhat soluble in carbon disulfide, benzene, chloroform.
Derivation: By reacting pentachlorobenzene with fuming nitric acid.
Grades: Dust; emulsion concentrate; wettable powder.
Containers: Casks or drums.
Hazard: Moderately toxic; skin irritant.
Uses: Intermediate; soil fungicide; slime prevention in industrial waters; herbicide.

pentachlorophenol (PCP) C_6Cl_5OH.
Properties: White powder or crystals; m.p. 190°C; b.p. 310°C with decomposition; sp. gr. 1.978 (22/4°C). Slightly soluble in water; soluble in dilute alkali, alcohol, acetone, ether, pine oil, benzene,

"Carbitol," and "Cellosolve"; slightly soluble in hydrocarbons.
Derivation: Chlorination of phenol.
Containers: Fiber drums; 50-lb bags; car lots.
Hazard: Toxic by ingestion, inhalation and skin absorption. Tolerance, 0.5 mg per cubic meter of air.
Uses: Fungicide; bactericide; algicide; herbicide; sodium pentachlorophenate; wood preservative (telephone poles, piling, etc.).

pentachlorothiophenol. See "RPA."

"Pentacite."[36] Trademark for resins based on pentaerythritol (q.v.).

pentadecane $C_{15}H_{32}$ and $CH_3(CH_2)_{13}CH_3$.
Properties: Colorless liquid; soluble in alcohol; insoluble in water. Sp. gr. 0.776; b.p. 270.5°C; m.p. 10°C. Combustible; low toxicity.
Grade: Technical.
Use: Organic synthesis.

n-**pentadecanoic acid** (pentadecylic acid) $CH_3(CH_2)_{13}COOH$. A saturated fatty acid normally not found in vegetable fats, but made synthetically.
Properties: Colorless crystals; sp. gr. 0.8423 (80/4°C); m.p. 51.8–52.8°C; b.p. 339.1°C, 212°C (16 mm); refractive index 1.4529 (60°C). Insoluble in water; soluble in alcohols and ethers.
Grades: 99% pure.
Uses: Organic synthesis; reference standard in gas chromatography.

pentadecanolide (15-hydroxypentadecanoic acid lactone; pentadecalactone) $CH_2(CH_2)_{13}C(O)O$.
Properties: Colorless liquid with a strong musky odor; congeals to white crystals at room temperature. Minimum congealing point 36°C. Soluble in equal volume of 90% ethanol. Combustible. Low toxicity.
Derivation: Angelica root oil.
Grade: 98% min.
Use: Perfumery.

pentadecenyl phenol. See "Cardanol."

pentadecylic acid. See n-pentadecanoic acid.

1,3-pentadiene. See piperylene.

pentaerythrite tetranitrate. Legal label name for pentaerythritol tetranitrate (q.v.).

pentaerythritol (PE; tetramethylomethane; monopentaerythritol) $C(CH_2OH)_4$.
Properties: White, crystalline powder. Readily esterified by common organic acids. Unaffected when boiled with diute caustic alkali. Soluble in water; slightly soluble in alcohol; insoluble in benzene, carbon tetrachloride, ether, and petroleum ether. B.p. 276°C (30 mm); m.p. 262°C; refractive index (20°C) 1.54 to 1.56. Sp. gr. 1.399 (25/4°C). Combustible.
Technical grade is approximately 88% monopentaerythritol and 12% dipentaerythritol.
Derivation: Reaction of acetaldehyde with an excess of formaldehyde in an alkaline medium.
Grades: Technical; nitration; C.P.
Containers: Fiber cartons; multiwall bags; drums; car lots.
Hazard: Classed as a nuisance particulate.
Uses: Alkyd resins; rosin and tall oil esters; special varnishes; pharmaceuticals; plasticizers; insecticides; synthetic lubricants; explosives; paint swelling agents.

pentaerythritol tetraacetate $C(CH_2OOCCH_3)_4$.
Properties: White, crystalline powder; extremely stable in sunlight. Soluble in water, alcohol and ether. M.p. 84°C; b.p. 225°C (30 m). Combustible.
Derivation: By the esterification of pentaerythritol with acetic acid.
Grade: Technical.

pentaerythritol tetrakis(diphenyl phosphite) (tetra(diphenylphosphito)pentaerythritol $C[CH_2OP(OC_6H_5)_2]_4$. A low-melting, white waxy solid with a slight phenolic odor; sp. gr. 1.24 (25/15.5°C); m.p. 30 to 60°C; refractive index (n 25/D) 1.5823. Combustible.
Use: Ingredient in stabilizer systems for resins.

pentaerythritol tetra(3-mercaptopropionate) $C(CH_2OOCCH_2CH_2SH)_4$.
Properties: Liquid; sp. gr. 1.28 (25°C); refractive index (n 25/D) 1.5300. Insoluble in water, alcohol and hexane; soluble in acetone and benzene. Combustible.
Uses: Curing or cross-linking agents for polymers, especially epoxy resins; intermediates for stabilizers and antioxidants.

pentaerythritol tetranitrate (PETN; penthrite) $C(CH_2ONO_2)_4$.
Properties: White crystalline material; sp. gr. 1.75; m.p. 138–140°C; decomposes above 150°C. Very soluble in acetone; slightly soluble in alcohol and ether; insoluble in water.
Derivation: Esterification of pentaerythritol with nitric acid.
Containers: Specially lined 300-lb steel drums.
Hazard: Highly shock-sensitive explosive. Detonates at 210°C.
Uses: Demolition explosive; blasting caps; detonating compositions.
Shipping regulations: (Rail) Consult regulations. Not accepted by express. (Air) Not acceptable. Legal label name: pentaerythrite tetranitrate.

pentaerythritol tetrastearate $C(CH_2OOCC_{17}H_{35})_4$. Hard, high-melting wax; ivory-colored and essentially neutal; acid number 1; softening point 67°C. Combustible. Used in polishes, coatings, textile finishes.

pentaerythritol tetrathioglycolate $C(CH_2OOCCH_2SH)_4$.
Properties: Liquid; sp. gr. 1.385 (25°C); refractive index (n 25/D) 1.5499. Insoluble in water, alcohol, and hexane; partially soluble in benzene; soluble in acetone. Combustible.
Uses: Curing or cross-linking agents for polymers, especially epoxy resins; intermediate for stabilizers and antioxidants.

pentaglycerine. See trimethylolethane.

pentahydroxycyclohexane. See quercitol.

2′,3,4′,5,7-pentahydroxyflavone. See morin.

pentalin. See pentachloroethane.

"Pentalyn."[266] Trademark for a series of nonreactive and heat-reactive pentaerythritol esters of rosin. Available in solid and/or flake form in various grades. Used in varnishes, floor polishes, inks, and adhesives.

1,1,3,3,5-pentamethyl-4,6-dinitroindane $C_{14}H_{18}N_2O_4$.
Properties: Pale yellow crystals with a musk-type odor; m.p. min 132°C. Slightly soluble in alcohol; soluble in diethyl phthalate.
Use: Perfumery.

pentamethylene. See cyclopentane.

pentamethyleneamine. See piperidine.

pentamethylene-1,1-bis(1-methylpyrrolidinium bitartrate). See pentolinium tartrate.

pentamethylenediamine. See cadaverine.

pentamethylene dibromide (1,5-dibromopentane) $BrCH_2(CH_2)_3CH_2Br$.
Properties: Colorless, aromatic liquid; f.p. −35°C; b.p. 224°C; insoluble in water.

pentamethylene glycol. See 1,5-pentanediol.

pentamethylpararosaniline chloride. See methyl violet.

pentanal. See n-valeraldehyde.

n-**pentane** (amyl hydride) $CH_3(CH_2)_3CH_3$.
Properties: Colorless liquid; pleasant odor; freezing point −129.723°C; b.p. 36.074°C; refractive index (n 20/D) 1.35748; sp. gr. (20°C) 0.62624; soluble in hydrocarbons, oils, and ether; very slightly soluble in alcohol; soluble in water. Flash point −57°F. Autoignition temp. 588°F.
Derivation: Fractional distillation from petroleum; purified by reactification.
Grades: Pure, technical; commercial.
Containers: Drums; tank cars.
Hazard: Highly flammable; dangerous fire and explosion risk. Explosive limits 1.4 to 8% in air. Tolerance, 500 ppm in air. Narcotic in high concentrations.
Uses: Artificial ice manufacture; low-temperature thermometers; solvent extraction processes; blowing agent in plastics, e.g., expandable polystyrene; pesticide.
Shipping regulations: (Rail) Red label. (Air) Flammable Liquid label.

pentanedinitrile. See glutaronitrile.

pentanedioic acid. See glutaric acid.

pentanedioic acid anhydride. See glutaric anhydride.

1,5-pentanediol (pentamethylene glycol) $HOCH_2(CH_2)_3CH_2OH$.
Properties: Colorless liquid; b.p. 260°C; f.p. −15.6°C; sp. gr. 0.9921 (20/20°C); wt/gal 8.2 lb (20°C); flash point (open cup) 265°F. Miscible with water and alcohol. Combustible; nontoxic. Autoignition temp. 633°F.
Grade: Technical.
Containers: Drums; tank cars.
Uses: Polyester and urethane resins; hydraulic fluid; lube oil additive; antifreeze.

2,3-pentanedione. See acetyl propionyl.

2,4-pentanedione. See acetylacetone.

pentanethiol (amyl mercaptan) $C_5H_{11}SH$. A mixture of isomers.
Properties: Water-white to light yellow liquid; sp. gr. (20°C) 0.83–0.84; mercaptan content at least 90.0%; initial b.p. not below 104.0°C; final b.p. not above 130°C; wt/gal 6.99 lb. Insoluble in water; soluble in alcohol. Flash point 65°F (open cup). These properties vary with proportions of isomers. Strong, offensive odor.
Derivation: Mixing amyl bromide and potassium hydrosulfide in alcohol.
Containers: Cans; drums.
Hazard: Flammable, dangerous fire risk. Toxic by inhalation.
Uses: Synthesis of organic sulfur compounds; chief constituent of odorant used in gas lines to locate leaks.
Shipping regulations: (Rail) Red label. (Air) Flammable Liquid label. Legal label name, amyl mercaptan.

pentanoic acid. See n-valeric acid.

1-pentanol. See n-amyl alcohol, primary.

2-pentanol (sec-n-amyl alcohol; sec-amyl alcohol, active; methyl propyl carbinol; 1-methylbutyl alcohol) $CH_3CH_2CH_2CHOHCH_3$.
Properties (racemic form): Colorless liquid; b.p. 119.3°C; sp. gr. (20/20°C) 0.811; wt/gal (20°C) 6.75 lb; refractive index (n 20/D) 1.4041; flash point (open cup) 105°F. Soluble in water; miscible with alcohol and ether. Autoignition temp. 657°F. Low toxicity. Combustible.
Derivation: Fractional distillation of the mixed alcohols resulting from the chlorination and hydrolysis of pentanes.
Containers: Drums; tank cars.
Hazard: Moderate fire risk.
Uses: Solvent for paints and lacquers; pharmaceutical intermediate.

3-pentanol (sec-n-amyl alcohol; 1-ethyl-1propanol; diethyl carbinol) $CH_3CH_2CHOHCH_2CH_3$.
Properties: Colorless liquid; sp. gr. 0.82 (20°C); freezing point less than −75°C; b.p. 115.6°C; wt/gal 6.81 lb; refractive index 1.41 (20°C). Flash point (C.C.) 94°F; autoignition temp. 650°F. Soluble in alcohol and ether; slightly soluble in water.
Containers: Drums; tank cars.
Hazard: Moderate fire risk.
Uses: Solvent; flotation agent; pharmaceuticals.

2-pentanone. See methyl propyl ketone.

3-pentanone. See diethyl ketone.

"Pentaphen."[204] Trademark for para-tert-amyl phenol (q.v.).

penta resin. Ester gum made from rosin and pentaerythritol.

pentasodium diethylenetriaminepentaacetate. A sodium salt of diethylenetriaminepentaacetic acid (q.v.).
Use: A chelating agent.

pentasodium triphosphate. See sodium tripolyphosphate.

n-**pentatriacontane** $C_{35}H_{72}$ or $CH_3(CH_2)_{33}CH_3$.
Properties: Crystals; sp. gr. 0.782 at 75°C; b.p. 331°C at 15 mm; m.p. 75°C. Combustible; low toxicity.
Use: Organic synthesis.

pentazocine $C_{19}H_{27}NO$ (2-dimethylallyl-5,9-dimethyl-2′-hydroxy-benzomorphan). A synthetic drug claimed to be as effective as morphine but without its addictive properties. Has a noncumulative effect. FDA approval.

"Pentecat L."[230] Trademark for a 50% aqueous solution of lithium naphthenate.
Use: As alcoholysis catalyst in alkyd varnish cooking.

"Pentek."[138] Trademark for pentaerythritol, technical.

1-pentene (alpha-n-amylene; propylethylene) $CH_3CH_2CH_2CH{:}CH_2$.
Properties: Colorless liquid; soluble in alcohol; insoluble in water; f.p. −165°C; b.p. 30°C; sp. gr. 0.6410 (20°C); flash point 0°F (open cup); autoignition temp. 523°F.
Derivation: Natural gasoline.
Hazard: Flammable, dangerous fire risk. Moderately toxic by ingestion, inhalation, and skin absorption.

Uses: Organic synthesis; blending agent for high octane motor fuel.
Shipping regulations: (Rail) Flammable liquid, n.o.s., Red label. (Air) Flammable Liquid label.

2-pentene (beta-n-amylene, sym-methylethylethylene) $CH_3CH_2CH{:}CHCH_3$. Mixed cis- and trans-isomers are available commercially.
cis-isomer.
Properties: B.p. 37°C; f.p. −152°C; sp. gr. 0.656 (20/4°C); flash point 0°F. Soluble in alcohol; insoluble in water.
Grades: Technical; 95.0 mole %.
trans-isomer.
Properties: B.p. 36.4°C; f.p. −139°C; sp. gr. 0.6482 (20°C); soluble in alcohol; insoluble in water. Flash point 0°F.
Derivation: Natural gasoline.
Hazard: Flammable; dangerous fire risk.
Uses: Polymerization inhibitor; organic synthesis.
Shipping regulations: See 1-pentene.

"Pentergent" T.[309] Trademark for a complete detergent containing alkyl, aryl, and fatty alcohol sulfonate with required alkali. Used in textile industry.

"Pentex."[248] Trademark for tetrabutylthiuram monosulfide (q.v.).
Uses: Rubber accelerator. When mixed with 87.5% clay, it is used for sponge rubbers and called "Pentex Flour."

"Pentids."[412] Trademark for crystalline penicillin G. See penicillin.

"Pentite."[62] Trademark for tetra(diphenylphosphito)-pentaerythritol. See pentaerythritol tetrakis(diphenyl phosphite).

pentlandite (Fe, Ni)S.
Properties: Light bronze-yellow mineral; metallic luster. Contains 35.57% of nickel. Soluble in nitric acid. Sp. gr. 4.6–5. Hardness 3.5–4.
Occurrence: Canada (Ontario), Norway.
Use: Nickel ore.

pentobarbital $C_{11}H_{18}O_3N_2$. 5-Ethyl-5-(1-methylbutyl)-barbituric acid. See barbiturate.

pentolinium tartrate $C_{23}H_{42}N_2O_{12}$. Pentamethylene-1,1-bis(1-methylpyrrolidinium bitartrate).
Properties: White to light cream-colored, crystalline powder; slightly soluble in alcohol; insoluble in ether, chloroform; very soluble in water; pH of 1% solution in water is 3.0–4.0. Decomposes 203°C.
Use: Medicine.

pentolite. A high explosive, consisting of equal parts of pentaerythritol tetranitrate and trinitrotoluene (q.v.).
Hazard: Dangerous; explodes on shock or heating.
Shipping regulations: (Rail) Consult regulations. (Air) Not acceptable.

"Penton."[266] Trademark for a thermoplastic resin derived from 3,3-bis(chloromethyl)oxetane

$(CH_2Cl)_2\overline{CCH_2OCH_2}$. A chlorinated polyether.
Properties: A linear polymer extremely resistant to chemicals, and to thermal degradation at molding and extrusion temperatures; sp. gr. 1.4; self-extinguishing; dimensionally stable; very low water absorption, outstanding chemical resistance. Natural, black or olive green molding powder. Finely divided powder for coatings.
Uses: Solid and lined valves, pumps, pipe and fittings; monofilament for filter supports and column packing.

pentosan. A complex carbohydrate (hemicellulose) present with the cellulose in many woody plant tissues, particularly cereal straws and brans; characterized by hydrolysis to give five-carbon-atom sugars (pentoses). Thus the pentosan xylan yields the sugar xylose ($HOH_2C \cdot CHOH \cdot CHOH \cdot CHOH \cdot CHO$) which is dehydrated with sulfuric acid to yield furfural ($C_5H_4O_2$).

pentose. General term for sugars with five carbon atoms per molecule.

"Pentothal."[3] Trademark for sodium thiopental; a barbiturate. See thiopental sodium.

"Pentox."[495] Trademark for a pale liquid, polyglycidyl ether which can be converted by suitable curing agents into a solid, insoluble state. An aliphatic liquid epoxy resin for air-drying solventless two-component finishes. Can also be used in formulating solventless one coat systems on concrete, wood, metal, etc.

"Pentoxone."[125] Trademark for 4-methoxy-4-methyl-pentanone-2 (q.v.).

pentyl. Synonym for the amyl group, C_5H_{11}-.

pentyl acetate. See amyl acetate.

pentylamine. See n-amylamine.

alpha-**pentylcinnamaldehyde.** See alpha-amylcinnamic aldehyde.

peppermint oil.
Properties: Colorless or slightly yellowish, volatile, essential oil; strong, aromatic, odor and taste, followed by a sensation of coolness. Odor varies with origin of oil. Sp. gr. 0.895–0.921; refractive index 1.4590–1.4650 (20°C); optical rotation −18 to −32°. Soluble in alcohol, ether and chloroform. Combustible; nontoxic.
Chief constituent: Menthol (F.C.C. and U.S.P. require not less than 50% menthol); esters of menthol, pinene, limonene, cineole, menthone, etc.
Derivation: By distilling the leaves and flowering tops of the peppermint plant (Washington, Oregon, Indiana).
Method of purification: Rectification.
Grades: Technical; U.S.P., F.C.C.
Containers: Drums.
Uses: Source of menthol; ingredient of mouth washes; flavoring of chewing gum, confectionery, liqueurs, etc.

pepsin (pepsinum).
Properties: A digestive enzyme. White or yellowish-white powder or lustrous transparent or translucent scales; should have no odor; converts proteins into albumoses and peptones. Soluble in water; insoluble in alcohol, chloroform and ether.
Derivation: From the glandular layer of fresh hogs' stomachs.
Grades: Technical; N.F.
Uses: Medicine (digestive ferment); substitute for rennet in cheese making.

pepsinogen. An inactive precursor of pepsin (q.v.).

peptidase. See protease.

peptide. See polypeptide.

peptization. A form of aggregation (q.v.) in which hydrophobic colloidal sols are stabilized by addition of electrolytes, which provide the necessary electric double layer of ionic charges around each particle. Such electrolytes are known as peptizing agents. The ions of the electrolyte are strongly adsorbed on the particle surfaces. Stable sols of non-ionizing substances acquire a charge in contact with water by preferential adsorption of the hydroxyl ions, which may be considered peptizing agents. The term is also loosely applied to the softening or liquefaction of one substance by trace quantities of another, analogous to the digestion of a protein by an enzyme (pepsin).

peptone.
Properties: (a) From albumin: White or pale yellow, amorphous powder. (b) From meat: Light brown, amorphous powder. (c) From milk; Light brown powder. Soluble in water; insoluble in alcohol or ether.
Derivation: (a) By digestion of egg albumin by pepsin and a small quantity of dilute hydrochloric acid at 38–40°C (body temperature). (b) By digestion of red meat with pancreatin at body temperature. (c) By digestion of casein.
Grades: Technical; reagent.
Uses: Preparation of nutrient media in bacteriology; nutrient.

per-. A prefix signifying complete or extreme, and specifically denoting: (1) a compound containing an element in its highest state of oxidation, as perchloric acid; (2) presence of the peroxy group, —O—O—, as peracetic and perchromic acids; (3) exhaustive substitution or addition, as perchloroethylene.

peracetic acid (peroxyacetic acid) CH_3COOOH.
Properties: Colorless liquid; strong odor; b.p. 105°C; f.p. (approx.) −30°C; sp. gr. (20°C) 1.15; flash point (open cup) 105°F. Solubility similar to acetic acid.
Derivation: (a) oxidation of acetaldehyde; (b) reaction of acetic acid and hydrogen peroxide with sulfuric acid catalyst.
Grades: Technical; 40% solution in acetic acid.
Containers: 65-lb glass carboys; 250-lb aluminum drums.
Hazard: Strong irritant. Oxidizing material; dangerous in contact with organic materials. Explodes at 110°C.
Uses: Bleaching textiles, paper, oils, waxes, starch; polymerization catalyst; bactericide and fungicide, especially in food processing; epoxidation of fatty acid esters and epoxy resin precursors; reagent in making caprolactam; synthetic glycerol.
Shipping regulations: (Rail) Yellow label. (Air) Solutions not over 40% by weight, Oxidizer label; over 40% by weight, Not acceptable.

per-acids. Derivatives of hydrogen peroxide, the molecules of which contain one or more directly linked pairs of oxygen atoms, —O—O—. See per- (2). Examples are persulfuric, perchromic, peracetic acids. Permanganic, perchloric, and periodic acids are not per-acids in this sense. See per-(1).

"Perapret."[440] Trademark for finishing agents and binders (polymer dispersions); additives for wash-and-wear finishing.

"Perazil."[301] Trademark for chlorcyclizine hydrochloride; an antihistamine.

"Perchlor."[177] Brand name for perchloroethylene.

percarbamide. See urea peroxide.

perchloric acid $HClO_4$.
Properties: Colorless, hygroscopic liquid. Unstable in concentrated form. Sp. gr. 1.764; f.p. −18°C; b.p. 203°C. Soluble in water and alcohol; forms a maximum boiling point mixture containing 71.6% acid. This is quite stable when other materials are absent.
Derivation: By distilling potassium perchlorate with strong sulfuric acid (96%), under reduced pressure in an oil bath at 140–190°C.
Method of purification: Rectification.
Grades: Technical; C.P., strength of solution 6 to 20%; 60%; 70–72%.
Containers: Glass bottles; carboys.
Hazard: Strong oxidizing agent; will explode in contact with organic materials, or by shock or heat. Strong irritant. Safety data sheet available from Manufacturing Chemists Assn., Washington, D.C.
Uses: Medicine; analytical chemistry; catalyst; manufacture of various esters; ingredient of the electrolytic bath in the deposition of lead; electro-polishing; explosives.
Shipping regulations: (not over 72%) (Rail) White label; (Air) Corrosive label. Not acceptable on passenger planes. (over 72%) (Rail) Consult regulations; (Air) Not acceptable.

perchlorobenzene. See hexachlorobenzene.

perchlorocyclopentadiene. See hexachlorocyclopentadiene.

perchloroethane. See hexachloroethane.

perchloroethylene ("per"; tetrachloroethylene) $Cl_2C{:}CCl_2$.
Properties: Colorless liquid; ether-like odor. Extremely stable. Resists hydrolysis. Sp. gr. (20/20°C) 1.625; b.p. 121°C; f.p. −22.4°C; weight 13.46 lb/gal (26°C); refractive index 1.5029 (25°C); flash point, none. Miscible with alcohol, ether, and oils, in all proportions. Insoluble in water. Nonflammable.
Derivation: (a) By chlorination of hydrocarbons, and pyrolysis of the carbon tetrachloride also formed; (b) from acetylene and chlorine via trichloroethylene.
Method of purification: Distillation.
Grades: Purified; technical; U.S.P., as tetrachloroethylene; spectrophotometric.
Containers: Drums; tank cars.
Hazard: Moderately toxic. Irritant to eyes and skin. Tolerance, 100 ppm in air. Safety data sheet available from Manufacturing Chemists Assn., Washington, D.C.
Uses: Dry-cleaning solvent; vapor-degreasing solvent; drying agent for metals and certain other solids; vermifuge; heat-transfer medium; mfg. of fluorocarbons.

perchloromethane. See carbon tetrachloride.

perchloromethyl mercaptan. See trichloromethylsulfenyl chloride.

perchloropentacyclodecane $C_{10}Cl_{12}$. White crystalline solid; m.p. (sealed tube) 485°C. Nonflammable.
Uses: Fire-retardant additive in elastomeric resin systems.

perchloropropylene. See hexachloropropylene.

perchloryl fluoride $ClFO_3$.
Properties: Colorless, noncorrosive gas with sweet odor; f.p. −146°C; b.p. −46.8°C; sp. gr. (liquid) 1.434 (20°C). Nonflammable, but supports combustion.
Containers: Steel cylinders as liquid under pressure. Stable under anhydrous conditions.

Hazard: Tolerance, 3 ppm in air. Dangerous in contact with organic materials; strong oxidizing agent.
Uses: Oxidant in rocket fuels; oxidizing and fluorinating agent in chemical reactions.
Shipping regulations: Oxidizing material, n.o.s., (Rail) Yellow label. (Air) Oxidizer label.

perchromic acid. Probably $(HO)_4Cr(OOH)_3$ or $H_3CrO_8 \cdot 2H_2O$.
Properties: Unstable acid formed when a solution of chromic acid is added to hydrogen peroxide. Below −15°C, forms deep blue crystals; the blue color can be extracted from solutions by ether. Decomposes in acid solution to form chromic salts and in alkaline solution to form chromates. The blue color can be used as a test for chloride or chromate.
Hazard: Probably toxic and irritant.

"Perclene."[28] Trademark for a drycleaning composition consisting of perchloroethylene and surfactant additives.

"Perdox."[28] Trademark for sodium borate perhydrate (q.v.).

"Peregal."[307] Trademark for a series of textile chemicals.
O. Polyoxyethylated fatty alcohol; nonionic.
OK. Methyl polyethanol quaternary amine; cationic.
ST. Polymeric dye complexing agent.
TW. Alkyl polyoxyethylene glycol amine.

"Perfecta."[45] Trademark for petrolatum, U.S.P., of high melting point and medium consistency.

perfect gas. See gas.

"Perflow."[134] Trademark for proprietary sulfur-free, semibright nickel plating additive.

perfluoro-2-butene. See octafluoro-2-butene.

perfluorobutyric acid. See heptafluorobutyric acid.

perfluorocarbon compound. A fluorocarbon in which the hydrogen directly attached to the carbon atoms is completely replaced by fluorine.

perfluorocyclobutane. See octafluorocyclobutane.

perfluorodimethylcyclobutane. An inert fluorocarbon liquid; b.p. 113°F; stable to 500°F. Evaporative coolant for electronic equipment.

perfluoroethylene. See tetrafluoroethylene.

perfluoropropane. Legal label name (Air) for octafluoropropane.

perfluoropropene. See hexafluoropropylene.

perfluorosulfonic acid. See "Nafion."

performic acid (peroxyformic acid) HCOOOH.
Properties: Colorless liquid. Miscible with water, alcohol, ether; soluble in benzene, chloroform. Solutions are unstable.
Derivation: Mixture of formic acid, peroxide and sulfuric acid is allowed to interact for 2 hours and then distilled.
Grades: 90% solution.
Hazard: Strong irritant and oxidizing agent. Explodes when shocked or heated, or in contact with reducing materials, metals, and metallic oxides.
Uses: Oxidation, epoxidation and hydroxylation reactions.
Shipping regulations: (Rail, Air) Not listed. Consult authorities.

perfume. A blend of pleasantly odorous substances (usually liquids) obtained from the essential oils of flowers, leaves, fruit, roots, or wood of a wide variety of plants, either by steam distillation or solvent extraction. Flower oils (rose, jasmine) are extracted with a non-polar solvent to give a waxy mixture called a concrete (q.v.); the wax is then removed by a second solvent (an alcohol), which is then in turn removed to form an absolute (q.v.). It is necessary that all the solvent be eliminated to obtain the finest perfumes. The center of this industry has long been in Grasse, France.

Perfume materials are also derived from animal sources (musk, ambergris), from resinous extracts (terpenes and balsams); they are also made synthetically. Cologne and toilet water are weak alcoholic solutions (5% or less) of perfumes. Fine perfumes may contain as many as 30 ingredients, and their blending is an art rather than a science. The largest volume use of perfumes is in soaps, lotions, shaving creams, and cosmetics.
See also fragrance; odorant.

"Perglow."[134] Trademark for proprietary bright nickel-plating additive.

"Pergut."[470] Trademark for a chlorinated rubber.

perhydrosqualene. See squalane.

peri acid. See 1-naphthylamine-8-sulfonic acid.

periclase MgO. A natural or calcined magnesium oxide used as a lining and maintenance material for basic oxygen steel-making furnaces and other refractories.

periodic acid $HIO_4 \cdot 2H_2O$.
Properties: White crystals; loses $2H_2O$ at about 100°C. Soluble in water, alcohol; slightly soluble in ether.
Derivation: By the interaction of iodine and concentrated perchloric acid; by low-temperature electrolysis of concentrated iodic acid.
Method of purification: Crystallization.
Hazard: Dangerous in contact with organic materials. Irritant and oxidizing material.
Uses: Oxidizing agent; increasing wet strength of paper; photographic paper.

periodic law. Originally stated in recognition of an empirical periodic variation of physical and chemical properties of the elements with atomic *weight*, this law is now understood to be based fundamentally on atomic *number* and atomic *structure*. A modern statement is: *The electronic configurations of the atoms of the elements vary periodically with their atomic number. Consequently all properties of the elements that depend on their atomic structure (electronic configuration) tend also to change with increasing atomic number in a periodic manner.*

R. T. Sanderson

periodic table. Any of a large number of possible arbitrary arrangements of the chemical elements by symbol in a geometric pattern designed to represent the periodic law by aligning the elements in periods so that the corresponding parts of the several periods are adjacent. When the elements are aligned in order of increasing atomic number, they constitute a succession of periods, each beginning with an alkali metal (one electron in the outermost principal quantum level) and ending with an element of the helium

IA	IIA	IIIB	IVB	VB	VIB	VIIB	VIII	VIII	VIII	IB	IIB	IIIA	IVA	VA	VIA	VIIA	NOBLE GASES	
1 H 1.00797 ±0.00001																	2 He 4.0026 ±0.00005	2
3 Li 6.939 ±0.0005	4 Be 9.0122 ±0.00005											5 B 10.811 ±0.003	6 C 12.01115 ±0.00005	7 N 14.0067 ±0.00005	8 O 15.9994 ±0.0001	9 F 18.9984 ±0.00005	10 Ne 20.183 ±0.0005	2 8
11 Na 22.9898 ±0.00005	12 Mg 24.312 ±0.0005											13 Al 26.9815 ±0.00005	14 Si 28.086 ±0.001	15 P 30.9738 ±0.00005	16 S 32.064 ±0.003	17 Cl 35.453 ±0.001	18 Ar 39.948 ±0.0005	2 8 8
19 K 39.102 ±0.0005	20 Ca 40.08 ±0.005	21 Sc 44.956 ±0.0005	22 Ti 47.90 ±0.005	23 V 50.942 ±0.0005	24 Cr 51.996 ±0.001	25 Mn 54.9380 ±0.00005	26 Fe 55.847 ±0.003	27 Co 58.9332 ±0.00005	28 Ni 58.71 ±0.005	29 Cu 63.54 ±0.005	30 Zn 65.37 ±0.005	31 Ga 69.72 ±0.005	32 Ge 72.59 ±0.005	33 As 74.9216 ±0.00005	34 Se 78.96 ±0.005	35 Br 79.909 ±0.002	36 Kr 83.80 ±0.005	2 8 18 8
37 Rb 85.47 ±0.005	38 Sr 87.62 ±0.005	39 Y 88.905 ±0.0005	40 Zr 91.22 ±0.005	41 Nb 92.906 ±0.0005	42 Mo 95.94 ±0.005	43 Tc (99)	44 Ru 101.07 ±0.005	45 Rh 102.905 ±0.0005	46 Pd 106.4 ±0.05	47 Ag 107.870 ±0.003	48 Cd 112.40 ±0.005	49 In 114.82 ±0.005	50 Sn 118.69 ±0.005	51 Sb 121.75 ±0.005	52 Te 127.60 ±0.005	53 I 126.9044 ±0.00005	54 Xe 131.30 ±0.005	2 8 18 18 8
55 Cs 132.905 ±0.0005	56 Ba 137.34 ±0.005	57 *La 138.91 ±0.005	72 Hf 178.49 ±0.005	73 Ta 180.948 ±0.0005	74 W 183.85 ±0.005	75 Re 186.2 ±0.05	76 Os 190.2 ±0.05	77 Ir 192.2 ±0.05	78 Pt 195.09 ±0.005	79 Au 196.967 ±0.0005	80 Hg 200.59 ±0.005	81 Tl 204.37 ±0.005	82 Pb 207.19 ±0.005	83 Bi 208.980 ±0.0005	84 Po (210)	85 At (210)	86 Rn (222)	2 8 18 32 18 8
87 Fr (223)	88 Ra (226)	89 †Ac (227)																

*Lanthanum Series

58 Ce 140.12 ±0.005	59 Pr 140.907 ±0.0005	60 Nd 144.24 ±0.005	61 Pm (147)	62 Sm 150.35 ±0.005	63 Eu 151.96 ±0.005	64 Gd 157.25 ±0.005	65 Tb 158.924 ±0.0005	66 Dy 162.50 ±0.005	67 Ho 164.930 ±0.0005	68 Er 167.26 ±0.005	69 Tm 168.934 ±0.0005	70 Yb 173.04 ±0.005	71 Lu 174.97 ±0.005	2 8 18 32 9 2

†Actinium Series

90 Th 232.038 ±0.0005	91 Pa (231)	92 U 238.03 ±0.005	93 Np (237)	94 Pu (242)	95 Am (243)	96 Cm (247)	97 Bk (247)	98 Cf (249)	99 Es (254)	100 Fm (253)	101 Md (256)	102 No (253)	103 Lw (257)	2 8 18 32 7 9 2

Based on chart published by Fisher Scientific Co.

Shown above is a form of the Periodic Table that has been generally accepted in recent years; it is by no means the only possible arrangement, nor even the best one, according to some authorities. [Ed.]

family, helium, neon, argon, krypton, xenon, and radon, (each having 8 electrons in the outermost principal quantum level, except for helium which is limited to 2). Each helium family element is followed directly in atomic number by an alkali metal, which begins a new period.

The advantage of placing each successive period so that it is adjacent to the preceding period in its corresponding parts is that similar elements are thus brought together in groups. For example, the alkali metals that begin each period have similar properties that correspond to their atomic structure, each having one outermost shell electron. In a periodic table, the elements exhibit a steady trend in properties from the beginning of each period, where they are metals, to the end where they are nonmetals. Within each group the elements are quite similar. A given element can be predicted, on the basis of its position in a periodic table, to resemble the other elements of its group and be intermediate in properties between its adjacent neighbors within its period. The chief function of a periodic table is to serve as a fundamental framework for the systematic organization of chemistry. See also Mendeleef.

R. T. Sanderson

Perkin, Sir William Henry (1838–1907). An English chemist who was the first to make a synthetic dyestuff (1856). He studied under Hofmann at the Royal College of London. Perkin's first dye was called mauveine, but he proceeded to synthetize alizarin and coumarin, the first man-made perfume. In 1907 he was awarded the first Perkin Medal, which has ever since been awarded by the American Division of the Society of Chemical Industry for distinguished work in chemistry. Notwithstanding the fact that Perkin patented and manufactured mauve dye in England, the center of the synthetic organic dye industry shifted to Germany, where it remained till 1914. See also Hofmann.

"Perluxe."[56] Trademark for a dry-cleaning solvent consisting of perchloroethylene or tetrachloroethylene.

"Perma-."[319] Trademark for a series of vitamin preparations; used in animal feeds.

"Permachlor Red."[141] Azo pigments made from para-chloro-ortho-nitraniline and beta-naphthol. Used in bright, light red and pink house paints.

"Permachrom Red."[141] Permanent red azo pigments derived from beta-hydroxynaphthoic acid.

Grades: Medium red; dark red; maroon.

Uses: Printing inks, paints, enamels, lacquers, plastics.

permafil. A mixture in which the liquid completely undergoes polymerization and hardens without the necessity of any evaporation. Anaerobic permafils harden out of contact with air.

"Permafresh."[42] Trademark for a series of resins used in the textile industry. They include cellulose reactants, melamine derivatives designed as a replacement for conventional trimethylol melamine and related products; partially condensed modified carbamide resins; cyclic nitrogenous reactant; bisketo-triazine; methylol imidazolidone; modified carbamide resin; urea-formaldehyde liquid; durable thermosetting urea-formaldehyde reactive resin.

Uses: Wash-and-wear fabrics; crease resistance; shrinkage control; durable finishes; bodying agent.

"Permagreen Gold."[141] Trademark for a yellow nickel azo pigment. Used in paints, printing inks, plastics.

"Permagrid."[155] Trademark for electronic tube alloy of 99.3% nickel, 0.4% titanium, and 0.3% magnesium. Used for grid wire.

"Permalloy." Trademark for an alloy containing 78.5% nickel and 21.5% iron. It has high magnetic permeability and electrical resistivity. Used in trans-ocean submarine cables.

"Permalume G."[288] Trademark for a semi-bright nickel electroplating process. The plating bath contains nickel sulfate, nickel chloride, boric acid, and organic addition agents.

"Permalux."[28] Trademark for a di-ortho-tolylguanidine salt of dicatechol borate, $(HOC_6H_4O)_2B \cdot HNC(NHC_6H_4CH_3)_2$.
Properties: Light grayish brown powder; sp. gr. 1.27; f.p. not lower than 165°C.
Uses: Rubber antioxidant; vulcanization of neoprene.

"Permalyn" 330[266] Trademark for a pale, hard resin, a glycerol ester of selected stabilized resin acids. Softening point 87°C; acid number 7; USDA color WW-X.

Permanent Orange. See dinitraniline orange.

permanent-press resin (durable-press resin). A thermosetting resin used as a textile impregnant or fiber coating to impart crease resistance and permanent hot-creasing to suitings, dress fabrics, etc. Chemicals such as formaldehyde and maleic anhydride are the basis of these products. The resin is "cured" after the fabric has been tailored into a garment. A permanent-press fabric that requires no resin has been developed (a blend of polyester with cotton or rayon).

Permanent Red 2B Amine. See 4-amino-2-chlorotoluene-5-sulfonic acid.

"Permanickel" 300.[283] Trademark for an alloy which contains 98.6% nickel and small quantities of other metals.

"Permanite."[250] Trademark for purified natural magnetic iron ore for use in radio cores, shields, etc.

"Permansa."[141] Trademark for a series of organic pigments.
Uses: Printing inks, paints, wallpaper, textiles.

"Permasep."[28] Trademark for a method of desalinating water by reverse osmosis using membranes composed of hollow nylon fibers, which provide maximum membrane surface. See also osmosis.

permenorm. Nickel-iron alloy produced by magnetic annealing and drastic cold reduction, and used for mechanical rectifiers and low-frequency amplifiers. This alloy has a rectangular hysteresis loop that eliminates arcing at the contacts of mechanical rectifiers, as well as other desirable properties.

permissible explosive. See explosive, permissible.

"Permolite."[300] Trademark for a solubilized alkyd resin used as a slip-proof finish for low-count viscose and acetate fabrics.

"Permolith."[141] Trademark for lithopone (q.v.) available in various grades and forms.

"Permox."[468] Trademark for a series of anti-corrosive pigments, primarily basic lead silicochromate.

"Permutit."[184] Trademark for a series of ion exchange agents. Commercial grades of cation exchangers include sulfonated polystyrene copolymer (designated by Q) and carboxylic acid (H-70); commercial grades of anion exchangers include aliphatic polyamide intermediate base (A and AB), phenolic intermediate base (CCG), strong base polystyrene alkyl quaternary ammonium (S-1), strong base polystyrene alkanol quaternary ammonium (S-2), strong base pyridinium quaternary amine (SK); nuclear grade ion exchangers include as cation exchanger, sulfonated polystyrene copolymer (organic); as anion exchanger, strong base polystyrene alkyl quaternary amine (organic). These are designated by a series of letters beginning with N.
Used in water purification and a wide variety of separation processes.

"Perone."[28] Trademark for hydrogen peroxide, 30, 35, and 50% solutions by weight. See hydrogen peroxide.

perovskite. A natural or synthetic crystalline mineral composed of calcium dioxide and titanium dioxide. The natural material was discovered in Russia about 1850; the synthetic type can be so made as to incorporate particles of catalysts (platinum or palladium) which are protected by the crystal structure from contamination by lead and other poisons, thus permitting use of leaded gasoline in cars equipped with emission-control devices.

peroxidase. An enzyme found in most plant cells and some animal cells which promote the oxidation of various substrates, such as phenols, aromatic amines, etc., by means of hydrogen peroxide.

peroxide. (1) Any compound containing a bivalent O-O group, i.e., the O atoms are univalent. Such compounds release atomic (nascent) oxygen readily. Thus they are strong oxidizing agents, and fire hazards when in contact with combustible materials, especially under high-temperature conditions. The chief industrial uses of peroxides are as oxidizing agents, bleaching agents, and initiators of polymerization. (2) Hydrogen peroxide.

"Peroxidol."[28] Trademark for a series of epoxy-type plasticizers for vinyl resins.

peroxyacetic acid. See peracetic acid.

peroxybenzoyl nitrate. A component of photochemical smog, found to be 200 times as irritating to the eyes as formaldehyde. A concentration of 0.02 ppm in air causes moderate to severe conjunctival irritation. See also smog.

peroxyformic acid. See performic acid.

peroxysulfuric acid. See Caro's acid.

perphenazine $C_{21}H_{26}ClN_3OS$.
Properties: Crystals; sensitive to light; m.p. 97–100°C; insoluble in water, soluble in ethanol and acetone.
Grade: N.D.
Use: Medicine.

perrhenic acid $HReO_4$. Exists only in solution; commercially available as aqueous syrup. Strong, very stable monobasic acid; extremely soluble in water and organic solvents.

"Persistol."[440] Trademark for agents for the wash-proof water-repellent finishing of textiles of natural and synthetic fibrous materials and fiber mixtures.

persulfuric acid. See Caro's acid.

"Perthane."[23] Trademark for an agricultural insecticide based on 1,1-dichloro-2,2-bis(para-ethylphenyl)-ethane. Supplied as a wettable powder or emulsifiable concentrate.

Peru balsam. See balsam.

pesticide. Any substance, organic or inorganic, used to destroy or inhibit the action of plant or animal pests; the term thus includes insecticides, herbicides, rodenticides, miticides, etc. See specific entry. Virtually all pesticides are toxic to man to a greater or lesser degree. Microencapsulated controlled-release forms are available. See note under methyl parathion; see also insecticide (Hazard). For additional information consult International Pesticide Institute, Syracuse, N.Y.

"Pestmaster" Soil Fumigant-1.[426] Trademark for a mixture of methyl bromide 98%, chloropicrin 2%.
Properties: Colorless liquefied gas with a sharp pungent odor.
Containers: 1-lb cans; steel cylinders to 400-lb net.
Hazard: See methyl bromide.
Uses: To control insects, weeds, nematodes, and certain fungi in the soil.

petalite $LiAl(Si_2O_5)_2$ or $Li_2O \cdot Al_2O_3 \cdot 8SiO_4$.
Properties: Colorless, white, gray or occasionally pink mineral, white streak, vitreous luster. Resembles spodumene in appearance. Contains up to 4.9% lithia, sometimes with partial replacement by sodium or, less often, by potassium. Insoluble in acids. Sp. gr. 2.39–2.46; Mohs hardness 6–6.5.
Occurrence: United States (Massachusetts, Maine); Sweden.
Uses: Source of lithium salts; in ceramics and glass.

petitgrain oil
Properties: Yellowish liquid; odor similar to neroli oil. Soluble in 70% alcohol, ether, chloroform and most fixed oils. Insoluble in glycerol. Sp. gr. 0.878–0.889 (25°C); refractive index 1.4580–1.4640 (20°C); angular rotation −4° to +1°C. Combustible; nontoxic.
Derivation: From leaves and twigs of the bitter orange tree, Citrus aurantium amara, especially in Paraguay.
Grades: Technical; F.C.C.
Uses: Perfumery (soaps; synthetic neroli; skin creams); flavoring.

PETN. Abbreviation for pentaerythritol tetranitrate (q.v.).

"Petrex."[266] Trademark for a series of terpene-derived alkyd resins.

"Petrobase."[25] Trademark for a series of emulsifying and rust-preventive bases. On dilution with suitable petroleum oils or light distillates the proper concentrate provides soluble cutting and grinding oils; solvent emulsion cleaner; preservative, slushing and household oils; and water-displacing fluids.

"Petro Bond."[236] Trademark for a bonding agent in the preparation of waterless foundry sands; used with oil and catalyst. Fine sands giving excellent reproduction of detail can be used since high permeability of the foundry sand is not required with this binder.

petrochemical. An organic compound for which petroleum or natural gas is the ultimate raw material. Thus, cracking of petroleum produces ethylene which is converted to ethylene glycol, the latter being a typical petrochemical. The term is also applied to substances such as ammonia, because the hydrogen used to form the ammonia is derived from natural gas. Thus synthetic fertilizers are considered to be petrochemicals. At least 175 substances are designated as petrochemicals including many paraffin, olefin, naphthene, and aromatic hydrocarbons (methane, ethane, propane, ethylene, propylene, butenes, cyclohexane, benzene, toluene, naphthalene, etc.) and their derivatives, even though some of their commercial production is from sources other than petroleum. The percentage of total hydrocarbon consumption represented by petrochemicals is steadily increasing; some authorities maintain that petroleum is too valuable to be used for fuel and that it should be conserved for future petrochemical development.

petrolatum (mineral wax; petroleum jelly; mineral jelly).
Properties: Colorless to amber, oily, transparent, amorphous mass whose consistency varies with the temperature. Soluble in chloroform, ether, carbon disulfide, benzene and oils; insoluble in water. Sp. gr. 0.815–0.880 at 60°C; m.p. 38–60°C. Combustible; nontoxic.
Derivation: By fractional distillation of still residues from the steam distillation of paraffin-base petroleum, or from steam-reduced crude oils from which the light fractions have been removed.
Grades: Natural, produced as above; artificial, made by mixing heavy petroleum lubricating oil with a low m.p. paraffin wax; U.S.P. (white petrolatum); N.F. (yellow petrolatum); F.C.C. (includes both types).
Containers: Glass bottles; tins; drums.
Uses: Protective dressing and substitute for fats in ointments; lubricating greases; metal polishes; leather grease; rust preventive; perfume extractor; insect repellents; in foods as defoaming agent, lubricant, release agent, protective coating; softener in white or colored rubber compounds.

petrolatum, liquid (white mineral oil)
Properties: Colorless, transparent oily liquid, a mixture of liquid hydrocarbons; almost tasteless and odorless; sp. gr. 0.828–0.880 (20°C) (light), 0.860–0.905 (25°C) (heavy); soluble in ether, chloroform, carbon disulfide, benzene, boiling alcohol, and fixed or volatile oils; insoluble in water, cold alcohol, glycerin. Combustible; nontoxic.
Derivation: Distillation of high-boiling (330–390°C) petroleum fractions.
Method of purifications: Treatment with sulfuric acid, caustic soda, filtration through decolorizing carbon, and crystallization to remove waxes.
Grades: Technical; U.S.P. (as mineral oil, formerly liquid petrolatum, heavy); N.F. (light); F.C.C. (as mineral oil, white).
Containers: Glass bottles; drums; tank cars.
Uses: Medicine (laxative); cosmetics; dispersants; diluents, etc. in white or colored rubber and plastic compounds; compressor and textile lubricants; dispersants for reactive compounds such as metal hydrides; catalyst carrier; in foods as binder, defoaming agent, lubricant, release agent, fermentation aid, protective coating.

petrolatum wax. A microcrystalline wax containing hydrocarbons from $C_{33}H_{70}$ to $C_{43}H_{88}$. Solidifying range 71 to 83°C. See also wax, microcrystalline.

"Petrolene."[200] Trademark for a petroleum solvent prepared by straight-run distillation.
Properties: Water-white; initial boiling point 140–145°F, 95% distills between 195 and 200°F; sp. gr.

0.701 (60°F); flash point (TCC) −16°F; mild; nonresidual odor.
Hazard: Highly flammable; dangerous fire risk.
Use: Rubber cements; sealers; fast-drying lacquers; lacquer dopes; roto inks used on high-speed presses.

petroleum (crude oil). A highly complex mixture of paraffinic, cycloparaffinic (naphthenic), and aromatic hydrocarbons, containing low percentages of sulfur and trace amounts of nitrogen and oxygen compounds. Said to have originated from both plant and animal sources from 10 to 20 million years ago. In the U.S., the chemical basis of petroleum varies with location; in general, Pennsylvania crudes are aliphatic (wax-base), Far Western crudes are aromatic (asphalt-base), and Mid-Continent crudes mixed-base. The most important petroleum fractions, obtained by cracking or distillation, are various hydrocarbon gases (butane, ethane, propane), naphtha of several grades, gasoline, kerosene, fuel oils, gas oil, lubricating oils, paraffin wax, and asphalt. From the hydrocarbon gases, ethylene, butylene and propylene are obtained; these are important industrial intermediates, being the source of alcohols, ethylene glycols, and monomers for a wide range of plastics, elastomers, and pharmaceuticals. Benzene, phenol, toluene, and xylene can be made from petroleum, and hundreds of other products, including biosynthetically produced proteins, are petroleum-derived. About 5% of the petroleum consumed in the U.S. is used as feedstocks by the chemical industries.
Occurrence: At present about two-thirds of the world's proven resources are in the Middle East and North Africa, the other third being divided among the U.S., Canada, Venezuela, and U.S.S.R., with minor sources in the North Sea area, Indonesia, Mexico, Rumania and Australia. Existence of sizable off-shore deposits on the eastern continental shelf has still to be proved.
Properties: Viscous dark-brown liquid; unpleasant odor; sp. gr. 0.78–0.97. Flash point 20–90°F.
Hazard: Flammable, moderate fire risk. Moderately toxic by ingestion; local skin irritant.
Shipping regulations: (Rail) Red label. (Air) Flammable Liquid label. For further information, refer to American Petroleum Institute, 1271 Ave. of the Americas, New York, N.Y.
See also natural gas; petrochemical.
Note: Authorities state that the world's petroleum resources will be exhausted before the year 2050 at present rate of consumption. The need for both conservation and full-scale development of alternative energy sources is an urgent necessity. See energy sources.

petroleum benzin. A special grade of ligroin (q.v.).

petroleum coke. See coke.

petroleum ether. This term is used synonymously with petroleum naphtha by MCA. It is also sometimes used as a synonym for ligroin or petroleum benzin. It is technically a misnomer, for it is not an ether in the chemical sense. For details about specified distillation ranges and other distinctive properties, consult ASTM and API specifications. Also see naphtha, petroleum.
Shipping regulations: (Rail) Red label. (Air) Flammable Liquid label.

petroleum gas, liquefied. See liquefied petroleum gas.

petroleum jelly. See petrolatum.

petroleum naphtha. See naphtha. (1)

petroleum spirits. See naphtha (1a). In Great Britain the term "petroleum spirits" refers to a volatile hydrocarbon mixture having a flash point below 32°F.
Hazard: Highly flammable, dangerous fire risk.
Shipping regulations: (Rail) Red label. (Air) Flammable Liquid label.
See also spirits.

petroleum, synthetic. See pyrolysis.

petroleum thinner. See naphtha (1a).

petroleum wax. A high molecular weight solid hydrocarbon derived from petroleum. There are three types: paraffin waxes, microcrystalline waxes, and petrolatum waxes. All three are made mostly by solvent dewaxing, although pressing and sweating processes are still used. See specific entry.

"Petrolite."[128] Trademark for a series of petroleum and synthetic microcrystalline waxes.

"Petromix."[45] Trademark for a soluble oil base, made from petroleum sulfonates.

"Petronates."[45] Trademark for salts of petroleum sulfonic acids varying in molecular weight and color.
Uses: Emulsifying agents; dispersing agents; wetting agents; corrosion-preventive.

"Petronauba."[128] Trademark for a series of emulsifiable microcrystalline waxes.

"Petrosul."[25] Trademark for a series of highly purified natural petroleum sulfonate products available in high, medium and low molecular weight ranges. Useful in applications requiring the surface active functions of foaming, detergency, emulsibility, dispersion, solubilization, spreading and rust protection.

"Petrotect."[25] Trademark for a series of rust-preventive and hydraulic fluids in general meeting military specifications. The rust preventives are classified as solvent cut backs, petrolatum barriers, general purpose preservatives, and engine preservative lubricants. Hydraulic fluids include both preservative and operational types.

"Petrothene."[192] Trademark for polypropylene resins for blown, cast, and water-quenched films; substrate coating; wire and cable coating; injection molding; blow molding; thermoforming; pipe extrusion, calendering. Available as solid cubes and pellets in natural and black.

"Petrotone."[236] Trademark for an organophilic clay that increases suspending properties of crude or refined oils and oil muds without appreciably increasing gels. Makes the use of weighted oil muds safer and more economical.

"Petrowet" R.[28] Trademark for a surface-active agent composed of saturated-hydrocarbon sodium sulfonate. A wetting and penetrating agent effective in high concentrations of electrolytes and acids, suitable for use in acidizing of oil wells.

pewter. Tin alloys with 5–15% antimony, 0–3% copper, and 0–15% lead. White metal (q.v.) and britannia metal are also of this general composition.

"Pexol."[266] Trademark for fortified rosin sizes based on rosin adducts. Available as dry products and pastes in both pale and dark types.

PF resins. Abbreviation for phenol-formaldehyde resins (q.v.).

PG. Abbreviation for polypropylene glycol (q.v.).

PGA. Abbreviation for pteroylglutamic acid. See folic acid.

pH

pH is a value taken to represent the acidity or alkalinity of an aqueous solution; it is defined as the logarithm of the reciprocal of the hydrogen-ion concentration of a solution:

$$pH = \ln \frac{1}{[H^+]}$$

Pure water is the standard used in arriving at this value. Under ordinary conditions water molecules dissociate into the ions H^+ and OH^-, with recombination at such a rate that with very pure water at 22°C there is a concentration of oppositely charged ions of 1/10,000,000, or 10^{-7}, mole per liter. This is commonly expressed by saying that pure water has a pH of 7, which means that its concentration of hydrogen ions is expressed by the exponent 7, without its minus sign.

When acids or hydroxyl-containing bases are in water solution they ionize more or less completely, furnishing varying concentrations of H^+ and OH^- ions, respectively, to the solution. Strong acids and bases ionize much more completely than weak acids and bases; thus strong acids give solutions of pH 1 to 3, while solutions of weak acids have a pH of about 6. Strong bases give solutions of pH 12 or 13, while weak bases give solutions of pH about 8. As the pH scale is logarithmic, the intervals are exponential, and thus represent far greater differences in concentration than the values themselves seem to indicate. (See table).

	pH Value	Ratio of H^+ or OH^- Concentration to That of Pure Water at 22°C
Acid side (excess of H^+ ions)	1	1,000,000
	2	100,000
	3	10,000
	4	1,000
	5	100
	6	10
Neutrality	7	1
Alkaline side (excess of OH^- ions)	8	10
	9	100
	10	1,000
	11	10,000
	12	100,000
	13	1,000,000

Liquid	pH Value
Pure water	7
Sea water	8.5–10
Electroplating bath	6.5–5
.01 *N* HCl	2
.01 *N* NaOH	12
.1 *N* acetic acid	3
.1 *N* NH_4OH	11
Gastric juices	2
Urine	5–7
Blood	7.3–7.5
Milk	6.5–7
Soil (optimum for crops)	6–7

In acid-base titrations, changes in pH can be detected by indicators (q.v.) such as methyl orange, phenolphthalein, etc. Litmus paper can also be used as a rough indication of acidity or alkalinity. In carrying out titrations the point at which the solution is found to be neutral (pH 7) is not always the correct end point.

pH control is of critical importance in a large number of industrial operations such as water purification, chrome tanning process for leather, in preservation of food products, in electroplating baths, dyeing, agriculture, and numerous other instances.

See also acid; base.

"Phaltan."[253] Trademark for fungicide formulations containing N-trichloromethylthiophthalimide.

"Phanadorn" Calcium.[162] Trademark for cyclobarbital calcium.

"Phantolid."[105] Trademark for a standardized mixture of structural isomers of 1,1,2,3,3,5-hexamethylindan methyl ketone.

Properties: Waxy, white solid with musk-like odor that begins to sinter at about 35°C and becomes liquid at about 40°C. Soluble in conventional solvents and essential oils.

Use: Perfumery.

pharmaceutical. A broad term that includes not only all types of drugs, medicinal and curative products but also such ancillary products as tonics, dietary supplements, vitamins, deodorants and the like. See also drug.

"Pharmasorb."[99] Trademark for a pharmaceutical grade of attapulgus clay.

"Pharmolin."[99] Trademark for a pharmaceutical grade of kaolin.

phase.

(1) One of the three states or conditions in which substances can exist, i.e., solid, liquid, or gas(vapor). The condition depends primarily on the concentration of atoms or molecules: solids are the most dense, gases the least, and liquids occupy the intermediate position. Solids are normally crystalline, liquids are amorphous, and gases are without structure. See also matter.

(2) A physically distinct and mechanically separable portion of a dispersion or solution. Phases may be either solid, liquid, or gaseous (vapor). In any mixture or solution the major component is called the continuous or external phase and the minor component the dispersed or internal phase. The latter may or may not be uniformly dispersed in the continuous phase. See colloid chemistry; solution.

phase rule. Propounded by J. Willard Gibbs (q.v.) in 1877, the phase rule is a general system F = n—r—2 stating the boundaries of thermodynamic equilibrium in a system of chemical reactants. The number of degrees of freedom (*F*) allowed in a given heterogeneous system may be examined by analysis or observation and plotted on a graph by proper choice of the components (*n*), the phases (*r*), and the independently variable factors of temperature and pressure.

The principles of the phase rule apply to all multicomponent systems, including solvent blends, glass, alloys and plastics.

Phe Abbreviation for phenylalanine.

alpha-phellandrene (4-isopropyl-1-methyl-1,5-cyclohexadiene) $CH_3C{:}CHCH_2CH[CH(CH_3)_2]CH{:}CH$. A monocyclic terpene occurring as (a) *d*- and (b) *l*-optical isomers.

Properties: Colorless oil; insoluble in water; soluble in ether. (a) Sp. gr. 0.8463 (25°C); b.p. 66–68°C (16 mm); refractive index 1.4777. (b) Sp. gr. 0.8324 (20°C); b.p. 58–59°C (16 mm); refractive index 1.4724. Combustible.
Derivation: (a) Found in ginger oil; Ceylon and Seychelles cinnamon oil. (b) Found in eucalyptus oil.
Hazard: Moderately toxic by ingestion and skin absorption; strong irritant.
Uses: Flavoring; perfumery.

beta-**phellandrene** (4-isopropyl-1-methylene-2-cyclohexene) $CH_2:CCH:CHCH[CH(CH_3)_2]CH_2CH_2$. A monocyclic terpene occurring as (a) *d*- and (b) *l*-optical isomers.
Properties: (a) Mobile oil with pleasant odor and a burning taste. Sp. gr. 0.8520 (20°C); b.p. 171-172°C (760 mm); refractive index 1.4788. (b) Mobile oil; sp. gr. 0.8497 (15°C); b.p. 178–179°; flash point 120°F (T.C.C.). Refractive index 1.4800. Both are insoluble in water and alcohol; soluble in ether. Combustible.
Derivation: (a) lemon oil. (b) Japanese peppermint oil.
Hazard: Moderate fire risk.

"Phemerol."[330] Trademark for benzethonium chloride (q.v.).

phenacaine hydrochloride
$C_2H_5OC_6H_4NCH(CH_3)NC_6H_4OC_2H_5 \cdot HCl \cdot H_2O$. N,N′-Bis(para-ethoxyphenyl) acetamidine hydrochloride.
Properties: Small, white crystals; odorless; faintly bitter taste. Incompatible with alkalies. M.p. 190°C. Soluble in alcohol, boiling water and chloroform; less so in cold water, insoluble in ether.
Grades: N.F.; technical.
Use: Medicine.

phenacemide (phenylacetylurea)
$C_6H_5CH_2CONHCONH_2$.
Properties: White to creamy white, odorless, tasteless crystalline solid; m.p. 212–216°C; slightly soluble in alcohol, benzene, chloroform and ether; very slightly soluble in water.
Use: Medicine.

phenacetin. U.S.P. name for acetophenetidin (q.v.).

phenacyl chloride. See chloroacetophenone.

phenacyl fluoride. See fluoroacetophenone.

"Phenamine."[307] Trademark for a series of direct dyestuffs, used for the dyeing of cotton and paper.

phenanthraquinone. See phenanthrenequinone.

phenanthrene $C_{14}H_{10}$.

Properties: Colorless, shining crystals. Soluble in alcohol, ether, benzene, carbon disulfide and acetic acid; insoluble in water. Sp. gr. 1.063; m.p. 100.35°C; b.p. 340°C. Combustible.
Derivation: Fractional distillation of high-boiling coal-tar oils, with subsequent recrystallization from alcohol.
Hazard: Carcinogenic agent. Tolerance, 0.2 mg per cubic meter in air.
Uses: Dyestuffs; explosives; synthesis of drugs; biochemical research; phenanthrenequinone.

phenanthrenequinone. (Erroneously: phenanthraquinone) $C_{14}H_8O_2$.
Properties: Yellow-orange, needle-like crystals. Soluble in sulfuric acid, benzene; glacial acetic acid and hot alcohol; slightly soluble in ether; insoluble in water. Sp. gr. 1.4045; m.p. 206–207°C; b.p. sublimes above 360°C.
Derivation: By oxidation of a boiling solution of phenanthrene in glacial acetic acid with chromic acid, solution in sodium disulfite, precipitation by means of hydrochloric acid and recrystallization.
Uses: Organic synthesis; dyes.

1,10-phenanthroline (4,5-phenanthroline; ortho-phenanthroline) $C_{12}H_8N_2 \cdot H_2O$. A heterotricyclic compound.
Properties: White crystalline powder; m.p. 93–94°C, anhydrous 117°C. Slightly soluble in water; soluble in alcohol, benzene.
Derivation: Made by heating ortho-phenylenediamine with glycerin, nitrobenzene and concentrated sulfuric acid; or in like manner from 8-aminoquinoline.
Uses: Forms a complex compound with ferrous ions used as an indicator; drier in coatings industry.

phenarsazine chloride. See diphenylaminechloroarsine.

phenazine (azophenylene) $C_6H_4N_2C_6H_4$. A tricyclic compound.
Properties: Yellow crystals; m.p. 170–171°C; b.p. >360°C; very slightly soluble in water; soluble in alcohol and ether. Combustible.
Uses: Organic synthesis; manufacture of dyes; larvicide.

phenethicillin. See potassium alpha-phenoxyethyl penicillin.

phenethyl acetate. See 2-phenylethyl acetate.

phenethyl alcohol (phenylethyl alcohol; 2 phenylethanol; benzyl carbinol) $C_6H_5CH_2CH_2OH$.
Properties: Colorless liquid; floral odor; sharp burning taste; sp. gr. 1.017–1.020 (25°C); refractive index (n 20/D) 1.5310–1.5340; f.p. −27°C; b.p. 219°C. Flash point 216°F. Soluble in 50% alcohol; soluble 1 part in 50 parts of water; soluble in fixed oils, alcohol, and glycerol; slightly soluble in mineral oil. Combustible.
Derivation: (a) By reduction of phenylacetic ethyl ester by sodium in absolute alcohol. (b) By the action of ethylene oxide on phenylmagnesium bromide and subsequent hydrolysis.
Grades: Technical; N.F.; F.C.C.
Containers: Tin cans and glass bottles; drums.
Uses: Organic synthesis; synthetic rose oil; soaps; flavors; antibacterial; perservative; medicine.

sec-**phenethyl alcohol.** See alpha-methylbenzyl alcohol.

phenethylamine. See 2-phenylethylamine.

phenethyl anthranilate. See 2-phenylethyl anthranilate.

phenethyl isobutyrate. See 2-phenylethyl isobutyrate.

phenethyl phenylacetate. See 2-phenylethyl phenylacetate.

phenethyl propionate. See 2-phenylethyl propionate.

phenethyl salicylate. See 2-phenylethyl salicylate.

ortho-**phenetidine** (2-aminophenetole) $NH_2C_6H_4OC_2H_5$.
Properties: Oily liquid; rapidly becomes brown on exposure to light or air. Solidifies below −20°C; b.p.

228–230°C. Soluble in alcohol and ether; insoluble in water. Combustible.
Derivation: Reduction of ortho-nitrophenetole with iron filings and hydrochloric acid.
Hazard: Toxic by ingestion and inhalation.
Use: Manufacture of dyes; laboratory reagent.

para-**phenetidine** (4-aminophenetole) $NH_2C_6H_4OC_2H_5$.
Properties: Colorless oily liquid; becomes red to brown on exposure to air and light. Sp. gr. 1.0613 (15°C); m.p. 2–4°C; b.p. 253–255°C. Insoluble in water; soluble in alcohol. Combustible.
Derivation: Ethylating para-nitrophenol with ethyl sulfate or chloride in presence of sodium hydroxide followed by reduction with iron filings and hydrochloric acid.
Hazard: Toxic by ingestion and inhalation.
Uses: Dyestuffs intermediate; pharmaceuticals; medicine; laboratory reagent.

phenetole (phenyl ethyl ether) $C_6H_5OC_2H_5$.
Properties: Colorless liquid; b.p. 172°C; f.p. −30°C; sp. gr. 0.967 (20/4°C); insoluble in water; soluble in alcohol and ether. Combustible; low toxicity.

phenetsal. See para-acetylaminophenyl salicylate.

"Phenex."[94] Trademark for alpha-ethyl-beta-propylacrylaniline. Used as an accelerator for natural and synthetic rubber and latexes.

"Phenidone."[219] Trademark for 1-phenyl-3-pyrazolidone (q.v.).

phenindione (2-phenyl-1,3-indanedione) $C_{15}H_{10}O_2$.
Properties: Pale yellow crystalline material; practically odorless; insoluble in water; soluble in methanol, alcohol, ether, acetone, benzene. Solutions in alkalies are red; in concentrated sulfuric acid blue.
Use: Medicine (blood anticoagulant).

pheniramine maleate (prophenpyridamine maleate) $C_{16}H_{20}N_2 \cdot C_4H_4O_4$. 1-Phenyl-1-(2-pyridyl)-3-dimethylaminopropane maleate.
Properties: White crystalline powder with faint amine-like odor. M.p. 104–108°C. Very soluble in alcohol and water, only slightly soluble in benzene and ether. 1% solution has pH between 4.5–5.5.
Grade: N.F.
Use: Medicine (antihistamine).

"Phenmad."[329] Trademark for a 10% phenylmercuric acetate aqueous solution. Used as a turf fungicide.
Hazard: Highly toxic.

phenobarbital (phenylbarbital; phenylethylmalonylurea; 5-ethyl-5-phenylbarbituric acid) $C_{12}H_{12}N_2O_3$.
Properties: White, shining, crystalline powder odorless, stable. M.p. 174–178°C. Soluble in alcohol, ether, chloroform, benzene, alkali hydroxides, alkali carbonate solutions; sparingly soluble in water.
Derivation: Condensation of phenylethylmalonic acid derivatives and urea.
Grade: U.S.P.
Containers: Glass bottles; fiber cans, drums.
Hazard: Toxic; may have damaging side effects. See barbiturate.
Use: Medicine (sedative); laboratory reagent. Also available as the sodium salt, which has good water-solubility.

phenocoll hydrochloride (aminoacetophenetidide hydrochloride; glycocoll-para-phenetidine hydrochloride) $C_2H_5OC_6H_4NHCOCH_2NH_2 \cdot HCl$.
Properties: Fine, white crystalline powder; soluble in water and warm alcohol; slightly soluble in chloroform, ether and benzene. M.p. 95°C.
Derivation: By the action of aminoacetic acid upon phenetidine and acidifying.
Use: Medicine.

"Pheno"Dyes.[57] Trademark for a group of direct dyes used for coloring paper.

"Phenoform."[307] Trademark for a series of dyestuffs and pigments used to color phenol-formaldehyde resins.

phenol
(1) A class of aromatic organic compounds in which one or more hydroxy groups are attached directly to the benzene ring. Examples are phenol itself (benzophenol), the cresols, xylenols, resorcinol, naphthols. Though technically alcohols, their properties are quite different. (2) Phenol (carbolic acid; phenylic acid; benzophenol; hydroxybenzene) C_6H_5OH.

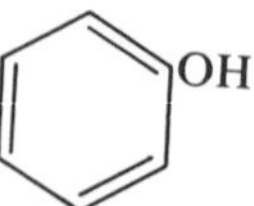

Properties: White, crystalline mass which turns pink or red if not perfectly pure or if under influence of light; absorbs water from the air and liquefies; distinctive odor; sharp burning taste. When in very weak solution it has a sweetish taste; sp. gr. 1.07; m.p. 42.5–43°C; b.p. 182°C; flash point 172.4°F. (C.C.). Soluble in alcohol, water, ether, chloroform, glycerol, carbon disulfide, petrolatum, fixed or volatile oils and alkalies. Combustible. Autoignition temp. 1319°F.
Derivation: Most of the phenol in the U.S. is made by the oxidation of cumene, yielding acetone as a by-product. The first step in the reaction yields cumene hydroperoxide, which decomposes with dilute sulfuric acid to the primary products, plus acetophenone and phenyl dimethyl carbinol. Other processes include sulfonation, chlorination of benzene, Raschig, and oxidation of benzene.
Method of purification: Rectification.
Grades: Fused, crystals or liquid, all as technical (82%, 90%, 95%, other components mostly cresols); C.P. and U.S.P.
Containers: 25- and 55-gal drums; tank cars; tank trucks.
Hazard: Toxic by ingestion, inhalation, and skin absorption. Strong irritant to tissue. Tolerance, 5 ppm in air. Safety data sheet available from Manufacturing Chemists Assn., Washington, D.C.
Uses: Phenolic resins; epoxy resins (bisphenol-A); nylon-6 (caprolactam); 2,4-D; selective solvent for refining lubricating oils; adipic acid; salicylic acid; phenolphthalein; pentachlorophenol; acetophenetidine; picric acid; germicidal paints; pharmaceuticals; laboratory reagent; dyes and indicators; slimicide.
Shipping regulations: (Rail, Air) Solid, or liquid if containing over 50% benzophenol: Poison label. Legal name: carbolic acid.
Note: High-boiling phenols are mixtures containing predominantly meta-substituted alkyl phenols. Their boiling point ranges from 238 to 288°C; they set to a glass below −30°C. They are used in phenolic resins, as fuel-oil sludge inhibitors, as solvents and as rubber chemicals.

phenolate process. A process for removing hydrogen sulfide from gas by the use of sodium phenolate, which reacts with the hydrogen sulfide to give sodium hydrosulfide and phenol. This can be reversed by steam heat to regenerate the sodium phenolate.

phenol coefficient. In determining the effectiveness of a disinfectant using phenol as a standard of comparison, the phenol coefficient is a value obtained by dividing the highest dilution of the test disinfectant by the highest dilution of phenol that sterilizes a given culture of bacteria under standard conditions of time and temperature.
See also disinfectant.

phenol-formaldehyde resin. The first synthetic thermosetting polymer, the reaction product of phenol with aqueous 37–50% formaldehyde at 50–100°C, with basic catalyst; discovered by Baekeland (q.v.) in 1907 and trademarked "Bakelite" in 1911. Polymerization is of the condensation type, proceeding through three stages. With an acid catalyst novolak resins (q.v.) are produced, which are thermoplastic.
Properties: Gray to black, hard, infusible solid when cured; resistant to moisture, solvents and heat up to 200°C; dimensionally stable; good electrical resistance; noncombustible; sound- and noise-absorbent; decomposed by oxidizing acids; fair resistance to alkalies. Cannot be successfully colored. Generally nontoxic.
Uses: Molded and cast articles; bonding powders; ion exchange; laminating and impregnating; plywood and glass-fiber composites; ablative coatings for aerospace use; binder for oil-well sands; paint and baked enamel coatings; thermal and acoustic insulation, brake linings, clutch facings, shell molds; chemical equipment, machine and instrument housings, machine parts, electrical devices. See also A-, B-, and C-stage resin; novolak; Baekeland; phenolic resin.

phenol-furfural resin. A phenolic resin that has a somewhat sharper transition than phenol-formaldehyde from the soft, thermoplastic stage to the cured, infusible state and can be fabricated by injection molding since it has little tendency to harden before curing conditions are reached.

phenolic resin. Any of several types of synthetic thermosetting resin obtained by the condensation of phenol or substituted phenols with aldehydes such as formaldehyde, acetaldehyde, and furfural. Phenol-formaldehyde resins (q.v.) are typical and constitute the chief class of phenolics.

phenolphthalein $(C_6H_4OH)_2C_2O_2C_6H_4$ (an approximation). 3,3-Bis(para-hydroxyphenyl)phthalide.
Properties: Pale yellow powder; forms an almost colorless solution in neutral or acid solution in presence of alkali, but colorless in the presence of large amounts of alkali. Soluble in alcohol, ether, and alkalies; insoluble in water. Sp. gr. 1.2765; m.p. 261°C.
Derivation: Interaction of phenol and phthalic anhydride in sulfuric acid.
Grades: Technical; pure reagent; N.F.
Containers: Bottles; cans; drums.
Uses: Dyes; acid-base indicator; laboratory reagent; medicine (laxative).

phenol red. See phenolsulfonephthalein.

phenolsulfonephthalein (phenol red)
$(C_6H_4OH)_2COSO_2C_6H_4$ (an approximation).
Properties: Bright to dark red crystalline powder. Stable in air. Slightly soluble in water, alcohol, and acetone; almost insoluble in chloroform and ether; soluble in alkali hydroxides and carbonates.
Derivation: Reaction of phenol with ortho-sulfobenzoic acid anhydride. Differs from phenolphthalein in containing an SO_2 group in place of CO.
Grades: Technical; reagent; U.S.P. The U.S.P. spelling is phenolsulfonphthalein.
Uses: Acid-base indicator; diagnostic reagent in medicine; laboratory reagent.

phenolsulfonic acid (sulfocarbolic acid) $HOC_6H_4SO_3H$.
Properties: Yellowish liquid, becoming brown on exposure to air. A mixture of ortho- and paraphenolsulfonic acids. Soluble in water and alcohol.
Derivation: Action of sulfuric acid on phenol.
Grades: Technical; reagent.
Hazard: Strong irritant to skin and tissue; moderately toxic.
Uses: Water analysis; laboratory reagent; electroplated tin coating baths; manufacture of intermediates and dyes; pharmaceuticals.
Shipping regulations: (Rail) Corrosive material, n.o.s., White label. (Air) Corrosive label.

phenol trinitrate. See picric acid.

"Phenoplast."[289] Trademark for a cold-setting liquid phenolic resin coating.

phenothiazine (thiodiphenylamine)

Properties: Grayish-green to greenish-yellow powder, granules or flakes. Tasteless; slight odor. Insoluble in ether and water; soluble in acetone. M.p. 175–185°C; b.p. 371°; sublimes 130°C (1 mm).
Derivation: By reaction of diphenylamine and sulfur in presence of an oxidizing catalyst.
Grades: Technical; N.F.
Hazard: Moderately toxic by ingestion. Absorbed by skin. Tolerance, 5 mg per cu meter of air.
Uses: Insecticide; manufacture of dyes; parent compound for chlorpromazine and related antipsychotic drugs; polymerization inhibitor; antioxidant.

phenoxyacetic acid $C_6H_5OCH_2COOH$.
Properties: Light tan powder; b.p. 285°C; m.p. 98°C; soluble in ether, water, methanol, carbon disulfide, glacial acetic acid. Combustible.
Uses: Intermediate for dyes, pharmaceuticals, pesticides, other organics; fungicides; flavoring; laboratory reagent.

phenoxybenzamine hydrochloride
$C_6H_5OCH_2CH(CH_3)N(CH_2C_6H_5)(CH_2)_2Cl \cdot HCl$.
Properties: Crystals; m.p. 137.5–140°C; soluble in alcohol, propylene glycol; slightly soluble in water.
Use: Medicine.

phenoxydihydroxypropane. See phenoxypropanediol.

2-phenoxyethanol. See ethylene glycol monophenyl ether.

alpha-**phenoxyethylpenicillin.** See penicillin.

phenoxymethylpenicillin (penicillin V) $C_{16}H_{18}N_2O_5S$.
Properties: White, odorless, crystalline powder; very slightly soluble in water; soluble in alcohol and acetone. Insoluble in fixed oils; pH of saturated solution is 2.5–4.0.
Grade: N.F.
Use: Medicine (antibiotic for oral use). Available also as potassium salt.

phenoxypropanediol (1-phenoxypropanediol-2,3) $C_6H_5OCH_2CHOHCH_2OH$.
Properties: White crystalline solid; m.p. 53°C; b.p. 150–155°C (4 mm); soluble in water, alcohol, glycerin, carbon tetrachloride, warm benzene; insoluble in gasoline. Combustible.
Derivation: Phenol and glycerol.
Uses: Medicine; plasticizer; resins; lacquers.

phenoxypropylene oxide $C_6H_5OCH_2\overline{CHCH_2O}$.
Properties: Practically colorless liquid with characteristic odor. Very slightly soluble in water; sp. gr. 1.1110 (20/20°C); b.p. 244.2°C (760 mm); vapor pressure less than 0.1 mm (20°C); f.p. 2.8°C; viscosity 6.93 cp (20°C).

phenoxy resin. A high molecular weight thermoplastic copolymer of bisphenol A and epichlorohydrin having the basic molecular structure $—[OC_6H_4C(CH_3)_2-C_6H_4OCH_2CH(OH)CH_2]_n—$ (n is about 100). It uses the same raw materials as epoxy resins, but contains no epoxy groups. It may be cured by reacting with polyisocyanates, anhydrides or other cross-linking agent capable of reacting with hydroxyl groups.

The ductility of phenoxy resins resembles that of metals. They are transparent and are also characterized by low mold shrinkage, good dimensional stability and moderately good resistance to temperature and corrosives. Phenoxy resins are soluble in methyl ethyl ketone and have been used for coatings and adhesives. A typical injection-molded specimen has a tensile strength of 9000 psi, heat distortion point 188°F at 264 psi load, and specific gravity 1.18.

They may be extruded or blow-molded. Parts may by thermally formed and heat- or solvent-welded. Some applications are blow-molded containers, pipe, ventilating ducts and molded parts.

"Phenurone."[3] Trademark for phenacemide (q.v.).

phenyl. The univalent C_6H_5 group derived from benzene, and characteristic of phenol and other derivatives. See following entries.

phenylacetaldehyde (alpha-toluic aldehyde) $C_6H_5CH_2CHO$.
Properties: Colorless liquid; strong hyacinth-like odor. Soluble in 2 parts of 80% alcohol; soluble in ether and most fixed oils; very slightly soluble in water. Sp. gr. 1.023 to 1.030 (25°C); f.p. below −10°C; b.p. 193 to 194°C; refractive index 1.520–1.530. Becomes more viscous on aging. Combustible.
Derivation: From phenyl-alpha-chloroacetic acid, by action of alkalies; oxidation of phenylethyl alcohol.
Grades: Technical; 50% solution in benzyl alcohol; F.C.C.
Uses: Perfumes; flavoring; laboratory reagent.

phenylacetaldehyde dimethylacetal (alpha-tolyl aldehyde dimethyl acetal) $C_6H_5CH_2CH(OCH_3)_2$.
Properties: Colorless liquid; strong odor; more stable than phenylacetaldehyde; not known to cause discoloration; sp. gr. 1.000–1.004 (25/25°C); refractive index 1.493–1.496 (20°C); soluble in 2 parts of 70% alcohol. Flash point 191°F (TCC). Combustible.
Grades: Technical; F.C.C.
Uses: Perfumery; flavoring; laboratory reagent.

phenylacetamide (alpha-phenylacetamide) $C_6H_5CH_2CONH_2$.
Properties: White crystals; b.p. 280–290°C (decomposes); m.p. 156–160°C; soluble in hot water and alcohol; very slightly soluble in cold water and ether. Combustible; low toxicity.
Derivation: Partial hydrolysis of benzyl cyanide; dehydration of ammonium phenyl acetate.
Containers: Fiber drums.
Uses: Organic synthesis; pharmaceuticals; penicillin precursors; laboratory reagent.

N-phenylacetamide. See acetanilide.

phenyl acetate (acetylphenol) $C_6H_5OOCCH_3$.
Properties: Water-white liquid; soluble in alcohol and ether; slightly soluble in water. Sp. gr. 1.073 (25/25°C); boiling point 195–196°C; flash point 176°F. Combustible. Low toxicity.
Derivation: (a) From phenol and acetyl chloride. (b) By heating triphenyl phosphate with potassium acetate and alcohol.
Containers: Glass bottles; carboys; drums.
Uses: Solvent; organic synthesis; laboratory reagent.

phenylacetic acid (alpha-toluic acid) $C_6H_5CH_2CO_2H$.
Properties: Shiny, white plate crystals; sp. gr. 1.0809; floral odor; f.p. 76–78°C; b.p. 262°C; soluble in alcohol and ether; slightly soluble in cold water. Combustible.
Derivation: From benzyl cyanide.
Grades: Technical; F.C.C.
Uses: Perfume; medicine; manufacture of penicillin; fungicide; plant hormones; flavoring; laboratory reagent.

phenylacetonitrile. See benzyl cyanide.

alpha-**phenylacetophenone.** See deoxybenzoin.

phenylacetyl chloride $C_6H_5CH_2COCl$.
Properties: Colorless liquid; refractive index 1.5320 (20°C).
Containers: Carboys, nickel drums.
Hazard: Strong irritant.
Uses: Acylating agent, including manufacture of esters for flavors; laboratory reagent.

phenylacetylurea. See phenacemide.

beta-**phenylacrylic acid.** See cinnamic acid.

phenylacrylyl chloride. See cinnamoyl chloride.

phenylalanine (alpha-amino-beta-phenylpropionic acid) $C_6H_5CH_2CH(CH_2)COOH$. An essential amino acid.
Properties: L(−)-phenylalanine: Plates and leaflets from concentrated aqueous solutions; hydrated needles from dilute aqueous solutions; decomposes 283°C; soluble in water; slightly soluble in methanol and ethanol.

D(+)-phenylalanine: Leaflets from water; decomposes 285°C; soluble in water; slightly soluble in methanol.

DL-phenylalanine: Leaflets or prisms from water or alcohol; sweet tasting; decomposes 318–320°C; soluble in water.

Sources: (L−)-Phenylalanine is isolated commercially from proteins (ovalbumin, lactalbumin, zein, and fibrin). DL-Phenylalanine is synthesized from alpha-acetaminocinnamic acid.
Grades: Technical; F.C.C.
Containers: Glass bottles; fiber cans; drums.
Uses: Biochemical research; dietary supplement; synthetic sweeteners; laboratory reagent. Available commercially as DL-dihydroxyphenylalanine and as DL-phenylalanine.

phenylallylic alcohol. See cinnamic alcohol.

phenylamine. See aniline.

phenylamino cadmium dilactate.
Hazard: Probably toxic by ingestion.
Use: Seed disinfectant.

phenyl-2-amino-5-naphthol-7-sulfonic acid (phenyl J acid; 6-anilino-1-naphthol-3-sulfonic acid) $HOC_{10}H_5(NHC_6H_5)(SO_3H)$.
Properties: Slate-colored crystals; soluble in alkali. Combustible.
Derivation: From H acid and aniline (condensation with heat).
Hazard: Probably toxic by ingestion.
Use: Dyes.

phenyl-2-amino-8-naphthol-6-sulfonic acid (phenyl gamma acid; 7-anilino-1-naphthol-3-sulfonic acid) $HOC_{10}H_5(NHC_6H_5)(SO_3H)$.
Properties: Gray crystals; soluble in alkali. Combustible.
Derivation: From gamma acid and aniline (condensation with heat).
Hazard: Probably toxic by ingestion.
Use: Dyes.

1-phenyl-2-aminopropane. See amphetamine.

N-phenylaniline. See diphenylamine.

ortho-**phenylaniline.** See ortho-aminobiphenyl.

phenylarsonic acid $C_6H_5AsO(OH)_2$.
Properties: Crystalline powder; m.p. 160°C with decomposition; soluble in water and alcohol; insoluble in chloroform.
Preparation: The Bart reaction between the diazonium salt and sodium arsenite.
Hazard: Highly toxic by ingestion.
Use: Analytical reagent for tin.
Shipping regulations: (Rail, Air) Arsenical compounds, n.o.s., Poison label.

phenylazoaniline. See aminoazobenzene.

1-(phenylazo)-2-naphthylamine. See yellow AB.

phenylbarbital. See phenobarbital.

pheneylbenzamide. See benzanilide.

3-phenyl-3-benzoborepin. A synthetic carbon-boron heterocyclic compound with aromatic properties. Stable in air. Key intermediate in its synthesis is the nonaromatic carbon-tin heterocycle 3,3-dimethyl-3-benzostannepin.

phenylbiphenyloxadiazole. See 1,3,4-phenylbiphenylyloxadiazole.

1,3,4-phenylbiphenylyloxadiazole (PBD; phenylbiphenyloxadiazole) $C_{20}H_{14}N_2O$.
Properties: Crystals; m.p. 166–168°C.
Grade: Purified.
Use: As primary fluors or as wave length shifters in solution scintillators.

phenylbis[1-(2-methyl)aziridinyl]phosphine oxide $C_6H_5(C_3H_6N)_2PO$.
Properties (technical): Straw-colored liquid; limited solubility in water; soluble in most organic solvents. The pure material is a low-melting solid. Combustible.
Hazard: Probably toxic.
Use: Polymerization additive.

phenyl bromide. See bromobenzene.

1-phenylbutane. See n-butylbenzene.

2-phenylbutane. See sec-butylbenzene.

phenylbutazone (4-n-butyl-1,2-diphenyl-3,5-pyrazolidinedione) $C_{19}H_{20}N_2O_2$. A synthetic pyrazolone derivative; also occurs naturally in certain herbs.
Properties: White or very light yellow powder; slightly bitter taste and very slight aromatic odor; m.p. 103–106°C; freely soluble in acetone, ether and ethyl acetate; very slightly soluble in water; stable if stored at room temperature in closed containers in absence of moisture. Also available as sodium salt.
Grade: N.F.
Hazard: Toxic; use by prescription only.
Use: Medicine. Antiinflammatory drug licensed for human use; veterinary medicine.

1-phenylbutene-2 $C_6H_5CH_2CH{:}CHCH_3$.
Properties: Colorless liquid. Boiling range 174–176°C; sp. gr. 0.888 (60/60°F); refractive index 1.511 (n 20/D); flash point 160°F. (TOC). Combustible.
Grades: Technical, 95 mole %.
Uses: Organic synthesis; laboratory reagent.

phenylbutynol (3-phenyl-1-butyn-3-ol) $HC{:}CC(C_6H_5)(OH)CH_3$.
Properties: Crystals; camphor odor; m.p. 51–52°C; b.p. 217–218°C; sp. gr. 1.0924 (20/20°C); slightly soluble in water; soluble in acetone benzene, most organic solvents. Combustible.
Uses: Acid inhibitor; organic synthesis.

2-phenylbutyric acid. See phenylethylacetic acid.

1-phenyl-3-carbethoxypyrazolone-5 $C_{12}H_{12}N_2O_3$.
Properties: White to light buff powder, stable in aqueous solution. M.p. 182–188°C. Combustible.
Containers: Drums.
Use: Dyestuff intermediate; laboratory reagent.

phenyl carbimide. See phenyl isocyanate.

phenylcarbinol. See benzyl alcohol.

phenylcarbylamine chloride $C_6H_5NCCl_2$.
Properties: Pale-yellow, oily liquid; onion-like odor. Mildly volatile. Soluble in alcohol, benzene, ether; insoluble in water. Sp. gr. 1.30 at 15°C; b.p. 208–210°C.
Derivation: Chlorination of phenylisothiocyanate.
Grade: Technical.
Hazard: Highly toxic by inhalation and skin absorption. Strong irritant to skin and mucous membranes.
Uses: Organic synthesis; military poison gas.
Shipping regulations: (Air) Not acceptable. (Rail) Poison Gas label. Not accepted by express.

phenyl "Cellosolove."[214] Trademark for ethylene glycol monophenyl ether (q.v.).

phenyl chloride. See chlorobenzene.

phenylchloroform. See benzotrichloride.

phenyl chloromethyl ketone. See chloroacetophenone.

1-phenyl-3-chloropropane. See phenylpropyl chloride.

phenyl cyanide. See benzonitrile.

phenylcyclohexane (cyclohexylbenzene) $C_6H_5C_6H_{11}$.
Properties: Colorless, oily liquid; pleasant odor; sp. gr. 0.938 (25/15°C); m.p. 5°C; b.p. 237.5°C; refractive index (n 25/D) 1.523; flash point 400°F (o.c.). Insoluble in water, glycerin; very soluble in alcohol, acetone, benzene, carbon tetrachloride, castor oil, hexane, xylene. Combustible; low toxicity.
Uses: High-boiling solvent; penetrating agent; intermediate; laboratory reagent.

2-phenylcyclohexanol $HOCH(CH_2)_4CH(C_6H_5)$.
Properties: Colorless to pale, straw-colored liquid; pour point −18°C; b.p. 276–281°C; sp. gr. 1.033

(25/25°); refractive index 1.536 (n 25/D); flash point 280°F; very slightly soluble in water; soluble in methanol, ether. Combustible; low toxicity.
Uses: Solvent; intermediate.

N-phenyl-N′-cyclohexyl-para-**phenylenediamine** $C_6H_5HNC_6H_4NHC_6H_{11}$.
Properties: Gray-violet powder; sp. gr. 1.16; m.p. 103–107°C; soluble in acetone, benzene, MEK, ethyl acetate, ethylene dichloride; insoluble in water.
Uses: Antioxidant-antiozonant for elastomers.

phenyldichloroarsine $C_6H_5AsCl_2$.
Properties: Liquid; microcrystalline mass at the f.p. Decomposed by water. Soluble in alcohol, benzene, and ether; insoluble in water. Sp. gr. 1.654 (20°C); b.p. 255–257°C; f.p. −20°C; vapor tension 0.014 mm (15°C); volatility 404 mg/cu m (20°C); coefficient of thermal expansion 0.00073.
Derivation: Arsenic trichloride and phenylmercuric chloride are heated together at 100°C.
Grade: Technical.
Containers: Steel bottles.
Hazard: Highly toxic by ingestion and inhalation. Strong irritant to eyes, skin, and tissue.
Uses: Lachrymator poison gas; solvent for diphenylcyanoarsine.
Shipping regulations: (Rail, Air) Poison label. Not accepted on passenger planes.

phenyldidecyl phosphite $C_6H_5OP(OC_{10}H_{21})_2$.
Properties: Nearly water-white liquid with an odor of alcohol; sp. gr. 0.940 (25/15.5°C); m.p. below 0°C; refractive index (n 25/D) 1.4785; flash point 425°F (COC). Combustible.
Containers: 55-gal steel drums; tank trucks.
Uses: Chemical intermediate; antioxidant; ingredient in stabilizer systems for resins.

phenyl diethanolamine $C_6H_5N(C_2H_4OH)_2$.
Constants: Colorless liquid. M.p. 58°C; b.p. 190°C (1 mm); vapor pressure <0.01 mm (20°C); wt 10.0 lb/gal (20°C); sp. gr. 1.1203 at 60/20°C; viscosity 1.19 poise (20°C). Slightly soluble in water; soluble in ethyl alcohol and acetone. Flash point 375°F (o.c.). Combustible. Low toxicity.
Grade: Technical.
Containers: 1-gal cans; 5-, 55-gal drums; tank cars.
Uses: Organic synthesis; dyestuff; laboratory reagent.

phenyl diglycol carbonate [diethyl glycol bis(phenyl carbonate)] $(C_6H_5OOCOCH_2CH_2)_2O$.
Properties: Colorless solid; sp. gr. 1.23 (20/4°C); m.p. 40°C; b.p. 225–229°C; refractive index (n 20/D) 1.525; evaporation rate 0.026 mg/sq cm/hr at 100°C; insoluble in water (very stable to hydrolysis); widely soluble in organic solvents; compatible with many resins and plastics. Combustible.
Use: Plasticizer.

N-phenyl-N′-(1,3-dimethyl butyl)-para-**phenylenediamine** $C_6H_5HNC_6H_4NHC_6H_{13}$.
Properties: Dark violet, staining, low-melting solid; m.p. 50°C; sp. gr. 1.07.
Uses: Antiozonant, antioxidant, and polymer stabilizer.

3-phenyl-1,1-dimethylurea. See fenuron.

meta-**phenylenediamine** (meta-diaminobenzene) $C_6H_4(NH_2)_2$.
Properties: Colorless needles; unstable in air; usually in the form of the stable hydrochloride; sp. gr. 1.1389; m.p. 63°C; b.p. 282–287°C; soluble in alcohol, ether, and water.
Derivation: Reduction of meta-dinitrobenzene or nitroaniline with iron and hydrochloric acid. Purified by crystallization.
Grades: Technical; 99% min purity.
Hazard: Strong irritant to skin. Toxic by inhalation.
Uses: Dyestuff manufacture; detection of nitrous acid; textile developing agent; laboratory reagent.

ortho-**phenylenediamine** (orthamine; ortho-diaminobenzene) $C_6H_4(NH_2)_2$.
Properties: Colorless monoclinic crystals; darkens in air; m.p. range 102–104°C; b.p. 252–258°C; soluble in alcohol, ether, water, and chloroform.
Derivation: Reduction of ortho-dinitrobenzene or nitroaniline with iron and hydrochloric acid. Purified by crystallization.
Grades: Technical; 99.0% min purity.
Uses: Manufacture of dyes; photographic developing agent; organic synthesis; laboratory reagent.

para-**phenylenediamine** (para-diaminobenzene) $C_6H_4(NH_2)_2$.
Properties: White to light purple crystals (oxidizes on standing in air to purple and black); m.p. about 147°C; b.p. 267°C; soluble in alcohol, ether; soluble in water and alcohol; affected by light. Flash point 312°F. Combustible.
Derivation: Reduction of para-dinitrobenzene or nitroaniline with iron and hydrochloric acid. Purified by crystallization.
Grades: Technical; 99% min purity.
Hazard: Toxic by inhalation; absorbed by skin. Tolerance, 0.1 mg per cubic meter of air.
Uses: Azo dye intermediate; photographic developing agent; photochemical measurements; intermediate in manufacture of antioxidants and accelerators for rubber; laboratory reagent; dyeing hair and fur.

phenylephrine hydrochloride (*l*-1-(meta-hydroxyphenyl-2-methylaminoethanol hydrochloride) $HOC_6H_4CH(OH)CH_2NHCH_3 \cdot HCl$.
Properties: White or nearly white crystals; odorless; has bitter taste; solutions are acid to litmus paper; freely soluble in water and in alcohol; m.p. 140–145°C; levorotatory in solution.
Grade: U.S.P.
Use: Medicine.

phenylethane. See ethylbenzene.

2-phenylethanol. See phenethyl alcohol.

phenylethanolamine $C_6H_5NHCH_2CH_2OH$.
Constants: Liquid; sp. gr. 1.0970 (20/20°C); b.p. 285.2°C; vapor pressure <0.01 mm (20°C); wt 9.1 lb/gal (20°C); f.p. 10.6°C; viscosity 1.01 poise (20°C). Flash point 305°F (o.c.). Combustible.
Hazard: Moderately toxic.
Uses: Organic synthesis; dyestuffs.

phenyl ether. See diphenyl oxide.

2-phenylethyl acetate (phenethyl acetate) $C_6H_5CH_2CH_2OOCCH_3$. (Not the same as sec-phenylethyl acetate).
Properties: Colorless liquid; peach-like odor. Soluble in alcohol, ether, and most fixed oils. Sp. gr. 1.030–1.034; refractive index 1.497–1.501 (20°C); b.p. 226°C; flash point 230°F; combustible.
Derivation: (a) Interaction of ethyl acetate and aluminumphenyl ethylate. (b) Interaction of acetic anhydride and phenylethyl alcohol in the presence of sodium acetate.
Grades: Technical; F.C.C.
Containers: Glass bottles.
Use: Perfumery; laboratory reagent.

sec-**phenylethyl acetate.** See alpha-methylbenzyl acetate.

phenylethylacetic acid (2-phenylbutyric acid) $C_2H_5CHC_6H_5COOH$.
Properties: White crystals with aromatic odor; m.p. 41.0°C (min); insoluble in water; soluble in alcohol, ketones, and esters. Combustible.
Use: Organic synthesis; laboratory reagent.

2-phenylethyl alcohol. See phenethyl alcohol.

2-phenylethylamine (phenethylamine; 1-amino-2-phenylethane) $C_6H_5C_2NH_2$.
Properties: Liquid with a fishy odor; absorbs carbon dioxide from the air; strong base; sp. gr. 0.9640; b.p. 194.5°C; soluble in water, alcohol, and ether. Combustible.
Derivation: From phenylethyl alcohol and ammonia under pressure.
Grades: Technical; scintillation.
Containers: Drums.
Hazard: Skin irritant.
Uses: Organic synthesis; laboratory reagent; scintillation counter (CO_2 absorber).

2-phenylethyl anthranilate (phenethyl anthranilate) $H_2NC_6H_4COOC_2H_4C_6H_5$.
Properties: Colorless liquid which yellows with age; odor of grape and orange; sp. gr. 1.14 (25/25°C). Combustible; nontoxic.
Uses: Perfumes; flavoring.

phenylethyl carbinol. See phenylpropyl alcohol.

phenylethylene. See styrene.

N-**phenylethylethanolamine** $C_6H_5N(C_2H_5)C_2H_4OH$.
Properties: Solid; m.p. 37.2°C; b.p. 268°C (740 mm); sp. gr. 1.04 (20/20°C); very slightly soluble in water. Flash point 270°F (COC). Soluble in alcohol, acetone, benzene. Combustible. Low toxicity.
Containers: Drums.
Uses: Solvents; chemical intermediates; preparation of dyes for acetate rayons; laboratory reagent.

phenyl ethyl ether. See phenetole.

5-phenyl-5-ethylhydantoin $(C_6H_5)(C_2H_5)\underline{CNHCONHC}O$.
Properties: Colorless, odorless crystalline powder; m.p. 199°C; insoluble in water.
Use: Medicine.

2-phenylethyl isobutyrate (phenethyl isobutyrate) $(CH_3)_2CHCOOC_2H_4C_6H_5$.
Properties: Colorless liquid; pleasant odor; sp. gr. 0.988 (25/25°C); refractive index (n 20/D) 1.488; soluble in alcohol and ether. Combustible; nontoxic.
Uses: Perfumes; flavoring.

phenylethylmalonylurea. See phenobarbital.

2-phenylethyl mercaptan $C_6H_5CH_2CH_2SH$.
Properties: Liquid. Boiling range 193–225°C; unpleasant odor; sp. gr. 1.0264 (60/60°F); refractive index 1.5582 (n 20/D); flash point 160°F. Combustible.
Containers: Bottles.
Hazard: Probably toxic.
Uses: Organic synthesis; laboratory reagent.

2-phenylethyl phenylacetate (phenethyl phenylacetate) $C_6H_5(CH_2)_2OOCCH_2C_6H_5$.
Properties: White crystals; hyacinth odor. Sp. gr. 1.080–1.082; congealing point 27°C. Combustible; low toxicity.
Containers: Bottles.
Uses: Perfumery; flavors.

2-phenylethyl propionate (phenethyl propionate) $C_2H_5COOC_2H_4C_6H_5$.
Properties: Synthetic colorless liquid; flower-fruit odor; miscible with alcohols and ether; sp. gr. 1.012 (25/25°C). Combustible; low toxicity.
Uses: Perfumes; flavors.

2-phenylethyl salicylate (phenethyl salicylate) $C_6H_5C_2H_4OOCC_6H_4OH$.
Properties: Snow-white crystals; very faint aromatic odor. Soluble in 14 parts of 95% alcohol. Congealing point 41.5°C. Combustible. Low toxicity.
Use: Flavors.

phenyl ferrocenyl ketone. See benzoylferrocene.

phenyl fluoride. See fluorobenzene.

phenyl fluoromethyl ketone. See fluoroacetophenone.

phenylformamide. See formanilide.

phenylformic acid. See benzoic acid.

phenyl gamma acid. See phenyl-2-amino-8-naphthol-6-sulfonic acid.

phenyl glycidyl ether (1,2-epoxy-3-phenoxypropane; PGE) $H_2COCHCH_2OC_6H_5$.
Properties: Colorless liquid; sp. gr. 1.11; b.p. 245°C; m.p. 3.5°C.
Hazard: Toxic by skin absorption; moderately irritating to eyes and skin. Tolerance, 10 ppm in air.

D(−)-alpha-**phenylglycine** $C_6H_5CH(NH_2)COOH$.
Properties: Crystals; m.p. 245–248°C; insoluble in water, ether, alcohol; soluble in acid.
Containers: Fiber drums.
Use: Intermediate.

phenylglycolic acid. See mandelic acid.

phenylhydrazine $C_6H_5NHNH_2$.
Properties: Pale yellow crystals or oily liquid; becomes red-brown on exposure to air. Soluble in alcohol, ether, chloroform, benzene, and dilute acids. Soluble in water, alcohol, and benzene. Sp. gr. 1.0978; m.p. 19.35°C; b.p. 243.5°C, with decomposition. Flash point 192°F (c.c.). Combustible. Autoignition temp. 345°F. Also available as the hydrochloride.
Derivation: Reduction of diazotized aniline; followed by reaction with sodium hydroxide.
Grades: Commercial; C.P.; reagent.
Containers: Glass bottles; drums.
Hazard: Toxic by inhalation, ingestion, and skin absorption. Tolerance, 5 ppm in air.
Uses: Analytical chemistry (reagent for detecting aldehydes, sugars, etc.); organic synthesis (intermediates, dyestuffs, pharmaceuticals).

alpha-**phenylhydroxyacetic acid.** See mandelic acid.

phenylic acid. See phenol.

phenylindan dicarboxylic acid. See 1,1,3-trimethyl-5-carboxy-3-(*p*-carboxyphenyl)indan.

phenyl isocyanate (phenyl carbimide; carbanil) C_6H_5NCO.
Properties: Liquid; b.p. 165°C; f.p. −30°C (approx); sp. gr. 1.095 (20/4°C); refractive index (n 19.6/D) 1.53684; decomposes in water and alcohol; very soluble in ether. Flash point 132°F (TOC). Combustible.

Hazard: Moderate fire risk. Toxic; strong irritant; a lachrymator.
Uses: Reagent for identifying alcohols and amines; intermediate.
Shipping regulations: (Air) Poison label. Not acceptable on passenger planes.

phenyl isothiocyanate. See phenyl mustard oil.

phenyl J acid. See phenyl-2-amino-5-naphthol-7-sulfonic acid.

phenyllithium C_6H_5Li. Available in a 20% by weight solution of a 70:30 volume % benzene-ether mixture. Used for organic synthesis.
Hazard: Flammable, dangerous fire risk.

phenylmagnesium bromide C_6H_5MgBr. A Grignard reagent available as a solution in ether; sp. gr. 1.14.
Derivation: From magnesium and bromobenzene.
Containers: Glass bottles; drums.
Hazard: Dangerous fire risk; solution highly flammable.
Use: Arylating agent in organic synthesis.

phenylmagnesium chloride C_6H_5MgCl. Available dissolved in tetrahydrofuran.
Hazard: Dangerous fire risk; solution highly flammable.
Use: Grignard reagent.

phenyl "MAPO."[293] Trademark for phenylbis[1-(2-methyl)aziridinyl]phosphine oxide (q.v.).

phenyl mercaptan C_6H_5SH.
Properties: Water-white liquid; offensive odor; oxidizes on exposure to air; supplied under nitrogen atmosphere. Sp. gr. 1.080 (15.5/15.5°C); refractive index (n 25/D) 1.5815; f.p. −15°C; b.p. 169°C. Insoluble in water; very soluble in aromatic and aliphatic hydrocarbons.
Hazard: Probably toxic; store out of contact with air and acids.
Uses: Chemical intermediate; mosquito larvicide.

phenylmercuric acetate $C_6H_5HgOCOCH_3$.
Properties: White to cream prisms; m.p. 148–150°C. Slightly soluble in water; soluble in alcohol, benzene, and glacial acetic acid. Slightly volatile at ordinary temperatures.
Derivation: Action of heat on benzene and mercuric acetate.
Grades: C.P.; technical; commercial.
Containers: Fiber drums, bottles.
Hazard: Highly toxic by ingestion, inhalation and skin absorption. Strong irritant. Tolerance, 0.05 mg per cubic meter of air.
Uses: Antiseptic, fungicide, herbicide; mildewcide for paints; slimicide in paper mills.
Shipping regulations: (Air) Poison label. (Rail) Mercury compounds, n.o.s., Poison label.

phenylmercuric benzoate $C_6H_5HgOOCC_6H_5$. Crystals; m.p. 94°C (min); used as a denaturant for alcohol; bactericide; fungicide.
Hazard: Highly toxic by ingestion, inhalation, and skin absorption. Tolerance, 0.05 mg per cubic meter in air. Solutions of 1% strength or greater are also poisonous.
Shipping reuglations: (Rail, Air) Mercury compounds, n.o.s., Poison label.

phenylmercuric borate $(C_5H_5Hg)_2HBO_3$.
Properties: White crystalline powder; m.p. 120–130°C. Slightly soluble in water; soluble in alcohol.
Derivation: Reaction of phenylmercuric acetate with boric acid.
Hazard: Highly toxic by ingestion, inhalation, and skin absorption. Tolerance, 0.05 mg per cubic meter in air. Solutions of 1% strength or greater are also poisonous.
Shipping regulations: (Rail, Air) Mercury compounds, n.o.s., Poison label.

phenylmercuric chloride C_6H_5HgCl.
Properties: White satiny crystals; m.p. 251°C. Insoluble in water; slightly soluble in hot alcohol; soluble in benzene, ether, pyridine.
Derivation: Reaction of phenylmercuric acetate and sodium chloride.
Hazard: Highly toxic by ingestion, inhalation, and skin absorption. Tolerance, 0.05 mg per cubic meter in air. Solutions of 1% strength or greater are also poisonous.
Uses: Antiseptic; fungicide, germicide.
Shipping regulations: (Rail, Air) Mercury compounds, n.o.s., Poison label.

phenylmercuric hydroxide C_6H_5HgOH.
Properties: Fine white to cream crystals; m.p. 197–205°C; slightly soluble in water; soluble in acetic acid; alcohol.
Grade: Technically pure.
Containers: Bottles; fiber drums.
Hazard: Highly toxic by ingestion, inhalation, and skin absorption. Tolerance, 0.05 mg per cubic meter in air. Solutions of 1% strength or greater are also poisonous.
Uses: Manufacture of phenylmercuric salts; fungicide and germicide. Principal compound in manufacturing organic mercury derivatives; denaturant for alcohol.
Shipping regulations: (Rail) Mercury compounds, n.o.s., Poison label. (Air) Poison label.

phenylmercuric lactate $C_6H_5HgOOCCHOHCH_3$. Used as bactericide and fungicide.
Hazard: Highly toxic by ingestion, inhalation, and skin absorption. Tolerance, 0.05 mg per cubic meter in air. Solutions of 1% strength or greater are also poisonous.
Shipping regulations: (Rail, Air) Mercury compounds, n.o.s., Poison label.

phenylmercuric naphthenate.
Prepared by interaction of phenylmercuric acetate and naphthenic acid, producing a colored solution. Used as a wood preservative and as a mildewproofing agent for paints and adhesives.
Hazard: Highly toxic by ingestion, inhalation, and skin absorption. Tolerance, 0.05 mg per cubic meter in air. Solutions of 1% strength or greater are also poisonous.
Shipping regulations: (Rail, Air) Mercury compounds, n.o.s., Poison label.

phenylmercuric nitrate (basic)
$C_6H_5HgNO_3 \cdot C_6H_5HgOH$.
Properties: Fine white crystals, or grayish powder; mercury content 63–65% (theory, 63.2%); melting range 175–185°C, with decomposition; ash 0.1% max; very slightly soluble in water; slightly soluble in alcohol; insoluble in ether; moderately soluble in glycerin.
Grade: N.F.
Containers: Bottles; fiber drums.
Hazard: Highly toxic by ingestion, inhalation, and skin absorption. Tolerance, 0.05 mg per cubic meter in air. Solutions of 1% strength or greater are also poisonous.
Shipping regulations: (Rail) Mercury compounds, n.o.s., Poison label. (Air) Poison label.

phenylmercuric oleate $C_6H_5HgOOC(CH_2)_7CH{:}CHC_8H_{17}$.
Properties: White crystalline powder; m.p. 45°C; insoluble in water; soluble in organic solvents and some oils.
Derivation: Reaction of phenylmercuric acetate with oleic acid.
Hazard: Highly toxic by ingestion, inhalation, and skin absorption. Tolerance, 0.05 mg per cubic meter in air. Solutions of 1% strength or greater are also poisonous.
Uses: Mildewproofing agent for paints; fungicide and germicide.
Shipping regulations: (Rail, Air) Mercury compounds, n.o.s., Poison label.

phenylmercuric propionate $C_6H_5HgOCOCH_2CH_3$.
Properties: Technical grades: White to off-white wax-like free flowing powder; m.p. 65–70°C; stable to 200°C for short periods; 57% min Hg content.
Containers: ½-lb container to 50-lb fiber drums.
Hazard: Highly toxic by ingestion, inhalation, and skin absorption. Tolerance 0.05 mg per cubic meter of air.
Uses: Fungicide and bactericide for paints and industrial finishes.
Shipping regulations: (Rail, Air) Mercury compounds, n.o.s., Poison label.

phenylmercuric salicylate $C_6H_4(OH)(COOHgC_6H_5)$.
Hazard: Highly toxic by ingestion, inhalation, and skin absorption. Tolerance 0.05 mg per cubic meter of air.
Uses: Seed disinfectant.
Shipping regulations: (Rail, Air) Mercury compounds, n.o.s., Poison label.

phenylmercuriethanolammonium acetate $[(HOC_2H_4)NH_2(C_6H_5Hg)]OOCCH_3$.
Properties: White crystalline solid; soluble in water.
Derivation: Reaction of phenylmercuric acetate with monoethanolamine.
Hazard: Highly toxic by ingestion, inhalation, and skin absorption. Tolerance 0.05 mg per cubic meter of air.
Uses: Insecticide and fungicide; may not be used on food crops.
Shipping regulations: (Rail, Air) Mercury compounds, n.o.s., Poison label.

phenylmercuritriethanolammonium lactate [tris(2-hydroxyethyl)(phenylmercuri)ammonium lactate] $[(HOC_2H_4)_3NHgC_6H_5]OOCCHOHCH_3$.
Properties: White crystalline solid; soluble in water.
Derivation: Reaction of phenylmercuric acetate with triethanolamine and lactic acid.
Hazard: Highly toxic by ingestion, inhalation, and skin absorption. Tolerance 0.05 mg per cubic meter of air.
Uses: Turf fungicide and eradicant fungicide for fruit trees.
Shipping regulations: (Rail, Air) Mercury compounds, n.o.s., Poison label.

phenyl mercury compounds. See phenylmercuric compounds.

phenylmercury formamide $HCONHHgC_6H_5$.
Hazard: Highly toxic by ingestion, inhalation, and skin absorption. Tolerance 0.05 mg per cubic meter of air.
Use: Seed disinfectant.
Shipping regulations: (Rail, Air) Mercury compounds, n.o.s., Poison label.

phenylmercury urea $C_6H_5HgNHCONH_2$.
Hazard: Highly toxic by ingestion, inhalation, and skin absorption. Tolerance 0.05 mg per cubic meter of air.
Uses: Disinfectant and fungicide for seed treatment.
Shipping regulations: (Rail, Air) Mercury compounds, n.o.s., Poison label.

phenylmethane. See toluene.

phenylmethanol. See benzyl alcohol.

phenylmethyl acetate. See benzyl acetate.

phenylmethylcarbinol. See alpha-methylbenzyl alcohol.

phenylmethylcarbinyl acetate. See alpha-methylbenzyl acetate.

N-**phenylmethylethanolamine** $C_6H_5N(CH_3)C_2H_4OH$.
Properties: Liquid which sets to a glass at −30°C; b.p. 192°C (100 mm); sp. gr. 1.0661 (20/20°C); slightly soluble in water; flash point 280°F (COC); combustible; low toxicity.
Uses: Chemical intermediate; solvent for dyes for acetate fibers.

phenyl methyl ketone. See acetophenone.

1-phenyl-3-methyl-5-pyrazolone. See 3-methyl-1-phenyl-2-pyrazolin-5-one.

phenyl alpha-**methylstyryl ketone.** See dypnone.

N-**phenylmorpholine** $C_6H_5\underline{NCH_2CH_2OCH_2CH_2}$.
Properties: White solid. Soluble in water. B.p. 268°C; m.p. 57°C; vapor pressure <0.1 mm (20°C). Flash point 220°F; sp. gr. 1.06 (57/20°C); combustible.
Grade: Technical.
Hazard: Probably toxic. Details unknown.
Uses: Chemical intermediate for dyestuffs, rubber accelerators, corrosion inhibitors and photographic developers; insecticide.

phenyl mustard oil (thiocarbanil; phenyl isothiocyanate; phenylthiocarbonimide) C_6H_5NCS.
Properties: Pale yellow or colorless liquid; penetrating, irritating odor; readily volatilzed with steam. Soluble in alcohol and ether; insoluble in water. Sp. gr. 1.1382; f.p. −21°C; b.p. 221°C. Combustible.
Derivation: (a) By action of concentrated hydrochloric acid on sulfocarbanilide; (b) by reaction of thiophosgene with aniline.
Hazard: Irritant to tissue; moderately toxic by ingestion and inhalation.
Uses: Medicine; organic synthesis.

N-**phenyl**-alpha-**naphthylamine** $C_{10}H_7NHC_6H_5$.
Properties: Crystallizes in prisms; white to slightly yellowish. Soluble in alcohol, ether, and benzene. M.p. 62°C; b.p. 335°C (260 mm). Combustible. Low toxicity.
Derivation: From alpha-naphthylamine and aniline. Purified by distillation.
Uses: Dyes and other organic chemials; rubber antioxidant.

N-**phenyl-beta-naphthylamine** $C_{10}H_7NHC_6H_5$.
Properties: Light gray powder; m.p. 107°C; b.p. 395°C; sp. gr. 1.24; insoluble in water; soluble in alcohol, acetone, benzene. Combustible; low toxicity.
Uses: Rubber antioxidant; lubricant; inhibitor (butadiene).

N-**phenyl-1-naphthylamine-8-sulfonic acid** (N-phenyl peri acid) $C_{16}H_{13}NO_3S$.

Properties: Greenish-gray needles. Insoluble in water; soluble in alcohol.
Derivation: Arylation of 1-naphthylamine-8-sulfonic acid with aniline.
Grades: Technical; mostly as sodium salt.
Containers: Barrels or steel drums.
Use: Azo dyes.

phenylneopentyl phosphite $(CH_3)_2C(CH_2O)_2POC_6H_5$.
Properties: Water-white liquid; m.p. 19°C; b.p. 138–140°C (10 mm); refractive index (n 25/D) 1.517; sp. gr. 1.135 (25/15°C).
Use: Chemical intermediate.

1-phenylnonane. See nonylbenzene.

1-phenylpentane. See n-amylbenzene.

2-phenylpentane. See sec-amylbenzene.

ortho-**phenylphenol** (ortho-hydroxydiphenyl; ortho-xenol) $C_6H_5C_6H_4OH$.
Properties: Nearly white or light buff crystals; m.p. 56–58°C; b.p. 280–284°C; sp. gr. 1.217 (25/25°C); flash point 255°F. Combustible. Soluble in alcohol, sodium hydroxide solution; insoluble in water.
Derivation: From reaction of chlorobenzene and caustic soda solution at elevated temperatures and pressure.
Hazard: Moderately toxic by ingestion.
Uses: Intermediate for dyes; germicide; fungicide; rubber chemicals; laboratory reagent; food packaging.

para-**phenylphenol** (para-hydroxydiphenyl; para-xenol) $C_6H_5C_6H_4OH$.
Properties: Nearly white crystals; m.p. 164–165°C; b.p. 308°C. Flash point 330°F. Soluble in alcohol, also in alkalies and most organic solvents; insoluble in water. Combustible.
Derivation: From reaction of chlorobenzene and caustic soda solution at elevated temperatures and pressure.
Hazard: Moderately toxic by ingestion.
Uses: Intermediate for dyes; resins; rubber chemicals; laboratory reagent; fungicide.

N-**phenyl**-para-**phenylenediamine.** See para-aminodiphenylamine.

phenylphosphine (phosphaniline). $C_6H_5PH_2$.
Properties: Colorless liquid; b.p. 160°C; sp. gr. 1.001 (15°C).
Hazard: Highly toxic by inhalation and ingestion. Tolerance, 0.05 ppm in air.
Shipping regulations: Not listed.

phenylphosphinic acid. See benzenephosphinic acid.

phenylphosphonic acid. See benzenephosphonic acid.

N-**phenylpiperazine** $C_6H_5NCH_2CH_2NHCH_2CH_2$.
Properties: Pale yellow oil; insoluble in water; soluble in alcohol and ether; sp. gr. 1.0621 (20/4°C); b.p. 286.5°C, 156–7°C (10 mm); m.p. 18.8°C; flash point 285°F. Combustible. Probably low toxicity.
Uses: Intermediate for pharmaceuticals, anthelmintics, surface-active agents, synthetic fibers.

3-phenyl-1-propanol. See phenylpropyl alcohol.

phenylpropanolamine hydrochloride $C_6H_5CH(OH)CH(CH_3)(NH_2) \cdot HCl$.
Properties: White crystalline powder with odor similar to benzoic acid; m.p. 198–199°C; freely soluble in alcohol and water; insoluble in benzene, chloroform, and ether; aqueous solution neutral to litmus.
Containers: Drums.
Use: Medicine.

1-phenylpropanone-1. See propiophenone.

3-phenylpropenal. See cinnamic aldehyde.

3-phenylpropenoic acid. See cinnamic acid.

3-phenylpropenol. See cinnamic alcohol.

1-(alpha-phenyl)-propenylveratrole. See benzyl isoeugenol.

3-phenylpropionaldehyde. See phenylpropyl aldehyde.

3-phenylpropionic acid. See hydrocinnamic acid.

phenylpropyl acetate (hydrocinnamyl acetate) $C_6H_5CH_2CH_2CH_2OOCCH_3$.
Properties: White crystals. Soluble in 70% alcohol. Sp. gr. 1.012–1.016; refractive index, 1.497. Combustible; low toxicity.
Uses: Perfumery; flavoring; laboratory reagent.

phenylpropyl alcohol (hydrocinnamic alcohol; 3-phenyl-1-propanol; phenylethyl carbinol) $C_6H_5CH_2CH_2CH_2OH$.
Properties: Colorless liquid; floral odor; b.p. 219°C. Soluble in 70% alcohol; insoluble in water. Sp. gr. 0.998–1.000; refractive index 1.524–1.528 (20°C). Combustible; low toxicity.
Grades: Technical; F.C.C.
Uses: Perfumery; flavoring; laboratory reagent.

phenylpropyl aldehyde (3-phenylpropionaldehyde; hydrocinnamic aldehyde) $C_6H_5CH_2CH_2CHO$.
Properties: Colorless liquid; floral odor. Soluble in 50% alcohol. Sp. gr. 1.010–1.020; refractive index 1.520–1.532. Combustible; low toxicity.
Grade: Chlorine-free.
Containers: Glass bottles; copper flasks.
Uses: Perfumery; flavors.

phenylpropyl chloride (hydrocinnamyl chloride; 1-phenyl-3-chloropropane) $C_6H_5CH_2CH_2CH_2Cl$.
Properties: Colorless to pale yellow liquid; b.p. 219–220°C; sp. gr. 1.056 (25/4°C); refractive index 1.5220 (20°C). Combustible; low toxicity.
Uses: Organic synthesis; laboratory reagent.

phenylpropylmethylamine (N,beta-dimethylphenylethylamine) $C_6H_5CH(CH_3)CH_2NHCH_3$.
Properties: Colorless to pale yellow liquid; 98% distils between 205–210°C; very soluble in alcohol, benzene, and ether; 1.2 parts dissolve in 100 parts water; aqueous solutions alkaline to litmus.
Uses: Medicine. Also available as hydrochloride.

1-phenyl-3-pyrazolidone (phenidone)

$C_6H_5NNHC(O)CH_2CH_2$.
Properties: Crystals; m.p. 121°C. Soluble in water. Combustible; low toxicity.
Uses: Photographic developer; laboratory reagent.

4-phenylpropylpyridine $C_6H_5(CH_2)_3C_5H_4N$.
Properties: Colorless liquid. B.p. 322°C; soluble in organic solvents.
Uses: Heat-transfer agent; chemical intermediate.

phenyl salicylate (salol) $C_6H_4OHCOOC_6H_5$.
Properties: White crystalline powder; faint aromatic odor and taste. Soluble in alcohol, ether, chloroform, benzene and fixed or volatile oils; sparingly soluble in water. Sp. gr. 1.2614; m.p. 41.9°C; b.p. 172–173°C. Absorbs light, especially at 290–330μ. Combustible; low toxicity.

Derivation: Heating salicylic acid and phenol with phosphorus pentachloride or other dehydrating agent.
Grades: N.F.; granular; powder.
Uses: Medicine; preservative; UV absorber in plastics, waxes, polishes, suntan oil; laboratory reagent.

phenylstearic acid. An organic fatty acid having a high degree of fluidity and no definite melting point. Pour point is −15°F. Its chief use is as a lubricant stabilizer; its potential uses include corrosion inhibitor, plasticizer and textile auxiliary.

Phenylstearic acid and its quaternary and ethoxylated derivatives are used in synthetic latices, as mineral oil emulsifiers, and in invert systems.

phenylsulfonic acid. See benzenesulfonic acid.

4-phenyl-1,4-thiazane $SCH_2CH_2N(C_6H_5)CH_2CH_2$.
Properties: White powder. Soluble in hot toluene. M.p. 108–111°C.
Derivation: Interaction of dichlorodiethyl sulfide and an aliphatic amine in the presence of alcohol and sodium carbonate.

phenylthiocarbonimide. See phenyl mustard oil.

phenyltrichlorosilane $C_6H_5SiCl_3$.
Properties: Colorless liquid. B.p. 201°C; sp. gr. 1.321 (25/25°C); refractive index (n 25/D) 1.5240; flash point (COC) 185°F. Soluble in benzene, ether, perchloroethylene. Readily hydrolyzed by moisture, with liberation of hydrochloric acid. Combustible.
Derivation: By Grignard reaction of silicon tetrachloride and phenylmagnesium chloride; reaction of benzene with trichlorosilane; of chlorobenzene, silicon and copper.
Hazard: Highly toxic. Strong irritant to tissue.
Uses: Intermediate for silicones; laboratory reagent.
Shipping regulations: (Rail) White label. (Air) Corrosive label. Not acceptable on passenger planes.

1-phenyltridecane. See tridecylbenzene.

phenyltrimethoxysilane $C_6H_5Si(OCH_3)_3$.
Properties: Liquid. Sp. gr. 1.063 (25°C); refractive index 1.4710 (25°C); b.p. 211°C. Soluble in acetone, benzene, perchloroethylene, methyl alcohol. Combustible.
Uses: In polymers to be applied to powders, glass, paper and fabrics.

phenylurethane. See ethyl phenylcarbamate.

phenyl valerate $C_4H_9COOC_6H_5$. Colorless liquid; slightly soluble in water; soluble in alcohol and ether. Used in flavors and odorants.

pheromone. A group of organic compounds produced by insects which function as communication means and as sex attractants. Synthetic pheromones are used to control insect pests by disrupting their mating behavior, e.g., 4-methyl-3-heptanone. See phoromone.

"Philprene."[303] Trademark for a series of styrene-butadiene rubbers, hot and cold, oil-extended, pigmented and non-pigmented, including special types for specific needs.
Containers: 70–90-lb bales.
Uses: Tire carcasses and treads, molded and extruded goods, sporting goods, footwear, coated fabrics, wire and cable jackets, hospital goods, floor tile, insulation.

"Phi-O-Sol WA."[328] Trademark for the sodium salt of a sulfonated fatty acid ester, for textile and industrial wetting applications. Removes wall paper and poultry feathers.

phloroglucinol (phloroglucine; 1,3,5-trihydroxybenzene) $C_6H_3(OH)_3 \cdot 2H_2O$.
Properties: White to yellowish crystals, odorless; m.p. 212–217°C if rapidly heated; 200–209°C, if slowly heated; b.p., sublimes with decomposition. Soluble in alcohol, ether, and pyridine; slightly soluble in water. Combustible.
Derivation: By fusion of resorcinol with caustic soda; by reduction of trinitrobenzene.
Containers: Glass bottles; fiber drums.
Uses: Analytical chemistry (reagent for pentoses and with vanillin for determining the presence of free hydrochloric acid); medicine; decalcifying agent for bones; preparation of pharmaceuticals and dyes, resins; preservative for cut flowers.

phorate. Generic name for O,O-diethyl S-[(ethylthio)methyl]phosphorodithioate, $(C_2H_5O)_2P(S)SCH_2SC_2H_5$.
Properties: Liquid; b.p. 118–120°C (0.8 mm). Insoluble in water; miscible with carbon tetrachloride, dioxane, xylene.
Hazard: Highly toxic by skin contact, inhalation, or ingestion. Rapidly absorbed through skin. Repeated inhalation or skin contact may, without symptoms, progressively increase susceptibility. Use may be restricted. Tolerance, 0.05 mg per cubic meter of air.
Use: Insecticide.
Shipping regulations: Organic phosphate, liquid, n.o.s., (Rail, Air) Poison label. Not acceptable on passenger planes.
Note: Approved by EPA as substitute for some uses of DDT.

phorbol (4,9,12-beta,13,20-pentahydroxy-1,6-tigliadien-3-on). The parent alcohol of tumor-producing compounds in croton oil (q.v.). A carcinogenic agent. Used in biochemical and medical research.

phoromone (7-ethyl-5-methyl-6,8-dioxabicyclo[3.2.1]-octane). Product excreted by bark beetles which acts as sex attractant. It has been isolated and synthesized for possible use in protection of forest timber. Details of use have not been developed.

phorone (diisopropylidene acetone). $(CH_3)_2CCHCOCHC(CH_3)_2$.
Properties: Yellow liquid or yellowish green prisms. Sp. gr. 0.8791 at 20/20°C; b.p. 197.9°C; freezing point 28.0°C; vapor pressure 0.38 mm (20°C); flash point 185°F; wt 7.3 lb/gal (20°C). Slightly soluble in water; soluble in alcohol. Combustible.
Uses: Solvent for nitrocellulose; lacquers; coating compositions; stains; intermediate (organic synthesis).

"Phorwite."[422] Trademark for a line of optical brighteners.
4205. Used for polyacrylonitrile, triacetate, acetate and polyester fibers.
BUP. A stilbene derivative used for brightening papers and other fibrous materials.
K 2002. Used for thermoplastic and thermosetting materials.

"Phos-Chek"[58] Trademark for fire-fighting solutions of phosphorus pentoxide: 201 and 202 contain 46.5% and 47% min P_2O_5, designed for application from aircraft; 258 contains 47.7% min P_2O_5, applied from ground equipment and helicopters. Said to be solutions of ammonium phosphate thickened with sodium methyl cellulose.

"Phosdrin."[125] Trademark for a mixture which contains not less than 60% of the alpha isomer of 2-carbomethoxy-1-methylvinyl dimethyl phosphate, $(CH_3O)_2P(O)OC(CH_3){:}CHCOOCH_3$, (generic name mevinphos, q.v.) and not more than 40% of insecticidally active related compounds; it is 100% active.
Properties: Yellow to orange liquid; b.p. 210–218°F (0.03 mm) miscible with water, alcohols and aromatic and chlorinated hydrocarbons. Slightly soluble in aliphatic hydrocarbons. Flash point 175°F. Combustible.
Hazard: Toxic by ingestion, inhalation, and skin absorption. Tolerance, 0.1 mg per cubic meter of air. Use may be restricted.
Use: Insecticide.
Shipping regulations: See mevinphos.

"Phos-Feed."[196] Trademark for a brand of dicalcium phosphate used as a mineral supplement of animal and poultry feeds.

"Phosflake."[177] Trademark for a uniform blend of caustic soda and trisodium phosphate prepared in flake form, especially for bottle-washing use.

"Phosfon."[40] Trademark for tributyl-2,4-dichlorobenzylphosphonium chloride (q.v.).

"Phosgard" C-22-R.[58] Trademark for a flame retardant organophosphorus compound with high efficiency through high phosphorus and chlorine content. Used in urethane foams, phenolics, polymethacrylates, polyester resins and epoxies.

phosgene (carbonyl chloride; carbon oxychloride; chloroformyl chloride) $COCl_2$.
Properties: Liquid, or easily liquefied gas. Colorless to light yellow; odor varies from strong and stifling when concentrated, to hay-like in dilute form. Sp. gr. 1.392 (19/4°C); f.p. −128°C; b.p. 8.2°C. Slightly soluble in water and slowly hydrolyzed by it; soluble in benzene and toluene; specific volume (70°F) 3.9 cu ft/lb. Noncombustible.
Derivation: By passing a mixture of carbon monoxide and chlorine over activated carbon.
Containers: Steel cylinders; special one-ton containers.
Hazard: Highly toxic by inhalation; strong irritant to eyes. Tolerance, 0.1 ppm in air. Safety data sheet available from Manufacturing Chemists Assn., Washington, D.C.
Uses: Organic synthesis, especially of isocyanates, polyurethane and polycarbonate resins, carbamates, organic carbonates and chloroformates; pesticides; herbicides; dye manufacture.
Shipping regulations: (Air) Not acceptable. (Rail) Consult regulations. Not accepted by express.

"Phos Kil."[55] Trademark for parathion-based insecticidal dusts and sprays.
Hazard: Highly toxic by ingestion and inhalation.

phosphamidon. Generic name for 2-chloro-2-diethylcarbamoyl-1-methylvinyl dimethyl phosphate, $(CH_3O)_2P(O)OC(CH_3){:}C(Cl)C(O)N(C_2H_5)_2$.
Properties: Colorless liquid; b.p. 162°C (1.5 mm); soluble in water and organic solvents.
Hazard: Toxic by ingestion, inhalation and skin absorption. Cholinesterase inhibitor. Use may be restricted.
Use: Insecticide.
Shipping regulations: Organic phosphate, liquid, n.o.s., (Rail, Air) Poison label. Not acceptable on passenger planes.

phosphatase, alkaline. An enzyme excreted into the bile by the liver and found in the blood. It is concerned with bone formation, probably being produced by osteoblasts. It hydrolyzes phosphoric acid esters at pH 7–8, liberating phosphate ions.
Use: Biochemical research.

phosphate, condensed. A phosphorus compound with two or more phosphorus atoms in the molecule. Examples are polyphosphates; pyrophosphates. See polyphosphoric acid.

phosphate glass. A type of glass containing phosphorus pentoxide. Aluminum meta-phosphate is frequently the basic material. Such glasses have properties not attainable in silicate glasses; e.g., resistance to hydrofluoric acid.

phosphate rock (phosphorite). A natural rock consisting largely of calcium phosphate and used as a raw material for manufacture of phosphate fertilizers, phosphoric acid, phosphorus, and animal feeds. A large amount is ground and applied directly to the soil. A primary source of superphosphate, prepared by treatment of the pulverized rock with sulfuric acid (superphosphate having 16–18% P_2O_5) or by acidifying with phosphoric acid (triple superphosphate having 40–48% P_2O_5). Nitric acid is sometimes used (see nitrophosphate). Defluorinated phosphate rock is the source of phosphate used in animal feeds and feed concentrates. Important deposits are in U.S. (Florida, North Carolina, Tennessee, California, Wyoming, Montana, Utah, Idaho), North Africa (Morocco, Tunisia, Algeria), U.S.S.R. and various islands in the Pacific.

phosphate slag. Glassy calcium silicate, by-product of electric furnace phosphorus manufacture.
Properties: Lumps; loose bulk density approximately 85 lb/cu ft.
Containers: Hopper cars, gondolas.

phosphatide. See phospholipid.

phosphatidyl choline. See lecithin.

phosphatidyl ethanolamine. See cephalin.

phosphatidyl serine. See cephalin.

phosphazene. A ring or chain polymer that contains alternating phosphorus and nitrogen atoms, with two substituents on each phosphorus atom. Characteristic structures are cyclic trimers, cyclic tetramers, and high polymers. The substituent can be a wide variety of organic groups, halogen, amino, etc. Most cyclic trimers are crystalline, solids, organosoluble and stable to weather conditions; the high polymers (polyphosphazenes) are elastomeric or thermoplastic.

"Phosphen."[233] Trademark for a group of aryl phosphates.

phosphinate. A derivative of the hypothetical phosphinic acid, $H_2P(O)OH$.

phosphine (hydrogen phosphide) PH_3.
Properties: Colorless gas; disagreeable, garlic-like odor. Soluble in alcohol, ether and cuprous chloride; slightly soluble in cold water; insoluble in hot water. Sp. gr. 1.185; f.p. −133.5°C; b.p. −85°C. Autoignition temp. 100°F.
Derivation: By action of nascent hydrogen or of caustic potash on phosphorus.

Hazard: Highly toxic by inhalation; strong irritant. Tolerance, 0.3 ppm in air. Spontaneously flammable.
Uses: Organic preparations; phosphonium halides; doping agent for solid state electronic components; polymerization initiator; condensation catalyst.
Shipping regulations: (Air) Not acceptable.
Note: A synthetic dye, chrysaniline yellow, is sometimes called phosphine.

phosphodiesterase. An enzyme which cleaves hydrolytically the carboxy- and phosphoesters of phosphatides (phospholipids).

"Phosphodust."[196] Trademark for a pesticide dust diluent derived from phosphate rock.

2-phosphoglyceric acid $HOCH_2CH[OPO(OH)_2]COOH$. An intermediate in the metabolism of carbohydrates in biological systems. See enolase.

phosphoglyceride. See phospholipid.

"Phospholeum."[58] Trademark for so-called superphosphoric acid.
Properties: Clear liquid containing 76% P_2O_5 and equivalent to 105% H_3PO_4. An azeotropic mixture of orthophosphoric and polyphosphoric acids which on dilution with water hydrolyze to orthophosphoric acid. Sp. gr. 1.90 (approx) (20.4°C); viscosity 800 cp (20.4°C); temperature rise, 15.5°C on dilution to 75% H_3PO_4; f.p. less than −50°C.
Grade: Technical.
Containers: Tank cars and tank trucks.
Uses: High-analysis liquid fertilizer; metal phosphatizing and aluminum bright dip bath component; desiccant; sequestrant for common trace minerals; catalyst or dehydrator in organic synthesis; rustproofing compounds; special bricks; steel pickling.

phospholipid (phosphatide). A group of lipid compounds that yield on hydrolysis phosphoric acid, an alcohol, fatty acid, and a nitrogenous base. They are widely distributed in nature and include such substances as lecithin, cephalin, and sphingomyelin.

phosphomolybdate. See heteromolybdate.

phosphomolybdic acid (phospho-12-molybdic acid; PMA; 12-molybdophosphoric acid) $H_3OMo_{12}O_{40} \cdot xH_2O$.
Properties: Yellowish crystals; soluble in water, alcohol, and ether; density, 3.15 g/cc; m.p. 78–90°C. Strong oxidizing agent in aqueous solution; strong acid in the free acid form.
Grades: Technical; C.P.; reagent.
Hazard: (solution) May ignite combustible materials.
Uses: Reagent for alkaloids; pigments; catalyst; fixing agent in photography; additive in plating processes; imparts water resistance to plastics, adhesives and cement.

phosphomolybdic pigment (molybdenum lake). A pigment made by precipitating a basic organic dye with phosphomolybdic acid or a mixture of phosphomolybdic and phosphotungstic acids. See also phosphotungstic pigment.

phosphonate. A derivative of the hypothetical phosphonic acid, $HP(O)(OH)_2$.

phosphonitrile. See phosphazene.

phosphonitrilic polymer. See fluoroelastomer.

phosphonium iodide (iodophosphonium) PH_4I.
Properties: Colorless or slightly yellowish crystals. Sp. gr. 2.86; m.p. 18.5°C; sublimes 61.8°C; b.p. 80°C; decomposed by water or alcohol.
Derivation: Action of phosphine upon hydrogen iodide.
Hazard: Probably toxic.
Use: Chemical synthesis.

"Phosphonol."[354] Trademark for polyphosphate esters used as surfactants, antistatic agents, plasticizers, and corrosion inhibitors.

phosphor (fluor). A substance, either organic or inorganic, liquid or crystalline, that is capable of luminescence, that is, of absorbing energy from sources such as x-rays, cathode rays, ultraviolet radiations, alpha particles and emitting a portion of the energy in the ultraviolet, visible or infrared. When the emission of the substance ceases immediately or in the order of 10^{-8} second after excitation, the material is said to be fluorescent. Material that continues to emit light for a period after the removal of the exciting energy is said to be phosphorescent The half-life of the afterglow varies with the substance and may range from 10^{-6} second to days.
Uses: Fluorescent light tubes; television, radar, and cathode ray tubes; instrument dials; scintillation counters.
See also fluorescence; phosphorescence.

phosphor bronze. A tin bronze (see brass and bronze) which has been deoxidized by the addition of up to 0.5% phosphorus. Relatively hard, strong, and corrosion-resistant. Has good cold work properties and high strength.
Grades: Grade A (5% tin), grade C (8% tin), grade D (10% tin), grade E (1.25% tin).
Uses: Springs, electrical switches, contact fingers, chains; fourdrinier wire.

phosphorescence. A type of luminescence (q.v.) in which the emission of radiation resulting from excitation of a crystalline or liquid material occurs *after* excitation has ceased, and may last from a fraction of a second to an hour or more. This phenomenon is characteristic of some organic compounds, as in the firefly, and also of a number of inorganic solid materials, both natural and synthetic. These are used industrially as phosphors (q.v.).
See also fluorescence.

phosphoric acid (orthophosphoric acid) H_3PO_4. Ninth highest-volume chemical produced in U.S. (1975).
Properties: Colorless, odorless, sparkling liquid or transparent crystalline solid, depending on concentration and temperature. At 20°C the 50 and 75% strengths are mobile liquids, the 85% is of a syrupy consistency, while the 100% acid is in the form of crystals; sp. gr. 1.834 (18°C); m.p. 42.35°C; loses ½H_2O at 213°C (to form pyrophosphoric acid); soluble in water and alcohol; corrosive to ferrous metals and alloys.
Derivation: (a) Action of sulfuric acid on pulverized phosphate rock; (b) Action of hydrochloric acid on phosphate rock, with extraction by tributylphosphate; (c) By heating phosphate rock, coke, and silica in an electric furnace, burning the elemental phosphorus produced, and then hydrating the phosphoric oxide (furnace acid).
Grades: Agricultural; technical (50, 75, 85, 90, 100%); food (50, 75, 85%); N.F. (85–88%); F.C.C. (75–85%). (Polyphosphoric acid is sometimes called 115% phosphoric acid).

Containers: Carboys; drums and barrels; tank trucks; tank cars.
Hazard: Toxic by ingestion and inhalation. Irritant to skin and eyes. Tolerance, 1 mg per cubic meter of air. Safety data sheet available from Manufacturing Chemists Assn., Washington, D.C.
Uses: Fertilizers; soaps and detergents; inorganic phosphates; pickling and rust-proofing metals; pharmaceuticals; sugar refining; gelatin manufacture; water treatment; animal feeds; electro-polishing; gasoline additive; conversion coatings for metals; catalyst for ethanol manufacture; lakes in cotton dyeing; yeasts; soil stabilizer; waxes and polishes; binder for ceramics; activated carbon; in foods and carbonated beverages as acidulant and sequestrant; laboratory reagent; source of uranium (experimental).
Shipping regulations: (Air) Corrosive label (solid or solution). (Rail) Not listed.

phosphoric acid, anhydrous. See phosphoric anhydride.

phosphoric acid, meta- (metaphosphoric acid; phosphoric acid, glacial) approximately $(HPO_3)_x$.
Properties: Transparent, highly deliquescent, glassy mass; sp. gr. 2.2–2.488. Soluble in water, slowly forming the ortho-acid; soluble in alcohol. Noncombustible.
Derivation: By heating orthophosphoric acid to redness; by treating phosphorus pentoxide with a calculated quantity of water; by heating diammonium phosphate.
Grades: Technical; C.P.
Containers: Glass bottles; carboys.
Uses: Phosphorylating agent; dehydrating agent; manufacture of dental cements; laboratory reagent.
See also polyphosphoric acid.

phosphoric anhydride (phosphorus pentoxide; phosphoric oxide; phosphoric acid, anhydrous) P_2O_5.
Properties: Soft, white powder. Absorbs moisture from the air with avidity, forming meta-, pyro-, or orthophosphoric acid, depending upon amount of water absorbed and upon conditions of absorption. Sp. gr. 2.387; m.p. 580–585°C; b.p., sublimes at 300°C.
Derivation: By burning phosphorus in dry air.
Grades: Technical; nitrogen-complex coated (for slow solution).
Containers: Glass bottles; cans; drums.
Hazard: Strong, caustic irritant. Reacts violently with water to evolve heat. Dangerous fire risk. Keep tightly sealed or stoppered. Safety data sheet available from Manufacturing Chemists Assn., Washington, D.C.
Uses: Preparation of phosphorus oxychloride and metaphosphoric acid; acrylate esters; surfactants; dehydrating agent; organic synthesis; medicine; sugar refining; laboratory reagent; fire extinguishing; special glasses.
Shipping regulations: (Rail) Yellow label. (Air) Flammable solid label. Not accepted on passenger planes.

phosphoric bromide. See phosphorus pentabromide.

phosphoric chloride. See phosphorus pentachloride.

phosphoric oxide. See phosphoric anhydride.

phosphoric perbromide. See phosphorus pentabromide.

phosphoric perchloride. See phosphorus pentachloride.

phosphoric sulfide. See phosphorus pentasulfide.

phosphorous acid, ortho- H_3PO_3.
Properties: White, or yellowish, crystalline mass. Very hygroscopic. Keep tightly sealed or stoppered. Absorbs oxygen very readily with formation of orthophosphoric acid. Soluble in alcohol, water. Sp. gr. 1.651; b.p. 200°C (dec); m.p. 70°C (approx).
Grades: Reagent; technical; 70%
Uses: Analysis (testing for mercury); chemical reducing agent); phosphite salts.

phosphorus P Nonmetallic element of atomic number 15, group VA of Periodic Table. Atomic weight 30.9738; valences 1, 3, 4, 5. Allotropes: white (or yellow), red, and black phosphorus. No stable isotopes. Several artificial radioactive isotopes from mass numbers 29 to 34.
Occurrence and derivation: Occurs in nature in phosphate rock [impure $Ca_3(PO_4)_2$], in apatite [$Ca_5(PO_4)_3F$], in bones, teeth, and in organic compounds of living tissue. Also as phosphorite nodules on ocean floor.
Derivation: (1) Produced in an electric furnace from phosphate rock, sand and coke. The phosphorus vapor is driven off and condensed under water. (2) By reaction of phosphate rock with sulfuric acid, the resulting $CaSO_4$ being removed by filtration and the phosphoric acid concentrated by evaporation (wet process).
Grades: Technical 99.9%; electronic grade, 99.9999%.
Containers: White (or yellow) (shipped under water): steel drums; tank cars. Red: tins and drums.

White (yellow) phosphorus:
Properties: Crystalline, wax-like, transparent solid; metastable with respect to red phosphorus, an impurity present in white allotrope. B.p. 280°C; vapor density corresponds to formula P_4; m.p. 44.1°C; sp. gr. (solid, 20°C) 1.82, (liquid, 44.5°C) 1.745. Mohs hardness 0.5; high electrical resistivity. Insoluble in water and alcohol; soluble in carbon disulfide. Exhibits phosphorescence at room temperature; an essential dietary nutrient.
Hazard: Ignites spontaneously in air at 86°F. Store under water and away from heat. Dangerous fire risk. Highly toxic by ingestion and inhalation. Tolerance, 0.1 mg per cubic meter of air. Safety data sheet available from Manufacturing Chemists Assn., Washington, D.C.
Uses: Manufacture of phosphoric acid (q.v.) and other phosphorus comounds; phosphor bronzes; metallic phosphides; additive to semiconductors; electroluminescent coatings; incendiaries, pyrotechnics, and rotenticides.
Shipping regulations: (dry) (Rail) Yellow label. Not accepted by express. (Air) Not acceptable. (in water) (Rail) Yellow label. (Air) Flammable Solid label. Not acceptable on passenger planes.

Red phosphorus:
Properties: Violet-red, amorphous or crystalline powder, obtained from white phosphorus by heating at 240°C with catalyst. Sublimes at 416°C; sp. gr. 2.34. Autoignition temp. 500°F. High electrical resistivity. Much less reactive than white phosphorus. Insoluble in most solvents.
Hazard: Large quantities ignite spontaneously, and on exposure to oxidizing materials. Reacts with oxygen and water vapor to evolve phosphine (q.v.). Extinguish with foam or dry chemical (not water).
Uses: Safety matches; phosphorus compounds.
Shipping regulations: (Rail) Yellow label. (Air) Flammable Solid label. Not acceptable on pasenger planes.

Black phosphorus:
Black crystalline or amorphous solid resembling graphite; sp. gr. 2.25–2.69. Obtained by heating white phosphorus under pressure. Insoluble in most solvents. Electrically conducting. No present uses.

phosphorus 32. Radioactive phosphorus of mass number 32.
Properties: Half-life, 14.3 days; radiation, beta.
Derivation: Pile irradiation of potassium dihydrogen phosphate or sulfur and sulfur compounds.
Forms available: Phosphate ion in weak hydrochloric acid solution; solid potassium dihydrogen phosphate; P-32 sterile solution; in tagged compounds such as hexaethyltetraphosphate, ribonucleic acid, triphenylphosphine, etc.
Grades: U.S.P. as sodium phosphate P-32 solution.
Hazard: Moderate.
Uses: Biochemical radioactive tracer studies; medical treatment of leukemia, skin lesions, etc.; industrial measurements, e.g., tire tread, wear, thickness of ink and paint films, lead detection.
Shipping regulations: Consult regulations.

phosphorus chloride. See phosphorus trichloride.

phosphorus heptasulfide (tetraphosphorus heptasulfide) P_4S_7. Light yellow crystals; sp. gr. 2.19; m.p. 310°C; b.p. 523°C. Slightly soluble in carbon disulfide.
Hazard: Flammable, dangerous fire risk.
Shipping regulations: (Air) Flammable Solid label. Not acceptable on passenger planes.

phosphorus nitride P_3N_5.
Properties: Amorphous white solid; insoluble in cold water; decomposes in hot water; soluble in common organic solvents. Nonhygroscopic and stable in air. Decomposes at 800°C.
Hazard: Probably toxic.
Use: Doping semiconductors.

phosphorus oxybromide $POBr_3$.
Properties: Colorless crystals; sp. gr. 2.82; m.p. 56°C; b.p. 189°C; decomposed by water; soluble in ether and benzene. Reacts strongly with organic matter.
Hazard: Highly toxic. Strong irritant to skin and tissue.
Use: Chemical intermediate.
Shipping regulations: (Rail) White label. (Air) Corrosive label. Not acceptable on passenger planes.

phosphorus oxychloride (phosphoryl chloride) $POCl_3$.
Properties: Colorless, fuming liquid; pungent odor. Sp. gr. 1.675 (20/20°C); m.p. 1.2°C; b.p. 107.2°C; refractive index 1.460 (25/D). Decomposed by water and alcohol.
Derivation: From phosphorus trichloride, phosphorus pentoxide and chlorine.
Grades: Technical; 99.999+%.
Containers: Steel-jacketed lead cylinders and carboys; nickel-lined drums and tank cars.
Hazard: Toxic by inhalation and ingestion; strong irritant to skin and tissue. Safety data sheet available from Manufacturing Chemists Assn., Washington, D.C.
Uses: Manufacture of cyclic and acyclic esters for plasticizers, gasoline additives, hydraulic fluids, and organophosphorus compounds; chlorinating agent and catalyst; dopant for semiconductor grade silicon; tricresyl phosphate and fire-retarding agents.
Shipping regulations: (Rail) White label. (Air) Corrosive label. Not acceptable on passenger planes.

phosphorus pentabromide (phosphoric bromide; phosphoric perbromide) PBr_5.
Properties: Yellow crystalline mass. Keep hermetically sealed! Soluble in water (dec); in carbon disulfide, carbon tetrachloride, benzene. B.p. 106°C (dec).
Grades: Technical.
Hazard: Toxic and irritant to skin and tissue.
Use: Organic synthesis.

phosphorus pentachloride (phosphoric chloride; phosphoric perchloride) PCl_5.
Properties: Slightly yellow, crystalline mass; irritating odor; fuming in moist air. Sp. gr. 3.60; m.p. (under pressure) 148°C. Ordinarily sublimes without melting at 160–165°C. Soluble in carbon disulfide and carbon tetrachloride.
Derivation: By action of chlorine on phosphorus or phosphorus trichloride.
Grades: Technical; reagent.
Hazard: Flammable, irritant to eyes and skin; tolerance, 1 mg per cubic meter of air. Reacts strongly with water. Use carbon dioxide or dry chemical to extinguish. Safety data sheet available from Manufacturing Chemists Assn., Washington, D.C.
Uses: Chlorinating and dehydrating agent; catalyst.
Shipping regulations: (Rail) Yellow label. (Air) Flammable Solid label. Not acceptable on passenger planes.

phosphorus pentafluoride PF_5. Colorless nonflammable gas; b.p. −84.8°C; decomposed by water. Available in small cylinders.
Grade: 99%.
Hazard: Toxic; strong irritant to eyes and skin. Tolerance (as F), 2.5 mg per cubic meter of air.
Shipping regulations: (Air) Nonflammable Gas label. Not acceptable on passenger planes.

phosphorus pentasulfide (phosphoric sulfide; phosphorus persulfide; thiophosphoric anhydride) P_2S_5.
Properties: Light-yellow or greenish-yellow crystalline masses. Odor similar to hydrogen sulfide. Keep in sealed containers. Very hygroscopic. Burns in air forming P_2O_5 and SO_2. Decomposed by moist air. Soluble in solutions of alkali hydroxides; slightly soluble in carbon disulfide. M.p. 286–290°C; b.p. 515°C; sp. gr. 2.03; vapor pressure 1 mm (300°C). Autoignition temp. 287°F.
Derivation: By reaction of phosphorus and sulfur.
Grades: Technical; distilled.
Containers: Fiber or steel drums; tote bins.
Hazard: Toxic by inhalation; strong irritant. Dangerous fire risk; ignites by friction. Contact with water or acids liberates poisonous and flammable hydrogen sulfide. Tolerance, 1 mg per cubic meter of air. Safety data sheet available from Manufacturing Chemists Assn., Washington, D.C.
Uses: Intermediate for lube oil additives, insecticides (chiefly parathion and malathion); flotation agents.
Shipping regulations: (Air) Flammable Solid label. Not acceptable on passenger planes.

phosphorus pentoxide. See phosphoric anhydride.

phosphorus persulfide. See phosphorus pentasulfide.

phosphorus salt. See sodium ammonium phosphate.

phosphorus sesquisulfide (tetraphosphorus trisulfide) P_4S_3.
Properties: Yellow, crystalline mass; soluble in carbon disulfide; insoluble in cold water; decomposed by hot water. Sp. gr. 2.00; m.p. 172°C; b.p. 407.8°C. Autoignition temp. 212°F.
Derivation: By gently heating phosphorus and sulfur.
Hazard: Dangerous fire risk; ignites by friction. Toxic and irritant.
Uses: Organic synthesis; manufacture of matches.

Shipping regulations: (Rail) Yellow label. (Air) Flammable Solid label. Not acceptable on passenger planes.

phosphorus sulfide. See phosphorus trisulfide.

phosphorus tribromide PBr_3.
Properties: Fuming, colorless liquid; penetrating odor; soluble in acetone, alcohol, carbon disulfide, hydrogen sulfide, water (decomposes). Sp. gr. 2.852 (15°C); b.p. 175°C; f.p. −40°C.
Hazard: Toxic; strong irritant to skin and tissue.
Uses: Analysis (testing for sugar and oxygen); catalyst; synthesis.
Shipping regulations: (Rail) White label. (Air) Corrosive label. Not acceptable on passenger planes.

phosphorus trichloride (phosphorus chloride) PCl_3.
Properties: Clear, colorless fuming liquid; decomposes rapidly in moist air. Soluble in ether, benzene, carbon disulfide and carbon tetrachloride. Sp. gr. 1.574; f.p. −111.8°C; b.p. 76°C.
Derivation: By passing a current of dry chlorine over gently heated phosphorus, which ignites. The trichloride, admixed with some pentachloride, distills over. A small amount of phosphorus is added and the whole distilled.
Grades: Technical; 99.9%.
Containers: Carboys; cylinders; tank cars.
Hazard: Highly toxic; strong irritant to skin and tissue. Dangerous fire risk in contact with water. Tolerance, 0.5 ppm in air. Safety data sheet available from Manufacturing Chemists Assn., Washington, D.C.
Uses: Making phosphorus oxychloride; intermediate for organophosphorus pesticides, surfactants, phosphites (reaction with alcohols and phenols), gasoline additives, plasticizers, dyestuffs; chlorinating agent; catalyst; preparing rubber surfaces for electrodeposition of metal; ingredient of textile finishing agents.
Shipping regulations: (Rail) White label. (Air) Corrosive label. Not accepted on passenger planes.

phosphorus triiodide PI_3.
Properties: Red crystals; hygroscopic. Soluble in alcohol, carbon disulfide, water (dec). M.p. 61°C (dec); sp. gr. 4.18.
Grades: Technical; reagent.
Hazard: Toxic and flammable; reacts with water. Irritant.
Use: Organic synthesis.

phosphorus trisulfide (phosphorus sulfide; tetraphosphorus hexasulfide; thiophosphorous anhydride) P_2S_3, or P_4S_6.
Properties: Grayish-yellow masses; tasteless; odorless. Keep well stoppered! Decomposes in moist air. Soluble in alcohol, carbon disulfide, ether. B.p. 490°C; m.p. 290°C.
Hazard: Highly toxic; flammable, dangerous fire risk; reacts with water.
Use: Organic chemistry (reagent).
Shipping regulations: (Air) Flammable Solid label. Not acceptable on passenger planes.

phosphorylase. An enzyme occurring in muscle and liver which catalyzes the conversion of glycogen into glucose-1-phosphate.

phosphorylation. A reaction in which phosphorus combines with an organic compound, usually in the form of the trivalent phosphoryl group $=P=O$. It occurs naturally in cellular metabolism and is of particular importance in vitamin activity and enzyme formation. It is also used to produce a modified cellulose (P-cellulose) for cation exchange in chromatographic separations.

phosphoryl chloride. See phosphorus oxychloride.

phosphotungstic acid (phospho-12-tungstic acid; phosphowolframic acid; 12-tungstophosphoric acid) $H_3PW_{12}O_{40} \cdot xH_2O$.
Properties: Yellowish-white solid; m.p. (for $24H_2O$ hydrate) 89°C. Very soluble in water, acetone, and diethyl ether. Relatively insoluble in nonpolar organic solvents. Strong oxidizing agent in aqueous solution; strong acid in the free acid form.
Derivation: Addition of phosphates to sodium tungstate in the presence of hydrochloric acid.
Grades: Reagent; technical.
Uses: Reagent in analytical chemistry and biology; manufacture of organic pigments; additive in plating industry; imparts water resistance to plastics, adhesives and cement; catalyst for organic reactions; photographic fixing agent; textile antistatic agent.

phosphotungstic acid, sodium salt. See sodium 12-tungstophosphate.

phosphotungstic pigment (tungsten lake). A green or blue pigment manufactured by precipitating basic dyestuffs such as malachite green or Victoria blue with solutions of phosphotungstic acid, or phosphomolybdic acid, or mixtures of both. See also phosphomolybdic pigment.
Uses: Printing inks; white paper; paints and enamels.

"Phostane" MDP.[58] Trademark for methyl diphenyl phosphate (q.v.).

"Phos-Trode."[407] Trademark for a phosphor bronze Grade C electrode and filler rod for joining both like and dissimilar metals and overlaying surfaces resistant to wear and corrosion.

photochemistry. The branch of chemistry concerned with the effect of absorption of radiant energy (light) in inducing or modifying chemical changes. Natural photosynthesis is the most important example of a photochemical reaction; others are the photosensitization of solids, applied in photography, photocells, etc. and occurring naturally in the formation of visual pigments; photochemical decomposition (photolysis); photo-induced polymerization, oxidation, and ionization; fluorescence and phosphorescence; and the reaction of chlorine with organic compounds. Free-radical chain mechanisms are usually involved. See also free radical.

photochromism. The ability of a transparent material to darken reversibly when exposed to light. See also glass, photochromic.

Plastics can be made light-sensitive by certain aromatic organic nitro compounds such as 2-(2,4-dinitrobenzyl)pyridine. Such chemicals are compatible with most transparent plastics and are either blended with the base resin or applied as coatings.

photo-glycin. See para-hydroxyphenylglycine.

photographic chemistry. In photographic films and papers the sensitive surface consists of microscopic grains of a silver halide, suspended in gelatin. Exposure to light renders the halide particles susceptible to reduction to metallic silver by developing agents (q.v.) containing a reducing agent, as well as an accelerator, preservative, and restrainer. The accelerator increases the activity of the reducing agent (due principally to ionization of the phenolic agents to their active form) and is usually an alkaline compound. The preservative, usually sodium sulfite, minimizes air oxidation. The restrainer helps to prevent "fog" (reduction of silver halide grains which have not been exposed to light) and is almost always potassium bromide.

Color sensitizers are dyes added to silver halide emulsions to broaden their response to various wavelengths. Unsensitized emulsions are most responsive in the blue region of the spectrum and thus do not correctly represent the light spectrum striking them. Widely used sensitizers include the cyanine dyes (q.v.), the merocyanines, the benzooxazoles, and the benzothiazoles. Cryptocyanine sensitizes the extreme red and infrared.

In color photography diethyl-para-phenylenediamine is an important developer, since its oxidation product readily couples with a large number of phenol and reactive methylene compounds to form indophenol and indoaniline dyes, which are the basis of most of the current color processes.

"Photogray."[20] Trademark for a photochromic glass that darkens in sunlight to 45% light transmittance. See glass, photochromic.

photolysis. Decomposition of a compound into simpler units as a result of absorbing one or more quanta of radiation; examples are splitting of hydrogen iodide by the reaction $2HI + h\nu \rightarrow H_2 + I_2$, and of ketene ($H_2C{=}CO$) into carbon monoxide and carbene (methylene) ($:CH_2$). Photodecomposition may also occur with aldehydes, ketones, azo compounds, and organometallic compounds. See also flash photolysis; photochemistry.

photon. The unit (quantum) of electromagnetic radiation. Light waves, gamma rays, x-rays, etc., consist of photons. Photons are concentrations of energy that behave partly like particles and partly like waves. They have no rest mass and exist only when moving at the speed of light. Their nature can be described only in mathematical terms. Photons are emitted when electrons move from one energy state to another, as in an excited atom. See also radiation.

photophor. See calcium phosphide.

photopolymer. A polymer or plastic so made that it undergoes a change on exposure to light. Such materials can be used for printing and lithography plates, for photographic prints and microfilm copying. The effect of the light may be to cause further polymerization or crosslinking, or may cause degradation. One application involves the use of esters of polyvinyl alcohol which crosslink and so become insoluble, whereas unexposed portions of the material remain soluble.

photosynthesis. The utilization of sunlight by plants to convert two inorganic substances (carbon dioxide and water) into carbohydrates. Chlorophyll (q.v.) acts as the energy-converter in this reaction, which is perhaps the most important on earth. The specific reaction is: $6CO_2 + 6H_2O + 672\ \text{kcal} \rightarrow C_6H_{12}O_6 + 6O_2$. This significance of this process lies in the conversion of energy from radiant to chemical form. The chemical energy a green plant stores by photosynthesis provides the total energy requirement of the plant. Directly or indirectly plants supply the primary organic nutrient for most other living organisms. Most fossil fuels are storehouses of the radiant energy transformed by photosynthesis in earlier geologic eras.

Photosynthesis is the principal source of atmospheric oxygen. At least two-thirds of the total photosynthetic activity of the earth takes place in the oceans (see algae). Its exact chemical mechanism is extremely complex, and even now has not been wholly elucidated. Essential features are the reduction of carbon dioxide and utilization of the hydrogen of water to form carbohydrates, the oxygen being liberated; the nucleotides uridine and adenosine triphosphate are involved in this conversion. Sugar (sucrose) is formed in the cytoplasm surrounding the chloroplasts.

Photosynthesis has been suggested as a possible source of fuels by the development of "energy plantations" (intensive growth of crops) and conversion of a small fraction of the solar energy to generation of electric power.

photovoltaic cell. See solar cell.

"Photox."[266] Trademark for photoconductive lead-free zinc oxides manufactured by the French process produced from zinc metal. Used in electrophotography.

"PH 990 Resin."[469] Trademark for a phosphonitrilic-modified phenolic resin. Stable, off-white, free-flowing powder. Soluble in most organic solvents. Flame retardant and retains electrical and structural properties at temperatures up to 500–800°F.

"Phthalamaquin."[342] Trademark for an aureoquin preparation, described as 4-(2-dimethylaminoethylamino)-6-methoxyquinoline diethylaminotetrahydrophthalate. Used in medicine.

phthalamide $C_6H_4(CONH_2)_2$. The double acid amide of phthalic acid.

Properties: Colorless crystals; m.p. 200–210°C (decomposes into phthalimide and ammonia). Very slightly soluble in water and alcohol; insoluble in ether.

Derivation: By stirring phthalimide with cold concentrated ammonia solution; by the reaction of phthalyl chloride and ammonia; or from the addition of ammonia to phthalic anhydride under pressure.

Use: Intermediate in organic synthesis; laboratory reagent.

phthalic acid (ortho-phthalic acid; ortho-benzene dicarboxylic acid) $C_6H_4(COOH)_2$.

Properties: Colorless crystals; soluble in alcohol; sparingly soluble in water and ether. Sp. gr. 1.585; m.p., decomposes at 191°C.

Derivation: From phthalic anhydrides.

Grades: Technical; reagent.

Uses: Dyes; medicine; phenolphthalein; phthalimide; anthranilic acid; synthetic perfumes; laboratory reagent.

para-**phthalic acid.** See terephthalic acid.

phthalic anhydride $C_6H_4(CO)_2O$.

Properties: White, crystalline needles; sublimes below b.p.; mild odor. Sp. gr. 1.527 (4°C); m.p. 131.16°C; b.p. 285°C. Flash point 305°F (C.C.). Combustible. Autoignition temp. 1083°F. Soluble in alcohol; slightly soluble in ether and hot water.

Derivation: Catalytic oxidation of naphthalene or ortho-xylene (catalyst is 10% vanadium pentoxide plus 20–30% potassium sulfate on silica gel carrier). Both fixed bed and fluidized bed processes are used.

Method of purification: Sublimation.

Grades: Pure; technical, 130.8°C min f.p.

Containers: Drums; paper bags; molten in tank trucks; tank cars.
Hazard: Toxic and skin irritant. Tolerance, 2 ppm in air. Safety data sheet available from Manufacturing Chemists Assn., Washington, D.C.
Uses: Alkyd resins; plasticizers; hardener for resins; polyesters; synthesis of phenolphthalein and other phthaleins, many other dyes; chlorinated products; pharmaceutical intermediates; insecticides; diethyl phthalate; dimethyl phthalate; laboratory reagent.

ortho-**phthalimide** $C_6H_4(CO)_2NH$.
Properties: White, crystalline leaflets. Slightly soluble in ether; insoluble in cold benzene; soluble in boiling benzene and in aqueous alkalies. M.p. 233–238°C; b.p., sublimes. Combustible. Low toxicity.
Derivation: By dissolving phthalic anhydride in ammonium hydroxide, evaporating to dryness and using the residue.
Uses: Synthetic indigo, via anthranilic acid; fungicide; organic synthesis; laboratory reagent.

phthalocyanine. Any of a group of benzoporphyrins which have strong pigmenting power, forming a family of dyes. The basic structure of the molecule comprises four isoindole groups, $(C_6H_4)C_2N$, joined by four nitrogen atoms. Four commercially important modifications are: (1) metal-free phthalocyanine $(C_6H_4C_2N)_4N_4$ having a blue-green color (structure shown below); (2) copper phthalocyanine, in which a copper atom is held by secondary valences of the isoindole nitrogen atoms; (see Pigment Blue 15); (3) chlorinated copper phthalocyanine, green, in which 15 to 16 hydrogen atoms are replaced by chlorine (see Pigment Green 7); and (4) sulfonated copper phthalocyanine, water-soluble, green, in which two hydrogen atoms are replaced by sulfonic acid (HSO_3) groups.

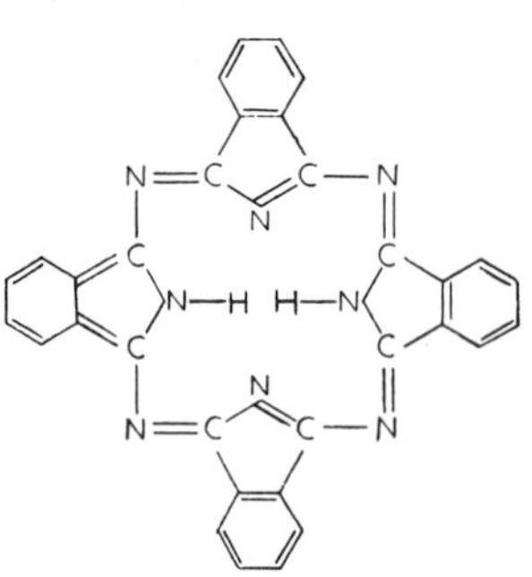

Uses: Decorative enamels, automotive finishes, linoleum plastics, roofing granules, printing inks, wallpaper, rubber goods, and similar applications where light fastness and chemical stability are required.

phthalonitrile (ortho-dicyanobenzene) $C_6H_4(CN)_2$.
Properties: Buff-colored crystals; m.p. 138°C; insoluble in water; soluble in acetone and benzene. Combustible.
Derivation: Vapor phase reaction of ammonia and phthalic anhydride over alumina catalyst at high temperature.
Hazard: May be toxic.
Uses: Intermediate in organic synthesis, especially pigments and dyes; base material for high-temperature lubricants and coatings; insecticide.

phthaloyl chloride (phthaloyl dichloride; phthalyl chloride) $C_6H_4(COCl)_2$.
Properties: Colorless, oily liquid; m.p. 16°C; b.p. 277°C; refractive index (n 20/D) 1.568. Decomposed by water or alcohol; soluble in ether. Combustible.
Derivation: By the action of phosphorus pentachloride on phthalic anhydride.
Hazard: Irritant by inhalation and skin contact.
Use: Chemical intermediate, especially for plasticizers and resins; laboratory reagent.
See also isophthaloyl chloride.

phycocolloid. One of several carbohydrate polymers (polysaccharides) occurring in algae (seaweed) (q.v.). They are hydrophilic colloids having a tendency to absorb water with swelling, and to form gels of varying strength and consistency. The chief types of phycocolloid are carrageenan from Irish moss, algin from brown algae (kelp), and agar from red algae. They contain complex galactose and mannose sugars, and are sometimes considered as seaweed mucilages. See specific entries for details.

"Phyllicin."[9] Trademark for theophylline-calcium salicylate.

physical chemistry. The science that concerns itself with application of the concepts and laws of physics to chemical phenomena in order to describe in quantitative (mathematical) terms a vast amount of qualitative (observational) information. A selection of only the most important concepts of physical chemistry would include: the electron wave equation and the quantum mechanical interpretation of atomic and molecular structure; the study of the subatomic fundamental particles of matter; application of thermodynamics to heats of formation of compounds and to heats of chemical reaction; the theory of rate processes and chemical equilibrium; orbital theory and chemical bonding; surface chemistry, including catalysis and finely divided particles; the principles of electrochemistry and ionization. Though physical chemistry is closely related to both inorganic and organic chemistry, it is considered a separate discipline.

physiological salt solution. A solution of sodium chloride and water (0.9%), which is identical with the concentration found in the body. Used in medicine to replace acute loss of water, as from burns, etc. Also called isotonic salt solution.

physostigmine (eserine; calabarine) $C_{15}H_{21}O_2N_3$. An alkaloid.
Properties: Colorless or pinkish crystals. Slightly soluble in water; soluble in alcohol and diluted acids. M.p. 86–87°C and 105–106°C (unstable and stable forms). Specific rotation −119 to −121°.
Derivation: By solvent extraction from the seeds of Physostigma venenosum.
Grade: U.S.P.
Hazard: Highly toxic by ingestion.
Use: Medicine. Available as the salicylate and sulfate.

phytane (2,6,10,14-tetramethylhexadecane) $C_{20}H_{42}$. A hydrocarbon. Found in rock specimens 2.5 to 3 billion years old. Is known to be synthesized only by living organisms (is a derivative of chlorophyll) and to withstand heat and pressure; so serves to date the existence of life on earth. See also pristane.

phytic acid (inositolhexaphosphoric acid) $C_6H_6[OPO(OH)_2]_6$. Occurs in nature in the seeds of many cereal grains, generally as the insoluble calcium-magnesium salt. It inhibits absorption of calcium in the intestine.
Properties: White to pale yellow liquid; odorless with

acid taste; pH less than 1.0 (in 1% solution); soluble in water and alcohol; sp. gr. 1.58; wt/gal 13.1 lb.
Derivation: From corn steep liquor.
Grade: Technical (as a 70% solution).
Uses: Chelation of heavy metals in processing of animal fats and vegetable oils; rust inhibitor; preparation of phytate salts; metal cleaning; treatment of hard water; nutrient.

phytochemistry. That branch of chemistry dealing with (1) plant growth and metabolism and (2) plant products. The former includes the absorption of inorganic nutrients (nitrogen, phosphorus, potassium, carbon dioxide, water, etc.) to form sugars, starches, proteins, fats, vitamins, etc., and is closely associated with photosynthesis (q.v.). Plant products comprise a vast group of natural materials and chemicals; besides those used directly as foods, these include alkaloids, cellulose, lignin, dyes, glucosides, essential oils, resins, gums, tannins, rubbers, terpene hydrocarbons, and glycerides (fats and oils). Some of these are basic raw materials for industry (paper, pharmaceutical, food, paint, perfume, flavoring, leather, rubber); there are also many miscellaneous plant products such as drugs, poisons, and vitamins. Phytochemistry also embraces the study of plant hormones or growth regulators (auxin, gibberellin, synthetic types).

phytol
$CH_3[CH(CH_3)CH_2CH_2CH_2]_3C(CH_3){:}CHCH_2OH$.
An alcohol obtained by the decomposition of chlorophyll.
Properties: Odorless liquid; b.p. 202–204°C (10 mm); sp. gr. 0.8497 (25/4°C); soluble in the common organic solvents; insoluble in water. Combustible; nontoxic.
Use: Synthesis of vitamins E and K.

phytonadione (2-methyl-3-phytyl-1,4-naphthoquinone; vitamin K_1) $CH_3C_{10}H_4O_2C_{20}H_{39}$.
Properties: Clear yellow, viscous, odorless liquid; sp. gr. 0.967 (25/25°C); refractive index 1.5230–1.5252 (25°C); stable in air. Protect from sunlight! Insoluble in water; soluble in benzene, chloroform and vegetable oils; slightly soluble in alcohol.
Derivation: Synthetically, from 2-methyl-1,4-naphthoquinone and phytol.
Grade: U.S.P.
Uses: Medicine; food supplement.

phytosterol. See sterol.

pi bond. A covalent bond formed between atoms by electrons moving in orbitals which extend above and below the plane of an organic molecule containing double bonds. A double bond consists of one pi and one sigma bond (q.v.) and a triple bond consists of one sigma and two pi bonds. See also metallocene; orbital theory.

"Picfume."[233] Trademark for fumigants containing chloropicrin.

pickle alum. See aluminum sulfate.

pickling (1) Removal of scale, oxides, and other impurities from metal surfaces by immersion in an inorganic acid, usually sulfuric, hydrochloric, or phosphoric. Rate of scale removal varies inversely with concentration and temperature; the usual concentration is 15%, at or above 212°F. The rate is also increased by electrolysis.

(2) A method of food preservation involving use of salt, sugar, spices, and organic acids (acetic).

(3) Preserving or preparing hides for tanning by immersion in a 6 to 12% salt solution, together with enough acid to maintain pH at 2.5 or less.

picloram (4-amino-3,5,6-trichloropicolinic acid). An herbicide used as defoliant in forest warfare. See also "Tordon." Claimed to leave toxic residue in soils. Toxic. Use may be restricted.

pico-. Prefix meaning 10^{-12} unit (symbol = p). 1 pg = 1 picogram = 10^{-12} gram.

alpha-**picoline** (2-methylpyridine; 2-picoline) $C_5H_4N(CH_3)$, or $NC(CH_3)CHCHCHCH$.

Properties: Colorless liquid; strong, unpleasant odor; sp. gr. 0.952; b.p. 129°C; f.p. −69.9°C; refractive index 1.4957 (n 20/D); miscible with water and alcohol. Flash point 102°F (O.C.). Autoignition temp. 1000°F. Combustible.
Derivation: Dry distillation of bones or coal.
Containers: Drums; tank cars.
Hazard: Moderate fire risk. Moderately toxic and irritant.
Uses: Organic intermediate for pharmaceuticals, dyes, rubber chemicals; solvent; source for vinyl pyridine; laboratory reagent.

beta-**picoline** (3-methylpyridine; 3-picoline)
$NCHC(CH_3)CHCHCH$.
Properties: Colorless liquid; unpleasant odor; b.p. 143.5°C; f.p. −18.3°C; sp. gr. 0.9613 (15/4°C); refractive index 1.5060 (n 20/D); soluble in water, alcohol, and ether. Combustible.
Derivation: Dry distillation of bones or coal.
Containers: Drums; tank cars.
Hazard: Moderate fire risk. Moderately toxic and irritant.
Uses: Solvent in synthesis of pharmaceuticals, resins, dyestuffs, rubber accelerators, insecticides; preparation of nicotinic acid, and nicotinic acid amide; waterproofing agents; laboratory reagent.

gamma-**picoline** (4-methylpyridine; 4-picoline)
$NCHCHC(CH_3)CHCH$.
Properties: Colorless, moderately volatile liquid; sp. gr. 0.957 (15/4°C); b.p. 144.9°C; refractive index 1.5050 (n 20/D); m.p. 3.8°C; soluble in water, alcohol, and ether. Flash point 134°F (O.C.). Combustible.
Derivation: Dry distillation of bones or coal.
Containers: Drums; tank cars.
Hazard: Moderately toxic and irritant. Moderate fire risk.
Uses: Solvent in synthesis of pharmaceuticals, resins, dyestuffs, rubber accelerators, pesticides, and waterproofing agents; laboratory reagent; making isoniazid; catalyst; curing agent.

picoline-N-oxide $N(O)C(CH_3)CHCHCHCH$ (formula for 2-picoline-N-oxide).
Properties: Crystals; very soluble in water. M.p. (2-isomer) 49.5°C; (3-isomer) 40.5°C; (4-isomer) 186.3°C. Combustible.
Use: Chemical synthesis.

4-picolylamine. $(CH_2NH_2)CHCHNCHCH$. A heterocyclic compound; a pyridine derivative. Highly reactive; strong base. Combustible.

Uses: Manufacture of polyamides, epoxy curing agents, carbinols, and amine polyols.

picramic acid (picraminic acid; 2-amino-4,6-dinitrophenol; dinitroaminophenol) $C_6H_2(NO_2)_2(NH_2)OH$.
Properties: Red crystals. Soluble in alcohol, benzene, glacial acetic acid, aniline, and ether; sparingly soluble in water. M.p. 168°C.
Derivation: By partial reduction of picric acid.
Hazard: May explode when shocked or heated. Dangerous fire risk.
Uses: Azo dyes; indicator; reagent for albumin.
Shipping regulations: (Rail) Consult regulations. (Air) Not listed. Consult authorities.

picramide. Legal label name (Air) for trinitroaniline (q.v.).

picraminic acid. See picramic acid.

picric acid (picronitric acid; trinitrophenol; nitroxanthic acid; carbazotic acid; phenoltrinitrate) $C_6H_2(NO_2)_3OH$.
Properties: Yellow crystals or liquid. Soluble in water, alcohol, chloroform, benzene, and ether. Very bitter taste. Sp. gr. 1.767; m.p. 122°C. B.p. explodes above 300°C. Flash point 302°F; autoignition temp. 572°F.
Derivation: Nitration of phenolsulfonic acid, obtained by heating phenol with concentrated sulfuric acid.
Grades: Technical paste; pure paste.
Containers: 1-, 5-lb bottles; 25-lb boxes; 100-lb kegs; 300-lb barrels.
Hazard: Severe explosion risk when shocked or heated; especially reactive with metals or metallic salts. Toxic by skin absorption. Tolerance, 0.1 mg per cubic meter of air.
Uses: Explosives; medicine (external); dyes; matches; electric batteries; etching copper; dyeing and printing textile fabrics with compound dyes which contain also such dyes as benzaldehyde green, methyl violet and indigo carmine; picrates.
Shipping regulations: (Rail) Consult regulations. (Air) Dry or wet with less than 10% water, Not acceptable. Wet with not less than 10% water, Flammable Solid label.

picrolonic acid $NO_2C_6H_4NNC(CH_3)C(NO_2)COH$. 3-Methyl-4-nitro-1-(para-nitrophenyl)-5-pyrazolone.
Properties: Yellow leaflets; m.p. 116–117°C; decomposes 125°C; slightly soluble in water; soluble in alcohol.
Uses: Reagent for alkaloid identifications, for tryptophan and phenylalanine; for the detection and estimation of calcium.

picronitric acid. See picric acid.

picrotoxin (cocculin) $C_{30}H_{34}O_{13}$. A glucoside.
Properties: Flexible shining, prismatic crystals or microcrystalline powder; odorless; very bitter taste; stable in air; affected by light; m.p. 200°C. Soluble in boiling water, boiling alcohol, diluted acids and alkalies; sparingly soluble in ether and chloroform.
Derivation: Derived from the fruit of Anamirta paniculata or cocculus indicus, fishberries.
Hazard: Highly toxic in overdose.
Use: Medicine, as CNS stimulant and antidote for barbiturate poisoning.
Shipping regulations: (Air) Cocculus, solid, Poison label.

picryl chloride (2-chloro-1,3,5-trinitrobenzene) $C_6H_2(NO_2)_3Cl$. A high explosive.
Hazard: Severe explosion and fire risk.
Shipping regulations: (Rail) Consult regulations. (Air) (wet with not less than 10% water) Flammable Solid label; (dry, or wet with less than 10% water) Not acceptable.

"Pictol."[329] Trademark for monomethyl para-aminophenol sulfate, photo-developer.

PIDA. Abbreviation for phenylindane dicarboxylic acid. See 1,1,3-trimethyl-5-carboxy-3-(p-carboxyphenyl) indane.

Pidgeon process (ferrosilicon process; silicothermic process). Process for the production of high-purity magnesium metal from dolomite or magnesium oxide by reduction with ferrosilicon at 1150°C under high vacuum.

piezochemistry. Study of reactions occurring at very high pressure, e.g., in interior of the earth's crust.

piezoelectricity. Opposite electric charges acquired by certain crystals on different surfaces as a result of mechanical stresses. Conversely, the property of expansion along one axis and contraction along another when subjected to an electric field.

pig iron. Product of blast-furnace reduction of iron oxide in presence of limestone. About half the ore is converted to iron. Average analysis is: 1% silicon, 0.03% sulfur, 0.27% phosphorus, 2.4% manganese, 4.6% carbon, balance iron. Pig iron is the basic raw material for steel (q.v.) and cast iron. In metal terminology, a "pig" is a bar or ingot of cooled metal.
See also iron.

pigment. Any substance, usually in the form of a dry powder, that imparts color to another substance or mixture. Most pigments are insoluble in organic solvents and water: exceptions are the natural organic pigments, such as chlorophyll, (q.v.) which are generally organosoluble. To qualify as a pigment, a material must have positive colorant value. This definition excludes whiting, barytes, clays, and talc (see fillers, extenders). Some pigments (zinc oxide, carbon black) are also reinforcing agents, but the two terms are not synonymous; in the parlance of the paint and rubber industries these distinctions are not always observed.
Pigments may be classified as follows:

I. Inorganic
 (a) metallic oxides (iron, titanium, zinc, cobalt chromium).
 (b) metal powder suspensions (gold, aluminum).
 (c) earth colors (siennas, ochers, umbers).
 (d) lead chromates

II. Organic
 (a) animal (rhodopsin, melanin).
 (b) vegetable (chlorophyll, xanthophyll, litmus, flavone, carotene). See also pigment, plant.
 (c) mineral (carbon black).
 (d) synthetic (phthalocyanine, lithols, toluidine, para red, toners, lakes, etc.).

See also dye, natural and synthetic.

"Pigmentar."[296] Trademark for tar products derived from the distillation and decomposition of oleoresinous southern pine. Produced to various viscosity grades. Used in rubber compounding and reclaiming, marine paints and roof coatings.

Pigment Blue 15 $C_{32}H_{16}N_8Cu$. A bright blue copper phthalocyanine pigment (q.v.). C.I. No. 74160.
Preparation: By heating phthalonitrile with cuprous chloride.
Uses: In paints; alkyd resin enamels; printing inks; lacquers; rubber; resins; papers; tinplate printing; colored chalks and pencils.

Pigment Blue 19 $C_{32}H_{28}N_3O_4SNa$. A bright blue to bright reddish navy triphenylmethane pigment (q.v.). C.I. No. 42750A.
Use: Coloring for candles.

Pigment Blue 24 $C_{37}H_{34}N_2O_9S_3Na_2$. A bright greenish blue triarylmethane pigment (q.v.). C.I. No. 42090.
Uses: In printing inks, especially for tinplate printing; in rubber; plastics; artist colors; lacquers.

pigment E. See barium potassium chromate pigment.

Pigment Green 7 $C_{32}O_{0-1}N_8Cl_{15-16}Cu$. A bright green chlorinated copper phthalocyanine pigment (q.v.). C.I. No. 74260.
Derivation: Heating copper phthalocyanine in sulfur dichloride under pressure.
Uses: Paints; printing inks; lacquers; leather and book cloth; paper surfacing; chalks; colored pencils.

pigment, plant. Any of a large number of organic natural colorants produced by living plants, with the exception of fungi and lichens. They may be classified into three groups; further information is given in specific entries.

(1) The chlorophylls (types a, b, and c): Green color. They are magnesium-containing porphyrins, and are technically considered to be microcrystalline waxes.

(2) The carotenoids: yellow and orange colors
 (a) Carotene (straight-chain hydrocarbon)
 (b) Xanthophyll (straight-chain hydrocarbon containing two oxygen molecules)

(3) The flavanoids: red, yellow, blue, orange, ivory colors. They are oxygen-containing heterocyclic compounds.
 (a) catechins
 (b) flavones, flavanols, anthocyanins
 (c) flavanones and leucoanthocyanidins
 (d) flavonols

Some of these pigments can be made synthetically. They have limited use as textile colorants and pharmaceutical products.

pigment, precipitated. See lake.

pigment volume concentration. See PVC (2).

Pigment Yellow 12 $C_{32}H_{26}Cl_2N_6O_4$. A yellow diazo pigment. C.I. No. 21090. See diazotization.
Preparation: Condensation of 3,3′-dichlorobenzidine di-diazotate with acetoacetanilide.
Uses: Printing inks; lacquers resistant to heat and solvents; in rubber and resins; in paper coloring, textile printing.

"Pilate" Fast Dyes.[440] Trademark for 1:1 metal complex dyes for dyeing and printing textiles of animal fibers and union materials of wool and nylon fibers.

pilchard oil. An oil expressed from the pilchard fish, a member of the herring family.
Properties: Pale yellow liquid; deposits stearin on long standing; sp. gr. 0.931–0.933; saponification value 186–189.6; refractive index 1.4751 (40°C). Combustible; nontoxic.
Uses: Potash soft soap; paints.

pilot plant. An operation intermediate between laboratory conditions (q.v.) and full-scale production. It is an essential step in chemial engineering development of a product. See also scale-up.

***l*-pimaric acid** (levopimaric acid) $C_{20}H_{30}O_2$.
Properties: Solid; m.p. 150°C; optical rotation [*a*] 20/D −280° (c = 0.7 in alcohol). Soluble in most organic solvents; insoluble in water. Combustible; low toxicity.
Derivation: From pine gum.
Use: Resins (see maleo-pimaric acid).

pimelic acid (1,7-heptanedioic acid)
 $HOOC[CH_2]_5COOH$.
Properties: Crystals; m.p. 105–106°C; slightly soluble in water; soluble in alcohol and ether; nearly insoluble in cold benzene. Combustible; low toxicity.
Derivation: From castor oil.
Use: Biochemical research; polymers; plasticizers.

pimelic ketone. See cyclohexanone.

pindone. Coined name for the insecticide 2-pivaloyl-1,3-indandione (q.v.).
Hazard: Highly toxic. Tolerance, 0.1 mg per cubic meter of air.
Shipping regulations: (Rail, Air) Poison label.

alpha-**pinene** $C_{10}H_{16}$. A terpene hydrocarbon derived from sulfate wood turpentine.
Properties: Colorless, transparent liquid; terpene odor; sp. gr. 0.8620–0.8645 (15.5/15.5°C); refractive index (n 20/D) 1.4655–1.4670; boiling range 95% between 156–160°C; f.p. −40°C; flash point 90°F (TCC); occurs in *d*-, *l*-, and racemic forms.
Containers: Tank cars and galvanized drums.
Hazard: Flammable, moderate fire risk.
Uses: Solvent for protective coatings, polishes, and waxes; synthesis of camphene, camphor, geraniol, terpin hydrate, terpineol, synthetic pine oil, terpene esters and ethers, lube oil additives, synthetic resins, and their derivatives; flavoring; odorant.

beta-**pinene** (nopinene) $C_{10}H_{16}$. A terpene hydrocarbon derived from sulfate wood turpentine.
Properties: Colorless, transparent liquid; terpene odor; sp. gr. 0.8740–0.8770 (15.5/15.5°C); refractive index (n 20/D) 1.4775–1.4790; boiling range 95% between 164–169°C; flash point 117°F; levorotatory. Combustible.
Containers: Tank cars and galvanized drums.
Hazard: Moderate fire risk.
Uses: Polyterpene resins; may be substituted for alpha-pinene; intermediate for perfumes and flavorings.

pinene hydrochloride. See bornyl chloride.

pine oil.
Properties: Colorless to light amber liquid having a strong, piny odor; miscible with alcohol in all proportions; sp. gr. 0.927–0.940; refractive index 1.4780–1.4820 (20°C) distilling range 200–225°C. Flash point 172°F (C.C.). Combustible; low toxicity.
Chief constituents: Tertiary and secondary terpene alcohols.
Derivation: From the wood of pinus palustris by extraction and fractionation, or by steam distillation; also from turpentine.
Containers: Bottles; drums.
Grade: N.F.
Uses: Odorant; disinfectant; penetrant; wetting agent; preservative (textile and paper industries); laboratory reagent.

pine tar
Properties: Sticky, viscous, dark brown to black liquid or semisolid with strong odor and sharp taste.

Translucent in thin layers; hardens with aging. Sp. gr. 1.03–1.07; boiling range 240–400°C. Flash point 130°F (C.C.). Low toxicity. Soluble in alcohol, acetone, fixed and volatile oils and in sodium hydroxide solution; slightly soluble in water. Combustible.
Chief constituents: Complex phenols; turpentine, rosin, toluene, xylene and other hydrocarbons.
Derivation: By destructive distillation of pine wood, especially Pinus palustris.
Grades: Kiln burnt; retort; N.F.
Containers: Tanks, drums, and barrels.
Hazard: Moderate fire risk. Subject to spontaneous heating.
Uses: Ore flotation; roofing compositions; paints and varnishes; softener in plastics and rubber processing; tar soaps; deKhotinsky cement; asphaltic compositions; marine preservative; medicine (cough syrups); laboratory reagent.

pine-tar oil. See tar oil, wood.

pine-tar pitch. The residue after distillation of practically all the volatile oils from pine tar. Similar to coal-tar pitch.

"Pipanol" Hydrochloride.[162] Trademark for trihexyphenidyl hydrochloride.

2-pipecoline. See 2-methylpiperidine.

alpha-**pipecoline.** See 2-methylpiperidine.

piperalin [3-(2-methylpiperidino)propyl-3,4-dichlorobenzoate] $CH_3C_5H_9N(CH_2)_3OC(O)C_6H_3Cl_2$.
Properties: Amber liquid; b.p. 156–157°C (0.2 mm). Slightly soluble in water; miscible in paraffin hydrocarbon, aromatic hydrocarbon, and chlorinated hydrocarbon solvents.
Hazard: May be toxic.
Use: Fungicide.

piperazidine. See piperazine.

piperazine (diethylenediamine; pyrazine hexahydride; piperazidine) $NHCH_2CH_2NHCH_2CH_2$.
Properties: Colorless, deliquescent, transparent, needle-like crystals, which absorb carbon dioxide from the air. Soluble in water, alcohol, glycerol, and glycols. M.p. 104–107°C; b.p. 145°C. Flash point 190°F. Combustible; low toxicity.
Derivation: Treatment of ethylene bromide or chloride with alcoholic ammonia at 100°C.
Containers: Glass bottles; drums.
Uses: Corrosion inhibitor; anthelmintic; insecticide. Accelerator for curing polychloroprene.

piperazine calcium edetate. USAN name for [dihydrogen(ethylenedinitrilo)tetraacetato]calcium piperazine salt; formerly piperazine calcium edathamil) $C_{14}H_{24}N_4O_8Ca$. A chelate compound produced by reacting EDTA with calcium carbonate and piperazine. Used in medicine.

piperazine dihydrochloride $C_4H_{10}N_2 \cdot 2HCl$.
Properties: White needles. Soluble in water.
Uses: Fibers; insecticides; pharmaceuticals. The monochloride, $C_4H_{10}N_2 \cdot HCl$, is also commercially available.

piperazine hexahydrate $C_4H_{10}N_2 \cdot 6H_2O$.
Properties: White crystals; m.p. 44°C. Soluble in water and alcohol.
Uses: Fibers; insecticides; pharmaceuticals; laboratory reagent.

"Pipersin."[188] Trademark for a substitute for oleoresin of black pepper. Officially recognized by USDA Meat Inspection Division for use under its supervision.

piperidine (hexahydropyridine; pentamethyleneamine) $CH_2CH_2CH_2CH_2CH_2NH$. Completely saturated ring compound.
Properties: Colorless liquid with odor of pepper; sp. gr. 0.862; b.p. 106°C; f.p. −7° to −9°C. Soluble in water, alcohol, and ether. It is a strong base. Combustible.
Derivation: By electrolytic reduction of pyridine.
Grades: 95% and 98% pure.
Containers: Drums.
Hazard: Strong irritant. Moderately toxic by ingestion.
Uses: Solvent and intermediate; curing agent for rubber and epoxy resins; catalyst for condensation reactions; ingredient in oils and fuels; complexing agent.

piperidine pentamethylene dithiocarbamate ("Pip-pip") $C_{11}H_{22}N_2S_2$.
Properties: White powder.
Hazard: Strong irritant to eyes and skin.
Use: Ultra-accelerator for rubber.

2-piperidinoethanol (N-2-hydroxyethylpiperidine) $C_5H_{10}NCH_2CH_2OH$.
Properties: Colorless liquid. Sp. gr. 0.972–0.974 (20/4°C); b.p. 115–117°C (45 mm); refractive index (n 20/D) 1.478–1.480. Miscible with water and most organic solvents in all proportions.
Use: Intermediate.

piperocaine hydrochloride $C_{16}H_{23}NO_2 \cdot HCl$. (3-(2-Methyl-1-piperidyl) propyl benzoate hydrochloride.
Properties: White crystalline powder; odorless and stable in air; m.p. 172–175°C; bitter taste. Solution (1 in 10) acid to litmus. Soluble in water, alcohol, and chloroform; almost insoluble in ether and fixed oils.
Use: Medicine.

piperonal (heliotropin; piperonyl aldehyde; 3,4-methylenedioxybenzaldehyde) $C_6H_3(CH_2OO)CHO$(bicyclic).
Properties: White, shining crystals; turns red-brown on exposure to light; floral odor; m.p. 35.5–37°C; b.p. 263°C; soluble in alcohol and ether; insoluble in water and glycerol. Combustible; low toxicity.
Derivation: By oxidation of isosafrole.
Grades: Technical; F.C.C.
Uses: Medicine; perfumery; suntan preparations; mosquito repellent; laboratory reagent; flavoring.

piperonyl butoxide. Generic name for alpha-[2-(2-butoxyethoxy)-ethoxy]-4,5-(methylenedioxy)-2-propyltoluene.
$C_3H_7C_6H_2(OCH_2O)CH_2OC_2H_4OC_2H_4OC_4H_9$.
Properties: Light-brown liquid; mild odor; insoluble in water; soluble in alcohol, benzene, petroleum hydrocarbons. Sp. gr. 1.06 (25°C); refractive index 1.50 (20°C); b.p. 180°C (1 mm). Flash point 340°F; combustible; low toxicity.
Containers: Glass bottles; tins; drums.
Use: Synergist in insecticides in combination with pyrethrins in oil solutions, emulsions, powders or aerosols; permissible animal food additive.

piperonyl cyclonene. Generic name for a mixture of 3-alkyl-6-carbethoxy-5-(3,4-methylenedioxyphenyl) 2-cyclohexen-1-one, and 3-alkyl-5-(3,4-methylenedioxyphenyl)-2-cyclohexen-1-one,
$(CH_2O_2)C_6H_3CHCH_2C(O)CH{:}CRCH_2$. R is usually C_6H_{13}.
Properties: Red liquid; insoluble in water, oils and

refrigerant 12. Flash point 290°F; combustible; low toxicity.
Use: Synergist in insecticides in combination with rotenone, pyrethrins or rotenone-pyrethrin mixtures in oil solutions, emulsions, or powders; permissible animal food additive.

piperylene (1,3-pentadiene) CH_2:CHCH:$CHCH_3$. Cis- and trans- forms.
Properties: Colorless liquid; sp. gr. 0.693 (60/60°F); f.p. cis −141°C, trans −87°C; b.p. cis −44°C, trans −42°C; refractive index (n 20/D) cis 1.43634; trans 1.43008. Insoluble in water; soluble in alcohol and ether. Flash point (mixture) −20°F.
Hazard: Highly flammable, dangerous fire risk.
Uses: Polymers; maleic anhydride adducts; intermediate.
Shipping regulations: Flammable liquid, n.o.s., (Rail) Red label. (Air) Flammable Liquid label.

"Pipron."[530] Trademark for piperalin; used for control of powdered mildew.

pitch. See coal-tar pitch; glance pitch; Burgundy pitch; fatty acid pitch; aslphalt; pine tar pitch.

pitchblende. A massive variety of uraninite (q.v.), or uranium oxide, found in metallic veins. Contains 55–75% UO_2; up to 30% UO_3; usually a little water, and varying amounts of other elements. Thorium and the rare earths are generally absent.
Properties: Color black; streak brownish black; luster pitchy to dull; Mohs hardness 5.5; sp. gr. 6.5–8.5.
Occurrence: Congo; Canada; Colorado; Europe.
Hazard: Radioactive material.
Uses: Most important ore of uranium; original source of radium.

"Pitocin."[330] Trademark for a sterile, aqueous solution of oxytocin injection.

"Pittabs."[177] Trademark for calcium hypochlorite (70% available chlorine) in tablet form.

"Pittchlor."[177] Trademark for granular calcium hypochlorite (70% available chlorine).

"Pitt-Consol."[545] Trademark for a group of phenols, cresols, cresylic acids, rubber chemicals, aryl mercaptans, and alkyl phenols.

pituitary. The extract obtained from the posterior lobe of the pituitary gland of hogs, sheep, etc.
Properties: Yellow or grayish amorphous powder with characteristic odor; partially soluble in water.
Grade: U.S.P.
Use: Medicine (hormone).

pivalic acid. See trimethylacetic acid.

2-pivaloyl-1,3-indandione (pivalyl-1,3-indandione; pindone) $C_9H_5O_2C(O)C(CH_3)_3$.
Properties: Bright yellow powder or crystals; m.p. 109°C. Insoluble in water; soluble in most organic solvents.
Hazard: Highly toxic by inhalation and ingestion. Tolerance, 0.1 mg per cubic meter of air.
Uses: Rodenticide; insecticide; pharmaceutical intermediate.
Shipping regulations: (Rail, Air) Poison label.

pK. A measurement of the completeness of an incomplete chemical reaction. It is defined as the negative logarithm (to the base 10) of the equilibrium constant, K, for the reaction in question. The pK is most frequently used to express the extent of dissociation or the strength of weak acids, particularly fatty acids, and amino acids, and also complex ions, or similar substances. The weaker an electrolyte the larger its pK. Thus, at 25°C for sulfuric acid (strong acid), pK is about −3.0; acetic acid (weak acid), pK = 4.76; boric acid (very weak acid), pK = 9.24. In a solution of a weak acid, if the concentration of undissociated acid is equal to the concentration of the anion of the acid, the pK will be equal to the pH (q.v.).

planetology, chemical. See chemical planetology.

plant. (1) Any of a broad group of living organisms comprised of all types of vegetation that contain chlorophyll (algae, mosses, grasses, vegetables, trees, etc., but excluding fungi). Their metabolic processes are vital to the maintenance of life on earth, and result in the following products: (1) oxygen (from respiration); (2) carbohydrates (from photosynthesis); (3) amino acids and proteins (from nitrates and nitrogen-fixing bacteria); (4) fats and oils; (5) vitamins; (6) natural fibers; (7) coal; (8) various other substances of value such as alkaloid drugs, rubber, etc. See also photosynthesis; phytochemistry.
(2) Any large-scale manufacturing unit, including pipelines, reaction equipment, machinery, etc.

plant growth regulator. An organic compound that modifies or controls one or more specific physiological processes within the plant. If the compound is produced by the plant it is called a plant hormone, e.g., auxin, which regulates the growth of longitudinal cells involved in bending of the stem one way or another. Substances applied externally also bring about modifications such as improved rooting of cuttings, increased rate of ripening (ethylene), and easier scission (separation of fruit from stem). A large number of chemicals tend to increase the yield of certain plants such as sugar cane, corn, etc. See dinitrobutylphenol; kinin; gibberellin; abscisin.

"Plaquenil" Sulfate.[162] Trademark for hydroxychloroquine sulfate.

"Plasdone."[307] Trademark for the pharmaceutical grade of polyvinylpyrrolidone.
Uses: Tablet binding and coating agent; detoxicant and demulcent lubricant in ophthalmic preparations; film forming agent in medical aerosols.

"Plaskon."[175] Trademark for plastics and resins including alkyd, urea, melamine, and nylon molding compounds; polyester, coating, foundry, bonding, impregnating, chlorotrifluoroethylene resins; adhesives; hardeners; phenolic laminating varnishes.
210-577A. A polyolefin designed for fabrication of tape for insulating computer cable; available in resin form or as specification product.
FR 1050. A flame-retardant polypropylene resin; continuous-use at 100°C; used as TV tube sockets, structural parts for appliances, etc.

"Plaslube."[539] Trademark for unreinforced thermoplastics with lubricating additives; includes nylons, polycarbonates, and acetals with various concentrations of molybdenum disulfide and/or "Teflon."

plasma. (1) The portion of the blood remaining after removal of the white and red cells and the platelets; it differs from serum in that it contains fibrinogen, which induces clotting by conversion into fibrin by activity of the enzyme thrombin. Plasma is made up of over 40 proteins, and also contains acids, lipids and metal ions. It is an amber, opalescent solution in which the proteins are in colloidal suspension and the solutes (electrolytes and nonelectrolytes) are either emulsified or in true solution. The proteins can be separated from each other and from the other solutes by ultrafiltration, ultracentrifugation, electrophoresis, and immunochemical techniques.

(2) The mixture of electrons and geseous ions formed when any substance is heated to the range of 10,000 to 50,000°F. A plasma may be formed by an electric arc of sufficient power, by sonic shock waves, or by other very sudden releases of very large quantities of energy, as in nuclear processes of fission or fusion. Uses are for spraying heat resistant coatings on missile cone surfaces and rocket nozzles, and high temperature research chemistry and physics. Hydrogen, deuterium, and tritium plasmas are involved in the development of fusion power, on which intensive research is in progress.

plasma volume expander. A substance used to partially or wholly replace blood plasma in treatment of the injured. Most important are gelatin, polyvinylpyrrolidone, and dextran.

plasmin. See fibrinolysin.

"Plastacele."[28] Trademark for cellulose acetate flake, a fine white powder used for molding powders, films, sheets, rods and tubes.
Containers: 50-lb paper bags.

"Plast-Alloy."[296] Trademark for ferro-alloy powder, such as ferroaluminum, ferrotitanium, etc. Used in sintered permanent magnets.

plaster of Paris. See calcium sulfate.

"Plastex."[160] Trademark for wires and cables with oilproof and flameproof polyvinyl chloride insulation which resists the action of oxygen, ozone, and sunlight; has high dielectric strength, high resistance to water, acids, and alkalies. It is firm, dense, and has a smooth finish. Supplied in several colors.

"Plastibase."[412] Trademark for plasticized hydrocarbon gel, a polyethylene and liquid petrolatum gel base.

plastic. (1) Capable of being shaped or molded, with or without the application of heat. Soft waxes and moist clay are good examples of this property. See also plasticity.

(2) A high polymer, usually synthetic, combined with other ingredients, such as curatives, fillers, reinforcing agents, colorants, plasticizers, etc.; the mixture can be formed or molded under heat and pressure in its raw state, and machined to high dimensional accuracy, trimmed and finished in its hardened state. The thermoplastic type can be resoftened to its original condition by heat; the thermosetting type cannot.

Plastics in general (including all forms) are sensitive to high temperatures, among the more resistant being fluorocarbon resins, nylon, phenolics, polyimides, and silicones, though even these soften or melt above 500°F. Other types are combustible when exposed to flame for a short time (polyethylene, acrylic polymers, polystyrene), and still others burn with evolution of toxic fumes (polyurethane).

Engineering plastics are those to which standard metal engineering equations can be applied; they are capable of sustaining high loads and stresses, and are machinable and dimensionally stable. They are used in construction, as machine parts, automobile components, etc. Among the more important are nylon, acetals, polycarbonates, ABS resins, PPO/styrene, and polybutylene terephthalate.

Fibers, films, and bristles are examples of extruded forms. Plastics may be shaped by either compression molding (direct pressure on solid material in a hydraulic press) or injection molding (ejection of a measured amount of material into a mold in liquid form). The latter is most generally used, and articles of considerable size can be produced. Because of their dielectric and non-toxic properties, plastics are essential components of electrical and electronic equipment (especially for use within the human body).

Plastics can be made into flexible and rigid foams by use of a blowing agent; they are light and strong, and the rigid type is machinable. These forms are collectively called cellular plastics (q.v.). Plastics can also be reinforced, usually with glass or metallic fibers for added strength. They are laminated to paper, cloth, wood, etc. for many uses in the packaging, electrical and furniture industries; they also can be metal-plated. Plastic pipe (q.v.) is widely used for underground transportation of gases and liquids over long distances, as well as intraplant.

Several natural materials (waxes, clays, and asphalts) have rheological properties similar to synthetic products, but as they are not polymeric, they are not considered true plastics. Certain proteins (casein, zein) are natural high polymers from which plastics are made (buttons, and other small items), but they are of decreasing importance.

Plastics have permeated industrial technology. Not only have they replaced and improved upon many materials formerly used, but also have made possible industrial and medical applications that would have been impracticable with older technologies. Use of plastics in the U.S. has been authoritatively estimated at close to 60 billion pounds a year by 1980, which is twice the 1970 consumption. Their major application areas are:

(1) automobile bodies and components; boat hulls
(2) building and construction (siding, piping, insulation, flooring)
(3) packaging (vapor-proof barriers; display cartons; bottles; drum linings)
(4) textiles (carpets, cordage, suiting, hosiery, drip-dry fabrics, etc.)
(5) organic coatings (paint and varnish vehicles)
(6) adhesives (plywood, reinforced plastics, laminated structures)
(7) pipelines
(8) electrical and electronic components.
(9) surgical implants
(10) miscellaneous (luggage, toys, tableware, brushes, furniture, etc.)

For additional information refer to Society of the Plastics Industry, 250 Park Ave., New York.

See also polymer, high; cellular plastic; reinforced plastic; foam, plastic; plastic pipe.

plastic flow. A type of rheological behavior in which a given material shows no deformation until the applied stress reaches a critical value, called the yield value. Most of the so-called plastics do not exhibit plastic flow. Common putty is an example of a material having plastic flow.

plastic foam. See foam, plastic; cellular plastic.

plastic, reinforced. See reinforced plastic.

plasticity. A rheological property of solid or semisolid materials expressed as the degree to which they will flow or deform under applied stress, and retain the shape so induced, either permanently or for a definite time interval. It may be considered the reverse of elasticity (q.v.). Application of heat and/or special additives is usually required for optimum results. See also thermoplastic; plasticizer.

plasticizer. An organic compound added to a high polymer both to facilitate processing and to increase the flexibility and toughness of the final product by internal modification (solvation) of the polymer molecule. The latter is held together by secondary valence bonds; the plasticizer replaces some of these with plasticizer-to-polymer bonds, thus aiding movement of the polymer chain segments. Plasticizers are classed as primary (high compatibility) and secondary (limited compatibility). Polyvinyl chloride and cellulose esters are the largest consumers of plasticizers; they are also used in rubber processing. Among the more important plasticizers are nonvolatile organic liquids and low-melting solids, e.g., phthalate, adipate and sebacate esters, polyols such as ethylene glycol and its derivatives, tricresyl phosphate, castor oil, etc. Camphor was used in the original modification of nitrocellulose to "Celluloid." See also plastisol; softener.

"Plasticone Red."[141] Trademark for light, medium, and deep shades of pyrazolone red pigments. Used in paints, enamels, lacquers, plastics, rubber, printing inks, textiles, and floor coverings.

plastic pipe. Tubes, cylinders, conduits and continuous length piping made (1) from thermoplastic polymers unreinforced (polyethylene, polyvinyl chloride, ABS polymers, polypropylene) or (2) from thermosetting polymers (polyesters, phenolics, epoxies) blended with 60–80% of such reinforcing materials as chopped asbestos or glass fibers to increase strength. The latter type is a reinforced plastic (q.v.). In general the properties of plastic tubing or pipe are those of the polymers that comprise it. Most have good resistance to chemicals, corrosion, weathering, etc., combined with flexibility, light weight, and high strength. They are combustible but generally slow-burning. The reinforced type is widely used as underground conduit for transportation of gases and fluids, including city water services, sewage disposal systems, etc. Its use in buildings is subject to local building codes. See Plastics Pipe Institute, 250 Park Avenue, New York, for further details.

"Plastic Steel."[445] Trademark for product composed of 80% steel and 20% plastic; used for repairing chemical and other equipment.

"Plastifix PC."[470] Trademark for a polychloroprene-based resin for corrosion-resistant, elastic metal coatings.

"Plast-Iron."[296] Trademark for high purity electrolytic iron powder and reduced iron oxide powder, annealed and unannealed.
Uses: Powder enrichment, catalyst, pole pieces, magnets, electronic cores, welding rod coatings, sintered structural parts and oil-less bearings.

plastisol. A dispersion of finely divided resin in a plasticizer. A typical composition is 100 parts resin and 50 parts plasticizer, forming a paste that gels when heated to about 300°F as a result of solvation of the resin particles by the plasticizer. If a volatile solvent is included, the plastisol is called an organosol (q.v.). Plastisols are used for molding thermoplastic resins, chiefly polyvinyl chloride. See also plasticizer.

"Plast-Manganese."[296] Trademark for electrolytic manganese powder. Used for welding rod coatings, pyrotechnics and fuses.

"Plast-Nickel."[296] Trademark for nickel powder. Used in welding rod coatings, sintered permanent magnets, filters and parts.

"Plasto."[243] Trademark for a series of liquid plasticizers used in lacquer coatings and PVC.

"Plastogen."[69,466] Trademark for a plasticizing agent.
Properties: Liquid, amber to mahogany; sp. gr. 0.81–0.84; acid number 1.0–1.1).
Uses: Plasticizer and softener in all elastomers; effective in sponge rubber.

"Plastolein."[242] Trademark for vinyl plasticizers, including di-n-hexyl azelate; diethylene glycol dipelargonate; diisooctyl azelate; di-2-ethylhexyl azelate; tetrahydrofurfuryl oleate; triethylene glycol dipelargonate.

"Plast-Silicon."[296] Trademark for silicon powder. Used in fuses and pyrotechnics.

platelet (thrombocyte). A proteinaceous cellular structure occurring in blood in the amount of $150–500 \times 10^3$ units per cubic millimeter. Platelets range from 2 to 4 microns in diameter, and contain no nuclei. They are rich in amine compounds which constrict the blood vessels at the site of an injury, to which the platelets adhere; on dissolution they release thromboplastin which initiates the coagulation mechanism. See also blood; fibrinogen; thrombin.

"Platfining."[416] Proprietary process for the treatment of hydrocarbon mixtures to remove deleterious materials such as sulfur and nitrogen compounds with no substantial decrease in the aromatic content, by treating the hydrocarbon mixtures with hydrogen in the presence of a solid catalyst containing platinum or platinum compounds.

"Platforming."[416] Proprietary process using special platinum-containing catalyst for making high-octane gasoline and/or a highly aromatic fraction for subsequent recovery of pure aromatics. Reactions include aromatization, dehydrogenation, cyclization, isomerization and hydrocracking. Reaction product may contain up to 60% aromatics. By-product hydrogen also is produced.

platinic ammonium chloride. See ammonium hexachloroplatinate.

platinic chloride. See chloroplatinic acid; platinum chloride.

platinic sal ammoniac. See ammonium hexachloroplatinate.

platinic sodium chloride. See sodium chloroplatinate.

platinic sulfate. See platinum sulfate.

platinous-ammonium chloride. See ammonium chloroplatinite.

platinous chloride. See platinum dichloride.

platinous iodide. See platinum iodide.

platinum Pt Metallic element of atomic number 78, group VIII of the periodic system. Atomic weight 195.09; valences 2, 4. There are 5 stable isotopes.

Properties: Silvery white ductile metal. Insoluble in mineral and organic acids; soluble in aqua regia. Attacked by fused alkalies. Sp. gr. 21.45; m.p. 1769°C; b.p. 3827 ± 100°C; Brinell hardness, 97; annealed (Vickers) 42. Nontoxic. Does not corrode or tarnish. Heated platinum absorbs large volumes of hydrogen. It is also a strong complexing agent.
Occurrence: Canada (Ontario), South Africa, U.S.S.R., Alaska. Usually mixed with ores of copper, nickel, etc.
Derivation: By dissolving the ore concentrate in aqua regia, precipitating the platinum by ammonium chloride as ammonium hexachloroplatinate, igniting the precipitate to form platinum sponge. This is then melted in an oxyhydrogen flame or in an electric furnace.
Grades: Physically pure (99.99%); chemically pure (99.9%); crucible platinum (99.5%); commercial (99.0%).
Forms available: Powder (platinum black); single crystals; wire (2″ by .05-.005 diam.); special compositions for electronics, metallizing, and decorating ceramics and metals.
Hazard: Flammable in powdered form. Soluble salts are toxic.
Uses: Catalyst (nitric acid, sulfuric acid, high-octane gasoline, automobile exhaust gas converters); laboratory ware; spinnerets for rayon and glass fiber manufacture; jewelry; dentistry; electrical contacts; thermocouples; surgical wire; bushings; electroplating; electric furnace windings; chemical reaction vessels; permanent magnets.

platinum-ammonium chloride. See ammonium hexachloroplatinate and ammonium chloroplatinate.

platinum barium cyanide. See barium cyanoplatinite.

platinum black. Finely divided metallic platinum.
Properties: Black powder; exhibits a metallic luster when rubbed. Soluble in aqua regia. Sp. gr. 15.8–17.6 (apparent).
Derivation: Reduction of solution of a platinum salt with zinc or magnesium.
Hazard: Flammable when dispersed in air.
Uses: Catalyst; absorbent of gases (hydrogen, oxygen, etc.) which it again liberates at red-heat; gas ignition apparatus.

platinum chloride (platinum tetrachloride; platinic chloride) (a) $PtCl_4$; (b) $PtCl_4 \cdot 5H_2O$. The platinum chloride of commerce is usually chloroplatinic acid (q.v.).
Properties: (a) Brown solid; (b) red crystals. Soluble in water and alcohol. (a) Sp. gr. 4.30 (25°C); decomposes at 370°C. (b) Sp. gr. 2.43; m.p., loses $4H_2O$ at 100°C.
Derivation: By solution of platinum in aqua regia and evaporation.
Hazard: Toxic. Tolerance (as Pt), 0.002 mg per cubic meter of air.
Use: Chemical reagent.

platinum-cobalt alloy. A 76.7 Pt/23.3 Co alloy forms a more powerful permanent magnet than any other known.

platinum dichloride (platinous chloride) $PtCl_2$.
Properties: Greenish gray powder which forms double salts with the chlorides of the alkali metals. Soluble in hydrochloric acid and ammonium hydroxide; insoluble in water. Sp. gr. 5.87; m.p., is decomposed at red heat, yielding platinum.
Derivation: (a) By heating platinum sponge in presence of dry chlorine; (b) by heating chloroplatinic acid to 200°C.
Hazard: May be toxic.
Use: Platinum salts.

platinum iodide (platinous iodide; platinum diiodide) PtI_2.
Properties: Heavy, black powder. Slightly soluble in hydriodic acid; insoluble in alkalies, water. Sp. gr. 6.4; m.p. 300–350°C (dec).

platinum-iridium alloy. The most important platinum alloy. Commercial alloys contain 1–30% iridium. As the iridium is increased the hardness of the alloy increases, as does the resistance to chemical attack. The m.p. of platinum is raised by the addition of iridium.
Uses: Jewelry ("medium" platinum is 95% Pt, 5% Ir and "hard" platinum is 90% Pt, 10% Ir); electrical contacts (10–25% Ir); fuse wire (10–20% Ir), hypodermic needles (20–30% Ir), and in general where high corrosion resistance is needed.
See also iridium.

platinum-lithium $LiPt_2$. Brittle, solid, nonreactive with water, made by direct combination at 540°C. If the lithium and platinum are combined at 200°C, the product can be decomposed by water, hydrolyzing and dissolving the lithium, and leaving unusually active platinum catalyst.

platinum metal. Any of a group of six metals, all members of group VIII of the periodic system: ruthenium, rhodium, palladium, osmium, iridium, and platinum. All of these are also transition metals.

platinum potassium chloride. See potassium chloroplatinate.

platinum-rhodium alloy. An alloy containing up to 40% rhodium, balance platinum. Such alloys are harder than platinum, but not as hard as the corresponding platinum-iridium alloys. The addition of rhodium to platinum increases the resistance to attack by aqua regia. The melting points of the alloys are higher than that of platinum.
Uses: Catalyst in nitric acid production; high temperature vessels; furnace resistors; thermocouples and resistance thermometers; spinneret nozzles.

platinum sodium chloride. See sodium chloroplatinate.

platinum sponge.
Properties: Grayish black, porous mass of finely divided platinum. Soluble in aqua regia.
Derivation: By ignition of ammonium hexachloroplatinate, or other salts.
Uses: Catalyst; ignition of hydrogen.
See also platinum black.

platinum sulfate (platinic sulfate) $Pt(SO_4)_2$.
Properties: Greenish black mass. Hygroscopic. Soluble in acids (dilute), alcohol, ether, water.
Use: Analysis (microtesting for bromine, chlorine, iodine).

platinum tetrachloride. See platinum chloride.

"Platreating."[416] Proprietary process for treatment of hydrocarbon mixtures without cracking to remove deleterious materials such as sulfur and nitrogen compounds and/or to effect an increase in the hydrogen content of such mixtures. The catalyst used contains platinum or platinum compounds. Commercial application has been directed to the cleanup of aromatic-rich streams for the production of high-purity benzene, toluene and xylene.

"Plexiglas."[23] Trademark for thermoplastic poly(methyl methacrylate)-type polymers. Available in bead or granule form and sheets.

Uses: Manufacture of lenses, ornaments, letters for signs, aircraft canopies and windows, light diffusers, industrial and architectural glazing, chalkboards, boat windshields, and similar products.

"Plexol."[23] Trademark for synthetic lubricants and additives for petroleum oils. Most grades are diesters of dibasic acids, some are polyesters or polyether alcohols. The ester lubricants have very low freezing points, high flash points, little change of viscosity with temperature.
Uses: Aircraft engine lubricants, hydraulic systems, instrument oils, petroleum-base lubricant formulation.

"Pliobond."[265] Trademark for general-purpose solvent-type thermoplastic adhesives which form good bonds with most like and unlike surfaces. Adheres well to wood glass, ceramics, metals, many plastics, leather, rubber, concrete, and plaster. May be employed as a wet adhesive, by solvent reactivation, or by the use of heat.

"Plioflex."[265] Trademark for a series of staining and nonstaining synthetic dry elastomers produced by emulsion polymerization. These rubbers are either styrene/butadiene or polybutadiene polymers. All are protected with antioxidants.

"Pliogrip."[265] A two-component adhesive for glass-reinforced plastics; it forms a highly cross-linked polymer on setting.

"Pliolite" Latices.[265] Trademark for a series of styrene/butadiene synthetic latices.

"Pliolite" Rubber Reinforcing Resins.[265] Trademark for a series of high-styrene/butadiene resins that provide effective reinforcement for SBR, NBR, neoprene and natural rubbers. Specifically developed for electrical applications.

"Pliolite" Solution Resins.[265] Trademark for a series of hard, thermoplastic resins for use in paint finishes, paper and packaging coatings, printing inks and adhesives.

"Pliopave."[265] Trademark for a series of rubber latices used specifically as modifiers of bituminous paving materials. The series is composed of polymers of styrene/butadiene, acrylonitrite/butadiene and others.

"Pliovic" Vinyl Resins.[265] Trademark for a group of thermoplastic resins composed of polymers and copolymers consisting of 50% or more vinyl chloride. Supplied in the form of fine white powders, the resins are easily compounded and formed into finished goods by extruding, calendering, compression molding, injection molding and blow molding.

plumbic acid, anhydrous. See lead dioxide.

plumbo-plumbic oxide. See lead oxide, red.

"Plumb-O-Sil" B and C.[304] Trademark for coprecipitates of lead orthosilicate and silica gel.
Properties: Soft white powders; sp. gr. (B) 3.9, (C) 3.1; refractive index, (B) 1.58–1.60, (C) 1.58.
Containers: Up to 150-lb fiberboard drums.
Uses: Translucent and colored vinyl film, sheeting, and upholstery stocks; as vinyl stabilizers.

plumbous oxide. See litharge.

plumbous sulfide. See lead sulfide.

plumbum. The Latin name for lead, hence the symbol Pb and the names plumbic and plumbous.

"Pluracol."[203] Trademark for a series of organic compounds used in hydraulic brake and other functional fluids; chemical intermediates; urethane foams, elastomers and coatings.

"Plurafac."[203] Trademark for a series of 100% active, nonionic biodegradable surfactants of straight chain, primary aliphatic, oxyethylated alcohols designed through advanced synthesis techniques. Available in liquid, paste, flake, and solid form.
Uses: Range from light-duty hand dishwashing formulations to heavy-duty industrial detergents, rinse aids, metal cleaners, etc.

"Pluronic."[203] Trademark for a nonionic series of 28 related difunctional block-polymers terminating in primary hydroxyl groups with molecular weights ranging from 1,000 to over 15,000. They are polyoxyalkylene derivatives of propylene glycol. Available in liquid, paste, flake, powder and cast-solid forms.
Uses: Defoaming agents, emulsifying and demulsifying agents, binders, stabilizers, dispersing agents, wetting agents, rinse aids, and chemical intermediates.

plutonium Pu Synthetic radioactive metallic element with atomic number 94; first prepared in 1941. Atomic weight 239.11. Valences 3, 4, 5, 6. Fifteen isotopes (from 232 to 246); six allotropic forms. Plutonium 239 (half-life 24,360 years) is produced in a nuclear reactor by neutron bombardment of the nonfissionable isotope U-238 according to the reaction:

$$U^{238}(n, \gamma)\ U^{239} \xrightarrow{\beta^-} Np^{239} \xrightarrow{\beta^-} Pu^{239}$$

Plutonium is readily fissionable with both slow and fast neutrons; though it can be used for nuclear weaponry, its most important application is as a controlled energy source for making electric power. One pound contains a heat energy equivalent of 10^6 kilowatt-hours. According to Glenn T. Seaborg, "in breeder reactors it is possible to create more new plutonium from U-238 than the plutonium consumed in sustaining the fission chain reaction. Because of this, plutonium is the key to unlocking the enormous energy reserves in the nonfissionable U-238." Reactor fuels containing plutonium can be either liquid or solid; since Pu forms low-melting alloys with a number of metals (gallium, bismuth, tin, iron, cobalt and nickel) these are often used as liquid reactor fuels. Cerium also may be a component.
Hazard: The most toxic of the elements and one of the most toxic substances known; dangerous ionizing radiation persists indefinitely; a powerful carcinogen. Must be handled by remote control and with adequate shielding.
See also breeder.

plutonium 242. An isotope of plutonium now available.
Hazard: Radioactive poison.
Uses: Tracer techniques; analytical chemistry.

"Plyac."[50] Trademark for polyethylene spreader-sticker; nonoily and nonionic.

"Plyacien."[36] Trademark for a protein base dust-free adhesive for use with the cold press no-clamp gluing process.
Use: Manufacture of interior-grade plywood.

"Plyamine."[36] Trademark for a group of liquid water-soluble urea-formaldehyde adhesive resins used as binders in the manufacture of plywood, furniture, wood particle products, etc.

"Plyamul."[36] Trademark for polyvinyl acetate adhesive bases.

"Plyophen."[36] Trademark for a water-soluble impregnating resin. Penetrates deeply and quickly into wood, canvas, asbestos, paper and other laminating and molding stocks. Can be diluted as much as 8–10 parts water to 1 part resin for spraying glass fiber or rock wool.

plywood. A composite composed of thin wood veneers (with grains placed at right angles to each other) bonded with a synthetic resin, usually phenol-formaldehyde or resorcinol-formaldehyde. It is superior to metals in strength-to-weight ratio, and has low thermal expansion, high heat capacity, and low water absorption. See also laminate; composite.

Pm Symbol for promethium.

PMA. Abbreviation for phosphomolybdic acid and for pyromellitic acid.

"PMA."[74] Trademark for a series of fungicides containing phenyl mercury acetate. Used in emulsion paint and other aqueous systems.
Hazard: See mercury compounds.

"PMD-10."[74] Trademark for a mineral spirits solution of phenyl mercury oleate. Used for mildew-resistant oil, oleoresinous and alkyd paints.
Hazard: See mercury compounds.

PMDA. Abbreviation for pyromellitic dianhydride (q.v.).

PMHP. Abbreviation for para-menthane hydroperoxide. A polymerization catalyst.

PMP. Abbreviation for 1-phenyl-3-methyl-5-pyrazolone (q.v.).

PMTA. Abbreviation for a mixture of phosphomolybdic and phosphotungstic acids, used in making pigments. See phosphotungstic pigment.

pn Abbreviation for propylenediamine, as used in formulas for coordination compounds. See also dien; en; py.

PNC. Abbreviation for phosphonitrilic chloride.

"PNF."[278] Trademark for a special-purpose synthetic rubber, a phosphonitrilic fluoroelastomer, said to be flexible at −70°F and serviceable up to 350°F. Resistant to oils over a wide temperature range. Can be processed on standard equipment.

Po Symbol for polonium.

"POCO.'[486] Trademark for a series of unique fine-grained, high strength, isotropic, formed graphite materials. Three basic density grades range between 1.50 and 1.88 gms/cm^3; the associated flexural strengths range from 4000 to 17,250 psi.
Uses: Jigs and fixtures for electronic components; electrodes for spectrographic use; electrical discharge machining; aerospace and nuclear fields.

POEMS. Abbreviation for polyoxyethylene monostearate; see polyoxyl 40 stearate.

POEOP. See polyoxyethyleneoxypropylene.

POF. See *dl*-alpha-lipoic acid.

poise. See centipoise.

poison. (1) Any substance that is harmful to living tissues when applied in relatively small doses. The most important factors involved in effective dosage are (a) quantity or concentration, (b) duration of exposure, (c) particle size or physical state of the substance, (d) its affinity for living tissue, (e) its solubility in tissue fluids, and (f) the sensitivity of the tissues or organs. Sharp distinction between poisons and non-poisons is not always possible, as many variables must be taken into consideration in each case. Poisons are divided into four classes by the shipping regulatory agencies as follows. Poison A: a gas or liquid so toxic that an extremely low percentage of the gas or the vapor formed by the liquid is dangerous to life. Poison B: Less toxic liquids and solids that are hazardous either by contact with the body (skin absorption) or by ingestion. Poison C: Liquids or solids that evolve toxic or strongly irritating fumes when heated or when exposed to air (excluding Class A poisons). Poison D: Radioactive materials.
See also toxicity; toxic substances.
Note: A computerized poison information center is operated by the FDA in Washington, D.C. The national clearinghouse for poison information centers is located at 5401 Westbard Ave., Bethesda, Md., 20016.

(2) In nuclear technology, any material with a high capture probability for neutrons that may divert an undesirable number of neutrons from the fission chain reaction.

(3) A substance that reduces or destroys the activity of a catalyst. Carbon monoxide, and phosphorus, arsenic, or sulfur compounds have this effect on the formation of ammonia from hydrogen and nitrogen gases, and the gases must be highly purified to avoid this. Another example is the poisoning of the platinum catalysts used in emission-control devices by organic lead compounds.

poison gas. A toxic or irritant gas or volatile liquid designed for use in chemical warfare or riot control. They vary in toxicity from nerve gases (q.v.), which are lethal, to tear gases (lachrymators) which cause only temporary disability. See also noxious gas; chemical warfare.

polar. Descriptive of a molecule in which the positive and negative electrical charges are permanently separated, as opposed to nonpolar molecules in which the charges coincide. Polar molecules ionize in solution and impart electrical conductivity. Water, alcohol, and sulfuric acid are polar in nature; most hydrocarbon liquids are not. Carboxyl and hydroxyl groups often exhibit an electric charge. The formation of emulsions and the action of detergents are dependent on this behavior. See also dipole moment.

polarimetry. Measurement of the degrees and direction of the plane of polarized light as it passes through an optically active compound. It is used in the investigation of optical isomers, especially in analysis of sugars.

"Polaris Red."[141] Trademark for precipitated azo pigments of a very bright, blue shade of red derived from beta-hydroxynaphthoic acid.
Uses: Printing inks, rubber, plastics.

polarized light. See nicol; polarimetry.

"Polectron."[307] Trademark for modified vinylpyrrolidone resins.
Properties: 40% active aqueous emulsion. Stable to intense mechanical sheer, freeze-thaw cycling. Compatible with other commercial latexes.
Uses: Binding agents for wood, cotton, paper, glass

fibers; stabilizer, opacifier, and dyestuff; medium precoat for photosensitive papers. Adhesive for metal, glass, cotton, paper and wood.

"Policarb."[155] Trademark for carbonized nickel in which one side is a "Duocarb" finish and the other side is "Radiocarb" finish.

"Polidase."[91] Trademark for cultured vegetable enzyme product.
Uses: Laboratory reagent; experimental nutrition; recovery of silver from photographic film; component of leather bates; conversion of starch in mashes used for alcohol production; low cost industrial enzyme.

"Polidene."[263] Trademark for a series of polyvinylidene chloride copolymer emulsions. Used in pigmented coating for textiles, paper, leather, etc.; fire retardant finishes.

polish. (1) A solid powder or a liquid or semi-liquid mixture that imparts smoothness, surface protection, or a decorative finish. The most widely used solid polishing agent is fine-ground red iron oxide (rouge), applied to the surface of plate glass, backs of mirrors, and optical glass. A wide variety of liquid and paste-like polishes are based on vegetable waxes (carnauba and candelilla), combined with softeners, fillers, and pigments or emulsified in alcohol or other solvent. Furniture polishes often contain red oil, lemon oil and petroleum solvent; most types of metal and wood polish contain organic solvents, and hence are flammable liquids. Nail polishes are nitrocellulose lacquers, usually with amyl acetate solvent. See also electro-polishing.
Hazard: (metal, furniture): May be toxic and flammable.

(2) The hard outer coating of cereal grains, especially rice, which is usually removed in processing. These coatings are rich in vitamin B_1. Their removal robs the cereal of much of its nutritive value.

"Politol."[228] Trademark for high molecular weight acidic wood products which are known to precipitate protein and similar substances.
Uses: Purifying and refining aid for fats and oils.

pollucite $Cs_4Al_4Si_9O_{26} \cdot H_2O$. A natural cesium aluminum silicate found in pegmatites. Colorless; Mohs hardness 6.5; sp. gr. 2.9.
Uses: Source of cesium; catalyst; fluxes; welding materials; ion propulsion; thermocouple units.

pollution. Introduction into any environment of substances that are not normally present therein and that are potentially toxic or otherwise objectionable. The most serious atmospheric contaminants have been (a) sulfur dioxide evolved from the fuels used in electric power production and industrial processing, and (b) automobile exhaust gases rich in carbon monoxide and tetraethyllead residues. The former is being alleviated by mandatory use of low-sulfur fuels, and the latter by eliminaton of tetraethyllead from high-octane gasoline and by use of catalytic converters.

Water pollution due to discharge of toxic chemical wastes is closely regulated by both EPA and FDA. Such substances are defined in the 1972 amendment of the Federal Water Pollution Control Act as those "which will cause death, disease, cancer, or genetic malfunctions in any organisms with which they come into contact." Substances added to water for purification purposes, (chlorine, aluminum sulfate, etc.) are excluded from the category of pollutants.
See also Environmental Protection Agency; air pollution; water pollution.

polonium Po Radioactive element of atomic number 84, member of Group VIA of the periodic Table. Atomic weight 210; valences 2, 4, 6; there are no stable isotopes. Polonium is a member of the uranium natural radioactive decay series, occurring naturally in uranium-bearing ores; it is produced artificially by bombarding bismuth with neutrons. It has been identified in cigarette smoke (see smoke, 4).
Properties: Similar to those of tellurium. M.p. 254°C; b.p. 962°C; sp. gr. 9.4. Dissolved by concentrated sulfuric and nitric acids and aqua regia, and by dilute hydrochloric acid.
Hazard: Dangerous radioactive poison.
Uses: Source of alpha radiation and neutrons; instrument calibration; oil-well logging; moisture determination; power source.
Shipping regulations: (Rail, Air) Consult regulations.

poly-. A prefix signifying many. For example, a polymer is an aggregate formed by combination of a number of single molecules. See polymer, high.

"Polyac."[28] Trademark for a butyl rubber conditioner containing 25% poly-para-dinitrosobenzene $[C_6H_4(NO)_2]_x$ with an inert wax. Dark brown waxy pellets; sp. gr. 0.96. Used as a processing aid and accelerator of vulcanization for butyl rubber.

polyacetal. See acetal resin.

polyacrylamide $(CH_2CHCONH_2)_x$. White solid; water-soluble high polymer.
Derivation: Polymerization of acrylamide with N,N′-methylene bisacrylamide.
Uses: Thickening agent; suspending agent; production of uranium; additive to adhesives. Permissible food additive.
See also acrylic resin.

polyacrylate. See acrylic resin.

polyacrylic acid. See acrylic acid; methacrylic acid.

polyacrylonitrile. See acrylonitrile; acrylic resin; acrylic fiber.

polyalcohol. See polyol.

"Polyall."[175] Trademark for a series of alkyd resin-based thermosetting compounds.
Properties: Good electrical resistance; high strength; flame resistance; good dimensional stability; very fast cure; resistance to solvents and weak acids; good colorfastness in sunlight and heat; fungus-resistant; sp. gr. 2.1; heat distortion temperature, 400°F.
Available in long glass fiber-filled and mineral glass-filled grades.

polyallomer. A copolymer which has a uniform crystal structure but a mixed chemical composition. It is prepared by anionic coordination catalysis using a Ziegler catalyst (q.v.). The best-known polyallomer is a copolymer of propylene and ethylene; it has the stereoregular crystalline structure of the homopolymers of these resins, but a variable chemical composition. Temperature range −40 to 210°F. The physical properties of polyallomers are generally intermediate between those of the homopolymers of the component resins, but give a better balance of properties than blends of the homopolymers.
Uses: Vacuum-formed, injection-molded, blow-molded, and extruded products; film, sheeting, wire cables.

polyamide. A high molecular weight polymer in which amide linkages (CONH) occur along the molecular chain. They may be either natural or synthetic. Important natural polyamides are casein, soybean and peanut proteins, and zein, the protein of corn, from all of which plastics, textile fibers and adhesive compositions can be made. Synthetic polyamides are typified by the numerous varieties of nylon, though some, e.g., "Versamide," are quite different from nylon. See also nylon; polypeptide; protein; aramid.

polyamine-methylene resin. A polyethylene polyamine methylene substituted resin of diphenylol dimethylmethane and formaldehyde in basic form.
Properties: Light amber, granular, freely flowing powder with appreciable odor. Insoluble in dilute acids and alkalies, alcohol, ether, and water.
Use: Medicine; ion-exchange resin.

polyaminotriazole (PAT). A synthetic polymer made from sebacic acid and hydrazine with small amounts of acetamide. Polyoctamethylene-aminotriazole is a specific example.
See monobasic.

"Poly B-D" Liquid Resins.[222] Trademark for low molecular weight liquid polymers based on butadiene containing controlled hydroxyl functionality. They consist of both homo- and terpolymers. Designed for general rubber products, coatings, adhesives, etc.

polybenzimidazole (PBI) $(C_7H_6N_2)_n$. A synthetic polymer designed for high-temperature space technology applications. Reputed to withstand temperatures up to 500°F for 1000 hours.
Derivation: Condensation of diphenyl isophthalate and 3,3′-diaminobenzidine.
Uses: Fibers; composites; adhesives (high adhesion to steel, titanium, beryllium and aluminum alloys); coatings; ablative materials.

polyblend. A mixture in any proportion of either (1) two homopolymers (natural or synthetic), (2) a homopolymer and a copolymer, or (3) two copolymers. An example of (1) is rubber-polystyrene, of (2) is rubber and butadiene-styrene, and of (3) is a mixture of butadiene-acrylonitrile and isobutylene-isoprene. A polyblend is a mixture that is made after its components have been polymerized, and thus is different from a copolymer, which is made by chemical combination of two monomers. See also homopolymer; copolymer.

polybutadiene. A synthetic thermoplastic polymer made by polymerizing 1,3-butadiene (q.v.) with a stereospecific organometallic catalyst (butyl lithium), though other catalysts such as titanium tetrachloride and aluminum iodide may be used. The *cis*-isomer, which is similar to natural rubber, is used in tire treads due to its abrasion and crack resistance and low heat build-up. Large quantities are also used as blends in SBR rubber (q.v.). The *trans*-isomer resembles gutta percha and has limited utility. Liquid polybutadiene, which is sodium-catalyzed, has specialty uses as a coating resin. It is cured with organic peroxides. Combustible; the liquid form is probably toxic by ingestion and inhalation, as well as a skin irritant. See also polymer, stereospecific.

polybutene (polybutylene; polyisobutylene; polyisobutene). Any of several thermoplastic isotactic (stereoregular) polymers of isobutene of varying molecular weight; also polymers of butene-1 and butene-2. Butyl rubber (q.v.) is a type of polyisobutene to which has been added about 2% of isoprene, which provides sulfur linkage sites for vulcanization. Isobutene can be homopolymerized to various degrees in chains containing from 10 to 1000 units, the viscosity increasing with molecular weight. It is combustible and essentially nontoxic.
See also "Vistanex."
Uses: Lubricating-oil additive; hot-melt adhesives; sealing tapes; special sealants; cable insulation; polymer modifier; viscosity index improvers.

polybutylene terephthalate. An engineering plastic derived from 1,4-butanediol.

"Polycarbafil."[539] Trademark for a glass fiber-reinforced polycarbonate (q.v.).

polycarbonate. A synthetic thermoplastic resin derived from bisphenol A and phosgene; a linear polyester of carbonic acid: $(COOC_6H_5C(CH_3)_2C_6H_5O)_n$. Can be formed from any dihydroxy compound and any carbonate diester, or by ester interchange. Polymerization may be in aqueous emulsion or in nonaqueous solution.
Properties: Transparent (90% light transmission); noncorrosive; weather and ozone-resistant; nontoxic; stain-resistant; combustible but self-extinguishing; low water absorption; high impact strength; heat-resistant; high dielectric strength; dimensionally stable; soluble in chlorinated hydrocarbons and attacked by strong alkalies and aromatic hydrocarbons; stable to mineral acids; insoluble in aliphatic alcohols. Excellent for all molding methods, extrusion, thermoforming etc.; easily fabricated by all methods including thermoforming and fluidized bed coating.
Uses: Molded products; solution-cast or extruded film; structural parts; tubes and piping; prosthetic devices; meter face plates; nonbreakable windows; street-light globes; household appliances; bicycles.

polycarboxylic acid. An organic acid containing two or more carboxyl (COOH) groups.

polychlor. General name for synthetic chlorinated hydrocarbons used as pesticides.

polychlorinated biphenyl (PCB). One of several aromatic compounds containing two benzene nuclei with two or more substituent chlorine atoms. They are colorless liquids with sp. gr. of 1.4 to 1.5. They are highly toxic. Their chief use is in heat-exchange and insulating fluids in closed systems. FDA has prohibited their use in plants manufacturing foods, animal feeds, and food-packaging materials, and has established strict tolerances on their presence in many food products (2.5 ppm).
Note: Because of their persistance and ecological damage from water pollution their manufacture has been discontinued in the U.S. (1976).

polychloroethylenesulfonyl chloride. See polyethylene, chlorosulfonated.

polychloroprene. See neoprene.

polychlorotrifluoroethylene (PCTFE). See chlorotrifluoroethylene resin.

"Polycin."[202] Trademark for (1) an elastic, tacky, gel-like solid resulting from the polymerization of castor oil; used in rubber compounding, floor tile manufacture and as a polymeric plasticizer; (2) a series of polyols used in the preparation and curing of urethane polymers for protective coatings, foamed insulation and elastomers.

"Polyclar."[307] Trademark for a series of beverage grades of polyvinylpyrrolidone. Used as a clarifier; stabilizer for flavor and color.

"Polyco."[65] Trademark for a series of thermoplastic polymers in the form of water emulsions or solvent solutions; applied to vinyl acetate polymers and copolymers, butadiene-styrene copolymer latics, polystyrenes, vinyl and vinylidene chloride copolymers, acrylic copolymers, and water-soluble polyacrylates.
Uses: Adhesives and coatings, in paint, leather, textiles, paper, cosmetics and construction fields.

polycondensation. See condensation (1); polymerization.

polycoumarone resin. See coumarone-indene resin.

"Polycryl."[65] Trademark for a series of acrylic polymers and copolymers polymerized in solvent medium and supplied as solutions in toluene, methyl ethyl ketone or other solvent.

polycyclic. An organic compound having four or more ring structures, which may be the same or different. See naphthacene; polynuclear.

poly(1-4-cyclohexylenedimethylene)terephthalate. A linear polyester film or fiber obtained by condensation of terephthalic acid with 1,4-cyclohexanedimethanol. It has good electrical resistivity and hydrolytic stability. Used principally for carpet fibers and chemically resistant films. Trademarked "Kodel." See also terephthalic acid.

"Polycyclol 1222."[214] Trademark for an intermediate for the preparation of alkyd-type resins used for coatings. These are known by the coined name "cyclyd" (q.v.).

poly-1,1-dihydroperfluorobutyl acrylate.
Properties: White, rubber-like polymer; sp. gr. 1.5; begins to degrade at 300°F; retains strength and elastomeric properties in contact with synthetic lubricants, solvents, hydraulic fluids, oils, etc., at temperatures in the range of 300–400°F; has limited flexibility at sub-zero temperatures. Nonflammable.
Uses: O-rings, seals, gaskets, diaphragms, hose, sheets, and coating for fabric and other surfaces.

poly-para-dinitrosobenzene. See "Polyac."

"Polydril."[233] Trademark for a synthetic water-soluble polymer; used as a flocculating agent in the oil industry.

polyelectrolyte. A high polymer substance, either natural (protein, gum arabic) or synthetic (polyethyleneimine, polyacrylic acid salts) containing ionic constituents; they may be either cationic or anionic. The former type is widely used for industrial applications. Water solutions of both types are electrically conducting; some are effective in concentrations as low as 1 ppm. In a given polyelectrolyte, ions of one sign are attached to the polymer chain, while those of opposite sign are free to diffuse into the solution. Major uses are flocculation of solids (especially dissolved phosphates) in potable water, dispersion of clays in oil well drilling muds, soil conditioning, antistatic agents, and treatment of paper-mill waste water. Ion-exchange resins (q.v.) are cross-linked (stabilized) polyelectrolytes.
See also flocculant; "Purifloc"; "Cat-Floc."

"Poly-Em" Emulsions.[533] Trademark for a series of polyethylene emulsions, many of which are designed for polishes.

polyene. Any unsaturated aliphatic or alicyclic compound containing more than four carbon atoms in the chain and having at least two double bonds. Examples are pentadiene, cyclooctatriene.

polyester fiber. Generic name for a manufactured fiber (either as staple or continuous filament) in which the fiber-forming substance is any long chain synthetic polymer composed of at least 85% by weight of an ester of a dihydric alcohol and terephthalic acid (Federal Trade Commission). See "Dacron"; polyethylene terephthalate.
Properties: Strength (staple), 2.2 to 4.0 g per denier; (continuous filament), up to 9.5 g per denier. M.p. 264°C; water absorption 0.5%. Nonflammable.
Uses: Tire fabric; seat belts; reinforcement of rubber hose for seawater cooling systems; as blend in clothing fabrics; fire-hose jackets.

polyester film. Continuously extruded polyester sheet of various thicknesses, especially useful in electrical equipment because of its high resistivity. Its tensile strength of 25,000 psi is much greater than that of other plastic films. Sensitized polyester film is used in magnetic tapes, in the photocopying technique known as reprography.

polyester resin. Any of a group of thermosetting synthetic resins, which are polycondensation products of dicarboxylic acids with dihydroxy alcohols. They are thus a special type of alkyd resin but, unlike other types, are not usually modified with fatty acids or drying oils. The outstanding characteristic of these resins is their ability, when catalyzed, to cure or harden at room temperature under little or no pressure. Most polyesters now produced contain ethylenic unsaturation, generally introduced by unsaturated acids. The unsaturated polyesters are usually cross-linked through their double bonds with a compatible monomer, also containing ethylenic unsaturation, and thus become thermosetting. Flame resistance is imparted by using either acid or glycol ingredients having a high content of halogens, e.g., HET acid.

The principal unsaturated acids used are maleic and fumaric. Saturated acids, usually phthalic and adipic, may also be included. The function of these acids is to reduce the amount of unsaturation in the final resin, making it tougher and more flexible. The acid anhydrides are often used if available and applicable. The dihydroxy alcohols most generally used are ethylene, propylene, diethylene, and dipropylene glycols. Styrene and diallyl phthalate are the most common cross-linking agents. Polyesters are resistant to corrosion, chemicals, solvents, etc.
Forms: Sheets; powder; chips.
Uses: Reinforced plastics; automotive parts; boat hulls; foams; encapsulation of electrical equipment; protective coatings; ducts, flues, and other structural applications; low-pressure laminates; magnetic tapes; piping; bottles.
See also alkyd resin; "Dacron"; "Mylar."

polyethenoid. Characterizing an aliphatic compound having more than one ethene group —CH:CH—. Linoleic acid is a polyethenoid fatty acid.

polyether. A polymer in which the repeating unit contains a carbon-oxygen bond derived from aldehydes or epoxides or similar materials. Examples are "Delrin" and "Celcon."

polyether, chlorinated. A highly crystalline material that is 46% chlorine. Outstanding corrosion-resistance. Good electrical resistance. Readily processed and fabricated.

Uses: Fluid-bed coating; tank linings; piping; valves; laboratory equipment; chemical processing equipment.

polyether foam. A polyurethane foam, either rigid or flexible, made by use of a polyether as distinct from a polyester or other resin component.
Hazard: See polyurethane.

polyether glycol. A compound with a structure skeleton such as HO-C-C-O-C-C-O-C-C-O-C-C-OH. The length of the chain can vary widely, and the number of consecutive carbon atoms may be greater than two. Examples are polyethylene glycol and polypropylene glycol.

polyethylene $(H_2C{:}CH_2)_x$
chlorosulfonated. See "Hypalon."
cross-linked (XLPE).
Properties: Thermosetting white solid; high temperature-resistant; excellent resistance to chemicals and to creep; high impact and tensile strength; high electrical resistivity; insoluble in organic solvents; does not stress-crack. Combustible. Nontoxic.
Derivation: (a) By irradiating linear polyethylene with electron beam or gamma radiation; cross-linking taking place through a primary valence bond, as shown:

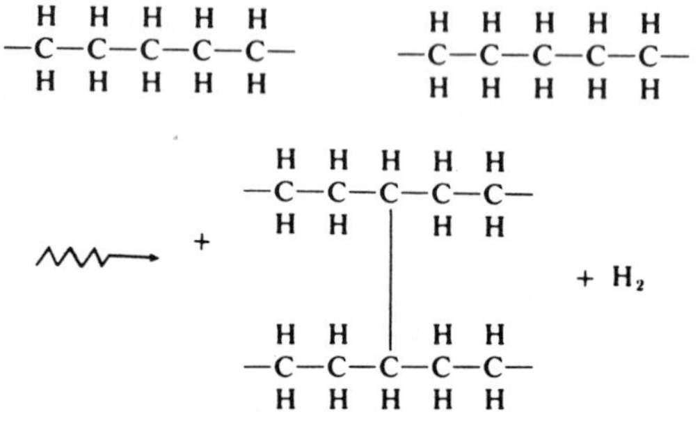

```
  H  H  H  H  H              H  H  H  H  H
—C—C—C—C—C—              —C—C—C—C—C—
  H  H  H  H  H              H  H  H  H  H

                  H  H  H  H  H
                —C—C—C—C—C—
                  H  H  |  H  H
/\/\/—>    +              |                  + H2
                          |
                  H  H  |  H  H
                —C—C—C—C—C—
                  H  H  H  H  H
```

(b) By chemical cross-linking agent such as an organic peroxide (e.g., benzoyl peroxide). All grades of polyethylene and most copolymers can be chemically cross-linked.
Uses: Wire and cable coatings and insulation (low-density grades); pipe and molded fittings (high-density grades). Special types having low electrical resistivity can be made; these can be regarded as semiconductors.
Note: In molding cross-linked polyethylene, the desired part must be formed before cross-linking is initiated, as material will not change its shape after cross-linking. Cross-linked polyethylene is a relatively new type of material, unlike other thermosets and thermplastics. The variations in composition and wide range of properties approach the ideal of a universal material more closely than most polymers.
linear (*low- and high-density*).
Properties: Thermoplastic white solid; molecular weight 6000 or more; high neutron absorption capacity; excellent barrier to water vapor and moisture; combustible; nontoxic.
Low-density: Sp. gr. 0.915; tensile strength 1500 psi; impact strength over 10 ft-lb/in./notch; thermal expansion 17×10^{-5} in./in./°C. Resistant to solvents and corrosive solutions; attacked by surfactants.
High-density: Sp. gr. 0.95; tensile strength 4000 psi; impact strength 8 ft-lb/in./notch; good electrical resistivity. Film is gas-permeable and hydrophobic; does not resist nitric acid.
Derivation: By polymerization of ethylene (a) with peroxide catalyst (amorphous); (b) by metal alkyl catalysts in a hydrocarbon solvent (crystalline). (See "Marlex"). Method (a) can be used at pressures as low as 500 to 600 psi; conventional pressures are 15,000 to 30,000 psi.
Forms: Rods, bars, sheets, tubes; thin films; molding powders; fibers.
Containers: Drums; hopper cars.
Hazard: Film is combustible and tends to disintegrate into burning globules.
Uses (high-density): Blow-molded products; injection-molded items; film and sheet; piping; gasoline and oil containers. (Low-density): packaging film, especially for food products; paper coating; liners for drums and other shipping containers; wire and cable coating; toys; cordage; refuse and waste bags; chewing-gum base; squeeze bottles; shipping drums; electrical insulation.
Note: Ethylene may be copolymerized with varying percentages of other materials, e.g., 2-butene or acrylic acid; a crystalline product results from copolymerization of ethylene and propylene. When butadiene is added to the copolymer blend, a vulcanizable elastomer is obtained. See also polyethylene, cross-linked.
low molecular weight.
Properties: Molecular weight from 2000 to 5000. Translucent white solids; excellent electrical resistance; abrasion-resistant; resistant to water and most chemicals; sp. gr. 0.92; slightly soluble in turpentine, petroleum naphtha, xylene, and toluene at room temperature; soluble in xylene, toluene, trichloroethylene, turpentine, and mineral oils at 180°F; practically insoluble in water; slightly soluble in methyl acetate, acetone, and ethanol up to the boiling point of these solvents. Available as emulsified and nonemulsified forms. Combustible.
Uses: Mold-release agent for rubber and plastics; paper and container coatings; liquid polishes and textile finishing agents.

polyethylene glycol (PEG; also called polyoxyethylene, polyglycol, or polyether glycol, q.v.). Any of several condensation polymers of ethylene glycol with the general formula $HOCH_2(CH_2OCH_2)_nCH_2OH$ or $H(OCH_2CH_2)_nOH$. Average molecular weights range from about 200 to 6000. Properties vary with molecular weight.
Properties: Clear, colorless, odorless, viscous liquids to waxy solids; soluble or miscible with water and for the most part with alcohol and other organic solvents; heat-stable; inert to many chemical agents; do not hydrolyze or deteriorate; have low vapor pressure. Combustible; nontoxic.
Derivation: By condensation of ethylene glycol, or of ethylene oxide and water.
Uses: Chemical intermediates (lower molecular weight varieties); plasticizers; softeners and humectants; lubricants; bases for cosmetics and pharmaceuticals; solvents; binders; metal and rubber processing; permissible additives to foods and animal feed; laboratory reagent. See also "Carbowax."

polyethylene glycol chloride $H(OCH_2CH_2)_nCl$. Any of a group of polymers, usually colorless liquids with very low vapor pressures at room temperature. Molecular weights from about 100 to about 600. Miscible with water; sp. gr. (20°C) for a low molecular weight polymer is 1.18, for a high molecular weight polymer 1.14. The former sets to a glass at −90°C, the latter sets to a wax-like solid at 20°C. Used as solvents for cleaning, extracting, and dewaxing. Combustible.

polyethylene glycol ester. A mono- or di- ester resulting from the interaction of an organic acid with one or both of the glycol ends of the polyethylene glycol polymer. These are also called polyoxyethylene esters, polyglycol esters, or by a coined generic name.

polyethyleneimine $(CH_2CH_2NH)_x$. Highly viscous hygroscopic liquid when anhydrous; completely miscible with water and lower alcohols; insoluble in benzene. Reactive toward cellulose. Combustible.
Uses: Adhesive and anchoring agent for paper and cellophane; dewatering agent and wet strength improver in paper manufacture; fixative; levelling agent in textile fibers; antiblocking agent on plastic films; flocculating agent; ion exchange resins; complexing agents; disinfectant for textiles, skins; photographic chemistry; absorbent for carbon dioxide; water purification; polyelectrolyte.

polyethylene oxide. A plastic reported to be dimensionally stable at high and low temperatures and designed as a substitute for phenolics.

polyethylene oxide sorbitan fatty acid esters. See polysorbate.

polyethylene terephthalate. A polyester formed from ethylene glycol by direct esterification or by catalyzed ester exchange between ethylene glycol and dimethyl terephthalate. Offered as oriented film or fiber. It melts at 265°C; tenacity is 2.2 to 4 g/denier (staple) and up to 9.0 g/denier as continuous filament. It has good electrical resistance and low moisture absorption. Resists combustion and is self-extinguishing.
Uses: Blended with cotton, for wash-and-wear fabrics; blended with wool, for worsteds and suitings; packaging films; recording tapes; soft-drink bottles.
See also "Dacron"; "Terylene"; "Mylar"; "Dalar."

polyethylene thiuram sulfide.
Derivation: Oxidation of diammonium ethylene bisdithiocarbamate with calcium hypochlorite.
Grades: 50% vegetable powder; 95% technical powder.
Containers: 50-lb multiwall paper bags.
Hazard: May be toxic.
Use: Fungicide.

"Poly-FBA."[158] Trademark for poly-1,1-dihydroperfluorobutyl acrylate (q.v.).

"Polyfilm."[233] Trademark for virgin polyethylene film specially formulated for converting and packaging.

"Polyflo."[416] Trademark for a series of all-organic inhibitor-dispersants designed for treatment of all fuel and diesel oils. Depending on precise inhibitor needs, this series contains products (1) to improve jet fuel thermal stability, (2) to stabilize No. 2 heating oils and diesel fuels against color deterioration and sludge deposition, (3) to eliminate fouling in feed exchangers and reboilers, (4) to eliminate tank bottoms in crude and residual oil storage tanks. Each member of the series exhibits a different degree of effectiveness in different types of stocks.

"Polyfon."[228] Trademark for a sodium lignosulfonate produced from pure, sugar-free, alkali lignin.
Properties: Nonhygroscopic, free-flowing brown amorphous powder. Completely free of wood sugars, hemicelluloses or other degradation products. Soluble in cold water. Yields alkaline solution.
Uses: Dispersant; thinner; plasticizer; binder; grinding aid; anticaking agent; protective colloid.

polyformaldehyde. See paraformaldehyde.

polyformaldehyde resin. See acetal resin.

"Poly-G."[84] Trademark for a series of polyethylene glycols, polypropylene glycols and polyoxypropylene adducts of glycerol.
G200, 300, 400, and 600 are liquid polyethylene glycols; G1000, 1500, GB-1530 and BG-2000 are waxy polyethylene glycols. The number indicates the approximate molecular weight.
G420P, 1020P, 2020P are propylene oxide condensation polymers of propylene glycol.
G1030PG, 3030PG, 4030PG are propylene oxide condensation polymers of glycerol.

"Polygard."[248] Trademark for a mixture of alkylated aryl phosphites.
Properties: Liquid; clear amber; sp. gr. 0.99; soluble in acetone, alcohol, benzene, carbon tetrachloride, solvent naphtha and ligroin; insoluble in water but can hydrolyze.
Use: Nondiscoloring stabilizer for rubber and plastics.

polyglycerol. One of several mixtures of ethers of glycerol with itself, ranging from diglycerol to triacontaglycerol. Some examples are: (a) diglycerol, possibly $(CH_2OHCHOHCH_2)_2O$, molecular weight 166, a liquid with 4 OH groups, viscosity of 287 cs at 150°F; (b) hexaglycerol, mol. wt. 462, a liquid with 8 OH groups, viscosity 1671 cs at 150°F; (c) decaglycerol, mol. wt. 758, a liquid with 12 OH groups, viscosity 3199 cs.
Properties: Viscous liquids to solids; soluble in water, alcohol, and other polar solvents. Act as humectants much like glycerin but have progressively higher molecular weights and boiling points. Combustible; low toxicity.
Derivation: Glycerol is heated with an alkaline catalyst (200–275°C) at normal or reduced pressure. A stream of inert gas may be used to blanket the reaction and help remove the water of reaction.
Uses: Surface-active agents, emulsifiers, plasticizers, adhesives, lubricants, and other compounds used for both edible and industrial applications.

polyglycerol ester. One of several partial or complete esters of saturated and unsaturated fatty acids with a variety of derivatives of polyglycerols ranging from diglycerol to triacontaglycerol. Prepared by (a) direct esterification and (b) transesterification reactions. Combustible; low toxicity.

Some examples of (a): decaglycerol monostearate, semisolid, sp. gr. 1.04; m.p. 51.9°C; decaglycerol monooleate, viscous liquid; sp. gr. 1.13, m.p. below zero; decaglycerol hexaoleate, liquid, sp. gr. 0.97, m.p. below zero.

Examples of (b); triglycerol monolinoleate, viscosity 322 cp (168°F) and triglycerol trilinoleate, viscosity 30.1 cp (168°F).
Uses: Lubricants, plasticizers, paint and varnish vehicles, gelling agents, urethane intermediates, adhesives, crosslinking agents, humectants, textile fiber finishes, functional fluids, surface-active agents, dis-

persants and emulsifiers in foods, pharmaceuticals, cosmetic preparations.

polyglycol. See polyethylene glycol.

polyglycol amine H-163 $HO[C_2H_4O]_2C_3H_6NH_2$.
Properties: Colorless liquid; sp. gr. 1.0556; b.p., decomposes; f.p. 14.5°C; soluble in water; wt/gal 8.8 lb; flash point 295°F. Combustible; low toxicity.
Containers: Drums.

polyglycol distearate (polyethylene glycol distearate) $C_{17}H_{35}COO(CH_2CH_2O)_xOCC_{17}H_{35}$. Distearate ester of polyglycol.
Properties: A soft, off-white solid; sp. gr. 1.04 (50°C); m.p. 43°C; pH of 10% dispersion 7.26; saponification number variable; soluble in chlorinated solvents, light esters and acetone. Slightly soluble in alcohols; insoluble in glycols, hydrocarbons and vegetable oils. Combustible; low toxicity.
Uses: Plasticizer for various resins; component of grinding and polishing pastes to promote easy removal in water.

polyglycol. See polyethylene glycol ester.

"Polygriptex." Trademark for an adhesive especially designed for bonding polyethylene sheeting to porous surfaces; used in making polyethylene-lined bags and multiwall kraft bags. Has good adhesion to waxed surfaces. Available in viscosities from 300 to 20,000 centipoises.

polyhalite $2CaSO_4 \cdot MgSO_4 \cdot K_2SO_4 \cdot 2H_2O$. A naturally occurring potash salt found in Germany, Texas, and New Mexico.
Use: Source of potash for fertilizer.

polyhexamethyleneadipamide. Same as nylon 66.

polyhexamethylene sebacamide. Same as nylon 610.

polyhydric alcohol. See polyol.

polyimide. Any of a group of high polymers which have an imide group (—CONHCO—) in the polymer chain.
Properties: Specific gravity 1.42, tensile strength 13,500 psi, water absorption (24 hrs at 77°K) 0.3%, heat distortion point above 500°F, dielectric constant at 2000 m.c. 3.2, and coefficient of linear expansion 28.4 × 10^{-6} in./in./°F. High-temperature stability (up to 700°F); excellent frictional characteristics; good wear resistance at high temperatures; resists radiation; exhibits low outgassing in high vacuum; resistant to organic materials at quite high temperatures; not resistant to strong alkalies and to long exposure to steam. Flame retardant.
Uses: High-temperature coatings; laminates and composites for aerospace vehicles; ablative materials, oil sealants and retainers, adhesives, semiconductors, valve seats, bearings; insulation for cables, printed circuits, magnetic tapes (high- and low-temperature-resistant); flame-resistant fibers; binders in abrasive wheels.

polyindene resin. See coumarone-indene resin.

"Poly-Ionic."[300] Trademark for a series of polyethylene emulsions. Used as softeners and lubricating agents for textiles.

polyisobutene. See polybutene.

polyisobutylene. See polybutene.

polyisocyanurate. See isocyanurate.

polyisoprene. $(C_5H_8)_n$. The major component of natural rubber; also made synthetically. Forms are stereospecific cis-1,4-, and trans-1,4-polyisoprene. Both can be produced synthetically by the effect of heat and pressure on isoprene in the presence of stereospecific catalysts. (See catalyst, stereospecific). Natural rubber is cis-1,4-; synthetic cis-1,4- is sometimes called synthetic natural rubber. Trans-1,4-polyisoprene resembles gutta-percha. Polyisoprene is thermoplastic until mixed with sulfur and vulcanized. Supports combustion; nontoxic. For further information on properties and uses, see rubber, natural and synthetic.

"Polylan."[493] Trademark for a polyunsaturated ester of linoleic acid and lanolin alcohols. An amber, viscous oily liquid; soluble in mineral oil; castor oil; anhydrous ethanol, isopropanol, ethyl acetate; insoluble in water. Used as a hydrophobic conditioner in cosmetics and pharmaceuticals.

"Poly-Lease."[175] Trademark for an aerosol mold release and parting agent for plastics and rubber materials, based on a low molecular weight polyethylene lubricant. The usual precautions for shipping and handling aerosol containers apply.

"Polylite."[36] Trademark for a group of 100% reactive alkyd resins, dissolved in styrene and other monomers. Highly diversified applications both alone and in combination with such materials as fibrous glass. This group also includes resins for use with diisocyanate to form rigid or flexible polyurethane foams.

"Polylite."[248] Trademark for nonylated diphenyl amine.

"Polymax."[332] Trademark for woven fabrics of polyolefin fibers for liquid and pneumatic filtration.

"Polymeg."[224] Trademark for polytetramethylene ether glycols. Available in three molecular weight ranges: 1000, 2000 and 3000.
Properties: Waxy solids which melt to clear, viscous liquids at about 100°F. On supercooling (or nucleation) the liquid resolidifies. Sp. gr. 0.985 (1000 MW) to 0.982 (3000 MW) at 95°F; soluble in aromatic and chlorinated hydrocarbons; slightly soluble in water, solubility decreasing with increasing molecular weight.
Use: Polyurethane technology.

polymer. A macromolecule formed by the chemical union of 5 or more identical combining units called monomers. In most cases the number of monomers is quite large (3500 for pure cellulose), and often is not precisely known. In synthetic polymers this number can be controlled to a predetermined extent, e.g., by shortstopping agents. (Combinations of 2, 3, or 4 monomers are called, respectively, dimers, trimers, and tetramers, and are known collectively as oligomers). A partial list of polymers by type is as follows:

I. Inorganic
   siloxane; sulfur chains; black phosphorus; boron-nitrogen; silicones.

II. Organic
   1. Natural
      (a) Polysaccharides
         starch; cellulose; pectin; seaweed gums (agar, etc.); vegetable gums (arabic, etc.)
      (b) Polypeptides (proteins)
         casein; albumin; globulin; keratin; insulin; DNA
      (c) Hydrocarbons
         rubber (polyisoprene); gutta percha; balata.
   2. Synthetic
      (a) Thermoplastic
         elastomers (unvulcanized); nylon; polyvinyl

chloride; polyethylene (linear); polystyrene; polypropylene; fluorocarbon resins; polyurethane; acrylate resins
(b) Thermosetting
elastomers (vulcanized); polyethylene (crosslinked); phenolics; alkyds; polyesters
3. Semisynthetic
cellulosics (rayon, methylcellulose, cellulose acetate); modified starches (starch acetate, etc.)

See also following entries.

polymer, addition. See addition polymer.

polymerase. An enzyme which catalyzes the formation of messenger DNA.

polymer, atactic. See atactic.

polymer, block. See block polymer.

polymer, condensation. A polymer formed by a condensation reacton (q.v.).

polymer, electroconductive. A polymer or elastomer made electrically conductive by incorporation of a substantial percentage of a suitable metal powder (e.g., aluminum) or acetylene carbon black; the proportion used must be high enough to permit the particles to be in contact with one another in the mixture. Polyelectrolytes such as ion-exchange resins, salts of polyacrylic acid and sulfonated polystyrene are electroconductive in the presence of water. Pyrolysis of polyacrylonitrile makes it electrically conductive without impairment of its structure.

polymer, graft. See graft polymer.

polymer, high. An organic macromolecule composed of a large number of monomers. The molecular weight may range from about 5000 into the millions (for some polypeptides). Natural high polymers are exemplified by cellulose $(C_6H_{10}O)_n$ and rubber $(C_5H_8)_n$. Proteins are natural high polymer combinations of amino acid monomers. The dividing line between low and high polymers is considered to be in the neighborhood of 5000 to 6000 molecular weight.

Synthetic high polymers (or "synthetic resins") include a wide variety of materials having properties ranging from hard and brittle to soft and elastic. Addition of such modifying agents as fillers, colorants, etc., yields an almost infinite number of products collectively called plastics (q.v.). High polymers are the primary constituents of synthetic fibers, coating materials (paints and varnishes), adhesives, sealants, etc. Polymers having special elastic properties are called rubbers, or elastomers (q.v.).

Synthetic polymers in general can be classified: (1) by thermal behavior, i.e., thermoplastic and thermosetting (q.v.); (2) by chemical nature, i.e., amino, alkyd, acrylic, vinyl, phenolic, cellulosic, epoxy, urethane, siloxane, etc., and (3) by molecular structure, i.e., atactic, stereospecific, linear, cross-linked, block, graft, ladder, etc. (see specific entries). Copolymers (q.v.) are products made by combining two or more polymers in one reaction (styrene-butadiene). See also cross-linking.

polymer, inorganic. A polymer in which the main chain contains no carbon atoms and in which behavior similar to that of an organic polymer can be developed, i.e., covalent bonding and cross-linking, as in silicone polymers. Here the element silicon replaces carbon in the straight chain; substituent groups are often present, forming highly useful polymers. Other inorganic high polymers are black phosphorus, boron and sulfur, all of which can form polymeric structures under special conditions. At present these have little or no commercial significance.

polymer, isotactic. A type of polymer structure in which groups of atoms which are not part of the backbone structure are located either all above or all below the atoms in the backbone chain, when the latter are all in one plane. See polymer, stereospecific.

```
   H2 H  H2 H  H2 H
   |  |  |  |  |  |
 — C——C——C——C——C——C —
      |     |     |
      O     O     O
      R     R     R
```

isotactic polymer

polymerization. A chemical reaction, usually carried out with a catalyst, heat or light, and often under high pressure, in which a large number of relatively simple molecules combine to form a chain-like macromolecule.

The polymerization reaction occurs spontaneously in nature; industrially it is performed by subjecting unsaturated or otherwise reactive substances to conditions that will bring about combination. This may occur by *addition*, in which free radicals are the initiating agents that react with the double bond of the monomer by adding to it on one side, at the same time producing a new free electron on the other:

$$R\cdot + CH_2{=}CHX \rightarrow R{-}CH_2{-}CHX\cdot$$

By this mechanism the chain becomes self-propagating. Polymerization may also occur by condensation, involving the splitting out of water molecules by two reacting monomers, and by so-called oxidative coupling (q.v.). The degree of polymerization (D.P.) is the number of monomer units in an average polymer unit of a given sample.

Polymerization techniques may be: (1) in the gas phase at high pressures and temperatures (200°C); (2) in solution at normal pressure and temperatures from −70° to +70°C; (3) bulk or batch polymerization at normal pressure at 150°C; (4) in suspension at normal pressure at 60° to 80°C; (5) in emulsion form at normal pressure at −20° to +60°C (used for copolymers). Catalysts of the peroxide type are necessary with some of these methods.

See also polymer, stereospecific; free radical.

polymer, ladder. See ladder polymer.

polymer, low. A polymer comprised of comparatively few monomer units and having a molecular weight from about 300 to 5000.

polymer, natural. See polymer.

polymer, stereoblock. See stereoblock polymer.

polymer, stereospecific (stereoregular). A polymer whose molecular structure has a definite spatial arrangement, i.e., a fixed position in geometrical space for the constituent atoms and atomic groups comprising the molecular chain, rather than the random and varying arrangement that characterizes an amorphous polymer. Achievement of this specific steric (three-dimensional) structure (also called tacticity, q.v.) requires use of special catalysts such as those developed by Ziegler and Natta (q.v.) about 1950.

Such polymers are wholly or partially crystalline. Synthetic natural rubber, cis-polyisoprene, is an example of a stereospecific polymer made possible by this means. There are five types of stereospecific (or stereoregular) structures; cis, trans, isotactic, syndiotactic, and tritactic. See also catalyst, stereospecific.

polymer, syndiotactic. See syndiotactic polymer.

polymer, synthetic. See polymer.

polymer, water-soluble. Any substance of high molecular weight that swells or dissolves in water at normal temperature. These fall into several groups, including natural, semisynthetic, and synthetic products. Their common property of water solubility makes them valuable for a wide variety of applications as thickeners, adhesives, coatings, food additives, textile sizing, etc. For further details see specific entries.

(1) *Natural.* This type is principally comprised of gums (q.v.), which are complex carbohydrates of the sugar group. They occur as exudations of hardened sap on the bark of various tropical species of trees. All are strongly hydrophilic. Examples are arabic, tragacanth, karaya, etc.

(2) *Semisynthetic.* This group (sometimes called water-soluble resins) includes such chemically treated natural polymers as carboxymethylcellulose, methylcellulose and other cellulose ethers, as well as various kinds of modified starches (ethers and acetates).

(3) *Synthetic.* The principal members of this class are polyvinyl alcohol (q.v.), ethylene oxide polymers, polyvinyl pyrrolidone, polyethyleneimine, etc.

polymethacrylate resin. See acrylic resin.

polymethylbenzene. See durene and pseudocumene, the two members of this group with some commercial production and use.

polymethylene polyphenol. See "Tetra-Flex."

polymethylene polyphenylisocyanate. A polymer of diphenylmethane 4,4′-diisocyanate (q.v.).

polymethylene wax. See wax, polymethylene.

poly-4-methylpentene-1.
Properties: High resistance to all chemicals except carbon tetrachloride and cyclohexane. Excellent heat resistance; high clarity and light transmittance. Temperature limit 170°C. Sp. gr. 0.83.
Uses: Laboratory ware (beakers, graduates, etc.); electronic and hospital equipment; food packaging, especially types subject to high temperature such as trays for TV dinners, etc.; light reflectors.

poly(methyl vinyl ether). See polyvinyl methyl ether.

polymorphism. See allotropy.

"Polymul MS-40."[309] Trademark for a polyethylene-wax dispersion for fabric finishing. Contains non-oxidized grade of polyethylene which is odor-free.

polymyxin. Generic term for a series of antibiotic substances produced by strains of Bacillus polymyxa. Various polymyxins are differentiated by the letters A, B, C, D and E. All are active against certain gram-negative bacteria. Polymyxin B (q.v.) is most used.
Properties: All are basic polypeptides, soluble in water; the hydrochlorides are soluble in water and methanol; insoluble in ether, acetone, chlorinated solvents and hydrocarbons. Permissible food additives. Toxicity low or negligible.
Uses: Medicine (antibiotic); beer production.

polynuclear. Descriptive of an aromatic compound containing four or more closed rings, usually of the benzenoid type. See also polycyclic; nucleus (3).

"Polynycel."[309] Trademark for aromatic amino condensates. Used as leveling agents for dye retardants for textiles, union dyeing of wool and nylon.

polyolefin. A class or group name for thermoplastic polymers derived from simple olefins (q.v.); among the more important are polyethylene, polypropylene, polybutenes, polyisoprene, and their co-polymers. Many are produced in the form of fibers. This group comprises the largest tonnage of all thermoplastics produced.

polyol. A polyhydric alcohol, i.e., one containing three or more hydroxyl groups. Those having three OH groups (trihydric) are glycerols; those with more than three are called sugar alcohols, with general formula $CH_2OH(CHOH)_nCH_2OH$, where n may be from 2 to 5. These react with aldehydes and ketones to form acetals and ketals. See also alcohol; glycerol.

polyorganosilicate graft polymer. An organoclay to which a monomer or an active polymer has been chemically bonded, often by the use of ionizing radiation. An example is the bonding of styrene to a polysilicate containing vinyl radicals, resulting in the growth of polystyrene chains from the surface of the silicate. Such complexes are stable to organic solvents. They have considerable use potential in the ion-exchange field, as ablative agents, reinforcing agents and hydraulic fluids.
See also: organoclay; graft polymer.

"Polyotic."[57] Trademark for tetracycline (q.v.).

"Polyox."[214] Trademark for a series of water-soluble ethylene oxide polymers with molecular weights in the 100,000 to several million range.
Uses: Textile warp size, paper coatings, detergents, hair spray, toothpastes, water-soluble packaging film, adhesives.

polyoxadiazole. A polymer of oxadiazole, cyclic C_2N_2O, prepared by cyclodehydration (ring formation from a chain with subsequent loss of water) of polyisophthalic hydrazide. Because of its high temperature tolerance (above 750°F) fibers made from it may be useful for space vehicles.

polyoxamide. A nylon-type material made from oxalic acid and diamines.

polyoxetane. See oxetane; "Penton."

polyoxyethylene. See polyethylene glycol.

polyoxyethyleneoxypropylene (POEOP). A polymer of ethylene and propylene glycols (ethylene oxidepropylene oxide). Used as a solvent.

polyoxyethylene fatty acid ester. See polysorbate.

polyoxyethylene (20) sorbitan monolaurate. See polysorbate.

polyoxyethylene (2) sorbitan monooleate. See polysorbate.

polyoxyethylene (20) sorbitan monostearate. See polysorbate.

polyoxyethylene (20) sorbitan tristearate. See polysorbate.

polyoxyethylene (8) stearate. A mixture of the mono- and diesters of stearic acid and mixed polyoxyethylene diols having an average polymer length of about 7.5 oxyethylene units.

Properties: Cream-colored, soft, waxy or pasty solid at 25°C; faint, fatty odor and a slightly bitter, fatty taste. Soluble in toluene, acetone, ether, and ethanol. Nontoxic.

Use: Emulsifier in bakery products.

polyoxyethylene (40) monostearate (polyethylene glycol stearate). A mixture of the mono- and distearate esters of mixed polyoxyethylene diols and corresponding free glycols. The monostearate can be represented as: $H(OCH_2CH_2)_nOCOC_{17}H_{35}$ (n is approximately 40).

Properties: Waxy, light tan, nearly odorless solid; congealing range 39–45°C; soluble in water, alcohol, ether, and acetone; insoluble in mineral oil and vegetable oils. Nontoxic.

Grade: U.S.P.

Uses: Ointments; emulsifier; surfactant; food additive.

polyoxymethylene. Any of several polymers of formaldehyde and trioxane. See acetal resin.

polyoxypropylene diamine (POPDA). Any of six high molecular weight amines of low viscosity and vapor pressure, high primary amine content, and light color. Used as cross-linking agents in epoxy coatings, imparting high flexibility and adhesion at low temperatures. Other possible uses are in polyamide and polyurethane coatings, adhesives, elastomers and foams, as intermediates for textile and paper treatment, and viscosity index improvers in lube oils.

polyoxypropylene ester. See polypropylene glycol ester.

polyoxypropylene-glycerol adduct. One of several condensation polymers of propylene oxide and glycerol, with molecular weights in the range 1000 to 4000. Clear, stable, almost colorless, noncorrosive liquids. Uses similar to those of polypropylene glycol (q.v.).

"Poly-pale."[266] Trademark for pale, hard, thermoplastic resins; 40% dimeric resin acids; acid number 145; USDA color WG; softening point 102°C. Available in solid and flake forms.

Uses: Adhesives, lacquers, varnishes, printing inks.

polypeptide (peptide). The class of compounds composed of acid units chemically bound together with amide linkages (CONH) with elimination of water. A polypeptide is thus a polymer of amino acids, forming chains that may consist of several thousand amino acid residues. A segment of such a chain is as follows:

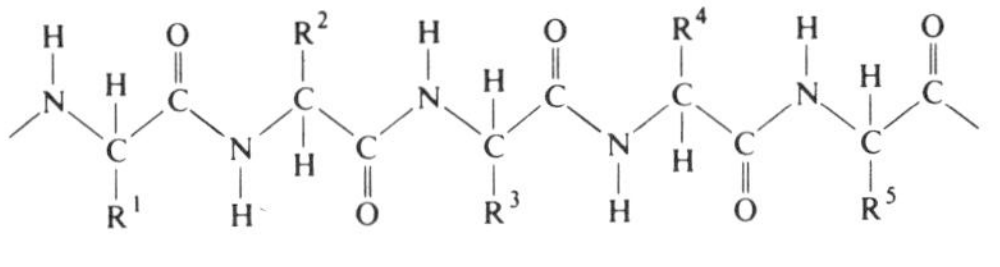

```
 H      O      R2     H      O      R4     H      O
 |      ‖      |      |      ‖      |      |      ‖
  N  H   C      C      N  H   C      C      N  H   C
/  \ | /  \   / | \  /  \ | /  \   / | \  /  \ | /  \
    C       N   H   C      C       N  H   C       C
    |       |       ‖      |       |      ‖       |
    R1      H       O      R3      H      O       R5
```

The sequence of amino acids in the chain is of critical importance in the biological functioning of the protein, and its determination is one of the most difficult problems in molecular biology. The chains may be relatively straight, or they may be coiled or helical. In the case of certain types of polypeptides, such as the keratins, they are cross-linked by the disulfide bonds of cystine. Linear polypeptides can be regarded as proteins. See also protein; polyamide; keratin.

polyphenylene oxide. See "PPO."

polyphenylene triazole $[-C_6H_4-C_2N_3(C_6H_5)-]_n$. A polymer stated to be serviceable up to 500°F for films, coatings, adhesives, and lamination.

"Polyphos."[84] Trademark for a water-soluble glassy sodium phosphate of standardized composition, $(Na_{12}P_{10}O_{31})$ analyzing 63.5% P_2O_5 (ratio of Na_2O:P_2O_5 is 1.2:1). It is closely similar to a sodium hexametaphosphate and sodium tetraphosphate; frequently the three names are used interchangeably.

Grades: Ground; walnut-size to pea-size lumps.

Containers: 100-lb bags; 100-, and 350-lb drums.

Uses: Boiler water compounds; detergents; textiles; leather tanning; photographic film developing; deflocculation of clays; flotation and desliming of minerals; dispersion of pigments; paper processing; industrial and municipal water treatment.

polyphosphazene. See phosphazene.

polyphosphoric acid $H_{n+2}P_nO_{3n+1}$, for $n > 1$. Any of a series of strong acids, from pyrophosphoric acid, $H_4P_2O_7$ (n = 2), through metaphosphoric acid (large values of n).

Properties: Viscous, water-white liquid; water-soluble; does not crystallize on standing. Hygroscopic. The commercial acid is a mixture of orthophosphoric acid with pyrophosphoric, triphosphoric and higher acids and is sold on the basis of its calculated content of H_3PO_4, as, for example 115%. Superphosphoric acid is a similar mixture sold at 105% H_3PO_4. These acids revert slowly to orthophosphoric acid on dilution with water.

Hazard: Moderately toxic by ingestion; strong irritant.

Uses: Dehydrating, catalytic and sequestering agents; for metal treating; many applications where a concentrated monoxidizing acid is needed; laboratory reagent.

See also phosphoric acid.

polypropylene $(C_3H_5)_n$. A synthetic crystalline thermoplastic polymer, with molecular weight of 40,000 or more. Note: low molecular weight polymers are also known which are amorphous in structure, and used as gasoline additives, detergent intermediates, greases, sealants, and lube oil additives: also available as a high-melting wax.

Derivation: Polymerization of propylene with a stereospecific catalyst (q.v.) such as aluminum alkyl.

Properties: Translucent white solid; specific gravity 0.90; m.p. 168–171°C; tensile strength 5000 psi; flexural strength 7000 psi; usable up to 250°F. Insoluble in cold organic solvents; softened by hot solvents. Maintains strength after repeated flexing. Degraded by heat and light unless protected by antioxidants. Readily colored; good electrical resistance; low water absorption and moixture permeability; poor impact strength below 15°F; not attacked by fungi or bacteria; resists strong acids and alkalies up to 140°F; but is attacked by chlorine, fuming nitric acid, and other strong oxidizing agents. Combustible, but slow-burning. Nontoxic. Fair abrasion and good heat resistance if properly modified. Can be chrome-plated, injection- and blow-molded, and extruded.

Forms: Molding powder; extruded sheet; cast film (1 to 10 mils); textile staple and continuous filament yarn; fibers with diameters from 0.05 to 1 micron and fiber webs down to 2 microns thick; low-density foam.

Uses: Packaging film; molded parts for automobiles, appliances, housewares, etc.; wire and cable coating;

food container closures; coated and laminated products; bottles (with PVC); printing plates; fibers for carpets and upholstery; cordage and bristles; storage battery cases; crates for soft-drink bottles; laboratory ware; toys; synthetic seaweed to encourage silt deposition; radiator grills; trays and containers for storing precision equipment; artificial grass and turfs; plastic pipe; wearing apparel (acid-dyed); fish nets; surgical casts; strapping; synthetic paper; reinforced plastics.

polypropylenebenzene. See dodecylbenzene.

polypropylene, chlorinated. White, odorless, nonflammable powder. A film-forming polymer used in coatings, inks, adhesives and paper coatings.

polypropylene glycol ester. Exactly analogous to polyethylene glycol ester (q.v.).

polypropylene glycol monobutyl ether. See butoxy polypropylene glycol.

polypropylene glycol (PG). $HO(C_3H_6O)_nH$. One of a group of compounds comparable to polyethylene glycols (q.v.), but more oil-soluble and substantially less water-soluble. Classified by approximate molecular weight, as 425, 1025, and 2025. Non-volatile, noncorrosive liquids; lower molecular weight members are soluble in water. Solvents for vegetable oils, waxes, resins. Combustible; low toxicity.
Uses: Hydraulic fluids; rubber lubricants; antifoam agents; intermediates in urethane foams, adhesives, coatings, elastomers; plasticizers; paint formulations; laboratory reagent.

polypropyleneimine. Polymeric form of propyleneimine (q.v.). Available in 50% aqueous solution.
Uses: Textile paper, and rubber industries.

polypropylene oxide $(C_3H_6O)_n$. A derivative of propylene used as intermediate for urethane foams.

polypyrrolidone. Synonym for nylon-4.

"Polyrad."[266] Trademark for reaction products of "Amine D" and ethylene oxide.
Grades: Various grades which differ in chain length of polyoxyethylene units and free amine content. Vary in viscosity at 25°C from 0.5 to 24.8 poises.
Uses: Corrosion inhibitors and detergents in petroleum processing equipment; wetting and emulsifying agents; inhibiting hydrochloric acid.

"Polyram."[55] Trademark for a wettable powder used as a fungicide and approved for many vegetables. Toxic by ingestion and inhalation.

polysaccharide. A combination of nine or more monosaccharides, linked together by glycosidic bonds. Examples: starch, cellulose, glycogen.
See also carbohydrate; phycocolloid.

polysiloxane. See siloxane.

"Poly-Solv."[84] Trademark for a series of glycol ether solvents for paints, varnishes, dry cleaning soaps, cutting oils, insecticides.
D2M. Diethylene glycol dimethyl ether. Used as anhydrous reaction medium for organometallic syntheses.

polysorbate (USAN name for a polyoxyethylene fatty acid ester). One of a group of nonionic surfactants obtained by esterification of sorbitol (q.v.) with one or three molecules of a fatty acid (stearic, lauric, oleic, palmitic) under conditions which cause splitting out of water from the sorbitol, leaving sorbitan. About 20 moles of ethylene oxide per mole of sorbitol are used in the condensation to effect water solubility.
Properties: Lemon to amber oily liquids; sp. gr. about 1.1; faint odor and bitter taste; most types are soluble in water, alcohol, and ethyl acetate. Combustible; nontoxic.
Grades: Polysorbate 20 (polyoxyethylene (20) sorbitan monolaurate). Polysorbate 60 (polyoxyethylene (20) sorbitan monostearate). Polysorbate 80 (polyoxyethylene (20) sorbitan monooleate). Polysorbate 65 (polyoxyethylene (20) sorbitan tristearate).
Uses: Surfactant; emulsifying agent; dispersing agents; shortenings and baked goods; pharmaceuticals; flavoring agents; foaming and defoaming agents.
See also sorbitan fatty acid ester.

polystyrene $(C_6H_5CHCH_2)_n$. Polymerized styrene (q.v.), a thermoplastic synthetic resin of variable molecular weight depending on degree of polymerization.
Properties: Transparent, hard solid; high strength and impact resistance; excellent electrical and thermal insulator. Attacked by hydrocarbon solvents but resists organic acids, alkalies, and alcohols. Not recommended for outdoor use; unmodified polymer yellows when exposed to light, but light-stable modified grades are available. Easily colored, molded and fabricated. Copolymerization with butadiene and acrylonitrile and blending with rubber or glass fiber increase impact strength and heat resistance. (See ABS; SAN). Nontoxic.
Forms: Sheet, plates, rods, rigid foam, expandable beads or spheres.
Hazard: Combustible; autoignition temperature about 800°F. See also under foam, plastic.
Uses: Refrigerator doors; air conditioner cases; containers and molded household wares; machine housings; electrical equipment; toys; packaging; clock and radio cabinets; phonograph records. (As foam); thermal insulation; light construction as in boats, etc.; ice buckets, water coolers; fillers in shipping containers; furniture construction. (As spheres): Radiator leak stopper.
See also "Styron"; "Styrofoam."

"Polysulfide."[28] Trademark for a mixture similar to sulfurated potash (q.v.), in which sodium replaces sufur. Yellow to yellow-green powder; sodium polysulfide content, 56% min; total sulfur, 50.7% min.
Containers: 10-lb tins (6/case); 100-lb drums.
Uses: Coloring copper and brass; stripping copper-plated deposits; purifying cyanide plating solutions.

polysulfide elastomer. A synthetic polymer in either solid or liquid form obtained by the reaction of sodium polysulfide with organic dichlorides such as dichlorodiethyl formal, alone or mixed with ethylene dichloride. Outstanding for resistance to oils and solvents and for impermeability to gases. Poor tensile strength and abrasion resistance but are resilient and have excellent low-temperature flexibility. Some grades have fairly strong odor, which is not objectionable in most applications. Sealant grades are furnished in two parts which cure at room temperature when blended.
Uses: Gasoline and oil-loading hose; sealants and adhesive compositions; binder in solid rocket propellants; gaskets; paint spray hose.
See also "Thiokol."

polysulfone. A synthetic thermoplastic polymer.
Properties: Hard, rigid transparent solid; tensile strength 10,000 psi; sp. gr. 1.24; flexural strength 15,000 psi; good electrical resistance; minimum creep; low expansion coefficient. Soluble in aromatic

hydrocarbons, ketones and chlorinated hydrocarbons; resistant to corrosive acids and alkalies, to heat and oxidation, and to detergents, oils, and alcohols. Dimensionally stable over wide temperature range −150 to 300°F). Tends to absorb moisture. Readily processed and fabricated. Combustible but self-extinguishing.

Derivation: Condensation of bis-phenol A and dichlorophenyl sulfone.

Uses: Power-tool housings; electrical equipment; extruded pipe and sheet; auto components; electronic parts; appliances; computer components; base matrix for stereotype printing plates.

"Poly-Tergent."[84] Trademark for a series of emulsifiers.

polyterpene resin. A class of thermoplastic resins or viscous liquids of amber color, obtained by polymerization of turpentine in the presence of catalysts such as aluminum chloride or mineral acids. The resins consist essentially of polymers of alpha- or beta-pinene and are soluble in most organic solvents.

Uses: Paints; rubber plasticizers; curing concrete; impregnating paper; adhesives; hot-melt coatings; pressure-sensitive tapes.

See also pinene.

polytetrafluoroethylene (PTFE: TFE) $(C_2F_4)_n$. A polymer of tetrafluoroethylene. It is essentially a straight chain of the repeating unit $[—CF_2—CF_2—]_n$. Soft and waxy, with a milk-white color. It can be molded by powder metallurgy techniques involving mixing with a diluent that is subsequently removed and sintering at 700°F.

Properties: Highly resistant to oxidation and action of chemicals, including strong acids, alkalies, oxidizing agents; resistant to nuclear radiation and ultraviolet rays, ozone, and weather. Halogenated solvents at high temperatures and pressures have some adverse effect. Retains useful properties from −450° to +550°F and is strong and tough. Low coefficient of friction (0.05), and antistick properties; excellent resistance to electricity. Coefficient of thermal expansion greater than other plastics and metals. Nonflammable.

Forms: Extrusion and molding powders, aqueous dispersion, film, multifilament fiber.

Uses: Gaskets, liners, seals, flexible hose; ablative coatings for rockets and space vehicles; chemical process equipment; coatings in aerospace; coaxial spacers, insulators, wire coating and tape in electrical and electronic fields; bearings, seals, piston rings for mechanical goods; antistick coatings for cooking vessels; felts, packings and bearings.

See also "Teflon"; "Halon."

polytetramethylene ether glycol. See "Polymeg."

"Poly-Tex."[474] Trademark for polyvinyl acetate emulsion resins and thermosetting acrylic resins.

polythene. Generic name for polyethylene (q.v.); no longer current in the United States, but is still used in England.

polythiadiazole $[—C_6H_4—C_2N_2S—]_n$. A polymer made from polyoxathiahydrazide. Can be converted to fibers. Stated to retain properties to 750°F, and to have resistance to thermal degradation such that it retains 60 or 70% of original tenacity after 32 hours at 750°F.

polythiazyl $(SN)_n$. An experimental polymer of sulfur nitride with covalent linkages, said to have the optical and electrical properties characteristic of metals. Thin films are reported to exhibit epitaxial growth.

polytrifluorochloroethylene resin. See chlorotrifluoroethylene resin.

polyunsaturated fat. A fat or oil based at least partly on fatty acids having two or more double bonds per molecule, such as linoleic or linolenic acids. Examples are corn oil and safflower oil.

Uses: Margarine and dietary foods; salad dressings, etc.

polyurethane. A thermoplastic polymer (which can be made thermosetting) produced by the condensation reaction of a polyisocyanate and a hydroxyl-containing material e.g., a polyol derived from propylene oxide or trichlorobutylene oxide. The basic polymer unit is formed as follows: $R_1NCO + R_2OH \rightarrow R_1NHCOOR_2$.

Fiber:

Properties: High elastic modulus, good electrical resistance; high moisture-proofness; crystalline structure. Combustible.

Derivation: Reaction of hexamethylene diisocyanate and 1,4-butanediol.

Uses: Chiefly in so-called spandex fibers (q.v.) for girdles and other textile structures requiring exceptional elasticity; bristles for brushes, etc.

Coatings:

Properties: Excellent hardness, gloss, flexibility, abrasion resistance, and adhesion; resistant to impact, weathering, acids and alkalies; attacked by aromatic and chlorinated solvents. Applied by brush, spray, or dipping. Combustible.

Derivation: Formed from "prepolymers" (q.v.) containing isocyanate groups (toluene and 4,4′-diphenyl methane diisocyanates) and hydroxyl-containing materials such as polyols and drying oils.

Uses: Baked coatings; two-component formulations; wire coatings; tank linings; maintenance paints; masonry coating.

Elastomers:

Properties: Good resistance to abrasion, weathering, and organic solvents; tend to harden and become brittle at low temperatures. Combustible.

Derivation: Reaction of polyisocyanates with linear polyesters or polyethers containing hydroxyl groups.

Uses: Sealants and caulking agents; adhesives; films and linings; shoe uppers and heels; encapsulation of electronc parts; binders for rocket propellants; abrasive wheels and other mechanical items; auto bumpers, fenders and other components.

Foams:

Properties: Both flexible and rigid foams are available. Density varies from 2 to 50 pounds per cubic foot; thermal conductivity as low as 0.11; excellent insulators.

Hazard: Combustible unless protected by effective thermal barrier. See also under foam, plastic.

Derivation: A polyether such as polypropylene glycol is treated with a diisocyanate in the presence of some water and a catalyst (amines, tin soaps, organic tin compounds). As the polymer forms, the water reacts with the isocyanate groups to cause cross linking, and also produces carbon dioxide which causes foaming. In other cases trifluoromethane or similar volatile material may be used as blowing agent.

Flexible foams are based on polyoxypropylenediols of 2000 molecular weight and triols up to 4000.

Rigid foams are based on polyethers made from sorbitol, methyl glucoside, or sucrose.

Uses: (Flexible) Furniture; mattresses; laminates and linings; building insulation; flooring leveling; seat cushions and other automotive accessories; carpet underlays; upholstery; absorbent of crude oil spills on seawater. (Rigid) Furniture; packaging and packing; marine flotation; auto components; cigarette filters; light structures, as boat hulls; soundproofing; salvaging operations; ship-building (for buoyancy); transportation insulation for box cars, refrigerated cars, tank and hopper cars, trucks and trailers (claimed to be twice as effective as glass fiber); insulation for storage tanks, ships' holds, and pipelines.

polyvinyl acetal. One of the family of vinyl resins resulting from the condensation of polyvinyl alcohol with an aldehyde; acetaldehyde, formaldehyde and butyraldehyde are commonly used. The three main groups are polyvinyl acetal itself, polyvinyl formal, and polyvinyl butyral. These are all thermoplastic, and can be processed by extruding, molding, coating and casting. They are used chiefly in adhesives, paints and lacquers, and films. Polyvinyl butyral is used in sheet form as an interlayer in safety glass and shatter-resistant acrylic protection in aircraft.

polyvinyl acetate (PVAc). $[—CH_2CH(OOCCH_3)—]_x$. A thermoplastic high polymer.

Properties: Colorless, odorless, transparent solid; sp. gr. 1.19 at 15°C; insoluble in water, gasoline, oils and fats; soluble in low molecular weight alcohols, esters, benzene, and chlorinated hydrocarbons. Resistant to weathering. Combustible; nontoxic.

Derivation: Polymerization of vinyl acetate with peroxide catalysts.

Uses: Latex water paints; adhesives for paper, wood, glass, metals, and porcelain; intermediate for conversion to polyvinyl alcohol and acetals; sealant; shatterproof photographic bulbs; paper coating and paperboard; bookbinding; textile finishing; nonwoven fabric binder; component of lacquers, inks, and plastic wood; strengthening agent for cements.

polyvinyl alcohol (PVA; PVOH) $(-CH_2CHOH-)_x$. A water-soluble synthetic polymer made by alcoholysis of polyvinyl acetate.

Properties: White to cream-colored powder; sp. gr. 1.27–1.31. Refractive index 1.49–1.53. Properties depend on degree of polymerization (q.v.) and the percentage of alcoholysis, both of which are controllable in processing. Water solubility increases as molecular weight decreases; strength, elongation, tear resistance, and flexibilty improve with increasing molecular weight decreases; strength, elongation, tear resistance, and flexibility improve with increasing molecular weight. Tensile strength up to 22,000 psi; decomposes at 200°C. PVA has high impermeability to gases; is unaffected by oils, greases, and petroleum hydrocarbons. Attacked by acids and alkalies. It forms films by evaporation from water solution. Combustible. Low toxicity.

Grades: Super-high viscosity (m.w. 250,000–300,000); high-viscosity (m.w. 170,000–220,000); medium-viscosity (m.w. 120,000–150,000); low viscosity (m.w. 25,000–35,000).

Uses: Textile warp and yarn size; laminating adhesives; binder for cosmetic preparations, ceramics, leather, cloth, nonwoven fabrics, and paper; paper coatings; greaseproofing paper; emulsifying agent; thickener and stabilizer; photosensitive films; cements and mortars; intermediate for other polyvinyls; component of mixtures for coloring shell eggs.

polyvinylbenzyltrimethyl ammonium chloride. An electrically conductive polymer, used to increase the conductivity of papers.

polyvinyl butyral. See polyvinyl acetal.

polyvinyl carbazole. A brown, thermoplastic resin obtained by the reaction of acetylene with carbazole. It softens about 150°C, has excellent dielectric properties, good heat and chemical stability, but poor mechanical strength. It is used principally as a substitute for mica in electrical equipment and as an impregnant for paper capacitors.

polyvinyl chloride (PVC) $(—H_2CCHCl—)_x$. A synthetic thermoplastic polymer.

Properties: White powder or colorless granules. Resistant to weathering and moisture; dimensionally stable; good dielectric properties; resistant to most acids, fats, petroleum hydrocarbons and fungus. Readily compounded into flexible and rigid forms by use of plasticizers, stabilizers, fillers, and other modifiers. Easily colored and processed by blow molding, extrusion, calendering, fluid bed coating, etc. Available as film, sheet, fiber and foam.

Derivation: Polymerization of vinyl chloride (q.v.) by suspension, emulsion, and solution methods. May be copolymerized with up to 15% of other vinyls.

Hazard: Decomposes at 300°F, evolving toxic fumes of hydrogen chloride. Tolerance, 1 ppm in air. See also vinyl chloride, and note.

Uses: Piping and conduits of all kinds; siding; gutters; window and door frames; officially approved for use in interior piping, plumbing, and other construction uses; raincoats, toys, gaskets, garden hose, electrical insulation, shoes, magnetic tape, film and sheeting; containers for toiletries, cosmetics, household chemicals and food products; fibers for athletic supports; sealant liners for ponds and reservoirs; adhesive and bonding agent; plastisols and organosols; tennis court playing surfaces; flooring; coating for paper and textiles; wire and cable protection; base for synthetic turf; phonograph records.

Note: Use of PVC in rigid and semirigid food containers, such as bottles, boxes, etc., is under restriction by FDA, as well as in coatings for fresh citrus fruits. Its use in thinner items such as films and package coatings is permissible. Possibility of migration of vinyl chloride monomer into food products is the critical factor; this tends to increase with the thickness of the material.

polyvinyl chloride-acetate. A vinyl chloride and vinyl acetate that is inherently more flexible than polyvinyl chloride. The copolymer usually contains 85 to 97% of the chloride. It is generally similar in properties and uses to polyvinyl chloride.

polyvinyl dichloride (PVDC). A chlorinated polyvinyl chloride. Has high strength and superior chemical resistance over a broad temperature range; self-extinguishing. Used for pipe and fittings for hot corrosive materials up to 212°F. Is immume to solvation or direct attack by inorganic reagents, aliphatic hydrocarbons and alcohols.

polyvinyl ether. See polyvinyl ethyl ether; polyvinyl isobutyl ether; polyvinyl methyl ether; polyvinyl methyl ether-maleic anhydride.

polyvinyl ethyl ether (PVE; polyvinyl ether) $[—CH(OC_2H_5)CH_2—]_n$.

Properties: Viscous gum to rubbery solid, depending on molecular weight. Colorless when pure; sp. gr. 0.97 (20°C); refractive index 1.45 (25°C). Insoluble in water; soluble in nearly all organic solvents. Sta-

ble toward dilute and concentrated alkalies and dilute acids. Compatible with a limited number of commercial resins, including rosin derivatives and some phenolics.
Derivation: Polymerization if vinyl ethyl ether.
Uses: Pressure-sensitive tape; to improve adhesion to porous surfaces, cellophane, cellulose acetate and vinyl sheet.

polyvinyl fluoride $(-H_2CCHF-)_n$. Polymer of vinyl fluoride. In film form it is characterized by superior resistance to weather, high strength, high dielectric constant, low permeability to air and water, as well as oil, chemical solvent and stain resistance.
Hazard: Not recommended for food packaging. Evolves toxic fumes on heating.
Uses: Protective material for outdoor use; packaging; electrical equipment.

polyvinyl formal. See polyvinyl acetal.

polyvinylidene chloride (saran). A stereoregular, thermoplastic polymer.
Properties: Tasteless, odorless, nontoxic; abrasion-resistant; low vapor transmission; impermeable to flavor. Highly inert to chemical attack; softened by chlorinated hydrocarbons and soluble in cyclohexanone and dioxane. Combustible but self-extinguishing.
Derivation: Polymerization of vinylidene chloride (q.v.) or copolymerization of vinylidene chloride with lesser amounts of other unsaturated compounds.
Forms: Extruded and molded products; oriented fibers; films.
Uses: Packaging of food products, especially meats and poultry; insecticide-impregnated multiwall paper bags; pipes for chemical processing equipment; seat covers, upholstery, fibers, bristles, latex coatings.
See also saran fiber; "Cryovac."

polyvinylidene fluoride $CH_2 = CF_2$. A thermoplastic fluorocarbon polymer suitable for compression and injection molding and extrusion.
Properties: Crystalline melting point 340°F. Thermally stable from −80 to 300°F. Self-extinguishing and nondripping. Tensile strength 7000 psi at 77°F; yield stress 5500 psi; elongation 300%; compression strength 10,000 psi; thermal conductivity 1.05 Btu/hr/sq ft/°F/in; water absorption 0.04% in 24 hrs; sp. gr. 1.76; refractive index 1.42. Resistant to oxidative degradation, electricity, acids, alkalies, oxidizers, halogens. Somewhat soluble in dimethylacetamide; attacked by hot conc. sulfuric acid or n-butylamine.
Forms: Powder, pellets, solution, and dispersion.
Uses: Insulation for high-temperature wire, tank linings, chemical tanks and tubing, protective paints and coatings with exceptional resistance (30 years) to weathering and U.V.; valve and impeller parts; shrinkage tubing to encapsulate resistors, diodes, and soldered joints; sealant.

polyvinyl isobutyl ether (PVI; polyvinyl ether) $[-CH(OCH_2CH(CH_3)_2)CH_2-]_n$.
Properties: White, opaque elastomer or viscous liquid depending on molecular weight; almost odorless. Sp. gr. 0.91–0.93 (20°C); refractive index 1.45–1.46 (25°C). Insoluble in water, ethanol, and acetone; soluble in most other organic solvents. Stable toward dilute and concentrated alkalies and dilute acids. Compatible with a limited number of commercial resins, including rosin derivatives and some phenolics. Combustible.
Derivation: Polymerization of vinyl isobutyl ether (q.v.) by peroxides or acid catalysts.
Grades: As 100% material in three molecular weight ranges.
Uses: Adhesives, waxes, tackifiers, plasticizers, surface coatings, laminating agents, cable filling compositions, lubricating oils.

polyvinyl methyl ether (PVM). $(-CH_2CHOCH_3-)_n$. A nonionic polymer of high molecular weight.
Properties: Colorless (when pure) tacky liquid; soluble in water below 32°C, above which it precipitates from water but redissolves on cooling. Soluble in most organic solvents except aliphatic hydrocarbons. Compatible with rubber latexes, rosin derivatives, and some phenolics. Sp. gr. 1.05; refractive index 1.47; solidifies near 0°C. Combustible; low toxicity.
Derivation: Polymerization of vinyl methyl ether (q.v.) with peroxides or acid catalysts.
Uses: Pressure-sensitive and hot-melt adhesives; bonding of paper to polyethylene; laminating adhesive; tackifier and plasticizer for coatings; heat sensitizer for rubber latex; pigment binder in textile finishing; printing inks; protective colloid in emulsions.

polyvinyl methyl ether-maleic anhydride (PVM/MA). Water-soluble copolymer of vinyl methyl ether and maleic anhydride $[-CH_2CHOCH_3CHCOOCOCH-]_n$.
Uses: Protective colloid, dispersing agent, thickener, binder, adhesive and size in coatings, detergents, emulsions, paper, textiles, leather, latex, rust preventive, foam stabilizer.

poly-2-vinylpyridine $(-CH(C_5H_4N)-CH_2-)_n$. A vinyl-type polymer suggested for use as a photographic dye mordant, tablet coating material, antistat for textiles and paper, and textile dye receptor.

polyvinylpyrrolidone (PVP) $(C_6H_9NO)_n$.
Properties: White, free-flowing amorphous powder or aqueous solution. Soluble in water and organic solvents. Combustible with wide range of hydrophilic and hydrophobic resins. Sp. gr. 1.23–1.29; bulk density 25 lb per cubic foot. Hygroscopic; nontoxic.
Grades: Various molecular weights: 10,000; 40,000; 160,000; 360,000.
Uses: Pharmaceuticals; blood plasma expander; cast films adherent to glass, metals, and plastics; complexing agent; detoxification of chemicals such as dyes, iodine, phenol, and poisonous drugs; tableting; photographic emulsions; cosmetics (hair sprays, shampoos, hand creams, skin lotions); dentifrices; dye-stripping; textile finishes; protective colloid; detergents; adhesives; beer and wine clarification.

polyvinyl resin (vinyl plastic). Any of a series of polymers (resins) derived by polymerization or copolymerization of vinyl monomers $(CH_2{=}CH-)$ including vinyl chloride and acetate, vinylidene chloride, methyl acrylate and methacrylate, acrylonitrile, styrene, the vinyl ethers, and numerous others. Specifically, polyvinyl chloride, acetate, alcohol, etc., and copolymers or closely related materials. See under both vinyl and polyvinyl.

"Polywood."[448] Trademark for polyester coatings for wood furniture and other wood products.

"Polyzime" P.[212] Trademark for a product containing diastatic and proteolytic enzymes.
Properties: Dry, fine white powder; fully water soluble; nontoxic; nonflammable; optimum pH for dia-

static reaction, 7.0–7.2, for proteolytic reaction 7.5–8.0; optimum temperature 45°C.
Use: Desizing of textile fabrics preparatory to dyeing, bleaching, mercerizing printing and finishing.

"Poly-Zole" AZDN.[511] Trademark for azodiisobutyronitrile, a blowing agent and catalyst for plastics.

"Pomalus."[243] Trademark for food grade of malic acid.

Ponceau 3R $(CH_3)_3C_6H_2NNC_{10}H_4(OH)(SO_3Na)_2$. 3-Hydroxy-4-[2,4,5-trimethylphenyl)azo]-2,7-naphthalenedisulfonic acid, disodium salt. A water-soluble red dye, C.I. No. 16155.
Properties: Dark-red powder. Soluble in water and acids to form a cherry-red solution; slightly soluble in alcohol; insoluble in alkalies.
Uses: Dyeing wool; biological stain.

"Ponolith."[28] Trademark for a type of highly dispersed pigments used primarily for the coloring and tinting of paper.

"Ponsol."[28] Trademark for a series of vat dyes derived from anthraquinone. Used for dyeing and printing of cotton and rayon.

"Ponstan."[330] Trademark for mefenamic acid.

"Pontachrome."[28] Trademark for a series of colors used on wool and nylon. Dyed by forming chromium lakes on the fiber.

"Pontacyl."[28] Trademark for a series of acid colors for wool, nylon, silk, leather, paper, and some types of "Orlon" acrylic fiber. Dyed in an acid bath. Produces bright shades.

"Pontamine."[28] Trademark for a series of dyes and developers for textile fibers, paper and leather.

pontianak. A type of Manila resin (q.v.) which is semi-fossil, hard, and soluble in oils and hydrocarbons.
Uses: Adhesives; paints, varnishes, and lacquers. See also copal resin.

"Pontocaine" Hydrochloride.[162] Trademark for tetracaine hydrochloride (q.v.).

POPDA. See polyoxypropylene diamine.

POPOP. Abbreviation for phenyl-oxazolyl-phenyl-oxazolyl-phenyl. See 1,4-bis[2-(5-phenyloxazolyl]-benzene.

porcelain. Potassium aluminum silicate $(4K_2O \cdot Al_2O_3 \cdot 3SiO_2)$. A mixture of clays, quartz and feldspar, usually containing at least 25% of alumina. Ball and china clays are ordinarily used. A slip or slurry is formed with water to form a plastic, moldable mass, which is then glazed and fired to a hard, smooth solid.
Properties: High impact strength; impermeable to liquids and gases; resistant to chemicals except hydrofluoric acid and hot, strong caustic solutions; usable up to 2000°F but subject to heat shock; sp. gr. 2.41, Mohs hardness 6–7, compression strength 100,000 psi.
Grades: Chemical and electrical.
Uses: Reaction vessels, sparkplugs, electrical resistors; electron tubes; corrosion-resistant equipment; ball mills and grinders; food-processing equipment; piping, valves, pumps, tower packing, lab ware.
See also ceramics; porcelain enamel; porcelain, zircon.

porcelain enamel. A substantially vitreous inorganic coating bonded to metal by fusion above 800°F (ASTM). Composed of various blends of low-sodium frit (q.v.), clay, feldspar and other silicates; ground in a ball mill and sprayed onto a metal surface (steel, iron, or aluminum) to which it bonds firmly after firing, giving a glass-like, fire-polished surface.
Properties: High hardness, abrasion and chemical resistance except to HF and hot, strong caustic; low expansion coefficient; corrosion-resistant; stable to heat-shock; easy to clean and decontaminate.
Uses: Chemical reaction equipment; light reflectors; lab bench tops; storage tanks and containers; high-temperature process equipment; sea-water treatment and marine engine parts; glazing frit.
See also porcelain; glaze.

porcelain, zircon. $(ZrO_2 \cdot SiO_2)$. A special high-temperature porcelain used for spark plugs and furnace trays because of its high mechanical strength and heat-shock resistance. Usable up to 1700°C. High dielectric strength, but rather low power factor at high frequencies.

"Porocel."[99] Trademark for an activated bauxite. Supplied in various standard meshes, moisture contents, and in regular low iron and low silica grade.
Use: Absorbent catalysts.

poromeric. A term coined to describe the microporosity, air-permeability, and water- and abrasion resistance of natural and synthetic leather. The pores decrease in diameter from the inner surface to the outer and thus permit air and water vapor to leave the material while excluding water from the outside. Polyester-reinforced urethane resins have poromeric properties, and vinyl resins have been used as leather substitutes with some success, primarily for shoe uppers. A recently introduced synthetic poromeric for this purpose is trademarked "Xylee."

porphyrin. Any of several physiologically active nitrogenous compounds occurring widely in nature. The parent structure is comprised of four pyrrole rings, shown as I, II, III, IV in the diagram below, together with four nitrogen atoms and two replaceable hydrogens, for which various metal atoms can be readily substituted. A metal-free porphyrin molecule has the structure:

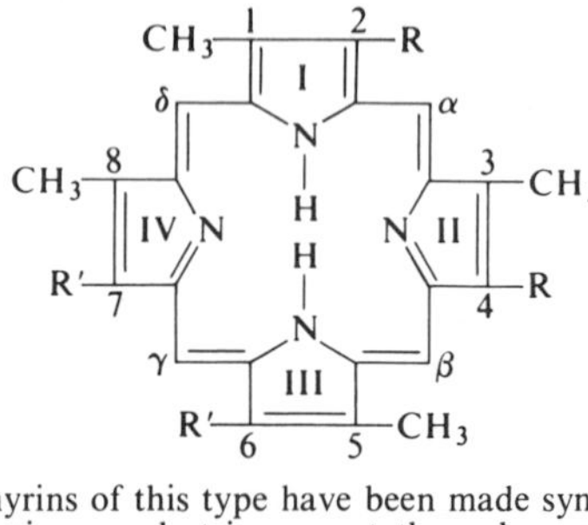

Porphyrins of this type have been made synthetically by passing an electric current through a mixture of ammonia, methane, and water vapor. Biochemists believe that this phenomenon may account for the original formation of chlorophyll and other porphyrins which have been essential factors in the origin of life.

The most important porphyrin derivatives are characterized by a central metal atom; hemin is the iron-containing porphyrin essential to mammalian blood, and chlorophyll is the magnesium-containing porphyrin that catalyzes photosynthesis. Other derivatives of importance are cytochromes, which function in cellular metabolism, and the phthalocyanine group of dyes.

"Poroplastic."[578] Trademark for a microporous plastic material reported to be effective for the controlled release of fragrances, flavors, medicinals, pesticides, etc. It has the mechanical properties of a polymer with pore size of molecular dimensions, which enable it to carry a "payload" of up to 98% of a liquid or solid immobilized within its structure. Compounds can be selectively admitted into or excluded from the material on a precisely controlled basis. Produced as microbeads, films, and granules.

porpoise oil (dolphin oil).
Properties: Pale yellow liquid; sp. gr. 0.926–0.929; saponification value about 290; iodine value 10 to 30. Soluble in ether, chloroform, benzene and carbon disulfide. The refined jaw and head oil has uniquely low pour point and high lubricity and is highly resistant to gumming, oxidation and evaporation.
Grades: Technical; also sold as body oil, jaw oil and junk oil (from the face blubber).
Containers: Steel drums; tins; glass bottles; vials.
Uses: Lubricant for watches and precision instruments.

"Portite."[326] Trademark for corrosion-resistant masonry sulfur-base cement.

Portland cement. See cement, Portland.

potash. See potassium carbonate. Also refer to American Potash Institute, 16th St. N.W., Washington, D.C.

potash alum. See aluminum potassium sulfate.

potash, caustic. See potassium hydroxide.

potash chrome alum. See chromium potassium sulfate.

potash feldspar. See feldspar.

potash magnesia double salt. A material containing potassium carbonate, magnesium sulfate and a low proportion of chloride, containing 20–30% K_2O and used as a fertilizer.

potash, sulfurated (potassium, sulfurated). A mixture composed chiefly of potassium polysulfides and potassium thiosulfate, containing not less than 12.8% sulfur in combination as sulfide.
Properties: Liver brown when freshly made, changing to a greenish yellow; odor of hydrogen sulfide; bitter, acrid, and alkaline taste; decomposes upon exposure to air; soluble in water leaving a residue.
Grade: U.S.P.
Uses: Medicine; decorative color effects on brass, bronze and nickel.

potassium K Element of atomic number 19, group IA of the periodic system; an alkali metal. Atomic weight 39.102; valence 1. Potassium 40 is a naturally occurring radioactive isotope. There are also two stable isotopes. The synthetic isotope, potassium 42, is used in tracer studies, primarily in medicine. An essential element in plant growth; occurs in all soils.
Properties: Soft, silvery metal; rapidly oxidized in moist air. Sp. gr. 0.862; m.p. 63°C; b.p. 770°C. Soluble in liquid ammonia, aniline, mercury, and sodium.
Derivation: Ores and deposits in Stassfurt (Germany), Carlsbad (New Mexico), Saskatchewan (Canada), Searles Lake (California), Great Salt Lake (Utah), Yorkshire (England), Ural Mts (USSR), Israel and eastern Mediterranean area. Most important ores are carnallite, sylvite, and polyhalite. Thermochemical distillation of potassium chloride with sodium is the chief U.S. method of production.
Grades: Technical; 99.95% pure.
Hazard: Dangerous fire risk; reacts with moisture to form potassium hydroxide and hydrogen. The reaction evolves much heat causing the potassium to melt and spatter. It also ignites the hydrogen. Burning potassium is difficult to extinguish; dry powdered soda ash or graphite or special mixtures of dry chemical is recommended. It can ignite spontaneously in moist air. Moderate explosion risk by chemical reaction. Potassium metal will form the peroxide and the superoxide at room temperature even when stored under mineral oil; may explode violently when handled or cut. Oxide-coated potassium should be destroyed by burning. Store in inert atmospheres, such as argon or nitrogen, or under liquids which are oxygen-free, such as toluene or kerosene, or in glass capsules which have been filled under vacuum or inert atmosphere.
Uses: Preparation of potassium peroxide; heat exchange alloys (See NaK); laboratory reagent; seeding of combustion gases in magnetohydrodynamic generators.
Shipping regulations: (Rail) Yellow label. (Air) Flammable Solid label. Not acceptable on passenger planes.

potassium abietate $KO_2CC_{19}H_{29}$. Water-soluble soap resulting from action of rosin and potassium hydroxide. Used as a pesticide.

potassium acetate $KC_2H_3O_2$.
Properties: White, crystalline deliquescent powder; saline taste. Soluble in water and alcohol; insoluble in ether. Solutions alkaline to litmus but not to phenolphthalein. M.p. 292°C; sp. gr. 1.57 (25°C). Low toxicity; combustible.
Derivation: By the action of acetic acid on potassium carbonate.
Grades: Pure; pure fused; C.P.; N.F. reagent.
Containers: Bottles; drums.
Uses: Dehydrating agent; textile conditioner; reagent in analytical chemistry; medicine; acetone; cacodylic derivatives; crystal glass; laboratory reagent.

potassium acid carbonate. See potassium bicarbonate.

potassium acid fluoride. See potassium bifluoride.

potassium acid oxalate. See potassium binoxalate.

potassium acid phosphate. See potassium phosphate, monobasic.

potassium acid saccharate $HOOC(CHOH)_4COOK$.
Properties: Light, off-white powder; pH of solution, 3.5; slightly soluble in cold water; soluble in hot water, acid, or alkaline solutions. Combustible.
Uses: Chelating agent; rubber formulations; metal plating; soaps and detergents.

potassium acid sulfate. See potassium bisulfate.

potassium acid sulfate, anhydrous. See potassium pyrosulfate.

potassium acid sulfite. See potassium bisulfite.

potassium acid tartrate. See potassium bitartrate.

potassium alginate (potassium polymannuronate) $(C_6H_7O_6K)_n$. Hydrophilic colloid having a molecular weight of 32,000 to 250,000.
Properties: Occurs in filamentous, grainy, granular,

and powdered forms. It is colorless or slightly yellow and may have a slight characteristic smell and taste. Slowly soluble in water forming a viscous solution; insoluble in alcohol. Combustible; nontoxic.
Grades: Technical; F.C.C.
Uses: Thickening agent and stabilizer in dairy products, canned fruits and sausage casings; emulsifier.
See also alginic acid.

potassium alum. See aluminum potassium sulfate.

potassium aluminate $K_2Al_2O_4 \cdot 3H_2O$.
Properties: Hard crystals, lustrous. Soluble in water with hydrolysis to form strongly alkaline solution; insoluble in alcohol.
Derivation: By fusing potassium hydroxide with aluminum oxide.
Grade: Technical.
Uses: Dyeing, printing (mordant); lakes; paper sizing.

potassium aluminosilicate. See feldspar.

potassium aluminum fluoride K_3AlK_6.
Properties: White powder; slightly soluble in water.
Derivation: Aluminum fluoride, ammonium fluoride, and potassium chloride.
Containers: Fiber cans.
Hazard: Toxic by ingestion and inhalation; strong irritant. Tolerance (as F), 2.5 mg per cubic meter of air.
Use: Insecticide.

potassium aluminum sulfate. See aluminum potassium sulfate.

potassium antimonyl tartrate. See antimony potassium tartrate.

potassium argentocyanide. See silver potassium cyanide.

potassium arsenate (Macquer's salt) KH_2AsO_4.
Properties: Colorless crystals; sp. gr. 2.867; m.p. 288°C; soluble in water; insoluble in alcohol.
Hazard: Highly toxic by ingestion and inhalation; strong irritant.
Uses: Manufacture of fly paper; insecticidal preparations; laboratory regent; preserving hides; printing textiles.
Shipping regulations: (Rail, Air) Poison label.

potassium arsenite (potassium metaarsenite) $KH(AsO_2)_2 \cdot H_2O$.
Properties: White powder; hygroscopic; decomposes slowly in air. Variable composition. Keep well stoppered! Soluble in water; slightly soluble in alcohol.
Grades: Technical; reagent.
Hazard: Highly toxic by ingestion and inhalation; strong irritant.
Uses: Medicine; mirrors.
Shipping regulations: (Rail, Air) Poison label.

potassium aurate $KAuO_2 \cdot 3H_2O$.
Properties: Yellow crystals; soluble in water, alcohol.
Derivation: Gold oxide dissolved in potassium hydroxide solution.
Use: To prepare other gold compounds.

potassium beryllium fluoride. See beryllium potassium fluoride.

potassium bicarbonate (potassium acid carbonate; baking soda) $KHCO_3$.
Properties: Colorless, odorless, transparent crystals or white powder; slightly alkaline, salty taste. Soluble in water and potassium carbonate solution; insoluble in alcohol. Sp. gr. 2.17; m.p., dec. between 100° and 120°C; refractive index 1.482.
Derivation: By passing carbon dioxide into a solution of potassium carbonate in water.
Grades: Commercial; highest purity; U.S.P.; reagent; F.C.C.
Containers: Drums; bags.
Uses: Baking; soft drinks; medicine (antacid); manufacture of pure potassium carbonate; fire-extinguishing agent; low pH liquid detergents; laboratory reagent.

potassium bichromate. See potassium dichromate.

potassium bifluoride (potassium acid fluoride; potassium hydrogen fluoride). KHF_2.
Properties: Colorless crystals; decomposed by heat. Soluble in alcohol (dilute), water; insoluble in alcohol (absolute). Sp. gr. 2.37; decomposes about 225°C.
Grade: Technical.
Hazard: Highly toxic and corrosive to tissue. Tolerance (as F), 2.5 mg per cubic meter of air.
Uses: Etching glass; production of fluorine; flux in metallurgy.
Shipping regulations: (Air) Solid: Poison label. Solution: Corrosive label.

potassium binoxalate (potassium acid oxalate; acid potassium oxalate; sorrel salt) $KHC_2O_4 \cdot \frac{1}{2}H_2O$.
Properties: White crystals; bitter, sharp taste; somewhat hygroscopic. Soluble in water; insoluble in alcohol; sp. gr. of the anhydrous salt 2.088; m.p. decomposes when heated.
Derivation: Neutral potassium oxalate and oxalic acid are dissolved in water and crystallized.
Hazard: Toxic, by ingestion.
Uses: Removing ink stains; scouring metals; cleaning wood; photography; bleaching; laboratory reagent.

potassium bismuth tartrate (bismuth potassium tartrate).
Properties: White, granular powder. Odorless; sweet taste; variable composition. Contains 60 to 64% bismuth. Decomposed by dilute mineral acid; darkens when exposed to light. Soluble in water; insoluble in alcohol, chloroform, ether.
Use: Medicine.

potassium bisulfate (acid potassium sulfate; potassium hydrogen sulfate; potassium acid sulfate) $KHSO_4$.
Properties: Colorless crystals; the fused salt is deliquescent. Soluble in water, yielding a solution with acid reaction; decomposes in alcohol. Sp. gr. 2.245; m.p. 214°C; b.p., decomposes. Low toxicity.
Derivation: Heating potassium sulfate with sulfuric acid.
Uses: Conversion of wine lees and tartrates into potassium bitartrate; flux; manufacture of mixed fertilizers; anisole; methyl acetate; ethyl acetate; laboratory reagent.

potassium bisulfide. See potassium hydrosulfide.

potassium bisulfite (potassium acid sulfite; potassium hydrogen sulfite). $KHSO_3$.
Properties: White, crystalline powder; sulfur dioxide odor. Soluble in water; insoluble in alcohol. M.p., decomposes at 190°C. Low toxicity.
Derivation: Sulfur dioxide is passed through a solution of potassium carbonate until no more carbon dioxide is given off; the solution is concentrated and allowed to crystallize.
Grades: Commercial; reagent; medicinal.
Uses: Reduction of various organic compounds; purification of aldehydes and ketones, iodine, sodium hydrosulfite; antiseptic; source of sulfurous acid, particularly in brewing; analytical chemistry; tan-

ning; bleaching straw and textile fibers; chemical preservative in foods, except meats and other sources of vitamin B_1.

potassium bitartrate (cream of tartar; potassium acid tartrate) $KHC_4H_4O_6$.
Properties: White crystals or powder; soluble in water; pleasant slightly acid taste; insoluble in alcohol. Sp. gr. 1.984 (18°C). Nontoxic.
Derivation: From wine lees by extraction with water and crystallization.
Grades: Technical; N.F.; F.C.C.
Containers: Barrels; multiwall paper bags.
Uses: Baking powder; preparation of other tartrates; medicine; galvanic tinning of metals; in foods as acid and buffer.

potassium borate. See potassium tetraborate.

potassium borofluoride. See potassium fluoborate.

potassium borohydride KBH_4.
Properties: White crystalline powder; soluble in water, alcohol, ammonia; insoluble in ethers and hydrocarbons; sp. gr. 1.18; stable in moist and dry air; stable in vacuum to 500°C. Decomposed by acids with evolution of hydrogen.
Derivation: By metathetical reaction of sodium borohydride and potassium hydroxide.
Grades: Technical; powder; pellets.
Containers: Glass bottles; polyethylene bags in drums.
Hazard: Flammable, dangerous fire risk.
Uses: Source of hydrogen and other borohydrides; reducing agent for aldehydes, ketones and acid chlorides.
Shipping regulations: (Rail) Flammable Solid, n.o.s., Yellow label. (Air) Flammable Solid label.

potassium bromate $KBrO_3$.
Properties: White crystals or crystalline powder. Soluble in water; slightly soluble in alcohol; sp. gr. 3.27; m.p. 434°C, with decomposition above 370°C. Strong oxidizing agent.
Derivation: By passing bromine into a solution of potassium hydroxide.
Grades: Pure; C.P.; F.C.C.
Hazard: Dangerous fire risk in contact with organic materials.
Uses: Laboratory reagent; oxidizing agent; permanent wave compounds; maturing agent in flour; dough conditioner; food additive.
Shipping regulations: (Rail) Yellow label; (Air) Oxidizer label.
Note: Food additive uses restricted as to proportions used.

potassium bromide KBr.
Properties: White, crystalline granules or powder; pungent, strong, bitter saline taste; somewhat hygroscopic. Soluble in water and glycerin; slightly soluble in alcohol and ether; sp. gr. 2.749; m.p. 730°C; b.p. 1435°C.
Derivation: Solutions of iron bromide and potassium carbonate are mixed and heated, the solution filtered and concentrated and the bromide crystallized out.
Grades: Technical; C.P.; N.F.; reagent; single crystals.
Hazard: Moderately toxic by ingestion and inhalation.
Uses: Medicine; photography (gelatin bromide papers and plates); process engraving and lithography; special soaps; spectroscopy; infrared transmission; laboratory reagent.

potassium tert-butoxide. See potassium tert-butylate.

potassium tert-butylate (potassium tert-butoxide) $(CH_3)_3COK$. White, hygroscopic powder; a strong organic base.
Hazard: Flammable; may cause severe caustic burns.

potassium carbonate (potash, pearl ash) (a) K_2CO_3; (b) $K_2CO_3 \cdot 1\frac{1}{2}H_2O$.
Properties: White, deliquescent, granular translucent powder; alkaline reaction. Soluble in water; insoluble in alcohol. (a) Sp. gr. 2.428 (19°C); m.p. 891°C; b.p., decomposes. Noncombustible.
Derivation: (1) Engel-Precht process uses magnesium oxide, potassium chloride, and carbon dioxide, separating the Engels salt ($MgCO_3 \cdot KHCO_3 \cdot 4H_2O$). Decomposition leaves potassium bicarbonate in solution which can be processed to potassium carbonate. (2) Alkyl amines or ion-exchange resins can be used with potassium chloride and carbon dioxide to yield potassium bicarbonate, which is calcined to the carbonate. This is analogous to the Solvay process for sodium carbonate (q.v.).
Grades: Crystals; technical; reagent; calcined 80–85%, 85–90%; 90–95%; 96–98%; hydrated 80–85%; F.C.C.
Containers: Drums; bags; kegs; barrels; unit trains.
Hazard: Solutions irritant to tissue.
Uses: Special glasses (optical and color TV tubes), potassium silicate; dehydrating agent; pigments; printing inks; laboratory reagent; soft soaps; raw wool washing; general-purpose food additive.

potassium chlorate $KClO_3$.
Properties: Transparent, colorless crystals or white powder; cooling, saline taste. Soluble in water, alkalies, and alcohol. Sp. gr. 2.337; m.p. 356°C; b.p., decomposes at 400°C, giving off oxygen.
Derivation: (a) By electrolyzing a hot concentrated alkaline solution of potassium chloride. Preferably (b) by interaction of solutions of potassium chloride and sodium chlorate or calcium chlorate.
Hazard: Moderately toxic; flammable; forms explosive mixtures with combustible materials (sulfur, sugar, etc.). Strong oxidizing agent.
Uses: Oxidizing agent; explosives; matches; source of oxygen; textile printing; pyrotechnics; percussion caps; medicines; dyes; paper manufacture; disinfectant; bleaching.
Shipping regulations: (Rail) Yellow label. (Air) Oxidizer label. Legal label name (Rail): chlorate of potash.

potassium chloride KC.
Properties: Colorless or white crystals; strong saline taste. Occurs naturally as sylvite. Soluble in water; slightly soluble in alcohol; insoluble in absolute alcohol. Sp. gr. 1.987; m.p. 772°C; sublimes at 1500°C; noncombustible; low toxicity.
Derivation: (a) Mined from sylvite deposits in New Mexico and Saskatchewan, purified by fractional crystallization or flotation; (b) extracted from salt lake brines, and purified by recrystallization.
Grades: Chemical (99.5 and 99.9%); agricultural grades sold as 60–62%, 48–52% and 22% K_2O; single crystals; U.S.P.; F.C.C.
Containers: Bags; drums; bulk.
Uses: Fertilizer; source of potassium salts; pharmaceutical preparations; photography; spectroscopy; plant nutrient; salt substitute; laboratory reagent.

potassium chloroaurate. See potassium gold chloride.

potassium chlorochromate (Peligot's salt) $KClCrO_3$.
Properties: Red crystals; on heating liberates chlorine. Soluble in water (dec). Sp. gr. 2.497.
Use: Oxidizing agent.

potassium chloroiridate. See iridium potassium chloride.

potassium chloroplatinate (platinum potassium chloride; potassium platinichloride) K_2PtCl_6.
Properties: Small, orange-yellow crystals or powder. Insoluble in alcohol; very slightly soluble in water; m.p. decomposes when heated. Sp. gr. 3.499 (24°C); decomposes at 250°C.
Derivation: By adding platinic chloride to a solution of a potassium salt and crystallizing.
Use: Photography; laboratory reagent.

potassium chromate (potassium chromate, yellow; neutral potassium chromate) K_2CrO_4.
Properties: Yellow crystals; soluble in water; insoluble in alcohol; sp. gr. 2.7319; m.p. 971°C.
Derivation: Roasting powdered chromite with potash and limestone, treating the cinder with hot potassium sulfate solution and leaching.
Grades: Technical; reagent.
Containers: Bottles; drums.
Hazard: Toxic by ingestion and inhalation.
Uses: Reagent in analytical chemistry; aniline black; textile mordant; enamels; chromate pigments; inks; medicine; fungicide; leather finishing; chromium compounds.

potassium chromate, red. See potassium dichromate.

potassium chromium sulfate. See chromium potassium sulfate.

potassium citrate $K_3C_6H_5O_7 \cdot H_2O$.
Properties: Colorless or white crystals or powder; deliquescent; cooling saline taste; odorless. Soluble in water and glycerol; almost insoluble in alcohol. Sp. gr. 1.98; m.p., decomposes when heated to about 230°C. Nontoxic.
Derivation: Action of citric acid on potassium carbonate.
Grades: Technical; C.P.; N.F.; F.C.C.
Containers: 25- to 250-lb drums.
Uses: Medicine; sequestrant, stabilizer, buffer in foods.

potassium cobaltinitrite. See cobalt potassium nitrite.

potassium columbate. Obsolete name for potassium niobate.

potassium copper cyanide (copper potassium cyanide; potassium cuprocyanide).
Properties: White, crystalline, double salt of copper cyanide and potassium cyanide. Copper content (Cu) min 25.8%; free KCN 1.25–3.0%.
Containers: 100-lb drums.
Hazard: Highly toxic.
Uses: For preparing and maintaining cyanide copper plating baths based on potassium cyanide.
Shipping regulations: (Air) Poison label. Legal label name potassium cuprocyanide.

potassium copper ferrocyanide. See copper potassium ferrocyanide.

potassium cupric ferrocyanide. See copper potassium ferrocyanide.

potassium cuprocyanide. Legal label name (Air) for potassium copper cyanide (q.v.).

potassium cyanate KOCN.
Properties: Colorless crystals; sp. gr. 2.05; decomposes 700–900°C. Soluble in water; insoluble in alcohol.
Derivation: Heating potassium cyanide with lead oxide.
Uses: Herbicide; manufacture of organic chemicals and drugs; treatment of sickle cell anemia.

potassium cyanide KCN.
Properties: White, amorphous, deliquescent lumps or crystalline mass; faint odor of bitter almonds. Soluble in water, alcohol and glycerol. Sp. gr. 1.52 (16°C); m.p. 634°C.
Derivation: Absorption of hydrocyanic acid in potassium hydroxide.
Grades: Commercial; pure; solution; reagent.
Containers: 25-, 100-lb drums.
Hazard: Highly toxic; absorbed by skin. Tolerance (as CN), 5 mg per cubic meter of air.
Uses: Extraction of gold and silver from ores; reagent in analytical chemistry; insecticide; fumigant; manufacture of various intermediate organic cyanogen derivatives.
Shipping regulations: (solid and solution) (Rail, Air) Poison label. Legal label name (Rail) Cyanide of potassium.

potassium cyanoargenate. See silver potassium cyanide.

potassium cyanoaurite. See potassium gold cyanide.

potassium cyclamate. See sodium cyclamate.

potassium di-n-butyl dithiocarbamate $(C_4H_9)_2NCSSK$. A 50% aqueous solution is a straw-colored liquid; sp. gr. 1.08–1.12. Used as an ultra-accelerator for natural and synthetic latexes.

potassium dichloroisocyanurate $Cl_2K(NCO)_3$. A cyclic compound. See also dichloroisocyanuric acid.
Properties: White; slightly hygroscopic; crystalline powder or granules; loose bulk density (approx) powder 37 lb/cu ft, granular 64 lb/cu ft. Active ingredient approx 59% available chlorine, decomp. at 240°C.
Containers: 200-lb fiber drums.
Hazard: Dangerous fire risk in contact with organic materials. Strong oxidizing agent. Toxic by ingestion.
Uses: Household dry bleaches, dishwashing compounds, scouring powders, detergent-sanitizers; replacement for calcium hypochlorite.
Shipping regulations: (Dry, containing more than 39% available chlorine) (Rail) Yellow label. (Air) Oxidizer label.

potassium dichromate (potassium bichromate; red potassium chromate) $K_2Cr_2O_7$.
Properties: Bright, yellowish red transparent crystals; bitter, metallic taste. Soluble in water; insoluble in alcohol. Sp. gr. 2.676 (25°C); m.p. 396°C; b.p., decomposes at 500°C.
Derivation: By adding sulfuric acid to crude potassium chromate solution.
Grades: Commercial; highest purity; highest purity fused; reagent.
Containers: 100-lb bags; 400-lb drums.
Hazard: Toxic by ingestion and inhalation. Dangerous fire risk in contact with organic materials. Strong oxidizing agent. Safety data sheet available from Manufacturing Chemists Assn., Washington, D.C.
Uses: Oxidizing agent (chemicals, dyes, intermediates); analytical reagent; brass pickling compositions; electroplating; pyrotechnics; explosives; safety matches; textiles; dyeing and printing; glass; chrome

glues and adhesives; chrome tanning leather; wood stains; poison fly paper; process engraving and lithography; photography; pharmaceutical preparations; synthetic perfumes; chrome alum manufacture; pigments; alloys; ceramic products.

potassium dihydrogen phosphate. See potassium phosphate, monobasic.

potassium dimethyldithiocarbamate. See sodium dimethyldithiocarbamate.

potassium diphosphate. See potassium phosphate, monobasic.

potassium dithionate (potassium hyposulfate) $K_2S_2O_6$.
Properties: Colorless crystals; soluble in water; insoluble in alcohol. Sp. gr. 2.2.78.
Use: Reagent.

potassium endothall (dipotassium salt of endothall, q.v.). The water solution is an amber liquid, used as a contact herbicide.

potassium ethyldithiocarbonate. See potassium xanthate.

potassium ethylxanthate. See potassium xanthate.

potassium ethylxanthogenate. See potassium xanthate.

potassium ferric oxalate $K_3Fe(C_2O_4)_3 \cdot 3H_2O$.
Properties: Monoclinic green crystals; loses 3 molecules water at 100°C; decomposes 230°C; soluble in water, acetic acid; incompatible with alkali and ammonia, since these react to precipitate ferric hydroxide.
Use: Photography and blue-printing.

potassium ferricyanide (red prussiate of potash; red potassium prussiate) $K_3Fe(CN)_6$.
Properties: Bright red, lustrous crystals or powder. Soluble in water; slightly soluble in alcohol. Sp. gr. 1.85 (25°C).
Derivation: Chlorine is passed into a solution of potassium ferrocyanide, the ferricyanide separating out.
Grades: Pure crystals; pure powder; reagent; technical.
Containers: 25-, 100-, 350-lb drums.
Hazard: Decomposes on strong heating to evolve highly toxic fumes, but the compound itself has low toxicity.
Uses: Tempering steel; etching liquid; production of pigments; electroplating; sensitive coatings on blueprint paper; fertilizer compositions; laboratory reagent.

potassium ferrocyanide (yellow prussiate of potash; yellow potassium prussiate) $K_4Fe(CN)_6 \cdot 3H_2O$.
Properties: Lemon yellow crystals or powder; mild saline taste; effloresces on exposure to air. Soluble in water; insoluble in alcohol. Sp. gr. 1.853 (17°C); m.p., loses its water of crystallization when heated to 70°C; b.p., decomposes.
Derivation: From nitrogenous waste products, iron filings and potassium carbonate.
Grades: Technical; C.P.
Containers: Drums; multiwall paper sacks.
Hazard: On heating to red heat it evolves highly toxic fumes, but the compound itself has low toxicity.
Uses: Medicine; potassium cyanide and ferricyanide; dry colors; tempering steel; dyeing; explosives; process engraving and lithography; laboratory reagent.

potassium fluoborate (potassium borofluoride) KBF_4.
Properties: Colorless crystals; sp. gr. 2.498 (20°C); decomp. 350°C; very slightly soluble in water; insoluble in alcohol, ether, alkalies.
Derivation: By the reaction of boric acid, hydrofluoric acid, and potassium hydroxide.
Containers: Fiber drums.
Uses: Sand-casting of aluminum and magnesium; grinding aid in resinoid grinding wheels; flux for soldering and brazing; electrochemical processes and chemical research.

potassium fluoride (a) KF; (b) $KF \cdot 2H_2O$.
Properties: White, crystalline, deliquescent powder; sharp saline taste. Soluble in water and hydrofluoric acid; insoluble in alcohol. Sp. gr. (a) 2.454, (b) 2.454. M.p. (a) 846°C; b.p. 1505°C.
Derivation: By saturation of hydrofluoric acid with potassium carbonate.
Grades: Technical; free of arsenic; C.P.
Containers: Polyethylene bottles; drums.
Hazard: Toxic by ingestion and inhalation; strong irritant to tissue. Tolerance (as F), 2.5 mg per cubic meter of air.
Uses: Etching glass; preservative; insecticide; production of fluorine.
Shipping regulations: (Air) (solid) No label required; (solution) Corrosive label.

potassium fluosilicate (potassium silicofluoride) K_2SiF_6.
Properties: White odorless, crystalline powder; sp. gr. 3.0; slightly soluble in water; soluble in hydrochloric acid.
Containers: 100-, 200-, 400-lb and in bulk.
Hazard: Toxic by ingestion and inhalation; strong irritant to tissue. Tolerance (as F), 2.5 mg per cubic meter of air.
Uses: Vitreous enamel frits; synthetic mica; metallurgy of aluminum and magnesium; ceramics; insecticide.

potassium fluotantalate. See tantalum potassium fluoride.

potassium fluozirconate. See zirconium potassium fluoride.

potassium germanium fluoride. See germanium potassium fluoride.

potassium gibberellate. Salt of gibberellic acid (q.v.). Used to promote and control development of malt in grain.

potassium gluconate $KC_6H_{11}O_7$.
Properties: Odorless, salty tasting, fine white crystalline powder. Soluble in water; insoluble in alcohol and benzene. Low toxicity.
Derivation: Reaction of potassium hydroxide or carbonate with gluconic acid.
Grade: Pharmaceutical.
Use: Medicine (vitamin tablets).

potassium glutamate (monopotassium L-glutamate) $KOOC(CH_2)_2CH(NH_2)COOH \cdot H_2O$.
Properties: White, practically odorless, free-flowing powder. Hygroscopic; freely soluble in water; slightly soluble in alcohol; pH of a 2% solution 6.9–7.1. Low toxicity.
Grade: F.C.C.
Uses: Flavor enhancer; salt substitute.

potassium glycerophosphate (potassium glycerinophosphate) $K_2C_3H_5O_2 \cdot H_2PO_4 \cdot 3H_2O$.
Properties: Pale yellow syrupy liquid; acid taste. Soluble in alcohol; miscible with water in all proportions. Nontoxic.
Derivation: Glycerol and phosphorus pentoxide or

metaphosphoric acid are mixed, warmed and exactly neutralized with potassium carbonate, warmed and concentrated.
Grades: Technical, 50 or 75% solution; F.C.C. (syrup or solution).
Uses: Food additive and dietary supplement.

potassium gold chloride (potassium aurichloride; potassium chloroaurate; gold potassium chloride) $KAuCl_4 \cdot 2H_2O$.
Properties: Yellow crystals. Soluble in water, alcohol and ether.
Derivation: By neutralizing chloroauric acid with potassium carbonate.
Uses: Photography; painting porcelain and glass; medicine.

potassium gold cyanide (potassium cyanoaurite; gold potassium cyanide) $KAu(CN)_2$.
Properties: White, crystalline powder. Soluble in water; slightly soluble in alcohol; insoluble in ether.
Derivation: Action of hydrocyanic acid on potassium aurate.
Hazard: Highly toxic.
Uses: Medicine; electrogilding.
Shipping regulations: (Rail, Air) Cyanides or cyanide mixtures, dry, Poison label.

potassium guaiacol sulfonate $C_6H_3OCH_3OHSO_3K \cdot \frac{1}{2}H_2O$.
Properties: White powder or crystals; gradually turns pink on exposure to air and light; bitter taste, afterward becoming sweetish; odorless. Contains approx. 60% guaiacol. Soluble in water; sparingly soluble in alcohol. Low toxicity.
Grade: N.F.
Containers: 25- to 300-lb drums.
Uses: Medicine; laboratory reagent.

potassium 2,4-hexadienoate. See potassium sorbate.

potassium hexafluorophosphate KPF_6.
Properties: Solid; m.p. 575°C (approx); b.p. decomposes. Soluble in water.
Grade: 98–100%.
Uses: Maintenance of fluoride atmospheres; preparation of bactericides and fungicides; laboratory reagent.

potassium hexanitrocobaltate III. See cobalt potassium nitrite.

potassium hexyl xanthane $C_6H_{13}OCSSK$. Used as a flotation agent.

potassium hydrate. See potassium hydroxide.

potassium hydride KH. Marketed as a semidispersion of gray powder in oil.
Properties: The solid decomposes on heating or in contact with moisture.
Hazard: Dangerous fire and explosion risk by reaction with water; evolves toxic and flammable gases on heating and on exposure to moisture.
Uses: Organic condensations and alkylations.
See also hydride.

potassium hydrogen fluoride. See potassium bifluoride.

potassium hydrogen phosphate. See potassium phosphate, dibasic.

potassium hydrogen phthalate. $HOOCC_6H_4COOK$.
Properties: Colorless crystals. Soluble in water; sp. gr. 1.636.
Derivation: Potassium hydroxide and phthalic anhydride.
Grades: C.P., analytical.
Use: Alkalimetric standard.

potassium hydrogen sulfate. See potassium bisulfate.

potassium hydrosulfide (potassium sulfhydrate; potassium bisufide) KHS.
Properties: Yellow crystals; hydrogen sulfide odor. When exposed to air, forms the polysulfide. Hygroscopic. Soluble in alcohol. Sp. gr. 1.69; m.p. 455°C. Decomposes in water. Low toxicity.
Grade: Technical.
Use: Separation of heavy metals.

potassium hydroxide (caustic potash; potassium hydrate; lye) KOH.
Properties: White, deliquescent pieces, lumps, sticks, pellets, or flakes having a crystalline fracture. Keep well stoppered; absorbs water and carbon dioxide from the air. Soluble in water, alcohol, glycerin; slightly soluble in ether. Sp. gr. 2.044; m.p. 405°C, but varies with water content.
Derivation: Electrolysis of concentrated potassium chloride solution.
Method of purification: Sulfur compounds are removed by the addition of potassium nitrate to the fused caustic. The purest form is obtained by solution in alcohol, filtration and evaporation.
Grades: Commercial; ground; flake; fused (88 to 92%); purified by alcohol (sticks, lumps and drops); reagent; highest purity; U.S.P.; liquid (45%); F.C.C.
Containers: Bottles; boxes; kegs; drums; tank cars.
Hazard: Toxic by ingestion and inhalation; strong irritant to tissue. Safety data sheet available from Manufacturing Chemists Assn., Washington, D.C.
Uses: Soap manufacture; bleaching; manufacture of potassium carbonate and tetrapotassium pyrophosphate; electrolyte in alkaline storage batteries and some fuel cells; absorbent for carbon dioxide and hydrogen sulfide; dyestuffs; liquid fertilizers; food additive; herbicides.
Shipping regulations: (solid and solution) (Air) Corrosive label. (Rail) White label. Legal label name, Caustic potash, liquid.

potassium hypophosphite (potassium hypophosphite, monobasic) KH_2PO_2.
Properties: White opaque crystals or powder with pungent saline taste, very deliquescent. Soluble in water and alcohol; decomposed by heat. Low toxicity.
Derivation: Interaction of calcium hypophosphite and potassium carbonate.
Containers: Drums.
Hazard: Moderate fire risk. May explode if ground with nitrates or other strong oxidizing agents.
Use: Medicine.

potassium hyposulfate. See potassium dithionate.

potassium hyposulfite. See potassium thiosulfate.

potassium iodate KIO_3.
Properties: White, crystalline powder. Odorless; soluble in water, sulfuric acid (dilute); insoluble in alcohol. Sp. gr. 3.9; m.p. 560°C (partial decomposition). Low toxicity.
Grades: Technical; C.P.; F.C.C.
Uses: Analysis (testing for zinc and arsenic); iodometry; medicine; reagent; feed additive; in foods as maturing agent and dough conditioner.

potassium iodide KI.
Properties: White crystals, granules or powder; strong bitter saline taste. Soluble in water, alcohol, acetone, and glycerol. Sp. gr. 3.123; m.p. 686°C; b.p. 1330°C. Low toxicity.
Grades: Reagent; U.S.P.; single crystals; F.C.C.
Containers: Bottles; drums.
Uses: Medicine; reagent in analytical chemistry; pho-

tography (precipitating silver); feed additive; spectroscopy, infrared transmission, scintillation; dietary supplement (up to 0.01% in table salt).

potassium iridium chloride. See iridium potassium chloride.

potassium laurate $KOOCC_{11}H_{23}$.
Properties: Light tan paste; soluble in water and alcohol.
Uses: Emulsifying agent; base for liquid soaps and shampoos.

potassium linoleate $KOOCC_{17}H_{31}$.
Properties: Light tan paste; soluble in water.
Use: Emulsifying agent.

potassium magnesium sulfate. $K_2SO_4 \cdot 2MgSO_4$.
Properties: White tetragonal crystals; sp. gr. 2.829; m.p. 927°C.
Use: Fertilizer.

potassium manganate K_2MnO_4.
Properties: Dark green powder or crystals; soluble in potassium hydroxide solution and water; decomposes in acid solution. M.p. 190°C (decomposes).
Derivation: Fusion of pyrolusite (q.v.) with potassium hydroxide.
Grade: Technical.
Hazard: Moderately toxic. Dangerous fire risk in contact with organic materials. Strong oxidizing agent.
Uses: Bleaching skins, fibers, oils; disinfectants; mordant (wool); batteries; photography; printing; source of oxygen (dyeing); water purification; oxidizing agent.

potassium mercuric iodide. See mercuric potassium iodide.

potassium metaarsenite. See potassium arsenite.

potassium metabisulfite (potassium pyrosulfite) $K_2S_2O_5$.
Properties: White granules or powder; pungent, sharp odor; sp. gr. 2.3; decomposes at 150–190°C; oxidizes in air and moisture to sulfate; slightly soluble in water and alcohol. Low toxicity.
Derivation: By heating potassium bisulfite until it loses water.
Grades: Technical; reagent; F.C.C.
Containers: Bottles; drums.
Uses: Antiseptic reagent in analytical chemistry; source of sulfurous acid; brewing (cleaning casks and vats); wine-making (kills only undesirable yeasts and bacteria); food preservative; developing agent (photography); process engraving and lithography; dyeing; antioxidant.

potassium, metallic, liquid alloy. A flammable mixture or alloy of potassium and another metal which is liquid at ordinary temperature. See, for example, NaK.
Shipping regulations: (Rail) Yellow label. (Air) Flammable Solid label. Not accepted on passenger planes.

potassium metaphosphate (monopotassium metaphosphate) KPO_3, or $(KPO_3)_6$.
Properties: White powder; soluble in acids (dilute); slightly soluble in water.

potassium molybdate K_2MoO_4.
Properties: White, microcrystalline powder; deliquescent; soluble in water; insoluble in alcohol; sp. gr. 2.91 (18°C); m.p. 919°C.
Use: Reagent.

potassium monophosphate. See potassium phosphate, dibasic.

potassium muriate. See potassium chloride.

potassium naphthenate.
Properties: Gray paste; soluble in water. Combustible.
Derivation: From naphthenic acids.
Uses: Driers; emulsifying agents.

potassium nickel sulfate. See nickel potassium sulfate.

potassium nitrate (niter; nitre; saltpeter) KNO_3.
Properties: Transparent, colorless or white crystalline powder or crystals; slightly hygroscopic; pungent, saline taste. Sp. gr. 2.1062; m.p. 337°C; b.p., decomposes at about 400°C. Soluble in water; slightly soluble in alcohol and glycerin. Low toxicity.
Grades: Commercial; C.P.; F.C.C.
Containers: Bottles; bags; barrels; drums.
Hazard: Dangerous fire and explosion risk when shocked or heated, or in contact with organic materials; strong oxidizing agent.
Uses: Pyrotechnics; explosives; matches; specialty fertilizer; reagent; to modify burning properties of tobacco; glass manufacture; metallurgy; oxidizer in solid rocket propellants; food preservative; color fixative in meats; anaphrodisiac.
Shipping regulations: (Rail) Yellow label. (Air) Oxidizer label.

potassium nitrite KNO_2.
Properties: White or slightly yellowish prisms or sticks; deliquescent. Sp. gr. 1.915; b.p., explodes at 1000°F; m.p. above 300°C. Soluble in water; insoluble in alcohol.
Grades: C.P.; technical; reagent; F.C.C.
Hazard: Dangerous fire and explosion risk when shocked or heated, or in contact with organic materials; strong oxidizing agent.
Uses: Analysis (testing for amino acids, cobalt, iodine, urea); medicine; organic synthesis.
Shipping regulations: (Rail) Yellow label. (Air) Oxidizer label.

potassium oleate $C_{17}H_{33}COOK$.
Properties: Gray-tan paste; soluble in water and alcohol. Combustible.
Uses: Textile soaps; emulsifying agent.

potassium orthophosphate. See potassium phosphate, monobasic, dibasic or tribasic.

potassium orthotungstate. See potassium tungstate.

potassium osmate (potassium perosmate) $K_2OsO_4 \cdot 2H_2O$.
Properties: Violet crystals. Hygroscopic. Soluble in water; insoluble in alcohol and ether.
Hazard: Moderately toxic.
Use: Determination of nitrogenous matter in water.

potassium oxalate $K_2C_2O_4 \cdot H_2O$.
Properties: Colorless transparent crystals. Odorless; soluble in water; efflorescent in warm dry air; sp. gr. 2.08; m.p., decomposes when heated.
Derivation: Potassium formate or carbonate mixed with a small quantity of oxalate and a slight excess of alkali is heated, the oxalate extracted with water and crystallized.
Grades: Technical; C.P.
Containers: Bottles; drums.
Hazard: Toxic by inhalation and ingestion.
Uses: Reagent in analytical chemistry; source of ox-

alic acid; bleaching and cleaning; removing stains from textiles; photography.

potassium oxide K_2O.
Properties: Gray, crystalline mass. Soluble in water; forms potassium hydroxide. Soluble in alcohol and ether. Sp. gr. 2.32 (0°C); decomposes 350°C.
Derivation: By heating potassium nitrate and metallic potassium.
Uses: Reagent and intermediate.

potassium oxyfluoniobate. See niobium potassium oxyfluoride.

potassium palladium chloride. See palladium potassium chloride.

potassium penicillin G (benzylpenicillin potassium) $C_{16}H_{17}KN_2O_4S$.
Properties: Colorless or white crystals or powder; odorless; moderately hygroscopic; solutions dextrorotatory; relatively stable to air and light; very soluble in water, in saline, and in dextrose solutions; moderately soluble in alcohol; pH of solution (30 mg/ml) is 5.0–7.5; m.p. 214–217°C (dec).
Grade: U.S.P.
Use: Medicine (antibiotic).

potassium penicillin V. See potassium phenoxymethylpenicillin.

potassium percarbonate $K_2C_2O_6 \cdot H_2O$.
Properties: Granular, white mass. Keep away from light and moisture. Soluble in water (liberates oxygen). M.P. 200–300°C.
Grade: Technical.
Hazard: Strong irritant to tissue. Strong oxidizing agent. Fire risk in contact with organic materials.
Uses: Analysis (testing for cerium, chromium, vanadium, titanium); microscopy; oxidizing agent; photography; textile printing.

potassium perchlorate $KClO_4$.
Properties: Colorless crystals or white, crystalline powder. Decomposed by concussion, organic matter, and agents subject to oxidation. More stable than potassium chlorate. Soluble in water; slightly soluble in alcohol. Sp. gr. 2.524; m.p. 610°C.
Grades: Technical; reagent; ordnance.
Containers: Drums.
Hazard: Strong oxidizing agent. Fire risk in contact with organic materials.
Uses: Explosives; medicine; oxidizing agent; photography; pyrotechnics and flares; reagent; oxidizer in solid rocket propellants; inflating agent in automobile safety air bags.
Shipping regulations: (Rail) Yellow label. (Air) Oxidizer label.

potassium periodate KIO_4.
Properties: Small, colorless crystals or white, granular powder; slightly soluble in water. Sp. gr. 3.168; m.p. 582°C (explodes at 1076°F).
Grades: Technical; C.P.; reagent.
Hazard: Toxic; strong irritant to tissue. Strong oxidizing agent. Fire risk in contact with organic materials.
Use: Analysis (oxidizing agent).

potassium permanganate $KMnO_4$.
Properties: Dark purple crystals; blue metallic sheen; sweetish, astringent taste; odorless. Soluble in water, acetone and methanol; decomposed by alcohol. Sp. gr. 2.7032; m.p., decomposes at 240°C. Oxidizing material.
Derivation: (a) By oxidation of the manganate in an alkaline electrolytic cell. (b) A hot solution of the manganate is treated with carbon dioxide; on cooling, the solution deposits crystals of the permanganate.
Grades: Technical; C.P.; U.S.P.
Containers: Bottles; drums.
Hazard: Toxic by ingestion and inhalation; strong irritant to tissue. Dangerous fire risk in contact with organic materials. Powerful oxidizing agent.
Uses: Oxidizer; disinfectant; deodorizer; bleach; dye; tanning; radioactive decontamination of skin; reagent in analytical chemistry; medicine; manufacture of organic chemicals; air and water purification.
Shipping regulations: (Rail) Yellow label. (Air) Oxidizer label.

potassium peroxide K_2O_2.
Properties: Yellow, amorphous mass; decomposes in water, evolving oxygen. M.p. 490°C.
Derivation: Oxidation of potassium in oxygen.
Hazard: Dangerous fire and explosion risk in contact with organic materials. Strong oxidizing agent. Irritant to skin and tissue.
Uses: Oxidizing agent; bleaching agent; oxygen-generating gas masks.
Shipping regulations: (Rail) Yellow label. (Air) Oxidizer label. Not acceptable on passenger planes.

potassium peroxydisulfate. See potassium persulfate.

potassium peroxymonosulfate. See "Oxone."

potassium persulfate (potassium peroxydisulfate) $K_2S_2O_8$.
Properties: White crystals; soluble in water; insoluble in alcohol; sp. gr. 2.477; m.p., decomposes below 100°C.
Derivation: By electrolysis of a saturated solution of potassium sulfate.
Containers: Glass bottles; drums; polyethylene-lined paper bags.
Hazard: Moderately toxic; strong irritant and oxidizing agent. Dangerous fire risk in contact with organic materials.
Uses: Bleaching; oxidizing agent; reducing agent in photography; antiseptic; soap manufacture; analytical reagent; polymerization promoter; pharmaceuticals; modification of starch; flour-maturing agent; desizing of textiles.
Shipping regulations: (Air) Oxidizer label.

potassium alpha-**phenoxyethylpenicillin** (potassium penicillin 152; potassium phenethicillin; phenethicillin) $KC_{17}H_{19}N_2O_5S$. Synthetically prepared; a mixture of two stereoisomers.
Properties: White, crystalline solid; moderately hygroscopic; decomposes above 220°C. Very soluble in water; resistant to acid decomposition.
Preparation: By N-acylation of alpha-phenoxypropionic acid and G-aminopenicillanic acid (produced by fermentation using Penicillium chrysogenum).
Grade: N.F.
Use: Antibiotic.

potassium phenoxymethylpenicillin (potassium penicillin V) $KC_{16}H_{17}N_2O_5S$.
Properties: White, odorless, crystalline powder; very soluble in water; slightly soluble in alcohol; insoluble in acetone.
Grade: U.S.P.
Use: Antibiotic.

potassium phosphate, dibasic (DKP; potassium hydrogen phosphate; potassium monophosphate; dipotassium orthophosphate) K_2HPO_4.
Properties: Deliquescent white crystals or powder.

Very soluble in water and alcohol; is converted to the pyrophosphate by ignition. Low toxicity.
Derivation: Action of phosphoric acid on potassium carbonate.
Grades: Commercial; pure; highest purity; N.F.; F.C.C.
Containers: Glass bottles; fiber drums.
Uses: Buffer in antifreezes; ingredient of "instant" fertilizers; nutrient for penicillin culturing; humectant; pharmaceuticals; in foods as a buffer, sequestrant, and yeast food; laboratory reagent.

potassium phosphate, monobasic (MKP; potassium acid phosphate; potassium diphosphate; potassium orthophosphate; potassium dihydrogen phosphate) KH_2PO_4.
Properties: Colorless crystals. Acid in reaction. Soluble in water; insoluble in alcohol; sp. gr. 2.338; m.p. 253°C. Low toxicity.
Derivation: Action of phosphoric acid on potassium carbonate.
Grades: Technical; C.P.; F.C.C.
Containers: Glass bottles; barrels; multiwall paper bags.
Uses: Medicine; baking powder; nutrient solutions; yeast foods; buffer and sequestrant in foods; laboratory reagent.

potassium phosphate, tribasic (potassium phosphate, neutral; potassium phosphate normal; tripotassium orthophosphate; potassium phosphate; tertiary; tripotassium phosphate) $K_3PO_4 \cdot H_2O$ or K_3PO_4.
Properties: Granular white powder. Hygroscopic. Soluble in water giving strongly basic solution; insoluble in alcohol. M.P. (anhydrous) 1340°C; sp. gr. (anhydrous) 2.564 (17°C).
Grades: Reagent; technical; F.C.C.
Containers: 275-, 300-, 400-lb drums.
Uses: Purification of gasoline; water-softening; liquid soaps; fertilizer; in foods as an emulsifier; laboratory reagent.

potassium phosphite, monobasic KH_2PO_3.
Properties: White powder; hygroscopic. Soluble in water; insoluble in alcohol. Slowly oxidized by air to phosphate.

potassium platinichloride. See potassium chloroplatinate.

potassium polymetaphosphate $(KPO_3)_n$. The molecular weight may be as high as 500,000. See sodium metaphosphate.
Properties: White odorless powder; insoluble in water; soluble in sodium salt solutions, which may have high viscosity. Low toxicity.
Derivation: Dehydration of monobasic potassium phosphate.
Grades: Technical; F.C.C.
Uses: Fat emulsifier and moisture-retaining agent in foods.

potassium polysulfide K_2S_x.
Properties: Crystals; soluble in water and alcohol.
Hazard: Toxic by ingestion. Irritant to skin and eyes. Moderate fire risk.
Use: Fungicide.

potassium prussiate, red. See potassium ferricyanide.

potassium prussiate, yellow. See potassium ferrocyanide.

potassium pyroantimonate $K_2H_2SbO_7 \cdot 4H_2O$ (approx).
Properties: White crystalline powder or granules; slightly soluble in cold water; readily soluble in hot water. Insoluble in alcohol.
Grades: Reagent; technical.
Uses: Starch sizes and flame-retarding compounds.

potassium pyroborate. See potassium tetraborate.

potassium pyrophosphate (TKPP; tetrapotassium pyrophosphate; potassium pyrophosphate, normal) $K_4P_2O_7 \cdot 3H_2O$.
Properties: Colorless crystals or white powder; somewhat hygroscopic in air (deliquesces above a relative humidity of 40–45%). Similar to tetrasodium pyrophosphate except for greater solubility; sp. gr. 2.33; dehydrated below 300°C; m.p. 1090°C. Soluble in water; insoluble in alcohol. Low toxicity.
Grades: Technical; 99.4%; 60% solution; F.C.C.
Containers: Fiber drums; multiwall paper bags.
Uses: Soap and detergent builder; sequestering agent; peptizing and dispersing agent.

potassium pyrosulfate (potassium acid sulfate, anhydrous) $K_2S_2O_7$.
Properties: Colorless needles or white, crystalline powder, or fused pieces. Soluble in water; converted to potassium bisulfate. Sp. gr. 2.512 (25/4°C); m.p. (approx) 325°C.
Use: Acid flux in analysis; laboratory reagent.

potassium pyrosulfite. See potassium metabisulfite.

potassium rhodanide. See potassium thiocyanate.

potassium ricinoleate $C_{17}H_{32}OHCOOK$.
Properties: White paste; soluble in water. Combustible.
Use: Emulsifying agent.

potassium silicate
Properties: (solid) Weight ratio $SiO_2:K_2O$ varies with grade, as 2.1:1; 2.5:1. Colorless, anhydrous lump, shattered or granular; soluble in water only at elevated temperatures and pressure. (Solution) Colorless liquid. Bé range 29–48°.
Derivation: Supercooled melt of potassium carbonate and pure silica sand.
Containers: (solid) Drums. (solution) Cans; drums; tank cars and tank trucks.
Uses: (solid) Manufacture of glass and refractory material; welding rods; potassium silicate solutions. (Solution) Nonefflorescing base for inorganic protective coatings; coating for roofing granules and welding rods; binder in the manufacture of carbon arc-light electrodes and for phosphors on television tubes; detergents; catalyst; adhesives.

potassium silicofluoride. See potassium fluosilicate.

potassium sodium carbonate. See sodium potassium carbonate.

potassium sodium ferricyanide $K_2NaFe(CN)_6$.
Properties: Red crystals, over 99% pure; m.p., decomposes; nonhygroscopic and stable; easily soluble in water.
Derivation: From ferrocyanides.
Containers: Fiber drums.
Uses: Blueprint paper and photography.

potassium-sodium phosphate. See sodium-potassium phosphate.

potassium-sodium tartrate (Rochelle salt; sodium potassium tartrate) $KNaC_4H_4O_6 \cdot 4H_2O$. It is a salt of L(+)-tartaric acid.
Properties: Colorless transparent efflorescent crystals

or white powder having cool, saline taste. Soluble in water; insoluble in alcohol. Loses water of crystallization at 140°C; unstable above about 225°C. Sp. gr. 1.77; m.p. 70–80°C. Nontoxic.
Derivation: Potassium acid tartrate is dissolved in water, the solution saturated with sodium carbonate, concentrated after purification and crystallized.
Grades: Highest purity; reagent; commercial crystals or powder; N.F.; F.C.C.
Containers: 25-, 225-, 250-lb drums.
Uses: Sequestrant and general-purpose food additive; baking powders; medicine (cathartic); component of Fehling's solution.

potassium sorbate (potassium 2,4-hexadienoate) $CH_3CH{:}CHCH{:}CHCOOK$.
Properties: White powder; m.p. 270°C (decomposes); 58.5% soluble in water (25°C); sp. gr. 1.36 (25/20°C). Nontoxic.
Containers: Cartons; fiber drums.
Grades: Technical; F.C.C.
Uses: Bacteriostat and antioxidant in foods; preservative in wines, sausage casings, margarine, etc.

potassium stannate $K_2SnO_3 \cdot 3H_2O$.
Properties: White to light tan crystals. Soluble in water; insoluble in alcohol. Sp. gr. 3.197.
Grade: Technical.
Containers: 100-, 350-lb drums.
Hazard: Toxic. Tolerance (as Sn), 2 mg per cubic meter of air.
Uses: Textiles (dyeing and printing); alkaline tin-plating bath.

potassium stearate $C_{17}H_{35}COOK$.
Properties: White crystalline powder; soluble in hot water; slightly soluble in alcohol; slight odor of fat. Nontoxic.
Grades: Commercial, contains considerable palmitate; F.C.C.
Uses: Base for textile softeners.

potassium strontium chlorate. See strontium potassium chlorate.

potassium styphnate $KC_6H_2N_3O_8 \cdot H_2O$.
Properties: Yellow prisms; m.p., loses water at 120°C; b.p., explodes.
Hazard: Explodes when shocked or heated.
Use: High explosive.
Shipping regulations: (Rail) Consult regulations. (Air) Not acceptable.

potassium sulfate K_2SO_4.
Properties: Colorless or white, hard crystals or powder; bitter, saline taste. Soluble in water; insoluble in alcohol. Sp. gr. 2.66; m.p. 1072°C.
Derivation: (a) By treatment of potassium chloride either with sulfuric acid or with sulfur dioxide, air, and water (Hargreaves process). (b) By fractional crystallization of a natural sulfate ore; (c) from salt lake brines.
Grades: Highest purity medicinal; commercial; crude; C.P.; agricultural.
Containers: Bags; drums.
Uses: Reagent in analytical chemistry; medicine; gypsum cements; fertilizer for chloride-sensitive crops such as tobacco and citrus; alum manufacture; glass manufacture; food additive.

potassium sulfhydrate. See potassium hydrosulfide.

potassium sulfide K_2S.
Properties: Red or yellow-red crystalline mass or fused solid; deliquescent in air. Soluble in water, alcohol, and glycerin; insoluble in ether. Sp. gr. 1.805 (20/4°C); m.p. 840°C.
Derivation: Potassium sulfate and carbon are heated in a tightly closed crucible to a moderate temperature.
Grade: Technical.
Containers: Cans; glass bottles; metal drums.
Hazard: Flammable, dangerous fire risk. May ignite spontaneously; explosive in form of dust or powder.
Uses: Reagent in analytical chemistry; depilatory; medicine.
Shipping regulations: (Rail) Yellow label. (Air) Flammable Solid label.

potassium sulfite $K_2SO_3 \cdot 2H_2O$.
Properties: White crystals or powder. Soluble in water; sparingly soluble in alcohol. Decomposes on heating and slowly oxidizes in air. Low toxicity.
Grades: Technical; C.P.; F.C.C.
Uses: Medicine; photography; preservative and antioxidant in foods.

potassium sulfocarbonate (potassium trithiocarbonate) K_2CS_3.
Properties: Yellowish-red crystals. Very hygroscopic. Soluble in alcohol and water.
Grade: Technical.
Uses: Analysis (testing for cobalt, nickel); medicine.

potassium sulfocyanate. See potassium thiocyanate.

potassium sulfocyanide. See potassium thiocyanate.

potassium tantalum fluoride. See tantalum potassium fluoride.

potassium tartrate $K_2C_4H_4O_6 \cdot \frac{1}{2} H_2O$.
Properties: Colorless, crystalline; soluble in water; insoluble in alcohol; decomposed by heat. Sp. gr. 1.98; decomposes 200–220°C. Nontoxic.
Grades: C.P.; technical.
Uses: Manufacture of potassium salts; medicine; laboratory reagent.

potassium tellurite K_2TeO_3.
Properties: Granular, white powder. Hygroscopic. Soluble in water. Decomposes at 460–470°C.
Use: Analysis (testing for bacteria).

potassium tetraborate (potassium pyroborate; potassium borate) $K_2B_4O_7 \cdot 5H_2O$.
Properties: White powder; alkaline taste. Soluble in water.

potassium tetrathiocyanodiammonochromate. See Reinecke salt.

potassium thiocyanate (potassium rhodanide; potassium sulfocyanate; potassium sulfocyanide) KCNS.
Properties: Colorless, transparent, hygroscopic, odorless crystals; soluble in water, alcohol and acetone. Saline, cooling taste. Sp. gr. 1.88; m.p. 173°C; turns brown, green, blue when fused, white again on cooling. B.p., decomposes at 500°C.
Derivation: By heating potassium cyanide with sulfur.
Grades: Commercial; pure; reagent; A.C.S.
Containers: Glass bottles; drums.
Hazard: Moderately toxic by ingestion.
Uses: Reagent in analytical chemistry; manufacture of sulfocyanides; thioureas; printing and dyeing textiles; photographic restrainer and intensifier; synthetic dyestuffs; medicine.

potassium thiosulfate (potassium hyposulfite) $K_2S_2O_3$ with varying proportions of water of crystallization.
Properties: Colorless crystals. Hygroscopic. Soluble in water.

Grades: Technical; C.P.
Use: Analytical reagent.

potassium titanate K_2TiO_3 (approx).
Properties: White salt; hydrolyzes in water to give a strongly alkaline solution.
Derivation: From titanic acid and potassium hydroxide.
Containers: Cartons.
Use: See potassium titanate fiber.

potassium titanate fiber. Approximate composition $K_2O \cdot (TiO_2)_n$ where n is 4 to 7.
Properties: M.p. 2500°F; high refractive index; can diffuse and reflect infrared radiation.
Uses: Rockets, missiles, nuclear-powered applications as an insulator, especially for the range 1300–2100°F.

potassium titanium fluoride. See titanium potassium fluoride.

potassium titanium oxalate. See titanium potassium oxalate.

potassium trichlorophenate $Cl_3C_6H_2OH$. Offered as a solution containing 47% potassium trichlorophenate, and 3% other potassium chlorophenates; sp. gr. 1.3; f.p. −9°C.
Use: Slime control agent for pulp and paper mill systems.

potassium tripolyphosphate (KTPP) $K_5P_3O_{10}$.
Properties: White crystalline solid; m.p. 620–640°C; sp. gr. 2.54; loose bulk density (approx) 67 lb/cu/ft; solubility in water (26°C), over 140 g/100 ml water.
Containers: Bags; fiber drums.
Uses: Water-treating compounds; cleaners; specialty fertilizers; sequestrant (including food preparation).

potassium trithiocarbonate. See potassium sulfocarbonate.

potassium tungstate (potassium orthotungstate; potassium wolframate) $K_2WO_4 \cdot 2H_2O$.
Properties: Heavy, crystalline powder; sp. gr. 3.1; m.p. 921°C. Deliquescent. Soluble in water; insoluble in alcohol.

potassium undecylenate $CH_2{:}CH(CH_2)_8COOK$.
Properties: Finely divided white powder; decomposes above 250°C; limited solubility in most organic solvents; soluble in water.
Hazard: May be toxic in high concentrations.
Uses: Bacteriostat and fungistat in cosmetics and pharmaceuticals.

potassium wolframate. See potassium tungstate.

potassium xanthate (potassium ethyldithiocarbonate; potassium xanthogenate; potassium ethyl xanthate; potassium ethylxanthogenate) $KS_2COC_2H_5$.
Properties: Colorless or light yellow crystals; soluble in water and alcohol; insoluble in ether; sp. gr. 1.558 (21.5°C).
Derivation: Reaction of potassium ethylate and carbon disulfide.
Containers: Glass bottles; fiber cans.
Uses: Fungicide for soil treatment; reagent in analytical chemistry.

potassium zinc iodide (zinc potassium iodide) $ZnI_2 \cdot KI$.
Properties: Colorless crystals; very hygroscopic.
Use: Analysis (testing for alkaloids).

potassium zinc sulfate. See zinc potassium sulfate.

potassium zirconifluoride. See zirconium potassium fluoride.

potassium zirconium chloride. See zirconium potassium chloride.

potassium zirconium sulfate. See zirconium potassium sulfate.

potentiator. A term used in the flavor and food industries to characterize a substance that intensifies the taste of a food product to a far greater extent than does an enhancer. The most important of these are the 5′-nucleotides, which appeared on the market in 1963. They are approved by FDA. Their effective concentration is measured in parts per *billion*, whereas that of enhancers such as MSG is in parts per thousand. The effect is thought to be due to synergism. Potentiators do not add any taste of their own, but intensify the taste response to substances already present in the food. See also enhancer; seasoning; flavor.

"Potter's" Compounds.[309] Trademark for drawing lubricants and compounds for nonferrous and space age materials. Basic formulations vary from water-soluble soaps through fortified synthetic oil types.

potting compound. See encapsulation.

pour point. (1) The lowest temperature at which a liquid will flow when a test container is inverted. (2) The temperature at which an alloy is cast.

pour point depressant. An additive for lubricating and automotive oils which lowers the pour point (or increases the flow point) by approximately 20°F. The agents now generally used are polymerized higher esters of acrylic acid derivatives. They are most effective with low-viscosity oils.

powder. Any solid, dry material of extremely small particle size ranging down to colloidal dimensions, prepared either by comminuting larger units (mechanical grinding), by combustion (carbon black, lampblack), or by precipitation from a chemical reaction (calcium carbonate, etc.). Powders that are so fine that the particles cannot be detected by rubbing between thumb and forefinger are called impalpable. Typical materials used in powder form are cosmetics, inorganic pigments, metals, plastics (molding powders), dehydrated dairy products, pharmaceuticals, and explosives. Metal powders are used to make specialized equipment by sintering and pressing (powder metallurgy), as well as for sprayed coatings and paint pigments (aluminum, bronze). Thermoplastic polymers in powder form are used in a technology known as powder molding, and thermosetting polymers are used in the sprayed coatings field for autos, machinery and other industrial applications, in which they have many advantages over sprayed solvent coatings.
See also metal, powdered; carbon black; black powder.

powder metallurgy. See metal, powdered; sintering.

"Poxeal."[116] Trademark for an insulation system for motor and generator coils whereby the electrical circuits are sealed in a tough flexible material (such as epoxy resin) capable of withstanding severe environmental and thermal shock conditions.

ppb. Abbreviation for parts per billion.

ppm. Abbreviation for parts per million, used to indicate extremely minute concentrations of particulates in liquids and gases.

"PPO."[245] Trademark for polyphenylene oxide.
Properties: Excellent engineering thermoplastic. Light beige, opaque; sp. gr. 1.06; Rockwell hardness (R scale), 118–120; tensile modulus at 73°F 3.6 to 3.8 × 10^5 psi; self-extinguishing; useful temperature range of more than 600°F; excellent mechanical properties and dielectric characteristics; soluble in most aromatic and chlorinated hydrocarbons; insoluble in alcohols, ketones, aliphatic hydrocarbons, and water. Highly resistant to hydrolysis, acids, bases, and detergents.
Derivation: Oxidative polymerization of 2,6-dimethylphenol in the presence of a copper-amine-complex catalyst.
Uses: Dielectric components; hospital and laboratory equipment; pump housings, impellers, pipe, valves and fittings required by chemical and food industries; substitute for die-cast metals; nose cones for space vehicles.

"PP-2000 Series."[210] Trademark for polyvinyl pyridines used as dye mordant and antistat tablet coating.

"PQ Soluble Silicates."[201] Trademark for a group of chemicals produced by varying the proportions of sodium or potassium oxide, silica and water. Includes sodium and potassium silicates, metasilicates, sesquisilicates, and orthosilicates.
Uses: Adhesives, soap builders, detergents, cleaning compounds, cements, binders, sizes, protective coatings and films, coagulant aids for raw and waste water, rust inhibitors, catalyst base and deflocculants.

Pr (1) Symbol for praseodymium.
(2) Informal abbreviation for propyl.

"Pramitol."[219] See "Prometone."

"Pranone."[321] Trademark for ethisterone.

"Prantal" Methylsulfate.[321] Trademark for a brand of diphemanil methylsulfate.

Prandtl number. For any substance, the ratio of the viscosity to the thermal conductivity. The lower the number, the higher is the convection capacity of the substance. This ratio is important in heat and chemical engineering calculations.

praseodymia. See praseodymium oxide; see also rare earths.

praseodymium Pr Element of atomic number 59; Group IIIB of the periodic table; one of the rare earth elements of the lanthanide group. Atomic weight 140.9077; valences 3,4. No stable isotopes.
Properties: Yellowish metal, tarnishes easily (color of salts green); sp. gr. 6.78–6.81; m.p. 930°C; b.p. 3020°C; ignites to oxide (200–400°C); liberates hydrogen from water; soluble in dilute acids; paramagnetic; nontoxic.
Forms and grades: Ingots; rods; sheets; 98.8 to 99.9+% pure.
Source: Monazite, cerite, and allonite; also fission product.
Derivation: Reduction of the trifluoride with an alkaline metal, or by electrolysis of the fused halides.
Uses: Praseodymium salts; ingredient of mischmetal (q.v.); core material for carbon arcs; colorant in glazes and glasses; catalyst; phosphors, lasers and masers.
See also didymium.

praseodymium oxalate $Pr_2(C_2O_4)_3 \cdot 10H_2O$. Green powder; insoluble in water; slightly soluble in acids.
Used in ceramics.
Containers: Glass bottles, fiber drums.

praseodymium oxide (praseodymia) Pr_2O_3. Yellow-green powder; sp. gr. 7.07. Insoluble in water; soluble in acids. Hygroscopic, absorbs carbon dioxide from air. Purities to 99.8% oxide. Combustible; low toxicity.
Uses: Glass and ceramic pigment; laboratory reagent.

precipitate. Small particles that have settled out of a liquid or gaseous suspension by gravity, or that result from a chemical reaction; precipitated compounds such as blanc fixe (barium sulfate) are prepared in this way, for example, by the reaction $BaCl + NaSO_4 \longrightarrow NaCl + BaSO_4$. A class of organic pigments called lakes are made by precipitating an organic dye onto an inorganic substrate. Colloidal particles dispersed in a gas, as flue dust in industrial stacks, can be precipitated by introducing an electric charge opposite to that which sustains the particles (see Cottrell).
See also sedimentation.

precision investment casting. See lost wax process.

precursor. In biochemistry, an intermediate compound or molecular complex present in a living organism; when activated physiochemically it is converted to a specific functional substance. The prefix "pro-" is usually used to indicate that the compound in question is a precursor. Examples are ergosterol (pro-vitamin D_2) which is activated by ultraviolet radiation to vitamin D; carotene (pro-vitamin A) is a precursor of Vitamin A; prothrombin forms thrombin upon activation in the blood-clotting mechanism.

prednisolone $C_{21}H_{28}O_5$. Delta^{1,4}-Pregnadiene-11β, 17α, 21-triol-3,20-dione. Generic name for an analog of hydrocortisone.
Properties: White to practically white, odorless, crystalline powder. Very slightly soluble in water; soluble in alcohol, chloroform, acetone, methanol, dioxane. M.p. about 235°C with some decomposition.
Grade: U.S.P.
Hazard: Causes sodium retention. May have side effects similar to cortisone.
Use: Medicine. Also available as acetate and sodium phosphate.

prednisone $C_{21}H_{26}O_5$. Delta^{1,4}-Pregnadiene-17α,21-diol-3,11,20-trione. Generic name for an analog of cortisone.

preferential. Descriptive of the selectivity of action, either chemical or physicochemical, exhibited by a substance when in contact with two other substances; it may be due either to chemical affinity or to surface phenomena. An example of a preferential chemical combination is that of hemoglobin with carbon monoxide, with which it unites 200 times as readily as with oxygen when exposed to a mixture of the two; such phenomena as adsorption, corrosion, and the wetting of dry powders by liquids are other examples.

pregnanediol $C_{21}H_{36}O_2$. 5β-Pregnane-3α,20α-diol. A steroid; the metabolic product of progesterone.
Properties: Crystallizes in plates from acetone; m.p. 238°C; dextrorotatory in solutions; sparingly soluble in organic solvents; not precipitated by digitonin.
Derivation: Isolation from pregnancy urine of women, cows, mares, and chimpanzees; by reduction of pregnanedione.
Uses: Synthesis of progesterone; medically as a pregnancy test.

pregnenedione. See progesterone.

pregneninolone. See ethisterone.

4-pregnen-21-ol-3,20-dione. See deoxycorticosterone.

pregnenolone $C_{21}H_{32}O_2$. delta^5-Pregnene-3β-ol-20-one. A steroid which is a biologically active hormone similar to progesterone and the adrenal steroid hormones.

Properties: Crystallizes in needles from dilute alcohol; m.p. 193°C; freely soluble in acetone, petroleum ether, benzene, and carbon tetrachloride.

Derivation: From stigmasterol or other steroids.

Uses: Medicine; biochemical research.

Also available as acetate salt.

"Preludin."[219] Trademark for a brand of phenmetrazine hydrochloride.

"Premerge."[233] Trademark for a dinitro weed killer.

premix molding. Mixtures of plastic ingredients prepared in advance of the molding or extruding operation and stored in bags or bins until required. They are made by mixing the components (resin, filler, fibrous materials such as glass, and necessary curatives) in a dough blender. Storage life may be from a few days to a year or more, depending on formulation. Such mixtures are then calendered or extruded after warming to suitable temperature.

prenitene (1,2,3,4-tetramethylbenzene; prenitol) $(CH_3)_4C_6H_2$.

Properties: Colorless liquid; soluble in alcohol; insoluble in water. Sp. gr. 0.901; b.p. 204°C; f.p. −7.7°C.

prepolymer. An adduct or reaction intermediate of a polyol and a monomeric isocyanate, in which either component is in considerable excess of the other. A polymer of medium molecular weight having reactive hydroxyl and —NCO groups. It is used in preparation of polyurethane coatings and foams.

prepreg. A term used in the reinforced plastics field to mean the reinforcing material containing or combined with the full complement of resin before molding.

"Preptone."[84] Trademark for sodium bromite desizing compound for textiles.

"Pre-San."[329] Trademark for a selective herbicide. See N-(2-mercaptoethyl)benzenesulfonamide S-(O,O-diisopropyl phosphorodithioate).

preservative. Any substance or mixture that prolongs the life of an organic material, either by admixture in low concentrations, by impregnating, by coating, or by immersion. Examples are such food additives as benzoic acid; antioxidants used in petroleum products, rubber and plastics; fungicides for textiles; paint on wood; and various alcohols and aldehydes used to prevent deterioration of biological specimens, e.g., formaldehyde. See also antioxidant; protective coating.

"Prestabit Oil" V. Trademark for an anionic textile chemical consisting of purified sulfated castor-oil fatty acids.

Uses: Dyeing assistant for cotton and wool fiber; in viscose manufacture, clarifying agent to prevent milkiness of the yarn; antistatic agent for acetate and polyacrylonitrile fibers.

"Prestone."[214] Trademark for a group of automotive service products which include an ethylene glycol base antifreeze, containing special inhibitors for prevention of corrosion, foaming, creepage and rust-loosening. See also antifreeze.

Priestley, Joseph (1733–1804). Born near Leeds, England, Priestley originally planned to enter the ministry. As a youth, he became interested in both physics and chemistry and his research soon established his position as a scientist. He was elected to the Royal Society in 1766. His greatest contribution to science was his discovery of oxygen in 1774. He emigrated from England to Northumberland, Pa., where he lived from 1784 to his death. His research in America resulted in the discovery of carbon monoxide (1799). In 1874, at a meeting at Priestley's old home convened to celebrate the centennial of the discovery of oxygen, plans were laid to organize the American Chemical Society, which later established the Priestley medal, one of the highest awards in chemistry. (See Avogadro).

prills. Small round or acicular aggregates of a material, usually a fertilizer, that are artificially prepared. In the explosives field prills-and-oil consists of 94% coarse, porous ammonium nitrate prills and 6% fuel oil. See also explosives, high.

"Primafloc."[23] Trademark for a series of organic polyelectrolyte products used for flocculating suspended solids in water, waste and process streams. See also polyelectrolyte.

"Primal."[23] Trademark for aqueous dispersions of acrylic resins, supplied in various grades that differ in hardness and flexibility, produce finishes which are water-insoluble, require no plasticizer for flexibility, unimpaired by aging, and adhere tenaciously to leather and lacquer coats.

primary. (1) In reference to monohydric alcohols, amines, and a few related compounds, this term, together with secondary and tertiary, describes the molecular structure of isomeric or chemically similar individuals. Monohydric alcohols are based on the methanol group, $-\overset{|}{\underset{|}{C}}-OH$, in which three of the bonds of the methanol carbon may be attached either to hydrogen atoms or to alkyl groups. A primary alcohol has one alkyl group and two hydrogens, e.g., $CH_3-\overset{H}{\underset{H}{C}}-OH$, except methyl alcohol, in which all three bonds are to hydrogen atoms. A secondary alcohol has two alkyl groups and one hydrogen, e.g., $CH_3-\overset{CH_3}{\underset{H}{C}}-OH$. A tertiary alcohol has three alkyl groups, e.g., $CH_3-\overset{CH_3}{\underset{CH_3}{C}}-OH$. The three types can be readily identified by the number of hydrogen atoms attached to the central (methanol) carbon atom: if it is 2 or more, the alcohol is primary; if 1, it is secondary; and if zero, it is tertiary. For example, CH_3CH_2OH is primary; $(CH_3)CHOH$ is secondary; $(CH_3)_3COH$ is tertiary.

Primary, secondary and tertiary amines are formed from ammonia (NH_3) when one, two, or three hydrogen atoms, respectively, are replaced by alkyl groups.

These terms are also used to name salts of orthophosphoric acid (H_3PO_4) in which one, two, or three of the hydrogen atoms have been replaced by metal or radicals: NaH_2PO_4 is primary sodium phosphate, Na_2HPO_4 is secondary sodium phosphate. The same system of names is used for salts of other acids containing three replaceable hydrogen atoms.

(2) A type of battery that is irreversible in respect to power output; a voltaic cell. A secondary battery (storage battery) is reversible and can be recharged.

(3) In the terminology of minerals, primary (in the case of metals) refers to direct production from the ore; in the case of petroleum it refers to production from wells by direct means. This meaning contrasts with the term "secondary," used to denote recovery of metal from scrap and, for petroleum, recovery by means of special techniques such as flooding and hydraulic pressure.

primary azo dyes. Azo dyes derived from primary amines.

primary calcium phosphate. See calcium phosphate, monobasic.

"Primol."[51] Trademark for high viscosity white mineral oils conforming to U.S.P. specifications for internal use.

primuline dye. See thiazole dye.

printing ink. A viscous to semisolid suspension of finely divided pigment in a drying oil such as heat-bodied linseed oil. Alkyd, phenol-formaldehyde or other synthetic resins are frequently used as binders; and cobalt, manganese, and lead soaps are added to catalyze the oxidative drying reaction. Some types of inks dry by evaporation of a volatile solvent rather than by oxidation and polymerization of a drying oil or resin. For further information refer to National Printing Ink Institute, Lehigh University, Bethlehem, Pa.

pristane (2,6,10,14-tetramethylpentadecane; norphytane) $C_{19}H_{40}$. Found in rock specimens 2.5 to 3 billion years old. Is known to be synthesized only by living organisms, and to withstand heat and pressure; thus it serves to date the existence of life on earth. See also phytane.

Properties: Colorless, faintly odored, transparent, stable, homogeneous liquid; b.p. 290°C; f.p. approx. −60°C; sp. gr. 0.775–0.795 (20°C); refractive index 1.435–1.440 (20°C); flash point 175°C. Soluble in most organic solvents. Combustible.

Grade: 90% purity.

Containers: 100-lb drums.

Uses: Precision lubricant; chromatographic oil; anticorrosion agent.

"Probrite."[428] Trademark for a line of zinc addition agents for both cyanide and acid plating solutions.

"Procadian."[197] Trademark for a solid urea product used as a supplement in animal feeding.

procaine hydrochloride (procaine) $C_6H_4NH_2COOCH_2CH_2N(C_2H_5)_2 \cdot HCl$.

Properties: Small, colorless crystals or white, crystalline powder; odorless; stable in air. Soluble in water and in alcohol at 25°C; slightly soluble in chloroform; almost insoluble in ether; solutions acid to litmus. M.p. 153–156°C.

Derivation: (a) By heating chloroethyl-para-nitrobenzoic ester with diethylamine for 24 hours under pressure at 120°C. The product is then reduced with tin and hydrochloric acid. (b) By condensation of ethylene chlorohydrin with diethylamine. The chloroethyldiethylamine formed is heated with sodium para-aminobenzoate.

Containers: Drums.

Grade: U.S.P.

Hazard: Moderately toxic by ingestion and inhalation.

Use: Medicine (local anesthetic).

procaine penicillin G $C_{16}H_{18}N_2O_4S \cdot C_{13}H_{20}N_2O_2 \cdot H_2O$.

Properties: White, fine crystals or powder; odorless; relatively stable to air and light; solutions dextrorotatory; sparingly soluble in water; slightly soluble in alcohol; fairly soluble in chloroform.

Grade: U.S.P.

Use: Antibiotic; animal feed additive.

process industry. See chemical process industry.

"Procinyl"[325] Trademark for reactive disperse dyes for nylon. Cover irregular-dyeing yarns, and produce dyeings have excellent wet fastness.

"Procion"[325] Trademark for a range of dyes which form a chemical linkage with cellulosic fiber molecules. Applied to cotton and rayon by dyeing and printing.

"Prodag."[46] Trademark for a dispersion of graphite in water. Used as a mold wash for aluminum permanent molds, ingot molds and molds for mechanical rubber goods; stop-off coating.

prodrug. A term applied in the pharmaceutical chemistry to a chemical compound that is converted into an active curative agent by metabolic processes within the body. See also precursor.

producer gas. A gas obtained by burning coal or coke with a restricted supply of air, or by passing air and steam through a bed of incandescent fuel under such conditions that the carbon dioxide formed is converted into carbon monoxide. The water vapor reacts to form carbon monoxide and hydrogen. Producer gas is cheap but has low Btu and is used where transportation is not required.

A typical gas from coke may analyze, in percent:

Carbon monoxide	25.3
Hydrogen	13.2
Methane	0.4
Carbon dioxide	5.4
Oxygen	0.6
Nitrogen	55.2

Hazard: Highly flammable and toxic. Explosive range 20 to 73% in air.

See also water gas; synthesis gas.

"Product BCO."[28] Trademark for a surface-active agent based on C-cetyl betaine. An amphoteric product showing cationic properties in acid solutions and anionic properties in alkaline solutions. Particularly effective in alkaline peroxide bleaching systems.

"Product BDO."[28] Trademark for a surface-active agent based on C-decyl betaine. Similar to "Product BCO." Effective in solutions containing high concentrations of electrolytes.

"Pro-fax."[266] Trademark for polypropylene molding powders. The 6000 Series is propylene homopolymer and 7000 Series, propylene copolymer. Can be precision-injection molded; blown; extended into film or sheet, into monofilament or multifilament, onto wire, or in profile forms; vacuum-formed, and machined. There are over 300 different grades and types.

"Proferm."[319] Trademark for vitamin B_{12} feed supplements.

"Profil."[539] Trademark for glass fiber-reinforced polypropylene.

profile. See soil.

proflavine sulfate (proflavine; 3, 6-diaminoacridinium hydrogen sulfate) $C_{13}H_{11}N_3 \cdot H_2SO_4$.

Properties: Reddish brown, odorless crystalline powder. Soluble in water and alcohol forming brownish solutions which fluoresce green on dilution, nearly insoluble in ether and chloroform.

Use: Medicine.

"Profume."[233] Trademark for an odorized methyl bromide product, used as a soil fumigant and insecticide.

progesterone (delta^4-pregnene-3,20-dione) $C_{21}H_{30}O_2$. The female sex hormone secreted in the body by the corpus luteum, by the adrenal cortex, or by the placenta during pregnancy. It is important in the preparation of the uterus for pregnancy, and for the maintenance of pregnancy. It exists in two crystalline forms (alpha- and beta-) of equal physiological activity. Progesterone is believed to be the precursor of the adrenal steroid hormones. Low toxicity.

Properties: White crystalline powder; odorless and stable in air but sensitive to light. M.p., alpha form 128–133°C, beta form, about 121°C. Practically insoluble in water, soluble in alcohol, acetone, and dioxane; sparingly soluble in vegetable oils.

Units: The international unit (IU) of progestational activity is expressed as 1 mg of progesterone.

Derivation: Isolation from corpus luteum of pregnant sows; synthesis from other steroids such as stigmasterol (q.v.).

Grade: N.F.

Use: Medicine; permissible food additive; oral contraceptive; laboratory reagent.

proguanyl (chlorguanide; 1-*p*-chlorophenyl-5-isopropyl biguanide). An antimalarial drug said to be less toxic than others.

"Progynon."[321] Trademark for estradiol.

prolactin. See luteotropin.

prolamin. See gliadin.

"Prolase."[173] Trademark for a proteolytic enzyme.

proline (2-pyrrolidinecarboxylic acid) C_4H_8NCOOH. A nonessential amino acid found naturally in the L(−) form.

Properties: Colorless crystals; soluble in water and alcohol; insoluble in ether; optically active; DL-proline: m.p. 205°C with decomposition.

D(+)-proline: m.p. 215–220°C with decomposition.

L(−)-proline: m.p. 220–222°C with decomposition.

Derivation: Hydrolysis of protein; also synthetically.

Uses: Biochemical and nutritional research; microbiological tests; culture media; dietary supplement; laboratory reagent. Available commercially as the L(−)-proline.

prolipin. A compound sterile solution of protein obtained from nonpathogenic bacteria, various animal fats and lipoids derived from bile.

"Proluton."[321] Trademark for progesterone.

"Promacetin."[330] Trademark for acetosulfone.

promazine hydrochloride (10-(3-dimethylaminopropyl)-phenothiazine hydrochloride) $C_{17}H_{20}N_2S \cdot HCl$. Isomeric with promethazine hydrochloride.

Properties: White to slightly yellow, practically odorless, crystalline powder. Oxidizes upon prolonged exposure to air and acquires a blue or pink color. Melts within a range of 3° between 172° and 182°C; pH of 5% solution between 4.2 and 5.2. Low toxicity.

Grade: N.F.

Use: Medicine; permissible food additive.

promethium Pm Radioactive rare-earth element, a member of the lanthanide series; atomic number 61; atomic weight 145. The 145 isotope has a half-life of 18 years; 147 half-life 2.64 years. The latter is the only form available.

Properties: Silvery-white metal; m.p. 1160°C; density 7.2 g/cc.

Derivation: The 147 isotope is recovered from spent uranium fission products; also by reduction of the chloride or fluoride with an alkali metal.

Hazard: Strong radioactive poison; requires shielding and glove-boxes.

Uses: (147 isotope) Nuclear auxiliary power generators; special semiconductor battery; luminescent paint for watch dials; x-ray source; thickness gauges; with tungsten as cermet for space power system (withstands 2000°C for 1000 hours).

Note: Some 30 compounds are known, but are not commercially available.

"Prometone."[219] Trademark for 2-methoxy-4,6-bis-(isopropylamino)-s-triazine. Available as "Pramitol" 25E, 25% emulsifiable solution, and "Pramitol" 5P, 5% pellets.

Hazard: Toxic by ingestion and inhalation.

Use: Herbicide.

"Promin."[330] Trademark for sodium glucosulfone injection.

promoter. (1) A substance which, when added in relatively small quantities to a catalyst, increases its activity, e.g., aluminum and potassium oxide are added as promoters to the iron catalyst used in facilitating combination of hydrogen and nitrogen to form ammonia. (2) In ore flotation, a substance which provides the minerals to be floated with a water-repellent surface that will adhere to air bubbles. Such reagents are generally more or less selective towards minerals of certain classes.

"Pro-Noxfish."[342] Trademark for fish-toxicant composition containing rotenone and sulfoxide in emulsifiable vehicles.

proof. The ethyl alcohol content of a liquid at 60°F, stated as twice the percent of ethyl alcohol by volume. One gallon of 95% alcohol is therefore equivalent to 1.9 gallons proof alcohol. In the U.S., the alcohol tax is based on the number of proof gallons.

propadiene. See allene.

"Propaloid-T."[236] Trademark for a creamy white organic-modified, beneficiated magnesium montmorillonite with a rapid hydration characteristic. Used as a viscosity control, stabilizer for emulsions and suspensions in aqueous systems. Primarily for use in paints, cosmetics and paper coatings.

propanal. See propionaldehyde.

propane (dimethylmethane) C_3H_8.
Properties: Colorless gas. Natural gas odor. Noncorrosive. B.p. −42.5°C; f.p. −189.9°C; density of liquid at 0°C 0.531; density of vapor at 0°C (air = 1) 1.56; flash point −156°F. Autoignition temp. 874°F. Soluble in ether, alcohol; slightly soluble in water. An asphyxiant gas.
Grades: Technical; research (99.9%).
Derivation: From petroleum and natural gas.
Containers: Cylinders; tank cars.
Hazard: Highly flammable; dangerous fire risk. Explosive limits in air 2.4 to 9.5%. For storage, see butane (note).
Uses: Organic synthesis; household and industrial fuel; manufacture of ethylene; extractant; solvent; refrigerant; gas enricher; aerosol propellant; mixture for bubble chambers.
Shipping regulations: (Rail) Red Gas label. (Air) Flammable Gas label. Not acceptable on passenger planes. Label name: Liquefied petroleum gas.

1,3-propanediamine. See 1,3-diaminopropane.

1,2-propanediol. See 1,2-propylene glycol.

1,3-propanediol. See trimethylene glycol.

propane hydrate. See gas hydrate.

propanenitrile. See ethyl cyanide.

1-propanethiol (n-propyl mercaptan). C_3H_7SH.
Properties: Offensive-smelling liquid; boiling range 67.73°C; sp. gr. (20/4°C) 0.8408; refractive index (20/D) 1.4380; flash point −5°F.
Grade: 95%.
Containers: Drums; tank cars.
Hazard: Highly flammable, dangerous fire risk.
Uses: Chemical intermediate; herbicide.
Shipping regulations: (Rail) Red label. (Air) Flammable Liquid label.

propanil. Generic name for 3,4-dichloropropionanilide $Cl_2C_6H_3NHCOCH_2CH_3$.
Properties (pure): Light brown solid; m.p. 85–89°C; (technical): Liquid, b.p. 91–91.5°C.
Hazard: May be toxic.
Use: Post-emergence herbicide, especially for rice culture.

propanoic acid. See propionic acid.

1-propanol. See propyl alcohol.

2-propanol. See isopropyl alcohol.

propanolamine. See 2-amino-1-propanol; 3-amino-1-propanol.

2-propanol nitrate. See isopropyl nitrate.

2-propanolpyridine $C_5NH_4C_3H_6OH$.
Properties: Colorless liquid. B.p. 260.2°C; sp. gr. (25°C) 1.060; refractive index 1.5298 (n 20/D); miscible with water in all proportions at 20°C.

4-propanolpyridine $C_5NH_4C_3H_6OH$.
Properties: Colorless liquid. B.p. 289.0°C; f.p. 36.7°C; sp. gr. (40°C) 1.053; soluble in water.

2-propanone. See acetone.

2-propanone oxime. See acetoxime.

propanoyl chloride. See propionyl chloride.

propargyl alcohol (2-propyn-1-ol) $HC{:}CCH_2OH$.
Properties: Colorless liquid; geranium-like odor; sp. gr. 0.9215; f.p. −17°C; b.p. 114°C; soluble in water, alcohol, and ether; immiscible with aliphatic hydrocarbons. Flash point (open cup) 97°F.
Derivation: From acetylene by high-pressure synthesis.
Grades: Technical; 75% solution.
Containers: Up to tank cars.
Hazard: Toxic by ingestion, inhalation, and skin absorption. Tolerance, 1 ppm in air. Flammable, moderate fire risk.
Uses: Chemical intermediate; corrosion inhibitor; laboratory reagent; solvent stabilizer; prevents hydrogen embrittlement of steel; soil fumigant.

propargyl bromide (3-bromo-1-propyne) $HC{:}CCH_2Br$.
Properties: Liquid; sp. gr. 1.520; b.p. 88–90°C. Flash point 65°F (COC). Sharp odor.
Derivation: From acetylene by high-pressure synthesis.
Containers: Up to tank cars.
Hazard: Flammable, dangerous fire risk. Toxic and irritant.
Uses: Chemical intermediate; soil fumigants.
Shipping regulations: (Rail, Air) Flammable Liquid label.

propargyl chloride (3-chloro-1-propyne) $HC{:}CCH_2Cl$.
Properties: Liquid; freezing point −76.9°C; b.p. range 56.0–57.1°C; refractive index (n 25/D) 1.4310; flash point 90–95°F.
Derivation: From acetylene by high-pressure synthesis.
Containers: Up to tank cars.
Hazard: Probably toxic. Flammable. Moderate fire risk.
Uses: Chemical intermediate; soil fumigant.

"Propasol"[214] Trademark for a series of solvents for water-based enamels.

propazine 80W. Generic name for a pre-emergence herbicide containing 80% 2-chloro-4,6-bis(isopropylamino)-s-triazine. Used especially for weed control in sorghum culture.

propellant. (1) A rocket fuel (q.v.). (2) A compressed gas used to expel the contents of containers in the form of aerosols. Until recently chlorofluorocarbons 11 and 12 have been widely used because of their nonflammability and nontoxicity. The strong possibility that they may contribute to depletion of the ozone layer of the upper atmosphere has resulted in reduction of their use for this purpose. Other propellants used are hydrocarbon gases such as butane and propane, carbon dioxide and nitrous oxide. The materials dispersed include insecticides, shaving cream, whipped cream and cosmetic preparations. See also ozone (note).

2-propenal. See acrolein.

propene. See propylene.

propenenitrile. See acrylonitrile.

2-propene-1-thiol. See allyl mercaptan.

propene-1,2,3-tricarboxylic acid. See aconitic acid.

propenoic acid. See acrylic acid.

2-propen-1-ol. See allyl alcohol.

propenyl alcohol. See allyl alcohol.

2-propenylamine. See allylamine.

para-**propenylanisole.** See anethole.

alpha-**propenyldichlorohydrin.** See alpha-dichlorohydrin.

propenyl guaethol (1-ethoxy-2-hydroxy-4-propenylbenzene) $C_2H_5OC_6H_3(OH)(C_3H_5)$.
Properties: Free-flowing white powder; odor and

taste similar to vanilla, but much more powerful; m.p. 85–86°C; very soluble in fats, edible solvents, and essential oils; very slightly soluble in water.
Uses: Artificial vanilla flavoring; flavor enhancer.

propenyl guaiacol. See methyl isoeugenol.

2-propenyl hexanoate. See allyl caproate.

2-propenyl isothiocyanate. See allyl isothiocyanate.

propiodal (1,3-bis(trimethylamino)-2-propanol diiodide; iodisan; hexamethyldiaminoisopropanol diiodide) $[CH_2N(CH_3)_3I]_2CHOH$.
Properties: White, crystalline solid; m.p. about 275°C (dec). Turns brown at 240°C. Freely soluble in water, slightly in alcohol; insoluble in ether, acetone.
Uses: Medicine (iodine therapy).

beta-**propiolactone** (USAN) (BPL) $\underline{OCH_2CH_2CO}$.

Properties: Colorless liquid; pungent odor; b.p. 155°C with rapid decomposition; f.p. −33.4°C; refractive index (n 20/D) 1.4131; sp. gr. (20/4°C) 1.1460; soluble in water; miscible with ethyl alcohol, acetone, ether, and chloroform at 25°C. Reacts with alcohol. Flash point (open cup) 167°F; stable when stored in glass at refrigeration temperature (+5° to +10°C). Combustible.
Derivation: Direct combination of ketene and formaldehyde.
Containers: Up to tank cars; tank trucks.
Hazard: Strong skin irritant. Causes cancer in experimental animals; worker exposure should be minimized.
Uses: Organic synthesis; vapor sterilant; disinfectant.

propionaldehyde (propanal; propyl aldehyde; propionic aldehyde) C_2H_5CHO.
Properties: Water-white liquid with suffocating odor; soluble in water. Flash point (open cup) 15°F; b.p. 48.8°C; f.p. −81°C; sp. gr. 0.807 (20/4°C); refractive index (n 20/D) 1.364. Soluble in water and alcohol. Autoignition temp. 405°F.
Derivation: (a) Oxidation of propyl alcohol with dichromate; (b) by passing propyl alcohol over copper at elevated temperatures.
Containers: Drums.
Hazard: Flammable, dangerous fire risk; explosive limits in air 3.7 to 16%. Moderately toxic and irritant.
Uses: Manufacture of propionic acid, polyvinyl and other plastics; synthesis of rubber chemicals; disinfectant; preservative.
Shipping regulations: (Rail) Red label. (Air) Flammable Liquid label.

propionamide nitrile. See cyanoacetamide.

propione. See diethyl ketone.

propionic acid (methylacetic acid; propanoic acid) $CH_3CH_2CO_2H$.
Properties: Colorless liquid; pungent odor; sp. gr. 0.9942 (20/4°C); f.p. −20.8°C; refractive index 1.3862 (20°C); b.p. 140.7°C; soluble in water, alcohol, chloroform and ether. Flash point 130°F. Combustible. Autoignition temp. 955°F.
Derivation: By reaction of ethyl alcohol with carbon monoxide, using a boron trifluoride catalyst; also by the reaction of carbon monoxide with hydrogen and olefins or alcohols.
Method of purification: Rectification.
Grades: Technical, 99.0%; F.C.C.
Containers: Drums; carboys; tank cars.
Hazard: Moderate fire risk; strong irritant.
Uses: Propionates, some of which are used as mold inhibitors in bread and fungicides in general; herbicides; preservative for grains and wood chips; emulsifying agents; solutions for electroplating nickel; perfume esters; artificial fruit flavors; pharmaceuticals; cellulose propionate plastics.
Shipping regulations: (Air) Corrosive label.

propionic aldehyde. See propionaldehyde.

propionic anhydride $(CH_3CH_2CO)_2O$.
Properties: Colorless liquid; pungent odor; f.p. −45°C; b.p. 167–169°C; sp. gr. 1.0119 at 20°C; vapor pressure 1 mm at 20°C; flash point 165°F; wt 8.4 lb/gal at 20°C; soluble in alcohol, ether, chloroform, and alkalies; insoluble in water. Combustible.
Hazard: Strong irritant to tissue.
Uses: Esterifying agent for fats, oils, cellulose; dehydrating medium for nitrations and sulfonations; alkyd resins, dyestuffs and pharmaceuticals.
Shipping regulations: (Air) Corrosive label.

propionitrile. See ethyl cyanide.

propionylbenzene. See propiophenone.

propionyl chloride (propanoyl chloride) CH_3CH_2COCl.
Properties: Colorless liquid with pungent odor; f.p. −94°C; b.p. 80°C; sp. gr. 1.065 (20/4°C). Decomposes in water; soluble in alcohol.
Use: Chemical intermediate.

propionyl peroxide $C_2H_5C(O)OOC(O)C_2H_5$. Available as a 25% solution in a high-boiling hydrocarbon; flash point 125°F. Used as an initiator in polymerization reactions, such as the high pressure polymerization of ethylene.
Hazard: Strong oxidizing agent. May explode if shocked or heated.
Shipping regulations: (Rail) Oxidizing material, n.o.s., Yellow label. (Air) Organic peroxide label.

propiophenone (ethyl phenyl ketone; propionylbenzene; 1-phenylpropanone-1) $C_6H_5COC_2H_5$.
Properties: Water-white to light amber liquid with strong persistent odor; sp. gr. 1.012 (20/20°C); refractive index (n 20/D) 1.527; congealing temperature 17.5–21°C; b.p. 218°C; flash point (TOC) 210°F. Insoluble in water, ethylene glycol, glycerine; miscible with ethyl ether, benzene, and toluene. Combustible.
Containers: Drums.
Uses: Fixative in perfumes; starting material for synthesis of ephedrine and other important pharmaceuticals and synthetic organic chemicals; laboratory reagent.

"Propi-Rhap."[248] Trademark for the 2-ethylhexyl ester of 2-(2,4-dichlorophenoxy)propionic acid.
Use: Herbicide.
Hazard: May be toxic.

propoxyphene (4-dimethylamino-3-methyl-1,2-diphenyl-2-diphenyl-2-butanol propionate) $C_{22}H_{29}NO_2$. The alpha-diastereoisomers are optically active and are preferred for their greater pain-relieving ability. The drug has about the same analgesic effect as codeine and is not considered addictive; it does have a narcotic effect which can be harmful in cases of overdosage. Recommendations for restriction of its use have been made by FDA.

n-propoxypropanol $C_6H_{14}O_2$.
Properties: Liquid; sp. gr. 0.8865 (20/20°C); b.p. 149.8°C; f.p. −80°C (sets to glass below this). Soluble in water. Flash point 128°F.
Hazard: Moderate fire risk.
Use: Solvent for water-based enamel.

n-**propyl acetate** $C_3H_7OOCCH_3$.
Properties: Colorless liquid; pleasant odor. Slightly soluble in water. Miscible with alcohols, ketones, esters, hydrocarbons Sp. gr. 0.887; flash point 58°F; boiling range 96.0–102.0°C; wt 7.36 lb/gal; autoignition temp. 842°F.
Derivation: Interaction of acetic acid and n-propyl alcohol in the presence of sulfuric acid.
Grade: Technical.
Hazard: Flammable, dangerous fire risk. Moderately toxic. Tolerance, 200 ppm in air. Explosive limits in air 2 to 8%.
Uses: Flavoring agents; perfumery; solvent for nitrocellulose and other cellulose derivatives; natural and synthetic resins; lacquers; plastics; organic synthesis; laboratory reagent.
Shipping regulations: (Rail) Red label. (Air) Flammable Liquid label.

propyl acetone. See methyl butyl ketone.

propyl alcohol (1-propanol; n-propyl alcohol) $CH_3CH_2CH_2OH$.
Properties: Colorless liquid; odor similar to ethyl alcohol; b.p. 97.2°C; f.p. −127.0°C; sp. gr. (20/4°C) 0.804; flash point (open cup) 77°F; autoignition temperature 700°F; refractive index (20°C) 1.385; viscosity 2.256 cp (20°C); soluble in water, alcohol, and ether. Low toxicity.
Derivation: From oxidation of natural gas hydrocarbons; also from fusel oil.
Containers: Up to 10,000 gal tank cars.
Hazard: Flammable, dangerous fire risk. Tolerance 200 ppm in air. Explosive limits in air 2 to 13%.
Uses: Organic synthesis and chemical intermediate; solvent for waxes, vegetable oils, natural and synthetic resins, cellulose esters and ethers; polishing compositions; brake fluids; solvent degreasing; antiseptic.
Shipping regulations: (Rail) Red label. (Air) Flammble Liquid label.

n-**propyl alcohol.** See propyl alcohol.

sec-**propyl alcohol.** See isopropyl alcohol.

propyl aldehyde. See propionaldehyde.

n-**propylamine** $C_3H_7NH_2$.
Properties: Colorless liquid; sp. gr. at 20°C, 0.7182; b.p. 47.8°C; vapor pressure 248 mm (20°C); f.p. −83°C; odor, amine; soluble in water, alcohol and ether. Flash point −35°F; autoignition temp. 604°F.
Containers: Drums; tank cars.
Hazard: Highly flammable, dangerous fire risk. Explosive limits in air 2 to 10%. Highly toxic; strong irritant to skin and tissue. Use alcohol foam to extinguish.
Use: Intermediate; laboratory reagent.
Shipping regulations: (Rail) Flammable Liquid, n.o.s., Red label. (Air) Flammable Liquid label.

propyl butyrate $C_3H_7OOCC_3H_7$.
Properties: Colorless liquid. Sp. gr. 0.8789 (15°C); b.p. 142.7°C; f.p. −95.2°C. Combustible. Slightly soluble in water.
Hazard: Irritant to mucous membranes; narcotic in high concentrations.
Use: Solvent mixtures for cellulose ethers.

propyl chloride (1-chloropropane) $CH_3CH_2CH_2Cl$. Liquid; f.p. −122.8°C; b.p. 46.6°C; refractive index (n 20/D) 1.3886. Soluble in alcohol and ether. Slightly soluble in water. Flash point 0°F.
Hazard: Highly flammable, dangerous fire risk. Explosive limits in air 2.5 to 11%. Irritant and narcotic.
See also isopropyl chloride.
Shipping regulations: (Rail) Flammable Liquid, n.o.s., Red label. (Air) Flammable Liquid label.

propyl chlorosulfonate $CH_3CH_2CH_2OSO_2Cl$.
Properties: Liquid. B.p. 70–72°C (20 mm).
Derivation: Interaction of n-propyl alcohol and sulfuryl chloride.
Hazard: Highly toxic by inhalation and ingestion; strong irritant to eyes.
Uses: Organic synthesis; military poison gas (lachrymator).
Shipping regulations: (Rail) Poisonous liquid or gas, n.o.s., Poison Gas label. Not accepted by express. (Air) Not acceptable.

n-**propyl cyanide.** See n-butyronitrile.

n-**propyl diethyl malonate.** See propylamalonic acid, diethyl ester.

propyl 3,5-diiodo-4-oxo-1(4H)pyridineacetate. See propyliodone.

propylene (propene) $CH_3CH{:}CH_2$. 14th highest-volume chemical produced in U.S. (1975).
Properties: Colorless gas; soluble in alcohol and ether; slightly soluble in water. B.p. −47.7°C; f.p. −185.2°C; sp. gr. (liquid) 0.5139 (20/4°C); density of vapor at 0°C (air = 1) 1.46. Flash point −162°F. Autoignition temp. 927°F. Low toxicity.
Derivation: Catalytic and thermal cracking of ethylene with zeolite catalyst; from naphtha.
Grades: 95%, 99% and research.
Containers: Cylinders; tank cars; pipeline; ocean tankers.
Hazard: Highly flammable, dangerous fire risk. Explosive limits in air 2 to 11%. Safety data sheet available from Manufacturing Chemists Assn., Washington, D.C.
Uses: Isopropyl alcohol, polypropylene, synthetic glycerol, acrylonitrile, propylene oxide, heptene, cumene, polymer gasoline; acrylic acid; vinyl resins; oxo chemicals.
Shipping regulations: (Rail) Red Gas label. (Air) Flammable Gas label. Not acceptable on passenger planes.

propylene carbonate CH_6CO_3, or
$\overline{OCOCH_2CH(CH_3)O}$.
Properties: Odorless, colorless liquid. Freezing point −49.2°C (easily super-cooled); b.p. 241.7°C; sp. gr. (20/4°C) 1.2057; wt/gal (20°C) 10 lbs; refractive index (n 20/d) 1.4209; flash point 270°F. Miscible with acetone, benzene, chloroform, ether, ethyl acetate. Moderately soluble in water and carbon tetrachloride. Combustible.
Uses: Solvent extraction; plasticizer; organic synthesis; natural gas purification; synthetic fiber spinning solvent.

propylene chloride. See propylene dichloride.

propylene chlorohydrin (chloro-isopropyl alcohol; 1-chloro-2-propanol) $CH_2ClCHOHCH_3$.
Properties: Colorless liquid. Mild odor; nonresidual. B.p. 127.5°C; vapor pressure 4.9 mm (20°C); flash point 125°F (closed cup); wt 9.3 lb/gal (20°C); sp.

gr. 1.1128 at 20/20°C; soluble in water (in all proportions). Low toxicity.
Grade: Technical.
Containers: Carboys; steel drums.
Hazard: Moderate fire risk.
Use: Organic synthesis (introducing hydroxypropyl group).

propylenediamine. Legal label name (Air) for 1,2-diaminopropane (q.v.).

propylene dichloride (1,2-dichloropropane; propylene chloride) $CH_3CHClCH_2Cl$.
Properties: Colorless, stable liquid. Chloroform-like odor. Boiling point: 96.3°C; sp. gr. 1.1583 at 20/20°C; wt/gal 9.6 lbs (20°C); refractive index 1.4068 (20°C); flash point 61°F; solubility in water 0.26% by wt. (20°C); freezing point −80°C. Miscible with most common solvents; insoluble in water. Autoignition temp. 1035°F.
Derivation: Action of chlorine on propylene.
Grade: Refined.
Containers: Drums; tank cars.
Hazard: Flammable, dangerous fire risk. Explosive limits in air 3.4 to 14.5%. Toxic by ingestion and inhalation. Tolerance, 75 ppm in air.
Uses: Intermediate for perchloroethylene and carbon tetrachloride; lead scavenger for antiknock fluids; solvents for fats, oils, waxes, gums, and resins; solvent mixtures for cellulose esters and ethers; scouring compounds; spotting agents; metal degreasing agents; soil fumigant for nematodes.
Shipping regulations: (Rail) Flammable liquid, n.o.s., Red label. (Air) Flammable Liquid label.

propylene disulfate $C_3H_8S_2O_8$. A known carcinogen.

1,2-propylene glycol (1,2-dihydroxypropane; 1,2-propanediol: methylene glycol; methyl glycol; 1,2-propanediol) $CH_3CHOHCH_2OH$.
Properties: Colorless, viscous, stable, hygroscopic liquid; practically odorless and tasteless. Miscible with water, alcohols, and many organic solvents. B.p. 187.3°C; sp. gr. 1.0381 at 20/20°C; wt/gal 8.64 lb (20°C); refractive index 1.4293 (27°C); surface tension 40.1 dynes/cm (25°C); viscosity 0.581 poise (20°C); vapor pressure 0.07 mm (20°C); specific heat 0.590 cal/g (20°C); latent heat of evaporation 168.6 cal/g at b.p.; flash point (open cup) 210°F; autoignition temp. 780°F; heat of combustion 431.0 kg cal/mole. Combustible; nontoxic.
Derivation: By hydration of propylene oxide.
Method of purification: By distillation.
Grades: Refined; technical; U.S.P.; F.C.C.; feed.
Containers: Glass bottles; drums; tank cars; tank trucks.
Uses: Organic synthesis, especially polypropylene glycol and polyester resins, cellophane; antifreeze solution; solvent for fats, oils, waxes, resins, flavoring extracts, perfumes, colors, soft-drink syrups, antioxidants; hygroscopic agent; coolant in refrigeration systems; plasticizers, hydraulic fluids, bactericide; textile conditioners; in foods as solvent, wetting agent, humectant; emulsifier; feed additive; anticaking agent; preservative (retards molds and fungi); cleansing creams; sun tan lotions; pharmaceuticals; brake fluids; deicing fluids for airport runways; tobacco.
See also polypropylene glycol.

1,3-propylene glycol. See trimethylene glycol.

propylene glycol alginate (hydroxypropyl alginate) $(C_9H_{14}O_7)$.
Properties: Vary with degree of esterification. White powder; practically tasteless and odorless. Soluble in water and dilute organic acids. Nontoxic.
Grade: F.C.C.
Uses: Stabilizer; thickener; emulsifier; food additive.

propylene glycol distearate. See propylene glycol monostearate.

propylene glycol monomethyl ether (polypropylene glycol methyl ether) $CH_3OCH_2CHOHCH_3$.
Properties: Colorless liquid; f.p. −95°C (sets to glass); b.p. 120.1°C; sp. gr. 0.9234 (20/20°C); lb/gal 7.65 (25°C); refractive index 1.402 (n 25/D); flash point 97°F. Soluble in water, methanol, ether.
Containers: Drums; tank cars.
Hazard: Flammable, moderate fire risk.
Uses: Solvent for celluloses, acrylics, dyes, inks, stains; solvent-sealing of cellophane.

propylene glycol monoricinoleate $C_{17}H_{32}(OH)COOCH_2CHOHCH_3$.
Properties: Pale yellow, moderately viscous oily liquid; mild odor; sp. gr. 0.960 (25/25°C); saponification value 160; hydroxyl value 285; solidifies at −26°C. Soluble in most organic solvents; insoluble in water. Combustible; low toxicity.
Derivation: Castor oil and propylene glycol.
Grade: Technical.
Containers: 5-gal cans; 55-gal drums.
Uses: Plasticizer; dye solvents; lubricant; cosmetics; urethane polymers and hydraulic fluids.

propylene glycol monostearate. The F.C.C. grade is a mixture of propylene glycol mono- and diesters of stearic and palmitic acids. White beads or flakes; bland odor and taste. Insoluble in water; soluble in alcohol, ethyl acetate, chloroform, and other chlorinated hydrocarbons. Combustible; nontoxic.
Uses: Emulsifier; stabilizer.

propylene glycol phenyl ether $C_6H_5OCH_2CHOHCH_3$.
Properties: Colorless liquid; sp. gr. 1.060–1.070 (25/25°C); boiling range 5.95%, 237–242°C. Flash point 275°F. Combustible; low toxicity.
Uses: High-boiling solvent; bactericidal agent; fixative for soaps and perfumes; intermediate for plasticizers.

propyleneimine (2-methylaziridine; propylenimine) $CH_3HCNHCH_2$.
Properties: Water-white liquid; b.p. 66–67°C; sp. gr. (25/25°C) 0.8039–0.8070; n (25/D) 1.4094–1.4109. Soluble in water and most organic solvents.
Hazard: Flammable, dangerous fire risk. Toxic by ingestion, inhalation, and skin absorption. Tolerance, 2 ppm in air.
Uses: Organic intermediate whose derivatives are used in the paper, textile, rubber and pharmaceutical industries.
Shipping regulations: (Inhibited) (Rail) Red label. (Air) Flammable Liquid label. Not acceptable on passenger planes. (Uninhibited) (Air) Not acceptable.

propylene oxide CH_3CHCH_2O.
Properties: Colorless liquid, ethereal odor. Sp. gr. 0.8304 at 20/20°C; b.p. 33.9°C; vapor pressure 445 mm (20°C); flash point −35°F; wt 6.9 lb/gal (20°C); freezing point −104.4°C; soluble in water, alcohol and ether.

Derivation: (1) Action of alkalies on (a) propylene glycol or (b) propylene chlorohydrin; (2) Oxidation of propylene; (3) Epoxidation of propylene by a hydroperoxide complex with molybdenum catalyst.
Containers: Drums; tank cars.
Hazard: Highly flammable, dangerous fire risk; explosive limits in air 2 to 22%. Tolerance, 100 ppm in air. Moderately toxic and irritant.
Uses: Polyols for urethane foams; propylene glycols; surfactants and detergents; isopropanol amines; fumigant; synthetic lubricants; synthetic elastomer (homopolymer).
Shipping regulations: (Rail) Red label. (Air) Flammable Liquid label.

propyl formate $CH_3CH_2CH_2OOCH$.
Properties: Liquid; sp. gr. 0.9006 (20/4°C); f.p. −92.9°C; b.p. 81.3°C; refractive index 1.3769 (20°C). Slightly soluble in water; miscible with alcohol and ether in all proportions. Flash point (COC) 27°F. Autoignition temp. 851°F. Low toxicity.
Hazard: Flammable, dangerous fire risk.
Use: Flavoring.
Shipping regulations: (Rail) Flammable liquid, n.o.s., Red label. (Air) Flammable Liquid label.

propylformic acid. See butyric acid.

n-**propyl furoate** $C_4H_3OCO_2C_3H_7$.
Properties: Colorless, fragrant liquid, becomes yellow in light. Practically insoluble in water; soluble in alcohol and ether. Sp. gr. 1.0745 (25.9°C/4°C); b.p. 210.9°C (corr); refractive index 1.4737 (25.9°C/D). Low toxicity. Combustible.
Uses: Flavoring.

propyl gallate $C_3H_7OOCC_6H_2(OH)_3$. Colorless crystals; m.p. 150°C. Almost insoluble in water; somewhat soluble in oils. Low toxicity.
Containers: Polyethylene-lined fiber drums.
Hazard: Use in foods restricted to 0.02% of fat content.
Uses: Antioxidant or synergist to retard or prevent rancidity in lard and other edible fats and oils; laboratory reagent.

propylhexedrine (1-cyclohexyl-2-methylaminopropane; N-dimethylcyclohexaneethylamine) $C_6H_{11}CH_2CH(CH_3)NHCH_3$.
Properties: Clear, colorless liquid with amine odor; b.p. 202–206°C; very slightly soluble in water; soluble in dilute acids; miscible with alcohol, chloroform, and ether. Sp. gr. 0.848–0.852. Volatilizes slowly at room temperature; absorbs carbon dioxide from the air. Solutions are alkaline to litmus.
Grade: N.F.
Use: Medicine.

propyl para-**hydroxybenzoate.** See propylparaben.

propyliodone (propyl 3,5-diiodo-4-oxo-1(4H)-pyridineacetate) $I_2(O)C_5H_2NCH_2COOC_3H_7$.
Properties: White, crystalline powder. Odorless or nearly so. Practically insoluble in water. Soluble in acetone, alcohol, and ether. M.p. 187–190°C.
Grade: U.S.P.
Use: Medicine (radiopaque medium).

propyl isome. Generic name for dipropyl-5,6,7,8-tetrahydro-7-methylnaphtho [2,3-d]-1,3-dioxole-5,6-dicarboxylate. $C_{20}H_{26}O_6$.
Properties: Orange liquid; insoluble in water; slightly soluble in oils; soluble in most other organic solvents.
Use: Insecticide synergist.

propylmagnesium bromide C_3H_7MgBr. Available as a solution in ether. A Grignard reagent.
Use: Alkylating agent in organic synthesis.

propylmalonic acid, diethyl ester (propyl diethyl malonate) $C_3H_7CH(COOC_2H_5)_2$.
Properties: Colorless liquid; fragrant odor; insoluble in water; soluble in alcohols, ethers, esters, and ketones; sp. gr. 0.9860 (25°C); b.p. 222°C.
Uses: Intermediate; tobacco flavoring.

n-propyl mercaptan. Legal label name (Air) for 1-propanethiol (q.v.).

2-n-propyl-4-methylpyrimidyl 6-N,N-dimethylcarbamate $C_3H_7C_4HN_2(CH_3)OOCN(CH_3)_2$. Liquid, miscible with water and most organic solvents.
Hazard: May be toxic.
Use: Insecticide.

n-**propyl nitrate** (NPN) $C_3H_7NO_3$.
Properties: White to straw-colored liquid; ethereal odor; sp. gr. 1.07 (20°C); boiling range 104–127°C; flash point 68°F. Autoignition temp. 350°F; freezing point −100°C; refractive index 1.3975 (n 20/D). Insoluble in water.
Grades: 96–98% pure.
Containers: 55-gal steel drums.
Hazard: Flammable, severe fire and explosion risk. Strong oxidizing material. Explosive limits in air 2 to 100%. Tolerance, 25 ppm in air.
Use: Rocket fuel (monopropellant).
Shipping regulations: (Air) Flammable Liquid label.

propylparaben (propyl para-hydroxybenzoate) $C_{10}H_{12}O_3$.
Properties: Colorless crystals or white powder; slightly soluble in water; soluble in alcohol, ether and acetone; m.p. 95–98°C. Low toxicity.
Grades: U.S.P., F.C.C.
Containers: Drums.
Uses: Medicine; food preservative; fungicide; mold control in sausage casings.

propyl pelargonate $C_3H_7OOCC_8H_{17}$.
Properties: Liquid; sp. gr. 0.870 (15/15°C); b.p. 237°C. Insoluble in water; soluble in alcohol and most organic solvents. Combustible.
Uses: Flavors and perfumes; bactericides and fungicides.

propylpiperidine. See coniine.

6-propylpiperonyl butyl diethylene glycol ether. See piperonyl butoxide.

n-**propyl propionate** $CH_3CH_2COOCH_2CH_2CH_3$.
Properties: Colorless liquid. Soluble in most organic solvents; very slightly soluble in water. Boiling range 122–124°C; f.p. −76°C; flash point 174°F (open cup); wt 7.31 lb/gal. Combustible. Low toxicity.
Grade: Technical.
Containers: 55-gal drums; tank cars.
Uses: Solvent for nitrocellulose; paints, varnishes, lacquers; coating agents.

4-n-propylpyridine $C_8H_{11}N$ or $C_3H_7C_5H_4N$. B.p. 188°C. Used as an intermediate. Not to be confused with coniine (propylpiperidine).

propylthiouracil (6-propyl-2-thiouracil) $C_7H_{10}N_2OS$.
Properties: White powdery crystalline substance; starchlike in appearance and to touch; bitter taste; m.p. 218–221°C. Sensitive to light. Very slightly soluble in water; sparingly soluble in alcohol; soluble in ammonia and alkali hydroxides.

Derivation: By the condensation of beta-oxocaproate with thiourea.
Grade: U.S.P.
Use: Medicine.

n-propyltrichlorosilane $C_3H_7SiCl_3$.
Properties: Colorless liquid. B.p. 123.5°C; sp. gr. 1.195 (25/25°C); refractive index (n 25/D) 1.4292; flash point (C.O.C.) 100°F. Readily hydrolyzed with liberation of hydrochloric acid.
Derivation: By Grignard reaction of silicon tetrachloride and propylmagnesium chloride.
Grade: Technical.
Hazard: Strong irritant; flammable, moderate fire risk.
Use: Intermediate for silicones.
Shipping regulations: (Rail) White label. (Air) Corrosive label. Not acceptable on passenger planes.

propyl xanthate. See xanthic acid.

propyne. See methylacetylene.

2-propyn-1-ol. See propargyl alcohol.

"Proquel."[428] Trademark for chrome bath mist suppressants used to control mist, fumes, and spray. Concentrated, viscous liquid which can produce a foam blanket.

"Prorex Oils."[331] Trademark for a brand of highly-refined, light-colored, light-bodied mineral oils.

"Proslik."[487] Trademark for a blended, complex fatty amide fabric softener.

prostaglandin. One of a group of physiologically active compounds derived from fatty acids having 20 carbon atoms; an approximate formula is $C_{20}H_{36}O_5$. Research on these compounds has been intense in recent years in view of their importance in various reproduction mechanisms and their effect on blood pressure. They are believed to have significant relationships to a number of hormones; they also affect the nervous system and inhibit production of gastric juice. They occur naturally in body tissues and biological fluids (especially semen). Both the chemical structure and metabolic functions of these compounds have been established with considerable accuracy, and several types have been synthesized; one pathway uses norbornadiene as a starting point followed by a ten-step sequence of conversions to a diol which serves as a precursor; another starts with cyclopentadiene, followed by hydroboration, yielding two intermediates from which prostaglandin can be derived. The most prolific source of natural prostaglandins is a marine organism called a gorgonian sea whip, which occurs in great numbers in coral reefs, especially in the Caribbean area. Harvesting of these has made the production of prostaglandins much less expensive. The intermediates and chemical analogs derived from this natural source of prostaglandins are known as syntons.

prosthetic group. A chemical grouping in which a metal ion is associated with a large molecule or molecular complex, for example, coenzymes and metal-porphyrin complexes such as chlorophyll and hemin. Such groups activate metabolic mechanisms such as phosphorylation, decarboxylation, etc. by coordination reactions with amino acids, proteins, enzymes, and nucleic acids. The behavior of certain vitamins and other metabolites is due in part to prosthetic groups; catalysis is also involved.

"Prostigmin."[190] Trademark for neostigmine U.S.P.

protactinium Pa A radioactive element of atomic number 91, a member of the actinide series. Atomic weight 231.0359; valences 4,5; 13 unstable isotopes, two of which occur naturally. Protactinium is a constituent of all uranium ores, about 340 mg being extracted from one ton. Protactinium may also be produced by irradiation of thorium-230. Purification is carried out by ion exchange and solvent extraction techniques. The longest lived isotope, Pa-231, decays by alpha emission and has a half-life of about 33,000 years. Protactinium may be precipitated as the double potassium fluoride K_2PaF_7 or the oxide Pa_2O_5. The metal may be prepared by reducing PaF_4 with barium or by heating PaI_4 in a vacuum. It is hard and white, melting near 1600°C. It is too rare for commercial use. Forms several compounds with halogens.
Hazard: Highly toxic radioactive material.

"Protan."[1] Trademark for sodium formate.
Uses: Masking agent in chrome tanning; neutralizer in wool dyeing; preparation of washable wallpapers; acidizing of oil wells; solubilizer and buffering agent for manufacture of chemicals.

"Protargol."[162] Trademark for strong silver protein for use as antibacterial. A different grade also available for use in staining nerve tissue.

protease. A proteolytic enzyme which weakens or breaks the peptide linkages in proteins. They include some of the more widely known enzymes such as pepsin, trypsin, ficin, bromelin, papain and rennin. Being water-soluble they solubilize proteins and are commercially used for meat tenderizers, bread baking and digestive aids.

protective coating. A film or thin layer of metal, glass or paint applied to a substrate primarily to inhibit corrosion and secondarily for decorative purposes. Metals such as nickel, chromium, copper and tin are electrodeposited on the base metal; paints may be sprayed or brushed on. Vitreous enamel coatings are also used, which require baking. Zinc coatings are applied by a continuous bath process, where a strip of ferrous metal is passed through molten zinc. See also galvanizing; terne plate; electroplating; paint; corrosion; cladding.

protective colloid. See colloid, protective.

"Protectol" 3.[307] Trademark for a protective colloid, sulfite cellulose liquor; 46% active.

protein. A complex high polymer containing carbon, hydrogen, oxygen, nitrogen, and usually sulfur, and comprised of chains of amino acids (q.v.) connected by peptide linkages (—CO.NH—). Proteins occur in the cells of all living organisms and in biological fluids (blood plasma, protoplasm). They are synthesized by plants largely because of the nitrogen-fixing ability of certain soil bacteria. Their molecular weight may be as high as 40 million (tobacco mosaic virus). They have many important functional forms: enzymes, hemoglobin, hormones, viruses, genes, antibodies, and nucleic acids. They also comprise the basic component of connective tissue (collagen), hair (keratin), nails, feathers, skin, etc. Some have been synthesized in the laboratory.

The sequence of amino acids in the polypeptide chain is of critical importance in genetics (see deoxyribonucleic acid). Proteins can be hydrolyzed to their constituent amino acids and can be broken down into simpler forms by proteolytic enzymes. They form colloidal solutions, and behave chemically as both acids and bases simultaneously (see amphoteric). They are denatured by changes in pH, by heat, ultraviolet radiation, and many organic solvents.

Simple proteins contain only amino acids; conjugated proteins contain amino acids plus nucleic acids, carbohydrates, lipids, etc. On the basis of solubility they can be classified as albumins (water-soluble), globulins (insoluble in water but soluble in aqueous salt solutions), and prolamines (soluble in alcohol-water mixtures but not in alcohol or water alone). A number of proteins have been synthesized, notably the hormone insulin. Proteins are an essential component of the diet, occurring chiefly in meat, eggs, milk and fish. Edible proteins suitable for human food as well as cattle feed can be produced from microorganisms grown in carbonaceous or nitrogenous media to form yeast-like materials. Paraffinic hydrocarbons (methane) and petroleum-derived ethyl alcohol can be used as growth media for protein biosynthesis.

Industrial applications of proteins include plastics, adhesives and fibers derived from casein and soybean protein, but these have been declining in recent years. Special forms in which proteins are commercially available include textured proteins for food products, protein hydrolyzate and liquid predigested protein, both for medical use. See also nutrition; RDA; amino acid; protein, textured; protein hydrolyzate; polypeptide; keratin.

protein hydrolyzate. Solution of protein hydrolyzed into its constituent amino acids. Used extensively in medicine and surgery. Usually administered by a stomach tube or intravenous injection.
Grades: U.S.P.

protein, single-cell. A coined name for any protein derived directly from microorganisms.

protein, spun. See protein, textured.

protein, textured. A meat extender or substitute made from defatted soybean flour or similar protein, usually by an extrusion process. Some types are used to fortify cereals and other food products. The filaments produced by extrusion are designed to simulate the fibrous structure (texture) of meats. The term "spinning" is also used, and the products are often called spun proteins.

"Protek-Sorb."[241] Trademark for a group of silica gels.

"Protobore."[342] Trademark for protoveratrines for medicine.

proteolysis. The structural breakdown of proteins, usually by hydrolysis, as a result of the action of an enzyme, e.g., trypsin, pepsin, papain, etc.

prothrombin. The precursor of the enzyme thrombin (q.v.); a proteinaceous component of blood plasma which is converted into thrombin by the blood-clotting mechanism when activated at the site of an open wound. See also coagulation.

protocatechuic aldehyde, methyl ether. See vanillin.

"Protolin."[23] Trademark for reducing agents based on zinc sulfoxylate and zinc formaldehyde sulfoxylate. Supplied as water-soluble white powder.
Uses: Stripping colors from fabrics; chemical synthesis.

proton. A fundamental unit of matter having a positive charge and a mass number of 1, equivalent to 1.67×10^{-24} gram. Its mass is 1837 times that of the negatively charged electron, but is almost identical with that of the uncharged neutron. Protons are constituents of all atomic nuclei, their number in each nucleus being the atomic number of the element. An atom of normal hydrogen contains one proton and one electron. In hydrogen compounds which readily ionize, such as HCl, the proton is represented as H^+ (hydrogen ion).

"Protopet."[45] Trademark for petrolatum of medium consistency and ranging in color from pure white to amber, but meeting U.S.P. or N.F. purity requirements for petrolatum.

protoplasm. The total contents of the living cell, including both nucleus and cytoplasm. Predominantly a mixture of proteins, protoplasm is the physical basis of life. Most of its components are in the colloidal size range. This term is falling into disuse among modern biochemists. See also cell (1).

protoveratrine. A substance isolated from the Veratrum album plant. It is a mixture of two alkaloids, designated protoveratrine A and protoveratrine B. Used in medicine to lower blood pressure.

"Protox."[268] Trademark for a series of lead-free zinc oxides manufactured by the American process; produced from zinc ore. Available in pellets and free flowing, nondusting, low bulk for easy handling. Nontoxic.
Uses: Reinforcing agent and accelerator activator for rubber.

"Protozyme."[309] Trademark for a series of diastatic, proteolytic, and pectolytic enzymes of fungal origin on a cereal base carrier for applicaton where subsequent filtration permits the removal of the inactive portion.

"Provinite."[304] Trademark for a two part barium-cadmium organic vinyl stabilizer.
Properties: (A) soft white powder; sp. gr. 1.4; refractive index 1.56; (B) clear straw-colored liquid; sp. gr. 0.91; refractive index 1.45.
Containers: Up to 375-lb metal drums.
Hazard: Probably toxic.
Uses: Heat and light stabilizer for vinyl film, sheeting and dispersion resin systems.

provitamin. The precursor (q.v.) of a vitamin. Examples are carotene and ergosterol, which upon activation become Vitamin A and Vitamin D, respectively. See also specific compounds.

"Proxocream."[487] Trademark for a self-emulsifiable, fatty ester-based nonionic softener and lubricant.

"Proxoft-N."[487] Trademark for a modified, self-emulsifiable, fatty ester-based, nonionic softener and lubricant.

"Proxol."[487] Trademark for a self-emulsifiable, mineral oil-based lubriant for wool fibers.

"Prunolide."[227] Trademark for gamma-nonyl lactone.

Prussian blue. The most common and best known name for blue iron ferrocyanide (iron blue) pigments made by a variety of procedures. See iron blue.

prussic acid. See hydrocyanic acid.

"Prym."[42] Trademark for a group of colorless syrups of thermosetting carbamide fiber reactants. Dissolve readily in water at 25°C.
Use: Textile finishes.

pseudobutylene glycol. See 2,3-butylene glycol.

pseudocumene (1,2,4-trimethylbenzene; uns-trimethylbenzene) $C_6H_3(CH_3)_3$.
Properties: Liquid; f.p. −43.91°C; b.p. 168.89°C; sp. gr. (20/4°C) 0.8758; refractive index (20/D) 1.5045; flash point 130°F. Insoluble in water; soluble in alcohol, benzene, and ether.
Derivation: From C_9 fraction of refinery reformate streams by fractional distillation.
Grades: 95%, 99%, and research.
Containers: Bottles; drums.
Hazard: Moderate fire risk. Moderately toxic.
Uses: Manufacture of trimellitic anhydride, dyes, pharmaceuticals and pseudocumidine.

pseudocumidine (2,4,5-trimethylaniline; 1,2,4-trimethyl-5-aminobenzene) $C_6H_2(CH_3)_3NH_2$.
Properties: White crystals; sp. gr. 0.957; m.p. 62°C; b.p. 236°C. Soluble in alcohol and ether; insoluble in water. Combustible.
Derivation: From pseudocumene.
Hazard: Moderately toxic.
Uses: Manufacture of dyes; organic synthesis.

***d*-pseudoephedrine hydrochloride** (USAN) $C_6H_5CHOHCH(NHCH_3)CH_3 \cdot HCl$.
Properties: Needles; m.p. 181–182°C. Soluble in water, alcohol, chloroform.
Use: Medicine.

pseudohexyl alcohol. See 2-ethylbutyl alcohol.

pseudoionone (6,10-dimethyl-3,5,9-undecatriene-2-one) $(CH_3)_2C{:}CH(CH_2)_2C(CH_3){:}CHCH{:}CHCOCH_3$.
Properties: Pale yellow liquid; sp. gr. 0.8984 (20°C); b.p. 143–145°C (12 mm). Soluble in alcohol and ether. Combustible.
Uses: Perfumery; cosmetics.

psi. Abbreviation for pounds per square inch.

psia. Abbreviation for pounds per square inch absolute.

psig. Abbreviation for pounds per square inch gauge.

psilocin $C_8H_5N(OH)C_2H_4N(CH_3)_2$. An indole derivative. An alkaloid from certain mushrooms; a hallucinogenic drug.
Hazard: Highly toxic.

psilocybin $C_8H_5N(OPO_3H_2)C_2H_4N(CH_3)_2$. An indole derivative. An alkaloid from certain mushrooms; hallucinogenic drug.
Hazard: Highly toxic.

psilomelane $BaMn_9O_{16}(OH)_4$. A natural oxide of variable composition. Calcium, nickel, cobalt and copper frequently are present. The name sometimes refers to mixtures of manganese minerals.
Properties: Color black; streak brownish black; luster submetallic; hardness 5–6; sp. gr. 3.7–4.7.
Occurrence: U.S.S.R.; India; South Africa; Cuba; Arkansas, Virginia, Georgia.
Use: Important ore of manganese.

psychotropic drug. Any of a number of therapeutic agents which affect the behavior, emotional state, or mental functioning of psychologically disturbed persons. They are widely known as "tranquilizers," but this term is no longer accepted as clinically accurate, as the minor tranquilizers (benzodiazepine and glycerol derivatives) act quite differently from the major tranquilizers. For this reason the latter are now classified as antipsychotics and antidepressants, and the term "antianxiety agent" is applied to the minor tranquilizers. Antipsychotic agents include phenothiazines (chlorpromazine), thioxanthenes, and butyrophenones; antidepressant agents have two major types, i.e., monoamine oxidase (MAO) inhibitors and several tricyclic compounds; antianxiety agents are glycerol derivatives (meprobamate) and benzodiazepine derivatives, e.g., oxazepam.

Pt Symbol for platinum.

PTA. Abbreviation for phosphotungstic acid.

pteroylglutamic acid. See folic acid.

PTFE. Abbreviation for polytetrafluoroethylene (q.v.).

PTMA. Abbreviation for a mixture of phosphotungstic and phosphomolybdic acids, used in making pigments. See phosphotungstic pigment.

PTMS. See para-toluidine-meta-sulfonic acid ($CH_3 = 1$).

ptomaine. A group of highly toxic substances (derivatives of ethers of polyhydric alcohols) resulting from the putrefaction or metabolic decomposition of animal proteins. Examples which have been isolated and prepared synthetically are cadaverine (1,5-diaminopentane) (q.v.), muscarine (hydroxyethyltrimethylammonium hydroxide), putrescine (tetraethylenediamine), and neurine (trimethylvinylammonium hydroxide) (q.v.).
Note: The term "ptomaine poisoning" is usually a misnomer for other types of food poisoning.

PTSA. Abbreviation for para-toluenesulfonamide (q.v.).

ptyalin. A salivary amylase which acts upon alpha-1,4-glycosidic linkages converting starch to various dextrins and maltose. It can act over a pH range of 4.0–9.0; optimum pH 5.6–6.5. It requires the presence of certain negative ions for activation; chlorides and bromides are the most effective.
Use: Biochemical research.

Pu Symbol for plutonium.

pulegium oil. See hedeoma oil.

pulegone (1-isopropylidene-4-methyl-2-cyclohexanone) $C_{10}H_{16}O$. A ketone found in pennyroyal and hedeoma oil. Combustible; low toxicity.
Properties: Oily liquid with pleasant odor; sp. gr. 0.9323 (20°C); b.p. 221°C; dextrorotatory; refractive index (n 20/D) 1.4894.
Uses: Chemical intermediate; flavoring.

pullulan. See starch polymer.

pulp, paper. Processed cellulosic fibers derived from hardwoods, softwoods, and other plants. There are two major types of pulp: (1) ground wood or mechanical pulp, which is merely finely divided wood without purification, and is made into newsprint, cheap manila papers, and nonpermanent tissues; (2) chemical pulp, of which there are three kinds: (a) soda process pulp, obtained from the digestion of wood chips (mostly poplar) by caustic soda; (b) sulfite process pulp (mostly spruce and other coniferous woods) obtained by digestion with a solution of magnesium, ammonium or calcium disulfite containing free sulfur dioxide; and (c) sulfate process (kraft) pulp, in which sodium sulfate is added to the caustic liquors but is reduced by the carbon present to the sulfide, which becomes a digesting agent. Sulfite and sulfate pulps (chiefly from softwoods, q.v.) comprise the bulk of paper pulps. Sulfate pulps are known as kraft pulps because of their strength, and are used for wrapping, packaging, container board, etc. A relatively new process called holopulping replaces sodium sulfate with oxidants. A synthetic pulp based on polyolefins (styrene copolymer fibers) has been developed to the production stage in Japan. See also holopulping; paper; digestion.

pumice. A highly porous igneous rock, usually containing 67–75% SiO_2 and 10–20% Al_2O_3; glassy texture. Potassium, sodium, and calcium are generally present. Insoluble in water; not attacked by acids.
Occurrence: Arizona, Oregon, California, Hawaii, New Mexico; Italy; New Zealand, Greece.
Grades: Lump; powdered coarse, medium, fine; N.F.; technical.
Containers: Bags; ton lots.
Uses: Concrete aggregate; heat and sound insulation; filtration; finishing glass and plastics; road construction; scouring preparations; paint fillers; absorbents; support for catalysts; dental abrasive; abherent for uncured rubber products.

"Purafil."[525] Trademark for an odoroxidant consisting of activated alumina impregnated with potassium permanganate in pellet form. Destroys odors by oxidation.

"Purecal."[203] Trademark for a series of calcium carbonates produced by a special precipitation process which results in finer particles of cubical shape.
Uses: Coating fine paper stocks; filler in rubber and plastics; extender for paints, printing inks; filler in dentifrices, cosmetics, pharmaceuticals; neutralizing agent in preparation of antibiotics; insecticide carrier; enamel-color extender in ceramics; calcium enrichment of food products.

purification. Removal of extraneous materials (impurities) from a substance or mixture by one or more separation techniques. A pure substance is one in which no impurity can be detected by any experimental procedure. Though absolute purity is impossible to attain, a number of standard procedures exist for approaching it to the extent of 1 part per million of impurity. The following fractionation techniques are widely used: crystallization, precipitation, distillation, adsorption (various types of chromatography), extraction, electrophoresis, and thermal diffusion. See also purity, chemical.

"Purifloc."[233] Trademark for a polyelectrolyte (q.v.) used to flocculate solids in water and industrial waste treatment.

purine (1) [imidazo (4,5-d) pyrimidine]

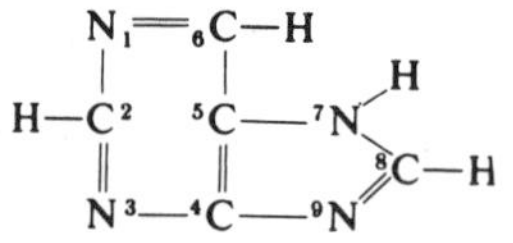

```
     N1==6C—H
     |     |      H
  H—C2   5C——7N/
     ||    ||     >8C—H
     N3—4C——9N/
```

Properties: Colorless crystals; m.p. 217°C; soluble in water, alcohol, toluene; slightly soluble in ether.
Derivation: Prepared from uric acid and regarded as the parent substance for compounds of the uric group, many of which occur naturally in animal waste products.
Uses: Organic synthesis; metabolism and biochemical research.

(2) One of a number of basic compounds found in living matter and having a purine-type molecular structure. See adenine, guanine, hypoxanthine, xanthine, uric acid, caffeine, and theobromine.

"Purinethol."[301] Trademark for 6-mercaptopurine (q.v.).

"Purite."[84] Trademark for a specially prepared fused soda ash furnished in the form of two-pound cast pigs and stated to contain over 98% sodium carbonate.
Uses: Cupola flux; refining and desulfuring iron, steel and other metals.

purity, chemical. A substance is said to be pure when its physical and chemical properties coincide with those previously established and recorded in the literature, and when no change in these properties occurs after application of the most selective fractionation techniques. In other words, purity exists when no impurity can be detected by any experimental procedure. There are a number of recognized standards of purity. See also grade.

"Purodigin."[24] Trademark for digitoxin (q.v.).

puromycin (USAN) $C_{22}H_{29}N_7O_5$. Crystals; m.p. 176°C. An antibiotic which inhibits protein synthesis, prevents transfer of amino acid from its carrier to the growing protein. Produced by Streptomyces alboniger; effective against bacteria, protozoa, parasitic worms, and cancerous tumors. Toxic to living cells of all kinds.

purple of Cassius. See gold-tin purple.

purpurin (1,2,4-trihydroxyanthraquinone) $C_{14}H_5O_2(OH)_3$.
Properties: Reddish needles; m.p. 256°C; slightly soluble in hot water; soluble in alcohol and ether.
Derivation: Occurs as a glucoside in madder root. Made synthetically by oxidation of alizarin.
Uses: Dye for cotton; stain for microscopy; reagent for calcium.

purpurin red. See anthrapurpurin.

"Purzaust."[416] Trademark for a catalytic converter developed for the removal of air pollutants normally present in exhaust gases from internal combustion engines. A fixed bed of solid catalyst promotes oxidation of combustible pollutants to carbon dioxide and water.

"PuTrol."[188] Trademark for a powerful aromatic compound used for masking the odors of putrefaction associated with the decomposition of proteins, as in fat-rendering operations, sewage disposal, and industrial wastes.

putty. A mixture of whiting (chalk) with from 12 to 18% of inseed oil, with or without white lead or other pigment. Containers must be air-tight.
Uses: Sealant; glass setting; caulking agent.

putty powder. A soft abrasive composed of tin oxide.

PVA. Abbreviation for polyvinyl alcohol.

PVAc. Abbreviation for polyvinyl acetate.

PVB. Abbreviation for polyvinyl butyral.

PVC. Abbreviation for (1) polyvinyl chloride, and (2) pigment volume concentration, a term used in paint technology to mean pigment volume divided by the sum of the pigment volume and the vehicle solids volume, multiplied by 100.

PVdC. See polyvinyl dichloride.

PVE. Abbreviation for polyvinyl ethyl ether.

PVI. Abbreviation for polyvinylisobutyl ether.

PVM. Abbreviation for polyvinyl methyl ether.

PVM/MA. See polyvinyl methyl ether/maleic anhydride.

PVOH. Abbreviation for polyvinyl alcohol.

PVP. Abbreviaton for polyvinylpyrrolidone.

PVT. Abbreviation for pressure-volume-temperature; used in chemical engineering.

"PX."[28] Trademark for special grade of pyroxylin-impregnated, washable book cloth for bookbindings.

py. An abbreviation for pyridine, used as in formulas for coordination compounds. See also dien; en; pn.

"Pycal."[89] Trademark for a series of plasticizers used for cellulose acetate, cellulose acetate butyrate, nitrocellulose, vinyls, and synthetic rubber.

"Pynol."[79] Trademark for a stream-distilled pine oil.
Uses: Mining-flotation; textile dyeing and cleaning; laundries; disinfectants; insecticides; deodorants; cleaning compounds; coated paper; paint and varnish; pharmaceuticals.

"Py-Ran."[58] Trademark for food grade of anhydrous monocalcium phosphate $CaH_4P_2O_8$ or $CaH_4(PO_4)_2$.

"Pyranol."[245] Trademark for a dielectric material, principally of the askarel type (q.v.).

"Pyratex."[248] Trademark for a vinylpyridine latex.
Properties: Total solids 40–42%, pH 10.4–11.5; sp. gr. 0.96.
Uses: To promote ashesion between rayon or nylon fibers and rubber, as in tire cord, belting, hose, etc.

"Pyrax."[69] Trademark for a ground pyrophyllite-aluminum silicate used as a filler in paints, plastics, and for dusting unvulcanized rubber.

pyrazine hexahydride. See piperazine.

pyrazole HNNCHCHCH.
Properties: Off-white crystalline solid with pyridine odor; m.p. 68–70°C; b.p. 186–188°C; soluble in water and alcohol.
Hazard: Toxic by ingestion and inhalation; irritant to skin and eyes.
Uses: Chemical intermediate; stabilizer for halogenated solvents and lubricating oils.

pyrazoline $HNNCHCH_2CH_2$. B.p. 144°C.
Use: Organic synthesis.

pyrazolone dye. A dye whose molecules contain both the —N=N— and the =C=C= chromophore groups in their structure. These are acid dyes most used for silk and wool, and to some extent for lakes. Tartrazine, C.I. 19140, is an important member of this group. See dye.

"Pyrefume."[342] Trademark for pyrethrin-extract insecticidal concentrates.

"Pyre-M.L."[28] Trademark for polyimide coated fabrics and laminates; used as class H electrical insulating materials.

pyrene $C_{16}H_{10}$. A condensed ring hydrocarbon.
Properties: Colorless solid (tetracene impurities give a yellow color); solutions have a slight blue fluorescence. M.p. 156°C; sp. gr. 1.271 (23°C); b.p. 404°C. Insoluble in water; fairly soluble in organic solvents.
Derivation: From coal tar.
Hazard: A carcinogenic agent. Tolerance, 0.2 mg per cubic meter in air.
Use: Biochemical research.

pyrethrin I $C_{21}H_{28}O_3$. Pyrethrolone ester of cyrysanthemummonocarboxylic acid. Most potent insecticidal ingredient of pyrethrum flowers. See also cinerin I and II and pyrethrin II.
Properties: Viscous liquid, oxidizes readily in air. Insoluble in water; soluble in other common solvents. Incompatible with alkalies.
Source: Pyrethrum extract (q.v.).
Hazard: Moderately toxic by ingestion and inhalation.
Uses: Household insecticide (flies, mosquitoes, garden insects, etc.).

pyrethrin II $C_{22}H_{28}O_5$. Pyrethrolone ester of chrysanthemumdicarboxylic acid. One of the four primary active insecticidal ingredients of pyrethrum flowers. See also pyrethrin I, cinerin I and II. Properties similar to those of pyrethrin I; less toxic than pyrethrin I.
Source: Pyrethrum extract (q.v.).
Hazard: Moderately toxic by ingestion and inhalation.
Uses: Household insecticide (flies, mosquitoes, garden insects, etc.).

pyrethrum extract. An extract obtained from powdered pyrethrum flowers by using a hydrocarbon of the kerosine type; also made synthetically. Not compatible with alkaline material. The chief constituents are pyrethrins I and II and cinerins I and II (see also allethrin). These compounds are nonvolatile and very slightly soluble in water.
Hazard: Moderately toxic by ingestion and inhalation.
Uses: Household insecticide (flies, mosquitoes, garden insects, etc.).

"Pyretox."[55] Trademark for an insecticide containing 1.0% by wt. of pyrethrin.

"Pyrexcel."[342] Trademark for synergized pyrethrin extracts for insecticides.

"Pyrex Glass Brand No. 7740."[20] Trademark for a borosilicate glass.
Properties: Linear coefficient of expansion, 32×10^{-7}/°C; elasticity coefficient 6.230 kg/sq mm; hardness (scleroscope) 120; sp. gr. 2.25; specific heat 0.20; refractive index 1.474 dispersion 0.00738; light and heat transmission higher than the best plate glass; dielectric constant 4.5 (25°C); upper working temperature (mechanical considerations only): annealed, extreme limit 490°C, normal service 230°C; thermal stress resistance 53°C; impact abrasion resistance 3.1; softening temp. 820°C.
Uses: Laboratory and pharmaceutical glassware and apparatus; electrochemical equipment; fiber manufacture; domestic ovenware apparatus and equipment for many processes.

"Pyribenzamine" Citrate.[305] Trademark for tripelennamine citrate (q.v.).

pyridine $N(CH)_4CH$.
Properties: Slightly yellow or colorless liquid; nauseating odor; burning taste; slightly alkaline in reaction. Soluble in water, alcohol, ether, benzene, ligroin and fatty oils. Sp. gr. 0.978; f.p. −42.0°C; b.p. 115.5°C; flash point (closed cup) 68°F. Autoignition temp. 900°F.
Derivation: (a) By coal carbonization, and recovery both from coke-oven gases and the coal tar middle oil. (b) Also synthetically from acetaldehyde and ammonia.

Grades: Technical, as 20°C, 2°C, etc. (meaning distillation range); medicinial; C.P.; spectrophotometric.
Containers: Bottles; drums; barrels; tank cars.
Hazard: Toxic by ingestion and inhalation. Tolerance, 5 ppm in air. Flammable, dangerous fire risk. Explosive limits in air 1.8 to 12.4%. Safety data sheet available from Manufacturing Chemists Assn., Washington, D.C.
Uses: Synthesis of vitamins and drugs; solvent; waterproofing; rubber chemicals; denaturant for alcohol and antifreeze mixtures; dyeing assistant in textiles; fungicides.
Shipping regulations: (Rail) Red label. (Air) Flammable Liquid label.

2-pyridine aldoxime methiodide (PAM) $C_5NH_4CHNOH \cdot ICH_3$. Antidote for nerve gas, because of its property of reactivating the enzyme cholinesterase by removal of phosphoryl groups introduced by the nerve gas.

pyridine-3-carboxylic acid. See niacin.

3-pyridine diazonium fluoborate. Intermediate in manufacture of 3-fluoropyridine. Explodes when dry.

2,5-pyridinedicarboxylic acid. See isocinchomeronic acid.

3-pyridinemethanol. See nicotinyl alcohol.

pyridine-N-oxide C_5H_5NO. F.p. 67.0°C. Soluble in water. Used as an intermediate.
Hazard: Probably toxic and flammable.

pyridine polymer. A polymer or copolymer of methylvinylpyridine and vinylpyridine; used for corrosion inhibitors and as intermediates for coatings and printing inks.

2-pyridinethiol-1-oxide. See 1-hydroxy-2-pyridine thione.

pyridinium bromide perbromide (PBPB) $C_5H_6NBr \cdot Br_2$.
Properties: Red prismatic crystals; m.p. 135–137°C (dec) with preliminary softening. The salt is stable in the dry state and can be used in glacial acetic acid, ethanol and related solvents. This compound has 45% to 50% available bromine.
Use: Brominating phenols, and addition to double bonds; mono- and polybromination of ketones, including aliphatic, alicyclic, steroid and amino carbonyls. Micro or semimicro quantitative analysis.

"Pyridium."[546] Trademark for phenazopyridine hydrochloride. Used in medicine.

pyridostigmine bromide $C_5H_4N(CH_3)OOCN(CH_3)_2Br$. 3-Hydroxy-1-methylpyridinium bromide dimethylcarbamate.
Properties: White or practically white, crystalline powder. Agreeable odor; hygroscopic. Freely soluble in water, alcohol, and chloroform. M.p. 154–157°C.
Grade: U.S.P.
Use: Medicine.

pyridoxal hydrochloride $C_8H_9NO_3 \cdot HCl$. An aldehyde derivative of pyridoxine, with vitamin B_6 activity.
Properties: Rhombic crystals; m.p. 165°C (dec); soluble in water and 95% ethyl alcohol. Low toxicity.
Use: Nutrition.

pyridoxal phosphate (2-methyl-3-hydroxy-4-formyl-5-pyridylmethylphosphoric acid) $CH_3C_5HN(OH)(CHO)(CH_2PO_4H_2)$. The coenzyme of amino acid metabolism, which also is the active group of various decarboxylases and other types of enzymes. It is closely related to pyridoxine (q.v.). Low toxicity.
Derivation: Synthesized (a) through the action of adenosine triphosphate, or phosphorus oxychloride, on pyridoxal, and (b) by phosphorylation of pyridoxamine followed by oxidation with 100% H_3PO_4.
Uses: Medicine; nutrition.

pyridoxine (vitamin B_6) $CH_3C_5HN(OH)(CH_2OH)_2$. 3-Hydroxy-4,5-dimethylol-2-methylpyridine. A group name designating the naturally occurring pyridine derivatives having vitamin B_6 activity. Essential for the dehydration and desulfhydration of amino acids and for the normal metabolism of tryptophan; appears to be related to fat metabolism. Pyridoxine is required in the nutrition of all species of animals. Nontoxic.
Sources: (Food): vegetable fats, whole grain cereals, legumes, yeast, muscle meats, liver, and fish. (Commercial): synthetic pyridoxine, pyridoxal, and pyridoxamine are produced by a complex series of reactions from isoquinoline.
Units: Amounts are expressed in micrograms.
Uses: Medicine; nutrition. (Available as pyridoxine hydrochloride).

alpha-**pyridylamine.** See 2-aminopyridine.

beta-**pyridylamine.** See 3-aminopyridine.

3-pyridylcarbinol (3-pyridine methanol) $C_5H_4NCH_2OH$.
Properties: Colorless liquid. B.p. 266°C; f.p. −6.5°C; density (20°C) 1.131 g/ml; refractive index (n 20°C) 1.5455. Soluble in water at 20°C in all proportions. Hygroscopic. Low toxicity. Combustible.
Use: Intermediate.

pyrimidine. One of a group of basic compounds found in living matter. They may be isolated following complete hydrolysis of nucleic acids. They include uracil, thymine, cytosine, and methylcytosine. Thiamine is also a pyrimidine derivative. Other pyrimidines such as alloxan and thiouracil are important in medicine and biochemical research. See also purine.

pyrite (iron pyrite; fool's gold). FeS_2, often with small amounts of copper, arsenic, nickel, cobalt, gold, selenium.
Properties: Brass-yellow or brown tarnished mineral, greenish or brownish-black streak, metallic luster. Sp. gr. 4.9–5.2; hardness 6–6.5.
Uses: Manufacture of sulfur, sulfuric acid and sulfur dioxide; ferrous sulfate; cheap jewelry; recovery of metals.
See also ferrous sulfide.

pyrite, magnetic. See pyrrhotite.

pyrithiamine (neopyrithiamine) $C_{14}H_{20}Br_2N_4O$. A thiamine antagonist.
Properties: Crystallizes from acetone; m.p. 219°C (dec); soluble in water.
Derivation: Synthetically from the condensation of 2-methyl-3-(beta-hydroxyethyl)pyridine with the pyrimidine moiety of thiamine.
Use: Biochemical research.

pyro-. A prefix indicating formation by heat; specifically, an inorganic acid derived by loss of one molecule of water from two molecules of an ortho acid, as pyrophosphoric acid.

"Pyrobor."[88] Trademark for dehydrated borax, $Na_2B_4O_7$.

pyroboric acid (tetraboric acid) $H_2B_4O_7$. Vitreous or white powder; soluble in water and in alcohol.

"Pyrobrite."[288] Trademark for a bright leveling, pyrophosphate copper electroplating process. The materials used are copper pyrophosphate trihydrate, potassium pyrophosphate, ammonium hydroxide, and addition agents.

pyrocatechol (ortho-dihydroxybenzene; catechol) $C_6H_4(OH)_2$.

Properties: Colorless crystals; discolors to brown on exposure to air and light, especially when moist; sp. gr. 1.371; m.p. 104°C; b.p. 245°C, sublimes; soluble in water, alcohol, ether, benzene and chloroform, also in pyridine and aqueous alkaline solutions. combustible. Flash point 261°F (C.C.).

Derivation: (a) By fusion of ortho-phenolsulfonic acid with caustic potash at 350°C. (b) By heating guaiacol with hydriodic acid.

Grades: Technical; C.P.; resublimed.

Containers: 25- to 200-lb drums.

Hazard: Strong irritant.

Uses: Antiseptic; photography; dyestuffs; electroplating; specialty inks; antioxidants and light stabilizers; organic synthesis.

pyrocatechol dimethyl ether. See veratrole.

pyrocatechol methyl ester. See guaiacol.

"Pyroceram" Brand Cement.[20] Trademark for powdered glasses which are thermosetting and utilized for sealing inorganic materials as in vacuum tubes. The resultant seals are crystalline and have service temperatures in excess of the sealing temperatures.

"Pyroceram" Brand 9606.[20] Trademark for a crystalline ceramic material made from glass by controlled nucleation (q.v.).

Properties: Hard, brittle solid. Sp. gr. 2.61; specific heat 0.185 (25°C); softening temp. 1260°C; thermal conductivity 0.0087 cgs; dielectric constant 5.5 (25°C); expansion coefficient 57×10^{-7} per °C; flexural strength 20,000 psi; Knoop hardness (100 g) 698.

Uses: Envelope material for vacuum tubes and other electronic devices.

"Pyroceram" Brand 9608.[20] Trademark for a crystalline ceramic material made from glass by controlled nucleation (q.v.).

Properties: Hard, brittle solid; specific gravity 2.5; softening temp. 1250°C; specific heat 0.19 (25°C); thermal conductivity 0.0047 cgs; dielectric constant (25°C) 6.54; flexural strength 14,000–23,000 psi; Knoop hardness (100 g) 703.

Uses: Telescope mirrors; special-purpose ceramic products.

pyrochlore ($NaCaNb_2O_6F$. A complex oxide of sodium, calcium and niobium. Tantalum, rare earth metals, and other elements may be present. Color brown to black, streak light brown; hardness 5–5.5; sp. gr. 4.2–6.4.

Occurrence: Canada (Quebec); Brazil; Africa.

Use: Ore of niobium.

pyrochroite. See manganous hydroxide.

pyrogallol (pyrogallic acid; 1,2,3-trihydroxybenzene) $C_6H_3(OH)_3$.

Properties: White, lustrous crystals; turn gray on exposure to light; sp. gr. 1.463; m.p. 132.5°C; b.p. 309°C; soluble in water, alcohol, and ether. A solution of pyrogallol acquires a brown color on exposure to air. This absorption of oxygen and change of color take place rapidly when the solution is made alkaline.

Derivation: By heating gallic acid with three times its weight of water in an autoclave.

Hazard: Toxic by ingestion and skin absorption.

Uses: Protective colloid in preparation of metallic colloidal solutions; photography; dyes; intermediates; synthetic drugs; medicine; process engraving; laboratory reagent; gas analysis; reducing agent. Antioxidant in lubricating oils.

"2-Pyrol"[307] Trademark for 2-pyrrolidone (q.v.).

pyroligneous acid (wood vinegar; pyroligneous liquor). Crude yellow to red liquid, a mixture of materials from wood distillation. Crude product contains methanol, acetic acid, acetone, furfural and various tars and related products; sp. gr. 1.018–1.030; miscible with water and alcohol.

pyrolusite (manganese dioxide, black) MnO_2.

Properties: Iron-black to dark steel-gray or bluish mineral; streak, black or bluish-black; luster, metallic or dull. Soluble in hydrochloric acid. Sp. gr. 4.73–4.86; Mohs hardness 2–2.5.

Occurrence: United States (Virginia, Georgia, Arkansas, Lake Superior region, Massachusetts, Vermont, New Mexico); Germany; Australia; India; Canada.

Uses: Manganese ore. See also manganese dioxide. For many uses the ore and synthetic material are interchangeable, but pyrolusite is not usable for batteries.

"Pyrolux Maroon."[141] Trademark for organic maroon pigments used in plastics.

pyrolysis. Transformation of a compound into one or more other substances by heat alone, i.e., without oxidation. It is thus similar to destructive distillation. Though the term implies decomposition into smaller fragments, pyrolytic change may also involve isomerization and formation of higher molecular weight compounds. Hydrocarbons are subject to pyrolysis, e.g., formation of carbon black and hydrogen from methane at 1300°C and decomposition of gaseous alkanes at 500 to 600°C. The latter is the basis of thermal cracking (pyrolysis gasoline).

One application of pyrolysis is conversion of acetone into ketenes by decomposition at about 700°C; the reaction is $CH_3COCH_3 \rightarrow H_2C{=}C{=}O + CH_4$. Pyrolysis of natural gas or methane at about 2000°C and 100 mm Hg pressure produces a unique form of graphite. Synthetic crude oil can be made by pyrolysis of coal, followed by hydrogenation of the resulting tar. Large-scale pyrolysis of solid wastes is being conducted for production of synthetic fuel oils and other products; the method is said to require only 30 minutes at about 1000°F (flash pyrolysis). A pyrolysis method for recovery of usable materials from scrap tires is under development.

See also destructive distillation.

pyrolysis gasoline. See under gasoline.

pyromellitic acid (PMA; 1,2,4,5-benzenetetracarboxylic acid) $C_6H_2(COOH)_4$.

Properties: White powder; sp. gr. 1.79; m.p. 257–265°C; b.p., converts to dianhydride; bulk density 32 lb/cu ft. Absorbs moisture slowly if exposed to atmosphere.

Grade: 99% purity.

Hazard: Skin irritant.

Uses: Intermediate for polyesters and polyamides used in electrical and nonfogging plasticizers, lubricants and waxes.

pyromellitic dianhydride (PMDA) $C_6H_2(C_2O_3)_2$.
Properties: White powder; sp. gr. 1.68; m.p. 286°C; b.p. 397–400°C, 305–310°C (30 mm); bulk density 21 lb/cu ft. Soluble in some organic solvents. Hydrolyzes to the acid when exposed to moisture.
Derivation: Oxidation of durene, either in wet process by nitric acid or a dichromate, or as direct air oxidation with catalyst.
Grade: 98+% purity.
Hazard: Skin irritant.
Uses: Curing agent for epoxy resins used in high temperature laminates, molds, and coatings; cross-linking agent for epoxy plasticizers in vinyls; alkyd resins; intermediate for pyromellitic acid.

pyrometric cone (Seger cone). A small pyramid composed of mixtures of oxides which melt at known temperatures; used to measure temperatures in autoclaves, curing ovens, etc. in the 1100–3700°F range.

pyromucamide. See furoamide.

pyromucic acid. See furoic acid.

pyromucic aldehyde. See furfural.

"Pyron."[433] Trademark for iron and steel powders.

"Pyronate."[45] Trademark for water-soluble petroleum sulfonates.

"Pyronil."[100] Trademark for pyrrobutamine phosphate (q.v.).

pyrophoric. Descriptive of a substance that will ignite in air at or below room temperature in the absence of added heat, shock or friction. Titanium dichloride and phosphorus are examples of pyrophoric solids; tributylaluminum and related compounds are pyrophoric liquids. Sodium, butyllithium and lithium hydride are spontaneously flammable in moist air, as they react exothermically with water. Such materials must be stored in an atmosphere of inert gas or under kerosene. Some alloys (barium, misch metal) are called pyrophoric because they spark when slight friction is applied; they are used as tips in pocket lighters and similar devices.
Hazard: Dangerous fire risk near combustible materials.
Shipping regulations: Pyrophoric liquids: (Rail) Red label. (Air) Not acceptable.

pyrophosphoric acid (also called diphosphoric acid) $H_4P_2O_7$. See also polyphosphoric acid.
Properties: A viscous, syrupy liquid which tends to solidify on long standing at room temperature. When diluted with water it is rapidly converted into orthophosphoric acid. Soluble in water. M.p. 54°C.
Derivation: By heating phosphoric acid at 250–260°C. Further heating produces metaphosphoric acid.
Grade: Technical.
Containers: Carboys; casks.
Hazard: Moderately toxic.
Uses: Catalyst; manufacture of organic phosphate esters; metal treatment; stabilizer for organic peroxides.

pyrophyllite (agalmatolite) $Al_2Si_4O_{10}(OH)$. A natural hydrous aluminum silicate, found in metamorphic rocks.
Properties: Color white, green, gray, brown; luster pearly to greasy; good micaceous cleavage; sp. gr. 2.8–2.9. Mohs hardness 1–2. Similar to talc. Low toxicity.
Occurrence: North Carolina, California; Newfoundland; Japan.
Uses: Ceramics; insecticides; slate pencils; substitute for talc; extender in paints; wallboard; buffer in high-pressure equipment; permissible animal feed additive.

pyroracemic acid. See pyruvic acid.

pyrosulfuric acid. See sulfuric acid, fuming.

pyrosulfuryl chloride (disulfuryl chloride) $S_2O_5Cl_2$.
Properties: Colorless, mobile; very refractive, fuming liquid; sp. gr. 1.819; b.p. 146°C; f.p. −38°C; refractive index (n 19/D) 1.449.
Containers: Carboys.
Hazard: Decomposes violently with water to sulfuric and hydrochloric acids. Toxic and irritant.
Use: Organic synthesis.
Shipping regulations: (Rail) White label. (Air) Corrosive label.

pyrotartaric acid (methylsuccinic acid) $HOOCCH(CH_3)CH_2COOH$.
Properties: White or yellowish crystals. Soluble in water, alcohol, and ether. Sp. gr. 1.4105; m.p. 111–112°C.
Derivation: By distilling tartaric acid.
Uses: Organic synthesis.

pyrotartaric acid, normal. See glutaric acid.

pyrotechnics. The formulation and manufacture of fireworks, signal flares and military warning devices; it involves the use of phosphorus; magnesium; potassium nitrate, chlorate, and dichromate; strontium compounds; and numerous other substances that emit a bright, colored light when deflagrated.

pyrovanadic acid. See vanadic acid.

"Pyrovatex" CP.[443] Trademark for a fiber-reactive phosphone alkyl amide.
Properties: Odorless, nontoxic, nonirritating; stable to laundering and dry-cleaning. Reduces tear and tensile strengths; no effect on "hand." Compatible with other finishing and proofing agents and precured durable press processes. From 25 to 35% by weight is required. Imparts self-extinguishing properties to cotton.
Uses: Flame retardance of 100% cotton fabrics, such as tenting, military uniforms, draperies, etc.

pyroxylin. See nitrocellulose.

pyrrhotite (magnetic pyrites, pyrrhotine) FeS. A natural iron sulfide. Frequently has a deficiency in iron. May contain small amounts of nickel, cobalt, manganese and copper.
Properties: Color brownish bronze; streak black; luster metallic; slightly magnetic; hardness 4; sp. gr. 4.6.
Occurrence: Tennessee, Pennsylvania; Europe; Canada.
Uses: Ore of iron; manufacture of sulfuric acid.

pyrrobutamine phosphate
$ClC_6H_4CH_2C(C_6H_5){:}CHCH_2NC_4H_8 \cdot 2H_3PO_4$. (1-[4-(para-Chlorophenyl)-3-phenyl-2-butenyl]pyrrolidine diphosphate).
Properties: Cream to off-white powder with a slight odor and a bitter taste. Soluble in water, slightly soluble in alcohol, almost insoluble in chloroform and ether. Melting range 127° to 131°C.
Grade: N.F.
Use: Medicine.

pyrrole. Pyrrole is regarded as a resonance hybrid, and no one structure adequately represents it. An approximation is:

```
HC —— CH
||     |
HC    CH
  \ + //
    N
    H
```

Properties: Yellowish or brown oil; burning, pungent taste; odor similar to chloroform; readily polymerizes by the action of light and turns brown. Soluble in alcohol, ether, and dilute acids; insoluble in water and dilute alkalies. B.p. 130–131°C; f.p. −24°C; sp. gr.0.968 (20/4°C); refractive index (n 20/D) 1.5091; flash point (TCC) 102°F; soluble in most organic chemicals. Combustible.
Derivation: Fractional distillation of bone-oil with sulfuric acid.
Method of purification: Conversion into the potassium compound (C_4H_4NK), washing with ether and treatment with water, followed by drying and distillation.
Grade: Technical.
Hazard: Moderate fire risk. Moderately toxic by ingestion and inhalation.
Use: Manufacture of pharmaceuticals.

pyrrolidine C_4H_9N.
Properties: Colorless to pale yellow liquid; penetrating amine-like odor. Sp. gr. 0.8660 (20/20°C); f.p. −60°C; b.p. 87°C; refractive index 1.4425 (n 20/D). Flash point 37°F (TCC). Soluble in water and alcohol.
Grades: 95% min purity.
Uses: Intermediate for pharmaceuticals, fungicides, insecticides, rubber accelerators; citrus decay control; curing agent for epoxy resins; inhibitor.
Containers: 1-, 5-gal cans; 55-gal drums.
Hazard: Flammable, dangerous fire risk. Moderately toxic by ingestion and inhalation.
Shipping regulations: (Rail) Flammable liquid, n.o.s., Red label. (Air) Flammable Liquid label. Not acceptable on passenger planes.

2-pyrrolidinecarboxylic acid. See proline.

2-pyrrolidone (2-pyrrolidinone, butyrolactam). $CH_2CH_2CH_2C(O)NH$.
Properties: Light yellow liquid; specific gravity 1.1; b.p. 245°C; soluble in water, ethanol, ethyl ether, chloroform, benzene, ethyl acetate, carbon disulfide. High-boiling, noncorrosive; flash point 265°F. Combustible. See polyvinylpyrrolidone.
Derivation: From acetylene and formaldehyde by high-pressure synthesis.
Uses: Plasticizer and coalescing agent for acrylic latexes in floor polishes; solvent for polymers, insecticides, polyhydroxylic alcohols, sugar, iodine, specialty inks, monomer for nylon-4.

pyrrone An aromatic heterocyclic polymer derived from a cyclic dianhydride and an aromatic orthotetramine or a derivative. Stable to about 900°F.
Uses: Films, coatings, adhesives, laminates and moldings.

pyruvaldehyde. See pyruvic aldehyde.

pyruvic acid (alpha-ketopropionic acid; acetylformic acid; pyroracemic acid) $CH_3COCOOH$. A fundamental intermediate in protein and carbohydrate metabolism in the cell.
Properties: Liquid with odor resembling acetic acid; sp. gr. 1.2272 (20/4°C); m.p. 13.6°C; b.p. 165°C; miscible with water, ether, and alcohol.
Derivation: Dehydration of tartaric acid by distilling with potassium acid sulfate.
Use: Biochemical research.

pyruvic alcohol. See hydroxy-2-propanone.

pyruvic aldehyde (pyruvaldehyde; methyl glyoxal) CH_3COHCO.
Properties: Supplied commercially as approximately 30% aqueous solution; sp. gr. 1.20 (20/20°C); 10 lb/gal (20°C). Low toxicity.
Containers: 1-gal cans; 5-, 55-gal drums.
Uses: Organic synthesis, as of complex chemical compounds such as pyrethrins; tanning leather; flavoring.

Q

"Quiana."[28] Trademark for a nylon-type textile fiber derived from polyamide.
Properties: Similar to silk in appearance, hand, drape, luster, and other physical properties; wash-and-wear characteristics superior to those of other synthetic fibers. Wrinkle- and crease-resistant. Self-extinguishing.
Form: Filament yarns of 30 and 60 denier; slubbed yarns of 130 denier.
Uses: Women's wearing apparel; decorative fabrics.

Q-lure. 4-(para-acetoxyphenyl)-2-butanone. A lure for the male melon fly.

"QO."[224] Trademark for a series of furan chemicals and derivatives.

"Q-Tac."[30] Trademark for a brand of starch used as a fiber-bonding agent and retention aid for pigments and filler material in papermaking.

"Quadrafos."[164] Trademark for a sodium phosphate glass commonly called sodium tetraphosphate; see sodium phosphate. A linear polymer containing a minimum of 63.5% P_2O_5.
Grades: Beads; granular; powder.
Uses: Sequestering, dispersing and deflocculating agent; to prevent scaling and corrosion of pipes; to soften water; to disperse clays in paper making and oil well drilling muds.

"Quadrol."[203] Trademark for N,N,N′,N′-tetrakis(2-hydroxypropyl)ethylenediamine (q.v.).

"Quaker."[319] Trademark for pure and denatured ethyl alcohols.

"Quakeral."[224] Trademark for furfural (q.v.).

"Quakersol."[319] Trademark for a solvent made from 190 proof alcohol. Mild odor. May be purchased without a Federal permit.
Hazard: Toxic by ingestion. Flammable, moderate fire risk. Carries a poison label in accordance with many states' statutes. Must not be taken internally.
Containers: Up to 55-gal drums.
Uses: Shellac solvent and thinner; nitrocellulose lacquers and other protective coatings; cleaning and drying metal parts; printing industry; glass cleaners; antifreeze; lamp and stove fuel; chemicals and chemical specialties.

"Qualex."[28] Trademark for a special grade of sodium carboxymethylcellulose developed for use as a fluid loss and viscosity control additive in drilling muds.

qualitative analysis. See analytical chemistry.

quantitative analysis. See analytical chemistry.

quantum number. The quantum is the basic unit of electromagnetic energy; it characterizes the wave properties of electrons, as distinct from their particulate properties. The quantum theory developed by Max Planck states that the energy associated with any quantum is proportional to the frequency of the radiation, that is, e (energy) $= h\nu$, where ν is the frequency and h is a universal constant.

An electron has four quantum numbers which define its properties. These are: (1) the principal quantum number is a constant that can be any positive integer ($n = 1,2,3\ldots$). It determines the principal energy level, or shell, of the electron, sometimes designated by letters such as K, L, or M, depending on the value of the principal quantum number. (2) The angular momentum constant l, also an integer, which is related to n as: $l = 0, 1\ldots, n-1$. Here again, letter designations are often used: in s electrons $l = 0$, in p electrons $l = 1$, in d electrons $l = 2$, and in f electrons $l = 3$. (3) The magnetic quantum number m is an integer related to l as: $m = -l\ldots, -1,0,+1\ldots,+l$. (4) The spin quantum number is independent of the other three and has a value of either $+\frac{1}{2}$ or $-\frac{1}{2}$, depending on the direction of rotation of the electron on its axis in the atomic frame of reference.

See also orbital theory; electron; photon; radiation; Pauli exclusion principle.

quartz SiO_2 Crystallized silicon dioxide.
Properties: Color white to reddish; luster vitreous; Mohs hardness 7; sp. gr. 2.65; insoluble in acids except hydrofluoric; only slightly attacked by solutions of caustic alkali; piezoelectric and pyroelectric; melting point 1713°C. Noncombustible.
Derivation: Synthetic crystals of good size and purity are grown by mass production methods under very carefully regulated conditions of temperature and concentration.
Hazard: Avoid inhalation (fine particles).
Uses: Electronic components; piezoelectric control in filters, oscillators, frequency standards, wave filters, radio and TV components; barrel-finishing abrasive. See also silica.

quartz, fused. Pure silica that has been melted to yield a glass-like material on cooling. Used for apparatus and equipment (such as vacuum tubes) where its high melting point, ability to withstand large and rapid temperature changes, chemical inertness and transparency (including ultraviolet light) are valuable. Produced as fibers and fabrics for heat resistance, low expansion coefficient and high dielectric strength. See also glass.

quassia (bitter ash; bitterwood).
Derivation: The wood or bark of Picrasma excelsa or Quassia amara; very bitter taste. White to bright yellow chips or shavings. Low toxicity.
Uses: Decoction or tincture as a fly poison; surrogate for hops; medicine (powerful bitter); hair lotion; flavoring.

quassin $C_{22}H_{30}O_6$. Extract of quassia.
Properties: Colorless crystals; m.p. 205°C; odorless but very bitter. Slightly soluble in water; soluble in organic solvents.
Hazard: Moderately toxic.
Use: To denature alcohol.

quaternary ammonium salt. A type of organic nitrogen compound in which the molecular structure includes a central nitrogen atom joined to four organic

groups as well as to an acid radical. Octadecyldimethylbenzyl ammonium chloride and hexamethonium chloride are examples. Pentavalent nitrogen ring compounds, such as lauryl pyridinium chloride, are also considered quaternary ammonium compounds. They are all cationic surface-active compounds and tend to be adsorbed on surfaces.
Uses: Disinfectant, cleanser and sterilizer; cosmetics (deodorants, dandruff removers, emulsion stabilizers); fungicides; mildew control; to increase affinity of dyes for film in photography; coating of pigment particles to improve dispersibility; to increase adhesion of road dressings and paints; antistatic additive (q.v.).
See also detergent, synthetic; coordination compound.

"Quaternary O."[219] Trademark for alkyl imidazolinium chloride.
$C_{17}H_{35}CNC_2H_4NR_1R_2Cl$.
Properties: 80% active viscous oil, soluble in water and most organic solvents; unstable at pHs over 7.
Uses: Cationic wetting agent, foamer, penetrant for strong acids, salt solutions, solvents; germicide. Corrosion inhibitor in acids; emulsion breaking.

para-**quaterphenyl** $C_6H_5C_6H_4C_6H_4C_6H_5$.
Properties: Crystals; m.p. 316–318°C; b.p. 428°C (18 mm).
Grade: Purified.
Use: As primary fluor or as wave-length shifter in solution scintillators.

quebrachine. See yohimbine.

quebracho extract.
Properties: A wood-derived tannin, the most important extract used in the American leather industry. Combustible.
Derivation: From Aspidosperma quebracho and Queracho lorentzi, imported as logs from Argentina.
Grades: Liquid: 35–37% tannin. Solid: 65% tannin.
Uses: Vegetable tanning; retanning of chrome-tanned upper leathers; dyeing; ore flotation; oil well drilling fluids.

"Questex."[1] Trademark for ethylenediaminetetraacetic acid (EDTA) and derivatives, a group of polyaminoacid-based organic sequestering agents which complex or chelate multivalent, metallic cations (such as calcium, magnesium, copper and iron) into stable, coordinated anionic complexes.

quick-. Prefix meaning alive or active, as in quicksilver (mercury), quicklime (unslaked lime), quicksand, quick (fingernail).

quicklime. See calcium oxide.

quicksilver. See mercury.

"Quik-Gel."[236] Trademark for a peptized high-swelling bentonite. Used as thickener and suspending agent in fluids or core hole drilling.

"Quilon."[28] Trademark for a Werner-type chromium complex in isopropanol; approximately a 30% solution of stearato chromic chloride.
Uses: Water repellent and sizing treatment of cellulosic materials; treatment of negatively charged surfaces; antiblocking agent; for insolubilizing various water-soluble or swellable coatings; improving grease-resistant coatings; treatment of feathers.

quinacridone. A light-fast pigment used in paints, printing inks, plastics, etc.

quinacrine hydrochloride $C_{23}H_{30}ClN_3O \cdot 2HCl \cdot 2H_2O$. 3-Chloro-7-methoxy-9-(1-methyl-4-diethylaminobutylamino)acridine dihydrochloride).
Properties: Bright yellow, crystalline powder; odorless and with a bitter taste. Decomposes at 248–250°C. Soluble in water and alcohol; pH of 1% water solution 4.5.
Derivation: Organic synthesis.
Grade: U.S.P.
Hazard: Moderately toxic.
Use: Medicine (antimalarial).

quinaldine (chinaldine; alpha-methylquinoline) $C_9H_6NCH_3$.
Properties: Colorless oily liquid; odor of quinoline; darkens to reddish brown in air. Soluble in alcohol, ether, and chloroform; slightly soluble in water. B.p. 246–247°C; m.p. −2°C. Sp. gr. 1.51.
Derivation: (a) By the treatment of aniline and paraldehyde with hydrochloric acid and heat. (b) From coal tar.
Uses: Manufacture of dyes, pharmaceuticals, fine organic chemicals, acid-base indicators.

quinaphthol (quinine-beta-naphthol alpha-sulfonate) $C_{20}H_{24}N_2O_2 \cdot (OHC_{10}H_6 \cdot SO_3H)_2$.
Properties: Yellow crystalline powder; contains 42% quinine; bitter taste. Moderately soluble in hot water or alcohol; insoluble in cold water. M.p. 185–186°C.
Derivation: Interaction of quinine and betanaphtholsulfonic acid.
Use: Medicine.

"Quindex."[74] Trademark for solubilized form of copper 8-quinolinolate. Contains 1.8% copper. Used when nonmercurial fungicide is required.

"Quindo."[438] Trademark for quinacridone pigments. Used in paints, printing inks, and plastics.

quinhydrone $C_6H_4O_2C_6H_4(OH)_2$.
Properties: Dark green crystals. Slightly soluble in water; soluble in alcohol, ether, hot water, ammonia. M.p. 171°C; sp. gr. 1.40.
Derivation: Oxidation of hydroquinone with sodium dichromate.
Grades: Reagent; technical.
Use: Electrode for pH determination.

quinic acid (chinic acid) $C_6H_7(OH)_4COOH \cdot H_2O$. Hexahydro-1,3,4,5-tetrahydroxybenzoic acid.
Properties: White, transparent crystals; very acid taste. Soluble in water, alcohol, and glacial acetic acid; insoluble in ether. Sp. gr. 1.637; m.p. 162°C; b.p., decomposes.
Derivation: From cinchona bark (see quinine).
Hazard: Moderately toxic and irritant.
Use: Medicine.

quinidine (chinidine; beta-quinine) $C_{20}H_{24}N_2O_2$.
Properties: Colorless; lustrous, crystalline alkaloid; efflorescing on exposure to air. Soluble in chloroform, alcohol, and ether; very slightly soluble in water. M.p. 171.5°C.
Derivation: Extraction of cinchona bark; also synthetic.
Hazard: Moderately toxic and irritant.
Use: Medicine (as the alkaloid or various salts).

quinine $C_{20}H_{24}N_2O_2 \cdot 3H_2O$. An alkaloid.
Properties: Bulky, white, amorphous powder or crystalline alkaloid; very bitter taste; odorless and levorotatory. Soluble in alcohol, ether, chloroform, carbon disulfide, ligroin, oils, glycerol, alkalies and

acids (with formation of salts); very slightly soluble in water.
Derivation: Finely ground cinchona bark mixed with lime is extracted with hot high-boiling paraffin oil. The solution is filtered, shaken with dilute sulfuric acid, the latter neutralized hot with sodium carbonate and on cooling quinine sulfate crystallizes out. The sulfate is treated with ammonia, the alkaloid being obtained.
Source: Indonesia; Bolovia.
Containers: Vials; up to 1000-oz drums.
Uses: Medicine (antimalarial), as the alkaloid or as numerous salts and derivatives; beverages.

beta-**quinine.** See quinidine.

quinine acid sulfate. See quinine.

quinine bisulfate. See quinine.

quinine carbacrylic resin. The quinine salt of a polyacrylic carboxylic acid resin. Contains 1.85% of quininium ion.
Properties: Buff, odorless, tasteless, free-flowing, amorphous, granular solid. Practically insoluble in dilute acids and alkalies, alcohol, ether and water.
Use: Medicine.

quinine sulfate $(C_{20}H_{24}N_2O_2)_2 \cdot H_2SO_4 \cdot 2H_2O$.
Properties: White needle-like crystals; odorless; very bitter taste; turns brown in the light; a saturated solution is neutral or slightly alkaline to litmus paper; m.p. 205°C; if exposed to damp air it may take up 5 H_2O. Loses water of crystallization at about 100°C. Soluble in hot water and in alcohol. Slightly soluble in chloroform and ether.
Grades: N.F.; F.C.C.
Containers: See quinine.
Uses: Medicine (antimalarial); flavoring agent.

quinine-urea hydrochloride (urea-quinine) $C_{20}H_{24}N_2O_2 \cdot HCl \cdot CO(NH_2)_2 \cdot HCl \cdot 5H_2O$.
Properties: White, granular powder, or colorless translucent prisms. Bitter taste, odorless. Contains not less than 58% anhydrous quinine. Soluble in water and strong alcohol.
Derivation: By dissolving quinine hydrochloride in hydrochloric acid, filtering, adding urea, heating and crystallizing.
Grade: U.S.P.
Use: Medicine.

quinizarin (1,4-dihydroxyanthraquinone) $C_{14}H_6O_2(OH)_2$.
Properties: Red or yellow-red crystals; m.p. 194–195°C. Soluble in hot water, alcohol, ether and benzene, and in potassium hydroxide and sulfuric acid.
Use: Antioxidant in synthetic lubricants.

quinolidine (chinoidine)
Properties: Brownish black lustrous mass; resinous appearance; conchoidal fracture; very bitter taste. It is a mixture of the amorphous alkaloids remaining in the solution from the extraction of cinchona bark, after the crystallizable alkaloids have been removed. Soluble in dilute acids, alcohol and chloroform. M.p., softens below 100°C.
Containers: 70-, 140-lb drums.
Hazard: Moderately toxic and irritant.
Uses: Medicine, either as such, or as the borate, citrate, hydrochloride, sulfate or tannate.

quinol. See hydroquinone.

quinoline (chinoline)
Properties: A basic heterocyclic nitrogenous compound occurring in coal tar and obtained from it, but more frequently by synthesis; highly refractive, colorless liquid; darkens with age; hygroscopic; peculiar odor. Soluble in water, alcohol, ether and carbon disulfide. Sp. gr. 1.0899; f.p. −15°C; b.p. 238°C. Combustible. Autoignition temp. 896°F.

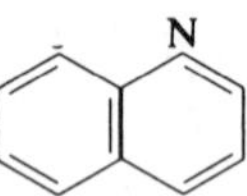

Derivation: By treatment of aniline and nitrobenzene with glycerol and sulfuric acid and heat.
Containers: Glass bottles; drums; tank cars.
Uses: Medicine; preserving anatomical specimens; manufacture of quinolinol sulfate, niacin, and copper-8-quinolinolate.

quinoline dye. See cyanine dye.

8-quinolinol. See 8-hydroxyquinoline.

8-quinolinol-7-iodo-5-sulfonic acid. See 7-iodo-8-hydroxyquinoline-5-sulfonic acid.

quinone (1,4-benzoquinone; chinone) $C_6H_4O_2$.
Properties: Yellow crystals; irritating odor. Soluble in alcohol, ether and alkalies; slightly soluble in hot water. Sp. gr. 1.307; m.p. 115.7°C; b.p. sublimes; volatile with steam, being in part decomposed. Combustible.
Derivation: By oxidation of aniline with chromic acid, extraction with ether and distillation.
Hazard: Toxic by inhalation; strong irritant to skin and mucous membranes. Tolerance, 0.1 ppm in air.
Uses: Manufacture of dyes and hydroquinone; fungicides.

para-**quinonedioxime.** See "G-M-F."

quinone oxime dye. See nitroso dye.

quinosol $C_9H_6NOSO_3K \cdot H_2O$. Yellow crystalline powder. Used as a fungicide. Combustible.

quinoxaline (1,4-benzodiazine; benzo-para-diazine) $C_8H_6N_2$ (bicyclic). An organic base.
Properties: Colorless crystalline powder; m.p. 30°C; b.p. 229°C; soluble in water and organic solvents.
Use: Organic synthesis.

quinoxalinedithiol cyclic trithiocarbonate. See thioquinox.

"Quotane."[71] Trademark for an ointment and lotion containing dimethisoquin hydrochloride.

"QUSO."[201] Trademark for a series of microfine precipitated amorphous silicas having a fully hydroxylated surface.
Uses: Thickening, flatting, anti-caking, and reinforcing agent, adsorptive carrier.

R

R (1) Symbol used to represent an organic group in a chemical formula, for example, CH_3, C_2H_5, C_6H_5, etc.
(2) A free radical (with superior dot, R˙).
(3) The universal gas constant, equal to $p_oV_o/273$.
(4) Abbreviation of Rankine temperature scale.

R & D. Abbreviation for Research and Development, usually referring to the department or division of a company whose major responsibility is applied research and creative development of new products and processes.

Ra Symbol for radium.

racemization. Conversion, by heat or by chemical reaction (e.g., enolization), of an optically active compound into an optically inactive form, in which half of the optically active substance becomes its mirror-image (enantiomer). This change results in a mixture of equal quantities of dextro- and levorotatory isomers, as a result of which the compound does not rotate plane-polarized light to either right or left since the two opposite rotations cancel each other. This is sometimes referred to as external compensation, as opposed to the internal compensation exhibited by meso- compounds. See also meso- (1); tartaric acid.

racephedrine (racemic ephedrine; *dl*-ephedrine)
$C_{10}H_{15}NO$. For optically active form, see ephedrine.
Properties: Crystals; m.p. 79°C; soluble in water, alcohol, chloroform, and oils.
Derivation: Synthetic.
Use: Medicine (also as hydrochloride and sulfate).

R acid. See 2-amino-3-naphthol-6-sulfonic acid; 2-naphthol-3,6-disulfonic acid.

2R acid. See 2-amino-8-naphthol-3,6-disulfonic acid.

rad. That quantity of ionizing radiation (q.v.) that results in the absorption of 100 ergs of energy per gram of irradiated material, regardless of the source of the radiation. The Federal radiation safety standard is 0.17 rad per person per year, and even this is considered too high by some authorities.

"Radapon."[233] Trademark for an herbicide based on dalapon.

"Radel."[214] Trademark for a water-soluble polyethylene oxide film.
Uses: Packaging powdered detergents, dyes, insecticides, fungicides and other household, industrial, and agricultural products that are dissolved in water prior to use.

radiation. Energy in the form of electromagnetic waves (also called radiant energy, or light). It is emitted from matter in the form of photons (quanta), each having an associated electromagnetic wave having frequency (ν) and wavelength (λ). The various forms of radiant energy are characterized by their wavelength, and together they comprise the electromagnetic spectrum, the components of which are as follows: (1) cosmic rays (highest energy, shortest wavelength); (2) gamma rays from radioactive disintegration of atomic nuclei; (3) x-rays; (4) ultraviolet rays; (5) visible light rays; (6) infrared; (7) microwave; and (8) radio (Hertzian) and electric rays. All these are identical in every way except wavelength, those having the shortest wavelength being the most penetrating. They are not electrically charged and have no mass; their velocity of propagation is the same; all display the properties characteristic of light and have a dual nature (wave-like and corpuscular). In a looser sense the term "radiation" also includes energy emitted in the form of particles which possess mass and may or may not be electrically charged, i.e., alpha (positive) and beta (negative); also neutrons. Beams of such particles may be considered as "rays." The charged particles may all be accelerated and high energy imparted to "beams" in particle accelerators such as cyclotrons, betatrons, synchrotrons and linear accelerators.

Radiation is used in medicine in the form of x-rays and radioactive isotopes; it is used in industry in many ways, e.g., as vitamin activator, sterilizing agent, and polymerization initiator; it is also the basis of all types of spectroscopic analysis.
See also following entries.

radiation biochemistry. The study of substances having the ability to protect cells and body tissue against the deleterious effects of ionizing radiation. As one of these effects is to deprive proteins of sulfhydryl (—SH) groups necessary for cell division, the injection of compounds rich in this radical (notably cysteine) has been successfully tried with laboratory animals. Thiourea has been found to protect DNA from depolymerization by x-rays; enzymes containing —SH groups inactivated by radiation are reactivated by addition of glutathione. Some of the other radiochemically induced reactions that adversely affect biochemical activity are (1) formation of hydrogen peroxide (a biological poison) by free radical mechanism; (2) denaturation of proteins; (3) change in substituent groups of amino acids; (4) oxidation of hemoglobin; (5) depolymerization of DNA. See also radiation, ionizing.

radiation, industrial. Chemical or physiochemical changes induced by exposure to various types and intensities of radiation. The reactions are of the following types: (1) synthesis of ethyl bromide from hydrogen bromide and ethylene, using gamma radiation from cobalt-60; (2) cross-linking of such polymers as polystyrene and polyethylene with either beta or gamma radiation; (3) vulcanization of rubber with ionizing radiation; (4) polymerization of methyl methacrylate monomer with cobalt-60 as source of gamma rays. Free radical formation is involved in both cross-linking and polymerization reactions. This technique is also being applied in the textile finishing field for grafting and cross-linking fibers with chemical agents for durable-press fabrics. (5) Irradiation of foodstuffs, milk, etc., by varying dosages of radiation for the purposes of (a) sterilization, (b) pasteurization, (c) increase of Vitamin D formation, (d) inhibition of sprouting, (e) prevention of infestation of stored wheat by insects. All these must be in accordance with FDA requirements. (6) Curing or hardening of paint films on plastic, wood, metal, etc. by use of low-voltage electron beam. See also "Electrocure."

radiation, ionizing. Extremely short-wavelength, highly energetic, penetrating rays of the following types: (a) gamma rays emitted by radioactive elements and isotopes (decay of atomic nucleus); (b) x-rays, generated by sudden stoppage of fast-moving electrons; (c) subatomic charged particles (electrons, protons, deuterons) when accelerated in a cyclotron or betatron. The term is restricted to electromagnetic radiation at least as energetic as x-rays, and to charged particles of similar energies. Neutrons also may induce ionization.

Such radiation is strong enough to remove electrons from any atoms in its path, leading to the formation of free radicals (q.v.). These short-lived but highly reactive particles initiate decomposition of many organic compounds. Thus ionizing radiation can cause mutations in DNA and in cell nuclei; adversely affect protein and amino acid mechanisms; impair or destroy body tissue; and attack bone marrow, the source of red blood cells. Exposure to ionizing radiation for even a short period is highly dangerous, and for an extended period may be lethal. The study of the chemical effects of such radiation is called radiation chemistry or (in the case of body reactions) radiation biochemistry (q.v.). See also radiation, industrial; x-radiation; rad.

radical. (1) An ionic group having one or more charges, either positive or negative, e.g., OH^-, NH_4^+, SO_4^{--}. See also group (2).

(2) See free radical.

radioactivity. Natural or artificial nuclear transformation; discovered by Becquerel in 1895. The energy of the process is emitted in the form of alpha, beta, or gamma rays. Thus radium-226 undergoes radioactive decay by the emission of an alpha particle, and the new product is radon-222. The decay series terminates in lead-206. Radioactivity is not affected by the physical state or chemical combination of the element. The radioactivity of a nuclide is characterized by the nature of the radiations, their energy, and the half-life of the process, i.e., the time required for the activity to decrease to one-half of the original. Half-lives vary from microseconds to millions of years. Some radioactive elements occur in nature (radium, uranium). Radioactivity can be caused artificially in many stable elements by irradiation with neutrons in a nuclear reactor, or by charged particles from an accelerator.

Amounts of radioactive material are usually expressed in units of activity, the rate of the radioactivity decay. The accepted unit is the curie and its metric fractions, the milli- and microcurie. A curie is 3.73×10^{10} disintegrations per second. A common unit is millicuries per millimole. Packaging and shipment of radioactive materials, which are highly toxic, must be in accord with official requirements.

"Radiocarb."[155] Trademark for carbonized nickel from which all the excess carbon has been removed without creating a sheen. Surface has a dull black appearance.

radiocarbon. See carbon 14.

radiochemistry. That subdivision of chemistry which deals with the properties and uses of radioactive materials in industry, biology, and medicine.

radioisotope (radionuclide). An isotopic form of an element (either natural or artificial) that exhibits radioactivity. Radioisotopes are used as curative agents in medicine, in biological tracer studies, and for many industrial purposes, from measurement of thickness to initiating polymerization.

See also tracer; isotope.

radium Ra Radioactive element of Group IIA of the Periodic Table; atomic number 88; atomic weight 226.0254; valence 2; 14 radioactive isotopes, but only Ra-226 with half-life of 1620 years is usable. Discovered by the Curies in 1898.

Properties: Brilliant white solid; m.p. 700°C; b.p. 1140°C; sp. gr. 5 (approx.). Luminescent; turns black on exposure to air; soluble in water with evolution of hydrogen. Forms water-soluble compounds. Decays by emission of alpha, beta, and gamma radiation; bone-seeking.

Occurrence: Colorado, Canada, Africa (Congo), France, USSR.

Derivation: Uranium ores (pitchblende and carnotite). The method used for isolating radium is similar to that of Mme. Curie, and involves coprecipitation with barium and lead, chemical separation with HCl, and further purification by repeated fractional crystallization. The metal is separated from its salts by electrolysis and subsequent distillation in hydrogen.

Containers: Dry salts are stored in sealed glass tubes, opened regularly by experienced workers to relieve pressure. The tubes are kept in lead containers.

Hazard: Highly toxic; emits ionizing radiation (q.v.). Lead shielding should be used in storage and handling; adequate protective clothing and remote control devices are essential. Destructive to living tissue.

Uses: Medical treatment for malignant growths; industrial radiography; source of neutrons and radon.

Shipping regulations: Consult regulations.

radium bromide $RaBr_2$.

Properties: White crystals, becoming yellow or pink; radioactive. Sp. gr. 5.79; m.p. 728°C; sublimes at 900°C; soluble in water and alcohol.

Derivation: Freed from the ores as a bromide mixed with barium bromide.

Method of purification: Fractional crystallization.

Impurities: Barium salts.

Grades: Technical; pure. The purity is determined by the strength of the ionizing power of the salt, i.e., the extent to which it causes air to conduct electricity.

Containers: Glass bottles; sealed tubes enclosed in sheet lead.

Hazard: See radium.

Uses: Medicine (cancer treatment); physical research.

Shipping regulations: Consult regulations.

radium carbonate $RaCO_3$.

Properties: Amorphous radioactive powder, white when pure, but sometimes yellow, orange, or pink due to impurities. Insoluble in water. Marketed as a mixture with barium carbonate.

Hazard: See radium.

Use: Medicine.

Shipping regulations: Consult regulations.

radium chloride $RaCl_2$.

Properties: Yellowish white crystals, becoming yellow

or pink on standing; radioactive. Soluble in water and alcohol. M.p. 1000°C; sp. gr. 4.91.
Derivation: Freed from the ores as a chloride mixed with barium chloride.
Method of purification: Fractional crystallization.
Grades: Technical; pure. The purity of radium salts is determined by the strength of their ionizing power, i.e., the extent to which they cause air to conduct electricity.
Containers: Glass bottles; sealed tubes enclosed in sheet lead.
Hazard: See radium.
Uses: Medicine (cancer treatment); physical research.
Shipping regulations: Consult regulations.

radium sulfate $RaSO_4$.
Properties: White crystals when pure, but sometimes yellow, orange, or pink due to impurities. Radioactive. Insoluble in acids and water.
Hazard: See radium.
Use: Medicine.

radon Rn Gaseous radioactive element. Atomic number 86; noble gas group of Periodic Table. Atomic weight 222.02; valences 2, 4, (6); 18 radioactive isotopes, all short-lived. The 222 isotope has a half-life of 3.8 days, emits alpha radiation.
Properties: Colorless gas; density 9.72 g/l (0°C); soluble in water; can be condensed to colorless transparent liquid (b.p. −61.8°C) and to an opaque, glowing solid. The heaviest gas known.
Derivation: Radioactive decay of radium. Radon is obtained by bubbling air through a radium salt solution and collecting the gas plus air.
Hazard: See radium.
Uses: Medicine (cancer treatment); tracer in leak detection; flow rate measurement; radiography; chemical research.
Shipping regulations: Consult regulations.

raffinate. The portion of an oil that is not dissolved in solvent refining of lubricating oil.

raffinose $C_{18}H_{32}O_{16} \cdot 5H_2O$. A trisaccharide composed of one molecule each of D(+)Galactose, D(+)-glucose, and D(−)-fructose.
Properties: White, crystalline powder; sweet taste. Soluble in water; very slightly soluble in alcohol. Sp. gr. 1.465; m.p. (anhydrous) 118° to 119°C; b.p. decomposes at about 130°C; optical rotation +104.5°. Split by invertase to melibiose and saccharose. Combustible; nontoxic.
Derivation: Hydrolysis of cottonseed meal; from sugar beet concentrates.
Uses: Bacteriology; preparation of other saccharides.

"Raldeine."[227] Trademark for methyl ionones, available in several grades.
Use: Perfumery (violet odor).

Raman spectra. See spectroscopy.

ramie. A natural fiber obtained from the stems of Boehmeria nivea, of the hemp family. High wet strength; absorbent but dries quickly; can be spun or woven. Wears well and has great rot and mildew resistance; tensile strength four times that of flax; elasticity 50% greater than flax.
Sources: Taiwan; Egypt; Brazil; Florida.
Hazard: Combustible, not self-extinguishing.
Uses: High-grade paper (Europe); fabrics (weaving apparel and car seat covers); stern-tube packing in ships; patching watermains (Great Britain).

"Ramrod."[58] Trademark for an herbicide available as a wettable powder (contains 65% 2-chloro-N-isopropylacetanilide) and granular form (contains 20% 2-chloro-N-isopropylacetanilide).

"Randox."[58] Trademark for a liquid herbicide containing 47.1% alpha-chloro-N,N-diallylacetamide.

Raney nickel. A spongy form of nickel.
Derivation: By leaching the aluminum from an alloy of 50% aluminum-50% nickel with 25% caustic soda solution.
Hazard: Ignites spontaneously in air; store under inert gas or under water. Dangerous fire risk.
Use: Catalyst for hydrogenation.
Shipping regulations: See under nickel.

Rankine. A scale of absolute temperature based on Fahrenheit degrees. Abbreviation: R. See absolute temperature.

Raoult's law. The vapor pressure of a substance, in equilibrium with a solution containing the substance, is equal to the product of the mole fraction of the substance in the solution, and the vapor pressure of the pure substance at the temperature of the solution. The law is not applicable to most solutions, but is often approximately applicable to mixtures of closely similar substances, particularly the substance present in high concentration.

rapeseed meal. The ground press-cake left from expression or rapeseed oil. Its major use is as an animal feed ingredient, but the presence of harmful glucosides, which may be goiter-inducing, limits its application for this purpose. Research efforts are being directed to removal of this constituent from meal produced in Canada. The meal also has some use as a fertilizer. If fed to animals, it should be blended with other feeds.

rapeseed oil. A vegetable oil derived from rape by expression or solvent extraction; it is now produced chiefly in Canada. It is a viscous, brownish liquid, though when refined it is yellow. Sp. gr. 0.913–0.916; solidifies about 0°C; flash point 325°F; autoignition temperature 836°F; subject to spontaneous heating. It is high in unsaturated acids, especially oleic, linoleic and erucic, the latter being 40% in older varieties, but recently introduced types have much lower erucic content. It is shipped in drums and tank cars.
Uses: Edible oil for salad dressings, margarine, etc.; lubricant additive; substitute for soybean oil.

"Rapidase."[173] Trademark for a bacterial amylase for removing starch sizings from textiles.

rare earth. One of a group of 15 chemically related elements in Group IIIB of the Periodic table (lanthanide series). Their names and atomic numbers are:

lanthanum	La	57
cerium	Ce	58
praseodymium	Pr	59
neodymium	Nd	60
promethium	Pm	61
samarium	Sm	62
europium	Eu	63
gadolinium	Gd	64
terbium	Tb	65
dysprosium	Dy	66
holmium	Ho	67

erbium	Er	68
thulium	Tm	69
ytterbium	Yb	70
lutetium	Lu	71

The elements 57–62 are known as the cerium subgroup, and 63–71 as the yttrium subgroup. Yttrium, atomic number 39, although not a rare earth element, is found associated with the rare earths and is only separated with difficulty. See also didymium.

Source: Monazite, bastnasite and related fluocarbonate minerals, as well as minerals of the yttrium group. These ores contain varying percentages of rare earth oxides, which are often loosely called rare earths. Rare earth elements also occur as fission products of uranium and plutonium, which is the only source of promethium.

Occurrence: (monazite) India, Brazil, Florida, Carolinas, Australia, South Africa; (bastnasite) California.

For properties and uses, see specific entry.

rare gas. Any of the six gases comprising the extreme right-hand group of the Periodic Table, namely, helium, neon, argon, krypton, xenon, and radon. They are preferably called noble gases or (less accurately) inert gases. The first three have a valence of 0 and are truly inert, but the others can form compounds to a limited extent.

rare metal. A loose term for the less common and more expensive metallic elements. They include the alkaline-earth metals (barium, calcium and strontium); beryllium, bismuth, cadmium, cobalt, gallium, germanium, hafnium, indium, lithium, boron, silicon, manganese, molybdenum, rhenium, selenium, tantalum, niobium, tellurium, thallium, thorium, titanium, tungsten, uranium, vanadium, zirconium, and the rare earths.

"Rareox."[241] Trademark for optical quality cerium oxide for high speed polishing.

Raschig process. See hydrazine.

"Rasorite."[441] Trademark for sodium borate concentrates.

"Raticate."[507] Trademark for 5-(α-hydroxy-α-2-pyridylbenzyl)-7-(α-2-pyridylbenzylidene)-5-norbornene-2,3-dicarboximide. A rodenticide specific for rats; nontoxic to other animals and to man (U.S. Dept. of Agriculture).

"Raudixin."[412] Trademark for rauwolfia serpentina whole root.

"Raunormine."[342] Trademark for deserpidine hydrochloride.

"Rau-Sed."[412] Trademark for reserpine (q.v.).

"Rauserfia."[342] Trademark for an extract of weakly basic alkaloids from Rauwolfia.

rauwolfia. The powdered whole root of Rauwolfia serpentina. The plant is of value as a source of alkaloids, especially reserpine (q.v.).

Properties: Light tan to light brown, bitter, fine, amorphous powder; slight aromatic odor. Sparingly soluble in alcohol and very slightly soluble in water.

Grade: N.F.

Containers: Barrels; drums.

Use: Medicine (antihypertensive agent).

rayon. Generic name for a semisynthetic fiber composed of regenerated cellulose, as well as manufactured fibers composed of regenerated cellulose in which substituents have replaced not more than 15% of the hydrogens of the hydroxyl groups. Rayon was first made by denitration of cellulose nitrate fibers (Chardonnet process), but most rayon is now made by the viscose and cuprammonium processes. "Regular" viscose tenacity = 2 g per denier; "high-tenacity" = 3 to 6 g per denier (tire cord). Elongation 15–30% (dry) and 20–40% (wet). Swells and weakens when wet. Moisture regain 11 to 13%; sp. gr. 1.50.

Modified rayon is made principally of regenerated cellulose and contains nonregenerated cellulose fiber-forming material; for example, a fiber spun from viscose containing casein or other protein (ASTM). This greatly increases both dry and wet strength and also permits mercerization. Rayon is readily dyed by standard methods.

Hazard: Flammable; not self-extinguishing. Moderate fire risk.

Uses: Nonwoven fabrics; surgical dressings; wearing apparel; mechanical rubber goods; coated fabrics; felts and blankets; blends with cotton for home furnishings, etc.; tire cord.

See also cellophane; acetate fiber.

rayon coning oil. An oil used to lubricate and reduce the static of yarns wound by a coning machine. Usually composed of mineral oils of low viscosity, so compounded as to emulsify in water.

raw material. Any ingredient or component of a mixture or product before mixing and processing take place. Bulk storage should be planned with due regard for flammable and toxic properties, possible emergencies due to spills or container breakage, etc.

"Rayox." Trademark for titanium dioxide (q.v.).

Rb Symbol for rubidium.

"RC."[52] Trademark for a series of plasticizers.

Uses: Electrical tapes, metal laminates, surgical tubing, gaskets, non-marring film and sheeting, organosols, plastisols, plasticizer for nitrocellulose and synthetic rubbers, mild lubricant.

"R-2" Crystals.[58] Trademark for a reaction product of carbon disulfide and 1,1′-methylenedipiperidine.

Uses: Vulcanization accelerator for natural and synthetic rubbers, mostly in cements and latex.

RDA (1) Abbreviation for Recommended Dietary Allowance of food requirements, including proteins, vitamins, and minerals, for infants, children, and adults, established by the Food and Nutrition Board of the National Academy of Sciences-National Research Council. They were revised in 1974, particularly in reference to certain vitamins (C, B_{12} and E) and proteins. Future revisions may be expected.

(2) Abbreviation for Recommended Daily Allowance for food and nutrition established by the Food and Drug Administration to serve as a basis for regulations on nutritional labeling of food products. These were effective as of January 1, 1975, and were based on the Recommended Dietary Allowance referred to in (1).

RDGE. See resorcinol diglycidyl ether.

RDX. See cyclonite.

Re Symbol for rhenium.

reaction, chemical. See chemical reaction.

reagent. Any substance used in a reaction for the purpose of detecting, measuring, examining or analyzing other substances. High purity and high sensitivity are essential requirements of laboratory reagents.

Over 8000 reagent chemicals are commercially available. See also grade.

rearrangement. A type of chemical reaction in which the atoms of a single compound recombine, usually under the influence of a catalyst, to form a new compound having the same molecular weight but different properties. Thus ammonium cyanate in solution will rearrange to form urea, in which the four hydrogen atoms are equally distributed between the two nitrogen atoms: $NH_4OCN \longrightarrow (NH_2)_2C{:}O$. Many such rearrangements have been named for their discoverers, e.g., Beckmann rearrangement (q.v.). See also Wöhler.

Reaumur. A temperature scale in which 0° is the freezing point of water and 80° is its boiling point.

recipe. A product formula; used primarily in the food industry.

reclaiming. See recycling (2).

recombinant DNA. Genetic material transferred from the genes of one species to those of another by uniting a portion of the DNA of one organism with extranuclear sections of DNA from another organism. Such "recombinant" molecules will replicate in the same way as in normal DNA behavior. This technique is being utilized in basic genetic research sponsored by the National Institutes of Health. Some concern has been expressed over the risks of producing abnormal and possibly dangerous species which might contaminate the ecosystem; in this regard, NIH has issued guidelines for such research. See also deoxyribonucleic acid; replication.

reconstitution. In food technology, restoration of a dehydrated food product to its original edible condition by adding water to it at the time of use. It is also called rehydration. See also dehydration.

rectification. The enrichment or purification of the vapor during the distillation process by contact and interaction with a countercurrent stream of liquid condensed from the vapor. See also reflux.

recycling. (1) As generally used in the chemical industries, this term refers to the practice of returning a portion of the reaction products to the start of the system, either for the purpose of more efficient conversion of unreacted components, or to reuse auxiliary materials that remain unchanged during processing. In the petroleum refining industry, some of the product stream may be recycled and blended with the fresh input materials to obtain a product of maximum value.

(2) In a less technical sense, the word is also used to describe the inclusion of a limited amount of comminuted waste stock in the manufacture of products such as plastics, paper, and glass. Steel from scrapped automobiles, iron from engine blocks, scrap aluminum, waste paper, etc. are recycled, though there is a definite maximum percentage that can be successfully used. If the scrap material requires chemical processing for this purpose, the more accurate term is reclaiming, for example, of rubber.

red acetate. See mordant rouge.

red arsenic. See arsenic disulfide.

"Redax."[69,265] Trademark for N-nitrosodiphenylamine; used as a pre-vulcanization retarder for natural and synthetic rubber compounds.

red brass. Copper-zinc (brass) alloys characterized by their red color and high copper content. The term is used for several different types of brass. One ASTM classification permits 2–8% zinc, tin less than zinc, and lead less than 0.5%. Other alloys referred to as red brass include those with 75–85% copper, up to 20% zinc, and usually very small amounts of lead and tin. In one alloy possessing good machining qualities, the lead content may be as high as 10%, the tin as high as 5%.

Red brasses are widely used for decorative purposes and in plumbing and piping because of their resistance to atmospheric corrosion and dezincification.

Red dye No. 2. See amaranth; FD&C colors.

red glass. A soda-zinc glass containing a small amount of cadmium and about 1% selenium. Red may also be obtained by using cuprous oxide or gold chloride, the latter usually in the form of purple of Cassius.

"Redicote."[15] Trademark for a series of compounded amines.

Containers: 55-gal drums; tank cars; tank trucks.

Uses: Cationic emulsifiers for asphalt.

red iron oxide. See iron oxide reds.

red iron trioxide. See ferric oxide.

red lake C pigment. One of a family of organic acid azo pigments prepared by coupling the diazonium salt of ortho-chloro-meta-toluidine-para-sulfonic acid with beta-naphthol. Both sodium and barium salts of the parent dye are used.

Uses: Plastics, rubber products, printing inks.

red lead. See lead oxide, red.

red liquor. See mordant rouge.

red ocher. See ocher.

red oil. A commercial grade of oleic acid comprised of 70% oleic and 15% each of linoleic and stearic acids.

redox. Short form of the term oxidation-reduction, as in redox reactions, redox conditions, etc. See also oxidation.

red oxide. See iron oxide red.

red phosphorus. See phosphorus.

reduced states. See corresponding states.

"Reducolite."[159] Trademark for a sodium zinc sulfoxylate formaldehyde compound used in stripping dyed colors from wool and synthetics.

reduction. (1) The opposite of oxidation. Reduction may occur in the following ways: (1) Acceptance of one or more electrons by an atom or ion; (2) removal of oxygen from a compound; (3) addition of hydrogen to a compound. See oxidation.

(2) See comminution.

refinery gas. A mixture of hydrocarbon gases (and often some sulfur compounds) produced in large-scale cracking and refining of petroleum. The usual components are hydrogen, methane, ethane, propane, butanes, pentanes, ethylene, propylene, butenes, pentenes, and small amounts of other components such as butadiene.

Use: Source of raw material for petrochemicals, high-octane gasoline, and organic synthesis of alcohols.

"Refinex."[98] Trademark for a specially processed grade of attapulgite clay.
Uses: Bleaching, neutralizing and deodorizing agent in the re-refining of engine and lubricating oils by the contact method.

refining. Essentially a separation process whereby undesirable components are removed from various types of mixtures to give a concentrated and purified product. Such separation may be effected (a) mechanically, by pressing, centrifuging, filtering, etc.; (b) by electrolysis; (c) by distillation, solvent extraction or evaporation; and (d) by chemical reaction. One or more of these operations is applied to (1) food products; (2) petroleum; (3) lubricating oils; and (4) metals. As regards petroleum, refining is generally understood to include not only fractional distillation of crude oil to naphthas, low-octane gasoline, kerosine, fuel oil, and asphaltic residues, but also the processes involved in thermal and catalytic cracking (hydroforming, reforming, etc.) for production of high-octane gasoline. See also specific entries.

reflux. In distillation processes in which a fractionating column is used, the term reflux refers to the liquid that has condensed from the rising vapor and allowed to flow back down the column toward the still. As it does so, it comes into intimate contact with the rising vapor, resulting in improved separtion of the components. The separation resulting from contact of the countercurrent streams of vapor and liquid is called rectification or fractionation.

reforming. Decomposition (cracking) of hydrocarbon gases or low-octane petroleum fractions by heat and pressure, either without a catalyst (thermoforming) or with a specific catalyst (molybdena, platinum). The latter method is the more efficient and is used almost exclusively in the U.S. The chief cracking reactions are (1) dehydrogenation of cyclohexanes to aromatic hydrocarbons; (2) dehydrocyclization of certain paraffins to aromatics; (3) isomerization, i.e., conversion of straight-chain to branched-chain structures, as octane to isooctane. These result in substantial increase in octane number. Steam reforming of natural gas is an important method of producing hydrogen; steam reforming of naphtha is used to produce synthetic natural gas. See also hydroforming; "Platforming."

"Refractaloy."[308] Trademark for a nickel-cobalt-chromium-molybdenum-iron alloy. Type 26 is a precipitation-hardenable material using titanium as the hardening agent, and having high strength, high ductility, and corrosion resistance up to 1450°F. Gas turbine discs, bolting and blading are typical applications.

refraction. The change in direction (apparent bending) of a light ray passing from one medium to another of different density, as from air to water or glass. The ratio of the sine of the angle of incidence to the sine of the angle of refraction is the index of refraction of the second medium. Index of refraction of a substance may also be expressed as the ratio of the velocity of light in a vacuum to its velocity in the substance. It varies with the wave length of the incident light, temperature and pressure. The usual light source is the D line of sodium, the standard temperature being 20°C; the expression of refractive index is n 20/D.

refractive index. See refraction.

refractory. (1) An earthy, ceramic material of low thermal conductivity that is capable of withstanding extremely high temperature (3000–4000°F) without essential change. There are three broad groups of these: (a) acidic (silica, fireclay); (b) basic (magnesite, dolomite); and (c) amphoteric (alumina, carbon, and silicon carbide). Their primary use is for lining steel furnaces, coke ovens, glass lehrs, and other continuous high-temperature applications. They are normally cast in the form of brick and are sometimes bonded to assure stability. The outstanding property of these materials is their ability to act as insulators. The most important are fireclay (aluminum silicates); silica; high alumina (70–80% Al_2O_3); mullite (clay-sand); magnesite (chiefly MgO); dolomite (CaO-MgO); forsterite (MgO-sand); carbon; chrome ore-magnesite; zirconia; and silicon carbide. See also specific entries.

(2) Characterizing the ability to withstand extremely high temperature, e.g., tungsten and tantalum are refractory metals; clay is a refractory earth; ceramics are refractory mixtures.

"Refrasil."[161] Trademark for a group of materials having outstanding high-temperature resistance. Composed of white, vitreous fibers having up to 99% silicon dioxide content. Available as bulk fiber, batt, cloth, tape, sleeving, yarn, cordage, and flakes.
Uses: Aircraft and industrial applications; missiles.

"Refrax."[280] Trademark for silicon nitride-bonded silicon carbide refractories. Available in brick and precision-formed shapes and parts.
Uses: Brazing and furnace fixtures; pumps and pump parts handling corrosive, abrasive slurries; rocket motor components; spray nozzles; burners; pyrometer protection tubes; sinker assemblies in wire aluminizing; bolts and nuts; valve parts; aluminum melting furnace linings and parts; conveyor parts.

"Refrex."[214] Trademark for a blend of alcohols for refrigerator car-heating fuels.

refrigerant. Any substance which, by undergoing a change of phase (solid to liquid or liquid to vapor), lowers the temperature of its environment because of its latent heat. Melting ice, with latent heat of 80 calories per gram, removes heat and exerts a considerable cooling effect. Most commercial refrigerants are liquids whose latent heat of vaporization results in cooling. Ammonia, sulfur dioxide, and ethyl or methyl chloride were once widely used. However, the flammabilty and toxicity of these compounds led to a search for safer refrigerants that resulted in the discovery of halogenated hydrocarbons, especially fluorocarbons, which are nonflammable. Under various trademarks, these are now generally used for domestic refrigeration and air-conditioning. Ice and circulating brine are still used for preservation of fish at sea, and ammonia systems are operated for seafood storage in warehouses. See also "Freon"; "Ucon"; "Genetron"; fluorocarbon.

"Regl."[275] Trademark for a group of oil furnace carbon blacks for rubber, paints, plastics, etc.

regeneration. (1) Restoration of a material to its original condition after it has undergone chemical modification necessary for manufacturing purposes. The most common instance is that of cellulose for rayon production. The wood pulp used must first be converted to a solution by reaction with sodium hydroxide and carbon disulfide; in this form (cellulose xanthate) it can be extruded through spinnerets.

After this operation it is regenerated to cellulose by passing it through acid (viscose process). Collagen can also be regenerated by acid treatment after it has been purified for use in food products by alkaline solution.

(2) Renewal or reactivation of a catalyst that has accumulated reaction residues such as coke, usually accomplished by passage of steam or reducing gases over the catalyst bed.

(3) Replenishing the sodium ions of a zeolite or similar ion-exchange agent by treatment with sodium chloride solution. Molecular sieves are regenerated by heat removal of the water (about 200°C), followed by treatment with an inert gas.

"Reagent yellows."[141] Trademark for lightfast chrome yellow pigments.
Uses: In paints for all exterior uses; in vinyl, particularly automotive or furniture upholstery; in inks for exterior use.

rehydration. See reconstitution.

Reichert-Meissl number. A measure of volatile soluble fatty acids derived under arbitrary conditions.

Reich process. A method for purifying carbon dioxide produced in fermentation. The small amounts of organic impurities are oxidized and absorbed and the gas is then dehydrated with chemicals.

Reimer-Tiemann reaction. Reaction for the formation of phenolic aldehydes by heating a phenol with chloroform in the presence of alkali.

reinecke salt (ammonium tetrathiocyanodiammonochromate; ammonium reineckate) $NH_4[Cr(NH_3)_2(SCN)_4] \cdot H_2O$.
Properties: Dark-red crystals or crystalline powder; moderately soluble in cold water; soluble in hot water and alcohol; decomposes in aqueous solutions.
Derivation: From fusion of ammonium thiocyanate with ammonium dichromate.
Uses: Reagent for organic bases, such as choline, amines, for certain amino acids, and for mercury.

reinforced plastic. A composite structure comprised of a thermosetting or thermoplastic resin and fibers, filaments, or whiskers of glass, metal, boron, or aluminum silicate. Unless otherwise indicated, this term refers to fiberglass-reinforced plastic (FRP).
Properties: Exceptionally high strength, good electrical resistivity, weather- and corrosion-resistance, low thermal conductivity and low flammability.
Derivation: Basic acid glass in the form of fiber (0.0005 inch), strands of 50–200 fibers, filaments, or woven fabric is coated by passing through a bath of molten resin, which acts as a binder. The assembly can be either compression- or injection- molded. Large, high-strength parts are made by filament winding (q.v.). Resins used are polyester, epoxy, phenolic, polypropylene, polystyrene, nylon, polycarbonate, and polyphenylene oxides.
Uses: Automotive body components; ablative coatings on rockets and space vehicles; appliances (air-conditioning and refrigerator cases); electrical equipment; oil-well piping and tubing; large-diameter pipe; industrial piping systems; chemical storage and mix tanks; unitized cargo containers; marine equipment; pressure vessels; prefabricated building panels and other structural components. See also glass fiber; composite; fiber; whiskers.

For further information consult the Reinforced Plastics/Composites Institute of the SPI, 250 Park Ave., New York.

reinforcing agent. (1) One of numerous fine powders used in rather high percentages to increase the strength, hardness, and abrasion resistance of rubber, plastics, and flooring compositions. The reinforcing effect is in general a function of the particle size of the powder. The finest of all is channel carbon black (q.v.) whose surface area may be as great as 18 acres per pound. Other widely used reinforcing agents are thermal and furnace blacks, magnesium carbonate, zinc oxide, hard clays (kaolin), and hydrated silicas. Though some reinforcing agents have positive coloring properties, the term should not be used as a general synonym for pigment. (See also pigment, filler).

(2) Fibers, fabric or metal insertions in plastics, rubber, flooring, etc. for the purpose of imparting impact strength and tear resistance. (See also whiskers).

relaxin. A polypeptide hormone.
Properties: Amorphous powder. Slightly soluble in water and alcohol; soluble in acid or alkaline solutions. Insoluble in absolute alcohol, ether, acetone, benzene. Decomposes above pH 9.0.
Derivation: Obtained commercially from pregnant sows' ovaries. Isolation procedure and purification by resin chromatography.
Use: Medicine.

"Rencal."[412] Trademark for sodium phytate granules (q.v.).

"Renex."[89] Trademark for a group of nonionic synthetic organic detergents, chiefly ethylene oxide esters and ethers.

"Renite."[577] Trademark for a series of high-temperature lubricants, swabbing compounds, release agents, and automatic spray equipment for hot-forming glass and hot-working metals. Used to improve the quality and production of glassware products, forgings, extrusions, and die castings.

rennet. See rennin.

rennin (rennase; chymosin). The enzyme secreted by the glands of the stomach which causes curdling of milk. It has the power of coagulating 25,000 times its own weight of milk.
Properties: Yellowish white powder or as yellow grains or scales. Characteristic and slightly salty taste and peculiar odor. Slightly hygroscopic. Partially soluble in water and dilute alcohol. Nontoxic.
Derivation: From the glandular layer (inner lining) of the true stomach of the calf.
Grade: Rennet is the dried commercial extract containing the rennin.
Containers: Glass bottles; fiber cans and drums.
Uses: Medicine; pharmacy; cheese-making; coagulation of casein for plastics; food additive.

"Renografin."[412] Trademark for meglumine diatrizoate injection.

"Reogen."[69,466] Trademark for a plasticizing agent for elastomers.

repellent. (1) A substance that causes an insect or animal to turn away from it or reject it as food. Repellents may be in the form of gases (olfactory), liquids, or solids (gustatory). Standard repellents for mosquitoes, ticks, etc. are citronella oil, dimethyl phthalate, n-butylmesityl oxide oxalate, and 2-ethyl hexanediol-1,3. Actidione is the most effective rodent repellent, but is too toxic and too costly to use. Thiuram disulfide, amino complexes with trinitrobenzene, and hexachlorophene are sucessfully used. Copper naphthenate and lime/sulfur mixtures protect vegetation against rabbits and deer. Shark repellents are copper acetate or formic acid mixed with ground asbestos. Bird repellents are chiefly based on taste, but this sense varies widely with the type of bird, so that generalization is impossible. Alpha-naphthol, naphthalene, sandalwood oil, quinine and ammonium compounds have been used, with no uniformity of result. See also fumigant.

(2) A substance which, becasuse of its physicochemical nature, will not mix or blend with another substance. All hydrophobic materials have water-repellent properties due largely to differences in surface tension or electric charges, e.g., oils, fats and waxes, and certain types of plastics. Silicone resin coatings can keep water from penetrating masonry by lining the pores, not by filling them; they will not exclude water under pressure.

replacement. See substitution.

replication. Making a reverse image of a surface by means of an impression on or in a receptive material; usually applied to microscopic techniques for obtaining plastic replicas of observed objects. In biochemistry the term refers to reproduction of the DNA molecule, which is composed of two interlocking chains of nucleotides (double helix). It reproduces itself by forming two identical daughter molecules, each of which receives one of the two chains of the original molecule, the other in each case being synthesized from nucleic acids by enzymes (DNA polymerases). In the drawing below, the solid lines are the strands of the original molecule and the broken lines are the synthesized strands:

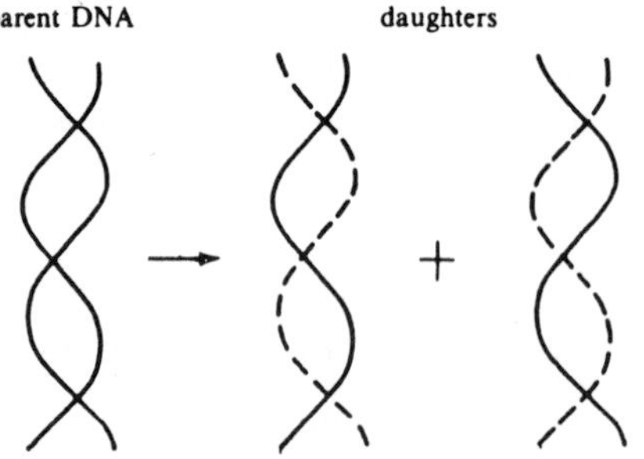

See also deoxyribonucleic acid; genetic code.

reprography. A coined name for the technique of reproducing drawings, blueprints and typographic matter by the use of photosensitized papers, or more recently polyester sheeting, which may be coated with diazo dyes. The process has broad potential in the photocopying field and in communications technology.

Reppe process. Any of several processes involving reaction of acetylene (a) with formaldehyde to produce 2-butyne-1,4-diol, which can be converted to butadiene; (b) with formaldehyde under different conditions to produce propargyl alcohol and from this, allyl alcohol; (c) with hydrocyanic acid to yield acrylonitrile; (d) with alcohols to give vinyl ethers; (e) with amines or phenols to give vinyl derivatives; (f) with carbon monoxide and alcohols to give esters of acrylic acid; (g) by polymerization to produce cyclooctatetraene; and (h) with phenols to make resins. The use of catalysts, of pressures up to 30 atmospheres, and of special techniques to avoid or contain explosions, are important factors in these processes. See also acetylene.

rescinnamine $C_{35}H_{42}N_2O_9$. Alkaloid from certain species of Rauwolfia.

Properties: White or pale buff to cream colored, odorless, crystalline powder. Darkens slowly on exposure to light, more rapidly when in solution. Soluble in chloroform and acetic acid. Slightly soluble in water; soluble in organic solvents. M.p. 238°C (in vacuum).
Grade: N.F.
Hazard: Moderately toxic.
Use: Medicine.

Research octane number. See octane number.

resene. The unsaponifiable component of rosin and other natural resins.

reserpine $C_{33}H_{40}N_2O_9$. An alkaloid.

Properties: White or pale buff to slightly yellowish, odorless powder. Darkens slowly on exposure to light, and darkens more rapidly in solution. Insoluble in water; very slightly soluble in alcohol. Soluble in chloroform and benzene. M.p. 264–265°C (decomposes).
Derivation: From Rauwolfia serpentina (q.v.).
Grade: U.S.P.
Uses: Medicine, often in the form of its salts; animal feeds; antihypertensive agent.

"Reserpoid."[327] Trademark for reserpine (q.v.).

residual oil. A liquid or semiliquid product resulting from the distillation of petroleum and containing largely asphaltic hydrocarbons. Also known as asphaltum oil, liquid asphalt, black oil, petroleum tailings, and residuum. Combustible.
Containers: Tank cars.
Uses: Roofing compounds; hot-melt adhesives; friction tape; sealants; heating oil for large buildings, factories, etc. See also fuel oil.

"Resilon."[326] Trademark for a bituminous base lining for chemical process equipment.
Use: To resist acids and alkalies up to 150°F.

"Resimene."[58] Trademark for melamine and urea-formaldehyde resins. Supplied in organic liquid solutions. The melamine is also available in water-alcohol, and soluble spray-dry powders.
Uses: Paint, varnish, lacquer for automobiles, machinery , appliances, construction; electronics, missiles; chemicals; pulp and paper.

resinate. A salt of the resin acids found in rosin (q.v.). They are mixtures rather than pure compounds. For uses, see soap (2).

"Resin C."[175] Trademark for a neutral synthetic coal-tar resin of high styrene content.
Properties: Light color; m.p. 115–123°C; sp. gr., approx 1.05; mineral oil cloud point, 130–150°C.
Containers: 55-gal steel drums.
Uses: To impart alkali- and grease-resistance to floor tile.

"Resinite."[65] Trademark for polyvinyl chloride film (q.v.).

resin, ion-exchange. See ion-exchange resins.

resin, liquid. An organic polymeric liquid which, when converted into its final state for use, becomes solid (ASTM). Example: linseed oil, raw or heat-bodied (partially polymerized). See also drying oil; resinoid.

resin, natural. (a) Vegetable-derived: amorphous mixtures of carboxylic acids, essential oils and terpenes occurring as exudations on the bark of many varieties of trees and shrubs. They are combustible; electrically nonconductive; hard and glassy with conchoidal fracture when cold, and soft and sticky below the glass transition point (q.v.). Most are soluble in alcohols, ethers and carbon disulfide, and insoluble in water. The best known of these are rosin and balsam, obtained from coniferous trees; these have a high acid content. Of more remote origin are such resins as kauri, congo, dammar, mastic, sandarac, and copal. Their use in varnishes, adhesives, and printing inks is still considerable, though diminishing in favor of synthetic products. (b) Miscellaneous types: Shellac (q.v.), obtained from the secretion of an Indian insect, is still in general use as a transparent coating. Amber (q.v.) is a hard, polymerized resin that occurs as a fossil. Ester gum (q.v.) is a modified rosin. Amorphous sulfur is considered an inorganic natural resin. Liquid resins, sometimes called resinoids (q.v.), are represented by linseed and similar drying oils. See also gum, natural (note); resin, synthetic (note).

resinography. The science of the morphology, structure, and related descriptive characteristics, as correlated with composition or treatment and properties or behavior, of resins, polymers, plastics, and their products.

resinoid. Any thermosetting synthetic resin, either in its initial temporarily fusible state or its final infusible state (ASTM). Heat-bodied linseed oil, partially condensed phenol-formaldehyde, and the like, are also considered resinoids.

resinol. A coal-tar distillation fraction containing phenols. It is the fraction soluble in benzene but insoluble in light petroleum, obtained by solvent extraction of low temperature tars or similar materials. Resinols are very sensitive to heat and oxidation.

"Resinox."[58] Trademark for a series of phenolic resins, supplied in various forms suitable for applications as bonding agents for shell molding and as core binders for metal casting; impregnants or bonding materials for grinding wheels, brake linings, insulation and similar industrial uses; as pipe linings, air conditioning equipment coatings; special primers; as laminating, bonding, impregnating resins for paper, fibers; for use in can, drum, and tank car linings requiring a high degree of chemical and solvent resistance; and for heavy duty products, such as equipment housings. Special formulations for high temperature use for space technology.

resin, synthetic. A man-made high polymer (q.v.) resulting from a chemical reaction between two (or more) substances, usually with heat or a catalyst. This definition includes synthetic rubbers, siloxanes, and silicones, but excludes modified, water-soluble polymers (often called resins). Distinction should be made between a synthetic resin and a plastic (q.v.): the former is the polymer itself, whereas the latter is the polymer plus such additives as filters, colorant, plasticizers, etc.

The first truly synthetic resin was developed by Baekeland (q.v.) in 1911 (phenol-formaldehyde). This was soon followed by a petroleum-derived product called coumarone-indene, which did indeed have the properties of a resin (see "Cumar"). The first synthetic elastomer was polychloroprene (1931) originated by Nieuwland, and later called neoprene. Since then many new types of synthetic polymers have been synthesized, perhaps the most sophisticated of which are nylon and its congeners (polyamides, by Carothers, q.v.) and the inorganic siloxane group (Kipping). Other important types are alkyds, acrylics, aminoplasts, polyvinyl halides, polyester, epoxies and polyolefins.

In addition to their many applications in plastics, textiles, and paints, special types of synthetic resins are useful as ion-exchange media (q.v.). See also plastic; paint; fiber; film; elastomer.

Note: Because the term "resin" is so broadly used as to be almost meaningless, it would be desirable to restrict its application to natural organo-soluble, hydrocarbon-based products derived from trees and shrubs. But in view of the tendency of inappropriate terminology to "gel" irreversibly, it seems like a losing battle to attempt to replace "synthetic resin" with the more precise "synthetic polymer."

See also note under gum, natural.

"Resipon."[300] Trademark for a series of textile resin finishes.

BC: Methylated urea-formaldehyde type.
C: Standard urea-formaldehyde type.
NC: Ethylene urea type.
NDC: Cyclic urea type, for wash and wear finishes.
ST: Triazone type.
V: Polyvinyl acetate emulsion.
RF: Solubilized alkyd type.

resist. A material which will prevent the fixation of dye on a fiber, thus making color designs and pattern prints possible. The resist may act mechanically, as a wax, resin or gel which prevents absorption of the dye, or it may act chemically to neutralize either the dye or its accompanying mordant. Citric acid, oxalic acid, and various alkalies are among the more common resists of the chemical type.

"Resistopen."[412] Trademark for sodium oxacillin.

resistor composition. A specially treated semiconducting metal powder compounded with glass binders and temporary organic carriers. Can be applied to glass or ceramic surfaces by stenciling, spraying, brushing or dipping; firing range 1300° to 1400°F. Compositions can be blended with members of the same series to produce intermediate resistance values. Fired resistors have good reproducibility, low temperature and voltage coefficients, and stability to abrasion, moisture, and relatively high (125°C) ambient temperature.

Containers: 1- to 32-oz jars.

Uses: To produce fired-on resistor components for electronic circuits.

"Resistox."[296] Trademark for stabilized grades of copper powder assaying a minimum of 99% copper with a specific gravity of 8.9 and apparent density range of 2.0–3.5 grams per cubic centimeter. Marketed in several grades of various particle sizes.

Uses: Fabrication of porous bearings; sintered ferrous machine parts; catalysts; magnesium chloride cements; metal friction surfaces; electric brushes; electrical contacts; metallic paints.

"Resitan."[300] Trademark for a synthetic tanning agent of the phenolic type with improved light fastness.

resite. See C-stage resin.

resitol. See B-stage resin.

"Resloom."[58] Trademark for a series of synthetic resins.
E-50 Resins. Supplied at 50% solid cyclic ethylene urea-formaldehyde resin in liquid form.
Melamine. A series of essentially monomeric melamine resins.
Use: Textile industry.

"Resmetal."[65] Trademark for a resin-metal composition which when catalyzed converts to metal-like solid. Recommended for mold making, patching, forming and general repair of metal surfaces and objects.

"Resodors."[12] Trademark for products for use in the neutralization of objectionable odors in plastics.

"Resoform."[307] Trademark for a line of dyestuffs used for coloring plastics.

resol. See A-stage resin.

"Resolube."[483] Trademark for a wide temperature high film strength lubricant (to −70°F) which does not have the plastic-attacking properties of most synthetic low-temperature oils.

resonance. (1) In chemistry, resonance (or mesomerism) is a mathematical concept based on quantum mechanical considerations (i.e., the wave functions of electrons); it is used to describe or express the true chemical structure of certain compounds which cannot be accurately represented by any one valence-bond structure. It was originally applied to aromatic compounds such as benzene, for which there are many possible approximate structures, none of which is completely satisfactory (see benzene). The resonance concept indicates that the actual molecular structure lies somewhere betewen these various approximations, but is not capable of objective representation. This idea can be applied to any molecule, organic or inorganic, in which an electron pair bond is present. The term "resonance hybrid" denotes a molecule which has this property. Such molecules do not vibrate back and forth between two or more structures, nor are they isotopes or mixtures: the resonance phenomenon is rather an idealized expression of an actual molecule which cannot be accurately pictured by any graphic device.

(2) In the terminology of spectroscopy, resonance is the condition in which the energy state of the incident radiation is identical with that of the absorbing atoms, molecules, or other chemical entities. Resonance is applied in various types of instrumental analysis such as nuclear resonance absorption and nuclear magnetic resonance. See also absorption spectroscopy.

Note: The multiple meanings of "nucleus" and "resonance" can be a source of confusion, especially when these terms are closely associated, as in nuclear magnetic resonance and resonance of a molecular nucleus. In the first of these expressions, nucleus is used in sense (1) under nucleus, and resonance in sense (2) under resonance. In the second expression, nucleus is used in sense (3) under nucleus and resonance in sense (1) under resonance.

resorcinol (resorcin; meta-dihydroxybenzene; 3-hydroxyphenol) $C_6H_4(OH)_2$.
Properties: White crystals, becoming pink on exposure to light when not perfectly pure; sweet taste; sp. gr. 1.2717; m.p. 110.7°C; b.p. 281°C; soluble in water, alcohol, ether, glycerol, cerol, benzene and amyl alcohol; slightly soluble in chloroform. Combustible. Flash point 261°F; autoignition temp. 1126°F.
Derivation: By fusing benzene-meta-disulfonic acid with sodium hydroxide.
Grades: U.S.P.; powder; resublimed; pure; reagent; technical; crude.
Containers: 25- to 350-lb fiber drums; multiwall paper bags.
Hazard: Moderately toxic. Irritant to skin and eyes.
Uses: Resorcinol-formaldehyde resins; dyes; pharmaceuticals; cross-linking agent for neoprene; rubber tackifier; adhesives for wood veneers and rubber-to-textile composites; medicine; manufacture of styphnic acid.

resorcinol acetate (resorcinol monoacetate) $HOC_6H_4OCOCH_3$.
Properties: Viscous, pale yellow or amber liquid with a faint odor and a burning taste. B.p. about 283°C (dec); boiling range (10 mm); 150–153°C. Saturated solution is acid to litmus. Soluble in alcohol and most organic solvents; sparingly soluble in water. Sp. gr. 1.203–1.207. Combustible. Low toxicity.
Derivation: Action of acetic anhydride on resorcinol.
Grades: C.P.; N.F.
Containers: Glass bottles; drums.
Uses: Medicine; cosmetics.

resorcinol blue. See lacmoid.

resorcinol diglycidyl ether (RDGE; 1,3-diglycidyloxybenzene; meta-bis(2,3-epoxypropoxybenzene) $C_6H_4(OCH_2CHOCH_2)_2$.
Properties: Straw-yellow liquid; sp. gr. 1.21 (25°C); b.p. 172°C (0.8 mm); refractive index 1.541 (n 25/D); viscosity 500 cp (25°C); flash point 350°F (COC). Combustible. Low toxicity. Miscible with most organic resins.
Use: Epoxy resins.

resorcinol dimethyl ether (dimethyl resorcinol; 1,3-dimethoxy benzene) $C_6H_4(OCH_3)_2$.
Properties: Pale straw-tint liquid; b.p. 204–212°C; sp. gr. 1.063–1.066 (25/25°C); refractive index (n 20/D) 1.523–1.527. Combustible. Low toxicity.
Uses: Organic intermediate; flavoring.

resorcinol-formaldehyde resin. A type of phenol-formaldehyde resin. Permanently fusible; soluble in water, ketones, and alcohols. By dissolving and adjusting the pH to 7, an adhesive base is formed. These adhesives can be used whenever phenolics are used and where fast or room-temperature cure is desired. An important use of these adhesives is in wood gluing, particularly marine plywood.

resorcinol monobenzoate $C_6H_5COOC_6H_4OH$.
Properties: White, crystalline solid; m.p. 132–135°C; b.p. 140°C (0.15 mm); insoluble in water, benzene, di-(2-ethylhexyl)phthalate. Soluble in acetone and ethanol. Combustible. Low toxicity.
Uses: Noncoloring ultraviolet inhibitor for various plastics; color stabilizer in cosmetic compositions.

resorcinolphthalein. See fluorescein.

resorcinolphthalein sodium. See uranine.

alpha-**resorcylic acid** (3,5-dihydroxybenzoic acid) $(OH)_2C_6H_3COOH$.
Properties: White crystals, m.p. 237°C. Soluble in water, ethanol and ether. Combustible. Low toxicity.
Grade: C.P.
Uses: Intermediate for dyes; pharmaceuticals; light stabilizers; resins.

beta-**resorcylic acid** (BRA; 2,4-dihydroxybenzene carboxylic acid; 2,4-dihydroxybenzoic acid; 4-hydroxysalicylic acid; 4-carboxyresorcinol) $(OH)_2C_6H_3COOH$.
Properties: White needles; m.p. (with decomposition) 219–220°C; b.p., decomposes; almost insoluble in water and benzene; soluble in alcohol, ethyl ether. The sodium, potassium, ammonium, calcium and barium salts are soluble in water; the silver, lead and copper salts are only slightly soluble. Combustible. Low toxicity.
Containers: 100-lb net nonreturnable fiber drums.
Uses: Dyestuff and pharmaceutical intermediate; chemical intermediate in synthesis of fine organic chemicals.

respiration. (1) In man and animals, inhalation of oxygen and exhalation of carbon dioxide; the oxygen supports the oxidation (combustion) of organic nutrients in the body, yielding energy, carbon dioxide and water. (2) In growing plants, respiration occurs in both the presence and the absence of light. Some of the energy produced in respiration is used to form adenosine triphosphate, pyruvic acid, and other metabolic intermediates. Fruits and vegetables continue to respire after harvest, a fact that must be taken into account in transportation and storage.

"Reswax."[65] Trademark for a series of wax-resin blends used as coatings and hot melt adhesives in paper conversion. The polymers used include butyl rubber, polyisobutylene, chlorinated rubber, polyethylene, and styrene co-polymers.

"Resyn."[53] Trademark for a series of synthetic resin-base adhesives and coatings.

ret. To reduce or digest fibers, especially linen, by enzymatic action.

retarder. See inhibitor.

"Retardion."[233] Trademark for an ion-exchange resin used in ion retardation, a separation process in which the flow of ions through a column is retarded by the ion-absorbing properties of the resin.

"Retardsol."[106] Trademark for a petroleum solvent (solvent naphtha).
Hazard: Highly flammable, dangerous fire risk.

"Reten" 205.[266] Trademark for a strongly cationic, high molecular weight, synthetic, water-soluble polymer. A finely divided white powder; dissolves in either hot or cold water to produce clear, smooth, viscous, nonthixotropic solutions. Available in a variety of viscosity grades and cationic functionality.
"Reten" 763 is an aqueous solution of a modified polyamide-epichlorohydrin resin.
Uses: Flocculant, binder and viscosifier.

retinal. Preferred name for retinene.

retinene (vitamin A aldehyde; retinal) $C_{20}H_{28}O$. A necessary component of rhodopsin, the light-sensitive pigment of the eye. Retinene is the aldehyde form of vitamin A, which is an alcohol.

retinol (1) $C_{20}H_{29}OH$. A component of vitamin A. See also carotene.
(2) A resin distillate similar to rosin oil (q.v.).

retorting. A process much used in the early years of chemistry for destructive distillation of heavy organic liquids and for laboratory separations. It involves the use of a cylindrical vessel made of glass (for laboratory work), fireclay or metal, with a neck bent at a downward angle to facilitate distillation. For gas manufacture the equipment is built on a heavier scale to handle destructive distillation of coal. Use of this term has been revived in current developments for processing shale oil (q.v.).

"Retrol."[217] Trademark for an acid-activated clay used as decolorizing adsorbent for refining used crankcase or other industrial oils. See also "Filtrol."

reverse osmosis. See osmosis; desalination.

reversible. (1) A chemical reaction which proceeds first to the right and then to the left when the ambient conditions are changed; the product of the first reaction decomposes to the original components as a result of different conditions of temperature or pressure. Examples are: $H_2O + CO_2 \rightleftharpoons H_2CO_3$, in which the carbonic acid reverts to water and carbon dioxide on heating; $NH_4Cl \rightleftharpoons NH_3 + HCl$, in which the ammonium chloride decomposes on heating to ammonia and hydrogen chloride, which recombine on cooling.
(2) A colloidal system such as a gel or suspension that can be changed back to its original liquid form by heating, addition of water, or other method. For example, evaporated egg white can be restored (reconstituted) by addition of water.

"Reversil."[425] Trademark for an inert hydrophobic chromatography powder, designed as a reversed-phase column support or thin-layer powder.

reversion. The softening and weakening of a natural rubber vulcanizate when the curing operation has been too long continued.

"Rexforming."[416] A proprietary process combining certain of the elements of the "Platforming" process (q.v.) and "Udex" extraction (q.v.) to convert a naphtha fraction into a highly aromatic motor fuel blending component of high octane rating.

"Rex" (Molybdate) Oranges and Reds.[206] Trademark for coprecipitations of normal lead chromate, lead sulfate, and lead molybdate. Used in pigments.

"R/Ex" Neutron Activation Foils and Flux Wires.[504] Trademark for about 50 types of foils and wires made of appropriate very pure metals and alloys for neutron activation and other uses requiring high-purity material.

"R/Ex" Neutron Gamma Shielding Materials.[504] Trademark for a group of materials composed of uniform mixtures of heavy metals in a hydrogen-containing base, to be used as shields against neutrons and gamma rays from nuclear reactors, particle accelerators, and other radiation sources. The hydrogen-containing base material is generally polyethylene, but epoxy, rubber or paraffin are also used. Boron may be added for neutron capture. Lead is generally used for gamma attenuation.

Reynold's number. The function DUP/μ used in fluid flow calculations to estimate whether flow through a pipe or conduit is streamline or turbulent in nature.

D is the inside pipe diameter, U is the average velocity of flow, P is density and μ is the viscosity of the fluid. Different systems of units give identical values of the Reynold's number, and values much below about 2100 correspond to streamline flow, while values above 3000 correspond to turbulent flow.

"Rezad."[11] Trademark for resorcinol derivatives and resins.

"Rezipas."[412] Trademark for aminosalicylic acid resin.

"Rezklad."[41] Trademark for epoxy-based flooring compound, grout, and cements. Gives corrosion-proof, strong, abrasion-resistant surfaces.

"Rezyl."[11] Trademark for short, medium, and long oil semioxidizing or oxidizing types of alkyd resins; also short nonoxidizing types; nondrying plasticizing types; specialty formulations.
Uses: Baking enamels, interior and exterior brushing enamels, air dry finishes, nitrocellulose lacquers.

Rf Symbol for rutherfordium (element 104).

RFNA. Abbreviation for red fuming nitric acid. See nitric acid, fuming.

Rh Symbol for rhodium.

rhamnose $C_6H_{12}O_5$. A deoxyhexose monosaccharide found combined in the form of glycosides in many plants. Two forms exist, alpha- and beta-, of which the alpha- is the more stable.
Properties: White crystalline powder. Very soluble in water and in methanol; soluble in absolute alcohol; m.p. 82–92°C (alpha- form converts partially to the beta- form on heating). For equilibrium mixture optical rotation +9.18°.
Uses: Synthetic sweetener research.

rhenium Re Metallic element, atomic number 75; group VIIB of the periodic table. Atomic weight 186.2; valences 1 to 7; 4, 6, and 7 are most common, the last being the most stable. There are two isotopes.
Properties: Silver-white solid or gray to black powder; sp. gr. 21.02 (20°C); m.p. 3180°C; b.p. 5630°C; tensile strength 80,000 psi; high modulus of elasticity. Attacked by strong oxidizing agents (nitric and sulfuric acids). Practically insoluble in hydrochloric acid. Retains its crystal structure all the way to its melting point; has widest range of valences of any element. Rhenium-molybdenum alloys are superconductive at 10°K. Not attacked by sea water. Nontoxic.
Source: Principally molybdenite.
Derivation: Solutions from refinery residues (molybdenum ore flue dust, copper ore treatment) are (a) concentrated by salting-out processes and reduced by hydrogen gas under pressure to give the metal or (b) passed through an anionic resin from which pure rhenium can be extracted by a strong mineral acid.
Forms: Powder which can be consolidated into rods, wires, or strips by powder metallurgy. Annealed metal is very ductile, and can be bent, coiled, or rolled. Single crystals 2 in. × .05–.005 in. dia.
Hazard: Flammable in powder form.
Uses: Additive to tungsten- and molybdenum-based alloys; electronic filaments; electrical contact material; high-temperature thermocouples; igniters for flash bulbs; refractory metal components of missiles; catalyst; plating of metals by electrolysis and vapor-phase deposition.

rhenium heptasulfide Re_2S_7.
Properties: Brown-black solid; sp. gr. 4.87; decomposes to ReS_2 at 600°C. Insoluble in water; dissolves in solutions of alkali sulfides.
Hazard: Ignites on heating in air.
Use: Catalyst.

rhenium heptoxide Re_2O_7.
Properties: Yellow crystalline solid; sp. gr. 6.103; m.p. approx. 297°C. Dissolves in water to form perrhenic acid $HReO_4$ (q.v.). Very soluble in alcohol.

rhenium pentachloride $ReCl_5$.
Properties: Dark green to black solid; sp. gr. 4.9; decomposes on heating. Decomposes in water; soluble in hydrochloric acid and alkalies.
Derivation: By reacting rhenium heptoxide with carbon tetrachloride at 400°C.

rhenium trichloride $ReCl_3$.
Properties: Dark red solid; on heating emits green vapor from which the metal may be deposited; nonelectrolyte in solution. Soluble in water and glacial acetic acid.
Derivation: Distillation of rhenium pentachloride.

rhenium trioxychloride ReO_3Cl.
Properties: Colorless liquid; sp. gr. 3.867 (20/4°C); m.p. 4°C; b.p. 131°C (760 mm). Decomposes in water; soluble in carbon tetrachloride; reacts readily with organic substances.

rheology. Science of the deformation and flow of matter in terms of stress, strain and time. See liquid, Newtonian; dilatancy, thixotropy, and viscosity. Has important bearing on the behavior of viscous liquids in plastic molding.

"Rheolube."[483] Trademark for a series of low-shear thixotropic greases and semifluid instrument lubricants for use where nonspreading properties are critical.

rhesus factor (Rh factor). A substance present in the red blood cells of the rhesus monkey, and of approximately 85% of an average white American population. Those whose red cells contain this factor are termed Rh-positive; others, Rh-negative. A negative individual may develop anti-Rh antibodies, if Rh-positive red cells enter his blood; such antibodies can then agglutinate Rh-positive cells. Hemolytic reactions may thus occur following transfusion of Rh-positive blood cells into a recipient previously sensitized and having Rh antibodies in the serum. Likewise, an Rh-positive fetus may give rise to antibodies in the blood of an Rh-negative mother; the antibodies, returning into the fetus, may then produce the disease erythroblastosis fetalis. There are many subtypes of the Rh factor; these can be distinguished by serologic tests, and the laws of their inheritance have been determined.

rhizobitoxin. A recently developed broad-spectrum herbicide, which attacks young growth and new leaves but has little effect on older growth. Has low toxicity to humans (USDA).

rho acid. See anthraquinone-1,5-disulfonic acid.

rhodamine B $C_{28}H_{31}ClN_2O_3$. A basic red fluorescent dye, C.I. No. 45170, structurally related to xanthene.
Properties: Green crystals or reddish violet powder. Very soluble in water and alcohol, forming bluish red, fluorescent solution; slightly soluble in acids or alkalies.

Derivation: By fusion of meta-diethylaminophenol and phthalic anhydride followed by acidification with hydrochloric acid.
Uses: Red dye for paper; also for wool and silk where brilliant fluorescent effects are desired and lightfastness is of secondary importance; analytical reagent for certain heavy metals; biological stain.

rhodamine toner. Red to maroon lakes of rhodamine dyes and phosphotungstic or phosphomolybdic acid. They have good lightfastness and are used principally in printing inks. See also phosphomolybdic pigment; phosphotungstic pigment.

rhodanates. Synonym for sulfocyanates or thiocyanates (preferred). See under the individual metals which form thiocyanates.

rhodanides. See rhodanates.

rhodanine (2-thio-4-keto-thiazolidine)
$\overline{SCH_2C(O)NHCS}$.
Properties: Finely crystalline; light yellow color; sp. gr. 0.868; bulk density 0.617; m.p. decomposes, often violently, 166°C; pure material, 167–168.5°C; soluble in methanol, ethyl ether and hot water.
Hazard: May explode when heated to 166°C, evolving toxic fumes.
Use: Organic synthesis (phenylalanine); laboratory reagent.

rhodinol. A mixture of terpene alcohols consisting principally of *l*-citronellol.
Properties: Colorless liquid with a pronounced rose-like odor; sp. gr. 0.860–0.880 (25°C); refractive index 1.4630–1.4730 (20°C); optical rotation −4 to −9°. Soluble in alcohol and mineral oil; insoluble in glycerin. Combustible; low toxicity.
Derivation: From Réunion geranium oil.
Grade: F.C.C.
Uses: Perfumery; flavoring agent.

rhodinyl acetate. A mixture of terpene alcohol acetates consisting primarily of *l*-citronellyl acetate.
Properties: Colorless to slightly yellow liquid with a rose-like odor; sp. gr. 0.895–0.908 (25°C); refractive index 1.4530–1.4580 (20°C); optical rotation −2° to −6°. Soluble in alcohol and mineral oil; insoluble in glycerin. Combustible; low toxicity.
Derivation: Action of acetic anhydride on rhodinol in the presence of sodium acetate.
Grades: Technical; F.C.C.
Uses: Perfumery; flavoring agent.

rhodium Rh Metallic element having atomic number 45, group VIII of the periodic system. Atomic weight 102.9055. No isotopes. Valence 3.
Properties: White solid of platinum group; sp. gr. 12.41 (20°C); m.p. 1966°C; b.p. 4500°C. Insoluble in acids and aqua regia; soluble in fused potassium bisulfate. Harder and higher-melting than platinum or palladium; highest in electrical and thermal conducitity of the platinum group. High surface reflectivity. A strong complexing agent.
Occurrence: Ontario, So. Africa, Siberia.
Derivation: Occurs with platinum, from which it is recovered during the purification process.
Forms: Produced as powder, which can be fabricated by casting or powder metallurgy techniques. Single crystals are available.
Hazard: Flammable in powder form. Tolerance (fumes and dusts) 0.1 mg per cubic meter of air; (soluble salts) 0.001 mg per cubic meter of air.
Uses: Alloy with platinum for high temperature thermocouples, furnace windings, laboratory crucibles; spinnerets in rayon industry; electrical contacts; jewelry; catalyst; optical instrument mirrors; electrodeposited coatings for metals; vacuum-deposited glass coatings; headlight reflectors.

rhodium chloride (rhodium trichloride) $RhCl_3$.
Properties: Brownish red powder. Soluble in solutions of alkalies and cyanides; insoluble in acids, water. B.p. 800°C (sublimes); m.p. 450-500°C (dec).

rhodochrosite $MnCO_3$ with partial replacement by Fe, Ca, Mg, Zn.
Properties: Light pink, rose red, brownish red or brown mineral; white streak; vitreous to pearly luster; photoluminescent. Found in veins with ores of silver, lead, copper, manganese.
Constants: Sp. gr. 3.3–3.6; Mohs hardness 3–4.
Occurrence: United States (Connecticut, New Jersey, Colorado, Montana, Nevada); Europe.
Use: Manganese ore.

rhodonite. An ore of manganese. See manganous silicate.

rhodopsin. The red-light-sensitive pigment of the eye, (visual purple) consisting of the protein, opsin, and retinene (vitamin A aldehyde). It occurs in land and marine vertebrates.

"Rhoduline."[307] Trademark for basic dyestuffs, used for dyeing of silk, cotton, rayon, paper, leather, lacquers, and plastics. Also used as phosphotungstic toners or lakes for printing inks, wall paper and coated paper. Characterized by exceptional brilliancy.

"Rhonite."[23] Trademark for thermosetting modified and unmodified urea-formaldehyde condensates. Supplied as water-clear solutions and aqueous pastes. Reactive with cotton, various grades producing shrink-resistance, crease-proofing or modification of hand.
Use: Finishing of natural and synthetic fabrics.

"Rhoplox."[23] Trademark for aqueous dispersions of acrylic copolymers. White opaque emulsions. Various grades differing in hardness, flexibility, adhesion and tack of film; some thermosetting. Produce colorless transparent films with outstanding permanence, durability, adhesion, and pigment-binding capacity.
Uses: Emulsion paints; paper coatings and saturation; floor sealers and wax emulsions; textile-backing and finishing; bonding fibers and pigments; clear and pigmented coatings on wood and metals.

"Rhosidol."[188] Trademark for a synthetic rhodinol compound.

"Rhothane."[23] Trademark for an agricultural insecticide based on 1,1-bis (chlorophenyl)-2,2-dichloroethane and supplied as a wettable powder or emulsion concentrate.

"Rhozyme."[23] Trademark for enzyme concentrates with diastatic or proteolytic activity. Buff-colored powders or liquids of fungal or bacterial origin which hydrolyze and solubilize proteins and starches.
Uses: Desizing textile fabrics; drycleaning; liquefaction of starch paste; fermentation processes; manufacture of corn syrup, fish solubles, septic tank formulations; animal feed; meat tenderizer.

"Rhulitol."[57] Trademark for tannic acid solution (q.v.).

"Ria."[511] Trademark for a finely ground surface-treated urea. Also available as paste. Used as cure accelerator and activator for resins.

Rib. Abbreviation for ribose.

"Rib-Bak."[515] Trademark for a latex compound that can be applied as a secondary carpet backing.

riboflavin (vitamin B_2) $C_{17}H_{20}N_4O_6$. 7,8-Dimethyl-10-(1^1-D-ribityl) isoalloxazine. A crystalline pigment, the principal growth-promoting factor of the vitamin B_2 complex. It functions as a flavoprotein in tissue respiration. A syndrome resembling pellagra is thought to be due to riboflavin deficiency.
Properties: Orange yellow crystals; bitter taste; m.p. 282°C (dec). Slightly soluble in water and alcohols, insoluble in lipid solvents, stable to heat in dry form and in acid solution; stable to ordinary oxidation; unstable in alkaline solution, and quite sensitive to light. In solution, riboflavin has an intense greenish yellow fluorescence. Nontoxic.
Units: Amounts are expressed in milligrams or micrograms of riboflavin.
Sources (Food): milk, green leafy vegetables, egg yolk, liver, enriched flour, yeast. (Commercial): distiller's residues, fermentation solubles; synthetic production (indirectly from dextrose, lactose, yeast, and whey.)
Grade: U.S.P.; F.C.C.
Uses: Medicine; nutrition; animal feed supplement; enriched flours; dietary supplement.

riboflavin 5′-phosphate (FMN; flavin mononucleotide). The phosphate ester of riboflavin (q.v.) in which the phosphate is esterified to the ribityl portion of riboflavin. It functions as a coenzyme for many flavine enzymes (q.v.). The riboflavin group has the ability to take up hydrogen atoms, thus oxidizing the substrate.
Properties (sodium salt): Yellow crystals. Much more soluble than riboflavin in water. Quite sensitive to ultraviolet light.
Derivation: By treating riboflavin with chlorophosphoric acid.
Containers (sodium salt): Fiber drums.
Use: Dietary supplement; flavor potentiator.

9-beta-D-**ribofuranosyladenine.** See adenosine.

ribonuclease. An enzyme which causes splitting of ribonucleic acid. Pancreatic ribonuclease, for example, cleaves only phosphodiester bonds that are linked to pyrimidine 3′-phosphates. It is a critical regulator of life processes in the cell. The first enzyme to be synthesized (1969), it is composed of 124 amino acid residues. It is one of the few proteins for which the sequence of amino acids has been elucidated (the order or sequence of amino acids is of critical importance in the functioning of enzymes, genes, and nucleotides.)

ribonucleic acid (ribose nucleic acid; RNA). Generic term for a group of natural polymers consisting of long chains of alternating phosphate and D-ribose units, with the bases adenine, guanine, cytosine, and uracil bonded to the 1-position of the ribose. Ribonucleic acid is universally present in living cells and has a functional genetic specificity due to the sequence of bases along the polyribonucleotide chain.
Four types are recognized:
(1) Messenger RNA, synthesized in the living cell by the action of an enzyme that carries out the polymerization of ribonucleotides on a DNA template region which carries the information for the primary sequence of amino acids in a structural protein. It is a ribonucleotide copy of the deoxynucleotide sequences in the primary genetic material.
(2) Ribosomal RNA, which exists as a part of a functional unit within living cells called the ribosome, a particle containing protein and ribosomal RNA in roughly 1:2 parts by weight, having a particle weight of about 3 million. (See ribosome). Messenger RNA combines with ribosomes to form polysomes containing several ribosome units, usually 5 (*e.g.*, during hemoglobin synthesis), complexed to the messenger RNA molecule. This aggregate structure is the active template for protein biosynthesis.
(3) Transfer RNA, the smallest and best characterized RNA class. Its molecules contain only about 80 nucleotides per chain. Within the class of transfer RNA molecules there are probably at least 20 separate kinds, correspondingly related to each of the 20 amino acids naturally occurring in proteins. Transfer RNA must have at least two kinds of specificity: (a) It must recognize (or be recognized by) the proper amino acid activating enzyme so that the proper amino acid will be transferred to its free 2′- or 3′-OH group. (b) It must recognize the proper triplet on the messenger RNA-ribosome aggregate. Having these properties, the transfer RNA accepts or forms an intermediate transfer RNA-amino acid that finds its way to the polysome, complexes at a triplet coding for the activated amino acid, and allows transfer of the amino acid into peptide linkage.
(4) Viral RNA isolated from a number of plant, animal and bacterial viruses, may be considered as a polycistronic messenger RNA. It has been shown to have molecular weights of 1 or 2 million. Generally speaking, there is one molecule of RNA per infective virus particle. The RNA of RNA virus can be separated from its protein component and is also infective, bringing about the formation of complete virus.
From article by F. J. Bollum in "Encyclopedia of Biochemistry."
See also deoxyribonucleic acid.

D-**ribose-5-phosphoric acid** $C_5H_9O_4 \cdot H_2PO_4$. A constituent of nucleotides and nucleic acids.
Properties: The barium salt (5½H_2O) is sparingly soluble in cold water and crystallizes in hexagonal plates.
Derivation: From inosinic acid.
Use: Biochemical research.

ribosome. A ribonucleoprotein, the smallest organized structure in the cell; ribosomes occur in all cells including bacteria, fungi, algae, and protozoa. They are the central point of protein synthesis (see deoxyribonucleic acid). They contain from 45 to 60% of ribonucleic acid (RNA), the balance being proteins. Ribosome crystals have been produced; in the electron microscope these appear as sheets of black dots, each sheet (one ribosome thick) containing hundreds of ribosomes in recurring groups of four. See also ribonucleic acid.

D-**ribosyl uracil.** See uridine.

"Ricilan."[493] Trademark for esters synthesized from castor oil and lanolin components. Amber viscous liquids. Used as hydrophobic emollients; pigment dispersants.

ricin. White powder. The albumin of the castor oil bean is the toxic principle. Extracted from the pressed seeds with 10% solution of sodium chloride followed by precipitation with magnesium sulfate.

Hazard: Highly toxic by ingestion; small particle in cut or abrasion, eye or nose, may prove fatal.
Use: Reagent for pepsin and trypsin.

ricinine $C_8H_8O_2N_2$. 1,2-Dihydro-4-methoxy-1-methyl-2-oxonicotinonitrile.
Properties: White crystalline alkaloid; soluble in water, chloroform and ether. M.p. 201.5°C; sublimes at 170–180°C under 20 mm pressure.
Derivation: From castor oil seeds and leaves.
Hazard: Highly toxic by ingestion.
Uses: Biochemical research.

ricinoleic acid (cis-12-hydroxyoctadec-9-enoic acid; 12-hydroxyoleic acid; castor oil acid) $CH_3(CH_2)CH(OH)CH_2CH{:}CH(CH_2)_7COOH$. A C_{18} unsaturated fatty acid which comprises approximately 80% of the fatty acid content of castor oil.
Properties: Colorless to yellow viscous liquid; soluble in most organic solvents; insoluble in water. Sp. gr. 0.940 (27.4/4°C); m.p. 5.5°C; b.p. 226°C (10 mm); refractive index 1.4697 (20°C). Dextrorotatory. Combustible. Low toxicity.
Derivation: Saponification of castor oil.
Uses: Soaps; Turkey red oils; textile finishing; source of sebacic acid and heptanol; ricinoleate salts; 12-hydroxystearic acid.

ricinoleyl alcohol (9-octadecen-1,12-diol) $CH_3(CH_2)_5CH(OH)CH_2CH{:}CH(CH_2)_7CH_2OH$. The fatty alcohol derived from ricinoleic acid. It has a long straight chain with one double bond and one hydroxyl group in a secondary position besides the primary group on the end. Available as a 90% product.
Properties: Colorless non-drying liquid at room temperature. Iodine value 91.8%; cloud point below 10°F; boiling range 170–328°C; viscosity 51 (SSU/21°C); combustible; low toxicity.
Derivation: Reduction of acid made from castor oil.
Impurities: Oleyl and linoleyl alcohols.
Uses: Protective coatings; polyesters; plasticizers; organic synthesis, pharmaceuticals, lubricants, surface active agents.

ricinus oil. See castor oil.

"Ridzlik."[15a] Trademark for an emulsifying agent developed for treatment of oil spills in both fresh and salt water. Claimed to have minimal toxicity to marine life and to be biodegradable.

"Rigortex."[11] Trademark for vinyl copolymer coating for severe corrosive conditions.

"Rila."[4] Trademark for a group of fluorescent pigments in red, green, yellow and orange. They are based on such phosphors as zinc sulfide-cadmium sulfide, cadmium sulfide-strontium sulfide.
Uses: Printing inks, fine coatings, paints, lacquers.

"Rilsan." Trademark for nylon 11. See nylon.

ring compound. See cyclic compounds; alicyclic; aromatic; heterocyclic.

"Rio Resin."[69,119] Trademark for a blend of resinous and protective materials.
Properties: Orange to dark red; sp. gr. 1.13 ±; softening point 54°C min.
Use: Compounding heat-resistant copolymers and corona-resistant neoprene.

ristocetin. An antibiotic produced by the fermentation of Nocardia lurida, a species of Actinomycetes. The antibiotic has two components, ristocetin A and ristocetin B. The commercial product is a lyophilized preparation representing a mixture of A and B, of which B comprises no more than 25%.
Properties: White or tan crystals or powder. Practically odorless; freely soluble in water; practically insoluble in organic solvents.
Grade: U.S.P.
Use: Medicine.

Rittinger's law. The energy required for reduction in particle size of a solid is directly proportional to the increase in surface area. See also Kick's law.

Rn Symbol for radon.

RNA. Abbreviation for ribonucleic acid (q.v.).

roasting. Heating in the presence of air or oxygen. Most commonly used in converting natural metal sulfide ores to oxides as a first step in recovery of metals such as zinc, lead, copper, etc. Roasting is an oxidation process. See also smelting.

"Robane."[415] Trademark for squalane; used for dermatologicals, topical pharmaceuticals and cosmetics.

Robison ester. See glucose 6-phosphate.

"Robuoy."[415] Trademark for pristane (q.v.).

"Roccal."[162] Trademark for a mixture of high molecular weight alkyl benzyl dimethyl ammonium chlorides in 10 and 50% aqueous solution. The alkyl groups are straight chain C_8 to C_{18} from coconut fatty acids. Used as a germicide in the food and beverage industry, and for similar applications.

Rochelle salt. See potassium-sodium tartrate.

rock crystal. See quartz.

rocket fuel (rocket propellant). A substance or mixture that has the capacity for extremely rapid but controlled combustion, which produces large volumes of gas at high pressure and temperature. They may be either liquid, solid, or combinations of both. Liquid monopropellants are hydrogen peroxide and hydrazine, catalyzed by finely divided metals to decompose them into gases. Liquid bipropellants consist of the fuel and an oxidizer; typical fuels of this type are hydrogen, hydrazine, ammonia, and boron hydride, the oxidizers being oxygen, nitric acid, ozone, hydrogen peroxide, and water.

Solid propellants include nitrocellulose, plasticized with nitroglycerin or various phthalates; inorganic salts suspended in a plastic or synthetic rubber (e.g., "Thiokol") and containing a finely divided metal. The inorganic oxidizers used are ammonium and potassium nitrates and perchlorates.

Rockwell hardness. See hardness.

"Rodar."[155] Trademark for an alloy composed of nickel, 29%; cobalt, 17%; manganese, 0.30%; iron, 53.7%.
Properties: Produces a permanent vacuum-tight seal with simple oxidation procedure; resists mercury corrosion; readily machined and fabricated; can be welded, soldered, or brazed.
Forms: Wire; strip; bar.
Use: Sealing metal to hard glass.

rodenticide. A chemical used to kill rats and other rodents. See warfarin; squill.

"Rodiform."[69] Trademark for rubber accelerators supplied in the form of extruded rods or pellets.

"Rodine."[342] Trademark for red squill liquid extract rodenticides.

"Rodo."[69] Trademark for a series of blended essential oils used to deodorize rubber.

Roentgen, W. K. (1845–1923). German physicist who discovered x-rays in 1895 for which he was awarded the Nobel Prize in 1901. Application of these to a number of important problems in analytical chemistry was developed by the Braggs, Moseley, von Laue, and Debye and Sherrer.

roentgen (r). The international unit of quantity or dose for both x-rays and gamma rays. It is defined as the quantity of x- or gamma rays which will produce as a result of ionization one electrostatic unit of electricity of either sign in 1 cc (0.001293 g) of dry air as measured at 0°C and standard atmospheric pressure. The use of the roentgen unit has been extended to include particle radiation such as alpha and beta particles and protons and neutrons. See also rad, curie.

Rohrbach solution.
Properties: Clear, yellow liquid. Very refractive; sp. gr. 3.5.
Derivation: An aqueous solution of mercuric barium iodide.
Hazard: Highly toxic by ingestion and inhalation.
Uses: Separating minerals by their specific gravity; microchemical detection of alkaloids.

"Romark."[448] Trademark for alkyd and chlorinated rubber type road-marking paints.

ronnel. Generic name for O,O-dimethyl O-(2,4,5 trichlorophenyl) phosphorothioate, $(CH_3O)_2P(S)OC_6H_2Cl_3$.
Properties: Powder or granules; m.p. 41°C. Insoluble in water; soluble in most organic solvents.
Hazard: Toxic by ingestion and inhalation. Tolerance, 10 mg per cubic meter of air. Cholinesterase inhibitor. Use may be restricted.
Use: Pesticide.
Shipping regulations: (Rail, Air) Organic phosphate, solid, n.o.s., Poison label. Not accepted on passenger planes.

"Ronopole" Oil.[165] Trademark for highly oxidized sulfonated castor oil.

room temperature. An interior temperature from 20 to 25°C (68 to 77°F).

"Roracyl."[28] Trademark for a group of soluble dyes that have good affinity and fastness properties on leather.

"Rosaldehyde."[188] Trademark for a synthetic floral perfume base.

rosaniline $HOC(C_6H_4NH_2)_2C_6H_3(CH_3)NH_2$. A triphenylmethane dye.
Properties: Reddish brown crystals; m.p. 186°C (dec). Soluble in acids and alcohol; slightly soluble in water.
Hazard: May be toxic.
Uses: Dye (usually as the hydrochloride); fungicide.

"Rosanlik."[188] Trademark for a synthetic replacement of otto of rose.

roscoelite $K_2V_4Al_2Si_6O_{20}(OH)_4$. A vanadium-bearing species of mica (q.v.). Formula variable, with V_2O_3 up to 28%. Occurs as minute scales with micaceous cleavage; dark green to brown in color; pearly luster. Mohs hardness 2.5; sp. gr. 3.0.
Occurrence: Colorado, California, Australia.
Use: Source of vanadium.

rose absolute. Pure oil of rose. The first filtrate obtained on separation of waxes from the cooled alcohol solution of rose concrete in perfume manufacture.

rose concrete. Semisolid residue, a mixture of essential oils and waxes, resulting from extraction of rose flower petals, leaves, seeds, fruit, roots, gums or bark by means of a volatile solvent.

rose oil (otto of rose oil; attar of roses; rose flower oil)
Properties: Pale yellow, pale green, or pale red, transparent, essential oil; mild, sweet taste; strong, fragrant odor; semi-solid at room temperature. Sp. gr. 0.845–0.865; solidifying point 18–37°C; saponification value 10–17; acid value 0.5–3; refractive index (n 30/D) 1.457–1.463. Combustible; nontoxic.
Chief constituents: Geraniol, citronellol and phenylethyl alcohol.
Derivation: By steam distillation of the fresh flowers of Rosa damascena, Rosa centifolia, Rosa galica and Rosa alba.
Grades: Bulgarian; French; Turkish; N.F.; F.C.C.
Uses: Perfumes; flavoring.

"Rosetone."[19] Trademark for trichloromethylphenylcarbinyl acetate (q.v.).

rosewood oil. See oil bois de rose.

rosin
Properties: Angular, translucent, amber-colored fragments; sp. gr. 1.08; m.p. 100–150°C; acid no. not less than 150. Flash point 370°F. Insoluble in water; freely soluble in alcohol, benzene, ether, glacial acetic acid, oils, carbon disulfide, dilute solutions of fixed alkali hydroxides. Low toxicity. Hard and friable at room temperature; soft and very sticky when warm. Combustible.
Chief constituents: Resin acids of the abietic and pimaric types, having the general formula $C_{19}H_{29}COOH$, and having a phenanthrene nucleus. See also turpentine.
Derivation: From pine trees, chiefly Pinus palustris and Pinus caribaea. (a) Gum rosin is the residue obtained after the distillation of turpentine oil from the oleoresin tapped from living trees. (b) Wood rosin is obtained by extracting pine stumps with naphtha and distilling off the volatile fraction. (c) Tall oil rosin is a byproduct of the fractionation of tall oil (q.v.).
Grades: Virgin; yellow dip; hard; N.F. Wood rosin is grades B, C, D, E, F, FF, G, H, I, J, K, L, M, N, W-G (window-glass), W-W (water-white). The grading is done by color, B being the darkest and W-W the lightest.
Containers: Drums; multi-wall paper bags.
Uses: Hot-melt and pressure-sensitive adhesives; mastics and sealants; varnishes; ester gum; soldering compounds; core oils; insulating compounds; soaps; paper sizing, printing inks; polyesters (formed by reaction of the conjugated acids of rosin with acrylic acid, followed by reaction with a glycol).
See also abietic acid.

rosin oil.
Properties: Water-white to brown liquid; viscous; odorless; strong, peculiar taste. Soluble in ether, chloroform, fatty oils and carbon disulfide; slightly soluble in alcohol; insoluble in water. Essentially decarboxylated rosin acids. Sp. gr. 0.980–1.110; iodine number 112–115.
Derivation: By fractional distillation of rosin, that portion distilling above 360°C being rosin oil.
Containers: Drums; tank cars.
Hazard: Spontaneous heating; fire risk when heated.

Uses: Lubricant; adulterant for boiled linseed oil; hot-melt adhesives; printing inks; impregnating paper for wrapping electric cables; core compounds.

rosin soap. See sodium abietate, and soap.

"Roskydal."[470] Trademark for a series of unsaturated polyesters curing with peroxides for coating applications.

rosolic acid. See aurin.

"Rotalin."[28] Trademark for a line of spirit-soluble printing colors.
Uses: Principally in the printing of glassine and other types of paper and cellophane.

"Rotax."[256] Trademark for a purified grade of 2-mercaptobenzothiazole.
Use: Primary accelerator for rubber.

rotenone (tubatoxin) $C_{23}H_{22}O_6$. A pentacyclic compound.
Properties: White, odorless crystals. Soluble in ether, alcohol, acetone, and other organic solvents; insoluble in water. Not compatible with alkalies. Sp. gr. 1.27 at 20°C; m.p. 163°C; strongly levorotatory in solution, specific rotation for D line, 230° in benzene, 62° in ethylene dichloride.
Grades: C. P. crystals; technical; also as extracts of derris and cube root.
Containers: Fiber drums; tins; multiwall paper sacks.
Hazard: Moderately toxic. Tolerance, 5 mg per cubic meter of air.
Uses: Insecticide (toxic to fish); flea powders; fly sprays; mothproofing agents.

"Rotoclor."[56] Trademark for ferric chloride solution with other additives specially prepared for rotogravure work. Contains 43% $FeCl_3$. Sold only in 155-lb carboys.

rouge. (1) A high-grade red pigment used as a polishing agent for glass, jewelry, etc. See also iron oxide red.
(2) A cosmetic prepared from dried flowers of the safflower.

"Rovana."[565] Trademark for a thermoplastic vinylidene chloride copolymer filament in the form of a folded flat tape, offered in 300, 400 and 550 deniers.

Roxel process. A process for increasing the flame resistance of cotton by chemically modifying the fiber, i.e., by applying a formulation consisting of tetrakis-(hydroxymethyl)phosphonium chloride (THPC) (q.v.), triethylolamine, and urea in aqueous solution. These substances crosslink with the hydroxyl groups of the cellulose (q.v.) to form an efficient and durable protective medium. This process is used in the manufacture of blankets, curtains, bed linen and industrial safety garments.

"Royalac."[248] Trademark for an accelerator for EPDM elastomers (q.v.).

"Royalene."[248] Trademark for a low-cost, high-performance ethylene-propylene terpolymer.
Properties: Resistance to ozone, heat, weather, sunlight aging, steam and chemicals. Resilient at low temperatures, excellent electrical properties and can be easily colored.
Uses: Tires, automotive parts, tank cars, hoses, and insulation.

royal jelly. A complex mixture secreted by "worker" honeybees, which comprises the sole nutrient of the queen bee. It contains 31% protein, 15% carbohydrate, 15% lipid, plus vitamins. The free fatty acid portion of the lipid is a mixture of C_{10} acids.

"Royal Methyl Violet."[141] Trademark for violet pigment produced by precipitation of the basic methyl violet dyestuff with phosphomolybdic acid.
Use: Printing inks.

"Royal Spectra."[133] Trademark for an impingement carbon black. Used in specialty application requiring highest blackness and reinforcing power.

"Royal Victoria Blue."[141] Trademark for blue pigment produced by precipitation of basic Victoria Blue dye with phosphomolybdic acid. Used in printing inks and some paints.

"RPA."[28] Trademark for a group of peptizing agents for natural and synthetic rubber.
No. 2. 2-Naphthalenethiol, $C_{10}H_7SH$, with inert wax.
No. 3. 36% or 71% xylenethiols, $(CH_3)_2C_6H_3SH$, in an inert hydrocarbon.
No. 6. Pentachlorothiophenol, C_6Cl_5SH.

"RR-10."[28] Trademark for a group of mixed dixylyl disulfides, $[(CH_3)_2C_6H_3]S_2$. Used to accelerate devulcanization of natural rubber, SBR, and mixed scrap for reclaiming purposes.

RR acid. See 2-amino-8-naphthol-3,6-disulfonic acid (q.v.).

"RSR."[173] Trademark for a proteolytic enzyme preparation for removal of stains; available in powder form specifically designed for removing albuminous spots and stains from garments.

"RTV."[245] Trademark for a family of silicone rubber compounds. RTV (room temperature-vulcanizing) rubbers have good physical properties and electrical properties similar to silicone rubber.
Uses: Sealing, caulking, encapsulating and flexible mold-making in electronic, aircraft, missile, and building industries.
See also silicone (uses).

Ru Symbol for ruthenium.

"Rubanox Red."[141] Trademark for lithol rubine pigments of bright bluish-red shades. Composed of calcium salts of azo pigments formed when 4-aminotoluene-3-sulfonic acid is coupled with beta-hydroxynaphthoic acid.
Uses: Printing inks, paints, enamels, lacquers, rubber, plastics, wallpaper, floor coverings.

rubber. Any of a number of natural or synthetic high polymers having unique properties of deformation (elongation or yield under stress) and elastic recovery after vulcanization (q.v.) with sulfur or other cross-linking agent, which in effect changes the polymer from thermoplastic to thermosetting. The yield or stretch of the vulcanized material ranges from a few hundred to over 1000 per cent. The deformation after break, called "permanent set," is usually taken as the index of recovery; it ranges from 5 to 10% for natural rubber to 50% or more for some synthetic elastomers, and varies considerably with the state of vulcanization and the pigment loading. See also elastomer; rubber, natural; rubber, synthetic.

rubber cement. See adhesive, rubber-based.

rubber, chlorinated. An elastomer (natural rubber or a polyolefin) to which 65% of chlorine has been added to give a solid, film-forming resin. White, amorphous powder available in viscosity grades from 5 to 125 centipoises, the figures indicating viscosity of a 20% solution in toluene. Decomposes at 125°C. Soluble in aromatics; insoluble in aliphatics and alcohols. Compatible with almost all natural and synthetic resins. Chief use is in maintenance paints (marine, swimming pool, traffic, masonry, etc.). See also "Parlon," "Hypalon," rubber hydrochloride.

Hazard: Do not dry-mill chlorinated rubber with zinc oxide; mixture reacts violently at 216°C. Do not use in baked enamels.

rubber, cold. See cold rubber.

rubber fiber. Generic name for a manufactured fiber in which the fiber-forming substance is comprised of natural or synthetic rubber (Federal Trade Commission). Often the rubber is a core around which cotton or other fibers are wrapped to make an elastic yarn used for girdles, swimwear, elastic bands and tapes.

rubber, hard. A rubber compounded with from 30 to 50% by weight of sulfur and cured until an extremely hard, brittle product is formed. Lime or magnesia is used as activator. The theoretical maximum of sulfur that can combine chemically with rubber hydrocarbon is 32%. Combustible; nontoxic.

Hazard: Flammable in form of dust.

Uses: Battery boxes; tank linings; acid- and alkali-resistant equipment; combs. As dust, filler for low-cost rubber products.

rubber hydrochloride. A hydrochloride derivative, as distinct from a chlorine derivative.

Properties: Thermoplastic white powder or clear film. Odorless, tasteless, nonflammable, nontoxic. Chlorine content 29–30.5%. Soluble in aromatic hydrocarbons. Softens at 110–120°C. Films are highly resistant to moisture, oils, acids and alkalies but tend to become brittle on exposure to sunlight. The life of such films is greatly extended by the incorporation of suitable stabilizers and plasticizers.

Derivation: A solution of natural rubber is treated with anhydrous hydrogen chloride under pressure and at low temperature. After neutralization of excess hydrogen chloride the product is precipitated by the addition of ethyl alcohol.

Uses: Protective coverings for machinery, raincloth-ing, shower curtains, food packaging.

"Rubber Lubricant RL 684."[233] Trademark for a polyoxyalkylene glycol formulation; used as a lubricant for mounting tires.

rubber latex. See latex.

rubber, liquid. Any of several proprietary products consisting of high polymers in liquid form for use as coatings, adhesives, etc. For manufacturers, see [29,222]

rubber, natural (polyisoprene, NR) $(C_5H_8)_x$ or

$$\left(CH_2{=}\overset{\overset{\displaystyle H}{|}}{C}{-}\overset{\overset{\displaystyle CH_3}{|}}{C}{=}CH_2\right)_x$$

(a) Crude (unvulcanized):

Properties: Sp. gr. 0.92. Chemically unsaturated. Amorphous when unstretched, but has oriented crystal structure on stretching. Not stable to temperature changes (thermoplastic); readily oxidizable by mastication. Soluble in acetone, carbon tetrachloride and most organic solvents. Refractive index 1.52; dielectric constant about 2.5 Combustible. Nontoxic. Processed by calenders and extruders; can be injection-molded with low sulfur and high accelerator. Cured by hot-molding or in open steam, at temperatures from 120 to 150°C after addition of about 3% sulfur, 1% organic accelerator, 3% zinc oxide, plus fillers or reinforcing agents. The only factors of significance in vulcanization are the time of exposure to heat and the temperature used.

Derivation: From latex (q.v.) obtained from Hevea trees; coagulated with acetic or formic acid. Also made synthetically. See "Coral," "Natsyn."

Occurrence: Malaya, Sumatra, Ceylon, Java, Brazil.

Grades: Ribbed smoked sheets; pale (yellow) crepe; brown crepe.

Containers: Burlap or polyethylene-wrapped bales; wooden boxes.

Uses: Cements, adhesives, electrical insulating tapes and cable wrapping.

(b) Cured (vulcanized, i.e., sulfur cross-linkages):

Properties: High tensile strength; relatively low permanent set; insensitive to temperature changes; attacked by heat, atmospheric oxygen, ozone, hydrocarbons and unsaturated fats and oils; insoluble in acetone; permeable to gases; supports combustion; abrasion resistance poor unless compounded with carbon black; dissipates vibration shock; high electrical resistivity.

Uses: Vehicle tires, hose, conveyor belt covers, footwear, specialized mechanical products, drug sundries, foam rubber, electric insulation, etc.

Note: Gutta percha and balata have similar chemical composition (isomeric) but have very different properties and few commercial uses.

See also latex.

rubber sponge (foam rubber; cellular rubber). A flexible foam produced by beating air into rubber latex with subsequent vulcanization, or by incorporating ammonium carbonate or sodium bicarbonate into a strongly masticated rubber stock. The heat of vulcanization releases NH_3 or CO_2, which inflates the rubber to a porous mass just before the onset of vulcanization.

Uses: Vibration damping pads and inserts; rug and carpet underlays; mattresses and upholstery; seat cushions.

rubber, synthetic. Any of a group of man-made elastomers which approximate one or more of the properties of natural rubber. Some of these are: sodium polysulfide ("Thiokol"); polychloroprene (neoprene); butadiene-styrene copolymers (SBR); acrylonitrile-butadiene copolymers (nitrile rubber); ethylene-propylene-diene (EPDM) rubbers; synthetic polyisoprene ("Coral," "Natsyn"), butyl rubber (copolymer of isobutylene and isoprene); polyacrylonitrile ("Hycar"); silicone (polysiloxane); epichlorohydrin; polyurethane ("Vulkollan").

The properties of these elastomers are widely different. All require vulcanization. In general, sulfur is used only for unsaturated polymers; peroxides, quinones, metallic oxides, or diisocyanates effect vulcanization with saturated types. Many are special-purpose rubbers; some can be used in tires when loaded with carbon black; others have high resistance to attack by heat and hydrocarbon oils and thus are superior to natural rubber for steam hose, gasoline and oil-loading hose. Most are available in latex form. For details, see specific type.

rubber, thermoplastic. Any of several block copolymers of styrene-butadiene or styrene-isoprene, reported to have the resilience of rubber and the processing ease of plastics. See also block polymer.

rubeanic acid. See dithiooxamide.

rubidium Rb Metallic element of atomic number 37; group IA of the periodic table; atomic weight 85.4678. Valence 1. Has only one stable form. Principal natural radioactive isotope is rubidium 87. It is the second most electropositive and the second most alkaline element; has lowest ionization potential. Highy reactive.
Properties: Soft, silvery white solid; easily oxidized in air. Soluble in acids and alcohol; decomposes water. Sp. gr. 1.532 (20°C, as solid); m.p. 39°C; b.p. 688°C. High heat capacity and heat transfer coefficient.
Derivation: (a) Thermochemical reduction of rubidium chloride with calcium; (b) electrolysis of fused cyanide or chloride.
Source: Lepidolite, carnallite, pollucite; mineral springs and natural brines.
Grades: Technical; 99+%.
Containers: Glass bottles under naphtha or kerosine.
Hazard: Reacts vigorously with air and water; must be stored under kerosine or similar liquid. Dangerous fire and explosion risk. Metal causes serious skin burns.
Uses: Photocells; vacuum tubes; catalyst or catalyst promotor. Suggested as heat transfer fluid for vapor turbines in space vehicles; plasmas in thermoelectric devices; grain-refining agents in metals; ultrasensitive magnetometers (Rb-87).
Shipping regulations: (Air) Flammable solid label. Not acceptable on passenger planes unless in cartridges.

rubidium alum. See aluminum rubidium sulfate.

rubidium carbonate Rb_2CO_3.
Properties: White powder; m.p. 837°C; extremely hygroscopic; soluble in water. Dissociates above 900°C.
Hazard: Strong irritant to tissue.
Use: Special glass formulations.

rubidium chloride RbCl.
Properties: White, crystalline powder. Lustrous. When heated it decrepitates, melts and volatizes. Soluble in water; almost insoluble in alcohols. Sp. gr. 2.76; m.p. 715°C; b.p. 1390°C.
Grades: Technical; C.P.; single crystals.
Use: Analysis (testing for perchloric acid); source of rubidium metal.

rubidium fluoride RbF.
Properties: White crystals; sp. gr. 3.557; m.p. 775°C; b.p. 1410°C. Soluble in water; insoluble in alcohol. Single large crystals available.
Hazard: Toxic; strong irritant to tissue. Tolerance (as F), 2.5 mg per cubic meter of air.

rubidium hexafluorogermanate Rb_2GeF_6.
Properties: White crystalline solid; m.p. approx. 696°C. Slightly soluble in cold water; very soluble in hot water.

ribidium hydroxide (rubidium hydrate) RbOH.
Properties: Grayish white mass; extremely hygroscopic; absorbs CO_2 from air; strong base. Soluble in alcohol, water. Sp. gr. 3.2; m.p. 300°C. Attacks glass at room temperature.
Hazard: Toxic; strong irritant to tissue.
Uses: Suggested for electrolyte in storage batteries for low-temperature use.

"Rubinol."[307] Trademark for fast alizarine direct red dyestuff.

"Rubramin."[412] Trademark for vitamin B_{12} (q.v.).

ruby. Synthetic rubies, usually in the form of rods, are made from aluminum oxide containing small amounts of other metals, such as 0.05% chromic oxide, by single crystal-growing technique. These are used extensively in masers and lasers.
See also corundum.

ruby glass. See glass, ruby.

"Ruelene."[233] Trademark for organophosphorus products and formulations for use as parasiticides, insecticides, veterinary drugs and animal health products.

"Rufert."[134] Trademark for nickel catalyst for hardening margarine.

"Ruflux."[337] Trademark for a series of titanium and zirconium products (potassium and sodium titanates, rutile, zirconium oxide and zircon in specially prepared forms). Used in metallurgical and welding products.

"Rulan."[28] Trademark for a flame-retardant plastic for electrical insulation.

rust. (1) The reddish corrosion product formed by electrochemical interaction between iron and atmospheric oxygen; ferric oxide, Fe_2O_3. The reaction occurs most rapidly in moist air, indicating the catalytic activity of water.
(2) A reddish-yellow fungus which attacks plants, especially cereal grains; it is also called smut. It can be controlled by treatment with formaldehyde.

"Rust-Ban."[29] Trademark for a series of protective coatings which include alkyd-based, epoxy ester-based, converted epoxy-based, oil-based, phenolic-based, silicone-based, vinyl-based and inorganic zinc coatings. The inorganic zinc coatings are primarily for heavy-duty protection of steel surfaces and are used on marine vessels and equipment, steel structures, offshore installations and heavy industrial equipment.

ruthenium Ru Metallic element of atomic number 44, group VIII of the periodic system; atomic weight 101.07; valences 3, 4, 5, 6, 8; there are 7 stable isotopes.
Properties: Silvery white solid of the platinum metal group. Sp. gr. 12.41 (20°C); m.p. 2310°C; b.p. 3900°C; Brinell hardness 220 (cast). Insoluble in acids and in aqua regia; attacked by alkaline oxidants (conc. NaOH solution), and by fused alkalies. Low toxicity.
Derivation: Occurs with platinum, from which it is recovered.
Grades: Technical; single crystals.
Uses: Hardener for platinum and palladium in jewelry; electrical contact alloys; catalyst; dental and medical instruments; corrosion-resistant alloys; electrodeposited coatings; nitrogen-fixing agent (experimental); the oxide is used to coat titanium anodes in electrolytic production of chlorine.

ruthenium chloride (ruthenic chloride; ruthenium sesquichloride) $RuCl_3$.
Properties: Black solid; deliquescent. Sp. gr. 3.11; decomposes above 500°C. Insoluble in cold water; decomposes in hot water. Slightly soluble in alcohol.

Grades: Technical; C.P.
Use: Analysis (testing for sulfur trioxide).

ruthenium red (ammoniated ruthenium oxychloride) $Ru_2(OH)_2Cl_4 \cdot 7NH_3 \cdot 3H_2O$.
Properties: Brownish red powder. Soluble in water.
Uses: Microscopic stain and reagent for pectin, plant mucin and gum.

ruthenium tetroxide RuO_4.
Properties: Yellow solid; m.p. 25°C; b.p. 100°C. Strong oxidizing agent.
Derivation: Acidic solutions of Ru compounds are heated with strong oxidizing agents.
Hazard: Fire risk in contact with organic materials.

ruthenocene $(C_5H_5)_2Ru$. Dicyclopentadienyl-ruthenium. A stable, light yellow solid, m.p. 199°C. Used as an intermediate for high-temperature compounds, and for ultraviolet radiation absorbers in paints.

Rutherford, Sir Ernest (1871–1937). Born in New Zealand, Rutherford studied under J. J. Thomson at the Cavendish Laboratory in England. His work constituted a notable landmark in the history of atomic research, as he developed Becquerel's discovery of radioactivity into an exact and documented proof that the atoms of the heavier elements, which had been thought to be immutable, actually disintegrate (decay) into various forms of radiation. Rutherford was the first to establish the theory of the nuclear atom and to carry out a transmutation reaction (1919) (formation of hydrogen and an oxygen isotope by bombardment of nitrogen with alpha particles). Uranium emanations were shown to consist of three types of rays: alpha (helium nuclei) of low penetrating powder; beta (electrons); and gamma, of exceedingly short wave length and great energy. Rutherford also discovered the half-life of radioactive elements and applied this to studies of age determination of rocks by measuring the decay period of radium to lead-206. (see Aston, Bohr).

rutherfordium. Rf Atomic number 104. Atomic wt. 261. Artificial element discovered at Lawrence Radiation Laboratory in 1969, and named after the English physicist, Sir Ernest Rutherford.

rutile TiO_2. Natural titanium dioxide. May contain up to 10% iron.
Properties: Color red, reddish brown, to black; luster adamantine to submetallic; streak pale brown; hardness 6–6.5; sp. gr. 4.3; m.p. 1640°C; refractive index 2.7. Insoluble in acids.
Occurrence: Virginia, West Africa, Australian beach sands.
Uses: Source of titanium and titanium compounds; ceramics; steel deoxidizer; welding rod coatings; pigment for paints, enamels and tiles.

"Rutile, Ceramic."[337] Trademark for 92% titanium oxide. Light brown powder; bulk density 98 lb/cu ft; m.p. 3260°F; particle size, less than 44 microns. Used for stable stains for ceramic glazes and bodies.

"Ryanicide."[342] Trademark for powdered ryania insecticide concentrates.

"Ryton."[303] Trademark for a polyphenylene sulfide plastic used for coatings and molded parts.

"RZ-50-A."[58] Trademark for a 50% solution of N,N-dimethylcyclohexylamine salt of dibutyldithiocarbamic acid. Also available as "RZ-50-B" with emulsifying agents for use in latex.
Use: Rubber accelerator.

S

S Symbol for sulfur.

D-saccharic acid (2,3,4,5-tetrahydroxyhexanedioic acid; tetrahydroxyadipic acid) $COOH(CHOH)_4COOH$. The 1,6-dicarboxylic acid formed by the oxidation of D-glucose.

Properties: White needles or syrup; very soluble in water, alcohol, or ether; deliquescent; m.p. 125–126°C with decomposition. Low toxicity. Combustible.

Derivation: Oxidation of cane sugar, glucose, starch by nitric acid.

saccharin (ortho-benzosulfimide; gluside; benzoylsulfonic imide). The anhydride of ortho-sulfimide benzoic acid. A non-nutritive sweetener.

(structure: benzene ring fused with CO–N–SO_2 ring)

Properties: White, crystalline powder; exceedingly sweet taste (500 times that of sucrose). M.p. 226–230°C. Soluble in amyl acetate, ethyl acetate, benzene and alcohol; slightly soluble in water, chloroform and ether.

Derivation: A mixture of toluenesulfonic acids is converted into the sodium salt, then distilled with phosphorus trichloride and chlorine to obtain the ortho-toluene sulfonyl chloride, which by means of ammonia is converted into ortho-toluenesulfamide. This is oxidized with permanganate, treated with acid and saccharin crystallized out. In food formulations saccharin is used in the form of its sodium and calcium salts. Sodium bicarbonate is added to provide water solubility.

Grades: Commercial; C.P.; U.S.P.; F.C.C.

Containers: Bottles; drums.

Uses: Manufacture of syrups; medicine (substitute for sugar); soft drinks, foods, dietetic foods, etc.

Note: Saccharin is no longer "generally recognized as safe" by FDA, which has placed limits on the percentages allowed in manufactured foods and beverages. Tests on experimental animals have not confirmed carcinogenicity.

saccharolactic acid. See mucic acid.

saccharose. See sucrose.

Sachsse process. See BASF process.

S-acid. See 1-amino-8-naphthol-4-sulfonic acid; 1-naphthylamine-8-sulfonic acid.

"Sacon."[173] Trademark for a concentrated water-soluble material used to restore sizing in dry-cleaned garments.

sacrificial protection. The preferential corrosion of a metal coating for the sake of protecting the substrate metal. For example, when zinc is in contact with a more noble (less reactive) metal in the electromotive series, such as steel, a galvanic cell is created in which electric curent will flow in the presence of an electrolyte. Atmospherically contaminated moisture constitutes the electrolyte. Under these conditions the zinc coating, rather than the steel is affected. Thus galvanic protection (zinc) is sometimes called sacrificial protection. See also galvanizing.

SAE. Abbreviation for Society of Automotive Engineers. The initials are applied to its specifications and tests for motor fuels, oils and steels.

SAE steel. A grade or type of steel indicated by a number system; they are principally plain carbon and of low to medium alloy content, used primarily in machine parts. The first two numbers designate either plain carbon or the alloy grouping and quantity, and the last two the mean carbon content in hundredths of 1%. Thus 10 are carbon steels; 13 manganese steels; 40 and 44 molybdenum steels (the latter of higher alloy content); 50 and 51 chromium steels (the latter of higher alloy content); 41 chromium-molybdenum; 61 chromium-vanadium; 92 silicon and silicon-chromium; 46 and 48 nickel-molybdenum (the latter of higher nickel content); and 81, 94, 86, 87, 88, 47, 43 and 93 nickel-chromium-molybdenum in the order of increasing alloy content. The letter B between the first two and last two numerals indicates the presence of boron in amounts of 0.0005 to 0.003% as a depth-hardening addition.

SAF black. Abbreviation for super-abrasion furance black.

safety glass (shatterproof glass). A composite or laminate consisting of two sheets of plate glass with an interlayer of polyvinyl butyral. The plastic is so bonded to the glass that shattering on impact is virtually eliminated. Required for automobile windshields; also used for bullet-proof glass, train windows, etc.

"Safety-Linen Sour."[244] Trademark for a laundering product consisting chiefly of acid fluorides.

Properties: White granular solid; soluble in water; neutralizing value 21.7 oz sodium bicarbonate per lb.

Containers: 150-lb, 300-lb net fiber drums.

Hazard: Probably toxic by ingestion.

Uses: Laundry sour, iron-removing type.

"Saffil." Trademark for a group of synthetic inorganic fibers made from alumina and zirconia.

safflower oil. Drying oil from safflower (carthamus) seed, somewhat similar to linseed oil. It is non-yellowing. Contents 78% linoleic acid (unsaturated fatty acid).

Properties: Straw-colored liquid; sp. gr. (25/25°C) 0.923–0.927; refractive index (25°C) 1.4740–1.4745; acid value 0.6–1.5; iodine no. 140–152; saponification

Superior numbers refer to Manufacturers of Trade Mark Products. For page number see Contents.

no. 186–193; unsaponifiable 0.3–1.0%. Combustible; nontoxic.
Derivation: Hydraulic press or solvent extraction of seeds.
Containers: Drums; tank cars.
Uses: Alkyd resins; paints; varnishes; medicine; dietetic foods; margarine; hydrogenated shortening.

"Saflex."[58] Trademark for polyvinyl butyral adhesive film supplied in clear, translucent, tinted, or gradated color, for the plastic interlayer in safety glass (q.v.).

safranine. A family of dyes, based on phenazine, having C.I. Nos. 50200–50375, some of which are useful as biological stains.

safrole (4-allyl-1,2-methylenedioxybenzene) $C_3H_5C_6H_3O_2CH_2$.
Properties: Colorless or pale yellow oil; the odor-giving constituent of sassafras, camphorwood and other oils; sp. gr. (15°C) 1.100–1.107; m.p. 11°C; b.p. 233°C; optical rotation (15°C) −0°30′; refractive index (n 20/D) 1.5363–1.5385; soluble in alcohol; slightly soluble in propylene glycol; insoluble in water and glycerin.
Derivation: From oil of sassafras or camphor oil.
Method of purification: Rectification or freezing.
Grade: Technical.
Containers: Bottles; drums.
Hazard: Toxic by ingestion. May not be used in food products (FDA).
Uses: Perfumery and soaps; manufacture of heliotropin; medicine; insecticides.

"SAG."[214] Trademark for silicone defoamers used in many industrial and chemical processes.

sage oil, clary.
Properties: Pale-yellow to light-yellow essential oil. Sp. gr. 0.886–0.929 (25/25°C); refractive index 1.4580–1.4730 (20°C).
Chief constituent: Linalyl acetate.
Derivation: Obtained by distillation of Salvia sclarea L. A floral absolute (solid) is obtained by volatile solvent extraction from the same plant.
Grade: F.C.C.
Uses: Perfumery; flavoring.

sage oil, Dalmatian (salvia oil).
Properties: Yellowish or greenish yellow liquid; has warm camphoraceous odor and flavor; sp. gr. 0.903–0.925 (25°C); refractive index 1.4570–1.4690 (20°C); optical rotation +2° to +29°. Soluble in alcohol and mineral oil; practically insoluble in glycerin.
Chief known constituents: Thujone, cineol, pinene.
Derivation: Steam distillation of partially dried leaves of Salvia officinalis.
Grades: F.C.C.; technical.
Uses: Medicine; perfumery; flavoring.

SAIB. Abbreviation for sucrose acetate isobutyrate.

sal ammoniac. See ammonium chloride.

salicin (salicyl alcohol glucoside) $HOCH_2C_6H_4OC_6H_{11}O_5$. A glucoside obtained from several species of Salix and Populus.
Properties: Colorless crystals or white powder Soluble in waer, alcohol, alkalies, glacial acetic acid; insoluble in ether. M.p. 199–202°C; sp. gr. 1.43. Nontoxic.
Uses: Medicine; reagent for nitric acid.

salicylal. See salicylaldehyde.

salicyl alcohol (ortho-hydroxybenzyl alcohol; alpha, 2-dihydroxytoluene; saligenin) $HOC_6H_4CH_2OH$.
Properties: White crystals; m.p. 86–87°C; pungent taste; sublimes at 100°C; somewhat soluble in water; very soluble in alcohol, chloroform, ether; soluble in propylene glycol, benzene and fixed oils; sparingly soluble in cold water; soluble in hot water. Nontoxic. Combustible. Sp. gr. 1.16.
Derivation: Hydrolysis of salicin; heating phenol and methylene chloride with caustic.
Use: Medicine (antipyretic and tonic); local anesthetic.

salicyl alcohol glucoside. See salicin.

salicylaldehyde (salicylal; salicylic aldehyde; ortho-hydroxybenzaldehyde) C_6H_4OHCHO.
Properties: Colorless, oily liquid or dark red oil; bitter almond-like odor; burning taste; sp. gr. 1.165–1.172; f.p. −7°C; b.p. 196°C. Soluble in alcohol, ether and benzene; slightly soluble in water. Flash point 172°F. Combustible; low toxicity.
Derivation: Interaction of phenol and chloroform in presence of aqueous alkali.
Uses: Analytical chemistry; perfumery (violet); synthesis of coumarin; gasoline additives; auxiliary fumigant.

salicylamide (ortho-hydroxybenzamide) $C_6H_4(OH)CONH_2$.
Properties: White or slightly pink crystals; m.p. 139–142°C; b.p. decomposes at 270°C; soluble in hot water, alcohol, ether, chloroform; slightly soluble in cold water, naphtha, and carbon tetrachloride. Low toxicity.
Derivation: Treatment of methyl salicylate with dry ammonia gas.
Grades: Technical; N.F.
Containers: Glass bottles; drums.
Use: Medicine (analgesic).

salicylanilide $HOC_6H_4CONHC_6H_5$.
Properties: Odorless, white or slightly pink crystals. M.p. 136–138°C; b.p., decomposes; stable in air; slightly soluble in water; freely soluble in alcohol, ether, chloroform, and benzene.
Grade: N.F.
Hazard: Moderately toxic by ingestion; irritant to skin.
Uses: Medicine; fungicide; slimicide; antimildew agent; intermediate.

salicylic acid (ortho-hydroxybenzoic acid) $C_6H_4(OH)(COOH)$.
Properties: White powder; acrid taste; stable in air but gradually discolored by light; sp. gr. 1.443 (20/4°C); m.p. 158–161°C; b.p. about 211°C at 20 mm; sublimes at 76°C; soluble in acetone, oil of turpentine, alcohol, ether, benzene; slightly soluble in water. Combustible; low toxicity.
Derivation: Reacting a hot solution of sodium phenolate with carbon dioxide, and acidifying the sodium salt thus formed.
Grades: Technical; U.S.P.; crude.
Containers: Fiber drums.
Hazard: Dust forms explosive mixtures in air.
Uses: Manufacture of aspirin and salicylates; resins; dyestuff intermediate; prevulcanization inhibitor; fungicide.

salicylic acid dipropylene glycol monoester. See dipropylene glycol monosalicylate.

salicylic aldehyde. See salicylaldehyde.

saline water. See brine; ocean water.

salmine. A protein specific to the salmon. Used in nutritional and biochemical research.

salol. See phenyl salicylate.

sal soda (washing soda; sodium carbonate decahydrate) $Na_2CO_3 \cdot 10H_2O$.
Properties: White crystals; sp. gr. 1.44; m.p. 32.5–34.5°C, loses water at this temperature. Easily soluble in water; insoluble in alcohol. A pure form of sodium carbonate (soda ash). Nontoxic.
Containers: Multiwall paper sacks; carloads.
Uses: Washing textiles; bleaching linen and cotton; general cleanser.

salt. (1) The compound formed when the hydrogen of an acid is replaced by a metal or its equivalent (e.g., an NH_4 radical). Example:

$$HCl + NaOH \longrightarrow NaCl + H_2O.$$

This is typical of the general rule that the reaction of an acid and a base yields a salt and water. See also soap. Most inorganic salts ionize in water solution. (2) Common salt, sodium chloride, occurs widely in nature, both as deposits left by ancient seas and in the ocean, where its average concentration is 2.6%. See sodium chloride.

salt bath. A molten mixture of sodium, potassium, barium and calcium chlorides or nitrates, to which sodium carbonate and sodium cyanide are sometimes added. Used for hardening and tempering of metals, and for annealing both ferrous and nonferrous metals. Temperatures used may be as high as 2400°F for hardening high-speed steels. Commercial mixtures are available for a variety of specifications. See also fused salt.

salt cake. Impure sodium sulfate (90–99%). For properties and derivation, see sodium sulfate.
Grades: Technical; glassmakers' (iron-free).
Containers: 200-lb bags; 500-lb barrels; carloads.
Uses: Paper pulp; detergents and soaps; plate and window glass; sodium salts; ceramic glazes; dyes.

salting out. The addition of salts (usually sodium chloride) to a mixture or production batch to precipitate or coagulate proteins, soaps, or simpler organic substances.

salt, molten. See fused salt.

saltpeter. See niter, and potassium nitrate.

salt, rock. See sodium chloride.

salt, fused. See fused salt.

salvia oil. See sage oil, Dalmatian.

"Salyrgan."[162] Trademark for mersalyl (q.v.).

samarium Sm A rare-earth metal of the lanthanide group (Group IIIB of the Periodic Table). Atomic number 62. Atomic weight 150.4. Valences 2 and 3. Seven stable isotopes.
Properties: Hard, brittle metal which quickly develops an oxide film in air. An active reducing agent. Ignites at 150°C; liberates hydrogen from water. Sp. gr. 7.53; m.p. 1072°C; b.p. 1900°C; hardness similar to iron. High neutron absorption capacity. Combustible.
Occurrence: Australia, Brazil, southeastern U.S. in monazite sands; South Africa. Also from bastnasite ore in California.
Derivation: Reduction of the oxide with barium or lanthanum.
Uses: Neutron absorber; dopant for laser crystals; metallurgical research; permanent magnets.

samarium chloride $SmCl_3 \cdot 6H_2O$.
Properties: Faintly yellow hygroscopic crystals; sp. gr. 2.383; loses $5H_2O$ at 110°C; soluble in water.
Derivation: By treating the carbonate or oxide with hydrochloric acid.

samarium oxide Sm_2O_3.
Properties: Cream-colored powder; sp. gr. 8.347; m.p. 2300°C. Insoluble in water; soluble in acids. Absorbs moisture and carbon dioxide from the air.
Uses: Catalyst in the dehydrogenation of ethyl alcohol; infrared-absorbing glass; neutron absorber; preparation of samarium salts.

sampling. The methods and the techniques used in obtaining representative test samples of quantity lots of raw materials, semiprocessed work, and finished product for production and quality control. Rules for sampling procedures for both solid and liquid materials have been established by the National Cottonseed Products Association, Memphis, Tenn., and by the National Institute of Oilseed Products, San Francisco, Cal. The techniques of physical sampling are only one application of statistical quality control.

SAN. Abbreviation for styrene-acrylonitrile polymer. See polystyrene.

sand. See silica.

sandalwood oil, East Indian (santalwood oil; santal oil; sandal oil).
Properties: Pale yellow, essential oil; faint, aromatic odor; harsh, resinous taste. Sp. gr. 0.965–0.980 (25°C); optical rotation −15° to −20° in 100 mm tube at 25°C; refractive index 1.50–1.51 (20°C); acid value 0.5–8.0; ester value 3–17, after acetylation, not less than 196; soluble in 3–5 vols of 70% alcohol; soluble in fixed oils; insoluble in glycerin. Combustible; nontoxic.
Derivation: Steam distillation of the wood of Santalum album of India.
Grades: F.C.C.; technical.
Uses: Medicine; perfumery; flavoring.

sandarac. A recent natural resin obtained from Morocco. Its commercial form is yellow, brittle, amorphous lumps or powder; soluble in alcohol; insoluble in benzene and water. Used in special types of varnishes and lacquers.

sand casting. See foundry sand.

"Sandril."[100] Trademark for reserpine (q.v.).

"Sandrol."[462] Trademark for a group of detergents, foam boosters, foam stabilizers, and thickeners. Made from aluric acid, coconut fatty acid, and coconut oil "Superamides"; coconut and lauric acid alkanolamides; hydrogenated coconut alkanolamide; and lauric acid isopropanolamide.

sandwich molecule. See metallocene.

"Sanforized."[7] Trademark used for fabrics treated by a proprietary process involving water spray and steam heat which limits shrinkage to not over 1% by government standard wash test. Fabrics include all-cottons, blends, and some synthetics.

"Sangamo."[492] Trademark for snow-white powder with a clean, neutral, sweet taste. Essentially, regular corn syrup with water removed. Used as a sweetener in a variety of food products.

"Sanikleen."[284] Trademark for a germicide containing detergents and detergent builders, wetting agents, water softeners and quaternary ammonium disinfectants. Used for cleaning and disinfecting floor surfaces, porcelain eating utensils, food-handling equipment, etc.

sanitizer. A special class of disinfectant designed for use on food-processing equipment; dairy utensils; dishes and glassware in restaurants. Among them are the hypochlorites; chloramines and other organic chlorine-liberating compounds; and quaternary ammonium compounds, many of which are proprietary. See also antiseptic; disinfectant.

santal oil. See sandalwood oil, East Indian.

santalol $C_{15}H_{24}O$. A sesquiterpene alcohol.
Properties: Colorless liquid; odor of oil of sandalwood. Soluble in 3 parts of 70% alcohol. Insoluble in water; sp. gr. 0.971–0.973; refractive index 1.504–1.508; b.p. about 300°C. Combustible; low toxicity.
Derivation: From sandalwood oil.
Use: Perfumery.

santalyl acetate. Acetic acid ester of a mixture of alpha- and beta-santalols.
Properties: Colorless liquid, light odor of sandalwood. Sp. gr. 0.982–0.985; refractive index 1.487–1.492.
Derivation: Treatment of sandalwood oil or santalol with acetic anhydride.
Use: Perfumery.

"Santicizer."[58] Trademark for a series of plasticizers, including various sulfonamides, phthalates, and glycollates.

"Santobrite."[58] Trademark for sodium pentachlorophenate, technical (q.v.).

"Santocel."[58] Trademark for a series of silica aerogels.
Uses: Anticaking flow conditioner; flatting agent; additive for drying emulsion paint films without flocculation; thixotropic and thickening of polyvinyl chloride plastisols and in various plasticizer dispersions.

"Santochlor."[58] Trademark for para-dichlorobenzene (q.v.).

"Santocure."[58] Trademark for a series of accelerators for natural and synthetic rubbers.

"Santoflex."[58] Trademark for a series of rubber antioxidants and antiozonants.

"Santolite."[58] Trademark for a group of sulfonamide resins.

"Santolube."[58] Trademark for a series of lubricating oil additives, e.g., zinc dialkyl dithiophosphate.

"Santomask" II.[58] Trademark for odor-masking ingredient for interior paints, printing inks and similar products having objectionable natural odors.

"Santomerse."[58] Trademark for a series of alkyl benzene sulfonates (biodegradable).

santonin $C_{15}H_{18}O_3$. A tricyclic structure.
Properties: White powder, turning yellow on exposure to light; odorless; tasteless at first, then bitter. Soluble in chloroform, alcohol, and most volatile and fatty oils; very slightly soluble in water. Solutions are levorotatory. Sp. gr. 1.187; m.p. 170–173°C; b.p. sublimes; specific rotation −170 to −175° (2 g/100 ml alcohol).
Derivation: By extraction from Artemisia.
Hazard: Moderately toxic by ingestion; affects color vision.
Grade: Technical.
Use: Medicine.

"Santophen" 1 Germicide.[58] Trademark for ortho-benzyl-para-chlorophenol.

"Santopoid."[58] Trademark for a series of gear oil additives.

"Santopour" C.[58] Trademark for pour point depressant for wax-bearing oils. See also pour point depressant.

"Santoquin."[58] Trademark for ethoxyquin (q.v.).
Uses: Antioxidant in animal feeds and dehydrated forage crops.

"Santosite."[58] Trademark for sodium sulfite anhydrous, technical.

"Santotan" KR.[58] Trademark for basic chrome sulfate containing approximately 50% $Cr_2(SO_4)_2(OH)_2$.
Properties: Dark green crystals; basicity (Cr_2O_3/SO_3) 101–105; basicity (Schorlemmer) 37.3–39.7; chromium as Cr_2O_3 23.9% min; clear solution in water (5:500).
Use: Chrome tanning of calf, cow and horse hides and sheepskins.

"Santovar A."[58] Trademark for 2,5-di-tert-amylhydroquinone (q.v.).

"Santowax" R.[58] Trademark for mixed terphenyls. Yellowish-white, noncrystalline flaked solid. Used as extender for polystyrene.

"Santowhite."[58] Trademark for a series of rubber antioxidants.

saponification. The chemical reaction in which an ester is heated with aqueous alkali such as sodium hydroxide, to form an alcohol (usually glycerol), and the sodium salt of the acid coresponding to the ester. The process is usually carried out on fats, (glyceryl esters of fatty acids). The sodium salt formed is called a soap (q.v.). A typical saponification reaction is:
$(C_{17}H_{35}COO)_3C_3H_5 + 3NaOH \longrightarrow 3C_{17}H_{35}COONa + C_3H_5(OH)_3$.

saponification number. The number of milligrams of potassium hydroxide required to hydrolyze 1 gram of a sample of an ester (glyceride; fat) or mixture.

saponin
(1) A general term applied to two groups of plant glucosides that on shaking with water form colloidal solutions giving soapy lathers. They also have the ability to hemolyze red blood corpuscles at very great dilutions. The two groups are triterpenoid and steroid saponins, the latter being used in research on sex hormones.
(2) Specific term; saponin derived from saponaria or quillaja.
Properties: White, amorphous glucoside; pungent; disagreeable taste and odor. It foams strongly when shaken with water. Soluble in water. Nontoxic except by injection.
Grades: Crude; purified; highest purity.
Hazard: Highly toxic by injection; destroys red blood cells.
Uses: Foam producer in fire extinguishers; detergent in textile industries; sizing; substitute for soap; emulsification agent for fats and oils.

SAPP. Abbreviation for sodium acid pyrophosphate. See sodium pyrophosphate, acid.

"SAPP #4."[1] Trademark for a sodium acid pyrophosphate.
Grade: F.C.C.
Uses: In leavening (has the slowest reaction rate of several phosphates which are otherwise chemically identical); in acid-type metal cleaners, oil well drilling muds, household and industrial cleaners.

sapphire (Al_2O_3). For natural material, see corundum. Synthetic sapphire is made by crystal-growing technique.
Properties: Hard, crystalline solid. Sp. gr. 3.98. Mohs hardness 9.0. M.p. 2040°C. Dielectric strength 480 kV/cm. Dielectric constant (20°C) 9.0. Coefficient of friction 0.05 micron. Inert to strong acids and alkalies. Excellent high-temperature stability. Can be sealed to glass. High transmission in infrared and ultraviolet.
Forms: Rods, spheres, disks, whiskers, single crystals.
Uses: Electron and microwave tubes; optical elements in radiation detectors; substrate for thin-film components and integrated circuits; abrasive; record needles; precision instrument bearings; aluminum composites; micromortars for hand-pulverizing chemicals.

"SAR."[182] Trademark applied to ion-exchange resin used in water treating and chemical process applications; Type II, strongly basic anion exchange resin, produced from styrene and divinylbenzene.

"Saraloy."[233] Trademark for a line of flexible thermoplastic flashing material based on a saran copolymer.

saran. See polyvinylidene chloride; saran fiber.

saran fiber. Generic name for a manufactured fiber in which the fiber-forming substance is any long-chain synthetic polymer composed of at least 80% by weight of vinylidene chloride units (—CH_2CCl_2—) (Federal Trade Commission).
Properties: Tenacity, 2 to 2.5 grams per denier; elongation 15–30%; moisture regain, none; sp. gr. 1.65–1.75; softens at 240–280°F; combustible, but self-extinguishing. Highly resistant to most chemicals and solvents, to weather, moths and mildew.
Uses: Screens; upholstery, curtain and drapery fabrics; rugs and carpets; awnings; filter cloth.
See also polyvinylidene chloride.

sarcolysin. See melphalan.

sarcosine (methyl glycocoll, methyl aminoacetic acid) CH_3NHCH_2COOH.
Properties: Deliquescent crystals with sweet taste; m.p. 210–215°C (dec). Very soluble in water; slightly soluble in alcohol. Combustible; low toxicity.
Derivation: Decomposition of creatine or caffeine.
Containers: Drums; tank cars.
Grade: Technical.
Use: Synthesis of foaming antienzyme compounds for toothpastes, cosmetics, and pharmaceuticals.

sardine oil. See fish oil.

sarin (methylphosphonofluoride acid, isopropyl ester) $[(CH_3)_2CHO](CH_3)FPO$. A nerve gas (q.v.).
Hazard: Highly toxic by inhalation and skin absorption. Cholinesterase inhibitor. Not manufactured.

"Sarkosyl."[219] Trademark for a series of surface-active N-acylated sarcosines.
Uses: Tooth pastes, cosmetics, and pharmaceuticals (anti-enzyme); corrosion inhibitors; lubricants and greases; inks.
See also sarcosine.

"Sarolex."[219] Trademark for nematocide and insecticide containing diazinon.
Hazard: May be toxic.

SAS. (1) Abbreviation for sodium aluminum sulfate. See aluminum sodium sulfate. (2) Abbreviation for sodium alkane sulfonate.

satin spar. See calcite.

satin white. A high-bulking filler used in paper coating formulations; a mixture of hydrated lime, potash alum, and aluminum sulfate. Readily hydrated. Particle size range. 0.2 to 2 microns.

saturation. (1) The state in which all available valence bonds of an atom (esp. carbon) are attached to other atoms. The straight-chain paraffins are typical saturated compounds. See also unsaturation. (2) The state of a solution when it holds the maximum equilibrium quantity of dissolved matter at a given temperature. See also solution, true; supersaturation.

"Savalux."[28] Trademark for a group of textile dyes selected to give outstanding fastness.

savory oil (summer savory oil)
Properties: Light yellow to dark brown essential oil; aromatic odor; sp. gr. 0.875–0.954 (25°C); refractive index 1.4860–1.5050 (20°C); optical rotation −5° to +4°. Soluble in most fixed oils and mineral oil, practically insoluble in glycerin. One ml dissolves in 2 ml 80% alcohol. Combustible; nontoxic.
Derivation: Steam distillation of the whole dried plant Satureia hortensis L., an aromatic mint of the United States and Europe.
Grade: F.C.C.
Use: Flavoring agent.

"Saxin."[301] Trademark for a noncaloric sweetening agent containing saccharin (q.v.).

saxitoxin $C_{10}H_{17}N_7O_4 \cdot 2HCl$. A toxic principle present in certain species of shellfish. It is a paralytic poison that attacks the central nervous system.

Saybolt Universal viscosity. The efflux time in seconds (SUS) of 60 ml of sample flowing through a calibrated Universal orifice in a Saybolt viscometer under specified conditions. See also viscosity.

Sb Symbol for antimony (from Latin stibium).

SBA. Abbreviation for sec-butyl alcohol (q.v.).

S.B.G. Abbreviation for standard battery grade; highly purified chemicals manufactured for use in the battery industry.

SBR. Abbreviation for styrene-butadiene rubber (q.v.).

"SBR."[182] Trademark applied to ion-exchange resins used in water treating and chemical process applications; Type I, strongly basic anion exchange resin, produced from styrene and divinylbenzene.

Sc Symbol for scandium.

"SC."[555] Trademark for triethylene glycol dicaprylate-caprate; used as a plasticizer.

scale. (1) A calcareous deposit in water tubes or steam boilers resulting from deposition of mineral compounds present in the water, e.g., calcium carbonate. See boiler scale. (2) A type of paraffin or petroleum wax from which all but a few per cent of oil has been removed by hydraulic pressing and subsequent processing. (3) A graduated standard of measurement in which the units (degrees) are defined in relation to some property of what is measured, e.g., temperature scale; Brix scale; Baumé scale.

scale-up. A term used in chemical engineering to describe the calculations and planning involved in carrying a chemical processing operation from the pilot plant to large-scale production stage.

scandium Sc Metallic element of atomic number 21; Group IIIB of the Periodic Table; atomic weight 44.9559; valence 3; no stable isotopes.
Properties: Silvery white solid; m.p. 1539°C; b.p. 2727°C; density 2.99 g/cc. Does not tarnish in air; reacts rapidly with acids; strongly electropositive. Not attacked by 1:1 mixture of nitric and 48% hydrofluoric acids. It is chemically similar to the rare earths.
Source: Chief ores are wolframite and thortveitite (Norway, Madagascar).
Derivation: Reduction of scandium fluoride with calcium, or with zinc or magnesium alloy. Purification by distillation gives purity of 99+%.
Uses: No major industrial uses. An artificial radioactive isotope has been used in tracer studies and leak detection, and there is some application in the semiconductor field.

scandium fluoride. ScF_3.
Derivation: Reaction of scandium oxide with ammonium hydrogen fluoride.
Uses: Preparation of scandium metal.

scandium oxide. Sc_2O_3.
Properties: White amorphous powder. Soluble in hot acids, less so in cold acids. Sp. gr. 3.864; specific heat 0.153. A weak base.
Source: Thortveitite.
Use: Preparation of scandium fluoride.

scavenger (getter). (1) In chemistry, any substance added to a system or mixture to consume or inactivate traces of impurities. (2) In metallurgy, an active metal added to a molten metal or alloy which combines with oxygen or nitrogen in the melt and causes its removal into the slag. Alloys including such metals as thorium, zirconium, and cerium, as well as carbon and misch metal, are used in vacuum tubes to absorb traces of residual gases.

"Scav-Ox."[84] Trademark for a hydrazine oxygen scavenger. Used to prevent corrosion in boilers and oil wheel casings.

Schaeffer's acid. See 2-naphthol-6-sulfonic acid.

Schaeffer's salt $C_{10}H_6OHSO_3Na$. Sodium salt of 2-naphthol-6-sulfonic acid (q.v.).
Uses: Intermediate for organic chemicals.

Schaffer's salt. See Schaeffer's salt.

Scheele, C. W. (1742–1786). One of the outstanding early chemical thinkers and experimenters, Scheele was a Swedish scientist who discovered a number of previously unknown substances, among which were tartaric acid, chlorine, manganese salts, arsine and copper arsenite (Scheele's green). He also noted the oxidation states of various metals; observed the nature of oxygen two years before Priestley's discovery; and discovered the chemical action of light on silver compounds, thus laying the foundation of photochemistry and photography. Scheelite, or natural calcium tungstate, is named after him.

Scheele's green. See copper arsenite.

scheelite $CaWO_4$. A natural calcium tungstate, found in igneous rocks, usually with granite. Some molybdenum may replace tungsten.
Uses: Ore of tungsten; phosphor.

Schiff base. A class of compounds derived by chemical reaction (condensation) of aldehydes or ketones with primary amines. The general formula is $RR'C = NR''$.
Properties: Usually colorless crystalline solids, although some are dyes. Very weakly basic and hydrolyzed by water and strong acids to carbonyl compounds and amines.
Uses: Rubber accelerators; dyes (phenylene blue, toluylene blue, and naphthol blue); chemical intermediate; liquid crystals in electronic display systems; perfume base.

Schoelkopf's acid. See 1-naphthylamine-8-sulfonic acid; and 1-naphthol-4,8-disulfonic acid.

schradan. Generic name for actamethyl pyrophosphoramide (OMPA), $[(CH_3)_2N]_2P(O)OP(O)[N(CH_3)_2]_2$.
Properties: Viscous liquid; sp. gr. 1.137; b.p. 120–125°C (0.5 mm); refractive index (n 25/D) 1.462. Miscible with water; soluble in most organic solvents; hydrolyzed in the presence of acids, but not by alkalies or water alone.
Hazard: Highly toxic by ingestion and ihhalation. A cholinesterase inhibitor. Use may be restricted.
Use: A systemic insecticide which is absorbed by the plant, which then becomes toxic to sucking and chewing insects.
Shipping regulations: (Rail, Air) Organic phosphate, liquid, n.o.s. Poison label. Not accepted on passenger planes.

Schweitzer's reagent. A solution of copper hydroxide in strong ammonia used in analytical chemistry as a test for wool. It dissolves cotton, silk and linen.

scintillator. See phosphor.

scission. (1) The rupture of a chemical bond, with production of about 5 electron-volts of energy.
(2) In agricultural technology, the separation of fruit or vegetable products from the tree or vine.

scleroprotein. Any of a large class of proteins that have a supporting or protective fucntion in tendons, bones, cartilages, ligaments and other hard or tough parts of the animal body. They include the collagens of skin, tendons and bones, as well as the elastic proteins known as elastins, and the keratins. Specific examples are the keratin of hair, hoofs and horns and fibroin from silk.

scleroscope hardness. See hardness.

scouring agent. A compound used to remove the natural oils and fats from raw wool; also used to remove lubricants applied to rayon yarns or fabrics during such operations as throwing, winding, weaving, knitting, etc.

"Scriptite."[58] Trademark for melamine and urea-formaldehyde resins; used in paper industry for wet strength improvement.

scrubbing. Process for removing one or more components from a mixture of gases and vapors by passing it upward and usually countercurrent to and in intimate contact with a stream of descending liquid, the latter being chosen so as to dissolve the desired components and not others. The gas or vapor may be broken into fine bubbles upon entering a tower filled with liquid, but more frequently the tower is filled with coke, broken stone or other packing, over which the liquid flows while exposing a relatively large surface to the rising gas or vapor. See also absorption (1).

"SD-75."[430] Trademark for alkenyl dimethyl ethyl ammonium bromide. Used primarily as an algicide in water towers and swimming pools.

SDA. Abbreviation for specially denatured alcohol.

SDA No. 1. Specially denatured alcohol (government regulation formula). It consists of 5 gal methyl alcohol added to 100 gal 95% ethyl alcohol.
Hazard: Highly toxic.

SDDC. Abbreviation for sodium dimethyldithiocarbamate.

Se Symbol for selenium.

Seaboard process. Method of removing hydrogen sulfide from a gas by absorption in sodium carbonate solution. Sodium bicarbonate and sodium hydrosulfide are formed. By blowing air through this solution, hydrogen sulfide is released and carried off, and the sodium carbonate is regenerated.

sealant. Any organic substance that is soft enough to pour or extrude and is capable of subsequent hardening to form a permanent bond with the substrate. Most sealants are synthetic polymers (silicones, urethanes, acrylics, polychloroprene) which are semisolid before application and later become elastomeric. A few of the best-known sealants are such natural products as linseed oil (putty), asphalt, and various waxes. See also adhesive.

"Sealol."[51] Trademark for "white" mineral oil used as lubricant for sealed refrigeration compressor units.

"Sealstix." Trademark for a cement of the deKhotinsky type (q.v.) which adheres to almost any surface. Insoluble in all common reagents except alcohol, strong caustics and chromic acid cleaning solution.

"Sealvar."[299] Trademark for an iron-nickel-cobalt alloy used for making hermetic seals with the harder glasses and ceramic materials.

"Sealz."[248] Trademark for a line of thermoplastic rubber compounds which are used as additives for asphalt and tar products to raise softening point, improve low temperature flexibility, enhance adhesive qualities and increase elasticity.

Searles Lake brine. See brine; Trona process.

seasoning. Any food additive used in low concentrations to contribute to the taste of a food. Best-known is sodium chloride, but the term also includes various spices, onion, mustard, etc. See also enhancer; potentiator.

"Sea Sorb."[55] Trademark for patented grades of magnesia, MgO, with specific adsorption characteristics.
Uses: Catalysts, chromatography, purification, decolorization, and dehumidification.

sea water. See ocean water.

seaweed. See algae; phycocolloid; carrageenan; kelp.

sebacic acid (1,8-octanedicarboxylic acid; sebacylic acid; decanedioic acid) $COOH(CH_2)_8COOH$.
Properties: White leaflets; m.p. 133°C; sp. gr. 1.110 (25°C); b.p. 295.0°C (100 mm); refractive index 1.422 (133.3°C); slightly soluble in water; soluble in alcohol and ether. Combustible.
Derivation: From butadiene via dichlorobutene and its nitrile derivatives; dry distillation of castor oil with alkali.
Containers: Fiber drums; bags.
Grades: C.P.; purified.
Uses: Stabilizer in alkyd resins, maleic and other polyesters, polyurethanes; fibers; paint products; candles and perfumes; low-temperature lubricants and hydraulic fluids; manufacture of nylon 610.

"Sebacol."[173] Trademark for a tanners' depilatory containing sodium hydrosulfite suitably stabilized for use in alkaline unhairing systems.

sebaconitrile $NC(CH_2)_8CN$.
Properties: Straw-colored oily liquid; b.p. 199°C (15 mm).
Use: Chemical intermediate for drugs, dyes and high polymers.

sebacoyl chloride (n-octane-1,8-dicarboxylic acid dichloride) $ClOC(CH_2)_8COCl$.
Properties: Liquid; b.p. 137–140°C (3 mm); 97% pure. Decomposes slowly in cold water; soluble in hydrocarbons and ethers.
Containers: Bottles; carboys; 55-gal drums.
Use: Organic intermediate.

sebacylic acid. See sebacic acid.

sec-. Abbreviation for secondary, as applied to names of organic compounds. See primary. The prefix is ignored in alphabetical arrangement.

"Seconal."[100] Trademark for secobarbital (q.v.).

secondary. See primary.

secondary calcium phosphate. See phosphate, dibasic.

sedanolic acid $C_{12}H_{20}O_3$. A constituent of celery seed oil. Its lactone, also present, which is the source of the characteristic odor, is sedanolide.

sedative. A natural or synthetic therapeutic agent having the property of inducing relaxation and varying degrees of depression of the central nervous system. The major types of sedatives are: (1) chlorine substitution products (chloral hydrate, chlorobutanol); (2) ethyl alcohol derivatives (ethyl carbamate, hedonal); (3) specific aldehydes and ketones (paraldehyde, amylene hydrate); barbituric acid derivatives (barbital, barbiturates, q.v.). Sedatives are almost always prescription drugs. See also tranquilizer.

sedimentation. The settling out by gravity of solid particles suspended in a liquid, the rate of settling being defined by Stoke's Law (q.v.). This method is used industrially in water purification. It is also an analytical procedure for separation of solids of different particle size, as well as for molecular weight determination. The *sedimentation* of large molecules in a strong centrifugal field permits determination of both average molecular weights and the distribution of molecular weights in certain systems. When a solution containing polymer or other large molecules is centrifuged at forces up to 250,000 times gravity, the molecules begin to settle, leaving pure solvent above a boundary which progressively moves toward the bottom of the cell. An optical system is provided for viewing this boundary, and a study as a function of the time of centrifuging yields the rate of sedimentation for the single component, or for each of many components of a polydisperse system. These sedimentation rates may then be related to the corresponding molecular weights of the species present after the diffusion coefficients for each species are determined by independent experiments. The result of this work is the distribution of molecular weights in the sample which is attainable by few other methods.

See also precipitate.

seed. (1) That part of a plant which includes the plant embryo itself, a quantity of stored food (fats, carbohydrates, and proteins in varying proportions), and the enclosing cellulosic coats. The food storage tissue is called the endosperm. Starch is an important food reserve in the endosperms of cereals and legumes; sugar in sweet corn; protein in soybean and wheat; and fats in such plants as coconut, cacao, castor bean, and most types of palms. Such growth substances as giberellin also occur in seeds; some also contain alkaloids (ricinine, hyoscine, and caffeine). The seeds or bulbs of some plants are highly toxic, e.g., hyacinth, larkspur, castor bean, cashew nut shells. Bitter almond contains as much as 10% of hydrocyanic acid, which is removed in processing. Seeds (which in many cases are equivalent to nuts) are an important source of vegetable oils, which are used as food components, in paints and other industrial products, and in medicine.

(2) Trace proportions of a material introduced to initiate a desired reaction, where the "seed" acts as a nucleating agent. An example is the seeding of a cloud with silver iodide to cause precipitation by providing means for the nucleation of droplets. See nucleation.

Seger cone. See pyrometric cone.

Seidlitz powder. A two-part dosage, to be dissolved in water and drunk while effervescing.

In the blue paper: 3 parts Rochelle salt with 1 part sodium bicarbonate.

In the white paper: Tartaric acid.

Grade: N.F.

Containers: Fiber drums.

Use: Medicine (antacid).

"Seismex."[28] Trademark for an ammonia dynamite with limited water resistance (for seismic prospecting).

Hazard: Dangerous fire risk; moderate explosion risk.

"Seismogel."[28] Trademark for a straight gelatin dynamite when water conditions are only moderately severe; for seismic prospecting.

Hazard: Dangerous fire risk; moderate explosion risk.

"Selectacel."[231] Trademark for ion-exchange cellulose products which are derived from purified wood pulp by chemical reaction and substitution. White fibrous powders packed in glass, plastic and fiber drums. Used for the isolation and purification of proteins.

selenic acid H_2SeO_4.

Properties: White hygroscopic solid; sp. gr. 3.004 (15/4°C); m.p. 58°C (easily undercools); decomposes at 260°C. Very soluble in water; decomposes in alcohol; usually available as a liquid, sp. gr. 2.609 (15°C).

Hazard: Highly toxic; strong irritant to skin and mucous membranes.

Shipping regulations: (Air) Liquid: Corrosive label. Not acceptable on passenger planes. Solid: Poison label.

selenious acid. See selenous acid.

selenium Se A nonmetal (q.v.). Atomic number 34; Group VIA of the Periodic Table. Atomic weight 78.96; valences 2, 4, 6. There are 6 stable isotopes.

Properties: Amorphous red powder, becoming black on standing and crystalline on heating; vitreous and colloidal forms may be prepared. Crystalline form has sp. gr. about 4.5; m.p. 217°C; b.p. 685°C. Amorphous form softens at 40°C and melts at 217°C. Crystalline selenium is a p-type semiconductor. Electrically it acts as a rectifier, and has marked photoconductive and photovoltaic action (converts radiant to electrical energy). The electrical conductivity increases with increasing light irradiation. Soluble in concentrated nitric acid and (in liquid form) in common alkalies. Forms binary alloys with silver, copper, zinc, lead, etc. Low toxicity.

Occurrence: Copper ores; soils; feed crops such as corn and soybeans. Se is a necessary nutritional factor in animals.

Grades: Commercial (powder or lumps); high-purity up to 99.999%.

Containers: Glass bottles; drums.

Uses: Electronics (xerographic plates, TV cameras, photocells, magnetic computer cores, solar batteries, rectifiers, relays; ceramics (colorant for glass); steel and copper (degasifier and machinability improver); rubber accelerator; catalyst; trace element in animal feeds.

For further information consult Selenium, Tellurium Development Ass'n. 11 Broadway, New York, N.Y.

selenium diethyldithiocarbamate $Se[SC(S)N(C_2H_5)_2]_4$.

Properties: Orange-yellow powder; sp. gr. (20/20°C) 1.32; melting range 63–71°C; characteristic odor. Soluble in carbon disulfide, benzene, chloroform; insoluble in water.

Containers: 5-, 50-, 100-lb drums.

Hazard: Toxic by inhalation; moderately toxic by ingestion and skin absorption. Tolerance (as Se), 0.2 mg per cubic meter of air.

Use: Vulicanization agent without added sulfur or as a primary or secondary accelerator with sulfur.

selenium dimethyldithiocarbamate. See "Methyl Selenac."

selenium dioxide (selenous acid anhydride) SeO_2.

Properties: White or yellowish white to slightly reddish, lustrous, crystalline powder or needles; sp. gr. 3.954 (15/15°C); m.p. 340–350°C (sublimes); soluble in alcohol, water.

Hazard: Toxic by inhalation; moderately toxic by ingestion and skin absorption. Tolerance (as Se), 0.2 mg per cubic meter of air.

Uses: Analysis (testing for alkaloids); medicine; oxidizing agent; antioxidant in lubricating oils; catalyst.

selenium sulfide (selenium disulfide) SeS_2.

Properties: Bright orange powder; faint odor at most; m.p. < 100°C. Practically insoluble in water and organic solvents.

Grade: U.S.P.

Hazard: Toxic by ingestion; strong irritant to eyes and skin. Tolerance, 0.2 mg per cubic meter of air.

Uses: Medicine (treatment of seborrhea; medicated shampoos).

selenous acid (selenious acid) H_2SeO_3.

Properties: Transparent, colorless deliquescent crys-

tals; soluble in water and alcohol; insoluble in ammonia. Sp. gr. 3.0066; m.p. 70°C, decomposes.
Derivation: Action of hot nitric acid on selenium.
Hazard: Toxic by inhalation; moderately toxic by ingestion and skin absorption. Tolerance (as Se), 0.2 mg per cubic meter of air.
Use: Chemical reagent.

selenous acid anhydride. See selenium dioxide.

"Sellogen."[309] Trademark for a series of detergent, emulsifying, and wetting agents used in the textile industry and in the manufacture of cosmetics.

semecarpus nut (acajou nut). The fruit of semecarpus anacardium; oriental cashew nut. Properties and uses similar to cashew nut. See cashew nutshell oil.

"Semesan."[28] Trademark for a wettable powder containing 25.3% hydroxymercurichlorophenol.
Hazard: See mercury compounds.

"Semesan Bel."[28] Trademark for a seed disinfectant containing 12.5% hydroxymercurinitrophenol and 3.8% hydroxymercurichlorophenol.
Hazard: See mercury compounds.

semicarbazide hydrochloride (carbamylhydrazine hydrochloride; aminourea hydrochloride) $H_2NCONHNH_2 \cdot HCl$.
Properties: Snow-white crystals; m.p. 172–175°C (dec); soluble in water; insoluble in absolute alcohol and ether; soluble in dilute alcohol.
Derivation: From hydrazine sulfate, potassium or sodium cyanate and sodium carbonate, or electrolytically by the reduction of nitrourea.
Grades: C.P.; technical.
Containers: Bottles; fiber drums.
Uses: Reagent for aldehydes and ketones; isolation of hormones and isolation of certain fractions from essential oils.

semiconductor. An element or compound having an electrical conductivity intermediate between that of conductors and non-conductors (insulators). Most metals have quite high conductivity, while substances like diamond and mica have very low conductivity (high resistance). Between these extremes lie the semiconductors, of which germanium, silicon, silicon carbide, and selenium are examples, with resistivities in the range of 10^{-2} to 10^{9} ohms/cm. Slight traces of impurities in the crystalline structure are essential for semiconduction; arsenic is a typical impurity in semiconductor crystals. These impurities function as electron donors or acceptors, and the semiconductor is designated n-type or p-type, depending on the electrical nature of the "holes," or energy deficits in the crystal lattice.

The functioning of semiconductors involves the science of solid state physics. Their discovery in the early 1940's made possible the development of transistors, with their manifold applications in electronic devices, in which they have largely replaced the vacuum tube.

There are a few organic semiconducting compounds which contain a significant amount of carbon-carbon bonding, and are also capable of supporting electronic conduction. Anthracene and Ziegler-catalyzed acetylene polymers (conjugated polyolefins) are examples. See also crystal; impurity; solid; solid state chemistry.

semimicrochemistry. Any chemical method (usually analytical) in which the weight of the sample used is from 10 to 100 mg.

semipermeable membrane. See membrane, semipermeable.

semisynthetic. A product involving a chemical reaction which is based on or derived from a natural material, but does not occur as such in nature. Examples are paper, soap, cement, glass, and the chemical modifications of cellulose (rayon, nitrocellulose, methylcellulose, cellulose acetate); similarly for the modified starches. See also synthesis.

"Sentry."[214] Trademark for sorbic acid and potassium sorbate (q.v.). Used in fungistats for the control of certain molds and yeasts in foods. Also trademark for propylene glycol, U.S.P. Used as solvent for flavors and colors; humectant for baked goods; and plasticizer for cork seals and crowns.

"Separan."[233] Trademark for a series of flocculating agents.
AP30. Synthetic high molecular weight anionic polymer.
C-90 and C-120. Synthetic, high-molecular-weight cationic polymers.
MGL. Similar to NP10. Used in production of uranium.
NP10. Synthetic, water-soluble nonionic, high-molecular-weight polymer of acrylamide. "Separan" NP10 potable water grade flocculant has been accepted, subject to maximum use concentration of one ppm by the U.S. Public Health Service.
NP20. Nonionic polyacrylamide polymer.
PG2. Similar to NP10. Used in paper mfg.

separation. A collective term including a number of unit operations which, in one way or another, isolate the various components of mixtures. Chief among these are evaporation, distillation, drying, gas absorption, sedimentation, solvent extraction, pressure extraction, adsorption, and filtration. Specialized methods include centrifugation, magnetic separation (mass spectrograph), gaseous diffusion, and various types of chromatography. For further information see specific entry.

"Sephadex."[485] Trademark for a dry, insoluble powder composed of macroscopic beads which are synthetic, organic compounds derived from the polysaccharide dextran. The dextran chains are cross-linked to give a three dimensional network and the functional ionic groups are attached to the glucose units of the polysaccharide chains by ether linkages. Available in various forms for use in many different phases of chromatography.

septiphene. See ortho-benzyl-para-chlorophenol.

"Septo-Sour."[244] Trademark for a product consisting chiefly of zinc salts of fluorine compounds.
Properties: White, dustless crystals; readily soluble in water; neutralizing value 25.6 oz sodium bicarbonate per lb.
Containers: 10-lb net corrugated paper drums; 60-lb net plywood drums; 150- and 300-lb net fiber drums.
Uses: Laundry sour, especially low temperatures.

sequestration. The formation of a coordination complex by certain phosphates with metallic ions in solution so that the usual precipitation reactions of the latter are prevented. Thus, calcium soap precipitates are not produced from hard water treated with certain polyphosphates and metaphosphates. (These polyphosphates are often improperly referred to as hexametaphosphates). The term sequestration may be used for any instance in which an ion is prevented from exhibiting its usual properties due to close combination with an added material. Two groups of organic sequestering agents (chelates in these examples) of economic importance are the aminopolycarboxylic acids such as ethylenediaminetetraacetic acid and the hydroxycarboxylic acids such as gluconic, citric, and tartaric acids. In the food industry sequestering agents aid in stabilizing color, flavor, and texture.

See also chelate; coordination compound; complex.

"Sequestrene."[219] Trademark for a series of complexing agents and metal complexes consisting of ethylenediaminetetraacetic acid and salts.

Ser Abbreviation for serine.

serendipity. An unexpected scientific discovery which turns out to be more important than the project being researched. One example is the discovery of the sodium sulfide polymer, later known as "Thiokol," while the researcher was seeking to develop an improved automotive coolant. Another is the discovery of the reinforcing effect of carbon black on rubber while technologists were using it as a black pigment to counteract the whitening effect of zinc oxide.

serine (beta-hydroxyalanine; alpha-amino-beta-hydroxypropionic acid. $HOCH_2CH(NH_2)COOH$. A nonessential amino acid occurring naturally in the L(−) form.

Properties: Colorless crystals; soluble in water; insoluble in alcohol and ether; optically active; DL-serine, m.p. 246°C with decomposition; D(+)-serine, m.p. 228°C with decomposition; L(−) serine, m.p. 228°C with decomposition. Available commercially in all three forms.

Derivation: Hydrolysis of protein (especially silk protein); synthesized from glycine.

Uses: Biochemical research; dietary supplement; culture media; microbiological tests; feed additive.

"Serizyme."[173] Trademark for a proteolytic enzyme preparation for desizing textiles.

serotonin (5-hydroxytryptamine; 5-hydroxy-3-(beta-aminoethyl)indole) $C_{10}H_{12}N_2O$. A powerful vasoconstrictor occurring in the brain and blood platelets; it is thought to play a part in regulation of blood pressure, and also has muscle-contracting properties. It has been isolated from beef serum and may be synthesized from 5-benzyloxyindole. Serotonin is similar to the hallucinogenic drugs mescaline, LSD, psilocin and psilocybin. Its formation is inhibited by *p*-chlorophenylalanine.

"Serpasil."[305] Trademark for reserpine (q.v.).

serum. (1) The continuous phase of a biocolloid after the solid or disperse phase has been removed by centrifugation, coagulation or similar means. In the case of milk, for example, the serum, or whey, is a true solution of sugars, proteins, and mineral compounds in water. (2) Specifically, blood from which corpuscles, platelets, etc. have been removed, especially when prepared with antigenic bacteria for inoculation to effect the cure of a disease.

SES. Abbreviation for sodium 2,4-dichlorophenoxyethyl sulfate. See sesone.

sesame oil (benne oil; teel oil)

Properties: A fixed, bland, yellow, almost odorless liquid. Does not readily become rancid; sp. gr. 0.9187 (25/25°C); solidifying point −5°C; m.p. 20–25°C (for free fatty acids); Hehner value 95.7; saponification value 188–193; iodine value 103–114; refractive index 1.4748–1.4762. Optically active. Soluble in chloroform, carbon disulfide, ether, and benzene; slightly soluble in alcohol. Combustible; nontoxic.

Chief known constituents: Olein (75%); stearin; palmitin; myristin; linolein, sesamin, and sesamolin.

Derivation: By pressing from Sesamum indicum (orientale).

Occurrence: China; Japan; East Indies; South America.

Grades: Edible, should contain less than 1% free fatty acids; semirefined; coast; U.S.P.

Containers: Drums.

Uses: Shortenings, salad oil, margarine and similar food products.

sesamex. Generic name for 2-(2-ethoxyethoxy)ethyl 3,4-(methylenedioxy)phenyl acetal of acetaldehyde. $CH_2O_2C_6H_3OCH(CH_3)OC_2H_4OC_2H_4OC_2H_5$.

Properties: Liquid; b.p. 137–141°C (0.08 mm). Soluble in non-polar solvents.

Hazard: May be toxic.

Use: Synergist for insecticides.

sesamin. Generic name for 2,6-bis[3,4-(methylenedioxy)phenyl]-3,7-dioxabicyclo[3.3.0]octane, $C_{20}H_{18}O_6$.

Properties: Solid; m.p. 122.5°C. Combustible.

Hazard: May be toxic.

Use: Synergist for insecticides and fungicides.

sesamolin. Generic name for 6-[3,4-(methylenedioxy) phenoxy]-2-[3,4-(methylenedioxy)phenyl]3, 7-dioxabicyclo[3.3.0]octane, $C_{22}H_{18}O_7$. Constituent of sesame oil which is a powerful synergist of pyrethrum. A 1:1 mixture of pyrethrum and sesamoline has 31 times the insecticidal activity of pyrethrum alone.

sesone. Generic name for sodium 2,4-dichlorophenoxy-ethyl sulfate (SES) $C_6H_3Cl_2OCH_2CH_2SO_4Na$.

Properties: Crystalline solid; m.p. 245°C; soluble in water.

Hazard: May be toxic.

Use: Pre-emergence herbicide.

sesqui. A prefix meaning one and a half. Often used for salts in which the proportions of metal oxide to acid anhydride are 2:3, or vice-versa, as in sodium sesquicarbonate.

sesquiterpene. A terpene having the formula $C_{15}H_{24}$, one and a half times the standard terpene formula of $C_{10}H_{16}$. An example is cadinene, a constituent of various wood oils (cade oil, cedarleaf oil, etc.).

"Setit."[337] Trademark for suspension aid used in the ceramic industry. Composed of a colloidal form of aluminum oxide.

"Setole."[42] Trademark for thermoplastic resin emulsions; used as finishing agents for all types of textile fabrics.

"Setrete."[48] Trademark for a seed disinfectant containing phenyl mercuric ammonium acetate. Used for cereal grains and cotton seed.

Hazard: See mercury compounds.

"Setsit."[69] Trademark for a series of activated dithiocarbamate liquid accelerators for natural and synthetic latex films and adhesives.

"Sevin."[214] Trademark for a carbaryl insecticide (1-naphthyl N-methyl carbamate).

"Sevron."[28] Trademark for a series of cationic dyes.

sewage sludge. A mixture of organic materials resulting from purification of municipal waste. There are two types:
(a) Imhoff sludge: A low grade sludge containing from 2 to 3% ammonia and about 1% phosphoric acid.
(b) Activated sludge: A high-grade sludge containing from 5.0 to 7.5% ammonia and from 2.5 to 4.0% phosphoric acid.
Derivation: (a) By running sewage through settling tanks without access of air. The sludge, or solid matter, is decomposed by anaerobic bacteria.
(b) By running sewage through settling tanks and pumping air (or oxygen) through porous plates at the bottom of the tanks. 20% of the current "make" is also added. The waste acts as nutrient for aerobic bacteria, which consume the polluting organic matter. The resulting solids are filtered and dried.
Use: Fertilizer.
See also milorganite.

"SF-100."[236] Trademark for a granular organic material used as the basic additive for preparing low filtrate casing packs, and for spotting fluid to free differentially stuck pipe. It can be prepared with density of 7.6 lb/gal for use where formation is too weak to support a full column of mud or heavier packs.

"S-Fatty Acids."[487] Trademark for fatty acids derived from soybean oil. S-210 is a soy fatty acid developed primarily for alkyd resins. S-230 is a soy-type fatty acid having a lower unsaturated acids content and higher titer than S-210.
Properties: S-210 is a light yellow liquid and S-230 a light yellow semi-solid at ambient temperature.
Containers: Tank cars and trucks.
Uses: Chemical intermediate, paints and varnishes, alkyd resins and soaps.

SFS. (1) Abbreviation for Saybolt Furol seconds. See Furol viscosity and Saybolt Universal viscosity.
(2)Abbreviation for sodium formaldehyde sulfoxylate.

"SGL" Activated Carbon.[323] Trademark for a decolorizing carbon also designed for use in fixed beds. Total surface area (N_2, BET Method) 950–1050 sq m/g; particle size 8 × 30 mesh; apparent density 30 lb/cu ft.

SGR. Abbreviation for steam gas recycle process for shale oil recovery. See also shale oil.

"Sale Ban."[236] Trademark for buffered lignosulfonate. Used to convert or maintain muds for drilling mud-making shales or for drilling and completing "dirty sand" formations.

shale oil. A crude oil obtained from oil shale (q.v.) by destructive distillation followed by hydrotreating. Extensive deposits of the shale are located in Colorado, Wyoming, and Utah. Its content or bitumen (kerogen) is reported to have the composition 75% carbon, 10% hydrogen, 2.5% nitrogen, 1.5% sulfur and 9% oxygen. After refining it yields about 18% gasoline, 30% kerosene, 25% gas oil, 15% light lube oil, and 10% heavy lube oil. Its chemical structure is aliphatic, the refining residues consisting of paraffin wax. Extraction may be effected by any of several retorting processes.
Note: The contribution of shale oil to energy requirements is likely to be slight in the short-range future, largely because of the low concentration of oil in the shale, which necessitates earth-moving operations on a tremendous scale for which satisfactory equipment is not available. There are also undesirable ecologic effects, as well as the high cost of extraction installations. A large scale extraction has been considered by a group of companies in Colorado in the Federal Naval Reserve area, based on scale-up of successful pilot-plant operation. Though production is technically feasible, the capital requirements are so great that significant production appears to be uneconomic for the indefinite future.

shape-selective catalyst. A catalytic solid (transition metal) introduced into a crystalline aluminosilicate (zeolite) having ports or openings of about 5 angstrom units. This permits the catalyst to exert a selective effect between molecules that differ in shape, rather than in the reactivity of their chemical groups. The cage structure of the zeolite effectively prevents contact of the catalyst with all molecules whose shape and size exclude their entry. The ability of a catalyst to discriminate among molecules on the basis of their shape is of great value in the cracking of straight-chain hydrocarbons, and has attractive possibilities in other types of catalytic reactions.

shark liver oil
Properties: Yellow to red-brown liquid; strong odor; sp. gr. 0.917–0.928. Soluble in ether, chloroform, benzene, and carbon disulfide.
Derivation: By expression from shark livers.
Method of purification: Chilling and filtration.
Grades: Crude; refined.
Uses: Source of vitamin A and of squalene; biochemical research.

"Sharstop."[204] Trademark for a series of short-stopping agents used in the polymerization of synthetic rubbers.

shell. (1) In physical chemistry, this term is applied to any of the several paths or orbits of the electrons in an atom as they revolve around its nucleus. They constitute a number of principal quantum paths representing successively higher energy levels. There may be from 1 to 7 shells, depending on the atomic number of the element and corresponding to the 7 periods of the Periodic Table. The shells are usually designated by number, though letter symbols have been used, i.e., K . . . Q. The laws of physics limit the number of electrons in the various shells as follows: 2 in the first (K), 8 in the second (L), 18 in the third (M), and 32 in the fourth (N). With the exception of hydrogen and helium, each shell contains two or more orbitals, each of which is capable of holding a maximum of 2 electrons. See also quantum number; orbital theory; Pauli exclusion principle.
(2) The hard integument of molluscs and crustaceans, consisting mostly of calcium carbonate, chitin, etc. See nacre; oyster shells.
(3) The brittle covering of avian eggs, chiefly calcium carbonate, lime, etc. The formation of proper shell structures in certain species of birds is said to be adversely affected by DDT and similar insecticidal contaminants of their food.
(4) The shells of nuts are cellulosic in character; some contain industrially useful oils.

"Shell A-A Weed Seed Killer."[125] Trademark for a preplant herbicide containing not less than 98 wt % allyl alcohol.

Hazard: Highly toxic by ingestion, inhalation, and skin absorption.

shellac (lac; garnet lac; gum lac; stick lac).
Derivation: A natural resin secreted by the insect Laccifer lacca (coccus lacca) and deposited on the twigs of trees in India. After collection, washing, and purification by melting and filtering it is formed into thin sheets, which are later fragmented into flakes of Orange Shellac; this may be dewaxed and bleached to a transparent product. Soluble in alcohol; insoluble in water.
Grades: (1) Orange: TN(impure), fine, superfine, heart, superior; (2) Bleached and dewaxed (colorless).
Containers: Glass bottles; terne-plated cans; bags.
Hazard: (alcohol solution) Flammable, dangerous fire risk.
Uses: Sealer coat under varnish; finish coat for floors, furniture, etc.; dielectric coatings; deKhotinsky cement (q.v.).

"Shellacol."[319] Trademark for an alcohol-type solvent; available both anhydrous and made from 190 proof alcohol. Mild odor, free from objectionable characteristics. May be purchased without a Federal permit.
Hazard: Flammable liquid.
Uses: Shellac solvent and thinner; nitrocellulose lacquers and other protective coatings; cleaning and drying metal parts; printing industry; glass cleaners; antifreeze; lamp and stove fuel; chemical and chemical specialties.

"Shellflex."[125] Trademark for a series of petroleum products largely composed of naphthenic and/or paraffinic hydrocarbons.
Properties: Colors ranging from near water-white to black; viscosities from 3 cp to 55 cp at 210°F; sp. gr. from 0.85 to 0.95; odor slight to none; very low volatilities.
Hazard: Probably flammable.
Uses: Rubber processing and extending oils; miscellaneous process oil uses.

"Shell" Isoprene Rubber.[125] Trademark for a synthetic elastomer having properties which approximate those of natural rubber. A stereospecific isoprene polymer which has been reacted to a high structural purity of about 92% cis-1,4-polyisoprene.

"Shell Sol."[125] Trademark for a series of aliphatic hydrocarbon solvents composed of mixtures of paraffinic, naphthenic and aromatic hydrocarbons.

"Shell"-S-Polymers.[125] Trademark for a series of general-purpose and specialty butadiene-styrene copolymers, including clear hot and cold copolymers, cold oil-extended copolymers, hot and cold black masterbatches, cold oil-black masterbatches, as well as high styrene resin/cold copolymer masterbatches and hot and cold latexes. Used in a wide variety of applications, including tires, tire repair materials, mechanical goods, etc.

"Shellwax."[125] Trademark for a series of paraffin waxes derived from distillate lube streams.

sherardizing. The process by which relatively small articles made of iron or steel are coated with zinc powder. The metal forms an alloy with the steel surface and produces a thin, tightly adherent corrosion-resistant coating.

"Sherbelizer."[322] Trademark for an algin derivative-vegetable gum composition.
Uses: Stabilizer for sherbets, water ices, frozen fruits, syrups, purees, chocolate ice cream.

"Sherflex."[141] Trademark for a series of plasticizers. "Sherflex DBP," dibutyl phthalate.

shielding. Protection of personnel from the harmful effects of ionizing radiation and/or neutrons by enclosure of the equipment (reactor, particle accelerator, x-ray generator, etc.) with an absorbing material. The most efficient of these are cadmium, lead, steel, and high-density concrete (3 feet of concrete equals about 1 foot of steel). Among the plastics, polyethylene affords reasonable protection in thick sheets. Water and paraffin wax are good neutron absorbers because of their high hydrogen content. Materials such as graphite and beryllium are able to deflect and retard neutrons and are used for this purpose as moderators in nuclear reactors. Heavy water is also used.

shortstopping agent. A material used in a polymerization reaction to cut off the reaction at a predetermined point. Example: use of diethylhydroxyl amine or sodium dimethyldithiocarbamate to control synthetic rubber polymerization; tetraethylsilanol is used in silicone polymers.

Si Symbol for silicon.

side-chain. See chain.

siderite (chalybite; spathic iron ore) $FeCO_3$, usually with some calcium, magnesium or manganese. The term siderite is also used for an iron alloy found in meteorites.
Properties: Gray, yellow, brown, green, white or brownish red mineral, vitreous inclining to pearly luster, white streak; sp. gr. 3.83–3.88; Mohs hardness 3.5–4.
Occurrence: United States (Vermont, Massachusetts, Connecticut, New York, North Carolina, Pennsylvania, Ohio); Europe.
Use: An ore of iron; when high in manganese, used in the manufacture of spiegeleisen.

sienna. A yellowish hydrated iron oxide. Raw sienna is a brown-tinted yellow ocher occurring in Alabama, California, Pennsylvania; Cyprus and Italy. Burnt sienna is an orange-brown pigment made by calcining raw sienna. See also ocher; iron oxide reds.
Containers (raw or burnt): Paper bags.
Uses: Colorant in oil paints, stains, pastels, etc.

"Sierralite."[38] Trademark for hydrous magnesium-aluminum silicate derived from the mineral prochlorite. Used in ceramics in the preparation of synthetic cordierite for high resistance to heat shock. Supplied as 200-mesh powder.

siglure. Generic name for sec-butyl 6-methyl-3-cyclohexane-1-carboxylate $CH_3C_6H_8COOCH(CH_3)C_2H_5$.
Properties: Liquid; b.p. 113–114°C (15 mm). Soluble in most organic solvents; insoluble in water. Combustible.
Use: Insect attractant.

sigma bond. A covalent bond directed along the line joining the centers of two atoms. They are the normal single bonds in organic molecules. See also pi bond.

silane (silicon tetrahydride). SiH_4.
Properties: A gas with repulsive odor. Solidifies at about −200°C, b.p. −112°C; insoluble in water. Sp. gr. 0.68.

Hazard: Highly toxic; strong irritant to tissue. Dangerous fire risk; ignites spontaneously in air. Tolerance, 0.5 ppm in air.
Use: Doping agent for solid-state devices.
Shipping regulations: (Air) Not acceptable.

"Silaneal."[149] Trademark for organopolysiloxanes used for making materials water-repellent.

silane compounds. Gaseous or liquid compounds of silicon and hydrogen (Si_nH_{2n+2}), analogous to alkanes or saturated hydrocarbons. SiH_3 is called silyl (analogous to methyl), and Si_2H_5 is disilanyl (analogous to ethyl). A cyclic silicon and hydrogen compound having the formula (SiH_2) is called a cyclosilane. Organo-functional silanes are noted for their ability to bond organic polymer systems to inorganic substrates.
Hazard: Highly toxic; dangerous fire risk.
See also silicone; siloxane.

"Silastic."[149] Trademark for compositions in physical character comparable to milled and compounded rubber prior to vulcanization but containing organosilicon polymers. Parts fabricated of "Silastic" are serviceable from −100 to 500°F; retain good physical and dielectric properties in such service; show excellent resistance to compression set, weathering, and corona. Thermal conductivity is high; water absorption low.
Uses: Diaphragms, gaskets and seals, O-rings, hose, coated fabrics, wire and cable, and insulating components for electrical and electronic parts.

"Silbond."[1] Trademark for ethyl silicate; available as pure, condensed, prehydrolyzed, and specialty formulations.

"Silco-Flex."[116] Trademark for a silicone-rubber insulation for use on large motors and generators.

"Silene."[177] Trademark for hydrated calcium silicates for rubber reinforcing and flow conditioning.

"Silflake."[31] Trademark for a commercially pure silver, containing about 1% of organic lubricant and traces of iron. Particle size, 2 to 10 microns.
Grades: 131, 135 and 850 (variable conductivities and covering power). "Wet" flake contains a flammable solvent.
Uses: Conductive coatings and adhesives.

"Silfrax."[280] Trademark for bonded refractories containing from 40% to 78% silicon carbide.
Properties: High refractoriness; great strength; high thermal conductivity; freedom from spalling; resistance to clinker adhesion; and resistance to mechanical and flame abrasion.
Uses: Bricks for boiler and furnace installations; kiln furniture in ceramic kilns; shapes for boiler furnaces, air-cooled furnace linings, glass lehrs, pit furnaces, and enameling furnace ware supports.

"Silgon."[1] Trademark for a group of silicone products.

silica (silicon dioxide) SiO_2. Occurs widely in nature as sand, quartz, flint, diatomite.
Properties: Colorless crystals or white powder; odorless and tasteless; sp. gr. 2.2–2.6; insoluble in water and acids except hydrofluoric; soluble in molten alkali when finely divided and amorphous. Combines chemically with most metallic oxides. Noncombustible; melts to a glass with lowest known coefficient of expansion (fused silica). Thermal conductivity about half that of glass. M.p. 1710°C; b.p. 2230°C. High dielectric constant; high heat and shock resistance.
Derivation: Can be made from a soluble silicate (water glass) by acidification, washing, and ignition. Arc silica is made from sand, vaporized in a 3000°C electric arc.
Grades: By purity and mesh size; silica aerogel; hydrated; precipitated.
Containers: Multiwall paper bags.
Hazard: Toxic by inhalation; chronic exposure to dust may cause silicosis.
Uses (powder): Manufacture of glass, water glass, ceramics; abrasives; water filtration; component of hydraulic cements; source of ferrosilicon and elemental silicon; filler in cosmetics, pharmaceuticals, paper, insecticides; hydrated and precipitated grades as rubber reinforcing agent, including silicone rubber; anticaking agent in foods; flatting agent in paints; thermal insulator. (Fused): Ablative material in rocket engines, spacecraft, etc.; fibers in reinforced plastics; special camera lenses.
See also quartz: silicic acid; silica gel.

silica, fumed. A colloidal form of silica made by combustion of silicon tetrachloride in hydrogen-oxygen furnaces. Fine white powder.
Uses: Thickener, thixotropic, and reinforcing agent in inks, resins, rubber, paints, cosmetics, etc.
See also "Aerosil."

silica, fused. See silica.

silica gel. A regenerative adsorbent consisting of amorphous silica. Nontoxic; noncombustible.
Derivation: From sodium silicate and sulfuric acid.
Grades: Commercial grades capable of withstanding temperatures up to 500-600°F are supplied in the following mesh sizes: 3–8, 6–16, 14–20, 14–42, 28–200 and through 325.
Containers: Air-tight metal containers.
Uses: Dehumidifying and dehydrating agent; air-conditioning; drying of compressed air and other gases, liquids, such as refrigerants, and oils containing water in suspension; recovery of natural gasoline from natural gas; bleaching of petroleum oils; catalyst and catalyst carrier; chromatography; anticaking agent in cosmetics and pharmaceuticals; in waxes to prevent slipping.
See also silicic acid.

"SilicAR."[329] Trademark for silica-gel-based formulations, suitable for various chromatographic applications. The numerical suffixes indicate the approximate pH of a 10% slurry. Letters F, G, or GF indicate that the product contains a fluorescent material, gypsum binder or both. TLC indicates suitability for thin layer chromatography.

silicate. Any of the widely occurring compounds containing silicon, oxygen, and one or more metals, with or without hydrogen. The silicon and oxygen may combine with organic groups to form silicate esters. Most rocks (except limestone and dolomite) and many mineral compounds are silicates. Typical natural silicates are gemstones (except diamond), beryl, asbestos, talc, clays, feldspar, mica, etc. Portland cement contains a high percentage of calcium silicates. Best known of the synthetic (soluble) silicates is sodium silicate (water glass) (q.v.).
Hazard (natural silicate dusts): Toxic by inhalation. Tolerances range from 5 to 50 million particles per cubic foot of air. They are also fire risks.

Uses: Fillers in plastics and rubber; paper coatings; antacids; anticaking agents; cements.

silicic acid (hydrated silica). $SiO_2 \cdot nH_2O$. The jelly-like precipitate obtained when sodium silcate solution is acidified. The proportion of water varies with the conditions of preparation and decreases gradually during drying and ignition, until relatively pure silica remains. During drying the jelly is converted to a white amorphous powder or lumps. Used as laboratory reagent and reinforcing agent in rubber. See silica gel.

"Siliclad."[239] Trademark for a water-soluble silicone concentrate, used for coating or treating laboratory glassware, ceramics, metal, rubber and plastic product.

silicochloroform. See trichlorosilane.

silicomanganese. Alloys consisting principally of manganese, silicon and carbon.
Use: Low-carbon steel in which silicon is not objectionable. Silicon manganese steels are used for springs and high-strength structural steels. See also manganese steels and ferromanganese.

silicomolybdic acid. See 12-molybdosilicic acid.

silcon Si A non-metal (q.v.). Atomic number 14; Group IVA of the Periodic Table. Atomic weight 28.086; valence 4; 3 stable isotopes. It is the second most abundant element (25% of the earth's crust).
Properties: Dark-colored crystals (the octahedral form in which the atoms have the diamond arrangement). The so-called amorphous form consists of minute crystals. Soluble in a mixture of nitric and hydrofluoric acids and in alkalies; insoluble in water, nitric and hydrochloric acid. Sp. gr. 2.33; m.p. 1410°C; b.p. 2355°C; Mohs hardness 7; dielectric constant 12.
Occurrence: Does not occur free in nature, but is a major portion of silica and silicates (rocks, quartz, sand, clays, etc.). See silicate.
Derivation: Heating sand with coke in an electric furnace. Single crystals of ultrahigh purity are made on a commercial scale.
Purification: Treatment with hydrochloric and hydrofluoric acids. High purity is achieved by distillation of silicon tetrachloride, tetraiodide or silane, reduction to silicon with ultrapure zinc, and vacuum or argon zone refining.
Grades: Ferrosilicon (50% Si); regular (97% Si); semiconductor or hyperpure (99.97% Si).
Hazard: Tolerance, 10 mg per cubic meter of air. Flammable in powder form.
Uses: Alloying agent for steels, aluminum, bronze, copper, and iron (ferrosilicon); production of halogenated silanes; organosilicon compounds (silicone resins); silicon carbide; spring steels; deoxidizer in steel manufacture; semiconductor in integrated circuits, rectifiers, transistors, diodes, solar batteries, other solid state devices (single crystals doped with boron or phosphorus); cermets and other special refractories.
Shipping regulations: (Powder) (Air) Flammable solid label.
See also silica; silicate; silicone; ferrosilicon; silica gel.

silicon bromide. See silicon tetrabromide.

silicon-bronze. An alloy of copper, tin and silicon used for telephone and telegraph wires.

silicon carbide SiC.
Properties: Bluish black iridescent crystals; Mohs hardness 9; sp. gr. 3.217; sublimes with decomposition at 2700°C. Insoluble in water and alcohol; soluble in fused alkalies and molten iron. Excellent thermal conductivity; electrically conductive; resists oxidation at high temperatures. Noncombustible; a nuisance particulate.
Derivation: Heating carbon and silica in an electric furnace at about 2000°C.
Forms available: Powder; filament; whiskers (3 million psi); single crystals.
Containers: Bags, barrels, drums.
Uses: Abrasive for cutting and grinding metals; grinding wheels; refractory in nonferrous metallurgy, ceramic industry, and boiler furnaces; composite tubes for steam reforming operations. Fibrous form used in filament-wound structures and heat-resistant, high-strength composites.
See also "Carborundum"; "Crystalon"; "Carbofrax."

silicon chloride. See silicon tetrachloride.

silicon-copper (copper silicide).
Properties: A hard, tough, bronze-like alloy containing 10–30% silicon.
Derivation: From silica and copper electrolytically.
Use: Manufacture of silicon-bronze.

silicon dioxide. See silica.

silicone. Any of a large group of siloxane polymers based on a structure consisting of alternate silicon and oxygen atoms with various organic radicals attached to the silicon:

$$\begin{array}{ccccc} CH_3 & & CH_3 & & CH_3 \\ | & & | & & | \\ -OSi & - & \left[O-Si \right]_n & - & OSi- \\ | & & | & & | \\ CH_3 & & CH_3 & & CH_3 \end{array}$$

Properties: Liquids, semisolids, or solids depending on molecular weight and degree of polymerization; viscosity ranges from less than 1 to over 1 million centistokes. Polymers may be straight-chain, or cross-linked with benzoyl peroxide or other free radical initiator, with or without catalyst. Stable over temperature range from −50 to +250°C. Very low surface tension; extreme water repellency; high lubricity; excellent dielectric properties; resistant to oxidation, weathering, and high temperature; permeable to gases. Soluble in most organic solvents; unhalogenated types are combustible. Nontoxic.
Derivation: (1) Silicon is heated in methyl chloride to yield methylchlorosilanes; these are separated and purified by distillation, and the desired compound mixed with water. A polymeric silicone results. (2) Reaction of silicon tetrachloride and a Grignard reagent (RMgCl), with subsequent hydrolysis and polymerization.
Forms: Fluids, powders, emulsions, solutions, resins, pastes, elastomers.
Uses: (Liquid) Adhesives; lubricants; protective coatings; coolant; mold-release agent; dielectric fluid; heat transfer; wetting agent and surfactant; foam stabilizer for polyurethanes; diffusion pumps; antifoaming agent for liquids; textile finishes; water repellent; weatherproofing concrete. (Resin) Coatings, molding compounds, laminates (with glass cloth); filament winding; sealants; room-temperature curing cements (see "RTV"); electrical insulation; impregnating electric coils; bonding agent; modifier for alkyd resins; vibration-damping devices. (Elastomer, or silicone rubber) Encapsulation of electronic parts; electrical insulation; gaskets; surgical membranes;

automobile engine components; flexible windows for face masks, air locks, etc.; medical devices used within the body; miscellaneous mechanical products.

silicone oil. See silicone (properties and uses).

silicone rubber. See silicone (properties and uses).

silicon fluoride. See silicon tetrafluoride.

silicon-gold alloy. See gold-silicon alloy.

silicon monoxide SiO.
Properties: Amorphous black solid; sp. gr. 2.15–2.18; sublimes at high temperature. Hard and abrasive. Noncombustible.
Grades: Lumps, powders, tablets; optical.
Uses: To form thin surface films for protection of aluminum coatings, optical parts, mirrors; dielectric or insulator.

silicon nitride Si_3N_4.
Properties: Gray powder (can be prepared as crystals); sublimes at 1900°C; density 3.44 g/cu m; bulk density 70–75 lb/cu ft, depending on mesh; Mohs hardness 9+; thermal conductivity 10.83 Btu/in/sq ft/hr/°F at 400–2400°F. Resistant to oxidation, various corrosive media, molten aluminum, zinc, lead and tin. Soluble in hydrofluoric acid.
Derivation: Reaction of powdered silicon and nitrogen in an electric furnace.
Uses: Refractory coatings; bonding silicon carbide; mortars; abrasives; thermocouple tubes in molten aluminum; crucibles for zone-refining germanium; rocket nozzles; high-strength fibers and whiskers; insulator and passivating agent in transistors and other solid-state devices.

silicon tetrabromide (silicon bromide) $SiBr_4$.
Properties: Fuming, colorless liquid. Turns yellow in air. Disagreeable odor. Decomposed by water with evolution of heat. Sp. gr. 2.82 (0°C); b.p. 153°C; m.p. 5°C. Noncombustible.
Containers: Sealed "Pyrex" or quartz ampules.
Purity: 99.999%.
Hazard: Strong irritant to tissue.

silicon tetrachloride (silicon chloride) $SiCl_4$.
Properties: Colorless exceedingly mobile, fuming liquid; suffocating odor. Corrosive to most metals when water is present; in the absence of water it has practically no action on iron, steel, or the common metals and alloys, and can be stored and handled in metal equipment without danger. Sp. gr. 1.483 (20°C); wt 12.4 lb/gal; f.p. −70°C; b.p. 57.6°C; refractive index (n 20/D) 1.412. Miscible in all proportions with carbon tetrachloride, tin tetrachloride, titanium tetrachloride, and sulfur mono- and dichlorides. Decomposed by water and alcohol, with evolution of hydrochloric acid. Noncombustible.
Derivation: Heating silicon dioxide and coke in a stream of chlorine.
Grades: Technical; 99.5%; C.P. (99.8%); semiconductor.
Containers: Iron drums; bottles; tank cars.
Hazard: Highly toxic by ingestion and inhalation. Strong irritant to tissue.
Uses: Smoke screens; manufacture of ethyl silicate and similar compounds; production of silicones; manufacture of high-purity silica and fused silica glass; source of silicon, silica, and hydrogen chloride; laboratory reagent.
Shipping regulations: (Rail) White label. (Air) Corrosive label.

silicon tetrafluoride (silicon fluoride) SiF_4.
Properties: Colorless gas; suffocating odor similar to hydrogen chloride; fumes strongly in air; sp. gr. (gas) (air = 1) 3.57 (15°C); f.p. −90°C; b.p. −86°C. Absorbed readily in large quantities by water with decomposition. Soluble in absolute alcohol. Noncombustible.
Derivation: (a) Action of hydrofluoric acid or concentrated sulfuric acid and a metallic fluoride on silica or silicates. (b) Direct synthesis.
Grade: Pure, 99.5% min.
Containers: Gas cylinders.
Hazard: Highly toxic by inhalation. Strong irritant to mucous membranes. Tolerance (as F), 2.5 mg per cubic meter of air.
Uses: Manufacture of fluosilicic acid; to seal water out of oil wells during drillings.
Shipping regulations: (Rail) Green label. (Air) Nonflammable Gas label. Not acceptable on passenger planes.

silicon tetrahydride. See silane.

silicon tetraiodide SiI_4.
Properties: White crystals; m.p. 120°C; b.p. 290°C. Noncombustible.
Containers: Sealed "Pyrex" or quartz ampoules.
Purity: Up to 99.999%.
Hazard: Toxic by ingestion and inhalation. Irritant to tissue.

silicotungstic acid (12-tungstosilicic acid; silicowolframic acid) $H_4SiW_{12}O_{40} \cdot 5H_2O$.
Properties: White crystalline powder; exceptionally soluble in water and polar organic solvents; relatively insoluble in non-polar organic solvents. Strong acid; noncombustible.
Grades: Reagent; technical.
Uses: Catalyst for organic synthesis; reagents; additives to plating processes; as precipitants and inorganic ion-exchangers in atomic energy research.
See also sodium 12-tungstosilicate and tungstosilicates.

silicowolframic acid. See silicotungstic acid.

"Silicure" I-816.[74] Trademark for 6% iron octoate; used as a silicone catalyst.

"Silimite."[116] Trademark for a high magnesium dolomitic lime used for silica reduction in hot-process water-softening equipment.

silk. A natural fiber secreted as a continuous filament by the silkworm, Bombyx mori. Silk consists essentially of the protein fibroin and, in the raw state, is coated with a gum, which is usually removed before spinning. Combustible, but self-extinguishing. Sp. gr. 1.25. Elongation at rupture about 20%; tenacity 3 to 5 grams/denier.

sillimanite An aluminum silicate. A high heat-resisting material containing a maximum amount of mullite, developed from the alteration of andalusite during firing. This necessitates firing above 1550°C for the development of a suitable crystalline structure.
Uses: Spark plugs; chemical laboratory ware; pyrometer tubes; special porcelain shapes; furnace patch and refractories.

"Sil-O-Cel."[247] Trademark for a type of diatomaceous earth used in filtration and insulation.

siloxane (oxosilane). A straight-chain compound (analogous to paraffin hydrocarbons) consisting of silicon atoms single-bonded to oxygen and so arranged that each silicon atom is linked with four oxygen atoms.

```
   |     |     |
   O     O     O
   |     |     |
 —Si—O—Si—O—Si—O—
   |     |     |
   O     O     O
   |     |     |
```

In some types, hydrogen may replace two or more of the oxygens. Disiloxane and trisiloxane are examples. See also silicone.

"Silpowder."[31] Trademark for a commercially pure silver, metallic impurities 0.1% max. Particle size, 5 to 125 microns depending on grade.
Derivation: By precipitation from solutions of silver salts and by atomization.
Grades: 120, 130, 140, 150 and 160 (galvanic and chemical precipitates of variable size and density).
Uses: Powder metallurgy parts such as electrical contacts, batteries and brushes; pigments for conductive coatings and adhesives.

"Sil-Temp."[349] Trademark for a substantially pure fibrous silica for use in rocket and missile constructions and for high temperature insulation of motor components and similar applications. As a construction material it is used in laminate form impregnated and bonded with high-temperature resins.

"Sil-Trode."[407] Trademark for silicon bronze electrodes and filler rod for use in inert gas welding.

"Silvacel."[129] Trademark for a series of wood fibers.
Uses: Insulation; pulp molding; special papers and boards where high bulk and absorbent qualities are desirable; filter aid and filter media; treatment of oil well drilling muds.

"Silvacon."[129] Trademark for a series of products made from Douglas fir (Pseudotsuga taxifolia) bark.
Uses: Phenolic adhesive extender; molding compound component; conditioner for fertilizer, insecticides, etc.; burnout material in foundry sands; rubber sponge manufacture; other vinyl and rubber compounds; treatment of oil well drilling fluids; replacement for cork.

silver Ag Metallic element, atomic number 47, Group I B of the periodic table. Atomic weight 107.868. Valence 1; two stable isotopes.
Properties: Soft, ductile, lustrous white solid. Highest electrical and thermal conductivity of all metals. Resists oxidation, but tarnishes in air through reaction with atmospheric sulfur compounds. Sp. gr. 10.53; m.p. 961°C; b.p. 2212°C. Thermal conductivity 1.01 cal/cm/sec/°C. Absorbs oxygen strongly at melting point. Soluble in nitric acid, hot sulfuric acid, and alkali cyanide solutions; insoluble in water and alkalies. Noncombustible, except as powder.
Derivation: By-product of operations on copper, zinc, lead, or gold ores, but some smelters still operate on native silver. The recovery ranges from 166 ounces to a few thousandths of an ounce per ton. See Parkes process; Pattinson process. The chief silver ores are native silver, argentite (silver sulfide) and cerargyrite (silver chloride).
Forms available: Pure ("fine"), sterling (7.5% Cu), various alloys, plate; ingot, bullion, moss, sheet, wire, tubing, castings; powder; high purity (impurities less than 100 ppm); single crystals; whiskers.
Hazard: Tolerance (metal and soluble compounds): 0.01 mg per cubic meter of air. Moderately toxic when absorbed in circulatory system.
Uses: Manufacture of silver nitrate, silver bromide, photographic chemicals; lining vats and other equipment for chemical reaction vessels, water distillation, etc.; catalyst in manufacture of ethylene; mirrors; electric conductors, such as bus bars; batteries; silver plating; electronic equipment; sterilant; low-temperature brazing alloys; table cutlery, jewelry; dental, medical and scientific equipment; electrical contacts; bearing metal; magnet windings; dental amalgams. Colloidal silver is used as a nucleating agent in photography, and in medicine, often combined with protein. See "Argyrol."

silver acetate CH_3COOAg.
Properties: White crystals or powder; sp. gr. 3.26. Moderately soluble in hot water; soluble in nitric acid.
Hazard: Toxic. Tolerance, 0.01 mg per cubic meter of air.
Use: Medicine (external); laboratory reagent.

silver acetylide Ag_2C_2.
Properties: White powder. A salt of acetylene.
Derivation: Reaction of acetylene with aqueous solution of argentous salts.
Hazard: Severe explosion risk when shocked or heated.
Uses: Detonators.
Shipping regulations: (Rail, Air) Explosives, n.o.s., Not acceptable.

silver arsenite Ag_3AsO_3.
Properties: Fine, yellow powder. Sensitive to light. Soluble in acetic acid, ammonium hydroxide, and nitric acid; insoluble in alcohol and water. M.p. 150°C (decomposes).
Hazard: Highly toxic. Tolerance, 0.01 mg per cubic meter of air.
Use: Medicine.
Shipping regulations: (Rail, Air) Poison label.

silver arsphenamine (silver diaminodihydroxyarsenobenzene, sodium salt).
Properties: Brownish black powder (containing approximately 20% arsenic, 15% silver) for which the exact molecular formula has not been established. Unstable in air. Soluble in water.
Derivation: Action of silver salts on arsphenamine, converting the product to the disodium salt and precipitating with alcohol, acetone, or ether.
Hazard: Highly toxic. Tolerance, 0.01 mg per cubic meter of air.
Use: Medicine.
Shipping regulations: (Rail, Air) Arsenical compounds, n.o.s., Poison label.

silver bromate $AgBrO_3$.
Properties: White powder. Sensitive to light. Keep in amber bottle. Soluble in ammonium hydroxide; slightly soluble in water (hot). Decomposed by heat. Sp. gr. 5.2.

silver bromide AgBr.
Properties: Pale yellow crystals or powder, darkening on exposure to light, finally turning black. Sp. gr. 6.473 (25°C); m.p. 432°C; b.p. decomposes at 700°C. Soluble in potassium bromide, potassium cyanide and sodium thiosulfate solutions; very slightly soluble in ammonia water; insoluble in water. Light-sensitive.
Derivation: Silver nitrate is dissolved in water and a

solution of alkali bromide added slowly. The precipitated silver bromide is washed repeatedly with hot water; the operation must be carried on in a darkroom under a ruby red light.
Containers: Amber or black glass bottles.
Use: Photograph film and plates; photochromic glass; laboratory reagent.

silver carbonate Ag_2CO_3.
Properties: Yellow to yellowish gray powder. Contains 78% (approx.) silver. Light-sensitive. Soluble in ammonium hydroxide, nitric acid; insoluble in alcohol and water. Sp. gr. 6.077; decomposes at 218°C.
Uses: Laboratory reagent.

silver chloride AgCl.
Properties: White granular powder, which darkens on exposure to light, finally turning black. Exists in several modifications differing in conduct toward light and also in their solubility in various solvents. Soluble in ammonium hydroxide, concentrated sulfuric acid and sodium thiosulfate and potassium bromide solutions; very slightly soluble in water. Can be melted, cast and fabricated like a metal. Sp. gr. 5.56; m.p. 445°C; b.p. 1550°C.
Derivation: Silver nitrate solution is heated and hydrochloric acid or salt solution added. The whole is boiled, then filtered, all in the dark or under a ruby-red light.
Method of purification: Re-solution in ammonium hydroxide and precipitation by hydrochloric acid.
Grades: Technical; C.P.; single pure crystals.
Containers: Amber or black glass bottles.
Uses: Photography; photometry and optics; batteries; photochromic glass; silver plating; production of pure silver; medicine. Single crystals are used for infrared absorption cells and lens elements; laboratory reagent.

silver chromate Ag_2CrO_4.
Properties: Dark, brownish red, powder. Soluble in acids, ammonium hydroxide, potassium cyanide, solutions of alkali chromates; insoluble in water. Sp. gr. 5.625.
Use: Laboratory reagent.

silver cyanide AgCN.
Properties: White, odorless, tasteless powder which darkens on exposure to light. Soluble in ammonium hydroxide, dilute boiling nitric acid and potassium cyanide and sodium thiosulfate solutions; insoluble in water. Sp. gr. 3.95; decomposes at 320°C.
Derivation: By adding sodium or potassium cyanide to a solution of silver nitrate.
Containers: Amber or black glass bottles.
Hazard: Highly toxic by ingestion or inhalation. Tolerance, 0.01 mg per cubic meter of air.
Uses: Medicine; silver plating.
Shipping regulations: (Rail) Cyanides, Poison label. (Air) Poison label.

silver dichromate (silver bichromate) $Ag_2Cr_2O_7$.
Properties: Dark red, almost black, crystalline powder; sp. gr. 4.770; soluble in ammonium hydroxide and nitric acid; slightly soluble in water.
Hazard: Tolerance, 0.01 mg per cubic meter of air.

silver fluoride $AgF \cdot H_2O$.
Properties: Yellow, or brownish, crystalline masses. Very hygroscopic. Becomes dark on exposure to light. Light-sensitive. Soluble in water. Sp. gr. 5.852 (anhydrous); m.p. 435°C; b.p. ca. 1159°C.
Hazard: Toxic; strong irritant. Tolerance, 0.01 mg per cubic meter of air.
Use: Medicine.

silver iodide AgI.
Properties: Pale yellow, odorless, tasteless powder, darkening on exposure to light. Soluble in hydriodic acid, potassium iodide, potassium cyanide, ammonium hydroxide, sodium chloride and sodium thiosulfate solutions; insoluble in water. Light-sensitive. Sp. gr. 5.675; m.p. 556°C.
Derivation: Silver nitrate solution is heated, alkali iodide solution added and the precipitate washed with boiling water in the dark or under ruby red illumination.
Containers: Amber or black glass bottles.
Hazard: Moderately toxic.
Uses: Medicine; photography; cloud seeding for artificial rain-making; laboratory reagent.

"Silver-Lume" A,B.[288] Trademark for a bright silver electroplating process for use by silversmiths and electronics manufacturers. The materials used are silver cyanide, potassium cyanide, potassium carbonate, and addition agents.

silver mercury iodide. See mercuric silver iodide.

silver methylarsonate (methanearsonic acid, disilver salt) $CH_3AsO_3Ag_2$.
Derivation: Reaction of disodium methylarsonate with silver salts.
Hazard: Highly toxic. Tolerance, 0.01 mg per cubic meter of air.
Use: Algicides.
Shipping regulations: (Rail, Air) Arsenical compounds, n.o.s., Poison label.

silver nitrate $AgNO_3$.
Properties: Colorless, transparent, tabular, rhombic crystals, becoming gray or grayish-black on exposure to light in the presence of organic matter; odorless; bitter, caustic metallic taste; strong oxidizing agent and caustic. Soluble in cold water; more soluble in hot water, glycerol and hot alcohol; slightly soluble in either. Sp. gr. 4.328; m.p. 212°C; b.p. decomposes.
Derivation: Silver is dissolved in dilute nitric acid and the solution evaporated. The residue is heated to a dull red heat to decompose any copper nitrate, dissolved in water, filtered and recrystallized.
Grades: Technical; C.P.; U.S.P.
Containers: Amber or black glass bottles.
Hazard: Tolerance, 0.01 mg per cubic meter of air. Highly toxic and strong irritant to skin and tissue.
Uses: Photography; mother of pearl; reagent; silver plating; indelible ink; silver salts; glass manufacture; silvering mirrors; germicide (as a wall spray); weak solutions are used in hair dyeing and as antiseptic; fused form to cauterize wounds; laboratory reagent.
Shipping regulations: (Rail) Yellow label. (Air) Oxidizer label.

silver nitride Ag_3N.
Properties: Colorless, solid.
Derivation: Reaction of silver compounds with ammonia, with or without additives.
Hazard: Severe explosion hazard when shocked. It is unusually sensitive to mechanical action of any kind, and can explode even in water suspension.
Shipping regulations: (Rail, Air) Not listed. Consult authorities.

silver nitrite $AgNO_2$.
Properties: Small, yellow, or grayish yellow needles. Become gray on exposure to light. Contain 70% (approx.) silver. Decomposes by acids. Soluble in (hot) water; insoluble in alcohol. Decomposes at 140°C. Sp. gr. 4.453 (26°C).
Hazard: Tolerance, 0.01 mg per cubic meter of air.
Uses: Organic synthesis; standardizing potassium permanganate solutions; water analysis; analysis (testing for alcohols).

silver orthophosphate. See silver phosphate.

silver oxide (argentous oxide) Ag_2O.
Properties: Dark brown or black odorless powder; metallic taste. Soluble in ammonium hydroxide, potassium cyanide solution, nitric acid and sodium thiosulfate solution; slightly soluble in water; insoluble in alcohol. Sp. gr. 7.14 (16°C); m.p. decomposes when heated above 300°C. Strong oxidizing agent.
Derivation: Silver nitrate and alkali hydroxide solutions are mixed, the precipitate filtered and washed.
Grades: Technical; up to 99.6% pure, particle size 2–3 microns.
Containers: Jars; 1-, 5-gal cans.
Hazard: Fire and explosion risk in contact with organic materials or ammonia. Moderately toxic.
Uses: Medicine; polishing glass; coloring glass yellow; catalyst; purifying drinking water; battery plates; laboratory reagent.

silver oxide battery. See zinc-silver oxide battery.

silver permanganate $AgMnO_4$.
Properties: Violet, crystalline powder. Sp. gr. 4.27 (25°C); m.p., decomposes. Contains 47.5% (approx.) silver. Decomposed by alcohol. Light-sensitive; use dark-colored bottles.
Grade: Technical.
Hazard: Dangerous explosion risk; may detonate if shocked or heated.
Uses: Gas masks; medicine.
Shipping regulations: Permanganates, n.o.s., (Rail) Yellow label. (Air) Oxidizer label.

silver peroxide (argentic oxide) Ag_2O_2.
Properties: Grayish powder; sp. gr. 7.44 (25°C); decomposes above 100°C; insoluble in water; soluble in sulfuric acid, nitric acid, and ammonium hydroxide.
Containers: 80-oz jars in metal outer-containers.
Hazard: Dangerous fire and explosion risk; strong oxidizing agent. Keep out of contact with organic materials.
Shipping regulations: (Rail, Air) Not listed. Consult authorities.

silver phosphate (silver orthophosphate) Ag_3PO_4.
Properties: Yellow powder; turns brown when heated or on exposure to light. Soluble in acids, potassium cyanide solution, ammonium hydroxide, ammonium carbonate, and acetic acid; very slightly soluble in water. Sp. gr. 6.370 (25°C); m.p. 849°C.
Derivation: Interaction of silver nitrate and sodium phosphate.
Containers: Amber or black glass bottles.
Hazard: Moderately toxic.
Uses: Photographic emulsions; catalyst; pharmaceuticals.

silver picrate $C_6H_2O(N_2)_3Ag \cdot H_2O$.
Properties: Yellow crystals containing 30% silver; soluble in water; slightly soluble in alcohol and acetone; insoluble in ether and chloroform.
Hazard: Severe explosion risk.
Use: Medicine.
Shipping regulations: (Air) Explosives, n.o.s., Not acceptable. (Rail) Consult regulations.

silver potassium cyanide (potassium cyanoargentate; potassium argentocyanide) $KAg(CN)_2$.
Properties: White crystals; sp. gr. 2.36 (25°C); sensitive to light. Soluble in water and alcohol; insoluble in acids.
Derivation: By adding silver chloride to a solution of potassium cyanide.
Containers: Glass bottles.
Hazard: Highly toxic. Tolerance, 0.01 mg per cubic meter in air.
Uses: Silver plating; bactericide; antiseptic.

silver sodium chloride (sodium silver chloride) $AgCl \cdot NaCl$.
Properties: Hard, white crystals. Decomposed by water. Soluble in solution of sodium chloride (conc.).

silver sodium thiosulfate (sodium silver thiosulfate) $Ag_2S_2O_3 \cdot 2Na_2S_2O_3 \cdot 2H_2O$.
Properties: White or gray, crystalline powder. Sweet taste; soluble in water.

silver, sterling. See sterling silver.

silver sulfate (silver sulfate, normal) Ag_2SO_4.
Properties: Small, colorless, lustrous crystals or crystalline powder. Contains 69% (approx.) silver. Turns gray on exposure to light. Soluble in ammonium hydroxide, nitric acid, sulfuric acid, (hot) water; insoluble in alcohol. Sp. gr. 5.45 (29°C); b.p. 1085°C (decomposes); m.p. 652°C.
Grades: Technical; C.P.
Hazard: Tolerance, 0.01 mg per cubic meter of air.
Use: Laboratory reagent.

silver sulfate, normal. See silver sulfate.

silver sulfide Ag_2S.
Properties: Grayish black powder. Soluble in concentrated sulfuric and nitric acids; insoluble in water. Sp. gr. 7.32 (25°C); b.p. decomposes; m.p. 825°C.
Derivation: By passing hydrogen sulfide gas into silver nitrate solution, washing and drying.
Uses: Inlaying in niello metal-work; ceramics.

silvex. Generic name for 2-(2,4,5-trichlorophenoxy)-propionic acid (fenoprop) $Cl_3C_6H_2OCH(CH_3)COOH$.
Properties: Solid; m.p. 180.4–181.6°C. Slightly soluble in water; freely soluble in acetone, methyl alcohol. Combustible.
Hazard: May be toxic.
Uses: Herbicide and plant growth regulator.
Also available in form of esters, as 2,4,5-trichlorophenoxypropionic acid, 2-ethylhexyl ester (see "Silvi-Rhap").

silvichemical. A chemical derived from wood, e.g., lignins, lignosulfonates (from spent sulfite liquor); vanillin; yeast (from fermentation of wood sugars); tall oil; sulfate turpentine; bark extracts; phenolic materials.

silvicide. A nonselective herbicide used to kill or defoliate bushes and small trees, e.g., ammonium sulfamate.

"Silvi-Rhap."[266] Trademark for 2-ethylhexyl ester of 2-(2,4,5-trichlorophenoxy)propionic acid. Used as a herbicide.
Hazard: May be toxic.

simazine. Generic name for 2-chloro-4,6-bis(ethylamino)-s-triazine $ClC_3N_3(NHC_2H_5)_2$.
Properties: White solid; m.p. 225°C. Insoluble in water; slightly soluble in organic solvents. Combustible.
Hazard: Moderately toxic by ingestion and inhalation; irritant to mucous membranes.
Use: Herbicide.

"Simconex."[160] Trademark for silicone-base insulation used in power cables.

Simons process. An electrochemical fluorination process which makes fluorocarbons by passing an electric current through a mixture of the organic starting compound and liquid anhydrous hydrogen fluoride. The products are hydrogen and the desired fluorocarbon.

simple distillation. Distillation in which no appreciable rectification of the vapor occurs, i.e., the vapor formed from the liquid in the still is completely condensed in the distillate receiver and does not undergo change in composition due to partial condensation or contact with previously condensed vapor.

"Singoserp."[305] Trademark for syrosingopine.

sintering. The agglomeration of metal or earthy powders at temperatures below the melting point. Occurs in both powder metallurgy and ceramic firing. While heat and pressure are essential, decrease in surface area is the critical factor. (see Rittinger's law). Sintering increases strength, conductivity, and density.

"Sipenol."[542] Trademark for ethoxylated fatty and short-chain amines. Used in cosmetics, textile industry, metal cleaning, agricultural emulsifiers, chemical and pharmaceutical intermediates.

"Sipex."[542] Trademark for industrial grade alcohol sulfates and alcohol ethoxylate sulfates. Used in detergents, emulsion polymerization, and textile industry. Available are principally ammonium, magnesium, sodium and triethanolamine lauryl sulfates, sodium tridecyl sulfate, sodium 2-ethylhexyl sulfate, and sodium lauryl ethoxylate sulfate. See also "Sipon."

"Sipomer."[542] Trademark for a group of speciality monomers; used in polymerization. They include dimethylaminoethyl methacrylate, hydroxyethyl methacrylate, dimethyl and diethyl maleates, and allyl glycolate.

"Sipon."[542] Trademark for a cosmetic grade of fatty alcohol sulfates and fatty alcohol ethoxylate sulfates. They include approximately the same materials as the "Sipex" grade, and also diethanolamine lauryl sulfate, sodium cetyl sulfate, and ammonium lauryl ethoxylate sulfate.

"Siponate."[542] Trademark for purified alkylarylsulfonates, including sodium dodecyl benzene sulfonate (branched or linear), and sodium lauroyl monoglyceride sulfate.

"Siponic."[542] Trademark for fatty alcohol ethoxylates. Used in detergents, cosmetics, finishes, and coatings. Available are ethoxylated lauryl alcohol, oleyl alcohol, and stearyl-cetyl alcohol.

SIPP. Abbreviation for sodium iron pyrophosphate.

"Sirlene."[233] Trademark for feed grade of propylene glycol.

sisal.
Properties: Hard, strong, light yellow to reddish fibers obtained from the leaves of Agave sisilana. Combustible; not self-extinguishing. Strength, 4.5 grams per denier; fineness ranges from 300 to 500 denier.
Source: Africa; Java; Haiti; Bahama.
Hazard: Moderate, by inhalation. Dust is flammable. May ignite spontaneously when wet.
Uses: Twine; rope; sacking; upholstery; life preservers.

"Sitol."[28] Trademark for an oxidizing agent in flake form used in discharge printing.

beta-**sitosterol** $C_{29}H_{50}O$. 22,23-Dihydrostigmasterol.
Properties: Waxy white solid; almost odorless and tasteless; insoluble in water, soluble in benzene, chloroform, carbon disulfide and ether. Can be crystallized from ether as anhydrous needles, or from aqueous alcohol as leaflets with one molecule of water.
Derivation: Soybeans.
Uses: Biochemical research. See also cholesterol.

"Sixide."[319] Trademark for insecticidal preparations containing benzene hexachloride.

size oil. See throwing oil.

sizing compound. (1) A material such as starch, gelatin, casein, gums, oils, waxes, asphalt emulsions, silicones, rosin, and water-soluble polymers applied to yarns, fabrics, paper, leather and other products to improve or increase their stiffness, strength, smoothness or weight. (2) A material used to modify the cooked starch solutions applied to warp ends prior to weaving. See also slashing compound.

"Skamex."[28] Trademark for a fluorocarbon plastic used as a metallurgical additive for the beneficiation of molten metal. Cakes and slugs of it are specifically designed for immersion in hot metal for removal of excessive quantities of dissolved hydrogen. Other forms are added to molds during casting to create a protective atmosphere against the effects of absorbed oxygen. Can be used for ferrous and nonferrous metals.
Containers: Powder, pellets, cakes and slugs: 50-lb drums. Mold wash: 100-lb 44% concentrate, 25-gal drums.

skatole (3-methylindole) C_9H_9N.
Properties: White crystalline substance, browning upon aging; fecal odor. M.p. 93° to 95°C; b.p. 265°C. Soluble in hot water, alcohol, benzene. Gives violet color in potassium ferrocyanide and sulfuric acid.
Derivation: Feces; African civet cat; Celtis reticulosa, a Javanese tree.
Use: Perfumery (fixative); artificial civet.

"Skellysolves."[409] Trademark for straight-run aliphatic naphthas having various boiling ranges, specific gravities, evaporation rates and other properties, which make them suitable for a number of industrial uses.
Hazard: Flammable; dangerous fire risk.

Skraup synthesis. Synthesis of quinoline or its derivatives by heating aniline or an aniline derivative, glycerol and nitrobenzene in the presence of sulfuric acid.

"Skydrol."[58] Trademark for a line of fire-resistant aircraft hydraulic fluids.

500-A. Used for hydraulic systems in turbo jet and turbo prop aircraft, which must operate at −65°F.
7000. Used in aircraft cabin superchargers, expansion turbines for air-conditioning systems and the aircraft hydraulic system itself.

slack. (1) Descriptive of a soft paraffin wax resulting from incomplete pressing of the settlings from the petroleum distillate. Though it has some applications in this form, it is actually an intermediate product between the liquid distillate and the scale wax made by expressing more of the oil. See also scale (2).
(2) Specifically, to react calcium oxide (lime) with water to form calcium hydroxide (slacked or hydrated lime); the reaction is $CaO + H_2O \longrightarrow Ca(OH)_2 +$ heat. The alternate spelling "slake" has the same meaning.

slaframine (1-acetoxy-8-aminooctahydroindolizine). An alkaloid derived from a fungus that infests clover. It is under research development for use as an agent in retarding cystic fibrosis.

slag. The fused agglomerate which separates in metal smelting and floats on the surface of the molten metal. Formed by combination of flux with gangue of ore, ash of fuel, and perhaps furnace lining. The slag is often the medium by means of which impurities may be separated from metal.
Uses: Railroad ballast; highway construction; cement and concrete aggregate; raw material for Portland cement; raw material for glass fibers (see mineral wool); fertilizer.

slake. See slack.

slaked lime. See calcium hydroxide; lime, hydrated.

slashing compound. A textile sizing material applied to cotton or rayon warp ends by a special machine (slasher).

slate. A fine-grained metamorphic rock which cleaves into thin slabs or sheets. Color usually gray to black, sometimes green, yellow, brown or red. Slates are composed of micas, chlorite, quartz, hematite, clays, and other minerals.
Occurrence: Pennsylvania, Vermont, Maine, Virginia, California, Colorado; Europe.
Uses: Roofing; blackboards; (as powder) filler in paint, rubber; abrasive.

slate black. See mineral black.

slate flour. Finely divided slate used as a filler and dusting agent in rubber, plastics, etc.

slate, green. See slate flour.

"S.L." Chemicals.[329] (Standard Luminescent). Trademark for chemicals specially developed and standardized to meet the exacting requirements of phosphor manufacture.

slimicide. A chemical which is toxic to the types of bacteria and fungi characteristic of aqueous slimes. Examples are chlorine and its compounds, organomercurial compounds, phenols, and related substances. Used largely in paper mills, and to some extent in textile and leather industries.

slip clay. A type of clay containing such a high percentage of fluxing impurities and of such a texture that it melts at a relatively low temperature to a greenish or brown glass, thus forming a natural glaze. It must be fine-grained, free from lumps or concretions, show a low air-shrinkage and mature in burning at as little above 1300°F as possible.

"Slip-Eze."[430] Trademark for a specially prepared fatty acid amide. Used as a slip additive and as an internal mold-release additive to polyethylene.

"Slipicone."[149] Trademark for fluid silicone compositions to prevent adhesion of materials to one another. Used on food-processing and packaging equipment.

sludge. A soft mud, slush, or mire; for example, the solid product of a filtration process before drying (filter cake). See also sewage sludge.

sludge acid. Waste or spent sulfuric acid, usually that from refining petroleum oils or crude benzenes.
Hazard: Highly toxic; strong irritant.
Shipping regulations: (Rail) White label. (Air) Corrosive label. Not acceptable on passenger planes.

"Sludge Conditioner."[108] Trademark for a series of polyelectrolytes.
Containers: Bags; drums.
Uses: Conditions sludge for dewatering and settling in municipal sewage treatment plants.

slurry. A thin watery suspension; for example, the feed to a filter press or to a fourdrinier machine; also a stream of pulverized metal ore. A special use of this term refers to a type of explosives called "slurry blasting agents" based on gelatinized aqueous ammonium nitrate, sensitized with various other explosives.

slushing compound. A nondrying oil, grease or similar compound which, when coated over a metal, affords temporary protection against corrosion.

Sm Symbol for samarium.

"SM-30."[241] Trademark for a catalyst designed for the production of maximum yields of gasoline and light cycle oil. Typical analysis: magnesia, 27.5% and silica, 68.8%.

smalt.
Properties: Blue powder.
Derivation: A potash-cobalt glass made by fusing pure sand and potash with cobalt oxide, grinding and powdering.
Containers: Wooden kegs; multiwall paper sacks.
Uses: Paint pigments; ceramic industries (pigment); coloring glass; bluing paper, starch and textiles; coloring rubber.

"SMA" Resins.[566] Trademark for a family of short-chain copolymers of styrene and maleic anhydride that offer typical anhydride functionality in a polymeric material.
Uses: Protective colloid in emulsion polymerization, a leveling resin in floor polishes, a dispersant in aqueous pigment systems, and a cross-linking agent for epoxy resins.

smectic. A phase of matter intermediate between solid and liquid, in which flow is not the normal liquid type. It is studied by polarized light, x-ray diffraction, and nuclear magnetic resonance. See also liquid crystal.

smelting. Heat treatment of an ore to separate the metallic portion, with subsequent reduction. See also roasting.

"Smentox."[236] Trademark for a chemical compound for reconditioning cement-contaminated drilling mud, or for preventing mud from becoming cement-contaminated, since contact with cement flocculates untreated drilling mud, rendering it unfit for use.

"Smite."[28] Trademark for emulsifiable insecticide containing 12.5% methoxychlor and 12.5% malathion.
Containers: 1-pt, 1-qt bottles.

"Smithol 25."[403] Trademark for a synthetic fatty alcohol ester of fatty acids to substitute for 45°NW sperm oil. May replace sperm oil, lard oil, olive oil, and other natural fatty oils for industrial uses. The lower cloud point and pour point and the fact that it does not have the objectionable odor of some natural oils makes it attractive for many applications where natural oils have limitations.

smog. A coined word denoting a persistent combination of smoke and fog occurring under appropriate meteorological conditions in large metropolitan or heavy industrial areas. The discomfort and danger of smog is increased by the action of sunlight on the combustion products in the air, especially sulfur dioxide, nitric oxide and exhaust gases (photochemical smog). Strongly irritating and even toxic substances may be present, e.g., peroxybenzoyl nitrate (q.v.). Fatalities have resulted from particularly severe photochemical smogs. See also air pollution.

smoke. A colloidal or microscopic dispersion of a solid in a gas; an aerosol.

(1) Coal smoke: A suspension of carbon particles in hydrocarbon gases or in air, generated by combustion. The larger particles can be removed by electrostatic precipitation in the stack (see Cottrell). Dark color, nauseating odor. See also smog; air pollution.

(2) Wood smoke: Light-colored particles of cellulose ash; pleasant aromatic odor. Smoke from special kinds of wood (e.g., hickory, maple) is used to cure ham, fish, etc., also to preserve crude rubber.

(3) Chemical smoke (q.v.): Generated by chemical means for military purposes (concealment, signaling, etc.).

(4) Metallic smoke (fume): An emanation from heated metals or metallic ores, the particles being of specific geometric shapes. Such smoke is particularly damaging to vegetation in the neighborhood of zinc and tin smelters.

(5) Cigarette smoke. There is rather conclusive evidence that the tars occurring in cigarette smoke can lead to lung cancer; chief factors are age of individual at initiation of smoking, extent of inhalation, and number smoked per day. Polonium, a radioactive element, is known to occur in cigarette smoke; more than 100 compounds have been identified including nicotine, cresol, pyridene, and benzopyrene, the latter a known carcinogen. See also cigarette tar.

smokeless powder. Nitrocellulose containing about 13.1% nitrogen, produced by blending material of somewhat lower (12.6%) and slightly higher (13.2%) nitrogen content, converting to a dough with alcohol-ether mixture, extruding, cutting and drying to a hard horny product. Small amounts of stabilizers, (amines) and plasticizers are usually present, as well as various modifying agents (nitrotoluene, nitroglycerin salts).
Hazard: A low explosive; dangerous fire and explosion risk when exposed to flame or impact.
Uses: Sports ammunition; military purposes.
Shipping regulations: (Air) Not acceptable. (Rail) Consult regulations.

"Smoothawal."[481] Trademark for a vinyl latex spackling paste.

smudge oil. An oil burned in fruit orchards to prevent frost from injuring the trees. No. 3 fuel oil is typical of oils used.

Sn Symbol for tin (from Latin stannum).

snake venom. There are two functional types: (1) those that bring about blood coagulation either by direct action on fibrogen or by converting prothrombin to thrombin; (2) neurotoxins that act on the central nervous system, e.g., by inactivation of acetylcholine. Rattlesnake and moccasin venom are examples of (1) and cobra venom of (2). The enzymes of snake venoms are thought to be the actual toxic principles. Solutions of cobra venom have found use in treatment of arthritis and cancer. The chemistry and pharmacological properties of these poisons are not well understood.
Note: A person bitten by a poisonous snake should be carried, *not walked*, to a hospital. *No alcohol* should be administered.

SNG. Abbreviation for synthetic natural gas (q.v.).

"Snodotte."[221] Trademark for a marine oil fatty acid; predominant chain length about 30% C_{16}.

snow, artificial. See artificial snow.

"Snowfine."[292] Trademark for a powdered form of sodium sesquicarbonate. See "Snowflake."

"Snowflake."[292] Trademark for sodium sesquicarbonate ($Na_2CO_3 \cdot NaHCO_3 \cdot 2H_2O$); small white sparkling needle-like crystals of uniform size which are free flowing and non-caking under normal conditions.
Uses: Base for household and industrial cleansers where mild alkaline action is desired, and as a base for both foaming and non-foaming bath salts.
Containers: 100-lb paper bags; 325-lb fiber drums; bulk.

SNTA. Abbreviation for sodium nitrilotriacetate.

soap. (1) The water-soluble reaction product of a fatty acid ester and an alkali (usually sodium hydroxide), with glycerol as by-product. For the reaction, see saponification. A soap is actually a specific type of salt (q.v.), the hydrogen of the fatty acid being replaced by a metal, which in common soaps is usually sodium. Soap lowers the surface tension of water and thus permits emulsification of fat-bearing soil particles. A typical commercial cleansing soap is made by reacting sodium hydroxide with a fatty acid. The lower the hydrogen content of the acid, the thinner the soap. The by-product of the reaction is glycerol (see saponification). Many different carboxyl-containing substances are used, including rosin, tall oil, vegetable and animal oils and fats (stearic, palmitic, and oleic acids). Olive oil is used for Castile soap; transparent soaps are made from decolorized fats. The specific gravity of soaps is slightly more than 1.0; inclusion of air gives a floating product. Water solutions of sodium soaps in bar, chip, or powder form are universally used as mild emulsifying detergents for washing textiles, skin, paint, etc. Liquid green soap is made with potassium hydroxide and a vegetable oil. For further information refer to Soap and Detergent Assn., 485 Madison Ave., New York.

(2) Heavy-metal soaps (loosely called metallic soaps) are those in which metals heavier than sodium are used (aluminum, calcium, cobalt, lead, zinc). These soaps are not water-soluble; specific types are

used in lubricating greases, gel thickeners, and in paints as driers and flatting agents. Napalm is an aluminum soap. See also saponification; detergent; drier.

soap builder. Any material mixed with soap to improve the cleaning properties, modify the alkali content, or impart water-softening characteristics. The commonest are some of the sodium phosphates, and rosin.

soap, heavy metal. See soap (2).

soap, soft. See green soap; soap.

soapstone. See talc.

Society of Chemical Industry. This society was founded in London in 1881 to advance applied chemistry in all its branches. The American Section was established in 1894. In 1906, on the anniversary of the "birth of the coal-tar industry," it founded the famous Perkin Medal, which has since been awarded annually for outstanding achievement in chemistry in the United States. It has also offered the Chemical Industry Medal since 1932, given for conspicuous service to applied chemistry. Perkin Medalists include L. H. Baekeland, F. G. Cottrell, Irving Langmuir, Herbert Dow, A. D. Little, R. R. Williams, Glenn T. Seaborg, and Roger Adams.

Society of Plastics Engineers (SPE). An incorporated technical organization devoted primarily to the application of sound engineering principles to the manufacture and use of plastics, this rapidly growing organization (founded in 1942) has contributed much to the correct evaluation of these versatile materials in many fields. It publishes a monthly journal and sponsors a series of technical books on both theoretical and applied aspects of plastics technology. Its headquarters is at Greenwich, Conn.

Society of the Plastics Industry (SPI). An incorporated technical organization serving the needs of the entire plastics industry in the United States. It establishes standards for the properties and selection of materials and for product design and engineering. Its two major publications are the "Plastics Engineering Handbook" (4th Edition) and the "Reinforced Plastics Handbook." With it are associated the Plastics Pipe Institute and the Reinforced Plastics/Composites Institute. Its offices are at 250 Park Ave., New York.

SOCMA. Abbreviation for Synthetic Organic Chemical Manufacturers Association.

soda. Any one of the forms of sodium carbonate (q.v.); also used loosely as equivalent to the word sodium in compounds.

soda alum. See aluminum sodium sulfate.

soda ash (soda, calcined; sodium carbonate, anhydrous). Na_2CO_3. The crude sodium carbonate of commerce. Tenth highest-volume chemical produced in the U.S. (1975).

Properties: Grayish white powder or lumps containing up to 99% sodium carbonate. Soluble in water; insoluble in alcohol. Noncombustible; nontoxic.

Derivation: For nearly a century most of the soda ash produced in the U.S. has been made synthetically by the ammonia soda (Solvay) process (q.v.). Largely because of high energy costs and strict pollution controls, most of the synthetic production is being abandoned in favor of the natural product obtained from deposits in Utah, California, Wyoming, etc. One of the largest producing sites of the natural product is in Green River, Wyoming.

Impurities: Sodium chloride, sodium sulfate, calcium carbonate and magnesium carbonate, sodium bicarbonate.

Grades: Dense 58%; light 58%; extra light; natural; refined.

Containers: Bags; barrels; drums; bulk.

Uses: Glass manufacture; chemicals; pulp and paper manufacture; sodium compounds; soaps and detergents; water treatment; aluminum production; textile processing; cleaning preparations; petroleum refining; sealing ponds from leakage (sodium ions bind to clay particles which swell to seal leaks).

soda, baking. See sodium bicarbonate.

soda, calcined. See soda ash.

soda, caustic. See sodium hydroxide.

soda, crystals. See sodium carbonate monohydrate.

soda lime. A mixture of calcium oxide with sodium or potassium hydroxide intended for the absorption of carbon dioxide gas and water vapor.

Properties: White or grayish white granules unless colored by a specified indicator. Must be kept in air-tight containers.

Grades: Technical; reagent. Usually % moisture and mesh size are stated.

Hazard: Toxic by ingestion and inhalation; strong irritant to tissue.

Uses: Drying agent and carbon dioxide absorbent; laboratory reagent.

Shipping regulations: (Air) Corrosive label.

soda lime glass. See glass.

sodamide. See sodium amide.

soda, modified (neutral soda). A combination of soda ash and bicarbonate of soda in definite proportions for purposes where an alkali is needed ranging in causticity between bicarbonate of soda and soda ash. White crystalline powders, water-soluble and possessing valuable cleansing and purifying properties. Prepared in various strengths.

Containers: See soda ash.

Uses: Washing powders; laundering; wool scouring powders; bottle cleansers; textile cleaners; mild detergents.

soda monohydrate. See sodium carbonate, monohydrate.

soda, natural. See trona.

soda niter. See sodium nitrate.

"Sodaphos."[55] Trademark for glassy sodium tetraphosphate (q.v.).

soda pulp. See pulp, paper.

soda, washing. See sal soda.

alpha-**sodio-sodium acetate** (sodium alpha-sodio-acetate) $NaCH_2COONa$.

Properties: Free-flowing powder; stable in dry air; decomposes slowly in moist air. Decomposes 280°C without melting. Insoluble in ethers and hydrocarbons; reacts mildly with water.

Grade: 80–85% pure. Impurities are sodium acetate, sodium amide and sodium hydroxide.

Hazard: Irritant to skin and mucous membranes. Toxic by inhalation.

Uses: Organic intermediate; drying agent for organic solvents.

"Sodite."[253] Trademark for an insecticide containing arsenic trioxide.

Hazard: Highly toxic.

sodium (natrium). Na Metallic element; atomic number 11; group IA of Periodic Table. Atomic weight 22.9898; valence 1; no stable isotopes, but several radioactive forms. Extremely reactive.
Properties: Soft, silver-white solid, oxidizing rapidly in air; wax-like at room temperature, brittle at low temperatures. Store in air-tight containers or in naphtha or similar liquid that does not contain water or free oxygen. Sp. gr. (25°C) 0.9674; m.p. 97.6°C; b.p. 892°C. Decomposes water on contact with evolution of hydrogen to form sodium hydroxide; insoluble in benzene, kerosine, and naphtha. Has excellent electrical conductivity.
Derivation: Electrolysis of a fused mixture of sodium chloride and calcium chloride in a Downs cell; a Castner cell may be used, but is less efficient.
Method of purification: Distillation.
Grades: Commercial; technical; brick, amalgam; coated powders; dispersions (see sodium dispersion); reactor (99.99% pure).
Hazard: Strong, caustic irritant to tissue. Severe fire risk in contact with water in any form. Ignites spontaneously in dry air when heated. To extinguish fires use *dry* soda ash, salt, or lime. Safety data sheet available from Manufacturing Chemists Assn., Washington, D.C.
Uses: Tetraethyl and tetramethyl lead; titanium reduction; sodium peroxide; sodium hydride; polymerization catalyst for synthetic rubber; laboratory reagent; coolant in nuclear reactors; electric power cable (encased in polyethylene); non-glare lighting for highways; radioactive forms in tracer studies and medicine.
Shipping regulations: (Rail) Yellow label. (Air) Flammable solid label. Not accepted on passenger planes.

sodium abietate (rosin soap; sodium resinate) $C_{19}H_{29}COONa$.
Properties: White powder. Soluble in water. Combustible.
Derivation: By leaching rosin with sodium hydroxide solution.
Uses: Medicine; soap making; paper coating.

sodium acetate (a) $NaC_2H_3O_2$; (b) $NaC_2H_3O_2 \cdot 3H_2O$.
Properties: Colorless, odorless crystals; efflorescent; soluble in water; slightly soluble in alcohol; soluble in ether. (a) Sp. gr. 1.528; m.p. 324°C; (b) sp. gr. 1.45; m.p. 58°C. Combustible. Low toxicity. Autoignition temp. 1125°F.
Grades: Highest purity; pure fused; C.P.; N.F.; technical; F.C.C.
Containers: Bags; drums.
Uses: Dye and color intermediate; pharmaceuticals; cinnamic acid; soaps; photography; purification of glucose; meat preservation; medicine; electroplating; tanning; dehydrating agent; buffer in foods; laboratory reagent.

sodium acetazolamide $C_4H_5N_4NaO_3S_2$.
Properties: Soluble in water; solution (1 in 10) has a pH range 9.0–10.0.
Grade: U.S.P.
Use: Medicine.

sodium N-acetoacetyl-para-sulfanilate $CH_3COCH_2CONHC_6H_4SO_3Na$.
Properties: Available in the form of a red-colored 40% aqueous solution. Sp. gr. 1.203.

sodium acetone bisulfite (acetone-sodium bisulfite) $(CH_3)_2CONaHSO_3$.
Properties: Crystalline material; soluble in water; decomposed by acids; slightly soluble in alcohol. Combustible.
Derivation: Interaction of sodium bisulfite and acetone.
Uses: Chemical (pure acetone); photography; textile (dyeing and printing).

sodium acetylformate. See sodium pyruvate.

sodium acid carbonate. See sodium bicarbonate.

sodium acid methanearsonate. See sodium methanearsonate.

sodium acid phosphate. See sodium phosphate, monobasic.

sodium acid pyrophosphate. See sodium pyrophosphate, acid.

sodium acid sulfate. See sodium bisulfate.

sodium acid sulfite. See sodium bisulfite.

sodium acid tartrate. See sodium bitartrate.

sodium alginate (sodium polymannuronate) $(C_6H_7O_6Na)$.
Properties: Colorless or slightly yellow solid occurring in filamentous, granular, and powdered forms. Forms a viscous colloidal solution with water; insoluble in alcohol, ether, and chloroform. Nontoxic; combustible.
Derivation: Extracted from brown seaweeds (see alginic acid).
Grades: N.F.; F.C.C.; technical.
Uses: Thickeners, stabilizers and emulsifiers in foods, especially ice cream; boiler compounds; medicine; experimental ocean-floor covering; textile printing; cement compositions; paper coating; food and pharmaceutical preparations; water-base paints.

sodium alkane sulfonate (SAS). RSO_3Na. The sodium salt of an alkane sulfonic acid of linear paraffins having chain lengths of from 14 to 18 carbon atoms.
Preparation: By reaction of paraffins with sulfur dioxide and oxygen in the presence of gamma radiation from a cobalt-60 source.
Use: Biodegradable detergent intermediate.

sodium alum. See aluminum sodium sulfate.

sodium aluminate $Na_2Al_2O_4$ or $NaAlO_2$.
Properties: White powder. Soluble in water; insoluble in alcohol; aqueous solution strongly alkaline. M.p. 1800°C.
Derivation: By heating bauxite with sodium carbonate and extracting the sodium aluminate with water.
Grades: Technical; reagent; also 27° Bé solution.
Containers: Bags; multiwall paper sacks; drums.
Hazard: (solution) Toxic; strong irritant to tissue.
Uses: Mordant; zeolites; water purification; sizing paper; manufacture of milk-glass, soap and cleaning compounds.
Shipping regulations (solution): (Rail) White label. (Air) Corrosive label.

sodium aluminosilicate (sodium silicoaluminate). A series of hydrated sodium aluminum silicates having a $Na_2O:Al_2O_3:SiO_2$ mole ratio of approximately 1:1:13.2.
Properties: Fine, white, amorphous powder or beads. Odorless and tasteless; insoluble in water and in alcohol and other organic solvents, but at 80 to 100°C is partially soluble in strong acids and solutions of

alkali hydroxides; pH of a 20% slurry is between 6.5 and 10.5. Low toxicity. Noncombustible.
Grades: Technical; F.C.C.
Uses: Anti-caking agent in food preparations (up to 2%).

sodium aluminum hydride $NaAlH_4$.
Properties: White crystalline material; sp. gr. 1.24. Stable in dry air at room temperature but very sensitive to moisture. Begins to melt at 183°C with decomposition to evolve hydrogen. Soluble in tetrahydrofuran, dimethyl "Cellosolve."
Derivation: By reaction of aluminum chloride with sodium hydride.
Containers: Glass bottles.
Hazard: Severe fire and explosion risk in contact with oxidizing agents and water; forms caustic and irritant compounds.
Use: Reducing agent similar to lithium aluminum hydride.
Shipping regulations: (Rail) Yellow label. (Air) Flammable Solid label. Not acceptable on passenger planes.

sodium aluminum phosphate (sodium aluminum phosphate, acidic) $NaAl_3H_{14}(PO_4)_8 \cdot 4H_2O$ or $Na_3Al_2H_{15}(PO_4)_8$.
Properties: White odorless powder; insoluble in water; soluble in hydrochloric acid. Low toxicity.
Grades: Technical; F.C.C.
Use: Food additive (baked products).

sodium aluminum silicofluoride $Na_5Al(SiF_6)_4$. Sodium aluminum fluosilicate.
Properties: White powder; somewhat soluble in cold water; corrosive to galvanized iron.
Hazard: Avoid prolonged skin contact. Tolerance (as F), 2.5 mg per cubic meter of air.
Uses: Moth proofing and insecticides; also to obtain acid medium in dyebath.

sodium aluminum sulfate. See aluminum sodium sulfate.

sodium amalgam Na_xHg_x.
Properties: A silver-white, porous, crystalline mass, containing 2–10% of metallic sodium. Decomposes water.
Derivation: Mercury is heated to about 200°C and sodium, in small pieces, added slowly. Also formed at one stage of process for making chlorine and sodium hydroxide by mercury cell process.
Grades: 2, 3, 4, 5, 6, 7, 8, 9, 10, and 20%.
Containers: Glass bottles.
Hazard: Flammable, dangerous fire risk. Toxic.
Uses: Preparation of hydrogen; reduction of metal halogen compounds and organic compounds; reagent in analytical chemistry.
Shipping regulations: (Rail) Flammable solids, n.o.s., Yellow label. (Air) Flammable Solid label.

sodium amide (sodamide) $NaNH_2$.
Properties: White crystalline powder with ammonia odor. Decomposes in water and hot alcohol. M.p. 210°C; b.p. 400°C.
Derivation: Dry ammonia gas is passed over metallic sodium at 350°C.
Hazard: Flammable, dangerous fire risk.
Uses: Manufacture of sodium cyanide; organic synthesis; laboratory reagent.
Shipping regulations: (Rail) Yellow label. (Air) Flammable Solid label. Not acceptable on passenger planes.

sodium para-**aminobenzoate** (PABA sodium) $NH_2C_6H_4COONa$. Sodium salt of para-aminobenzoic acid.
Properties: Crystals; soluble in water; slightly soluble in alcohol; nearly insoluble in ether.
Use: Medicine.

sodium aminophenylarsonate. See sodium arsanilate.

sodium para-**aminosalicylate** (PAS sodium) $NaOOCC_6H_3(OH)NH_3$.
Properties: White to pale yellow crystalline powder. Practically odorless, with sweet saline taste. Freely soluble in water; sparingly soluble in alcohol; very slightly soluble in ether and chloroform; 2% solution is clear and colorless and has pH 6.5 to 8.5. Solutions decompose slowly and darken in color.
Grade: U.S.P.
Containers: Drums.
Use: Medicine.

sodium ammonium hydrogen phosphate. See sodium ammonium phosphate.

sodium ammonium phosphate (microcosmic salt; sodium ammonium hydrogen phosphate; phosphorus salt) $NaNH_4HPO_4 \cdot 4H_2O$.
Properties: Transparent, colorless, odorless, efflorescent, monoclinic crystals. Gives off water and ammonia on heating, leaving $NaPO_3$. Soluble in water, insoluble in alcohol. Sp. gr. 1.57; m.p. about 79°C with decomposition.
Derivation: Mixing solutions of sodium phosphate and ammonium chloride.
Grades: Granular; C.P.; technical.
Use: Analytical reagent.

sodium anilinearsonate. See sodium arsanilate.

sodium antimonate (antimony sodiate) $NaSbO_3$. Other forms are sodium metaantimonate $2NaSbO_3 \cdot 7H_2O$ and sodium pyroantimonate $Na_2H_2Sb_2O_7 \cdot H_2O$.
Properties: White, granular powder; slightly soluble in water and alcohol. Insoluble in dilute alkalies, mineral acids, but soluble in tartaric acid. Noncombustible.
Grades: Technical; glassmakers' grade.
Hazard: Toxic by ingestion and inhalation. Tolerance, as Sb, 0.5 mg per cubic meter of air.
Uses: Opacifier in enamels, for cast iron and glass; ingredient of acid resisting sheet steel enamels; high-temperature oxidizing agent.

sodium antimony biscatechol-2,4-disulfonate. See stibophen.

sodium arsanilate (sodium anilinearsonate; sodium aminophenylarsonate) $C_6H_4NH_2(AsO \cdot OH \cdot ONa)$, often with 1 or more H_2O.
Properties: White, crystalline, odorless powder; faint salty taste. Soluble in water; slightly soluble in alcohol.
Derivation: By dissolving arsanilic acid in sodium carbonate solution and crystallizing.
Grades: Technical; medicinal.
Hazard: Highly toxic by ingestion and inhalation. May cause blindness.
Uses: Medicine; veterinary medicine; organic synthesis.
Shiping regulations: (Rail, Air) Poison label.

sodium arsenate $Na_3AsO_4 \cdot 12H_2O$.
Properties: Clear, colorless crystals; mild alkaline taste. Soluble in water; slightly soluble in alcohol and glycerol; insoluble in ether. Sp. gr. 1.7593; m.p. 86°C.
Derivation: Reaction of arsenic trioxide and sodium nitrate.

Grades: Pure crystals; C.P.; technical (60% arsenic pentoxide).
Containers: Bottles; multiwall paper sacks; drums.
Hazard: Highly toxic by ingestion and inhalation.
Uses: Medicine; insecticides; dry colors; textiles (mordant and assist in dyeing and printing); other arsenates; germicide.
Shipping regulations: (Rail, Air) Poison label.

sodium arsenite (sodium metaarsenite) $NaAsO_2$.
Properties: Grayish white powder, which absorbs carbon dioxide from the air. Soluble in water; slightly soluble in alcohol. Sp. gr. 1.87.
Derivation: Arsenic trioxide is dissolved in a solution of sodium carbonate or hydroxide and boiled.
Grades: Crude; pure; 75% arsenious oxide (94% soluble).
Hazard: Highly toxic by ingestion and inhalation.
Uses: Manufacture of arsenical soaps for taxidermists; antiseptic; dyeing; insecticides; hide preservation; herbicide.
Shipping regulations: (Rail, Air) Poison label.

sodium arsphenamine. See arsphenamine.

sodium ascorbate $C_6H_7NaO_6$. The sodium salt of ascorbic acid (q.v.).

sodium azide NaN_3.
Properties: Colorless, hexagonal crystals; decomposes at about 300°C; sp. gr. 1.846 (20°C); soluble in water and in liquid ammonia; slightly soluble in alcohol; insoluble in ether.
Hazard: Highly toxic. Severe explosion risk when shocked or heated.
Uses: Explosive; medicinals.
Shipping regulations: (Rail, Air) Poison label.

sodium bacitracin methylene disalicylate. See bacitracin methylene disalicylate.

sodium barbital (barbital, soluble) $C_8H_{11}N_2NaO_3$.
Properties: White powder, stable, odorless; bitter taste. Soluble in water (solution alkaline to litmus); slightly soluble in alcohol; insoluble in ether.
Hazard: Toxic by ingestion. See barbiturate.
Use: Medicine.

sodium barbiturate $C_4H_3N_2O_3Na$.
Properties: White to yellow tinted powder; pH of a 1% aqueous solution 10.5. Soluble in water and dilute mineral acid.
Containers: Fiber drums.
Hazard: Toxic by ingestion. See barbiturate.
Uses: Synthetic intermediate; catalyst in ammonium nitrate propellants; wood impregnating solutions.

sodium benzenesulfonchloramine. See chloramine-B.

sodium benzoate C_6H_5COONa.
Properties: White, crystalline or granular, odorless powder; sweetish, astringent taste. Soluble in water and alcohol. Combustible. Low toxicity.
Derivation: Benzoic acid is neutralized with sodium bicarbonate solution, the solution filtered, concentrated and allowed to crystallize.
Grades: U.S.P.; F.C.C.; technical.
Containers: Cartons; bottles; boxes; drums.
Hazard: Use in foods limited to 0.1%.
Uses: Food preservative; antiseptic; medicine; tobacco; pharmaceutical preparations; intermediate for manufacture of dyes; rust and mildew inhibitor.

sodium benzosulfimide. See saccharin.

sodium benzylpenicillin. See penicillin.

sodium benzyl succinate $C_6H_5CH_2O_2C(CH_2)_2COONa$.
Properties: White amorphous or crystalline powder having a slightl benzyl odor and cool salty taste. Soluble in hot and cold water.

sodium beryllium fluoride. See beryllium sodium fluoride.

sodium bicarbonate (baking soda; sodium acid carbonate) $NaHCO_3$.
Properties: White powder or crystalline lumps; cooling, slightly alkaline taste. Soluble in water; insoluble in alcohol. Stable in dry air, but slowly decomposes in moist air. Sp. gr. 2.159; m.p. loses carbon dioxide at 270°C. Low toxicity. Noncombustible.
Derivation: Principally, by treating a saturated solution of soda ash with carbon dioxide to precipitate the less soluble bicarbonate; also by purifying the crude product from the Solvay process.
Grades: Commercial; pure; highest purity; C.P.; U.S.P.; F.C.C.
Containers: Bottles and cartons; bags; drums; bulk.
Uses: Manufacture of effervescent salts and beverages, artificial mineral water, baking powder; other sodium salts; pharmaceuticals; sponge rubber; gold and platinum plating; treating wool and silk; fire extinguishers; ceramics; prevention of timber mold; laboratory reagent; antacid; mouthwash.

sodium bichromate. See sodium dichromate.

sodium bifluoride (sodium acid fluoride) $NaHF_2$.
Properties: White, crystalline powder. Soluble in water. Decomposed on heating. Sp. gr. 2.08.
Hazard: Highly toxic; strong irritant to tissue. Tolerance (as F), 2.5 mg per cubic meter of air.
Uses: Tin plate production; neutralizer in laundry rinsing operations; preservative for zoological and anatomical specimens; etching glass; antiseptic.
Shipping regulations: (Air) (solid) Poison label; (solution) Corrosive label. (Rail) Not listed.

sodium biphosphate. See sodium phosphate, monobasic.

sodium bis(2-methoxyethoxy)-aluminohydride. An organometallic metal hydride, the alkoxy adduct of sodium aluminum hydride; soluble in benzene and other hydrocarbons. Reacts less strongly with water than lithium aluminum hydride. Used for reduction of organic compounds.

sodium bismuthate $NaBiO_3$.
Properties: Yellow or brown, amorphous powder. Slightly hygroscopic. Insoluble in cold waer; decomposes in hot water and acids. Low toxicity.
Uses: Analysis (testing for manganese in iron, steel and ores); reagent; pharmaceuticals.

sodium bisulfate (sodium acid sulfate; niter cake; sodium hydrogen sulfate) (a) $NaHSO_4$ or (b) $NaHSO_4 \cdot H_2O$.
Properties: Colorless crystals or white, fused lumps; aqueous solution is strongly acid. Soluble in water. (a) Sp. gr. 2.435 (13°C); m.p. >315°C. (b) Sp. gr. 2.103 (13°C); m.p. 58.5°C. Noncombustible.
Derivation: A by-product in the manufacture of hydrochloric and nitric acids.
Method of purification: Recrystallization.
Grades: Pure crystals; pure fused; pure dry; reagent; crude; C.P.; technical; F.C.C.
Forms available: Cakes; powder; prills; pearls.

Containers: Various sizes of drums; bulk in cars.
Hazard: Strong irritant to tissue.
Uses: Flux for decomposing minerals; substitute for sulfuric acid in dyeing; disinfectant; manufacture of sodium hydrosulfide, sodium sulfate and soda alum; liberating carbon dioxide in CO_2 baths; in thermophores; carbonizing wool; manufacture of magnesia cements, paper, soap, perfumes; foods; industrial cleaners; metal pickling compounds; laboratory reagent.
Shipping regulations: (Air) (solid and solution) Corrosive label. (Rail) Not listed.

sodium bisulfite (sodium acid sulfite; sodium hydrogen sulfite). $NaHSO_3$.
Properties: White crystals or crystalline powder; slight sulfurous odor and taste; soluble in water; insoluble in alcohol. Unstable in air. Sp. gr. 1.48; m.p. decomposes. Low toxicity. Noncombustible.
Derivation: Sodium carbonate solution is saturated with sulfur dioxide gas and the solution crystallized.
Grades: Crystals; pure dry; commercial dry; reagent; commercial solution 35° Bé; powder (67% SO_2); C.P.; U.S.P.; F.C.C.
Hazard: Not permitted in meats or other sources of vitamin B_1. Strong irritant to skin and tissue.
Uses: Chemicals (sodium salts, cream of tartar); dyes; intermediates; organic chemicals; perfumery; wood pulp (digestion); leather (depilatory, tanning); textiles (antichlor, mordant, discharge); food preservative; photographic reducing agent; fermentation industries; medicine; glucose and sugar syrups; brewing (cask sterilization); copper and brass plating; color preservative for pale crepe rubber; general antiseptic; analytical reagent; pesticides; source of sulfur dioxide.
Shipping regulations: (Air) Corrosive label. (Rail) Not listed.

sodium bitartrate (acid sodium tartrate) $NaHC_4H_5O_6 \cdot H_2O$.
Properties: White, crystalline powder. Soluble in water (aqueous solution is acid); slightly soluble in alcohol. M.p., loses H_2O at 100°C; b.p. 219°C (decomposes). Low toxicity. Combustible.
Grades: C.P.; technical; reagent.
Uses: Analysis (testing for potassium); effervescing mixtures.

sodium bithionolate (USAN) (thiobis[4,6-dichloro-ortho-phenylene)oxy] disodium; disodium 2,2′-thiobis-(4,6-dichlorophenoxide)) $C_{12}H_4Cl_4Na_2O_2S$.
Properties: Solid; soluble in water.
Uses: Medicine; in self-sanitizing polishes, waxes and cleaners; in shoe polish and leather conditioners.

sodium borate (sodium tetraborate; sodium borate (2,4,7); sodium pyroborate; borax) $Na_2B_4O_7 \cdot 10H_2O$.
See also borax, anhydrous; borax, pentahydrate.
Properties: White crystals or powder; odorless; sp. gr. 1.73 (20/4°C); loss of water of crystallization when heated, with melting, between 75 and 320°C; fuses to a glassy mass at red heat (borax glass); effloresces slightly in warm, dry air. Soluble in water and glycerol; insoluble in alcohol. Low toxicity, noncombustible.
Derivation: Fractional crystallization from Searles Lake brine; solution of kernite ore followed by crystallization. Also from colemanite, natural borax, uxelite, and other borates.
Grades: Crystal; granulated; powdered (refined, U.S.P.); C.P.; technical.
Containers: Barrels; bulk (granulated); kegs; bags.
Uses: Heat-resistant glass; porcelain enamel; starch and adhesives; detergents; herbicides; fertilizers; rust inhibitors; pharmaceuticals; leather; photography; bleaches; paint; boron compounds; flux for smelting; antifreeze; adhesives; insulation materials; laboratory reagent.

sodium borate, anhydrous. See borax, anhydrous.

sodium borate perhydrate $NaBO_2 \cdot H_2O_2 \cdot 3H_2O$.
Properties: Free-flowing, white powder. Dissolves readily in water to produce mildly alkaline peroxide solution. M.p. 63°C; b.p., loses H_2O at 130–150°C.
Containers: 125-lb fiber drums.
Hazard: Oxidizing agent; fire risk near organic materials.
Uses: Dentrifices; hair wave preparations; bleaching; oxidation of dyestuffs; stain remover for plastic dishes.
See also sodium perborate.

sodium boroformate $NaH_2BO_3 \cdot 2HCOOH \cdot 2H_2O$.
Properties: White, crystalline solid. Soluble in water; m.p. 110°C.
Uses: Buffering agent toward both acid and alkali in the range of pH 8.5; textile treating and tanning baths.

sodium borohydride $NaBH_4$.
Properties: White crystalline powder; soluble in water, ammonia, amines, pyridine, and dimethyl ether of diethylene glycol; insoluble in other ethers, hydrocarbons and alkyl chlorides. Sp. gr. 1.07; hygroscopic; stable in dry air to 300°C; decomposes slowly in moist air or in vacuum at 400°C.
Derivation: By reaction of sodium hydride and trimethyl borate.
Grades: Technical; powdered; pellets.
Containers: Glass bottles; polyethylene bags in metal containers.
Hazard: Reacts with water to evolve hydrogen and sodium hydroxide. Flammable, dangerous fire risk. Store out of contact with moisture.
Uses: Source of hydrogen and other borohydrides. Reduces aldehydes, ketones and acid chlorides; bleaching wood pulp; blowing agent for plastics; precipitation of mercury from waste effluent (by reduction); decolorizer for plasticizers; recycling of gold and Pt group metals.
Shipping regulations: (Air) Flammable Solid label. (Rail) Not listed.

sodium bromate $NaBrO_3$.
Properties: White odorless crystals or powder. Soluble in water; insoluble in alcohol. Sp. gr. 3.339; m.p. 381°C (dec).
Derivation: By passing bromine into a solution of sodium carbonate, sodium bromide and sodium bromate being formed.
Hazard: Moderately toxic. Oxidizing material. Dangerous fire risk near organic materials.
Use: Analytical reagent.
Shipping regulations: (Rail) Yellow label. (Air) Oxidizer label.

sodium bromide (a) NaBr, (b) $NaBr \cdot 2H_2O$.
Properties: White, crystalline powder or granules; saline and somewhat bitter taste; absorbs moisture from the air, becoming very hard. Soluble in water; moderately soluble in alcohol. Sp. gr. (a) 3.208, (b) 2.176; m.p. (a) 757.7°C; b.p. 1390°C.
Derivation: Occurs naturally in some salt deposits. Made synthetically by first causing iron to react with bromine and water. The resulting ferroso-ferric bromide is dissolved in water, sodium carbonate added, the solution filtered and evaporated.

Grades: C.P.; crystalline; powdered; commercial; pure; highest purity; N.F.
Containers: Bottles; boxes; barrels; drums.
Hazard: Moderately toxic by inhalation and ingestion.
Uses: Photography; medicine; preparation of bromides; organic chemicals; source of bromine.

sodium bromite $NaBrO_2$. Used as a desizing compound for textiles.

sodium cacodylate (sodium dimethylarsenate) $(CH_3)_2AsOONa \cdot 3H_2O$.
Properties: White, amorphous crystals or powder; deliquescent. Melts about 60°C. Loses H_2O at 120°C. Soluble in water and alcohol.
Derivation: Oxidation and neutralization of cacodyl oxide.
Hazard: Highly toxic by inhalation and ingestion.
Uses: Medicine; herbicide.
Shipping regulations: (Solid) (Rail, Air) Poison label.

sodium carbolate. See sodium phenate.

sodium carbonate (soda). See soda ash; sal soda; sodium bicarbonate; sodium carbonate, monohydrate; sodium sesquicarbonate.

sodium carbonate, monohydrate (crystal, carbonate; soda monohydrate; soda crystals) $Na_2CO_3 \cdot H_2O$.
Properties: White; odorless; small crystals or crystalline powder; alkaline taste; sp. gr. 1.55. Soluble in water and glycerol; insoluble in alcohol. M.p. 109°C (loses water), 851°C. Noncombustible. Low toxicity.
Grades: U.S.P.; technical; F.C.C.
Uses: Medicine; photography; cleaning and boiler compounds; pH control of water; food additive.

sodium carbonate peroxide $2Na_2CO_3 \cdot 3H_2O_2$ or $Na_2CO_3 \cdot H_2O_2 \cdot \frac{1}{2}H_2O$.
Properties: White crystalline powder. Soluble in water. Stable at room temperature when dry; decomposes rapidly at 100°C, evolving oxygen. Active oxygen content 14% min.
Derivation: Crystallization from solution of soda ash and hydrogen peroxide.
Grade: Technical.
Containers: 100-lb fiber drums.
Hazard: Oxidizing agent; dangerous near organic materials.
Uses: Source of hydrogen peroxide; household detergents; dental cleansers; bleaching and dyeing; modification of starch.
Shipping regulations: (Rail) Oxidizing materials, n.o.s., Yellow label. (Air) Oxidizer label.

sodium carboxymethylcellulose. See carboxymethylcellulose.

sodium caseinate.
Properties: White, coarse powder. Odorless, tasteless. Contains 65% proteins. Soluble in water (usually with turbidity).
Derivation: By dissolving casein in sodium hydroxide and evaporating.
Grades: Edible.
Uses: Medicine; foods; as emulsifier and stabilizer.

sodium catechol disulfonate. See disodium 1,2-dihydroxybenzene-3,5-disulfonate.

sodium cellulosate.
Properties: A cellulosic fiber.
Derivation: Reaction of sodium methoxide with cotton or rayon swollen in methyl alcohol.
Uses: Intermediate in preparing grafted fibers when combined with polyacrylonitrile; there is an average substitution of one polyacrylonitrile chain for every 100 to 300 anhyroglucose units of the cellulose.

sodium chlorate $NaClO_3$.
Properties: Colorless, odorless crystals; cooling, saline taste; must not be triturated with any combustible substance. Soluble in water and alcohol. Sp. gr. 2.490; m.p. 255°C; b.p., decomposes.
Derivation: A concentrated acid solution of sodium chloride is heated and electrolyzed, the chlorate crystallizing out.
Grades: Technical; C.P.; crystals; powder.
Containers: Drums; stainless steel tank cars.
Hazard: Dangerous fire risk; strong oxidant; contact with organic materials may cause fire. Safety data sheet available from Manufacturing Chemists Assn., Washington, D.C.
Uses: Oxidizing agent and bleach (especially to make chlorine dioxide) for paper pulps; ore processing; herbicide and defoliant; medicine; substitute for potassium chlorate, being more soluble in water; matches; explosives; flares and pyrotechnics; recovery of bromine from natural brines; leather tanning and finishing; textile mordant; to make perchlorates.
Shipping regulations: (Rail) Yellow label. (Air) Oxidizer label.

sodium chloraurate. See sodium gold chloride.

sodum chloride (table salt; sea salt; halite; rock salt) NaCl.
Properties: Colorless, transparent crystals or white crystalline powder; somewhat hygroscopic; soluble in water and glycerol; very slightly soluble in alcohol. Sp. gr. 2.165; m.p. 801°C; b.p. 1413°C. Noncombustible; low toxicity. Essential in diet to maintain chloride balance in body.
Occurrence: Deposits in central New York, Southern Michigan, Gulf Coast, Great Salt Lake, Newfoundland.
Derivation: (a) evaporation and crystallization of natural brines. (b) solar evaporation of sea water. (c) direct mining from underground or surface deposits.
Method of purification: Recrystallization.
Impurities: Sulfates; heavy metals; alkaline earths; magnesium salts; ammonium salts.
Grades: Highest purity medicinal, crystals; highest purity, dried; highest purity, fine powder; highest purity, fused; reagent; reagent, fused; sea evaporated; ground; microsized; powdered; table salt; rock salt; C.P.; U.S.P.; F.C.C.; single pure crystals.
Uses: Chemical (sodium hydroxide, soda ash, hydrochloric acid, chlorine, metallic sodium); ceramic glazes; metallurgy; curing of hides; food preservative; mineral waters; soap manufacture (salting out); home water softeners; highway deicing; regeneration of ion-exchange resins; photography; food seasoning; herbicide; fire extinguishing; nuclear reactors; mouthwash; medicine (heat exhaustion); salting out dyestuffs; supercooled solutions. Single crystals are used for spectroscopy, ultraviolet and infrared transmission.
See also fused salt.

sodium chlorite $NaClO_2$.
Properties: White crystals or crystalline powder; slightly hygroscopic; soluble in water. M.p. 180–200°C (dec).
Grades: Technical; reagent.
Containers: Drums.

Hazard: (Solid) Flammable; strong oxidizing agent; dangerous fire and moderate explosion risk; (solution) strong irritant to skin and tissue.
Uses: For improving taste and odor of potable water (as an oxidizing agent); bleaching agent for textiles, paper pulp, edible and inedible oils, shellacs, varnishes, waxes and straw products; oxidizing agent; reagent.
Shipping regulations: (Solid) (Rail) Yellow label. (Air) Oxidizer label. Not acceptable on passenger planes. (Solution not over 42%) (Rail) White label. (Solution not over 40%) (Air) Corrosive label. (Solution over 40%) Not acceptable.

sodium chloroacetate $ClCH_2COONa$.
Properties: White nonhyrogscopic powder, odorless, easier to handle than chloroacetic acid. Soluble in water; slightly soluble in methanol; insoluble in acetone; benzene, ether, and carbon tetrachloride. Low toxicity. Noncombustible.
Uses: Manufacture of weed killers, dyes, vitamins, pharmaceuticals; also a defoliant.

sodium chloroasurate. See sodum gold chloride.

sodium para-**chloro**-meta-**cresolate.** A water-soluble preservative for cutting oils.

sodium 4-chlorophthalate (monosodium 4-chlorophthalate) $ClC_6H_3(COOH)COONa$.
Properties: White to light gray powder.
Grade: Commercial.
Containers: Fiber drums.
Use: Modifying agent in the manufacture of phthalocyanine pigments.

sodium chloroplatinate (platinic sodium chloride; platinum sodium chloride; sodium platinichloride) $Na_2PtCl_6 \cdot 4H_2O$.
Properties: Yellow powder; soluble in alcohol, water. Low toxicity; noncombustible.
Grades: Technical; C.P.
Containers: Glass bottles.
Uses: Etching on zinc; ink (indelible); microscopy; mirrors; medicine; photography; plating; catalyst; determination of potassium.

sodium ortho-**chlorotoluene**-para-**sulfonate** $NaSO_3C_6H_3(CH_3)Cl$.
Properties: Gray to light tan powder; soluble in water and organic solvents.
Derivation: Sulfonation of ortho-chlorotoluene, neutralized to form the sodium salt.
Uses: Synthesis of dyes, intermediates and drugs.

sodium chromate $Na_2CrO_4 \cdot 10H_2O$.
Properties: Yellow, translucent, efflorescent crystals. Soluble in water; slightly soluble in alcohol. Sp. gr. 1.483; m.p. 19.92°C. Anhydrous sodium chromate is also available commercially.
Derivation: Chrome iron ore is melted in a reverberatory furnace with lime and soda, in presence of air. The melt is dissolved in water, a small amount of sodium carbonate added, the solution decanted, acidified with acetic acid, concentrated and crystallized.
Grades: Pure neutral; highest purity; technical; C.P.; reagent.
Containers: Bottles; drums; barrels.
Uses: Inks; dyeing; paint pigment; leather tanning; other chromates; protection of iron against corrosion; wood preservative.

sodium chromate, tetrahydrate $Na_2CrO_4 \cdot 4H_2O$.
Properties: Yellow crystals; deliquescent; soluble in water.
Grade: Technical.
Containers: Multiwall paper bags, 100 lb net; fiber drums.
Uses: Pigment manufacture; corrosion inhibition; leather tanning; other chromium compounds.

sodium citrate (trisodium citrate) $C_6H_5O_7Na_3 \cdot 2H_2O$.
Properties: White crystals or granular powder; odorless; stable in air; pleasant acid taste. Soluble in water; insoluble in alcohol. M.p. loses $2H_2O$ at 150°C; b.p. decomposes at red heat. Combustible; nontoxic.
Derivation: Sodium sulfate solution is treated with calcium citrate, filtered, concentrated and crystallized.
Grades: Highest purity, medicinal; pure; commercial; C.P.; U.S.P.; F.C.C.
Containers: Bottles; drums; bags.
Uses: Medicine; soft drinks; photography; frozen desserts; meat products; detergents; special cheeses; electroplating; sequestrant and buffer; nutrient for cultured buttermilk; removal of sulfur dioxide from smelter waste gases; blood anticoagulant.

sodium copper chloride. See copper sodium chloride.

sodium copper cyanide (copper sodium cyanide; sodium cyanocuprate) $NaCu(CN)_2$.
Properties: White, crystalline, double salt of copper cyanide and sodium cyanide. Sp. gr. 1.013 (20°C); decomposes at 100°C. Soluble in water.
Containers: 100-lb drums.
Hazard: Tolerance (as CN), 5 mg per cubic meter of air.
Use: For preparing and maintaining cyanide copper plating baths based on sodium cyanide.

sodium cyanate NaOCN.
Properties: White crystalline powder; soluble in water; insoluble in alcohol and ether. Sp. gr. 1.937 (20°C); m.p. 700°C.
Containers: Fiber drums; bottles.
Uses: Organic synthesis; heat treating of steel; intermediate for manufacture of medicinals; treatment of sickle cell anemia.

sodium cyanide NaCN.
Properties: White deliquescent, crystalline powder. Soluble in water; slightly soluble in alcohol; m.p. 563°C; b.p. 149°C. The aqueous solution is strongly alkaline and decomposes rapidly on standing.
Derivation: By absorption of hydrocyanic acid in a solution of sodium hydroxide, with subsequent vacuum evaporation.
Grades: 30% solution; 73 to 75%; 96 to 98%; reagent; technical; briquettes granular.
Containers: 25-lb packages; 100-, 160-, 200-lb drums.
Hazard: Toxic by ingestion and inhalation. Tolerance (as CN), 5 mg per cubic meter of air. Safety data sheet available from manufacturing Chemists Assn., Washington, D.C.
Use: Extraction of gold and silver from ores; electroplating; heat treatment of metals; making hydrocyanic acid; insecticide; cleaning metals; fumigation; manufacture of dyes and pigments; nylon intermediates; chelating compounds; ore flotation.
Shipping regulations: (Rail, Air) Poison label.

sodium cyanoaurite. See sodium gold cyanide.

sodium cyanocuprate. See sodium copper cyanide.

sodium cyclamate (sodium cyclohexylsulfamate) $C_6H_{11}NHSO_3Na$.
Properties: White, crystalline, practically odorless powder; sweet taste. Freely soluble in water; practically insoluble in alcohol, benzene, chloroform and

ether; pH (10% solution) 5.5–7.5. Sweetening power approximately 30 times that of sucrose.
Grades: N.F.; F.C.C.
Containers: 100-lb drums.
Hazard: Claimed to cause cancer in laboratory animals. Prohibited by FDA for food use.
Use: Nonnutritive sweetener.

sodium decametaphosphate. See sodium metaphosphate.

sodium dehydroacetate $C_8H_7NaO_4 \cdot H_2O$.
Properties: Tasteless white powder. Soluble in water and propylene glycol. Insoluble in most organic solvents. See also dehydroacetic acid.
Grade: F.C.C.
Uses: Fungicide; plasticizer; toothpaste; pharmaceutical; preservative in food; mold inhibitor for strawberries and similar fruits.

sodium deoxycholate. See deoxycholic acid.

sodium dextran sulfate (dextran sulfate).
Properties: Solid; soluble in water.
Derivation: Derivatives of dextran (q.v.) having a molecular weight of 500 thousand to 2 million.
Uses: Fractionation and separation of biological preparations.

sodium 3,5-diacetamido-2,4,6-triiodobenzoate. See sodium diatrizoate.

sodium diacetate $CH_3COONa \cdot x(CH_3COOH)$, anhydrous, or $CH_3COONa \cdot x(CH_3COOH) \cdot yH_2O$, technical.
Properties: White crystals with an acetic acid odor; soluble in water; slightly soluble in alcohol; insoluble in ether. Decomposes above 150°C. Combustible. Low toxicity.
Containers: 5- to 250-lb drums.
Grade: F.C.C.
Uses: Buffer; mold inhibitor; souring agent; intermediate for acid salts; mordants, varnish hardeners; antitarnishing agents; sequestrant and preservative in foods.

sodium diatrizoate (sodium 3,5-diacetamido-2,4,6-triiodobenzoate) $C_6I_3(COONa)(NHCOCH_3)_2$.
Properties: White crystals; soluble in water. Solutions are radiopaque.
Grade: U.S.P. (as solution for injection).
Use: Radiopaque medium; medicine.

sodium 1-diazo-2-naphthol-4-sulfonate. See 1-diazo-2-naphthol-4-sulfonic acid.

sodium dibutyldithiocarbamate. See "Tepidone."

sodium dibutyl naphthalene sulfonate. See "Sorbit" AC.

sodium alpha, beta-**dichloroisobutyrate.** A plant growth regulator.

sodium dichloroisocyanurate (sodium salt of dichloro-s-triazine-2,4,6-trione) NaNC(O)NClC(O)NClCO.
Properties: White, slightly hygroscopic, crystalline powder; loose bulk density (approx.) powder 37 lb/cu ft, granulate 57 lb/cu ft. Active ingredient; approx 60% available chlorine; decomp. at 230°C.
Containers: 200-lb fiber drums.
Hazard: Strong oxidizing material; fire risk near organic materials. Toxic by ingestion.
Uses: Active ingredient in dry bleaches, dishwashing compounds, scouring powders; detergent-sanitizers, swimming pool disinfectants, water and sewage treatment; replacement of calcium hypochlorite.
Shipping regulations: (Dry, containing more than 39% available chlorine) (Rail) Yellow label. (Air) Oxidizer label.

sodium 2,4-dichlorophenoxyacetate (2,4-D, sodium salt) $C_6H_3(OCH_2COONa)Cl_2$.
Properties: Crystalline solid. Decomposes at 215°C. Slightly soluble in water.
Hazard: Toxic and irritant by inhalation.
Use: Herbicide. See 2,4-D.

sodium 2,4-dichlorophenoxyethyl sulfate. See sesone.

sodium 2,2-dichloropropionate. See dalapon.

sodium dichromate (sodium bichromate). $Na_2Cr_2O_7 \cdot 2H_2O$.
Properties: Red or red-orange deliquescent crystals. Sp. gr. 2.52 (13°C); m.p. 357°C; decomposes at 400°C; loses $2H_2O$ on prolonged heating at 100°C. Soluble in water; insoluble in alcohol. Noncombustible.
Derivation: (a) From chromite ore by alkaline roasting and subsequent leaching; (b) action of sulfuric acid on sodium chromate.
Grades: Technical crystalline; technical liquor containing 69–70% $Na_2Cr_2O_7 \cdot 2H_2O$; anhydrous.
Containers: Crystals: multiwall paper bags; fiber drums. Liquor: tank cars or tank trucks.
Hazard: Toxic by inhalation and ingestion; strong irritant. Safety data sheet available from Manufacturing Chemists Assn., Washington, D.C.
Uses: Chemical reactant for oxidation reactions; chromic acid; corrosion inhibitor; manufacture of pigments; tanning of leather; electroplating; mordanting; defoliating agent; catalyst; wood preservative.

sodium diethyldithiocarbamate. $(C_2H_5)_2NCS_2Na$. Oxidation inhibitor in ethyl ether; in trace quantities it prevents peroxide formation.

sodium dihydrogen phosphate. See sodium phosphate, monobasic.

sodium dihydroxyethylglycine (N,N-bis(2-hydroxyethyl)-glycine) $NaOOCCH_2N(CH_2CH_2OH)_2$.
Properties: Clear straw-colored liquid; sp. gr. 1.204 (25°C); f.p. below −10°C.
Use: Complexing agent for the transition metals.

sodium dimethylarsenate. See sodium cacodylate.

sodium dimethyldithiocarbamate (SDDC) $(CH_3)_2NCS_2Na$.
Properties: 40% solution is amber to light green; sp. gr. 1.17–1.20 (25/25°C).
Derivation: Reaction of dimethylamine, carbon disulfide and sodium hydroxide.
Containers: 5-gal cans; 55-gal drums; tanks (solution).
Hazard: Moderately toxic.
Uses: Fungicide; corrosion inhibitor; rubber accelerator; intermediate; polymerization shortstop.

sodium dinitro-ortho-**cresylate** $CH_3C_6H_2(NO_2)_2ONa$.
Properties: Brilliant orange-yellow dye which may stain clothing and wood.
Derivation: By treating 4,6-dinitro-ortho-cresol with sodium hydroxide.
Hazard: Toxic by ingestion and inhalation.
Uses: Herbicide (control of mustard and other susceptible weeds); fungicide.
See also 4,6-dinitro-ortho-cresol.

sodium dioctyl sulfosuccinate. See dioctyl sodium sulfosuccinate.

sodium dioxide. NaO_2. Exists as impurity (about 10%) in sodium peroxide; obtained by heating sodium peroxide in oxygen; reacts with water to yield hydrogen peroxide, oxygen, and sodium hydroxide.

sodium diphenylhydantoin (phenytoin, soluble) $C_{15}H_{11}N_2O_2Na$. Sodium 5,5-diphenylhydantoinate.
Properties: White, odorless powder; freely soluble in water; soluble in alcohol; practically insoluble in ether and in chloroform; somewhat hygroscopic; on exposure to air diphenylhydantoin is liberated.
Containers: Up to 100-lb drums.
Grades: U.S.P.
Use: Medicine.

sodium dispersion. A stable suspension of microscopic sodium particles in a hydrocarbon or other medium boiling at temperatures above the melting point of sodium (97.5°C), e.g., heptane, n-octane, toluene, xylene, naphtha, kerosine, mineral oil, n-butyl ether, etc. Particles range in size from submicron to 30 microns depending on the method of prepartion. Dispersions contain up to 50% (by weight) of sodium metal.
Hazard: Flammable, dangerous fire risk. To extinguish use soda ash, sodium chloride or graphite, followed by carbon dioxide. Do not use carbon tetrachloride.
Uses: For chemical reactions where advantages of controlled reaction rate, lower reaction temperature, increased yields, or substitution for more expensive reagent can be achieved; as, removal of sulfur from hydrocarbons and petroleum; metal powders; sodium hydride, alcohol-free alcoholates, phenylsodium.
Shipping regulations: (Rail) Yellow label. (Air) Flammable Solid label. Not acceptable on passenger planes.

sodium dithionate (sodium hyposulfate) $Na_2S_2O_6 \cdot 2H_2O$.
Properties: Large transparent crystals; bitter taste. Soluble in water; insoluble in alcohol and concentrated hydrochloric acid. Sp. gr. 2.189 (25°C); loses $2H_2O$ at 110°C; decomposes at 267°C. Combustible.
Use: Chemical reagent.

sodium dithionite (sodium hydrosulfite) $Na_2S_2O_4$ or $Na_2S_2O_4 \cdot 2H_2O$.
Properties: Light lemon-colored solid in powder or flake form; m.p. 55° (decomp.). Also available in liquid form. Soluble in water; insoluble in alcohol.
Derivation: (1) Zinc is dissolved in a solution of sodium bisulfite, the zinc-sodium sulfite precipitated by milk of lime leaving the hydrosulfite in solution. (2) Reaction of sodium formate with sodium hydroxide and sulfur dioxide.
Grades: Technical; reagent.
Containers: Drums; carlots; refrigerated tanks (liquid).
Hazard: Fire risk in contact with moisture. Strong reducing agent. Use dry sand, soda ash or carbon dioxide to extinguish fires.
Uses: Vat dyeing of fibers and textiles; stripping agent for dyes; reagent; bleaching sugar, soap, oils, minerals; oxygen scavenger for synthetic rubbers.
Shipping regulations: (Rail) Yellow label. (Air) Flammable Solid label.

sodium diuranate (uranium yellow) $Na_2U_2O_7 \cdot 6H_2O$.
Properties: Yellow-orange solid, insoluble in water; soluble in dilute acids.
Derivation: By treating a solution of uranyl salt with sodium hydroxide.
Hazard: Probably toxic.
Uses: Ceramics, to produce colored glazes; manufacture of fluorescent uranium glass.

sodium dodecylbenzenesulfonate $C_{12}H_{25}C_6H_4SO_3Na$.
Properties: White to light yellow flakes, granules, or powder; biodegradable. The dodecyl radical may have many isomers, and the benzene may be attached to it in many positions. Combustible.
Derivation: Benzene is alkylated with dodecene, to which it attaches itself in any secondary position; the resulting dodecylbenzene is sulfonated with sulfuric acid and neutralized with caustic soda. For ABS (branched-chain alkyl) the dodecene is usually a propylene tetramer, made by catalytic polymerization of propylene. For LAS (straight-chain alkyl), the dodecene may be removed from kerosine or crudes by molecular sieve, may be formed by Ziegler polymerization of ethylene, or by cracking wax paraffins to alpha-olefins. See also biodegradability; alkyl sulfonate, linear; detergent; alkylate (3).
Containers: Carlots.
Use: Most widely used synthetic detergent.

sodium dodecyldiphenyl oxide disulfonate $C_{24}H_{34}O(SO_3)_2Na_2$. M.p. 150°C (dec); dry form 90% min active solution (45%), sp. gr. 1.1. Very soluble in water, strong acids, bases and electrolytes; stable to oxidation.

sodium erythorbate (sodium isoascorbate) $NaC_6H_7O_6$. White, free-flowing crystals. Soluble in water.
Grades: Technical; F.C.C.
Uses: Antioxidant and preservative.

sodium ethoxide. See sodium ethylate.

sodium ethylate (sodium ethoxide; caustic alcohol) C_2H_5ONa.
Properties: White powder, sometimes having brownish tinge. Readily hydrolyzes to alcohol and sodium hydroxide.
Derivation: By carefully adding small amounts of sodium to absolute alcohol kept at a temperature of 10°C, heating carefully to 37.7°C, again carefully adding sodium, cooling to 10°C, and adding the same quantity of absolute alcohol as was used originally.
Hazard: See sodium hydroxide, which is formed when exposed to moisture.
Use: Organic synthesis.

sodium ethylenebisdithiocarbamate. See nabam.

sodium 2-ethylhexyl sulfoacetate $C_8H_{17}OOCCH_2SO_3Na$.
Properties: Light cream-colored flakes; water-soluble. Good foaming properties and excellent resistance to hard water. Solutions practically neutral and stable to mineral acids. Low toxicity. Combustible.
Uses: Solubilizing agent, particularly for soapless shampoo compositions; electroplating detergent.

sodium ethylmercurithiosalicylate. See thimerosal.

sodium ethyl oxalacetate. See ethyl sodium oxalacetate.

sodium ethylxanthate (sodium xanthogenate; sodium xanthate) $C_2H_5OC(S)SNa$. See also xanthic acids.
Properties: Yellowish powder; soluble in water, alcohol.
Use: Ore flotation agent.

sodium ferricyanide (red prussiate of soda) $Na_3Fe(CN)_6 \cdot H_2O$.

Properties: Ruby-red, deliquescent crystals. Soluble in water; insoluble in alcohol. Low toxicity.
Derivation: Chlorine is passed into sodium ferrocyanide solution.
Grades: Technical; C.P.
Uses: Production of pigments; dyeing; printing.

sodium ferrocyanide (yellow prussiate of soda) $Na_4Fe(CN)_6 \cdot 10H_2O$.
Properties: Yellow, semitransparent crystals. Soluble in water; insoluble in alcohol. Sp. gr. 1.458. Low toxicity.
Grades: Technical; F.C.C.
Containers: Bags.
Uses: Photography; manufacture of sodium ferricyanide; blue pigments; blueprint paper; metallurgy; tanning; dyes; anticaking agent for sodium chloride.

sodium fluoborate $NaBF_4$.
Properties: White powder; bitter acid taste; slowly decomposed by heat; sp. gr. 2.47 (20°C); m.p. 384°C. Soluble in water; slightly soluble in alcohol.
Derivation: By heating sodium fluoride and hydrofluoboric acid and cooling slowly.
Uses: Sand casting of aluminum and magnesium; electrochemical processes; oxidatoin inhibitor; fluxes for nonferrous metals; fluorinating agent.

sodium fluophosphate. See sodium fluorophosphate.

sodium fluorescein. U.S.P. name for uranine.

sodium fluoride NaF.
Properties: Clear, lustrous crystals or white powder. Insecticide grade frequently dyed blue. Soluble in water; very slightly soluble in alcohol. Sp. gr. 2.558 (41°C); m.p. 988°C; b.p. 1695°C.
Derivation: By adding sodium carbonate to hydrofluoric acid.
Grades: Pure; C.P.; U.S.P.; insecticide; technical; single pure crystals.
Containers: 100-lb multiwall paper bags; 125-lb drums; 400-lb carlots.
Hazard: Toxic by ingestion and inhalation; strong irritant to tissue. Tolerance (as F), 2.5 mg per cubic meter of air.
Uses: Fluoridation of municipal water (1 ppm); treatment of steel; wood preservative; insecticide (not to be used on living plants); fungicide, and rodenticide; chemical cleaning; electroplating; glass manufacture; vitreous enamels; preservative for adhesives; toothpastes; single crystals used as windows in ultraviolet and infrared radiation detecting systems.
Shipping regulations: (solution) (Air) Corrosive label.

sodium fluoroacetate (also known as 1080) FCH_2COONa.
Properties: Fine, white, odorless powder; soluble in water. Nonvolatile. Decomposes at 200°C; soluble in water; insoluble in most organic solvents.
Derivation: Ethyl chloroacetate and potassium fluoride form ethyl fluoroacetate, which is then treated with a methanol solution of sodium hydroxide.
Hazard: Highly toxic by ingestion, inhalation, and skin absorption. Tolerance, 0.05 mg per cubic meter of air. For use by trained operators only.
Use: Rodenticide.
Shipping regulations: (Rail, Air) Not listed.

sodium fluorophosphate (sodium fluophosphate) Na_2PO_3F.
Properties: Colorless crystals; m.p. 625°C (approx.); soluble in water.
Grade: 97%.
Hazard: Toxic; strong irritant.
Uses: Preparation of bactericides and fungicides.

sodium fluorosilicate (sodium silicofluoride, sodium fluosilicate). Na_2SiF_6.
Properties: White, odorless, tasteless, free-flowing powder; sp. gr. 2.7; m.p., decomposes at red heat; very slightly soluble in cold water; more soluble in hot water; insoluble in alcohol.
Derivation: From fluosilicic acid and sodium carbonate, or sodium chloride.
Grades: Technical; C.P.
Containers: Multiwall bags; fiber drums; various sized barrels.
Hazard: Toxic by ingestion and inhalation; strong irritant to tissue. Tolerance (as F), 2.5 mg per cubic meter of air.
Uses: Fluoridation of drinking water; laundry sours; opalescent glass; vitreous enamel frits; metallurgy of aluminum and beryllium; insecticides and rodenticides; chemical intermediate; glue, leather and wood preservative.

sodium formaldehyde bisulfite $HOCH_2SO_3Na$.
Properties: White water-soluble solid.
Derivation: Action of sodium bisulfite, formaldehyde and water.
Use: Textile stripping agent.
See also hydrosulfite-formaldehyde compounds.

sodium formaldehyde hydrosulfite. See hydrosulfite-formaldehyde compounds. Also used in synthetic rubber polymerization, and textile dyeing and stripping agent.

sodium formaldehyde sulfoxylate (sodium sulfoxylate; sodium sulfoxylate formaldehyde; SFS) $HCHO \cdot HSO_2Na \cdot 2H_2O$.
Properties: White solid; m.p. 64°C; soluble in water and alcohol.
Purity: Usually admixed with a sulfite.
Containers: Drums; truckloads.
Hazard: Probably toxic.
Uses: Stripping and discharge agent for textiles; bleaching agent for molasses, soap.
See also hydrosulfite-formaldehyde preparations.

sodium formate HCOONa.
Properties: White, slightly hygroscopic, crystalline powder, slight odor of formic acid; soluble in water; slightly soluble in alcohol; insoluble in ether. Sp. gr. 1.919; m.p. 253°C.
Derivation: Carbon monoxide and sodium hydroxide are heated under pressure; also from pentaerythritol manufacture.
Grades: Technical; C.P.
Containers: Bags; carlots.
Uses: Reducing agent; manufacture of formic acid and oxalic acid; organic chemicals; mordant; manufacture of sodium dithionite; complexing agent; analytical reagent.

sodium gentisate $C_7H_5O_4Na$. The sodium salt of gentisic acid.
Proerties: Crystallizes with 5.5 H_2O from water; loses $3H_2O$ rapidly upon exposure to air; soluble in water.
Use: Medicine.

sodium glucoheptonate $HOCH_2(CHOH)_5COONa$. Light tan crystalline powder. A sequstering agent for polyvalent metals.
Uses: Metal cleaning; bottle washing; kier boiling;

mercerizing; caustic boiloff; paint stripping; aluminum etching.

sodium gluconate $NaC_6H_{11}O_7$.
Properties: White to yellowish, crystalline powder; readily soluble in water; sparingly soluble in alcohol. Low toxicity.
Derivation: From glucose by fermentation.
Grades: Purified; technical; F.C.C.
Containers: Fiber drums; bags.
Use: Foods and pharmaceuticals; sequestering agent; metal cleaners; paint stripper; aluminum deoxidizer; bottle-washing preparations; rust removal.

sodium glucosulfone $C_{24}H_{34}N_2Na_2O_{18}S_3$. para′-diaminodiphenylsulfone-N,N′-di-(dextrose sodium sulfonate). The U.S.P. grade is an aqueous solution (for injection), clear and pale yellow; pH 5.0–6.5. Used in medicine.

sodium glutamate (monosodium glutamate; MSG) $COOH(CH_2)_2CH(NH_2)COONa$. Sodium salt of glutamic acid, one of the common naturally occurring amino acids.
Properties: White crystalline powder; m.p., decomposes; soluble in water and alcohol. Shows optical activity. Most effective between pH 6 and 8. Nontoxic.
Derivation: (a) Alkaline hydrolysis of the waste liquor from beet sugar refining; (b) a similar hydrolysis of wheat or corn gluten; (c) organic synthesis based on acrylonitrile.
Grades: Technical, 99%; N.D.; F.C.C.
Containers: Bottles, cans, fiber drums; retail shaker cans; paper bags; bulk.
Uses: Flavor enhancer for foods, in concentrations of about 0.3%. See also flavor; glutamic acid.

sodium glycolate (sodium hydroxyacetate) $NaOOCCH_2OH$.
Properties: White powder.
Uses: Buffer in electroless plating and textile finishing.

sodium gold chloride (sodium aurichloride; sodium chloraurate; sodium chloroaurate; gold sodium chloride; gold salts) $NaAuCl_4 \cdot 2H_2O$.
Properties: Yellow crystals; soluble in water and alcohol.
Derivation: By neutralizing chloroauric acid with sodium carbonate.
Uses: Photography; staining fine glass; decorating porcelain; medicine.

sodium gold cyanide (sodium cyanoaurite; sodium aurocyanide; gold sodium cyanide) $NaAu(CN)_2$.
Properties: Yellow powder; soluble in water; contains 46% gold (min).
Hazard: Toxic. Tolerance (as CN), 5 mg per cubic meter of air.
Use: For gold-plating radar and electric parts, razor holders, lamps, clocks, jewelry and tableware.

sodium guanylate (GMP; disodium gyanylate) $Na_2C_{10}H_{12}N_5O_8P \cdot 2H_2O$. A 5′-nucleotide. See also guanylic acid.
Properties: Crystals; soluble in cold water; very soluble in hot water.
Derivation: From a seaweed or from dried fish.
Use: Flavor potentiator in foods.

sodium gynocardate (sodium hydnocarpate). Sodium salt of the fatty acids in chaulmoogra oil.
Properties: Yellow powder. Soluble in alcohol, water. Aqueous solution slightly alkaline.
Use: Medicine.

sodium heparin. See heparin.

sodium heptametaphosphate. See sodium metaphosphate.

sodium hexachloroosmate (osmium-sodium chloride; sodium-osmium chloride) Na_2OsCl_6.
Properties: Orange, rhombic prisms. Contains 40.3% osmium. Unstable. Soluble in alcohol, water.
Grade: Technical.
Use: Catalyst (oxidation).

sodium hexametaphosphate. See "Calgon."

sodium hexylene glycol monoborate $C_6H_{12}O_3BNa$.
Properties: Amorphous white solid; bulk density 0.25 g/cc; m.p. 426°C. Soluble in nonpolar solvents.
Purity: Minimum 98%.
Uses: Corrosion inhibitor in organic systems; additive to lubricating oils; flame-retardant; siloxane resin additive.

sodium hydnocarpate. See sodium gynocardate.

sodium hydrate. See sodium hydroxide.

sodium hydride NaH.
Properties: Practically odorless powder; sp. gr. 0.92; m.p. 800°C (dec). Must be kept cool and dry. Particle size range 5-50μ. Starts to decompose with evolution of hydrogen at about 255°C.
Preparation: Reaction of sodium metal with hydrogen. A microcrystalline dispersion of gray powder in oil, concentration 50 or 25% by weight.
Containers: The dispersion is packed in polyethylene bags in a metal can. Packaging ranges from 1 pint to 55-gal drums.
Hazard: Dangerous fire risk. Reacts violently with water evolving hydrogen.
Uses: Condensing or alkylating agent, especially for amines; descaling metals.
Shipping regulations: (Rail) Yellow label. (Air) Flammable Solid label. Not acceptable on passenger planes.

sodium hydrosulfide (sodium sulfhydrate) $NaSH \cdot 2H_2O$.
Properties: Colorless needles to lemon-colored flakes. Soluble in water, alcohol, and ether. 70–72% NaSH; m.p. 55°C; water of crystallization 26–28%.
Derivation: From calcium sulfide by treating it in the cold with sodium bisulfate.
Grades: Technical; flake, 70–72%; solution, 40–44%.
Containers: Drums; tank cars.
Hazard: Contact with acids causes evolution of toxic gases.
Uses: Paper pulping; dyestuffs processing; rayon and cellophane desulfurizing; unhairing hides; bleaching reagent.

sodium hydrosulfite. See sodium dithionite.

sodium hydroxide (caustic soda; sodium hydrate; lye; white caustic) NaOH. The most important commercial caustic. Sixth highest-volume chemical produced in U.S. (1975).
Properties: White deliquescent solid, now chiefly in form of beads or pellets; also 50% and 73% aqueous solution. Absorbs water and carbon dioxide from the air. Sp. gr. 2.13; m.p. 318°C; b.p. 1390°C. Soluble in water, alcohol, and glycerol.
Derivation: Electrolysis of sodium chloride (brines) (there are several variations).
Grades: Commercial; ground; flake; beads; F.C.C.; granulated (60% and 76% Na_2O); rayon (low in iron, copper, and manganese); purified by alcohol (sticks, lumps, and drops); reagent; highest purity: C.P., U.S.P.
Containers: U.S.P., C.P.: 1-lb bottles; 5-, 10-lb cans;

commercial; drums and barrels; solution; drums; tank trucks and cars; barges.
Hazard: Corrosive to tissue in presence of moisture; strong irritant to tissue (eyes, skin, mucous membranes). Tolerance, 2 mg per cubic meter of air. Safety data sheet available from Manufacturing Chemists Assn., Washington, D.C.
Uses: Chemical manufacture; rayon and cellophane; petroleum refining; pulp and paper; aluminum; detergents; soap; textile processing; vegetable oil refining; reclaiming rubber; regenerating ion exchange resins; organic fusions; peeling of fruits and vegetables in food industry; laboratory reagent; etching and electroplating.
Shipping regulations: (Rail) (solution) White label. (Air) (solid and solution) Corrosive label.

sodium hypochlorite $NaOCl \cdot 5H_2O$.
Properties: Unstable in air unless mixed with sodium hydroxide. Strong oxidizing agent, usually stored and used in solution; disagreeable, sweetish odor and pale greenish color. Soluble in cold water; decomposed by hot water. M.p. 18°C.
Derivation: Addition of chlorine to cold dilute solution of sodium hydroxide.
Grade: Technical.
Containers: Carboys; drums.
Hazard: Strong irritant to tissue. Fire risk in contact with organic materials.
Uses: Bleaching paper pulp, textiles, etc.; intermediate; organic chemicals; water purification; medicine; fungicides; swimming pools; household bleach; laundering; reagent.
Shipping regulations: (Rail) Hypochlorite solutions containing more than 7% available chlorine: White label. (Air) Hypochlorite solutions: Corrosive label.

sodium hypophosphite $NaH_2PO_2 \cdot H_2O$.
Properties: Colorless, pearly, crystalline plates or white, granular powder; saline taste. Deliquescent. Soluble in water, partially in alcohol.
Derivation: By neutralizing hypophosphoric acid with sodium carbonate.
Grades: Technical; C.P.
Containers: Glass bottles; drums; car lots; truck loads.
Hazard: Explosion risk when mixed with strong oxidizing agents; decomposes to phosphine on heating. Store in cool, dry place, away from oxidizing materials.
Uses: Medicine; reducing agent in electroless nickel plating of plastics and metals; laboratory reagent.
Shipping regulations: Consult authorities.

sodium inosinate $C_{10}H_{11}Na_2N_4O_8P$. A 5′-nucleotide derived from seaweed or dried fish. Sodium guanylate (q.v.) is a byproduct.
Use: Flavor potentiator in foods.
See also inosinic acid.

sodium iodate $NaIO_3$.
Properties: White crystals; sp. gr. 4.28. Soluble in water and acetone; insoluble in alcohol. Low toxicity.
Derivation: Interaction of sodium chlorate and iodine in presence of nitric acid.
Grades: C.P.; reagent; technical.
Hazard: Oxidizing agent; fire risk near organic materials.
Uses: Medicine; disinfectant; feed additive; reagent.

sodium iodide (a) NaI (b) $NaI \cdot 2H_2O$.
Properties: White cubical crystals or powder, or colorless, odorless crystals; slowly becomes brown in air; deliquescent; saline, somewhat bitter taste. Soluble in water, alcohol, and glycerin. Sp. gr. (a) 3.665; (b) 2.448 (21°C). M.p. (a) 653°C; b.p. (a) 1304°C.
Grades: Technical; C.P.; U.S.P.; single crystals.
Containers: 25-, 100-, 300-lb drums.
Hazard: Moderately toxic.
Uses: Photography; solvent for iodine; organic chemicals; reagent; medicine; feed additive; artificial rainmaking; scintillation (thallium-activated form).

sodium iodide I 131 (sodium radio-iodide). A radioactive form of sodium iodide containing iodine 131 which can be used as a tracer. See iodine 131.
Grade: U.S.P. (as capsules or solution).

sodium iodipamide $C_{20}H_{12}I_6N_2Na_2O_6$. N,N′-adipolybis-(3-amino-2,4,6-triiodobenzoic acid) disodium salt.
Properties: Radiopaque; water-soluble. Available as a 20% solution for injection as a clear colorless to pale yellow, slightly viscous liquid.
Derivation: By dissolving the free acid in dilute sodium hydroxide and buffering to pH 6.5–7.7
Grade: U.S.P.
Use: X-ray contrast medium.

sodium iodomethanesulfonate. See sodium methiodal.

sodium iothalamate (sodium 5-acetamido-2,4,6-triiodo-N-methylisophthalamate) $C_6I_3(CONHCH_3)_2COONa$.
Grade: U.S.P. (as injection).
Use: Medicine (radiopaque medium).

sodium ipodate (USAN) (sodium 3-(dimethylaminomethyleneamino)-2,4,6-triiodohydrocinnamate) $NaOOCCH_2CH_2C_6H(I_3)N{:}CHN(CH_3)_2$.
Uses: Radiopaque agent in medicine.

sodium iron pyrophosphate (SIPP) $Na_8Fe_4(P_2O_7)_5 \cdot xH_2O$.
Properties: 300 Mesh powder; tan in color; insoluble in water but soluble in dilute acid. Minimum 14.5% iron. Iron is in complex form and will not catalyze oxidation reactions.
Derivations: By reacting tetrasodium pyrophosphate with a soluble iron salt.
Containers: Drums.
Use: For iron enrichment, particularly in flours and cereals.

sodium isoascorbate. See sodium erythorbate.

sodium isobutylxanthate $CH(CH_3)_2CH_2OC(S)SNa$. See xanthic acid.
Use: Ore flotation agent.

sodium isopropylnaphthalenesulfonic acid. See "Vatsol."

sodium isopropylxanthate $(CH_3)_2CHOC(S)SNa$.
Properties: Light-yellow crystals soluble in water; deliquescent; decomposes 150°C. Low toxicity.
Hazard: Irritant to skin and mucous membranes. Moderate fire risk.
Uses: Chemical weed killer of seeded crops; fortifying agent for certain oils; flotation reagent for base and precious metal ores.

sodium isovalerate. See sodium valerate.

sodium lactate $CH_3CHOHCOONa$.
Properties: Colorless or yellowish syrupy liquid, very hygroscopic. Soluble in water. M.p. 17°C; decomposes 140°C. Combustible; low toxicity.

Grades: Technical; U.S.P. (solution with pH 6.0 to 7.3).
Uses: Medicine; hygroscopic agent; glycerin substitute; plasticizer; corrosion inhibitor in alcohol antifreeze.

sodium N-lauroyl sarcosinate $C_{15}H_{28}NO_4Na$. Available as 30% aqueous solution in 55-gal drums.
Uses: Dentrifices, hair shampoos, rug shampoos.

sodium lauryl sulfate $NaC_{12}H_{25}SO_4$.
Properties: Small, white or light yellow crystals; slight characteristic odor. Soluble in water, forming an opalescent solution. Low toxicity.
Grades: U.S.P.; technical; F.C.C.
Containers: 55-gal drums; tank cars.
Uses: Wetting agent in textiles; detergent in toothpaste; food additive and surfactant.

sodium lead alloy. One of several alloys, as follows: (1) usually containing 10% sodium and 90% lead, used in the manufacture of lead tetraethyl; (2) containing 2% sodium used as a deoxidizer and homogenizer in nonferrous metals where lead is a component; (3) used as a stabilizer and deoxidizer for lead in cable sheathing.
Hazard: Toxic; moderate fire and explosion risk; reacts with moisture, acids, and oxidizing agents.

sodium lead hyposulfite. See lead sodium thiosulfate.

sodium lead thiosulfate. See lead sodium thiosulfate.

sodium lignosulfonate.
Properties: Tan, free-flowing spray-dried powder, containing 70–80% total lignin sulfonates, balance wood sugars. Combustible; low toxicity.
Containers: 50-lb bags.
Uses: Dispersant, emulsion stabilizer, chelating agent. See also lignin sulfonate.

sodium liothyronine
$NaOOCCH(NH_2)CH_2C_6H_2I_2OC_6H_3IOH$. Sodium L-3[4-(4-hydroxy-3-iodophenoxy)-3,5-diiodophenyl]-alanine.
Properties: Light-tan, odorless, crystalline powder. Very slightly soluble in water. Slightly soluble in alcohol, insoluble in most other organic solvents.
Grade: U.S.P.
Use: Medicine (a thyroid hormone).

sodium MBT (NaMBT) $C_7H_4NS_2Na$. A 50% aqueous solution of sodium mercaptobenzothiazole. Light-amber liquid; density 10.5 lb/gal.
Uses: Corrosion inhibitor for nonferrous metals; antifreeze, paper mill systems.

sodium mercaptoacetate. See sodium thioglycolate.

sodum 2-mercaptobenzothiazole. See sodium MBT.

sodium metabisulfite (sodium pyrosulfite) $Na_2S_2O_5$. Chief constituent of commercial dry sodium bisulfite, with which most of its properties and uses are practically identical.
Grade: F.C.C.
Uses: In foods, as preservative; laboratory reagent.

sodium metaborate $NaBO_2$.
Properties: White lumps; sp. gr. 2.464. Soluble in water. M.p. 966°C; b.p. 1434°C. Low toxicity. Noncombustible.
Derivation: By fusing sodium carbonate and borax.
Use: Herbicide.
Also available commercially as octahydrate and tetrahydrate.

sodium metanilate $NaSO_3C_6H_4NH_2$.
Derivation: The meta-sodium sulfonate of aniline sold as a solid or 20% aqueous solution prepared by neutralizing metanilic acid.
Grades: Technical; 99%; also 20% solution.
Containers: Barrels; drums.
Use: Manufacture of synthetic dyestuffs and drugs.

sodium metaperiodate. See sodium periodate.

sodium metaphosphate $(NaPO_3)_n$, where *n* ranges from 3 to 10 (cyclic molecules) or may be a much larger number (polymers). Cyclic sodium metaphosphate, based on rings of alternating phosphorus and oxygen atoms, range from the trimetaphosphate, $(NaPO_3)_3$, to at least the decametaphosphate. So-called sodium hexametaphosphate is probably a polymer where *n* is between 10 and 20 (see "Calgon").

The vitreous sodium phosphates having a Na_2O/P_2O_5 mole ratio near unity are classified as sodium metaphosphates (Graham's salts). The average number of phosphorus atoms per molecule in these glasses ranges from about 25 to infinity. The term sodium metaphosphate has also been extended to short-chain vitreous compositions, the molecules of which exhibit the polyphosphate formula $Na_{n+2}P_nO_{3n+1}$ with n as low as 4–5. These materials are more correctly called sodium polyphosphates (q.v.).
Uses: Dental polishing agents; detergent builders; water softening; sequestrants; emulsifiers; food additives; textile processing; laundering.

sodum metasilicate, anhydrous. Na_2SiO_3. A crystalline silicate.
Properties: Dustless white granules; m.p. 1089°C; total Na_2O content 51.5%; total Na_2O in active form 48.6%; density 2.61 or 75 lb/cu ft. Soluble in water; precipitated by acids and by alkaline earth and heavy metal ions; pH of 1% solution 12.6. Nontoxic; noncombustible.
Derivation: Crystallized from a melt of Na_2O and SiO_2 below 1089°C.
Containers: Bags; fiber drums; carlots.
Uses: Laundry, dairy and metal cleaning; floor cleaning; base for detergent formulations; bleaching aid; deinking paper.
Also available as the pentahydrate, whose properties are m.p. 72.2°C; total Na_2O content 29.3%; total Na_2O in active form 27.8%; density 1.75, or 55 lb per cu ft.

sodium metavanadate $NaVO_3$, often with $4H_2O$.
Properties: Colorless, monoclinic, prismatic crystals, or pale green crystalline powder. Soluble in water. M.p. 630°C. Noncombustible; low toxicity.
Derivation: Sodium hydrate and vanadium pentoxide in water solution.
Grades: Technical; C.P.
Containers: Glass bottles; fiber drums.
Uses: Inks; fur dyeing; photography; inoculation of plant life; mordants and fixers; corrosion inhibitor in gas-scrubbing systems.

sodium methacrylate $CH_2{:}C(CH_3)COONa$.
Properties: Water-soluble monomer.
Containers: Drums.
Uses: Resins; chemical intermediate.

sodium methanearsonate.
(1) See disodium methylarsonate.
(2) (monosodium methanearsonate; sodium acid methanearsonate; sodium methylarsonate) $CH_3AsO(OH)(ONa)$.
Properties: White solid; m.p. 130–140°C; very soluble in water.
Hazard: Highly toxic by ingestion and inhalation.

Use: Post-emergence herbicide for grassy weeds.
Shipping regulations: (Rail, Air) Arsenical compounds, n.o.s., Poison label.

sodium methiodal (sodium iodomethanesulfonate) ICH_2SO_3Na.
Properties: A white crystalline powder; odorless with slight salty taste followed by sweetish aftertaste. Decomposes on exposure to light; solutions are neutral to litmus; soluble in water; very soluble in methanol; slightly soluble in alcohol; practically insoluble in acetone, ether and benzene.
Derivation: From sodium sulfite and methylene iodide.
Grade: N.F.
Use: Medicine (as radiopaque contrast medium).

sodium methylate (sodium methoxide) CH_3ONa.
Properties: White, free-flowing powder; sensitive to oxygen; decomposed by water. Soluble in methyl and ethyl alcohol. Decomposes in air above 260° F.
Containers: Steel drums; also 25% solution in methanol in drums. Bulk density 4.15 lb/gal.
Hazard: (Solid) Flammable when exposed to heat or flame. (Solution). Flammable, moderate fire risk. Solution contains methyl alcohol, and is toxic and corrosive to tissue.
Uses: Condensation reactions; catalyst for treatment of edible fats and oils, especially lard; intermediate for pharmaceuticals; preparation of sodium cellulosate; analytical reagent.
Shipping regulations: (Rail) (dry) Yellow label; (alcohol mixture) Red label. (Air) (dry) Flammable Solid label; (alcohol mixture) Flammable Liquid label; Corrosive label.

sodium methyl carbonate $CH_3OCOONa$.
Properties: White powder; m.p. 330° C (decomposes); sp. gr. 1.66; purity 90% min.

sodium N-methyldithiocarbamate dihydrate $CH_3NHC(S)SNa \cdot 2H_2O$.
Properties: White crystalline solid; readily soluble in water; moderately soluble in alcohol; stable in concentrated aqueous solution but decomposed in dilute aqueous solution; unstable in moist soil.
Hazard: Irritant to tissue; may be toxic to plants and vegetation.
Uses: Fungicide, nematocide, weed killer, and insecticide; soil fumigant.
See also "Vapam."

sodium N-methyl-N-oleoyl taurate (oleyl methyl tauride)$(CH_3)(CH_2)_7CHCH(CH_2)_7CON(CH_3)CH_2CH_2SO_2ONa$.
Properties: Fine white powder. Sweet odor. Non-toxic.
Grades: Technical, 32% purity (remainder is mainly sodium sulfate).
Uses: Detergent; pesticide aid.

sodium methyl siliconate. Most effective water-repellent and cleaner for limestone, concrete and similar masonry. Reaction product of aqueous sodium hydroxide and a resinous silicone. Total solids about 30%.

sodium N-methyltaurate. See N-methyltaurine.

sodium molybdate. Commercially, the normal molybdate Na_2MoO_4 or its dihydrate (called sodium molybdate crystals). Chemically, a wide variety of complex molybdates with sodium are known.
Properties: Small, lustrous, crystalline plates; soluble in water; m.p. 687°C; sp. gr. 3.28 (18°C). Noncombustible.
Derivation: By the action of sodium hydroxide on molybdenum trioxide. Complex molybdates are prepared by dissolving large amounts of molybdenum trioxide in solutions of normal molybdates.
Grades: Anhydrous; crystals.
Containers: 100-, 150-, 200-, 385-lb drums.
Hazard: Moderately toxic and irritant.
Uses: Reagent in analytical chemistry; medicine; paint pigment; production of molybdated toners and lakes; metal finishing; fertilizer; brightening agent for zinc plating; corrosion inhibitor; catalyst in dye and pigment production; additive for fertilizers and feeds.

sodium 12-molybdophosphate (sodium phospho-12-molybdate) $Na_3PMo_{12}O_{40}$.
Properties: Yellow crystals; sp. gr. 2.83. Soluble in water, although less soluble than the free acid; strong oxidizing action in aqueous solution.
Grade: Technical.
Uses: Analysis; neuromicroscopy; catalysts; additives in photographic processes; imparting water resistance to plastics adhesives, and cements; pigments.
See also heteromolybdate.

sodium 12-molybdosilicate (sodium silico-12-molybdate) $Na_4SiMo_{12}O_{40} \cdot xH_2O$.
Properties: Yellow crystals; density 3.44 g/cc; soluble in water, acetone, alcohol, ethyl acetate. Insoluble in ether, benzene, and cyclohexane.
Grade: Reagent.
Containers: 24-gal drums.
Hazard: Water solution has strong oxidizing action; store away from organic materials.
Uses: Catalysts; reagents; fixing and oxidizing agents in photography; precipitants and ion exchangers in atomic energy; plating processes; imparting water resistance to plastics, adhesives, and cement.

sodium monofluorophosphate (MFP). Tooth decay preventive in dental cream.

sodium monoxide (sodium oxide) Na_2O.
Properties: White powder, sp. gr. 2.27; sublimes 1275°C. Soluble in molten caustic soda or potash. Converted to sodium hydroxide by water.
Containers: 60-lb pails; 300-lb drums.
Hazard: Caustic and strong irritant when wet with water.
Uses: Condensing or polymerizing agent in organic reactions; dehydrating agent; strong base.
Shipping regulations: (Air) Corrosive label.

sodium myristyl sulfate $NaC_{14}H_{29}SO$. Anionic detergent.
Uses: Foaming, wetting, and emulsifying in cosmetic, household and industrial uses.

sodium naphthalenesulfonate $C_{10}H_7SO_3Na$.
Properties: Yellowish, crystalline plates, or white, odorless scales. Soluble in water; insoluble in alcohol. Could be either the 1- or 2-(alpha or beta) sulfonate. Combustible.
Grade: Technical.
Containers: 360-, 500-lb barrels.
Hazard: May be toxic by ingestion and inhalation.
Uses: Organic preparations; liquefying agent in animal glue preparations; naphthols.

sodium naphthenate. A white paste. The most important of the naphthenic acid salts. Commercial sam-

ples have consistency of grease, but this will vary with source and manner of processing. Excellent emulsifying and foam-producing properties; low hydrolytic dissociation.
Uses: Detergent; emulsifier; disinfectant; manufacture of paint driers. See also naphthenic acid.

sodium naphthionate (sodium alpha-naphthylaminesulfonate) $NH_2C_{10}H_6SO_3Na \cdot 4H_2O$.
Properties: White crystals, become violet on exposure to light. Soluble in water; insoluble in ether. Combustible.
Grades: Technical (paste, crystals).
Uses: Riegler's reagent for nitrous acid; manufacture of dyestuffs.

sodium naphtholsulfonates. See various naphtholsulfonic acids, sodium salts.

sodium naphthylaminesulfonate. See sodium naphthionate.

sodium niobate (sodium columbate) $Na_2Nb_2O_6 \cdot 7H_2O$. Important in the purification of niobium materials. The crystalline compound forms when a niobium compound is treated with hot concentrated sodium hydroxide. It is sparingly soluble in water.

sodium nitrate (soda niter) $NaNO_3$. Chile saltpeter (caliche) is impure natural sodium nitrate.
Properties: Colorless, transparent, odorless crystals; saline, slightly bitter taste; sp. gr. 2.267; m.p. 308°C; explodes at 1000°F; decomp. at 380°C. Soluble in water and glycerol; slightly soluble in alcohol.
Derivation: From nitric acid and sodium carbonate; and from Chile saltpeter.
Grades: Granular, sticks, powder; crude; 99.5%; double refined; recrystallized; C.P.; technical; reagent; diuretic; F.C.C.
Containers: Tins; glass bottles; multiwall paper sacks; bulk.
Hazard: Fire risk near organic materials. Ignites on friction and explodes when shocked or heated to 1000°F. Content in cured meats, fish, and other food products restricted to 500 ppm. USDA has proposed its elimination.
Uses: Oxidizing agent; solid rocket propellants; fertilizer; flux; glass manufacture; pyrotechnics; reagent; medicine; refrigerant; matches; dynamites; black powders; manufacturing sodium salts and nitrates; dyes; pharmaceuticals; anaphrodisiac; color fixative and preservative in cured meats, fish, etc.; enamel for pottery; modifying burning properties of tobacco.
Shipping regulations: (Rail) Yellow label. (Air) Oxidizer label.

sodium nitrilotriacetate. See nitrilotriacetic acid.

sodium nitrite $NaNO_2$.
Properties: Slightly yellowish or white crystals, pellets, sticks or powder. Oxidizes on exposure to air. Soluble in water; slightly soluble in alcohol and ether. Sp. gr. 2.157; m.p. 271°C; b.p. explodes at 1000°F; decomp. at 320°C.
Grades: Reagent; technical; U.S.P.; F.C.C.
Containers: 25-, 100-lb drums; 150-lb kegs; 400-lb barrels.
Hazard: Dangerous fire and explosion risk when heated to 1000°F or in contact with organic materials. Strong oxidizing agent. Content in meats and fish products restricted to 200 ppm, and may be reduced to 100 ppm.
Uses: Diazotization (by reaction with HCl to form nitrous acid); rubber accelerators; color fixative and preservative in cured meats, meat products, fish, etc.; pharmaceuticals; photographic and analytical reagent; dye manufacture; preparation of nitric oxide.
Shipping regulations: (Rail) Yellow label. (Air) Oxidizer label.

sodium nitroferricyanide (sodium nitroprussiate; sodium nitroprusside) $Na_2Fe(CN)_5NO \cdot 2H_2O$.
Properties: Red, transparent crystals; sp. gr. 1.72. Soluble in water, with slow decomposition; slightly soluble in alcohol.
Grades: Reagent; technical.
Use: Analytical reagent.

sodium para-nitrophenolate. See para-nitrophenol, sodium salt.

sodium novobiocin. See novobiocin.

sodium nucleinate (sodium nucleate).
Properties: Yellowish-white almost odorless powder containing approximately 4.5% sodium. Soluble in water; insoluble in alcohol.
Derivation: From yeast.
Use: Medicine.

sodium octyl sulfate $C_8H_{17}OSO_3Na$. Anionic detergent. Available commercially as a 35% solution.
Uses: Wetting, dispersing, and emulsifying agent.

sodium oleate $C_{17}H_{33}COONa$.
Properties: White powder; slight tallow-like odor. M.p. 232–235°C. Soluble in water with partial decomposition; soluble in alcohol. Combustible; low toxicity.
Derivation: Action of alcoholic sodium hydroxide on oleic acid.
Containers: Bags; barrels; drums.
Uses: Ore flotation; waterproofing textiles; emulsifier of oil/water systems.

sodium orthophosphate. See sodium phosphate, mono-, di, and tribasic.

sodium orthosilicate $Na_2SiO_3 \cdot 2NaOH$ or other proportions such as $2\ Na_2O \cdot SiO_2$ (anhydrous) or $2Na_2O \cdot SiO_2 \cdot 5.4H_2O$.
Properties: (Composition $2Na_2O \cdot SiO_2$). Dustless white, flaked product; density 75 lb/cu ft; total Na_2O content 60.8%; percent of total Na_2O in active form 59.0%. Soluble in water; pH of a 1% solution 13.0.
Containers: Bags; fiber drums; carlots.
Hazard: Strong irritant to skin, eyes, and mucous membranes.
Uses: Commercial laundries; metal cleaning; heavy-duty cleaning.

sodium oxalate $Na_2C_2O_4$.
Properties: White, crystalline powder; sp. gr. 2.34; m.p. 250–270°C (decomposes). Soluble in water; insoluble in alcohol.
Grades: Reagent; technical, 88%, 99%.
Containers: 100-lb bags; 225-lb barrels.
Hazard: Moderately toxic by ingestion.
Uses: Reagent; textile finishing; pyrotechnics; leather finishing; blue printing.

sodium oxide. See sodium monoxide.

sodium palconate. The sodium salt of an acid that may be extracted with alkali from redwood dust. The dark reddish brown material consists mainly of a partially methylated phenolic acid containing aliphatic hydroxyls, phenolic hydroxyls and carboxyl groups in the ratio of 2:4:3. The viscosity of aqueous solutions rises rapidly with concentration.
Uses: To control viscosity and water loss in drilling muds, and as a dispersing agent.

sodium palladium chloride. See palladium sodium chloride.

sodium palmitate $CH_3(CH_2)_{14}COONa$. The sodium salt of palmitic acid. Combustible. Low toxicity.
Uses: Polymerization catalyst for synthetic rubbers; laundry and toilet soaps; detergents; cosmetics; pharmaceuticals; printing inks; emulsifier.

sodium paraperiodate (sodium triparaperiodate) $Na_3H_2IO_6$.
Properties: White crystalline solid; very slightly soluble in water; soluble in concentrated sodium hydroxide solutions.
Uses: Selectively oxides specific carbohydrates and amino acids; wet-strengthens paper; aids combustion of tobacco.

sodium pentaborate $Na_2B_{10}O_{16} \cdot 10H_2O$.
Properties: White crystals; free flowing; stable under ordinary conditions; solubility in water, 15.40% (20°C), increasing with temperature; sp. gr. 1.72; pH of solution approx 7.5. Noncombustible; low toxicity.
Containers: 100-lb paper bags.
Uses: Weed killer; cotton defoliant; fireproofing compositions; glass manufacture; boron supplement for tree fruit and truck crops.

sodium pentachlorophenate C_6Cl_5ONa.
Properties: White or tan powder; soluble in water, ethanol, and acetone; insoluble in benzene.
Containers: Bags; drums; carlots.
Grades: Technical; powder, pellets, or briquettes.
Hazard: Avoid skin contact and dust inhalation.
Uses: Fungicide, herbicide, slimicide; fermentation disinfectant, especially in finishes and papers.

sodium pentobarbital. See barbiturate.

"Sodium Pentothal." See "Pentothal," thiopental.

sodium perborate (sodium metaborate peroxyhydrate) $NaBO_2 \cdot 3H_2O$ or $1H_2O$. Often described as sodium peroborate tetrahydrate, $NaBO_3 \cdot 4H_2O$; see also sodium borate perhydrate.
Properties: White, odorless crystals or powder; salty taste. M.p. 63°C; loses H_2O at 130–150°C. Stable in cool, dry air but decomposes with evolution of oxygen in warm or moist air. Moderately soluble in water (with decomposition) and glycerol. pH of aqueous solutions 10.0 to 10.3. Active oxygen content 10% min.
Derivation: (a) Electrolysis of a solution of borax and soda ash; (b) crystallization from solution of borax or boric acid, sodium peroxide, and hydrogen peroxide.
Grades: Technical; C.P.; N.F. (as $NaBO_3 \cdot 4H_2O$).
Containers: Cartons; boxes; bags.
Hazard: Moderately toxic by ingestion. Fire risk in contact with organic materials. Strong oxidizing agent.
Uses: Developing vat dyes; textile bleaching; synthetic detergents; neutralizing cold wave preparations; dental compositions; electroplating; laboratory reagent; germicide; deodorant; mouthwash.

sodium percarbonate $Na_2C_2O_6$ or Na_2CO_4. Decomposes in aqueous solution to hydrogen peroxide and sodium carbonate. A stable form has been developed and is commercially available.
Hazard: Moderately toxic by ingestion. Fire risk near organic materials. Strong oxidizing agent.
Uses: Substitute for sodium perborate in detergents.

sodium perchlorate $NaClO_4$, sometimes with $1H_2O$.
Properties: White deliquescent crystals. Soluble in water and alcohol. M.p. 482°C; b.p., decomposes; sp. gr. 2.02.
Derivation: (a) Sodium chlorate and sodium chloride are mixed and heated until fused. The unchanged chloride is leached out. (b) A cold solution of sodium chlorate is electrolyzed, the solution concentrated and crystallized.
Hazard: Dangerous fire and explosion risk in contact with organic materials and sulfuric acid. Moderately toxic and irritant.
Uses: Explosives; jet fuel; analytical reagent.
Shipping regulations: (Rail) Perchlorates, n.o.s., Yellow label. (Air) Oxidizer label.

sodium periodate (sodium metaperiodate) (a) $NaIO_4$; (b) $NaIO_4 \cdot 3H_2O$. See also sodium paraperiodate.
Properties: Colorless crystals; sp. gr. (a) 3.865 (16°C); (b) 3.219 (18°C); m.p. (a) 300°C (dec), (b) 175°C (dec). Very soluble in water.
Containers: Glass bottles; steel pails.
Hazard: Toxic by ingestion; fire risk in contact with organic materials.
Uses: Source of periodic acid; analytical reagent; oxidizing agent.

sodium permanganate $NaMnO_4 \cdot 3H_2O$.
Properties: Purple to reddish-black crystals or powder; soluble in water. Sp. gr. 2.47; m.p. 170°C (dec).
Derivation: Sodium manganate is dissolved in water and chlorine or ozone passed in. The solution is concentrated and crystallized.
Grades: Technical; sold commercially in solution.
Containers: Wooden barrels; steel drums; glass bottles.
Hazard: Toxic; dangerous fire risk in contact with organic materials. Strong oxidizing agent.
Uses: Oxidizing agent; disinfectant; bactericide; manufacture of saccharin; antidote for poisoning by morphine, curare and phosphorus.
Shipping regulations: (Rail) Yellow label. (Air) Oxidizer label.

sodium peroxide Na_2O_2.
Properties: Yellowish white powder, turning yellow when heated. Absorbs water and carbon dioxide from air. Active oxygen content approximately 20% by weight; sp. gr. 2.805; m.p. 460°C (dec); b.p. 657°C (dec). Soluble in cold water, with evolution of heat.
Derivation: Metallic sodium is heated at 300°C in aluminum trays in a retort in a current of dry air, from which the carbon dioxide has been removed.
Grades: Technical; reagent.
Containers: Drums; carlots; truckloads.
Hazard: Dangerous fire and explosion risk in contact with water, alcohols, acids, powdered metals, and organic materials. Keep dry. Toxic by ingestion. Strong oxidizing agent.
Uses: Oxidizing agent; bleaching of miscellaneous materials including paper and textiles; deodorant; antiseptic; medicine; organic chemicals; water purification; pharmaceuticals; oxygen generation for diving bells, submarines, etc.; textile dyeing and printing; ore processing; analytical reagent; calorimetry; germicidal soaps.
Shipping regulations: (Rail) Yellow label. (Air) Oxidizer label. Not acceptable on passenger planes.

sodium persulfate (sodium peroxydisulfate) $Na_2S_2O_8$.
Properties: White, crystalline powder; soluble in water; decomposed by alcohol; decomposes in moist air.
Uses: Bleaching agent (fats, oils, fabrics, soap); battery depolarizers; medicine.

sodium phenate (sodium phenolate; sodium carbolate) C_6H_5ONa.
Properties: White, deliquescent crystals. Soluble in water and alcohol; decomposed by carbon dioxide in the air.
Derivation: Phenol is dissolved in caustic soda solution, concentrated and crystallized.
Hazard: Strong irritant to skin and tissue.
Uses: Antiseptic; salicylic acid; organic synthesis.
Shipping regulations: (Air) Corrosive label.

sodium phenobarbital (phenobarbital, soluble). See barbiturate.

sodium phenolate. Legal label name (Air) for sodium phenate.

sodium phenolsulfonate (sodium sulfocarbolate) $HOC_6H_4SO_3Na \cdot 2H_2O$.
Properties: Colorless crystals or granules, slightly efflorescent; chars at high temperature, evolving phenol. Soluble in water, hot alcohol, and glycerol.

sodium phenylacetate (sodium alpha-toluate) $C_6H_5CH_2 \cdot COONa$.
Properties: Soluble in water; insoluble in alcohol, ether, and ketones; 50% aqueous solution has pH 7.0 to 8.5 and is pale yellow. Solution tends to crystallize at 15°C.
Containers: 3000-gal tanks; 500-lb drums.
Grades: 50% solution; dry salt.
Uses: Precursor in production of penicillin G; intermediate for producing heavy metal salts which act as fungicides.

sodium N-phenylglycinamide-para-**arsonate.** See tryparsamide.

sodium ortho-**phenylphenate** (sodium ortho-phenylphenolate) $C_6H_4(C_6H_5)ONa \cdot 4H_2O$.
Properties: Practically white flakes. Bulk density 38–43 lb/cu ft; pH of saturated solution in water 12.0–13.5. Soluble in water, methanol, acetone.
Hazard: Probably toxic.
Use: Industrial preservative (bactericide and fungicide); mold inhibitor for apples and other fruit (postharvest).

sodium phenylphosphinate $C_6H_5PH(O)(ONa)$.
Properties: Crystals; m.p. 355°C (decomposes to give phenylphosphine); stable at room temperature; soluble in water.
Hazard: May be toxic.
Uses: Antioxidant; heat and light stabilizer.

sodium phosphate. See sodium metaphosphate; sodium phosphate, dibasic; sodium phosphate, monobasic; sodium phosphate P-32; sodium phosphate, tribasic; sodium polyphosphate; sodium pyrophosphate; sodium pyrophosphate, acid; sodium tripolyphosphate.

sodium phosphate, dibasic (DSP; disodium phosphate; sodium orthophosphate, secondary; disodium orthophosphate; disodium hydrogen phosphate) (a) Na_2HPO_4; (b) $Na_2HPO_4 \cdot 2H_2O$; (c) $Na_2HPO_4 \cdot 7H_2O$; (d) $Na_2HPO_4 \cdot 12H_2O$. The dihydrate (b) is also marketed as the duohydrate.
Properties: Colorless, translucent crystals or white powder; saline taste. Soluble in water; very soluble in alcohol. (a) Hygroscopic; converted to sodium pyrophosphate at about 240°C. (b) M.p. loses H_2O at 92.5°C; sp. gr. (15°C) 2.066; (c) Sp. gr. 1.679; loses $5H_2O$ at 48°C; (d) M.p. 35°C; sp. gr. 1.5235. Readily loses $5H_2O$ on exposure to air at room temperature; loses $12H_2O$ at 100°C. pH of 1% solution 8.0–8.8. Low toxicity. Nonflammable.
Derivation: (1) By treating phosphoric acid with a slight excess of soda ash, boiling the solution to drive off carbon dioxide, and cooling to permit the dodecahydrate to crystallize; (2) by precipitating calcium carbonate from a solution of dicalcium phosphate with soda ash.
Grades: Commercial; N.F. (a) and (c); F.C.C. (a) or (b).
Containers: Paper bags; fiber drums; barrels.
Uses: Chemicals; dyes; fertilizers; pharmaceuticals; medicine; textiles (weighting silk, dyeing and printing); fireproofing wood, paper and ceramic glazes; tanning; paint pigments; baking powders; galvanoplastics; soldering enamels; analytical reagent; cheese; detergents; water treatment; dietary supplement; buffer; sequestrant in foods.

sodium phosphate, monobasic (sodium acid phosphate; sodium biphosphate; sodium orthophosphate, primary; MSP; sodium dihydrogen phosphate) (a) NaH_2PO_4, (b) $NaH_2PO_4 \cdot H_2O$.
Properties: (a) White crystalline powder; slightly hygroscopic; very soluble in water; has acid reaction; forms sodium acid pyrophosphate at 225–250°C and sodium metaphosphate at 350–400°C; (b) large, transparent crystals; m.p. loses H_2O at 100°C; sp. gr. 2.040; very soluble in water; insoluble in alcohol. pH of 1% solution 4.4–4.5. Low toxicity; nonflammable.
Derivation: By treating disodium phosphate with proper proportion of phosphoric acid.
Grades: Commercial; food; (b) N.F.; (a) F.C.C.
Containers: Paper bags; drums; barrels.
Uses: Boiler water treatment; electroplating; dyeing; acid cleansers; baking powders; cattle food supplement; buffer, emulsifier, nutrient supplement in food; laboratory reagent.

sodium phosphate P 32 (sodium radio-phosphate). A radioactive form of sodium phosphate (which phosphate is not specified) containing phosphorus 32 which can be used as a tracer. See phosphorus 32.
Grade: U.S.P., as solution.
Use: Medicine; biochemical research.

sodium phosphate, tribasic (TSP; trisodium orthophosphate; trisodium phosphate; tertiary sodium phosphate; sodium orthophosphate, tertiary) $Na_3PO_4 \cdot 12H_2O$.
Properties: Colorless crystals; soluble in water. Sp. gr. 1.62 (20°C); m.p. 75°C (dec); loses $12H_2O$ at 100°C. pH of 1% solution is 11.8–12.0. Nonflammable.
Derivation: By mixing soda ash and phosphoric acid in proper proportions to form disodium phosphate, and then adding caustic soda.
Grades: Commercial; high purity; C.P.; F.C.C. (anhydrous). Anhydrous salt also available.
Containers: Barrels; bags.
Hazard: Moderately toxic by ingestion; irritant to tissue.
Uses: Water softeners; boiler compounds; detergent; metal cleaner; textiles; manufacture of paper; laundering; tanning; sugar purification; photographic developers; medicine; paint removers; industrial cleaners; dietary supplement; buffer; emulsifier.

sodium phosphide Na_3P.
Properties: Red solid; decomposes on heating and in water, forming phosphine.
Hazard: Dangerous fire risk; reacts with water and acids to form phosphine (q.v.).
Shipping regulations: (Air) Flammable Solid label. Not acceptable on passenger planes. (Rail) Not listed.

sodium phosphite $Na_2HPO_3 \cdot 5H_2O$.
Properties: White, crystalline powder. Hygroscopic. Soluble in water; insoluble in alcohol. M.p. 53°C; b.p. 200–250°C (dec).
Use: Medicine; antidote in mercuric chloride poisoning.

sodium phosphoaluminate. White powder composed primarily of sodium aluminate (hydrated), sodium phosphate (ortho) and small amounts of sodium carbonate and sodium silicate. Used primarily in the paper industry as a sizing adjunct, as an aid in retention of filler and fiber and in pH control. Also used in boiler feed water treatment and as a food additive.

sodium phospho-12-molybdate. See sodium 12-molybdophosphate.

sodium phospho-12-tungstate. See sodium 12-tungstophosphate.

sodium phytate (USAN) (inositol hexaphosphoric ester, sodium salt) $C_6H_9O_{24}P_6Na_9$.
Properties: Hygroscopic powder, water-soluble.
Uses: Chelating agent for trace (heavy) metals; color improvement; medicine.

sodium picramate $NaOC_6H_2(NO_2)_2NH_2$.
Derivation: Yellow, water-soluble salt resulting from neutralization of picramic acid with caustic soda.
Containers: Drums.
Hazard: Dangerous fire and explosion hazard when dry. Toxic by ingestion and skin absorption.
Uses: Manufacture of dye intermediates; organic synthesis.
Shipping regulations: Wet with not less than 20% water: (Rail) Yellow label. (Air) Flammable Solid label. Not accepted on passenger planes. Dry, or wet with less than 20% water: Not acceptable.

sodium platinichloride. See sodium chloroplatinate.

sodium plumbate $Na_2PbO_3 \cdot 3H_2O$.
Properties: Fused, light yellow lumps. Hygroscopic. Decomposed by water and acids; soluble in alkalies.

sodium plumbite Na_2PbO_2.
Derivation: Solution of PbO (litharge) in sodium hydroxide.
Hazard: Highly toxic and corrosive.
Use: Doctor solution (q.v.) for improving the odor of gasoline and other petroleum distillates.

sodium polyphosphate $Na_{n+2}P_nO_{3n+1}$. The two prominent crystalline sodium polyphosphates are the pyrophosphate (n = 2) and the tripolyphosphate, also known as the triphosphate (n = 3). The term sodium polyphosphate also includes the system of vitreous sodium phosphates for which the mole ratio of Na_2O/P_2O_5 is between 1 and 2. See also sodium metaphosphate.
Uses: Sequestering and deflocculating agents, primarily in water treatment, food processing and cleaning compounds; heavy-set detergent builders. See sodium meta-, pyro-, and tripolyphosphates.

sodium polystyrene sulfonate. A cation-exchange resin prepared in the sodium form which can exchange about 12% of its weight of potassium.
Properties: Golden brown, fine powder; odorless and tasteless; insoluble in water.
Grade: U.S.P.
Use: Medicine.

sodium polysulfide Na_2S_x.
Properties: Yellow-brown granular free-flowing polymer; density 56 lb/cu ft. Combustible.
Containers: 100-lb, 400-lb drums.
Uses: Manufacture of sulfur dyes and colors, insecticides, oil-resistant synthetic rubber ("Thiokol"), petroleum additives; electroplating.

sodium-potassium alloy. Legal label name for NaK (q.v.).

sodium-potassium carbonate (potassium-sodium carbonate) $NaKCO_3 \cdot 6H_2O$.
Properties: Colorless crystals. The double salt fuses more readily than the single salts; sp. gr. 1.6344; m.p. 135°C (dec). Soluble in water.
Derivation: Mixture of potassium and sodium carbonates.
Use: Analysis (flux).

sodium-potassium phosphate (potassium-sodium phosphate) $NaKHPO_4 \cdot 7H_2O$.
Properties: White powder; stable in air. Soluble in water.

sodium-potassium tartrate. See potassium-sodium tartrate.

sodium propionate CH_3CH_2COONa or $C_2H_5COONa \cdot xH_2O$.
Properties: Transparent crystals or granules; almost odorless; deliquescent in moist air; soluble in water and alcohol. Combustible; low toxicity.
Grades: N.F.; F.C.C.
Containers: 25-, 100-, 250-lb drums.
Uses: Fungicide; mold-preventive; food preservative (bread and other bakery products).

sodium prussiate, red. See sodium ferricyanide.

sodium prussiate, yellow. See sodium ferrocyanide.

sodium pyroantimonate. See sodium antimonate.

sodium pyroborate. See sodium borate.

sodium pyrophosphate (tetrasodium pyrophosphate; sodium pyrophosphate, normal; TSPP) (a) $Na_4P_2O_7$; (b) $Na_4P_2O_7 \cdot 10H_2O$. One of the sodium polyphosphates (q.v.).
Properties: Colorless, transparent crystals or white powder. (a) M.p. 880°C; sp. gr. 2.45; soluble in water; decomposes in alcohol; (b) m.p. 94°C (loses H_2O); sp. gr. 1.8 soluble in water; insoluble in alcohol and ammonia. Low toxicity.
Derivation: By fusing sodium phosphate, dibasic.
Grades: Pure crystals; dried; fused; C.P.; F.C.C.
Containers: Bottles; paper bags; drums; bulk cars.
Uses: Water softener; soap and synthetic detergent builder; dispersing and emulsifying agent; metal cleaner; boiler water treatment; viscosity control of drilling muds; de-inking news print; synthetic rubber manufacture; textile dyeing; scouring of wool; buffer; sequestrant; nutrient supplement.

sodium pyrophosphate, acid (disodium pyrophosphate; sodium acid pyrophosphate; disodium diphosphate;

disodium dihydrogen pyrophosphate; SAPP) $Na_2H_2P_2O_7 \cdot 6H_2O$.
Properties: White crystalline powder; m.p. (dec) 220°C; sp. gr. 1.862; soluble in water. Low toxicity.
Derivation: Incomplete decomposition of monobasic sodium phosphate.
Grades: Technical; food; F.C.C.
Containers: 100-lb bags; 125-, 350-lb drums.
Uses: Electroplating; metal cleaning and phosphatizing; drilling muds; baking powders and leavening agent; buffer; sequestrant; peptizing agent in cheese and meat products; frozen desserts.

sodium pyrophosphate, normal. See sodium pyrophosphate.

sodium pyrophosphate peroxide $Na_4P_2O_7 \cdot 2H_2O_2$.
Properties: White powder; bulk density 73 lb/cu ft. Active oxygen minimum 9.0% by wt. Water-soluble; mildly alkaline.
Containers: Fiber drums.
Hazard: Fire risk in contact with organic materials. Oxidizing agent.
Uses: Denture cleansers, dentrifices, household and laundry detergents; antiseptic.

sodium pyroracemate. See sodium pyruvate.

sodium pyrosulfite. See sodium metabisulfite.

sodium pyrovanadate $Na_4V_2O_7 \cdot 8H_2O$.
Properties: Colorless six-sided plates. Soluble in water; insoluble in alcohol; m.p. (anhydrous) 654°C.
Derivation: Sodium hydroxide and vanadium pentoxide in water solution.

sodium pyruvate (sodium pyroracemate; sodium acetylformate) $NaOOCCOCH_3$. White powder; apparent melting point 205°C. Very soluble in water.
Use: Biochemical research.

sodium resinate. See sodium abietate.

sodium rhodanate. See sodium thiocyanate.

sodium rhodanide. See sodium thiocyanate.

sodium ricinoleate $C_{17}H_{32}OHCOONa$.
Properties: White or slightly yellow, nearly odorless powder; soluble in water or alcohol. Combustible; low toxicity.
Derivation: Sodium salt of the fatty acids from castor oil.
Uses: Emulsifying agent in special soaps; medicine.

sodium saccharin (sodium benzosulfimide; gluside, soluble; soluble saccharin) $C_7H_4NNaO_3S \cdot 2H_2O$. The sodium salt of saccharin (q.v.).
Properties: White crystals or crystalline powder; odorless or with a faint aromatic odor; in dilute solutions has an intensely sweet taste (500 times as sweet as sugar). Very soluble in water; slightly soluble in alcohol.
Grades: N.F.; F.C.C.
Uses: Medicine; foods (non-nutritive sweetener).

sodium salicylate HOC_6H_4COONa.
Properties: Lustrous, white, crystalline scales or amorphous powder; saline taste. Soluble in water, alcohol and glycerol. Combustible. Low toxicity.
Grades: Technical; C.P.; U.S.P.
Containers: Drums; barrels.
Uses: Medicine; production of salicylic acid; preservative for paste, mucilage, glue and hides.

sodium sarcosinate (sodium sarcosine) CH_3NHCH_2COONa.
Grade: 33% aqueous solution.
Uses: Intermediate; stabilizer for diazonium salts; chelating agent.

sodium secobarbital. See barbiturate.

sodium selenate $Na_2SeO_4 \cdot 10H_2O$.
Properties: White crystals. Sp. gr. 1.603–1.620. Soluble in water.
Hazard: Toxic by ingestion. Tolerance (as Se), 0.2 mg per cu meter.
Uses: Reagent; insecticide for nonedible plants.

sodium selenite $Na_2SeO_3 \cdot 5H_2O$.
Properties: White crystals. Soluble in water; insoluble in alcohol.
Derivation: By neutralizing selenious acid with sodium carbonate and crystallizing.
Hazard: Toxic by ingestion. Tolerance (as Se), 0.2 mg per cu meter.
Uses: Glass manufacture; reagent in bacteriology; testing germination of seeds; decorating porcelain.

sodium sesquicarbonate (sesqui) $Na_2CO_3 \cdot NaHCO_3 \cdot 2H_2O$.
Properties: White needle-shaped crystals; sp. gr. 2.112; m.p., decomposes. Soluble in water. Less alkaline than sodium carbonate. Noncombustible. Nontoxic.
Derivation: Crystallization of a solution containing equimolar quantities of sodium carbonate and sodium bicarbonate; also occurs native (as trona) in desert areas and in Searles Lake brine (q.v.).
Grades: Technical; F.C.C.
Containers: Barrels and kegs; paper or burlap bags; fiber drums; bulk.
Uses: Detergent and soap builder; mild alkaline agent for general cleaning and water softening; bath crystals; alkaline agent in leather tanning; food additive.

sodium sesquisilicate. $Na_6Si_2O_7$ (anhydrous).
Properties: White; granular powder; soluble in water; pH of 1% solution 12.7. Noncombustible.
Derivation: Crystallization from solutions obtained by heating silica or sodium metasilicate with sodium hydroxide. Intermediate in composition between ortho- and metasilicates; less alkaline than sodium orthosilicate.
Containers: Bags; barrels; fiber drums; bulk.
Uses: Heavy-duty cleaning (metals, laundries); textile processing.

sodium silicate (water glass). See also the other soluble sodium silicates: sodium metasilicate anhydrous, sodium metasilicate pentahydrate, sodium sesquisilicate, sodium orthosilicate. The simplest form of glass.
Formulas: Vary from $Na_2O \cdot 3.75\ SiO_2$ to $2Na_2O \cdot SiO_2$, and with various proportions of water.
Properties: Lumps of greenish glass soluble in steam under pressure; white powders of varying degrees of solubility; or liquids cloudy or clear and varying from highly fluid to extreme viscosity; viscosity range from 0.4 to 600,000 poises; f.p. slightly lower than water; miscible with some polyhydric alcohols; partially miscible with primary alcohols and ketones. Gels form with acids between pH 3 to 9; coagulated by brine; precipitated by alkaline earth and heavy metal ions. Noncombustible. Nontoxic.
Derivation: By fusing sand and soda ash.
Grades (liquid): 40,47,52° Be.
Containers: Barrels; drums; tank trucks; carlots; bulk.
Uses: Catalysts and silica gels; soaps and detergents; adhesives, especially sealing and laminating paper board containers; water treatment; bleaching and sizing of textiles and paper pulp; ore treatment; soil

solidification; glass foam; pigments; drilling fluids; binder for foundry cores and molds; waterproofing mortars and cements; impregnating wood.

sodium silicoaluminate. See sodium aluminosilicate.

sodium silicofluoride. See sodium fluorosilicate.

sodium silico-12-molybdate. See sodium 12-mlybdosilicate.

sodium 12-silicotungstate. See sodium 12-tungstosilicate.

sodium silver chloride. See silver sodium chloride.

sodium silver thiosulfate. See silver sodium thiosulfate.

sodium alpha-**sodioacetate.** See alpha-sodiosodium acetate.

sodium sorbate $CH_3CH{:}CHCH{:}CHCOONa$. Combustible. Nontoxic.
Uses: Food preservative.

sodium stannate $Na_2SnO_3 \cdot 3H_2O$, or $Na_2Sn(OH)_6$.
Properties: White to light tan crystals; soluble in water; insoluble in alcohol; decomposes in air. Aqueous solution slightly alkaline. Loses $3H_2O$ in 140°C.
Derivation: (a) By fusion of metastannic acid and sodium hydroxide. (b) By boiling tin scrap and sodium plumbate solution.
Hazard: Toxic. Tolerance, 2 mg per cubic meter of air.
Uses: Mordant in dyeing; ceramics; glass; source of tin for electroplating and immersion plating; textile fireproofing; stabilizer for hydrogen peroxide; blueprint paper; laboratory reagent.

sodium stearate $NaOOCC_{17}H_{35}$.
Properties: White powder with fatty odor. Soluble in hot water and hot alcohol; slowly soluble in cold water and cold alcohol; insoluble in many organic solvents.
Impurities: Varying quantities of sodium palmitate.
Grade: Technical.
Containers: 150-lb drums; 200-lb barrels.
Uses: Waterproofing and gelling agent; toothpaste and cosmetics; stabilizer in plastics.

sodium stearoyl 2-lactylate.
Properties: White powder. Melting range 46–52°C. Nontoxic.
Derivation: Sodium salt of reaction product of lactic and stearic acids.
Uses: Emulsifier; dough conditioner; whipping agent in baked products, desserts, and mixes; complexing agent for starches and proteins.

sodium styrenesulfonate $CH_2{:}CH_2C_6H_4SO_3Na$. White, free-flowing powder.
Use: Reactive monomer. See sodium polystyrenesulfonate.

sodium subsulfite. See sodium thiosulfate.

sodium succinate $Na_2C_4H_4O_4 \cdot 6H_2O$.
Properties: White crystals or odorless granules; soluble in water. Loses $6H_2O$ at 120°C.
Use: Medicine.

sodium sulfate, anhydrous Na_2SO_4. See also salt cake.
Properties: White crystals or powder; odorless; bitter saline taste; sp. gr. 2.671; m.p. 888°C; soluble in water and glycerol; insoluble in alcohol. Noncombustible; nontoxic.
Derivation: (a) By-product of hydrochloric acid production from salt and sulfuric acid. (b) Purification of natural sodium sulfate from deposits or brines. (c) By-product of phenol manufacture (caustic fusion process); (d) Hargreaves process (q.v.).
Grades: Technical; C.P.; detergent; rayon; glass makers.
Containers: Bags; drums.
Uses: Manufacture of kraft paper, paperboard, and glass; filler in synthetic detergents; sodium salts; ceramic glazes; processing textile fibers; dyes; tanning; glass; pharmaceuticals; freezing mixtures; laboratory reagent; food additive.

sodium sulfate decahydrate (sodium sulfate, crystals; Glauber's salt) $Na_2SO_4 \cdot 10H_2O$.
Properties: Large transparent crystals, small needles, or granular powder; sp. gr. 1.464 (crystals); m.p. 33°C (liquefies); loses water of hydration at 100°C. Soluble in water and glycerin; insoluble in alcohol; solutions neutral to litmus. Nontoxic; nonflammable.
Derivation: Crystallization of sodium sulfate from water solutions. (Glauber's salt); also occurs in nature as mirabilite (q.v.).
Grades: Technical; N.F.
Uses: See under anhydrous form.

sodium sulfhydrate. See sodium hydrosulfide.

sodium sulfide (a) Na_2S; (b) $Na_2S \cdot 9H_2O$.
Properties: Yellow or brick red lumps or flakes or deliquescent crystals; (a) sp. gr. 1.856 (14°C); m.p. 1180°C; (b) sp. gr. 1.427 (16°C); decomposes at 920°C. Soluble in water; slightly soluble in alcohol; insoluble in ether; largely hydrolyzed to sodium acid sulfide and sodium hydroxide.
Derivation: By heating sodium acid sulfate with salt and coal to above 950°C, extraction with water, and crystallization.
Grades: Flake; fused; chip sulfide (60% Na_2S), 60% fused and broken; 30% crystals; liquid.
Containers: Barrels; drums; bulk.
Hazard: Flammable, dangerous fire risk. Strong irritant to skin and tissue. Liberates toxic hydrogen sulfide on contact with acids.
Uses: Organic chemicals; dyes (sulfur); intermediates; rayon (denitrating); leather (depilatory); paper pulp; solvent for gold in hydrometallurgy of gold ores; sulfiding oxidized lead and copper ores preparatory to flotation; sheep dips; photographic reagent; engraving and lithography; analytical reagent.
Shipping regulations: (Rail) Yellow label. (Air) Flammable Solid label.

sodium sulfite (a) Na_2SO_3; (b) $Na_2SO_3 \cdot 7H_2O$.
Properties: White crystals or powder; saline, sulfurous taste. Soluble in water; sparingly soluble in alcohol. Sp. gr.: (a) 2.633; (b) 1.5939. M.p.: (a) decomposes; (b) loses $7H_2O$ at 150°C.
Derivation: (a) Sulfur dioxide is reacted with soda ash and water, and a solution of the resulting sodium bisulfite is treated with additional soda ash; (b) by-product of the caustic fusion process for phenol.
Grades: Reagent; technical; F.C.C.
Containers: Bags; drums.
Hazard: Use prohibited in meats and other sources of Vitamin B_1.
Uses: Paper industry (semichemical pulp); water treatment; photographic developer; food preservative and antioxidant; textile bleaching (antichlor); dietary supplements.

sodium sulfobromophthalein $C_{20}H_8Br_4O_{10}S_2Na_2$.
Properties: White, crystalline powder; odorless with a bitter taste; hygroscopic. Soluble in water; insoluble in alcohol and acetone.
Derivation: From phenol and tetrabromophthalic acid or anhydride.
Grades: U.S.P.; technical.
Use: Medicine (diagnostic aid).

sodium sulfocarbolate. See sodium phenolsulfonate.

sodium sulfocyanate. See sodium thiocyanate.

sodium sulfocyanide. See sodium thiocyanate.

sodium sulfonate. Class name for various sulfonates derived from petroleum, e.g., sodium dodecylbenzenesulfonate; sodium xylenesulfonate; sodium toluenesulfonate. See also sodium alkanesulfonate.
Uses: Textile processing oils; oils for metal working (emulsifying and antirust agents); lubricating oils; emulsifiers for insecticides, herbicides, fungicides; preparation of dyes and intermediates; hydrotropic solvent; coatings in food packaging.

sodium sulforicinoleate.
Derivation: Product of successive sulfonation (partial) and saponification of castor oil. Composition indefinite.
Use: Emulsifying and wetting agent.

sodium sulfoxylate. See sodium formaldehyde sulfoxylate.

sodium tartrate (sal tartar; disodium tartrate) $Na_2C_4H_4O_6 \cdot 2H_2O$.
Properties: White crystals or granules. Soluble in water; insoluble in alcohol. Sp. gr. 1.794; loses $2H_2O$ at 150°C. Nontoxic.
Derivation: Neutralization of tartaric acid with sodium carbonate, concentration and crystallization.
Grades: Technical; C.P.; reagent; F.C.C.
Uses: Medicine; chemical reagent; food additive, as sequestrant and stabilizer.

sodium tartrate, acid. See sodium bitartrate.

sodium TCA. See sodium trichloroacetate.

sodium tellurite Na_2TeO_3.
Properties: White powder; soluble in water.
Hazard: Toxic by ingestion.
Use: Bacteriology; medicine.

sodium tetraborate. See sodium borate.

sodium 2,3,4,6-tetrachlorophenate $C_6HCl_4ONa \cdot H_2O$.
Properties: Buff to light brown flakes; bulk density 26–29 lb/cu ft; pH of water-saturated solution 9.0–13.0. Soluble in water, methanol, acetone.
Hazard: Toxic by ingestion.
Use: Industrial preservative (bactericide and fungicide).

sodium tetradecyl sulfate (sodium 7-ethyl-2-methyl-4-hendecanol sulfate). $C_{14}H_{29}SO_4Na$.
Properties: White, waxy, odorless solid. Soluble in alcohol, ether, and water. 5% solution is clear and colorless. pH (5% solution) 6.5–9.0.
Uses: Medicine; wetting agent.

sodium tetraphosphate. See sodium polyphosphate.

sodium tetrasulfide Na_2S_4.
Properties: Yellow, hygroscopic crystals or clear dark red liquid; m.p. of crystals 275°C.
Grade: Aqueous solution containing 40% by weight of compound.
Containers: Glass bottles; carboys; drums.
Hazard: Moderate fire risk when exposed to flame; irritant to skin and tissue.
Uses: Reducing organic nitro compounds; manufacture of sulfur dyes; insecticides and fungicides; ore flotation; soaking hides and skins; preparation of metal sulfide finishes.

sodium thiocyanate (sodium sulfocyanate; sodium sulfocyanide; sodium rhodanate; sodium rhodanide) NaSCN.
Properties: Colorless, deliquescent crystals or white powder; m.p. 287°C. Soluble in water and alcohol. Hygroscopic and affected by light.
Derivation: By boiling sodium cyanide with sulfur.
Grades: Technical; pure, crystal or dried; C.P.; reagent; A.C.S.
Containers: Bottles; tins; drums.
Hazard: May be toxic.
Uses: Analytical reagent; dyeing and printing textiles; medicine; black nickel plating; manufacturing other thiocyanate salts and artificial mustard oil; solvent for polyacrylates; medicine.

sodium thioglycolate (sodium mercaptoacetate) $HSCH_2COONa$. The sodium salt of thioglycolic acid.
Properties: Crystals; characteristic odor; hygroscopic; discolors on exposure to air or iron; soluble in water; slightly soluble in alcohol. Combustible.
Containers: Bottles; carboys.
Hazard: Yields toxic hydrogen sulfide on decomposition. May be toxic by skin absorption.
Uses: Bacteriology; cold waving of hair; depilatory; analytical reagent.

sodium thiopental (thiopentone soluble) $C_{11}H_{17}N_2O_2SNa$. Sodium 5-ethyl-5-(1-methylbutyl)-2-thiobarbiturate.
Properties: Yellowish white, hygroscopic powder with a disagreeable odor. Soluble in water and alcohol; insoluble in absolute ether, benzene and ligroin. Solution decomposes on standing. Precipitation occurs on boiling of solution.
Grade: U.S.P.
Use: Medicine.
See also barbiturate.

sodium thiosulfate (sodium subsulfite; hypo) $Na_2S_2O_3 \cdot 5H_2O$. The anhydrous salt is also commercially available.
Properties: White, translucent crystals or powder; cooling taste and bitter aftertaste. Soluble in water and oil of turpentine; insoluble in alcohol; deliquescent in moist air; efflorescent above 33°C in dry air. Sp. gr. 1.729 (17°C); m.p. 48°C; b.p. decomposes.
Derivation: Heating a solution of sodium sulfite with powdered sulfur.
Grades: Technical; crystals; granulated; photographic; C.P.; pure; U.S.P.; F.C.C.
Containers: Drums; bulk.
Hazard: Use in foods restricted to 0.1%.
Uses: Photography (fixing agent to dissolve unchanged silver salts from exposed negatives); chrome tanning; removing chlorine in bleaching and papermaking; extraction of silver from its ores; dechlorination of water; mordant; reagent; medicine; bleaching bone, straw, ivory; reducing agent in chrome dyeing; sequestrant in foods.

sodium titanate (sodium tritanate) $Na_2Ti_3O_7$.
Properties: White crystals. Insoluble in water; sp. gr. 3.35–3.50; m.p. 1128°C. Combustible.
Containers: Cartons, drums; carlots.
Use: Welding.

sodium alpha-**toluate.** See sodium phenylacetate.

sodium toluenesulfonate (para-toluenesulfonic acid, sodium salt) $CH_3C_6H_4SO_3Na$.
Properties: Crystals; soluble in water.
Uses: Dye chemistry; hydrotropic solvent.

sodium para-**toluenesulfonchloramine.** See chloramine

sodium trichloroacetate (sodium TCA) CCl_3COONa.
Containers: Cans; drum; carlots.
Hazard: Toxic by ingestion. Irritant to skin and eyes.
Uses: Herbicide; pesticide.

sodium 2,4,5-trichlorophenate $C_6H_2Cl_3ONa \cdot 1\frac{1}{2} H_2O$.
Properties: Buff to light brown flakes; bulk density 28–33 lb/cu ft; pH of water-saturated solution 11.0–13.0. Soluble in water, methanol, acetone.
Use: Industrial preservative (bactericide and fungicide).

sodium tridecylbenzenesulfonate. Not a true compound, but a mixture of C_{12} and C_{15} alkyl benzene sulfonates which approximates C_{13}. For derivation and use, see sodium dodecylbenzene sulfonate.

sodium trimetaphosphate. See sodium metaphosphate.

sodium triparaperiodate. See sodium paraperiodate.

sodium triphenyl-para-**rosaniline sulfonate.** See methyl blue.

sodium triphosphate. See sodium tripolyphosphate.

sodium tripolyphosphate (STPP; sodium triphosphate; tripoly; pentasodium triphosphate) $Na_5P_3O_{10}$. (The sodium phosphate produced in largest volume in the U.S.).
Properties: White powder; two crystalline forms of anhydrous salt (transition pt. 417°C; m.p. 622°C to give melt and sodium pyrophosphate), and a hexahydrate. Low toxicity.
Derivation: Controlled calcination of sodium orthophosphate mixture from sodium carbonate and phosphoric acid. May contain up to 10% pyrophosphate and up to 5% trimetaphosphate.
Grades: Powdered and granular, F.C.C.; food.
Containers: Multiwall paper bags, drums, bulk carlots or trucklots.
Uses: Water softening; sequestering, peptizing, or deflocculating agent; food additive and texturizer.

sodium trititanate. See sodium titanate.

sodium tungstate (sodium wolframate) $Na_2WO_4 \cdot 2H_2O$.
Properties: Colorless crystals; soluble in water; insoluble in alcohol and acids. Sp. gr. 3.245; m.p., loses $2H_2O$ at 100°C and then melts at 692°C. Low toxicity; noncombustible.
Derivation: By dissolving tungsten trioxide or the ground ore in caustic soda solution, concentration and crystallization.
Grades: Technical; C.P.; crystalline.
Containers: 1-, 5-lb bottles; 25-, 50-lb cans; 100-lb kegs.
Uses: Intermediate in preparation of tungsten and its compounds; reagent; fireproofing fabrics and cellulose.

sodium 12-tungstophosphate (sodium phosphotungstate; phosphotungstic acid, sodium salt) $Na_3[P(W_3O_4)] \cdot xH_2O$.
Properties: Yellowish-white powder; very soluble in water and alcohols. An oxidizing agent.
Grades: Reagent; technical.
Uses: Reagent; manufacture of organic pigments; treatment of furs; antistatic agent for textiles; leather tanning; making plastic films, cements, and adhesives water-resistant.

sodium 12-tungstosilicate $Na_4SiW_{12}O_{40} \cdot 5H_2O$.
Properties: White crystalline powder; soluble in water and alcohols, although less so than the free acid. Low toxicity. Noncombustible.
Grades: Reagent; technical.
Uses: Catalyst for organic synthesis; precipitant and inorganic ion-exchanger; additive in plating processes.

sodium undecylenate $CH_2{:}CH(CH_2)_8COONa$.
Properties: White powder; decomposes above 200°C; limited solubility in most organic solvents; soluble in water. Combustible.
Uses: Bacteriostat and fungistat in cosmetics and pharmaceuticals.

sodium uranate. See sodium diuranate.

sodium vanadate. See sodium orthovanadate; sodium pyrovanadate; sodium metavanadate.

sodium para-**vinylbenzene sulfonate.** See sodium styrene sulfonate.

sodium warfarin (sodium(3-alpha-acetonylbenzyl)-4-hydroxycoumarin) $C_{19}H_{15}NaO_4$. See warfarin.

sodium xanthate. See sodium ethylxanthate.

sodium xanthogenate. See sodium ethylxanthate.

sodium xylenesulfonate (dimethylbenzenesulfonic acid, sodium salt) $(CH_3)_2C_6H_3SO_3Na \cdot H_2O$.
Use: Hydrotropic solvent, used in detergents.

sodium zinc hexametaphosphate. See "Calgon Composition TG." See also sodium metaphosphate.

sodium zirconium glycolate $NaH_3ZrO(H_2COCOO)_3$.
Properties: Clear, light straw-colored solution, sp. gr. 1.28–1.30, containing 35.7–38.6% solids; 12.5–13.5% ZrO_2.
Uses: Deodorant; astringent; germicide; sequestrant; fire retardant.

sodium zirconium lactate $NaH_3ZrO(CH_3CHOCOO)_3$.
Properties: Clear, straw colored solution, sp. gr. 1.28–1.30, containing 12.5–13.5% ZrO_2, equivalent to 42.5–45.9% sodium zirconium lactate; pH 7.5–8.0.
Uses: Deodorant and antiperspirant.

sodium zirconium sulfate. See zirconium sodium sulfate.

softener (1) A substance used when dry powders are added to a polymeric material (e.g., rubber or plastic) to reduce the friction of mechanical mixing and to facilitate subsequent processing. They exert both lubricating and dispersing action, often by means of emulsification. Examples are vegetable oils, asphaltic materials, and stearic acid, the latter being especially effective with carbon black. It is difficult to distinguish precisely between softeners and plasticizers (q.v.); in general, softeners do not enter into chemical combination with the polymer, and their softening effect tends to be temporary.
(2) A fatliquoring agent (q.v.) used to soften leather.
(3) A sulfonated oil, fatty alcohol, or quaternary ammonium compound used in textile finishing to impart superior "hand" to the fabric and facilitate mechanical processing.
(4) A substance that reduces the hardness of water by removing or sequestering calcium and magnesium ions; among those used are various sodium phosphates and zeolites (q.v.). See water, hard.

"Softener 64."[73] Trademark for a fatty amide used as a cationic softener for textiles.

softwood. In papermaking terminology, an arbitrary name for the wood from coniferous trees (pine, spruce, fir, hemlock), regardless of the hardness or softness of the wood itself. Softwoods are used for almost all commercial grades of paper. See also paper pulp.

soil. (1) A mixture of inorganic matter derived from weathered rocks (about 50%) and organic components resulting from decay of prior vegetation (5%). Eight elements are present in the inorganic component in excess of 1% (oxygen, silicon, aluminum, iron, calcium, potassium, magnesium and sodium), most of which are in the ionized state. Water and air are also present, either in the voids between the particles or adsorbed on their surfaces. Many other elements occur in lower percentages including trace elements (q.v.) in concentrations of less than 1000 ppm. Some of these, e.g., boron (about 20 ppm) are essential for plant nutrition. Both nitrogen and phosphorus are associated with the organic content. The concentration of these is a fraction of 1%, respectively, but they play a vital part in plant and animal life. The pH of soils varies widely with location; some are as low as pH 4.5 (very acid) and others as high as pH 10 (strongly alkaline). For most crops the pH ranges around the neutral point (6.5–7.5). Texturally, soils are classified on the basis of their content of sand, silt, and clay. Those having 45–50% sand and 20–28% clay are called loams; those with over 50% sand are called sandy; and those with over 28% clay are in the clay group. Technologists consider soil as being made up of layers, known as horizons, each having a characteristic composition and physical properties; the spectrum of these horizons is called the soil profile. Organic matter is usually excluded from the profile.
(2) In textile literature any foreign matter present in or on fiber or fabric, i.e., dirt, oil, grease, etc. These are usually removable by the action of soap, synthetic detergents, or organic solvents.

soil conditioner. (1) A synthetic long-chain organic molecule having carboxyl groups along the chain whose charges react with the positive charges on the soil particles (aluminum and iron). The conditioners affect the anion exchange capacity of a soil.
(2) Loosely, any material added to topsoil to reduce acidity (lime) and promote growth (bone meal).

"Soilfume."[55] Trademark for a soil fumigant whose active ingredient is ethylene dibromide.
Hazard: See ethylene dibromide.

sol. See solution, colloidal.

solan. Generic name for 3′-chloro-2-methyl-para-valerotoluidide; (N-(3-chloro-4-methylphenyl)-2-methylpentanamide)
$H_3CC_6H_3(Cl)NHCOCH(CH_3)CH_2CH_2CH_3$.
Properties: Solid; m.p. 86° C; insoluble in water; soluble in pine oil, diisobutyl ketone, isophorone, and xylene. Low toxicity; combustible.
Use: Herbicide.

solanine. A toxic alkaloid formed in potatoes and some other root crops on long exposure to light during storage. Chlorophyll may also be formed.

"Solar."[307] Trademark for a series of phosphotungstic and phosphomolybdic lakes in dispersed powder form.
Use: Coloring various grades of paper.

solar cell. A device for converting the radiant energy of the sun into electrical or thermal energy for use in power generation and heating and cooling of buildings. There are two basic types. One type utilizes a p-type semiconductor, such as a ribbon of pure silicon, the energy conversion being electrical in nature. Selenium and cadmium sulfide can also be used. The other type is a heat-collecting cell which may be 3 feet wide by 6 or more feet long, utilizing either a circulating gas or glass plates and an aluminum absorber plate. A variation of this is the parabolic concentrator developed at Argonne National Laboratory, consisting of channels made of epoxy resin with vapor-deposited aluminum surface; each wall of the channels is a portion of a parabolic surface. Prototypes of these are under construction. See also energy converter.

solar energy. See solar cell; energy converter; chlorophyll.

"Solastral."[243] Trademark for phthalocyanine pigments.

"Solcut."[320] Trademark for a self-emulsifying superfatted soluble cutting fluid for difficult nonferrous metal machining operations.

solder. A low-melting alloy, usually of the lead-tin type, used for joining metals at temperatures below 800° F. The solder acts as an adhesive, and does not form an intermetallic solution with the metals being joined. See also brazing; welding.

"Soledon."[325] Trademark for a series of leuco vat esters.

"Solfast."[141] Trademark for a series of pigments used in paints, printing inks, textiles, paper, etc. They are stable to sunlight. The colors offered are blue, violet, green, red and yellow.

"Solganal."[321] Trademark for aurothioglucose (q.v.).

solid. Matter in its most highly concentrated form, i.e., the atoms or molecules are much more closely packed than in gases or liquids, and thus more resistant to deformation. The normal condition of the solid state is crystalline structure—the orderly arrangement of the constituent atoms of a substance in a framework called a lattice. (See crystal). Crystals are of many types, and normally have defects and impurities that profoundly affect their applications, as in semiconductors (q.v.). The geometric structure of solids is determined by x-rays, which are reflected at characteristic angles from the crystal lattices, which act as diffraction gratings. This is the science of crystallography (q.v.).

Some materials that are physically rigid, such as glass, are regarded as highly viscous liquids because they lack crystalline structure. All solids can be melted (i.e., the attractive forces acting between the crystals are disrupted) by heat, and are thus converted to liquids. For ice, this occurs at 32° F; for some metals the melting point may be as high as 6000° F.

solids. The nonliquid portion of a suspension or emulsion, e.g., a paint of low solids content.

solid state chemistry. Study of the exact arrangement of atoms in solids, especially crystals, with particular emphasis on imperfections and irregularities in the electronic and atomic patterns in a crystal, and the effects of these on electrical and other properties. See also crystal; semiconductor; impurity.

"Solinox."[64] Trademark for a series of chemically treated soybean oils.
Uses: Lacquers; coated fabrics; artificial leather.

"Solithane."[27] Trademark for liquid urethane prepolymers which can be cured to produce solid materials of varying degrees of hardness and flexibility.
Uses: For casting and molding mechanical components, sheets, tubing, other shapes; coatings; potting and encapsulation.

"Solka."[231] Trademark for a purified wood cellulose.

"Solka-Floc."[231] Trademark for powdered cellulose products derived by mechanical comminution of purified wood pulp. Available in various fiber lengths including dense free-flowing powders. Combustible.
Properties: White; brightness up to 92; particle size 40 to 165 microns; moisture 5–7%; sp. gr. 1.58; apparent density 9–34 lbs/cu ft; relatively inert to acids, alkalies and solvents; practically ashless; bone dry, 99.5+% cellulose. Soft, adsorbent and non-abrasive.
Containers: Multiwall bags.
Uses: Filter aid; raw material for cellulose derivatives, filler in rubber; component in welding rod coatings; inert bulking agent in food products.

"Solo."[79] Trademark for petroleum-insoluble resin. Available in solid, flake and pulverized forms.
Uses: Adhesives; asphalt emulsions; core binders; electrical insulating compounds, fiberboard panels, insulation batt manufacture; oil emulsions; sulfur grinding.

"Solox."[192] Trademark for a general-purpose solvent formulation comprised of specially denatured alcohol with low percentages of solvent modifiers. Available in regular and anhydrous grades.

"Solozone."[28] Trademark for a technical grade of sodium peroxide.

"Solprene."[303] Trademark for a solution of polymers of butadiene/styrene.
Uses: Footwear, wire and cable, sponge, floor tile and cove base, and other molded and extruded goods.

"Solricin" 135.[202] Trademark for 35% aqueous solution of potassium ricinoleate. Mild germicide; synergizes phenol coefficients of disinfectants.

"Solricin" 285.[202] Trademark for an 85% aqueous solution of ammonium ricinoleate; used as a rust-proofing agent.

"Solros."[79] Trademark for a heat-treated "FF" wood rosin.
Uses: Adhesive tape; artificial Burgundy pitch; belt dressings; branding paint; core oil; electric insulating compounds; pitch; printing ink; rock wool; roofing cement; rubber cement; shellac diluent; smoking molds; spirit varnishes; sticky fly paper; synthetic rosin oil; tree banding; Venice turpentine; wire coating compounds.

"Soltrol"[303] Trademark for a series of odorless mineral spirits.

solubility. The ability or tendency of one substance to blend uniformly with another, e.g., solid in liquid, liquid in liquid, gas in liquid, gas in gas. Solids vary from 0 to 100% in their degree of solubility in liquids, depending on the chemical nature of the substances; to the extent that they are soluble, they lose their crystalline form and become molecularly or ionically dispersed in the solvent to form a true solution. Examples are sugar/water, salt/water. Liquids and gases are often said to be miscible in other liquids and gases, rather than soluble. Thus nitrogen, oxygen, and carbon dioxide are freely miscible in each other, and air is a solution (uniform mixture) of these gases.

The physical chemistry of solubility is an extremely complex mathematical subject in which the principles of electrolytic dissociation, diffusion and thermodynamics play a controlling part. Raoult's law and Henry's law are also involved. See also miscibility; solution, true.

soluble oil. An oil (also called emulsifying oil) which, when mixed with water, produces milky emulsions. In some soluble oils the emulsion is so fine that instead of milky solutions in water, amber colored transparent solutions are formed. Typical examples are sodium and potassium petroleum sulfonates.
Uses: Metal cutting lubricants; textile lubricants; metal boring lubricants; emulsifying agents.

soluble starch. See starch, modified.

"Solulan."[493] Trademark for ethoxylated derivatives of lanolin and lanolin components. Some are also acetylated. "Solulan" C-24 is a polyethoxylated cholesterol.
Uses: Pharmaceuticals and cosmetics.

"Solu-Rez."[170] Trademark for a modified polyvinyl resin emulsion designed as multi-purpose packaging adhesive.

solute. One or more substances dissolved in another substance, called the solvent; the solute is uniformly dispersed in the solvent in the form of either molecules (sugar) or ions (salt), the resulting mixture comprising a solution. See solution (true); solvent.

solution (true). A uniformly dispersed mixture, at the molecular or ionic level, of one or more substances (the solute) in one or more other substances (the solvent). These two parts of a solution are called phases (q.v.). Common types are:
liquid/liquid: alcohol/water.
solid/liquid: salt/water.
solid/solid: carbon/iron.

Solutions that exhibit no change of internal energy on mixing and complete uniformity of cohesive forces are called *ideal;* their behavior is described by Raoult's Law. Solutions are involved in most chemical reactions, refining and purification, industrial processing and biological phenomena.

The proportion of substances in a solution depends on their limits of solubility. The solubility of one substance in another is the maximum amount that can be dissolved at a given temperature and pressure. A solution containing such a maximum amount is *saturated.* A state of supersaturation can be created, but such solutions are unstable and may precipitate spontaneously.

solution, colloidal. A liquid colloidal dispersion, often called a sol. Since colloidal particles are larger than molecules, it is strictly incorrect to call such dispersions solutions; however, this term is widely used in the literature.
Note: Wolfgang Ostwald stated, ". . There are no sharp differences between mechanical suspensions, colloidal solutions, and molecular [true] solutions. There is a gradual and continuous transition from the first through the second to the third." See also colloid chemistry.

solutrope. A ternary mixture having two liquid phases between which one component is distributed in an apparent ratio varying with concentration from less than one to more than one. In other words, the solute may be selectively dissolved in one or the other of the phases or solvents depending on the concentration. This phenomenon has been compared to azeotropic behavior.

"Solvat."[243] Trademark for leuco esters of vat dyes used for wool, cellulose and synthetic fibers.

solvation. In the parlance of colloid chemistry, the adsorption of a microlayer or film of water or other solvent on individual dispersed particles of a solution or dispersion. The term "solvated hulls" has been used to describe such particles. It is also applied to the action of plasticizers on resin dispersions in plastisols (q.v.).

"Solvatone."[214] Trademark for mixture of low boiling alcohols and ketones, including isopropanol and acetone.

Solvay process (ammonia soda process). Manufacture of sodium carbonate (soda ash, Na_2CO_3) from salt, ammonia, carbon dioxide and limestone by an ingenious sequence of reactions involving recovery and reuse of practically all the ammonia and part of the carbon dioxide. Limestone is heated to produce lime and carbon dioxide. The latter is dissolved in water containing the ammonia and salt, with resultant precipitation of sodium bicarbonate. This is separated by filtration, dried and heated to form normal sodium carbonate. The liquor from the bicarbonate filtration is heated and treated with lime to regenerate the ammonia. Calcium chloride is a major byproduct.

Note: Because this process requires much energy and pollutes streams and rivers with chloride effluent, its continued use in its present form is questionable. Many plants using it have closed, future production being obtained from the natural deposits in western U.S.

"Solvenol."[266] Trademark for a group of monocyclic terpene hydrocarbons with minor amounts of terpene alcohols and ketones.

Uses: General solvent; rubber reclaiming.

solvent. A substance capable of dissolving another substance (solute) to form a uniformly dispersed mixture (solution) at the molecular or ionic size level. Solvents are either polar (high dielectric constant) or non-polar (low dielectric constant). Water, the most common of all solvents, is strongly polar (dielectric constant 81), but hydrocarbon solvents are non-polar. Aromatic hydrocarbons have higher solvent power than aliphatics (alcohols). Other organic solvent groups are esters, ethers, ketones, amines and nitrated and chlorinated hydrocarbons.

The chief uses of organic solvents are in the coatings field (paints, varnishes and lacquers), industrial cleaners, printing inks, extractive processes, and pharmaceuticals. Since many solvents are flammable and toxic to varying degrees, they contribute to air pollution and fire hazards. For this reason their use in coatings and cleaners is expected to decline rather sharply, perhaps by as much as 75% by 1980. For specific properties and uses, see individual compound.

solvent, aprotic. A solvent that cannot act as a proton acceptor or donor, i.e., as an acid or base.

solvent drying. Removal of water from metal surfaces by means of a solvent that displaces it preferentially, as on precision equipment, electronic components, etc. Examples of solvents used are acetone, 1,1,2,-trichloro-1,2,2,-trifluoroethane, 1,1,1-trichloro-ethane.

solvent dyeing. See dyeing, solvent.

solvent extraction. A separation operation which may involve three types of mixtures: (a) a mixture composed of two or more solids, such as a metallic ore; (b) a mixture composed of a solid and a liquid; (c) a mixture of two or more liquids. One or more components of such mixtures are removed (extracted) by exposing the mixture to the action of a solvent in which the component to be removed is soluble. If the mixture consists of two or more solids, extraction is performed by percolation of an appropriate solvent through it. This procedure is also called leaching, especially if the solvent is water; coffee-making is an example.

In liquid-liquid extraction one or more components are removed from a liquid mixture by intimate contact with a second liquid which is itself nearly insoluble in the first liquid and dissolves the impurities and not the substance that is to be purified. In other cases the second liquid may dissolve, i.e. extract, from the first liquid, the component that is to be purified, and leave associated impurities in the first liquid. Liquid-liquid extraction may be carried out by simply mixing the two liquids with agitation, and then allowing them to separate by standing. It is often economical to use countercurrent extraction, in which the two immiscible liquids are caused to flow past or through one another in opposite directions. Thus fine droplets of heavier liquid can be caused to pass downward through the lighter liquid in a vertical tube or tower.

The solvents used vary with the nature of the products involved. Widely used are water, hexane, acetone, isopropyl alcohol, furfural, xylene, liquid sulfur dioxide, and tributyl phosphate. Solvent extraction is an important method of both producing and purifying such products as lubricating and vegetable oils, pharmaceuticals, and nonferrous metals.

solvent, latent (co-solvent). An organic liquid that will dissolve nitrocellulose in combination with an active solvent. Latent solvents are usually alcohols and are used widely in nitrocellulose lacquers in a ratio of 1 part alcohol to 2 parts active solvent.

solvent naphtha. See naphtha (2b).

solvent refining. See solvent extraction.

Solvent Yellow 3. See ortho-aminoazotoluene.

"Solvesso."[51] Trademark for aromatic petroleum solvents. Grades available include "Solvesso Toluene," "Solvesso Xylene," "Solvesso 100," and "Solvesso 150."

"Solvofen" HM.[307] Trademark for a substituted pyrrolidone.

Properties: Light yellow liquid, b.p. 202°C.

Uses: Cleaning agent for rubber and metal padder or print rollers in textile processing; woolen felts and other paper mill equipment. Dispersing aid for organic and inorganic pigments for plastic and rubber systems.

solvolysis. A reaction involving substances in solution, in which the solvent reacts with the dissolved substance (solute) to form a new substance. Intermediate compounds are usually formed in this process. See also hydrolysis.

"Solwyte."[79] Trademark for a heat-treated tall oil rosin.

Uses: Rosin esters; rosin-modified synthetic resins; paper size.

soman (methylphosphonofluoridic acid, 1,2,2-trimethylpropyl ester) $(CH_3)_3CCH(CH_3)OPF(O)CH_3$. A nerve gas.

Properties: Colorless liquid; evolves odorless gas; b.p. 167°C; f.p. −70°C; sp. gr. 1.026 (20°C).
Hazard: Highly toxic by ingestion, inhalation, and skin absorption. May be fatal on short exposure. Cholinesterase inhibitor. No longer manufactured.

somatotropic hormone (STH; somatotropin). Hormone secreted by the anterior lobe of the pituitary. It causes an increase in general body growth and also affects carbohydrate and lipid metabolism.

"Sono-Jell."[45] Trademark for a balanced blend of white mineral oils and waxes of U.S.P. purity.
Uses: Waterless cleaning creams; ointments.

sonolysis. The breaking-up (molecular fragmentation) of molecules by ultrasonic radiation. Examples: sonolysis in pure water produces hydrogen atoms, hydroxyl radicals, molecular hydrogen, oxygen and hydrogen peroxide; acetonitrile, in an argon atmosphere, produces molecular hydrogen, nitrogen, and methane.

"Sonostat."[498] Trademark for a series of antistatic agents based on nitrogenous compounds for use on natural and synthetic fibers.

"Sonowax."[45] Trademark for a series of wax emulsions based on microcrystalline wax for use with resins or as top finish on various textile fabrics.

"Sopanox."[48] Trademark for ortho-tolyl biguanide (q.v.).

"Sorapon" SF-78.[307] Trademark for an anionic surfactant, sodium alkylaryl sulfonate; 85% active.

sorb. See sorption.

sorbic acid (2,4-hexadienoic acid) $CH_3CH{:}CHCH{:}CHCOOH$.
Properties: White, crystalline solid. M.p. 134.5°C; b.p. 228°C (dec), 153°C (50 mm); flash point (open cup) 260°F. Slightly soluble in water; soluble in many organic solvents. Combustible; low toxicity.
Derivation: Trimerization of acetaldehyde and catalytic air oxidation of the resulting hexadienal. Found in berries of mountain ash.
Grade: F.C.C.; technical.
Containers: Glass bottles; fiber cans.
Uses: Fungicide; food preservative; copolymerization; upgrading of drying oils; cold rubber additive; intermediate for plasticizers and lubricants.

sorbide (dianhydrosorbitol) $C_6H_8O_2(OH)_2$. Generic name for anhydrides (dicyclic ether dihydric alcohols) derivable from sorbitol (q.v.) by removal of two molecules of water. The name is also applied to specific commercial varieties.

"Sorbit."[219] Trademark for sodium dibutyl naphthalene sulfonate $(C_4H_9)_2C_{10}H_5SO_3Na$, 65% active.
Properties: Light tan flakes. Soluble in water, polar organic solvents, strong electrolytes. Stable to acids and alkalies.
Uses: Wetting agent and penetrant; hydrotrope; dispersant and thinning agent; emulsifier; detergents.

sorbitan (monoanhydrosorbitol; sorbitol anhydride) $C_6H_8O(OH)_4$. Generic name for anhydrides (cyclic ether tetrahydric alcohols) derivable from sorbitol (q.v.) by removal of one molecule of water.

sorbitan fatty acid esters. Mixtures of partial esters of sorbitol and its anhydrides with fatty acids.
Properties: Sorbitan monolaurate and sorbitan monooleate are amber liquids; sorbitan monostearate, sorbitan monopalmitate and sorbitan tristearate are cream-colored waxy solids with slight odor and bland taste. Sp. gr. of both liquid and solid esters is approx. 1.0 (25°C); m.p. of solid esters is approx. 54°C. They are insoluble in water; somewhat soluble in organic solvents. Low toxicity; combustible. See also "Span"; "Tween."
Derivation: Esterification of sorbitol with fatty acids.
Grades: Technical; F.C.C. (for sorbitan monostearate).
Uses: Emulsifiers and stabilizers in foods, cosmetics, drugs, textiles; plastics; argicultural chemicals.
See also polysorbate.

sorbitol (D-sorbite; D-sorbitol; hexahydric alcohol) $C_6H_8(OH)_6$.
Properties: White, odorless, crystalline powder; hygroscopic; faint, sweet taste. Soluble in water, glycerol, and propylene glycol; slightly soluble in methanol, ethanol, acetic acid, phenol and acetamide. Almost insoluble in most other organic solvents. Sp. gr. 1.47 (−5°C); m.p. (metastable form) 93°C; (stable form) 97.5°C. Approved by FDA for food use.
Derivation: By pressure hydrogenation of dextrose with nickel catalyst. Occurs in small amounts in various fruits and berries.
Grades: Crystals; technical; 70% aqueous solution (U.S.P.); resin; powder; F.C.C. (solid and solution).
Containers: Drums; tank cars; tank trucks.
Uses: Ascorbic acid fermentation. In solution form, for moisture-conditioning of cosmetic creams and lotions, toothpaste, tobacco, gelatin; bodying agent for paper, textiles, and liquid pharmaceuticals; softener for candy; sugar crystallization inhibitor; surfactants; urethane resins and rigid foams; plasticizer; stabilizer for vinyl resins; varnishes and lacquers; food additive (sweetener, humectant, emulsifier, thickener, anticaking agent).

sorbitol anhydride. See sorbitan.

"Sorbo."[89] Trademark for 70% sorbitol solution.

"Sorbo-Cell."[247] Trademark for a chemical-coated diatomite filter aid used for selectively removing traces of oil from oil-in-water emulsions. Effective for removing other trace components from free or emulsified systems.
Use: Conditioning of boiler feed water.

L(−)-sorbose $HOCH_2CO(CHOH)_3CH_2OH$.
Properties: White crystalline powder; sweet taste; m.p. 159–161°C; soluble in water; slightly soluble in ethyl or isopropyl alcohol, insoluble in ether, acetone, benzene, chloroform. Combustible; nontoxic.
Derivation: From sorbitol by submerged culture aerobic fermentation.
Grades: Technical; reagent.
Uses: Manufacture of ascorbic acid (vitamin C); preparation of special diets and media for the study of metabolism in animals and microorganisms.

Sorel cement. See magnesium oxychloride cement.

sorghum. A cereal plant cultivated in the temperate zones; its stems are rich in sucrose and can be processed in much the same way as sugar-cane and used as a source of sugar. It is grown in the U.S. in the Midwest, especially Texas.

"Soromine."[307] Trademark for a series of fiber softening agents.
AT. Complex fatty amido compound; 20% active; amphoteric.

CT-75. Dimethyldistearyl quaternary ammonium chloride; cationic.
CAZ-75. Complex polyalkyl amido imidazolinium sulfate; cationic.

sorption. A surface phenomenon which may be either absorption or adsorption, or a combination of the two. The term is often used when the specific mechanism is not known.

sorrel salt. See potassium binoxalate.

sour (1) Any substance used in textile or laundry operations to neutralize residual alkali or decompose residual hypochlorite bleach. The commonly used sours are sodium bifluoride and sodium fluosilicate.
(2) Contaminated with sulfur compounds, e.g., gasoline or natural gas. See doctor treatment.
(3) Taste characteristic of an acidic or fermented substance.

"Source."[523] Trademark for a biconstituent fiber composed of about 70% nylon and 30% polyester by weight. The fiber is a dispersion of polyester fibrils within and oriented to the axis of the nylon fiber. Chief use is for luxury carpet fiber.

"Soyagel."[221] Trademark for a catalytically polymerized soybean oil, for use in paints, enamels, primers and sealers.

soybean flour. A fine-ground powder having a particle size of 100-mesh or less, made by steaming soybeans to inactive enzymes, followed by removal of hulls and mechanical grinding. It contains from 40 to 50% protein, about 20% fat and 5% moisture. Defatted flours are made by extraction with hexane to remove the oil. The flakes produced are used chiefly for animal feeds, but the full-fat flour has become an increasingly important factor in high-protein food products, especially textured proteins and meat analogs.

soybean meal. The crushed residue from the extraction of soybeans. Extraction by the hydraulic or expeller process produces normally a meal with approximately 6% residual oils while from the solvent process approximately 1% residual oil. Typical analysis; crude protein 43%; crude fiber 5.5%; nitrogen-free extract 30%; ash 6%; oil content between 1 and 6%. Total digestible nutrients approximate 75%.
Containers: Bulk or bags.
Use: Animal feeds; adhesives; medium for bacitracin production.

soybean oil (soya bean oil; Chinese bean oil; soy oil).
Properties: Pale yellow, fixed drying oil. Soluble in alcohol, ether, chloroform and carbon disulfide. Sp. gr. 0.924–0.929; m.p. 22–31°C; refractive index 1.4760–1.4775; solidifying point −15 to −8°C; Hehner value 94–96; saponification value 190–193; iodine value 137–143. Flash point 540°F; combustible; nontoxic. Moderate spontaneous heating. Autoignition temp. 833°F.
Derivation: Expression and solvent extraction of soybeans.
Method of purification: Oil to be used for edible purposes is bleached with fuller's earth; oil for technical use is purified with chemicals.
Grades: Coast; refined (salad); crude; foots (for soapstock); clarified.
Containers: Drums; tank cars.
Uses: Soap manufacture; high-protein foods; paints and varnishes; cattle feeds; margarine and salad dressings; printing inks; source of nylon 9; plasticizer (epoxidized); alkyd resins.
Note: Soybean oil is the most widely used vegetable oil for both edible and industrial use in the U.S.

space velocity. The volume of gas or liquid, measured at specified temperature and pressure, (usually standard conditions) passing through unit volume in unit time. Used in comparing flow processes involving different conditions, rates of flow, and sizes or shape of containers.

spalling. Chipping an ore for crushing; or the cracking, breaking or splintering of materials due to heat.

"Span."[89] Trademark for each member of a series of general-purpose emulsifiers and surface-active agents. They are fatty acid partial esters of sorbitol anhydrides (or sorbitan). Generally insoluble in water and soluble in most organic solvents.

spandex. Generic name for a manufactured fiber in which the fiber-forming substance is a long chain synthetic polymer comprised of at least 85% of a segmented polyurethane (Federal Trade Commission). Imparts elasticity to garments such as girdles, socks, special hosiery. See, for example, "Lycra."

spar. (1) A type of crystalline material such as Iceland spar or feldspar, usually containing calcium carbonate or an aluminum silicate; fluorspar is calcium fluoride. Iceland spar has unique optical properties.
(2) A weather-resistant varnish originally used for coating wooden decks of ships, which may be the reason for its name. See spar varnish.

spar, Greenland. See cryolite.

spar, heavy. See barite.

spar, Iceland. See calcite.

"Sparine" Hydrochloride.[24] Trademark for promazine hydrochloride (q.v.).

sparking metal. See pyrophoric alloy.

"Spark-L."[212] Trademark for a liquid pectinase enzyme used to depectinize apples at low temperatures.

spar, satin. See calcite; gypsum.

spar varnish. A durable, water-resistant varnish for severe service on exterior exposure. It consists of one or more drying oils (linseed, tung or dehydrated castor oil); one or more resins (rosin, ester gum, phenolic resin or modified phenolic resin); one or more volatile thinners (turpentine or petroleum spirits), and driers (linoleates, resinates or naphthenates of lead, manganese and cobalt). It is classed as a long-oil varnish and generally consists of 45–50 gallons of oil to each 100 lb of resin.
See also varnish.

SPE. Abbreviation for Society of Plastics Engineers (q.v.).

spearmint oil.
Properties: Colorless to pale yellowish liquid; characteristic odor and taste. Soluble in alcohol, ether, and chloroform. Sp. gr. 0.930–0.940; optical rotation −48° to −59°; refractive index about 1.4910 (20°C). Combustible; nontoxic.
Chief known constituents: Carvone (40–60%); linalool; pinene.
Derivation: By distillation of spearmint leaves.
Grades: Technical; N.F.; F.C.C.
Containers: Bottles; tins; drums.
Uses: Flavoring (foods, medicine, chewing gum); source of carvone.

specific gravity. The weight of a given volume of any substance compared with the weight of an equal volume of a reference material, namely, water for solids and liquids and air for gases. The weight of one cubic centimeter of water is taken as 1 gram; hence, if the density of a given substance is 5 grams per cc, its specific gravity is 5/1, or 5. Since weights vary with temperature, it is necessary to specify both temperatures involved, except for rough or approximate values. Thus the specific gravity of alcohol should be given as 0.7893 at 20/4°C, the first temperature referring to the alcohol and the latter to the water. At 15.56°C the specific gravity of alcohol is 0.816. See also API gravity; density.

specific heat. The ratio of the heat capacity of a substance to the heat capacity of water; or the quantity of heat required for a one degree temperature change in a unit weight of material. Commonly expressed in Btu/lb/°F or in cal/g/°F. For a gas the specific heat at constant pressure is greater than that at constant volume by the amount of heat needed for expansion.

specific volume. The volume of unit weight of a substance, as cubic feet per pound, or gallons per pound, but more frequently milliliters per gram. The reciprocal of density.

specific weight. The weight per unit volume of a substance.

"SpectrAR."[329] Trademark for a group of spectrophotometric-grade solvents that have been developed for the extremely high purity required by modern analytical techniques.

spectrophotometry. See absorption spectroscopy.

spectroscopy (instrumental analysis). A branch of analytical chemistry devoted to identification of elements and elucidation of atomic and molecular structure by measurement of the radiant energy absorbed or emitted by a substance in any of the wavelengths of the electromagnetic spectrum in response to excitation by an external energy source. The types of absorption and emission spectroscopy are usually identified by the wavelength involved, namely, gamma ray, x-ray, ultraviolet, visible, infrared and microwave. The technique of spectroscopic analysis was originated by Fraunhofer who in 1814 discovered certain dark (D) lines in the solar spectrum, which were later identified as characterizing the element sodium. In 1861 Kirchhoff and Bunsen produced emission spectra and showed their relationship to Fraunhofer lines. X-ray spectroscopy was utilized by Moseley (1912) to determine the precise location of elements in the Periodic system. Since then, still more sophisticated techniques have been developed, e.g., nuclear magnetic resonance (NMR) and dynamic reflectance spectroscopy. See also absorption spectroscopy; emission spectroscopy; infrared spectroscopy.

"Spectrosil."[220] Trademark for a synthetic fused silica used in laboratory ware.

"Speedflow."[218] Trademark for diatomite filteraids, used in food, pharmaceutical, and industrial processing.

spelter. Relatively pure zinc as encountered in industrial operations such as galvanizing. Lead and/or iron are common impurities.

"Spen-Amm."[533] Trademark for ammonia used for direct application.

"Spensol Green."[533] Trademark for ammoniating solutions.

spent mixed acid. Mixed acid (q.v.) which has given up part of its nitric acid.
Hazard: Dangerous fire risk; strong irritant to tissue.
Shipping regulations: (Rail) White label. (Air) Corrosive label.

spent oxide. See iron sponge, spent.

"Spergon."[248] Trademark for a series of fungicides and seed protectants based on tetrachloro-para-benzoquinone.

spermaceti
Properties: White, unctuous, semitransparent, waxy solid; almost odorless and tasteless; becomes rancid on exposure. Soluble in ether, chloroform, carbon disulfide and hot alcohol; insoluble in water and cold alcohol. Sp. gr. 0.945; m.p. 42–50°C; refractive index about 1.4330; saponification no. 120–136; iodine no. 3–4.4. Combustible; Low toxicity.
Chief constituents: Cetyl palmitate, esters of lauric, myristic and stearic acids.
Derivation: Found in the head of the spermwhale; filtered under pressure to remove stearin, boiled with water and a small amount of caustic soda, following by repeated washings. Also made synthetically.
Grades: Technical; U.S.P.; as blocks or cakes.
Containers: 50-, 60-lb cases; bags.
Uses: Base for ointments, cerates and emulsions; manufacture of candles, soaps, cosmetics, laundry wax; finishing and lustering linens.
Note: See note under sperm oil.

sperm oil
Properties: Light yellow liquid; sp. gr. 0.8781–0.8835; flash point 428°F saponification number 123–147; iodine number 79.5–84; very low in free fatty acids. Soluble in chloroform, ether, and benzene. A monohydric fatty alcohol ester of fatty acids—not a triglyceride. One of the best natural lubricants known. Combustible; low toxicity.
Derivation: From the head cavity of the sperm whale. (Not the same as whale oil, q.v.).
Grades: Bleached winter; natural winter.
Containers: Drums, tank cars and trucks.
Uses: High-grade lubricating oil for precision machinery (watches, clocks and scientific instruments); fatliquoring leather; rust proofing; extreme pressure lubricants; automatic transmission fluids.
See also porpoise oil; jojoba oil.
Note: For animal conservation reasons, use of sperm oil has been prohibited in the U.S. since 1971.

Sperry process. An electrolytic process for the manufacture of lead carbonate, basic (white lead) from desilverized lead containing some bismuth. The impure lead forms the anode. A diaphragm separates anode and cathode compartments, and carbon dioxide is passed into the solution. Impurities, including bismuth, remain on the anode as a slime blanket.

sphalerite (blende; zinc blende) ZnS. Natural zinc sulfide, usually containing some cadmium, iron, and manganese.
Properties: Color yellow, brown, black, or red; luster resinous; hardness 3.5–4; sp. gr. 3.9–4.1; good cleavage; soluble in hydrochloric acid.

Occurrence: Missouri, Kansas, Oklahoma, Colorado, Montana, Wisconsin, Idaho; Australia; Canada; Mexico.
Uses: Most important ore of zinc; also a source of cadmium; phosphor; source of sulfur dioxide for production of sulfuric acid.

"Spheron."[275] Trademark for pelleted channel carbon blacks for rubber.

sphingomyelin. Diaminophosphatides occurring primarily in nervous tissue and containing a fatty acid, phosphoric acid, choline, and sphingosine. They are soluble in hot absolute alcohol, and insoluble in ether, acetone, and water.

sphingosine (1,3-dihydroxy-2-amino-4-octadecene) $CH_3(CH_2)_{12}CH{:}CHCHOHCH(NH_2)CH_2OH$. Forms part of certain phosphatides, such as cerebrosides and sphingomyelins.
Properties: Waxy crystals; m.p. 67°C; soluble in ether.

SPI. Abbreviation for Society of the Plastics Industry (q.v.).

"Spiceal."[342] Trademark for water-miscible flavor concentrates from essential oils and oleoresin bases.

spike oil (lavender-spike oil; Spanish lavender oil; Spanish spike oil). See also lavandin oil, abrial.
Properties: Pale yellow essential oil; lavender-like odor; sp. gr. 0.893–0.909 (25°C); refractive index 1.4630–1.4680 (20°C); angular rotation −5° to +5°. Soluble in some dilution between one and three volumes of 70% alcohol. Soluble in most fixed oils and propylene glycol; slightly soluble in glycerin and mineral oil.
Derivation: Steam distillation of flowers of Lavandula latifolia.
Grade: F.C.C., which requires not less than 40% and not more than 50% of total alcohols, calculated as linalool, and not less than 1.5% and not more than 3% esters, calculated as linalyl acetate.
Uses: Soaps; bath preparations; masking odor in sprays and disinfectants; flavoring agent.

spindle oil. Low-viscosity lubricating oil for lubrication of textile and other high-speed machinery.

spinel $MgAl_2O_4$. A natural oxide of magnesium and aluminum, with replacement of magnesium by iron, zinc and manganese, and of aluminum by iron and chromium. There are also synthetic spinels, as magnesia-alumina or magnesia-chromia. Their structure is similar to ferrites (q.v.).
Properties: Color, various shades of red, grading to green, brown, and black; luster vitreous; hardness 8. There are many varieties.
Occurrence: New York, New Jersey, Massachusetts, Virginia, North Carolina; Ceylon; Thailand; Madagascar.
Uses: Crystallography; synthetic spinel is used as a refractory, and in electronic applications.

"Spiralloy."[266] Trademark for glass- and other filament-wound, resin-bonded internal and external pressure vessels.
Uses: Rocket motor cases, underwater and space structures, pressure vessels, radomes, torpedo cases, booms and tubes.

spirits. An obsolescent and ambiguous term usually taken to mean the distilled essence of a substance. Mineral or petroleum spirits is a volatile hydrocarbon distillate similar to naphtha; in pharmacy, the term refers to an alcoholic solution of volatile principles, e.g., spirits of ammonia, niter, camphor, etc. See also tincture.

spirocyclane. See spiropentane.

spironolactone $C_{24}H_{32}O_4S$. 17-Hydroxy-7α-mercapto-3-oxo-17α-pregn-4-ene-21-carboxylic acid γ-lactone 7-acetate.
Properties: Light cream-colored to light tan crystalline powder. Mild mercaptan-like odor; stable in air. Practically insoluble in water; soluble in ethyl acetate and alcohol; slightly soluble in methanol. M.p. 198–207°C (dec).
Grade: U.S.P.
Use: Medicine.

spiropentane (spirocyclane; cyclopropanespirocyclopropane $H_2CH_2CCCH_2CH_2$.
Properties: Colorless liquid; refractive index (n 20/D) 1.41220; sp. gr. 0.7551 (20/4°C); freezing point −107.05°C; b.p. 39.03°C.
Derivation: Heating pentaerythrityl tetrabromide in ethanol with zinc dust.

spiro system. A structural formula consisting of two rings having one atom in common. Most bicyclic compounds, such as naphthalene, have two atoms in common. See, for example, spiropentane.

splitting, molecular, see hydrolysis; atomic, see fission.

spodumene $LiAl(SiO_3)_2$.
Properties: White, pale green, emerald green, pink or purple mineral; white streak; vitreous luster. Contains up to 8% lithium oxide with some replacement by sodium. Insoluble in acids. Sp. gr. 3.13–3.20; Mohs hardness 6.5–7.
Occurrence: United States (North Carolina, California, Massachusetts, South Dakota); Brazil; Madagascar.
Uses: Source of lithium; in ceramics and glass as a source of lithia and alumina.

sponge (1) Metal: A finely divided and porous form of a metal such as iron, platinum, or titanium. May be used in sponge form as a catalyst, or pressed into metal ingots. (2) Plastic: See cellular plastic; foam; rubber sponge. (3) Natural: Siliceous cells of the Porifera group of sessile sea animals.

sponge iron. See iron sponge.

"Spotleak."[204] Trademark for a group of natural gas odorants.
1001. A mercaptan-sulfide blend that combines the high odor impact of tert-butyl mercaptan with the chemical stabilty of alkyl sulfides.
1003. A 100% synthetic mercaptan of tert-btyl mescaptan and isopropyl mercaptan.
1009. A 100% synthetic mercaptan mixture containing approximately 80% tert-butyl mercaptan.

"Spot-Lite."[529] Trademark for nonradioactive phosphorescent paint, including an aerosol type and pressure-sensitive vinyl film. Both can be activated by any ambient light (sunlight, incandescent or fluorescent lamp) and are characterized by a high initial glow and a long and low afterglow; the dominant wavelength of both is about 5600Å.

sputtered coating. A protective metallic coating applied in a vacuum tube and consisting of metal ions emanating from a cathode and deposited as a film on objects within the tube. The process involves three phases: generation of metal vapor, diffusion of the vapor, and its condensation. Paper, plastics, and similar materials can be coated in this way.

"Sprout Nip."[177] Trademark for potato sprout inhibiting formulation of isopropyl N-(3-chlorophenyl)-carbamate.
Grades: Aerosol; emulsifiable concentrate; 5% dust.

spun protein. See protein, textured.

"SQ Phosphate."[58] A glassy polyphosphate with high calcium-sequestering properties. Used in formulating industrial cleaning compounds.

squalane (perhydrosqualene; 2,6,10,15,19,23-hexamethyltetracosane; spinacane; dodecahydrosqualane) $C_{30}H_{62}$. A saturated hydrocarbon.
Properties: Colorless, odorless, tasteless liquid; miscible with vegetable and mineral oils, organic solvents and lipophilic substances. Nondrying, nonoxidizing, noncongealing. Sp. gr. 0.805–0.812 (20°C); b.p. approx. 350°C; f.p. approx. −38°C; refractive index 1.4520–1.4525 (n 20/D). Combustible.
Derivation: Hydrogenation of squalene; may occur naturally in sebum.
Containers: Bottles; drums.
Uses: High-grade lubricating oil; vehicle for externally applied pharmaceuticals and cosmetics; perfume fixative; gas chromatographic analysis.

squalene (spinacene; 2,6,10,15,19,23-hexamethyl-2,6,10,14,18,22-tetracosahexaene) $C_{30}H_{50}$. A natural raw material found in human sebum (5%), and in shark liver oil. An unsaturated aliphatic hydrocarbon (carotenoid) with six unconjugated double bonds.
Properties: Oil with faint odor; sp. gr. 0.858–0.860 (20°C); b.p. ca. 225°C; f.p. −60°C; refractive index 1.49–1.50; iodine no. 360–380; saponification value 0–5. Insoluble in water; slightly soluble in alcohol; soluble in lipids and organic solvents. Combustible.
Grade: 90% min.
Uses: Biochemical and pharmaceutical research; a precursor of cholesterol in biosynthesis.

Sr Symbol for strontium.

"SR-173."[245] Trademark for a silicone intermediate resin for reaction with alkyd coating resin constituents for air-dry and baking co-polymer high-performance coatings. Also available with special solvent to comply with air pollution requirements.

SRF black. Abbreviation for semireinforcing furnace black. See furnace black.

SS acid. See 8-amino-1-naphthol-5,7-disulfonic acid.

SSU. Abbreviation sometimes used for standard Saybolt Universal viscosity. The preferred abbreviation is now SUS. See Saybolt Universal viscosity.

"Stabilide."[329] Trademark for potassium iodide stabilized with calcium stearate. Used in animal feeds.

"Stabilisal C."[230] Trademark for an organic compound used as an antigellation agent for nitrocellulose-based bronzing lacquers and for stabilizing dipping lacquers.

"Stabilite."[94] Trademark for a series of antioxidants for natural and synthetic rubbers and latices.
N,N′-diphenylethylenediamine.
Alba. Di-o-tolyl ethylenediamine.
FLX. A mixture of N,N′-diphenylethylenediamine and diphenyl-p-phenylenediamine.
L. N,N′-diphenyl propylenediamine.
White. Polyalkylated phenol.

stabilizer. Any substance which tends to keep a compound, mixture, or solution from changing its form or chemical nature. Stabilizers may retard a reaction rate, preserve a chemical equilibrium, act as antioxidants, keep pigments and other components in emulsion form, or prevent the particles in a colloidal suspension from precipitating. See also inhibitor.

"Staclipse" Starches.[492] Trademark for specially processed acid-modified thin-boiling starches. Highly converted products having relatively high percentage of cold water-soluble fractions.
Uses: Finishes and binders in textile wet processing, adhesives, and paper industry for surface sizing, calender sizing and coating.

"Stacolloid" Gums.[492] Trademark for a group of non-congealing sizing and finishing agents derived from corn starch. Used in the sizing of spun synthetic, combed cotton, worsted yarns, and paper.

"Stacrylic" 115 Warp Size.[492] A pale, yellow, cloudy solution of an acrylate terpolymer used for sizing polyester filament yarns of all types.

"Stadex" Dextrins.[492] Trademark for products of the partial hydrolysis of corn starch produced by heating the starch in a dry atmosphere in the presence of acid.
Uses: Adhesives, binders, thickeners, sizing ingredients, stabilizers, bodying agents and in edible products as carriers for color and flavor additives.

"Stafast."[50] Trademark for a naphthaleneacetic acid product (q.v.).

"Staflex."[36] Trademark for a series of vinyl plasticizers.

"Sta-fresh."[50] Trademark for a sodium bisulfite powder for preserving grass silage.

stain. (1) An organic protective coating similar to a paint, but with much lower solids content (pigment loading). Used for both exterior and interior coating of wood, furniture, flooring, etc. (2) Any compound used to color bacteria for microscopic examination, e.g., osmium tetroxide, phosphotungstic acid, uranyl acetate, and certain chromium compounds.

stainless steel. See steel, stainless.

"Staleydex" Dextrose.[492] Trademark for a crystalline dextrose produced by an enzyme conversion process. Several types of dextrose available varying from coarse particles to very fine dextrose monohydrate crystals, as well as liquid dextrose.
Uses: Canning, brewing, confections, beverages, prepared mixes, pharmaceuticals, chemicals, and other industrial uses.

"Stamere."[406] Trademark for a series of edible hydrocolloids extracted from seaweeds. Various products are available which are standardized and recommended for particular applications:

standard cell. See Weston cell.

"Standfast Metal."[60] Trademark for a bismuth, lead, tin, and cadmium alloy. M.P. 158°F.
Uses: Liquid heat and pressure transfer medium.

stannic acid. See stannic oxide.

stannic anhydride. See stannic oxide.

stannic bromide (tin bromide; tin tetrabromide) $SnBr_4$.
Properties: White, crystalline mass. Fumes when exposed to air. Soluble in water, alcohol, and carbon tetrachloride. Sp. gr. 3.3; b.p. 203°C; m.p. 31°C.

Hazard: Irritant to skin and eyes. Tolerance (as Sn), 2 mg per cubic meter of air.
Use: Mineral separations.

stannic chloride (tin chloride; tin tetrachloride) $SnCl_4$. Often sold in the form of the double salt with sodium chloride, $Na_2SnCl_6 \cdot H_2O$.
Properties: Colorless, fuming, caustic liquid, which water converts into a crystalline solid, $SnCl_4 \cdot 5H_2O$. Keep well stoppered! Sp. gr. 2.2788; f.p. −33°C; b.p. 114°C. Soluble in cold water, alcohol, carbon disulfide and oil of turpentine; decomposed by hot water.
Derivation: Treatment of tin or stannous chloride with chlorine.
Grades: Technical; C.P.
Containers: Bottles; drums.
Hazard: Toxic, corrosive liquid. Tolerance (as Sn), 2 mg per cubic meter of air. Evolves heat on contact with moisture.
Uses: Electroconductive and electroluminescent coatings; textiles; perfume stabilization; manufacture of fuchsin; color lakes; ceramics; bleaching agent for sugar; stabilizer for certain resins; manufacture of blue print and other sensitized papers; other tin salts; bacteria and fungi control in soaps.
Shipping regulations: (Rail) White label. (Air) Corrosive label. Legal label name, tin tetrachloride.

stannic chloride pentahydrate $SnCl_4 \cdot 5H_2O$.
Properties: White solid, m.p. 56°C; soluble in water or alcohol.
Hazard: Toxic. Tolerance (as Sn), 2 mg per cubic meter of air.
Use: Substitute for anhydrous stannic chloride where the presence of water is not objectionable.

stannic chromate (tin chromate) $Sn(CrO_4)_2$.
Properties: Brownish yellow, crystalline powder. Partially soluble in water.
Derivation: Action of chromic acid on stannic hydroxide.
Hazard: Toxic. Tolerance (as Sn) 2 mg per cubic meter of air.
Use: Coloring porcelain and china.

stannic oxide (stannic anhydride; tin peroxide; tin dioxide; stannic acid) SnO_2, or $SnO_2 \cdot nH_2O$.
Properties: White powder, anhydrous or containing variable amounts of water; sp. gr. 6.6–6.9; m.p. 1127°C; sublimes at 1800–1900°C. Soluble in concentrated sulfuric acid, hydrochloric acid; insoluble in water. Low toxicity. Noncombustible.
Derivation: (a) Found in nature as in the mineral cassiterite (q.v.). (b) Precipitated from stannic chloride solution by ammonium hydroxide.
Grades: White, pure; white; gray; C.P.
Containers: Bottles; drums.
Uses: Tin salts; catalyst; ceramic glazes and colors; putty; perfume preparations and cosmetic; textiles (mordant, weighting); polishing powder for steel, glass, etc.; manufacture of special glasses.

stannic phosphide (tin phosphide) Sn_2P_2, or SnP.
Properties: Silver white, hard mass or lumps. Soluble in acids. Sp. gr. 6.56; m.p., forms Sn_4P_3 at 415°C.
Derivation: By heating tin and phosphorus.
Hazard: Flammable, dangerous fire risk. Tolerance (as Sn), 2 mg per cubic meter of air.
Shipping regulations: (Rail) Flammable solids, n.o.s., Yellow label. (Air) Not acceptable.

stannic sulfide (artificial gold; mosaic gold; tin bronze; tin disulfide) SnS_2.
Properties: Yellow to brown powder. Soluble in concentrated hydrochloric acid and alkaline sulfides; insoluble in water. Sp. gr. 4.42–4.60; m.p., decomposes at 600°C.
Derivation: (a) Action of sulfide on a solution of stannic chloride. (b) By heating tin amalgam with sulfur and ammonium chloride, distilling off the mercury sulfide and ammonium chloride.
Grades: Technical; reagent.
Containers: Glass bottles; boxes; drums.
Hazard: Irritant to skin and eyes. Tolerance (as Sn), 2 mg per cubic meter of air.
Uses: Imitation gilding; pigment.

"Stannochlor."[288] Trademark for anhydrous stannous chloride.
Uses: Immersion tinning; liquor finishing of wire; major ingredient in acid electro-tin and tin-nickel plating solutions; reducing agent; catalyst; analytical sensitizing of glass and plastics before metallizing; tanning agent; soldering flux; intermediate for tin chemicals; manufacture of color pigments, pharmaceuticals and sensitized paper.

stannous bromide (tin bromide; tin dibromide) $SnBr_2$.
Properties: Yellow powder. Soluble in hydrochloric acid (dilute); soluble in water, alcohol, ether, and acetone. Oxidizes and turns brown in air. Sp. gr. 5.117 (17°C); b.p. 619°C; m.p. 215°C.
Hazard: Skin irritant. Tolerance (as Sn), 2 mg per cubic meter of air.

stannous chloride (tin crystals; tin salt; tin dichloride; tin protochloride) (a) $SnCl_2$ (b) $SnCl_2 \cdot 2H_2O$.
Properties: White, crystalline mass, which absorbs oxygen from the air, being converted into the insoluble oxychloride. Soluble in water, alkalies, tartaric acid, and alcohol. (a) Sp. gr. 3.95 (25/4°C); M.p. 246.8°C; b.p. 652°C. (b) Sp. gr. 2.71; m.p. 37.7°C; b.p., decomposes.
Derivation: By dissolving tin in hydrochloric acid.
Grades: Technical; C.P.; reagent; anhydrous; hydrated.
Containers: Bottles; drums.
Hazard: Toxic; irritant to skin. Tolerance (as Sn), 2 mg per cubic meter of air. Use in foods restricted to 0.0015%, calculated as tin.
Uses: Reducing agent in manufacture of chemicals, intermediates, dyes, polymers, phosphors; manufacture of lakes; textiles (reducing agent in dyeing, discharge in printing); tin galvanizing; reagent in analytical chemistry; medicine; silvering mirrors; revivification of yeast sown in must (accelerator); antisludging agent for lubricating oils; food preservative; stabilizer for perfume in soaps.

stannous chromate (tin chromate) $SnCrO_4$.
Properties: Brown powder. Almost insoluble in water.
Derivation: Interaction of stannous chloride and sodium chromate.
Hazard: Toxic. Tolerance (as Sn) 2 mg per cubic meter of air.
Use: Decorating porcelain.

stannous 2-ethylhexoate (stannous octoate; tin octoate) $Sn(C_8H_{15}O_2)_2$.
Properties: Light yellow liquid; insoluble in water, methanol; soluble in benzene, toluene, petroleum ether; hydrolyzed by acids and bases. Sp. gr. 1.25; Gardner color, 3 (max).
Hazard: Toxic. Tolerance (as Sn), 0.1 mg per cubic meter of air.
Uses: Polymerization catalyst for urethane foams; lubricant; addition agent; stabilizer for transformer oils.

stannous fluoride (tin fluoride; tin difluoride) SnF_2.
Properties: White, lustrous, crystalline powder with bitter salty taste; m.p. 212–214°C. Practically in-

soluble in alcohol, ether, and chloroform; slightly soluble in water.
Grade: N.F.
Hazard: Toxic by ingestion; strong irritant to skin and tissue. Tolerance (as Sn), 2 mg per cubic meter of air.
Uses: Fluoride source in toothpastes; medicine.
Note: Stannous hexafluorozirconate is said to be more effective than the fluoride in preventing dental caries.

stannous octoate. See stannous 2-ethylhexoate.

stannous oleate (tin oleate) $Sn(C_{18}H_{33}O_2)_2$.
Properties: Light yellow liquid; insoluble in water and methanol; soluble in benzene, toluene, petroleum ether; hydrolyzed by acids and bases.
Hazard: Toxic. Tolerance (as Sn), 0.1 mg per cubic meter of air. Absorbed by skin.
Uses: Polymerization catalyst; inhibitor.

stannous oxalate (tin oxalate) SnC_2O_4.
Properties: Heavy, white, crystalline powder; sp. gr. 3.56; m.p., decomposes at ca. 280°C; soluble in acids; insoluble in water and acetone.
Derivation: By the action of oxalic acid on stannous oxide.
Grades: Technical; C.P.; reagent.
Hazard: Toxic. Tolerance (as Sn), 0.1 mg per cubic meter of air. Absorbed by skin.
Uses: Dyeing and printing textiles; catalyst for esterification reactions.

stannous oxide (tin oxide; tin protoxide) SnO.
Properties: Brownish black powder; unstable in air. Reacts with acids and strong bases; insoluble in water. Sp. gr. 6.3; m.p. 1080°C (600 mm), decomposes. A nuisance particulate. Low toxicity.
Derivation: By heating stannous hydroxide in a current of carbon dioxide.
Grades: Technical; C.P.
Containers: 1-, 5-lb bottles; wooden kegs.
Uses: Reducing agent; intermediate in preparation of stannous salts as used in plating and glass industries; pharmaceuticals; soft abrasive (putty powder).

stannous pyrophosphate $Sn_2P_2O_7$.
Properties: White, free-flowing crystals insoluble in water. Sp. gr. 4.009 (16°C).
Uses: Toothpaste additive.

stannous sulfate (tin sulfate) $SnSO_4$.
Properties: Heavy, white or yellowish, crystals; soluble in water and sulfuric acid. Water solution decomposes rapidly. M.p. loses sulfur dioxide at 360°C.
Derivation: Action of sulfuric acid on stannous oxide.
Containers: Fiber drums.
Hazard: Toxic. Tolerance (as Sn), 2 mg per cubic meter in air.
Uses: Dyeing; tin-plating; particularly for plating automobile pistons and steel wire.

stannous sulfide (tin monosulfide; tin protosulfide; tin sulfide) SnS.
Properties: Dark gray or black crystalline powder; sp. gr. 5.080; b.p. 1230°C; m.p. 880°C. Soluble in concentrated hydrochloric acid (decomposes); insoluble in dilute acids and water.
Hazard: Toxic. Tolerance (as Sn), 2 mg per cubic meter in air.
Uses: Making bearing material; catalyst in polymerization of hydrocarbons; chemical reagent.

stannous tartrate (tin tartrate) $SnC_4H_4O_6$.
Properties: Heavy, white, crystalline powder; soluble in water, dilute hydrochloric acid.
Derivation: Action of tartaric acid on stannous oxide.
Hazard: Toxic. Tolerance (as Sn), 0.1 mg per cubic meter of air.
Uses: Dyeing and printing fabrics.

stannum. The Latin name for tin, hence the symbol Sn in chemical nomenclature.

staple. A cotton fiber, usually in reference to length, i.e., short or long staple cotton.

starch. A carbohydrate polymer having the repeating unit:

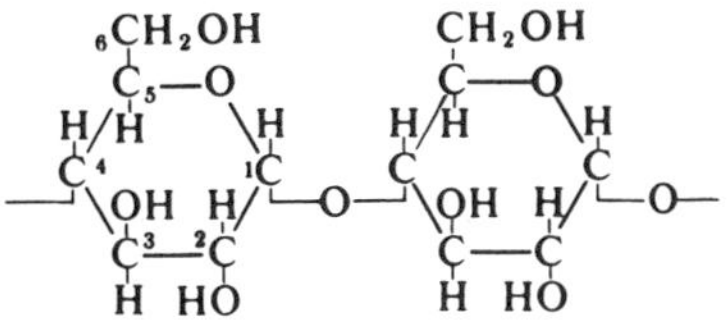

```
       6CH2OH              CH2OH
         C5—O               C——O
      H/H    \H          H/H    \H
      C4      1C          C       C
    —  \OH  H/  —O—        \OH  H/  —O—
        C3——2C             C——C
        H   HO             H   HO
```

It is composed of about 25% amylose (anhydroglucopyranose units joined by glucosidic bonds) and 75% amylopectin, a branched-chain structure.
Properties: White, amorphous, tasteless powder or granules; various crystalline forms may be obtained, including microcrystalline. Irreversible gel formation occurs in hot water; swelling of granules can be induced at room temperature with such compounds as formamide, formic acid, and strong bases and metallic salts.
Occurrence: Starch is a reserve polysaccharide in plants (corn, potatoes, tapioca, rice, and wheat are commercial sources).
Grades: Commercial; powdered; pearl; laundry; technical; reagent; edible; U.S.P.
Uses: Adhesive (gummed paper and tapes, cartons, bags, etc.); machine-coated paper; textile filler and sizing agent; beater additive in papermaking; gelling agent and thickener in food products (gravies, custards, confectionery); oil-well drilling fluids; filler in baking powders (cornstarch); fabric stiffener in laundering; urea-formaldehyde resin adhesives for particle board and fiberboard; explosives (nitrostarch); dextrin (starch gum); chelating and sequestering agent in foods; indicator in analytical chemistry; anticaking agent in sugar; face powders; abherent and mold-release agent; polymer base (see next entry).

starch-based polymer. (1) A reactive polyol derived from a mixture of a starch with dibasic acids, hydrogen-donating compounds, and catalysts dissolved in water; the slurry is subjected to high temperatures and pressures, yielding a low-viscosity polymer in a 50% solids aqueous solution. A molecular rearrangement takes place, and the polymer formed is completely different from starch in structure and properties. It can be further reacted with acids, bases and cross-linking agents. Suggested uses are in high wet-strength papers, as binders in paper coatings, as moisture barriers in packaging, and as water-resistant adhesives.

(2) Yeast fermentation of starch to form a biodegradable plastic called pullulan (a trigluco polysaccharide) has been reported to be commercially feasible.

starch dialdehyde. Starch in which the original anhydroglucose units have been partially oxidized to dialdehyde form by oxidation, for example, the product of the oxidation of cornstarch by periodic acid.

Available in cationic dispersions up to 15% solids for mixing with paper pulp.

Uses: Thickening agent; tanning agent; binder for leaf tobacco; adhesives; wet-strength additive in paper.

starch, modified. Any of several water-soluble polymers derived from a starch (corn, potato, tapioca) by acetylation, chlorination, acid hydrolysis, or enzymatic action. These reactions yield starch acetates, esters and ethers in the form of stable and fluid solutions and films. Modified starches are used as textile sizing agents and paper coatings. Thin-boiling starches have high gel strength; oxidized starches made with sodium hypochlorite have low gelling tendency. Introduction of carboxyl, sulfonate, or sulfate groups into starch gives sodium or ammonium salts of anionic starches, yielding clear, non-gelling dispersions of high viscosity. Cationic starches result from addition of amino groups.

The glucose units of starch can be crosslinked with such agents as formaldehyde, soluble metaphosphates, and epichlorohydrin; this increases viscosity and thickening power for adhesives, canned foods, etc.

starch phosphate. An ester made from the reaction of a mixture of orthophosphate salts (sodium dihydrogen phosphate and disodium hydrogen phosphate) with starch.

Properties: Soluble in cold water (unlike regular starch), and has high thickening power. Can be frozen and thawed repeatedly without change in physical properties.

Uses: Thickener for frozen foods; taconite ore binder; in adhesives, drugs, cosmetics; substitute for arabic gum, locust bean gum, and carboxymethyl cellulose.

starch syrup. See glucose.

starch, thin-boiling. See starch, modified.

starch xanthate. A water-insoluble synthetic polysaccharide made by reacting starch with sodium hydroxide and carbon disulfide. Biodegradable. Used to encapsulate pesticides; the coating, though insoluble, is permeable to water, thus slowly releasing the pesticide. Also used as a rubber reinforcing agent.

starch xanthide. Starch xanthate which has been cross-linked with oxygen; used as strengthening agent in paper and manufacture of powdered rubbers.

"Starex."[79] Trademark for a heat-treated tall oil rosin.

Uses: Rosin esters; rosin-modified synthetic resins; limed rosin; paper size.

"Starez."[170] Trademark for a modified polyvinyl acetate monopolymer resin emulsion. Used as a heatseal coating.

"Starfol."[221] Trademark for various grades of glyceryl monostearate and butyl stearate. Most grades have been approved for food additives.

"Starwax" 100.[128] Trademark for a hard petroleum microcrystalline wax; minimum m.p. 180° F.

"Sta-set."[50] Trademark for a trichlorophenoxypropionic acid product for use on apples to reduce preharvest drop.

"Sta-Sol" Lecithin Concentrate.[492] Trademark for mixtures of naturally occurring phosphatides or phospholipids, derived from soybean oil. They contain some residual oil as a solvent or carrier. Available in regular (plastic) or fluid (pourable liquid) form, bleached and unbleached. They act as emulsifiers, stabilizers, antioxidants.

Uses: Foods, especially chocolate and compound coatings, candies, bakery products, margarine; industrial uses; paints and printing inks, soaps and cosmetics, textile compounds, leather tanning, petroleum lubricants; animal feeds and pet foods.

"Statex."[133] Trademark for furnace carbon blacks used in rubber, printing inks and protective coatings. There are numerous grades. See furnace black.

"Stat-Eze."[430] Trademark for a specially prepared quaternary ammonium compound, used as an externally applied antistatic agent on plastics, textiles, paper and carpeting.

"Sta-Thik" Starch.[492] Trademark for a modified corn starch of high viscosity.

Uses: Wet-end additive in paper industry; binder for both cellulose and mineral wool building materials; textile finishing; adhesives; and briquetting.

Staudinger, Hermann (1881–1965). A German chemist, winner of Nobel Prize in 1953 for his pioneer work on the structure of macromolecules and polymerization. A large part of modern high-polymer chemistry is based on his original research.

"Staybelite."[266] Trademark for hydrogenated rosin, a pale, thermoplastic resin. Acid number 165; USDA color X; softening point 75°C; saponification number 167.

Uses: Adhesives and protective coatings.

"Stayco" Starches.[492] Trademark for a family of oxidized starches. Available in five viscosities: S, A, G, C, and M. Used by the paper industry for wet-end addition, in tub, press, and calender sizing, and as coating adhesives.

"Stayrite."[104] Trademark for stabilizers used to provide increased heat and light stability to vinyl resins. Available in range of products, both liquid and solid.

"Staysize" 109 Starch.[492] A white, highly uniform, chemically modified corn starch in pearl form developed for surface sizing.

"Stayzyme" Starch.[492] A thick-boiling corn starch buffered to a pH of 6.3 to 7.0 to produce the proper pH for enzyme conversion in the paper mill.

steam. An allotropic form of water (H_2O) formed at 212°F (100°C) and having a latent heat of condensation of 540 calories per gram. It has a number of industrial uses, one of the most important being the production of hydrogen by the steam-hydrocarbon gas process (reforming), by the steam-water gas process, the steam-iron process, and the steam-methanol process. It is also used in steam cracking of gas oil and naphtha, in food processing, as a cleaning agent, in rubber vulcanization, as a source of heat and power, and in distillation of plants for production of essential oils and perfumes.

Steam from geothermal sources such as hot springs and fumaroles is being utilized as an energy source. See geothermal energy, latent heat.

steam distillation. See hydrodistillation.

steam reforming. See reforming.

steapsin. A lipase in the pancreatic juice. See enzyme.

stearamide $CH_3(CH_2)_{16}CONH_2$. Octadecanamide.

Properties: Colorless leaflets; m.p. 109°C; b.p. 251°C (12 mm); insoluble in water; slightly soluble in alcohol and ether.

Use: Corrosion inhibitor in oil wells.

stearato chromic chloride. A polynuclear complex in the form of a six-membered ring. The two chromium atoms are bridged on one side by a hydroxyl group, and on the other side by the carboxyl oxygens of the stearic acid. The water-soluble complex results from the neutralization of stearic acid with basic chromic chloride. It acts as a water repellent and abherent.

stearic acid (n-octadecanoic acid) $CH_3(CH_2)_{16}COOH$. The most common fatty acid occurring in natural animal and vegetable fats. Most commercial stearic acid is about 45% palmitic acid, 50% stearic acid, and 5% oleic acid, but purer grades are increasingly used.
Properties: Colorless, wax-like solid. Odor and taste slight, suggesting tallow. Soluble in alcohol, ether, chloroform, carbon disulfide, carbon tetrachloride; insoluble in water. Sp. gr. 0.8390 (80/4°C); m.p. 69.6°C; b.p. 361.1°C; refractive index 1.4299 (80°C). Flash point 385°F; autoignition temp. 743°F. Combustible; nontoxic.
Derivation: (a) From high-grade tallows and yellow grease stearin by washing, hydrolysis with the Twitchell or similar reagent, boiling, distilling, cooling and pressing; (b) from oleic acid by hydrogenation.
Grades: Saponified; distilled; single-pressed; double-pressed; triple-pressed; U.S.P.; F.C.C.; 90% stearic with low oleic; grade free from chick edema factor; 99.8% pure.
Containers: Cans; barrels; bags (in slab form); tank cars and trucks.
Uses: Chemicals, especially stearates and stearate driers; lubricants; soaps; pharmaceuticals and cosmetics; accelerator activator, dispersing agent, and softener in rubber compounds; shoe and metal polishes; coatings; food packaging.

stearin (tristearin; glyceryl tristearate) $C_3H_5(C_{18}H_{35}O_2)_3$.
Properties: Colorless crystals or powder; odorless; tasteless; m.p. 71.6°C; sp. gr. 0.943 (65°C); insoluble in water; soluble in alcohol, chloroform, carbon disulfide. Combustible; nontoxic.
Derivation: Constituent of most fats.
Grades: Technical; also grades as to source.
Uses: Soap; candles; candies; adhesive pastes; metal polishes; water-proofing paper; textile sizes; leather stuffing; mfg. of stearic acid.

stearin pitch. See fatty acid pitch.

"Stearite."[104] Trademark for stearic acid produced by selective hydrogenation process. Used in rubber compounding and other processing.

stearone. An aliphatic ketone; insoluble in water; stable to high temperatures, acids, and alkalies. Compatible with high-melting vegetable waxes, paraffins, and fatty acids. Incompatible with resins, polymers, and organic solvents at room temperature, but compatible with them at high temperature. Used as an antiblocking agent. Combustible. Low toxicity.

N-**stearoyl**-para-**aminophenol** $HO(C_6H_4)NHCOCH_2(CH_2)_{15}CH_3$.
Properties: White to off-white powder; m.p. 131–134°C; insoluble in water; soluble in polar organic solvents (especially when hot) such as alcohol, dioxane, acetone, and dimethylformamide. Combustible.
Use: Antioxidant.

stearoyl chloride (n-octadecanoyl chloride) $CH_3(CH_2)_{16}COCl$.
Properties: M.p. 23°C; b.p. 174–178°C (2 mm); soluble in hydrocarbons and ethers. Combustible.
Containers: Bottles, carboys, steel drums.
Uses: Preparation of substituted amines and amides, acid anhydrides, esterification of alcohols, synthesis of other organic compounds.

stearyl alcohol (1-octadecanol; octadecyl alcohol) $CH_3(CH_2)_{16}CH_2OH$.
Properties: Unctuous white flakes or granules with faint odor and bland taste; sp. gr. 0.8124 (59/4°C); b.p. 210.5°C (15 mm); m.p. 59°C. Soluble in alcohol, acetone, and ether; insoluble in water. Combustible; nontoxic.
Derivation: Reduction of stearic acid.
Grades: Commercial; technical; U.S.P.
Containers: 50-lb bags; 200-lb fiber drums; steam-coiled tank cars.
Uses: Perfumery; cosmetics; intermediate; surface active agents; lubricants; resins.

stearyl mercaptan (octadecyl mercaptan) $CH_3(CH_2)_{17}SH$.
Properties: Solid; M.p. 25°C; b.p. 205–209°C (11 mm); sp. gr. 0.8420 (25/4°C); refractive index (n 34/D) 1.4591.
Grade: 95% (min) purity.
Hazard: Toxic; strong irritant.
Uses: Organic intermediate; synthetic rubber processing.

stearyl methacrylate. Group name for $CH_2{:}C(CH_3)COO(CH_2)_nCH_3$(n = 13 to 17).
Properties: Density 0.864 g/ml (25/15°C); boiling range 310–370°C (760 mm).
Containers: Drums.
Uses: Monomer for plastics, molding powders, solvent coatings, adhesives, oil additives; emulsions for textile, leather, paper finishing.

stearyl monoglyceridyl citrate.
Properties: A soft, practically tasteless, off-white to tan, waxy solid having a lard-like consistency. Insoluble in water but is soluble in chloroform and ethylene glycol. Low toxicity.
Derivation: Reaction of citric acid, monoglycerides of fatty acids (obtained by the glycerolysis of edible fats and oils or derived from fatty acids) and stearyl alcohol.
Grade: F.C.C.
Use: Emulsion stabilizer in foods (not over 0.15%).

stearylurea. See "Warcobase A."

steatite. A mixture of talc, clay, and alkaline-earth oxides; used chiefly as a ceramic insulator in electronic devices.

"Stedbac."[430] Trademark for stearyl dimethyl benzyl ammonium chloride, a hair conditioner used primarily in after-shampoo hair rinses.

steel. An alloy of iron and 0.02 to 1.5% carbon; it is made from molten pig iron by oxidizing out the excess carbon and other impurities. The open-hearth process, which uses air for this purpose has been supplanted by the basic oxygen process in which pure oxygen is injected into molten iron. A small percentage of steel is also made in the electric furnace, with iron ore as oxidant. Low-carbon (mild) steels contain 0.02 to 0.3% carbon; medium-carbon grades from 0.3 to 0.7% carbon; and high-carbon grades 0.7 to 1.5%.

There are many special-purpose types of steel in which one or more alloying metals are used, with or without special heat treatment. The most common additives are chromium and nickel, as in the 18–8 stainless steels; these add greatly to corrosion resistance. High-speed and tool steels, designed primarily for efficient cutting, contain such alloying metals as tungsten, molybdenum, manganese, and vanadium, as well as chromium. Cobalt and zirconium are used for construction steels. For further information, refer to American Iron and Steel Institute, New York, N.Y.

Forms available: Rods, bars, sheet, strips, wire, wool.

Uses: Construction; ship hulls; auto bodies; machinery and machine parts; cables; abrasive; chemical equipment; belts for tire reinforcement.

steel, stainless. Alloy steels containing high percentages of chromium, from less than 10 to more than 25%. There are three groups: (1) Austenitic, which contain both chromium (16% min.) and nickel (7% min.); a stress-corrosion resistant type contains 2% silicon. (2) Ferritic, which contain chromium only and cannot be hardened by heat treatment. (3) Martensitic, which contain chromium and can be hardened by heat treatment; these are ferromagnetic. A subgroup of the martensitic steels comprises the precipitation-hardening types. See also SAE steel.

Steffen process. A process used in beet sugar manufacture to separate residual sugar from molasses. Based on the formation of insoluble tricalcium saccharate and its subsequent decomposition to sugar in the presence of a weak acid such as carbonic.

Steinbuhl yellow. See barium chromate.

"Stellite."[214] Trademark for a series of cobalt-chromium-tungsten alloys. There are nine alloys available in the wear-resistant group.

Uses: Special castings, hard-facing rods, and metal-cutting tools. High-temperature grades are also produced by the investment casting process for jet engine turbine blades and vanes.

stendomycin salicylate. A fungicide.

Stengel process. A method of making ammonium nitrate fertilizer from anhydrous ammonia and nitric acid. The fertilizer particles can be made in different sizes according to the use.

"Stepan."[451] Trademark for a series of detergent and detergent intermediates, including lauryl sulfates, alkylolamides, fatty alcohol ether sulfates, linear alkylate sulfonates and ethoxylated nonionics.

"Sterane."[299] Trademark for prednisolone.

stereoblock polymer. A polymer whose molecule is made of comparatively long sections of identical stereospecific structure, these sections being separated from one another by segments of different structure, as for example, blocks of an isotactic polymer interspersed with blocks of the same polymer with a syndiotactic structure. See polymer, stereospecific; block polymer.

$$\left(\begin{array}{cccc} H_2 & H & H_2 & H \\ | & | & | & | \\ -C- & C- & C- & C- \\ & | & & | \\ & O & & O \\ & R & & R \end{array}\right)_n \left(\begin{array}{cccc} & R & & R \\ H_2 & O & H_2 & O \\ | & | & | & | \\ -C- & C- & C- & C- \\ & | & & | \\ & H & & H \end{array}\right)_n$$

stereoblock polymer

stereochemistry. A subdiscipline of organic chemistry devoted to study of the three-dimensional spatial configurations of molecules. One aspect of the subject deals with stereoisomers—compounds which have identical chemical constitution, but differ as regards the arrangement of the atoms or groups in space. Stereoisomers fall into two broad classes: optical isomers and geometric (cis-trans) isomers. Another aspect of stereochemistry concerns control of the molecular configuration of high polymer substances by use of appropriate (stereospecific) catalysts. See also optical isomer; enantiomer; asymmetry; polymer, stereospecific; Ziegler catalyst; geometric isomer.

stereoisomer. See stereochemistry.

stereoregular polymer. See polymer, stereospecific.

stereospecific catalyst. See catalyst, stereospecific.

stereospecific polymer. See polymer, stereospecific.

steric hindrance. A characteristic of molecular structure in which the molecules have a spatial arrangement of their atoms such that a given reaction with another molecule is prevented or retarded.

sterilization. (1) Complete destruction of all bacteria and other infectious organisms in an industrial, food, or medical product; it must be followed by aseptic packaging to prevent recontamination, usually by hermetic sealing. The methods used involve either wet or dry heat, use of chemicals such as formaldehyde and ethylene oxide filtration (pharmaceutical products), and irradiation by ultraviolet or gamma radiation. Milk is sterilized by heating for 30 seconds at 135°C, followed by in-can heating at 115°C for several minutes and aseptic packaging.

(2) Rendering an insect incapable of reproduction by radiation or chemical treatment. See chemosterilant.

"Sterimine."[282] Trademark for trichloromelamine compounded with salt, wetting agent, and acid phosphate buffer. Available chlorine: 50%.

Properties: White, free-flowing powder; slight chlorine odor.

Containers: 10-, 25-, 100-lb fiber drums.

Uses: Sanitizer of ion exchange beds, and of surfaces of equipment and utensils.

sterling silver. According to U.S. law this term is restricted to silver alloys containing at least 92.5% pure silver and 7.5% max. other metal (usually copper). See also silver.

steroid. One of a group of polycyclic compounds closely related biochemically to terpenes. They include cholesterol, numerous hormones, precursors of certain vitamins, bile acids, alcohols (sterols), and certain natural drugs and poisons (such as the digitalis derivatives). Steroids have as a common nucleus a fused reduced 17-carbon-atom ring system, cyclopentanoperhydrophenanthrene. Most steroids also have two methyl groups and an aliphatic side-chain attached to the nucleus. The length of the side-chain varies and generally contains 8, 9 or 10 carbon atoms in the sterols, 5 carbon atoms in the bile acids, 2 in the adrenal cortical steroids, and none in the estrogens and androgens. Steroids are classed as lipids because of their solubility in organic solvents and insolubility in water.

Most of the naturally occurring steroids have been synthesized and many new steroids unknown in nature have been synthesized for use in medicine, such as the fluorosteroids (dexamethasone).

sterol. A steroid alcohol. Such alcohols contain the common steroid nucleus, plus an 8 to 10-carbon-atom side-chain and a hydroxyl group. Sterols are widely distributed in plants and animals, both in the free form and esterified to fatty acids. Cholesterol (q.v.) is the most important animal sterol; ergosterol is an important plant sterol (phytosterol).

"Sterosan."[219] Trademark for a brand of chlorquinaldol (q.v.).

"Sterox."[58] Trademark for a series of nonionic surface-active agents including polyoxyethylene ethers, and polyoxyethylene thioethers.

STH. See somatotropic hormone.

stibic anhydride. See antimony pentoxide.

stibine. See antimony hydride.

stibium. The Latin name for the element antimony; hence the symbol Sb.

stibnite (gray antimony; antimony glance; antimonite) Sb_2S_3.
Properties: Lead gray mineral; subject to blackish tarnish; metallic luster. Soluble in concentrated boiling hydrochloric acid with evolution of H_2S. Sp. gr. 4.52–4.62; Mohs hardness 2.
Occurrence: Japan; China; Mexico; Bolivia, Peru, South Africa.
Use: The most important ore of antimony.

stibophen $C_{12}H_4Na_5O_{16}S_4Sb \cdot 7H_2O$. Sodium antimony II bis(catechol-2,4-disulfonate) heptahydrate.
Properties: White, crystalline, odorless powder. Affected by light. Freely soluble in water; nearly insoluble in alcohol, ether, and chloroform.
Derivation: Reaction of sodium pyrocatechol-3,5-disulfonate with antimony trioxide and precipitating with alcohol.
Grade: U.S.P.
Use: Medicine.

stigmasterol $C_{29}H_{48}O \cdot H_2O$. A plant sterol.
Properties: Anhydrous form has m.p. of 170°C. Insoluble in water; soluble in usual organic solvents. Combustible.
Derivation: From soy or calabar beans.
Uses: Preparation of progesterone and other important steroids.

stilbene (toluylene; trans form of alpha, beta-diphenylethylene) $C_6H_5CH{:}CHC_6H_5$.
Properties: Colorless or slightly yellow crystals; sp. gr. 0.9707; m.p. 124–125°C; b.p. 306–307°C. Soluble in benzene and ether; slightly soluble in alcohol; insoluble in water. Combustible.
Derivation: By passing toluene over hot lead oxide.
Method of purification: Crystallization; zone melting used for very pure crystals.
Grades: Technical; pure.
Uses: Manufacture of dyes and optical bleaches; crystals are used as phosphors and scintillators.
Note: The cis form of alpha, beta-diphenylethylene (isostilbene), is a yellow oil; b.p. 145°C (13 mm); m.p. 1°C.

stilbene dye. A dye whose molecules contain both the —N=N— and the >C=C< chromophore groups in their structure and whose CI numbers range from 40000 to 40999. These are direct cotton dyes.

stilbestrol. See diethylstilbestrol.

stillage. The grain residue from alcohol production, used in feeds and feed supplements.

Stock system. See chemical nomenclature.

Stoddard solvent. A widely used dry-cleaning solvent. U.S. Bureau of Standards and ASTM D—484—52 define it as a petroleum distillate clear and free from suspended matter and undissolved water, and free from rancid and objectionable odor. The minimum flash point is 100°F. Distillation range: not less than 50% over at 350°F (177°C), 90% over at 375°F (190°C), and the end point not higher than 410°F (210°C). Autoignition temp. 450°F. Combustible.
Containers: 55-gal steel drums; tank cars; tank trucks.
Hazard: Moderate fire risk. Moderately toxic by ingestion. Tolerance, 200 ppm in air.

"Stod-Sol."[200] Trademark for a petroleum solvent prepared by straight-run distillation.
Properties: Water-white; initial boiling point 308–316°F, 95% distills between 363–373°F; sp. gr. 0.780 (60°F); flash point (TCC) 103°F; mild, nonresidual odor. Combustible.
Hazard: Moderate fire risk.
Use: Dry cleaning.

stoichiometry. The branch of chemistry and chemical engineering that deals with the quantities of substances that enter into and are produced by chemical reactions. For example, when methane unites with oxygen in complete combustion, 16 grams of methane require 64 grams of oxygen. At the same time 44 grams of carbon dioxide and 36 grams of water are formed as reaction products. Every chemical reaction has its characteristic proportions. The methods of obtaining these from chemical formulas, equations, atomic weights and molecular weights, and determination of what and how much is used and produced in chemical processes is the major concern of stoichiometry.

Stokes' law. (1) The rate at which a spherical particle will rise or fall when suspended in a liquid medium varies as the square of its radius; the density of the particle and the density and viscosity of the liquid are essential factors. Stokes' law is used in determining sedimentation of solids, creaming rate of fat particles in milk, etc. The equation is:

$$v = \frac{h}{t} = \frac{gD^2(\rho_s - \rho_l)}{18\eta}$$

where v is the terminal velocity, h the height of fall, t the time, g the gravitational constant, D, the particle diameter, ρ_s, ρ are the densities of the solid and the suspending medium, respectively, and η is the viscosity.

(2) In atomic processes, the wavelength of fluorescent radiation is always longer than that of the exciting radiation.

storage. Methods of keeping raw materials, finished goods, or food products while awaiting use, shipment, or consumption. Normally materials are stored in suitably protected interior areas, though in some cases outdoor storage is practicable, e.g., logs for wood pulp, materials received in metal drums, and certain bulk solids.

Raw materials. These should be stored in well-ventilated rooms at temperatures ranging from about 35 to 100°F. Materials which may react to form hazardous products in case of spillage should be kept well separated; oxidizing agents (nitrates, peroxides, etc.) should not be stored near reducing or combustible materials. Heat-sensitive materials should be kept away from hot pipes or other heat source, especially in the case of flammable liquids. Textiles, paper and other hygroscopic materials require a humidity-controlled environment.

Finished goods. Combustible products that may build up internal heat on long standing at high temperatures should be stored in well-ventilated or air-conditioned areas, e.g., paper (as in telephone books), bulk wool and cellulosics.

Food Products. Long shelf-life at or near room temperature is highly desirable for processed foods. This is achieved partly by use of antioxidants and other preservatives, and partly by processing techniques. Much experimental work has been done in this field. Refrigerated storage at temperatures near 40°F is used for meats, eggs and other dairy products. Meats and quick-frozen foods can be stored indefinitely at or below 0°F. Controlled-atmosphere storage to retard post-harvest ripening has long been used for unprocessed fruits and vegetables (see atmosphere, controlled).

storage battery. A secondary battery, so called because the conversion of chemical to electrical energy is reversible, and the battery is thus rechargable. An automobile battery usually consists of from 12 to 17 cells with plates (electrodes) made of sponge lead (negative plate or anode) and lead dioxide (positive plate or cathode), which is in the form of a paste. The electrolyte is sulfuric acid. The chemical reaction that yields electric current is $Pb + PbO_2 + 2H_2SO_4 \longleftrightarrow 2PbSO_4 + 2H_2O + 2e$. More complicated and expensive types have nickel-iron, nickel-cadmium, silver-zinc and silver-cadmium as electrode materials. A sodium-liquid sulfur battery for high-temperature operation as well as a chlorine-zinc type using titanium electrodes have also been developed. See also battery.

storax. U.S.P. name for styrax (q.v.).

"Stoxil."[71] Trademark for a brand of idoxuridine (q.v.).

STP. Abbreviation for standard temperature and pressure, i.e., 0°C and 1 atmosphere pressure.

STPP. Abbreviation for sodium tripolyphosphate.

straight-chain. See aliphatic.

straight-run. See gasoline.

stramonium (jimson weed; thorn apple; stinkweed).
Derivation: Dried leaves and flowering tops of Datura stramonium or its variety tutula.
Constituents: Atropine, scopolamine, proteins, albumin.
Occurrence: Europe; Asia; America.
Hazard: Toxic; skin irritant.
Uses: Medicine (antispasmodic); source of alkaloids.

strangeness. See antiparticle.

"Stratabond."[526] Trademark for organic phosphate coatings used for the preparation of steel, galvanized steel, and aluminum for painting. Application by spray, dip, flow-coat or roller coat.

streaming potential. A potential resulting from the friction of a liquid passing through a tube or pipe; its strength depends on the viscosity of the liquid, diameter of the pipe, and rate of flow. This can be a factor in pipelines conveying oil and chemicals.

"Stren."[28] Trademark for a high-strength, and low-elongation nylon monofilament for fishing lines.

streptolin.
Properties: An antibiotic, isolated as the hydrochloride. Gummy mass; soluble in water; most stable at pH 3.0–3.5.
Derivation: Produced by Streptomyces no. 11.
Hazard: May be toxic.
Uses: Antibiotic; possible rodenticide.

streptomycin $C_{21}H_{39}N_7O_{12}$. A specific antibiotic, but the term is also used loosely to designate several chemically related antibiotics produced by actinomycetes. Streptomycin is produced by Streptomyces griseus and consists of streptidine attached in glycosidic linkage to the disaccharide, streptobiosamine. It is active against gram-negative bacteria and the tubercle bacillus.
Properties: A base which readily forms salts with anions. Quite stable but very hygroscopic.
Units: One unit equals one microgram of pure crystalline streptomycin base.
Derivation: From streptomyces griseus by submerged culture. The streptomycin is then concentrated by adsorption on activated carbon and purified.
Hazard: Possible damage to nerves and kidneys may result. Use restricted by FDA.
Use: Medicine (usually as sulfate).
See also dihydrostreptomycin.

streptonigrin (USAN) $C_{25}H_{22}N_4O_8$. An antibiotic derived from Streptomyces flocculus. Dark brown rectangular crystals.

"Stripkolex."[28] Trademark for a dynamite for coal-stripping operations.

"Stripolite."[159] Trademark for a modified sodium sulfoxylate formaldehyde. Used as a powerful reducing agent to strip dyed colors from wool and synthetics.

stripping.
(1) Removal of relatively volatile components from a gasoline or other liquid mixture by distillation, evaporation, or by passage of steam, air or other gas through the liquid mixture.
(2) Rapid removal of color from an improperly dyed fabric or fiber by a chemical reaction. Compounds used for this purpose in vat dyeing or in discharge printing are termed discharging agents. Substances commonly used as strippers are sodium hydrosulfite, titanous sulfate, sodium and zinc formaldehyde sulfoxylates.

strontia. See strontium oxide.

strontianite $SrCO_3$. Natural strontium carbonate.
Properties: Color white, gray, yellow, green; luster vitreous; Mohs hardness 3.5–4; sp. gr. 3.7.

Occurrence: California, New York, Washington; Germany; Mexico.
Use: Source of strontium chemicals.

strontium Sr Metallic element, of atomic number 38, Group IIA of Periodic Table. Atomic weight 87.62. Valence 2. Radioactive isotopes Sr-89 and Sr-90. There are 4 stable isotopes.
Properties: Pale yellow, soft metal, chemically similar to calcium. Soluble in alcohol and acids; decomposes water on contact. Occurs in nature in the minerals strontianite (carbonate) and celestite (sulfate). Sp. gr. 2.54; m.p. 752°C; b.p. 1390°C. Low toxicity.
Derivation: (a) Electrolysis of molten strontium chloride in a graphite crucible with cooling of the upper, cathodic space; (b) thermal reduction of the oxide with metallic aluminum (strontium aluminum alloy formed), and distilling the strontium in a vacuum.
Grade: Technical.
Containers: Glass bottles containing sufficient naphtha to cover the metal.
Hazard: Spontaneously flammable in powder form; ignites when heated above melting point. Reacts with water to evolve hydrogen. Store under naphtha.
Uses: Alloys; "getter" in electron tubes.
Shipping regulations: (Air) (alloys) Flammable Solid label; (metal in organic solvent), (Air) Not acceptable on passenger planes.

strontium 90. Radioactive strontium isotope.
Properties: Half-life, 38 years; radiation, beta.
Derivation: From the fission products of nuclear reactor fuels.
Forms available: A mixture containing strontium 90, yttrium 90, and strontium 89 chlorides in hydrochloric acid solution; also as the carbonate and titanate.
Hazard: Highly toxic radioactive poison; present in fallout from nuclear explosions. Absorbed by growing plants; when ingested attacks bone marrow with possibly fatal results. It may be partially removed from milk by treatment with vermiculite (q.v.).
Uses: Radiation source in industrial thickness gauges; elimination of static charge; treatment of eye diseases; in radio-autography to determine the uniformity of material distribution; in electronics for studying strontium oxide in vacuum tubes; activation of phosphors; source of ionizing radiation in luminous paint; cigarette density control; measuring silt density; atomic batteries, etc.
Shipping regulations: (Rail, Air) Consult regulations.

strontium acetate $Sr(C_2H_3O_2)_2 \cdot \frac{1}{2} H_2O$.
Properties: White, crystalline powder; soluble in water. Loses $\frac{1}{2} H_2O$ at 150°C.
Derivation: Interaction of strontium hydroxide and acetic acid, followed by crystallization.
Grades: Technical; reagent.
Use: Medicine; intermediate for strontium compounds; catalyst production.

strontium arsenite $Sr(AsO_2)_2 \cdot 4H_2O$ (approx.).
Properties: White powder; slightly soluble in water and alcohol. Odorless; soluble in dilute acids.
Hazard: Highly toxic.
Use: Medicine.
Shipping regulations: (Rail, Air) Poison label.

strontium bromate $Sr(BrO_3)_2 \cdot H_2O$.
Properties: Colorless or yellowish, crystalline, lustrous powder. Hygroscopic. Soluble in water. Sp. gr. 3.773; loses H_2O at 120°C; decomposes 240°C.
Hazard: Moderately toxic. Strong oxidant; fire risk in contact with organic materials.

strontium bromide $SrBr_2 \cdot 6H_2O$.
Properties: White, hygroscopic crystals or powder. Soluble in water, alcohol, and amyl alcohol. Insoluble in ether. Sp. gr. 2.386 (25/4°C); loses $4H_2O$ at 89°C, remaining H_2O by 180°C; m.p. anhydrous salt 643°C.
Derivation: Strontium carbonate is treated with bromine or hydrobromic acid.
Grades: Anhydrous powder; crystal; technical; C.P.
Containers: Bottles; 150-lb drums.
Hazard: Moderately toxic by ingestion and inhalation.
Uses: Medicine; laboratory reagent.

strontium carbonate $SrCO_3$.
Properties: White, impalpable powder. Soluble in acids, carbonated water and solutions of ammonium salts; slightly soluble in water. Sp. gr. 3.62; loses CO_2 at 1340°C. Low toxicity.
Derivation: Celestite is boiled with a solution of ammonium carbonate or is fused with sodium carbonate.
Grades: Precipitated; technical; natural; reagent.
Containers: 50-lb bags; drums.
Uses: Catalyst; in radiation-resistant glass for color television tubes; ceramic ferrites; pyrotechnics.

strontium chlorate (a) $Sr(ClO_3)_2$; (b) $Sr(ClO_3)_2 \cdot 8H_2O$.
Properties: White, crystalline powder. Soluble in water; slightly soluble in alcohol. (a) Sp. gr. 3.152; m.p., decomposes at 120°C.
Derivation: Strontium hydroxide solution is warmed and chlorine passed in, with subsequent crystallization.
Grades: Technical; reagent.
Containers: Tins; glass bottles.
Hazard: Dangerous explosion risk in contact with organic materials; highly sensitive to shock, heat and friction. Moderately toxic by ingestion.
Use: Manufacture of red-fire and other pyrotechnics; tracer bullets.
Shipping regulations: (wet and dry) (Rail) Yellow label. (Air) Oxidizer label.

strontium chloride (a) $SrCl_2$; (b) $SrCl_2 \cdot 6H_2O$.
Properties: White, crystalline needles; odorless; sharp, bitter taste; soluble in water and alcohol. (a) Sp. gr. 3.054; m.p. 872°C; b.p. 1250°C. (b) Sp. gr. 1.964; m.p. loses $6H_2O$ at 112°C.
Derivation: Strontium carbonate is fused with calcium chloride, the melt extracted with water, the solution concentrated and crystallized.
Grades: Reagent; technical; anhydrous.
Containers: Bottles; wooden kegs; fiber drums; carloads.
Uses: Strontium salts; pyrotechnics; medicine; electron tubes.

strontium chromate $SrCrO_4$.
Properties: Light yellow pigment; sp. gr. 3.84; rust-inhibiting and corrosion-resistant properties. Has good heat and light resistance and low reactivity in high acid vehicles.
Containers: Fiber drums.
Uses: Metal protective coatings to prevent corrosion; polyvinyl chloride resins to produce pastel primrose yellows; pyrotechnics.

strontium dioxide. See strontium peroxide.

strontium fluoride SrF_2.
Properties: White powder; soluble in hydrochloric acid and hydrofluoric acid. Insoluble in water. Sp. gr. 4.2; m.p. 1463°C; b.p. 2489°C.

Grades: Crystal, 99.2% pure.
Hazard: Toxic. Tolerance (as F), 2.5 mg per cubic meter of air.
Uses: Medicine; substitute for other fluorides; electronic and optical applications; single crystals for lasers; high-temperature dry film lubricants.

strontium hydrate. See strontium hydroxide.

strontium hydroxide (strontium hydrate) (a) $Sr(OH)_2$; (b) $Sr(OH)_2 \cdot 8H_2O$.
Properties: Colorless, deliquescent crystals. Soluble in acids and hot water; slightly soluble in cold water. Absorbs carbon dioxide from air. Sp. gr. (a) 3.625; m.p. 375° C; decomposes 710° C. (b) Sp. gr. 1.90. Low toxicity.
Derivation: Heating the carbonate or sulfide in steam.
Grades: Technical; C.P.; reagent.
Containers: Bottles; drums.
Uses: Extraction of sugar from beet sugar molasses; lubricant soaps and greases; stabilizer for plastics; glass; adhesives.

strontium hyposulfite. See strontium thiosulfate.

strontium iodide (a) $SrI_2 \cdot 6H_2O$.
Properties: (a) White, crystalline plates; decomposes in moist air. Becomes yellow on exposure to air or light. (b) White crystals; soluble in water and alcohol. (a) Sp. gr. 4.549; m.p. 515° C; b.p. decomposes. (b) Sp. gr. 2.67 (25° C); decomposes 90° C.
Derivation: By treating strontium carbonate with hydriodic acid.
Hazard: Moderately toxic by ingestion and inhalation.
Uses: Medicine; intermediate.

strontium lactate $Sr(C_3H_5O_3)_2 \cdot 3H_2O$.
Properties: White crystals or granular powder. Soluble in water, alcohol, ether. Odorless.
Use: Medicine.

strontium molybdate $SrMoO_4$.
Properties: Crystalline powder; density 4 g/cc; m.p. 1600° C (approx); insoluble in water. Low toxicity.
Grades: 99.39% pure; single crystals.
Uses: Anticorrosion pigments; electronic and optical applications; single-crystal, solid-state lasers.

strontium monosulfide. See strontium sulfide.

strontium nitrate $Sr(NO_3)_2$.
Properties: White powder. Soluble in water; slightly soluble in absolute alcohol. Sp. gr. 2.98; m.p. 570° C; b.p. 645° C.
Derivation: A concentrated solution of strontium chloride is treated with a solution of sodium nitrate.
Grades: Technical; reagent.
Containers: Bags; barrels; drums.
Hazard: Moderately toxic. Strong oxidizing agent. Fire risk in contact with organic materials. May explode when shocked or heated.
Uses: Pyrotechnics; marine signals, railroad flares; matches; medicine; intermediate; electron tubes.
Shipping regulations: (Rail) Yellow label. (Air) Oxidizer label.

strontium nitrite $Sr(NO_2)_2$.
Properties: White or yellowish powder, or hygroscopic needles. Soluble in water; insoluble in alcohol. Sp. gr. 2.8. Decomp. at 240° C.

strontium oxalate $SrC_2O_4 \cdot H_2O$.
Properties: White, odorless powder; loses H_2O at 150° C; insoluble in cold water.
Hazard: May be toxic.
Uses: Pyrotechnics; catalyst manufacture; tanning.

strontium oxide (strontia) SrO
Properties: Grayish white powder; sp. gr. 4.7; m.p. 2430° C; b.p. approx 3000° C. Soluble in fused potassium hydroxide; converted to hydroxide by water. Combustible.
Derivation: Decomposition of strontium carbonate or hydroxide.
Grades: Lump; powder; porous carbide-free; reagent.
Containers: Sealed cans; drums.
Uses: Manufacture of strontium salts; pyrotechnics; pigments; medicine; greases and soaps; a major desiccant.

strontium perchlorate $Sr(ClO_4)_2$.
Properties: Colorless crystals; very soluble in water and alcohol.
Hazard: Moderately toxic. Fire and explosion risk in contact with organic materials.
Uses: Oxidizing agent; pyrotechnics.
Shipping regulations: (Rail) Yellow label. (Air) Oxidizer label.

strontium peroxide (strontium dioxide) (a) SrO_2 (b) $SrO_2 \cdot 8H_2O$.
Properties: White powder; odorless and tasteless; sp. gr. (a) 4.56; m.p., decomposes at 215° C; (b) sp. gr. 1.951; loses $8H_2O$ at 100° C and decomposes when heated to a higher temperature. Soluble in ammonium chloride solution and alcohol; decomposes in hot water. Slightly soluble in cold water.
Derivation: (a) By passing oxygen over heated strontium oxide. (b) Reaction of strontium hydroxide and hydrogen peroxide.
Hazard: Dangerous fire and explosion risk in contact with organic materials; sensitive to heat, shock, catalysts, reducing agents.
Uses: Bleaching; medicine; fireworks.
Shipping regulations: (Rail) Yellow label. (Air) Oxidizer label

strontium-potassium chlorate (potassium-strontium chlorate) $Sr(ClO_3)_2 \cdot 2KClO_3$.
Properties: White, crystalline powder. Soluble in water.
Grade: Technical.
Hazard: Dangerous fire and explosion risk in contact with organic materials; sensitive to heat, shock, catalysts, reducing agents.
Use: Pyrotechnics.
Shipping regulations: (Rail, Air) Chlorates n.o.s. Oxidizer label.

strontium salicylate $Sr(C_7H_5O_3)_2 \cdot 2H_2O$.
Properties: White crystals or powder; odorless with sweet saline taste. Soluble in water and alcohol; decomposes when heated. Protect from light.
Derivation: Interaction of strontium carbonate and salicylic acid.
Uses: Medicine; pharmaceutical and fine chemical manufacture.

strontium stearate $Sr(OOCC_{17}H_{35})_2$.
Properties: White powder; m.p. 130–140° C. Insoluble in alcohol; soluble (forms gel) in aliphatic and aromatic hydrocarbons.
Use: Grease and wax compounding.

strontium sulfate $SrSO_4$.
Properties: White precipitate or crystals of the mineral celestite. Odorless. Slightly soluble in concentrated acids; very slightly soluble in water; insoluble in alcohol and dilute sulfuric acid. Sp. gr. 3.71–3.97; m.p. 1605° C.
Derivation: (a) Celestite is ground. (b) Precipitation of any soluble strontium salt with sodium sulfate.

Grades: Natural; precipitated; air-floated, 90%, 325 mesh; free from sodium salts; C.P.
Uses: Pyrotechnics; ceramics and glass; paper manufacture.

strontium sulfide (strontium monosulfide) SrS.
Properties: Gray powder. Has hydrogen sulfide odor when in presence of moist air. Soluble in acids (decomposes); slightly soluble in water. Sp. gr. 3.7; m.p. above 2000°C.
Derivation: Reduction of celestite with coke at high temperatures.
Grades: Technical; high purity.
Containers: Glass bottles; sealed drums.
Hazard: Irritant to skin and tissue. Moderate fire and explosion risk.
Uses: Depilatory; luminous paints; source of hydrogen sulfide; manufacture of strontium chemicals.

strontium tartrate $SrC_4H_4O_6 \cdot 4H_2O$.
Properties: White crystals; sp. gr. 1.966; slightly soluble in water. Combustible. Low toxicity.
Use: Pyrotechnics.

strontium thiosulfate (strontium hyposulfite) $SrS_2O_3 \cdot 5H_2O$.
Properties: Fine needles. Soluble in water; insoluble in alcohol; sp. gr. 2.17; m.p. loses $4H_2O$ at 100°C.

strontium titanate $SrTiO_3$.
Properties: Powder, sp. gr. 4.81; m.p. ca 2060°C. Insoluble in water and most solvents. Low toxicity.
Grades: Technical; single crystals.
Use: Electronics; electrical insulation.

strontium zirconate $SrZrO_3$.
Properties: Powder; m.p. ca 2600°C. Low toxicity.
Use: Electronics.

"Structo-Foam."[1] Trademark for expanded polystyrene foam products used for insulation in buildings, refrigerator cars and trucks and for bouyancy in floats, moorings, etc.

structural antagonist. See antagonist, structural.

structural formula. See formula, chemical.

"Struex-80."[471] Trademark for a fine particle size grade of hydrated aluminum silicate produced from kaolin. Used in paint systems.

"Struflow."[471] Trademark for large, thin, platey particles of hydrated aluminum silicates produced from kaolin. Particle size: 10 to 20 microns in diameter, 1.0 micron thick. Used in paint systems.

strychnidine $C_{21}H_{24}ON_2$.
Properties: Colorless crystals; m.p. 256°C.
Use: Reagent for nitrate determination.

strychnine $C_{21}H_{22}N_2O_2$. An alkaloid.
Properties: Hard, white, crystals or powder; bitter taste. M.p. 268–290°C; b.p. 270°C (5 mm). Soluble in chloroform; slightly soluble in alcohol and benzene; slightly soluble in water and ether.
Derivation: Extraction of the seeds of Nux vomica with acetic acid, filtration, precipitation by alkali and filtration.
Grades: Crystal; powder; technical.
Hazard: Highly toxic by ingestion and inhalation; tolerance, 0.15 mg per cubic meter of air.
Uses: In medicine as such, or as phosphates; rodent poison.
Shipping regulations: (Rail, Air) Poison label.

"Strycin."[412] Trademark for streptomycin syrup.

stuff. Term used by paper makers to refer to the aqueous pulp suspension fed onto the fourdrinier wire from the headbox. See also furnish.

"Sty-Grade."[576] Trade mark for an additive to polystyrene which imparts biodegradability, primarily for such disposable products as plastic bottles.

"Stymer" Vinyl Styrene.[58] Trademark for resins used as sizes for filament acetates.
LF. A vinyl resin, soluble with ammonium hydroxide.
S. Styrene copolymer resin, soluble in water.

S-type synthetic elastomer. See styrene-butadiene rubber.

"Styphen."[233] Trademark for a mixture of styrenated phenols.
Use: Antioxidant.

styphnic acid (2,4,6-trinitroresorcinol) $C_6H(OH)_2(NO_2)_3$.
Properties: Yellow crystals; astringent taste; an initiating explosive; m.p. 179–180°C; forms additive compounds with many hydrocarbons. Soluble in ethyl alcohol and ether; slightly soluble in water.
Derivation: Nitration of resorcinol.
Hazard: Severe explosion risk when heated. Probably toxic.
Use: Priming agents in the explosives industry.
Shipping regulations: (Rail) Consult authorities. (Air) (dry or wet with less than 10% water) Not acceptable; (wet with not less than 10% water) Flammable Solid label.
Legal label name: trinitroresorcinol.

styracin. See cinnamyl cinnamate.

"Styrafil."[539] Trademark for polystyrene with glass fiber reinforcement.

styralyl acetate. See alpha-methylbenzyl acetate.

styralyl alcohol. See alpha-methylbenzyl alcohol.

styrax. A type of balsam found in Central America and the Near East. See balsam.

styrenated oil. A drying oil whose drying and hardening characteristics have been modified by incorporation of styrene or a similar monomer.

styrene. See polystyrene; styrene monomer.

styrene monomer (vinylbenzene; phenylethylene; cinnamene) $C_6H_5CH{:}CH_2$.
Properties: Colorless, oily liquid; aromatic odor. F.p. −30.63°C; b.p. 145.2°C; sp. gr. (25/25°C) 0.9045; wt/gal (20°C) 7.55 lbs; flash point 88°F. Autoignition temp 914°F. Insoluble in water; soluble in alcohol and ether. Readily undergoes polymerization when heated or exposed to light or a peroxide catalyst. The polymerization releases heat and may become explosive.
Derivation: From ethylene and benzene in the presence of aluminum chloride to yield ethylbenzene, which is catalytically dehydrogenated at about 630°C to form styrene.
Grades: Technical 99.2%; polymer 99.6%.
Containers: Glass bottles; carboys; steel drums; tank cars and tank trucks.
Hazard: Moderately toxic by ingestion and inhalation. Tolerance, 100 ppm in air. Flammable, moderate fire risk. Explosive limits in air 1.1 to 6.1%. Must be

inhibited during storage. Safety data sheet available from Manufacturing Chemists Assn., Washington, D.C.

Uses: Polystyrene plastics; SBR, ABS and SAN resins; protective coatings (Styrene-butadiene latex; alkyds); styrenated polyesters; rubber-modified polystyrene; copolymer resins; intermediate

Shipping regulations: (Rail) Not listed. (Air) (inhibited) No label required; (uninhibited) Not acceptable.

styrene-acrylonitrile. See polystyrene.

styrene-butadiene rubber (SBR). By far the most widely used type of synthetic rubber; its consumption for all applications is about four times that of polybutadiene, its nearest competitor, and 1½ times that of all other elastomers combined. Its manufacture involves copolymerization of about 3 parts butadiene with 1 part styrene. These materials are suspended in finely divided emulsion form in a large proportion of water, in the presence of a soap or detergent. Also present in small amounts are an initiator or catalyst which is usually a peroxide, and a chain-modifying agent such as dodecyl mercaptan.

Uses: Tires, footwear, mechanical goods; coatings; adhesives; solvent-release sealants; carpet backing.

See also rubber, synthetic; polymerization; free radical.

styrene nitrosite. A compound resulting from the reaction between styrene and nitrogen dioxide and used as a qualitative or quantitative specific test for monomeric styrene in mixtures with other hydrocarbons.

styrene oxide $C_6H_5\overline{CHOCH_2}$.

Properties: Colorless to pale straw-colored liquid. Boiling range (5 to 95%) 194.2–195°C; f.p. −36.6°C; flash point 180°F (COC); refractive index (n 25/D) 1.5328; sp. gr. (25/4°C) 1.0469; miscible with benzene, acetone, ether, and methanol. Combustible.

Hazard: Moderately toxic and irritant.

Uses: Highly reactive organic intermediate.

"Styresol."[36] Trademark for a group of styrenated alkyd resins with air-drying and baking properties and high resistance to gasoline, alkalies, acids, and water.

"Styrocrete."[233] Trademark for latex formulation used as an additive for cement mortar to bond plastic foam to various surfaces.

"Styretex."[474] Trademark for styrenated alkyd resins.

"Styrofoam."[233] Trademark for expanded, cellular polystyrene (available in colors).

Used: Insulating material; light-weight materials for boats, toys, etc.; separators in packing containers; airport runways; highway construction; battery cases.

"Styron."[233] Trademark for polystyrene resins; general purpose, medium and hi impact, heat and impact-heat resistant, and light-stabilized resins ("Styron Verelite"). Available in wide range of translucent and opaque colors, as well as natural and crystal.

Uses: Packaging, toys, appliance parts, bottle closures and containers, hot and cold drinking cups, television cabinet backs, automotive components and machine housings, lighting equipment.

styryl carbinol. See cinnamic alcohol.

suberane. See cycloheptane.

suberic acid (octanedioic acid) $HOOC(CH_2)_6COOH$.

Properties: Colorless crystals from water; m.p. 143°C; b.p. 279°C at 100 mm. Sparingly soluble in ether; soluble in alcohol and hot water; slightly soluble in cold water. Combustible.

Derivation: Oxidation of cyclooctene or cyclooctane.

Uses: Intermediate for the synthesis of drugs, dyes and high polymers.

suberone. See cycloheptanone.

sublimation. The direct passage of a substance from solid to vapor without appearing in the intermediate (liquid) state. An example is solid carbon dioxide which vaporizes at room temperature; the conversion may also be from vapor to solid under appropriate conditions of temperature.

subnuclear particle. A particle either found in the nucleus or observed coming from the nucleus as the result of nuclear reaction or rearrangement, i.e., neutrons, mesons, etc.

substance. Any chemical element or compound. All substances are characterized by a unique and identical constitution, and are thus homogeneous (q.v.). "A material of which every part is like every other part is said to be homogeneous and is called a substance." (Black and Conant, "Practical Chemistry.")

See also homogeneous.

substantive dye. See direct dye.

substituent. An atom or radical that replaces another in a molecule as the result of a reaction. See substitution.

substitution. The replacement of one element or radical by another as a result of a chemical reaction. Chlorination of benzene to produce chlorobenzene is a typical example; in this case a chlorine atom replaces a hydrogen atom in the benzene molecule.

substrate. (1) A substance upon which an enzyme or ferment acts. (2) Any solid surface on which a coating or layer of a different material is deposited.

subtillin. An antibiotic produced by the metabolic processes of a strain of Bacillus subtilis. It is a cyclic polypeptide similar to bacitracin in chemical structure and antibiotic activity, but not as important clinically. Subtilin is active against many gram-positive bacteria, some gram-negative cocci, and some species of fungi. It is a surface tension depressant, and its antibiotic action is increased by use of wetting agents.

Properties: Soluble in water in pH range 2.0–6.0; soluble in methanol and ethanol (up to 80%); insoluble in dry ethanol or other common organic solvents. Relatively stable in acid solutions. Inactivated by pepsin and trypsin, and destroyed by light.

Uses: Medicine; seed disinfectant.

succinaldehyde (butanedial) $OHCCH_2CH_2CHO$.

Properties: Liquid; sp. gr. 1.064 (20/4°C); b.p. 169-170°C. Refractive index 1.4254. Soluble in water, alcohol, and ether. The name succinaldehyde is often incorrectly used in commerce as a synonym for succinic anhydride.

succinic acid (butanedioic acid) $CO_2H(CH_2)_2CO_2H$.

Properties: Colorless crystals; slightly soluble in water; soluble in alcohol and ether; odorless; acid taste. Sp. gr. 1.552; m.p. 185°C; b.p. 235°C. Combustible. Low toxicity.

Derivation: Fermentation of ammonium tartrate.

Grades: Technical; C.P.; F.C.C.

Containers: Bottles, barrels; kegs; fiber drums.

Uses: Medicine; organic synthesis; manufacture of lacquers, dyes, esters for perfumes, succinates; photography; in foods as a sequestrant, buffer, neutralizing agent.

succinic acid, 2,2-dimethylhydrazide $(CH_3)_2NNHCOCH_2Ch_2COOH$.

Properties: White crystals; m.p. 155°C; pH 3.8 (500

ppm); soluble in water; insoluble in simple hydrocarbons.
Use: Growth retardant used in greenhouses; retards premature fruit drop.

succinic acid peroxide $(HOOCCH_2CH_2CO)_2O_2$.
Properties: Fine white, odorless powder; m.p. 125°C (dec). Moderately soluble in water; insoluble in petroleum solvents and benzene.
Hazard: Moderately toxic by ingestion and inhalation; irritant to skin. Fire risk in contact with combustible materials.
Uses: Polymerization catalyst; deodorants; antiseptics.
Shipping regulations: (Rail) (Air) Yellow label. Organic Peroxide label.

succinic anhydride (2,5-diketotetrahydrofurane; succinyl oxide; butanedioic anhydride) $H_2CC(O)OC(O)CH_2$.
Properties: Colorless or light-colored needles or flakes; sp. gr. 1.104 (20/4°C); m.p. 120°C; b.p. 261°C. Soluble in alcohol and chloroform; insoluble in water. Sublimes at 115°C at 5 mm pressure. Combustible.
Grade: Distilled.
Containers: 250-lb drums; carlots.
Hazard: Moderately toxic and irritant.
Uses: Manufacture of chemicals, pharmaceuticals, esters; hardener for resins.

succinimide (2,5-diketopyrrolidine)
$H_2CC(O)NHC(O)CH_2$ or $C_4H_5O_2N \cdot H_2O$.
Properties: Colorless crystals or thin light tan flakes; nearly odorless; sweet taste; m.p. 125–127°C; b.p. 287–288°C; sp. gr. 1.41. Soluble in hot and cold water and in sodium hydroxide solution; slightly soluble in alcohol; insoluble in ether and chloroform.
Derivation: Conversion of succinic acid to succinamide, followed by heating; ammonia splits off to give a diacyl-substituted derivative (succinimide).
Grades: Purified; technical.
Uses: Growth stimulants for plants; organic synthesis.

succiniodimide. See N-iodosuccinimide.

succinonitrile. See ethylene cyanide.

succinyl oxide. See succinic anhydride.

sucrase. See invertase.

sucroblanc. A mixture used to defecate and bleach sugar solution in one operation. Contains high-test calcium hypochlorite, calcium superphosphate, lime and Filtercel (a grade of "Celite.")

sucrose (table sugar; saccharose) $C_{12}H_{22}O_{11}$.
Properties: Hard, white, dry crystals, lumps or powder; sweet taste; odorless. Soluble in water; slightly soluble in alcohol. Solutions are neutral to litmus. Sp. gr. 1.5877; decomposes 160–186°C. Combustible; nontoxic; optical rotation +33.6°.
Derivation: By crushing and extraction of sugar cane (Saccharum officinarum) with water or extraction of the sugar-beet (Beta vulgaris) with water, evaporating and purifying with lime, carbon and various liquids. Also obtainable form sorghum by conventional methods. Occurs in low percentages in honey and maple sap.
Grades: Reagent; U.S.P.; technical; refined.
Containers: Multiwall paper sacks; bulk.
Uses: Sweetener in foods and soft drinks; manufacture of syrups; source of invert sugar; confectionery; preserves and jams; demulcent; pharmaceutical products; caramel; chemical intermediate for detergents; emulsifying agents and other sucrose derivatives.
See also sugar.

sucrose acetate isobutyrate (SAIB) $(CH_3COO)_2C_{12}H_{14}O_3[OOCCH(CH_3)_2]_6$.
Properties: Clear sucrose derivative available either as a semi-solid (100%), or as a 90% solution in ethyl alcohol. Sp. gr. 1.146 (25/25°C). Combustible; nontoxic.
Derivation: By controlled esterification of sucrose with acetic and isobutyric anhydrides.
Containers: 55-gal drums; tanks.
Uses: Modifier for lacquers, hot-melt coating formulations, extrudable plastics.

sucrose monostearate.
Properties: Odorless, tasteless white powder.
Derivation: By reaction of sugar and methyl stearate in a suitable solvent and with potassium carbonate catalyst.
Uses: Low foam nonionic detergent; surfactants.

sucrose octaacetate $C_{12}H_{14}O_3(OOCCH_3)_8$.
Properties: White, hygroscopic, crystalline material. Bitter taste. When once melted does not recrystallize on cooling but becomes a transparent film. Molten film is very adhesive but this property is lost on cooling. Rate of hydrolysis, practically nil. Gives no action with Fehling's solution. Soluble in acetic acid, acetone, benzene; slightly soluble in water. Sp. gr. 1.28 (fused) (20/20°C); b.p. 260°C (0.1 mm); m.p. 79–84°C; refractive index 1.4660 (20°C); decomposes above 285°C; viscosity (100°C) 29.5 poises; specific rotation (CCl_4) +54.96°C. Combustible; low toxicity.
Grades: Technical; reagent; denaturing.
Containers: Bags; drums; 200-lb barrels.
Uses: Plasticizer for cellulose esters and synthetic resins; adhesive compositions; coating compositions; insecticide; termite repellent; denaturant in rubbing alcohol formulas; paper; plastics; lacquers.

"Sudafed."[301] Trademark for pseudoephedrine hydrochloride (q.v.).

"Sudan."[307] Trademark for a series of dyestuffs; soluble in hydrocabons; used for coloring fats, oils, waxes, etc.

"Suganilla."[342] Trademark for concentrated extractives of vanilla beans adsorbed on sugar for food-flavoring.

sugar. A carbohydrate (q.v.) product of photosynthesis comprised of one, two, or more saccharose groups. The monosaccharide sugars (often called simple sugars) are composed of chains of from 2 to 7 carbon atoms. One of the carbons carries aldehydic or ketonic oxygen which may be combined in acetal or ketal forms. The remaining carbons usually have hydrogen atoms and hydroxyl groups. Chief among the monosaccharides are glucose (dextrose) and fructose (levulose). These are optical isomers of formula $C_6H_{10}O_5$, i.e., their crystals have the property of rotating to either left or right under polarized light. Hence the alternate names dextrose and levulose. (See also D, and optical rotation).

Among the disaccharides are sucrose (cane or beet sugar) (q.v.); lactose, found in milk; maltose, obtained by hydrolysis of starch; and cellobiose from partial hydrolysis of cellulose. High-polymer sugars occur as water-soluble gums such as arabic, tragacanth, etc. Hydrolysis of sucrose yields invert sugar, composed of equal parts fructose and glucose. If on a scale of relative sweetness sucrose has an assigned value of 100, ten per cent solutions have the following values: fructose 120, glycerol 77, glucose 69, and lactose 39. Sugar is an important source of metabolic energy in foods, and its formation in plants is an essential factor in the life process.

sugar-cane wax. A hard wax varying form dark green to tan and brown; produced by solvent extraction of cane. M.p. 76–79°C. Combustible; nontoxic.
Grades: Domestic refined, in slabs.
Containers: Cartons.
Uses: Carbon paper; pigment disperser, castings, lubricant for plastics in food wrappers.

sugar, corn. See dextrose.

sugar, grape. See dextrose.

sugar, invert. See invert sugar; sugar.

sugar of lead. See lead acetate.

sugar of milk. See lactose.

sugar, reducing. Sugars that will reduce Fehling's solution or similar test liquids, with conversion of the blue soluble copper salt to a red, orange or yellow precipitate of cuprous oxide. Glucose and maltose are typical examples of reducing sugars, their molecules containng an aldehyde group that is the basis for this type of reaction.

sugar substitute. See sweetener, nonnutritive.

sulfa drug. Any of a group of compounds characterized by the presence of both sulfur and nitrogen, with high specificity for certain bacteria. Because of their toxicity and often serious side effects, their use in treating disease is limited, since the less toxic antibiotics are generally available. Many are used in veterinary medicine. Among the best known are sulfanilimide, sulfadiazine, sulfapyridine, sulfamerazine, sulfasoxazole, and gantricin.

"Sulfads."[69] Trademark for a preparation of dipentamethylene thiuram tetrasulfide. Used as a curing agent for rubber.

sulfamic acid HSO_3NH_2.
Properties: White crystalline solid; non-volatile; nonhydroscopic; sp. gr. 2.1; m.p. 205°C (decomposes). Moderately soluble in water; slightly soluble in organic solvents. Odorless. Aqueous solutions are highly ionized, giving pH values lower than solutions of formic, phosphoric, and oxalic acids. All the common salts (including calcium, barium, and lead) are extremely soluble in water. Combustible; low toxicity.
Grades: Reagent; crystalline; granular.
Containers: Fiber drums; carlots.
Uses: Metal and ceramic cleaning; nitrite removal in azo dye operations; gas-liberating compositions; organic synthesis; analytical acidimetric standard; amine sulfamates used as plasticizers and fire retardants for paper and other cellulosics; stabilizing agent for chlorine and hypochlorite in swimimng pools, closed water systems; bleaching paper pulp and textiles; catalyst for urea-formaldehyde resins.

"Sulfan."[50] Trademark for stabilized sulfuric anhydride. A colorless liquid; m.p. 17°C; b.p. 45°C. An efficient sulfonating agent used in the production of water-soluble and oil-soluble detergents. Also used in the preparation of organic intermediates for pharmaceuticals, dyestuffs and other products.
Containers: Glass ampules, steel drums, tank cars.
Grades: 99.5% of SO_3 min, stabilized to retain 17°C melting point.

sulfanilic acid (para-aminobenzenesulfonic acid; para-anilinesulfonic acid) $H_2NC_6H_4SO_3H \cdot H_2O$.
Properties: Grayish white, flat crystals. Soluble in fuming hydrochloric acid; slightly soluble in water; very slightly soluble in alcohol and ether. M.p., chars at 280–300°C. Combustible; low toxicity.
Derivation: By heating aniline with weak fuming sulfuric acid and pouring the reaction product into water.
Grades: Technical; pure; reagent.
Containers: Bottles; drums.
Uses: Dyestuffs; organic synthesis; medicine.

meta-**sulfanilic acid.** See metanilic acid.

sulfanilylbutylurea. See 1-butyl-3-sulfanilylurea.

"Sulfanole."[42] Trademark for a group of sulfonated detergents used in the textile industry.

"Sulfanthrene."[28] Trademark for a group of thioindigoid vat dyes characterized by good fastness properties. Used principally for the dyeing and printing of cotton and rayon.

"Sulfasan" R.[58] Trademark for 4,4'-dithiodimorpholine (q.v.).

"Sulfatate B-1."[73] Trademark for a hydrocarbon sulfonate used as a wetting agent.

sulfate pulp. See pulp, paper.

sulfenamide. A compound having the typical structure $RSNH_2$.

sulfide dye (sulfur dye). One of a group of dyes produced by heating various organic compounds with sulfur. The characteristic chromophore groupings are $\equiv C—S—C\equiv$ and $\equiv C—S—S—C\equiv$. CI numbers range from 53000–54999. Application is usually to cotton from a sodium sulfide bath. Sulfur black (CI No. 53185) is an important example.

sulfite acid liquor. An aqueous solution of calcium bisulfite or calcium and magnesium busulfites containing a large amount of free sulfur dioxide. It is prepared from sulfur dioxide and limestone, dolomite or lime by countercurrent extraction.
Use: Manufacture of sulfite pulp for paper.

sulfite pulp. See pulp, paper.

sulfite waste liquor. A waste liquor produced in the sulfite paper process. Sold in variations from a dilute solution to a solid.
Uses: Foam producer; emulsifier; adhesives; tanning agent; binder for briquets, cores, unpaved roads; source of torula yeast.
See also lignin sulfonate; waste disposal.

meta-**sulfobenzoic acid** $HO_3SC_6H_4COOH \cdot 2H_2O$.
Properties: Grayish white solid; stable but hygroscopic in air; m.p. 98°C, anhydrous form melts at 141°C. Soluble in water, alcohol; insoluble in benzene.
Derivation: Direct sulfonation of benzoic acid with sulfur trioxide.
Use: Derivative for surface-active agents.

ortho-**sulfobenzoic acid** $HO_3SC_6H_4COOH$.
Properties: White needles. Soluble in water and alcohol; insoluble in ether. M.p. 68–69°C (with $3H_2O$ of crystallization); m.p. 134°C (dry).

Derivation: (a) From saccharin and concentrated hydrochloric acid; (b) by the oxidation of thiosalicylic acid, with potassium permanganate in alkaline solution.
Uses: Manufacture of sulfonaphthalein indicators; dyes.

ortho-**sulfobenzoic anhydride** $C_6H_4COOSO_2$.

Properties: Solid; m.p. 129.5°C; b.p. 184–186°C (18 mm). Soluble in hot water, ether, and benzene. Used as polymerization inhibitor.

"Sulfobetaines."[560] Trademark for a group of ionic neutral surfactants, compatible with all other classes of surface-active agents. The "zwitterion" of the unusual series of compounds contains a quaternary ammonium ion and a sulfonic acid ion, arranged in a betaine-like inner salt structure.
Uses: Cosmetic soap bars, liquid hand soap, soap-based shampoos, germicidal soap preparations, and soap-based laundry preparations.

sulfocarbanilide. See thiocarbanilide.

sulfocarbolic acid. See phenolsulfonic acid.

1-(4-sulfo-2,3-dichlorophenyl)-3-methylpyrazolone $(Cl_2C_6H_2SO_3H)NNC(CH_3)CH_2CO$.

Properties: White or yellowish powder or crystals. Very soluble in water; soluble in alkalies.
Derivation: By condensation of dichlorophenylhydrazine sulfonic acid with ethylacetoacetate.
Use: Intermediate for dyes.

"Sulfogene."[28] Trademark for a series of sulfur colors. Used extensively on cotton work-clothing and similar fabrics and to a limited extent on rayon and other materials.

sulfolane (tetrahydrothiophene 1,1-dioxide; tetramethylene sulfone) $CH_2CH_2CH_2CH_2SO_2$.
Properties: Usually a liquid; sp. gr. 1.261 (30/4°C); m.p. 27°C; b.p. 285°C; refractive index 1.481 (30°C); flash point 350°F (open cup). Miscible with water, acetone, and toluene. A highly polar compound with outstanding solvent properties. Combustible.
Derivation: By the reaction of sulfur dioxide with butadiene to form sulfolene, which is then hydrogenated.
Uses: Extraction of aromatic hydrocarbons from oil refinery streams; fractionation of wood tars, tall oil, and other fatty acids; polymerization solvent; plasticizer; component of hydraulic fluid; textile finishing.

"Sulfole."[303] Trademark for group of high-quality thiols ranging from C_4 to C_{16}. Derived by the catalytic addition of hydrogen sulfide to petroleum olefins.
Uses: Rubber and plastics industries as modifiers for "hot" and "cold" rubber, nitrile rubber, and certain types of plastics.

sulfolene. See 2,5-dihydrothiophene-1,1-dioxide.

sulfonamide. A compound having the typical structure RSO_2NH_2. Used as anti-infective agents in medicine. Their use is restricted by FDA. See sulfa drug.

"Sul-fon-ate."[93] Trademark for a series of anionic wetting agents.
AA9 and AA10. Crisp, white flakes of sodium dodecyl benzene sulfonate.
OA-5. Light brown liquid of sulfonated oleic acid, sodium salt.

sulfonated castor oil. See Turkey red oil.

sulfonated oil. A vegetable or animal oil which has been treated with sulfuric acid and the excess of acid washed out and the oil neutralized with a small amount of caustic soda or ammonia. These oils are soluble (emulsifiable) in water and are claimed to be capable of: (1) supplying lubrication for various purposes; (2) emulsifying (dispersing) in water various other materials; (3) plasticizing; (4) defoaming; (5) dispersing; (6) penetrating; (7) scouring; (8) wetting out; and (9) softening. For some of these operations specially formulated oils may be necessary.

sulfonation. The formation of a sulfonic acid, i.e., a compound containng the $—SO_2OH$ group. The conversion of benzene (C_6H_6) into benzenesulfonic acid ($C_6H_5SO_2OH$) is an example. Common sulfonating agents are: concentrated sulfuric acid, fuming sulfuric acid, sulfur trioxide, alkali disulfates, pyrosulfates, chlorosulfonic acid and a mixture of manganese dioxide and sulfurous acid.

para-**sulfondichloraminobenzoic acid.** See halazone.

sulfonyl bisphenol. See dihydroxydiphenyl sulfone.

sulfonyl chloride. See sulfuryl chloride.

4,4′-sulfonyldianiline (DDS; dapsone; 4,4′-diaminodiphenyl sulfone) $C_6H_4NH_2)_2SO_2$.
Properties: White or creamy white, crystalline powder; odorless; slightly bitter taste. Very slightly soluble in water; freely soluble in alcohol; soluble in acetone and dilute mineral acids. Melting range between 175° and 181°C.
Grades: U.S.P. (as dapsone, the USAN name); technical.
Hazard: Toxic by ingestion.
Uses: Curing agent for epoxy resins; medicine.

para-**1-sulfophenyl-3-methyl-5-pyrazolone** $(C_6H_4SO_3H)NNC(CH_3)CH_2CO$.

Properties: White or yellowish powder. Very slightly soluble in water; soluble in alkalies.
Derivation: By condensation of phenylhydrazine sulfonic acid with ethylacetoacetate.

4-sulfophthalic anhydride $HO_3SC_6H_3(CO)_2O$.
Properties: Reddish brown syrup; crystallizes partially on long standing; hygroscopic; fluorescent in solution under ultraviolet radiation; sp. gr. 1.62 (25°C). Very soluble in water, alcohol; insoluble in ether, benzene. Combustible.
Uses: Esters of 4-sulfophthalic acid used in wetting, cleansing, emulsifying, softening and equalizing agents with textiles. Derivatives have application as surface-active agents.

"Sulforon X."[28] Trademark for a microfine sulfur used as a fungicide in a water spray.

5-sulfosalicylic acid $C_6H_3OH(SO_2OH)COOH \cdot 2H_2O$.
Properties: Colorless crystals; colored pink by traces of iron. Very soluble in water. M.p. 120°C; decomposes at higher temperatures.
Derivation: Action of sulfuric acid on salicylic acid.
Uses: Reagent for albumin; colorimetric reagent for ferric iron; intermediate for dyes, surfactants, catalysts, grease additives.

sulfotepp. Generic name for ethyl thiopyrophosphate (tetraethyl dithiopyrophosphate) $(C_2H_5O)_2P(S)OP(S)(OC_2H_5)_2$.
Properties: Yellow liquid; b.p. 136–139°C (2 mm). Slightly soluble in water; soluble in most organic solvents.
Hazard: Highly toxic by ingestion, inhalation, and skin absorption. Cholinesterase inhibitor. Use may be restricted.
Use: Insecticide.
Shipping regulations: (Rail, Air) Poison label. Not accepted on passenger planes. Legal label name: tetraethyl dithiopyrophosphate. Consult regulations for details.

sulfoxide. Generic name for 1,2-(methylenedioxy)-4-[2-(octylsulfinyl)propyl]benzene (n-octyl sulfoxide isosafrole) $C_{18}H_{28}O_3S$.
Properties: Brown liquid; insoluble in water; slightly soluble in oils; miscible with most organic solvents.
Use: Insect synergist.

"Sulframin."[449] Trademark for a series of alkylaryl sulfonates.

"Sulfricin."[202] Trademark for a modified castor oil of increased hydroxyl content used for sulfonation.

sulfur S Nonmetallic element, atomic number 16, group VIA of periodic system. Atomic weight 32.06. Valances 2,4,6. Four stable isotopes.
Properties: Pure sulfur exists in two stable crystalline forms (a) and (b) and at least two amorphous (liquid) forms.
(a) Alpha-sulfur, rhombic, octahedral yellow crystals stable at room temperature. Sp. gr. 2.06; transition to beta form 94.5°C; m.p. (rapid heating) 112.8°C; refractive index 1.957.
(b) Beta-sulfur, monoclinic, prismatic pale yellow crystals slowly changing to alpha form below 94.5°C. Sp. gr. 1.96; m.p. 119°C; b.p. 444.6°C; flash point 405°F; autoignition temp. 450°F. Index of refraction 2.038. Both forms are insoluble in water; slightly soluble in alcohol and ether; soluble in carbon disulfide, carbon tetrachloride, and benzene. Combustible; nontoxic.
Occurrence: Native in Texas; Louisiana; Sicily; Canada (Alberta); Poland; Saudi Arabia; Mexico; Iraq; offshore deposits in Gulf of Mexico; U.S.S.R.; Japan.
Derivation: Direct mining by Frasch process (q.v.); low-grade ores by Chemico process (q.v.); smelter waste gas; sour natural gas; iron pyrites; gypsum; solvent extraction of volcanic ash; petroleum; coke oven gas by Thylox process (q.v.).
Grades: Technical (lumps, roll, flour); rubber makers; N.F. (sublimed); crude; refined; precipitated (milk of sulfur); high purity (impurities less than 10 ppm).
Containers: Bags; barrels; bulk in carloads and barges; molten, in steam-heated pipelines.
Hazard: Fire and explosion risk in finely divided form. Safety data sheet available from Manufacturing Chemists Assn., Washington, D.C.
Uses: Sulfuric acid manufacture; pulp and paper manufacture; carbon disulfide; rubber vulcanization; dyes and chemicals; drugs and pharmaceuticals; explosives; insecticides; rodent repellents; soil conditioner; fungicide; coating for controlled-release fertilizers; nucleating agent for photographic film; cement sealant (crack repair in airport runways); road paving compositions (experimental).
For further information refer to Sulfur Institute, 1725 K St. N.W., Washington, D.C.

sulfur 35. Radioactive sulfur of mass number 35.
Properties: Half-life 87.1 days; radiation, beta; radiotoxicity, moderately hazardous.
Derivation: By pile irradiation of elemental sulfur or of various chlorides.
Forms available: Solid elemental sulfur; sulfate in weak hydrochloric acid; barium sulfide in barium hydroxide solution; elemental sulfur in benzene solution; in tagged compounds such as carbon disulfide, chlorosulfonic acid, thiourea, sulfanilamide, thiamine, heparin, insulin, "Sucaryl," etc.
Hazard: Radioactive poison.
Uses: Research tool in studying mechanism of rubber vulcanization and polymerization of synthetic rubber; role of sulfur in the coking process and in steel; effect of sulfur on engine wear; sulfur removal in the viscose process; behavior of detergents during washing; sulfur deposition in diesel engines; action of sulfur in silver plating solutions; protein metabolism; surface active agents and surface phenomena; drug actions; etc.
Shipping regulations: (Rail, Air) Consult regulations.

sulfurated lime. See lime, sulfurated.

sulfurated potash. See potash, sulfurated.

sulfur bromide (sulfur monobromide) S_2Br_2.
Properties: Yellow liquid. Becomes red when exposed to air. Decomposed by water. Soluble in carbon disulfide. Sp. gr. 2.6 (15°C); f.p. −40°C; b.p. 54°C (0.2 mm).
Hazard: Avoid inhalation of fumes. Strong irritant to tissue.

sulfur chloride (sulfur subchloride; sulfur monochloride) S_2Cl_2.
Properties: Amber to yellowish red, oily, fuming liquid; penetrating odor. Soluble in alcohol, ether, benzene, carbon disulfide and amyl acetate; decomposes on contact with water. Sp. gr. 1.690 (15.5°C); f.p. −80°C; b.p. 138°C; flash point 266°F. Combustible.
Derivation: By passing chlorine over molten sulfur.
Method of purification: Distillation.
Containers: Carboys; drums; tank cars.
Hazard: Highly toxic by inhalation and ingestion; strong irritant to tissue. Reacts violently with water in a closed vessel. Tolerance, 1 ppm in air. Safety data sheet available from Manufacturing Chemists Assn., Washington, D.C.
Uses: Chemicals (sulfur solvent, acetic anhydride, thionyl chloride, carbon tetrachloride from carbon disulfide, various chlorohydrins from glycerol, glycol, etc.); analytical reagent; rubber vulcanizing; vulcanized oils; purifying sugar juices; military poison gas; insecticide; hardening soft woods (by treatment with sulfur chloride dissolved in carbon disulfide); pharmaceuticals; textile finishing and dyeing; extraction of gold from its ores.
Shipping regulations: (Rail) White label. (Air) Corrosive label. Not acceptable on passenger planes.

sulfur dichloride SCl_2.
Properties: Reddish brown fuming liquid with pungent chlorine odor. Sp. gr. 1.638 (15.5°C); f.p. −78°C; b.p. decomposes above 59°C; on rapid heating, boils near 60°C; decomposes in water, and alcohol; soluble in benzene; refractive index 1.567 (n 20/D); noncombustible.
Derivation: Chlorine is passed into sulfur monochloride to saturation, at 6° to 10°C, followed by carbon dioxide to drive off the excess chlorine.
Grade: Technical.
Containers: Drums; tank cars.

Hazard: Highly toxic by inhalation and ingestion; strong irritant to tissue. Safety data sheet available from Manufacturing Chemists Assn., Washington, D.C.
Uses: In general, as a chlorine carrier or chlorinating agent; rubber vulcanizing; vulcanized oils; purifying sugar juices; sulfur solvent; chloridizing agent in metallurgy; manufacture of organic chemicals and insecticides.
Shipping regulations: (Rail) White label. (Air) Corrosive label. Not acceptable on passenger planes.

sulfur dioxide SO_2.
Properties: Colorless gas or liquid with sharp pungent odor. Soluble in water, alcohol, and ether. Forms sulfurous acid H_2SO_3. Sp. gr. 1.4337, liquid at 0°C; f.p. −76.1°C; b.p. −10°C; vapor pressure 3.2 atmosphere at 68°F; refractive index (liquid) 1.410 (n 24/D). An outstanding oxidizing and reducing agent. Noncombustible.
Derivation: (a) By roasting pyrites in special furnaces. The gas is readily liquefied by cooling with ice and salt, or at a pressure of 3 atm. (b) By purifying and compressing sulfur dioxide gas from smelting operations. (3) By burning sulfur.
Graded: Commercial; U.S.P.; technical; refrigeration; anhydrous 99.98% min.
Containers: 2- to 300-lb pressure cylinders; ton drums on multi-unit cars; 40,000-lb tank cars.
Hazard: Toxic by inhalation; strong irritant to eyes and mucous membranes, especially under pressure. Dangerous air contaminant and constituent of smog. Tolerance, (for industrial air spaces), 5 ppm; U.S. atmospheric standard, 0.140 ppm. Not permitted in meats and other sources of vitamin B_1. Safety data sheet available from Manufacturing Chemists Assn., Washington, D.C.
Uses: Chemicals (sulfuric acid, salt cake, sulfites, hydrosulfites of potassium and sodium, thiosulfates, alum from shale, recovery of volatile substances); sulfite paper pulp; ore and metal refining; soybean protein; intermediates; solvent extraction of lubricating oils; bleaching agent for oils, foods; sulfonation of oils; disinfecting and fumigating; food additive (inhibition of browning, of enzyme-catalyzed reactions, bacterial growth); preservation of grapes; reducing agent; antioxidant.
Shipping regulations: (Rail) Green label. (Air) Nonflammable Gas label. Not accepted on passenger planes.

sulfur dye. See sulfide dye.

sulfuretted hydrogen. See hydrogen sulfide.

sulfur hexafluoride SF_6.
Properties: Colorless gas; f.p. −64°C (sublimes); density 6.5 grams per liter; sp. gr. liquid 1.67; specific volume 2.5 cu ft/lb (70°F); slightly soluble in water; soluble in alcohol and ether. Odorless, noncombustible, low toxicity.
Containers: 60-lb cylinders.
Hazard: Tolerance, 1000 ppm in air.
Use: Dielectric (gaseous insulator for electrical equipment and radar wave guides).
Shipping regulations: (Rail) Green label. (Air) Nonflammable Gas label.

sulfuric acid (hydrogen sulfate, battery acid) H_2SO_4. By far the most widely used industrial chemical; its production is approximately 60 billion pounds a year in U.S. alone. Highest-volume chemical produced in U.S. (1975).
Properties: Strongly corrosive, dense, oily liquid; colorless to dark brown depending on purity. Miscible with water in all proportions. Very reactive, dissolves most metals; concentrated acid oxidizes, dehydrates, or sulfonates most organic compounds, often causes charring. Sp. gr. of pure material 1.84; m.p. 10.4°C; b.p. varies over range 315–338°C due to loss of sulfur trioxide during heating to 300°C or higher.
Note: Use great caution in mixing with water due to heat evolution that causes explosive spattering. Always add the acid to water, *never* the reverse.
Derivation: From sulfur, pyrite (FeS_2), hydrogen sulfide and sulfur-containing smelter gases by the contact process (vanadium pentoxide catalyst). It can also be made from sulfur dioxide by the "Cat-Ox" process (q.v.) and from gypsum (calcium sulfate).
Grades: Commercial; 60° Bé (sp. gr. 1.71, 77.7% H_2SO_4); 66° Bé (sp. gr. 1.84, 93.2% H_2SO_4); 98% (sp. gr. 1.84); 99% (sp. gr. 1.84); 100% (sp. gr. 1.84) depending on supplier; reagent A.C.S.; C.P.
Containers: Bottles, carboys, drums, tank trucks, tank transports, tank cars, tank barges. Lined equipment used for chemically pure shipments.
Hazard: Highly toxic; strong irritant to tissue. Tolerance, 1 mg per cubic meter of air. See note above. Safety data sheet available from Manufacturing Chemists Assn., Washington, D.C.
Uses: Fertilizers; chemicals; inorganic pigments; petroleum refining; etchant; alkylation catalyst; electroplating baths; iron and steel; rayon and film; industrial explosives; laboratory reagent; nonferrous metallurgy.
Shipping regulations: (Rail) White label. (Air) Corrosive label.
See also sulfuric acid, fuming.

sulfuric acid, fuming (oleum; $xH_2SO_4 \cdot ySO_3$). A solution of sulfur trioxide in sulfuric acid.
Properties: Heavy, oily liquid. Colorless to dark brown depending on purity. Fumes strongly in moist air. Extremely hygroscopic.
Derivation: Sulfur trioxide produced by the contact process (q.v.) is absorbed in concentrated sulfuric acid.
Grades: Commercial (10% to 70% free SO_3, depending on supplier); C.P.
Containers: Bottles, drums, tank cars, tank trucks, tank transports, tank barges.
Hazard: Highly toxic; strong irritant to tissue. (Reacts violently with water). Safety data sheet available from Manufacturing Chemists Assn., Washington, D.C.
Uses: Sulfating and sulfonating agent; dehydrating agent in nitrations; dyes; explosives; petroleum refining; laboratory reagent.
Shipping regulations: (Rail) White label. (Air) Corrosive label. Not acceptable on passenger planes.

sulfuric anhydride. See sulfur trioxide.

sulfuric chloride. See sulfuryl chloride.

sulfuric oxychloride. See sulfuryl chloride.

sulfur monobromide. See sulfur bromide.

sulfur monochloride. See sulfur chloride.

sulfurous acid. A solution of sulfur dioxide in water. The formula H_2SO_3 is used, but the acid is known only through its salts.
Properties: Colorless liquid; suffocating sulfur odor. Sp. gr. about 1.03; unstable. Soluble in water.
Derivation: Absorption of sulfur dioxide in water.
Grades: Technical; C.P.
Containers: Carboys; drums.
Hazard: Highly toxic by ingestion and inhalation; strong irritant to tissue.
Uses: Organic synthesis; bleaching straw and textiles, etc.; paper manufacture; wine manufacture; brewing; metallurgy; ore flotation; medicine; reagent in analytical chemistry; sulfites; as preservative for fruits, nuts, foods, wines; disinfecting ships; refining crude oils and paraffins.

sulfur oxychloride. See thionyl chloride.

sulfur subchloride. See sulfur chloride.

sulfur tetrafluoride SF_4.
Properties: A gas; f.p. −124°C; b.p. −40°C; decomposes in water. Noncombustible.
Grade: Available in cylinders at 90–94% purity.
Hazard: Highly toxic by inhalation; strong irritant to eyes and mucous membranes. Tolerance, 0.1 ppm in air.
Uses: Fluorinating agent for making water and oil repellents, and lubricity improvers.
Shipping regulations: (Air) Nonflammable Gas label. Not acceptable on passenger planes.

sulfur trioxide (sulfuric anhydride) SO_3; $(SO_3)_n$.
Properties: Exists in three soliid modifications; alpha, m.p. 62°C; beta, m.p. 32.5°C; gamma, m.p. 16.8°C. The alpha form appears to be the stable form but the solid transitions are commonly slow; a given sample may be a mixture of the various forms, and its m.p. not constant. The solids sublime easily. All three forms boil at 45°C.
Derivation: Passing a mixture of SO_2 and O_2 over a heated catalyst such as platinum or vanadium pentoxide.
Hazard: Highly toxic; strong irritant to tissue. Oxidizing agent. Fire risk in contact with organic materials. An explosive increase in vapor pressure occurs when the alpha form melts. The anhydride combines with water, forming sulfuric acid and evolving a large amount of heat.
Containers: (Stabilized, liquid) 750-lb drums; tank cars.
Uses: Sulfonation of organic compounds, especially nonionic detergents, solar energy collectors. It is usually generated in the plant where it is to be used.
Shipping regulations: (Stabilized) (Rail) White label. (Air) (Beta or gamma, stabilized) Corrosive label. (Beta or gamma, not stabilized, and alpha with or without stabilizer) Not acceptable.

sulfuryl chloride (chlorosulfuric acid; sulfonyl chloride; sulfuric chloride; sulfuric oxychloride) SO_2Cl_2.
Properties: Colorless liquid. Pungent odor. Rapidly decomposed by alkalies and by hot water. Soluble in glacial acetic acid. Sp. gr. 1.667 at 20°C; b.p. 69.2°C; f.p. −54°C; vapor density 4.6.
Derivation: (a) By heating chlorosulfonic acid in the presence of catalysts. (b) From sulfur dioxide and chlorine in the presence of either activated carbon or camphor.
Grades: Technical.
Containers: 5-gal carboys; 55-gal drums; 725-lb drums.
Hazard: Highly toxic; strong irritant to tissue. Safety data sheet available from Manufacturing Chemists Assn., Washington, D.C.
Uses: Organic synthesis (chlorinating agent, dehydrating agent; acylating agent) pharmaceuticals; dyestuffs; rubber-base plastics; rayon; poison gases; solvent; catalyst.
Shipping regulations: (Rail) White label. (Air) Corrosive label.

sulfuryl fluoride SO_2F_2.
Properties: Colorless gas; f.p. −136.7°C; b.p. −55.4°C. Slightly soluble in cold water and most organic solvents. Noncombustible.
Containers: Steel cylinders.
Hazard: Toxic by inhalation. Tolerance, 5 ppm in air.
Uses: Insecticide; fumigant.
Shipping regulations: (Rail) Green label. (Air) Nonflammable Gas label.

"Sulphon."[307] Trademark for acid dyestuffs used on wool and silk; fair to good fastness to light, good fastness to washing, etc. Can be used on leather and paper.

"Sulphonated Olevene."[309] Trademark for a sulfonated synthetic olive oil; used for finishing textiles as a softener and plasticizer.

sultam acid. See 1,8-naphthosultam-2,4-disulfonic acid.

"Sumstar" 190.[212] Trademark for a polymeric dialdehyde. A fine, nonvolatile, virtually odorless powder. Produced by highly specific periodate oxidation of the anhydroglucose units of starch. Bulk density 40 to 42 lb/cu ft.
Uses: Film-forming agents, adhesives, leather, and resins.

"Sunaptic."[494] Trademark for three grades of naphthenic acids having an average molecular weight of 310, 340, and 415.
Uses: Plasticizers; corrosion and rust inhibitors; cleaning, defoaming, foaming, dispersing, emulsifying and wetting agents; metallic naphthenates.

"Sun Colors."[134] Trademark for a series of lightfast pigments resistant to heat and chemicals based on heavy metal modification of titanium dioxide.

"Sundex" Oils.[494] Trademark for a series of aromatic process and extender oils. Recommended for rubber processing where color is not a problem.
Uses: Plasticizers, softeners.

sunflower cake. The presscake resulting from hydraulic pressure expression of sunflower seeds. Typical analysis; proteins 21.0%; fats 8.5%; fiber 48.9%; water 10.2%; ash 11.4%.
Containers: Bags; bulk.
Uses: Cattle food; fertilizer ingredient.

sunflower meal. The mealy form assumed by sunflower seeds after crushing and heating. If the oil cake is ground the product again is in this mealy form. It contains 55–60% protein.
Uses: Animal feed; fertilizers.

sunflower oil.
Properties: Pale yellow semidrying oil; mild taste; pleasant odor. Soluble in alcohol, ether, chloroform and carbon disulfide. Sp. gr. 0.924–0.926; iodine value 130–135; refractive index 1.4611. Nontoxic. Combustible.
Chief constituents: Mixed triglycerides of linoleic, oleic and other fatty acids.
Derivation: By expression from the seeds of Helianthus annus.
Method of purification: Filtration.
Grades: Crude; refined.
Containers: Barrels; steel drums; tank cars.

Uses: Modified alkyd resins; soap; edible oil; margarine; shortening.

"Sunolite."[104] Trademark for waxes used to prevent disintegration of rubber stocks which are constantly exposed to the atmosphere.

"Sunolith."[296] Trademark for a pigment consisting of various types of lithopone.

"Sunolox."[204] Trademark for a formulation designed to promote the stability of hydrogen peroxide bleach solutions.
Use: Bleaching of textiles near the neutral point; softness is improved by low silicate content.

"Sunpar."[494] Trademark for a series of nonstaining paraffinic oils. Recommended for use with ethylene-propylene copolymers. Used as plasticizers and softeners.

"Sunproof."[248] Trademark for a series of protective waxes to prevent static atmospheric cracking in all types of synthetic and natural rubbers.

"Sunsize" 536A.[42] Trademark for a complex of fiber-reactive, alkyl-substituted, synthetic nitrogen derivatives. Used as an internal size in textile industry in place of rosin-alum or rosin-alum-wax.

"Sunthene."[494] Trademark for a series of nonstaining naphthenic oils. Used as plasticizers, softeners, and extenders.

"Superaid."[218] Trademark for diatomite filteraids; used in food and industrial processing.

"Super-Alkali."[292] Trademark for combinations of caustic soda and soda ash with varying percentages of the two basic ingredients.
Containers: 400-lb fiber drums; 400-lb steel drums.
Uses: Heavy-duty cleaning such as in metal cleaning and meat-packing houses.

superalloy. An iron-base, cobalt-base, or nickel-base alloy which combines high-temperature mechanical properties, oxidation resistance, and creep resistance to an unusual degree. Constitutions of these alloys are as follows: Iron-base: 10–45% nickel, 13–19% chromium, 1.3–6% molybdenum, balance iron. Cobalt-base: 0–26% nickel, 0–26% chromium, 0–15% tungsten, balance cobalt. Nickel-base: 55–75% nickel, 10–20% chromium, 0–6% aluminum, 0–5% titanium. All contain less than 0.5% carbon, plus other special ingredients. Superalloys can be used up to 2500°F.
Uses: Jet engine parts; turbo-superchargers; extreme high-temperature applications.

"Super-Amide."[328] Trademark for high-activity fatty acid diethanolamines. Used as foam stabilizers, emulsifiers, thickeners.

"Superba."[133] Trademark for a medium high-color carbon black for automotive finishes, locomotive paints, high grade industrial work. Available in standard form and beads.

"Super-Beckacite."[36] Trademark for a series of oil-soluble, oil-reactive, heat-hardening pure phenolic synthetic resins. Available in both foaming and non-foaming types.
Uses: Varnishes and enamels of the spar and marine types.

"Super-Beckamine."[36] Trademark for a group of melamine-formaldehyde resins.

"Super-Beckosol."[36] Trademark for a group of isophthalic acid alkyd resins for drying, flexibility, and high exterior durability paints and printing inks.

"Super Cat."[74] Trademark for a series of hydrogenation catalysts available in three stabilized flake forms (100, 200, 300). Used in edible oils, inedible oils and fatty acids.

"Super-CE."[88] Trademark for a cerium polish, consisting of cerium oxide and other ingredients.

"Superchrome."[243] Trademark of acid mordant dyes for wool and synthetics.

superconductivity. The phenomenon that causes certain metals, alloys, and compounds near absolute zero to lose both electrical resistance and magnetic permeability, i.e., to have infinite electrical conductivity. Depending upon the substance, the maximum temperature (transition temperature) for the behavior is 0.5–18°K. Superconductivity does not occur in alkali metals, noble metals, ferro- and antiferromagnetic metals. It is well-known in elements having 3, 5, or 7 valence electrons per atom, and is associated with high room-temperature resistivity. A system for transmitting electric current underground by means of superconducting cables has been developed.
See also cryogenics.

"Super Dylan."[11] Trademark for high-density polyethylene, made by Ziegler or low pressure process.
Uses: Molded items, extruded sheets, film, pipe, bottles, wire insulation, filaments, etc.

"Superresins."[188] Trademark for a series of standardized essential oil/oleoresin mixtures, designed to replace natural spices under prescribed conditions.

"Super-Fast."[241] Trademark for premium quality cerium oxide for ultrafast glass polishing.

"Superfine."[1] Trademark for a flour sulfur air-classified for delivery in two particle sizes, one 93% and the other 98% through a 325 U.S. Sieve.
Uses: Manufacture of glass, matches, chrome oxide pigments, sulfur cements, stock foods, pyrotechnics and explosives; refining of petroleum and casting magnesium and aluminum.

"Super-Foil."[524] Trademark for aluminum and aluminum alloys in foil form.

"Superglo."[108] Trademark for a completely neutral, liquid synthetic detergent. Combination of highly active wetting agents and foam stabilizers. Soluble in hot and cold water.

"Superinone."[162] Trademark for tyloxapol (q.v.).

"Superlith."[223] Trademark for imported zinc sulfide pigments.
Grades: Available as pure zinc sulfide, 60% zinc sulfide.

"Superloid."[322] Trademark for ammonium alginate, a hydrophilic colloid.
Uses: Suspending, thickening, emulsifying and stabilizing agent in creaming and bodying of rubber latex products; protective colloid in resin emulsion paints; adhesives; fire-retarding compositions; ceramics, etc.

"Superlose."[553] Trademark for amylose fraction of potato starch.

"Superlume."[288] Trademark for a super-leveling bright nickel electroplating process on steel stampings, brass, copper, zinc die castings, etc. The materials used are nickel sulfate, nickel chloride, boric acid and addition agents.

supernatant. A liquid or fluid forming a layer on the surface of another liquid.

superoxide. A compound characterized by the presence in its structure of the O_2^- ion. The O_2^- ion has an odd number of electrons (13) and, as a result, all superoxide compounds are paramagnetic. At room temperature they have a yellowish color. At low temperature many of them undergo reversible phase transitions accompanied by a color change to white. The stable superoxides are:

Sodium superoxide	NaO_2
Potassium superoxide	KO_2
Rubidium superoxide	RbO_2
Cesium superoxide	CsO_2
Calcium superoxide	$Ca(O_2)_2$
Strontium superoxide	$Sr(O_2)_2$
Barium superoxide	$Ba(O_2)_2$
Tetramethylammonium superoxide	$(CH_3)_4NO_2$

In these compounds each oxygen atom has an oxidation number of $-\frac{1}{2}$ instead of -2, as a normal oxide.

"Supernilla."[342] Trademark for natural vanilla concentrate in liquid form.

"Superoxol."[123] Trademark for hydrogen peroxide (q.v.).

"Superpax."[337] Trademark for 92 to 94.5% zirconium silicate with bulk density 68 lbs/cu ft; average particle size 5 microns max. Used in ceramic glazes and as a filler for resins and rubbers.

superphosphate (acid phosphate). The most important phosphorus fertilizer, made by the action of sulfuric acid on insoluble phosphate rock (essentially calcium phosphate, tribasic) to form a mixture of gypsum and calcium phosphate, monobasic. A typical composition is $CaH_4(PO_4)_2 \cdot H_2O$ 30%; $CaHPO_4$ 10%; $CaSO_4$ 45%; iron oxide, alumina, silica 10%; water 5%.
Typical analysis: Moisture 10–15%; available phosphoric acid (as P_2O_5) 18–21%; insoluble phosphoric acid 0.3–2%; total phosphoric acid (as P_2O_5) 19–23%.
Grades: Based on available P_2O_5.
Containers: Bags; bulk; multiwall paper sacks; carloads.
Use: Fertilizer.
See also triple superphosphate and nitrophosphate.

superphosphoric acid. See polyphosphoric acid.

supersaturation. The condition in which a solvent contains more dissolved matter (solute) than is present in a saturated solution of the same components at equivalent temperature. Such solutions may occur, or can be made, when a saturated solution cools gradually so that nucleating crystals do not form. They are extremely unstable and will precipitate upon addition of even one or two crystals of the solute, or upon shaking or other slight agitation. Supersaturated solutions occur in the confectionery industry, e.g., in fudges, maple sugar, etc.

"Supersheen."[292] Trademark for caustic soda solution containing chelating agent and wetting agent.
Containers: Tank cars and tank trucks.
Hazard: Strong irritant to skin and tissue.
Uses: Bottle washing and food plant sanitation.

"Supersil."[436] Trademark for ground silica. Available in standard grades from 100 to 325 mesh.
Uses: Manufacture of ceramics, porcelain enamel, scouring powders and buffing compounds, fiber glass, autoclave concrete products, chemicals, asbestos products and for cementing deep oil wells.

"Supersilicate."[244] Trademark for a compound consisting essentially of the formula, $1.5\ Na_2O \cdot SiO_2 \cdot 5.5\ H_2O$.
Uses: Laundry and metal cleaning; paint remover; concrete floor cleaner; base for cleaning compounds.

"Supersize."[553] Trademark for hydroxyethyl corn starches.

"Super-sol."[25] Trademark for an odorless petroleum naphtha; a rapid-drying highly purified solvent.
Uses: Carrier for insecticides; preparation of odorless paints; cleaning compositions.

"Super Spectra."[133] Trademark for an impingement carbon black used for jet black enamel and lacquers requiring satin finish. Containers 6¼-lb bags.

"Super Stod-Sol."[200] Trademark for a petroleum solvent.
Properties: Water-white; boiling range 310–353°F; sp. gr. 0.779 (60°F); wt/gal 6.49 lbs (60°F); flash point 102°F. Combustible.
Containers: Drums, tank wagon, tank cars.
Hazard: Moderate fire risk.
Uses: Dry-cleaning solvent.
See also Stoddard Solvent.

"Supralan."[307] Trademark for metallized acid colors of good fastness and level dyeing properties.

"Supramine" XA.[307] Trademark for a leather chemical, solubilized sulfur phenol condensate; 75% active.

"Supranol."[307] Trademark for dyestuffs used on wool and silk. Good fastness to light, washing, and sea water. Can also be used on leather.

"Suprarenin."[162] Trademark for synthetic epinephrine.

"Suprex."[285] Trademark for a group of clays (sedimentary kaolins) from South Carolina.

"Surfacaine."[100] Trademark for cyclomethycaine (q.v.).

surface. In physical chemistry, the area of contact between two different phases or states of matter, e.g., finely divided solid particles and air or other gas (solid-gas); liquids and air (liquid-gas); insoluble particles and liquid (solid-liquid). Surfaces are the sites of the physicochemical activity between the phases that is responsible for such phenomena as adsorption, reactivity, and catalysis. The depth of a surface is of molecular order of magnitude. The term interface is approximately synonymous with surface, but it also includes dispersions involving only one phase of matter, i.e., solid-solid or liquid-liquid. See also interface, surface area, surface chemistry.

surface-active agent (surfactant). Any compound that reduces surface tension (q.v.) when dissolved in water or water solutions, or which reduces interfacial tension between two liquids, or between a liquid and a solid. There are three categories of surface active agents: detergents, wetting agents, and emulsifiers; all have the same basic chemical mechanism and differ chiefly in the nature of the surfaces involved. For further information see under these three entries; also see interface; surface chemistry.

surface area. The total area of exposed surface of a finely divided solid (powder, fiber, etc.) including irregularities of all types. Since activity is greatest at the surface, that is, the boundary between the particle and its environment, the larger the surface area of a given substance the more reactive it becomes. Thus reduction to small particles is a means of increasing the efficiency of both chemical and physical reactions; for example, the coloring effect of pigments is increased by maximum size reduction. Carbon black is notable among solids for its huge surface area (as much as 18 acres a pound for some types); the activity of its surface accounts for its outstanding ability to increase the strength and abrasion resistance of rubber. The capacity of activated carbon to adsorb molecules of gases is due to this factor. Surface area is measured most accurately by nitrogen adsorption techniques.

surface chemistry. The observation and measurement of forces acting at the surfaces of gases, liquids, and solids or at the interfaces between them. This includes the surface tension of liquids (vapor pressure, solubility); emulsions (liquid/liquid interfaces); finely divided solid particles (adsorption, catalysis); permeable membranes and microporous materials; and biochemical phenomena such as osmosis, cell function and metabolic mechanisms in plants and animals. Surface chemistry has many industrial applications, a few of which are air pollution; soaps and synthetic detergents; reinforcement of rubber and plastics; behavior of catalysts; color and optical properties of paints; aerosol sprays of all types; monolayers and thin films, both metallic and organic. Outstanding names in the development of this science are Graham, Freundlich and W. Ostwald in the 19th century, and Harkins, Langmuir, LaMer, and McBain in the 20th. See also colloid chemistry.

surface tension. In any liquid, the attractive force exerted by the molecules below the surface upon those at the surface/air interface, resulting from the high molecular concentration of a liquid compared to the low molecular concentration of a gas. An inward pull, or internal pressure, is thus created which tends to restrain the liquid from flowing. Its strength varies with the chemical nature of the liquid. Polar liquids have high surface tension (water = 73 dynes/cm at 20°C); nonpolar liquids have much lower values (benzene = 29 dynes/cm, ethyl alcohol = 22.3 dynes/cm); thus they flow more readily than water. Mercury, with the highest surface tension of any liquid, (480 dynes/cm), does not flow, but disintegrates into droplets. See also interface; surface-active agent.

surfactant. See surface-active agent.

"Surfactol."[202] Trademark for a series of castor oil-derived nonionic surfactants.
Uses: Emulsifiers; defoamers; plasticizers; solubilizers for oils, dyes; lubricants; in emulsion paints, pigment dispersions, cosmetics and polishes.

"Surfaseptic."[233] Trademark for synthetic molding resins containing germicides.

"Surfex."[244] Trademark for a high purity precipitated calcium carbonate (surface-coated).
Uses: Paints, plastics, rubber, inks.

"Surflo."[236] Trademark for a series of bactericides and scale inhibitors. "Surflo-B11" is a film-forming amine that acts as a corrosion inhibitor and bactericide in low pH water-base drilling muds. "Surflo-B33" is a bactericide for treating water-base packer fluids or oil field waters.

"Surfonic."[137] Trademark for a series of nonionic, biodegradable, surface-active agents, described as alkoxy-polyethoxy-ethanols, the ethylene oxide adducts of primary or straight-chain alcohols.
Uses: Wetting agent; detergent; emulsifier; biodegradable intermediate for sulfation purposes.

"Surfpac."[233] Trademark for biological oxidation media, based on saran; used in aerobic treatment of wastewaters.

"Surfy-nol."[144] Trademark for a group of organic surface-active agents.
Properties: Acetylenic alcohols or glycols or their ethoxylated derivatives; waxy or powdered solids, or liquids; nonfoaming, nonionic. For example, "Surfynol" 61 is 3,5-dimethyl-1-hexyne-3-ol.
Containers: 55-gal phosphatized steel drums, except 82 grade (44-gal fiber drum), 5-gal and 1-gal cans.
Uses: Wetting and foam suppression; rinsing aids; viscosity reduction; detergent formulations; penetrating agents.

"Surlyn."[28] Trademark for a group of ionomer resins (q.v.).
Properties: ("Surlyn" A): Thermoplastic produced as a granular material; flexible, transparent, grease-resistant; very light-weight but tough. Izod impact strength 5.7–14.6 ft-lb/in (higher than any other polyolefin); tensile strength 3,500–5,500 psi; elongation 300–400%; softening point 160°F. Insoluble in any commercial solvent; subject to slow swelling by hydrocarbons, to slow attack by acids.
Uses: Coatings; packaging films; products made by injection or blow molding, or by thermoforming.

SUS. Abbreviation for Saybolt Universal seconds. See Saybolt Universal viscosity.

suspension. A system in which very small particles (solid, semisolid, or liquid) are more or less uniformly dispersed in a liquid or gaseous medium. If the particles are small enough to pass through filter membranes, the system is a colloidal suspension (or solution). Examples of solid-in-liquid suspensions are comminuted wood pulp in water, which becomes paper on filtration; the fat particles in milk; and the red corpuscles in blood. A liquid-in-gas suspension is represented by fog or by an aerosol spray. If the particles are larger than colloidal dimensions they will tend to precipitate, if heavier than the suspending medium, or to agglomerate and rise to the surface, if lighter. This can be prevented by incorporation of protective colloids. Polymerization is often carried out in suspension, the product being in the form of spheres or beads. See also solution, colloidal; dispersion; emulsion; colloid chemistry.

"Sustane."[416,505] Trademark for a group of nontoxic formulations based on butylated hydroxyanisole.
Forms available: 3 and 6, liquid formulations; BHA, white tablets; E, water emulsifiable formulation; 1-F and 3-F, flakes.
Uses: Antioxidants for food products, especially animal fats and vegetable oils; in cooked food products such as potato chips, baked goods, nuts, etc.; stabilizer for petroleum wax coatings for food packaging; petroleum products.

sweeten. (1) To add sugar or a synthetic product to foods or beverages to provide a sweet taste (see flavor). (2) To deodorize and purify petroleum products by removing sulfur compounds (see doctor treatment). (3) In industrial slang, to increase the quality of a low-cost product by adding more expensive ingredients.

sweetener, nonnutritive. A food additive, either natural or synthetic, usually having much greater sweetness intensity than sugar (sucrose), but without its caloric value. In some cases they act as enhancers or potentiators of sweetness. Chief among them at present is saccharin, a benzoic acid derivative. The cyclamate group was removed from the market by FDA in 1970 because of animal carcinogenicity, though the evidence in respect to human toxicity is controversial. Efforts to change this ruling have been unsuccessful.

Increasing research in the last decade has developed several new noncaloric sweeteners: the dihydrochalcone group of disaccharides; glycyrrhizin (licorice extract), especially its ammoniated derivative; and the glycoprotein "miracle fruit", indigenous to West Africa but now being cultivated in the U.S., the product to be known as miracle fruit concentrate (MFC). See also flavanone; dihydrochalcone; glycyrrhizin.

sweet oil. See olive oil.

sweet water. (1) The glycerin and water mixture obtained when fats are split (or hydrolyzed) with water to give fatty acid and glycerin. (2) The washings from char used in sugar refining. (3) In engineering terminology, plain water cooled to just above the freezing point and used to preserve milk and other food products.

swep. Generic name for methyl 3,4-dichlorocarbanilate (methyl N-3,4-dichlorophenylcarbamate) $CH_3OOCNHC_6H_3Cl_2$.
Properties: Crystals; m.p. 113°C. Insoluble in water and kerosine; soluble in acetone and dimethylformamide.
Hazard: May be toxic.
Use: Herbicide.

"Sylfat."[296] Trademark for high-purity tall oil fatty acids used in protective coatings, soaps, detergents, disinfectants, intermediates, and flotation chemicals.

"Sylflex."[149] Trademark for compositions used in the preparation of leather to impart permanent water repellence.

"Sylgard."[149] Trademark for a series of silicone resin encapsulants used in electronic assemblies.

"Sylkyd."[149] Trademark for organosiloxy compositions for use as intermediates in the production of resins and formulating varnishes, paints, and enamels.

"Sylmer."[149] Trademark for silicone compositions used in the finishing textile fabrics, and in the preparation and treatment of leather to impart permanent water repellence.

"Syl-off."[149] Trademark for a series of stick-proof silicone coatings for paper. Coatings are unaffected by melts up to 275°F.

"Syloid."[241] Trademark for a series of micron-sized silicas. Used as flatting agents, lacquer, baking finishes, inks, paper reproduction, adhesives, pharmaceuticals, insulation.

"Sylvamix."[296] Trademark for an internal additive for concrete masonry. A noncorrosive, nonflammable resin derivative of tall oil, developed principally to reduce water absorption, improve surface texture and inhibit efflorescence in all types of concrete masonry units and in mortar joints.

"Sylvaros."[296] Trademark for high purity tall oil rosins used in sizing of paper and manufacture of resins for protective coatings and food containers.

"Sylvatal."[296] Trademark for a refined tall oil.

sylvic acid. See abietic acid.

sylvite (sylvine) KCl. A natural potassium chloride. Contains about 43% potassium chloride, 57% sodium chloride, sometimes with up to 0.26% bromide.
Properties: Colorless or white, bluish or reddish in color; streak, white; vitreous luster. Resembles rock salt in appearance. Sp. gr. 1.97–1.99; Mohs hardness 2.
Occurrence: West Texas; New Mexico; Europe.
Use: Major source of potassium compounds in the United States; used for fertilizers.

symmetrical compound. A compound derived from benzene in which the hydrogen atom on three alternate carbon atoms has been replaced with another element or group, for example, at the 1, 3, and 5 positions.

symmetry. Arrangement of the constituents of a molecule in a definite and continuously repeated space pattern or coordinate system. It is described in terms of three parameters called elements of symmetry, namely: (1) The center of symmetry, around which the constituent atoms are located in an ordered arrangement; there is only one such center in a molecule, which may or may not be an atom. (2) Planes of symmetry, which represent division of a molecule into mirror-image segments. (3) Axes of symmetry, represented by lines passing through the center of symmetry; if the molecule is rotated it will have the same position in space more than once in a complete 360° turn, e.g., the benzene molecule, with six axes of symmetry, requires a 60° rotation to return to its identical position. See also stereochemistry; asymmetry.

"Symmetrel"[28] (amantadine hydrochloride). Trademark for a synthetic antiviral drug specific for A_2 type virus infection (one type of Asian flu). It is taken orally, and acts by preventing the virus from entering cells. It is inactive against B type viruses. Reported to be effective in treatment of Parkinson's disease.

"Sympatol" Tartrate.[162] Trademark for synephrine tartrate.

synaptase. See emulsin.

"Synasol."[214] Trademark for solvent composed of 100 gal of S.D. 1 ethanol denatured with 1 gal of methyl isobutyl ketone, 1 gal of ethyl acetate (87%), and 1 gal of aviation gasoline.
Uses: Wherever an alcohol-type solvent is required, except for anti-freeze, where its use is prohibited by the Bureau of Internal Revenue.

syndet. Abbreviated form for synthetic detergent. See detergent.

syndiotactic. A type of polymer molecule in which groups of atoms that are not part of the backbone structure are located in some symmetrical and recurring fashion above and below the atoms in the backbone chain, when the latter are arranged so as to be in a single plane. See polymer, stereospecific.

```
                    R
   H2  H   H2  O   H2  H
   |   |   |   |   |   |
 — C — C — C — C — C — C —
       |       |       |
       O       H       O
       R               R
```

Syndiotactic polymer

"Synektan."[309] Trademark for a series of synthetic tanning materials.

synephrine tartrate (para-methylaminoethanolphenol tartrate; 1-(para-hydroxyphenyl)-2-methylaminoethanol tartrate) $(HOC_6H_4CHOHCH_2NHCH_3)_2C_4H_6O_6$.
Properties: White crystals; freely soluble in water and alcohol; m.p. 182–85°C.
Use: Medicine.

syneresis. The contraction of a gel on standing, with exudation of liquid. The separation of serum from a blood clot, or of the whey in milk souring or cheese making are examples.

synergism. A chemical phenomenon (the opposite of antagonism) in which the effect of two active components of a mixture is more than additive, i.e., it is greater than the equivalent volume or concentration of either component alone. A substance that induces this result when added to another substance is called a synergist.

"Synkamin."[330] Trademark for 4-amino-2-methyl-1-naphthol, as the hydrochloride, $NH_2(CH_3)C_{10}H_5OH \cdot HCl$. Used in medicine.

"Synoca."[108] Trademark for a group of cleaning compounds, including wetting agents, detergents or alkaline hexametaphosphate.
Containers: 100 and 300 lb. drums; 55 gal. drums.
Uses: Continuous conditioning and batch washing of paper machine felts. Paper mill system boil-outs. Deinking of secondary fiber.

"Synotol."[555] Trademark for a series of alkanolamides of fatty acids. Used as foam builders, viscosity improvers, wetting agents, detergents, and opacifiers.

syntan. A synthetic organic tanning material made from phenolsulfonic acids and formaldehyde. Used for both chrome and vegetable tanned leathers. See also tanning.

synthane process. See gasification.

"Synthenol."[64] Trademark for a series of dehydrated castor oils with fast, controllable rate of bodying, excellent color retention, and water resistance.
Uses: Varnishes, enamels, house paints.

synthesis. Creation of a substance which either duplicates a natural product or is a unique material not found in nature, by means of one or more chemical reactions, or (for elements) by a nuclear change. High temperatures and pressures as well as a catalyst are usually required. Though syntheses are more readily achieved with organic compounds because of the great versatility of the carbon atom, extremely important syntheses of inorganic compounds were made in the early years of chemistry. A list of noteworthy syntheses, with approximate dates, would include the following:

Inorganic	
sulfuric acid	
(chamber)	1746
(contact)	1890
soda ash (LeBlanc)	1800
(Solvay)	1861
ammonia	
(Haber-Bosch)	1912
plutonium	1940

Organic	
urea (Wohler)	1828
mauveine (Perkin)	1857
Celluloid (Hyatt)	1869
ethylbenzene	
(Friedel-Crafts)	1877
rayon	
(Chardonnet)	1910
phenolic resins	
(Baekeland)	1910
neoarsphenamine	
(Ehrlich)	1910
aldehydes, alcohols	
(Oxo synthesis)	1920
insulin (Banting)	1925
methanol	1927
neoprene	
(Nieuwland)	1930
nylon (Carothers)	1935
SBR rubber	1940
polyisoprene	1950

The tremendous proliferation of synthetic materials in recent years, especially in the high-polymer field, was made possible by the increasingly sophisticated use of catalysts, particularly the Ziegler and Natta stereospecific type.

Synthesis of elements has also occurred; since 1940 all the transuranic elements from 93 to 105, as well as a vast array of radioisotopes of natural elements, have been created by nuclear bombardment of various types.

See also following entries; resin, synthetic; semisynthetic.

synthesis gas. Any of several gaseous mixtures used for synthesizing a wide range of compounds, both organic and inorganic, especially ammonia. Such mixtures result from reacting carbon-rich substances with steam (steam reforming) or steam and oxygen (partial oxidation); they contain chiefly carbon monoxide and hydrogen, plus low percentages of carbon dioxide and usually of nitrogen (less than 2.0%). The organic source materials may be natural gas, methane, naphtha, heavy petroleum oils, or coke (coal). The reactions are nickel-catalyzed steam-cracking (reforming) of methane or natural gas ($CH_4 + H_2O \rightarrow CO + 3H_2$); partial oxidation of methane, naphtha, or heavy oils; and (especially in view of the petroleum shortage) the water-gas reaction with coke ($C + H_2O \rightarrow CO + H_2$). With transition-metal catalysts synthesis gas yields alcohols, aldehydes, acrylic acid, etc., and has many useful reactions with acetylene; it is the basis of the Oxo and Fischer-Tropsch reactions. Synthesis gas from which the CO has been removed and adjusted to a ratio of 3 parts hydrogen to 1 part nitrogen is used for ammonia synthesis (nitrogen fixation). Processing with a catalyst at high temperatures and pressures yields ammonia. See also ammonia, anhydrous; Oxo process; Haber; water gas; gasification.
Note: Do not confuse with synthetic natural gas.

synthetic natural gas (SNG). Any manufactured fuel gas of approximately the same composition and Btu value as that obtained naturally from oil fields (about 85% methane and 15% ethane). There are two major methods of synthesis involving catalysts, high temperature and high pressure: (1) direct hydrogenation of coal and (2) methanation of synthesis gas (q.v.), which is obtained by hydrogenolysis of coal or steam-reforming of naphtha or similar petroleum distillate. Other methods involve pyrolysis of solid wastes and extraction from oil shale and manures. See also gasification; hydrogenolysis.

synethetic resin. See resin, synthetic.

synthine process. See Fischer-Tropsch process.

"Synthrapol."[325] Trademark for a series of ethylene oxide condensation products. Used as nonionic all-purpose surface-active agents.

"Synthravon."[325] Trademark for a series of nonsubstantive softeners used as textile napping assistants, lubricants, and resin-finishing plasticizers.

synton. Any of several isomers of prostaglandin from which prostaglandin analogs and intermediates may be derived.

"Synvarac."[495] Trademark for a group of resinous compositions specifically designed for use in the construction of chemically resistant equipment. Principal components are graphite and a thermosetting resin.

"Synvaren."[495] Trademark for phenol-formaldehyde resins in liquid form.

"Synvarex."[495] Trademark for specialty resins derived from various forms of soya, linseed, fish oil, lauric, coconut, fatty acid type flat, medium and short copolymer alkyds and from PE ester of tall oil, and nonphenolic spar.
Uses: Paints and enamels.

syrosingopine $C_{35}H_{42}N_2O_{11}$. Methyl reserpate ester of syringic acid ethyl carbonate. An analog of reserpine.
Properties: White or slightly yellowish powder. Practically insoluble in water; slightly soluble in methanol; soluble in chloroform and acetic acid.
Grade: N.F.
Use: Medicine.

syrup. Commercial name for an aqueous solution of cane or beet sugar (sucrose) sold in tank car lots to manufacturers of candy, soft drinks, soda-fountain goods, etc. U.S.P. grade is an aqueous solution of cane sugar (85g/100 ml). A viscous liquid with sp. gr. about 1.313.

"Systox."[181] Trademark for demeton (q.v.).

"Syton."[58] Trademark for a series of colloidal silicas dispersed in water. Used as anti-soil and anti-slip agents.

T

T Symbol for tritium (q.v.).

2,4,5-T. Abbreviation for 2,4,5-trichlorophenoxyacetic acid.

2,4,6,-T. Abbreviation for 2,4,6-trichlorophenol.

Ta Symbol for tantalum.

table salt. See sodium chloride.

tabun (dimethylphosphoramidocyanidic acid, ethyl ester) $(CH_3)_2NP(O)(C_2H_5O)(CN)$.
Properties: Liquid, f.p. −50°C; b.p. 240°C; flash point 172°F; sp. gr. 1.4250 (20/4°C); readily soluble in organic solvents; miscible with water but readily hydrolyzed. Destroyed by bleaching powder, which, however, generates cyanogen chloride. Combustible.
Hazard: Highly toxic by inhalation; cholinesterase inhibitor. Not manufactured.

"TAC."[329] Trademark for tested additive chemical items, satisfactory for food additives and medical uses.

tachysterol $C_{28}H_4O$. A sterol.
Properties: Oil; levorotatory; insoluble in water, soluble in most organic solvents. Protect from air.
Use: Medicine, as the dihydrotachysterol.

tackiness (tack). Property of being sticky or adhesive.

taconite. A low-grade iron ore consisting essentially of a mixture of hematite and silica. It contains about 25% iron. Found in the Lake Superior district and western states.

tacticity. The regularity or symmetry in the molecular arrangement or structure of a polymer molecule. Contrasts with random positioning of substituent groups along the polymer backbone, or random position with respect to one another of successive atoms in the backbone chain of a polymer molecule.
See polymer, stereospecific; isotactic.

"Tag."[253] Trademark for a type of fungicide containing phenyl mercuric acetate.
Hazard: See mercury compounds.

"Tagathen."[315] Trademark for chlorothen citrate (q.v.).

tagged atom. A radioactive isotope used in tracing the behavior of a substance in both biochemical and engineering research, e.g. carbon-14 or iodine-131. See tracer; label (2).

Tag Open Cup. See TOC.

tailings. (1) In flour-milling, the product left after grinding and bolting middlings (q.v.). (2) Impurities remaining after the extraction of useful minerals from an ore. (3) In general, any residue from a mechanical refining or separation process.

"Tajmir."[575] Trademark for a nylon-4 resin and fiber, a synthetic based on 2-pyrrolidone. It has excellent dyeability, high moisture regain, and water absorption capability exceeding that of other synthetic fibers and equivalent to cotton. It is also antistatic and abrasion-resistant. It is used in all types of garments and clothing, as well as athletic shoes and leather substitutes.

"Takalab TLM."[212] Trademark for a product containing diastatic and proteolytic enzymes.
Properties: Dry, fine, white powder, fully water-soluble, nonhazardous, nonflammable; optimum pH for diastatic reaction 6.5–7.0; for proteolytic reaction 7.0–8.4; optimum temperature 45°C.
Use: Digestion of albumin and starch-containing stains in commercial drycleaning plants.

"Takamine" Pectinase.[212] Trademark for an enzyme system which hydrolyzes and depolymerizes pectins over a wide range of conditions. Used to produce clear fruit juices, wine, and jellies.

"Takatabs."[212] Trademark for sodium *d*-erythorbate in tablet form.

"Take-Hold."[1] Trademark for a completely water-soluble mixture of ammonium and potassium phosphates giving a substantially neutral solution.
Use: Agricultural starter solution mix for transplanting set-outs.

"Takimerse."[212] Trademark for a product containing diastatic and proteolytic enzymes.
Properties: Dry, fine white powder, water-soluble; nonhazardous, nonflammable. Optimum pH 7.0–8.0; optimum temperature 40–45°C.
Uses: Digestion of albumin and starch-containing stains by immersion, in commercial drycleaning plants.

"Talase."[212] Trademark for a special enzyme formulation having both amylase and protease activity. Used in textile industry.

talc (talcum; mineral graphite; steatite) $Mg_3Si_4O_{10}(OH)_2$ or $3MgO \cdot 4SiO_2 \cdot H_2O$. A natural hydrous magnesium silicate. Compact massive varieties may be called steatite in distinction from the foliated varieties, which are called talc. Soapstone is an impure variety of steatite.
Properties: White, apple green, gray powder; luster pearly or greasy; feel greasy; Mohs hardness 1–1.5 (may be harder when impure). High resistance to acids, alkalies and heat. Sp. gr. 2.7–2.8.
Grades: Crude; washed; air-floated; U.S.P. fibrous (99.5%, 99.95%).
Containers: 50-lb paper bags; 200-lb multiwall bags; bulk.
Hazard: Moderate by inhalation. Tolerance, 20 million particles per cubic foot in air.
Uses: Ceramics; cosmetics and pharmaceuticals; filler in rubber, paints, soap, putty, plaster, oilcloth; abherent; dusting agent; lubricant; paper; slate pencils and crayons; electrical insulation.
See also magnesium silicate.

tall oil (tallol; liquid rosin). A mixture of rosin acids, fatty acids, and other materials obtained by acid treatment of the alkaline liquors from the digesting (pulping) of pine wood. Combustible; low toxicity. Flash point 360°F.
Derivation: The spent black liquor from the pulping process is concentrated until the sodium salts (soaps) of the various acids separate out and are skimmed off. These are acidified by sulfuric acid. Composition and properties vary widely, but average 35–40% rosin acids, 50–60% fatty acids.
Grades: Crude; refined.
Containers: Drums; tank cars; tank trucks.
Uses: Paint vehicles; source of rosin; alkyd resins; soaps, cutting oils and emulsifiers; driers; flotation agents; oil-well drilling muds; core oils; lubricants and greases; asphalt derivatives; rubber reclaiming; synthesis of cortisone and sex hormones.

tallow. An animal fat containing C_{16} to C_{18}.
Properties: The solidifying points of the different tallows are as follows; from 20–45°C for horse fat; 27–38°C for beef tallow; 54–56°C for stearin and oleo; 32–41°C for mutton tallow. Combustible; nontoxic. Sp. gr. 0.86; refractive index (40°C) 46–49 (Zeiss); iodine value 193–202; flash point 509°F. Combustible.
Derivation: Extracted from the solid fat or "suet" of cattle, sheep or horses by dry or wet rendering.
Chief constituents: Stearin, palmitin and olein.
Grades: Edible; inedible; beef tallow; mutton tallow; horse fats, acidless; edible, extra.
Containers: 50-lb tierces; 375-lb barrels; tank trucks; tank cars.
Uses: Soap stock; leather dressing; candles; greases; manufacture of stearic and oleic acids; animal feeds; abherent in tire molds.

"Talwin."[162] Trademark for pentazocine (q.v.).

"Tamarax."[342] Trademark for extract of tamarind rind for flavoring.

"Tamax."[408] Trademark for a premium grade, bonded mullite refractory available as brick and special shapes. Manufactured from calcined Indian kyanite or other high purity aluminum silicate materials. Used in the construction of glass-melting furnaces.

"Tamed Iodine."[284] Trademark for iodophors (iodine complexed with surface-active agents that have detergent properties).

"Tamol."[23] Trademark for anionic, polymer-type dispersing agents. Supplied as light-colored powders or aqueous solutions. Effective dispersant for aqueous suspensions of insoluble dyestuffs, polymers, clays, tanning agents and pigments.
Use: Manufacture of dyestuff pastes; textile printing and dyeing; pigment dispersion in textile backings, latex paints and paper coatings; retanning and bleaching of leather; dye resist in leather dyeing; dispersion of pitch in paper manufacture; pre-floc prevention in the manufacture of synthetic rubber.

"Tamul."[408] Trademark for a synthetic mullite produced by sintering selected grades of bauxite. Used in the manufacture of a wide variety of special refractory materials. Also the brand name of some of the products manufactured from "Tamul" synthetic mullite grain.

"Tanak" MRX.[57] Trademark for melamineformaldehyde resin tanning agent used to make pure white leather and for bleaching and filling chrome leather.

"Tanamer."[57] Trademark for sodium polyacrylate adhesive for use during the drying of leather. See acrylate.

"Tanarc."[250] Trademark for a material which provides a source of titanium dioxide in welding electrode coatings. Also serves as a ceramic colorant.

"Tanasol."[309] Trademark for a series of synthetic, organic sulfonic acid condensation products used in tanning of hides and skins.

"Tandearil."[219] Trademark for a brand of oxyphenbutazone.

tangerine oil. See citrus peel oil.

tankage (animal tankage; tankage, rough ammoniate). The product obtained in abattoir by-product plants from meat scraps and bones, which are boiled under pressure and allowed to settle. The grease is removed from the top and the liquor drawn off. The scrap is then pressed, dried and sold for fertilizer.
Grades: Based on per cent of ammonia and bone phosphate. A medium grade has 10% ammonia and 20% bone phosphate. Concentrated tankage has had the boiled down tank liquor and press water added to it before drying, and runs about 15–17% ammonia.
Hazard: Flammable; may ignite spontaneously.
Shipping regulations: (Air) Not acceptable. (Rail) Yellow label. Not accepted by express.

tankage, garbage (tankage fertilizer). Garbage treated with steam under pressure, the water and some of the grease removed by pressing and further grease removed by solvent extraction. Contains from 3–4% ammonia, 2–5% phosphoric acid and 0.50–1.00% potash.
Hazard: Flammable; may ignite spontaneously.
Use: Fertilizer.
Shipping regulations: (Air) Not acceptable. (Rail) Yellow label. Not accepted by express.

"Tannex."[236] Trademark for a full substitute for quebracho in the thinning of drilling muds. Used in mud thinning when exposed to the high bottom-hole temperatures of deep wells. May be added directly to the mud system, but gives better results when dissolved in caustic soda solution before adding to the mud system.

tannic acid (gallotannic acid; tannin) $C_{76}H_{52}O_{46}$, described as a penta-(m-digalloyl)-glucose. Natural substance widely found in nutgalls, tree barks, and other plant parts. Tannins are known to be gallic acid derivatives. A solution of tannic acid will precipitate albumin.

Tannins are classified according to their behavior on dry distillation into two groups, (1) condensed tannins, which yield catechol, and (2) hydrolyzable tannins, which yield pyrogallol. Group (2) comprises two groups on the basis of its products of hydrolysis, glucose and (a) ellagic acid or (b) gallic acid.
Properties: Lustrous, faintly yellowish, amorphous powder, glistening scales or spongy mass; darkens on exposure to air; odorless; strong astringent taste; m.p., decomposes at 210°C; soluble in water, alcohol, and acetone; almost insoluble in benzene, chloroform, ether, and petroleum ether. Combustible. Flash point 390°F, autoignition temp. 980°F.
Derivation: Extraction of powdered nutgalls with water and alcohol.
Grades: Technical; C.P.; N.F.; fluffy; F.C.C.
Containers: Barrels; drums.

Hazard: Moderately toxic by ingestion and inhalation.
Uses: Chemicals (tannates, gallic acid, pyrogallic acid, hydrosols of the noble metals); alcohol denaturant; tanning; textiles (mordant and fixative); electroplating; galvanoplastics (gelatin precipitant); clarification agent in wine manufacture, brewing and foods; writing inks; pharmaceuticals; deodorization of crude oil; photography; paper (sizing, mordant for colored papers); treatment of minor burns.

tanning. The preservation of hides or skins by use of a chemical, which (1) makes them immune to bacterial attack, (2) raises the shrinkage temperature, and (3) prevents the collagen fibers from sticking together on drying, so that the material remains porous, soft, and flexible. Vegetable tanning is used mostly for sole and heavy-duty leathers. The chief vegetable tannins are water extracts of special types of wood or bark, especially quebracho (q.v.) and wattle (q.v.). The main active constituent is tannic acid (q.v.). The tannins penetrate the skin or hide after long periods of soaking, during which the molecular aggregates of the tannin form cross-links between the polypeptide chains of the skin proteins; hydrogen bonding is an important factor.

In mineral or chrome tanning, the sulfates of chromium, aluminum and zirconium are used (the last two for white leather); here the reaction is of a coordination nature between the carboxyl groups of the skin collagen and the metal atom. Syntans are also used; these are sulfonated phenol or naphthol condensed with formaldehyde. Condensation products other than phenol having strong hydrogen-bonding power have been developed. Tannage by any method is a time-consuming and exacting process, requiring careful control of pH, temperature, humidity, and concentration factors. See also leather. For further information refer to Tanners' Council, University of Cincinnati, Cincinnati, Ohio.

tannyl acetate. See acetyltannic acid.

"Tanolin."[582] Trademark for basic chromium sulfate.
Uses: Chrome tanning of leather; colors; compounding.

"Tansul 7" Clay.[236] Trademark for beneficiated magnesium montmorillonite. Used as a flocculating agent and selective absorbent primarily in the beverage industry.

"Tansul" Enzyme.[236] Trademark for a mixture of plant-derived proteolytic enzymes in white powder form or clear liquid form with a hydrolysis potency of 60–75 Activity Units/gram. Used as a protein hydrolyzing agent in beverages.

tantalic acid anhydride. See tantalum oxide.

tantalic chloride. See tantalum chloride.

tantalite (Fe, Mn) $(TaCb)_2O_6$. The most important ore of tantalum, found in Canada, Africa, Brazil, Malaysia. When Cb content exceeds that of Ta, the ore is called columbite.

tantalum Ta Element of atomic number 73 in group VB of the periodic system. Atomic weight 180.9479; valences 2, 3, 5. No stable isotopes.
Properties: (a) Black powder. (b) Steel-blue-colored metal when unpolished; nearly a platinum-white color when polished. Soluble in fused alkalies; insoluble in acids except hydrofluoric and fuming sulfuric acids. Sp. gr. (a) 14.491; (b) 16.6 (worked metal); m.p. 2996°C; b.p. 5425 ± 100°C. Tensile strength of drawn wire may be as high as 130,000 psi. Refractive index 2.05. Expansion coefficient 8×10^{-6} over range 20–1500°C. Electrical resistance 13.6 microhm-cm (0°C), 32.0 (500°C).
Derivation: From tantalum potassium fluoride, by heating in an electric furnace, by sodium reduction, or by fused salt electrolysis. The powdered metal is converted to a massive metal by sintering in a vacuum. Foot-long crystals can be grown by arc-fusion.
Corrosion resistance: 99.5% pure tantalum is resistant to all concentrations of hot and cold sulfuric (except concentrated boiling), hydrochloric, nitric and acetic acids, hot and cold dilute sodium hydroxide, all dilutions of hot and cold ammonium hydroxide, mine and sea waters, moist sulfurous atmospheres, aqueous solutions of chlorine.
Forms and grades available: Powder 99.6% pure; sheet; rods; wire; ultrapure; single crystals.
Hazard: Dust or powder may be flammable. Moderately toxic by inhalation. Tolerance, 5 mg per cubic meter of air.
Uses: Capacitors; chemical equipment; dental and surgical instruments; rectifiers; vacuum tubes; furnace components; high-speed tools; catalyst; getter alloys; sutures and body implants; electronic circuitry; thin-film components.

tantalum alcoholate (pentaethoxytantalum) $(C_2H_5O)_5Ta$.
Uses: Catalyst; intermediate for pure tantalates; preparing thin dielectric films.

tantalum carbide TaC. Hard, heavy, refractory crystalline solid; m.p. 3875°C; b.p. 5500°C; sp. gr. 14.5; hardness 1800 kg/sq mm; resistivity 30 microohm cm (room temp); extremely resistant to chemical action except at elevated temperatures.
Derivation: Tantalum oxide and carbon heated at high temperatures.
Uses: Cutting tools and dies; cemented carbide tools.

tantalum chloride (tantalic chloride; tantalum pentachloride) $TaCl_5$.
Properties: Pale yellow, crystalline powder. Highly reactive. Decomposed by moist air. Keep well stoppered! Sp. gr. 3.7; b.p. 242°C; m.p. 221°C. Soluble in alcohol and potassium hydroxide.
Grade: Technical.
Uses: Chlorination of organic substances; intermediate; production of pure metal.

tantalum disulfide TaS_2. Black powder or crystals; m.p. above 3000°C; insoluble in water. Available in 40 micron size. Used as a solid lubricant.

tantalum nitride TaN.
Properties: Hexagonal brown, bronze or black crystals; sp. gr. 16.3; m.p. 3310–3410°C; insoluble in water; slightly soluble in aqua regia, nitric acid, hydrofluoric acid.
Grade: Technical; powder.

tantalum oxide (tantalic acid anhydride; tantalum pentoxide) Ta_2O_5.
Properties: Rhombic, crystalline prisms; sp. gr. 7.6; m.p. 1800°C; insoluble in water and acids except hydrofluoric.
Derivation: From tantalite, by removal of other metals.
Grades: Technical; optical (99.995% pure); single crystals.
Uses: Production of tantalum; "rare-element" optical glass; intermediate in preparation of tantalum car-

bide; piezoelectric, maser and laser applications; dielectric layers in electronic circuits.

tantalum pentachloride. See tantalum chloride.

tantalum pentoxide. See tantalum oxide.

tantalum potassium fluoride (potassium tantalum fluoride; potassium fluotantalate; K_2TaF_7.

Properties: White, silky needles. Slightly soluble in cold water, quite soluble in hot water.

Hazard: Toxic by inhalation. Tolerance (as F), 2.5 mg per cubic meter of air.

Uses: Intermediate in preparation of pure tantalum.

tantiron. A ferrous alloy containing 84.87% iron, 13.5% silicon, 1% carbon, 0.4% manganese, 0.18% phosphorus, 0.05% sulfur. It is resistant to acids; used for chemical equipment.

"Tanz."[48] Trademark for an ammonium lignin sulfonate used as an extender for vegetable tannin extracts.

"Tapazole."[100] Trademark for methimazole, U.S.P.

tapioca. A product rich in starch, obtained from the tuberous roots of the cassava plant. It is used both directly as a food and as a thickening agent in emulsions for food-products and industrial applications (lithographic inks). In many uses it is a satisfactory substitute for gum arabic.

TAPPI. Abbreviation for Technical Association of the Pulp and Paper Industry (q.v.).

tar. See coal tar; cigarette tar; pine tar; tar, wood.

tar acid. Any mixture of phenols present in tars or tar distillates and extractable by caustic soda solutions. Usually refers to tar acids from coal tar and includes phenol, cresols, and xylenols. When applied to the products from other tars it should be qualified by the appropriate prefix, e.g., wood-tar acid, lignite tar acid, etc.

Properties: Soluble in alcohol and coal-tar hydrocarbons. Combustible.

Grades: 15–18%, 25–28% and 50–53% phenol.

Containers: Drums; tank cars.

Hazard: Toxic by ingestion, inhalation, and skin absorption. Strong irritant to tissue.

Uses: Wood preservative; insecticide for cattle and sheep dipping, (dip oils or sheep dip); manufacture of disinfectants.

tar base. A basic nitrogen compound derived from coal tar, such as pyridine, picoline, lutidine, and quinoline.

tar camphor. See naphthalene.

tar camphor, chlorinated. See chloronaphthalene.

"Tar-Cel."[235] Trademark for 72% "Dry Liquid" concentrate of "Tarene," a pine tar type plasticizer.

Properties: Light brown, free-flowing powder; sp. gr. 1.41.

Uses: Plasticizer and softener for plastics and rubber.

tar, dehydrated (tar, refined).

Properties: Dark brown, thick, viscid liquid; combustible.

Derivation: Tar from which the water has been driven off.

Grade: Technical.

Containers: Barrels; tank cars.

Hazard: Toxic; strong irritant; absorbed by skin.

Uses: Water-proofing compounds; roads; medicine.

"Tarene."[235] Trademark for mixed naval stores (pine tar type).

Properties: Non-toxic, dark brown viscous liquid; sp. gr. 0.99–1.02; ash 0.5%, benzene-soluble. Combustible.

Uses: Plasticizer and softener.

tarnish. A reaction product that occurs readily at room temperature between metallic silver and sulfur in any form. The well-known black film that appears on silverware results from reaction between atmospheric sulfur dioxide and metallic silver, forming silver sulfide. It is easily removable with a cleanig compound and is not a true form of corrosion. Plating with a mixture of silver and indium will increase tarnish resistance. Gold will also tarnish in the presence of a high concentration of sulfur in the environment.

tar oil. See creosote, coal-tar.

tar-oil, wood (pine-tar oil).

Properties: Almost colorless liquid when freshly distilled; turns dark reddish-brown; strong odor and taste. Sp. gr. 0.862–0.872. Soluble in ether, chloroform, alcohol and carbon disulfide. Flash point 144°F (closed cup). Combustible. Low toxicity.

Derivation: Obtained by the destructive distillation of the wood of Pinus palustris.

Method of purification: Rectification.

Grades: Technical; rectified; refined.

Containers: Drums; glass bottles; tank cars.

Uses: Paints; waterproofing paper; rubber reclaiming; varnishes; stains; ore flotation; cattle dips; insecticides.

"Tarpine."[79] Trademark for a pine tar used as a tackifier in rubber compounding and in rubber reclaiming.

tarragon oil. See estragon oil.

tar, refined. See tar, dehydrated.

tar sands. See oil sands.

"Tartan."[158] Trademark for a synthetic turf or surfacing material for athletic fields. Composed of specially compounded synthetic resins on asphalt base.

tartar, cream of. See potassium bitartrate.

tartar emetic. See antimony potassium tartrate.

tartaric acid (dihydroxysuccinic acid).

$HOOC(CHOH)_2COOH$.

Properties: Colorless, transparent crystals or white crystalline powder; odorless; acidic taste; stable in air. Soluble in water, alcohol, and ether. Sp. gr. 1.76; m.p. 170°C. Nontoxic. It has unusual optical properties. The molecule has two asymmetric carbons which result in three isomeric forms; the natural *D*-form and the corresponding *L*-form, which are optically active, as well as the inactive or meso form, as shown:

```
     COOH          COOH          COOH
      |             |             |
   H—C—OH       HO—C—H        H—C—OH
      |             |             |
  HO—C—H         H—C—OH       H—C—OH
      |             |             |
     COOH          COOH          COOH
      L             D            Meso
```

The occurrence of some DL-tartaric acid along with natural D-tartaric acid in the wine industry is explained on the basis of partial racemization.

Derivation: Occurs naturally in wine lees; made synthetically from maleic anhydride and hydrogen peroxide. A new process involving enzymatic reaction with a succinic acid derivative has been developed.
Grades: Technical; C.P.; crystalline; powder; granular; N.F.; F.C.C
Containers: Bags; drums; carlots.
Uses: Chemicals (cream of tartar, tartar emetic, acetaldehyde); sequestrant; tanning; effervescent beverages; baking powder; fruit esters; ceramics; galvanoplastics; photography (printing and developing; light-sensitive iron salts); textile industry; silvering mirrors; coloring metals.

tartrazine (FD & C Yellow No. 5) $C_{16}H_9N_4O_9S_2$. 3-Carboxy-5-hydroxy-1-para-sulfophenyl-4-para-sulfophenylazopyrazole trisodium salt. C.I. No. 19140.
Properties: Bright orange-yellow powder. Soluble in water.
Use: Dye, especially for foods, drugs, and cosmetics.

tar, wood (tar, hardwood).
Properties: Black, syrup-like, viscous fluid.
Derivation: A byproduct of the destructive distillation of wood.
Grades: Technical.
Containers: Tank cars.
Hazard: Flammable, moderate fire risk.
Use: Hardwood pitch; wood creosote; heavy, high-boiling wood oils; wood preserving oils; paint thinners.

"Tasil."[408] Trademark for a refractory brick, usually manufactured of calcined Indian kyanite; used in the construction of glass and metallurgical furnaces.

taurine (2-aminoethanesulfonic acid) $NH_2CH_2CH_2SO_3H$. A crystallizable amino acid found in combination with bile acids; its combination with cholic acid is called taurocholic acid.
Properties: Solid; decomposes 300°C; soluble in water; insoluble in alcohol.
Derivation: Isolated from ox bile; organic synthesis.
Uses: Biochemical research; pharmaceuticals; organic synthesis; wetting agents.

taurocholic acid (cholaic acid; cholyltaurine) $C_{26}H_{45}NO_7S$. Occurs as sodium salt in bile. It is formed by the combination of the sulfur-containing amino acid, taurine, and cholic acid (q.v.) as the sodium salt. It aids in digestion and absorption of fats.
Properties: Crystals; stable in air. M.p. 125°C (dec). Freely soluble in water; soluble in alcohol; almost insoluble in ether and ethyl acetate.
Derivation: Isolation from bile.
Uses: Biochemical research; emulsifying agent in foods (not over 0.1%).

tautomerism. A type of isomerism in which migration of a hydrogen atom results in two or more structures, called tautomers. The two tautomers are in equilibrium. For example, acetoacetic ester has the properties of both an unsaturated alcohol and a ketone. The tautomers are called enol and keto. See also enol; isomer (1).

"Taycor."[408] Trademark for a group of corundum base refractory products. Manufactured of high purity alumina which has been sintered to form tabular corundum.

Tb Symbol for terbium.

TBH. Abbreviation for technical benzene hexachloride. See 1,2,3,4,5,6-hexachlorocyclohexane. Used as an insecticide.

TBP. Abbreviation for tributyl phosphate (q.v.).

TBT. Abbreviation for tetrabutyl titanate (q.v.).

TBTO. Abbreviation for bis(tributyltin oxide) (q.v.).

Tc Symbol for technetium.

TC. Abbreviation for trichloroacetic acid (q.v.) or its sodium salt.

TCA cycle (tricarboxylic acid cycle; Krebs cycle; citric acid cycle). A series of enzymatic reactions occurring in living cells of aerobic organisms, the net result of which is the conversion of pyruvic acid, formed by anaerobic metabolism of carbohydrates, into carbon dioxide and water. The metabolic intermediates are degraded by a combination of decarboxylation and dehydrogenation. It is the major terminal pathway of oxidation in animal, bacterial, and plant cells. Recent research indicates that the TCA cycle may have predated life on earth and may have provided the pathway for formation of amino acids.

TCB. Abbreviation for tetracarboxybutane (q.v.).

TCC. Abbreviation for Tagliabue Closed Cup, a standard method of determining flash points.

TCDD. See dioxin.

TCP. Abbreviation for tricresyl phosphate (q.v.).

TDE. Generic name for 1,1-dichloro-2,2-bis(para-chlorophenyl)ethane. (tetrachlorodiphenylethane; DDD) $(ClC_6H_4)_2CHCHCl_2$.
Properties: Colorless crystals; m.p. 109–110°C; soluble in organic solvents; insoluble in water; not compatible with alkalies. Combustible.
Derivation: Chlorination of ethanol and condensation with chlorobenzene.
Grade: Technical.
Containers: Fiber drums.
Hazard: Toxic by ingestion, inhalation and skin absorption. Use restricted in some states.
Uses: Similar to DDT; as dusts, emulsions and wettable powders for contact control of leaf rollers and other insects.

TDI. Abbreviation for toluene diisocyanate (q.v.).

TDQP. Abbreviation for trimethyldihydroquinoline polymer (q.v.).

Te Symbol for tellurium.

TEA. (1) Abbreviation for triethanolamine. (2) Abbreviation for triethylaluminum.

TEAC. Abbreviation for tetraethylammonium chloride (q.v.).

technetium Tc Element with atomic number 43 of group VIIB of the periodic system. Atomic weight 98.9062; valences 4, 5, 6, 7; 3 radioactive isotopes of half-life more than 105 years; also several of relatively short half-life, some of which are gamma emitters. Technetium was first obtained by the deuteron bombardment of molybdenum, but since has been found in the fission products of uranium and plutonium.

The chemistry of technetium has been studied by tracer techniques and is similar to that of rhenium and manganese. The free metal is obtained from reactor fission products by solvent extraction followed by crystallization as ammonium pertechnetate, which is reduced with hydrogen. The metal is silver gray in appearance, and melts at 2200°C (4000°F); sp. gr. ca. 11.5. It is slightly magnetic. Compounds of the types

TcO_2, Tc_2O_7, NH_4TcO_4, etc, have been prepared. The pertechnetate ion has strong anticorrosive properties. Technetium and its alloys are superconductors and can be used to create high-strength magnetic fields at low temperature.

Uses: Metallurgical tracer; cryochemistry; corrosion resistance; medicine (radiology).

Technical Association of the Pulp and Paper Industry (TAPPI). A professional group of scientists devoted to the interests of pulp and paper chemistry and technology. Founded in 1915, it has seven sections, each concerned with a specific phase of the industry. It also has eleven local sections which hold monthly meetings. The Association publishes its own journal, as well as industry data sheets, bibliographies, technical monographs on subjects relating to the paper industry, etc. Its headquarters is at 155 E. 44th Street, New York, 10017.

"Technoscents."[188] Trademark for odor modifiers designed to cover a disagreeable odor associated with solvents, cutting oils, paints, lubricating oils, detergents, scrub soaps and powders, fuel oils.

"Tecmangam."[256] Trademark for free-flowing granular solid containing 75–78% manganese sulfate. Used in fertilizers and feeds.

"Tecquinol."[256] Trademark for technical hydroquinone (98.5% min).

"Tecsol."[256] Trademark for solvents consisting of denatured anhydrous and denatured 95% ethyl alcohol.

"Tedlar."[28] Trademark for polyvinylfluoride film.

TEDP. Abbreviation for tetraethyl dithionopyrophosphate (q.v.).

"Teepol" 610.[125] Trademark for a detergent and wetting agent in liquid form based on the sodium salt of secondary alkyl sulfates, of the type $R_1R_2CHOSO_2ONa$, where R_1 is a large alkyl group and R_2 is a relatively small alkyl group.

"Teflon."[28] Trademark for tetrafluoroethylene (TFE) fluorocarbon polymers available as molding and extrusion powders, aqueous dispersion, film, finishes, and multifilament yarn or fiber. The name also applies to fluorinated ethylene-propylene (FEP) resins (q.v.) available in the same forms. The no-stick cookware finishes may be of either type. Fibers are monofilaments made from copolymer of TFE and FEP.

Uses: Packing; bearings; filters; electrical insulation; high-temperature industrial plastics; cooking utensils; plumbing sealants.

See also fluorocarbon polymer.

TEG. Abbreviation for tetraethylene glycol and triethylene glycol.

"Tego."[23] Trademark for thin tissue impregnated with heat-convertible phenol-formaldehyde resin, supplied in rolls. Produces waterproof bond with plywood veneers.

Use: Hot-press bonding of furniture veneers; premium wall panelling.

TEL. Abbreviation for tetraethyl lead (q.v.).

"Teldrin."[71] Trademark for chlorpheniramine maleate (q.v.).

"Telepaque."[162] Trademark for iopanoic acid.

"Telloy."[69] Trademark for finely ground tellurium. Used as a secondary vulcanizing agent for rubber.

"Tellurac."[69] Trademark for tellurium diethyldithiocarbamate $[(C_2H_5)_2NC(S)S]_4Te$.

Properties: Orange-yellow powder; sp. gr. 1.44 ± .03; melting range 108–118°C; soluble in benzene, carbon disulfide, chloroform; slightly soluble in alcohol, gasoline; insoluble in water.

Uses: Primary or secondary (with thiazoles) rubber accelerator.

telluric acid (hydrogen tellurate) $H_2TeO_4 \cdot 2H_2O$, or H_6TeO_6.

Properties: White heavy crystals. Soluble in hot water and alkalies; slightly soluble in cold water. Sp. gr. 3.07; m.p. 136°C.

Derivation: Action of sulfuric acid on barium tellurate.

Hazard: Toxic by ingestion.

Uses: Chemical reagent.

telluric bromide. See tellurium tetrabromide.

tellurium Te A nonmetallic element with many properties similar to selenium and sulfur. Atomic number 52; Group VIA of the periodic system. Atomic weight 127.60; valences 2, 4, 6. Eight stable isotopes. Low toxicity.

Properties: Silvery white, lustrous solid with metal characteristics. Density 6.24 g/cc (30°C); Mohs hardness 2.3; m.p. 450°C; b.p. 990°C. Soluble in sulfuric acid, nitric acid, potassium hydroxide and potassium cyanide solution; insoluble in water. Imparts garlic-like odor to breath; can be depilatory. It is a p-type semiconductor and its conductivity is sensitive to light exposure.

Source: From anode slime produced in the electrolytic refining of copper and lead.

Derivation: Reduction of telluric oxide with sulfur dioxide; by dissolving the oxide in a caustic soda solution and plating out the metal.

Grades: Powder, sticks, slabs and tablets, 99.5% pure; crystals, up to 99.999% pure.

Uses: Alloys (tellurium lead, stainless steel, iron castings); secondary rubber vulcanizing agent; manufacture of iron and stainless steel castings; coloring agent in glass and ceramics; thermoelectric devices; catalysts; with lithium, in storage batteries for space craft.

For further information refer to Selenium-Tellurium Development Assoc., 11 Broadway, New York.

tellurium bromide. See tellurium dibromide and tellurium tetrabromide.

tellurium chloride. See tellurium dichloride.

tellurium dibromide (tellurium bromide; tellurous bromide) $TeBr_2$.

Properties: Blackish green, crystalline mass, or gray to black needles. Very hygroscopic. Decomposed by water. Soluble in ether. Violet vapor. M.p. 210°C; b.p. 339°C.

tellurium dichloride (tellurium chloride; tellurous chloride) $TeCl_2$.

Properties: Amorphous, black mass. Greenish yellow when powdered. Decomposed by water. Sp. gr. 6.9; b.p. 327°C; m.p. 209°C.

tellurium diethyldithiocarbamate. See "Tellurac."

tellurium dioxide (tellurous acid anhydride) TeO_2.

Properties: Heavy, white, crystalline powder, odorless. Soluble in acids (conc.), alkalies; slightly soluble in acids (dilute), water. Sp. gr. 5.89; m.p. 733°C; b.p. 1245°C.

Hazard: Ingestion causes persistant garlic breath.

tellurium disulfide (tellurium sulfide) TeS_2.
Properties: Red powder. Turns in time to a dark brown, amorphous powder. Fuses to gray, lustrous mass. Soluble in alkali sulfides; insoluble in acids, water.

tellurium lead. See lead, tellurium.

tellurium sulfide. See tellurium disulfide.

tellurium tetrabromide (telluric bromide; tellurium bromide) $TeBr_4$.
Properties: Yellow crystals. Soluble in a little water (decomposes in excess water). Sp. gr. 4.3; m.p. 363°C; b.p. approx 420°C, with decomposition into bromine and dibromide.

tellurous acid H_2TeO_3.
Properties: White, crystalline powder. Soluble in acids (dilute), alkalies; slightly soluble in water, alcohol. Sp. gr. 3.053; m.p. 40°C (decomposes).

tellurous acid anhydride. See tellurium dioxide.

tellurous bromide. See tellurium dibromide.

tellurous chloride. See tellurium dichloride.

"Telone."[233] Trademark for fumigants containing 1,3-dichloropropene and related C_3 hydrocarbons.

"Tel-Tale."[241] Trademark for silica gel which is impregnated with cobalt chloride and turns from blue to pink as the relative humidity increases.

"Telvar."[28] Trademark for wettable powder containing 80% monuron, for weed control.
Containers: 4-lb bags; 50-lb fiber drums and bags.
"Telvar ML" is a stable liquid suspension containing 28% monuron, used for pre-emergence weed control in cotton.
Containers: ½-gal cans.

TEM. Abbreviation for triethylene melamine (q.v.).

"Temasept."[430] Trademark for series of antimicrobial agents.
"Temasept I." 1:1 mixture of 4′,5-dibromosalicylanilide and 3,4′,5-tribromosalicylanilide. Used in fabric treatment and other sanitizing applications.
"Temasept II." Essentially pure 3,4′,5-tribromosalicylanilide; used in deodorants, antibacterial toilet soaps, and other skin contact applications.

"Temex" 15.[304] Trademark for a barium-cadmium organic vinyl stabilizer. A fine white powder with a sp. gr. of 1.3. Good dispersion characteristics.
Uses: Vinyl phonograph record and other rigid stocks, based on vinyl copolymer resins.

temper. To increase the hardness and strength of a metal by quenching or heat treatment.

temperature. The thermal state of a body considered with reference to its ability to communicate heat to other bodies (J. C. Maxwell). There is a distinction between temperature and heat, as is evidenced by Helmholtz' definition of heat as "energy that is transferred from one body to another by a thermal process," where by a thermal process is meant radiation, conduction, and/or convection.

Temperature is measured by such instruments as thermometers, pyrometers, thermocouples, etc.; and by scales such as centigrade (Celsius), Fahrenheit, Réaumur, and absolute (Kelvin). See also absolute temperature.

tenacity. Strength per unit weight of a fiber or filament, expressed as grams per denier. It is the rupture load divided by the linear density of the fiber. See also tensile strength; denier.

"Tenamene."[256] Trademark for a group of additives for gasoline and other petroleum products.

"Ten-Cem."[480] Trademark for a group of neodecanoates of cobalt, copper, calcium, lead, manganese and zinc.
Uses: Driers for printing inks; activators for protective coatings and plastics.

tenderization. (1) Treatment of meats with ultraviolet radiation or with certain enzyme preparations to accelerate softening of the collagen fibers and thus reduce the time necessary to "hang" the meat. (2) The degradation and mechanical weakening undergone by textile fibers under excessive wet abrasion during laundering.

"Tenex."[79] Trademark for an extremely pale heat-treated wood rosin.
Uses: Adhesive tape; artificial Burgundy pitch; core oil; cosmetics; disinfectants; electric insulating compounds; emulsions; ester gum; printing ink; rubber cement; shellac diluent; solder; spirit varnishes; synthetic resins; varnishes; Venice turpentine.

"Tenite."[256] Trademark for cellulose esters and polyolefin thermoplastics. "Tenite" acetate, "Tenite" butyrate, "Tenite" propionate are made from cellulose acetate, cellulose acetate butyrate, and cellulose acetate propionate respectively. "Tenite" polyethylene and "Tenite" polypropylene are made by the polymerization of ethylene and propylene. "Tenite" polyallomer (q.v.) is a block copolymer of propylene and ethylene; the lightest solid plastic offered commercially. Available in formulations meeting FDA requirements for use in contact with food.

"Tenlo" 70.[309] Trademark for a blend of fatty amides and polyethylene glycol ester. Used as a pigment grinding oil.

"Tenox."[256] Trademark for food grade antioxidants containing one or more of the following ingredients; butylated hydroxyanisole; butylated hydroxytoluene and/or propyl gallate with or without citric acid. Some formulas are supplied in solvents such as propylene glycol.

"Teox 120."[84] Trademark for a nonionic surface-active agent composed of polyethenoxy tallate.
Properties: Light straw-colored liquid; bland odor; sp. gr. 1.065–1.070 (77°F); refractive index 1.4762 (77°F); viscosity 190 cp (77°F); heat-stable to 644°F; very slightly hygroscopic; pH (0.2% soln) 7.2; soluble in water, acetone, benzene, ethyl ether, carbon tetrachloride, ethanol, methanol, and xylene.
Containers: 450-lb drums; tank trucks; tank cars.
Uses: Detergents; surface cleaners; textile cleaning and dyeing; emulsion formulations.

tensile strength. The rupture strength (stress-strain product at break) per unit area of a material subjected to a specified dynamic load; it is usually expressed in pounds per square inch (psi). This definition applies to elastomeric materials as well as to certain metals. For strength of textile fibers, see tenacity.

tepa. Generic name for tris(1-aziridinyl)phosphine oxide, $(CH_2CH_2N)_3POH$. See triethylenephosphoramide.
Hazard: Highly toxic; strong irritant to skin and tissue.
Use: Insect chemosterilant.

Shipping regulations: (Rail) White label. (Air) Corrosive label. Legal label name for tris(1-aziridinyl)phosphine oxide.

"Tepidone."[28] Trademark for a water solution of sodium dibutyldithiocarbamate, $(C_4H_9)_2NC(S)SNa$. An amber liquid; sp. gr. 1.08.
Use: To accelerate vulcanization of natural and synthetic rubber and latex compounds.

tepp. Generic name for ethyl pyrophosphate (tetraethyl pyrophosphate) $(C_2H_5)_4P_2O_7$.
Properties: Water-white to amber liquid depending on purity; hygroscopic; miscible in all proportions with water and all organic solvents except aliphatic hydrocarbons; hydrolyzed in water with formation of mono-, di and triethyl ortho-phosphates; water solutions attack metals. Commercial material contains 40% tepp; sp. gr. approx 1.20; refractive index 1.420; b.p. (pure compound) 135–138°C (1 mm).
Derivation: From phosphorus oxychloride and ethanol or phosphorus oxychloride and triethyl phosphate.
Grade: 40%.
Containers: 5-gal cans; 50- and 55-gal drums.
Hazard: Highly toxic by skin contact, inhalation or ingestion. Rapidly absorbed through skin. Repeated exposure may, without symptoms, be increasingly hazardous. Tolerance, 0.05 mg per cubic meter of air. Use may be restricted.
Uses: Insecticides for aphids and mites, rodenticide.
Shipping regulations: (Rail, Air) Poison label. Not acceptable on passenger planes. Legal label name: tetraethyl pyrophosphate. Consult regulations for details.

tera-. Prefix meaning 10^{12} units (symbol T). 1 Tg = 1 teragram = 10^{12} grams.

"Terbec."[233] Trademark for a wet strength improver for soils; based on 4-tert-butylcatechol.

terbia. See terbium oxide.

terbium Tb Atomic number 65; Group IIIB of the periodic table; a rare-earth element of the yttrium subgroup (lanthanide series). Atomic weight 158.9254. Valences 3, 4. No stable isotopes.
Properties: Metallic luster. Reacts slowly with water, and is soluble in dilute acids. Sp. gr. 8.332; m.p. 1356°C; b.p. 2800°C; salts colorless. Highly reactive; handled in inert atmosphere or vacuum.
Source: See rare-earth.
Derivation: Reduction of fluoride with calcium.
Grades: Regular; 99.9+% purity (ingots, lumps); single crystals.
Uses: Phosphor activator; dope for solid state devices.

terbium chloride $TbCl_3 \cdot 6H_2O$.
Properties: Transparent, colorless, prismatic crystals; readily soluble in water or alcohol; sp. gr. 4.35; m.p. (anhydrous) 588°C; very hygroscopic.
Derivation: By treatment of carbonate or oxide with hydrochloric acid in an atmosphere of dry hydrogen chloride.

terbium fluoride $TbF_3 \cdot 2H_2O$.
Properties: Solid; m.p. 1172°C; b.p. 2280°C; insoluble in water.
Hazard: Toxic; strong irritant.
Use: Source of terbium.

terbium nitrate $Tb(NO_3)_3 \cdot 6H_2O$.
Properties: Colorless monoclinic needles or white powder, soluble in water; m.p. 89.3°C.
Derivation: By treatment of oxide, carbonate or hydroxide with nitric acid.
Hazard: Strong oxidant. Fire risk in contact with organic materials.
Shipping regulations: (Rail) Nitrates, n.o.s. Yellow label. (Air) Oxidizer label.

terbium oxide (terbia) Tb_2O_3.
Properties: Dark brown powder; soluble in dilute acids; slightly hygroscopic; absorbs carbon dioxide from air.
Derivation: By ignition of hydroxides or salts of oxyacids.
Grades: 98–99%.
Containers: Glass bottles.
See also rare earth.

terbium sulfate $Tb_2(SO_4)_3 \cdot 8H_2O$.
Properties: Colorless crystals which lose $8H_2O$ at 360°C. Soluble in water.

terebene. A mixture of terpenes, chiefly dipentene and terpinene.
Properties: Colorless liquid; sp. gr. 0.862–0.866; optical rotation, inactive; b.p. 160–172°C. Soluble in alcohol; insoluble in water.
Derivation: From oil of turpentine.
Hazard: Flammable, moderate fire risk. Moderately toxic by ingestion and inhalation.
Use: Medicine.

terephthalic acid (para-phthalic acid; TPA; benzene-para-dicarboxylic acid) $C_6H_4(COOH)_2$.
Properties: White crystals or powder; insoluble in water, chloroform, ether, acetic acid; slightly soluble in alcohol; soluble in alkalies. Sp. gr. 1.51; sublimes above 300°C. Low toxicity. Combustible.
Derivation: (1) Oxidation of para-xylene or of mixed xylenes and other alkyl aromatics (phthalic anhydride). (2) reacting benzene and potassium carbonate over a cadmium catalyst.
Grades: Commercial; fiber.
Containers: Up to car and truck loads.
Uses: Production of linear, crystalline polyester resins, fibers and films by combination with glycols. Also used as a reagent for alkali in wool; additive to poultry feeds.

2-terephthaloylbenzoic acid
$HOOCC_6H_4COC_6H_4COOH$.
Properties: White to gray powder; m.p. 233–237°C.
Use: Chemical intermediate.

terephthaloyl chloride $C_6H_4(COCl)_2$. 1,4-Benzenedicarbonyl chloride.
Properties: Colorless needles; m.p. 82–84°C; b.p. 259°C; decomposes in water and alcohol; soluble in ether. Flash point 356°F; combustible.
Hazard: Skin irritant.
Uses: Dye manufacture; synthetic fibers, resins, films; ultraviolet absorption; pharmaceuticals; rubber chemicals; cross-linking agent for polyurethanes and polysulfides.

"Tergitol."[214] Trademark for a series of nonionic and anionic surfactants.
Uses: Detergents, wetting agents, emulsifiers in water systems; leveling and spreading agents.

"Ternalloy."[453] Trademark for a series of four aluminum-zinc-magnesium alloys. Ternalloys 5 and 6 are of self-aging, while Ternalloys 7 and 8 are amenable to heat treatment. Nominal compositions are 0.5% Mn, 0.3% Cr, and 3.0–4.8% Zn, 1.6–2.3% Mg, dependent on particular alloy. Alloys have good corrosion resistance and excellent machinability in both sand and permanent mold type castings. Supplied in ingot form.

ternary. Descriptive of a solution or alloy having three components, or of a chemical compound having three constituent elements or groups.

terneplate (roofing tin). A lead-tin alloy used for coating iron or steel; its composition is approximately 75% lead and 25% tin. It has a dull finish. Used chiefly for roofing; also for deep stamping, gasoline tanks, etc.

1,4(8)-terpadiene. See terpinolene.

terpene. An unsaturated hydrocarbon having the formula $C_{10}H_{16}$, occurring in most essential oils and oleoresins of plants. The terpenes are based on the isoprene unit C_5H_8, and may be either acyclic or cyclic with one or more benzenoid groups. They are classified as monocyclic (dipentene), dicyclic (pinene), or acyclic (myrcene), according to the molecular structure. Many terpenes exhibit optical activity. Terpene derivatives (camphor, menthol, terpineol, borneol, geraniol, etc.) are called terpenoids; many are alcohols. See also polyterpene resin.

terpeneless oil. An essential oil from which the terpene components have been removed by extraction and fractionation, either alone or in combination. The optical activity of the oil is thus reduced. The terpeneless grades are much more highly concentrated than the original oil (from 15 to 30 times). Removal of terpenes is necessary to inhibit spoilage, particularly of oils derived from citrus sources; on atmospheric oxidation the specific terpenes form compounds that impair the value of the oil; for example, *d*-limonene oxidizes to carvone and gamma-terpinene to *p*-cymene. Terpeneless grades of citrus oils are commercially available.

terpenoid. See terpene.

"Terpex."[296] Trademark for a polymerized turpentine modified with various additives. Used as a component of coatings, especially aluminum paints.

"Terpex" Extra.[206] Trademark for a low molecular weight liquid terpene resin.
Uses: Rubber tackifier and plasticizer; adhesives; resin and oil extender.

para-**terphenyl** (1,4-diphenylbenzene) $(C_6H_5)_2C_6H_4$.
Properties: Liquid; sp. gr. 1.234 (0°C); m.p. 213°C; b.p. 405°C. Flash point 405°F. Combustible.
Derivation: From para-dibromobenzene, or bromobenzene, and sodium.
Method of purification: Zone-melting.
Grades: Technical; scintillation.
Containers: Glass bottles; fiber drums.
Hazard: Toxic by ingestion and inhalation. Tolerance, 1 ppm in air.
Uses: Polymerized with styrene to make a plastic phosphor. Single crystals used as scintillation counters.

terpilenol. See terpineol.

terpinene $C_{10}H_{16}$. A mixture of three isomeric cyclic terpenes, alpha-, beta-, and gamma-terpinene.
alpha-Terpinene has a b.p. of 180–182°C and sp. gr. 0.8484 (14°C). It is found in cardamom, marjoram and coriander oils.
beta-Terpinene occurs naturally only in association with alpha-terpinene.
gamma-Terpinene has a b.p. of 183°C and sp. gr. 0.853 (15°C). It is found in coriander, lemon, cumin, and ajawan oils.

terpineol (alpha-terpineol; beta-terpineol; gamma-terpineol; terpilenol) $C_{10}H_{17}OH$.
Properties: Colorless liquid or low melting transparent crystals; lilac odor. Usually sold commercially as a mixture of the three isomers. Sp. gr. 0.930–0.936 (25°C); solidification point 2°C; optical rotation between −0° 10′ and +0° 10′; boiling range between 214 and 224°C, 90% within 5°C; refractive index (n 20/D) 1.4825–1.4850; soluble in 2 vols 70% alcohol; slightly soluble in water, glycerin. Combustible.
Derivation: By heating terpin hydrate with phosphoric acid and distilling, or with dilute sulfuric, using an azeotropic separation; fractional distillation of pine oil. Occurs naturally in several essential oils.
Grades: Technical; perfumery; extra; prime; F.C.C.
Containers: Bottles, tins; drums.
Uses: Solvent for hydrocarbon materials; mutual solvent for resins and cellulose esters and ethers; perfumes; soaps; disinfectant; antioxidant; medicine; flavoring agent.

"Terpineol 318."[266] Trademark for a highly refined mixture of tertiary terpene alcohols, predominantly alpha-terpineol; 97% total terpene alcohols. Colorless liquid; f.p. less than −10°C; sp. gr. 0.937 (15.6/15.6°C); ASTM distillation range 5–95%, 216–220°C.

terpin hydrate (dipentene glycol) A cyclohexane derivative. $CH_3(OH)C_6H_9C(CH_3)_2OH \cdot H_2O$.
Properties: Colorless, lustrous, rhombic crystalline prisms, or white powder; slight characteristic odor; slightly bitter taste; efflorescent; m.p. 115–117°C; anhydrous, m.p. 102–105°C; b.p. 258°C; soluble in alcohol and ether; slightly soluble in water. Combustible.
Derivation: Action of nitric acid and dilute sulfuric acid, or dilute sulfuric alone, on pine oil boiling between 200–220°C.
Grade: Technical: N.F.
Containers: 1-lb cans and cartons: 100-lb net fiber drums.
Uses: Pharmaceuticals; source of terpineol; veterinary medicine.

terpinolene (1,4(8)-terpadiene) $C_{10}H_{16}$.
Properties: Water-white to pale amber liquid. Insoluble in water; soluble in alcohol, ether, glycol. Sp. gr. (15.5/15.5°C) 0.864; b.p. 183–185°C; flash point (closed cup) 99°F; wt/gal 7.2 lbs (15.5°C).
Derivation: Fractionation of wood turpentine.
Hazard: Flammable, moderate fire risk.
Uses: Solvent for resins; essential oils; manufacture of synthetic resins, chemical derivatives.

terpinyl acetate $C_{10}H_{17}OOCCH_3$.
Properties: Colorless liquid; odor suggestive of bergamot and lavender; sp. gr. 0.958–0.968 (15°C); refractive index (n 20/D) 1.4640–1.4660; optical rotation varies around 0°; f.p. −50°C; b.p. 220°C. Soluble in 5 or more vols of 70% alcohol; slightly soluble in water and glycerol. Combustible.
Derivation: By heating terpineol with acetic acid or anhydride in the presence of sulfuric acid, and subsequent distillation.
Grades: Technical; prime; extra; F.C.C.
Containers: Cans; drums.
Uses: Perfumes; flavoring agent.

terpolymer. A polymer made from three monomers, e.g., ABS polymers (q.v.).

"Terraclor."[84] Trademark for pentachloronitrobenzene (q.v.). Available as technical grade and formulations of 75% wettable powder, 2 lb concentrate, 20% dust, and 10% granular.
Uses: Soil sterilant and fungicide.

"Tersan" 75.[28] Trademark for a turf fungicide containing 75% thiram.
"Tersan" OM contains 45% thiram and 10% hydroxymercurichlorophenol.
Hazard: Toxic by ingestion and inhalation.

tert-. Abbreviation for tertiary. See primary.

tertiary calcium phosphate. See calcium phosphate, tribasic.

tertiary sodium phosphate. See sodium phosphate, tribasic.

"Tervan."[51] Trademark for wax-polymer blends used primarily for coating food cartons.

"Terylene."[479] Trademark for a synthetic polyester textile fiber based on terephthalic acid. Used in wash-and-wear fabrics.

"Tessalon."[305] Trademark for benzonatate.

testosterone $C_{19}H_{28}O_2$. An androgenic steroid; the male sex hormone produced by the testis. It has six times the androgenic activity of its metabolic product, androsterone.
Properties: White or slightly cream-white crystals or as crystalline powder; odorless and stable in air; m.p. 153–157°C; dextrorotatory in dioxane solution. Very soluble in chloroform; soluble in alcohol, dioxane and vegetable oils; slightly soluble in ether; insoluble in water.
Derivation: Isolation from extract of testis; synthesis from cholesterol or from the plant steroid diosgenin.
Grade: N.F.
Containers: Bottles.
Uses: Medicine; biochemical research.
See also methyltestosterone.

"Testryl."[412] Trademark for testosterone (q.v.).

TETD. Abbreviation for tetraethylthiuram disulfide.

tetraaminoditolylmethane $[CH_3C_6H_2(NH_2)_2]_2Ch_2$.
Properties: White crystalline powder. Insoluble in water. M.p. 195°C.
Derivation: Condensation of meta-tolylenediamine and formaldehyde.

tetraamylbenzene $(C_5H_{11})_4C_6H_2$.
Properties: Colorless liquid. Sp. gr. at 20°C 0.89; boiling range 320–350°C; odor faintly aromatic; flash point 295°F. Combustible.
Uses: Solvent; chemical intermediate.

tetraazobenzene-beta-**naphthol.** See benzeneazo-para-benzeneazo-beta-naphthol.

tetra base. See tetramethyldiaminodiphenylmethane.

tetraboric acid. See pyroboric acid.

tetrabromobisphenol A. $(C_6H_2Br_2OH)_2C(CH_3)_2$.
Properties: Off-white powder; m.p. 180–184°C; soluble in methanol and ether; insoluble in water.
Use: Flame retardant for plastics, paper, and textiles.

sym-**tetrabromoethane.** See acetylene tetrabromide.

tetrabromoethylene C_2Br_4.
Properties: Colorless crystals; m.p. 55–56°C; b.p. 227°C.
Derivation: Bromination of dibromoacetylene.
Use: Organic synthesis.

tetrabromofluorescein. See eosin.

tetrabromomethane. See carbon tetrabromide.

tetrabromophthalic anhydride $C_6Br_4C_2O_3$.
Properties: Pale yellow crystalline solid; m.p. 280°C.
Use: Flame retardant for plastics, paper, and textiles.

tetra-n-butylammonium chloride $(C_4H_9)_4NCl$.
Properties: Deliquescent, light tan powder; m.p. 50°C.
Uses: Substitute in solution for glass-calomel electrode systems; coagulant for silver iodide solutions; stereospecific catalyst.

tetrabutylthiuram disulfide $[(C_4H_9)_2NCH]_2S_2$.
Properties: Liquid; amber color; sp. gr. (20/20°C) 1.03–1.06°C; solidifies approx. −30°C; slight sweet odor. Soluble in carbon disulfide, benzene, chloroform, and gasoline; insoluble in water and 10% caustic. Combustible. Low toxicity.
Uses: Vulcanizing and accelerating agent.

tetrabutylthiuram monosulfide $[(C_4H_9)_2NCS]_2S$.
Properties: Brown, free flowing liquid; sp. gr. 0.99; soluble in acetone, benzene, gasoline, and ethylene dichloride; insoluble in water. Combustible. Low toxicity.
Use: Rubber accelerator.

tetrabutyltin $(C_4H_9)_4Sn$.
Properties: Colorless or slightly yellow oily liquid; sp. gr. 1.0572 (20/4°C); f.p. −97°C; b.p. 145°C (10 mm); decomposes at 265°C. Insoluble in water; soluble in most common organic solvents. Combustible.
Derivation: Reaction of tin tetrachloride with butyl magnesium chloride.
Hazard: Toxic and irritant. Tolerance (as Sn), 0.1 mg per cubic meter of air.
Uses: Stabilizing and rust-inhibiting agent for silicones; lubricant and fuel additive; polymerization catalyst; hydrochloric acid scavenger.

tetrabutyl titanate (TBT; butyl titanate; titanium butylate) $Ti(OC_4H_9)_4$.
Properties: Colorless to light yellow liquid. B.p. 310–314°C; forms a glass below −55°C; sp. gr. 0.996; refractive index 1.486; flash point 170°F; decomposes in water; soluble in most organic solvents except ketones. Combustible.
Derivation: Reaction of titanium tetrachloride with butyl alcohol.
Uses: Ester exchange reactions; heat-resistant paints (up to 500°C); improving adhesion of paints, rubber, and plastics to metal surfaces; cross-linking agent; condensation catalyst.

tetrabutyl urea $(C_4H_9)_2NCON(C_4H_9)_2$.
Properties: Liquid; sp. gr. 0.880; refractive index 1.4535; vapor pressure less than 0.01 mm; b.p. 305°C; f.p. less than −60°C; flash point 200°F; insoluble in water. Combustible, low toxicity.
Use: Plasticizer.

tetrabutyl zirconate $(C_4H_9O)_4Zr$. White solid from reaction of zirconium tetrachloride with butyl alcohol.
Uses: Condensation catalyst and cross-linking agent.

tetracaine $CH_3(CH_2)_3NHC_6H_4COOCH_2CH_2N(CH_3)_2$.
2-Dimethylaminoethyl para-butylaminobenzoate.
Properties: White or light yellow, waxy solid. Very slightly soluble in water; soluble in alcohol, ether, benzene, chloroform. M.p. 41–46°C.
Grade: U.S.P.
Use: Medicine. (Also available as the hydrochloride).

tetracalcium aluminoferrate. An ingredient of Portland cement.

1,2,3,4-tetracarboxybutane (TCB; 1,2,3,4-butanetetracarboxylic acid) $HOOCCH_2CH(COOH)CH(COOH)CH_2COOH$.
Uses: Alkyd resins; epoxy curing agent; sequestrant.

tetracene. See naphthacene.

tetracine. See tetrazene.

1,2,3,4-tetrachlorobenzene $C_6H_2Cl_4$.
Properties: White crystals; m.p. 46.6°C; b.p. 254°C; insoluble in water. Low toxicity. Combustible.
Uses: Component of dielectric fluids; synthesis.

1,2,4,5-tetrachlorobenzene $C_6H_2Cl_4$.
Properties: White flakes; m.p. 137.5–140°C; flash point 311°F; distillation range 240–246°C. Low toxicity. Combustible.
Uses: Intermediate for herbicides and defoliants; insecticide; impregnant for moisture resistance; electrical insulation; temporary protection in packing.

tetrachloro-para-**benzoquinone.** See chloranil.

tetrachlorobisphenol A $(C_6H_2Cl_2OH)_2C(CH_3)_2$. A monomer for flame-retardant epoxy, polyester and polycarbonate resins.

2,3,7,8-tetrachlorodibenzo-*p*-dioxin. See dioxin.

sym-**tetrachlorodifluoroethane** CCl_2FCCl_2F.
Properties: White solid or colorless liquid with slightly camphor-like odor when concentrated. B.p. 92.8°C; m.p. 26°C; critical temperature 278°C; sp. gr. (25°C) 1.6447; refractive index (25°C) 1.413; wt/gal 13.8 lbs. Nonflammable. Insoluble in water; soluble in alcohol.
Grades: Purified; solvent.
Hazard: Moderately toxic by inhalation. Tolerance, 500 ppm in air.
Use: Degreasing solvent.

tetrachlorodinitroethane $O_2NCCl_2CCl_2NO_2$.
Properties: White crystals; decomposes at 130°C to nitrogen peroxide.
Hazard: Highly toxic by ingestion and inhalation. Strong irritant.
Shipping regulations: (Air) Not acceptable. (Rail) Poisonous solids, n.o.s., Poison label.

tetrachlorodiphenylethane. See TDE.

tetrachlorodiphenyl sulfone. See tetradifon.

sym-**tetrachloroethane** (acetylene tetrachloride) $CHCl_2CHCL_2$.
Properties: Heavy, colorless, corrosive liquid. Chloroform-like odor. Soluble in alcohol and ether; slightly soluble in water. Sp. gr. 1.593 (25/25°C); b.p. 146.5°C; f.p. −43°C; weight 13.25 lbs/gal (25°C); refractive index 1.4918 (25°C); flash point none. Nonflammable.
Derivation: Reaction of acetylene and chlorine, and subsequent distillation.
Grade: Technical.
Containers: 55-gal drums.
Hazard: Toxic by ingestion, inhalation, and skin absorption. Tolerance, 5 ppm in air. Safety data sheet available from Manufacturing Chemists Assn., Washington, D.C.
Uses: Solvent; cleansing and degreasing metals; paint removers, varnishes, lacquers, photographic film; resins and waxes; extraction of oils and fats; alcohol denaturant; organic synthesis; insecticides; weed killer; fumigant.
Shipping regulations: (Rail) Poisonous liquid, n.o.s., Poison label. (Air) Poison label.

tetrachloroethylene. See perchloroethylene.

cis-**N-[(1,1,2,2-tetrachloroethyl)thio]-4-cyclohexene-1,2-dicarboximide** ("Difolatan"). $C_{10}H_9Cl_4NO_2S$.
Properties: White solid; m.p. 160°C; slightly soluble in most organic solvents; insoluble in water.
Use: Fungicide.

tetrachloromethane. See carbon tetrachloride.

tetrachloronaphthalene. See chlorinated naphthalene.

2,3,4,6-tetrachlorophenol C_6HCl_4OH.
Properties: Brown flakes or sublimed mass with a strong odor; m.p. 69–70°C; b.p. 164°C (23 mm): sp. gr. 25°/4°C 1.839; soluble in acetone, benzene, ether, and alcohol. Nonflammable.
Hazard: Toxic by ingestion and inhalation; strong irritant.
Use: Fungicide.

2,4,5,6-tetrachlorophenol C_6HCl_4OH.
Properties: Brown solid; phenol odor. Sp. gr. 1.65 at 60/4°C; m.p. > 50°C. Soluble in sodium hydroxide solutions and most organic solvents; insoluble in water.
Hazard: Toxic by ingestion and inhalation; strong irritant.
Uses: Fungicide; wood preservative.

tetrachlorophthalic acid $C_6Cl_4(CO_2H)_2$.
Properties: Colorless, crystalline plates. Soluble in hot water; sparingly soluble in cold water.
Derivation: By passing a stream of chlorine through a mixture of phthalic anhydride and antimony pentachloride.
Uses: Dyes; intermediates.

tetrachlorophthalic anhydride $C_6Cl_4(CO)_2O$.
Properties: White, odorless, free-flowing, nonhygroscopic powder; slightly soluble in water; m.p. 254–255°C; b.p. 371°C.
Containers: Fiber drums.
Uses: Intermediate in dyes, pharmaceuticals, plasticizers, and other organic materials; flame-retardant in epoxy resins.

tetrachloroquinone. See chloranil.

tetrachlorothiophene CClCClCClCClS (ring).
Properties: Liquid. M.p. 29–30°C; b.p. 104°C (10 mm). Soluble in benzene, hexane, alcohols, chlorocarbons.
Hazard: May be toxic.
Uses: Agricultural chemicals; lubricants.

tetracosane $C_{24}H_{50}$ and $CH_3(CH_2)_{22}CH_3$.
Properties: Crystals; soluble in alcohol; insoluble in water. Sp. gr. 0.779 (51°/4°C); b.p. 324.1°C; m.p. 51.5°C. Combustible; low toxicity.
Use: Organic synthesis.

n-**tetracosanoic acid.** See lignoceric acid.

tetracyanoethylene $(CN)_2C{:}C(CN)_2$. The first member of a class of compounds called cyanocarbons.
Properties: Colorless crystals; sublimes above 120°C; m.p. 198–200°C; b.p. 223°C. High thermal stability; burns in oxygen with a hotter flame than acetylene.
Hazard: Hydrolyzes in moist air to hydrogen cyanide.
Uses: Organic synthesis; dyes; to make colored solutions with aromatics.

tetracycline $C_{22}H_{24}N_2O_8 \cdot 0$ to $6H_2O$. An antibiotic obtained from certain Streptomyces species. It can also be prepared by catalytic hydrogenation of chlortetracycline or oxytetracycline, which it resembles in its

actions and uses. Its chemical structure is that of a modified naphthacene molecule. See naphthacene.
Properties: Yellow, odorless, crystalline powder. Stable in air; affected by strong sunlight. Potency affected in solutions with pH below 2 and destroyed in alkali hydroxide solutions. Practically insoluble in chloroform and ether; very slightly soluble in water; slightly in alcohol; very soluble in dilute hydrochloric acid and alkali hydroxide solutions; pH (saturated solution) 3.0–7.0. Induces fluorescence in mitochondria of living cells in tissue culture or in fresh preparations from various organs.
Grade: U.S.P.
Hazard: May be toxic side effects.
Use: Medicine (also available as the hydrochloride).

"Tetracyn."[299] Trademark for tetracycline hydrochloride.

n-**tetradecane** $C_{14}H_{30}$ and $CH_3(CH_2)_{12}CH_3$.
Properties: Colorless liquid; sp. gr. (20/4°C) 0.7653; m.p. 5.5°C; b.p. 253.5°C; refractive index (n 20/D) 1.4302; flash point 212°F. Autoignition temp. 396°F. Soluble in alcohol; insoluble in water. Combustible.
Grades: 95%, 99%.
Containers: Bottles; small drums.
Hazard: Lower flammable limit in air 0.5%.
Uses: Organic synthesis; solvent; standardized hydrocarbon; distillation chaser.

tetradecanoic acid. See myristic acid.

tetradecanol. See myristyl alcohol; see also 7-ethyl-2-methyl-4-undecanol.

1-tetradecene (alpha-tetradecylene) $CH_2{:}CH(CH_2)_{11}CH_3$.
Properties: Colorless liquid; density 0.775 g/ml (20/4°C); f.p. −12°C; b.p. 256°C; flash point 230°F. Insoluble in water; very slightly soluble in alcohol and ether. Combustible.
Uses: Solvent in perfumes, flavors, medicines, dyes, oils, resins.

cis-**tetradec-9-enoic acid.** See myristoleic acid.

tetradecylamine $C_{14}H_{29}NH_2$.
Properties: White solid with odor of ammonia; m.p. 37°C; b.p. 291.2°C; insoluble in water; soluble in alcohol and ether. Combustible.
Grade: 90% purity.
Containers: Drums; tank cars.
Uses: Intermediate for manufacture of cationic surface-active agents; germicides.

tetradecylbenzyldimethylammonium chloride monohydrate $[C_{14}H_{29}N(CH_3)_2CH_2C_6H_5]Cl \cdot H_2O$. A quaternary ammonium salt.
Properties: (Of a 50% solution in aqueous isopropanol). Viscous liquid which may gelatinize on standing; sp. gr. 0.978 (16°C); miscible with water, alcohol, glycerin and acetone; pH of a 10% solution in distilled water 7–8.
Grade: 50% solution in aqueous isopropanol.
Use: Production of bacteriostatic and fungistatic paper.

tetradecyl chloride (myristyl chloride) $CH_3(CH_2)_{13}Cl$.
Properties: Water-white distilled liquid, mild odor. Sp. gr. 0.8590; f.p. −0.2°C; b.p. 154–155°C (15 mm); 15.2% chloride; subject to mild hydrolysis on standing.
Grade: 97% min.

alpha-**tetradecylene.** See 1-tetradecene.

tetradecyl thiol (myristyl mercaptan) $CH_3(CH_2)_{13}SH$.
Properties: Liquid; m.p. 6.5°C; b.p. 176–180°C (22 mm); sp. gr. 0.8398 (25/4°C); refractive index 1.4612 (n 20/D); strong odor; combustible.
Grade: 95% (min) purity.
Hazard: Moderately toxic by inhalation; strong irritant.
Uses: Organic intermediate; synthetic rubber processing.

tetradifon. Generic name for 4-chlorophenyl 2,4,5-trichlorophenyl sulfone (2,4,4′,5-tetrachlorodiphenylsulfone), $Cl_3C_6H_2SO_2C_6H_4Cl$.
Properties: White crystalline powder; m.p. 147°C. Soluble in chloroform and aromatic hydrocarbons; insoluble in water.
Hazard: Probably toxic.
Use: Insecticide; acaricide.

tetra(diphenylphosphito) pentaerythritol. See pentaerythritol tetrakis(diphenyl phosphite).

tetraethanolammonium hydroxide $(HOCH_2CH_2)_4NOH$.
Properties: White, crystalline solid; m.p. 123°C; vapor pressure less than 0.01 mm (20°C); completely soluble in water. A strong base, approaching sodium hydroxide in alkalinity. Aqueous solutions are stable at room temperature but decompose on heating to weakly basic polyethanolamines.
Grades: Commercial grade is a 40% water solution.
Containers: 1-gal cans; 5-, 55-gal drums.
Hazard: Strong irritant to skin and tissue.
Uses: Alkaline catalyst; solvent for certain types of dyes; screen printing of dyes; metal-plating solutions.

1,1,3,3-tetraethoxypropane $[(C_2H_5O)_2CH]_2CH_2$.
Properties: Liquid; b.p. 105°C; refractive index 1.410 (n 20/D); sp. gr. 0.920 (20/20°C); slightly soluble in water; soluble in ether and alcohol; flash point 190°F. Combustible; low toxicity.
Containers: 55-gal drums.
Use: Organic synthesis.

tetraethylammonium chloride (TEAC; TEA chloride) $(C_2H_5)_4NCl$.
Properties: Anhydrous: Colorless, odorless, hygroscopic crystals; sp. gr. 1.080; freely soluble in water, alcohol, chloroform, acetone; slightly soluble in benzene and ether. Tetrahydrate: $(C_2H_5)_4NCl \cdot 4H_2O$, crystals; m.p. 37.5°C; sp. gr. 1.084.
Use: Medicine.

tetraethylammonium hexafluorophosphate $(C_2H_5)_4NPF_6$.
Properties: Solid; m.p. 255°C; nonhygroscopic, but soluble in water. Stable to heat; can be stored in solution without decomposition of the PF_6 ion.
Hazard: Toxic and irritant to skin.
Uses: Maintenance of fluoride atmospheres; preparation of bactericides and fungicides.

tetra-(2-ethylbutyl) silicate $[(C_2H_5)C_4H_8O]_4Si$.
Properties: Colorless liquid. Sp. gr. (20/20°C) 0.8920–0.9018; f.p. below −100°C; b.p. 238°C (50 mm); insoluble in water; slightly soluble in methanol; miscible with most organic solvents.
Uses: Heat-transfer medium; hydraulic fluid; wide temperature-range lubricant.

tetraethyl dithionopyrophosphate (TEDP). $(C_2H_5O)_4P_2OS_2$.
Properties: Liquid; sp. gr. 1.19; b.p. 138°C; insoluble in water; soluble in alcohol. Combustible.
Hazard: Highly toxic. Absorbed by skin. Cholinesterase inhibitor. Use may be restricted.
Use: Insecticide.
Shipping regulations: (Rail, Air) Organic phosphate, liquid, n.o.s., Poison label. Not accepted on passenger planes.

tetraethyl dithiopyrophosphate. See sulfotepp.

tetraethylene glycol (TEG) $HO(C_2H_4O)_3C_2H_4OH$.
Properties: Colorless liquid; hygroscopic. Soluble in water; insoluble in benzene, toluene, or gasoline. Sp. gr. 1.1248 at 20/20°C; f.p. −4°C; b.p. 327.3°C; vapor pressure >0.001 mm (20°C); refractive index 1.4577 (n 20/D); flash point 345°F; wt/gal 9.4 lb (20°C). Combustible. Low toxicity.
Containers: Cans; drums.
Uses: Solvent for nitrocellulose; plasticizer; lacquers; coating compositions.

tetraethylene glycol dibutyl ether. See dibutoxytetraglycol.

tetraethylene glycol dimethacrylate
Properties: Water-white to pale straw liquid; b.p. (1 mm) 200°C; sp. gr. (20/20°C) 1.075; refractive index (20°C) 1.4620; viscosity 12 cp; insoluble in water; soluble in styrene, many esters and aromatics; limited solubility in aliphatic hydrocarbons. Combustible.
Hazard: Irritant to skin and eyes.
Use: Plasticizer.

tetraethylene glycol dimethyl ether. See dimethoxytetraglycol.

tetraethylene glycol distearate
$(C_{17}H_{35}COOCH_2CH_2OCH_2CH_2)_2O$.
Properties: Liquid. M.p. 32–33°C; insoluble in water. Combustible.
Use: Plasticizer.

tetraethylene glycol monostearate
$C_{17}H_{35}COO(CH_2CH_2O)_4H$.
Properties: Liquid. Sp. gr. 0.971; m.p. 30–31°C; insoluble in water. Combustible.
Use: Plasticizer.

tetraethylenepentamine
$NH_2(CH_2CH_2NH)_3CH_2CH_2NH_2$.
Properties: Viscous, hygroscopic liquid. Sp. gr. 0.9980 at 20/20°C; f.p. −30°C; b.p. 333°C; vapor pressure >0.01 mm (20°C); wt/gal 8.3 lb (20°C). Flash point 325°F. Soluble in most organic solvents and water. Combustible.
Containers: 1-gal cans; 5-, 55-gal drums; tank cars.
Hazard: Toxic and strong irritant to eyes and skin.
Uses: Solvent for sulfur, acid gases, various resins and dyes; saponifying agent for acidic materials; manufacture of synthetic rubber; dispersant in motor oils; intermediate for oil additives.

tetra-(2-ethylhexyl) silicate $[C_4H_9CH(C_2H_5)CH_2O]_4Si$.
Properties: Colorless liquid. Sp. gr. 0.8838; b.p. 350–370°C; f.p. −90°C; solubility in water <0.01; pounds/gallon 7.4; flash point 390°F. Combustible.
Uses: Synthetic lubricants and functional fluids.

tetra-(2-ethylhexyl) titanate (tetrakis(2-ethylhexyl) titanate $[C_4H_9CH(C_2H_5)CH_2O]_4Ti$. Light yellow, viscous liquid from the transesterification of isopropyl titanate with 2-ethylhexanol. Combustible; low toxicity.
Uses: Crosslinking agent; condensation catalyst; adhesion promoter; water repellents.

tetraethyllead (TEL) $Pb(C_2H_5)_4$.
Properties: Colorless, oily liquid; pleasant odor. Soluble in all organic solvents. Insoluble in water and dilute acids or alkalies. Sp. gr. 1.65; b.p. 198–202°C; 75–85°C (13–14 mm); f.p. −136°C; decomposes slowly at room temperature, rapidly at 125–150°C. Combustible.
Derivation: (a) Alkylation of lead-sodium alloy with ethyl chloride; (b) electrolysis of an ethyl Grignard reagent, with an anode of lead pellets.
Grade: One grade only—approximately 98% pure.
Hazard: Toxic by ingestion, inhalation, and skin absorption. Tolerance (as Pb), 0.1 mg per cubic meter of air.
Uses: Aviation gasoline; other automotive gasoline; ethylation operations.
Note: While organic lead compounds are unquestionably toxic, the extent to which their presence in automobile exhaust fumes is a public health risk remains controversial. The consumption of TEL in gasolines has been reduced considerably in recent years, partly because of their contribution to air pollution and partly because they poison the platinum catalysts used in emission control devices. Even so-called lead-free gasolines contain trace quantities of lead compounds. See also antiknock agent.
Shipping regulations: (Rail, Air) Poison label. Not accepted on passenger planes.

tetraethyl orthosilicate. See ethyl silicate.

tetraethyl pyrophosphate. See tepp.

tetraethylthiuram disulfide [disulfiram; TTD; TETD; bis(diethylthiocarbamyl) disulfide] $[(C_2H_5)_2NCS]_2S_2$.
Properties: Light gray powder; sp. gr. 1.27; melting range 65–70°C; slight odor. Soluble in carbon disulfide, benzene and chloroform; insoluble in water.
Containers: 5-, 50-, 100-, 150-lb drums.
Hazard: Produces toxic symptoms only when ingested in presence of alcohol.
Uses: Fungicide and insecticide; treatment of alcoholism; ultra-accelerator for rubber.

tetraethylthiuram sulfide [bis-(diethylthiocarbamyl) sulfide] $[(C_2H_5)_2NCS]_2S$.
Properties: Dark brown powder; slight odor; sp. gr. (20/20°C) 1.12; boiling range 225–240°C (3 mm).
Hazard: Moderately toxic by ingestion and inhalation.
Uses: Pharmaceutical ointments; fungicide; insecticide.

tetraethyltin $Sn(C_2H_5)_4$.
Properties: Colorless liquid; sp. gr. (23°C) 1.187; b.p. 181°C; f.p. −112°C; insoluble in water; soluble in alcohol and ether.
Hazard: Toxic. Tolerance (as Sn), 0.1 mg per cubic meter of air.

"Tetra-Flex."[511] Trademark for internally plasticized resins based on polymethylene polyphenol. Permanent flexibility makes them suitable for industrial and electrical laminates.

tetrafluorodichloroethane. See dichlorotetrafluoroethane.

tetrafluoroethylene (perfluoroethylene; TFE) $F_2C:CF_2$. Monomer for polytetrafluoroethylene polymers.
Properties: Colorless gas; f.p. −142.5°C; b.p. −78.4°C; insoluble in water. Much heavier than air.
Derivation: By passing chlorodifluoromethane through a hot tube.
Hazard: Flammable, dangerous fire risk.
Shipping regulations: Inhibited: (Rail) Red Gas label. (Air) Flammable Gas label. Not acceptable on passenger planes. Uninhibited: (Air) Not acceptable.

tetrafluoroethylene epoxide (TFEO). $FCFCF_2$ (epoxide, O bridging the two carbons).
Derivation: Oxidation of tetrafluoroethylene at 120° C with ultraviolet light. Reaction proceeds by free-radical mechanism.
Use: Monomer for products ranging from dimers to polymers of d.p. 35. See also "Freon E" and "Krytox."

tetrafluorohydrazine F_2NNF_2.
Properties: Colorless mobile liquid or colorless gas; b.p. (calc) −73°C; heat of vaporization 3170 cal/mole; critical temperature 36°C.
Hazard: Toxic and irritant. Explodes in contact with reducing agents and at high pressures.
Uses: Organic synthesis; oxidizer in fuels for rockets, missiles, etc.
Shipping regulations: (Rail, Air) Not listed. Consult authorities.

tetrafluoromethane (fluorocarbon 14; carbon tetrafluoride) CF_4.
Properties: Colorless gas; f.p. −184°C; b.p. −128°C; slightly soluble in water. Nonflammable. Density of liquid 1.96 g/ml at −184°C; sp. vol. (70° F) 4.4 cu ft/lb.
Grade: 95% min purity.
Containers: Steel cylinders.
Hazard: Moderately toxic by inhalation.
Uses: Refrigerant; gaseous insulator.
Shipping regulations: (Air) Nonflammable Gas label.

tetraglycol dichloride $(ClCH_2CH_2OCH_2CH_2)_2O$.
Properties: Colorless liquid. Slightly miscible with water. Sp. gr. 1.186; b.p. (2 mm) 114°C. Combustible; low toxicity.
Uses: High-boiling solvent and extractant for oils, fats, waxes and greases; chemical intermediate.

tetragrammaton. A compound formula which can be represented by four capital letters, e.g., HCCH.

1,2,3,6-tetrahydrobenzaldehyde. See 3-cyclohexene-1-carboxaldehyde.

1,2,3,4-tetrahydrobenzene. See cyclohexene.

tetrahydrocannibol. The active principle of marijuana; a hallucinatory drug. It has been synthesized and is available in laboratory quantities, subject to legal restrictions. It is comparatively nontoxic; animal tests have indicated that it can retard cancer growth and also may promote the acceptance of organ transplants in the human body.

tetrahydrofuran (THF) $CH_2CH_2CH_2CH_2O$ (ring).
Properties: Water-white liquid with ethereal odor; sp. gr. (20°C) 0.888; refractive index (n 20/D) 1.4070; f.p. −65°C; b.p. 66°C; flash point (open cup) 5°F. Autoignition temp. 610°F. Soluble in water and organic solvents. Low toxicity.
Derivation: (1) Catalytic hydrogenation of furan with nickel catalyst. (2) Acid-catalyzed dehydration of 1,4-butanediol.
Grades: Technical; spectrophotometric.
Containers: Drums; tank cars.
Hazard: Flammable, dangerous fire risk. Tolerance, 200 ppm in air. Flammable limits in air 2 to 11.8%.
Uses: Solvent for natural and synthetic resins, particularly vinyls, in topcoating solutions, polymer coating, cellophane, protective coatings, adhesives, magnetic tapes, printing inks, etc. Grignard reactions, lithium aluminum hydride reductions, and polymerizations; chemical intermediate and monomer.
Shipping regulations: (Rail) Flammable liquid, n.o.s., Red label. (Air) Flammable Liquid label.

tetrahydrofurfuryl acetate $C_4H_7OCH_2OOCCH_3$.
Properties: Colorless liquid. Soluble in water, alcohol, ether, and chloroform. Sp. gr. 1.061 (20/0°C); b.p. 194–195°C (753 mm). Low toxicity. Combustible.
Derivation: By treatment of tetrahydrofurfuryl alcohol with acetic anhydride.
Use: Flavoring.

tetrahydrofurfuryl alcohol ("THFA"; tetrahydrofuryl carbinol) $C_4H_7OCH_2OH$.
Properties: Colorless liquid; mild odor; hygroscopic; miscible with water; sp. gr. 1.0543 (20/20°C); b.p. 178°C; refractive index 1.4520 (20/D); flash point (open cup) 167°F; viscosity 5.49 cp (25°C). Autoignition temp. 540°F. Combustible. Low toxicity.
Derivation: Catalytic hydrogenation of furfural.
Grades: Commercial; industrial (about 80%).
Containers: Cans; 55-gal drums; tank cars; tank trucks.
Uses: Solvent for vinyl resins, dyes for leather; chlorinated rubber; cellulose esters; solvent-softener for nylon; vegetable oils; coupling agent.

tetrahydrofurfurylamine $C_4H_7OCH_2NH_2$.
Properties: Colorless to light yellow liquid; refractive index 1.4520–1.4535 (25°C); distilling range −150 to −156°C. Combustible. Low toxicity.
Uses: Chemical intermediate; fine grain photographic development; vulcanization accelerator.

tetrahydrofurfuryl benzoate $C_4H_7OCH_2OOCC_6H_5$.
Properties: Colorless liquid; insoluble in water; soluble in alcohol, ether, and chloroform. Sp. gr. 1.137 (20/0°C); b.p. 300–302°C, 138–140°C (2 mm). Combustible. Low toxicity.
Derivation: Tetrahydrofurfuryl alcohol and benzoic acid by esterification.
Uses: Chemical intermediate.

tetrahydrofurfuryl laurate $C_4H_7OCH_2OOCC_{11}H_{23}$.
Properties: Colorless liquid. Sp. gr. (25°C) 0.930; insoluble in water. Combustible.
Use: Plasticizer.

tetrahydrofurfuryl levulinate $CH_3CO(CH_2)_2COOCH_2C_4H_7O$.
Properties: Colorless liquid. M.p. 59–61°C; soluble in water. Combustible.
Use: Plasticizer.

tetrahydrofurfuryl oleate $C_{17}H_{33}COOCH_2C_4H_7O$.
Properties: Colorless liquid; sp. gr. (25°C) 0.923; b.p. (5 mm) 240°C; f.p. −30°C; insoluble in water. Flash point 329°F. Combustible.
Containers: 1-, 5-, 55-gal drums.
Use: Plasticizer.

tetrahydrofurfuryl phthalate $C_6H_4(COOCH_2C_4H_7O)_2$.
Properties: Colorless liquid. Sp. gr. (25°C) 1.194; m.p. less than 15°C; insoluble in water. Combustible.
Use: Plasticizer.

tetrahydrofuryl carbinol. See tetrahydrofurfuryl alcohol.

tetrahydrogeraniol. See 3,7-dimethyl-1-octanol.

tetrahydrolinalool (3,7-dimethyl-3-octanol) $C_{10}H_{21}OH$.
Properties: Colorless liquid; floral odor. Sp. gr. 0.832–0.837; optically inactive. Combustible.
Containers: Up to 55-gal drums.
Uses: Perfumery; flavoring.

1,2,3,4-tetrahydro-6-methylquinoline $C_{10}H_{13}N$.
Properties: Yellowish crystals; strong civet-like odor;

m.p. 33°C; soluble in 2 parts 80% alcohol. Combustible.
Use: Perfumery.

tetrahydronaphthalene $C_{10}H_{12}$.
Properties: Colorless liquid; pungent odor. Miscible with most solvents, and compatible with natural and synthetic vehicles; insoluble in water. Sp. gr. 0.981 at 13°C; b.p. 206°C; refractive index 1.540–1.547; flash point 160°F; freezing point −25°C; moisture content none; residue on evaporation none; acidity neutral; wt/gal approx. 8 lbs. Autoignition temp. 723°F; combustible.
Derivation: Hydrogenation of naphthalene in the presence of a catalyst at 150°C.
Grade: Technical.
Containers: 50-gal drums.
Hazard: Moderately toxic and irritant. Narcotic in high concentrations.
Uses: Solvent; chemical intermediate.

tetrahydrophthalic anhydride $C_6H_8(CO)_2O$.
Properties: White crystalline powder, solidification point 99–101°C; sp. gr. (105°C); 1.20; slightly soluble in petroleum ether and ethyl ether; soluble in benzene. Flash point 315°F (open cup); combustible.
Derivation: Diels-Alder reaction of butadiene and maleic anhydride.
Containers: Fiber drums.
Uses: Chemical intermediate for light-colored alkyds, polyesters, plasticizers and adhesives; intermediate for pesticides; hardener for resins.

tetrahydropyran-2-methanol
$OCH_2CH_2CH_2CH_2CHCH_2OH$.
Properties: Liquid; sp. gr. (20°C) 1.0272; b.p. 187.2°C; f.p., sets to glass below −70°C; miscible with water. Flash point 200°F (open cup). Combustible.
Containers: Cans; drums.
Use: Chemical intermediate.

1,2,5,6-tetrahydropyridine C_5H_9N.
Properties: Colorless liquid. Sp. gr. (20/4°C) 0.912–0.914; f.p. −44°C; b.p. 115.5–120.0°C. Combustible.
Purity: 96% min.
Use: Organic intermediate.

tetrahydrothiophene (thiophane) $CH_2CH_2CH_2CH_2S$.
Properties: Water-white liquid; sp. gr. 1.00 (15.6/15.6°C); boiling range 115–124.4°C. Combustible.
Uses: Solvent; intermediate; fuel gas odorant.

tetrahydrothiophene 1,1-dioxide. See sulfolane.

tetrahydroxyadipic acid. For two of fourteen possible isomers, see mucic acid; saccharic acid.

tetrahydroxybutane. See erythritol.

tetrahydroxydiphenyl. See diresorcinol.

tetrahydroxyethylethylenediamine [N,N,N′,N′-tetrakis-(2-hydroxyethyl)ethylenediamine] $(HOCH_2CH_2)_2NCH_2CH_2N(CH_2CH_2OH)_2$.
Properties: Clear, viscous liquid; good heat stability; combustible; low toxicity.
Uses: Organic intermediate; crosslinking of rigid polyurethane foams; chelating agent; humectant; gas absorbent; resin formation; detergent processing.

2,3,4,5-tetrahydroxyhexanedioic acid. See D-saccharic acid.

tetraiodoethylene (iodoethylene) $I_2C{:}CI_2$.
Properties: Light-yellow, odorless crystals; on exposure to light turns brown; m.p. 187°C; sp. gr. 2.98. Insoluble in water; soluble in most organic solvents.
Derivation: Action of iodine on diiodoacetylene obtained from calcium carbide and iodine.
Uses: Surgical dusting powder; antiseptic ointment; fungicide.

tetraiodofluorescein. See iodeosin.

tetraisopropylthiuram disulfide $[(CH_3CH_3CH)_2NCS]S_2$.
Properties: Tan powder; sp. gr. (20/20°C) 1.12; melting range 95–99°C; amine odor. Soluble in benzene, chloroform, gasoline; insoluble in water, 10% caustic, carbon disulfide.
Use: Rubber accelerator.

tetraisopropyl titanate (TPT; titanium isopropylate; isopropyl titanate) $Ti[OCH(CH_3)_2]_4$.
Properties: Light-yellow liquid which fumes in moist air. B.p. (10 mm) 102–104°C; m.p. 14.8°C; sp. gr. 0.954; index of refraction 1.46; apparent viscosity (25°C) 2.11 cp. Decomposes rapidly in water; soluble in most organic solvents.
Derivation: Reaction of titanium tetrachloride with isopropanol.
Uses: Ester exchange reactions; adhesion of paints, rubber, and plastics to metals, condensation catalyst.

tetraisopropyl zirconate $Zr[OCH(CH_3)_2]_4$.
Properties: White solid; decomposes before melting.
Derivation: By reaction of zirconium tetrachloride with isopropanol.
Uses: Condensation catalyst; crosslinking agent.

tetrakis(hydroxymethyl)phosphonium chloride (THPC) $(HOCH_2)_4PCl$. A crystalline compound made by the reaction of phosphine, formaldehyde and hydrochloric acid.
Uses: Flame-retarding agent for cotton fabrics. May be used in combination with triethylolamine and urea (Roxel process) or with triethanolamine and tris(1-aziridinyl) phosphine oxide.

N,N,N′N′-tetrakis(2-hydroxypropyl)ethylenediamine (ethylenedinitrilotetra-2-propanol) $[-CH_2N(CH_2CHOHCH_3)_2]_2$.
Properties: Viscous, water-white liquid; infinitely soluble in water; soluble in ethyl alcohol, toluene, ethylene glycol and perchloroethylene. Combustible.
Uses: Cross-linking agent and catalyst in urethane foams; epoxy resin curing; metal complexes; intermediate.

"Tetralin."[28] Trademark for tetrahydronaphthalene (q.v.).

tetralite. See tetryl.

1-tetralone $CHCHCHCHCCCH_2CH_2CH_2CO$. A ketone of tetrahydronaphthalene.
Properties: Liquid; sp. gr. 1.090–1.095 (20/20°C); b.p. 120–125°C (10 mm); vapor-pressure 0.02 mm (20°C); m.p. 5.3–6.0°C. Insoluble in water. Flash point 265°F. Combustible.
Grade: Solvent and intermediate.

tetram (O,O-Diethyl-S-(β-diethylamino)-ethyl phosphorothioate hydrogen oxalate) $(C_2H_5O)_2POSCH_3CH_2N(C_2H_5)_2$.
Hazard: Toxic by ingestion and inhalation. Cholinesterase inhibitor. Use may be restricted.
Use: Insecticide.

Shipping regulations: (Rail, Air) Organic phosphate, n.o.s., Poison label. Not accepted on passenger planes.

tetramer. An oligomer whose molecule is composed of four molecules of the same chemical composition. See also polymer.

1,1,3,3-tetramethoxypropane $[(CH_3O)_2CH]_2CH_2$.
Properties: Liquid; b.p. 183°C; refractive index 1.408 (n 20/D); sp. gr. 0.995 (20/20°C); soluble in water, hexane, ether and alcohol. Combustible.
Containers: 55-gal drums.
Use: Organic synthesis.

tetramethylammonium chloride $(CH_3)_4NCl$. A quaternary ammonium compound.
Properties: White crystalline solid; sp. gr. 1.1690 (20/4°C); m.p., decomposes. Soluble in water and alcohol; insoluble in ether. Also available as a 50% solution.
Uses: Chemical intermediate; catalyst; inhibitor.

tetramethylammonium chlorodibromide $(CH_3)_4NClBr_2$.
Properties: Powder, M.p. 118–126°C. Soluble in water and other polar solvents.
Hazard: Evolves bromine on contact with water.
Uses: Dry brominating agent; ingredient in formulation of sanitizers.

tetramethylammonium hydroxide $(CH_3)_4NOH$.
Hazard: Highly toxic; strong irritant to skin and tissue.
Shipping regulations: (Rail) Corrosive liquid, n.o.s., White label. Corrosive label. (Air).

1,2,3,5-tetramethylbenzene. See isodurene.

sym-**tetramethylbenzene.** See durene.

N,N,N',N'-**tetramethyl-1,3-butanediamine** $CH_3CHN(CH_3)_2CH_2CH_2N(CH_3)_2$.
Properties: Colorless, stable liquid, f.p. below −100°C; b.p. 165.0°C; sp. gr. 0.8020 (20/20°C); vapor pressure 1.64 mm (20°C); miscible in all proportions with water; viscosity 1.0 cp (20°C). Flash point 114°F (TOC). Combustible.
Uses: Catalyst for polyurethane foams and epoxy resins; high-energy fuels.

2,2,4,4-tetramethyl-1,3-cyclobutanediol $(CH_3)_4C_4H_2(OH)_2$.
Properties: White solid; m.p. 124–135°C; b.p. 220–225°C. Isomer composition (approx) 50% cis, 50% trans. Flash point 125°F.
Hazard: Moderate fire risk; moderately toxic and irritant.
Uses: Chemical intermediate; lubricants.

tetramethyldiamidophosphoric fluoride. See dimefox.

tetramethyldiaminobenzhydrol (tetramethyldiaminodiphenylcarbinol; Michler's hydrol; hydrol) $(CH_3)_2NC_6H_4CH(OH)C_6H_4N(CH_3)_2$.
Properties: Colorless prisms; forms a colorless solution in ether or benzene and a blue solution in alcohol or acetic acid. Soluble in alcohol, ether, benzene, and acetic acid. M.p. 96°C. Combustible.
Derivation: Reaction of tetramethyldiaminodiphenylmethane, hydrochloric acid and glacial acetic acid; oxidized with lead peroxide.
Grade: Technical.
Containers: Fiber drums; tank trucks.
Uses: Dye intermediate; organic synthesis.

tetramethyldiaminobenzophenone (Michler's ketone) $CO[C_6H_4N(CH_3)_2]_2$. (4,4'-bis(dimethylamino) benzophenone).
Properties: Crystalline leaflets; m.p. 172°C; b.p., decomposes at 360°C; soluble in alcohol, ether, and water. Combustible.
Derivation: From dimethylaniline by reaction with phosgene.
Containers: Barrels; fiber drums.
Use: Synthesis of dyestuffs, especially auramine derivatives.

tetramethyldiaminodiphenylmethane (tetra base) $H_2C[C_6H_4N(CH_3)_2]_2$.
Properties: Yellowish leaflets or glistening plates; m.p. 90–91°C; sublimes with decomposition; b.p. 390°C; insoluble in water; soluble in benzene, ether, carbon disulfide and acids.
Derivation: By heating dimethylaniline with hydrochloric acid and formaldehyde.
Use: Dye intermediate.

tetramethyldiaminodiphenylsulfone (4,4'-bis(dimethylamino)diphenylsulfone) $[(CH_3)_2NC_6H_4]_2SO_2$. Solid; m.p. 259–260°C. Combustible.
Grades: Technical; reagent.
Uses: Intermediate in making dyestuffs and medicinal chemicals; analytical reagent for lead.

tetramethylene. See cyclobutane.

tetramethylene bismethanesulfonate. See busulfan.

tetramethylenediamine $H_2N(CH_2)_4NH_2$.
Properties: Colorless crystals with strong odor; m.p. 27°C; b.p. 158–159°C. Soluble in water with strongly basic reaction. Combustible.
Use: Chemical intermediate.

tetramethylene dichloride. See 1,4-dichlorobutane.

tetramethylene glycol. See 1,4-butylene glycol.

tetramethylene sulfone. See sulfolane.

1,1,4,4-tetramethyl-6-ethyl-7-acetyl-1,2,3,4-tetrahydronaphthalene $C_{18}H_{26}O$. A polycyclic musk.
Properties: Colorless crystals; m.p. 45°C; b.p. (2 mm) 130°C; insoluble in water; soluble in alcohol.
Uses: Perfumes; cosmetics; soaps.

tetramethylethylenediamine (TMEDA; N,N,N',N'-tetramethylethylenediamine) $(CH_3)_2NCH_2CH_2N(CH_3)_2$.
Properties: Colorless, liquid with slight ammoniacal odor. Soluble in water and most organic solvents. B.p. 121–122°C; sp. gr. 0.7765 (20/4°C); refractive index (n 25/D) 1.4170; f.p. −55.1°C. Combustible.
Grade: Anhydrous.
Containers: Up to 55-gal steel drums.
Uses: Preparation of epoxy curing agents; polyurethane formation; corrosion inhibitor; textile finishing agents; intermediate for quaternary ammonium compounds.

tetramethylguanidine $(CH_3)_2NC(NH)N(CH_3)_2$.
Properties: Liquid with slight ammoniacal odor; b.p. 159–160°C; soluble in both water and organic solvents. A strong base. Combustible.
Containers: Drums.

2,6,10,14-tetramethylhexadecane. See phytane.

tetramethyllead (TML) $(CH_3)_4Pb$.
Properties: Colorless liquid. Sp. gr. 1.995; f.p. −27.5°C; b.p. 110°C (10 mm). Insoluble in water; slightly soluble in benzene, petroleum ether, alcohol. Flash point 100°F.
Derivation, grades, and containers: See tetraethyl lead. Methyl reagents used instead of ethyl.
Hazard: Highly toxic by ingestion, inhalation, and skin absorption. Flammable, moderate fire risk. Tolerance

(as Pb), 0.15 mg. per cubic meter of air. Lower explosion level, 1.8%.
Uses: Antiknock agent in aviation gasoline.
Shipping regulations: (Air) Poison label. Not accepted on passenger planes.
See note under tetraethyllead.

tetramethylmethane. See neopentane.

N,N,N′,N′-tetramethylmethylenediamine $(CH_3)_2NCH_2N(CH_3)_2$.
Properties: Liquid; refractive index 1.4005 (n 20/D).

3,3′-tetramethylnonyl thiodipropionate. See ditridecyl thiodipropionate.

tetramethylsilane $(CH_3)_4Si$.
Properties: Colorless volatile liquid. B.p. 26.5° C; sp. gr. 0.646 (20/4° C). Insoluble in water and cold, concentrated sulfuric acid; soluble in most organic solvents. Low toxicity. Flash pt. 0° F.
Derivation: By Grignard reaction of silicon tetrachloride and methylmagnesium chloride.
Grades: Technical; purified.
Containers: Bottles; 50-lb drums.
Hazard: Flammable; high fire risk.
Use: Aviation fuel; internal standard for NMR analytical instruments.
Shipping regulations: (Rail, Air) Flammable liquid, n.o.s., Flammable Liquid label.

tetramethylthiuram disulfide. See thiram.

tetramethylthiuram monosulfide (bisdimethylthiocarbamyl sulfide) $[(CH_3)_2NCH]_2S$.
Properties: Yellow powder; sp. gr. 1.40; m.p. 104–107° C; soluble in acetone, benzene, and ethylene dichloride; insoluble in water and gasoline. Combustible.
Hazard: Moderately toxic.
Uses: Ultra-accelerator for rubber; fungicide; insecticide.

"Tetramix."[28] Trademark for a liquid antiknock compound, the active component of which is a redistribution mixture of tetraethyl and tetramethyl lead. Several redistribution mixtures are made by changing the molar percentages of the different lead alkyls. These mixtures are identified by the molar percentage of the TML used in the reaction.
Hazard: See tetraethyl lead.
Containers: Drums; tank cars.

"Tetran."[204] Trademark for polytetrafluoroethylene (q.v.).

tetranitroaniline (TNA) $C_6H(NO_2)_4NH_2$. A nitration product of aniline which melts at 170° C and explodes at 237° C.
Hazard: Toxic. Dangerous fire and explosion risk.
Use: Manufacture of detonators and primers.
Shipping regulations: (Rail, Air) Consult regulations.

tetranitromethane $C(NO_2)_4$.
Properties: Colorless liquid. Pungent odor. B.p. 125.7° C; m.p. 12.5° C; sp. gr. (13° C) 1.650. Miscible with alcohol and ether; insoluble in water. Decomposed by alcoholic solution of potassium hydroxide. Powerful oxidizing agent.
Derivation: By action of fuming nitric acid on benzene, acetic anhydride, or acetylene.
Hazard: Toxic by ingestion, inhalation, and skin absorption. Dangerous fire and explosion risk. Tolerance, 1 ppm in air.
Uses: Rocket fuel, as an oxidant or monopropellant; qualitative test for unsaturated compounds.
Shipping regulations: (Rail) Yellow label. (Air) Not acceptable.

"Tetranol."[300] Trademark for a highly sulfated fatty ester of the oleic type, with wetting and dye-leveling properties.

tetra(octylene glycol) titanate. See octylene glycol titanate.

"Tetra-Phen."[511] Trademark for a series of straight or modified phenolics, solid or liquid. Special resins for industrial use.

1,1,4,4-tetraphenylbutadiene (TPB) $(C_6H_5)_2C{:}CHCH{:}C(C_6H_5)_2$.
Properties: White crystals; two forms, m.p. 194–196° C and m.p. 202–204° C; insoluble in water, soluble in most organic solvents. Combustible.
Grade: Purified.
Use: Primary fluor or wave length shifter in solution scintillators.

tetraphenylsilane $(C_6H_5)_4Si$.
Properties: White solid; m.p. 237° C; b.p. 428° C. Very stable and inert. Combustible.
Derivation: By Grignard reaction of silicon tetrachloride and phenylmagnesium chloride.
Grade: Technical.
Use: Heat-transfer medium; polymers.

tetraphenyltin $(C_6H_5)_4Sn$.
Properties: White powder. S. gr. 1.490; m.p. 225–228° C; b.p. above 420° C. Insoluble in water; soluble in hot benzene, toluene, xylene.
Derivation: Reaction of tin tetrachloride with phenylmagnesium bromide.
Hazard: Toxic; skin irritant. Tolerance, 0.1 mg per cubic meter of air.
Uses: Stabilizer in chlorinated transformer oils; mothproofing agent; scavenger in dielectric fluids; intermediate.

tetraphosphoric acid. See polyphosphoric acids.

tetraphosphorus heptasulfide. See phosphorus heptasulfide.

tetraphosphorus hexasulfide. See phosphorus trisulfide.

tetraphosphorus trisulfide. See phosphorus sesquisulfide.

tetrapotassium EDTA. See ethylenediaminetetraacetic acid salts.

tetrapotassium ethylenediaminetetraacetate. See ethylenediaminetetraacetic acid (note).

tetrapotassium pyrophosphate (TKPP). See potassium pyrophosphate.

tetrapropenylsuccinic anhydride. See dodecenylsuccinic anhydride. Many isomers are possible.

tetra-n-propyl dithionopyrophosphate $(C_3H_7O)_2P(S)OP(S)(OC_3H_7)_2$.
Properties: Amber liquid; b.p. 148° C (2 mm). Miscible with most organic solvents; insoluble in water.
Hazard: Highly toxic.
Use: Insecticide; acaricide.
Shipping regulations: (Rail, Air) Organic phosphates, liquid, n.o.s., Poison label. Not accepted on passenger planes.

tetrapropylene (dodecene; propylene tetramer) $C_{12}H_{24}$. Mixture of C_{12} monoolefins.

Properties: Liquid; sp. gr. (20/20°C) 0.770; boiling range 183°-218°C; weight 6.44 lb/gal (60°F). Combustible.
Derivation: Olefin fraction obtained from catalytic polymerization of propylene.
Containers: Barges; tank cars.
Uses: Detergents (dodecylbenzene); lubricant additives; plasticizers.

tetrapropylthiuram disulfide $[(C_3H_7)_2NCS]_2S_2$.
Properties: Light cream color; sp. gr. (20/20°C) 1.13; melting range 49–51.5°C; musty odor. Soluble in carbon disulfide, benzene, chloroform and gasoline; insoluble in water and 10% caustic. Combustible.

"Tetraquinone."[227] Trademark for 1,2,3,4-tetrahydro-6-methylquinoline (q.v.).

tetrasilane Si_4H_{10}.
Properties: Colorless liquid; f.p. −93.5°C; b.p. 109°C; sp. gr. 0.825 (0°C).
Hazard: Severe fire and explosion risk; can ignite or explode in air.
Shipping regulations: (Rail, Air) Consult authorities.

tetrasodium diphosphate. See sodium pyrophosphate and sodium polyphosphates.

tetrasodium EDTA (ethylenediaminetetraacetic acid, tetrasodium salt; EDTA Na_4) $C_{10}H_{12}N_2Na_4O_8$, anhydrous or $2H_2O$. White powder; freely soluble in water. Used as a general-purpose chelating agent.
See also ethylenediaminetetraacetic acid (note).

tetrasodium monopotassium tripolyphosphate $Na_4KP_3O_{10}$.
Properties: White crystalline solid; m.p. 580–600°C; sp. gr. 2.55; solubility in water (26°C) 30 g/100 ml.
Use: Sequestrant.

tetrasodium pyrophosphate. See sodium pyrophosphate and sodium polyphosphate.

tetrastearyl titanate. Organic intermediate, adhesion promoter, pigment dispersant.

tetrazene (4-amidino-1-(nitrosamino-amidino)-1-tetrazene) $H_2NC(:NH)NHNHN:NC(:NH)NHNHNO$.
Properties: Colorless or pale yellow, fluffy solid; apparent density 0.45 but yields a pellet of density 1.05 under pressure of 3000 psi. Practically insoluble in water, alcohol, ether, benzene, and carbon tetrachloride. Slightly hygroscopic.
Derivation: Interaction of an aminoguanidine salt with sodium nitrite in absence of free mineral acid.
Use: Initiating explosive. Noncorrosive type.
Shipping regulations: (Rail) Consult regulations. Not accepted by express. (Air) Not acceptable.

tetrazolium chloride (tetrazolium salt; TTC; 2,4,5-triphenyltetrazolium chloride) $CN_4Cl(C_6H_5)_3$.
Properties: White to pale-yellow crystalline powder which darkens on exposure to light. Readily soluble in water. M.p. (with decomposition) 245°C.
Uses: In germination and viability tests. Viable parts of seed are stained red by deposition of red insoluble triphenyl formazan.

"Tetrine."[73] Trademark for a series of organic chelating, sequestering, and complexing agents consisting of ethylenediaminetetraacetic acid (EDTA) and its salts.

tetraodotoxin. A highly toxic venom occurring in puffer-like fish.

tetrol. See furan.

"Tetron."[88] Trademark for tetraethyl pyrophosphate (tepp), technical, and in liquid formulations for insecticidal use.
Hazard: Highly toxic.

"Tetrone" A.[28] Trademark for dipentamethylenethiuram tetrasulfide rubber accelerator.

"Tetronic."[203] Trademark for a nonionic tetrafunctional series of polyether block-polymers ranging in physical form from liquids through pastes to flakable solids. They are polyoxyalkylene derivatives of ethylenediamine. Physical state varies with molecular weight and oxyethylene content; 100% active.
Uses: Low-foaming detergent formulations; defoaming agents; flexible and rigid polyurethane foams; emulsifying and demulsifying agents; textile processing.

"Tetrosan 3,4D."[328] Trademark for alkyldimethyl 3,4-dichlorobenzylammonium chloride. Used as disinfectant, deodorant, germicide and fungicide.

tetryl (tetralite; nitramine). Common commercial names for trinitrophenylmethylnitramine.
$(NO_2)_3C_6H_2N(NO_2)CH_3$.
Properties: Yellow crystals; m.p. 130–132°C; sp. gr. 1.57 (19°C); explodes at 187°C. Insoluble in water; soluble in alcohol, ether, benzene, glacial acetic acid.
Hazard: Dangerous fire and explosion risk; toxic; skin irritant. Tolerance, 1.5 mg per cubic meter of air. Absorbed by skin.
Uses: Detonating agent for less sensitive high explosives.
Shipping regulations: (Rail) Consult regulations. Not accepted by express. (Air) Not acceptable.

"Texanol."[256] Trademark for 2,2,4-trimethyl-1,3-pentanediol monoisobutyrate.

"Texicote."[263] Trademark for a series of vinyl, acrylic and styrene polymer and copolymer emulsions, plasticized and unplasticized. Available in many grades.
Uses: Adhesives; paint; carpet backsizing; pigment binder in clay coatings; self-polishing finishes.

"Texicryl."[263] Trademark for a series of acrylic and methacrylic polymer and copolymer emulsions.
Uses: Paper coating, paint, leather finishes, textiles, adhesives, and wall paper.

"Texigel."[263] Trademark for a series of polyacrylic thickening agents, water-soluble anionic colloids. Used as stabilizers, protective colloid and thickeners for natural and synthetic latices and other aqueous dispersions, binders, and flocculants.

"Texilac."[263] Trademark for a series of vinyl and acrylic polymer and copolymer solutions in various solvents, plasticized and unplasticized.
Uses: Wet-bond and heat-seal adhesives, impregnants, specialty lacquers, paper coating, nitrocellulose finishes, and printing inks for plastic films.

"Texiprint."[263] Trademark for a series of synthetic resin pigment pastes and thickeners for screen and roller printing on textiles.

textile assistant. See assistant.

textile auxiliary. See auxiliary.

textile oil. Any of various specially compounded oils used to condition raw fibers, yarns or fabric for manufacturing, bleaching, dyeing and finishing operations.

"Textiline."[498] Trademark for a series of synthetic fiber lubricants based on nitrogen derivatives, esters and white mineral oils.

"Textolite."[245] Trademark for laminated plastic sheets, tubes, and rods used as insulating materials. Includes laminations of various combinations of phenolic, melamine, silicone, and epoxy resins with such base materials as paper, asbestos, cotton, linen, and nylon. Also some of the above materials with copper foil cladding for printed circuit applications.

"Textone."[84] Trademark for a sodium chlorite preparation that bleaches fabrics (cotton, linen, rayon, nylon) to high whiteness without degradation.

textured protein. See protein, textured.

textryl. Generic name for nonwoven structures which may be manufactured by wet-processing from staple fibers and fibrid binder.

"Texzyme."[114] Trademark for a proteolytic enzyme which digests blood, albumin, food, beverage, urine, perspiration, etc. Used in silk and filament rayon fabric to accelerate the rate of desizing and insure good results in dyeing, bleaching, and finishing.

"T-Fatty Acids."[487] Trademark for fatty acids derived from animal fats. T-10 is a completely hydrogenated tallow fatty acid. T-11 is a tallow fatty acid which has been partially hydrogenated by a special process designed to maximize reduction of polyunsaturated fatty acids. T-18, T-22, and T-33 are grades of distilled tallow and animal fatty acids.
Properties: White to yellow waxy solids which liquefy from about 43° to 64°C.
Containers: Tank cars and trucks.
Uses: Chemical intermediates, rubber compounding, grease manufacture, cosmetic ingredients, buffing compounds, candles, lubricants and emulsifiers.

TFE. Abbreviation for tetrafluoroethylene. See also polytetrafluoroethylene.

TFEO. Abbreviation for tetrafluoroethylene epoxide (q.v.).

TGA. Abbreviation for triglycollamic acid. See nitrilotriacetic acid.

Th Symbol for thorium.

"Thalan."[58] Trademark for 98% phthalic anhydride (q.v.).

thallic oxide. See thallium peroxide.

thallium Tl Metallic element of atomic number 81, Group IIIA of the periodic table. Atomic weight 204.37; valences 1,3; 2 stable isotopes.
Properties: Bluish-white, lead-like solid. Sp. gr. 11.85; m.p. 302°C; b.p. 1457°C. Oxidizes in air at room temperature. Soluble in nitric and sulfuric acids; insoluble in water, but readily forms soluble compounds when exposed to air or water.
Derivation: Flue dusts from lead and zinc smelting. The thallium compounds recovered are treated to obtain the metal by electrolysis, precipitation, or reduction.
Grades: Technical; high-purity.
Containers: Glass bottles.
Hazard: Forms toxic compounds on contact with moisture; keep from skin contact. Tolerance (as Tl) for soluble salts, 0.1 mg per cubic meter of air.
Uses: Thallium salts; alloys; low-melting glasses; pesticides; photoelectric applications; electrodes in dissolved oxygen analyzers.

thallium acetate (thallous acetate) $TlOCOCH_3$.
Properties: White, deliquescent crystals. Soluble in water, alcohol.
Constants: M.p. 131°C; sp. gr. 3.68.
Derivation: Interaction of acetic acid and thallium carbonate.
Hazard: Toxic. See under thallium.
Uses: Medicine; high specific gravity solutions used to separate ore constituents by flotation.
Shipping regulations: (Rail, Air) Poison label.

thallium bromide (thallous bromide) TlBr.
Properties: Yellowish white, crystalline powder. Soluble in alcohol; slightly soluble in water; insoluble in acetone. Sp. gr. 7.557; b.p. 815°C; m.p. (approx) 460°C.
Hazard: Toxic. See under thallium.
Use: Mixed crystals with thallium iodide for infrared radiation transmitters used in military detection devices.
Shipping regulations: (Rail, Air) Poison label.

thallium carbonate (thallous carbonate) Tl_2CO_3.
Properties: Heavy shiny, colorless or white crystals. Highly refractive. Melts to dark gray mass. Slightly alkaline taste. Soluble in water; insoluble in alcohol. Sp. gr. 7; m.p. 272°C.
Hazard: Toxic. See under thallium.
Uses: Analysis (testing for carbon disulfide); artificial diamonds.
Shipping regulations: (Rail, Air) Poison label.

thallium chloride (thallous chloride)TlCl.
Properties: White, crystalline powder. Becomes violet on exposure to light. Slightly soluble in water; insoluble in alcohol, ammonium hydroxide. Sp. gr. 7.004 (30/4°C); m.p. 430°C; b.p. 720°C.
Containers: Bottles.
Hazard: Toxic. See under thallium.
Uses: Catalyst (chlorination); medicine; sun-tan lamps.
Shipping regulations: (Rail, Air) Poison label.

thallium hydroxide (thallous hydroxide) $TlOH \cdot H_2O$.
Properties: Yellow needles. Soluble in alcohol, water. B.p. (dehydrated) 139°C, decomposes.
Hazard: Toxic. See under thallium.
Use: Analysis (testing for ozone; indicator).
Shipping regulations: (Rail, Air) Poison label.

thallium iodide (thallous iodide) TlI.
Properties: Yellow powder. Insoluble in alcohol; slightly soluble in water; soluble in aqua regia. Sp. gr. 7.09; b.p. 824°C; m.p. 440°C. Becomes red at 170°C.
Hazard: Toxic. See under thallium.
Uses: Medicine; mixed crystals with thallium bromide for infrared radiation transmitters.
Shipping regulations: (Rail, Air) Poison label.

thallium-mercury alloy. Reported to have freezing point −60°C, about 20°C below that of mercury.
Hazard: Toxic. See under thallium.
Uses: Substitute for mercury in electrical switches, thermometers, etc. for extreme low-temperature service.

thallium monoxide (thallium oxide; thallous oxide) Tl_2O.
Properties: Black powder. Oxidizes when exposed to air. Keep well stoppered! Soluble in alcohol, water (decomposes). M.p. 300°C; b.p. 1080°C; sp. gr. 9.52 (16°C).
Hazard: Toxic. See under thallium.

Uses: Analysis (testing for ozone); artificial gem; optical glass of high refractive index.
Shipping regulations: (Rail, Air) Poison label.

thallium nitrate (thallous nitrate) $TlNO_3$.
Properties: Colorless crystals. Soluble in hot water; insoluble in alcohol. Sp. gr. 5.5; m.p. 206°C (solidifies to a glass-like solid); decomposes at 450°C.
Grade: Technical.
Hazard: Highly toxic. Tolerance, 0.1 mg per cubic meter of air. Strong oxidizing agent. Fire and explosion risk.
Uses: Analysis; pyrotechnics; green fire.
Shipping regulations: (Rail, Air) Poison label.

thallium sesquichloride (thallo-thallic chloride) $TlCl_3 \cdot 3TlCl$ or Tl_2Cl_3.
Properties: Yellow, crystalline powder. Slightly soluble in water. Sp. gr. 5.9; m.p. 400–500 °C.
Hazard: Toxic. See under thallium.
Shipping regulations: (Rail, Air) Poison label.

thallium sulfate (thallous sulfate) Tl_2SO_4.
Properties: Colorless crystals. Soluble in water. Sp. gr. 6.77; m.p. 632°C.
Grades: Technical, 99%.
Hazard: Toxic. See under thallium.
Uses: Analysis (testing for iodine in the presence of chlorine); medicine; ozonometry; rodenticides; pesticide.
Shipping regulations: (Rail, Air) Poison label.

thallium sulfide (thallous sulfide) Tl_2S.
Properties: Blue-black, lustrous, microscopic crystals or amorphous powder. Soluble in mineral acids; insoluble in water, alcohol or ether. Sp. gr. 8.46; m.p. 448°C.
Hazard: Toxic. See under thallium.
Uses: Infrared-sensitive photocells.
Shipping regulations: (Rail, Air) Poison label.

thallous compound. See corresponding thallium compound.

THAM. See tris(hydroxymethyl)aminomethane.

"Thancat."[137] Trademark for a series of catalysts for use in urethane resins. The three letters that follow the trademark most nearly identify the chemical name. Some examples are:
DME. N,N-Dimethylethanolamine.
DMP. N,N'-Dimethylpiperazine.
NEM. N-Ethylmorpholine.
NMM. N-Methylmorpholine.

"Thanite."[266] Trademark for a technical grade of isobornyl thiocyanoacetate.
Use: Insecticide.

"Thanol."[137] Trademark for a series of polyols for use in polyurethane resins. "Thanol" is followed by letters denoting the broad use or composition, or by numbers indicating the molecular range.

thebaine (para-morphine) $C_{19}H_{21}NO_3$.
Properties: White, crystalline alkaloid. Slightly soluble in water; soluble in alcohol and ether. M.p. 193°C. Sp. gr. 1.30. It is non-addictive; extracted from poppies of a different species from morphine-producing opium poppies. It is a precursor of cocaine.

"THEIC."[50] Trademark for tris(2-hydroxyethyl) isocyanurate (q.v.).

theine. See caffeine.

Thenard's blue. See cobalt blue.

thenyl. The radical $C_4H_3SCH_2$—or $\underline{CHCHSCHC}CH_2$—, based on methylthiophene.

thenyl alcohol (2-thienylmethanol; 2-hydroxymethylthiophene; 2-thiophenecarbinol). $C_4H_4SCH_2OH$.
Properties: Colorless liquid; b.p. 207°C; insoluble in water; soluble in alcohol and ether.
Derivation: A heterocyclic alcohol made by reaction of 2-thienylmagnesium iodide and formaldehyde.
Uses: No commercially developed applications.

thenyldiamine. Coined name for 2-[2-dimethylaminoethyl]-3-thenylaminopyridine. $(C_4H_3SCH_2)N(C_5H_4N)CH_2CH_2N(CH_3)_2$.
Properties: Liquid; b.p. (1.0 mm) 169–172°C.
Derivation: Condensing N,N-dimethylaminoethyl-alpha-aminopyridine with 3-thenyl bromide.
Uses: Medicine (as base for various salts, especially the hydrochloride).

theobroma oil (cacao butter; cocoa butter).
Properties: Yellowish-white solid with chocolate-like taste and odor. Sp. gr. 0.858–0.864 (100/25°C); m.p. 30–35°C; refractive index (n 40/D) 1.4537–1.4585; saponification number 188–195; iodine number 35–43. Insoluble in water; slightly soluble in alcohol; soluble in boiling dehydrated alcohol; freely soluble in ether and chloroform. Combustible. Low toxicity.
Derivation: From the cacao bean, by expression, decoction, or extraction by solvent.
Chief constituents: Glycerides of stearic, palmitic, and lauric acids.
Grades: Crude; refined; U.S.P.
Containers: Tins; drums.
Uses: Confectionery; suppositories and pharmaceuticals; soaps; cosmetics.

theobromine (3,7-dimethylxanthine) $C_7H_8N_4O_2$. The alkaloid found in cocoa and chocolate products. A purine base closely related to caffeine. Also occurs in tea and cola nuts.
Hazard: Toxic by ingestion.

theophylline (1,3-dimethylxanthine) $C_7H_8N_4)_2 \cdot H_2O$.
Properties: White, crystalline alkaloid; odorless; bitter taste; m.p. 270–274°C; slightly soluble in water and alcohol; more soluble in hot water; soluble in alkaline solutions.
Derivation: (a) By extraction from tea leaves; (b) synthetically from ethyl cyanoacetate.
Grades: Technical; N.F.
Containers: Glass bottles; 1-, 5-lb tins; 25-, 100-lb drums.
Use: Medicine.

theoretical plate. See HETP.

therm. A unit of heat equal to 100,000 Btu. It has also been used to mean 1 Btu or 1 small calorie, but these uses have been abandoned in the U.S.

thermal black. Carbon black made from natural gas by the thermatomic process (q.v.). Production by pyrolysis of bituminous coal is an alternative method.
Grades: Fine thermal black (FT); medium thermal black (MT).
Hazard: Dust may be fire risk.
Uses: Rubber compounding; reclaimed rubber; pigment. See also furnace black.

thermal cracking. See cracking.

thermal decomposition See decomposition (2); pyrolysis.

thermal expansion coefficient. The change in volume per unit volume per degree change in temperature (cubical coefficient). For isotropic solids the expansion is equal in all directions; and the cubical coefficient is approximately three times the linear coefficient of expansion. These coefficients vary with temperature, but for gases at constant pressure the coefficient of volume expansion is nearly constant and equals 0.00367 for 1°C at any temperature.

thermal neutron. A slow neutron. See neutron.

thermal pollution. Heat introduced into rivers or estuaries by power plants or other industrial cooling waters or chemical wastes, which has adverse effects on estuarine ecology. See also water pollution.

"Therma-Tite."[170] Trademark for a specially formulated adhesive for high-speed bottle label machines to withstand the dangers of thermoshock. Material will adhere to cold glass without crystallizing.

thermatomic process. Methane or natural gas is cracked over hot bricks at a temperature of 1600°F to form amorphous carbon (carbon black) and hydrogen. See thermal black; carbon black.

"Thermax."[69] Trademark for a medium thermal carbon black used for compounding with rubber.

"Therminol" FR.[58] Trademark for heat-transfer liquids which are thermally and oxidatively stable to 600°F and extremely fire-resistant.

thermite. A mixture of ferric oxide and powdered aluminum, usually enclosed in a metal cylinder and used as an incendiary bomb invented by a German chemist Hans Goldschmidt about 1900. On ignition by a ribbon of magnesium, the reaction produces a temperature of about 4000°F, which is sufficient to soften steel. This is typical of some oxide/metal reactions which provide their own oxygen supply and thus are very difficult to stop.
Hazard: Dangerous fire risk.
Shipping regulations: (Rail, Air) Consult authorities.

thermochemistry. That branch of chemistry comprising the measurement and interpretation of heat changes that accompany changes of state and chemical reactions. It is closely related to chemical thermodynamics (q.v.). The heat of formation of a compound is the heat absorbed when it is formed from its elements in their standard states. An exothermic reaction evolves heat; an endothermic reaction requires heat for initiation. Application of thermochemical principles to generation of hydrogen by water splitting is a future possibility.

thermodynamics. A rigorously mathematical analysis of energy relationships (heat, work, temperature, and equilibrium) the principles of which were first elaborated by J. Willard Gibbs (q.v.) in the mid-19th century. It describes systems whose states are determined by thermal parameters, such as temperature, in addition to mechanical and electromagnetic parameters. A *system* is a geometric section of the universe whose boundaries may be fixed or varied, and which may contain matter or energy or both. The *state* of a system is a reproducible condition, defined by assigning fixed numerical values to the measurable attributes of the system. These attributes may be wholly reproduced as soon as a fraction of them have been reproduced. In this case the fractional number of attributes determines the state, and is referred to as the *number of variables of state* or the *number of degrees of freedom* of the system.

The concept of *temperature* can be evolved as soon as a means is available for determining when a body is "hotter" or "colder." Such means might involve the measurement of a physical parameter such as the volume of a given mass of the body. When a "hotter" body, A, is placed in contact with a "colder" body, B, it is observed that A becomes "colder" and B "hotter." When no further changes occur, and the joint system involving the two bodies has come to equilibrium, the two bodies are said to have the same temperature. Thus temperature can only be measured at equilibrium. Therefore thermodynamics is a science of equilibrium, and a thermodynamic state is necessarily an equilibrium state. Thermodynamics is a macroscopic discipline, dealing only with the properties of matter and energy in bulk, and does not recognize atomic and molecular structure. Although severely limited in this respect, it has the advantage of being completely insensitive to any change in our ideas concerning molecular phenomena, so that its laws have broad and permanent generality. Its chief service is to provide mathematical relations between the measurable parameters of a system in equilibrium so that, for example, a variable like pressure may be computed when the temperature is known, and *vice versa*.

The three laws of thermodynamics involve concepts too involved to be described here. For a more detailed explanation, see the article by Howard Reiss in "Encyclopedia of Chemistry", from which this entry was adapted.

thermodynamics, chemical. See chemical thermodynamics.

thermoelectricity. Electricity produced directly by applying a temperature difference to various parts of electrically conducting or semiconducting materials. Usually two dissimilar materials are used, and the points of contact are kept at different temperatures (Peltier effect). Many temperature-measuring devices (thermocouples, thermopiles) work on this principle, since the voltage is proportional to the temperature difference. Metallic conductors are usually used for these "thermometers," which produce a rather small current. A newer use for the effect is as a source of electrical energy, i.e., a means of direct conversion of heat into electricity (or vice versa) without the use of a generator (or motor). The materials used for these thermoelectric couples are semiconductors (e.g., tellurium; zinc antimonide; lead, bismuth, and germanium tellurides; samarium sulfide) or thermoelectric alloys, all of which produce relatively large currents. Several of these "cells" are then hooked in series much like the cells of a battery.

"Thermoflex" A.[28] Trademark for a rubber antioxidant containing 25% di-para-methoxydiphenylamine $(CH_3OC_6H_4)_2NH$, 25% diphenyl-para-phenylenediamine $C_6H_4(NHC_6H_5)_2$, and 50% phenyl-beta-naphthylamine $C_{10}H_7NHC_6H_5$.
Properties: Dark gray pellets; sp. gr. 1.21; f.p. not lower than 67°C.
Uses: Tire carcasses, transmission belts, etc., to promote resistance to flexing at operating temperatures. See also antioxidant.

thermofor. A heat-transfer medjum. See coolant.

thermoforming. (1) See reforming. (2) Forming or shaping a thermoplastic sheet by heating the sheet above its melting point, fitting it along the contours of a mold with pressure supplied by vacuum or other force, and removing it from the mold after cooling below its softening point. The method is applied to polystyrenes, acrylics, vinyls, polyolefins, cellulosics, etc.

Thermofor process. Catalytic cracking process in which petroleum vapor is passed up through a reactor countercurrent to a flow of small beads of catalyst. The deactivated catalyst then passes through a regenerator and is recirculated.

"Thermoguard."[288] Trademark for antimony-based materials for incorporation in PVC and other chlorine-containing plastics for flameproofing properties.

"Thermolastic."[125] Trademark for a copolymer of butadiene and styrene which does not require vulcanization. Has the elastic and resilient properties of rubber while offering the techniques of thermoplastic processing. The desired end product may be formed by thermoplastic injection molding, extrusion, blow molding, or vacuum forming.

"Thermolite."[288] Trademark for a series of stabilizers for vinyl resins.

thermonuclear reaction. A nuclear fusion reaction initiated by a temperature in the neighborhood of 15 million degrees Kelvin, which can be attained in practice only by means of a fission reaction, and experimentally (on a minute scale) by high energy input from a laser. This phenomenon, which on the sun releases in one second energy equivalent to about 10^{11} megatons of TNT, occurs in a hydrogen bomb explosion in less than a millionth of a second. The problem of maintaining and controlling this energy release in a useful manner, as has been done successfully with fission, is of almost inconceivable magnitude. A controlled thermonuclear reactor would have to operate electromagnetically at temperatures of 10 million degrees Kelvin, far higher than any known materials can endure. Fuels would be elements fom hydrogen through lithium (hydrogen and its isotopes, deuterium and tritium, helium and lithium-7), irradiated by laser or electron-beam pulses. Intensive research is under way, in view of the importance of this reaction as a possible future energy source. See also nuclear reaction; fusion.

thermoplastic. A high polymer that softens when exposed to heat and returns to its original condition when cooled to room temperature. Natural substances that exhibit this behavior are crude rubber and a number of waxes; however, the term is usually applied to synthetics such as polyvinyl chloride, nylons, fluorocarbons, linear polyethylene, polyurethane prepolymer, polystyrene, polypropylene, and cellulosic and acrylic resins. See also thermoset.

thermoset. A high polymer that solidifies or "sets" irreversibly when heated. This property is usually associated with a cross-linking seaction of the molecular constituents induced by heat or radiation, as with proteins, and in the baking of doughs. In many cases, it is necessary to add "curing" agents such as organic peroxides or (in the case of rubber) sulfur. For example, linear polyethylene can be cross-linked to a thermosetting material either by radiation or by chemical reaction. Phenolics, alkyds, amino resins, polyesters, epoxides, and silicones are usually considered to be thermosetting; but the term also applies to materials where additive-induced cross-linking is possible, e.g., natural rubber.

THF. Abbreviation for tetrahydrofuran (q.v.).

"THFA."[224] Trademark for tetrahydrofurfuryl alcohol (q.v.).

thia-. Prefix indicating the presence of sulfur in a heterocyclic ring.

"Thiacryl."[27] Trademark for millable polyacrylate base synthetic rubbers.

Uses: Seals, gaskets, O-rings for automotive power train components and hydraulic assemblies; also hose, belting.

thiamine (vitamin B_1) $C_{12}H_{17}ClN_4OS$. 3-(4-Amino-2-methylpyrimidyl-5-methyl)-4-methyl-5, beta-hydroxyethylthiazolium chloride. The antineuritic vitamin, essential for growth and the prevention of beriberi. It functions in intermediate carbohydrate metabolism in coenzyme form in the decarboxylation of alpha-keto acids. Deficiency symptoms: emotional hypersensitivity, loss of appetite, susceptibility to fatigue, muscular weakness and polyneuritis.

Sources: Enriched and whole grain cereals, milk, legumes, meats, yeast. Most of the thiamine commercially available is synthetic.

Uses: Medicine; nutrition; enriched flours. Isolated usually as the chloride (see formula above). Available as thiamine hydrochloride and thiamine mononitrate.

"Thiate."[69] Trademark for a series of accelerators for neoprene.

"Thiate A." (thiohydropyrimidine).

Properties: White crystalline powder; sp. gr. 1.12 ± .03; melts at 250°C min; moderately soluble in chloroform, ethyl alcohol; insoluble in water, dilute caustic, carbon disulfide, gasoline, benzene. Used for neoprenes to be cured in open steam. Used alone or with thiurams or guanidines plus sulfur.

"Thiate B." (trialkyl thiourea).

Properties: Reddish brown liquid; sp. gr. 1.05 ± .12; very soluble in benzene, carbon disulfide, chloroform, acetone, ethyl alcohol; soluble in water; insoluble in dilute caustic, gasoline. Used where low compression set is required.

"Thiate E." (trimethyl thiourea).

Properties: Light tan flakes; sp. gr. 1.23 ± .03; melting range 68–78°C; very soluble in water; soluble in toluene; insoluble in petroleum ether. Used where low compression set is required.

1,4-thiazane $CH_2SCH_2CH_1NHCH_2$.

Properties: Colorless liquid. Pyridine-like odor. Fumes in air. Absorbs carbon dioxide from the air. Soluble in alcohol, benzene, ether, water. S.p. 169°C (758 mm). combustible.

Derivation: Interaction of alcoholic ammonia and dichlorodiethyl sulfide.

Grade: Technical.

Use: Organic synthesis.

thiazole SCH:NCH:CH.

Properties: Colorless or pale yellow liquid; sp. gr. 1.198; b.p. 116.8°C; soluble in alcohol and ether; slightly soluble in water; odor resembles that of pyridine.

Uses: Organic synthesis of fungicides, dyes, and rubber accelerators.

thiazole dye. A dye whose molecular structure contains the thiazole ring (see thiazole). The chromophore groups are =C=N—, —S—C=, but the conjugated double bonds are also of importance. The members of the class are mainly used as direct or developed dyes for cotton, though some find use as union dyes. One example is primuline, CI No. 49000.

2-thiazylamine. See 2-aminothiazole.

"Thibenzole."[123] Trademark for thiabendazole, a veterinary anthelmintic.

"Thibetolide."[227] Trademark for pentadecanolide (q.v.).

thickening agent. Any of a variety of substances used to increase the viscosity of liquid mixtures and solutions and to aid in maintaining stability by their emulsifying properties. Four classifications are recognized: (1) starches, gums, casein, gelatin and phycocolloids, (2) semisynthetic cellulose derivatives (carboxymethylcellulose, etc.), (3) polyvinyl alcohol and carboxyvinylates (synthetic), and (4) bentonite, silicates and colloidal silica. The first group is widely used in the food industries, especially in ice creams, confectionery, gravies, etc.; other major consumers are the paper, adhesives, textiles, and detergent fields. See also emulsion.

thimerosal (sodium ethylmercurithiosalicylate; ethylmercurithiosalicylic acid, sodium salt) $NaOOCC_6H_4SHgC_2H_5$.
Properties: Light cream-colored crystalline powder with slight characteristic odor. pH (1% solution) about 6.7. Affected by light. Soluble in water and alcohol; almost insoluble in ether and benzene.
Derivation: Reaction between ethylmercuric chloride and thiosalicylic acid in alcoholic sodium hydroxide.
Grade: N.F.
Hazard: Toxic by ingestion.
Uses: Medicine; bacteriostat, fungistat.

"Thimet."[57] See phorate.

thin. A nontechnical word used by scientists, with a variety of meanings. (1) In electronic metallurgy, a thin film is a vapor-deposited coating having a thickness of only a single atom; such *monatomic* films, e.g., thorium on tungsten, are used in electronic devices such as cathodes. (2) A coating or film of a fatty acid on water which is one molecule thick (about 200 angstroms) is called a *monomolecular* film. (3) In thin-layer chromatography the term applies to a specially prepared mixture of adsorbents spread on a glass slide to a thickness of about 1/100 inch. (4) The word is also used in the sense of a liquid of low viscosity, as in paint thinner and thin-boiling starch.

thin-boiling starch. See starch, modified; thin.

thin film. In electronic engineering, a film having a "thickness" of a single atom and consisting of a metal deposited on a metallic substrate either externally by vapor deposition or internally by diffusion. The base metal is usually tungsten (for a cathode), the film being any of a number of other metals (thorium, cesium, zirconium, barium, or cerium). "The greatest benefit is obtained when the films are of a monatomic nature; in the case of thorium on tungsten, the optimum coverage is 0.67 monatomic layer, i.e., the thorium atoms do not completely cover the tungsten surface. Such films are very tightly bound to the base metal by atomic forces." (W.H. Kohl). See also thin; film.

thin-layer chromatography (TLC). A micro type of chromatography (q.v.). The thin layer (about 0.01 inch) is the adsorbent, usually a special silica gel spread on glass, or incorporated in a plastic film. Single drops of the solutions to be investigated are placed along one edge of the glass plate, and this edge then dipped into a solvent. The solvent carries the constituents of the original test drops up the thin layer in a selective separation, so that a comparison with known standards, and various identifying tests may be made on the spots formed. See also thin.

thinner. A hydrocarbon (naphtha) or oleoresinous solvent (turpentine) used to reduce the viscosity of paints to appropriate working consistency usually just prior to application. In this sense a thinner is a liquid diluent, except that it has active solvent power on the dissolved resin.

thio-. A prefix used in chemical nomenclature to indicate the presence of sulfur in a compound, usually as a substitute for oxygen. See also thiol.

thioacetamide CH_3CSNH_2.
Properties: Colorless leaflets; stable in solution; m.p. 115°C; soluble in water, alcohol, ether, benzene. Combustible.
Hazard: Moderately toxic by ingestion and inhalation.
Use: To replace gaseous hydrogen sulfide in qualitative analysis.

thioacetic acid (thiacetic acid; ethanethiolic acid) CH_3COSH.
Properties: Clear yellow liquid; strong, unpleasant odor; sp. gr. 1.05 (25°C); f.p. −17°C; b.p. 81.8°C (630 mm); soluble in water, alcohol and ether. Combustible.
Derivation: By heating glacial acetic acid and phosphorus pentasulfide, with subsequent distillation.
Hazard: Moderately toxic by ingestion and inhalation.
Uses: Chemical reagent; lachrymator.

thioallyl ether. See allyl sulfide.

thioanisole $C_6H_5SCH_3$.
Properties: Colorless liquid with strong, unpleasant odor; sp. gr. 1.053 (25°C); f.p. −15.5°C; b.p. 188°C; refractive index 1.5842 (n 25/D). Insoluble in water; soluble in most organic solvents. Combustible.
Uses: Intermediate; solvent for polymeric systems.

thiobenzoic acid C_6H_5COSH.
Properties: Yellow oil or crystals. Sp. gr. (20/4°C) 1.1825–1.1835; m.p. 24°C; b.p. 77.5°C (5 mm), 122°C (30 mm); refractive index (n 20/D) 1.602–1.604. Insoluble in water; miscible in all proportions with organic solvents. Combustible.
Grade: 95% min.
Uses: Organic intermediate; in zinc thiobenzoate.

4,4′-thiobis(6-tert-butyl-meta-cresol).
Properties: Light gray to tan powder; m.p. 150°C min; sp. gr. 1.10 (25°C).
Uses: Protection of light-colored rubber from oxidation and of non-staining neoprene compounds against deterioration.

2,2′-thiobis(chlorophenol) $[ClC_6H_3(OH)]_2S$.
Properties: White crystalline solid; m.p. 175.8–186.8°C; odorless; insoluble in water.
Hazard: May be toxic by ingestion.
Uses: Bacteriostat for cosmetics; fungicide.

2,2′-thiobis(4,6-di-sec-amylphenol) (2,2′-thiobis[4,6-bis-(1-methylbutyl)phenyl]) $[(CH_3[CH_2]_2CH[CH_3])_2OHC_6{:}H_2]_2S$. A dark viscous liquid; softening point 0°C; sp. gr. 0.99 (50°C).
Uses: Rubber antioxidant.

thiocarbamide. See thiourea.

thiocarbanil. See phenyl mustard oil.

thiocarbanilide (N,N′-diphenylthiourea; sulfocarbanilide) $CS(NHC_6H_5)_2$.
Properties: Gray powder; m.p. 148°C; sp. gr. 1.32. Soluble in alcohol and ether; insoluble in water. Combustible. Low toxicity.

Derivation: Interaction of aniline and carbon disulfide and alcohol in the presence of sulfur.
Containers: Bags; drums; carlots.
Uses: Intermediates; dyes (sulfur colors, indigo, methyl indigo); vulcanization accelerator; synthetic organic pharmaceuticals; flotation agent; acid inhibitor.

thiocarbonyl chloride. See thiophosgene.

thioctic acid. See dl-alpha-lipoic acid.

thiodiethylene glycol. See thiodiglycol.

thiodiglycol (thiodeithylene glycol; beta-bis-hydroxyethyl sulfide; dihydroxyethyl sulfide) $(CH_2CH_2OH)_2S$.
Properties: Syrupy colorless liquid. Characteristic odor. Sp. gr. 1.1852 at 20°C; b.p. 283°C; freezing point −10°C; viscosity 0.652 poise (20°C); flash point 320°F; weight per gallon 9.85 lbs; refractive index (20/D) 1.5217. Soluble in acetone, alcohol, chloroform, water; slightly soluble in benzene, carbon tetrachloride, and ether. Combustible. Low toxicity.
Derivation: Hydrolysis of dichloroethyl sulfide; interaction of ethylene chlorohydrin and sodium sulfide.
Containers: Cans; drums; tank cars.
Hazard: Do not use with hydrochloric acid.
Uses: Organic synthesis; solvent for dyes in textile printing; antioxidant.

thiodiglycolic acid $HOOCCH_2SCH_2COOH$. A dicarboxylic acid. Colorless crystals; m.p. 128°C. Soluble in water and alcohol. Combustible.
Hazard: Moderately toxic.
Uses: Analytical reagent.

4,4-thiodiphenol (TDP) $(C_6H_5OH)_2S$.
Uses: Engineering plastics; antioxdant, flame retardant intermediate.

thiodiphenylamine. See phenothiazine.

thiodipropionic acid $HOOCCH_2 \cdot S \cdot CH_2CH_2COOH$. A dicarboxylic acid.
Uses: Preservative and antioxidant for edible fats and oils; pharmaceuticals; food packaging ingredient.
Hazard: Use in foods restricted to 0.02% of fat and oil content, including essential oils.

beta,beta-**thiodipropionitrile** $S(Ch_2CH_2CN)_2$.
Properties: White crystals or light yellow liquid; sp. gr. (30°C) 1.1095; m.p. 28.65°C; slightly soluble in water and alcohol; soluble in acetone, chloroform and benzene.
Hazard: May be toxic.
Uses: Preservative; selective solvent; chromatography.

thio-1,3-dithio[4,5-b]**quinoxaline.** See thioquinox.

thioethanolamine. See 2-aminoethanethiol.

"Thiofast."[438] Trademark for a thio-indigo deep maroon pigment with a red-violet undertone. Used in paints, printing inks, and plastics.

"Thiofide."[58] Trademark for 2-2′-dithiobis(benzothiazole) (q.v.).

thioflavine T $CH_3C_6H_3N(HCl)SCC_6H_4N(CH_3)_2$.
Properties: A yellow basic dye of the thiazole class. CI No. 49005. Fluoresces yellow to yellowish green when excited by ultraviolet.
Derivation: By heating para-toluidine with sulfur in the presence of lead oxide.
Uses: Textile dyeing; fluorescent sign paints; in combination with green or blue pigments to produce brilliant greens; phosphotungstic pigments.

thiofuran. See thiophene.

"Thiogel."[91] Trademark for gelatin into which thiol or -SH groups have been introduced by an improved thiolation process.
Uses: Vehicle for gradual release of active compounds; enzyme activity measurement; tissue culture medium; photography and graphic arts; gel filtration; surface coatings.

1-thioglycerol $CH_2(OH)CH(OH)CH_2SH$.
Properties: Water-white liquid; b.p. 118°C at 5 mm; sp. gr. 1.295 (14.4°C). Soluble in water, alcohol and ether. Combustible; low toxicity.
Uses: Reducing agent for cystine molecule in human hair and wool; for stabilization of acrylonitrile polymers; medicine.

thioglycolic acid (mercaptoacetic acid) $HSCH_2COOH$.
Properties: Colorless liquid; strong, unpleasant odor; sp. gr. 1.325; f.p. −16.5°C; b.p. 105°C. Miscible with water, alcohol or ether. Combustible.
Derivation: Heating chloroacetic acid with potassium hydrogen sulfide.
Containers: Carboys; drums.
Hazard: Toxic by ingestion and inhalation; strong irritant to tissue.
Uses: Reagent for iron; manufacture of thioglycolates, permanent wave solutions and depilatories; vinyl stabilizer; manufacture of pharmaceuticals.
Shipping regulations: (Rail) Corrosive liquids, n.o.s., (Air) White label. Corrosive label.

2-thiohydantoin (glycolylthiourea) $NHC(S)NHC(O)CH_2$.
Properties: Crystals or tan powder. M.p. 230°C. Slightly soluble in water; insoluble in alcohols and ethers.
Purity: 99% min.
Use: Intermediate for pharmaceuticals, rubber accelerators, copper plating brighteners and dyestuffs.

2-thio-4-keto-thiazolidine. See rhodanine.

"Thiokol."[27] Trademark for a series of polysulfide elastomers in the form of liquids, water dispersions and millable (dry) rubbers. Combustible. Characterized by almost 100% resistance to solvents, particularly hydrocarbons, but has relatively low tensile strength.
Uses: Sealants for gasoline tanks; sealer adhesive for machine components; potting and sealing electrical parts; caulking ship decks and buildings; flexibilizing constituent of resin-based adhesives; paint spray and gasoline hose; oil suction and discharge hose; rocket propellant binder.

thiol (mercaptan). Any of a group of organic compounds resembling alcohols, but having the oxygen of the hydroxyl group replaced by sulfur, as in ethanethiol (C_2H_5SH). Many thiols are characterized by strong and repulsive odors.
Hazard: Aliphatic thiols are flammable and toxic by inhalation.
Uses: Warning agents in fuel gas lines; chemical intermediates. See also specific compound.
Shipping regulations: (Mercaptan mixtures, aliphatic) (Rail) Red label. (Air) Flammable Liquid label.
Note: Adoption of the name "thiol" to replace "mercaptan" has been officially recommended as being more consistent with the molecular constitution of these compounds. The older term, which literally means "mercury-seizing", is inappropriate.

thiomalic acid (mercaptosuccinic acid) $HOOCCH(SH)CH_2COOH$.
Properties: White crystals or powder; sulfuric odor; m.p. 149–150°C; soluble in water, alcohol, acetone,

and ether; slightly soluble in benzene. Low toxicity; combustible.
Use: Biochemical research; intermediate; rust inhibitor; antidarkening agent for crepe rubber; tackifier for synthetic rubber.

thionazin. Generic name for O,O-diethyl O-2-pyrazinyl phosphorothioate
$\overline{\text{NCHCHNCHC}}\text{OPS}(C_2H_5O)_2$.
Properties: Amber liquid; m.p. −1.7°C; b.p. 80°C (.001 mm). Slightly soluble in water; miscible with most organic solvents.
Hazard: Toxic by ingestion, inhalation and skin absorption. Cholinesterase inhibitor.
Uses: Insecticide; fungicide; nematocide.
Shipping regulations: (Rail, Air) Organic phosphates, liquid, n.o.s. Poison label. Not accepted on passenger planes.

"Thionex."[28] Trademark for tetramethylthiuram monosulfide. $[(CH_3)_2NCH]_2S$.
Properties: Lemon yellow powder or grains; sp. gr. 1.39; m.p. not lower than 105°C.
Use: Ultra-accelerator for rubber.
See also "Monex."

"Thionone."[232] Trademark for a series of sulfur dyestuffs.

thionyl chloride (sulfurous oxychloride; sulfur oxychloride) $SOCl_2$.
Properties: Pale yellow to red liquid with sharp odor; sp. gr. 1.638; f.p. −105°C; b.p. 79°C; decomposes at 140°C. decomposes (fumes) in water; soluble in benzene.
Grades: 93%, 97.5%.
Containers: Glass carboys; drums.
Hazard: Toxic; strong irritant to skin and tissue. Safety data sheet available from Manufacturing Chemists Assn., Washington, D.C.
Uses: Pesticides; engineering plastics; chlorinating agent; catalyst.
Shipping regulations: (Rail) White label. (Air) Corrosive label. Not acceptable on passenger planes.

2-thio-4-oxypyrimidine. See thiouracil.

thiopental sodium ("Pentothal"; "Sodium Pentothal"). A rapidly acting barbiturate administered intravenously for general anesthesia and hypnosis. Commonly known as "truth serum". Its chemical name is sodium 5-ethyl-5(1-methylbutyl)-2-thiobarbiturate ($C_{11}H_{17}N_2O_2SNa$). May cause respiratory failure; should be used only with physician in attendance.

thiophane. See tetrahydrothiophene.

thiophene (thiofuran) $\underline{\text{CHCHCHCHS}}$. A cyclic organosulfur; highly reactive.
Properties: Colorless liquid; refractive index (n 20/D) 1.5285; sp. gr. 1.0644 (20°/4°C); f.p. −38.5°C; b.p. 84°C; flash point 30°F. Soluble in alcohol and ether; insoluble in water.
Derivation: From coal tar (benzene fraction) and petroleum; synthetically from heating sodium succinate with phosphorus trisulfide.
Hazard: Flammable, dangerous fire risk. Moderately toxic.
Use: Organic synthesis (condenses with phenol and formaldehyde; copolymerizes with maleic anhydride).
Shipping regulations: (Rail) Flammable liquid, n.o.s., Red label. (Air) Flammable Liquid label.

alpha-**thiophenealdehyde** C_4H_3SCHO. (2-Thiophenecarboxaldehyde).
Properties: Oily liquid with almond-like odor; b.p. 198°C, 90°C (20 mm); sp. gr. 1.210–1.220; very soluble in alcohol, benzene, ether; slightly soluble in water. Combustible.
Grade: 95%.
Containers: Drums.
Uses: Thiophene derivatives; introducing thenyl group into organic compounds.

thiophenol (phenyl mercaptan) C_6H_5SH.
Properties: Water-white liquid, repulsive odor; b.p. 168.3°C; b.p. (15 mm) 71°C; refractive index 1.5891; sp. gr. (25/25°C) 1.075; insoluble in water; soluble in alcohol and ether. Combustible.
Derivation: Reduction of benzenesulfonyl chloride with zinc dust in sulfuric acid.
Grade: 99%.
Hazard: Skin irritant, moderately toxic.
Uses: Pharmaceutical synthesis.

"Thiophos."[57] Trademark for parathion, technical (q.v.).

thiophosgene (thiocarbonyl chloride) $CSCl_2$.
Properties: Reddish liquid; sp. gr. 1.5085 (15°C); b.p. 73.5°C. Decomposes in water and alcohol. Soluble in ether.
Hazard: Highly toxic by ingestion and inhalation.
Use: Organic synthesis.
Shipping regulations: (Rail, Air) Poison label. Not acceptable on passenger planes.

thiophosphoryl chloride $PSCl_3$.
Properties: Colorless liquid; penetrating odor; sp. gr. 1.635; b.p. 126°C; f.p. −35°C; decomposed by water; soluble in carbon disulfide, carbon tetrachloride. Flash point, none. Nonflammable.
Hazard: Highly toxic; strong irritant to skin and tissue.
Shipping regulations: (Rail) White label. (Air) Corrosive label. Not acceptable on passenger planes.

thioquinox. Generic name for 2,3-quinoxalinedithiol cyclic trithiocarbonate. $C_6H_4N_2C_2S_2CS$ (tricyclic).
Properties: Yellow solid; m.p. 180°C. Insoluble in water. slightly soluble in acetone, kerosine and alcohol.
Hazard: Toxic by ingestion.

thioridazine (USAN) $C_{21}H_{26}N_2S_2$. 2-Methylthio-10-[2-(N-methyl-2-piperidyl)-ethyl]phenothiazine.
Properties: Colorless crystals; m.p. 158°C; soluble in water and alcohol.
Grade: N.D. (as the hydrochloride).
Use: Medicine.

thiosalicylic acid (2-mercaptobenzoic acid) $HOOCC_6H_4SH$.
Properties: Yellow solid; m.p. 164–165°C; sublimes; slightly soluble in hot water; soluble in alcohol, ether and acetic acid. Combustible.
Grades: Reagent; technical, 80%; pharmaceutical.
Containers: Drums.
Uses: Dyes; reagent for iron determination; intermediate.

thiosemicarbazide (aminothiourea) $NH_2CSNHNH_2$.
Properties: White crystalline powder, no odor, soluble in water and alcohol. Melting point 180–184°C. Low toxicity.
Derivation: From potassium thiocyanate and hydrazine salts.
Grades: Technical and pure.
Containers: Bottles; drums.

Uses: Reagent for ketones and certain metals; photography; rodenticide.

thiosinamine. See allyl thiourea.

"Thiostat-B."[248] Trademark for 40% aqueous solution of sodium dimethyldithiocarbamate.
Uses: Bactericide in oil-well water flooding; paper slimicide; specialty bactericide.

"Thiostop."[248] Trademark for aqueous solutions of the sodium and potassium salts of dimethyl dithiocarbamate.
Properties: Clear yellow to amber liquid; sp. gr. 1.18–1.23; good storage stability; potassium salt will crystallize out at 20° F and the sodium at 32° F. Avoid long storage in partially filled containers.
Uses: Shortstop in SBR polymerization.

"Thiotax."[58] Trademark for 2-mercaptobenzothiazole (q.v.).

thiotepa. U.S.P. name for triethylenethiophosphoramide.

thiouracil (2-thio-4-oxypyrimidine; 2-mercapto-4-hydroxypyrimidine) HOCNC(SH)NCHCH.
Properties: White, odorless, powder; bitter taste; melts with decomposition at about 340° C; characteristic ultraviolet absorption spectrum; slightly soluble in water, alcohol, and ether. Combustible.
Hazard: Moderately toxic by ingestion.
Uses: Medicine; veterinary medicine.

thiourea (thiocarbamide) $(NH_2)_2CS$.
Properties: White, lustrous crystals; bitter taste; sp. gr. 1.406; m.p. 180–182° C; b.p. sublimes in vacuo at 150–160° C; soluble in cold water, ammonium thiocyanate solution, and alcohol; nearly insoluble in ether.
Derivation: (1) By heating dry ammonium thiocyanate, extraction with a concentrated solution of ammonium thiocyanate with subsequent crystallization; (2) action of hydrogen sulfide on cyanamide.
Grades: Technical; reagent.
Containers: Bottles; bags.
Hazard: Toxic. May not be used in food products (FDA). Skin irritant (allergenic).
Uses: Photography and photocopying papers; organic synthesis (intermediates, dyes, drugs, hair preparations); rubber accelerator; analytical reagent; medicine (external); amino resins; mold inhibitor.

"Thiovanic Acid."[312] Trademark for thioglycolic acid (q.v.).

"Thiovanol."[312] Trademark for vacuum-distilled alpha-monothioglycerol. See 1-thioglycerol.

thioxylenol (mixed isomers) $C_6H_3(CH_3)_2SH$.
Properties: Colorless liquid; sp. gr. 1.03 (15.6/15.6° C); b.p. 122–133° C (50 mm). Oxidizes on exposure to air; supplied under nitrogen atmosphere. Combustible.
Uses: Chemical intermediate; peptizer for natural and SBR rubbers.

thiram (USAN) (tetramethylthiuram disulfide; bis-(dimethylthiocarbamyl) disulfide; thiuram; TMTD) $[(CH_3)_2NCH]_2S_2$.
Properties: White crystalline powder with a characteristic odor; soluble in alcohol, benzene, chloroform, carbon disulfide; insoluble in water, dilute alkali, gasoline; sp. gr. 1.29 (20° C); melting range 155–156° C.
Grades: 75% wettable powder; 95% technical powder; N.D.
Containers: 50-lb multiwall paper bags.
Hazard: Moderately toxic by ingestion and inhalation; irritant to skin and eyes. Tolerance, 5 mg per cubic meter of air.
Uses: Vulcanizing agent for rubber, especially for steam hose and other heat-resistant uses; fungicide; insecticide; seed disinfectant; lube oil additive; bacteriostat; animal repellent.

"Thiramad."[329] Trademark for a 75% thiram formulation. A turf fungicide.
Hazard: See thiram.

"Thiram-75W."[248] Trademark for a tetramethylthiuram disulfide formulation; used in seed treatment.
Hazard: See thiram.

"Thiurad."[58] Trademark for tetramethylthiuram disulfide.
Hazard: See thiram.

thiuram. A compound containing the group R_2NCS. Most are disulfides. See thiram. The most common monosulfide is tetramethylthiuram monosulfide (trademarked "Thionex" and "Monex").

"Thixatrol" ST.[202] Trademark for a colloidal thixotropic agent, designed especially for use in the heat developing types of dispersers.

"Thixcin."[202] Trademark for a colloidal thixotropic agent used in paints, inks, cosmetics, calks, sealants, epoxy and polyester resins, pharmaceuticals, etc. to prevent pigment settling, syneresis, excess flow and penetration.

"Thixo-Jel."[471] Trademark for a Wyoming sodium bentonite. Used where a granular bentonite with high gelling and bonding characteristics is required.

"Thixon."[465] Trademark for a family of adhesives or cements composed of mixtures of rubber and other bonding agents in solvents. Used in bonding natural and synthetic rubber to metals or plastics and for splicing various elastomers to themselves or each other.

thixotropy. The ability of certain colloidal gels to liquefy when agitated (as by shaking or ultrasonic vibration) and to return to the gel form when at rest. This is observed in some clays, paints, and printing inks which flow freely on application of slight pressure, as by brushing or rolling. Suspensions of bentonite clay in water display this property, which is desirable in oil-well drilling fluids. See also rheology; dilatancy.

Thomson, Sir J. J. (1856–1940). A native of England, Thomson entered Cambridge University in 1876 and remained there permanently as a professor of physics, especially in the field of electrical phenomena. His observations and calculations of cathode ray experiments led to proof of the existence of the electron as the lightest particle of matter (1896). This proof was announced at the Royal Institution in the following year. This was the keystone of the theory of atomic structure and one of the most notable discoveries in the history of science. (See Rutherford).

thomsonite $NaCa_2Al_5(SiO_4)_5 \cdot 6H_2O$. A mineral; one of the zeolites (q.v.).

thonzylamine hydrochloride $C_{16}H_{22}N_4O \cdot HCl$; 2-[(2-dimethylaminoethyl)(para-methoxybenzyl)amino]-pyrimidinehydrochloride.
$CH_3OC_6H_4Ch_2N(C_4H_3N_2)CH_2CH_2N(CH_3)_2 \cdot HCl$.
Properties: White, crystalline powder with faint odor. M.p. 173–176° C. Very soluble in water; freely soluble

in alcohol and chloroform; practically insoluble in ether and benzene; pH (2% solution) 5.0–6.0.
Grade: N.F.
Use: Medicine (antihistamine).

"Thorazine."[71] Trademark for a brand of chlorpromazine (q.v.).

thoria. See thorium dioxide.

thorin $HOC_{10}H_4(SO_3H)_2NNC_6H_4AsO_3H_2$. A reagent for the colorimetric determination of microgram quantities of thorium.
Hazard: Toxic by ingestion.

thorite $ThSiO_4$. A natural thorium silicate, usually impure, found in pegmatites.
Properties: Color black to orange; luster vitreous to resinous; hardness 4.5–5; sp. gr. 4.4–5.2; radioactive.
Occurrence: Norway, Ceylon.
Use: Source of thorium.

thorium Th Metallic element of atomic number 90; a member of the actinide series (Group IIIB of Periodic Table). Atomic weight 232.0381; valence 4; radioactive. No stable isotopes.
Properties: Soft metal with bright silvery luster when freshly cut; similar to lead in hardness when pure. Can be cold-rolled, extruded, drawn and welded. Sp. gr. about 11.7; m.p. about 1700°C; b.p. about 4500° C. Soluble in acids; insoluble in alkalies and water. Low toxicity. Some alloys may ignite spontaneously. The metal in massive form is not flammable. See Hazard.
Source: Monazite, thorite. It is about as abundant as lead.
Derivation: (a) Reduction of thorium dioxide with calcium; (b) fused salt electrolysis of the double fluoride $ThF_4 \cdot KF$. The product of both processes is thorium powder, fabricated into the metal by powder metallurgy techniques. Hot surface decomposition to the iodide produces crystalbar thorium.
Forms available: Powder; unsintered bars; sintered bars; sheets.
Hazard: Flammable and explosive in powder form. Dusts of thorium have very low ignition points and may ignite at room temperature. Radioactive decay isotopes are dangerous when ingested.
Uses: Nuclear fuel (thorium 232 is converted to uranium 233 on neutron bombardment after several decay steps); sun lamps; photoelectric cells; target in x-ray tubes; alloys.
Shipping regulations: (powder) (Rail) Yellow label. (Air) Flammable Solid label. Not acceptable on passenger planes.
Note: As of 1974, U.S. reserves of thorium were about 3 billion tons.

thorium acetylacetonate $Th[OC(CH_3):CHCOCH_3]_4$ Crystalline powder. Slightly soluble in water. Resistant to hydrolysis. A chelating, nonionizing compound.

thorium anhydride. See thorium dioxide.

thorium carbide ThC_2.
Properties: Yellow solid; sp. gr. 8.96 (18°C); m.p. 2630–2680°C; b.p. approx. 5000°C. Decomposes in water.
Use: Nuclear fuel.

thorium chloride (thorium tetrachloride) $ThCl_4$.
Properties: Colorless or white, lustrous needles (light yellow color caused by iron trace). Hygroscopic. Partially volatile. Crystallizes with variable water of crystallization. Soluble in alcohol, water. Sp. gr. 4.59; b.p. 928°C (decomposes); m.p. 820°C.
Grade: Technical; as 50% THO_2.
Containers: Glass bottles; fiber drums.
Hazard: Moderately toxic.
Use: Incandescent lighting.

thorium decay series. The series of radioactive elements produced as successive intermediate products when the element thorium undergoes its spontaneous natural radioactive disintegration into lead. Many of these are severe radioactive poisons when ingested or inhaled in the form of thorium dust particles.

thorium dioxide (thorium anhydride; thorium oxide; thoria) ThO_2.
Properties: Heavy, white powder; sp. gr. 9.7; m.p. about 3300°C (highest of all oxides); b.p. 4400°C; hardness 6.5 (Mohs). Very refractory. Soluble in sulfuric acid; insoluble in water.
Derivation: Reduction of thorium nitrate.
Grades: Technical and purities to 99.8% ThO_2; granular particles, crystals.
Containers: Lined drums.
Uses: Ceramics (high temperature); gas mantles; nuclear fuel; flame spraying; crucibles; medicine; nonsilica optical glass; catalyst; thoriated tungsten filaments in incandescent lamps, cathodes in electron tubes, and arc-melting electrodes.

thorium disulfide ThS_2. Dark brown-crystals; sp. gr. 7.30 (25/4°C); m.p. 1875–1975°C (vacuum); insoluble in water.
Use: Solid lubricant.

thorium fluoride ThF_4.
Properties: White powder having approximate formula $ThF_4 \cdot 1.4H_2O$. Dehydrated between 200 and 300°C; sp. gr. (anhydrous) 6.32 (24°C); m.p. 1111°C. Above 500°C reacts with atmospheric moisture to form thorium oxyfluoride, $ThOF_2$, and finally the oxide ThO_2. Forms a series of compounds with other metallic fluorides such as NaF and KF.
Grades: 79–80% ThO_2.
Containers: Glass bottles; fiber drums.
Hazard: Toxic. Tolerance (as F), 2.5 mg per cubic meter of air.
Use: Production of thorium metal and magnesium-thorium alloys; high temperature ceramics. $ThOF_2$ is used as a protective coating on reflective surfaces.

thorium nitrate $Th(NO_3)_4 \cdot 4H_2O$.
Properties: White, crystalline mass. Soluble in water and alcohol. Strong oxidizing agent. M.p. 500°C (decomposes).
Grades: Technical; C.P.
Containers: Glass bottles; fiber drums.
Hazard: Dangerous fire and explosion risk in contact with organic materials; radioactive.
Uses: Medicine, reagent for determination of fluorine; thoriated tungsten.
Shipping regulations: (Rail) Yellow label. (Air) Oxidizer label. Consult regulations for radioactive label requirement.

thorium oxalate $Th(C_2O_4)_2 \cdot 2H_2O$.
Properties: White powder, insoluble in water and most acids. Sp. gr. (anhydrous) 4.637 (16°C). Soluble in solutions of alkali and of ammonium oxalates. Above 300–400°C decomposes to thorium oxide, ThO_2.
Grades: Purities to 99.9%; as 59% ThO_2.
Containers: Glass bottles; fiber drums.
Use: Ceramics.

thorium oxide. See thorium dioxide.

thorium sulfate (thorium sulfate, normal) $Th(SO_4)_2 \cdot 8H_2O$.
Properties: White, crystalline powder; sp. gr. 2.8; m.p. loses $4H_2O$ at 42°C, remainder at 400°C. Slightly soluble in water; soluble in ice water.
Grades: As 43% ThO_2.
Containers: Glass bottles; fiber drums.

thorium tetrachloride. See thorium chloride.

"Thornel."[214] Trademark for a graphite yarn.
Properties: Tensile strength 200,000 psi; elastic modulus 25 million psi; sp. gr. about 1.5.
Uses: Structural components for high-performance aircraft; solid-propellant-rocket motor casings; reentry vehicles and deep submergence vessels; high-strength composites. See also graphite.

thortveitite. An ore containing 37–42% scandium oxide. A basic material in production of scandium.

Thr Abbreviation of threonine.

threonine (alpha-amino-beta-hydroxybutyric acid) $CH_3CH(OH)CH(NH_2)COOH$. An essential amino acid.
Properties: Colorless crystals, soluble in water; optically active.
DL-threonine, m.p. 228–229°C with decomposition;
L(−)-threonine (naturally occurring), m.p. 255–257°C with decomposition;
DL-allo-threonine, m.p. 250–252°C.
Derivation: Hydrolysis of protein (casein); organic synthesis.
Containers: Bottles; drums.
Uses: Nutrition and biochemical research, dietary supplement.

threshold limit value (TLV). A set of standards established by the American Conference of Governmental Industrial Hygienists for concentrations of airborne substances in workroom air; they are time-weighted averages based on conditions which it is believed that workers may be repeatedly exposed to day after day without adverse effects. The TLV values are intended to serve as guides in control of health hazards, rather than definitive marks between safe and dangerous concentrations. In this book, these are indicated by the word "tolerance." See also air pollution.

thrombin. A proteolytic enzyme which catalyzes conversion of fibrinogen to fibrin and thus is essential in the clotting mechanism of blood. It is present in the blood in the form of prothrombin (q.v.) under normal conditions; when bleeding begins, the prothrombin is converted to thrombin, which in turn activates the formation of fibrin.

throwing oil. An oil applied to prepare raw silk and filament rayon for "throwing," the operation by which the filaments are twisted into threads. Applied by a bath, the oils condition the filaments and yarns for subsequent weaving or knitting. Usually compounded to be self-emulsifying and may contain a sizing agent such as dextrin, gelatin, etc. See also slashing compound.

throwing power. A term denoting the effectiveness of an electrolytic cell for depositing metal uniformly over the surface being electroplated particularly in irregular and recessed areas. The throwing power is the weight of deposition per unit distance between the electrodes.

THPC. See tetrakis(hydroxymethyl)phosphonium chloride.

thuja oil (arbor vitae oil).
Properties: Pale yellow essential oil; camphor-like odor; sp. gr. 0.910–0.920; refractive index (n 20/D) 1.459; optical rotation −10° to −13° in 100 mm tube. Soluble in alcohol, ether, chloroform and carbon disulfide; in fixed oils and mineral oil. Combustible; low toxicity.
Chief known constituents: Dextro-pinene; levofenchone; thujone; should contain not less than 60% ketones calculated as thujone.
Derivation: Distilled from the leaves of the white cedar, Thuja occidentalis.
Grades: Technical; F.C.C. (as cedarleaf oil).
Containers: Glass bottles.
Uses: Medicine; perfumery; flavoring.

thujone $C_{10}H_{16}O$. A terpene-type ketone contained in thuja oil and the oils of sage, tansy and wormwood.
Properties: Colorless liquid; sp. gr. 0.915–0.919 (20°/20°C); b.p. 203°C. Insoluble in water; soluble in alcohol. Combustible; low toxicity.
Uses: Solvent.

thulia. See thulium oxide.

thulium Tm Atomic number 69; Group IIIB of the periodic table; a rare earth element of the lanthanide group. Atomic weight 168.9342. Valence 3. No stable isotopes.
Properties: Metallic luster; reacts slowly with water; soluble in dilute acids; salts colored green; sp. gr. 9.318; m.p. 1550°C; b.p. 1727°C. Low toxicity.
Derivation: For source see rare earth minerals. Isolated by reduction of the fluoride with calcium.
Grades: Regular high purity (ingots, lumps).
Hazard: Fire risk in form of dust.
Uses: Ferrites; x-ray source.

thulium 170. Radioactive thulium of mass number 170, used as x-ray source in portable units.
Shipping regulations: Consult regulations.

thulium chloride $TmCl_3 \cdot 7H_2O$.
Properties: Green deliquescent crystals; m.p. 824°C; b.p. 1440°C. Very soluble in water and alcohol.

thulium oxalate. $Tm_2(C_2O_4)_3 \cdot 6H_2O$.
Properties: Greenish white precipitate. Loses $1H_2O$ at 50°C. Soluble in aqueous alkali oxalates.
Derivation: Precipitation for a solution containing a thulium salt and a mineral acid by addition of oxalic acid.
Use: Analytical separation of thulium (and other rare-earth metals) from the common metals.

thulium oxide (thulia) Tm_2O_3.
Properties: Dense white powder with greenish tinge; slightly hygroscopic; absorbs water and CO_2 from the air; sp. gr. 8.6. Exhibits a reddish incandescence on heating, changing to yellow and then white on prolonged heating. Slowly soluble in strong acids.
Derivation: By ignition of thulium oxalate, salt of other oxyacids or hydroxide.
Grades: 99–99.9%.
Containers: Glass bottles.
Use: Source of thulium metal.

"Thylate."[28] Trademark for a wettable off-white powder containing 65% thiram (q.v.).

Thylox process. A process whereby hydrogen sulfide from coke-oven gas is converted to elemental sulfur by absorption in a solution of arsenious oxide and soda ash (or sodium thioarsenate) in water. Air blown through the solution liberates sulfur, which can be skimmed off.

thyme oil. Essential oil distilled from flowering plant Thymus vulgaris or Thymus zygis.

Properties: Colorless to reddish brown liquid; pleasant odor and sharp taste; sp. gr. 0.910–0.935 (25/25°C); refractive index 1.4950–1.5050 (20°C); levorotatory, but no more than −3°; very slightly soluble in water; soluble in alcohol. Combustible; low toxicity.
Chief constituents: Not less than 40% by volume of phenols, and including thymol, carvacrol, cymene, pinene, linalool, and bornyl acetate.
Grades: N.F.; F.C.C.; red; white.
Containers: Cans; drums.
Uses: Medicine, perfumery, cosmetics, toilet soaps, flavoring.

thymic acid. See thymol.

thymidine (thymine-2-deoxyriboside) $C_{10}H_{14}N_2O_5$. The nucleoside (deoxyriboside) of thymine. Occurs in DNA.
Properties: Crystalline needles; m.p. 185°C; dextrorotatory in solution; soluble in water, methanol, hot alcohol, hot acetone and hot ethyl acetate; sparingly soluble in hot chloroform; soluble in pyridine and glacial acetic acid.
Use: Biochemical research.
Also available as trityl thymidine, and as tritiated thymidine in a radioactive form.

thymidylic acid. The nucleotide of thymine; i.e. the phosphate ester of thymidine.

thymine (5-methyluracil) $CH_3CC(O)NHC(O)NHCH$.
5-Methyl-2,4-dioxypyrimidine. One of the pyrimidine bases of living matter.
Properties: White crystalline powder; decomposes at 335–337°C; slightly soluble in hot water; insoluble in cold water, alcohol; sparingly soluble in ether; readily soluble in alkalies.
Derivation: Hydrolysis of deoxyribonucleic acids; from methylcyanoacetylurea by catalytic reduction.
Use: Biochemical research.

thymine-2-deoxyriboside. See thymidine.

thymol (isopropyl-meta-cresol; thyme camphor; thymic acid) $(CH_3)_2CHC_6H_3(CH_3)OH$.
Properties: White crystals with aromatic odor and taste. Soluble in alcohol, carbon disulfide, chloroform, glacial acetic acid, ether and fixed or volatile oils; slightly soluble in water and glycerol. Sp. gr. 0.979; m.p. 48–51°C; b.p. 233°C. Combustible.
Derivation: From thyme oil or other oils; synthetically from meta-cresol and isopropyl chloride by the Friedel-Crafts method at −10°C.
Grades: Technical; N.F.; reagent.
Containers: Bottles; tins; drums.
Hazard: Moderately toxic by ingestion and inhalation.
Uses: Perfumery; thymol compounds; microscopy; preservative; antioxidant; flavoring; laboratory reagent; synthetic menthol.

thymol blue (thymolsulfonphthalein)
$C_6H_4SO_2OC[C_6H_2(CH_3)(OH)CH(CH_3)_2]_2$.
Properties: Brown green powder or crystals; insoluble in water, soluble in alcohol or dilute alkali. M.p. 223°C (dec.).
Use: Acid base indicator in pH range 1.5 (pink) to 2.8 to 8 (yellow) to 9.6 (blue). See indicator.

thymol iodide. Principally dithymol diiodide, $[C_6H_2(CH_3)(OI)(C_3H_7)]_2$.
Properties: Red-brown powder or crystals, slight aromatic odor; affected by light. Soluble in ether, chloroform and fixed or volatile oils; slightly soluble in alcohol, insoluble in water. Low toxicity. Combustible.
Derivation: Interaction of thymol and potassium iodide in alkaline solution.
Grade: Technical.
Containers: Bottles; boxes; drums.
Uses: Medicine; feed additive.

thymolphthalein
$C_6H_4COOC[C_6H_2(CH_3)(OH)CH(CH_3)_2]_2$.
Properties: White powder; m.p. 245°C; insoluble in water, soluble in alcohol and acetone and in dilute alkali and acids.
Uses: Medicine; acid base indicator in pH range 9.3 (colorless) to 10.5 (blue). See indicator.

thymolsulfonphthalein. See thymol blue.

para-**thymoquinone** (2-isopropyl-5-methylbenzoquinone) $C_6H_2O_2(CH_3)CH(CH_3)_2$.
Properties: Bright yellow crystals; penetrating odor. M.p. 45.5°C; b.p. 232°C. Slightly soluble in water; soluble in alcohol and ether. Combustible.
Derivation: From diazonium salt of aminothymol and nitrous acid.
Hazard: May be toxic.
Use: Fungicide.

"Thyrite."[245] Trademark for composite dielectric or resistance material in molded form.

thyrocalcitonin. A thyroid hormone having significant effect on the calcium content of bone and blood. It can retard bone damage and may be involved in the incidence of dental caries and the form of dental calculus known as tartar.

thyroid.
Properties: Yellow, amorphous powder with slight odor, saline taste.
Derivation: Dried, cleansed, powdered thyroid gland obtained from domesticated animals.
Grade: U.S.P.
Use: Medicine (hormone). See thyroxine.

thyronine (desiodothyroxine)
$HOC_6H_4OC_6H_4CH_2CH(NH_2)COOH$. The para-hydroxyphenyl ether of tyrosine. Thyronine and its iodinated derivatives are used in biochemical research on the thyroid gland and its activity.

thyrotropic hormone (TSH; thyrotropin). A hormone secreted by the anterior lobe of the pituitary gland. It increases the rate of removal of iodine from the blood by the thyroid gland, of synthesis of the thyroid hormone, and of its release into the bloodstream. The thyrotropic hormone is a protein which has a low molecular weight (approximately 10,000) and which contains some carbohydrate.

thyroxine $C_{15}H_{11}I_4NO_4$ or
$HOC_6H_2I_2OC_6H_2I_2CH_2CH(NH_2)COOH$. 3,5,3′,5′-Tetraiodothyronine. The hormone produced by the thyroid gland. (See also triiodothyronine). It is an amino acid and a derivative of tyrosine. It increases the metabolic rate and oxygen consumption of animal tissues.
Properties: Optically active; the L-isomer is the natural and physiologically active form.
DL-thyroxine: Needles; decompose 231–233°C; insoluble in water, alcohol, and the common organic solvents; soluble in alcohol in the presence of mineral acids or alkalies.

L-thyroxine: Crystals; decompose 235–236° C.
D-thyroxine: Crystals; decompose 237° C.
Derivation: Obtained from the thyroid glands of animals; also made synthetically.
Uses: Medicine; biochemical research.

Ti Symbol for titanium.

TIBAL. Abbreviation for triisobutylaluminum (q.v.).

"Ticon."[337] Trademark for a series of metal salts used as additives in ferroelectric and piezoelectric devices. They include barium, bismuth, calcium, cerium, lead, magnesium, strontium and zinc titanates; barium, bismuth, calcium, lead, magnesium, and strontium zirconates; barium, calcium and strontium stannates.

tiglic acid (methylcrotonic acid; crotonolic acid; trans-2-methyl-2-butenoic acid) $CH_3CH{:}C(CH_3)COOH$. The trans isomer of angelic acid.
Properties: Thick, syrupy liquid or colorless crystals; spicy odor. Soluble in alcohol and ether; slightly soluble in water. Sp. gr. 0.9641; m.p. 65° C; b.p. 198.5° C. Combustible.
Derivation: From croton oil. Also occurs in English chamomile oil.
Hazard: Toxic by ingestion.
Use: Medicine.

tiglium oil. See croton oil.

"Ti-Kote."[537] Trademark for bonded coating of a silicon carbide material completely impermeable to all gases. Has unusual resistance to abrasion, thermal shock, oxidation, mineral acid corrosion. Used for mechanical seals of pumps, ejector nozzles, thermocouple protection tubes, molds and mandrels, rocket nozzle inserts, nuclear fuel elements, missile leading edges, spacecraft re-entry surfaces, and deflection elements.

tin (stannum) Sn Metallic element of atomic number 50; group IVA of the periodic system. Atomic weight 118.69; valences 2, 4. There are ten isotopes.
Properties: Silver-white, ductile solid (beta form); sp. gr. (20°C) 7.29; m.p. 232°C; b.p. 2260°C; changes to brittle grey (alpha) tin at temperatures below 18° C but the transition is normally very slow. Soluble in acids, and hot potassium hydroxide solution; insoluble in water. Elemental tin has low toxicity, but organic compounds of tin are toxic.
Derivation: By roasting the ore (cassiterite) to oxidize sulfates and to remove arsine, then reducing with coal in a reverberatory furnace, or by smelting in an electric furnace. Secondary recovery from tin plate.
Occurrence: Malaysia (called Straits tin), Indonesia, Thailand, Bolivia.
Grades: By % purity. Spectrographic grade is 99.9999%. Block tin is a common designation for pure tin.
Forms: Sheet; wire; tape; pipe; bar; ingot; powder; single crystals.
Hazard: All organic compounds of tin are toxic. Tolerance (as Sn), 0.1 mg per cubic meter of air.
Uses: Tin plate; terneplate; Babbitt metal; pewter; bronze; packaging and wrapping foil; collapsible tubes; anodes for electrotin plating; hot-dipped coatings; cladding; solders; low-melting alloys for fire control, organ pipes, dental amalgams, diecasting, white, type and casting metal; manufacture of chemicals; tinned wire (all copper wire which is to be rubber covered). Block tin is used in coating copper cooking utensils and lead sheet, or lining lead pipe for distilled water, beer, carbonated beverages, and some chemicals.
Note: In speaking of fabricated articles "tin" is often used when tinplate (thin sheets of steel coated with tin) is meant; e.g. "a tin can." To distinguish, articles (such as condenser coils) made of solid tin are said to be made of "block tin." For further information refer to Tin Research Institute, Inc., 483 West Sixth Ave., Columbus, Ohio.

tin anhydride. See stannic oxide.

tin ash. See stannic oxide.

tin bisulfide. See stannic sulfide.

tin bronze. See stannic sulfide.

tincal. See borax.

tincture. An alcoholic or aqueous alcoholic solution of an animal or vegetable drug or a chemical substance. The tincture of potent drugs is essentially a 10% solution. Tinctures are more dilute than fluid extracts and less volatile than spirits (q.v.).

"Tinopal."[219] Trademark for a group of optical brighteners, which absorb ultraviolet light in the near visible range and re-emit the energy as visible light.
Uses: In heavy duty detergents and detergent specialties, to whiten fabrics; in soap bars as product brighteners; starch products; plastic (e.g molded grade nylon); fibers; coatings, waxes, etc.

tin oxide. See stannous oxide.

tin plating. The process of covering steel, iron or other metal with a layer of tin by dipping in the molten metal, by electroplating, or by immersion in solutions which deposit tin by chemical action of their components. Ingredients for the chemical process are cream of tartar and stannous chloride. The metal being plated also takes part in the process. The objective is to utilize the superior corrosion resistance of tin and in some cases to improve appearance. See also terneplate.

tin protochloride. See stannous chloride.

tin protosulfide. See stannous sulfide.

tin protoxide. See stannous oxide.

tin resinate. See soap (2).

tin salt. See stannous chloride.

tin spirits. Solutions of tin salts used in dyeing.

tin tetrabromide. See stannic bromide.

tin tetrachloride. See stannic chloride.

tin tetraiodide. See stannic iodide.

"Tinuvin."[219] Trademark for a group of ultraviolet absorbers; substituted hydroxyphenyl benzotriazoles. Example: See 2-(2′-hydroxy-5′-methylphenyl)benzotriazole.
Properties: Off-white to pale yellow powders. One or more are soluble in styrene, methyl methacrylate, cyclohexane, toluene, xylene. Characterized by high UV absorptivity below 400 mμ, high light transmission in the visible range. Stable to temperatures normally encountered in plastics production. Also stable to strong acid, peroxide catalysts, reducing agents.
Uses: Protection of plastics (including polyolefins), coatings, rubber, adhesives, waxes, oils.
See also ultraviolet absorber.

"Tipersul."[28] Trademark for fibrous potassium titanate. Crystalline fibers 1 micron in diameter melting at 2500° F; useful to 2200° F. Used for high temperature thermal, acoustical and electrical insulation; also for filter media. Available as lumps, blocks, and loose fibers.

"Ti-Pure."[28] Trademark for titanium dioxide pigments available in two types, rutile and anatase. Both are fine, dry, white powders, available in various grades. Rutile; sp. gr. 4.2; refractive index 2.71. Anatase; sp. gr. 3.88; refractive index 2.53.
Containers: 50-lb bags.
Use: Paints; lacquers; paper; leather; inks; rubber; plastics.

"Tirex."[160] Trademark for flexible, lead-cured cord and cable.

"Tiron."[169] Trademark for disodium-1,2-dihydroxybenzene-3,5-disulfonate.
Uses: Colorimetric determination of ferric iron, titanium, or molybdenum.

Tischenko reaction. Reaction for the formation of esters by the condensation of two molecules of aldehyde catalyzed by aluminum alcoholate in the presence of a halide.

titanellow. See titanium trioxide (Not to be confused with titan yellow, an organic dye containing no titanium).

titania. See titanium dioxide.

titanic acid (titanic hydroxide; metatitanic acid) H_2TiO_3 or $Ti(OH)_4$. Water content variable.
Properties: White powder; insoluble in mineral acids and alkalies except when freshly precipitated; insoluble in water.
Derivation: From hydrochloric acid solution of titanates by treating with ammonia and then drying over concentrated sulfuric acid or by boiling titanium sulfate solution.
Grades: Technical.
Dontainers: Bottles; fiber drums; multiwall paper bags.
Use: Mordant.

titanium Ti Metallic element of atomic number 22; Group IVB of Periodic Table. Atomic weight 47.90; valences 2, 3, 4. There are 5 isotopes.
Properties: Silvery solid or dark gray amorphous powder) density 4.5 g/cc (20°C); m.p. 1675°C; b.p. 3260°C; specific heat 0.13 Btu/lb/°F; thermal conductivity 105 Btu/ft^2/°F/hour; as strong as steel but 45% lighter; Vickers hardness 80–100; excellent resistance to atmospheric and seawater corrosion, and to corrosion by chlorine, chlorinated solvents and sulfur compounds; reactive when hot or molten. Insoluble in water. Inert to nitric acid but attacked by concentrated sulfuric acid and hydrochloric acid. Unaffected by strong alkalies. Low toxicity.
Sources: Ilmenite, rutile, titanite, titanium slag from certain iron ores.
Derivation: (a) Reduction of titanium tetrachloride with magnesium (Kroll process) or sodium in an inert atmosphere of helium or argon. The titanium sponge is consolidated by melting. (b) Electrolysis of titanium tetrachloride in a bath of fused salts (alkali or alkaline-earth chlorides).
Grades: Technical (powder); commercially pure (sheets, bars, tubes, rods, wire and sponge); single crystals.
Containers: Wet powder in cans, boxes or 10-, 100-lb kegs, sponge in steel pails.
Hazard: (dry powder) Flammable, dangerous fire and explosion risk. (Metal) Ignites in air at 1200°C; will burn in atmosphere of nitrogen. Do not use water or carbon dioxide to extinguish.
Uses: Alloys (especially ferrotitanium); as structural material in aircraft, jet engines, missiles, marine equipment, textile machinery, chemical equipment (especially as anode in chlorine production), desalination equipment, surgical instruments, orthopedic appliances, foodhandling equipment; x-ray tube targets; abrasives; cermets; metal-ceramic brazing, especially in nickel-cadmium batteries for space vehicles; electrodeposited and dipped coatings on metals and ceramics; electrodes in chlorine battery.
Shipping regulations: Powder, wet (with not less than 20% water) or dry. (Rail) Yellow label. (Air) Flammable Solid label. Not acceptable on passenger planes.

titanium acetylacetonate. See tianyl acetylacetonate.

titanium boride (titanium diboride) TiB_2.
Properties: Extremely hard solid with oxidative resistance up to 1400°C. M.p. 2980°C; sp. gr. 4.50; hardness 9+ (Mohs); low electrical resistivity.
Uses: Metallurgical additive; high-temperature electrical conductor; refractory; cermet component; coatings resistant to attack by molten metals; aluminum manufacture; super alloys; nuclear steels.

titanium butylate. See tetrabutyl titanate.

titanium carbide TiC.
Properties: Extremely hard crystalline solid with gray metallic color; m.p. 3140° ± 90°C; b.p. 4300°C; sp. gr. 4.93; resistivity 60 micron-ohm-cm (room temperature). Insoluble in water; soluble in nitric acid and aqua regia.
Uses: Additive (with tungsten carbide) in making cutting tools and other parts subjected to thermal shock; arc-melting electrodes; cermets; coating dies for metal extrusion (.0002 inch film by vapor deposition).

titanium chelate. A titanium compound whose general formula is $(HOYO)_2Ti(OR)_2$ or $(H_2NYO)_2Ti(OR)_2$ where Y and R are hydrocarbon groups, e.g., octylene glycol titanate; triethanolamine titanate.
Uses: Surface-active agents; corrosion inhibitors; cross-linking agents.

titanium diboride. See titanium boride.

titanium dichloride $TiCl_2$.
Properties: Black powder; sp. gr. 3.13. Decomposed by water; decomposes at 475° in vacuum. Hygroscopic; soluble in alcohol; insoluble in chloroform, ether, carbon disulfide.
Derivation: Reduction of titanium tetrachloride with sodium amalgam; dissolving titanium metal in hydrochloric acid.
Hazard: Flammable, dangerous fire risk. Ignites in air. Keep under water or inert gas.
Shipping regulations: (Rail, Air) Not listed. Consult authorities.

titanium dioxide (titanic anhydride; titanic acid anhydride; titanic oxide; titanium white; titania) TiO_2.
Properties: White to black powder, depending on purity; is also prepared in two crystalline forms anatase (q.v.) and rutile (q.v.). It has the greatest hiding power of all white pigments. Noncombustible. Nontoxic.
Derivation: From ilmenite or rutile. (a) Ilmenite is treated with sulfuric acid and the titanium sulfate further processed. The product is primarily the anatase form. (b) Rutile is chlorinated and the titanium tetrachloride converted to the rutile form by vapor-phase oxidation.
Grades: Technical, of many variations; pure; U.S.P., single crystals; whiskers.
Containers: Fiber drums; multiwall paper bags.

Uses: White pigment in paints, paper, rubber, plastics, etc.; opacifying agent; cosmetics; radioactive decontamination of skin; floor coverings; glassware and ceramics; enamel frits; delustering synthetic fibers; printing inks; welding rods. Single crystals are high-temperature transducers.

titanium disilicide. Ti_2Si. Used for special alloy applications, as a flame or blast impingement-resistant coating material.

titanium disulfide TiS_2. Yellow solid. Sp. gr. 3.22 (20°C). Decomposed by steam.
Use: Solid lubricant.

titanium ester. A titanium compound whose general formula is $Ti(OR)_4$ where R is a hydrocarbon group, e.g., tetraisopropyl titanate. See also tetrabutyl titanate, tetra(2-ethylhexyl) titanate.
Uses: Adhesion promoters, ester exchange catalysts, cross-linking agents; heat-resistant paints.

titanium ferrocene. See titanocene dichloride.

titanium hydride TiH_2.
Properties: Black metallic powder, which dissociates above 288°C. The evolution of hydrogen is gradual and practically complete at 650°C. Sp. gr. 3.8. Attacked by strong oxidizing agents. Low toxicity.
Derivation: Direct combination of titanium with hydrogen; reduction of titanium oxide with calcium hydride in the presence of hydrogen above 600°C.
Containers: Polyethylene bags in metal drums.
Hazard: Flammable, dangerous fire risk. Dust may explode in presence of oxidizing agents.
Uses: Powder metallurgy; production of pure hydrogen (contains approx 1800 cc (STP) hydrogen per cc of hydride); production of foamed metals; solder for metal-glass; electronic getter; reducing atmosphere for furnaces; hydrogenation agent; refractories.
Shipping regulations: (Air) Flammable Solid label. Not acceptable on passenger planes.

titanium isopropylate. See tetraisopropyl titanate.

titanium monoxide TiO. A weakly basic oxide of titanium, prepared by reduction of the dioxide at 1500°C. It has no important industrial uses.

titanium nitride TiN.
Properties: Golden-brown, hard, brittle plates; m.p. 2927°C; sp. gr. 5.24; specific heat 8.86 cal/mole at 25°C; electrical resistivity 21.7 micro-ohm-cm.
Uses: High-temperature bodies, cermets, alloys, rectifiers, semiconductor devices.

titanium ore. See rutile and ilmenite.

titanium oxalate (titanous oxalate) $Ti_2(C_2O_4)_3 \cdot 10H_2O$.
Properties: Yellow prisms. Soluble in water; insoluble in alcohol and ether.
Derivation: Action of oxalic acid on titanous chloride.

titanium peroxide. See titanium trioxide.

titanium potassium fluoride (potassium titanium fluoride) TiK_2F_6.
Properties: White leaflets. Soluble in hot water.
Hazard: Toxic. Tolerance (as F), 2.5 mg per cubic meter of air.
Uses: Titanic acid; titanium metal.

titanium potassium oxalate (potassium titanium oxalate) $TiO(C_2O_4K)_2 \cdot 2H_2O$.
Properties: Colorless, lustrous crystals. Soluble in water.
Derivation: By treating titanium hydroxide with potassium oxalate and oxalic acid.
Grades: Technical; pure; 22% TiO_2 (min).
Containers: Kegs; barrels.
Uses: Mordant in cotton and leather dyeing; sensitization of aluminum for photography.

titanium sesquisulfate. See titanous sulfate.

titanium sponge. The metal in crude form as reduced from the tetrachloride; ingots are made from it by consumable-electrode refining (arc-melting). See titanium.

titanium sulfate (Titanic sulfate; basic titanium sulfate; titanyl sulfate) $Ti(SO_4)_2 \cdot 9H_2O$; also $TiOSO_4 \cdot H_2SO_4 \cdot 8H_2O$. A commercial material, possibly a mixture.
Properties: White cake-like solid; highly acidic, similar to 50% sulfuric acid; typical composition 20% TiO_2, 50% H_2SO_4, 30% H_2O; hygroscopic; sp. gr. about 1.47; soluble in water; solutions hydrolyze readily unless protected from heat and dilution.
Derivation: Action of sulfuric acid on ilmenite ore.
Hazard: Toxic and irritant to skin and tissue.
Uses: Treatment of chrome yellow and other colors; production of titanous sulfate used as reducing agent or stripper for dyes; laundry chemical.
Shipping regulations: (Rail) Solution containing not more than 45% sulfuric acid: Corrosive liquid. White label.

titanium tetrachloride (titanic chloride) $TiCl_4$.
Properties: Colorless liquid. Fumes strongly when exposed to moist air forming a dense and persistent white cloud. Pure: Sp. gr. 1.7609 at 0°C; b.p. 136.4°C; freezing point −30°C. Soluble in dilute hydrochloric acid; soluble in water with evolution of heat; concentrated aqueous solutions are stable and corrosive; dilute solutions precipitate insoluble basic chlorides.
Derivation: By heating titanium dioxide or the ores and carbon to redness in a current of chlorine.
Grades: Technical; C.P.
Containers: Glass bottles; steel drums; tank cars.
Hazard: Toxic by inhalation; strong irritant to skin and tissue.
Uses: Pure titanium and titanium salts; mordant dye; iridescent effects in glass; smoke screens; titanium pigments; polymerization catalyst.
Shipping regulations: (Rail) White label. (Air) Corrosive label.

titanium trichloride (titanous chloride) $TiCl_3$.
Properties: Dark violet anhydrous deliquescent crystals. Sp. gr. 2.6; decomposes above 440°C; decomposes in air and water with heat evolution. Soluble in alcohol, acetonitrile, certain amines; slightly soluble in chloroform. Insoluble in ether and hydrocarbons.
Hazard: Fire risk in presence of oxidizing materials. Irritant to skin and tissue. MCA warning: Open container only in oxygen-free or inert atmosphere.
Uses: Reducing agent; organic synthesis; cocatalyst for polyolefin polymerization; organometallic synthesis involving titanium; laundry stripping agent.
Shipping regulations: (Rail, Air) Consult authorities.

titanium trioxide (titanium peroxide; pertitanic acid). TiO_3.
Properties: Yellow powder; soluble in acids.
Containers: 1-, 5-lb bottles; fiber containers.
Uses: Dental porcelain and cements; yellow tile.

titanocene dichloride (dicyclopentadienyltitanium dichloride; titanium ferrocene) $(C_5H_5)_2TiCl_2$.
Properties: Red crystals; m.p. 287–289°C; moderately soluble in toluene, chloroform, sparingly soluble in ether and water; stable in dry air; slowly hydrolyzed in moist air.
Containers: Bottles; fiber drums.

Hazard: Toxic by inhalation; irritant to skin and mucous membranes.
See also metallocene.

titanous chloride. See titanium trichloride.

titanous oxalate. See titanium oxalate.

titanous sulfate (titanium sesquisulfate) $Ti_2(SO_4)_3$.
Properties: Green crystalline powder; insoluble in water, alcohol, concentrated sulfuric acid; but soluble in dilute hydrochloric or sulfuric acids giving violet solutions.
Grades: Commercial grade made and supplied as a dark purple solution containing about 15% $Ti_2(SO_4)_3$.
Containers: Glass bottles; carboys.
Use: Textile industry as reducing agent for stripping or discharging colors.

"Titanox."[336] Trademark of a series of white pigments comprising titanium dioxide in both anatase and rutile crystal forms, and also extended with calcium sulfate (titanium dioxide-calcium pigment). Noncombustible; nontoxic.
Containers: Multiwall paper bags.
Uses: Paints, paper, rubber, plastics, leather and leather finishes, inks, coated textiles, textile delustering, ceramics, roofing granules, welding rod coatings, and floor coverings.

titanyl acetylacetonate (titanium acetylacetonate) $TiO[OC(CH_3):CHCOCH_3]_2$.
Properties: Crystalline powder; slightly soluble in water. Resistant to hydrolysis. A chelating, nonionizing compound.
Derivation: Reaction of titanium oxychloride with acetylacetone and sodium carbonate.
Use: Crosslinking agent for cellulosic lacquers.

titanyl sulfate. See titanium sulfate.

titer. In solutions (1) the concentration of a dissolved substance as determined by titration; (2) the minimum amount or volume needed to bring about a given result in titration; or (3) the solidification point of fatty acids which have been liberated from the fat by hydrolysis.

titration. Any of a number of methods for determining volumetrically the concentration of a desired substance in solution by adding a standard solution of known volume and strength until the reaction is completed, usually as indicated by a change in color due to an indicator. Organic solvents are used in titrating water-insoluble substances (non-aqueous titration). Several types of electrical measurement techniques are used, among which are amperometric, conductometric, and coulometric titration methods, and spectrophotometry is a further possible procedure. Titration has been successfully automated, the extent depending on the type of measurement used.

TKP. Abbreviation for tripotassium phosphate. See potassium phosphate, tribasic.

TKPP. Abbreviation for tetrapotassium pyrophosphate. See potassium pyrophosphate.

Tl Symbol for thallium.

"TLA."[569] Trademark for a series of lubricant additives.

TLV. See threshold limit value.

Tm Symbol for thulium.

TMA. Abbreviation for trimethylamine (q.v.) and trimellitic anhydride (q.v.).

TMEDA. Abbreviation for tetramethylethylenediamine (q.v.).

TML. Abbreviation for tetramethyllead (q.v.).

TMTD. Abbreviation for tetramethylthiuram disulfide. See thiram.

TNA. Abbreviation for tetranitroaniline (q.v.).

TNB. Abbreviation for trinitrobenzene (q.v.).

TNT. Abbreviation for trinitrotoluene (q.v.).

tobacco. Cured leaves of the plant Solanaceae, genus Nicotiniana; the species N. tabacum is the most important domestic source. Curing consists of drying and long ageing. The leaves are dehydrated by hanging in warm air to terminal moisture content of 20–30%, during which time starches reduce to sugars, the chlorophyll is discharged, and the color darkens. The product is then aged from one to five years to remove unpleasant odor; cigar tobacco is "fermented" with water, with resulting hydrolysis, deamination and decarboxylation. Aging and fermentation improve taste, aroma and smoking qualities. Humectants and flavoring agents are added to some cigarette blends.

Scores of chemical compounds have been identified in tobacco. Basically it is cellulose. The cured product contains acids (citric, oxalic, formic); alkaloids (nicotine, anabasine, myosmine); carbohydrates (lignin, pentosans, starch, sucrose); as well as tannin, ammonia, glutamine, and micro amounts of zinc, iodine, copper, and manganese. For contents of smoke, see cigarette tar; smoke (4).

A nicotine-free tobacco substitute trademarked "Cytrel" has been developed. See "Cytrel".
Hazard: May have toxic (possibly carcinogenic) effects on chronic inhalation.

tobacco mosaic virus. The first virus to be obtained in crystalline form (from diseased tobacco plants). The protein portion of this virus (95% of each particle) contains about 2300 peptide chains, each having about 150 amino acids and ending with threonine. Molecular weight of each chain is about 17,000; total m.w. about 40 million. The complete sequence of the amino acids has been determined.

tobermorite gel $3CaO \cdot 2SiO_2 \cdot 3H_2O$. A calcium silicate hydrate; a main cementing ingredient of concrete because of its great surface area.

Tobias acid. See 2-naphthylamine-1-sulfonic acid.

"Tobin Bronze-4641."[324] Trademark for a high strength, corrosion-resistant rod alloy, developed originally for marine use. Nominal composition is copper 60%, zinc 39.25%, tin 0.75%.

TOC. Abbreviation for Tagliabue Open Cup, a standard method for determining flash points.

"Tocopherex."[412] Trademark for *d*-alpha-tocopheryl acetate (q.v.).

dl-alpha-**tocopherol.** $C_{29}H_{50}O_2$. *dl*-2,5,7,8-Tetra-methyl-2-(4′,8′,12′-trimethyltridecyl)-6-chromanol. One of the vitamin E group.
Properties: Clear yellow viscous oil; sp. gr. 0.947–0.958 (25/25°C); refractive index 1.5030–1.5070 (20°C). Nontoxic. See also tocopherol.
Derivation: Synthetic.
Grades: N.F.; F.C.C.
Uses: Biological antioxidant; dietary supplement; nutrient; medicine. Also available as the acetate.

tocopherol (vitamin E). Any of a group of related substances (alpha-, beta-, gamma-, and delta-tocopherol) which constitute vitamin E. The alpha-form, $C_{29}H_{50}O_2$, (which occurs naturally as the *d*-isomer), is the most potent. Occur naturally in plants, especially wheat germ. All are derivatives of dihydrobenzo-gamma-pyran, and differ from each other only in the number and position of methyl groups. Vitamin E is required by certain rodents for normal reproduction. Muscular and central nervous system depletion along with generalized edema are deficiency symptoms in all animals. It is not required as a dietary supplement for humans.
Properties: Viscous oils; soluble in fats; insoluble in water; stable to heat in the absence of oxygen, to strong acids, and to visible light; unstable to ultraviolet light, alkalies and oxidation. Nontoxic.
Uses: Medicine; nutrition; antioxidants for fats; animal feed additive.

tocophersolan (USAN) (*d*-alpha-tocopheryl polyethylene glycol 1000 succinate)
$C_{29}H_{49}O \cdot OOC(CH_2)_2COO(CH_2CH_2O)$ H. n = ca 22. A water-miscible vitamin E.
Use: Dietary food supplement; medicine.

"Tofaxin."[162] Trademark for tocopherol (q.v.).

toiletries. Liquid cosmetic preparations usually consisting of an essential oil-alcohol mixture used chiefly for application to hair and scalp, as after-shave lotions, etc.

"Tok."[23] Trademark for 2,4-dichlorophenyl-4-nitrophenyl ether product, used as a herbicide.

tokamak. A magnetic fusion device developed in USSR and at Princeton. The latter is large enough to simulate conditions of plasma suitable for a fusion reactor. It will be used to study the laws of plasma physics. See plasma; fusion.

tolan (diphenylacetylene) $C_6H_5C{:}CC_6H_5$.
Properties: Monoclinic crystals; m.p. 59-61°C; b.p. 300°C; 170°C (19 mm); sp. gr. 0.966 (100/4°C). Insoluble in water; soluble in ether or hot alcohol.
Grades: Technical; purified.
Uses: Organic synthesis; purified grade, primary fluor or wave length shifter in solution scintillators.

"Toleron."[329] Trademark for ferrous fumarate (q.v.).

"Tolex."[457] Trademark for an aromatic solvent, proposed as an alternate for toluene. Distillation range (ASTM D86-59) 215–235°F.

ortho-**tolidine** (3,3′-dimethylbenzidine; diaminoditolyl) $[C_6H_3(CH_3)NH_2]_2$.
Properties: Glistening plates, white to reddish; m.p. 129–131°C. Soluble in alcohol and ether; sparingly soluble in water. Combustible. Affected by light. Low toxicity.
Derivation: Reduction of ortho-nitrotoluene with zinc dust and caustic soda and conversion of the hydrazotoluene by boiling with hydrochloric acid.
Grades: Technical, dry or paste.
Uses: Dyes; sensitive reagent for gold (1:10 million detectable); and for free chlorine in water; curing agent for urethane resins. (Also available as the dihydrochloride).

tolnaftate. USAN name for ortho-2-naphthyl meta, N-dimethylthiocarbanilate,
$C_{10}H_7OC(S)N(CH_3)C_6H_4CH_3$.
Use: Antifungal treatment (medicine).

3-ortho-**toloxy-1,2-propanediol.** See mephenesin.

"Tolpit."[79] Trademark for a tall oil pitch used in roofing cements, caulking compounds; shoemaker's wax; asphalt emulsions and anti-stripping agents; coal dust briquetting.

tolualdehyde. See tolyl aldehydes.

toluazotoluidine. See ortho-aminoazotoluene.

tolu balsam. See balsam.

toluene methylbenzene; phenylmethane). 17th highest-volume chemical produced in U.S. (1975).

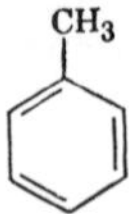

Properties: Colorless liquid; benzene-like odor. Sp. gr. 0.866 (20/4°C); f.p. −94.5°C; b.p. 110.7°C; refractive index 1.497 (20°C); aniline equivalent 15; flash point 40°F (C.C.). Autoignition temp. 997°F. Soluble in alcohol, benzene, and ether; insoluble in water.
Derivation: (a) By catalytic reforming of petroleum. (b) By fractional distillation of coal-tar light oil.
Method of purification: Rectification.
Grades (usually defined in terms of boiling ranges): Pure; commercial; straw-colored; nitration; scintillation; industrial.
Containers: Drums; tank cars.
Hazard: Flammable, dangerous fire risk. Explosive limits in air 1.27 to 7%. Toxic by ingestion, inhalation, and skin absorption. Tolerance, 100 ppm in air. Safety data sheet available from Manufacturing Chemists Assn., Washington, D.C.
Uses: Aviation gasoline and high-octane blending stock; benzene, phenol, and caprolactam; solvent for paints and coatings, gums, resins, most oils, rubber, vinyl organosols; diluent and thinner in nitrocellulose lacquers; adhesive solvent in plastic toys and model airplanes; chemicals (benzoic acid, benzyl and benzoyl derivatives, saccharin, medicines, dyes, perfumes); source of toluenediisocyanates (polyurethane resins); explosives (TNT); toluene sulfonates (detergents); scintillation counter.
Shipping regulations: (Rail) Red label. (Air) FLammable Liquid label. Legal label name (Rail): toluol.

toluene-2,4-diamine (meta-tolylenediamine; MTD: meta-toluylenediamine; diaminotoluene)
$CH_3C_6H_3(NH_2)_2$, (CH_3 = 1).
Properties: Colorless crystals. Soluble in water, alcohol and ether. M.p. 99°C; b.p. 280°C.
Derivation: Reduction of meta-dinitrotoluene with iron and hydrochloric acid.
Hazard: Toxic by ingestion and inhalation. Skin irritant.
Uses: Dye intermediate; direct oxidation black for furs and hair; source for toluene-2,4-diisocyanate.

toluene-2,4-diisocyanate (2,4-tolylene diisocyanate; meta-tolylene diisocyanate; TDI) $CH_3C_6H_3(NCO)_2$. The 2,6-isomer is an impurity.
Properties: Water-white to pale yellow liquid; sharp, pungent odor; b.p. 251°C, 120°C (10 mm); flash point 270°F; m.p. (pure isomer) 19.5–21.5°C; sp. gr. 1.22 (25°/15.5°C); vapor pressure (approx) 0.01 mm at 20°C. Reacts with water producing CO_2; reacts with compounds containing active hydrogen (may be violent). Soluble in ether, acetone, and other organic solvents. Combustible.

Derivation: Reaction of 2,4-diaminotoluene with phosgene.
Method of purification: Distillation to remove hydrochloric acid.
Grade: 100% 2,4-isomer; 80% and 65% 2,4-isomer both mixed with 2,6-isomer.
Containers: Drums; tank cars.
Hazard: Toxic by ingestion and inhalation; strong irritant to skin and tissue, especially to eyes. Tolerance, 0.02 ppm in air. Safety data sheet available from Manufacturing Chemists Assn., Washington, D.C.
Uses: Polyurethane foams, elastomers and coatings; cross-linking agent for nylon 6.

toluene-2,6-diisocyanate. See toluene-2,4-diisocyanate.

para-**toluenesulfamine.** See para-toluenesulfonamide.

para-**toluenesulfanilide** $CH_3C_6H_4SO_2C_6H_4NH_2$.
Properties: White to pink crystalline solid. M.p. 103°C. Soluble in most lacquer solvents. Combustible; low toxicity.
Derivation: Para-toluene sulfonchloride treated with aniline in presence of lime or carefully regulated amounts of alkalies.
Grade: Technical.
Uses: Softener for acetylcellulose in proportions up to 50%; dyestuff intermediate.

ortho-**toluenesulfonamide** $CH_3C_6H_4SO_2NH_2$.
Properties: Colorless crystals; m.p. 156.3°C; soluble in alcohol; slightly soluble in water and ether. Combustible.
Use: Plasticizer.

para-**toluenesulfonamide** (para-toluenesulfamine; PTSA). $CH_3C_6H_4SO_2NH_2$.
Properties: White leaflets. Soluble in alcohol; very slightly soluble in water. M.p. 137°C. Combustible.
Derivation: By amination of para-toluene sulfonchloride.
Uses: Organic synthesis; plasticizers and resins; fungicide and mildewicide in paints and coatings.

para-**toluenesulfondichloroamide.** See dichloramine-T.

ortho-**toluenesulfonic acid** (ortho-toluenesulfonate); $C_6H_4(SO_3H)(CH_3)$.
Properties: Colorless crystals; m.p. 67.5°C; b.p. 129°C; soluble in alcohol, water, and ether. Combustible.
Derivation: By sulfonating toluene with concentrated sulfuric acid below 100°C.
Grades: Anhydrous; monohydrate; 40% aqueous solution.
Containers: 55-gal drums.
Hazard: Toxic by ingestion and inhalation; strong irritant to tissue.
Uses: Dyes; organic synthesis; acid catalyst.
Shipping regulations: (solution) (Air) Corrosive label.

para-**toluenesulfonic acid** (para-toluenesulfonate) $C_6H_4(SO_3H)(CH_3)$.
Properties: Colorless leaflets; m.p. 107°C; b.p. 140°C (20 mm). Soluble in alcohol, ether, and water. Combustible.
Derivation: By action of chlorosulfonic acid on toluene at a low temperature.
Grades: Anhydrous, monohydrate; 40% aqueous solution.
Containers: 55-gal drums.
Hazard: Moderately toxic; skin irritant.
Uses: Dyes; organic synthesis; organic catalyst.

ortho-**toluenesulfonyl chloride** (ortho-toluenesulfochloride; ortho-toluenesulfonchloride) $H_3CC_6H_4SO_2Cl$.
Properties: Oily liquid; sp. gr. 1.3383 (20/4°C); m.p. 10.2°C; b.p. 154°C (36 mm). Insoluble in water; soluble in hot alcohol and in ether and benzene. Combustible.
Derivation: Action of chlorosulfonic acid on toluene.
Use: Intermediate in the synthesis of saccharin and dyestuffs.

para-**toluenesulfonyl chloride** (para-toluenesulfochloride; para-toluenesulfonchloride) $H_3CC_6H_4SO_2Cl$.
Properties: Solid; m.p. 71°C; b.p. 145–146°C (15 mm). Insoluble in water; soluble in alcohol, ether, and benzene. Combustible.
Use: Organic synthesis.

para-**toluenesulfonyl semicarbazide** $C_8H_{11}N_3O_3S$.
Properties: Fine white powder; sp. gr. 1.438; decomposition temperature, 440°F dry, 415–430°F, compounded.
Use: Blowing agent for polyolefins, impact polystyrene, polypropylene, ABS, etc.

toluenethiol (thiocresol, tolyl mercaptan). $CH_3C_6H_4SH$.
Properties: Cream to white moist crystals; musty odor; b.p. about 195°C. Insoluble in water; soluble in alcohol or ether. There are three isomers with different boiling points.
Hazard: May be toxic and skin irritant.
Uses: Medicine; intermediate; bacteriostat.

alpha-**toluenethiol.** See benzyl thiol.

toluene trichloride. See benzotrichloride.

toluene trifluoride. See benzotrifluoride.

toluhydroquinone $CH_3C_6H_3(OH)_2$.
Properties: Pink to white crystals; m.p. 126–127°C; slightly soluble in water; soluble in alcohol and acetone.
Containers: Fiber drums.
Uses: Antioxidant; polymerization inhibitor.

alpha-**toluic acid.** See phenylacetic acid.

meta-**toluic acid** (meta-toluylic acid; 3-methylbenzoic acid) $C_6H_4CH_3COOH$.
Properties: White to yellowish crystals; slightly soluble in water; soluble in alcohol and ether. Sp. gr. 1.0543; m.p. 109°C; b.p. 263°C; ionization constant 5.3×10^{-5}. Combustible; low toxicity.
Derivation: Oxidation of meta-xylene with nitric acid.
Uses: Organic synthesis; to form N,N-diethyl-meta-toluamide, a broad-spectrum insect repellent.

ortho-**toluic acid** (ortho-toluylic acid; 2-methylbenzoic acid) $C_6H_4CH_3COOH$.
Properties: White crystals; slightly soluble in water; soluble in alcohol and chloroform. Sp. gr. 1.0621; m.p. 103.5–104°C; b.p. 259°C; refractive index (114.6°C) 1.512; ionization constant 1.2×10^{-5}. Combustible; low toxicity.
Derivation: Oxidation of ortho-xylene with dilute nitric acid.
Uses: Bacteriostat.

para-**toluic acid** (para-toluylic acid; 4-methylbenzoic acid) $C_6H_4CH_3COOH$.
Properties: Transparent crystals; slightly soluble in water; soluble in alcohol and ether. M.p. 180°C; b.p. 275°C; ionization constant 4.3×10^{-5}. Combustible; low toxicity.

Derivation: By treating cymene or turpentine with nitric acid.
Uses: Agricultural chemicals; animal feed supplement.

alpha-**toluic aldehyde.** See phenylacetaldehyde.

meta-**toluidine** (meta-aminotoluene) $CH_3C_6H_4NH_2$.
Properties: Colorless liquid; sp. gr. 0.980; f.p. −31.5°C; b.p. 203.3°C; slightly soluble in water; soluble in alcohol or ether. Flash point 187°F. Combustible.
Derivation: Reduction of meta-nitrotoluene.
Containers: Drums; tank cars.
Hazard: Toxic by inhalation and ingestion; absorbed by skin. Safety data sheet available from Manufacturing Chemists Assn., Washington, D.C.
Uses: Dyes; manufacture of organic chemicals.
Shipping regulations: (Air) Poison label.

ortho-**toluidine** (ortho-aminotoluene) $CH_3C_6H_4NH_2$.
Properties: Light yellow liquid; becomes reddish brown on exposure to air and light; volatile with steam. Sp. gr. 1.008 (20/20°C); f.p. −16°C; b.p. 200°C; flash point 185°F. Soluble in alcohol and ether; very slightly soluble in water. Combustible.
Derivation: By the reduction of ortho-nitrotoluene or obtained mixed with para-toluidine by the reduction of crude nitrotoluene.
Grade: Technical.
Containers: 450-, 900-lb drums; tank cars.
Hazard: Toxic by inhalation and ingestion; absorbed by skin. Tolerance, 5 ppm in air. Safety data sheet available from Manufacturing Chemists Assn., Washington, D.C.
Uses: Dyes; printing textiles blue-black and making various colors fast to acid; vulcanization accelerators; organic synthesis.
Shipping regulations: (Air) Poison label.

para-**toluidine** (para-aminotoluene) $CH_3C_6H_4NH_2$.
Properties: White, lustrous plates or leaflets. Sp. gr. 1.046 (20/4°C); m.p. 45°C; b.p. 200.3°C. Soluble in alcohol and ether; very slightly soluble in water. Flash point 189°F. Combustible.
Derivation: By the reduction of para-nitrotoluene with iron and hydrochloric acid.
Grades: Technical, flake or cast.
Containers: 450-, 900-lb drums.
Hazard: Toxic by inhalation and ingestion; absorbed by skin. Safety data sheet available from Manufacturing Chemists Assn., Washington, D.C.
Uses: Dyes; organic synthesis; test reagent for lignin, nitrite, phloroglucinol.

toluidine maroon
$CH_3C_6H_3NO_2N_2C_{10}H_5OHCONHC_6H_4NO_2$. An organic azo pigment obtained by the azo coupling of meta-nitro-para-toluidine with the meta-nitroanilide of beta-hydroxynaphthoic acid.
Properties: Good lightfastness and weather resistance; excellent acid and alkali resistance; poor resistance to bleeding.
Hazard: May be toxic.
Uses: Air-dried and baked enamels; truck body finishes.

toluidine red. Any of a class of pigments based on couplings of beta-naphthol and meta-nitro-para-toluidine. See toluidine maroon.

tolu oil (tolu balsam oil).
Properties: Yellow liquid; hyacinth-like odor. Soluble in alcohol, ether, chloroform and carbon disulfide. Sp. gr. 0.945–1.09.
Chief known constituents: A terpene, $C_{10}H_{16}$, and esters of cinnamic and benzoic acid.
Derivation: From tolu balsam by distillation.
Grade: Technical.
Use: Medicine (expectorant); cough syrups. See also balsam.

toluol. Obsolete name for toluene.

toluquinone (2-methylquinone; para-toluquinone) $CH_3C_6H_3O_2$.
Properties: Yellow leaflets or needles; m.p. 65–67°C; soluble in hot water; very soluble in alcohol, ether, acetone, ethyl acetate, and benzene.

"Tolu-Sol."[125] Trademark for a series of solvents, predominantly C_7 hydrocarbons, low in naphthenic hydrocarbons and containing from 3% to 50% aromatics, the balance being essentially paraffinic. Used mainly as lacquer diluents and gravure ink solvents.
Hazard: May be flammable.

toluyl aldehyde. See tolyl aldehyde.

toluylene. See stilbene.

meta-**toluylenediamine.** See toluene-2,4-diamine.

toluylene red. See neutral red.

meta-, ortho-, and para-**toluylic acid.** See corresponding toluic acid.

tolyl acetate. See cresyl acetate.

para-**tolylaldehyde** (para-toluyl aldehyde; para-tolualdehyde; para-methylbenzaldehyde) $CH_3C_6H_4CHO$.
Properties: Colorless liquid; refractive index (n 16.6/D) 1.54693; sp. gr. 1.020; b.p. 204°C; slightly soluble in water; soluble in alcohol and ether. There are also ortho and meta isomers. Combustible.
Grades: Technical; pure.
Containers: Tins; drums.
Uses: Perfumes; pharmaceutical and dyestuff intermediate; flavoring agent.

alpha-**tolylaldehyde dimethylacetal.** See phenylacetaldehyde dimethylacetal.

4-ortho-**tolylazo**-ortho-**diacetotoluide.** See diacetylaminoazotoluene.

ortho-**tolyl biguanide** $NH_2(CNHNH)_2C_6H_4CH_3$.
Properties: White to off-white powder; m.p. 138°C (min). Combustible.
Use: Antioxidant for soaps produced from animal or vegetable oil.

para-**tolyldiethanolamine** $(HOC_2H_4)_2NC_6H_4CH_3$.
Properties: Crystals. M.p. 62°C; b.p. 297.1°C; vapor pressure <0.1 (20°C); sp. gr. 1.0723 (20/20°C); solubility in water 1.67% by weight (20°C); viscosity 155 cp (20°C). Very soluble in acetone, ethanol, ethyl acetate, benzene. Flash point 385°F. Combustible. Low toxicity.
Containers: 500-lb drums.
Uses: Emulsifier, dyestuff intermediate.

meta-**tolylenediamine.** See toluene-2,4-diamine.

meta-**tolylenediaminesulfonic acid** [4,6-diamino-meta-toluenesulfonic acid (SO_3H = 1)] $CH_3C_6H_2(NH_2)_2SO_3H$.
Properties: White crystalline solid; soluble in alkalies.
Derivation: By addition of meta-toluylenediamine sulfate to oleum and heating.
Hazard: May be toxic.
Use: Dyes.

meta-**tolylenediisocyanate.** See toluene-2,4-diisocyanate.

para-**tolyl isobutyrate.** See para-cresyl isobutyrate.

para-**tolyl-1-naphthylamine-8-sulfonic acid** (tolylperi acid) $C_{17}H_{15}NO_3S$.
Properties: Greenish gray needles. Soluble in alcohol; rather insoluble in water. Combustible.
Derivation: Arylation of 1-naphthylamine-8-sulfonic acid with para-toluidine.
Grades: Technical; mostly as sodium salt.
Hazard: May be toxic.
Use: Azo colors.

para-**tolyl phenylacetate.** See para-cresyl phenylacetate.

N-meta-**tolylphthalamic acid.** See "Duraset."

tolyl-para-**toluenesulfonate.** See cresyl-para-toluenesulfonate.

tomatidine. A steroid secondary amine; the nitrogenous aglycone of tomatine. Isolated from the roots of the Rutgers tomato plant as the hydrochloride, $C_{27}H_{45}NO_2 \cdot HCl$. Crystals decompose at 275–280°C.

tomatine. A glycosidal alkaloid prepared from the dried leaves and stems of the tomato plant. White crystals, used as plant fungicide and as a specific precipitating agent for cholesterol. The crude extract is referred to as tomatin.

"Tona."[173] Trademark for a proteolytic enzyme meat-tenderizer. Supplied as powder and used as liquid.

"Tonalid."[105] Trademark for 1,1,2,2,3,3,5-heptamethylindan-6-methyl ketone. Readily soluble in all common perfumery raw materials, solvents and aerosol propellants. Stable and non-discolorizing in soaps and cosmetics. Used as a musk odorant.

"Tonco."[518] Trademark for octadecyl isocyanate, technical.

toner. An organic pigment which does not contain inorganic pigment or inorganic carrying base. (ASTM) See also lake. Important toners are Pigment Green 7, Pigment Blue 15, Pigment Yellow 12, and Pigment Blue 19.

tonka (tonka bean; coumarouna bean; dipteryx).
Properties: Black-brownish seeds with wrinkled surface and brittle shining or fatty skins; aromatic, bitterish taste; balsamic, vanilla-like odor; efflorences of coumarin often appear on the surface. Combustible; nontoxic.
Derivation: Bean of Dipteryx oppositifolia and other species of the Dipteryx.
Occurrence: Brazil, Guiana and Venezuela.
Grades: Angostura; Brazilian.
Containers: Casks; cases.
Uses: Production of natural coumarin; medicine; flavoring extracts; toilet powders.

tonka bean camphor. See coumarin.

"Tonox."[248] Trademark for a blend of aromatic primary amines, the main component of which is para, para-diaminodiphenylmethane.
Uses: Epoxy and urethane curative.

topochemical reaction. Any chemical reaction that is not expressible in stoichiometric relationships. Such reactions are characteristic of cellulose; they can take place only at certain sites on the molecule where reactive groups are available, i.e., in the amorphous areas or on the surfaces of the crystalline areas.

"Tophel."[155] Trademark for a nickel-base thermocouple alloy of approximately 10% chromium. Contains a number of minor elements to give proper E.M.F. against pure platinum and produces a black oxide whose structure matches the alloy.
Properties: M.p. 2600°F; density 8.63 g/cc; sp. heat at 20°C, 0.11 cal/gm; tensile strength at 20°C 88,000 psi.

"Tophet."[155] Trademark for a group of alloys composed of nickel and chromium.

TOPO. Abbreviation for trioctylphosphine oxide (q.v.).

"Toranil."[168] Trademark for a desugared calcium lignosulfonate produced from a desugared extract of coniferous woods.
Uses: Adhesives, binding agents, dispersants, refractory materials, boiler water treating compounds, oil well drilling muds, dye mordants, emulsifiers and foaming agents, concrete, leather, agricultural sprays.

"Tordon."[233] Trademark for herbicides and defoliants containing 4-amino-3,5,6-trichloropicolinic acid (q.v.) as the potassium, isopropanolamine, etc. salts. Relatively nontoxic.

torr. A pressure unit used chiefly in vacuum technology; it is the pressure required to support 1 millimeter of mercury at 0°C.

torula yeast. A yeast that utilizes fermentable sugar in industrial wastes, such as fruit cannery refuse and sulfite liquor from pulp mills. The dried yeast is high in protein and vitamin content, enabling it to be used for enriching animal feeds. The enzymes present are destroyed during drying. It is now being made by a new process utilizing petroleum-derived ethyl alcohol.

tosyl (Ts). The para-toluenesulfonyl radical, $CH_3C_6H_4SO_2$—. Esters of para-toluenesulfonic acid are known as tosylates.

Toth process. A comparatively recent process for production of aluminum metal, which utilizes kaolin and other high-alumina clays as well as bauxite. The clay is chlorinated after calcination and the aluminum chloride resulting is reacted with metallic manganese to yield aluminum and manganese chloride. The reaction occurs at the comparatively low temperature of 500°F. The manganese chloride is recovered as manganese metal and chlorine by oxidation and subsequent reduction, the manganese being recyled. This is a much cheaper and more efficient method than the present Hall process, and if successful it may be used extensively, especially in view of its much lower electric energy requirement.

tourmaline $(Na,Ca)(Al,Fe)B_3Al_3(AlSi_2O_9)(O,OH,F)_4$. A complex borosilicate of aluminum.
Uses: Pressure gauges; optical equipment; oscillator plates; source of boric acid.

"Toval."[28] Trademark for a dense, economical gelatine dynamite used in construction and quarry work.

"Tovex."[28] Trademark for a series of non-nitroglycerin water-compatible explosives with varying densities. Sensitized with an amine nitrate, and manufactured as a cartridged water gel, it is much more resistant than dynamite to accidental detonation.

toxaphene. Generic name for technical chlorinated camphene with the approximate formula $C_{10}H_{10}Cl_8$. Contains 67–69% chlorine.
Properties: Amber, waxy solid with a mild odor of chlorine and camphor; melting range 65–90°C; sp. gr. 1.66 (27°C). Soluble in common organic solvents.
Containers: Drums.
Hazard: Tolerance, 0.5 mg per cubic meter of air.

Use: Insecticide (primarily for cotton and early growth stage of vegetables).

toxicity. The ability of a substance to cause damage to living tissue, impairment of the central nervous system, severe illness or, in extreme cases, death when ingested, inhaled, or absorbed by the skin. The amounts required to produce these results vary widely with the nature of the substance and the time of exposure to it. "Acute" toxicity refers to exposure of short duration, i.e., a single brief exposure; "chronic" toxicity refers to exposure of long duration, i.e., repeated or prolonged exposures.

The toxic hazard of a material may depend on its physical state and on its solubility in water and acids. Some metals that are harmless in solid or bulk form are quite toxic as fume, powder or dust. Many substances that are intensely poisonous are actually beneficial when administered in micro amounts, as in prescription drugs, e.g., strychnine.

Toxicity is objectively evaluated on the basis of test dosages made on experimental animals under controlled conditions. Most important of these are the LD_{50} (lethal dose, 50%) and the LC_{50} (lethal concentration, 50%) tests, which include exposure of the animal to (a) oral ingestion), (b) extended skin contact, and (c) inhalation of the material under test. Such studies are the basis for toxicity ratings determined by the Manufacturing Chemists Association and by the Dept. of Health, Education and Welfare. Reference to these sources should be made for detailed clinical information on toxicity of materials. See also poison (1); toxic materials.

toxic materials. The following list includes a number of chemical individuals and groups that are generally regarded as having toxic properties by either ingestion, inhalation, or absorption via the skin. There is considerable variation in the degree of toxicity among these, and the listing is by no means exhaustive. See also toxicity.

Individuals

aniline
arsenic and cpds.
asbestos (carcinogen)
barium and soluble cpds.
beryllium and soluble cpds.
benzidine (carcinogen)
cadmium and cpds.
benzpyrene (carcinogen)
carbon monoxide
chlorine
chromium (hexavalent cpds.)
coal-tar pitch (carcinogen)
cresol
fluorine and cpds.
hydrogen peroxide
hydrogen sulfide
hydrogen selenide
lead compounds
mercury (soluble cpds.)
methyl alcohol
nickel carbonyl
osmium tetroxide
oxalic acid
ozone
phenol
pyrene
pyridine
phosphine
selenium cpds.
stibine
sulfur dioxide
thallium (soluble cpds.)
tin (organic cpds.)
vinyl chloride monomer (carcinogen)

Groups

aldehydes
alkaloids
allyl cpds.
barbiturates
chlorinated hydrocarbons
corrosive materials (q.v.)
cyanides
fluorides
organic phosphate esters
radioactive elements

Note: An International Register of Potentially Toxic Chemicals (IRPTC) is under development by a group of toxicologists in Geneva. The data will be available to UN member countries.

toxicology. The branch of medical science devoted to the study of poisons, including their mode of action, effects, detection, and countermeasures. The subject may be subdivided into (1) clinical, (2) environmental, (3) forensic, and (4) occupational (industrial). The Institute of Chemical Toxicology was formed in 1975, its members comprising a number of the larger chemical companies. See also toxicity; poison (1).

TPA. Abbreviation for terephthalic acid (q.v.).

TPB. Abbreviation for tetraphenylbutadiene (q.v.).

TPG. Abbreviation for triphenylguanidine (q.v.).

TPN. Abbreviation for triphosphopyridine nucleotide. See nicontinamide adenine dinucleotide phosphate.

Tpp. (1) Abbreviation for triphenyl phosphate (q.v.).
(2) Abbreviation for thiamine pyrophosphate. See cocarboxylase.

"TPSA."[58] Trademark for tetrapropenylsuccinic anhydride. See dodecenylsuccinic anhydride.

"TPS-PG."[1] Trademark for a polyester resin stabilizer/accelerator.

TPT. (1) Abbreviation for triphenyltetrazolium chloride. See tetrazolium chloride.
(2) Abbreviation for tetraisopropyl titanate (q.v.).

trace element (micronutrient). An element essential to plant and animal nutrition in trace concentrations, i.e., minute fractions of 1 per cent (1000 ppm or less). Plants require iron, copper, boron, zinc, manganese, molybdenum, sodium, and chlorine. Animals require iron, copper, iodine, manganese, cobalt, and selenium. Such elements are also called micronutrients. Do not confuse with "tracer" (q.v.).

tracer. A chemical entity (almost invariably radioactive and usually an isotope) added to the reacting elements or compounds in a chemical process, which can be traced through the process by appropriate detection methods, e.g., Geiger counter. Compounds containing tracers are often said to be "tagged" or "labelled." Carbon-14 (q.v.) is a commonly used tracer, and radioactive forms of iodine and sodium are also used. Many complex biochemical reactions have been examined in this way (e.g., photosynthesis). Nonradioactive dueterium (hydrogen isotope) is sometimes used, the detection being by molecular weight determination. Radioactive enzymes are also available for tracer studies, e.g., ribonuclease, pepsin, trypsin, and others. See also labeling (2).

trade mark. A word, symbol, or insignia designating one or more proprietary products, or the manufacturer of such products, which has been officially registered with the government trademark agency. The accepted designation is a superior capital R enclosed in a circle; however, quotation marks may also be used, as in this Dictionary. The term "trade name", though widely used, is not applicable to such products; according to the U.S. Trade Mark Association, a trade name is the name under which a company does business, e.g., the Blank Chemical Company. Use of a trade mark without proper indication of its proprietary nature places the name in jeopardy; a number of trade marks have been invalidated as a result of this practice.

trade name. See trade mark.

trade sales. In the paint industry this term is applied to paints intended for sales to the general public, as in hardware stores and similar outlets.

tragacanth gum.

Properties: Dull white, translucent plates or yellowish powder. Soluble in alkaline solutions, aqueous hydrogen peroxide solution; strongly hydrophilic; insoluble in alcohol. Combustible; nontoxic.

Constituents: Polysaccharides of galactose, fructose xylose and arabinose with glucuronic acid.

Occurrence: Southwestern Europe, Greece, Turkey, Iran.

Grades: U.S.P.; F.C.C.; No. 1, 2, 3.

Containers: Barrels; cases.

Uses: Pharmacy (emulsions), adhesives; leather dressing; textile printing and sizing; thickener and emulsifier; dyes; food products (ice cream, desserts); toothpastes; coating soap chips and powders; hair wave preparations.

tranquilizer. See psychotropic drug.

trans-. See cis-.

"Trans-4."[303] Trademark for trans-1,4-polybutadiene rubber.

transalkylation. A type of disproportionation reaction (q.v.) by which toluene is hydrogenated to benzene and mixed xylene isomers free from ethylbenzene, avoiding the formation of methane resulting from the conventional hydrodealkylation process (q.v.). Transalkylation of toluene to benzene involves the use of an undisclosed catalyst; the yield is claimed to be 97%, based on toluene feed.

transferase. An enzyme whose activity causes a transfer of a radical from one molecule to another. Examples are transaminases, transacetylases, and transmethylases, which effect the transfer of amino, acetyl, methyl groups respectively. See enzyme.

transfer RNA. See ribonucleic acid.

transference number. That portion of the total current carried by any species of ion in an electrolyte in the fluid state. The symbol t^+ is usually used for a positive ion and t^- for a negative ion.

transformer oil. A liquid having the property of insulating the coils of transformers, both electrically and thermally. There are two broad classes, (1) natural and (2) synthetic. The natural type includes refined mineral oils (petroleum fractions), which have low viscosity and high chemical and oxidative stability. The synthetic types are (a) chlorinated aromatics (chlorinated biphenyls and trichlorobenzene) known collectively as askarels (q.v.); (b) silicone oils; and (c) ester liquids such as dibutyl sebacate. All these types are nonflammable, but are combustible. Flash points are from 250 to 300°F. See also dielectric.

Note: Use of PCB (chlorinated biphenyls) has been discontinued because of their ecologically damaging effects.

"TransistAR."[329] A grade of reagent quality chemicals, specially controlled for use in the manufacture of semiconductors and other electronic devices and precision instruments.

transition element (transition metal) Any of a number of elements in which the filling of the outermost shell to 8 electrons within a period is interrupted to bring the penultimate shell from 8 to 18 or 32 electrons. Only these elements can use penultimate shell orbitals as well as outermost shell orbitals in bonding. All other elements, called "major group" elements, can use only outermost shell orbitals in bonding. Transition elements include elements 21 through 29 (scandium through copper), 39 through 47 (yttrium through silver), 57 through 79 (lanthanum through gold), and all known elements from 89 (actinium) on. All are metals. Many are noted for exhibiting a variety of oxidation states and forming numerous complex ions, as well as possessing extremely valuable properties in the metallic state. See also orbital theory.

R. T. Sanderson.

transmutation. The natural or artificial transformation of atoms of one element into atoms of a different element as the result of a nuclear reaction. The reaction may be one in which two nuclei interact, as in the formation of oxygen from nitrogen and helium nuclei (alpha particles), or one in which a nucleus reacts with an elementary particle such as a neutron or a proton. Thus a sodium atom and a proton form a magnesium atom. Radioactive decay e.g., of uranium, can be regarded as a type of transmutation.

"Transphalt."[140] Trademark for a series of dark thermoplastic resins which are polymeric polynuclear hydrocarbons. Available as 100°C softening point material.

Containers: Steel drums, tank trucks and cars.

Use: Paint, saturants for paper and wall board, pipe coatings, joint sealers, floor tile, extenders in epoxy systems, calendered and extruded rubber products, and secondary plasticizers.

transportation label. See label (1).

transuranic element. An element of higher atomic number than uranium, not found naturally, and produced by nuclear bombardment. The most recently discovered is 105. They are all radioactive. See actinide.

"Trebo-Phos."[57] Trademark for triple superphosphate.

"Treflan."[530] Trademark for trifluralin.

"Trefmid."[530] Trademark for selective herbicide composed of trifluralin and diphenamid.

"Trek."[214] Trademark for a concentrated synthetic methanol base antifreeze containing special corrosion inhibitors.

Properties: Deep violet (dye added) liquid; practically odorless. B.p. 149°F; sp. gr. (61/60°F) 0.800; vol concentration of antifreeze to lower freezing point of water solution to 0°F, 27%; −20°F, 37%; wt/gal at 68°F, 6.6 lbs.

Hazard: Highly toxic by ingestion.

Use: Automotive antifreeze.

tremolite $Ca_2Mg_5Si_8O_{22}(OH)_2$. A variety of asbestos (q.v.). Some tremolite is sold as "fibrous talc."

Properties: Color white to light green; luster vitreous to silky; hardness 5–6; sp. gr. 3.0–3.3. Resistant to acids. Noncombustible.

Occurrence: New York, California, Maryland; South Africa.
Hazard: Inhalation of dust or fine particles is dangerous. Tolerance 177 particles per cc. of air. Carcinogenic.
Uses: As asbestos, particularly in acid-resisting applications; ceramics; paint.

"Treopax."[337] Trademark for 90% minimum purity form of zirconium oxide. Contains 5% maximum silicon dioxide. M.p. 4500°F. Used as a source for zirconium oxide for welding rod coatings and to stabilize color in titanium enamels.

tretamine. Generic name for 2,4,6-tris(1-aziridinyl)-s-triazine. See under triethylenemelamine.

"Tret-O-Lite." Trademark for an oil-field demulsifying agent designed especially for breaking crude oil/water emulsions.

tri. Prefix meaning three.

triacetate fiber. See acetate fiber.

triacetin (glyceryl triacetate) $C_3H_5(CH_2CH_3)_3$.
Properties: Colorless liquid with slight fatty odor and a bitter taste; sp. gr. 1.160 (20°C); b.p. 258–260°C; sets to a glass below −37°C; refractive index 1.4312 (20°C); flash point 300°F; wt/gal 9.7 lbs. Slightly soluble in water; very soluble in alcohol, ether, and other organic solvents. Combustible. Low toxicity.
Derivation: Action of acetic acid on glycerol.
Method of purification: Vacuum distillation.
Grades: Technical; C.P.; N.D.; F.C.C.
Containers: Tins; drums; tank cars.
Uses: Plasticizer; fixative in perfumery; manufacture of cosmetics; specialty solvent; to remove carbon dioxide from natural gas; medicine (external); food additive.

triacontanoic acid. See melissic acid.

1-triacontanol $CH_3(CH_2)_{28}CH_2OH$. A 30-carbon straight-chain fatty alcohol.
Properties: Colorless needles from ether; m.p. about 85–88°C; soluble in most organic solvents; insoluble in water. Combustible. Low toxicity. Sp. gr. at melting point 0.777.
Occurrence: Beeswax; carnauba wax.
Use: Biochemical research.

"Triafil."[470] Trademark for electric insulating film based on cellulose triacetate.

trialkyl boranes R_3B. See tributylborane; triethylborane.

trialkylsilanol. An alcohol derivative of silane.
Hazard: May be toxic and flammable.
Uses: Short stopping agent for silicone polymers.

triallylamine $(H_2C{:}CHCH_2)_3N$.
Properties: Liquid; sp. gr. 0.800 (20/4°C); f.p. below −70°C; b.p. 150–151°C; refractive index 1.4501 (n 20/D); flash point 103°F (TOC).
Hazard: Moderate fire risk; probably toxic and irritant.
Use: Intermediate.

triallyl cyanurate $(CH_2{:}CHCH_2OC)_3N_3$.
Properties: Colorless liquid or solid; m.p. 27.32°C; flash point >176°F (TOC); sp. gr. 1.1133 (30°C); refractive index n 25/D 1.5049. Miscible with acetone, benzene, chloroform, dioxane, ethyl acetate, ethyl alcohol, and xylene. Combustible.
Hazard: Moderately toxic by ingestion and inhalation.
Uses: Polymers as monomer and modifier; organic intermediate.

triallyl phosphate $(CH_2{:}CHCH_2O)_3PO$.
Properties: Water-white liquid; f.p. below −50°C; b.p. 80°C (0.5 mm); refractive index 1.448 (n 25/D); sp. gr. 1.064 (25/15°C). Combustible.
Containers: 55-gal drums; tank trailers.
Hazard: Probably toxic.
Use: Intermediate.

triamcinolene (9α-fluoro-16α-hydroxyprednisolone) $C_{21}H_{27}FO_6$.
Properties: White crystalline powder; m.p. 264–268°C; insoluble in water; slightly soluble in usual organic solvents; soluble in dimethylformamide.
Grade: N.D.
Hazard: May be toxic.
Use: Medicine. Also available as the acetonide.

1,3,5-triaminobenzene $C_6H_3(NH_2)_3$.
Properties: Solid; m.p., anhydrous, 129°C; hydrate, 84–86°C (1.5 moles water). Soluble in water, acetone, and alcohol. Insoluble in ether, cold benzene, carbon tetrachloride, and petroleum ether. Combustible.
Containers: Bottles; fiber drums. Supplied as hydrochloride.
Hazard: May be toxic.
Uses: Ion exchange resin intermediate; wetting and frothing agents; photographic developers; organic reactions.

2,4,6-triaminotoluene trihydrochloride $C_6H_2(NH_2HCl)_3CH_3 \cdot H_2O$.
Properties: Light tan to cream crystals; very soluble in water; soluble in alcohol and acetone; insoluble in benzene. Melting point 119°C (free base).
Containers: Bottles; fiber drums.
Uses: Nongelatin photographic emulsion with ethylenediamine for fixation; ion exchange resins; wetting and frothing agents; photographic developers; intermediate varnishes and rubber chemicals.

2,4,6-triamino-syn**-triazine.** See melamine.

triamylamine $(C_5H_{11})_3N$.
Properties: Colorless to yellow liquid; sp. gr. (20°C) 0.79–0.80; trimylamine content at least 98.0%; 95% boils between 225 and 260°C; flash point 215°F; refractive index (18°C) 1.4374; insoluble in water; soluble in gasoline. Combustible.
Derivation: Reaction of amyl chloride and ammonia.
Containers: 1-gal, 5-gal cans; 55-gal drums.
Hazard: Toxic and irritant.
Uses: Corrosion inhibitor; insecticidal preparations.

triamylbenzene $(C_5H_{11})_3C_6H_3$.
Properties: Colorless liquid. Sp. gr. (20°C) 0.87; boiling range 300–320°C; odor faintly aromatic. Flash point 270°F. Combustible.
Uses: Chemical intermediate.

triamyl borate $(C_5H_{11})_3BO_3$.
Properties: Colorless liquid; sp. gr. (20°C) 0.845; boiling range 220–280°C; flash point 180°F; odor faintly alcoholic. Soluble in alcohol and ether. Combustible. Low toxicity.
Derivation: Direct heating of boric acid and amyl alcohol.
Use: Varnishes.

tri-para-tert-**amylphenyl phosphate** $(C_5H_{11}C_6H_4)_3PO$.
Properties: Liquid. Boiling range 305–345°C at 5mm; m.p. 62–63°C; white solid; odorless; insoluble in water. Combustible.
Hazard: Probably toxic.
Use Plasticizer.

tri-para-**anisylchloroethylene.** See chlorotrianisene.

triarylmethane dye. Any of a group of dyes whose molecular structure involves a central carbon atom joined to three aromatic nuclei. C.I. numbers range from 42000 to 44999. The color is due in part to the aromatic rings and to the chromophore groups =C=NH and =C=N—. The members of this class function as basic dyes for cotton, using tannin as a mordant, or, if they contain sulfonic acid groups, as acid dyes for wool and silk. Examples are malachite green and methyl violet. See also triphenylmethane dye.

s-**triazine derivatives.** See ammelide; ammeline; melamine.

s-**triazine-3,5(2H,4H)dione riboside.** See 6-azauridine.

s-**triazine-2,4,6-triol.** See cyanuric acid.

triazole. A five-membered ring compound containing three nitrogens in the ring: $C_2H_3N_3$. Suggested as photoconductors in copying systems.

triazone resin. One of a class of amino resins produced from urea, formaldehyde and a primary amine. Used in textile and fabric treatment. See dimethylolethyltriazone.

"Tribase."[304] Trademark for lead sulfate, tribasic.
"Tribase-E." Basic lead silicate sulfate.
Uses: Stabilizer for vinyl resins.

tribasic. See monobasic.

tribromoacetaldehyde (bromal) CBr_3CHO.
Properties: Oily yellowish liquid; sp. gr. 2.66; b.p. 174°C. Soluble in water, alcohol, or ether. Combustible.
Derivation: (a) By adding bromine to a solution of paraldehyde in ethylacetate. (b) By adding bromine to absolute alcohol, fractionating, treating the fraction boiling at 165°C with water and distilling.
Hazard: Probably toxic. See bromine.
Uses: Medicine; organic synthesis.

tribromoacetic acid CBr_3COOH.
Properties: Colorless crystals; soluble in water, alcohol, or ether. M.p. 135°C; b.p. 245°C. to 250°C. Combustible.
Derivation: By oxidizing tribromoacetaldehyde with nitric acid.
Hazard: Probably toxic. See bromine.
Use: Organic synthesis.

tribromo-tert-**butyl alcohol** (acetone-bromoform) $CBr_3C(CH_3)_2OH$.
Properties: Fine white prismatic crystals; camphor odor and taste; m.p. 176°C. Slightly soluble in water; soluble in alcohol and ether. Combustible.
Derivation: Reaction of acetone and bromoform with solid potassium hydroxide.
Hazard: Probably toxic. See bromine.
Use: Medicine.

tribromoethanol (1,1,1-tribromoethyl alcohol) CBr_3CH_2OH.
Properties: White crystals or powder with slight aromatic odor and taste; m.p. 79–82°C; b.p. 94°C (11 mm); unstable in air and light; slightly soluble in water; soluble in alcohol, ether, benzene, and amylene hydrate; aqueous and alcoholic solutions decompose on exposure to light. Combustible.
Grade: N.F.
Derivation: By reduction of tribromoacetaldehyde with aluminum isopropylate.
Hazard: Moderately toxic.
Use: Medicine (basal anesthetic).

tribromomethane. See bromoform.

1,1,1-tribromo-2-methyl-2-propanol $CBr_3C(CH_3)_2OH$.
Properties: Fine white crystals; m.p. 176–177°C; soluble in water, methanol, ether. Combustible.
Hazard: May be toxic.
Use: Organic synthesis.

tribromonitromethane. See bromopicrin.

tribromophenol. See bromol.

tribromophenol-bismuth. See bismuth tribromophenate.

3,4′5-tribromosalicylanilide $Br_2C_6H_2(OH)C(O)NHC_6H_4Br$. An active antiseptic; used in soaps.

tributoxyethyl phosphate $[CH_3(CH_2)_3O(CH_2)_2O]_3PO$.
Properties: Slightly yellow oily liquid. Insoluble or limited solubility in glycerin, glycols and certain amines. Soluble in most organic liquids. Sp. gr. 1.020 (20°C); f.p. −70°C (viscous liquid); boiling range 215–228°C (4 mm); flash point 435°F; refractive index 1.434 (25°C). Combustible.
Containers: 5-gal cans; 55-gal steel drums.
Hazard: May be toxic.
Uses: Primary plasticizer for most resins and elastomers; floor finishes and waxes; flame-retarding agent.

tri-n-butyl aconitate $C_3H_3(COOC_4H_9)_3$.
Properties: Colorless, odorless liquid; sp. gr. 1.018 (20°C); refractive index 1.4500–1.4530 (25°C); b.p. 190°C (3 mm); insoluble in water; soluble in organic solvents. Combustible.
Containers: Drums; tank cars.
Uses: Plasticizer-stabilizer for vinylidene chloride polymers, synthetic rubbers, and cellulosic lacquers; insecticides.

tri-n-butylaluminum $(CH_3CH_2CH_2Ch_2)_3Al$.
Properties: Colorless pyrophoric liquid; density 0.823 g/ml (20°C); f.p. −26.7°C.
Derivation: Exchange reaction of butene-1 and isobutyl aluminum.
Hazard: Highly flammable, dangerous fire risk. Ignites spontaneously.
Use: Production of organo-tin compounds.
Shipping regulations: (Rail) Pyrophoric liquid, n.o.s., Red label. (Air) Not acceptable.

tri-n-butylamine $(C_4H_9)_3N$.
Properties: Pale yellow liquid with amine odor. B.p. 214°C; f.p. −70°C; sp. gr. (20/20°C) 0.8; wt/gal 6.5 lbs; refractive index 1.4297 (n 20/D); flash point (open cup) 185°F. Slightly soluble in water; soluble in most organic solvents. Combustible. Low toxicity.
Derivation: By reaction of butanol or butyl chloride with ammonia.
Grade: Technical.
Containers: 5-, 55-gal drums; tank cars.
Uses: Solvent; inhibitor in hydraulic fluids; intermediate.

tri-n-butylborane (tri-n-butylborine) $(CH_3CH_2CH_2CH_2)_3B$.
Properties: Colorless pyrophoric liquid; f.p. −34°C; b.p. +170°C (222 mm); density (25°C) 0.747 g/ml; vapor pressure (20°C) 0.1 mm; refractive index (n 20/D) 1.4285; insoluble in water; soluble in most organic solvents. Flash point −32°F.
Hazard: Highly flammable, dangerous fire risk; store and use in inert atmosphere; ignites spontaneously in air.

Uses: Petrochemical industry; organic reactions; catalyst.
Shipping regulations: (Rail) Pyrophoric liquid n.o.s., Red label. (Air) Not acceptable.

tributyl borate (butyl borate) $(C_4H_9)_3BO_3$.
Properties: Water-white liquid; sp. gr. 0.8550–0.8570; b.p. 232.4°C; distillation range, 85% distills between 135°C and 140°C (40 mm); refractive index (n 25/D) 1.4071; f.p. less than −70°C; viscosity 1.601 cp (25°C); flash point (open cup) 200°F; hydrolyzes rapidly; miscible with common organic liquids. Combustible.
Derivation: From butyl alcohol and boric acid.
Uses: Welding fluxes; intermediate in preparation of borohydrides; flame retardant for textiles (with boric acid).

tri-n-butylborine. See tri-n-butylborane.

tri-n-butylchlorostannane. See tributyltin chloride.

tributyl citrate (butyl citrate) $C_3H_5O(COOC_4H_9)_3$.
Properties: Colorless or pale yellow, stable, odorless nonvolatile liquid. Practically insoluble in water. F.p. −20°C; b.p. approximately 233.5°C at 22.5 mm; flash point 315°F (COC); refractive index 1.4453 at 20°C; sp. gr. (25/25°C) 1.042; wt/gal 8.7 lbs at 68°F; pour point −80°F; viscosity (25°C) 31.9 cp. Combustible. Low toxicity.
Grade: Technical.
Containers: Cans; drums; tank cars.
Uses: Plasticizer; antifoam agent; solvent for nitrocellulose.

tributyl(2,4-dichlorobenzyl)phosphonium chloride $Cl_2C_6H_3CH_2P(C_4H_9)_3Cl$.
Properties: White, crystalline solid with a mild aromatic odor; soluble in water, acetone, ethanol, isopropanol and hot benzene; insoluble in hexane and ether; technical grade melts 114–120°C.
Use: Growth retardant for ornamental plants.

2,4,6-tri-tert-butylphenol $[(CH_3)_3C]_3C_6H_2OH$.
Properties: Solid; m.p. 131°C; insoluble in water; soluble in most organic solvents. Combustible.
Hazard: Moderately toxic.
Use: Permissible antioxidant for aviation gasolines (when mixed with other butylphenols) (ASTM).

tri-para-tert-butylphenyl phosphate $[(CH_3)_3CC_6H_4O]_3PO$.
Properties: Solid; b.p. (5 mm) 320°C; m.p. 102–105°C; flash point 545°F; insoluble in water. Combustible.
Hazard: May be toxic.
Use: Plasticizer.

tributyl phosphate (TBP) $(C_4H_9)_3PO_4$.
Properties: Stable, colorless, odorless liquid. Miscible with most solvents and diluents. Soluble in water. Refractive index 1.4226 at 25°C; b.p. 292°C; latent heat of vaporization 55.1 cals/gm at 289°C; f.p. below −80°C; flash point 295°F (COC); Saybolt viscosity 38.6 seconds at 85°F; wt/gal 8.19 lbs; sp. gr. 0.978 (20/20°C). Combustible.
Grade: Technical.
Containers: 1-gal cans; 5-, 55-gal steel drums; tank cars.
Hazard: Toxic by ingestion and inhalation. Irritant to skin. Tolerance 5 mg per cubic meter of air.
Uses: Heat-exchange medium; solvent extraction of metal ions from solution of reactor products; solvent for nitrocellulose, cellulose acetate; plasticizer; pigment grinding assistant; antifoam agent; dielectric.

tri-n-butyl phosphine $(CH_3CH_2CH_2CH_2)_3P$.
Properties: Colorless liquid with garlic odor. Sp. gr. 0.8100 (min at 25/4°C); f.p. −60 to −65°C; b.p. 240°C; flash point 104°F; autoignition point 392°F; refractive index 1.463 (20°C); almost insoluble in water; miscible with ether, methanol, ethanol, benzene. An organic base and strong reducing agent. Combustible.
Uses: Polymerization cross-linking catalyst; organic intermediate; fuels.

tributyl phosphite $(C_4H_9O)_3P$.
Properties: Water-white liquid; b.p. 120°C (8 mm); sp. gr. 0.911 (25°C); refractive index (n 25/D) 1.4301; flash point 250°F. Decomposes in water. Soluble in common organic solvents. Combustible; low toxicity.
Containers: Carboys.
Uses: Additive for greases and extreme-pressure lubricants; stabilizer for fuel oils and polyamides; gasoline additive.

O,O,O-tributyl phosphorothioate (tributyl thiophosphate) $(C_4H_9O)_3PS$.
Properties: Colorless liquid with characteristic odor. B.p. (4.5 mm) 142–145°C; sp. gr. 0.987; flash point (COC) 295°F. Insoluble in water; soluble in most organic solvents. Combustible.
Containers: Bottles; drums.
Hazard: Highly toxic.
Uses: Plasticizer; lubricant additive; antifoam agent; hydraulic fluid; intermediate.
Shipping regulations: (Rail, Air) Organic phosphates, n.o.s.,Poison label. Not accepted on passenger planes.

S,S,S-tributyl phosphorotrithioate ("DEF") $(C_4H_9S)_3PO$.
Properties: Liquid; boiling point 150°C (0.3 mm). Insoluble in water; soluble in aliphatic, aromatic, and chlorinated hydrocarbons.
Hazard: Toxic. Possible cholinesterase inhibitor.
Use: Cotton defoliant.

tributylphosphorotrithioite $(C_4H_9S)_3P$.
Properties: Nearly colorless liquid; insoluble in water; soluble in a variety of organic solvents; b.p. (0.08 mm) 115–134°C; sp. gr. 1.02 (20°C); refractive index 1.542 (25°C).
Derivation: Reaction of butyl mercaptan with phosphorus trichloride.
Hazard: Probably toxic.
Use: Cotton defoliant.

tributyl thiophosphate. See tributyl phosphorothioate.

tributyltin acetate $(C_4H_9)_3SnOOCCH_3$.
Properties: White crystalline solid.
Derivation: Reaction of sodium acetate with tributyltin chloride.
Hazard: Toxic. Tolerance (as Sn), 0.1 mg per cubic meter of air.
Uses: Fungicide and bactericide.

tributyltin chloride (tri-n-butylchlorostannate) $(C_4H_9)_2SnCl$.
Properties: Colorless liquid; soluble in the common organic solvents, including alcohol, heptane, benzene and toluene; insoluble in cold water, but hydrolyzes in hot water. Sp. gr. 1.20 (20/4°C); refractive index 1.4903 (n 25/D); b.p. 145–147°C (5 mm).
Derivation: Reaction of tetrabutyltin with dibutyltin chloride.
Hazard: Toxic. Tolerance (as Sn), 0.1 mg per cubic meter of air.
Uses: Rodenticide; intermediate; rodent-repellent cable coatings.

tributyltin oxide $(C_4H_9)_3SnOSn(C_4H_9)_3$.
Properties: Colorless to pale yellow liquid; soluble in

many organic solvents; practically insoluble in water; b.p. 180°C (2 mm)
Hazard: Toxic. Tolerance (as Sn), 0.1 mg per cubic meter of air.
Uses: Bactericide, fungicide; intermediate.

tri-n-butyl tricarballylate $(C_4H_9OCOCH_2)_2CHCOOC_4H_9$.
Properties: Liquid. Sp. gr. (24°C) 1.004; refractive index (26.5°C) 1.4388; b.p. 305°C; insoluble in water. Combustible.
Use: Plasticizer.

tributyrin. See glyceryl tributyrate.

tricalcium aluminate. See calcium aluminate.

tricalcium citrate. See calcium citrate.

tricalcium orthoarsenate. See calcium arsenate.

tricalcium orthophosphate. See calcium phosphate, tribasic.

tricalcium phosphate. See calcium phosphate, tribasic.

tricalcium silicate. See cement, Portland. Also used as an anticaking agent in foods (up to 2%).

tricamba. Generic name for 3,5,6-trichloro-ortho-anisic acid, $C_6HCl_3(COOH)(OCH_3)$.
Properties: Crystals; m.p. 137–139°C. Very slightly soluble in water; moderately soluble in xylene; freely soluble in alcohol.
Hazard: May be toxic.
Use: Herbicide.

tricarbimide. See cyanuric acid.

tricarboxylic acid cycle. See TCA cycle.

"Trichlor."[177] Trademark for trichloroethylene.

trichlorfon (dylox). Generic name for O,O-dimethyl (2,2,2-tri-chloro-1-hydroxyethyl)phosphonate, $(CH_3O)_2P(O)CH(OH)CCl_3$.
Properties: White crystalline solid; m.p. 83–84°C; b.p. 100°C (1 mm); sp. gr. 1.73 (20/4°C). Soluble in water, benzene, chloroform, ether; insoluble in oils.
Hazard: Highly toxic; cholinesterase inhibitor. Absorbed by skin. Use may be restricted.
Use: Systemic insecticide.
Shipping regulations: (Rail, Air) Organic phosphates, n.o.s. Poison label. Not acceptable on passenger planes.

trichloroacetaldehyde. See chloral.

trichloroacetaldehyde, hydrated. See chloral hydrate.

trichloroacetic acid (TCA) CCl_3COOH.
Properties: Deliquescent colorless crystals; sharp pungent odor; sp. gr. 1.6298; m.p. 57.5°C; b.p. 197.5°C; soluble in water, alcohol, and ether. Flash point, none. Nonflammable.
Derivation: (a) Treating chloral hydrate with fuming nitric acid; (b) from glacial acetic acid by the action of chlorine in presence of sunlight, ultraviolet radiation or catalysts.
Grades: Technical; C.P.; U.S.P.
Containers: Tightly stoppered glass bottles; drums; carlots.
Hazard: Toxic by ingestion and inhalation; strong irritant to skin and tissue.
Uses: Organic synthesis; reagent for detection of albumin; medicine; pharmacy; herbicides.
Shipping regulations: (solid and solution) (Air) Corrosive label. (Rail) Not listed.

trichloroacetyl chloride CCl_3COCl. Liquid; sp. gr. 1.654 (0/4°C); b.p. 118°C. Decomposes in water; soluble in alcohol.
Hazard: Highly toxic by ingestion and inhalation; strong irritant to skin and tissue.
Shipping regulations: (Rail) Not listed. (Air) Not acceptable.

S-2,3,3-trichloroallyl-N,N-diisopropylthiolcarbamate $[(CH_3)_2CH]_2NC(O)SCH_2CCl{:}CCl_2$.
Properties: Oily liquid; b.p. 148–149°C (9 mm); m.p. 29–30°C. Practically insoluble in water; soluble in alcohol, acetone, ether, and heptane. Combustible.
Hazard: Probably toxic.
Use: Herbicide.

trichloro-ortho-**anisic acid.** See tricamba.

1,2,3-trichlorobenzene $C_6H_3Cl_3$.
Properties: White crystals; sp. gr. (solid) 1.69; refractive index (19°C) 1.5776; b.p. 221°C; m.p. 52.6°C; insoluble in water; slightly soluble in alcohol; soluble in ether. Flash point 235°F (closed cup); combustible.
Hazard: Moderately toxic by ingestion and inhalation.
Use: Organic intermediate.

1,2,4-trichlorobenzene $C_6H_3Cl_3$.
Properties: Colorless, stable liquid. Odor similar to that of ortho-dichlorobenzene. Miscible with most organic solvents and oils. Insoluble in water. Sp. gr. 1.4634 (25°C); b.p. 213°C; m.p. 17°C; flash point 210°F. Combustible.
Derivation: Chlorination of monochlorobenzene.
Grades: Technical; 99%; mixture of 1,2,4- and 1,2,3-isomers distilling at 213–219°C.
Containers: Cans; drums; tank cars.
Hazard: Moderately toxic by ingestion and inhalation.
Uses: Solvent in chemical manufacturing; dyes and intermediates; dielectric fluid; synthetic transformer oils; lubricants; heat-transfer medium; insecticides.

2,3,6-trichlorobenzyloxypropanol $(Cl)_3C_6H_2Ch_2OCH_2CH(CH_3)OH$.
Properties: Liquid; b.p. 121–124°C (0.1 mm). Almost insoluble in water; soluble in most organic solvents. Combustible.
Hazard: Moderately toxic by ingestion and inhalation.
Use: Herbicide.

1,1,1-trichloro-2,2-bis(chlorophenyl)ethane. See DDT.

1,1,1-trichloro-2,2-bis(para**-methoxyphenyl)ethane.** See methoxychlor.

B-trichloroborazole BClNHBClNHBClNH.
Properties: White crystalline solid; m.p. 84.5–85.5°C; b.p. 96.5–98°C/37 mm. Soluble in many organic solvents. Highly reactive.
Uses: Intermediate; gelling agent; catalyst; complexing agent.

trichlorobromomethane. See bromotrichloromethane.

trichlorobutyraldehyde hydrate. See butyl chloral hydrate.

3,4,4′-trichlorocarbanilide $C_6H_3Cl_2NHCONHC_6H_4Cl$. Colorless, heat-resistant, white powder; m.p. 250°C. Used as a bacteriostat in soaps and detergents, plastics.

1,1,1-trichloroethane (methyl chloroform) CH_3CCl_3.
Properties: Colorless liquid; sp. gr. 1.325; b.p. 75°C; f.p. −38°C. Insoluble in water; soluble in alcohol and ether. Flash point, none. Nonflammable.
Containers: Drums; tank cars.

Hazard: Moderately toxic by inhalation. Tolerance, 350 ppm in air. Safety data sheet available from Manufacturing Chemists Assn., Washington, D.C.
Uses: Solvent for cleaning precision instruments; metal degreasing; pesticide.

1,1,2-trichloroethane (vinyl trichloride; beta-trichloroethane) $CHCl_2CH_2Cl$.
Properties: Clear, colorless liquid. Characteristic sweet odor. B.p. 113.7°C; sp. gr. 1.4432 (20°C/4°C); refractive index 1.4458; vapor pressure 16.7 mm (20°C); wt/gal 12.0 lbs (20°C); f.p. −36.4°C; flash point, none; nonflammable. Miscible with alcohols, ethers, esters, and ketones; insoluble in water.
Grade: Technical.
Containers: 55-gal drums; tank cars.
Hazard: Toxic and irritant. Absorbed by skin. Tolerance, 10 ppm in air.
Uses: Solvent for fats, oils, waxes, resins, other products; organic synthesis.

trichloroethanol CCl_3Ch_2OH.
Properties: Viscous liquid; ether-like odor. Slightly soluble in water; miscible with alcohol, ether, and carbon tetrachloride. B.p. 150°C; f.p. (approx) 13°C; sp. gr. (25/4°C) 1.541. Combustible.
Hazard: Moderately toxic.
Use: Intermediate; anesthetic.

trichloroethylene (tri) $CHCl:CCl_2$.
Properties: Stable, low-boiling, colorless, photoreactive liquid. Chloroform-like odor. Will not attack the common metals even in the presence of moisture; b.p. 86.7°C; f.p. −73°C; sp. gr. 1.456–1.462 (25/25°C); refractive index 1.4735 (27°C). Miscible with common organic solvents; slightly soluble in water. Nonflammable.
Derivation: From tetrachloroethane by treatment with lime or alkali in the presence of water, or by thermal decomposition, followed by steam distillation.
Grades: U.S.P.; technical; high purity; electronic; metal degreasing; extraction.
Containers: Cans; drums; tank trucks; tank cars.
Hazard: Toxic by inhalation; causes cancer in mice. Use may be restricted. Tolerance, 100 ppm in air. Use prohibited in some states. Safety data sheet available from Manufacturing Chemists Assn., Washington, D.C.
Uses: Metal degreasing; extraction solvent for oils, fats, waxes; solvent dyeing; dry cleaning; refrigerant and heat exchange liquid; fumigant; cleaning and drying electronic parts; diluent in paints and adhesives; textile processing; chemical intermediate; aerospace operations (flushing liquid oxygen).

trichlorofluoromethane (fluorotrichloromethane; fluorocarbon-11) CCl_3F.
Properties: Colorless, nearly odorless, volatile liquid. B.p. 23.7°C; f.p. −111°C; sp. gr. 1.494 (17.2°C); critical pressure 43.2 atm. Noncombustible.
Derivation: From carbon tetrachloride and hydrogen fluoride, in the presence of fluorinating agents such as antimony tri- and pentafluoride.
Grades: Technical; 99.9% min.
Containers: Drums; cylinders.
Hazard: Tolerance, 1000 ppm in air.
Uses: Solvent; fire extinguishers; refrigerant; air conditioning; chemical intermediate; blowing agent.

trichloroisocyanuric acid (1,3,5-trichloro-s-triazine-2,4,6-trione) OCNClCONClCONCl.
Properties: White, slightly hygroscopic, crystalline powder or granules; loose bulk density (approx) powder 31 lbs/cu ft, granular 60 lbs/cu ft. Available chlorine 85%; decomp. 225°C.
Containers: 200-lb fiber drums.
Hazard: Moderately toxic by ingestion. Fire risk in contact with organic materials, strong oxidizing agent.
Uses: Active ingredient in household dry bleaches, dishwashing compounds, scouring powders, detergent-sanitizers, commercial laundry bleaches; swimming pool disinfectant; bactericide, algicide, bleach and deodorant.
Shipping regulations: Dry, containing more than 39% available chlorine: (Rail) Yellow label. (Air) Oxidizer label.

trichloroisopropyl alcohol. See isopral.

trichloromelamine (N,N′,N″-trichloro-2,4,6-triamine-1,3,5-triazine) NC(NHCl)NC(NHCl)NC(NHCl).
Properties: Fine white powder; slightly soluble in water and glacial acetic acid; insoluble in carbon tetrachloride and benzene; pH saturated aqueous solution 4. Low toxicity.
Derivation: By chlorination of melamine.
Grades: 89% available chlorine; see "Sterimine."
Containers: Polyethylene bags in fiber drums.
Hazard: Dangerous fire risk; can ignite spontaneously in contact with reactive organic materials. Autoignition temp. 320°F.
Uses: Chlorine bleach and bactericide.
Shipping regulations: (Rail, Air) Not listed.

trichloromethane. See chloroform.

alpha-**(trichloromethyl)benzyl acetate.** See trichloromethylphenylcarbinyl acetate.

trichloromethyl chloroformate (diphosgene) $ClCOOCCl_3$.
Properties: Colorless liquid. Odor similar to phosgene (newmown hay). Decomposed by heat, porous substances, activated carbons (with evolution of phosgene); also by alkalies, hot water. Soluble in alcohol, benzene, and ether. Sp. gr. 1.65 (15°C); b.p. 127–128°C; f.p. −57°C; vapor density 6.9 (air = 1); refractive index 1.45664 (22°C). Noncombustible.
Derivation: (a) By chlorinating methyl formate. (b) By chlorinating methyl chloroformate. In both methods the mixture of chloro-derivatives is then separated by fractionation.
Grade: Technical.
Hazard: Highly toxic by inhalation and ingestion. Strong irritant to tissue.
Uses: Organic synthesis; military poison gas.
Shipping regulations: (Rail) Poison gas label. Not accepted by express. (Air) Not acceptable. Legal label name: diphosgene.

trichloromethyl ether $CHCl_2OCH_2Cl$.
Properties: Liquid. Pungent odor. Sp. gr. 1.5066 (10°C); b.p. 130–132°C. Soluble in alcohol, benzene, and ether; insoluble in water.
Hazard: Strong irritant to eyes and skin; evolves lachrymatory fumes.

N-(trichloromethylmercapto)-tetrahydrophthalimide. See captan.

trichloromethylphenylcarbinyl acetate (alpha-(trichloromethyl)benzyl acetate) $C_6H_5CH(CCl_3)OOCCH_3$.
Properties: White crystalline solid; intense rose odor; m.p. 86–88°C. Clearly soluble in 18 parts of 95% alcohol.
Containers: Fiber drums.
Uses: Perfumes; fixative for essential oils and perfumes.

trichloromethylphosphonic acid $CCl_3PO(OH)_2$. Strong dibasic acid, soluble in water and alcohol; insoluble in benzene and hexane.
Uses: Catalyst and condensation agent.

1,1,1-trichloro-2-methyl-2-propanol. See chlorobutanol.

trichloromethylsulfenyl chloride (perchloromethyl mercaptan). $ClSCCl_3$.
Properties: Yellow, oil liquid. Disagreeable odor. Mildly decomposed by moist air. Subject to the action of oxidizing agents, reducing agents, chlorine, etc. sp. gr. 1.722 (0°C); b.p. 148–149°C (decomposes); vapor density 6.414; volatility 18,000 mg/cu m (20°C). Insoluble in water. Nonflammable, but supports combustion.
Derivation: Chlorination of carbon disulfide, thiophosgene, or methyl thiocyanate.
Grade: Technical.
Containers: Steel bottles.
Hazard: Highly toxic by ingestion and inhalation; strong irritant to eyes and skin. Tolerance, 0.1 ppm in air.
Uses: Organic synthesis; dye intermediate; fumigant.
Shipping regulations: (Rail, Air) Poison label. Not acceptable on passenger planes. Legal label name perchloromethyl mercaptan.

N-[(trichloromethyl)thio]-4-cyclohexene-1,2-dicarboximide. See captan.

trichloronaphthalene. See chlorinated naphthalene.

trichloronitromethane. See chloropicrin.

thrichloronitrosomethane CCl_3NO.
Properties: Dark blue liquid. Unpleasant odor. Slowly decomposes, but is more stable in solution. Soluble in alcohol, benzene, ether; insoluble in water. Sp. gr. 1.5 (20°C); b.p. 5°C (70 mm).
Derivation: Interaction of sulfuric acid, sodium trichloromethylsulfinate, and sodium nitrate.
Grade: Technical.
Hazard: Strong irritant to eyes and tissue.
Uses: Organic synthesis; military poison gas (lachrymator).

trichlorononylsilane. See nonyl trichlorosilane.

trichloroöctadecylsilane. See octadecyltrichlorosilane.

trichloroöctylsilane. See octyl trichlorosilane.

2,4,5-trichlorophenol $C_6H_2Cl_3OH$.
Properties: Gray flakes in sublimed mass with a strong phenolic odor; sp. gr. (25/4°C) 1.678; b.p. 252°C; m.p. 68–70°C; no flash point. Soluble in alcohol, ether, and acetone. Nonflammable.
Hazard: May cause skin irritation.
Uses: Fungicide, bactericide.

2,4,6-trichlorophenol $C_6H_2Cl_3OH$. (2,4,6-T).
Properties: Yellow flakes with strong phenolic odor; sp. gr. (25/4°C) 1.675; f.p. 61°C; b.p. 248–249°C; no flash point. Soluble in acetone, alcohol, and ether. Nonflammable.
Hazard: May cause skin irritation.
Use: Fungicide, herbicide, defoliant.

2,4,5-trichlorophenoxyacetic acid (2,4,5-T) $C_6H_2Cl_3OCH_2CO_2H$.
Properties: Light tan solid; m.p. 151–153°C; soluble in alcohol; insoluble in water; available as sodium and amine salts.
Containers: 50-, and 200-lb drums.
Hazard: Toxic. Tolerance, 10 mg per cubic meter of air. Permitted by EPA for use on rice, range land and rights of way; all other uses cancelled. See also dioxin.
Uses: Plant hormone; herbicide; defoliant.

2-(2,4,5-trichlorophenoxy)ethyl 2,2-dichloropropionate. See erbon.

2-(2,4,5-trichlorophenoxy)propionic acid. See silvex.

2,4,5-trichlorophenyl acetate $C_6H_2Cl_3OOCCH_3$.
Hazard: May be toxic.
Use: As fungicide, especially on cotton seed.

1,2,3-trichloropropane $CH_2ClCHClCH_2Cl$.
Properties: Colorless liquid; sp. gr. 1.3888 (20/4°C); f.p. −15°C; b.p. 156.17°C; refractive index (n 20/D) 1.4822. Flash point (COC) 180°F. Slightly soluble in water; dissolves oils, fats, waxes, chlorinated rubber and numerous resins. Combustible. Autoignition temp. 580°F.
Derivation: Chlorination of propylene.
Hazard: Toxic by inhalation, and skin absorption. Strong irritant. Tolerance, 50 ppm in air.
Uses: Paint and varnish remover; solvent; degreasing agent.

trichlorosilane
(1) $SiHCl_3$ (silicochloroform).
Properties: Colorless, volatile liquid; sp. gr. 1.336; f.p. −127°C; b.p. 32°C; refractive index 1.3990. Soluble in benzene, ether, heptane, perchloroethylene; decomposed by water. Flash point 7°F.
Containers: To 55-gal drums.
Hazard: Flammable, dangerous fire risk. Moderately toxic.
Use: Laboratory synthesis of organic silanes (trimethyl silane and triphenyl silane).
Shipping regulations: (Rail) Red label. (Air) Flammable Liquid label. Not acceptable on passenger planes.
(2) Generic name for compounds of the formula $RSiCl_3$ of which methyl trichlorosilane, CH_3SiCl_3, is most important.

N,N′,N″-trichloro-2,4,6-triamine-1,3,5-triazine. See trichloromelamine.

2,4,6-trichloro-1,3,5-triazine. See cyanuric chloride.

1,3,5-trichloro-s-triazine-2,4,6-trione. See trichloroisocyanuric acid.

trichlorotrifluoroacetone (1,1,3-trichloro-1,3,3-trifluoroacetone) $CCl_2FCOCClF_2$.
Properties: Colorless liquid; b.p. 84.5°C; f.p. below −78°C. Soluble in all proportions with water and most organic solvents. Stable to acid but not alkalies. Nonflammable.
Hazard: Strong irritant to eyes.
Uses: Solvent in acid media; complexing agent.

1,1,2-trichloro-1,2,2-trifluoroethane (trifluorotrichloroethane) CCl_2FCClF_2.
Properties: Colorless, nearly odorless, volatile liquid. B.p. 47.6°C; f.p. −35°C; critical pressure 33.7 atm; sp. gr. 1.42 (25°C). Noncombustible.
Derivation: From perchloroethylene and hydrofluoric acid.
Grades: Technical; spectrophotometric.
Containers: Drums.
Hazard: Tolerance, 1000 ppm in air.
Uses: Dry-cleaning solvent; fire extinguishers; refrigerant; air-conditioning units; to make chlorotrifluoro-

ethylene; blowing agent; polymer intermediate; solvent drying; drying electronic parts and precision equipment.

tricholine citrate (tris(2-hydroxyethyl)trimethylammonium citrate) $[(CH_3)_3NCH_2CH_2OH]_3 \cdot C_6H_5O_7$.
Containers: Carboys (65% solution); drums.
Uses: Medicine; nutrition.

"Triclene."[28] Trademark for trichloroethylene (q.v.). Available in five grades.

tricobalt tetraoxide. See cobalto-cobaltic oxide.

"Tricoloid."[301] Trademark for tricyclamol chloride (q.v.).

tricosane $CH_3(CH_2)_{21}CH_3$.
Properties: Glittering leaflets. Soluble in alcohol; insoluble in water. Sp. gr. 0.779 (48°C); b.p. 234°C (15 mm); m.p. 48°C. Combustible.
Grade: Technical.
Containers: Bottles; fiber containers.
Use: Organic synthesis.

n-**tricosanoic acid** $CH_3(CH_2)_{21}COOH$. A saturated fatty acid not normally found in natural fats or oils. Synthetic compound is a white crystalline solid; m.p. 79.1°C. Purified product is used in medical research and as reference standard for gas chromatography.

tri-meta, para-**cresyl borate** $(CH_3C_6H_4)_3BO_3$.
Properties: Light amber liquid; sp. gr. 1.065 (25°C); b.p. 385–395°C; refractive index 1.5480 (24°C); flash point 240°F (COC). Soluble in all proportions in acetone, benzene, chloroform, hydrolyzes on contact with H_2O. Combustible.
Containers: 55-gal drums.
Hazard: Moderately toxic.
Uses: Plasticizer; organic synthesis.

tricresyl phosphate (tritolyl phosphate; TCP) $(CH_3C_6H_4O)_3PO$. A mixture of isomers.
Properties: Practically colorless, odorless liquid. Stable, nonvolatile. B.p. 420°C; refractive index 1.556 (25°C); sp. gr. 1.162 (25/25°C); wt/gal 9.7 lb; crystallizing point below −35°C. Miscible with all the common solvents and thinners, also with vegetable oils; insoluble in water. Flash point 437°F; autoignition temp. 770°F. Combustible.
Derivation: From cresol and phosphorus oxychloride.
Containers: 1-, 5-, 55-gal drums; tank cars.
Hazard: Moderately toxic by ingestion and skin absorption. The ortho isomer is highly toxic; its tolerance is 0.1 mg per cubic meter of air.
Uses: Plasticizer for polyvinyl chloride, polystyrene, nitrocellulose; fire retardant for plastics; air filter medium; solvent mixtures; waterproofing; additive to extreme pressure lubricants; hydraulic fluid and heat exchange medium.

tricresyl phosphite $(CH_3C_6H_4O)_3P$.
Properties: Colorless liquid; slight phenolic odor. B.p. (0.11 mm) 191°C; sp. gr. (20/4°C) 1.115; flash point (open cup) 440°F. Insoluble in water; miscible with acetone, alcohol, benzene, ether, and kerosine. Combustible.
Grade: Technical.
Uses: Stabilizer and plasticizer for plastics and resins.

tricyanic acid. See cyanuric acid.

tricyclamol chloride
$C_6H_{11}C(C_6H_5)(OH)CH_2CH_2C_4H_8N \cdot CH_3Cl$. 1-Cyclohexyl-1-phenyl-3-pyrrolidino-1-propanol methylchloride.
Properties: White, extremely bitter, crystalline powder; faint characteristic odor; soluble in alcohol and in water.
Use: Medicine.

tricyclic. An organic compound comprised of three (only) ring structures, which may be the same or different, e.g., anthracene.

sym-**tricyclodecane.** See adamantane.

tricyclohexyl borate. See boric acid ester.

n-**tridecane** $CH_3(CH_2)_{11}CH_3$.
Properties: Colorles liquid. Soluble in alcohol; insoluble in water. Sp. gr. 0.755 (20/4°C); b.p. 225.5°C; f.p. −5.45°C; refractive index 1.4250 (20/D); flash point 175°F. Combustible; low toxicity.
Grades: 95%; 99%; research.
Containers: Glass bottles; 1-, 5-gal drums.
Uses: Organic synthesis; distillation chaser.

n-**tridecanoic acid** (tridecylic acid; tridecoic acid) $CH_3(CH_2)_{11}COOH$. A saturated fatty acid usually prepared synthetically.
Properties: Colorless crystals; m.p. 44.5°C; sp. gr. 0.8458 (80/4°C); b.p. 312.4°C, 192.2°C (16 mm); refractive index 1.4328 (50°C). Slightly soluble in water; soluble in alcohol and ether. Combustible; low toxicity.
Grade: 99% pure.
Uses: Organic synthesis; medical research.

tridecanol. See tridecyl alcohol.

tridecoic acid. See n-tridecanoic acid.

tridecyl alcohol (tridecanol). A commercial mixture of isomers of the formula $C_{12}H_{25}CH_2OH$.
Properties: Low-melting white solid with pleasant odor; b.p. 274°C; m.p. 31°C; sp. gr. (20/20°C) 0.845; wt/gal 7.0 lb; flash point (TOC) 180°F. Combustible; low toxicity.
Derivation: Oxo process (q.v.) from C_{15} hydrocarbons.
Grade: Technical.
Containers: 55-gal drums; tank cars.
Uses: Esters for synthetic lubricants; detergents; antifoam agent; other tridecyl compounds; perfumery.

tridecylbenzene (1-phenyltridecane) $C_6H_5(CH_2)_{12}CH_3$.
Properties: Colorless liquid; sp. gr. 0.85–0.86 (60/60°F); refractive index 1.4815–1.4830. Combustible.
Use: Detergent intemediate.

tridecylic acid. See n-tridecanoic acid.

tri(decyl) orthoformate $CH(OC_{10}H_{21})_3$.
Properties: Liquid; b.p. 194°C; f.p. −15 to −20°C; refractive index 1.448; insoluble in water; soluble in benzene, naphtha, ether, and alcohol.
Use: To remove small quantities of water from ethers or other solvents where acid catalysts can be employed.

tri(decyl) phosphite $(C_{10}H_{21}O)_3P$.
Properties: Water-white liquid; decyl alcohol odor; sp. gr. 0.892 (25/15.5°C); m.p. less than 0°C; refractive index 1.4565 (25°C). Flash point 455°F. Combustible.
Containers: 55-gal drums.
Uses: Chemical intermediate; stabilizer for polyvinyl and polyolefin resins.

tridihexethyl chloride
$C_6H_{11}C(C_6H_5)(OH)CH_2CH_2N(C_2H_5)_2 \cdot C_2H_5Cl$. (3-Diethylamino-1-phenyl-1-cyclohexylpropanol ethochloride).
Properties: White, odorless, crystalline powder, bitter taste. Freely soluble in water, in methanol, in chloro-

form, and in alcohol. Practically insoluble in ether and in acetone. Melting range 198–202°C.
Grade: N.F.
Use: Medicine.

2,4,6-tri(dimethylaminomethyl)phenol $[(CH_3)_2NCH_2]_3C_6H_2OH$.
Properties: Liquid; refractive index 1.5181. Combustible.
Hazard: May be toxic.
Uses: Antioxidants, acid neutralizers, stabilizers, and catalysts for epoxy and polyurethane resins.

tri(dimethylphenyl)phosphite (trixylenyl phosphate) $[(CH_3)_2C_6H_3O]_3PO$.
Properties: Liquid. Sp. gr. 1.155; refractive index 1.5535; b.p. (10 mm), 243–265°C; flash point 450°F; solubility in water (85°C), 0.002% by weight. Combustible.
Use: Plasticizer.

"Tridione."[3] Trademark for trimethadione, U.S.P.

tridodecyl amine. See trilauryl amine.

tridodecyl borate. See boric acid ester.

tridymite SiO_2. A vitreous, colorless or white, native form of pure silica. Found variously but not so commonly as quartz (q.v.). Quartz will change into tridymite with a 16.2% increase in volume at 870°C. Unlike quartz, it is soluble in boiling sodium carbonate solution. Sp. gr. 2.28–2.3; Mohs hardness 7.

trietazine. Generic name for 2-chloro-4-diethylamino-6-ethylamino-s-triazine $ClC_3N_3[N(C_2H_5)_2]NHC_2H_5$.
Properties: Solid; practically insoluble in water.
Uses: Herbicide; plant growth regulator.

"Tri-Ethane."[177] Trademark for 1,1,1-trichloroethane (q.v.).

triethanolamine (TEA; tri(2-hydroxyethyl)amine) $(HOCH_2CH_2)_3N$.
Properties: Colorless, viscous, hygroscopic liquid with slight ammoniacal odor; m.p. 21.2°C; b.p. 335°C (dec); vapor pressure <0.01 mm (20°C); sp. gr. 1.126; flash point (open cup) 375°F; wt/gal 9.4 lb; miscible with water, alcohol; soluble in chloroform; slightly soluble in benzene and ether; slightly less alkaline than ammonia. Commercial product contains up to 25% diethanolamine and up to 5% monoethanolamine. Combustible; low toxicity.
Derivation: Reaction of ethylene oxide and ammonia.
Grades: Technical; regular; 98%; U.S.P.
Containers: Drums; tank cars.
Uses: Fatty acid soaps used in drycleaning, cosmetics, household detergents, and emulsions; wool scouring; textile antifume agent and water-repellent; dispersion agent; corrosion inhibitor; softening agent, humectant, and plasticizer; insecticide; chelating agent; rubber accelerator.

triethanolamine lauryl sulfate $(HOC_2H_4)_3NOS(O)_2OC_{12}H_{25}$. A liquid or paste.
Containers: Drums, tank cars; tank trucks.
Uses: Detergent; wetting, foaming and dispersing agent for industrial, cosmetic and pharmaceutical applications, especially shampoos.

triethanolamine methanearsonate $CH_3As(O)[ONH(C_2H_4OH)_3]_2$.
Hazard: Highly toxic by ingestion.
Use: Herbicide.
Shipping regulations: (Rail, Air) Arsenical compounds, n.o.s., Poison label.

triethanolamine oleate. See trihydroxyethylamine oleate.

triethanolamine stearate. See trihydroxyethylamine stearate.

triethanolamine titanate. See titanium chelate.

1,1,3-triethoxyhexane $CH(OC_2H_5)_2CH_2CH(OC_2H_5)C_3H_7$.
Properties: Colorless liquid; sp. gr. 0.8746 (20/20°C); b.p. 133°C (50 mm); f.p. −100°C; wt/gal 7.3 lb; flash point 210°F. Insoluble in water. Combustible. Low toxicity.
Use: Synthesis of aldehydes, acids, esters, chloride, amines, etc.

triethoxymethane. See triethyl orthoformate.

1,1,3-triethoxy-3-methoxypropane (triethylmethyl malonaldehyde diacetal) $(CH_3O)(C_2H_5O)CHCH_2CH(OC_2H_5)_2$.
Properties: Colorless liquid. Sp. gr. (25/4°C), 0.9300; b.p. (6 mm) 86°C. Combustible.
Grade: 99%.
Uses: Intermediate; crosslinking and insolubilizing agent.

triethylaconitate $C_2H_5OOCCHC(COOC_2H_5)CH_2COOC_2H_5$.
Properties: Liquid. Sp. gr. (25°C) 1.096; refractive index (26°C) 1.4517; b.p. (5 mm) 154–156°C. Combustible.
Use: Plasticizer.

triethylaluminum (ATE; TEA; aluminum triethyl). $(C_2H_5)_3Al$.
Properties: Colorless liquid; sp. gr. 0.837; f.p. −52.5°C; b.p. 194°C; specific heat 0.527 (91.4°F). Miscible with saturated hydrocarbons. Flash point −63°F.
Derivation: By introduction of ethylene and hydrogen into an autoclave containing aluminum. The reaction proceeds under moderate temperature and varying pressures.
Grades: 88–94%.
Containers: Cylinders.
Hazard: Highly toxic; destructive to tissue. Flammable, dangerous fire risk; ignites spontaneously in air; reacts violently with water, acids, alcohols, halogens, and amines.
Uses: Catalyst intermediate for polymerization of olefins, especially ethylene; pyrophoric fuels; production of alpha-olefins and long chain alcohols; gas plating of aluminum.
Shipping regulations: (Rail) Red label. (Air) Not acceptable. (Rail) Legal label name; aluminum triethyl.

triethylamine $(C_2H_5)_3N$.
Properties: Colorless liquid; strong ammoniacal odor. B.p. 89.7°C; f.p. −115.3°C; sp. gr. (20/20°C) 0.7293; wt/gal (20°C) 6.1 lb; flash point (open cup) 20°F. Soluble in water and alcohol.
Derivation: From ethyl chloride and ammonia under heat and pressure.
Containers: 1-, 5-gal cans; 55-gal drums; tank cars.
Hazard: Flammable, dangerous fire risk. Explosive limits in air 1.2 to 8.0%. Tolerance, 25 ppm in air. Toxic by ingestion and inhalation; strong irritant to tissue.
Uses: Catalytic solvent in chemical synthesis; accelerator activators for rubber; wetting, penetrating and waterproofing agents of quaternary ammonium

types; curing and hardening of polymers (e.g., core-binding resins); corrosion inhibitor; propellant.
Shipping regulations: (Rail) Flammable liquids, n.o.s., Red label. (Air) Flammable Liquid label.

triethylborane (triethylborine; boron triethyl). $(C_2H_5)_3B$.
Properties: Colorless liquid. Sp. gr. (25°C); flash point −32°F; f.p. −93°C; b.p. 95°C; refractive index 1.3971; heat of combustion 20,000 Btu/lb. Miscible with most organic solvents; immiscible with water.
Derivation: Reaction of triethylaluminum and boron halide, or diborane and ethylene.
Hazard: Flammable, dangerous fire risk. Ignites spontaneously in air. Highly toxic by inhalation; strong irritant. Reacts violently with water and oxidizing materials.
Uses: Igniter or fuel for jet and rocket engines; fuel additive; olefin polymerization catalyst; intermediate.
Shipping regulations: (Rail) Pyrophoric liquid, n.o.s., Red label. (Air) Not acceptable.

triethyl borate (ethyl borate) $(C_2H_5)_3BO_3$.
Properties: Colorless liquid; mild odor. Hydrolyzes rapidly, depositing boric acid in finely divided crystalline form. B.p. 120°C; sp. gr. 0.863–0.864 (20/20°C); flash point 51.8°F (C.C.); wt/gal 7.20 lb. (20°C); refractive index 1.37311 (20°C).
Uses: Antiseptics; disinfectants; antiknock agent.
Hazard: Flammable, dangerous fire risk. Moderately toxic.
Shipping regulations: (Rail) Flammable liquid, n.o.s., Red label. (Air) Flammable Liquid label. (Air) Legal label name, ethyl borate.

triethylborine. See triethylborane.

triethyl citrate (ethyl citrate) $C_3H_5(COOC_2H_5)_3$.
Properties: Colorless, mobile liquid. Bitter taste; b.p. 294°C; b.p. (1 mm) 126–127°C; sp. gr. 1.136 (25°C); refractive index 1.4405 (24.5°C); pour point −50°F; solubility in water 6.5 g/100 cc; solubility in oil 0.8 g/100 cc. Flash point 303°F (COC); combustible. Low toxicity.
Derivation: Esterification of citric acid.
Grades: Technical; refined; F.C.C.
Containers: Metal drums and cans; tank cars.
Uses: Solvent and plasticizer for nitrocellulose and natural resins; softener; paint removers; agglutinant; perfume base; food additive (not over 0.25%).

triethylenediamine $N(CH_2CH_2)_3N$. Catalyst used in the production of polyurethanes. Combustible.

triethylene glycol (TEG) $HO(C_2H_4O)_3H$.
Properties: Colorless, hygroscopic, practically odorless liquid. Sp. gr. 1.1254 (20/20°C); b.p. 287.4°C; vapor pressure less than 0.01 mm (20°C); flash point 350°F (C.C.); wt/gal 9.4 lb (20°C); freezing point −7.2°C; viscosity 0.478 poise (20°C). Autoignition temp. 700°F. Soluble in water; immiscible with benzene, toluene and gasoline. Combustible; low toxicity.
Derivation: From ethylene and oxygen, as a by-product of ethylene glycol manufacture.
Grades: Technical; C.P.
Containers: 1-, 5-gal cans; 55-gal drums; tank cars.
Uses: Solvent and plasticizer in vinyl, polyester and polyurethane resins; dehydration of natural gas; humectant in printing inks; extraction solvent ("Udex" process).

triethylene glycol diacetate
$CH_3COOCH_2CH_2OCH_2CH_2OCH_2CH_2OOCCH_3$.
Properties: Colorless liquid. Sp. gr. (25°C) 1.112; refractive index n (25°C) 1.437; b.p. 300°C; f.p. less than −60°C. Combustible; low toxicity.
Use: Plasticizer.

triethylene glycol dibenzoate
$C_6H_5CO(OCH_2CH_2)_3OOCC_6H_5$.
Properties: Crystals; b.p. 210–223°C; m.p. 46°C; flash point 457°F (TOC); sp. gr. 1.168. Combustible. Low toxicity.
Uses: Plasticizer for vinyl resins; adhesives.

triethylene glycol dicaprylate (triethylene glycol dioctoate) $C_7H_{15}COO(CH_2CH_2O)_3OCC_7H_{15}$.
Properties: Clear liquid; sp. gr. 0.973 (20°C); acidity 0.3% max. (caprylic); moisture 0.05% max; f.p. −3°C; b.p. 243°C (5 mm). Soluble in most organic solvents. Combustible; low toxicity.
Uses: Low-temperature plasticizer for elastomers.

triethylene glycol dichloride. See triglycol dichloride.

triethylene glycol didecanoate
$C_9H_{19}COO(C_2H_4O)_3OCC_9H_{19}$.
Properties: Colorless liquid. B.p. 237°C (2.0 mm); sp. gr. 0.9584 (20/20°C); viscosity 28.6 cp (20°C). Combustible. Low toxicity.
Use: Plasticizer.

triethylene glycol di(2-ethylbutyrate)
$C_5H_{11}OCOCH_2(CH_2OCH_2)_2CH_2OCOC_5H_{11}$.
Properties: Light-colored liquid; sp. gr. 0.9946 (20/20°C); 8.3 lb/gal (20°C); b.p. 196°C (5 mm); vapor pressure 5.8 mm Hg (200°C); solubility in water 0.02% by wt (20°C); viscosity 10.3 cp (20°C). Flash point 385°F. Combustible. Low toxicity.
Use: Plasticizer.

triethylene glycol di(2-ethylhexoate)
$C_7H_{15}OCOCH_2(CH_2OCH_2)_2CH_2OCOC_7H_{15}$.
Properties: Light-colored liquid; sp. gr. 0.9679 (20/20°C); 8.1 lb/gal (20°C); b.p. 219°C (5 mm); vapor pressure 1.8 mm Hg (200°C); insoluble in water; viscosity 15.8 cp (20°C). Flash point 405°F. Combustible. Low toxicity.
Use: Plasticizer.

triethylene glycol dihydroabietate
$C_{19}H_{31}COO(C_2H_4O)_3OCC_{19}H_{31}$.
Properties: Liquid. Sp. gr. (25°C) 1.080–1.090; refractive index (20°C) 1.5180; vapor pressure (225°C) 2.5; flash point 226°C; insoluble in water. Combustible. Low toxicity.
Use: Plasticizer.

triethylene glycol dimethyl ether
$CH_3(OCH_2CH_2)_3OCH_3$.
Properties: Water-white liquid; mild ether odor; sp. gr. (20/20°C) 0.9862; refractive index 1.4233 (n 20/D); flash point 232°F; b.p. (760 mm) 216.0°C; (100 mm) 153.6°C; f.p. −46°C. Autoignition temp. 1166°F. Completely soluble in water and hydrocarbons at 20°C. May contain peroxides. Combustible. Low toxicity.
Containers: Glass bottles; cans; 55-gal drums.
Uses: Solvent for gases; coupling immiscible liquids.

triethylene glycol dioctoate. See triethylene glycol dicaprylate.

triethylene glycol dipelargonate
$C_8H_{17}COO(C_2H_4O)_3OCC_8H_{17}$.
Properties: Clear liquid; sp. gr. 0.964 (20/20°C); b.p. 251°C (5 mm); f.p. +1 to −4°C; refractive index 1.4470 (23°C); flash point 410°F. Almost insoluble in water; soluble in most organic solvents. Combustible. Low toxicity.
Use: Plasticizer.

triethylene glycol dipropionate $C_2H_5CO(OCH_2CH_2)_3OOCC_2H_5$.
Properties: Colorless liquid. Sp. gr. (25°C) 1.066; refractive index (25°C) 1.436; b.p. (2 mm) 138–142°C; f.p. less than −60°C; solubility in water, 6.70% by weight. Combustible. Low toxicity.
Use: Plasticizer.

triethylene glycol monobutyl ether. See butoxytriglycol.

triethylenemelamine (tretamine; TEM; 2,4,6-tris(1-aziridinyl)-s-triazine) $NC[N(CH_2)_2]NC[N(CH_2)_2]NC[N(CH_2)_2]$.
Properties: White, crystalline, odorless powder; m.p. 160°C (polymerizes); polymerizes readily with heat or moisture; soluble in alcohol, water, methanol, chloroform, and acetone.
Grade: N.F.
Hazard: Highly toxic.
Uses: Medicine (see nitrogen mustards); insecticide; chemosterilant.

triethylenephosphoramide (tepa; tris-(1-aziridinyl)phosphine oxide; APO) $(NCH_2CH_2)_3PO$.
Properties: Colorless crystals; m.p. 41°C; soluble in water, alcohol and ether. Combustible.
Derivation: From ethyleneimine.
Hazard: Highly toxic. Strong irritant to skin and tissue.
Uses: Medicine (see nitrogen mustards); insect sterilant. Also used with tetrakis(hydroxymethyl)phosphonium chloride (THPC) to form a condensation polymer suitable for flameproofing cotton. See also tris[1-(2-methyl)aziridinyl]phosphine oxide.
Shipping regulations: (Rail) White label. (Air) Corrosive label. Legal label name: tris(1-aziridinyl)-phosphine oxide.
See also tepa.

triethylenetetramine $NH_2(C_2H_4NH)_2C_2H_4NH_2$.
Properties: Moderately viscous yellowish liquid. Less volatile than diethylenetriamine but resembles it in many other properties. Soluble in water. B.p. 277.5°C; sp. gr. 0.9818 (20/20°C); m.p. 12°C; flash point 275°F (C.C.); wt 8.2 lb/gal (20°C). Combustible. Autoignition temp. 640°F.
Grades: Technical; anhydrous.
Containers: Cans; drums; tank cars.
Hazard: Strong irritant to tissue. Causes skin burns and eye damage.
Uses: Detergents and softening agents; synthesis of dyestuffs, pharmaceuticals and rubber accelerators.
Shipping regulations: (Air) Corrosive label.

tri(2-ethylhexyl)phosphate $[C_4H_9CH(C_2H_5)CH_2]_3PO_4$.
Properties: Light-colored liquid; sp. gr. 0.9260 (20/20°C); 7.70 lb/gal (20°C); b.p. 220°C (5 mm); vapor pressure 1.9 mm Hg (200°C); insoluble in water; viscosity 14.1 cp (20°C); pour point −74°C; flash point 405°F. Combustible; low toxicity.
Use: Plasticizer.

tri(2-ethylhexyl) phosphite $(C_8H_{17}O)_3P$.
Properties: Straw-colored liquid; sp. gr. 0.897 (25/15°C); m.p., glass at low temperature; refractive index 1.451 (n 25/D); flash point 340°F (COC). Combustible; low toxicity.
Uses: Plasticizer; intermediate.

tri(2-ethylhexyl) trimellitate $C_6H_3(COOC_8H_{17})_3$.
Properties: Clear liquid, mild odor; sp. gr. 0.992 (20/20°C); distillation range at 3 mm, 278–284°C (5–95%); f.p., a gel at −35°C; refractive index 1.4846 (23°C); wt/gal 8.26 lb (20°C). Combustible; low toxicity.
Use: Plasticizer.

triethylmethane. See 3-ethylpentane.

triethylmethyl malonaldehyde diacetal. See 1,1,3-triethoxy-3-methoxypropane.

triethyl orthoformate (triethoxymethane) $CH(OC_2H_5)_3$.
Properties: Colorless liquid; pungent odor; b.p. 145.9°C; refractive index 1.39218 (18.8°C); sp. gr. 0.895 (20/20°C). Soluble in alcohol, ether; decomp. in water. Flash point 86°F (C.C.). Low toxicity.
Derivation: Reaction of sodium ethylate with chloroform or reaction of hydrochloric acid with hydrogen cyanide in ethyl alcohol solution.
Containers: 55-gal steel drums.
Hazard: Flammable, moderate fire risk.
Uses: Organic synthesis; pharmaceuticals.

triethyl phosphate (TEP) $(C_2H_5)_3PO_4$.
Properties: Colorless, high-boiling liquid. Mild odor; very stable at ordinary temperatures. Compatible with many gums and resins. Soluble in most organic solvents; completely miscible in water. When mixed with water is quite stable at room temperature, but at elevated temperatures it hydrolyzes slowly. F.p. −56.4°C; b.p. 216°C, flash point 240°F; refractive index 1.4055 (20°C); wt/gal 8.90 lb (68°F). Combustible.
Grades: Technical; 97%.
Containers: Drums; tank cars; tank trucks.
Hazard: May cause nerve damage but to less extent than other cholinesterase-inhibiting compounds.
Uses: Solvent; plasticizer for resins, plastics, gums; manufacture of pesticides; catalyst; lacquer remover.
Shipping regulations: (Rail, Air) Organic phosphate, liquid, n.o.s., Poison label. Not acceptable on passenger planes.

triethyl phosphite $(C_2H_5)_3PO_3$.
Properties: Colorless liquid; sp. gr. 0.9687 (20°C); b.p. 156.6°C; refractive index 1.413 (n 25/D); flash point 130°F; insoluble in water; soluble in alcohol and ether. Combustible.
Containers: Glass bottles; 5-, 55-gal drums.
Hazard: Moderate fire risk. May be toxic.
Uses: Synthesis; plasticizers; stabilizers; lube and grease additives.

O,O,O-**triethyl phosphorothioate** (triethyl thiophosphate) $(C_2H_5O)_3PS$.
Properties: Colorless liquid with characteristic odor. B.p. (10 mm) 93.5–94°C; sp. gr. 1.074; flash point 225°F (COC); combustible.
Containers: Bottles; drums.
Hazard: Toxic by ingestion; cholinesterase inhibitor.
Uses: Plasticizer; lubricant additive; antifoam agent; hydraulic fluid; intermediate.
Shipping regulations: (Rail, Air) Organic phosphate, liquid, n.o.s., Poison label. Not acceptable on passenger planes.

triethyl tricarballylate $(C_2H_5OCOCH_2)_2CHCOOC_2H_5$.
Properties: Colorless liquid. Sp. gr. (20°C) 1.087; refractive index (26°C) 1.4234; b.p. (5 mm) 158–160°C; solubility in water (20°C) 0.62% by weight. Combustible.
Use: Plasticizer.

"Triexcel."[342] Trademark for rotenone and isomesynergized pyrethrin extracts in concentrate form for insecticidal formulations.

trifluoroacetic acid CF_3COOH.
Properties: Colorless, fuming liquid; hygroscopic; pungent odor. B.p. 72.4°C; sp. gr. 1.535; f.p. −15.25°C; index of refraction (n 20/D) 1.2850; very soluble in water. Nonflammable.
Containers: Custom-packed.
Hazard: Probably toxic and irritant to skin.
Uses: Strong non-oxidizing acid; laboratory reagent; solvent; catalyst.

trifluoroamine oxide (F_3NO). A perfluorated amine, obtained by fluorination of nitrosyl fluoride with ultraviolet light or high temperature or pressure. An alternative process is by burning nitric oxide and fluorine, rapidly quenching the gaseous mixture as it leaves the flame zone.
Hazard: Probably toxic.

trifluorochloroethylene. Legal label name (Rail) for chlorotrifluoroethylene (q.v.).

trifluorochloromethane. See chlorotrifluoromethane.

alpha, alpha, alpha-**trifluoro-2,6-dinitro**-N,N-**dipropyl**-para-**toluidine.** See trifluralin.

trifluoromethane. See fluoroform.

trifluoromethylbenzene. See benzotrifluoride.

trifluoromethylhydrothiazide. See hydroflumethiazide.

3-trifluoromethyl-4-nitrophenol (alpha, alpha, alpha-trifluoro-4-nitro-meta-cresol) $CF_3C_6H_3(NO_2)OH$.
Properties: Crystals; m.p. 74–76°C.
Use: To exterminate lampreys, especially in the Great Lakes. It is placed in tributary streams where it kills the lamprey larvae.

trifluoronitrosomethane CF_3NO.
Properties: Bright blue, fairly stable gas. Disagreeable odor. B.p. −84°C; f.p. −150°C. Nonflammable.
Derivation: (a) Interaction of fluorine and silver cyanide in the presence of silver nitrate; (b) from nitric oxide and iodotrifluoromethane or bromotrifluoromethane in the presence of ultraviolet light.
Hazard: Strong irritant to mucous membranes and tissue.
Uses: Monomer for nitroso rubber; military poison.

trifluorostyrene $C_8H_5F_3$. A monomer designed for the production of polytrifluorostyrene and for copolymerization with vinyl monomers.
Properties: Liquid; boiling point 68°C; freezing pt. −23°C; refractive index 1.474; sp. gr. 1.22; dipole moment 1.98 (D). The polymer is soluble in toluene, chloroform, and methyl ethyl ketone and has dielectric constant of 2.56. Available in laboratory quantities only. Nonflammable.
Uses: Membranes for fuel cells and water purification.

trifluorotrichloroethane. See trichlorotrifluoroethane.

trifluorovinylchloride. See chlorotrifluoroethylene.

triflupromazine hydrochloride (10-(3-dimethylaminopropyl)-2-(trifluoromethyl)phenothiazine hydrochloride) $C_{18}H_{19}F_3N_2S \cdot HCl$.
Properties: White crystalline powder; decomposes 173–174°C. Soluble in water, alcohol, acetone; solutions are sensitive to air and light.
Grade: N.D.
Use: Medicine.

trifluralin. Generic name for 1,1,1-trifluoro-2,6-dinitro-N,N-dipropyl-para-toluidine, $F_3C(NO_2)_2C_6H_2N(C_3H_7)_2$.
Properties: Yellowish orange solid; m.p. 48.5–49°C; b.p. 139–140°C (4.2 mm). Insoluble in water; soluble in xylene, acetone, and ethanol. Probably low toxicity.
Use: Herbicide; especially for cotton plant.

triformol. See sym-trioxane.

"Trigamine."[73] Trademark for buffered aliphatic amine.
Properties: Water-white to pale yellow viscous liquid with a pleasant odor. Soluble in water, ethyl alcohol (50%), glycerin and diethylene glycol. Insoluble in oils and hydrocarbon solvents. Sp. gr. (25°C) 1.17; pH (10% solution) 9.5; neutralization value 208–210.
Uses: Solvent and plasticizer for aqueous solutions of casein and shellac; emulsifier for waxes, polishes, etc.

triglyceride. Any naturally occurring ester of a normal acid (fatty acid) and glycerol. The chief constituents of fats and oils, they have the general formula: $CH_2(OOCR_1)CH(OOCR_2)CH_2(OOR_3)$, where R_1, R_2, and R_3 are usually of different chain length. Refining processes often yield commercial products in which the R chain lengths are the same.
Derivation: Extraction from animal, vegetable and marine matter.
Uses: Fatty acids and derivatives; manufacture of edible oils and fats; manufacture of monoglycerides.

triglycerol monolinolenate. See polyglycerol ester.

triglycerol trilinoleate. See polyglycerol ester.

triglycine. See nitrilotriacetic acid.

triglycol dichloride (triethylene glycol dichloride) $Cl(C_2H_4O)_2C_2H_4Cl$.
Properties: Colorless liquid; sp. gr. 1.1974 (20/20°C); b.p. 241.3°C; flash point 250°F; wt/gal 10.0 lb (20°C); freezing point −31.5°C. Insoluble in water. Combustible. Low toxicity.
Grade: Technical.
Containers: Cans; drums.
Uses: Solvent for hydrocarbons, oils, etc; extractant; intermediate for resins and insecticides; organic synthesis.

triglycollamic acid. See nitrilotriacetic acid.

trigonelline (coffearine; caffearine; gynesine) $C_5H_4NCOOCH_3 \cdot H_2O$. N-Methylnicotinic acid betaine. A base formed in the seeds of many plants.
Properties: Colorless prisms; m.p. 218°C (dec); very soluble in water; soluble in alcohol; nearly insoluble in ether, benzene, and chloroform.
Derivation: Plant seeds; coffee beans; synthetically by heating nicotinic acid with methyl iodine and treatment with silver oxide.
Use: Biochemical research.

tri-n-**hexylaluminum** $(C_6H_{13})_3Al$.
Properties: Colorless pyrophoric liquid; b.p. (0.001 mm) 105°C.
Derivation: Exchange reaction between hexene and isobutyl aluminum.
Hazard: Toxic; ignites at room temperature.
Use: Polyolefin catalyst.
Shipping regulations: Pyrophoric liquids, n.o.s., Red label. (Air) Not acceptable.

trihexylene glycol biborate $(C_6H_{12}O_2)_3B_2$. A cyclic borate.
Properties: Colorless liquid; sp. gr. 0.982 (21°C); boiling range 314–326°C; refractive index 1.4375 (25°C); flash point 345°F. Soluble in most organic

solvents; hydrolyzes slowly in water. Combustible. Low toxicity.
Derivation: Reaction of hexylene glycol with boric oxide.
Containers: 55-gal drums.
Uses: Gasoline additive; chemical intermediate.

trihexyl phosphite $(C_6H_{13}O)_3P$.
Properties: Mobile, colorless liquid with characteristic odor; sp. gr. 0.897 (20/4°C); b.p. 135–141°C (0.2 mm); flash point 320°F (COC). Miscible with most common organic solvents; insoluble in water; hydrolyzes very slowly. High degree of thermal stability. Exposure to air should be minimum. Combustible. Low toxicity.
Containers: 5-gal, 55-gal steel drums.
Uses: Intermediate for insecticides; component of vinyl stabilizers; lubricant additive; specialty solvent.

trihydric. Any alcohol in which three hydroxyl groups are present. See also polyol; glycerol.

1,2,3-trihydroxyanthraquinone. See anthragallol.

1,2,4-trihydroxyanthraquinone. See purpurin.

1,2,7-trihydroxyanthraquinone. See anthrapurpurin.

1,2,3-trihydroxybenzene. See pyrogallol.

1,3,5-trihydroxybenzene. See phloroglucinol.

3,4,5-trihydroxybenzoic acid. See gallic acid.

2,4,5-trihydroxybutyrophenone $C_6H_2(OH)_3COC_3H_7$.
Properties: Yellow-tan crystals; m.p. 149–153°C; density 6.0 lb/gal (20°C). Very slightly soluble in water; soluble in alcohol and propylene glycol. Nontoxic.
Use: Antioxidant for polyolefins and paraffin waxes; food additive.

tri(2-hydroxyethyl)amine. See triethanolamine.

trihydroxyethylamine oleate (triethanolamine oleate) $(HOCH_2CH_2)_3N \cdot HOOCC_{17}H_{33}$. Surfactant made by reaction of triethanolamine with oleic acid. Combustible.
Use: Emulsifying agent.

trihydroxyethylamine stearate (triethanolamine stearate) $(HOCH_2CH_2)_3N \cdot HOOCC_{17}H_{35}$.
Properties: Cream-colored, wax-like solid. Faint fatty odor. Soluble in methyl alcohol, ethyl alcohol, mineral oil, vegetable oil. Dispersible in hot water. Sp. gr. 0.968; pH (25°C) 8.8–9.2 (5% aqueous dispersion); m.p. 42–44°C. Combustible; low toxicity.
Containers: 1-, 5-gal cans; 50-gal drums.
Uses: Emulsifying agent for cosmetic and pharmaceutical industries.

tri(hydroxymethyl)aminomethane. See tris(hydroxymethyl)aminomethane.

1,3,8-trihydroxy-6-methylanthraquinone. See emodin.

2,4,6-trihydroxytoluene. See methylphloroglucinol.

triiodebenzoic acid. A hormone-like synthetic used to increase yields of plants when applied in the flowering stage.

triiodomethane. See iodoform.

triiodothyronine (liothyronine) $HOC_6H_3IOC_6H_2I_2CH_2CH(NH_2)COOH$.
3,5,3′-Triiodothyronine. Either a derivative or precursor of thyroxine (q.v.). Triiodothyronine increases the metabolic rate and oxygen consumption of animal tissues. Low toxicity at normal dosage.
Use: Biochemical research; medicine (metabolic insuffiency).

triisobutylaluminum (TIBAL) $[(CH_3)_2CHCH_2]_3Al$.
Properties: Colorless liquid; sp. gr. 0.7876 (20°C); f.p. −5.6°C; b.p. 114°C (30 mm). Flash point below 32°F; autoignition temp. below 39°F.
Derivation: Reaction of isobutylene and hydrogen with aluminum under moderate temperature and varying pressures.
Hazard: Flammable, dangerous fire risk; ignites spontaneously in air; reacts violently with water, acids, alcohols, amines, and halogens. Highly toxic; destroys tissue.
Uses: Polyolefin catalyst; manufacture of primary alcohols and olefins; pyrophoric fuel.
Shipping regulations: (Rail) Pyrophoric liquid, n.o.s., Red label. (Air) Not acceptable.

triisobutylene. A mixture of isomers of the formula $(C_4H_8)_3$ readily prepared by polymerizing isobutylene. A typical mixture is 2,2,4,6,6-pentamethylheptane-3 and 2-neopentyl-4,4-dimethylpentene-1. May be depolymerized to simpler isobutylene derivatives.
Properties: Liquid. Sp. gr. 0.764 (60°F); boiling range 348–354°F. Combustible; low toxicity.
Containers: 55-gal drums; tank cars.
Uses: Synthesis of resins, rubbers, and intermediate organic compounds; lubricating oil additive, raw material for alkylation in producing high octane motor fuels.

triisooctyl phosphite $(C_8H_{17}O)_3P$.
Properties: Colorless liquid with characteristic odor. Miscible with most common organic solvents; insoluble in water. Hydrolyzes very slowly. Exposure to air should be minimum. High thermal stability. Sp. gr. 0.891 (20/4°C); b.p. 161–164°C (0.3 mm); flash point 385°F (COC). Combustible.
Containers: 5-gal, 55-gal steel drums.
Uses: Intermediate for insecticides; component of vinyl stabilizers; lubricant additive; specialty solvent.

O,O,O-triisooctyl phosphorothioate (triisooctyl thiophosphate) $(C_8H_{17}O)_3PS$.
Properties: Colorless liquid with characteristic odor. B.p. (0.2 mm) 160–170°C; sp. gr. 0.933; flash point (COC) 410°F. Insoluble in water; soluble in most organic solvents. Combustible.
Containers: Bottles; drums.
Hazard: Toxic, with nerve damage. Cholinesterase inhibitor.
Uses: Plasticizer; lubricant additive; hydraulic fluid; intermediate.
Shipping regulations: (Rail, Air) Organic phosphates, liquid, n.o.s., Poison label. Not acceptable on passenger planes.

triisooctyl trimellitate $C_6H_3(COOC_8H_{17})_3$.
Properties: Clear liquid; mild odor; sp. gr. 0.992 (20/20°C); distillation range 272–286°C (5–95%); f.p. gel at −45°C; refractive index 1.4852 (23°C). Practically insoluble in water. Combustible; low toxicity.
Use: Plasticizer.

triisopropanolamine $N(C_3H_6OH)_3$.
Properties: Crystalline white solid. Mild base. (A mixture of isopropanolamines which has sp. gr. of 1.004–1.010 and is liquid at room temperature is also marketed). Sp. gr. 0.9996 (50/20°C); m.p. 45°C;

b.p. 305°C; vapor pressure < 0.01 mm (20°C); melting point 58°C; viscosity 1.38 poise (60°C). Soluble in water. Flash point 320°F. (O.C.). Combustible.
Grade: Technical.
Containers: Drums; tank cars.
Hazard: Moderately irritant to skin and eyes.
Use: Emulsifying agents.

triisopropyl borate $[(CH_3)_2CH]_3BO_3$.
Properties: Colorless liquid; b.p. 138–140°C. f.p. −59°C; sp. gr. 0.8138; flash point 82°F (TCC).
Derivation: Reaction of isopropyl alcohol with boric oxide.
Hazard: Flammable; moderate fire risk. Moderately toxic.

triisopropyl phosphite $[(CH_3)_2CH]_3PO_3$.
Properties: Colorless liquid, characteristic odor; sp. gr. 0.914 (20/4°C); b.p. 94–96°C (50 mm); flash point 165°F (COC). Miscible with most common organic solvents; insoluble in water. Hydrolyzes slowly in water. Exposure to air should be minimum. High thermal stability. Combustible.
Containers: 55-gal, 5-gal steel drums.
Uses: Intermediate for insecticides; component of vinyl stabilizers; lubricant additive; specialty solvent.

triketohydrindene hydrate. See ninhydrin.

"Trilafon."[321] Trademark for perphenazine (q.v.).

trilaurin. The glyceride of lauric acid; glyceryl trilaurate.

trilauryl amine (tridodecyl amine) $(C_{12}H_{25})_3N$. Colorless liquid; sp. gr. 0.82; m.p. 14°C; soluble in organic solvents; insoluble in water. Combustible. Toxicity probably low.
Uses: Chemical intermediate; metal complexes.

trilauryl phosphite $(C_{12}H_{25}O)_3P$.
Properties: Water-white liquid; sp. gr. 0.866 (25/15°C); refractive index 1.456 (n 25/D); m.p. below 10°C. Combustible. Toxicity probably low.
Uses: Stabilizer in polymers; chemical intermediate.

trilauryl trithiophosphite $(C_{12}H_{25}S)_3P$.
Properties: Pale yellow liquid; sp. gr. 0.915 (25/15°C); m.p. 20°C; refractive index 1.502 (n 25/D); flash point 430°F (COC). Combustible. Toxicity probably low.
Uses: Stabilizer; lubricant; chemical intermediate.

"Triluxe."[56] Trademark for dry-cleaning solvent consisting of trichloroethylene.
Hazard: See trichloroethylene.

trimagnesium phosphate. See magnesium phosphate, tribasic.

"Tri-Mal."[304] Trademark for tribasic lead maleate vinyl stabilizer.

trimedlure. Generic name for tert-butyl 4(or 5)-chloro-2-methylcyclohexanecarboxylate, $H_3C(Cl)C_6H_9COOC(CH_3)_3$.
Properties: Liquid; b.p. 90–92°C (0.6 mm). Soluble in most organic solvents; insoluble in water.
Use: Insect attractant.

trimellitic anhydride (TMA) $HOCOC_6H_3COOCO$ (ring closure between CO groups)
1,2,4-Benzenetricarboxylic acid, 1,2-anhydride.
Properties: Solid; m.p. 164–166°C. Combustible.
Derivation: From pseudocumene.
Uses: Plasticizer for polyvinylchloride; alkyd coating resins; high-temperature plastics; wire insulation; gaskets; automotive upholstery.

"Trimene Base."[248] Trademark for a reaction product of ethyl chloride, formaldehyde, and ammonia.

Properties: Dark brown, viscous liquid; sp. gr. 1.10; soluble in water and acetone; insoluble in gasoline and benzene.
Uses: Rubber accelerator.

trimer. An oligomer whose molecule is comprised of three molecules of the same chemical composition. Examples are trioxane and tripropylene. See also polymer; dimer.

trimercuric orthophosphate. See mercuric phosphate.

trimercurous orthophosphate. See mercurous phosphate.

trimesoyl trichloride (benzene-1,3,5-tricarboxylic acid chloride) $C_6H_3(COCl)_3$.
Use: Specialty organic.

"Trimet."[319] Trademark for a trihydric alcohol. See trimethylolethane.

trimethadione (3,5,5-trimethyl-2,4-oxazolidinedione)
C_6H_9NO or $OC(O)N(CH_3)C(O)C(CH_3)_2$ (ring closure between O and C).
Properties: White, granular, crystalline substance. Camphor-like odor. M.p. 45–47°C; soluble in water; freely soluble in alcohol, chloroform, and ether; pH (5% solution) about 6.0.
Grade: U.S.P. Toxic in overdose.
Hazard: May have adverse side effects.
Use: Medicine (anticonvulsant).

trimethoxyborine. See trimethyl borate.

trimethoxyboroxine (methyl metaborate) $(CH_3O)_3B_3O_3$. A cyclic compound.
Properties: Colorless liquid; m.p. 10–11°C; b.p. dissociates; sp. gr. 1.216 (25°C); refractive index 1.3986. Nonflammable.
Derivation: Reaction of methyl borate with boric acid.
Grade: 99%.
Use: Metal-fire extinguishing fluid.

trimethoxymethane. See methyl orthoformate.

3,4,5-trimethoxyphenethylamine. See mescaline.

2,4,5-trimethoxy-1-propenylbenzene (asarone) $(CH_3O)_3C_6H_2CH{:}CHCH_3$. Crystals; m.p. 67°C; insoluble in water; soluble in alcohol. A component of calamus oil and other essential oils. Combustible.

trimethylacethydrazide ammonium chloride. See Girard's T Reagent.

trimethylacetic acid (pivalic acid; neopentanoic acid) $(CH_3)_3CCOOH$.
Properties: Colored crystals; sp. gr. 0.905 (50°C); refractive index 1.3931 (36.5°C); m.p. 35.5°C; b.p. 163.8°C; soluble in water, alcohol and ether. Combustible. Low toxicity.
Uses: Intermediate, as a replacement for some natural materials.

trimethylaluminum (aluminum trimethyl; ATM) $(CH_3)_3Al$.
Properties: Colorless pyrophoric liquid; b.p. 126°C; m.p. 15.4°C; sp. gr. 0.752.
Derivation: By sodium reduction of dimethylaluminum chloride.
Hazard: Highly flammable; dangerous fire risk. Flames instantly on contact with air; reacts violently with water, acids, halogens, alcohols and amines.
Uses: Catalyst for olefin polymerization; pyrophoric fuel; manufacture of straight-chain primary alcohols and olefins; to produce luminous trails in upper atmosphere to track rockets.
Shipping regulations: (Rail) Red label. Legal label name; aluminum trimethyl. (Air) Not acceptable.

trimethyladipic acid. $C_9H_{16}O_4$.
Properties: Powder. Low toxicity. Combustible.
Uses: Esterification agent for plasticizers, lubricants, alkyd resins, polyurethane, polyester, special polyamides; intermediate for the production of glycols.

trimethylamine (TMA) $(CH_3)_3N$.
Properties: Colorless gas at room temperature; readily liquefied. Anhydrous form shipped as liquefied compressed gas. Fishy, ammoniacal odor; sp. gr. 0.662 (−5°C); b.p. −4°C; f.p. −117.1°C; flash point of 25% solution 38°F. (TOC). Soluble in water, alcohol, and ether. Autoignition temp. 374°F. Flash point 10°F (C.C.)
Derivation: Interaction of methanol and ammonia over a catalyst at high temperature. The mono-, di, and trimethylamines are produced, and yields are regulated by conditions.
Method of separation: Azeotropic or extractive distillation.
Grades: Anhydrous 99% min; aqueous solution 25, 30, 40%.
Containers (Solution): Glass bottles; drums; tank cars (Anhydrous) Cylinders; tank cars.
Hazard: Flammable, dangerous fire risk. Explosive limits in air, 2 to 11.6%. Toxic by inhalation; vapor highly irritating. Safety data sheet available from Manufacturing Chemists Assn., Washington, D.C.
Uses: Organic synthesis, especially of choline salts; warning agent for natural gas; manufacture of disinfectants; flotation agent; insect attractant; quaternary ammonium compounds; plastics.
Shipping regulations: (Anhydrous) (Rail) Red gas label. (Air) Flammable Gas label. Not acceptable on passenger planes. (Aqueous solution) (Rail) Red label. (Air) Flammable Liquid label.

trimethylamine sulfur trioxide $(CH_3)_3N \cdot SO_3$.
Properties: White powder; m.p. 232–238°C (decomp.) soluble in hot water, ethanol. Soluble with difficulty in cold water and acetone. Not dissociated in benzene and chloroform solutions. Distinctly different from the isomeric adduct of trimethylamine oxide and sulfur dioxide.
Uses: Separation of isomers; soil sterilant; catalyst for thermosetting resins.

2,4,5-trimethylaniline. See pseudocumidine.

1,2,3-trimethylbenzene. See hemimellitene.

1,2,4-trimethylbenzene. See pseudocumene.

1,3,5-trimethylbenzene. See mesitylene.

(trimethylbenzyl)dodecyldimethylammonium chloride $[(CH_3)_3C_6H_2CH_2N(CH_3)_2C_{12}H_{25}]Cl$. A quaternary ammonium salt.
Properties: White to slightly yellow crystalline powder; mild odor and taste. M.p. 162–163°C; density 4.62 lbs/gal. Soluble in water, alcohol, glycerin, acetone; pH of 10% solution in distilled water, 4.4.
Use: Germicide.

trimethyl borate (methyl borate; trimethoxyborine) $(Ch_3O)_3B$.
Properties: Water-white liquid. B.p. 67–68°C; sp. gr. 0.915; f.p. −29°C. Miscible with ether, methanol, hexane, tetrahydrofuran; decomposes in presence of water. Flash point about 80°F; low toxicity.
Derivation: Reaction of boric acid and methanol.
Containers: Truck and car lots.
Hazard: Flammable, fire risk; reacts with water and oxidizing agents.
Uses: Solvent; dehydrating agent; fungicide for citrus fruit; neutron scintillation counters; brazing flux; boron compounds; intermediate.

2,2,3-trimethylbutane (isopropyltrimethylmethane; triptane) $CH_3C(CH_3)_2C(CH_3)CH_3$.
Properties: Colorless liquid. Soluble in alcohol; insoluble in water. Sp. gr. 0.691; b.p. 81.0°C; f.p. −24.96°C; refractive index 1.3895 (20°C).
Hazard: Flammable, moderate fire risk.
Uses: Organic synthesis; aviation fuel.

trimethyl carbinol. See tert-butyl alcohol.

1,1,3-trimethyl-5-carboxy-3-(*p*-carboxyphenyl) indane. An aromatic di-acid used as intermediate in manufacture of polyester fibers, polyamides, and alkyd resins; hot-melt adhesives; engineering thermoplastics.

trimethylchlorosilane $(CH_3)_3SiCl$.
Properties: Colorless liquid. B.p. 57°C; sp. gr. 0.854 (25/25°C); refractive index (n 25/D) 1.3893; flash point −18°F. Readily hydrolyzed with liberation of hydrochloric acid; soluble in benzene, ether, and perchloroethylene.
Derivation: By Grignard reaction of silicon tetrachloride and methylmagnesium chloride.
Containers: Up to 55-gal drums.
Hazard: Highly flammable; dangerous fire risk. Strong irritant to tissue.
Uses: Intermediate for silicone fluids; as a chain terminating agent; imparting water repellency.
Shipping regulations: (Rail) Red label. (Air) Flammable Liquid label. Not acceptable on passenger planes.

trimethylcyclododecatriene (TMCDT). A cyclic hydrocarbon. Available in 1, 5, and 55-gallon containers. Intermediate in making derivatives useful in the perfume and pharmaceutical industries, as well as catalysts.

3,3,5-trimethylcyclohexanol-1 $C_6H_8(CH_3)_3OH$.
Properties: Colorless liquid. Sp. gr. 0.878 (40/20°C); m.p. 35.7°C; b.p. 198°C. Soluble in most organic solvents, hydrocarbons, oils; insoluble in water. Flash point 165°F (O.C.) Combustible.
Hazard: Toxic by inhalation. Strong irritant. Safety data sheet available from Manufacturing Chemists Assn., Washington, D.C.
Uses: Menthol and camphor substitute; antifoaming agent; hydraulic fluids and textile soaps; odor masking; esterification agent; pharmaceuticals; wax additive; printing inks.

3,5,5-trimethyl-2-cyclohexen-1-one. See isophorone.

3,3,5-trimethylcyclohexyl salicylate. See homomenthyl salicylate.

trimethyl dihydroquinoline polymer (TDQP) $(C_{12}H_{15}N)_n$ (probably three or more quinoline groups).
Properties: Amber pellets; sp. gr. 1.08; softening point 75°C. Insoluble in water; miscible with ethanol, acetone, benzene, monochlorobenzene, isopropyl acetate and gasoline.
Uses: Antioxidant; stabilizer or polymerization inhibitor.

3,7,11-trimethyl-1,6,10-dodecatrien-3-ol. See nerolidol.

3,7,11-trimethyl-2,6,10-dodecatrien-1-ol. See farnesol.

trimethylene. See cyclopropane.

trimethylene bromide (1,3-dibromopropane) $CH_2BrCH_2CH_2Br$.
Properties: Colorless liquid; sweet odor; sp. gr. 1.979 (20/4°C); b.p. 166°C; insoluble in water; soluble in organic solvents; f.p. −34.4°C. Combustible. Low toxicity.
Grades: Technical; C.P.
Containers: Carboys, drums.
Uses: Intermediate for dyestuff and pharmaceutical industries, cyclopropane manufacture.

trimethylene chlorobromide. See 1-bromo-3-chloropropane.

trimethylene chlorohydrin (3-chloro-1-propanol) $ClCH_2CH_2CH_2OH$.
Properties: Colorless to pale yellow liquid; characteristic odor; sp. gr. 1.130–1.150 (25/25°C); refractive index 1.445–1.447 (n 25/D). Soluble in water, alcohols and ethers; insoluble in hydrocarbons. Combustible.
Use: Intermediate.

trimethylenedicyanide. See glutaronitrile.

trimethylene glycol (1,3-propylene glycol; 1,3-propanediol) $CH_2OHCH_2CH_2OH$.
properties: Colorless, odorless liquid; sp. gr. 1.0537 (25°C); b.p. 210–211°C; soluble in water, alcohol and ether. Combustible. Low toxicity. Autoignition temp. 752°F.
Derivation: From acrolein.
Grades: Technical, 95%; pure, 99%.
Use: Intermediate, especially for polyesters.

trimethylene oxide. See oxetane.

sym-**trimethylene trinitramine.** See cyclonite.

trimethylethylene. See 3-methyl-2-butene.

trimethylglycine. See betaine.

trimethylheptanoic acid. See isodecanoic acid.

2,2,5-trimethylhexane $(CH_3)_3CCH_2CH_2CH(CH_3)_2$.
Properties: Colorless liquid; f.p. −105.84°C; b.p. 124.06°C; sp. gr. 0.711 (60/60°F); refractive index 1.399 (20/D); flash point 55°F.
Grades: 95%, 99%; research.
Containers: Bottles; drums.
Hazard: Flammable, dangerous fire risk. The 2,3,3-, 2,3,4-, and 3,3,4-isomers are less flammable.
Uses: Synthesis; motor fuel additive.
Shipping regulations: (Rail) Flammable liquids, n.o.s., Red label. (Air) Flammable Liquid label.

3,5,5-trimethylhexan-1-ol $C_9H_{20}O$.
Properties: Colorless liquid of mild odor; b.p. 194°C; sp. gr. 0.8236 (25/4°C); wt/gal 6.86 lbs (25°C); refractive index (n 25/D) 1.4300; flash point (open cup) 200°F. Insoluble in water. Combustible. Low toxicity.
Derivation: High-pressure synthesis.
Uses: Synthetic lubricants; additives to lubricating oils; wetting agent; softener in manufacture of various plastics; disinfectants and germicides.

trimethylmethane. See isobutane.

trimethyl nitrilotripropionate $N(CH_2CH_2COOCH_3)_3$
Properties: Weakly basic tertiary amine and organic ester.
Uses: Plasticizer for PVC; intermediate for pharmaceutical and agricultural chemicals; high-boiling solvents; low-temperature lubricants.

trimethylnonanol. See 2,6,8-trimethylnonyl-4 alcohol.

2,6,8-trimethyl-4-nonanone (isobutyl heptyl ketone) $(CH_3)_2CHCH_2COCH_2CH(CH_3)CH_2CH(CH_3)_2$.
Properties: Water-white liquid with pleasant odor; high solvent power for vinyl resins, the cellulose esters and ethers, and many difficultly soluble substances; insoluble in water; sp. gr. 0.8165 (20/20°C); wt/gal 6.8 lbs (20°C); f.p. −75°C; b.p. 211–219°C; viscosity 1.91 cp (20°C); flash point 195°F (COC). Combustible. Low toxicity.
Uses: Solvent; dispersant; intermediate; lube oil dewaxing.

2,6,8-trimethylnonyl-4 alcohol (trimethylnonanol) $(CH_3)_2CHCH_2CHOHCH_2CH(CH_3)CH_2CH(CH_3)_2$.
Properties: Colorless liquid with characteristic odor; sp. gr. 0.8913 (20/20°C); wt/gal 6.9 lbs (20°C); b.p. 225.2°C; f.p. −60°C (sets to a glass); viscosity 21.4 cp; flash point 200°F (COC). Insoluble in water. Combustible. Low toxicity.
Derivation: Oxo process.
Uses: Surface-active and flotation agents; lube additives; rubber chemicals.

trimethylolethane (pentaglycerine; methyltrimethylolmethane) $CH_3C(CH_2OH)_3$.
Properties: Colorless hygroscopic crystals. Soluble in water and alcohol. Combustible.
Containers: 200-lb drums.
Uses: Conditioning agent; manufacture of varnishes, alkyd and polyester resins, synthetic drying oils.

trimethylolmelamine $C_3N_3(NHCH_2OH)_3$. The first stage in making melamine resins.

trimethylolpropane (hexaglycerol) $C_2H_5C(CH_2OH)_3$.
Properties: Colorless hygroscopic crystals. Soluble in water and alcohol. Combustible.
Containers: 200-lb drums.
Uses: Conditioning agent; manufacture of varnishes, alkyd resins; synthetic drying oils; urethane foams and coatings; silicone lube oils; lactone plasticizers; textile finishes; surfactants; epoxidation products.

trimethylolpropane monooleate. Theoretically $C_2H_5C(CH_2OH)_2Ch_2OOCC_{17}H_{33}$. The commercial product is a mixture of mono-, di, and tri- esters, free polyol and free oleic acid. Combustible.
Properties: Oily liquid; sp. gr. 0.954 (25°C); f.p. less than −20°C. Insoluble in water; soluble in most organic solvents.
Uses: Water-in-oil emulsifier; corrosion inhibitor; low-temperature plasticizer; deicing agent for gasoline.

trimethylopropane tris(mercaptopropionate) $C_2H_5C(CH_2OOCCH_2CH_2SH)_3$.
Properties: Liquid; sp. gr. 1.21 (25°C); refractive index 1.5151 (n 25/D). Insoluble in water and hexane; soluble in acetone, benzene and alcohol. Combustible.
Uses: Curing or cross-linking agents for polymers, especially epoxy resins; intermediate for stabilizers and antioxidants.

trimethylolpropane trithioglycolate $CH_3CH_2C(CH_2OOCCH_2SH)_3$.
Properties: Liquid; sp. gr. 1.28; refractive index 1.5292 (n 25/D). Combustible.
Uses: Curing or cross-linking agent for polymeric systems, especially epoxy resins; intermediate for stabilizers and antioxidants.

3,5,5-trimethyl-2,4-oxazolidinedione. See trimethadione.

2,2,4-trimethylpentane. See isooctane.

2,3,4-trimethylpentane $(CH_3)_2CHCH(CH_3)CH(CH_3)CH_3$.
Properties: Liquid; f.p. −109.43°C; b.p. 113°C; sp. gr. 0.723 (60/60°F); refractive index 1.4042 (20/D); flash point 41°F.

Grades: 95%, 99%; research.
Containers: Bottles; drums.
Hazard: Flammable, dangerous fire risk.
Uses: Intermediate; azeotropic distillation entrainer.
Shipping regulations: (Rail) Flammable liquid, n.o.s., Red label. (Air) Flammable Liquid label.

2,2,4-trimethyl-1,3-pentanediol $(CH_3)_2CHCH(OH)C(CH_3)_2CH_2OH$.
Properties: (96% pure). White solid; sp. gr. 0.928 (55/15°C); m.p. 46–55°C; b.p. 215–235°C; flash point 235°F. Slightly soluble in water; soluble in alcohol, acetone, ether, and benzene. Combustible.
Containers: Molten in tank cars and trucks.
Uses: Polyester resins, plasticizers, lubricants, surface coatings, and printing inks; insect repellent.

2,2,4-trimethyl-1,3-pentanediol monoisobutyrate $(CH_3)_2CHCH(OH)C(CH_3)_2CH_2OOCCH(CH_3)_2$.
Properties: Liquid; sp. gr. 0.945–0.955 (20/20°C); b.p. 180–182°C (125 mm); pour point −57°C; refractive index 1.4423 (n 20/D); flash point 245°F (COC). Insoluble in water; soluble in benzene, alcohol, acetone, and carbon tetrachloride. Combustible.
Use: Intermediate in the manufacture of plasticizers, surfactants, pesticides, and resins.

2,4,4-trimethylpentene-1 (alpha-diisobutylene) $H_2C:C(CH_3)CH_2C(CH_3)_3$.
Properties: Colorless liquid; b.p. 101.44°C; f.p. −93.5°C; refractive index 1.4086 (20°C); sp. gr. 0.7150 (20°C); flash point 35°F.
Derivation: Polymerization of isobutene.
Grades: 95%; 99%; research.
Containers: Bottles; drums.
Hazard: Flammable, dangerous fire risk.
Uses: Organic synthesis; motor fuel synthesis, particularly isooctane; peroxide reactions.
Shipping regulations: (Rail) Flammable liquid, n.o.s., Red label. (Air) Flammable Liquid label.

2,4,4-trimethylpentene-2 (beta-diisobutylene) $H_3CC(CH_3):CHC(CH_3)_3$.
Properties: Colorless liquid; b.p. 104.55°C; sp. gr. 0.724 (60/60°F); f.p. −106.4°C; refractive index 1.416 (20/D); flash point 35°F (TOC).
Grade: 95%.
Containers: Bottles; drums.
Hazard: Flammable, dangerous fire risk. May be irritant and narcotic in high concentrations.
Use: Organic synthesis.
Shipping regulations: (Rail) Flammable liquid, n.o.s., Red label. (Air) Flammable Liquid label.
See diisobutylene.

tri-2-methylpentylaluminum $[(CH_3)_2CH(CH_2)_3]_3Al$.
Properties: Colorless liquid.
Derivation: Reaction of 2-methylpentene and isobutylaluminum.
Hazard: Flammable and toxic.
Uses: Polyolefinn catalyst.

N,alpha,alpha-**trimethylphenethylamine.** See mephentermine.

trimethyl phosphate $(CH_3O)_3PO$.
Properties: Colorless liquid; 22.1% phosphorus; density 1.210 g/ml at 68°F; flash point above 300°F; b.p. 193°C; refractive index (n 20/D) 1.397; pour point −46°C. Soluble in both gasoline and water. Combustible.
Hazard: Toxic by ingestion and inhalation; strong irritant to skin and eyes.
Uses: For controlling spark plug fouling, surface ignition and rumble in gasoline engines.

trimethyl phosphite $(CH_3O)_3P$.
Properties: Colorless liquid; b.p. 108–108.5°C; pour point, less than −60°C; sp. gr. (20/4°C) 1.046; flash point (COC) 100°F. Insoluble in water, soluble in hexane, benzene, acetone, alcohol, ether, carbon tetrachloride and kerosine.
Hazard: Flammable, moderate fire risk.
Uses: Chemical intermediate, especially for insecticides.

trimethyl phosphorothionate $(CH_3O)_3PS$.
Properties: Water-white liquid; b.p. 78°C (12 mm).
Containers: 55-gal drums.
Hazard: Probably toxic.
Uses: Extraction of mineral salts from alkyl acid phosphate solvent solutions; plasticizer.

1,2,4-trimethylpiperazine $(CH_3)_3C_4H_7N_2$.
Properties: Colorless liquid; sp. gr. 0.851 (25/25°C); f.p. −44°C; b.p. 151°C; refractive index 1.4480 (n 20/D); flash point 257°F (open cup); miscible in water. Combustible.
Use: Polymerization catalyst.

trimethylpropylmethane. See 2,2-dimethylpentane.

2,4,6-trimethylpyridine. See 2,4,6-collidine.

2,4,6-trimethyl-1,3,5-trioxane. See paraldehyde.

1,3,5-trimethyl-2,4,6-tris(3,5-di-*tert*-butyl-4-hydroxybenzyl)benzene.
Properties: Free-flowing white crystalline powder; m.p. 244°C; no odor; partially soluble in benzene and methylene chloride; insoluble in water. Low toxicity. Combustible. Permissible in contact with food products.
Containers: Fiber drums.
Uses: Antioxidant for polypropylene, high-density polyethylene, spandex fibers, polyamides, and specialty rubbers.

2,6,10-trimethyl-9-undecen-1-al $(CH_3)_2C:CHC_2H_4CH(CH_3)C_3H_6CH(CH_3)CHO$.
Properties: Clear yellow liquid; strong, pungent ozone-like odor; sp. gr. 0.850–0.860 (25/25°C); refractive index 1.4530–1.4630 (20°C). Clearly soluble in 1 part of 90% alcohol; flash point above 212°F (TCC). Combustible; low toxicity.
Use: Perfume.

trimethylvinylammonium hydroxide. See neurine.

trimethylxanthine. See caffeine.

"Trimeton Maleate."[321] Trademark for pheniramine maleate (q.v.).

"Trimulso."[236] Trademark for a liquid synthetic surfactant used in the preparation of oil-in-water emulsion drilling muds; effective in both fresh water and brine muds.

trimyristin. The glyceride of myristic acid; glyceryl trimyristate.

trinickelous orthophosphate. See nickel phosphate.

trinitroaniline (picramide) $C_6H_2NH_2(NO_2)_3$.
Properties: Orange-red crystals; m.p. 188°C; b.p. explodes; sp. gr. 1.762.
Derivation: Nitrating aniline in glacial acetic acid solution or by the use of mixed nitric-sulfuric acid in limited amounts.

Hazard: Dangerous; explodes by heat or shock.
Use: Explosive.
Shipping regulations: (Rail) Consult regulations. (Air) Not acceptable dry, or wet with less than 10% water; Flammable Solid label when wet with not less than 10% water.

trinitroanisole (methyl picrate; 2,4,6-trinitrophenyl methyl ether) $CH_3OC_6H_2(NO_2)_3$.
Properties: Crystals; m.p. 68.4°C; sp. gr. 1.408 (20/4°C).
Hazard: Dangerous; explodes by heat or shock. Highly toxic.
Derivation: Interaction of methyl iodide and silver picrate; nitration of anisic acid.
Shipping regulations: (Rail) Consult regulations. (Air) Not acceptable.

1,3,5-trinitrobenzene (TNB) $C_6H_3(NO_2)_3$.
Properties: Yellow crystals with a sp. gr. of 1.688 (20/4°C); m.p. 122°C; soluble in alcohol and ether; insoluble in water.
Derivation: From trinitrotoluene by removal of methyl group.
Hazard: Dangerous; explodes by heat or shock.
Use: Explosive.
Shipping regulations: (Rail) Consult regulations. (Air) Dry, or wet with less than 10% water, Not acceptable. Wet with not less than 10% water, Flammable Solid label.

2,4,6-trinitrobenzoic acid (trinitrobenzoic acid) $C_6H_2(NO_2)_3COOH$.
Properties: Orthorhombic crystals; m.p. 228.7°C; sublimes with decomposition forming carbon dioxide and trinitrobenzene; slightly soluble in water and benzene; soluble in alcohol, ether, and acetone.
Derivation: Oxidation of 2,4,6-trinitrotoluene with chromic acid.
Hazard: Dangerous; explodes by heat or shock.
Use: Synthesis.
Shipping regulations: (Rail) Consult regulations. (Air) Dry, or wet with less than 10% water, Not acceptable. Wet with not less than 10% water, Flammable Solid label.

2,4,6-trinitro-meta-**cresol** (cresolite; cresylite) $(NO_2)_3C_6H(Ch_3)OH$.
Properties: Yellow crystals; m.p. 106°C; readily soluble in alcohol, ether, and acetone. When heated to decomposition it emits highly toxic fumes of oxides of nitrogen.
Derivation: Prepared from meta-cresol by a process similar to which picric acid is prepared from phenol.
Hazard: Explodes at 300°F. Severe explosion risk when shocked or heated.
Uses: Bursting charges and other high explosive uses.
Shipping regulations: (Rail) Consult regulations (Air) Not acceptable.

trinitroglycerin. See nitroglycerin.

trinitronaphthalene (naphtite) $C_{10}H_5(NO_2)_3$. Commercial preparation is a mixture of isomers which melts about 110°C.
Hazard: Explosion risk when shocked or heated.
Uses: Explosive; stabilizer for nitrocellulose.
Shipping regulations: (Rail) Consult regulations (Air) Not acceptable.

trinitrophenol. See picric acid.

2,4,6-trinitrophenyl methyl ether. See trinitroanisole.

trinitrophenylmethylnitramine. See tetryl.

2,4,6-trinitroresorcinol. See styphnic acid.

2,4,6-trinitroltoluene (TNT; methyltrinitrobenzene) $CH_3C_6H_2(NO_2)_3$.
Properties: Yellow, monoclinic needles; sp. gr. 1.654; m.p. 80.9°C; soluble in alcohol and ether; insoluble in water.
Derivation: Nitration of toluene with mixed acid. Small amounts of the 2,3,4-and 2,4,5-isomers are produced which may be removed by washing with aqueous sodium sulfite solution.
Grade: Technical.
Containers: Wooden cases or kegs; multiwall paper sacks.
Hazard: Toxic by ingestion, inhalation, and skin absorption. Tolerance, 1.5 mg per cubic meter of air. Flammable, dangerous fire risk; moderate explosion risk; will detonate only if vigorously shocked or heated to 450°F.
Uses: Explosive; intermediate in dyestuffs and photographic chemicals.
Shipping regulations: (Rail) (Dry) See regulations for high explosives (Wet, not to exceed 16 oz.) Flammable solid. (Air) Dry, or wet with less than 10% water, Not acceptable. Wet with not less than 10% water, Flammable solid label.

trinitrotrimethylenetriamine. See cyclonite.

trioctadecyl phosphite $(C_{18}H_{37}O)_3P$. White waxy solid; m.p. 45–47°C; sp. gr. 0.940 (25/15°C). Used as a stabilizer in polymers and an intermediate.

tri-n-octylaluminum $(C_8H_{17})_3Al$.
Properties: Colorless pyrophoric liquid.
Derivation: Reaction between octene and isobutylaluminum.
Hazard: Toxic and flammable. Ignites in air.
Use: Polyolefin catalyst.
Shipping regulations: (Rail) Pyrophoric liquid, n.o.s., Red label. (Air) Not acceptable.

trioctyl phosphate (octyl phosphate) $(C_8H_{17})_3PO_4$.
Properties: Liquid. Sp. gr. 0.924 (26°C); b.p. 220–30 (8 mm); soluble in alcohol, acetone, and ether. Combustible; low toxicity.
Uses: Solvent, antifoaming agent; plasticizer.

trioctylphosphinic oxide (TOPO). $(C_8H_{17})_3PO$
Properties: Solid; m.p. 55°C; min. purity 95%.
Containers: 25 and 100-lb. fiber drums.
Uses: Reagent for extraction of metals from aqueous and nonaqueous solutions, including fissionable actinide elements.

trioctyl phosphite. See tris-2-ethyl-hexyl phosphite.

"Triodine."[284] Trademark for a cleaner-sanitizer-disinfectant particularly formulated for use in the bottling industry. Contains nonionic-iodine complexes.

triolein. See olein.

triorthocresylphosphate. See tricresyl phosphate (hazard).

"Triox."[253] Trademark for a weed killer containing sodium arsenite.
Hazard: Highly toxic by ingestion.

sym-**trioxane** (triformol; trioxin; metaformaldehyde) $(CH_2O)_3$ or $\underline{CH_2OCH_2OCH_2O}$. A trimer of formaldehyde; not to be confused with paraformaldehyde (q.v.) which consists of 8 or more formaldehyde units.
Properties: White crystals, formaldehyde odor; m.p. 62°C; sublimes at 115°C; flash point (open cup) 113°F. Soluble in water; soluble in alcohol and ether. Autoignition temp. 777°F.

Derivation: By distillation of formaldehyde with an acid catalyst, and extraction with solvent.
Containers: Drums; solid blocks in polyethylene containers.
Hazard: Moderate fire risk; explosive limits in air 3.6 to 29%.
Uses: Organic synthesis; disinfectant; nonluminous, odorless fuel.
See also formaldehyde.

2,6,8-trioxypurine. See uric acid.

tripalmitin (palmitin; glyceryl tripalmitate) $C_3H_5(OOCC_{15}H_{31})_3$.
Properties: White, crystalline powder. Soluble in ether and chloroform. Insoluble in water. M.p. 65.5°C; sp. gr. 0.866 (80/4°C). Combustible. Low toxicity.
Derivation: From glycerol and palmitic acid.
Grade: Technical.
Containers: Tins; fiber drums.
Uses: Medicine; soap; leather dressing.

tripelennamine citrate $C_{16}H_{21}N_3 \cdot C_6H_8O_7$. 2-[Benzyl(2-dimethylaminoethly)amino]pyridine dihydrogen citrate.
Properties: White, bitter crystalline powder. Solutions are acid to litmus. M.p. 107°C. Soluble in water and alcohol; very slightly soluble in ether; practically insoluble in chloroform and benzene. 1% solution in water has a pH of about 4.3.
Grade: U.S.P.
Hazard: Moderately toxic by ingestion.
Use: Medicine (antihistamine; sunburn treatment).

tripentaerythritol $(CH_2OH)_3CCH_2OCH_2C(CH_2OH)_2 \cdot CH_2OCH_2C(CH_2OH)_3$.
Properties: White to ivory powder; has eight primary hydroxyl groups, all esterifiable. Melting range 225–240°C. Combustible. Low toxicity.
Uses: Hard resins, varnishes, and fast-drying tall oil vehicles.

"Tripentek."[138] Trademark for tripentaerythritol, technical.

triphenol P. See 1,1,3-tris(hydroxyphenyl)-propane.

triphenylantimony (triphenylstibine) $Sb(C_6H_5)_3$.
Properties: White crystalline solid; sp. gr. 1.434 (25°C); m.p. 46–53°C; b.p. less than 360°C. Insoluble in water; slightly soluble in alcohol; soluble in most organic solvents. Combustible. Low toxicity.
Derivation: Reaction of antimony trichloride with phenyl magnesium bromide or phenyl sodium.
Containers: 10-, 50-, 200-lb drums.
Uses: Stibonium salts; co-catalyst in converting trienes to aromatics and hydroaromatics; reacts with nitric-sulfuric acid to give trinitro derivatives; polymerization inhibitor; lubricating oil additive.

triphenylcarbinol, bis(4-dimethylamino)-. $C_6H_5C(OH)[C_6H_4N(CH_3)_2]_2$.
Properties: Solid; m.p. 121–123°C; very soluble in ether and hot benzene; soluble in acids.
Use: Dyestuffs (see malachite green).

triphenyl formazan $CN_4H(C_6H_5)_3$. Red insoluble derivative of tetrazolium chloride; formed when the latter comes into contact with viable portions of a seed. Used in germination and viability tests.

triphenylguanidine (TPG) $C_6H_5NC(C_6H_5NH)_2$.
Properties: White crystalline powder; soluble in alcohol. Sp. gr. 1.10; m.p. 144°C. Combustible; low toxicity.
Derivation: Desulfurization of thiocarbanilide in presence of aniline.
Use: Accelerator for vulcanization of rubber.

triphenylmethane dye. Any of a group of dyes whose molecular structure is basically derived from $(C_6H_5)_3CH$, usually by substitution of NH_2, OH, HSO_3, or other groups or atoms for some of the hydrogen of the C_6H_5 groups. Many coal tar and synthetic dyes are of this class, including rosaniline, fuchsin, malachite green, and crystal violet. See also triarylmethane dye.

triphenylmethane triisocyanate. Available as a brown 20% solution in methylene chloride. Used for bonding uncured rubber to metal or other surfaces.

triphenylmethylhexafluorophosphate.
Properties: Orange-colored, free-flowing crystalline powder.
Hazard: May be toxic.
Uses: Catalyst in manufacture of polyoxymethylenes (polymers of formaldehyde and trioxane).
See also acetal resin.

triphenyl phosphate (TPP) $PO(OC_6H_5)_3$.
Properties: Colorless, odorless, crystalline powder. Soluble in most lacquers, solvents, thinners, oils. Insoluble in water. M.p. 50°C; b.p. 245°C (11 mm Hg); sp. gr. 1.268 (60°C); wt/gal 10.5 lbs; refractive index 1.550 (60°C). Combustible. Flash point 428°F. (C.C.).
Derivation: Interaction of phenol and phosphorus oxychloride.
Grade: Technical.
Containers: 200-lb fiber drums; barrels; multiwall paper sacks.
Hazard: Toxic by inhalation. Tolerance, 3 mg per cubic meter in air.
Uses: Fire-retarding agent; plasticizer for cellulose acetate and nitrocellulose.

triphenylphosphine. See triphenylphosphorus.

triphenyl phosphite $(C_6H_5O)_3P$.
Properties: Water-white to pale yellow solid or oily liquid; pleasant odor; sp. gr. 1.184 (25/25°C); m.p. 22–25°C; b.p. 155–160°C (0.1 mm); refractive index 1.589 (25°C); flash point 425°F (COC). Combustible.
Containers: 55-gal drums.
Uses: Chemical intermediate; stabilizer systems for resins; metal scavenger; diluent for epoxy resins.

triphenylphosphorus (triphenylphosphine) $(C_6H_5)_3P$.
Properties: White, crystalline solid; m.p. 79–82°C; b.p. above 360°C; sp. gr. 1.132 (25°C). Insoluble in water; slightly soluble in alcohol; soluble in benzene, acetone, carbon tetrachloride. Flash point 356°F. (O.C.) Combustible.
Derivation: By a modified Grignard synthesis.
Containers: 10-, 50-, 150-lb drums.
Uses: Synthesis of organic compounds, phosphonium salts, other phosphorus compounds.

triphenylstibine. See triphenylantimony.

triphenyltetrazolium chloride. See tetrazolium chloride.

triphenyltin acetate $(C_6H_5)_3SnOOCCH_3$. An agricultural biocide. White crystalline solid, made by reaction of sodium acetate with triphenyltin chloride.
Hazard: Toxic and irritant to skin. Tolerance (as Sn), 0.1 mg per cubic meter in air.

triphenyltin chloride $(C_6H_5)_3SnCl$.
Properties: White crystalline solid; m.p. 106°C; b.p. 240°C (13.5 mm). Insoluble in water; soluble in organic solvents.
Derivation: Reaction of tin tetrachloride with phenylmagnesium bromide.
Hazard: Toxic and irritant to skin. Tolerance (as Sn), 0.1 mg per cubic meter in air.
Use: Biocidal intermediate.

triphenyltin hydroxide $(C_5H_5)_3SnOH$.
Properties: White solid; m.p. 118–120°C. Insoluble in water; soluble in ether, benzene, and alcohol.
Hazard: Toxic and irritant to skin. Tolerance (as Sn), 0.1 mg per cubic meter in air.
Uses: Insect chemisterilant; fungicide.

"Triphosaden."[91] Trademark for adenosine triphosphate used for biochemical and clinical research, and medicine.

triphosgene. See hexachloromethylcarbonate.

triphosphoric acid. See polyphosphoric acid.

triple bond. A highly unsaturated linkage between the two carbon atoms of acetylenic compounds (alkynes), typified by acetylene (HC≡CH). It is somewhat less reactive than a double bond in a comparable structure toward electrophilic reagents, but far more reactive than the single bonds in paraffinic compounds. See also double bond; chemical bonding.

triple point. The temperature and pressure at which the solid, liquid, and vapor of a substance can exist simultaneously. Also applied to similar equilibrium between any three phases, i.e., two solids and a liquid, etc. The triple point of water is of special importance because it is the fixed point (273.16°K) for the absolute scale of temperature; it is 0.00°C at 4.6 mm Hg.

triple superphosphate. A dry, granular, free-flowing product, gray in color; produced by addition of phosphoric acid to phosphate rock, thus avoiding formation of insoluble gypsum, as in superphosphate, and achieving about three times the amount of available phosphate (as P_2O_5).
Typical analysis: Moisture 2%; available P_2O_5 50%; water soluble P_2O_5 45%; free phosphoric acid 1%; also minor ingredients.
Containers: Bags; bulk; carloads.
Use: Fertilizer.
See also superphosphate and nitrophosphate.

tripoli (rottenstone).
Derivation: A porous, siliceous rock resulting from the natural decomposition of siliceous sandstone.
Grades: Various grades according to fineness for polishing; rose; cream; white.
Uses: Abrasive; polishing powder; filtering material; absorbent for insecticidal chemicals; paints (inert filler, wood filler); rubber filler; base for scouring soaps and powders.

tripolyphosphate. See sodium tripolyphosphate.

tripotassium orthophosphate. See potassium phosphate, tribasic.

tripotassium phosphate. See potassium phosphate, tribasic.

tripropionin. See glyceryl tripropionate.

tri-n-propylaluminum $(C_3H_7)_3Al$.
Properties: Colorless pyrophoric liquid; density 0.820 g/ml; f.p. −84°C.
Derivation: Reaction of propylene and isobutylaluminum.
Hazard: Toxic and flammable; dangerous fire risk; ignites spontaneously in air.
Use: Polyolefin catalyst.
Shipping regulations: Pyrophoric liquids, n.o.s., (Air) Not acceptable. (Rail) Red label.

tripropylamine $(CH_3CH_2CH_2)_3N$.
Properties: Water white liquid; amine odor; f.p. −94°C; boiling range 150–156°C; sp. gr. (20/20°C) 0.754; refractive index 1.417 (20°C); flash point 105°F (O.C.). Probably low toxicity. Combustible.
Containers: 5-gal cans; 55-gal drums; tank cars.
Hazard: Moderate fire risk.

tripropylene (propylene trimer) $(C_3H_6)_3$.
Properties: Colorless liquid; sp. gr. (20/20°C) 0.738; boiling range 133.3°-141.7°C). Weight 6.17 lb/gal (60°F); flash point 75°F (TOC). Probably low toxicity.
Derivation: Catalytic polymerization of propylene.
Containers: Drums; barges; tank cars.
Hazard: Flammable, dangerous fire risk.
Uses: Oxo feed stock; lubricant additive; plasticizers; nonyl phenol.
Shipping regulations: Flammable liquids, n.o.s. (Rail, Air) Flammable Liquid label.

tripropylene glycol $HO(C_3H_6O)_2C_3H_6OH$.
Properties: Colorless liquid; supercools instead of freezing; b.p. 268°C; sp. gr. 1.019 (25/25°C); lb/gal 8.51; refractive index 1.442 (n 25/D); flash point 285°F. Soluble in water, methanol, ether. Combustible. Low toxicity.
Containers: Drums; tank cars; tank trucks.
Uses: Intermediate in resins, plasticizers, pharmaceuticals; insecticides; dyestuffs; mold lubricants.

tripropylene glycol monomethyl ether $HO(C_3H_6O)_2C_3H_6OCH_3$.
Properties: Colorless liquid. Sp. gr. 0.961 (25°C); b.p. 242°C; 116°C (10 mm); viscosity 5.5 cp (25°C); refractive index 1.427 (25°C); flash point 250°F. Completely miscible with water, VMP naphtha, acetone, ethanol, benzene, carbon tetrachloride, ether, methanol, monochlorobenzene, and petroleum ether. Combustible. Low toxicity.
Use: Ingredient in hydraulic fluids.

triptane. See 2,2,3-trimethylbutane.

"Triptide."[91] Trademark for lyophilized monosodium glutathione used for biochemical and clinical research, and medicine.

tris-. A prefix indicating that a certain chemical grouping occurs three times in a molecule, e.g., (tris(hydroxymethyl)aminomethane $(CH_2OH)_3CNH_2$. See also bis-.

tris amine buffer. See tris(hydroxymethyl)aminomethane.

"Tris Amino."[319] Trademark for pharmaceutical grade of tris(hydroxymethyl)aminomethane.

tris(1-aziridinyl)phosphine oxide. See triethylenephosphoramide.

2,4,6-tris(1-aziridinyl)-s-triazine. See triethylenemelamine.

trisazo dye. One of the four kinds of azo dyes, characterized by the presence of three azo couplings (—N≡N—) in each molecule. (CI 30000 to 34999).

tris(2-chloroethyl) phosphate $(ClC_2H_4O)_3PO$.
Properties: Clear, transparent liquid; sp. gr. 1.425

(20/20°C); b.p. 214°C (25 mm); refractive index 1.4721 (n 20/D). Flash point 421°F (COC); combustible.
Use: Flame-retardant plasticizer.

tris(2-chloroethyl) phosphite $(ClC_2H_4O)_3P$.
Properties: Colorless liquid with characteristic odor. Sp. gr. 1.353 (20/4°C); b.p. 119°C (0.15 mm); flash point 280°F (O.C.). Miscible with most common organic solvents. Undergoes intramolecular isomerization at higher temperatures. Exposure to air should be minimum. Insoluble in water, and hydrolyzes in water. Combustible.
Containers: 5-gal, 55-gal steel drums.
Uses: Intermediate; component of vinyl stabilizers; grease additives; flameproofing compositions.

tris(2-chloroisopropyl) thionophosphate $[CH_3(CH_2Cl)CHO]_3PS$.
Properties: Liquid. Phosphorus content 9.0%; density 1.282 g/ml at 68°F; open cup flash point above 347°F; pour point less than −50°C. Readily soluble in gasoline; insoluble in water. Combustible.
Uses: To extend spark plug life, to control deposit-induced knocking in gasoline engines.

tris(2,3-dibromopropyl) phosphate $(CH_2BrCHBrCH_2O)_3PO$.
Properties: Viscous, pale yellow liquid; density 18.5 lb/gal; refractive index 1.5772 (20°C). Combustible.
Containers: Steel drums.
Use: Flame retardant for plastics.

tris(2,4-dichlorophenoxyethyl) phosphite. See "Falone."

tris(2,3-dichloropropyl) phosphate $(CH_2ClCHClCH_2O)_3PO$. Used as a flame retardant in plastics, and as a secondary plasticizer. Combustible.

tris(diethylene glycol monoethyl ether) citrate $HOC[CH_2COO(CH_2CH_2O)_2C_2H_5]_2COO-(CH_2CH_2O)_2C_2H_5$.
Properties: Solid. Sp. gr. (25°C) 1.28; m.p. 16–19°C; soluble in water. Used as a plasticizer. Combustible.

2,4,6-tris-(ethyleneimino)-s-triazine. See triethylenemelamine.

tris(2-ethylhexyl) phosphate $[C_4H_9CH(C_2H_5)CH_2O]_3PO$.
Properties: Colorless liquid; sp. gr. 0.9260 (20/20°C); f.p. below −90°C (sets to glass); pour point −74°C; mid-boiling point 216°C (4 mm); refractive index 1.441 (25°C). Insoluble in water; soluble in mineral oil and gasoline. Flash point 420°F. Combustible.
Use: Low-temperature plasticizer for PVC resins; imparting flame and fungus resistance.

tris-2-ethylhexyl phosphite (trioctyl phosphite) $[C_4H_9CH(C_2H_5)CH_2O]_3P$.
Properties: Colorless liquid; sp. gr. 0.902; b.p. (0.3 mm) 163–164°C; insoluble in water; soluble in alcohol and ether. Flash point 340°F. Combustible. Low toxicity.
Containers: Glass bottles; 5-, 55-gal drums.
Uses: Synthesis; plasticizers; stabilizers; lube and grease additives; flameproofing compositions.

tris(2-hydroxyethyl) isocyanurate (THEIC; tris(2-hydroxyethyl)-s-triazine-2,4,6-trione) $C_3N_3O_3(CH_2CH_2OH)_3$.
Properties: White solid; m.p. 135°C; b.p., dissociates at 180°C (3 mm). Very soluble in water; somewhat soluble in alcohol and acetone; insoluble in chloroform and benzene. Combustible.
Use: Additive to plastics, especially to impart thermal stability.

tris(hydroxymethyl)acetic acid $HOCH_2C(CH_2OH)_2COOH$. A photographic chemical made by bacterial oxidation of pentaerythritol.

tris(hydroxymethyl)aminomethane [Tri(hydroxymethyl)-aminomethane; THAM; 2-amino-2-hydroxymethyl-1,3-propanediol; tris amine buffer] $(CH_2OH)_3CNH_2$.
Properties: White crystalline solid. Solubility in water, 80 g/100 cc at 20°C. M.p. 171–172°C; b.p. (10 mm) 219–220°C; pH 0.1M aqueous solution 10.36. Corrosive to copper, brass, aluminum. Combustible.
Containers: Fiberpak boxes; drums.
Hazard: Irritant to skin and eyes.
Uses: Emulsifying agent (in soap form) for oils, fats, and waxes; absorbent for acidic gases; chemical synthesis; buffer; medicine.

tris(hydroxymethyl)nitromethane (2-hydroxymethyl-2-nitro-1,3-propanediol) $(CH_2OH)_3CNO_2$.
Properties: White crystals or amorphous solid; m.p. 175°C; decomposes. Soluble in water and alcohol.
Hazard: Moderate fire risk; probably irritant to skin and eyes.
Uses: Bactericide and slimicide for aqueous systems; cutting oil emulsions; pulp and paper; industrial water systems; drilling muds.

1,1,3-tris(hydroxyphenyl)propane (triphenol P) $(C_6H_4OH)_2CHCH_2CH_2C_6H_4OH$.
Properties: White solid; m.p. 84°C; sp. gr. 1.226 (20/20°C); f.p. 90–110°C; sets to glass below this temperature. Combustible.
Uses: Antioxidant; intermediate for polyester and alkyd resins.

"Tri-Sil."[534] Trademark for a ready mixed, single solution of solvent, reagent, and catalyst used for rapidly converting samples of polar materials (e.g., carbohydrates, phenols, steroids, etc.) into volatile trimethylsilyl ethers for gas chromatographic analyses.

trisiloxane. See siloxane.

tris[1-(2-methyl)aziridinyl]phosphine oxide $(C_3H_6N)_3PO$. See metepa.

"Tris Nitro."[319] Trademark for tris(hydroxymethyl)-nitromethane (q.v.).

trisodium citrate. See sodium citrate.

trisodium dipotassium tripolyphosphate $Na_3K_2P_3O_{10}$.
Properties: White crystalline solid; m.p. 620–640°C; density 2.48; solubility in water (26°C) 80 g/100 ml.
Use: Sequestrant.

trisodium EDTA (ethylenediaminetetraacetic acid trisodium salt) $C_{10}H_{13}N_2Na_3O_8 \cdot H_2O$. White powder; freely soluble in water. Used as an alkaline chelating agent.

trisodium hydroxyethylethylenediaminetriacetate $(NaOOCCH_2)_2NC_2H_4N(CH_2COONa)C_2H_4OH$.
Properties: Liquid; sp. gr. 1.285; f.p. below −5°C.
Use: Chelating agent.

trisodium nitrilotriacetate $N(CH_2COONa)_3$. A sodium salt of nitrilotriacetic acid.
Use: Sequestrant.

trisodium orthophosphate. See sodium phosphate, tribasic.

trisodium phosphate. See sodium phosphate, tribasic.

trisodium phosphate, chlorinated $4(Na_3PO_4 \cdot 12H_2O) \cdot NaOCl$. Active ingredients 3.25% min sodium hypochlorite and 91.75% min trisodium phosphate dodecahydrate. Inert ingredient, less than 5% NaCl.
Properties: White crystalline water-soluble solid; stable under normal storage conditions. In solution has the properties of both trisodium phosphate and sodium hypochlorite.
Derivation: By reacting sodium phosphate, caustic soda and sodium hypochlorite.
Containers: Bags; drums; and bulk.
Hazard: Moderately toxic; irritant to skin and eyes.
Uses: Cleaner and bactericide in dairies, food plants; dish washing compounds and scouring powders.

trisodium phosphate monohydrate. See sodium phosphate, tribasic, monohydrate.

tristearin. See stearin.

tris(tetrahydrofurfuryl) phosphate $(C_4H_7OCH_2O)_3PO$.
Properties: Yellow liquid; f.p. −75°C; refractive index 1.4759 (n 20/D); soluble in water and aromatic solvents.
Hazard: May be toxic.
Uses: Plasticizer; solvent.

"Tritac."[62] Trademark for a group of herbicides whose active ingredient is 2,3,6-trichlorobenzyloxypropanol.

"Trithion."[1] Trademark for an organic phosphate insecticide-acaricide available as aqueous emulsion, dust, and wettable powder containing various percentages of carbophenothion.
Hazard: See carbophenothion.

"Tritiac."[529] Trademark for tritium-activated self-luminous compounds and paints. Available in varying grades of luminosity up to 100 micro-lamberts.

tritium T Radioactive isotope of hydrogen; mass number 3; isotopic weight 3.017 (two neutrons and one proton in the nucleus).
Properties: Half-life 12.5 years; radiation, beta; radiotoxicity, very low.
Derivation: Bombardment of lithium with low-energy neutrons.
Forms available: Gas packaged in ampules and in tagged compounds such as water, streptomycin, cortisone, epinephrine, octadecane, stearic acid, etc.
Uses: Bombarding particle in cyclotrons; activator in self-luminous phosphors; in cold cathode tubes; tracer in biochemical research and various special problems in chemical analysis; luminous instrument dials.

tritolyl phosphate. See tricresyl phosphate (ortho isomer).

triton. The nucleus of the tritium atom (one proton, two neutrons).

"Triton."[23] Trademark for surfactants based on alkylaryl polyether alcohols, sulfonates and sulfates; nonionic, cationic and anionic types; oil-soluble and water-soluble types. Supplied as viscous liquids, pastes, or aqueous solutions. Surface activity includes detergency, emulsification, wetting, spreading, dispersing action.
Uses: Processing of textile fabrics and yarns; leather tannery operations; emulsification of pesticides; sanitizing and cleansing formulations; cosmetic preparations; medicinal soaps; paper manufacture; oil-well flooding; polymer manufacture; petroleum additives.

tritopine. See laudanidine.

"Tri-Una-Sol."[58] Trademark for nitrogen fertilizer solutions; varying analysis of 28, 30 and 32% nitrogen; containing urea, ammonium nitrate and water.

triuranium octoxide (uranous-uranic oxide; uranyl uranate) U_3O_8.
Properties: Olive-green to black solid; crystals or granules; insoluble in water; soluble in nitric acid and sulfuric acid. Sp. gr. 8.39; decomposes when heated to 1300°C to uranium dioxide.
Source: The naturally occurring uranium oxide found in pitchblende.
Derivation: (a) As one of the forms of uranium produced from the ores, often by a solvent extraction process. The solvent used is dodecylphosphoric acid. (b) A common form of triuranium octoxide is yellow cake, the powder obtained by evaporating an ammonia solution of the oxide.
Hazard: Radioactive poison. Use appropriate protection in handling.
Uses: Nuclear technology; preparation of other uranium compounds.
Shipping regulations: (Rail, Air) Consult regulations.

trixylenyl phosphate. See tri(dimethylphenyl)phosphate.

"Trizone."[233] Trademark for fumigants containing methyl bromide, chloropicrin, and 3-bromo-1-propyne.

"Trolene."[253] Trademark for insecticides containing O,O-dimethyl-O-(2,4,5-trichlorophenyl)phosphorothioate.
Hazard: Toxic; cholinesterase inhibitor.

"Troluoil."[200] Trademark for a petroleum solvent prepared by straight-run distillation.
Properties: Water-white, initial boiling point 194–205°F, 95% distills between 235 and 247°F; sp. gr. 0.741 (60°F); flash point (TCC) 25°F; mild, nonresidual odor.
Hazard: Flammable, dangerous fire risk.
Uses: Lacquers, rubber cements, and gravure inks.

"Tromexan."[219] Trademark for ethyl biscoumacetate.

trona (urao) $Na_2CO_3 \cdot NaHCO_3 \cdot 2H_2O$. A natural sodium sesquicarbonate and the most important of the natural sodas.
Properties: White, gray or yellow with vitreous, glistening luster. Noncombustible. Low toxicity. See Trona process.
Occurrence: Hungary, Egypt, Africa, Venezuela, and United States (Wyoming; California, especially Searles Lake, Owens Lake).
Use: Source of sodium compounds, especially the sodium carbonates.

"Tronabor."[88] Trademark for crude borax pentahydrate, from Searles Lake brines.

Trona process. The method used for separation and purification of soda ash, anhydrous sodium sulfate, boric acid, borax, potassium sulfate, bromine and potassium chloride from Searles Lake (California) brine.

"Tronox."[84] Trademark for chloride process titanium dioxide; used as a white pigment.

tropacocaine hydrochloride $C_{15}H_{19}NO_2 \cdot HCl$. An alkaloidal salt.

Properties: White crystals. Soluble in water, alcohol, and ether. M.p. 271°C.
Derivation: From a variety of Erythroxylon coca.
Hazard: Toxic by ingestion and inhalation.
Use: Medicine.

3-tropanol. See torpine.

tropeolin D. See methyl orange.

tropeolin OO (Orange IV) $NaSO_3C_6H_4NNC_6H_4NHC_6H_5$. para-Diphenylamino-azobenzene-sodium sulfonate. A biological stain and acid-base indicator, red at pH 1.4, yellow at pH 2.6. CI No. 13080.

tropine (3-tropanol) $C_8H_{15}NO$.
Properties: White, crystalline alkaloid; very hygroscopic. Soluble in water; alcohol, ether, and chloroform. M.p. 61.2–63°C; b.p. 233°C.
Derivation: By heating atropine or hyoscyamine with barium hydroxide.
Hazard: Toxic by ingestion and inhalation.
Use: Medicine.

Trouton's rule. The molal heat of vaporization of normal liquids, at the boiling point and under atmospheric pressure, divided by the absolute boiling temperature is a constant, approximately 22.

Trp. Abbreviation for tryptophan.

true solution. See solution (true).

truttine. A protein obtained from fish of the trout family.

"Trycite."[233] Trademark for an oriented polystyrene film. Used for packaging and in envelope windows.

tryparsamide (sodium N-phenylglycineamide-para-arsonate) $NaOAs(OOH)C_6H_4NHCH_2CONH_2 \cdot \frac{1}{2}H_2O$.
Properties: White crystalline powder, odorless. Contains 24.6% arsenic. May affect eyes. Soluble in water; almost insoluble in alcohol; insoluble in ether, chloroform, benzene.
Grades: Medicinal; U.S.P.
Hazard: Highly toxic by ingestion and inhalation.
Use: Medicine (treatment of syphilis).
Shipping regulations: (Rail, Air) Arsenical compounds, n.o.s., Poison label.

trypsin. The proteolytic enzyme of pancreatic juice. Yellow to grayish-powder; soluble in water; insoluble in alcohol or glycerin. It acts on albuminoid material producing amino acids. The maximum result is obtained in a neutral or slightly alkaline medium. Trypsins or similar materials are found not only in the pancreas but also in the spleen, leucocytes and urine, as well as in beer yeast, molds and bacteria.
Grade: N.F. (crystallized).
Hazard: Irritant to skin.
Uses: Medicine; dehairing of hides.

trypsinogen. An inactive precursor of trypsin (q.v.).

tryptophan (indole-alpha-aminopropionic acid; 1-alpha-amino-3-indolepropionic acid)

$C_6H_4NHCHCCH_2CHNH_2COOH$. One of the essential amino acids occurring naturally in the L-(−)-form.
Properties:
DL-: White crystals; slightly soluble in water; stable in alkaline solution; decomposed by strong acids.
D(+)-: Characteristic sweet taste; m.p. 275–290°C (dec); soluble in water, hot alcohol, alkali hydroxides; insoluble in chloroform.
L(−)-: Flat taste (other properties identical with D(+)-tryptophan).
Derivation: (a) Synthetic tryptophan can be made by the conversion of indole to gramine, followed by methylation, interaction with acetylaminomalonic ester and hydrolysis; (b) hydrolysis of proteins.
Grades: Reagent; technical; F.C.C.
Containers: 1-, 5-lb bottles; drums.
Uses: Nutrition and research; medicine; dietary supplement; cereal enrichment.
Available commercially in all three forms, as well as acetyl-DL-tryptophan.

"Trysben" 200.[28] Trademark for a weed killer based on aqueous solution of the dimethylamine salt of trichlorobenzoic acid, containing 2 lb of acid equivalent per gallon.
Use: Control of broadleaf weeds.

Ts. Abbreviation for tosyl.

TSA. Abbreviation for toluenesulfonic acid (q.v.).

TSH. See thyrotropic hormone.

TSP. Abbreviation for trisodium phosphate. See sodium phosphate, tribasic.

TSPA. Abbreviation for triethylenethiophosphoramide.

TSPP. Abbreviation for tetrasodium pyrophosphate. See sodium pyrophosphate.

TTC. Abbreviation for tetrazolium chloride (q.v.).

TTD. Abbreviation for tetraethylthiuram disulfide.

"Tuads." See "Ethyl Tuads"; "Methyl Tuads."

"Tuamine."[100] Trademark for tuaminoheptane sulfate.

tuaminoheptane. See 2-aminoheptane.

tuaminoheptane sulfate $[CH_3(CH_2)_4CHNH_2CH_3]_2 \cdot H_2SO_4$.
Properties: White odorless powder; readily soluble in water and alcohol.
Grade: N.F.
Use: Medicine.

"Tuasal."[233] Trademark for a brominated salicylanilide mixture; used as a bacteriostat.

"Tubarine."[301] Trademark for *d*-tubocurarine chloride (q.v.).

tubatoxin. See rotenone.

tuberose oil.
Properties: Colorless to very light-colored oil; sp. gr. 1.007–1.035 at 15°C; taken from Polianthes tuberosa, by enfleurage. Nontoxic.
Uses: Perfume; flavoring.

***d*-tubocurarine chloride** $C_{38}H_{44}Cl_2N_2O_6 \cdot 5H_2O$.
Properties: White to light tan, odorless, crystalline alkaloid. M.p. about 270°C with decomposition; soluble in water and alcohol; insoluble in acetone, chloroform, and ether. Aqueous solution is strongly dextrorotatory. (Specific rotation for 1% solution of anhydrous −208° to +218°.)
Grade: U.S.P.
Hazard: Toxic in high concentrations.
Use: Medicine (muscle relaxant).

"Tuex."[248] Trademark for tetramethylthiuram disulfide. Available in pellets as "Tuex Naugets." See thiram.

"Tuf-Foil."[524] Trademark for aluminum and aluminum alloys in foil form.

"Tumerol."[342] Trademark for oleoresin of tumeric for food coloring and flavoring.

tung oil (China-wood oil).
Properties: Yellow drying oil. Soluble in chloroform, ether, carbon disulfide, and oils. Sp. gr. 0.9360–0.9432; saponification value 193; iodine value 150–165; refractive index 1.5030. Flash point 552°F. Autoignition temp. 855°F. Low toxicity. Combustible.
Derivation: From the seeds of Aleurites cordata, a tree indigenous to China. It is now produced in China, Argentina and Paraguay; U.S. production has been virtually discontinued.
Chief constituent: Eleostearic acid (80%).
Uses: Exterior paint; varnishes.

"Tung-Spray."[456] Trademark for a suspension of pure, submicron tungsten disulfide in a fast-drying isopropanol carrier. Stable in air to 950°F.
Containers: 16-oz. aerosol cans.
Hazard: Flammable, dangerous fire risk.

tungstated pigment. See phosphotungstic pigment.

tungstate white. See barium tungstate.

tungsten (wolfram) W Metallic element; atomic number 74; group VIB of periodic system. Atomic weight 183.85. Valences 2, 4, 5, 6. Five isotopes.
Properties: Hard, brittle, gray solid. Not found native. The ores are scheelite and wolframite. Sp. gr. 19.3 (20°C); m.p. 3410°C (highest of all the metals); b.p. 5927°C. High electrical conductivity. Soluble in a mixture of nitric acid and hydrofluoric acid. Low toxicity. Corroded by seawater. Oxidizes in air at 400°C, the rate increasing rapidly with temperature.
Derivation: (a) Aluminothermic reduction of tungstic oxide; (b) hydrogen reduction of tungstic acid or its anhydride.
Fabrication: The metal can be plated onto objects by vapor deposition from tungsten hexafluoride or hexacarbonyl and can bond metal parts together. Tungsten powder (produced by carbon reduction) is converted into solid metal by powder metallurgical techniques. Large single crystals are grown by an arc-fusion process. Granules obtained by reduction of the hexafluoride.
Occurrence: Korea, Mexico, U.S.S.R., China, Spain and United States (Arizona, California, Colorado, Nebraska, Nevada, New Mexico, and Texas).
Grades: Technical; powder; single crystals; granules, ultrapure of 50–600 microns.
Containers: Cans, barrels, drums (powder).
Hazard: Finely divided form is highly flammable and may ignite spontaneously. Tolerance (as W) (insoluble compounds): 5 mg per cubic meter of air; soluble compounds, 1 mg per cubic meter of air.
Uses: High-speed tool steel; ferrous and nonferrous alloys; ferrotungsten; filaments for electric light bulbs; contact points; dentistry; x-ray and electron tubes; welding electrodes; heating elements in furnaces and vacuum-metallizing equipment; rocket nozzles and other aerospace applications; shell steel; chemical apparatus; high speed rotors as in gyroscopes; solar energy devices (as vapor-deposited film, which retains heat at about 500°C).

tungsten boride WB_2.
Properties: Silvery solid; sp. gr. 10.77; m.p. ca 2900°C; insoluble in water; soluble in aqua regia and concentrated acids; decomposed by chlorine at 212°F.
Derivation: Heating tungsten and boron in electric furnace.
Use: Refractory.

tungsten carbide WC. A ditungsten carbide, W_2C, with similar physical properties, is also known. WC is said to be the strongest of all structural materials.
Properties: Gray powder; sp. gr. 15.6; m.p. 2780°C; b.p. 6000°C; hardness of 9+ (Mohs) in solid form; insoluble in water but readily attacked by a nitric acid-hydrofluoric acid mixture. Stable to 400°C with chlorine; burns with fluorine at room temperature; oxidizes on heating with air.
Derivation: Heating tungsten and lamp black at 1500–1600°C.
Hazard: Toxic by inhalation. Tolerance (as W), 5 mg per cubic meter of air.
Uses: Cemented carbide tools; dies; wear-resistant parts; cermets; electrical resistors. An abrasive in liquids.

tungsten carbide, cemented. A mixture consisting of tungsten carbide 85–95% and cobalt 5–15%.
Properties: Sp. gr. 12–16; hardness about that of corundum; not affected by high temperatures.
Derivation: Ball milling of powdered tungsten carbide with metallic cobalt, followed by sintering.
Hazard: See tungsten carbide.
Uses: Machine tools and abrasives for machining and grinding metals, rocks, molded products, porcelain and glass; in gages, knife edges, blast nozzles.

tungsten carbonyl. See tungsten hexacarbonyl.

tungsten diselenide WSe_2. Lamellar-structured, dry solid lubricant with exceptional high temperature and high vacuum stability; retains its lubricity to temperatures as low as −423°F.
Grades: 1–2 micron and 40 micron.
Hazard: Toxic. Tolerance (as W), 5 mg per cubic meter of air.

tungsten disulfide WS_2.
Properties: Grayish black solid; m.p. above 1480°C; can lubricate at temperatures above 2400°F (1316°C). Attacked by fluorine and hot sulfuric and hydrofluoric acids.
Grades: 0.40, 0.70 and 1–2 micron grades.
Hazard: Toxic. Tolerance (as W), 5 mg per cubic meter of air.
Use: Solid lubricant for many applications, including use as an aerosol.

tungsten hexacarbonyl (tungsten carbonyl) $W(CO)_6$.
Properties: White, volatile, high refractive, crystalline solid; decomposes without melting at 150°C. One of the more stable carbonyls. Sp. gr. 2.65; vapor pressure 0.1 mm (20°C). Insoluble in water; soluble in organic solvents.
Derivation: Reaction of tungsten with carbon monoxide at high pressures; reduction of tungsten hexachloride with iron alloy powders in CO atmosphere.
Hazard: Toxic. Tolerance (as W), 5 mg per cubic meter of air.
Use: Tungsten coatings on base metals by deposition and decomposition of the carbonyl.

tungsten hexachloride WCl_6.
Properties: Dark blue or violet hexagonal crystals; volatile; m.p. 275°C; b.p. 347°C; sp. gr. 3.52; vapor pressure 43 mm (215°C); electrical conductivity (fused state) poor. Soluble in organic solvents including ligroin and ethanol. Decomposed by moist air and water; reduced by hydrogen to the metal.
Derivation: Heating tungsten with dry chlorine at red heat.

Hazard: Toxic. Tolerance (as W), 5 mg per cubic meter of air.
Uses: Tungsten coatings on base metals; vapor deposition for bonding metals; single crystal tungsten wire; additive to tin oxide to produce electrically conducting coating for glass; catalyst for olefin polymers.

tungsten hexafluoride WF_6.
Properties: Colorless gas or light yellow liquid. Sp. gr. (liquid) 3.44; m.p. 2.5°C; b.p. 19.5°C. Decomposes in water.
Derivation: Direct fluorination of powdered tungsten. Purified by distillation under pressure.
Hazard: Toxic by ingestion and inhalation. Strong irritant to tissue. Tolerance (as W), 1 mg per cubic meter of air.
Uses: Vapor phase deposition of tungsten; fluorinating agent.

tungsten lake. See phosphotungstic pigment.

tungsten oxychloride $WOCl_4$.
Properties: Dark red, acicular crystals. Decomposed by water and moist air. Keep in sealed glass containers! B.p. (approx) 227.5°C; m.p. (approx) 211°C; sp. gr. 11.92; soluble in carbon disulfide.
Derivation: By the action of chlorine on tungsten or tungstic oxide at elevated temperatures.
Purification: Vacuum distillation.
Grade: Technical.
Hazard: Toxic and irritant. Tolerance, 1 mg per cubic meter of air (as W).
Use: Incandescent lamps.

tungsten silicide. A ceramic. Probably WSi_2.
Properties: Blue-gray, very hard solid; sp. gr. 9.4; m.p. above 900°C. Insoluble in water; attacked by fused alkalies and mixtures of nitric and hydrofluoric acids.
Grades: Cylindrical shapes, lumps, standard sieve sizes.
Hazard: Dust flammable. Tolerance, 5 mg per cubic meter of air (as W).
Uses: Oxidation-resistant coatings; electrical resistance and refractory applications.

tungsten steel. In many of its alloying effects, tungsten is similar to molybdenum. Tungsten increases the density of alloys to which it is added. It is used to obtain steels with great wear resistance and special resistance to tempering, as in high-speed steels, hot-work steels, finishing steels that maintain keen cutting edge and great wear resistance, creep-resisting steels, and oxidation-resistant, high-temperature, high-strength alloys.

tungsten trioxide. See tungstic oxide.

tungstic acid (wolframic acid; orthotungstic acid) H_2WO_4.
Properties: Yellow powder; sp. gr. 5.5. Insoluble in water; soluble in hydrofluoric acid and alkalies. A white form of tungstic acid exists, having the formula $H_2WO_4 \cdot H_2O$. This is formed by acidifying tungstate solutions in the cold.
Derivation: Decomposition of sodium tungstate with hot sulfuric acid.
Grades: Technical; C.P.; reagent.
Containers: Drums; polyethylene bags; tonnage lots.
Hazard: Toxic. Tolerance (as W), 5 mg per cubic meter of air.
Uses: Textiles (mordant, color resist); plastics; tungsten metal, wire, etc.

tungstic acid anhydride. See tungstic oxide.

tungstic anhydride. See tungstic oxide.

tungstic oxide (tungstic acid anhydride; tungstic anhydride; tungsten trioxide; wolframic acid, anhydrous) WO_3.
Properties: Canary yellow, heavy powder; dark orange when heated and regains original color on cooling; m.p. 1473°C; sp. gr. 7.16. Insoluble in water; soluble in caustic alkalies; soluble with difficulty in acids. Noncombustible.
Derivation: Scheelite ore is treated with hydrochloric acid and the resulting product dissolved out with ammonia. The complex ammonium tungstate can then be ignited to tungstic oxide.
Hazard: Toxic. Tolerance (as W), 5 mg per cubic meter of air.
Uses: To form metal by reduction; alloys; preparation of tungstates for x-ray screens; fireproofing fabrics; yellow pigment in ceramics.

"Tungstide."[289] Trademark for a series of coatings consisting of metallic tungsten of near-colloidal particle-size suspended in a liquid plastic and incorporated with a form of "Liqui-Moly," (a molybdenum disulfide product) to produce a hard, abrasion-resistant self-lubricating coating.

tungstophosphate. Structure and properties are similar to molybdophosphate (q.v.).

12-tungstophosphoric acid. See phosphotungstic acid.

tungstosilicate. One of a group of complex inorganic compounds of high molecular weight, containing a central silicon atom surrounded by tungsten oxide octahedra. They have high molecular weight, high degree of hydration, strong oxidizing action in aqueous solution; they decompose in strongly basic aqueous solutions to give simple tungstate solutions; they form highly colored anions or reaction products. See silicotungstic acid and sodium 12-tungstosilicate.

12-tungstosilicic acid. See silicotungstic acid.

"Turgum S."[285] Trademark for a resin acid; sp. gr. 1.06–1.07; an odorless tackifier, retarder, and processing aid. Slightly retarding in cure.

turkey brown (turkey umber). Natural earth used as permanent pigment. Contains iron and manganese oxides with some clay.

Turkey red. See iron oxide red.

Turkey red oil (castor oil, sulfonated; castor oil, soluble). It is also known as alizarin assistant and alizarin oil because of its use in dyeing with alizarin.
Properties: Viscous liquid. Sp. gr. 0.95; iodine no. 82.1; acid no. 174.3; saponification no. 189.3. Soluble in water. Combustible; low toxicity. Autoignition temp. 833°F.
Derivation: By sulfonating castor oil with sulfuric acid and washing.
Grades: Sulfonated castor oil graded as to moisture and color.
Containers: 55-gal barrels and drums; 500-lb barrels.
Uses: Textiles; leather; manufacture of soaps; alizarin dye assistant; paper coatings.

Turnbull's Blue. An inorganic blue pigment made by the reaction of a ferrous salt and potassium ferricyanide $[Fe_3Fe(CN)_6]_2$. One of its important uses is in making blueprints, in which sensitized paper containing ferric ammonium citrate and potassium ferricyanide is exposed to light, the ferric ions being thus reduced to ferrous ions. See also Iron Blue.

turpentine (gum). The oleoresin or pitch obtained from living pine trees. Sticky, viscous balsamic liquid comprising a mixture of rosin and turpentine oil. Strong piny odor. Soluble in alcohol, ether, chloroform and glacial acetic acid. Combustible; low toxicity.

Uses: Source of turpentine oil and gum rosin.

See also rosin.

turpentine (oil) $C_{10}H_{16}$. A volatile essential oil whose chief constituents are pinene and diterpene.

Derivation: Steam-distillation of the turpentine gum (q.v.) exuded from living pine trees (gum turpentine), or naphtha-extraction of pine stumps (wood turpentine); destructive distillation of pine wood. See also rosin; rosin oil.

Properties: Colorless liquid; penetrating odor. Immiscible with water; lighter than water. Considerable variation appears in constants reported. The following are based on tests made by the Forest Products Laboratory: Sp. gr. (15°C) 0.860–0.875; refractive index 1.463–1.483 (20°C); flash point (closed cup) 90 to 115°F; acidity none; wt in lb/gal 7.18; autoignition temp. 488°F.

Containers: Bottles; cans; drums.

Hazard: Moderate fire risk. Toxic by ingestion. Tolerance, 100 ppm in air.

Uses: Solvent; thinner for paints, varnishes and lacquers; rubber solvent and reclaiming agent; insecticide; synthesis of camphor and menthol; wax-based polishes; medicine (liniments); perfumery.

"Tutane."[530] Trademark for 2-aminobutane; used as a fungicide.

Tw. Abbreviation for Twaddell, used in reporting specific gravities for densities greater than water, as °Tw. A Twaddell reading, multiplied by five and added to 1000, gives specific gravity with reference to water as 1000.

"Tween."[89] Trademark for a series of general-purpose emulsifiers and surface active agents. They are polyoxyethylene derivatives of fatty acid partial esters of sorbitol anhydrides. Generally soluble or dispersible in water, and differ widely in organic solubilities.

"Twitchell."[242] Trademark for a group of textile processing oils, fat-splitting reagents, and bases for soluble oils.

Twitchell reagent. Catalyst for the Twitchell process (acid hydrolysis of fats). It is a sulfonated addition product of naphthalene and oleic acid, a naphthalenestearosulfonic acid.

"TX."[247] Trademark for a hydrous magnesium oxide fiber which has an ultimate temperature resistance of 5070°F. Can be fabricated into reinforced plastic parts for high temperature conditions (e.g., rocket exhaust nozzles, re-entry nose cones).

"Tybek."[28] Trademark for a high-density polyethylene fiber.

"Tybrene."[233] Trademark for a series of acrylonitrile-butadiene-styrene (ABS) resins.

"Tydex."[233] Trademark for polyethyleneimine products.

"Tygobond."[326] Trademark for a series of vinyl- and rubber-based adhesives for cementing porous and semiporous materials to each other.

"Tygofil."[326] Trademark for a modified epoxy base metal filler.

"Tygoflex."[326] Trademark for a plastisol vinyl compound derived from "Tygon." For specialty corrosion protection.

"Tygon."[326] Trademark for a series of vinyl compounds used as linings, coatings, adhesives, tubing, and extruded shapes applied to chemical process equipment as corrosion protection.

"Tygorust."[326] Trademark for a vinyl-based primer for application to damp or dry rusted steel.

"Tygoweld."[326] Trademark for a modified epoxy-base structural adhesive for bonding both similar and dissimilar materials.

"Tygozinc."[326] Trademark for a series of inorganic zinc-rich protective coatings of both self-curing and post-cured types.

"Tylac."[513] Trademark for a series of synthetic latexes and elastomers.

NBL. High-strength, film-forming nitrile rubber latexes and rubbers and carboxylated polymers characterized by exceptional oil, solvent, and abrasion resistance.

SBL. Butadiene/styrene and carboxylated butadiene/styrene and high styrene latexes of various monomer ratios. Modified versions available.

Uses: Paper, textiles, adhesives, Portland cement, molded products, polyvinyl chloride and phenolic resin blends.

"Tylan."[530] Trademark for tylosin phosphate; used as an antibiotic in veterinary medicine.

"Tylose."[450] Trademark for a wide range of water-soluble cellulose ethers. See methylcellulose, carboxymethylcellulose. Numerous grades and viscosities are available.

Uses: Thickeners, binders, dispersing agent, emulsifiers, protective colloids, lubricants and film forming materials; drilling muds; detergents; textile, paper, print, and varnish industries; ceramics; cosmetics and pharmaceuticals.

tylosin. Antibiotic substance isolated from a strain of Streptomyces fradiae.

Properties: Crystals; m.p. 128–130°C; solutions are stable between pH 4 to 9.

Uses: Veterinary medicine.

tyloxapol (USAN) An oxyethylated tert-octylphenol-polymethylene polymer, and essentially nontoxic, nonionic surfactant.

Tyndall effect. A colloidal phenomenon in which particles too small to be resolved in an optical microscope suspended in a gas or liquid reveal their presence by scattering a beam of light as it passes through the suspension, the extent of reflection being dependent of the position of the irregularly shaped particles relative to the incident light. The effect causes the appearance of a visible cone of light through the suspension. This principle is utilized in the ultramicroscope.

"Tynex."[28] Trademark for nylon filament. Available tapered with an essentially uniform taper from butt to tip, and also level, i.e., in a wide range of constant diameters. The tapered form used primarily in paint brushes; the level form in other brushes.

type metal. Alloy of 75–95% lead, 2½–18% antimony with a little tin and sometimes copper, which expands slightly upon solidification and produces sharp castings.

"Ty-Ply."[525] Trademark for a family of nontacky vulcanizing adhesives for bonding natural and synthetic rubber compounds.

Tyr Abbreviation for tyrosine.

tyramine (tyrosamine; para-beta-amino-ethylphenol) $HOC_6H_4CH_2CH_2NH_2$. A base found in mistletoe, putrefied animal tissues, certain cheeses, and ergot. It is usually made synthetically.
Properties: Colorless crystals; m.p. 164–165°C; soluble in boiling alcohol; slightly soluble in water, benzene, and xylene.
Use: Medicine.

"Tyril."[233] Trademark for a group of styrene-acrylonitrile copolymers.

"Tyrilfoam" 80.[233] Trademark for expanded styrene-acrylonitrile for flotation uses under conditions of gasoline spillage, petroleum scum from outboard motors, and stagnant water.

tyrocidine. Antibiotic produced by the metabolic processes of the bacteria, Bacillus brevis. It is a cyclic polypeptide which is active against most gram-positive pathogenic bacteria. It is one of the two antibiotic components of tyrothricin (q.v.) but has been isolated and used alone.
Properties (probably the hydrochloride): Fine crystalline needles which decompose at 240°C. Soluble in 95% alcohol, acetic acid and pyridine; slightly soluble in water, acetone and absolute alcohol; insoluble in ether, chloroform and hydrocarbons. Depresses surface tension; forms fairly stable colloidal emulsion in distilled water.
Use: Medicine (usually as component of tyrothricin); possible fungistat and bacteriostat.

tyrosamine. See tyramine.

tyrosinase. An enzyme containing copper which occurs in plant and animal tissue and is responsible for turning peeled potatoes black when exposed to air.
Use: Medicine.

tyrosine (beta-para-hydroxyphenylalanine; alpha-amino-beta-para-hydroxyphenylpropionic acid) $C_6H_4OHCH_2CHNH_2COOH$. A nonessential amino acid.
Properties: White crystals, readily oxidized by the animal organism; soluble in water; slightly soluble in alcohol; insoluble in ether; optically active.
DL-tyrosine m.p. 316°C;
D(+)-tyrosine m.p. 310–314°C;
L(−)-tyrosine m.p. 295°C with decomposition; sp. gr. 1.456 (20/4°C).
Derivation: Hydrolysis of protein (casein); organic synthesis.
Grade: F.C.C.
Uses: Growth factor in nutrition; biochemical research; dietary supplement. Available commercially as DL-tyrosine.

tyrothricin. An antibiotic produced by growth of Bacillus brevis. It consists of a mixture of antibiotics, principally gramicidin and tyrocidine. Gramicidin is the more active component. Use is generally limited to local external applications. It is active against some gram-positive bacteria, including species of pneumococci, streptococci and staphylococci.

"Tyzine."[299] Trademark for tetrahydrozoline hydrochloride.

"Tyzor."[28] Trademark for a group of simple and chelated esters of orthotitanic acid such as tetrabutyltitanate of varying reactivity.
Uses: Chemical intermediates; primers for adhesion promotion in extrusion coating; dispersants; catalysts; scratch-resistant finishes on glass; masonry water repellents; cross-linking agents.

U

U Symbol for uranium.

"U-A-S."[197] Trademark for a nitrogen fertilizer solution containing urea, water, and ammonia.
Use: Solid nitrogen-containing fertilizers.

"Ubatol."[22] Trademark for a series of fine particle size styrene- and acrylic-based polymer and copolymer emulsions, some less than 0.01 micron. Films deposited from compounded polymer emulsions exhibit high gloss, increased water resistance and durability. Film characteristics of straight polymers range from hard non-film forming to soft-flexible.
Uses: Coatings for paper, leather, textiles, tapes; vehicle for gloss latex paints; self-polishing floor wax, latex compounding, detackifying agents, detergent opacifying agent.

"Ucane."[214] Trademark for soft detergent alkylates which are anionic in character after sulfonation; derived from the chlorination of n-paraffins and the subsequent alkylation of benzene with these n-paraffin chlorides. Show almost complete biodegradability.

"Ucar."[214] Trademark for various synthetic organic chemicals including butylene oxide, butylphenol, nonylphenol, triphenol P, bisphenol A, para-tert-amylphenol, ethylene and propylene glycols; also applied to synthetic latexes and water-soluble polymers.

"Ucet."[214] Trademark for epoxy resin type of wrinkle-resistant finishes for cotton and rayon, and for a type of acetylene black.

"Ucon."[214] (1) Trademark for a series of nonflammable, low-toxicity fluorocarbon solvents and solvent blends with high chemical stability and extremely low residue levels. Used for cleaning electronic and mechanical instruments and controls, degreasing motors, cleaning liquid oxygen equipment, and motion picture and television film and magnetic tape.

(2) Trademark for polyalkylene glycols and diesters. Available as water-soluble or insoluble products.
Grades: LB-Series, 50-HB Series, 75-H Series, DLB Series and Hydrolube Series.
Uses: High-temperature lubricants, low temperature fluids, compressor lubricants, hydraulic brake fluids, quenchant fluid, heat transfer fluids, textile lubricants, rock drill lubricants, leather and paper-treating compounds, rubber lubricants, plasticizers and solvents, chemical intermediate; stationary phase in gas chromatography.
See fluorocarbon.

"Udel."[214] Trademark for biaxially oriented polypropylene film used for overwraps, laminations, box windows, and shrink wraps; also for polysulfone resins.

"Udex."[233,416] Proprietary process for extracting aromatic hydrocarbons by using a glycol-water mixture as solvent, utilizing a unique countercurrent contacting technique. The aromatic extract is distilled from the solvent and extremely high purity individual aromatics separated from one another by further distillation. The composition of the solvent is readjusted by removing water or other impurities and continually returning solvent to process.

UDP. Abbreviation for uridine diphosphate. See uridine phosphate.

UDPG. Abbreviation for uridine diphosphate glucose.

"UF 71."[319] Trademark for an aqueous concentrate of 51% formaldehyde by weight stabilized by the addition of 20% urea.
Uses: Urea-formaldehyde resins and nitrogen fertilizers.

"Uformite."[23] Trademark for synthetic resins based on urea-formaldehyde, melamine-formaldehyde, and triazine condensates. Supplied as colorless or light-colored aqueous solutions or solutions in volatile solvents. Solvent type produces hard, alkali-resistant, colorless coatings on curing, with adhesion to a variety of surfaces.
Uses: With alkyd resins in coatings; industrial finishes on appliances, automobiles, etc.; adhesives for paperboard boxes; paper coatings; wet-strength paper; textile pigment binding.

ulexite (cotton balls) $NaCaB_5O_9 \cdot 8H_2O$. A natural hydrated borate of sodium and calcium.
Properties: Color white; luster silky; Mohs hardness 1–2.5; sp. gr. 1.96; usually found as rounded, loose-textured masses of fine crystals.
Occurrence: Chile; Argentina; California, Nevada.
Use: Source of borax.

Ullman reaction. A modification of the Fittig synthesis (q.v.) in which copper powder is used instead of sodium.

ulmin brown. See Van Dyke brown.

ulmin. One of a class of amorphous substances resulting from the decomposition of the cellulose and lignite tissues of plants. Ulmins represent one of the initial changes by which vegetable matter is converted into coal.

"Ultex."[94] Trademark for a dithiocarbamate accelerator for natural and synthetic rubbers; suitable for white and light-colored goods.

ultra-accelerator. An unusually powerful accelerator of rubber vulcanization, typified by thiuram sulfides and dithiocarbamates.

"Ultra-Clor."[329] Trademark for a turf fungicide whose active ingredients are mercuric dimethyldithiocarbamate, potassium chromate, and cadmium succinate.
Hazard: Highly toxic by ingestion.

ultramarine blue.
Properties: Inorganic pigment; blue powder; good alkali and heat resistance; low hiding power; poor acid resistance; poor outdoor durability; noncombustible; low toxicity.

Derivation: Heating a mixture of sulfur, clay, alkali, and a reducing agent at high temperatures.
Containers: Barrels; fiber drums; multiwall paper bags.
Uses: Colorant for machinery and toy enamels; white baking enamels; printing inks; rubber products; soaps and laundry blues; cosmetics; textile printing. *Note:* Used in very low percentages to intensify whiteness of white enamels, rubber compounds, laundered clothing, etc. by offsetting yellowish undertones; gives a "blue" rather than a "yellow" white.

"Ultramid."[440] Trademark for nylon 6, nylon 6,6 and nylon 6,10; used in filament and engineering plastics. See nylon.

"Ultrapoles."[449] Trademark for a series of alkanolamine condensates, detergent base and surface active materials used as detergents and ingredients of detergents, wetting and foaming agents, foam stabilizers, ingredients for cosmetic preparations, emollients, thickening agents.

"UltraSeal."[506] Trademark for synthetic rubber industrial sealants.

ultrasonics. The science of effects of sound vibrations beyond the limit of audible frequencies. Used for dust, smoke and mist precipitation; preparation of colloidal dispersions; cleaning of metal parts, precision machinery, fabrics, etc.; friction welding; formation of catalysts; degassing and solidification of molten metals; extracting flavor oils in brewing; electroplating; drilling hard materials; fluxless soldering; nondestructive testing. Also used for investigation of physical properties, determination of molecular weights of liquid polymers, degree of association of water, and for inducing chemical reaction. Biological effects are also under study.

"Ultrathene."[192] Trademark for a series of ethylene-vinyl acetate copolymer resins for adhesives, conversion coatings, and thermoplastic modifiers. Wide range of melt indexes. Improves specific adhesion of hot-melt, solvent-based, and pressure-sensitive adhesives.

ultraviolet. Radiation in the region of the electromagnetic spectrum including wavelengths from 100 to 3900 A. See radiation.
Hazard: Dangerous to eyes; overexposure may cause severe skin burns (sunburn).
Uses: Irradiation of milk; air sterilization in hospitals; microscopy.

ultraviolet absorber. A substance which absorbs radiant energy in the wavelength of ultraviolet (q.v.). The radiant energy absorbed is converted to heat (thermal energy). Ultraviolet absorbers are added to unsaturated substances (plastics, rubbers, etc.) to decrease light sensitivity and consequent discoloring and degradation. Among compounds used are benzophenones, benzotriazoles, substituted acrylonitriles, and phenol-nickel complexes. See also absorption (2).

"Ultrawet."[136] Trademark for a series of biodegradable linear alkylate sulfonate (LAS) anionic detergents or surface-active agents (linear dodecylbenzene type). Available forms: Sodium or triethanolamine salts in liquid, slurry, flake or bead form.

"Ultra-X Foam Liquid."[270] Trademark for a compound used at 1% to 3% in water to produce ultra-high expansion fire fighting foam. When used in appropriate foam generators, foam expansions as high as 1000 times the liquid volume result. Used for combatting confined flammable liquid and structural fires, for total flooding of basements, mines, warehouses, etc.

umber. A naturally occurring brown earth containing ferric oxide together with silica, alumina, manganese oxides and lime. Raw umber is umber which is ground and then levigated. Burnt umber is umber calcined at low heat.
Uses: Paint pigment; lithographic inks; wall paper (pigment); artists' color.

"Umbrex Fatty Alcohol."[487] See "CO Fatty Alcohols."

UMP. Abbreviation for uridine monophosphate. See uridine phosphates and also uridylic acid.

"Unads."[69] Trademark for tetramethylthiuram monosulfide (q.v.).
Use: Ultra-accelerator for rubber.

uncertainty principle. The conclusion of Heisenberg based on quantum mechanical theory that the precise position of a specific electron in an atomic orbit cannot be determined, and that consequently the ultimate nature of matter is not susceptible to objective measurement. The result of this concept was development of the orbital theory, in which electron behavior is dealt with on a statistical basis. Its validity was confirmed by 1930 by the work of other mathematical physicists such as DeBroglie, Fermi and Schrodinger. See also orbital theory; Heisenberg.

gamma-**undecalactone** (peach aldehyde; gamma-undecyl lactone; 4-hydroxy-undecanoic acid, gamma lactone). $CH_3(CH_2)_6\underline{CHCH_2CH_2COO}$.

Properties: Colorless to light-yellow liquid; peach-like odor. Sp. gr. 0.941–0.944; refractive index 1.450–1.454. Soluble in 4 to 5 vols. of 60% alcohol; soluble in benzyl alcohol, benzyl benzoate, and most fixed oils. Combustible.
Derivation: By heating undecylenic acid in the presence of sulfuric acid.
Grades: Chlorine-free; F.C.C.
Uses: Perfumery; flavoring agent.

undecanal (n-undecylic aldehyde; hendecanal) $CH_3(CH_2)_9CHO$.
Properties: Colorless liquid; sweet odor; sp. gr. 0.825–0.832 (25°C); refractive index 1.4310–1.4350 (20°C); soluble in oils and alcohol; insoluble in glycerol and water. Flash point 235°F; combustible.
Derivation: By oxidation of 1-undecanol or reduction of undecanoic acid.
Grade: F.C.C.
Hazard: Moderately toxic by ingestion and inhalation; irritant to tissue.
Uses: Perfumery; flavors.

n-**undecane** (hendecane) $CH_3(CH_2)_9CH_3$.
Properties: Colorless liquid; sp. gr. 0.7402 (20/4°C); f.p. −25.75°C; b.p. 195.6°C; refractive index 1.41725 (n 20/D); flash point 149°F. Combustible.
Grades: 95%; 99%; research.
Containers: Bottles and drums.
Uses: Petroleum research; organic synthesis; distillation chaser.

undecanoic acid (n-undecylic acid; hendecanoic acid) $CH_3(CH_2)_9COOH$. Small amounts occur in castor oil. It is best derived from undecylenic acid by hydrogenation.
Properties: Colorless crystals. Sp. gr. 0.8505 (80/4°C); m.p. 28.5°C; b.p. 284.0°C; 222.2°C (128 mm); refractive index 1.4319 (40°C); insoluble in water; soluble in alcohol and ether.
Grades: Technical; 99%.
Use: Organic synthesis.

1-undecanol (n-undecyl alcohol; decyl carbinol; 1-hendecanol) $CH_3(CH_2)_9CH_2OH$.
Properties: Colorless liquid with a citrus odor. Sp. gr. 0.829–0.834; refractive index 1.435–1.443; m.p. 19°C. Soluble in 60% alcohol. Flash point about 200°F. Combustible. Low toxicity.
Uses: Perfumery; flavoring.

2-undecanol (2-hendecanol) $CH_3(CH_2)_8CHOHCH_3$.
Properties: Colorless liquid; sp. gr. 0.8363 (20°C); m.p. 12°C; b.p. 228–229°C; insoluble in water; soluble in alcohol and ether. Flash point 235°F. Combustible.
Containers: Drums.
Uses: Antifoaming agent; intermediate; perfume fixatives; plasticizer.

2-undecanone. See methyl nonyl ketone.

undecenal (undecylenic aldehyde; hendecen-1-al). Listed by different authorities as the 10-undecenal, $CH_2{:}CH(CH_2)$ CHO, and the 9-undecenal, $CH_3CH{:}CH(CH_2)_7CHO$.
Properties: Colorless liquid; strong odor suggesting rose; sp. gr. 0.840–0.850 (25/25°C); refractive index 1.4410–1.4470 (20°C). Soluble in 80% alcohol. Combustible. Low toxicity.
Uses: Perfumery; flavoring.

10-undecenoic acid. See undecylenic acid.

10-undecen-1-ol. See undecylenic alcohol.

n-**undecyl alcohol.** See 1-undecanol.

undecylenic acid (10-undecenoic acid) $CH_2{:}CH(CH_2)_8COOH$.
Properties: Light-colored liquid; fruity-rosy odor. Almost insoluble in water; miscible with alcohol, chloroform, ether, benzene, and with fixed and volatile oils. Congealing point 21°C; sp. gr. (25/25°C) 0.910–0.913; refractive index (25°C) 1.4475–1.4485. Flash point 295°F; combustible; low toxicity.
Derivation: Destructive distillation of castor oil.
Grades: Technical; N.F.
Containers: 6-, 13-gal carboys; 5-, 10-, 50-gal drums.
Uses: Perfumery; flavoring; medicinals; plastics; modifying agent (plasticizer, lubricant additive, etc.).

undecylenic alcohol (n-undecylenic alcohol; 10-undecen-1-ol; alcohol C-11) $CH_2{:}CH(CH_2)_8CH_2OH$.
Properties: Colorless liquid; citrus odor; sp. gr. 0.842–0.847; refractive index n 20 1.449–1.454; f.p. −3.0°C; soluble in 70% alcohol. Combustible; low toxicity.
Use: Perfumes.

undecylenic aldehyde. See undecenal.

undecylenyl acetate (10-hendecenyl acetate) $CH_3COO(CH_2)_9CH{:}CH_2$.
Properties: Colorless liquid; floral-fruity odor. Sp. gr. 0.876–0.883; refractive index 1.438–1.442. Soluble in 80% alcohol. Combustible; low toxicity.
Uses: Perfumery; flavoring.

n-**undecylic acid.** See undecanoic acid.

u-**undecylic aldehyde.** See undecanal.

gamma-**undecyl lactone.** See gamma-undecalactone.

UNH. Abbreviation for uranyl nitrate hydrated. See uranyl nitrate.

unhairing (dehairing). Removal of hair from hides and skins as practised on a commercial scale in the leather industry. Several methods are used, involving application of hydrated lime, dimethylamine, trypsin and other enzymes.

"Unicel."[28] Trademark for blowing agents for natural and synthetic rubber sponge.

"Unicor."[416] Trademark for ash-free, oil-soluble, surface active film-former. Used as a corrosion inhibitor in various refinery operations.
"Unicor-LHS" is a film-forming organic phosphate base used also as an anti-icer and carburetor detergent.

"Unisize."[144] Trademark for sizing agents for paper and paperboard.
Properties: Modified polyvinyl alcohol; free-flowing powder; medium-viscosity or high-viscosity solutions; water-soluble.
Containers: 50-lb multiwall bags.
Uses: Single-station surface-sizing agents for paper and paperboard.

"Unisoil."[99] Trademark for a colloidal clay. Used as a soil conditioner in sandy or light soils.

"Unisol."[136,416] Proprietary process for removing mercaptans from gasoline and burning oils by use of caustic methanol solutions. The caustic is separated by decantation and the mercaptans separated by settling; the methanol is distilled from the hydrocarbons. Both chemicals are continuously returned to process.

"Unisorb."[416] Proprietary process for separating hydrocarbon types from each other by adsorption from the liquid phase in a fixed bed of adsorbent at moderate temperatures and pressures.

"Unitane."[57] Trademark for titanium dioxide (TiO_2) pigment available in both anatase and rutile forms.

"Unithane."[27] Trademark for urethane-base compounds.
Uses: Adhesives for bonding or laminating textiles, flocking, textile coating and finishing treatments.

"Unitol."[420] Trademark for a series of tall oil fatty acids, distilled tall oil, tall oil rosin, acid-refined tall oils, synthetic crude tall oils, and tall oil pitch.
Uses: Alkyd resins, oleoresinous vehicles, esters, maleic-modified resins, fumaric-modified resins, gloss oils, epoxy resins, polyurethanes, soaps, synthetic detergents, ore flotation, stabilizer-plasticizers, metallic soaps, paper size, rubber processing aids and others.

unit operation. A particular kind of physical change used in the industrial production of various chemicals and related materials. Filtration, evaporation, distillation, fluid flow and heat transfer are examples. See also chemical engineering.

unit process. A process characterized by a particular kind of chemical reaction; oxidation, hydrolysis, esterification, and nitration are examples. See also kinetics, chemical; chemical engineering.

"Univis."[51] Trademark for a series of power transmission or hydraulic oils. They have viscosity indexes of 150 or higher and pour points of −50°F or lower, permitting wide temperature ranges in operation.

"Univolt."[51] Trademark for oils used as electrical insulating mediums in transformers, switches and some electrical cables. Suitable for transformers, whether used indoors or outdoors or under low-temperature conditions. See also transformer oil.

"Unox" Epoxides.[214] Trademark for a series of stabilizers and intermediates for plasticizers and resins.

"Unox" Fire-Fighting Penetrant.[214] Trademark for a wetting agent. Increases the effectiveness of water in extinguishing fires by aiding penetration of dense burning materials, putting out the fire quickly with less smoke, less water damage, and less overhaul.

"Unox Process."[214] Trademark for an improved method of oxidizing organic wastes by use of oxygen instead of air.

uns- (unsym). Abbreviation for unsymmetrical. A prefix denoting the structure of organic compounds in which substituents are disposed unsymmetrically with respect to the carbon skeleton or to a functional group, such as a double bond. For example, unsdichloroethane is CH_3CHCl_2.

unsaturation. Of a chemical compound, the state in which not all the available valence bonds are satisfied; in such compounds the extra bonds usually form double or triple bonds (chiefly with carbon). Thus unsaturated compounds are more reactive than saturated compounds, as other elements readily add to the unsaturated linkage. An unsaturated compound (ethylene, C_2H_4, butadiene, C_4H_6, benzene, C_6H_6) has fewer hydrogen atoms or equivalent groups than the corresponding saturated compound (ethane, C_2H_6, butane, C_4H_{10}, cyclohexane, C_6H_{12}).

In structural formulas unsaturation may be represented by parallel lines joining the carbon atoms (ethylene, $H_2C{=}CH_2$, butadiene $H_2C{=}CHCH{=}CH_2$) or by colons or triple dots, $H_2C{:}CH_2$ $H_2C{:}CH_2$ (ethylene) and HC⋮CH (acetylene).

"UOP."[416] Designation for a broad group of antioxidants and related products especially designed for use in motor fuels, lubricating oils and greases. Among other things they inhibit gum formation and tetraethyllead decomposition in high-octane gasolines.

"Urab."[50] Trademark for a complex of fenuron and TCA, available in liquid concentrate, granular, and pelleted formulations. A brush and weed killer for control of woody plants and deep-rooted weeds in noncrop land.

"Urac."[57] Trademark for products based on urea-formaldehyde condensates used mainly as adhesives for the production of moisture-proof bonds in plywood manufacture, plywood assembly, and furniture manufacture.

uracil HNC(O)NHC(O)CHCH (2,4-Dioxypyrimidine). A pyrimidine that is a constituent of ribonucleic acids and the coenzyme, uridine diphosphate glucose (q.v.).

Properties: Crystalline needles; m.p. 335°C (dec). Soluble in hot water, ammonium hydroxide and other alkalies; insoluble in alcohol and ether.

Derivation: Hydrolysis of nucleic acids; precipitation from urea and ethyl formylacetate. Radioactive forms available.

Use: Biochemical research.

uracil-6-carboxylic acid. See orotic acid.

"Uracil Mustard."[327] Trademark for 5-[bis(2-chloroethyl)amino]uracil.

uracil, D-ribosyl. See uridine.

"Urafil."[539] Trademark for a glass fiber-reinforced urethane.

"Ura-Green."[533] Trademark for non-pressure nitrogen solutions.

"Uramite."[28] Trademark for a fertilizer derived from urea-formaldehyde containing 38% nitrogen.

"Uramon."[28] Trademark for solutions of urea in aqueous ammonia; used in manufacture of mixed fertilizers.

"Uran."[197] Trademark for a series of nitrogen solutions containing water, ammonium nitrate, and urea. They are used as solutions for direct application to the soil and in manufacture of complete liquid fertilizers.

"Urana."[197] Trademark for a series of nitrogen solutions containing water, ammonium nitrate, urea, and ammonia. These solutions are used in the manufacture of solid fertilizers having special handling properties.

urania. See uranium dioxide.

urania-thoria. Crystals of the mixed oxides of uranium and thorium are available. Used as nuclear fuel. The crystals are denser and cheaper than the pellet form.

uranic chloride. See uranium tetrachloride.

uranic oxide. See uranium dioxide.

uranine (uranine yellow; sodium fluorescein; resorcinolphthalein sodium) $Na_2C_{20}H_{10}O_5$. Colour Index No. 45350.

Properties: Orange red, odorless powder; hygroscopic; soluble in water and sparingly soluble in alcohol.

Derivation: By treatment of fluorescein with sodium carbonate solution and crystallizing.

Method of purification: Recrystallization.

Grades: Technical; U.S.P. (as sodium fluorescein).

Containers: Custom-packed.

Uses: Dyeing silk and wool yellow; tracing subterranean waters; marking water for air-sea rescues; clinical test solution.

uraninite UP_2. A natural oxide of uranium usually partly oxidized to UO_3, with variable amounts of lead, radium, thorium, rare earth metals, helium, argon, and nitrogen. Pitchblende is an important variety.

Occurrence: Colorado, Utah, South Africa, Canada, Europe, Congo, U.S.S.R.; Australia.

Use: Source of uranium and radium.

uranium U Metallic element number 92, a member of the actinide series. Atomic weight 238.029; valences 3, 4, 6. Three natural radioactive isotopes: U-234 (0.006%), U-235 (0.7%), and U-238 (99%). (See separate entries for details).

Properties: Dense, silvery solid; strongly electropositive; ductile and malleable; poor conductor of electricity; sp. gr. 19.0; m.p. 1132°C; b.p. 3818°C; heat of fusion 4.7 kcal/mole; heat capacity 6.6 cal/mole/°C. Forms solid solutions (for nuclear reactors) with molybdenum, niobium, titanium and zirconium. The metal reacts with nearly all nonmetals. It is attacked by water, acids, and peroxides; but is inert towards alkalies. Green tetravalent uranium and yellow uranyl ion (UO_2^{++}) are the only species which are stable in solution.

Occurrence: Pitchblende (essentially UO_2), a variety of uraninite, coffinite ($USiO_4$) and carnotite (Colorado, Canada, Congo, So. Africa, Australia, USSR).

Derivation: Finely ground ore is leached under oxidizing conditions to give uranyl nitrate solution. The uranyl nitrate, purified by solvent extraction (ether, alkyl phosphate esters), is then reduced with hydrogen to uranium dioxide. This is treated with hydrogen fluoride to obtain uranium tetrafluoride, followed by either electrolysis in fused salts or by reduction with calcium or magnesium. Uranium can also be recovered from phosphoric acid by a recently developed process (1973). U.S. reserves are estimated to be about 500,000 tons.

Forms available: Solid pure metal; alloys; powder (99.7%).

Hazard: Highly toxic radioactive material; source of ionizing radiation. Tolerance (including metal and all compounds, as U), 0.2 mg per cubic meter of air. (Powder) Dangerous fire risk; ignites spontaneously in air.

Uses: Nuclear fuel; source of fissionable isotope 235; source of plutonium by neutron capture; electric power generation.

See also enrichment (2); uranium compounds.

uranium-233 (U-233). A fissionable isotope of uranium produced artificially by bombarding thorium with neutrons. Used as an atomic fuel in molten salt reactor and is a possible fuel in breeder reactors. Half-life 1.62×10^5 years.

uranium-234 (U-234): A natural isotope of uranium with half-life of 2.48×10^5 years; it is separated by extraction with trioctylphosphine oxide. Used in nuclear research, with potential use in fission detectors for counting fast neutrons.

uranium-235 (U-235). The readily fissionable isotope of uranium used to enrich natural uranium in nuclear fuels. It is present in uranium only to the extent of 0.7%, and can be separated from it by any of several methods: the gaseous diffusion process using uranium hexafluoride, the gas centrifuge process, and the electromagnetic separation method. Its half-life is 7.13×10^8 years. It was the energy source used in the original atom bomb. See also fission; uranium.

uranium-238 (U-238). The abundant isotope of uranium, of which it comprises 99%. It is not fissionable, but will form plutonium-239 as a result of bombardment by neutrons in a reactor. Its half-life is 4.51×10^9 years. It will be used in breeder reactors, together with plutonium, where its energy potential will be exploited by transmuting it to fissionable plutonium. See also breeder.

uranium compounds. Before the advent of nuclear energy, uranium had very limited uses. It has been suggested for filaments of lamps. A small tube of uranium dioxide, UO_2, connected in series with the tungsten filaments of large incandescent lamps used for photography and motion pictures, tends to eliminate the sudden surge of current through the bulbs when the light is turned on, thereby extending their life. Compounds of uranium have been used in photography for toning, and in the leather and wood industries uranium compounds have been used for stains and dyes. Uranium salts are mordants of silk or wool. In making special steels, a little ferrouranium has been utilized, but its value is questionable in this connection. Such alloys have not proved commercially attractive. In the production of ceramics, sodium and ammonium diuranates have been used to produce colored glazes.

Uranium carbide has been suggested as a good catalyst for the production of synthetic ammonia. Uranium salts in small quantities are claimed to stimulate plant growth, but large quantities are clearly poisonous to plants.

By far the most important use of uranium lies in its application for nuclear (or atomic) energy. This use, in fact, has so increased the value of uranium as to eliminate its use for many of the purposes mentioned above.

Glenn T. Seaborg
("Encyclopedia of Chemical Elements," ed. by C. A. Hampel)

uraniumn decay series (uranium-radium series). The series of elements produced as successive intermediate products when the element uranium undergoes spontaneous natural radioactive disintegration into lead. Radium and radon are members of this series.

uranium, depleted. Uranium from which most of the U-235 isotope has been removed. See uranium-238.

uranium dicarbide (uranium carbide) UC_2.

Properties: Gray crystals; sp. gr. 11.28 (18°C); m.p. 2350°C; b.p. 4370°C. Decomposes in water; slightly soluble in alcohol.

Hazard: Highly toxic; radiation risk.

Use: As crystals, pellets, or microspheres for nuclear reactor fuel.

uranium dioxide (uranium oxide; uranic oxide; urania) UO_2.

Properties: Black crystals, insoluble in water, soluble in nitric acid and concentrated sulfuric acid; sp. gr. 10.9; m.p. 3000 ± 200°C.

Derivation of pure oxide: Powdered uranium ore is digested with hot nitric-sulfuric acid mixture and filtered to remove the insoluble portion. Sulfate is precipitated from the solution with barium carbonate and uranyl nitrate is extracted with ether. After re-extraction into water, it is heated to drive off nitric acid, leaving uranium trioxide. The latter is reduced with hydrogen to the dioxide. Can also be prepared from uranium hexafluoride by treating with ammonia and subsequent heating of the ammonium diuranate.

Hazard: Highly toxic; radiation risk. Ignites spontaneously in finely divided form.

Uses: A crystalline (or pellet) form is used to pack nuclear fuel rods.

uranium, enriched. Natural uranium to which a few percent of the fissionable 235 isotope has been added. See also enrichment.

uranium hexafluoride UF_6.

Properties: Colorless volatile crystals; sublimes; triple point 64.0°C (1134 mm); m.p. 64.5°C (2 atm); sp. gr. 5.06 (25°C); soluble in liquid bromine, chlorine, carbon tetrachloride, sym-tetrachloroethane and fluorocarbons. Reacts vigorously with water, alcohol, ether, and most metals. Vapor behaves as nearly perfect gas.

Derivation: (1) Triuranium octoxide (U_3O_8) and nitric acid react to form solution of uranyl nitrate; this is decomposed to UO_3 and reduced to the dioxide

with hydrogen. The dioxide as a fluidized bed is reacted with hydrogen fluoride. The resulting tetrafluoride is fluorinated to the hexafluoride. (2) Triuranium octoxide is converted directly to the hexafluoride with hydrogen fluoride and fluorine, then purified by fractional distillation.
Hazard: Highly toxic and corrosive. Radiation risk.
Use: Gaseous diffusion process for separating isotopes of uranium.

uranium hydride UH_3.
Properties: Brown gray to black powder; sp. gr. 10.92; conductor of electricity.
Derivation: Action of hydrogen on hot uranium.
Hazard: Highly toxic; ignites spontaneously in air.
Uses: Preparation of finely divided uranium metal by decomposition; separation of hydrogen isotopes; reducing agent; laboratory source of pure hydrogen.

uranium monocarbide (uranium carbide) UC.
Properties: Lumps or powder that can be formed into desired shapes by powder metallurgy or arc-melt casting; m.p. 2375°C, density 13.63 g/cc; thermal conductivity 0.08 cal/sec/cm_2/°C/cm; must be stored in inert atmosphere.
Hazard: Radioactive poison.
Use: Nuclear reactor fuel.

uranium tetrafluoride (green salt) UF_4.
Properties: Green, nonvolatile crystalline powder; sp. gr. 6.70; m.p. 1036°C; insoluble in water.
Derivation: Treatment of uranium dioxide with hydrogen fluoride. See uranium hexafluoride.
Hazard: Highly toxic and corrosive; radioactive poison.
Use: Intermediate in preparation of uranium metal.

uranocene (bis-cyclooctatetraenyluranium). A synthetic metallocene "sandwich" molecule whose unique feature is that the pi-molecular orbitals of the large organic rings share electrons with f-atomic orbitals of the uranium. See also metallocene.

uranyl nitrate (uranium nitrate; UNH; yellow salt) $UO_2(NO_3)_2 \cdot 6H_2O$.
Properties: Yellow, rhombic crystals. Sp. gr. 2.807; m.p. 60.2°C; b.p. 118°C. Soluble in water, alcohol, and ether.
Derivation: Action of nitric acid on uranium octoxide (U_3O_8).
Hazard: Highly toxic; severe fire and explosion risk when shocked or heated, and in contact with organic materials.
Uses: Source of uranium dioxide; extraction of uranium into non-aqueous solvents.
Shipping regulations: (Rail) Radioactive and Yellow label. (Air) Oxidizer label. Consult regulations for radioactive label.

urban waste. As conventionally used, this term includes garbage, cellulosics, glass, etc., but not sewage (municipal waste). In some localities the organic portion, mixed with oil, is being used for fuel.

urea (carbamide) $CO(NH_2)_2$. Occurs in urine and other body fluids. The first organic compound to be synthesized (Wohler, 1824).
Properties: White crystals or powder, almost odorless; saline taste; sp. gr. 1.335; m.p. 132.7°C; decomposes before boiling. Soluble in water, alcohol and benzene; slightly soluble in ether, almost insoluble in chloroform. Low toxicity. Noncombustible.
Derivation: Liquid ammonia and liquid carbon dioxide at 1750–3000 psi and 160–200°C react to form ammonium carbamate, $NH_4CO_2NH_2$, which decomposes at lower pressure (about 80 psi) to urea and water. Several variations of the process include once-through, partial recycle, and total recycle.
Method of purification: Crystallization.
Grades: Technical; C.P.; U.S.P.; fertilizer (45–46% nitrogen); feed grade (about 42% nitrogen).
Containers: 80-, 100-lb bags; 100-, 225-lb drums. Solution in tank cars and tank trucks.
Uses: Fertilizer; animal feed; plastics; chemical intermediate; stabilizer in explosives; medicine; adhesives; separation of hydrocarbons (as urea adducts); sulfamic acid production; flameproofing agents; viscosity modifier for starch or casein-based paper coatings; reported helpful in treating sickle-cell anemia.
See also urea-formaldehyde resin.

urea adduct. See inclusion complex.

urea ammonia liquor. A solution of crude urea in aqueous ammonia containing ammonium carbamate.
Containers: Tank cars.
Use: Reaction with superphosphate in preparation of fertilizers, furnishing combined nitrogen.

urea-ammonium orthophosphate. A fertilizer developed especially for food-deficient regions, especially rice-dependent areas. Several grades contain all three primary plant nutrients (nitrogen, phosphorus, and potassium). Contains up to 60% nitrogen, phosphoric anhydride, and potassium oxide.

urea-ammonium polyphosphate. A fertilizer similar to urea-ammonium orthosphosphate (q.v.), except that about half the phosphorus is in polyphosphate form, which gives improved sequestering action and solubility. It is excellent for use as a liquid fertilizer.

ureaform. A urea-formaldehyde reaction product that contains more than one molecule of urea per molecule of formaldehyde. It can be used as a fertilizer because of its high nitrogen content, its insolubility in water and its gradual decomposition in the soil during the growing season to yield soluble nitrogen.

urea-formaldehyde resin. An important class of amino resin. Urea and formaldehyde are united in a two-stage process in the presence of pyridine, ammonia, or certain alcohols with heat and control of pH to form intermediates (methylolurea, dimethylolurea) that are mixed with fillers to produce molding powders. These are converted to thermosetting resins by further controlled heating and pressure in the presence of catalysts. These were the first plastics that could be made in white, pastel, and colored products; widely used for scale housings, lamp shades, dinnerware and other decorative household applications; also plywood for interior use; foundry core binder.
Uses: See amino resin; melamine resin; "Plaskon."

urea half-chloride $(NH_2CONH_2)_2 \cdot HCl$.
Properties: Practically white, odorless powder. Very soluble in water.
Use: Catalyst.

urea hydrogen peroxide. See urea peroxide.

urea nitrate $CO(NH_2)_2 \cdot HNO_3$.
Properties: Colorless crystals. Decomposes 152°C. Slightly soluble in water, soluble in alcohol. Low toxicity.

Derivation: By adding an excess of nitric acid to a strong aqueous solution of urea.
Hazard: Dangerous fire and explosion risk.
Use: Explosives; manufacture of urethane (q.v.).
Shipping regulations: Dry or wet with less than 10% water: (Rail) Consult regulations. (Air) Not acceptable. Wet with not less than 10% water: (Rail) Yellow label. (Air) Flammable Solid label.

urea peroxide (urea hydrogen peroxide; percarbamide; carbamide peroxide) $CO(NH_2)_2 \cdot H_2O_2$.
Properties: White crystals or crystalline powder; m.p. (dec) 75–85°C. Decomposed by moisture at temperatures above 40°C. Soluble in water, alcohol, and ethylene glycol. Solvents such as ether and acetone extract the hydrogen peroxide and may form explosive solutions. Active oxygen (min) 16%.
Grades: Technical; sometimes compounded with waxes in pellet form.
Containers: 100-lb fiber drums.
Hazard: Dangerous fire risk in contact with organic materials; strong oxidizing agent. Moderately irritant.
Uses: Source of water-free hydrogen peroxide; bleaching; disinfectant; cosmetics; pharmaceuticals; blue print developer; modification of starches.
Shipping regulations: (Rail) Yellow label. (Air) Organic peroxide label.

urea phosphoric acid. See carbamide phosphoric acid.

urea-quinine. See quinine-urea hydrochloride.

urease. Enzyme present in the soy bean and in urine, and secreted by certain microorganisms. Its principal use is in the determination of urea in urine and in blood. It splits urea into ammonia and carbon dioxide or ammonium carbonate.

"Urecholine."[123] Trademark for bethanechol chloride.

"Urepan."[470] Trademark for urethane rubbers which have to be crosslinked with isocyanates or peroxides.

urethane (ethyl carbamate; ethyl urethane) $CO(NH_2)OC_2H_5$. Its structure is typical of the repeating unit in polyurethane resins.
Properties: Colorless crystals or white powder; odorless; saltpeter-like taste; solutions neutral to litmus. Soluble in water, alcohol, ether, glycerol and chloroform; slightly soluble in olive oil. Sp. gr. 0.9862; m.p. 49°C; b.p. 180°C. Combustible.
Derivation: (a) By heating ethyl alcohol and urea nitrate at 120–130°C; (b) by action of ammonia on ethyl carbonate or ethyl chloroformate.
Grades: Technical; N.F.
Containers: 55-gal steel drums; 250-lb fiber drums.
Hazard: Toxic by ingestion.
Uses: Intermediate for pharmaceuticals, pesticides and fungicides; biochemical research; medicine; veterinary medicine.
See also polyurethane.

urethane alkyd (uralkyd). A urethane resin modified with a drying oil (linseed or safflower) for use in high-quality paints.

uric acid (lithic acid; uric oxide; 2,6,8-trioxypurine)

OCNHC(O)NHCCNHC(O)NH (keto form). May also be written in phenolic form. The end-product of purine metabolism in man and other primates, birds and some dogs and reptiles.
Properties: Odorless, tasteless, white crystals. Soluble in hot concentrated sulfuric acid; very slightly soluble in water; insoluble in alcohol and ether; soluble in glycerol, solutions of alkali hydroxides, sodium acetate, and sodium phosphate. Sp. gr. 1.855–1.893; m.p., decomposes.
Derivation: From guano.
Grades: Technical; reagent.
Hazard: Evolves highly toxic hydrogen cyanide when heated.
Use: Organic synthesis.
See also pyrine.

uridine (D-ribosyl uracil) $C_9H_{12}N_2O_6$. The nucleoside of uracil. It is a constituent of ribonucleic acid and some coenzymes (such as uridine diphosphate glucose).
Properties: White, odorless powder, of slightly acrid and faintly sweet taste; m.p. 165°C; soluble in water, acid, and base; slightly soluble in dilute alcohol; insoluble in strong alcohol.
Derivation: From nucleic acid hydrolyzates from yeast. Radioactive forms available.
Use: Biochemical research.

uridine diphosphate glucose (UDPG). A coenzyme which acts in the transfer of glucose from the coenzyme to another chemical compound during the reaction for which the coenzyme is a catalyst. Used in biochemical research.

uridine monophosphate. See uridylic acid.

uridine phosphate. A nucleotide used by the body in growth processes; important in biochemical and physiological research. Those isolated and commercially available (as sodium salts) are the monophosphate (UMP), the diphosphate (UDP), and the triphosphate (UTP).
See also uridine diphosphate glucose (UDPG).

uridine-phosphoric acid. See uridylic acid.

uridylic acid (uridine phosphoric acid; UMP; uridine monophosphate) ($C_9H_{13}N_2O_9P$). The monophosphoric ester of uracil; i.e., the nucleotide containing uracil, D-ribose and phosphoric acid. The phosphate may be esterified to either the 2,3, or 5 carbon of ribose, yielding uridine-2′-phosphate, uridine-3′-phosphate, and uridine-5′-phosphate, respectively.
Properties (uridine-3′-phosphate): Crystallizes in prisms from methanol. M.p. 202°C (dec). Freely soluble in water and alcohol. Dextrorotatory in solution.
Derivation of commercial product: From yeast ribonucleic acid. Also made synthetically. Radioactive forms available.
Use: Biochemical research.

"Uritone."[330] Trademark for methenamine. See hexamethylenetetramine.

uronic acid. Any of a class of compounds similar to sugars but differing from them in that the terminal carbon has been oxidized from an alcohol to a carboxyl group. The most common are galacturonic acid and glucuronic acid.

"Urox."[50] Trademark for a soil sterilant herbicide developed for general weed control in non-crop areas. Contains monuron trichloroacetate.
Properties: Low toxicity; miscible with kerosine, fuel oil, diesel oil, aromatic weed oil, liquid asphalts and tars. Available in granular and liquid oil concentrate form.

ursin. See arbutin.

urushiol. Mixture of catechol derivatives.
Properties: Pale yellow liquid; sp. gr. 0.968; b.p. 200°C. Soluble in alcohol, ether, and benzene.
Derivation: Poison ivy (Rhus toxicodendron).
Hazard: The toxic principle of poison ivy. Causes severe allergenic dermatitis.

USAN. Abbreviation for United States Adopted Name, a nonproprietary name approved by the American Pharmaceutical Association, American Medical Association, and the U.S. Pharmacopeia. Such names applied to pharmaceutical products do not imply endorsement; their use in advertising and labeling is required by law.

USDA. Abbreviation for United States Department of Agriculture, the Federal regulatory authority for meats and meat products. This department maintains Regional Laboratories for agricultural and food research and development located in Philadelphia, Peoria, New Orleans, and Albany (Cal.) Its central office is in Washington, D.C.

usnic acid (usninic acid) $C_{18}H_{16}O_7$. A tricyclic compound. A constituent of many lichens. Known in *d*-, *l*-, and *dl*-forms.
Properties: Crystalline yellow solid; melting range 192–203°C. Insoluble in water; slightly soluble in alcohol and ether.
Derivation: From Usnea barbata, a lichen growing on trees.
Use: Medicine (antibiotic).

U.S.P. Abbreviation for United States Pharmacopeia, the official publication for drug product standards. The 18th edition appeared in 1970.

UTP. See uridine phosphate.

"Uversoft."[134] Trademark for softening agents for fabrics and paper products. Based on a quaternary ammonium salt containing two long straight chain hydrocarbon chains attached to the nitrogen atom. Are cationic and substantive to cellulose.

"Uversol."[134] Trademark for metal salts of naphthenic acids. Available as solids and liquids of most of the common metals. Used as paint driers, wetting agents, catalysts, etc.

"Uvinul."[307] Trademark for ultraviolet light absorbers (q.v.).
Uses: Effective in the range of 200–400 millimicrons. They do not darken or decompose upon prolonged exposure to an intense ultraviolet source. The absorbed energy is not re-emitted in the visible spectrum. Typical applications include protection of plastics, oils, cosmetics, paper, wood, and leather.

"Uvitex."[443] Trademark for optical brighteners for textile fibers, plastics, etc.

V Symbol for vanadium.

"V-25."[1] Trademark for a dicalcium phosphate dihydrate, finely milled and air-floated so that 99.5% passes a 325-mesh screen.
Grade: F.C.C.

vacancy. See hole.

vacuum deposition. The process of coating a base material by evaporating a metal under high vacuum and condensing it on the surface of the material to be coated, which is usually another metal or a plastic. Aluminum is most commonly used for this purpose. The coatings obtained range in thickness from 0.01 to as much as 3 mils. A vacuum of about one-millionth atmosphere is necessary. The process is used for jewelry, electronic components, decorative plastics, etc. Thermally evaporated metals and dielectric coatings can be effectively applied to glass by this method. It is also called vacuum coating and vacuum metallizing.

vacuum distillation. Distillation at a pressure less than atmospheric but not so low that it would be classed as molecular distillation. Since lowering the pressure also lowers the boiling point, vacuum distillation is useful for distilling high-boiling and heat-sensitive materials such as heavy distillates in petroleum, fatty acids, vitamins, etc.

vacuum forming (thermoforming). Shaping a thermoplastic sheet to the exact contours of a mold by application of vacuum. The sheet is first heated above its softening point (250 to 450°F); it is then placed in a male or female mold. Vacuum is then applied, which acts through channels in the base of the mold; the force thus exerted causes the sheet to assume the shape of the mold. After cooling below its softening point, the formed sheet is removed. Plastics most widely used for this process are polystyrene, acrylics, vinyls, polyethylene, polypropylene and cellulosics.

Val Abbreviation for valine.

"Valcar 85."[496] Trademark for polyester dye carriers.

"Valchrome CA."[496] Trademark for a chromium acetate solution; used as a mordant for dyeing and printing.

"Valclene."[28] Trademark for clear fluorocarbon formulations with slight ethereal odor; nonflammable. Used as drycleaning fluids.

"Valdet."[496] Trademark for a series of detergents. 561. Ethylene oxide condensate of nonyl phenol. AC-40, CC. Amine condensate.
Uses: Wetting agent; emulsifying; leveling; dispersing; antistatic agent; multipurpose detergent for dyeing, bleaching, scouring, washing after printing; anionic synthetic detergents; kier boiling aids.

valence. A whole number which represents or denotes the combining power of one element with another. By balancing these integral valence numbers in a given compound, the relative proportions of the elements present can be accounted for. If hydrogen and chlorine both have a valence of 1, oxygen 2, and nitrogen 3, the valence-balancing principle gives the formulas HCl, H_2O,NH_3, Cl_2O, NCl_3, and N_2O_3, which indicate the relative numbers of atoms of these elements in compounds which they form with each other. In inorganic compounds it is necessary to assign either a positive or negative value to each valence number, so that valence-balancing will give a zero sum by algebraic addition. Negative numbers are called polar valence numbers (−1 and −2). The valence of chlorine may be −1, +1, +3, +5, and +7, depending on the type of compound in which it occurs. In organic chemistry only nonpolar valence numbers are used. See also chemical bonding; oxidation; oxidation number; coordination number.

valentinite (antimony trioxide, ortho-rhombic; white antimony) Sb_2O_3. White or gray mineral, sometimes pale red. White streak and adamantine or silky luster. Sp. gr. 5.57–5.76; Mohs hardness 2-3.
Occurrence: Algeria; Yugoslavia; Italy; Germany.
Use: Ore of antimony.

valeral. See n-valeraldehyde.

n-**valeraldehyde** (valeric aldehyde; valeral; amyl aldehyde; pentanal) $CH_3(Ch_2)_3CHO$.
Properties: Colorless liquid; sp. gr. 0.8095; (20/4°C); f.p. −91°C; b.p. 102–103°C; refractive index (n 20/D) 1.3944; flash point (closed cup) 54°F. Slightly soluble in water; soluble in alcohol and ether. Low toxicity. See also isovaleraldehyde.
Derivation: Oxidation of amyl alcohol; also by the Oxo process.
Hazard: Flammable, dangerous fire risk.
Use: Flavoring; rubber accelerators.
Shipping regulations: (Rail) Flammable liquid, n.o.s., Red label. (Air) Flammable liquid label.

valerianic acid. See n-valeric acid.

valerian oil.
Properties: Yellowish or brownish liquid; penetrating odor. Soluble in alcohol, ether, chloroform, acetone, benzene, and carbon disulfide. Sp. gr. 0.903–0.960; refractive index (n 20/D) about 1.486. Combustible. Low toxicity.
Chief constituents: Pinene, camphene, borneol, and esters of borneol and valeric acid.
Derivation: Distilled from roots and rhizome of Valeriana officinalis.
Uses: Medicine; tobacco perfume; industrial odorant; flavors.

valeric acid (valerianic acid; n-pentanoic acid) $CH_3(CH_2)_3COOH$.
Properties: Colorless liquid; penetrating odor and taste; sp. gr. 0.9394 (20/4°C); b.p. 185.4°C; refractive index 1.4081 (20°C); b.p. 185.4°C; refractive index 1.4081 (20°C); vapor pressure 0.08 mm (20°C); f.p. −34°C; flash point 205°F (open cup); soluble in water; soluble in alcohol and ether. Undergoes reactions typical of normal monobasic organic acids. Combustible. See also isovaleric acid.

Derivation: With other C_5 acids by distillation from valerian; by oxidation of n-amyl alcohol; numerous essential oils.
Grades: Technical; reagent.
Containers: Drums; tank cars.
Hazard: Strong irritant to skin and tissue.
Uses: Intermediate for flavors and perfumes; ester-type lubricants; plasticizers; pharmaceuticals; vinyl stabilizers.
Shipping regulations: Corrosive label. (Air)

valeric aldehyde. See n-valeraldehyde.

gamma-**valerolactone** $CH_3\overline{CHCH_2CH_2C(O)O}$.
Properties: Colorless liquid. Sp. gr. (25/25°C) 1.0518; b.p. 205–206.5°C; crystallizing point −37°C; flash point (COC) 205°F; refractive index (25°C) 1.4301. Surface tension (25°C) 39 dynes/cm; viscosity (25°C) 2.18 cp; pH, anhydrous 7.0; pH 10% solution in distilled water 4.2; miscible with water and most organic solvents, resins, waxes, etc. Slightly miscible with zein, beeswax, petrolatum. Not miscible with anhydrous glycerin, glue, casein, arabic gum, and soybean protein. Combustible; low toxicity.
Uses: In dye baths (coupling agent), brake fluids, cutting oils, and as solvent for adhesives, insecticides and lacquers.

valine (alpha-aminoisovaleric acid) $(CH_3)_2CHCH(NH_2)COOH$. An essential amino acid.
Properties: White crystalline solid; soluble in water; very slightly soluble in alcohol; insoluble in ether. Shows the following optical isomers:
DL-valine: M.p. 298°C with decomposition.
D-valine (natural isomer): M.p. 315°C with decomposition.
L-valine: M.p. 293°C with decomposition.
Derivation: Hydrolysis of proteins; synthesized by the reaction of ammonia with alpha-chloroisovaleric acid. Available commercially as D-, L-, or DL-valine.
Containers: Drums.
Uses: Medicine; dietary supplement; culture media; biochemical and nutritional investigations.

"Valium."[190] Trademark for diazepam (q.v.).

"Valkleer 5722."[496] Trademark for emulsifiable resin solvent blend for printing with pigment colors.

"Vallestril."[70] Trademark for methallenestril (q.v.).

"Valmine SA."[496] Trademark for an organic amine salt of a volatile acid. Used as a protective agent to keep fabrics from tendering when printing with aniline black.

"Valpen NC."[496] Trademark for an organic sodium sulfonate; used as a mercerizing agent.

"Valrez."[496] Trademark for a series of wash and wear resins, many of which are urea derivatives or modified urea-formaldehyde resins.

"Valron."[28] Trademark for a solution of urea and formaldehyde in water; used for making fertilizers.

"Valsof."[496] Trademark for a series of polyethylene emulsions; used as fabric softeners.

"Valspex" A-115, N-123.[496] Trademark for polyethylene emulsions with 1% emulsifier content; used to increase abrasion resistance, tear and tensile strength of textiles.

"Valstat" E.[496] Trademark for a modified fatty condensate; used as an antistatic agent for synthetics, cotton and as a napping assistant.

"Valsyl."[496] Trademark for nylon and polyester sizing agents.

"Valwax."[496] Trademark for a series of textile lubricants.

"Valwet."[496] Trademark for a series of wetting agents for textile industry.
092. Fortified alkylarylsulfonate blend.
RW. Sulfonated alkyl ester; used in rewetting and compressive shrinking.

"Valzopon ON."[496] Trademark for a polyoxyethylated fatty alcohol. Used as a dispersing agent to improve crockfastness of naphthol dyeings; stabilizer for color salts.

vanadic acid (a) metal-HVO_3; (b) ortho-H_3VO_4; (c) pyro-$H_4V_2O_7$ These acids apparently do not exist in the pure state, but are represented in the various alkali and other metal vanadates. Ordinarily, vanadic acid implies vanadium pentoxide (vanadic acid anhydride).

vanadic acid anhydride. See vanadium pentoxide.

vanadic sulfate. See vanadyl sulfate.

vanadic sulfide. See vanadium sulfide.

vanadinite $Pb_5Cl(VO_3)$. A natural chlorovanadate of lead.
Properties: Color ruby red, orange red, brown, yellow; luster resinous to adamantine; Mohs hardness 3; sp. gr. 6.7–7.1. Soluble in strong nitric acid.
Occurrence: New Mexico, Arizona; Africa; Scotland; U.S.S.R.
Use: Ore of vanadium and lead.

vanadium V Metallic element having atomic number 23; Group VB of the periodic system. Atomic weight 50.94; valences 2, 3, 4, 5. Two natural isotopes.
Properties: Silvery-white ductile solid. Insoluble in water; resistant to corrosion, but soluble in nitric, hydrofluoric, and concentrated sulfuric acids; attacked by alkali, forming water-soluble vanadates. Sp. gr. 6.11; m.p. 1900 ± 25°C; b.p. ca. 3000°C. Acts as either a metal or a nonmetal, and forms a variety of complex compounds. Nontoxic as metal.
Source: Not found native. Principal ores are patronite, roscoelite, carnotite and vanadinite. Also from phosphate rock (Idaho, Montana, Arkansas).
Occurrence: Colorado, Utah, New Mexico, Arizona; Mexico and Peru.
Derivation: (a) Calcium reduction of vanadium pentoxide yields 99.8+% pure ductile vanadium. (b) Aluminum, cerium, etc. reduction produces a less pure product. (c) Solvent extraction of petroleum ash or ferrophosphorus slag from phosphorus production. (d) Electrolytic refining using a molten salt electrolyte containing vanadium chloride.
Grades: 99.99% pure (electrolytic process); single crystals.
Uses: Target material for x-rays; manufacture of alloy steels (see ferrovanadium); vanadium compounds, especially catalysts for sulfuric acid and synthetic rubber.

vanadium acetylacetonate.
Properties: Blue to blue-green crystals. Decomposes before melting.

Derivation: Reaction of vanadyl sulfate with acetylacetone and sodium carbonate.
Use: Catalyst.

vanadium carbide.
Properties: Crystals with hardness 2800 kg/sq mm; sp. gr. 5.77; m.p. 2800°C; b.p. 3900°C; resistivity 150 micro-ohm cm (room temp.).
Uses: Alloys for cutting tools; steel additive.

vanadium dichloride (vanadous chloride) VCl_2.
Properties: Apple-green hexagonal plates. Soluble in alcohol and ether; decomposes in hot water. Sp. gr. 3.23 (18°C).
Derivation: From vanadium trichloride by heating in atmosphere of nitrogen.
Method of purification: Sublimation in nitrogen.
Grade: C.P.
Containers: Sealed glass bottles.
Hazard: Toxic. Strong irritant to tissue.
Uses: Strong reducing agent; purification of hydrogen chloride from arsenic.

vanadium disulfide V_2S_2. Solid; sp. gr. 4.20; m.p. decomposes. Soluble in hot sulfuric or nitric acids; insoluble in alkalies. Used as a solid lubricant.

vanadium ethylate $(C_2H_5O)_4V$.
Properties: Dark reddish-brown solid.
Derivation: Reaction of vanadium chloride with sodium ethylate.
Use: Polymerization catalyst.

vanadium hexacarbonyl $V(CO)_6$.
Properties: Blue-green powder; sublimes easily at 50°C (15 mm); paramagnetic; decomposes without melting at 60–70°C.
Hazard: Toxic; strong irritant to tissue. Store under inert gas.
Use: Chemical intermediate; production of plating compounds and fuel additives.

vanadium hexacarbonyl, sodium salt. A convenient, though incomplete, name for the complex compound $Na(C_6H_{14}O_3)_2V(CO)_6$, which has also been named bis-diglymesodium hexacarbonyl vanadate.
Properties: Yellow solid; m.p. 173–176°C (with decomposition); soluble in water, alcohol and ether; slightly soluble in hydrocarbons. Relatively inert to air, but is stored and shipped under nitrogen.
Hazard: Toxic; strong irritant to tissue.
Use: Source of vanadium hexacarbonyl by treatment with phosphoric acid under special conditions.

vanadium nitride VN.
Properties: Black solid; sp. gr. 6.13; m.p. 2320°C. Insoluble in water; slightly soluble in aqua regia.
Use: Refractory.

vanadium oxydichloride. See vanadyl chloride.

vanadium oxytrichloride $VOCl_3$.
Properties: Lemon-yellow liquid; sp. gr. 1.811 (32°C); f.p. −78.9°C; b.p. 125–127°C; nonionizing solvent; dissolves most nonmetals; dissolves and/or reacts with many organic compounds. Hydrolyzes in moisture.
Hazard: Highly toxic; strong irritant to tissue.
Uses: Catalyst in olefin polymerization, (ethylene-propylene rubber); organovanadium synthesis.
Shipping regulations: (Rail) white label. (Air) Corrosive label. Not accepted on passenger planes.

vanadium pentoxide (vanadic acid anhydride) V_2O_5.
Properties: Yellow to red crystalline powder; sp. gr. 3.357 (18°C); m.p. 690°C; b.p. decomposes at 1750°C. Soluble in acids and alkalies; slightly soluble in water.
Derivation: (a) Alkali or acid extraction from vanadium minerals. (b) By igniting ammonium metavanadate. (c) From concentrated ferrophosphorus slag by roasting with sodium chloride, leaching with water, and purification by solvent extraction followed by precipitation and heating.
Method of purification: Alkali solution, precipitation as ammonium metavanadate and ignition to V_2O_5.
Grades: Commercial air-dried; commercial fused; C.P. air-dried; C.P. fused.
Containers: Drums; multiwall paper sacks.
Hazard: Toxic by inhalation. Tolerance (as V), (dust) 0.5 (fume) 0.1 mg per cubic meter of air.
Uses: Catalyst for oxidation of sulfur dioxide in sulfuric acid manufacture (see also contact process): ferrovanadium (q.v.); catalyst for many organic reactions; ceramic coloring material; vanadium salts; inhibiting ultraviolet transmission in glass; photographic developer; dyeing textiles.

vanadium sesquioxide. See vanadium trioxide.

vanadium sulfate. See vanadyl sulfate.

vanadium sulfide (vanadium pentasulfide; vanadic sulfide) V_2S_5.
Properties: Black-green powder. Soluble in acids, alkali-metal sulfides and alkalies; insoluble in water. Sp. gr. 3.0. Decomposes on heating.
Derivation: Action of hydrogen sulfide on vanadium chloride solution.
Hazard: May be toxic by inhalation (especially to animals).
Use: Vanadium compounds.

vanadium tetrachloride VCl_4.
Properties: Red liquid. Soluble in absolute alcohol and ether; decomposes slowly to vanadium trichloride and chlorine below 63°C. Sp. gr. 1.816 (20°C); f.p. −28°C; b.p. 154°C. Nonflammable.
Derivation: Chlorination of ferrovanadium.
Method of purification: Distillation and fractionation.
Hazard: Highly toxic by ingestion, inhalation and skin absorption. Open containers only in dry oxygen-free atmosphere or inert gas; wear goggles and protective clothing (MCA).
Uses: Preparation of vanadium trichloride, vanadium dichloride, and organovanadium compounds.
Shipping regulations: (Rail) white label. (Air) Corrosive label. Not acceptable on passenger planes.

vanadium tetraoxide V_2O_4.
Properties: Blue-black powder; sp. gr. 4.339; m.p. 1967°C; insoluble in water; soluble in alkalies and acids.
Derivation: (1) From vanadium pentoxide by oxalic acid reduction. (2) From vanadium pentoxide by carbon reduction.
Hazard: May be toxic and irritant.
Use: Catalyst at high temperature.

vanadium trichloride VCl_3.
Properties: Pink deliquescent crystals; soluble in absolute alcohol and ether. Decomposes in water. Sp. gr. 3.0 (18°C); decomposes on heating.
Derivation: From vanadium tetrachloride boiling under reflux condenser.
Hazard: Toxic and irritant.
Use: Preparation of vanadium dichloride and organovanadium compounds.
Shipping regulations: (Air) Poison label.

vanadium trioxide (vanadium sesquioxide) V_2O_3.
Properties: Black crytals; soluble in alkalies and hydro-

fluoric acid; slightly soluble in water. Sp. gr. 4.87 (18°C); m.p. 1970°C.
Derivation: From vanadium pentoxide by either hydrogen or carbon reduction.
Hazard: May be toxic and irritant.
Use: Catalyst for conversion of ethylene to ethyl alcohol.

vanadous chloride. See vanadium dichloride.

vanadyl chloride (vanadium oxydichloride; vanadyl dichloride; divanadyl tetrachloride) $V_2O_2Cl_4 \cdot 5H_2O$.
Properties: Green, very deliquescent crystals. Slowly decomposed by water. Usual technical product is a dark green syrupy mass 76–82% pure, or a solution. Soluble in water, alcohol, and acetic acid.
Grade: Technical.
Hazard: Toxic and irritant.
Use: Mordanting textiles.

vanadyl sulfate (vanadic sulfate; vanadium sulfate) $VOSO_4 \cdot 2H_2O$.
Properties: Blue crystals; soluble in water.
Derivation: Reduction of cold solution of concentrated sulfuric acid and vanadium pentoxide by sulfur dioxide gas.
Containers: Glass bottles; kegs.
Hazard: May be toxic and irritant.
Uses: Mordant; catalyst; aniline black preparation; reducing agent; colorant in glasses and ceramics.

"Vancide 89."[69] Trademark for N-(trichloromethylmercapto)-4-cyclohexene 1,2-dicarboximide. See captan.
Use: Fungicide for vinyl compositions.

"Vancide 26EC."[69] Trademark for lauryl pyridinium 5-chloro-2-benzothiazyl sulfide.
Use: Preservative for cotton fabrics used in rubber structures. Applied from an aqueous solution.

"Vancide 51Z."[69] Trademark for zinc dimethyldithiocarbamate $[(CH_3)_2NC(S)S]_2Zn$, with a small proportion of zinc 2-mercaptobenzothiazole.
Use: Fungicide for neoprene compositions; ultra-accelerator for rubber.

"Vancocin."[100] Trademark for vancomycin hydrochloride (q.v.).

vancomycin hydrochloride.
Properties: Tan to brown powder, odorless, with bitter taste. Soluble in water; moderately soluble in dilute methanol; insoluble in higher alcohols, acetone, ether.
Derivation: Produced by Streptomyces orientalis from Indonesian and Indian soil.
Grade: U.S.P.
Use: Medicine (antibiotic).

van der Waals' forces. Weak attractive forces acting between molecules. They are somewhat weaker than hydrogen bonds (q.v.) and far weaker than interatomic valences. They are involved in the van der Waals equation of state for gases which compensates for the actual volume of the molecules and the forces acting between them. "Information regarding the numerical values of van der Waals forces is mostly semiempirical, derived with the aid of theory from an analysis of chemical or physical data. Attempts to calculate the forces from first principles have had a measure of success only for the simplest systems, such as H—H, He—He and a few others. When judging the difficulties of such calculations, one must bear in mind that the energies sought are of the same order of magnitude as the *errors* in the best atomic energy calculations." (Henry Margenau) See also hydrogen bond; chemical bonding.

"Vandex."[69] Trademark for a finely ground selenium.
Properties: Dark gray metallic powder; sp. gr. 4.80±.03; fineness through 200 mesh 99%; melting range above 217°C; slightly soluble in benzene; soluble in carbon disulfide; insoluble in water.
Uses: Rubber vulcanization.

Van Dyke brown (Cassel brown; Cologne brown; Cologne earth; ulmin brown). A naturally occurring pigment.
Derivation: Indefinite mixtures of iron oxide and organic matter. Obtained from bog-earth, peat deposits or from ochers containing bituminous matter.
Containers: Barrels or fiber drums.
Use: Pigment for artists' colors and stains.

Van Dyke red. A brownish red pigment consisting of copper ferrocyanide; sometimes used to refer to red varieties of ferric oxide. See iron oxide red.
Use: Pigment.

"Vanfre."[69] Trademark for a series of release agents for various plastics.

vanillin (3-methoxy-4-hydroxybenzaldehyde; vanillic aldehyde) $(CH_3O)(OH)C_6H_3CHO$. The methyl ether of protocatechuic aldehyde.
Properties: White crystalline needles; sweetish smell. Sp. gr. 1.056; m.p. 81–83°C; b.p. 285°C. Soluble in 125 parts water; in 20 parts glycerol and in 2 parts 95% alcohol; soluble in chloroform and ether. Combustible; low toxicity.
Derivation: (a) By extraction of the vanilla bean; (b) from lignin contained in sulfite waste pulp liquor.
Method of purification: Crystallization.
Grades: Technical; U.S.P.; F.C.C.
Containers: Cartons, bottles; cans; drums.
Uses: Perfumes; flavoring; pharmaceuticals; laboratory reagent; source of L-dopa.

"Vanplast 125."[69] Trademark for cresyl diphenyl phosphate; used as a flame-retardant plasticizer for vinyl formulations, including foam.

"Vanstay."[69] Trademark for a series of heat and light stabilizers for polyvinyl chloride resins.

"Vantoc" DD-50.[325] Trademark for a blend of mineral oil with sulfonated oils and esters. Used as textile softener and shrink-proofing oil.

"Vapam."[1] Trademark for a soil fumigant composed of sodium N-methyldithiocarbamate dihydrate (q.v.).

"Vapona."[125] Trademark for an insecticide which contains not less than 93% 2,2-dichlorovinyl dimethyl phosphate (see DDVP) and not more than 7% active, related compounds.
Hazard: See dichlorovos.

vapor. An air dispersion of molecules of a substance that is liquid or solid in its normal state, i.e., at standard temperature and pressure. Examples are water vapor and benzene vapor. Vapors of organic liquids are also loosely called fumes. See also evaporation; gas.

vapor-phase chromatography. See gas chromatography.

vapor pressure. The pressure (usually expressed in millimeters of mercury) characteristic at any given temperature of a vapor in equilibrium with its liquid or solid form.

vapor tension. See vapor pressure.

"Vapotone."[253] Trademark for a series of insecticides containing tetraethylpyrophosphate.
Hazard: See tepp.

"Variamine."[307] Trademark for an azoic composition for printing blues on cotton and rayon.

"Varidase."[315] Trademark for streptokinase-streptodornase enzymes.

"Varkon."[300] Trademark for sequestering agents of the ethylenediaminetetraacetate type. Used in kier boiling, peroxide bleaching, dyeing and stripping operations. Available as both powder and liquid.

varnish. (1) An organic protective coating, similar to a paint, except that it does not contain a colorant. It may be comprised of a vegetable oil (linseed, tung, etc.) and solvent, or of a synthetic or natural resin and solvent. In the first case the formation of the film is due to polymerization of the oil, and in the second to evaporation of the solvent. "Long-oil" varnishes such as spar varnish (q.v.) have a high proportion of drying oil: "short-oil" types have a lower proportion, i.e., furniture varnishes. Spirit varnishes contain such solvents as methanol, toluene, ketones, etc., and often also thinners such as naphtha or other light hydrocarbon. Flammable. (2) A hard, tightly adherent deposit on the metal surfaces of automobile engines resulting from resinous oxidation products of gasoline and lubricating oils.

varnish remover. See paint remover.

"Varnon."[446] Trademark for hard-fired super-duty fireclay brick which resists the destructive action of carbon monoxide and other reducing gases. Used as glass tank regenerator checkers; also to line various metallurgical furnaces, rotary kilns, shaft kilns, carbon baking furnaces, and incinerators.

"Varox."[69] Trademark for a 50% active blend of 2,5-bis(tert-butylperoxy)-2,5-dimethylhexane with an inert mineral carrier.
Properties: White powder; sp. gr. 1.43 ± .03.
Uses: Cross-linking agent for polymers such as polyethylene.

"Varsol."[51] Trademark for straight petroleum aliphatic solvents used as paint and varnish thinners, for dry cleaning and for general plant machinery cleaning. Conform to CS3-40, the U.S. Dep't of Commerce commercial standard for Stoddard Solvent and have minimum TCC flash points of 100°F. Combustible.

"Varsoy."[221] Trademark for a prebodied chemically modified soybean oil for varnish cooking or bodied drying oil.

"Vaseline" Petroleum Jelly.[97] Trademark for a commercial product of petroleum, largely employed in pharmacy, alone and as a vehicle for external applications of medicinal agents, especially when local action rather than absorption is desired; as a protective coating for metallic surfaces, and for other purposes. Consists of a semisolid mixture of hydrocarbons, having a m.p. usually ranging between 38° and 60°C. Colorless or pale yellow, translucent, semisolid unctuous mass. Does not readily oxidize on exposure to the air and is not readily acted on by chemical reagents. Soluble in chloroform, benzene, carbon disulfide, oil of turpentine, warm ether; slightly soluble in hot alcohol, but separates from the latter in flakes on cooling.

vasopressin (beta-hypophamine; antidiuretic hormone). One of the hormones secreted by the posterior lobe of the pituitary gland. It causes an increase in blood pressure and an increase in water retention by the kidney. Vasopressin is an octapeptide consisting of eight different amino acids.
Derivation: Synthetic, or from the posterior lobe of the pituitary of food animals.
Grade: U.S.P., as an aqueous solution for injection.
Use: Medicine.

vat dye. A class of dyes that can be easily reduced, i.e., vatted, to a soluble and usually colorless leuco form in which they can readily impregnate fibers. Subsequent oxidation then produces the insoluble colored dyestuff in a form that is remarkably fast to washing, light and chemicals. Examples are indigo (CI No. 73000), and Indanthrene Blue BFP (CI No. 69825). The reducing agents are usually an alkaline solution of sodium hydrosulfite ($Na_2S_2O_4$) or some derivative of the latter. Oxidation is by air, perborate, dichromate, etc.

vat printing assistant. A mixture of gums, reducing and wetting agents used to carry the dye in printing fabrics with vat dyes. They assist in securing penetration of the fabric and in converting the dyes from a semi-leuco to a leuco state.

"Vatrolite."[159] Trademark for a group of concentrated sodium hydrosulfite preparations.
Uses: Bleaching textiles and paper; oxygen scavenger in synthetic rubber production.

"Vatsol."[57] Trademark for a series of wetting agents made in several different grades and types: OS, sodium isopropyl naphthalene sulfonic acid; OT, sodium dioctyl sulfosuccinate.

"Vazo."[28] Trademark for azobisisobutyronitrile.

"V-Bor."[58] Trademark for refined borax pentahydrate $Na_{32}B_4O_7 \cdot 5H_2O$; 99.8% purity. Obtained from Searles Lake brines.

VC. Abbreviation for vinyl chloride or vinylidene chloride.

"V-C 13 Nemacide."[40] Trademark for a 75% emulsifiable concentrate whose active ingredient is O-2,4-dichlorophenyl-O,O-diethylphosphorothioate.
Hazard: Highly toxic.
Uses: Insecticide and nematocide.

vector. In biochemistry, an organism that carries or transports an infectious disease.

"Vectra."[29] Trademark for a polypropylene fiber.

"Vegadex" Herbicide.[58] Trademark for a herbicide containing 2-chloroallyl diethyldithiocarbamate as the active ingredient.
Hazard: May be toxic.

vegetable black. In general any form of more or less pure carbon produced by incomplete combustion or destructive distillation of vegetable matter, wood, vines, wine lees.

vegetable dye. A colorant derived from a vegetable source, i.e., logwood, indigo, madder, etc.

vegetable gum. See dextrin.

vegetable ivory. A mannose polysaccharide obtained from the hardened seeds of the ivory-nut (Corozo-nut) tree (Phytelephas macrocarpa). The latter must be distinguished from the cohune palm (Attalea cohume) whose seeds are also sometimes referred to as corozo nuts.
Uses: Button manufacture; mannose production.

vegetable oil. An oil extracted from the seeds, fruit or nuts of plants and generally considered to be mixtures of mixed glycerides (e.g., cottonseed, linseed, corn, coconut, babassu, olive, tung, peanut, perilla, oiticica, etc.). Many types are edible.
Containers: Bottles; drums; tank cars; tank trucks.
Uses: Paints (as drying oils); shortenings; salad dressings, margarine; soaps; rubber softeners; medicine; dietary supplements; pesticide carriers.
See also specific entry.

vegetable tanning. The tanning of leather by plant extracts. See tannic acid; tanning; wattle bark; quebracho.

vehicle. A term used in paint technology to indicate the liquid portion of a paint, comprised of drying oil or resin, solvent, and thinner, in which the solid components are dissolved or dispersed. See also paint.

"Velban."[100] Trademark for vinblastine sulfate.

"Velsicol" 2-1.[316] Trademark for a mercuric turf fungicide.
Hazard: See mercury compounds.

"Velsicol" Emmi.[316] Trademark for a mercuric fungicide.
Hazard: See mercury compounds.

"Velvetex."[133] Trademark for carbon black in the high structure, coarse particle size range; gives improved processing and physical properties of various elastomers.

Venetian red. A high-grade ferric oxide pigment of a pure red hue. It is obtained either native as a variety of hematite red (q.v.) or more often artificially, by calcining copperas (ferrous sulfate) in the presence of lime. The composition ranges from 15 to 40% ferric oxide and from 60 to 80% calcium sulfate. The 40% ferric oxide is the "pure" grade and has a sp. gr. of 3.45.
Grades: 20 to 40% ferric oxide.
See also iron oxide red.

"Veon."[233] Trademark for herbicides consisting of amine salts of 2,4-D and 2,4,5-T.
Hazard: See 2,4-D and 2,4,5-T.

"Vera Blanc."[148] Trademark for a filler or extender composed of calcium carbonate, whiting, finely ground limestone, primer. Water-ground, water-floated and silk-bolted. Sp. gr. 2.72; wt. per solid gal. 22.66 lbs.
Chemical and screen analysis: Calcium carbonate 99.5%; 100% 325 mesh; average particle size 4 microns.
Uses: Oil and water paints, colors, rubber compounds, adhesives, oilcloth, linoleum, textile fabrics, fur dressing, putty, plastics, ceramics, etc.

"Verabore."[342] Trademark for extract of antihypertensive alkaloids from Veratrum species.

"Verafleurs."[188] Trademark for a series of synthetic replacements for natural floral absolutes.

veratrole (1,2-dimethoxybenzene; pyrocatechol dimethyl ether) $C_6H_4(OCH_3)_2$.
Properties: Colorless crystals or liquid; m.p. 21–22°C; b.p. 206–207°C; sp. gr. 1.084 (25/25°C). Soluble in alcohol and ether; slightly soluble in water.
Derivation: Treatment of catechol in methyl alcohol with dimethyl sulfate and caustic.
Use: Medicine (antiseptic).

veratrum alkaloid. One of a group of alkaloids used in medicine to relieve hypertension. They include veratrum viride (American hellebore).
Hazard: May have severely toxic side effects. See also alkaloid.

"Verel."[256] Trademark for a modified acrylic fiber.

"Verelite."[233] See "Styron."

verdigris. See copper acetate, basic.

vermiculite. Hydrated magnesium-iron-aluminum silicate capable of expanding six to twenty times when heated to about 2000°F. The platelets exhibit an active curling movement when heated; hence the name.
Occurrence: Montana, North Carolina, South Carolina, Wyoming, Colorado; South Africa.
Properties: Platelet-type crystalline structure; high porosity; high void volume to surface area ratio; low density; large range of particle size; insoluble in water and organic solvents; soluble in hot concentrated sulfuric acid; water vapor adsorption capacity of expanded vermiculite less than 1%, liquid adsorption dependent on conditions and particle size, ranges 200–500%. Noncombustible; nontoxic.
Grides: Unexpanded (ore concentrate); expanded (also called exfoliated); flake; activated.
Containers: Multiwall paper bags, 4 cu ft.
Uses: Lightweight concrete aggregate; insulation; sound conditioning; fireproofing; plaster; soil conditioner; additive for fertilizers; seed bed for plants; refractory; lubricant; oil well drilling mud; filler in rubber, paint, plastics; wall paper printing; removal of strontium-90 from milk; absorption of oil spills on seawater; animal feel additive; packing; carrier for insecticides; catalyst and catalyst support; litter for hatcheries; adsorbent.
See also verxite.

vermilion, natural. See cinnabar.

vernolepin. A sesquiterpene dilactone extracted from leaves of an African plant *Veronia hymenolepis*.
Uses: Biochemical research (inhibits tumor growth in rats and reversibly retards plant growth).

"Veronal."[162] Trademark for barbital. See barbiturate.

"Versadyme."[259] Trademark for dimerized vegetable fatty acids.
Uses: Production of polymers, petroleum additives, corrosion inhibitors, paint additives and surfactants.

"Versalide."[227] Trademark for 1,1,4,4-tetramethyl-6-ethyl-7-acetyl-1,2,3,4-tetrahydronaphthalene, a polycyclic musk.

"Versalon."[259] Trademark for thermoplastic polyamide resins of high strength, good adhesion and flexibility.
Uses: Hot-melt adhesives, heat-seal coatings, electrical encapsulation.

"Versamid."[259] Trademark for thermoplastic and reactive polyamide resins. The reactive resins copolymerize with epoxy resins.
Uses: Heat-seal coatings; hot-melt adhesives; varnishes, paints, and inks; coatings; structural adhesives.

"Versatic" 911 Acid.[125] Trademark for a saturated synthetic tertiary monocarboxylic acid having C_9, C_{10} and C_{11} chain length.
Derivation: From olefins, water and carbon monoxide in the presence of a strong acid catalyst.
Uses: Chemical intermediate to make extremely stable esters, metal salts, nitrogen derivatives, and hindered alcohols.

"Versatyl."[141] Trademark for a series of printing ink vehicles and pigment colors flushed in the vehicles.
Flushed Colors: Wide range of yellows, reds, blues and greens.
Uses: Letterpress, lithographic, offset, tin lithography; quick-set, gloss, multicolor and snap-dry inks.

"Versenate."[233] Trademark for disodium salt of ethylenediaminetetraacetic acid and related compounds.

"Versene."[233] Trademark for a series of chelating agents based on ethylenediaminetetraacetic acid (q.v.).

"Versene Fe-3 Specific."[233] Trademark for an iron specialty chelating agent. Active ingredient is sodium dihydroxyethylglycine. Available as straw-colored liquid or white powder.

"Versenex" 80.[233] Trademark for pentasodium salt of diethylenetriaminepentaacetic acid.

"Versenol" 120.[233] Trademark for the trisodium salt of N-hydroxyethylethylenediaminetriacetic acid, ($C_{10}H_{15}O_7N_2Na_3$). Used as an organic chelating agent.

"Versilad."[244] Trademark for a silicate-based adhesive designed specifically for the corrugated boxboard industry.

"Versilate."[244] Trademark for a composition of fortified sodium silicate.
Properties: Viscous liquid; opalescent.
Containers: Tank trucks and tank cars.
Use: Base for "Versilad" adhesives.

"Versilube."[245] Trademark for a series of silicone hydraulic fluids, lubricating fluids and greases.

"Vertifume."[233] Trademark for fumigants containing carbon tetrachloride and carbon disulfide.

verxite (exfoliated hydrobiotite).
Properties: Thermally expanded (exfoliated) magnesium-iron-aluminum silicate having a minimum of 98% purity and a bulk density of 5 to 7 lb per cubic foot. Expansion occurs by heating at about 1400°F. Sponge-like structure which absorbs liquids and permits reexpansion after compression to 70–80% of the original heat-expanded volume.
Uses: Poultry feed in quantities not greater than 5% as a non-nutritive bulking agent; pelleting or anticaking agent and nutrient carrier in dog and ruminant feeds. For dog feeds the maximum permitted is 1.5%. Labeling must state content when in excess of 1%. See also vermiculite.

"Vespel."[28] Trademark for fabricated parts, based on polyimide resin, and diamond abrasive wheels formulated with a high-temperature polyimide binder.

"Vesprin."[412] Trademark for triflupromazine hydrochloride (q.v.).

"Vetalog."[412] Trademark for triamcinolone acetonide; used in veterinary medicine.

"Vetame."[412] Trademark for triflupromazine hydrochloride.

vetiver oil (cuscus oil; vetivert).
Properties: Viscid essential oil; violet-like odor; sp. gr. 0.990–1.040 (15°C); optical rotation +15 to +45°; refractive index (n 20/D) 1.5200–1.5280; saponification value 14–45; soluble in 1–3 volumes of 80% alcohol, in fixed oils; insoluble in glycerin and propylene glycol. Combustible; nontoxic.
Derivation: Steam distillation of partially dried roots of East Indian grass, Vetiveria zizanioides. Purified by rectification.
Grades: Characterized by its geographical origin (Java, Haiti, East Indian, Bourbon Reunion, French).
Containers: Bottles; cans.
Use: Perfumery.

"Vestrep."[123] Trademark for a veterinary preparation containing streptomycin sulfate.

"Vexar."[28] Trademark for both low and high density polyethylene and polypropylene plastic netting. Available in wide variety of forms and colors for use in packaging.

"V-G-B."[248] Trademark for reaction product of acetaldehyde and aniline.
Properties: Brown resinous powder; sp. gr. 1.152; m.p. 60–80°C; soluble in acetone, benzene, and ethylene dichloride; insoluble in water and gasoline.
Use: Rubber antioxidant.

"Vialon."[440] Trademark for a series of dyes for nylon fibers and furs.

"Vibrathane."[248] Trademark for a group of polyurethane raw materials for manufacture of foam and elastomers. Includes isocyanates, polyesters, polyether glycols, polyester and polyether prepolymers, liquid casting resins and gums and catalysts.

"Vibrin."[248] Trademark for resin compositions of polyesters and cross-linking monomers which, when catalyzed, will polymerize to infusible solid resins without evolving water or other by-products.
Uses: Molding; laminating; impregnating; casting; automotive and aircraft structural parts; wall panels, table tops; coating for paper; boat hulls; chemically inert tanks; large-diameter pipe.

vic-. Prefix meaning vicinal (q.v.).

"Vicalloy."[155] Trademark for a magnetic alloy composed of iron, 38%; cobalt, 52%, and vanadium, 10%.
Properties: Density, 8.09 8.09 g/cm^3; tensile strength, 100,000 psi.
Uses: Magnetic recording tape; machinable permanent magnets.

vicinal (abbreviated as vic-). Neighboring or adjoining positions on a carbon ring or chain; the term is used in naming derivatives with substituting groups in such locations in a structural formula or molecule. For example, vicinal locations in the molecule shown are occupied by the hydrogen atoms and the hydroxyl groups:

```
  H   H
  |   |
—C—C—
  |   |
 HO  OH
```

"Victacid 105."[1] Trademark for a high-strength technical grade of phosphoric acid composed of approxi-

mately equal quantities of orthophosphoric and polyphosphoric acids.
Uses: Dehydrating agent for organic reactions; sequestrant for heavy metals and control of P_2O_5 content of bath process for electropolishing and bright-dipping metals.

"Victamide."[1] Trademark for exceedingly fine particles of ammonium salt of an amido polyphosphate, practically all less than five microns. Slowly soluble in cold water; more rapidly in hot water.
Uses: Sequestering agent for metallic ions; flameproofing agent; deflocculating agent for oil-drilling muds, paint pigments, and clay slips.

"Victamines."[1] Trademark for tan, waxy solids; cationic surface-active phosphorus compounds which disperse in water.
Uses: Wetting and softening agents for textiles and leather; oil additives.

"Victamuls."[1] Trademark for nonionic, surface-active phosphorus compounds.
Uses: Emulsifying, penetrating, solubilizing, or dispersing agents.

"Victawets."[1] Trademark for surface-active phosphorus compounds. Anionic and nonionic wetting agents, used as penetrants, dye carriers, and dispersing agents.

"Victor Cream."[1] Trademark for sodium acid pyrophosphate. Purity meets all requirements of federal and state pure food laws; F.C.C. grade.
Uses: Baking acid in doughnut and prepared flours; manufacture of commercial baking powders and instant puddings; for conditioning oil-well drilling muds; formulation of acid-type metal cleaners.

Victoria blue $C_{33}H_{31}N_3 \cdot HCl$. CI No. 44045.
Properties: Crystalline powder, bronze colored. Soluble in hot water, alcohol or ether.
Derivation: Michler's ketone is condensed with phenyl-alpha-naphthylamine.
Uses: Dyeing silk, wool, and cotton; biological stain; dye intermediate for complex acid pigment toners.

Victoria green. See malachite green.

vicuna. A soft wool-like fiber obtained from a South American animal similar to the llama. Used for specialty high-grade coats, sweaters, etc. Combustible.

"Vidden D."[233] Trademark for 1,3-dichloropropene and 1,2-dichloropropane mixture.

"Vigofac 6."[299] Trademark for an unidentified growth factor for addition to animal feeds. Derived from dried streptomyces fermentation solubles.

"Vikane."[233] Trademark for sulfuryl fluoride (q.v.).

"Vim-Thane."[320] Trademark for polyurethane-impregnated leather packings.

"Vinac."[144] Trademark for polyvinyl acetate emulsions, powders, beads and solutions.
Emulsions: Properties (range): 48% to 57% solids; 80 to 4200 cp; 4.0 to 7.0 pH; 0.15μ to 2.5μ average particle size; nonionic.
Uses: Binder in pigmented paper coatings; adhesive bases; concrete adhesive; textile finishes.
Powders: Properties: White, free-flowing powder; 33 lbs per cu ft bulk density; 98% through 100 mesh; viscosity of 5000 to 7000 cp in 50% dispersion; particle size of 2μ to 6μ in dispersion.
Use: Adhesive for concrete and joint cements.
Beads: Properties (range): Glass-like spheres; low, medium, and high molecular weight grades; soluble in organic solvents except aliphatic hydrocarbons; ASB-grade is soluble in alkalies in addition to organic solvents.
Uses: Hot melt and solvent adhesives; solvent paints; specialty paper coatings; binders in inks.
Solutions: Properties: Methanol solvent; 50% to 51% polymer; 900–1100 cp.
Hazard: Toxic and flammable.
Use: Heat-seal adhesives.

"Vinactane" Sulfate.[305] Trademark for viomycin sulfate.

vinal fiber. Generic name for a manufactured fiber in which the fiber-forming substance is any long-chain synthetic polymer composed of at least 50% by weight of vinyl alcohol units, $—CH_2CHOH—$, and in which the total of the vinyl alcohol units and any one or more of the various acetal units is at least 85% by weight of the fiber (Federal Trade Commission). It has good chemical resistance, low affinity for water; good resistance to mildew and fungi. Combustible.
Uses: Fishing nets; stockings, gloves; hats; rainwear; swimsuits.

"Vincel."[332] Trademark for a group of fabrics of polyvinyl chloride fibers for liquid and pneumatic filtration.

"Vin-Clad."[326] Trademark for a series of dry, free-flowing polyvinyl chloride powders formulated for application by the fluidized bed process to provide protective and decorative coatings.

vinegar.
Properties: Brownish or colorless liquid; dilute aqueous solution containing from 4 to 8% acetic acid (q.v.), depending on source. Legal minimum is 4%. Also contains low percentages of alcohols and mineral salts. Nontoxic; nonflammable.
Derivation: (a) Bacterial fermentaton of apple cider, wine or other fruit juice. (b) Fermentation of malt or barley. The fermenting agent is usually a mold (e.g., Mycoderma aceti), generally known as "mother" (q.v.). Either type can be distilled to remove color and other impurities, and is then called white vinegar.
Containers: Bottles; carboys.
Uses: Mayonnaise; salad dressings; pickled foods; preservative; medicine (antifungal agent); latex coagulant.

"Vinethene."[123] Trademark for an anesthetic preparation consisting of vinyl ether.

"Vinoflex."[440] Trademark for vinyl chloride homo- and copolymers.

"Vinol."[144] Trademark for polyvinyl alcohol (q.v.).
Properties: White, free-flowing powders; water-soluble; films are impervious to oxygen and other gases; films resist greases and solvents and range from water-soluble to water resistant, depending on grade.
Uses: Adhesives; finishing and sizing of textiles; protective coatings; films; paper coating; protective colloid in emulsion polymerization; molded and foamed products.

"Vinrez."[553] Trademark for vinyl acetate polymer and copolymer emulsions.

"Vinsol."[266] Trademark for a series of low-cost, dark, brittle, thermoplastic resins, ruby red by transmitted light, dark brown by reflected light. Available in solid form, flakes, fine powder, and aqueous dispersion.
Uses: Adhesives; emulsions; electrical insulation; inks; platics.

"Vinycol."[223] Trademark for pigments of titanium dioxide ("Vinycol" No. 100) or high-color carbon black ("Vinycol" No. 200) dispersed in vinyl resins and plasticizers. Used in high-gloss vinyl finishes.

vinyl. See vinyl compound.

"Vinylac."[65] Trademark for a series of resinous tackifier dispersions for modification of polyvinyl acetate adhesive systems.

vinyl acetate $CH_3COOCH:CH_2$. Raw material for polyvinyl resins (q.v.). See also polyvinyl acetal resins; polyvinyl acetate; polyvinyl alcohol.
Properties: Colorless liquid, stabilized with either hydroquinone or diphenylamine inhibitors. The HQ-stabilized material can be polymerized without redistillation. The DPA-stabilized material must be distilled prior to polymerization. Sp. gr. 0.9345 (20/20°C; f.p. −100.2°C; b.p. 73°C; refractive index 1.3941; wt/gal 7.79 lb; flash point 30°F (TOC). Soluble in most organic solvents including chlorinated solvents; insoluble in water. Autoignition temp. 800°F.
Derivation: (a) Vapor-phase reaction of ethylene and acetic acid, with a palladium catalyst. (b) Vapor-phase reaction of acetylene and acetic acid, with zinc acetate catalyst.
Grade: Technical.
Containers: 1-, 5-gal cans; 55-gal steel drums; tank cars.
Hazard: Flammable, dangerous fire risk. Flammable limits in air 2.6 to 13.4%. Toxic by inhalation and ingestion. Tolerance, 10 ppm in air. Safety data sheet available from Manufacturing Chemists Assn., Washington, D.C.
Uses: Polyvinyl acetate, polyvinyl alcohol, polyvinyl butyral, and polyvinyl chloride-acetate resins (q.v.). These are used particularly in latex paints; paper coating; adhesives; textile finishing; safety glass interlayers.
Shipping regulations: (Rail) Red label. (Air) Flammable Liquid label.

vinyl acetate-ethylene copolymer. See "Aquace."

vinylacetonitrile. See allyl cyanide.

vinylacetylene C_4H_4 or $H_2C:CHC:CH$. The dimer of acetylene, formed by passing it into a solution of cuprous and ammonium chlorides in hydrochloric acid.
Properties: Colorless gas or liquid; sp. gr. 0.6867 (0/20°C); b.p. 5°C. Combustible.
Use: Intermediate in manufacture of neoprene and for various organic syntheses.
See also divinyl acetylene.

vinyl alcohol (ethenol) $CH_2:CHOH$. Unstable liquid. Isolated only in form of its esters, or the polymer, polyvinyl alcohol (q.v.).

vinylation. The formation of a vinyl derivative by reaction with acetylene. Thus vinylation of alcohols yields vinyl ethers such as vinyl ethyl ether, $C_2H_5OC_2H_3$. See also Reppe processes.

vinylbenzene. See styrene.

vinyl bromide. CH_2CHBr.
Properties: Gas; f.p. −138°C; b.p. 15.6°C; sp. gr. 1.51.
Hazard: Moderately toxic. Tolerance, 250 ppm in air.
Uses: Flame-retarding agent for acrylic fibers.

vinyl n-butyl ether (n-butyl vinyl ether; BVE) $CH_2:CHOC_4H_9$.
Properties: Liquid; sp. gr. 0.7803 (20°C); b.p. 94.1°C; f.p. −113°C; refractive index 1.3997; flash point 15°F (open cup); wt/gal 7.45 lb (20°C). Slightly soluble in water; soluble in alcohol and ether.
Derivation: Reaction of acetylene with n-butyl alcohol.
Grade: Technical (98%).
Containers: Glass bottles; drums; tank cars.
Hazard: Flammable, dangerous fire risk.
Uses: Synthesis; copolymerization.
Shipping regulations: (Rail) Flammable liquid, n.o.s., Red label. (Air) Flammable Liquid label.

vinyl butyrate $CH_2:CHOOCC_3H_7$.
Properties: Liquid; sp. gr. 0.9022 (20/20°C); b.p. 116.7°C; f.p. −86.8°C. Very slightly soluble in water. Flash point 68°F (open cup). Low toxicity.
Containers: 55-gal drums.
Hazard: Flammable, dangerous fire risk.
Uses: Polymers; emulsion paints.
Shipping regulations: (Rail) Flammable liquid, n.o.s., Red label. (Air) Flammable Liquid label.

N-vinylcarbazole $C_{12}H_8NHC:CH_2$.
Properties: Liquid. Combustible; low toxicity.
Derivation: From acetylene and carbazole.
Use: Polymerizes to form heat-resistant and insulating resins somewhat similar to mica in dielectric properties. See polyvinyl carbazole.

vinyl cetyl ether. See cetyl vinyl ether.

vinyl chloride (VC; chloroethene; chloroethylene) $CH_2:CHCl$. The most important vinyl monomer.
Properties: Compressed gas; easily liquefied; ethereal odor. Usually handled as liquid; phenol is added as a polymerization inhibitor. Sp. gr. 0.9121 (liquid, at 20/20°C); b.p. −13.9°C; f.p. −159.7°C; vapor pressure 2300 mm (20°C); flash point −108°F. Autoignition temp. 882°F. Slightly soluble in water; soluble in alcohol and ether.
Derivation: (a) Oxychlorination of ethylene; (b) reaction of acetylene and hydrochloric acid, either as liquids or gases; (c) pyrolysis of ethylene dichloride (see Wulff process). Method of purification: Distillation.
Grades: Technical; pure, 99.9%.
Containers: Cylinders; tank cars; ocean tankers.
Hazard: Highly flammable; severe explosion risk at 30,000 ppm. Probably carcinogenic. Tolerance for workroom exposure established by OSHA is 1 ppm (average over an 8-hr period); Exposure times up to 15 minutes are permissible up to 5 ppm. Respirators must be used in cases where this concentration is exceeded. Use in aerosol sprays prohibited. Safety Data Sheet available from Manufacturing Chemists Assoc., Washington, D.C.
Uses: Polyvinyl chloride and copolymers; organic synthesis; adhesives for plastics.
Shipping regulations: (Rail) Red Gas label. (Air) (inhibited) Flammable Gas label. Not acceptable on passenger planes; (uninhibited) Not acceptable.

vinyl 2-chloroethyl ether $CH_2:CHOCH_2CH_2Cl$.
Properties: Liquid; sp. gr. 1.0498 (20°C); b.p. 109.1°C; f.p. −69.7°C; very slightly soluble in water; flash point (open cup) 80°F.
Hazard: Flammable, dangerous fire risk.
Use: Polymerization monomer.

Shipping regulations: (Rail) Flammable liquid, n.o.s., Red label. (Air) Flammable Liquid label.

vinyl compound. A compound having the vinyl grouping ($CH_2{=}CH{-}$), specifically vinyl chloride, vinyl acetate and similar esters, but also referring more generally to compounds such as styrene $C_6H_5CH{:}CH_2$, methyl methacrylate $CH_2{:}C(CH_3)COOCH_3$ and acrylonitrile $CH_2{:}CHCN$. The vinyl compounds are highly reactive, polymerize easily, and are the basis of a number of important plastics.

vinyl cyanide. See acrylonitrile.

vinylcyclohexene (1-vinylcyclohexene-3; 4-vinylcyclohexene-1; cyclohexenylethylene)

$CH_2{:}CHCHCH_2CH{:}CHCH_2CH_2$. A butadiene dimer.

Properties: Liquid; sp. gr. 0.8303 (20/4°C); f.p. −108.9°C; b.p. 128°C; refractive index 1.464 (n 20/D); flash point 70°F (TOC). Autoignition temp. 517°F. Temperatures above 80°F and prolonged exposure to oxygen-containing gases should be avoided as these conditions lead to discoloration and gum formation.

Grades: Technical, 95%; pure, 99%; research.

Containers: 1-, 5-gal cans; 55-gal drums; tank trucks; tank cars.

Hazard: Flammable, dangerous fire risk. May be narcotic in high concentrations.

Uses: Polymers; organic synthesis.

Shipping regulations: (Rail) Flammable liquid, n.o.s., Red label. (Air) Flammable Liquid label.

vinylcyclohexene dioxide $CH_2CHOC_6H_9O$. Also called vinylcyclohexane dioxide.

Properties: Colorless liquid; sp. gr. 1.098 (20/20°C); b.p. 227°C; refractive index 1.4782 (n 20/D); viscosity 7.77 cp (20°C); flash point 230°F. Combustible.

Hazard: Strong irritant to skin and tissue. Moderately toxic by ingestion and skin absorption.

Uses: Polymers; organic synthesis.

vinylcyclohexene monoxide $CH_2{:}CHC_6H_9O$. Also called vinylcyclohexane monoxide.

Properties: Liquid; sp. gr. 0.9598 (20/20°C); b.p. 169°C; f.p. −100°C; flash point 136°F; viscosity 1.69 cp (20°C). Very slightly soluble in water. Combustible.

Hazard: Moderately toxic and irritant. Moderate fire risk.

Uses: Polymers; organic synthesis.

vinyl ether (divinyl ether; divinyl oxide)
$CH_2{:}CHOCH{:}CH_2$.

Properties: Colorless liquid with characteristic odor; sp. gr. 0.769; b.p. 39°C; refractive index (20/D) 1.3989. Slightly soluble in water; miscible with alcohol, acetone, chloroform, and ether. Must be protected from light. Flash point below −22°F; autoignition temp. 680°F.

Derivation: Treatment of dichloroethyl ether with alkali.

Grade: N.F. (contains 96–97% vinyl ether, remainder dehydrated alcohol).

Containers: 50- and 75-cc bottles.

Hazard: Highly flammable; severe fire and explosion risk. Explosive limits in air 1.7 to 27%. Highly toxic by inhalation; overexposure may be fatal.

Uses: Copolymer with 3 to 5% polyvinyl chloride for plastic products such as clear plastic bottles; medicine (anesthetic, for brief operations only).

Shipping regulations: (Rail) Flammable liquid, n.o.s., Red label. (Air) Flammable Liquid label. Legal label name (Air), divinyl ether.

vinyl beta-ethoxyethyl sulfide
$CH_2{:}CHSCH_2CH_2OC_2H_5$.

Properties: Colorless liquid. Pungent, camphor-like odor. Sp. gr. 0.9532 (15°C); b.p. 65°C (8 mm).

Use: Organic synthesis.

vinylethylene. See butadiene.

vinyl ethyl ether (ethyl vinyl ether; EVE)
$CH_2{:}CHOC_2H_5$.

Properties: Colorless liquid. Extremely reactive; can be polymerized in liquid or vapor phase; slightly soluble in water: 0.9% by wt; sp. gr. 0.754 (20/20°C); 6.28 lb/gal (20°C); b.p. 35.5°C; vapor pressure 428 mm (20°C); f.p. −115.0°C; viscosity 0.22° cp (20.0°C); refractive index 1.3739; flash point −50°F. Autoignition temp. 395°F. Commercial material contains inhibitor to prevent premature polymerization. Often stored underground to minimize vapor losses.

Derivation: Reaction of acetylene with ethyl alcohol.

Grade: Technical.

Containers: Drums; tank cars.

Hazard: Highly flammable; severe fire and explosion risk. Explosive limits in air 1.7 to 28%.

Uses: Copolymerization; intermediate.

Shipping regulations: (Rail) Flammable liquid, n.o.s., Red label. (Air) (Inhibited) Flammable Liquid label. (Uninhibited) Not acceptable.

vinyl 2-ethylhexoate $CH_2{:}CHOOCCH(C_2H_5)C_4H_9$.

Properties: Liquid; sp. gr. 0.8751 (20/20°C); b.p. 185.2°C; f.p. −90°C; flash point 165°F (O.C.). Insoluble in water. Combustible.

Containers: 55-gal drums.

Uses: Polymers; emulsion paints.

vinyl 2-ethylhexyl ether $CH_2{:}CHOCH_2CH(C_2H_5)C_4H_9$.

Properties: Liquid; sp. gr. 0.8102 (20/20°C); b.p. 177.7°C; f.p. −100°C; flash point 135°F (open cup). Autoignition temp. 395°F. Insoluble in water. Combustible.

Containers: 55-gal drums.

Hazard: Moderate fire and explosion risk.

Uses: Intermediate for pharmaceuticals, insecticides, adhesives; viscosity index improver.

2-vinyl-5-ethylpyridine $(CH_2{:}CH)C_5H_3N(C_2H_5)$.

Properties: Liquid. Sp. gr. (20/20°C) 0.9449; b.p. (100 mm) 138°C; vapor pressure (20°C) 0.2 mm; f.p. −50.9°C; insoluble in water; flash point 200°F (COC). Combustible.

Hazard: Probably toxic.

Uses: Copolymer; synthesis.

vinyl fluoride (fluoroethylene) $CH_2{:}CHF$.

Properties: Colorless gas; b.p. −72°C; insoluble in water; soluble in alcohol and ether.

Containers: Cylinders.

Hazard: Flammable, dangerous fire and explosion risk. Probably toxic by inhalation. Tolerance (as F), 2.5 mg per cubic meter of air.

Use: Monomer. See polyvinyl fluoride.

Shipping regulations: (Inhibited) (Rail) Red gas label. (Air) Flammable Gas label. Not acceptable on passenger planes. (Uninhibited) Not acceptable.

vinylidene chloride (VC) $CH_2:CCl_2$.
Properties: Colorless liquid; f.p. −122.53°C; b.p. 37°C; flash point 14°F (open cup); Insoluble in water. Autoignition point 856°F. Readily polymerizes; commercially contains small proportion of inhibitor.
Hazard: Flammable, dangerous fire risk. Explosive limits in air 5.6 to 11.4%.
Uses: Copolymerized with vinyl chloride or acrylonitrile to form various kinds of saran. Other copolymers are also made. Adhesives; component of synthetic fibers. See also saran.
Shipping regulations: (Inhibited) (Rail) Red label. (Air) Flammable Liquid label. Uninhibited Not acceptable.

vinylidene fluoride (1,1-difluoroethylene) $H_2C:CF_2$. A monomer.
Properties: Colorless gas with faint ethereal odor; b.p. −83°C (1 atm); f.p. −144°C (1 atm); density (liquid) 0.617 g/cc (24°C). Slightly soluble in water; soluble in alcohol and ether.
Derivation: Interaction of hydrogen with dichlorodifluoroethane.
Grade: 99% min purity.
Containers: Steel cylinders.
Hazard: Flammable, dangerous fire risk. Toxic by inhalation. Explosive limits in air 5.5 to 21%. Tolerance (as F), 2.5 mg per cubic meter of air.
Uses: Polymers and copolymers; chemical intermediate. See polyvinylidene fluoride.
Shipping regulations: (Rail) Red Gas label. (Air) Flammable Gas label. Not acceptable on passenger planes. Legal label name (Air) difluoroethylene.

vinylidene resin (polyvinylidene resin). A polymer in which the unit structure is ($—H_2CCX_2—$), in which X is usually chlorine, fluorine, or cyanide radical. Examples are saran and "Viton" A.

vinyl isobutyl ether (isobutyl vinyl ether; IVE) $CH_2:CHOCH_2CH(CH_3)_2$.
Properties: Colorless liquid; sp. gr. 0.7706 (20/20°C); b.p. 83.3°C; vapor pressure 68 mm (20°C); f.p. −132°C; refractive index 1.3938; flash point 15°F (open cup). Very slightly soluble in water; soluble in alcohol and ether; easily polymerized.
Derivation: Catalytic union of acetylene and isobutyl alcohol.
Method of purification: Washing with water, drying in presence of alkali, and distillation from metallic sodium.
Grade: Technical.
Containers: Drums; tank cars.
Hazard: Flammable, dangerous fire risk.
Uses: Polymer and copolymers used in surgical adhesives, coatings and lacquers; modifier for alkyd and polystyrene resins; plasticizer for nitrocellulose and other plastics; chemical intermediate.
Shipping regulations: (Rail) Flammable liquid, n.o.s., Red label. (Air) (Inhibited) Flammable Liquid label. (Uninhibited) Not acceptable.

vinylmagnesium chloride $CH_2:CHMgCl$, dissolved in tetrahydrofuran. Available in up to 55-gal drums. Used as a Grignard reagent.

vinyl methyl ether (methyl vinyl ether; methoxyethylene; MVE) $CH_2:CHOCH_3$.
Properties: Colorless, compressed gas, or colorless liquid; sp. gr. 0.7500 (20/20°C); b.p. 6.0°C; vapor pressure 1052 mm (20°C); flash point −60°F; f.p. −121.6°C. Slightly soluble in water; soluble in alcohol and ether; easily polymerized; commercial material contains a small proportion of polymerization inhibitor.
Derivation: Catalytic reaction of acetylene and methyl alcohol.
Grades: Technical (95% min); pure.
Containers: 150-lb cylinders; single-unit tank cars and ton multi-unit (TMU) tank cars.
Hazard: Highly flammable; severe fire and explosion risk. Explosive limits in air 2.6 to 39%.
Uses: Copolymers used in coatings and lacquers; modifier for alkyl, polystyrene and ionomer resins; plasticizer for nitrocellulose and adhesives. See polyvinyl methyl ether.
Shipping regulations: (Inhibited) (Rail) Red gas label. (Air) Flammable Gas label. Not acceptable on passenger planes. (Uninhibited) Not acceptable.

vinyl methyl ketone (3-buten-2-one; methyl vinyl ketone) $CH_3COCH:CH_2$.
Properties: Colorless liquid; sp. gr. 0.8636 (20/4°C); b.p. 80°C; miscible with water; soluble in alcohol. Flash point 20°F (closed cup).
Hazard: Flammable, dangerous fire risk. Skin and eye irritant.
Use: Monomer for polymerization; component of ionomer resins.
Shipping regulations: (Rail) Red label. (Air) (Inhibited) Flammable Liquid label. (Uninhibited) Not acceptable. Legal label name methyl vinyl ketone.

vinyl plastics. See polyvinyl resins.

vinyl propionate $CH_2:CHOOCC_2H_5$.
Properties: Liquid; sp. gr. 0.9173 (20/20°C); b.p. 95.0°C; f.p. −81.1°C; flash point 34°F (open cup). Almost insoluble in water. Low toxicity.
Containers: 55-gal drums; tank cars.
Hazard: Flammable, dangerous fire risk.
Uses: Polymer; emulsion paints.
Shipping regulations: (Rail) Flammable liquid, n.o.s., Red label. (Air) Flammable Liquid label.

vinylpyridine $C_5H_4NCH:CH_2$.
Properties: Colorless liquid, boils with resinification at about 159°C; sp. gr. 0.9746 (20°C); refractive index 1.5509 (n 20/D). Dissolves in water to extent of 2.5%; water dissolves in it to about 15%; soluble in dilute acids, hydrocarbons, alcohols, ketones, esters. Commercial material contains inhibitor. Combustible.
Containers: Drums; tank cars.
Hazard: Probably toxic and irritant to skin and eyes.
Uses: Elastomers and pharmaceuticals. Latex used as tire cord binder (41% solids). The latex is copolymerized with butadiene-styrene.

N-vinyl-2-pyrrolidone $CH_2:CH\overline{NCH_2CH_2CH_2C}O$.
Properties: Colorless liquid; b.p. 148°C (100 mm); m.p. 13.5°C; flash point 209°F (COC); sp. gr. 1.04. Combustible.
Derivation: From acetylene and formaldehyde by high-pressure synthesis.
Hazard: Probably irritant and narcotic in high concentrations.
Uses: Polyvinylpyrrolidone; organic synthesis.

vinyl stabilizer. A substance added to vinyl chloride resins during compounding, to retard the rate of deterioration due to formation of hydrogen chloride. Many combine readily with hydrogen chloride but do not otherwise interfere with the properties and uses of the final product. Amines, basic oxides, and metallic soaps are commonly used.

vinyl stearate $CH_3(CH_2)_{16}COOCH{:}CH_2$.
Properties: White, waxy solid. M.p. 28–30°C; b.p. 175°C (3 mm); sp. gr. 0.9037 (20/20); refractive index 1.4355–1.4362 (n 55/D); iodine no. 80–82. Insoluble in water and alcohol; moderately soluble in ketones and vegetable oils; soluble in most hydrocarbon and chlorinated solvents. Combustible. Low toxicity.
Containers: 35-, 210-, and 390-lb drums.
Uses: Plasticizer (copolymerizer); lubricant.

vinylstyrene. See divinyl benzene.

vinyltoluene (methyl styrene) $CH_2{:}CHC_6H_4CH_3$.
Properties: Colorless liquid; f.p. −76.8°C; b.p. 170–171°C; sp. gr. 0.890 (25/25°C); lb/gal 7.41; refractive index 1.534 (n 34/D); flash point 130°F (C.C.). Autoignition temp. 921°F. Very slightly soluble in water. Soluble in methanol, ether. Combustible.
Containers: Drums; tank cars.
Hazard: Moderate fire risk. Moderately toxic. Tolerance, 100 ppm in air.
Uses: Solvent; intermediate.

vinyl trichloride. See 1,1,2-trichloroethane.

vinyltrichlorosilane $CH_2CHSiCl_3$.
Properties: Colorless or pale yellow liquid. B.p. 90.6°C; sp. gr. 1.265 (25/25°C); refractive index (n 20/D) 1.432; flash point 16°F. Readily hydrolyzed, with liberation of hydrochloric acid; polymerizes easily. Soluble in most organic solvents; reacts with alcohol.
Derivation: Reaction of acetylene and trichlorosilane (peroxide catalyst); reaction of trichlorosilane with vinyl chloride.
Grade: Technical.
Hazard: Flammable, dangerous fire risk. Highly toxic by ingestion and inhalation. Strong irritant to tissue.
Uses: Intermediate for silicones; coupling agent in adhesives and bonds.
Shipping regulations: (Rail) Red label. (Air) Flammable Liquid label. Not acceptable on passenger planes.

"Vinymul."[170] Trademark for a textile finisher for all fabrics or fibers where flexibilty is desired at low temperatures, such as back-coating or immersion finishing of automobile fabrics. Also as primer with adhesives on vinyl pressure sensitive tapes or as a prime coat on metal for adhesion to metal and improved adhesion for second protective coating.

vinyon. Generic name for a manufactured fiber in which the fiber-forming substance is any long chain synthetic polymer composed of at least 85% by weight of vinyl chloride units, $—CH_2CHCl—$ (Federal Trade Commission). It has good resistance to chemicals, bacteria, moths; is unaffected by water and sunlight; low softening point. Tenacity 3.1 grams per denier; difficult to ignite; self-extinguishing. See polyvinyl chloride.

"Vinyzene."[8] Trademark for series of fungicides and bactericides formulated as additives to vinyls. Typical formulation is a condensation product of epoxidized soybean oil and 10,10′-oxybisphenoxarsine, functioning as both primary plasticizer-stabilizer and biocide.
Hazard: Probably toxic by ingestion and inhalation.

"Viocin."[299] Trademark for viomycin sulfate.

"Vioform."[305] Trademark for iodochlorhydroxyquin.

"Violamine."[307] Trademark for a series of acid dyestuffs. Used on wool and silk. Fairly good fastness to light, washing, fulling, etc. Can also be used on paper and leather.

violanthrone. See dibenzanthrone.

Violet No. 1. An FD&C color used for meat grading, cosmetics, beverages, etc. It is a triphenylmethane dye. Banned by the FDA in 1973 due to possible carcinogenic risk.

violet gentian. See methyl violet.

"Violite."[4] Trademark for a series of fluorescent and phosphorescent pigments, based on such phosphors as zinc sulfide-cadmium sulfide and cadmium sulfide-strontium sulfide.
Uses: Printing inks, paints, lacquers, watercolors, paper, marking material, etc.

viomycin. An antibiotic produced by Streptomyces puniceus. Unique among antibiotics in that it is more active against acid-fast organisms than against other groups. Mycobacteria are most sensitive to viomycin, and the antibiotic is active against strains of Mycobacterium tuberculosis which are resistant to other antibiotics. Available commercially as sulfate.

"Vircol."[40] Trademark for neutral phosphorus polyols, used as flame retardant for urethane foams and plasticizer for polyvinyl alcohol.

"Virco-Pet."[40] Trademark for neutral organic phosphorus compounds for use as a corrosion inhibitor for steel and aluminum. Tan, viscous liquids, miscible with a variety or organic solvents; sp. gr. 0.96–1.06 (20°C).

"Virgin Rock."[1] Trademark for a 99.8% pure sulfur cast in solid barrels.
Uses: For applications in which a clean, pure sulfur is needed for melting, or for burning to form sulfur dioxide (fumigation, sugar and starch refining, bleaching of wool and straw, "sulfuring" of hops and grouting of machinery foundations).

"Virgo" Salts.[62] Trademark for salts for fused salt baths, for descaling, desanding, and otherwise cleaning metal products.

"Viridine."[227] Trademark for phenylacetaldehyde dimethylacetal (q.v.).

virus. An infectious agent composed almost entirely of protein and nucleic acids (nucleoprotein, q.v.). Viruses can reproduce only within living cells, and are so small they they can be resolved only with an electron microscope. Since they pass through filters that retain bacteria, they are often called filtrable viruses. Tobacco mosaic was the first virus to be crystallized and isolated (Dr. W. M. Stanley, 1935); it contains some 2000 protein molecules in a sequence of 158 amino acids (m. w. 40,000,000). Bushy stunt virus, found in tomato plants, has a m.w. of 7,600,000. First synthesis of a virus was reported in 1967.

"Viruses differ from organisms in that they are only half alive; they lack metabolism, are unable to utilize oxygen, to synthesize macromolecules, to grow or to die. They are parasites, relying on a living host cell." They account for many diseases, including mumps, measles, scarlet fever, smallpox, influenza, and possibly the common cold. Their shapes are similar to those of bacteria (rods, spheres, filaments). They have the ability to mutate; they are also antigenic and thus initiate formation of antibodies. Some act as bacteriophages (q.v.). A direct relation between virus and cancer has been shown, the DNA of the virus becoming irreversibly bound to the DNA of the affected cells. See also bacteria; deoxyribonucleic acid.

"Viscarin."[124] Trademark for carrageenan extractives from Irish moss. Dry, free-flowing powder packed in fiber drums.
Uses: Gelling, suspending, binding and viscosity-building in food products.

"Viscasil."[245] Trademark for a series of polish grade dimethyl silicone fluids. Available in viscosities from 50 to 1,000 centistokes at 25°C. Are not compatible with waxes and act as lubricants, producing a polish which is more easily rubbed out and gives a higher gloss.

"Viscolan."[493] Trademark for liquid lanolin consisting of low melting point esters isolated by physical fractionation. A crystal-clear amber viscous liquid; anhydrous; soluble in mineral and vegetable oils.
Uses: Nonionic w/o emulsifier; pigment dispersant; plasticier; emollient.

viscose process. The best-known process for making regenerated cellulose (rayon) by converting cellulose to the soluble xanthate, which can be spun into fibers and then reconverted to cellulose by treatment with acid. Wood pulp is steeped with 17–20% caustic soda; the resulting alkali cellulose is pressed to remove excess liquor and the soluble beta- and gamma-cellulose, and then shredded and aged. It is then treated with carbon disulfide and sodium hydroxide to form an orange, viscous solution of cellulose xanthate. After filtration and deaeration, this solution (viscose) is forced through minute spinneret openings (or long slit dies in the case of cellophane) into a bath containing sulfuric acid and various salts such as sodium and zinc sulfate. The salts cause the viscose to gel immediately, forming a fiber or film of sufficient strength to permit it to be drawn through the bath under tension. At the same time the sulfuric acid decomposes the xanthate, converting the fibers to cellulose, in which form they are washed and dried. See also rayon; cellophane.

viscosity. The internal resistance to flow exhibited by a fluid. A liquid has a viscosity of one poise if a force of one dyne per square centimeter causes two parallel liquid surfaces one square centimeter in area and one centimeter apart to move past one another at a velocity of one centimeter per second. One poise equals 100 centipoises (cp). Viscosity in centipoises divided by the liquid density at the same temperature gives kinematic viscosity in centistokes (cs). One hundred centistokes equal 1 stoke. To determine kinematic viscosity, the time is measured for an exact quantity of liquid to flow by gravity through a standard capillary.

Water is the primary viscosity standard with an accepted viscosity at 20°C of 0.01002 poise. Hydrocarbon liquids such as hexane are less viscous. Molasses may have a viscosity of several hundred centistokes, while for a very heavy lubrication oil the viscosity may be 100 centistokes. There are many empirical methods for measuring viscosity which generally involve measurement of the time of flow or movement of a ball, ring or other object in a specially shaped or sized apparatus. See Saybolt Universal Viscosity; liquid; liquid, Newtonian.

viscosity index improver. A lubricating oil additive that has the effect of increasing the viscosity of the oil in such a way that it is greater at high temperature than at low temperature. Agents used for this purpose are polymers of alkyl esters of methacrylic acid, polyisobutylenes, etc.

viscosity, kinematic. See viscosity.

"Viscotone."[271] Trademark for a mineral by-product used for oil-well drilling muds, tanning agents, ore dressing, dispersants and emulsifiers.

"VisKing."[313] Trademark for plastic film and sheets for wrapping, packaging and containers; rods, bars and tubing.

"Vistac."[230] Trademark for a series of synthetic hydrocarbon polymers used in rubber-base adhesives and cements, waxes and asphalts, paints and caulking compounds, and in latex and asphalt emulsions.

"Vistalon." Trademark for an ethylene-propylene rubber, claimed to have excellent resistance to ozone and weathering, heat resistance, low-temperature flexibility and resistance to acids and detergents.
Uses: Automotive appliances; industrial hose; cable insulation; electrical equipment.

"Vistanex."[29] Trademark for polyisobutylene (q.v.) a thermoplastic saturated polymer unvulcanizable by itself, but frequently added to impart special properties to other elastomers. Also used in unvulcanized compositions, alone and in combinations. Soluble in hydrocarbon solvents.
Uses: Cable insulation compounds; pressure-sensitive tapes and adhesives; caulking and sealing compounds; additives to wax, asphalt, and polyethylene; viscosity index improver.

visual purple. See rhodopsin.

vitamin. Any of a number of complex organic compounds, present in natural products or made synthetically, which are essential in small proportions in the diet of animals and man. Some are fat-soluble (A, D, K); others are water-soluble (B complex, C). Their precursors are called provitamins (q.v.). A normal diet usually contains sufficient vitamins for health; their use in medicine is restricted to correction of specific metabolic deficiencies. Some authorities believe that habitual intake of standard vitamin preparations readily available on the market is of little if any nutritional benefit. The following list of cross-references will serve to locate technical information about the various vitamin entries in this book:

Vitamin A. See carotene, cryptoxanthin, retinol, 3-dehydroretinol; provitamin.

Vitamin B. See Vitamin B complex; also thiamine, riboflavin, niacin, pantothenic acid, biotin, cobalamin, pyridoxine, folic acid, inositol.

Vitamin C. See ascorbic acid.

Vitamin D. See ergosterol, ergocalciferol, cholecalciferol.

Vitamin E. See tocopherol.

Vitamin K. See phytonadione, menadione, phthiocol.

vitamin B complex. A group of closely interrelated vitamins found in rice bran, yeast, wheat germ, etc., originally thought to be one substance. Studies carried out by R. J. Williams and associates later revealed the astonishing complexity of this group. He states as follows: "The physiological activity originally observed was due to the *additive* effect of a considerable number of substances, *each one* of which is of itself essential. If and when the designations B_1, B_2, B_3, etc., are used, they have an entirely different meaning from the parallel use of D_1, D_2, D_3, or K_1 and K_2, because in the case of the D and K vitamins one form can replace another. In the case of the B vitamins each form is a distinctly different substance with different functions, and each member of the family is separately indispensable. No one B vitamin can replace any other."

"Vitel."[265] Trademark for a series of polyester resins; for spinning into polyester staple and high tenacity fibers for wearing apparel, industrial fabrics, tire cord; for extrusion of high strength monofilaments and film; for extrusion coating on paper, textiles, foils, plastics; also used as hot melt resins.

"Vithane."[265] Trademark for a cured or curable polyurethane coating resin of high molecular weight.
Uses: Coatings for tent fabrics, ponchos, fire hose, industrial belting, magnetic recording tape, balloons and other air-filled structures.

"Viton."[28] Trademark for a series of fluoroelastomers based on the copolymer of vinylidene fluoride and hexafluoropropylene, with the repeating structure possibly $—CF_2—CH_2—CF_2—CF(CF_3)—$.
Properties: White transparent solid; sp. gr. 1.72–1.86. Resistant to corrosive liquids and chemicals up to 600°F. Useful continuous service at 400–450°F. Resistant to ozone, weather, flame, oils, fuels, lubricants, many solvents. Radiation resistance good. Nonflammable.
Uses: Gaskets, seals, diaphragms, tubing; aerospace and automotive components; high vacuum equipment; low-temperature and radiation equipment.

"Vitrafos."[1] Trademark for a clear, glassy, granular sodium phosphate.
Properties: Phoshorus pentoxide, 63% min; sieve size, 10- to 80-mesh screen; pH (1% soln), 7.8; bulk density 79 lb/cu ft.
Uses: Detergent builder; deflocculating agent in oil-well drilling muds; sequestering agent in the textile industry.

"Vitresil."[220] Trademark for a high-quality natural silica, fused into a pure and uniform woven fabric containing over 99.8% SiO_2. Available in translucent and transparent forms. Used as filter medium.

vitreous. Descriptive of a material having the appearance and properties of a glass, i.e., a hard, amorphous, brittle structure, as in porcelain enamel. See vitrification; glass.

vitreous enamel. See porcelain enamel.

"Vitrex."[41] Trademark for an acid-proof silicate cement which sets by chemical action. Inert to acids, except hydrofluoric, up to 2100°F.

"Vitric 10."[326] Trademark for a silicate base acid proof, chemical setting cement suitable for use with strong oxidizing acids such as nitric and chromic.

vitrification. The process of converting a siliceous material into glass or a glassy substance by heat, fusion, and slow cooling.

vitriol. An obsolete term once used to refer to a number of sulfates (lead, copper, zinc) because of their glassy appearance. Sulfuric acid was called oil of vitriol. Derived from vitrum (glass).

"Vitrobond."[41] Trademark for a plasticized sulfur cement, resistant to nonoxidizing acids, oxidizing acids, acid and neutral salts, up to 200°F.

"Vitroplast."[41] Trademark for a polyester resin cement, resistant to nonoxidizing acids, acid and neutral salts, some organic solvents and weak alkalies, up to 275°F.

"Vituf."[265] Trademark for polyester extrusion resins designed for jacket applications over electrical wire insulations as a replacement for nylon over vinyl and polyethylene.
Uses: Aircraft, auto and electronic hook-up wire, building wire, missile ground cable.

"Vivana."[565] Trademark for fibers, filaments and yarn of polymers and interpolymers of vinylidene chloride.

"Vix-Syn."[320] Trademark for synthetic rubber products. Made of duck or asbestos with neoprene or butyl polymers.

"Viz-Thin."[48] Trademark for a ferro-chrome lignin sulfonate, used as a thinner for gypsum base oil well drilling muds.

V.M.&P. naphtha. See under naphtha.

voids. Empty spaces of molecular dimensions occurring between closely packed solid particles, as in powder metallurgy. Their presence permits barriers made by powder metallurgy techniques to act as diffusion membranes for separation of uranium isotopes in the gaseous diffusion process. See diffusion, gaseous.

"Volan."[28] Trademark for a Werner type chromium complex (methacrylatochromic chloride, q.v.) in isopropanol.
Used as a bonding agent for glass fibers, paper, wood, and polymeric coatings.

volatility. The tendency of a solid or liquid material to pass into the vapor state at a given temperature. Specifically the vapor pressure of a component divided by its mole fraction in the liquid or solid.

"Volck."[253] Trademark for refined petroleum oil used as a phytonomic contact insecticide.

Volhard's solution. A solution of potassium thiocyanate used in analytical chemistry.

voltaic cell. Two conductive metals of different potentials in contact with an electrolyte which generate an electric current. The original voltaic cell was comprised of silver and zinc, with brine-moistened paper as electrolyte. Semisolid pastes are now used; electrodes may be lead, nickel, zinc or cadmium.

volumetric analysis. See titration.

"Voranate."[233] Trademark for a series of uretnane intermediates which are the reaction products of polyols and isocyanates. They are adducts or quasi-prepolymers to be used in combination with "Voranol" products to obtain rigid urethane foams.

"Vorane."[233] Trademark for a group of polyurethane chemicals, raw materials for polyurethane elastomers, coatings and foams.

"Voranol."[233] Trademark for a series of polyols used as intermediates in urethane elastomers, coatings, flexible and rigid foams.

"Voraspan."[233] Trademark for urethane resins.

"Vorite."[202] Trademark for a series of ricinoleate urethane prepolymers for use in protective coatings, foams, elastomers and adhesives. Cured with polyols (see "Polycin") or catalyst.

VPC. Abbreviation for vapor-phase chromatography. See gas chromatography.

"VPM."[28] Trademark for a solution containing 32.7% sodium methyldithiocarbamate used as a soil fumigant.
Containers: 1-gal bottles; 5- and 30-gal drums.

"V-S-P."[160] Trademark for thermoplastic voltage-stabilized polyethylene; used in power cables.

"Vulca."[53] Trademark for a series of ether derivatives of ungelatinized starch; graded by resistance to swelling in boiling water. "Vulca" 100 is completely nonswelling, resistant to agents that gelatinize starch, retains stability during autoclaving and pressure cooking, and may be steam sterilized without any appreciable change in its powdery appearance.
Uses: Thickeners; cosmetics; dusting powders.

"Vulcalock."[119] Trademark for rubber-based adhesive especially designed for bonding rubber to metals.

"Vulcan."[275] Trademark for oil furnace carbon blacks for rubber and plastics.

vulcanization. A physicochemical change resulting from cross-linking of the unsaturated hydrocarbon chain of polyisoprene (rubber) with sulfur, usually with application of heat. The precise mechanism which produces the network structure of the cross-linked molecules is still incompletely known. Sulfur is also used with unsaturated types of synthetic rubbers; some types require use of peroxides, metallic oxides, chlorinated quinones, or nitrobenzenes. Natural rubber can be vulcanized with selenium, organic peroxides and quinone derivatives, but these have limited industrial use; high-energy radiation curing is an important innovation.

Vulcanization can be effected with sulfur alone in high percentage, but the time required is too long to be economic and the properties obtained are poor. Inorganic accelerators and metallic oxides (usually zinc) are essential for satisfactory cure. Organic accelerators (q.v.), introduced in the early 1920's, notably shortened vulcanization time.

Three factors affect the properties of a vulcanizate: (1) the percentage of sulfur and accelerator used; (2) the temperature; and (3) time of cure. Sulfur percentage is usually from 1 to 3%; with strong acceleration the time can be as short as three minutes at high temperature (150°C). Vulcanization can also occur at room temperature with specific formulations (self-curing cements).

Vulcanization was discovered in 1846 by Charles Goodyear in the U.S. and simultaneously by Thomas Hancock in England. Its over-all effect is to convert rubber hydrocarbon from a soft, tacky, thermoplastic to a strong, temperature-stable thermoset having unique elastic modulus and yield properties. See also rubber; rubber, synthetic.

"Vulcoid."[281] Trademark for a resin-impregnated vulcanized fiber. A light-weight electrical insulator of great mechanical adaptability which has arc and moisture resistance and can be drawn and formed. Approved by Underwriters' Laboratories as Class A Insulation.
Forms: Sheets; tubes; rods; fabricated parts.
Uses: Contact and instrument panels; knife switch guides; arc deflector spacer bushings; electrical insulation; motor and transformer lead bushings and terminal blocks.

"Vulcosal."[233] Trademark for the industrial grade of salicylic acid; used as a stabilizer and retarder of vulcanization.

"Vulklor."[248] Trademark for tetrachloro-para-benzoquinone; used as a vulcanizing agent.

"Vulkollan."[470] Trademark for urethane elastomers. See polyurethane.

"Vultac."[204] Trademark for a series of vulcanizing agents composed chiefly of alkyl phenol disulfides.

"Vultex."[572] Trademark for a vulcanized rubber or synthetic rubber latex. Used especially for surgeons' gloves, drug sundries, dipped products, adhesives.

"Vycor" Brand Glass No. 7900.[20] Trademark for a glass made by a process in which an article fabricated by conventional methods is chemically leached to remove substantially all of the ingredients except silica. When fired at high temperatures a transparent glass of high softening point and extremely low expansion coefficient is produced.
Physical properties: Softening point approximately 1500°C. Temperature limit in service, 900°C. Linear coefficient of expansion per °C, 0.0000008; sp. gr. 2.18; refractive index 1.458. (A similar glass, No. 7910, will transmit over 60% radiation at 254 millimicrons in 2 mm section.)
Uses: Laboratory and industrial glassware, including beakers, crucibles, flasks, dishes, tubes, cylinders, containers, flat glass rods.

"Vycron."[287] Trademark for a polyester fiber (q.v.).

"Vygen."[179] Trademark for a complete series of polyvinyl chloride polymers.
Uses: Film, sheeting, coated fabrics, luggage, wall covering, extruded goods, gaskets, wire insulation, toys, and automotive parts.

"Vynex."[482] Trademark for an extruded rigid vinyl sheet with exceptional clarity and gloss finish. Available in sheet or roll form for a wide range of applications such as transparent film, book bindings, dictating records, identification cards, etc.

"Vyn-Eze."[430] Trademark for a specially prepared fatty acid amide used as an internal mold release agent in PVC molded products.

"Vynolour."[438] Trademark for pigments dispersed on vinyl copolymer for use in vinyl chloride and vinyl chloride-vinylidene chloride types for inks, calendering solution resin coatings, extrusion and molding.

"Vyron."[41] Trademark for a porous plastic of high density polyethylene of the Ziegler type. Its permeability is uniform at 50 microns, permitting retention of particles down to 25 microns. Used for filtering, pneumatic silencing and fluidizing applications.

W Symbol for tungsten.

"W-545."[308] Trademark for an austenitic iron-base precipitation-hardening alloy containing nickel, chromium, and small proportions of molybdenum, titanium, boron, silicon, and manganese. Developed primarily for improved gas turbine discs, critical components of jet engines.

"Wallpol."[36] Trademark for colloidal dispersions of solid polyvinyl acetate resin particles in water. Used in interior and exterior paints and primer sealers.

"Warcobase A."[42] Trademark for a mixed stearylurea (60% mono, 40% di-stearyl); off-white granules; m.p. 107–110°C.
Uses: Antistatic agent in plastics; chemical intermediate.

"Warcocide."[42] Trademark for a series of germicides.
1400. Methylnaphthyl-dodecyl-dimethyl-ammonium chloride.
1450. A 50% solution in aqueous isopropanol of tetradecylbenzyl-dimethylammonium chloride monohydrate. Used in the paper industry.
TMB. Trimethylbenzyl-dodecyl-dimethyl ammonium chloride.

"Warcoflex."[42] Trademark for a three-component polyurethane system: BR-233, base resin; CL-239, crosslinking agent; A-240, accelerator. When mixed they produce a system highly adhesive, flexible, resistant to solvents, with good properties over a wide temperature range.

"Warco" GFC, GFI, GND.[42] Trademark for a series of gas-fading inhibitors for acetate fabrics. High molecular weight amino compounds and blend of organic amines.

"Warcosol" NF3.[42] Trademark for a complex of surface-active sulfonated polyester and phosphorated alcohol.
Uses: Dish- and bottle-washing compounds; adhesives and asphalt emulsions; alkaline and acid metal-cleaning; package- and pad-dyeing of textiles with vat and sulfur colors; acid-dyeing of wool.

warfarin (3-(alpna-acetonylbenzyl)-4-hydroxycoumarin) $C_6H_4C_3O(OH)(O)CH(CH_2COCH_3)C_6H_5$.
Properties: Colorless, odorless, tasteless crystals; m.p. 161°C. Soluble in alcohol and ether; insoluble in water. Also available as sodium salt.
Derivation: Condensation of benzylideneacetone and 4-hydroxycoumarin.
Grade: Technical.
Containers: Drums.
Hazard: Highly toxic by ingestion. Tolerance, 0.1 mg per cubic meter in air.
Use: Rodenticide.
Shipping regulations: (Air) Poison label.

warning label. See label (1).

wash-and-wear fabric. See aminoplast resin.

waste treatment. Waste materials can be classified by type as gaseous, liquid, solid, and radioactive, and by source as industrial, agricultural, urban, and nuclear. Many methods of treating wastes, either by converting them to useful by-products or to dispose of them, are in operation or under experimentation. Dumping in streams and rivers has long been illegal, and ocean dumping has been prohibited since 1973. Older treatment methods include incineration (garbage, paper, plastics); precipitation (smoke, solid-in-liquid suspensions); adsorption (gases and vapors); chemical treatment (neutralization, ion-exchange, chlorination); reclamation of sulfite liquors (paper industry); recycling; bacterial digestion (sewage); comminution and melting (glass, metals). Some cases may involve a combination of these.

Several new techniques have recently been researched and some are now in limited use. (1) Flash pyrolysis of certain solid wastes yields synthetic fuel oil and other useful products (see pyrolysis). (2) Urban, animal and agricultural wastes can be fermented by anaerobic bacteria to yield proteins, fuel oil, etc.; the possibilities of converting cannery and other food-processing wastes into protein-rich foods and feeds are being investigated. (3) Catalytic oxidation of exhaust emission gases; devices have been installed in cars since 1973 for air-pollution control. (4) Recovery of methane from manures. (5) Incorporation of special additives to certain plastics (polystyrene bottles) to render them biodegradable. (6) High-pressure hydrogenation of garbage to yield a low-sulfur combustible oil. (7) Deactivating radioactive wastes by adsorption or ion exchange, as well as by solidification and hydraulic fracturing; high-activity wastes are buried, but no longer sea-dumped. (8) Catalytic oxidation of waste chlorinated hydrocarbons, with partial recovery of chlorine. See also air pollution; water pollution.

waste paper, wet. Flammable solid. DOT requires a Yellow label; not accepted by express.

waste textile, wet. See waste paper, wet.

waste wool, wet. See wool waste.

"Watchung."[28] Trademark for precipitated diazo red pigments.

water (ice, steam) HOH
Properties: Colorless, odorless, tasteless liquid; allotropic forms are ice (solid) and steam (vapor). Water is a polar liquid with high dielectric constant (81 at 17°C); this largely accounts for its solvent power. It is a weak electrolyte, ionizing as H_3O^+ and OH^-. At atmospheric pressure it has sp. gr. 1.00 (4°C); freezing pt. 0°C (32°F), and expands about 10% on freezing; viscosity 0.01002 poise (20°C); sp. heat 1 calorie per gram; vapor pressure (100°C) 760 mm Hg. Triple point 0.00°C at 4.6 mm Hg. Surface tension 73 dynes/cm at 20°C. Latent heat of fusion (ice) 80 cal/gram; latent heat of condensation (steam), 540 cal/gram. Wt. per gallon (15°C 8.337 lbs; wt. per cu ft

62.3 lbs. Refractive index 1.333. Water may be superheated by enclosing in an autoclave and increasing pressure; it may be supercooled by adding sodium chloride or other ionizing compound. It has definite catalytic activity, especially of metal oxidation. Physiologically water is classed as a nutrient substance.
Derivation: (1) Oxidation of hydrogen; (2) end product of combustion; (3) end product of acid-base reaction; (4) end product of condensation reaction.
Purification: (1) Distillation; (2) ion exchange reaction (zeolite); (3) chlorination; (4) filtration.
Containers: Glass carboys; tank cars; pipelines.
Uses: Paper making; textile processing; solvent and diluent; industrial coolant; filtration; washing and scouring; sulfur mining; hydrolysis; Portland cement; hydraulic systems; power source; steam generation; food industry; source of hydrogen by electrolysis and thermochemical decomposition (see under hydrogen).
See also ice; steam; heavy water; ocean water; mineral water; water, hard.

water, bound. See bound water.

water gas (blue gas). A mixture of gases made from coke, air, and steam. The steam is decomposed by passing it over a bed of incandescent coke, or by high-temperature reaction with natural gas or similar hydrocarbons. Carbon monoxide about 40%, hydrogen about 50%; carbon dioxide about 3% and nitrogen about 3%.
Hazard: Flammable, dangerous fire and explosion risk. Explosive limits 7 to 72% in air. Toxic by inhalation.
Uses: Organic synthesis; fuel gas; formerly for ammonia synthesis.
See also synthesis gas.

water glass. See sodium silicate.

water, hard. Water containing low percentages of calcium and magnesium carbonates, bicarbonates, sulfates or chlorides, as a result of long contact with rocky substrates and soils. Degree of hardness is expressed either as grains per gallon or parts per million (ppm) of calcium carbonate. (1 grain of $CaCO_3$ per gallon is equivalent to 17.1 ppm). Up to 5 grains is considered soft; over 30 grains is very hard. Hardness may be temporary (carbonates and bicarbonates) or permanent (sulfates, chlorides). Treatment with zeolites (q.v.) is necessary to soften permanently hard water. Temporary hardness can be reduced by boiling. These impurities are responsible for boiler scale (q.v.) and corrosion of metals on long contact. Hard waters require use of synthetic detergents for satisfactory "sudsing." See also zeolite.

water of crystallization. Water chemically combined in many crystallized substances; it can be removed at or near 100°C, usually with loss of crystalline properties.

water pollution. Contamination of fresh or salt water with materials that are toxic, noxious, or otherwise harmful to fish and other animals, and to man, including thermal pollution (q.v.). Disposal of untreated chemical and municipal wastes in streams and rivers has been illegal since the early 1900's; in 1973 the EPA prohibited dumping of all types of wastes into the ocean. Unintentional pollution results from run-off containing toxic insecticidal residues. Oil spills at sea are a continual problem and probably will remain so. See also waste treatment; pollution; toxic material; oil spill treatment.

waterproofing agent. (1) Any film-forming substance which coats a substrate with a water-repellent layer, such as a paint, a rubber or plastic film, a wax, or an asphaltic compound. These are used on a wide variety of surfaces, including cement, masonry, metals, textiles, etc.

(2) Any metal salt or other chemical which impregnates textile fibers in such a way as to give an air-permeable, water-resistant product. There are three types of these: renewable, semidurable, and durable. Renewable repellents (e.g., "Aridex") are water dispersions containing aluminum acetate or formate, emulsifying agents and protective colloids in the continuous phase, and a blend of waxes in the disperse phase. Semidurable repellents involve precipitation of insoluble metal salts on the fibers; water-soluble soaps and waxes are usually added to the mixture, which is especially effective on synthetic fibers. Durable repellents coat each fiber with a protective film without bonding them together or sealing the apertures. Several trademarked products are used (see, e.g., "Zelan," "Norane"), some containing fatty pyridinium salts.

water purification (water conditioning). Any process whereby water is treated in such a way as to remove or reduce undesirable impurities. The following methods are used: (1) sedimentation, in which coarse suspended matter is allowed to settle by gravity in special tanks or reservoirs; (2) coagulation of aggregates (called "floc") by means of aluminum sulfate, ferric sulfate, or sodium aluminate (the aggregates are formed from colloidally dispersed impurities activated by the coagulant); (3) filtration through a bed of fine sand, either by gravity flow or by pressure, to remove suspended particles; (4) chlorination, which is effective in sterilizing potable water, swimming pools, etc.; (5) adsorption on activated carbon for removal of organic contaminants causing unpleasant taste and odor; (6) hardness removal by ion-exchange or zeolite process.

USDA reported (1973) that mercury can be removed from water by treatment with low percentages of black liquor from kraft papermaking.

water-soluble gum. See gum, natural.

water-soluble oil. Ammonia, potash, or sodium soaps of oleic, rosin, or naphthenic acids dissolved in mineral oils.
Uses: Boring; lathe-cutting; milling; polishing lubricants; dressing textile fibers; dust laying.
See also soluble oil.

water-soluble resin. See polymer, water-soluble.

wattle bark (Australian bark, mimosa bark).
Derivation: From the Australian wattles and South African acacias. Bark contains 25–35% tannin.
Grades: Based on tannin content.
Use: Source of wattle bark extract, used in vegetable tanning of leather, especially retannage of upper leathers and production of heavy leathers.

wax. A low-melting organic mixture or compound of high molecular weight, solid at room temperature and generally similar in composition to fats and oils except that it contains no glycerides. Some are hydrocarbons; others are esters of fatty acids and alcohols. They are classed among the lipids. Waxes are thermoplastic, but since they are not high polymers, they are not considered in the family of plastics. Common properties are water repellency; smooth texture; nontoxicity; freedom from objectionable odor and color. They are combustible, and have good dielectric properties. Soluble in most organic solvents; insoluble in water. The major types are as follows:

I. Natural
   1. Animal (beeswax, spermaceti, lanolin, shellac wax)
   2. Vegetable (carnauba, candelilla, bayberry, sugarcane)

3. Mineral
   (a) Fossil or earth waxes (ozocerite, ceresin, montan)
   (b) Petroleum waxes (paraffin, microcrystalline petrolatum) (slack or scale wax)

II. Synthetic
   1. Ethylenic polymers and polyol ether-esters ("Carbowax," sorbitol)
   2. Chlorinated naphthalenes ("Halowax")
   3. Hydrocarbon type via Fischer-Tropsch synthesis

Uses: Polishes; candles; crayons; sealants; sun-cracking protection of rubber and plastic products; cosmetics; paper coating; packaging food products; electrical insulation; waterproofing and cleaning compounds; carbon paper; precision investment casting.

See also following entries.

wax, chloronaphthalene

Properties: Translucent, black, light, and varied colors. Sp. gr. 1.40–1.7 (300°F); m.p. 190–265°F; b.p. 550–700°F. Soluble in many organic solvent liquids and oils (when heated together).

Derivation: By chlorinating naphthalene.

Containers: 130-lb drums.

Hazard: Toxic by ingestion and skin contact.

Uses: Condenser impregnation; moisture-, flame-, acid-, insect-proofing of wood, fabric and other fibrous bodies; moisture- and flame-proofing covered wire and cable; solvent (for rubber, aniline and other dyes, mineral and vegetable oils, varnish gums and resins, and other waxes when mixed in the molten state).

wax, microcrystalline. A wax, usually branched-chain paraffins, characterized by a crystal structure much smaller than that of normal wax and also by much higher viscosity. Obtained by dewaxing tank bottoms, refinery residues and other petroleum waste products, they have an average molecular weight of 500–800 (about twice that of paraffin). Viscosity 45–120 seconds (SUS at 210°F); penetration value from 3 to 33. Petroleum-derived products are used for adhesives, paper coating, cosmetic creams, floor wax, electrical insulation, heat-sealing, glass fabric impregnation, leather treatment, emulsions, etc. Some natural products, notably chlorophyll, are classed as microcrystalline waxes.

wax, polymethylene. White, odorless solid with congealing point of 205°F. Offered in flaked form. Approved by FDA.

"Waxine" Sizes.[57] Trademark for acid-stable, aqueous emulsions of various petroleum waxes, combined with rosin. Rosin is present entirely in free state (not saponified).

Use: Paper coating.

"Waxolan."[493] Trademark for higher melting fraction of lanolin esters isolated by physical fractionation. A light tan waxy solid; anhydrous; oil miscible.

Uses: Nonionic w/o emulsifier; auxiliary emulsifier and stabilizer for o/w systems.

wax tailings. Brown, sticky, semiasphalt product obtained in the destructive distillation of petroleum tar just prior to formation of coke.

Uses: Wood preservative; roofing paper.

"WBR."[182] Trademark applied to ion-exchange resin used in water treating and chemical process applications; weakly basic anion exchange resin produced from styrene and divinylbenzene.

weed-killer. See herbicide.

weight. See mass.

weighting agent. (1) In soft drink technology, an oil or oil-soluble compound of high specific gravity, such as a brominated olive oil, which is added to citrus flavoring oils to raise the specific gravity of the mixture to about 1.00, so that stable emulsions with water can be made for flavoring. (2) In the textile industry a compound used both to deluster and lower the cost of a fabric, at the same time improving its "hand" or feeling. Zinc acetylacetonate, clays, chalk, etc. are used.

"Weldal."[142] Trademark for chromate-containing composition designed for preparing aluminum for resistance welding. Supplied as a yellow powder dissolved in water.

welding. A metal-joining process by which coalescence is produced by heating to suitable temperatures with or without the use of filler metal. (American Welding Society). Brazing means welding above 1000°F with a nonferrous filler metal; soldering implies use of a lower melting filler (solder) (below 1000°F) the joint depending on adhesion rather than alloying. Aluminum can be cold-welded by pressure alone. Ultrasonic welding needs no heat and only light pressure.

Werner, A. (1866–1919). A native of Switzerland, Werner was awarded the Nobel Prize for his development of the concept of the coordination theory of valence, which he advanced in 1893. His ideas revolutionized the approach to the structure of inorganic compounds and in recent years have permeated this entire area of chemistry. The tern "Werner complex" has largely been replaced by "coordination compound". A noteworthy book by Dr. John Bailar (about 1945) admirably presented this important concept.

"W.E.S. Oil."[175] Trademark for a semi-refined coal tar distillate.

Properties: Straw to dark amber-colored liquid; distils 10% to not below 165°C; 70% to not above 190°C and 95% to not above 235°C; sp. gr. (15.5/15.5°C) 0.939–0.950; approx wt/gal 7.75 lb.

Use: High-boiling, slow-drying solvent for use in wire enamels and metal coatings.

"Westasept."[284] Trademark for a vegetable and oil soap formulated with a high percentage of coconut oil containing 3% hexachlorophene. Used as a liquid antiseptic surgical hand soap.

Weston cell. An electrical cell used as a standard which consists of an amalgamated cadmium anode covered with crystals of cadmium sulfate dipping into a saturated solution of the salt, and a mercury cathode covered with solid mercury sulfate.

"Wetanol."[73] Trademark for a modified sodium lauryl sulfate.

Properties: White, practically odorless powder, pH of 1% aqueous dispersion at 25°C, 5. Soluble in water, insoluble in alcohol, hydrocarbons, mineral oil, vegetable oils.

Uses: Wetting, penetrating, scouring agent.

"Wet-Ege spirits."[200] Trademark for a petroleum solvent prepared by straight-run distillation.

Properties: Water-white; initial boiling point 320–328°F, 95% distills between 380 and 388°F; sp. gr. 1.784 (60°F); flash point (TCC) 118°F; mild, nonresidual odor; holds a wet edge longer than mineral spirits.

Hazard: Flammable, moderate fire risk.
Uses: Brushing enamels, fly sprays, polishes, cleaning garments, degreasing hides, washing metal before painting, and as a carrier for chemicals in application.

"Wetsit."[309] Trademark for modified, blended alkyl sulfonates.
Uses: Textile wetting agents, scouring agents, general-purpose liquid detergent bases.

wetting agent. A surface-active agent (q.v.) which, when added to water, causes it to penetrate more easily into, or to spread over the surface of, another material by reducing the surface tension of the water. Soaps and alcohols are examples. See also detergent.

WFNA. Abbreviation for white fuming nitric acid. See nitric acid, fuming.

"WGR."[182] Trademark for an ion-exchange resin used in water treating and chemical processing; produced from an epoxy polymer.

whale oil (blubber oil).
Properties: Yellowish brown, nondrying, fixed oil; strong fishy odor. Soluble in alcohol, ether, benzene, chloroform, and carbon disulfide. Sp. gr. 0.925–0.939; saponification value 188–193; iodine value 120. Combustible; nontoxic.
Derivation: By boiling the blubber of whales, and skimming off the oil.
Containers: Barrels; tank cars.
Uses: Leather dressing; lubrication; tempering steel; soapmaking; fat manufacture (by hydrogenation); plant insecticide; sheep dips.
See also sperm oil (note); porpoise oil.

wheat germ oil. Light-yellow fat-soluble oil extracted from wheat germ.
Containers: 5-gal drums.
Uses: Source of Vitamin E; medicine; dietary supplement.
See also tocopherol.

whey. The serum remaining after removal of the solids (fat and casein) from milk. Dried whey contains about 13% protein, 71% lactose, 2.3% lactic acid, 4.5% water, and 8% ash, including a low concentration of phosphoric anydride. Besides its value as an inexpensive source of protein, whey is used as a source of lactose and lactic acid, as well as for the synthesis of riboflavin, acetone, and butyl alcohol by fermentation processes. Some types of cheese are made from whey, and it is also a possible culture medium. Dried whey may be used to replace up to 75% of the polyol component of rigid polyurethane foams. See also lactose.

whiskers. Single axially oriented crystal filaments of metals (iron, cobalt, aluminum, tungsten, rhenium, nickel, etc), refractory materials (sapphire, aluminum oxide, silicon carbide), carbon, boron, etc. They have tensile strengths of from 3 to 6 million psi and fantastically high elastic moduli. Their upper temperature limit in oxidizing atmospheres may be as high as 1700°C, and in inert atmospheres up to 2000°C. Length may be up to 2 inches, with diameter up to 10 microns. Their chief use is in the manufacture of composite structures with plastics, glass, or graphite which have many applications in the aircraft and space vehicle field, where their high heat capacity and tremendous strength are invaluable, especially as ablative agents.

whiskey.
Properties: Light yellow to amber liquid; sp. gr. 0.923–0.935 (15.56°C); 47–53% alcohol by volume. Flash point about 80°F (closed cup).
Derivation: Distillation of fermented malted grains, (corn, rye, or barley). After distillation whiskey is aged in wooden containers for several years. The characteristics of various brands and grades are due to slight "impurities," and also to the principal grain used.
Hazard: Flammable, moderate fire risk. A noncumulative poison, usually harmless in moderate amounts, but may be toxic when habitually taken in large amounts.
Containers: Wooden barrels and casks; bottles.
Uses: Beverage; medicine (stimulant; antiseptic; vasodilator). See also ethyl alcohol; proof.

white acid. A mixture of ammonium bifluoride and hydrofluoric acid used for etching glass.
Hazard: Strong irritant to skin and tissue.
Shipping regulations: (Air) Corrosive label.

white arsenic. See arsenic trioxide.

white copperas. See zinc sulfate.

white dye. An optical bleach (q.v.) or, in general, any substance, such as bluing, which may be added to a white article to increase its apparent whiteness.

white gasoline. See under gasoline.

"Whitekote."[139] Trademark for a dolomitic, normal, hydrated lime having a neutralizing value reported as 166% in terms of calcium carbonate; used in neutralizing and soil stabilization and other processes requiring an air-floated lime high in magnesium.

white lead. Name primarily applied to lead carbonate, basic (q.v.), but also used for lead sulfate, basic (white lead sulfate) and lead silicate, basic (white lead silicate).
Hazard: Toxic by ingestion and skin absorption.
Uses: Paint pigment.

white liquor. See liquor.

white metal. (1) Any of a group of alloys having relatively low melting points. They usually contain tin, lead, or antimony as the chief component. Type metal, Babbitt, pewter, and Britannia metal are of this group. (2) Copper matte containing about 75% copper, as obtained in copper smelting operations. See copper.

whitener. (1) Any of several oil-in-water emulsions in powder form that are dried and concentrated; when added to an aqueous medium they form stable emulsions, giving a white color and a cream-like body to coffee. A typical formulation contains vegetable fat, protein, sugar, corn syrup solids, plus emulsifier and stabilizers. (2) A white pigment or colorant used in the paper and textile industries.

white oil. Any of several derivatives of paraffinic hydrocarbons having moderate viscosity, low volatility and high flash point. Combustible; low toxicity. Available in U.S.P. grade.
Uses: Textile lubricant and finishing agent; plastics modifier.

white phosphorus. See phosphorus.

white precipitate. See mercury, ammoniated.

whitewash. A suspension of hydrated lime or calcium carbonate in water used for temporary antiseptic coatings of interiors of chicken houses and the like. See also calcimine.

whiting. Finely ground, naturally occurring calcium carbonate, $CaCO_3$, about 98% pure, contaminated with silica, iron, aluminum, or magnesium. See also calcium carbonate; chalk.
Properties: White or off-white powder; sp. gr. 2.7; in-

soluble in water, soluble in acids. It has no tinctorial power and hence is not a pigment.
Derivation: Traditionally from chalk (q.v.), obtained from England, France, Belgium. A pure limestone or calcite is the principal commercial source. Crude chalk or limestone is ground dry or wet, air- or water-floated and sieved. Grades are based on particle size, softness and light reflectance. Dry ground, air-floated limestone whiting can be as fine as 99% through 300 mesh.
Grades: Various. Paris white is the finest; coarser grades are extra gilders whiting, gilders whiting, and commercial, the last being quite coarse and of poor color. A putty grade is also sold.
Uses: Filler in rubber, plastics, and paper coatings; putty (with linseed oil); whitewashes; sealants.

willemite Zn_2SiO_4. Natural zinc orthosilicate. Troostite is a manganese-bearing variety.
Properties: Color yellow, green, red, brown, white; luster vitreous to resinous; sometimes fluoresces in ultraviolet light; Mohs hardness 5.5; sp. gr. 3.3.
Occurrence: New Jersey; New Mexico; Africa; Greenland.
Uses: Ore of zinc; a phosphor.

Williamson synthesis. An organic method for preparing ethers by the interaction of an alkyl halide with a sodium alcoholate (or phenolate).

"Wilmar."[497] Trademark for a series of fatty acids, fatty acid esters and polyesters used in foods, drugs, cosmetics, textiles, polishes, waxes and plastics.

"Wiltrol."[511] Trademark for a series of scorch inhibitors; used in the rubber industry, e.g., nitrosodiphenylamine and surface-treated phthalic anhydride.

wine. The fermented juice of grapes or other fruits or plants. Contains from 7 to 20% alcohol (by volume). The higher percentages are obtained by addition of pure alcohol (fortifying). Coloring matter, sugars, and small amounts of acetic acid, salts, higher fatty acids, etc. give wines their distinctive appearance and flavor. Low toxicity.
Hazard: Flammable; moderate fire risk.

wine ether. See ethyl pelargonate.

wine gallon. Same as U.S. gallon.

wine lees. A deposit or sediment formed in the bottom of wine casks during fermentation. Wine lees vary greatly in quality, but usually contain from 20–35% potassium acid tartrate and up to 20% calcium tartrate. They also contain yeast cells, proteins, and other solid matter which was suspended in the grape juice.
Use: Source of tartaric acid and tartrates.

"Wing-Stay."[265] Trademark for a series of para-phenylenediamine derivatives used as stabilizers, antioxidants and antiozonants for natural and synthetic rubbers and latices.

"Wing-Stop."[265] Trademark for sodium dimethyldithiocarbamate used in terminating the polymerization reactions of SBR elastomers, in "hot" and "cold" reactions.

wintergreen oil. See methyl salicylate.

winterize. A process of refrigerating edible and lubricating oils to crystallize the saturated glycerides, which are then removed by filter pressing.

"Witall."[104] Trademark for a series of driers manufactured from highly refined tall oil acids available in usual metallic salts including cobalt, lead and manganese. Equals naphthenate driers in action.

"Witcarb."[104] Trademark for a series of finely precipitated calcium carbonates.

"Witcizer."[104] Trademark for plasticizers used in the plastics, paint and lacquer industries.

"Witcoblak."[104] Trademark for pigment carbon blacks. Suitable for use in inks, paints, plastics, paper, caulking compounds, etc.

"Witcogard."[104] Trademark for a coating system used to protect new and used guardrails. A 2-component system, the base is a bituminous adhesive consisting of refined asphalts, petroleum resins, plasticizers, asbestos, mica, wetting agent, copolymer and petroleum solvent. The second component is colored ceramic granules which are opaque and self-cleaning.

witherite $BaCO_3$. A natural barium carbonate usually found in veins with lead ores.
Properties: White, yellowish or grayish. Vitreous, inclining to resinous luster. Sp. gr. 4.27–4.35; Mohs hardness 3–3.75.
Occurrence: United States (Kentucky, Lake Superior region), England (most important source).
Hazard: Toxic by ingestion; skin irritant.
Uses: Chemicals (barium dioxide, barium hydroxide, blanc fixe); plate glass and porcelain; brick making; rodenticide.

"Wittox."[104] Trademark for copper and zinc naphthenate fungicides.

Wöhler, Friedrich (1800–1882). A native of Germany, Wöhler, working with Berzelius, placed the qualitative analysis of minerals on a firm foundation. In the early 19th century chemists still thought it impossible to synthesize organic compounds. In 1828, Wöhler publicized his laboratory synthesis of urea, which he had obtained four years previously by heating the inorganic substance ammonium cyanate; the synthesis is a result of intramolecular rearrangement:

$$(NH_4)OCN \qquad NH_2{-}\overset{\overset{\displaystyle O}{\|}}{C}{-}NH_2$$

This was the beginning of the science of synthetic organic chemistry.

Wohlwill process. The official process of the U.S. mints for refining gold. It consists in subjecting gold anodes to electrolysis in a hot solution of hydrochloric acid containing gold chloride, the solution being continuously agitated with compressed air.

wolfram. See tungsten. Wolfram is the official international alternate name for tungsten. The latter is preferred in the U.S.

wolframic acid, anhydrous. See tungstic oxide.

wolframite $(Fe,Mn)WO_4$. A natural tungstate of iron and manganese. Ferberite is the iron-rich member of the series, and huebnerite is the manganese-rich member.
Properties: Color black to brown; luster sub-metallic to resinous; streak black to brown; Mohs hardness 5–5.5; sp. gr. 7.0–7.5.
Occurrence: Colorado, South Dakota, Nevada, Australia, Bolivia, Europe.
Use: Chief ore of tungsten.

wolfram white. See barium tungstate.

wollastonite $CaSiO_3$. A natural calcium silicate, found in metamorphic rocks.
Properties: Color white to brown, red, gray, yellow; luster vitreous to pearly; Mohs hardness 4.5–5; sp. gr. 2.8–2.9.
Occurrence: New York; California.
Grades: Fine, medium paint grades.
Uses: Ceramics; paint extender; welding rod coatings; rubber filler; silica gels; paper coating; filler in plastics, cements, and wallboard; mineral wool; soil conditioner.

"Wolman."[11] Trademark for a wood preservative comprising a mixture of chlorinated arsenate, fluoride and phenolic salts used in aqueous solutions to impregnate lumber to protect it against decay and termites.

wood. Wood is composed principally of 40–60% cellulose and 20–40% lignin, together with gums, resins, a variable amount of water, and potassium compounds. Its fuel value varies widely around 3000–6000 Btu per lb according to variety, moisture, etc. Combustible.
Uses: Pulp and paper; construction; packaging and cooperage; furniture; destructive distillation products (charcoal); extraction products (turpentine, rosin, tall oil, pine oil, etc.); methyl alcohol; plywood; fuel; rayon and cellophane; flock.

wood alcohol. See methyl alcohol.

wood ash. The inorganic residue of wood combustion.
Use: Fertilizer for its potash content, which averages around 4% K_2O.

wood flour (wood flock). Pulverized dried wood from either soft or hardwood wastes. Graded according to color and fineness.
Grades: Domestic standard, domestic fine, imported 40–60 mesh, 70–80 mesh, etc.
Hazard: Flammable; dangerous fire risk, especially suspended in air.
Uses: Manufacture of dynamite; filler for plastics, rubber, paperboard; fur cleaning; polishing agents; Sorel cement.

wood, indurated. A wood hardened by impregnation with a phenol-formaldehyde product. Used in storage batteries.

wood, petrified. Wood in which the original chemical components have been replaced by silica. The change occurs in such a way that the original form and structure of the wood are preserved. The most famous instance is the Petrified Forest of Arizona.

wood pulp. See pulp, paper.

wood rosin. See rosin.

Wood's metal. A four-component fusible alloy used largely in sprinkler systems; it melts at 70°C, the composition being bismuth 50%, cadmium 10%, tin 13.3%, lead 26.7%. See also alloy, fusible; eutectic.

wood sugar. See D(+)-xylose.

wood tar. See pine tar; creosote, wood-tar.

wood turpentine. See turpentine (oil).

wool. Staple fibers, usually 2-8 in. long, obtained from the fleece of sheep (and also alpaca, vicuna, and certain goats). Physically, wool differs from hair in fineness and by the presence of prominent cortical scales and a natural crimp. The latter properties are responsible for the felting properties of wool and the ability of the fibers to cling together when spun into yarns. Chemically, wool consists essentially of protein chains (keratin) bound together by disulfide cross linkages.
Properties: Tenacity ranges from 1 to 2 g per denier; elongation 25–50%; sp. gr. 1.32; moisture regain 16% (70°F, 65% relative humidity); decomposes about 260°F, scorches at 400°F. Resistant to most acids except hot sulfuric; destroyed by alkalies and chlorine bleach; resistant to mildew but attacked by insects. Combustible. Amphoteric to dyes.
Sources: Australia, Argentina, U.S., New Zealand, Uruguay, U.S.S.R., England.
Uses: Outerwear; blankets; carpets; upholstery; felt; clothing; source of lanolin.
For additional information refer to Wool Bureau, 360 Lexington Ave., N.Y.

wool fat. See lanolin.

wool waste. Wool from scrap materials cut up for remaking into cloth. Used as a fertilizer. Wool waste usually contains from 4–7% ammonia. Pure wool shoddy may contain as much as 15% nitrogen and is particularly valued by hop growers.
Hazard: Wet wool waste is flammable and a dangerous fire risk.
Shipping regulations: (Wet) (Rail) Yellow label. Not accepted by express.

wort. A clear infusion of grain extract (usually malt) used as the basis of fermentation in the brewing industry. See also brewing.

"Worthite."[269] Trademark for an austenitic stainless steel, containing 3% silicon, 20% chromium, 24% nickel, 2.75% molybdenum, 1.75% copper. Available in castings and wrought bars.
Uses: Pumps for corrosive environments, (paper industry, citrus fruit canners, sea water, sulfuric acid plants, etc.).

writing ink. A solution of colorant in water, usually containing also low percentages of tannic or gallic acid. Washable inks contain glycerol. Various water-soluble dyes are used for colored inks; dispersions of carbon black stabilized with a protective colloid are used for drawing inks. Fountain pen inks retain the fluidity of water; for ball-point pens the mixture is of a paste-like consistency.

wrought iron. A ferrous material, aggregated from a solidifying mass of pasty particles of highly refined metallic iron into which from 1 to 4% of slag is uniformly dispersed without subsequent melting. Wrought iron is distinguished by its low carbon and manganese contents. Carbon seldom exceeds 0.035%, and manganese content is held at 0.06% max. Phosphorus usually ranges from 0.10% to 0.15%; it adds about 1000 psi for each 0.01% above 0.10%. Sulfur content is normally low, ranging from 0.006% to below 0.015%. Silicon content ranges from 0.075% to 0.15%; silicon content of base metals is 0.015% or less. Residuals such as Cr, Ni, Co, Cu and Mo are generally low, totaling less than 0.05%.
Wrought iron is readily fabricated by standard methods and is quite corrosion-resistant.

wulfenite $PbMoO_4$, sometimes with Ca, Cr, V. Yellow, orange or bright orange-red mineral of resinous

luster. Found in veins with ores of lead. Sp. gr. 6.7–7.0; Mohs hardness 2.75–3.
Occurrence: United States (Massachusetts, New York, Pennsylvania, Nevada, Utah, New Mexico, Arizona); Europe; Australia.
Uses: Ore of molybdenum.

Wulff process. A process for producing acetylene by treating a hydrocarbon gas such as butane with superheated steam in a regenerative type of refractory furnace at 2100–2500°F. Contact times are very short. Acetylene and ethylene are produced in 1 to 1 ratio; used for vinyl chloride. See also acetylene.

X

xanthan gum. A synthetic biopolymer made by fermentation of carbohydrates; it is a thickening and suspending agent that is heat-stable, with good tolerance for strongly acid and basic solutions. A relatively new product, its expected uses are in drilling fluids, ore flotation and the food and pharmaceutical fields, especially in prepared mixes and high-protein fast foods. Viscosity remains stable over wide temperature range.

xanthate. A salt (usually potassium or sodium) of a xanthic acid (q.v.). Available as yellow pelletized solids having a pungent odor; soluble in water. Examples are potassium amyl xanthate, potassium ethyl xanthate, sodium isobutyl xanthate, sodium isopropyl xanthate. See also cellulose xanthate; viscose process.
Use: Collector agents in the flotation of sulfide minerals, metallic elements such as copper, silver, and gold, and some oxidized minerals of lead and copper.

xanthene (dibenzopyran, tricyclic). $CH_2(C_6H_4)_2O$. The central structure of the fluoroscein, eosin and rhodamine dyes.
Properties: Yellowish, crystalline leaflets. M.p. 100.5°C; b.p. 315°C. Soluble in ether; slightly soluble in alcohol; very slightly soluble in water.
Derivation: By the condensation of phenol and orthocresol by means of aluminum chloride.
Uses: Organic synthesis; fungicide.

xanthene dye. A group of dyes whose molecular structure is related to that of xanthene. The aromatic (C_6H_4) groups constitute the chromophore. Colour Index number ranges from 45000 to 45999. The dyes are closely related structurally to diaryl methane dyes. Eosin (CI No. 45380) is an example.

xanthenol. See xanthydrol.

xanthic acid (xanthogenic acid). A substituted dithiocarbonic acid of the type ROC(S)SH, in which R is ordinarily an alkyl radical. Unless otherwise designated, xanthic acid is understood to be the ethyl derivative, also called ethylxanthic acid, $C_2H_5OC(S)SH$, a liquid melting at −53°C and decomposing at room temperature. Xanthic acid salts are called xanthates (q.v.).

xanthine (dioxopurine) $C_5H_4N_4O_2$. A purine base occurring in the blood and urine and in some plants. Theophylline and theobromine, the alkaloids of tea and cocoa, respectively, are both dimethylxanthines; caffein, in coffee, is a trimethylxanthine.
Properties: Yellowish white powder; sublimes with partial decomposition. Soluble in potassium hydroxide; insoluble in water and acid.
Derivation: By the action of nitrous acid on guanine.
Grades: Technical, C.P.; monohydrate; sodium salt. Radioactive forms available.
Hazard: Moderately toxic by ingestion.
Uses: Organic synthesis; medicine.

xanthine oxidase. An enzyme found in animal tissues which acts upon hypoxanthine, xanthine, aldehydes, reduced coenzyme I, etc., producing, respectively, xanthine, uric acid, acids, oxidized coenzyme I, etc.
Use: Biochemical research.

xanthogenic acid. See xanthic acid.

xanthone (benzophenone oxide, dibenzopyrone, xanthene ketone) $CO(C_6H_4)_2O$. Tricyclic. Occurs in some plant pigments.
Properties: White needles or crystalline powder; m.p. 173–4°C; b.p. 350°C; sublimes. Insoluble in water, soluble in alcohol, chloroform, and benzene, especially when hot.
Uses: Larvicide; intermediate for dyes, perfumes and pharmaceuticals.

xanthophyll $C_{40}H_{56}O_2$.
Properties: Yellow pigment; an oxygenated carotenoid occurring in green vegetation and in some animal products, notably egg yolk. M.p. 190–3°C; insoluble in water, slightly soluble in alcohol and ether.
See also lutein; carotenoid.

xanthopterin $C_6H_5N_5O_2 \cdot H_2O$. 2-Amino-4,6-dihydroxypteridine. Pigment found in the wings of butterflies. Can be converted by yeast into folic acid.
Properties: Orange-yellow crystals; sinters around 360°C; decomposes above 410°C. Practically insoluble in water; freely soluble in dilute ammonium or sodium hydroxide giving yellow solutions, and in 2N hydrochloric acid giving colorless solutions.
Use: Biochemical research.

xanthydrol (xanthenol) $HCOH(C_6H_4)_2O$. A derivative of xanthene.
Properties: White powder; m.p. 123°C. Insoluble in water; soluble in alcohol.
Derivation: Reduction of xanthone with alcohol and sodium.
Grades: C.P. (analytical).
Use: Determination of urea and DDT.

"X-Cor."[236] Trademark for an ethoxylated diamine liquid. Added to drilling and packer fluids to protect drill pipe, casing and tubing from hydrogen sulfide attack, oxidation or the corrosive effects of electrolytes. Chemically stable, nonfluorescing; does not adversely affect drilling mud properties. Not affected by calcium or chloride contaminants; effective over a wide range of pH and temperature.

Xe Symbol for xenon.

xenon Xe Element of atomic number 54; noble gas group of the periodic table. Atomic weight 131.30. Valences 2, 4, 6, 8. There are 9 stable isotopes.
Properties: Colorless, odorless gas or liquid. Gas (at STP) has density of 5.8971 g/l and dielectric constant (25°C), 1 atm) of 1.0012. Liquid has b.p. of −108.12°C at 1 mm; density (at b.p.) 2.987 g/cc. Liquefaction temp. −106.9°C. Noncombustible; nontoxic. Chemically unreactive, but not completely inert.
Derivation: Fractional distillation of liquid air.
Containers: Hermetically sealed glass flasks; pressurized steel cylinders.

Uses: Luminescent tubes; flash lamps in photography; fluorimetry; lasers; tracer studies; anesthesia.
Shipping regulations: (Air) Nonflammable Gas label.
See also noble gas; xenon compounds.

xenon compounds.
Xenon tetrafluoride, XeF_4, is easily prepared by mixing fluorine and xenon in gaseous form, heating in a nickel vessel to 400°C, and cooling. The product forms large colorless crystals. The difluoride and hexafluoride, XeF_2 and XeF_6, also colorless crystals, can be obtained somewhat similarly. The hexafluoride melts to a yellow liquid at 50°C and boils at 75°C. Many XeF complexes with other compounds are also known. Xenon oxytetrafluoride, $XeOF_4$, a volatile liquid at room temperature, is obtained from the reaction of xenon hexafluoride and silica. Gram amounts have been isolated and studied.

All these fluorides must be protected from moisture to avoid formation of xenon trioxide, XeO_3, a colorless, nonvolatile solid which is dangerously explosive when dry. Its solution, the so-called xenic acid, is a stable weak acid but a strong oxidizing agent, which will even liberate chlorine from hydrochloric acid.

In an alkaline solution, xenon trioxide reacts to give free xenon and a perxenate, such as $Na_4XeO_6 \cdot 8H_2O$. Perxenates are probably the most powerful oxiding agents known, just as the xenon fluorides are extremely effective fluorinating agents.

Complex compounds containing nitrogen bonded to xenon have also been prepared.
Hazard: Highly toxic by inhalation; oxidizing agents; strong irritants.

xenyl. The biphenyl group, $C_6H_5C_6H_4$-.

xerography. A "dry" method of photography or photocopying. A metal plate is covered with a layer of photoconductive powder, such as selenium; the surface of this plate is given an electric charge by passing it under a series of charged wires. An image of the material to be photographed is projected onto the charged plate through a camera lens. The electric charges disappear in the areas exposed to light, but elsewhere the surface retains its charge. A powder consisting of a coarse carrier and a fine developing resin is then spread over the plate. Adhesion between powder and plate occurs only at the charged areas. Elsewhere developing resin and carrier are not retained on the plate, which thus has become a negative of the original image. A positive is obtained by placing a piece of paper against the plate, and applying an electric charge as in the first stage of the process. This causes adhesion of developing resin and its carrier to the paper. This positive print is fixed by heating in a press for a few seconds to melt the developing resin and fuse it to the paper. Colored prints are possible by use of suitable developing resins. Various materials other than paper can be printed.

XLPE. Trade abbreviation for cross-linked polyethylene (q.v.).

x-radiation (rotengen rays; x-rays). Electromagnetic radiation of extremely short wavelength (0.06 to 120 A), emitted as the result of electron transitions in the inner orbits of heavy atoms bombarded by cathode rays in a vacuum tube. Those of the shortest wavelength have the highest intensity, and are called "hard" x-rays. X-radiation was discovered by Roentgen in 1898. Its properties are: (1) Penetration of solids of moderate density, such as human tissue; they are retarded by bone, barium sulfate, lead, and other dense materials. (2) Action on photographic plates and fluorescent screens. (3) Ionization of the gases through which they pass. (4) Ability to damage or destroy diseased tissue; there is also a cumulative deleterious effect on healthy tissue.
Hazard: Overexposure can permanently damage cells and tissue structures; effect is cumulative.
Uses: Spectrometry; structure determination of molecules; cancer therapy; diagnostic medicine; nondestructive testing of metals; identification of original paintings.
See also radiation, ionizing; diffraction, x-ray; Roentgen.

xylene (dimethylbenzene) $C_6H_4(CH_3)_2$. A commercial mixture of the three isomers, ortho-, meta-, and para-xylene. The last two predominate.
Properties: Clear liquid; soluble in alcohol and ether; insoluble in water. Sp. gr. approximately 0.86; see under Grades for boiling range. Flash point from 81 to 115°F (TOC).
Derivation: (a) Fractional distillation from petroleum (90%), coal tar or coal gas; (b) by catalytic reforming from petroleum, followed by separation of para-xylene by continuous crystallization; (c) from toluene by transalkylation (q.v.).
Grades: Nitration (b.p. range 137.2–140.5°C); 3° (b.p. range 138–141°C); 5° (b.p. range 137–142°C, high in meta isomer); 10° (b.p. range 135–145°C); industrial (b.p. 90% below 150°C, complete below 160°C). Also other grades depending upon use. In some cases one or another of the industrial isomers are partially removed for use in chemical production.
Containers: Glass bottles, 5-, 55-, 110-gal drums; tank cars, tank trucks.
Hazard: Flammable, moderate fire risk; toxic by ingestion and inhalation. Tolerance 100 ppm in air.
Uses: Aviation gasoline; protective coatings; solvent for alkyd resins, lacquers, enamels, rubber cements; synthesis of organic chemicals.
Shipping regulations: (Rail) Red label. (Air) Flammable Liquid label.
See also following entries.

meta-**xylene** (1,3-dimethylbenzene) 1,3-$C_6H_4(CH_3)_2$.
Properties: Clear, colorless liquid, soluble in alcohol and ether; insoluble in water. Sp. gr. (15°C) 0.8684; f.p. −47.4°C; b.p. 138.8°C; refractive index (20°C) 1.4973. Flash point 85°F. Autoignition temp. 982°F.
Derivation: Selective crystallization or solvent extraction of meta-para mixtures.
Grades: 95% (technical); 99%; 99.9% (research).
Hazard: Flammable, moderate fire risk.
Uses: Solvent; intermediate for dyes and organic synthesis, especially isophthalic acid; insecticides; aviation fuel.

ortho-**xylene** (1,2-dimethylbenzene) 1,2-$C_6H_4(CH_3)_2$.
Properties: Clear, colorless liquid; soluble in alcohol and ether; insoluble in water. Sp. gr. (20/4°C) 0.880; f.p. −25°C; b.p. 144°C; refractive index (20°C) 1.505. Flash point 115°F (TOC). Autoignition temp. 867°F. Combustible.
Grades: 99%, free of hydrogen sulfide and sulfur dioxide; technical 95%; research 99.9%.
Hazard: Moderate fire risk.
Uses: Manufacture of phthalic anhydride; vitamin and pharmaceutical syntheses; dyes; insecticides; motor fuels.

para-**xylene** (1,4-dimethylbenzene) 1,4-$C_6H_4(CH_3)_2$.
Properties: Colorless liquid; crystals at low temperature; soluble in alcohol and ether; insoluble in water. Sp. gr. (20°C) 0.8611; m.p. 13.2°C; b.p. 138.5°C; refractive index (21°C) 1.5004. Flash point 81°F (TOC).
Derivation: Selective crystallization or solvent extraction of meta-para mixtures; separation from mixed-xylene feedstocks by adsorption (Parex process[416]).
Hazard: Flammable, dangerous fire risk.
Uses: Synthesis of terephthalic acid for polyester resins and fibers ("Dacron," "Mylar," "Terylene"); vitamin and pharmaceutical syntheses; insecticides.
Shipping regulations: See xylene.

para-**xylene**-alpha,alpha'-**diol** $C_6H_4(CH_2OH)_2$.
Properties: White crystalline solid. M.p. 118°C; b.p. 138–144°C (0.8–1.0 mm). Slightly soluble in water (25°C).
Purity: Approximately 98%.
Uses: Polyester resins.

xylenol (dimethylphenol, hydroxydimethylbenzene; dimethylhydroxybenzene) $(CH_3)_2C_6H_3OH$. There are five isomers (2,4-; 2,5-; 2,6-; 3,4-; 3,5-). This entry describes commercially offered mixtures.
Properties: White, crystalline solid; sp. gr. (15°C) 1.02–1.03; m.p. 20–76°C; b.p. 203–225°C. Only slightly soluble in water, soluble in most organic solvents and in caustic soda solution. Combustible.
Derivation: Cresylic acid or tar acid fraction of coal tar.
Containers: Fiber drums.
Hazard: Toxic by ingestion and skin absorption.
Uses: Disinfectants; solvents, pharmaceuticals, insecticides and fungicides; plasticizers; rubber chemicals; additives to lubricants and gasolines; manufacture of polyphenylene oxide (2,6-isomer only); wetting agents; dyestuffs.

"Xylex 780."[277] Trademark for a chemical peptizing agent or softener to facilitate processing of rubbers; reclaiming natural and/or synthetic rubbers.
Containers: Up to tank cars.

xylidine (aminodimethylbenzene; aminoxylene; dimethylaniline) $(CH_3)_2C_6H_3NH_2$. A varying mixture of isomers (2,3-; 2,4-; 2,5-; 2,6-).
Properties: Liquid; sp. gr. 0.97–0.99; b.p. 213–226°C; slightly soluble in water; soluble in alcohol and ether. Flash point 206°F (C.C.). Combustible.
Derivation: Nitration of xylene and subsequent reduction.
Containers: 425- and 800-lb drums; tank cars.
Hazard: Toxic by ingestion, inhalation, and skin absorption. Tolerance, 5 ppm in air.
Uses: Dye intermediates; organic syntheses; pharmaceuticals.
Shipping regulations: (Air) Poison label.

D(+)-**xylose** (wood sugar) $C_5H_{10}O_5$. (not to be confused with phenylosazone of the same name.)
Properties: White crystalline dextrorotatory powder, sweet taste; sp. gr. (20°C) 1.525; m.p. 144°C (also given as 153°C); soluble in water and alcohol. Nontoxic; combustible.
Derivation: Hydrolysis with hot dilute acids of wood, straw, corn cobs, etc.; wood pulp wastes.
Grades: Reagent, technical.
Uses: Dyeing; tanning; diabetic food; nonnutritive sweetener.

xylyl bromide (alpha-bromoxylene) $CH_3C_6H_4CH_2Br$. Mixed ortho-, meta-, and para-isomers.
Properties: Colorless liquid; pleasant aromatic odor. Decomposed slowly by water. Sp. gr. 1.4; b.p. 210–220°C. Combustible.
Derivation: Bromination of xylene.
Grade: Technical.
Containers: Lead-lined drums.
Hazard: Highly toxic by inhalation, and ingestion; strong irritant to eyes, skin, and tissue.
Uses: Organic synthesis; tear gas.
Shipping regulations: (Rail) Tear Gas label. (Air) Poison label. Not acceptable on passenger planes.

m-**xylyl chloride** $CH_3C_6H_4CH_2Cl$.
Properties: Colorless liquid; b.p. 196°C; sp. gr. 1.064. Combustible.
Hazard: Toxic by ingestion and inhalation. Strong irritant to eyes and skin.
Use: Intermediate.

para-**xylylene** $CH_2{:}C_6H_4{:}CH_2$. A monomer. See parylene.

Y Symbol for yttrium.

"Yarmor."[266] Trademark for series of pine oils for widely varied uses. Total terpene alcohol contents from 55% up to 91%.
Uses: Disinfectants, textile specialties, wetting and flotation agents, special solvents, household and industrial cleaners; odorants.

Yb Symbol for ytterbium.

yeast (barm). Unicellular organisms known as saccharomycetaceae. The following description applies to the cultured commercial product, and not to various wild varieties.
Properties: Yellowish-white viscid liquid or soft mass, flakes, or granules, consisting of cells and spores of Saccharomyces cerevisiae. Nontoxic.
Derivation: A ferment obtained in brewing. Yeasts induce fermentation by enzymes (zymases) which convert glucose and other carbohydrates into carbon dioxide and water in the presence of oxygen, or into alcohol and carbon dioxide (or lactic acid) in the absence of oxygen.
Grades: Technical; brewers'; cooking; compressed (contains about 74% moisture); dried; N.F. (contains no starch or filler, not more than 7% moisture nor more than 8% ash). Also graded according to vitamin B_1 content.
Containers: Tins, boxes, drums, tank trucks.
Uses: Fermentation of sugars, molasses, and cereals for alcohol; brewing; medicine; baking; food supplement; protein biosynthesis from many carbonaceous and nitrogenous materials, including petroleum; source of vitamins, enzymes, nucleic acids, etc.; biochemical research.
See also bacteria; fermentation.

yeast adenylic acid. See adenylic acid.

yellow AB (1-(phenylazo)-2-naphthylamine) $C_6H_5N_2C_{10}H_6NH_2$. Colour Index No. 11380.
Properties: Orange or red platelets; m.p. 102–104°C; insoluble in water; soluble in alcohol and oils.
Hazard: Consult regulations before using in food products.
Use: Biological stain.

yellow brass. A brass containing from 34 to 37% zinc; it has excellent fabrication properties and corrosion resistance. Used for structural and decorative purposes. See also brass.

yellow enzyme. See flavin enzyme.

yellow gentian. See gentian.

yellow glass. A soda-lime glass.

yellow lake. Any of several pigments made by precipitating soluble yellow dyes on an aluminum hydrate base. They are transparent in oil and lacquer vehicles and are used for metal decorating.

yellow OB (1-ortho-tolueneazonaphthylamine-2) $CH_3C_6H_4N_2C_{10}H_6NH_2$. Colour Index No. 11390.
Properties: Orange or yellow powder; m.p. 122–125°C; insoluble in water; soluble in alcohol and oils.
Hazard: Consult regulations before using in food products.
Use: Biological stain.

yellow phosphorus. See phosphorus.

yellow precipitate. See mercuric oxide,yellow.

yellow prussiate of potash. See potassium ferrocyanide.

yellow prussiate of soda. See sodium ferrocyanide.

yellow salt. See uranyl nitrate.

ylang ylang oil. An essential yellowish volatile oil distilled from the flowers of Cananga odorata, the principal sources of which are the Philippines and Réunion (Bourbon variety). It is often confused with cananga oil. The major components are linalool, geraniol and their esters, pinene, with small amounts of para-cresol methyl ether, and various other substances. Nontoxic.
Properties: (Manila) Sp. gr. 0.911–0.958 (30/4°C); optical rotation −27 to −49° 7′ (mostly −32 to −45°); refractive index 1.4747–1.4940 (rarely over 1.4900); ester value 90–150 (usually 100 or more). This is the first fraction to distill over. Later fractions are sometimes called cananga oil.
(Réunion) Sp gr. 0.932–0.962 (15°C); optical rotation −34 to −64°; ester value 96–134.
Containers: Bottles.
Uses: Perfumery; flavoring.

ylem. The original substance from which atoms are believed to have been derived, thought to have consisted of neutrons and electromagnetic radiation. According to one theory, most of the elements were formed in only a few minutes' time. See also element; abundance.

ylid. A substance in which a carbanion (q.v.) is attached to a heteratom with a high degree of positive charge, i.e., $>C^-{-}X^+$. It is similar to a zwitterion and related to the Wittig reaction.

yohimbine (aphrodine; corynine; quebrachine) $C_{21}H_{26}O_3N_2$.
Properties: Glistening, needle-like alkaloid; m.p. 234°C; soluble in alcohol and ether; very slightly soluble in water.
Derivation: By extraction from the bark of Corynanthe yohimbe, found in the Cameroons.
Hazard: Toxic by ingestion.

Young's modulus. See modulus of elasticity.

ytterbium Yb A rare-earth metal (q.v.) of yttrium subgroup; atomic number 70; atomic weight 173.04; valence 2,3. Exists in alpha and beta forms, the lat-

ter being semiconductive at pressures above 16,000 atm. There are 7 natural isotopes.
Properties: Metallic luster; quite malleable; m.p. 824°C; b.p. 1427°C; sp. gr. 7.01; reacts slowly with water; soluble in dilute acids and liquid ammonia. Low toxicity.
Source: See rare earth metals.
Derivation: Reduction of the oxide with lanthanum or misch metal.
Grades: Regular high purity (ingots and lumps).
Uses: Lasers; dopant for garnets; portable x-ray source; chemical research.

ytterbium chloride $YbCl_3 \cdot 6H_2O$.
Properties: Green crystals; sp. gr. 2.575; loses $6H_2O$ at 180°C; m.p. 865°C; very soluble in water; hygroscopic.

ytterbium fluoride YbF_3.
Properties: Solid; m.p. 1157°C; b.p. 2200°C; insoluble in water; hygroscopic.
Hazard: Toxic. Tolerance (as F), 2.5 mg per cubic meter of air.

ytterbium oxide (ytterbia) Yb_2O_3.
Properties: Colorless mass when free of thulia but tinted brown or yellow when containing thulia. The weakest base of the yttrium group with the exception of scandia and lutetia. Slightly hygroscopic, absorbs water and carbon dioxide from the air; sp. gr. 9.2; melting point 2346°C. Soluble in hot, dilute acids, less so in cold acids.
Containers: Glass bottles.
Uses: Special alloys; dielectric ceramics; carbon rods for industrial lighting; catalyst; special glasses.

ytterbium sulfate $Yb_2(SO_4)_3 \cdot 8H_2O$.
Properties: Crystalline solid; sp. gr. 3.286; soluble in water.

yttria. See yttrium oxide.

yttrium Y Element of atomic number 39, group IIIB of the peridic table. Atomic weight 88.9059. Valence 3. No stable isotopes.
Properties: Dark gray metal; sp. gr. 4.47; m.p. 1500°C; b.p. 2927°C. Soluble in dilute acids and potassium hydroxide solution; decomposes water; known only in the tripositive state; low neutron capture cross-section.
Source: See rare-earth metals.
Derivation: Reduction of the fluoride with calcium.
Grades: Regular high purity (ingots, lumps, turnings); metallurgical; low-oxygen; crystal sponge; powder.
Containers: Wooden kegs or fiber drums.
Hazard: Tolerance, 1 mg per cubic meter in air. Flammable in finely divided form.
Uses: Nuclear technology; iron and other alloys; incandescent gas mantles; deoxidizer for vanadium and other nonferrous metals; microwave ferrites; coating on high-temperature alloys; special semiconductors.

yttrium acetate $Y(C_2H_3O_2)_3 \cdot 8H_2O$.
Properties: Colorless crystals. Soluble in water.
Derivation: Action of acetic acid on yttrium oxide.
Use: Analytical chemistry.

yttrium antimonide YSb. A high-purity binary semiconductor.

yttrium arsenide YAs. A high-purity binary semiconductor.
Hazard: Highly toxic.

yttrium bromide $YBr_3 \cdot 9H_2O$.
Properties: Colorless crystals. Hygroscopic. M.p. (anhydrous) 904°C. Soluble in water; slightly soluble in alcohol; insoluble in ether.

yttrium chloride $YCl_3 \cdot 6H_2O$.
Properties: Reddish white, transparent, deliquescent prisms. Soluble in water and alcohol; insoluble in ether. Sp. gr 2.18; decomposes at 100°C.
Derivation: By the action of hydrochloric acid on yttrium oxide.
Grades: Purities to 99%.
Use: Analytical chemistry.

yttrium oxide (yttria) Y_2O_3.
Properties: Yellowish-white powder. Soluble in dilute acids; insoluble in water. Sp. gr. 4.84; m.p. 2410°C.
Derivation: By the ignition of yttrium nitrate.
Grades: Purities to 99.8%; electronic grade 99.999%.
Containers: Wooden kegs or fiber drums.
Uses: Phosphors for color TV tubes (alloy with europium); yttrium-iron garnets for microwave filters; stabilizer for zirconia refractories.

yttrium phosphide YP. Used as a high-purity binary semiconductor.

yttrium sulfate $Y_2(SO_4)_3 \cdot 8H_2O$.
Properties: Small reddish-white, monosymmetric crystals. Soluble in concentrated sulfuric acid; sparingly soluble in water; insoluble in alkalies. Sp. gr. 2.558; loses $8H_2O$ at 120°C; decomposes 700°C.
Derivation: Action of sulfuric acid on monazite sand.
Method of purification: Fractional crystallization.
Use: Reagent.

yttrium vanadate YVO_4.
Properties: White crystalline solid.
Use: With europium vanadate as red phosphor in color television tubes.

Z

"Zaclon."[28] Trademark for a series of fluxes based on active zinc ammonium chloride combined with additives; available in various forms.

"Zactane" Citrate.[24] Trademark for ethoheptazine citrate.

"Zalba."[28] Trademark for a rubber antioxidant containing a hindered phenol. Yellow cream-colored powder; sp. gr. 1.30.
Uses: Nondiscoloring antioxidant for natural and synthetic rubbers and latex. Stabilizer in SBR manufacture.

"Zamak."[268] Trademark for zinc die-casting alloys.

Zanzibar gum. A hard, usually fossil type of copal.
Source: Found on the island of Zanzibar and the adjacent African mainland.
Properties: Sp. gr. 1.062–1.068; m.p. 240–250°C; insoluble in most solvents. Combustible.
Use: Varnishes.

"Zarontin."[330] Trademark for ethosuximide (q.v.).

"Z-B-X."[248] Trademark for zinc butylxanthate (q.v.).

ZDP. Abbreviation for zinc dithiophosphate.

"Zectran."[233] Trademark for insecticides containing 4-dimethylamino-3,5-xylyl N-methylcarbamate.

"Zeecon."[48] Trademark for sodium lignin sulfonate with controlled wood sugar content. Used as a cement dispersant in the manufacture of concrete.

"Zee-Mill."[48] Trademark for a lignin base grinding aid for Portland cement.

"Zefkrome."[565] Trademark for acrylic fiber; available in brilliant shades which are colorfast, uniform, and consistent.

"Zefran."[565] Trademark for an acrylic fiber in white staple form, based on polyacrylonitrile and supplemented with a dye-receptive component. Has higher strength than other acrylics and different dyeing properties.

zein. The protein of corn.
Properties: White to slightly yellow powder. Odorless, nontoxic protein of the prolamine class, derived from corn. Contains 17 amino acids. Tasteless. Free of cystine, lysine, and tryptophan. A resinous material dispersible in water with neutral sulfonated castor oil. Soluble in dilute alcohol; insoluble in water, dilute acids, anhydrous alcohols, turpentine, esters, oils, fats. Sp. gr. 1.226. Combustible.
Derivation: By-product of corn processing, by extraction of gluten meal with 85% isopropyl alcohol, extraction of the zein from the extract with hexane, precipitation by water, and spray drying.
Containers: Bags; carloads; truckloads.
Uses: Paper coating; grease-resistant coating; label varnishes; laminated board; solid color prints; printing inks; food coatings; microencapsulation; fibers.

"Zelan."[28] Trademark for a line of durable water-repellent textile finishes.

"Zelcon" SL.[28] Trademark for a quaternized long-chain complex amine condensation product used as a fabric softener and conditioner for compounding into home laundry softeners.

"Zelec."[28] Trademark for a series of antistatic agents, both durable and nondurable.

"Zendel."[214] Trademark for polyethylene films. Available both cast and blown. Used in packaging products and netting for agricultural and construction industries.

"Zenite."[28] Trademark for a group of rubber accelerators based on zinc salt of 2-mercaptobenzothiazole, with or without various modifying agents.

"Zeo."[285] Trademark for a series of hydrous silicas produced by the precipitation process.
Uses: Reinforcing agent for rubber compounds; adsorbent carrier for insecticides; flatting agent for lacquers and varnishes; anticaking agent in foods, feeds, and chemicals.

"Zeodur."[184] Trademark for a processed greensand used as a low capacity inorganic cation exchanger suitable for salt cycle, primarily water softening.

"Zeogel."[236] Trademark for a special clay used in drilling fluids to give an equally high yield and stable viscosity and gel characteristics in either fresh or salt water, regardless of the concentration of salt in the latter.

"Zeokarb."[184] Trademark for a commercial grade cation exchanger of sulfonated coal. Available in two ionic forms, Na and H. Intermediate acidity; used for metal recovery, deionization, removal of cations, purification and esterification catalyst.

zeolite. A natural hydrated silicate of aluminum and either sodium or calcium or both, of the type $Na_2O \cdot Al_2O_3 \cdot nSiO_2 \cdot xH_2O$; or an artificial ion-exchange resin. Both natural and artificial zeolites are used extensively for water softening. For this purpose the sodium or potassium compounds are required, since their usefulness depends on the cationic exchange of the sodium of the zeolite for the calcium or magnesium of hard water. When the zeolite has become saturated with calcium or magnesium ion, it is flooded with strong salt solution; a reverse exchange of cations takes place, and the material is regenerated. The natural zeolites are analcite, chabazite, heulandite, natrolite, stilbite, and thomsonite.

Artificial zeolites are made in a variety of forms ranging from gelatinous to porous and sandlike, and are used as gas adsorbents, drying agents, and catalysts as well as water softeners. They include such diverse groups of compounds of sulfonated organics or basic resins, which act in a similar manner to effect either cation or anion exchange. See ion exchange resin; molecular sieve; water purification.

"Zeotone."[108] Trademark for a powdered, corrosion-inhibiting, hexametaphosphate compound specially formulated for cleaning and disinfecting domestic and industrial water softeners.

"Zepar" BP.[28] Trademark for a reducing agent based on sodium hydrosulfite; used as a bleaching agent for ground wood pulp, unbleached pulp, and old papers.

"Zepel."[28] Trademark for a fluorocarbon textile finish used as a durable oil- and water-repellent.

"Zephiran" Chloride.[162] Trademark for benzalkonium chloride.

Zerewitinoff reagent. Solution of methylmagnesium iodide in purified n-butyl ether. A clear, light, colored liquid which reacts rapidly with moisture and oxygen.
Uses: Analytical reagent for active hydrogen atoms in organic compounds; also to determine water, alcohols, and amines in inert solvents.

"Zerice."[51] Trademark for light lubricating oils suitable for many general machinery applications, particularly at low temperatures. Used in refrigerating machines having low evaporator temperatures.

"Zerlate."[28] Trademark for a wettable fungicide powder containing 76% ziram.
Hazard: Irritant to eyes and mucous membranes.

"Zerlon."[233] Trademark for a methyl methacrylate-styrene copolymer used as a plastic molding material.

zero group. See group (1).

"Zerok" 101.[41] Trademark for a synthetic resin coating of the vinyl type which is resistant to oxidizing acids and useful for protection against fumes and splashing, up to 150°F.

"Zero-Lite."[247] Trademark for a low temperature expanded polystyrene pipe insulation.

"Zerostat."[443] Trademark for antistatic finishes for hydrophobic synthetic fibers.

zeroth law (of thermodynamics). Two bodies which have been shown to be individually in equilibrium with a third body will be in equilibrium when placed in contact with each other, that is, they will have the same temperature. (H. Reiss).

"Zeset."[28] Trademark for a fiber-reactant resin for crease resistance and dimensional stabilization of textiles.

"Zetafin."[233] Trademark for a series of ethylene-ethyl acrylate resins which are soft, highly flexible copolymers with end-use characteristics similar to vinyl.

zeta potential (electrokinetic potential). The potential across the interface of all solids and liquids. Specifically, the potential across the diffuse layer of ions surrounding a charged colloidal particle, which is largely responsible for colloidal stability. Discharge of the zeta potential, accompanied by precipitation of the colloid, occurs by addition of polyvalent ions of sign opposite to that of the colloidal particles. Zeta potentials can be calculated from electrophoretic mobilities, i.e., the rates at which colloidal particles travel between charged electrodes placed in the solution.

"Zetax."[265] Trademark for zinc 2-mercaptobenzothiazole.
Properties: Pale yellow powder; sp. gr. 1.70 ± .03; melting point above 300°C; zinc content 15–18%.
Use: Rubber accelerator.

Ziegler catalyst. A type of stereospecific catalyst, usually a chemical complex derived from a transition metal halide and a metal hydride or a metal alkyl. The transition metal may be any of those in groups IV to VIII of the periodic table; the hydride or alkyl metals are those of groups I, II, and III. Typically, titanium chloride is added to aluminum alkyl in a hydrocarbon solvent to form a dispersion or precipitate of the catalyst complex. These catalysts usually operate at atmospheric pressure and are used to convert ethylene to linear polyethylene, and also in stereospecific polymerization of propylene to crystalline polypropylene (Ziegler process). See also Natta catalyst.

"Zilloy."[268] Trademark for a zinc-base alloy.

"Zimate." See "Ethyl Zimate"; "Methyl Zimate."

"Zinar."[79] Trademark for a high melting, pale colored zinc resinate which has a negative acid number (slightly basic).
Uses: Paints; varnishes; adhesives; linoleum print-paint; ethyl cellulose lacquers.

zinc Zn Metallic element of atomic number 30, Group IIB of periodic table. Atomic weight 65.37. Valence 2. Five stable isotopes.
Properties: Shining white metal with bluish gray luster (called spelter). Not found native. Soluble in acids and alkalies. Insoluble in water. Sp. gr. 7.14; m.p. 419°C; b.p. 907°C. Malleable at 100–150°C; strongly electropositive. Zinc foil will ignite in presence of moisture. Low toxicity.
Ores and minerals: See calamine, franklinite, hydrozincite, smithsonite, sphalerite, willemite, wurtzite.
Sources: British Columbia; Japan; Mexico; U.S. (Colorado); Peru; Australia.
Derivation: Extracted from ores by two distinct methods, both starting with zinc oxide formed by roasting the ores: (1) the pyrometallurgical or distillation process wherein the zinc oxide is reduced with carbon in retorts from which the resultant zinc is distilled and condensed; and (2) the hydrometallurgical or electrolytic process wherein the zinc oxide is leached from the roasted or calcined material with sulfuric acid to form zinc sulfate solution which is electrolyzed in cells to deposit zinc on cathodes.
Grades: Special high-grade (99.990%); high-grade (89.90%); intermediate (99.5%); brass special (99%); prime western (98%).
Forms available: Slab, rolled (strip, sheet, rod, tubing), wire, mossy zinc, zinc dust powder (99% pure); single crystals; zinc anodes.
Hazard (dust): Flammable, dangerous fire and explosion risk. See also zinc dust.
Uses: Alloys (brass, bronze, and die-casting alloys); galvanizing iron and other metals; electroplating; metal spraying; automotive parts; electrical fuses, anodes; dry cell batteries; fungicides; nutrition (essential growth element). Roofing, gutters; engravers' plates; cable wrappings; organ pipes.
Shipping regulations: (Dust or powder) (Air) Flammable Solid label.
For further information refer to American Zinc Institute, 292 Madison Ave., New York.

zinc 65. Radioactive isotope of mass number 65.
Properties: Half-life, 250 days; radiation, beta and gamma.
Derivation: Pile irradiation of zinc metal and, in the cyclotron, by bombarding copper 65 with deuterons.
Forms available: Zinc metal and zinc chloride in hydrochloric acid solution.
Hazard: A radioactive poison.
Uses: Tracer nuclide in study of wear in alloys, the

nature of phosphor activators, galvanizing, the function of traces of zinc in body metabolism, the functions of oil additives in lubricating oils, etc.
Shipping regulations: Consult regulations.

zinc abietate. See zinc resinate.

zinc acetate $Zn(C_2H_3O_2)_2 \cdot 2H_2O$.
Properties: White, monoclinic, crystalline plates; pearly luster; faint acetous odor; astringent taste. Soluble in water and alcohol. Sp. gr. 1.735; loses $2H_2O$ at 100°C; m.p. 200°C (decomp.). Low toxicity.
Derivation: Action of acetic acid on zinc oxide.
Containers: Bottles; cartons; fiber drums; carlots. Also available in solution.
Uses: Medicine; preserving wood; textile dyeing (mordant and resist); zinc chromate; laboratory reagent; feed additive; cross-linking agent for polymers; ingredient of dietary supplements (up to 1 mg daily).

zinc acetylacetonate $Zn[OC(CH_3):CHCO(CH_3)]_2$.
Properties: Crystalline solid; m.p. 138°C; b.p. sublimes. Very soluble in benzene and acetone; decomposes in water. Low toxicity.
Uses: Catalyst in synthesis of long-chain alcohols and aldehydes; textile weighting agent.

"Zincalume."[288] Trademark for a bright zinc electroplating process for consumer's goods and military materials. The bath contains zinc cyanide, sodium cyanide, sodium hydroxide, and addition agents.

zinc ammonium chloride $ZnCl_2 \cdot 2NH_4Cl$. A complex salt. Double salts with 3 or 6 molecules ammonium chloride have also been prepared.
Properties: White powder or crystals; soluble in water; sp. gr. 1.8. Low toxicity.
Grade: Technical (foaming and nonfoaming).
Containers: Bags; barrels; carlots.
Uses: Welding; soldering flux; dry batteries; galvanizing.

zinc ammonium nitrite. $ZnNH_4(NO_2)_3$.
Properties: White powder. Strong oxidizing agent.
Hazard: Dangerous fire risk in contact with organic materials.
Shipping regulations: (Rail) Yellow label.

zinc antimonide Zn_3Sb_2. Silvery-white crystals; sp. gr. 6.33; m.p. 570°C. Decomposes in water. Used in thermoelectric devices.
Hazard: May be toxic and irritant to skin.

zinc arsenate. Various forms are known. The description following is for zinc orthoarsenate, $Zn_3(AsO_4)_2 \cdot 8H_2O$.
Properties: White, odorless powder; sp. gr. 3.31 (15°C); loses $1H_2O$ at 100°C. Insoluble in water; soluble in acids and alkalies.
Derivation: (a) Occurs in nature as mineral koettigite; (b) by reactoin of a solution of sodium arsenate and a soluble zinc salt.
Hazard: Highly toxic by ingestion and inhalation.
Use: Insecticide; wood preservative.
Shipping regulations: (Rail, Air) Poison label.

zinc arsenite (zinc metaarsenite; ZMA) $Zn(AsO_2)_2$.
Properties: Colorless powder, soluble in acids, insoluble in water. Federal Specification TT-W-581 describes the composition of the solution used for wood preservation.
Hazard: Highly toxic by ingestion and inhalation.
Uses: Timber preservative; insecticide.
Shipping regulations: (Rail, Air) Poison label.

zinc bacitracin.
Properties: Creamy-white powder; slightly soluble in water; good thermal stability. Usually has 50–60 units/mg of bacitracin activity. Nontoxic.
Derivation: Action of zinc salts on bacitracin broth.
Grade: U.S.P.
Use: Preserving silage for feed; medicine.

zinc borate. Typical composition: ZnO 45%, B_2O_3 34%. May have 20% water of hydration.
Properties: White, amorphous powder; soluble in dilute acids; slightly soluble in water. Zinc borate of composition $3ZnO \cdot 2B_2O_3$ has a sp. gr. 3.64; m.p. 980°C. Nonflammable; low toxicity.
Derivatoin: Interaction of the oxides at 500–1000°C, or of zinc oxide slurries with solutions of boric acid or borax.
Uses: Medicine; fireproofing textiles; fungistat and mildew inhibitor; flux in ceramics.

zinc bromate $Zn(BrO_3)_2 \cdot 6H_2O$.
Properties: White solid; sp. gr. 2.566; m.p. 100°C; deliquescent; loses $6H_2O$ at 200°C. Very soluble in water.
Hazard: Dangerous fire risk in contact with organic materials; strong oxidizing agent.
Shipping regulations: (Air) Oxidizer label.

zinc bromide $ZnBr_2$.
Properties: White, hygroscopic, crystalline powder. Soluble in water, alcohol and ether. Sp. gr. 4.219; m.p. 394°C; b.p. 650°C. Low toxicity.
Derivation: Interaction of solutions of barium bromide and zinc sulfate, with subsequent crystallization.
Uses: Medicine; photography (plates, papers); manufacture of rayon. A solution of 80% zinc bromide is used as a radiation viewing shield.

zinc butylxanthate $Zn(C_4H_9OCS_2)_2$.
Properties: White powder; sp. gr. 1.45; decomposes when heated; moderately soluble in benzene and ethylene dichloride; slightly soluble in acetone; insoluble in water and gasoline.
Use: Ultra-accelerator used in self-curing rubber cements.
See also xanthate.

zinc cadmium sulfide. A fluorescent pigment; a phosphor.
Hazard: Probably toxic. See cadmium.

zinc caprylate (zinc octanoate) $Zn(C_8H_{15}O_2)_2$.
Properties: Lustrous scales; slightly soluble in boiling water, fairly soluble in boiling alcohol. M.p. 136°C. Decomposes in moist amosphere giving off caprylic acid.
Derivation: By precipitation from a solution of ammonium caprylate with zinc sulfate.
Use: Fungicide.

zinc carbolate. See zinc phenate.

zinc carbonate $ZnCO_3$.
Properties: White, crystalline powder. Soluble in acids, alkalies and ammonium salt solutions; insoluble in water. Sp. gr. 4.42–4.45; evolves carbon dioxide at 300°C. Nontoxic.
Derivation: (a) Grinding the mineral smithsonite; (b) Action of sodium bicarbonate on a solution of a zinc salt.
Uses: Ceramics; fire-proofing filler for rubber and plastic compositions exposed to flame temperature;

cosmetics and lotions; pharmaceuticals (ointments, dusting powders); zinc salts; feed additive.

zinc chlorate $Zn(ClO_3)_2 \cdot 4H_2O$.
Properties: Colorless to yellowish crystals; deliquescent; sp. gr. 2.15; decomposes at 60°C. Soluble in water and alcohol; soluble in glycerin and ether.
Hazard: Dangerous fire risk in contact with organic materials. Strong oxidizing agent.
Shipping regulations: (Rail) Yellow label. (Air) Oxidizer label.

zinc chloride $ZnCl_2$.
Properties: White granular deliquescent crystals or crystalline powder; soluble in water, alcohol, glycerine and ether. Sp. gr. 2.91; m.p. 290°C; b.p. 732°C. A 10% solution is acid to litmus.
Derivation: Action of hydrochloric acid on zinc or zinc oxide.
Method of purification: Recrystallization.
Grades: C.P.; technical. Fused; crystal; granulated; 62½% solution; 50% solution; U.S.P.
Containers: Bottles; drums; tank cars; tank trucks.
Hazard: (Solid) Skin irritant. (Solution) Severe irritant to skin and tissue. Tolerance, (as fume) 1 mg per cubic meter of air.
Uses: Galvanizing iron; catalyst; dehydrating and condensing agent in organic synthesis; wood preservative; soldering fluxes; burnishing and polishing compounds for steel; electroplating; antiseptic and deodorant preparations (up to 2% solution); textiles (mordant, carbonizing agent, mercerizing, sizing and weighting compositions, resist for sulfur colors, albumin colors and para red); adhesives; dental cements; glass etching; petroleum refining; parchment; dentrifices; embalming and taxidermists' fluids; medicine; dyestuffs; pigments; antistatic; feed additive; denaturant for alcohol; dietary supplement.
Shipping regulations: (Air) Corrosive label.

zinc chloride, chromated. A mixture of zinc chloride and sodium dichromate used as a wood preservative. Federal Specification TT-W-551 requires that it contain not less than 77.5% zinc chloride and 17.5% sodium dichromate dihydrate.
Hazard: Toxic by ingestion.

zinc chloroiodide. Mixture of zinc chloride and iodide.
Properties: White powder. Soluble in water.
Uses: Disinfectant; pharmaceutical preparations.

zinc chromate $ZnCrO_4$. There is a series of compounds in which the ratio of ZnO to CrO_3 is 5:1, 4:1, 2:1, 1:1 and 1:2, usually with some water of crystallization. The basic compounds containing the highest zinc content are more insoluble and stable than the acidic compounds. The following is a description of $ZnCrO_4$.
Properties: Yellow powder; sp. gr. 3.40; insoluble in water; soluble in acids.
Derivation: Action of chromic acid on slurries of zinc oxide, or on zinc hydroxide.
Grades: Technical; C.P.
Containers: Bottles, drums.
Hazard: Toxic by ingestion and inhalation; strong irritant. A carcinogen. Tolerance, 0.1 mg per cubic meter of air.
Uses: Artists' color; varnishes; pigment in rust-resistant primer and automotive paints; to impart corrosion resistance to epoxy laminates.
See also zinc yellow (principally zinc potassium chromate).

"Zincrometal."[579] Trademark for a precoated, highly corrosion-resistant steel designed chiefly for use in automobile bodies. The coating is stated to be comprised of two parts: an undercoat of zinc-chromium solution and a topcoat of zinc-rich epoxy-based resin. The combination of coatings is applied to the steel before shaping and other mechanical processing. The effectiveness of this product is said to be greater than that of other anticorrosion systems.

zinc cyanide $Zn(CN)_2$.
Properties: White powder. Sp. gr. 1.852; m.p. 800°C (decomposes); soluble in dilute mineral acids with production of hydrogen cyanide; soluble in alkalies; insoluble in water and alcohol.
Derivation: By precipitation of a solution of zinc sulfate or chloride with potassium cyanide.
Grade: Technical.
Containers: Glass bottles; kegs; drums.
Hazard: Highly toxic by ingestion and inhalation.
Uses: Medicine; metal plating; chemical reagent; insecticide.
Shipping regulations: (Air) Poison label. (Rail) Consult regulations.

zinc dialkyldithiophosphate. A lube oil additive for corrosion resistance, wear resistance, antioxidant.

zinc dibenzyldithiocarbamate $Zn[SCSN(C_7H_7)_2]_2$.
Properties: White powder; sp. gr. 1.41; melting range 165–175°C; moderately soluble in benzene and ethylene dichloride; insoluble in acetone, gasoline, and water.
Uses: Accelerator for latex dispersions and cements.

zinc dibutyldithiocarbamate $Zn[SC(S)N(C_4H_9)_2]_2$.
Properties: White powder; sp. gr. (20/20°C) 1.24; melting range 104–108°C; pleasant odor. Soluble in carbon disulfide; benzene, and chloroform; insoluble in water.
Uses: Accelerator for latex dispersions, cements, etc.; ultra-accelerator for lubricating oil additive.

zinc dichromate $ZnCr_2O_7 \cdot 3H_2O$.
Properties: Orange-yellow powder; soluble in acids and hot water; insoluble in alcohol and ether.
Derivation: Action of chromic acid on zinc hydroxide.
Hazard: Toxic by ingestion and inhalation.
Use: Pigment.

zinc diethyl. See diethylzinc.

zinc diethyldithiocarbamate $Zn[SC(S)N(C_2H_5)_2]_2$.
Properties: White powder; sp. gr. (20/20°C) 1.47; melting range 172–176°C. Soluble in carbon disulfide, benzene and chloroform; insoluble in water.
Containers: Drums.
Hazard: Strong irritant to eyes and mucous membranes.
Uses: Rubber vulcanization accelerator, especially latex foam; heat stabilizer for polyethylene.

zinc dimethyldithiocarbamate. See ziram.

zinc dimethyldithiocarbamate cyclohexylamine complex (zinc dithioamine complex). White powder or slurry of low solubility.
Hazard: Toxic by ingestion.
Uses: Fungicide; rodent poison; deer and rabbit repellent.

zinc dioxide. See zinc peroxide.

zinc dithionite. See zinc hydrosulfite.

zinc dust. A gray powder.
Grades: Commercial; pigment.
Containers: Drums; carlots.

Hazard: Dangerous fire risk. May form explosive mixtures with air; in bulk when damp may heat and ignite spontaneously on exposure to air.
Uses: Zinc salts and other zinc compounds; reducing agent, precipitating agent, purifier, catalyst; rust-resistant paints; bleaches; pyrotechnics; soot-removal; pipe-thread compounds; sherardizing; decorative effect in resins; autobody coatings.
Shipping regulations: (Air) Flammable Solid label.

zinc ethyl. See diethylzinc.

zinc ethylenebisdithiocarbamate. See zineb.

zinc 2-ethylhexoate. See zinc octoate.

zinc ethylsulfate $Zn(C_2H_5SO_4)_2 \cdot 2H_2O$.
Properties: Colorless, hygroscopic, crystalline leaflets. Soluble in water and alcohol.
Derivation: Interaction of zinc hydroxide and diethyl sulfate.
Use: Organic synthesis.

zinc fluoride ZnF_2.
Properties: White powder. Soluble in hot acids; slightly soluble in water; insoluble in alcohol. Sp. gr. (15°C) 4.84; m.p. 872°C; b.p. about 1500°C.
Derivation: (a) Action of hydrofluoric acid on zinc hydroxide; (b) Addition of sodium fluoride to a solution of zinc acetate.
Grade: Technical, about 95% pure.
Containers: Drums, barrels.
Hazard: Toxic. Tolerance (as F), 2.5 mg. per cubic meter of air.
Uses: Ceramic glazes and enamels; impregnating lumber; galvanizing.

zinc fluoroborate $Zn(BF_4)_2$. Colorless liquid, handled as 40 or 48% solution.
Uses: Plating and bonderizing; resin curing.

zinc fluorosilicate (zinc silicofluoride) $ZnSiF_6 \cdot 6H_2O$.
Properties: White crystals; sp. gr. 2.104. Decomposes on heating; soluble in water.
Derivation: Reaction of zinc oxide and fluosilicic acid.
Uses: Concrete hardener; laundry sour; preservative; mothproofing agents.

zinc formaldehyde sulfoxylate $Zn(HSO_2 \cdot CH_2O)_2$ (normal); $Zn(OH)(HSO_2 \cdot CH_2O)$ (basic).
Properties: Rhombic prisms. Very soluble in water (normal) (basic form is insoluble in water); insoluble in alcohol. Decomposes in acid.
Derivation: Reaction of formaldehyde and zinc sulfoxylate.
Grades: Basic; normal.
Containers: 250-lb and 300-lb drums.
Hazard: Probably toxic by ingestion.
Uses: Stripping and discharging agent for textiles.
See also hydrosulfite-formaldehyde preparations.

zinc formate $Zn(CHO_2)_2 \cdot 2H_2O$.
Properties: White crystals; sp. gr. (20°C) 2.207; loses $2H_2O$ at 140°C. Soluble in water; insoluble in alcohol.
Derivation: Action of formic acid on zinc hydroxide.
Hazard: May be toxic by ingestion.
Uses: Catalyst for production of methyl alcohol; waterproofing agent; textiles; antiseptic.

zinc gluconate. Dietary supplement and food additive; vitamin tablets.

zinc glycerophosphate (zinc glycerinophosphate) $C_3H_5(OH)_2OPO_3Zn$.
Properties: White, amorphous powder. Soluble in water; insoluble in alcohol and ether.
Derivation: Action of glycerophosphoric acid on zinc hydroxide.
Use: Medicine.

zinc green. One of a group of brilliant green pigments consisting of mixtures of Prussian blue and zinc yellow. They are permanent to light but not to alkali or water. Used for flat wall paints and interior work.

zinc hydrosulfite (zinc dithionite) (ZnS_2O_4).
Properties: White, amorphous solid; soluble in water.
Grade: Technical.
Containers: Cartons, drums; carlots.
Uses: Brightening groundwood, kraft, and other paper pulps; treatment of beet and cane sugar juices; depressant in mining flotations; bleaching textiles, vegetable oils, straw, hemp, vegetable tannins, animal glue, etc.

zinc hydroxide $Zn(OH)_2$.
Properties: Colorless crystals; sp. gr. 3.053; decomposes 125°C. Very soluble in water. Forms both zinc salts and zincates.
Derivation: Addition of a strong alkali to a solution of a zinc salt.
Uses: Intermediate; absorbent in surgical dressings; rubber compounding.

zinc hypophosphite $Zn(H_2PO_2)_2 \cdot H_2O$.
Properties: White hygroscopic crystals. Soluble in water and alkalies.
Derivation: Action of hypophosphorous acid on zinc hydroxide.
Use: Medicine.

zinc iodide ZnI_2.
Properties: Hygroscopic, white, crystalline powder; sharp, saline taste. Turns brown on exposure to light or air. Soluble in water, alcohol and alkalies. Sp. gr. 4.67; m.p. 446°C; b.p. 625°C (decomposes).
Derivation: Interaction of barium iodide and zinc sulfate, with subsequent crystallization.
Uses: Medicine (topical antiseptic); analytical reagent.

zinc lactate $Zn(C_3H_5O_3)_2 \cdot 3H_2O$.
Properties: White crystals. Soluble in water. Combustible. Low toxicity.
Derivation: Action of lactic acid on zinc hydroxide.
Use: Medicine.

zinc laurate $Zn(C_{12}H_{23}O_2)_2$.
Properties: White powder; m.p. 128°C; almost insoluble in water and alcohol. Combustible. Low toxicity.
Derivation: Precipitation of a soluble coconut oil soap with a solution of a zinc salt.
Containers: Barrels.
Uses: Paints; varnishes; rubber compounding (softener and activator).

zinc linoleate $Zn(C_{17}H_{31}COO)_2$. Brown solid containing 8.5 to 9.5% zinc. Combustible. Low toxicity.
Derivation: Precipitation from solutions of sodium linoleate and soluble zinc salt, or by fusion of the fatty acid and zinc oxide.
Uses: Paint drier, especially with cobalt and manganese soaps.

zinc malate $Zn(OOCCH_2CHOHCOO) \cdot 3H_2O$.
Properties: White, crystalline powder. Soluble in water. Combustible. Low toxicity.
Derivation: Action of malic acid on zinc hydroxide.
Use: Medicine.

zinc 2-mercaptobenzothiazole $Zn(C_7H_4NS_2)_2$. Rubber accelerator; fungicide.

zinc molybdate $ZnMoO_4 \cdot 2H_2O$.
Properties: Solid; sp. gr. 3.3; m.p. 1650°C (approx); insoluble in water.
Uses: Anticorrosion agent; starting material for growing single crystals.

zinc naphthenate $Zn(C_6H_5COO)_2$.
Properties: Amber, viscous, basic liquid or basic solid. The liquid contains 8–10% Zn, the solid contains 16% Zn. Very soluble in acetone. Combustible. Low toxicity.
Derivation: Fusion of zinc oxide or hydroxide and naphthenic acid, or precipitation from mixture of soluble zinc salts and sodium naphthenate.
Containers: Small metal drums; fiber drums.
Uses: Drier and wetting agent in paints, varnishes and resins; insecticide, fungicide, and mildew preventive; wood preservative; waterproofing textiles; insulating materials.

zinc nitrate $Zn(NO_3)_2 \cdot 6H_2O$.
Properties: Colorless lumps or crystals. Soluble in water and alcohol. Sp. gr. (13°C) 2.065; m.p. 36.4°C; loses water of crystallization between 105 and 131°C.
Derivation: Action of nitric acid on zinc or zinc oxide.
Containers: Drums; carlots.
Hazard: Dangerous fire and explosion risk; strong oxidizing agent. Moderately toxic.
Uses: Acidic catalyst; latex coagulant; medicine; reagent and intermediate; mordant.
Shipping regulations: (Rail) Nitrates, n.o.s., Yellow label. (Air) Oxidizer label.

zinc octanoate. See zinc caprylate.

zinc octoate (zinc 2-ethylhexoate) $Zn(OOCCH(C_2H_5)C_4H_9)_2$.
Properties: Light straw-colored viscous liquid. Sp. gr. 1.16; insoluble in water, soluble in common organic hydrocarbon solvents. Combustible. Low toxicity.
Use: Catalyst.

zinc oleate $Zn(C_{17}H_{33}COO)_2$.
Properties: Dry, white to tan, greasy, granular powder, containing from 8.5 to 10.5% zinc; m.p. 70°C. Soluble in alcohol, ether, carbon disulfide and ligroin; insoluble in water. Combustible. Low toxicity.
Derivation: Interaction of solutions of zinc acetate and sodium oleate, or by fusion of zinc oxide and oleic acid.
Uses: Paints, resins and varnishes (drier).

"Zincon."[169] Trademark for 2-carboxy-2′-hydroxy-5′-sulfoformazylbenzene used in colorimetric determination of zinc and copper.

zinc orthoarsenate. See zinc arsenate.

zinc orthophosphate. See zinc phosphate.

zinc orthosilicate. See zinc silicate.

zinc oxalate $ZnC_2O_4 \cdot 2H_2O$.
Properties: White powder. Soluble in acids and alkalies; slightly soluble in water. Sp. gr. (24°C) 2.562; m.p. (dec) 100°C. Combustible.
Derivation: Interaction of zinc sulfate and sodium oxalate.
Uses: Zinc oxide; organic synthesis.

zinc oxide (Chinese white; zinc white) ZnO.
Properties: Coarse white or grayish powder; odorless; bitter taste; absorbs carbon dioxide from the air; has greatest ultraviolet absorption of all commercial pigments; sp. gr. 5.47; m.p. 1975°C; soluble in acids and alkalies; insoluble in water and alcohol. Noncombustible; nontoxic as powder.
Derivation: (a) Oxidation of vaporized pure zinc (French process) or (b) roasting of zinc oxide ore (franklinite) with coal and subsequent oxidation with air; (c) similar treatment starting with other ores; (d) oxidation of vapor-fractionated die castings.
Grades: American process, lead-free; French process, lead-free, green seal, red seal, white seal (according to fineness); leaded (with lead sulfate); U.S.P., single crystals.
Containers: Boxes; drums; multiwall paper bags.
Hazard: ZnO fume is harmful by inhalation. Tolerance (fume), 5 mg per cubic meter in air. ZnO powder reacts violently with chlorinated rubber at 215°C.
Use: Accelerator activator, pigment, and reinforcing agent in rubber; ointments; pigment and mold-growth inhibitor in paints; ceramics; floor tile; glass; zinc salts; feed additive; dietary supplement; seed treatment; cosmetics; semiconductor in electronic devices; photoconductor in office copying machines and in color photography; piezoelectric devices.

zinc oxychloride. A saturated solution of zinc chloride and zinc oxide. Used in dentistry.

zinc palmitate $Zn(C_{16}H_{31}O_2)_2$.
Properties: White amorphous powder; sp. gr. 1.121; m.p. 100°C; insoluble in water and alcohol; slightly soluble in benzene and toluene. Combustible; low toxicity.
Containers: Cartons, bags.
Uses: Flatting agent in lacquer; pigment suspending agent for paints; rubber compounding; lubricant in plastics.

zinc perborate $Zn(BO_3)_2$ with water of hydration.
Properties: Amorphous, white powder; insoluble in water but slowly decomposed by it, liberating hydrogen peroxide.
Derivation: Interaction of sodium peroxide, boric acid and zinc salt, or of boric acid and zinc peroxide.
Containers: Tins, glass bottles.
Hazard: Fire risk when wet, in contact with organic materials.
Uses: Medicine; oxidizing agent.

zinc permanganate $Zn(MnO_4)_2 \cdot 6H_2O$.
Properties: Violet-brown or black, hygroscopic crystals; sp. gr. 2.47; loses $5H_2O$ at 100°C. Decomposes on exposure to light and air. Soluble in water and acids; decomposes in alcohol.
Grade: Technical (about 95% pure).
Containers: Glass bottles, tins.
Hazard: Dangerous fire risk in contact witih organic materials; strong oxidizing agent.
Uses: Oxidizing agent; medicine (antiseptic).
Shipping regulations: (Rail) Yellow label. (Air) Oxidizer label.

zinc peroxide (zinc dioxide) ZnO_2.
Properties: White powder containing 45–60% ZnO_2, balance ZnO; sp. gr. 1.571; decomposes rapidly above 150°C. Decomposes in acids, alcohol, acetone; insoluble in water but decomposed by it.
Derivation: Action of barium peroxide on zinc sulfate solution, followed by filtration.
Grades: U.S.P. (mixture of peroxide, carbonate and hydroxide); technical, 50–60%.
Containers: Glass bottles; drums.
Hazard: Severe explosion risk when heated; explosive range from 190°C to 212°C. Fire risk in contact with organic materials. Strong oxidizing agent.
Uses: Cosmetics; medicine; high-temperature oxidation reactions.
Shipping regulations: (Rail) Yellow label. (Air) Oxidizer label.

zinc phenate (zinc carbolate; zinc phenolate) $Zn(C_6H_5O)_2$. (May be only a mixture of zinc oxide and phenol.)
Properties: White powder. Soluble in alcohol; slightly soluble in water. Combustible.
Derivation: By heating zinc hydroxide with phenol and extracting with alcohol.
Hazard: Probably toxic by ingestion.
Uses: Medicine; insecticide.

zinc 1,4-phenolsulfonate (zinc sulfophenate; zinc sulfocarbolate) $Zn(SO_3C_6H_4OH)_2 \cdot 8H_2O$.
Properties: Colorless, transparent crystals or white granular powder; odorless; astringent metallic taste; effloresces in air; turns pink on exposure to air and light; loses water of crystallization at 120°C; soluble in water and alcohol.
Derivation: By heating zinc hydroxide with para-phenolsulfonic acid.
Grade: Technical.
Containers: Glass bottles; drums.
Hazard: Probably toxic by ingestion.
Use: Insecticide; medicine (antiseptic).

zinc phosphate (zinc orthophosphate; zinc phosphate, tribasic) $Zn_3(PO_4)_2$.
Properties: White powder. Soluble in acids and ammonium hydroxide; insoluble in water. Sp. gr. 3.998 (15°C); m.p. 900°C. Low toxicity.
Derivation: Interaction of zinc sulfate and trisodium phosphate.
Grade: Technical, about 98% pure.
Containers: Boxes; wooden kegs; glass bottles.
Uses: Dental cements; phosphors; conversion coating of steel, aluminum, and other metal surfaces.

zinc phosphide Zn_3P_2.
Properties: Dark gray, gritty powder; sp. gr. 4.55 (15°C); m.p. >420°C; stable if kept dry; insoluble in alcohol; soluble in acids; decomposes in water.
Derivation: By passing phosphine into a solution of zinc sulfate.
Grade: Technical, about 80–85% pure.
Containers: Tins; glass bottles.
Hazard: Highly toxic by ingestion. Reacts violently with oxidizing agents; produces toxic and flammable phosphine by reaction with acids.
Uses: Medicine; rat poisons.
Shipping regulations: (Air) Poison label. (Rail) Poisonous solids, n.o.s., Poison label.

zinc potassium chromate. See zinc yellow.

zinc potassium iodide. See potassium zinc iodide.

zinc propionate $Zn(OOCC_2H_5)_2$.
Properties: Platelets, tablets, or needlelike crystals. Fairly soluble in water, slightly soluble in alcohol. Decomposes in moist atmosphere, liberating propionic acid. Combustible; low toxicity.
Derivation: By dissolving zinc oxide in dilute propionic acid and concentrating the solution.
Use: Fungicide on adhesive tape.

zinc pyrophosphate $Zn_2P_2O_7$.
Properties: White powder; sp. gr. 3.756 (23°C); soluble in acids and alkalies; insoluble in water.
Containers: Wooden kegs or fiber drums.
Use: Pigment.

zinc resinate.
Properties: Powder; clear amber lumps, or yellowish liquid. May be acid, basic or neutral. Soluble in some organic solvents (ether, amyl alcohol). Combustible.
Chief constituent: Zinc abietate.
Derivation: By fusion of zinc oxide and rosin, or by precipitation from solutions of zinc salts and sodium resinate.
Containers: 55-gal drums.
Uses: Wetting, dispersing and hardening agent; drier in paints, varnishes and resins.

zinc rhodanide. See zinc thiocyanate.

zinc ricinoleate $Zn[CH_3(CH_2)_5CHOHCH_2CH{:}CH(CH_2)_7CO_2]_2$.
Properties: Fine white powder with faint fatty acid odor. M.p. 92–95°C; sp. gr. (25/25°C) 1.10. Combustible. Low toxicity.
Containers: 50-lb bags.
Uses: Fungicide; emulsifier; greases; lubricants; waterproofing; lubricating-oil additive; stabilizer in vinyl compounds.

"Zincrometal."[579] Trademark for a precoated, highly corrosion-resistant steel designed chiefly for use in automobile bodies. The coating is stated to be comprised of two parts: an undercoat of zinc-chromium solution and a topcoat of a zinc-rich epoxy-based resin. The combination of coatings is applied to the steel before shaping and other mechanical processing. The effectiveness of this product is said to be greater than that of other anticorrosion systems.

zinc salicylate $Zn[C_6H_4(OH)COO]_2 \cdot 3H_2O$.
Properties: White, crystalline needles or powder; soluble in water and alcohol. Combustible; low toxicity.
Derivation: By heating zinc hydroxide and salicylic acid.
Use: Medicine (antiseptic).

zinc selenide ZnSe.
Properties: Yellowish to reddish solid; sp. gr. 5.42 (15/4°C); m.p. above 1100°C. Insoluble in water.
Hazard: Moderately toxic; fire risk in contact with water or acids.
Use: Windows in infrared optical equipment.

zinc silicate (zinc orthosilicate) Zn_2SiO_4. See also willemite.
Properties: White crystals; sp. gr. 4.103; m.p. 1509°C. Insoluble in water.
Uses: Phosphors; spray ingredients; to remove traces of copper from gasoline.

zinc silicofluoride. See zinc fluorosilicate.

zinc-silver oxide battery. Primary or secondary battery used where space and weight are critical, i.e., in missiles. The battery has large energy output for its weight, but the components are expensive and the cycle life is short. To avoid deterioration, potassium hydroxide electrolyte is added just before use.
See also battery.

zinc stearate $Zn(C_{18}H_{35}O_2)_2$. Percentage of zinc may vary according to intended use, some products being more basic than others.
Properties: (pure substance) White, hydrophobic powder, free from grittiness; faint odor; sp. gr. 1.095; m.p. about 130°C. Soluble in acids; soluble in common solvents when hot; insoluble in water, alcohol and ether. Nontoxic. Combustible.
Derivation: Action of sodium stearate on solution of zinc sulfate.
Grades: U.S.P.; technical; available free from chick edema factor.

Containers: Glass bottles, bags, cartons, carlots.
Uses: Cosmetics, lacquers, plastics, powder metallurgy; lubricant; mold-release agent; filler; antifoamer; heat and light stabilizer; medicine (dermatitis); tablet manufacture; dietary supplement.

zinc subcarbonate. See zinc carbonate, precipitated.

zinc sulfate (white vitriol; white copperas; zinc vitriol) $ZnSO_4 \cdot 7H_2O$.
Properties: Colorless crystals, small needles or granular crystalline powder, without odor; astringent, metallic taste; efflorescent in air. Solutions acid to litmus. Sp. gr. 1.957 (25/4°C); m.p. 100°C; loses $7H_2O$ at 280°C; soluble in water and glycerol; insoluble in alcohol. Low toxicity.
Derivation: (a) Roasting zinc blende and lixiviating, with subsequent purification; (b) action of sulfuric acid on zinc or zinc oxide.
Grades: Technical; U.S.P.; reagent.
Containers: Glass bottles; barrels; fiber drums; multiwall paper sacks.
Uses: Precipitating agent for viscose rayon; trace nutrient in animal feeds; fertilizer ingredient; fungicide; ore flotation; water purification; galvanizing.

zinc sulfate monohydrate $ZnSO_4 \cdot H_2O$.
Properties: White, free-flowing powder; soluble in water; insoluble in alcohol.
Uses: Rayon manufacture; agricultural sprays; chemical intermediate; dyestuffs; electroplating.

zinc sulfide ZnS. Exists in two crystalline forms, alpha (wurtzite) and beta (sphalerite).
Properties: Yellowish-white powder, stable if kept dry. Alpha: sp. gr. 3.98. Beta: sp. gr. 4.102; changes to alpha form at 1020°C; sublimes at 1180°C. Soluble in acids; insoluble in water. Low toxicity.
Derivation: By passing hydrogen sulfide gas into a solution of a zinc salt.
Grades: Technical; C.P., fluorescent or luminous; single crystals.
Containers: 1-lb bottles; barrels; bags.
Uses: Pigment; white and opaque glass; base for color lakes; rubber; plastics; dyeing (hydrosulfite process); ingredient of lithopone; phosphor in x-ray and television screens; luminous watch faces; fungicide.

zinc sulfite $ZnSO_3 \cdot 2H_2O$.
Properties: White, crystalline powder; absorbs oxygen from the air to form sulfate. Loses $2H_2O$ at 100°C; decomposes at 200°C. Soluble in sulfurous acid; insoluble in cold water and alcohol; decomposes in hot water. Low toxicity.
Derivation: Action of sulfurous acid on zinc hydroxide.
Containers: Glass bottles; tins.
Uses: Medicine; preservative for anatomical specimens.

zinc sulfocarbolate. See zinc phenolsulfonate.

zinc sulfocyanate. See zinc thiocyanate.

zinc sulfophenate. See zinc phenolsulfonate.

zinc sulfoxylate $ZnSO_2$.
Properties: White crystalline material; decomposed by heat; salt of unstable sulfoxylic acid, H_2SO_2; a strong reducing agent.
Derivation: Action of zinc and sulfuryl chloride in ethereal solution, or of sulfur dioxide on granulated zinc in absolute alcohol.
Containers: Fiber drums; multiwall paper bags.
Use: Stripping agent in dyeing.

zinc telluride ZnTe. Reddish crystals; sp. gr. 6.34 (15°C); m.p. 1238°C. Decomposes in water. Single crystals available for phosphors.
Hazard: Moderately toxic.

zinc thiocyanate (zinc rhodanide; zinc sulfocyanate) $Zn(CNS)_2$.
Properties: White hygroscopic powder or crystals. Soluble in water, alcohol and ammonium hydroxide. Low toxicity.
Derivation: Interaction of zinc hydroxide and ammonium thiocyanate.
Grades: Technical; solution; reagent; A.C.S.
Containers: Glass bottles; drums.
Uses: Analytical chemistry; swelling agent for cellulose esters; dyeing assistant.

zinc thiophenate. A coined name for a class of peptizing agents for natural and synthetic rubbers. A typical example is the tert-butylphenyl sulfide, $[(CH_3)_3CC_6H_4S]_2Zn$.

zinc undecylenate $[CH_2{:}CH(CH_2)_8COO]_2Zn$.
Properties: White amorphous powder; nearly insoluble in water and alcohol; m.p. 115–116°C. Combustible.
Grade: N.F.
Containers: Drums.
Uses: Medicine (fungistat); cosmetics; chemical intermediate.

zinc white. Zinc oxide paste used by commercial artists. See also zinc oxide.

zinc yellow (citron yellow, buttercup yellow, zinc potassium chromate, zinc chrome).
Properties: Greenish-yellow pigment of comparatively low tinting strength. Partially water-soluble. Consists principally of zinc potassium chromate, about $4ZnO \cdot K_2O \cdot 4Cr_2O_3 \cdot 3H_2O$.
Derivation: Reaction of a solution of potassium dichromate with zinc oxide and sulfuric acid.
Containers: 250-lb bags.
Hazard: Toxic by ingestion.
Uses: Rust-inhibiting paints; artists' color.

zinc zirconium silicate $ZnO \cdot ZrO_2 \cdot SiO_2$.
Properties: White powder; sp. gr. 4.8; density 115 lbs/cu ft; m.p. 2100°C; soluble in hydrofluoric acid; insoluble in water and alkalies; slightly soluble in mineral acids and hot conc. sulfuric acid. Noncombustible.
Containers: Bags, barrels, carloads.
Uses: Opacifier for ceramic glazes.

zineb. Generic name for zinc ethylenebis(dithiocarbamate), $Zn(CS_2NHCH_2)_2$.
Properties: Light-tan solid, insoluble in water; soluble in pyridine. Decomposes on heating.
Derivation: Reaction of sodium ethylenebisdithiocarbamate with zinc sulfate or other zinc salts. In practical application as a fungicide these reactants are mixed in the presence of lime, and the zineb is not formed until after reaction of the carbon dioxide of the air with the film of the other chemicals on the leaf or fruit.
Grades: Commercial dusts and wettable powders usually contain 65% active material.
Hazard: Moderately toxic by inhalation and ingestion; irritant to eyes and muscous membranes.
Uses: Insecticide and fungicide.

zingerone (4,(4-hydroxy-3-methoxyphenyl)-2-butanone) $HOC_6H_3(OCH_3)CH_2CH_2COCH_3$.
Properties: Crystals; m.p. 40–41°C. Soluble in ether; sparingly soluble in water and petroleum ether.
Use: Flavoring.

"Zinol."[79] Trademark for a special zinc resinate in solution in mineral spirits.
Hazard: Flammable; moderate fire risk.
Uses: Printing ink; paint and varnish.

"Zin-O-Lyte."[28] Trademark for a series of electroplating products for use in zinc cyanide plating baths.

"Zinophos."[57] Trademark for a soil insecticide and nematocide whose active ingredient is thionazin (q.v.).

"Zinros."[79] Trademark for a pale, high-melting zinc resinate used in adhesives; printing ink; rubber compounding.

ziram. Generic name for zinc dimethyldithiocarbamate, $Zn(SCSNCH_3CH_3)_2$.
Properties: White and odorless when pure; sp. gr. 1.71; m.p. 246°C; almost insoluble in water; soluble in acetone, carbon disulfide, chloroform, in dilute alkalies , and in concentrated hydrochloric acid.
Derivation: Reaction of sodium dimethyldithiocarbamate with a soluble zinc salt in aqueous solution.
Grades: 76% wettable powder; 90% technical powder.
Containers: Multiwall bags; drums.
Hazard: Moderately toxic; strong irritant to eyes and mucous membranes.
Uses: Fungicide; rubber accelerator.

"Zirberk."[81] Trademark for ziram (q.v.).

"Zircaloy." Trademark for alloys of zirconium with low percentages of tin, iron, chromium, and nickel, used as cladding for nuclear fuel elements and other reactor applications.

"Zirco."[230] Trademark for an oil-soluble polymeric zirconyl complex in odorless mineral spirits. Not a paint drier in itself; has synergistic action on metallic driers.

"Zircofrax."[280] Trademark for super-refractory products from zirconium oxide and zirconium silicate.
Properties: High heat resistance; great strength; high thermal conductivity; high resistance to attack by acids and acid slags; porosity about 25%; permeability low.
Uses: Bricks and special shapes for ceramic kiln furniture and in chemical and metallurgical furnaces.

zircon $ZrSiO_4$ or $ZrO_2 \cdot SiO_2$. A natural zirconium silicate; represents 60 to 70% of all zirconium used.
Properties: Color brown, gray, red, colorless; luster adamantine; hardness 7.5; sp. gr. 4.68. Insoluble in acids.
Occurrence: Georgia; Florida; Australia; Brazil.
Uses: Source of zirconium oxide, metallic zirconium, and hafnium; abrasive; refractories; enamels; refractory porcelain.

Zircon Granular "G."[337] Trademark for zirconium silicate of 98% minimum purity. Used in refractories, enamel frits, and electrical insulation.

"Zircon H-W."[446] Trademark for a compound made from the purified mineral by impact pressing, or by air ramming or slip casting.
Properties: High density (230 lbs/cu ft) helps resist the wetting and penetration of molten glass. Resistant also to thermal spalling, fluxing conditions; has constancy of volume under soaking heat in excess of 2900°F.
Uses: To pave glass tank bottoms and floors; line sodium metaphosphate and sodium silicate furnaces; tap hole blocks for aluminum and nonferrous melting; nozzles for continuous steel casting.

zirconia. See zirconium oxide.

zirconic anhydride. See zirconium oxide.

"Zirconite."[337] Trademark for foundry materials containing 97–98% zircon minimum. M.p. 4050–4100°F. Available as sand and flour. Used in foundries for cores, molds, and facings; reduces cleaning costs and produces smoother castings.

zirconium Zr Metallic element of atomic number 40, group IVB of the periodic table. Atomic weight 91.22; valences 2,3 (halogens only), 4. There are 5 stable isotopes.
Properties: Hard, lustrous, grayish, crystalline scales or gray amorphous powder; sp. gr. 6.4; m.p. about 1850°C; b.p. 4377°C; soluble in hot, very concentrated acids; insoluble in water and cold acids. Corrosion-resistant; low neutron absorption. Low toxicity.
Sources: Zircon; baddeleyite (zirconia).
Derivation: The ore is converted to a cyanonitride, which is chlorinated to obtain zirconium tetrachloride. This is reduced with magnesium (Kroll process) in inert atmosphere. The metal can be prepared in a highly pure and ductile form by vapor-phase decompostion of the tetraiodide. Hafnium must be removed for uses in nuclear reactors (see hafnium).
Grades: Plate, strip, bars, wire; sponge and briquettes; powder; foil; technical; pure (hafnium-free); single crystals.
Hazard: Flammable and explosive as dust or powder, and in form of borings, shavings, etc. Tolerance (as Zr): 5 mg per cubic meter of air (applies to all Zr compounds). Powder should be kept wet in storage, and protective clothing should be worn. Safety data sheet available from Manufacturing Chemists Assn., Washington, D.C.
Uses: Nuclear technology; corrosion-resistant alloys; photoflash bulbs (foil); pyrotechnics; metal-to-glass seals; special welding fluxes; getter in vacuum tubes; explosive primers; acid manufacturing plants; deoxidizer and scavenger in steel manufacture.
Shipping regulations: Many particulate forms require Yellow label (Rail) and are Not acceptable by air. Consult regulations for details.
Note: Use of zirconium compounds in aerosol sprays as an antiperspirant has been restricted by FDA due to suspected toxicity.

zirconium 95. Radioactive zirconium of mass number 95.
Properties: Half-life, 63 days; radiation, beta and gamma.
Derivation: Obtained in a mixture with niobium from the fission products of nuclear reactor fuels.
Forms available: Zirconium oxalate complex in oxalic acid solution.
Hazard: Radioactive poison.
Uses: To trace the flow of petroleum products in pipelines; to measure the rate of catalyst circulation in petroleum cracking plants; to study the cracking and polymerization of hydrocarbons with various catalysts; etc.
Shipping regulations: Consult regulations.

zirconium acetate $H_2ZrO_2(C_2H_3O_2)_2$
Properties: (a) Available as aqueous solution, 22% ZrO_2. Clear to pale amber liquid; sp. gr. 1.46 (approx); pH 3.8–4.2 (20°C); f.p. −7°C; stable at room temperature. (b) Available as 13% ZrO_2 (aqueous solution); sp. gr. 1.20 (approx); pH 3.3–4.0 (20°C); stable at room temperature, but temperature of hydrolysis decreases with pH; undergoes exchange with anion exchange resins, but not with cation exchangers.
Containers: Polyethylene-lined drums; carloads.
Hazard: Tolerance (as Zr), 5 mg per cubic meter of air.
Uses: Preparation of water repellents, pharmaceuticals, other chemicals.

zirconium acetylacetonate. See zirconium tetraacetylacetonate.

zirconium ammonium floride (ammonium zirconifluoride) $Zr(NH_4)_2F_6$.
Properties: White crystals. Soluble in water.
Hazard: Toxic and irritant. Tolerance (as Zr), 5 mg per cubic meter of air.

zirconium anhydride. See zirconium oxide.

zirconium boride (zirconium diboride) ZrB_2.
Properties: Gray metallic crystals or powders; sp. gr. 6.085; m.p. 3000°C; Mohs hardness 8; electrical resistivity 9.2 micro-ohm-cm at 20°C; excellent thermal shock resistance; poor oxidation resistance above 1100°C.
Hazard: Tolerance (as Zr), 5 mg per cubic meter of air.
Uses: Refractory for aircraft and rocket applications; thermocouple protection tubes; high temperature electrical conductor; cutting tool component; coating tantalum; cathode in high-temperature electrochemical systems.

zirconium carbide ZrC.
Properties: Gray crystalline solid; sp. gr. 6.78; hardness Mohs 8+; m.p. 3400°C; b.p. 5100°C; insoluble in water and hydrochloric acid; soluble in oxidizing acids and attacked by oxidizers.
Derivation: By heating zirconium oxide and coke in an electric furnace.
Grade: Technical.
Containers: Iron drums.
Hazard: Powder or dust will ignite spontaneously. Tolerance (as Zr), 5 mg per cubic meter of air.
Uses: Incandescent filaments; abrasive; cermet component; high temperature electrical conductor; refractory; metal cladding; cutting tool component.

zirconium carbonate, basic (zirconyl carbonate; zirconium carbonate) $ZrOCO_3$ or $ZrOCO_3 \cdot xH_2O$.
Properties: White, amorphous powder. Soluble in acids; insoluble in water.
Derivation: By adding sodium carbonate to a solution of zirconium salt.
Hazard: Tolerance (as Zr), 5 mg per cubic meter of air.
Use: Preparation of zirconium oxide.

zirconium chloride. See zirconium tetrachloride.

zirconium chloride, basic. See zirconium oxychloride.

zirconium diboride. See zirconium boride.

zirconium dioxide. See zironcium oxide.

zirconium disilicide (zirconium silicide) $ZrSi_2$. Gray solid; sp. gr. 4.88 (22°C); soluble in hydrofluoric acid; insoluble in water and aqua regia.
Hazard: Tolerance (as Zr), 5 mg per cubic meter of air.
Uses: Coatings resistant to flame or blast impingement; special alloys.

zirconium disulfide ZrS_2. Gray crystalline solid; sp. gr. 3.87; m.p. ca. 1550°C; insoluble in water. Used as a solid lubricant.
Hazard: Tolerance (as Zr), 5 mg per cubic meter of air.

zirconium fluoride. See zirconium tetrafluoride.

zirconium glycolate $H_2ZrO(C_2H_2O_3)_3$.
Properties: Solid; decomposes without melting on heating to about 220°C; insoluble in water and organic solvents; soluble in alkali and sulfuric acid solutions. One or more of the acidic hydrogens may be replaced by alkali metals or ammonium to give water-soluble salts.
Containers: 250-lb (polyethylene-lined) fiber drums (41 gals).
Hazard: Tolerance (as Zr), 5 mg per cubic meter of air.
Uses: Cosmetics (deodorant); medicine; sequestrant.

zirconium hydride ZrH_2. Contains 1.7 to 2.1% combined hydrogen which can driven off in a vacuum above 600°C.
Properties: Gray-black metallic powder. Stable toward air and water. Sp. gr. 5.6. Autoignition temp. 518°F.
Derivation: Reduction of zirconia with calcium hydride or magnesium in the presence of hydrogen; direct combination of hydrogen and zirconium.
Grades: Commercial (contains hafnium); reactor (hafnium-free); electronic.
Hazard: Flammable, dangerous fire risk, especially in presence of oxidizing agents. Tolerance (as Zr), 5 mg per cubic meter of air.
Uses: Vacuum-tube getter; powder metallurgy; source of hydrogen; metal-foaming agent; nuclear moderator; reducing agent; hydrogenation catalyst.
Shipping regulations: (Air) Flammable Solid label. Not acceptable on passenger planes.

zirconium hydroxide $Zr(OH)_4$.
Properties: White, bulky, amorphous powder. Soluble in dilute mineral acids; insoluble in water and alkalies. Sp. gr. 3.25; decomposes to ZrO_2 at 550°C.
Derivation: Action of a solution of sodium hydroxide on a solution of a zirconium salt.
Hazard: Tolerance (as Zr), 5 mg per cubic meter of air.
Use: Zirconium compounds; pigments; dyes; glass.

zirconium hydroxychloride solution. See zirconyl hydroxychloride solution.

zirconium lactate $H_4ZrO(CH_3CHOCO_2)_3$.
Properties: White, slightly moist pulp; decomposes without melting; very slightly soluble in water and the common organic solvents; soluble in aqueous alkalies with formation of salts; decomposes to hydrous zirconia above pH 10.5. Efficient odor absorber. Combustible.
Grade: Zirconia 25% (min).
Containers: 250-lb (polyethylene-lined) fiber drums.
Hazard: Tolerance (as Zr), 5 mg per cubic meter of air.
Uses: Body deodorants; source of zirconia.

zirconium naphthenate.
Properties: Amber-colored, heavy transparent liquid. Viscosity equivalent to that of heavy lubricating oil. Very stable. Unlike other metallic naphthenates, possesses no drying properties. Soluble in all common solvents. Sp. gr. 1.05. Combustible.
Derivation: By heating a mixture of naphthenic acid and zirconium sulfate.
Hazard: Tolerance (as Zr), 5 mg per cubic meter of air.
Uses: Ceramics (enamels, glazes); lubricants; paints and varnishes (anti-chalking agent, minimizer of moisture and solar radiation effects).

zirconium nitrate $Zr(NO_3)_4 \cdot 5H_2O$.
Properties: White hygroscopic crystals. Soluble in water and alcohol. Decomposes at 100°C.
Derivation: Action of nitric acid on zirconium oxide.
Containers: Kegs, drums.
Hazard: Dangerous fire and explosion risk in contact with organic materials; strong oxidizing agent. Tolerance, 5 mg per cubic meter of air.
Use: Preservative.
Shipping regulations: (Rail) Nitrates, n.o.s., Yellow label. (Air) Oxidizer label.

zirconium nitride ZrN. A brassy-colored powder produced by heating the metal in nitrogen.
Properties: Sp. gr. 7.09; hardness Mohs 8+; m.p. 2930°C; slightly soluble in dilute hydrochloric or

sulfuric acid; soluble in concentrated acids.
Hazard: Tolerance (as Zr), 5 mg per cubic meter of air.
Uses: Special crucibles; cermets; refractories.

zirconium orthophosphate. See zirconium phosphate.

zirconium oxide (zirconia; zirconium dioxide; zirconic anhydride; zirconium anhydride) ZrO_2. Occurs in nature as baddeleyite.
Properties: Heavy white amorphous powder; sp. gr. 5.73; m.p. 2700°C; Mohs hardness 6.5; refractive index 2.2; insoluble in water and most acids or alkalies at room temperature; soluble in nitric acid and hot concentrated hydrochloric, hydrofluoric, and sulfuric acids. Most heat-resistant of commercial refractories; dielectric.
Derivation: By heating zirconium hydroxide or zirconium carbonate.
Grades: Reagent; technical; crystals; fused; whiskers; C.P. (99% zirconia); hydrous. The fused grade is reported to be harder than diamond (11 on Mohs scale).
Containers: Bags; drums; carloads.
Hazard: Tolerance (as Zr), 5 mg per cubic meter of air.
Uses: (Unstabilized) Production of piezoelectric crystals; high-frequency induction coils; colored ceramic glazes; special glasses; source of zirconium metal; heat-resistant fibers; (Hydrous) odor absorbent; to cure dermatitis caused by poison ivy. (Stabilized with CaO) Refractory furnace linings, crucibles, etc.; solid electrolyte for batteries operating at 1000°C or more.

"Zirconium Oxide, E.F."[337] Trademark designation for zirconium oxide available in lump or milled form. Lump contains 97–98% zirconium oxide and milled, 94.9%. Melting point, 4900°F. Used in massive packing of high temperature installations and refractory uses up to 4000°F not requiring stabilized zirconia.

zirconium oxychloride (zirconium chloride, basic; zirconyl chloride) $ZrOCl_2 \cdot 8H_2O$.
Properties: White, silky crystals; loses $6H_2O$ at 150°C, $8H_2O$ at 210°C; density 44 lb/cu ft. Soluble in water, methanol, and ethanol; insoluble in other organic solvents. Aqueous solutions are acidic.
Derivation: Action of hydrochloric acid on zirconium oxide.
Grades: Technical; 36% ZrO_2; H.P.
Containers: Barrels; 250-lb (plastic-lined) fiber drums; carloads.
Uses: Textile, cosmetic, and grease additive; antiperspirant; water repellents; chemical reagent; zirconium salts; in lakes and toners of acid and basic dyes; oil-field acidizing aid.

zirconium phosphate (zirconium phosphate, basic; zirconium orthophosphate) $ZrO(H_2PO_4)_2 \cdot 3H_2O$.
Properties: White, dense, amorphous powder. Decomposes on heating. Soluble in acids; insoluble in water and organic solvents. Extensively hydrolyzed in basic solution.
Derivation: Action of phosphoric acid on zirconium hydroxide.
Containers: Glass bottles; fiber drums; carloads.
Hazard: Tolerance (as Zr), 5 mg per cubic meter of air.
Uses: Chemical reagent; cation scavenger; coagulant; carrier for radioactive phosphorus.

zirconium potassium chloride (potassium zirconium chloride) $ZrCl_4 \cdot KCl$. A source of zirconium for magnesium alloys; to remove iron in an insoluble form.

zirconium potassium fluoride ZrK_2F_6. (potassium fluozirconate, potassium zirconifluoride).
Properties: White crystals. Soluble in water (hot).
Hazard: Toxic and irritant. Tolerance (as F), 2.5 mg per cubic meter in air.
Uses: Grain refiner in magnesium and aluminum; welding fluxes; catalyst; optical glass.

zirconium potassium sulfate (potassium zirconium sulfate) $2K_2SO_4 \cdot Zr(SO_4)_2 \cdot 3H_2O$.
Properties: White, crystalline powder. Slightly soluble in water.
Hazard: Tolerance, 5 mg per cubic meter in air.

zirconium pyrophosphate ZrP_2O_7.
Properties: White solid; stable to about 1550°C; insoluble in water and dilute acids other than hydrofluoric acid; coefficient of thermal expansion 5×10^{-6} (approx) at 1000°C.
Containers: 250-lb fiber drums; carloads.
Hazard: Tolerance (as Zr), 5 mg per cubic meter in air.
Uses: Refractory; olefin polymerization catalyst; phosphor.

zirconium silicate. See zircon.

zirconium silicide. See zirconium disilicide.

"Zirconium Spinel."[337] Trade designation for a synthetic complex containig 39–41% zirconium oxide, 20–22% silicon dioxide, 18.5–20.5% aluminum oxide, and 17–21% zinc oxide. M.p. 3100°F. Used as a glaze opacifier in the ceramic industry.

zirconium sulfate $Zr(SO_4)_2 \cdot 4H_2O$.
Properties: White, crystalline powder; bulk density 70 lb/cu ft; decomposes to monohydrate at about 100°C. Soluble in water; slightly soluble in alcohol; insoluble in hydrocarbons. Aqueous solutions are strongly acidic; will precipitate potassium ions and amino acids from solution; are decomposed by bases and heat.
Derivation: Action of sulfuric acid on zirconium hydroxide.
Containers: Glass bottles; steel drums.
Hazard: Tolerance (as Zr), 5 mg per cubic meter in air.
Uses: Chemical reagent; lubricants; catalyst; protein precipitation; tanning of white leather.

zirconium sulfate, basic (zirconyl sulfate) $Zr_5O_8(SO_4)_2 \cdot xH_2O$ (approximately). Similar in properties to the oxychloride, and is prepared in a similar fashion, the end result being in cake form. Used in textile treatment and white leather tanning and retannage.

zirconium tetraacetylacetonate (zirconium acetylacetonate) $Zn[OC(CH_3){:}CHCO(CH_3)]_4$.
Properties: A colorless, crystalline tetrachelate; sp. gr. 1.415; m.p. 194–5° C (decomposition) begins at 125°C). Soluble in pyridine, acetone, benzene, and other organic solvents having some polarity; slightly soluble in water.
Derivation: Reaction between zirconyl chloride, acetylacetone and sodium carbonate.
Hazard: Tolerance (as Zr), 5 mg per cubic meter in air.
Uses: Crosslinking agent for polyol, polyester, and polyalkyoxy resins; lubricant and grease additive; reagent; catalyst.

zirconium tetrachloride (zirconium chloride) $ZrCl_4$.
Properties: White, lustrous crystals. Soluble in alco-

hol; decomposes in water. Sp. gr. 2.8; sublimes above 300°C.
Derivation: Action of hydrochloric acid on zirconium hydroxide.
Grade: Technical.
Containers: Glass bottles; drums.
Hazard: Toxic and irritant. Tolerance (as Zr), 5 mg per cubic meter in air.
Uses: Source of the pure metal (formed as intermediate in process); analytical chemistry; water repellents for textiles; tanning agent; catalysis; zirconium compounds.
Shipping regulations: (Air) Corrosive label.

zirconium tetrafluoride (zirconium fluoride) ZrF_4.
Properties: White powder; sp. gr. 4.43; m.p. ca. 600°C (sublimes). Slightly soluble in water and hydrogen fluoride.
Hazard: Toxic; strong irritant. Tolerance (as F), 2.5 mg per cubic meter of air.
Use: Component of molten salts for nuclear reactors.

Zircon Milled "G."[337] Trademark for a form of zirconium silicate of 97% minimum purity. Used in electrical and chemical porcelains and refractories.

zirconocene dichloride (dicyclopentadienylzirconium dichloride) $(C_5H_5)_2ZrCl_2$.
Properties: White crystals; m.p. 244°C; soluble in polar organic solvents; stable in dry air, very slowly hydrolyzes in moist air.
Containers: Bottles and fiber drums.
Hazard: Toxic by inhalation and skin contact; irritant to eyes and mucous membranes.
Uses: Rubber accelerator; component of a catalyst system for polymerization of vinyl monomers; curing agent for water-repellent silicone materials; agent for plating with zirconium.
See also metallocene.

zircon, porcelain. See porcelain, zircon.

zircon sand. A sand containing considerable zirconium, titanium and related metals. Used as a source of these elements, and also as a heat-resistant sand for casting of alloys.

zirconyl carbonate. See zirconium carbonate, basic.

zirconyl chloride. See zirconium oxychloride.

zirconyl hydroxychloride $ZrOOHCl \cdot nH_2O$.
Properties: Colorless or slightly amber liquid (aqueous solution); sp. gr. 1.26; forms a soluble glass on evaporation; pH of solution 0.8 (approx); reacts with alkalies to form hydrous zirconia. Contains 20% zirconia.
Containers: 500-lb (plastic-lined) steel drums; carloads.
Uses: Pharmaceuticals; deodorants; precipitation of acid dyes; zirconium compounds; water repellents for textiles.

zirconyl nitrate (basic) (zirconyl hydroxynitrate) $ZrO(OH)NO_3$.
Properties: Aqueous solution; Sp. gr. (25°C) 1.35.
Hazard: Fire risk in contact with organic materials.
Uses: Gelatins and improving lamination bonds of polyvinyl alcohol.

zirconyl sulfate. See zirconium sulfate, basic.

"Zircopax."[337] Trademark for zirconium silicate, 94–96.5% min. purity. Used as an opacifier for ceramic glazes.

"Zircotan."[23] Trademark for zirconium tanning agents which produce through-white leather.
Uses: Tannage of white kid suede, glove leathers; retannage of chrome leather.

"Zirex."[79] Trademark for a special zinc resinate having a high melting point; high zinc content (twice as high as "Zitro") with a pale color and a negative acid number (slightly basic).
Uses: Paints; varnishes; adhesives; ethyl cellulose lacquers.

"Zirmul."[408] Trademark for a bonded alumina-zirconia-silica refractory containing a minimum of glassy phase. It has broad applications in construction of glass melting furnaces, both above and below the glass line.

"Zirox" B.[337] Trademark for a polishing compound of zirconium oxide. Used for ophthalmic lenses, precision glass, marble, and granite.

"Zitro."[79] Trademark for a zinc resinate having a high melting point, pale color and low acid number.
Uses: Paints; varnishes; adhesives; ethyl cellulose lacquers.

ZMA. Abbreviation for zinc metaarsenite. See zinc arsenite.

Zn Symbol for zinc.

zoalene. Generic name for 3,5-dinitro-ortho-toluamide $(O_2N)_2C_6H_2(CH_3)CONH_2$.
Properties: Yellowish solid; m.p. 177°C. Very slightly soluble in water; soluble in acetone, acetonitrile, dioxane, and dimethylformamide.
Use: Coccidiostat; permissible food additive.

"Zoamix."[233] Trademark for poultry coccidiostats containing 3,5-dinitro-o-toluamide.

"Zobar."[28] Trademark for a weed killer based on an aqueous solution of the dimethylamine salts of polychlorobenzoic acids, containing 4 lb of acid equivalent per gallon.

"Zolex."[457] Trademark for an aromatic solvent, proposed as an alternate for xylene. Distillation range (ASTM) 270-325°F.

"Zonarez."[252] Trademark for a series of polyterpene resins. Thermoplastic polymers produced by the polymerization of terpene hydrocarbons consisting primarily of beta-pinene and dipentene. Available in two series of products—Zonarez B and Zonarez 7000. The B series is produced by the polymerization of beta-pinene, whle the 7000 series is based primarily on dipentene. Products of comparable softening points in the two series are differentiated mainly by molecular weight, molecular weight distribution and polymer structure. These resins are bright, clear and pale-colored, low molecular weight polymers which impart high levels of tack and adhesion to many elastomeric and polymeric materials. They have excellent stability in adhesive systems whereby variations in color, viscosity, holding power and tack are minimized. They are neutral in nature, have good resistance to attack by acids, alkalies, salts and water, and exhibit excellent compatability with numerous organic solvents including aliphatic and aromatic hydrocarbons, chlorinated hydrocarbons, rubbers and elastomers, drying oils, polyethylenes, ethylene copolymers, waxes, esters, terpenes, rosin and rosin derivatives.
Uses: Manufacture of pressure-sensitive adhesives for tapes and labels, rubber cements, solvent-based adhesives, emulsion adhesives, hot-melt adhesives and coatings, can sealants, caulking and general

sealant compounds, inks, investment casting waxes, paints, concrete waterproofing agents, varnishes, chewing and bubble gum bases, moisture resistant soft gelatin capsules and powders of ascorbic acid and its salts, and other applications which require low molecular weight linear polymers that promote tack and adhesion and have excellent aging properties.

zone refining. A purification process that involves repeated melting and crystallization. The sample to be purified is placed in a relatively long narrow tube and then passed slowly through a furnace having short, alternate hot and cold zones. Melting occurs opposite the hot zones and crystallization opposite the cold zones. As the rod moves through the furnace the zones move along the rod. Impurities remain in the molten zones and so are carried to one end of the rod. The process has been most used for relatively high cost materials used in small quantities at very high purities, as for solid state electronic purposes.

"Zonolite."[241] Trademark for verxite (expanded hydrobiotite) (q.v.).

"Zonyl."[28] Trademark for a series of surface-active agents and lubricants.

A. Modified polyethylene glycol type nonionic.

E-7. A fluoroalkyl ester, the condensation product of pyromellitic anhydride and a mixture of C-5 and C-7 trihydrofluoroalcohols.

E-91. A fluoroalcohol camphorate.

S-13. The free acid of a fluoroalkyl phosphate.

"Zopaque."[296] Trademark for pure titanium dioxide (q.v.), manufactured from ilmenite and specially processed to control crystal growth.

Uses: White pigment in rubber, plastics, paints, etc.

"Zoron" WSP.[28] Trademark for a resin finish used to insolubilize starch, polyvinyl alcohol and other pigments. Used by paper industry to impart a water resistant finish.

zoxazolamine (2-amino-5-chlorobenzoxazole) $C_7H_5ClN_2O$.

Properties: White to creamy white, odorless powder or glistening crystals; freely soluble in alcohol; nearly insoluble in water; m.p. 183–188°C.

Use: Medicine.

Zsigmondy, Richard (1865–1929). A native of Austria, Zsigmondy received the Nobel Prize in chemistry in 1925 for his work in the field of colloid chemistry, which was initiated by his interest in ruby glass (a colloidal gold suspension). His most important contribution to chemistry was his invention of the ultramicroscope (with Siedentopf) in 1903. By illuminating the sample with a light beam at right angles to the incident light in a compound microscope, the presence of particles in a suspension as small as 3 millimicrons could be detected due to the light-scattering effect of the particles. This phenomenon had been discovered by John Tyndall a few years earlier, but Zsigmondy was the first to apply it to microscopy. The ultramicroscope became of immeasurable value in the study of colloidal suspensions.

Zr Symbol for zirconium.

zwitterion. See isoelectric point.

zymase. The enzyme present in yeast which converts sugars to alcohol and carbon dioxide.

zymohexase. See aldolase.

"Zymol."[221] Trademark for a chemically modified linseed oil for part or complete replacement for tung oil in varnish manufacturing.

"Zyrol."[487] Trademark for a concentrated amine condensate surfactant particuliarly useful for woolens.

"Zytel."[28] Trademark for a nylon resin available as molding powder, extrusion powder and soluble resin.

"Zytron."[233] Trademark for O-[2,4-dichlorophenyl]-O-methyl isopropylphosphoramidothioate (q.v.).

Hazard: May be toxic.

APPENDIX I

Manufacturers of Trade marked Products

Numerical List

(for addresses, see Alphabetical List, page 952)

1. Stauffer Chemical Co.
3. Abbott Laboratories
4. Rhode Island Laboratories, Inc.
5. Firestone Tire and Rubber Co.
7. Sanforized Co.
8. Scientific Chemicals, Inc.
9. Knoll Pharmaceutical Co.
10. Van Dyk & Co., Inc.
11. Koppers Co., Inc.
12. Sindar Corporation
13. Formica Corporation
15. Armour Industrial Chemical Co.
15a. Reheis Chemical Co.
16. Thomas, Arthur H. Co.
17. Heresite & Chemical Co.
19. Inmont Corp.
19a. Interpace Corp.
20. Corning Glass Works
21. Neville Chemical Co.
22. Staley Chemical Co.
23. Rohm & Haas Co.
24. Wyeth Laboratories
25. Pennsylvania Refining Co.
27. Thiokol Chemical Corporation
28. Du Pont de Nemours, E. I. & Co.
29. Enjay Chemical Co.
30. Corn Products Co.
31. Handy & Harman
35. Firestone Plastics Co.
36. Reichhold Chemicals, Inc.
38. Sierra Talc & Chemical Corporation
40. V-C Chemical Co.
41. Atlas Minerals and Chemicals Division, Electric Storage Battery Co.
42. Sun Chemical Corporation
45. See 104
46. Acheson Colloids Co.
47. Duriron Company, Inc.
48. Cleary, W. A. Corporation
50. Allied Chemical Corp.; (all divisions).
51. Humble Oil & Refining Co.
52. Rubber Corporation of America
53. National Starch and Chemical Corporation
54. Carrier Corporation
55. FMC Corporation
56. Canadian Industries Ltd.
57. American Cyanamid Co.
58. Monsanto Industrial Chemicals Co.
60. Cerro Sales Corporation
62. Hooker Chemical Corporation
63. Richardson Co., The
64. Spencer Kellogg Division, Textron, Inc.
65. Borden Chemical
67. Climax Molybdenum Co.
69. Vanderbilt Co., R. T., Inc.
70. Searle Chemicals, Inc.
71. Smith Kline & French Laboratories
73. Glyco Chemicals, Inc.
74. Tenneco Chemicals, Inc., Intermediates Division
79. Tenneco Chemicals, Inc., Newport Division
81. Ventron Corp.
82. Graphite Metallizing Corp.
84. Olin Corp.
85. Shulton Inc.
88. American Potash & Chemical Corp.
89. ICI United States, Inc.
91. Schwarz/Mann
92. Masonite Corp.
93. Tennessee Corp.
94. Hall, C. P., Co.
97. Chesebrough-Pond's, Inc.
98. Floridin Co.
99. Minerals & Chemicals Div., Engelhard Minerals & Chemicals Corp.
100. Lilly, Eli, and Co.
103. Hoyt, Arthur S., Co., Inc.
104. Witco Chemical Co, Inc.
106. American Mineral Spirits Co.
107. American Norit Co., Inc.
108. See 323.
109. Arapahoe Chemicals, Inc.
110. United Carbon Co.
114. Premier Malt Products, Inc.
115. Eastman Kodak Co.
116. Allis-Chalmers Corp.
117. Alox Corp.
118. California Industrial Minerals Co.
119. Goodrich, B. F., Chemical Co.
121. American Can Co.
122. G & A Laboratories, Inc.
123. Merck & Co., Inc.
124. Marine Colloids, Inc.
125. Shell Chemical Co.
126. Pacific Lumber Co.
128. Petrolite Corp., Bareco Division
129. Weyerhaeuser Co.
133. Columbian Carbon Co., Inc.
134. Harshaw Chemical Co.
136. Atlantic Refining Co.
137. Jefferson Chemical Co., Inc.
138. Tenneco Chemicals, Inc., Heyden Div.
139. Moores Lime Co.
140. Pennsylvania Industrial Chemical Corp.
141. Sherwin-Williams Chemicals

142. Enthone Inc.
144. Airco Chemicals and Plastics
145. Lithcote Corp.
147. Chipman Chemical Co.
148. National Lead Co., DeLore Division
149. Dow Corning Corp.
151. Chevron Chemical, Co.
152. Swift Edible Oil Co.
154. Wallace & Tiernan, Inc., Lucidol Division
155. Driver, Wilbur B., Co.
157. Nulomoline Division, SuCrest Corp.
158. 3M Company
159. Royce Chemical Co.
160. Simplex Wire & Cable Co.
161. Hitco
162. Winthrop Laboratories, Sterling Drug Co.
164. Rumford Chemical Works
165. Apex Chemical Co., Inc.
166. Hoskins Mfg. Co.
167. Kendall Refining Co.
168. Lake States Chemical Div., St. Regis Paper Co.
169. LaMotte Chemical Products Co.
170. Paisley Products
173. Wallerstein Co.
175. See 50.
177. PPG Industries, Inc.
179. General Tire & Rubber Co.
181. Chemagro Agricultural Div., Mobay Chemical Corp.
182. Nalco Chemical Co.
183. Natural Gas Odorizing, Inc.
184. Permutit Co.
188. Fritzsche Dodge & Olcutt, Inc.
189. Wallace & Tiernan, Inc., Harchem Div.
190. Hoffmann-LaRoche, Inc.
191. Owens-Corning Fiberglas Corp.
192. U. S. Industrial Chemicals Co.
196. American Agricultural Chemical Co.
197. See 50.
200. Apco Oil Corp.
201. Philadelphia Quartz Co.
202. Baker Castor Oil Co.
203. BASF-Wyandotte Corp.
204. Pennsalt Chemicals Corp.
205. Grace & Co., W. R. Construction Products Div.
206. See 89.
206a. Chemical Mfg. Co., Inc.
210. Ionac Chemical Co.
212. Miles Laboratories, Inc.
214. Union Carbide Corp.
216. Amoco Chemicals Corp.
217. Filtrol Corp.
218. General Refractories Co.
219. Ciba-Geigy Corp.
220. Thermal American Fused Quartz Co.
221. Archer Daniels Midland Co.
222. Arco Chemical Co.
223. Osborn, C. J., Chemicals, Inc.
224. Quaker Oats Co.
225. Bromine Producers Div., Drug Research, Inc.
226. Aluminum Co. of America
227. Givaudan Corp.
228. West Virginia Pulp & Paper Co.
230. Carlisle Chemical Works, Inc., Advance Division
231. Brown Co.
232. Holliday, L. B., & Co., Ltd.
233. Dow Chemical Co.
235. National Rosin Oil Products, Inc.
236. National Lead Co., Baroid Div.
238. Chemical Development Corp.
239. Clay-Adams, Inc.
241. Grace, W. R., & Co., Davison Chemical Division
242. Emery Industries, Inc.
243. See 50.
244. Diamond Alkali Co.
245. General Electric Co., Polymers Product Dept.
247. Johns-Manville Products Corp.
248. Uniroyal Chemical
249. Norton Co., Metals Division
250. Foote Mineral Co.
251. Republic Steel Corp.
252. Arizona Chemical Co.
253. See 151.
256. Eastman Chemical Products, Inc.
259. General Mills Chemicals, Inc.
261. FMC Corp., Avicel Dept.
263. Scott Bader & Co., Ltd.
265. Goodyear Tire & Rubber Co.
266. Hercules, Inc.
267. Kenrich Petrochemicals, Inc.
268. New Jersey Zinc
269. Worthington Corp.
270. Mearl Corp.
271. American Lignite Products Co.
272. Ames Laboratories, Inc.
274. Beckman Instruments, Inc.
275. Cabot Corp.
277. Xylos Rubber Co.
278. Firestone Synthetic Rubber and Latex Co.
280. See 442.
281. Budd Co.
282. See 154, 189.
283. Huntington Alloy Products Div., International Nickel Co., Inc.
284. West Chemical Products, Inc.
285. Huber, J. M., Corp.
287. Beaunit Corp.
288. M & T Chemicals, Inc.
289. Lockrey Co.
292. See 50.
293. Inmont Corp.
296. Glidden-Durkee Div., SCM Corp.
299. Pfizer, Inc., Chemicals Division
300. Arkansas Co., Inc.
301. Burroughs Wellcome & Co., Inc.
303. Phillips Petroleum Co.
304. National Lead Co.
305. Ciba Pharmaceutical Co.
307. GAF Corp.
308. Westinghouse Electric Corp.
309. Nopco Division, Diamond Shamrock Chemical Co.
310. Norac Co., Inc.
311. Grace, W. R., & Co., Organic Chemicals Div.
312. Evans Chemetics, Inc.
312a. Fabric Research Laboratories
313. Ethyl Corp.
315. American Cyanamid Co., Fine Chemicals Dept.
316. Velsicol Chemical Corp.
319. Commercial Solvents Corp.
320. Houghton, E. F., & Co.
321. Schering Corp.
322. Kelco Co.
323. Calgon Corp., Pittsburgh Activated Carbon Division
324. Anaconda American Brass Co.
325. I.C.I. Organics, Inc.
326. U. S. Stoneware Co.
327. Upjohn Co.
328. Onyx Chemical Co.
329. Mallinckrodt, Inc.
330. Parke, Davis & Co.
331. Mobil Chemical Co.
332. National Filter Media Corp.
336. See 304.

337. See 304.
342. Penick, S. B., & Co.
343. Hooker Chemical Corp., Parker Rustproof Div.
345. Hammond, W. A., Drierite Co.
346. Consumer Products Division, Sterling Drug Co.
347. Kennametal, Inc.
348. Hynson, Westcott & Dunning, Inc.
349. Haveg Industries, Inc.
350. Driver-Harris Co.
351. Celotex Corp.
352. Celanese Chemical Co.
353. Catalin Corp.
354. See 346.
355. Wisconsin Alumni Research Fdn.
400. Aceto Chemical Co., Inc.
401. Morton Chemical Co.
403. Smith, Werner G., Inc.
405. Mona Industries, Inc.
406. Meer Corp.
407. Ampco Metal, Inc.
408. Taylor Sons, Charles, Co.
409. Skelly Oil Co.
410. Gross, A., & Co.
412. Squibb, E. R., & Sons, Inc.
415. Robeco Chemicals, Inc.
416. Universal Oil Products Co.
418. Roussel Corp.
419. Cadet Chemical Corp.
420. Union Camp Corp.
422. Verona Division, Baychem Corp.
424. Johnson, S. C. & Son, Inc.
425. Applied Science Laboratories, Inc.
426. Michigan Chemical Corp.
427. Albermar Co.
428. Cowles Chemical Co.
429. Avisun Corp.
430. Fine Organics, Inc.
431. Biddle, James G., Co.
433. American Metal Climax, Inc.
434. National Lead Co., Evans Lead Div.
436. Pennsylvania Glass Sand Corp.
438. See 50.
439. See 304.
440. See 203.
440a. Baird Chemical Industries, Inc.
441. U. S. Borax & Chemical Corp.
442. Harbison-Carborundum Corp.
443. See 219.
445. Devcon Corp.
446. Harbison-Walker Refractories Co.
448. See 296.
449. See 104.
450. American Hoeschst Corp.
451. Stepan Chemical Co.
453. Apex Smelting Co.
454. American Rubber & Chemical Co.
455. Babcock & Wilcox Co.
457. Brown, R. J., Co.
458. Carlisle Chemical Works, Inc.
459. Carus Chemical Co.
460. Chemactants, Inc.
461. Chemtree Corp.
463. Interplastic Corp.
464. Consolidated Astronautics, Inc.
465. Dayton Chemical Products Laboratories
466. King, Robert J., Co.
467. Eastman Kodak Co.
468. Eagle-Picher Industries, Inc.
469. El Monte Chemical Co.
470. Farbenfabriken Bayer AG
471. Georgia Kaolin Co.
472. Air Products & Chemicals, Inc.
473. Humko-Sheffield Chemical
474. Jones Dabney Div., Devoe & Raynolds Co., Inc.
475. Kaiser Chemicals Div., Kaiser Aluminum
476. Lignosol Chemicals Ltd.
477. Loctite Corp.
478. Marblette Co.
479. Millhaven Fibres, Ltd.
480. Mooney Chemicals, Inc.
481. National Gypsum Co.
482. Nixon-Baldwin Chemicals, Inc.
483. Nye, William F., Inc.
484. Ozark-Mahoning Co.
485. Pharmacia Fine Chemicals, Inc.
486. Poco Graphite, Inc.
487. Procter & Gamble Co.
488. Puerto Rico Chemical Co., Inc.
489. Reynolds Metal Co.
491. Stainless Foundry & Engineering, Inc.
492. Staley, A. E., Mfg. Co.
493. American Cholesterol Products Inc.
494. Sun Oil Co.
495. Synvar Corp.
496. Valchem, Chemical Division
497. Wilson Pharmaceutical & Chemical Corp.
498. See 104.
499. See 104.
500. See 104.
502. Dawe's Laboratories, Inc.
503. Douglas Laboratories
504. Reactor Experiments, Inc.
505. See 416.
506. General Adhesives & Chemical Co.
507. McDanel Refractory Porcelain Co.
508. Breon Laboratories
510. Kay-Fries Chemicals, Inc.
511. National Polychemicals, Inc.
512. Hodag Chemical Corp.
513. Tylac, Standard Brands Chemical Industries, Inc.
515. Southern Latex Corp.
516. Polyvinyl Chemical Industries
517. Grace, W. R., & Co., Hampshire Div.
518–520. Upjohn Co.
521. High Temperature Materials, Inc.
522. Pfaudler Co.
523. See 50.
524. Hunter Engineering Co.
525. Borg-Warner Chemicals
526. International Rustproof Co.
528. See 188.
529. Canrad Precision Industries, Inc.
530. Elanco Products Co.
531. Amerace Corp.
532. Escambia Chemical Corp.
532a. Essex Chemical Corp.
533. Gulf Oil Chemicals Co.
535. Selas Flotronics
536. Nease Chemical Co., Inc.
538. Mallinckrodt, Inc., Washine Div.
539. Fiberfil, Inc.
540. Stevens, J.P., & Co., Inc.
541. Aldrich Chemical Co., Inc.
542. Alcolac Inc.
544. Continental Oil Co.
545. Pitt-Consol Chemical Co.
546. Warner-Chilcott Laboratories
547. Hughson Chemical Co.
548. Ansul Co.
550. Woburn Chemical Corp.
551. Huyck Metals Co.
552. Fairmount Chemical Co.
553. Stein, Hall & Co., Inc.
554. Barium and Chemicals, Inc.

555. Drew Chemical Corp.
557. Kolene Corp.
559. Nalge Co., Inc.
560. Textilana Corp.
561. Ameripol, Inc.
562. Degussa, Inc.
563. Humphrey Chemical Co.
564. See 151.
565. Dow Badische Co.
566. Sinclair Petrochemicals, Inc.
567. Mobay Chemical Co.
569. Texaco Inc.
570. Kohnstamm, H., & Co., Inc.
572. General Latex & Chemical Co.
573. See 311.
574. Amchem Products, Inc.
575. Alrac Corp.
576. Biodegradable Plastics, Inc.
577. Renite Co.
578. Moleculon Research Corp.
579. See 309.
580. Christy Chemical Corp.
581. R. W. Barker & Co.
582. Hamblet & Hayes Co.
583. Napp Chemicals

APPENDIX II

Manufacturers of Trademarked Products

Alphabetical List

A

Abbott Laboratories
Chemical Division
1400 Sheridan Rd.
North Chicago, Ill. 60064

Aceto Chemical Co., Inc.
126–02 Northern Blvd.
Flushing, N.Y. 11368

Acheson Colloids Co.
Port Huron, Mich.

Airco Chemicals & Plastics
150 East 42nd St.
New York, N.Y. 10017

Air Products & Chemicals, Inc.
Chemical Intermediates
Box 538
Allentown, Pa. 18105

Albermar Co.
51 Madison Ave.
New York, N.Y. 10010

Alcolac Inc.
3440 Fairfield Rd.
Baltimore, Md. 21226

Aldrich Chemical Co., Inc.
940 W. St. Paul Ave.
Milwaukee, Wis. 53233

Amchem Products, Inc.
Ambler, Pa. 19002

Allied Chemical Corp.
Morristown, N.J. 07960

Allis-Chalmers Corp.
Milwaukee, Wis. 53201

Alox Corp.
Niagara Falls, N.Y. 14302

Alrac Corp.
649 Hope St.
Stamford, Conn. 06907

Aluminum Co. of America
Pittsburgh, Pa. 15219

Amerace Corp., Penetone Div.,
74 Hudson Ave.
Tenafly, N.J. 07670

American Agricultural
Chemical Co.
100 Church St.
New York, N.Y. 10007

American Can Co.
247 American Lane
Greenwich, Conn. 06830

American Cholesterol Products,
Inc.
Edison, N.J. 08817

American Cyanamid Co.
Fine Chemicals Dept.
Pearl River, N.Y. 10965

American Cyanamid Co.
Pigments Division
Bound Brook, N.J. 08805

American Cyanamid Co.
Plastics Division
Berdan Ave.
Wayne, N.J. 07470

American Hoechst Corp.
Dyes & Pigments Div.
Box 2500
Somerville, N.J. 08876

American Lignite Products Co.
Ione, California

American Metal Climax, Inc.
1270 Ave. of Americas
New York, N.Y. 10020

American Mineral Spirits Co.
Division of Union Oil Co.
of California
Palatine, Ill. 60067

American Norit Co., Inc.
6301 Glidden Way
Jacksonville, Fla. 32208

American Potash & Chemical
Corp.
3000 West 6th St.
Los Angeles, Cal. 90054

American Rubber & Chemical Co.
Box 1034
Louisville, Ky. 40201

Ameripol, Inc.
3135 Euclid Ave.
Cleveland, O. 44115

Ames Laboratories, Inc.
200 Rock Lane
Milford, Conn. 06460

Amoco Chemicals Corp.
200 E. Randolph Drive
Chicago, Ill. 60601

Ampco Metal, Inc.
Box 2004
Milwaukee, Wis. 53201

Anaconda American Brass Co.
Waterbury, Conn. 06720

Ansul Co.
Marinette, Wis. 54143

Apco Oil Corp.
Box 1841
Oklahoma City, Okla. 73101

Apex Chemical Co., Inc.
200 So. First St.
Elizabethport, N.J. 07206

Apex Smelting Co.
Division of American Metal
Climax, Inc.
6700 Grant Ave.
Cleveland, O. 44105

Applied Science Laboratories, Inc.
Box 440
State College, Pa. 16801

Arapahoe Chemicals, Inc.
2075 No. 55th St.
Boulder, Colorado 80301

Archer Danials Midland Co.
Box 1470 Decatur, Ill. 62525

Arco Chemical Co.
1500 Market St.
Philadelphia, Pa. 19101

Arizona Chemical Co.
Wayne, N.J. 07470

Arkansas Co., Inc.
Box 210
Newark, N.J. 07101

Armour Industrial Chemical Co.
Box 1805
Chicago, Ill. 60690

Atlantic Refining Co.
Box 7258
Philadelphia, Pa. 19101

Atlas Minerals and Chemicals Co.
Mertztown, Pa. 19539

Avisun Corp.
21 South 12th St.
Philadelphia, Pa. 19107

B

Babcock & Wilcox Co.
Refractories Div.
Augusta, Ga. 30903

Baird Chemical Industries, Inc.
185 Madison Ave.
New York, N.Y. 10016

Baker Castor Oil Co.
Bayonne, N.J. 07002

Barium and Chemicals, Inc.
Box 218
Steubenville, O. 43952

Barker, R. W. & Co.
13 Charterhouse Sq.
London, England

BASF Wyandotte Corp.
Intermediate Chemicals Dept.
Box 181
Parsippany, N.J. 07054

Beaunit Corp.
261 Fifth Ave.
New York, N.Y. 10016

Beckman Instruments, Inc.
Fullerton, Cal. 92634

Biddle, James G. Co.
Plymouth Meeting, Pa. 19462

Biodegradable Plastics, Inc.
Boise, Idaho

Borden Chemical
Division Borden, Inc.
180 E. Broad St.
Columbus, O. 43215

Borg-Warner Chemicals
International Center
Parkersburg, W. Virginia 26101

Breon Laboratories
90 Park Ave.
New York, N.Y. 10016

Bromine Producers
Div. Drug Research, Inc.
Adrian, Michigan 49221

Brown Co.
555 Fifth Ave.
New York, N.Y. 10017

Brown, R. J., Co.
1430 So. Vandeventer Ave.
St. Louis, Mo. 63110

Budd Co., Polychemical Div.
Newark, Del. 19711

Burroughs Wellcome & Co.
Tuckahoe, N.Y. 10707

C

Cabot Corp.
125 High St.
Boston, Mass. 02110

Cadet Chemical Corp.
Burt, N.Y. 14028

Calgon Corp.
Pittsburgh Activated Carbon Div.
Box 1346
Pittsburgh, Pa. 15230

California Industrial Minerals Co.
Friant, California 93626

Canadian Industries, Ltd.
Box 10
Montreal, P.Q.

Canrad Precision Industries, Inc.
630 Fifth Ave.
New York, N.Y. 10020

Carlisle Chemical Works, Inc.
Reading, Ohio 45215

Carlisle Chemical Works, Inc.
Advance Division
New Brunswick, N.J. 08903

Carrier Corp.
Syracuse, N.Y. 13201

Carus Chemical Co.
1500 Eighth St.
LaSalle, Ill. 61301

Catalin Corp.
1 Park Ave.
New York, N.Y. 10016

Celanese Chemical Co.
1211 Ave. of Americas
New York, N.Y. 10036

Celotex Corp.
120 No. Florida Ave.
Tampa, Fla. 33602

Cerro Sales Corp.
300 Park Ave.
New York, N.Y. 10022

Chemactants Inc.
Edison, N.J. 08817

Chemagro Agricultural Div.,
Mobay Chemical Corp.
Box 4913
Kansas City, Mo. 64120

Chemical Development Corp.
Danvers, Mass. 09123

Chemical Mfg. Co., Inc.
Subsidiary ICI America
151 South St.
Stamford, Conn. 06904

Chemtree Corp.
Central Valley, N.Y. 10917

Chesebrough-Ponds Inc.
485 Lexington Ave.
New York, N.Y. 10017

Chevron Chemical Co.
575 Market St.
San Francisco, Cal. 94105

Chipman Chemical Co.
Box 1065
Burlingame, Cal. 94011

Christy Chemical Corp.
37 Harvard St.
Worcester, Mass.

Ciba-Geigy Corp.
Dyestuffs and Chemicals Div.
Greensboro, No. Carolina 27409

Ciba Pharmaceutical Co.
556 Morris Ave.
Summit, N.J. 07901

Clay-Adams Inc.
141 East 25th St.
New York, N.Y. 10010

Cleary, W. A., Corp.
Box 10
Somerset, N.J. 08873

Climax Molybdenum Co.
1 Greenwich Plaza
Greenwich, Conn. 06830

Columbian Carbon Co.
380 Madison Ave.
New York, N.Y. 10017

Commercial Solvents Corp.
245 Park Ave.
New York, N.Y. 10017

Consolidated Astronautics, Inc.
41–45 Crescent St.
Long Island City, N.Y. 11101

Consumer Products Division
Sterling Drug Co.
Montvale, N.J. 07645

Continental Oil Co.
Petrochemicals Dept.
Saddle Brook, N.J. 07662

Corning Glass Works
Corning, N.Y. 14832

Corn Products Co.
Englewood Cliffs, N.J. 07632

Cowles Chemical Co.
12000 Shaker Blvd.
Cleveland, O. 44120

Crown Zellerbach Corp.
Chemical Products Div.
Camas, Washington 98607

D

Dawe's Laboratories, Inc.
450 State St.
Chicago Heights, Ill. 60411

Dayton Chemical Products Labs.
West Alexandria, O. 45381

Degussa, Inc. Chemicals Div.
2 Penn. Plaza
New York, N.Y. 10001

Devcon Corp.
Danvers, Mass. 01923

Diamond Alkali Co.
Union Commerce Bldg.
Cleveland, O. 44114

Diamond Shamrock Chemical Co.
Chemetals Division
711 Pittman Rd.
Baltimore, Md. 21226

Diamond Shamrock Chemical Co.
Nopco Division
Morristown, N.J. 07960

Douglas Laboratories
2945 West 48th St.
Chicago, Ill. 60632

Dow Badische Co.
Williamsburg, Va. 23185

Dow Chemical Co.
Midland, Michigan 48640

Dow-Corning Corp.
Midland, Mich. 48640

Drew Chemical Corp.
701 Jefferson Rd.
Parsippany, N.J. 07054

Driver-Harris Co.
Harrison, N.J. 07029

Driver, Wilbur B., Co.
1875 McCarter Hwy.
Newark, N.J. 07104

DuPont de Nemours, E. I., Co.
Wilmington, Del. 19898

Duriron Co., Inc.
Dayton, O. 45401

E

Eagle-Picher Industries, Inc.
580 Walnut St.
Cincinnati, O. 45202

Eastman Chemical Products, Inc.
Box 431
Kingsport, Tenn. 37662

Eastman Kodak Co., Organic
Chemicals Dept.
343 State St.
Rochester, N.Y. 14650

Elanco Products Co.
Box 1750
Indianapolis, Ind. 46206

El Monte Chemical Co.
521 East Green St.
Pasedena, Cal. 91101

Emery Industries, Inc.
4300 Carew Tower
Cincinnati, O. 45202

Enjay Chemical Co.
Division Humble Oil & Refining Co.
60 W. 49th St.
New York, N.Y.

Enthone Inc.
Box 1900
New Haven, Conn. 06508

Escambia Chemical Corp.
Wilton, Conn.
(261 Madison Ave., N.Y. 10016)

Essex Chemical Corp.
1401 Broad St.
Clifton, N.J. 07015

Ethyl Corp.
Industrial Chemicals Div.
451 Florida St.
Baton Rouge, La. 70801

Evans Chemetics, Inc.
90 Tokeneke Rd.
Darien, Conn. 06820

F

FMC Corp., Industrial Chemical Div.
2000 Market St.
Philadelphia, Pa. 19103

FMC Corp., Avicel Dept.
Marcus Hook, Pa. 19061

FMC Corp., Agricultural Chemicals Div.
Middleport, N.Y. 14105

Fabric Research Laboratories
Dedham, Mass.

Fairmount Chemical Co.
117 Blanchard St.
Newark, N.J. 07105

Farbenfabriken Bayer AG
425 Park Ave.
New York, N.Y. 10022

Fiberfil, Inc.
Evansville, Ind. 47711

Filtrol Corp.
5959 W. Century Blvd.
Los Angeles, Cal. 90045

Fine Organics, Inc.
205 Main St.
Lodi, N.J. 07644

Firestone Plastics Co.
Box 699
Pottstown, Pa. 19464

Firestone Synthetic Rubber & Latex Co.
381 West Wilbeth Rd.
Akron, O. 44301

Firestone Tire & Rubber Co.
Akron, O. 44301

Floridin Co.
3 Penn Center
Pittsburgh, Pa. 15235

Foote Mineral Co.
Exton, Pa. 19341

Ford Motor Co.
Dearborn, Michigan

Formica Corp.
4614 Spring Grove Ave.
Cincinnati, O. 45232

Fritzsche Dodge & Olcutt, Inc.
76 Ninth Ave.
New York, N.Y. 10011

G

G & A Laboratories, Inc.
Box 1217
Savannah, Ga. 31402

GAF Corp. Chemical Products
140 W. 51st St.
New York, N.Y. 10020

General Adhesives & Chemical Co.
6100 Centennial Blvd.
Nashville, Tenn. 37202

General Electric Co.
Polymers Product Dept.
Pittsfield, Mass. 01201

General Latex & Chemical Corp.
666 Main St.
Cambridge, Mass. 02139

General Mills Chemicals, Inc.
4620 W. 77th St.
Minneapolis, Minn. 55435

General Refractories Co.
3450 Wilshire Blvd.
Los Angeles, Cal. 90010

General Tire & Rubber Co.
Chemical Division
Akron, O. 44329

Georgia Kaolin Co.
433 No. Broad St.
Elizabeth, N.J. 07207

Givaudan Corp.
100 Delawanna Ave.
Clifton, N.J. 07014

Glidden-Durkee Div., SCM Corp.
Organic Chemicals Dept.
Box 389
Jacksonville, Fla. 32201

Glyco Chemicals, Inc.
Box 700
Greenwich, Conn. 06830

Goodrich, B. F. Chemical Co.
6100 Oak Tree Blvd.
Cleveland, O. 44131

Goodyear Tire & Rubber Co.
Akron, O. 44316

Grace, W. R., & Co.
Davison Chemical Div.
Charles and Baltimore Sts.
Baltimore, Md. 21203

Grace, W. R., & Co.
Construction Products Div.
62 Whittemore Ave.
Cambridge, Mass. 02140

Grace, W. R., & Co.
Organic Chemicals Div.
55 Hayden Ave.
Lexington, Mass. 02173

Grace, W. R., & Co.
Hampshire Chemical Div.
Nashua, N.H. 03060

Graphite Metallizing Corp.
1050 Nepperhan Ave.
Yonkers, N.Y. 10702

Gross, A. & Co.
Box 818
Newark, N.J. 07101

Gulf Oil Chemicals Co.
Industrial & Specialty Chemicals Div.
Box 2100
Houston, Texas 77001

H

Hall, C. P. Co.
7300 So. Central Ave.
Chicago, Ill. 60638

Hamblet & Hayes Co.
Colonial Rd.
Salem, Mass. 01970

Hammond, W. A., Drierite Co.
Xenia, O. 45385

Handy & Harman
Stamford, Conn.
(850 Third Ave., N.Y.)

Harbison-Carborundum Corp.
Box 337
Niagara Falls, N.Y. 14302

Harbison-Walker Refractories Co.
2 Penn Center
Pittsburgh, Pa. 15222

Harshaw Chemical Co.
Division Kewanee Oil Co.
1945 E. 97th St.
Cleveland, O. 44106

Haveg Industries, Inc.
Sub. Hercules Inc.
900 Greenbank Rd.
Wilmington, Del. 19898

Hercules Inc.
910 Market St.
Wilmington, Del. 19899

Heresite & Chemical Co.
Manitowoc, Wis. 54221

High Temperature Materials, Inc.
Lowell, Mass. 01852

Hitco
Box 1097
Gardena, Cal. 90249

Hodag Chemical Corp.
7247 No. Central Park Ave.
Skokie, Ill. 60076

Hoffmann-LaRoche Inc.
340 Kingsland St.
Nutley, N.J. 07110

Holliday, L. B., & Co., Ltd.
Huddersfield, England

Hooker Chemical Corp.
Industrial Chemicals Div.
Box 344
Niagara Falls, N.Y. 14302

Hooker Chemical Corp.
Parker Rustproof Div.
2177 E. Milwaukee Dr.
Detroit, Mich. 48211

Hoskins Mfg. Co.
4445 Lawton Ave.
Detroit, Mich. 48208

Houghton, E. F., & Co.
303 West Lehigh Ave.
Philadelphia, Pa. 19133

Hoyt, Arthur S., Co., Inc.
Box 24
Hicksville, N.Y. 11802

Huber, J. M., Corp.
Box 831
Borger, Texas 79007

Hughson Chemical Co.
Erie, Pa. 16512

Humble Oil & Refining Co.
Box 2180
Houston, Texas 77001

Humko Sheffield Chemical
Div. Kraftco Corp.
Box 398
Memphis, Tenn. 38101
Humphrey Chemical Co.
7261 Devine St.
North Haven, Conn. 06473
Hunter Engineering Co.
Div. American Metal Climax, Inc.
Box 5437
Riverside, Cal. 92507
Huntington Alloy Products Div.
International Nickel Co., Inc.
Huntington, West Virginia 25720
Huyck Metals Co.
Box 30
Milford, Conn. 06461
Hynson, Westcott & Dunning, Inc.
Baltimore, Md. 21201

I

I.C.I. Organics, Inc.
55 Canal St.
Providence, R.I. 02901
ICI United States, Inc.
Wilmington, Del. 19897
Inmont Corp.
150 Wagaraw Rd.
Hawthorne, N.J. 07506
International Rustproof Co.
Div. of Lubrizol Corp.
Box 17100
Cleveland, O. 44117
Interpace Corp.
Box 785
Ione, Cal. 95640
Interplastic Corp.
2015 N.E. Broadway
Minneapolis, Minn. 55413
Ionac Chemical Co.
Div. Pfaudler Permutit, Inc.
Birmingham, N.J. 08011

J

Jefferson Chemical Co., Inc.
Box 53300
Houston, Texas 77052
Johns-Manville Products Corp.
Celite Division
Manville, N.J.
(22 E. 40 St., N.Y. 10016)
Johnson, S. C., and Son, Inc.
1525 Howe St.
Racine, Wis. 53403
Jones-Dabney Division
Devoe & Raynolds Co., Inc.
Box 1863
Louisville, Ky. 40201

K

Kaiser Chemicals, Div. Kaiser Aluminum
300 Lakeside Dr.
Oakland, Cal. 94643
Kay-Fries Chemicals, Inc.
360 Lexington Ave.
New York, N.Y. 10017
Kelco Co.
Box 23076
San Diego, Cal. 92123
Kendall Refining Co.
Bradford, Pa. 16701
Kennametal Inc.
Latrobe, Pa. 15650
Kenrich Petrochemicals, Inc.
Bayonne, N.J. 07002
King, Robert J., Co.
Box 588
Norwalk, Conn. 06856
Knoll Fine Chemicals, Inc.
120 E. 56 St.
New York, N.Y. 10022
Kohnstamm, H., & Co., Inc.
161 Ave. of Americas
New York, N.Y. 10013
Kolene Corp.
12890 Westwood Ave.
Detroit, Mich. 48223
Koppers Co., Inc.
Organic Materials Div.
1900 Koppers Bldg.
Pittsburgh, Pa. 15219

L

Lake States Chemical Div.
St. Regis Paper Co.
Rhinelander, Wis. 54501
La Motte Chemical Products Co.
Chestertown, Md. 21620
Lignosol Chemicals
Box 2025
Quebec 6, P.Q.
Lilly, Eli, & Co.
Box 618
Indianapolis, Ind. 46206
Lithcote Corp.
5000 West Lake St.
Melrose Park, Ill. 60161
Lockrey Co.
Southampton, N.Y. 11968
Loctite Corp.
705 No. Mountain Rd.
Newington, Conn. 06111

M

3M Co., Chemical Division
2501 Hudson Rd.
St. Paul, Minn. 55119
M & T Chemicals, Inc.
Woodbridge Rd.
Rahway, N.J. 07065
Mallinckrodt Inc.
Box 5439
St. Louis, Mo. 63147
Mallinckrodt Inc.
Washine Div.
165 Main St.
Lodi, N.J. 07644
Marblette Co.
Div. Allied Products Corp.
37-31 Thirtieth St.
Long Island City, N.Y. 11101
Marine Colloids, Inc.
Rockland, Maine
Masonite Corp.
29 No. Wacker Dr.
Chicago, Ill. 60606
McDanel Refractory Porcelain Co.
Box 560
Beaver Falls, Pa. 15010
Mearl Corp.
41 East 42 St.
New York, N.Y. 10017
Meer Corp.
North Bergen, N.J. 07047
Merck & Co., Inc.
126 E. Lincoln Ave.
Rahway, N.J. 07065
Michigan Chemical Corp.
2 No. Riverside Plaza
Chicago, Ill. 60606
Miles Laboratories, Inc.
1127 Myrtle St.
Elkhart, Ind. 46514
Millhaven Fibres Ltd.
Box 10
Montreal, P.Q.
Minerals & Chemicals Div.
Englehard Minerals & Chemicals Corp.
Menlo Park, N.J. 08817
Mobay Chemical Corp.
Penn-Lincoln Pky. West
Pittsburgh, Pa. 15205
Mobil Chemical Co.
Petrochemicals Div.
120 E. 42 St.
New York, N.Y. 10017
Mona Industries, Inc.
Paterson, N.J. 07524
Monsanto Industrial Chemicals Co.
800 No. Lindbergh Blvd.
St. Louis, Mo. 63166
Mooney Chemicals, Inc.
2301 Scranton Rd.
Cleveland, O. 44113
Moores Lime Co.
Box 878
Springfield, O. 45501
Morton Chemical Co.
Box 66655
Baton Rouge, La. 70806

N

Nalco Chemical Co.
Oakbrook, Ill. 60521
Nalge Co., Inc.
Rochester, N.Y. 14602
Napp Chemicals
199 Main St.
Lodi, N.J. 07644
National Filter Media Corp.
New Haven, Conn. 06514
National Gypsum Co.
Buffalo, N.Y. 14202
National Lead Co.
Baroid Division
Box 1675
Houston, Texas 77001
National Lead Co.
111 Broadway
New York, N.Y. 10006
National Lead Co.
DeLore Division
St. Louis, Mo. 63111

National Lead Co.
Evans Lead Division
Box 1467
Charleston, West Virginia 25325
National Polychemicals, Inc.
Wilmington, Mass. 01887
National Rosin Oil Products, Inc.
Box 1217
Savannah, Ga. 31402
National Starch & Chemical Corp.
750 Third Ave.
New York, N.Y. 10017
Natural Gas Odorizing, Inc.
Box 15252
Houston, Texas 77020
Nease Chemical Co., Inc.
Box 221
State College, Pa. 16801
Neville Chemical Co.
Neville Island
Pittsburgh, Pa. 15225
New Jersey Zinc
2045 City Line Rd.
Bethlehem, Pa. 18017
Nixon-Baldwin Chemicals, Inc.
Nixon, N.J. 08818
Norac Co., Inc.
Azusa, Cal. 91702
Norton Co., Metals Div.
45 Industrial Pl
Newton, Mass.
Nulomoline Division
SuCrest Corp.
New York, N.Y. 10005
Nye, William F., Inc.
Box 927
New Bedford, Mass. 02742

O

Olin Corp.
120 Long Ridge Rd.
Stamford, Conn. 06904
Onyx Chemical Co.
190 Warren St.
Jersey City, N.J. 07302
Osborn, C. J. Chemicals, Inc.
Pennsauken, N.J. 08109
Owens-Corning Fiberglas Corp.
Toledo, O. 43601
Ozark-Mahoning Co.
1870 So. Boulder
Tulsa, Okla. 74119

P

Pacific Lumber Co.
1111 Columbus Ave.
San Francisco, Cal. 94133
Paisley Products
Div. of International Latex & Chemical Corp.
1770 Canalport Ave.
Chicago, Ill. 60616
Parke, Davis & Co.
Box 118
Detroit, Mich. 48232
Penick, S. B., & Co.
1050 Wall St.
W. Lyndhurst, N.J. 07071
Pennsalt Chemicals Corp.
3 Penn Center
Philadelphia, Pa. 19102
Pennsylvania Glass Sand Corp.
3 Penn Center
Pittsburgh, Pa. 15235
Pennsylvania Industrial Chemical Corp.
131 State St.
Clairton, Pa. 15025
Pennsylvania Refining Co.
Butler, Pa. 16001
Permutit Co.
Div. of Pfaudler-Permutit, Inc.
Paramus, N.J. 07632
Petrolite Corp., Bareco Div.
6910 East 14 St.
Tulsa, Okla. 74115
Pfaudler Co.
Div. of Pfaudler-Permutit, Inc.
Rochester, N.Y. 14603
Pfizer Inc., Chemicals Div.
235 E. 42 St.
New York, N.Y. 10017
Pharmacia Fine Chemicals, Inc.
New Market, N.J. 08854
Philadelphia Quartz Co.
Box 840
Valley Forge, Pa. 19482
Phillips Petroleum Co.
Rubber Chemicals Div.
Bartlesville, Okla. 74004
Pitt-Consol Chemical Co.
Teterboro, N.J. 07608
Poco Graphite, Inc.
Box 1524
Garland, Texas 75041
Polyvinyl Chemical Industries
Wilmington, Mass. 01887
PPG Industries, Chemical Div.
1 Gateway Center
Pittsburgh, Pa. 15222
Premier Malt Products, Inc.
1137 No. Eighth St.
Milwaukee, Wis. 53201
Procter & Gamble Co.
Industrial Chemicals Div.
Box 599
Cincinnati, O. 45201
Puerto Rico Chemical Co., Inc.
277 Park Ave.
New York, N.Y. 10017

Q

Quaker Oats Co.
Merchandise Mart Plaza
Chicago, Ill. 60654

R

Reactor Experiments, Inc.
Belmont, Cal. 94002
Reheis Chemical Co.
Div. of Armour Pharmaceutical Co.
Phoenix, Arix. 85077
Reichhold Chemicals, Inc.
Newport Div.
Box 1433
Pensacola, Fla. 32596
Renite Co.
2500 East Fifth Ave.
Columbus, O. 43219
Republic Steel Corp.
1441 Republic Bldg.
Cleveland, O. 44101
Reynolds Metals Co.
6601 West Broad St.
Richmond, Va. 23261
Rhode Island Laboratories, Inc.
West Warwick, R.I. 02893
Richardson Co.
Organic Chemicals Div.
2400 E. Devon St.
Desplaines, Ill. 60018
Robeco Chemicals, Inc.
51 Madison Ave.
New York, N.Y. 10010
Rohm & Haas Co.
Independence Mall West
Philadelphia, Pa. 19105
Rorer-Amchem Products, Inc.
Ambler, Pa. 19002
Roussel Corp.
155 E. 44th St.
New York, N.Y. 10017
Royce Chemical Co.
Box 237
E. Rutherford, N.J. 07073
Rubber Corp. of America
Hicksville, N.Y. 11802
Rumford Chemical Works
Rumford, R.I. 02916

S

Sanforized Co.
530 Fifth Ave.
New York, N.Y. 10036
Schering Corp.
60 Orange St.
Bloomfield, N.J. 07003
Schwarz/Mann
Orangeburg, N.Y. 10962
Scientific Chemicals, Inc.
1637 So. Kilbourn Ave.
Chicago, Ill. 60623
Scott Bader & Co., Ltd.
Wollaston, Wellingborough
Northamptonshire, England
Searle Chemicals, Inc.
Box 8526
Chicago, Ill. 60680
Selas Flotronics
Box 300
Spring House, Pa. 19477
Shell Chemical Co.
1 Shell Plaza
Houston, Texas 77002
Sherwin Williams Chemicals
Box 6520
Cleveland, O. 44101
Shulton, Inc.
Clifton, N.J. 07015
Sierra Talc & Chemical Corp.
Box 370
So. Pasadena, Cal. 91031
Simplex Wire & Cable Co.
No. Berwick, Maine
Sinclair Petrochemicals, Inc.
600 Fifth Ave.
New York, N.Y. 10020

Sindar Corp.
321 W. 44th St.
New York, N.Y. 10036
Skelly Oil Co.
1437 Boulder St.
Tulsa, Okla. 74102
Smith Kline & French Labs.
1500 Spring Garden St.
Philadelphia, Pa. 19101
Smith, Werner G., Inc.
1730 Train Ave.
Cleveland, O. 44113
Southern Latex Corp.
Austell, Ga. 30001
Spencer Kellogg Div.
Textron, Inc.
120 Delaware Ave.
Buffalo, N.Y. 14240
Squibb & Sons, E. R., Inc.
745 Fifth Ave.
New York, N.Y. 10022
Stainless Foundry & Engineering, Inc.
5132 No. 35th St.
Milwaukee, Wis. 53209
Staley Chemical Co.
Div. A. E. Staley Mfg. Co.
96 Western Ave.
Allston, Mass.
Staley, A. E., Mfg. Co.
Box 151
Decatur, Ill. 62525
Stauffer Chemical Co.
Industrial Chemical Div.
Westport, Conn. 06880
Stein, Hall & Co., Inc.
605 Third Ave.
New York, N.Y. 10016
Stepan Chemical Co.
Industrial Chemicals Div.
Northfield, Ill. 60093
Stevens, J. P., & Co. Inc.
1460 Broadway
New York, N.Y. 10036
Sun Chemical Corp.
Box 70
Chester, So. Carolina 29706
Sun Oil Co.
1608 Walnut St.
Philadelphia, Pa. 19103
Swift Edible Oil Co.
115 W. Jackson Blvd.
Chicago, Ill. 60604
Synvar Corp.
917 Washington St.
Wilmington, Del. 19899

T

Taylor Sons Co., Charles
Box 14058
Cincinnati, O. 45214
Tenneco Chemicals, Inc.
Intermediates Div.
Box 2
Piscataway, N.J. 08854
Tenneco Chemicals, Inc.
Heyden Division
280 Park Ave.
New York, N.Y. 10017
Tenneco Chemicals, Inc.
Newport Div.
Pensacola, Fla. 32502
Tennessee Corp.
Box 2205
Atlanta, Ga. 30303
Texaco Inc.
Petrochemical Dept.
Box 52332
Houston, Texas 77052
Textilana Corp.
12607 Cerise Ave.
Hawthorne, Cal. 90250
Thermal American Fused Quartz Co.
Montville, N.J. 07045
Thiokol Chemical Corp.
780 No. Clinton Ave.
Trenton, N.J. 08638
Thomas, Arthur H., Co.
Box 779
Philadelphia, Pa. 19105
Tylac, Standard Brands Chemical Industries, Inc.
Dover, Del. 19901

U

Union Camp Corp.
Box 6170
Jacksonville, Fla. 32205
Union Carbide Corp.
270 Park Ave.
New York, N.Y. 10017
Uniroyal Chemical
Div. Uniroyal Inc.
Naugatuck, Conn. 06770
United Carbon Co.
Box 1503
Houston, Texas 77001
Universal Oil Products Co.
Chemical Div.
East Rutherford, N.J. 07073
UOP Chemical Co.
Division Universal Oil Products Co.
Des Plaines, Ill. 60016
Upjohn Co.
Kalamazoo, Mich. 49001
U.S. Borax & Chemical Corp.
3075 Wilshire Blvd.
Los Angeles, Cal. 90010
U.S. Industrial Chemicals Co.
99 Park Ave.
New York, N.Y. 10016
U.S. Stoneware Co.
5300 E. Tallmadge Ave.
Akron, O. 44310

V

Valchem, Chemical Div.
Box 38
Langley, So. Carolina 29834
Vanderbilt, R. T., & Co.
Norwalk, Conn. 06855
(230 Park Ave, N.Y.)
Van Dyk & Co., Inc.
Main and William Sts.
Belleville, N.J. 07109
V-C Chemical Co.
401 East Main St.
Richmond, Va. 23208
Velsicol Chemical Corp.
341 East Ohio St.
Chicago, Ill. 60611
Ventron Corp.
Beverly, Mass. 01915
Verona Division
Baychem Corp.
Box 385
Union, N.J. 07083

W

Wallace & Tiernan, Inc.
Harchem Div.
25 Main St.
Belleville, N.J. 07101
Wallace & Tiernan, Inc.
Lucidol Div.
1740 Military Rd.
Buffalo, N.Y. 14240
Wallerstein Co.
Div. Travenol Laboratories
Morton Grove, Ill. 60053
Warner-Chilcott Laboratories
Div. Warner-Lambert Pharmaceutical Co.
Morris Plains, N.J. 07950
West Chemical Products, Inc.
42-16 West St.
Long Island City, N.Y. 11101
Westinghouse Electric Corp.
E. Pittsburgh, Pa. 15235
West Virginia Pulp & Paper Co.
Chemical Division
230 Park Ave.
New York, N.Y. 10017
Weyerhaeuser Co.
Tacoma, Wash. 98401
Wilson Pharmaceutical & Chemical Corp.
Jackson and Swanson Sts.
Philadelphia, Pa. 19148
Winthrop Laboratories
Div. Sterling Drug Co.
90 Park Ave.
New York, N.Y. 10016
Wisconsin Alumni Research Fdn.
Box 37
Madison, Wis. 53701
Witco Chemical Corp.
Organics Division
277 Park Ave.
New York, N.Y. 10017
Woburn Chemical Corp.
1200 Harrison Ave.
Harrison, N.J. 07029
Worthington Corp.
Harrison, N.J. 07029
Wyeth Laboratories
Div. American Home Products Corp.
Box 8299
Philadelphia, Pa. 19101

X

Xylos Rubber Co.
Div. Firestone Synthetic Rubber & Latex Co.
1300 Emerling Ave.
Akron, O. 44301